MECHATRONICS

A MULTIDISCIPLINARY APPROACH

Fifth Edition

William Bolton

PEARSON

Harlow, England • London • New York • Boston • San Francisco • Toronto • Sydney • Auckland • Singapore • Hong Kong
Tokyo • Seoul • Taipei • New Delhi • Cape Town • São Paulo • Mexico City • Madrid • Amsterdam • Munich • Paris • Milan

Pearson Education Limited
Edinburgh Gate
Harlow
Essex CM20 2JE
England

and Associated Companies throughout the world

Visit us on the World Wide Web at:
www.pearson.com/uk

First published 1995
Second edition published 1999
Third edition published 2003
Fourth edition published 2008
Fifth edition published 2011

ISBN: 978-0-273-74286-9

British Library Cataloguing-in-Publication Data
A catalogue record for this book is available from the British Library

10 9 8 7 6 5 4 3 2 1
15 14 13 12 11

Typeset in 10/11pt, Ehrhardt MT Std by 73
Printed and bound by Ashford Colour Press Ltd., Gosport

Contents

Supporting resources

Visit **www.pearsoned.co.uk/bolton** to find valuable online resources

For instructors

- Instructor's Manual
- PowerPoint Slides

For more information please contact your local Pearson Education sales representative or visit **www.pearsoned.co.uk/bolton**

Preface

The term **mechatronics** was 'invented' by a Japanese engineer in 1969, as a combination of 'mecha' from mechanisms and 'tronics' from electronics. The word now has a wider meaning, being used to describe a philosophy in engineering technology in which there is a co-ordinated, and concurrently developed, integration of mechanical engineering with electronics and intelligent computer control in the design and manufacture of products and processes. As a result, many products which used to have mechanical functions have had many replaced with ones involving microprocessors. This has resulted in much greater flexibility, easier redesign and reprogramming, and the ability to carry out automated data collection and reporting.

A consequence of this approach is the need for engineers and technicians to adopt an interdisciplinary and integrated approach to engineering. Thus engineers and technicians need skills and knowledge that are not confined to a single subject area. They need to be capable of operating and communicating across a range of engineering disciplines and linking with those having more specialised skills. This book is an attempt to provide a basic background to mechatronics and provide links through to more specialised skills.

The first edition was designed to cover the Business and Technology Education Council (BTEC) Mechatronics units for Higher National Certificate/Diploma courses for technicians and designed to fit alongside more specialist units such as those for design, manufacture and maintenance determined by the application area of the course. The book was widely used for such courses and has also found use in undergraduate courses in both Britain and in the United States. Following feedback from lecturers in both Britain and the United States, the second edition was considerably extended and with its extra depth it was not only still relevant for its original readership but also suitable for undergraduate courses. The third edition involved refinements of some explanations, more discussion of microcontrollers and programming, increased use of models for mechatronics systems, and the grouping together of key facts in the Appendices. The fourth edition was a complete reconsideration of all aspects of the text, both layout and content, with some regrouping of topics, movement of more material into Appendices to avoid disrupting the flow of the text, new material – in particular an introduction to artificial intelligence, more case studies and a refinement of some topics to improve clarity. Also, objectives and key point summaries were included with each chapter. The fifth edition has kept the same structure but, after consultation with many users of the book, many aspects have had extra detail and refinement added. Chapter 1 has been made a better introduction to the subject, and chapters 2, 3, 5, 6, 8, 9, 10, 15, 21, and 22 have had additions. There is also a new Appendix on Electrical Circuit Analysis to make easily accessible to students the basic methods used for both d.c. and a.c. circuit analysis.

The overall aim of the book is to give a comprehensive coverage of mechatronics which can be used with courses for both technicians and undergraduates in engineering, and hence, to help the reader:

- Acquire a mix of skills in mechanical engineering, electronics and computing which is necessary if he/she is to be able to comprehend and design mechatronics systems.
- Become capable of operating and communicating across the range of engineering disciplines necessary in mechatronics.
- Be capable of designing mechatronic systems.

Each chapter of the book includes objectives, a summary, is copiously illustrated and contains problems, answers to which are supplied at the end of the book. With Chapter 24 research and design assignments are also included, clues as to their possible answers also being given.

The structure of the book is:

- Chapter 1 is a general introduction to mechatronics.
- Chapters 2 to 6 form a coherent block on sensors and signal conditioning.
- Chapters 7 to 9 cover actuators.
- Chapters 10 to 16 are concerned with system models.
- Chapters 17 to 23 deal with microprocessor systems.
- Chapter 24 provides an overall conclusion in considering the design of mechatronic systems.

A large debt is owed to the publications of the manufacturers of the equipment referred to in the text. I would also like to thank those reviewers in Britain, Canada and the United States who painstakingly read through the fourth edition and made suggestions for improvements.

W. Bolton

Part I
Introduction

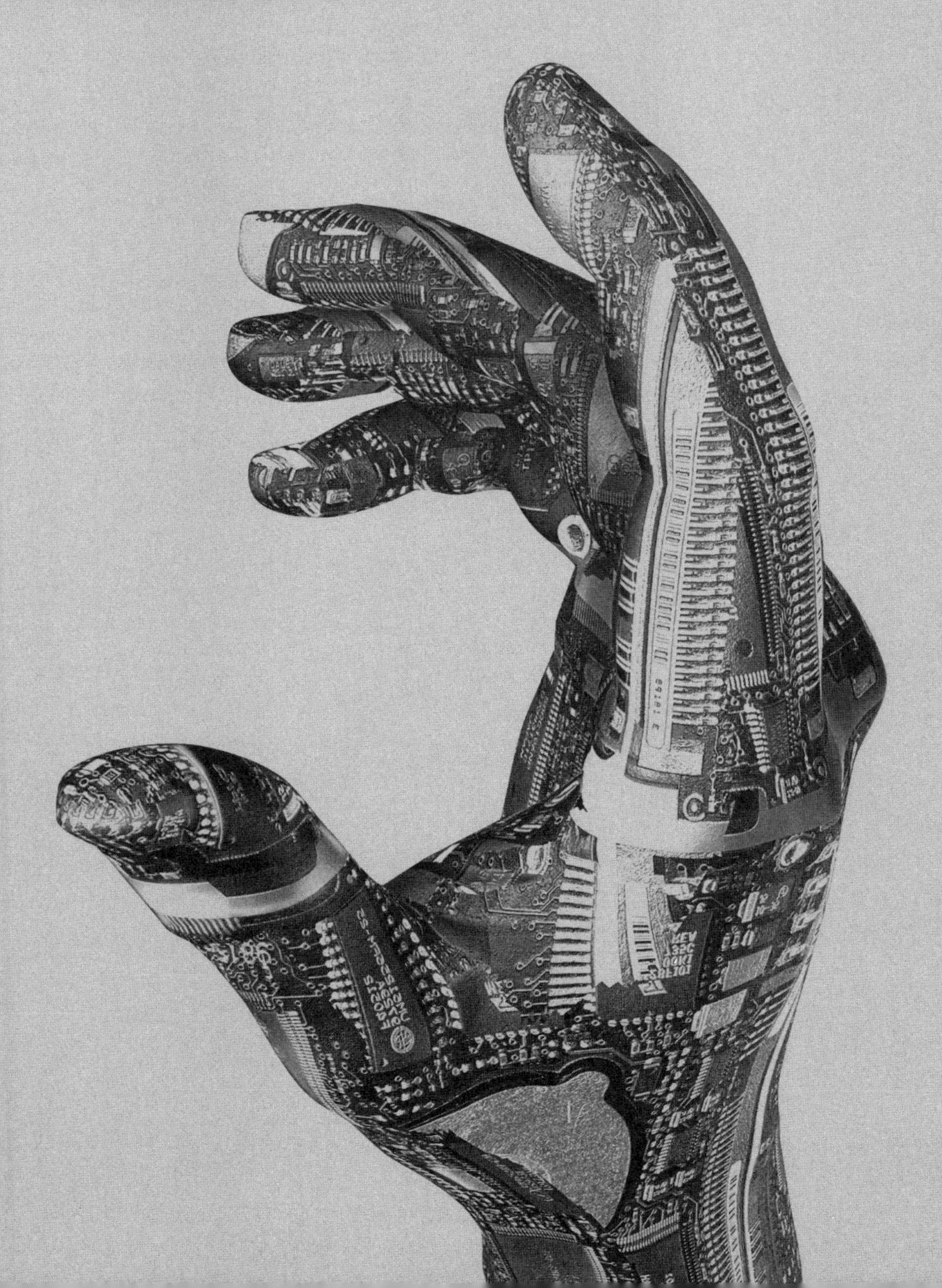

Chapter one Introducing Mechatronics

Objectives

The objectives of this chapter are that, after studying it, the reader should be able to:

- **Explain what is meant by mechatronics and appreciate its relevance in engineering design.**
- **Explain what is meant by a system and define the elements of measurement systems.**
- **Describe the various forms and elements of open-loop and closed-loop control systems.**
- **Recognise the need for models of systems in order to predict their behaviour.**

1.1 What is mechatronics?

The term **mechatronics** was 'invented' by a Japanese engineer in 1969, as a combination of 'mecha' from mechanisms and 'tronics' from electronics. The word now has a wider meaning, being used to describe a philosophy in engineering technology in which there is a co-ordinated, and concurrently developed, integration of mechanical engineering with electronics and intelligent computer control in the design and manufacture of products and processes. As a result, mechatronic products have many mechanical functions replaced with electronic ones. This results in much greater flexibility, easy redesign and reprogramming, and the ability to carry out automated data collection and reporting.

A mechatronic system is not just a marriage of electrical and mechanical systems and is more than just a control system; it is a complete integration of all of them in which there is a concurrent approach to the design. In the design of cars, robots, machine tools, washing machines, cameras, and very many other machines, such an integrated and interdisciplinary approach to engineering design is increasingly being adopted. The integration across the traditional boundaries of mechanical engineering, electrical engineering, electronics and control engineering has to occur at the earliest stages of the design process if cheaper, more reliable, more flexible systems are to be developed. Mechatronics has to involve a concurrent approach to these disciplines rather than a sequential approach of developing, say, a mechanical system, then designing the electrical part and the microprocessor part. Thus mechatronics is a design philosophy, an integrating approach to engineering.

Mechatronics brings together areas of technology involving sensors and measurement systems, drive and actuation systems, and microprocessor systems (Figure 1.1), together with the analysis of the behaviour of systems and control systems. That essentially is a summary of this book. This chapter is an introduction to the topic, developing some of the basic concepts in order to give a framework for the rest of the book in which the details will be developed.

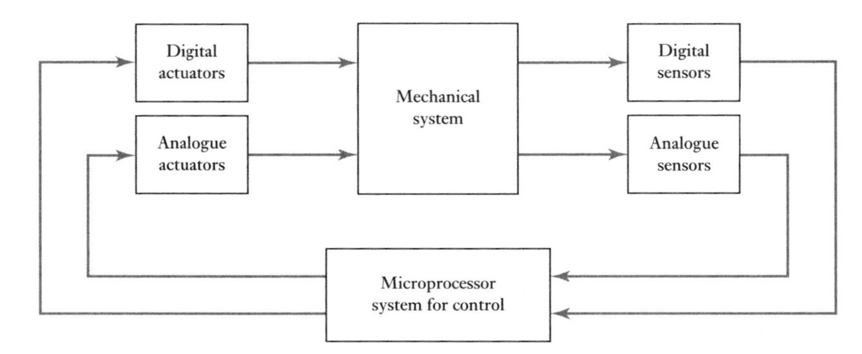

Figure 1.1 The basic elements of a mechatronic system.

1.1.1 Examples of mechatronic systems

Consider the modern autofocus, auto-exposure camera. To use the camera all you need to do is point it at the subject and press the button to take the picture. The camera can automatically adjust the focus so that the subject is in focus and automatically adjust the aperture and shutter speed so that the correct exposure is given. You do not have to manually adjust focusing and aperture or shutter speed controls. Consider a truck smart suspension. Such a suspension adjusts to uneven loading to maintain a level platform, adjusts to cornering, moving across rough ground, etc., to maintain a smooth ride. Consider an automated production line. Such a line may involve a number of production processes which are all automatically carried out in the correct sequence and in the correct way with a reporting of the outcomes at each stage in the process. The automatic camera, the truck suspension and the automatic production line are examples of a marriage between electronics, control systems and mechanical engineering.

1.1.2 Embedded systems

The term **embedded system** is used where microprocessors are embedded into systems and it is this type of system we are generally concerned with in mechatronics. A microprocessor may be considered as being essentially a collection of logic gates and memory elements that are not wired up as individual components but whose logical functions are implemented by means of software. As an illustration of what is meant by a logic gate, we might want an output if input A AND input B are both giving on signals. This could be implemented by what is termed an AND logic gate. An OR logic gate would give an output when either input A OR input B is on. A microprocessor is thus concerned with looking at inputs to see if they are on or off, processing the results of such an interrogation according to how it is programmed, and then giving outputs which are either on or off. See Chapter 17 for a more detailed discussion of microprocessors.

For a microprocessor to be used in a control system, it needs additional chips to give memory for data storage and for input/output ports to enable it to process signals from and to the outside world. **Microcontrollers** are microprocessors with these extra facilities all integrated together on a single chip.

An embedded system is a microprocessor-based system that is designed to control a range of functions and is not designed to be programmed by the end user in the same way that a computer is. Thus, with an embedded system, the user cannot change what the system does by adding or replacing software.

As an illustration of the use of microcontrollers in a control system, a modern washing machine will have a microprocessor-based control system to control the washing cycle, pumps, motor and water temperature. A modern car will have microprocessors controlling such functions as anti-lock brakes and engine management. Other examples of embedded systems are autofocus, auto-exposure cameras, camcorders, cell phones, DVD players, electronic card readers, photocopiers, printers, scanners, televisions and temperature controllers.

1.2 The design process

The design process for any system can be considered as involving a number of stages:

1 *The need*
The design process begins with a need from, perhaps, a customer or client. This may be identified by market research being used to establish the needs of potential customers.

2 *Analysis of the problem*
The first stage in developing a design is to find out the true nature of the problem, i.e. analysing it. This is an important stage in that not defining the problem accurately can lead to wasted time on designs that will not fulfil the need.

3 *Preparation of a specification*
Following the analysis, a specification of the requirements can be prepared. This will state the problem, any constraints placed on the solution, and the criteria which may be used to judge the quality of the design. In stating the problem, all the functions required of the design, together with any desirable features, should be specified. Thus there might be a statement of mass, dimensions, types and range of motion required, accuracy, input and output requirements of elements, interfaces, power requirements, operating environment, relevant standards and codes of practice, etc.

4 *Generation of possible solutions*
This is often termed the **conceptual stage.** Outline solutions are prepared which are worked out in sufficient detail to indicate the means of obtaining each of the required functions, e.g. approximate sizes, shapes, materials and costs. It also means finding out what has been done before for similar problems; there is no sense in reinventing the wheel.

5 *Selections of a suitable solution*
The various solutions are evaluated and the most suitable one selected. Evaluation will often involve the representation of a system by a model and then simulation to establish how it might react to inputs.

6 *Production of a detailed design*
The detail of the selected design has now to be worked out. This might require the production of prototypes or mock-ups in order to determine the optimum details of a design.

7 *Production of working drawings*
The selected design is then translated into working drawings, circuit diagrams, etc., so that the item can be made.

It should not be considered that each stage of the design process just flows on stage by stage. There will often be the need to return to an earlier stage and give it further consideration. Thus when at the stage of generating possible solutions there might be a need to go back and reconsider the analysis of the problem.

1.2.1 Traditional and mechatronics designs

Engineering design is a complex process involving interactions between many skills and disciplines. With traditional design, the approach was for the mechanical engineer to design the mechanical elements, then the control engineer to come along and design the control system. This gives what might be termed a sequential approach to the design. However, the basis of the mechatronics approach is considered to lie in the concurrent inclusion of the disciplines of mechanical engineering, electronics, computer technology and control engineering in the approach to design. The inherent concurrency of this approach depends very much on system modelling and then simulation of how the model reacts to inputs and hence how the actual system might react to inputs.

As an illustration of how a multidisciplinary approach can aid in the solution of a problem, consider the design of bathroom scales. Such scales might be considered only in terms of the compression of springs and a mechanism used to convert the motion into rotation of a shaft and hence movement of a pointer across a scale; a problem that has to be taken into account in the design is that the weight indicated should not depend on the person's position on the scales. However, other possibilities can be considered if we look beyond a purely mechanical design. For example, the springs might be replaced by load cells with strain gauges and the output from them used with a microprocessor to provide a digital readout of the weight on an LED display. The resulting scales might be mechanically simpler, involving fewer components and moving parts. The complexity has, however, been transferred to the software.

As a further illustration, the traditional design of the temperature control for a domestic central heating system has been the bimetallic thermostat in a closed-loop control system. The bending of the bimetallic strip changes as the temperature changes and is used to operate an on/off switch for the heating system. However, a multidisciplinary solution to the problem might be to use a microprocessor-controlled system employing perhaps a thermodiode as the sensor. Such a system has many advantages over the bimetallic thermostat system. The bimetallic thermostat is comparatively crude and the temperature is not accurately controlled; also devising a method for having different temperatures at different times of the day is complex and not easily achieved. The microprocessor-controlled system can, however, cope easily with giving precision and programmed control. The system is much more flexible. This improvement in flexibility is a common characteristic of mechatronics systems when compared with traditional systems.

1.3 Systems

In designing mechatronic systems, one of the steps involved is the creation of a model of the system so that predictions can be made regarding its behaviour when inputs occur. Such models involve drawing block diagrams to represent systems. A **system** can be thought of as a box or block diagram which

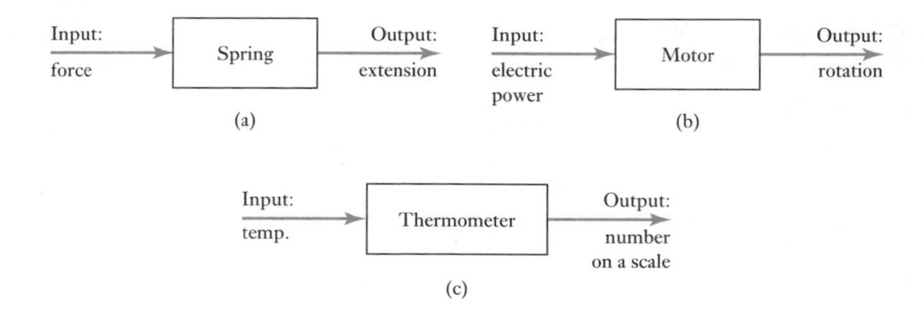

Figure 1.2 Examples of systems: (a) spring, (b) motor, (c) thermometer.

has an input and an output and where we are concerned not with what goes on inside the box but with only the relationship between the output and the input. The term **modelling** is used when we represent the behaviour of a real system by mathematical equations, such equations representing the relationship between the inputs and outputs from the system. For example, a spring can be considered as a system to have an input of a force F and an output of an extension x (Figure 1.2(a)). The equation used to model the relationship between the input and output might be $F = kx$, where k is a constant. As another example, a motor may be thought of as a system which has as its input electric power and as output the rotation of a shaft (Figure 1.2(b)).

A **measurement system** can be thought of as a box which is used for making measurements. It has as its input the quantity being measured and its output the value of that quantity. For example, a temperature measurement system, i.e. a thermometer, has an input of temperature and an output of a number on a scale (Figure 1.2(c)).

1.3.1 Modelling systems

The response of any system to an input is not instantaneous. For example, for the spring system described by Figure 1.2(a), though the relationship between the input, force F, and output, extension x, was given as $F = kx$, this only describes the relationship when steady-state conditions occur. When the force is applied it is likely that oscillations will occur before the spring settles down to its steady-state extension value (Figure 1.3). The responses of systems are functions of time. Thus, in order to know how systems behave when there are inputs to them, we need to devise models for systems which relate the output to the input so that we can work out, for a given input, how the output will vary with time and what it will settle down to.

As another example, if you switch on a kettle it takes some time for the water in the kettle to reach boiling point (Figure 1.4). Likewise, when a microprocessor controller gives a signal to, say, move the lens for focusing

Figure 1.3 The response to an input for a spring.

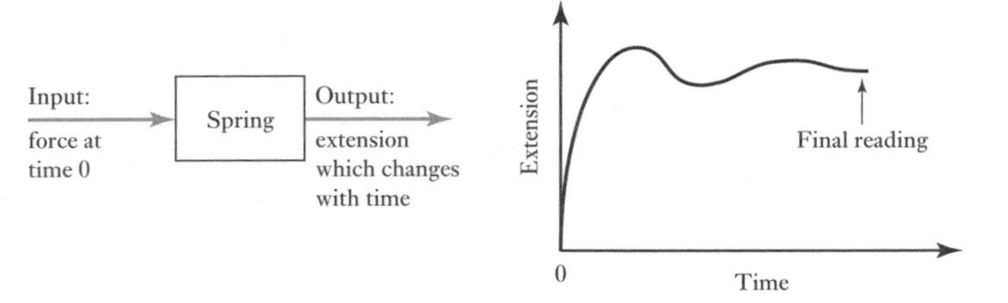

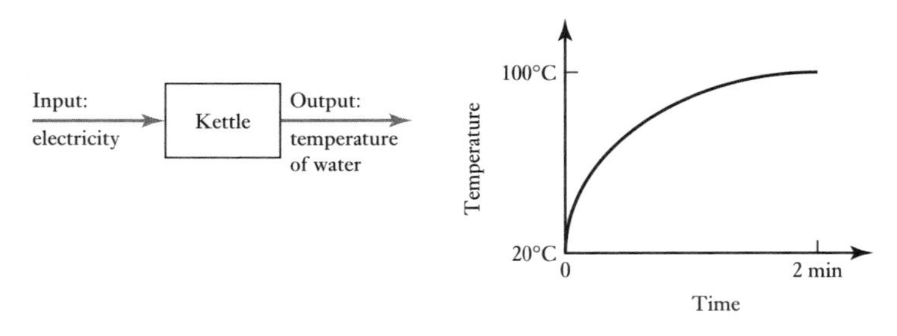

Figure 1.4 The response to an input for a kettle system.

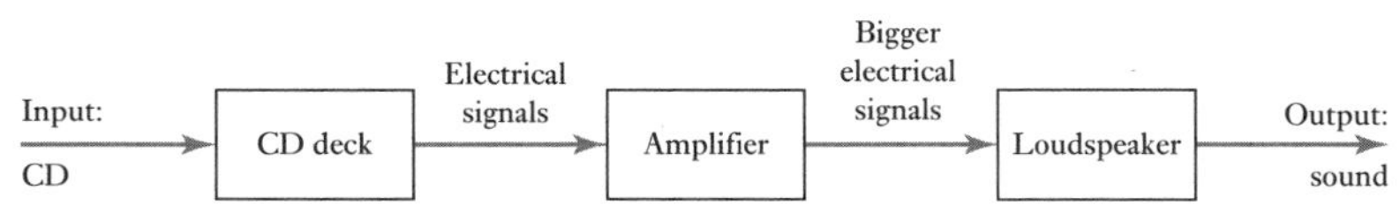

Figure 1.5 A CD player.

in an automatic camera then it takes time before the lens reaches its position for correct focusing.

Often the relationship between the input and output for a system is described by a differential equation. Such equations and systems are discussed in Chapter 10.

1.3.2 Connected systems

In other than the simplest system, it is generally useful to consider it as a series of interconnected blocks, each such block having a specific function. We then have the output from one block becoming the input to the next block in the system. In drawing a system in this way, it is necessary to recognise that lines drawn to connect boxes indicate a flow of information in the direction indicated by an arrow and not necessarily physical connections. An example of such a connected system is a CD player. We can think of there being three interconnected blocks: the CD deck which has an input of a CD and an output of electrical signals, an amplifier which has an input of these electrical signals, and an output of bigger electrical signals, and a speaker which has an input of the electrical signals and an output of sound (Figure 1.5). Another example of such a set of connected blocks is given in the next section on measurement systems.

1.4 Measurement systems

Of particular importance in any discussion of mechatronics are measurement systems. **Measurement systems** can, in general, be considered to be made up of three basic elements (as illustrated in Figure 1.6):

1 A **sensor** which responds to the quantity being measured by giving as its output a signal which is related to the quantity. For example, a thermocouple is a temperature sensor. The input to the sensor is a temperature and the output is an e.m.f. which is related to the temperature value.
2 A **signal conditioner** takes the signal from the sensor and manipulates it into a condition which is suitable either for display, or, in the case of a control system, for use to exercise control. Thus, for example, the output

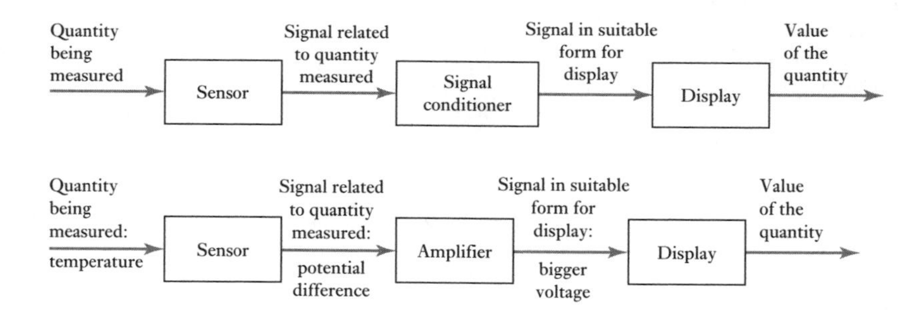

Figure 1.6 A measurement system and its constituent elements.

Figure 1.7 A digital thermometer system.

from a thermocouple is a rather small e.m.f. and might be fed through an amplifier to obtain a bigger signal. The amplifier is the signal conditioner.

3 A **display system** where the output from the signal conditioner is displayed. This might, for example, be a pointer moving across a scale or a digital readout.

As an example, consider a digital thermometer (Figure 1.7). This has an input of temperature to a sensor, probably a semiconductor diode. The potential difference across the sensor is, at constant current, a measure of the temperature. This potential difference is then amplified by an operational amplifier to give a voltage which can directly drive a display. The sensor and operational amplifier may be incorporated on the same silicon chip.

Sensors are discussed in Chapter 2 and signal conditioners in Chapter 3. Measurement systems involving all elements are discussed in Chapter 6.

1.5 Control systems

A **control system** can be thought of as a system which can be used to:

1 Control some variable to some particular value, e.g. a central heating system where the temperature is controlled to a particular value.
2 Control the sequence of events, e.g. a washing machine where when the dials are set to, say, 'white' and the machine is then controlled to a particular washing cycle, i.e. sequence of events, appropriate to that type of clothing.
3 Control whether an event occurs or not, e.g. a safety lock on a machine where it cannot be operated until a guard is in position.

1.5.1 Feedback

Consider an example of a control system with which we are all individually involved. Your body temperature, unless you are ill, remains almost constant regardless of whether you are in a cold or hot environment. To maintain this constancy your body has a temperature control system. If your temperature begins to increase above the normal you sweat, if it decreases you shiver. Both these are mechanisms which are used to restore the body temperature back to its normal value. The control system is maintaining constancy of temperature. The system has an input from sensors which tell it what the temperature is and then compare this data with what the temperature should be and provide the appropriate response in order to obtain the required

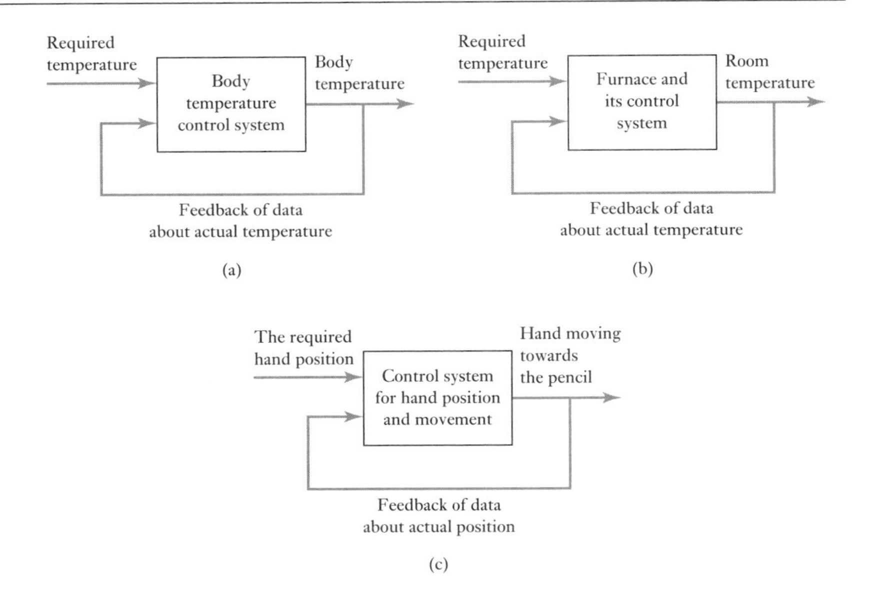

Figure 1.8 Feedback control: (a) human body temperature, (b) room temperature with central heating, (c) picking up a pencil.

temperature. This is an example of **feedback control**: signals are fed back from the output, i.e. the actual temperature, in order to modify the reaction of the body to enable it to restore the temperature to the 'normal' value. Feedback control is exercised by the control system comparing the fed-back actual output of the system with what is required and adjusting its output accordingly. Figure 1.8(a) illustrates this feedback control system.

One way to control the temperature of a centrally heated house is for a human to stand near the furnace on/off switch with a thermometer and switch the furnace on or off according to the thermometer reading. That is a crude form of feedback control using a human as a control element. The term feedback is used because signals are fed back from the output in order to modify the input. The more usual feedback control system has a thermostat or controller which automatically switches the furnace on or off according to the difference between the set temperature and the actual temperature (Figure 1.8(b)). This control system is maintaining constancy of temperature.

If you go to pick up a pencil from a bench there is a need for you to use a control system to ensure that your hand actually ends up at the pencil. This is done by your observing the position of your hand relative to the pencil and making adjustments in its position as it moves towards the pencil. There is a feedback of information about your actual hand position so that you can modify your reactions to give the required hand position and movement (Figure 1.8(c)). This control system is controlling the positioning and movement of your hand.

Feedback control systems are widespread, not only in nature and the home but also in industry. There are many industrial processes and machines where control, whether by humans or automatically, is required. For example, there is process control where such things as temperature, liquid level, fluid flow, pressure, etc. are maintained constant. Thus in a chemical process there may be a need to maintain the level of a liquid in a tank to a particular level or to a particular temperature. There are also control systems which involve consistently and accurately positioning a moving part or maintaining a constant speed. This might be, for example, a motor designed to run at a constant speed or perhaps a machining operation in which the position, speed and operation of a tool are automatically controlled.

1.5.2 Open- and closed-loop systems

There are two basic forms of control system, one being called **open loop** and the other **closed loop**. The difference between these can be illustrated by a simple example. Consider an electric fire which has a selection switch which allows a 1 kW or a 2 kW heating element to be selected. If a person used the heating element to heat a room, he or she might just switch on the 1 kW element if the room is not required to be at too high a temperature. The room will heat up and reach a temperature which is only determined by the fact that the 1 kW element was switched on and not the 2 kW element. If there are changes in the conditions, perhaps someone opening a window, there is no way the heat output is adjusted to compensate. This is an example of open-loop control in that there is no information fed back to the element to adjust it and maintain a constant temperature. The heating system with the heating element could be made a closed-loop system if the person has a thermometer and switches the 1 kW and 2 kW elements on or off, according to the difference between the actual temperature and the required temperature, to maintain the temperature of the room constant. In this situation there is feedback, the input to the system being adjusted according to whether its output is the required temperature. This means that the input to the switch depends on the deviation of the actual temperature from the required temperature, the difference between them being determined by a comparison element – the person in this case. Figure 1.9 illustrates these two types of system.

An example of an everyday open-loop control system is the domestic toaster. Control is exercised by setting a timer which determines the length of time for which the bread is toasted. The brownness of the resulting toast is determined solely by this preset time. There is no feedback to control the degree of browning to a required brownness.

To illustrate further the differences between open- and closed-loop systems, consider a motor. With an open-loop system the speed of rotation of the shaft might be determined solely by the initial setting of a knob which

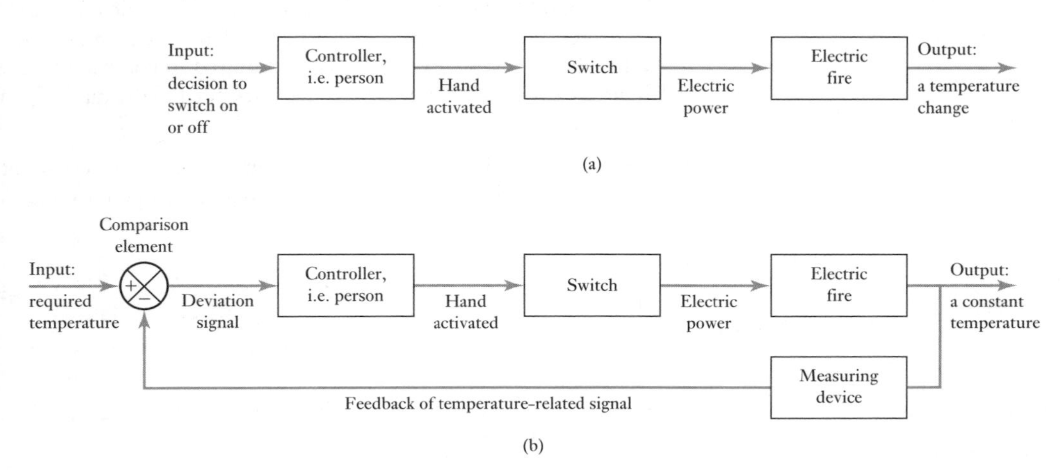

Figure 1.9 Heating a room: (a) an open-loop system, (b) a closed-loop system.

affects the voltage applied to the motor. Any changes in the supply voltage, the characteristics of the motor as a result of temperature changes, or the shaft load will change the shaft speed but not be compensated for. There is no feedback loop. With a closed-loop system, however, the initial setting of the control knob will be for a particular shaft speed and this will be maintained by feedback, regardless of any changes in supply voltage, motor characteristics or load. In an open-loop control system the output from the system has no effect on the input signal. In a closed-loop control system the output does have an effect on the input signal, modifying it to maintain an output signal at the required value.

Open-loop systems have the advantage of being relatively simple and consequently low cost with generally good reliability. However, they are often inaccurate since there is no correction for error. Closed-loop systems have the advantage of being relatively accurate in matching the actual to the required values. They are, however, more complex and so more costly with a greater chance of breakdown as a consequence of the greater number of components.

1.5.3 Basic elements of a closed-loop system

Figure 1.10 shows the general form of a basic closed-loop system. It consists of the following elements:

1 *Comparison element*
   This compares the required or reference value of the variable condition being controlled with the measured value of what is being achieved and produces an error signal. It can be regarded as adding the reference signal, which is positive, to the measured value signal, which is negative in this case:

$$\text{error signal} = \text{reference value signal} - \text{measured value signal}$$

   The symbol used, in general, for an element at which signals are summed is a segmented circle, inputs going into segments. The inputs are all added, hence the feedback input is marked as negative and the reference signal positive so that the sum gives the difference between the signals. A **feedback loop** is a means whereby a signal related to the actual condition being achieved is fed back to modify the input signal to a process. The feedback is said to be **negative feedback** when the signal which is fed back subtracts from the input value. It is negative feedback that is required to control a system. **Positive feedback** occurs when the signal fed back adds to the input signal.

2 *Control element*
   This decides what action to take when it receives an error signal. It may be, for example, a signal to operate a switch or open a valve. The control

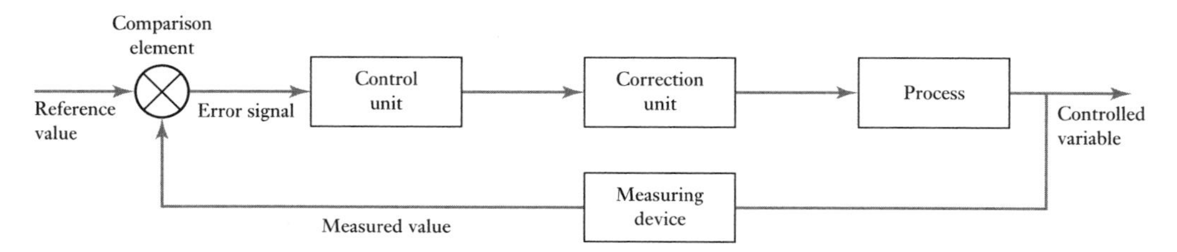

Figure 1.10 The elements of a closed-loop control system.

plan being used by the element may be just to supply a signal which switches on or off when there is an error, as in a room thermostat, or perhaps a signal which proportionally opens or closes a valve according to the size of the error. Control plans may be **hard-wired systems** in which the control plan is permanently fixed by the way the elements are connected together, or **programmable systems** where the control plan is stored within a memory unit and may be altered by reprogramming it. Controllers are discussed in Chapter 17.

3 *Correction element*
The correction element produces a change in the process to correct or change the controlled condition. Thus it might be a switch which switches on a heater and so increases the temperature of the process or a valve which opens and allows more liquid to enter the process. The term **actuator** is used for the element of a correction unit that provides the power to carry out the control action. Correction units are discussed in Chapters 7, 8 and 9.

4 *Process element*
The process is what is being controlled. It could be a room in a house with its temperature being controlled or a tank of water with its level being controlled.

5 *Measurement element*
The measurement element produces a signal related to the variable condition of the process that is being controlled. It might be, for example, a switch which is switched on when a particular position is reached or a thermocouple which gives an e.m.f. related to the temperature.

With the closed-loop system illustrated in Figure 1.10 for a person controlling the temperature of a room, the various elements are:

Controlled variable	–	the room temperature
Reference value	–	the required room temperature
Comparison element	–	the person comparing the measured value with the required value of temperature
Error signal	–	the difference between the measured and required temperatures
Control unit	–	the person
Correction unit	–	the switch on the fire
Process	–	the heating by the fire
Measuring device	–	a thermometer

An automatic control system for the control of the room temperature could involve a thermostatic element which is sensitive to temperature and switches on when the temperature falls below the set value and off when it reaches it (Figure 1.11). This temperature-sensitive switch is then used to switch on the heater. The thermostatic element has the combined functions of comparing the required temperature value with that occurring and then controlling the operation of a switch. It is often the case that elements in control systems are able to combine a number of functions.

Figure 1.12 shows an example of a simple control system used to maintain a constant water level in a tank. The reference value is the initial setting of the lever arm arrangement so that it just cuts off the water supply at the

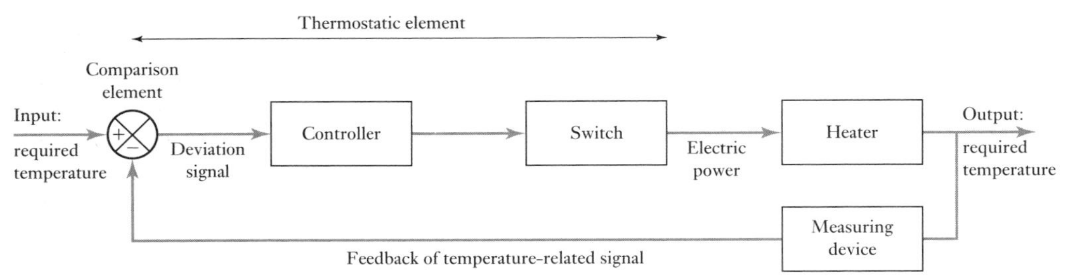

Figure 1.11 Heating a room: a closed-loop system.

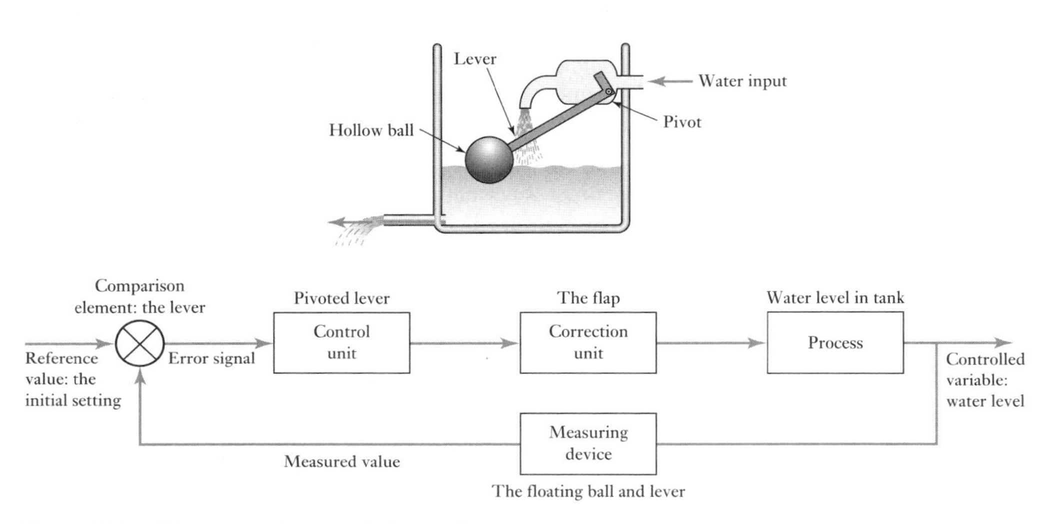

Figure 1.12 The automatic control of water level.

required level. When water is drawn from the tank the float moves downwards with the water level. This causes the lever arrangement to rotate and so allows water to enter the tank. This flow continues until the ball has risen to such a height that it has moved the lever arrangement to cut off the water supply. It is a closed-loop control system with the elements being:

Controlled variable	–	water level in tank
Reference value	–	initial setting of the float and lever position
Comparison element	–	the lever
Error signal	–	the difference between the actual and initial settings of the lever positions
Control unit	–	the pivoted lever
Correction unit	–	the flap opening or closing the water supply
Process	–	the water level in the tank
Measuring device	–	the floating ball and lever

The above is an example of a closed-loop control system involving just mechanical elements. We could, however, have controlled the liquid level by means of an electronic control system. We thus might have had a level

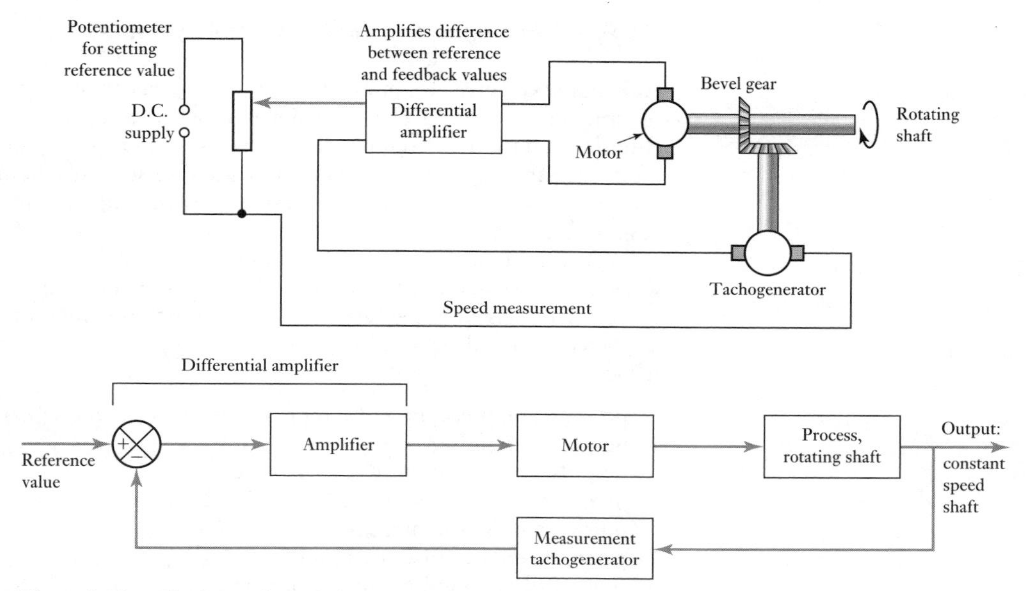

Figure 1.13 Shaft speed control.

sensor supplying an electrical signal which is used, after suitable signal conditioning, as an input to a computer where it is compared with a set value signal and the difference between them, the error signal, then used to give an appropriate response from the computer output. This is then, after suitable signal conditioning, used to control the movement of an actuator in a flow control valve and so determine the amount of water fed into the tank.

Figure 1.13 shows a simple automatic control system for the speed of rotation of a shaft. A potentiometer is used to set the reference value, i.e. what voltage is supplied to the differential amplifier as the reference value for the required speed of rotation. The differential amplifier is used both to compare and amplify the difference between the reference and feedback values, i.e. it amplifies the error signal. The amplified error signal is then fed to a motor which in turn adjusts the speed of the rotating shaft. The speed of the rotating shaft is measured using a tachogenerator, connected to the rotating shaft by means of a pair of bevel gears. The signal from the tachogenerator is then fed back to the differential amplifier:

Controlled variable	–	speed of rotation of shaft
Reference value	–	setting of slider on potentiometer
Comparison element	–	differential amplifier
Error signal	–	the difference between the output from the potentiometer and that from the tachogenerator system
Control unit	–	the differential amplifier
Correction unit	–	the motor
Process	–	the rotating shaft
Measuring device	–	the tachogenerator

1.5.4 Analogue and digital control systems

Analogue systems are ones where all the signals are continuous functions of time and it is the size of the signal which is a measure of the variable (Figure 1.14(a)). The examples so far discussed in this chapter are such systems. **Digital signals** can be considered to be a sequence of on/off signals, the value of the variable being represented by the sequence of on/off pulses (Figure 1.14(b)).

Where a digital signal is used to represent a continuous analogue signal, the analogue signal is sampled at particular instants of time and the sample values each then converted into effectively a digital number, i.e. a particular sequence of digital signals. For example, we might have for a three-digit signal the digital sequence of:

no pulse, no pulse, no pulse representing an analogue signal of 0 V,
no pulse, no pulse, a pulse representing 1 V,
no pulse, pulse, no pulse representing 2 V,
no pulse, pulse, pulse representing 3 V,
pulse, no pulse, no pulse representing 4 V,
pulse, no pulse, pulse representing 5 V,
pulse, pulse, no pulse representing 6 V,
pulse, pulse, pulse representing 7 V

Because most of the situations being controlled are analogue in nature and it is these that are the inputs and outputs of control systems, e.g. an input of temperature and an output from a heater, a necessary feature of a digital control system is that the real-world analogue inputs have to be converted to digital forms and the digital outputs back to real-world analogue forms. This involves the uses of analogue-to-digital converters (ADC) for inputs and digital-to-analogue converters (DAC) for the outputs.

Figure 1.15(a) shows the basic elements of a digital closed-loop control system; compare it with the analogue closed-loop system in Figure 1.10. The reference value, or set point, might be an input from a keyboard. Analogue-to-digital (ADC) and digital-to-analogue (DAC) elements are included in the loop in order that the digital controller can be supplied with digital

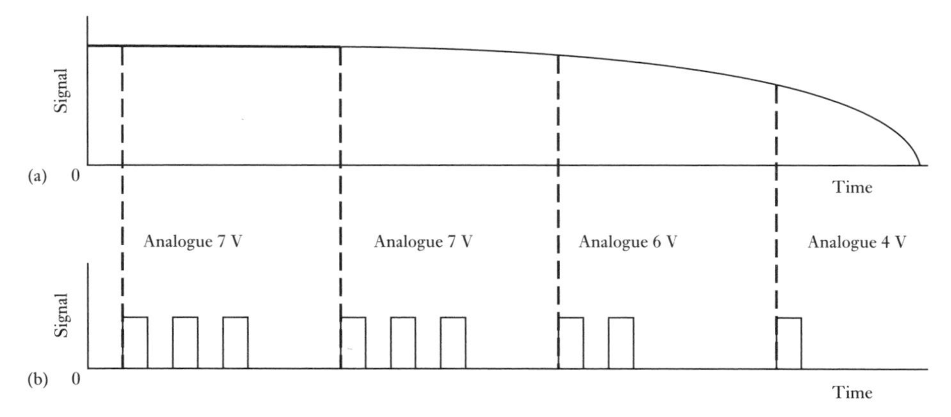

Figure 1.14 Signals: (a) analogue, and (b) the digital version of the analogue signal showing the stream of sampled signals.

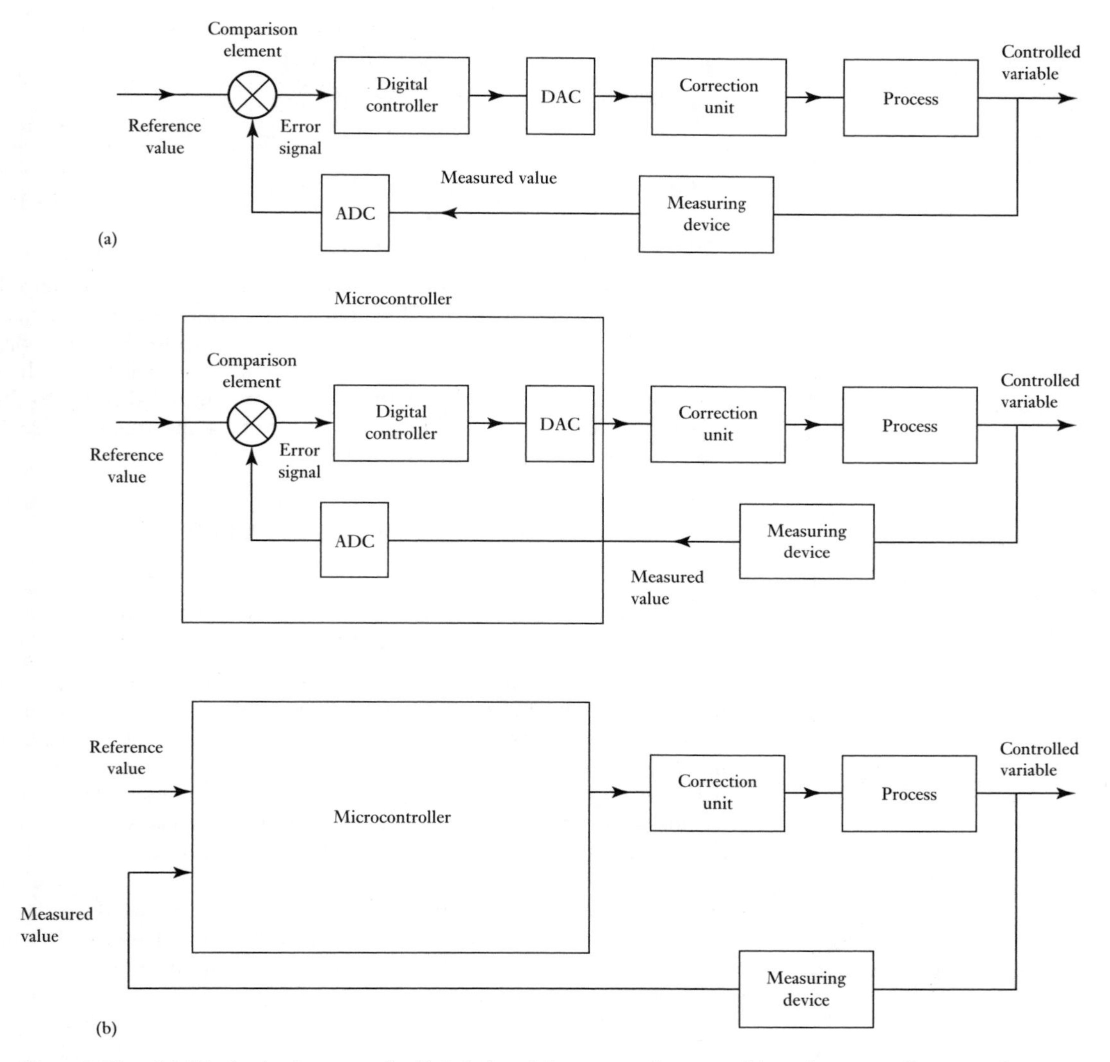

Figure 1.15 (a) The basic elements of a digital closed-loop control system, (b) a microcontroller control system.

signals from analogue measurement systems and its output of digital signals can be converted to analogue form to operate the correction units. It might seem to be adding a degree of complexity to the control system to have this analogue-to-digital conversion and digital-to-analogue conversion, but there are some very important advantages: digital operations can be controlled by a program, i.e. a set of stored instructions; information storage is easier, accuracy can be greater; digital circuits are less affected by noise and also are generally easier to design.

The digital controller could be a digital computer which is running a program, i.e. a piece of software, to implement the required actions. The term control algorithm is used to describe the sequence of steps needed to solve the control problem. The control algorithm that might be used for digital control could be described by the following steps:

Read the reference value, i.e. the desired value.
Read the actual plant output from the ADC.
Calculate the error signal.
Calculate the required controller output.
Send the controller output to the DAC.
Wait for the next sampling interval.

However, many applications do not need the expense of a computer and just a microchip will suffice. Thus, in mechatronics applications a microcontroller is often used for digital control. A microcontroller is a microprocessor with added integrated elements such as memory and analogue-to-digital and digital-to-analogue converters; these can be connected directly to the plant being controlled so the arrangement could be as shown in Figure 1.15(b). The control algorithm then might be:

Read the reference value, i.e. the desired value.
Read the actual plant output to its ADC input port.
Calculate the error signal.
Calculate the required controller output.
Send the controller output to its DAC output port.
Wait for the next sampling interval.

An example of a digital control system might be an automatic control system for the control of the room temperature involving a temperature sensor giving an analogue signal which, after suitable signal conditioning to make it a digital signal, is inputted to the digital controller where it is compared with the set value and an error signal generated. This is then acted on by the digital controller to give at its output a digital signal which, after suitable signal conditioning to give an analogue equivalent, might be used to control a heater and hence the room temperature. Such a system can readily be programmed to give different temperatures at different times of the day.

As a further illustration of a digital control system, Figure 1.16 shows one form of a digital control system for the speed a motor might take. Compare this with the analogue system in Figure 1.13.

The software used with a digital controller needs to be able to:

Read data from its input ports.
Carry out internal data transfer and mathematical operations.
Send data to its output ports.

In addition it will have:

Facilities to determine at what times the control program will be implemented.

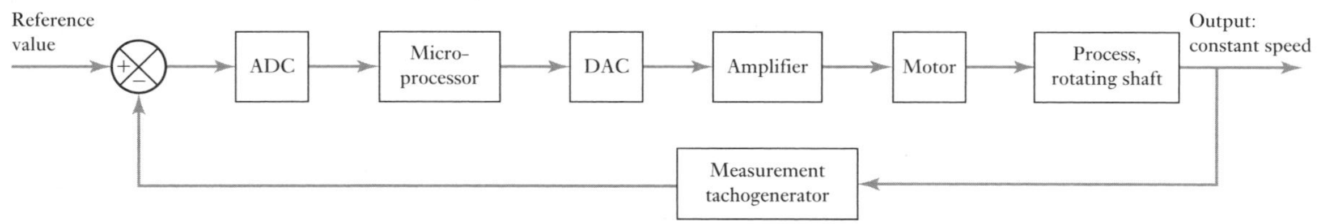

Figure 1.16 Shaft speed control.

Thus we might have the program just waiting for the ADC sampling time to occur and then spring into action when there is an input of a sample. The term **polling** is used for such a situation, the program repeatedly checking the input ports for such sampling events. So we might have:

Check the input ports for input signals.
No signals so do nothing.
Check the input ports for input signals.
No signals so do nothing.
Check the input ports for input signals.
Signal so read data from its input ports.
Carry out internal data transfer and mathematical operations.
Send data to its output ports.
Check the input ports for input signals.
No signals so do nothing.
And so on.

An alternative to polling is to use **interrupt control**. The program does not keep checking its input ports but receives a signal when an input is due. This signal may come from an external clock which gives a signal every time the ADC takes a sample.

No signal from external clock.
Do nothing.
Signal from external clock that an input is due.
Read data from its input ports.
Carry out internal data transfer and mathematical operations.
Send data to its output ports.
Wait for next signal from external clock.

1.5.5 Sequential controllers

There are many situations where control is exercised by items being switched on or off at particular preset times or values in order to control processes and give a step sequence of operations. For example, after step 1 is complete then step 2 starts. When step 2 is complete then step 3 starts, etc.

The term **sequential control** is used when control is such that actions are strictly ordered in a time- or event-driven sequence. Such control could be obtained by an electric circuit with sets of relays or cam-operated switches which are wired up in such a way as to give the required sequence. Such hard-wired circuits are now more likely to have been replaced by a microprocessor-controlled system, with the sequencing being controlled by means of a software program.

As an illustration of sequential control, consider the domestic washing machine. A number of operations have to be carried out in the correct sequence. These may involve a pre-wash cycle when the clothes in the drum are given a wash in cold water, followed by a main wash cycle when they are washed in hot water, then a rinse cycle when they are rinsed with cold water a number of times, followed by spinning to remove water from the clothes. Each of these operations involves a number of steps. For example, a pre-wash cycle involves opening a valve to fill the machine drum to the required level, closing the valve, switching on the drum motor to rotate the drum for a specific time, and operating the pump to empty the water from the drum. The operating sequence is called a **program,** the sequence of instructions in each program being predefined and 'built' into the controller used.

Figure 1.17 shows the basic washing machine system and gives a rough idea of its constituent elements. The system that used to be used for the washing machine controller was a mechanical system which involved a set of cam-operated switches, i.e. mechanical switches, a system which is readily adjustable to give a greater variety of programs.

Figure 1.18 shows the basic principle of one such switch. When the machine is switched on, a small electric motor slowly rotates its shaft, giving an amount of rotation proportional to time. Its rotation turns the controller cams so that each in turn operates electrical switches and so switches on circuits in the correct sequence. The contour of a cam determines the time at which it operates a switch. Thus the contours of the cams are the means by which the program is specified and stored in the machine. The sequence of instructions and the instructions used in a particular washing program are determined by

Figure 1.17 Washing machine system.

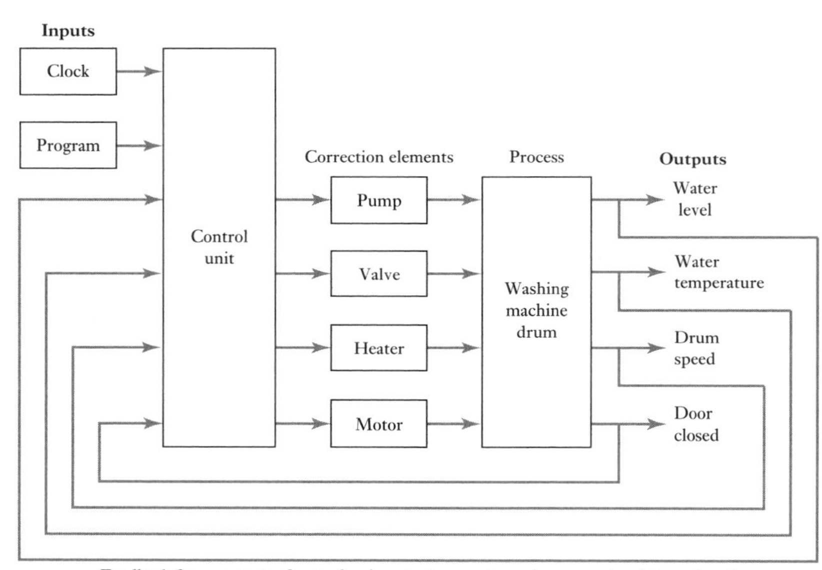

Figure 1.18 Cam-operated switch.

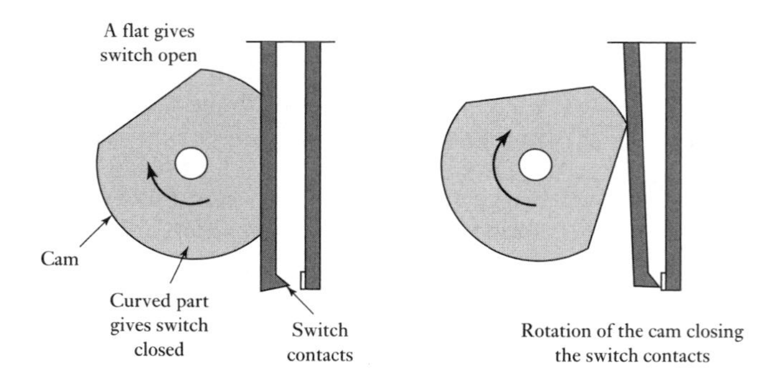

the set of cams chosen. With modern washing machines the controller is a microprocessor and the program is not supplied by the mechanical arrangement of cams but by a software program. The microprocessor-controlled washing machine can be considered an example of a mechatronics approach in that a mechanical system has become integrated with electronic controls. As a consequence, a bulky mechanical system is replaced by a much more compact microprocessor.

For the pre-wash cycle an electrically operated valve is opened when a current is supplied and switched off when it ceases. This valve allows cold water into the drum for a period of time determined by the profile of the cam or the output from the microprocessor used to operate its switch. However, since the requirement is a specific level of water in the washing machine drum, there needs to be another mechanism which will stop the water going into the tank, during the permitted time, when it reaches the required level. A sensor is used to give a signal when the water level has reached the preset level and give an output from the microprocessor which is used to switch off the current to the valve. In the case of a cam-controlled valve, the sensor actuates a switch which closes the valve admitting water to the washing machine drum. When this event is completed the microprocessor, or the rotation of the cams, initiates a pump to empty the drum.

For the main wash cycle, the microprocessor gives an output which starts when the pre-wash part of the program is completed; in the case of the cam-operated system the cam has a profile such that it starts in operation when the pre-wash cycle is completed. It switches a current into a circuit to open a valve to allow cold water into the drum. This level is sensed and the water shut off when the required level is reached. The microprocessor or cams then supply a current to activate a switch which applies a larger current to an electric heater to heat the water. A temperature sensor is used to switch off the current when the water temperature reaches the preset value. The microprocessor or cams then switch on the drum motor to rotate the drum. This will continue for the time determined by the microprocessor or cam profile before switching off. Then the microprocessor or a cam switches on the current to a discharge pump to empty the water from the drum.

The rinse part of the operation is now switched as a sequence of signals to open valves which allow cold water into the machine, switch it off, operate the motor to rotate the drum, operate a pump to empty the water from the drum, and repeat this sequence a number of times.

The final part of the operation is when the microprocessor or a cam switches on just the motor, at a higher speed than for the rinsing, to spin the clothes.

1.6 Programmable logic controller

In many simple systems there might be just an embedded microcontroller, this being a microprocessor with memory all integrated on one chip, which has been specifically programmed for the task concerned. A more adaptable form is the **programmable logic controller (PLC)**. This is a microprocessor-based controller which uses programmable memory to store instructions and to implement functions such as logic, sequence, timing, counting and arithmetic to control events and can be readily reprogrammed for different tasks. Figure 1.19 shows the control action of a programmable logic controller, the inputs being signals from, say, switches being closed and the

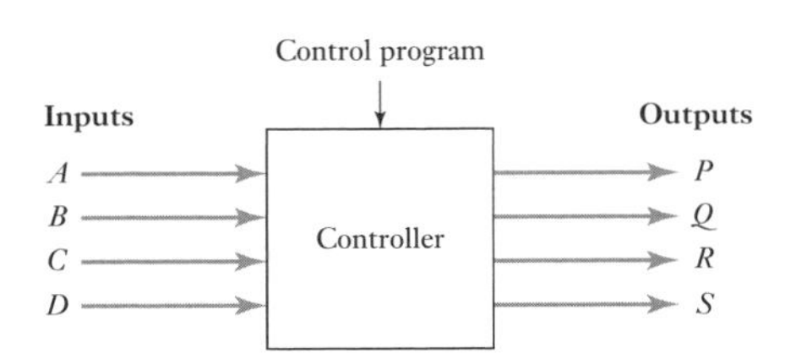

Figure 1.19 Programmable logic controller.

program used to determine how the controller should respond to the inputs and the output it should then give.

Programmable logic controllers are widely used in industry where on/off control is required. For example, they might be used in process control where a tank of liquid is to be filled and then heated to a specific temperature before being emptied. The control sequence might thus be:

1 Switch on pump to move liquid into the tank.
2 Switch off pump when a level detector gives the on signal, so indicating that the liquid has reached the required level.
3 Switch on heater.
4 Switch off heater when a temperature sensor gives the on signal to indicate the required temperature has been reached.
5 Switch on pump to empty the liquid from the container.
6 Switch off pump when a level detector gives an on signal to indicate that the tank is empty.

See Chapter 21 for a more detailed discussion of programmable logic controllers and examples of their use.

1.7 Examples of mechatronic systems

Mechatronics brings together the technology of sensors and measurements systems, embedded microprocessor systems, actuators and engineering design. The following are examples of mechatronic systems and illustrate how microprocessor-based systems have been able not only to carry out tasks that previously were done 'mechanically' but also to do tasks that were not easily automated before.

1.7.1 The digital camera and autofocus

A digital camera is likely to have an autofocus control system. A basic system used with less expensive cameras is an open-loop system (Figure 1.20(a)). When the photographer presses the shutter button, a transducer on the front of the camera sends pulses of infrared (IR) light towards the subject of the photograph. The infrared pulses bounce off the subject and are reflected back to the camera where the same transducer picks them up. For each metre the subject is distant from the camera, the round-trip is about 6 ms. The time difference between the output and return pulses is detected and fed to a microprocessor. This has a set of values stored in its memory and so gives an output which rotates the lens housing and moves the lens to a position where the object is in focus. This type of autofocus can only be used for distances up to about 10 m as the returning infrared pulses are too weak at greater distances. Thus for greater distances the microprocessor gives an output which moves the lens to an infinity setting.

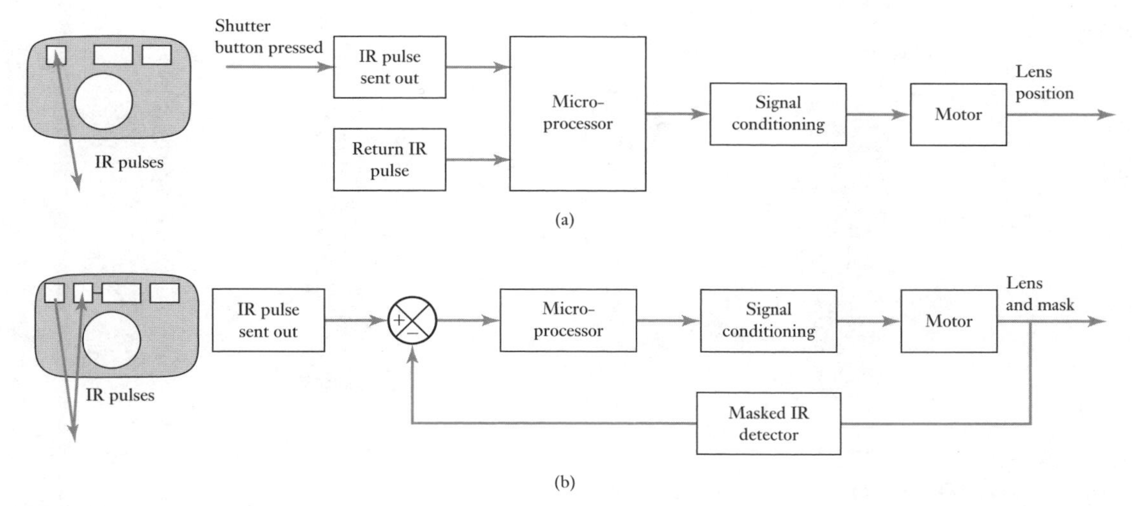

Figure 1.20 Autofocus.

A system used with more expensive cameras involves triangulation (Figure 1.20(b)). Pulses of infrared radiation are sent out and the reflected pulses are detected not by the same transducer that was responsible for the transmission but by another transducer. However, initially this transducer has a mask across it. The microprocessor thus gives an output which causes the lens to move and simultaneously the mask to move across the transducer. The mask contains a slot which is moved across the face of the transducer. The movement of the lens and the slot continues until the returning pulses are able to pass through the slot and impact on the transducer. There is then an output from the transducer which leads the microprocessor to stop the movement of the lens, and so give the in-focus position.

1.7.2 The engine management system

The engine management system of an automobile is responsible for managing the ignition and fuelling requirements of the engine. With a four-stroke internal combustion engine there are several cylinders, each of which has a piston connected to a common crankshaft and each of which carries out a four-stroke sequence of operations (Figure 1.21).

When the piston moves down a valve opens and the air–fuel mixture is drawn into the cylinder. When the piston moves up again the valve closes and the air–fuel mixture is compressed. When the piston is near the top of the cylinder the spark plug ignites the mixture with a resulting expansion of the hot gases. This expansion causes the piston to move back down again and so the cycle is repeated. The pistons of each cylinder are connected to a common crankshaft and their power strokes occur at different times so that there is continuous power for rotating the crankshaft.

The power and speed of the engine are controlled by varying the ignition timing and the air–fuel mixture. With modern automobile engines this is done by a microprocessor. Figure 1.22 shows the basic elements of a microprocessor control system. For ignition timing, the crankshaft drives a

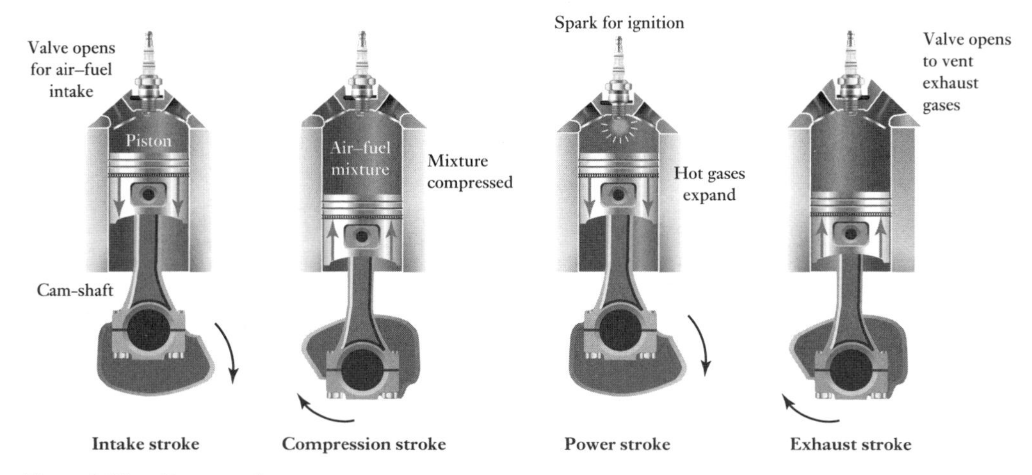

Figure 1.21 Four-stroke sequence.

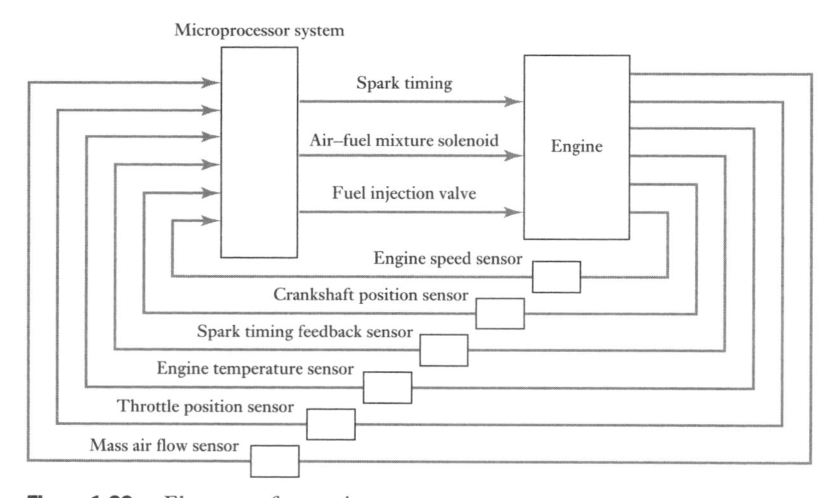

Figure 1.22 Elements of an engine management system.

distributor which makes electrical contacts for each spark plug in turn and a timing wheel. This timing wheel generates pulses to indicate the crankshaft position. The microprocessor then adjusts the timing at which high-voltage pulses are sent to the distributor so they occur at the 'right' moments of time. To control the amount of air–fuel mixture entering a cylinder during the intake strokes, the microprocessor varies the time for which a solenoid is activated to open the intake valve on the basis of inputs received of the engine temperature and the throttle position. The amount of fuel to be injected into the air stream can be determined by an input from a sensor of the mass rate of air flow, or computed from other measurements, and the microprocessor then gives an output to control a fuel injection valve. Note that the above is a very simplistic indication of engine management.

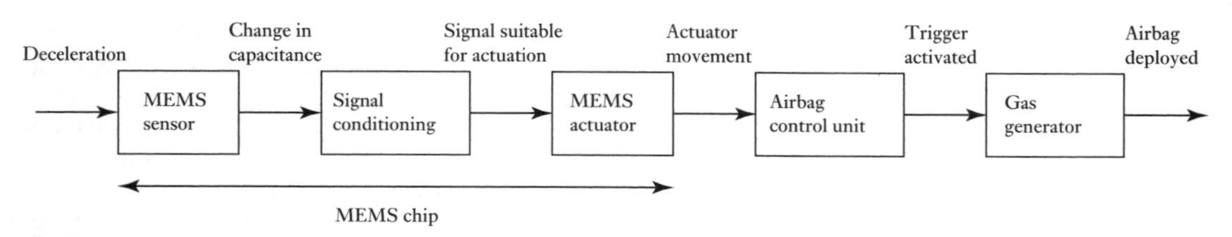

Figure 1.23 Airbag control system.

1.7.3 MEMS and the automobile airbag

Microelectromechanical systems (MEMS) are mechanical devices that are built onto semiconductor chips, generally ranging in size from about 20 micrometres to a millimetre and made up of components 0.001 to 0.1 mm in size. They usually consist of a microprocessor and components such as microsensors and microactuators. MEMS can sense, control and activate mechanical processes on the micro scale. Such MEMS chips are becoming increasingly widely used, and the following is an illustration.

Airbags in automobiles are designed to inflate in the event of a crash and so cushion the impact effects on the vehicle occupant. The airbag sensor is a MEMS accelerometer with an integrated micromechanical element which moves in response to rapid deceleration. See Figure 2.9 for basic details of the ADXL-50 device which is widely used. The rapid deceleration causes a change in capacitance in the MEMS accelerometer, which is detected by the electronics on the MEMS chip and actuates the airbag control unit to fire the airbag. The airbag control unit then triggers the ignition of a gas generator propellant to rapidly inflate a nylon fabric bag (Figure 1.23). As the vehicle occupant's body collides with and squeezes the inflated bag, the gas escapes in a controlled manner through small vent holes and so cushions the impact. From the onset of the crash, the entire deployment and inflation process of the airbag is about 60 to 80 milliseconds.

Summary

Mechatronics is a co-ordinated, and concurrently developed, integration of mechanical engineering with electronics and intelligent computer control in the design and manufacture of products and processes. It involves the bringing together of a number of technologies: mechanical engineering, electronic engineering, electrical engineering, computer technology and control engineering. Mechatronics provides an opportunity to take a new look at problems, with engineers not just seeing a problem in terms of mechanical principles but having to see it in terms of a range of technologies. The electronics, etc., should not be seen as a bolt-on item to existing mechanical hardware. A mechatronics approach needs to be adopted right from the design phase.

Microprocessors are generally involved in mechatronics systems and these are **embedded.** An embedded system is one that is designed to control a range of functions and is not designed to be programmed by the end user in the same way that a computer is. Thus, with an embedded system, the user cannot change what the system does by adding or replacing software.

A **system** can be thought of as a box or block diagram which has an input and an output and where we are concerned not with what goes on inside the box but with only the relationship between the output and the input.

In order to predict how systems behave when there are inputs to them, we need to devise **models** which relate the output to the input so that we can work out, for a given input, how the output will vary with time.

Measurement systems can, in general, be considered to be made up of three basic elements: sensor, signal conditioner and display.

There are two basic forms of **control system: open loop** and **closed loop.** With closed loop there is feedback, a system containing a comparison element, a control element, correction element, process element and the feedback involving a measurement element.

Problems

1.1 Identify the sensor, signal conditioner and display elements in the measurement systems of (a) a mercury-in-glass thermometer, (b) a Bourdon pressure gauge.

1.2 Explain the difference between open- and closed-loop control.

1.3 Identify the various elements that might be present in a control system involving a thermostatically controlled electric heater.

1.4 The automatic control system for the temperature of a bath of liquid consists of a reference voltage fed into a differential amplifier. This is connected to a relay which then switches on or off the electrical power to a heater in the liquid. Negative feedback is provided by a measurement system which feeds a voltage into the differential amplifier. Sketch a block diagram of the system and explain how the error signal is produced.

1.5 Explain the function of a programmable logic controller.

1.6 Explain what is meant by sequential control and illustrate your answer by an example.

1.7 State steps that might be present in the sequential control of a dishwasher.

1.8 Compare and contrast the traditional design of a watch with that of the mechatronics-designed product involving a microprocessor.

1.9 Compare and contrast the control system for the domestic central heating system involving a bimetallic thermostat and that involving a microprocessor.

Part II
Sensors and Signal Conditioning

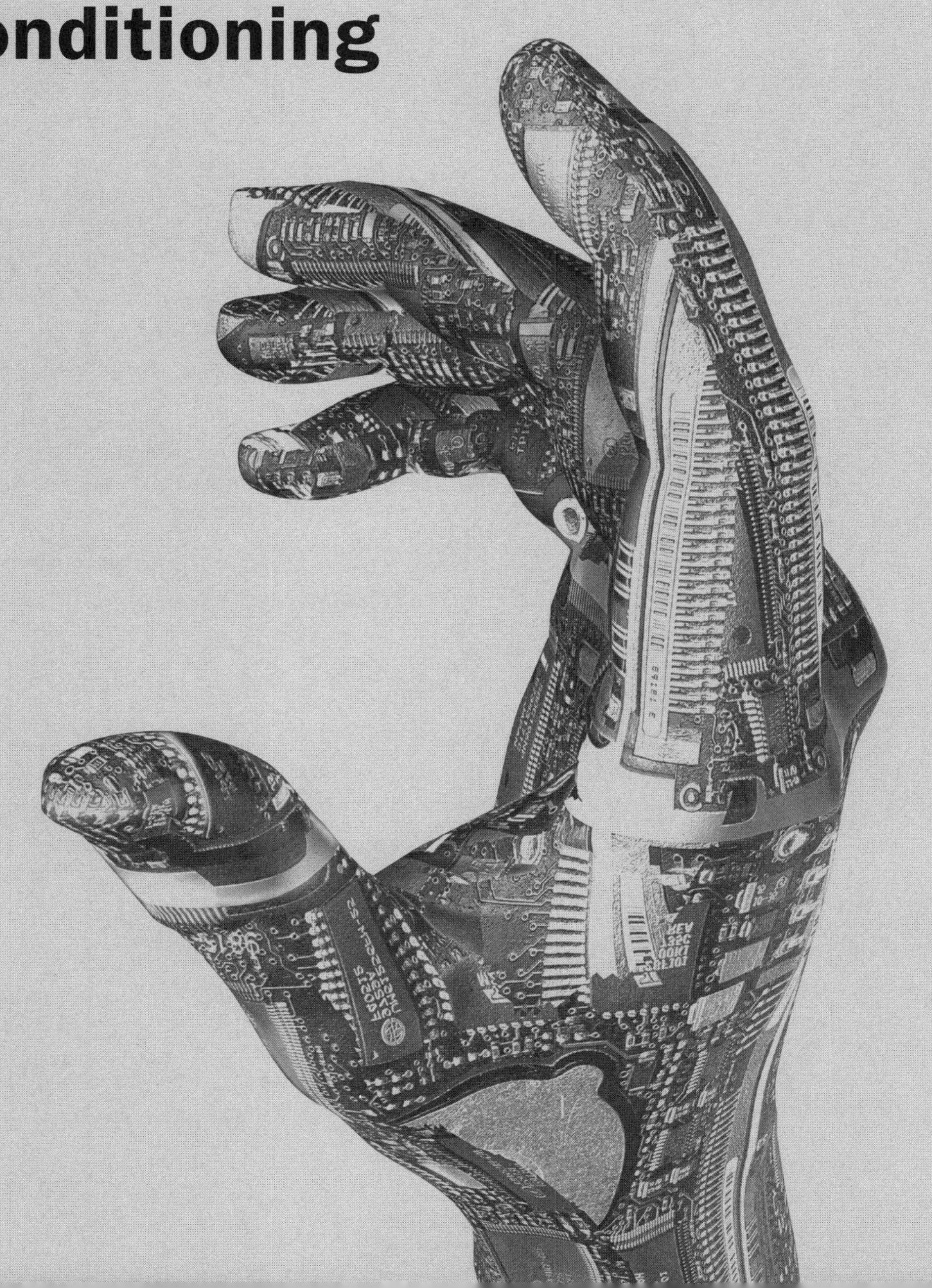

Chapter two Sensors and Transducers

Objectives

The objectives of this chapter are that, after studying it, the reader should be able to:

- **Describe the performance of commonly used sensors using terms such as range, span, error, accuracy, sensitivity, hysteresis and non-linearity error, repeatability, stability, dead band, resolution, output impedance, response time, time constant, rise time and settling time.**
- **Evaluate sensors used in the measurement of displacement, position and proximity, velocity and motion, force, fluid pressure, liquid flow, liquid level, temperature, and light intensity.**
- **Explain the problem of bouncing when mechanical switches are used for inputting data and how it might be overcome.**

2.1 Sensors and transducers

The term **sensor** is used for an element which produces a signal relating to the quantity being measured. Thus in the case of, say, an electrical resistance temperature element, the quantity being measured is temperature and the sensor transforms an input of temperature into a change in resistance. The term **transducer** is often used in place of the term sensor. Transducers are defined as elements that when subject to some physical change experience a related change. Thus sensors are transducers. However, a measurement system may use transducers, in addition to the sensor, in other parts of the system to convert signals in one form to another form. A sensor/transducer is said to be **analogue** if it gives an output which is analogue and so changes in a continuous way and typically has an output whose size is proportional to the size of the variable being measured. The term **digital** is used if the systems give outputs which are digital in nature, i.e. a sequence of essentially on/off signals which spell out a number whose value is related to the size of the variable being measured.

This chapter is about transducers and in particular those used as sensors. The terminology that is used to specify the performance characteristics of transducers is defined and examples of transducers commonly used in engineering are discussed.

2.1.1 Smart sensors

Some sensors come combined with their signal conditioning all in the same package. Such an integrated sensor does still, however, require further data processing. However, it is possible to have the sensor and signal conditioning combined with a microprocessor all in the same package. Such an arrangement is termed a **smart sensor**. A smart sensor is able to have such functions as the ability to compensate for random errors, to adapt to changes in the environment, give an automatic calculation of measurement accuracy, adjust for non-linearities to give a linear output, self-calibrate and give self-diagnosis of faults.

Such sensors have their own standard, IEEE 1451, so that smart sensors conforming to this standard can be used in a 'plug-and-play' manner, holding and communicating data in a standard way. Information is stored in the form of a TEDS (Transducer Electronic Datasheet), generally in EEPROM, and identifies each device and gives calibration data.

2.2 Performance terminology

The following terms are used to define the performance of transducers, and often measurement systems as a whole.

1 *Range and span*
The range of a transducer defines the limits between which the input can vary. The span is the maximum value of the input minus the minimum value. Thus, for example, a load cell for the measurement of forces might have a range of 0 to 50 kN and a span of 50 kN.

2 *Error*
Error is the difference between the result of the measurement and the true value of the quantity being measured:

$$\text{error} = \text{measured value} - \text{true value}$$

Thus if a measurement system gives a temperature reading of 25°C when the actual temperature is 24°C, then the error is +1°C. If the actual temperature had been 26°C then the error would have been −1°C. A sensor might give a resistance change of 10.2 Ω when the true change should have been 10.5 Ω. The error is −0.3 Ω.

3 *Accuracy*
Accuracy is the extent to which the value indicated by a measurement system might be wrong. It is thus the summation of all the possible errors that are likely to occur, as well as the accuracy to which the transducer has been calibrated. A temperature-measuring instrument might, for example, be specified as having an accuracy of ±2°C. This would mean that the reading given by the instrument can be expected to lie within plus or minus 2°C of the true value. Accuracy is often expressed as a percentage of the full range output or full-scale deflection. The percentage of full-scale deflection term results from when the outputs of measuring systems were displayed almost exclusively on a circular or linear scale. A sensor might, for example, be specified as having an accuracy of ±5% of full range output. Thus if the range of the sensor was, say, 0 to 200°C, then the reading given can be expected to be within plus or minus 10°C of the true reading.

4 *Sensitivity*
The sensitivity is the relationship indicating how much output there is per unit input, i.e. output/input. For example, a resistance thermometer may have a sensitivity of 0.5 Ω/°C. This term is also frequently used to indicate the sensitivity to inputs other than that being measured,

i.e. environmental changes. Thus there can be the sensitivity of the transducer to temperature changes in the environment or perhaps fluctuations in the mains voltage supply. A transducer for the measurement of pressure might be quoted as having a temperature sensitivity of ±0.1% of the reading per °C change in temperature.

5 *Hysteresis error*
Transducers can give different outputs from the same value of quantity being measured according to whether that value has been reached by a continuously increasing change or a continuously decreasing change. This effect is called hysteresis. Figure 2.1 shows such an output with the hysteresis error as the maximum difference in output for increasing and decreasing values.

Figure 2.1 Hysteresis.

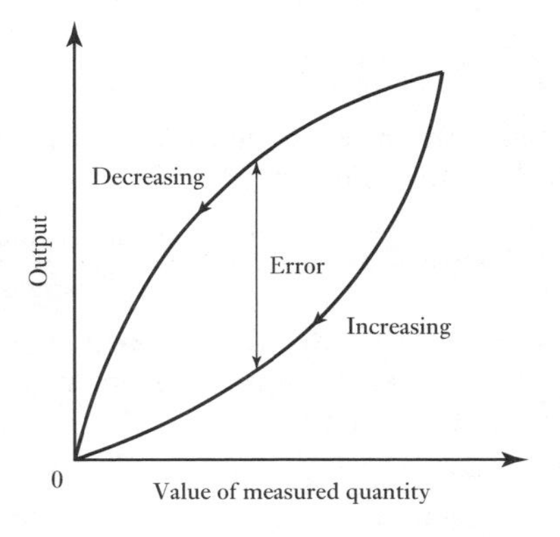

6 *Non-linearity error*
For many transducers a linear relationship between the input and output is assumed over the working range, i.e. a graph of output plotted against input is assumed to give a straight line. Few transducers, however, have a truly linear relationship and thus errors occur as a result of the assumption of linearity. The error is defined as the maximum difference from the straight line. Various methods are used for the numerical expression of the non-linearity error. The differences occur in determining the straight line relationship against which the error is specified. One method is to draw the straight line joining the output values at the end points of the range; another is to find the straight line by using the method of least squares to determine the best fit line when all data values are considered equally likely to be in error; another is to find the straight line by using the method of least squares to determine the best fit line which passes through the zero point. Figure 2.2 illustrates these three methods and how they can affect the non-linearity error quoted. The error is generally quoted as a percentage of the full range output. For example, a transducer for the measurement of pressure might be quoted as having a non-linearity error of ±0.5% of the full range.

7 *Repeatability/reproducibility*
The terms repeatability and reproducibility of a transducer are used to describe its ability to give the same output for repeated applications of the same input value. The error resulting from the same output not being

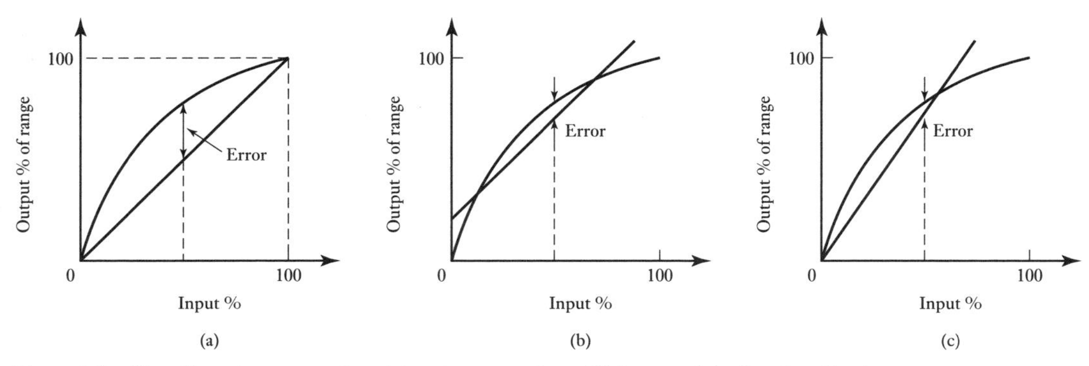

Figure 2.2 Non-linearity error using: (a) end-range values, (b) best straight line for all values, (c) best straight line through zero point.

given with repeated applications is usually expressed as a percentage of the full range output:

$$\text{repeatability} = \frac{\text{max.} - \text{min. values given}}{\text{full range}} \times 100$$

A transducer for the measurement of angular velocity typically might be quoted as having a repeatability of $\pm 0.01\%$ of the full range at a particular angular velocity.

8 *Stability*

The stability of a transducer is its ability to give the same output when used to measure a constant input over a period of time. The term **drift** is often used to describe the change in output that occurs over time. The drift may be expressed as a percentage of the full range output. The term **zero drift** is used for the changes that occur in output when there is zero input.

9 *Dead band/time*

The dead band or dead space of a transducer is the range of input values for which there is no output. For example, bearing friction in a flowmeter using a rotor might mean that there is no output until the input has reached a particular velocity threshold. The dead time is the length of time from the application of an input until the output begins to respond and change.

10 *Resolution*

When the input varies continuously over the range, the output signals for some sensors may change in small steps. A wire-wound potentiometer is an example of such a sensor, the output going up in steps as the potentiometer slider moves from one wire turn to the next. The resolution is the smallest change in the input value that will produce an observable change in the output. For a wire-wound potentiometer the resolution might be specified as, say, 0.5° or perhaps a percentage of the full-scale deflection. For a sensor giving a digital output the smallest change in output signal is 1 bit. Thus for a sensor giving a data word of N bits, i.e. a total of 2^N bits, the resolution is generally expressed as $1/2^N$.

11 *Output impedance*
When a sensor giving an electrical output is interfaced with an electronic circuit it is necessary to know the output impedance since this impedance is being connected in either series or parallel with that circuit. The inclusion of the sensor can thus significantly modify the behaviour of the system to which it is connected. See Section 6.1.1 for a discussion of loading.

To illustrate the above, consider the significance of the terms in the following specification of a strain gauge pressure transducer:

Ranges: 70 to 1000 kPa, 2000 to 70 000 kPa
Supply voltage: 10 V d.c. or a.c. r.m.s.
Full range output: 40 mV
Non-linearity and hysteresis: ±0.5% full range output
Temperature range: −54°C to +120°C when operating
Thermal zero shift: 0.030% full range output/°C

The range indicates that the transducer can be used to measure pressures between 70 and 1000 kPa or 2000 and 70 000 kPa. It requires a supply of 10 V d.c. or a.c. r.m.s. for its operation and will give an output of 40 mV when the pressure on the lower range is 1000 kPa and on the upper range 70 000 kPa. Non-linearity and hysteresis will lead to errors of ±0.5% of 1000, i.e. ±5 kPa on the lower range and ±0.5% of 70 000, namely ±350 kPa, on the upper range. The transducer can be used between the temperatures of −54 and +120°C. When the temperature changes by 1°C the output of the transducer for zero input will change by 0.030% of 1000 = 0.3 kPa on the lower range and 0.030% of 70 000 = 21 kPa on the upper range.

2.2.1 Static and dynamic characteristics

The **static characteristics** are the values given when steady-state conditions occur, i.e. the values given when the transducer has settled down after having received some input. The terminology defined above refers to such a state. The **dynamic characteristics** refer to the behaviour between the time that the input value changes and the time that the value given by the transducer settles down to the steady-state value. Dynamic characteristics are stated in terms of the response of the transducer to inputs in particular forms. For example, this might be a step input when the input is suddenly changed from zero to a constant value, or a ramp input when the input is changed at a steady rate, or a sinusoidal input of a specified frequency. Thus we might find the following terms (see Chapter 12 for a more detailed discussion of dynamic systems):

1 *Response time*
This is the time which elapses after a constant input, a step input, is applied to the transducer up to the point at which the transducer gives an output corresponding to some specified percentage, e.g. 95%, of the value of the input (Figure 2.3). For example, if a mercury-in-glass thermometer is put into a hot liquid there can be quite an appreciable time lapse, perhaps as much as 100 s or more, before the thermometer indicates 95% of the actual temperature of the liquid.

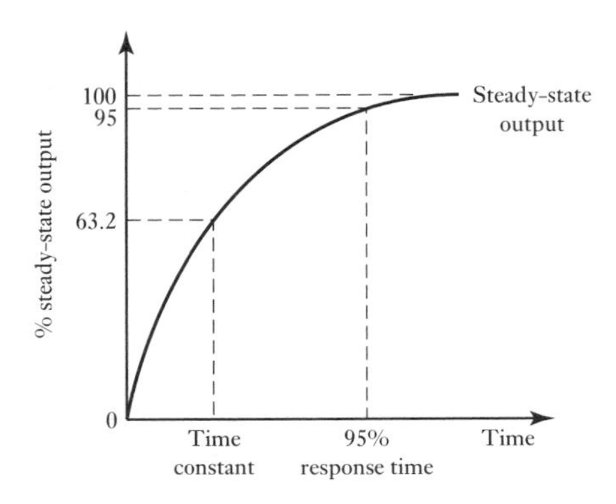

Figure 2.3 Response to a step input.

2 *Time constant*
This is the 63.2% response time. A thermocouple in air might have a time constant of perhaps 40 to 100 s. The time constant is a measure of the inertia of the sensor and so how fast it will react to changes in its input: the bigger the time constant, the slower the reaction to a changing input signal. See Section 12.3.4 for a mathematical discussion of the time constant in terms of the behaviour of a system when subject to a step input.

3 *Rise time*
This is the time taken for the output to rise to some specified percentage of the steady-state output. Often the rise time refers to the time taken for the output to rise from 10% of the steady-state value to 90 or 95% of the steady-state value.

4 *Settling time*
This is the time taken for the output to settle to within some percentage, e.g. 2%, of the steady-state value.

To illustrate the above, consider the graph in Figure 2.4 which indicates how an instrument reading changed with time, being obtained from a thermometer plunged into a liquid at time $t = 0$. The steady-state value is 55°C and so, since 95% of 55 is 52.25°C, the 95% response time is about 228 s.

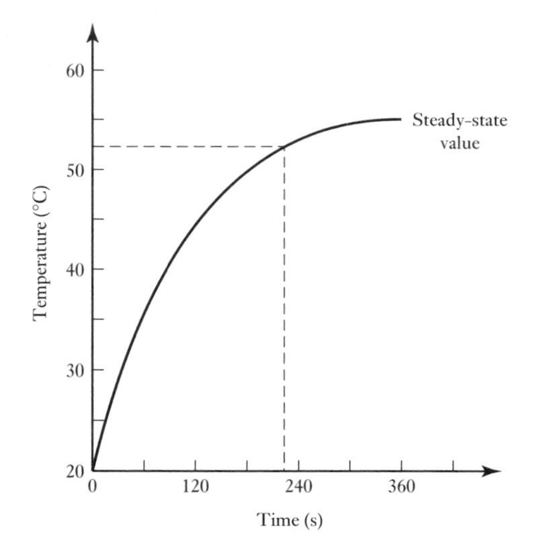

Figure 2.4 Thermometer in liquid.

The following sections give examples of transducers grouped according to what they are being used to measure. The measurements considered are those frequently encountered in mechanical engineering, namely: displacement, proximity, velocity, force, pressure, fluid flow, liquid level, temperature, and light intensity.

2.3 Displacement, position and proximity

Displacement sensors are concerned with the measurement of the amount by which some object has been moved; **position** sensors are concerned with the determination of the position of some object in relation to some reference point. **Proximity** sensors are a form of position sensor and are used to determine when an object has moved to within some particular critical distance of the sensor. They are essentially devices which give on/off outputs.

Displacement and position sensors can be grouped into two basic types: contact sensors in which the measured object comes into mechanical contact with the sensor, or non-contacting where there is no physical contact between the measured object and the sensor. For those linear displacement methods involving contact, there is usually a sensing shaft which is in direct contact with the object being monitored. The displacement of this shaft is then monitored by a sensor. The movement of the shaft may be used to cause changes in electrical voltage, resistance, capacitance or mutual inductance. For angular displacement methods involving mechanical connection, the rotation of a shaft might directly drive, through gears, the rotation of the transducer element. Non-contacting sensors might involve the presence in the vicinity of the measured object causing a change in the air pressure in the sensor, or perhaps a change in inductance or capacitance. The following are examples of commonly used displacement sensors.

2.3.1 Potentiometer sensor

A **potentiometer** consists of a resistance element with a sliding contact which can be moved over the length of the element. Such elements can be used for linear or rotary displacements, the displacement being converted into a potential difference. The rotary potentiometer consists of a circular wire-wound track or a film of conductive plastic over which a rotatable sliding contact can be rotated (Figure 2.5). The track may be a single turn or helical. With a constant input voltage V_s, between terminals 1 and 3, the output voltage V_o between terminals 2 and 3 is a fraction of the input voltage,

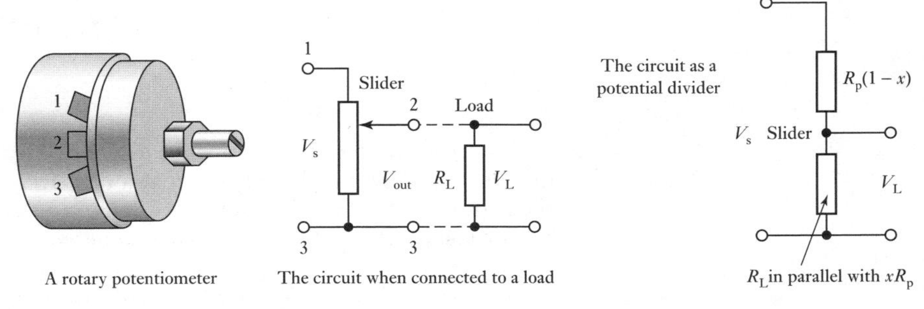

Figure 2.5 Rotary potentiometer.

the fraction depending on the ratio of the resistance R_{23} between terminals 2 and 3 compared with the total resistance R_{13} between terminals 1 and 3, i.e. $V_o/V_s = R_{23}/R_{13}$. If the track has a constant resistance per unit length, i.e. per unit angle, then the output is proportional to the angle through which the slider has rotated. Hence an angular displacement can be converted into a potential difference.

With a wire-wound track the slider in moving from one turn to the other will change the voltage output in steps, each step being a movement of one turn. If the potentiometer has N turns then the resolution, as a percentage, is $100/N$. Thus the resolution of a wire track is limited by the diameter of the wire used and typically ranges from about 1.5 mm for a coarsely wound track to 0.5 mm for a finely wound one. Errors due to non-linearity of the track tend to range from less than 0.1% to about 1%. The track resistance tends to range from about 20 Ω to 200 kΩ. Conductive plastic has ideally infinite resolution, errors due to non-linearity of the track of the order of 0.05% and resistance values from about 500 to 80 kΩ. The conductive plastic has a higher temperature coefficient of resistance than the wire and so temperature changes have a greater effect on accuracy.

An important effect to be considered with a potentiometer is the effect of a load R_L connected across the output. The potential difference across the load V_L is only directly proportional to V_o if the load resistance is infinite. For finite loads, however, the effect of the load is to transform what was a linear relationship between output voltage and angle into a non-linear relationship. The resistance R_L is in parallel with the fraction x of the potentiometer resistance R_p. This combined resistance is $R_L x R_p/(R_L + xR_p)$. The total resistance across the source voltage is thus

$$\text{total resistance} = R_p(1 - x) + R_L x R_p/(R_L + xR_p)$$

The circuit is a potential divider circuit and thus the voltage across the load is the fraction that the resistance across the load is of the total resistance across which the applied voltage is connected:

$$\frac{V_L}{V_s} = \frac{xR_L R_p/(R_L + xR_p)}{R_p(1 - x) + xR_L R_p/(R_L + xR_p)}$$

$$= \frac{x}{(R_p/R_L)x(1 - x) + 1}$$

If the load is of infinite resistance then we have $V_L = xV_s$. Thus the error introduced by the load having a finite resistance is

$$\text{error} = xV_s - V_L = xV_s - \frac{xV_s}{(R_p/R_L)x(1 - x) + 1}$$

$$= V_s \frac{R_p}{R_L}(x^2 - x^3)$$

To illustrate the above, consider the non-linearity error with a potentiometer of resistance 500 Ω, when at a displacement of half its maximum slider travel, which results from there being a load of resistance 10 kΩ. The supply voltage is 4 V. Using the equation derived above

$$\text{error} = 4 \times \frac{500}{10\,000}(0.5^2 - 0.5^3) = 0.025 \text{ V}$$

As a percentage of the full range reading, this is 0.625%.

Potentiometers are used as sensors with the electronic systems in cars, being used for such things as the accelerator pedal position and throttle position.

2.3.2 Strain-gauged element

The electrical resistance strain gauge (Figure 2.6) is a metal wire, metal foil strip, or a strip of semiconductor material which is wafer-like and can be stuck onto surfaces like a postage stamp. When subject to strain, its resistance R changes, the fractional change in resistance $\Delta R/R$ being proportional to the strain ε, i.e.

$$\frac{\Delta R}{R} = G\varepsilon$$

where G, the constant of proportionality, is termed the gauge factor.

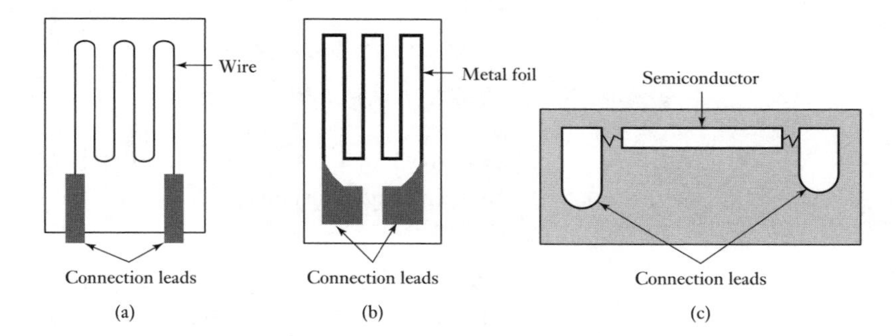

Figure 2.6 Strain gauges: (a) metal wire, (b) metal foil, (c) semiconductor.

Since strain is the ratio (change in length/original length) then the resistance change of a strain gauge is a measurement of the change in length of the element to which the strain gauge is attached. The gauge factor of metal wire or foil strain gauges with the metals generally used is about 2.0 and resistances are generally of the order of about 100 Ω. Silicon p- and n-type semiconductor strain gauges have gauge factors of about +100 or more for p-type silicon and −100 or more for n-type silicon and resistances of the order of 1000 to 5000 Ω. The gauge factor is normally supplied by the manufacturer of the strain gauges from a calibration made of a sample of strain gauges taken from a batch. The calibration involves subjecting the sample gauges to known strains and measuring their changes in resistance. A problem with all strain gauges is that their resistance changes not only with strain but also with temperature. Ways of eliminating the temperature effect have to be used and are discussed in Chapter 3. Semiconductor strain gauges have a much greater sensitivity to temperature than metal strain gauges.

To illustrate the above, consider an electrical resistance strain gauge with a resistance of 100 Ω and a gauge factor of 2.0. What is the change in resistance of the gauge when it is subject to a strain of 0.001? The fractional change in resistance is equal to the gauge factor multiplied by the strain, thus

$$\text{change in resistance} = 2.0 \times 0.001 \times 100 = 0.2\ \Omega$$

One form of displacement sensor has strain gauges attached to flexible elements in the form of cantilevers (Figure 2.7(a)), rings (Figure 2.7(b)) or

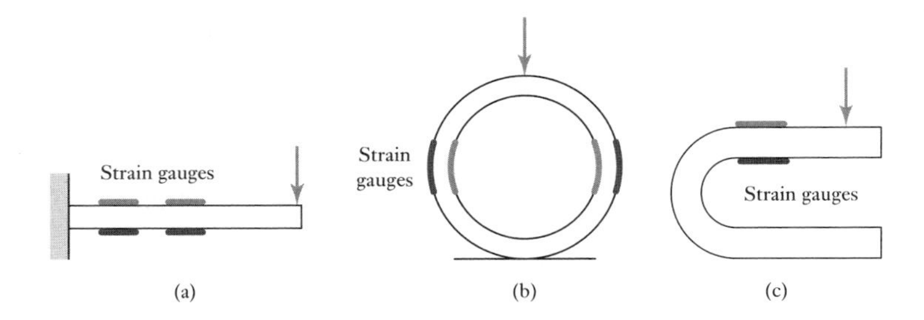

Figure 2.7 Strain-gauged element.

U-shapes (Figure 2.7(c)). When the flexible element is bent or deformed as a result of forces being applied by a contact point being displaced, then the electrical resistance strain gauges mounted on the element are strained and so give a resistance change which can be monitored. The change in resistance is thus a measure of the displacement or deformation of the flexible element. Such arrangements are typically used for linear displacements of the order of 1 to 30 mm and have a non-linearity error of about $\pm 1\%$ of full range.

2.3.3 Capacitive element

The capacitance C of a parallel plate capacitor is given by

$$C = \frac{\varepsilon_r \varepsilon_0 A}{d}$$

where ε_r is the relative permittivity of the dielectric between the plates, ε_0 a constant called the permittivity of free space, A the area of overlap between the two plates and d the plate separation. Capacitive sensors for the monitoring of linear displacements might thus take the forms shown in Figure 2.8. In (a) one of the plates is moved by the displacement so that the plate separation changes; in (b) the displacement causes the area of overlap to change; in (c) the displacement causes the dielectric between the plates to change.

For the displacement changing the plate separation (Figure 2.8(a)), if the separation d is increased by a displacement x then the capacitance becomes

$$C - \Delta C = \frac{\varepsilon_0 \varepsilon_r A}{d + x}$$

Hence the change in capacitance ΔC as a fraction of the initial capacitance is given by

$$\frac{\Delta C}{C} = -\frac{d}{d + x} - 1 = -\frac{x/d}{1 + (x/d)}$$

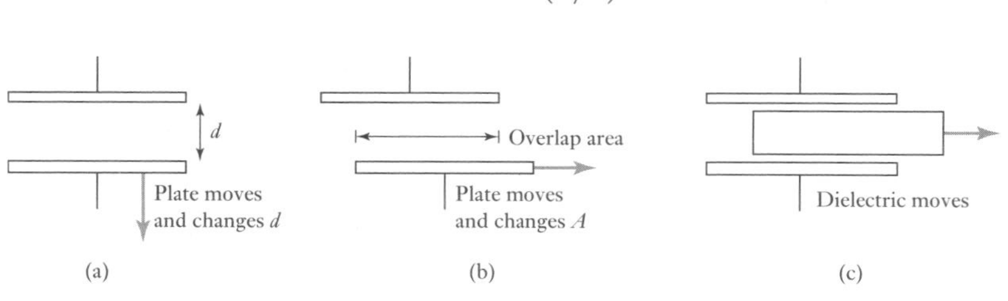

Figure 2.8 Forms of capacitive sensing element.

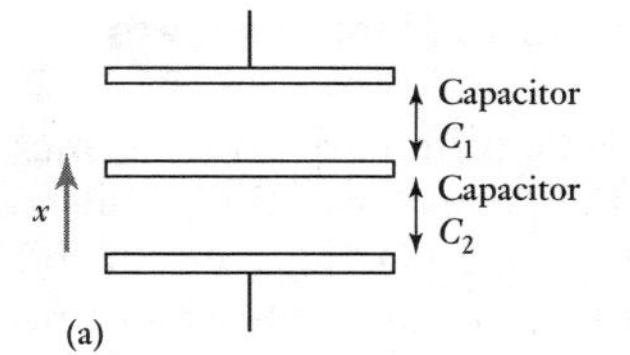

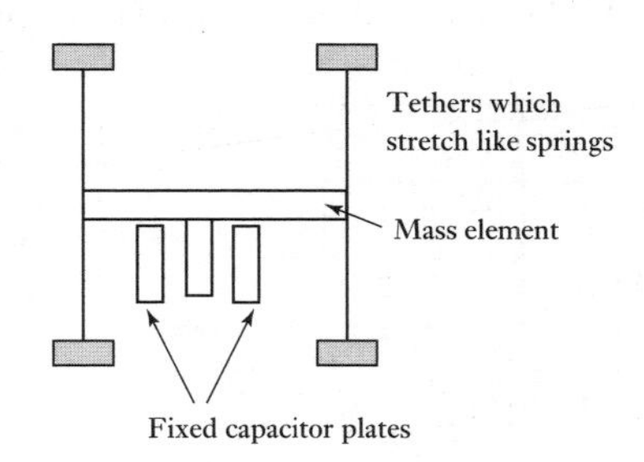

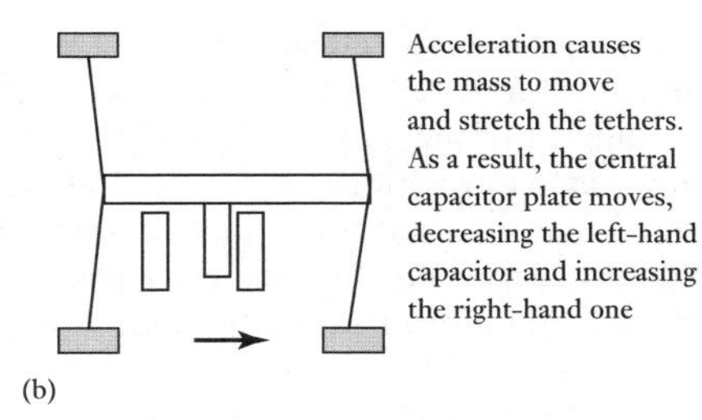

Figure 2.9 (a) Push–pull sensor, (b) such a sensor used as an element in the ADXL-50 MEMS accelerometer. The Analog Devices ADXL50 consists of a mass spring system as well as a system to measure displacement and the appropriate signal conditioning circuitry.

There is thus a non-linear relationship between the change in capacitance ΔC and the displacement x. This non-linearity can be overcome by using what is termed a **push–pull displacement sensor** (Figure 2.9(a)). Figure 2.9(b) shows how this can be realised in practice. This has three plates with the upper pair forming one capacitor and the lower pair another capacitor. The displacement moves the central plate between the two other plates. The result of, for example, the central plate moving downwards is to increase the plate separation of the upper capacitor and decrease the separation of the lower capacitor. We thus have

$$C_1 = \frac{\varepsilon_0 \varepsilon_r A}{d + x}$$

$$C_2 = \frac{\varepsilon_0 \varepsilon_r A}{d - x}$$

When C_1 is in one arm of an a.c. bridge and C_2 in the other, then the resulting out-of-balance voltage is proportional to x. Such a sensor is typically used for monitoring displacements from a few millimetres to hundreds of millimetres. Non-linearity and hysteresis are about $\pm 0.01\%$ of full range.

One form of capacitive proximity sensor consists of a single capacitor plate probe with the other plate being formed by the object, which has to be metallic and earthed (Figure 2.10). As the object approaches so the 'plate separation' of the capacitor changes, becoming significant and detectable when the object is close to the probe.

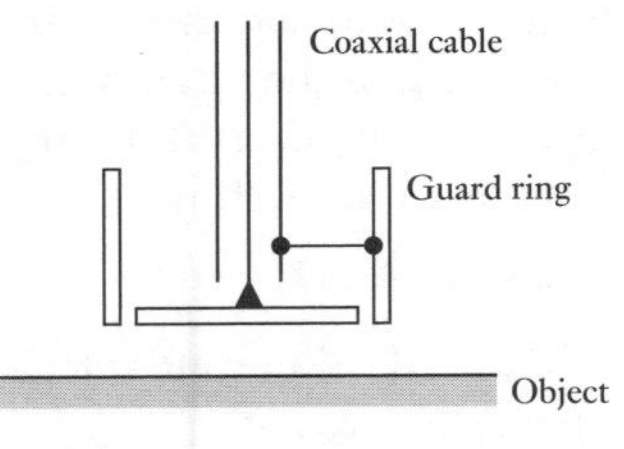

Figure 2.10 Capacitive proximity sensor.

2.3.4 Differential transformers

The linear variable differential transformer, generally referred to by the abbreviation LVDT, consists of three coils symmetrically spaced along an insulated tube (Figure 2.11). The central coil is the primary coil and the other two are identical secondary coils which are connected in series in such a way that their outputs oppose each other. A magnetic core is moved through the central tube as a result of the displacement being monitored.

Figure 2.11 LVDT.

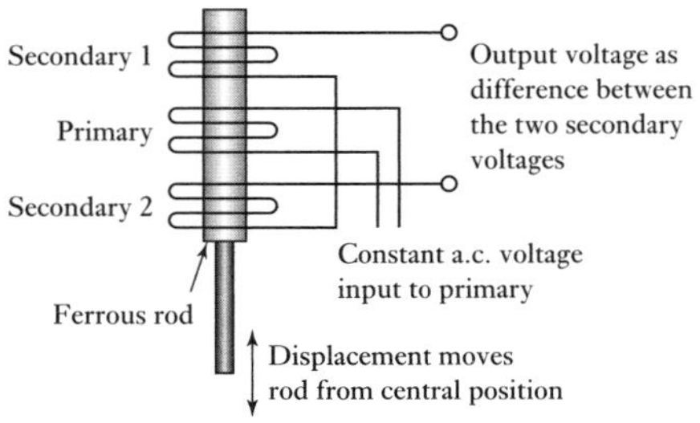

Secondary 1
Primary
Secondary 2
Ferrous rod
Output voltage as difference between the two secondary voltages
Constant a.c. voltage input to primary
Displacement moves rod from central position

When there is an alternating voltage input to the primary coil, alternating e.m.f.s are induced in the secondary coils. With the magnetic core central, the amount of magnetic material in each of the secondary coils is the same. Thus the e.m.f.s induced in each coil are the same. Since they are so connected that their outputs oppose each other, the net result is zero output.

However, when the core is displaced from the central position there is a greater amount of magnetic core in one coil than the other, e.g. more in secondary coil 2 than coil 1. The result is that a greater e.m.f. is induced in one coil than the other. There is then a net output from the two coils. Since a greater displacement means even more core in one coil than the other, the output, the difference between the two e.m.f.s increases the greater the displacement being monitored (Figure 2.12).

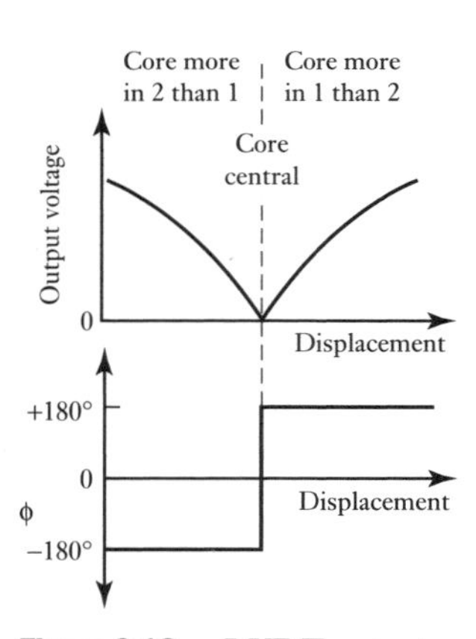

Figure 2.12 LVDT output.

The e.m.f. induced in a secondary coil by a changing current i in the primary coil is given by

$$e = M\frac{di}{dt}$$

where M is the mutual inductance, its value depending on the number of turns on the coils and the ferromagnetic core. Thus, for a sinusoidal input current of $i = I \sin \omega t$ to the primary coil, the e.m.f.s induced in the two secondary coils 1 and 2 can be represented by

$$v_1 = k_1 \sin(\omega t - \phi) \text{ and } v_2 = k_2 \sin(\omega t - \phi)$$

where the values of k_1, k_2 and ϕ depend on the degree of coupling between the primary and secondary coils for a particular core position. ϕ is the phase difference between the primary alternating voltage and the secondary alternating voltages. Because the two outputs are in series, their difference is the output

$$\text{output voltage} = v_1 - v_2 = (k_1 - k_2)\sin(\omega t - \phi)$$

When the core is equally in both coils, k_1 equals k_2 and so the output voltage is zero. When the core is more in 1 than in 2 we have $k_1 > k_2$ and

$$\text{output voltage} = (k_1 - k_2)\sin(\omega t - \phi)$$

When the core is more in 2 than in 1 we have $k_1 < k_2$. A consequence of k_1 being less than k_2 is that there is a phase change of 180° in the output when the core moves from more in 1 to more in 2. Thus

$$\text{output voltage} = -(k_1 - k_2)\sin(\omega t - \phi)$$
$$= (k_2 - k_1)\sin[\omega t + (\pi - \phi)]$$

Figure 2.12 shows how the size and phase of the output change with the displacement of the core.

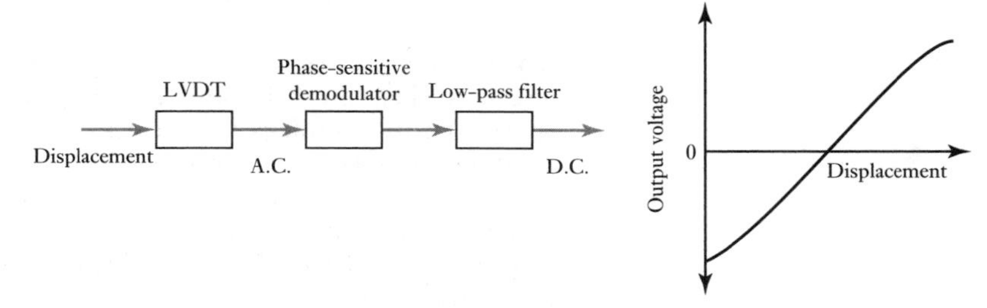

Figure 2.13 LVDT d.c. output.

With this form of output, the same amplitude output voltage is produced for two different displacements. To give an output voltage which is unique to each value of displacement we need to distinguish between where the amplitudes are the same but there is a phase difference of 180°. A phase-sensitive demodulator, with a low-pass filter, is used to convert the output into a d.c. voltage which gives a unique value for each displacement (Figure 2.13). Such circuits are available as integrated circuits.

Typically, LVDTs have operating ranges from about ±2 to ±400 mm with non-linearity errors of about ±0.25%. LVDTs are very widely used as primary transducers for monitoring displacements. The free end of the core may be spring loaded for contact with the surface being monitored, or threaded for mechanical connection. They are also used as secondary transducers in the measurement of force, weight and pressure; these variables are transformed into displacements which can then be monitored by LVDTs.

A rotary variable differential transformer (RVDT) can be used for the measurement of rotation (Figure 2.14); it operates on the same principle as the LVDT. The core is a cardioid-shaped piece of magnetic material and rotation causes more of it to pass into one secondary coil than the other. The range of operation is typically ±40° with a linearity error of about ±0.5% of the range.

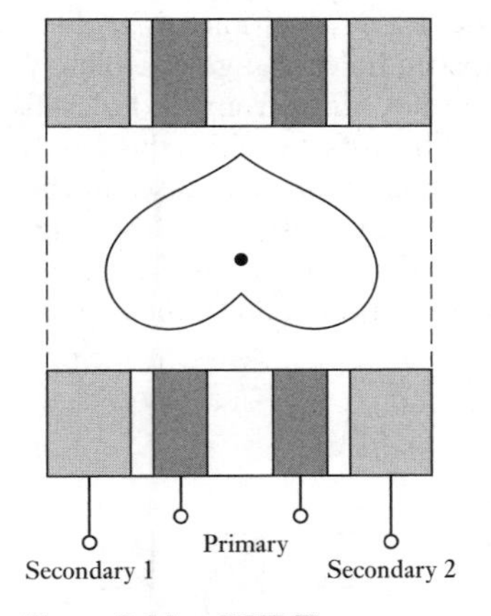

Figure 2.14 RVDT.

2.3.5 Eddy current proximity sensors

If a coil is supplied with an alternating current, an alternating magnetic field is produced. If there is a metal object in close proximity to this alternating magnetic field, then eddy currents are induced in it. The eddy currents themselves produce a magnetic field. This distorts the magnetic field responsible for their production. As a result, the impedance of the coil changes and so the amplitude of the alternating current. At some preset level, this change can be used to trigger a switch. Figure 2.15 shows the basic form of such a sensor; it is used for the detection of non-magnetic but conductive materials. They have the advantages of being relatively inexpensive, small in size, with high reliability, and can have high sensitivity to small displacements.

Reference coil
Sensor coil
Conducting object

Figure 2.15 Eddy current sensor.

2.3.6 Inductive proximity switch

This consists of a coil wound round a core. When the end of the coil is close to a metal object its inductance changes. This change can be monitored by its effect on a resonant circuit and the change used to trigger a switch. It can only be used for the detection of metal objects and is best with ferrous metals.

2.3.7 Optical encoders

An **encoder** is a device that provides a digital output as a result of a linear or angular displacement. Position encoders can be grouped into two categories: **incremental encoders** that detect changes in rotation from some datum position and **absolute encoders** which give the actual angular position.

Figure 2.16(a) shows the basic form of an incremental encoder for the measurement of angular displacement. A beam of light passes through slots in a disc and is detected by a suitable light sensor. When the disc is rotated, a pulsed output is produced by the sensor with the number of pulses being proportional to the angle through which the disc rotates. Thus the angular position of the disc, and hence the shaft rotating it, can be determined by the number of pulses produced since some datum position. In practice three concentric tracks with three sensors are used (Figure 2.16(b)). The inner track has just one hole and is used to locate the 'home' position of the disc. The other two tracks have a series of equally spaced holes that go completely round the disc but with the holes in the middle track offset from the holes in the outer track by one-half the width of a hole. This offset enables the direction of rotation to be determined. In a clockwise direction the pulses in the outer track lead those in the inner; in the anti-clockwise direction they lag. The resolution is determined by the number of slots on the disc. With 60 slots in 1 revolution then, since 1 revolution is a rotation of 360°, the resolution is $360/60 = 6°$.

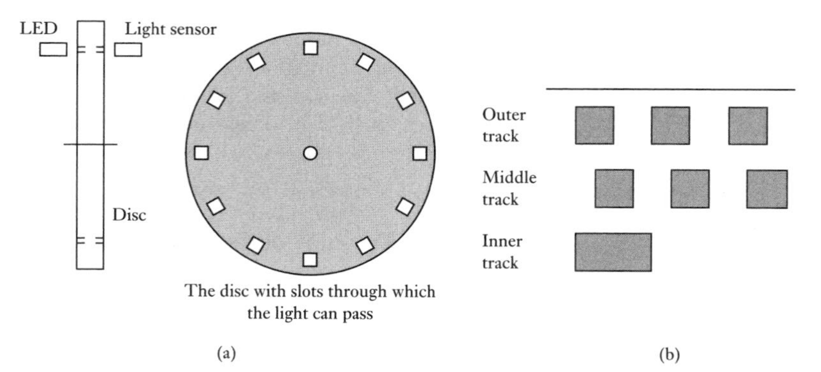

Figure 2.16 Incremental encoder: (a) the basic principle, (b) concentric tracks.

Figure 2.17 shows the basic form of an absolute encoder for the measurement of angular displacement. This gives an output in the form of a binary number of several digits, each such number representing a particular angular position. The rotating disc has three concentric circles of slots and three sensors to detect the light pulses. The slots are arranged in such a way that the sequential output from the sensors is a number in the binary code. Typical encoders tend to have up to 10 or 12 tracks. The number of bits in the binary

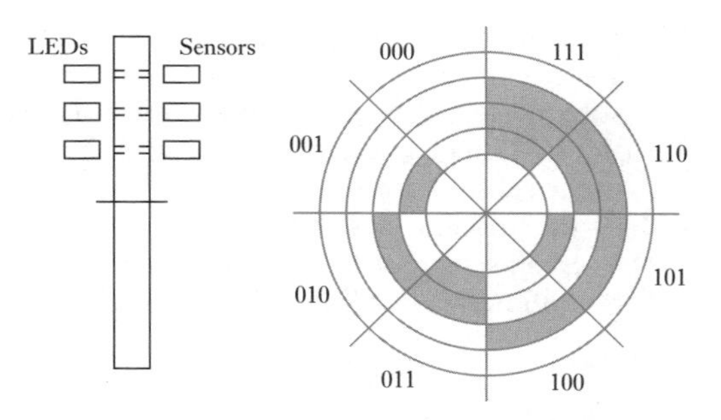

Figure 2.17 A 3-bit absolute encoder.

number will be equal to the number of tracks. Thus with 10 tracks there will be 10 bits and so the number of positions that can be detected is 2^{10}, i.e. 1024, a resolution of $360/1024 = 0.35°$.

The normal form of binary code is generally not used because changing from one binary number to the next can result in more than one bit changing, e.g. in changing from 001 to 110 we have two bits changing, and if, through some misalignment, one of the bits changes fractionally before the others then an intermediate binary number is momentarily indicated and so can lead to false counting. To overcome this the **Gray code** is generally used (see Appendix B). With this code only one bit changes in moving from one number to the next. Figure 2.18 shows the track arrangements with normal binary code and the Gray code.

	Normal binary	Gray code
0	0000	0000
1	0001	0001
2	0010	0011
3	0011	0010
4	0100	0110
5	0101	0111
6	0110	0101
7	0111	0100
8	1000	1100
9	1001	1101
10	1010	1111

Figure 2.18 Binary and Gray codes.

Optical encoders, e.g. HEDS-5000 from Hewlett Packard, are supplied for mounting on shafts and contain an LED light source and a code wheel. Interface integrated circuits are also available to decode the encoder and convert from Gray code to give a binary output suitable for a microprocessor. For an absolute encoder with seven tracks on its code disc, each track will give one of the bits in the binary number and thus we have 2^7 positions specified, i.e. 128. With eight tracks we have 2^8 positions, i.e. 256.

2.3.8 Pneumatic sensors

Pneumatic sensors involve the use of compressed air, displacement or the proximity of an object being transformed into a change in air pressure. Figure 2.19 shows the basic form of such a sensor. Low-pressure air is allowed

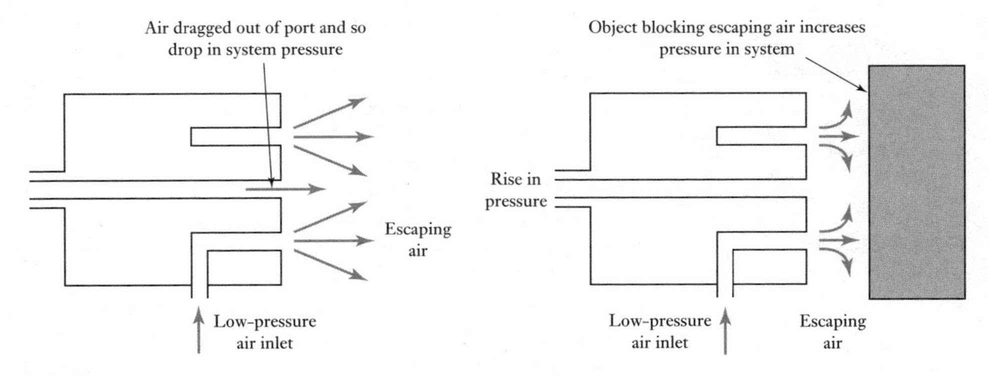

Figure 2.19 Pneumatic proximity sensor.

to escape through a port in the front of the sensor. This escaping air, in the absence of any close-by object, escapes and in doing so also reduces the pressure in the nearby sensor output port. However, if there is a close-by object, the air cannot so readily escape and the result is that the pressure increases in the sensor output port. The output pressure from the sensor thus depends on the proximity of objects.

Such sensors are used for the measurement of displacements of fractions of millimetres in ranges which typically are about 3 to 12 mm.

2.3.9 Proximity switches

There are a number of forms of switch which can be activated by the presence of an object in order to give a proximity sensor with an output which is either on or off.

The **microswitch** is a small electrical switch which requires physical contact and a small operating force to close the contacts. For example, in the case of determining the presence of an item on a conveyor belt, this might be actuated by the weight of the item on the belt depressing the belt and hence a spring-loaded platform under it, with the movement of this platform then closing the switch. Figure 2.20 shows examples of ways such switches can be actuated.

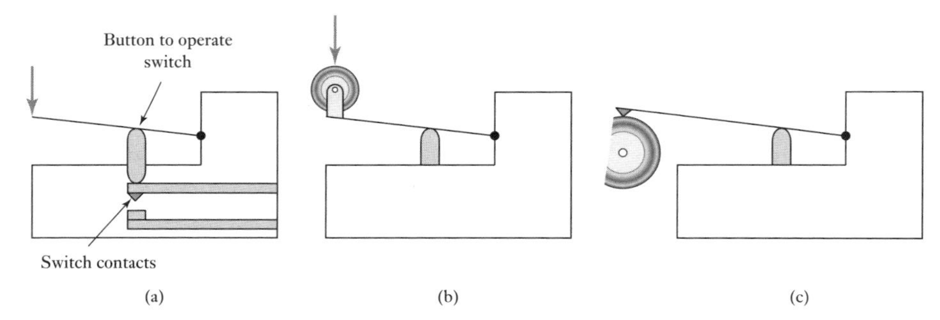

Figure 2.20 (a) Lever-operated, (b) roller-operated, (c) cam-operated switches.

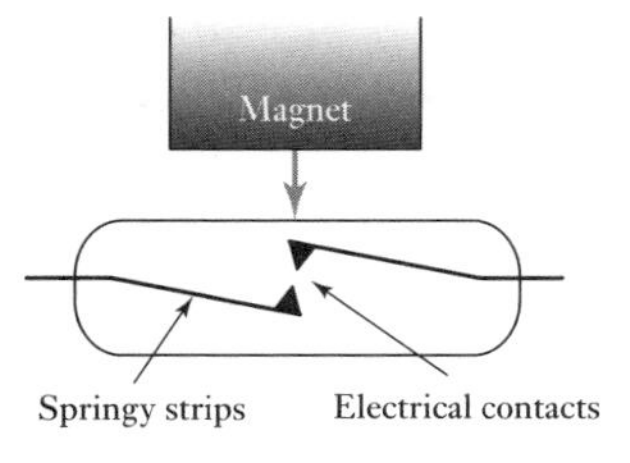

Figure 2.21 Reed switch.

Figure 2.21 shows the basic form of a **reed switch.** It consists of two magnetic switch contacts sealed in a glass tube. When a magnet is brought close to the switch, the magnetic reeds are attracted to each other and close the switch contacts. It is a non-contact proximity switch. Such a switch is very widely used for checking the closure of doors. It is also used with such devices as tachometers, which involve the rotation of a toothed wheel past the reed switch. If one of the teeth has a magnet attached to it, then every time it passes the switch it will momentarily close the contacts and hence produce a current/voltage pulse in the associated electrical circuit.

Photosensitive devices can be used to detect the presence of an opaque object by its breaking a beam of light, or infrared radiation, falling on such a device or by detecting the light reflected back by the object (Figure 2.22).

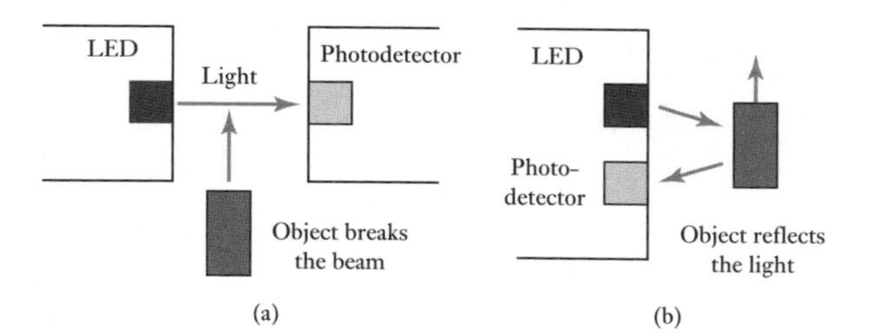

Figure 2.22 Using photoelectric sensors to detect objects by (a) the object breaking the beam, (b) the object reflecting light.

2.3.10 Hall effect sensors

When a beam of charged particles passes through a magnetic field, forces act on the particles and the beam is deflected from its straight line path. A current flowing in a conductor is like a beam of moving charges and thus can be deflected by a magnetic field. This effect was discovered by E.R. Hall in 1879 and is called the **Hall effect.** Consider electrons moving in a conductive plate with a magnetic field applied at right angles to the plane of the plate (Figure 2.23). As a consequence of the magnetic field, the moving electrons are deflected to one side of the plate and thus that side becomes negatively charged, while the opposite side becomes positively charged since the electrons are directed away from it. This charge separation produces an electric field in the material. The charge separation continues until the forces on the charged particles from the electric field just balance the forces produced by the magnetic field. The result is a transverse potential difference V given by

$$V = K_H \frac{BI}{t}$$

where B is the magnetic flux density at right angles to the plate, I the current through it, t the plate thickness and K_H a constant called the **Hall coefficient**. Thus if a constant current source is used with a particular sensor, the Hall voltage is a measure of the magnetic flux density.

Figure 2.23 Hall effect.

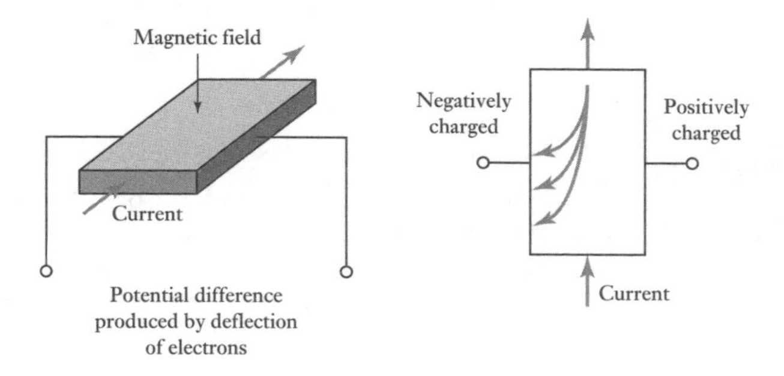

Hall effect sensors are generally supplied as an integrated circuit with the necessary signal processing circuitry. There are two basic forms of such sensor: linear, where the output varies in a reasonably linear manner with the magnetic flux density (Figure 2.24(a)); and threshold, where the output shows a sharp drop at a particular magnetic flux density (Figure 2.24(b)). The linear output Hall effect sensor 634SS2 gives an output which is fairly

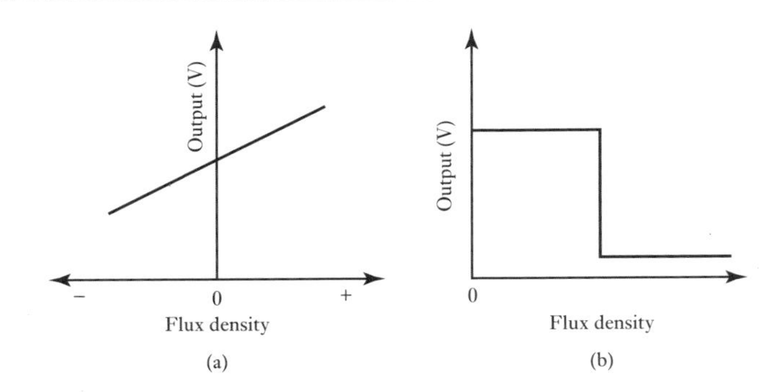

Figure 2.24 Hall effect sensors: (a) linear, (b) threshold.

linear over the range −40 to +40 mT (−400 to +400 gauss) at about 10 mV per mT (1 mV per gauss) when there is a supply voltage of 5 V. The threshold Hall effect sensor Allegro UGN3132U gives an output which switches from virtually zero to about 145 mV when the magnetic flux density is about 3 mT (30 gauss). Hall effect sensors have the advantages of being able to operate as switches which can operate up to 100 kHz repetition rate, cost less than electromechanical switches and do not suffer from the problems associated with such switches of contact bounce occurring and hence a sequence of contacts rather than a single clear contact. The Hall effect sensor is immune to environmental contaminants and can be used under severe service conditions.

Such sensors can be used as position, displacement and proximity sensors if the object being sensed is fitted with a small permanent magnet. As an illustration, such a sensor can be used to determine the level of fuel in an automobile fuel tank. A magnet is attached to a float and as the level of fuel changes so the float distance from the Hall sensor changes (Figure 2.25). The result is a Hall voltage output which is a measure of the distance of the float from the sensor and hence the level of fuel in the tank.

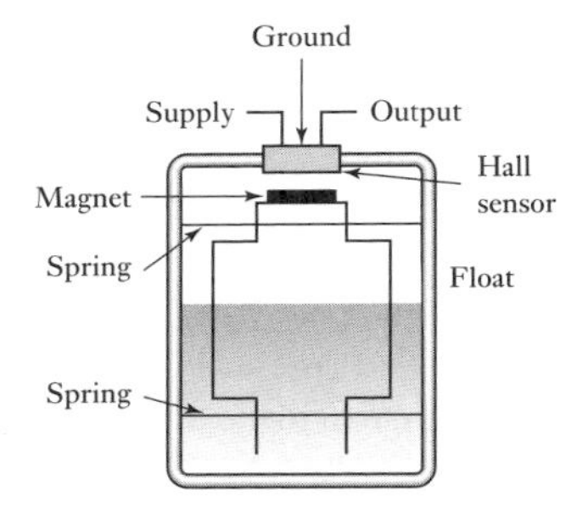

Figure 2.25 Fluid-level detector.

Another application of Hall effect sensors is in brushless d.c. motors. With such motors it is necessary to determine when the permanent magnet rotor is correctly aligned with the windings on the stator so that the current through the windings can be switched on at the right instant to maintain the rotor rotation. Hall effect sensors are used to detect when the alignment is right.

2.4 Velocity and motion

The following are examples of sensors that can be used to monitor linear and angular velocities and detect motion. The application of motion detectors includes security systems used to detect intruders and interactive toys and appliances, e.g. the cash machine screen which becomes active when you get near to it.

2.4.1 Incremental encoder

The incremental encoder described in Section 2.3.7 can be used for the measurement of angular velocity, the number of pulses produced per second being determined.

2.4.2 Tachogenerator

The tachogenerator is used to measure angular velocity. One form, the **variable reluctance tachogenerator**, consists of a toothed wheel of ferromagnetic material which is attached to the rotating shaft (Figure 2.26). A pick-up coil is wound on a permanent magnet. As the wheel rotates, so the teeth move past the coil and the air gap between the coil and the ferromagnetic material changes. We have a magnetic circuit with an air gap which periodically changes. Thus the flux linked by a pick-up coil changes. The resulting cyclic change in the flux linked produces an alternating e.m.f. in the coil.

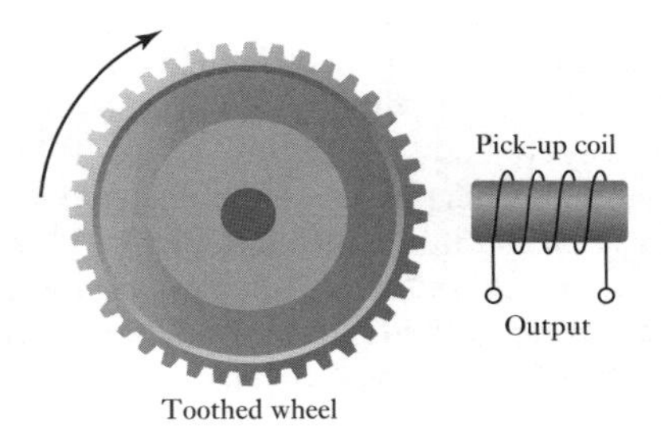

Figure 2.26 Variable reluctance tachogenerator.

If the wheel contains n teeth and rotates with an angular velocity ω, then the flux change with time for the coil can be considered to be of the form

$$\Phi = \Phi_0 + \Phi_a \cos n\omega t$$

where Φ_0 is the mean value of the flux and Φ_a the amplitude of the flux variation. The induced e.m.f. e in the N turns of the pick-up coil is $-N\,d\Phi/dt$ and thus

$$e = N\Phi_a n\omega \sin \omega t$$

and so we can write

$$e = E_{max} \sin \omega t$$

where the maximum value of the induced e.m.f. E_{max} is $N\Phi_a n\omega$ and so is a measure of the angular velocity.

Instead of using the maximum value of the e.m.f. as a measure of the angular velocity, a pulse-shaping signal conditioner can be used to transform the output into a sequence of pulses which can be counted by a counter, the number counted in a particular time interval being a measure of the angular velocity.

Another form of tachogenerator is essentially an **a.c. generator**. It consists of a coil, termed the rotor, which rotates with the rotating shaft. This coil rotates in the magnetic field produced by a stationary permanent magnet or electromagnet (Figure 2.27) and so an alternating e.m.f. is induced in it. The amplitude or frequency of this alternating e.m.f. can be used as a measure of the angular velocity of the rotor. The output may be rectified to give a d.c. voltage with a size which is proportional to the angular velocity. Non-linearity for such sensors is typically of the order of $\pm 0.15\%$ of the full range and the sensors are typically used for rotations up to about 10 000 rev/min.

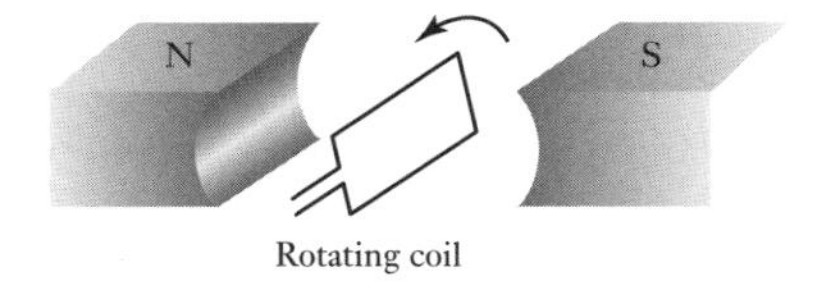

Figure 2.27 A.C. generator form of tachogenerator.

2.4.3 Pyroelectric sensors

Pyroelectric materials, e.g. lithium tantalate, are crystalline materials which generate charge in response to heat flow. When such a material is heated to a temperature just below the Curie temperature, this being about 610°C for lithium tantalate, in an electric field and the material cooled while remaining in the field, electric dipoles within the material line up and it becomes polarised (Figure 2.28, (a) leading to (b)). When the field is then removed, the material retains its polarisation; the effect is rather like magnetising a piece of iron by exposing it to a magnetic field. When the pyroelectric material is exposed to infrared radiation, its temperature rises and this reduces the amount of polarisation in the material, the dipoles being shaken up more and losing their alignment (Figure 2.28(c)).

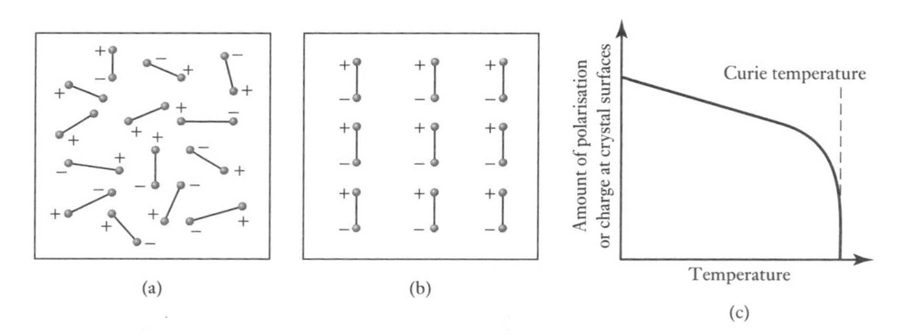

Figure 2.28 (a), (b) Polarising a pyroelectric material, (c) the effect of temperature on the amount of polarisation.

A pyroelectric sensor consists of a polarised pyroelectric crystal with thin metal film electrodes on opposite faces. Because the crystal is polarised with charged surfaces, ions are drawn from the surrounding air and electrons from any measurement circuit connected to the sensor to balance the surface charge (Figure 2.29(a)). If infrared radiation is incident on the crystal and changes its temperature, the polarisation in the crystal is reduced and consequently that is a reduction in the charge at the surfaces of the crystal. There is then an excess of charge on the metal electrodes over that needed to balance the charge on the crystal surfaces (Figure 2.29(b)). This charge leaks away through the measurement circuit until the charge on the crystal once

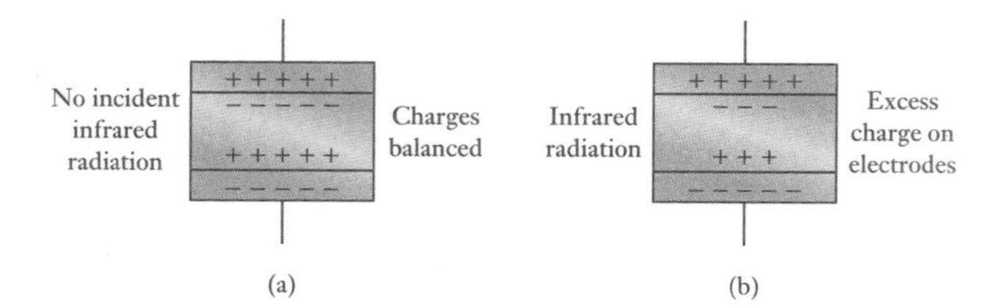

Figure 2.29 Pyroelectric sensor.

again is balanced by the charge on the electrodes. The pyroelectric sensor thus behaves as a charge generator which generates charge when there is a change in its temperature as a result of the incidence of infrared radiation. For the linear part of the graph in Figure 2.28(c), when there is a temperature change the change in charge Δq is proportional to the change in temperature Δt:

$$\Delta q = k_{\text{p}} \Delta t$$

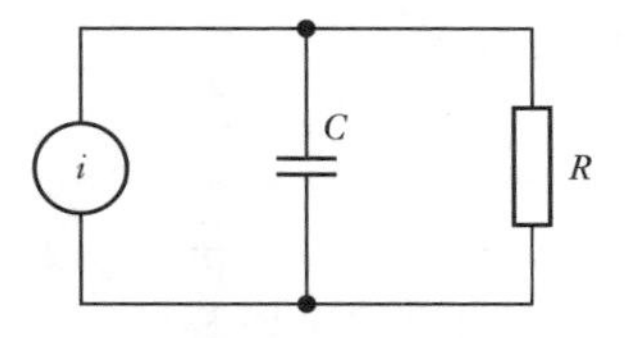

Figure 2.30 Equivalent circuit.

where k_{p} is a sensitivity constant for the crystal. Figure 2.30 shows the equivalent circuit of a pyroelectric sensor, it effectively being a capacitor charged by the excess charge with a resistance R to represent either internal leakage resistance or that combined with the input resistance of an external circuit.

To detect the motion of a human or other heat source, the sensing element has to distinguish between general background heat radiation and that given by a moving heat source. A single pyroelectric sensor would not be capable of this and so a dual element is used (Figure 2.31). One form has the sensing element with a single front electrode but two, separated, back electrodes. The result is two sensors which can be connected so that when both receive the same heat signal their outputs cancel. When a heat source moves so that the heat radiation moves from one of the sensing elements to the other, then the resulting current through the resistor alternates from being first in one direction and then reversed to the other direction. Typically a moving human gives an alternating current of the order of 10^{-12} A. The resistance R has thus to be very high to give a significant voltage. For example, 50 GΩ with such a current gives 50 mV. For this reason a transistor is included in the circuit as a voltage follower to bring the output impedance down to a few kilo-ohms.

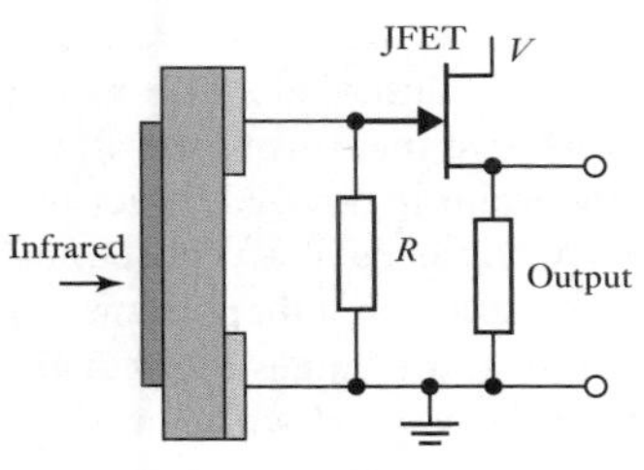

Figure 2.31 Dual pyroelectric sensor.

A focusing device is needed to direct the infrared radiation onto the sensor. While parabolic mirrors can be used, a more commonly used method is a Fresnel plastic lens. Such a lens also protects the front surface of the sensor and is the form commonly used for sensors to trigger intruder alarms or switch on a light when someone approaches.

2.5 Force

A spring balance is an example of a force sensor in which a force, a weight, is applied to the scale pan and causes a displacement, i.e. the spring stretches. The displacement is then a measure of the force. Forces are commonly measured by the measurement of displacements, the following method illustrating this.

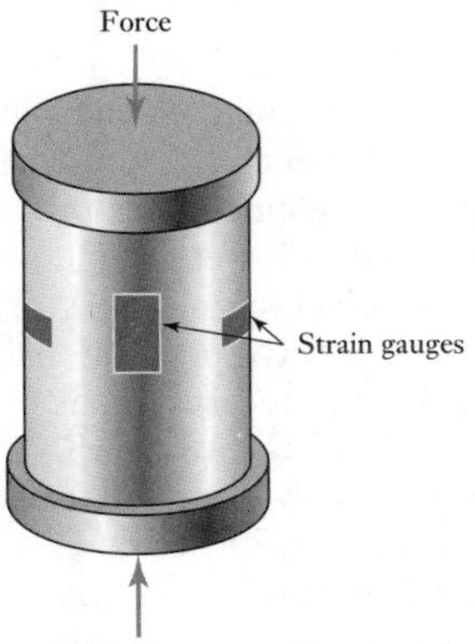

2.5.1 Strain gauge load cell

Figure 2.32 Strain gauge load cell.

A very commonly used form of force-measuring transducer is based on the use of electrical resistance strain gauges to monitor the strain produced in some member when stretched, compressed or bent by the application of the force. The arrangement is generally referred to as a **load cell.** Figure 2.32 shows an example of such a cell. This is a cylindrical tube to which strain gauges have been attached. When forces are applied to the cylinder to compress it, then the strain gauges give a resistance change which is a measure of the strain and hence the applied forces. Since temperature also produces a resistance change, the signal conditioning circuit used has to be able to eliminate the effects due to temperature (see Section 3.5.1). Typically such load cells are used for forces up to about 10 MN, the non-linearity error being about ±0.03% of full range, hysteresis error ±0.02% of full range and repeatability error ±0.02% of full range. Strain gauge load cells based on the bending of a strain-gauged metal element tend to be used for smaller forces, e.g. with ranges varying from 0 to 5 N up to 0 to 50 kN. Errors are typically a non-linearity error of about ±0.03% of full range, hysteresis error ±0.02% of full range and repeatability error ±0.02% of full range.

2.6 Fluid pressure

Many of the devices used to monitor fluid pressure in industrial processes involve the monitoring of the elastic deformation of diaphragms, capsules, bellows and tubes. The types of pressure measurements that can be required are: absolute pressure where the pressure is measured relative to zero pressure, i.e. a vacuum, differential pressure where a pressure difference is measured and gauge pressure where the pressure is measured relative to the barometric pressure.

For a diaphragm (Figure 2.33(a) and (b)), when there is a difference in pressure between the two sides then the centre of the diaphragm becomes displaced. Corrugations in the diaphragm result in a greater sensitivity. This movement can be monitored by some form of displacement sensor, e.g. a strain gauge, as illustrated in Figure 2.34. A specially designed strain gauge is often used, consisting of four strain gauges with two measuring the strain in a circumferential direction while two measure strain in a radial direction. The four strain gauges are then connected to form the arms of a Wheatstone bridge (see Chapter 3). While strain gauges can be stuck on a diaphragm, an alternative is to create a silicon diaphragm with the strain gauges as specially doped areas of the diaphragm. Such an arrangement is used with the electronic systems for cars to monitor the inlet manifold pressure.

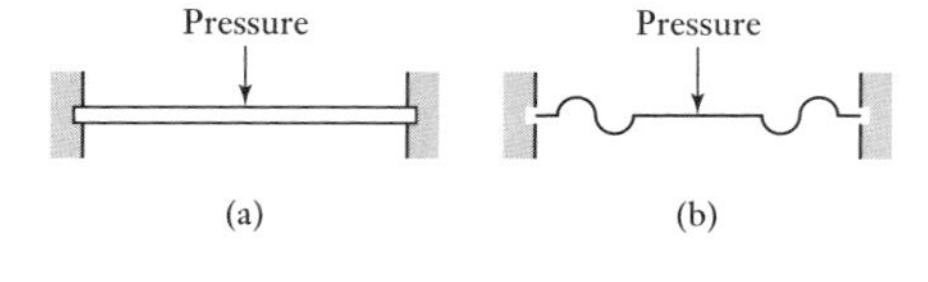

Figure 2.33 Diaphragms: (a) flat, (b) corrugated.

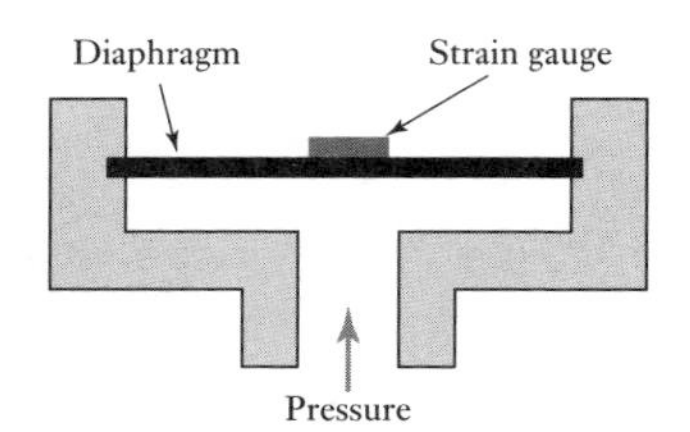

Figure 2.34 Diaphragm pressure gauge.

With the Motorola MPX pressure sensors, the strain gauge element is integrated, together with a resistive network, in a single silicon diaphragm chip. When a current is passed through the strain gauge element and pressure applied at right angles to it, a voltage is produced. This element, together with signal conditioning and temperature compensation circuitry, is packaged as the MPX sensor. The output voltage is directly proportional to the pressure. Such sensors are available for use for the measurement of absolute pressure (the MX numbering system ends with A, AP, AS or ASX), differential pressure (the MX numbering system ends with D or DP) and gauge pressure (the MX numbering system ends with GP, GVP, GS, GVS, GSV or GVSX). For example, the MPX2100 series has a pressure range of 100 kPa and with a supply voltage of 16 V d.c. gives in the absolute pressure and differential pressure forms a voltage output over the full range of 40 mV. The response time, 10 to 90%, for a step change from 0 to 100 kPa is about 1.0 ms and the output impedance is of the order of 1.4 to 3.0 kΩ. The absolute pressure sensors are used for such applications as altimeters and barometers, the differential pressure sensors for air flow measurements and the gauge pressure sensors for engine pressure and tyre pressure.

Capsules (Figure 2.35(a)) can be considered to be just two corrugated diaphragms combined and give even greater sensitivity. A stack of capsules is just a bellows (Figure 2.35(b)) and even more sensitive. Figure 2.36 shows how a bellows can be combined with an LVDT to give a pressure sensor with an electrical output. Diaphragms, capsules and bellows are made from such materials as stainless steel, phosphor bronze and nickel, with rubber and nylon also being used for some diaphragms. Pressures in the range of about 10^3 to 10^8 Pa can be monitored with such sensors.

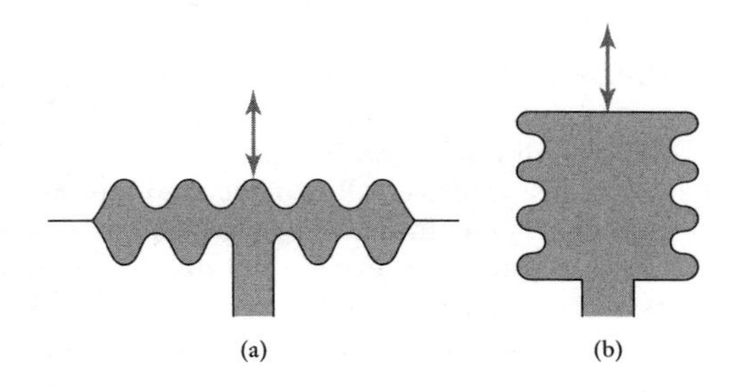

Figure 2.35 (a) Capsule, (b) bellows.

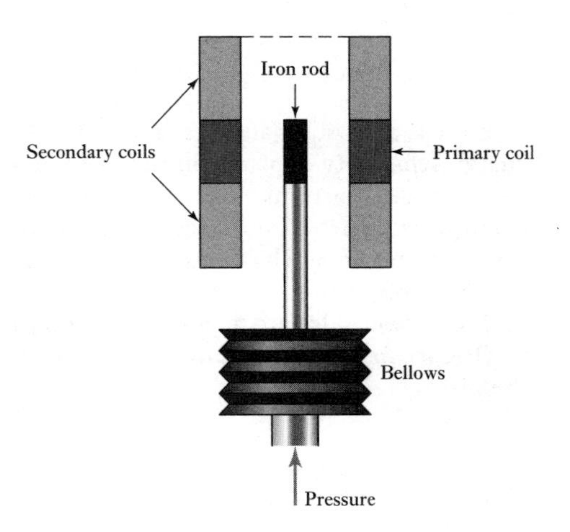

Figure 2.36 LVDT with bellows.

A different form of deformation is obtained using a tube with an elliptical cross-section (Figure 2.37(a)). Increasing the pressure in such a tube causes it to tend to a more circular cross-section. When such a tube is in the form of a C-shaped tube (Figure 2.37(b)), this being generally known as a **Bourdon tube**, the C opens up to some extent when the pressure in the tube increases. A helical form of such a tube (Figure 2.37(c)) gives a greater sensitivity. The tubes are made from such materials as stainless steel and phosphor bronze and are used for pressures in the range 10^3 to 10^8 Pa.

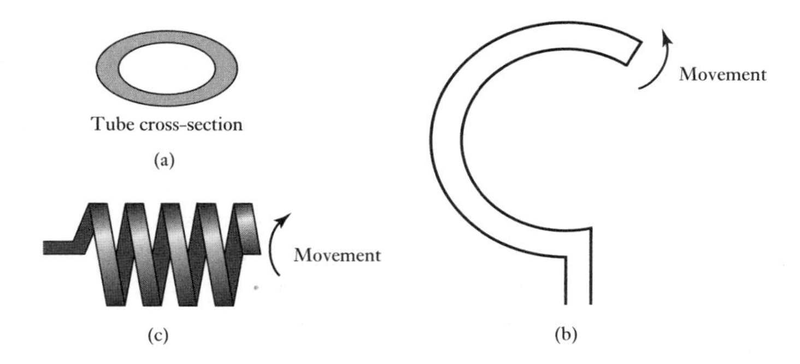

Figure 2.37 Tube pressure sensors.

2.6.1 Piezoelectric sensors

Piezoelectric materials when stretched or compressed generate electric charges with one face of the material becoming positively charged and the opposite face negatively charged (Figure 2.38(a)). As a result, a voltage is produced. Piezoelectric materials are ionic crystals which when stretched or compressed result in the charge distribution in the crystal changing so that there is a net displacement of charge with one face of the material becoming positively charged and the other negatively charged. The net charge q on a surface is proportional to the amount x by which the charges have been displaced, and since the displacement is proportional to the applied force F:

$$q = kx = SF$$

where k is a constant and S a constant termed the **charge sensitivity**. The charge sensitivity depends on the material concerned and the orientation of its crystals. Quartz has a charge sensitivity of 2.2 pC/N when the crystal is cut in one particular direction and the forces applied in a specific direction; barium titanate has a much higher charge sensitivity of the order of 130 pC/N and lead zirconate–titanate about 265 pC/N.

Metal electrodes are deposited on opposite faces of the piezoelectric crystal (Figure 2.38(b)). The capacitance C of the piezoelectric material between the plates is

$$C = \frac{\varepsilon_0 \varepsilon_r A}{t}$$

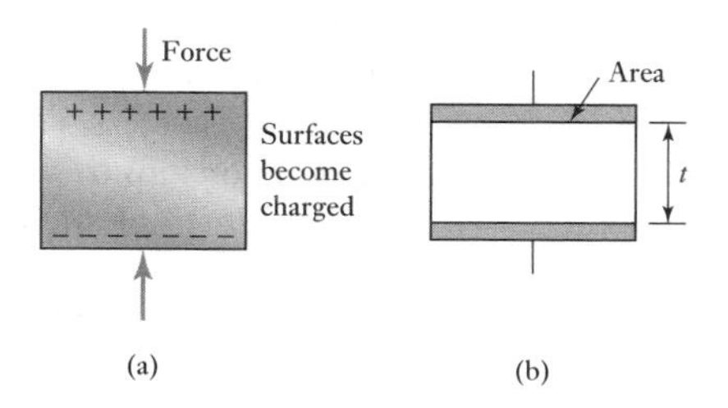

Figure 2.38 (a) Piezoelectricity, (b) piezoelectric capacitor.

where ε_r is the relative permittivity of the material, A is area and t its thickness. Since the charge $q = Cv$, where v is the potential difference produced across a capacitor, then

$$v = \frac{St}{\varepsilon_0 \varepsilon_r A} F$$

The force F is applied over an area A and so the applied pressure p is F/A and if we write $S_v = (S/\varepsilon_0\varepsilon_r)$, this being termed the **voltage sensitivity factor**, then

$$v = S_v t p$$

The voltage is proportional to the applied pressure. The voltage sensitivity for quartz is about 0.055 V/m Pa. For barium titanate it is about 0.011 V/m Pa.

Piezoelectric sensors are used for the measurement of pressure, force and acceleration. The applications have, however, to be such that the charge produced by the pressure does not have much time to leak off and thus tends to be used mainly for transient rather than steady pressures.

The equivalent electric circuit for a piezoelectric sensor is a charge generator in parallel with capacitance C_s and in parallel with the resistance R_s arising from leakage through the dielectric (Figure 2.39(a)). When the sensor is connected via a cable, of capacitance C_c, to an amplifier of input capacitance C_A and resistance R_A, we have effectively the circuit shown in Figure 2.39(b) and a total circuit capacitance of $C_s + C_c + C_A$ in parallel with a resistance of $R_A R_s/(R_A + R_s)$. When the sensor is subject to pressure it becomes charged, but because of the resistance the capacitor will discharge with time. The time taken for the discharge will depend on the time constant of the circuit.

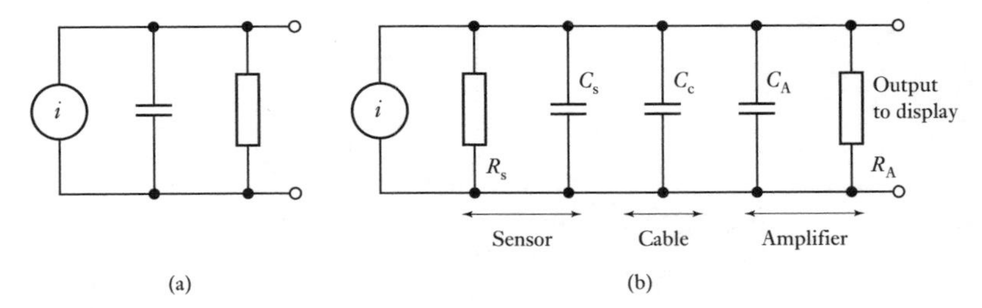

Figure 2.39 (a) Sensor equivalent circuit, (b) sensor connected to charge amplifier.

2.6.2 Tactile sensor

A tactile sensor is a particular form of pressure sensor. Such a sensor is used on the 'fingertips' of robotic 'hands' to determine when a 'hand' has come into contact with an object. They are also used for 'touch display' screens where a physical contact has to be sensed. One form of tactile sensor uses piezoelectric polyvinylidene fluoride (PVDF) film. Two layers of the film are used and are separated by a soft film which transmits vibrations (Figure 2.40). The lower PVDF film has an alternating voltage applied to it and this results in mechanical oscillations of the film (the piezoelectric effect described above in reverse). The intermediate film transmits these vibrations to the upper PVDF film. As a consequence of the piezoelectric effect, these

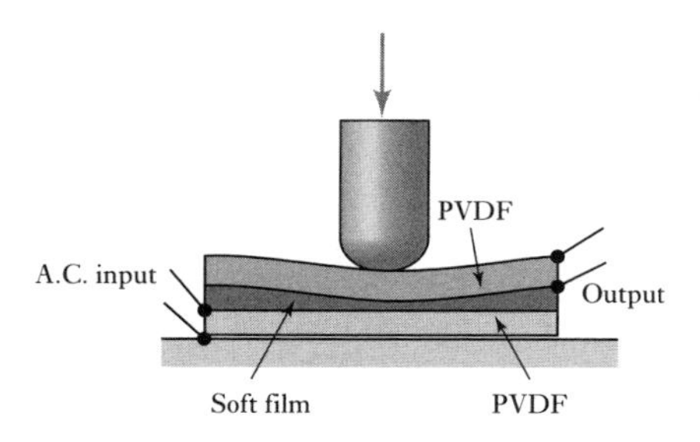

Figure 2.40 PVDF tactile sensor.

vibrations cause an alternating voltage to be produced across the upper film. When pressure is applied to the upper PVDF film its vibrations are affected and the output alternating voltage is changed.

2.7 Liquid flow

The traditional method of measuring the flow rate of liquids involves devices based on the measurement of the pressure drop occurring when the fluid flows through a constriction (Figure 2.41). For a horizontal tube, where v_1 is the fluid velocity, p_1 the pressure and A_1 the cross-sectional area of the tube prior to the constriction, v_2 the velocity, p_2 the pressure and A_2 the cross-sectional area at the constriction, with ρ the fluid density, then Bernoulli's equation gives

$$\frac{v_1^2}{2g} + \frac{p_1}{\rho g} = \frac{v_2^2}{2g} + \frac{p_2}{\rho g}$$

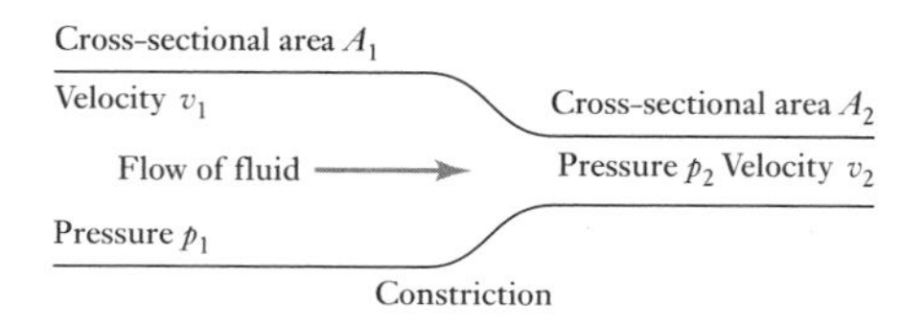

Figure 2.41 Fluid flow through a constriction.

Since the mass of liquid passing per second through the tube prior to the constriction must equal that passing through the tube at the constriction, we have $A_1 v_1 \rho = A_2 v_2 \rho$. But the quantity Q of liquid passing through the tube per second is $A_1 v_1 = A_2 v_2$. Hence

$$Q = \frac{A}{\sqrt{1 - (A_2/A_1)^2}} \sqrt{\frac{2(p_1 - p_2)}{\rho}}$$

Thus the quantity of fluid flowing through the pipe per second is proportional to $\sqrt{}$(pressure difference). Measurements of the pressure difference can thus be used to give a measure of the rate of flow. There are many devices based on this principle, and the following example of the orifice plate is probably one of the commonest.

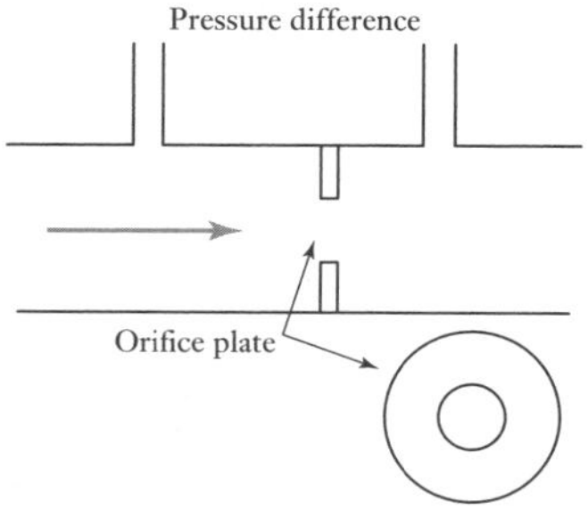

Pressure difference

Orifice plate

Figure 2.42 Orifice plate.

2.7.1 Orifice plate

The orifice plate (Figure 2.42) is simply a disc, with a central hole, which is placed in the tube through which the fluid is flowing. The pressure difference

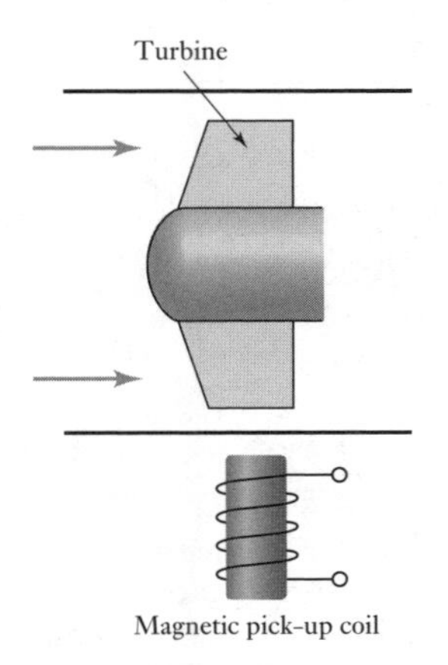

Figure 2.43 Turbine flowmeter.

is measured between a point equal to the diameter of the tube upstream and a point equal to half the diameter downstream. The orifice plate is simple, cheap, with no moving parts, and is widely used. It does not, however, work well with slurries. The accuracy is typically about ±1.5% of full range, it is non-linear, and it does produce quite an appreciable pressure loss in the system to which it is connected.

2.7.2 Turbine meter

The turbine flowmeter (Figure 2.43) consists of a multi-bladed rotor that is supported centrally in the pipe along which the flow occurs. The fluid flow results in rotation of the rotor, the angular velocity being approximately proportional to the flow rate. The rate of revolution of the rotor can be determined using a magnetic pick-up. The pulses are counted and so the number of revolutions of the rotor can be determined. The meter is expensive with an accuracy of typically about ±0.3%.

2.8 Liquid level

The level of liquid in a vessel can be measured directly by monitoring the position of the liquid surface or indirectly by measuring some variable related to the height. Direct methods can involve floats; indirect methods include the monitoring of the weight of the vessel by, perhaps, load cells. The weight of the liquid is $Ah\rho g$, where A is the cross-sectional area of the vessel, h the height of liquid, ρ its density and g the acceleration due to gravity. Thus changes in the height of liquid give weight changes. More commonly, indirect methods involve the measurement of the pressure at some point in the liquid, the pressure due to a column of liquid of height h being $h\rho g$, where ρ is the liquid density.

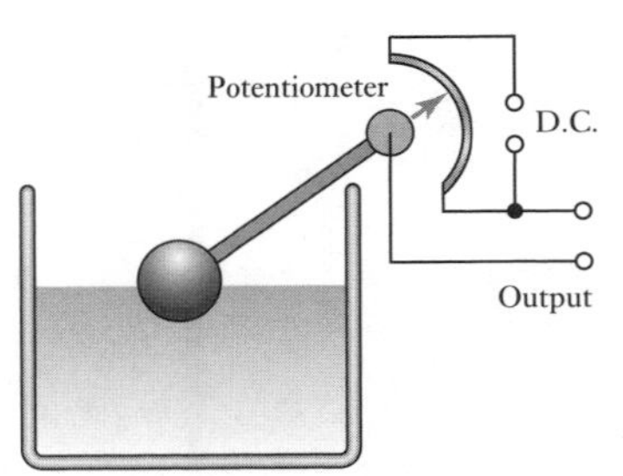

Figure 2.44 Float system.

2.8.1 Floats

A direct method of monitoring the level of liquid in a vessel is by monitoring the movement of a float. Figure 2.44 illustrates this with a simple float system. The displacement of the float causes a lever arm to rotate and so move a slider across a potentiometer. The result is an output of a voltage related to the height of liquid. Other forms of this involve the lever causing the core in an LVDT to become displaced, or stretch or compress a strain-gauged element.

2.8.2 Differential pressure

Figure 2.45 shows two forms of level measurement based on the measurement of differential pressure. In Figure 2.45(a), the differential pressure cell determines the pressure difference between the liquid at the base of the vessel and atmospheric pressure, the vessel being open to atmospheric pressure. With a closed or open vessel the system illustrated in (b) can be used. The differential pressure cell monitors the difference in pressure between the base of the vessel and the air or gas above the surface of the liquid.

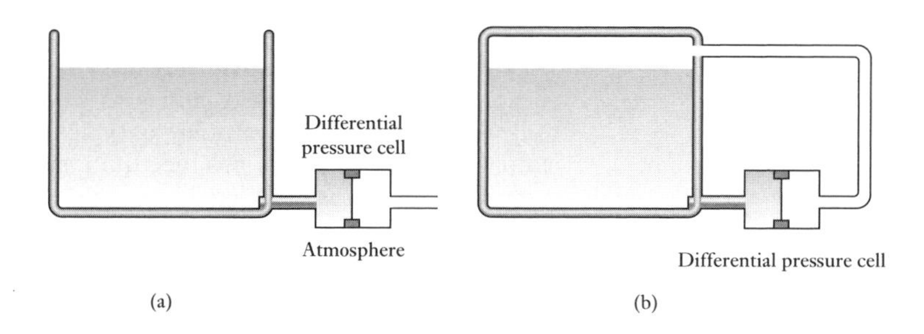

Figure 2.45 Using a differential pressure sensor.

2.9 Temperature

Changes that are commonly used to monitor temperature are the expansion or contraction of solids, liquids or gases, the change in electrical resistance of conductors and semiconductors and thermoelectric e.m.f.s. The following are some of the methods that are commonly used with temperature control systems.

2.9.1 Bimetallic strips

This device consists of two different metal strips bonded together. The metals have different coefficients of expansion and when the temperature changes the composite strip bends into a curved strip, with the higher coefficient metal on the outside of the curve. This deformation may be used as a temperature-controlled switch, as in the simple thermostat which was commonly used with domestic heating systems (Figure 2.46). The small magnet enables the sensor to exhibit hysteresis, meaning that the switch contacts close at a different temperature from that at which they open.

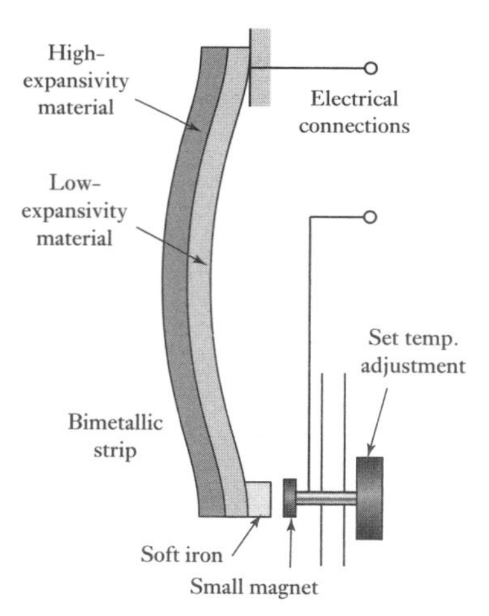

Figure 2.46 Bimetallic thermostat.

2.9.2 Resistance temperature detectors (RTDs)

The resistance of most metals increases, over a limited temperature range, in a reasonably linear way with temperature (Figure 2.47). For such a linear relationship:

$$R_t = R_0(1 + \alpha t)$$

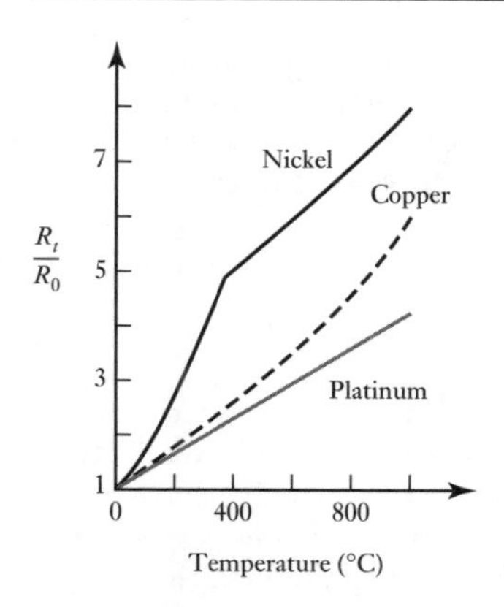

Figure 2.47 Variation of resistance with temperature for metals.

where R_t is the resistance at a temperature t(°C), R_0 the resistance at 0°C and α a constant for the metal termed the temperature coefficient of resistance. Resistance temperature detectors (RTDs) are simple resistive elements in the form of coils of wire of such metals as platinum, nickel or nickel–copper alloys; platinum is the most widely used. Thin-film platinum elements are often made by depositing the metal on a suitable substrate, wire-wound elements involving a platinum wire held by a high-temperature glass adhesive inside a ceramic tube. Such detectors are highly stable and give reproducible responses over long periods of time. They tend to have response times of the order of 0.5 to 5 s or more.

2.9.3 Thermistors

Thermistors are small pieces of material made from mixtures of metal oxides, such as those of chromium, cobalt, iron, manganese and nickel. These oxides are semiconductors. The material is formed into various forms of element, such as beads, discs and rods (Figure 2.48(a)).

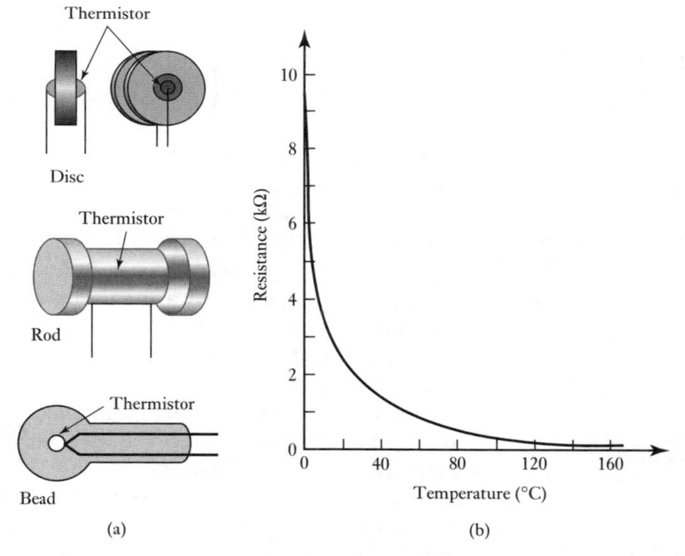

Figure 2.48 Thermistors: (a) common forms, (b) typical variation of resistance with temperature.

The resistance of conventional metal-oxide thermistors decreases in a very non-linear manner with an increase in temperature, as illustrated in Figure 2.48(b). Such thermistors have negative temperature coefficients (NTCs). Positive temperature coefficient (PTC) thermistors are, however, available. The change in resistance per degree change in temperature is considerably larger than that which occurs with metals. The resistance–temperature relationship for a thermistor can be described by an equation of the form

$$R_t = Ke^{\beta/t}$$

where R_t is the resistance at temperature t, with K and β being constants. Thermistors have many advantages when compared with other temperature sensors. They are rugged and can be very small, so enabling temperatures to be monitored at virtually a point. Because of their small size they respond very rapidly to changes in temperature. They give very large changes in resistance per degree change in temperature. Their main disadvantage is their non-linearity. Thermistors are used with the electronic systems for cars to monitor such variables as air temperature and coolant air temperature.

2.9.4 Thermodiodes and transistors

A junction semiconductor diode is widely used as a temperature sensor. When the temperature of doped semiconductors changes, the mobility of their charge carriers changes and this affects the rate at which electrons and holes can diffuse across a p–n junction. Thus when a p–n junction has a potential difference V across it, the current I through the junction is a function of the temperature, being given by

$$I = I_0\left(e^{eV/kT} - 1\right)$$

where T is the temperature on the Kelvin scale, e the charge on an electron, and k and I_0 are constants. By taking logarithms we can write the equation in terms of the voltage as

$$V = \left(\frac{kT}{e}\right)\ln\left(\frac{I}{I_0} + 1\right)$$

Thus, for a constant current, we have V proportional to the temperature on the Kelvin scale and so a measurement of the potential difference across a diode at constant current can be used as a measure of the temperature. Such a sensor is compact like a thermistor but has the great advantage of giving a response which is a linear function of temperature. Diodes for use as temperature sensors, together with the necessary signal conditioning, are supplied as integrated circuits, e.g. LM3911, and give a very small compact sensor. The output voltage from LM3911 is proportional to the temperature at the rate of 10 mV/°C.

In a similar manner to the thermodiode, for a thermotransistor the voltage across the junction between the base and the emitter depends on the temperature and can be used as a measure of temperature. A common method is to use two transistors with different collector currents and determine the difference in the base–emitter voltages between them, this difference being directly proportional to the temperature on the Kelvin scale. Such transistors can be combined with other circuit components on a single chip to give a temperature sensor with its associated signal conditioning, e.g. LM35 (Figure 2.49). This sensor can be used in the range −40 to 110°C and gives an output of 10 mV/°C.

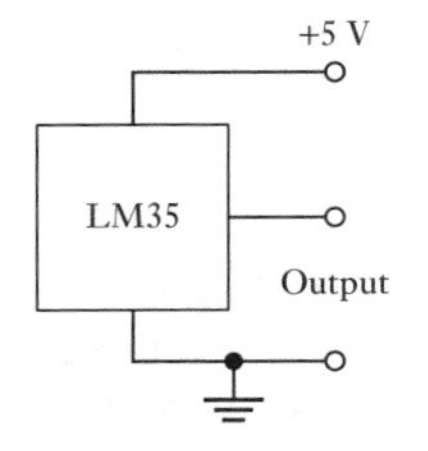

Figure 2.49 LM35.

2.9.5 Thermocouples

If two different metals are joined together, a potential difference occurs across the junction. The potential difference depends on the metals used and the temperature of the junction. A thermocouple is a complete circuit involving two such junctions (Figure 2.50(a)).

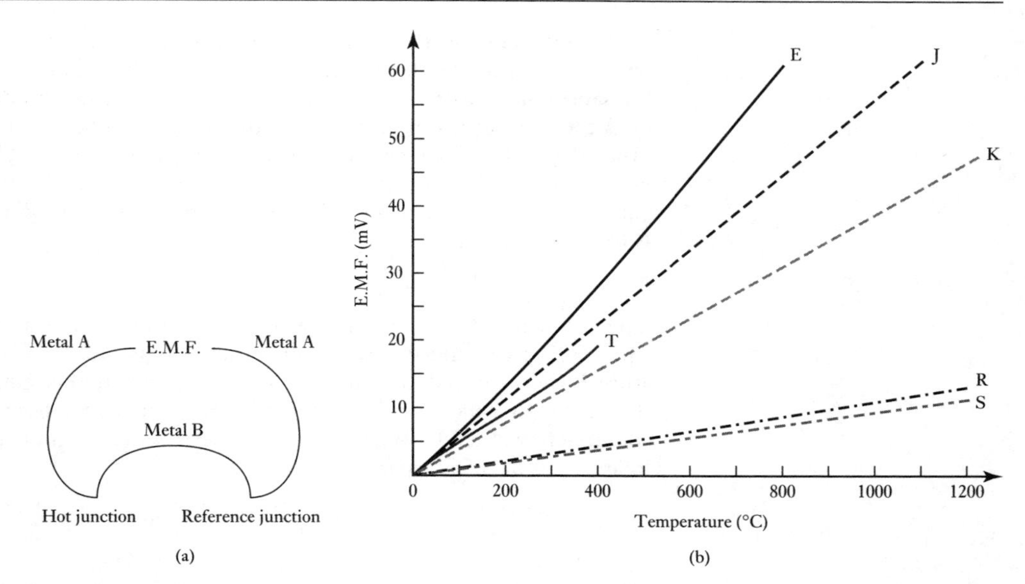

Figure 2.50 (a) A thermocouple, (b) thermoelectric e.m.f.–temperature graphs.

If both junctions are at the same temperature there is no net e.m.f. If, however, there is a difference in temperature between the two junctions, there is an e.m.f. The value of this e.m.f. E depends on the two metals concerned and the temperatures t of both junctions. Usually one junction is held at 0°C and then, to a reasonable extent, the following relationship holds:

$$E = at + bt^2$$

where a and b are constants for the metals concerned. Commonly used thermocouples are shown in Table 2.1, with the temperature ranges over which they are generally used and typical sensitivities. These commonly used thermocouples are given reference letters. For example, the iron–constantan thermocouple is called a type J thermocouple. Figure 2.50(b) shows how the e.m.f. varies with temperature for a number of commonly used pairs of metals.

Table 2.1 Thermocouples.

Ref.	Materials	Range (°C)	(μV/°C)
B	Platinum 30% rhodium/platinum 6% rhodium	0 to 1800	3
E	Chromel/constantan	−200 to 1000	63
J	Iron/constantan	−200 to 900	53
K	Chromel/alumel	−200 to 1300	41
N	Nirosil/nisil	−200 to 1300	28
R	Platinum/platinum 13% rhodium	0 to 1400	6
S	Platinum/platinum 10% rhodium	0 to 1400	6
T	Copper/constantan	−200 to 400	43

A thermocouple circuit can have other metals in the circuit and they will have no effect on the thermoelectric e.m.f. provided all their junctions are at the same temperature. This is known as the **law of intermediate metals.**

A thermocouple can be used with the reference junction at a temperature other than 0°C. The standard tables, however, assume a 0°C junction and hence a correction has to be applied before the tables can be used. The correction is applied using what is known as the **law of intermediate temperatures,** namely

$$E_{t,0} = E_{t,I} + E_{I,0}$$

The e.m.f. $E_{t,0}$ at temperature t when the cold junction is at 0°C equals the e.m.f. $E_{t,I}$ at the intermediate temperature I plus the e.m.f. $E_{I,0}$ at temperature I when the cold junction is at 0°C. To illustrate this, consider a type E thermocouple which is to be used for the measurement of temperature with a cold junction at 20°C. What will be the thermoelectric e.m.f. at 200°C? The following is data from standard tables:

Temp. (°C)	0	20	200
E.M.F. (mV)	0	1.192	13.419

Using the law of intermediate temperatures

$$E_{200,0} = E_{200,20} + E_{20,0} = 13.419 - 1.192 = 12.227 \text{ mV}$$

Note that this is not the e.m.f. given by the tables for a temperature of 180°C with a cold junction at 0°C, namely 11.949 mV.

To maintain one junction of a thermocouple at 0°C, i.e. have it immersed in a mixture of ice and water, is often not convenient. A compensation circuit can, however, be used to provide an e.m.f. which varies with the temperature of the cold junction in such a way that when it is added to the thermocouple e.m.f. it generates a combined e.m.f. which is the same as would have been generated if the cold junction had been at 0°C (Figure 2.51). The compensating e.m.f. can be provided by the voltage drop across a resistance thermometer element.

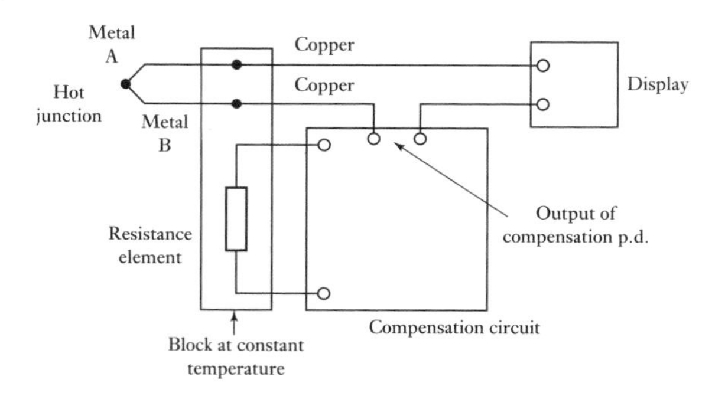

Figure 2.51 Cold junction compensation.

The base-metal thermocouples, E, J, K and T, are relatively cheap but deteriorate with age. They have accuracies which are typically about ±1 to 3%. Noble-metal thermocouples, e.g. R, are more expensive but are more stable with longer life. They have accuracies of the order of ±1% or better.

Thermocouples are generally mounted in a sheath to give them mechanical and chemical protection. The type of sheath used depends on the temperatures at which the thermocouple is to be used. In some cases the sheath is packed with a mineral which is a good conductor of heat and a good electrical insulator. The response time of an unsheathed thermocouple is very fast. With a sheath this may be increased to as much as a few seconds if a large sheath is used. In some instances a group of thermocouples are connected in series so that there are perhaps 10 or more hot junctions sensing the temperature. The e.m.f. produced by each is added together. Such an arrangement is known as a **thermopile**.

2.10 Light sensors

Photodiodes are semiconductor junction diodes (see Section 9.3.1 for a discussion of diodes) which are connected into a circuit in reverse bias, so giving a very high resistance (Figure 2.52(a)). With no incident light, the reverse current is almost negligible and is termed the dark current. When light falls on the junction, extra hole–electron pairs are produced and there is an increase in the reverse current and the diode resistance drops (Figure 2.52(b)). The reverse current is very nearly proportional to the intensity of the light. For example, the current in the absence of light with a reverse bias of 3 V might be 25 μA and when illuminated by 25 000 lumens/m^2 the current rises to 375 μA. The resistance of the device with no light is $3/(25 \times 10^{-6}) = 120\ k\Omega$ and with light is $3/(375 \times 10^{-6}) = 8\ k\Omega$. A photodiode can thus be used as a variable resistance device controlled by the light incident on it. Photodiodes have a very fast response to light.

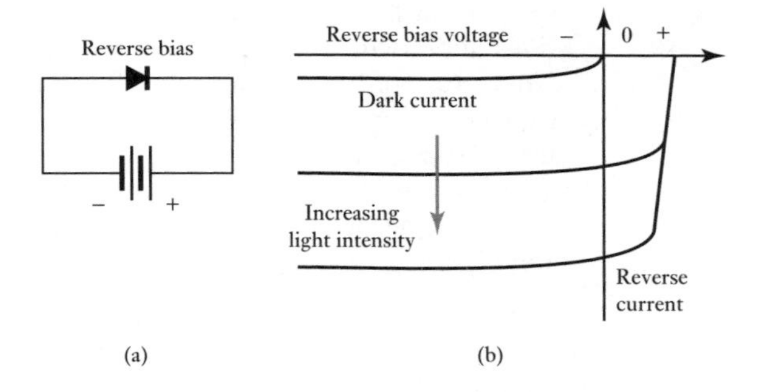

Figure 2.52 Photodiode.

The **phototransistors** (see Section 9.3.3 for a discussion of transistors) have a light-sensitive collector–base p–n junction. When there is no incident light there is a very small collector-to-emitter current. When light is incident, a base current is produced that is directly proportional to the light intensity. This leads to the production of a collector current which is then a measure of the light intensity. Phototransistors are often available as integrated packages with the phototransistor connected in a Darlington arrangement with a conventional transistor (Figure 2.53). Because this arrangement gives a higher current gain, the device gives a much greater collector current for a given light intensity.

A **photoresistor** has a resistance which depends on the intensity of the light falling on it, decreasing linearly as the intensity increases. The cadmium sulphide photoresistor is most responsive to light having wavelengths shorter than about 515 nm and the cadmium selinide photoresistor for wavelengths less than about 700 nm.

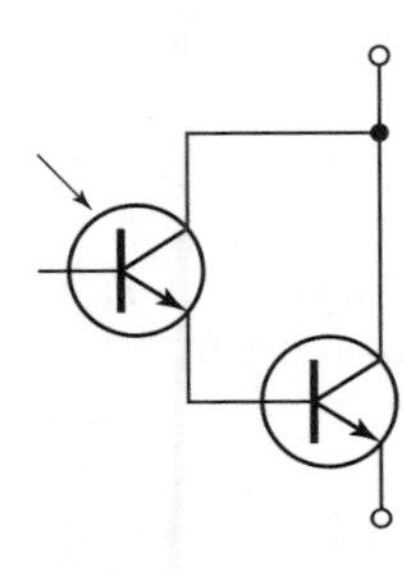

Figure 2.53 Photo Darlington.

An array of light sensors is often required in a small space in order to determine the variations of light intensity across that space. An example of this is in the digital camera to capture the image being photographed and convert it into a digital form. For this purpose a **charge-coupled device** (CCD) is often used. A CCD is a light-sensitive arrangement of many small light-sensitive cells termed pixels. These cells are basically a p-layer of silicon, separated by a depletion layer from an n-type silicon layer. When exposed to light, a cell becomes electrically charged and this charge is then converted by electronic circuitry into an 8-bit digital number. In taking a photograph the digital camera electronic circuitry discharges the light-sensitive cells, activates an electromechanical shutter to expose the cells to the image, then reads the 8-bit charge value for each cell and so captures the image. Since the pn cells are colour blind and we need colour photographs, the light passes through a colour filter matrix before striking the cells. This allows just green light to fall on some cells, blue on others and red light on others. Then, by later taking account of the output from neighbouring cells, a colour image can be created.

2.11 Selection of sensors

In selecting a sensor for a particular application there are a number of factors that need to be considered:

1. The nature of the measurement required, e.g. the variable to be measured, its nominal value, the range of values, the accuracy required, the required speed of measurement, the reliability required, the environmental conditions under which the measurement is to be made.
2. The nature of the output required from the sensor, this determining the signal conditioning requirements in order to give suitable output signals from the measurement.
3. Then possible sensors can be identified, taking into account such factors as their range, accuracy, linearity, speed of response, reliability, maintainability, life, power supply requirements, ruggedness, availability, cost.

The selection of sensors cannot be taken in isolation from a consideration of the form of output that is required from the system after signal conditioning, and thus there has to be a suitable marriage between sensor and signal conditioner.

To illustrate the above, consider the selection of a sensor for the measurement of the level of a corrosive acid in a vessel. The level can vary from 0 to 2 m in a circular vessel which has a diameter of 1 m. The empty vessel has a weight of 100 kg. The minimum variation in level to be detected is 10 cm. The acid has a density of 1050 kg/m^3. The output from the sensor is to be electrical.

Because of the corrosive nature of the acid an indirect method of determining the level seems appropriate. Thus it is possible to use a load cell, or load cells, to monitor the weight of the vessel. Such cells would give an electrical output. The weight of the liquid changes from 0 when empty to, when full, $1050 \times 2 \times \pi(1^2/4) \times 9.8 = 16.2$ kN. Adding this to the weight of the empty vessel gives a weight that varies from about 1 to 17 kN. The resolution required is for a change of level of 10 cm, i.e. a change in weight of $0.10 \times 1050 \times \pi(1^2/4) \times 9.8 = 0.8$ kN. If three load cells are used to support the tank then each will require a range of about 0 to 6 kN with a resolution of 0.27 kN. Manufacturers' catalogues can then be consulted to see if such load cells can be obtained.

2.12 Inputting data by switches

Mechanical switches consist of one or more pairs of contacts which can be mechanically closed or opened and in doing so make or break electrical circuits. Thus 0 or 1 signals can be transmitted by the act of opening or closing a switch. The term **limit switch** is used when the switches are opened or closed by the displacement of an object and used to indicate the limit of its displacement before action has to be initiated.

Mechanical switches are specified in terms of their number of poles and throws. **Poles** are the number of separate circuits that can be completed by the same switching action and **throws** are the number of individual contacts for each pole. Figure 2.54(a) shows a single pole–single throw (SPST) switch, Figure 2.54(b) a single pole–double throw (SPDT) switch and Figure 2.54(c) a double pole–double throw (DPDT) switch.

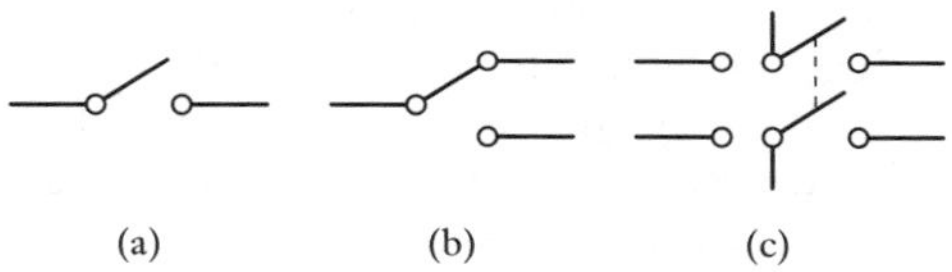

Figure 2.54 Switches: (a) SPST, (b) SPDT, (c) DPDT.

2.12.1 Debouncing

A problem that occurs with mechanical switches is **switch bounce**. When a mechanical switch is switched to close the contacts, we have one contact being moved towards the other. It hits the other and, because the contacting elements are elastic, bounces. It may bounce a number of times (Figure 2.55(a)) before finally settling to its closed state after, typically, some 20 ms. Each of the contacts during this bouncing time can register as a separate contact. Thus, to a microprocessor, it might appear that perhaps two or more separate switch actions have occurred. Similarly, when a mechanical switch is opened, bouncing can occur. To overcome this problem either hardware or software can be used.

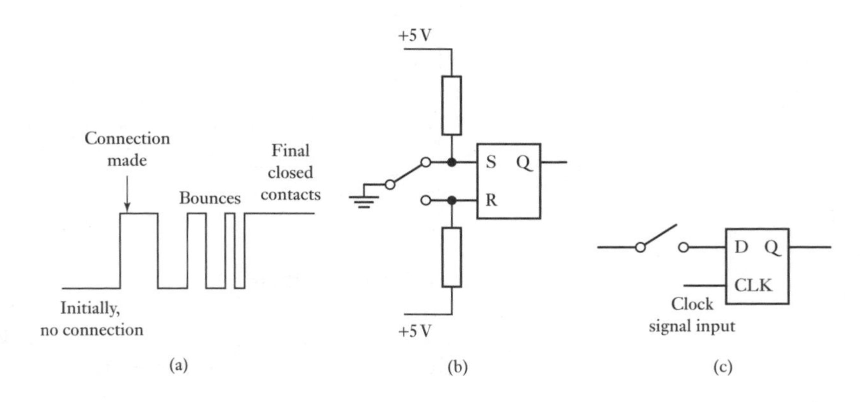

Figure 2.55 (a) Switch bounce on closing a switch, (b) debouncing using an SR flip-flop, (c) debouncing using a D flip-flop.

With software, the microprocessor is programmed to detect if the switch is closed and then wait, say, 20 ms. After checking that bouncing has ceased and the switch is in the same closed position, the next part of the program can take place. The hardware solution to the bounce problem is based on the use of a flip-flop. Figure 2.55(b) shows a circuit for debouncing an SPDT switch which is based on the use of an SR flip-flop (see Section 5.4.1). As shown, we have S at 0 and R at 1 with an output of 0. When the switch is moved to its lower position, initially S becomes 1 and R becomes 0. This gives an output of 1. Bouncing in changing S from 1 to 0 to 1 to 0, etc. gives no change in the output. Such a flip-flop can be derived from two NOR or two NAND gates. An SPDT switch can be debounced by the use of a D flip-flop (see Section 5.4.4). Figure 2.55(c) shows the circuit. The output from such a flip-flop only changes when the clock signal changes. Thus by choosing a clock period which is greater than the time for which the bounces last, say, 20 ms, the bounce signals will be ignored.

An alternative method of debouncing using hardware is to use a **Schmitt trigger**. This device has the 'hysteresis' characteristic shown in Figure 2.56(a). When the input voltage is beyond an upper switching threshold and giving a low output, then the input voltage needs to fall below the lower threshold before the output can switch to high. Conversely, when the input voltage is below the lower switching threshold and giving a high, then the input needs to rise above the upper threshold before the output can switch to low. Such a device can be used to sharpen slowly changing signals: when the signal passes the switching threshold it becomes a sharply defined edge between two well-defined logic levels. The circuit shown in Figure 2.56(b) can be used for debouncing; note the circuit symbol for a Schmitt trigger. With the switch open, the capacitor becomes charged and the voltage applied to the Schmitt trigger becomes high and so it gives a low output. When the switch is closed, the capacitor discharges very rapidly and so the first bounce discharges the capacitor; the Schmitt trigger thus switches to give a high output. Successive switch bounces do not have time to recharge the capacitor to the required threshold value and so further bounces do not switch the Schmitt trigger.

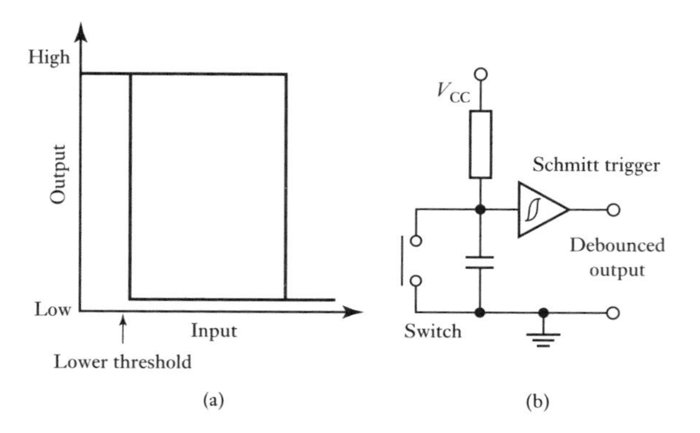

Figure 2.56 Schmitt trigger: (a) characteristic, (b) used for debouncing a switch.

2.12.2 Keypads

A keypad is an array of switches, perhaps the keyboard of a computer or the touch input membrane pad for some device such as a microwave oven. A

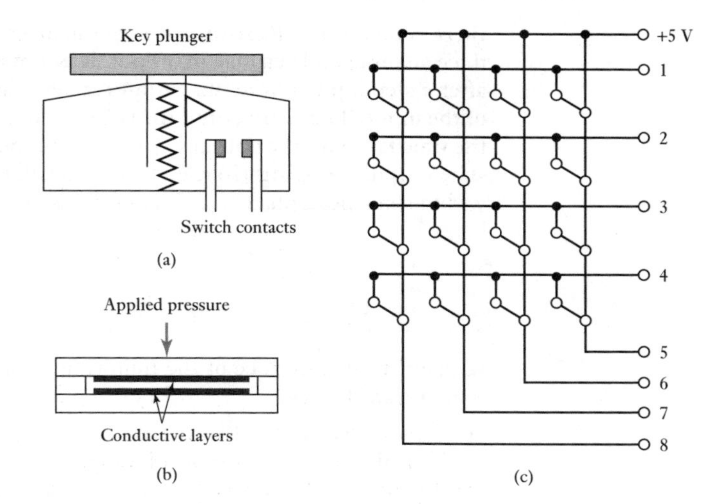

Figure 2.57 (a) Contact key, (b) membrane key, (c) 16-way keypad.

contact-type key of the form generally used with a keyboard is shown in Figure 2.57(a). Depressing the key plunger forces the contacts together with the spring returning the key to the off position when the key is released. A typical membrane switch (Figure 2.57(b)) is built up from two wafer-thin plastic films on which conductive layers have been printed. These layers are separated by a spacer layer. When the switch area of the membrane is pressed, the top contact layer closes with the bottom one to make the connection and then opens when the pressure is released.

While each switch in such arrays could be connected individually to give signals when closed, a more economical method is to connect them in an array where an individual output is not needed for each key but each key gives a unique row–column combination. Figure 2.57(c) shows the connections for a 16-way keypad.

Summary

A **sensor** is an element which produces a signal relating to the quantity being measured. A **transducer** is an element that, when subject to some physical change, experiences a related change. Thus sensors are transducers. However, a measurement system may use transducers, in addition to the sensor, in other parts of the system to convert signals in one form to another form.

The **range** of a transducer defines the limits between which the input can vary. **Span** is the maximum value of the input minus the minimum value. The **error** is the difference between the result of a measurement and its true value. **Accuracy** is the extent to which the measured value might be wrong. **Sensitivity** indicates how much output there is per unit input. **Hysteresis error** is the difference between the values obtained when reached by a continuously increasing and a continuously decreasing change. **Non-linearity error** is the error given by assuming a linear relationship. **Repeatability/reproducibility** is a measure of the ability to give the same output for repeated applications of the same input. **Stability** is the ability to give the same output for a constant input. **Dead band** is the range of input values for which there is no output. **Resolution** is the smallest change in input that will produce an observable change in output. **Response time** is the time that elapses after a step input before the output reaches a specified percentage, e.g. 95%, of the input. The **time constant** is 63.2% of the response time. **Rise time** is the time taken for the output to rise to some specified percentage of its steady-state output. **Settling time** is the time taken for the output to settle down to within some percentage, e.g. 2%, of the steady-state value.

Problems

2.1 Explain the significance of the following information given in the specification of transducers:

(a) A piezoelectric accelerometer.
Non-linearity: ±0.5% of full range.
(b) A capacitive linear displacement transducer.
Non-linearity and hysteresis: ±0.01% of full range.
(c) A resistance strain gauge force measurement transducer.
Temperature sensitivity: ±1% of full range over normal environmental temperatures.
(d) A capacitance fluid pressure transducer.
Accuracy: ±1% of displayed reading.
(e) Thermocouple.
Sensitivity: nickel chromium/nickel aluminium thermocouple: 0.039 mV/°C when the cold junction is at 0°C.
(f) Gyroscope for angular velocity measurement.
Repeatability: ±0.01% of full range.
(g) Inductive displacement transducer.
Linearity: ±1% of rated load.
(h) Load cell.
Total error due to non-linearity, hysteresis and non-repeatability: ±0.1%.

2.2 A copper–constantan thermocouple is to be used to measure temperatures between 0 and 200°C. The e.m.f. at 0°C is 0 mV, at 100°C it is 4.277 mV and at 200°C it is 9.286 mV. What will be the non-linearity error at 100°C as a percentage of the full range output if a linear relationship is assumed between e.m.f. and temperature over the full range?

2.3 A thermocouple element when taken from a liquid at 50°C and plunged into a liquid at 100°C at time $t = 0$ gave the following e.m.f. values. Determine the 95% response time.

Time (s)	0	20	40	60	80	100	120
E.M.F. (mV)	2.5	3.8	4.5	4.8	4.9	5.0	5.0

2.4 What is the non-linearity error, as a percentage of full range, produced when a 1 kΩ potentiometer has a load of 10 kΩ and is at one-third of its maximum displacement?

2.5 What will be the change in resistance of an electrical resistance strain gauge with a gauge factor of 2.1 and resistance 50 Ω if it is subject to a strain of 0.001?

2.6 You are offered a choice of an incremental shaft encoder or an absolute shaft encoder for the measurement of an angular displacement. What is the principal difference between the results that can be obtained by these methods?

2.7 A shaft encoder is to be used with a 50 mm radius tracking wheel to monitor linear displacement. If the encoder produces 256 pulses per revolution, what will be the number of pulses produced by a linear displacement of 200 mm?

2.8 A rotary variable differential transformer has a specification which includes the following information:

Ranges: $\pm 30°$, linearity error $\pm 0.5\%$ full range
$\pm 60°$, linearity error $\pm 2.0\%$ full range
Sensitivity: 1.1 (mV/V input)/degree
Impedance: primary 750 Ω, secondary 2000 Ω

What will be (a) the error in a reading of 40° due to non-linearity when the RDVT is used on the $\pm 60°$ range, and (b) the output voltage change that occurs per degree if there is an input voltage of 3 V?

2.9 What are the advantages and disadvantages of the plastic film type of potentiometer when compared with the wire-wound potentiometer?

2.10 A pressure sensor consisting of a diaphragm with strain gauges bonded to its surface has the following information in its specification:

Ranges: 0 to 1400 kPa, 0 to 35 000 kPa
Non-linearity error: $\pm 0.15\%$ of full range
Hysteresis error: $\pm 0.05\%$ of full range

What is the total error due to non-linearity and hysteresis for a reading of 1000 kPa on the 0 to 1400 kPa range?

2.11 The water level in an open vessel is to be monitored by a differential pressure cell responding to the difference in pressure between that at the base of the vessel and the atmosphere. Determine the range of differential pressures the cell will have to respond to if the water level can vary between zero height above the cell measurement point and 2 m above it.

2.12 An iron–constantan thermocouple is to be used to measure temperatures between 0 and 400°C. What will be the non-linearity error as a percentage of the full-scale reading at 100°C if a linear relationship is assumed between e.m.f. and temperature?

E.M.F. at 100°C = 5.268 mV; e.m.f. at 400°C = 21.846 mV

2.13 A platinum resistance temperature detector has a resistance of 100.00 Ω at 0°C, 138.50 Ω at 100°C and 175.83 Ω at 200°C. What will be the non-linearity error in °C at 100°C if the detector is assumed to have a linear relationship between 0 and 200°C?

2.14 A strain gauge pressure sensor has the following specification. Will it be suitable for the measurement of pressure of the order of 100 kPa to an accuracy of ± 5 kPa in an environment where the temperature is reasonably constant at about 20°C?

Ranges: 2 to 70 MPa, 70 kPa to 1 MPa
Excitation: 10 V d.c. or a.c. (r.m.s.)
Full range output: 40 mV
Non-linearity and hysteresis errors: $\pm 0.5\%$
Temperature range: -54 to $+120$°C

Thermal shift zero: 0.030% full range output/°C
Thermal shift sensitivity: 0.030% full range output/°C

2.15 A float sensor for the determination of the level of water in a vessel has a cylindrical float of mass 2.0 kg, cross-sectional area 20 cm^2 and a length of 1.5 m. It floats vertically in the water and presses upwards against a beam attached to its upward end. What will be the minimum and maximum upthrust forces exerted by the float on the beam? Suggest a means by which the deformation of the beam under the action of the upthrust force could be monitored.

2.16 Suggest a sensor that could be used as part of the control system for a furnace to monitor the rate at which the heating oil flows along a pipe. The output from the measurement system is to be an electrical signal which can be used to adjust the speed of the oil pump. The system must be capable of operating continuously and automatically, without adjustment, for long periods of time.

2.17 Suggest a sensor that could be used, as part of a control system, to determine the difference in levels between liquids in two containers. The output is to provide an electrical signal for the control system.

2.18 Suggest a sensor that could be used as part of a system to control the thickness of rolled sheet by monitoring its thickness as it emerges from rollers. The sheet metal is in continuous motion and the measurement needs to be made quickly to enable corrective action to be taken quickly. The measurement system has to supply an electrical signal.

Chapter three Signal Conditioning

Objectives

The objectives of this chapter are that, after studying it, the reader should be able to:

- **Explain the requirements for signal conditioning.**
- **Explain how operational amplifiers can be used.**
- **Explain the requirements for protection and filtering.**
- **Explain the principles of the Wheatstone bridge and, in particular, how it is used with strain gauges.**
- **Explain the principle of pulse modulation.**
- **Explain the problems that can occur with ground loops and interference and suggest possible solutions to these problems.**
- **State the requirements for maximum power transfer between electrical components.**

3.1 Signal conditioning

The output signal from the sensor of a measurement system has generally to be processed in some way to make it suitable for the next stage of the operation. The signal may be, for example, too small and have to be amplified, contain interference which has to be removed, be non-linear and require linearisation, be analogue and have to be made digital, be digital and have to be made analogue, be a resistance change and have to be made into a current change, be a voltage change and have to be made into a suitable size current change, etc. All these changes can be referred to as **signal conditioning**. For example, the output from a thermocouple is a small voltage, a few millivolts. A signal conditioning module might then be used to convert this into a suitable size current signal, provide noise rejection, linearisation and cold junction compensation (i.e. compensating for the cold junction not being at 0°C).

Chapter 4 continues with a discussion of signal conditioning involving digital signals.

3.1.1 Signal conditioning processes

The following are some of the processes that can occur in conditioning a signal:

1. *Protection* to prevent damage to the next element, e.g. a microprocessor, as a result of high current or voltage. Thus there can be series current-limiting resistors, fuses to break if the current is too high, polarity protection and voltage limitation circuits (see Section 3.3).
2. Getting the signal into the *right type of signal.* This can mean making the signal into a d.c. voltage or current. Thus, for example, the resistance change

of a strain gauge has to be converted into a voltage change. This can be done by the use of a Wheatstone bridge and using the out-of-balance voltage (see Section 3.5). It can mean making the signal digital or analogue (see Section 4.3 for analogue-to-digital and analogue-to-digital converters).

3 Getting the *level* of the signal right. The signal from a thermocouple might be just a few millivolts. If the signal is to be fed into an analogue-to-digital converter for inputting to a microprocessor then it needs to be made much larger, volts rather than millivolts. Operational amplifiers are widely used for amplification (see Section 3.2).
4 Eliminating or reducing *noise*. For example, filters might be used to eliminate mains noise from a signal (see Section 3.4).
5 Signal *manipulation*, e.g. making it a linear function of some variable. The signals from some sensors, e.g. a flowmeter, are non-linear and thus a signal conditioner might be used so that the signal fed on to the next element is linear (see Section 3.2.6).

The following sections outline some of the elements that might be used in signal conditioning.

3.2 The operational amplifier

An amplifier can be considered to be essentially a system which has an input and an output (Figure 3.1), the **voltage gain** of the amplifier being the ratio of the output and input voltages when each is measured relative to the earth. The **input impedance** of an amplifier is defined as the input voltage divided by the input current, the **output impedance** being the output voltage divided by the output current.

The basis of many signal conditioning modules is the **operational amplifier.** The operational amplifier is a high-gain d.c. amplifier, the gain typically being of the order of 100 000 or more, that is supplied as an integrated circuit on a silicon chip. It has two inputs, known as the inverting input (−) and the non-inverting input (+). The output depends on the connections made to these inputs. There are other inputs to the operational amplifier, namely a negative voltage supply, a positive voltage supply and two inputs termed offset null, these being to enable corrections to be made for the non-ideal behaviour of the amplifier (see Section 3.2.8). Figure 3.2 shows the pin connections for a 741-type operational amplifier.

An ideal model for an operational amplifier is as an amplifier with an infinite gain, infinite input impedance and zero output impedance, i.e. the output voltage is independent of the load.

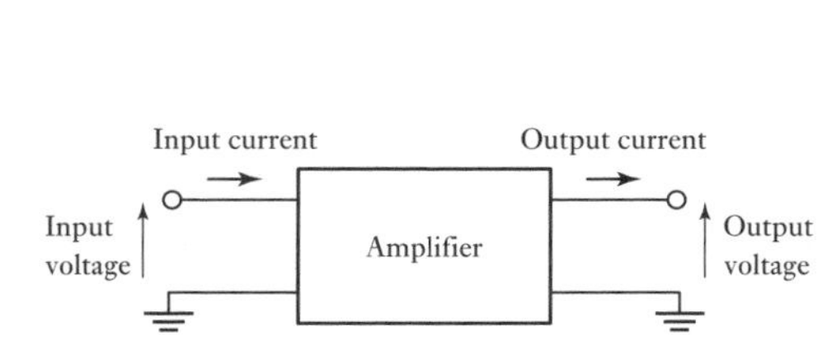

Figure 3.1 Amplifier.

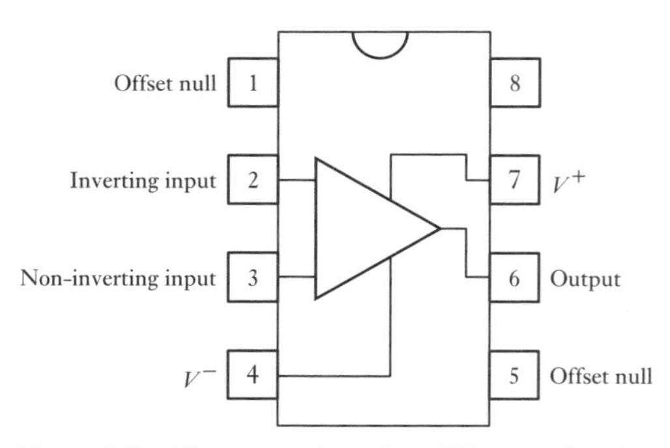

Figure 3.2 Pin connections for a 741 operational amplifier.

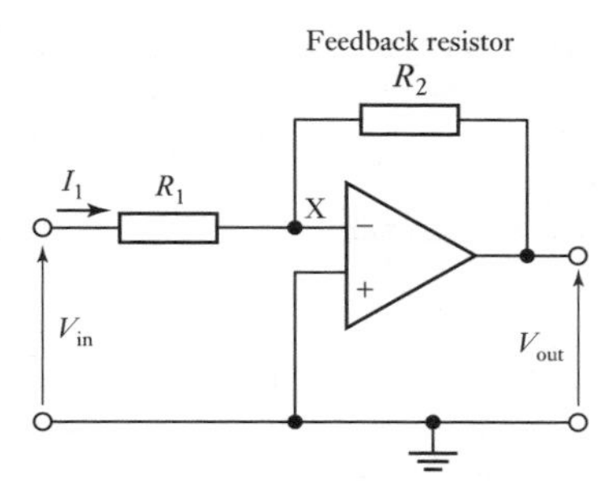

Figure 3.3 Inverting amplifier.

The following indicates the types of circuits that might be used with operational amplifiers when used as signal conditioners.

3.2.1 Inverting amplifier

Figure 3.3 shows the connections made to the amplifier when used as an **inverting amplifier.** The input is taken to the inverting input through a resistor R_1 with the non-inverting input being connected to ground. A feedback path is provided from the output, via the resistor R_2 to the inverting input. The operational amplifier has a voltage gain of about 100 000 and the change in output voltage is typically limited to about ± 10 V. The input voltage must then be between $+0.0001$ and -0.0001 V. This is virtually zero and so point X is at virtually earth potential. For this reason it is called a **virtual earth.** The potential difference across R_1 is $(V_{in} - V_X)$. Hence, for an ideal operational amplifier with an infinite gain, and hence $V_X = 0$, the input potential V_{in} can be considered to be across R_1. Thus

$$V_{in} = I_1 R_1$$

The operational amplifier has a very high impedance between its input terminals; for a 741 about 2 MΩ. Thus virtually no current flows through X into it. For an ideal operational amplifier the input impedance is taken to be infinite and so there is no current flow through X. Hence the current I_1 through R_1 must be the current through R_2. The potential difference across R_2 is $(V_X - V_{out})$ and thus, since V_X is zero for the ideal amplifier, the potential difference across R_2 is $-V_{out}$. Thus

$$-V_{out} = I_1 R_2$$

Dividing these two equations,

$$\text{voltage gain of circuit} = \frac{V_{out}}{V_{in}} = -\frac{R_2}{R_1}$$

Thus the voltage gain of the circuit is determined solely by the relative values of R_2 and R_1. The negative sign indicates that the output is inverted, i.e. 180° out of phase, with respect to the input.

To illustrate the above, consider an inverting operational amplifier circuit which has a resistance of 1 MΩ in the inverting input line and a feedback resistance of 10 MΩ. What is the voltage gain of the circuit?

$$\text{Voltage gain of circuit} = \frac{V_{out}}{V_{in}} = -\frac{R_2}{R_1} = -\frac{10}{1} = -10$$

As an example of the use of the inverting amplifier circuit, photodiodes are widely used sensors (see Section 2.10) and give small currents on exposure to light. The inverting amplifier circuit can be used with such a sensor to give a current to voltage converter, the photodiode being reverse bias connected in place of resistor R_1, and so enable the output to be used as input to a microcontroller.

3.2.2 Non-inverting amplifier

Figure 3.4(a) shows the operational amplifier connected as a non-inverting amplifier. The output can be considered to be taken from across a potential divider circuit consisting of R_1 in series with R_2. The voltage V_X is then the fraction $R_1/(R_1 + R_2)$ of the output voltage, i.e.

$$V_X = \frac{R_1}{R_1 + R_2} V_{out}$$

Since there is virtually no current through the operational amplifier between the two inputs there can be virtually no potential difference between them. Thus, with the ideal operational amplifier, we must have $V_X = V_{in}$. Hence

$$\text{voltage gain of circuit} = \frac{V_{out}}{V_{in}} = \frac{R_1 + R_2}{R_1} = 1 + \frac{R_2}{R_1}$$

A particular form of this amplifier is when the feedback loop is a short circuit, i.e. $R_2 = 0$. Then the voltage gain is 1. The input to the circuit is into a large resistance, the input resistance typically being 2 MΩ. The output resistance, i.e. the resistance between the output terminal and the ground line, is, however, much smaller, e.g. 75 Ω. Thus the resistance in the circuit that follows is a relatively small one and is less likely to load that circuit. Such an amplifier is referred to as a **voltage follower**; Figure 3.4(b) shows the basic circuit.

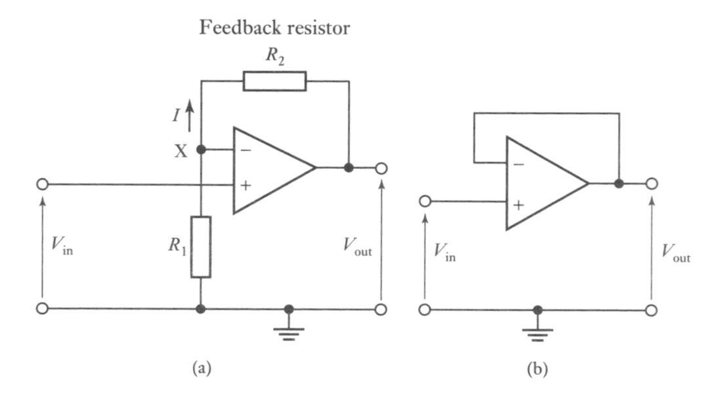

Figure 3.4 (a) Non-inverting amplifier, (b) voltage follower.

3.2.3 Summing amplifier

Figure 3.5 shows the circuit of a summing amplifier. As with the inverting amplifier (Section 3.2.1), X is a virtual earth. Thus the sum of the currents entering X must equal that leaving it. Hence

$$I = I_A + I_B + I_C$$

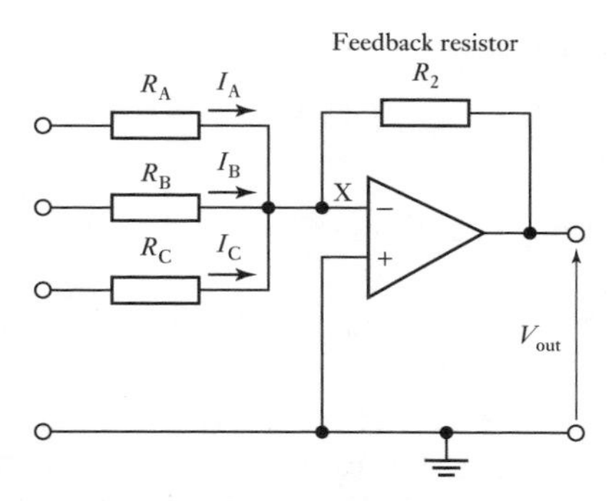

Figure 3.5 Summing amplifier.

But $I_A = V_A/R_A$, $I_B = V_B/R_B$ and $I_C = V_C/R_C$. Also we must have the same current I passing through the feedback resistor. The potential difference across R_2 is $(V_X - V_{out})$. Hence, since V_X can be assumed to be zero, it is $-V_{out}$ and so $I = -V_{out}/R_2$. Thus

$$-\frac{V_{out}}{R_2} = \frac{V_A}{R_A} + \frac{V_B}{R_B} + \frac{V_C}{R_C}$$

The output is thus the scaled sum of the inputs, i.e.

$$V_{out} = -\left(\frac{R_2}{R_A}V_A + \frac{R_2}{R_B}V_B + \frac{R_2}{R_C}V_C\right)$$

If $R_A = R_B = R_C = R_1$ then

$$V_{out} = -\frac{R_2}{R_1}(V_A + V_B + V_C)$$

To illustrate the above, consider the design of a circuit that can be used to produce an output voltage which is the average of the input voltages from three sensors. Assuming that an inverted output is acceptable, a circuit of the form shown in Figure 3.5 can be used. Each of the three inputs must be scaled to 1/3 to give an output of the average. Thus a voltage gain of the circuit of 1/3 for each of the input signals is required. Hence, if the feedback resistance is 4 kΩ the resistors in each input arm will be 12 kΩ.

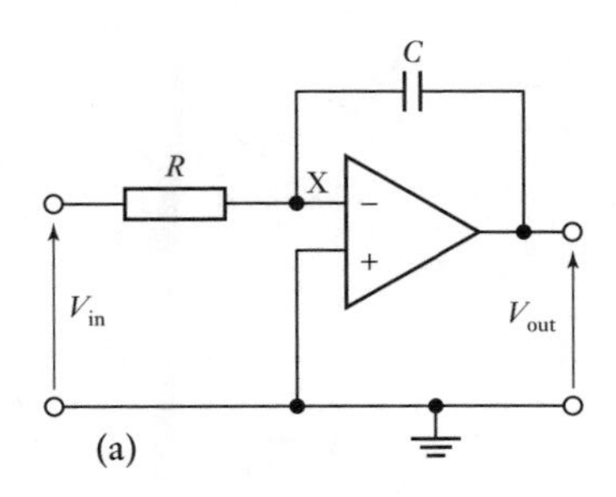

(b)

Figure 3.6 (a) Integrating amplifier, (b) differentiator amplifier.

3.2.4 Integrating and differentiating amplifiers

Consider an inverting operational amplifier circuit with the feedback being via a capacitor, as illustrated in Figure 3.6(a).

The current is the rate of movement of charge q and since for a capacitor the charge $q = Cv$, where v is the voltage across it, then the current through the capacitor $i = dq/dt = C\,dv/dt$. The potential difference across C is $(v_X - v_{out})$ and since v_X is effectively zero, being the virtual earth, it is $-v_{out}$. Thus the current through the capacitor is $-C\,dv_{out}/dt$. But this is also the current through the input resistance R. Hence

$$\frac{v_{in}}{R} = -C\frac{dv_{out}}{dt}$$

Rearranging this gives

$$\mathrm{d}v_{\text{out}} = -\left(\frac{1}{RC}\right)v_{\text{in}}\,\mathrm{d}t$$

Integrating both sides gives

$$v_{\text{out}}(t_2) - v_{\text{out}}(t_1) = -\frac{1}{RC}\int_{t_1}^{t_2} v_{\text{in}}\,\mathrm{d}t$$

$v_{\text{out}}(t_2)$ is the output voltage at time t_2 and $v_{\text{out}}(t_1)$ is the output voltage at time t_1. The output is proportional to the integral of the input voltage, i.e. the area under a graph of input voltage with time.

A differentiation circuit can be produced if the capacitor and resistor are interchanged in the circuit for the integrating amplifier. Figure 3.6(b) shows the circuit. The input current i_{in} to capacitor C is $\mathrm{d}q/\mathrm{d}t = C\,\mathrm{d}v/\mathrm{d}t$. With the ideal case of zero op-amp current, this is also the current through the feedback resistor R, i.e. $-v_{\text{out}}/R$ and so

$$\frac{v_{\text{out}}}{R} = -C\frac{\mathrm{d}v_{\text{in}}}{\mathrm{d}t}$$

$$v_{\text{out}} = -RC\frac{\mathrm{d}v_{\text{in}}}{\mathrm{d}t}$$

At high frequencies the differentiator circuit is susceptible to stability and noise problems. A solution is to add an input resistor R_{in} to limit the gain at high frequencies and so reduce the problem.

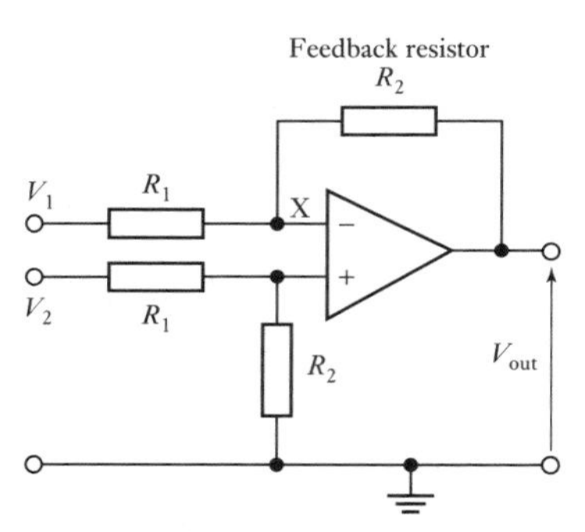

Figure 3.7 Difference amplifier.

3.2.5 Difference amplifier

A difference amplifier is one that amplifies the difference between two input voltages. Figure 3.7 shows the circuit. Since there is virtually no current through the high resistance in the operational amplifier between the two input terminals, there is no potential drop and thus both the inputs X will be at the same potential. The voltage V_2 is across resistors R_1 and R_2 in series. Thus the potential V_X at X is

$$\frac{V_X}{V_2} = \frac{R_2}{R_1 + R_2}$$

The current through the feedback resistance must be equal to that from V_1 through R_1. Hence

$$\frac{V_1 - V_X}{R_1} = \frac{V_X - V_{out}}{R_2}$$

This can be rearranged to give

$$\frac{V_{out}}{R_2} = V_X\left(\frac{1}{R_2} + \frac{1}{R_1}\right) - \frac{V_1}{R_1}$$

Hence substituting for V_X using the earlier equation,

$$V_{out} = \frac{R_2}{R_1}(V_2 - V_1)$$

The output is thus a measure of the difference between the two input voltages.

As an illustration of the use of such a circuit with a sensor, Figure 3.8 shows it used with a thermocouple. The difference in voltage between the e.m.f.s of the two junctions of the thermocouple is being amplified. The values of R_1 and R_2 can, for example, be chosen to give a circuit with an output of 10 mV for a temperature difference between the thermocouple junctions of 10°C if such a temperature difference produces an e.m.f. difference between the junctions of 530 μV. For the circuit we have

$$V_{out} = \frac{R_2}{R_1}(V_2 - V_1)$$

$$10 \times 10^{-3} = \frac{R_2}{R_1} \times 530 \times 10^{-6}$$

Hence $R_2/R_1 = 18.9$. Thus if we take for R_1 a resistance of 10 kΩ then R_2 must be 189 kΩ.

A difference amplifier might be used with a Wheatstone bridge (see Section 3.5), perhaps one with strain gauge sensors in its arms, to amplify the out-of-balance potential difference that occurs when the resistance in one or more arms changes. When the bridge is balanced, both the output terminals of the bridge are at the same potential; there is thus no output potential difference. The output terminals from the bridge might both be at, say, 5.00 V. Thus the differential amplifier has both its inputs at 5.00 V. When the bridge is no

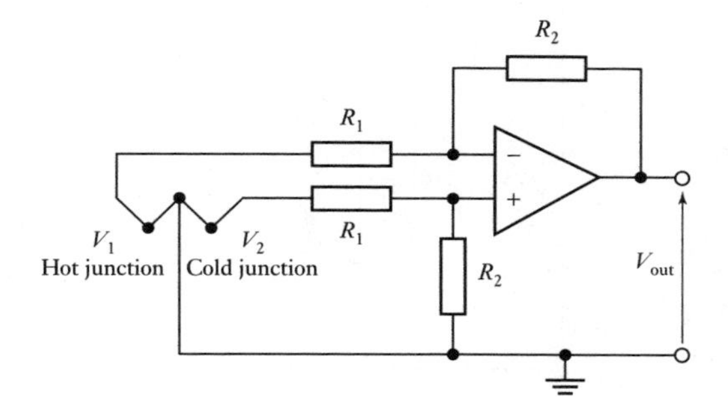

Figure 3.8 Difference amplifier with a thermocouple.

longer balanced we might have one output terminal at 5.01 V and the other at 4.99 V and so the inputs to the differential amplifier are 5.01 and 4.99 V. The amplifier amplifies this difference in the voltages of 0.02 V. The original 5.00 V signal which is common to both inputs is termed the **common mode voltage** V_{CM}. For the amplifier only to amplify the difference between the two signals assumes that the two input channels are perfectly matched and the operational amplifier has the same, high, gain for both of them. In practice this is not perfectly achieved and thus the output is not perfectly proportional to the difference between the two input voltages. Thus we write for the output

$$V_{out} = G_d \Delta V + G_{CM} V_{CM}$$

where G_d is the gain for the voltage difference ΔV, G_{CM} the gain for the common mode voltage V_{CM}. The smaller the value of G_{CM}, the smaller the effect of the common mode voltage on the output. The extent to which an operational amplifier deviates from the ideal situation is specified by the **common mode rejection ratio** (CMRR):

$$\text{CMRR} = \frac{G_d}{G_{CM}}$$

To minimise the effect of the common mode voltage on the output, a high CMRR is required. CMRRs are generally specified in decibels (dB). Thus, on the decibel scale a CMRR of, say, 10 000 would be 20 lg 10 000 = 80 dB. A typical operational amplifier might have a CMRR between about 80 and 100 dB.

A common form of **instrumentation amplifier** involves three operational amplifiers (Figure 3.9), rather than just a single difference amplifier, and is available as a single integrated circuit. Such a circuit is designed to have a high input impedance, typically about 300 MΩ, a high voltage gain and excellent CMRR, typically more than 100 dB. The first stage involves the amplifiers A_1 and A_2, one being connected as an inverting amplifier and the other as a non-inverting amplifier. Amplifier A_3 is a differential amplifier with inputs from A_1 and A_2.

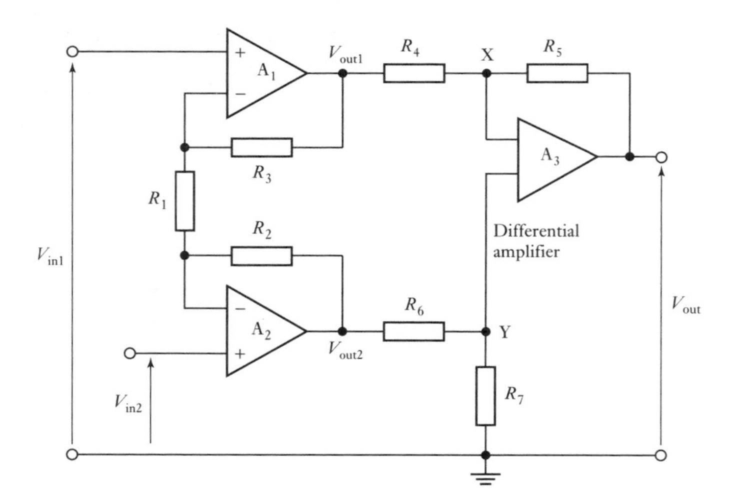

Figure 3.9 Instrumentation amplifier.

Because virtually no current passes through A_3, the current through R_4 will be the same as that through R_5. Hence

$$\frac{V_{out1} - V_X}{R_4} = \frac{V_X - V_{out}}{R_5}$$

The differential input to A_3 is virtually zero, so $V_Y = V_X$. Hence the above equation can be written as

$$V_{out} = \left(1 + \frac{R_5}{R_4}\right)V_Y - \frac{R_5}{R_4}V_{out1}$$

R_6 and R_7 form a potential divider for the voltage V_{out2} so that

$$V_Y = \frac{R_6}{R_6 + R_7}V_{out2}$$

Hence we can write

$$V_{out} = \frac{1 + \dfrac{R_5}{R_4}}{1 + \dfrac{R_7}{R_6}}V_{out2} - \frac{R_5}{R_4}V_{out1}$$

Hence by suitable choice of resistance values we obtain equal multiplying factors for the two inputs to the difference amplifier. This requires

$$1 + \frac{R_5}{R_4} = \left(1 + \frac{R_7}{R_6}\right)\frac{R_5}{R_4}$$

and hence $R_4/R_5 = R_6/R_7$.

We can apply the **principle of superposition**, i.e. we can consider the output produced by each source acting alone and then add them to obtain the overall response. Amplifier A_1 has an input of the difference signal V_{in1} on its non-inverting input and amplifies this with a gain of $1 + R_3/R_1$. It also has an input of V_{in2} on its inverting input and this is amplified to give a gain of $-R_3/R_1$. Also the common mode voltage V_{cm} on the non-inverting input is amplified by A_1. Thus the output of A_1 is

$$V_{out1} = \left(1 + \frac{R_3}{R_1}\right)V_{in1} - \left(\frac{R_3}{R_1}\right)V_{in2} + \left(1 + \frac{R_3}{R_1}\right)V_{cm}$$

Amplifier A_2 likewise gives

$$V_{out2} = \left(1 + \frac{R_2}{R_1}\right)V_{in2} - \left(\frac{R_2}{R_1}\right)V_{in1} + \left(1 + \frac{R_2}{R_1}\right)V_{cm}$$

The difference input to A_3 is $V_{out1} - V_{out2}$ and so

$$V_{out2} - V_{out1} = \left(1 + \frac{R_3}{R_1} + \frac{R_2}{R_1}\right)V_{in1} - \left(1 + \frac{R_2}{R_3} + \frac{R_3}{R_1}\right)V_{in2} + \left(\frac{R_3}{R_1} - \frac{R_2}{R_1}\right)V_{cm}$$

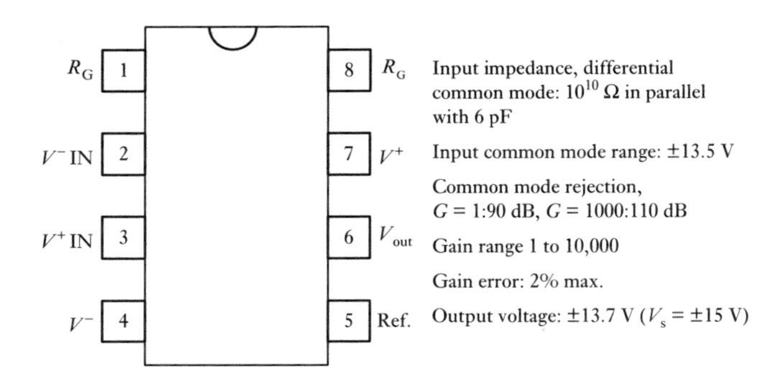

Figure 3.10 INA114.

With $R_2 = R_3$ the common mode voltage term disappears, thus

$$V_{out2} - V_{out1} = \left(1 + \frac{2R_2}{R_1}\right)(V_{in1} - V_{in2})$$

The overall gain is thus $(1 + 2R_2/R_1)$ and is generally set by varying R_1.

Figure 3.10 shows the pin connections and some specification details for a low-cost, general-purpose instrumentation amplifier (Burr-Brown INA114) using this three op-amps form of design. The gain is set by connecting an external resistor R_G between pins 1 and 8, the gain then being $1 + 50/R_G$ when R_G is in kΩ. The 50 kΩ term arises from the sum of the two internal feedback resistors.

3.2.6 Logarithmic amplifier

Some sensors have outputs which are non-linear. For example, the output from a thermocouple is not a perfectly linear function of the temperature difference between its junctions. A signal conditioner might then be used to linearise the output from such a sensor. This can be done using an operational amplifier circuit which is designed to have a non-linear relationship between its input and output so that when its input is non-linear, the output is linear. This is achieved by a suitable choice of component for the feedback loop.

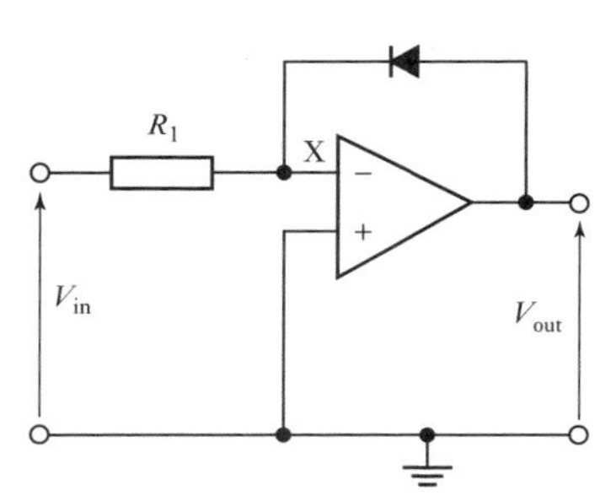

Figure 3.11 Logarithmic amplifier.

The logarithmic amplifier shown in Figure 3.11 is an example of such a signal conditioner. The feedback loop contains a diode (or a transistor with a grounded base). The diode has a non-linear characteristic. It might be represented by $V = C \ln I$, where C is a constant. Then, since the current through the feedback loop is the same as the current through the input resistance and the potential difference across the diode is $-V_{out}$, we have

$$V_{out} = -C \ln(V_{in}/R) = K \ln V_{in}$$

where K is some constant. However, if the input V_{in} is provided by a sensor with an input t, where $V_{in} = A\,e^{at}$, with A and a being constants, then

$$V_{out} = K \ln V_{in} = K \ln(A\,e^{at}) = K \ln A + Kat$$

The result is a linear relationship between V_{out} and t.

3.2.7 Comparator

A comparator indicates which of two voltages is the larger. An operational amplifier used with no feedback or other components can be used as a comparator. One of the voltages is applied to the inverting input and the other to the non-inverting input (Figure 3.12(a)). Figure 3.12(b) shows the relationship between the output voltage and the difference between the two input voltages. When the two inputs are equal there is no output. However, when the non-inverting input is greater than the inverting input by more than a small fraction of a volt then the output jumps to a steady positive saturation voltage of typically +10 V. When the inverting input is greater than the non-inverting input then the output jumps to a steady negative saturation voltage of typically −10 V. Such a circuit can be used to determine when a voltage exceeds a certain level, the output then being used to perhaps initiate some action.

Figure 3.12 Comparator.

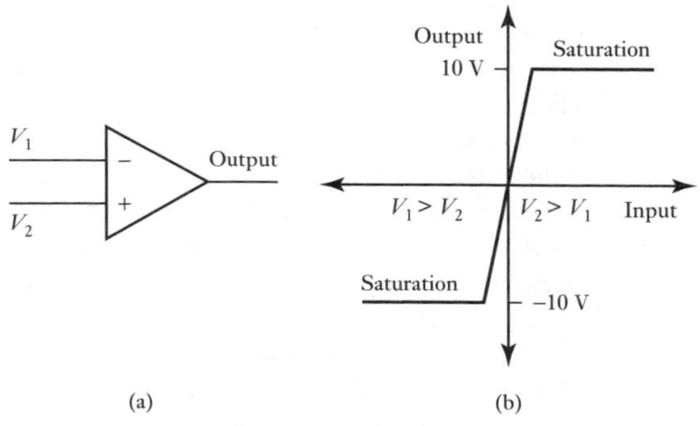

As an illustration of such a use, consider the circuit shown in Figure 3.13. This is designed so that when a critical temperature is reached a relay is activated and initiates some response. The circuit has a Wheatstone bridge with a thermistor in one arm. The resistors in the bridge have their resistances selected so that at the critical temperature the bridge will be balanced. When the temperature is below this value the thermistor resistance R_1 is more than R_2 and the bridge is out of balance. As a consequence there is a voltage difference between the inputs to the operational amplifier and it gives an output at its lower saturated level. This keeps the transistor off, i.e. both the base–emitter and base–collector junctions are reverse biased, and so no current passes through the relay coil. When the temperature rises and the resistance of the thermistor falls, the bridge becomes balanced and the operational

Figure 3.13 Temperature switch circuit.

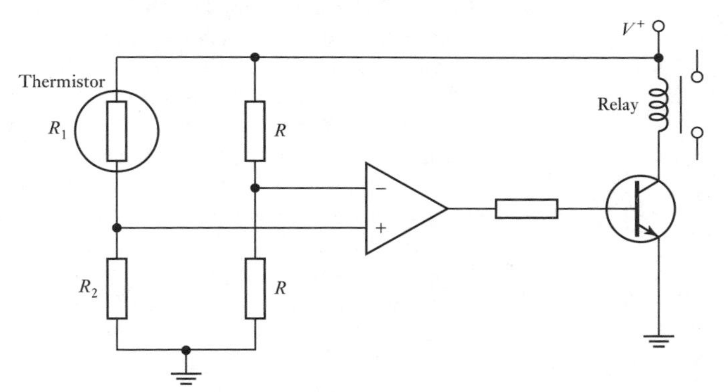

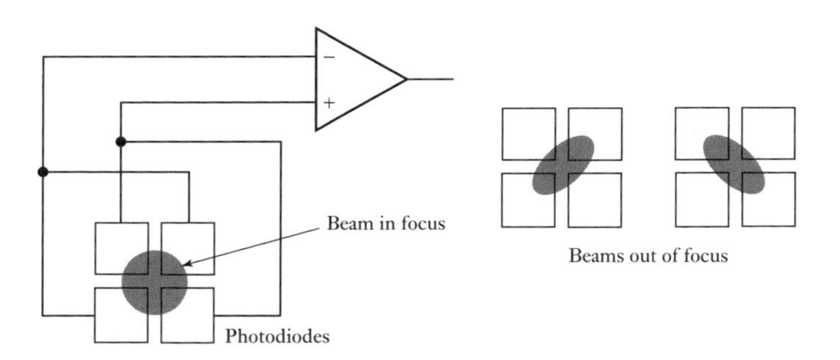

Figure 3.14 Focusing system for a CD player.

amplifier then switches to its upper saturation level. Consequently the transistor is switched on, i.e. its junctions become forward biased, and the relay energised.

As another illustration of the use of a comparator, consider the system used to ensure that in a compact disc player the laser beam is focused on the disc surface. With a CD player, lenses are used to focus a laser beam onto a CD, this having the audio information stored as a sequence of microscopic pits and flats. The light is reflected back from the disc to an array of four photodiodes (Figure 3.14). The output from these photodiodes is then used to reproduce the sound. The reason for having four photodiodes is that the array can also be used to determine whether the beam of laser light is in focus. When the beam is in focus on the disc then a circular spot of light falls on the photodiode array with equal amounts of light falling on each photodiode. As a result the output from the operational amplifier, which is connected as a comparator, is zero. When the beam is out of focus an elliptical spot of light is produced. This results in different amounts of light falling on each of the photocells. The outputs from the two diagonal sets of cells are compared and, because they are different, the comparator gives an output which indicates that the beam is out of focus and in which direction it is out of focus. The output can then be used to initiate correcting action by adjusting the lenses focusing the beam onto the disc.

3.2.8 Real amplifiers

Operational amplifiers are not in the real world the perfect (ideal) element discussed in the previous sections of this chapter. A particularly significant problem is that of the **offset voltage**.

An operational amplifier is a high-gain amplifier which amplifies the difference between its two inputs. Thus if the two inputs are shorted we might expect to obtain no output. However, in practice this does not occur and quite a large output voltage might be detected. This effect is produced by imbalances in the internal circuitry in the operational amplifier. The output voltage can be made zero by applying a suitable voltage between the input terminals. This is known as the **offset voltage**. Many operational amplifiers are provided with arrangements for applying such an offset voltage via a potentiometer. With the 741 this is done by connecting a 10 kΩ potentiometer between pins 1 and 5 (see Figure 3.2) and connecting the sliding contact of the potentiometer to a negative voltage supply (Figure 3.15). The imbalances within the operational amplifier are corrected by adjusting the

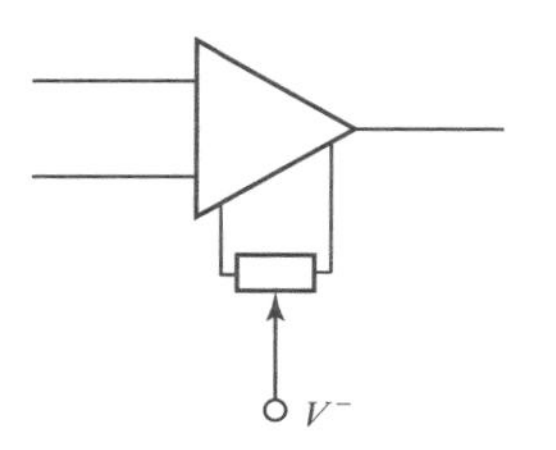

Figure 3.15 Correcting the offset voltage.

position of the slider until with no input to the amplifier there is no output. Typically a general-purpose amplifier will have an offset voltage between 1 and 5 mV.

Operational amplifiers draw small currents at the input terminals in order to bias the input transistors. The bias current flowing through the source resistance at each terminal generates a voltage in series with the input. Ideally, the bias currents at the two inputs will be equal; however, in practice this will not be the case. Thus the effect of these bias currents is to produce an output voltage when there is no input signal and the output should be zero. This is particularly a problem when the amplifier is operating with d.c. voltages. The average value of the two bias currents is termed the **input bias current**. For a general-purpose operational amplifier, a typical value is about 100 nA. The difference between the two bias currents is termed the **input-offset current**. Ideally this would be zero but for a typical general-purpose amplifier it is likely to be about 10 nA, about 10 to 25% of the input bias current.

An important parameter which affects the use of an operational amplifier with alternating current applications is the **slew rate**. This is the maximum rate of change at which the output voltage can change with time in response to a perfect step-function input. Typical values range from 0.2 V/μs to over 20 V/μs. With high frequencies, the large-signal operation of an amplifier is determined by how fast the output can swing from one voltage to another. Thus for use with high-frequency inputs a high value of slew rate is required.

As an illustration of the above, the general-purpose amplifier LM348 with an open-loop voltage gain of 96 dB has an input bias current of 30 nA and a slew rate of 0.6 V/μs. The wide-band amplifier AD711 with an open-loop gain of 100 has an input bias current of 25 pA and a slew rate of 20 V/μs.

3.3 Protection

There are many situations where the connection of a sensor to the next unit, e.g. a microprocessor, can lead to the possibility of damage as a result of perhaps a high current or high voltage. A high current can be protected against by the incorporation in the input line of a series resistor to limit the current to an acceptable level and a fuse to break if the current does exceed a safe level. High voltages, and wrong polarity, may be protected against by the use of a Zener diode circuit (Figure 3.16). Zener diodes behave like ordinary diodes up to some breakdown voltage when they become conducting. Thus to allow a maximum voltage of 5 V but stop voltages above 5.1 V getting through, a Zener diode with a voltage rating of 5.1 V might be chosen. When the voltage rises to 5.1 V the Zener diode breakdown and its resistance drop to a very low value. The result is that the voltage across the diode, and hence that outputted to the next circuit, drops. Because the Zener diode is a diode with a low resistance for current in one direction through it and a high resistance for the opposite direction, it also provides protection against wrong polarity. It is connected with the correct polarity to give a high resistance across the output and so a high voltage drop. When the supply polarity is reversed, the diode has low resistance and so little voltage drop occurs across the output.

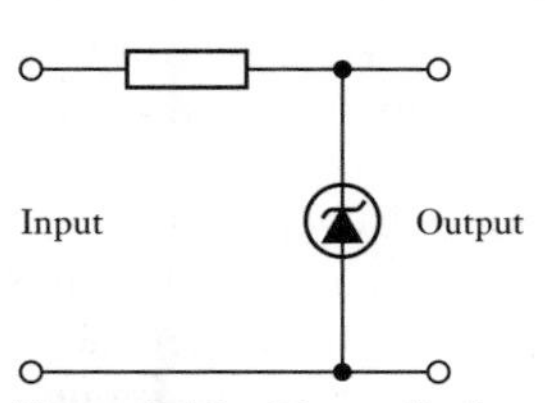

Figure 3.16 Zener diode protection circuit.

In some situations it is desirable to isolate circuits completely and remove all electrical connections between them. This can be done using an

optoisolator. Thus we might have the output from a microprocessor applied to a light-emitting diode (LED) which emits infrared radiation. This radiation is detected by a phototransistor or triac and gives rise to a current which replicates the changes occurring in the voltage applied to the LED. Figure 3.17 shows a number of forms of optoisolator. The term **transfer ratio** is used to specify the ratio of the output current to the input current. Typically, a simple transistor optoisolator (Figure 3.17(a)) gives an output current which is smaller than the input current and a transfer ratio of perhaps 30% with a maximum value of 7 mA. However, the Darlington form (Figure 3.17(b)) gives an output current larger than the input current, e.g. the Siemens 6N139 gives a transfer ratio of 800% with a maximum output value of 60 mA. Another form of optoisolator (Figure 3.17(c)) uses the triac and so can be used with alternating current, a typical triac optoisolator being able to operate with the mains voltage. Yet another form (Figure 3.17(d)) uses a triac with a zero-crossing unit, e.g. Motorola MOC3011, to reduce transients and electromagnetic interference.

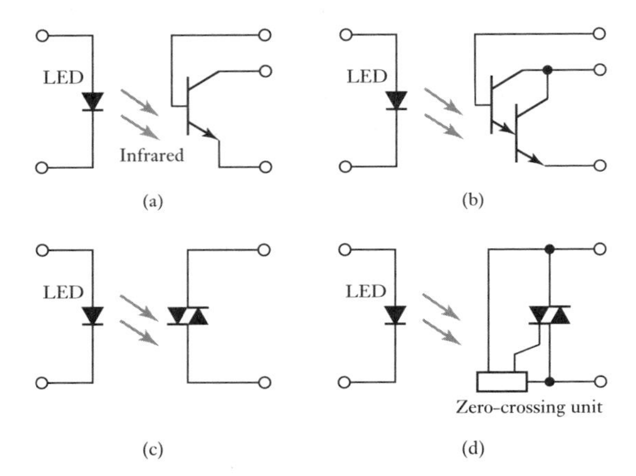

Figure 3.17 Optoisolators: (a) transistor, (b) Darlington, (c) triac, (d) triac with zero-crossing unit.

Optoisolator outputs can be used directly to switch low-power load circuits. Thus a Darlington optoisolator might be used as the interface between a microprocessor and lamps or relays. To switch a high-power circuit, an optocoupler might be used to operate a relay and so use the relay to switch the high-power device.

A protection circuit for a microprocessor input is thus likely to be like that shown in Figure 3.18; to prevent the LED having the wrong polarity or too high an applied voltage, it is also likely to be protected by the Zener diode circuit shown in Figure 3.16 and if there is alternating signal in the input a diode would be put in the input line to rectify it.

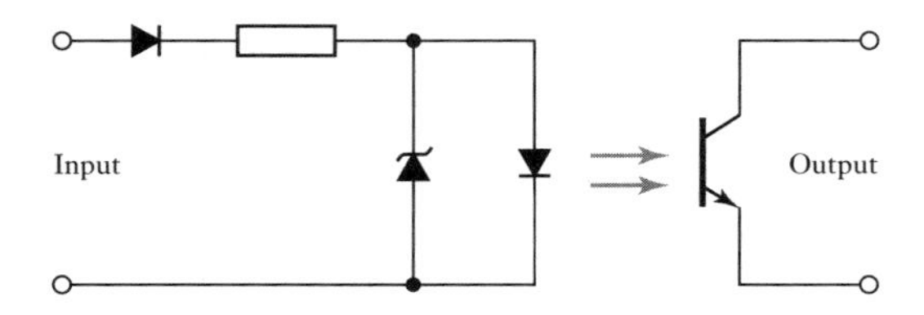

Figure 3.18 Protection circuit.

3.4 Filtering

The term **filtering** is used to describe the process of removing a certain band of frequencies from a signal and permitting others to be transmitted. The range of frequencies passed by a filter is known as the **pass band**, the range not passed as the **stop band** and the boundary between stopping and passing as the **cut-off frequency**. Filters are classified according to the frequency ranges they transmit or reject. A **low-pass filter** (Figure 3.19(a)) has a pass band which allows all frequencies from zero up to some frequency to be transmitted. A **high-pass filter** (Figure 3.19(b)) has a pass band which allows all frequencies from some value up to infinity to be transmitted. A **band-pass filter** (Figure 3.19(c)) allows all the frequencies within a specified band to be transmitted. A **band-stop filter** (Figure 3.19(d)) stops all frequencies with a particular band from being transmitted. In all cases the cut-off frequency is defined as being that at which the output voltage is 70.7% of that in the pass band. The term **attenuation** is used for the ratio of input and output powers, this being written as the ratio of the logarithm of the ratio and so gives the attenuation in units of bels. Since this is a rather large unit, decibels (dB) are used and then attenuation in dB = 10 lg(input power/output power). Since the power through an impedance is proportional to the square of the voltage, the attenuation in dB = 20 lg(input voltage/output voltage). The output voltage of 70.7% of that in the pass band is thus an attenuation of 3 dB.

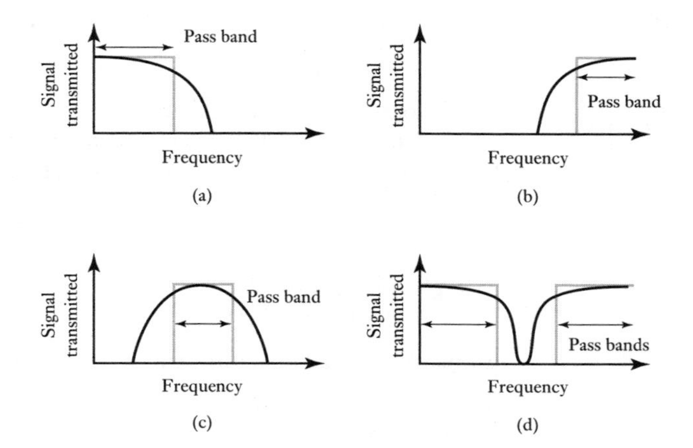

Figure 3.19 Characteristics of ideal filters: (a) low-pass, (b) high-pass, (c) band-pass, (d) band-stop.

The term **passive** is used to describe a filter made up using only resistors, capacitors and inductors, the term **active** being used when the filter also involves an operational amplifier. Passive filters have the disadvantage that the current that is drawn by the item that follows can change the frequency characteristic of the filter. This problem does not occur with an active filter.

Low-pass filters are very commonly used as part of signal conditioning. This is because most of the useful information being transmitted is low frequency. Since noise tends to occur at higher frequencies, a low-pass filter can be used to block it off. Thus a low-pass filter might be selected with a cut-off frequency of 40 Hz, thus blocking off any interference signals from the a.c. mains supply and noise in general. Figure 3.20 shows the basic form that can be used for a low-pass filter.

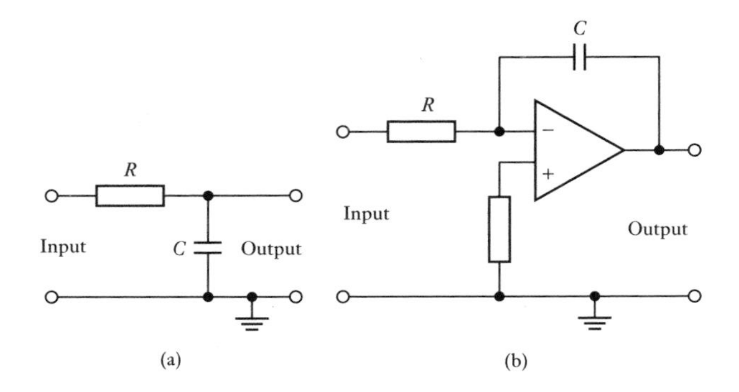

Figure 3.20 Low-pass filter: (a) passive, (b) active using an operational amplifier.

3.5 Wheatstone bridge

The **Wheatstone bridge** can be used to convert a resistance change to a voltage change. Figure 3.21 shows the basic form of such a bridge. When the output voltage V_o is zero, then the potential at B must equal that at D. The potential difference across R_1, i.e. V_{AB}, must then equal that across R_3, i.e. V_{AD}. Thus $I_1R_1 = I_2R_2$. It also means that the potential difference across R_2, i.e. V_{BC}, must equal that across R_4, i.e. V_{DC}. Since there is no current through BD, then the current through R_2 must be the same as that through R_1 and the current through R_4 the same as that through R_3. Thus $I_1R_2 = I_2R_4$. Dividing these two equations gives

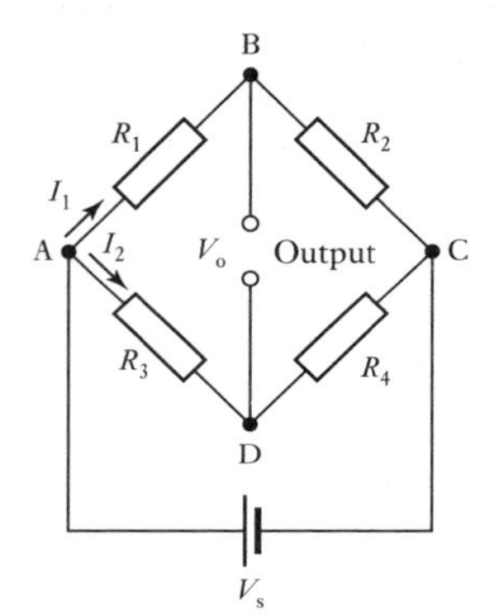

Figure 3.21 Wheatstone bridge.

$$\frac{R_1}{R_2} = \frac{R_3}{R_4}$$

The bridge is said to be **balanced**.

Now consider what happens when one of the resistances changes from this balanced condition. The supply voltage V_s is connected between points A and C and thus the potential drop across the resistor R_1 is the fraction $R_1/(R_1 + R_2)$ of the supply voltage. Hence

$$V_{AB} = \frac{V_sR_1}{R_1 + R_2}$$

Similarly, the potential difference across R_3 is

$$V_{AD} = \frac{V_sR_3}{R_3 + R_4}$$

Thus the difference in potential between B and D, i.e. the output potential difference V_o, is

$$V_o = V_{AB} - V_{AD} = V_s\left(\frac{R_1}{R_1 + R_2} - \frac{R_3}{R_3 + R_4}\right)$$

This equation gives the balanced condition when $V_o = 0$.

Consider resistance R_1 to be a sensor which has a resistance change. A change in resistance from R_1 to $R_1 + \delta R_1$ gives a change in output from V_o to $V_o + \delta V_o$, where

$$V_o + \delta V_o = V_s\left(\frac{R_1 + \delta R_1}{R_1 + \delta R_1 + R_2} - \frac{R_3}{R_3 + R_4}\right)$$

Hence

$$(V_o + \delta V_o) - V_o = V_s\left(\frac{R_1 + \delta R_1}{R_1 + \delta R_1 + R_2} - \frac{R_1}{R_1 + R_2}\right)$$

If δR_1 is much smaller than R_1 then the above equation approximates to

$$\delta V_o \approx V_s\left(\frac{\delta R_1}{R_1 + R_2}\right)$$

With this approximation, the change in output voltage is thus proportional to the change in the resistance of the sensor. This gives the output voltage when there is no load resistance across the output. If there is such a resistance then the loading effect has to be considered.

To illustrate the above, consider a platinum resistance temperature sensor which has a resistance at 0°C of 100 Ω and forms one arm of a Wheatstone bridge. The bridge is balanced, at this temperature, with each of the other arms also being 100 Ω. If the temperature coefficient of resistance of platinum is 0.0039/K, what will be the output voltage from the bridge per degree change in temperature if the load across the output can be assumed to be infinite? The supply voltage, with negligible internal resistance, is 6.0 V. The variation of the resistance of the platinum with temperature can be represented by

$$R_t = R_0(1 + \alpha t)$$

where R_t is the resistance at t (°C), R_0 the resistance at 0°C and α the temperature coefficient of resistance. Thus

$$\text{change in resistance} = R_t - R_0 = R_0\alpha t$$
$$= 100 \times 0.0039 \times 1 = 0.39\ \Omega/\text{K}$$

Since this resistance change is small compared with the 100 Ω, the approximate equation can be used. Hence

$$\delta V_o \approx V_s\left(\frac{\delta R_1}{R_1 + R_2}\right) = \frac{6.0 \times 0.39}{100 + 100} = 0.012\ \text{V}$$

3.5.1 Temperature compensation

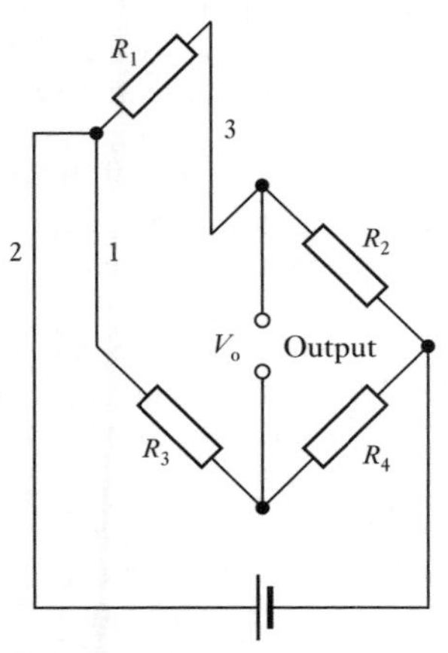

Figure 3.22 Compensation for leads.

In many measurements involving a resistive sensor the actual sensing element may have to be at the end of long leads. Not only the sensor but the resistance of these leads will be affected by changes in temperature. For example, a platinum resistance temperature sensor consists of a platinum coil at the ends of leads. When the temperature changes, not only will the resistance of the coil change but so also will the resistance of the leads. What is required is just the resistance of the coil and so some means has to be employed to compensate for the resistance of the leads to the coil. One method of doing this is to use three leads to the coil, as shown in Figure 3.22. The coil is connected into the Wheatstone bridge in such a way that lead 1 is in series with the R_3 resistor while lead 3 is in series with the platinum resistance coil R_1. Lead 2 is the connection to the power supply. Any change in lead resistance is likely to affect all three leads equally, since they are of the same material, diameter and length and held close together. The result is that changes in lead resistance occur equally in two arms of the bridge and cancel out if R_1 and R_3 are the same resistance.

The electrical resistance strain gauge is another sensor where compensation has to be made for temperature effects. The strain gauge changes resistance when the strain applied to it changes. Unfortunately, it also changes if the temperature changes. One way of eliminating the temperature effect is to use a **dummy strain gauge.** This is a strain gauge which is identical to the one under strain, the active gauge, and is mounted on the same material but is not subject to the strain. It is positioned close to the active gauge so that it suffers the same temperature changes. Thus a temperature change will cause both gauges to change resistance by the same amount. The active gauge is mounted in one arm of a Wheatstone bridge (Figure 3.23(a)) and the dummy gauge in another arm so that the effects of temperature-induced resistance changes cancel out.

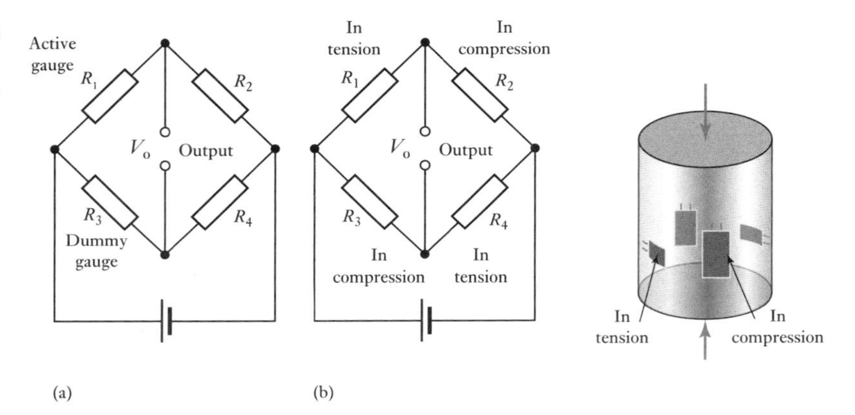

Figure 3.23 Compensation with strain gauges: (a) use of dummy gauge, (b) four active arm bridge.

Strain gauges are often used with other sensors such as load cells or diaphragm pressure gauges to give measures of the amount of displacement occurring. In such situations, temperature compensation is still required. While dummy gauges could be used, a better solution is to use four strain gauges. Two of them are attached so that when forces are applied they are in tension and the other two in compression. The load cell in Figure 3.23(b) shows such a mounting. The gauges that are in tension will increase in resistance while those in compression will decrease in resistance. As the gauges are connected as the four arms of a Wheatstone bridge (Figure 3.23(b)), then since all will be equally affected by any temperature changes, the arrangement is temperature compensated. It also gives a much greater output voltage than would occur with just a single active gauge.

To illustrate this, consider a load cell with four strain gauges arranged as shown in Figure 3.23 which is to be used with a four active arm strain gauge bridge. The gauges have a gauge factor of 2.1 and a resistance of 100 Ω. When the load cell is subject to a compressive force, the vertical gauges show compression and, because when an item is squashed there is a consequential sideways extension, the horizontal gauges are subject to tensile strain (the ratio of the transverse to longitudinal strains is called Poisson's ratio and is usually about 0.3). Thus, if the compressive gauges suffer a strain of -1.0×10^{-5} and the tensile gauges $+0.3 \times 10^{-5}$, the supply voltage for the bridge is 6 V and the output voltage from the bridge is amplified by a differential operational amplifier circuit, what will be the ratio of the feedback resistance to that of the input resistances in the two inputs of the amplifier if the load is to produce an output of 1 mV?

The change in resistance of a gauge subject to the compressive strain is given by $\Delta R/R = G\varepsilon$:

$$\text{change in resistance} = G\varepsilon R = -2.1 \times 1.0 \times 10^{-5} \times 100 = -2.1 \times 10^{-3}\,\Omega$$

For a gauge subject to tension we have

$$\text{change in resistance} = G\varepsilon R = 2.1 \times 0.3 \times 10^{-5} \times 100 = 6.3 \times 10^{-4}\,\Omega$$

The out-of-balance potential difference is given by (see earlier in Section 3.5)

$$V_o = V_s\left(\frac{R_1}{R_1 + R_2} - \frac{R_3}{R_3 + R_4}\right) = V_s\left(\frac{R_1(R_3 + R_4) - R_3(R_1 + R_2)}{(R_1 + R_2)(R_3 + R_4)}\right) = V_s\left(\frac{R_1R_4 - R_2R_3}{(R_1 + R_2)(R_3 + R_4)}\right)$$

We now have each of the resistors changing. We can, however, neglect the changes in relation to the denominators where the effect of the changes on the sum of the two resistances is insignificant. Thus

$$V_o = V_s\left(\frac{(R_1 + \delta R_1)(R_4 + \delta R_4) - (R_2 + \delta R_2)(R_3 + \delta R_3)}{(R_1 + R_2)(R_3 + R_4)}\right)$$

Neglecting products of δ terms and since we have an initially balanced bridge with $R_1R_4 = R_2R_3$, then

$$V_o = \frac{V_sR_1R_4}{(R_1 + R_2)(R_3 + R_4)} = \left(\frac{\delta R_1}{R_1} - \frac{\delta R_2}{R_2} - \frac{\delta R_3}{R_3} + \frac{\delta R_4}{R_4}\right)$$

Hence

$$V_o = \frac{6 \times 100 \times 100}{200 \times 200}\left(\frac{2 \times 6.3 \times 10^{-4} + 2 \times 2.1 \times 10^{-3}}{100}\right)$$

The output is thus 3.6×10^{-5} V. This becomes the input to the differential amplifier; hence, using the equation developed in Section 3.2.5,

$$V_o = \frac{R_2}{R_1}(V_2 - V_1)$$

$$1.0 \times 10^{-3} = \frac{R_2}{R_1} \times 3.6 \times 10^{-5}$$

Thus $R_2/R_1 = 27.8$.

3.5.2 Thermocouple compensation

A thermocouple gives an e.m.f. which depends on the temperature of its two junctions (see Section 2.9.5). Ideally, if one junction is kept at 0°C, then the temperature relating to the e.m.f. can be directly read from tables. However, this is not always feasible and the cold junction is often allowed to be at the ambient temperature. To compensate for this a potential difference has to be added to the thermocouple. This must be the same as the e.m.f. that would be generated by the thermocouple with one junction at 0°C and the other at the ambient temperature. Such a potential difference can be produced by using a resistance temperature sensor in a Wheatstone bridge. The bridge is balanced at 0°C and the output voltage from the bridge provides the correction potential difference at other temperatures.

The resistance of a metal resistance temperature sensor can be described by the relationship

$$R_t = R_0(1 + \alpha t)$$

where R_t is the resistance at t (°C), R_0 the resistance at 0°C and α the temperature coefficient of resistance. Thus

$$\text{change in resistance} = R_t - R_0 = R_0\alpha t$$

The output voltage for the bridge, taking R_1 to be the resistance temperature sensor, is given by

$$\delta V_o \approx V_s\left(\frac{\delta R_1}{R_1 + R_2}\right) = \frac{V_s R_0 \alpha t}{R_0 + R_2}$$

The thermocouple e.m.f. e is likely to vary with temperature t in a reasonably linear manner over the small temperature range being considered – from 0°C to the ambient temperature. Thus $e = kt$, where k is a constant, i.e. the e.m.f. produced per degree change in temperature. Hence for compensation we must have

$$kt = \frac{V_s R_0 \alpha t}{R_0 + R_2}$$

and so

$$kR_2 = R_0(V_s\alpha - k)$$

For an iron-constantan thermocouple giving 51 μV/°C, compensation can be provided by a nickel resistance element with a resistance of 10 Ω at 0°C and a temperature coefficient of resistance of 0.0067/K, a supply voltage for the bridge of 1.0 V and R_2 as 1304 Ω.

3.6 Pulse modulation

A problem that is often encountered with dealing with the transmission of low-level d.c. signals from sensors is that the gain of an operational amplifier used to amplify them may drift and so the output drifts. This problem can be overcome if the signal is a sequence of pulses rather than a continuous-time signal.

One way this conversion can be achieved is by chopping the d.c. signal in the way suggested in Figure 3.24. The output from the chopper is a chain of

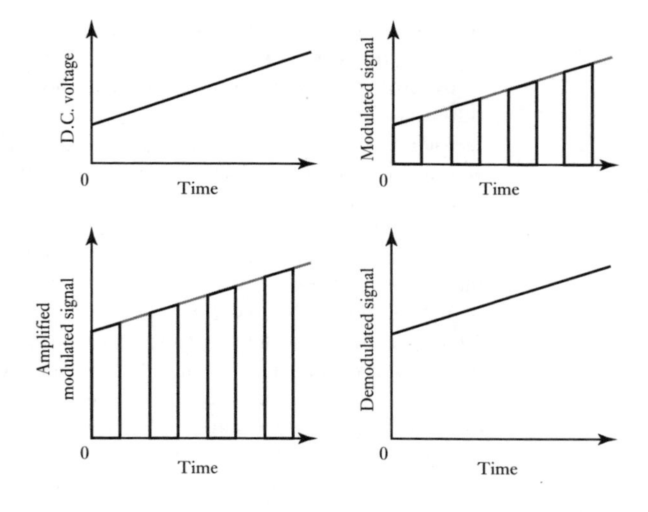

Figure 3.24 Pulse amplitude modulation.

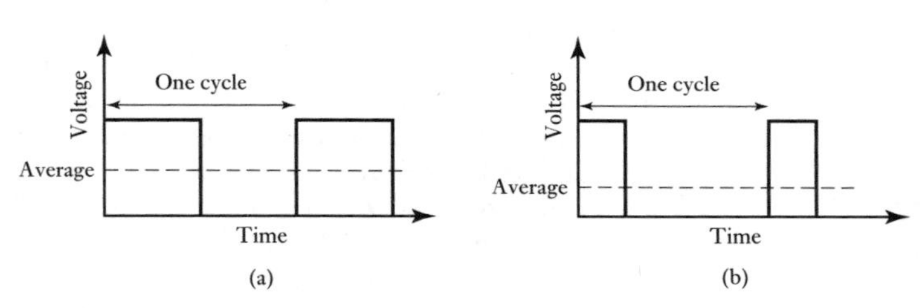

Figure 3.25 PWM for voltage control: (a) duty cycle 50%, (b) duty cycle 25%.

pulses, the heights of which are related to the d.c. level of the input signal. This process is called **pulse amplitude modulation.** After amplification and any other signal conditioning, the modulated signal can be demodulated to give a d.c. output. With pulse amplitude modulation, the height of the pulses is related to the size of the d.c. voltage.

Pulse width modulation (PWM) is widely used with control systems as a means of controlling the average value of a d.c. voltage. Thus if there is a constant analogue voltage and it is chopped into pulses, by varying the width of the pulses so the average value of the voltage can be changed. Figure 3.25 illustrates this. The term **duty cycle** is used for the fraction of each cycle for which the voltage is high. Thus for a PWM signal where the signal is high for half of each cycle, the duty cycle is ½ or 50%. If it is only on for a quarter of each cycle then the duty cycle is ¼ or 25%.

3.7 Problems with signals

When connecting sensors to signal conditioning equipment and controllers, problems can occur with signals as a result of grounding and electromagnetic interference.

3.7.1 Grounding

Generally the signals from sensors and signal conditioning equipment are transmitted as voltages to the controller. Such voltages are the potential differences between two points. If one of the points is earthed it is said to be a **grounded signal source**. If neither point is grounded then it is said to be a **floating signal source**. With a grounded source the voltage output is the potential difference between the system ground and the positive signal lead of the source. If it is a floating source, the signal source is not referenced to any absolute value and each of the voltage lines may have a potential relative to the ground.

Differential systems, e.g. a differential amplifier, are concerned with the potential difference between two input lines. If each has a voltage referred to a common ground, V_A and V_B, then the **common mode voltage** is the average of the two, i.e. $\frac{1}{2}(V_A + V_B)$. Thus if we have one input line at 10 V and the other at 12 V, the potential difference will be 2 V and the common mode voltage 11 V. The differential measurement system is concerned with the difference between the two inputs ($V_A - V_B$) and not the common mode voltage. Unfortunately, the common mode voltage can have an effect on the indicated potential difference value, and the extent to which it affects the difference is described by the **common mode rejection ratio** (CMRR) (see Section 3.2.5). This is the ratio of differential gain of the system to the common-mode gain or, when expressed in decibels, 20 lg (differential gain/common mode gain). The higher the CMRR the greater the differential gain when compared with the common mode gain and the less significance is attached to the common mode voltage. A CMRR of 10 000, or 80 dB, for a differential amplifier would mean that, if the desired difference signal was the same size as the common mode voltage, it will appear at the output 10 000 times greater in size that the common mode.

Problems can arise with systems when a circuit has several grounding points. For example, it might be that both the sensor and the signal conditioner are grounded. In a large system, multiple grounding is largely inevitable. Unfortunately, there may be a potential difference between the two grounding points and thus significant currents can flow between the grounding points through the low but finite ground resistance (Figure 3.26). These are termed **ground-loop currents**. This potential difference between two grounding points is not necessarily just d.c. but can also be a.c., e.g. a.c. mains hum. There is also the problem that we have a loop in which currents can be induced by magnetic coupling with other nearby circuits. Thus a consequence of having a ground loop can be to make remote measurements difficult.

Ground loops from multiple point grounding can be minimised if the multiple earth connections are made close together and the common ground has a

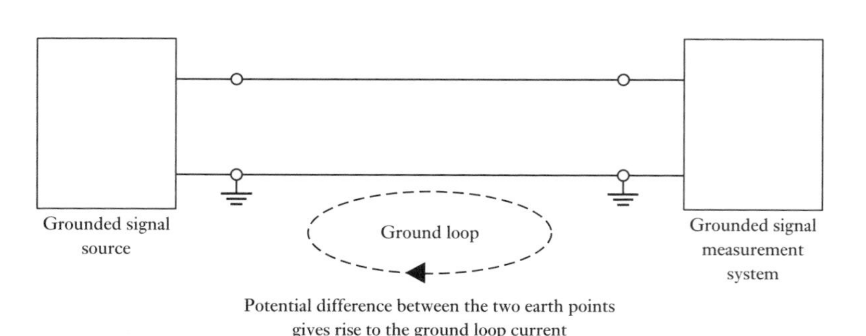

Figure 3.26 A ground loop.

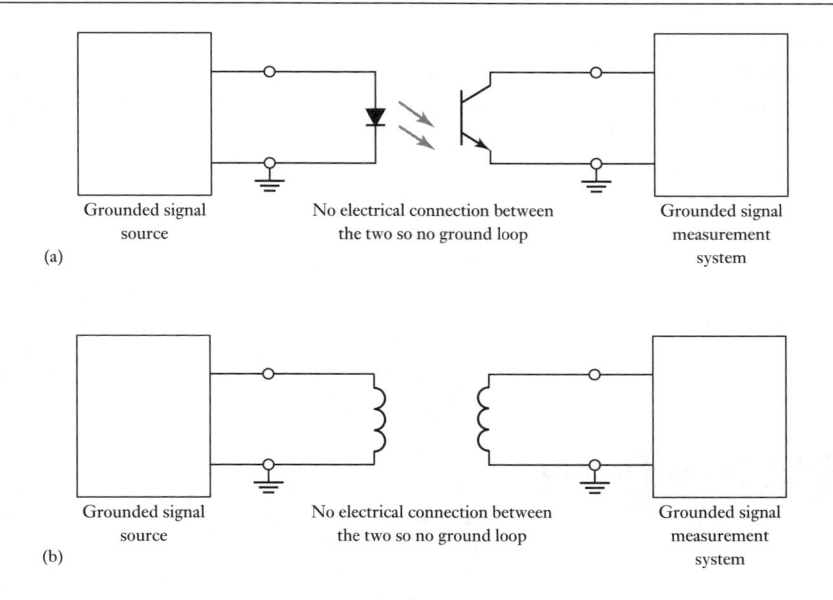

Figure 3.27 Isolation using (a) an optoisolator, (b) a transformer.

resistance small enough to make the voltage drops between the earthing points negligible. Ground loops can be eliminated if there is electrical isolation of the signal source system from the measurement system. This can be achieved by using an optoisolator (see Section 3.3) or a transformer (Figure 3.27).

3.7.2 Electromagnetic interference

Electromagnetic interference is an undesirable effect on circuits resulting from time-varying electric and magnetic fields. Common sources of such interference are fluorescent lamps, d.c. motors, relay coils, household appliances and the electrics of motor cars.

Electrostatic interference occurs as a result of mutual capacitance between neighbouring conductors. Electric shielding can guard against this. This is a shield of electrically conductive material, e.g. copper or aluminium, that is used to enclose a conductor or circuit. Thus a screened cable might be used to connect a sensor to its measurement system. If the sensor is earthed then the screen should be connected to the same point where the sensor is earthed, thus minimising the ground loop (Figure 3.28).

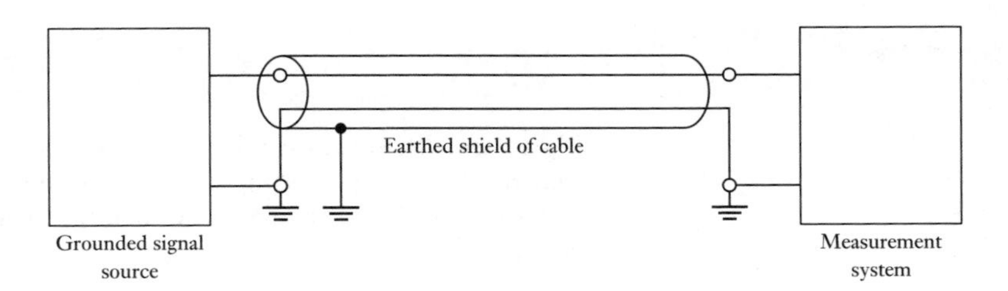

Figure 3.28 Use of a shielded cable to minimise electrostatic interference.

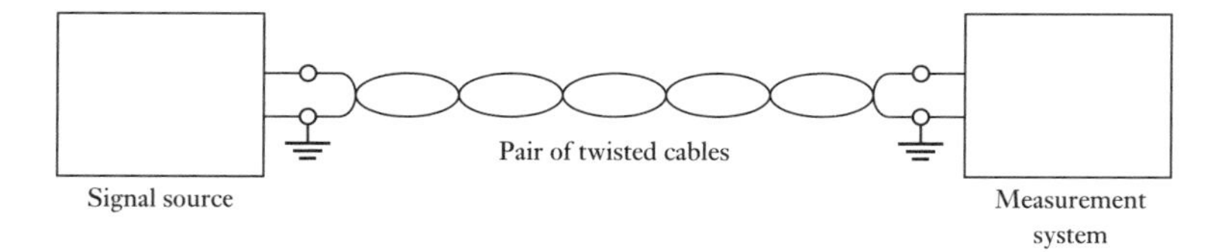

Figure 3.29 Twisted pair of cables to minimise electromagnetic interference.

Interference also occurs when there is a changing magnetic field which induces voltages in the measurement system. Protection against this can be achieved by such methods as placing components as far as possible from sources of interference and minimising the area of any loops in the system and by the use of twisted pairs of wires for interconnections (Figure 3.29). With twisted wire, the coupling alternates in phase between adjacent twists and so leads to cancellation of the effect.

3.8 Power transfer

There are many control system situations when components are interconnected. Thus with electrical components we might have a sensor system connected to an amplifier. With a mechanical system we might have a motor rotating a load. A concern is the condition for maximum power transfer between the two elements.

As an introduction, consider a d.c. source of e.m.f. E and internal resistance r supplying a load of resistance R (Figure 3.30). The current supplied to the load is $I = E/(R + r)$ and so the power supplied to the load is

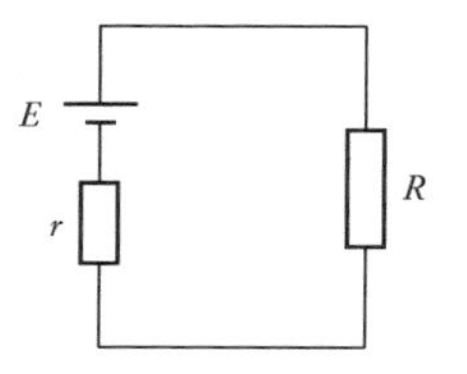

Figure 3.30 D.C. source supplying a load.

$$P = I^2R = \frac{E^2R}{(R + r)^2}$$

The maximum power supplied to the load will when $dP/dt = 0$.

$$\frac{dP}{dt} = \frac{(R + r)^2E^2 - E^2R^2(R + r)}{(R + r)^3}$$

When this is zero then $(R + r) = 2R$ and so the condition for maximum power transfer is $R = r$, i.e. when the source and load resistances are matched.

With an alternating current source having an internal impedance supplying a load impedance, the condition for maximum power transfer can similarly be derived and is when the source and load impedances are matched. With, for example, a high impedance sensor being matched to an electronic system, an impedance matching amplifier might be used between the source and the load in order to achieve this maximum power transmission. Such an amplifier is typically a high gain amplifier with a high input impedance and a low output impedance.

Summary

Signal conditioning can involve protection to prevent damage to the next element in a system, getting a signal into the form required, getting the level of a signal right, reducing noise, manipulating a signal to perhaps make it linear.

Commonly used signal conditioning elements are **operational amplifiers**, these being high-gain d.c. amplifiers with gains of the order of 100 000 or more.

Protection against perhaps a high voltage or current can involve the use of resistors and fuses; Zener diodes can be used to protect against wrong polarity and high voltages. Optoisolators are used to isolate circuits completely, removing all electrical connections between them.

Filters can be used to remove a particular band of frequencies from a signal and permit others to be transmitted.

The **Wheatstone bridge** can be used to convert an electrical resistance change to a voltage change.

When connecting sensors to signal conditioning equipment and controllers, problems can occur with signals when a circuit has several **grounding** points and **electromagnetic interference** as a result of time-varying electric and magnetic fields.

For **maximum power transfer** between electrical components the impedances must match.

Problems

3.1 Design an operational amplifier circuit that can be used to produce an output that ranges from 0 to −5 V when the input goes from 0 to 100 mV.

3.2 An inverting amplifier has an input resistance of 2 kΩ. Determine the feedback resistance needed to give a voltage gain of 100.

3.3 Design a summing amplifier circuit that can be used to produce an output that ranges from −1 to −5 V when the input goes from 0 to 100 mV.

3.4 A differential amplifier is used with a thermocouple sensor in the way shown in Figure 3.8. What values of R_1 and R_2 would give a circuit which has an output of 10 mV for a temperature difference between the thermocouple junctions of 100°C with a copper–constantan thermocouple if the thermocouple is assumed to have a constant sensitivity of 43 μV/°C?

3.5 The output from the differential pressure sensor used with an orifice plate for the measurement of flow rate is non-linear, the output voltage being proportional to the square of the flow rate. Determine the form of characteristic required for the element in the feedback loop of an operational amplifier signal conditioner circuit in order to linearise this output.

3.6 A differential amplifier is to have a voltage gain of 100. What will be the feedback resistance required if the input resistances are both 1 kΩ?

3.7 A differential amplifier has a differential voltage gain of 2000 and a common mode gain of 0.2. What is the common mode rejection ratio in dB?

3.8 Digital signals from a sensor are polluted by noise and mains interference and are typically of the order of 100 V or more. Explain how protection can be afforded for a microprocessor to which these signals are to be inputted.

3.9 A platinum resistance temperature sensor has a resistance of 120 Ω at 0°C and forms one arm of a Wheatstone bridge. At this temperature the bridge is balanced with each of the other arms being 120Ω. The temperature coefficient of resistance of the platinum is 0.0039/K. What will be the output voltage from the bridge for a change in temperature of 20°C? The loading across the output is effectively open circuit and the supply voltage to the bridge is from a source of 6.0 V with negligible internal resistance.

3.10 A diaphragm pressure gauge employs four strain gauges to monitor the displacement of the diaphragm. The four active gauges form the arms of a Wheatstone bridge, in the way shown in Figure 3.23. The gauges have a gauge factor of 2.1 and resistance 120 Ω. A differential pressure applied to the diaphragm results in two of the gauges on one side of the diaphragm being subject to a tensile strain of 1.0×10^{-5} and the two on the other side a compressive strain of 1.0×10^{-5}. The supply voltage for the bridge is 10 V. What will be the voltage output from the bridge?

3.11 A Wheatstone bridge has a single strain gauge in one arm and the other arms are resistors with each having the same resistance as the unstrained gauge. Show that the output voltage from the bridge is given by $\frac{1}{4}V_s G\varepsilon$, where V_s is the supply voltage to the bridge, G the gauge factor of the strain gauge and ε the strain acting on it.

Chapter four Digital Signals

Objectives

The objectives of this chapter are that, after studying it, the reader should be able to:
- **Explain the principles and main methods of analogue-to-digital and digital-to-analogue converters.**
- **Explain the principles and uses of multiplexers.**
- **Explain the principles of digital signal processing.**

4.1 Digital signals

The output from most sensors tends to be in analogue form, the size of the output being related to the size of the input. Where a microprocessor is used as part of the measurement or control system, the analogue output from the sensor has to be converted into a *digital* form before it can be used as an input to the microprocessor. Likewise, most actuators operate with analogue inputs and so the digital output from a microprocessor has to be converted into an analogue form before it can be used as an input by the actuator.

4.1.1 Binary numbers

The **binary system** is based on just the two symbols or states 0 and 1, these possibly being 0 V and 5 V signals. These are termed *bi*nary dig*its* or **bits**. When a number is represented by this system, the digit position in the number indicates the weight attached to each digit, the weight increasing by a factor of 2 as we proceed from right to left:

...	2^3	2^2	2^1	2^0
	bit 3	bit 2	bit 1	bit 0

For example, the decimal number 15 is $2^0 + 2^1 + 2^2 + 2^3 = 1111$ in the binary system. In a binary number the bit 0 is termed the **least significant bit** (LSB) and the highest bit the **most significant bit** (MSB). The combination of bits to represent a number is termed a **word**. Thus 1111 is a 4-bit word. Such a word could be used to represent the size of a signal. The term **byte** is used for a group of 8 bits. See Appendix B for more discussion of binary numbers.

4.2 Analogue and digital signals

Analogue-to-digital conversion involves converting analogue signals into binary words. Figure 4.1(a) shows the basic elements of analogue-to-digital conversion.

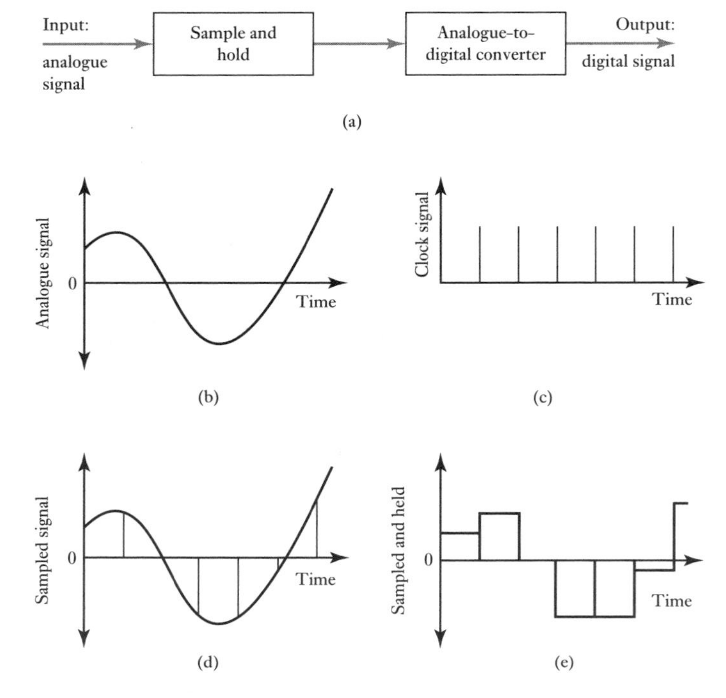

Figure 4.1 (a) Analogue-to-digital conversion, (b) analogue input, (c) clock signal, (d) sampled signal, (e) sampled and held signal.

The procedure used is that a clock supplies regular time signal pulses to the analogue-to-digital converter (ADC) and every time it receives a pulse it samples the analogue signal. Figure 4.1 illustrates this analogue-to-digital conversion by showing the types of signals involved at the various stages. Figure 4.1(b) shows the analogue signal and Figure 4.1(c) the clock signal which supplies the time signals at which the sampling occurs. The result of the sampling is a series of narrow pulses (Figure 4.1(d)). A **sample and hold** unit is then used to hold each sampled value until the next pulse occurs, with the result shown in Figure 4.1(e). The sample and hold unit is necessary because the ADC requires a finite amount of time, termed the **conversion time**, to convert the analogue signal into a digital one.

The relationship between the sampled and held input and the output for an ADC is illustrated by the graph shown in Figure 4.2 for a digital output

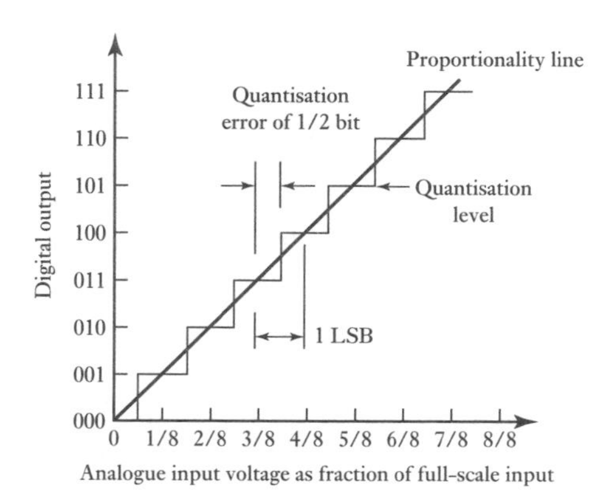

Figure 4.2 Input/output for an ADC.

which is restricted to 3 bits. With 3 bits there are $2^3 = 8$ possible output levels. Thus, since the output of the ADC to represent the analogue input can be only one of these eight possible levels, there is a range of inputs for which the output does not change. The eight possible output levels are termed **quantisation levels** and the difference in analogue voltage between two adjacent levels is termed the **quantisation interval**. Thus for the ADC given in Figure 4.2, the quantisation interval is 1 V. Because of the step-like nature of the relationship, the digital output is not always proportional to the analogue input and thus there will be error, this being termed the **quantisation error**. When the input is centred over the interval, the quantisation error is zero, the maximum error being equal to one-half of the interval or $\pm\frac{1}{2}$ bit.

The word length possible determines the **resolution** of the element, i.e. the smallest change in input which will result in a change in the digital output. The smallest change in digital output is 1 bit in the least significant bit position in the word, i.e. the far right bit. Thus with a word length of n bits the full-scale analogue input V_{FS} is divided into 2^n pieces and so the minimum change in input that can be detected, i.e. the resolution, is $V_{FS}/2^n$.

Thus if we have an ADC with a word length of 10 bits and the analogue signal input range is 10 V, then the number of levels with a 10-bit word is $2^{10} = 1024$ and thus the resolution is $10/1024 = 9.8$ mV.

Consider a thermocouple giving an output of 0.5 mV/°C. What will be the word length required when its output passes through an ADC if temperatures from 0 to 200°C are to be measured with a resolution of 0.5°C? The full-scale output from the sensor is $200 \times 0.5 = 100$ mV. With a word length n, this voltage will be divided into $100/2^n$ mV steps. For a resolution of 0.5°C we must be able to detect a signal from the sensor of $0.5 \times 0.5 = 0.25$ mV. Thus we require

$$0.25 = \frac{100}{2^n}$$

Hence $n = 8.6$. Thus a 9-bit word length is required.

4.2.1 Sampling theorem

ADCs sample analogue signals at regular intervals and convert these values to binary words. How often should an analogue signal be sampled in order to give an output which is representative of the analogue signal?

Figure 4.3 illustrates the problem with different sampling rates being used for the same analogue signal. When the signal is reconstructed from the samples, it is only when the sampling rate is at least twice that of the highest frequency in the analogue signal that the sample gives the original form of signal. This criterion is known as the **Nyquist criterion** or **Shannon's sampling theorem**. When the sampling rate is less than twice the highest frequency, the reconstruction can represent some other analogue signal and we obtain a false image of the real signal. This is termed **aliasing**. In Figure 4.3(c) this could be an analogue signal with a much smaller frequency than that of the analogue signal that was sampled.

Whenever a signal is sampled too slowly, there can be a false interpretation of high-frequency components as arising from lower frequency aliases.

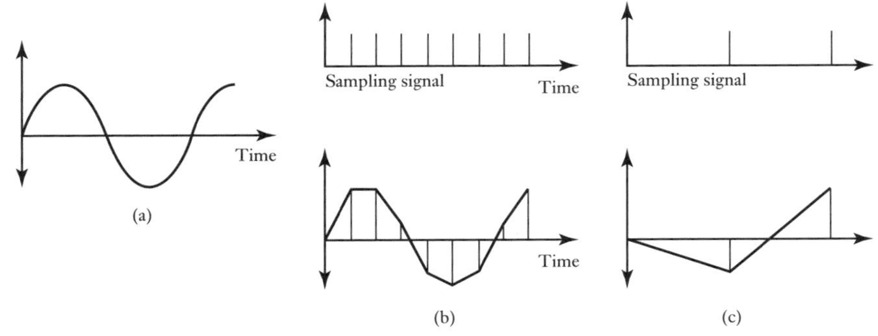

Figure 4.3 Effect of sampling frequency: (a) analogue signal, (b) sampled signal, (c) sampled signal.

High-frequency noise can also create errors in the conversion process. To minimise errors due to both aliasing and high-frequency noise, a low-pass filter is used to precede the ADC, the filter having a bandwidth such that it passes only low frequencies for which the sampling rate will not give aliasing errors. Such a filter is termed an **anti-aliasing filter**.

4.2.2 Digital-to-analogue conversion

The input to a digital-to-analogue converter (DAC) is a binary word; the output is an analogue signal that represents the weighted sum of the non-zero bits represented by the word. Thus, for example, an input of 0010 must give an analogue output which is twice that given by an input of 0001. Figure 4.4 illustrates this for an input to a DAC with a resolution of 1 V for unsigned binary words. Each additional bit increases the output voltage by 1 V.

Consider the situation where a microprocessor gives an output of an 8-bit word. This is fed through an 8-bit digital-to-analogue converter to a control valve. The control valve requires 6.0 V to be fully open. If the fully open state is indicated by 11111111 what will be the output to the valve for a change of 1 bit?

The full-scale output voltage of 6.0 V will be divided into 2^8 intervals. A change of 1 bit is thus a change in the output voltage of $6.0/2^8 = 0.023$ V.

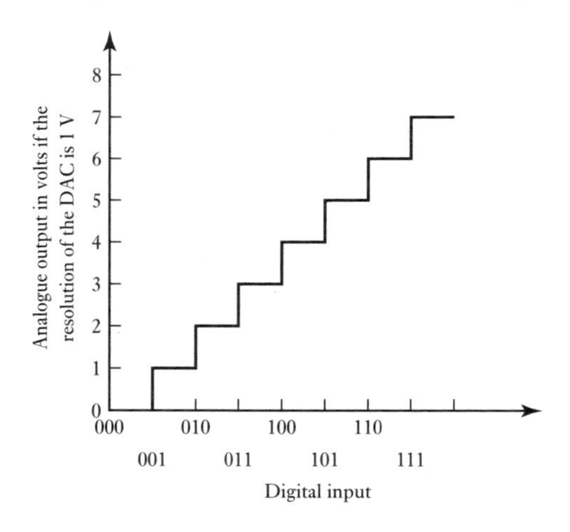

Figure 4.4 Input/output for a DAC.

4.3 Digital-to-analogue and analogue-to-digital converters

The following are commonly encountered forms of DACs and ADCs.

4.3.1 DACs

A simple form of DAC uses a summing amplifier (see Section 3.2.3) to form the weighted sum of all the non-zero bits in the input word (Figure 4.5). The reference voltage is connected to the resistors by means of electronic switches which respond to binary 1. The values of the input resistances depend on which bit in the word a switch is responding to, the value of the resistor for successive bits from the LSB being halved. Hence the sum of the voltages is a weighted sum of the digits in the word. Such a system is referred to as a **weighted-resistor network.** The function of the op-amp circuit is to act as a buffer to ensure that the current out of the resistor network is not affected by the output load and also so that the gain can be adjusted to give an output range of voltages appropriate to a particular application.

Figure 4.5 Weighted-resistor DAC.

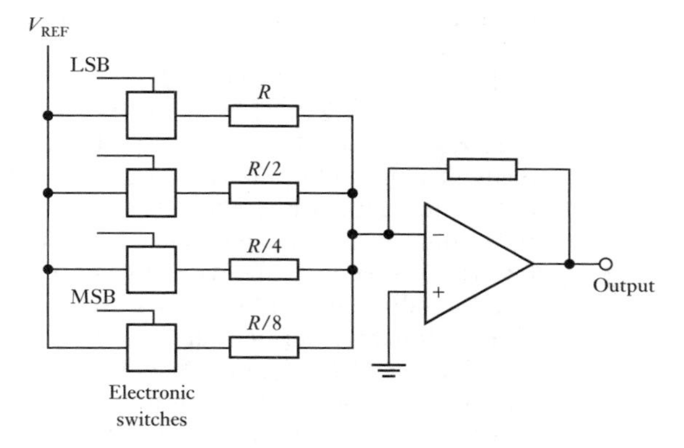

A problem with the weighted-resistor network is that accurate resistances have to be used for each of the resistors and it is difficult to obtain such resistors over the wide range needed. As a result this form of DAC tends be limited to 4-bit conversions.

Another, more commonly used, version uses an ***R*–2*R* ladder network** (Figure 4.6). This overcomes the problem of obtaining accurate resistances

Figure 4.6 *R*–2*R* ladder DAC.

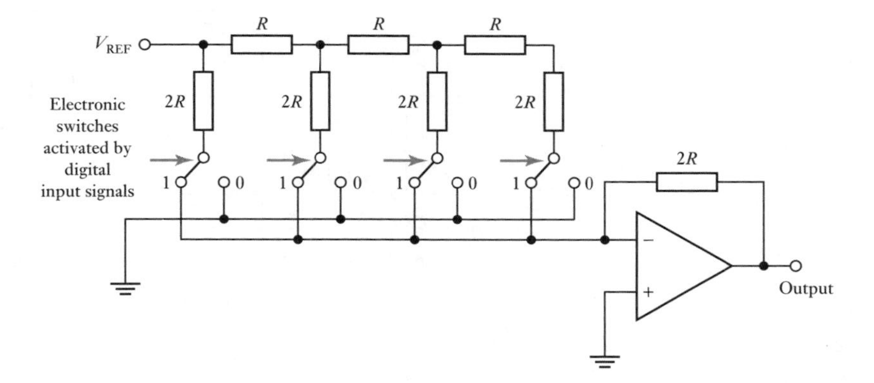

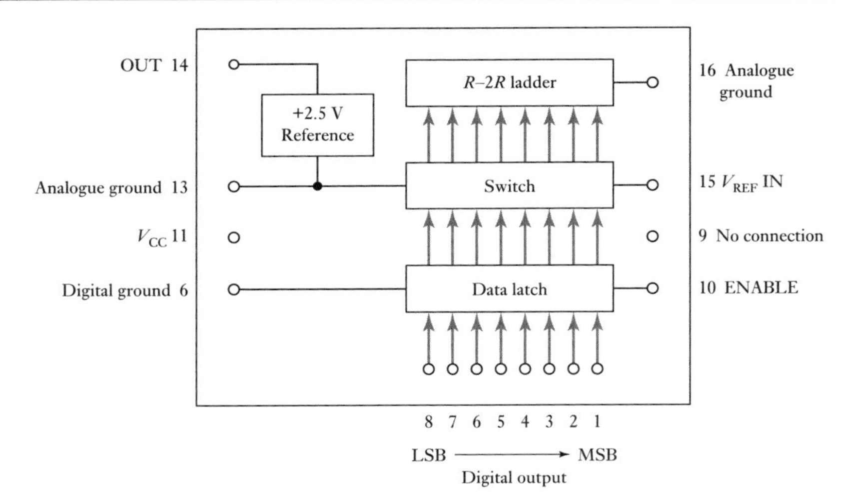

Figure 4.7 ZN558D DAC.

over a wide range of values, only two values being required. The output voltage is generated by switching sections of the ladder to either the reference voltage or 0 V according to whether there is a 1 or 0 in the digital input.

Figure 4.7 shows details of the GEC Plessey ZN558D 8-bit latched input DAC using a *R*-2*R* ladder network. After the conversion is complete, the 8-bit result is placed in an internal latch until the next conversion is complete. Data is held in the latch when ENABLE is high, the latch being said to be transparent when ENABLE is low. A **latch** is just a device to retain the output until a new one replaces it. When a DAC has a latch it may be interfaced directly to the data bus of a microprocessor and treated by it as just an address to send data. A DAC without a latch would be connected via a peripheral interface adapter (PIA), such a device providing latching (see Section 20.4). Figure 4.8 shows how the ZN558D might be used with a microprocessor when the output is required to be a voltage which varies between zero and the reference voltage, this being termed **unipolar operation**. With $V_{ref\ in} = 2.5$ V, the output range is +5 V when $R_1 = 8$ kΩ and $R_2 = 8$ kΩ and the range is +10 V when $R_1 = 16$ kΩ and $R_2 = 5.33$ kΩ.

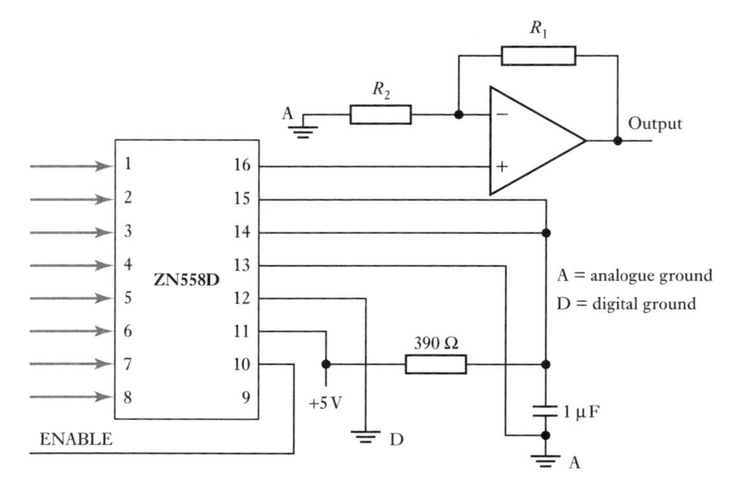

Figure 4.8 Unipolar operation.

The specifications of DACs include such terms as:

1 The *full-scale output*, i.e. the output when the input word is all ones. For the ZN558D this is typically 2.550 V.
2 The *resolution*, 8-bit DACS generally being suitable for most microprocessor control systems. The ZN558D is 8-bit.
3 The *settling time*, this being the time taken by the DAC to reach within $\frac{1}{2}$ LSB of its new voltage after a binary change. This is 800 ns for the ZN558D.
4 The *linearity*, this being the maximum deviation from the straight line through zero and the full range of the output. This is a maximum of ±0.5 LSB for the ZN558D.

4.3.2 ADCs

The input to an ADC is an analogue signal and the output is a binary word that represents the level of the input signal. There are a number of forms of ADC, the commonest being successive approximations, ramp, dual ramp and flash.

Successive approximations is probably the most commonly used method. Figure 4.9 illustrates the subsystems involved. A voltage is generated by a clock emitting a regular sequence of pulses which are counted, in a binary manner, and the resulting binary word converted into an analogue voltage by a DAC. This voltage rises in steps and is compared with the analogue input voltage from the sensor. When the clock-generated voltage passes the input analogue voltage, the pulses from the clock are stopped from being counted by a gate being closed. The output from the counter at that time is then a digital representation of the analogue voltage. While the comparison could be accomplished by starting the count at 1, the LSB, and then proceeding bit by bit upwards, a faster method is by successive approximations. This involves selecting the MSB that is less than the analogue value, then adding successive lesser bits for which the total does not exceed the analogue value. For example, we might start the comparison with 1000. If this is too large we try 0100. If this is too small we then try 0110. If this is too large we try 0101. Because each of the bits in the word is tried in sequence, with an n-bit word it only takes n steps to make the comparison. Thus if the clock has a frequency f, the time between pulses is $1/f$. Hence the time taken to generate the word, i.e. the conversion time, is n/f.

Figure 4.10 shows the typical form of an 8-bit ADC (GEC Plessey ZN439) designed for use with microprocessors and using the successive approximations method. Figure 4.11 shows how it can be connected so that it is controlled by a microprocessor and sends its digital output to the

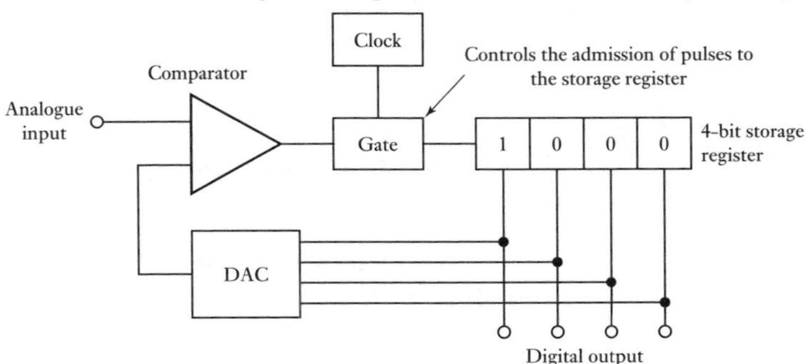

Figure 4.9 Successive approximations ADC.

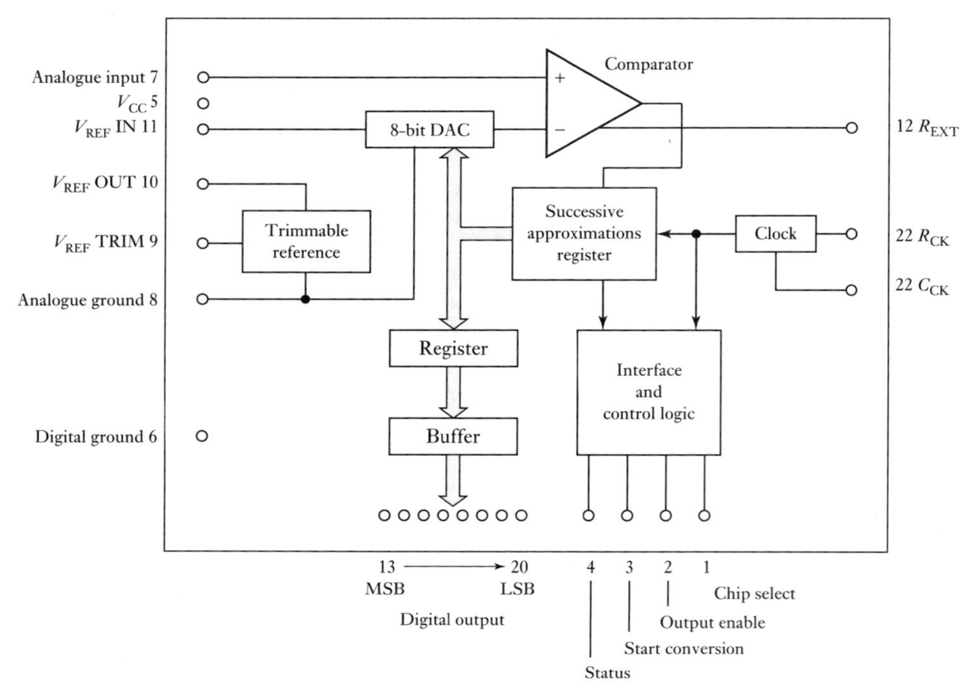

Figure 4.10 ZN439 ADC.

Figure 4.11 ZN439 connected to a microprocessor.

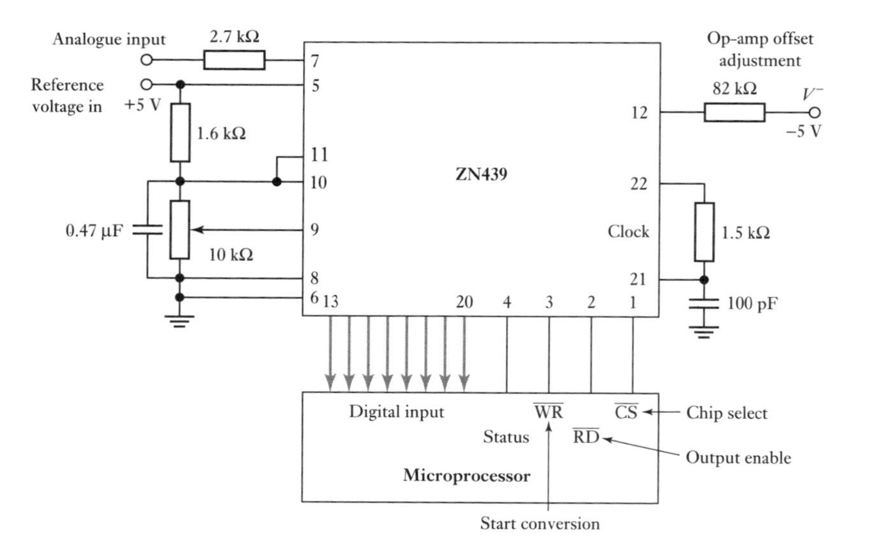

microprocessor. All the active circuitry, including the clock, is contained on a single chip. The ADC is first selected by taking the chip select pin low. When the start conversion pin receives a negative-going pulse the conversion starts. At the end of the conversion the status pin goes low. The digital output is sent to an internal buffer where it is held until read as a result of the output enable pin being taken low.

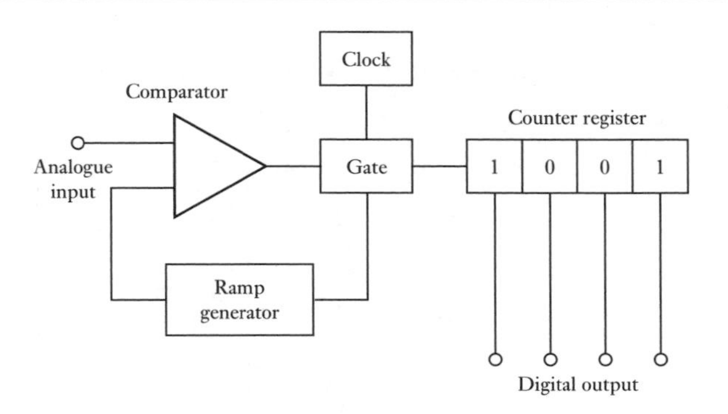

Figure 4.12 Ramp ADC.

The **ramp** form of ADC involves an analogue voltage which is increased at a constant rate, a so-called ramp voltage, and applied to a comparator where it is compared with the analogue voltage from the sensor. The time taken for the ramp voltage to increase to the value of the sensor voltage will depend on the size of the sampled analogue voltage. When the ramp voltage starts, a gate is opened which starts a binary counter counting the regular pulses from a clock. When the two voltages are equal, the gate closes and the word indicated by the counter is the digital representation of the sampled analogue voltage. Figure 4.12 indicates the subsystems involved in the ramp form of ADC.

The **dual ramp converter** is more common than the single ramp. Figure 4.13 shows the basic circuit. The analogue voltage is applied to an integrator which drives a comparator. The output from the comparator goes high as soon as the integrator output is more than a few millivolts. When the comparator output is high, an AND gate passes pulses to a binary counter. The counter counts pulses until it overflows. The counter then resets to zero, sends a signal to a switch which disconnects the unknown voltage and connects a reference voltage, and starts counting again. The polarity of the reference voltage is opposite to that of the input voltage. The integrator voltage then decreases at a rate proportional to the reference voltage. When the integrator output reaches zero, the comparator goes low, bringing the AND gate low and so switching the clock off. The count is then a measure of the analogue input voltage. Dual ramp ADCs have excellent noise rejection

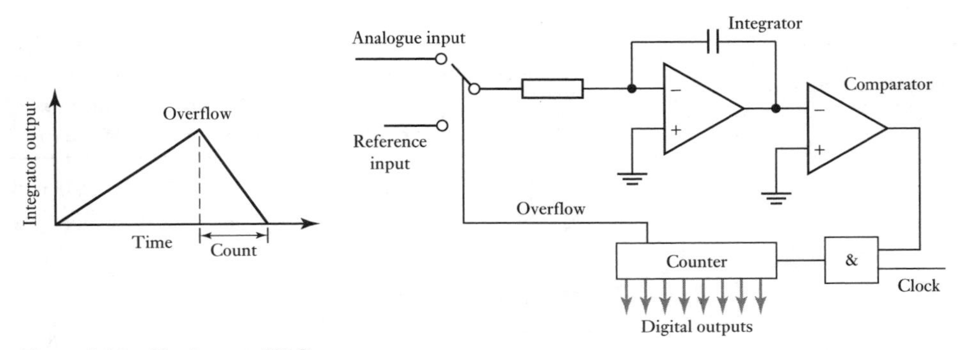

Figure 4.13 Dual ramp ADC.

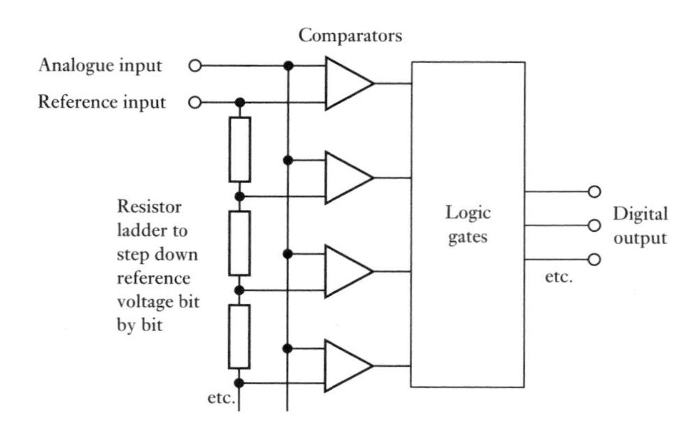

Figure 4.14 Flash ADC.

because the integral action averages out random negative and positive contributions over the sampling period. They are, however, very slow.

The **flash ADC** is very fast. For an n-bit converter, $2^n - 1$ separate voltage comparators are used in parallel, with each having the analogue input voltage as one input (Figure 4.14). A reference voltage is applied to a ladder of resistors so that the voltage applied as the other input to each comparator is 1 bit larger in size than the voltage applied to the previous comparator in the ladder. Thus when the analogue voltage is applied to the ADC, all those comparators for which the analogue voltage is greater than the reference voltage of a comparator will give a high output and those for which it is less will be low. The resulting outputs are fed in parallel to a logic gate system which translates them into a digital word.

In considering the specifications of ADCs the following terms will be encountered:

1 *Conversion time*, i.e. the time required to complete a conversion of the input signal. It establishes the upper signal frequency that can be sampled without aliasing; the maximum frequency is $1/(2 \times \text{conversion time})$.
2 *Resolution*, this being the full-scale signal divided by 2^n, where n is the number of bits. It is often just specified by a statement of the number of bits.
3 *Linearity error*, this being the deviation from a straight line drawn through zero and full-scale. It is a maximum of $\pm\frac{1}{2}$ LSB.

Table 4.1 shows some specification details of commonly used ADCs.

Table 4.1 ADCs.

ADC	Type	Resolution (bits)	Conversion time (ns)	Linearity error (LSB)
ZN439	SA	8	5 000	±1/2
ZN448E	SA	8	9 000	±1/2
ADS7806	SA	12	20 000	±1/2
ADS7078C	SA	16	20 000	±1/2
ADC302	F	8	20	±1/2

SA = successive approximations, F = flash.

4.3.3 Sample and hold amplifiers

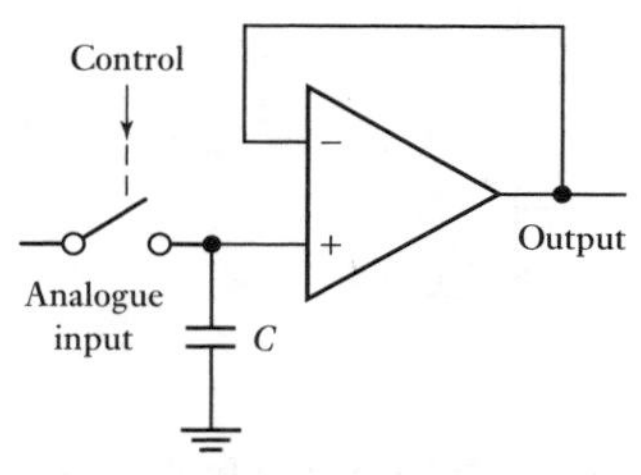

Figure 4.15 Sample and hold.

It takes a finite time for an ADC to convert an analogue signal to digital and problems can arise if the analogue signal changes during the conversion time. To overcome this, a sample and hold system is used to sample the analogue signal and hold it while the conversion takes place.

The basic circuit (Figure 4.15) consists of an electronic switch to take the sample, with a capacitor for the hold and an operational amplifier voltage follower. The electronic switch is controlled so that the sample is taken at the instant dictated by the control input. When the switch closes, the input voltage is applied across the capacitor and the output voltage becomes the same as the input voltage. If the input voltage changes while the switch is closed, the voltage across the capacitor and the output voltage change accordingly. When the switch opens, the capacitor retains its charge and the output voltage remains equal to the input voltage at the instant the switch was opened. The voltage is thus held until such time as the switch closes again. The time required for the capacitor to charge to a new sample of the input analogue voltage is called the **acquisition time** and depends on the value of the capacitance and the circuit resistance when the switch is on. Typical values are of the order of 4 μs.

4.4 Multiplexers

A **multiplexer** is a circuit that is able to have inputs of data from a number of sources and then, by selecting an input channel, give an output from just one of them. In applications where there is a need for measurements to be made at a number of different locations, rather than use a separate ADC and microprocessor for each measurement, a multiplexer can be used to select each input in turn and switch it through a single ADC and microprocessor (Figure 4.16). The multiplexer is essentially an electronic switching device which enables each of the inputs to be sampled in turn.

Figure 4.16 Multiplexer.

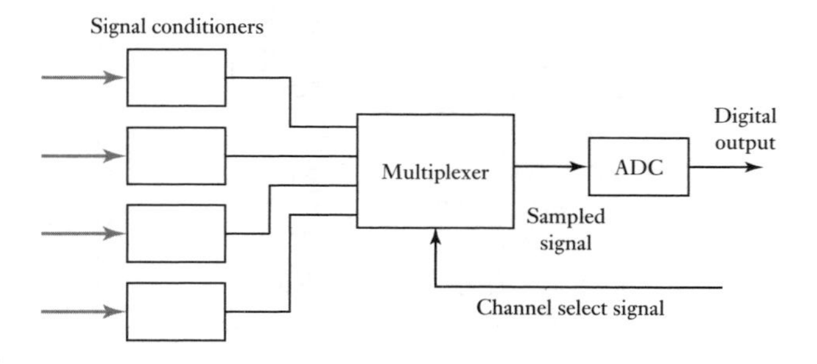

As an illustration of the types of analogue multiplexers available, the DG508ACJ has eight input channels with each channel having a 3-bit binary address for selection purposes. The transition time between taking samples is 0.6 μs.

4.4.1 Digital multiplexer

Figure 4.17 shows the basic principle of a multiplexer which can be used to select digital data inputs; for simplicity only a two-input channel system is

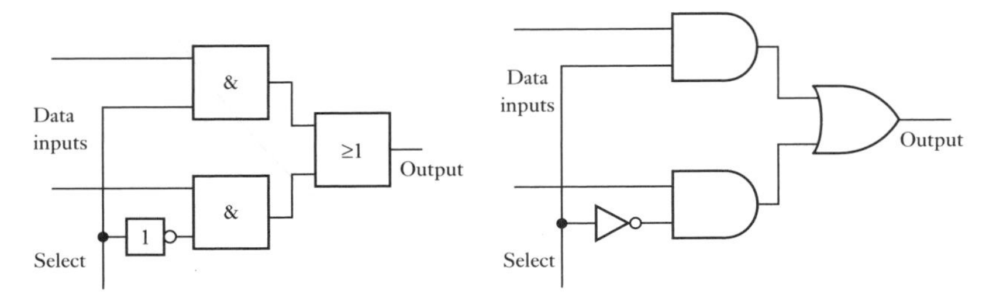

Figure 4.17 Two-channel multiplexer.

shown. The logic level applied to the select input determines which AND gate is enabled so that its data input passes through the OR gate to the output (see the next chapter for a discussion of such gates). A number of forms of multiplexers are available in integrated packages. The 151 types enable one line from eight to be selected, the 153 types one line from four inputs which are supplied as data on two lines each, the 157 types one line from two inputs which are supplied as data on four lines.

4.4.2 Time division multiplexing

Often there is a need for a number of peripheral devices to share the same input/output lines from a microprocessor. So that each peripheral can be supplied with different data it is necessary to allocate each a particular time slot during which data is transmitted. This is termed **time division multiplexing.** Figure 4.18 illustrates how this can be used to drive two display devices. In Figure 4.18(a) the system is not time multiplexed, in (b) it is.

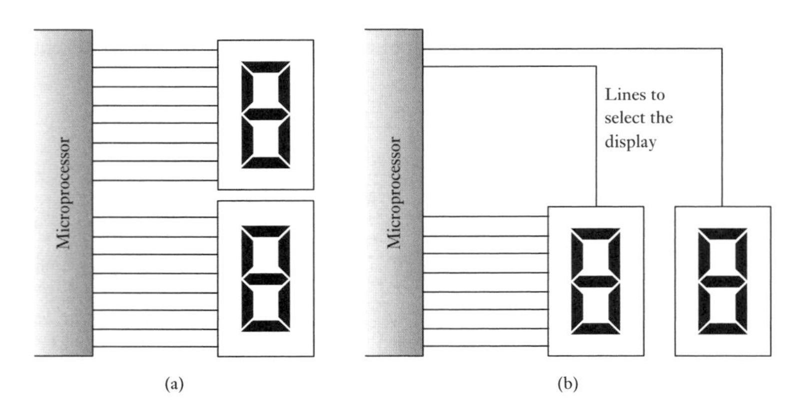

Figure 4.18 Time division multiplexing.

4.5 Data acquisition

The term **data acquisition**, or **DAQ**, is used for the process of taking data from sensors and inputting that data into a computer for processing. The sensors are connected, generally via some signal conditioning, to a data acquisition board which is plugged into the back of a computer (Figure 4.19(a)). The DAQ board is a printed circuit board that, for analogue inputs, basically provides a multiplexer, amplification, analogue-to-digital conversion, registers and control circuitry so that sampled digital signals are applied to the computer system. Figure 4.19(b) shows the basic elements of such a board.

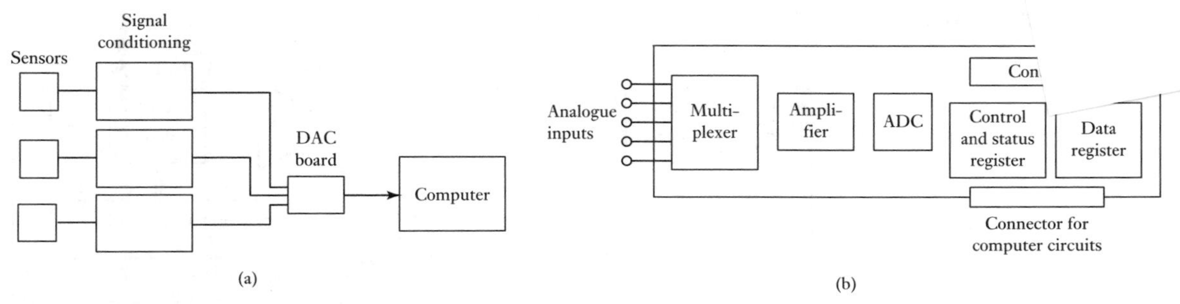

Figure 4.19 DAQ system.

Computer software is used to control the acquisition of data via the DAQ board. When the program requires an input from a particular sensor, it activates the board by sending a control word to the control and status register. Such a word indicates the type of operation that the board has to carry out. As a consequence the board switches the multiplexer to the appropriate input channel. The input from the sensor connected to that input channel is then passed via an amplifier to the ADC. After conversion the resulting digital signal is passed to the data register and the word in the control and status register changes to indicate that the signal has arrived. Following that signal, the computer then issues a signal for the data to be read and taken into the computer for processing. This signal is necessary to ensure that the computer does not wait doing nothing while the board carries out its acquisition of data, but uses this to signal to the computer when the acquisition is complete and then the computer can interrupt any program it is implementing, read the data from the DAQ and then continue with its program. A faster system does not involve the computer in the transfer of the data into memory but transfers the acquired data directly from the board to memory without involving the computer, this being termed **direct memory address** (DMA).

The specifications for a DAQ board include the sampling rate for analogue inputs, which might be 100 kS/s (100 000 samples per second). The Nyquist criteria for sampling indicate that the maximum frequency of analogue signal that can be sampled with such a board is 50 kHz, the sample rate having to be twice the maximum frequency component. In addition to the above basic functions of a DAQ board, it may also supply analogue outputs, timers and counters which can be used to provide triggers for the sensor system.

As an example of a low-cost multifunction board for use with an IBM computer, Figure 4.20 shows the basic structure of the National Instruments DAQ board PC-LPM-16. This board has 16 analogue input channels, a sampling rate of 50 kS/s, an 8-bit digital input and an 8-bit digital output, and a counter/timer which can give an output. Channels can be scanned in sequence, taking one reading from each channel in turn, or there can be continuous scanning of a single channel.

4.5.1 Data accuracy

An advantage of digital signal processing is that two voltage ranges are used rather than two exact voltage levels to distinguish between the two binary states for each bit. Thus data accuracy is less affected by noise, drift,

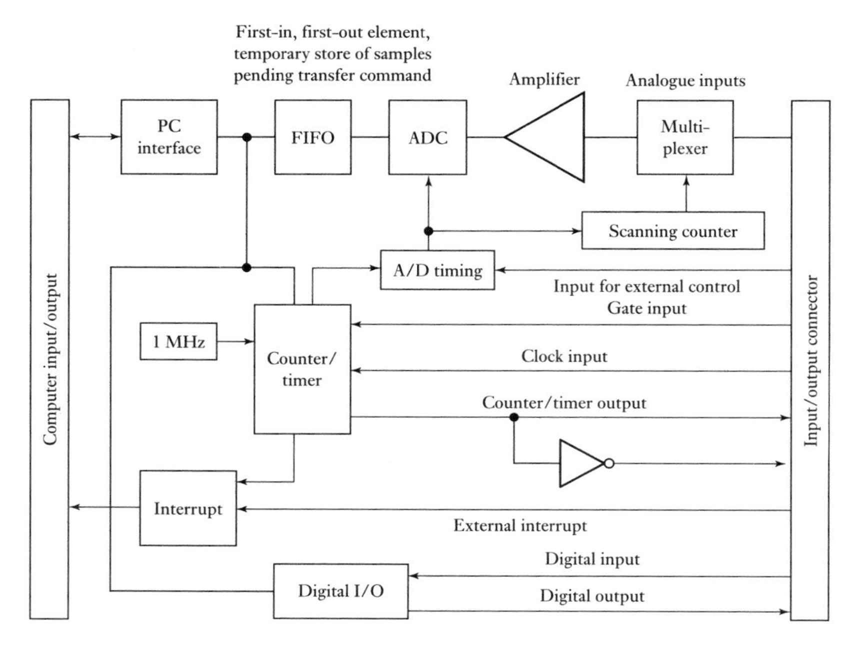

Figure 4.20 PC-LPM-16 DAQ board.

component tolerances, and other factors causing fluctuations in voltages which would be critical for transmission as analogue voltages. For example, in a 5 V system, the difference between the two binary states is typically a minimum of 3 V. So two signals could be 0 and 5 V or 1 V and 4 V and still be distinguished as 0 and 1.

4.5.2 Parity method for error detection

The movement of digital data from one location to another can result in transmission errors, the receiver not receiving the same signal as transmitted by the transmitter as a result of electrical noise in the transmission process. Sometimes a noise pulse may be large enough at some point to alter the logic level of the signal. For example, the sequence 1001 may be transmitted and be received as though 1101. In order to detect such errors a **parity bit** is often used. A parity bit is an extra 0 or 1 bit attached to a code group at transmission. In **even parity** the value of the bit is chosen so that the total number of ones in the code group, including the parity bit, is an even number. For example, in transmitting 1001 the parity bit used would be 0 to give 01001 and so an even number of ones. In transmitting 1101 the parity bit used would be 1 to give 11101 and so an even number of ones. With **odd parity** the parity bit is chosen so that the total number of ones, including the parity bit, is odd. Thus if at the receiver the number of ones in a code group does not give the required parity, the receiver will know that there is an error and can request the code group be retransmitted.

An extension of the parity check is the **sum check** in which blocks of code may be checked by sending a series of bits representing their binary sum. Parity and sum checks can only detect single errors in blocks of code; double errors go undetected. Also the error is not located so that correction by the receiver can be made. Multiple error detection techniques and methods to pinpoint errors have been devised.

4.6 Digital signal processing

The term **digital signal processing or discrete-time signal processing** is used for the processing applied to a signal by a microprocessor. Digital signals are discrete-time signals in that they are not continuous functions of time but exist at only discrete times. Whereas signal conditioning of analogue signals requires components such as amplifiers and filter circuits, digital signal conditioning can be carried out by a program applied to a microprocessor, i.e. processing the signal. To change the characteristics of a filter used with analogue signals it is necessary to change hardware components, whereas to change the characteristics of a digital filter all that is necessary is to change the software, i.e. the program of instructions given to a microprocessor.

With a digital signal processing system there is an input of a word representing the size of a pulse and an output of another word. The output pulse at a particular instant is computed by the system as a result of processing the present input pulse, together with previous pulse inputs and possibly previous system outputs.

For example, the program used by the microprocessor might read the value of the present input and add to it the previous output value to give the new output. If we consider the present input to be the kth pulse in the input sequence of pulses we can represent this pulse as $x[k]$. The kth output of a sequence of pulses can be represented by $y[k]$. The previous output, i.e. the $(k - 1)$th pulse, can be represented by $y[k - 1]$. Thus we can describe the program which gives an output obtained by adding to the value of the present input the previous output by

$$y[k] = x[k] + y[k - 1]$$

Such an equation is called a **difference equation.** It gives the relationship between the output and input for a discrete-time system and is comparable with a differential equation which is used to describe the relationship between the output and input for a system having inputs and outputs which vary continuously with time.

For the above difference equation, suppose we have an input of a sampled sine wave signal which gives a sequence of pulses of

$$0.5, 1.0, 0.5, -0.5, -1.0, -0.5, 0.5, 1.0, \ldots$$

The $k = 1$ input pulse has a size of 0.5. If we assume that previously the output was zero then $y[k - 1] = 0$ and so $y[1] = 0.5 + 0 = 0.5$. The $k = 2$ input pulse has a size of 1.0 and so $y[2] = x[2] + y[2 - 1] = 1.0 + 0.5 = 1.5$. The $k = 3$ input pulse has a size of 0.5 and so $y[3] = x[3] + y[3 - 1] = 0.5 + 1.5 = 2.0$. The $k = 4$ input pulse has a size of -0.5 and so $y[4] = x[4] + y[4 - 1] = -0.5 + 2.0 = 1.5$. The $k = 5$ input pulse has a size of -1.0 and so $y[5] = x[5] + y[5 - 1] = -1.0 + 1.5 = 0.5$. The output is thus the pulses

$$0.5, 1.5, 2.0, 1.5, 0.5, \ldots$$

We can continue in this way to obtain the output for all the pulses.

As another example of a difference equation we might have

$$y[k] = x[k] + ay[k - 1] - by[k - 2]$$

The output is the value of the current input plus a times the previous output and minus b times the last but one output. If we have $a = 1$ and $b = 0.5$ and consider the input to be the sampled sine wave signal considered above, then the output now becomes

$$0.5, 1.5, 1.75, 0.5, -1.37, \ldots$$

We can have a difference equation which produces an output which is similar to that which would have been obtained by integrating a continuous-time signal. Integration of a continuous-time signal between two times can be considered to be the area under the continuous-time function between those times. Thus if we consider two discrete-time signals $x[k]$ and $x[k-1]$ occurring with a time interval of T between them (Figure 4.21), the change in area is $\frac{1}{2}T(x[k] + x[k-1])$. Thus if the output is to be the sum of the previous area and this change in area, the difference equation is

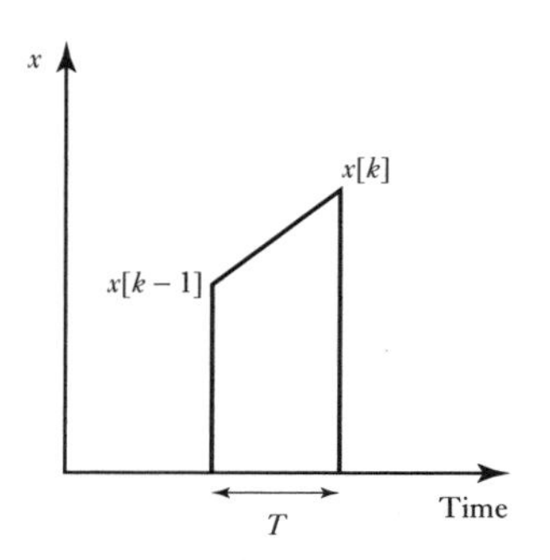

Figure 4.21 Integration.

$$y[\kappa] = y[k-1] + \tfrac{1}{2}T(x[k] + x[k-1])$$

This is known as *Tustin's approximation* for integration.

Differentiation can be approximated by determining the rate at which the input changes. Thus when the input changes from $x[k-1]$ to $x[k]$ in time T the output is

$$y[k] = (x[k] - x[k-1])/T$$

Summary

Analogue-to-digital conversion involves converting analogue signals into binary words. A clock supplies a regular time signal to the analogue-to-digital converter (ADC) and it samples the analogue signal at each clock pulse. A sample and hold unit then holds each sampled value until the next pulse occurs. Forms of ADC are successive approximations, ramp, dual ramp and flash.

Digital-to-analogue conversion involves converting a binary word into an analogue signal. Forms of digital-to-analogue converters (DACs) are weighted-resistor and R–$2R$ ladder.

A **multiplexer** is a circuit that is able to have inputs of data from a number of sources and then, by selecting an input channel, give an output for just one of them.

The term **data acquisition,** or DAQ, is used for the process of taking data from sensors and inputting that data into a computer for processing.

The term **digital signal processing** or **discrete-time signal processing** is used for the processing applied to a signal by a microprocessor.

Problems

4.1 What is the resolution of an ADC with a word length of 12 bits and an analogue signal input range of 100 V?

4.2 A sensor gives a maximum analogue output of 5 V. What word length is required for an ADC if there is to be a resolution of 10 mV?

4.3 An R–$2R$ DAC ladder of resistors has its output fed through an inverting operational amplifier with a feedback resistance of $2R$. If the reference voltage is 5 V, determine the resolution of the converter.

4.4 For a binary weighted-resistor DAC how should the values of the input resistances be weighted for a 4-bit DAC?

4.5 What is the conversion time for a 12-bit ADC with a clock frequency of 1 MHz?

4.6 In monitoring the inputs from a number of thermocouples the following sequence of modules is used for each thermocouple in its interface with a microprocessor

Protection, cold junction compensation, amplification, linearisation, sample and hold, analogue-to-digital converter, buffer, multiplexer.

Explain the function of each of the modules.

4.7 Suggest the modules that might be needed to interface the output of a microprocessor with an actuator.

4.8 For the 4-bit weighted-resistor DAC shown in Figure 4.5, determine the output from the resistors to the amplifier for inputs of 0001, 0010, 0100 and 1000 if the inputs are 0 V for a logic 0 and 5 V for a logic 1.

4.9 If the smallest resistor in a 16-bit weighted-resistor DAC is R, how big would the largest resistor need to be?

4.10 A 10-bit ramp ADC has a full-scale input of 10 V. How long will it take to convert such a full-scale input if the clock period is 15 μs?

4.11 For a 12-bit ADC with full-scale input, how much faster will a successive approximations ADC be than a ramp ADC?

Chapter five Digital Logic

Objectives

The objectives of this chapter are that, after studying it, the reader should be able to:

- **Recognise the symbols used for the logic gates AND, OR, NOT, NAND, NOR and XOR, and use such gates in applications, recognising the significance of logic families.**
- **Explain how SR, JK and D flip-flops can be used in control systems.**
- **Explain the operation of decoders and the 555 timer.**

5.1 Digital logic

Many control systems are concerned with setting events in motion or stopping them when certain conditions are met. For example, with the domestic washing machine, the heater is only switched on when there is water in the drum and it is to the prescribed level. Such control involves *digital* signals where there are only two possible signal levels. Digital circuitry is the basis of digital computers and microprocessor controlled systems.

With **digital control** we might, for example, have the water input to the domestic washing machine switched on if we have both the door to the machine closed and a particular time in the operating cycle has been reached. There are two input signals which are either yes or no signals and an output signal which is a yes or no signal. The controller is here programmed to only give a yes output if both the input signals are yes, i.e. if input A and input B are both 1 then there is an output of 1. Such an operation is said to be controlled by a **logic gate**, in this example an AND gate. There are many machines and processes which are controlled in this way. The term **combinational logic** is used for the combining of two or more basic logic gates to form a required function. For example, a requirement might be that a buzzer sounds in a car if the key is in the ignition and a door is opened or if the headlights are on and the driver's door is opened. Combinational logic depends only on the values of the inputs at a particular instant of time.

In addition to a discussion of combinational logic, this chapter also includes a discussion of **sequential logic**. Such digital circuitry is used to exercise control in a specific sequence dictated by a control clock or enable–disable control signals. These are combinational logic circuits with memory. Thus the timing or sequencing history of the input signals plays a part in determining the output.

5.2 Logic gates

Logic gates are the basic building blocks for digital electronic circuits.

5.2.1 AND gate

Suppose we have a gate giving a high output only when both input A and input B are high; for all other conditions it gives a low output. This is an AND logic gate. We can visualise the AND gate as an electric circuit involving two switches in series (Figure 5.1(a)). Only when switch A and switch B are closed is there a current. Different sets of standard circuit symbols for logic gates have been used, with the main form being that originated in the United States. An international standard form (IEEE/ANSI), however, has now been developed; this removes the distinctive shape and uses a rectangle with the logic function written inside it. Figure 5.1(b) shows the US form of symbol used for an AND gate and (c) shows the new standardised form, the & symbol indicating AND. Both forms will be used in this book. As illustrated in the figure, we can express the relationship between the inputs and the outputs of an AND gate in the form of an equation, termed a **Boolean equation** (see Appendix C). The Boolean equation for the AND gate is written as

$$A \cdot B = Q$$

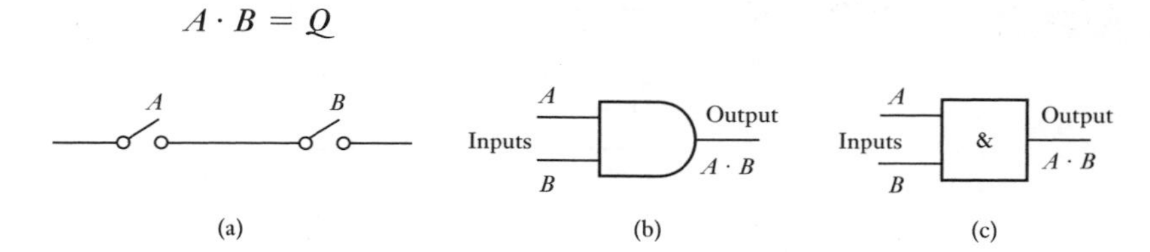

Figure 5.1 AND gate: (a) represented by switches, (b) US symbols, (c) new standardised symbols.

An example of an AND gate is an interlock control system for a machine tool such that if the safety guard is in place and gives a 1 signal and the power is on, giving a 1 signal, then there can be an output, a 1 signal, and the machine operates. Another example is a burglar alarm in which it gives an output, the alarm sounding, when the alarm is switched on and when a door is opened to activate a sensor.

The relationships between inputs to a logic gate and the outputs can be tabulated in a form known as a **truth table**. This specifies the relationships between the inputs and outputs. Thus for an AND gate with inputs A and B and a single output Q, we will have a 1 output when, and only when, $A = 1$ and $B = 1$. All other combinations of A and B will generate a 0 output. We can thus write the truth table as

Inputs		Output
A	B	Q
0	0	0
0	1	0
1	0	0
1	1	1

Consider what happens when we have two digital inputs which are functions of time, as in Figure 5.2. Such a figure is termed an AND gate timing diagram. There will only be an output from the AND gate when each of the inputs is high and thus the output is as shown in the figure.

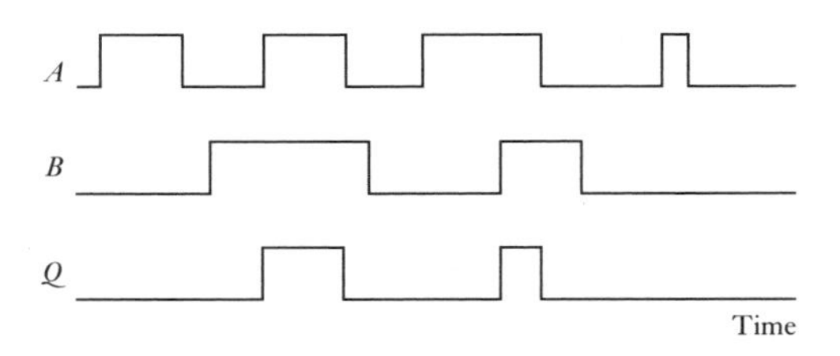

Figure 5.2 AND gate.

5.2.2 OR gate

An OR gate with inputs A and B gives an output of a 1 when A or B is 1. We can visualise such a gate as an electric circuit involving two switches in parallel (Figure 5.3(a)). When switch A or B is closed, then there is a current. OR gates can also have more than two inputs. The truth table for the gate is

Inputs		Output
A	B	Q
0	0	0
0	1	1
1	0	1
1	1	1

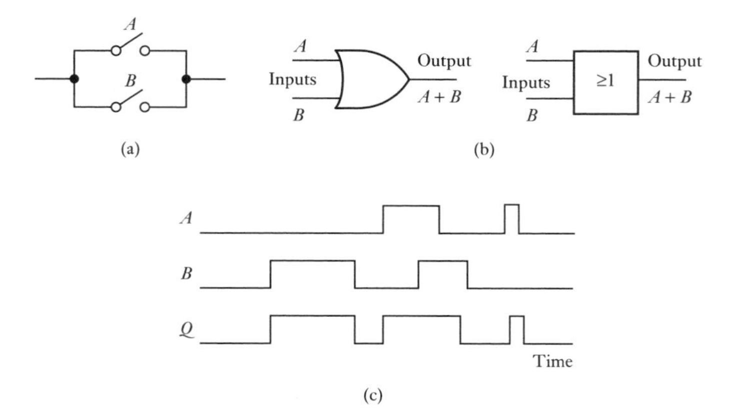

Figure 5.3 OR gate: (a) representation by switches, (b) symbols, (c) timing diagram.

We can write the Boolean equation for an OR gate as

$$A + B = Q$$

The symbols used for an OR gate are shown in Figure 5.3(b); the use of a greater than or equal to 1 sign to depict OR arises from the OR function being true if at least more than one input is true. Figure 5.3(c) shows a timing diagram.

5.2.3 NOT gate

A NOT gate has just one input and one output, giving a 1 output when the input is 0 and a 0 output when the input is 1. The NOT gate gives an output which is the inversion of the input and is called an **inverter**. Figure 5.4(a) shows the symbols used for a NOT gate. The 1 representing NOT actually symbolises logic identity, i.e. no operation, and the inversion is depicted by the circle on the output. Thus if we have a digital input which varies with time, as in Figure 5.4(b), the out variation with time is the inverse.

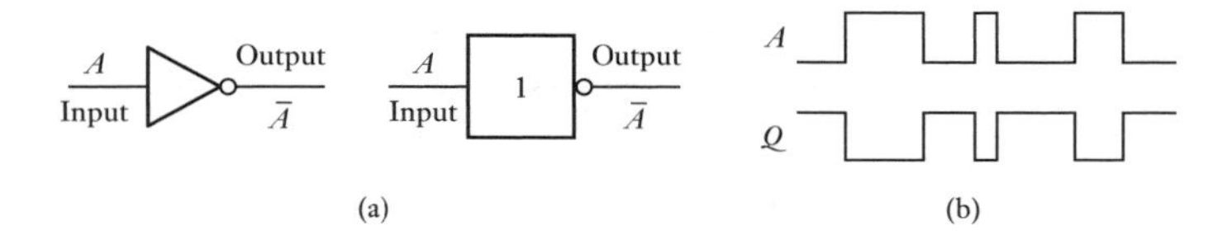

Figure 5.4 NOT gate.

The following is the truth table for the NOT gate:

Input A	Output Q
0	1
1	0

The Boolean equation describing the NOT gate is

$$\bar{A} = Q$$

A bar over a symbol is used to indicate that the inverse, or complement, is being taken; thus the bar over the A indicates that the output Q is the inverse value of A.

5.2.4 NAND gate

The NAND gate can be considered as a combination of an AND gate followed by a NOT gate (Figure 5.5(a)). Thus when input A is 1 and input B is 1, there is an output of 0, all other inputs giving an output of 1.

The NAND gate is just the AND gate truth table with the outputs inverted. An alternative way of considering the gate is as an AND gate with a NOT gate applied to invert both the inputs before they reach the AND gate. Figure 5.5(b) shows the symbols used for the NAND gate, being the

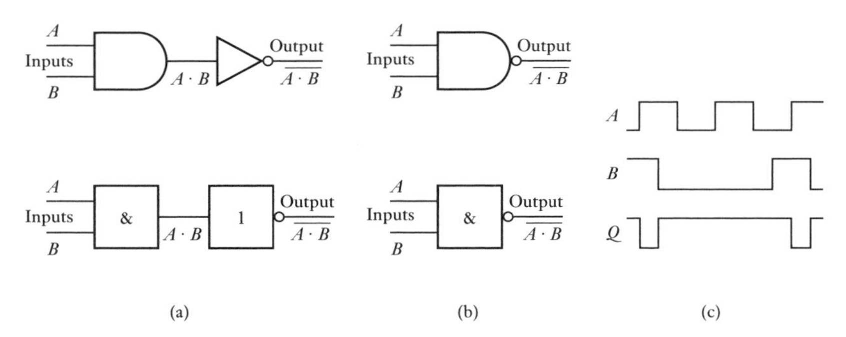

Figure 5.5 NAND gate.

AND symbol followed by the circle to indicate inversion. The following is the truth table:

Inputs		Output
A	*B*	*Q*
0	0	1
0	1	1
1	0	1
1	1	0

The Boolean equation describing the NAND gate is

$$\overline{A \cdot B} = Q$$

Figure 5.5(c) shows the output that occurs for a NAND gate when its two inputs are digital signals which vary with time. There is only a low output when both the inputs are high.

5.2.5 NOR gate

The NOR gate can be considered as a combination of an OR gate followed by a NOT gate (Figure 5.6(a)). Thus when input *A* or input *B* is 1 there is an output of 0. It is just the OR gate with the outputs inverted. An alternative way of considering the gate is as an OR gate with a NOT gate applied to invert both the inputs before they reach the OR gate. Figure 5.6(b) shows the symbols used for the NOR gate; it is the OR symbol followed by the circle to

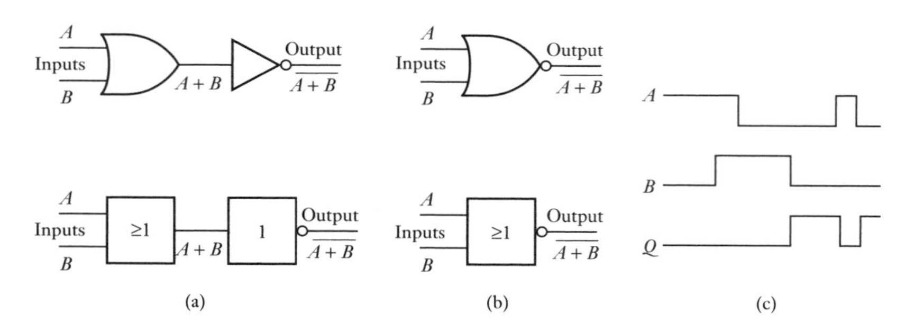

Figure 5.6 NOR gate.

indicate inversion. The Boolean equation for the NOR gate is

$$\overline{A + B} = Q$$

The following is the truth table for the NOR gate and Figure 5.6(c) shows its timing diagram:

Inputs		Output
A	B	Q
0	0	1
0	1	0
1	0	0
1	1	0

5.2.6 XOR gate

The EXCLUSIVE-OR gate (XOR) can be considered to be an OR gate with a NOT gate applied to one of the inputs to invert it before the inputs reach the OR gate (Figure 5.7(a)). Alternatively it can be considered as an AND gate with a NOT gate applied to one of the inputs to invert it before the inputs reach the AND gate. The symbols are shown in Figure 5.7(b); the =1 depicts that the output is true if only one input is true. The following is the truth table and Figure 5.7(c) shows a timing diagram:

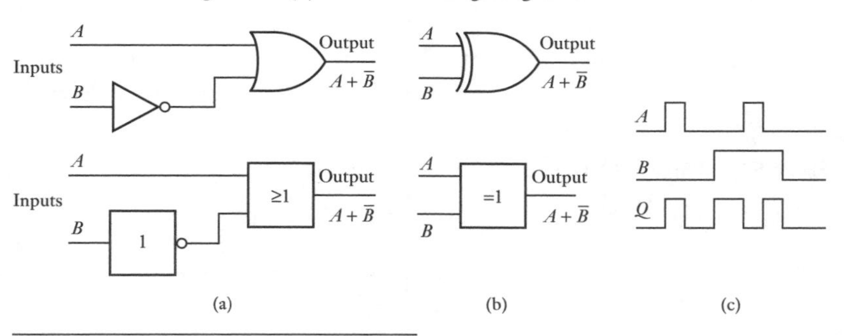

Figure 5.7 XOR gate.

Inputs		Output
A	B	Q
0	0	0
0	1	1
1	0	1
1	1	0

5.2.7 Combining gates

It might seem that to make logic systems we require a range of gates. However, as the following shows, we can make up all the gates from just one. Consider the combination of three NOR gates shown in Figure 5.8.

Figure 5.8 Three NOR gates.

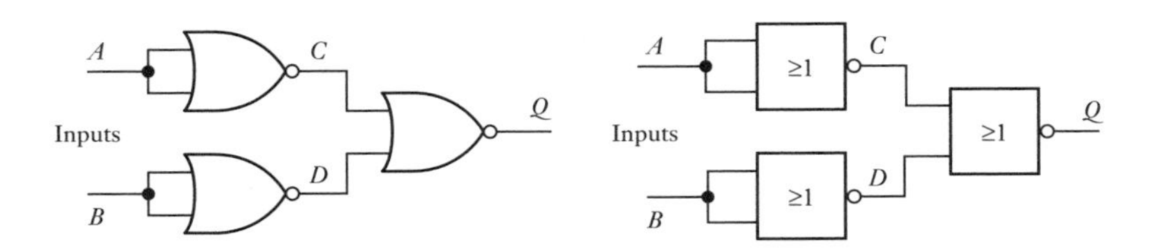

The truth table, with the intermediate and final outputs, is as follows:

A	*B*	*C*	*D*	*Q*
0	0	1	1	0
0	1	1	0	0
1	0	0	1	0
1	1	0	0	1

The result is the same as an AND gate. If we followed this assembly of gates by a NOT gate then we would obtain a truth table the same as a NAND gate.

A combination of three NAND gates is shown in Figure 5.9. The truth table, with the intermediate and final outputs, is as follows:

A	*B*	*C*	*D*	*Q*
0	0	1	1	0
0	1	1	0	1
1	0	0	1	1
1	1	0	0	1

Figure 5.9 Three NAND gates.

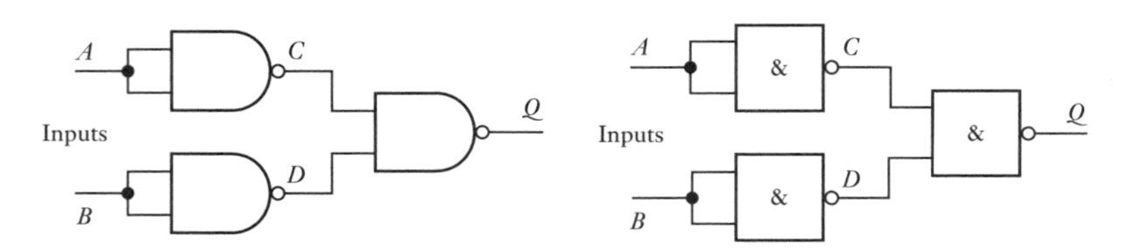

The result is the same as an OR gate. If we followed this assembly of gates by a NOT gate then we would obtain a truth table the same as a NOR gate.

The above two illustrations of gate combinations show how one type of gate, a NOR or a NAND, can be used to substitute for other gates, provided we use more than one gate. Gates can also be combined to make complex gating circuits and sequential circuits.

Logic gates are available as integrated circuits. The different manufacturers have standardised their numbering schemes so that the basic part numbers are the same regardless of the manufacturer. For example, Figure 5.10(a) shows the gate systems available in integrated circuit 7408; it has four two-input AND gates and is supplied in a 14-pin package. Power supply connections are made to pins 7 and 14, these supplying the operating voltage for all the four AND gates. In order to indicate at which end of the package pin 1 starts, a notch is cut between pins 1 and 14. Integrated circuit

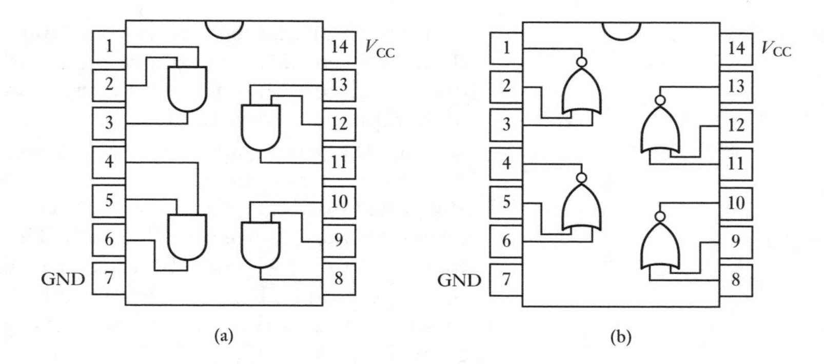

Figure 5.10 Integrated circuits: (a) 7408, (b) 7402.

7411 has three AND gates with each having three inputs; integrated circuit 7421 has two AND gates with each having four inputs. Figure 5.10(b) shows the gate systems available in integrated circuit 7402. This has four two-input NOR gates in a 14-pin package, power connections being to pins 7 and 14. Integrated circuit 7427 has three gates with each having three inputs; integrated circuit 7425 has two gates with each having four inputs.

For a discussion of how Boolean algebra and techniques such as De Morgan's law and Karnaugh maps can be used to generate the required logic functions from logic gates, see Appendix C.

5.2.8 Logic families and integrated circuits

In order to implement digital logic designs, it is necessary to understand the significance of logic families and their different operating principles. Integrated circuits made with the same technology and electrical characteristics comprise a **logic family**. Commonly encountered families are transistor–transistor logic (TTL), complementary metal-oxide semiconductor (CMOS) and emitter-coupled logic (ECL). The general parameters are:

1 **Logic level**, i.e. the range of voltage levels that can correspond to the binary 1 and 0 states. For the standard 74XX TTL series, the typical voltage guaranteed to register as binary 0 is between 0 and 0.4 V and for binary 1 between 2.4 V and 5.0 V. For CMOS, the levels depend on the supply voltage V_{DD} used. This can be from +3 V to +15 V and the maximum voltage for logic 1 is $0.3V_{DD}$ while the minimum for logic 1 is $0.7V_{DD}$.
2 **Noise immunity** or **noise margin**, i.e. the circuit's ability to tolerate noise without causing spurious changes in the output voltage. For the standard 74XX TTL series, the noise margin is 0.4 V. Thus 0.4 V is the leeway that can be accepted on the logic 0 and logic 1 inputs and they still register as 0 and 1. For CMOS the noise margin depends on the supply voltage and is $0.3V_{DD}$.
3 **Fan-out**, i.e. the number of gate inputs that can be driven by a standard gate output while maintaining the desired LOW or HIGH levels. This is determined by how much current a gate can supply and how much is needed to drive a gate. For a standard TTL gate, the fan-out is 10, for CMOS it is 50 and for ECL 25. If more gates are connected to the driver gate then it will not supply enough current to drive them.

4 **Current-sourcing** or **current-sinking action**, i.e. how the current flows between the output of one logic gate and the input of another. For one gate driving another, with current-sourcing the driving gate when high supplies a current to the input of the next gate. With current-sinking, the driver gate when low receives a current back to it from the driven gate. TTL gates operate as current sinking.
5 **Propagation delay time**, i.e. how fast a digital circuit responds to a change in the input level. Typically TTL gates have delay times of 2 to 40 ns, this being generally about 5 to 10 times faster than CMOS gates but slower than ECL gates which typically have propagation delays of 2 ns.
6 **Power consumption**, i.e. the amount of power the logic gate will drain from the power supply. TTL consumes about 10 mW per gate while CSMOS draws no power unless it is in the act of switching. ECL consumes about 25 to 60 mW per gate.

The main criteria generally involved in determining which logic family to use are propagation delay and power consumption. The principal advantage of CMOS over TTL is the low power consumption which makes it ideal for battery-operated equipment. It is possible for integrated circuits from different logic families to be connected together, but special interfacing techniques must be used.

The TTL family is widely used, the family being identified as the 74XX series. There are a number of forms. Typically, the standard TTL is 7400 with a power dissipation of 10 mW and a propagation delay of 10 ns. The low-power Schottky TTL (LS) is 74LS00 with a power dissipation of 2 mW and the same propagation delay. The advanced low power Schottky TTL (ALS) is 74ALS00 and is faster and dissipates even lower power, the propagation delay being 4 ns and the power dissipation 1 mW. The fast TTL(F) is 74F00 and has a propagation delay of 3 ns and power dissipation of 6 mW.

The CMOS family includes the 4000 series which had the low power dissipation advantage over the TTL series, but unfortunately was much slower. The 40H00 series was faster but still slower than TTL (LS). The 74C00 series was developed to be pin-compatible with the TTL family, using the same numbering system but beginning with 74C. While it has a power advantage over the TTL family, it is still slower. The 74HC00 and 74HCT00 are faster with speeds comparable with the TTL (LS) series.

5.3 Applications of logic gates

The following are some examples of the uses of logic gates for a number of simple applications.

5.3.1 Parity generators

In the previous chapter the use of parity bits as an error detection method was discussed. A single bit is added to each code block to force the number of ones in the block, including the parity bit, to be an odd number if odd parity is being used or an even number if even parity is being used.

Figure 5.11 shows a logic gate circuit that could be used to determine and add the appropriate parity bit. The system employs XOR gates; with an XOR gate if all the inputs are 0 or all are 1 the output is 0, and if the inputs are not equal the output is a 1. Pairs of bits are checked and an output of 1 given if they are not equal. If odd parity is required the bias bit is 0; if

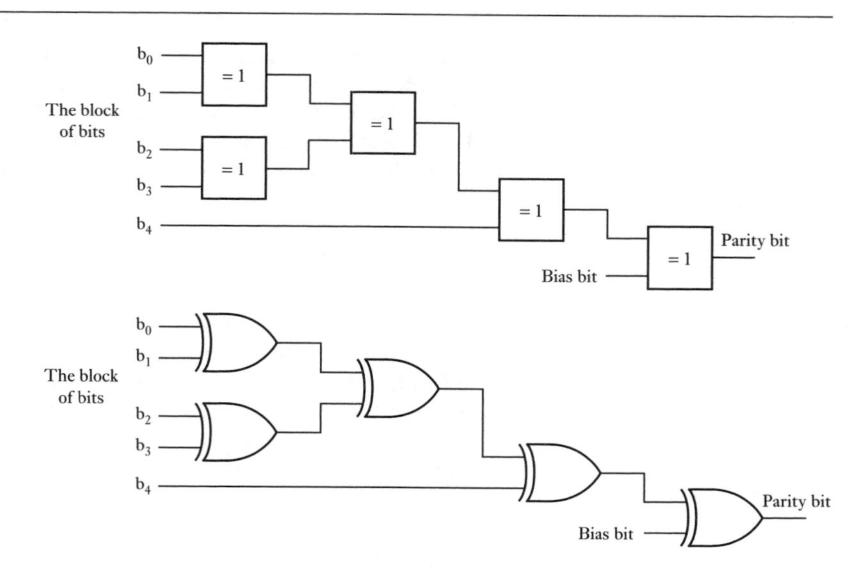

Figure 5.11 Parity bit generator.

even parity it is 1. The appropriate bias bit can then be added to the signal for transmission. The same circuit can be used to check the parity at the receiver, with the final output being a 1 when there is an error. Such circuits are available as integrated circuits.

5.3.2 Digital comparator

A digital comparator is used to compare two digital words to determine if they are exactly equal. The two words are compared bit by bit and a 1 output given if the words are equal. To compare the equality of two bits, an XOR gate can be used; if the bits are both 0 or both 1 the output is 0, and if they are not equal the output is a 1. To obtain a 1 output when the bits are the same we need to add a NOT gate, this combination of XOR and NOT being termed an XNOR gate. To compare each of the pairs of bits in two words we need an XNOR gate for each pair. If the pairs are made up of the same bits then the output from each XNOR gate is a 1. We can then use an AND gate to give a 1 output when all the XNOR outputs are ones. Figure 5.12 shows the system.

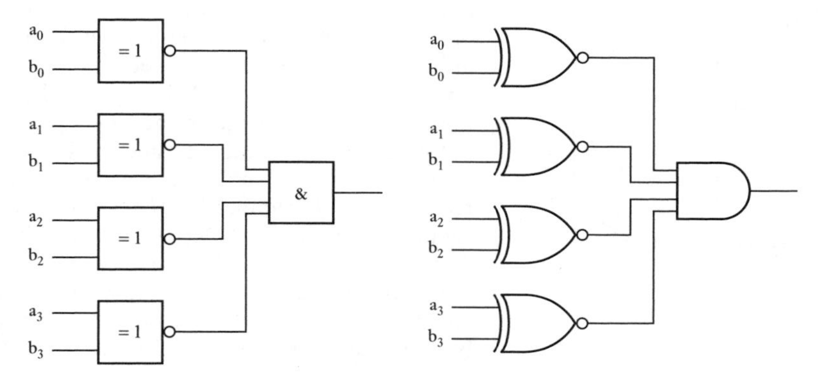

Figure 5.12 Comparator.

Digital comparators are available as integrated circuits and can generally determine not only if two words are equal but which one is greater than the other. For example, the 7485 4-bit magnitude comparator compares two 4-bit words A and B, giving a 1 output from pin 5 if A is greater than B, a 1 output from pin 6 if A equals B and a 1 output from pin 7 if A is less than B.

5.3.3 Coder

Figure 5.13 shows a simple system by which a controller can send a coded digital signal to a set of traffic lights so that the code determines which light, red, amber or green, will be turned on. To illuminate the red light we might use the transmitted signal $A = 0$, $B = 0$, for the amber light $A = 0$, $B = 1$ and for the green light $A = 1$, $B = 0$. We can switch on the lights using these codes by using three AND gates and two NOT gates.

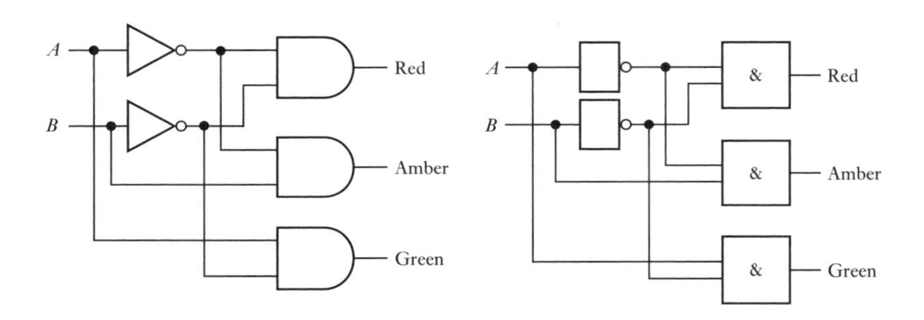

Figure 5.13 Traffic lights.

5.3.4 Code converter

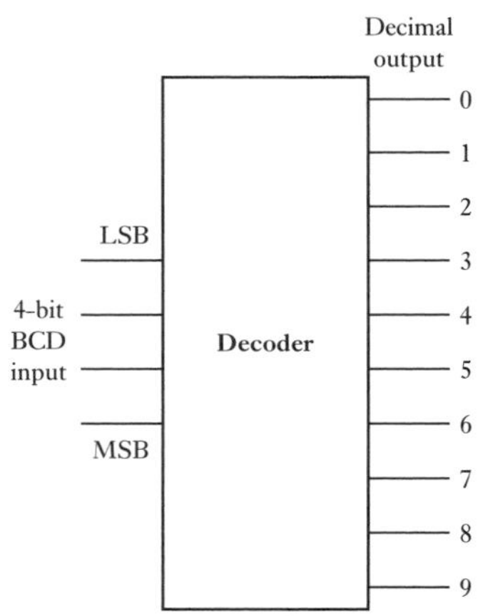

Figure 5.14 Decoder.

In many applications there is a need to change data from one type of code to another. For example, the output from a microprocessor system might be BCD (Binary-Coded Decimal) and need to be transformed into a suitable code to drive a seven-segment display. The term **data decoding** is used for the process of converting some code group, e.g. BCD, binary, hex, into an individual active output representing that group. A decoder has n binary input lines for the coded input of an n-bit word and gives m output lines such that only one line is activated for one possible combination of inputs, i.e. only one output line gives an output for a particular word input code. For example, a BCD-to-decimal decoder has a 4-bit input code and 10 output lines so that a particular BCD input will give rise to just one of the output lines being activated and so indicate a particular decimal number with each output line corresponding to a decimal number (Figure 5.14).

Thus, in general, a **decoder** is a logic circuit that looks at its inputs, determines which number is there, and activates the one output that corresponds to that number. Decoders are widely used in microprocessor circuits.

Decoders can have the active output high and the inactive ones low or the active output low and the inactive ones high. For active-high output a decoder can be assembled from AND gates, while for active-low output NAND gates can be used. Figure 5.15 shows how a BCD-to-decimal

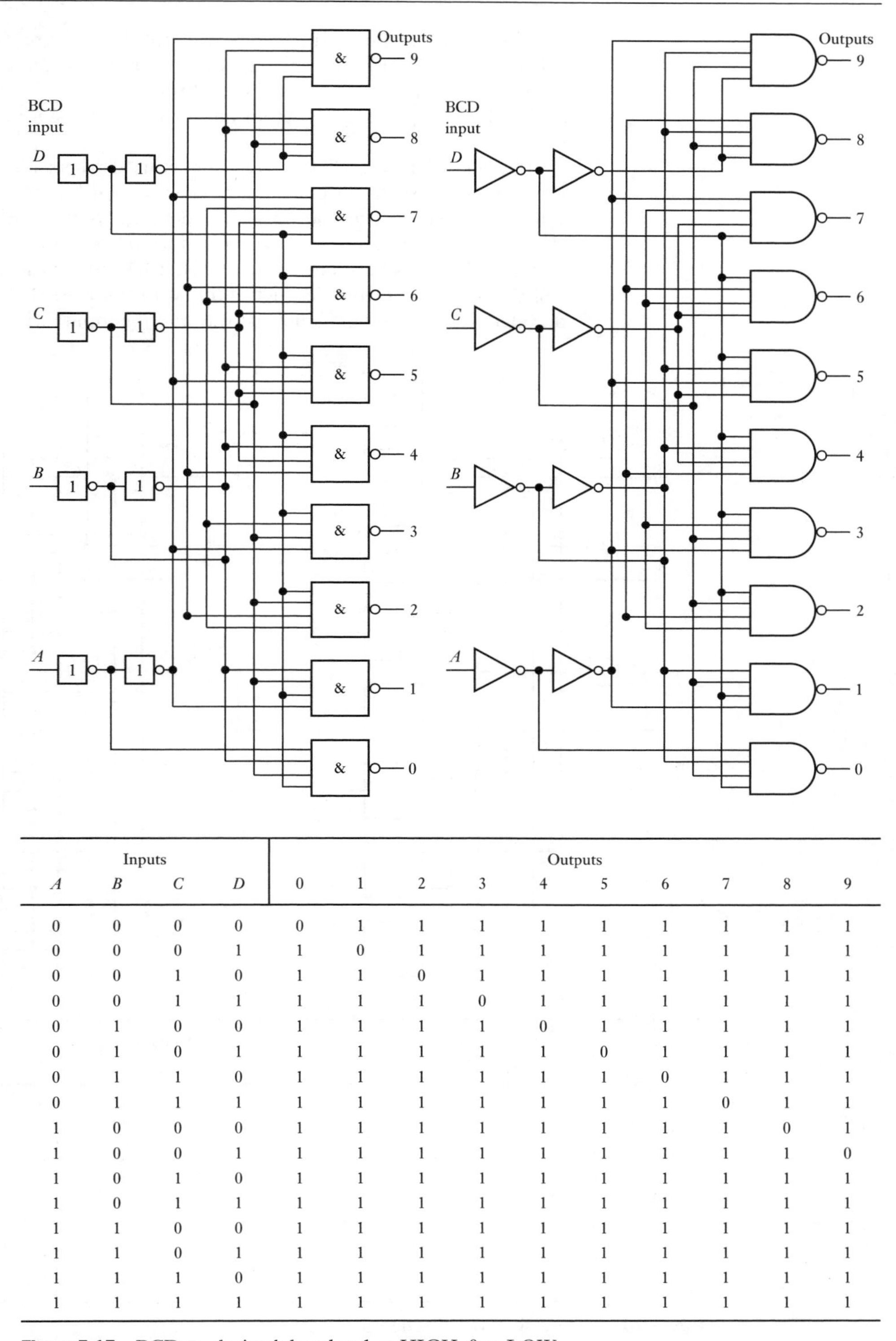

Inputs				Outputs									
A	*B*	*C*	*D*	0	1	2	3	4	5	6	7	8	9
0	0	0	0	0	1	1	1	1	1	1	1	1	1
0	0	0	1	1	0	1	1	1	1	1	1	1	1
0	0	1	0	1	1	0	1	1	1	1	1	1	1
0	0	1	1	1	1	1	0	1	1	1	1	1	1
0	1	0	0	1	1	1	1	0	1	1	1	1	1
0	1	0	1	1	1	1	1	1	0	1	1	1	1
0	1	1	0	1	1	1	1	1	1	0	1	1	1
0	1	1	1	1	1	1	1	1	1	1	0	1	1
1	0	0	0	1	1	1	1	1	1	1	1	0	1
1	0	0	1	1	1	1	1	1	1	1	1	1	0
1	0	1	0	1	1	1	1	1	1	1	1	1	1
1	0	1	1	1	1	1	1	1	1	1	1	1	1
1	1	0	0	1	1	1	1	1	1	1	1	1	1
1	1	0	1	1	1	1	1	1	1	1	1	1	1
1	1	1	0	1	1	1	1	1	1	1	1	1	1
1	1	1	1	1	1	1	1	1	1	1	1	1	1

Figure 5.15 BCD-to-decimal decoder: 1 = HIGH, 0 = LOW.

decoder for active-low output can be assembled and the resulting truth table. Such a decoder is readily available as an integrated circuit, e.g. 74LS145.

A decoder that is widely used is BCD-to-seven, e.g. 74LS244, for taking a 4-bit BCD input and giving an output to drive the seven segments of a display.

The term **3-line-to-8-line decoder** is used where a decoder has three input lines and eight output lines. It takes a 3-bit binary number and activates the one of the eight outputs corresponding to that number. Figure 5.16 shows how such a decoder can be realised from logic gates and its truth table.

Some decoders have one or more ENABLE inputs that are used to control the operation of the decoder. Thus with the ENABLE line HIGH the decoder will function in its normal way and the inputs will determine which

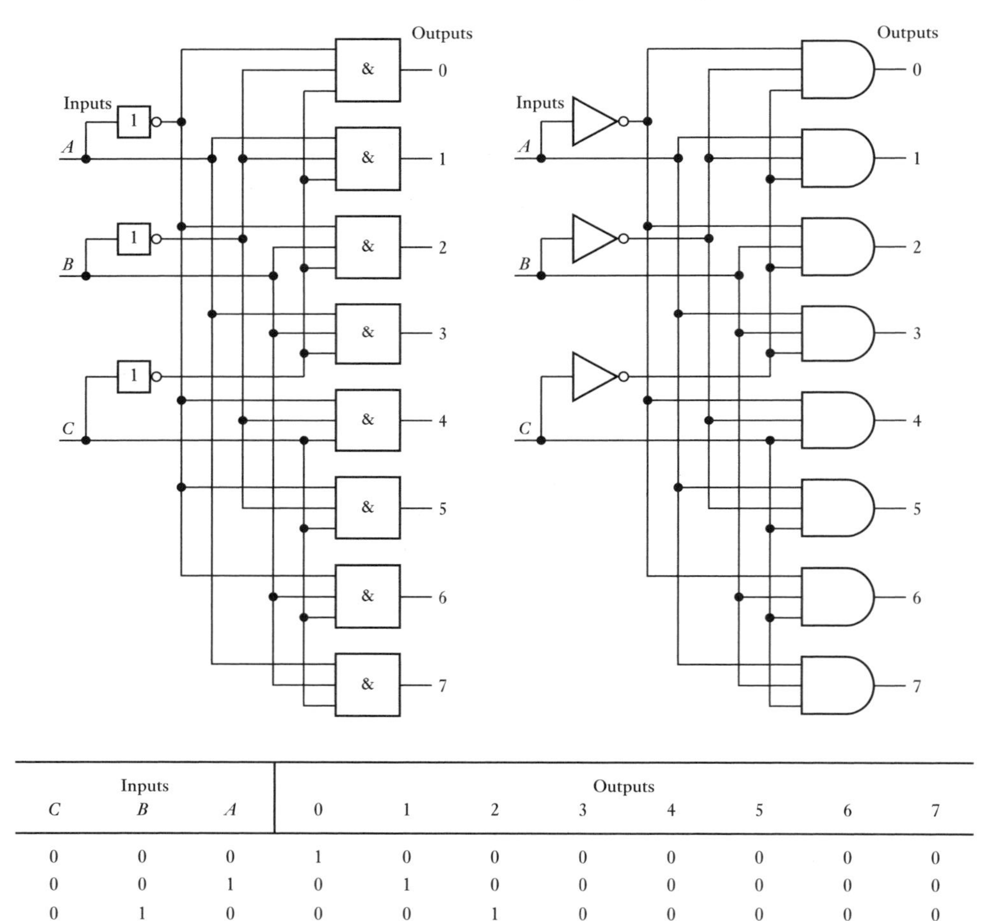

Inputs			Outputs							
C	B	A	0	1	2	3	4	5	6	7
0	0	0	1	0	0	0	0	0	0	0
0	0	1	0	1	0	0	0	0	0	0
0	1	0	0	0	1	0	0	0	0	0
0	1	1	0	0	0	1	0	0	0	0
1	0	0	0	0	0	0	1	0	0	0
1	0	1	0	0	0	0	0	1	0	0
1	1	0	0	0	0	0	0	0	1	0
1	1	1	0	0	0	0	0	0	0	1

Figure 5.16 The 3-line-to-8-line decoder.

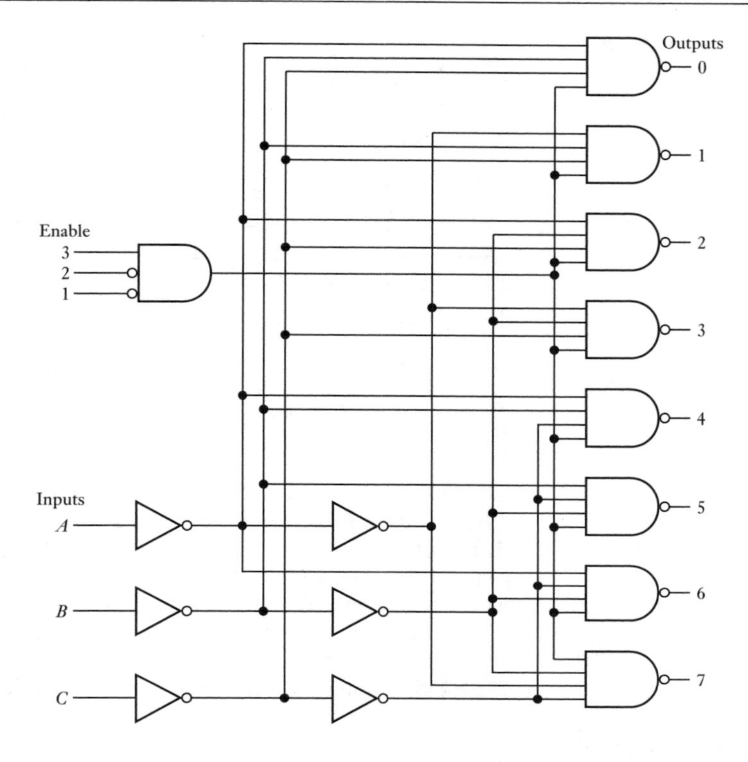

Enable			Inputs			Outputs							
E1	E2	E3	C	B	A	0	1	2	3	4	5	6	7
1	X	X	X	X	X	1	1	1	1	1	1	1	1
X	1	X	X	X	X	1	1	1	1	1	1	1	1
X	X	0	X	X	X	1	1	1	1	1	1	1	1
0	0	1	0	0	0	0	1	1	1	1	1	1	1
0	0	1	0	0	1	1	0	1	1	1	1	1	1
0	0	1	0	1	0	1	1	0	1	1	1	1	1
0	0	1	0	1	1	1	1	1	0	1	1	1	1
0	0	1	1	0	0	1	1	1	1	0	1	1	1
0	0	1	1	0	1	1	1	1	1	1	0	1	1
0	0	1	1	1	0	1	1	1	1	1	1	0	1
0	0	1	1	1	1	1	1	1	1	1	1	1	0

Figure 5.17 The 74LS138: 1 = HIGH, 0 = LOW, X = does not matter.

output is HIGH; with the ENABLE line LOW all the outputs are held LOW regardless of the inputs. Figure 5.17 shows a commonly used 3-line-to-8-line decoder with this facility, the 74LS138. Note that the outputs are active-LOW rather than the active-HIGH of Figure 5.16, and that the decoder has three ENABLE lines with the requirement for normal functioning that E1 and E3 are LOW and E3 is HIGH. All other variations result in the decoder being disabled and just a HIGH output.

Figure 5.18 illustrates the type of response we can get from a 74LS138 decoder for different inputs.

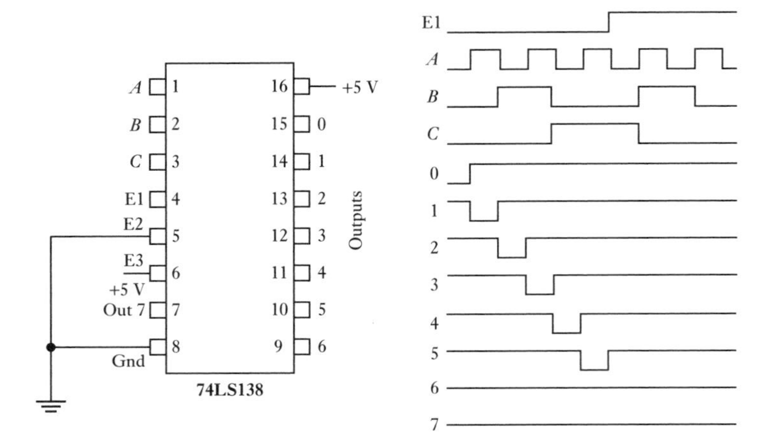

Figure 5.18 The 74LS138.

A 74LS138 decoder might be used with a microprocessor with the ENABLE used to switch on the decoder and then depending on the output from three output lines from the microprocessor so one of the eight decoder outputs receives the LOW output with all the others remaining HIGH. Thus, we can consider each output device to have an address, i.e. a unique binary output number, so that when a microprocessor sends an address to the decoder it activates the device which has been allocated that address. The 74LS138 can then be referred to as an address decoder.

5.4 Sequential logic

The logic circuits considered in earlier sections of this chapter are all examples of combinational logic systems. With such systems the output is determined by the combination of the input variables at a particular instant of time. For example, if input *A* and input *B* occur at the same time then an AND gate gives an output. The output does not depend on what the inputs previously were. Where a system requires an output which depends on earlier values of the inputs, a **sequential logic** system is required. The main difference between a combinational logic system and a sequential logic system is that the sequential logic system must have some form of memory.

Figure 5.19 shows the basic form of a sequential logic system. The combinational part of the system accepts logic signals from external inputs and from outputs from the memory. The combinational system then operates on these inputs to produce its outputs. The outputs are thus a function of both its external inputs and the information stored in its memory.

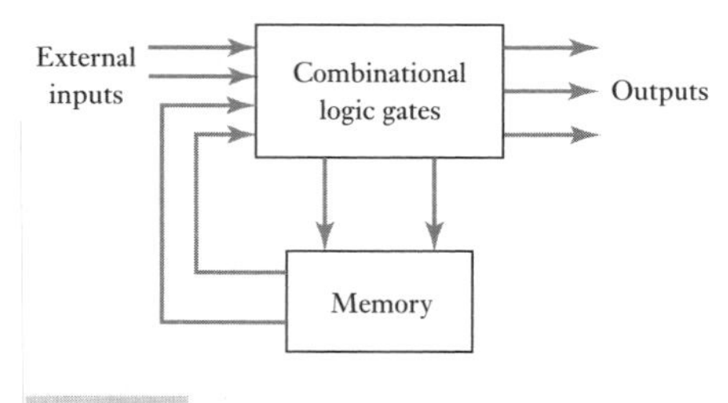

Figure 5.19 Sequential logic system.

5.4.1 The flip-flop

The **flip-flop** is a basic memory element which is made up of an assembly of logic gates and is a sequential logic device. There are a number of forms of flip-flops. Figure 5.20(a) shows one form, the SR (set–reset) flip-flop, involving NOR gates. If initially we have both outputs 0 and $S = 0$ and $R = 0$, then when we set and have S change from 0 to 1, the output from NOR gate 2 will become 0. This will then result in both the inputs to NOR gate 1 becoming 0 and so its output becomes 1. This feedback acts as an input to NOR gate 2 which then has both its inputs at 1 and results in no further change.

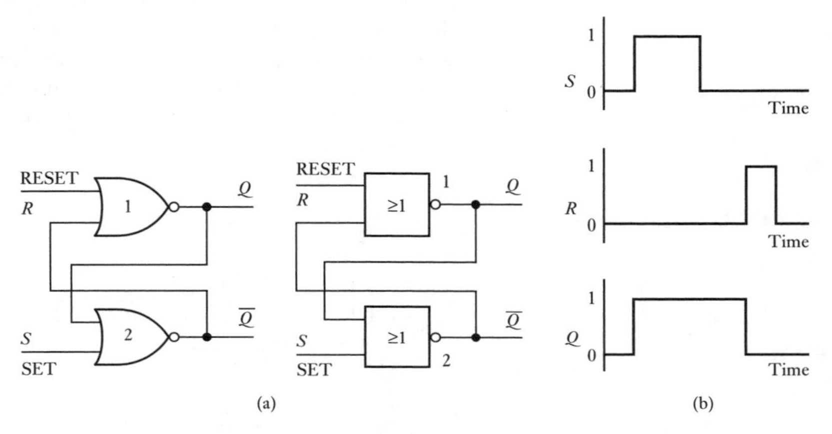

Figure 5.20 SR flip-flop.

Now if S changes from 1 to 0, the output from NOR gate 1 remains at 1 and the output from NOR gate 2 remains at 0. There is no change in the outputs when the input S changes from 1 to 0. It will remain in this state indefinitely if the only changes are to S. It 'remembers' the state it was set to. Figure 5.20(b) illustrates this with a timing diagram in which a rectangular pulse is used as the input S.

If we change R from 0 to 1 when S is 0, the output from NOR gate 1 changes to 0 and hence the output from NOR gate 2 changes to 1. The flip-flop has been reset. A change then of R to 0 will have no effect on these outputs.

Thus when S is set to 1 and R made 0, the output Q will change to 1 if it was previously 0, remaining at 1 it was previously 1. This is the set condition and it will remain in this condition even when S changes to 0. When S is 0 and R is made 1, the output Q is reset to 0 if it was previously 1, remaining at 0 if it was previously 0. This is the rest condition. The output Q that occurs at a particular time will depend on the inputs S and R and also the last value of the output. The following state table illustrates this:

S	R	$Q_t \rightarrow Q_{t+1}$	$\overline{Q}_t \rightarrow \overline{Q}_{t+1}$
0	0	$0 \rightarrow 0$	$1 \rightarrow 1$
0	0	$1 \rightarrow 1$	$0 \rightarrow 0$
0	1	$0 \rightarrow 0$	$1 \rightarrow 1$
0	1	$1 \rightarrow 0$	$0 \rightarrow 0$
1	0	$0 \rightarrow 1$	$1 \rightarrow 0$
1	0	$1 \rightarrow 1$	$0 \rightarrow 0$
1	1	Not allowed	
1	1	Not allowed	

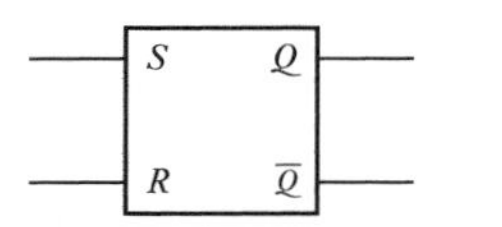

Figure 5.21 SR flip-flop.

Note that if S and R are simultaneously made equal to 1, no stable state can occur and so this input condition is not allowed. Figure 5.21 shows the simplified block symbol used for the SR flip-flop.

As a simple illustration of the use of a flip-flop, consider a simple alarm system in which an alarm is to sound when a beam of light is interrupted and remain sounding even when the beam is no longer interrupted. Figure 5.22 shows a possible system. A phototransistor might be used as the sensor and so connected that when it is illuminated it gives a virtually 0 V input to S, but when the illumination ceases it gives about 5 V input to S. When the light beam is interrupted, S becomes 1 and the output from the flip-flop becomes 1 and the alarm sounds. The output will remain as 1 even when S changes to 0. The alarm can only be stopped if the reset switch is momentarily opened to produce a 5 V input to R.

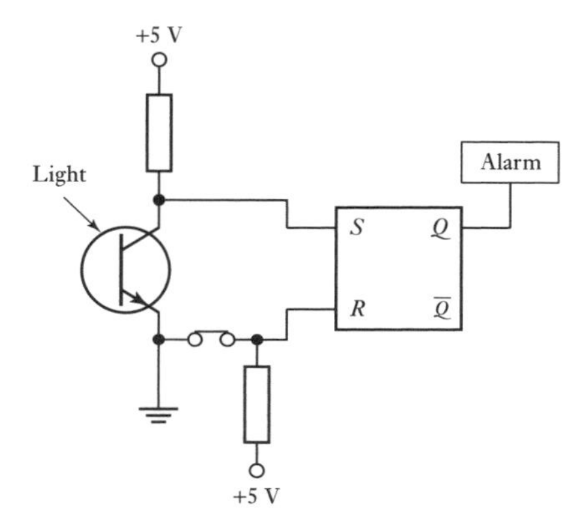

Figure 5.22 Alarm circuit.

5.4.2 Synchronous systems

It is often necessary for set and reset operations to occur at particular times. With an unclocked or **asynchronous system** the outputs of logic gates can change state at any time when any one or more of the inputs change. With a clocked or **synchronous system** the exact times at which any output can change state are determined by a signal termed the clock signal. This is generally a rectangular pulse train and when the same clock signal is used for all parts of the system, outputs are synchronised. Figure 5.23(a) shows the principle of a **gated SR flip-flop**. The set and clock signal are supplied through an AND gate to the S input of the flip-flop. Thus the set signal only arrives at the flip-flop when both it and the clock signal are 1. Likewise the reset signal is supplied with the clock signal to the R input via another AND gate. As a consequence, setting and resetting can only occur at the time determined by the clock. Figure 5.23(b) shows the timing diagram.

5.4.3 JK flip-flop

For many applications the indeterminate state that occurs with the SR flip-flop when $S = 1$ and $R = 1$ is not acceptable and another form of flip-flop

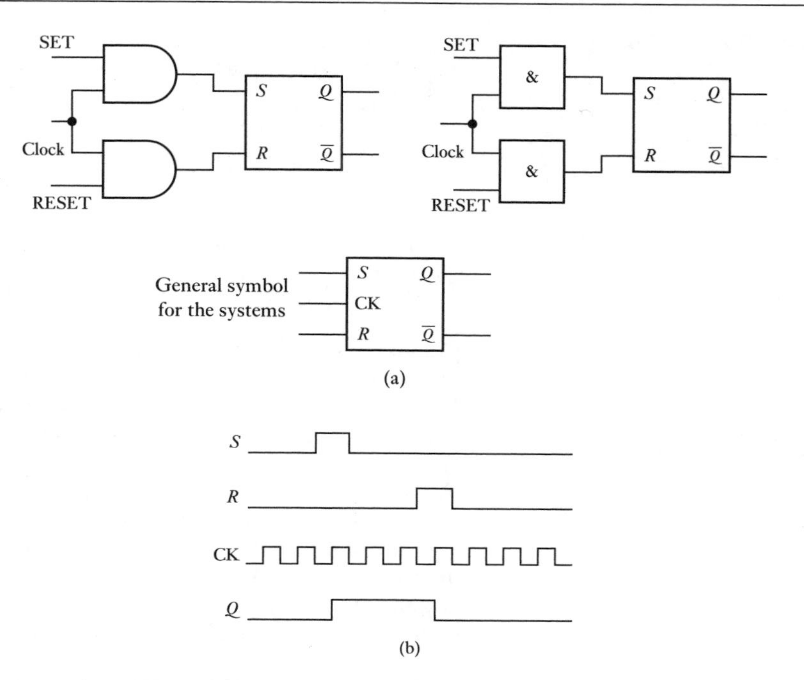

Figure 5.23 Clocked SR flip-flop.

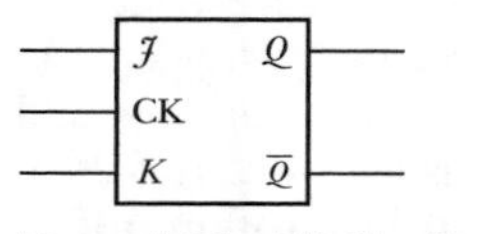

Figure 5.24 JK flip-flop.

is used, the **JK flip-flop** (Figure 5.24). This has become a very widely used flip-flop device.

The following is the truth table for this flip-flop; note that the only changes from the state table for the SR flip-flop are the entries when both inputs are 1:

J	K	$Q_t \rightarrow Q_{t+1}$	$\overline{Q}_t \rightarrow \overline{Q}_{t+1}$
0	0	0 → 0	1 → 1
0	0	1 → 1	0 → 0
0	1	0 → 0	1 → 1
0	1	1 → 0	0 → 0
1	0	0 → 1	1 → 0
1	0	1 → 1	0 → 0
1	1	0 → 1	1 → 0
1	1	1 → 0	0 → 1

As an illustration of the use of such a flip-flop, consider the requirement for a high output when input A goes high and then some time later B goes high. An AND gate can be used to determine whether two inputs are both high, but its output will be high regardless of which input goes high first. However, if the inputs A and B are used with a JK flip-flop, then A must be high first in order for the output to go high when B subsequently goes high.

5.4.4 D flip-flop

The data or **D flip-flop** is basically a clocked SR flip-flop or a JK flip-flop with the D input being connected directly to the S or J inputs and via a NOT

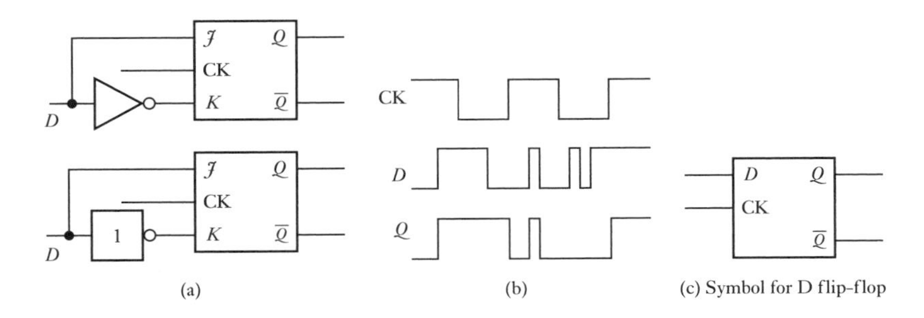

Figure 5.25 D flip-flop.

gate to the R or K inputs (Figure 5.25(a)); in the symbol for the D flip-flop this joined R and K input is labelled D. This arrangement means that a 0 or a 1 input will then switch the outputs to follow the input when the clock pulse is 1 (Figure 5.25(b)). A particular use of the D flip-flop is to ensure that the output will only take on the value of the D input at precisely defined times. Figure 5.25(c) shows the symbol used for a D flip-flop.

With the above form of D flip-flop, when the clock or enable input goes high, the output follows the data presented at input D. The flip-flop is said to be transparent. When there is a high-to-low transition at the enable input, output Q is held at the data level just prior to the transition. The data at transition is said to be **latched**. D flip-flops are available as integrated circuits. The 7475 is an example; it contains four transparent D latches.

The 7474 D flip-flop differs from the 7475 in being an edge-triggered device; there are two such flip-flops in the package. With an edge-triggered D flip-flop, transitions in Q only occur at the edge of the input clock pulse and with the 7474 it is the positive edge, i.e. low-to-high transition. Figure 5.26(a) illustrates this. The basic symbol for an edge-triggered D flip-flop differs from that of a D flip-flop by a small triangle being included on the CK input (Figure 5.26(b)). There are also two other inputs called preset and clear. A low on the preset sets the output Q to 1 while a low on clear clears the output, setting Q to 0.

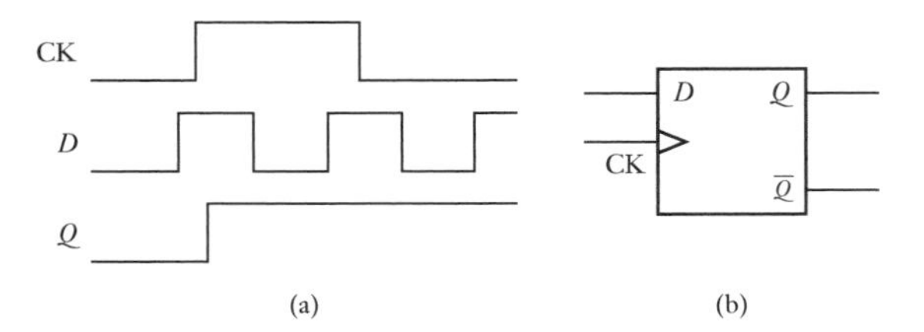

Figure 5.26 (a) Positive edge-triggered, (b) symbol for edge-triggered D flip-flop.

As an illustration of a simple application for such a flip-flop, Figure 5.27 shows a system that could be used to show a green light when the sensor input is low and a red light when it goes high and sound an alarm. The red light is to remain on as long as the sensor input is high but the alarm can be switched off. This might be a monitoring system for the temperature in some process, the sensor and signal conditioning giving a low signal when the temperature is below the safe level and a high signal when it is above. The flip-flop has a high input. When a low input is applied to the CK input and the sensor input is low, the green light is on. When the sensor input changes to high, the green light goes out, the red light on and the alarm sounds. The

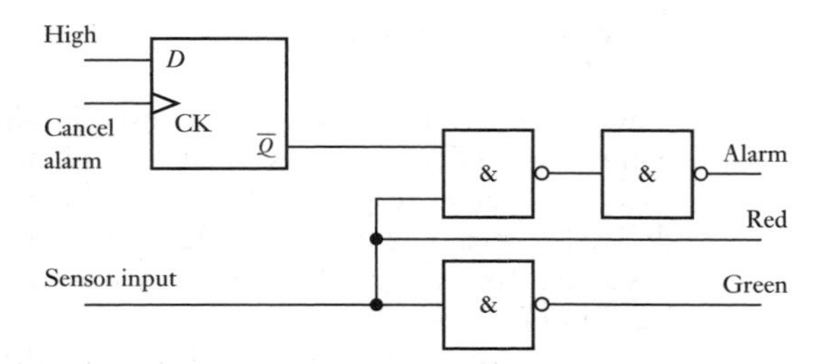

Figure 5.27 Alarm system.

alarm can be cancelled by applying a high signal to the CK input, but the red light remains on as long as the sensor input is high. Such a system could be constructed using a 7474 and an integrated circuit or circuits giving three NAND gates.

5.4.5 Registers

A **register** is a set of memory elements and is used to hold information until it is needed. It can be implemented by a set of flip-flops. Each flip-flop stores a binary signal, i.e. a 0 or a 1. Figure 5.28 shows the form a 4-bit register can take when using D flip-flops.

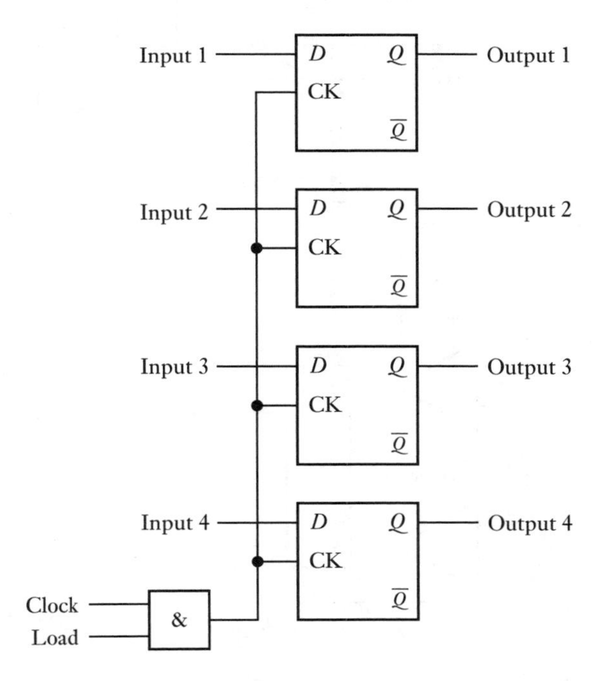

Figure 5.28 Register.

When the load signal is 0, no clock input occurs to the D flip-flops and so no change occurs to the states of the flip-flops. When the load signal is 1, then the inputs can change the states of the flip-flops. As long as the load signal is 0, the flip-flops will hold their old state values.

5.4.6 The 555 timer

The 555 timer chip is very widely used in digital circuits as it can provide a wide variety of timing tasks. It consists of an SR flip-flop with inputs fed by two comparators (Figure 5.29). The comparators each have an input voltage derived from a potentiometric chain of equal size resistors. So comparator A has an non-inverting voltage input of $V_{CC}/3$ and comparator B has an inverting input of $2V_{CC}/3$.

One use of a 555 timer is as a **monostable multivibrator**, this being a circuit which will generate a single pulse of the desired time duration when it receives a trigger signal. Figure 5.30(a) shows how the time is connected for such a use. Initially, the output will be low with the transistor shorting the capacitor and the outputs of both comparators low (Figure 5.30(b)).

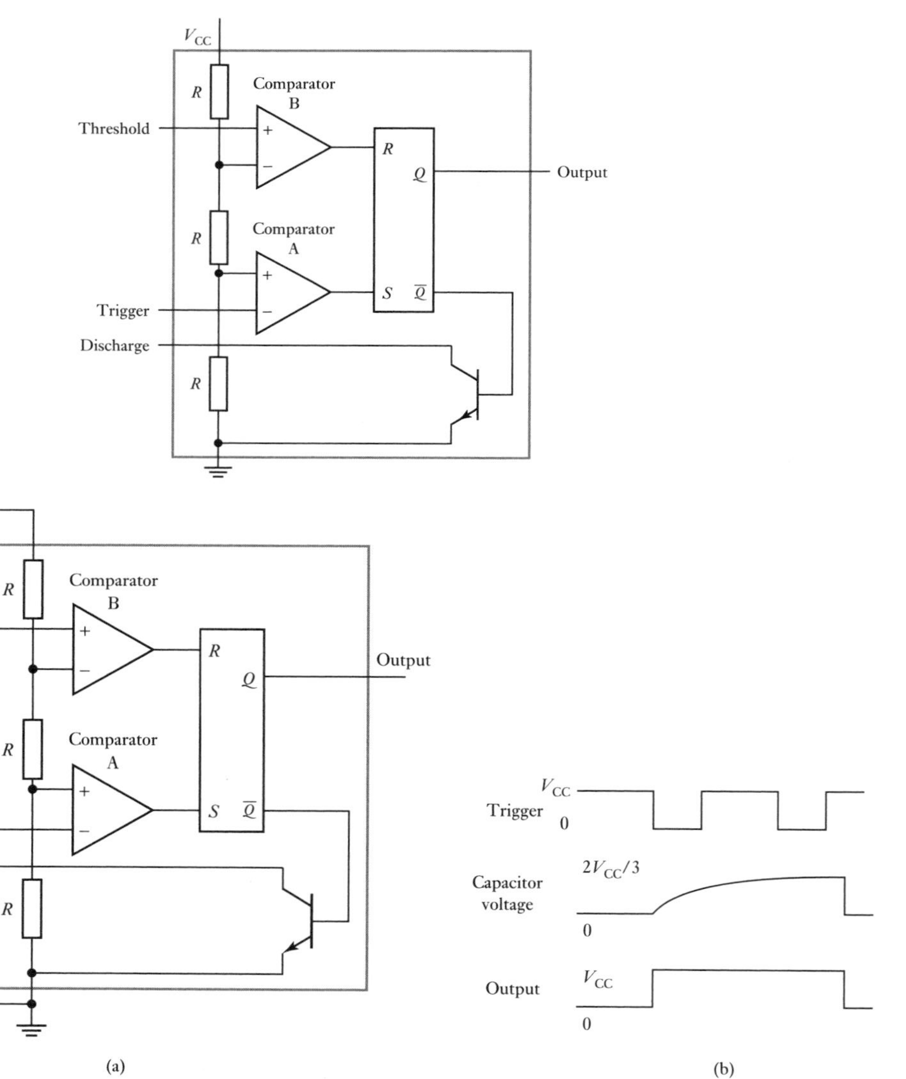

Figure 5.29 The 555 timer.

Figure 5.30 Monostable multivibrator.

When the trigger pulse goes below $V_{CC}/3$, the trigger comparator goes high and sets the flip-flop. The output is then high and the transistor cuts off and the capacitor begins to charge. When the capacitor reaches $2V_{CC}/3$, the threshold comparator resets the flip-flop and thus resets the output to low and discharges the capacitor. If the trigger is pulsed while the output is high it has no effect. The length of the pulse is thus the time taken for the capacitor to charge up to $2V_{CC}/3$ and this depends on its time constant, i.e. its value of R_tC, and is given by the normal relationship for the charging up of a capacitor though a resistance as $1.1R_tC$. As an illustration, consider the situation where a burglar alarm is to sound if a door is opened and the rightful householder does not enter the requisite number on a keypad within 30 s. If the circuit of Figure 5.30 is used with a capacitor of 1 μF then R_t would need to have a value of $30/(1.1 \times 1 \times 10^{-6}) = 27.3$ MΩ.

Summary

With **combinational logic systems,** the output is determined by the combination of the input variables at a particular instant of time. The output does not depend on what the inputs previously were. Where a system requires an output which depends on earlier values of the inputs, a **sequential logic system** is required. The main difference between a combinational logic system and a sequential logic system is that the sequential logic system must have some form of memory.

Commonly encountered logic families are **transistor–transistor logic (TTL), complementary metal-oxide semiconductor (CMOS)** and **emitter-coupled logic (ECL)**, being distinguished by their logic levels, noise immunity, fan-out, current-sourcing or current-sinking action, propagation delay time and power dissipation.

A **decoder** is a logic circuit that looks at its inputs, determines which number is there, and activates the one output that corresponds to that number.

The **flip-flop** is a basic memory element which is made up of an assembly of logic gates and is a sequential logic device.

A **register** is a set of memory elements and is used to hold information until it is needed.

The **555 timer** chip consists of an SR flip-flop with inputs fed by two comparators.

Problems

5.1 Explain what logic gates might be used to control the following situations:

(a) The issue of tickets at an automatic ticket machine at a railway station.

(b) A safety lock system for the operation of a machine tool.

(c) A boiler shut-down switch when the temperature reaches, say, 60°C and the circulating pump is off.

(d) A signal to start a lift moving when the lift door is closed and a button has been pressed to select the floor.

5.2 For the time signals shown as A and B in Figure 5.31, which will be the output signal if A and B are inputs to (a) an AND gate, (b) an OR gate?

Figure 5.31 Problem 5.2.

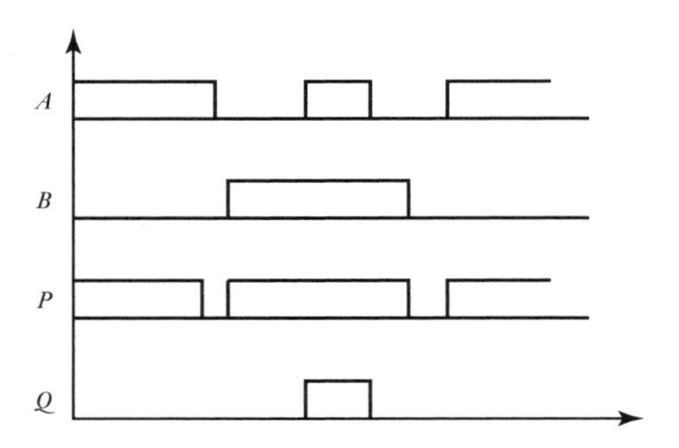

5.3 A clock signal as a continuous sequence of pulses is applied to a logic gate and is to be outputted only when an enable signal is also applied to the gate. What logic gate can be used?

5.4 Input A is applied directly to a two-input AND gate. Input B is applied to a NOT gate and then to the AND gate. What condition of inputs A and B will result in a 1 output from the AND gate?

5.5 Figure 5.32(a) shows the input signals A and B applied to the gate system shown in Figure 5.32(b). Draw the resulting output waveforms P and Q.

Figure 5.32 Problem 5.5.

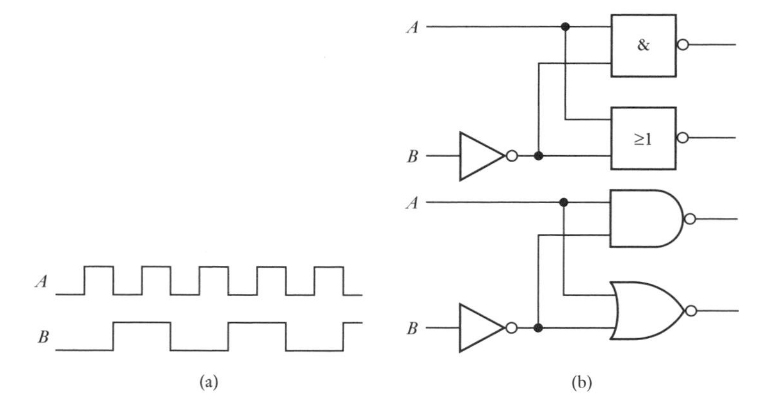

5.6 Figure 5.33 shows the timing diagram for the S and R inputs for an SR flip-flop. Complete the diagram by adding the Q output.

Figure 5.33 Problem 5.6.

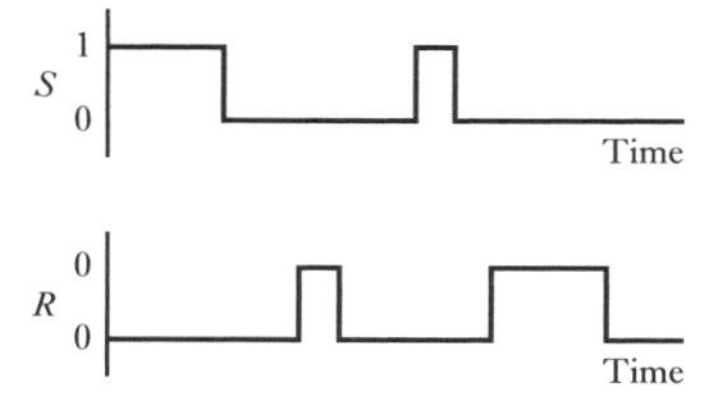

5.7 Explain how the arrangement of gates shown in Figure 5.34 gives an SR flip-flop.

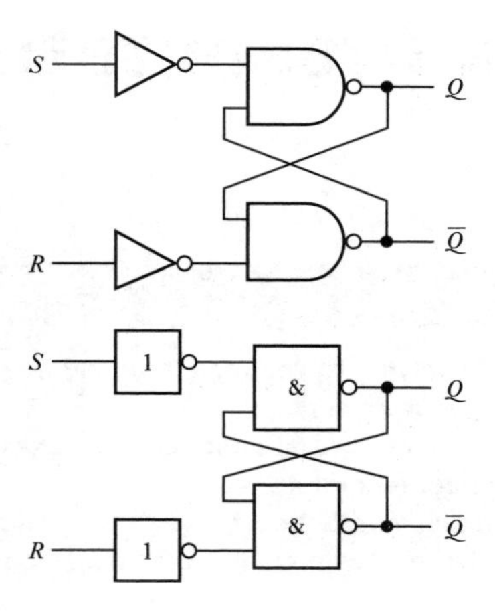

Figure 5.34 Problem 5.7.

Chapter six Data Presentation Systems

Objectives

The objectives of this chapter are that, after studying it, the reader should be able to:

- **Explain the problem of loading in measurement systems.**
- **Identify commonly used data presentation elements and describe their characteristics.**
- **Explain the principles of magnetic and optical recording.**
- **Explain the principles of displays, in particular the LED seven-segment and dot matrix displays.**
- **Describe the basic elements of data acquisition systems and virtual instruments.**

6.1 Displays

This chapter is about how data can be displayed, e.g. as digits on an LED display or as a display on a computer screen, and stored, e.g. on a computer hard disk or a CD.

Measurement systems consist of three elements: sensor, signal conditioner and display or data presentation element (see Section 1.4). There are a very wide range of elements that can be used for the presentation of data. Traditionally they have been classified into two groups: indicators and recorders. **Indicators** give an instant visual indication of the sensed variable while **recorders** record the output signal over a period of time and give automatically a permanent record.

This chapter can also be considered as the completion of the group of chapters concerned with measurement systems, i.e. sensors, signal conditioning and now display, and so the chapter is used to bring the items together in a consideration of examples of complete measurement systems.

6.1.1 Loading

A general point that has to be taken account of when putting together any measurement system is **loading**, i.e. the effect of connecting a load across the output terminals of any element of a measurement system.

Connecting an ammeter into a circuit to make a measurement of the current changes the resistance of the circuit and so changes the current. The act of attempting to make such a measurement has modified the current that was being measured. When a voltmeter is connected across a resistor then we effectively have put two resistances in parallel, and if the resistance of the voltmeter is not considerably higher than that of the resistor the current through the resistor is markedly changed and so the voltage being measured

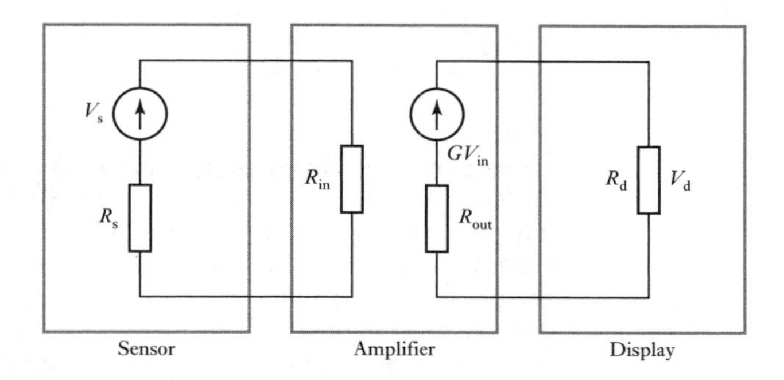

Figure 6.1 Measurement system loading.

is changed. The act of attempting to make the measurement has modified the voltage that was being measured. Such acts are termed loading.

Loading can also occur within a measurement system when the connection of one element to another modifies the characteristics of the preceding element. Consider, for example, a measurement system consisting of a sensor, an amplifier and a display element (Figure 6.1). The sensor has an open-circuit output voltage of V_s and a resistance R_s. The amplifier has an input resistance of R_{in}. This is thus the load across the sensor. Hence the input voltage from the sensor is divided so that the potential difference across this load, and thus the input voltage V_{in} to the amplifier, is

$$V_{in} = \frac{V_s R_{in}}{R_s + R_{in}}$$

If the amplifier has a voltage gain of G then the open-circuit voltage output from it will be GV_{in}. If the amplifier has an output resistance of R_{out} then the output voltage from the amplifier is divided so that the potential difference V_d across the display element, resistance R_d, is

$$V_d = \frac{GV_{in}R_d}{R_{out} + R_d} = \frac{GV_s R_{in} R_d}{(R_{out} + R_d)(R_s + R_{in})}$$

$$= \frac{GV_s}{\left(\frac{R_{out}}{R_d} + 1\right)\left(\frac{R_s}{R_{in}} + 1\right)}$$

Thus if loading effects are to be negligible we require $R_{out} \gg R_d$ and $R_s \gg R_{in}$.

6.2 Data presentation elements

This section is a brief overview of commonly used examples of data presentation elements.

6.2.1 Analogue and digital meters

The **moving-coil meter** is an analogue indicator with a pointer moving across a scale. The basic instrument movement is a d.c. microammeter with shunts, multipliers and rectifiers being used to convert it to other ranges of direct current and measurement of alternating current, direct voltage and alternating voltage. With alternating current and voltages, the instrument is restricted to between about 50 Hz and 10 kHz. The accuracy of such a meter depends on a number of factors, among which are temperature, the presence nearby of magnetic fields or ferrous materials, the way the meter is mounted, bearing friction, inaccuracies in scale marking during manufacture, etc. In addition there are errors involved in reading the meter, e.g. parallax errors when the position of the pointer against the scale is read from an angle other than directly at right angles to the scale and errors arising from estimating the position of the pointer between scale markings. The overall accuracy is generally of the order of ± 0.1 to $\pm 5\%$. The time taken for a moving-coil meter to reach a steady deflection is typically in the region of a few seconds. The low resistance of the meter can present loading problems.

The **digital voltmeter** gives its reading in the form of a sequence of digits. Such a form of display eliminates parallax and interpolation errors and can give accuracies as high as $\pm 0.005\%$. The digital voltmeter is essentially just a sample and hold unit feeding an analogue-to-digital converter with its output counted by a counter (Figure 6.2). It has a high resistance, of the order of 10 MΩ, and so loading effects are less likely than with the moving-coil meter with its lower resistance. Thus, if a digital voltmeter specification includes the statement 'sample rate approximately 5 readings per second' then this means that every 0.2 s the input voltage is sampled. It is the time taken for the instrument to process the signal and give a reading. Thus, if the input voltage is changing at a rate which results in significant changes during 0.2 s then the voltmeter reading can be in error. A low-cost digital voltmeter has typically a sample rate of 3 per second and an input impedance of 100 MΩ.

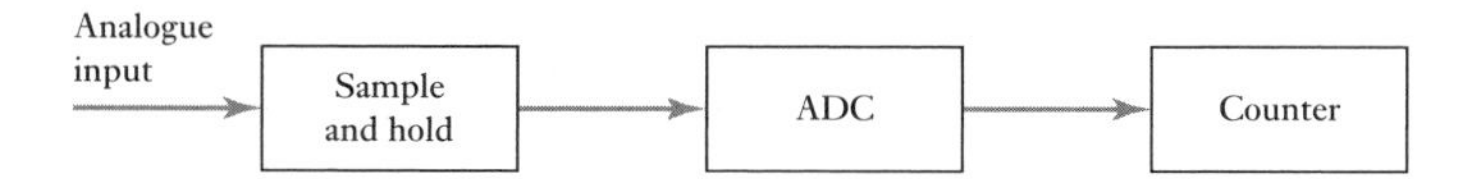

Figure 6.2 Principle of digital voltmeter.

6.2.2 Analogue chart recorders

Analogue chart recorders have data recorded on paper by fibre-tipped ink pens, by the impact of a pointer pressing a carbon ribbon against the paper, by the use of thermally sensitive paper which changes colour when a heated pointer moves across it, by a beam of ultraviolet light falling on paper sensitive to it and by a tungsten-wire stylus moving across the surface of specially coated paper, a thin layer of aluminium over coloured dye, and the electrical discharge removing the aluminium and exposing the dye. In many applications they have been superseded by virtual instruments (see later this chapter).

6.2.3 Cathode-ray oscilloscope

The cathode-ray oscilloscope is a voltage-measuring instrument which is capable of displaying extremely high-frequency signals. A general-purpose instrument can respond to signals up to about 10 MHz while more specialist instruments can respond up to about 1 GHz. Double-beam oscilloscopes enable two separate traces to be observed simultaneously on the screen while storage oscilloscopes enable the trace to remain on the screen after the input signal has ceased, only being removed by a deliberate action of erasure. Digital storage oscilloscopes digitise the input signal and store the digital signal in a memory. The signal can then be analysed and manipulated and the analogue display on the oscilloscope screen obtained from reconstructing the analogue signal. Permanent records of traces can be made with special-purpose cameras that attach directly to the oscilloscope.

A general-purpose oscilloscope is likely to have vertical deflection, i.e. Y-deflection, sensitivities which vary between 5 mV per scale division and 20 V per scale division. In order that a.c. components can be viewed in the presence of high d.c. voltages, a blocking capacitor can be switched into the input line. When the amplifier is in its a.c. mode, its bandwidth typically extends from about 2 Hz to 10 MHz and when in the d.c. mode, from d.c. to 10 MHz. The Y-input impedance is typically about 1 MΩ shunted with about 20 pF capacitance. When an external circuit is connected to the Y-input, problems due to loading and interference can distort the input signal. While interference can be reduced by the use of coaxial cable, the capacitance of the coaxial cable and any probe attached to it can be enough, particularly at low frequencies, to introduce a relatively low impedance across the input impedance of the oscilloscope and so introduce significant loading. A number of probes exist for connection to the input cable and which are designed to increase the input impedance and avoid this loading problem. A passive voltage probe that is often used is a 10-to-1 attenuator (Figure 6.3). This has a 9 MΩ resistor and variable capacitor in the probe tip. However, this reduces not only the capacitive loading but also the voltage sensitivity and so an active voltage probe using an FET is often used.

Figure 6.3 Passive voltage probe.

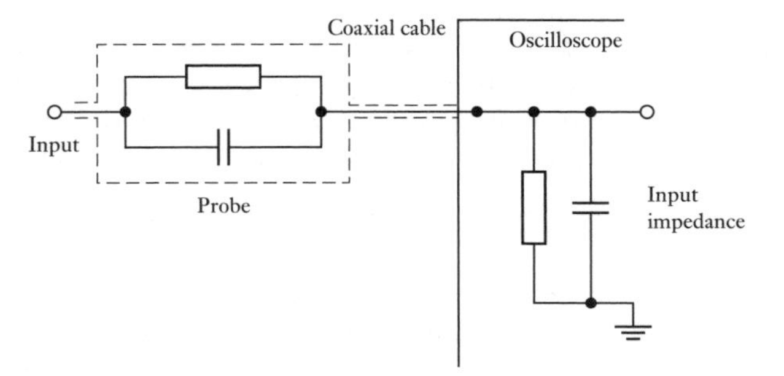

6.2.4 Visual display unit

Output data is increasingly being presented using a visual display unit (VDU). With a cathode-ray tube screen, the picture on the screen is built up by moving the spot formed by an electron beam in a series of horizontal scan lines, one after another down the screen. The image is built up by varying the intensity of the spot on the screen as each line is scanned. This raster form of display is termed **non-interlaced** (Figure 6.4(a)). To reduce the effects of flicker two scans down the screen are used to trace a complete picture. On the first scan all the odd-numbered lines are traced out and on the second the even-numbered lines are traced. This technique is called **interlaced scanning** (Figure 6.4(b)).

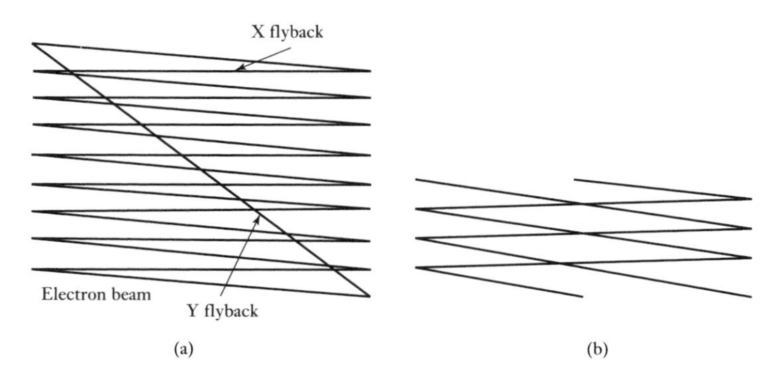

Figure 6.4 (a) Non-interlaced, (b) interlaced displays.

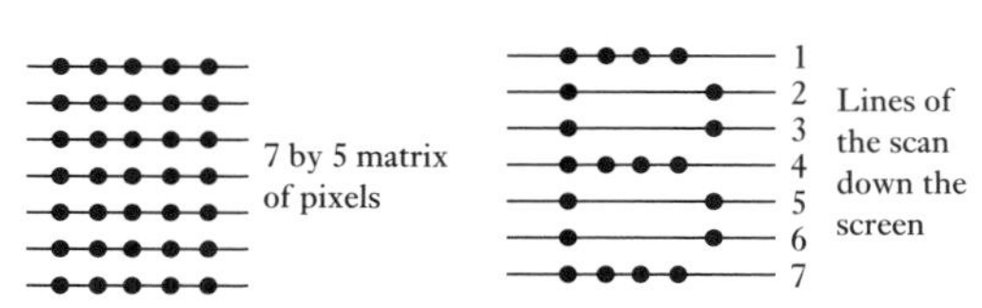

Figure 6.5 Character build-up by selective lighting.

The screen of the visual display unit is coated with a large number of phosphor dots, these dots forming the **pixels**. The term pixel is used for the smallest addressable dot on any display device. A text character or a diagram is produced on the screen by selectively lighting these dots. Figure 6.5 shows how, for a 7 by 5 matrix, characters are built up by the electron beam moving in its zigzag path down the screen. The input data to the VDU is usually in digital **ASCII (American Standard Code for Information Interchange)** format. This is a 7-bit code and so can be used to represent $2^7 = 128$ characters. It enables all the standard keyboard characters to be covered, as well as some control functions such as RETURN which is used to indicate the return from the end of a line to the start of the next line. Table 6.1 gives an abridged list of the code.

Table 6.1 ASCII code.

Character	ASCII	Character	ASCII	Character	ASCII
A	100 0001	N	100 1110	0	011 0000
B	100 0010	O	100 1111	1	011 0001
C	100 0011	P	101 0000	2	011 0010
D	100 0100	Q	101 0001	3	011 0011
E	100 0101	R	101 0010	4	011 0100
F	100 0110	S	101 0011	5	011 0101
G	100 0111	T	101 0100	6	011 0110
H	100 1000	U	101 0101	7	011 0111
I	100 1001	V	101 0110	8	011 1000
J	100 1010	W	101 0111	9	011 1001
K	100 1011	X	101 1000		
L	100 1100	Y	101 1001		
M	100 1101	Z	101 1010		

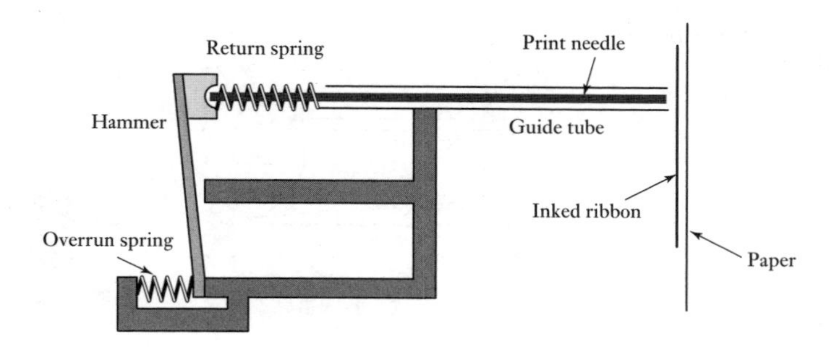

Figure 6.6 Dot matrix print head mechanism.

6.2.5 Printers

Printers provide a record of data on paper. There are a number of versions of such printers: the dot matrix printer, the ink/bubble jet printer and the laser printer.

The **dot matrix printer** has a print head (Figure 6.6) which consists of either 9 or 24 pins in a vertical line. Each pin is controlled by an electromagnet which when turned on propels the pin onto the inking ribbon. This transfers a small blob of ink onto the paper behind the ribbon. A character is formed by moving the print head in horizontal lines back and forth across the paper and firing the appropriate pins.

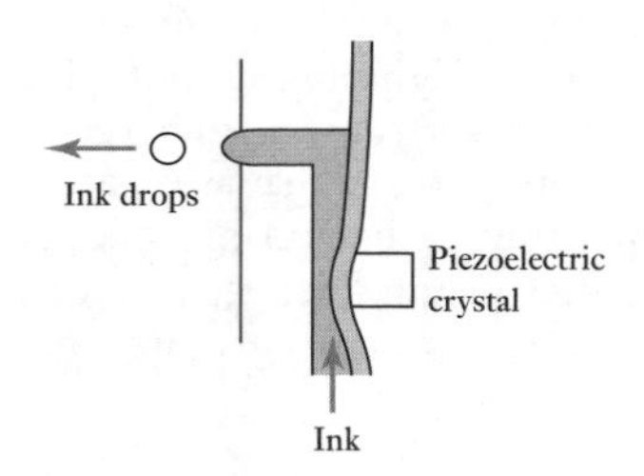

Figure 6.7 Producing a stream of drops.

The **ink jet printer** uses a conductive ink which is forced through a small nozzle to produce a jet of very small drops of ink of constant diameter at a constant frequency. With one form a constant stream of ink passes along a tube and is pulsed to form fine drops by a piezoelectric crystal which vibrates at a frequency of about 100 kHz (Figure 6.7). Another form uses a small heater in the print head with vaporised ink in a capillary tube, so producing gas bubbles which push out drops of ink (Figure 6.8). In one printer version each drop of ink is given a charge as a result of passing through a charging

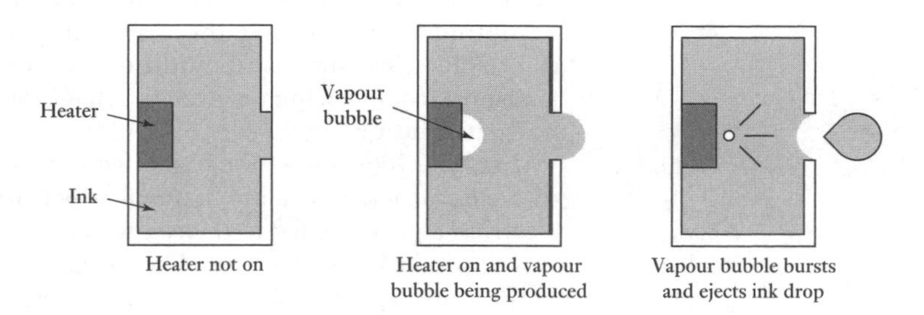

Figure 6.8 Principle of the bubble jet.

electrode and the charged drops are deflected by passing between plates between which an electric field is maintained; in another version a vertical stack of nozzles is used and each jet is just switched on or off on demand. Ink jet printers can give colour prints by the use of three different colour ink jet systems. The fineness of the drops is such that prints can be produced with more than 600 dots per inch.

The **laser printer** has a photosensitive drum which is coated with a selenium-based light-sensitive material (Figure 6.9). In the dark the selenium has a high resistance and consequently becomes charged as it passes close to the charging wire; this is a wire at a high voltage and off which charge leaks.

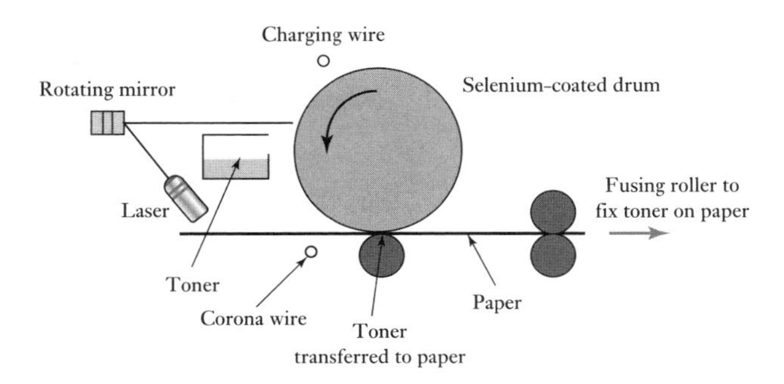

Figure 6.9 Basic elements of a laser printer.

A light beam is made to scan along the length of the drum by a small rotating eight-sided mirror. When light strikes the selenium its resistance drops and it can no longer remain charged. By controlling the brightness of the beam of light, so points on the drum can be discharged or left charged. As the drum passes the toner reservoir, the charged areas attract particles of toner which thus stick to the areas that have not been exposed to light and do not stick on the areas that have been exposed to light. The paper is given a charge as it passes another charging wire, the so-called corona wire, so that as it passes close to the drum, it attracts the toner off the drum. A hot fusing roller is then used to melt the toner particles so that, after passing between rollers, they firmly adhere to the paper. General-use laser printers are currently able to produce 1200 dots per inch.

6.3 Magnetic recording

Magnetic recording is used for the storage of data on the floppy disks and hard disks of computers. The basic principles are that a recording head, which responds to the input signal, produces corresponding magnetic patterns on a thin layer of magnetic material and a read head gives an output by converting the magnetic patterns on the magnetic material to electrical signals. In addition to these heads the systems require a transport system which moves the magnetic material in a controlled way under the heads.

Figure 6.10(a) shows the basic elements of the recording head; it consists of a core of ferromagnetic material which has a non-magnetic gap. When electrical signals are fed to the coil which is wound round the core, magnetic flux is produced in the core. The proximity of the magnetic coated plastic to the non-magnetic gap means that the magnetic flux readily follows a path through the core and that part of the magnetic coating in the region of the gap. When there is magnetic flux passing through a region of the magnetic coating, it becomes permanently magnetised. Hence a magnetic record is produced of the electrical input signal. Reversing the direction of the current reverses the flux direction.

The replay head (Figure 6.10(b)) has a similar construction to that of the recording head. When a piece of magnetised coating bridges the non-magnetised gap, then magnetic flux is induced in the core. Flux changes in the core induce e.m.f.s in the coil wound round the core. Thus the output from the coil is an electrical signal which is related to the magnetic record on the coating.

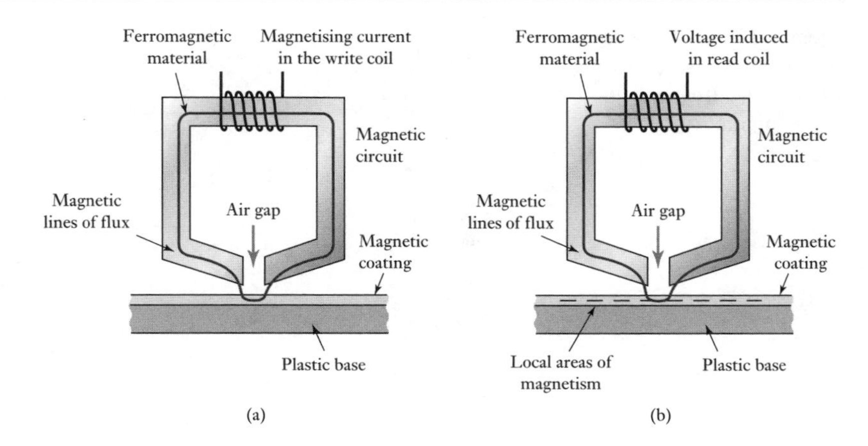

Figure 6.10 Basis of magnetic (a) recording, (b) replay head.

6.3.1 Magnetic recording codes

Digital recording involves the recording of signals as a coded combination of bits. A bit cell is the element of the magnetic coating where the magnetism is either completely saturated in one direction or completely saturated in the reverse direction. Saturation is when the magnetising field has been increased to such an extent that the magnetic material has reached its maximum amount of magnetic flux and further increases in magnetising current produce no further change.

The bit cells on the magnetic surface might then appear in the form shown in Figure 6.11. An obvious method of putting data on the magnetic material might seem to be to use the magnetic flux in one direction to represent a 0 and in the reverse direction a 1. However, it is necessary to read each cell and thus accurate timing points are needed in order to indicate clearly when sampling should take place. Problems can arise if some external clock is used to give the timing signals, as a small mismatch between the timing signals and the rate at which the magnetic surface is moving under the read head can result in perhaps a cell being missed or even read twice. Synchronisation is essential. Such synchronisation is achieved by using the bit cells themselves to generate the signals for taking samples. One method is to use transitions on the magnetic surface from saturation in one direction to saturation in the other, i.e. where the demarcation between two bits is clearly evident, to give feedback to the timing signal generation in order to adjust it so that it is in synchronisation with the bit cells.

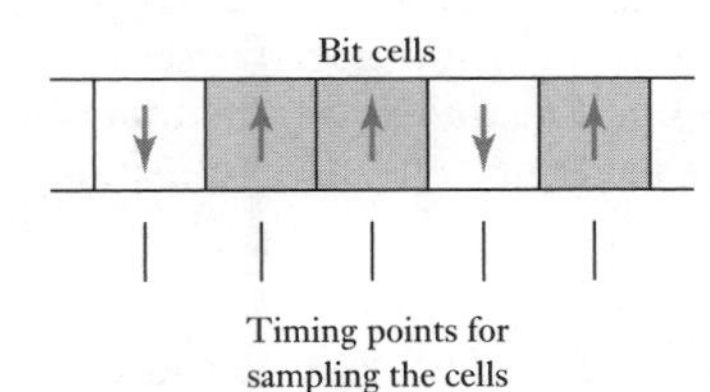

Figure 6.11 Bit cells.

If the flux reversals do not occur sufficiently frequently, this method of synchronisation can still result in errors occurring. One way of overcoming this problem is to use a form of encoding. The following are some of the methods commonly used:

1 *Non-return-to-zero* (NRZ)
With this system the flux is recorded on the tape such that no change in flux represents 0 and a change in flux 1 (Figure 6.12(a)). It is, however, not self-clocking.

2 *Phase encoding* (PE)
Phase encoding has the advantage of being self-clocking with no external clock signals being required. Each cell is split in two with one half having

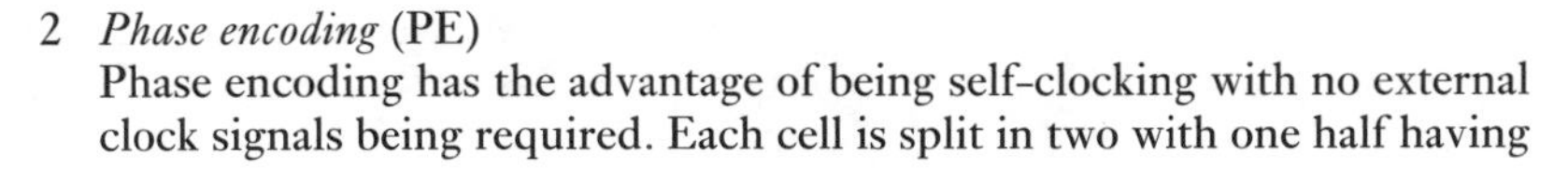

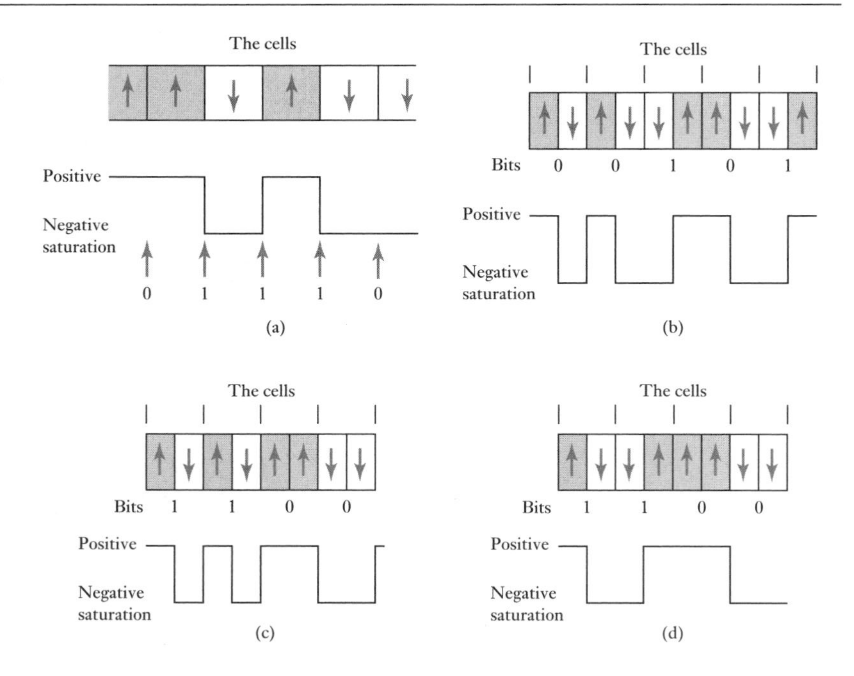

Figure 6.12 (a) Non-return-to-zero, (b) phase, (c) frequency, (d) modified frequency modulation.

positive saturation flux and the other a negative saturation flux. A digit 0 is then recorded as a half-bit positive saturation followed by a half-bit negative saturation; a 1 digit is represented by a half-bit negative saturation followed by a half-bit positive saturation. The mid-cell transition of positive to negative thus indicates a 0 and a negative to positive transition a 1 (Figure 6.12(b)).

3 *Frequency modulation* (FM)
This is self-clocking and similar to phase encoding but there is always a flux direction reversal at the beginning of each cell (Figure 6.12(c)). For a 0 bit there is then no additional flux reversal during the cell but for a 1 there is an additional flux reversal during the cell.

4 *Modified frequency modulation* (MFM)
This is a modification of the frequency modulation code, the difference being that the flux reversal at the beginning of each bit code is only present if the current and previous bit were 0 (Figure 6.12(d)). This means that only one flux reversal is required for each bit. This and the run length limited code are the codes generally used for magnetic discs.

5 *Run length limited* (RLL)
This is a group of self-clocking codes which specify a minimum and maximum distance, i.e. run, between flux reversals. The maximum run is short enough to ensure that the flux reversals are sufficiently frequent for the code to be self-clocking. A commonly used form of this code is $RLL_{2,7}$, the 2,7 indicating that the minimum distance between flux reversals is to be 2 bits and the maximum is to be 7. The sequence of codes is described as a sequence of S-codes and R-codes. An S-code, a space code, has no reversals of flux while an R-code, a reversal code, has a reversal during the bit. Two S/R-codes are used to represent each bit. The bits

are grouped into sequences of 2, 3 and 4 bits and a code assigned to each group, the codes being:

Bit sequence	Code sequence
10	SRSS
11	RSSS
000	SSSRSS
010	RSSRSS
011	SSRSSS
0010	SSRSSRSS
0011	SSSSRSSS

Figure 6.13 shows the coding for the sequence 0110010, it being broken into groups 011 and 0010 and so represented by SSRSSSSRSSRSS. There are at least two S-codes between R-codes and there can be no more than seven S-codes between R-codes.

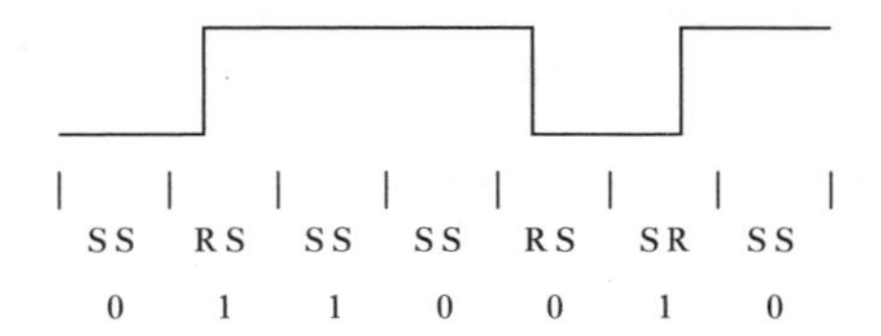

Figure 6.13 RLL code.

The optimum code is the one that allows the bits to be packed as close as possible and which can be read without error. The read heads can locate reversals quite easily but they must not be too close together. The RLL code has the advantage of being more compact than the other codes, PE and FM taking up the most space. MFM and NRZ take up the same amount of space. NRZ has the disadvantage of, unlike the other codes, not being self-clocking.

6.3.2 Magnetic disks

Digital recording is very frequently to a hard disk. The digital data is stored on the disk surface along concentric circles called tracks, a single disk having many such tracks. A single read/write head is used for each disk surface and the heads are moved, by means of a mechanical actuator, backwards and forwards to access different tracks. The disk is spun by the drive and the read/write heads read or write data into a track. Hard disks (Figure 6.14(a)) are sealed units with data stored on the disk surface along concentric circles. A hard disk assembly has more than one such disk and the data is stored on magnetic coatings on both sides of the disks. The disks are rotated at high speeds and the tracks accessed by moving the read/write heads. Large amounts of data can be stored on such assemblies of disks; storages of hundreds of gigabytes are now common.

The disk surface is divided into sectors (Figure 6.14(b)) and so a unit of information on a disk has an address consisting of a track number and

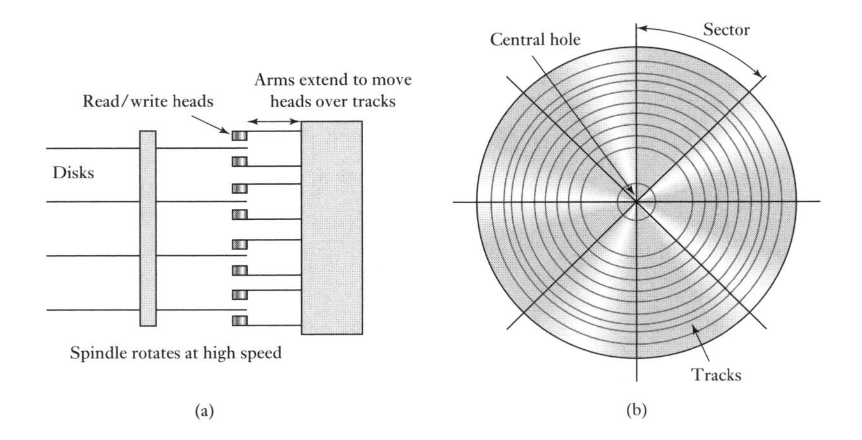

Figure 6.14 Hard disk: (a) arrangement of disks, (b) tracks and sectors.

a sector number. A floppy disk has normally between 8 and 18 sectors and about 100 tracks. A hard disk might have about 2000 tracks per surface and 32 sectors. To seek data the head must be moved to over the required track, the time this takes being termed the **seek time**, and then wait there until the required segment moves under it, this time being termed the **latency**. In order that an address can be identified, it is necessary for information to have been recorded on the disk to identify segments and tracks. The writing of this information is called **formatting** and has to be carried out before data can be stored on a disk. The technique usually used is to store this location information on the tracks so that when data is stored the sequence of information on a track becomes:

index marker,
sector 0 header, sector 0 data, sector 0 trailer,
sector 1 header, sector 1 data, sector 1 trailer,
sector 2 header, sector 2 data, sector 2 trailer,
etc.

The index marker contains the track number with the sector header identifying the sector. The sector trailer contains information, e.g. a cyclic redundancy check, which can be used to check that a sector was read correctly.

6.4 Optical recording

Like magnetic disks, CD-ROMs store the data along their tracks. Unlike a magnetic disk which has a series of concentric tracks, a CD-ROM has a spiral track. The recording surface is coated with aluminium and is highly reflective. The information is then stored in a track about 0.6 μm wide as a series of pits, etched into the surface by focused light from a laser in a beam about 1 μm in diameter, and these result in light being strongly reflected or not reflected according to whether it strikes a pit or a non-depressed area. Data is thus read as a sequence of reflected and non-reflected light pulses.

Optical recording uses similar coding methods to that used with magnetic recording, the RLL form of coding being generally used. Because optical recordings can be very easily corrupted by scratches or dust obstructing the laser beams used to read them, methods have to be used to detect and correct for errors. One method is **parity checking**. With this method, groups of bits are augmented with an extra, parity, bit which is either set to 0 or 1 so that the total number of 1 bits in a group is either an odd or an even number. When the information is read, if one of the bits has been corrupted, then the number of bits will have changed and this will be detected as an error.

6.5 Displays

Many display systems use light indicators to indicate on/off status or give alphanumeric displays. The term **alphanumeric** is a contraction of the terms alphabetic and numeric and describes displays of the letters of the alphabet and numbers 0 to 9 with decimal points. One form of such a display involves seven 'light' segments to generate the alphabetic and numeric characters. Figure 6.15 shows the segments and Table 6.2 shows how a 4-bit binary code input can be used to generate inputs to switch on the various segments.

Another format involves a 7 by 5 or 9 by 7 dot matrix (Figure 6.16). The characters are then generated by the excitation of appropriate dots.

The light indicators for such displays might be neon lamps, incandescent lamps, **light-emitting diodes** (LEDs) or **liquid crystal displays** (LCDs). **Neon lamps** need high voltages and low currents and can be powered directly from the mains voltage but can only be used to give a red light. **Incandescent lamps** can be used with a wide range of voltages but need a comparatively high current. They emit white light and so use lenses to generate any required colour. Their main advantage is their brightness.

Figure 6.15 Seven-segment display.

Table 6.2 Seven-segment display.

Binary input				Segments activated							Number displayed
				a	b	c	d	e	f	g	
0	0	0	0	1	1	1	1	1	1	0	0
0	0	0	1	0	1	1	0	0	0	0	1
0	0	1	0	1	1	0	1	1	0	1	2
0	0	1	0	1	1	1	1	0	0	1	3
0	1	0	0	0	1	1	0	0	1	1	4
0	1	0	1	1	0	1	1	0	1	1	5
0	1	1	0	0	0	1	1	1	1	1	6
0	1	1	1	1	1	1	0	0	0	0	7
1	0	0	0	1	1	1	1	1	1	1	8
1	0	0	1	1	1	1	0	0	1	1	9

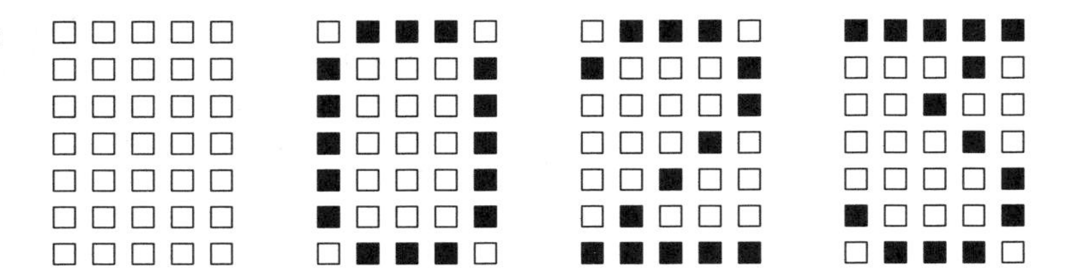

Figure 6.16 A 7 by 5 dot matrix display.

6.5.1 Light-emitting diodes

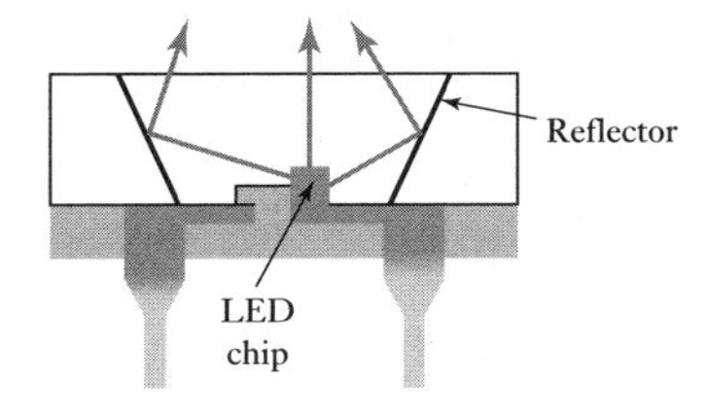

Figure 6.17 LED.

LEDs require low voltages and low currents and are cheap. These diodes when forward biased emit light over a certain band of wavelengths. Figure 6.17 shows the basic form of a LED, the light emitted from the diode being enhanced in one direction by means of reflectors. Commonly used LED materials are gallium arsenide, gallium phosphide and alloys of gallium arsenide with gallium phosphide. The most commonly used LEDs can give red, yellow or green colours. With microprocessor-based systems, LEDs are the most common form of indicator used.

A current-limiting resistor is generally required with an LED in order to limit the current to below the maximum rated current of about 10 to 30 mA. Typically an LED might give a voltage drop across it of 2.1 V when the current is limited to 20 mA. Thus when, say, a 5 V output is applied, 2.9 V has to be dropped across a series resistor. This means that a resistance of $2.9/0.020 = 145\ \Omega$ is required and so a standard resistor of 150 Ω is likely to be used. Some LEDs are supplied with built-in resistors so they can be directly connected to microprocessor systems.

LEDs are available as single light displays, seven- and sixteen-segment alphanumeric displays, in dot matrix format and bar graph form.

Figure 6.18(a) shows how seven LEDs, to give the seven segments of a display of the form shown in Figure 6.16, might be connected to a driver so that when a line is driven low, a voltage is applied and the LED in that line is switched on. The voltage has to be above a 'turn-on' value before the LED emits significant light; typical turn-on voltages are about 1.5 V. Such

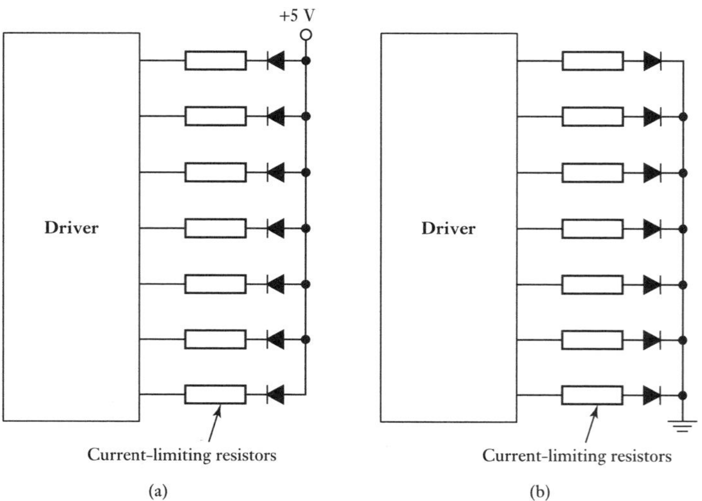

Figure 6.18 (a) Common a node connection for LEDs, (b) common cathode.

an arrangement is known as the **common anode** form of connection since all the LED anodes are connected together. An alternative arrangement is the **common cathode** (Figure 6.18(b)). The elements in common anode form are made active by the input going low, in the common cathode type by going high. Common anode is the usual choice since the direction of current flow and the size of current involved are usually most appropriate.

Examples of such types of display are the seven-segment 7.6 mm and 10.9 mm high-intensity displays of Hewlett Packard which are available as either common anode or common cathode form. In addition to the seven segments to form the characters, there is either a left-hand or right-hand decimal point. By illuminating different segments of the display, the full range of numbers and a small range of alphabetical characters can be formed.

Often the output from the driver is not in the normal binary form but in **Binary-Coded Decimal** (BCD) (see Appendix B). With BCD, each decimal digit is coded separately in binary. For example, the decimal number 15 has the 1 coded as 0001 and the 5 as 0101 to give the BCD code of 0001 0101. The driver output has then to be decoded into the required format for the LED display. The 7447 is a commonly used decoder for driving displays (Figure 6.19).

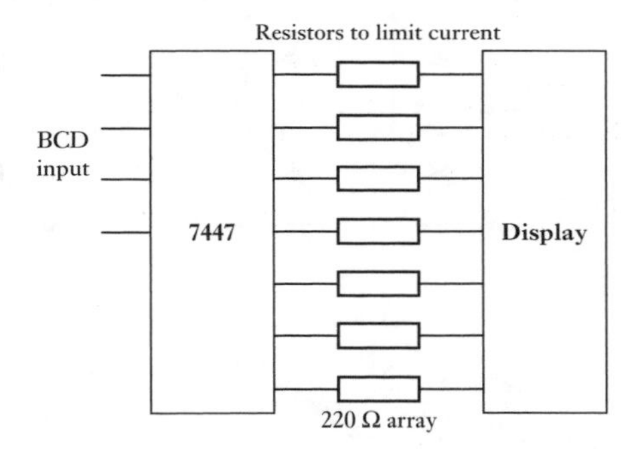

Figure 6.19 Decoder with seven-segment display.

Figure 6.20 shows the basic form used for a 5 by 7 dot matrix LED display. The array consists of five column connectors, each connecting the anodes of seven LEDs. Each row connects to the cathodes of five LEDs. To turn on a particular LED, power is applied to its column and its row is grounded. Such a display enables all the ASCII characters to be produced.

6.5.2 Liquid crystal displays

Liquid crystal displays do not produce any light of their own but rely on reflected light or transmitted light. The liquid crystal material is a compound with long rod-shaped molecules which is sandwiched between two sheets of polymer containing microscopic grooves. The upper and lower sheets are grooved in directions at 90° to each other. The molecules of the liquid crystal material align with the grooves in the polymer and adopt a smooth 90° twist between them (Figure 6.21).

When plane polarised light is incident on the liquid crystal material its plane of polarisation is rotated as it passes through the material. Thus if it is sandwiched between two sheets of polariser with their transmission directions at right angles, the rotation allows the light to be transmitted and so the material appears light.

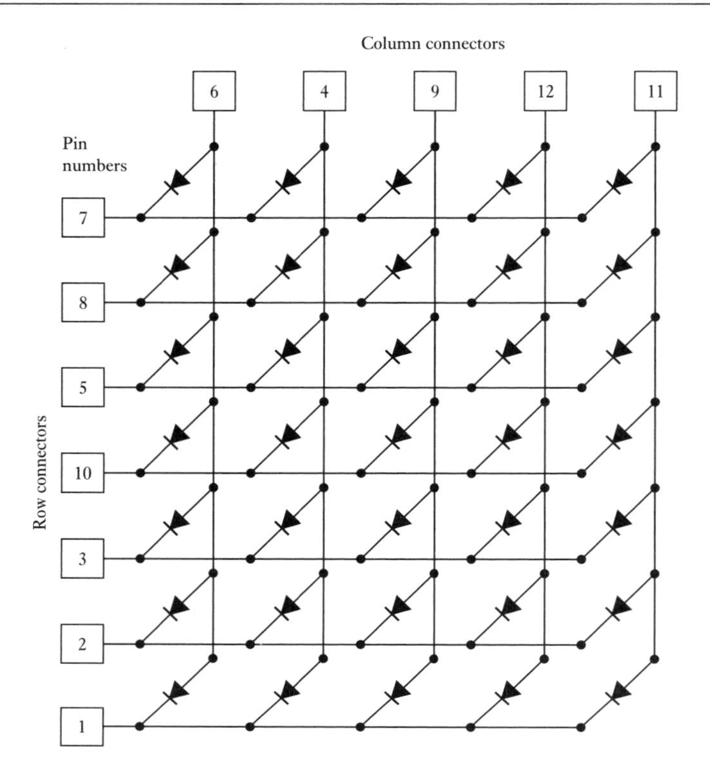

Figure 6.20 Dot matrix display.

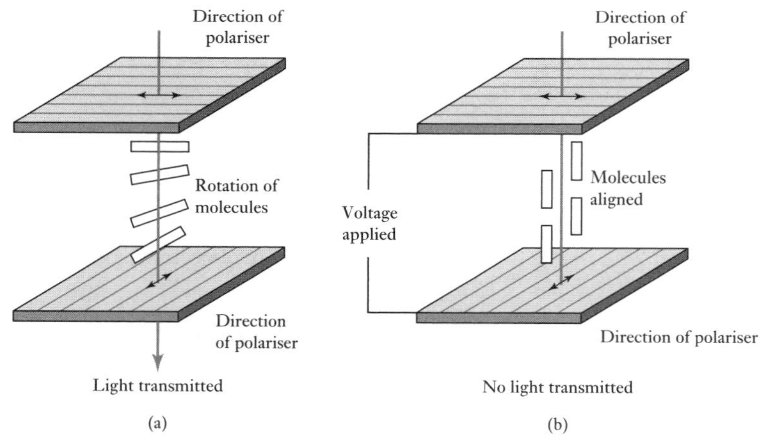

Figure 6.21 Liquid crystal: (a) no electric field, (b) electric field.

However, if an electric field is applied across the material, the molecules become aligned with the field and the light passing through the top polariser is not rotated and cannot pass through the lower polariser but becomes absorbed. The material then appears dark.

The arrangement is put between two sheets of glass on which are transparent electrodes in the form of the required display. An LED display can be transmissive or reflective. With the transmissive display, the display is back-lit. When the rotation of the plane of polarisation allows light to be transmitted, the display is light, otherwise it is dark. With the reflective light display, there is a reflective surface behind the crystals. Thus when the incident light passes through the display, it is reflected back and so the display appears light. When the incident light cannot pass through the display it appears dark.

LCDs are available in many segment layouts, including a seven-segment display similar to the seven-segment LED display. The application of voltages to the various display elements results in them appearing black against the lighter display where there is no electric field. To turn on a segment, an a.c. electric field of about 3 to 12 V is used. The drive voltage must not be d.c. but a.c. since d.c. voltages generate reactions which destroy the crystals. LCDs have a relatively slow response time, typically about 100 to 150 ms. Power consumption is low.

LCDs are also available as dot matrix displays. LCD modules are also available with displays with one or more rows of characters. For example, a two row 40 character display is available.

Integrated circuit drivers are available to drive LEDs. Thus the MC14543B can be used for a seven-segment LCD display. Drivers are available for when the input is in BCD code. A 5 × 8-dot matrix display can be driven by the MC145000 driver. Displays are available combined with drivers. For example, the Hitachi LM018L is a 40 character × 2 line reflective-type LCD module with a built-in driver HD44780 which provides a range of features, including 192 5 × 7-dot characters plus 8 user-defined characters and thus can be directly interfaced to a 4-bit or 8-bit microprocessor.

LEDs are the form of display used in battery-operated devices such as cell phones, watches and calculators.

6.6 Data acquisition systems

The term **data acquisition** (DAQ) tends to be frequently used for systems in which inputs from sensors are converted into a digital form for processing, analysis and display by a computer. The systems thus contain: sensors, wiring to connect the sensors to signal conditioning to carry out perhaps filtering and amplification, data acquisition hardware to carry out such functions as conversion of input to digital format and conversion of output signals to analogue format for control systems, the computer and data acquisition software. The software carries out analysis of the digital input signals. Such systems are also often designed to exercise control functions as well.

6.6.1 Computer with plug-in boards

Figure 6.22 shows the basic elements of a data acquisition system using plug-in boards with a computer for the data acquisition hardware. The signal conditioning prior to the inputs to the board depends on the sensors concerned, e.g. it might be for thermocouples: amplification, cold junction compensation and

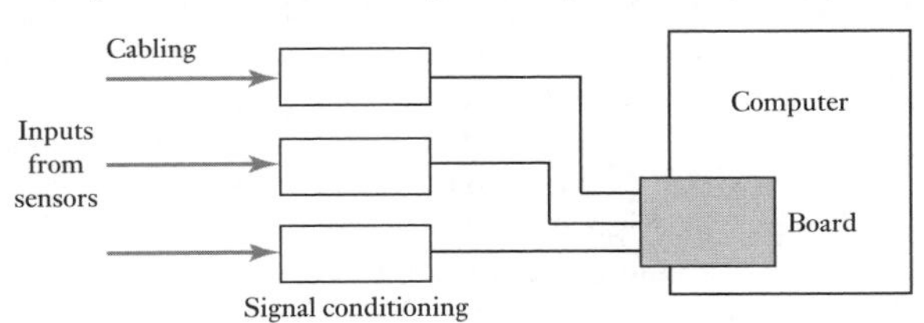

Figure 6.22 Data acquisition system.

linearisation; for strain gauges: Wheatstone bridge, voltage supply for bridge and linearisation; for RTDs: current supply, circuitry and linearisation.

In selecting the DAQ board to be used the following criteria have to be considered:

1 What type of computer software system is being used, e.g. Windows, MacOS?
2 What type of connector is the board to be plugged into, e.g. PCMCIA for laptops, NuBus for MacOS, PCI?
3 How many analogue inputs will be required and what are their ranges?
4 How many digital inputs will be required?
5 What resolution will be required?
6 What is the minimum sampling rate required?
7 Are any timing or counting signals required?

Figure 6.23 shows the basic elements of a DAQ board. Some boards will be designed only to handle analogue inputs/outputs and others digital inputs/outputs.

Figure 6.23 DAQ board elements.

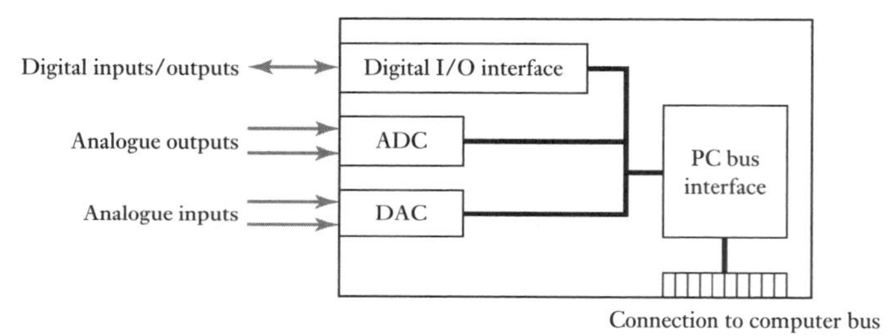

All DAQ boards use **drivers**, software generally supplied by the board manufacturer with a board, to communicate with the computer and tell it what has been inserted and how the computer can communicate with the board. Before a board can be used, three parameters have to be set. These are the addresses of the input and output channels, the interrupt level and the channel to be used for direct memory access. With 'plug-and-play' boards for use with Windows software, these parameters are set by the software; otherwise microswitches have to be set on the card in accordance with the instructions in the manual supplied with the board.

Application software can be used to assist in the designing of measurement systems and the analysis of the data. As an illustration of the type of application software available, LabVIEW is a graphical programming software package that has been developed by National Instruments for data acquisition and instrument control. LabVIEW programs are called **virtual instruments** because in appearance and operation they imitate actual instruments. A virtual instrument has three parts: a front panel which is the interactive user interface and simulates the front panel of an instrument by containing control knobs, push-buttons and graphical displays; a block diagram which is the source code for the program with the programming being done graphically by drawing lines between connection points on selected icons on the computer screen; and representation as an icon and connector which can provide a graphical representation of the virtual instrument if it is wanted for use in other block diagrams.

Figure 6.24(a) shows the icon selected for a virtual instrument where one analogue sample is obtained from a specified input channel, the icon having been selected from the Analog Input palette. The 'Device' is the device

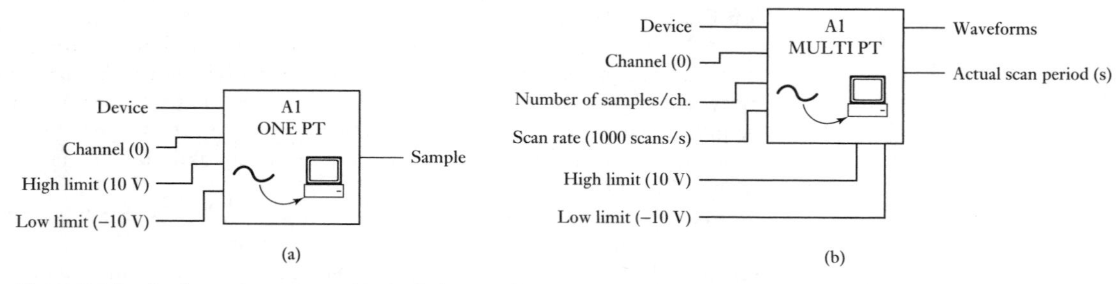

Figure 6.24 Analogue input icon: (a) single input, (b) for sampling from a number of channels.

number assigned to the DAQ board, the 'Channel' is the source of the data, a 'Sample' is one analogue-to-digital conversion, and 'High limit' and 'Low limit' are the voltage limits expected for the signal (the default is +10 V and −10 V and changing these values automatically changes the gain of the amplifier on the DAQ board).

If we want a waveform from each channel in a designated channel string then the icon shown in Figure 6.24(b) can be selected. For each input channel a set of samples is acquired over a period of time, at a specified sampling rate, and gives a waveform output showing how the analogue quantity varies with time.

By connecting other icons to, say, the above icon, a block diagram can be built up which might take the inputs from a number of analogue channels, sample them in sequence and display the results as a sequence of graphs. The type of front panel display we might have for a simple DAQ acquisition of samples and display is shown in Figure 6.25. By using the up and down arrows the parameters can be changed and the resulting display viewed.

Virtual instruments have a great advantage over traditional instruments in that the vendor of a traditional instrument determines its characteristics and interface while with a virtual instrument these can all be defined by the user and readily changed.

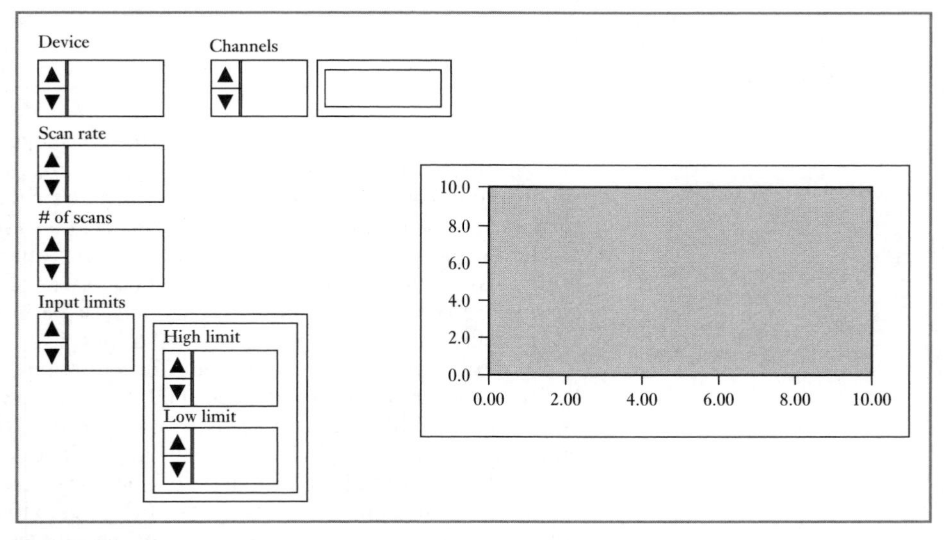

Figure 6.25 Front panel.

6.6.2 Data loggers

The term **data logger** is used for DAQ systems which are able to be used away from a computer. Once the program has been set by a computer, it can be put onto a memory card which can be inserted into the logger or have the program downloaded to it from a computer, so enabling it to carry out the required DAQ functions.

Figure 6.26 shows the basic elements of a data logger. Such a unit can monitor the inputs from a large number of sensors. Inputs from individual sensors, after suitable signal conditioning, are fed into the multiplexer. The multiplexer is used to select one signal which is then fed, after amplification, to the analogue-to-digital converter. The digital signal is then processed by a microprocessor. The microprocessor is able to carry out simple arithmetic operations, perhaps taking the average of a number of measurements. The output from the system might be displayed on a digital meter that indicates the output and channel number, used to give a permanent record with a printer, stored on a floppy disk or transferred to perhaps a computer for analysis.

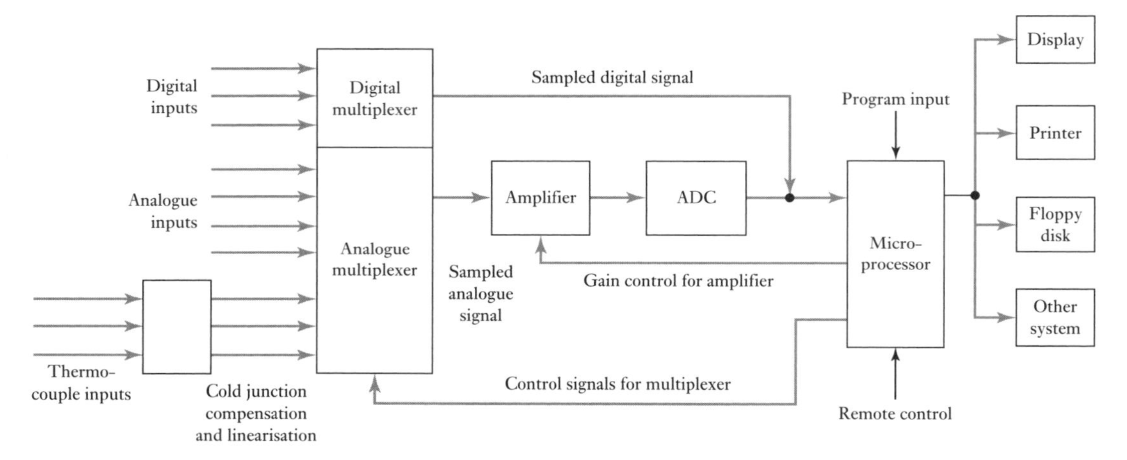

Figure 6.26 Data logger system.

Because data loggers are often used with thermocouples, there are often special inputs for thermocouples, these providing cold junction compensation and linearisation. The multiplexer can be switched to each sensor in turn and so the output consists of a sequence of samples. Scanning of the inputs can be selected by programming the microprocessor to switch the multiplexer to sample just a single channel, carry out a single scan of all channels, a continuous scan of all channels, or perhaps a periodic scan of all channels, say every 1, 5, 15, 30 or 60 minutes.

Typically a data logger may handle 20 to 100 inputs, though some may handle considerably more, perhaps 1000. It might have a sample and conversion time of 10 μs and be used to make perhaps 1000 readings per second. The accuracy is typically about 0.01% of full-scale input and linearity is about ±0.005% of full-scale input. Cross-talk is typically 0.01% of full-scale input on any one input. The term **cross-talk** is used to describe the interference that can occur when one sensor is being sampled as a result of signals from other sensors.

6.7 Measurement systems

The following examples illustrate some of the points involved in the design of measurement systems for particular applications.

6.7.1 Load cell for use as a link to detect load lifted

A link-type load cell, of the form shown in Figure 6.27, has four strain gauges attached to its surface and can be inserted in between the cable lifting a load and the load to give a measure of the load being lifted. Two of the strain gauges are in the longitudinal axis direction and two in a transverse direction. When the link is subject to tensile forces, the axial gauges will be in tension and the transverse gauges in compression. Suppose we have the design criteria for the load cell of a sensitivity such that there is an output of about 30 mV when the stress applied to the link is 500 MPa. We will assume that the strain gauges may have gauge factors of 2.0 and resistances of 100 Ω.

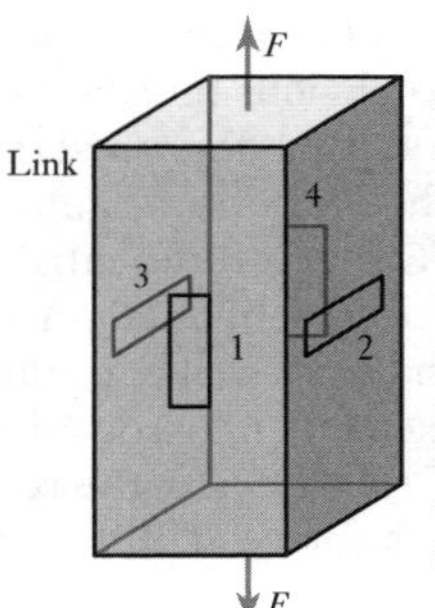

Figure 6.27 Load cell.

When a load F is applied to the link then, since the elastic modulus E is stress/strain and stress is force per unit area, the longitudinal axis strain ε_l is F/AE and the transverse strain ε_t is $-\nu F/AE$, where A is the cross-sectional area and ν is Poisson's ratio for the link material. The responses of the strain gauges (see Section 2.3.2) to these strains are

$$\frac{\delta R_1}{R_1} = \frac{\delta R_4}{R_4} = G\varepsilon_l = \frac{GF}{AE}$$

$$\frac{\delta R_3}{R_3} = \frac{\delta R_2}{R_2} = G\varepsilon_t = -\frac{\nu GF}{AE}$$

The output voltage from the Wheatstone bridge (see Section 3.5) is given by

$$V_o = \frac{V_s R_1 R_4}{(R_1 + R_2)(R_3 + R_4)}\left(\frac{\delta R_1}{R_1} - \frac{\delta R_2}{R_2} - \frac{\delta R_3}{R_3} + \frac{\delta R_4}{R_4}\right)$$

With $R_1 = R_2 = R_3 = R_4 = R$, and with $\delta R_1 = \delta R_4$ and $\delta R_2 = \delta R_3$, then

$$V_o = \frac{V_s}{2R}(\delta R_1 - \delta R_2) = \frac{V_s GF}{2AE}(1 + \nu)$$

Suppose we consider steel for the link. Then tables give E as about 210 GPa and ν about 0.30. Thus with a stress $(= F/A)$ of 500 MPa we have, for strain gauges with a gauge factor of 2.0,

$$V_o = 3.09 \times 10^{-3} V_s$$

For a bridge voltage with a supply voltage of 10 V this would be an output voltage of 30.9 mV. No amplification is required if this is the only load value required; if, however, this is a maximum value and we want to determine loads below this level then we might use a differential amplifier. The output can be displayed on a high-resistance voltmeter – high resistance to avoid loading problems. A digital voltmeter might thus be suitable.

6.7.2 Temperature alarm system

A measurement system is required which will set off an alarm when the temperature of a liquid rises above 40°C. The liquid is normally at 30°C. The output from the system must be a 1 V signal to operate the alarm.

Since the output is to be electrical and a reasonable speed of response is likely to be required, an obvious possibility is an electrical resistance element. To generate a voltage output the resistance element could be used with a Wheatstone bridge. The output voltage will probably be less than 1 V for a change from 30 to 40°C, but a differential amplifier could be used to enable the required voltage to be obtained. A comparator can then be used to compare the value with the set value for the alarm.

Suppose a nickel element is used. Nickel has a temperature coefficient of resistance of 0.0067/K. Thus if the resistance element is taken as being 100 Ω at 0°C then its resistance at 30°C will be

$$R_{30} = R_0(1 + \alpha t) = 100(1 + 0.0067 \times 30) = 120.1\ \Omega$$

and at 40°C

$$R_{40} = 100(1 + 0.0067 \times 40) = 126.8\ \Omega$$

Thus there is a change in resistance of 6.7 Ω. If this element forms one arm of a Wheatstone bridge which is balanced at 30°C, then the output voltage V_o is given by (see Section 3.5)

$$\delta V_o = \frac{V_s \delta R_1}{R_1 + R_2}$$

With the bridge balanced at 30°C and, say, all the arms have the same value and a supply voltage of 4 V, then

$$\delta V_o = \frac{4 \times 6.7}{126.8 + 120.1} = 0.109\ \text{V}$$

To amplify this to 1 V we can use a difference amplifier (see Section 3.2.5)

$$V_o = \frac{R_2}{R_1}(V_2 - V_1)$$

$$1 = \frac{R_2}{R_1} \times 0.109$$

Hence $R_2/R_1 = 9.17$ and so if we use an input resistance of 1 kΩ the feedback resistance must be 9.17 kΩ.

6.7.3 Angular position of a pulley wheel

A potentiometer is to be used to monitor the angular position of a pulley wheel. Consider the items that might be needed to enable there to be an output to a recorder of 10 mV per degree if the potentiometer has a full-scale angular rotation of 320°.

When the supply voltage V_s is connected across the potentiometer we will need to safeguard it and the wiring against possible high currents and so a resistance R_s can be put in series with the potentiometer R_p. The total voltage drop across the potentiometer is thus $V_sR_p/(R_s + R_p)$. For an angle θ with a potentiometer having a full-scale angular deflection of θ_F we will obtain an output from the potentiometer of

$$V_\theta = \frac{\theta}{\theta_F}\frac{V_sR_p}{R_s + R_p}$$

Suppose we consider a potentiometer with a resistance of 4 kΩ and let R_s be 2 kΩ. Then for 1 mV per degree we have

$$0.01 = \frac{1}{320}\frac{4V_s}{4 + 2}$$

Hence we would need a supply voltage of 4.8 V. To prevent loading of the potentiometer by the resistance of the recorder, a voltage follower circuit can be used. Thus the circuit might be of the form shown in Figure 6.28.

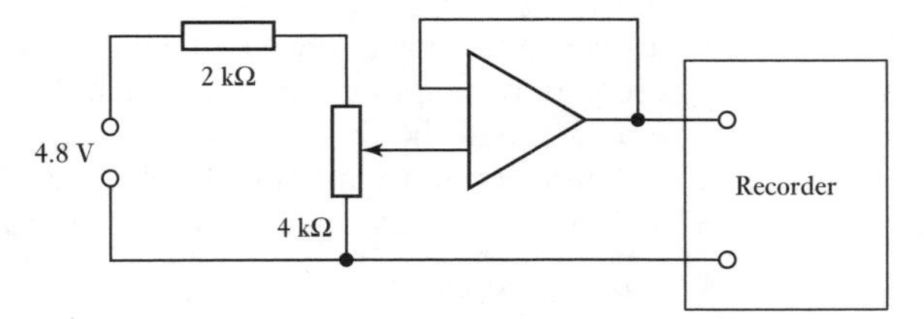

Figure 6.28 Pulley wheel monitor.

6.7.4 Temperature measurement to give a binary output

Consider the requirement for a temperature measurement system for temperatures in the range 0 to 100°C and which will give an 8-bit binary output with a change in 1 bit corresponding to a temperature change of 1°C. The output is intended for inputting to a microprocessor as part of a temperature control system.

A linear temperature sensor is required and so the thermotransistor LM35 can be used (see Section 2.9.4). LM35 gives an output of 10 mV/°C when it has a supply voltage of 5 V. If we apply the output from LM35 to an 8-bit analogue-to-digital converter then a digital output can be obtained. We need the resolution of the ADC to be 10 mV so that each step of 10 mV will generate a change in output of 1 bit. Suppose we use a successive approximations ADC, e.g. ADC0801; then this requires an input of a reference voltage which when subdivided into $2^8 = 256$ bits gives 10 mV per bit. Thus a reference voltage of 2.56 V is required. For this to be obtained the reference voltage

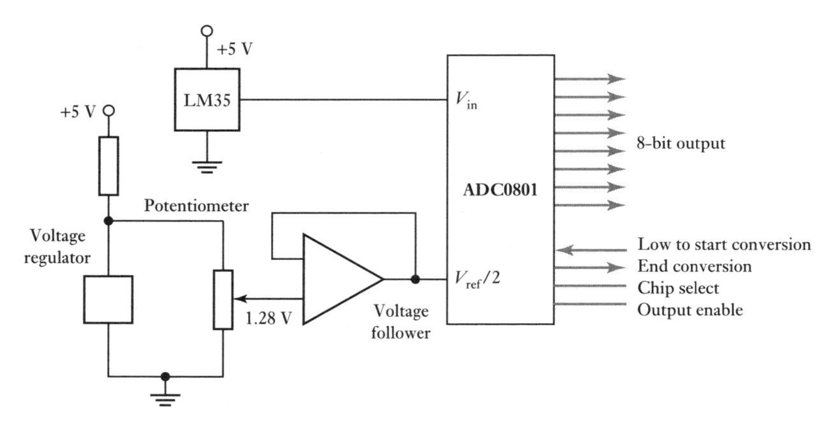

Figure 6.29 Temperature sensor.

input to the ADC0801 has to be $V_{ref}/2$ and so an accurate input voltage of 1.28 V is required. Such a voltage can be obtained by using a potentiometer circuit across the 5 V supply with a voltage follower to avoid loading problems. Because the voltage has to remain steady at 1.28 V, even if the 5 V supply voltage fluctuates, a voltage regulator is likely to be used, e.g. a 2.45 V voltage regulator ZN458/B. Thus the circuit might be as in Figure 6.29.

6.8 Testing and calibration

Testing a measurement system installation falls into three stages:

1 *Pre-installation testing*
This is the testing of each instrument for correct calibration and operation prior to it being installed.

2 *Piping and cabling testing*
In the case of pneumatic lines this involves, prior to the connection of the instruments, blowing through with clean, dry air prior to connection and pressure testing to ensure they are leak free. With process piping, all the piping should be flushed through and tested prior to the connection of instruments. With instrument cables, all should be checked for continuity and insulation resistance prior to the connection of any instruments.

3 *Pre-commissioning*
This involves testing that the installation is complete, all instrument components are in full operational order when interconnected and all control room panels or displays function.

6.8.1 Calibration

Calibration consists of comparing the output of a measurement system and its subsystems against standards of known accuracy. The standards may be other instruments which are kept specially for calibration duties or some means of defining standard values. In many companies some instruments and items such as standard resistors and cells are kept in a company standards department and used solely for calibration purposes. The relationship between the calibration of an instrument in everyday use and national standards is likely to be:

1 National standards are used to calibrate standards for calibration centres.
2 Calibration centre standards are used to calibrate standards for instrument manufacturers.

3 Standardised instruments from instrument manufacturers are used to provide in-company standards.
4 In-company standards are used to calibrate process instruments.

There is a simple traceability chain from the instrument used in a process back to national standards. The following are some examples of calibration procedures that might be used in-company:

1 *Voltmeters*
These can be checked against standard voltmeters or standard cells giving standard e.m.f.s.

2 *Ammeters*
These can be checked against standard ammeters.

3 *Gauge factor of strain gauges*
This can be checked by taking a sample of gauges from a batch and applying measured strains to them when mounted on some test piece. The resistance changes can be measured and hence the gauge factor computed.

4 *Wheatstone bridge circuits*
The output from a Wheatstone bridge can be checked when a standard resistance is introduced into one of the arms.

5 *Load cells*
For low-capacity load cells, dead-weight loads using standard weights can be used.

6 *Pressure sensors*
Pressure sensors can be calibrated by using a dead-weight tester (Figure 6.30). The calibration pressures are generated by adding standard weights W to the piston tray. After the weights are placed on the tray, a screw-driven plunger is forced into the hydraulic oil in the chamber to lift the piston–weight assembly. The calibration pressure is then W/A, where A is the cross-sectional area of the piston. Alternatively the dead-weight tester can be used to calibrate a pressure gauge and this gauge can be used for the calibration of other gauges.

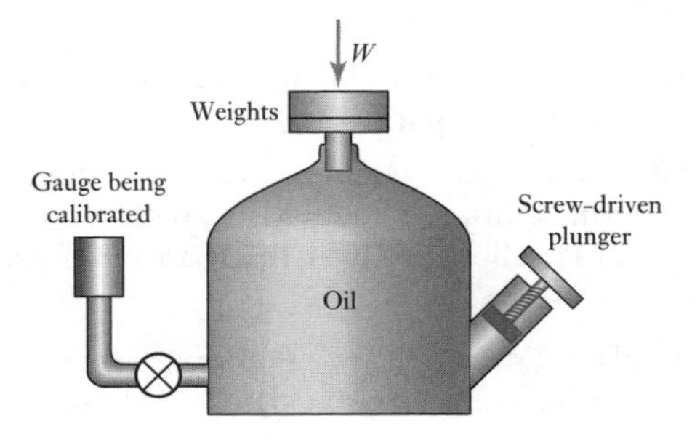

Figure 6.30 Dead-weight calibration for pressure gauges.

7 *Temperature sensors*
These can be calibrated by immersion in a melt of a pure metal or water. The temperature of the substance is then slowly reduced and a temperature–time record obtained. When the substance changes state from liquid to solid, the temperature remains constant. Its value can be looked up from tables and hence an accurate reference temperature for calibration obtained. Alternatively, the temperature at which a liquid boils can be

used. However, the boiling point depends on the atmospheric pressure and corrections have to be applied if it differs from the standard atmospheric pressure. Alternatively, in-company the readings given by the measurement system can be compared with those of a standard thermometer.

Summary

A general point that has to be taken account of when putting together any measurement system is **loading**, i.e. the effect of connecting a load across the output terminals of any element of a measurement system.

Indicators give an instant visual indication of the sensed variable while **recorders** record the output signal over a period of time and give automatically a permanent record.

The term **data acquisition** (DAQ) tends to be frequently used for systems in which inputs from sensors are converted into a digital form for processing, analysis and display by a computer. The term **data logger** is used for DAQ systems which are able to be used away from a computer.

Virtual instruments are software-generated instruments; in appearance and operation they imitate actual instruments.

Testing a measurement system installation falls into three stages: pre-installation testing, piping and cabling testing, pre-commissioning.

Calibration consists of comparing the output of a measurement system and its subsystems against standards of known accuracy.

Problems

6.1 Explain the significance of the following terms taken from the specifications of display systems:

(a) Recorder: dead band ±0.2% of span.

(b) The hard disk has two disks with four read/write heads, one for each surface of the disks. Each surface has 614 tracks and each track 32 sectors.

(c) Data logger: number of inputs 100, cross-talk on any one input 0.01% of full-scale input.

(d) Double-beam oscilloscope: vertical deflection with two identical channels, bandwidth d.c. to 15 MHz, deflection factor of 10 mV/div to 20 V/div in 11 calibrated steps, time base of 0.5 μs/div to 0.5 s/div in 19 calibrated steps.

6.2 Explain the problems of loading when a measurement system is being assembled from a sensor, signal conditioner and display.

6.3 Suggest a display unit that could be used to give:

(a) A permanent record of the output from a thermocouple.

(b) A display which enables the oil pressure in a system to be observed.

(c) A record to be kept of the digital output from a microprocessor.

(d) The transient voltages resulting from monitoring of the loads on an aircraft during simulated wind turbulence.

6.4 A cylindrical load cell, of the form shown in Figure 2.32, has four strain gauges attached to its surface. Two of the gauges are in the circumferential direction and two in the longitudinal axis direction. When the cylinder is subject to a compressive load, the axial gauges will be in compression while the circumferential ones will be in tension. If the material of the cylinder has a cross-sectional area A and an elastic modulus E, then a force F acting on the cylinder will give a strain acting on the axial gauges of $-F/AE$ and on the circumferential gauges of $+\nu F/AE$, where ν is Poisson's ratio for the material. Design a complete measurement system, using load cells, which could be used to monitor the mass of water in a tank. The tank itself has a mass of 20 kg and the water when at the required level 40 kg. The mass is to be monitored to an accuracy of ± 0.5 kg. The strain gauges have a gauge factor of 2.1 and are all of the same resistance of 120.0 Ω. For all other items, specify what your design requires. If you use mild steel for the load cell material, then the tensile modulus may be taken as 210 GPa and Poisson's ratio 0.30.

6.5 Design a complete measurement system involving the use of a thermocouple to determine the temperature of the water in a boiler and give a visual indication on a meter. The temperature will be in the range 0 to 100°C and is required to an accuracy of $\pm 1\%$ of full-scale reading. Specify the materials to be used for the thermocouple and all other items necessary. In advocating your design you must consider the problems of cold junction and non-linearity. You will probably need to consult thermocouple tables. The following data is taken from such tables, the cold junction being at 0°C, and may be used as a guide:

	E.M.F. in mV at				
Materials	**20°C**	**40°C**	**60°C**	**80°C**	**100°C**
Copper–constantan	0.789	1.611	2.467	3.357	4.277
Chromel–constantan	1.192	2.419	3.683	4.983	6.317
Iron–constantan	1.019	2.058	3.115	4.186	5.268
Chromel–alumel	0.798	1.611	2.436	3.266	4.095
Platinum–10% Rh, Pt	0.113	0.235	0.365	0.502	0.645

6.6 Design a measurement system which could be used to monitor the temperatures, of the order of 100°C, in positions scattered over a number of points in a plant and present the results on a control panel.

6.7 A suggested design for the measurement of liquid level in a vessel involves a float which in its vertical motion bends a cantilever. The degree of bending of the cantilever is then taken as a measure of the liquid level. When a force F is applied to the free end of a cantilever of length L, the strain on its surface a distance x from the clamped end is given by

$$\text{strain} = \frac{6(L - x)}{wt^2E}$$

where w is the width of the cantilever, t its thickness and E the elastic modulus of the material. Strain gauges are to be used to monitor the bending of the cantilever with two strain gauges being attached longitudinally to the upper surface and two longitudinally to the lower surface. The gauges are then to be incorporated into a four-gauge Wheatstone bridge and the output voltage, after possible amplification, then taken as a measure of the liquid level. Determine the specifications required for the components of this system if there is to be an output of 10 mV per 10 cm change in level.

6.8 Design a static pressure measurement system based on a sensor involving a 40 mm diameter diaphragm across which there is to be a maximum pressure difference of 500 MPa. For a diaphragm where the central deflection y is much smaller than the thickness t of the diaphragm,

$$y \approx \frac{3r^2P(1 - \nu^2)}{16Et^3}$$

where r is the radius of the diaphragm, P the pressure difference, E the modulus of elasticity and ν Poisson's ratio. Explain how the deflection y will be converted into a signal that can be displayed on a meter.

8.9 Suggest the elements that might be considered for the measurement systems to be used to:

(a) Monitor the pressure in an air pressure line and present the result on a dial, no great accuracy being required.

(b) Monitor continuously and record the temperature of a room with an accuracy of ± 1°C.

(c) Monitor the weight of lorries passing over a weighing platform.

(d) Monitor the angular speed of rotation of a shaft.

Part III
Actuation

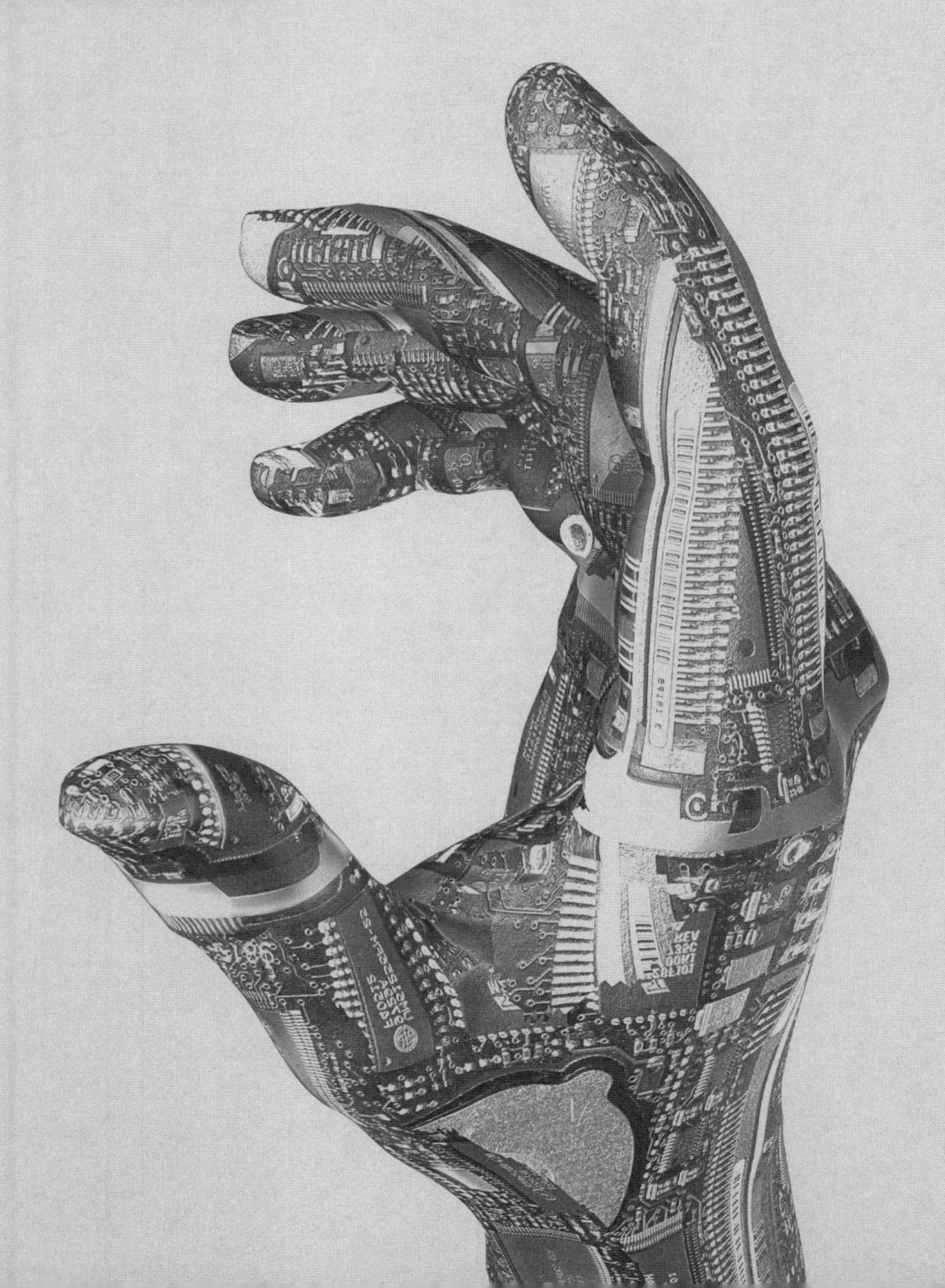

Chapter seven Pneumatic and Hydraulic Actuation Systems

Objectives

The objectives of this chapter are that, after studying it, the reader should be able to:

- **Interpret system drawings, and design simple systems, for sequential control systems involving hydraulic/pneumatic directional control valves and cylinders.**
- **Explain the principles of process control valves, their characteristics and sizing.**

7.1 Actuation systems

Actuation systems are the elements of control systems which are responsible for transforming the output of a microprocessor or control system into a controlling action on a machine or device. Thus, for example, we might have an electrical output from the controller which has to be transformed into a linear motion to move a load. Another example might be where an electrical output from the controller has to be transformed into an action which controls the amount of liquid passing along a pipe.

In this chapter fluid power systems, namely pneumatic and hydraulic actuation systems, are discussed. **Pneumatics** is the term used when compressed air is used and **hydraulics** when a liquid, typically oil. In Chapter 8 mechanical actuator systems are discussed and in Chapter 9 electrical actuation systems.

7.2 Pneumatic and hydraulic systems

Pneumatic signals are often used to control final control elements, even when the control system is otherwise electrical. This is because such signals can be used to actuate large valves and other high-power control devices and so move significant loads. The main drawback with pneumatic systems is, however, the compressibility of air. Hydraulic systems can be used for even higher power control devices but are more expensive than pneumatic systems and there are hazards associated with oil leaks which do not occur with air leaks.

The atmospheric pressure varies with both location and time but in pneumatics is generally taken to be 10^5 Pa, such a pressure being termed 1 bar.

7.2.1 Hydraulic systems

With a hydraulic system, pressurised oil is provided by a pump driven by an electric motor. The pump pumps oil from a sump through a non-return

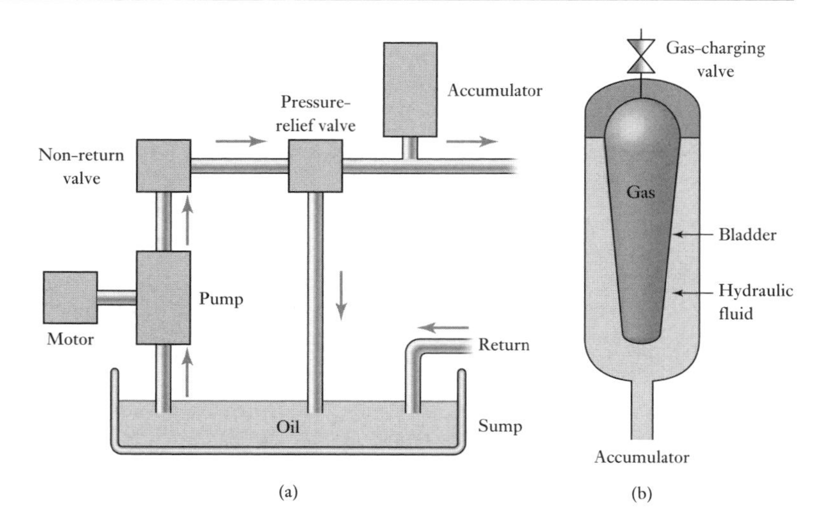

Figure 7.1 (a) Hydraulic power supply, (b) accumulator.

valve and an accumulator to the system, from which it returns to the sump. Figure 7.1(a) illustrates the arrangement. A pressure-relief valve is included, this being to release the pressure if it rises above a safe level, the non-return valve is to prevent the oil being back driven to the pump and the accumulator is to smooth out any short-term fluctuations in the output oil pressure. Essentially the accumulator is just a container in which the oil is held under pressure against an external force, Figure 7.1(b) showing the most commonly used form which is gas pressurised and involves gas within a bladder in the chamber containing the hydraulic fluid; an older type involved a spring-loaded piston. If the oil pressure rises then the bladder contracts, increases the volume the oil can occupy and so reduces the pressure. If the oil pressure falls, the bladder expands to reduce the volume occupied by the oil and so increases its pressure.

Commonly used hydraulic pumps are the gear pump, the vane pump and the piston pump. The **gear pump** consists of two close-meshing gear wheels which rotate in opposite directions (Figure 7.2(a)). Fluid is forced through the pump as it becomes trapped between the rotating gear teeth and the housing and so is transferred from the inlet port to be discharged at the outlet port. Such pumps are widely used, being low cost and robust. They generally operate at pressures below about 15 MPa and at 2400 rotations per minute. The maximum flow capacity is about 0.5 m^3/min. However, leakage occurs between the teeth and the casing and between the interlocking teeth, and this limits the efficiency. The **vane pump** has spring-loaded sliding vanes slotted in a driven rotor (Figure 7.2(b)). As the rotor rotates, the vanes follow the contours of the casing. This results in fluid becoming trapped between successive vanes and the casing and transported round from the inlet port to outlet port. The leakage is less than with the gear pump. **Piston pumps** used in hydraulics can take a number of forms. With the **radial piston pump** (Figure 7.2(c)), a cylinder block rotates round the stationary cam and this causes hollow pistons, with spring return, to move in and out. The result is that fluid is drawn in from the inlet port and transported round for ejection from the discharge port. The **axial piston pump** (Figure 7.2(d)) has pistons which move axially rather than radially. The pistons are arranged

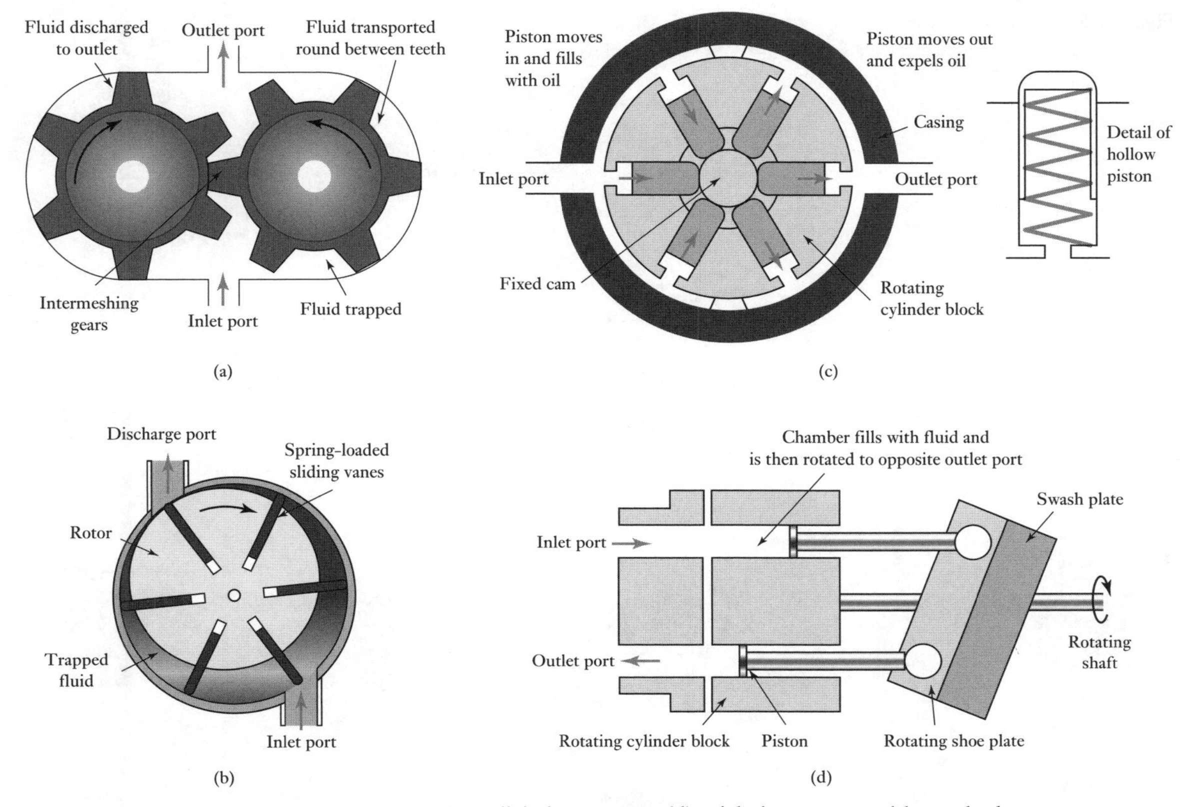

Figure 7.2 (a) Gear pump, (b) vane pump, (c) radial piston pump, (d) axial piston pump with swash plate.

axially in a rotating cylinder block and made to move by contact with the swash plate. This plate is at an angle to the drive shaft and thus as the shaft rotates they move the pistons so that air is sucked in when a piston is opposite the inlet port and expelled when it is opposite the discharge port. Piston pumps have a high efficiency and can be used at higher hydraulic pressures than gear or vane pumps.

7.2.2 Pneumatic systems

With a **pneumatic** power supply (Figure 7.3) an electric motor drives an air compressor. The air inlet to the compressor is likely to be filtered and via a silencer to reduce the noise level. A pressure-relief valve provides protection against the pressure in the system rising above a safe level. Since the air compressor increases the temperature of the air, there is likely to be a cooling system and to remove contamination and water from the air a filter with a water trap. An air receiver increases the volume of air in the system and smoothes out any short-term pressure fluctuations.

Commonly used air compressors are ones in which successive volumes of air are isolated and then compressed. Figure 7.4(a) shows the basic form of

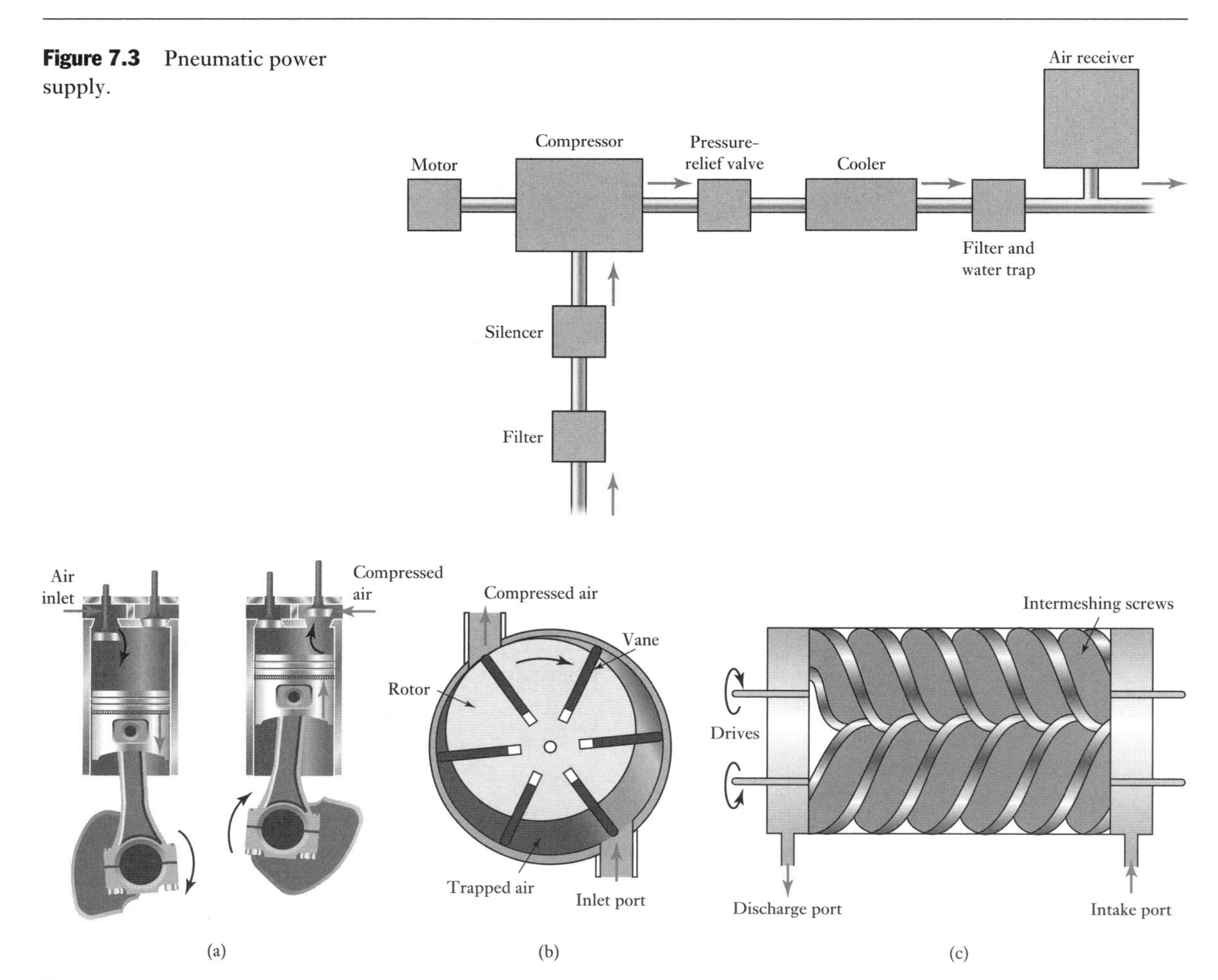

Figure 7.3 Pneumatic power supply.

Figure 7.4 (a) Single-acting, single stage, vertical, reciprocating compressor, (b) rotary vane compressor, (c) screw compressor.

a single-acting, single-stage, vertical, reciprocating compressor. On the air intake stroke, the descending piston causes air to be sucked into the chamber through the spring-loaded inlet valve and when the piston starts to rise again, the trapped air forces the inlet valve to close and so becomes compressed. When the air pressure has risen sufficiently, the spring-loaded outlet valve opens and the trapped air flows into the compressed-air system. After the piston has reached the top dead centre it then begins to descend and the cycle repeats itself. Such a compressor is termed **single-acting** because one pulse of air is produced per piston stroke; **double-acting** compressors are designed to produce pulses of air on both the up and down strokes of the piston. It is also termed **single-stage** because the compressor goes directly from atmospheric pressure to the required pressure in a single operation. For the production of compressed air at more than a few bars, two or more stages are generally used. Normally two stages are used for pressures up to about 10 to 15 bar and more stages for higher pressures. Thus with a two-stage compressor we might have the first stage taking air at atmospheric pressure and compressing it to, say, 2 bar and then the second stage compressing this air to, say, 7 bar. Reciprocating piston compressors can be used as a single-stage compressor to produce air pressures up to about 12 bar and as a multistage compressor up to about 140 bar. Typically, air flow deliveries tend to range from about 0.02 m^3/min free air delivery to about 600 m^3/min free air delivery; free air is the term used for air at normal atmospheric pressure. Another form of compressor is the **rotary vane compressor**. This has a rotor mounted eccentrically in a cylindrical chamber (Figure 7.4(b)). The rotor has blades, the vanes, which are free to slide in radial slots with rotation causing the vanes to be driven outwards against the walls of the cylinder. As the rotor rotates, air is trapped in pockets formed by the vanes and as the rotor rotates so the pockets become smaller and the air is compressed. Compressed packets of air are thus discharged from the discharge port. Single-stage, rotary vane compressors typically can be used for pressures up to about 800 kPa with flow rates of the order of 0.3 m^3/min to 30 m^3/min free air delivery. Another form of compressor is the **rotary screw compressor** (Figure 7.4(c)). This has two intermeshing rotary screws which rotate in opposite directions. As the screws rotate, air is drawn into the casing through the inlet port and into the space between the screws. Then this trapped air is moved along the length of the screws and compressed as the space becomes progressively smaller, emerging from the discharge port. Typically, single-stage, rotary screw compressors can be used for pressures up to about 1000 kPa with flow rates of between 1.4 m^3/min and 60 m^3/min free air delivery.

7.2.3 Valves

Valves are used with hydraulic and pneumatic systems to direct and regulate the fluid flow. There are basically just two forms of valve, the **finite position** and the **infinite position** valves. The finite position valves are ones where the action is just to allow or block fluid flow and so can be used to switch actuators on or off. They can be used for directional control to switch the flow from one path to another and so from one actuator to another. The infinite position valves are able to control flow anywhere between fully on and fully off and so are used to control varying actuator forces or the rate of fluid flow for a process control situation.

7.3 Directional control valves

Pneumatic and hydraulic systems use directional control valves to direct the flow of fluid through a system. They are not intended to vary the rate of flow of fluid but are either completely open or completely closed, i.e. on/off devices. Such on/off valves are widely used to develop sequenced control systems (see later in this chapter). They might be activated to switch the fluid flow direction by means of mechanical, electrical or fluid pressure signals.

A common type of directional control valve is the **spool valve**. A spool moves horizontally within the valve body to control the flow. Figure 7.5

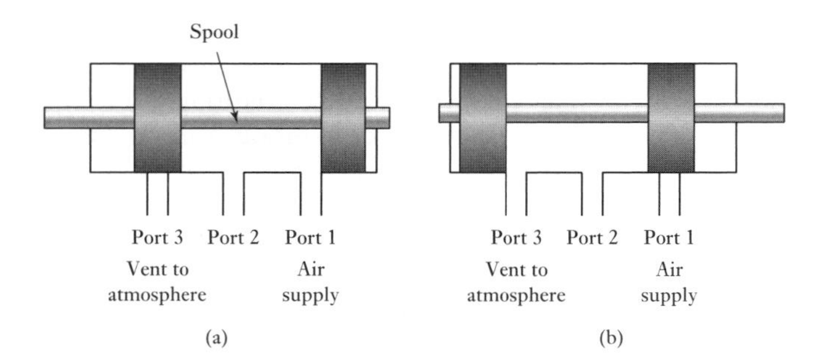

Figure 7.5 Spool valve.

shows a particular form. In (a) the air supply is connected to port 1 and port 3 is closed. Thus the device connected to port 2 can be pressurised. When the spool is moved to the left (Figure 7.5(b)) the air supply is cut off and port 2 is connected to port 3. Port 3 is a vent to the atmosphere and so the air pressure in the system attached to port 2 is vented. Thus the movement of the spool has allowed the air firstly to flow into the system and then be reversed and flow out of the system. **Rotary spool valves** have a rotating spool which, when it rotates, opens and closes ports in a similar way.

Another common form of directional control valve is the **poppet valve**. Figure 7.6 shows one form. This valve is normally in the closed condition, there being no connection between port 1 to which the pressure supply is connected and port 2 to which the system is connected. In poppet valves, balls, discs or cones are used in conjunction with valve seats to control the flow. In the figure a ball is shown. When the push-button is depressed, the ball is pushed out of its seat and flow occurs as a result of port 1 being connected to port 2. When the button is released, the spring forces the ball back up against its seat and so closes off the flow.

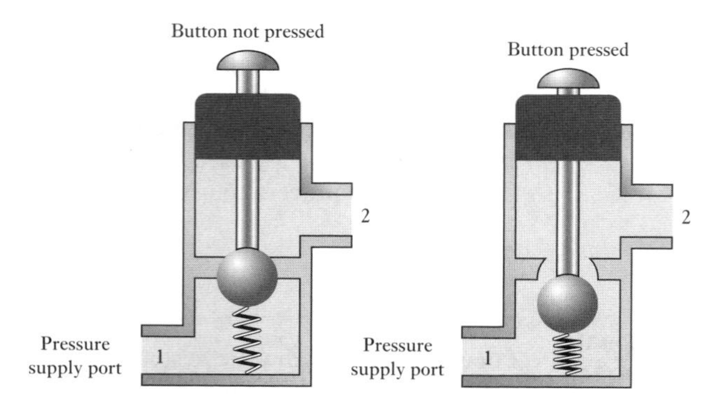

Figure 7.6 Poppet valve.

7.3.1 Valve symbols

The symbol used for a control valve consists of a square for each of its switching positions. Thus for the poppet valve shown in Figure 7.6, there are two positions: one with the button not pressed and one with it pressed. Thus a two-position valve will have two squares, a three-position valve three squares. Arrow-headed lines (Figure 7.7(a)) are used to indicate the directions of flow in each of the positions, with blocked-off lines indicating closed flow lines (Figure 7.7(b)). The initial position of the valve has the connections (Figure 7.7(c)) to the ports shown; in Figure 7.7(c) the valve has four ports. Ports are labelled by a number or a letter according to their function. The ports are labelled 1 (or P) for pressure supply, 3 (or T) for hydraulic return port, 3 or 5 (or R or S) for pneumatic exhaust ports, and 2 or 5 (or B or A) for output ports.

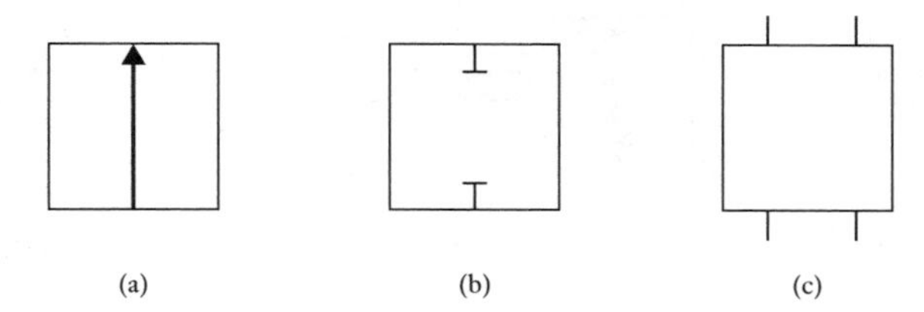

Figure 7.7 (a) Flow path, (b) flow shut-off, (c) initial connections.

Figure 7.8(a) shows examples of some of the symbols which are used to indicate the various ways the valves can be actuated. More than one of these symbols might be used with the valve symbol. As an illustration, Figure 7.8(b) shows the symbol for the two-port, two-position poppet valve of Figure 7.6. Note that a two-port, two-position valve would be described as a 2/2 valve, the first number indicating the number of ports and the second number the number of positions. The valve actuation is by a push-button and a spring.

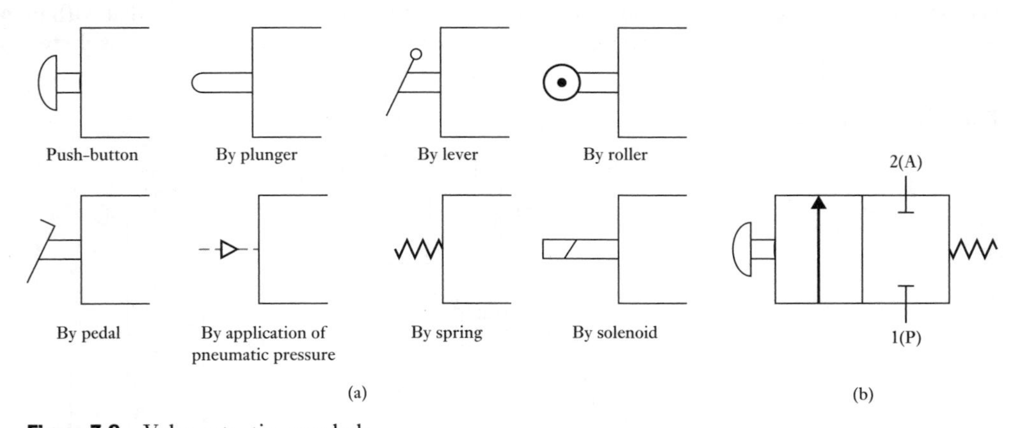

Figure 7.8 Valve actuation symbols.

As a further illustration, Figure 7.9 shows a solenoid-operated spool valve and its symbol. The valve is actuated by a current passing through a solenoid and returned to its original position by a spring.

Figure 7.10 shows the symbol for a 4/2 valve. The connections are shown for the initial state, i.e. 1(P) is connected to 2(A) and 3(R) closed. When the

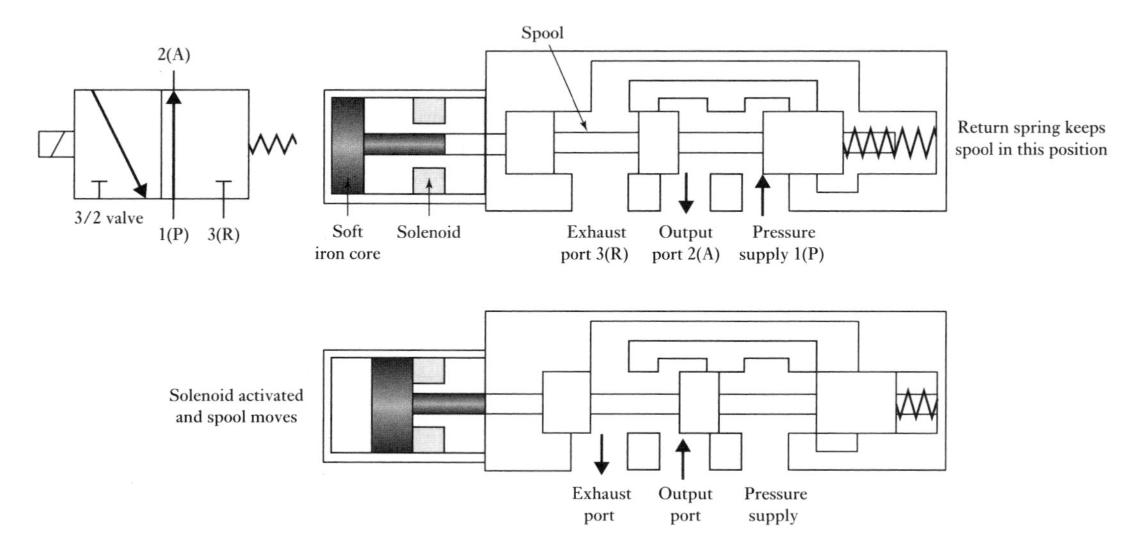

Figure 7.9 Single-solenoid valve.

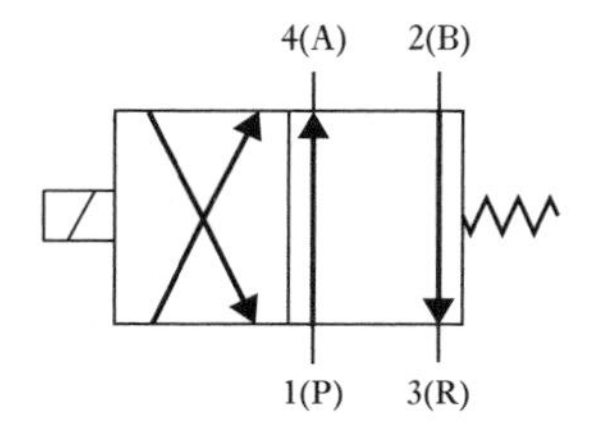

Figure 7.10 A 4/2 valve.

solenoid is activated it gives the state indicated by the symbols used in the square to which it is attached, i.e. we now have 1(P) closed and 2(A) connected to 3(R). When the current through the solenoid ceases, the spring pushes the valve back to its initial position. The spring movement gives the state indicated by the symbols used in the square to which it is attached.

Figure 7.11 shows a simple example of an application of valves in a pneumatic lift system. Two push-button 2/2 valves are used. When the button on the up valve is pressed, the load is lifted. When the button on the down valve is pressed, the load is lowered. Note that with pneumatic systems an open arrow is used to indicate a vent to the atmosphere.

Figure 7.11 Lift system.

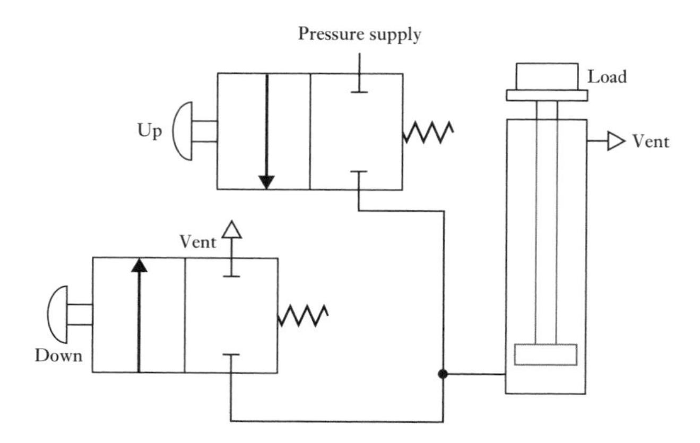

7.3.2 Pilot-operated valves

The force required to move the ball or shuttle in a valve can often be too large for manual or solenoid operation. To overcome this problem a **pilot-operated**

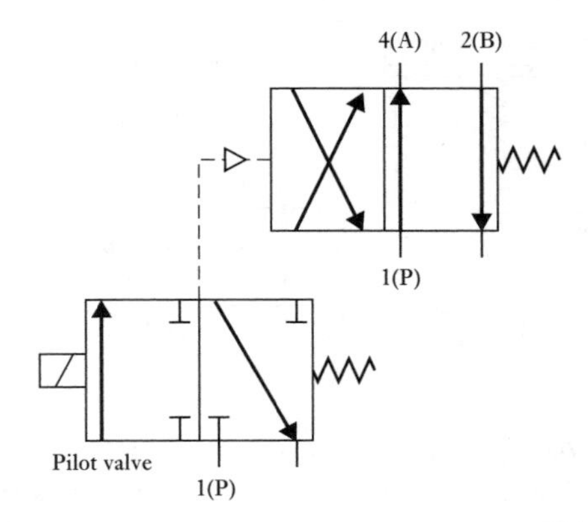

Figure 7.12 Pilot-operated system.

system is used where one valve is used to control a second valve. Figure 7.12 illustrates this. The pilot valve is small capacity and can be operated manually or by a solenoid. It is used to allow the main valve to be operated by the system pressure. The pilot pressure line is indicated by dashes. The pilot and main valves can be operated by two separate valves but they are often combined in a single housing.

7.3.3 Directional valves

Figure 7.13 shows a simple **directional valve** and its symbol. Free flow can only occur in one direction through the valve: that which results in the ball being pressed against the spring. Flow in the other direction is blocked by the spring forcing the ball against its seat.

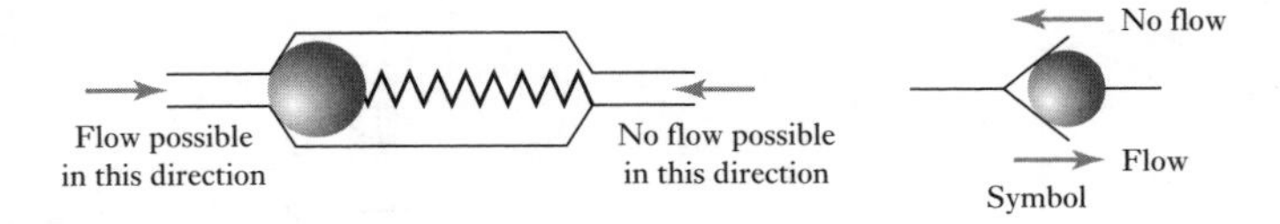

Figure 7.13 Directional valve.

7.4 Pressure control valves

There are three main types of pressure control valves:

1 *Pressure-regulating valves*
These are used to control the operating pressure in a circuit and maintain it at a constant value.

2 *Pressure-limiting valves*
These are used as safety devices to limit the pressure in a circuit to below some safe value. The valve opens and vents to the atmosphere, or back to the sump, if the pressure rises above the set safe value. Figure 7.14 shows a **pressure-limiting/relief valve** which has one orifice which is normally closed. When the inlet pressure overcomes the force exerted by the spring, the valve opens and vents to the atmosphere, or back to the sump.

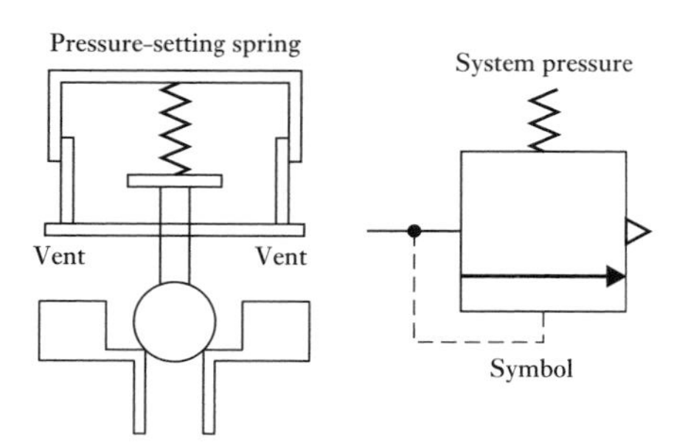

Figure 7.14 Pressure-limiting valve.

3 *Pressure sequence valves*
These valves are used to sense the pressure of an external line and give a signal when it reaches some preset value. With the pressure-limiting valve of Figure 7.15, the limiting pressure is set by the pressure at the inlet to the valve. We can adapt such a valve to give a sequence valve. This can be used to allow flow to occur to some part of the system when the pressure has risen to the required level. For example, in an automatic machine we might require some operation to start when the clamping pressure applied to a workpiece is at some particular value. Figure 7.15(a) shows the symbol for a sequence valve, the valve switching on when the inlet pressure reaches a particular value and allowing the pressure to be applied to the system that follows. Figure 7.15(b) shows a system where such a sequential valve is used. When the 4/3 valve first operates, the pressure is applied to cylinder 1 and its ram moves to the right. While this is happening the pressure is too low to operate the sequence valve and so no pressure is applied to cylinder 2. When the ram of cylinder 1 reaches the end stop, then the pressure in the system rises and, at an appropriate level, triggers the sequence valve to open and so apply pressure to cylinder 2 to start its ram in motion.

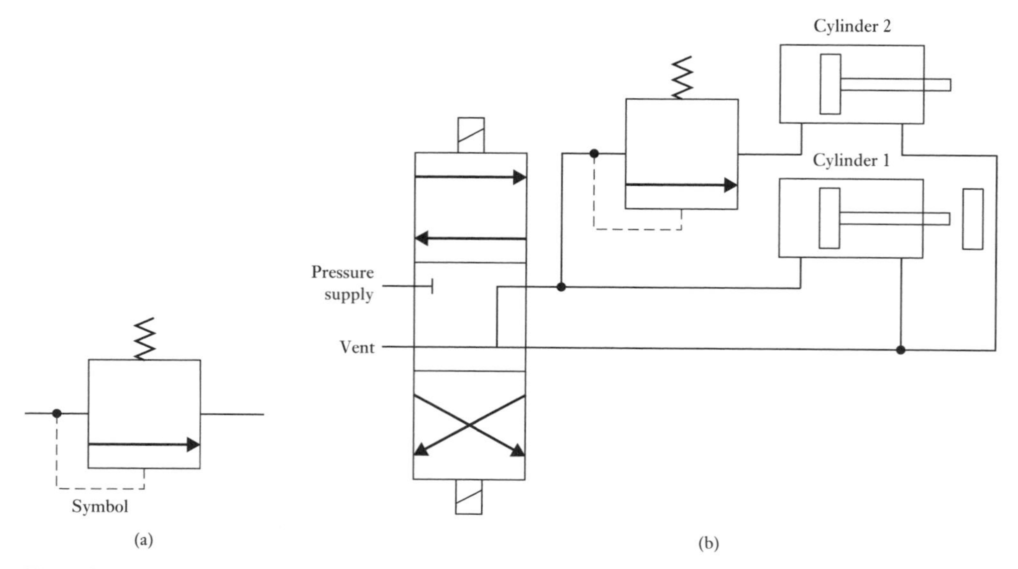

Figure 7.15 (a) Pressure sequence valve symbol, (b) a sequential system.

7.5 Cylinders

The **hydraulic** or **pneumatic cylinder** is an example of a linear actuator. The principles and form are the same for both hydraulic and pneumatic versions, differences being purely a matter of size as a consequence of the higher pressures used with hydraulics. The cylinder consists of a cylindrical tube along which a piston/ram can slide. There are two basic types, single-acting cylinders and double-acting cylinders.

The term single acting is used when the control pressure is applied to just one side of the piston, a spring often being used to provide the opposition to the movement of the piston. The other side of the piston is open to the atmosphere. Figure 7.16 shows one such cylinder with a spring return. The fluid is applied to one side of the piston at a gauge pressure p with the other side being at atmospheric pressure and so produces a force on the piston of pA, where A is the area of the piston. The actual force acting on the piston rod will be less than this because of friction.

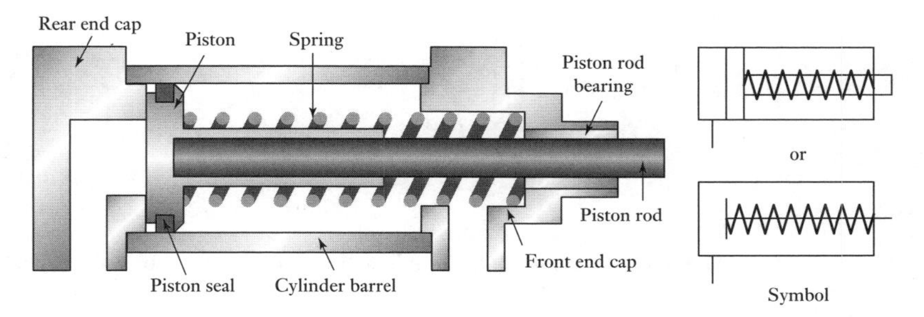

Figure 7.16 Single-acting cylinder.

For the single-acting cylinder shown in Figure 7.17, when a current passes through the solenoid, the valve switches position and pressure is applied to move the piston along the cylinder. When the current through the solenoid ceases, the valve reverts to its initial position and the air is vented from the cylinder. As a consequence the spring returns the piston back along the cylinder.

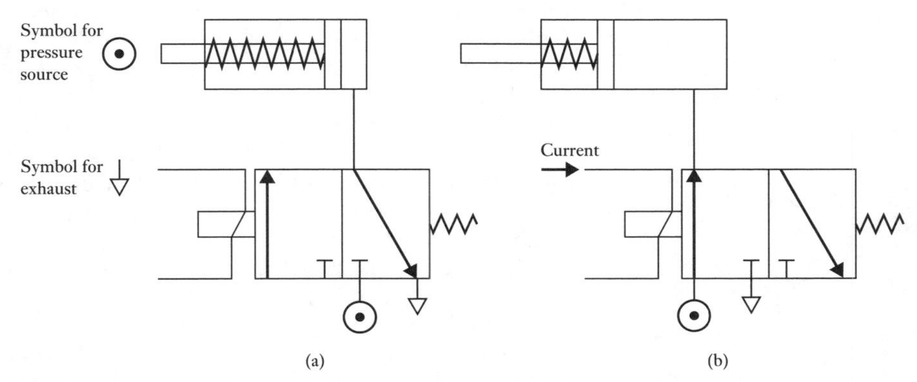

Figure 7.17 Control of a single-acting cylinder with (a) no current through solenoid, (b) a current through the solenoid.

The term double acting is used when the control pressures are applied to each side of the piston (Figure 7.18). A difference in pressure between the two sides then results in motion of the piston, the piston being able to move in either direction along the cylinder as a result of high-pressure signals. For

Figure 7.18 Double-acting cylinder.

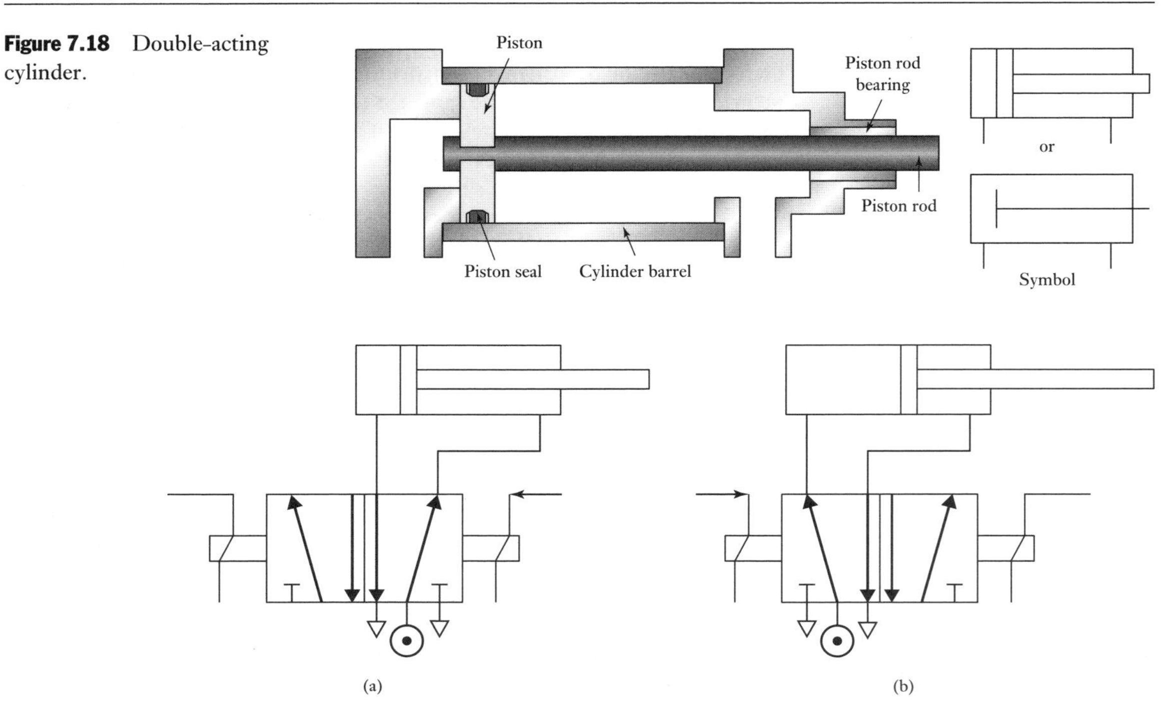

Figure 7.19 Control of a double-acting cylinder with solenoid, (a) not activated, (b) activated.

the double-acting cylinder shown in Figure 7.19, current through one solenoid causes the piston to move in one direction with current through the other solenoid reversing the direction of motion.

The choice of cylinder is determined by the force required to move the load and the speed required. Hydraulic cylinders are capable of much larger forces than pneumatic cylinders. However, pneumatic cylinders are capable of greater speeds. The force produced by a cylinder is equal to the cross-sectional area of the cylinder multiplied by the working pressure, i.e. the pressure difference between the two sides of the piston, in the cylinder. A cylinder for use with a working pneumatic pressure of 500 kPa and having a diameter of 50 mm will thus give a force of 982 N. A hydraulic cylinder with the same diameter and a working pressure of 15 000 kPa will give a force of 29.5 kN.

If the flow rate of hydraulic liquid into a cylinder is a volume of Q per second, then the volume swept out by the piston in a time of 1 s must be Q. But for a piston of cross-sectional area A this is a movement through a distance of v in 1 s, where we have $Q = Av$. Thus the speed v of a hydraulic cylinder is equal to the flow rate of liquid Q through the cylinder divided by the cross-sectional area A of the cylinder. Thus for a hydraulic cylinder of diameter 50 mm and a hydraulic fluid flow of 7.5×10^{-3} m^3/s, the speed is 3.8 m/s. The speed of a pneumatic cylinder cannot be calculated in this way since its speed depends on the rate at which air can be vented ahead of the advancing piston. A valve to adjust this can be used to regulate the speed.

To illustrate the above, consider the problem of a hydraulic cylinder to be used to move a workpiece in a manufacturing operation through a distance of 250 mm in 15 s. If a force of 50 kN is required to move the workpiece, what is the required working pressure and hydraulic liquid flow rate if a cylinder with a piston diameter of 150 mm is available? The cross-sectional area of the piston is $\frac{1}{4}\pi \times 0.150^2 = 0.0177\ \text{m}^2$. The force produced by the cylinder is equal to the product of the cross-sectional area of the cylinder and the working pressure. Thus the working pressure is $50 \times 10^3/0.0177 = 2.8$ MPa. The speed of a hydraulic cylinder is equal to the flow rate of liquid through the cylinder divided by the cross-sectional area of the cylinder. Thus the required flow rate is $(0.250/15) \times 0.0177 = 2.95 \times 10^{-4}\ \text{m}^3/\text{s}$.

7.5.1 Cylinder sequencing

Many control systems employ pneumatic or hydraulic cylinders as the actuating elements and require a sequence of extensions and retractions of the cylinders to occur. For example, we might have two cylinders A and B and require that when the start button is pressed, the piston of cylinder A extends and then, when it is fully extended, the piston of cylinder B extends. When this has happened and both are extended, we might need the piston of cylinder A to retract, and when it is fully retracted we might then have the piston of B retract. In discussions of sequential control with cylinders it is common practice to give each cylinder a reference letter A, B, C, D, etc., and to indicate the state of each cylinder by using a + sign if it is extended or a − sign if retracted. Thus the above required sequence of operations is A+, B+, A−, B−. Figure 7.20 shows a circuit that could be used to generate this sequence.

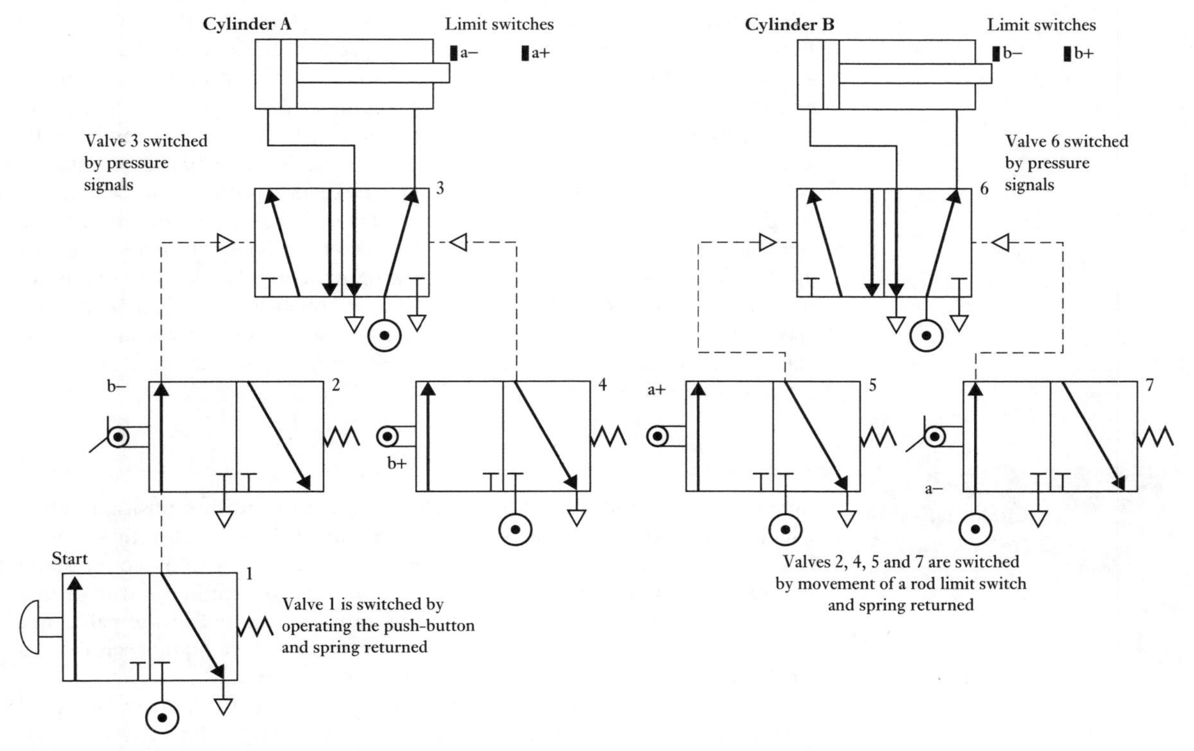

Figure 7.20 Two-actuator sequential operation.

The sequence of operations is:

1 Initially both the cylinders have retracted pistons. Start push-button on valve 1 is pressed. This applies pressure to valve 2, as initially limit switch b− is activated, hence valve 3 is switched to apply pressure to cylinder A for extension.
2 Cylinder A extends, releasing limit switch a−. When cylinder A is fully extended, limit switch a+ operates. This switches valve 5 and causes pressure to be applied to valve 6 to switch it and so apply pressure to cylinder B to cause its piston to extend.
3 Cylinder B extends, releasing limit switch b−. When cylinder B is fully extended, limit switch b+ operates. This switches valve 4 and causes pressure to be applied to valve 3 and so applies pressure to cylinder A to start its piston retracting.
4 Cylinder A retracts, releasing limit switch a+. When cylinder A is fully retracted, limit switch a− operates. This switches valve 7 and causes pressure to be applied to valve 5 and so applies pressure to cylinder B to start its piston retracting.
5 Cylinder B retracts, releasing limit switch b+. When cylinder B is fully retracted, limit switch b− operates to complete the cycle.

The cycle can be started again by pushing the start button. If we wanted the system to run continuously then the last movement in the sequence would have to trigger the first movement.

An alternative way of realising the above sequence involves the air supply being switched on and off to valves in groups and is termed **cascade control**. This avoids a problem that can occur with circuits, formed in the way shown in Figure 7.20, of air becoming trapped in the pressure line to control a valve and so preventing the valve from switching. With cascade control, the sequence of operations is divided into groups with no cylinder letter appearing more than once in each group. Thus for the sequence A+, B+, B−, A− we can have the groups A+, B+ and A−, B−. A valve is then used to switch the air supply between the two groups, i.e. air to the group A+B+ and then air switched to the group with A−B−. A start/stop valve is included in the line that selects the first group, and if the sequence is to be continuously repeated, the last operation has to supply a signal to start the sequence over again. The first function in each group is initiated by that group supply being switched on; further actions within the group are controlled by switch-operated valves, and the last valve operation initiates the next group to be selected. Figure 7.21 shows the pneumatic circuit.

7.6 Servo and proportional control valves

Servo and **proportional control valves** are both infinite position valves which give a valve spool displacement proportional to the current supplied to a solenoid. Basically, servo valves have a torque motor to move the spool within a valve (Figure 7.22). By varying the current supplied to the torque motor, an armature is deflected and this moves the spool in the valve and hence gives a flow related to the current. Servo valves are high precision and costly and generally used in a closed-loop control system.

Proportional control valves are less expensive and basically have the spool position directly controlled by the size of the current to the valve solenoid. They are often used in open-loop control systems.

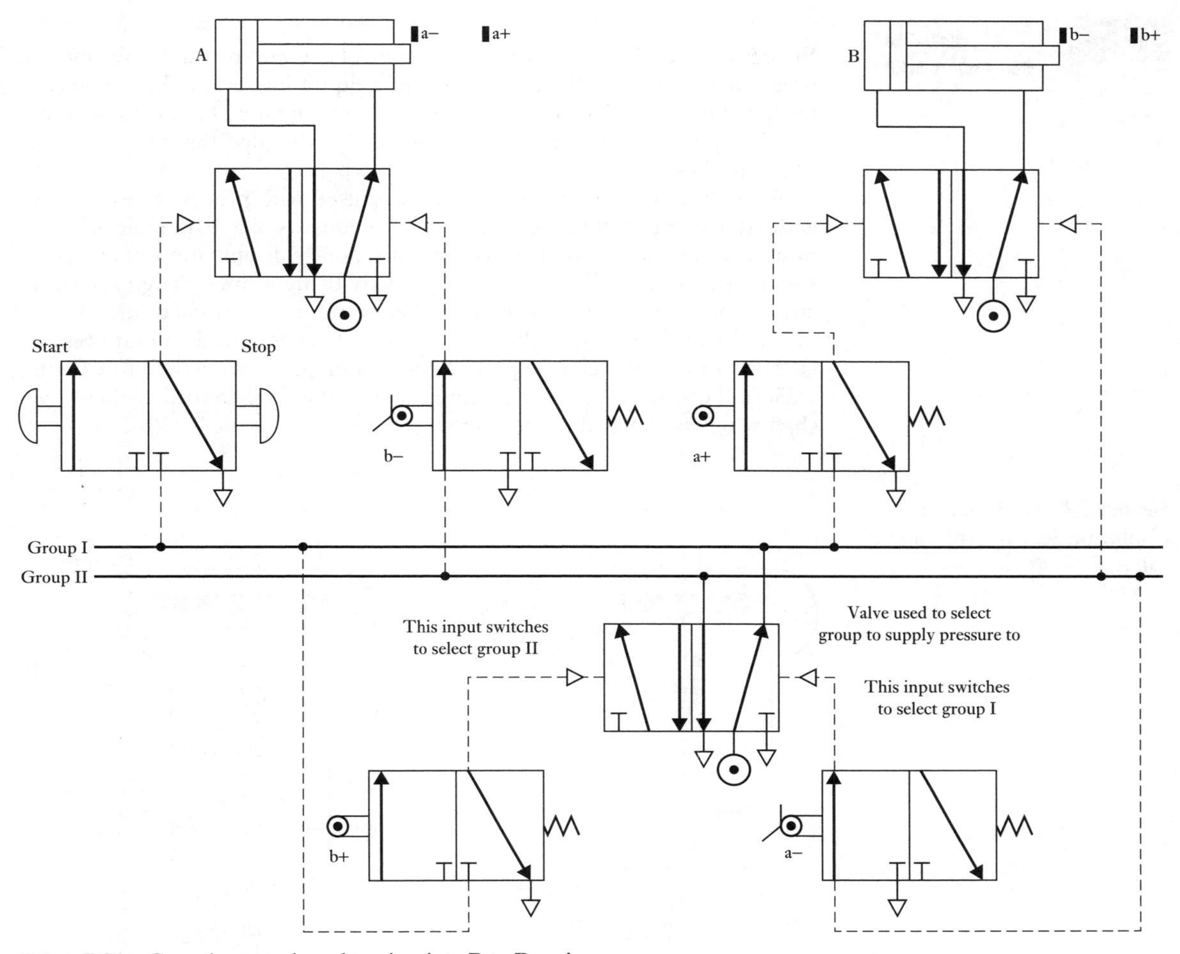

Figure 7.21 Cascade control used to give A+, B+, B−, A−.

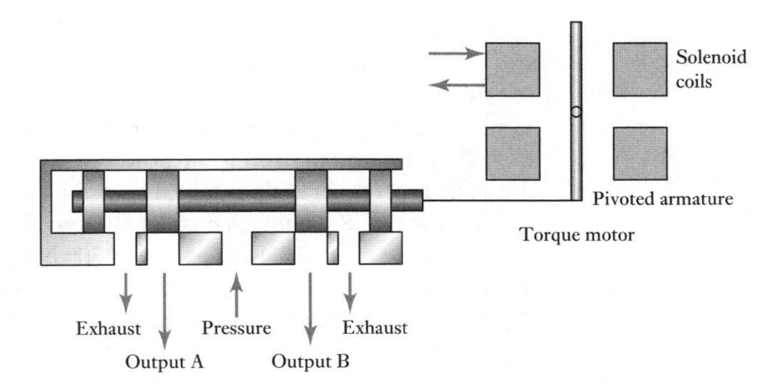

Figure 7.22 The basic form of a servo valve.

7.7 Process control valves

Process control valves are used to control the rate of fluid flow and are used where, perhaps, the rate of flow of a liquid into a tank has to be controlled. The basis of such valves is an actuator being used to move a plug into the flow pipe and so alter the cross-section of the pipe through which the fluid can flow.

A common form of pneumatic actuator used with process control valves is the **diaphragm actuator**. Essentially it consists of a diaphragm with the input pressure signal from the controller on one side and atmospheric pressure on the other, this difference in pressure being termed the **gauge pressure**. The diaphragm is made of rubber which is sandwiched in its centre between two circular steel discs. The effect of changes in the input pressure is thus to move the central part of the diaphragm, as illustrated in Figure 7.23(a). This movement is communicated to the final control element by a shaft which is attached to the diaphragm, e.g. as in Figure 7.23(b).

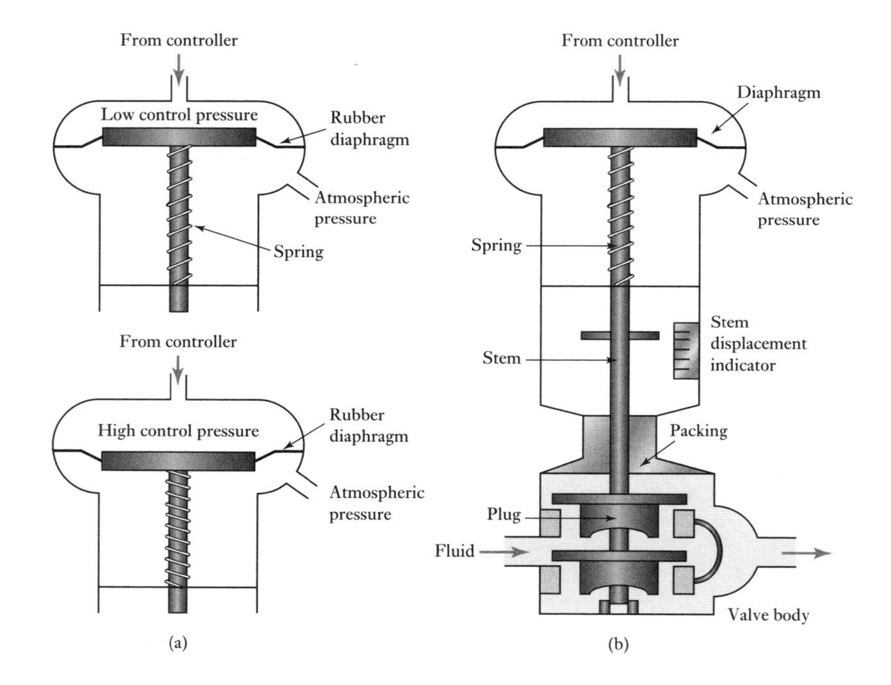

Figure 7.23 (a) Pneumatic diaphragm actuator, (b) control valve.

The force F acting on the shaft is the force that is acting on the diaphragm and is thus the gauge pressure P multiplied by the diaphragm area A. A restoring force is provided by a spring. Thus if the shaft moves through a distance x, and assuming the compression of the spring is proportional to the force, i.e. $F = kx$ with k being a constant, then $kx = PA$ and thus the displacement of the shaft is proportional to the gauge pressure.

To illustrate the above, consider the problem of a diaphragm actuator to be used to open a control valve if a force of 500 N must be applied to the valve. What diaphragm area is required for a control gauge pressure of 100 kPa? The force F applied to the diaphragm of area A by a pressure P is given by $P = F/A$. Hence $A = 500/(100 \times 10^3) = 0.005\ \text{m}^2$.

7.7.1 Valve bodies and plugs

Figure 7.23(b) shows a cross-section of a valve for the control of rate of flow of a fluid. The pressure change in the actuator causes the diaphragm to move and so consequently the valve stem. The result of this is a movement of the inner-valve plug within the valve body. The plug restricts the fluid flow and so its position determines the flow rate.

There are many forms of valve body and plug. Figure 7.24 shows some forms of valve bodies. The term **single-seated** is used for a valve where there is just one path for the fluid through the valve and so just one plug is needed to control the flow. The term **double-seated** is used for a valve where the fluid on entering the valve splits into two streams, as in Figure 7.23, with each stream passing through an orifice controlled by a plug. There are thus two plugs with such a valve.

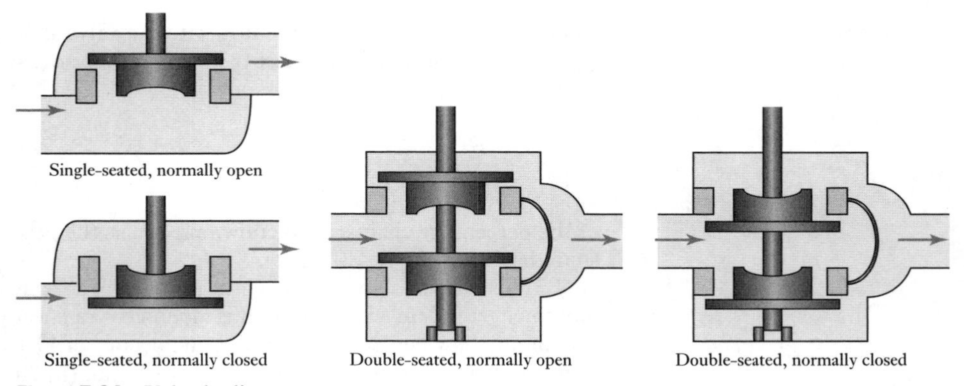

Figure 7.24 Valve bodies.

A single-seated valve has the advantage that it can be closed more tightly than a double-seated one, but the disadvantage that the force on the plug due to the flow is much higher and so the diaphragm in the actuator has to exert considerably higher forces on the stem. This can result in problems in accurately positioning the plug. Double-seated valves thus have an advantage here. The form of the body also determines whether an increasing air pressure will result in the valve opening or closing.

The shape of the plug determines the relationship between the stem movement and the effect on the flow rate. Figure 7.25(a) shows three commonly used types and Figure 7.25(b) how the percentage by which the volumetric rate of flow is related to the percentage displacement of the valve stem.

With the **quick-opening** type a large change in flow rate occurs for a small movement of the valve stem. Such a plug is used where on/off control of flow rate is required.

With the **linear-contoured** type, the change in flow rate is proportional to the change in displacement of the valve stem, i.e.

$$\text{change in flow rate} = k(\text{change in stem displacement})$$

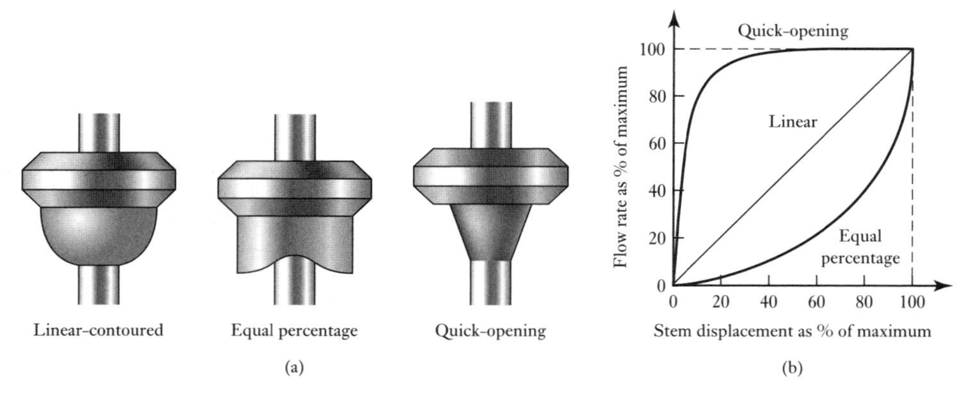

Figure 7.25 (a) Plug shapes, (b) flow characteristics.

where k is a constant. If Q is the flow rate at a valve stem displacement S and Q_{max} is the maximum flow rate at the maximum stem displacement S_{max}, then we have

$$\frac{Q}{Q_{max}} = \frac{S}{S_{max}}$$

or the percentage change in the flow rate equals the percentage change in the stem displacement.

To illustrate the above, consider the problem of an actuator which has a stem movement at full travel of 30 mm. It is mounted on a linear plug valve which has a minimum flow rate of 0 and a maximum flow rate of 40 m^3/s. What will be the flow rate when the stem movement is (a) 10 mm, (b) 20 mm? Since the percentage flow rate is the same as the percentage stem displacement, then: (a) a percentage stem displacement of 33% gives a percentage flow rate of 33%, i.e. 13 m^3/s; (b) a percentage stem displacement of 67% gives a percentage flow rate of 67%, i.e. 27 m^3/s.

With the **equal percentage** type of plug, equal percentage changes in flow rate occur for equal changes in the valve stem position, i.e.

$$\frac{\Delta Q}{Q} = k\Delta S$$

where ΔQ is the change in flow rate at a flow rate of Q and ΔS the change in valve position resulting from this change. If we write this expression for small changes and then integrate it we obtain

$$\int_{Q_{min}}^{Q} \frac{1}{Q} \mathrm{d}Q = k \int_{S_{min}}^{S} \mathrm{d}S$$

$$\ln Q - \ln Q_{min} = k(S - S_{min})$$

If we consider the flow rate Q_{max} which is given by S_{max} then

$$\ln Q_{max} - \ln Q_{min} = k(S_{max} - S_{min})$$

Eliminating k from these two equations gives

$$\frac{\ln Q - \ln Q_{\text{min}}}{\ln Q_{\text{max}} - \ln Q_{\text{min}}} = \frac{S - S_{\text{min}}}{S_{\text{max}} - S_{\text{min}}}$$

$$\ln \frac{Q}{Q_{\text{min}}} = \frac{S - S_{\text{min}}}{S_{\text{max}} - S_{\text{min}}} \ln \frac{Q_{\text{max}}}{Q_{\text{min}}}$$

and so

$$\frac{Q}{Q_{\text{min}}} = \left(\frac{Q_{\text{max}}}{Q_{\text{min}}}\right)^{(S - S_{\text{min}})/(S_{\text{max}} - S_{\text{min}})}$$

The term **rangeability** R is used for the ratio $Q_{\text{max}}/Q_{\text{min}}$.

To illustrate the above, consider the problem of an actuator which has a stem movement at full travel of 30 mm. It is mounted with a control valve having an equal percentage plug and which has a minimum flow rate of 2 m^3/s and a maximum flow rate of 24 m^3/s. What will be the flow rate when the stem movement is (a) 10 mm, (b) 20 mm? Using the equation

$$\frac{Q}{Q_{\text{min}}} = \left(\frac{Q_{\text{max}}}{Q_{\text{min}}}\right)^{(S - S_{\text{min}})/(S_{\text{max}} - S_{\text{min}})}$$

we have for (a) $Q = 2 \times (24/2)^{10/30} = 4.6\ \text{m}^3/\text{s}$ and for (b) $Q = 2 \times (24/2)^{20/30} = 10.5\ \text{m}^3/\text{s}$.

The relationship between the flow rate and the stem displacement is the inherent characteristic of a valve. It is only realised in practice if the pressure losses in the rest of the pipework, etc., are negligible compared with the pressure drop across the valve itself. If there are large pressure drops in the pipework so that, for example, less than half the pressure drop occurs across the valve, then a linear characteristic might become almost a quick-opening characteristic. The linear characteristic is thus widely used when a linear response is required and most of the system pressure is dropped across the valve. The effect of large pressure drops in the pipework with an equal percentage valve is to make it more like a linear characteristic. For this reason, if a linear response is required when only a small proportion of the system pressure is dropped across the valve, then an equal percentage value might be used.

7.7.2 Control valve sizing

The term control valve sizing is used for the procedure of determining the correct size of valve body. The equation relating the rate of flow of liquid Q through a wide open valve to its size is

$$Q = A_{\text{V}} \sqrt{\frac{\Delta P}{\rho}}$$

where A_{V} is the valve flow coefficient, ΔP the pressure drop across the valve and ρ the density of the fluid. This equation is sometimes written, with the quantities in SI units, as

$$Q = 2.37 \times 10^{-5}\, C_{\text{V}} \sqrt{\frac{\Delta P}{\rho}}$$

where C_V is the valve flow coefficient. Alternatively it may be found written as

$$Q = 0.75 \times 10^{-6} C_V \sqrt{\frac{\Delta P}{G}}$$

where G is the specific gravity or relative density. These last two forms of the equation derive from its original specification in terms of US gallons. Table 7.1 shows some typical values of A_V, C_V and valve size.

Table 7.1 Flow coefficients and valve sizes.

Flow coefficients	Valve size (mm)							
	480	640	800	960	1260	1600	1920	2560
C_V	8	14	22	30	50	75	110	200
$A_V \times 10^{-5}$	19	33	52	71	119	178	261	474

To illustrate the above, consider the problem of determining the valve size for a valve that is required to control the flow of water when the maximum flow required is 0.012 m^3/s and the permissible pressure drop across the valve at this flow rate is 300 kPa. Using the equation

$$Q = A_V \sqrt{\frac{\Delta P}{\rho}}$$

then, since the density of water is 1000 kg/m^3,

$$A_V = Q\sqrt{\frac{\rho}{\Delta P}} = 0.012\sqrt{\frac{1000}{300 \times 10^3}} = 69.3 \times 10^{-5}$$

Thus, using Table 7.1, the valve size is 960 mm.

7.7.3 Example of fluid control system

Figure 7.26(a) shows the essential features of a system for the control of a variable such as the level of a liquid in a container by controlling the rate at which liquid enters it. The output from the liquid-level sensor, after signal conditioning, is transmitted to the current-to-pressure converter as a current of 4 to 20 mA. It is then converted into a gauge pressure of 20 to 100 kPa which then actuates a pneumatic control valve and so controls the rate at which liquid is allowed to flow into the container.

Figure 7.26(b) shows the basic form of a current-to-pressure converter for such a system. The input current passes through coils mounted on a core which is attracted towards a magnet, the extent of the attraction depending on the size of the current. The movement of the core causes movement of the lever about its pivot and so the movement of a flapper above the nozzle. The position of the flapper in relation to the nozzle determines the rate at which air can escape from the system and hence the air pressure in the system. Springs on the flapper are used to adjust the sensitivity of the converter so that currents of 4 to 20 mA produce gauge pressures of 20 to 100 kPa. These are the standard values that are generally used in such systems.

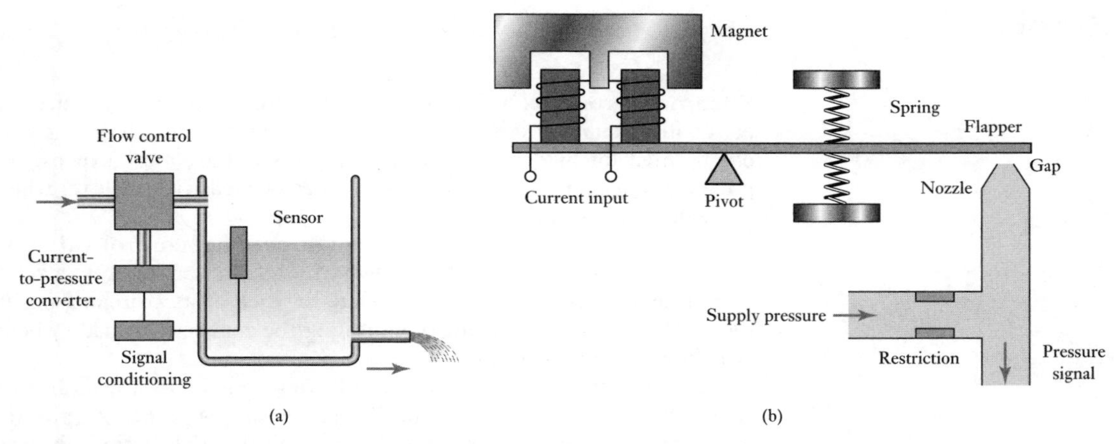

Figure 7.26 (a) Fluid control system, (b) current-to-pressure converter.

7.8 Rotary actuators

A linear cylinder can, with suitable mechanical linkages, be used to produce rotary movement through angles less than 360°, Figure 7.27(a) illustrating such an arrangement. Another alternative is a **semi-rotary actuator** involving a vane (Figure 7.27(b)). A pressure difference between the two ports causes the vane to rotate and so give a shaft rotation which is a measure of the pressure difference. Depending on the pressures, so the vane can be rotated clockwise or anti-clockwise.

For rotation through angles greater than 360° a pneumatic motor can be used; one such form is the **vane motor** (Figure 7.27(c)). An eccentric rotor has slots in which vanes are forced outwards against the walls of the cylinder by the rotation. The vanes divide the chamber into separate compartments which increase in size from the inlet port round to the exhaust port. The air entering such a compartment exerts a force on a vane and causes the rotor to rotate. The motor can be made to reverse its direction of rotation by using a different inlet port.

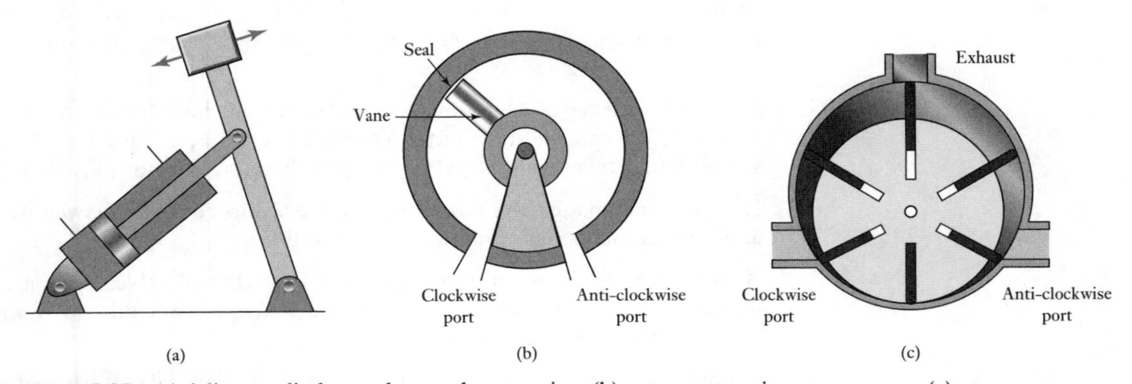

Figure 7.27 (a) A linear cylinder used to produce rotation, (b) vane-type semi-rotary actuator, (c) vane motor.

Summary

Pneumatic systems use air, **hydraulic systems** use oil. The main drawback with pneumatic systems is the compressibility of air. Hydraulic systems can be used for higher power control devices but are more expensive than pneumatic systems and there are hazards associated with oil leaks which do not occur with air leaks.

Pneumatic and hydraulic systems use **directional control valves** to direct the flow of fluid through a system. Such valves are on/off valves. The symbol used for such a valve is a square for each of its switching positions, the symbols used in each square indicating the connections made when that switching position is activated.

The **hydraulic** or **pneumatic cylinder** consists of a cylindrical tube along which a piston/ram can slide. There are two basic types, **single-acting cylinders** and **double-acting cylinders**. With single-acting, the control pressure is applied to just one side of the piston, a spring often being used to provide the opposition to the movement of the piston. The other side of the piston is open to the atmosphere. The term double-acting is used when the control pressures are applied to each side of the piston.

Servo and **proportional control valves** are both infinite position valves which give a valve spool displacement proportional to the current supplied to a solenoid.

Process control valves are used to control the rate of fluid flow. The basis of such valves is an actuator being used to move a plug into the flow pipe and so alter the cross-section of the pipe through which the fluid can flow. There are many forms of valve body and plug, these determining how the valve controls the fluid flow.

Problems

7.1 Describe the basic details of (a) a poppet valve, (b) a shuttle valve.

7.2 Explain the principle of a pilot-operated valve.

7.3 Explain how a sequential valve can be used to initiate an operation only when another operation has been completed.

7.4 Draw the symbols for (a) a pressure-relief valve, (b) a 2/2 valve which has actuators of a push-button and a spring, (c) a 4/2 valve, (d) a directional valve.

7.5 State the sequence of operations that will occur for the cylinders A and B in Figure 7.28 when the start button is pressed. a−, a+, b− and b+ are limit switches to detect when the cylinders are fully retracted and fully extended.

7.6 Design a pneumatic valve circuit to give the sequence A+, followed by B+ and then simultaneously followed by A− and B−.

7.7 A force of 400 N is required to open a process control valve. What area of diaphragm will be needed with a diaphragm actuator to open the valve with a control gauge pressure of 70 kPa?

7.8 A pneumatic system is operated at a pressure of 1000 kPa. What diameter cylinder will be required to move a load requiring a force of 12 kN?

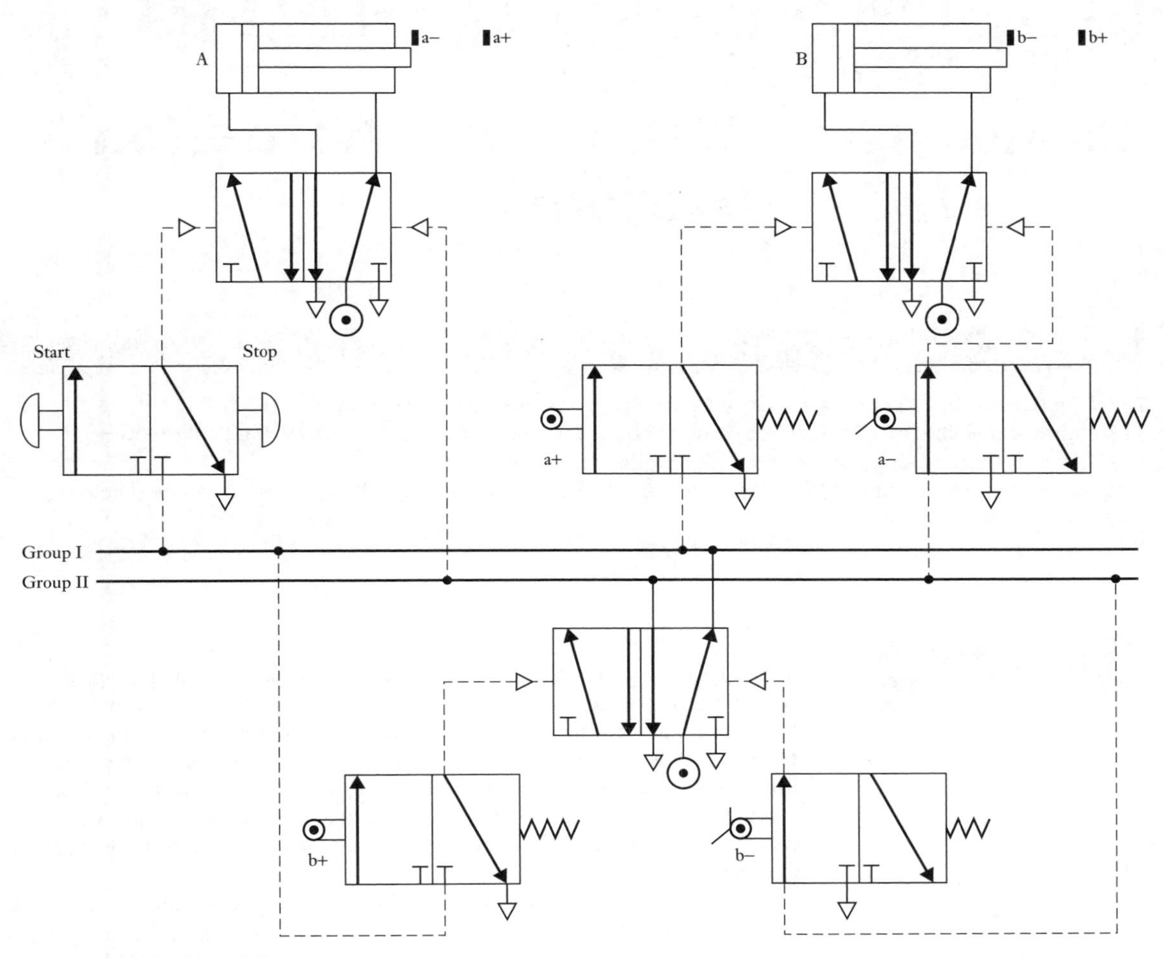

Figure 7.28 Problem 7.5.

7.9 A hydraulic cylinder is to be used to move a workpiece in a manufacturing operation through a distance of 50 mm in 10 s. A force of 10 kN is required to move the workpiece. Determine the required working pressure and hydraulic liquid flow rate if a cylinder with a piston diameter of 100 mm is available.

7.10 An actuator has a stem movement which at full travel is 40 mm. It is mounted with a linear plug process control valve which has a minimum flow rate of 0 and a maximum flow rate of 0.20 m^3/s. What will be the flow rate when the stem movement is (a) 10 mm, (b) 20 mm?

7.11 An actuator has a stem movement which at full travel is 40 mm. It is mounted on a process control valve with an equal percentage plug and which has a minimum flow rate of 0.2 m^3/s and a maximum flow rate of 4.0 m^3/s. What will be the flow rate when the stem movement is (a) 10 mm, (b) 20 mm?

7.12 What is the process control valve size for a valve that is required to control the flow of water when the maximum flow required is 0.002 m^3/s and the permissible pressure drop across the valve at this flow rate is 100 kPa? The density of water is 1000 kg/m^3.

Chapter eight Mechanical Actuation Systems

Objectives

The objectives of this chapter are that, after studying it, the reader should be able to:

- **Determine possible mechanical actuation systems for motion transmission involving linear-to-rotary, rotary-to-rotary, rotary-to-linear and cyclic motion transmission.**
- **Evaluate the capabilities of linkages, cams, gears, ratchet-and-pawl, belt and chain drives and bearings for actuation systems.**

8.1 Mechanical systems

This chapter is a consideration of **mechanisms**: mechanisms are devices which can be considered to be motion converters in that they transform motion from one form to some other required form. They might, for example, transform linear motion into rotational motion, or motion in one direction into a motion in a direction at right angles, or perhaps a linear reciprocating motion into rotary motion, as in the internal combustion engine where the reciprocating motion of the pistons is converted into rotation of the crank and hence the drive shaft.

Mechanical elements can include the use of linkages, cams, gears, rack-and-pinion, chains, belt drives, etc. For example, the rack-and-pinion can be used to convert rotational motion to linear motion. Parallel shaft gears might be used to reduce a shaft speed. Bevel gears might be used for the transmission of rotary motion through 90°. A toothed belt or chain drive might be used to transform rotary motion about one axis to motion about another. Cams and linkages can be used to obtain motions which are prescribed to vary in a particular manner. This chapter is a consideration of the basic characteristics of a range of such mechanisms.

Many of the actions which previously were obtained by the use of mechanisms are, however, often nowadays being obtained, as a result of a mechatronics approach, by the use of microprocessor systems. For example, cams on a rotating shaft were previously used for domestic washing machines in order to give a timed sequence of actions such as opening a valve to let water into the drum, switching the water off, switching a heater on, etc. Modern washing machines use a microprocessor-based system with the microprocessor programmed to switch on outputs in the required sequence. Another example is the hairspring balance wheel and gears and pointer of a watch which have now largely been replaced by an integrated circuit with perhaps a liquid crystal display. The mechatronics approach has resulted in a simplification, and often a reduction in cost.

Mechanisms still, however, have a role in mechatronics systems. For example, the mechatronics system in use in an automatic camera for adjusting the aperture for correct exposures involves a mechanism for adjusting the size of the diaphragm.

While electronics might now be used often for many functions that previously were fulfilled by mechanisms, mechanisms might still be used to provide such functions as:

1 Force amplification, e.g. that given by levers.
2 Change of speed, e.g. that given by gears.
3 Transfer of rotation about one axis to rotation about another, e.g. a timing belt.
4 Particular types of motion, e.g. that given by a quick-return mechanism.

The term **kinematics** is used for the study of motion without regard to forces. When we consider just the motions without any consideration of the forces or energy involved then we are carrying out a kinematic analysis of the mechanism. This chapter is an introduction to such a consideration.

8.2 Types of motion

The motion of any rigid body can be considered to be a combination of translational and rotational motions. By considering the three dimensions of space, a **translation motion** can be considered to be a movement which can be resolved into components along one or more of the three axes (Figure 8.1(a)). A **rotational motion** can be considered as a rotation which has components rotating about one or more of the axes (Figure 8.1(b)).

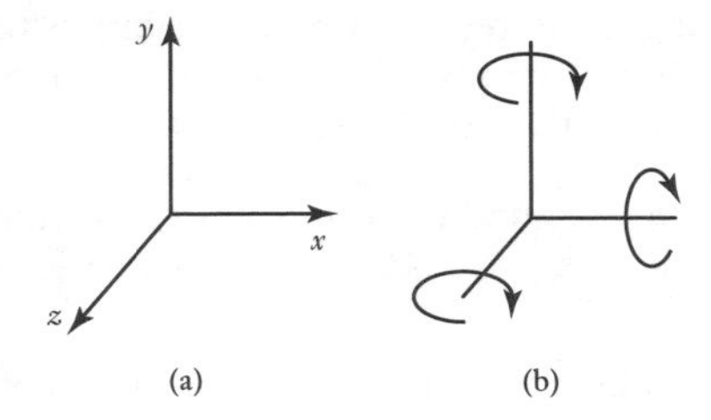

Figure 8.1 Types of motion: (a) translational, (b) rotational.

A complex motion may be a combination of translational and rotational motions. For example, think of the motion required for you to pick up a pencil from a table. This might involve your hand moving at a particular angle towards the table, rotation of the hand, and then all the movement associated with opening your fingers and moving them to the required positions to grasp the pencil. This is a sequence of quite complex motions. However, we can break down all these motions into combinations of translational and rotational motions. Such an analysis is particularly relevant if we are not moving a human hand to pick up the pencil but instructing a robot to carry out the task. Then it really is necessary to break down the motion into combinations of translational and rotational motions so that we can design mechanisms to carry out each of these components of the motion. For example, among the sequence of control signals sent to a mechanism might be such groupings of signals as those to instruct joint 1 to rotate by 20° and link 2 to be extended by 4 mm for translational motion.

8.2.1 Freedom and constraints

An important aspect in the design of mechanical elements is the orientation and arrangement of the elements and parts. A body that is free in space can move in three, independent, mutually perpendicular directions and rotate in three ways about those directions (Figure 8.1). It is said to have six degrees of freedom. The number of **degrees of freedom** is the number of components of motion that are required in order to generate the motion. If a joint is constrained to move along a line then its translational degrees of freedom are reduced to one. Figure 8.2(a) shows a joint with just this one translational degree of freedom. If a joint is constrained to move on a plane then it has two translational degrees of freedom. Figure 8.2(b) shows a joint which has one translational degree of freedom and one rotational degree of freedom.

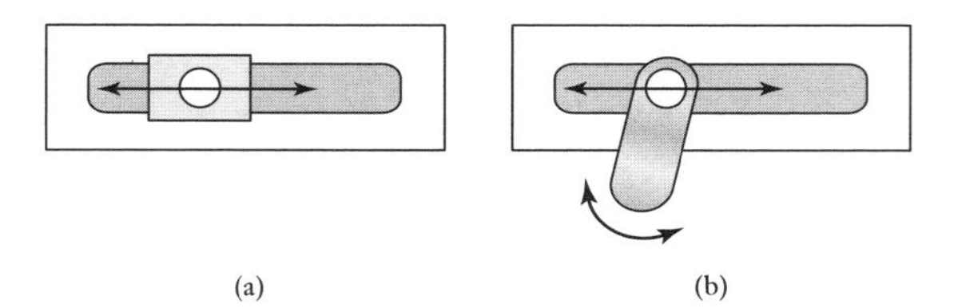

Figure 8.2 Joints with: (a) one, (b) two degrees of freedom.

The problem in design is often to reduce the number of degrees of freedom and this then requires an appropriate number and orientation of constraints. Without any constraints a body would have six degrees of freedom. A constraint is needed for each degree of freedom that is to be prevented from occurring. Provided we have no redundant constraints then the number of degrees of freedom would be 6 minus the number of constraints. However, redundant constraints often occur and so for constraints on a single rigid body we have the basic rule

$$6 - \text{number of constraints} = \text{number of degrees of freedom} - \text{number of redundancies}$$

Thus if a body is required to be fixed, i.e. have zero degrees of freedom, then if no redundant constraints are introduced the number of constraints required is 6.

A concept that is used in design is that of the **principle of least constraint**. This states that in fixing a body or guiding it to a particular type of motion, the minimum number of constraints should be used, i.e. there should be no redundancies. This is often referred to as **kinematic design**.

For example, to have a shaft which only rotates about one axis with no translational motions, we have to reduce the number of degrees of freedom to 1. Thus the minimum number of constraints to do this is 5. Any more constraints than this will give redundancies. The mounting that might be used to mount the shaft has a ball bearing at one end and a roller bearing at the other (Figure 8.3). The pair of bearings together prevent translational motion at right angles to the shaft, the y-axis, and rotations about the z-axis and the y-axis. The ball bearing prevents translational motion along the x-axis and along the z-axis. Thus there is a total of five constraints. This leaves just one degree of freedom, the required rotation about the x-axis. If

Figure 8.3 Shaft with no redundancies.

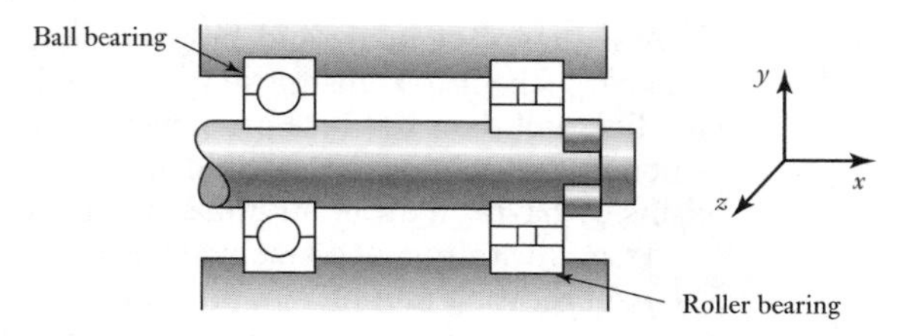

there had been a roller bearing at each end of the shaft then both the bearings could have prevented translational motion along the x-axis and the z-axis and thus there would have been redundancy. Such redundancy might cause damage. If ball bearings are used at both ends of the shaft, then in order to prevent redundancy one of the bearings would have its outer race not fixed in its housing so that it could slide to some extent in an axial direction.

8.2.2 Loading

Mechanisms are structures and as such transmit and support loads. Analysis is thus necessary to determine the loads to be carried by individual elements. Then consideration can be given to the dimensions of the element so that it might, for example, have sufficient strength and perhaps stiffness under such loading.

8.3 Kinematic chains

When we consider the movements of a mechanism without any reference to the forces involved, we can treat the mechanism as being composed of a series of individual links. Each part of a mechanism which has motion relative to some other part is termed a **link.** A link need not necessarily be a rigid body but it must be a resistant body which is capable of transmitting the required force with negligible deformation. For this reason it is usually taken as being represented by a rigid body which has two or more points of attachment to other links, these being termed **nodes.** Each link is capable of moving relative to its neighbouring links. Figure 8.4 shows examples of links with two, three and four nodes. A **joint** is a connection between two or more links at their nodes and which allows some motion between the connected links. Levers, cranks, connecting rods and pistons, sliders, pulleys, belts and shafts are all examples of links.

Figure 8.4 Links: (a) with two nodes, (b) with three nodes, (c) with four nodes.

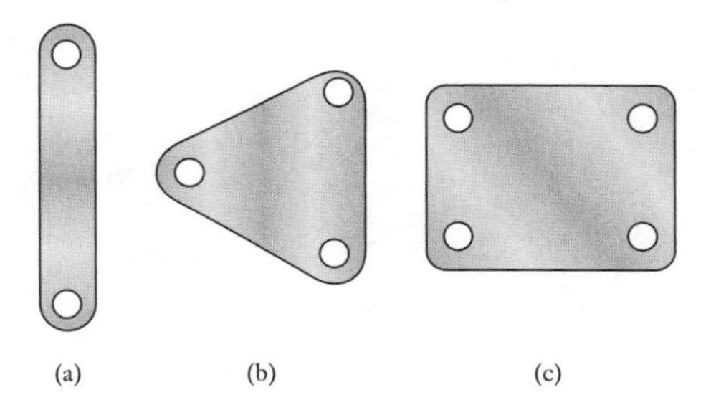

A sequence of joints and links is known as a **kinematic chain.** For a kinematic chain to transmit motion, one link must be fixed. Movement of one link will then produce predictable relative movements of the others. It is possible to obtain from one kinematic chain a number of different mechanisms by having a different link as the fixed one.

As an illustration of a kinematic chain, consider a motor car engine where the reciprocating motion of a piston is transformed into rotational motion of a crankshaft on bearings mounted in a fixed frame (Figure 8.5(a)). We can represent this as being four connected links (Figure 8.5(b)). Link 1 is the crankshaft, link 2 the connecting rod, link 3 the fixed frame and link 4 the slider, i.e. piston, which moves relative to the fixed frame (see Section 8.3.2 for further discussion).

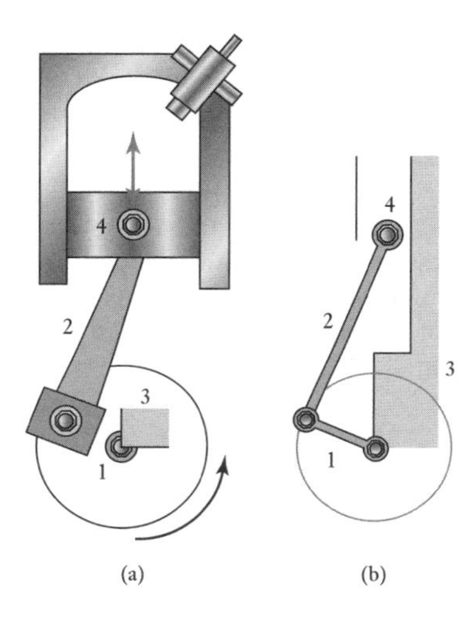

Figure 8.5 Simple engine mechanism.

The designs of many mechanisms are based on two basic forms of kinematic chains, the four-bar chain and the slider–crank chain. The following illustrates some of the forms such chains can take.

8.3.1 The four-bar chain

The **four-bar chain** consists of four links connected to give four joints about which turning can occur. Figure 8.6 shows a number of forms of the four-bar chain produced by altering the relative lengths of the links. If the

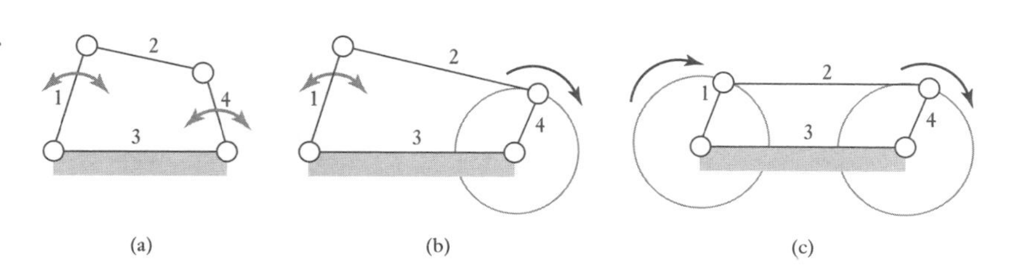

Figure 8.6 Examples of four-bar chains.

sum of the length of the shortest link plus the length of the longest link is less than or equal to the sum of the lengths of the other two links then at least one link will be capable of making a full revolution with respect to the fixed link. If this condition is not met then no link is capable of a complete revolution. This is known as the Grashof condition. In Figure 8.6(a), link 3 is fixed and the relative lengths of the links are such that links 1 and 4 can oscillate but not rotate. The result is a **double-lever mechanism**. By shortening link 4 relative to link 1, then link 4 can rotate (Figure 8.6(b)) with link 1 oscillating and the result is termed a **lever–crank mechanism**. With links 1 and 4 the same length and both able to rotate (Figure 8.6(c)), then the result is a **double-crank mechanism**. By altering which link is fixed, other forms of mechanism can be produced.

Figure 8.7 illustrates how such a mechanism can be used to advance the film in a cine camera. As link 1 rotates so the end of link 2 locks into a sprocket of the film, pulls it forward before releasing and moving up and back to lock into the next sprocket.

Figure 8.7 Cine film advance mechanism.

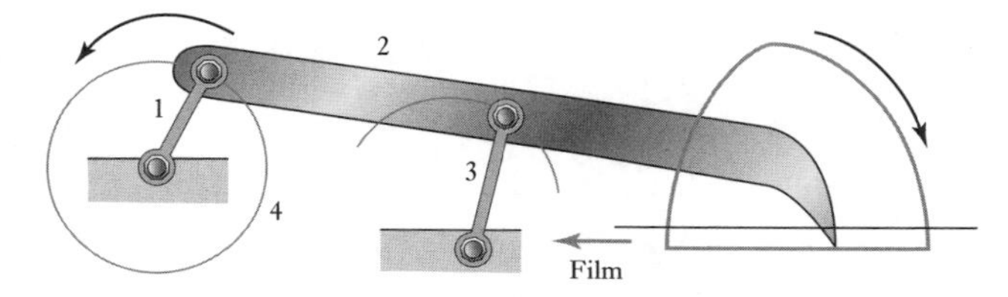

Some linkages may have **toggle positions**. These are positions where the linkage will not react to any input from one of its links. Figure 8.8 illustrates such a toggle, being the linkage used to control the movement of the tailgate of a truck so that when link 2 reaches the horizontal position no further load on link 2 will cause any further movement. There is another toggle position for the linkage and that is when links 3 and 4 are both vertical and the tailgate is vertical.

Figure 8.8 Toggle linkage.

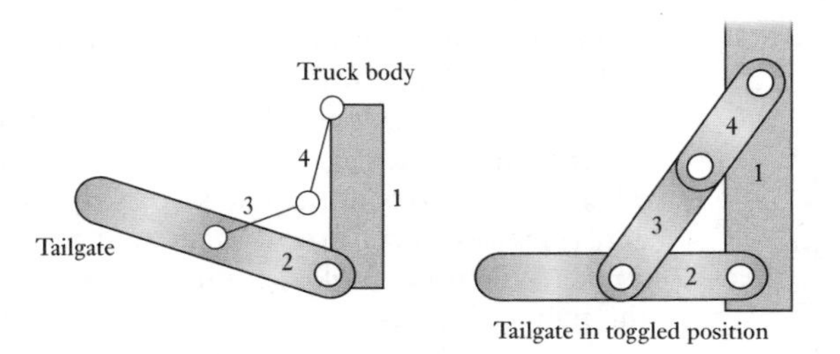

8.3.2 The slider-crank mechanism

This form of mechanism consists of a crank, a connecting rod and a slider and is the type of mechanism described in Figure 8.5 which showed the simple engine mechanism. With that configuration, link 3 is fixed, i.e. there is no relative movement between the centre of rotation of the crank and the housing in which the piston slides. Link 1 is the crank that rotates, link 2 the connecting rod and link 4 the slider which moves relative to the fixed link. When the piston moves backwards and forwards, i.e. link 4 moves

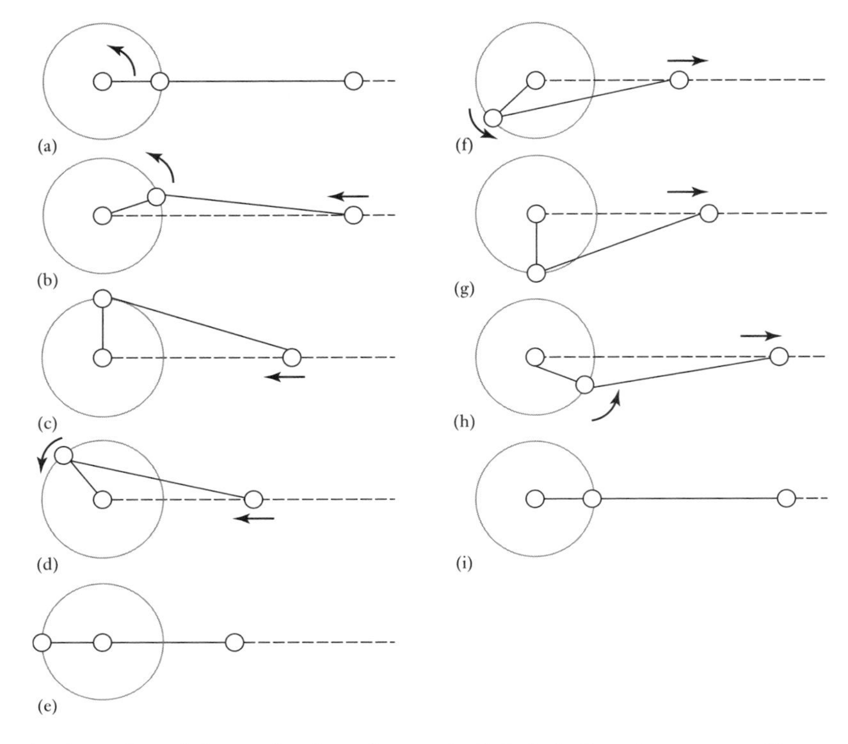

Figure 8.9 The position sequence for the links in a slider-crank mechanism.

backwards and forwards, then the crank, link 1, is forced to rotate. Hence the mechanism transforms an input of backwards and forwards motion into rotational motion. Figure 8.9 shows a number of stages in this motion. A useful way of seeing how any mechanism might behave is to construct, to scale, a cardboard model and move the links. Changing the length of a link then enables the changes in behaviour of the mechanism to be determined.

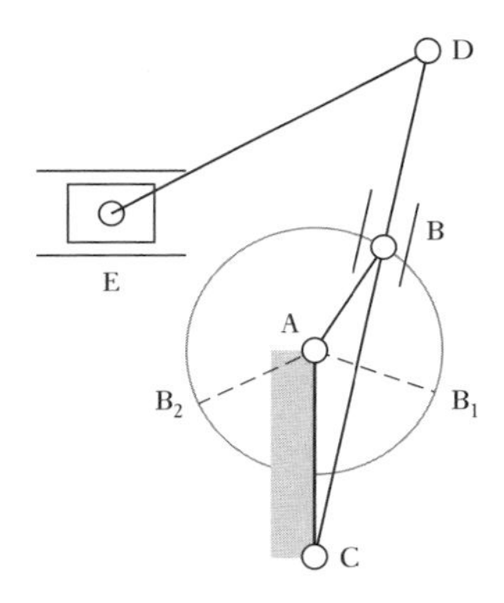

Figure 8.10 Quick-return mechanism.

Figure 8.10 shows another form of this type of mechanism, a **quick-return mechanism**. It consists of a rotating crank, link AB, which rotates round a fixed centre, an oscillating lever CD, which is caused to oscillate about C by the sliding of the block at B along CD as AB rotates, and a link DE which causes E to move backwards and forwards. E might be the ram of a machine and have a cutting tool attached to it. The ram will be at the extremes of its movement when the positions of the crank are AB_1 and AB_2. Thus, as the crank moves anti-clockwise from B_1 to B_2, the ram makes a complete stroke, the cutting stroke. When the crank continues its movement from B_2 anti-clockwise to B_1 the ram again makes a complete stroke in the opposite direction, the return stroke. With the crank rotating at constant speed, because the angle of crank rotation required for the cutting stroke is greater than the angle for the return stroke, the cutting stroke takes more time than the return stroke hence the term quick-return for the mechanism. Similar diagrams, and a cardboard model, can be constructed in the same way as that shown in Figure 8.9.

8.4 Cams

A **cam** is a body which rotates or oscillates and in doing so imparts a reciprocating or oscillatory motion to a second body, called the **follower**, with which it is in contact (Figure 8.11). As the cam rotates so the follower is made to rise, dwell and fall, the lengths of times spent at each of these positions

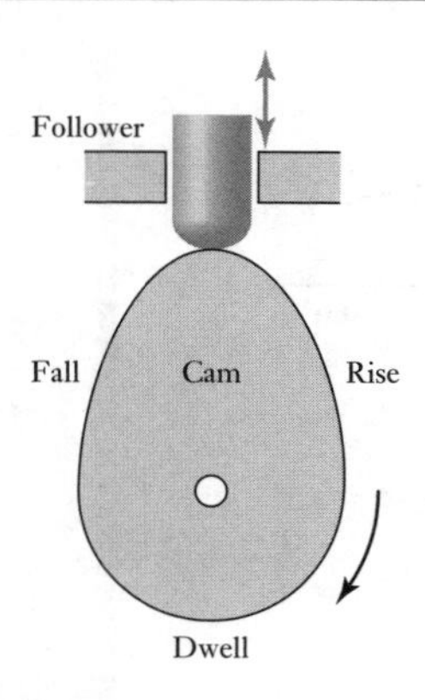

Figure 8.11 Cam and cam follower.

depending on the shape of the cam. The rise section of the cam is the part that drives the follower upwards, its profile determining how quickly the cam follower will be lifted. The fall section of the cam is the part that lowers the follower, its profile determining how quickly the cam follower will fall. The dwell section of the cam is the part that allows the follower to remain at the same level for a significant period of time. The dwell section of the cam is where it is circular with a radius that does not change.

The cam shape required to produce a particular motion of the follower will depend on the shape of the cam and the type of follower used. Figure 8.12 shows the type of follower displacement diagram that can be produced by an eccentric cam with a point-shaped follower. This is a circular cam with an offset centre of rotation. It produces an oscillation of the follower which is simple harmonic motion and is often used with pumps. The radial distance from the axis of rotation of the cam to the point of contact of the cam with the follower gives the displacement of the follower with reference to the axis of rotation of the cam. The figure shows how these radial distances, and hence follower displacements, vary with the angle of rotation of the cam. The vertical displacement diagram is obtained by taking the radial distance of the cam surface from the point of rotation at different angles and projecting them round to give the displacement at those angles.

Figure 8.12 Displacement diagram for an eccentric cam.

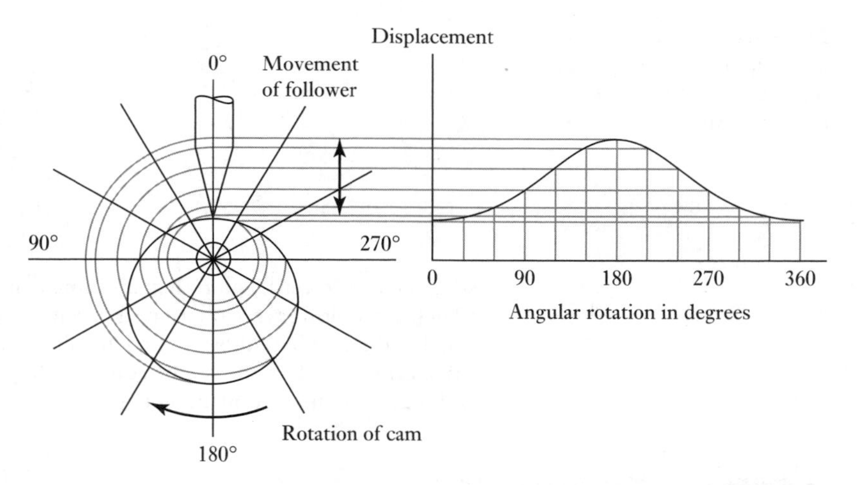

Figure 8.13 shows the types of follower displacement diagrams that can be produced with two other different shaped cams and either point or knife followers.

The heart-shaped cam (Figure 8.13(a)) gives a follower displacement which increases at a constant rate with time before decreasing at a constant rate with time, hence a uniform speed for the follower. The pear-shaped cam (Figure 8.13(b)) gives a follower motion which is stationary for about half a revolution of the cam and rises and falls symmetrically in each of the remaining quarter revolutions. Such a pear-shaped cam is used for engine valve control. The dwell holds the valve open while the petrol/air mixture passes into the cylinder. The longer the dwell, i.e. the greater the length of the cam surface with a constant radius, the more time is allowed for the cylinder to be completely charged with flammable vapour.

Figure 8.14 shows a number of examples of different types of cam followers. Roller followers are essentially ball or roller bearings. They have the

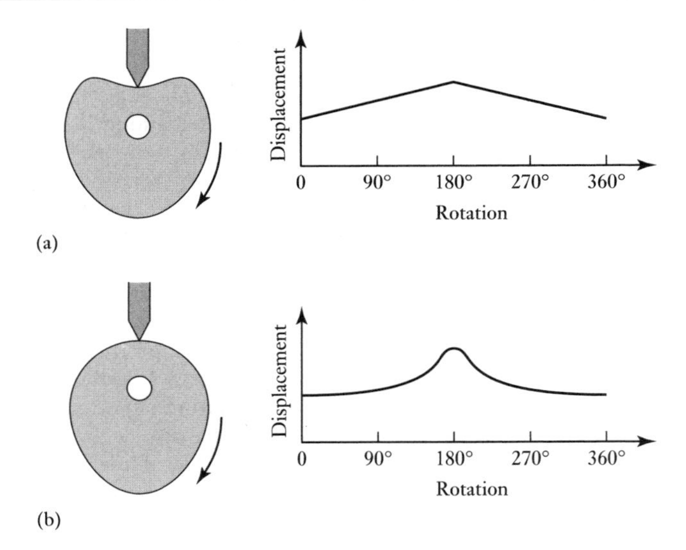

Figure 8.13 Cams: (a) heart-shaped, (b) pear-shaped.

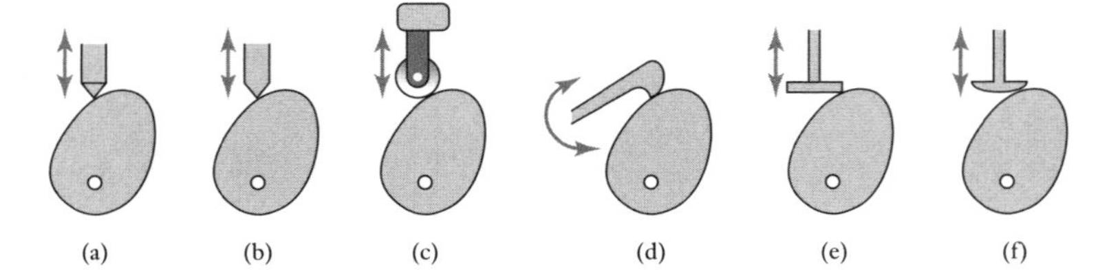

Figure 8.14 Cam followers: (a) point, (b) knife, (c) roller, (d) sliding and oscillating, (e) flat, (f) mushroom.

advantage of lower friction than a sliding contact but can be more expensive. Flat-faced followers are often used because they are cheaper and can be made smaller than roller followers. Such followers are widely used with engine valve cams. While cams can be run dry, they are often used with lubrication and may be immersed in an oil bath.

8.5 Gears

Gear trains are mechanisms which are very widely used to transfer and transform rotational motion. They are used when a change in speed or torque of a rotating device is needed. For example, the car gearbox enables the driver to match the speed and torque requirements of the terrain with the engine power available.

Gears can be used for the transmission of rotary motion between parallel shafts (Figure 8.15(a)) and for shafts which have axes inclined to one another (Figure 8.15(b)). The term **bevel gears** is used when the lines of the shafts intersect, as illustrated in Figure 8.15(b). When two gears are in mesh, the larger gear wheel is often called the **spur** or **crown wheel** and the smaller one the **pinion.** Gears for use with parallel shafts may have axial teeth with the teeth cut along axial lines parallel to the axis of the shaft (Figure 8.15(c)). Such gears are then termed **spur gears.** Alternatively they may have helical teeth with the teeth being cut on a helix (Figure 8.15(d)) and are then termed

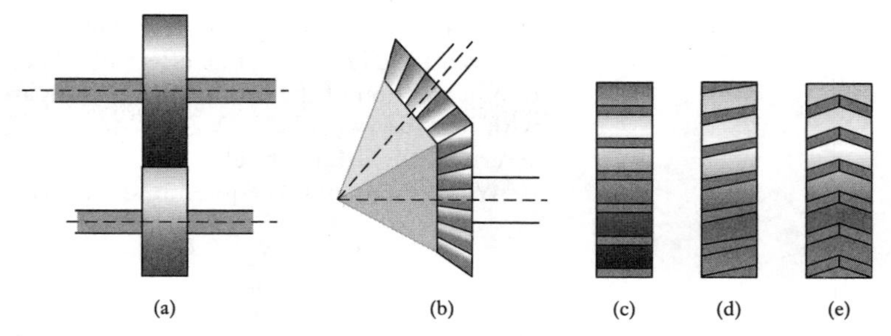

Figure 8.15 (a) Parallel gear axes, (b) axes inclined to one another, (c) axial teeth, (d) helical teeth, (e) double helical teeth.

helical gears. Helical gears have the advantage that there is a gradual engagement of any individual tooth and consequently there is a smoother drive and generally prolonged life of the gears. However, the inclination of the teeth to the axis of the shaft results in an axial force component on the shaft bearing. This can be overcome by using double helical teeth (Figure 8.15(e)).

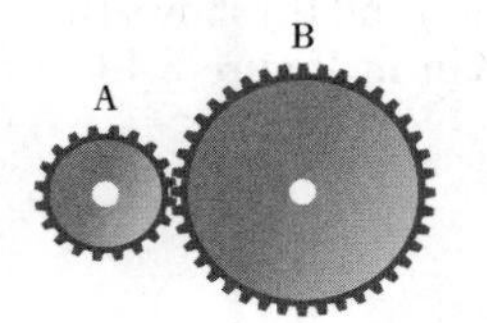

Figure 8.16 Two meshed gears.

Consider two meshed gear wheels A and B (Figure 8.16). If there are 40 teeth on wheel A and 80 teeth on wheel B, then wheel A must rotate through two revolutions in the same time as wheel B rotates through one. Thus the angular velocity ω_A of wheel A must be twice that ω_B of wheel B, i.e.

$$\frac{\omega_A}{\omega_B} = \frac{\text{number of teeth on B}}{\text{number of teeth on A}} = \frac{80}{40} = 2$$

Since the number of teeth on a wheel is proportional to its diameter, we can write

$$\frac{\omega_A}{\omega_B} = \frac{\text{number of teeth on B}}{\text{number of teeth on A}} = \frac{d_B}{d_A}$$

Thus for the data we have been considering, wheel B must have twice the diameter of wheel A. The term **gear ratio** is used for the ratio of the angular speeds of a pair of intermeshed gear wheels. Thus the gear ratio for this example is 2.

8.5.1 Gear trains

The term **gear train** is used to describe a series of intermeshed gear wheels. The term **simple gear train** is used for a system where each shaft carries only one gear wheel, as in Figure 8.17. For such a gear train, the overall gear ratio is the ratio of the angular velocities at the input and output shafts and is thus ω_A/ω_C, i.e.

$$G = \frac{\omega_A}{\omega_C}$$

Consider a simple gear train consisting of wheels A, B and C, as in Figure 8.17, with A having 9 teeth and C having 27 teeth. Then, as the angular velocity

Figure 8.17 Simple gear train.

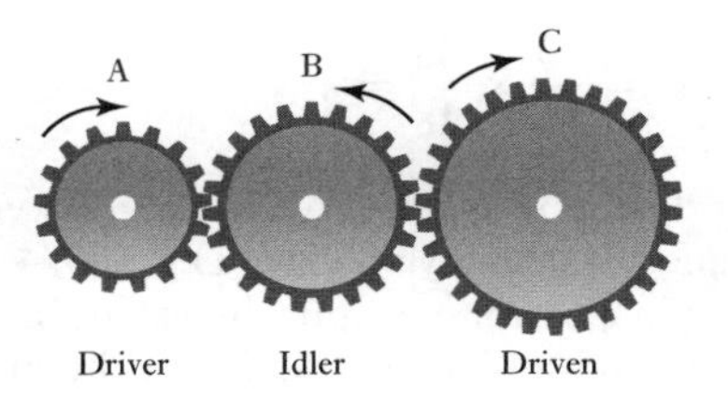

of a wheel is inversely proportional to the number of teeth on the wheel, the gear ratio is $27/9 = 3$. The effect of wheel B is purely to change the direction of rotation of the output wheel compared with what it would have been with just the two wheels A and C intermeshed. The intermediate wheel, B, is termed the **idler wheel.**

We can rewrite this equation for the overall gear ratio G as

$$G = \frac{\omega_A}{\omega_C} = \frac{\omega_A}{\omega_B} \times \frac{\omega_B}{\omega_C}$$

But ω_A/ω_B is the gear ratio for the first pair of gears and ω_B/ω_C the gear ratio for the second pair of gears. Thus the overall gear ratio for a simple gear train is the product of the gear ratios for each successive pair of gears.

The term **compound gear train** is used to describe a gear train when two wheels are mounted on a common shaft. Figure 8.18(a) and (b) shows two examples of such a compound gear train. The gear train in Figure 8.18(b) enables the input and output shafts to be in line.

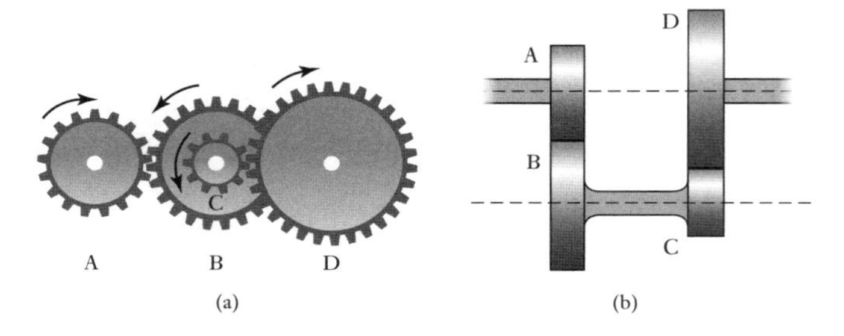

Figure 8.18 Compound gear trains.

When two gear wheels are mounted on the same shaft they have the same angular velocity. Thus, for both of the compound gear trains in Figure 8.18, $\omega_B = \omega_C$. The overall gear ratio G is thus

$$G = \frac{\omega_A}{\omega_D} = \frac{\omega_A}{\omega_B} \times \frac{\omega_B}{\omega_C} \times \frac{\omega_C}{\omega_D} = \frac{\omega_A}{\omega_B} \times \frac{\omega_C}{\omega_D}$$

For the arrangement shown in Figure 8.18(b), for the input and output shafts to be in line we must also have for the radii of the gears

$$r_A + r_B = r_D + r_C$$

Consider a compound gear train of the form shown in Figure 8.18(a), with A, the first driver, having 15 teeth, B 30 teeth, C 18 teeth and D, the final driven wheel, 36 teeth. Since the angular velocity of a wheel is inversely proportional to the number of teeth on the wheel, the overall gear ratio is

$$G = \frac{30}{15} \times \frac{36}{18} = 4$$

Thus, if the input to wheel A is an angular velocity of 160 rev/min, then the output angular velocity of wheel D is $160/4 = 40$ rev/min.

A simple gear train of spur, helical or bevel gears is usually limited to an overall gear ratio of about 10. This is because of the need to keep the gear train down to a manageable size if the number of teeth on the pinion is to be kept above a minimum number which is usually about 10 to 20. Higher gear ratios can, however, be obtained with compound gear trains. This is because the gear ratio is the product of the individual gear ratios of parallel gear sets.

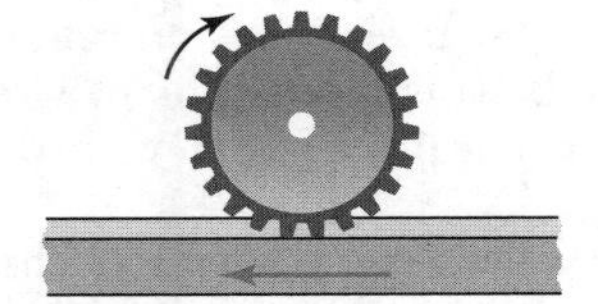

Figure 8.19 Rack-and-pinion.

8.5.2 Rotational to translational motion

The **rack-and-pinion** (Figure 8.19) is another form of gear, being essentially two intermeshed gears with one having a base circle of infinite radius. Such gears can be used to transform either linear motion to rotational motion or rotational motion to linear motion.

Another method that has been used for converting rotary to translational motion is the **screw and nut system**. With the conventional form of screw and nut, the nut is rotated and moved along the stationary screw. However, if the screw is rotated then a nut, which is attached to the part to be driven, moves along the screw thread. Such an arrangement is termed a **lead screw**. The lead L is the distance moved parallel to the screw axis when the nut is given one turn; for a single thread the lead is equal to the pitch. In n revolutions the distance moved parallel to the screw axis will be nL. If n revolutions are completed in a time t, the linear velocity v parallel to the screw axis is nL/t. As n/t is the number of revolutions per second f for the screw then:

$$v = \frac{nL}{t} = fL$$

There are, however, problems with using such an arrangement for converting rotational motion to linear motion. There are the high friction forces involved in the direct sliding contact between the screw and the nut and also the lack of rigidity. Friction can be overcome by using a **ball screw**. Such a screw is identical in principle to the lead screw, but ball bearings are located in the thread of the nut. Such an arrangement has been used with robots with the arm being driven by a ball screw powered by a geared d.c. motor (Figure 8.20). The motor rotates the screw which moved the nut up or down its thread. The movement of the nut is conveyed to the arm by means of a linkage.

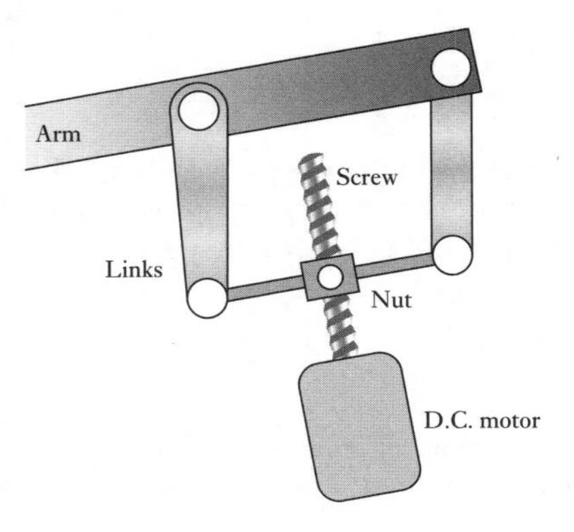

Figure 8.20 Ball screw and links used to move a robot arm.

8.6 Ratchet and pawl

Ratchets can be used to lock a mechanism when it is holding a load. Figure 8.21 shows a ratchet and pawl. The mechanism consists of a wheel, called a **ratchet**, with saw-shaped teeth which engage with an arm called **a pawl**. The arm is pivoted and can move back and forth to engage the wheel. The shape of the teeth is such that rotation can occur in only one direction. Rotation of the ratchet wheel in a clockwise direction is prevented by the pawl and can only take place when the pawl is lifted. The pawl is normally spring loaded to ensure that it automatically engages with the ratchet teeth.

Thus a winch used to wind up a cable on a drum may have a ratchet and pawl to prevent the cable unwinding from the drum when the handle is released.

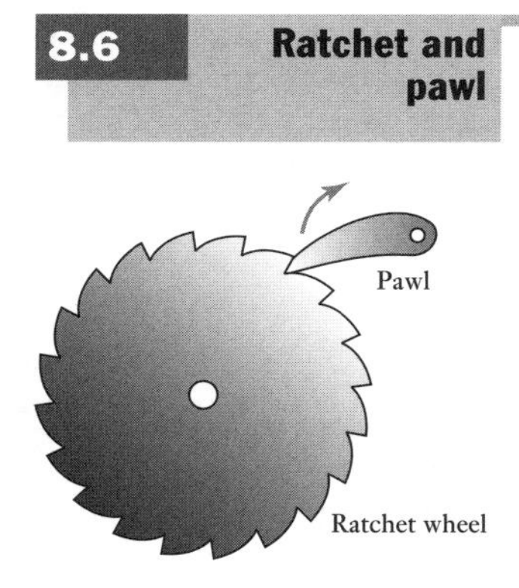

Figure 8.21 Ratchet and pawl.

8.7 Belt and chain drives

Belt drives are essentially just a pair of rolling cylinders with the motion of one cylinder being transferred to the other by a belt (Figure 8.22). Belt drives use the friction that develops between the pulleys attached to the shafts and the belt around the arc of contact in order to transmit a torque. Since the transfer relies on frictional forces then slip can occur. The transmitted torque is due to the differences in tension that occur in the belt during operation. This difference results in a tight side and a slack side for the belt. If the tension on the tight side is T_1, and that on the slack side T_2, then with pulley A in Figure 8.22 as the driver

$$\text{torque on A} = (T_1 - T_2)r_A$$

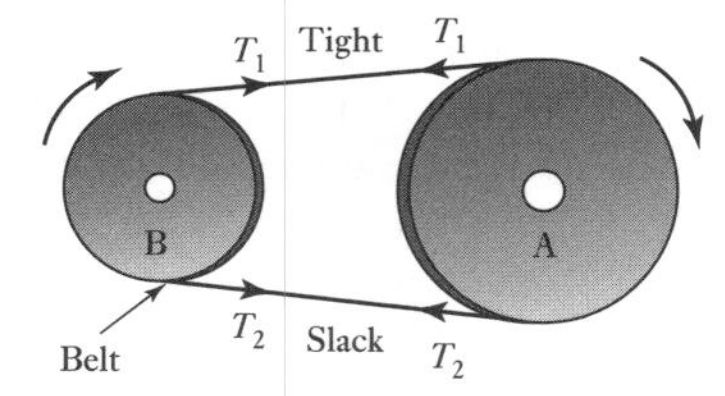

Figure 8.22 Belt drive.

where r_A is the radius of pulley A. For the driven pulley B we have

$$\text{torque on B} = (T_1 - T_2)r_B$$

where r_B is the radius of pulley B. Since the power transmitted is the product of the torque and the angular velocity, and since the angular velocity is v/r_A for pulley A and v/r_B for pulley B, where v is the belt speed, then for either pulley we have

$$\text{power} = (T_1 - T_2)v$$

As a method of transmitting power between two shafts, belt drives have the advantage that the length of the belt can easily be adjusted to suit a wide range of shaft-to-shaft distances and the system is automatically protected against overload because slipping occurs if the loading exceeds the maximum tension that can be sustained by frictional forces. If the distances between shafts is large, a belt drive is more suitable than gears, but over small distances gears are to be preferred. Different-size pulleys can be used to give a gearing effect. However, the gear ratio is limited to about 3 because of the need to maintain an adequate arc of contact between the belt and the pulleys.

The belt drive shown in Figure 8.22 gives the driven wheel rotating in the same direction as the driver wheel. Figure 8.23 shows two types of reversing drives. With both forms of drive, both sides of the belt come into contact with the wheels and so V-belts or timing belts cannot be used.

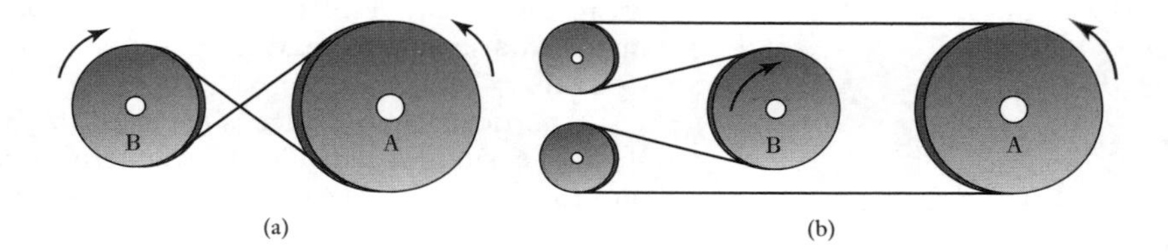

Figure 8.23 Reversed belt drives: (a) crossed belt, (b) open belt.

8.7.1 Types of belts

The four main types of belts (Figure 8.24) are:

1 *Flat*
The belt has a rectangular cross-section. Such a drive has an efficiency of about 98% and produces little noise. They can transmit power over long distances between pulley centres. Crowned pulleys are used to keep the belts from running off the pulleys.

2 *Round*
The belt has a circular cross-section and is used with grooved pulleys.

3 *V*
V-belts are used with grooved pulleys and are less efficient than flat belts but a number of them can be used on a single wheel and so give a multiple drive.

4 *Timing*
Timing belts require toothed wheels, having teeth which fit into the grooves on the wheels. The timing belt, unlike the other belts, does not stretch or slip and consequently transmits power at a constant angular velocity ratio. The teeth make it possible for the belt to be run at slow or fast speeds.

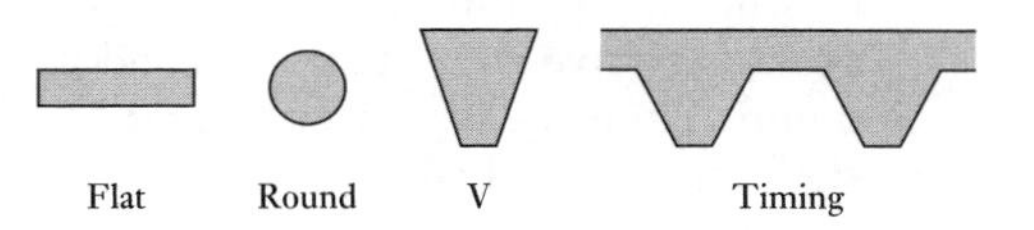

Figure 8.24 Types of belt.

8.7.2 Chains

Slip can be prevented by the use of chains which lock into teeth on the rotating cylinders to give the equivalent of a pair of intermeshing gear wheels. A chain drive has the same relationship for gear ratio as a simple gear train. The drive mechanism used with a bicycle is an example of a chain drive. Chains enable a number of shafts to be driven by a single wheel and so give a multiple drive. They are not as quiet as timing belts but can be used for larger torques.

8.8 Bearings

Whenever there is relative motion of one surface in contact with another, either by rotating or sliding, the resulting frictional forces generate heat which wastes energy and results in wear. The function of a **bearing** is to guide with minimum friction and maximum accuracy the movement of one part relative to another.

Of particular importance is the need to give suitable support to rotating shafts, i.e. support radial loads. The term **thrust bearing** is used for bearings that are designed to withstand forces along the axis of a shaft when the relative motion is primarily rotation. The following sections outline the characteristics of commonly used forms of bearings.

8.8.1 Plain journal bearings

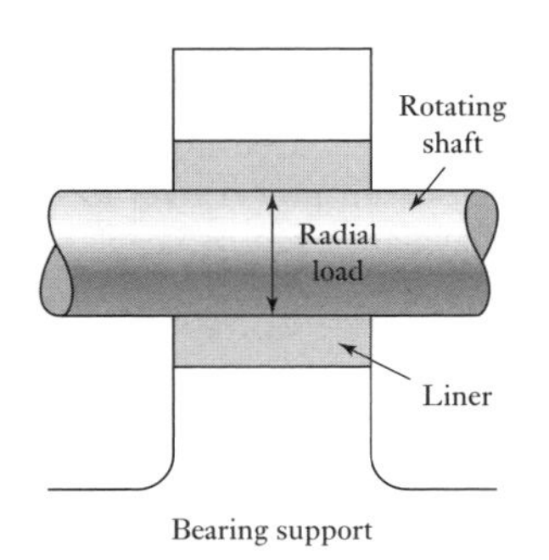

Figure 8.25 Plain journal bearing.

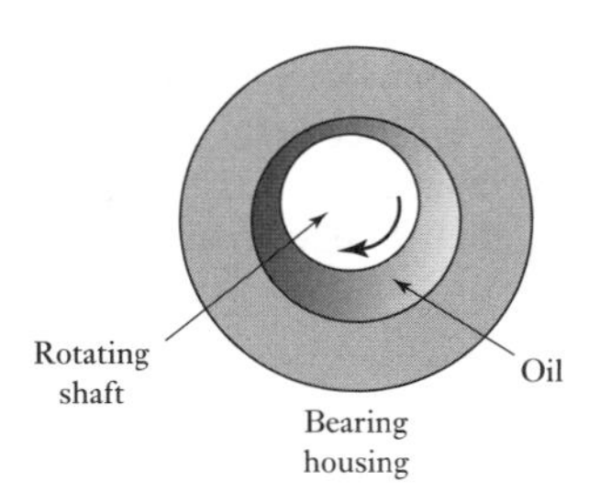

Figure 8.26 Hydrodynamic journal bearing.

Journal bearings are used to support rotating shafts which are loaded in a radial direction; the term **journal** is used for a shaft. The bearing basically consists of an insert of some suitable material which is fitted between the shaft and the support (Figure 8.25). Rotation of the shaft results in its surface sliding over that of the bearing surface. The insert may be a white metal, aluminium alloy, copper alloy, bronze or a polymer such as nylon or PTFE. The insert provides lower friction and less wear than if the shaft just rotated in a hole in the support. The bearing may be a dry rubbing bearing or lubricated. Plastics such as nylon and PTFE are generally used without lubrication, the coefficient of friction with such materials being exceptionally low. A widely used bearing material is sintered bronze; this is bronze with a porous structure which allows it to be impregnated with oil and so the bearing has a 'built-in' lubricant.

The lubricant may be:

1 *Hydrodynamic*
The **hydrodynamic journal bearing** consists of the shaft rotating continuously in oil in such a way that it rides on oil and is not supported by metal (Figure 8.26). The load is carried by the pressure generated in the oil as a result of the shaft rotating.

2 *Hydrostatic*
A problem with hydrodynamic lubrication is that the shaft only rides on oil when it is rotating and when at rest there is metal-to-metal contact. To avoid excessive wear at start-up and when there is only a low load, oil is pumped into the load-bearing area at a high-enough pressure to lift the shaft off the metal when at rest.

3 *Solid-film*
This is a coating of a solid material such as graphite or molybdenum disulphide.

4 *Boundary layer*
This is a thin layer of lubricant which adheres to the surface of the bearing.

8.8.2 Ball and roller bearings

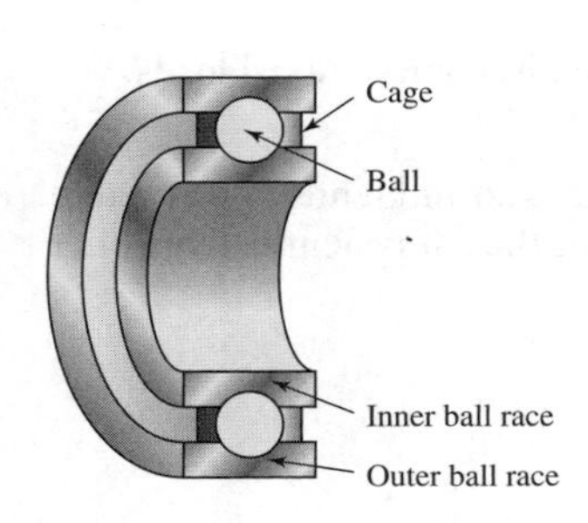

Figure 8.27 Basic elements of a ball bearing.

With this type of bearing, the main load is transferred from the rotating shaft to its support by rolling contact rather than sliding contact. A rolling element bearing consists of four main elements: an inner race, an outer race, the rolling element of either balls or rollers, and a cage to keep the rolling elements apart (Figure 8.27). The inner and outer races contain hardened tracks in which the rolling elements roll.

There are a number of forms of ball bearings:

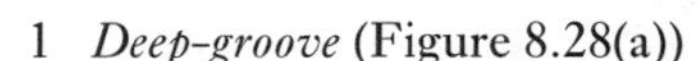

1 *Deep-groove* (Figure 8.28(a))
This is good at withstanding radial loads but is only moderately good for axial loads. It is a versatile bearing which can be used with a wide range of load and speed.

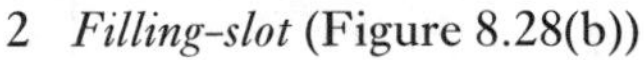

2 *Filling-slot* (Figure 8.28(b))
This is able to withstand higher radial loads than the deep-groove equivalent but cannot be used when there are axial loads.

3 *Angular contact* (Figure 8.28(c))
This is good for both radial and axial loads and is better for axial loads than the deep-groove equivalent.

4 *Double-row* (Figure 8.28(d))
Double-row ball bearings are made in a number of types and are able to withstand higher radial loads than their single-row equivalents. The figure shows a double-row deep-groove ball bearing, there being double-row versions of each of the above single-row types.

5 *Self-aligning* (Figure 8.28(e))
Single-row bearings can withstand a small amount of shaft misalignment but where there can be severe misalignment a self-aligning bearing is used. This is able to withstand only moderate radial loads and is fairly poor for axial loads.

6 *Thrust, grooved race* (Figure 8.28(f))
These are designed to withstand axial loads but are not suitable for radial loads.

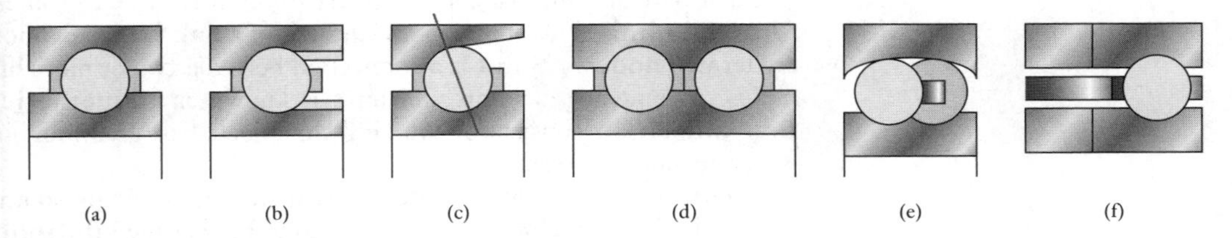

Figure 8.28 Types of ball bearings.

There are also a number of forms of roller bearing, the following being common examples:

1 *Straight roller* (Figure 8.29(a))
This is better for radial loads than the equivalent ball bearing but is not generally suitable for axial loads. They will carry a greater load than ball bearings of the same size because of their greater contact area. However, they are not tolerant of misalignment.

2 *Taper roller* (Figure 8.29(b))
This is good for radial loads and good in one direction for axial loads.

3 *Needle roller* (Figure 8.29(c))
This has a roller with a high length/diameter ratio and tends to be used in situations where there is insufficient space for the equivalent ball or roller bearing.

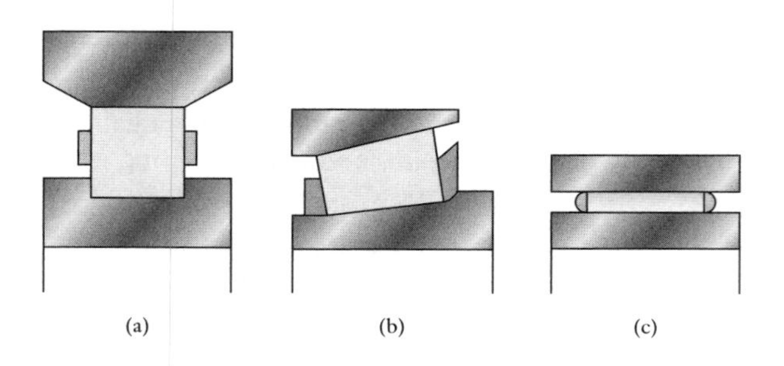

Figure 8.29 Roller bearings.

8.8.3 Selection of bearings

In general, dry sliding bearings tend to be used only for small-diameter shafts with low-load and low-speed situations, ball and roller bearings, i.e. bearings involving rolling, with a much wider range of diameter shafts and higher load and higher speed, and hydrodynamic bearings for the high loads with large-diameter shafts.

Summary

Mechanisms are devices which can be considered to be motion converters in that they transform motion from one form to some other required form.

The motion of a body can be considered to be a combination of translational and rotational motions. The number of **degrees of freedom** is the number of components to motion that are required to generate the motion.

Each part of a mechanism which has motion relative to some other part is termed a **link.** The points of attachment of a link to its neighbouring links are termed **nodes.** A **joint** is a connection between two or more links at their nodes. A sequence of joints and links is known as a **kinematic chain.** A **four-bar chain** consists of four links connected to give four joints about which turning can occur.

A **cam** is a body which rotates or oscillates and in doing so imparts a reciprocating or oscillatory motion to a second body, called the **follower**, with which it is in contact.

Gears can be used for the transmission of rotary motion between parallel shafts and for shafts which have axes inclined to one another.

The **rack-and-pinion** and the **screw-and-nut** systems can be used to convert rotational motion to translational motion.

Ratchets can be used to lock a mechanism when it is holding a load.

Belt and **chain drives** can be used to transmit rotational motion between shafts which are parallel and some distance apart.

Bearings are used to guide with minimum friction and maximum accuracy the movement of one part relative to another.

Problems

8.1 Explain the terms (a) mechanism, (b) kinematic chain.

8.2 Explain what is meant by the four-bar chain.

8.3 By examining the following mechanisms, state the number of degrees of freedom each has.

(a) A car hood hinge mechanism.
(b) An estate car tailgate mechanism.
(c) A windscreen wiper mechanism.
(d) Your knee.
(e) Your ankle.

8.4 Analyse the motions of the following mechanisms and state whether they involve pure rotation, pure translation or are a mixture of rotation and translation components.

(a) The keys on a computer keyboard.
(b) The pen in an XY plotter.
(c) The hour hand of a clock.
(d) The pointer on a moving-coil ammeter.
(e) An automatic screwdriver.

8.5 For the mechanism shown in Figure 8.30, the arm AB rotates at a constant rate. B and F are sliders moving along CD and AF. Describe the behaviour of this mechanism.

Figure 8.30 Problem 8.5.

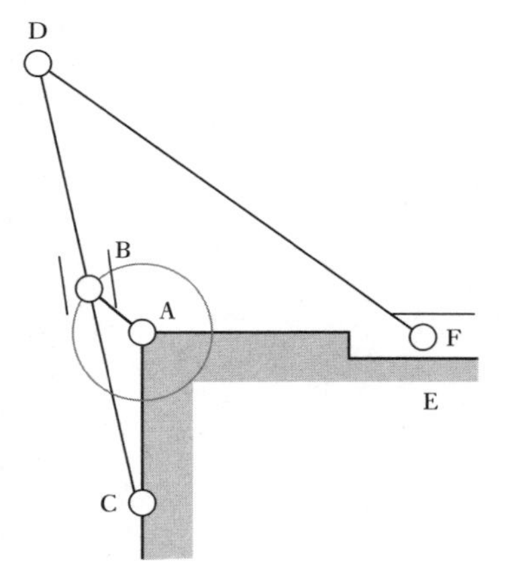

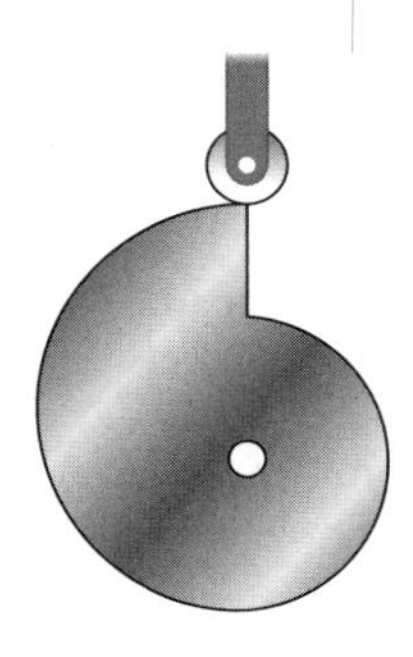

Figure 8.31
Problem 8.6.

8.6 Describe how the displacement of the cam follower shown in Figure 8.31 will vary with the angle of rotation of the cam.

8.7 A circular cam of diameter 100 mm has an eccentric axis of rotation which is offset 30 mm from the centre. When used with a knife follower with its line of action passing through the centre of rotation, what will be the difference between the maximum and minimum displacements of the follower?

8.8 Design a cam follower system to give constant follower speeds over follower displacements varying from 40 to 100 mm.

8.9 Design a mechanical system which can be used to:

(a) Operate a sequence of microswitches in a timed sequence.

(b) Move a tool at a steady rate in one direction and then quickly move it back to the beginning of the path.

(c) Transform a rotation into a linear back-and-forth movement with simple harmonic motion.

(d) Transform a rotation through some angle into a linear displacement.

(e) Transform a rotation of a shaft into rotation of another, parallel shaft some distance away.

(f) Transform a rotation of one shaft into rotation of another, close shaft which is at right angles to it.

8.10 A compound gear train consists of the final driven wheel with 15 teeth which meshes with a second wheel with 90 teeth. On the same shaft as the second wheel is a wheel with 15 teeth. This meshes with a fourth wheel, the first driver, with 60 teeth. What is the overall gear ratio?

Chapter nine Electrical Actuation Systems

Objectives

The objectives of this chapter are that, after studying it, the reader should be able to:

- **Evaluate the operational characteristics of electrical actuation systems: relays, solid-state switches (thyristors, bipolar transistors and MOSFETs), solenoids, d.c. motors, a.c. motors and steppers.**
- **Explain the principles of d.c. motors, including the d.c. permanent magnet motor and how it can have its speed controlled.**
- **Explain the principle of the brushless permanent magnet d.c. motor.**
- **Explain the principles of the variable reluctance, permanent magnet and hybrid forms of stepper motor and how step sequences can be generated.**
- **Explain the requirements in selecting motors of inertia matching and torque and power requirements.**

9.1 Electrical systems

In any discussion of electrical systems used as actuators for control, the discussion has to include:

1. *Switching devices* such as mechanical switches, e.g. relays, and solid-state switches, e.g. diodes, thyristors, and transistors, where the control signal switches on or off some electrical device, perhaps a heater or a motor.
2. *Solenoid-type devices* where a current through a solenoid is used to actuate a soft iron core, as, for example, the solenoid-operated hydraulic/pneumatic valve where a control current through a solenoid is used to actuate a hydraulic/pneumatic flow.
3. *Drive systems*, such as d.c. and a.c. motors, where a current through a motor is used to produce rotation.

This chapter is an overview of such devices and their characteristics.

9.2 Mechanical switches

Mechanical switches are elements which are often used as sensors to give inputs to systems (see Section 2.12), e.g. keyboards. In this chapter we are concerned with their use as actuators to switch on perhaps electric motors or heating elements, or switch on the current to actuate solenoid valves controlling hydraulic or pneumatic cylinders. The electrical **relay** is an example of a mechanical switch used in control systems as an actuator.

9.2.1 Relays

Relays are electrically operated switches in which changing a current in one electric circuit switches a current on or off in another circuit. For the relay shown in Figure 9.1(a), when there is a current through the solenoid of the relay, a magnetic field is produced which attracts the iron armature, moves the push rod, and so closes the normally open (NO) switch contacts and opens the normally closed (NC) switch contacts.

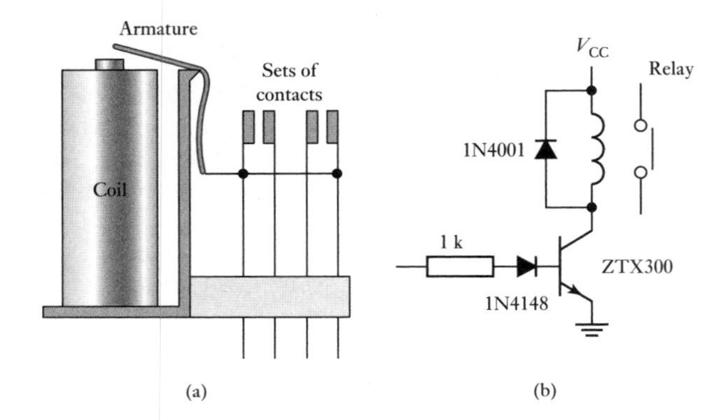

Figure 9.1 (a) A relay and (b) a driver circuit.

Relays are often used in control systems. The output from a controller is a relatively small current and so it is often used in conjunction with a transistor to switch on the current through the relay solenoid and hence use the relay to switch on the much larger current needed to switch on or off a final correction element such as an electric heater in a temperature control system or a motor. Figure 9.1(b) shows the type of circuit that might be used. Because relays are inductances, they can generate a back voltage when the energising current is switched off or when their input switches from a high to low signal. As a result, damage can occur in the connecting circuit. To overcome this problem, a diode is connected across the relay. When the back e.m.f. occurs, the diode conducts and shorts it out. Such a diode is termed a **free-wheeling** or **flyback** diode.

As an illustration of the ways relays can be used in control systems, Figure 9.2 shows how two relays might be used to control the operation of pneumatic valves which in turn control the movement of pistons in three cylinders A, B and C. The sequence of operation is:

1 When the start switch is closed, current is applied to the A and B solenoids and results in both A and B extending, i.e. A+ and B+.
2 The limit switches a+ and b+ are then closed; the a+ closure results in a current flowing through relay coil 1 which then closes its contacts and so supplies current to the C solenoid and results in it extending, i.e. C+.
3 Its extension causes limit switch c+ to close and so current to switch the A and B control valves and hence retraction of cylinders A and B, i.e. A− and B−.
4 Closing limit switch a− passes a current through relay coil 2; its contacts close and allow a current to valve C and cylinder C to retract, i.e. C−.

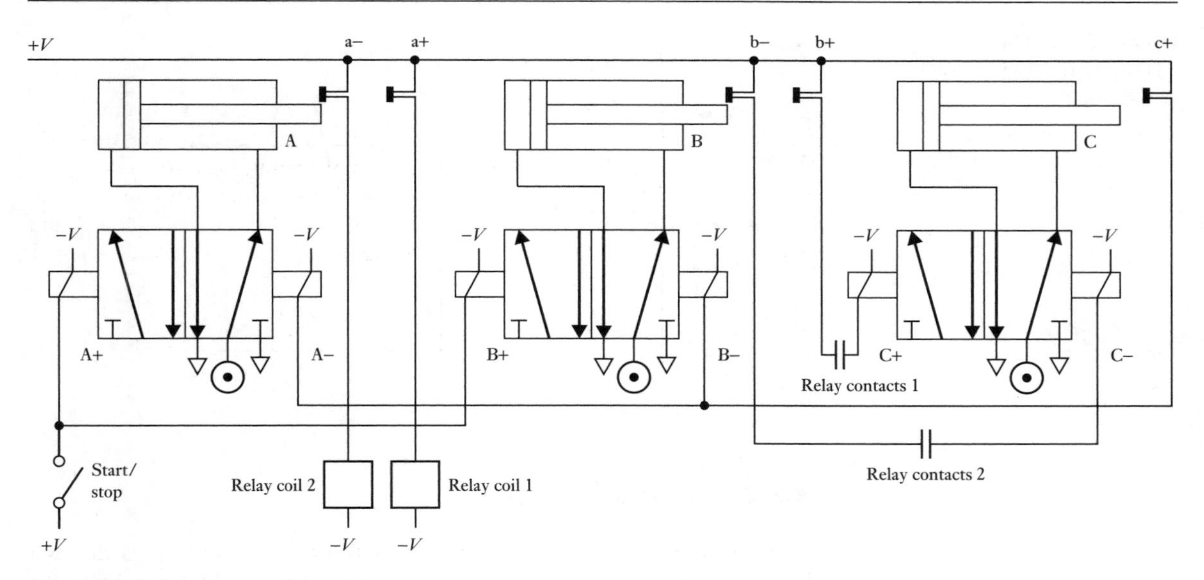

Figure 9.2 Relay-controlled system.

The sequence thus given by this system is A+ and B+ concurrently, then C+, followed by A− and B− concurrently and finally C−.

Time-delay relays are control relays that have a delayed switching action. The time delay is usually adjustable and can be initiated when a current flows through the relay coil or when it ceases to flow through the coil.

9.3 Solid-state switches

There are a number of solid-state devices which can be used electronically to switch circuits. These include:

1 Diodes.
2 Thyristors and triacs.
3 Bipolar transistors.
4 Power MOSFETs.

9.3.1 Diodes

The **diode** has the characteristic shown in Figure 9.3(a), only passing a current when forward biased, i.e. with the anode being positive with respect to the cathode. If the diode is sufficiently reverse biased, i.e. a very high voltage, it will break down. If an alternating voltage is applied across a diode, it can be regarded as only switching on when the direction of the voltage is such as to forward-bias it and being off in the reverse-biased direction. The result is that the current through the diode is half-rectified to become just the current due to the positive halves of the input voltage (Figure 9.3(b)), i.e. the circuit only 'switches on' for the positive half cycle.

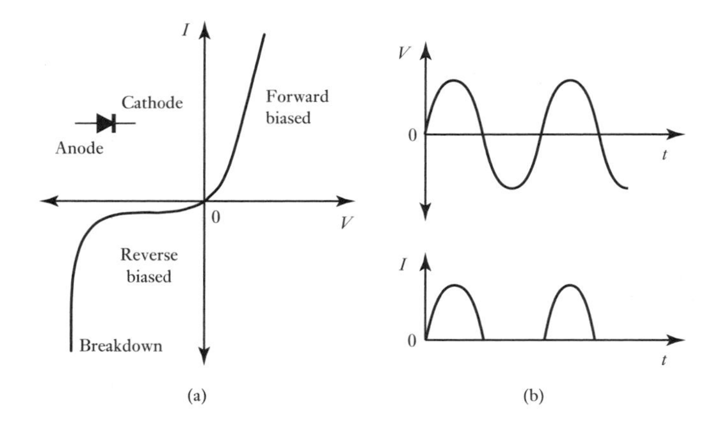

Figure 9.3 (a) Diode characteristic, (b) half-wave rectification.

9.3.2 Thyristors and triacs

The **thyristor**, or **silicon-controlled rectifier** (SCR), can be regarded as a diode which has a gate controlling the conditions under which the diode can be switched on. Figure 9.4(a) shows the thyristor characteristic. With the gate current zero, the thyristor passes negligible current when reverse biased (unless sufficiently reverse biased, hundreds of volts, when it breaks down). When forward biased the current is also negligible until the forward breakdown voltage is exceeded. When this occurs the voltage across the diode falls to a low level, about 1 to 2 V, and the current is then only limited by the external resistance in a circuit. Thus, for example, if the forward breakdown is at 300 V then when this voltage is reached the thyristor switches on and the voltage across it drops to 1 or 2 V. If the thyristor is in series with a resistance of, say, 20 Ω (Figure 9.4(b)) then before breakdown we have a very high resistance in series with the 20 Ω and so virtually all the 300 V is across the thyristor and there is negligible current. When forward breakdown occurs, the voltage across the thyristor drops to, say, 2 V and so there is now $300 - 2 = 298$ V across the 20 Ω resistor, hence the current rises to $298/20 = 14.9$ A. Once switched on, the thyristor remains on until the forward current is reduced to below a level of a few milliamps. The voltage at which forward breakdown occurs is determined by the current entering the gate: the higher the current, the lower the breakdown voltage. The power-handling capability of a thyristor is high and thus it is widely used for switching high-power applications. As an example, the Texas Instruments

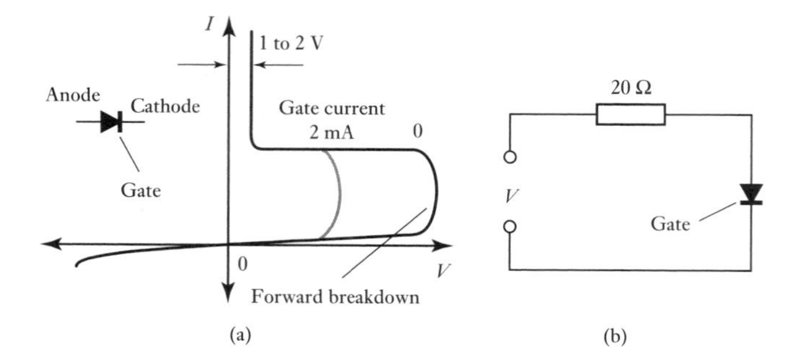

Figure 9.4 (a) Thyristor characteristic, (b) thyristor circuit.

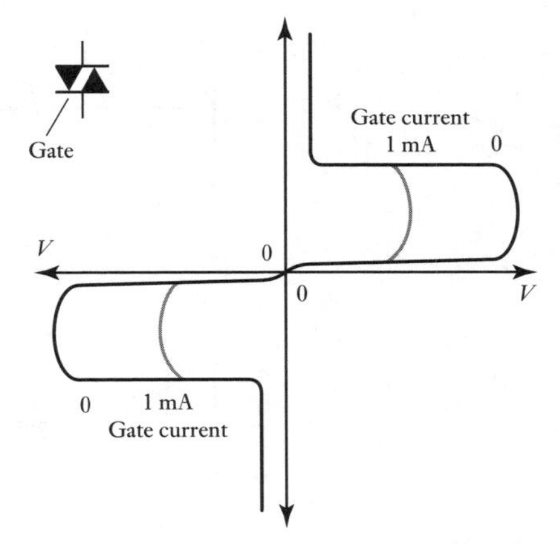

Figure 9.5 Triac characteristic.

CF106D has a maximum off-state voltage of 400 V and a maximum gate trigger current of 0.2 mA.

The triac is similar to the thyristor and is equivalent to a pair of thyristors connected in reverse parallel on the same chip. The triac can be turned on in either the forward or reverse direction. Figure 9.5 shows the characteristic. As an example, the Motorola MAC212-4 triac has a maximum off-state voltage of 200 V and a maximum on-state current of 12 A r.m.s. Triacs are simple, relatively inexpensive, methods of controlling a.c. power.

Figure 9.6 shows the type of effect that occurs when a sinusoidal alternating voltage is applied across (a) a thyristor and (b) a triac. Forward breakdown occurs when the voltage reaches the breakdown value and then the voltage across the device remains low.

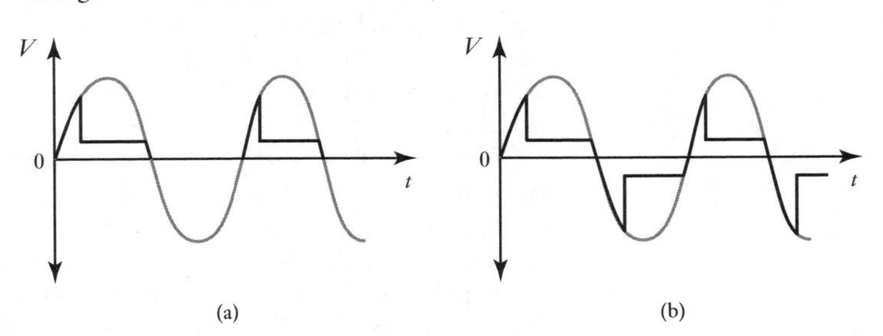

Figure 9.6 Voltage control: (a) thyristor, (b) triac.

As an example of how such devices can be used for control purposes, Figure 9.7 illustrates how a thyristor could be used to control a steady d.c. voltage V. In this the thyristor is operated as a switch by using the gate to switch the device on or off. By using an alternating signal to the gate, the supply voltage can be chopped and an intermittent voltage produced. The average value of the output d.c. voltage is thus varied and hence controlled by the alternating signal to the gate.

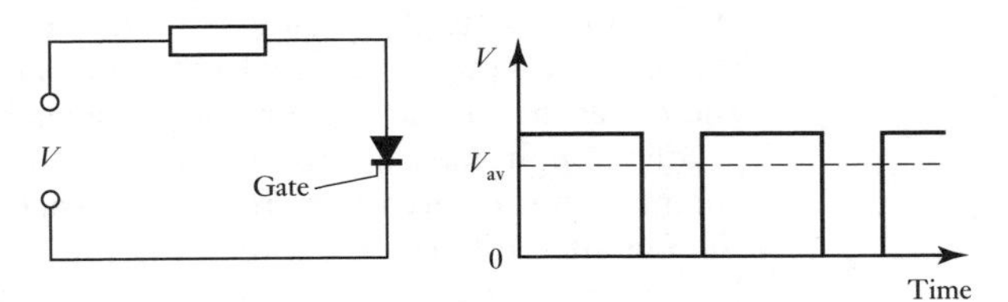

Figure 9.7 Thyristor d.c. control.

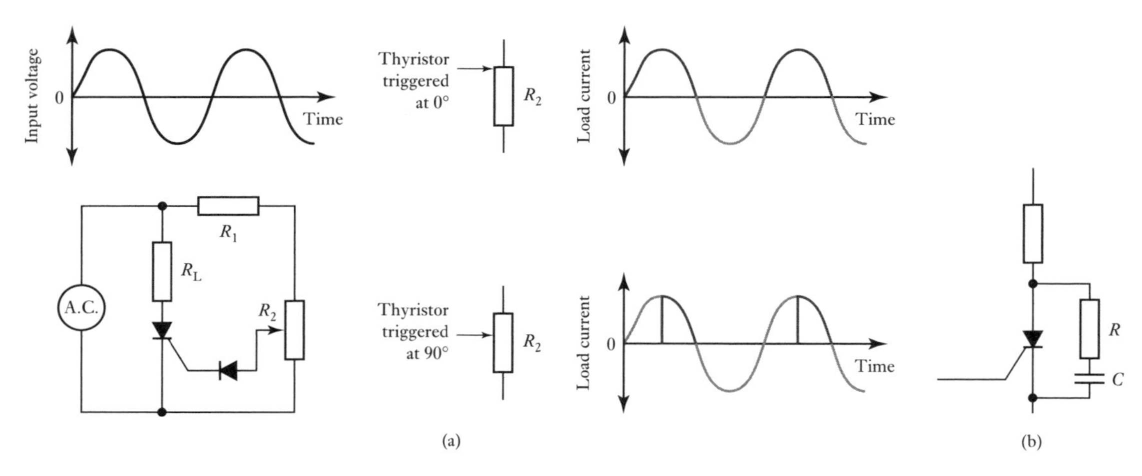

Figure 9.8 (a) Phase control, (b) snubber circuit.

Another example of control is that of alternating current for electric heaters, electric motors or lamp dimmers. Figure 9.8(a) shows a half-wave, variable resistance, phase control circuit. The alternating current is applied across the load, e.g. the lamp for the lamp dimming circuit, in series with a thyristor. R_1 is a current-limiting resistor and R_2 is a potentiometer which sets the level at which the thyristor is triggered. The diode is to prevent the negative part of the alternating voltage cycle being applied to the gate. By adjusting R_2 the thyristor can be made to trigger at any point between 0° and 90° in the positive half cycle of the applied alternating voltage. When the thyristor is triggered near the beginning of the cycle, i.e. 0°, it conducts for the entire positive half cycle and the maximum power is delivered to the load. As the triggering of the thyristor is delayed to later in the cycle, so the power delivered to the load is reduced.

When a source voltage is suddenly applied to a thyristor, or a triac, with the gate off, the thyristor may switch from off to on. A typical rate of voltage change that would produce this effect is of the order of 50 V/μs. If the source is a d.c. voltage the thyristor can remain in this conducting state until there is a circuit interruption. In order to prevent this sudden change in source voltage producing this effect, the rate at which the voltage changes with time, i.e. dV/dt, is controlled by using a **snubber circuit**. This is a resistor in series with a capacitor and is placed in parallel with the thyristor (Figure 9.8(b)).

9.3.3 Bipolar transistors

Bipolar transistors come in two forms, the npn and the pnp. Figure 9.9(a) shows the symbol for each. For the npn transistor, the main current flows in at the collector and out at the emitter, a controlling signal being applied to the base. The pnp transistor has the main current flowing in at the emitter and out at the collector, a controlling signal being applied to the base.

For an npn transistor connected as shown in Figure 9.9(b), the so-termed common emitter circuit, the relationship between the collector current I_C and the potential difference between the collector and emitter V_{CE} is described by the series of graphs shown in Figure 9.9(c). When the base current I_B

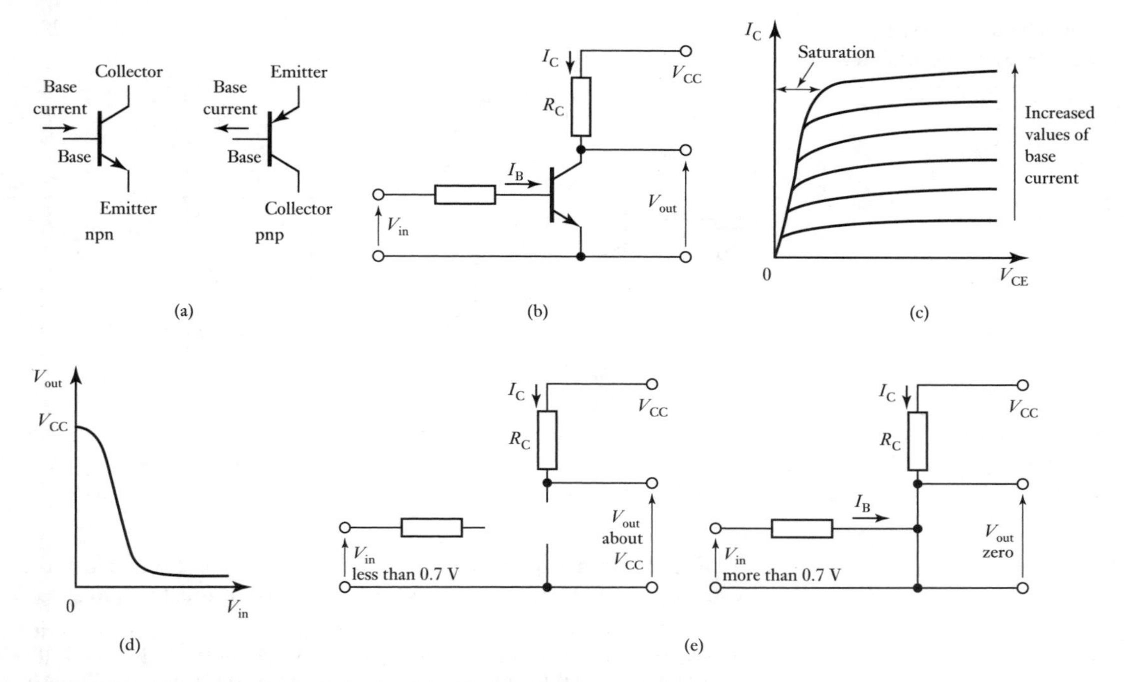

Figure 9.9 (a) Transistor symbols, (b), (c), (d), (e) transistor switch.

is zero, the transistor is cut off; in this state both the base–emitter and the base–collector junctions are reverse biased. When the base current is increased, the collector current increases and V_{CE} decreases as a result of more of the voltage being dropped across R_C. When V_{CE} reaches a value $V_{CE(sat)}$, the base–collector junction becomes forward biased and the collector current can increase no further, even if the base current is further increased. This is termed **saturation**. By switching the base current between 0 and a value that drives the transistor into saturation, bipolar transistors can be used as switches. When there is no input voltage V_{in} then virtually the entire V_{CC} voltage appears at the output. When the input voltage is made sufficiently high, the transistor switches so that very little of the V_{CC} voltage appears at the output (Figure 9.9(d)). Figure 9.9(e) summarises this switching behaviour of a typical transistor.

The relationship between collector current and the base current I_B at values below that which drives the transistor into saturation is

$$I_C = h_{FE} I_B$$

where h_{FE} is the **current gain**. At saturation the collector current $I_{C(sat)}$ is

$$I_{C(sat)} = \frac{V_{CC} - V_{CE(sat)}}{R_C}$$

To ensure that the transistor is driven into saturation the base current must thus rise to at least

$$I_{B(sat)} = \frac{I_{C(sat)}}{h_{FE}}$$

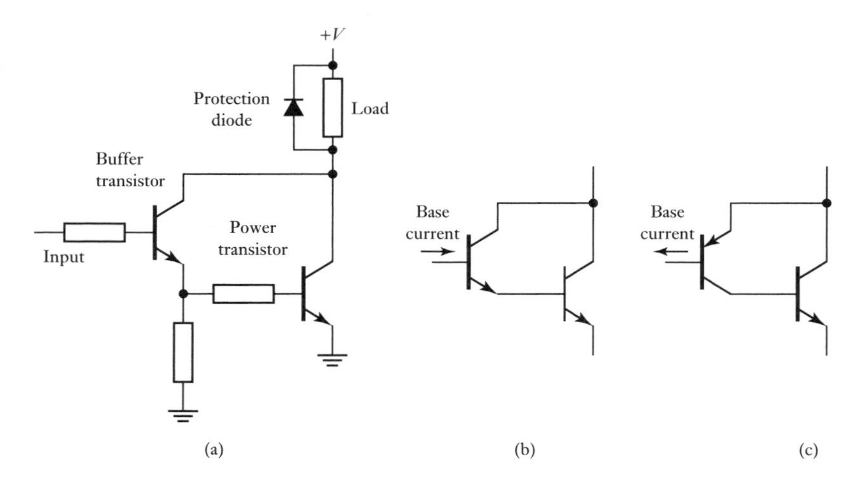

Figure 9.10 (a) Switching a load, (b) and (c) Darlington pairs.

Thus for a transistor with h_{FE} of 50 and $V_{CE(sat)}$ of 1 V, then for a circuit with $R_C = 10\ \Omega$ and $V_{CC} = 5$ V, the base current must rise to at least about 8 mA.

Because the base current needed to drive a bipolar power transistor is fairly large, a second transistor is often needed to enable switching to be obtained with relatively small currents, e.g. that supplied by a microprocessor. Thus the switching circuit can be of the form shown in Figure 9.10(a). Such a combination of a pair of transistors to enable a high current to be switched with a small input current is termed a **Darlington pair** and they are available as single-chip devices. A **protection diode** (free-wheeling diode) is generally connected in parallel with the load to prevent damage when the transistor is switched off since it is generally used with inductive loads and large transient voltages can occur. The integrated circuit ULN2001N from SGS-Thomson contains seven separate Darlington pairs, each pair being provided with a protection diode. Each pair is rated as 500 mA continuous and can withstand surges up to 600 mA.

Figure 9.10(b) shows the Darlington connections when a small npn transistor is combined with a large npn transistor, the result being equivalent to a large npn transistor with a large amplification factor. Figure 9.10(c) shows the Darlington connections for a small pnp transistor with a large npn transistor, the result being equivalent to a single large pnp transistor.

In using transistor-switched actuators with a microprocessor, attention has to be given to the size of the base current required and its direction. The base current required can be too high and so a **buffer** might be used. The buffer increases the drive current to the required value. It might also be used to invert. Figure 9.11 illustrates how a buffer might be used when transistor switching is used to control a d.c. motor by on/off switching. Type 240 buffer is inverting while types 241 and 244 are non-inverting. Buffer 74LS240 has a high-level maximum output current of 15 mA and a low-level maximum output current of 24 mA.

Bipolar transistor switching is implemented by base currents and higher frequencies of switching are possible than with thyristors. The power-handling capability is less than that of thyristors.

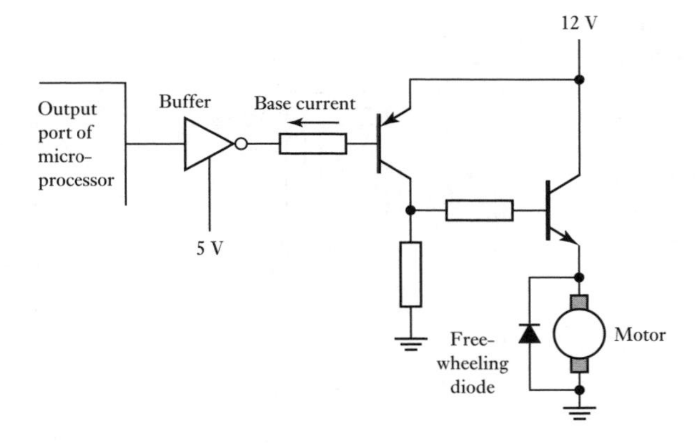

Figure 9.11 Control of d.c. motor.

9.3.4 MOSFETs

MOSFETs (metal-oxide field-effect transistors) come in two types, the n-channel and the p-channel. Figure 9.12(a) and (b) shows the symbols. The main difference between the use of a MOSFET for switching and a bipolar transistor is that no current flows into the gate to exercise the control. The gate voltage is the controlling signal. Thus drive circuitry can be simplified in that there is no need to be concerned about the size of the current.

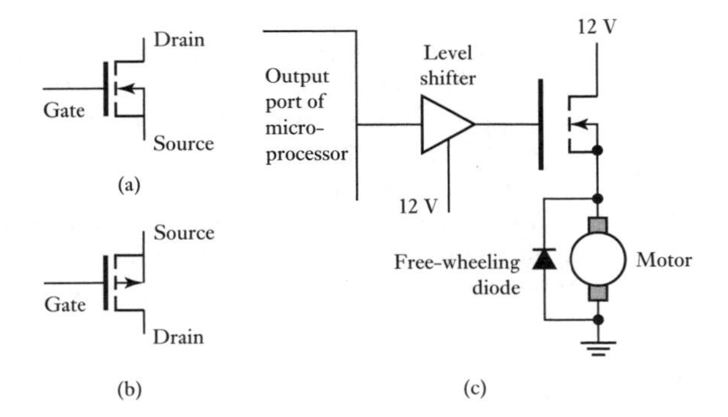

Figure 9.12 MOSFETs: (a) n-channel, (b) p-channel, (c) used to control a d.c. motor.

Figure 9.12(c) illustrates the use of a MOSFET as an on/off switch for a motor; compare the circuit with that in Figure 9.11 where bipolar transistors are used. A level shifter buffer is indicated, this being to raise the voltage level to that required for the MOSFET.

With MOSFETs, very high-frequency switching is possible, up to 1 MHz, and interfacing with a microprocessor is simpler than with bipolar transistors.

9.4 Solenoids

Essentially, solenoids consist of a coil of electrical wire with an armature which is attracted to the coil when a current passes through it and produces a magnetic field. The movement of the armature contracts a return spring which then allows the armature to return to its original position when the current ceases. The solenoids can be linear or rotary, on/off or variable positioning and operated by d.c. or a.c. Such an arrangement can be used to

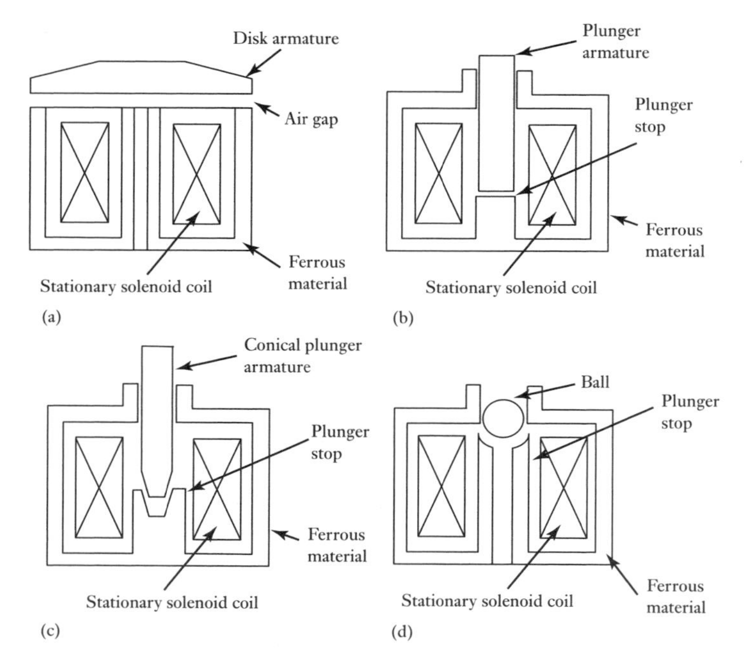

Figure 9.13 The basic forms of linear solenoids with (a) disk, (b) plunger, (c) conical plunger, (d) ball forms of armature. Not shown in the figures are the springs required to return the armature back to its original position when the current through the solenoid ceases.

provide electrically operated actuators which are widely used for short stroke devices, typically up to 25 mm.

Figure 9.13 shows four examples of linear solenoids with different forms of armature. The form of the armature, the pole pieces and the central tube will depend on the use for which the actuator is designed. Disk armatures are useful where small distances of travel and fast action are required. Plunger armatures are widely used for applications requiring small distances of travel and fast action. Conical armatures are used for long-stroke applications, a typical application being for an automotive door lock mechanism. Ball armatures are used with fluid control applications, a typical application being an air bag deployment mechanism.

For a simple on/off device, there is no necessity for the design to give a linear characteristic. Where a proportional actuator is required, careful design is needed to give a movement of the armature proportional to the solenoid current. A simple example of the use of an on/off solenoid actuator is as a door lock with the lock either being actuated by the passage of a current through the solenoid or the reverse case when the passage of the current unlocks the door.

Solenoid valves are another example of such devices, being used to control fluid flow in hydraulic or pneumatic systems (see Figure 7.9). When a current passes through a coil, a soft iron plunger form of armature is pulled into the coil and, in doing so, can open or close ports to allow the flow of a fluid. The force exerted by the solenoid on the armature is a function of the current in the coil and the length of the armature within the coil. With on/off valves, i.e. those used for directional control, the current in the coil is controlled to be either on or off and the core is consequently in one of two positions. With proportional control valves, the current in the coil is controlled to give a plunger movement which is proportional to the size of the current.

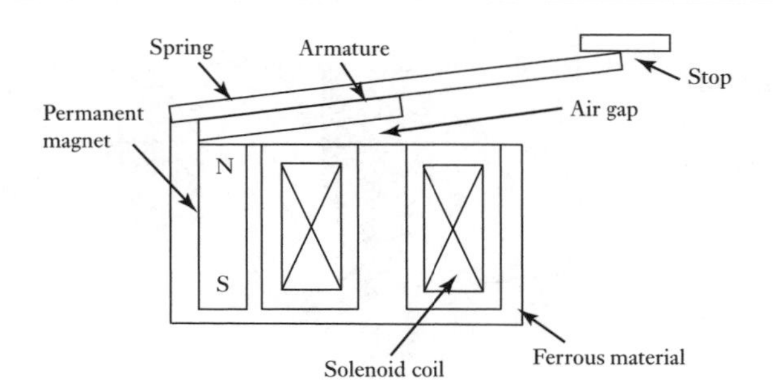

Figure 9.14 A latching solenoid actuator.

Solenoid actuators can be made to be latching, i.e. retain their actuated position when the solenoid current is switched off. Figure 9.14 illustrates this. A permanent magnet is added so that when there is no current through the solenoid it is not strong enough to pull the armature against its retaining spring into the closed position. However, when there is a current through the solenoid to give a magnetic field in the same direction as the permanent magnet then the armature is pulled into the closed position. When the current through the solenoid is switched off, the permanent magnet is strong enough to retain the armature in its closed position. To open it, the current through the solenoid has to be reversed to give a magnetic field in the opposite direction to that of the permanent magnet. Such a solenoid actuator can thus be used to switch on some device and leave it switched on until the reverse current signal is received.

9.5 D.C. motors

Electric motors are frequently used as the final control element in positional or speed control systems. Motors can be classified into two main categories: d.c. motors and a.c. motors, most motors used in modern control systems being d.c. motors. D.C. motors can be divided into two main groups, those using brushes to make contact with a commutator ring assembly on the rotor to switch the current from one rotor winding to another and the brushless type. With the brush type of motor, the rotor has the coil winding and the stator can be either a permanent magnet or an electromagnet. With the brushless type, the arrangement is reversed in that the rotor is a permanent magnet and the stator has the coil winding.

9.5.1 Brush-type d.c. motor

A **brush-type d.c. motor** is essentially a coil of wire which is free to rotate, and so termed the rotor, in the field of a permanent magnet or an electromagnet, the magnet being termed the stator since it is stationary (Figure 9.15(a)). When a current is passed through the coil, the resulting forces acting on its sides at right angles to the field cause forces to act on those sides to give rotation. However, for the rotation to continue, when the coil passes through the vertical position the current direction through the coil has to be reversed and this is achieved by the use of brushes making contact with a split-ring commutator, the commutator rotating with the coil.

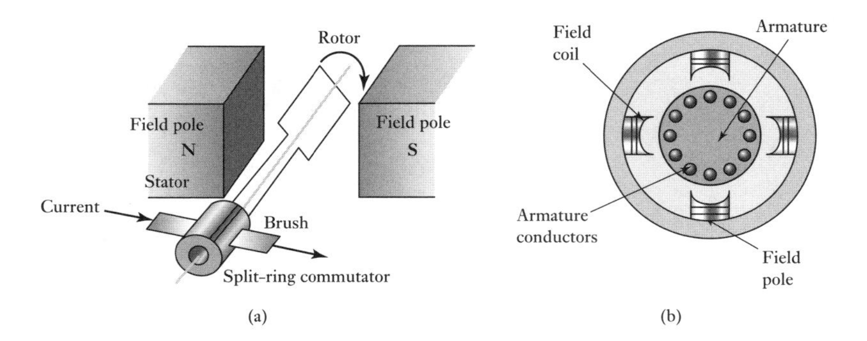

Figure 9.15 D.C. motor: (a) basics, (b) with two sets of poles.

In the conventional d.c. motor, coils of wire are mounted in slots on a cylinder of magnetic material called the **armature**. The armature is mounted on bearings and is free to rotate. It is mounted in the magnetic field produced by **field poles**. These may be, for small motors, permanent magnets or electromagnets with their magnetism produced by a current through the **field coils**. Figure 9.15(b) shows the basic principle of a four-pole d.c. motor with the magnetic field produced by current-carrying coils. The ends of each armature coil are connected to adjacent segments of a segmented ring called the commutator with electrical contacts made to the segments through carbon contacts called brushes. As the armature rotates, the commutator reverses the current in each coil as it moves between the field poles. This is necessary if the forces acting on the coil are to remain acting in the same direction and so the rotation continue. The direction of rotation of the d.c. motor can be reversed by reversing either the armature current or the field current.

Consider a permanent magnet d.c. motor, the permanent magnet giving a constant value of flux density. For an armature conductor of length L and carrying a current i, the force resulting from a magnetic flux density B at right angles to the conductor is BiL (Fig. 9.16(a)). The forces result in a torque T about the coil axis of Fb, with b being the breadth of the coil. Thus:

$$\text{Torque on a armature turn } T = BbLi = \Phi i$$

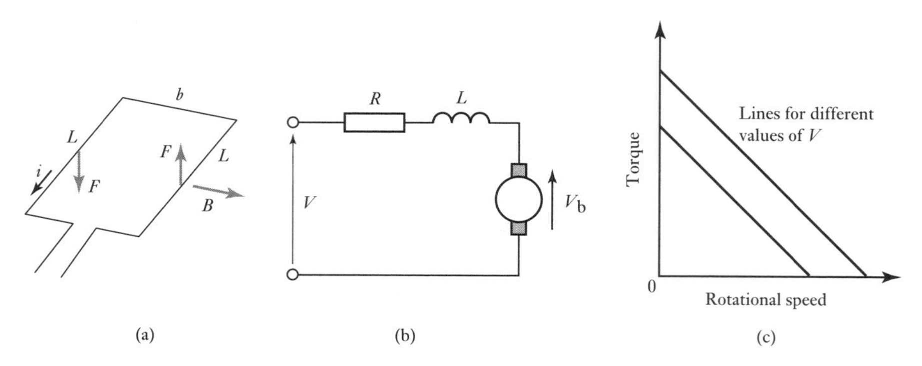

Figure 9.16 D.C. motor: (a) forces on armature, (b) equivalent circuit, (c) torque–speed characteristic.

where Φ is the flux linked per armature turn. In practice there will be more than one armature turn and more than one set of poles, so we can write

$$\text{torque } T = k_t \Phi i$$

and k_t is a constant. The equation can be written as $T = K_t i$ where K_t is termed the torque constant for a motor. Since an armature coil is rotating in a magnetic field, electromagnetic induction will occur and a back e.m.f. will be induced. The back e.m.f. v_b is proportional to the rate at which the flux linked by the coil changes and hence, for a constant magnetic field, is proportional to the angular velocity ω of the rotation. Thus:

$$\text{back e.m.f. } v_b = k_v \Phi \omega$$

where k_v is a constant. The equation can be written as $v_b = K_v \omega$ where K_v is the back e.m.f. constant for a motor.

We can consider a d.c. motor to have the equivalent circuit shown in Figure 9.16(b), i.e. the armature coil being represented by a resistor R in series with an inductance L in series with a source of back e.m.f. If we neglect the inductance of the armature coil then the voltage providing the current i through the resistance is the applied voltage V minus the back e.m.f., i.e. $V - v_b$. Hence:

$$i = \frac{V - v_b}{R} = \frac{V - k_v \Phi \omega}{R} = \frac{V - K_v \omega}{R}$$

The torque T is thus:

$$T = k_t \Phi i = \frac{k_t \Phi}{R}(V - k_v \Phi \omega) = \frac{K_t}{R}(V - K_v \omega)$$

Graphs of the torque against the rotational speed ω are a series of straight lines for different voltage values (Fig. 9.16(c)). The starting torque, i.e. the torque when $\omega = 0$, is, when putting this zero value in the derived equations, $K_t V/R$ and is thus proportional to the applied voltage, and the starting current is V/R. The torque decreases with increasing speed. If a permanent magnet motor developed a torque of 6 N m with an armature current of 2 A, then, as $T = K_t i$, the torque developed with a current of 1 A would be 3 N m.

The speed of a permanent magnet motor depends on the current through the armature coil and thus can be controlled by changing the armature current. The electrical power converted to mechanical power developed by a motor when operating under steady-state conditions is the product of the torque and the angular velocity. The power delivered to the motor in steady-state conditions is the sum of the power loss through the resistance of the armature coil and the mechanical power developed.

As an example, a small permanent magnet motor S6M41 by PMI Motors has $K_t = 3.01$ N cm/A, $K_V = 3.15$ V per thousand rev/min, a terminal resistance of 1.207 Ω and an armature resistance of 0.940 Ω.

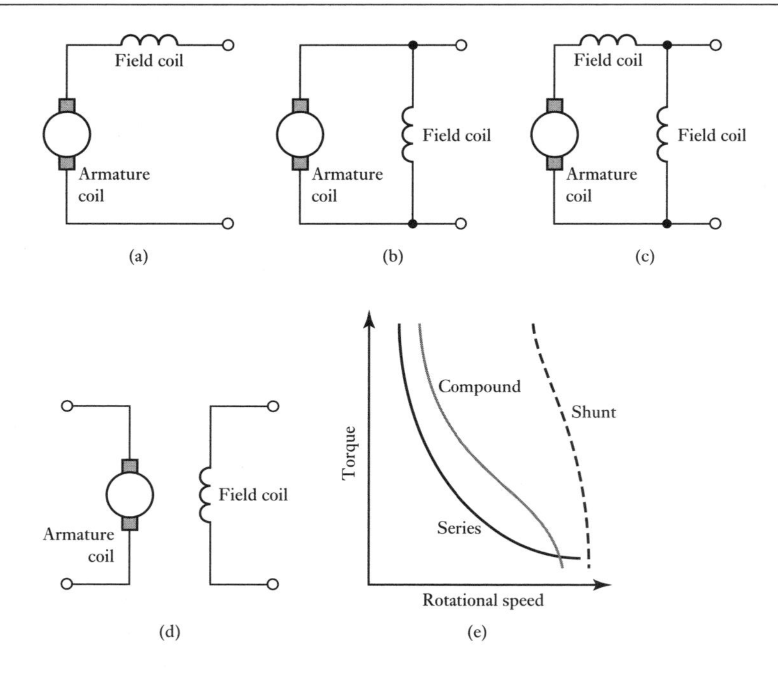

Figure 9.17 D.C. motors: (a) series, (b) shunt, (c) compound, (d) separately wound, (e) torque–speed characteristics.

9.5.2 Brush-type d.c. motors with field coils

D.C. motors with field coils are classified as series, shunt, compound and separately excited according to how the field windings and armature windings are connected (Figure 9.17).

1 *Series wound motor* (Figure 9.17(a))
With the series-wound motor, the armature and field coils are in series and thus carry the same current. The flux Φ depends on the armature current i_a and so the torque acting on the armature is $k_t\Phi i_a = ki_a^2$. At start-up, when $\omega = 0$, $i_a = V/R$ and so the starting torque $= k(V/R)^2$. As such motors have a low resistance, they have a high starting torque and high no-load speed. As the speed is increased, so the torque decreases. Since Ri is small, $V = v_b + Ri \simeq v_b$ and so, since $v_b = k_v\Phi\omega$ and Φ is proportional to i, we have V proportional to $i\omega$. To a reasonable approximation V is constant and so the speed is inversely proportional to the current. The speed thus drops quite markedly when the load is increased. Reversing the polarity of the supply to the coils has no effect on the direction of rotation of the motor; it will continue rotating in the same direction since both the field and armature currents have been reversed. Such d.c. motors are used where large starting torques are required. With light loads there is a danger that a series-wound motor might run at too high a speed.

2 *Shunt-wound motor* (Figure 9.17(b))
With the shunt-wound motor, the armature and field coils are in parallel. It provides the lowest starting torque and a much lower no-load speed and has good speed regulation. The field coil is wound with many turns of fine wire and so has a much larger resistance than the armature coil. Thus, with a constant supply voltage, the field current is virtually constant. The

torque at start-up is $k_t V/R$ and thus it provides a low starting torque and a low no-load speed. With V virtually constant, the motor gives almost constant speed regardless of load and such motors are very widely used because of this characteristic. To reverse the direction of rotation, either the armature or field supplied must be reversed.

3 *Compound motor* (Figure 9.17(c))
The compound motor has two field windings, one in series with the armature and one in parallel. The aim is to get the best features of the series- and shunt-wound motors, namely a high starting torque and good speed regulation.

4 *Separately excited motor* (Figure 9.17(d))
The separately excited motor has separate control of the armature and field currents and can be considered to be a special case of the shunt-wound motor.

The speed of such d.c. motors can be changed by changing either the armature current or the field current. Generally it is the armature current that is varied. This can be done by a series resistor. However, this method is very inefficient since the controller resistor consumes large amounts of power. An alternative is to control the armature voltage (see Section 9.5.3). D.C. motors develop a torque at standstill and so are self-starting. They can, however, require a starting resistance to limit the starting current as the starting current $i = (V - v_b)/R$. Since there is initially no back e.m.f. v_b to limit the current, the starting current can be very large.

The choice of motor will depend on its application. For example, with a robot manipulator, the robot wrist might use a series-wound motor because the speed decreases as the load increases. A shunt-wound motor would be used where a constant speed was required, regardless of the load.

9.5.3 Control of brush-type d.c. motors

The speed of a permanent magnet motor depends on the current through the armature coil. With a field coil motor the speed can be changed by varying either the armature current or the field current; generally it is the armature current that is varied. Thus speed control can be obtained by controlling the voltage applied to the armature. However, because fixed voltage supplies are often used, a variable voltage is obtained by an electronic circuit.

With an a.c. supply, the thyristor circuit of Figure 9.4(b) can be used to control the average voltage applied to the armature. However, we are often concerned with the control of d.c. motors by means of control signals emanating from microprocessors. In such cases the technique known as **pulse width modulation** (PWM) is generally used. This basically involves taking a constant d.c. supply voltage and chopping it so that the average value is varied (Figure 9.18).

Figure 9.19(a) shows how PWM can be obtained by means of a basic transistor circuit. The transistor is switched on or off by means of a signal applied to its base. The diode is to provide a path for current which arises when the transistor is off as a result of the motor acting as a generator. Such a circuit can only be used to drive the motor in one direction; a circuit (Figure 9.19(b))

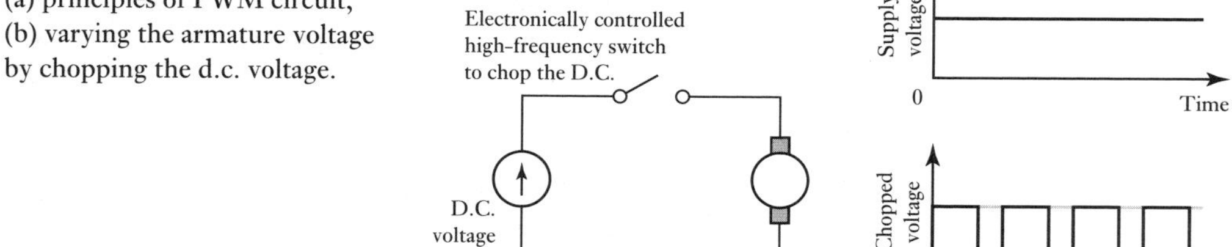

Figure 9.18 PWM:
(a) principles of PWM circuit,
(b) varying the armature voltage by chopping the d.c. voltage.

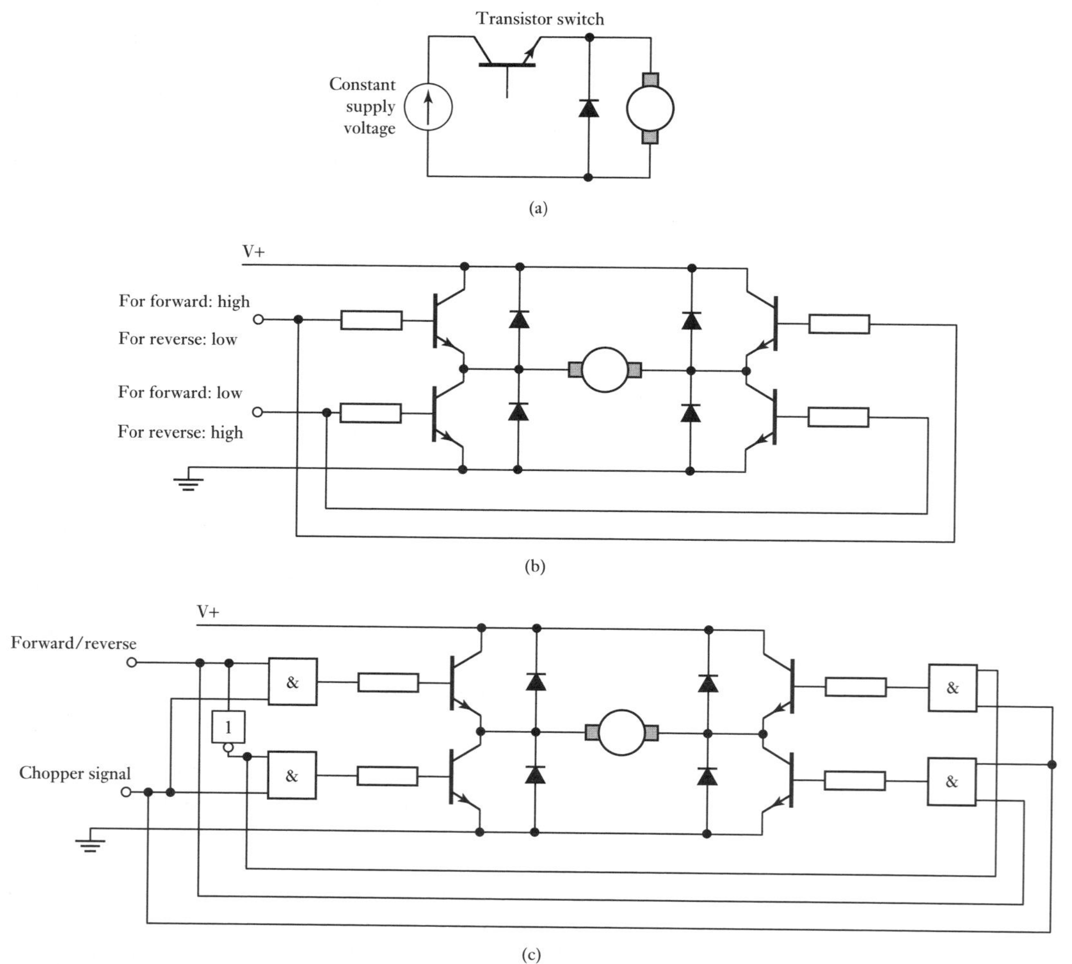

Figure 9.19 (a) Basic transistor circuit, (b) H-circuit, (c) H-circuit with logic gates.

involving four transistors, termed an H circuit, can be used to enable the motor to be operated in forward and reverse directions. This circuit can be modified by the use of logic gates so that one input controls the switching and one the direction of rotation (Figure 9.19(c)).

The above are examples of open-loop control; this assumes that conditions will remain constant, e.g. the supply voltage and the load driven by the motor. Closed-loop control systems use feedback to modify the motor speed if conditions change. Figure 9.20 shows some of the methods that might be employed.

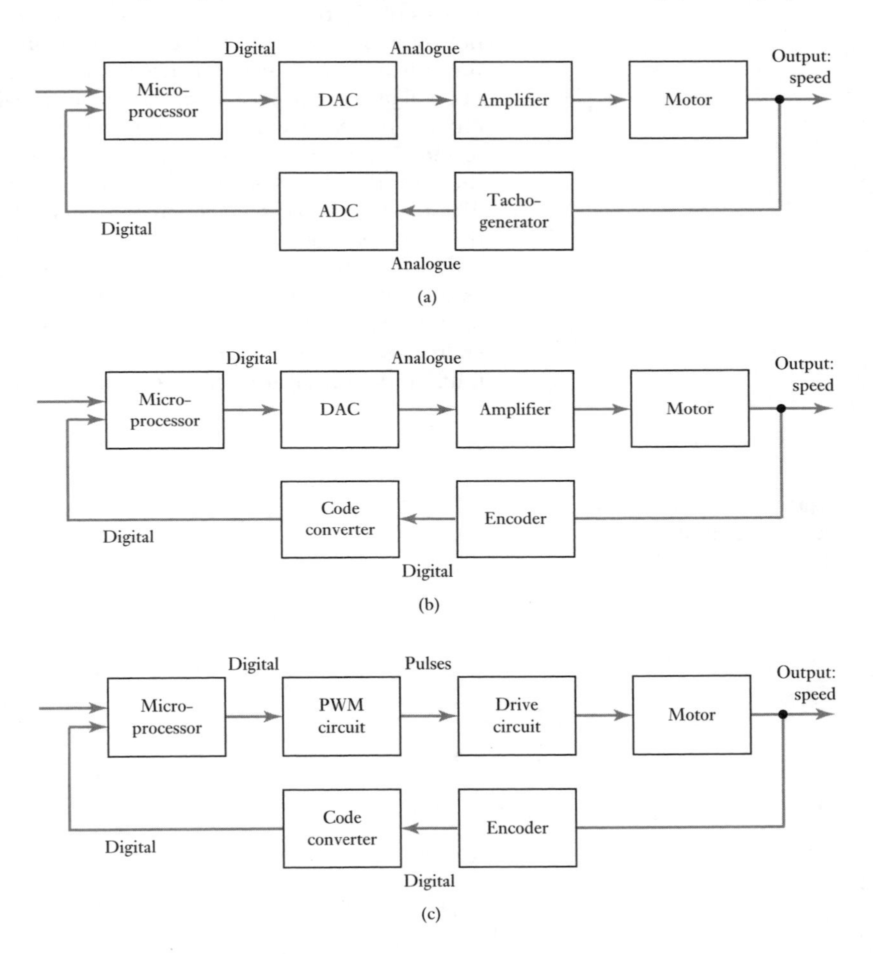

Figure 9.20 Speed control with feedback.

In Figure 9.20(a) the feedback signal is provided by a tachogenerator, this giving an analogue signal which has to be converted to a digital signal by an ADC for input to the microprocessor. The output from the microprocessor is converted to an analogue signal by an ADC and used to vary the voltage applied to the armature of the d.c. motor. In Figure 9.20(b) the feedback signal is provided by an encoder, this giving a digital signal which after code conversion can be directly inputted to the microprocessor. As in (a), the system shows an analogue voltage being varied to control the motor speed. In Figure 9.20(c) the system is completely digital and PWM is used to control the average voltage applied to the armature.

9.5.4 Brushless permanent magnet d.c. motors

A problem with d.c. motors is that they require a commutator and brushes in order periodically to reverse the current through each armature coil. The brushes make sliding contacts with the commutator and as a consequence sparks jump between the two and they suffer wear. Brushes thus have to be periodically changed and the commutator resurfaced. To avoid such problems **brushless motors** have been designed.

Essentially they consist of a sequence of stator coils and a permanent magnet rotor. A current-carrying conductor in a magnetic field experiences a force; likewise, as a consequence of Newton's third law of motion, the magnet will also experience an opposite and equal force. With the conventional d.c. motor, the magnet is fixed and the current-carrying conductors made to move. With the brushless permanent magnet d.c. motor, the reverse is the case: the current-carrying conductors are fixed and the magnet moves. The rotor is a ferrite or ceramic permanent magnet. Figure 9.21(a) shows the basic form of such a motor. The current to the stator coils is electronically switched by transistors in sequence round the coils, the switching being controlled by the position of the rotor so that there are always forces acting on the magnet causing it to rotate in the same direction. Hall sensors are generally used to sense the position of the rotor and initiate the switching by the transistors, the sensors being positioned around the stator.

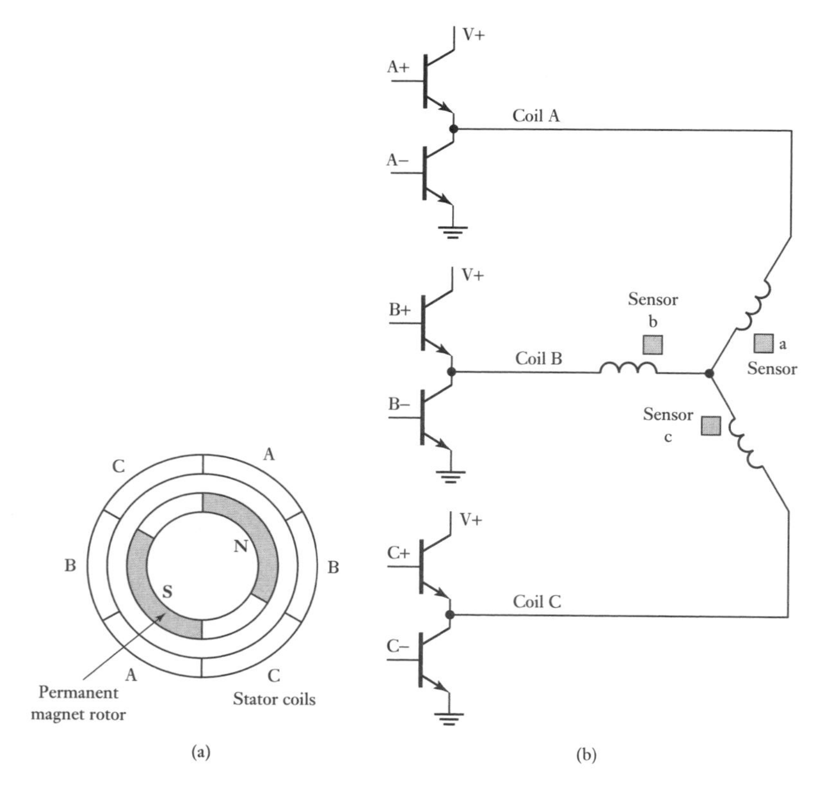

Figure 9.21 (a) Brushless permanent magnet motor, (b) transistor switching.

Figure 9.21(b) shows the transistor switching circuits that might be used with the motor shown in Figure 9.21(a). To switch the coils in sequence we need to supply signals to switch the transistors on in the right sequence. This is provided by the outputs from the three sensors operating through a decoder circuit to give the appropriate base currents. Thus when the rotor is in the vertical position, i.e. 0°, there is an output from sensor c but none from a and b. This is used to switch on transistors A+ and B−. For the rotor in the 60° position there are signals from the sensors b and c and transistors A+ and C− are switched on. Table 9.1 shows the entire switching sequence. The entire circuit for controlling such a motor is available on a single integrated circuit.

Table 9.1 Switching sequence.

Rotor position	Sensor signals			Transistors on	
	a	b	c		
0°	0	0	1	A+	B−
60°	0	1	1	A+	C−
120°	0	1	0	B+	C−
180°	1	1	0	B+	A−
240°	1	0	0	C+	A−
360°	1	0	1	C+	B−

Brushless permanent magnet d.c. motors are becoming increasingly used in situations where high performance coupled with reliability and low maintenance are essential. Because of their lack of brushes, they are quiet and capable of high speeds.

9.6 A.C. motors

A.C. motors can be classified into two groups, single-phase and polyphase, with each group being further subdivided into induction and synchronous motors. Single-phase motors tend to be used for low-power requirements while polyphase motors are used for higher powers. Induction motors tend to be cheaper than synchronous motors and are thus very widely used.

The **single-phase squirrel-cage induction motor** consists of a squirrel-cage rotor, this being copper or aluminium bars that fit into slots in end rings to form complete electric circuits (Figure 9.22(a)). There are no external electrical connections to the rotor. The basic motor consists of this rotor with a stator having a set of windings. When an alternating current passes through the stator windings an alternating magnetic field is produced. As a result of electromagnetic induction, e.m.f.s are induced in the conductors of the rotor and currents flow in the rotor. Initially, when the rotor is stationary, the forces on the current-carrying conductors of the rotor in the magnetic field of the stator are such as to result in no net torque. The motor is not self-starting. A number of methods are used to make the motor self-starting and give this initial impetus to start it; one is to use an auxiliary starting winding to give the rotor an initial push. The rotor rotates at a speed determined by the frequency of the alternating current applied to the stator. For a constant frequency supply to a two-pole single-phase motor the

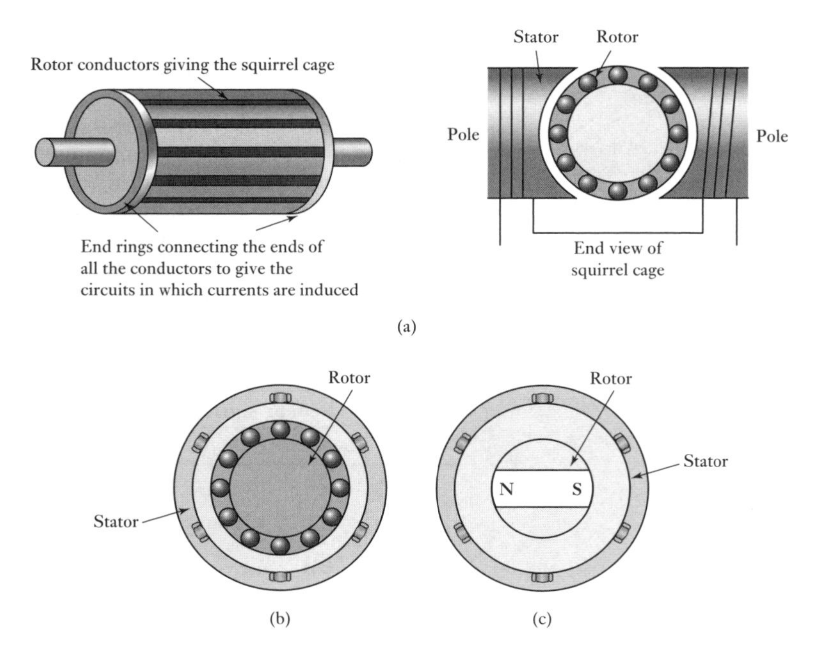

Figure 9.22 (a) Single-phase induction motor, (b) three-phase induction motor, (c) three-phase synchronous motor.

magnetic field will alternate at this frequency. This speed of rotation of the magnetic field is termed the **synchronous speed**. The rotor will never quite match this frequency of rotation, typically differing from it by about 1 to 3%. This difference is termed **slip**. Thus for a 50 Hz supply the speed of rotation of the rotor will be almost 50 revolutions per second.

The **three-phase induction motor** (Figure 9.22(b)) is similar to the single-phase induction motor but has a stator with three windings located 120° apart, each winding being connected to one of the three lines of the supply. Because the three phases reach their maximum currents at different times, the magnetic field can be considered to rotate round the stator poles, completing one rotation in one full cycle of the current. The rotation of the field is much smoother than with the single-phase motor. The three-phase motor has a great advantage over the single-phase motor of being self-starting. The direction of rotation is reversed by interchanging any two of the line connections, this changing the direction of rotation of the magnetic field.

Synchronous motors have stators similar to those described above for induction motors but a rotor which is a permanent magnet (Figure 9.22(c)). The magnetic field produced by the stator rotates and so the magnet rotates with it. With one pair of poles per phase of the supply, the magnetic field rotates through 360° in one cycle of the supply and so the frequency of rotation with this arrangement is the same as the frequency of the supply. Synchronous motors are used when a precise speed is required. They are not self-starting and some system has to be employed to start them.

A.C. motors have the great advantage over d.c. motors of being cheaper, more rugged, reliable and maintenance free. However, speed control is generally more complex than with d.c. motors and as a consequence a speed-controlled d.c. drive generally works out cheaper than a speed-controlled a.c. drive, though the price difference is steadily dropping as a result of technological developments and the reduction in price of solid-state devices. Speed control of a.c. motors is based around the provision of a variable frequency supply, since the speed of such motors is determined by the frequency of the supply. The torque developed by an a.c. motor is constant when the ratio of the applied stator voltage to frequency is constant. Thus to maintain a constant torque at the different speeds when the frequency is varied, the voltage applied to the stator has also to be varied. With one method, the alternating current is first rectified to direct current by a **converter** and then **inverted** back to alternating current again but at a frequency that can be selected (Figure 9.23). Another method that is often used for operating slow-speed motors is the **cycloconverter**. This converts alternating current at one frequency directly to alternating current at another frequency without the intermediate d.c. conversion.

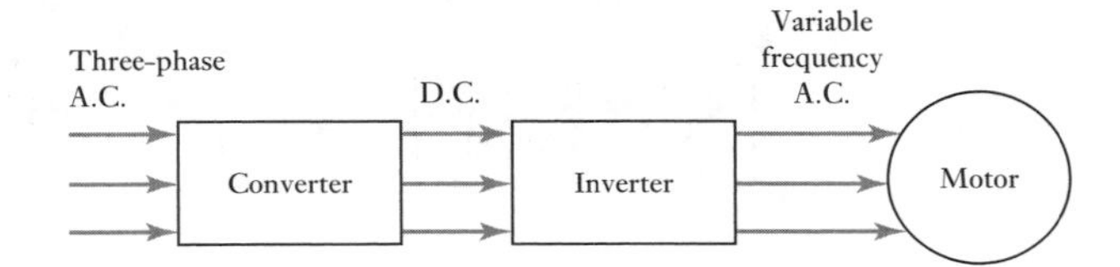

Figure 9.23 Variable speed a.c. motor.

9.7 Stepper motors

The **stepper motor** is a device that produces rotation through equal angles, the so-called **steps**, for each digital pulse supplied to its input. Thus, for example, if with such a motor 1 pulse produces a rotation of 6° then 60 pulses will produce a rotation through 360°. There are a number of forms of stepper motor:

1 *Variable reluctance stepper*
   Figure 9.24 shows the basic form of the variable reluctance stepper motor. With this form the rotor is made of soft steel and is cylindrical with four poles, i.e. fewer poles than on the stator. When an opposite pair of windings has current switched to them, a magnetic field is produced with lines of force which pass from the stator poles through the nearest set of poles on the rotor. Since lines of force can be considered to

Figure 9.24 Variable reluctance stepper motor.

This pair of poles energised by current being switched to them and rotor rotates to next position

Stator

S

N

Rotor

N

S

This pair of poles energised by current being switched to them to give next step

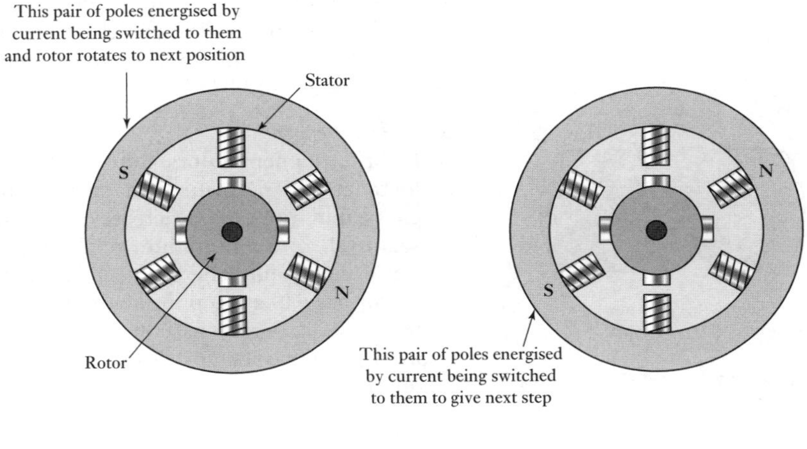

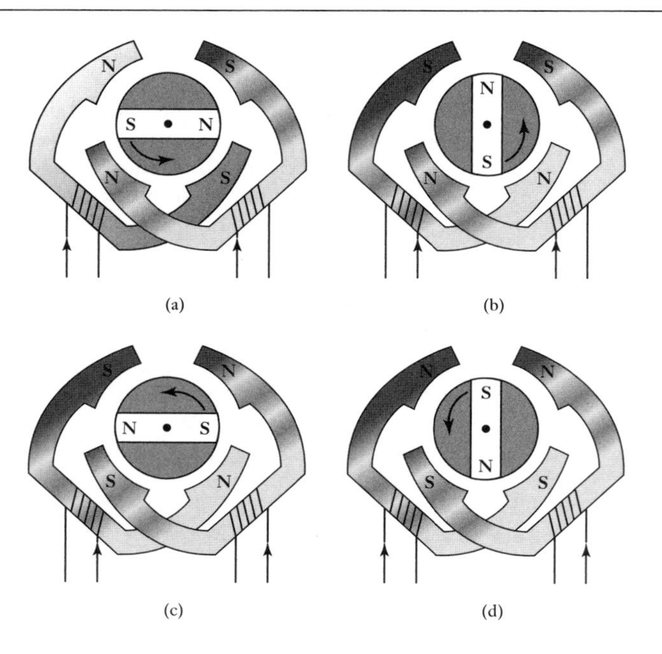

Figure 9.25 Permanent magnet two-phase stepper motor with 90° steps. (a), (b), (c) and (d) show the positions of the magnet rotor as the coils are energised in different directions.

be rather like elastic thread and always trying to shorten themselves, the rotor will move until the rotor and stator poles line up. This is termed the position of minimum reluctance. This form of stepper generally gives step angles of 7.5° or 15°.

2 *Permanent magnet stepper*
Figure 9.25 shows the basic form of the permanent magnet motor. The motor shown has a stator with four poles. Each pole is wound with a field winding, the coils on opposite pairs of poles being in series. Current is supplied from a d.c. source to the windings through switches. The rotor is a permanent magnet and thus when a pair of stator poles has a current switched to it, the rotor will move to line up with it. Thus for the currents giving the situation shown in the figure, the rotor moves to the 45° position. If the current is then switched so that the polarities are reversed, the rotor will move a further 45° in order to line up again. Thus by switching the currents through the coils, the rotor rotates in 45° steps. With this type of motor, step angles are commonly 1.8°, 7.5°, 15°, 30°, 34° or 90°.

3 *Hybrid stepper*
Hybrid stepper motors combine the features of both the variable reluctance and permanent magnet motors, having a permanent magnet encased in iron caps which are cut to have teeth (Figure 9.26). The rotor sets itself in the minimum reluctance position in response to a pair of stator coils being energised. Typical step angles are 0.9° and 1.8°. If a motor has n phases on the stator and m teeth on the rotor, the total number of steps per revolution is nm. Such stepper motors are extensively used in high-accuracy positioning applications, e.g. in computer hard disk drives.

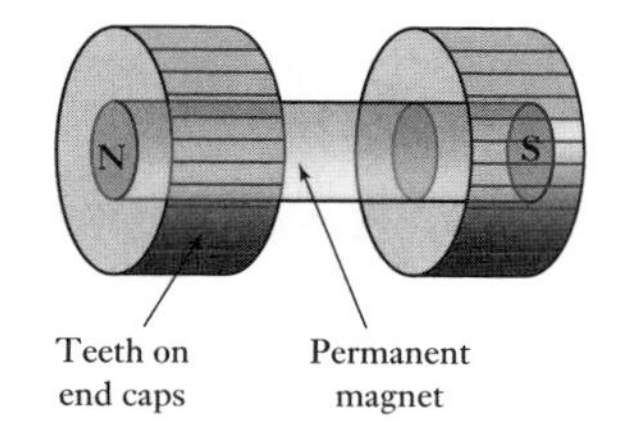

Figure 9.26 Hybrid motor rotor.

9.7.1 Stepper motor specifications

The following are some of the terms commonly used in specifying stepper motors:

1 *Phase*
This term refers to the number of independent windings on the stator, e.g. a four-phase motor. The current required per phase and its resistance and inductance will be specified so that the controller switching output is specified. Two-phase motors, e.g. Figure 9.25, tend to be used in light-duty applications, three-phase motors tend to be variable reluctance steppers, e.g. Figure 9.24, and four-phase motors tend to be used for higher power applications.

2 *Step angle*
This is the angle through which the rotor rotates for one switching change for the stator coils.

3 *Holding torque*
This is the maximum torque that can be applied to a powered motor without moving it from its rest position and causing spindle rotation.

4 *Pull-in torque*
This is the maximum torque against which a motor will start, for a given pulse rate, and reach synchronism without losing a step.

5 *Pull-out torque*
This is the maximum torque that can be applied to a motor, running at a given stepping rate, without losing synchronism.

6 *Pull-in rate*
This is the maximum switching rate at which a loaded motor can start without losing a step.

7 *Pull-out rate*
This is the switching rate at which a loaded motor will remain in synchronism as the switching rate is reduced.

8 *Slew range*
This is the range of switching rates between pull-in and pull-out within which the motor runs in synchronism but cannot start up or reverse.

Figure 9.27 shows the general characteristics of a stepper motor.

9.7.2 Stepper motor control

Solid-state electronics is used to switch the d.c. supply between the pairs of stator windings. Two-phase motors, e.g. Figure 9.25, are termed **bipolar motors** when they have four connecting wires for signals to generate the switching sequence (Figure 9.28(a)). Such a motor can be driven by H-circuits (see Figure 9.19 and the associated discussion); Figure 9.28(b) shows the circuit and Table 9.2 shows the switching sequence required for the transistors to carry out the four steps, the sequence then being repeated for

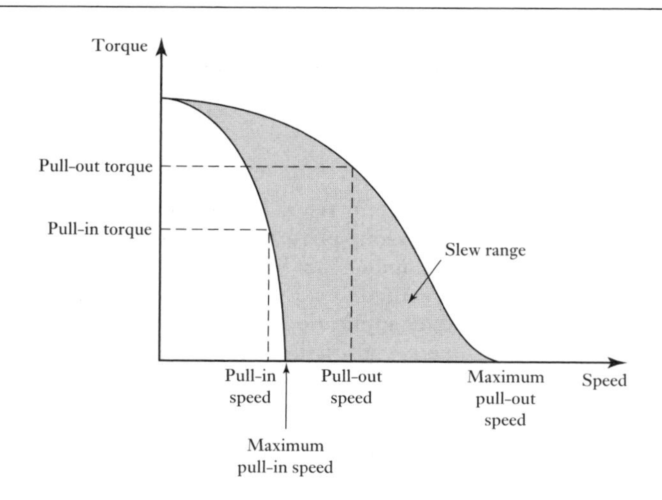

Figure 9.27 Stepper motor characteristics.

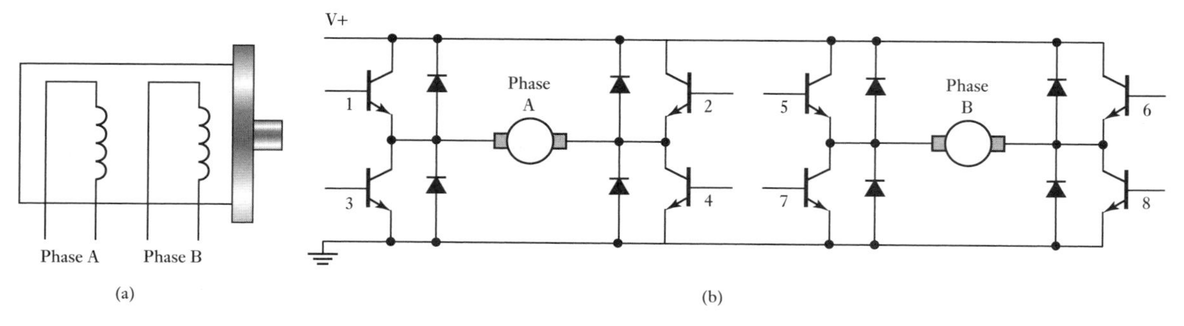

Figure 9.28 (a) Bipolar motor, (b) H-circuit.

Table 9.2 Switching sequence for full-stepping bipolar stepper.

	Transistors			
Step	1 and 4	2 and 3	5 and 8	6 and 7
1	On	Off	On	Off
2	On	Off	Off	On
3	Off	On	Off	On
4	Off	On	On	Off

further steps. The sequence gives a clockwise rotation; for an anti-clockwise rotation the sequence is reversed.

Half-steps, and hence finer resolution, are obtainable if instead of the full-stepping sequence needed to implement a pole reversal to get from one step to the next, the coils are switched so that the rotor stops at a position halfway to the next full step. Table 9.3 shows the sequence for half-stepping with the bipolar stepper.

Two-phase motors are termed **unipolar** when they have six connecting wires for the generation of the switching sequence (Figure 9.29). Each of the coils has a centre-tap. With the centre-taps of the phase coils connected together, such a form of stepper motor can be switched with just

Table 9.3 Half-steps for bipolar stepper.

Step	Transistors 1 and 4	2 and 3	5 and 8	6 and 7
1	On	Off	On	Off
2	On	Off	Off	Off
3	On	Off	Off	On
4	Off	Off	Off	On
5	Off	On	Off	On
6	Off	On	Off	Off
7	Off	On	On	Off
8	Off	Off	On	Off

Table 9.4 Switching sequence for full-stepping unipolar stepper.

Step	Transistors 1	2	3	4
1	On	Off	On	Off
2	On	Off	Off	On
3	Off	On	Off	On
4	Off	On	On	Off

Table 9.5 Half-steps for unipolar stepper.

Step	Transistors 1	2	3	4
1	On	Off	On	Off
2	On	Off	Off	Off
3	On	Off	Off	On
4	Off	Off	Off	On
5	Off	On	Off	On
6	Off	On	Off	Off
7	Off	On	On	Off
8	Off	Off	On	Off

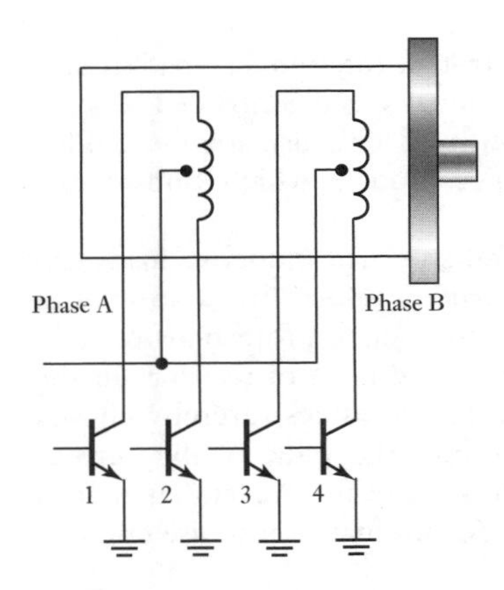

Figure 9.29 Unipolar motor.

four transistors. Table 9.4 gives the switching sequence for the transistors in order to produce the steps for clockwise rotation, the sequence then being repeated for further steps. For anti-clockwise rotation the sequence is reversed. Table 9.5 shows the sequence when the unipolar is half-stepping.

Integrated circuits are available to provide the drive circuitry. Figure 9.30 shows the connections with the integrated circuit SAA 1027 for a four-phase stepper. The three inputs are controlled by applying high or low signals to them. When the set terminal is held high, the output from the integrated circuit changes state each time the trigger terminal goes from low to high. The sequence repeats itself at four-step intervals but can be reset to the zero condition at any time by applying a low signal to the trigger terminal. When the rotation input is held low there is clockwise rotation, when high it is anti-clockwise.

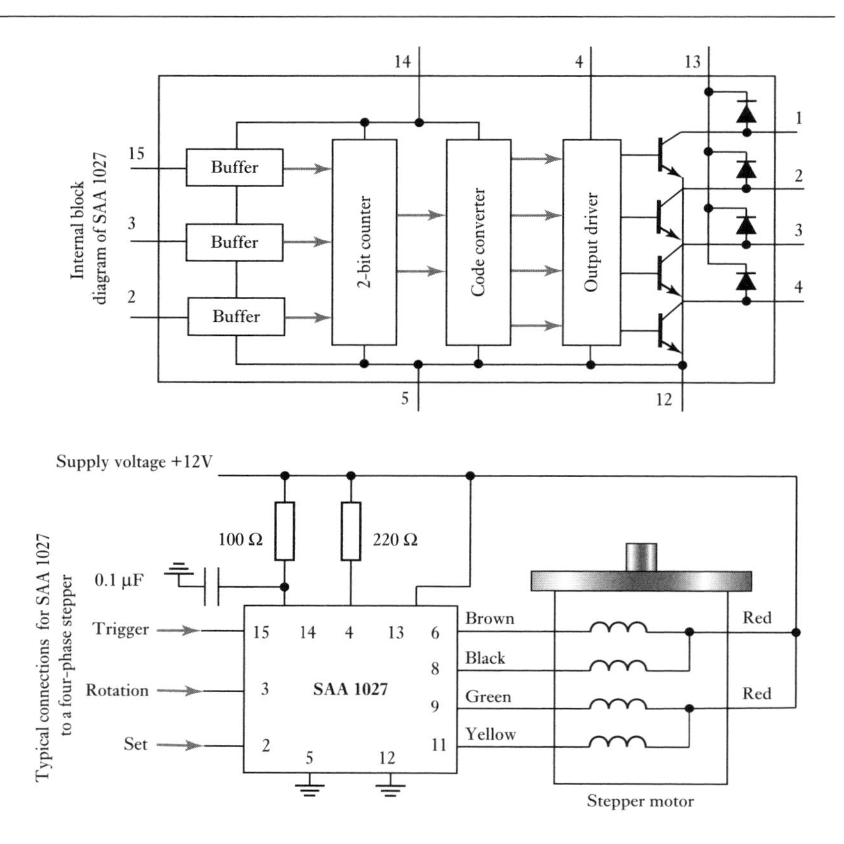

Figure 9.30 Integrated circuit SAA 1027 for stepper motor.

Some applications require very small step angles. While the step angle can be made small by increasing the number of rotor teeth and/or the number of phases, generally more than four phases and 50 to 100 teeth are not used. Instead a technique known as **mini-stepping** is used. This involves dividing each step into a number of equal size substeps. This is done by using different currents to the coils so that the rotor moves to intermediate positions between normal step positions. Thus, for example, a step of 1.8° might be subdivided into 10 equal steps.

Stepper motors can be used to give controlled rotational steps but also can give continuous rotation with their rotational speed controlled by controlling the rate at which pulses are applied to it to cause stepping. This gives a very useful controlled variable speed motor which finds many applications.

Because stepper coils have inductance and switched inductive loads can generate large back e.m.f.s when switched, when steppers are connected to microprocessor output ports it is necessary to include protection to avoid damage to the microprocessor. This may take the form of resistors in the lines to limit the current, though these must have values carefully chosen both to provide the protection and not to limit the value of the current needed to switch the transistors. Diodes across the coils prevent current in the reverse direction and so give protection. An alternative is to use optoisolators (see Section 3.3).

9.7.3 Selection of a stepper motor

Selection of a stepper motor should be based on a consideration of:

1 The operating torque requirements of the application. The rating torque must be high enough to accommodate the torque and slew range requirement. Also the torque-speed characteristics must be appropriate.
2 The step angle must be of high enough resolution to provide the required output motion increments.
3 Cost.

This will require looking at the data specifications for stepper motors. The following are some typical values taken from a manufacturer's data sheet for a unipolar stepper motor (Canon 42M048C1U-N):

D.C. operating voltage	5 V
Resistance per winding	9.1 Ω
Inductance per winding	8.1 mH
Holding torque	66.2 mNm/9.4 oz. in
Rotor moment of inertia	12.5×10^{-4} gm^2
Detent torque	12.7 mNm/1.8 oz. in
Step angle	7.5°
Step angle tolerance	± 0.5°
Steps per revolution	48

The detent torque is the torque required to rotate the stepper motor when the motor windings are not energised.

Once a motor has been selected, a drive system will need to be found which is compatible with the motor. For example, for use with a unipolar motor the Cybernetics CY512 might be used if a maximum input voltage of 7 V and maximum current per phase of 80 mA is acceptable. The SAA1027 by Signetics is a widely used driver for small unipolar stepper motors with a maximum input voltage of 18 V and a maximum current per phase of 350 mA. For a two-phase bipolar or four-phase unipolar motor, the SCS-Thomson L297/L298 could be considered, it being a two-chip logic driver set. The L297 chip generates the motor phase sequences of four-phase TTL logic signals for two-phase and four-phase unipolar motors and the L298 is a bridge driver designed to accept such signals and drive inductive loads, in this case a stepper. A bipolar motor can be driven with winding currents up to 2 A.

When a pulse is supplied to a stepper motor we have essentially an input to an inductor–resistor circuit and the resulting torque is applied to the load, resulting in angular acceleration. As a consequence, the system will have a natural frequency; it will not go directly to the next step position but will generally have damped oscillations about it before settling down to the steady-state value (Figure 9.31). See Section 24.1.2 for a discussion of this and a derivation of the natural frequency and damping factor.

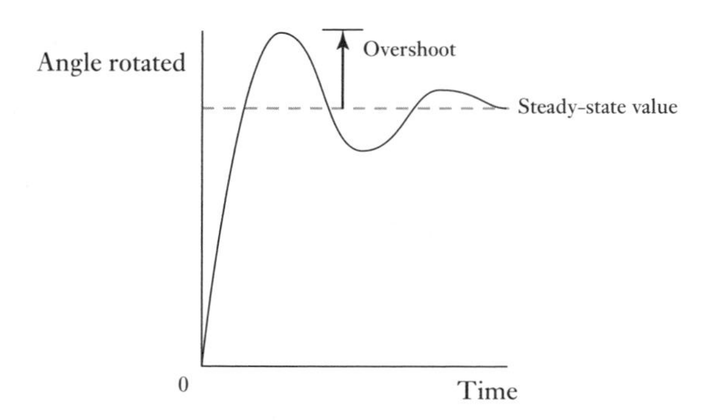

Figure 9.31 Oscillations about the steady-state angle.

9.8 Motor selection

When selecting a motor for a particular application, factors that need to be taken into account include:

1 Inertia matching
2 Torque requirements
3 Power requirements

9.8.1 Inertia matching

The concept of impedance matching introduced in Section 3.8 for electrical impedances can be extended to mechanical systems, and an analogous situation to that described there for electrical circuits is that of a motor, a torque source, directly rotating a load (Figure 9.32(a)). The torque required to give a load with moment of inertia I_L and angular acceleration α is $I_L\alpha$. The torque required to accelerate the motor shaft is $T_M = I_M\alpha_M$ and that required to accelerate the load is $T_L = I_L\alpha_L$. The motor shaft will, in the absence of gearing, have the same angular acceleration and the same angular velocity. The power needed to accelerate the system as a whole is $T_M\omega + T_L\omega$, where ω is the angular velocity. Thus:

$$\text{power} = (I_M + I_L)\alpha\omega$$

This power is produced by the motor torque T_M and thus the power must equal $T_M\omega$. Hence:

$$T = (I_M + I_L)\alpha$$

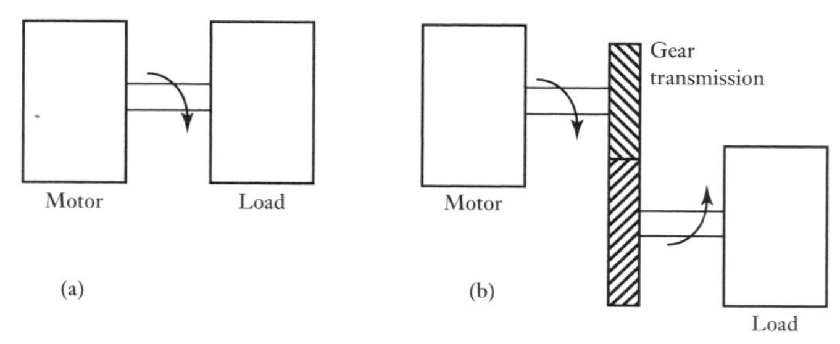

Figure 9.32 (a) Motor rotating load, (b) motor with gear transmission rotating load.

The torque to obtain a given angular acceleration will be minimised when $I_M = I_L$. Thus, for maximum power transfer, the moment of inertia of the load should be similar to that of the motor.

For the situation where the motor rotates the load through a gear transmission (Figure 9.32(b), the condition for maximum power transfer is that the moment of inertia of the motor equals the reflected moment of inertia of the load, this being $n^2 I_L$, where n is the gear ratio and I_L the moment of inertia of the load (see Section 10.2.2).

Thus for maximum power transfer, the moment of inertia of the motor should match that of the load or the reflected load when gears are used. This will mean that the torque to obtain a given acceleration will be minimised. This is particularly useful if the motor is being used for fast positioning. With a geared system, adjustment of the gear ratio can be used to enable matching to be obtained.

9.8.2 Torque requirements

Figure 9.33 shows the operating curves for a typical motor. For continuous running, the stall torque value should not be exceeded. This is the maximum torque value at which overheating will not occur. For intermittent use, greater torques are possible. As the angular speed is increased so the ability of the motor to deliver torque diminishes. Thus if higher speeds and torques are required than can be given by a particular motor, a more powerful motor needs to be selected.

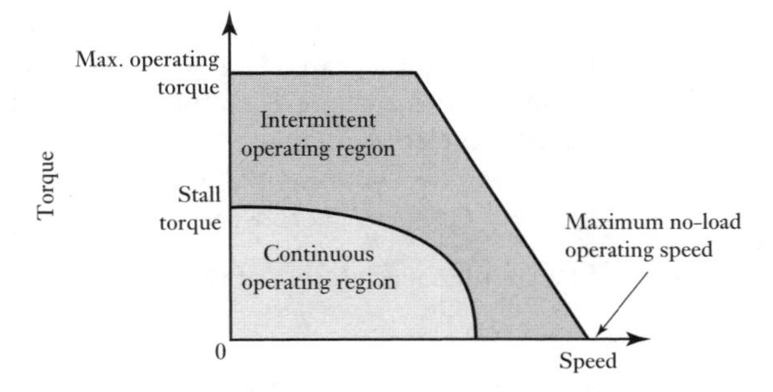

Figure 9.33 Torque–speed graph

Suppose we require a motor to operate a drum-type hoist and lift a load (Figure 9.34). With a drum diameter of, say, 0.5 m and a maximum load m of 1000 kg, then the tension in the cable will be $mg = 1000 \times 9.81 = 9810$ N. The torque at the drum will be $9810 \times 0.25 = 24\,525$ Nm or about 2.5 kNm. If the hoist is operating at a constant speed v of 0.5 m/s then the drum angular velocity ω is $v/r = 0.5/0.25 = 2$ rad/s or $2/2\pi = 0.32$ revs/s. The motor is driving the shaft through a gear. We might decide that the gear ratio should be such that the maximum motor speed should be about 1500 rev/min or 25 rev/s. This means a gear ratio n of 25/0.32 or near enough 80:1.The load torque on the motor will be reduced by a factor of 80 from that on the drum and so will be $2500/80 = 31.25$ Nm. If we allow for some friction in the gear then the maximum torque on the motor which we should allow for is about 35 Nm.

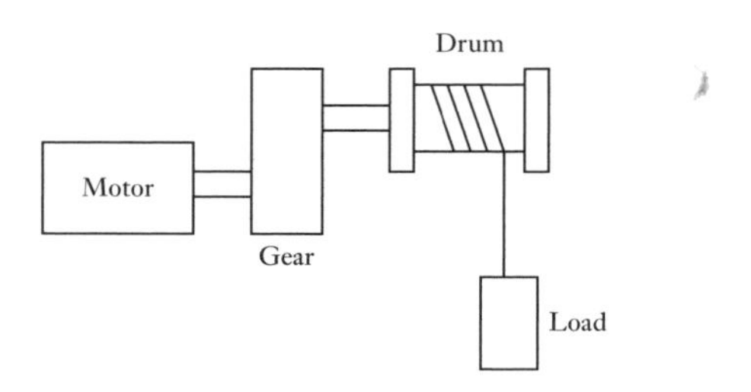

Figure 9.34 A motor lifting a load.

However, this is only the maximum torque when the load is being lifted at a constant speed. We need to add to this the torque needed to accelerate the load from rest to the speed of 0.5 m/s. If, say, we want to reach this speed from rest in 1 s then the accelerating torque needed is $I\alpha$, where I is the moment of inertia and α the angular acceleration. The effective moment of inertia of the load as seen by the motor via the gear is $(1/n^2) \times$ the moment of inertia of the load mr^2 and so is $(1/80)^2 \times 1000 \times 0.25^2 = 0.0098$ kg m^2 or about 0.01 kg m^2. The referred moment of inertia of the drum and gear might add 0.02 kg m^2. To find the total moment of inertia involved in the lifting we also have to add to this the moment of inertia of the motor. Manufacturers' data sheets might give a value of, say, 0.02 kg m^2 and so the total moment of inertia involved in the lifting might be $0.01 + 0.02 + 0.02 = 0.05$ kg m^2. The motor speed is required to rise from 0 to 25 rev/s in 1 s so the angular acceleration is $(25 \times 2\pi)/1 = 157$ rad/s^2 or about 160 rad/s^2. The accelerating torque required is thus $0.05 \times 160 = 8$ N m. Hence the maximum torque we have to allow for is that required to lift the load at a constant velocity plus that need to accelerate it to this velocity from rest and so is $35 + 8 = 43$ Nm.

We can write the arguments involved in the above example algebraically as follows. The torque T_m required from a motor is that needed by the load T_L, or T_L/n for a geared load with gear ratio n, and that needed to accelerate the motor $I_m\alpha_m$, where I_m is the moment of inertia of the motor and α_m its angular acceleration:

$$T_m = \frac{T_L}{n} + I_m\alpha_m$$

The angular acceleration of the load α_L is given by

$$\alpha_m = n\alpha_L$$

Because there will be torque T_f required to overcome the load friction, the torque used to accelerate the load will be $(T_L - T_f)$ and so

$$T_L - T_f = I_L\alpha_L$$

Thus we can write

$$T_m = \frac{1}{n}[T_f + \alpha_L(I_L + n^2 I_m)]$$

9.8.3 Power requirements

The motor needs to be able to run at the maximum required velocity without overheating. The total power P required is the sum of the power required to overcome friction and that needed to accelerate the load. As power is the product of torque and angular speed, then the power required to overcome the frictional torque T_f is $T_f\omega$ and that required to accelerate the load with angular acceleration α is $(I_L\alpha)\omega$, where I_L is the moment of inertia of the load. Thus:

$$P = T_f\omega + I_L\alpha\omega$$

Summary

Relays are electrically operated switches in which changing a current in one electric circuit switches a current on or off in another circuit.

A **diode** can be regarded as only passing a current in one direction, the other direction being very high resistance.

A **thyristor** can be regarded as a diode which has a gate controlling the conditions under which the diode can be switched on. A **triac** is similar to the thyristor and is equivalent to a pair of thyristors connected in reverse parallel on the same chip.

Bipolar transistors can be used as switches by switching the base current between zero and a value that drives the transistor into saturation. **MOSFETs** are similar and can also be used for switching.

The basic principle of a **d.c. motor** is of a loop of wire, the armature, which is free to rotate in the field of a magnet as a result of a current passing through the loop. The magnetic field may be provided by a permanent magnet or an electromagnet, i.e. a field coil. The speed of a permanent magnet motor depends on the current through the armature coil; with a field coil motor it depends on either the current through the armature coil or that through the field coil. Such d.c. motors require a commutator and brushes in order periodically to reverse the current through each armature coil. The **brushless permanent magnet d.c. motor** has a permanent magnet rotor and a sequence of stator coils through which the current is switched in sequence.

A.C. motors can be classified into two groups, single-phase and polyphase, with each group being further subdivided into induction and synchronous motors. Single-phase motors tend to be used for low-power requirements while polyphase motors are used for higher powers. Induction motors tend to be cheaper than synchronous motors and are thus very widely used.

The **stepper motor** is a motor that produces rotation through equal angles, the so-called **steps**, for each digital pulse supplied to its input.

Motor selection has to take into account **inertia matching**, and the **torque** and **power requirements**.

Problems

9.1 Explain how the circuit shown in Figure 9.35 can be used to debounce a switch.

9.2 Explain how a thyristor can be used to control the level of a d.c. voltage by chopping the output from a constant voltage supply.

9.3 A d.c. motor is required to have (a) a high torque at low speeds for the movement of large loads, (b) a torque which is almost constant regardless of speed. Suggest suitable forms of motor.

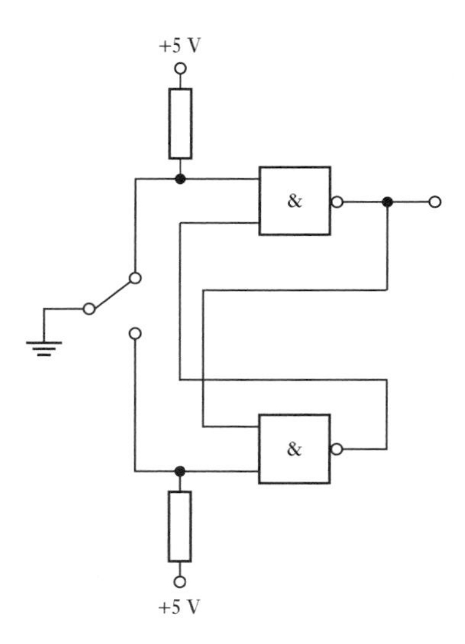

Figure 9.35 Problem 9.1.

9.4 Suggest possible motors, d.c. or a.c., which can be considered for applications where (a) cheap, constant torque operation is required, (b) high controlled speeds are required, (c) low speeds are required, (d) maintenance requirements have to be minimised.

9.5 Explain the principle of the brushless d.c. permanent magnet motor.

9.6 Explain the principles of operation of the variable reluctance stepper motor.

9.7 If a stepper motor has a step angle of 7.5°, what digital input rate is required to produce a rotation of 10 rev/s?

9.8 What will be the step angle for a hybrid stepper motor with eight stator windings and ten rotor teeth?

9.9 A permanent magnet d.c. motor has an armature resistance of 0.5 Ω and when a voltage of 120 V is applied to the motor it reaches a steady-state speed of rotation of 20 rev/s and draws 40 A. What will be (a) the power input to the motor, (b) the power loss in the armature, (c) the torque generated at that speed?

9.10 If a d.c. motor produces a torque of 2.6 N m when the armature current is 2 A, what will be the torque with a current of 0.5 A?

9.11 How many steps/pulses per second will a microprocessor need to output per second to a stepper motor if the motor is to give an output of 0.25 rev/s and has a step angle of 7.5°?

9.12 A stepper motor is used to rotate a pulley of diameter 240 mm and hence a belt which is moving a mass of 200 kg. If this mass is to be accelerated uniformly from rest to 100 mm/s in 2 s and there is a constant frictional force of 20 N, what will be the required pull-in torque for the motor?

Part IV
System Models

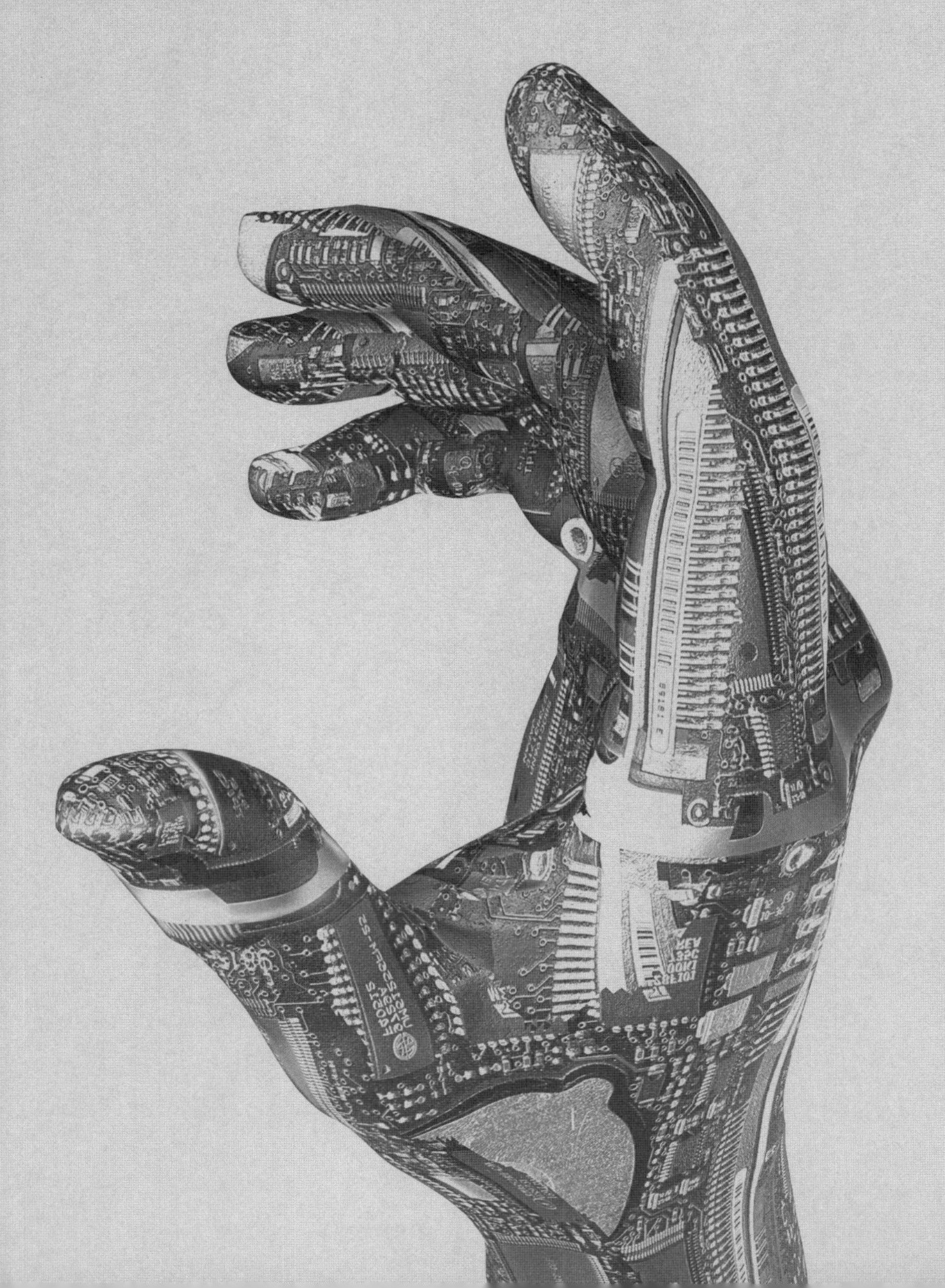

Chapter ten Basic System Models

Objectives

The objectives of this chapter are that, after studying it, the reader should be able to:

- **Explain the importance of models in predicting the behaviour of systems.**
- **Devise models from basic building blocks for mechanical, electrical, fluid and thermal systems.**
- **Recognise analogies between mechanical, electrical, fluid and thermal systems.**

10.1 Mathematical models

Consider the following situation. A microprocessor switches on a motor. How will the rotation of the motor shaft vary with time? The speed will not immediately assume the full-speed value but will only attain that speed after some time. Consider another situation. A hydraulic system is used to open a valve which allows water into a tank to restore the water level to that required. How will the water level vary with time? The water level will not immediately assume the required level but will only attain that level after some time.

In order to understand the behaviour of systems, **mathematical models** are needed. These are simplified representations of certain aspects of a real system. Such a model is created using equations to describe the relationship between the input and output of a system and can then be used to enable predictions to be made of the behaviour of a system under specific conditions, e.g. the outputs for a given set of inputs, or the outputs if a particular parameter is changed. In devising a mathematical model of a system it is necessary to make assumptions and simplifications and a balance has to be chosen between simplicity of the model and the need for it to represent the actual real-world behaviour. For example, we might form a mathematical model for a spring by assuming that the extension x is proportional to the applied force F, i.e. $F = kx$. This simplified model might not accurately predict the behaviour of a real spring where the extension might not be precisely proportional to the force and where we cannot apply this model regardless of the size of the force, since large forces will permanently deform the spring and might even break it and this is not predicted by the simple model.

The basis for any mathematical model is provided by the fundamental physical laws that govern the behaviour of the system. In this chapter a range of systems will be considered, including mechanical, electrical, thermal and fluid examples.

Like a child building houses, cars, cranes, etc., from a number of basic building blocks, systems can be made up from a range of building blocks.

Each building block is considered to have a single property or function. Thus, to take a simple example, an electric circuit system may be made up from building blocks which represent the behaviour of resistors, capacitors and inductors. The resistor building block is assumed to have purely the property of resistance, the capacitor purely that of capacitance and the inductor purely that of inductance. By combining these building blocks in different ways, a variety of electric circuit systems can be built up and the overall input/output relationships obtained for the system by combining in an appropriate way the relationships for the building blocks. Thus a mathematical model for the system can be obtained. A system built up in this way is called a **lumped parameter** system. This is because each parameter, i.e. property or function, is considered independently.

There are similarities in the behaviour of building blocks used in mechanical, electrical, thermal and fluid systems. This chapter is about the basic building blocks and their combination to produce mathematical models for physical, real, systems. Chapter 11 looks at more complex models. It needs to be emphasised that such models are only aids in system design. Real systems often exhibit non-linear characteristics and can depart from the ideal models developed in these chapters. This matter is touched on in Chapter 11.

10.2 Mechanical system building blocks

The models used to represent mechanical systems have the basic building blocks of springs, dashpots and masses. **Springs** represent the stiffness of a system, **dashpots** the forces opposing motion, i.e. frictional or damping effects, and **masses** the inertia or resistance to acceleration (Figure 10.1). The mechanical system does not have to be really made up of springs, dashpots and masses but have the properties of stiffness, damping and inertia. All these building blocks can be considered to have a force as an input and a displacement as an output.

The stiffness of a spring is described by the relationship between the forces F used to extend or compress a spring and the resulting extension or compression x (Figure 10.1(a)). In the case of a spring where the extension or compression is proportional to the applied forces, i.e. a linear spring,

$$F = kx$$

where k is a constant. The bigger the value of k, the greater the forces have to be to stretch or compress the spring and so the greater the stiffness. The

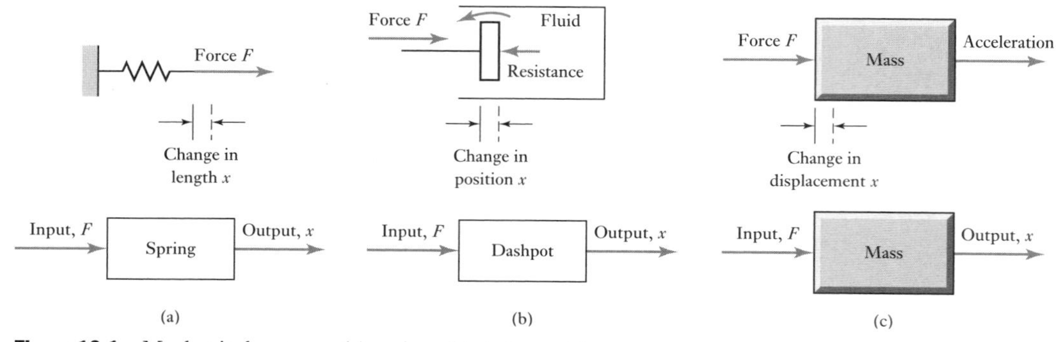

Figure 10.1 Mechanical systems: (a) spring, (b) dashpot, (c) mass.

object applying the force to stretch the spring is also acted on by a force, the force being that exerted by the stretched spring (Newton's third law). This force will be in the opposite direction and equal in size to the force used to stretch the spring, i.e. kx.

The dashpot building block represents the types of forces experienced when we endeavour to push an object through a fluid or move an object against frictional forces. The faster the object is pushed, the greater the opposing forces become. The dashpot which is used pictorially to represent these damping forces which slow down moving objects consists of a piston moving in a closed cylinder (Figure 10.1(b)). Movement of the piston requires the fluid on one side of the piston to flow through or past the piston. This flow produces a resistive force. In the ideal case, the damping or resistive force F is proportional to the velocity v of the piston. Thus

$$F = cv$$

where c is a constant. The larger the value of c, the greater the damping force at a particular velocity. Since velocity is the rate of change of displacement x of the piston, i.e. $v = \mathrm{d}x/\mathrm{d}t$, then

$$F = c\frac{\mathrm{d}x}{\mathrm{d}t}$$

Thus the relationship between the displacement x of the piston, i.e. the output, and the force as the input is a relationship depending on the rate of change of the output.

The mass building block (Figure 10.1(c)) exhibits the property that the bigger the mass, the greater the force required to give it a specific acceleration. The relationship between the force F and the acceleration a is (Newton's second law) $F = ma$, where the constant of proportionality between the force and the acceleration is the constant called the mass m. Acceleration is the rate of change of velocity, i.e. $\mathrm{d}v/\mathrm{d}t$, and velocity v is the rate of change of displacement x, i.e. $v = \mathrm{d}x/\mathrm{d}t$. Thus

$$F = ma = m\frac{\mathrm{d}v}{\mathrm{d}t} = m\frac{\mathrm{d}(\mathrm{d}x/\mathrm{d}t)}{\mathrm{d}t} = m\frac{\mathrm{d}^2x}{\mathrm{d}t^2}$$

Energy is needed to stretch the spring, accelerate the mass and move the piston in the dashpot. However, in the case of the spring and the mass we can get the energy back but with the dashpot we cannot. The spring when stretched stores energy, the energy being released when the spring springs back to its original length. The energy stored when there is an extension x is $\frac{1}{2}kx^2$. Since $F = kx$ this can be written as

$$E = \frac{1}{2}\frac{F^2}{k}$$

There is also energy stored in the mass when it is moving with a velocity v, the energy being referred to as kinetic energy, and released when it stops moving:

$$E = \frac{1}{2}mv^2$$

However, there is no energy stored in the dashpot. It does not return to its original position when there is no force input. The dashpot dissipates energy rather than storing it, the power P dissipated depending on the velocity v and being given by

$$P = cv^2$$

10.2.1 Rotational systems

The spring, dashpot and mass are the basic building blocks for mechanical systems where forces and straight line displacements are involved without any rotation. If there is rotation then the equivalent three building blocks are a **torsional spring**, a **rotary damper** and the **moment of inertia**, i.e. the inertia of a rotating mass. With such building blocks the inputs are torque and the outputs angle rotated. With a torsional spring the angle θ rotated is proportional to the torque T. Hence

$$T = k\theta$$

With the rotary damper a disc is rotated in a fluid and the resistive torque T is proportional to the angular velocity ω, and since angular velocity is the rate at which angle changes, i.e. $\mathrm{d}\theta/\mathrm{d}t$,

$$T = c\omega = c\frac{\mathrm{d}\theta}{\mathrm{d}t}$$

The moment of inertia building block has the property that the greater the moment of inertia I, the greater the torque needed to produce an angular acceleration α:

$$T = I\alpha$$

Thus, since angular acceleration is the rate of change of angular velocity, i.e. $\mathrm{d}\omega/\mathrm{d}t$, and angular velocity is the rate of change of angular displacement, then

$$T = I\frac{\mathrm{d}\omega}{\mathrm{d}t} = I\frac{\mathrm{d}(\mathrm{d}\theta/\mathrm{d}t)}{\mathrm{d}t} = I\frac{\mathrm{d}^2\theta}{\mathrm{d}t^2}$$

The torsional spring and the rotating mass store energy; the rotary damper just dissipates energy. The energy stored by a torsional spring when twisted through an angle θ is $\frac{1}{2}k\theta^2$ and since $T = k\theta$ this can be written as

$$E = \frac{1}{2}\frac{T^2}{k}$$

The energy stored by a mass rotating with an angular velocity ω is the kinetic energy E, where

$$E = \frac{1}{2}I\omega^2$$

The power P dissipated by the rotatory damper when rotating with an angular velocity ω is

$$P = c\omega^2$$

Table 10.1 summarises the equations defining the characteristics of the mechanical building blocks when there is, in the case of straight line

Table 10.1 Mechanical building blocks.

Building block	Describing equation	Energy stored or power dissipated
Translational		
Spring	$F = kx$	$E = \frac{1}{2}\frac{F^2}{k}$
Dashpot	$F = c\frac{dx}{dt} = cv$	$P = cv^2$
Mass	$F = m\frac{d^2x}{dt^2} = m\frac{dv}{dt}$	$E = \frac{1}{2}mv^2$
Rotational		
Spring	$T = k\theta$	$E = \frac{1}{2}\frac{T^2}{k}$
Rotational damper	$T = c\frac{d\theta}{dt} = c\omega$	$P = c\omega^2$
Moment of inertia	$T = I\frac{d^2\theta}{dt^2} = I\frac{d\omega}{dt}$	$E = \frac{1}{2}I\omega^2$

displacements (termed translational), a force input F and a displacement x output and, in the case of rotation, a torque T and angular displacement θ.

10.2.2 Building up a mechanical system

Many systems can be considered to be essentially a mass, a spring and dashpot combined in the way shown in Figure 10.2(a) and having an input of a force F and an output of displacement x (Figure 10.2(b)). To evaluate the relationship between the force and displacement for the system, the procedure to be adopted is to consider just one mass, and just the forces acting on that body. A diagram of the mass and just the forces acting on it is called a **free-body diagram** (Figure 10.2(c)).

When several forces act concurrently on a body, their single equivalent resultant can be found by vector addition. If the forces are all acting along the same line or parallel lines, this means that the resultant or net force acting on the block is the algebraic sum. Thus for the mass in Figure 10.2(c), if we consider just the forces acting on that block then the net force applied to the

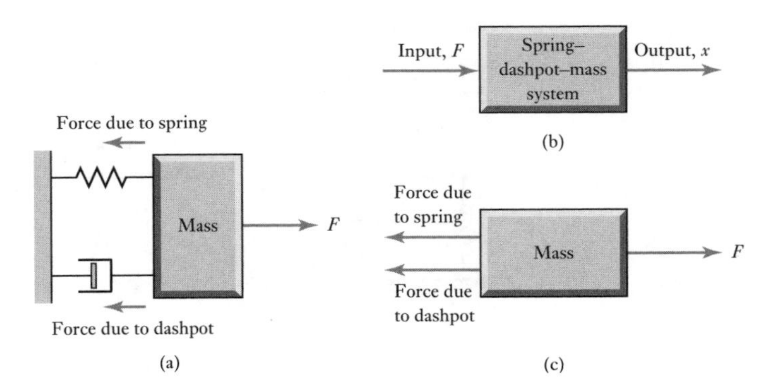

Figure 10.2 (a) Spring–dashpot–mass, (b) system, (c) free-body diagram.

mass is the applied force F minus the force resulting from the stretching or compressing of the spring and minus the force from the damper. Thus

$$\text{net force applied to mass } m = F - kx - cv$$

where v is the velocity with which the piston in the dashpot, and hence the mass, is moving. This net force is the force applied to the mass to cause it to accelerate. Thus

$$\text{net force applied to mass} = ma$$

Hence

$$F - kx - c\frac{dx}{dt} = m\frac{d^2x}{dt^2}$$

or, when rearranged,

$$m\frac{d^2x}{dt^2} + c\frac{dx}{dt} + kx = F$$

This equation, called a **differential equation**, describes the relationship between the input of force F to the system and the output of displacement x. Because of the d^2x/dt^2 term, it is a second-order differential equation; a first-order differential equation would only have dx/dt.

There are many systems which can be built up from suitable combinations of the spring, dashpot and mass building blocks. Figure 10.3 illustrates some.

Figure 10.3(a) shows the model for a machine mounted on the ground and could be used as a basis for studying the effects of ground disturbances on the displacements of a machine bed. Figure 10.3(b) shows a model for the wheel and its suspension for a car or truck and can be used for the study of

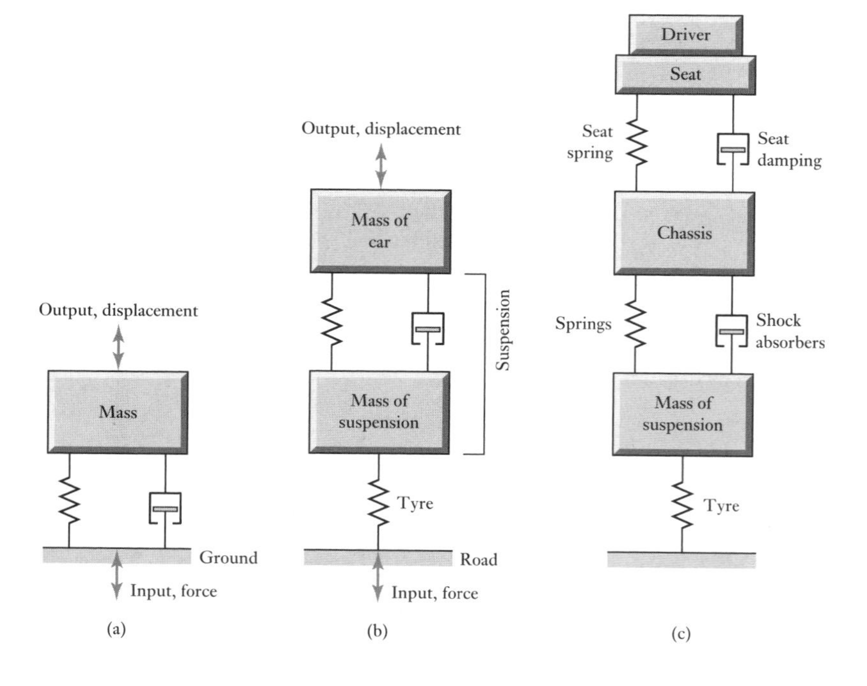

Figure 10.3 Model for (a) a machine mounted on the ground, (b) the chassis of a car as a result of a wheel moving along a road, (c) the driver of a car as it is driven along a road.

the behaviour that could be expected of the vehicle when driven over a rough road and hence as a basis for the design of the vehicle suspension. Figure 10.3(c) shows how this model can be used as part of a larger model to predict how the driver might feel when driven along a road. The procedure to be adopted for the analysis of such models is just the same as outlined above for the simple spring–dashpot–mass model. A free-body diagram is drawn for each mass in the system, such diagrams showing each mass independently and just the forces acting on it. Then for each mass the resultant of the forces acting on it is equated to the product of the mass and the acceleration of the mass.

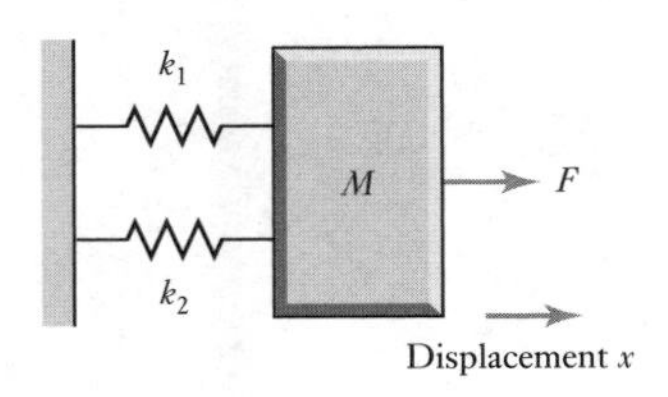

Figure 10.4 Example.

To illustrate the above, consider the derivation of the differential equation describing the relationship between the input of the force F and the output of displacement x for the system shown in Figure 10.4.

The net force applied to the mass is F minus the resisting forces exerted by each of the springs. Since these are k_1x and k_2x, then

$$\text{net force} = F - k_1x - k_2x$$

Since the net force causes the mass to accelerate, then

$$\text{net force} = m\frac{d^2x}{dt^2}$$

Hence

$$m\frac{d^2x}{dt^2} + (k_1 + k_2)x = F$$

The procedure for obtaining the differential equation relating the inputs and outputs for a mechanical system consisting of a number of components can be summarised as:

1 Isolate the various components in the system and draw free-body diagrams for each.
2 Hence, with the forces identified for a component, write the modelling equation for it.
3 Combine the equations for the various system components to obtain the system differential equation.

As an illustration, consider the derivation of the differential equation describing the motion of the mass m_1 in Figure 10.5(a) when a force F is applied. Consider the free-body diagrams (Figure 10.5(b)). For mass m_2 these are the force F and the force exerted by the upper spring. The force exerted by the upper spring is due to its being stretched by $(x_2 - x_3)$ and so is $k_2(x_3 - x_2)$. Thus the net force acting on the mass is

$$\text{net force} = F - k_2(x_3 - x_2)$$

This force will cause the mass to accelerate and so

$$F - k_2(x_3 - x_2) = m_2\frac{d^2x_3}{dt}$$

For the free-body diagram for mass m_1, the force exerted by the upper spring is $k_2(x_3 - x_2)$ and that by the lower spring is $k_1(x_1 - x_2)$. Thus the net force acting on the mass is

$$\text{net force} = k_1(x_2 - x_1) - k_2(x_3 - x_2)$$

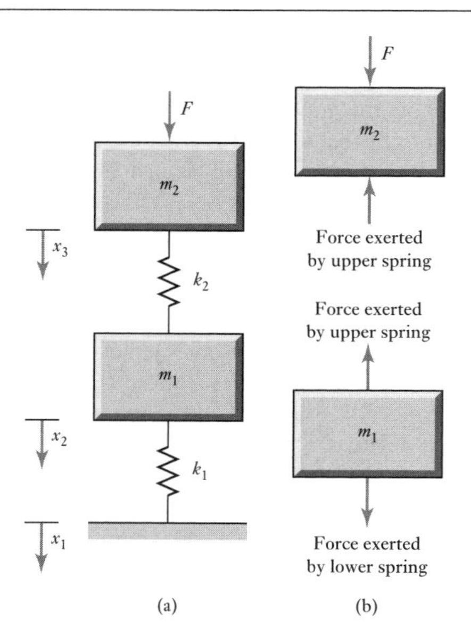

Figure 10.5 Mass–spring system.

This force will cause the mass to accelerate and so

$$k_1(x_2 - x_1) - k_2(x_3 - x_2) = m_1\frac{d^2x_2}{dt}$$

We thus have two simultaneous second-order differential equations to describe the behaviours of the system.

Similar models can be constructed for rotating systems. To evaluate the relationship between the torque and angular displacement for the system the procedure to be adopted is to consider just one rotational mass block, and just the torques acting on that body. When several torques act on a body simultaneously, their single equivalent resultant can be found by addition in which the direction of the torques is taken into account. Thus a system involving a torque being used to rotate a mass on the end of a shaft (Figure 10.6(a)) can be considered to be represented by the rotational building blocks shown in Figure 10.6(b). This is a comparable situation with that analysed above (Figure 10.2) for linear displacements and yields a similar equation

$$I\frac{d^2\theta}{dt^2} + c\frac{d\theta}{dt} + k\theta = T$$

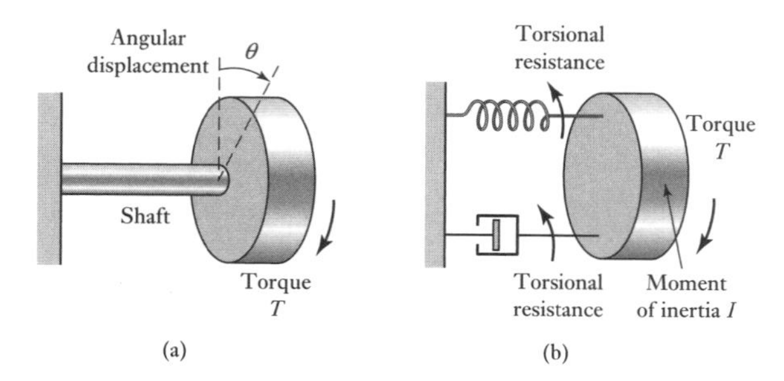

Figure 10.6 Rotating a mass on the end of a shaft: (a) physical situation, (b) building block model.

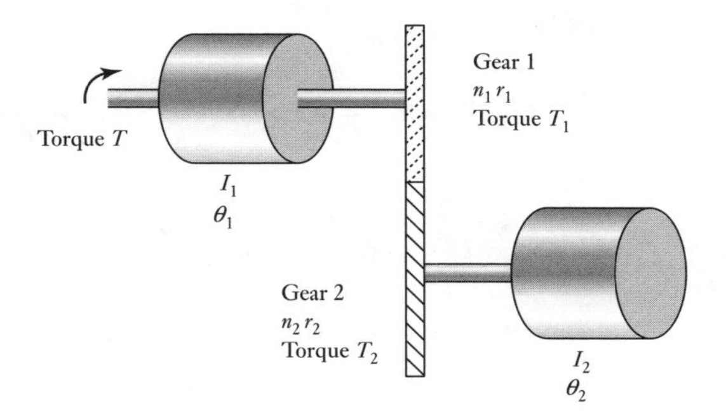

Figure 10.7 A two-gear train system.

Motors operating through gear trains to rotate loads are a feature of many control systems. Figure 10.7 shows a simple model of such a system. It consists of a mass of moment of inertia I_1 connected to gear 1 having n_1 teeth and a radius r_1 and a mass of moment of inertia I_2 connected to a gear 2 with n_2 teeth and a radius r_2. We will assume that the gears have negligible moments of inertia and also that rotational damping can be ignored.

If gear 1 is rotated through an angle θ_1 then gear 2 will rotate through an angle θ_2 where

$$r_1\theta_1 = r_2\theta_2$$

The ratio of the gear teeth numbers is equal to the ratio n of the gear radii:

$$\frac{r_1}{r_2} = \frac{n_1}{n_2} = n$$

If a torque T is applied to the system and torque T_1 is applied to gear 1 then the net torque is $T - T_1$ and so

$$T - T_1 = I_1\frac{\mathrm{d}^2\theta_1}{\mathrm{d}t^2}$$

If the torque T_2 occurs at gear 2 then

$$T_2 = I_2\frac{\mathrm{d}^2\theta_2}{\mathrm{d}t^2}$$

We will assume that the power transmitted by gear 1 is equal to that transmitted by gear 2 and so, as the power transmitted is the product of the torque and angular velocity, we have

$$T_1\frac{\mathrm{d}\theta_1}{\mathrm{d}t} = T_2\frac{\mathrm{d}\theta_2}{\mathrm{d}t}$$

Since $r_2\theta_1 = r_2\theta_2$ it follows that

$$r_1\frac{\mathrm{d}\theta_1}{\mathrm{d}t} = r_2\frac{\mathrm{d}\theta_2}{\mathrm{d}t^2}$$

and so

$$\frac{T_1}{T_2} = \frac{r_1}{r_2} = n$$

Thus we can write

$$T - T_1 = T - nT_2 = T - n\left(I_2\frac{d^2\theta_2}{dt^2}\right)$$

and so

$$T - n\left(I_2\frac{d^2\theta_2}{dt^2}\right) = I_1\frac{d^2\theta_1}{dt^2}$$

Since $\theta_2 = n\theta_1$, $d\theta_2/dt = nd\theta_1/dt$ and $d^2\theta_2/dt^2 = nd^2\theta_1/dt^2$ and so

$$T - n^2\left(I_2\frac{d^2\theta_1}{dt^2}\right) = I_1\frac{d^2\theta_1}{dt^2}$$

$$(I_1 + n^2 I_2)\frac{d^2\theta_1}{dt^2} = T$$

Without the gear train we would have had simply

$$I_1\frac{d^2\theta_1}{dt^2} = T$$

Thus the moment of inertia of the load is reflected back to the other side of the gear train as an additional moment of inertia term $n^2 I_2$.

10.3 Electrical system building blocks

The basic building blocks of electrical systems are inductors, capacitors and resistors (Figure 10.8).

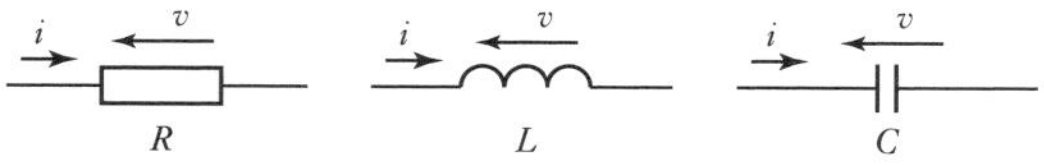

Figure 10.8 Electrical building blocks.

For an **inductor** the potential difference v across it at any instant depends on the rate of change of current (di/dt) through it:

$$v = L\frac{di}{dt}$$

where L is the inductance. The direction of the potential difference is in the opposite direction to the potential difference used to drive the current through the inductor, hence the term back e.m.f. The equation can be rearranged to give

$$i = \frac{1}{L}\int v\,dt$$

For a **capacitor**, the potential difference across it depends on the charge q on the capacitor plates at the instant concerned:

$$v = \frac{q}{C}$$

where C is the capacitance. Since the current i to or from the capacitor is the rate at which charge moves to or from the capacitor plates, i.e. $i = \mathrm{d}q/\mathrm{d}t$, then the total charge q on the plates is given by

$$q = \int i \, \mathrm{d}t$$

and so

$$v = \frac{1}{C} \int i \, \mathrm{d}t$$

Alternatively, since $v = q/C$ then

$$\frac{\mathrm{d}v}{\mathrm{d}t} = \frac{1}{C}\frac{\mathrm{d}q}{\mathrm{d}t} = \frac{1}{C}i$$

and so

$$i = C\frac{\mathrm{d}v}{\mathrm{d}t}$$

For a **resistor**, the potential difference v across it at any instant depends on the current i through it

$$v = Ri$$

where R is the resistance.

Both the inductor and capacitor store energy which can then be released at a later time. A resistor does not store energy but just dissipates it. The energy stored by an inductor when there is a current i is

$$E = \frac{1}{2}Li^2$$

The energy stored by a capacitor when there is a potential difference v across it is

$$E = \frac{1}{2}Cv^2$$

The power P dissipated by a resistor when there is a potential difference v across it is

$$P = iv = \frac{v^2}{R}$$

Table 10.2 summarises the equations defining the characteristics of the electrical building blocks when the input is current and the output is potential difference. Compare them with the equations given in Table 10.1 for the mechanical system building blocks.

Table 10.2 Electrical building blocks.

Building block	Describing equation	Energy stored or power dissipated
Inductor	$i = \frac{1}{L}\int v \, \mathrm{d}t$	$E = \frac{1}{2}Li^2$
	$v = L\frac{\mathrm{d}i}{\mathrm{d}t}$	
Capacitor	$i = C\frac{\mathrm{d}v}{\mathrm{d}t}$	$E = \frac{1}{2}Cv^2$
Resistor	$i = \frac{v}{R}$	$P = \frac{v^2}{R}$

10.3.1 Building up a model for an electrical system

The equations describing how the electrical building blocks can be combined are **Kirchhoff 's laws**. These can be expressed as:

Law 1: the total current flowing towards a junction is equal to the total current flowing from that junction, i.e. the algebraic sum of the currents at the junction is zero.

Law 2: in a closed circuit or loop, the algebraic sum of the potential differences across each part of the circuit is equal to the applied e.m.f.

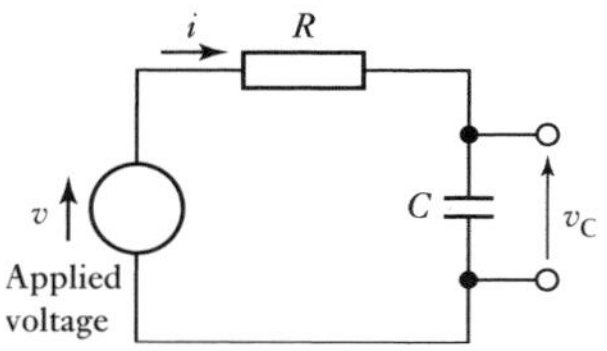

Figure 10.9
Resistor–capacitor system.

Now consider a simple electrical system consisting of a resistor and capacitor in series, as shown in Figure 10.9. Applying Kirchhoff's second law to the circuit loop gives

$$v = v_R + v_C$$

where v_R is the potential difference across the resistor and v_C that across the capacitor. Since this is just a single loop, the current i through all the circuit elements will be the same. If the output from the circuit is the potential difference across the capacitor, v_C, then since $v_R = iR$ and $i = C(dv_C/dt)$,

$$v = RC\frac{dv_C}{dt} + v_C$$

This gives the relationship between the output v_C and the input v and is a first-order differential equation.

Figure 10.10 shows a resistor–inductor–capacitor system. If Kirchhoff's second law is applied to this circuit loop,

$$v = v_R + v_L + v_C$$

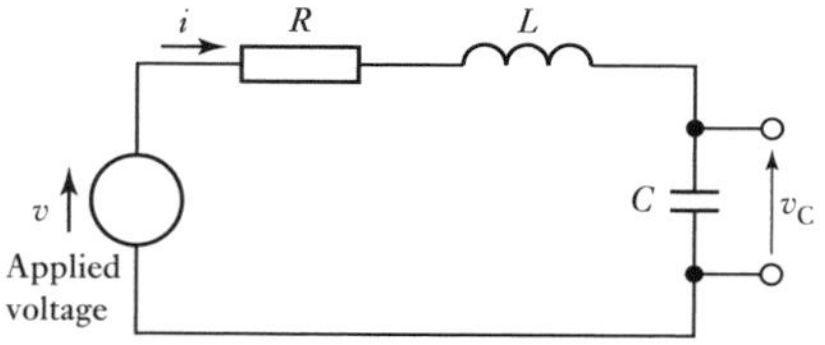

Figure 10.10
Resistor–inductor–capacitor system.

where v_R is the potential difference across the resistor, v_L that across the inductor and v_C that across the capacitor. Since there is just a single loop, the current i will be the same through all circuit elements. If the output from the circuit is the potential difference across the capacitor, v_C, then since $v_R = iR$ and $v_L = L(di/dt)$

$$v = iR + L\frac{di}{dt} + v_C$$

But $i = C(dv_C/dt)$ and so

$$\frac{di}{dt} = C\frac{d(dv_C/dt)}{dt} = C\frac{d^2v_C}{dt^2}$$

Hence

$$v = RC\frac{dv_C}{dt} + LC\frac{d^2v_C}{dt^2} + v_C$$

This is a second-order differential equation.

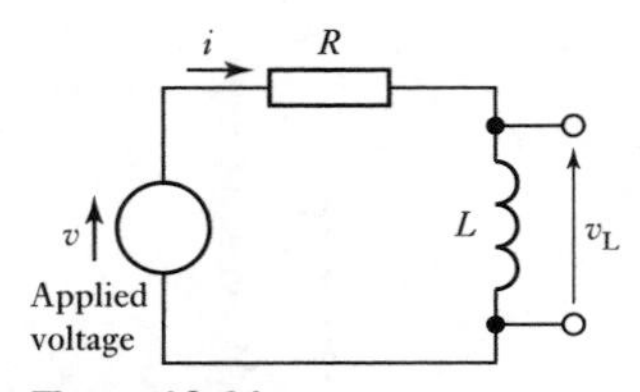

Figure 10.11
Resistor–inductor system.

As a further illustration, consider the relationship between the output, the potential difference across the inductor of v_L, and the input v for the circuit shown in Figure 10.11. Applying Kirchhoff's second law to the circuit loop gives

$$v = v_R + v_L$$

where v_R is the potential difference across the resistor R and v_L that across the inductor. Since $v_R = iR$,

$$v = iR + v_L$$

Since

$$i = \frac{1}{L}\int v_L \mathrm{d}t$$

then the relationship between the input and output is

$$v = \frac{R}{L}\int v_L \,\mathrm{d}t + v_L$$

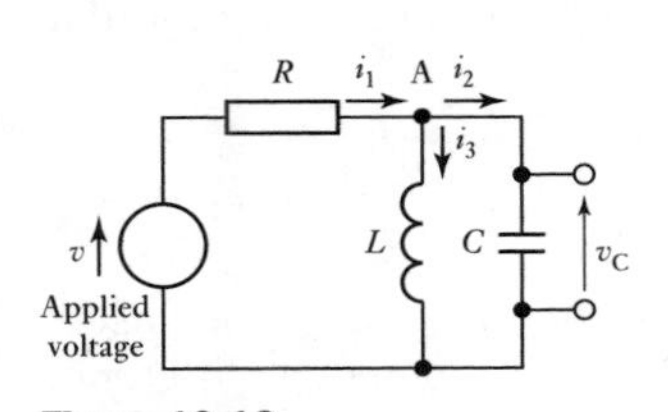

Figure 10.12
Resistor–capacitor–inductor system.

As another example, consider the relationship between the output, the potential difference v_C across the capacitor, and the input v for the circuit shown in Figure 10.12. Applying Kirchhoff's law 1 to node A gives

$$i_1 = i_2 + i_3$$

But

$$i_1 = \frac{v - v_A}{R}$$

$$i_2 = \frac{1}{L}\int v_A \,\mathrm{d}t$$

$$i_3 = C\frac{\mathrm{d}v_A}{\mathrm{d}t}$$

Hence

$$\frac{v - v_A}{R} = \frac{1}{L}\int v_A \,\mathrm{d}t + C\frac{\mathrm{d}v_A}{\mathrm{d}t}$$

But $v_C = v_A$. Hence, with some rearrangement,

$$v = RC\frac{\mathrm{d}v_C}{\mathrm{d}t} + v_C + \frac{R}{L}\int v_C \,\mathrm{d}t$$

10.3.2 Electrical and mechanical analogies

The building blocks for electrical and mechanical systems have many similarities (Figure 10.13). For example, the electrical resistor does not store energy but dissipates it, with the current i through the resistor being given by $i = v/R$, where R is a constant, and the power P dissipated by $P = v^2/R$. The mechanical analogue of the resistor is the dashpot. It also does not store energy but dissipates it, with the force F being related to the velocity v by $F = cv$, where c is a constant, and the power P dissipated by $P = cv^2$. Both these sets of equations have similar forms. Comparing them, and taking the

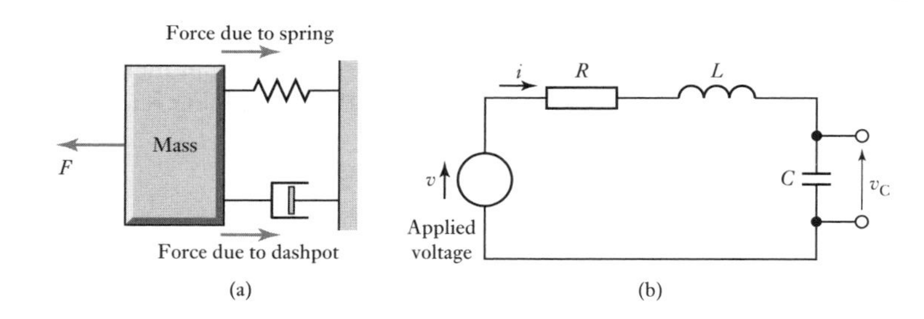

Figure 10.13 Analogous systems.

current as being analogous to the force, then the potential difference is analogous to the velocity and the dashpot constant c to the reciprocal of the resistance, i.e. $(1/R)$. These analogies between current and force, potential difference and velocity, hold for the other building blocks with the spring being analogous to inductance and mass to capacitance.

The mechanical system in Figure 10.1(a) and the electrical system in Figure 10.1(b) have input/output relationships described by similar differential equations:

$$m\frac{d^2x}{dt^2} + c\frac{dx}{dt} + kx = F \quad \text{and} \quad RC\frac{dv_C}{dt} + LC\frac{d^2v_C}{dt^2} + v_C = v$$

The analogy between current and force is the one most often used. However, another set of analogies can be drawn between potential difference and force.

10.4 Fluid system building blocks

In fluid flow systems there are three basic building blocks which can be considered to be the equivalent of electrical resistance, capacitance and inductance. Fluid systems can be considered to fall into two categories: hydraulic, where the fluid is a liquid and is deemed to be incompressible; and pneumatic, where it is a gas which can be compressed and consequently shows a density change.

Hydraulic resistance is the resistance to flow which occurs as a result of a liquid flowing through valves or changes in a pipe diameter (Figure 10.14(a)). The relationship between the volume rate of flow of liquid q through the resistance element and the resulting pressure difference $(p_1 - p_2)$ is

$$p_1 - p_2 = Rq$$

where R is a constant called the hydraulic resistance. The bigger the resistance, the bigger the pressure difference for a given rate of flow. This equation, like that for the electrical resistance and Ohm's law, assumes a linear relationship. Such hydraulic linear resistances occur with orderly flow through capillary tubes and porous plugs but non-linear resistances occur with flow through sharp-edged orifices or if flow is turbulent.

Hydraulic capacitance is the term used to describe energy storage with a liquid where it is stored in the form of potential energy. A height of liquid in a container (Figure 10.14(b)), i.e. a so-called pressure head, is one form of such a storage. For such a capacitance, the rate of change of volume V in the

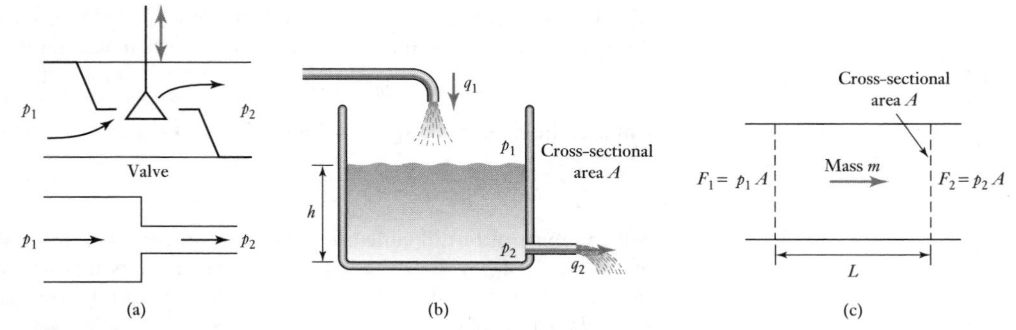

Figure 10.14 Hydraulic examples: (a) resistance, (b) capacitance, (c) inertance.

container, i.e. dV/dt, is equal to the difference between the volumetric rate at which liquid enters the container q_1 and the rate at which it leaves q_2,

$$q_1 - q_2 = \frac{dV}{dt}$$

But $V = Ah$, where A is the cross-sectional area of the container and h the height of liquid in it. Hence

$$q_1 - q_2 = \frac{d(Ah)}{dt} = A\frac{dh}{dt}$$

But the pressure difference between the input and output is p, where $p = h\rho g$ with ρ being the liquid density and g the acceleration due to gravity. Thus, if the liquid is assumed to be incompressible, i.e. its density does not change with pressure,

$$q_1 - q_2 = A\frac{d(p/\rho g)}{dt} = \frac{A}{\rho g}\frac{dp}{dt}$$

The hydraulic capacitance C is defined as being

$$C = \frac{A}{\rho g}$$

Thus

$$q_1 - q_2 = C\frac{dp}{dt}$$

Integration of this equation gives

$$p = \frac{1}{C}\int (q_1 - q_2)\,dt$$

Hydraulic inertance is the equivalent of inductance in electrical systems or a spring in mechanical systems. To accelerate a fluid and so increase its velocity, a force is required. Consider a block of liquid of mass m (Figure 10.14(c)). The net force acting on the liquid is

$$F_1 - F_2 = p_1A - p_2A = (p_1 - p_2)A$$

where $(p_1 - p_2)$ is the pressure difference and A the cross-sectional area. This net force causes the mass to accelerate with an acceleration a, and so

$$(p_1 - p_2)A = ma$$

But a is the rate of change of velocity dv/dt, hence

$$(p_1 - p_2)A = m\frac{dv}{dt}$$

But the mass of liquid concerned has a volume of AL, where L is the length of the block of liquid or the distance between the points in the liquid where the pressures p_1 and p_2 are measured. If the liquid has a density ρ then $m = AL\rho$ and so

$$(p_1 - p_2)A = AL\rho\frac{dv}{dt}$$

But the volume rate of flow $q = Av$, hence

$$(p_1 - p_2)A = L\rho\frac{dq}{dt}$$

$$p_1 - p_2 = I\frac{dq}{dt}$$

where the hydraulic inertance I is defined as

$$I = \frac{L\rho}{A}$$

With pneumatic systems the three basic building blocks are, as with hydraulic systems, resistance, capacitance and inertance. However, gases differ from liquids in being compressible, i.e. a change in pressure causes a change in volume and hence density. **Pneumatic resistance** R is defined in terms of the mass rate of flow dm/dt (note that this is often written as an m with a dot above it to indicate that the symbol refers to the mass rate of flow and not just the mass) and the pressure difference $(p_1 - p_2)$ as

$$p_1 - p_2 = R\frac{dm}{dt} = R\dot{m}$$

Pneumatic capacitance C is due to the compressibility of the gas, and is comparable with the way in which the compression of a spring stores energy. If there is a mass rate of flow dm_1/dt entering a container of volume V and a mass rate of flow of dm_2/dt leaving it, then the rate at which the mass in the container is changing is $(dm_1/dt - dm_2/dt)$. If the gas in the container has a density ρ then the rate of change of mass in the container is

$$\text{rate of change of mass in container} = \frac{d(\rho V)}{dt}$$

But, because a gas can be compressed, both ρ and V can vary with time. Hence

$$\text{rate of change of mass in container} = \rho\frac{dV}{dt} + V\frac{d\rho}{dt}$$

Since $(\mathrm{d}V/\mathrm{d}t) = (\mathrm{d}V/\mathrm{d}p)(\mathrm{d}p/\mathrm{d}t)$ and, for an ideal gas, $pV = mRT$ with consequently $p = (m/V)RT = \rho RT$ and $\mathrm{d}\rho/\mathrm{d}t = (1/RT)(\mathrm{d}p/\mathrm{d}t)$, then

$$\text{rate of change of mass in container} = \rho\frac{\mathrm{d}V}{\mathrm{d}p}\frac{\mathrm{d}p}{\mathrm{d}t} + \frac{V}{RT}\frac{\mathrm{d}p}{\mathrm{d}t}$$

where R is the gas constant and T the temperature, assumed to be constant, on the Kelvin scale. Thus

$$\frac{\mathrm{d}m_1}{\mathrm{d}t} - \frac{\mathrm{d}m_2}{\mathrm{d}t} = \left(\rho\frac{\mathrm{d}V}{\mathrm{d}p} + \frac{V}{RT}\right)\frac{\mathrm{d}p}{\mathrm{d}t}$$

The pneumatic capacitance due to the change in volume of the container C_1 is defined as

$$C_1 = \rho\frac{\mathrm{d}V}{\mathrm{d}p}$$

and the pneumatic capacitance due to the compressibility of the gas C_2 as

$$C_2 = \frac{V}{RT}$$

Hence

$$\frac{\mathrm{d}m_1}{\mathrm{d}t} - \frac{\mathrm{d}m_2}{\mathrm{d}t} = (C_1 + C_2)\frac{\mathrm{d}p}{\mathrm{d}t}$$

or

$$p_1 - p_2 = \frac{1}{C_1 + C_2}\int(\dot{m}_1 - \dot{m}_2)\,\mathrm{d}t$$

Pneumatic inertance is due to the pressure drop necessary to accelerate a block of gas. According to Newton's second law, the net force is $ma = \mathrm{d}(mv)/\mathrm{d}t$. Since the force is provided by the pressure difference $(p_1 - p_2)$, then if A is the cross-sectional area of the block of gas being accelerated

$$(p_1 - p_2)A = \frac{\mathrm{d}(mv)}{\mathrm{d}t}$$

But m, the mass of the gas being accelerated, equals ρLA with ρ being the gas density and L the length of the block of gas being accelerated. And the volume rate of flow $q = Av$, where v is the velocity. Thus

$$mv = \rho LA\frac{q}{A} = \rho Lq$$

and so

$$(p_1 - p_2)A = L\frac{\mathrm{d}(\rho q)}{\mathrm{d}t}$$

But $\dot{m} = \rho q$ and so

$$p_1 - p_2 = \frac{L}{A}\frac{\mathrm{d}\dot{m}}{\mathrm{d}t}$$

$$p_1 - p_2 = I\frac{\mathrm{d}\dot{m}}{\mathrm{d}t}$$

with the pneumatic inertance I being $I = L/A$.

Table 10.3 shows the basic characteristics of the fluid building blocks, both hydraulic and pneumatic.

For hydraulics the volumetric rate of flow and for pneumatics the mass rate of flow are analogous to the electric current in an electrical system. For both hydraulics and pneumatics the pressure difference is analogous to the potential difference in electrical systems. Compare Table 10.3 with Table 10.2. Hydraulic and pneumatic inertance and capacitance are both energy storage elements; hydraulic and pneumatic resistance are both energy dissipaters.

10.4.1 Building up a model for a fluid system

Figure 10.15 shows a simple hydraulic system, a liquid entering and leaving a container. Such a system can be considered to consist of a capacitor, the liquid in the container, with a resistor, the valve.

Table 10.3 Hydraulic and pneumatic building blocks

Building block	Describing equation	Energy stored or power dissipated
Hydraulic		
Inertance	$q = \frac{1}{L}\int (p_1 - p_2)\,dt$ $p = L\frac{dq}{dt}$	$E = \frac{1}{2}Iq^2$
Capacitance	$q = C\frac{d(p_1 - p_2)}{dt}$	$E = \frac{1}{2}C(p_1 - p_2)^2$
Resistance	$q = \frac{p_1 - p_2}{R}$	$P = \frac{1}{R}(p_1 - p_2)^2$
Pneumatic		
Inertance	$\dot{m} = \frac{1}{L}\int (p_1 - p_2)\,dt$	$E = \frac{1}{2}I\dot{m}^2$
Capacitance	$\dot{m} = C\frac{d(p_1 - p_2)}{dt}$	$E = \frac{1}{2}C(p_1 - p_2)^2$
Resistance	$\dot{m} = \frac{p_1 - p_2}{R}$	$P = \frac{1}{R}(p_1 - p_2)^2$

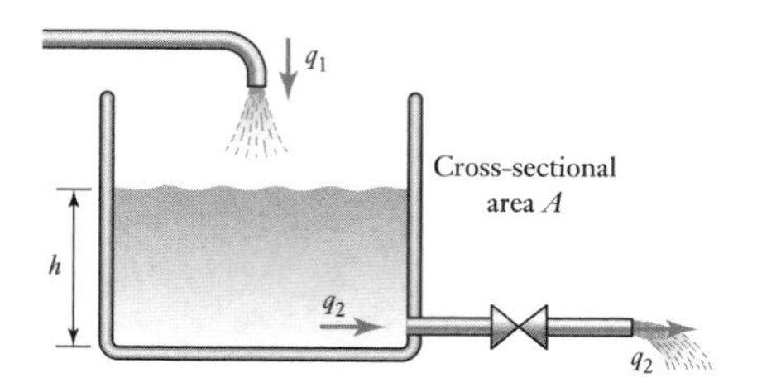

Figure 10.15 A fluid system.

Inertance can be neglected since flow rates change only very slowly. For the capacitor we can write

$$q_1 - q_2 = C\frac{dp}{dt}$$

The rate at which liquid leaves the container q_2 equals the rate at which it leaves the valve. Thus for the resistor

$$p_1 - p_2 = Rq_2$$

The pressure difference $(p_1 - p_2)$ is the pressure due to the height of liquid in the container and is thus $h\rho g$. Thus $q_2 = h\rho g/R$ and so substituting for q_2 in the first equation gives

$$q_1 - \frac{h\rho g}{R} = C\frac{d(h\rho g)}{dt}$$

and, since $C = A/\rho g$,

$$q_1 = A\frac{dh}{dt} + \frac{\rho gh}{R}$$

This equation describes how the height of liquid in the container depends on the rate of input of liquid into the container.

A bellows is an example of a simple pneumatic system (Figure 10.16). Resistance is provided by a constriction which restricts the rate of flow of gas into the bellows and capacitance is provided by the bellows itself. Inertance can be neglected since the flow rate changes only slowly.

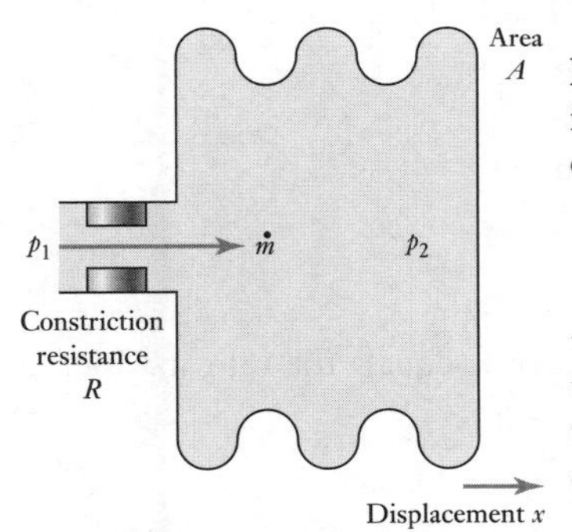

Figure 10.16 A pneumatic system.

The mass flow rate into the bellows is given by

$$p_1 - p_2 = R\dot{m}$$

where p_1 is the pressure prior to the constriction and p_2 the pressure after the constriction, i.e. the pressure in the bellows. All the gas that flows into the bellows remains in the bellows, there being no exit from the bellows. The capacitance of the bellows is given by

$$\dot{m}_1 - \dot{m}_2 = (C_1 + C_2)\frac{dp_2}{dt}$$

The mass flow rate entering the bellows is given by the equation for the resistance and the mass leaving the bellows is zero. Thus

$$\frac{p_1 - p_2}{R} = (C_1 + C_2)\frac{dp_2}{dt}$$

Hence

$$p_1 = R(C_1 + C_2)\frac{dp_2}{dt} + p_2$$

This equation describes how the pressure in the bellows p_2 varies with time when there is an input of a pressure p_1.

The bellows expands or contracts as a result of pressure changes inside it. Bellows are just a form of spring and so we can write $F = kx$ for the relationship between the force F causing an expansion or contraction and the resulting displacement x, where k is the spring constant for the bellows. But the force F depends on the pressure p_2, with $p_2 = F/A$ where A is the cross-sectional area of the bellows. Thus $p_2A = F = kx$. Hence substituting for p_2 in the above equation gives

$$p_1 = R(C_1 + C_2)\frac{k}{A}\frac{dx}{dt} + \frac{k}{A}x$$

This equation, a first-order differential equation, describes how the extension or contraction x of the bellows changes with time when there is an input of a pressure p_1. The pneumatic capacitance due to the change in volume of the container C_1 is $\rho dV/dp_2$ and since $V = Ax$, C_1 is $\rho A\, dx/dp_2$. But for the bellows $p_2A = kx$, thus

$$C_1 = \rho A\frac{dx}{d(kx/A)} = \frac{\rho A^2}{k}$$

C_2, the pneumatic capacitance due to the compressibility of the air, is $V/RT = Ax/RT$.

The following illustrates how, for the hydraulic system shown in Figure 10.17, relationships can be derived which describe how the heights of the liquids in the two containers will change with time. With this model inertance is neglected.

Container 1 is a capacitor and thus

$$q_1 - q_2 = C_1\frac{dp}{dt}$$

where $p = h_1\rho g$ and $C_1 = A_1/\rho g$ and so

$$q_1 - q_2 = A_1\frac{dh_1}{dt}$$

The rate at which liquid leaves the container q_2 equals the rate at which it leaves the valve R_1. Thus for the resistor,

$$p_1 - p_2 = R_1q_2$$

The pressures are $h_1\rho g$ and $h_2\rho g$. Thus

$$(h_1 - h_2)\rho g = R_1q_2$$

Using the value of q_2 given by this equation and substituting it into the earlier equation gives

$$q_1 - \frac{(h_1 - h_2)\rho g}{R_1} = A_1\frac{dh_1}{dt}$$

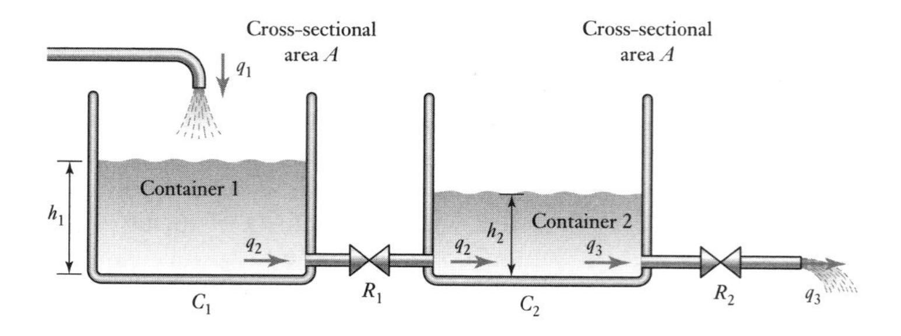

Figure 10.17 A fluid system.

This equation describes how the height of the liquid in container 1 depends on the input rate of flow.

For container 2 a similar set of equations can be derived. Thus for the capacitor C_2,

$$q_2 - q_3 = C_2 \frac{dp}{dt}$$

where $p = h_2 \rho g$ and $C_2 = A_2/\rho g$ and so

$$q_2 - q_3 = A_2 \frac{dh_2}{dt}$$

The rate at which liquid leaves the container q_3 equals the rate at which it leaves the valve R_2. Thus for the resistor,

$$p_2 - 0 = R_2 q_3$$

This assumes that the liquid exits into the atmosphere. Thus, using the value of q_3 given by this equation and substituting it into the earlier equation gives

$$q_2 - \frac{h_2 \rho g}{R_2} = A_2 \frac{dh_2}{dt}$$

Substituting for q_2 in this equation using the value given by the equation derived for the first container gives

$$\frac{(h_1 - h_2)\rho g}{R_1} - \frac{h_2 \rho g}{R_2} = A_2 \frac{dh_2}{dt}$$

This equation describes how the height of liquid in container 2 changes.

10.5 Thermal system building blocks

There are only two basic building blocks for thermal systems: resistance and capacitance. There is a net flow of heat between two points if there is a temperature difference between them. The electrical equivalent of this is that there is only a net current i between two points if there is a potential difference v between them, the relationship between the current and potential difference being $i = v/R$, where R is the electrical resistance between the points. A similar relationship can be used to define **thermal resistance R**. If q is the rate of flow of heat and $(T_1 - T_2)$ the temperature difference, then

$$q = \frac{T_2 - T_1}{R}$$

The value of the resistance depends on the mode of heat transfer. In the case of conduction through a solid, for unidirectional conduction

$$q = Ak \frac{T_1 - T_2}{L}$$

where A is the cross-sectional area of the material through which the heat is being conducted and L the length of material between the points at which the temperatures are T_1 and T_2; k is the thermal conductivity. Hence, with this mode of heat transfer,

$$R = \frac{L}{Ak}$$

When the mode of heat transfer is convection, as with liquids and gases, then

$$q = Ah(T_2 - T_1)$$

where A is the surface area across which there is the temperature difference and h is the coefficient of heat transfer. Thus, with this mode of heat transfer,

$$R = \frac{1}{Ah}$$

Thermal capacitance is a measure of the store of internal energy in a system. Thus, if the rate of flow of heat into a system is q_1 and the rate of flow out is q_2, then

$$\text{rate of change of internal energy} = q_1 - q_2$$

An increase in internal energy means an increase in temperature. Since

$$\text{internal energy change} = mc \times \text{change in temperature}$$

where m is the mass and c the specific heat capacity, then

$$\text{rate of change of internal energy} = mc \times \text{rate of change of temperature}$$

Thus

$$q_1 - q_2 = mc\frac{dT}{dt}$$

where dT/dt is the rate of change of temperature. This equation can be written as

$$q_1 - q_2 = C\frac{dT}{dt}$$

where C is the thermal capacitance and so $C = mc$. Table 10.4 gives a summary of the thermal building blocks.

Table 10.4 Thermal building blocks.

Building block	Describing equation	Energy stored
Capacitance	$q_1 - q_2 = C\frac{dT}{dt}$	$E = CT$
Resistance	$q = \frac{T_1 - T_2}{R}$	

10.5.1 Building up a model for a thermal system

Consider a thermometer at temperature T which has just been inserted into a liquid at temperature T_L (Figure 10.18).

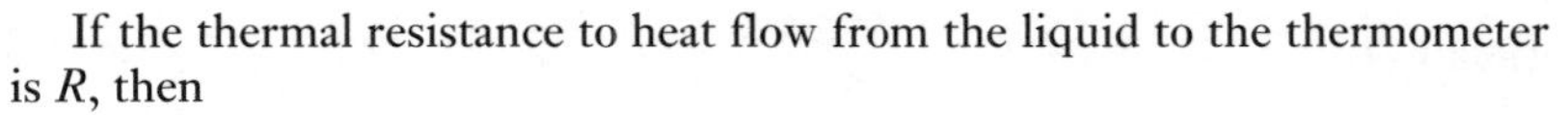

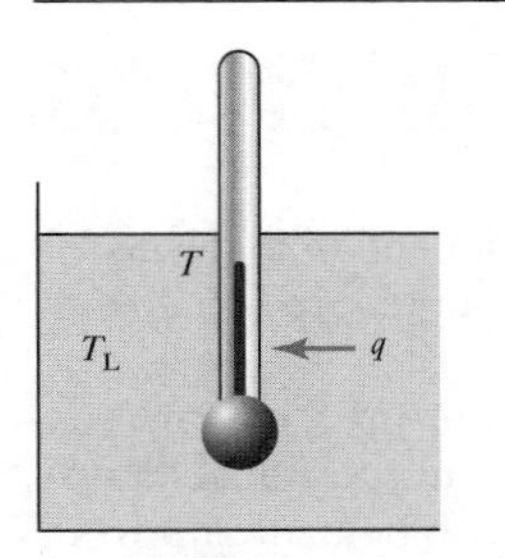

Figure 10.18 A thermal system.

If the thermal resistance to heat flow from the liquid to the thermometer is R, then

$$q = \frac{T_L - T}{R}$$

where q is the net rate of heat flow from liquid to thermometer. The thermal capacitance C of the thermometer is given by the equation

$$q_1 - q_2 = C\frac{dT}{dt}$$

Since there is only a net flow of heat from the liquid to the thermometer, $q_1 = q$ and $q_2 = 0$. Thus

$$q = C\frac{dT}{dt}$$

Substituting this value of q in the earlier equation gives

$$C\frac{dT}{dt} = \frac{T_L - T}{R}$$

Rearranging this equation gives

$$RC\frac{dT}{dt} + T = T_L$$

This equation, a first-order differential equation, describes how the temperature indicated by the thermometer T will vary with time when the thermometer is inserted into a hot liquid.

In the above thermal system the parameters have been considered to be lumped. This means, for example, that there has been assumed to be just one temperature for the thermometer and just one for the liquid, i.e. the temperatures are only functions of time and not position within a body.

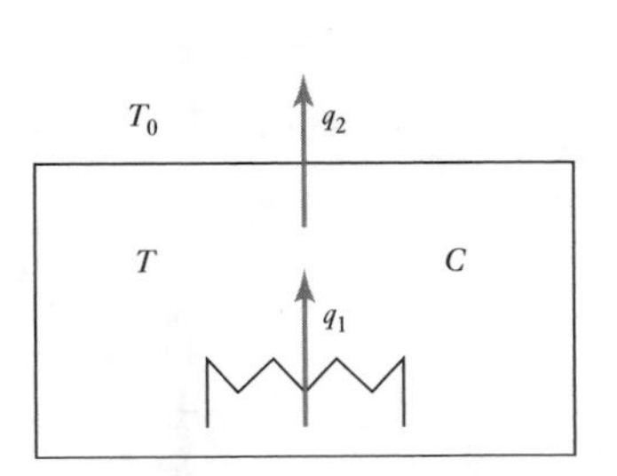

Figure 10.19 Thermal system.

To illustrate the above consider Figure 10.19 which shows a thermal system consisting of an electric fire in a room. The fire emits heat at the rate q_1 and the room loses heat at the rate q_2. Assuming that the air in the room is at a uniform temperature T and that there is no heat storage in the walls of the room, derive an equation describing how the room temperature will change with time.

If the air in the room has a thermal capacity C then

$$q_1 - q_2 = C\frac{dT}{dt}$$

If the temperature inside the room is T and that outside the room T_0 then

$$q_2 = \frac{T - T_0}{R}$$

where R is the resistivity of the walls. Substituting for q_2 gives

$$q_1 - \frac{T - T_0}{R} = C\frac{dT}{dt}$$

Hence

$$RC\frac{dT}{dt} + T = Rq_1 + T_0$$

Summary

A **mathematical model** of a system is a description of it in terms of equations relating inputs and outputs so that outputs can be predicted from inputs.

Mechanical systems can be considered to be made up from masses, springs and dashpots, or moments of inertia, springs and rotational dampers if rotational. Electrical systems can be considered to be made up from resistors, capacitors and inductors, hydraulic and pneumatic systems from resistance, capacitance and inertance, and thermal systems from resistance and capacitance.

There are many elements in mechanical, electrical, fluid and thermal systems which have similar behaviours. Thus, for example, mass in mechanical systems has similar properties to capacitance in electrical systems, capacitance in fluid systems and capacitance in thermal systems. Table 10.5 shows a comparison of the elements in each of these systems and their defining equations.

Table 10.5 System elements.

	Mechanical (translational)	**Mechanical (rotational)**	**Electrical**	**Fluid (hydraulic)**	**Thermal**
Element	Mass	Moment of inertia	Capacitor	Capacitor	Capacitor
Equation	$F = m\frac{d^2x}{dt^2}$	$T = I\frac{d^2\theta}{dt^2}$			
	$F = m\frac{dv}{dt}$	$T = I\frac{d\omega}{dt}$	$i = C\frac{dv}{dt}$	$q = C\frac{d(p_1 - p_2)}{dt}$	$q_1 - q_2 = C\frac{dT}{dt}$
Energy	$E = \frac{1}{2}mv^2$	$E = \frac{1}{2}I\omega^2$	$E = \frac{1}{2}Cv^2$	$E = \frac{1}{2}C(p_1 - p_2)^2$	$E = CT$
Element	Spring	Spring	Inductor	Inertance	None
Equation	$F = kx$	$T = k\theta$	$v = L\frac{di}{dt}$	$p = L\frac{dq}{dt}$	
Energy	$E = \frac{1}{2}\frac{F^2}{k}$	$E = \frac{1}{2}\frac{T^2}{k}$	$E = \frac{1}{2}Li^2$	$E = \frac{1}{2}Iq^2$	
Element	Dashpot	Rotational damper	Resistor	Resistance	Resistance
Equation	$F = c\frac{dx}{dt} = cv$	$T = c\frac{d\theta}{dt} = c\omega$	$i = \frac{v}{R}$	$q = \frac{p_1 - p_2}{R}$	$q = \frac{T_1 - T_2}{R}$
Power	$P = cv^2$	$P = c\omega^2$	$P = \frac{v^2}{R}$	$P = \frac{1}{R}(p_1 - p_2)^2$	

Problems

10.1 Derive an equation relating the input, force F, with the output, displacement x, for the systems described by Figure 10.20.

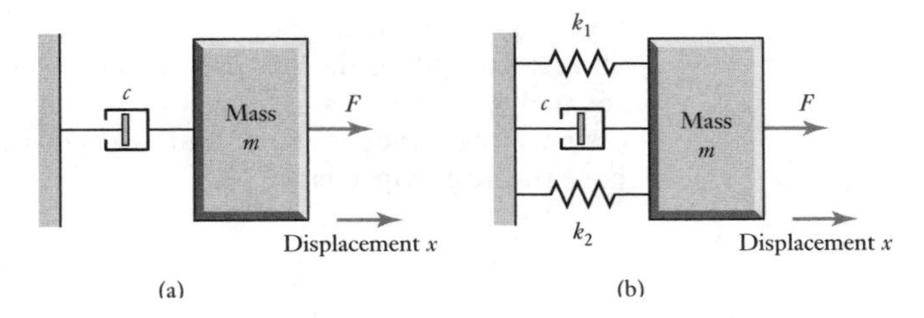

Figure 10.20 Problem 10.1.

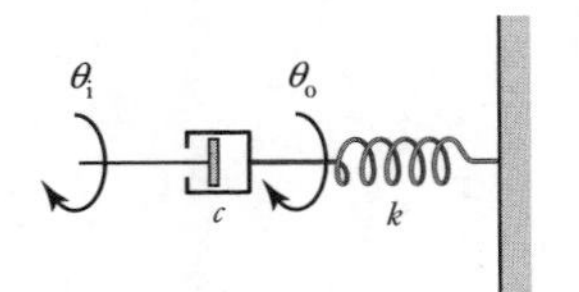

Figure 10.21 Problem 10.3.

10.2 Propose a model for the metal wheel of a railway carriage running on a metal track.

10.3 Derive an equation relating the input angular displacement θ_i with the output angular displacement θ_o for the rotational system shown in Figure 10.21.

10.4 Propose a model for a stepped shaft (i.e. a shaft where there is a step change in diameter) used to rotate a mass and derive an equation relating the input torque and the angular rotation. You may neglect damping.

10.5 Derive the relationship between the output, the potential difference across the resistor R of v_R, and the input v for the circuit shown in Figure 10.22 which has a resistor in series with a capacitor.

10.6 Derive the relationship between the output, the potential difference across the resistor R of v_R, and the input v for the series LCR circuit shown in Figure 10.23.

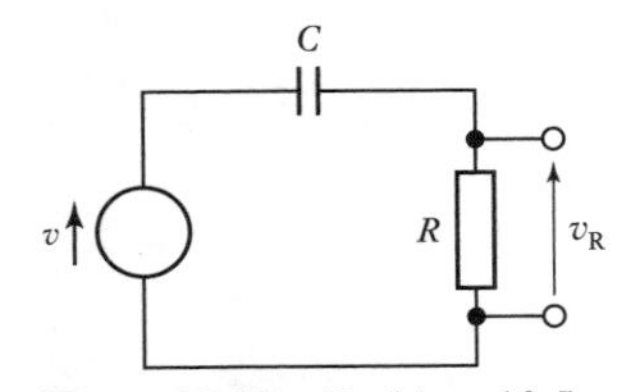

Figure 10.22 Problem 10.5.

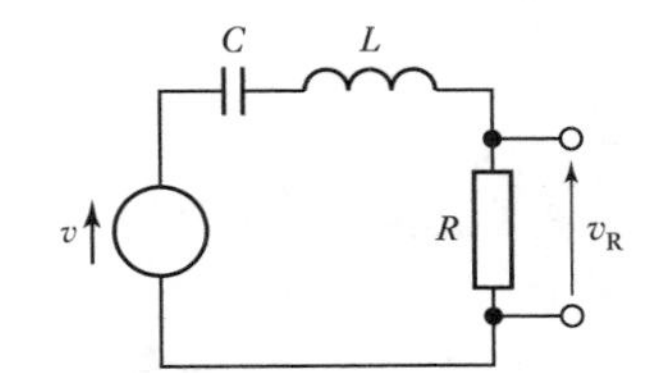

Figure 10.23 Problem 10.6.

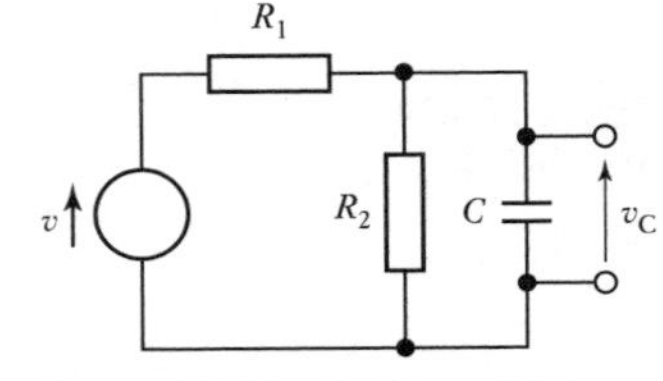

Figure 10.24 Problem 10.7.

10.7 Derive the relationship between the output, the potential difference across the capacitor C of v_C, and the input v for the circuit shown in Figure 10.24.

10.8 Derive the relationship between the height h_2 and time for the hydraulic system shown in Figure 10.25. Neglect inertance.

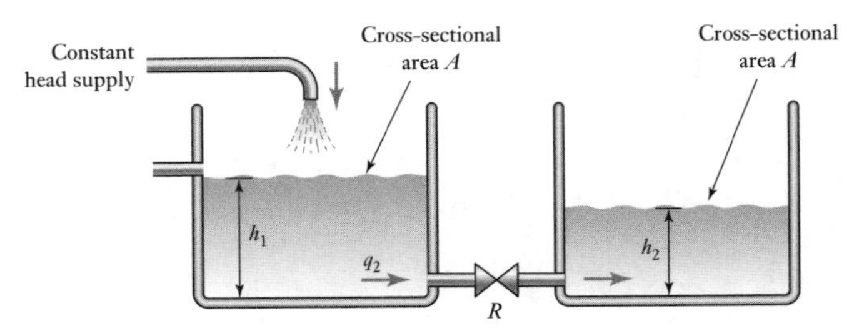

Figure 10.25 Problem 10.8.

10.9 A hot object, capacitance C and temperature T, cools in a large room at temperature T_r. If the thermal system has a resistance R, derive an equation describing how the temperature of the hot object changes with time and give an electrical analogue of the system.

10.10 Figure 10.26 shows a thermal system involving two compartments, with one containing a heater. If the temperature of the compartment containing the heater is T_1, the temperature of the other compartment T_2 and the temperature surrounding the compartments T_3, develop equations describing how the temperatures T_1 and T_2 will vary with time. All the walls of the containers have the same resistance and negligible capacitance. The two containers have the same capacitance C.

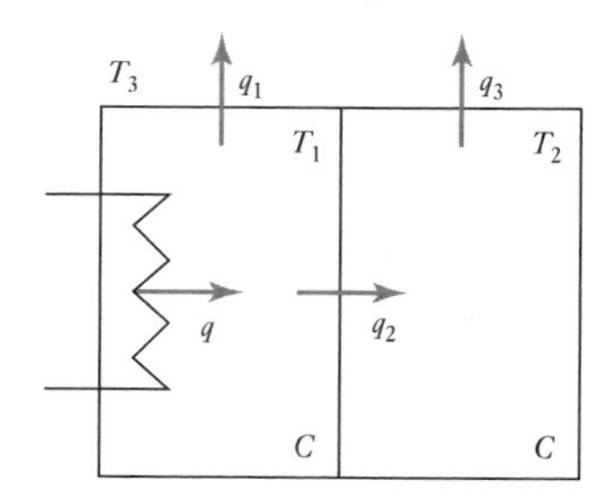

Figure 10.26 Problem 10.10.

10.11 Derive the differential equation relating the pressure input p to a diaphragm actuator (as in Figure 7.23) to the displacement x of the stem.

10.12 Derive the differential equation for a motor driving a load through a gear system (Figure 10.27) which relates the angular displacement of the load with time.

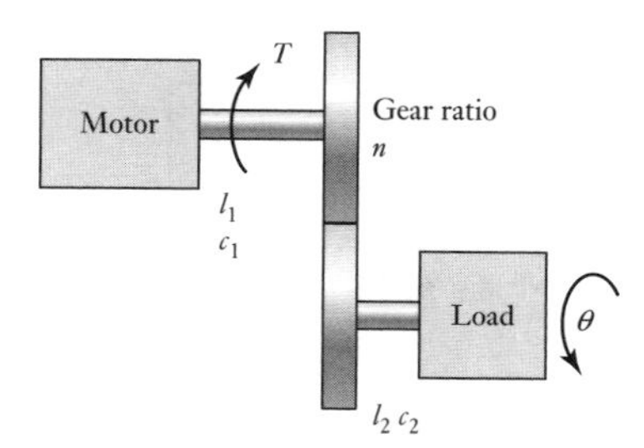

Figure 10.27 Problem 10.12.

Chapter eleven System Models

Objectives

The objectives of this chapter are that, after studying it, the reader should be able to:

- **Devise models for rotational–translational, electromechanical and hydraulic—mechanical systems.**
- **Linearise non-linear relationships in order to generate linear models.**

11.1 Engineering systems

In Chapter 10 the basic building blocks of translational mechanical, rotational mechanical, electrical, fluid and thermal systems were separately considered. However, many systems encountered in engineering involve aspects of more than one of these disciplines. For example, an electric motor involves both electrical and mechanical elements. This chapter looks at how single-discipline building blocks can be combined to give models for such multidiscipline systems and also addresses the issue that often real components are not linear. For example, in considering a spring the simple model assumes that the force and extension are proportional, regardless of how large the force was. The mathematical model might thus be a simplification of a real spring. Non-linear models are, however, much more difficult to deal with and so engineers try to avoid them and a non-linear system might be approximated by a linear model.

11.2 Rotational–translational systems

There are many mechanisms which involve the conversion of rotational motion to translational motion or vice versa. For example, there are rack-and-pinion, shafts with lead screws, pulley and cable systems, etc.

To illustrate how such systems can be analysed, consider a rack-and-pinion system (Figure 11.1). The rotational motion of the pinion is transformed into translational motion of the rack. Consider first the pinion element. The net torque acting on it is $(T_{in} - T_{out})$. Thus, considering the moment of inertia element, and assuming negligible damping,

$$T_{in} - T_{out} = I\frac{d\omega}{dt}$$

where I is the moment of inertia of the pinion and ω its angular velocity. The rotation of the pinion will result in a translational velocity v of the rack. If the

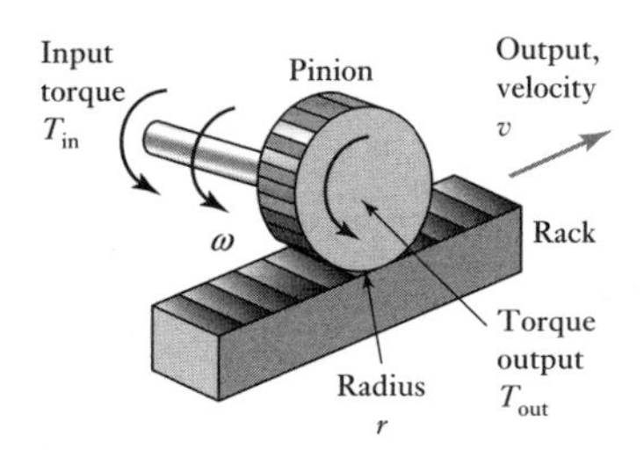

Figure 11.1 Rack-and-pinion.

pinion has a radius r, then $v = r\omega$. Hence we can write

$$T_{in} - T_{out} = \frac{I}{r}\frac{dv}{dt}$$

Now consider the rack element. There will be a force of T/r acting on it due to the movement of the pinion. If there is a frictional force of cv then the net force is

$$\frac{T_{out}}{r} - cv = m\frac{dv}{dt}$$

Eliminating T_{out} from the two equations gives

$$T_{in} - rcv = \left(\frac{I}{r} + mr\right)\frac{dv}{dt}$$

and so

$$\frac{dv}{dt} = \left(\frac{r}{1 + mr^2}\right)(T_{in} - rcv)$$

The result is a first-order differential equation describing how the output is related to the input.

11.3 Electromechanical systems

Electromechanical devices, such as potentiometers, motors and generators, transform electrical signals to rotational motion or vice versa. This section is a discussion of how we can derive models for such systems. A potentiometer has an input of a rotation and an output of a potential difference. An electric motor has an input of a potential difference and an output of rotation of a shaft. A generator has an input of rotation of a shaft and an output of a potential difference.

11.3.1 Potentiometer

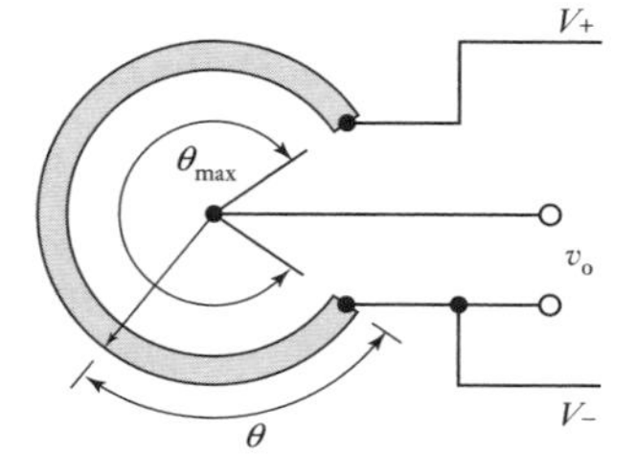

Figure 11.2 Rotary potentiometer.

The **rotary potentiometer** (Figure 11.2) is a potential divider and thus

$$\frac{v_0}{V} = \frac{\theta}{\theta_{max}}$$

where V is the potential difference across the full length of the potentiometer track and θ_{max} is the total angle swept out by the slider in being rotated from one end of the track to the other. The output is v_o for the input θ.

11.3.2 D.C. motor

The d.c. motor is used to convert an electrical input signal into a mechanical output signal, a current through the armature coil of the motor resulting in a shaft being rotated and hence the load rotated (Figure 11.3).

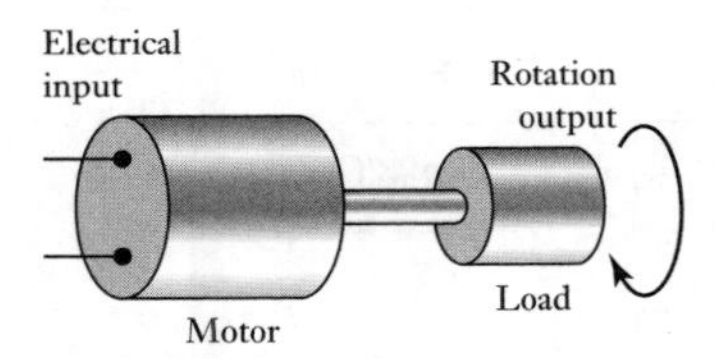

Figure 11.3 Motor driving a load.

The motor basically consists of a coil, the armature coil, which is free to rotate. This coil is located in the magnetic field provided by a current through field coils or a permanent magnet. When a current i_a flows through the armature coil, then, because it is in a magnetic field, forces act on the coil and cause it to rotate (Figure 11.4). The force F acting on a wire carrying a current i_a and of length L in a magnetic field of flux density B at right angles to the wire is given by $F = Bi_aL$ and with N wires is $F = Nbi_aL$. The forces on the armature coil wires result in a torque T, where $T = Fb$, with b being the breadth of the coil. Thus

$$T = NBi_aLb$$

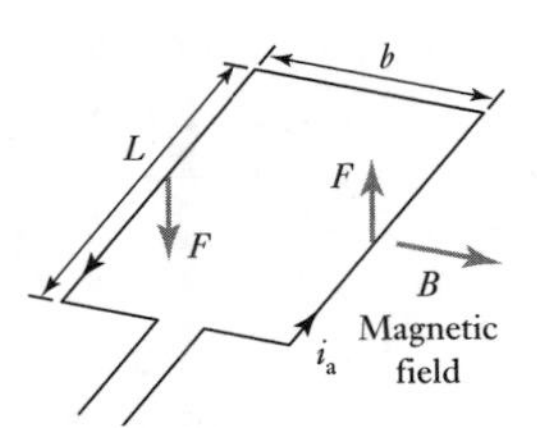

Figure 11.4 One wire of armature coil.

The resulting torque is thus proportional to (Bi_a), the other factors all being constants. Hence we can write

$$T = k_1Bi_a$$

Since the armature is a coil rotating in a magnetic field, a voltage will be induced in it as a consequence of electromagnetic induction. This voltage will be in such a direction as to oppose the change producing it and is called the back e.m.f. This back e.m.f. v_b is proportional to the rate or rotation of the armature and the flux linked by the coil, hence the flux density B. Thus

$$v_b = k_2B\omega$$

where ω is the shaft angular velocity and k_2 a constant.

Consider a d.c. motor which has the armature and field coils separately excited. With a so-called **armature-controlled motor** the field current i_f is held constant and the motor controlled by adjusting the armature voltage v_a. A constant field current means a constant magnetic flux density B for the armature coil. Thus

$$v_b = k_2B\omega = k_3\omega$$

where k_3 is a constant. The armature circuit can be considered to be a resistance R_a in series with an inductance L_a (Figure 11.5).

If v_a is the voltage applied to the armature circuit then, since there is a back e.m.f. of v_b, we have

$$v_a - v_b = L_a\frac{di_a}{dt} + R_ai_a$$

Figure 11.5 D.C. motor circuits.

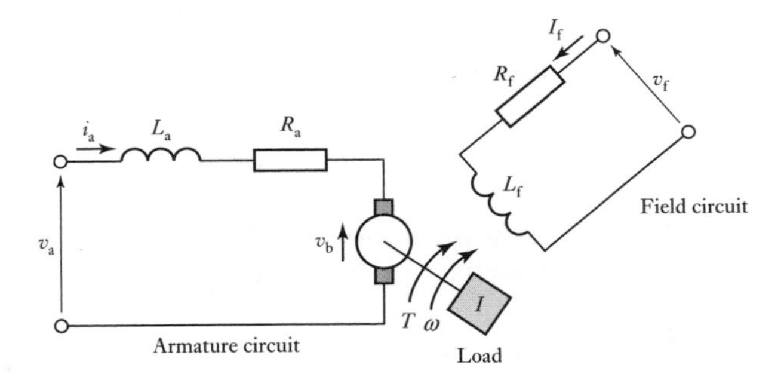

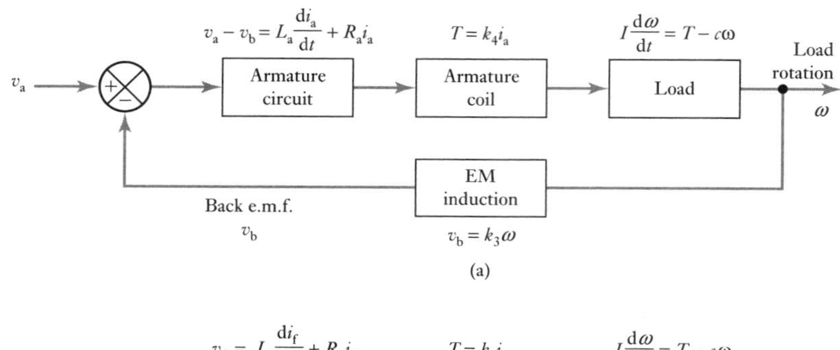

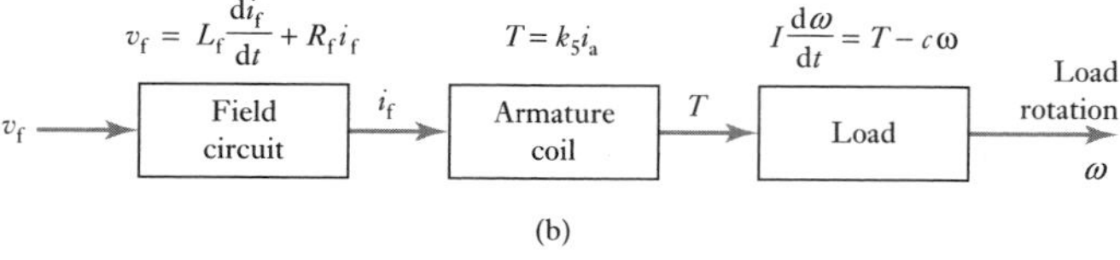

Figure 11.6 D.C. motors: (a) armature-controlled, (b) field-controlled.

We can think of this equation in terms of the block diagram shown in Figure 11.6(a). The input to the motor part of the system is v_a and this is summed with the feedback signal of the back e.m.f. v_b to give an error signal which is the input to the armature circuit. The above equation thus describes the relationship between the input of the error signal to the armature coil and the output of the armature current i_a. Substituting for v_b,

$$v_a - k_3\omega = L_a \frac{di_a}{dt} + R_a i_a$$

The current i_a in the armature generates a torque T. Since, for the armature-controlled motor, B is constant we have

$$T = k_1 B i_a = k_4 i_a$$

where k_4 is a constant. This torque then becomes the input to the load system. The net torque acting on the load will be

$$\text{net torque} = T - \text{damping torque}$$

The damping torque is $c\omega$, where c is a constant. Hence, if any effects due to the torsional springiness of the shaft are neglected,

$$\text{net torque} = k_4 i_a - c\omega$$

This will cause an angular acceleration of $d\omega/dt$, hence

$$I \frac{d\omega}{dt} = k_4 i_a - c\omega$$

We thus have two equations that describe the conditions occurring for an armature-controlled motor, namely

$$v_a - k_3\omega = L_a \frac{di_a}{dt} + R_a i_a \quad \text{and} \quad I \frac{d\omega}{dt} = k_4 i_a - c\omega$$

We can thus obtain the equation relating the output ω with the input v_a to the system by eliminating i_a. See the brief discussion of the Laplace transform in Chapter 13, or that in Appendix A, for details of how this might be done.

With a so-called **field-controlled motor** the armature current is held constant and the motor controlled by varying the field voltage. For the field circuit (Figure 11.5) there is essentially just inductance L_f in series with a resistance R_f. Thus for that circuit

$$v_f = R_f i_f + L_f \frac{di_f}{dt}$$

We can think of the field-controlled motor in terms of the block diagram shown in Figure 11.6(b). The input to the system is v_f. The field circuit converts this into a current i_f, the relationship between v_f and i_f being the above equation. This current leads to the production of a magnetic field and hence a torque acting on the armature coil, as given by $T = k_1 B i_a$. But the flux density B is proportional to the field current i_f and i_a is constant, hence

$$T = k_1 B i_a = k_5 i_f$$

where k_5 is a constant. This torque output is then converted by the load system into an angular velocity ω. As earlier, the net torque acting on the load will be

$$\text{net torque} = T - \text{damping torque}$$

The damping torque is $c\omega$, where c is a constant. Hence, if any effects due to the torsional springiness of the shaft are neglected,

$$\text{net torque} = k_5 i_f - c\omega$$

This will cause an angular acceleration of $d\omega/dt$, hence

$$I\frac{d\omega}{dt} = k_5 i_f - c\omega$$

The conditions occurring for a field-controlled motor are thus described by the equations

$$v_f = R_f i_f + L_f \frac{di_f}{dt} \quad \text{and} \quad I\frac{d\omega}{dt} = k_5 i_f - c\omega$$

We can thus obtain the equation relating the output ω with the input v_f to the system by eliminating i_f. See the brief discussion of the Laplace transform in Chapter 13, or that in Appendix A, for details of how this might be done.

11.4 Linearity

In combining blocks to create models of systems we are assuming that the relationship for each block is linear. The following is a brief discussion of linearity and how, because many real engineering items are non-linear, we need to make a linear approximation for a non-linear item.

The relationship between the force F and the extension x produced for an ideal spring is linear, being given by $F = kx$. This means that if force F_1 produces an extension x_1 and force F_2 produces an extension x_2, a force equal to $(F_1 + F_2)$ will produce an extension $(x_1 + x_2)$. This is called the **principle of superposition** and is a necessary condition for a system that can be termed a **linear system.** Another condition for a linear system is that if an input

F_1 produces an extension x_1, then an input cF_1 will produce an output cx_1, where c is a constant multiplier.

A graph of the force F plotted against the extension x is a straight line passing through the origin when the relationship is linear (Figure 11.7(a)). Real springs, like any other real components, are not perfectly linear (Figure 11.7(b)). However, there is often a range of operation for which linearity can be assumed. Thus for the spring giving the graph in Figure 11.7(b), linearity can be assumed provided the spring is only used over the central part of its graph. For many system components, linearity can be assumed for operations within a range of values of the variable about some operating point.

Figure 11.7 Springs: (a) ideal, (b) real.

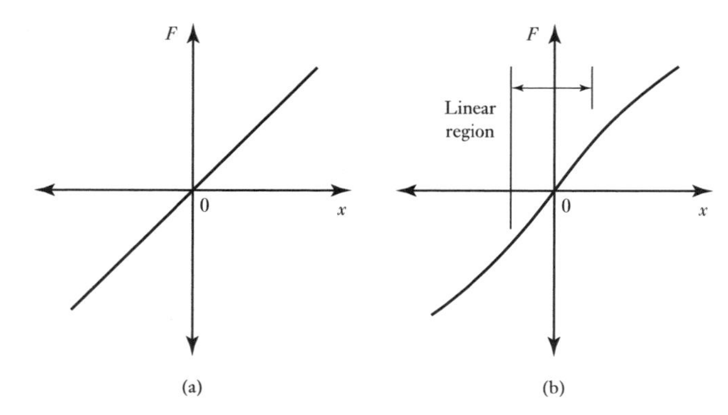

For some system components (Figure 11.8(a)) the relationship is non-linear. For such components the best that can be done to obtain a linear relationship is just to work with the straight line which is the slope of the graph at the operating point.

Figure 11.8 A non-linear relationship.

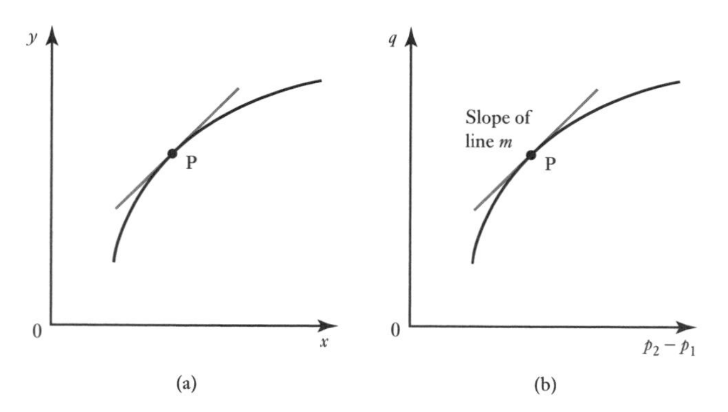

Thus for the relationship between y and x in Figure 11.8(a), at the operating point P where the slope has the value m

$$\Delta y = m\,\Delta x$$

where Δy and Δx are small changes in input and output signals at the operating point.

For example, the rate of flow of liquid q through an orifice is given by

$$q = c_d A \sqrt{\frac{2(p_1 - p_2)}{\rho}}$$

where c_d is a constant called the discharge coefficient, A the cross-sectional area of the orifice, ρ the fluid density and $(p_1 - p_2)$ the pressure difference. For a constant cross-sectional area and density the equation can be written as

$$q = C\sqrt{p_1 - p_2}$$

where C is a constant. This is a non-linear relationship between the rate of flow and the pressure difference. We can obtain a linear relationship by considering the straight line representing the slope of the rate of flow/pressure difference graph (Figure 11.8(b)) at the operating point. The slope m is $\mathrm{d}q/\mathrm{d}(p_1 - p_2)$ and has the value

$$m = \frac{\mathrm{d}q}{\mathrm{d}(p_1 - p_2)} = \frac{C}{2\sqrt{p_{o1} - p_{o2}}}$$

where $(p_{o1} - p_{o2})$ is the value at the operating point. For small changes about the operating point we will assume that we can replace the non-linear graph by the straight line of slope m and therefore can write $m = \Delta q/\Delta(p_1 - p_2)$ and hence

$$\Delta q = m\,\Delta(p_1 - p_2)$$

Hence, if we had $C = 2\ \mathrm{m^3/s}$ per kPa, i.e. $q = 2(p_1 - p_2)$, then for an operating point of $(p_1 - p_2) = 4$ kPa with $m = 2/(2\sqrt{4}) = 0.5$, the linearised version of the equation would be

$$\Delta q = 0.5\,\Delta(p_1 - p_2)$$

Linearised mathematical models are used because most of the techniques of control systems are based on there being linear relationships for the elements of such systems. Also, because most control systems are maintaining an output equal to some reference value, the variations from this value tend to be rather small and so the linearised model is perfectly appropriate.

11.5 Hydraulic–mechanical systems

Hydraulic–mechanical converters involve the transformation of hydraulic signals to translational or rotational motion, or vice versa. Thus, for example, the movement of a piston in a cylinder as a result of hydraulic pressure involves the transformation of a hydraulic pressure input to the system to a translational motion output.

Figure 11.9 shows a hydraulic system in which an input of displacement x_i is, after passing through the system, transformed into a displacement x_o of a load. The system consists of a spool valve and a cylinder. The input displacement x_i to the left results in the hydraulic fluid supply pressure p_s causing fluid to flow into the left-hand side of the cylinder. This pushes the piston in the cylinder to the right and expels the fluid in the right-hand side of the chamber through the exit port at the right-hand end of the spool valve.

The rate of flow of fluid to and from the chamber depends on the extent to which the input motion has uncovered the ports allowing the fluid to enter or leave the spool valve. When the input displacement x_i is to the right, the

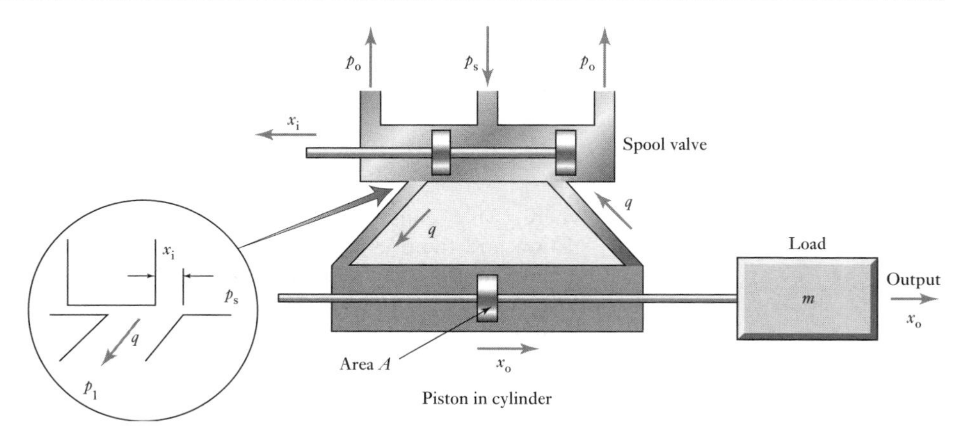

Figure 11.9 Hydraulic system and load.

spool valve allows fluid to move to the right-hand end of the cylinder and so results in a movement of the piston in the cylinder to the left.

The rate of flow of fluid q through an orifice, which is what the ports in the spool valve are, is a non-linear relationship depending on the pressure difference between the two sides of the orifice and its cross-sectional area A. However, a linearised version of the equation can be used (see the previous section for its derivation)

$$\Delta q = m_1 \Delta A + m_2 \Delta \text{ (pressure difference)}$$

where m_1 and m_2 are constants at the operating point. For the fluid entering the chamber, the pressure difference is $(p_s - p_1)$ and for the exit $(p_2 - p_o)$. If the operating point about which the equation is linearised is taken to be the point at which the spool valve is central and the ports connecting it to the cylinder are both closed, then for this condition q is zero, and so $\Delta q = q$, A is proportional to x_s if x_s is measured from this central position, and the change in pressure on the inlet side of the piston is $-\Delta p_1$ relative to p_s and on the exit side Δp_2 relative to p_o. Thus, for the inlet port the equation can be written as

$$q = m_1 x_i + m_2(-\Delta p_1)$$

and for the exit port

$$q = m_1 x_i + m_2 \Delta p_2$$

Adding the two equations gives

$$2q = 2m_1 x_i - m_2(\Delta p_1 - \Delta p_2)$$

$$q = m_1 x_i - m_3(\Delta p_1 - \Delta p_2)$$

where $m_3 = m_2/2$.

For the cylinder, the change in the volume of fluid entering the left-hand side of the chamber, or leaving the right-hand side, when the piston moves a distance x_o is Ax_o, where A is the cross-sectional area of the piston. Thus the rate at which the volume is changing is $A(\mathrm{d}x_o/\mathrm{d}t)$. The rate at which fluid is entering the left-hand side of the cylinder is q. However, since there is some leakage flow of fluid from one side of the piston to the other,

$$q = A\frac{dx_o}{dt} + q_L$$

where q_L is the rate of leakage. Substituting for q gives

$$m_1x_i - m_3(\Delta p_1 - \Delta p_2) = A\frac{dx_o}{dt} + q_L$$

The rate of leakage flow q_L is a flow through an orifice, the gap between the piston and the cylinder. This is of constant cross-section and has a pressure difference $(\Delta p_1 - \Delta p_2)$. Hence, using the linearised equation for such a flow,

$$q_L = m_4(\Delta p_1 - \Delta p_2)$$

Thus, using this equation to substitute for q_L,

$$m_1x_i - m_3(\Delta p_1 - \Delta p_2) = A\frac{dx_o}{dt} + m_4(\Delta p_1 - \Delta p_2)$$

$$m_1x_i - (m_3 + m_4)(\Delta p_1 - \Delta p_2) = A\frac{dx_o}{dt}$$

The pressure difference across the piston results in a force being exerted on the load, the force exerted being $(\Delta p_1 - \Delta p_2)A$. There is, however, some damping of motion, i.e. friction, of the mass. This is proportional to the velocity of the mass, i.e. (dx_o/dt). Hence the net force acting on the load is

$$\text{net force} = (\Delta p_1 - \Delta p_2)A - c\frac{dx_o}{dt}$$

This net force causes the mass to accelerate, the acceleration being (d^2x_o/dt^2). Hence

$$m\frac{d^2x_o}{dt^2} = (\Delta p_1 - \Delta p_2)A - c\frac{dx_o}{dt}$$

Rearranging this equation gives

$$\Delta p_1 - \Delta p_2 = \frac{m}{A}\frac{d^2x_o}{dt^2} + \frac{c}{A}\frac{dx_o}{dt}$$

Using this equation to substitute for the pressure difference in the earlier equation,

$$m_1x_i - (m_3 + m_4)\left(\frac{m}{A}\frac{d^2x_o}{dt^2} + \frac{c}{A}\frac{dx_o}{dt}\right) = A\frac{dx_o}{dt}$$

Rearranging gives

$$\frac{(m_3 + m_4)m}{A}\frac{d^2x_o}{dt^2} + \left(A + \frac{c(m_3 + m_4)}{A}\right)\frac{dx_o}{dt} = m_1x_i$$

and rearranging this equation leads to

$$\frac{(m_3 + m_4)m}{A^2 + c(m_3 + m_4)}\frac{d^2x_o}{dt^2} + \frac{dx_o}{dt} = \frac{Am_1}{A^2 + c(m_3 + m_4)}x_i$$

This equation can be simplified by introducing two constants k and τ, the latter constant being called the time constant (see Chapter 12). Hence

$$\tau\frac{d^2x_o}{dt^2} + \frac{dx_o}{dt} = kx_i$$

Thus the relationship between input and output is described by a second-order differential equation.

Summary

Many systems encountered in engineering involve aspects of more than one discipline and these can be considered by examining how the system can be built up from single-discipline building blocks.

A system is said to be linear when its basic equations, whether algebraic or differential, are such that the magnitude of the output produced is directly proportional to the input. For an algebraic equation, this means that the graph of output plotted against input is a straight line passing through the origin. So doubling the input doubles the output. For a linear system we can obtain the output of the system to a number of inputs by adding the outputs of the system to each individual input considered separately. This is called the **principle of superposition**.

Problems

11.1 Derive a differential equation relating the input voltage to a d.c. servo motor and the output angular velocity, assuming that the motor is armature controlled and the equivalent circuit for the motor has an armature with just resistance, its inductance being neglected.

11.2 Derive differential equations for a d.c. generator. The generator may be assumed to have a constant magnetic field. The armature circuit has the armature coil, having both resistance and inductance, in series with the load. Assume that the load has both resistance and inductance.

11.3 Derive differential equations for a permanent magnet d.c. motor.

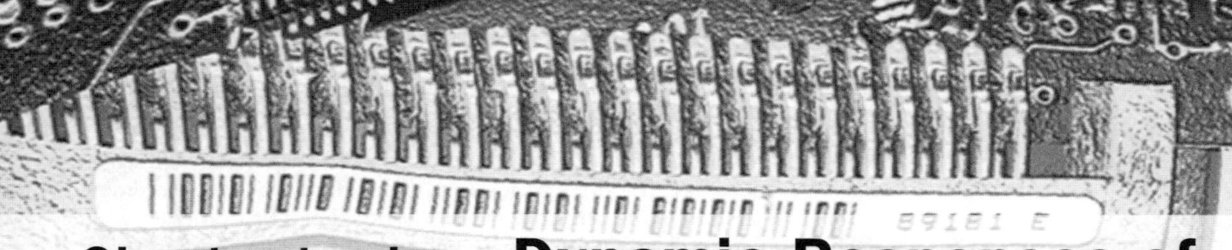

Chapter twelve Dynamic Responses of Systems

Objectives

The objectives of this chapter are that, after studying it, the reader should be able to:

- **Model dynamic systems by means of differential equations.**
- **Determine the outputs of first-order systems to inputs and determine time constants.**
- **Determine the outputs of second-order systems to inputs and identify the under-damped, critically damped and over-damped conditions.**
- **Describe the characteristics of second-order system responses in terms of rise time, overshoot, subsidence ratio, decrement and settling time.**

12.1 Modelling dynamic systems

The most important function of a model devised for measurement or control systems is to be able to predict what the output will be for a particular input. We are not just concerned with a static situation, i.e. that after some time when the steady state has been reached an output of x corresponds to an input of y. We have to consider how the output will change with time when there is a change of input or when the input changes with time. For example, how will the temperature of a temperature-controlled system change with time when the thermostat is set to a new temperature? For a control system, how will the output of the system change with time when the set value is set to a new value or perhaps increased at a steady rate?

Chapters 10 and 11 were concerned with models of systems when the inputs varied with time, with the results being expressed in terms of differential equations. This chapter is about how we can use such models to make predictions about how outputs will change with time when the input changes with time.

12.1.1 Differential equations

To describe the relationship between the input to a system and its output we must describe the relationship between inputs and outputs which are both possible functions of time. We thus need a form of equation which will indicate how the system output will vary with time when the input is varying with time. This can be done by the use of a differential equation. Such an equation includes derivatives with respect to time and so gives information about how the response of a system varies with time. A derivative $\mathrm{d}x/\mathrm{d}t$ describes the rate at which x varies with time; the derivative d^2x/dt^2 states how dx/dt varies with time. Differential equations can be classed as first-order, second-order, third-order, etc., according to the highest order of the derivative in the equation. For a first-order equation the highest order will be dx/dt, with a second-order d^2x/dt^2, with a third-order d^3x/dt^3,with nth-order d^nx/dt^n.

This chapter is about the types of responses we can expect from first-order and second-order systems and the solution of such differential equations in order that the response of the system to different types of input can be obtained. This chapter uses the 'try a solution' approach in order to find a solution; the Laplace transformation method is introduced in Chapter 13.

12.2 Terminology

In this section we look at some of the terms that are used in describing the dynamic responses of systems.

12.2.1 Natural and forced responses

The term **natural response** is used for a system when there is no input to the system forcing the variable to change but it is just changing naturally. As an illustration, consider the first-order system of water being allowed naturally to flow out of a tank (Figure 12.1(a)).

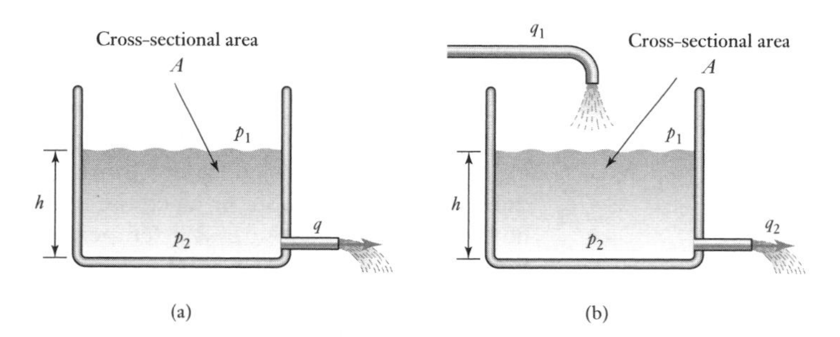

Figure 12.1 Water flowing out of a tank: (a) naturally with no input, (b) with forcing input.

For such a system we have

$$p_1 - p_2 = Rq$$

where R is the hydraulic resistance. But $p_1 - p_2 = h\rho g$, where ρ is the density of the water, and q is the rate at which water leaves the tank and so is $-dV/dt$, with V being the volume of water in the tank and so being Ah. Thus $q = -d(Ah)/dt = -Adh/dt$ and so the above equation can be written as

$$h\rho g = -RA\frac{dh}{dt}$$

This is the natural response in that there is no input to the system forcing the variable h to change; it is just naturally changing with time. We can draw attention to this by writing the differential equation with all the output terms, i.e. h, on the same side of the equals sign and the input term of zero on the right, i.e.

$$RA\frac{dh}{dt} + (\rho g)h = 0$$

In Section 10.4.1 the differential equation was derived for a water tank from which water was flowing but also into which there was a flow of water (Figure 12.1(b)). This equation has a forcing input function of q_1 and can be written as

$$RA\frac{dh}{dt} + (\rho g)h = q_1$$

As another example, consider a thermometer being placed in a hot liquid at some temperature T_L. The rate at which the reading of the thermometer T changes with time was derived in Section 10.5.1 as being given by the differential equation

$$RC\frac{dT}{dt} + T = T_L$$

Such a differential equation has a forcing input of T_L.

12.2.2 Transient and steady-state responses

The total response of a control system, or element of a system, can be considered to be made up of two aspects, the steady-state response and the transient response. The **transient response** is that part of a system response which occurs as a result of a change in input and which dies away after a short interval of time. The **steady-state response** is the response that remains after all transient responses have died down.

To give a simple illustration of this, consider a vertically suspended spring (Figure 12.2) and what happens when a weight is suddenly suspended from it. The deflection of the spring abruptly increases and then may well oscillate until after some time it settles down to a steady value. The steady value is the steady-state response of the spring system; the oscillation that occurs prior to this steady state is the transient response.

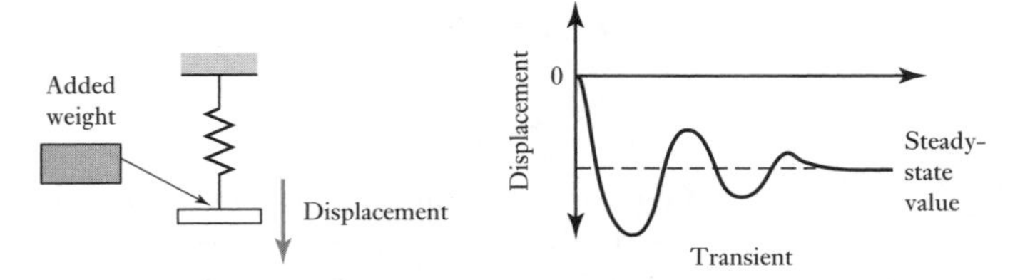

Figure 12.2 Transient and steady-state responses of a spring system.

12.2.3 Forms of inputs

The input to the above spring system, the weight, is a quantity which varies with time. Up to some particular time there is no added weight, i.e. no input, then after that time there is an input which remains constant for the rest of the time. This type of input is known as a **step input** and is of the form shown in Figure 12.3(a).

The input signal to systems can take other forms, e.g. impulse, ramp and sinusoidal signals. An **impulse** is a very short-duration input (Figure 12.3(b)); a **ramp** is a steadily increasing input (Figure 12.3(c)) and can be described by

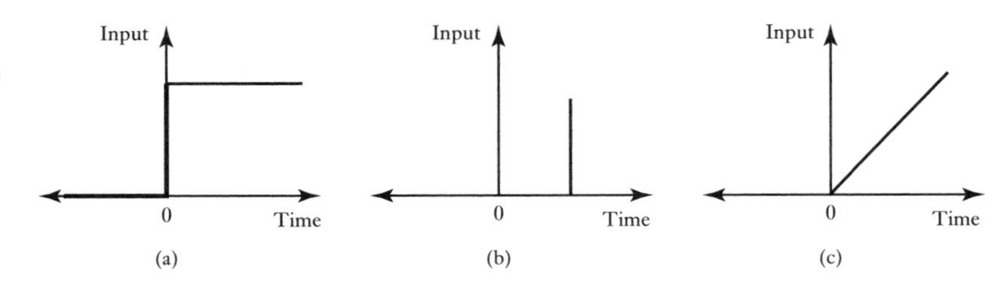

Figure 12.3 Inputs: (a) step at time 0, (b) impulse at some time, (c) ramp starting at time 0.

an equation of the form $y = kt$, where k is a constant; and a **sinusoidal** input can be described by an equation of the form $y = k \sin \omega t$, with ω being the so-called angular frequency and equal to $2\pi f$ where f is the frequency.

Both the input and the output are functions of time. One way of indicating this is to write them in the form $f(t)$, where f is the function and (t) indicates that its value depends on time t. Thus for the weight W input to the spring system we could write $W(t)$ and for the deflection d output $d(t)$. $y(t)$ is commonly used for an input and $x(t)$ for an output.

12.3 First-order systems

Consider a first-order system with $y(t)$ as the input to the system and $x(t)$ the output and which has a forcing input $b_0 y$ and can be described by a differential equation of the form

$$a_1 \frac{dx}{dt} + a_0 x = b_0 y$$

where a_1, a_0 and b_0 are constants.

12.3.1 Natural response

The input $y(t)$ can take many forms. Consider first the situation when the input is zero. Because there is no input to the system we have no signal forcing the system to respond in any way other than its natural response with no input. The differential equation is then

$$a_1 \frac{dx}{dt} + a_0 x = 0$$

We can solve this equation by using the technique called **separation of variables.** The equation can be written with all the x variables on one side and all the t variables on the other:

$$\frac{dx}{x} = -\frac{a_0}{a_1} dt$$

Integrating this between the initial value of $x = 1$ at $t = 0$, i.e. a unit step input, and x at t gives

$$\ln x = -\frac{a_0}{a_1} t$$

and so we have

$$x = e^{-a_0 t / a_1}$$

We could, however, have recognised that the differential equation would have a solution of the form $x = Ae^{st}$, where A and s are constants. We then have $dx/dt = sAe^{st}$ and so when these values are substituted in the differential equation we obtain

$$a_1 sAe^{st} + a_0 Ae^{st} = 0$$

and so $a_1 s + a_0 = 0$ and $s = -a_0/a_1$. Thus the solution is

$$x = Ae^{-a_0 t/a_1}$$

This is termed the natural response since there is no forcing function. We can determine the value of the constant A given some initial (boundary) condition. Thus if $x = 1$ when $t = 0$ then $A = 1$. Figure 12.4 shows the natural response, i.e. an exponential decay:

$$x = e^{-a_0 t/a_1}$$

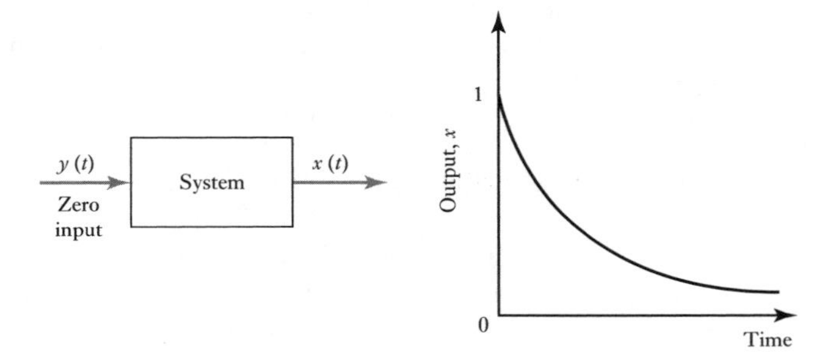

Figure 12.4 Natural response of a first-order system.

12.3.2 Response with a forcing input

Now consider the differential equation when there is a **forcing function**, i.e.

$$a_1 \frac{dx}{dt} + a_0 x = b_0 y$$

Consider the solution to this equation to be made up of two parts, i.e. $x = u + v$. One part represents the transient part of the solution and the other the steady-state part. Substituting this into the differential equation gives

$$a_1 \frac{d(u + v)}{dt} + a_0(u + v) = b_0 y$$

Rearranging this gives

$$\left(a_1 \frac{du}{dt} + a_0 u\right) + \left(a_1 \frac{dv}{dt} + a_0 v\right) = b_0 y$$

If we let

$$a_1 \frac{dv}{dt} + a_0 v = b_0 y$$

then we must have

$$a_1 \frac{du}{dt} + a_0 u = 0$$

and so two differential equations, one of which contains a forcing function and one which is just the natural response equation. This last equation is just the natural equation which we solved earlier in this section and so has a solution of the form

$$u = Ae^{-a_0 t/a_1}$$

The other differential equation contains the forcing function y. For this differential equation the form of solution we try depends on the form of the input signal y. For a step input when y is constant for all times greater than 0, i.e. $y = k$, we can try a solution $v = A$, where A is a constant. If we have an input signal of the form $y = a + bt + ct^2 + \dots$, where a, b and c are constants which can be zero, then we can try a solution which is of the form $v = A + Bt + Ct^2 + \dots$. For a sinusoidal signal we can try a solution of the form $v = A \cos \omega t + B \sin \omega t$.

To illustrate this, assume there is a step input at a time of $t = 0$ with the size of the step being k (Figure 12.5(a)). Then we try a solution of the form $v = A$. Differentiating a constant gives zero; thus when this solution is substituted into the differential equation we obtain $a_0 A = b_0 k$ and so $v = (b_0/a_0)k$.

The full solution will be given by $x = u + v$ and so will be

$$x = Ae^{-a_0 t/a_1} + \frac{b_0}{a_0}k$$

We can determine the value of the constant A given some initial (boundary) conditions. Thus if the output $x = 0$ when $t = 0$ then

$$0 = A + \frac{b_0}{a_0}k$$

Thus $A = -(b_0/a_0)k$. The solution then becomes

$$x = \frac{b_0}{a_0}k(1 - e^{-a_0 t/a_1})$$

When $t \to \infty$ the exponential term tends to zero. The exponential term thus gives that part of the response which is the transient solution. The steady-state response is the value of x when $t \to \infty$ and so is $(b_0/a_0)k$. Thus the equation can be written as

$$x = \text{steady-state value} \times (1 - e^{-a_0 t/a_1})$$

Figure 12.5(b) shows a graph of how the output x varies with time for the step input.

12.3.3 Examples of first-order systems

As a further illustration of the above, consider the following examples of first-order systems.

An electrical transducer system consists of a resistance in series with a capacitor and when subject to a step input of size V gives an output of a potential difference across the capacitor v which is given by the differential equation

$$RC\frac{dv}{dt} + v = V$$

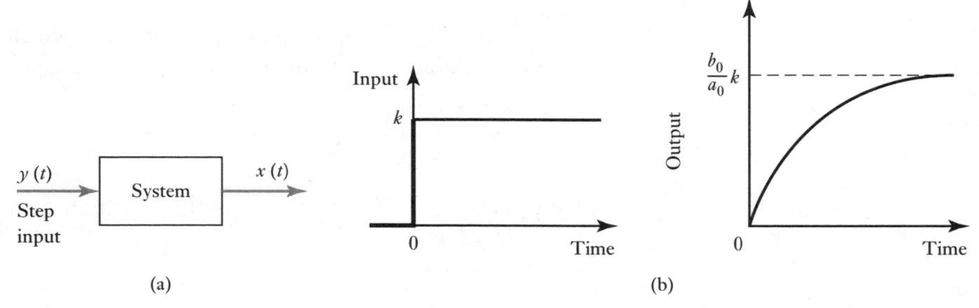

Figure 12.5 (a) Step input, (b) resulting output.

Comparing the differential equation with the equation solved earlier: $a_1 = RC$, $a_0 = 1$ and $b_0 = 1$. Then the solution is of the form

$$v = V(1 - e^{-t/RC})$$

Now consider an electric circuit consisting of a 1 MΩ resistance in series with a 2 μF capacitance. At a time $t = 0$ the circuit is subject to a ramp voltage of $4t$ V, i.e. the voltage increases at the rate of 4 V every 1 s. The differential equation will be of a similar form to that given in the previous example but with the step voltage V of that example replaced by the ramp voltage $4t$, i.e.

$$RC\frac{dv}{dt} + v = 4t$$

Thus, using the values given above,

$$2\frac{dv}{dt} + v = 4t$$

Taking $v = v_n + v_f$, i.e. the sum of the natural and forced responses, we have for the natural response

$$2\frac{dv_n}{dt} + v_n = 0$$

and for the forced response

$$2\frac{dv_f}{dt} + v_f = 4t$$

For the natural response differential equation we can try a solution of the form $v_n = Ae^{st}$. Hence, using this value

$$2Ase^{st} + Ae^{st} = 0$$

Thus $s = -\frac{1}{2}$ and so $v_n = Ae^{-t/2}$. For the forced response differential equation, since the right-hand side of the equation is $4t$ we can try a solution of the form $v_f = A + Bt$. Using this value gives $2B + A + Bt = 4t$. Thus we must have $B = 4$ and $A = -2B = -8$. Hence the solution is $v_f = -8 + 4t$. Thus the full solution is

$$v = v_n + v_f = Ae^{-t/2} - 8 + 4t$$

Since $v = 0$ when $t = 0$ we must have $A = 8$. Hence

$$v = 8e^{-t/2} - 8 + 4t$$

As a further example, consider a motor when the relationship between the output angular velocity ω and the input voltage v for the motor is given by

$$\frac{IR}{k_1 k_2}\frac{d\omega}{dt} + \omega = \frac{1}{k_1} v$$

Comparing the differential equation with the equation solved earlier, then $a_1 = IR/k_1 k_2$, $a_0 = 1$ and $b_0 = 1/k_1$. The steady-state value for a step input of size 1 V is thus $(b_0/a_0) = 1/k_1$.

12.3.4 The time constant

For a first-order system subject to a step input of size k we have an output y which varies with time t according to

$$x = \frac{b_0}{a_0} k(1 - e^{-a_0 t/a_1})$$

or

$$x = \text{steady-state value} \times (1 - e^{-a_0 t/a_1})$$

When the time $t = (a_1/a_0)$ then the exponential term has the value $e^{-1} = 0.37$ and

$$x = \text{steady-state value} \times (1 - 0.37)$$

In this time the output has risen to 0.63 of its steady-state value. This time is called the **time constant** τ:

$$\tau = \frac{a_1}{a_0}$$

In a time of $2(a_1/a_0) = 2\tau$, the exponential term becomes $e^{-2} = 0.14$ and so

$$x = \text{steady-state value} \times (1 - 0.14)$$

In this time the output has risen to 0.86 of its steady-state value. In a similar way, values can be calculated for the output after 3τ, 4τ, 5τ, etc. Table 12.1 shows the results of such calculations and Figure 12.6 the graph of how the output varies with time for a unit step input.

Table 12.1 Response of a first-order system to a step input.

Time t	Fraction of steady-state output
0	0
1τ	0.63
2τ	0.86
3τ	0.95
4τ	0.98
5τ	0.99
∞	1

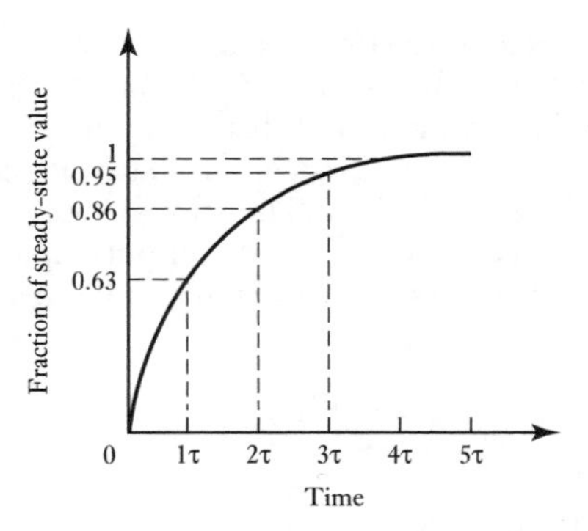

Figure 12.6 Response of a first-order system to a step input.

In terms of the time constant τ, we can write the equation describing the response of a first-order system as

$$x = \text{steady-state value} \times (1 - e^{-t/\tau})$$

The time constant τ is (a_1/a_0), thus we can write our general form of the first-order differential equation

$$a_1 \frac{dx}{dt} + a_0 x = b_0 y$$

as

$$\tau \frac{dx}{dt} + x = \frac{b_0}{a_0} y$$

But b_0/a_0 is the factor by which the input y is multiplied to give the steady-state value. We can term it the **steady-state gain** since it is the factor stating by how much bigger the output is than the input under steady-state conditions. Thus if we denote this by G_{SS} then the differential equation can be written in the form

$$\tau \frac{dx}{dt} + x = G_{SS} y$$

To illustrate this, consider Figure 12.7 which shows how the output v_o of a first-order system varies with time when subject to a step input of 5 V. The

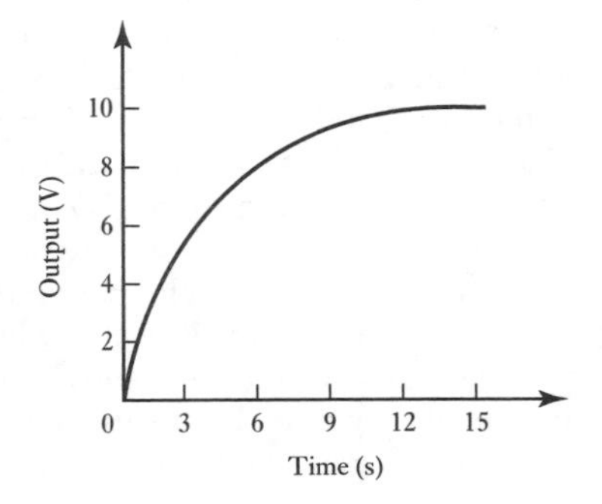

Figure 12.7 Example.

time constant is the time taken for a first-order system output to change from 0 to 0.63 of its final steady-state value. In this case this time is about 3 s. We can check this value, and that the system is first order, by finding the value at 2, i.e. 6 s. With a first-order system it should be 0.86 of the steady-state value. In this case it is. The steady-state output is 10 V. Thus the steady-state gain G_{SS} is (steady-state output/input) = 10/5 = 2. The differential equation for a first-order system can be written as

$$\tau \frac{dx}{dt} + x = G_{SS}y$$

Thus, for this system, we have

$$3 \frac{dv_o}{dt} + v_o = 2v_i$$

12.4 Second-order systems

Many second-order systems can be considered to be analogous to essentially just a stretched spring with a mass and some means of providing damping. Figure 12.8 shows the basis of such a system.

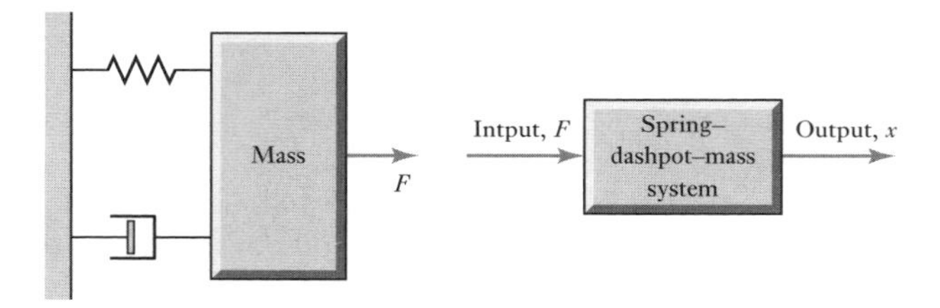

Figure 12.8 Spring–dashpot–mass system.

Such a system was analysed in Section 10.2.2. The equation describing the relationship between the input of force F and the output of a displacement x is

$$m \frac{d^2x}{dt^2} + c \frac{dx}{dt} + kx = F$$

where m is the mass, c the damping constant and k the spring constant.

The way in which the resulting displacement x will vary with time will depend on the amount of damping in the system. Thus if the force was applied as a step input and there was no damping at all then the mass would freely oscillate on the spring and the oscillations would continue indefinitely. No damping means $c = 0$ and so the dx/dt term is zero. However, damping will cause the oscillations to die away until a steady displacement of the mass is obtained. If the damping is high enough there will be no oscillations and the displacement of the mass will just slowly increase with time and gradually the mass will move towards its steady displacement position. Figure 12.9 shows the general way that the displacements, for a step input, vary with time with different degrees of damping.

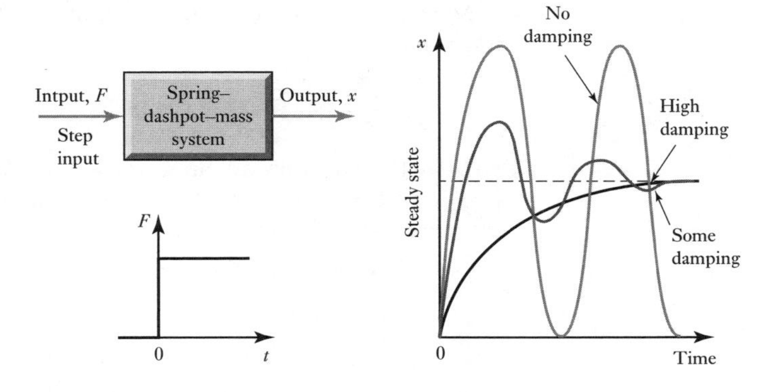

Figure 12.9 Effect of damping with a second-order system.

12.4.1 Natural response

Consider a mass on the end of a spring. In the absence of any damping and left to oscillate freely without being forced, the output of the second-order system is a continuous oscillation (simple harmonic motion). Thus, suppose we describe this oscillation by the equation

$$x = A \sin \omega_n t$$

where x is the displacement at a time t, A the amplitude of the oscillation and ω_n the angular frequency of the free undamped oscillations. Differentiating this gives

$$\frac{dx}{dt} = \omega_n A \cos \omega_n t$$

Differentiating a second time gives

$$\frac{d^2x}{dt^2} = -\omega_n^2 A \sin \omega_n t = -\omega_n^2 x$$

This can be reorganised to give the differential equation

$$\frac{d^2x}{dt^2} + \omega_n^2 x = 0$$

But for a mass m on a spring of stiffness k we have a restoring force of kx and thus

$$m\frac{d^2x}{dt^2} = -kx$$

This can be written as

$$\frac{d^2x}{dt^2} + \frac{k}{m}x = 0$$

Thus, comparing the two differential equations, we must have

$$\omega_n^2 = \frac{k}{m}$$

and $x = A \sin \omega_n t$ is the solution to the differential equation.

Now consider when we have damping. The motion of the mass is then described by

$$m\frac{d^2x}{dt^2} + c\frac{dx}{dt} + kx = 0$$

To solve this equation we can try a solution of the form $x_n = Ae^{st}$. This gives $dx_n/dt = Ase^{st}$ and $d^2x_n/dt^2 = As^2e^{st}$. Thus, substituting these values in the differential equation gives

$$mAs^2e^{st} + cAse^{st} + kAe^{st} = 0$$

$$ms^2 + cs + k = 0$$

Thus $x_n = Ae^{st}$ can only be a solution provided the above equation equals zero. This equation is called the **auxiliary equation**. The roots of the equation can be obtained by factoring or using the formula for the roots of a quadratic equation. Thus

$$s = \frac{-c \pm \sqrt{c^2 - 4mk}}{2m} = -\frac{c}{2m} \pm \sqrt{\left(\frac{c}{2m}\right)^2 - \frac{k}{m}}$$

$$= -\frac{c}{2m} \pm \sqrt{\frac{k}{m}\left(\frac{c^2}{4mk}\right) - \frac{k}{m}}$$

But $\omega_n^2 = k/m$ and so, if we let $\zeta^2 = c^2/4mk$, we can write the above equation as

$$s = -\zeta\omega_n \pm \omega_n\sqrt{\zeta^2 - 1}$$

ζ is termed the **damping factor**.

The value of s obtained from the above equation depends very much on the value of the square root term. Thus when ζ^2 is greater than 1 the square root term gives a square root of a positive number, and when ζ^2 is less than 1 we have the square root of a negative number. The damping factor determines whether the square root term is a positive or negative number and so the form of the output from the system:

1 *Over-damped*

With $\zeta > 1$ there are two different real roots s_1 and s_2:

$$s_1 = -\zeta\omega_n + \omega_n\sqrt{\zeta^2 - 1}$$

$$s_2 = -\zeta\omega_n - \omega_n\sqrt{\zeta^2 - 1}$$

and so the general solution for x_n is

$$x_n = Ae^{s_1t} + Be^{s_2t}$$

For such conditions the system is said to be **over-damped**.

2 *Critically damped*

When $\zeta = 1$ there are two equal roots with $s_1 = s_2 = -\omega_n$. For this condition, which is called **critically damped**,

$$x_n = (At + B)\,e^{-\omega_n t}$$

It may seem that the solution for this case should be $x_n = Ae^{st}$, but two constants are required and so the solution is of this form.

3 *Under-damped*
With $\zeta < 1$ there are two complex roots since the roots both involve the square root of (-1):

$$s = -\zeta\omega_n \pm \omega_n\sqrt{\zeta^2 - 1} = -\zeta\omega_n \pm \omega_n\sqrt{-1}\sqrt{1 - \zeta^2}$$

and so writing j for $\sqrt{-1}$,

$$s = -\zeta\omega_n \pm j\omega_n\sqrt{1 - \zeta^2}$$

If we let

$$\omega = \omega_n\sqrt{1 - \zeta^2}$$

then we can write $s = -\zeta\omega_d \pm j\omega$ and so the two roots are

$$s_1 = -\zeta\omega_d + j\omega \text{ and } s_2 = -\zeta\omega_d - j\omega$$

The term ω is the angular frequency of the motion when it is in the damped condition specified by ζ. The solution under these conditions is thus

$$x_n = Ae^{(\zeta\omega_n + j\omega)t} + Be^{(-\zeta\omega_n - j\omega)t} = e^{-\zeta\omega_n t}(Ae^{j\omega t} + Be^{-j\omega t})$$

But $e^{j\omega t} = \cos\omega t + j\sin\omega t$ and $e^{-j\omega t} = \cos\omega t - j\sin\omega t$. Hence

$$x_n = e^{-\zeta\omega_n t}(A\cos\omega t + jA\sin\omega t + B\cos\omega t - jB\sin\omega t)$$

$$= e^{-\zeta\omega_n t}[(A + B)\cos\omega t + j(A - B)\sin\omega t)]$$

If we substitute constants P and Q for $(A + B)$ and $j(A - B)$, then

$$x_n = e^{-\zeta\omega_n t}(P\cos\omega t + Q\sin\omega t)$$

For such conditions the system is said to be **under-damped**.

12.4.2 Response with a forcing input

When we have a forcing input F the differential equation becomes

$$m\frac{d^2x}{dt^2} + c\frac{dx}{dt} + kx = F$$

We can solve this second-order differential equation by the same method used earlier for the first-order differential equation and consider the solution to be made up of two elements, a transient (natural) response and a forced response, i.e. $x = x_n + x_f$. Substituting for x in the above equation then gives

$$m\frac{d^2(x_n + x_f)}{dt^2} + c\frac{d(x_n + x_f)}{dt} + k(x_n + x_f) = F$$

If we let

$$m\frac{d^2x_n}{dt^2} + c\frac{dx_n}{dt} + kx_n = 0$$

then we must have

$$m\frac{d^2x_f}{dt^2} + c\frac{dx_f}{dt} + kx_f = F$$

The previous section gave the solutions for the natural part of the solution. To solve the forcing equation,

$$m\frac{d^2x_f}{dt^2} + c\frac{dx_f}{dt} + kx_f = F$$

we need to consider a particular form of input signal and then try a solution. Thus for a step input of size F at time $t = 0$ we can try a solution $x_f = A$, where A is a constant (see Section 12.3.2 on first-order differential equations for a discussion of the choice of solutions). Then $dx_f/dt = 0$ and $d^2x_f/dt^2 = 0$. Thus, when these are substituted in the differential equation, $0 + 0 + kA = F$ and so $A = F/k$ and $x_f = F/k$. The complete solution, the sum of natural and forced solutions, is thus for the over-damped system

$$x = Ae^{s_1t} + Be^{s_2t} + \frac{F}{k}$$

for the critically damped system

$$x = (At + B)e^{-\omega_n t} + \frac{F}{k}$$

and for the under-damped system

$$x = e^{-\zeta\omega_n t}(P\cos\omega t + Q\sin\omega t) + \frac{F}{k}$$

When $t \to \infty$ the above three equations all lead to the solution $x = F/k$. This is the **steady-state condition.**

Thus a second-order differential equation in the form

$$a_2\frac{d^2x}{dt^2} + a_1\frac{dx}{dt} + a_0x = b_0y$$

has a natural frequency given by

$$\omega_n^2 = \frac{a_0}{a_2}$$

and a damping factor given by

$$\zeta^2 = \frac{a_1^2}{4a_2a_0}$$

12.4.3 Examples of second-order systems

The following examples illustrate the points made above.

Consider a series RLC circuit (Figure 12.10) with $R = 100\ \Omega$, $L = 2.0$ H and $C = 20\mu$F. When there is a step input V, the current i in the circuit is given by (see the text associated with Figure 10.9)

$$\frac{d^2i}{dt^2} + \frac{R}{L}\frac{di}{dt} + \frac{1}{LC}i = \frac{V}{LC}$$

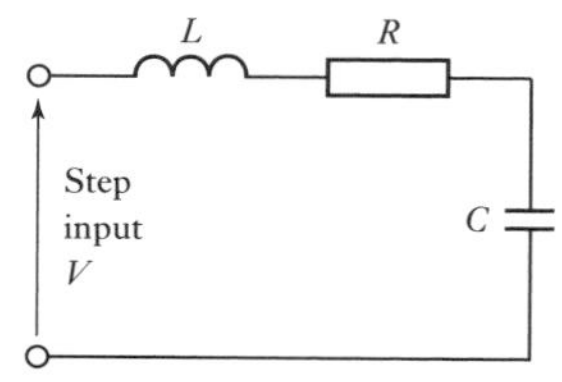

Figure 12.10 RLC system.

If we compare the equation with the general second-order differential equation of

$$a_2 \frac{d^2x}{dt^2} + a_1 \frac{dx}{dt} + a_0 x = b_0 y$$

then the natural angular frequency is given by

$$\omega_n^2 = \frac{1}{LC} = \frac{1}{2.0 \times 20 \times 10^{-6}}$$

and so $\omega_n = 158$ Hz. Comparison with the general second-order equation also gives

$$\zeta^2 = \frac{(R/L)^2}{4 \times (1/LC)} = \frac{R^2C}{4L} = \frac{100^2 \times 20 \times 10^{-6}}{4 \times 2.0}$$

Thus $\zeta = 0.16$. Since ζ is less than 1 the system is under-damped. The damped oscillation frequency ω is

$$\omega = \omega_n\sqrt{1 - \zeta^2} = 158\sqrt{1 - 0.16^2} = 156 \text{ Hz}$$

Because the system is under-damped the solution will be of the same form as

$$x = e^{-\zeta\omega_n t}(P \cos \omega t + Q \sin \omega t) + \frac{F}{k}$$

and so

$$i = e^{-0.16 \times 158t}(P \cos 156t + Q \sin 156t) + V$$

Since $i = 0$ when $t = 0$, then $0 = 1(P + 0) + V$. Thus $P = -V$. Since $di/dt = 0$ when $t = 0$, then differentiating the above equation and equating it to zero gives

$$\frac{di}{dt} = e^{-\zeta\omega_n t}(\omega P \sin \omega t - \omega Q \cos \omega t) - \zeta\omega_n e^{-\zeta\omega_n t}(P \cos \omega t + Q \cos \omega t)$$

Thus $0 = 1(0 - \omega Q) - \zeta\omega_n(P + 0)$ and so

$$Q = \frac{\zeta\omega_n P}{\omega} = -\frac{\zeta\omega_n V}{\omega} = -\frac{0.16 \times 158V}{156} \approx -0.16 \text{ V}$$

Thus the solution of the differential equation is

$$i = V - Ve^{-25.3t}(\cos 156t + 0.16 \sin 156t)$$

Now consider the system shown in Figure 12.11. The input, a torque T, is applied to a disc with a moment of inertia I about the axis of the shaft. The shaft is free to rotate at the disc end but is fixed at its far end. The shaft rotation is opposed by the torsional stiffness of the shaft, an opposing torque of $k\theta_o$ occurring for an input rotation of θ_o. k is a constant. Frictional forces damp the rotation of the shaft and provide an opposing torque of $c\ d\theta_o/dt$, where c is a constant. Suppose we need to determine the condition for this system to be critically damped.

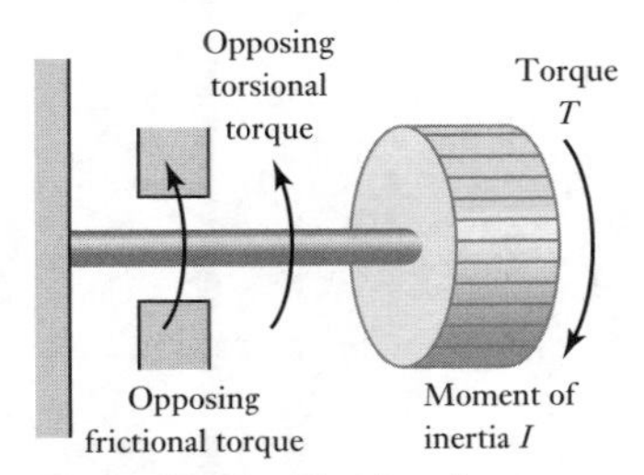

Figure 12.11 Torsional system.

We first need to obtain the differential equation for the system. The net torque is

$$\text{net torque} = T - c\frac{d\theta_o}{dt} - k\theta_o$$

The net torque is $I\,\mathrm{d}^2\theta_o/\mathrm{d}t^2$, hence

$$I\frac{\mathrm{d}^2\theta_o}{\mathrm{d}t^2} = T - c\frac{\mathrm{d}\theta_o}{\mathrm{d}t} - k\theta_o$$

$$I\frac{\mathrm{d}^2\theta_o}{\mathrm{d}t^2} + c\frac{\mathrm{d}\theta_o}{\mathrm{d}t} + k\theta_o = T$$

The condition for critical damping is given when the damping ratio ζ equals 1. Comparing the above differential equation with the general form of the second-order differential equation, then

$$\zeta^2 = \frac{a_1^2}{4a_2a_0} = \frac{c^2}{4Ik}$$

Thus for critical damping we must have $c = \sqrt{(Ik)}$.

12.5 Performance measures for second-order systems

Figure 12.12 shows the typical form of the response of an under-damped second-order system to a step input. Certain terms are used to specify such a performance.

The **rise time** t_r is the time taken for the response x to rise from 0 to the steady-state value x_{SS} and is a measure of how fast a system responds to the input. This is the time for the oscillating response to complete a quarter of a cycle, i.e. $\frac{1}{2}\pi$. Thus

$$\omega t_r = \frac{1}{2}\pi$$

The rise time is sometimes specified as the time taken for the response to rise from some specified percentage of the steady-state value, e.g. 10%, to another specified percentage, e.g. 90%.

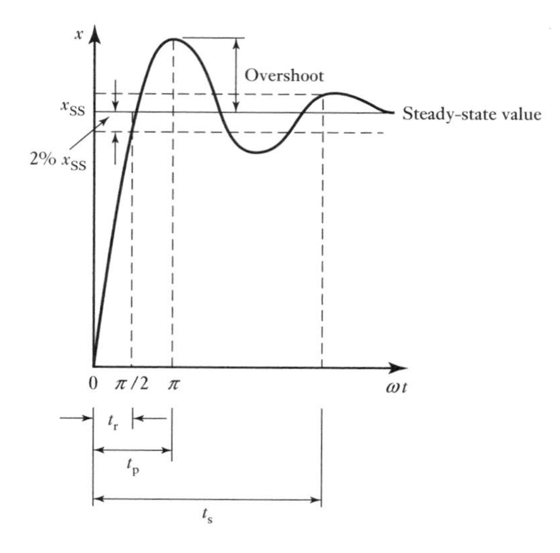

Figure 12.12 Step response of an under-damped system.

The **peak time** t_p is the time taken for the response to rise from 0 to the first peak value. This is the time for the oscillating response to complete one half cycle, i.e. π. Thus

$$\omega t_p = \pi$$

The **overshoot** is the maximum amount by which the response overshoots the steady-state value. It is thus the amplitude of the first peak. The overshoot is often written as a percentage of the steady-state value. For the under-damped oscillations of a system we can write

$$x = e^{-\zeta\omega_n t}(P\cos\omega t + Q\sin\omega t) + \text{steady-state value}$$

Since $x = 0$ when $t = 0$, then $0 = 1(P + 0) + x_{SS}$ and so $P = -x_{SS}$. The overshoot occurs at $\omega t = \pi$ and thus

$$x = e^{-\zeta\omega_n\pi/\omega}(P + 0) + x_{SS}$$

The overshoot is the difference between the output at that time and the steady-state value. Hence

$$\text{overshoot} = x_{SS}\, e^{-\zeta\omega_n\pi/\omega}$$

Since $\omega = \omega_n\sqrt{(1-\zeta^2)}$ then we can write

$$\text{overshoot} = x_{SS}\exp\left(\frac{-\zeta\omega_n\pi}{\omega_n\sqrt{1-\zeta^2}}\right) = x_{SS}\exp\left(\frac{-\zeta\pi}{\sqrt{1-\zeta^2}}\right)$$

Expressed as a percentage of x_{SS},

$$\text{percentage overshoot} = \exp\left(\frac{-\zeta\pi}{\sqrt{1-\zeta^2}}\right)\times 100\%$$

Table 12.2 gives values of the percentage overshoot for particular damping ratios.

Table 12.2 Percentage peak overshoot.

Damping ratio	Percentage overshoot
0.2	52.7
0.4	25.4
0.6	9.5
0.8	1.5

An indication of how fast oscillations decay is provided by the **subsidence ratio or decrement.** This is the amplitude of the second overshoot divided by that of the first overshoot. The first overshoot occurs when we have $\omega t = \pi$, the second overshoot when $\omega t = 3\pi$. Thus,

$$\text{first overshoot} = x_{SS}\exp\left(\frac{-\zeta\pi}{\sqrt{1-\zeta^2}}\right)$$

$$\text{second overshoot} = x_{SS}\exp\left(\frac{-3\zeta\pi}{\sqrt{1-\zeta^2}}\right)$$

and so

$$\text{subsidence ratio} = \frac{\text{second overshoot}}{\text{first overshoot}} = exp\left(\frac{-2\zeta\pi}{\sqrt{1-\zeta^2}}\right)$$

The **settling time** t_s is used as a measure of the time taken for the oscillations to die away. It is the time taken for the response to fall within and remain within some specified percentage, e.g. 2%, of the steady-state value (see Figure 12.12). This means that the amplitude of the oscillation should be less than 2% of x_{SS}. We have

$$x = e^{-\zeta\omega_n t}(P\cos\omega t + Q\sin\omega t) + \text{steady-state value}$$

and, as derived earlier, $P = -x_{SS}$. The amplitude of the oscillation is $(x - x_{SS})$ when x is a maximum value. The maximum values occur when ωt is some multiple of π and thus we have $\cos\omega t = 1$ and $\sin\omega t = 0$. For the 2% settling time, the settling time t_s is when the maximum amplitude is 2% of x_{SS}, i.e. $0.02x_{SS}$. Thus

$$0.02x_{SS} = e^{-\zeta\omega_n t_s}(x_{SS} \times 1 + 0)$$

Taking logarithms gives $\ln 0.02 = -\zeta\omega_n t_s$ and since $\ln 0.02 = -3.9$ or approximately -4, then

$$t_s = \frac{4}{\zeta\omega_n}$$

The above is the value of the settling time if the specified percentage is 2%. If the percentage is 5% the equation becomes

$$t_s = \frac{3}{\zeta\omega_n}$$

Since the time taken to complete one cycle, i.e. the periodic time, is $1/f$, where f is the frequency, and since $\omega = 2\pi f$, then the time to complete one cycle is $2\pi/f$. In a settling time of t_s the number of oscillations that occur is

$$\text{number of oscillations} = \frac{\text{settling time}}{\text{periodic time}}$$

and thus for a settling time defined for 2% of the steady-state value,

$$\text{number of oscillations} = \frac{4/\zeta\omega_n}{2\pi/\omega}$$

Since $\omega = \omega_n\sqrt{(1-\zeta^2)}$, then

$$\text{number of oscillations} = \frac{2\omega_n\sqrt{1-\zeta^2}}{\pi\zeta\omega_n} = \frac{2}{\pi}\sqrt{\frac{1}{\zeta^2} - 1}$$

To illustrate the above, consider a second-order system which has a natural angular frequency of 2.0 Hz and a damped frequency of 1.8 Hz. Since $\omega = \omega_n\sqrt{(1-\zeta^2)}$, then the damping factor is given by

$$1.8 = 2.0\sqrt{1-\zeta^2}$$

and $\zeta = 0.44$. Since $\omega t_r = \frac{1}{2}\pi$, then the 100% rise time is given by

$$t_r = \frac{\pi}{2 \times 1.8} = 0.87 \text{ s}$$

The percentage overshoot is given by

$$\text{percentage overshoot} = \exp\left(\frac{-\zeta\pi}{\sqrt{1-\zeta^2}}\right) \times 100\%$$
$$= \exp\left(\frac{-0.44\pi}{\sqrt{1-0.44^2}}\right) \times 100\%$$

The percentage overshoot is thus 21%. The 2% settling time is given by

$$t_s = \frac{4}{\zeta\omega_n} = \frac{4}{0.44 \times 2.0} = 4.5 \text{ s}$$

The number of oscillations occurring within the 2% settling time is given by

$$\text{number of oscillations} = \frac{2}{\pi}\sqrt{\frac{1}{\zeta^2} - 1} = \frac{2}{\pi}\sqrt{\frac{1}{0.44^2} - 1} = 1.3$$

12.6 System identification

In Chapters 10 and 11 models were devised for systems by considering them to be made up of simple elements. An alternative way of developing a model for a real system is to use tests to determine its response to some input, e.g. a step input, and then find the model that fits the response. This process of determining a mathematical model is known as **system identification.** Thus if we obtain a response to a step input of the form shown in Figure 12.5 then we might assume that it is a first-order system and determine the time constant from the response curve. For example, suppose the response takes a time of 1.5 s to reach 0.63 of its final height and the final height of the signal is five times the size of the step input. Table 12.1 indicates a time constant of 1.5 s and so the differential equation describing the model is

$$1.5\frac{dx}{dt} + x = 5y$$

An under-damped second-order system will give a response to a step input of the form shown in Figure 12.12. The damping ratio can be determined from measurements of the first and second overshoots with the ratio of these overshoots, i.e. the subsidence ratio, giving the damping ratio. The natural frequency can be determined from the time between successive overshoots. We can then use these values to determine the constants in the second-order differential equation.

Summary

The **natural response** of a system is when there is no input to the system forcing the variable to change but it is just changing naturally. The **forced response** of a system is when there is an input to the system forcing it to change.

A first-order system with no forcing input has a differential equation of the form

$$a_1\frac{dx}{dt} + a_0 x = 0$$

and this has the solution $x = e^{-a_0 t/a_1}$.

When there is a **forcing function** the differential equation is of the form

$$a_1\frac{dx}{dt} + a_0x = b_0y$$

and the solution is x = steady-state value $\times\ (1 - e^{-a_0t/a_1})$.

The **time constant** τ is the time the output takes to rise to 0.63 of its steady-state value and is (a_1/a_0).

A second-order system with no forcing input has a differential equation of the form

$$m\frac{d^2x}{dt^2} + c\frac{dx}{dt} + kx = 0$$

The natural angular frequency is given by $\omega_n^2 = k/m$ and the damping constant by $\zeta^2 = c^2/4mk$. The system is **over-damped** when we have $\zeta > 1$ and the general solution for x_n is

$$x_n = Ae^{s_1t} + Be^{s_2t} \text{ with } s = -\zeta\omega_n \pm \omega_n\sqrt{\zeta^2 - 1}$$

When $\zeta = 1$ the system is **critically damped** and

$$x_n = (At + B)\,e^{-\omega_n t}$$

and with $\zeta < 1$ the system is **under-damped** and

$$x_n = e^{-\zeta\omega_n t}(P\cos\omega t + Q\sin\omega t)$$

When we have a forcing input F the second-order differential equation becomes

$$m\frac{d^2x}{dt^2} + c\frac{dx}{dt} + kx = F$$

and for the over-damped system

$$x = Ae^{s_1t} + Be^{s_2t} + \frac{F}{k}$$

for the critically damped system

$$x = (At + B)\,e^{-\omega_n t} + \frac{F}{k}$$

and for the under-damped system

$$x = e^{-\zeta\omega_n t}(P\cos\omega t + Q\sin\omega t) + \frac{F}{k}$$

The **rise time** t_r is the time taken for the response x to rise from 0 to the steady-state value x_{SS} and is a measure of how fast a system responds to the input and is given by $\omega t_r = \frac{1}{2}\pi$. The **peak time** t_p is the time taken for the response to rise from 0 to the first peak value and is given by $\omega t_p = \pi$. The **overshoot** is the maximum amount by which the response overshoots the steady-state value and is

$$\text{overshoot} = x_{SS}\exp\left(\frac{-\zeta\pi}{\sqrt{1 - \zeta^2}}\right)$$

The **subsidence ratio** or **decrement** is the amplitude of the second overshoot divided by that of the first overshoot and is

$$\text{subsidence ratio} = \exp\left(\frac{-2\zeta\pi}{\sqrt{1-\zeta^2}}\right)$$

The **settling time** t_s is the time taken for the response to fall within and remain within some specified percentage, e.g. 2%, of the steady-state value, this being given by

$$t_s = \frac{4}{\zeta\omega_n}$$

Problems

12.1 A first-order system has a time constant of 4 s and a steady-state transfer function of 6. What is the form of the differential equation for this system?

12.2 A mercury-in-glass thermometer has a time constant of 10 s. If it is suddenly taken from being at 20°C and plunged into hot water at 80°C, what will be the temperature indicated by the thermometer after (a) 10 s, (b) 20 s?

12.3 A circuit consists of a resistor R in series with an inductor L. When subject to a step input voltage V at time $t = 0$ the differential equation for the system is

$$\frac{di}{dt} + \frac{R}{L}i = \frac{V}{L}$$

What is (a) the solution for this differential equation, (b) the time constant, (c) the steady-state current i?

12.4 Describe the form of the output variation with time for a step input to a second-order system with a damping factor of (a) 0, (b) 0.5, (c) 1.0, (d) 1.5.

12.5 An RLC circuit has a current i which varies with time t when subject to a step input of V and is described by

$$\frac{d^2i}{dt^2} + 10\frac{di}{dt} + 16i = 16V$$

What is (a) the undamped frequency, (b) the damping ratio, (c) the solution to the equation if $i = 0$ when $t = 0$ and $di/dt = 0$ when $t = 0$?

12.6 A system has an output x which varies with time t when subject to a step input of y and is described by

$$\frac{d^2x}{dt^2} + 10\frac{dx}{dt} + 25x = 50y$$

What is (a) the undamped frequency, (b) the damping ratio, (c) the solution to the equation if $x = 0$ when $t = 0$ and $dx/dt = -2$ when $t = 0$ and there is a step input of size 3 units?

12.7 An accelerometer (an instrument for measuring acceleration) has an undamped angular frequency of 100 Hz and a damping factor of 0.6. What will be (a) the maximum percentage overshoot and (b) the rise time when there is a sudden change in acceleration?

12.8 What will be (a) the undamped angular frequency, (b) the damping factor, (c) the damped angular frequency, (d) the rise time, (e) the percentage maximum overshoot and (f) the 0.2% settling time for a system which gave the following differential equation for a step input y?

$$\frac{d^2x}{dt^2} + 5\frac{dx}{dt} + 16x = 16y$$

12.9 When a voltage of 10 V is suddenly applied to a moving-coil voltmeter it is observed that the pointer of the instrument rises to 11 V before eventually settling down to read 10 V. What is (a) the damping factor and (b) the number of oscillations the pointer will make before it is within 0.2% of its steady-state value?

12.10 A second order system is described by the differential equation

$$\frac{d^2x}{dt^2} + c\frac{dx}{dt} + 4x = F$$

What value of damping constant c will be needed if the percentage overshoot is to be less than 9.5%?

12.11 Observation of the oscillations of a damped system when responding to an input indicates that the maximum displacement during the second cycle is 75% of the first displacement. What is the damping factor of the system?

12.12 A second-order system is found to have a time of 1.6 s between the first overshoot and the second overshoot. What is the natural frequency of the system?

Chapter thirteen System Transfer Functions

Objectives

The objectives of this chapter are that, after studying it, the reader should be able to:

- **Define the transfer function and determine it from differential equations for first- and second-order systems.**
- **Determine the transfer functions for systems with feedback loops.**
- **Determine, using Laplace transforms, the responses of first- and second-order systems to simple inputs.**
- **Determine the effect of pole location on the responses of systems.**

13.1 The transfer function

For an amplifier system it is customary to talk of the **gain** of the amplifier. This states how much bigger the output signal will be when compared with the input signal. It enables the output to be determined for specific inputs. Thus, for example, an amplifier with a voltage gain of 10 will give, for an input voltage of 2 mV, an output of 20 mV; or if the input is 1 V an output of 10 V. The gain states the mathematical relationship between the output and the input for the block. We can indicate when a signal is in the time domain, i.e. is a function of time, by writing it as $f(t)$. Thus, for an input of $y(t)$ and an output of $x(t)$ (Figure 13.1(a)),

$$\text{gain} = \frac{\text{output}}{\text{input}} = \frac{x(t)}{y(t)}$$

However, for many systems the relationship between the output and the input is in the form of a differential equation and so a statement of the function as just a simple number like the gain of 10 is not possible. We cannot just divide the output by the input because the relationship is a differential equation and not a simple algebraic equation. We can, however, transform a differential equation into an algebraic equation by using what is termed the **Laplace transform**. Differential equations describe how systems behave with time and are transformed by means of the Laplace transform into simple algebraic equations, not involving time, where we can carry out normal algebraic manipulations of the quantities. We talk of behaviour in the **time domain** being transformed to the ***s*-domain**. When in the *s*-domain a function is written, since it is a function of s, as $F(s)$. It is usual to use a capital letter F for the Laplace transform and a lower case letter f for the time-varying function $f(t)$.

We then define the relationship between output and input in terms of a **transfer function**, this stating the relationship between the Laplace transform of the output and the Laplace transform of the input. Suppose the input to a linear system has a Laplace transform of $Y(s)$ and the Laplace transform of the output is $X(s)$. The transfer function $G(s)$ of the system is then defined as

$$\text{transfer function} = \frac{\text{Laplace transform of output}}{\text{Laplace transform of input}}$$

$$G(s) = \frac{X(s)}{Y(s)}$$

with all the initial conditions being zero, i.e. we assume zero output when zero input and zero rate of change of output with time when zero rate of change of input with time. Thus the output transform is $X(s) = G(s)Y(s)$, i.e. the product of the input transform and the transfer function. If we represent a system in the s-domain by a block diagram (Figure 13.1(b)) then $G(s)$ is the function in the box which takes an input of $Y(s)$ and converts it to an output of $X(s)$.

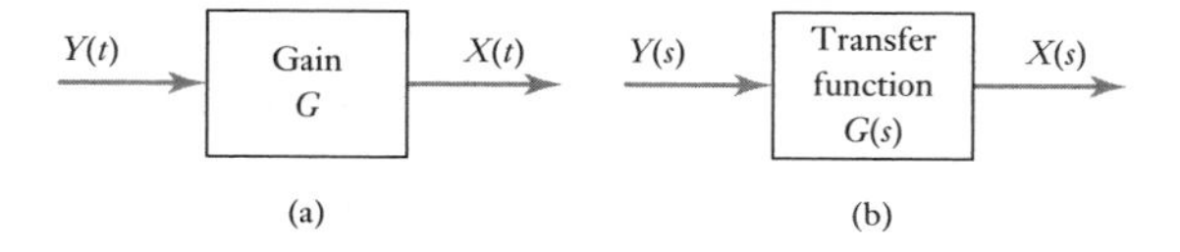

Figure 13.1 Block diagrams: (a) in time domain, (b) in s-domain.

13.1.1 Laplace transforms

To obtain the Laplace transform of a differential equation which includes quantities which are functions of time we can use tables coupled with a few basic rules (Appendix A includes such a table and gives details of the rules). Figure 13.2 shows basic transforms for common forms of inputs.

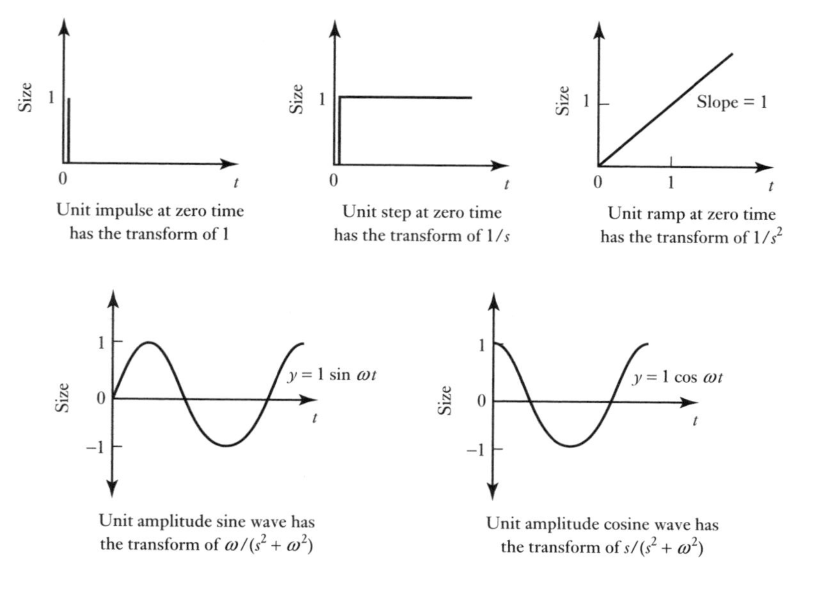

Figure 13.2 Laplace transforms for common inputs.

The following are some of the basic rules involved in working with Laplace transforms:

1 If a function of time is multiplied by a constant then the Laplace transform is multiplied by the same constant, i.e.

$$af(t) \text{ has the transform of } aF(s)$$

For example, the Laplace transform of a step input of 6 V to an electrical system is just six times the transform for a unit step and thus $6s$.

2 If an equation includes the sum of, say, two separate quantities which are functions of time, then the transform of the equation is the sum of the two separate Laplace transforms, i.e.

$$f(t) + g(t) \text{ has the transform } F(s) + G(s)$$

3 The Laplace transform of a first derivative of a function is

$$\text{transform of } \left\{ \frac{\mathrm{d}}{\mathrm{d}t} f(t) \right\} = sF(s) - f(0)$$

where $f(0)$ is the initial value of $f(t)$ when $t = 0$. However, when we are dealing with a transfer function we have all initial conditions zero.

4 The Laplace transform for the second derivative of a function is

$$\text{transform of } \left\{ \frac{\mathrm{d}^2}{\mathrm{d}t^2} f(t) \right\} = s^2F(s) - sf(0) - \frac{\mathrm{d}}{\mathrm{d}t} f(0)$$

where $\mathrm{d}f(0)/\mathrm{d}t$ is the initial value of the first derivative of $f(t)$ when we have $t = 0$. However, when we are dealing with a transfer function we have all initial conditions zero.

5 The Laplace transform of an integral of a function is

$$\text{transform of } \left\{ \int_0^t f(t)\, \mathrm{d}t \right\} = \frac{1}{s} F(s)$$

Thus, in obtaining the transforms of differential or integral equations when all the initial conditions are zero, we:

replace a function of time $f(t)$ *by* $F(s)$,
replace a first derivative $\mathrm{d}f(t)/\mathrm{d}t$ *by* $sF(s)$,
replace a second derivative $\mathrm{d}^2f(t)/\mathrm{d}t^2$ *by* $s^2F(s)$,
replace an integral $\int f(t)\, \mathrm{d}t$ *by* $F(s)/s$.

When algebraic manipulations have occurred in the s-domain, then the outcome can be transformed back to the time domain by using the table of transforms in the inverse manner, i.e. finding the time domain function which fits the s-domain result. Often the transform has to be rearranged to be put into a form given in the table. The following are some useful such inversions; for more inversions see Appendix A:

	Laplace transform	*Function of time*
1	$\dfrac{1}{s + a}$	e^{-at}
2	$\dfrac{a}{s(s + a)}$	$(1 - \mathrm{e}^{-at})$

3 $\dfrac{b-a}{(s+a)(s+b)}$ $\qquad$ $e^{-at} - e^{-bt}$

4 $\dfrac{s}{(s+a)^2}$ $\qquad$ $(1-at)e^{-at}$

5 $\dfrac{a}{s^2(s+a)}$ $\qquad$ $t - \dfrac{1-e^{-at}}{a}$

The following sections illustrate the application of the above to first-order and second-order systems.

13.2 First-order systems

Consider a system where the relationship between the input and the output is in the form of a first-order differential equation. The differential equation of a first-order system is of the form

$$a_1\frac{dx}{dt} + a_0x = b_0y$$

where a_1, a_0, b_0 are constants, y is the input and x the output, both being functions of time. The Laplace transform of this, with all initial conditions zero, is

$$a_1sX(s) + a_0X(s) = b_0Y(s)$$

and so we can write the transfer function $G(s)$ as

$$G(s) = \frac{X(s)}{Y(s)} = \frac{b_0}{a_1s + a_0}$$

This can be rearranged to give

$$G(s) = \frac{b_0/a_0}{(a_1/a_0)s + 1} = \frac{G}{\tau s + 1}$$

where G is the gain of the system when there are steady-state conditions, i.e. there is no dx/dt term. (a_1/a_0) is the time constant τ of the system (see Section 12.3.4).

13.2.1 First-order system with step input

When a first-order system is subject to a unit step input, then $Y(s) = 1/s$ and the output transform $X(s)$ is

$$X(s) = G(s)Y(s) = \frac{G}{s(\tau s + 1)} = G\frac{(1/\tau)}{s(s + 1/\tau)}$$

Hence, since we have the transform in the form $a/s(s+a)$, using the inverse transformation listed as item 2 in the previous section gives

$$x = G(1 - e^{-t/\tau})$$

13.2.2 Examples of first-order systems

The following examples illustrate the above points in the consideration of the transfer function of a first-order system and its behaviour when subject to a step input:

1 Consider a circuit which has a resistance R in series with a capacitance C. The input to the circuit is v and the output is the potential difference v_C across the capacitor. The differential equation relating the input and output is

$$v = RC\frac{dv_C}{dt} + v_C$$

Determine the transfer function.

Taking the Laplace transform, with all initial conditions zero, then

$$V(s) = RCsV_C(s) + V_C(s)$$

Hence the transfer function is

$$G(s) = \frac{V_C(s)}{V(s)} = \frac{1}{RCs + 1}$$

2 Consider a thermocouple which has a transfer function linking its voltage output V and temperature input of

$$G(s) = \frac{30 \times 10^{-6}}{10s + 1} \text{ V/°C}$$

Determine the response of the system when subject to a step input of size 100°C and hence the time taken to reach 95% of the steady-state value.

Since the transform of the output is equal to the product of the transfer function and the transform of the input, then

$$V(s) = G(s) \times \text{input}(s)$$

A step input of size 100°C, i.e. the temperature of the thermocouple is abruptly increased by 100°C, is $100/s$. Thus

$$V(s) = \frac{30 \times 10^{-6}}{10s + 1} \times \frac{100}{s} = \frac{30 \times 10^{-4}}{10s(s + 0.1)}$$

$$= 30 \times 10^{-4}\frac{0.1}{s(s + 0.1)}$$

The fraction element is of the form $a/s(s + a)$ and so the inverse transform is

$$V = 30 \times 10^{-4}(1 \times e^{-0.1t}) \text{ V}$$

The final value, i.e. the steady-state value, is when $t \rightarrow \infty$ and so is when the exponential term is zero. The final value is therefore 30×10^{-4} V. Thus the time taken to reach, say, 95% of this is given by

$$0.95 \times 30 \times 10^{-4} = 30 \times 10^{-4}(1 \times e^{-0.1t})$$

Thus $0.05 = e^{-0.1t}$ and $\ln 0.05 = -0.1t$. The time is thus 30 s.

3 Consider a ramp input to the above thermocouple system of $5t$ °C/s, i.e. the temperature is raised by 5°C every second. Determine how the voltage of the thermocouple varies with time and hence the voltage after 12 s.

The transform of the ramp signal is $5/s^2$. Thus

$$V(s) = \frac{30 \times 10^{-6}}{10s + 1} \times \frac{5}{s^2} = 150 \times 10^{-6} \frac{0.1}{s^2(s + 0.1)}$$

The inverse transform can be obtained using item 5 in the list given in the previous section. Thus

$$V = 150 \times 10^{-6}\left(t - \frac{1 - e^{-0.1t}}{0.1}\right)$$

After a time of 12 s we would have $V = 7.5 \times 10^{-4}$ V.

4 Consider an impulse input of size 100°C, i.e. the thermocouple is subject to a momentary temperature increase of 100°C. Determine how the voltage of the thermocouple varies with time and hence the voltage after 2 s.

The impulse has a transform of 100. Hence

$$V(s) = \frac{30 \times 10^{-6}}{10s + 1} \times 100 = 3 \times 10^{-4} \frac{1}{s + 0.1}$$

Hence $V = 3 \times 10^{-4} e^{-0.1t}$ V. After 2 s, the thermocouple voltage $V = 1.8 \times 10^{-4}$ V.

13.3 Second-order systems

For a second-order system, the relationship between the input y and the output x is described by a differential equation of the form

$$a_2 \frac{d^2x}{dt^2} + a_1 \frac{dx}{dt} + a_0 x = b_0 y$$

where a_2, a_1, a_0 and b_0 are constants. The Laplace transform of this equation, with all initial conditions zero, is

$$a_2 s^2 X(s) + a_1 s X(s) + a_0 X(s) = b_0 Y(s)$$

Hence

$$G(s) = \frac{X(s)}{Y(s)} = \frac{b_0}{a_2 s^2 + a_1 s + a_0}$$

An alternative way of writing the differential equation for a second-order system is

$$\frac{d^2x}{dt^2} + 2\zeta\omega_n \frac{dx}{dt} + \omega_n^2 x = b_0 \omega_n^2 y$$

where ω_n is the natural angular frequency with which the system oscillates and ζ is the damping ratio. The Laplace transform of this equation gives

$$G(s) = \frac{X(s)}{Y(s)} = \frac{b_0 \omega_n^2}{s^2 + 2\zeta\omega_n s + \omega_n^2}$$

The above are the general forms taken by the transfer function for a second-order system.

13.3.1 Second-order system with step input

When a second-order system is subject to a unit step input, i.e. $Y(s) = 1/s$, then the output transform is

$$X(s) = G(s)Y(s) = \frac{b_0\omega_n^2}{s(s^2 + 2\zeta\omega_n s + \omega_n)}$$

This can be rearranged as

$$X(s) = \frac{b_0\omega_n^2}{s(s + p_1)(s + p_2)}$$

where p_1 and p_2 are the roots of the equation

$$s^2 + 2\zeta\omega_n s + \omega_n^2 = 0$$

Hence, using the equation for the roots of a quadratic equation,

$$p = \frac{-2\zeta\omega_n \pm \sqrt{4\zeta^2\omega_n^2 - 4\omega_n^2}}{2}$$

and so the two roots p_1 and p_2 are

$$p_1 = -\zeta\omega_n + \omega_n\sqrt{\zeta^2 - 1} \qquad p_2 = -\zeta\omega_n - \omega_n\sqrt{\zeta^2 - 1}$$

With $\zeta > 1$ the square root term is real and the system is over-damped. To find the inverse transform we can either use partial fractions (see Appendix A) to break the expression down into a number of simple fractions or use item 14 in the table of transforms in Appendix A; the result in either case is

$$x = \frac{b_0\omega_n^2}{p_1p_2}\left[1 - \frac{p_2}{p_2 - p_1}e^{-p_2t} + \frac{p_1}{p_2 - p_1}e^{-p_1t}\right]$$

With $\zeta = 1$ the square root term is zero and so $p_1 = p_2 = -\omega_n$. The system is critically damped. The equation then becomes

$$X(s) = \frac{b_0\omega_n^2}{s(s + \omega_n)^2}$$

This equation can be expanded by means of partial fractions (see Appendix A) to give

$$Y(s) = b_0\left[\frac{1}{s} - \frac{1}{s + \omega_n} - \frac{\omega_n}{(s + \omega_n)^2}\right]$$

Hence

$$x = b_0[1 - e^{-\omega_nt} - \omega_nte^{-\omega_nt}]$$

With $\zeta < 1$, then

$$x = b_0\left[1 - \frac{e^{-\zeta\omega_nt}}{\sqrt{1 - \zeta^2}}\sin(\omega_n\sqrt{(1 - \zeta^2)}t + \phi)\right]$$

where $\cos\phi = \zeta$. This is an under-damped oscillation.

13.3.2 Examples of second-order systems

The following examples illustrate the above:

1 What will be the state of damping of a system having the following transfer function and subject to a unit step input?

$$G(s) = \frac{1}{s^2 + 8s + 16}$$

For a unit step input $Y(s) = 1/s$ and so the output transform is

$$X(s) = G(s)Y(s) = \frac{1}{s(s^2 + 8s + 16)} = \frac{1}{s(s + 4)(s + 4)}$$

The roots of $s^2 + 8s + 16$ are thus $p_1 = p_2 = -4$. Both the roots are real and the same and so the system is critically damped.

2 A robot arm having the following transfer function is subject to a unit ramp input. What will be the output?

$$G(s) = \frac{K}{(s + 3)^2}$$

The output transform $X(s)$ is

$$X(s) = G(s)Y(s) = \frac{K}{(s + 3)^2} \times \frac{1}{s^2}$$

Using partial fractions (see Appendix A) this becomes

$$X(s) = \frac{K}{9s^2} - \frac{2K}{9(s + 3)} + \frac{K}{9(s + 3)^2}$$

Hence the inverse transform is

$$x = \frac{1}{9}Kt - \frac{2}{9}Ke^{-3t} + \frac{1}{9}Kte^{-3t}$$

13.4 Systems in series

If a system consists of a number of subsystems in series, as in Figure 13.3, then the transfer function $G(s)$ of the system is given by

$$G(s) = \frac{X(s)}{Y(s)} = \frac{X_1(s)}{Y(s)} \times \frac{X_2(s)}{X_1(s)} \times \frac{X(s)}{X_2(s)}$$
$$= G_1(s) \times G_2(s) \times G_3(s)$$

The transfer function of the system as a whole is the product of the transfer functions of the series elements.

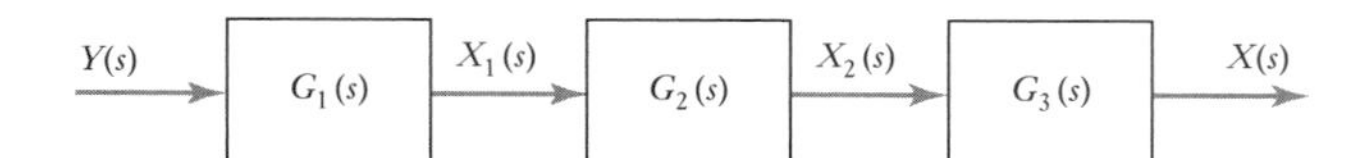

Figure 13.3 Systems in series.

13.4.1 Examples of systems in series

The following examples illustrate this. It has been assumed that when subsystems are linked together no interaction occurs between the blocks which would result in changes in their transfer functions, e.g. with electric circuits there can be problems when subsystem circuits interact and load each other.

1 What will be the transfer function for a system consisting of three elements in series, the transfer functions of the elements being 10, $2/s$ and $4/(s + 3)$?

Using the equation developed above,

$$G(s) = 10 \times \frac{2}{s} \times \frac{4}{s+3} = \frac{80}{s(s+3)}$$

2 A field-controlled d.c. motor consists of three subsystems in series: the field circuit, the armature coil and the load. Figure 13.4 illustrates the arrangement and the transfer functions of the subsystems. Determine the overall transfer function.

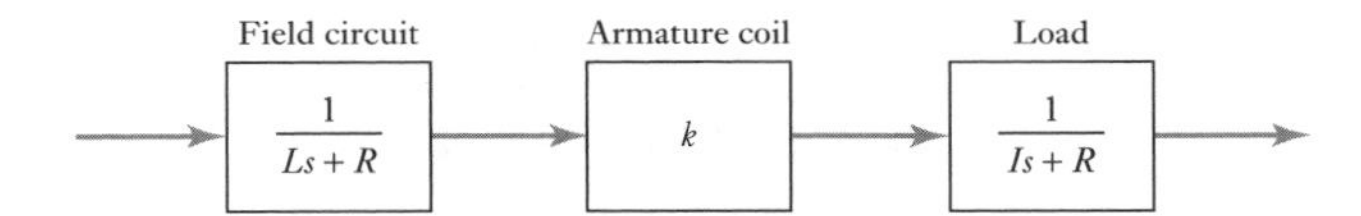

Figure 13.4 Field-controlled d.c. motor.

The overall transfer function is the product of the transfer functions of the series elements. Thus

$$G(s) = \frac{1}{Ls + R} \times k \times \frac{1}{Is + c} = \frac{k}{(Ls + R)(Is + c)}$$

13.5 Systems with feedback loops

Figure 13.5 shows a simple system having negative feedback. With **negative feedback** the system input and the feedback signals are subtracted at the summing point. The term **forward path** is used for the path having the transfer function $G(s)$ in the figure and **feedback path** for the one having $H(s)$. The entire system is referred to as a **closed-loop system**.

For the negative feedback system, the input to the subsystem having the forward-path transfer function $G(s)$ is $Y(s)$ minus the feedback signal. The feedback loop has a transfer function of $H(s)$ and has as its input $X(s)$, thus the feedback signal is $H(s)X(s)$. Thus the $G(s)$ element has an input of $Y(s) - H(s)X(s)$ and an output of $X(s)$ and so

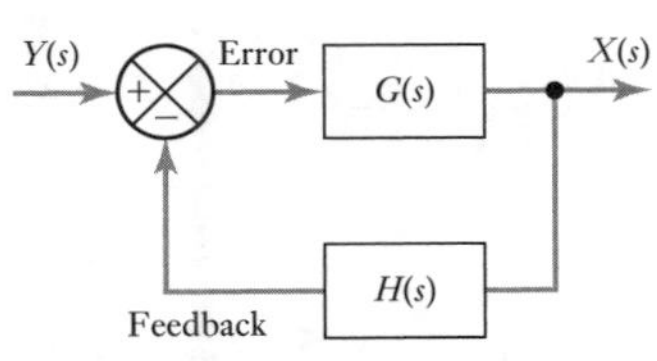

Figure 13.5 Negative feedback system.

$$G(s) = \frac{X(s)}{Y(s) - H(s)X(s)}$$

This can be rearranged to give

$$\frac{X(s)}{Y(s)} = \frac{G(s)}{1 + G(s)H(s)}$$

Hence the overall transfer function for the negative feedback system $T(s)$ is

$$T(s) = \frac{X(s)}{Y(s)} = \frac{G(s)}{1 + G(s)H(s)}$$

13.5.1 Examples of systems with negative feedback

The following examples illustrate the above:

1 What will be the overall transfer function for a closed-loop system having a forward-path transfer function of $2/(s+1)$ and a negative feedback-path transfer function of $5s$?

Using the equation developed above,

$$T(s) = \frac{G(s)}{1 + G(s)H(s)} = \frac{2/(s+1)}{1 + [2/(s+1)]5s} = \frac{2}{11s + 1}$$

2 Consider an armature-controlled d.c. motor (Figure 13.6). This has a forward path consisting of three elements: the armature circuit with a transfer function $1/(Ls + R)$, the armature coil with a transfer function k and the load with a transfer function $1/(Is + c)$. There is a negative feedback path with a transfer function K. Determine the overall transfer function for the system.

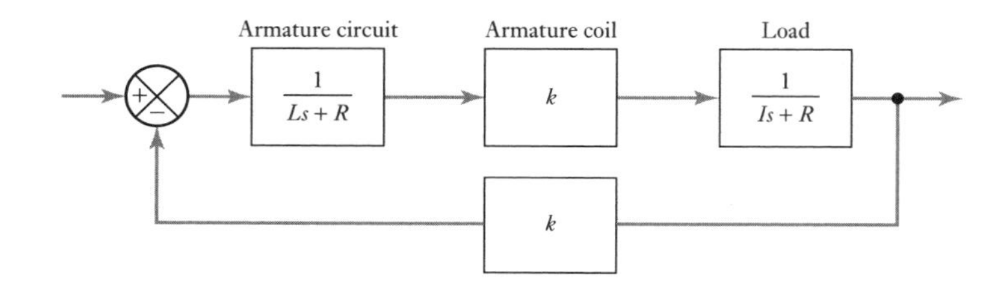

Figure 13.6 Armature-controlled d.c. motor.

The forward-path transfer function for the series elements is the product of the transfer functions of the series elements, i.e.

$$G(s) = \frac{1}{Ls + R} \times k \times \frac{1}{Is + c} = \frac{k}{(Ls + R)(Is + c)}$$

The feedback path has a transfer function of K. Thus the overall transfer function is

$$T(s) = \frac{G(s)}{1 + G(s)H(s)} = \frac{\dfrac{k}{(Ls + R)(Is + c)}}{1 + \dfrac{kK}{(Ls + R)(Is + c)}}$$

$$= \frac{k}{(Ls + R)(Is + c) + kK}$$

13.6 Effect of pole location on transient response

We can define a system as being **stable** if, when it is given an input, it has transients which die away with time and leave the system in its steady-state condition. A system is said to be **unstable** if the transients do not die away with time but increase in size and so the steady-state condition is never attained.

Consider an input of a unit impulse to a first-order system with a transfer function of $G(s) = 1/(s + 1)$. The system output $X(s)$ is

$$X(s) = \frac{1}{s + 1} \times 1$$

and thus $x = e^{-t}$. As the time t increases so the output dies away eventually to become zero. Now consider the unit impulse input to a system with the transfer function $G(s) = 1/(s-1)$. The output is

$$X(s) = \frac{1}{s-1} \times 1$$

and so $x = e^{t}$. As t increases, so the output increases with time. Thus a momentary impulse to the system results in an ever increasing output; this system is unstable.

For a transfer function, the values of s which make the transfer function infinite are termed its **poles**; they are the roots of the characteristic equation. Thus for $G(s) = 1/(s + 1)$, there is a pole of $s = -1$. For $G(s) = 1/(s-1)$, there is a pole of $s = +1$. Thus, for a first-order system the system is stable if the pole is negative, and **unstable** if the pole is positive (Figure 13.7).

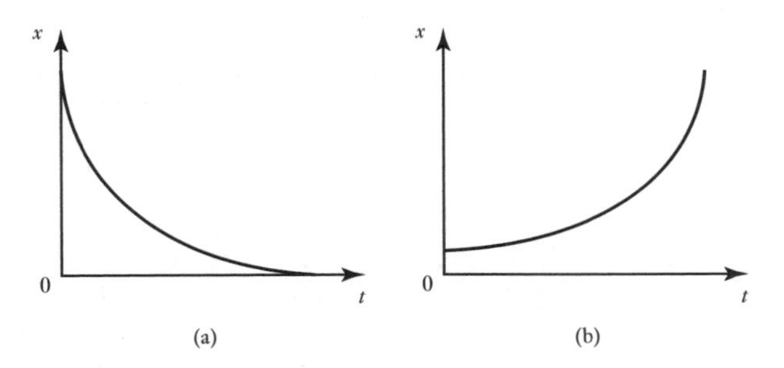

Figure 13.7 First-order systems: (a) negative pole, (b) positive pole.

For a second-order system with transfer function

$$G(s) = \frac{b_0\omega_n^2}{s^2 + 2\zeta\omega_n s + \omega_n^2}$$

when subject to a unit impulse input,

$$X(s) = \frac{b_0\omega_n^2}{(s + p_1)(s + p_2)}$$

where p_1 and p_2 are the roots of the equation

$$s^2 + 2\zeta\omega_n s + \omega_n = 0$$

Using the equation for the roots of a quadratic equation,

$$p = \frac{-2\zeta\omega_n \pm \sqrt{4\zeta^2\omega_n^2 - 4\omega_n^2}}{2} = -\zeta\omega_n \pm \omega_n\sqrt{\zeta^2 - 1}$$

Depending on the value of the damping factor, the term under the square root sign can be real or imaginary. When there is an imaginary term the output involves an oscillation. For example, suppose we have a second-order system with transfer function

$$G(s) = \frac{1}{[s - (-2 + j1)][s - (-2 - j1)]}$$

i.e. $p = -2 \pm j1$. When subject to a unit impulse input the output is $e^{-2t} \sin t$. The amplitude of the oscillation, i.e. e^{-2t}, dies away as the time increases and so the effect of the impulse is a gradually decaying oscillation (Figure 13.8(a)). The system is stable.

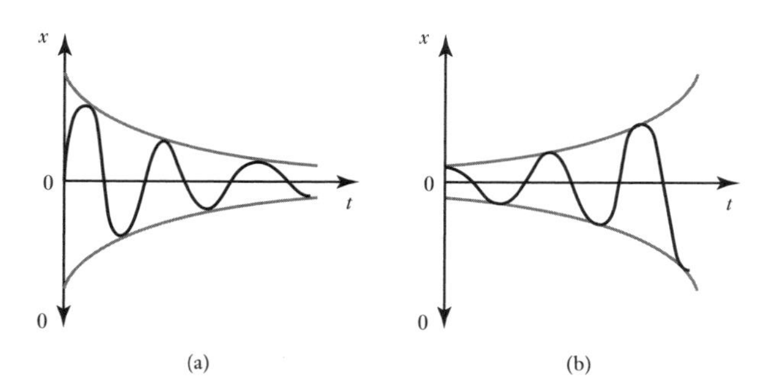

Figure 13.8 Second-order systems.

Suppose, however, we have a system with transfer function

$$G(s) = \frac{1}{[s - (2 + j1)][s - (2 - j1)]}$$

i.e. $p = +2 \pm j1$. When subject to a unit impulse input, the output is e^{2t} sin t. The amplitude of the oscillation, i.e. e^{2t}, increases as the time increases (Figure 13.8(b)). The system is unstable.

In general, when an impulse is applied to a system, the output is in the form of the sum of a number of exponential terms. If just one of these terms is of exponential growth then the output continues to grow and the system is unstable. When there are pairs of poles involving plus or minus imaginary terms then the output is an oscillation.

A system is stable if the real part of all its poles is negative.
A system is unstable if the real part of any of its poles is positive.

13.6.1 The s-plane

We can plot the positions of the poles of a system on a graph with the x-axis being the real parts and the y-axis the imaginary parts. Such a graph is termed the **s-plane**. The location of the poles on the plane determines the stability of a system. Figure 13.9 shows such a plane and how the location of roots affects the response of a system.

13.6.2 Compensation

The output from a system might be unstable or perhaps the response is too slow or there is too much overshoot. Systems can have their responses to

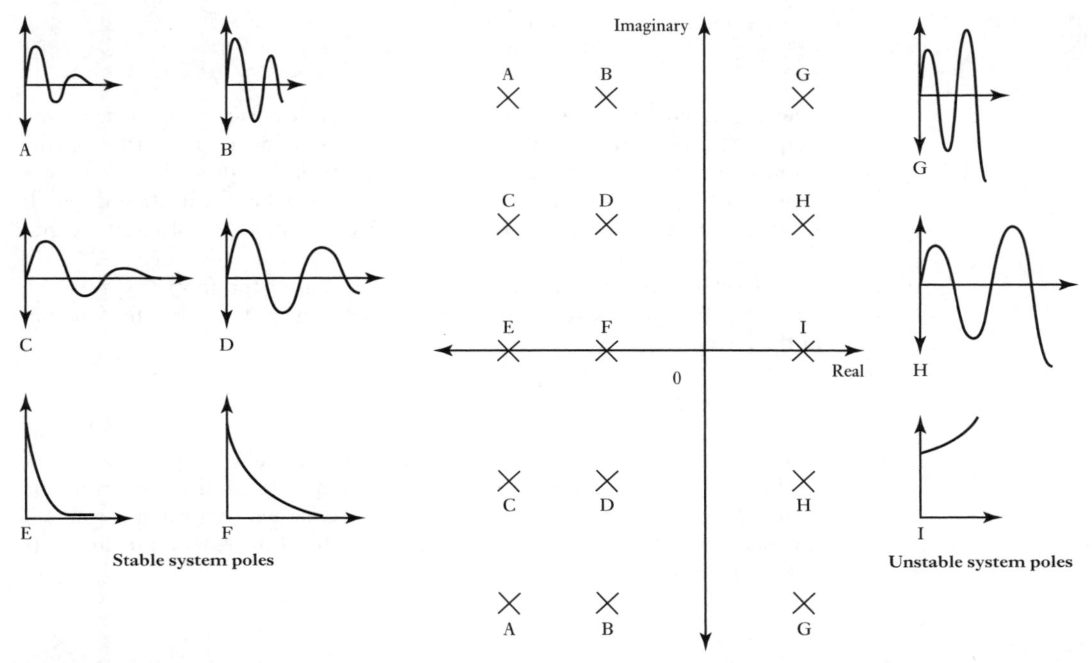

Figure 13.9 The s-plane.

inputs altered by including **compensators**. A compensator is a block which is incorporated in the system so that it alters the overall transfer function of the system in such a way as to obtain the required characteristics.

As an illustration of the use of a compensator, consider a position control system which has a negative feedback path with a transfer function of 1 and two subsystems in its forward path: a compensator with a transfer function of K and a motor/drive system with a transfer function of $1/s(s + 1)$. What value of K is necessary for the system to be critically damped? The forward path has a transfer function of $K/s(s + 1)$ and the feedback path a transfer function of 1. Thus the overall transfer function of the system is

$$T(s) = \frac{G(s)}{1 + G(s)H(s)} = \frac{\dfrac{K}{s(s+1)}}{1 + \dfrac{K}{s(s+1)}} = \frac{K}{s(s+1) + K}$$

The denominator is thus $s^2 + s + K$. This will have the roots

$$s = \frac{-1 \pm \sqrt{1 - 4K}}{2}$$

To be critically damped we must have $1 - 4K = 0$ and hence the compensator must have the proportional gain of $K = \frac{1}{4}$.

Summary

The **transfer function** $G(s)$ of a system is (Laplace transform of the output)/(Laplace transform of the input). To obtain the transforms of differential or integral equations when all the initial conditions are zero we: replace a function of time $f(t)$ by $F(s)$, replace a first derivative $\mathrm{d}f(t)/\mathrm{d}t$ by $sF(s)$, replace a second derivative $\mathrm{d}^2f(t)/\mathrm{d}t^2$ by $s^2F(s)$, replace an integral $\int f(t)\,\mathrm{d}t$ by $F(s)/s$.

A **first-order system** has a transfer function of the form $G/(\tau s + 1)$, where τ is the time constant. A **second-order system** has a transfer function of the form

$$G(s) = \frac{b_0\omega_n^2}{s^2 + 2\zeta\omega_n s + \omega_n^2}$$

where ζ is the damping factor and ω_n the natural angular frequency.

The values of s which make the transfer function infinite are termed its **poles**; they are the roots of the characteristic equation. A system is stable if the real part of all its poles is negative and unstable if the real part of any of its poles is positive.

Problems

13.1 What are the transfer functions for systems giving the following input/output relationships?

(a) A hydraulic system has an input q and an output h where

$$q = A\frac{\mathrm{d}h}{\mathrm{d}t} + \frac{\rho g h}{R}$$

(b) A spring-dashpot-mass system with an input F and an output x, where

$$m\frac{\mathrm{d}^2x}{\mathrm{d}t^2} + c\frac{\mathrm{d}x}{\mathrm{d}t} + kx = F$$

(c) An RLC circuit with an input v and output v_C, where

$$v = RC\frac{\mathrm{d}v_C}{\mathrm{d}t} + LC\frac{\mathrm{d}^2v_C}{\mathrm{d}t^2} + v_C$$

13.2 What are the time constants of the systems giving the transfer functions (a) $G(s) = 5/(3s + 1)$, (b) $G(s) = 2/(2s + 3)$?

13.3 Determine how the outputs of the following systems vary with time when subject to a unit step input at time $t = 0$: (a) $G(s) = 2/(s + 2)$ (b) $G(s) = 10/(s + 5)$.

13.4 What is the state of the damping for the systems having the following transfer functions?

(a) $G(s) = \dfrac{5}{s^2 - 6s + 16}$, (b) $G(s) = \dfrac{10}{s^2 + s + 100}$,

(c) $G(s) = \dfrac{2s + 1}{s^2 + 2s + 1}$, (d) $G(s) = \dfrac{3s + 20}{s^2 + 2s + 20}$

13.5 What is the output of a system with the transfer function $s/(s + 3)^2$ and subject to a unit step input at time $t = 0$?

13.6 What is the output of a system having the transfer function $G = 2/[(s + 3) \times (s + 4)]$ and subject to a unit impulse?

13.7 What are the overall transfer functions of the following negative feedback systems?

Forward path	*Feedback path*
(a) $G(s) = \dfrac{4}{s(s + 1)}$	$H(s) = \dfrac{1}{s}$
(b) $G(s) = \dfrac{2}{s + 1}$	$H(s) = \dfrac{1}{s + 2}$
(c) $G(s) = \dfrac{4}{(s + 2)(s + 3)}$	$H(s) = 5$
(d) two series elements $G_1(s) = 2/(s + 2)$ and $G_2(s) = 1/s$	$H(s) = 10$

13.8 What is the overall transfer function for a closed-loop system having a forward-path transfer function of $5/(s + 3)$ and a negative feedback-path transfer function of 10?

13.9 A closed-loop system has a forward path having two series elements with transfer functions 5 and $1/(s + 1)$. If the feedback path has a transfer function $2/s$, what is the overall transfer function of the system?

13.10 A closed-loop system has a forward path having two series elements with transfer functions of 2 and $1/(s + 1)$. If the feedback path has a transfer function of s, what is the overall transfer function of the system?

13.11 A system has a transfer function of $1/[(s + 1)(s + 2)]$. What are its poles?

13.12 Which of the following systems are stable or unstable?

(a) $G(s) = 1/[(s + 5)(s + 2)]$,
(b) $G(s) = 1/[(s - 5)(s + 2)]$,
(c) $G(s) = 1/[(s - 5)(s - 5)]$,
(d) $G(s) = 1/(s^2 + s + 1)$,
(e) $G(s) = 1/(s^2 - 2s + 3)$.

Chapter fourteen Frequency Response

Objectives

The objectives of this chapter are that, after studying it, the reader should be able to:

- **Explain the meaning of the frequency-response function.**
- **Analyse the frequency response of systems subject to sinusoidal inputs.**
- **Plot and interpret Bode plots.**
- **Use Bode plots for system identification.**
- **Explain the term bandwidth.**
- **Explain how the gain margin and phase margin can be used to indicate the stability of a system.**

14.1 Sinusoidal input

In the previous two chapters, the response of systems to step, impulse and ramp inputs has been considered. This chapter extends this to when there is a sinusoidal input. While for many control systems a sinusoidal input might not be encountered normally, it is a useful testing input since the way the system responds to such an input is a very useful source of information to aid the design and analysis of systems. It is also useful because many other signals can be considered as the sum of a number of sinusoidal signals. In 1822 Jean Baptiste Fourier proposed that any periodic waveform, e.g. a square waveform, can be made up of a combination of sinusoidal waveforms and by considering the behaviour of a system to each individual sinusoidal waveform it is possible to determine the response to the more complex waveform.

14.1.1 Response of a system to a sinusoidal input

Consider a first-order system which is described by the differential equation

$$a_1 \frac{dx}{dt} + a_0 x = b_0 y$$

where y is the input and x the output. Suppose we have the unit amplitude sinusoidal input of $y = \sin \omega t$. What will the output be? Well we must end up with the sinusoid $b_0 \sin \omega t$ when we add $a_1\, dx/dt$ and $a_0 x$. But sinusoids have the property that when differentiated, the result is also a sinusoid and with the same frequency (a cosine is a sinusoid, being just $\sin(\omega t + 90°)$). This applies no matter how many times we carry out the differentiation. Thus we should expect that the steady-state response x will also be sinusoidal and with the same frequency. The output may, however, differ in amplitude and phase from the input.

14.2 Phasors

In discussing sinusoidal signals it is convenient to use **phasors**. Consider a sinusoid described by the equation $v = V\sin(\omega t + \phi)$, where V is the amplitude, ω the angular frequency and ϕ the phase angle. The phasor can be represented by a line of length $|V|$ making an initial angle of ϕ with the phase reference axis (Figure 14.1). The | | lines are used to indicate that we are only concerned with the magnitude or size of the quantity when specifying its length. To completely specify a phasor quantity requires a magnitude and angle to be stated. The convention generally adopted is to write a phasor in bold, non-italic, print, e.g. **V**. When such a symbol is seen it implies a quantity having both a magnitude and an angle.

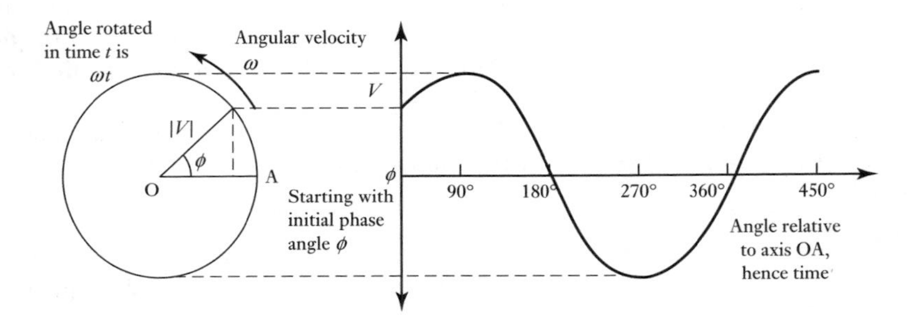

Figure 14.1 Representing a sinusoidal signal by a phasor.

Such a phasor can be described by means of complex number notation. A complex quantity can be represented by $(x + jy)$, where x is the real part and y the imaginary part of the complex number. On a graph with the imaginary component as the y-axis and the real part as the x-axis, x and y are Cartesian co-ordinates of the point representing the complex number (Figure 14.2(a)).

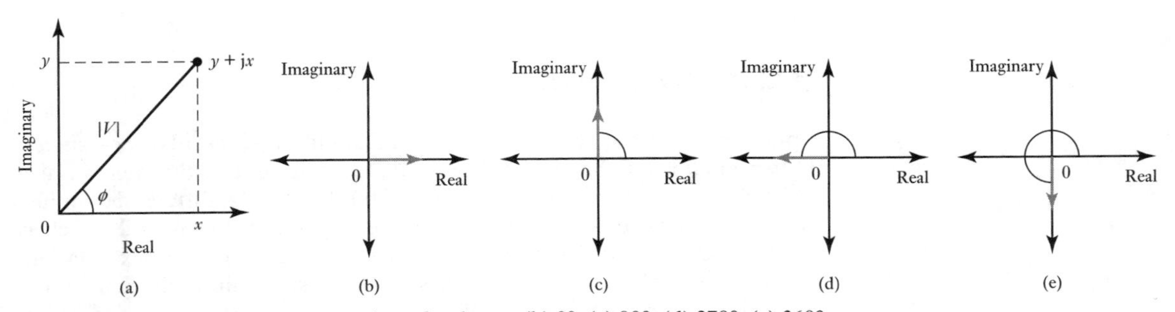

Figure 14.2 (a) Complex representation of a phasor, (b) 0°, (c) 90°, (d) 270°, (e) 360°.

If we take the line joining this point to the graph origin to represent a phasor, then we have the phase angle ϕ of the phasor represented by

$$\tan\phi = \frac{y}{x}$$

and its length by the use of Pythagoras's theorem as

$$\text{length of phasor } |V| = \sqrt{x^2 + y^2}$$

Since $x = |V|\cos\phi$ and $y = |V|\sin\phi$, then we can write

$$\mathbf{V} = x + \mathrm{j}y = |V|\cos\theta + \mathrm{j}|V|\sin\theta = |V|(\cos\theta + \mathrm{j}\sin\theta)$$

Thus a specification of the real and imaginary parts of a complex number enables a phasor to be specified.

Consider a phasor of length 1 and phase angle 0° (Figure 14.2(b)). It will have a complex representation of $1 + \mathrm{j}0$. Now consider the same length phasor but with a phase angle of 90° (Figure 14.2(c)). It will have a complex representation of $0 + \mathrm{j}1$. Thus rotation of a phasor anti-clockwise by 90° corresponds to multiplication of the phasor by j. If we now rotate this phasor by a further 90° (Figure 14.2(d)), then following the same multiplication rule we have the original phasor multiplied by j^2. But the phasor is just the original phasor in the opposite direction, i.e. just multiplied by -1. Hence $\mathrm{j}^2 = -1$ and so $\mathrm{j} = \sqrt{(-1)}$. Rotation of the original phasor through a total of 270°, i.e. $3 \times 90°$, is equivalent to multiplying the original phasor by $\mathrm{j}^3 = \mathrm{j}(\mathrm{j}^2) = -\mathrm{j}$.

To illustrate the above, consider a voltage v which varies sinusoidally with time according to the equation

$$v = 10\sin(\omega t + 30°)\ \mathrm{V}$$

When represented by a phasor, what are (a) its length, (b) its angle relative to the reference axis, (c) its real and imaginary parts when represented by a complex number?

(a) The phasor will have a length scaled to represent the amplitude of the sinusoid and so is 10 V.
(b) The angle of the phasor relative to the reference axis is equal to the phase angle and so is 30°.
(c) The real part is given by the equation $x = 10\cos 30° = 8.7$ V and the imaginary part by $y = 10\sin 30° = 5.0$ V. Thus the phasor is specified by $8.7 + \mathrm{j}5.0$ V.

14.2.1 Phasor equations

Consider a phasor as representing the unit amplitude sinusoid of $x = \sin\omega t$. Differentiation of the sinusoid gives $\mathrm{d}x/\mathrm{d}t = \omega\cos\omega t$. But we can also write this as $\mathrm{d}x/\mathrm{d}t = \omega\sin(\omega t + 90°)$. In other words, differentiation just results in a phasor with a length increased by a factor of ω and which is rotated round by 90° from the original phasor. Thus, in complex notation, we have multiplied the original phasor by $\mathrm{j}\omega$, since multiplication by j is equivalent to a rotation through 90°.

Thus the differential equation

$$a_1\frac{\mathrm{d}x}{\mathrm{d}t} + a_0x = b_0y$$

can be written, in complex notation, as a **phasor equation**

$$\mathrm{j}\omega a_1\mathbf{X} + a_0\mathbf{X} = b_0\mathbf{Y}$$

where the bold, non-italic, letters indicate that the data refers to phasors. We can say that the differential equation, which was an equation in the time domain, has been transformed into an equation in the **frequency domain**. The frequency domain equation can be rewritten as

$$(j\omega a_1 + a_0)\mathbf{X} = b_0\mathbf{Y}$$

$$\frac{\mathbf{X}}{\mathbf{Y}} = \frac{b_0}{j\omega a_1 + a_0}$$

But, in Section 13.2, when the same differential equation was written in the s-domain, we had

$$G(s) = \frac{X(s)}{Y(s)} = \frac{b_0}{a_1 s + a_0}$$

If we replace s by $j\omega$ we have the same equation. It turns out that we can always do this to convert from the s-domain to the frequency domain. This thus leads to a definition of a **frequency-response function** or **frequency transfer function** $G(j\omega)$, for the steady state, as

$$G(j\omega) = \frac{\text{output phasor}}{\text{input phasor}}$$

To illustrate the above, consider the determination of the frequency-response function for a system having a transfer function of

$$G(s) = \frac{1}{s + 1}$$

The frequency-response function is obtained by replacing s by $j\omega$. Thus

$$G(j\omega) = \frac{1}{j\omega + 1}$$

14.3 Frequency response

The procedure for determining the frequency response of a system is thus:

1 Replace s in the transfer function by $j\omega$ to give the frequency-response function.
2 The amplitude ratio between the output and the input is then the magnitude of the complex frequency-response function, i.e. $\sqrt{(x^2 + y^2)}$.
3 The phase angle between the output and the input is given by $\tan \phi = y/x$ or the ratio of the imaginary and real parts of the complex number representing the frequency-response function.

14.3.1 Frequency response for a first-order system

A first-order system has a transfer function which can be written as

$$G(s) = \frac{1}{1 + \tau s}$$

where τ is the time constant of the system (see Section 13.2). The frequency-response function $G(j\omega)$ can be obtained by replacing s by $j\omega$. Hence

$$G(j\omega) = \frac{1}{1 + j\omega\tau}$$

We can put this into a more convenient form by multiplying the top and bottom of the expression by $(1 - j\omega\tau)$ to give

$$G(j\omega) = \frac{1}{1 + j\omega\tau} \times \frac{1 - j\omega\tau}{1 - j\omega\tau} = \frac{1 - j\omega\tau}{1 + j^2\omega^2\tau^2}$$

But $j^2 = -1$, thus

$$G(j\omega) = \frac{1}{1 + \omega^2\tau^2} - j\frac{\omega\tau}{1 + \omega^2\tau^2}$$

This is of the form $x + jy$ and so, since $G(j\omega)$ is the output phasor divided by the input phasor, we have the size of the output phasor bigger than that of the input phasor by a factor which can be written as $|G(j\omega)|$, with

$$|G(j\omega)| = \sqrt{x^2 + y^2} = \sqrt{\left(\frac{1}{1 + \omega^2\tau^2}\right)^2 + \left(\frac{\omega\tau}{1 + \omega^2\tau^2}\right)^2} = \frac{1}{\sqrt{1 + \omega^2\tau^2}}$$

$|G(j\omega)|$ tells us how much bigger the amplitude of the output is than the amplitude of the input. It is generally referred to as the **magnitude** or **gain**. The phase difference ϕ between the output phasor and the input phasor is given by

$$\tan\phi = \frac{y}{x} = -\omega\tau$$

The negative sign indicates that the output phasor lags behind the input phasor by this angle.

The following examples illustrate the above:

1 Determine the frequency-response function, the magnitude and phase of a system (an electric circuit with a resistor in series with a capacitor across which the output is taken) that has a transfer function of

$$G(s) = \frac{1}{RCs + 1}$$

The frequency-response function can be obtained by substituting $j\omega$ for s and so

$$G(j\omega) = \frac{1}{j\omega RC + 1}$$

We can multiply the top and bottom of the above equation by $1 - j\omega RC$ and then rearrange the result to give

$$G(j\omega) = \frac{1}{1 + \omega^2(RC)^2} - j\frac{\omega(RC)}{1 + \omega^2(RC)^2}$$

Hence

$$|G(j\omega)| = \frac{1}{\sqrt{1 + \omega^2(RC)^2}}$$

and $\tan\phi = -\omega RC$.

2 Determine the magnitude and phase of the output from a system when subject to a sinusoidal input of 2 sin(3t + 60°) if it has a transfer function of

$$G(s) = \frac{4}{s + 1}$$

The frequency-response function is obtained by replacing s by jω. Thus

$$G(j\omega) = \frac{4}{j\omega + 1}$$

Multiplying the top and bottom of the equation by $(-j\omega + 1)$,

$$G(j\omega) = \frac{-j4\omega + 4}{\omega^2 + 1} = \frac{4}{\omega^2 + 1} - j\frac{4\omega}{\omega^2 + 1}$$

The magnitude is thus

$$|G(j\omega)| = \sqrt{x^2 + y^2} = \sqrt{\frac{4^2}{(\omega^2 + 1)^2} + \frac{4^2\omega^2}{(\omega^2 + 1)^2}} = \frac{4}{\sqrt{\omega^2 + 1}}$$

and the phase angle is given by $\tan\phi = y/x$ and so

$$\tan\phi = -\omega$$

For the specified input we have $\omega = 3$ rad/s. The magnitude is thus

$$|G(j\omega)| = \frac{4}{\sqrt{3^2 + 1}} = 1.3$$

and the phase is given by $\tan\phi = -3$. Thus $\phi = -72°$. This is the phase angle between the input and the output. Thus the output is 2.6 sin(3t − 12°).

14.3.2 Frequency response for a second-order system

Consider a second-order system with the transfer function (see Section 13.3)

$$G(s) = \frac{\omega_n^2}{s^2 + 2\zeta\omega_n s + \omega_n^2}$$

where ω_n is the natural angular frequency and ζ the damping ratio. The frequency-response function is obtained by replacing s by jω. Thus

$$G(j\omega) = \frac{\omega_n^2}{-\omega^2 + j2\zeta\omega\omega_n + \omega_n^2} = \frac{\omega_n^2}{(\omega_n^2 - \omega^2) + j2\zeta\omega_n}$$

$$= \frac{1}{\left[1 - \left(\frac{\omega}{\omega_n}\right)^2\right] + j2\zeta\left(\frac{\omega}{\omega_n}\right)}$$

Multiplying the top and bottom of the expression by

$$\left[1 - \left(\frac{\omega}{\omega_n}\right)^2\right] - j2\zeta\left(\frac{\omega}{\omega_n}\right)$$

gives

$$G(j\omega) = \frac{\left[1 - \left(\dfrac{\omega}{\omega_n}\right)^2\right] - j2\zeta\left(\dfrac{\omega}{\omega_n}\right)}{\left[1 - \left(\dfrac{\omega}{\omega_n}\right)^2\right]^2 + \left[2\zeta\left(\dfrac{\omega}{\omega_n}\right)\right]^2}$$

This is of the form $x + jy$ and so, since $G(j\omega)$ is the output phasor divided by the input phasor, we have the size or magnitude of the output phasor bigger than that of the input phasor by a factor which is given by $\sqrt{(x^2 + y^2)}$ as

$$|G(j\omega)| = \frac{1}{\sqrt{\left[1 - \left(\dfrac{\omega}{\omega_n}\right)^2\right]^2 + \left[2\zeta\left(\dfrac{\omega}{\omega_n}\right)\right]^2}}$$

The phase ϕ difference between the input and output is given by $\tan\phi = x/y$ and so

$$\tan\phi = -\frac{2\zeta\left(\dfrac{\omega}{\omega_n}\right)}{1 - \left(\dfrac{\omega}{\omega_n}\right)^2}$$

The minus sign is because the output phase lags behind the input.

14.4 Bode plots

The frequency response of a system is the set of values of the magnitude $|G(j\omega)|$ and phase angle ϕ that occur when a sinusoidal input signal is varied over a range of frequencies. This can be expressed as two graphs, one of the magnitude $|G(j\omega)|$ plotted against the angular frequency ω and the other of the phase ϕ plotted against ω. The magnitude and angular frequency are plotted using logarithmic scales. Such a pair of graphs is referred to as a **Bode plot**.

The magnitude is expressed in decibel units (dB):

$$|G(j\omega)| \text{ in dB} = 20 \lg_{10} |G(j\omega)|$$

Thus, for example, a magnitude of 20 dB means that

$$20 = 20 \lg_{10}|G(j\omega)|$$

so $1 = \lg_{10}|G(j\omega)|$ and $10^1 = |G(j\omega)|$. Thus a magnitude of 20 dB means that the magnitude is 10, and therefore the amplitude of the output is 10 times that of the input. A magnitude of 40 dB would mean a magnitude of 100 and so the amplitude of the output is 100 times that of the input.

14.4.1 Bode plot for $G(s) = K$

Consider the Bode plot for a system having the transfer function $G(s) = K$, where K is a constant. The frequency-response function is thus $G(j\omega) = K$. The magnitude $|G(j\omega)| = K$ and so, in decibels, $|G(j\omega)| = 20 \lg K$. The

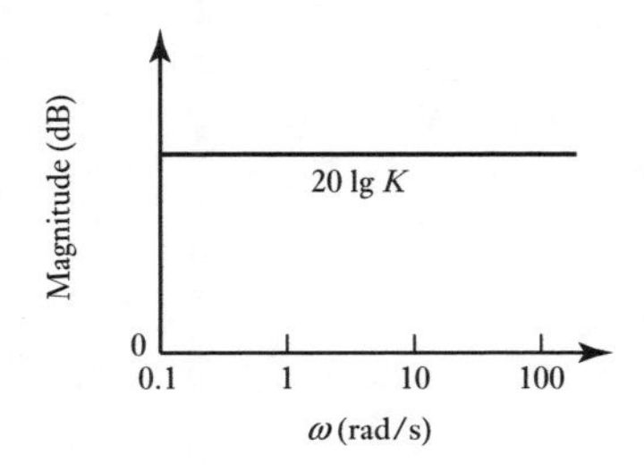

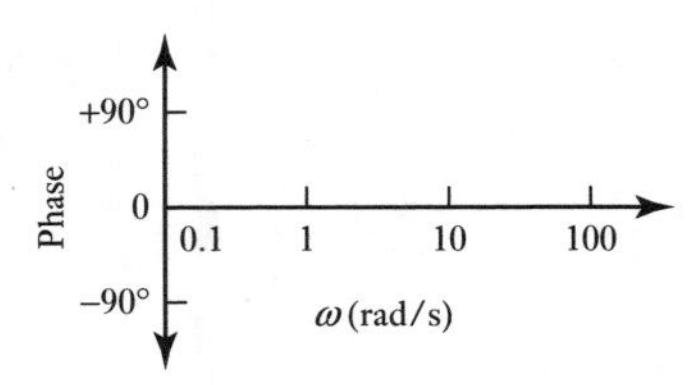

Figure 14.3 Bode plot for $G(s) = K$.

magnitude plot is thus a line of constant magnitude; changing K merely shifts the magnitude line up or down by a certain number of decibels. The phase is zero. Figure 14.3 shows the Bode plot.

14.4.2 Bode plot for $G(s) = 1/s$

Consider the Bode plot for a system having a transfer function $G(s) = 1/s$. The frequency-response function $G(j\omega)$ is thus $1/j\omega$. Multiplying this by j/j gives $G(j\omega) = -j/\omega$. The magnitude $|G(j\omega)|$ is thus $1/\omega$. In decibels this is $20 \lg(1/\omega) = -20 \lg \omega$. When $\omega = 1$ rad/s the magnitude is 0. When $\omega = 10$ rad/s it is -20 dB. When $\omega = 100$ rad/s it is -40 dB. For each 10-fold increase in angular frequency the magnitude drops by -20 dB. The magnitude plot is thus a straight line of slope -20 dB per decade of frequency which passes through 0 dB at $\omega = 1$ rad/s. The phase of such a system is given by

$$\tan \phi = \frac{-\dfrac{1}{\omega}}{0} = -\infty$$

Hence $\phi = -90°$ for all frequencies. Figure 14.4 shows the Bode plot.

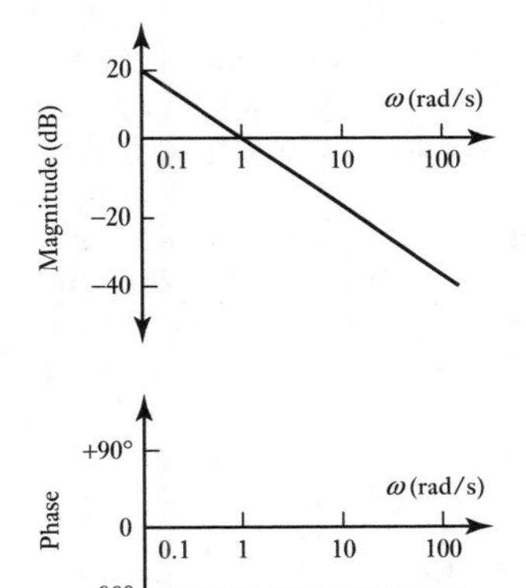

Figure 14.4 Bode plot for $G(s) = 1/s$.

14.4.3 Bode plot for a first-order system

Consider the Bode plot for a first-order system for which the transfer function is given by

$$G(s) = \frac{1}{\tau s + 1}$$

The frequency-response function is then

$$G(j\omega) = \frac{1}{j\omega t + 1}$$

The magnitude (see Section 14.2.1) is then

$$|G(j\omega)| = \frac{1}{\sqrt{1 + \omega^2\tau^2}}$$

In decibels this is

$$20 \lg\left(\frac{1}{\sqrt{1 + \omega^2\tau^2}}\right)$$

When $\omega \ll 1/\tau$, then $\omega^2\tau^2$ is negligible compared with 1 and so the magnitude is $20 \lg 1 = 0$ dB. Hence at low frequencies there is a straight line magnitude plot at a constant value of 0 dB. For higher frequencies, when $\omega \gg 1/\tau$, $\omega^2\tau^2$ is much greater than 1 and so the 1 can be neglected. The magnitude is then $20 \lg(1/\omega\tau)$, i.e. $-20 \lg \omega\tau$. This is a straight line of slope -20 dB per decade of frequency which intersects the 0 dB line when $\omega\tau = 1$, i.e. when $\omega = 1/\tau$. Figure 14.5 shows these lines for low and high frequencies

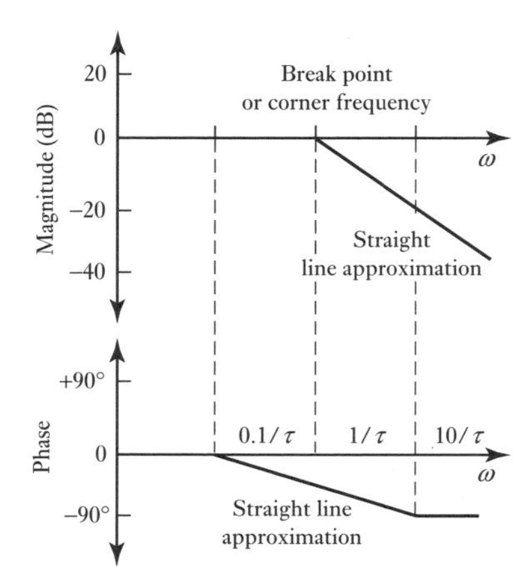

Figure 14.5 Bode plot for first-order system.

with their intersection, or so-called **break point** or **corner frequency**, at $\omega = 1/\tau$. The two straight lines are called the asymptotic approximation to the true plot. The true plot rounds off the intersection of the two lines. The difference between the true plot and the approximation is a maximum of 3 dB at the break point.

The phase for the first-order system (see Section 14.2.1) is given by $\tan\phi = -\omega\tau$. At low frequencies, when ω is less than about $0.1/\tau$, the phase is virtually 0°. At high frequencies, when ω is more than about $10/\tau$, the phase is virtually −90°. Between these two extremes the phase angle can be considered to give a reasonable straight line on the Bode plot (Figure 14.5). The maximum error in assuming a straight line is 5.5°.

An example of such a system is an *RC* filter (see Section 13.2.2), i.e. a resistance *R* in series with a capacitance *C* with the output being the voltage across the capacitor. It has a transfer function of $1/(RCs + 1)$ and so a frequency-response function of $1/(j\omega\tau + 1)$ where $\tau = RC$. The Bode plot is thus as shown in Figure 14.5.

14.4.4 Bode plot for a second-order system

Consider a second-order system with a transfer function of

$$G(s) = \frac{\omega_n^2}{s^2 + 2\zeta\omega_n s + \omega_n^2}$$

The frequency-response function is obtained by replacing *s* by $j\omega$:

$$G(j\omega) = \frac{\omega_n^2}{-\omega^2 + j2\zeta\omega_n\omega + \omega_n^2}$$

The magnitude is then (see Section 14.3.2)

$$|G(j\omega)| = \frac{1}{\sqrt{\left[1 - \left(\frac{\omega}{\omega_n}\right)^2\right]^2 + \left[2\zeta\left(\frac{\omega}{\omega_n}\right)\right]^2}}$$

Thus, in decibels, the magnitude is

$$20 \lg \frac{1}{\sqrt{\left[1-\left(\dfrac{\omega}{\omega_n}\right)^2\right]^2+\left[2\zeta\left(\dfrac{\omega}{\omega_n}\right)\right]^2}}$$

$$= -20 \lg \sqrt{\left[1-\left(\dfrac{\omega}{\omega_n}\right)^2\right]^2+\left[2\zeta\left(\dfrac{\omega}{\omega_n}\right)\right]^2}$$

For $(\omega/\omega_n) \ll 1$ the magnitude approximates to $-20 \lg 1$ or 0 dB and for $(\omega/\omega_n) \gg 1$ the magnitude approximates to $-20 \lg(\omega/\omega_n)^2$. Thus when ω increases by a factor of 10 the magnitude increases by a factor of $-20 \lg 100$ or -40 dB. Thus at low frequencies the magnitude plot is a straight line at 0 dB, while at high frequencies it is a straight line of -40 dB per decade of frequency. The intersection of these two lines, i.e. the break point, is at $\omega = \omega_n$. The magnitude plot is thus approximately given by these two asymptotic lines. The true value, however, depends on the damping ratio ζ. Figure 14.6 shows the two asymptotic lines and the true plots for a number of damping ratios.

Figure 14.6 Bode plot for a second-order system.

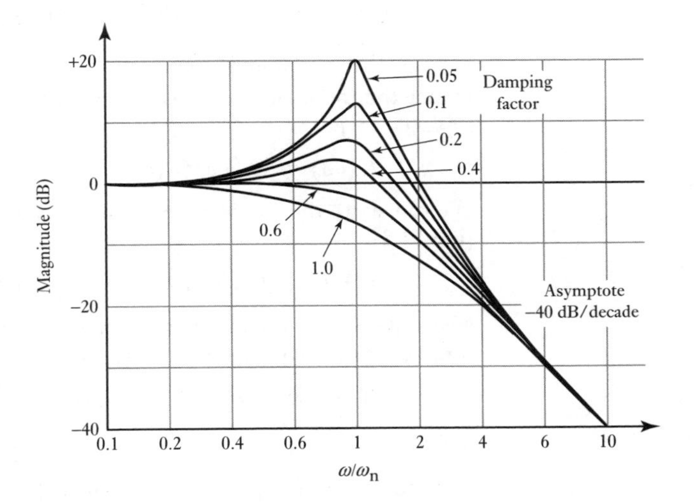

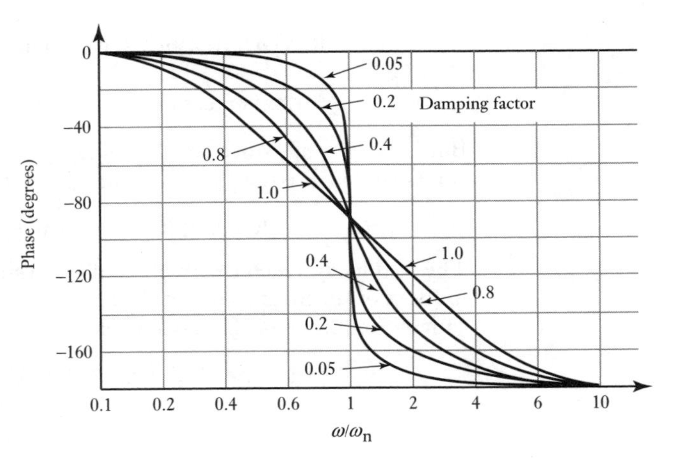

The phase is given by (see Section 14.3.2)

$$\tan\phi = -\frac{2\zeta\left(\dfrac{\omega}{\omega_n}\right)}{1-\left(\dfrac{\omega}{\omega_n}\right)^2}$$

For $(\omega/\omega_n) \ll 1$, e.g. $(\omega/\omega_n) = 0.2$, then $\tan\phi$ is approximately 0 and so $\phi = 0°$. For $(\omega/\omega_n) \gg 1$, e.g. $(\omega/\omega_n) = 5$, $\tan\phi$ is approximately $-(-\infty)$ and so $\phi = -180°$. When $\omega = \omega_n$ then we have $\tan\phi = -\infty$ and so $\phi = -90°$. A reasonable approximation is given by a straight line drawn through $-90°$ at $\omega = \omega_n$ and the points $0°$ at $(\omega/\omega_n) = 0.2$ and $-180°$ at $(\omega/\omega_n) = 5$. Figure 14.6 shows the graph.

14.4.5 Building up Bode plots

Consider a system involving a number of elements in series. The transfer function of the system as a whole is given by (see Section 13.4)

$$G(s) = G_1(s)G_2(s)G_3(s)\ldots$$

Hence the frequency-response function for a two-element system, when s is replaced by $j\omega$, is

$$G(j\omega) = G_1(j\omega)G_2(j\omega)$$

We can write the transfer function $G_1(j\omega)$ as a complex number (see Section 14.2)

$$x + jy = |G_1(j\omega)|\,(\cos\phi_1 + j\sin\phi_1)$$

where $|G(j\omega)|$ is the magnitude and ϕ_1 the phase of the frequency-response function. Similarly we can write $G_2(j\omega)$ as

$$|G_2(j\omega)|\,(\cos\phi_2 + j\sin\phi_2)$$

Thus

$$G(j\omega) = |G_1(j\omega)|\,(\cos\phi_1 + j\sin\phi_1) \times |G_2(j\omega)|\,(\cos\phi_2 + j\sin\phi_2)$$

$$= |G_1(j\omega)|\,|G_2(j\omega)|\,[\cos\phi_1\cos\phi_2$$

$$+ j(\sin\phi_1\cos\phi_2 + \cos\phi_1\sin\phi_2) + j^2\sin\phi_1\sin\phi_2]$$

But $j^2 = -1$ and, since $\cos\phi_1\cos\phi_2 - \sin\phi_1\sin\phi_2 = \cos(\phi_1+\phi_2)$ and $\sin\phi_1\cos\phi_2 + \cos\phi_1\sin\phi_2 = \sin(\phi_1+\phi_2)$, then

$$G(j\omega) = |G_1(j\omega)|\,|G_2(j\omega)|\,[\cos(\phi_1+\phi_2) + j\sin(\phi_1+\phi_2)]$$

The frequency-response function of the system has a magnitude which is the product of the magnitudes of the separate elements and a phase which is the sum of the phases of the separate elements, i.e.

$$|G(j\omega)| = |G_1(j\omega)|\,|G_2(j\omega)|\,|G_3(j\omega)|\ldots$$

$$\phi = \phi_1 + \phi_2 + \phi_3 + \ldots$$

Now, considering the Bode plot where the logarithms of the magnitudes are plotted,

$$\lg |G(j\omega)| = \lg |G_1(j\omega)| + \lg |G_2(j\omega)| + \lg |G_3(j\omega)| + \ \ldots$$

Thus we can obtain the Bode plot of a system by adding together the Bode plots of the magnitudes of the constituent elements. Likewise the phase plot is obtained by adding together the phases of the constituent elements.

By using a number of basic elements, the Bode plots for a wide range of systems can be readily obtained. The basic elements used are:

1 $G(s) = K$ gives the Bode plot shown in Figure 14.3.
2 $G(s) = 1/s$ gives the Bode plot shown in Figure 14.4.
3 $G(s) = s$ gives a Bode plot which is a mirror image of that in Figure 14.4. $|G(j\omega)| = 20$ dB per decade of frequency, passing through 0 dB at $\omega = 1$ rad/s. ϕ is constant at 90°.
4 $G(s) = 1/(\tau s + 1)$ gives the Bode plot shown in Figure 14.5.
5 $G(s) = \tau s + 1$ gives a Bode plot which is a mirror image of that in Figure 14.5. For the magnitude plot, the break point is at $1/\tau$ with the line prior to it being at 0 dB and after it at a slope of 20 dB per decade of frequency. The phase is zero at $0.1/\tau$ and rises to +90° at $10/\tau$.
6 $G(s) = \omega_n^2/(s^2 + 2\zeta\omega_n s + \omega_n^2)$ gives the Bode plot shown in Figure 14.6.
7 $G(s) = (s^2 + 2\zeta\omega_n s + \omega_n^2)/\omega_n^2$ gives a Bode plot which is a mirror image of that in Figure 14.6.

To illustrate the above, consider the drawing of the asymptotes of the Bode plot for a system having a transfer function of

$$G(s) = \frac{10}{2s + 1}$$

The transfer function is made up of two elements, one with a transfer function of 10 and one with transfer function $1/(2s + 1)$. The Bode plots can be drawn for each of these and then added together to give the required plot. The Bode plot for transfer function 10 will be of the form given in Figure 14.3 with $K = 10$ and that for $1/(2s + 1)$ like that given in Figure 14.5 with $\tau = 2$. The result is shown in Figure 14.7.

As another example, consider the drawing of the asymptotes of the Bode plot for a system having a transfer function of

$$G(s) = \frac{2.5}{s(s^2 + 3s + 25)}$$

The transfer function is made up of three components: one with a transfer function of 0.1, one with transfer function $1/s$ and one with transfer function $25/(s^2 + 3s + 25)$. The transfer function of 0.1 will give a Bode plot like that of Figure 14.3 with $K = 0.1$. The transfer function of $1/s$ will give a Bode plot like that of Figure 14.4. The transfer function of $25/(s^2 + 3s + 25)$ can be represented as $\omega_n^2/(s^2 + 2\zeta\omega_n s + \omega_n^2)$ with $\omega_n = 5$ rad/s and $\zeta = 0.3$. The break point will be when $\omega = \omega_n = 5$ rad/s. The asymptote for the phase passes through −90° at the break point, and is 0° when we have $(\omega/\omega_n) = 0.2$ and −180° when $(\omega/\omega_n) = 5$. Figure 14.8 shows the resulting plot.

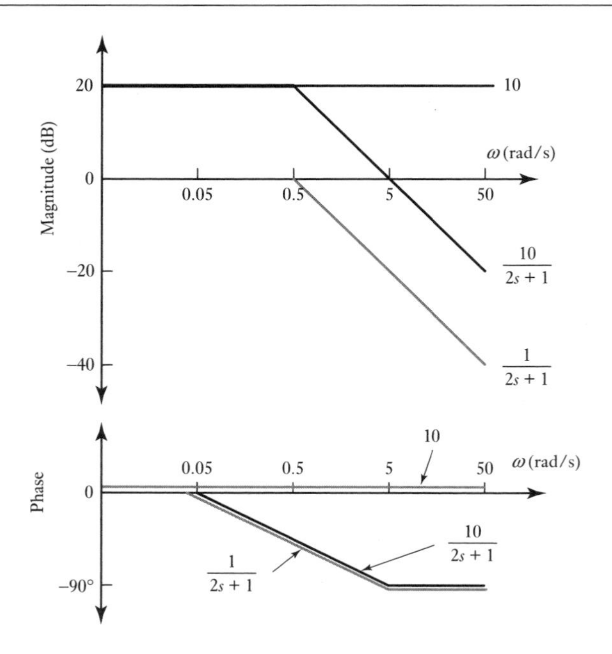

Figure 14.7 Building up a Bode diagram.

The above method of obtaining a Bode plot by building it up from its constituent elements, using the straight line approximations, was widely used but is now less necessary in the age of computers.

14.4.6 System identification

If we experimentally determine the Bode diagram for a system by considering its response to a sinusoidal input, then we can obtain the transfer function for the system. Basically we draw the asymptotes on the magnitude Bode plot and consider their gradients. The phase angle curve is used to check the results obtained from the magnitude analysis.

1 If the gradient at low frequencies prior to the first corner frequency is zero then there is no s or $1/s$ element in the transfer function. The K element in the numerator of the transfer function can be obtained from the value of the low-frequency magnitude; the magnitude in dB $= 20 \lg K$.
2 If the initial gradient at low frequencies is -20 dB/decade then the transfer function has a $1/s$ element.
3 If the gradient becomes more negative at a corner frequency by 20 dB/decade, there is a $(1 + s/\omega_c)$ term in the denominator of the transfer function, with ω_c being the corner frequency at which the change occurs. Such terms can occur for more than one corner frequency.
4 If the gradient becomes more positive at a corner frequency by 20 dB/decade, there is a $(1 + s/\omega_c)$ term in the numerator of the transfer function, with ω_c being the frequency at which the change occurs. Such terms can occur for more than one corner frequency.
5 If the gradient at a corner frequency becomes more negative by 40 dB/decade, there is a $(s^2/\omega_c^2 + 2\zeta s/\omega_c + 1)$ term in the denominator of the

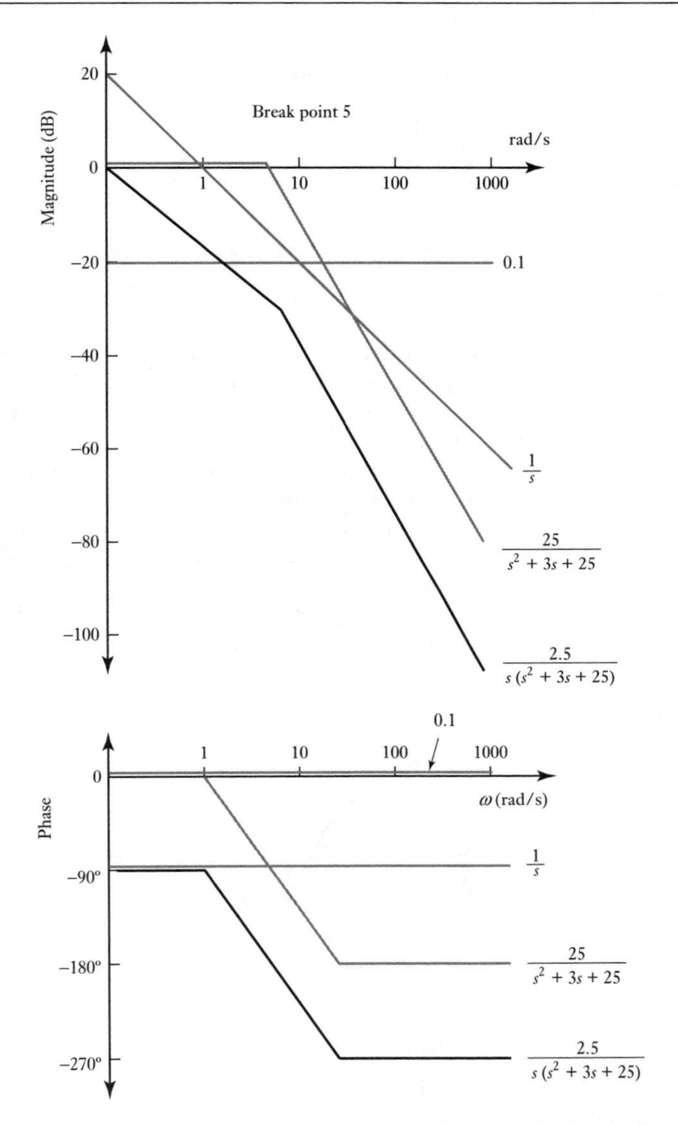

Figure 14.8 Building up a Bode plot.

transfer function. The damping ratio ζ can be found by considering the detail of the Bode plot at a corner frequency, as in Figure 14.6.

6 If the gradient at a corner frequency becomes more positive by 40 dB/decade, there is a $(s^2/\omega_c^2 + 2\zeta s/\omega_c + 1)$ term in the numerator of the transfer function. The damping ratio ζ can be found by considering the detail of the Bode plot at a corner frequency, as in Figure 14.6.

7 If the low-frequency gradient is not zero, the K term in the numerator of the transfer function can be determined by considering the value of the low-frequency asymptote. At low frequencies, many terms in transfer functions can be neglected and the gain in dB approximates to $20 \lg(K/\omega^2)$. Thus, at $\omega = 1$ the gain in dB approximates to $20 \lg K$.

As an illustration of the above, consider the Bode magnitude plot shown in Figure 14.9. The initial gradient is 0 and so there is no $1/s$ or s term in

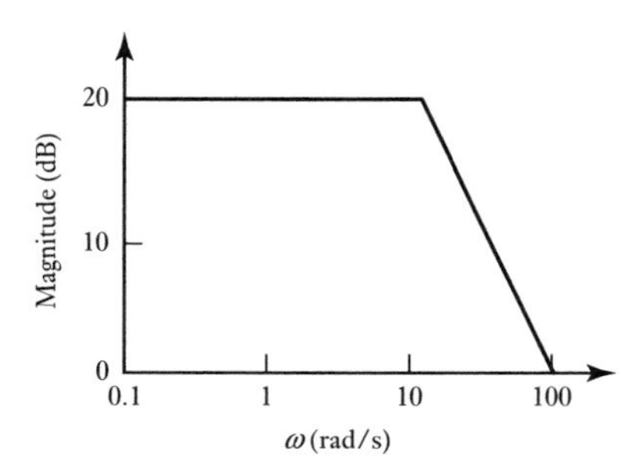

Figure 14.9 Bode plot.

the transfer function. The initial gain is 20 and so $20 = 20 \lg K$ and $K = 10$. The gradient changes by -20 dB/decade at a frequency of 10 rad/s. Hence there is a $(1 + s/10)$ term in the denominator. The transfer function is thus $10/(1 + 0.1s)$.

As a further illustration, consider Figure 14.10. There is an initial slope of -20 dB/decade and so a $1/s$ term. At the corner frequency 1.0 rad/s there is a -20 dB/decade change in gradient and so a $1/(1 + s/1)$ term. At the corner frequency 10 rad/s there is a further -20 dB/decade change in gradient and so a $1/(1 + s/10)$ term. At $\omega = 1$ the magnitude is 6 dB and so $6 = 20 \lg K$ and $K = 10^{6/20} = 2.0$. The transfer function is thus $2.0/s(1 + s)(1 + 0.1s)$.

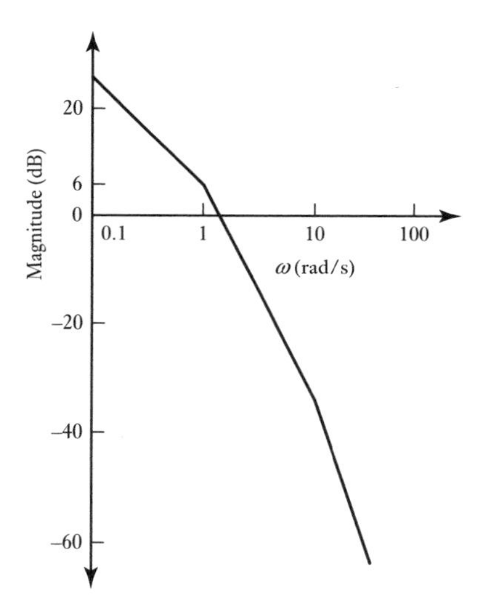

Figure 14.10 Bode plot.

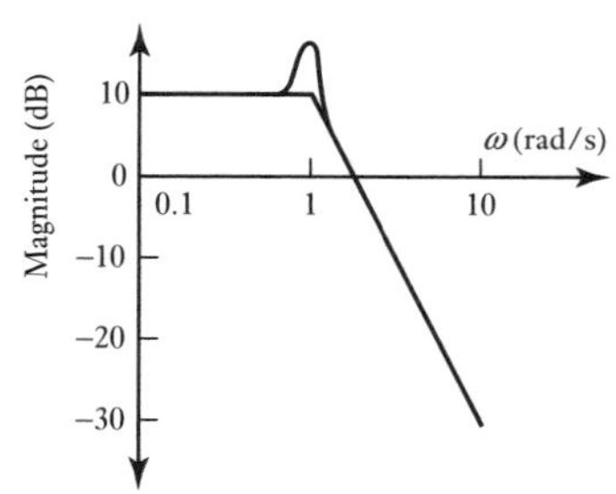

Figure 14.11 Bode plot.

As a further illustration, Figure 14.11 shows a Bode plot which has an initial zero gradient which changes by -40 dB/decade at 10 rad/s. The initial magnitude is 10 dB and so $10 = 20 \lg K$ and $K = 10^{0.5} = 3.2$. The change of -40 dB/decade at 10 rad/s means that there is a $(s^2/10^2 + 2\zeta s/10 + 1)$ term in the denominator. The transfer function is thus $3.2/(0.01s^2 + 0.2\zeta s + 1)$. The damping factor can be obtained by comparison of the Bode plot at the corner frequencies with Figure 14.6. It rises by about 6 dB above the corner and this corresponds to a damping factor of about 0.2. The transfer function is thus $3.2/(0.01s^2 + 0.04s + 1)$.

14.5 Performance specifications

The terms used to describe the performance of a system when subject to a sinusoidal input are peak resonance and bandwidth. The **peak resonance** M_p is defined as being the maximum value of the magnitude (Figure 14.12). A large value of the peak resonance corresponds to a large value of the maximum overshoot of a system. For a second-order system it can be directly related to the damping ratio by comparison of the response with the Bode plot of Figure 14.6, a low damping ratio corresponding to a high peak resonance.

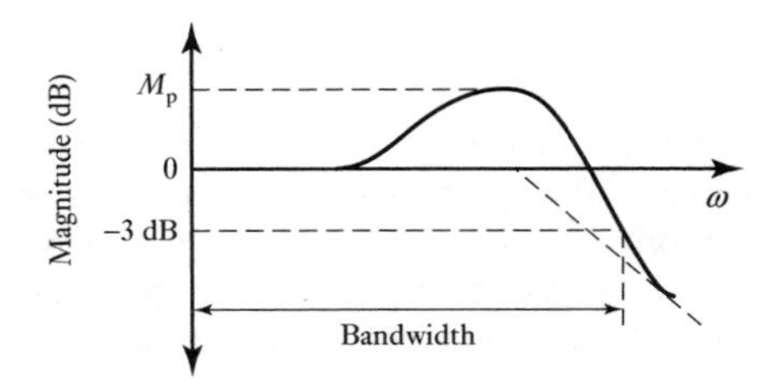

Figure 14.12 Performance specifications.

The **bandwidth** is defined as the frequency band between which the magnitude does not fall below −3 dB, the frequencies at which this occurs being termed the cut-off frequencies. With the magnitude expressed in decibel units (dB),

$$|G(j\omega)| \text{ in dB} = 20\,\lg_{10}|G(j\omega)|$$

and so

$$-3 = 20\,\log_{10}|G(j\omega)|$$

and $|G(j\omega)| = 0.707$ so the amplitude has dropped to 0.707 of its initial value. Since the power of a sinusoidal waveform is the square of its amplitude, then the power has dropped to $0.707^2 = 0.5$ of its initial value. Thus, the −3 dB cut-off is the decibel value at which the power of the input signal is attenuated to half the input value. For the system giving the Bode plot in Figure 14.12, the bandwidth is the spread between zero frequency and the frequency at which the magnitude drops below −3 dB. This is typical of measurement systems; they often exhibit no attenuation at low frequencies and the magnitude only degrades at high frequencies.

As an illustration, for the example described in Section 13.2.2, item 1, the magnitude of a system (an electric circuit with a resistor in series with a capacitor across which the output is taken) with a transfer function of

$$G(s) = \frac{1}{RCs + 1}$$

was determined as

$$|G(j\omega)| = \frac{1}{\sqrt{1 + \omega^2(RC)^2}}$$

For this magnitude ratio to be 0.707, the cut-off frequency ω_c must be given by

$$0.707 = \frac{1}{\sqrt{1 + \omega_c^2(RC)^2}}$$

$$1 + \omega_c^2(RC)^2 = (1/0.707)^2 = 2$$

Hence $\omega_c = 1/RC$. Such a circuit is called a low-pass filter since lower frequencies are passed to the output with little attenuation and higher frequencies are attenuated.

14.6 Stability

When there is a sinusoidal input to a system, the output from that system is sinusoidal with the same angular frequency but can have an output with an amplitude and phase which differ from that of the input. Consider a closed-loop system with negative feedback (Figure 14.13) and no input to the system. Suppose, somehow, we have a half-rectified sinusoidal pulse as the error signal in the system and that it passes through to the output and is fed back to arrive at the comparator element with amplitude unchanged but delayed by just half a cycle, i.e. a phase change of 180° as shown in the figure. When this signal is subtracted from the input signal, we have a resulting error signal which just continues the initial half-rectified pulse. This pulse then goes back round the feedback loop and once again arrives just in time to continue the signal. Thus we have a self-sustaining oscillation.

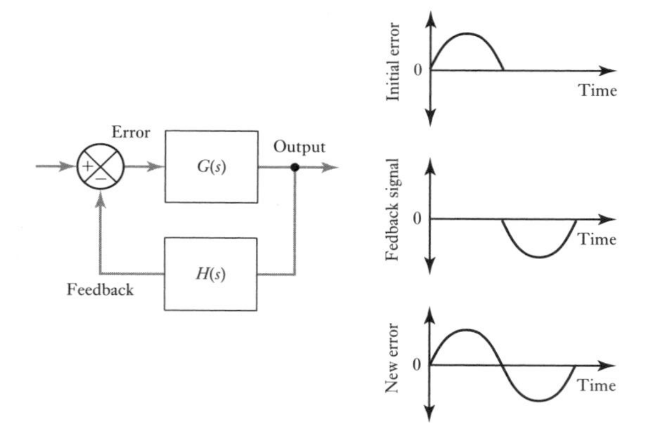

Figure 14.13 Self-sustaining oscillations.

For self-sustained oscillations to occur we must have a system which has a frequency-response function with a magnitude of 1 and a phase of −180°. The system through which the signal passes is $G(s)$ in series with $H(s)$. If the magnitude is less than 1 then each succeeding half-wave pulse is smaller in size and so the oscillation dies away. If the magnitude is greater than 1 then each succeeding pulse is larger than the previous one and so the wave builds up and the system is unstable.

1 A control system will oscillate with a constant amplitude if the magnitude resulting from the system $G(s)$ in series with $H(s)$ is 1 and the phase is $-180°$.
2 A control system will oscillate with a diminishing amplitude if the magnitude resulting from the system $G(s)$ in series with $H(s)$ is less than 1 and the phase is $-180°$.
3 A control system will oscillate with an increasing amplitude, and so is unstable, if the magnitude resulting from the system $G(s)$ in series with $H(s)$ is greater than 1 and the phase is $-180°$.

A good, stable control system usually requires that the magnitude of $G(s)H(s)$ should be significantly less than 1. Typically a value between 0.4 and 0.5 is used. In addition, the phase angle should be between about $-115°$ and $-125°$. Such values produce a slightly under-damped control system which gives, with a step input, about a 20 to 30% overshoot with a subsidence ratio of about 3 to 1 (see Section 12.5 for an explanation of these terms).

A concern with a control system is how stable it is and thus not likely to oscillate as a result of some small disturbance. The term **gain margin** is used for the factor by which the magnitude ratio must be multiplied when the phase is $-180°$ to make it have the value 1 and so be on the verge of instability. The term **phase margin** is used for the number of degrees by which the phase angle is numerically smaller than $-180°$ when the magnitude is 1. These rules mean a gain margin of between 2 and 2.5 and a phase margin between 45° and 65° for a good, stable control system.

Summary

We can convert from the s-domain to the **frequency domain** by replacing s by $j\omega$. The **frequency-response function** is the transfer function when transformed into the frequency domain.

The frequency response of a system is the set of values of the magnitude $|G(j\omega)|$ and phase angle ϕ that occur when a sinusoidal input signal is varied over a range of frequencies. This can be expressed as two graphs, one of the magnitude $|G(j\omega)|$ plotted against the angular frequency ω and the other of the phase ϕ plotted against ω. The magnitude and angular frequency are plotted using logarithmic scales. Such a pair of graphs is referred to as a **Bode plot**.

We can obtain the Bode plot of a system by adding together the Bode plots of the magnitudes of the constituent elements. Likewise the phase plot is obtained by adding together the phases of the constituent elements.

The **peak resonance** M_p is the maximum value of the magnitude. The **bandwidth** is the frequency band between which the magnitude does not fall below -3 dB, the frequencies at which this occurs being termed the cut-off frequencies.

For self-sustained oscillations to occur with a feedback system, i.e. it is on the verge of **instability**, we must have a system which has a frequency-response function with a magnitude of 1 and a phase of $-180°$. The **gain margin** is the factor by which the magnitude ratio must be multiplied when the phase is $-180°$ to make it have the value 1 and so be on the verge of instability. The **phase margin** is the number of degrees by which the phase angle is numerically smaller than $-180°$ when the magnitude is 1.

Problems

14.1 What are the magnitudes and phases of the systems having the following transfer functions?

(a) $\dfrac{5}{s + 2}$, (b) $\dfrac{2}{s(s + 1)}$, (c) $\dfrac{1}{(2s + 1)(s^2 + s + 1)}$

14.2 What will be the steady-state response of a system with a transfer function $1/(s + 2)$ when subject to the sinusoidal input $3 \sin(5t + 30°)$?

14.3 What will be the steady-state response of a system with a transfer function $5/(s^2 + 3s + 10)$ when subject to the input $2 \sin(2t + 70°)$?

14.4 Determine the values of the magnitudes and phase at angular frequencies of (i) 0 rad/s, (ii) 1 rad/s, (iii) 2 rad/s, (iv) ∞ rad/s for systems with the transfer functions (a) $1/[s(2s + 1)]$, (b) $1/(3s + 1)$.

14.5 Draw Bode plot asymptotes for systems having the transfer functions (a) $10/[s(0.1s + 1)]$, (b) $1/[(2s + 1)(0.5s + 1)]$.

14.6 Obtain the transfer functions of the systems giving the Bode plots in Figure 14.14.

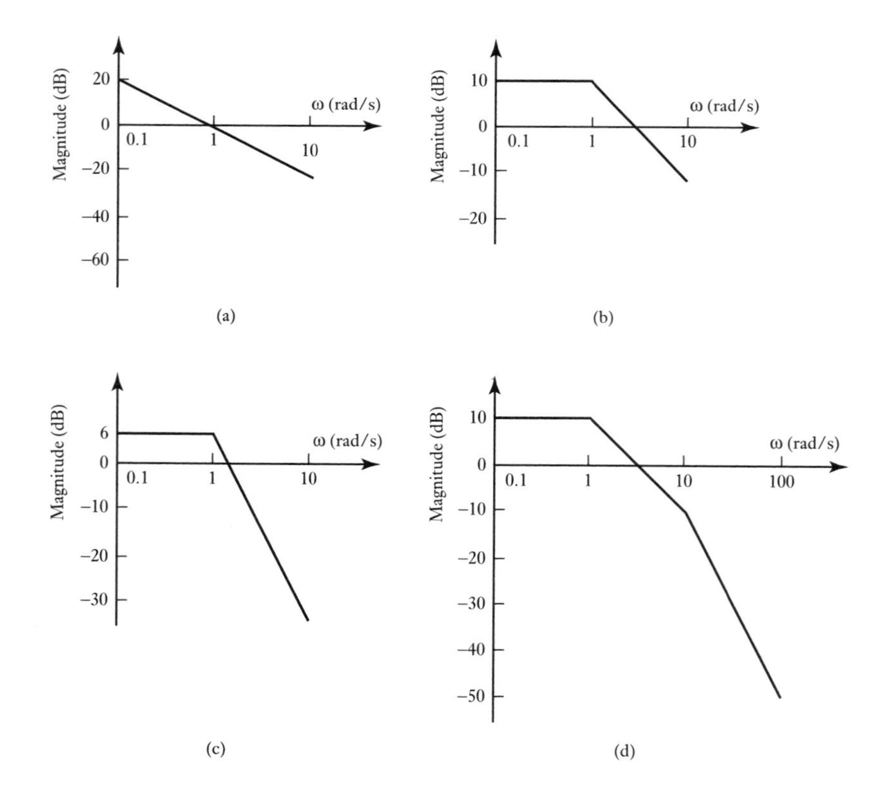

Figure 14.14 Problem 14.6.

Chapter fifteen Closed-loop Controllers

Objectives

The objectives of this chapter are that, after studying it, the reader should be able to:

- **Explain the term steady-state error.**
- **Explain the operation of the two-step mode of control.**
- **Predict the behaviour of systems with proportional, integral, derivative, proportional plus integral, proportional plus derivative and PID control.**
- **Describe how digital controllers operate.**
- **Explain how controllers can be tuned.**

15.1 Continuous and discrete control processes

Open-loop control is often just a switch on–switch off form of control, e.g. an electric fire is either switched on or off in order to heat a room. With **closed-loop control systems**, a controller is used to compare continuously the output of a system with the required condition and convert the error into a control action designed to reduce the error. The error might arise as a result of some change in the conditions being controlled or because the set value is changed, e.g. there is a step input to the system to change the set value to a new value. In this chapter we are concerned with the ways in which controllers can react to error signals, i.e. the **control modes** as they are termed, which occur with continuous processes. Such controllers might, for example, be pneumatic systems or operational amplifier systems. However, computer systems are rapidly replacing many of these. The term **direct digital control** is used when the computer is in the feedback loop and exercising control in this way. This chapter is about such closed-loop control.

Many processes not only involve controlling some variable, e.g. temperature, to a required value, but also involve the sequencing of operations. A domestic washing machine (see Section 1.5.5) where a number of actions have to be carried out in a predetermined sequence is an example. Another example is the manufacture of a product which involves the assembly of a number of discrete parts in a specific sequence by some controlled system. The sequence of operations might be **clock-based** or **event-based** or a combination of the two. With a clock-based system the actions are carried out at specific times; with an event-based system the actions are carried out when there is feedback to indicate that a particular event has occurred.

In many processes there can be a mixture of continuous and discrete control. For example, in the domestic washing machine there will be sequence control for the various parts of the washing cycle with feedback loop control of the temperature of the hot water and the level of the water.

15.1.1 Open- and closed-loop systems

Closed-loop systems differ from open-loop systems in having feedback. An open-loop system is one where the input signal does not automatically depend on the actual process output. With a closed-loop system, there is a feedback from the output to modify the input so that the system maintains the required output.

One consequence of having feedback is that there is a reduction of the effects of disturbance signals on the system. A disturbance signal is an unwanted signal which affects the output signal of a system. All physical systems are subject to some forms of extraneous signals during their operation. With an electric motor, this might be brush or commutator noise.

Consider the effect of external disturbances on the overall gain of an open-loop system. Figure 15.1 shows a two-element open-loop system with a disturbance which gives an input between the two elements. For a reference input $R(s)$ to the system, the first element gives an output of $G_1(s)R(s)$. To this is added the disturbance $D(s)$ to give an input of $G_1(s)R(s) + D(s)$. The overall system output $X(s)$ will then be

$$X(s) = G_2(s)[G_1(s)R(s) + D(s)] = G_1(s)G_2(s)R(s) + G_2(s)D(s)$$

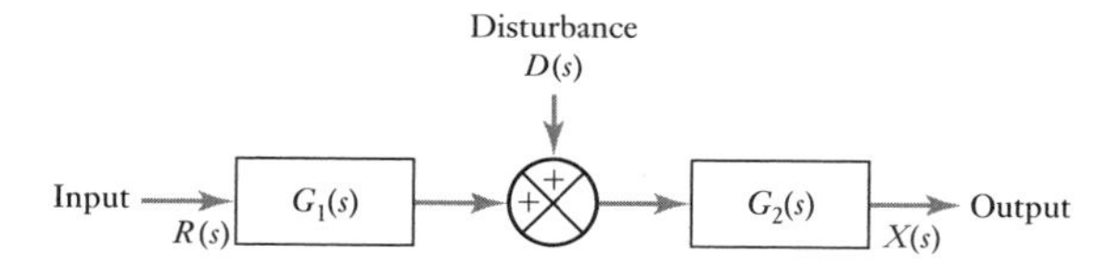

Figure 15.1 Disturbance with an open-loop system.

For the comparable system with negative feedback (Figure 15.2), the input to the first forward element $G_1(s)$ is $R(s)-H(s)X(s)$ and so its output is $G_1(s)[R(s)-H(s)X(s)]$. The input to $G_2(s)$ is $G_1(s)[R(s)-H(s)X(s)] + D(s)$ and so its output is

$$X(s) = G_2(s)\{G_1(s)[R(s)-H(s)X(s)] + D(s)\}$$

Thus

$$X(s) = \frac{G_1(s)G_2(s)}{1 + G_1(s)G_2(s)H(s)}R(s) + \frac{G_2(s)}{1 + G_1(s)G_2(s)H(s)}D(s)$$

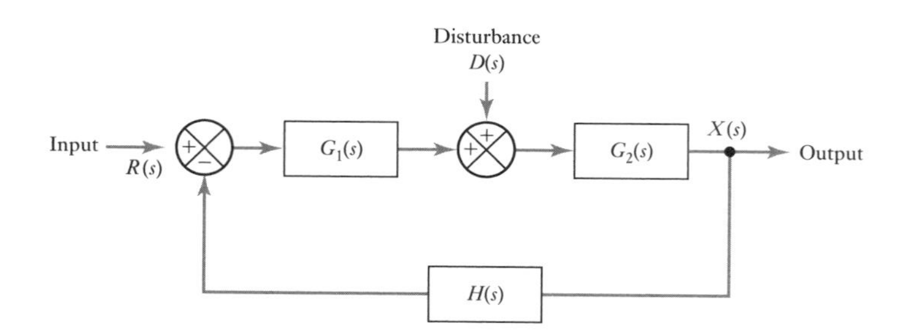

Figure 15.2 Disturbance with closed-loop system.

Comparing this with the equation for the open-loop system indicates that with the closed-loop system the effect of the disturbance on the output of the system has been reduced by a factor of $[1 + G_1(s)G_2(s)H(s)]$. The effect of a disturbance is reduced when there is feedback.

15.2 Terminology

The following are terms commonly used in discussing closed-loop controllers.

15.2.1 Lag

In any control system there are **lags**. Thus, for example, a change in the condition being controlled does not immediately produce a correcting response from the control system. This is because time is required for the system to make the necessary responses. For example, in the control of the temperature in a room by means of a central heating system, a lag will occur between the room temperature falling below the required temperature and the control system responding and switching on the heater. This is not the only lag. Even when the control system has responded there is a lag in the room temperature responding as time is taken for the heat to transfer from the heater to the air in the room.

15.2.2 Steady-state error

A closed-loop control system uses a measurement of the system output and a comparison of its value with the desired output to generate an error signal. We might get an error signal to the controller occurring as a result of the controlled variable changing or a change in the set value input. For example, we might have a ramp input to the system with the aim that the controlled variable increases steadily with time. When a change occurs, there are likely to be some transient effects; these, however, die away with time. The term **steady-state error** is used for the difference between the desired set value input and the output after all transients have died away. It is thus a measure of the accuracy of the control system in tracking the set value input. Whenever there is an error, the output is *not* at the desired output.

Consider a control system which has unity feedback (Figure 15.3). When there is a reference input of $R(s)$, there is an output of $X(s)$. The feedback signal is $X(s)$ and so the error signal is $E(s) = R(s) - X(s)$. If $G(s)$ is the forward-path transfer function, then for the unity feedback system as a whole:

$$\frac{X(s)}{R(s)} = \frac{G(s)}{1 + G(s)H(s)} = \frac{G(s)}{1 + G(s)}$$

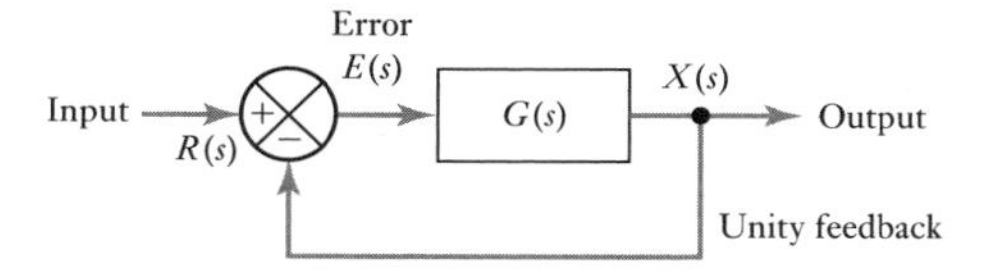

Figure 15.3 Unity feedback.

Hence

$$E(s) = R(s) - X(s) = R(s) - \frac{G(s)R(s)}{1 + G(s)} = \frac{1}{1 + G(s)}R(s)$$

The error thus depends on $G(s)$.

In order to determine the steady-state error we can determine the error e as a function of time and then determine the value of the error when all transients have died down and so the error as the time t tends to an infinite value. While we could determine the inverse of $E(s)$ and then determine its value when $t \to \infty$, there is a simpler method using the **final-value theorem** (see Appendix A); this involves finding the value of $sE(s)$ as s tends to a zero value:

$$e_{SS} = \lim_{t \to \infty} e(t) = \lim_{s \to 0} sE(s)$$

To illustrate the above, consider a unity feedback system with a forward-path transfer function of $k/(\tau s + 1)$ and subject to a unit step input of $1/s$:

$$e_{SS} = \lim_{s \to 0} sE(s) = \lim_{s \to 0} \left[s \frac{1}{1 + k/(\tau s + 1)} \frac{1}{s} \right] = \frac{1}{1 + k}$$

There is thus a steady-state error; the output from the system will never attain the set value. By increasing the gain k of the system then the steady-state error can be reduced.

The forward path might be a controller with a gain of k and a system with a transfer function $1/(\tau s + 1)$. Such a controller gain is termed a proportional controller. The steady-state error in this case is commonly termed **offset**. It can be minimised by increasing the gain.

However, if the unity feedback system had a forward-path transfer function of $k/s(\tau s + 1)$ and was subject to a step input, then the steady-state error would be

$$e_{SS} = \lim_{s \to 0} sE(s) = \lim_{s \to 0} \left[s \frac{1}{1 + k/s(\tau s + 1)} \frac{1}{s} \right] = 0$$

There is no steady-state error with this system. In this case, the forward path might be a controller with a gain of k/s and a system with a transfer function $1/(\tau s + 1)$. Such a controller gain is termed an integral controller and it gives no offset. Thus by combining an integral and a proportional controller it is possible to eliminate offset. Adding a derivative controller enables the controller to respond more rapidly to changes.

15.2.3 Control modes

There are a number of ways by which a control unit can react to an error signal and supply an output for correcting elements:

1. The *two-step mode* in which the controller is essentially just a switch which is activated by the error signal and supplies just an on/off correcting signal.
2. The *proportional mode* (P) which produces a control action that is proportional to the error. The correcting signal thus becomes bigger, the bigger the error. Thus as the error is reduced the amount of correction is reduced and the correcting process slows down.
3. The *derivative mode* (D) which produces a control action that is proportional to the rate at which the error is changing. When there is a sudden change in the error signal the controller gives a large correcting signal; when there is a gradual change only a small correcting signal is produced. Derivative control can be considered to be a form of anticipatory control in that the existing rate of change of error is measured, a coming larger error is anticipated and correction is applied before the larger error has arrived. Derivative control is not used alone but always in conjunction with proportional control and, often, integral control.
4. The *integral mode* (I) which produces a control action that is proportional to the integral of the error with time. Thus a constant error signal will produce an increasing correcting signal. The correction continues to increase as long as the error persists. The integral controller can be considered to be 'looking back', summing all the errors and thus responding to changes that have occurred.
5. *Combinations of modes*: proportional plus derivative (PD) modes, proportional plus integral (PI) modes, proportional plus integral plus derivative (PID) modes. The term **three-term controller** is used for PID control.

These five modes of control are discussed in the following sections of the chapter. A controller can achieve these modes by means of pneumatic circuits, analogue electronic circuits involving operational amplifiers or by the programming of a microprocessor or computer.

15.3 Two-step mode

An example of the **two-step mode** of control is the bimetallic thermostat (see Figure 2.46) that might be used with a simple temperature control system. This is just a switch which is switched on or off according to the temperature. If the room temperature is above the required temperature then the bimetallic strip is in an off position and the heater is off. If the room temperature falls below the required temperature then the bimetallic strip moves into an on position and the heater is switched fully on. The controller in this case can be in only two positions, on or off, as indicated by Figure 15.4(a).

With the two-step mode the control action is discontinuous. A consequence of this is that oscillations of the controlled variable occur about the required condition. This is because of lags in the time that the control system and the process take to respond. For example, in the case of the temperature control for a domestic central heating system, when the room temperature drops below the required level, the time that elapses before the control system responds and switches the heater on might be very small in comparison with the time that

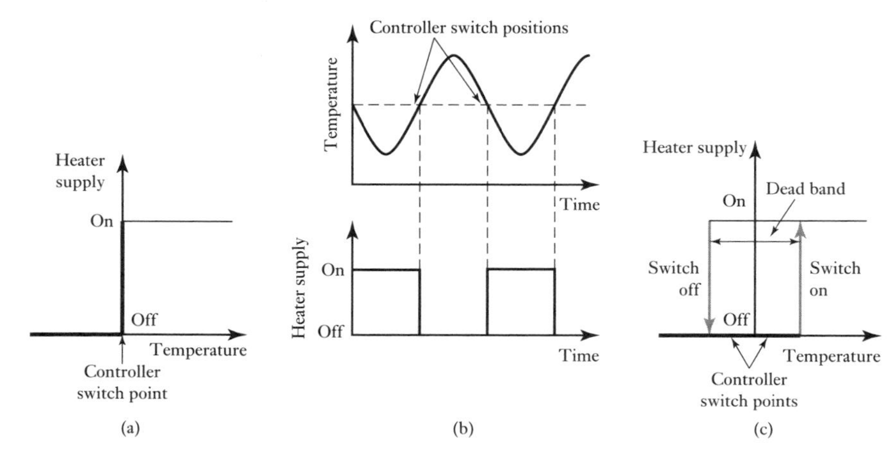

Figure 15.4 Two-step control.

elapses before the heater begins to have an effect on the room temperature. In the meantime the temperature has fallen even more. The reverse situation occurs when the temperature has risen to the required temperature. Since time elapses before the control system reacts and switches the heater off, and yet more time while the heater cools and stops heating the room, the room temperature goes beyond the required value. The result is that the room temperature oscillates above and below the required temperature (Figure 15.4(b)).

With the simple two-step system described above, there is the problem that when the room temperature is hovering about the set value the thermostat might be almost continually switching on or off, reacting to very slight changes in temperature. This can be avoided if, instead of just a single temperature value at which the controller switches the heater on or off, two values are used and the heater is switched on at a lower temperature than the one at which it is switched off (Figure 15.4(c)). The term **dead band** is used for the values between the on and off values. A large dead band results in large fluctuations of the temperature about the set temperature; a small dead band will result in an increased frequency of switching. The bimetallic element shown in Figure 2.46 has a permanent magnet for a switch contact; this has the effect of producing a dead band.

Two-step control action tends to be used where changes are taking place very slowly, i.e. with a process with a large capacitance. Thus, in the case of heating a room, the effect on the room temperature of switching the heater on or off is only a slow change. The result of this is an oscillation with a long periodic time. Two-step control is thus not very precise, but it does involve simple devices and is thus fairly cheap. On/off control is not restricted to mechanical switches such as bimetallic strips or relays; rapid switching can be achieved with the use of thyristor circuits (see Section 9.3.2); such a circuit might be used for controlling the speed of a motor, and operational amplifiers.

15.4 Proportional mode

With the two-step method of control, the controller output is either an on or an off signal, regardless of the magnitude of the error. With the **proportional mode,** the size of the controller output is proportional to the size of the error: the bigger the error, the bigger the output from the controller. This means that the correction element of the control system, e.g. a valve, will receive a signal which is proportional to the size of the correction required. Thus

$$\text{controller output} = K_P e$$

where e is the error and K_P a constant. Thus taking Laplace transforms,

$$\text{controller output }(s) = K_P E(s)$$

and so K_P is the transfer function of the controller.

15.4.1 Electronic proportional controller

A summing operational amplifier with an inverter can be used as a proportional controller (Figure 15.5). For a summing amplifier we have (see Section 3.2.3)

$$V_{out} = -R_f\left(\frac{V_0}{R_2} + \frac{V_e}{R_1}\right)$$

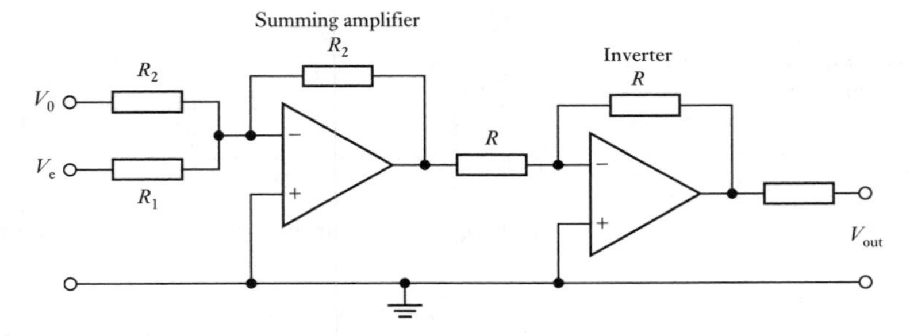

Figure 15.5 Proportional controller.

The input to the summing amplifier through R_2 is the zero error voltage value V_0, i.e. the set value, and the input through R_1 is the error signal V_e. But when the feedback resistor $R_f = R_2$, then the equation becomes

$$V_{out} = -\frac{R_2}{R_1}V_e - V_0$$

If the output from the summing amplifier is then passed through an inverter, i.e. an operational amplifier with a feedback resistance equal to the input resistance, then

$$V_{out} = \frac{R_2}{R_1}V_e + V_0$$

$$V_{out} = K_P V_e + V_0$$

where K_P is the proportionality constant. The result is a proportional controller.

As an illustration, Figure 15.6 shows an example of a proportional control system for the control of the temperature of a liquid in a container as liquid is pumped through it.

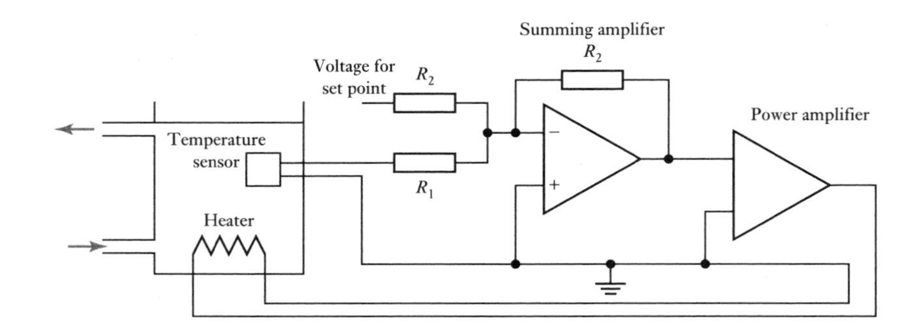

Figure 15.6 Proportional controller for temperature control.

15.4.2 System response

With proportional control we have a gain element with transfer function K_P in series with the forward-path element $G(s)$ (Figure 15.7). The error is thus

$$E(s) = \frac{K_P G(s)}{1 + K_P G(s)} R(s)$$

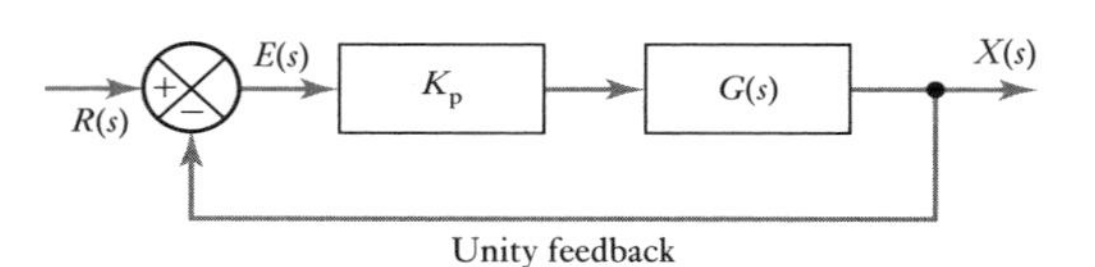

Figure 15.7 System with proportional control.

and so, for a step input, the steady-state error is

$$e_{SS} = \lim_{s \to 0} sE(s) = \lim_{s \to 0} \left[s \frac{1}{1 + 1/K_P G(s)} \frac{1}{s} \right]$$

This will have a finite value and so there is always a steady-state error. Low values of K_P give large steady-state errors but stable responses. High values of K_P give smaller steady-state errors but a greater tendency to instability.

15.5 Derivative control

With the **derivative mode** of control the controller output is proportional to the rate of change with time of the error signal. This can be represented by the equation

$$\text{controller output} = K_D \frac{de}{dt}$$

K_D is the constant of proportionality. The transfer function is obtained by taking Laplace transforms, thus

$$\text{controller output } (s) = K_D s E(s)$$

Hence the transfer function is $K_D s$.

With derivative control, as soon as the error signal begins to change, there can be quite a large controller output since it is proportional to the rate of

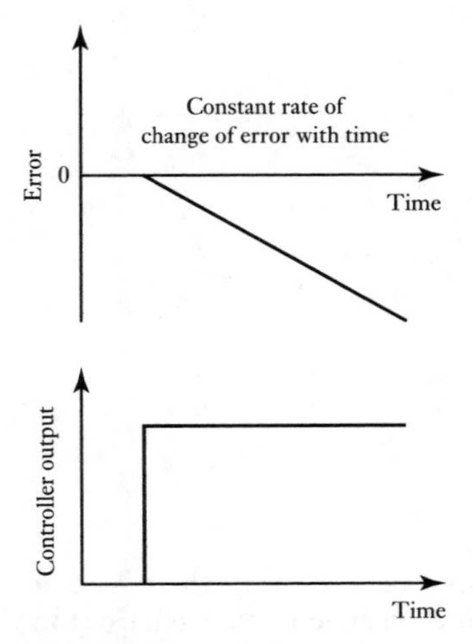

Figure 15.8 Derivative control.

change of the error signal and not its value. Rapid initial responses to error signals thus occur. Figure 15.8 shows the controller output that results when there is a constant rate of change of error signal with time. The controller output is constant because the rate of change is constant and occurs immediately the deviation occurs. Derivative controllers do not, however, respond to steady-state error signals, since with a steady error the rate of change of error with time is zero. Because of this, derivative control is always combined with proportional control; the proportional part gives a response to all error signals, including steady signals, while the derivative part responds to the rate of change. Derivative action can also be a problem if the measurement of the process variable gives a noisy signal, the rapid fluctuations of the noise resulting in outputs which will be seen by the controller as rapid changes in error and so give rise to significant outputs from the controller.

Figure 15.9 shows the form of an electronic derivative controller circuit, the circuit involving an operational amplifier connected as a differentiator circuit followed by another operational amplifier connected as an inverter. The derivative time K_D is R_2C.

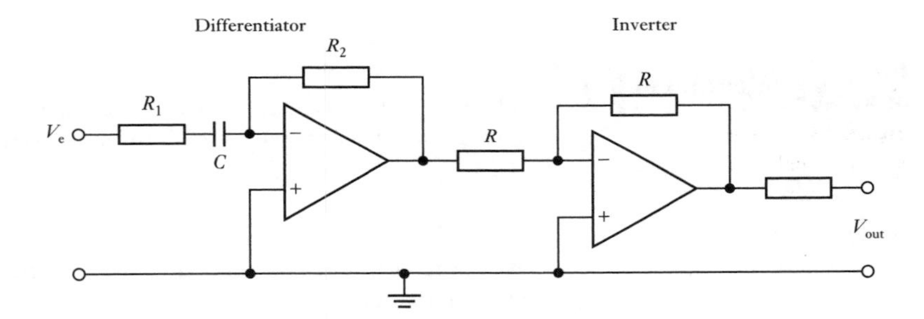

Figure 15.9 Derivative controller.

15.5.1 Proportional plus derivative (PD) control

Derivative control is never used alone because it is not capable of giving an output when there is a steady error signal and so no correction is possible. It is thus invariably used in conjunction with proportional control so that this problem can be resolved.

With proportional plus derivative control the controller output is given by

$$\text{controller output} = K_P e + K_D \frac{de}{dt}$$

K_P is the proportionality constant and K_D the derivative constant, de/dt is the rate of change of error. The system has a transfer function given by

$$\text{controller output } (s) = K_P E(s) + K_D s E(s)$$

Hence the transfer function is $K_P + K_D s$. This is often written as

$$\text{transfer function} = K_D\left(s + \frac{1}{T_D}\right)$$

where $T_D = K_D/K_P$ and is called the **derivative time constant**.

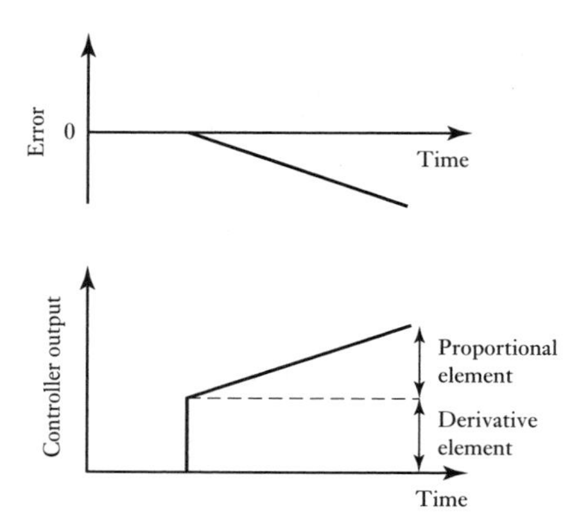

Figure 15.10 PD control.

Figure 15.10 shows how the controller output can vary when there is a constantly changing error. There is an initial quick change in controller output because of the derivative action followed by the gradual change due to proportional action. This form of control can thus deal with fast process changes.

15.6 Integral control

The **integral mode** of control is one where the rate of change of the control output I is proportional to the input error signal e:

$$\frac{\mathrm{d}I}{\mathrm{d}t} = K_I e$$

K_I is the constant of proportionality and has units of $1/s$. Integrating the above equation gives

$$\int_{I_0}^{I_{out}} \mathrm{d}I = \int_0^t K_I e \, \mathrm{d}t$$

$$I_{out} - I_0 = \int_0^t K_I e \, \mathrm{d}t$$

I_0 is the controller output at zero time, I_{out} is the output at time t.

The transfer function is obtained by taking the Laplace transform. Thus

$$(I_{out} - I_0)(s) = \frac{1}{s} K_I E(s)$$

and so

$$\text{transfer function} = \frac{1}{s} K_I$$

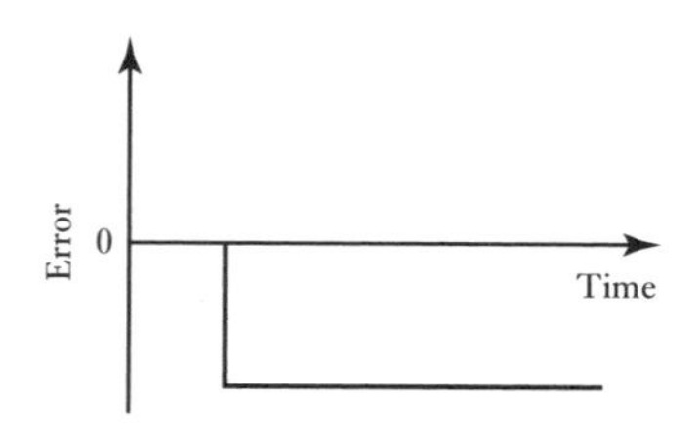

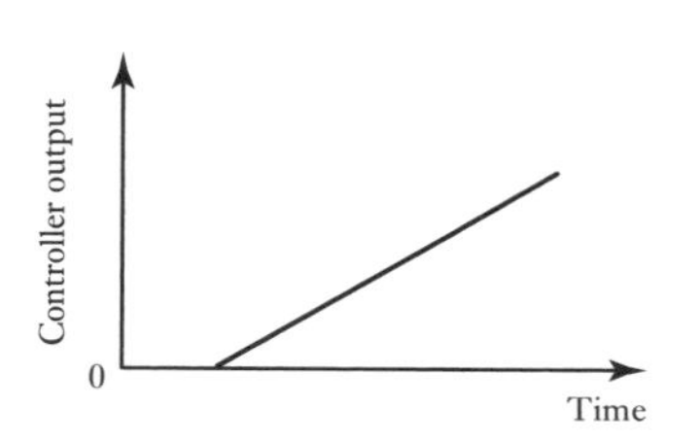

Figure 15.11 Integral control.

Figure 15.11 illustrates the action of an integral controller when there is a constant error input to the controller. We can consider the graphs in two ways. When the controller output is constant, the error is zero; when the controller output varies at a constant rate, the error has a constant value. The alternative way of considering the graphs is in terms of the area under the error graph:

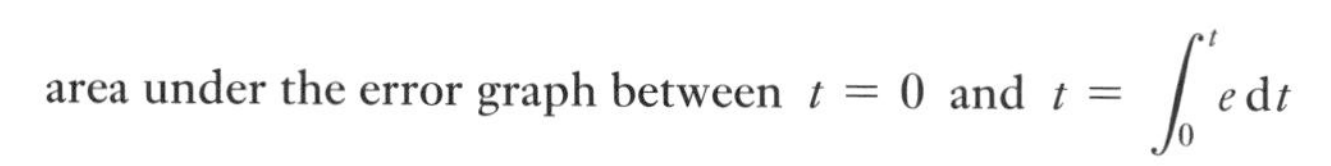

$$\text{area under the error graph between } t = 0 \text{ and } t = \int_0^t e \, \mathrm{d}t$$

Thus up to the time when the error occurs the value of the integral is zero. Hence $I_{out} = I_0$. When the error occurs it maintains a constant value. Thus the area under the graph is increasing as the time increases. Since the area increases at a constant rate the controller output increases at a constant rate.

Figure 15.12 shows the form of the circuit used for an electronic integral controller. It consists of an operational amplifier connected as an integrator and followed by another operational amplifier connected as a summer to add the integrator output to that of the controller output at zero time. K_I is $1/R_I C$.

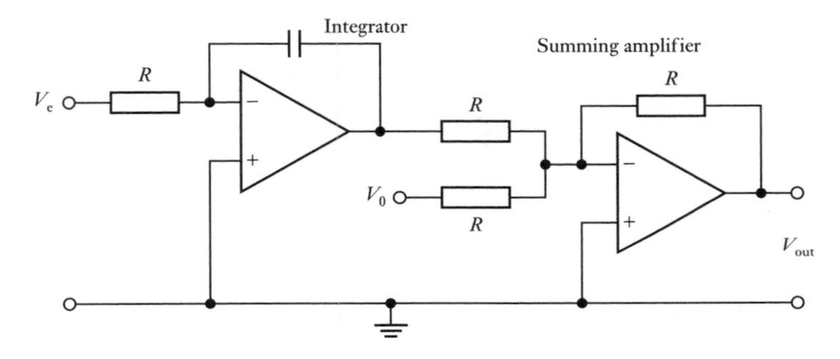

Figure 15.12 Integral controller.

15.6.1 Proportional plus integral (PI) control

The integral mode of control is not usually used alone but is frequently used in conjunction with the proportional mode. When integral action is added to a proportional control system the controller output is given by

$$\text{controller output} = K_P e + K_I \int e\,dt$$

where K_P is the proportional control constant, K_I the integral control constant and e the error e. The transfer function is thus

$$\text{transfer function} = K_P + \frac{K_I}{s} = \frac{K_P}{s}\left(s + \frac{1}{T_I}\right)$$

where $T_I = K_P/K_I$ and is the **integral time constant**.

Figure 15.13(a) shows how the system reacts when there is an abrupt change to a constant error. The error gives rise to a proportional controller output which remains constant since the error does not change. There is then superimposed on this a steadily increasing controller output due to the integral action. Figure 15.13(b) shows the effects of the proportional action and the integral action if we create an error signal which is increased from the zero value and then decreased back to it again. With proportional action alone the controller mirrors the change and ends up back at its original set point value. With the integral action the controller output increases in proportion to the way the area under the error–time graph increases and since, even when the error has reverted back to zero, there is still a value for the area, there is a change in controller output which continues after the error has ceased.

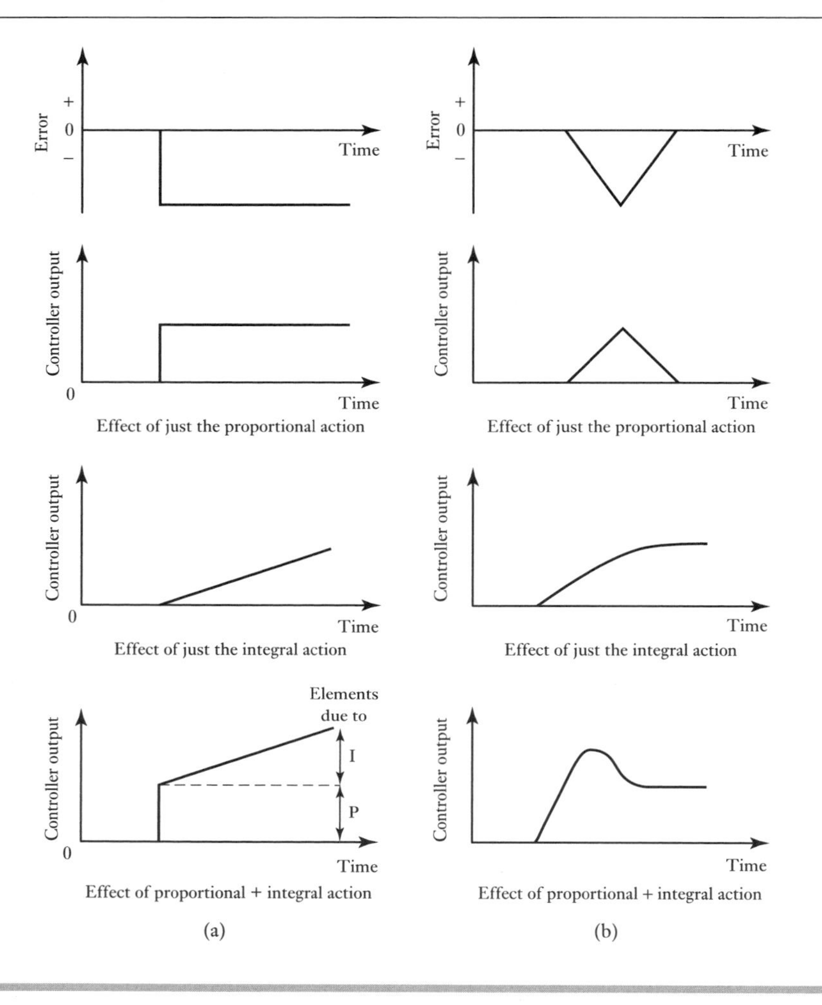

Figure 15.13 PI control.

15.7 PID controller

Combining all three modes of control (proportional, integral and derivative) gives a controller known as a **three-mode controller** or **PID controller**. The equation describing its action can be written as

$$\text{controller output} = K_{P}e + K_{I}\int e\,dt + K_{D}\frac{de}{dt}$$

where K_P is the proportionality constant, K_I the integral constant and K_D the derivative constant. Taking the Laplace transform gives

$$\text{controller output }(s) = K_{P}E(s) + \frac{1}{s}K_{I}E(s) + sK_{D}(s)$$

and so

$$\text{transfer function} = K_{P}e + \frac{1}{s}K_{I} + sK_{D} = K_{P}\left(1 + \frac{1}{T_{I}s} + T_{D}s\right)$$

15.7.1 Operational amplifier PID circuits

A three-mode controller can be produced by combining the various circuits described earlier in this chapter for the separate proportional, derivative and integral modes. A more practical controller can, however, be produced with a single operational amplifier. Figure 15.14 shows one such circuit. The proportional constant K_P is $R_I/(R + R_D)$, the derivative constant K_D is $R_D C_D$ and the integral constant K_I is $1/R_I C_I$.

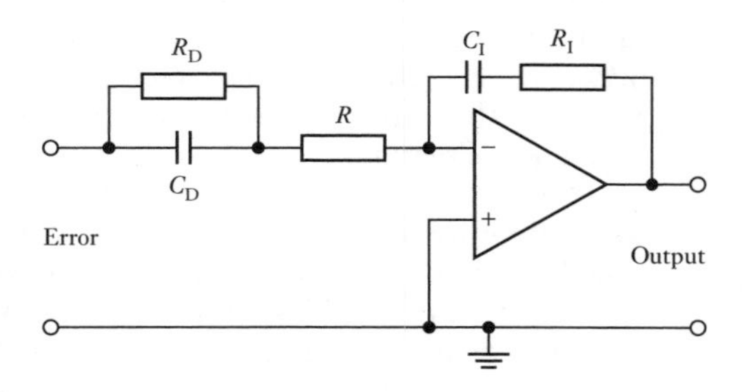

Figure 15.14 PID circuit.

15.8 Digital controllers

Figure 15.15 shows the basis of a direct digital control system that can be used with a continuous process; the term **direct digital control** is used when the digital controller, basically a microprocessor, is in control of the closed-loop control system. The controller receives inputs from sensors, executes control programs and provides the output to the correction elements. Such controllers require inputs which are digital, process the information in digital form and give an output in digital form. Since many control systems have analogue measurements an analogue-to-digital converter (ADC) is used for the inputs. A clock supplies a pulse at regular time intervals and dictates when samples of the controlled variable are taken by the ADC. These samples are then converted to digital signals which are compared by the microprocessor with the set point value to give the error signal. The microprocessor can then initiate a control mode to process the error signal and give a digital output. The control mode used by the microprocessor is determined by the program of instructions used by the microprocessor for processing the digital signals, i.e. the *software*. The digital output, generally after processing by a digital-to-analogue converter since correcting elements generally require analogue signals, can be used to initiate the correcting action.

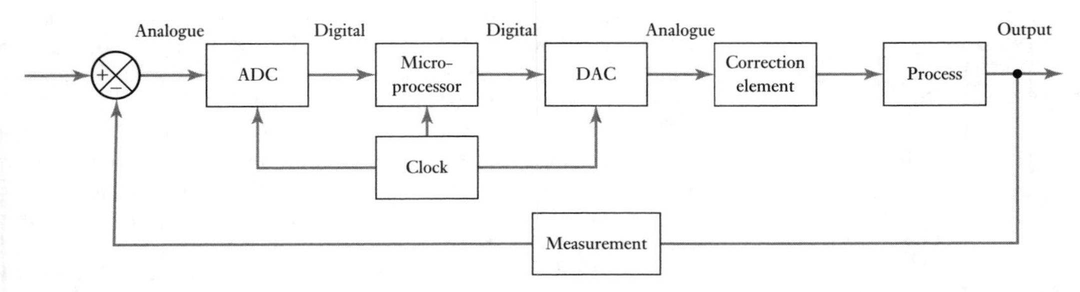

Figure 15.15 Digital closed-loop control system.

A digital controller basically operates the following cycle of events:

1 Samples the measured value.
2 Compares it with the set value and establishes the error.
3 Carries out calculations based on the error value and stored values of previous inputs and outputs to obtain the output signal.
4 Sends the output signal to the DAC.
5 Waits until the next sample time before repeating the cycle.

Microprocessors as controllers have the advantage over analogue controllers that the form of the controlling action, e.g. proportional or three-mode, can be altered by purely a change in the computer software. No change in hardware or electrical wiring is required. Indeed the control strategy can be altered by the computer program during the control action in response to the developing situation.

They also have other advantages. With analogue control, separate controllers are required for each process being controlled. With a microprocessor many separate processes can be controlled by sampling processes with a multiplexer (see Section 4.4). Digital control gives better accuracy than analogue control because the amplifiers and other components used with analogue systems change their characteristics with time and temperature and so show drift, while digital control, because it operates on signals in only the on/off mode, does not suffer from drift in the same way.

15.8.1 Implementing control modes

In order to produce a digital controller which will give a particular mode of control it is necessary to produce a suitable program for the controller. The program has to indicate how the digital error signal at a particular instant is to be processed in order to arrive at the required output for the following correction element. The processing can involve the present input together with previous inputs and previous outputs. The program is thus asking the controller to carry out a difference equation (see Section 4.6).

The transfer function for a PID analogue controller is

$$\text{transfer function} = K_P + \frac{1}{s}K_I + sK_D$$

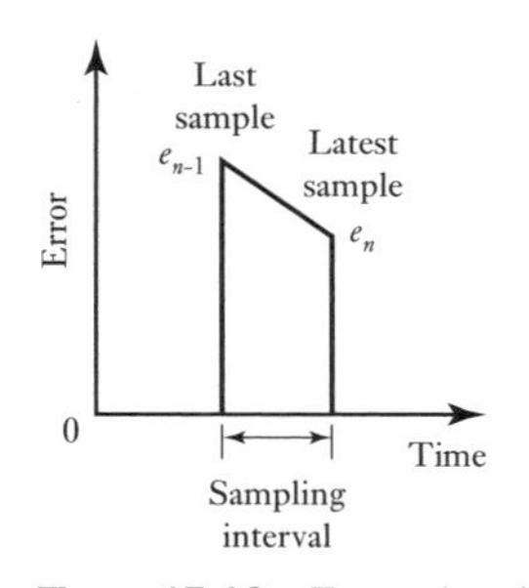

Figure 15.16 Error signals.

Multiplication by s is equivalent to differentiation. We can, however, consider the gradient of the time response for the error signal at the present instant of time as being (latest sample of the error e_n minus the last sample of the error e_{n-1})/(sampling interval T_s) (Figure 15.16).

Division by s is equivalent to integration. We can, however, consider the integral of the error at the end of a sampling period as being the area under the error–time graph during the last sampling interval plus the sum of the areas under the graph for all previous samples (Int_{prev}). If the sampling period is short relative to the times involved then the area during the last sampling interval is approximately $\frac{1}{2}(e_n + e_{n-1})/T_s$ (see Section 4.6 for another approximation known as Tustin's approximation). Thus we can write for the controller output x_n at a particular instant the equivalent of the transfer function as

$$x_n = K_P e_n + K_I\left(\frac{(e_n + e_{n-1})T_s}{2} + \text{Int}_{\text{prev}}\right) + K_D\frac{e_n - e_{n-1}}{T_s}$$

We can rearrange this equation to give

$$x_n = Ae_n + Be_{n-1} + C(\text{Int}_{\text{prev}})$$

where $A = K_P + 0.5K_I T_s + K_D/T_s$, $B = 0.5K_I T_s - K_D/T_s$ and $C = K_I$.

The program for PID control thus becomes:

1 Set the values of K_P, K_I and K_D.
2 Set the initial values of e_{n-1}, Int_{prev} and the sample time T_s.
3 Reset the sample interval timer.
4 Input the error e_n.
5 Calculate y_n using the above equation.
6 Update, ready for the next calculation, the value of the previous area to $\text{Int}_{\text{prev}} + 0.5(e_n + e_{n-1})T_s$.
7 Update, ready for the next calculation, the value of the error by setting e_{n-1} equal to e_n.
8 Wait for the sampling interval to elapse.
9 Go to step 3 and repeat the loop.

15.8.2 Sampling rate

When a continuous signal is sampled, for the sample values to reflect accurately the continuous signal, they have to be sufficiently close together in time for the signal not to fluctuate significantly between samples. During a sampling interval, no information is fed back to the controller about changes in the output. In practice this is taken to mean that the samples have to be taken at a rate greater than twice the highest frequency component in the continuous signal. This is termed Shannon's sampling theorem (see Section 4.2.1). In digital control systems, the sampling rate is generally much higher than this.

15.8.3 A computer control system

Typically a computer control system consists of the elements shown in Figure 15.15 with set points and control parameters being entered from a keyboard. The software for use with the system will provide the program of instructions needed, for example, for the computer to implement the PID control mode, provide the operator display, recognise and process the instructions inputted by the operator, provide information about the system, provide start-up and shut-down instructions, and supply clock/calendar information. An operator display is likely to show such information as the set point value, the actual measured value, the sampling interval, the error, the controller settings and the state of the correction element. The display is likely to be updated every few seconds.

15.9 Control system performance

The transfer function of a control system is affected by the mode chosen for the controller. Hence the response of the system to, say, a step input is affected. Consider the simple system shown in Figure 15.17.

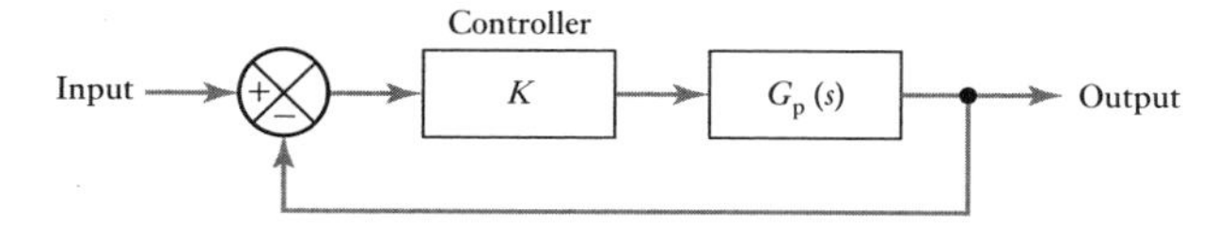

Figure 15.17 Control system.

With proportional control the transfer function of the forward path is $K_P G(s)$ and so the transfer function of the feedback system $G(s)$ is

$$G(s) = \frac{K_P G_p(s)}{1 + K_P G_p(s)}$$

Suppose we have a process which is first order with a transfer function of $1/(\tau s + 1)$ where τ is the time constant (it might represent a d.c. motor, often modelled as a first-order system – see Section 13.5.1). With proportional control, and unity feedback, the transfer function of the control system becomes

$$G(s) = \frac{K_P/(\tau s + 1)}{1 + K_P/(\tau s + 1)} = \frac{K_P}{\tau s + 1 + K_P}$$

The control system remains a first-order system. The proportional control has had the effect of just changing the form of the first-order response of the process. Without the controller, the response to a unit step input was (see Section 13.2.1)

$$y = 1 - e^{-t/\tau}$$

Now it is

$$y = K_P(1 - e^{-t/(\tau/1 + K_P)})$$

The effect of the proportional control has been to reduce the time constant from τ to $\tau/(1 + K_P)$, making it faster responding to the higher value of K_P. It also decreases the steady-state error.

With integral control we have a forward-path transfer function of $K_I G_p(s)/s$ and so the system transfer function is

$$G(s) = \frac{K_I G_p(s)}{s + K_I G_p(s)}$$

Thus, if we now have a process which is first order with a transfer function of $1/(\tau s + 1)$, with proportional control and unity feedback the transfer function of the control system becomes

$$G(s) = \frac{K_I/(\tau s + 1))}{s + K_I/(\tau s + 1)} = \frac{K_I}{s(\tau s + 1) + K_I} = \frac{K_I}{\tau s^2 + s + K_I}$$

The control system is now a second-order system. With a step input, the system will give a second-order response instead of a first-order response.

With a system having derivative control the forward-path transfer function is $sK_DG(s)$ and so, with unity feedback, the system transfer function is

$$G(s) = \frac{sK_DG_p(s)}{1 + K_DG_p(s)}$$

With a process which is first order with a transfer function of $1/(\tau s + 1)$, derivative control gives an overall transfer function of

$$G(s) = \frac{sK_D/(\tau s + 1)}{1 + sK_D/(\tau s + 1)} = \frac{sK_D}{\tau s + 1 + sK_D}$$

15.10 Controller tuning

The term **tuning** is used to describe the process of selecting the best controller settings. With a proportional controller this means selecting the value of K_P; with a PID controller the three constants K_P, K_I and K_D have to be selected. There are a number of methods of doing this; here just two methods will be discussed, both by Ziegler and Nichols. They assumed that when the controlled system is open loop a reasonable approximation to its behaviour is a first-order system with a built-in time delay. Based on this, they then derived parameters for optimum performance. This was taken to be settings which gave an under-damped transient response with a decay (subsidence) ratio of ¼., i.e. the second overshoot is ¼ of the first overshoot (see Section 12.5). This overshoot criterion gives a good compromise of small rise time, small settling time and a reasonable margin of stability.

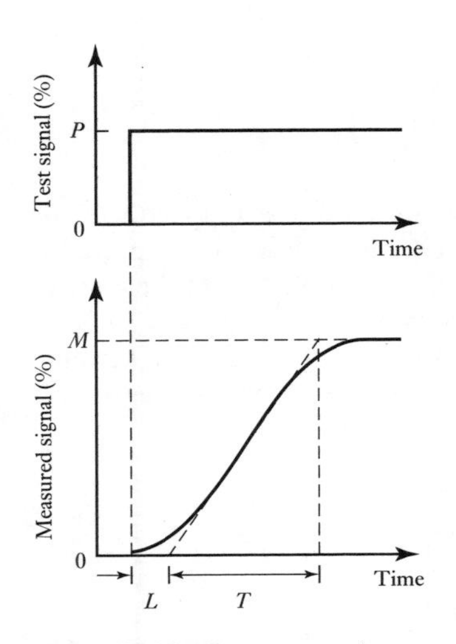

Figure 15.18 Process reaction curve.

15.10.1 Process reaction method

The process control loop is opened, generally between the controller and the correction unit, so that no control action occurs. A test input signal is then applied to the correction unit and the response of the controlled variable determined. The test signal should be as small as possible. Figure 15.18 shows the form of test signal and a typical response. The test signal is a step signal with a step size expressed as the percentage change P in the correction unit. The graph of the measured variable plotted against time is called the **process reaction curve**. The measured variable is expressed as the percentage of the full-scale range.

A tangent is drawn to give the maximum gradient of the graph. For Figure 15.18 the maximum gradient R is M/T. The time between the start of the test signal and the point at which this tangent intersects the graph time axis is termed the lag L. Table 15.1 gives the criteria recommended by Ziegler and Nichols for control settings based on the values of P, R and L.

Table 15.1 Process reaction curve criteria.

Control mode	K_P	T_I	T_D
P	P/RL		
PI	$0.9P/RL$	$3.33L$	
PID	$1.2P/RL$	$2L$	$0.5L$

Consider the following example. Determine the settings required for a three-mode controller which gave the process reaction curve shown in Figure 15.19 when the test signal was a 6% change in the control valve position. Drawing a tangent to the maximum gradient part of the graph gives a lag L of 150 s and a gradient R of $5/300 = 0.017/\text{s}$. Hence

$$K_P = \frac{1.2P}{RL} = \frac{1.2 \times 6}{0.017 \times 150} = 2.82$$

$$T_I = 2L = 300\,\text{s}$$

$$T_D = 0.5L = 0.5 \times 150 = 75\,\text{s}$$

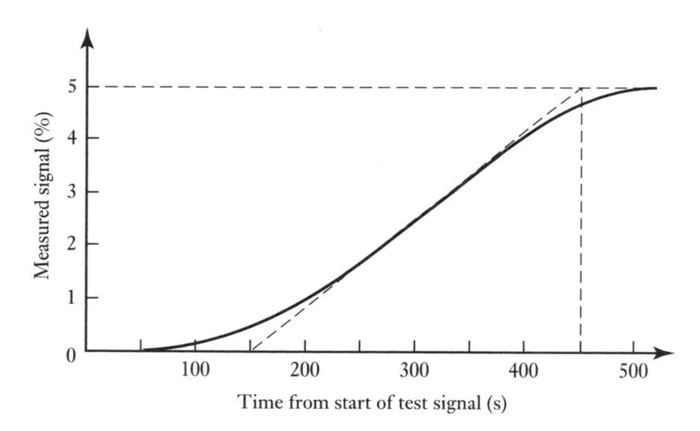

Figure 15.19 Process curve example.

15.10.2 Ultimate cycle method

With this method, integral and derivative actions are first reduced to their minimum values. The proportional constant K_P is set low and then gradually increased. While doing this small disturbances are applied to the system. This is continued until continuous oscillations occur. The critical value of the proportional constant K_{Pc} at which this occurs is noted and the periodic time of the oscillations T_c measured. Table 15.2 shows how the Ziegler and

Table 15.2 Ultimate cycle criteria.

Control mode	K_P	T_I	T_D
P	$0.5K_{Pc}$		
PI	$0.45K_{Pc}$	$T_c/1.2$	
PID	$0.6K_{Pc}$	$T_c/2.0$	$T_c/8$

Nichols recommended criteria for controller settings are related to this value of K_{Pc}. The critical proportional band is $100/K_{Pc}$.

Consider the following example. When tuning a three-mode control system by the ultimate cycle method it was found that oscillations begin when K_{Pc} is 3.33. The oscillations have a periodic time of 500 s. What are the suitable settings for the controller? Using the criteria given in Table 15.2, then $K_P = 0.6K_{Pc} = 0.6 \times 3.33 = 2.0$, $T_I = T_c/2.0 = 500/2 = 2.5$ s, $T_D = T_c/8 = 500/8 = 62.5$ s.

15.11 Velocity control

Consider the problem of controlling the movement of a load by means of a motor. Because the motor system is likely to be of second order, proportional control will lead to the system output taking time to reach the required displacement when there is, say, a step input to the system and may oscillate for a while about the required value. Time will thus be taken for the system to respond to an input signal. A higher speed of response, with fewer oscillations, can be obtained by using PD rather than just P control. There is, however, an alternative of achieving the same effect and this is by the use of a second feedback loop which gives a measurement related to the rate at which the displacement is changing. This is termed **velocity feedback**. Figure 15.20 shows such a system; the velocity feedback might involve the use of a tachogenerator giving a signal proportional to the rotational speed of the motor shaft, and hence the rate at which the displacement is changing, and the displacement might be monitored using a rotary potentiometer.

15.12 Adaptive control

There are many control situations where the parameters of the plant change with time or, perhaps, load, e.g. a robot manipulator being used to move loads when the load is changed. If the transfer function of the plant changes then retuning of the system is desirable for the optimum values to be determined for proportional, derivative and integral constants. For the control systems so far considered, it has been assumed that the system once tuned retains its values of proportional, derivative and integral constants until the operator decides to retune. The alternative to this is an **adaptive control system** which 'adapts' to changes and changes its parameters to fit the circumstances prevailing.

The adaptive control system is based on the use of a microprocessor as the controller. Such a device enables the control mode and the control parameters used to be adapted to fit the circumstances, modifying them as the circumstances change.

An adaptive control system can be considered to have three stages of operation:

1. Starts to operate with controller conditions set on the basis of an assumed condition.
2. The desired performance is continuously compared with the actual system performance.
3. The control system mode and parameters are automatically and continuously adjusted in order to minimise the difference between the desired and actual system performance.

For example, with a control system operating in the proportional mode, the proportional constant K_P may be automatically adjusted to fit the

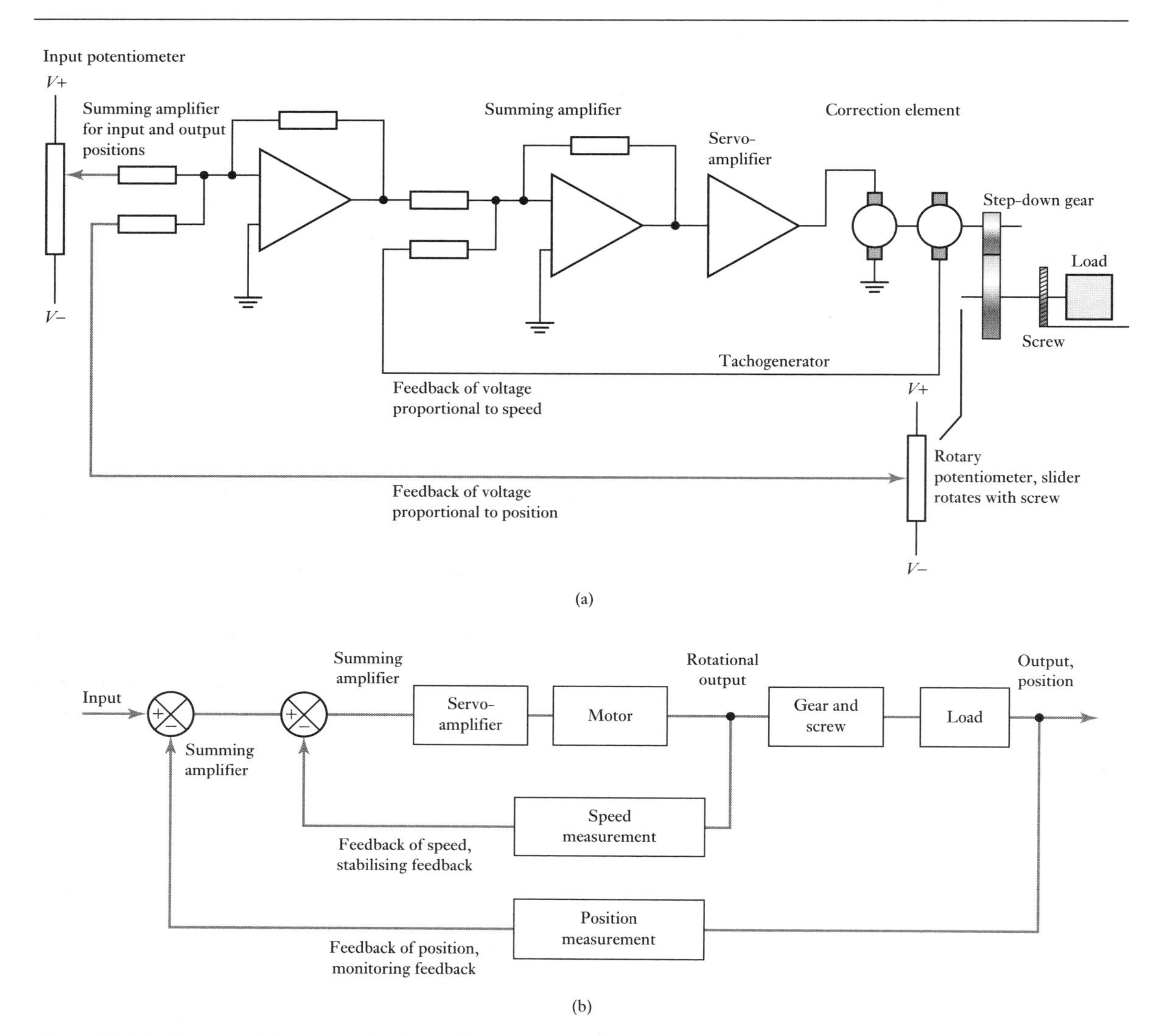

Figure 15.20 System with velocity feedback: (a) descriptive diagram of the system, (b) block diagram of the system.

circumstances, changing as they do. Adaptive control systems can take a number of forms. Three commonly used forms are:

1 Gain-scheduled control.
2 Self-tuning.
3 Model-reference adaptive systems.

15.12.1 Gain-scheduled control

With **gain-scheduled control** or, as it is sometimes referred to, **pre-programmed adaptive control**, preset changes in the parameters of the controller are made on the basis of some auxiliary measurement of some

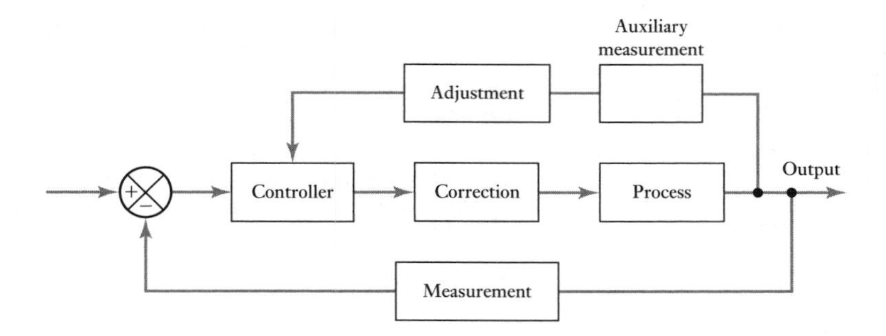

Figure 15.21 Gain-scheduled control.

process variable. Figure 15.21 illustrates this method. The term gain-scheduled control was used because the only parameter originally adjusted was the gain, i.e. the proportionality constant K_P.

For example, for a control system used to control the positioning of some load, the system parameters could be worked out for a number of different load values and a table of values loaded into the memory of the controller. A load cell might then be used to measure the actual load and give a signal to the controller indicating a mass value which is then used by the controller to select the appropriate parameters.

A disadvantage of this system is that the control parameters have to be determined for many operating conditions so that the controller can select the one to fit the prevailing conditions. An advantage, however, is that the changes in the parameters can be made quickly when the conditions change.

15.12.2 Self-tuning

With **self-tuning control** the system continuously tunes its own parameters based on monitoring the variable that the system is controlling and the output from the controller. Figure 15.22 illustrates the features of this system.

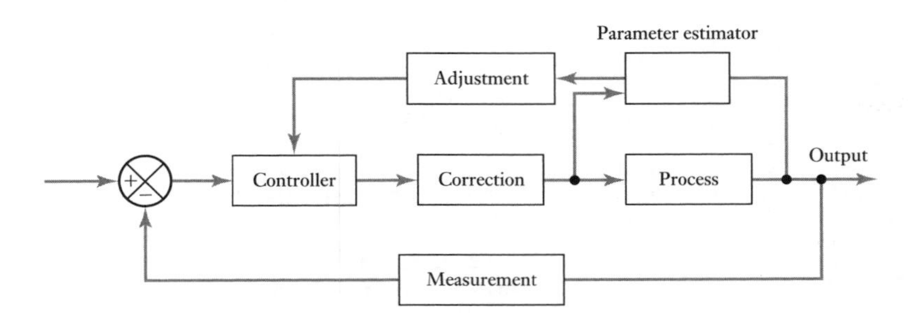

Figure 15.22 Self-tuning.

Self-tuning is often found in commercial PID controllers, generally then being referred to as **auto-tuning**. When the operator presses a button, the controller injects a small disturbance into the system and measures the response. This response is compared to the desired response and the control parameters adjusted, by a modified Ziegler–Nichols rule, to bring the actual response closer to the desired response.

15.12.3 Model-reference adaptive systems

With the **model-reference adaptive system** an accurate model of the system is developed. The set value is then used as an input to both the actual and the model systems and the difference between the actual output and the output from the model compared. The difference in these signals is then used to adjust the parameters of the controller to minimise the difference. Figure 15.23 illustrates the features of the system.

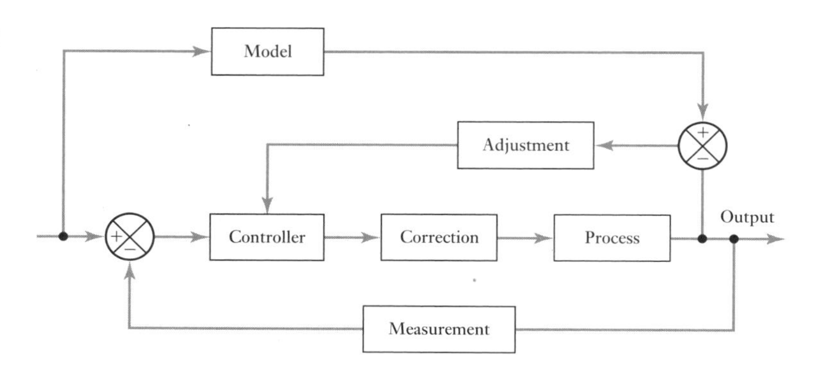

Figure 15.23 Model-referenced control.

Summary

The **steady-state error** is the difference between the desired set value input and the output after all transients have died away.

Control modes are **two-step** in which the controller supplies an on/off correcting signal, **proportional** (P) in which the correcting signal is proportional to the error, **derivative** (D) in which the correcting signal is proportional to the rate at which the error is changing, and **integral** (I) in which the correcting signal is proportional to the integral of the error with time. The transfer function for a PID system is

$$\text{transfer function} = K_{P}e + \frac{1}{s}K_{I} + sK_{D} = K_{P}\left(1 + \frac{1}{T_{I}s} + T_{D}s\right)$$

A **digital controller** basically operates by sampling the measured value, comparing it with the set value and establishing the error, carrying out calculations based on the error value and stored values of previous inputs and outputs to obtain the output signal, outputting and then waiting for the next sample.

The term **tuning** is used to describe the process of selecting the best controller settings, i.e. the values of K_P, K_I and K_D.

The term **adaptive control** is used for systems which 'adapt' to changes and change their parameters to fit the circumstances prevailing. Three commonly used forms are gain-scheduled control, self-tuning and model-reference adaptive systems.

Problems

15.1 What are the limitations of two-step (on/off) control and in what situation is such a control system commonly used?

15.2 A two-position mode controller switches on a room heater when the temperature falls to 20°C and off when it reaches 24°C. When the heater is on, the air in the room increases in temperature at the rate of 0.5°C per minute; when the heater is off, it cools at 0.2°C per minute. If the time lags in the control system are negligible, what will be the times taken for (a) the heater switching on to off, (b) the heater switching off to on?

15.3 A two-position mode controller is used to control the water level in a tank by opening or closing a valve which in the open position allows water at the rate of 0.4 m^3/s to enter the tank. The tank has a cross-sectional area of 12 m^2 and water leaves it at the constant rate of 0.2 m^3/s. The valve opens when the water level reaches 4.0 m and closes at 4.4 m. What will be the times taken for (a) the valve opening to closing, (b) the valve closing to opening?

15.4 A proportional controller is used to control the height of water in a tank where the water level can vary from 0 to 4.0 m. The required height of water is 3.5 m and the controller is to close a valve fully when the water rises to 3.9 m and open it fully when the water falls to 3.1 m. What transfer function will be required?

15.5 Describe and compare the characteristics of (a) proportional control, (b) proportional plus integral control, (c) proportional plus integral plus derivative control.

15.6 Determine the settings of K_P, T_I and T_D required for a three-mode controller which gave a process reaction curve with a lag L of 200 s and a gradient R of 0.010%/s when the test signal was a 5% change in the control valve position.

15.7 When tuning a three-mode control system by the ultimate cycle method it was found that oscillations began when the proportional critical value was 5. The oscillations had a periodic time of 200 s. What are the suitable values of K_P, T_I and T_D?

15.8 Explain the basis on which the following forms of adaptive control systems function: (a) gain-scheduled, (b) self-tuning, (c) model-reference.

15.9 A d.c. motor behaves like a first-order system with a transfer function of relating output position to which it has rotated a load with input signal of $1/s(1 + s\tau)$. If the time constant τ is 1 s and the motor is to be used in a closed-loop control system with unity feedback and a proportional controller, determine the value of the proportionality constant which will give a closed-loop response with a 25% overshoot.

15.10 The small ultrasonic motor used to move the lens for automatic focusing with a camera (see Section 24.2.3) drives the ring with so little inertia that the transfer function relating angular position with input signal is represented by $1/cs$, where c is the constant of proportionality relating the frictional torque and angular velocity. If the motor is to be controlled by a closed-loop system with unity feedback, what type of behaviour can be expected if proportional control is used?

Chapter sixteen Artificial Intelligence

Objectives

The objectives of this chapter are that, after studying it, the reader should be able to:

- **Explain what is meant by an intelligent machine and the capabilities of such machines.**
- **Explain the meaning of neural networks and their relevance to pattern recognition.**
- **Explain the term fuzzy logic.**

16.1 What is meant by artificial intelligence?

What constitutes an intelligent machine? A dictionary definition of intelligent might be 'endowed with the ability to reason'. The more intelligent we think a person is, the more we consider he or she is able to learn, generalise from this acquired knowledge, be capable of reasoning and able to make predictions by considering what is possible, learning from any mistakes. We can apply the same criteria to a machine: an **intelligent machine** is one endowed with the ability to reason.

A central heating system makes decisions about its actions. For example, should the boiler switch on or off as a result of information from a thermostat? It is not, however, considered to be intelligent because it is not capable of reasoning and making decisions under a wide range of conditions. For example, it cannot recognise a pattern in inputs from a thermostat and so make predictions about whether to switch the boiler on or off. It just does what it is told to do. It does not 'think for itself'.

In this chapter we take a brief look at basic concepts associated with intelligent machines.

16.1.1 Self-regulation

We can consider the closed-loop feedback systems discussed in earlier chapters as being self-regulation systems in that they regulate the output of a system to a required value. Thus a thermostatically controlled central heating system is used to maintain the room temperature at the value set for the thermostat. Such systems cannot, however, be considered intelligent, they merely do what they were told to do.

16.2 Perception and cognition

Perception with an intelligent system is the collecting of information using sensors and the organising of the gathered information so that decisions can be made. For example, a control system used with a production line might have a video camera to observe components on a conveyor belt.

The signals received from the camera enable a computed representation of the components to be made so that features can be identified. This will contain information about critical elements of the components. These can then be compared with representations of the components so that decisions can be made by the control system as to whether the component is correctly assembled or perhaps which component it is. Then action can be taken by the control system perhaps to reject faulty components or divert particular components to particular boxes.

Thus, with a mechatronics system, perception involves sensors gathering appropriate information about a system and its environment, decoding it and processing it to give useful information which can be used elsewhere in the system to make decisions.

16.2.1 Cognition

Once a machine has collected and organised information, it has to make decisions about what to do as a consequence of the information gathered. This is termed **cognition**. Vital to this perception and cognition is **pattern recognition**. What are the patterns in the data gathered?

Humans are very good at pattern recognition. Think of a security guard observing television monitors. He or she is able to look at the monitors and recognise unusual patterns, e.g. a person where there should be no person, an object having been moved, etc. This is the facility required of intelligent machines. An autopilot system on an aircraft monitors a lot of information and, on the basis of the patterns perceived in that data, makes decisions as to how to adjust the aircraft controls.

Machine pattern recognition can be achieved by the machine having a set of patterns in its memory and gathered patterns are then compared with these and a match sought. The patterns in its memory may arise from models or a process of training in which data is gathered for a range of objects or situations and these given identification codes. For example, for recognising coins, information may be gathered about diameter and colour. Thus a particular one-pound coin might be classified as having a diameter of 2.25 cm and a colour which is a particular degree of redness (it is a bronze coin). However, an intelligent machine will need to take account of worn and dirty coins and still be able to recognise the one-pound coin.

16.2.2 Neural networks

In the above example of the coins, only two dimensions were considered, namely diameter and colour. In more complex situations there may be many more dimensions. The human brain is faced with sorting and classifying multidimensional information and does this using **neural networks**. Artificial neural networks are now being used with intelligent machines. Such networks do not need to be programmed but can learn and generalise from examples and training. A neural network (Figure 16.1) is composed of a large number of interconnected processing units, the outputs from some units being inputs to others. Each processor in the network receives information at its inputs, and multiplies each by a weighting factor. If operating as AND, it

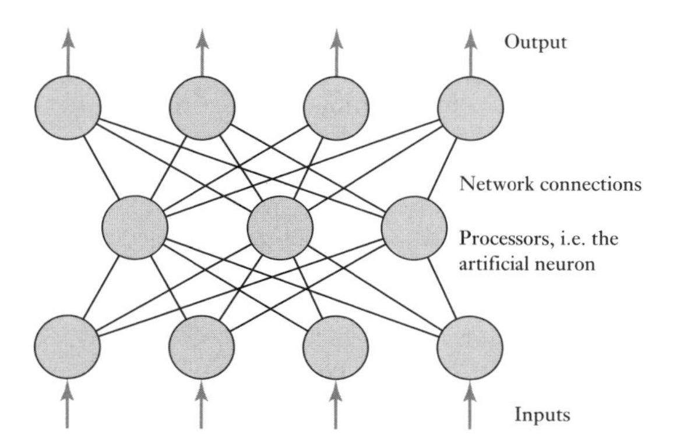

Figure 16.1 Neural network.

then sums these weighted inputs and gives an output of a 1 if the sum exceeds a certain value or is positive. For example, we might have an input of 1 with a weighting factor of -1.5 to give -1.5, another input of 1 with a weighting factor of 1.0 to give 1.0 and a third input of 1 with a weighting factor of 1.0 to give 1.0. The sum of these weighted inputs is thus $-1.5 + 1.0 + 1.0 = 0.5$ and so an output of 1 if the values are to be positive for an output. With these inputs as 1×-1.5, 0×1.0 and 0×1.0, the weighted sum is -1.5 and so an output of 0. The network can be programmed by learning from examples and so be capable of learning.

16.3 Reasoning

Reasoning is the process of going from what is known to what is not known. There are a number of mechanisms for carrying out reasoning.

16.3.1 Reasoning mechanisms

An example of **deterministic reasoning** is to use the 'if–then' rule. Thus, we might deduce that *if* a coin has a diameter of 1.25 cm *then* it is a pound coin. If the first part of the statement is true then the second part of the statement is true; if it is false then the second part is not true. In this form of reasoning we have a *true–false* situation and it is assumed that there is no default knowledge so that when the deduction is made there are no exceptions. Thus, in the above example, we cannot take account of there being a coin from another country with the same diameter.

Non-deterministic reasoning allows us to make predictions based on probability. If you toss a coin, there are two ways it can land: face upwards or face downwards. Out of the two ways there is just one way which will give face upwards. Hence, the probability of its landing face upwards is 1 in 2 or 1/2. An alternative way of arriving at this value is to toss a coin repeatedly and, in the long run, in 1/2 of the times it will end up with face upwards. Figure 16.2(a) shows how we can represent this as a probability tree. If we throw a six-sided die then the probability of its landing with a six uppermost is 1/6. Figure 16.2(b) shows how we can represent this as a probability tree. On each limb of the tree the probability is written. The chance of a coin landing with either heads or tails is 1. Thus, for a tree, the total probability will be 1.

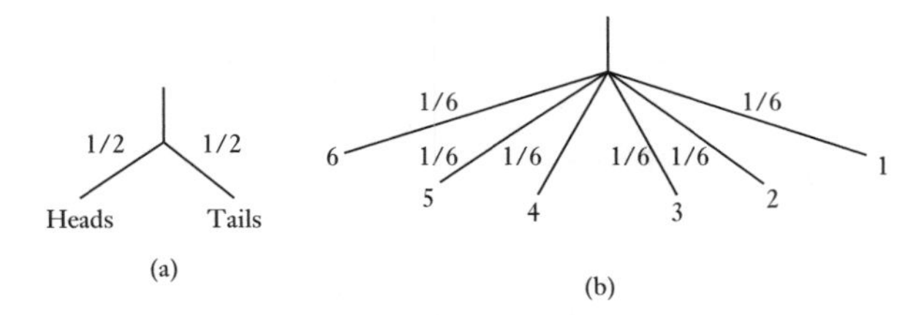

Figure 16.2 Probability trees: (a) a coin, (b) a die.

Thus in the example of the pound coin we might want to consider that there is a probability of 0.9 of a coin with diameter 1.25 cm being a pound coin. In the case of a mechatronics system we might monitor it for, say, 1000 hours and during that time the number of hours that the temperature has been high was found to be 3 hours. We can then say that the probability of the temperature being high is $3/1000 = 0.003$.

Sometimes we might know the probability of an event occurring and want to establish the probability that it will result in some other event. Thus, in a mechatronics system we might want to know what are the chances of, say, when a sensor detects a low pressure that the system will overheat, bearing in mind that there could be other reasons for a high temperature. We might represent this as the tree shown in Figure 16.3.

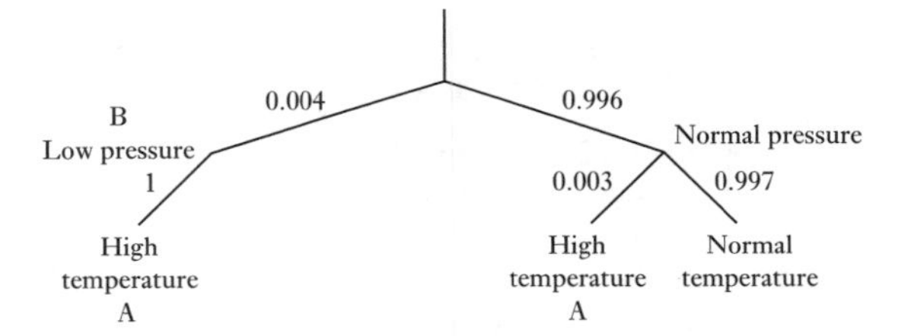

Figure 16.3 A conditional probability tree.

Bayes' rule can be used to solve this problem. This can be stated as

$$p(A|B) = \frac{p(B|A) \times p(A)}{p(B)}$$

$p(A|B)$ is the probability of A happening given that B has happened, $p(B|A)$ is the probability of B happening given that A has happened, $p|A|$ is the probability of A happening, $p|B|$ is the probability of B happening. Thus, if the probability for the system of a high temperature occurring $p|A|$ is 0.003, i.e. in 3 times in 1000 a high temperature occurs, and the probability of there being a low pressure $p|B|$ is 0.004, i.e. in 4 times in 1000 a low pressure occurs, then as we might be certain that the system will overheat if the pressure is low, i.e. $p(B|A)$ is 1, we must have a conditional probability of $(1 \times 0.003)/0.004 = 0.75$ that the system will overheat when a low pressure is detected.

16.3.2 Rule-based reasoning

At the heart of a **rule-based system** are a set of rules. These, when combined with facts, i.e. in mechatronics this could be inputs from sensors and users, enable inferences to be made which are then used to actuate actuators and control outputs. Figure 16.4 illustrates this sequence. The combination

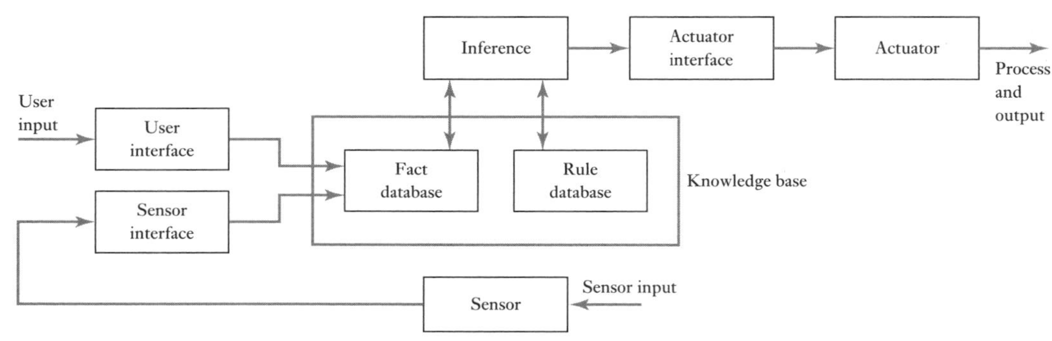

Figure 16.4 Rule-based system.

of fact and rule databases is known as the knowledge base for a machine. Inference is where the reasoning takes place as a result of the input facts being combined with the rules and decisions made which are then fed to actuators.

The rules used are often in the form of 'if–then' statements. Thus we might have a group of rules for a central heating system of the form:

If boiler on
Then pump is on

If pump on AND room temperature less than 20°C
Then open valve

If boiler not on
Then pump not on

etc.

The fact database with such a system would contain the facts:

Room temperature < 20°C
Timer On
Valve Open
Boiler On
Pump On

The rules can also be in the form of propositions involving probability statements or fuzzy logic.

Lotfi Zadeh proposed in 1965 a form of reasoning which has become known as **fuzzy logic**. One of its main ideas is that propositions need not be classified as true or false, but their truth or falsehood can be weighted so that it can be classified between the two on a scale. A **membership function** is defined for a value as being whether it is a member of a particular set. Thus we might define one set of temperature values as being 0 to 20°C and another as 20 to 40°C. If the temperature is, say, 18°C then membership of the 0 to 20°C set is 1 and that of the 20 to 40°C set 0. However, with fuzzy logic we define overlapping sets, e.g. cold 0 to 20°C, warm 10 to 30°C and hot 20 to 40°C. A temperature of 18°C is thus a member of two sets. If the fuzzy set membership functions are defined as shown in Figure 16.5, then 18°C has a cold membership of 0.2, a warm membership of 0.8 and a hot membership of 0. On the basis of data such as this, rules can be devised to trigger appropriate action. For example, a cold membership of 0.2 might have heating switched on low, but a cold membership of 0.6 might have it switched on high.

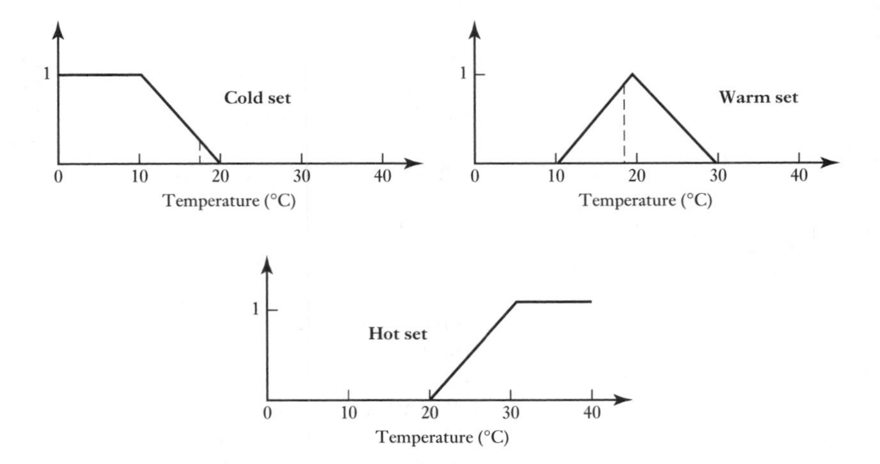

Figure 16.5 Fuzzy set membership functions.

Fuzzy logic is now being used in many commonly encountered products. For example, domestic washing machines can sense the fabric type, dirtiness and load size and adjust the wash cycle accordingly.

16.4 Learning

Machines that can learn and extend their knowledge base have great advantages compared with machines that cannot learn. **Learning** can be thought of as adapting to the environment based on experience. With machines, learning can be accomplished in a number of ways.

A simple method of learning is by new data being inputted and accumulated in memory. Machines can also learn by the data they receive being used to modify parameters in the machine.

Another method of learning that can be used is when reasoning is defined in terms of probabilities and that is to update the probabilities used in the light of what happens. We can think of this in terms of a simple example. Say we have a bag containing 10 coloured balls, all being red apart from one black one. When we first draw a ball from the bag, the probability of pulling the black ball out is 1/10. If we find it is a red ball, then the next time we draw a ball out the probability that it will be a black ball is 1/9. Our 'machine' can learn from the first ball being red by adjusting its probability value for a black ball being drawn. Bayes' rule given in Section 16.3.1 can be used to update a machine, being now written as

$$p(H\,|\,E) = \frac{p(E|H) \times p(H)}{p(E)}$$

where H is the hypothesis that we start with and E the example now encountered. Then $p(H|E)$ is the probability of the hypothesis H being true given that the example E has happened, $p(E\,|\,H)$ is the probability of the example E happening given that the hypothesis H is true, $p|\,E\,|$ is the probability of an example E happening, $p|\,H|$ is the probability of the hypothesis H being true. This allows the machine to update the probability of H every time new information comes in.

Yet another method a machine can learn is from examples. This is when a machine generalises from a set of examples. These may be the result of training with examples being supplied to the machine so that is can build up its rules or as a consequence of events it has encountered. Pattern recognition generally involves this form of learning. Thus, given an example of the number 2 in an array of pixels, the machine can learn to recognise the number 2. Neural networks (Section 16.2.2) also involve learning by example.

A machine may also learn by drawing analogies between a problem it has solved before and a new problem.

Summary

An **intelligent machine** is one endowed with the ability to reason. **Perception** with an intelligent system is the collecting of information using sensors and the organising of the gathered information so that decisions can be made. **Reasoning** is the process of going from what is known to what is not known. An example of **deterministic reasoning** is to use the 'if–then' rule. **Non-deterministic reasoning** allows us to make predictions based on probability. With **fuzzy logic** propositions need not be classified as true or false, but their truth or falsehood can be weighted so that it can be classified between the two on a scale. **Learning** can be thought of as adapting to the environment based on experience.

Problems

16.1 Examine a range of coins of your country and produce a pattern recognition table.

16.2 What is the probability of (a) throwing a six with a single six-sided die, (b) throwing two dice and one of them giving a six, (c) taking a black ball out of a bag containing nine red balls and one black one?

16.3 If the probability of a mechatronics system showing a high temperature is 0.01, what is the probability it will not show a high temperature?

16.4 A machine has been monitored for 2000 hours and during that time the cooling system has only shown leaks for 4 hours. What is the probability of leaks occurring?

16.5 The probability of a cooling system of a machine leaking has been found to be 0.005 and the probability of the system showing a high temperature 0.008. If a leakage will certainly cause a high temperature, what is the probability that a high temperature will be caused by a cooling system leak?

16.6 The probability of there being a malfunction with a machine consisting of three elements A, B and C is 0.46. If the probability of element A being active is 0.50 and the probability a malfunction occurs with A is 0.70, what is the probability that A was responsible for a malfunction?

16.7 Propose 'if–then' rules for a temperature controller that is used to operate a boiler and has a valve allowing water to circulate round central heating radiators when it only operates at a certain time period.

Part V
Microprocessor Systems

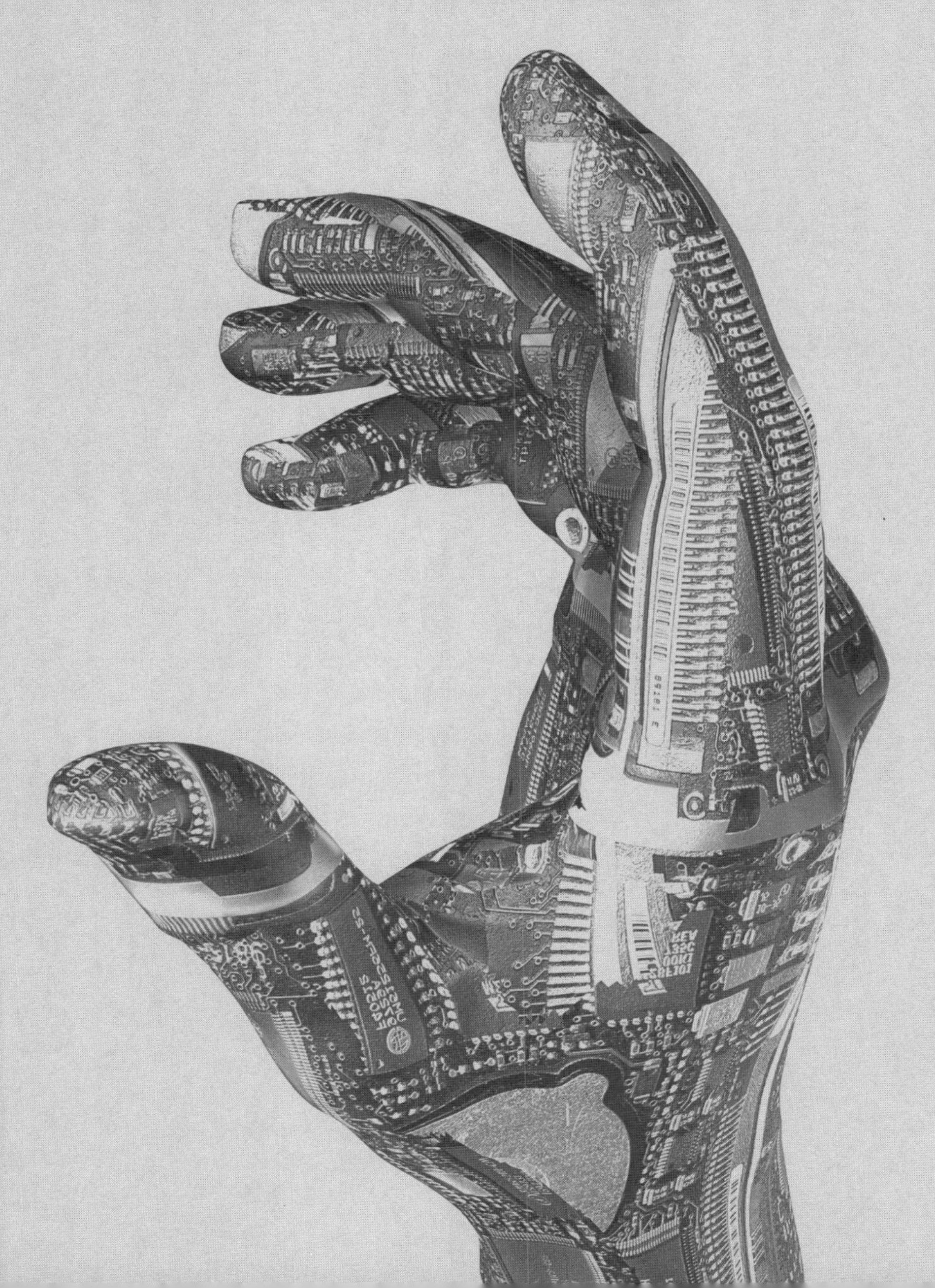

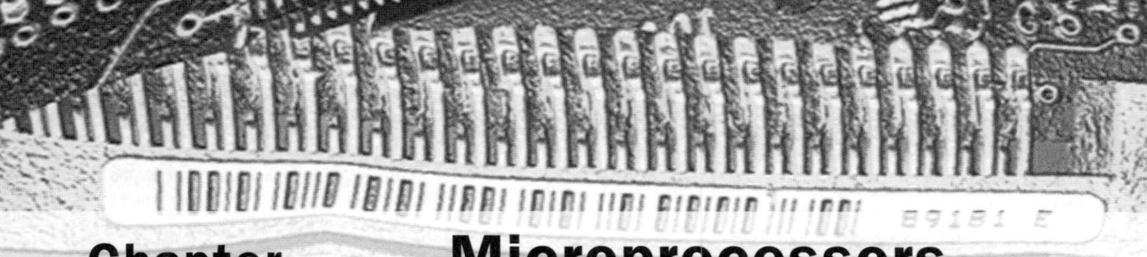

Chapter seventeen Microprocessors

Objectives

The objectives of this chapter are that, after studying it, the reader should be able to:

- **Describe the basic structure of a microprocessor system.**
- **Describe the architecture of common microprocessors and how they can be incorporated in microprocessor systems.**
- **Describe the basic structure of microcontrollers and how their registers can be set up to carry out tasks.**
- **Explain how programs can be developed using flow charts and pseudocode.**

17.1 Control

If we take a simple control problem, e.g. the sequencing of the red, amber, green lights at a traffic crossing, then it should be possible to solve it by an electronic control system involving combinational and sequential logic integrated circuits. However, with a more complex situation there might be many more variables to control in a more complex control sequence. The simplest solution now becomes not one of constructing a system based on hard-wired connections of combinational and sequential logic integrated circuits but of using a microprocessor and using software to make the 'interconnections'.

The microprocessor systems that we are concerned with in this book are for use as control systems and are termed **embedded microprocessors.** This is because such a microprocessor is dedicated to controlling a specific function and is self-starting, requiring no human intervention and completely self-contained with its own operating program. For the human, it is not apparent that the system is a microprocessor one. Thus, a modern washing machine contains a microprocessor but all that the operator has to do to operate it is to select the type of wash required by pressing the appropriate button or rotating a switch and then press the button to start.

This chapter is an overview of the structure of microprocessors and microcontrollers with the next two chapters discussing programming and Chapter 20 interfacing.

17.2 Microprocessor systems

Systems using microprocessors basically have three parts: a **central processing unit** (CPU) to recognise and carry out program instructions (this is the part which uses the microprocessor), **input and output interfaces** to handle communications between the microprocessor and the outside world (the term **port** is used for the interface), and **memory** to hold the program

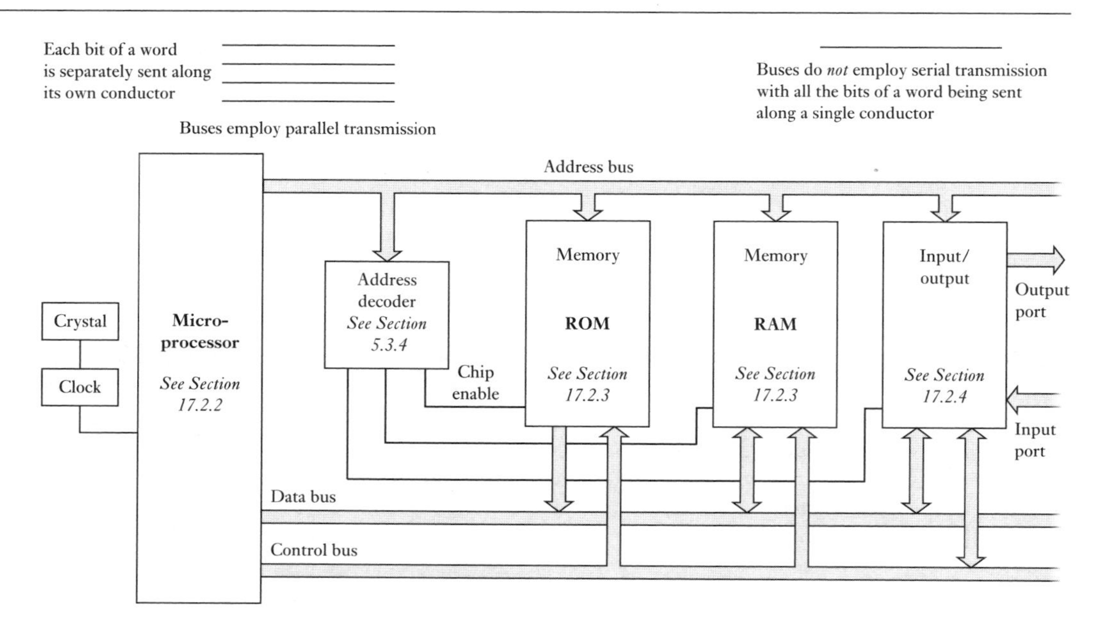

Figure 17.1 General form of a microprocessor system and its buses.

instructions and data. Figure 17.1 illustrates the general arrangement of a microprocessor system.

Microprocessors which have memory and various input/output arrangements all on the same chip are called **microcontrollers**.

17.2.1 Buses

Digital signals move from one section to another along paths called **buses**. A bus, in the physical sense, is just a number of parallel conductors along which electrical signals can be carried and are paths which can be shared by all the chips in the system. This is because if separate connections were used between the chips, there would be a very large number of connecting conductors. Using shared connection buses does mean that when one chip puts data on the bus, the other chips have to wait their turn until the data transfer is complete before one of them can put its data on the bus. Typically a bus has 16 or 32 parallel connections so that each can carry 1 bit of a data word simultaneously. This gives faster transmission than having a serial connection in which an entire word is sent in a sequence of bits along a single conductor.

There are three forms of bus in a microprocessor system:

1 *Data bus*
The data associated with the processing function of the CPU is carried by the **data bus**. Thus, it is used to transport a word to or from the CPU and the memory or the input/output interfaces. Each wire in the bus carries a binary signal, i.e. a 0 or a 1. Thus with a four-wire bus we might have the

word 1010 being carried, each bit being carried by a separate wire in the bus, as:

Word	Bus wire
0 (least significant bit)	First data bus wire
1	Second data bus wire
0	Third data bus wire
1 (most significant bit)	Fourth data bus wire

The more wires the data bus has, the longer the word length that can be used. The range of values which a single item of data can have is restricted to that which can be represented by the word length. Thus with a word of length 4 bits the number of values is $2^4 = 16$. Thus if the data is to represent, say, a temperature, then the range of possible temperatures must be divided into 16 segments if we are to represent that range by a 4-bit word. The earliest microprocessors were 4-bit (word length) devices, and such 4-bit microprocessors are still widely used in such devices as toys, washing machines and domestic central heating controllers. They were followed by 8-bit microprocessors, e.g. the Motorola 6800, the Intel 8085A and the Zilog Z80. Now, 16-bit, 32-bit and 64-bit microprocessors are available; however, 8-bit microprocessors are still widely used for controllers.

2 *Address bus*
The **address bus** carries signals which indicate where data is to be found and so the selection of certain memory locations or input or output ports. Each storage location within a memory device has a unique identification, termed its address, so that the system is able to select a particular instruction or data item in the memory. Each input/output interface also has an address. When a particular address is selected by its address being placed on the address bus, only that location is open to the communications from the CPU. The CPU is thus able to communicate with just one location at a time. A computer with an 8-bit data bus has typically a 16-bit-wide address bus, i.e. 16 wires. This size of address bus enables 2^{16} locations to be addressed; 2^{16} is 65 536 locations and is usually written as 64K, where K is equal to 1024. The more memory that can be addressed, the greater the volume of data that can be stored and the larger and more sophisticated the programs that can be used.

3 *Control bus*
The signals relating to control actions are carried by the **control bus**. For example, it is necessary for the microprocessor to inform memory devices whether they are to read data from an input device or write data to an output device. The term READ is used for receiving a signal and WRITE for sending a signal. The control bus is also used to carry the system clock signals; these are to synchronise all the actions of the microprocessor system. The clock is a crystal-controlled oscillator and produces pulses at regular intervals.

17.2.2 The microprocessor

The microprocessor is generally referred to as the central processing unit (CPU). It is that part of the processor system which processes the data,

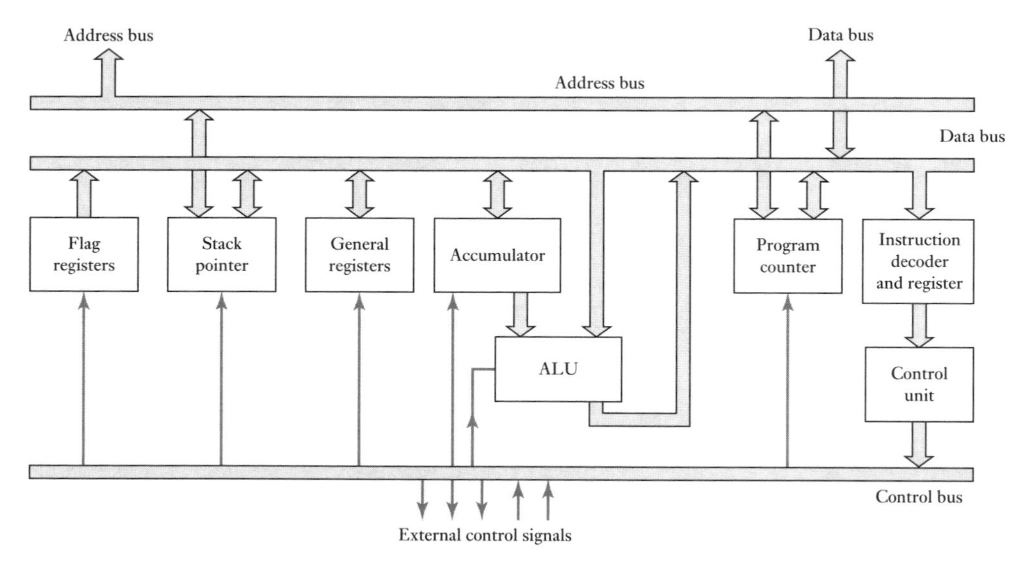

Figure 17.2 General internal architecture of a microprocessor.

fetching instructions from memory, decoding them and executing them. The internal structure – the term **architecture** is used – depends on the microprocessor concerned. Figure 17.2 indicates, in a simplified manner, the general architecture of a microprocessor.

The following are the functions of the constituent parts of a microprocessor:

1 *Arithmetic and logic unit* (ALU)
The arithmetic and logic unit is responsible for performing the data manipulation.

2 *Registers*
Internal data that the CPU is currently using is temporarily held in a group of **registers** while instructions are being executed. These are memory locations within the microprocessor and are used to store information involved in program execution. A microprocessor will contain a group of registers, each type of register having a different function.

3 *Control unit*
The **control unit** determines the timing and sequence of operations. It generates the timing signals used to fetch a program instruction from memory and execute it. The Motorola 6800 uses a clock with a maximum frequency of 1 MHz, i.e. a clock period of 1 μs, and instructions require between two and twelve clock cycles. Operations involving the microprocessor are reckoned in terms of the number of cycles they take.

There are a number of types of register, the number, size and types of registers varying from one microprocessor to another. The following are common types of registers:

1 *Accumulator register*
The accumulator register (A) is where data for an input to the arithmetic and logic unit is temporarily stored. In order for the CPU to be able to access, i.e. read, instructions or data in the memory, it has to supply the

address of the required memory word using the address bus. When this has been done, the required instructions or data can be read into the CPU using the data bus. Since only one memory location can be addressed at once, temporary storage has to be used when, for example, numbers are combined. For example, in the addition of two numbers, one of the numbers is fetched from one address and placed in the accumulator register while the CPU fetches the other number from the other memory address. Then the two numbers can be processed by the arithmetic and logic section of the CPU. The result is then transferred back into the accumulator register. The accumulator register is thus a temporary holding register for data to be operated on by the arithmetic and logic unit and also, after the operation, the register for holding the results. It is thus involved in all data transfers associated with the execution of arithmetic and logic operations.

2 *Status register, or condition code registeror flag register*
This contains information concerning the result of the latest process carried out in the arithmetic and logic unit. It contains individual bits with each bit having special significance. The bits are called **flags**. The status of the latest operation is indicated by each flag with each flag being set or reset to indicate a specific status. For example, they can be used to indicate whether the last operation resulted in a negative result, a zero result, a carry output occurs (e.g. the sum of two binary numbers such as 1010 and 1100 is (1)0110 which might be bigger than the microprocessor's word size and thus there is a carry of a 1), an overflow occurs or the program is to be allowed to be interrupted to allow an external event to occur. The following are common flags:

Flag	Set, i.e. 1	Reset, i.e. 0
Z	Result is zero	Result is not zero
N	Result is negative	Result is not negative
C	Carry is generated	Carry is not generated
V	Overflow occurs	Overflow does not occur
I	Interrupt is ignored	Interrupt is processed normally

As an illustration, consider the state of the Z, N, C and V flags for the operation of adding the hex numbers 02 and 06. The result is 08. Since it is not zero, then Z is 0. The result is positive, so N is 0. There is no carry, so C is 0. The unsigned result is within the range -128 to $+127$ and so there is no overflow and V is 0. Now consider the flags when the hex numbers added are F9 and 08. The result is (1)01. The result is not zero, so Z is 0. Since it is positive, then N is 0. The unsigned result has a carry and so C is 1. The unsigned result is within the range -128 to $+127$ and so V is 0.

3 *Program counter register* (PC) *or instruction pointer* (IP)
This is the register used to allow the CPU to keep track of its position in a program. This register contains the address of the memory location that contains the next program instruction. As each instruction is executed, the program counter register is updated so that it contains the address of the memory location where the next instruction to be executed is stored. The program counter is incremented each time so that the CPU executes instructions sequentially unless an instruction, such as JUMP or BRANCH, changes the program counter out of that sequence.

4 *Memory address register* (MAR)
This contains the address of data. Thus, for example, in the summing of two numbers the memory address register is loaded with the address of the first number. The data at the address is then moved to the accumulator. The memory address of the second number is then loaded into the memory address register. The data at this address is then added to the data in the accumulator. The result is then stored in a memory location addressed by the memory address register.

5 *Instruction register* (IR)
This stores an instruction. After fetching an instruction from the memory via the data bus, the CPU stores it in the instruction register. After each such fetch, the microprocessor increments the program counter by one with the result that the program counter points to the next instruction waiting to be fetched. The instruction can then be decoded and used to execute an operation. This sequence is known as the **fetch–execute cycle**.

6 *General-purpose registers*
These may serve as temporary storage for data or addresses and be used in operations involving transfers between other registers.

7 *Stack pointer register* (SP)
The contents of this register form an address which defines the top of the stack in RAM. The **stack** is a special area of the memory in which program counter values can be stored when a subroutine part of a program is being used.

The number and form of the registers depends on the microprocessor concerned. For example, the Motorola 6800 microprocessor (Figure 17.3) has two accumulator registers, a status register, an index register, a stack pointer register and a program counter register. The status register has flag bits to show negative, zero, carry, overflow, half-carry and interrupt. The Motorola 6802 is similar but includes a small amount of RAM and a built-in clock generator.

The Intel 8085A microprocessor is a development of the earlier 8080 processor; the 8080 required an external clock generator, whereas the 8085A has an in-built clock generator. Programs written for the 8080 can be run on the 8085A. The 8085A has six general-purpose registers B, C, D, E, H and L, a stack pointer, a program counter, a flag register and two temporary registers. The general-purpose registers may be used as six 8-bit registers or in pairs BC, DE and HL as three 16-bit registers. Figure 17.4 shows a block diagram representation of the architecture.

As will be apparent from Figure 17.3 and Figure 17.4, microprocessors have a range of timing and control inputs and outputs. These provide outputs when a microprocessor is carrying out certain operations and inputs to influence control operations. In addition there are inputs related to interrupt controls. These are designed to allow program operation to be interrupted as a result of some external event.

17.2.3 Memory

The memory unit in a microprocessor system stores binary data and takes the form of one or more integrated circuits. The data may be program instruction codes or numbers being operated on.

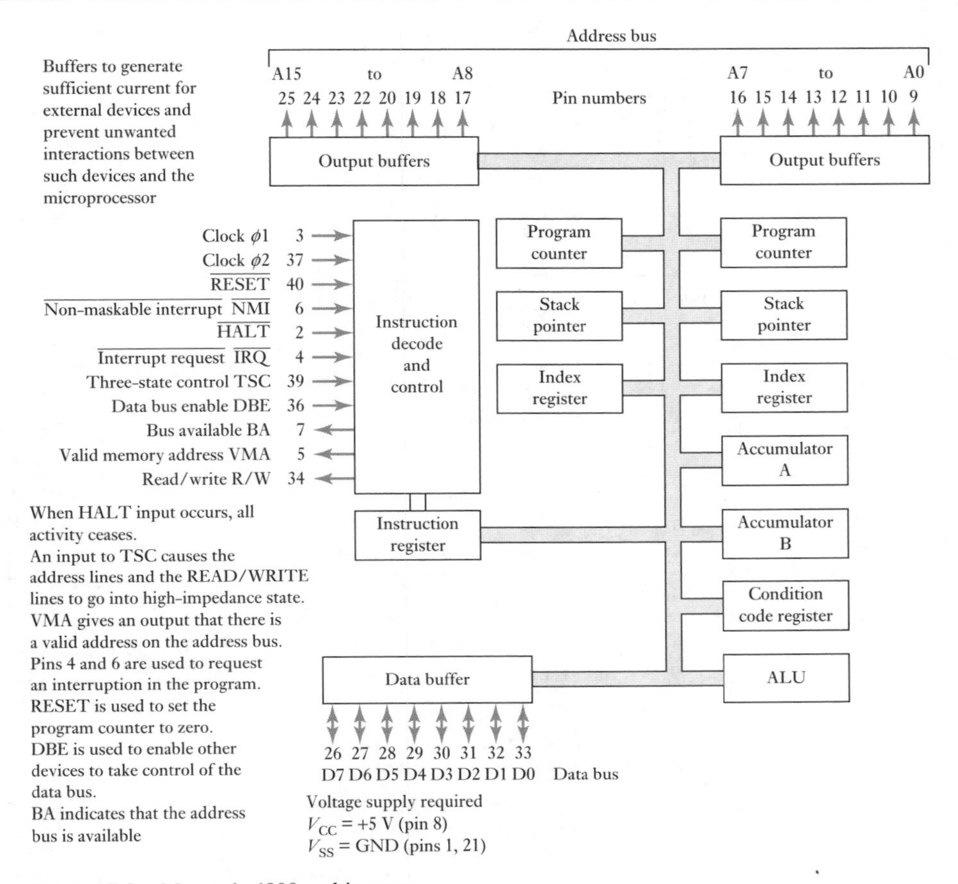

Figure 17.3 Motorola 6800 architecture.

The size of the memory is determined by the number of wires in the address bus. The memory elements in a unit consist essentially of large numbers of storage cells with each cell capable of storing either a 0 or a 1 bit. The storage cells are grouped in locations with each location capable of storing one word. In order to access the stored word, each location is identified by a unique address. Thus with a 4-bit address bus we can have 16 different addresses with each, perhaps, capable of storing 1 byte, i.e. a group of 8 bits (Figure 17.5).

The size of a memory unit is specified in terms of the number of storage locations available; 1K is 2^{10} = 1024 locations and thus a 4K memory has 4096 locations.

There are a number of forms of memory unit:

1 *ROM*
For data that is stored permanently a memory device called a **read-only memory** (ROM) is used. ROMs are programmed with the required contents during the manufacture of the integrated circuit. No data can then be written into this memory while the memory chip is in the computer. The data can only be read and is used for fixed programs such as computer operating systems and programs for dedicated microprocessor applications. They do not lose their memory when power is removed.

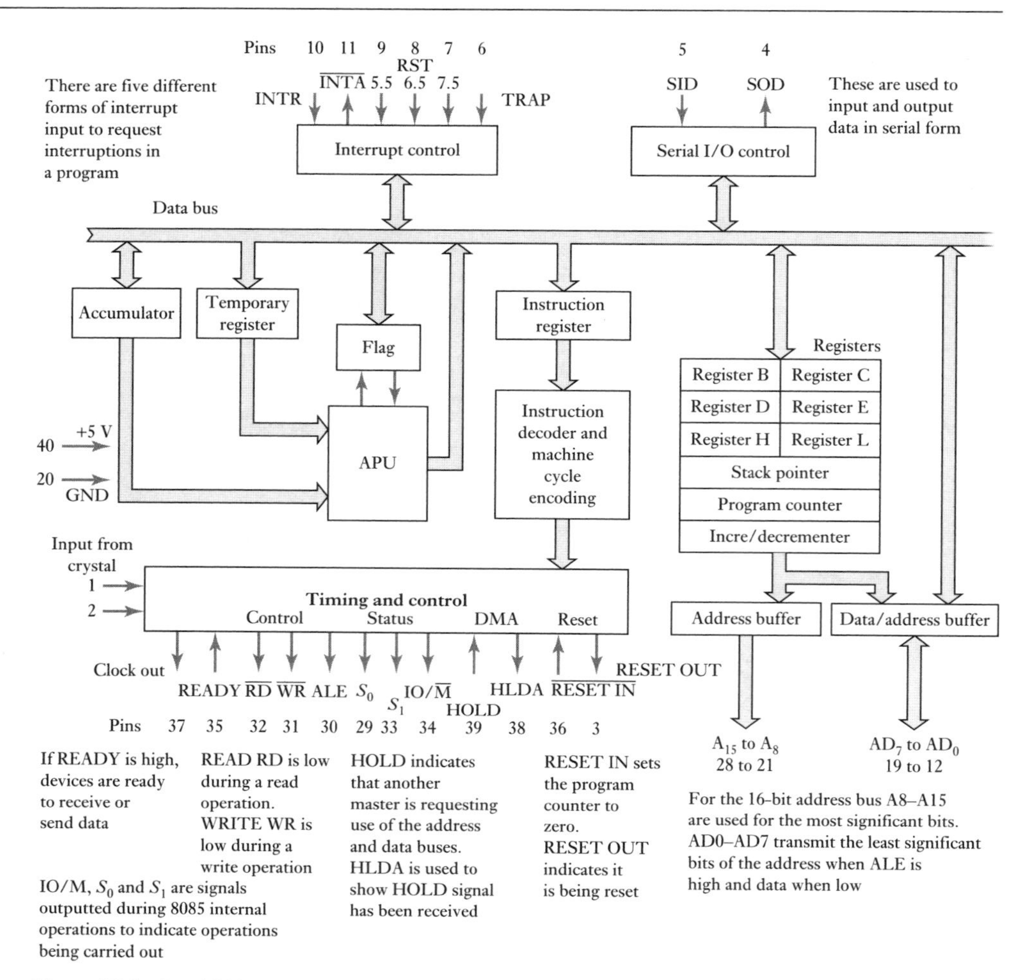

Figure 17.4 Intel 8085A architecture.

Address	Data contents
0000	
0001	
0010	
0011	
0100	
etc.	
1111	

Figure 17.5 Address bus size.

Figure 17.6(a) shows the pin connections of a typical ROM chip which is capable of storing 1K × 8 bits.

2 *PROM*

The term **programmable ROM** (PROM) is used for ROM chips that can be programmed by the user. Initially every memory cell has a fusible link which keeps its memory at 0. The 0 is permanently changed to 1 by sending a current through the fuse to open it permanently. Once the fusible link has been opened the data is permanently stored in the memory and cannot be further changed.

3 *EPROM*

The term **erasable and programmable ROM** (EPROM) is used for ROMs that can be programmed and their contents altered. A typical EPROM chip contains a series of small electronic circuits, cells, which can store charge. The program is stored by applying voltages to the integrated circuit connection pins and producing a pattern of charged

outputs a peripheral interface adapter (PIA) (see Section 20.4) is used and for serial inputs/outputs an asynchronous interface adapter (ACIA) (see Section 20.5) is used. These can be programmed to deal with both inputs and outputs and give the required buffering.

1 *RAM*
   Address lines A14 and A15 are connected to the enable inputs of the RAM chip. When both these lines are low then the RAM chip will be conversing with the microprocessor.
2 *ROM*
   Address lines A14 and A15 are connected to the enable inputs of the ROM chip and when the signals on both these lines are high then the ROM chip is addressed.
3 *Inputs/outputs*
   Address lines A14 and A15 are connected to the enable inputs of the PIA and ACIA. When the signal on line 15 is low and the signal on A14 high then the input/output interfaces are addressed. In order to indicate which of the devices is being enabled, address line A2 is taken high for the PIA and address line A3 is taken high for the ACIA.

17.3 Microcontrollers

For a microprocessor to give a system which can be used for control, additional chips are necessary, e.g. memory devices for program and data storage and input/output ports to allow it to communicate with the external world and receive signals from it. The microcontroller is the integration of a microprocessor with memory and input/output interfaces, and other peripherals such as timers, on a single chip. Figure 17.9 shows the general block diagram of a microcontroller.

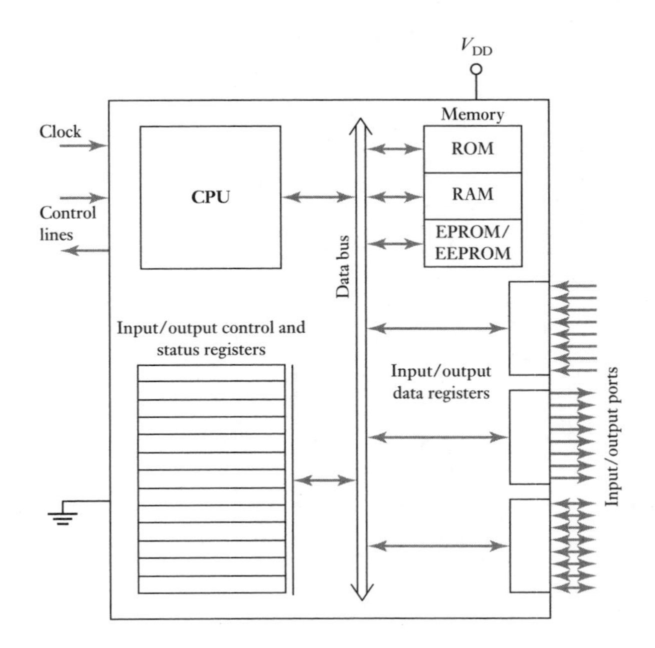

Figure 17.9 Block diagram of a microcontroller.

The general microcontroller has pins for external connections of inputs and outputs, power, clock and control signals. The pins for the inputs and outputs are grouped into units called input/output ports. Usually such ports have eight lines in order to be able to transfer an 8-bit word of data. Two ports may be used for a 16-bit word, one to transmit the lower 8 bits and the other the upper 8 bits. The ports can be input only, output only or programmable to be either input or output.

The Motorola 68HC11, the Intel 8051 and the PIC16C6x/7x are examples of 8-bit microcontrollers in that the data path is 8 bits wide. The Motorola 68HC16 is an example of a 16-bit microcontroller and the Motorola 68300 a 32-bit microcontroller. Microcontrollers have limited amounts of ROM and RAM and are widely used for embedded control systems. A microprocessor system with separate memory and input/output chips is more suited to processing information in a computer system.

17.3.1 Motorola M68HC11

Motorola offers two basic 8-bit families of microcontrollers, the 68HC05 being the inexpensive core and the 68HC11 the higher performance core. The Motorola M68HC11 family (Figure 17.10), this being based on the Motorola 6800 microprocessor, is very widely used for control systems.

There are a number of versions of this family, the differences being due to differences in the RAM, ROM, EPROM, EEPROM and configuration register features. For example, one version (68HC11A8) has 8K ROM,

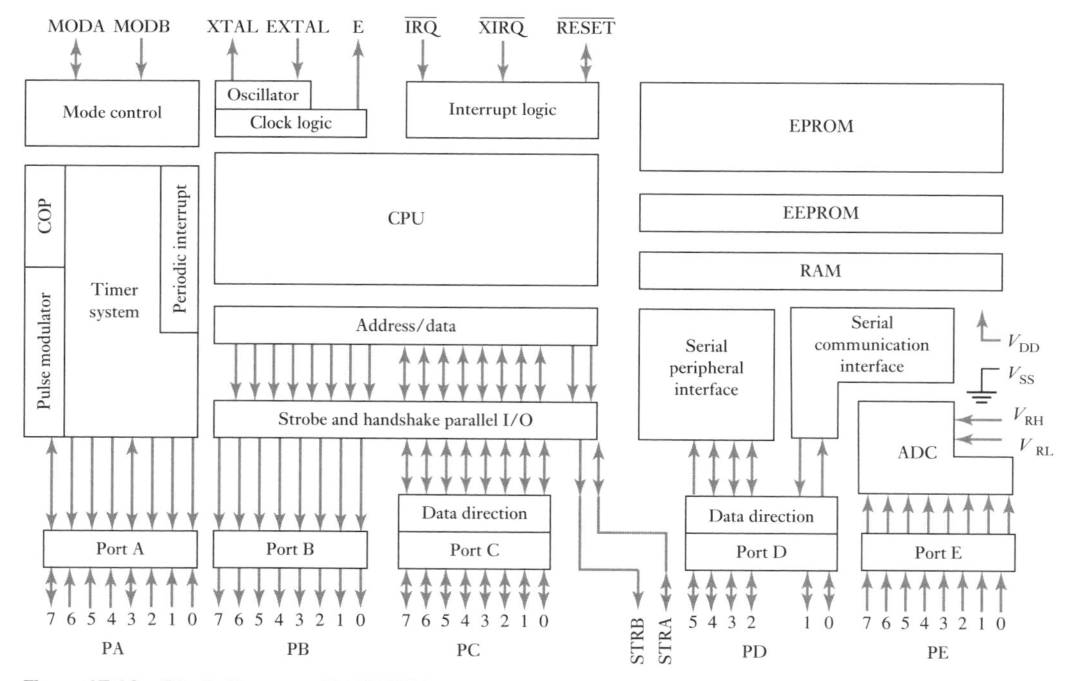

Figure 17.10 Block diagram of M68HC11.

512-byte EEPROM, 256-byte RAM, a 16-bit timer system, a synchronous serial peripheral interface, an asynchronous non-return-to-zero serial communication interface, an 8-channel, 8-bit analogue-to-digital converter for analogue inputs, and five ports A, B, C, D and E.

1 *Port A*
Port A has three input lines only, four output lines only and one line that can serve as either input or output. The port A data register is at address \$1000 (Figure 17.11) with a pulse accumulator control register (Figure 17.12) at address \$1026 being used to control the function of each bit in port A. This port also provides access to the internal timer of the microcontroller, the PAMOD, PEDGE, RTR1 and RTRO bits controlling the pulse accumulator and clock.

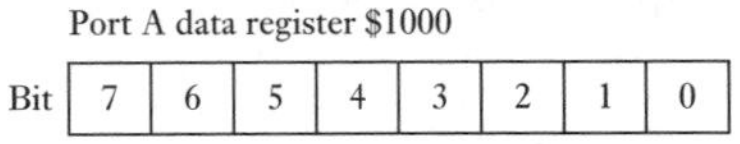

Port A data register \$1000

Bit	7	6	5	4	3	2	1	0

Figure 17.11 Port A register.

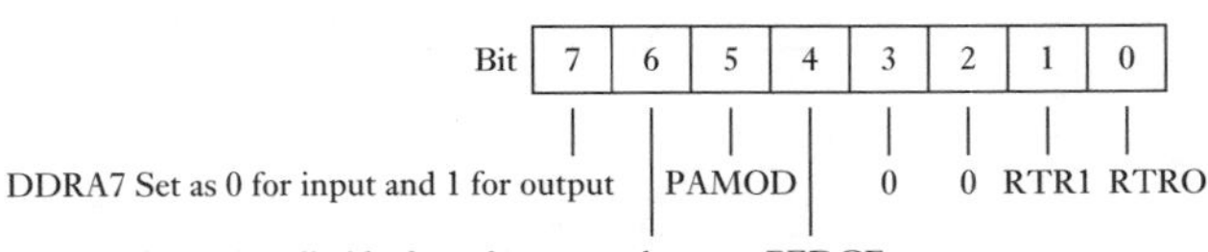

Figure 17.12 Pulse accumulator control register.

2 *Port B*
Port B is output only and has eight output lines (Figure 17.13). Input data cannot be put on port B pins. Its data register is at the address \$1004 and to output data it has to be written to this memory location.

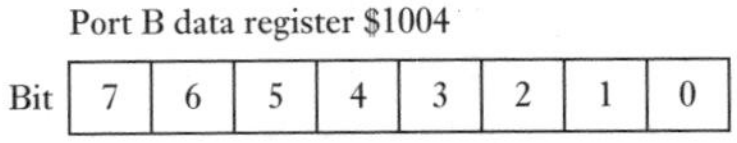

Port B data register \$1004

Bit	7	6	5	4	3	2	1	0

Figure 17.13 Port B register.

3 *Port C*
Port C can be either input or output, data being written or read from its data register at address \$1003 (Figure 17.14). Its direction is controlled by the port data direction register at address \$1007. The 8 bits in this register correspond to the individual bits in port C and determine whether the lines are inputs or outputs; when the data direction register bit is set to 0 it is an input and when set to 1 it is an output. The lines STRA and STRB (when operating in single-chip mode) are associated with ports B and C and are used for handshake signals with those ports. Such lines control

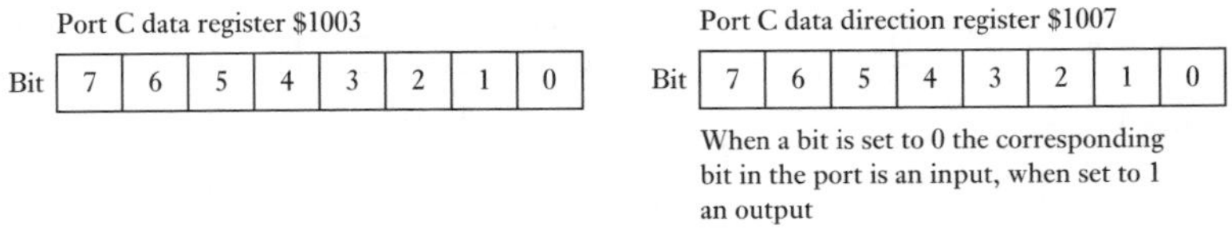

Port C data register \$1003

Bit	7	6	5	4	3	2	1	0

Port C data direction register \$1007

Bit	7	6	5	4	3	2	1	0

When a bit is set to 0 the corresponding bit in the port is an input, when set to 1 an output

Figure 17.14 Port C registers.

the timing of data transfers. The parallel I/O control register PIOC at address $1002 contains bits to control the handshaking mode and the polarity and active edges of the handshaking signals.

4 *Port D*

Port D contains just six lines; these can be either input or output and have a data register at address $1008 (Figure 17.15); the directions are controlled by a port data direction register at address $1009 with the corresponding bit being set to 0 for an input and 1 for an output. Port D also serves as the connection to the two serial subsystems of the microcontroller. The serial communication interface is an asynchronous system that provides serial communication that is compatible with modems and terminals. The serial peripheral interface is a high-speed synchronous system which is designed to communicate between the microcontroller and peripheral components that can access at such rates.

Figure 17.15 Port D registers.

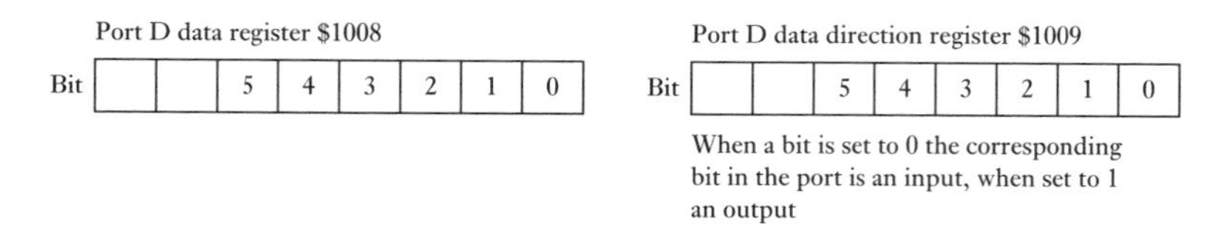

5 *Port E*

Port E is an 8-bit input-only port (Figure 17.16) which can be used as a general-purpose input port or for analogue inputs to the internal analogue-to-digital converter. The two inputs V_{RH} and V_{RL} provide reference voltages to the ADC. The port E data register is at address $1002.

Figure 17.16 Port E register.

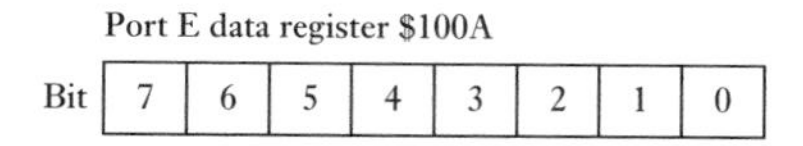

The 68HC11 has an internal analogue-to-digital (A/D) converter; port E bits 0, 1, 2, 3, 4, 5, 6 and 7 are the analogue input pins. Two lines V_{RH} and V_{LH} provide the reference voltages used by the ADC; the high reference voltage V_{RH} should not be lower than V_{DD}, i.e. 5 V, and the low reference voltage V_{LH} should not be lower than V_{SS}, i.e. 0 V. The ADC must be enabled before it can be used. This is done by setting the A/D power-up (ADPU) control bit in the OPTION register (Figure 17.17), this being bit 7. Bit 6 selects the clock source for the ADC. A delay of at least 100 μs is required after powering up to allow the system to stabilise.

Figure 17.17 OPTION register.

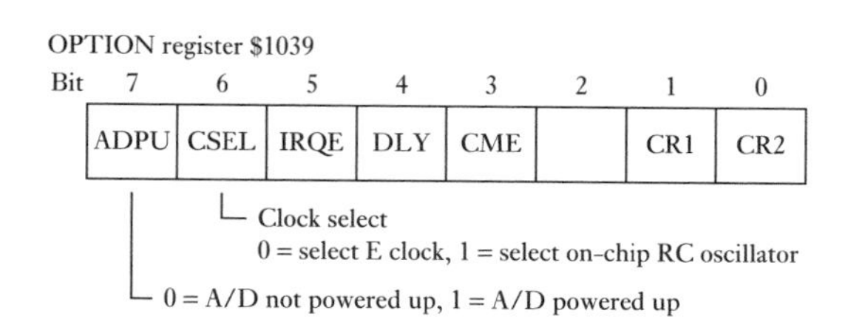

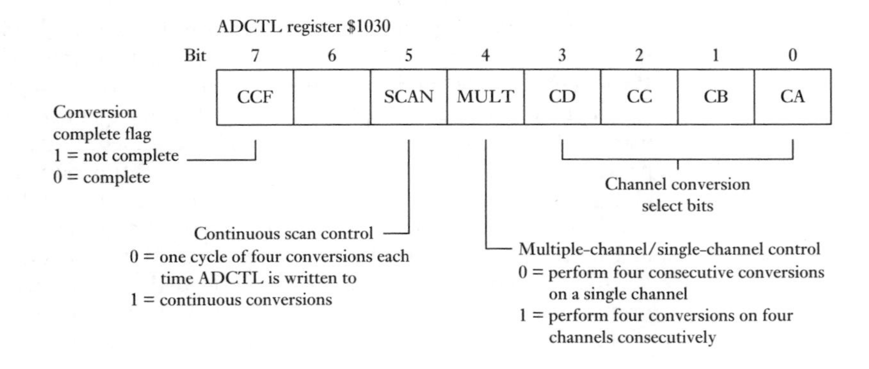

MULT = 0

CD	*CC*	*CB*	*CA*	*Channel converted*
0	0	0	0	PE0
0	0	0	1	PE1
0	0	1	0	PE2
0	0	1	1	PE3
0	1	0	0	PE4
0	1	0	1	PE5
0	1	1	0	PE6
0	1	1	1	PE7

MULT = 1

				A/D result register			
CD	*CC*	*CB*	*CA*	*ADR1*	*ADR2*	*ADR3*	*ADR4*
0	0	×	×	PE0	PE1	PE2	PE3
0	1	×	×	PE4	PE5	PE6	PE7

Figure 17.18 ADCTL register.

The analogue-to-digital conversion is initiated by writing to the A/D control/status register (ADCTL) after powering up and the stabilisation delay (Figure 17.18). This involves selecting the channels and operation modes. The conversion will then start one clock cycle later. For example, if single-channel mode is selected by setting MULT = 0 then four successive A/D conversions will occur of the channel selected by the CD–CA bits. The results of the conversion are placed in the A/D result registers ADR1–ADR4.

6 *Modes*

MODA and MODB are two pins that can be used to force the microcontroller into one of four modes at power-up, these modes being special bootstrap, special test, single chip and expanded:

MODB	**MODA**	**Mode**
0	1	Special bootstrap
0	1	Special test
1	0	Single chip
1	1	Expanded

In the single-chip mode the microcontroller is completely self-contained with the exception of an external clock source and a reset circuit. With such a mode the microcontroller may not have enough resources, e.g. memory, for some applications and so the expanded mode can be used so that the number of addresses can be increased. Ports B and C then provide address, data and control buses. Port B functions as the upper eight address pins and port C as the multiplexed data and low address pins. The bootstrap mode allows a manufacturer to load special programs in a special ROM for an M68HC11 customer. When the microcontroller is set in this mode, the special program is loaded. The special test mode is primarily used during Motorola's internal production testing.

After the mode has been selected, the MODA pin becomes a pin which can be used to determine if an instruction is starting to execute. The MODB pin has the other function of giving a means by which the internal RAM of the chip can be powered when the regular power is removed.

7 *Oscillator pins*
The oscillator system pins XTAL and EXTAL are the connections to access the internal oscillator. Figure 17.19 shows an external circuit that might be used. E is the bus clock and runs at one-quarter of the oscillator frequency and can be used to synchronise external events.

Figure 17.19 Oscillator output.

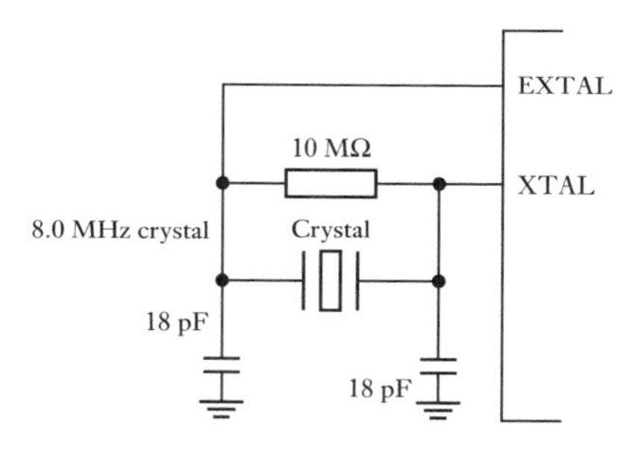

8 *Interrupt controller*
The interrupt controller is to enable the microcontroller to interrupt a program (see Section 20.3.3). An interrupt is an event that requires the CPU to stop normal program execution and perform some service related to the event. The two lines IRQ and XIRQ are for the external inputs of interrupt signals. RESET is for resetting the microcontroller and allowing an orderly system start-up. The state of the pin can be set either internally or externally. When a reset condition is detected, the pin signal is set low for four clock cycles. If after a further two cycles it is found to be still low, then an external reset is considered to have occurred. If a positive transition is detected on the power input V_{DD} a power-on reset occurs. This provides a 4064 cycle time delay. If the reset pin is low at the end of the power-on delay time, the microcontroller remains in the rest condition until it goes high.

9 *Timer*
The M68HC11 contains a timer system. This has a free-running counter, five-output compare function, the ability to capture the time when an external event occurs, a real-time periodic interrupt and a counter, called

the pulse accumulator, for external events. The free-running counter, called TCNT, is a 16-bit counter which starts counting at 0000 when the CPU is reset and continues thereafter continuously running and cannot be reset by the program. Its value can be read at any time. The source for the counter is the system bus clock and its output can be prescaled by setting the PR0 and PR1 bits as bits 0 and 1 in the TMSK2 register at address $1024 (Figure 17.20).

Figure 17.20 TMSK2 register.

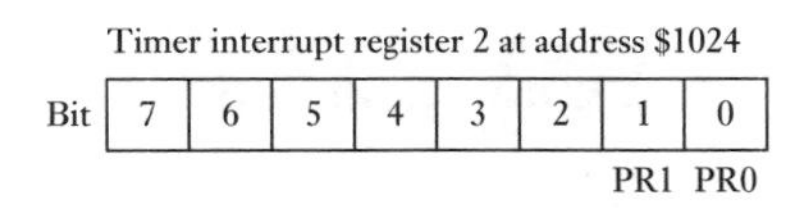

Prescale factors

			One count	
		Prescale	*Bus frequency*	
PR1	*PR0*	*factor*	*2 MHz*	*1 MHz*
0	0	1	0.5 ms	1 ms
0	1	4	2 ms	4 ms
1	0	8	4 ms	8 ms
1	1	16	8 ms	16 ms

The output compare functions allow times, i.e. timer counts, to be specified at which an output will occur when the preset count is reached. The input capture system captures the value of the counter when an input occurs and so the exact time at which the input occurs is captured. The pulse accumulator can be configured to operate as an event counter and count external clock pulses or as a gated time accumulator and store the number of pulses occurring in a particular time interval as a result of the counter being enabled and then, at some later time, disabled. The pulse accumulator control register PACTL (see Figure 17.12) at address $1026 is used to select the mode of operation. The bit PAEN is set to 0 to disable the pulse accumulator and to 1 to enable it, the bit PAMOD to 0 for the event counter mode and 1 for the gated time mode, and the bit PEDGE to 0 for the pulse accumulator to respond to a falling edge when in the event counter mode and 1 to respond to a rising edge in that mode. In gated mode, bit PEDGE set to 0 causes the count to be disabled when port A, bit 7 is 0 and to accumulate when 1, and when bit PEDGE is set to 1 in this mode, the count is disabled when port A, bit 7 is 1 and enabled when it is 0.

10 *COP*

Another timer function is COP, the computer operating properly function. This is a timer which times out and resets the system if an operation is not concluded in what has been deemed to be a reasonable time (see Section 23.2). This is often termed a **watchdog timer**.

11 *PWM*

Pulse width modulation (PWM) is used for the control of the speed of d.c. motors (see Section 3.6 and 9.5.3) by using a square wave signal and, by varying the amount of time for which the signal is on, changing the average value of the signal. A square wave can be generated by a microcontroller by arranging for an output to come on every half-period.

However, some versions of M68HC11 have a PWM module and so, after the PWM module has been initialised and enabled, the PWM waveforms can be automatically outputted.

As will be apparent from the above, before a microcontroller can be used it is necessary to initialise it, i.e. set the bits in appropriate registers, so that it will perform as required.

17.3.2 Intel 8051

Another common family of microcontrollers is the Intel 8051, Figure 17.21 showing the pin connections and the architecture. The 8051 has four parallel input/output ports, ports 0, 1, 2 and 3. Ports 0, 2 and 3 also have alternative functions. The 8051AH version has 4K ROM bytes, 128-byte RAM, two timers and interrupt control for five interrupt sources.

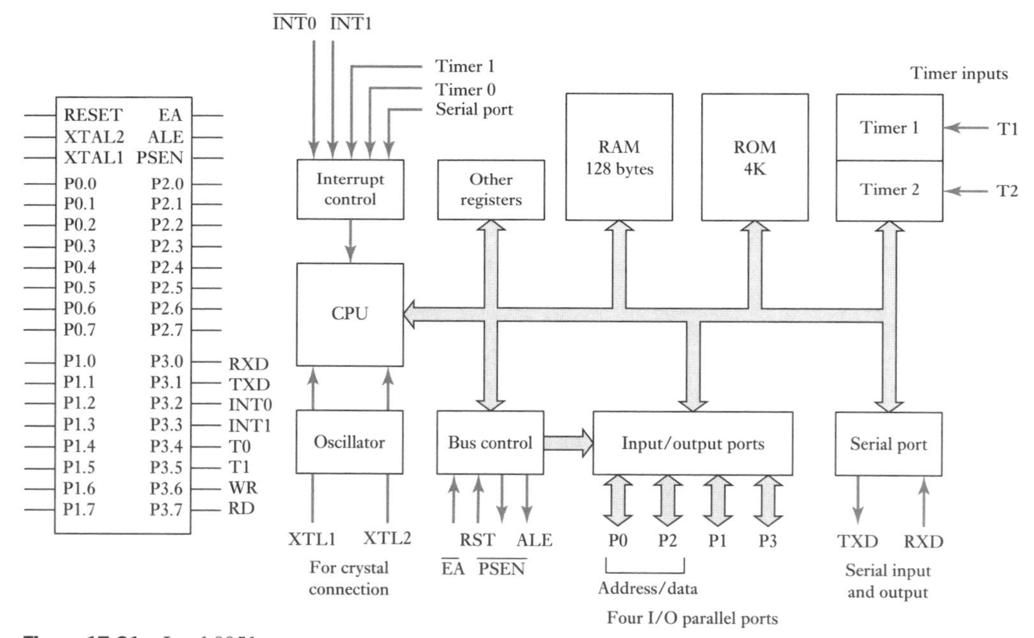

Figure 17.21 Intel 8051.

1 *Input/output ports*
Port 0 is at address 80H, port 1 at address 90H, port 2 at address A0H and port 3 at address B0H (note the use of H, or h, after the address with Intel to indicate that it is in hex). When a port is to be used as an output port, the data is put into the corresponding special function register. When a port is to be used as an input port, the value FFH must first be written to it. All the ports are bit addressable. Thus we might, for example, use just bit 6 in port 0 to switch a motor on or off and perhaps bit 7 to switch a pump on or off.

Port 0 can be used as an input port or an output port. Alternatively it can be used as a multiplexed address and data bus to access external memory. Port 1 can be used as an input port or an output port. Port 2 can be used as an input port or an output port. Alternatively it can be used for the high address bus to access external memory. Port 3 can be used as an input port or an output port. Alternatively, it can be used as a special-purpose input/output port. The alternative functions of port 3 include interrupt and timer outputs, serial port input and output and control signals for interfacing with external memory. RXD is the serial input port, TXD is the serial output port, INT0 is the external interrupt 0, INT1 is the external interrupt 1, T0 is the timer/counter 0 external input, T1 is the timer/counter 1 external input, WR is the external memory write strobe and RD is the external memory read strobe. The term **strobe** describes a connection used to enable or disable a particular function. Port 0 can be used as either an input or output port. Alternatively it can be used to access external memory.

2 *ALE*

The address latch enable (ALE) pin provides an output pulse for latching the low-order byte of the address during access to external memory. This allows 16-bit addresses to be used. Figure 17.22 illustrates this.

Figure 17.22 Use of ALE.

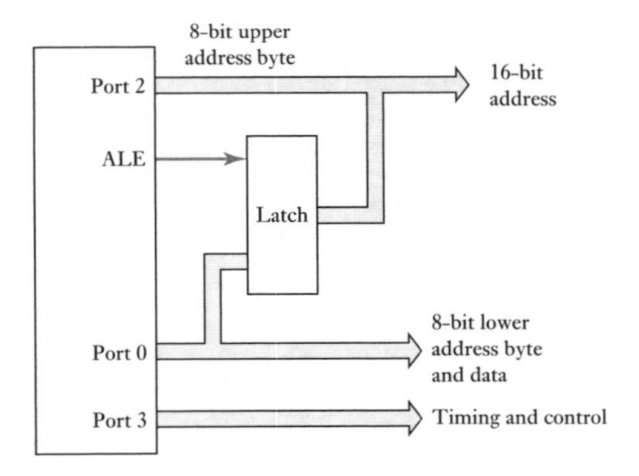

3 *PSEN*

The program store enable (PSEN) pin is the read signal pin for external program memory and is active when low. It is connected to the output enable pin of external ROM or EPROM.

4 *EA*

The external access (EA) pin is taken low for the microprocessor to access only external program code; when high it automatically accesses internal or external code depending on the address. Thus when the 8051 is first reset, the program counter starts at $0000 and points to the first program instruction in the internal code memory unless EA is tied low. Then the CPU issues a low on PSEN to enable the external code memory to be used. This pin is also used on a microcontroller with EPROM to receive the programming supply voltage for programming the EPROM.

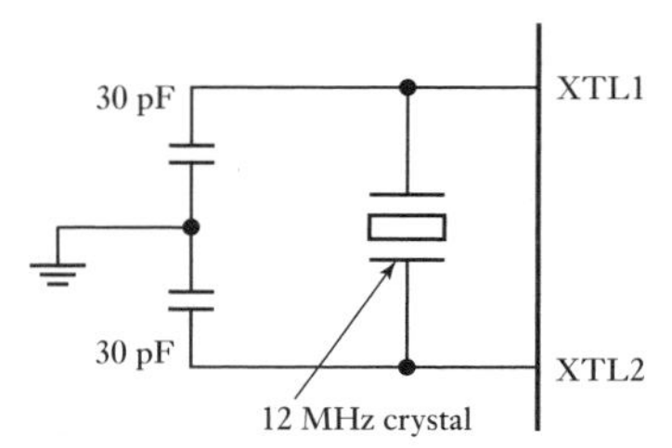

Figure 17.23 Crystal.

5 *XTAL1, XTAL2*
These are the connecting pins for a crystal or external oscillator. Figure 17.23 illustrates how they are used with a crystal. The most commonly used crystal frequency is 12 MHz.

6 *RESET*
A high signal on this pin for at least two machine cycles resets the microcontroller, i.e. puts in a condition to allow an orderly system start-up.

7 *Serial input/output*
Writing to the serial data buffer SBUF at address 99H loads data for transmission; reading SBUF accesses received data. The bit addressable serial port control register SCON at address 98H is used to control the various modes of operation.

8 *Timing*
The timer mode register TMOD at address 89H is used to set the operating mode for timer 0 and timer 1 (Figure 17.24). It is loaded at an entity and is not individually bit addressable. The timer control register TCON (Figure 17.25) contains status and control bits for timer 0 and timer 1. The upper 4 bits are used to turn the timers on and off or to signal a timer overflow. The lower 4 bits have nothing to do with timers but are used to detect and initiate external interrupts.

Figure 17.24 TMOD register.

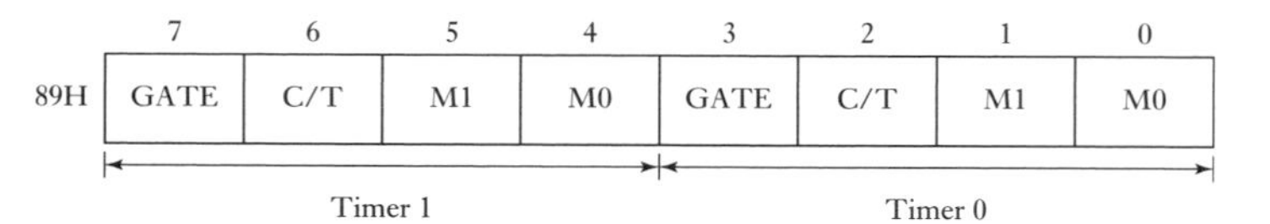

Gate: 0 = timer runs whenever TR0/TR1 set
1 = timer runs only when INT0/INT1 is high along with TR0/TR1

C/T: counter/timer select
0 = input from system clock, 1 = input from TX0/TX1

M0 and M1 set the mode

M1	M0	Mode	
0	0	0	13-bit counter, lower 5 bits of TL0 and all 8 bits of TH0
0	1	1	16-bit counter
1	0	2	8-bit auto-reload timer/counter
1	1	3	TL0 is an 8-bit timer/counter controlled by timer 0 control bits. TH0 is an 8-bit timer controlled by timer 1 control bits. Timer 1 is off

Figure 17.25 TCON register.

	7	6	5	4	3	2	1	0
88H	TF1	TR1	TF0	TR0	IE1	IT1	IE0	IT0

TF0, TF1	Timer overflow flag; set by hardware when time overflows and cleared by hardware when the processor calls the interrupt routine
TR0, TR1	Timer run control bits: 1 = timer on, 0 = timer off
IE0, IE1	Interrupt edge flag set by hardware when external interrupt edge or low level detected and cleared when interrupt processed
IT0, IT1	Interrupt type set by software: 1 = falling-edge-triggered interrupt, 0 = low-level-triggered interrupt

The source of the bits counted by each timer is set by the C/T bit; if the bit is low the source is the system clock divided by 12, otherwise if high it is set to count an input from an external source. The timers can be started by setting TR0 or TR1 to 1 and stopping by making it 0. Another method of controlling a timer is by setting the GATE to 1 and so allowing a timer to be controlled by the INT0 or INT1 pin on the microcontroller going to 1. In this way an external device connected to one of these pins can control the counter on/off.

9 *Interrupts*
Interrupts force the program to call a subroutine located at a specified address in memory; they are enabled by writing to the interrupt enable register IE at address A8H (Figure 17.26).

Figure 17.26 IE register.

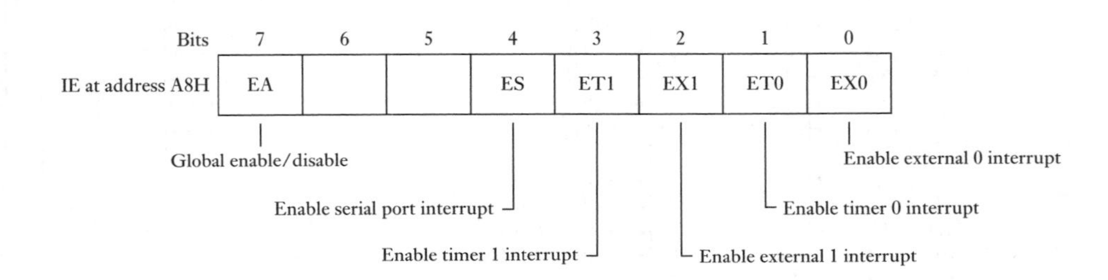

Address	Register	Address	Register
8D	TH1	F0	B
8C	TH0	E0	ACC
8B	TL1	D0	PSW
8A	TL0	B8	IP
89	TMOD	B0	P3
88	TCON	A8	IE
87	PCON	A0	P2
83	DPH	99	SBUF
82	DPL	98	SCON
81	SP	90	P1
80	P0		

Figure 17.27 Registers.

The term **special function registers** is used for the input/output control registers (Figure 17.27), like IE above, and these are located at addresses 80 to FF. Accumulator A (ACC) is the major register used for data operations; the B register is used for multiplication and division. P0, P1, P2 and P3 are the latch registers for ports 0, 1, 2 and 3.

17.3.3 Microchip™ microcontrollers

Another widely used family of 8-bit microcontrollers is that provided by Microchip™. The term PIC (Peripheral Interface Controller) is used for its single-chip microcontrollers. These use a form of architecture termed **Harvard architecture**. With this architecture, instructions are fetched from program memory using buses that are distinct from the buses used for accessing variables (Figure 17.28). In the other microcontrollers discussed in this chapter, separate buses are not used and thus program data fetches

Figure 17.28 Harvard architecture.

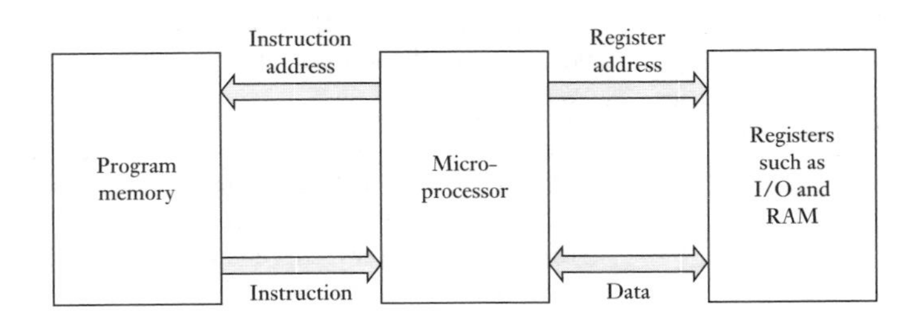

have to wait for read/write and input/output operations to be completed before the next instruction can be received from memory. With Harvard architecture, instructions can be fetched every cycle without waiting, each instruction being executed during the cycle following its fetch. Harvard architecture enables faster execution speeds to be achieved for a given clock frequency. Figure 17.29 shows the pin connections for one version of the PIC16C74A and the 16F84 microcontroller, and Figure 17.30 the basic form of the architecture.

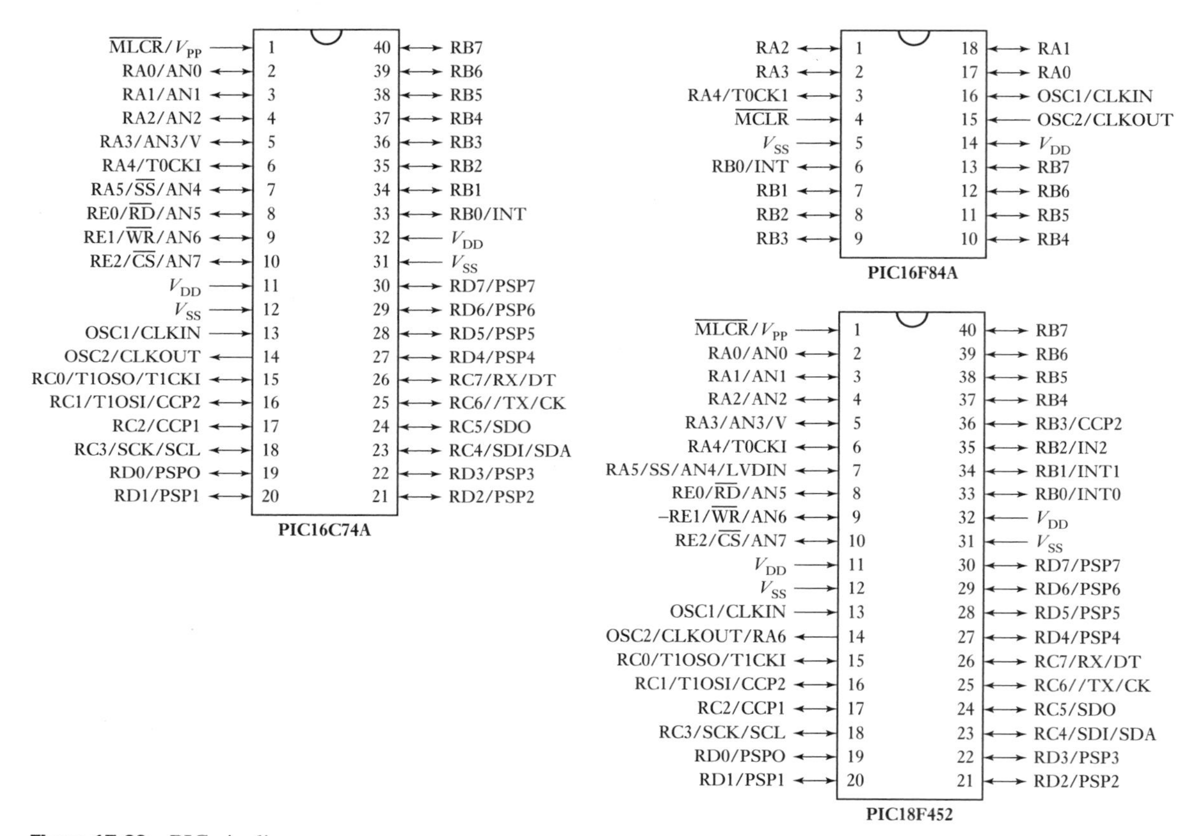

Figure 17.29 PIC pin diagrams.

The basic features of the 16C74 microcontroller, other PIC microcontrollers having similar functions, are:

1 *Input/output ports*
Pins 2, 3, 4, 5, 6 and 7 are for the bidirectional input/output port A. As with the other bidirectional ports, signals are read from and written to via port registers. The direction of the signals is controlled by the TRIS direction registers; there is a TRIS register for each port. TRIS is set as 1 for read and 0 for write (Figure 17.31).

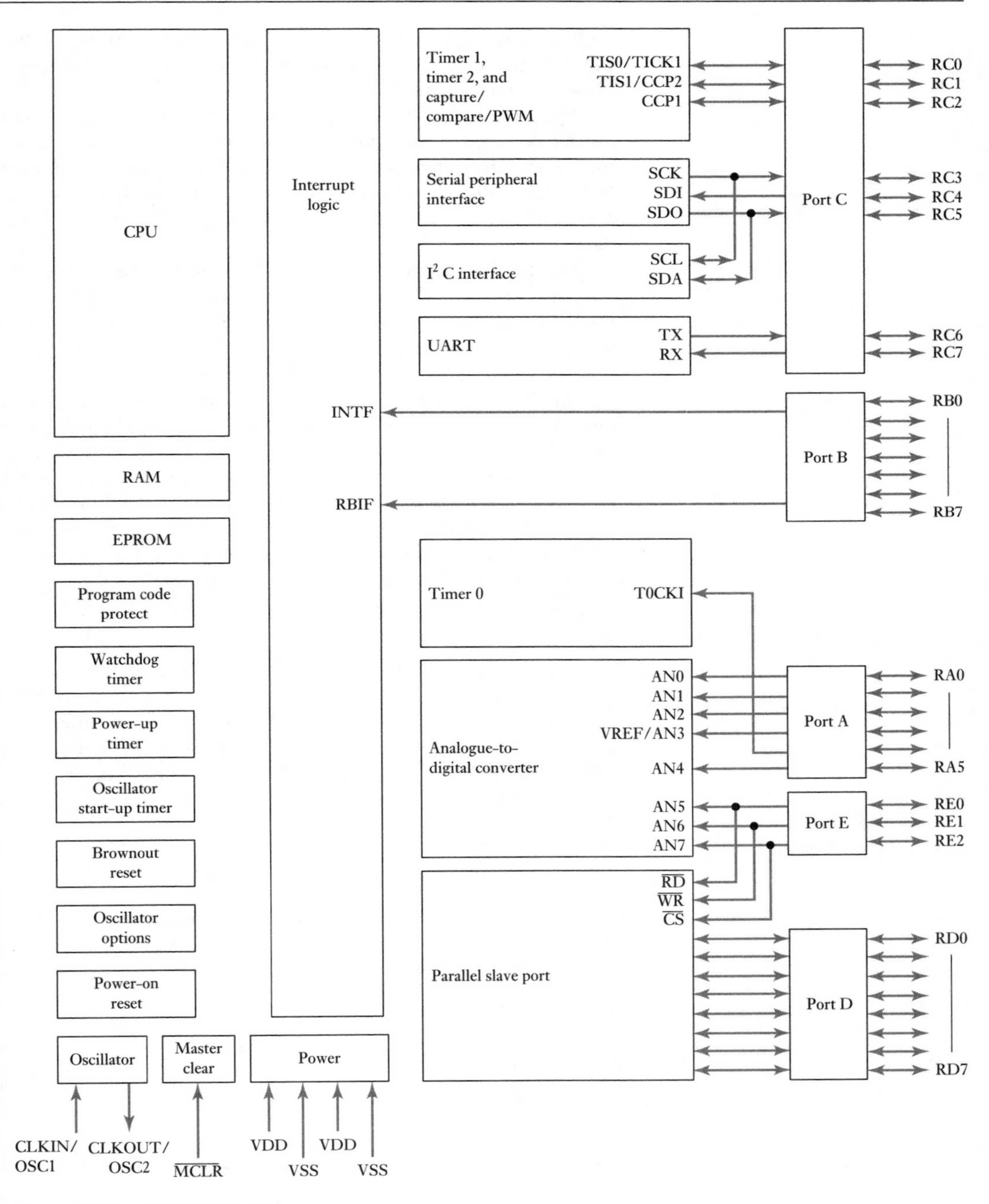

Figure 17.30 PIC 16C74/74A.

Figure 17.31 Port direction.

	7	6	5	4	3	2	1	0
Port bits								
TRIS for 5 outputs and 3 inputs	1	1	1	0	0	0	0	0

Pins 2, 3, 4 and 5 can also be used for analogue inputs, pin 6 for a clock input to timer 0; pin 7 can also be the slave select for the synchronous serial port (see later in this section).

Pins 33, 34, 35, 36, 37, 38, 39 and 40 are for the bidirectional input/output port B; the direction of the signals is controlled by a corresponding TRIS direction register. Pin 33 can also be the external interrupt pin. Pins 37, 38, 39 and 40 can also be the interrupt on change pins. Pin 39 can also be the serial programming clock and pin 40 the serial programming data.

Pins 15, 16, 17, 18, 23, 24, 25 and 26 are for the bidirectional input/output port C; the direction of the signals is controlled by a corresponding TRIS direction register. Pin 15 can also be the timer 1 output or the timer 1 clock input. Pin 16 can also be the timer 1 oscillator input or Capture 2 input/Compare 2 output/PWM2 output.

Pins 19, 20, 21, 22, 27, 28, 29 and 30 are for the bidirectional input/output port D; the direction of the signals is controlled by a corresponding TRIS direction register.

Pins 8, 9 and 10 are for the bidirectional input/output port E; the direction of the signals is controlled by a corresponding TRIS direction register. Pin 8 can also be the read control for the parallel slave port or analogue input 5. The parallel slave port is a feature that facilitates the design of personal computer interface circuitry; when in use the pins of ports D and E are dedicated to this operation.

2 *Analogue inputs*
Pins 2, 3, 4, 5 and 7 of port A and pins 8, 9 and 10 of port E can also be used for analogue inputs, feeding through an internal ADC. Registers ADCON1 and TRISA for port A (TRISE for port E) must be initialised to select the reference voltage to be used for the conversion and select channels as inputs. Then ADCON0 has to be initialised using the settings indicated below:

ADCON0 bits			
5	4	3	For analogue input on
0	0	0	Port A, bit 0
0	0	1	Port A, bit 1
0	1	0	Port A, bit 2
0	1	1	Port A, bit 3
1	0	0	Port A, bit 5
1	0	1	Port E, bit 0
1	1	0	Port E, bit 1
1	1	1	Port E, bit 2

3 *Timers*
The microcontroller has three timers: timer 0, timer 1 and timer 2. Timer 0 is an 8-bit counter which can be written to or read from and can be used to count external signal transitions, generating an interrupt when the required number of events have occurred. The source of the count can

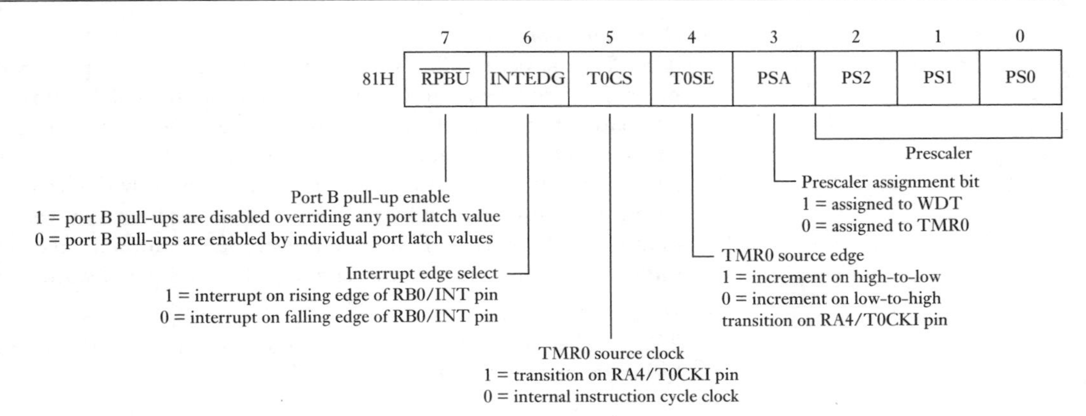

Figure 17.32 OPTION register.

be either the internal bus clock signal or an external digital signal. The choice of count source is made by the TOCS bit in the OPTION register (Figure 17.32).

If the prescaler is not selected then the count is incremented after every two cycles of the input source. A prescaler can be used so that signals are only passed to the counter after some other fixed number of clock cycles. The following shows the scaling rates possible. WDT gives the scaling factors selected when the watchdog timer is enabled. It is used to time out and reset the system if an operation is not concluded in a reasonable time; the default time is nominally 18 ms.

Prescalar bit values			TMR0	WDT
PS2	**PS1**	**PS0**	**rate**	**rate**
0	0	0	1 : 2	1 : 1
0	0	1	1 : 4	1 : 2
0	1	0	1 : 8	1 : 4
0	1	1	1 : 16	1 : 8
1	0	0	1 : 32	1 : 16
1	0	1	1 : 64	1 : 32
1	1	0	1 : 128	1 : 64
1	1	1	1 : 256	1 : 128

Timer 1 is the most versatile of the timers and can be used to monitor the time between signal transitions on an input pin or control the precise time of transitions on an output pin. When used with the capture or compare modes, it enables the microcontroller to control the timing of an output on pin 17.

Timer 2 can be used to control the period of a PWM output. PWM outputs are supplied at pins 16 and 17.

4 *Serial input/output*
The PIC microcontroller includes a synchronous serial port (SSP) module and a serial communications interface module (SCI). Pin 18 has the alternative functions of the synchronous serial clock input or output for SPI serial peripheral interface mode and I^2C mode. The I^2C bus provides a two-wire bidirectional interface that can be used with a range of other chips; it can also be used for connecting a master microcontroller to slave microcontrollers. UART, i.e. the universal asynchronous receiver transmitter, can be used to create a serial interface to a personal computer.

5 *Parallel slave port*
The parallel slave port uses ports D and E and enables the microcontroller to provide an interface with a PC.

6 *Crystal input*
Pin 13 is for the oscillator crystal input or external clock source input; pin 14 is for the oscillator crystal output. Figure 17.33(a) shows the arrangement that might be used for accurate frequency control, Figure 17.33 (b) that which might be used for a low-cost frequency control; for a frequency of 4 MHz we can have R = 4.7 kΩ and C = 33 pF. The internal clock rate is the oscillator frequency divided by 4.

Figure 17.33 Frequency control.

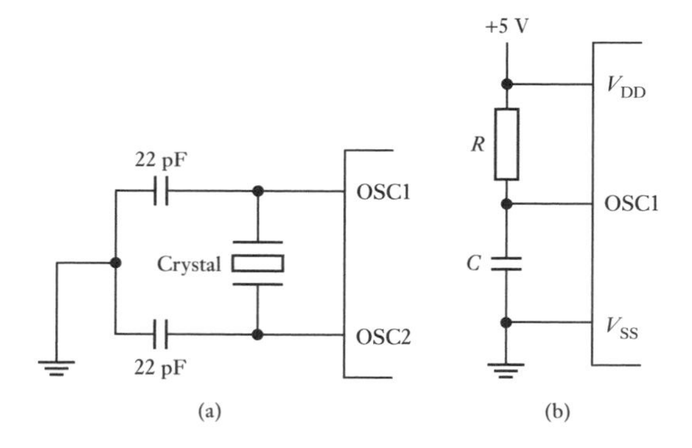

7 *Master clear and resets*
Pin 1 is the master clear MCLR, i.e. reset input, and is taken low to reset the device on demand and give an orderly start-up. When a V_{DD} rise is detected, a power-on-rest (POR) pulse is generated to provide a fixed time-out delay and keep the processor in the reset state. If the V_{DD} voltage goes below a specified voltage level for more than a certain amount of time, a brownout reset is activated. The watchdog timer is another way reset can occur. This is a timer which times out and rests the microcontroller if an operation is not concluded in what has been deemed a reasonable time.

The **special-purpose registers** (Figure 17.34) are used for input/output control, as illustrated above in relation to a few of these registers. The registers for the PIC16C73/74 are arranged in two banks and before a particular register can be selected, the bank has to be chosen by setting a bit in the status register (Figure 17.35).

Figure 17.34 Special-purpose registers.

File address	*Bank 0*	*Bank 1*	*File address*
00h	INDF	INDF	80h
01h	TMR0	OPTION	81h
02h	PCL	PCL	82h
03h	STATUS	STATUS	83h
04h	FSR	FSR	84h
05h	PORTA	TRISA	85h
06h	PORTB	TRISB	86h
07h	PORTC	TRISC	87h
08h	PORTD	TRISD	88h
09h	PORTE	TRISE	89h
0Ah	PCLATH	PCLATH	8Ah
0Bh	INTCON	INTCON	8Bh
0Ch	PIR1	PIE1	8Ch
0Dh	PIR2	PIE2	8Dh
0Eh	TMR1L	PCON	8Eh
0Fh	TMR1H		8Fh
10h	T1CON		90h
11h	TMR2		91h
12h	T2CON	PR2	92h
13h	SSPBUF	SSPADD	93h
14h	SSPCON	SSPSTAT	94h
15h	CCPR1L		95h
16h	CCPR1H		96h
17h	CCP1CON		97h
18h	RCSTA	TXSTA	98h
19h	TXREG	SPBRG	99h
1Ah	RCREG		9Ah
1Bh	CCPR2L		9Bh
1Ch	CCPR2H		9Ch
1Dh	CCPR2CON		9Dh
1Eh	ADRES		9Eh
1Fh	ADCON0	ADCON1	9Fh
20h 7Fh	General-purpose registers	General-purpose registers	A0h FFh

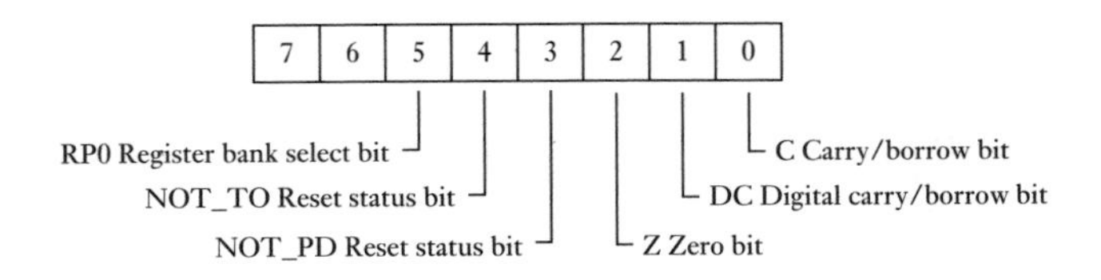

Figure 17.35 STATUS register.

17.3.4 Selecting a microcontroller

In selecting a microcontroller the following factors need to be considered:

1 *Number of input/output pins*
How many input/output pins are going to be needed for the task concerned?

2 *Interfaces required*
What interfaces are going to be required? For example, is PWM required? Many microcontrollers have PWM outputs, e.g. the PIC17C42 has two.

3 *Memory requirements*
What size memory is required for the task?

4 *The number of interrupts required*
How many events will need interrupts?

5 *Processing speed required*
The microprocessor takes time to execute instructions (see Section 18.2.2), this time being determined by the processor clock.

As an illustration of the variation of microcontrollers available, Table 17.1 shows details of members of the Intel 8051 family, Table 17.2 the PIC16Cxx family and Table 17.3 the M68HC11 family.

Table 17.1 Intel 8051 family members.

	ROM	EPROM	RAM	Timers	I/O ports	Interrupts
8031AH	0	0	128	2	4	5
8051AH	4K	0	128	2	4	5
8052AH	8K	0	256	3	4	6
8751H	0	4K	128	2	4	5

Table 17.2 PIC16C family members.

	I/O	EPROM	RAM	ADC channels	USART	CCP modules
PIC16C62A	22	2K	128	0	0	1
PIC16C63	22	4K	192	0	1	2
PIC16C64A	33	2K	128	0	0	1
PIC16C65A	33	4K	192	0	1	2
PIC16C72	22	2K	128	5	0	1
PIC16C73A	22	4K	192	5	1	2
PIC16C74A	33	4K	192	8	1	3

17.4 Applications

The following are two examples of how microcontrollers are used; more case studies are given in Chapter 24.

17.4.1 Temperature measurement system

As a brief indication of how a microcontroller might be used, Figure 17.36 shows the main elements of a temperature measurement system using an

Table 17.3 M68HC11 family members.

	ROM	EEPROM	RAM	ADC	Timer	PWM	I/O	Serial	E-clock MHz
68HC11AO	0	0	256	8 ch., 8-bit	(1)	0	22	SCI, SPI	2
68HC11A1	0	512	256	8 ch., 8-bit	(1)	0	22	SCI, SPI	2
68HC11A7	8K	0	256	8 ch., 8-bit	(1)	0	38	SCI, SPI	3
68HC11A8	8K	512	256	8 ch., 8-bit	(1)	0	38	SCI, SPI	3
68HC11C0	0	512	256	4 ch., 4-bit	(2)	2 ch., 8-bit	36	SCI, SPI	2
68HC11D0	0	0	192	None	(2)	0	14	SCI, SPI	2

Timer: (1) is three-input capture, five-output compare, real-time interrupt, watchdog timer, pulse accumulator, (2) is three- or four-input capture, five- or four-output compare, real-time interrupt, watchdog timer, pulse accumulator. Serial: SCI is asynchronous serial communication interface, SPI is synchronous serial peripheral interface.

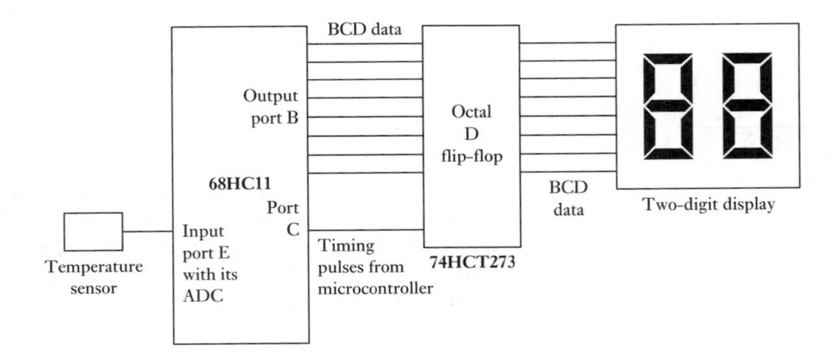

Figure 17.36 Temperature measurement system.

MC68HC11. The temperature sensor gives a voltage proportional to the temperature (e.g. a thermotransistor such as LM35; see Section 2.9.4). The output from the temperature sensor is connected to an ADC input line of the microcontroller. The microcontroller is programmed to convert the temperature into a BCD output which can be used to switch on the elements of a two-digit, seven-element display. However, because the temperature may be fluctuating it is necessary to use a storage register which can hold data long enough for the display to be read. The storage register, 74HCT273, is an octal D-type flip-flop which is reset on the next positive-going edge of the clock input from the microcontroller.

17.4.2 Domestic washing machine

Figure 17.37 shows how a microcontroller might be used as the controller for a domestic washing machine. The microcontroller often used is the Motorola M68HC05B6; this is simpler and cheaper than the Motorola M68HC11 microcontroller discussed earlier in this chapter and is widely used for low-cost applications.

The inputs from the sensors for water temperature and motor speed are via the analogue-to-digital input port. Port A provides the outputs for the

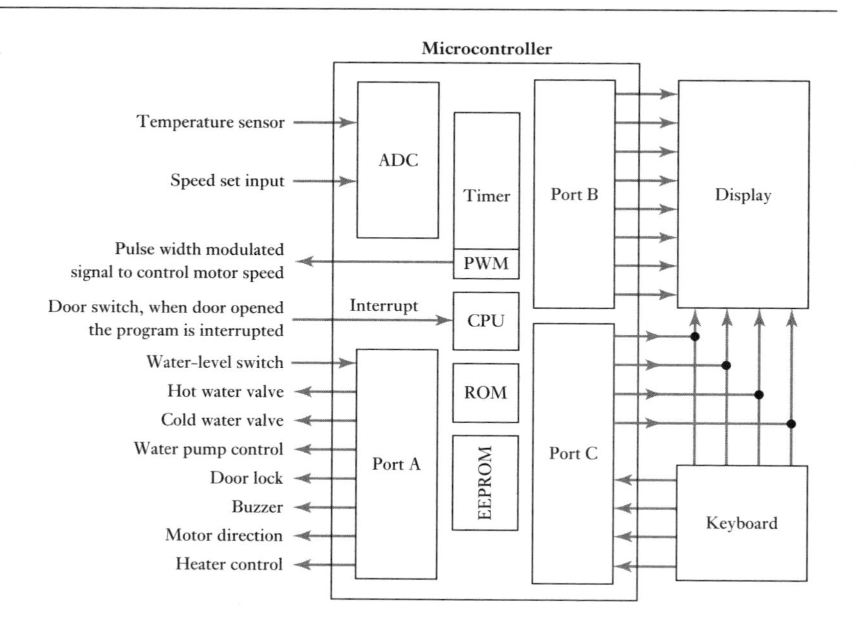

Figure 17.37 Washing machine.

various actuators used to control the machine and also the input for the water-level switch. Port B gives outputs to the display. Port C gives outputs to the display and also receives inputs from the keyboard used to input to the machine the various program selections. The PWM section of the timer provides a PWM signal to control the motor speed. The entire machine program is interrupted and stopped if the door of the washing machine is opened.

17.5 Programming

A commonly used method for the development of programs follows the following steps:

1 Define the problem, stating quite clearly what function the program is to perform, the inputs and outputs required, any constraints regarding speed of operation, accuracy, memory size, etc.
2 Define the algorithm to be used. An **algorithm** is the sequence of steps which define a method of solving the problem.
3 For systems with fewer than thousands of instructions a useful aid is to represent the algorithm by means of a **flow chart**. Figure 17.38(a) shows the standard symbols used in the preparation of flow charts. Each step of an algorithm is represented by one or more of these symbols and linked together by lines to represent the program flow. Figure 17.38(b) shows part of a flow chart where, following the program start, there is operation A, followed by a branch to either operation B or operation C depending on whether the decision to the query is yes or no. Another useful design tool is **pseudocode**. Pseudocode is a way of describing the steps in an algorithm in an informal way which can later be translated into a program (see the next section).
4 Translate the flow chart/algorithm into instructions which the microprocessor can execute. This can be done by writing the instructions in some

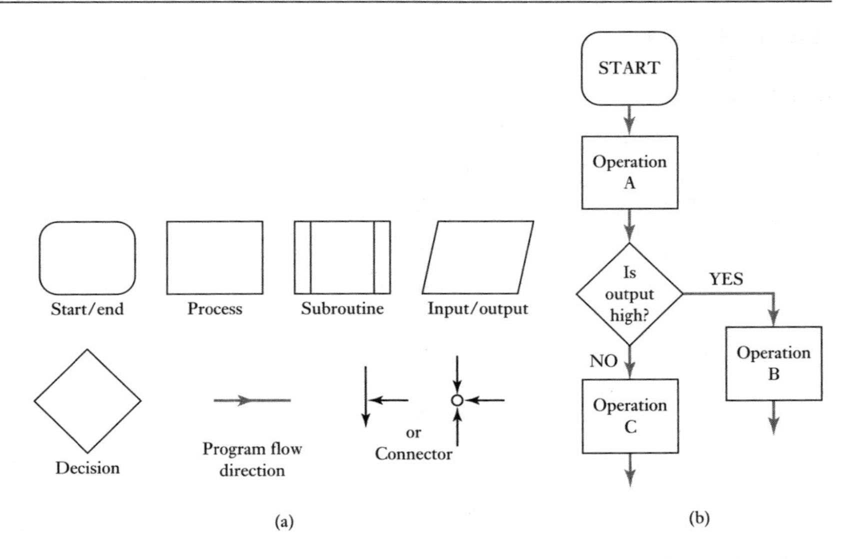

Figure 17.38 Flow chart: (a) symbols, (b) example.

language, e.g. assembly language or perhaps C, and then converting these, either manually or by means of an assembler computer program, into a code which is acceptable to the microprocessor, i.e. machine code.

5 Test and debug the program. Errors in programs are referred to as **bugs** and the process of tracking them down and eliminating them as **debugging**.

17.5.1 Pseudocode

Pseudocode is rather like drawing a flow chart and involves writing a program as a sequence of functions or operations with the decision element IF–THEN–ELSE and the repetition element WHILE–DO.

A sequence (Figure 17.39(a)) would be written as:

```
BEGIN A
  ...
END A
  ...
BEGIN B
  ...
END B
```

and a decision as:

```
IF X
THEN
  BEGIN A
    ...
  END A
ELSE
  BEGIN B
    ...
  END B
ENDIF X
```

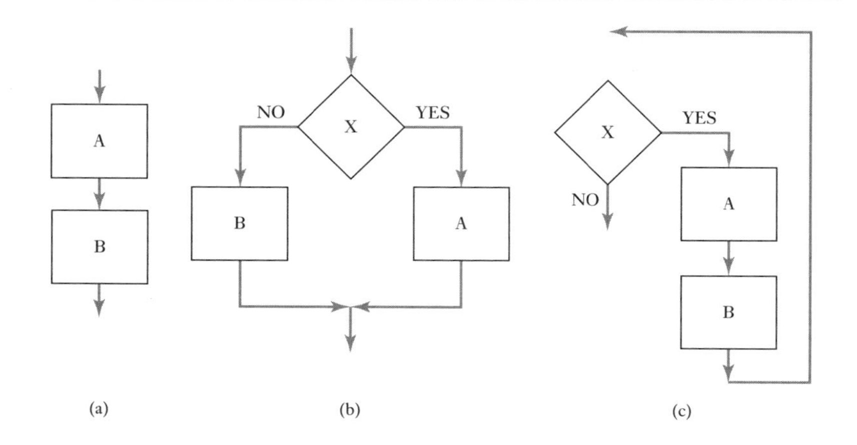

Figure 17.39 (a) Sequence, (b) IF–THEN–ELSE, (c) WHILE–DO.

Figure 17.39(b) shows such a decision in a flow chart. A repetition is written as:

```
WHILE X
DO
  BEGIN A
  ...
  END A
  BEGIN B
  ...
  END B
ENDO WHILE X
```

Figure 17.39(c) shows the WHILE–DO as a flow chart. A program written in this way might then appear as:

```
BEGIN PROGRAM
  BEGIN A
    IF X
      BEGIN B
      END B
    ELSE
      BEGIN C
      END C
    ENDIF X
  END A
  BEGIN D
    IF Z
      BEGIN E
      END E
    ENDIF Z
  END D
```

Chapter 18 shows how programs can be written in assembly language and Chapter 19 in C.

Summary

Basically, systems involving **microprocessors** have three parts: a central processing unit (CPU), input and output interfaces, and memory. Within a microprocessor, digital signals move along **buses**, these being parallel tracks for transmission of parallel rather than serial data.

Microcontrollers are the integration on a single chip of a microprocessor with memory, input/output interfaces and other peripherals such as timers.

An **algorithm** is the sequence of steps which define a method of solving a problem. **Flow charts** and **pseudocode** are two methods of describing such steps.

Problems

17.1 Explain, for a microprocessor, the roles of (a) accumulator, (b) status, (c) memory address, (d) program counter registers.

17.2 A microprocessor uses eight address lines for accessing memory. What is the maximum number of memory locations that can be addressed?

17.3 A memory chip has 8 data lines and 16 address lines. What will be its size?

17.4 How does a microcontroller differ from a microprocessor?

17.5 Draw a block diagram of a basic microcontroller and explain the function of each subsystem.

17.6 Which of the M68HC11 ports is used for (a) the ADC, (b) a bidirectional port, (c) serial input/output, (d) as just an 8-bit output-only port?

17.7 How many bytes of memory does the M68HC11A7 have for data memory?

17.8 For the Motorola M68HC11, port C is bidirectional. How is it configured to be (a) an input, (b) an output?

17.9 The Motorola M68HC11 can be operated in single-chip and in extended mode. Why these modes?

17.10 What is the purpose of the ALE pin connection with the Intel 8051?

17.11 What input is required to reset an Intel 8051 microcontroller?

17.12 Write pseudocode to represent the following:

(a) If A is yes then B, else C.

(b) While A is yes do B.

Chapter eighteen Assembly Language

Objectives

The objectives of this chapter are that, after studying it, the reader should be able to:

- **Use assembly language to write programs involving data transfers, arithmetic, logic, jumps, branches, subroutines, delays and look-up tables.**

18.1 Languages

Software is the term used for the **instructions** that tell a microprocessor or microcontroller what to do. The collection of instructions that a microprocessor will recognise is its **instruction set**. The form of the instruction set depends on the microprocessor concerned. The series of instructions that is needed to carry out a particular task is called a **program**.

Microprocessors work in binary code. Instructions written in binary code are referred to as being in **machine code**. Writing a program in such a code is a skilled and very tedious process. It is prone to errors because the program is just a series of zeros and ones and the instructions are not easily comprehended by just looking at the pattern. An alternative is to use an easily comprehended form of shorthand code for the patterns of zeros and ones. For example, the operation of adding data to an accumulator might be represented by just ADDA. Such a shorthand code is referred to as a **mnemonic code**, a mnemonic code being a 'memory-aiding' code. The term **assembly language** is used for such a code. Writing a program using mnemonics is easier because they are an abbreviated version of the operation performed by the instruction. Also, because the instructions describe the program operations and can easily be comprehended, they are less likely to be used in error than the binary patterns of machine code programming. The assembler program has still, however, to be converted into machine code since this is all the microprocessor will recognise. This conversion can be done by hand using the manufacturer's data sheets which list the binary code for each mnemonic. However, computer programs are available to do the conversion, such programs being referred to as **assembler programs**.

High-level languages, e.g. BASIC, C, FORTRAN and PASCAL, are available which provide a type of programming language which is even closer to describing in easily comprehended language the types of operations required. Such languages have still, however, to be converted into machine code, by a computer program, for the microprocessor to be able to use. This chapter is an outline of how programs might be written in assembly language, Chapter 19 using C.

18.2 Instruction sets

The following are commonly used instructions that may be given to a microprocessor, the entire list of such instructions being termed the instruction set. Appendix C gives instruction sets for three commonly encountered types of microcontroller. The instruction set differs from one microprocessor to another. The following are some of the commonly encountered instructions:

Data transfer/movement

1 *Load*
This instruction reads the contents of a specified memory location and copies it to a specified register location in the CPU and is typically used with Motorola microprocessors, e.g. LDAA $0010:

Before instruction	*After instruction*
Data in location 0010	Data in location 0010 Data from 0010 in accumulator A

2 *Store*
This instruction copies the current contents of a specified register into a specified memory location and is typically used with Motorola microprocessors, e.g. STA $0011:

Before instruction	*After instruction*
Data in accumulator A	Data in accumulator A Data copied to location 0011

3 *Move*
This instruction is used to move data into a register or copy data from one register to another and is used with PIC and Intel microprocessors, e.g. with PIC, MOV R5,A:

Before instruction	*After instruction*
Data in register A	Data in register A Data copied to register R5

4 *Clear*
This instruction resets all bits to zero, e.g. with Motorola, CLRA to clear accumulator A; with PIC, CLRF 06 to clear file register 06.

Arithmetic

5 *Add*
This instruction adds a number to the data in some register, e.g. with Intel, ADD A, #10h:

Before instruction	*After instruction*
Accumulator A with data	Accumulator A plus 10 hex

and with Motorola, ADDD $0020:

Before instruction	*After instruction*
Accumulator D with data	Accumulator D plus contents of location 0020

or the contents of a register to the data in a register, e.g. with Intel, ADD A, @R1:

Before instruction	*After instruction*
Accumulator A with data	Accumulator A plus contents of location R1

and with PIC, addwf 0C:

Before instruction	*After instruction*
Register 0C with data	Register 0C plus contents of location w

6 *Decrement*
This instruction subtracts 1 from the contents of a specified location. For example, we might have register 3 as the specified location and so, with Intel, DEC R3:

Before instruction	*After instruction*
Register R3 with data 0011	Register R3 with data 0010

7 *Increment*
This instruction adds 1 to the contents of a specified location, e.g. INCA with Motorola to increment the data in accumulator A by 1, incf 06 with PIC to increment the data in register 06 by 1.

8 *Compare*
This instruction indicates whether the contents of a register are greater than, less than or the same as the contents of a specified memory location. The result appears in the status register as a flag.

Logical

9 *AND*
This instruction carries out the logical AND operation with the contents of a specified memory location and the data in some register. Numbers are ANDed bit by bit, e.g. with Motorola, ANDA %1001:

Before instruction	*After instruction*
Accumulator A with data 0011 Memory location with data 1001	Accumulator A with data 0001

Only in the least significant bit in the above data do we have a 1 in both sets of data and the AND operation only gives a 1 in the least significant bit of the result. With PIC, ANDLW 01 adds the binary number 01 to the number in W and if the least significant bit is, say, 0 then the result is 0.

10 *OR*
This instruction carries out the logical OR operation with the contents of a specified memory location and the data in some register, bit by bit, e.g. with Intel, ORL A,#3Fh will OR the contents of register A with the hex number 3F.

11 *EXCLUSIVE-OR*
This instruction carries out the logical EXCLUSIVE-OR operation with the contents of a specified memory location and the data in some register,

bit by bit, e.g. with PIC, xorlw 81h (in binary 10000001):

Before instruction	*After instruction*
Register w with 10001110	Register w with 00001111

XORing with a 0 leaves a data bit unchanged whilst with a 1 the data bit is inverted.

12 *Logical shift (left or right)*
Logical shift instructions involve moving the pattern of bits in the register one place to the left or right by moving a 0 into the end of the number. For example, for logical shift right a 0 is shifted into the most significant bit and the least significant bit is moved to the carry flag in the status register. With Motorola the instruction might be LSRA for shift to the right and LSLA for a shift to the left.

Before instruction	*After instruction*
Accumulator with 0011	Accumulator with 0001 Status register indicates Carry 1

13 *Arithmetic shift (left or right)*
Arithmetic shift instructions involve moving the pattern of bits in the register one place to the left or right but preserve the sign bit at the left end of the number, e.g. for an arithmetic shift right with the Motorola instruction ASRA:

Before instruction	*After instruction*
Accumulator with 1011	Accumulator with 1001 Status register indicates Carry 1

14 *Rotate (left or right)*
Rotate instructions involve moving the pattern of bits in the register one place to the left or right and the bit that spills out is written back into the other end, e.g. for a rotate right, Intel instruction RR A:

Before instruction	*After instruction*
Accumulator with 0011	Accumulator with 1001

Program control

15 *Jump or branch*
This instruction changes the sequence in which the program steps are carried out. Normally the program counter causes the program to be carried out sequentially in strict numerical sequence. However, the jump instruction causes the program counter to jump to some other specified location in the program (Figure 18.1(a)). Unconditional jumps occur without the program testing for some condition to occur. Thus with Intel we can have

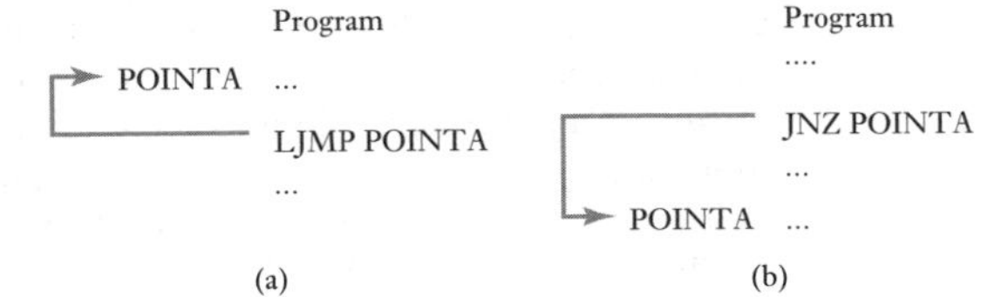

Figure 18.1 (a) Unconditional jump, (b) conditional jump.

LJMP POINTA for the program to jump to the line in the program labelled POINTA, with Motorola the instruction would be JMP POINTA and with PIC it would be GOTO POINTA. Conditional jumps occur if some condition is realised (Figure 18.1(b)). With Intel we can have JNZ POINTA for the program to jump to the line in the program labelled POINTA if any bits in the accumulator are not zero, otherwise it continues with the next line. JZ POINTA is all the bits in the accumulator are zero. With PIC, a conditional jump can involve two lines of code: BTFC 05,1 to 'bit test and skip', i.e. test if bit 1 of file register 5, and if the result is 0 then it jumps the next program line, if 1 it executes it. The next line is GOTO POINTA. With Motorola branch is a conditional jump instruction for the program to determine which branch of a program will be followed if the specified conditions occur. For example, Motorola uses BEQ to branch if equal to zero, BGE for branch if greater than or equal to, BLE for branch if less than or equal to.

16 *Halt/stop*
This instruction stops all further microprocessor activity.

Numerical data may be binary, octal, hex or decimal. Generally in the absence of any indicator the assembler assumes that the number is decimal. With Motorola, a number is indicated by the prefix #; a binary number is preceded by % or followed by B; an octal number is preceded by @ or followed by O; a hex number is preceded by $ or followed by H; and a decimal number requires no indicating letter or symbol. With Intel, numerical values must be preceded by # to indicate a number and by B for binary, O or Q for octal, H or h for hex and D or nothing for decimal. With PIC microcontrollers the header file has R = DEC for decimal to be the default. Then for binary the number is enclosed in quotation marks and preceded by B and for hex by H.

18.2.1 Addressing

When a mnemonic, such as LDA, is used to specify an instruction it will be followed by additional information to specify the source and destination of the data required by the instruction. The data following the instruction is referred to as the **operand.**

There are several different methods that are used for specifying data locations, i.e. addressing, and hence the way in which the program causes the microprocessor to obtain its instructions or data. Different microprocessors have different addressing modes. The Motorola 68HC11 has the six addressing modes of immediate, direct, extended, indexed, inherent and relative; the Intel 8051 has the five modes of immediate, direct, register, indirect and indexed; the PIC microcontroller has the three modes of immediate, direct and indirect with the indirect mode allowing indexing. The following are commonly used addressing modes:

1 *Immediate addressing*
The data immediately following the mnemonic is the value to be operated on and is used for the loading of a predetermined value into a register or memory location. For example, with the Motorola code, LDA B #$25 means load the number 25 into accumulator B. The # signifies immediate mode and a number, the $ that the number is in hexadecimal notation. With the Intel code we might have MOV A,#25H to move the number 25 to the accumulator A. The # indicates a number and the H indicates a hex

number. With the PIC code we might have movlw H'25' to load the number 25 into the working register w, the H indicating it is a hex number.

2 *Direct, absolute, extended or zero-page addressing*
With this form of addressing, the data byte that follows the operation code directly gives an address that defines the location of the data to be used in the instruction. With Motorola the term **direct addressing** is used when the address given is only 8 bits wide; the term **extended addressing** is used when it is 16 bits wide. For example, with Motorola code, LDAA $25 means load the accumulator with the contents of memory location 0025; the 00 is assumed. With Intel code, for the same operation, we can have the direct address instruction MOV A,20H to copy the data at address 20 to the accumulator A. With the PIC code we might have movwf Reg1 to copy the contents of Reg1 into the working register, the address of Reg1 having been previously defined.

3 *Implied addressing, or inherent addressing*
With this mode of addressing, the address is implied in the instruction. For example, with Motorola code and Intel code, CLR A means clear accumulator A. With PIC code clrw means clear the working register.

4 *Register*
With this form of addressing, the operand is specified as the contents of one of the internal registers. For example, with Intel ADD R7,A to add the contents of the accumulator to register R7.

5 *Indirect*
This form of addressing means that that the data is to be found in a memory location whose address is given by the instruction. For example, with the PIC system the INDF and FSR registers are used. The address is first written to the FSR register and then this serves as an address pointer. A subsequent direct access of INDF with the instruction movf INDF,w will load the working register w using the contents of FSR as a pointer to the data location.

6 *Indexed addressing*
Indexed addressing means that the data is in a memory location whose address is held in an index register. The first byte of the instruction contains the operation code and the second byte contains the offset; the offset is added to the contents of the index register to determine the address of the operand. A Motorola instruction might thus appear as LDA A $FF,X; this means load accumulator A with data at the address given by adding the contents of the index register and FF. Another example is STA A $05,X; this means store the contents of accumulator A at the address given by the index register plus 05.

7 *Relative addressing*
This is used with branch instructions. The operation code is followed with a byte called the relative address. This indicates the displacement in address that has to be added to the program counter if the branch occurs. For example, Motorola code BEQ $F1 indicates that if the data is equal to zero then the next address in the program is F1 further on. The relative address of F1 is added to the address of the next instruction.

As an illustration, Table 18.1 shows some instructions with the modes of addressing used in Motorola systems.

Table 18.1 Examples of addressing.

Address mode	Instruction	
Immediate	LDA A #$F0	Load accumulator A with data F0
Direct	LDA A $50	Load accumulator A with data at address 0050
Extended	LDA A$0F01	Load accumulator A with data at address 0F01
Indexed	LDA A $CF,X	Load accumulator with data at the address given by the index register plus CF
Inherent	CLR A	Clear accumulator A
Extended	CLR $2020	Clear address 2020, i.e. store all zeros at address 2020
Indexed	CLR $10,X	Clear the address given by the index register plus 10, i.e. store all zeros at that address

18.2.2 Data movement

The following is an example of the type of information that will be found in a manufacturer's (6800) instruction set sheet:

		Addressing modes					
		IMMED			DIRECT		
Operation	Mnemonic	OP	~	#	OP	~	#
Add	ADDA	8B	2	2	9B	3	2

~ is the number of microprocessor cycles required and # is the number of program bytes required.

This means that when using the immediate mode of addressing with this processor the Add operation is represented by the mnemonic ADDA. When the immediate form of addressing is used, the machine code for it is 8B and it will take two cycles to be fully expressed. The operation will require 2 bytes in the program. The **op-code** or **operation code** is the term used for the instruction that the microprocessor will act on and is expressed in hexa-decimal form. A byte is a group of eight binary digits recognised by the microprocessor as a word. Thus two words are required. With the direct mode of addressing, the machine code is 9B and takes three cycles and two program bytes.

To illustrate how information passes between memory and microprocessor, consider the following tasks. The addresses in RAM where a new program may be placed are just a matter of convenience. For the following examples, the addresses starting at 0010 have been used. For direct addressing to be used, the addresses must be on the zero page, i.e. addresses between 0000 and 00FF. The examples are based on the use of the instruction set for the M6800 microprocessor.

Task: Enter all zeros in accumulator A.

Memory address	*Op-code*	
0010	8F	CLR A

The next memory address that can be used is 0011 because CLR A only occupies one program byte. This is the inherent mode of addressing.

Task: Add to the contents of accumulator A the data 20.

Memory address	*Op-code*	
0010	8B 20	ADD A #$20

This uses the immediate form of addressing. The next memory address that can be used is 0012 because, in this form of addressing, ADD A occupies two program bytes.

Task: Load accumulator A with the data contained in memory address 00AF.

Memory address	*Op-code*	
0010	B6 00AF	LDA A $00AF

This uses the absolute form of addressing. The next memory address that can be used is 0013 because, in this form of addressing, LDA A occupies three program bytes.

Task: Rotate left the data contained in memory location 00AF.

Memory address	*Op-code*	
0010	79 00AF	ROL $00AF

This uses the absolute form of addressing. The next memory address that can be used is 0013 since ROL, in this mode, occupies three program bytes.

Task: Store the data contained in accumulator A into memory location 0021.

Memory address	*Op-code*	
0010	D7 21	STA A $21

This uses the direct mode of addressing. The next memory address that can be used is 0012 since STA A, in this mode, occupies two program bytes.

Task: Branch forward four places if the result of the previous instruction is zero.

Memory address	*Op-code*	
0010	27 04	BEQ $04

This uses the relative mode of addressing. The next memory address if the result is not zero is 0012 since BEQ, in this mode, occupies two program bytes. If the result is zero then the next address is 0012 + 4 = 0016.

18.3 Assembly language programs

An assembly language program should be considered as a series of instructions to an assembler which will then produce the machine code program. A program written in assembly language consists of a sequence of statements, one statement per line. A statement contains from one to four sections or **fields**, these being:

Label Op-code Operand Comment

A special symbol is used to indicate the beginning or end of a field, the symbols used depending on the microprocessor machine code assembler concerned. With the Motorola 6800 spaces are used. With the Intel 8080 there are a colon after a label, a space after the op-code, commas between entries in the address field and a semicolon before a comment. In general, a semicolon is used to separate comments from the operand.

The **label** is the name by which a particular entry in the memory is referred to. Labels can consist of letters, numbers and some other characters. With the Motorola 6800, labels are restricted to one to six characters, the first of which must be a letter, and cannot consist of a single letter A, B or X since these are reserved for reference to the accumulator or index register. With the Intel 8080, five characters are permitted with the first character a letter, @ or ?. The label must not use any of the names reserved for registers, instruction codes or pseudo-operations (see later in this section). All labels within a program must be unique. If there is no label then a space must be included in the label field. With the Motorola 6800, an asterisk (*) in the label field indicates that the entire statement is a comment, i.e. a comment included to make the purpose of the program clearer. As such, the comment will be ignored by the assembler during the assembly process for the machine code program.

The op-code specifies how data is to be manipulated and is specified by its mnemonic, e.g. LDA A. It is the only field that can never be empty. In addition, the op-code field may contain directives to the assembler. These are termed **pseudo-operations** since they appear in the op-code field but are not translated into instructions in machine code. They may define symbols, assign programs and data to certain areas of memory, generate fixed tables and data, indicate the end of the program, etc. Common assembly directives are:

Set program counter

ORG	This defines the starting memory address of the part of the program that follows. A program may have several origins.

Define symbols

EQU, SET, DEF	Equates/sets/defines a symbol for a numerical value, another symbol or an expression.

Reserve memory locations

RMB, RES	Reserves memory bytes/space.

Define constant in memory

FCB	Forms constant byte.
FCC	Forms constant character string.
FDB	Forms double byte constant.
BSW	Block storage of zeros.

The information included in the **operand** field depends on the mnemonic preceding it and the addressing mode used. It gives the address of the data to be operated on by the process specified by the op-code. It is thus often referred to as the **address field**. This field may be empty if the instructions given by the op-code do not need any data or address. Numerical data in this field may be hexadecimal, decimal, octal or binary. The assembler assumes that numbers are decimal unless otherwise specified. With the Motorola 6800 a hexadecimal number is preceded by $ or followed by H, an octal number preceded by @ or followed by O or Q, a binary number preceded by % or followed by B. With the Intel 8080, a hexadecimal number is followed by H, an octal number by O or Q, a binary number by B. Hexadecimal numbers must begin with a decimal digit, i.e. 0 to 9, to avoid confusion with names. With the Motorola 6800, the immediate mode of address can be indicated by preceding the operand by #, the indexed mode of address involving the operand being followed by X. No special symbols are used for direct or extended addressing modes. If the address is on the zero page, i.e. FF or less, then the assembler automatically assigns the direct mode. If the address is greater than FF, the assembler assigns the extended mode.

The comment field is optional and is there to allow the programmer to include any comments which may make the program more understandable by the reader. The comment field is ignored by the assembler during the production of the machine code program.

18.3.1 Examples of assembly language programs

The following examples illustrate how some simple programs can be developed.

Problem: The addition of the two 8-bit numbers located in different memory addresses and the storage of the result back into memory.

The algorithm is:

1 Start.
2 Load the accumulator with the first number. The accumulator is where the results of arithmetic operations are accumulated. It is the working register, i.e. like a notepad on which the calculations are carried out before the result is transferred elsewhere. Thus we have to copy data to the accumulator before we can carry out the arithmetic. With PIC the term working (w) register is used.
3 Add on the second number.
4 Store the sum in the designated memory location.
5 Stop.

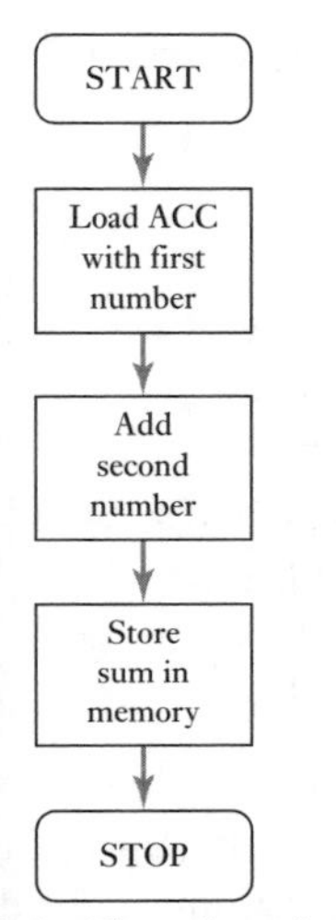

Figure 18.2 Flow chart for the addition of two numbers.

Figure 18.2 shows these steps represented as a flow chart.

The following are the programs written for three different microcontrollers. In each, the first column gives the label, the second the op-code, the third the operand and the fourth the comment. Note that all comments are preceded by a semicolon.

M68HC11 program

```
                ; Addition of two numbers

NUM1      EQU     $00       ; location of number 1
NUM2      EQU     $01       ; location of number 2
SUM       EQU     $02       ; location for the sum

          ORG     $C000     ; address of start of user RAM
START     LDAA    $NUM1     ; load number 1 into acc. A
          ADDA    $NUM2     ; add number 2 to A
          STAA    SUM       ; save the sum to $02
          END
```

The first line in the program specifies the address of the first of the numbers to be added. The second line specifies the address of the number to be added to the first number. The third line specifies where the sum is to be put. The fourth line specifies the memory address at which the program should start. The use of the labels means that the operand involving that data does not have to specify the addresses but merely the labels.

The same program for an Intel 8051 would be as follows.

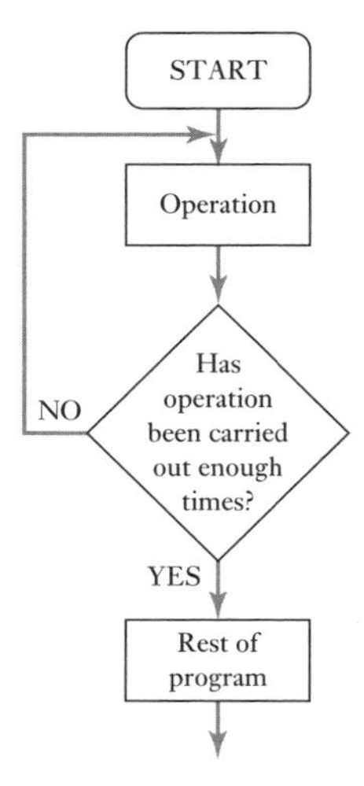

Figure 18.3 A loop.

8051 program

```
                ; Addition of two numbers

NUM1      EQU    20H        ; location of number 1
NUM2      EQU    21H        ; location of number 2
SUM       EQU    22H        ; location for the sum
          ORG    8000H      ; address of start of user RAM
START     MOV    A,NUM1     ; load number 1 into acc. A
          ADD    A,NUM2     ; add number 2 to A
          MOV    SUM,A      ; save the sum to address 22H
          END
```

The same program for a PIC microcontroller would be as follows.

PIC program

```
                ; Addition of two numbers

Num1      equ    H'20'      ; location of number 1
Num2      equ    H'21'      ; location of number 2
Sum       equ    H'22'      ; location for the sum
          org    H'000'     ; address of start of user RAM
Start     movlw  Num1       ; load number 1 into w
          addlw  Num2       ; add number 2 to w
          movwf  Sum        ; save the sum H'22'
          End
```

In many programs there can be a requirement for a task to be carried out a number of times in succession. In such cases the program can be designed so that the operation passes through the same section a number of times. This is termed **looping**, a **loop** being a section of a program that is repeated a number of times. Figure 18.3 shows a flow diagram of a loop. With such a loop, a certain operation has to be performed a number of times before the program proceeds. Only when the number of such operations is completed can the program proceed. The following problem indicates such a looping.

Problem: the addition of numbers located at 10 different addresses (these might be the results of inputs from 10 different sensors which have each to be sampled).

The algorithm could be:

1 Start.
2 Set the count as 10.
3 Point to location of bottom address number.
4 Add bottom address number.
5 Decrease the count number by 1.
6 Add 1 to the address location pointer.
7 Is count 0? If not branch to 4. If yes proceed.
8 Store sum.
9 Stop.

Figure 18.4 shows the flow chart.

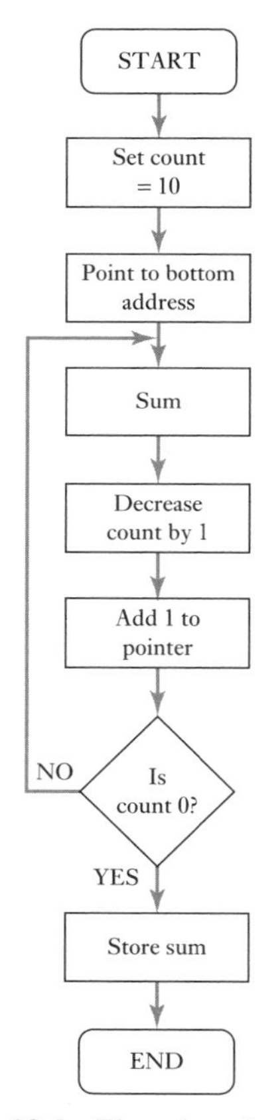

Figure 18.4 Flow chart for adding 10 numbers.

The program is:

```
COUNT    EQU     $0010
POINT    EQU     $0020
RESULT   EQU     $0050
         ORG     $0001
         LDA B   COUNT     ; Load count
         LDX     POINT     ; Initialise index register at start of
                           ; numbers
SUM      ADD A   X         ; Add addend
         INX               ; Add 1 to index register
         DEC B             ; Subtract 1 from accumulator B
         BNE     SUM       ; Branch to sum
         STA A   RESULT    ; Store
         WAI               ; Stop program
```

The count number of 10 is loaded into accumulator B. The index register gives the initial address of the data being added. The first summation step is to add the contents of the memory location addressed by the index register to the contents of the accumulator, initially assumed zero (a CLR A instruction could be used to clear it first). The instruction INX adds 1 to the index register so that the next address that will be addressed is 0021. DEC B subtracts 1 from the contents of accumulator B and so indicates that a count of 9 remains. BNE is then the instruction to branch to SUM if not equal to zero, i.e. if the Z flag has a 0 value. The program then loops and repeats the loop until ACC B is zero.

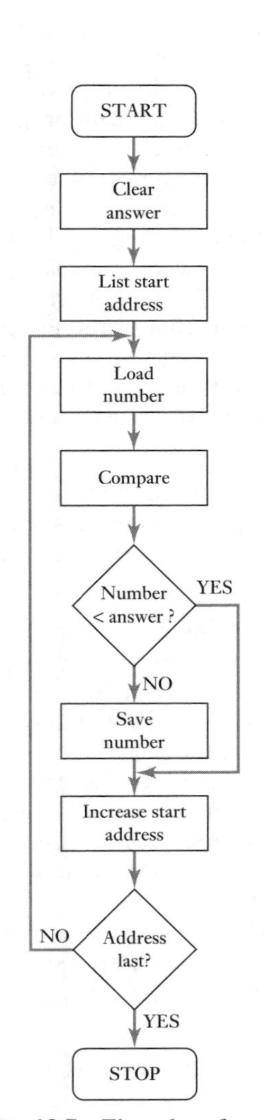

Figure 18.5 Flow chart for biggest number.

Problem: the determination of the biggest number in a list of numbers (it could be the determination of, say, the largest temperature value inputted from a number of temperature sensors).

The algorithm could be:

1 Clear answer address.
2 List starting address.
3 Load the number from the starting address.
4 Compare the number with the number in the answer address.
5 Store answer if bigger.
6 Otherwise save number.
7 Increase starting address by 1.
8 Branch to 3 if the address is not the last address.
9 Stop.

Figure 18.5 shows the flow chart. The program is:

```
FIRST    EQU     $0030
LAST     EQU     $0040
ANSW     EQU     $0041
         ORG     $0000
         CLR     ANSW      ; Clear answer
         LDX     FIRST     ; Load first address
NUM      LDA A   $30,X     ; Load number
         CMP A   ANSW      ; Compare with answer
         BLS     NEXT      ; Branch to NEXT if lower or same
         STA A   ANSW      ; Store answer
```

```
NEXT    INX             ; Increment index register
        CPX     LAST    ; Compare index register with LAST
        BNE     NUM     ; Branch if not equal to zero
        WAI             ; Stop program
```

The procedure is that first the answer address is cleared. The first address is then loaded and the number in that address put into accumulator A. LDA A $30,X means load accumulator A with data at address given by the index register plus 30. Compare the number with the answer, keeping the number if it is greater than the number already in the accumulator, otherwise branch to repeat the loop with the next number.

18.4 Subroutines

It is often the case that a block of programming, a subroutine, might be required a number of times in a program. For example, a block of programming might be needed to produce a time delay. It would be possible to duplicate the subroutine program a number of times in the main program. This, however, is an inefficient use of memory. Alternatively we could have a single copy of it in memory and branch or jump to it every time the subroutine was required. This, however, presents the problem of knowing, after completion of the subroutine, the point at which to resume in the main program. What is required is a mechanism for getting back to the main program and continuing with it from the point at which it was left to carry out the subroutine. To do this we need to store the contents of the program counter at the time of branching to the subroutine so that this value can be reloaded into the program counter when the subroutine is complete. The two instructions which are provided with most microprocessors to enable a subroutine to be implemented in this way are:

1 JSR (jump to routine), or CALL, which enables a subroutine to be called.
2 RTS (return from subroutine), or RET (return), which is used as the last instruction in a subroutine and returns it to the correct point in the calling program.

Subroutines may be called from many different points in a program. It is thus necessary to store the program counter contents in such a way that we have a last-in–first-out store (LIFO). Such a register is referred to as a **stack**. It is like a stack of plates in that the last plate is always added to the top of the pile of plates and the first plate that is removed from the stack is always the top plate and hence the last plate that was added to the stack. The stack may be a block of registers within a microprocessor or, more commonly, using a section of RAM. A special register within the microprocessor, called the stack pointer register, is then used to point to the next free address in the area of RAM being used for the stack.

In addition to the automatic use of the stack when subroutines are used, a programmer can write a program which involves the use of the stack for the temporary storage of data. The two instructions that are likely to be involved are:

1 PUSH, which causes data in specified registers to be saved to the next free location in the stack.
2 PULL, or POP, which causes data to be retrieved from the last used location in the stack and transferred to a specified register.

For example, prior to some subroutine, data in some registers may have to be saved and then, after the subroutine, the data restored. The program elements might thus be, with the Motorola 6800:

```
SAVE        PSH A   ; Save accumulator A to stack
            PSH B   ; Save accumulator B to stack
            TPA     ; Transfer status register to accumulator A
            PSH A   ; Save status register to stack
; Subroutine
RESTORE     PUL A   ; Restore condition code from stack to accumulator A
            TAP     ; Restore condition code from A to status register
            PUL B   ; Restore accumulator B from stack
            PUL A   ; Restore accumulator A from stack
```

18.4.1 Delay subroutine

Delay loops are often required when the microprocessor has an input from a device such as an analogue-to-digital converter. The requirement is often to signal to the converter to begin its conversion and then wait a fixed time before reading the data from the converter. This can be done by providing a loop which makes the microprocessor carry out a number of instructions before proceeding with the rest of the program. A simple delay program might be:

```
DELAY    LDA A    #$05     ; Load accumulator A with 05
LOOP     DEC A             ; Decrement accumulator A by 1
         BNE      LOOP     ; Branch if not equal to zero
         RTS               ; Return from subroutine
```

For each movement through the loop a number of machine cycles are involved. The delay program, when going through the loop five times, thus takes:

Instruction	Cycles	Total cycles
LDA A	2	2
DEC A	2	10
BNE	4	20
RTS	1	1

The total delay is thus 33 machine cycles. If each machine cycle takes, say, 1 μs then the total delay is 33 μs. For a longer delay a bigger number can be initially put into accumulator A.

An example of a time-delay subroutine for a PIC microcontroller is:

```
         movlw    Value    ; load count value required
         movwf    Count    ; loop counter
Delay    decfsz   Count    ; decrement counter
         goto     Delay    ; loop
```

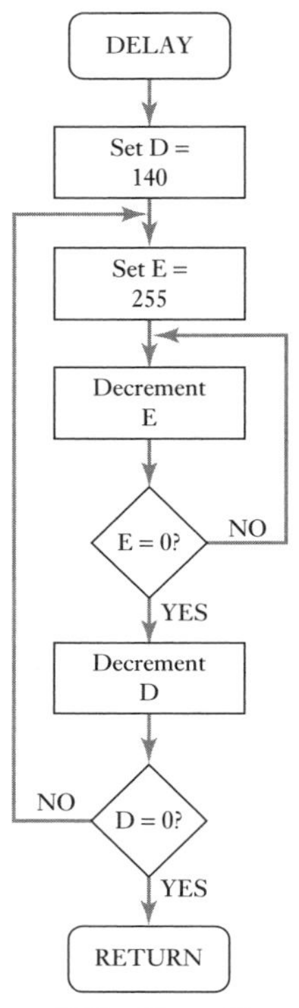

Figure 18.6 Nested loop delay.

The decfsz instruction takes one cycle and the goto instruction takes two cycles. This loop will be repeated (count − 1) times. In addition we have the movlw and movwf instructions, each taking one cycle, and when the count equals 1 we have decfsz which gives a further two cycles. Thus the total number of cycles is:

$$\text{number of instruction cycles} = 3(\text{count} - 1) + 4$$

Each instruction cycle takes four clock cycles and so the number of delay cycles introduced by this subroutine is:

$$\text{number of clock cycles} = 4[2(\text{count} - 1) + 4]$$

With a 4 MHz clock each clock cycle takes $1/(4 \times 10^6)$ s.

Often the delay obtained by using just the single loop described above is not enough. One way to obtain a longer delay is to use a nested loop. Figure 18.6 shows the flow chart for a nested loop delay. The inner loop is the same as the single-loop program described earlier. It will decrement register E 255 times before the looping is completed and the zero flag set. The outer loop causes the inner-loop routine to be repeatedly executed as register D is decremented down to zero. Thus with register D initially set with a loop count of, say, 140 then the time delay will be $140 \times 2.298 = 321.71$ ms.

The program is thus:

```
DELAY    MOV    D,8CH     ; set D to 8CH, i.e. 140
OLOOP    MOV    E,FFH     ; set E to FFH, i.e. 255
ILOOP    DEC    E         ; decrement E, i.e. inner-loop counter
         JNZ    ILOOP     ; repeat ILOOP 255 times
         DEC    D         ; decrement D, i.e. outer-loop counter
         JNZ    OLOOP     ; repeat OLOOP 140 times
```

The following are some examples of programs where time-delay subroutines are involved:

1 *Problem*: Switch an LED on and off repeatedly.
With this problem a subroutine DELAY is used with loops to provide time delays; the microprocessor takes a finite amount of time to process the instructions in a loop and thus to complete the loop. The structure of the program is:

```
1  If LED on
       Turn LED off
       While LED off, do subroutine TIME_DELAY
2  ELSE
       Turn LED on
       Do subroutine TIME_DELAY

Subroutine: TIME_DELAY
   Do an instruction, or instructions, or a loop, or a double loop depending
   on the length of time delay required.
```

Because of the length of time delay required, a double loop is likely to be used for the time delay. With Intel 8051 programming, it is possible to use the instruction DJNZ, decrement and jump if the result is not zero. It decrements the location indicated by the first operand and jumps to the second operand if the resulting value is not zero. The LED is connected to bit 0 of port 1 of the microcontroller. The program with Intel 8051 assembly instructions might thus be:

```
FLAG      EQU    0FH              ; flag set when LED is on
          ORG    8000H
START     JB     FLAG,LED_OFF     ; jump if LED_OFF bit set, i.e. LED is on
          SETB   FLAG             ; else set FLAG bit
          CLR    P1.0             ; turn LED on
          LCALL  DELAY            ; call up delay subroutine
          SJMP   START            ; jump to START
LED_OFF   CLR    FLAG             ; clear the LED on flag to indicate the
                                    LED is off
          SETB   P1.0             ; turn LED off
          LCALL  DELAY            ; call up delay subroutine
          LJMP   START            ; jump to START
DELAY     MOV    R0,#0FFH         ; outer-loop delay value
ILOOP     MOV    R1,#0FFH         ; inner-loop delay value
OLOOP     DJNZ   R1,ILOOP         ; wait through inner loop
          DJNZ   R0,OLOOP         ; wait through outer loop
          RET                     ; return from subroutine
          END
```

2 *Problem*: Switch on in sequence eight LEDS.
The rotate instruction can be used successively to turn on LEDs so that we have initially the bit pattern 0000 0001 which is then rotated to give 0000 0011, then 0000 0111, and so on. The following is a program in Motorola 68HC11 assembly language that can be used, the LEDs being connected to port B; a short delay is incorporated in the program:

```
COUNT    EQU    8           ; the count gives the number of loops
                            ; required, i.e. the number of
                            ; bits to be switched on
FIRST    EQU    %00000001   ; turn on 0 bit
PORTB    EQU    $1004       ; address of port B
         ORG    $C000
         LDAA   #FIRST      ; load initial value
         LDAB   #COUNT      ; load count
LOOP     STAA   PORTB       ; turn on bit 1 and so LED 1
         JSR    DELAY       ; jump to delay subroutine
         SEC                ; set carry bit to rotate into least significant
                            ; bit to maintain bit as 1
         ROLA               ; rotate left
         DECB               ; decrement count
         BNE    LOOP        ; branch to loop eight times
DELAY    RTS                ; simple short delay
         END
```

18.5 Look-up tables

Indexed addressing can be used to enable a program to look up values in a table. For example, in determining the squares of integers, a possible method is to look up the value corresponding to a particular integer in a table of squares, instead of doing the arithmetic to determine the square. Look-up tables are particularly useful when the relationship is non-linear and not described by a simple arithmetic equation, e.g. the engine management system described in Section 1.7.2 where ignition timing settings are a function of the angle of the crankshaft and the inlet manifold pressure. Here the microcontroller has to give a timing signal that depends on the input signals from the speed sensor and crankshaft sensors.

To illustrate how look-up tables are used, consider the problem of determining the squares of integers. We can place a table of squares of the integers 0, 1, 2, 3, 4, 5, 6, . . . in program memory and have the entries of the squares 0, 1, 4, 9, 16, 25, 36, . . . at successive addresses. If the number to be squared is 4 then this becomes the index for the index address of the data in the table, the first entry being index 0. The program adds the index to the base address of the table to find the address of the entry corresponding to the integer. Thus we have:

Index	0	1	2	3	4	5	6
Table entry	0	1	4	9	16	25	36

For example, with the Motorola 68HC11 microcontroller we might have the following look-up program to determine the squares:

```
REGBAS  EQU   $B600      ; base address for the table
        ORG   $E000
        LDAB  $20        ; load acc. B with the integer to be squared
        LDX   #REGBAS    ; point to table
        ABX              ; add contents of acc. B to index register X
        LDAA  $00,X      ; load acc. A with the indexed value
```

and we could have loaded the table into memory by using the pseudo-operation FDB:

```
ORG    $B600
FDB    $00,$01,$04,$09    ; giving values to the reserved memory
                          ; block
```

With the Intel 8051 microcontroller the instruction MOVC A,@A+DPTR fetches data from the memory location pointed to by the sum of DPTR and the accumulator A and stores it in the accumulator. This instruction can be used to look up data in a table where the data pointer DPTR is initialised to the beginning of the table. As an illustration, suppose we want to use a table for the conversion of temperatures on the Celsius scale to the Fahrenheit scale. The program involves parameter passing of the temperature requiring conversion to a subroutine, so it might include the following instructions:

```
         MOV    A,#NUM          ; load the value to be converted
         CALL   LOOK_UP         ; call the LOOK_UP subroutine

LOOK_UP  MOV    DPTR,#TEMP      ; point to table
         MOVC   A,@A+DPTR       ; get the value from the table
         RET                    ; return from the subroutine
TMP      DB     32, 34, 36, 37, ; giving values to the table
                39, 41, 43, 45
```

Another example of the use of a table is to sequence a number of outputs. This might be a sequence to operate traffic lights to give the sequence red, red plus amber, green, amber. The red light is illuminated when there is an output from RD0, the amber is illuminated from RD1, and the green is illuminated from RD2. The data table might then be:

	Red	Red + amber	Green	Amber
Index	0	1	2	3
	0000 0001	0000 0011	0000 0100	0000 0010

18.5.1 Delay with a stepper motor

In operating a stepper motor delays have to be used between each instruction to advance by a step to allow time for that step to occur before the next program instruction. A program to generate a continuous sequence of step pulses could thus have the algorithm:

1 Start.
2 State sequence of outputs needed to obtain the required step sequence.
3 Set to initial step position.
4 Advance a step.
5 Jump to delay routine to give time for the step to be completed.
6 Is this the last step in the step sequence for one complete rotation? If not, continue to next step; if yes, loop back to step 3.
7 Continue until infinity.

The following is a possible program for a stepper in the full-step configuration and controlled by the microcontroller M68HC11 using outputs from PB0, PB1, PB2 and PB3. A 'look-up' table is used for the output code sequence necessary from the outputs to drive the stepper in the step sequence. The following is the table used.

The code sequence required for full-step stepper operation is thus A, 9, 5, 6, A and so these values constitute the sequence that the pointer has to look up in the table. FCB is the op-code for 'form constant byte' and is used to initialise data bytes for the table.

	The outputs required from Port B				
Step	**PB0**	**PB1**	**PB2**	**PB3**	**Code**
1	1	0	1	0	A
2	1	0	0	1	9
3	0	1	0	1	5
4	0	1	1	0	6
1	1	0	1	0	4

```
BASE     EQU   $1000
PORTB    EQU   $4                ; Output port
TFLG1    EQU   $23               ; Timer interrupt flag register 1
TCNT     EQU   $0E               ; Timer counter register
TOC2     EQU   $18               ; Output compare 2 register
TEN_MS   EQU   20000             ; 10 ms on clock
```

```
          ORG    $0000
STTBL     FCB    $A                     ; This is the look-up table
          FCB    $9
          FCB    $5
          FCB    $6
ENDTBL    FCB    $A                     ; End of look-up table

          ORG    $C000
          LDX    #BASE
          LDAA   #$80
          STAA   TFLG1,X                ; Clear flag
START     LDY    #STTBL
BEG       LDAA   0,Y                    ; Start with first position in table
          STAA   PORTB,X
          JSR    DELAY                  ; Jump to delay
          INY                           ; Increment in table
          CPY    #ENTBL                 ; Is it end of table?
          BNE    BEG                    ; If not branch to BEG
          BRA    START                  ; If yes, go to start again

DELAY     LDD    TCNT,X
          ADDD   #TEN_MS                ; Add a 10 ms time delay
          STD    TOC2,X
HERE      BRCLR  TFLG1, X, $80, HERE    ; Wait till time delay elapsed
          LDAA   #$80
          STAA   TFLG1,X                ; Clear flag
          RTS
```

Note that, in the above program, the label TEN_MS has the space underlined to indicate that both TEN and MS are part of the same label.

The delay in the above program is obtained by using the timer block in the microcontroller. A time delay of 10 ms is considered. For a microcontroller system with a 2 MHz clock a 10 ms delay is 20 000 clock cycles. Thus to obtain such a delay, the current value of the TCNT register is found and then 20 000 cycles added to it and the TOC2 register loaded with this value.

18.6 Embedded systems

Microprocessors and microcontrollers are often ‘embedded’ in systems so that control can be exercised. For example, a modern domestic washing machine has an embedded microcontroller which has been programmed with the different washing programs; all that the machine operator has to do is select the required washing program by means of a switch and the required program is implemented. The operator does not have to program the microcontroller. The term **embedded system** is used for a microprocessor-based system that is designed to control a function or range of functions and is not designed to be programmed by the system user. The programming has been done by the manufacturer and has been ‘burnt’ into the memory system and cannot be changed by the system user.

18.6.1 Embedding programs

In an embedded system the manufacturer makes a ROM containing the program. This is only economical if there is a need for a large number of these chips. Alternatively, for prototyping or low-volume applications, a program could be loaded into the EPROM/EEPROM of the application hardware. The following illustrates how the EPROM/EEPROM of microcontrollers can be programmed.

For example, to program the EPROM of the Intel 8051 microcontroller, the arrangement shown in Figure 18.7(a) is required. There must be a 4–6 MHz oscillator input. The procedure is:

1 The address of an EPROM location, to be programmed in the range 0000H to 0FFFH, is applied to port 1 and pins P2.0 and P2.1 of port 2; at the same time, the code byte to be programmed into that address is applied to port 0.
2 Pins P2.7, RST and ALE should be held high, pins P2.6 and PSEN low. For pins P2.4 and P2.5 it does not matter whether they are high or low.
3 Pin EA/V_{pp} is held at a logic high until just before ALE is to be pulsed, then it is raised to +21 V, ALE is pulsed low for 50 ms to program the code byte into the addressed location, and then EA is returned to a logic high.

Verification of the program, i.e. reading out of the program, is achieved by the arrangement shown in Figure 18.7(b).

1 The address of the program location to be read is applied to port 1 and pins P2.0 to P2.3 of port 2.
2 Pins EA/V_{pp}, RST and ALE should be held high, pins P2.7, P2.6 and PSEN low. For pins P2.4 and P2.5 it does not matter whether they are high or low.
3 The contents of the addressed location come out on port 0.

A security bit can be programmed to deny electrical access by any external means to the on-chip program memory. Once this bit has been programmed, it can only be cleared by the full erasure of the program memory. The same arrangement is used as for programming (Figure 18.7(a)) but P2.6 is held

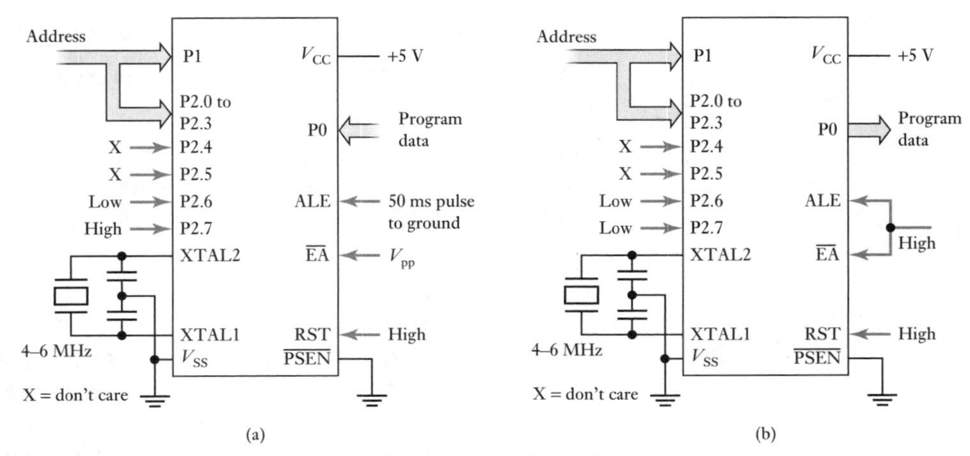

Figure 18.7 The Intel 8051: (a) programming, (b) verification.

high. Erasure is by exposure to ultraviolet light. Since sunlight and fluorescent lighting contain some ultraviolet light, prolonged exposure (about 1 week in sunlight or 3 years in room-level fluorescent lighting) should be avoided and the chip window should be shielded by an opaque label.

The Motorola 68HC11 microcontroller is available with an internal electrically erasable programmable read-only memory (EEPROM). The EEPROM is located at addresses \$B600 to \$B7FF. Like an EPROM, a byte is erased when all the bits are 1 and programming involves making particular bits 0. The EEPROM is enabled by setting the EEON bit in the CONFIG register (Figure 18.8) to 1 and disabled by setting it to 0. Programming is controlled by the EEPROM programming register (PPROG) (Figure 18.8).

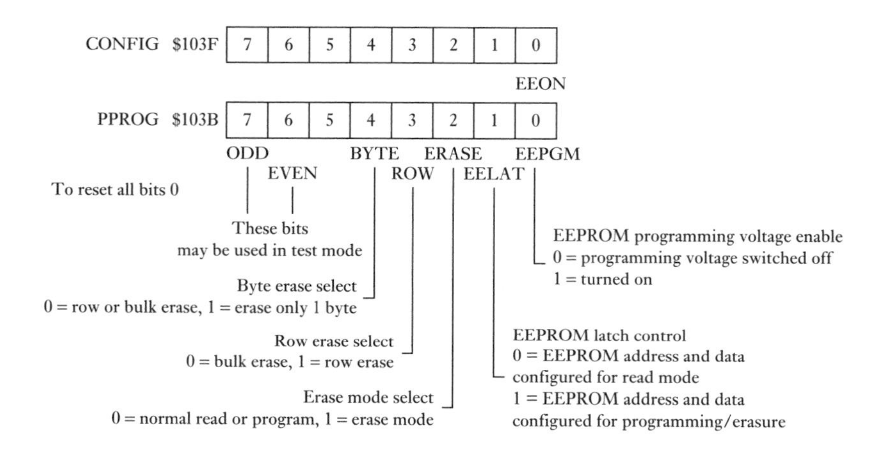

Figure 18.8 CONFIG and PPROG.

The procedure for programming is:

1 Write to the PPROG register to set the EELAT bit to 1 for programming.
2 Write data to the EEPROM address selected. This latches in the address and data to be programmed.
3 Write to the PPROG register to set the EEPGM bit to 1 to turn on the programming voltage.
4 Delay for 10 ms.
5 Write to the PPROG register to turn off, i.e. to 0, all the bits.

Here, in assembly language, is a programming subroutine for use with the MC68HC11:

```
EELAT    EQU    %00000010    ; EELAT bit
EEPGM    EQU    %00000001    ; EEPGM bit
PPROG    EQU    $1028        ; address of PPROG register

EEPROG
         PSHB
         LDAB   #EELAT
         STAB   PPROG        ; set EELAT = 1 and EEPGM = 0
         STAA   0,X          ; store data X to EEPROM address
         LDAB   #%00000011
         STAB   PPROG        ; set EELAT = 1 and EEPGM = 1
         JSR    DELAY_10     ; jump to delay 10 ms subroutine
```

```
            CLR     PPROG           ; clear all the PPROG bits and return
                                      to the read mode
            PULB
            RTS

; Subroutine for approximately 10 ms delay
DELAY_10
            PSHX
            LDX     #2500           ; count for 20 000 cycles
DELAY       DEX
            BNE     DELAY
            PULX
            RTS
```

The procedure for erasure is:

1 Write to the PPROG register to select for erasure of a byte, row or the entire EEPROM.
2 Write to an EEPROM address within the range to be erased.
3 Write a 1 to the PPROG register to turn on the EEPGM bit and hence the erase voltage.
4 Delay for 10 ms.
5 Write zeros to the PPROG register to turn off all the bits.

With the built-in EEPROM with a PIC microcontroller, a program to write data into it is (Figure 18.9):

```
        bcf     STATUS, RP0     ; Change to Bank 0 for the data
        mov.f   Data, w         ; Load data to be written
        movwf   EEDATA
        movf    Addr, w         ; Load address of write data
        movwf   EEADR
        bsf     STATUS, RP0     ; Change to Bank 1
        bcf     INTCON, GIE     ; Disable interrupts
        bsf     EECON1, WREN    ; Enable for writing
        movlw   55h             ; Special sequence to enable
                                ; writing
        movwf   EECON2
        movlw   0AAh
        movwf   EECON2
        bsf     EECON1, WR      ; Initiate write cycle
        bsf     INTCON, GIE     ; Re-enable interrupts
EE_EXIT btfsc   EECON, WR       ; Check that the write is completed
        goto    EE_EXIT         ; If not, retry
        bsf     EECON, WREN     ; EEPROM write is complete
```

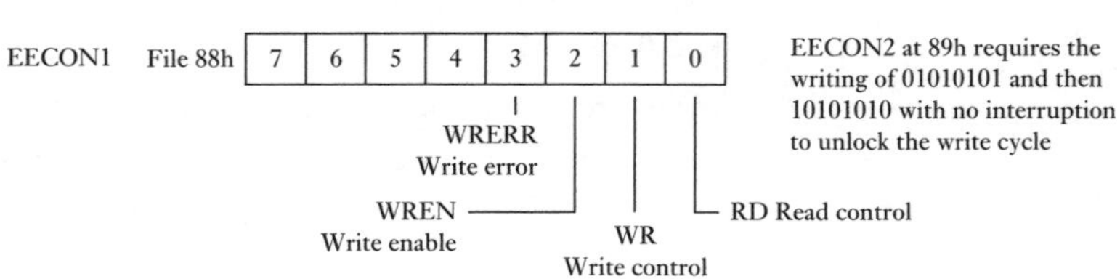

Figure 18.9 EECON registers.

Summary

The collection of instructions that a microprocessor will recognise is its **instruction set**. The series of instructions that is needed to carry out a particular task is called a **program**.

Microprocessors work in binary code. Instructions written in binary code are referred to as being in **machine code**. A shorthand code using simple, identifiable, terms rather than binary code is referred to as a **mnemonic code**, a mnemonic code being a 'memory-aiding' code. Such a code is termed **assembly language**. Assembly language programs consist of a sequence of statements, one per line, with each statement containing from one to four fields: label, op-code, operand and comment. The **label** is the name by which a particular entry in the memory is referred to. The **op-code** specifies how data is to be manipulated. The **operand** contains the address of the data to be operated on. The **comment** field is to allow the programmer to include comments which may make the program more understandable to the reader.

Problems

18.1 Using the following extract from a manufacturer's instruction set (6800), determine the machine codes required for the operation of adding with carry in (a) the immediate address mode, (b) the direct address mode.

		Addressing modes					
		IMMED			DIRECT		
Operation	Mnemonic	OP	~	#	OP	~	#
Add with carry	ADC A	89	2	2	99	3	2

18.2 The clear operation with the Motorola 6800 processor instruction set has an entry only in the implied addressing mode column. What is the significance of this?

18.3 What are mnemonics for, say, the Motorola 6800 for (a) clear register A, (b) store accumulator A, (c) load accumulator A, (d) compare accumulators, (e) load index register?

18.4 Write a line of assembler program for (a) load the accumulator with 20 (hex), (b) decrement the accumulator A, (c) clear the address $0020, (d) ADD to accumulator A the number at address $0020.

18.5 Explain the operations specified by the following instructions: (a) STA B $35, (b) LDA A #$F2, (c) CLC, (d) INC A, (e) CMP A #$C5, (f) CLR $2000, (g) JMP 05,X.

18.6 Write programs in assembly language to:

(a) Subtract a hexadecimal number in memory address 0050 from a hexadecimal number in memory location 0060 and store the result in location 0070.

(b) Multiply two 8-bit numbers, located at addresses 0020 and 0021, and store the product, an 8-bit number, in location 0022.

(c) Store the hexadecimal numbers 0 to 10 in memory locations starting at 0020.

(d) Move the block of 32 numbers starting at address $2000 to a new start address of $3000.

18.7 Write, in assembly language, a subroutine that can be used to produce a time delay and which can be set to any value.

18.8 Write, in assembly language, a routine that can be used so that if the input from a sensor to address 2000 is high the program jumps to one routine starting at address 3000, and if low the program continues.

Chapter nineteen C Language

Objectives

The objectives of this chapter are that, after studying it, the reader should be able to:

- **Comprehend the main features of C programs.**
- **Use C to write simple programs for microcontrollers.**

19.1 Why C?

This chapter is intended to give an introduction to the C language and the writing of programs. C is a high-level language that is often used in place of assembly language (see Chapter 18) for the programming of microprocessors. It has the advantages when compared with the assembly language of being easier to use and that the same program can be used with different microprocessors; all that is necessary for this is that the appropriate compiler is used to translate the C program into the relevant machine language for the microprocessor concerned. Assembly language is different for the different microprocessors while C language is standardised, the standard being that of the American National Standards Institute (ANSI).

19.2 Program structure

Figure 19.1 gives an overview of the main elements of a C program. There is a pre-processor command that calls up a standard file, followed by the main function. Within this main function there are other functions which are called up as subroutines. Each function contains a number of statements.

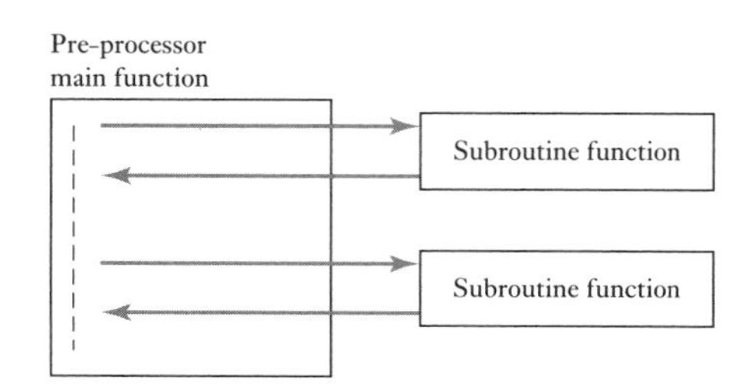

Figure 19.1 Structure of a C program.

19.2.1 Key features

The following are key features of programs written in C language. Note that in C programs spaces and carriage returns are ignored by the compiler and purely used for the convenience of the programmer to make it easier to read the program.

1 *Keywords*
In C certain words are reserved as keywords with specific meanings. For example, *int* is used to indicate that integer values are concerned; *if* is used for when a program can change direction based on whether a decision is true or false. C requires that all keywords are in lower case letters. Such words should not be used for any other purpose in a C program. The following are the ANSI C standard keywords:

auto	double	int	struct
break	else	long	switch
case	enum	register	typedef
char	extern	return	union
const	float	short	unsigned
continue	for	signed	void
default	goto	sizeof	volatile
do	if	static	while

2 *Statements*
These are the entries which make up a program, every statement being terminated by a semicolon. Statements can be grouped together in blocks by putting them between braces, i.e. { }. Thus for a two-statement group we have:

```
{
    statement 1;
    statement 2;
}
```

3 *Functions*
The term **function** is used for a self-contained block of program code which performs a specific set of actions and has a name by which it can be referred (it is like a subroutine in an assembly language program). A function is written as a name followed by brackets, i.e. name( ). The brackets enclose arguments; a function's argument is a value that is passed to the function when the function is called. A function is executed by calling it up by its name in the program statement. For example, we might have the statement:

```
printf("Mechatronics");
```

This would mean that the word Mechatronics is passed to the function printf( ), a pre-written function which is called up by the pre-processor command, and, as a result, the word is displayed on screen. In order to indicate that characters form a sequence, e.g. those making up the word Mechatronics, they are enclosed within double quotes.

4 *Return*

A function may return a value to the calling routine. The **return type** appears in front of the function name, this specifying the type of value to be returned to the calling function when execution of the function is completed. For example, int main( ) is used for an integer return from the main function. The return type may be specified as void if the function does not return a value, e.g. void main(void). Often a header file will contain this return information and so it will not need to be specified for functions defined by the header file.

To return a value from a function back to the calling point, the keyword return is used, e.g. to return the result:

```
return result;
```

The return statement terminates the function.

5 *Standard library functions*

C packages are supplied with libraries containing a large number of predefined functions containing C code that have already been written and so saving you the effort of having to write them. These can be called up by naming them. In order to use the contents of any particular library, that library has to be specified in a header file. Examples of such library files are:

math.h for mathematical functions
stdio.h for input and output functions
time.h for date and time functions

For example, the function printf( ) is a function that can be called up from the stdio.h library and is the function for printing to the screen of the monitor. Another function is scanf( ) which can be used to read data from a keyboard.

6 *Pre-processor*

The **pre-processor** is a program that is identified by **pre-processor commands** so that it is executed prior to the compilation. All such commands are identified by having # at the beginning of the line. Thus we might have:

```
# include < >
```

to include the file named between the angle brackets < >. When this command is reached, the specified file will be inserted into the program. It is frequently used to add the contents of standard header files, these giving a number of declarations and definitions to enable standard library functions to be used. The entry would then be:

```
# include <stdio.h>
```

As an illustration, consider the simple program:

```
# include <stdio.h>
main( )
{
  printf("Mechatronics");
}
```

Before starting the main program the file stdio.h is added. Thus when the main program starts we are able to access the function printf( ) which results in the word Mechatronics being displayed on the screen.

Another type of pre-processor command is:

```
# define pi 3.14
```

and this can be used to define values that will be inserted whenever a particular symbol is encountered in the program. Thus whenever pi is encountered the value 3.14 will be used.

```
# define square(x) (x)*(x)
```

will replace the term square in the program by (x)*(x).

7 *Main function*
Every C program must have a function called main( ). This function is the one that exercises control when the program is executed and is the first function to be called up. Execution starts with its first statement. Other functions may be called up within statements, each one in turn being executed and control returned to the main function. The statement:

```
void main(void)
```

indicates that no result is to be returned to the main program and there is no argument. By convention a return value of 0 from main( ) is used to indicate normal program termination, i.e. the entry:

```
return 0;
```

8 *Comments*
/* and */ are used to enclose comments. Thus we might have an entry such as:

```
/* Main program follows */
```

Comments are ignored by the compiler and are just used to enable a programmer more easily to comprehend the program. Comments can span more than one line, e.g.

```
/* An example of a program used to
illustrate programming */
```

9 *Variables*
A **variable** is a named memory location that can hold various values. Variables that can hold a character are specified using the keyword *char*, such a variable being 8 bits long and generally used to store a single character. Signed integers, i.e. numbers with no fractional parts and which are signed to indicate positive or negative, are specified using the keyword *int*. The keyword *float* is used for floating-point numbers, these being numbers which have a fractional part. The keyword *double* is also used for floating-point numbers but provides about twice the number of

significant digits as *float*. To declare a variable the type is inserted before the variable name, e.g.:

```
int counter;
```

This declares the variable 'counter' to be of the integer type. As another example we might have:

```
float x, y;
```

This indicates that the variables x and y are both floating-point numbers.

10 *Assignments*
An assignment statement assigns the value of the expression to the right of the = sign to the variable on its left. For example, $a = 2$ assigns the value 2 to the variable a.

11 *Arithmetic operators*
The arithmetic operators used are: addition +, subtraction −, multiplication *, division /, modulus %, increment + + and decrement− −. Increment operators increase the value of a variable by 1, decrement operators decrease it by 1. The normal rules of arithmetic hold for the precedence of operations. For example, 2*4 + 6/2 gives 11. An example of a program involving arithmetic operators is:

```
/*program to determine area of a circle*/

#include <stdio.h> /*identifies the function library*/

int radius, area /*variables radius and area are integers*/

int main(void) /*starts main program, the int specifies
    that an integer value is returned, the void indicates
    that main( ) has no parameters*/
{
    printf("Enter radius:"); /*"Enter radius" on screen*/
    scanf("%d", &radius); /*Reads an integer from
    keyboard and assigns it to the variable radius*/
    area = 3.14 * radius * radius; /*Calculates area*/
    printf("\nArea = %d", area); /*On new line prints Area
    = and puts in numerical value of the area*/
    return 0; /*returns to the calling point*/
}
```

12 *Relational operators*
Relational operators are used to compare expressions, asking questions such as 'Is x equal to y?' or 'Is x greater than 10?' The relational operators are: is equal to ==, is not equal to !=, less than < , less than or equal to < =, greater than > , greater than or equal to > =. Note that == has to be used when asking if two variables are the same, = is used for assignment when you are stating that they are the same. For example, we might have the question 'Is x equal to 2?' and represent this by (a == 2).

13 *Logical operators*
The logical operators are:

<table>
<tr><th>Operator</th><th>Symbol</th></tr>
<tr><td>AND</td><td>&&</td></tr>
<tr><td>OR</td><td>||</td></tr>
<tr><td>NOT</td><td>!</td></tr>
</table>

Note that in C the outcome is equal to 1 if true and 0 if false.

14 *Bitwise operations*
The bitwise operators treat their operands as a series of individual bits rather than a numerical value, comparing corresponding bits in each operand, and only work with integer variables. The operators are:

<table>
<tr><th>Bitwise operation</th><th>Symbol</th></tr>
<tr><td>AND</td><td>&</td></tr>
<tr><td>OR</td><td>|</td></tr>
<tr><td>EXCLUSIVE-OR</td><td>^</td></tr>
<tr><td>NOT</td><td>~</td></tr>
<tr><td>Shift right</td><td>≫</td></tr>
<tr><td>Shift left</td><td>≪</td></tr>
</table>

Thus, for example, we might have the statement:

```
portA = portA | 0x0c;
```

The prefix 0x is used to indicate that the 0c is a hex value, being 0000 1100 in binary. The value ORed with port A is thus a binary number that forces bits 2 and 3 on, all the other bits remaining unchanged.

```
portA = portA ^ 1;
```

This statement causes all the bits except for bit 1 of port A to remain unchanged. If bit 0 is 1 in port A the XOR will force it to 0 and if it is 0 it will force it to 1.

15 *String*
A sequence of characters enclosed within double quotes, i.e. " ", is termed a string. As the term implies, the characters within the double quotes are treated as a linked entity. For example, we might have:

```
printf("Sum = %d", x)
```

The argument in ( ) specifies what is passed to the print function. There are two arguments, the two being separated by a comma. The first argument is the string between the double quotes and specifies how the output is to be presented, the %d specifying that the variable is to be displayed as a decimal integer. Other format specifiers are:

%c	character
%d	signed decimal integer
%e	scientific notation
%f	decimal floating point
%o	unsigned octal
%s	string of characters
%u	unsigned decimal integer
%x	unsigned hexadecimal
%%	prints a % sign

The other argument x specifies the value that is to be displayed.

As another example, the statement:

```
scanf("%d", &x);
```

reads a decimal integer from the keyboard and assigns it to the integer variable x. The & symbol in front of x is the 'address of' operator. When placed before the name of a variable, it returns the address of the variable. The command thus scans for data and stores the item using the address given.

16 *Escape sequences*
Escape sequences are characters that 'escape' from the standard interpretation of characters and are used to control the location of output on a display by moving the screen cursor or indicating special treatments. Thus we might have:

```
printf("\nSum = %d", d)
```

with the \n indicating that a new line is to be used when it is printed on the screen. Escape sequences commonly used are:

\a	sound a beep
\b	backspace
\n	new line
\t	horizontal tab
\\	backslash
\?	question mark
\'	single quotation

19.2.2 Example of a C program

An example of a simple program to illustrate the use of some of the above terms is:

```
/*A simple program in C*/

# include <stdio.h>
void main(void)
{
```

```
    int a, b, c, d; /*a, b, c and d are integers*/
    a = 4; /*a is assigned the value 4*/
    b = 3; /*b is assigned the value 3*/
    c = 5; /*c is assigned the value 5*/
    d = a * b * c; /*d is assigned the value of a * b * c*/
    printf("a * b * c = %d\n", d);
}
```

The statement int a, b, c, d; declares the variables a, b, c and d to be integer types. The statements a = 4, b = 3, c = 5 assign initial values to the variables, the = sign being used to indicate assignment. The statement d = a * b * c directs that a is to be multiplied by b and then by c and stored as d. The printf in the statement printf("a * b * c = %d\n", d) is the display on screen function. The argument contains %d and this indicates that it is to be converted to a decimal value for display. Thus it will print a * b * c = 60. The character \n at the end of the string is to indicate that a new line is to be inserted at that point.

19.3 Branches and loops

Statements to enable branching and looping in programs include *if*, *if/else*, *for*, *while* and *switch*.

1 *If*

The *if* statement allows branching (Figure 19.2(a)). For example, if an expression is true then the statement is executed, if not true it is not, and the program proceeds to the next statement. Thus we might have statements of the form:

```
if (condition 1 = = condition 2);
printf ("\nCondition is OK.");
```

Figure 19.2 (a) If, (b) if/else.

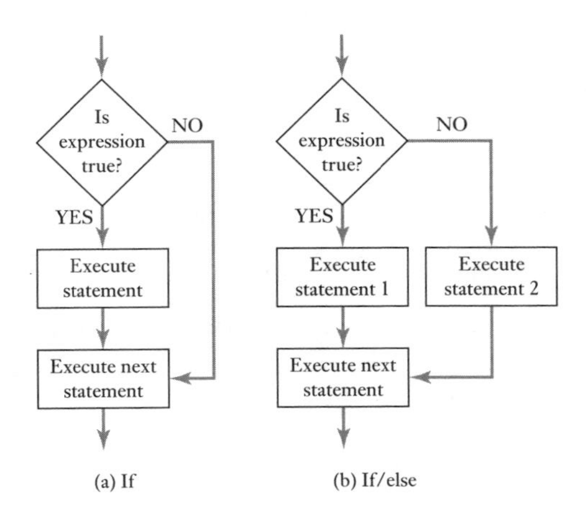

An example of a program involving if statements is:

```
#include <studio.h>

int x, y;
main( )
```

```
{
    printf("\nInput an integer value for x: ");
    scanf("%d", &x);
    printf("\nInput an integer value for y: ");
    scanf(%d", &y);
    if( x = = y)
      printf("x is equal to y");
    if(x > y)
      printf("x is greater than y");
    if(x < y)
      printf("x is less than y");
      return 0;
}
```

The screen shows Input an integer value for x: and then a value is to be keyed in. The screen then shows Input a value for y: and then a value is to be keyed in. The if sequence then determines whether the keyed-in values are equal or which is greater than the other and displays the result on the screen.

2 *If/else*
The *if* statement can be combined with the *else* statement. This allows one statement to be executed if the result is yes and another if it is no (Figure 19.2(b)). Thus we might have:

```
#include <studio.h>

main( )
{
  int temp;
  if(temp > 50)
    printf("Warning");
  else
   printf("System OK");
}
```

3 *For*
The term *loop* is used for the execution of a sequence of statements until a particular condition reaches the required condition of being true, or false. Figure 19.3(a) illustrates this. One way of writing statements for a loop is to use the function *for*. The general form of the statement is:

```
for(initialising expression; test expression; increment  expression)
loop statement;
```

Thus we might have:

```
#include <studio.h>

int count

main( )
{
  for(count = 0; count < 7; count ++)
  printf("\n%d", count);
}
```

Figure 19.3 (a) For, (b) while.

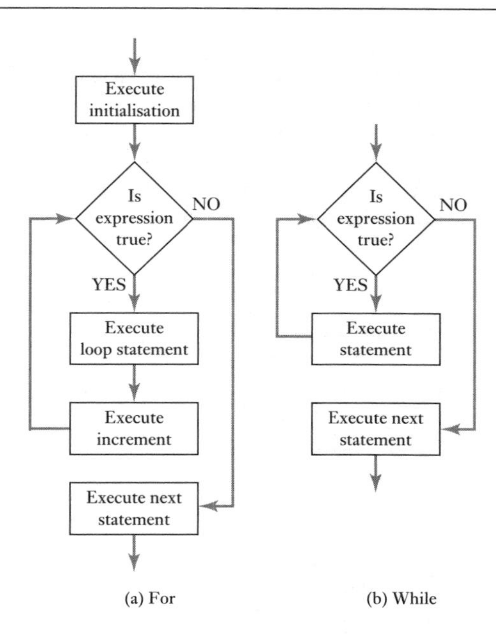

Initially the count is 0 and will be incremented by 1 and then looped to repeat the for statement as long as the count is less than 7. The result is that the screen shows 0 1 2 3 4 5 6 with each number being on a separate line.

4 *While*
The *while* statement allows for a loop to be continually repeated as long as the expression is true (Figure 19.3(b)). When the expression becomes false then the program continues with the statement following the loop. As an illustration we could have the following program where the while statement is used to count as long as the number is less than 7, displaying the results:

```
#include <studio.h>

int count;
int main( );
{
   count = 1;
   while(count < 7)
      {
         printf("\n%d", count);
         count ++;
      }
      return 0;
}
```

The display on the screen is 1 2 3 4 5 6 with each number on a separate line.

5 *Switch*
The *switch* statement allows for the selection between several alternatives, the test condition being in brackets. The possible choices are identified by

case labels, these identifying the expected values of the test condition. For example, we might have the situation where if case 1 occurs we execute statement 1, if case 2 occurs we execute statement 2, etc. If the expression is not equal to any of the cases then the default statement is executed. After each case statement there is normally a break statement to transfer execution to the statement after the switch and stop the switch continuing down the list of cases. The sequence is thus (Figure 19.4):

```
switch(expression)
{
  case 1;
    statement 1;
  break
  case 2;
    statement 2;
    break;
  case 3;
    statement 3;
    break;
  default;
    default statement;
}
next statement
```

Figure 19.4 Switch.

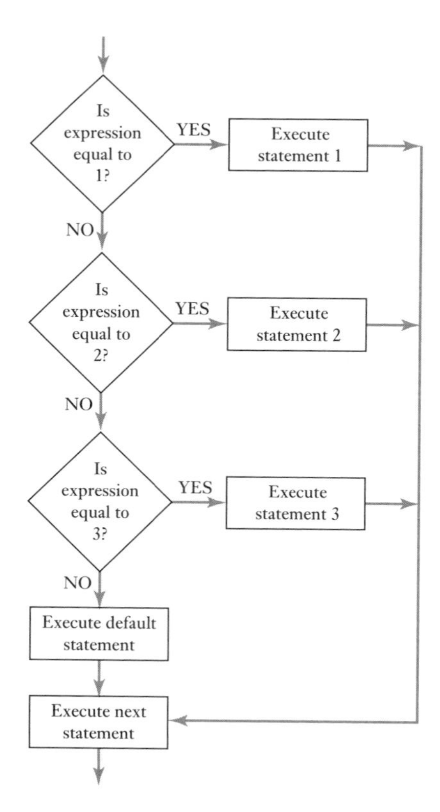

The following is an example of a program which recognises the numbers 1, 2 and 3 and will display whichever one is entered from the keyboard:

```
#include <stdio.h>

int main ( );
{
  int x;

  printf("Enter a number 0, 1, 2 or 3:  ");
  scanf("%d", &x);

  switch (x)
  {
    case 1:
        printf("One");
        break;
    case 2:
        printf("Two");
        break;
    case 3:
        printf("Three");
        break;
    default;
        printf("Not 1, 2 or 3");
  }
  return 0;
}
```

19.4 Arrays

Suppose we want to record the mid-day temperature for each day for a week and then later be able to find the temperature corresponding to any one particular day. This can be done using an array. An **array** is a collection of data storage locations with each having the same data type and referenced through the same name. To declare an array with the name Temperature to store values of type float we use the statement:

```
float Temperature[7];
```

The size of the array is indicated between square brackets [ ] immediately after the array name. In this case 7 has been used for the data for the seven days of the week. Individual elements in an array are referenced by an index value. The first element has the number 0, the second 1, and so on to the last element in an n sequence which will be $n - 1$. Figure 19.5(a) shows the form of a sequential array. To store values in the array we can write:

```
temperature[0] = 22.1;
temperature [1] = 20.4;
etc.
```

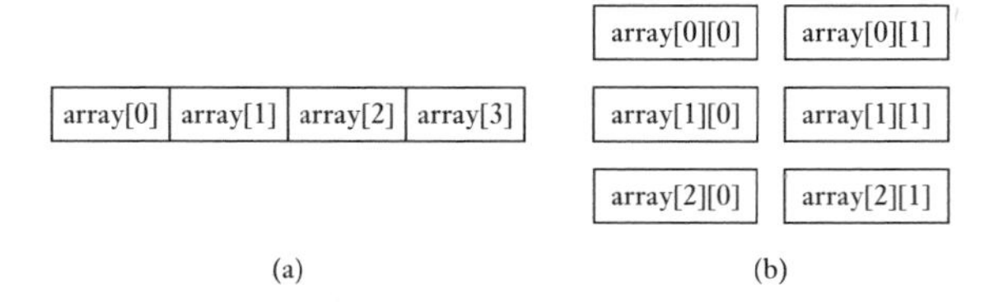

Figure 19.5 (a) A four-element sequential array, (b) a two-dimensional array.

If you want to use scanf( ) to input a value into an array element, put & in front of the array name, e.g.:

```
scanf("%d", &temperature [3]);
```

The following is an example of a simple program to store and display the squares of the numbers 0, 1, 2, 3 and 4:

```
#include <stdio.h>

int main(void)
{
  int sqrs[5];
  int x;

  for(x = 1; x<5; x++)
    sqrs[x – 1] = x * x;
  for(x = 0; x < 4; x++)
    printf("%d", sqrs[x]);

  return 0;
}
```

Arrays can be given initial values when first declared, e.g.:

```
int array[7] = {10, 12, 15, 11, 10, 14, 12};
```

If you omit the array size, the compiler will create an array just large enough to hold the initialisation values:

```
int array[ ] = {10, 12, 15, 11, 10, 14, 12};
```

Multidimensional arrays can be used. For example, a table of data is a two-dimensional array (Figure 19.5(b)), where x represents the row and y the column, and is written as:

```
array[x][y];
```

19.5 Pointers

Each memory location has a unique address and this provides the means by which data stored at a location can be accessed. A **pointer** is a special kind of variable that can store the address of another variable. Thus if a variable called p contains the address of another variable called x, then p is said to **point** to x. Thus if x is at the address 100 in the memory then p would have the value 100. A pointer is a variable and, as with all other variables, has to be declared before it can be used. A pointer declaration takes the form:

```
type *name;
```

The * indicates that the name refers to a pointer. Often names used for pointers are written with the prefix p, i.e. in the form pname. Thus we might have:

```
int *pnumber;
```

To initialise a pointer and give it an address to point to, we can use &, which is the address-of operator, in a statement of the form:

```
pointer = &variable;
```

The following short program illustrates the above:

```
#include <stdio.h>

int main(void)
{
  int *p, x;
  x = 12;
  p = &x;   /*assigns p the address of x*/
  printf("%d", *p);   /*displays the value of x using pointer*/
  return 0;
{
```

The program thus displays the number 12 on the screen. Accessing the contents of a variable by using a pointer, as above, is called **indirect access**. The process of accessing the data in the variable pointed to by a pointer is termed **dereferencing** the pointer.

19.5.1 Pointer arithmetic

Pointer variables can have the arithmetic operators +, −, + + and − − applied to them. Incrementing or decrementing a pointer results in its pointing to the next or previous element in an array. Thus to increment a pointer to the next item in an array we can use:

```
pa++; /*using the increment by 1 operator*/
```

or:

```
pa = pa + 1; /*adding 1*/
```

19.5.2 Pointers and arrays

Pointers can be used to access individual elements in an array, the following program showing such access:

```
#include <stdio.h>

int main(void)
{
```

```
    int x[5] = (0, 2, 4, 6, 8);
    int *p;
    p = x; /*assigns to p the address of the start of x*/
    printf("%d %d", x[0], x[2]);

    return 0;
}
```

The statement printf("%d %d", x[0], x[2]); results in pointing to the address given by x and hence the values at the addresses [0] and [2] being displayed, i.e. 0 and 4, on separate lines.

19.6 Program development

In developing programs the aim is to end up with a set of machine language instructions which can be used to operate a microprocessor/microcontroller system. These instructions are called the **executable file**. In order to arrive at such a file the following sequence of events occurs:

1. *Creating the source code*
   This is the writing of the sequence of statements in C that will constitute the program. Many compilers come with an editor and so the programmer can simply type in the source code from the keyboard. Otherwise a program such as Notepad with Microsoft Windows can be used. Using a word processor can present problems in that additional formatting information is included which can prevent compilation unless the file is saved without formatting information.
2. *Compiling the source code*
   Once the source code has been written, the programmer can direct the compiler to translate it into machine code. Before the compilation process starts, all the pre-processor commands are executed. The compiler can detect several different forms of error during the translation and generate messages indicating the errors. Sometimes a single error may result in the cascading sequence of errors all following from that single first error. Errors usually involve going back to the editor stage and re-editing the source code. The compiler then stores the resulting machine code in another file.
3. *Linking to create an executable file*
   Then the compiler is used to bring together, i.e. link, the generated code with library functions to give a single executable file. The program is then stored as the executable file.

19.6.1 Header files

Pre-processor commands are used at the beginning of a program to define the functions used in that program; this is so they can be referred to by simple labels. However, to save having to write long lists of standard functions for each program, a pre-processor instruction can be used to indicate that a file should be used which includes the relevant standard functions. All that is necessary is to indicate which file of standard functions should be used by the compiler; this file is a **header** since it comes at the head of the program. For example, <stdio.h> contains standard input and output functions such as gets (inputs, i.e. reads a line from a device), puts (outputs, i.e. writes a line to a device) and scanf (reads data); <math.h> contains mathematical functions such as cos, sin, tan, exp (exponential) and sqrt (square root).

Header files are also available to define the registers and ports of microcontrollers and save the programmer having to define each register and port by writing pre-processor lines for each. Thus for an Intel 8051 microcontroller we might have the header <reg.51.h>; this defines registers, e.g. the ports P0, P1, P2 and P3, and individual bits in bit-addressable registers, e.g. bits TF1, TR1, TF0, TR0, IE1, IT1, IE0 and IT0 in register TCON. Thus we can write instructions referring to port 0 inputs/outputs by purely using the label P0 or TF1 to refer to the TF1 bit in register TCON. Similarly, for a Motorola M68HC11E9 the header <hc11e9.h.> defines registers, e.g. PORTA, PORTB, PORTC and PORTD, and individual bits in bit-addressable registers, e.g. bits STAF, STAI, CWOM, HNDS, OIN, PLS, EGA and INVB in register PIOC. Thus, for example, we can write instructions referring to port A inputs/outputs by purely using the label PORTA. Libraries might also supply routines to help with the use of hardware peripheral devices such as keypads and liquid crystal displays.

The main program written for perhaps one specific microcontroller can, as a result of changing the header file, be easily adapted for running with another microcontroller. Libraries thus make C programs highly portable.

19.7 Examples of programs

The following are examples of programs written in C for microcontroller systems.

19.7.1 Switching a motor on and off

Consider the programming of a microcontroller, M68HC11, to start and stop a d.c. motor. Port C is used for the inputs and port B for the output to the motor, via a suitable driver (Figure 19.6). The start button is connected to PC0 to switch from a 1 to a 0 input when the motor is to be started. The stop button is connected to PC1 to switch from a 1 to a 0 input when the motor is to be stopped. The port C data direction register DDRC has to be set to 0 so that port C is set for inputs.

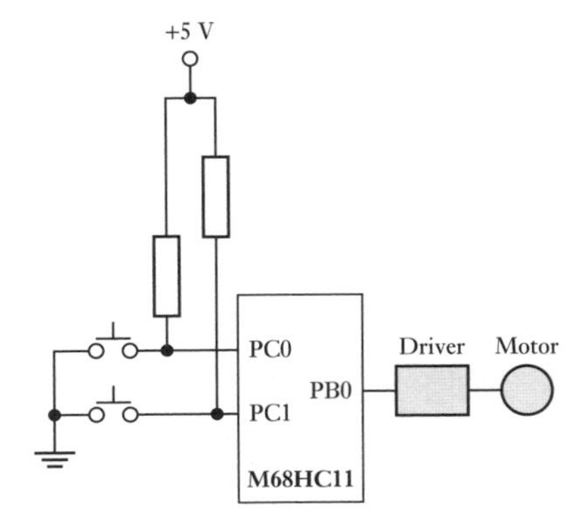

Figure 19.6 Motor control.

A program might be:

```
#include <hc11e9.h>  /*includes the header file*/

void main(void)
{
  PORTB.PB0 &=0; /*initially ensures motor off*/
  DDRC = 0; /*sets port C to be input*/
  while (1) /*repeats while this condition holds*/
  {
    if (PORTC.PC0 = =0)   /*is start button pressed?*/
      PORTB.PB0 |=1;  /*start output if pressed*/
    else if(PORTC.PC1 = =0)   /*is stop button pressed?*/
      PORTB.PB0 &=0;   /*stop output if pressed*/
  }
}
```

Note that | is the OR operator and sets a bit in the result to 0 only if the corresponding bits in both operands are 0, otherwise it sets to 1. It is used to turn on, or set, one or more bits in a value. Thus Port B.PB0 |=1 has 1 ORed with the value in PB0 and thus switches the motor on. It is a useful way of switching a number of bits in a port simultaneously. The & in PORTB.PB0 &=0 is used to AND the PB0 bit with 0 and so, since PB0 is already 1, assign to PORTB.PB0 the value 0.

19.7.2 Reading an ADC channel

Consider the task of programming a microcontroller (M68HC11) so that a single channel of its ADC can be read. The M68HC11 contains an eight-channel multiplexed, 8-bit, successive approximations ADC with inputs via port E (Figure 19.7). The ADC control/status register ADCTL contains the conversion complete flag CCF at bit 7 and other bits to control the multiplexer and the channel scanning. When CCF = 0 the conversion is not complete and when 1 it is complete. The analogue-to-digital conversion is initiated by writing a 1 to the ADPU bit in the OPTION register. However, the ADC must have been turned on for at least 100 μs prior to reading a value.

To convert the analogue input to PE0, the first 4 bits in the ADCTL register, i.e. CA, CB, CC and CD, have to be all set to 0. When operating to convert just a single channel, bit 5 SCAN might be set to 0 and bit 4 MULT to 0. A simple program to read a particular channel might thus involve, after powering up the ADC, turning all the bits in the ADCTL register to zeros, putting in the channel number and then reading the input while CCF is 0.

The program might thus be as follows:

```
#include <hc11e9.h> /*the header file*/

  void main(void)
  {
    unsigned int k; /*this enters the channel number*/

    OPTION=0; /*this and following line turns the ADC on*/
    OPTION.ADPU=1;
```

Figure 19.7 ADC.

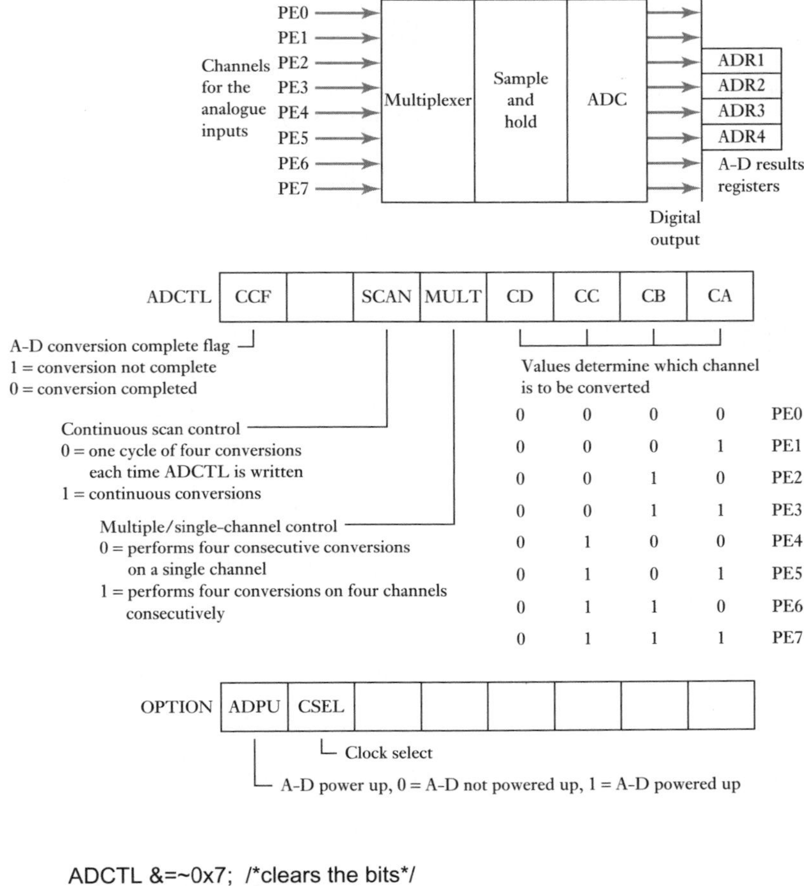


```
    ADCTL &=~0x7; /*clears the bits*/
    ADCTL |=k; /*puts the channel number to be read*/
    while (ADCTL.CCF==0);
    return ADR1; /*returns converted value to address 1*/
}
```

Note that ~ is the complement operator and its action is to reverse every bit in its operand, changing all the zeros to ones and vice versa. Thus bit 7 is set. | is the OR operator and sets a bit in the result to 0 only if the corresponding bits in both operands are 0, otherwise it sets to 1. It is used to turn on, or set, one or more bits in a value. In this case with k = 1 it just sets CA to 1. A delay subroutine can be included to ensure that after power-up the value is not read too quickly.

Summary

C is a high-level language which has the advantages when compared with the assembly language of being easier to use and that the same program can be used with different microprocessors; all that is necessary for this is that the appropriate compiler is used to translate the C program into the relevant machine language for the microprocessor concerned. Assembly language is different for the different microprocessors while C language is standardised.

C packages are supplied with libraries containing a large number of predefined functions containing C code that have already been written. In order to use the contents of any particular library, that library has to be specified in a header file. Every C program must have a function called main( ); this exercises control when the program is executed and is the first function to be called up. A program is made up of statements, every statement being terminated by a semicolon. Statements can be grouped together in blocks by putting them between braces, i.e. { }.

Problems

19.1 The following questions are all concerned with components of programs.

(a) State what is indicated by int in the statement:

```
nt counter;
```

(b) State what the following statement indicates:

```
num = 10
```

(c) State what the result of the following statement will be:

```
printf("Name");
```

(d) State what the result of the following statement will be:

```
printf("Number %d", 12);
```

(e) State what the effect of the following is:

```
#include <stdio.h>
```

19.2 For the following program, what are the reasons for including the line (a) #include <stdio.h>, (b) the { and }, (c) the /d, and (d) what will appear on the screen when the program is executed?

```
#include <stdio.h>

main( )
{
   printf(/d"problem 3");
}
```

19.3 For the following program, what will be displayed on the screen?

```
#include <stdio.h>

int main(void);
```

```
{
  int num;
  num = 20;

printf("The number is %d", num);
return 0;
}
```

19.4 Write a program to compute the area of a rectangle given its length and width at screen prompts for the length and width and then display the answer preceded by the words 'The area is'.

19.5 Write a program that displays the numbers 1 to 15, each on a separate line.

19.6 Explain the reasons for the statements in the following program for the division of two numbers:

```
#include <stdio.h>

int main(void);
{
  int num1, num2;

  printf("Enter first number:");
  scanf("%d", &num1);

  printf("Enter second number: ");
  scanf("%d", &num2);

  if(num2 = = 0)
    print f("Cannot divide by zero")
  else
    printf("Answer is: %d", num1/num2);
return 0;
}
```

Chapter twenty Input/Output Systems

Objectives

The objectives of this chapter are that, after studying it, the reader should be able to:

- **Identify interface requirements and how they can be realised: buffers, handshaking, polling and serial interfacing.**
- **Explain how interrupts are used with microcontrollers.**
- **Explain the function of peripheral interface adapters and be able to program them for particular situations.**
- **Explain the function of asynchronous communication interface adapters.**

20.1 Interfacing

When a microprocessor is used to control some system it has to accept input information, respond to it and produce output signals to implement the required control action. Thus there can be inputs from sensors to feed data in and outputs to such external devices as relays and motors. The term **peripheral** is used for a device, such as a sensor, keyboard, actuator, etc., which is connected to a microprocessor. It is, however, not normally possible to connect directly such peripheral devices to a microprocessor bus system due to a lack of compatibility in signal forms and levels. Because of such incompatibility, a circuit known as an interface is used between the peripheral items and the microprocessor. Figure 20.1 illustrates the arrangement. The interface is where this incompatibility is resolved.

Figure 20.1 The interfaces.

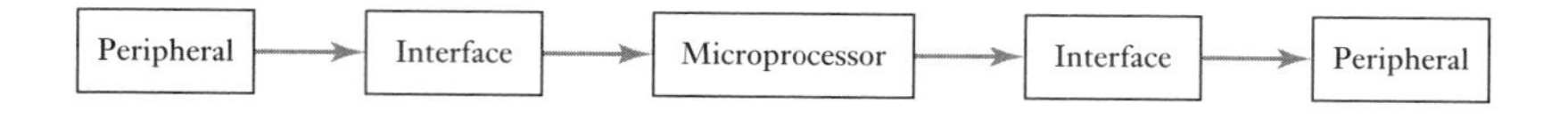

This chapter discusses the requirements of such interfaces and the very commonly used Motorola MC6820 Peripheral Interface Adapter and Motorola MC6850 Asynchronous Communications Interface Adapter.

20.2 Input/output addressing

There are two ways that microprocessors can select input/output devices. Some microprocessors, e.g. the Zilog Z80, have **isolated input/output,** and special input instructions such as IN are used to read from an input device and special output instructions such as OUT are used to output to an output device. For example, with the Z80 we might have:

```
IN A,(B2)
```

to read input device B2 and put the data in the accumulator A. An output instruction might be:

```
OUT (C), A
```

to write the data in accumulator A to port C.

More commonly, microprocessors do not have separate instructions for input and output but use the same instructions as they use for reading from or writing to memory. This is termed **memory-mapped input/output.** With this method, each input/output device has an address, just like a memory location. The Motorola 68HC11, Intel 8051 and PIC microcontrollers have no separate input/output instructions and use memory mapping. Thus, with memory mapping we might use:

```
LDAA $1003
```

to read the data input at address $1003 and:

```
STAA $1004
```

to write data to the output at address $1004.

Microprocessors use parallel ports to input or output bytes of data. Many peripherals often require several input/output ports. This can be because the data word of the peripheral is longer than that of the CPU. The CPU must then transfer the data in segments. For example, if we require a 16-bit output with an 8-bit CPU the procedure is:

1 The CPU prepares the eight most significant bits of the data.
2 The CPU sends the eight most significant bits of the data to the first port.
3 The CPU prepares the eight least significant bits of the data.
4 The CPU sends the eight least significant bits of the data to the second port.
5 Thus, after some delay, all the 16 bits are available to the peripheral.

20.2.1 Input/output registers

The Motorola 68HC11 microcontroller has five ports A, B, C, D and E (see Section 17.3.1). Ports A, C and D are bidirectional and can be used for either input or output. Port B is output only and port E input only. Whether a bidirectional port is used for input or output depends on the setting of a bit in its control register. For example, port A at address $1000 is controlled by the pulse accumulator control register PACTL at address $1026. To set port A for input requires bit 7 to be 0; output requires bit 7 to be 1 (see Figure 17.12). Port C is bidirectional and the 8 bits in its register at address $1003 are controlled by the corresponding bits in its port data direction register at address $1007. When the corresponding data direction bit is set to 0 it is an input, when set to 1 it is an output. Port D is bidirectional and contains just six input/output lines at address $1008. It is controlled by a port direction register at address $1009. The direction of each line is controlled by the corresponding bit in the control register; it is set to 0 for an input and 1 for an output. Some of the ports can also be set to carry out other functions by setting other bits in their control registers.

For a fixed-direction port, e.g. port B in the Motorola 68HC11 is output only, the instructions needed to output some value, e.g. $FF, are simply those needed to load the data to that address. The instructions might be:

```
REGBAS    EQU     $1000      ; base address of I/O registers
PORTB     EQU     $04        ; offset of PORTB from REGBAS
          LDX     #REGBAS    ; load index register X
          LDAA    #$FF       ; load $FF into accumulator
          STAA    PORTB,X    ; store value at PORTB address
```

For the fixed-direction port E, which is input only, the instruction to read a byte from it might be:

```
REGBAS    EQU     $1000      ; base address of I/O registers
PORTE     EQU     $0A        ; offset of PORTE from REGBAS
          LDAA    PORTE,X    ; load value at PORTE into the
                             ; accumulator
```

For a bidirectional port such as C, before we can use it for an input we have to configure the port so that it acts as an input. This means setting all the bits to 0. Thus we might have:

```
REGBAS    EQU     $1000      ; base address of I/O registers
PORTC     EQU     $03        ; offset of PORTC from REGBAS
DDRC      EQU     $07        ; offset of data direction register from
                             ; REGBAS
          CLR     DDRC,X     ; set DDRS to all 0
```

For the Intel 8051 microcontroller (see Section 17.3.2) there are four parallel bidirectional input/output ports. When a port bit is to be used for output, the data is just put into the corresponding special function register bit; when it is used for input a 1 must be written to each bit concerned, thus FFH might be written for an entire port to be written to. Consider an example of Intel 8051 instructions to light an LED when a push-button is pressed. The push-button provides an input to P3.1 and an output to P3.0; the push-button pulls the input low when it is pressed:

```
          SETB    P3.1       ; make bit P3.1 a 1 and so an input
LOOP      MOV     C,P3.1     ; read the state of the push-button
                             ; and store it in the carry flag
          CPL     C          ; complement the carry flag
          MOV     P3.0, C    ; copy state of carry flag to output
          SJMP    LOOP       ; keep on repeating the sequence
```

With the PIC microcontroller the direction of the signals at its bidirectional ports is set by the TRIS direction registers (see Section 17.3.3). TRIS is set as 1 for read and 0 for write. The registers for the PIC16C73/74 are arranged in two banks and before a particular register can be selected the bank has to be chosen by setting bit 5 in the STATUS register. This register is in both banks and so we do not have to select the bank in order to use this register. The TRIS registers are in bank 1 and the PORT registers in bank 0. Thus to set port B as output we have first to select bank 1 and then set TRISB to 0. We can then select bank 0 and write the output to PORTB. The bank is selected by setting a bit in the STATUS register. The instructions to select port B as an output are thus:

```
Output  clrf  PORTB        ; clear all the bits in port B
        bsf   STATUS,RP0   ; use status register to select bank 1 by
                           ; setting RP0 to 1
        clrf  TRISB        ; clear bits so output
        bcf   STATUS,RP0   ; use status register to select bank 0
                           ; port B is now an output set to 0
```

20.3 Interface requirements

The following are some of the actions that are often required of an interface circuit:

1 *Electrical buffering/isolation*
This is needed when the peripheral operates at a different voltage or current to that on the microprocessor bus system or there are different ground references. The term **buffer** is used for a device that provides isolation and current or voltage amplification. For example, if the output of a microprocessor is connected to the base of a transistor, the base current required to switch the transistor is greater than that supplied by the microprocessor and so a buffer is used to step up the current. There also has often to be isolation between the microprocessor and the higher power system.

2 *Timing control*
Timing control is needed when the data transfer rates of the peripheral and the microprocessor are different, e.g. when interfacing a microprocessor to a slower peripheral. This can be achieved by using special lines between the microprocessor and the peripheral to control the timing of data transfers. Such lines are referred to as **handshake lines** and the process as **handshaking**.

3 *Code conversion*
This is needed when the codes used by the peripherals differ from those used by the microprocessor. For example, an LED display might require a decoder to convert the BCD output from a microprocessor into the code required to operate the seven display elements.

4 *Changing the number of lines*
Microprocessors operate on a fixed word length of 4 bits, 8 bits or 16 bits. This determines the number of lines in the microprocessor data bus. Peripheral equipment may have a different number of lines, perhaps requiring a longer word than that of the microprocessor.

5 *Serial-to-parallel, and vice versa, data transfer*
Within an 8-bit microprocessor, data is generally manipulated 8 bits at a time. To transfer 8 bits simultaneously to a peripheral thus requires eight data paths. Such a form of transfer is termed **parallel data transfer**. It is, however, not always possible to transfer data in this way. For example,

data transfer over the public telephone system can only involve one data path. The data has thus to be transferred sequentially 1 bit at a time. Such a form of transfer is termed **serial data transfer**. Serial data transfer is a slower method of data transfer than parallel data transfer. Thus, if serial data transfer is used, there will be a need to convert incoming serial data into parallel data for the microprocessor and vice versa for outputs from the microprocessor.

6 *Conversion from analogue to digital and vice versa*
The output from sensors is generally analogue and this requires conversion to digital signals for the microprocessor. The output from a microprocessor is digital and this might require conversion to an analogue signal in order to operate some actuator. Many microcontrollers have built-in analogue-to-digital converters, e.g. PIC 16C74/74A (see Figure 17.30) and Motorola M68HC11 (see Figure 17.10), so can handle analogue inputs. However, where required to give analogue outputs, the microcontroller output has generally to pass through an external digital-to-analogue converter (see Section 20.6.2 for an example).

20.3.1 Buffers

A **buffer** is a device that is connected between two parts of a system to prevent unwanted interference between the two parts. An important use of a buffer is in the microprocessor input port to isolate input data from the microprocessor data bus until the microprocessor requests it. The commonly used buffer is a **tristate buffer**. The tristate buffer is enabled by a control signal to provide logic 0 or 1 outputs, when not enabled it has a high impedance and so effectively disconnects circuits. Figure 20.2 shows the symbols for tristate buffers and the conditions under which each is enabled. Figure 20.2(a) and (b) shows the symbol for buffers that does not change the logic of the input and Figure 20.2(c) and (d) for ones that do.

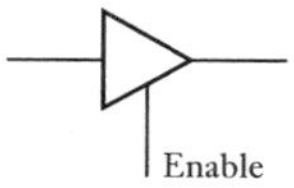

Enable

Enable	Input	Output
0	0	High impedance
0	1	High impedance
1	0	0
1	1	1

(a)

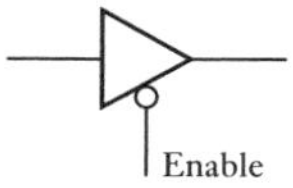

Enable

Enable	Input	Output
0	0	0
0	1	1
1	0	High impedance
1	1	High impedance

(b)

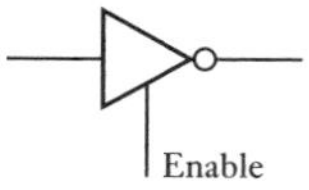

Enable

Enable	Input	Output
0	0	High impedance
0	1	High impedance
1	0	1
1	1	0

(c)

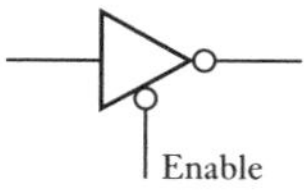

Enable

Enable	Input	Output
0	0	1
0	1	0
1	0	High impedance
1	1	High impedance

(d)

Figure 20.2 Buffers: (a) no logic change, enabled by 1, (b) no logic change, enabled by 0, (c) logic change, enabled by 1, (d) logic change, enabled by 0.

With PIC microcontrollers (see Section 17.3.3), the TRIS bit is connected to the enable input of a tristate buffer. If the bit is 0, the tristate buffer is enabled and simply passes its input value to its output, if it is 1 the tristate buffer is disabled and the output becomes high impedance (as in Figure 20.2(b)).

Such tristate buffers are used when a number of peripheral devices have to share the same data lines from the microprocessor, i.e. they are connected to the data bus, and thus there is a need for the microprocessor to be able to enable just one of the devices at a time with the others disabled. Figure 20.3 shows how such buffers can be used. Such buffers are available as integrated circuits, e.g. the 74125 with four non-inverting, active-low buffers and the 74126 with four non-inverting, active-high buffers.

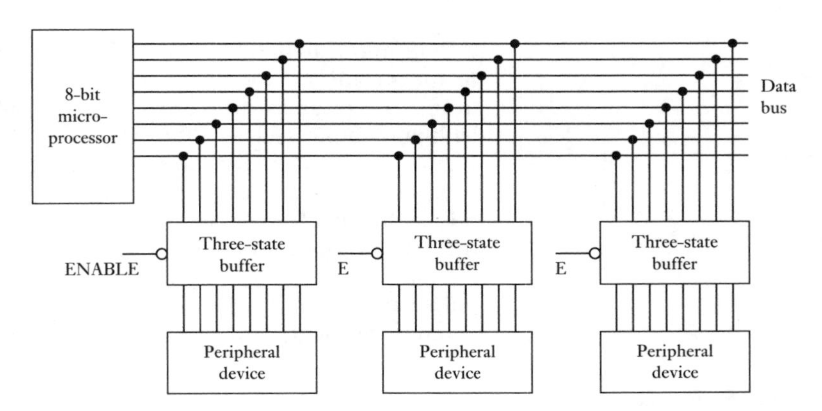

Figure 20.3 Three-state buffer.

20.3.2 Handshaking

Unless two devices can send and receive data at identical rates, handshaking is necessary to exchange data. With handshaking the slower device controls the rate at which the data is transferred. For parallel data transfer **strobe-and-acknowledge** is the commonly used form of handshaking. The peripheral sends a DATA READY signal to the input/output section. The CPU then determines that the DATA READY signal is active. The CPU then reads the data from the input/output section and sends an INPUT ACKNOWLEDGED signal to the peripheral. This signal indicates that the transfer has been completed and thus the peripheral can send more data. For an output, the peripheral sends an OUTPUT REQUEST or PERIPHERAL READY signal to the input/output section. The CPU determines that the PERIPHERAL READY signal is active and sends the data to the peripheral. The next PERIPHERAL READY signal may be used to inform the CPU that the transfer has been completed.

With the microcontroller MC68HC11, the basic strobed input/output operates as follows. The handshaking control signals use pins STRA and STRB (Figure 20.4(a), also see Figure 17.10 for the full block model), port C is used for the strobed input and port B for the strobed output. When data is ready to be sent by the microcontroller a pulse is produced at STRA and sent to the peripheral device. When the microcontroller receives either a rising or falling edge to a signal on STRB, then the relevant output port of the microcontroller sends the data to the peripheral. When data is ready to

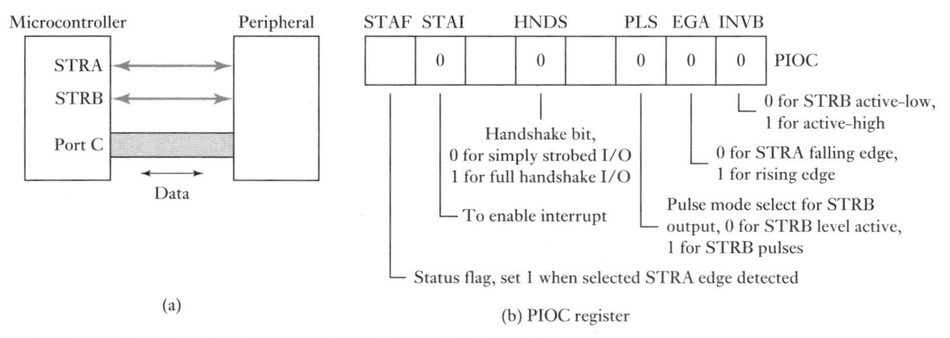

Figure 20.4 Handshaking control: strobe-and-acknowledge.

be transmitted to the microcontroller, the peripheral sends a signal to STRA that it is ready and then a rising or falling edge to a signal on STRB is used to indicate readiness to receive. Before handshaking can occur, the parallel input/output register PIOC at address $1002 has to be first configured. Figure 20.4(b) shows the states required of the relevant bits in that register.

Full handshake input/output involves two signals being sent along STRB, the first being to indicate ready to receive data and the next one that the data has been read. This mode of operation requires that in PIOC the HNDS bit is set to 1 and if PLS is set to 0 the full handshake is said to be pulsed and if to 1 it is interlocked. With pulsed operation a pulse is sent as acknowledgement; with interlocked STRB there is a reset (Figure 20.5).

Figure 20.5 Full handshaking: (a) pulsed, (b) interlocked.

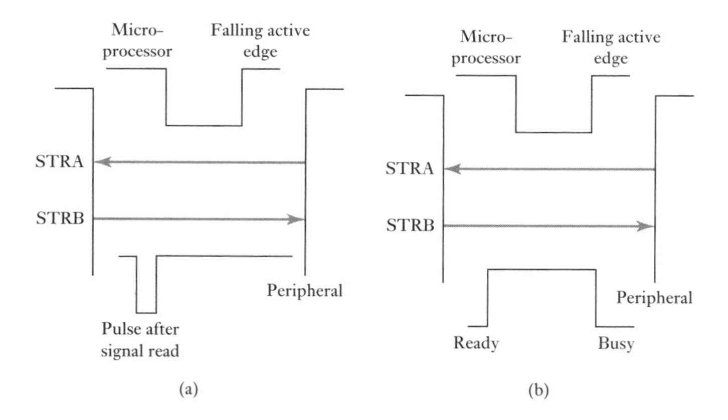

20.3.3 Polling and interrupts

Consider the situation where all input/output transfers of data are controlled by the program. When peripherals need attention they signal the microprocessor by changing the voltage level of an input line. The microprocessor can then respond by jumping to a program service routine for the device. On completion of the routine, a return to the main program occurs. Program control of inputs/outputs is thus a loop to read inputs and update outputs continuously, with jumps to service routines as required. This process of repeatedly checking each peripheral device to see if it is ready to send or accept a new byte of data is called **polling**.

An alternative to program control is **interrupt control**. An interrupt involves a peripheral device activating a separate interrupt request line. The reception of an interrupt results in the microprocessor suspending execution of its main program and jumping to the service routine for the peripheral. The interrupt must not lead to a loss of data and an interrupt handling routine has to be incorporated in the software so that the state of processor registers and the last address accessed in the main program are stored in dedicated locations in memory. After the interrupt service routine, the contents of the memory are restored and the microprocessor can continue executing the main program from where it was interrupted (Figure 20.6).

Figure 20.6 Interrupt control.

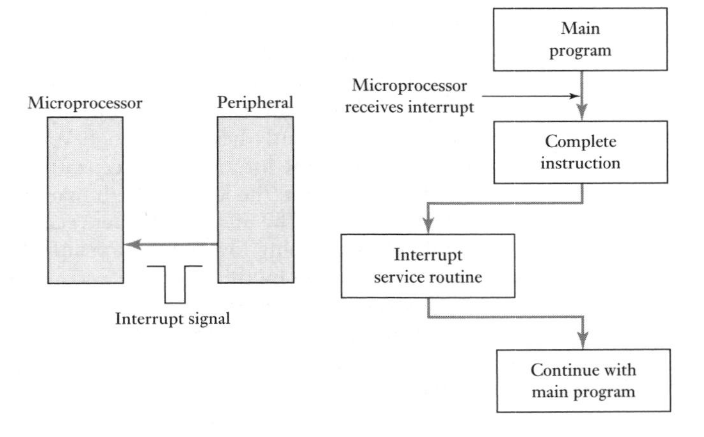

Thus, when an interrupt occurs:

1 The CPU waits until the end of the instruction it is currently executing before dealing with the interrupt.
2 All the CPU registers are pushed onto the stack and a bit set to stop further interrupts occurring during this interruption. The stack is a special area of memory in which program counter values can be stored when a subroutine is to be executed. The program counter gives the address of the next program instruction in a program and thus storing this value enables the program to be resumed at the place where it broke off to execute the interrupt.
3 The CPU then determines the address of the interrupt service routine to be executed. Some microprocessors have dedicated interrupt pins and the pin that is chosen determines which address is to be used. Other microprocessors have only one interrupt pin and the interrupting device must then supply data that tells the microprocessor where the interrupt service is located. Some microprocessors have both kinds of interrupt inputs. The starting address of an interrupt service routine is called an **interrupt vector**. The block of memory assigned to store the vectors is known as the **vector table**. Vector addresses are fixed by the chip manufacturer.
4 The CPU branches to the interrupt service routine.
5 After completion of this routine, the CPU registers are returned from the stack and the main program continues from the point it left off.

Unlike a subroutine call, which is located at a specific point in a program, an interrupt can be called from any point in the program. Note that the program does not control when an interrupt occurs; control lies with the interrupting event.

Input/output operations frequently use interrupts since often the hardware cannot wait. For example, a keyboard may generate an interrupt input signal when a key is pressed. The microprocessor then suspends the main program to handle the input from the keyboard; it processes the information and then returns to the main program to continue from where it left off. This ability to code a task as an interrupt service routine and tie it to an external signal simplifies many control tasks, enabling them to be handled without delay. For some interrupts it is possible to program the microprocessor to ignore the interrupt request signal unless an enable bit has been set. Such interrupts are termed **maskable**.

The Motorola 68HC1 has two external interrupt request inputs. XIRQ is a non-maskable interrupt and will always be executed on completion of the instruction currently being executed. When the XIRQ interrupt occurs, the CPU jumps to the interrupt service routine whose interrupt vector is held at address $FFF4/5 (the low and high bytes of the address). IRQ is a maskable interrupt. When the microcontroller receives a signal at the interrupt request pin IRQ by its going low, the microcontroller jumps to the interrupt service routine indicated by the interrupt vectors $FFF2/3. IRQ can be masked by the instruction set interrupt mask SEI and unmasked by the instruction clear interrupt mask CLI. At the end of an interrupt service routine the instruction RTI is used to return to the main program.

With the Intel 8051, interrupt sources are individually enabled or disabled through the bit-addressable register IE (interrupt enable) at address 0A8H (see Figure 17.26), a 0 disabling an interrupt and a 1 enabling it. In addition there is a global enable/disable bit in the IE register that is set to enable all external interrupts or cleared to disable all external interrupts. The TCON register (Figure 17.25) is used to determine the type of interrupt input signal that will initiate an interrupt.

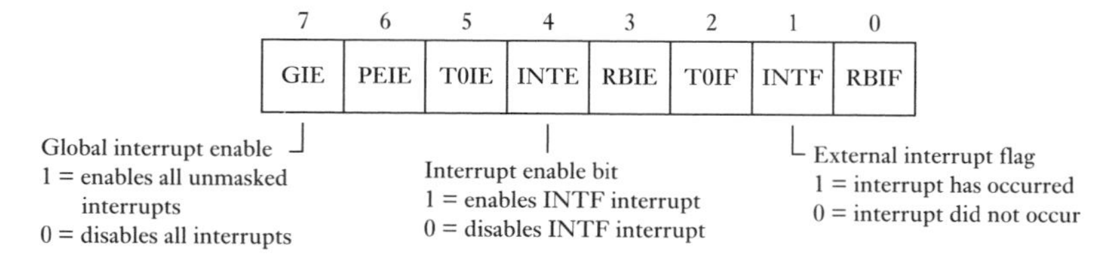

Figure 20.7 INTCON.

With the PIC microcontrollers, interrupts are controlled by the INTCON register (Figure 20.7). To use bit 0 of port B as an interrupt, it must be set as an input and the INTCON register must be initialised with a 1 in INTE and a 1 in GIE. If the interrupt is to occur on a rising edge then INTEDG (bit 6) in the OPTION register (see Figure 17.32) must be set to 1; if on a falling edge it must be set to 0. When an interrupt occurs, INTF is set. It can be cleared by the instruction bcf INTCON,INTF.

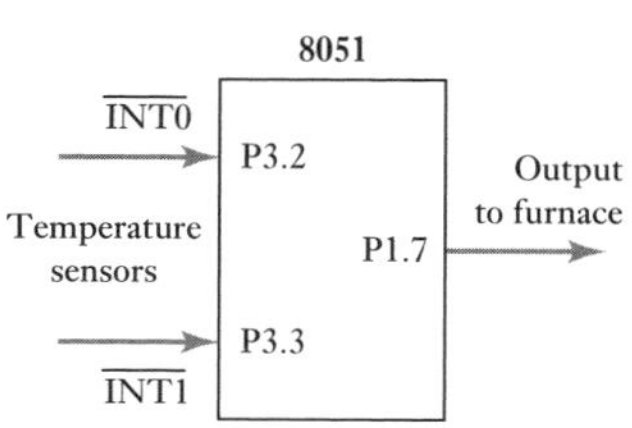

Figure 20.8 Central heating system.

As an illustration of a program involving external interrupts, consider a simple on/off control program for a central heating system involving an Intel 8051 microcontroller (Figure 20.8). The central heating furnace is controlled by an output from P1.7 and two temperature sensors are used, one to determine when the temperature falls below, say, 20.5°C and the other when it rises above 21.0°C. The sensor for the 21.0°C temperature is connected to interrupt INT0, port 3.2, and the sensor for the 20.5°C temperature is connected to interrupt INT1, port 3.3. By selecting the IT1 bit to be 1 in the TCON register, the external interrupts are edge triggered, i.e. activated when there is a change from 1 to 0. When the temperature rises to 21.0°C the external interrupt INT0 has an input which changes from 1 to 0 and the interrupt is activated to give the instruction CLR P1.7 for a 0 output to turn the furnace off. When the temperature falls to 20.5°C the external interrupt INT1 has an input which changes from 0 to 1 and the interrupt is activated to give the instruction SETB P1.7 for a 1 output to turn the furnace on. The MAIN program is just a set of instructions to configure and enable the interrupts, establish the initial condition of the furnace to be on if the temperature is less than 21.0° or off if above, and then to wait doing nothing until an interrupt occurs. With the program, a header file has been assumed:

```
        ORG     0
        LJMP    MAIN

        ORG     0003H       ; gives the entry address for ISR0
ISR0    CLR     P1.7        ; interrupt service routine to turn the
                            ; furnace off
        RETI                ; return from interrupt

        ORG     0013H       ; gives the entry address for ISR1
ISR1    SETB    P1.7        ; interrupt service routine to turn furnace off
        RETI                ; return from interrupt

        ORG     30H
MAIN    SETB    EX0         ; to enable external interrupt 0
        SETB    EX1         ; to enable external interrupt 1
        SETB    IT0         ; set to trigger when change from 1 to 0
        SETB    IT1         ; set to trigger when change from to 0
        SETB    P1.7        ; turn the furnace on
        JB      P3.2,HERE   ; if temperature greater than 21.0°C jump to
                            ; HERE and leave furnace on
        CLR     P1.7        ; turn the furnace off
HERE    SJMP    HERE        ; just doing nothing until an interrupt occurs
        END
```

Microcontrollers, in addition to the interrupt request, have a reset interrupt and a non-maskable interrupt. The reset interrupt is a special type of interrupt and when this occurs the system resets; thus when this is activated, all activity in the system stops, the starting address of the main program is loaded and the start-up routine is executed. The microcontroller M68HC11 has a computer operating properly (COP) watchdog timer. This is intended to detect software processing errors when the CPU is not executing certain sections of code within an allotted time. When this occurs the COP timer times out and a system reset is initiated.

The non-maskable interrupt cannot be masked and so there is no method of preventing the interrupt service routine being executed when it is connected to this line. An interrupt of this type is usually reserved for emergency routines such as those required when there is a power failure, e.g. switching to a back-up power supply.

20.3.4 Serial interfacing

With the parallel transmission of data, one line is used for each bit; serial systems, however, use a single line to transmit data in sequential bits. There are two basic types of serial data transfer: asynchronous and synchronous.

With **asynchronous transmission**, the receiver and the transmitter each use their own clock signals so it is not possible for a receiver to know when a word starts or stops. Thus it is necessary for each transmitted data word to carry its own start and stop bits so that it is possible for the receiver to tell where one word stops and another starts (Figure 20.9). With such a mode of transmission, the transmitter and receiver are typically remote (see Chapter 22 for details of standard interfaces). With **synchronous transmission**, the transmitter and receiver have a common clock signal and thus transmission and reception can be synchronised.

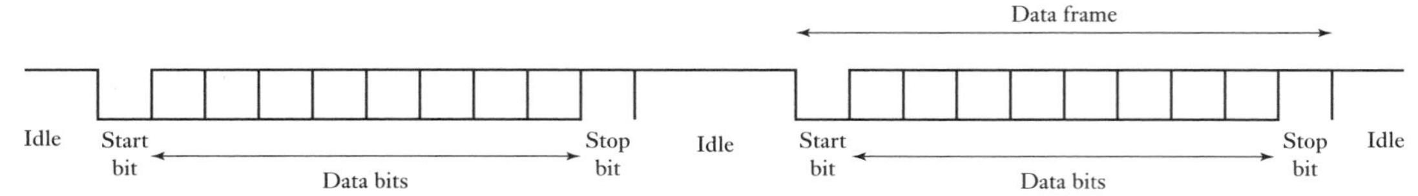

Figure 20.9 Asynchronous transmission.

The microcontroller MC68HC11 (see Figure 17.10) has a serial communications interface (SCI) which can be used for asynchronous transmission and thus can be used to communicate with remote peripheral devices. The SCI uses port D pin PD1 as a transmit line and port PD0 as a receive line. These lines can be enabled or disabled by the SCI control register. The microcontroller also has a serial peripheral interface (SPI) for synchronous transmission. This can be used for local serial communication, local meaning essentially inside the machine in which the chip is located.

20.4 Peripheral interface adapters

Interfaces can be specifically designed for particular inputs/outputs; however, programmable input/output interface devices are available which permit various different input and output options to be selected by means of software. Such devices are known as **peripheral interface adapters** (PIAs).

A commonly used PIA parallel interface is the Motorola MC 6821. It is part of the MC6800 family and thus can be directly attached to Motorola MC6800 and MC68HC11 buses. The device can be considered to be essentially just two parallel input/output ports, with their control logic, to link up with the host microprocessor. Figure 20.10 shows the basic structure of the MC6821 PIA and the pin connections.

The PIA contains two 8-bit parallel data ports, termed A and B. Each port has:

1 A *peripheral interface register*. An output port has to operate in a different way to an input port because the data must be held for the peripheral. Thus for output a register is used to store data temporarily. The register is said to be **latched**, i.e. connected, when a port is used for output and unlatched when used for input.
2 A *data direction register* that determines whether the input/output lines are inputs or outputs.

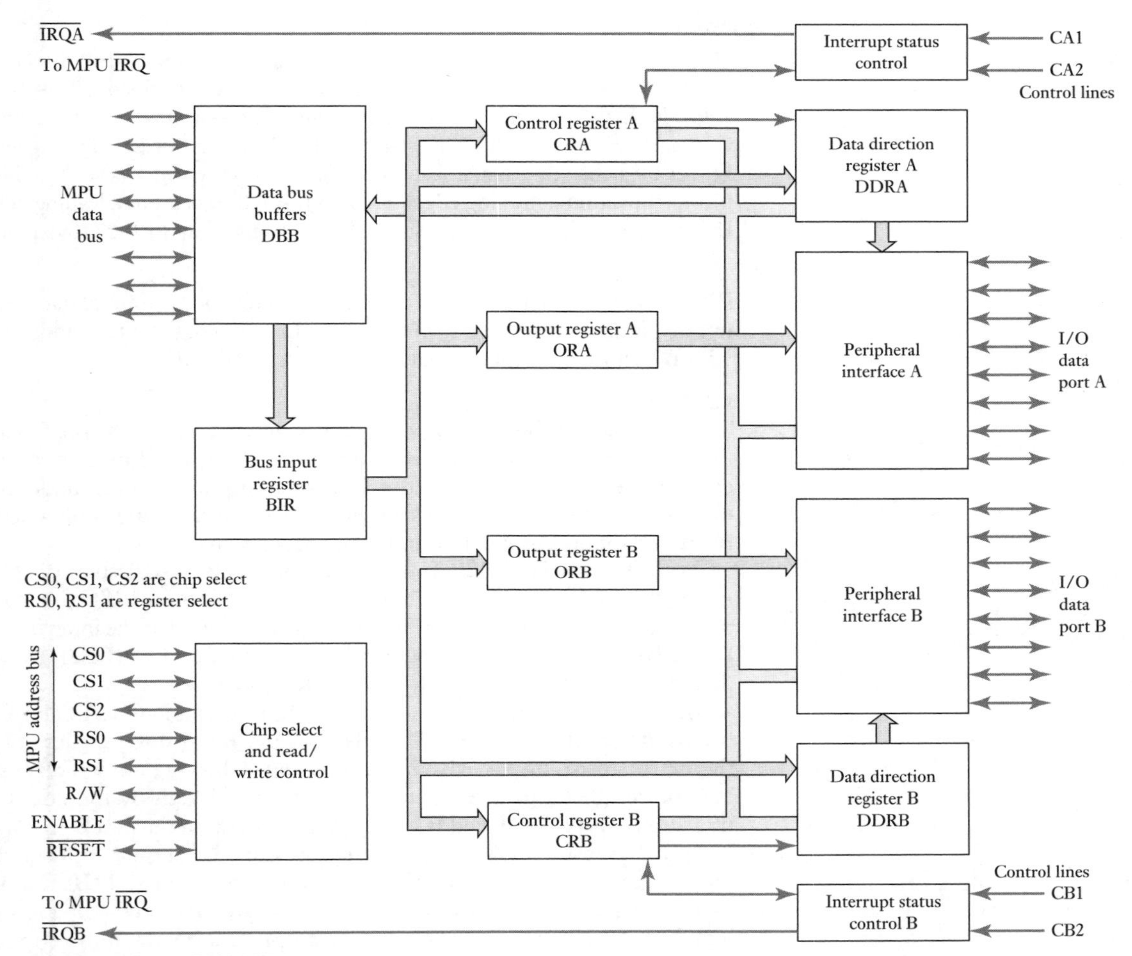

Figure 20.10 MC6821 PIA.

3 A *control register* that determines the active logical connections in the peripheral.
4 Two *control lines*, CA1 and CA2 or CB1 and CB2.

Two microprocessor address lines connect the PIA directly through the two register select lines RS0 and RS1. This gives the PIA four addresses for the six registers. When RS1 is low, side A is addressed and when it is high, side B. RS0 addresses registers on a particular side, i.e. A or B. When RS0 is high, the control register is addressed, when low the data register or the data direction register. For a particular side, the data register and the data direction register have the same address. Which of them is addressed is determined by bit 2 of the control register (see below).

Each of the bits in the A and B control registers is concerned with some features of the operation of the ports. Thus for the A control register we have the bits shown in Figure 20.11. A similar pattern is used for the B control register.

Figure 20.11 Control register.

B7	B6	B5	B4	B3	B2	B1	B0
IRQA1	IRQA2	CA2 control			DDRA access	CA1 control	

Bits 0 and 1

The first two bits control the way that CA1 or CB1 input control lines operate. Bit 0 determines whether the interrupt output is enabled. B0 = 0 disables the IRQA(B) microprocessor interrupt, B0 = 1 enables the interrupt. CA1 and CB1 are not set by the static level of the input but are edge triggered, i.e. set by a changing signal. Bit 1 determines whether bit 7 is set by a high-to-low transition (a trailing edge) or a low-to-high transition (a leading edge). B1 = 0 sets a high-to-low transition, B1 = 1 sets a low-to-high transition.

Bit 2

Bit 2 determines whether data direction registers or peripheral data registers are addressed. With B2 set to 0, data direction registers are addressed, with B2 set to 1, peripheral data registers are selected.

Bits 3, 4 and 5

These bits allow the PIA to perform a variety of functions. Bit 5 determines whether control line 2 is an input or an output. If bit 5 is set to 0, control line 2 is an input, if set to 1, it is an output. In input mode, both CA2 and CB2 operate in the same way. Bits 3 and 4 determine whether the interrupt output is active and which transitions set bit 6.

With B5 = 0, i.e. CA2(CB2) set as an input: B3 = 0 disables IRQA(B) microprocessor interrupt by CA2(CB2); B3 = 1 enables IRQA(B) microprocessor interrupt by CA2(CB2); B4 = 0 determines that the interrupt flag IRQA(B), bit B6, is set by a high-to-low transition on CA2(CB2); B4 = 1 determines that it is set by a low-to-high transition.

B5 = 1 sets CA2(CB2) as an output. In output mode CA2 and CB2 behave differently. For CA2: with B4 = 0 and B3 = 0, CA2 goes low on the first high-to-low ENABLE (E) transition following a microprocessor read of peripheral data register A and is returned high by the next CA1 transition; with B4 = 0 and B3 = 1, CA2 goes low on the first high-to-low ENABLE transition following a microprocessor read of the peripheral data register A and is returned to high by the next high-to-low ENABLE transition. For CB2: with B4 = 0 and B3 = 0, CB2 goes low on the first low-to-high ENABLE transition following a microprocessor write into peripheral data register B and is returned to high by the next CB1 transition; with B4 = 0 and B3 = 1, CB2 goes low on the first low-to-high ENABLE transition following a microprocessor write into peripheral data register B and is returned high by the next low-to-high ENABLE transition. With B4 = 1 and B3 = 0, CA2(CB2) goes low as the microprocessor writes B3 = 0 into the control register. With B4 = 0 and B3 = 1, CA2(CB2) goes high as the microprocessor writes B3 = 1 into the control register.

Bit 6

This is the CA2(CB2) interrupt flag, being set by transitions on CA2(CB2). With CA2(CB2) as an input (B5 = 0), it is cleared by a microprocessor read of the data register A(B). With CA2(CB2) as an output (B5 = 1), the flag is 0 and is not affected by CA2(CB2) transitions.

Bit 7

This is the CA1(CB1) interrupt flag, being cleared by a microprocessor read of data register A(B).

The process of selecting which options are to be used is termed **configuring** or **initialising** the PIA. The RESET connection is used to clear all the registers of the PIA. The PIA must then be initialised.

20.4.1 Initialising the PIA

Before the PIA can be used, a program has to be written and used so that the conditions are set for the desired peripheral data flow. The PIA program is placed at the beginning of the main program so that, thereafter, the microprocessor can read peripheral data. The initialisation program is thus only run once.

The initialisation program to set which port is to be input and which is to be output can have the following steps:

1 Clear bit 2 of each control register by a reset, so that data direction registers are addressed. Data direction register A is addressed as XXX0 and data direction register B as XXX2.
2 For A to be an input port, load all zeros into direction register A.
3 For B to be an output port, load all ones into direction register B.
4 Load 1 into bit 2 of both control registers. Data register A is now addressed as XXX0 and data register B as XXX2.

Thus an initialisation program in assembly language to make side A an input and side B an output could be, following a reset:

```
INIT    LDAA    #$00     ; Loads zeros
        STAA    $2000    ; Make side A input port
        LDAA    #$FF     ; Load ones
        STAA    $2000    ; Make side B output port
        LDAA    #$04     ; Load 1 into bit 2, all other bits 0
        STAA    $2000    ; Select port A data register
        STAA    $2002    ; Select port B data register
```

Peripheral data can now be read from input port A with the instruction LDAA 2000 and the microprocessor can write peripheral data to output port B with the instruction STAA 2002.

20.4.2 Connecting interrupt signals via a PIA

The Motorola MC6821 PIA (Figure 20.12) has two connections IRQA and IRQB through which interrupt signals can be sent to the microprocessor so that an interrupt request from CA1, CA2 or CB1, CB2 can drive the IRQ pin of the microprocessor to the active-low state. When the initialisation program for a PIA was considered in the previous section, only bit 2 of the control register was set as 1, the other bits being 0. These zeros disabled interrupt inputs. In order to use interrupts, the initialisation step which stores $04 into the control register must be modified. The form of the modification will depend on the type of change in the input which is required to initiate the interrupt.

Suppose, for example, we want CA1 to enable an interrupt when there is a high-to-low transition, with CA2 and CB1 not used and CB2 enabled and used for a set/reset output. The control register format to meet this specification is, for CA:

B0 is 1 to enable interrupt on CA1.
B1 is 0 so that the interrupt flag IRQA1 is set by a high-to-low transition on CA1.

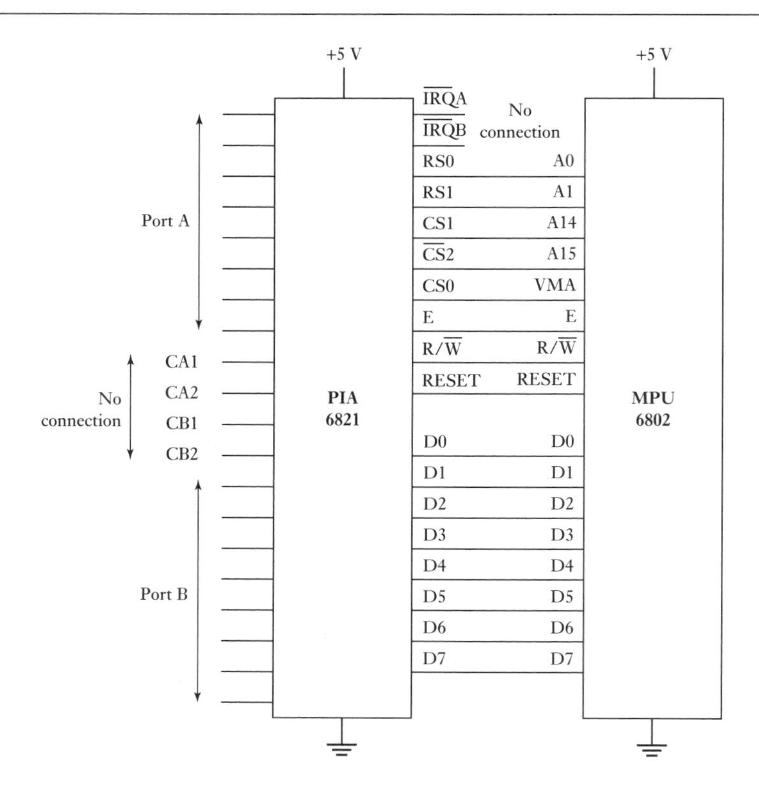

Figure 20.12 Interfacing with a PIA.

B2 is 1 to give access to the data register.
B3, B4, B5 are 0 because CA2 is disabled.
B6, B7 are read-only flags and thus a 0 or 1 may be used.

Hence the format for CA1 can be 00000101 which is 05 in hexadecimal notation. The control register format for CB2 is:

B0 is 0 to disable CB1.
B1 may be 0 or 1 since CB1 is disabled.
B2 is 1 to give access to the data register.
B3 is 0, B4 is 1 and B5 is 1, to select the set/reset.
B6, B7 are read-only flags and thus a 0 or 1 may be used.

Hence the format for CA1 can be 00110100 which is 34 in hexadecimal notation. The initialisation program might then read:

```
INIT    LDAA    #$00     ; Load zeros
        STAA    $2000    ; Make side A input port
        LDAA    #$FF     ; Load ones
        STAA    $2000    ; Make side B output port
        LDAA    #$05     ; Load the required control register format
        STAA    $2000    ; Select port A data register
        LDAA    #$34     ; Load the required control register format
        STAA    $2002    ; Select port B data register
```

20.4.3 An example of interfacing with a PIA

As an example of interfacing with a PIA, Figure 20.13 shows a circuit that can be used with a unipolar stepper motor (see Section 9.7.2). The inductive windings can generate a large back e.m.f. when switched so some way of isolating the windings from the PIA is required. Optoisolators, diodes or resistors might be used. Diodes give a cheap and simple interface, resistors not completely isolating the PIA.

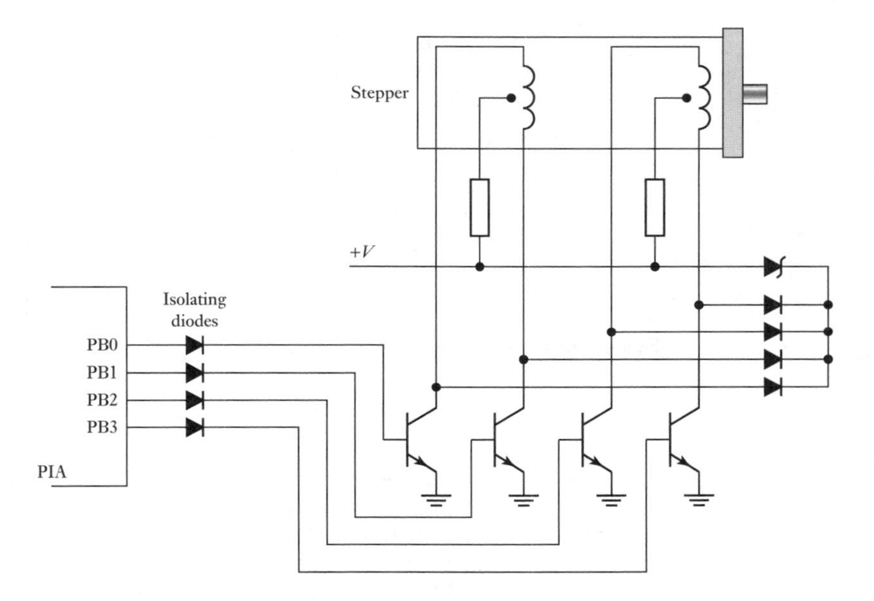

Figure 20.13 Interfacing a stepper.

20.5 Serial communications interface

The **universal asynchronous receiver/transmitter** (UART) is the essential element of a serial communication system, the function being to change serial data to parallel for input and parallel data to serial for output. A common programmable form of a UART is the **asynchronous communications interface adapter** (ACIA) from Motorola MC6850; Figure 20.14 shows a block diagram of the constituent elements.

Data flow between the microprocessor and the ACIA is via eight bidirectional lines D0 to D7. The direction of the data flow is controlled by the microprocessor through the read/write input to the ACIA. The three chip select lines are used for addressing a particular ACIA. The register select line is used to select particular registers within the ACIA; if the register select line is high then the data transmit and data receive registers are selected, if low then the control and status registers are selected. The status register contains information on the status of serial data transfers as they occur and is used to read data carrier detect and clear-to-send lines. The control register is initially used to reset the ACIA and subsequently to define the serial data transfer rate and data format.

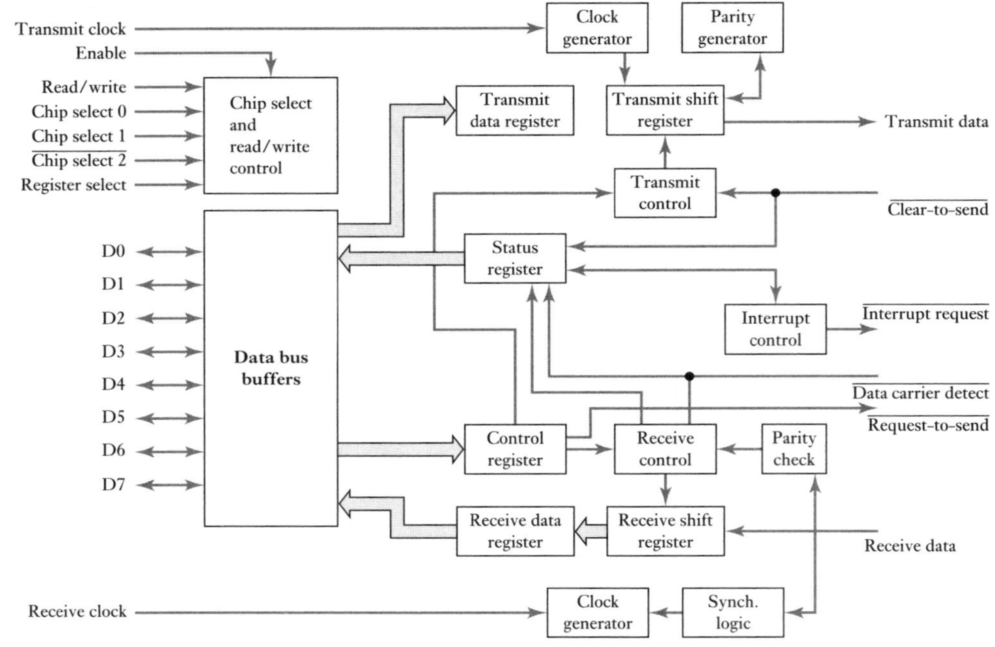

Figure 20.14 MC6850 ACIA.

The peripheral side of the ACIA includes two serial data lines and three control lines. Data is sent by the transmit data line and received by the receive data line. Control signals are provided by clear-to-send, data carrier detect and request-to-send. Figure 20.15 shows the bit formats of the control and Figure 20.16 the status registers.

Asynchronous serial data transfer is generally used for communications between two computers, with or without a modem, or a computer and a printer (see Chapter 22 for further discussion).

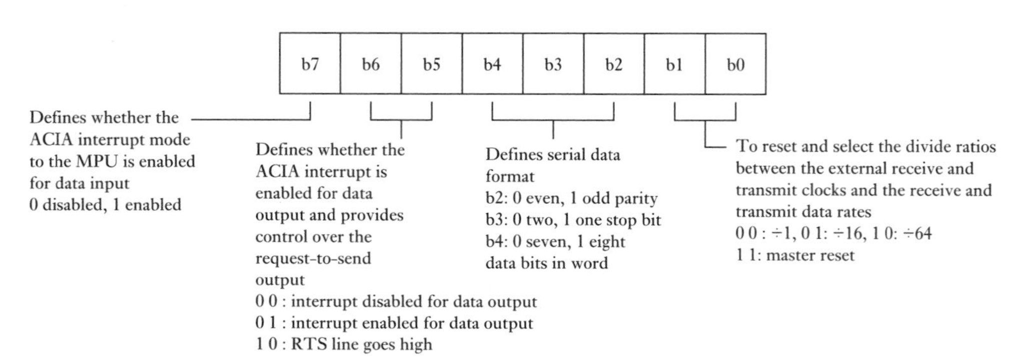

Figure 20.15 Control register.

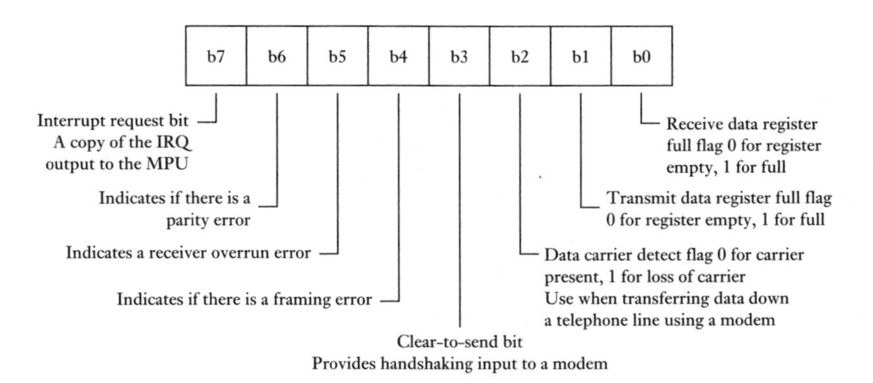

Figure 20.16 Status register.

20.5.1 Serial interfaces of microcontrollers

Many microcontrollers have serial interfaces, i.e. built-in UARTs. For example, the Motorola M68HC11 has a serial peripheral interface (SPI), a synchronous interface, and a serial communication interface (SCI), an asynchronous interface (see Figure 17.10). The SPI requires the same clock signal to be used by the microcontroller and the externally connected device or devices (Figure 20.17(a)). The SPI allows several microcontrollers, with this facility, to be interconnected. The SCI is an asynchronous interface and so allows different clock signals to be used by the SCI system and the externally connected device (Figure 20.17(b)). General-purpose microprocessors do not have an SCI so a UART, e.g. Motorola MC6850, has then to be used to enable serial communication to take place. In some situations more than one SCI is required and thus the microcontroller M68HC11 requires supplementing with a UART.

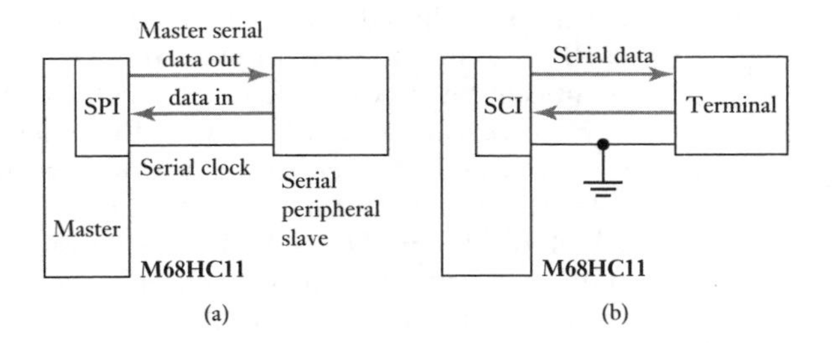

Figure 20.17 (a) SPI, (b) SCI.

The SPI is initialised by bits in the SPI control register (SPCR) and the port D data direction control register (DDRD). The SPI status register contains status and error bits. The SCI is initialised by using the SCI control register 1, the SCI control register 2 and the baud rate control register. Status flags are in the SCI status register.

The Intel 8051 has a built-in serial interface with four modes of operation, these being selected by writing ones or zeros into SMO and SM1 bits in the SCON (serial port control) register at address 98H (Figure 20.18) (Table 20.1).

In mode 0, serial data enters and leaves by RXD. Pin TXD outputs the shift clock and this is then used to synchronise the data transmission and

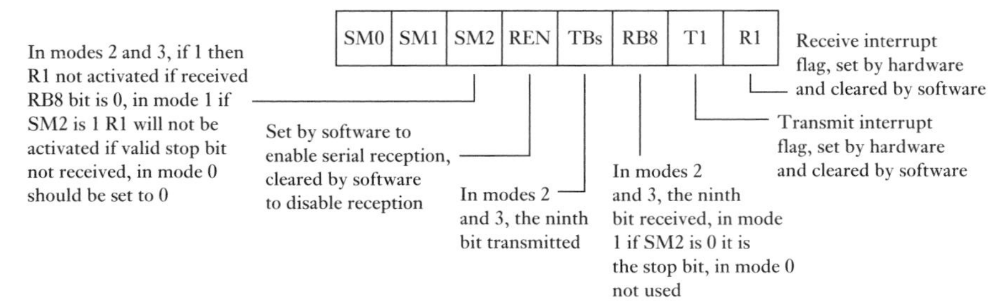

Figure 20.18 SCON register.

Table 20.1 Intel 8051 serial port modes.

SMO	SM1	Mode	Description	Baud rate
0	0	0	Shift register	Osc. freq./12
0	1	1	8-bit UART	Variable
1	0	2	9-bit UART	Osc. freq./12 or 64
1	1	3	9-bit UART	Variable

reception. Reception is initiated when REN is 1 and R1 is 0. Transmission is initiated when any data is written to SBUF, this being the serial port buffer at address 99H. In mode 1, 10 bits are transmitted on TXD or received on RXD; these are the start bit of 0, eight data bits and a stop bit of 1. Transmission is initiated by writing to SBUF and reception initiated by a 1 to 0 transition on RXD. In modes 2 and 3, 11 bits are transmitted on TXD or received on RXD.

PIC microcontrollers have an SPI (see Figure 17.30) which can be used for synchronous serial communications. When data is written to the SSBUF register it is shifted out of the SDO pin in synchronism with a clock signal on SCK and outputted through pin RC5 as a serial signal with the most significant bit appearing first and a clock signal though RC3. Input into the SSBUF register is via RC4. Many PIC microcontrollers also have a UART to create a serial interface for use with asynchronously transmitted serial data. When transmitting, each 8-bit byte is framed by a START bit and a STOP bit. When the START bit is transmitted, the RX line drops to a low and the receiver can then synchronise on this high-to-low transition. The receiver then reads the 8 bits of serial data.

20.6 Examples of interfacing

The following are examples of interfacing.

20.6.1 Interfacing a seven-segment display with a decoder

Consider where the output from a microcontroller is used to drive a seven-segment LED display unit (see Section 6.5). A single LED is an on/off indicator and thus the display number indicated will depend on which LEDs are on. Figure 20.19 shows how we can use a microcontroller to drive a common

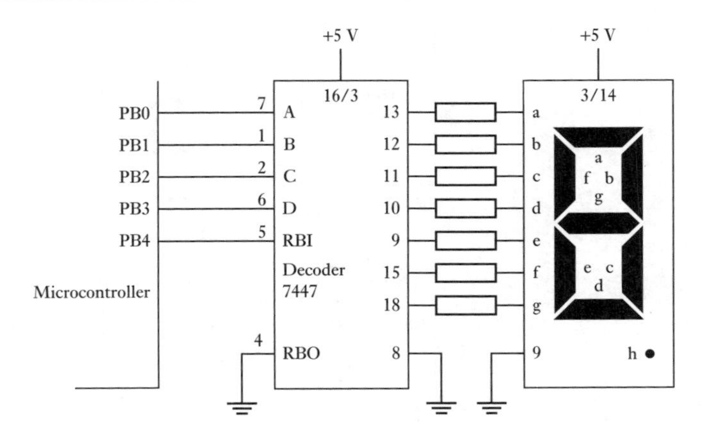

Figure 20.19 Driving a display.

anode display using a decoder driver, this being able to take in a BCD input and convert it to the appropriate code for the display.

For the 7447 decoder, pins 7, 1, 2 and 6 are the input pins of the decoder for the BCD input with pins 13, 12, 11, 10, 9, 15 and 14 being the outputs for the segments of the display. Pin 9 of the display is the decimal point. Table 20.2 shows the input and output signals for the decoder.

Table 20.2 The 7447 BCD decoder for a seven-segment display.

	Input pins				Output pins						
Display	6	2	1	7	13	12	11	10	9	15	14
0	L	L	L	L	ON	ON	ON	ON	ON	ON	OFF
1	L	L	L	H	OFF	ON	ON	OFF	OFF	OFF	OFF
2	L	L	H	L	ON	ON	OFF	ON	ON	OFF	ON
3	L	L	H	H	ON	ON	ON	ON	OFF	OFF	ON
4	L	H	L	L	OFF	ON	ON	OFF	OFF	ON	ON
5	L	H	H	L	ON	OFF	ON	ON	OFF	ON	ON
6	L	H	H	L	OFF	OFF	ON	ON	ON	ON	ON
7	L	H	H	H	ON	ON	ON	OFF	OFF	OFF	OFF
8	H	L	H	H	ON	ON	ON	OFF	OFF	OFF	OFF
9	H	L	H	L	ON	ON	ON	OFF	OFF	OFF	OFF

Blanking is when none of the segments are lit. This is used to prevent a leading 0 occurring when we have, say, three display units and want to display just 10 rather than 010 and so blank out the leading 0 and prevent it being illuminated. This is achieved by the ripple blanking input (RBI) being set low. When RBI is low and the BCD inputs A, B, C and D are low then the output is blanked. If the input is not zero the ripple blanking output (RBO) is high regardless of the RBI status. The RBO of the first digit in the display can be connected to the RBI of the second digit and the RBO of the second connected to the RBI of the third digit, thus allowing only the final 0 to be blanked (Figure 20.20).

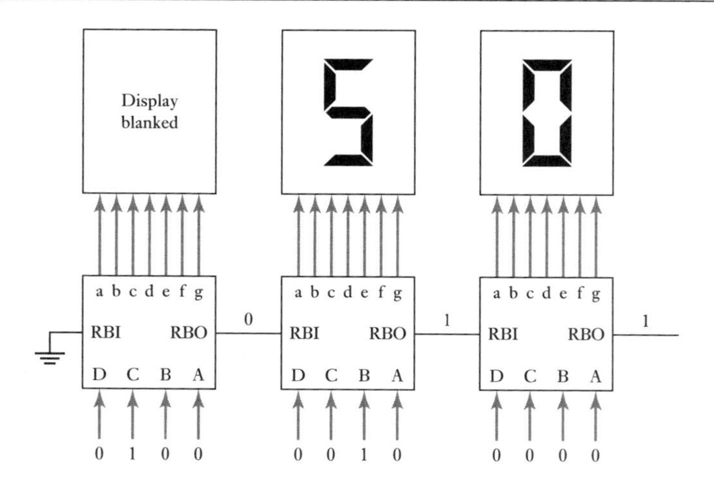

Figure 20.20 Ripple blanking.

With displays having many display elements, rather than use a decoder for each element, multiplexing is used with a single decoder. Figure 20.21 shows the circuit for multiplexing a four-element common cathode type of display. The BCD data is outputted from port A and the decoder presents the decoder output to all the displays. Each display has its common cathode connected to ground through a transistor. The display cannot light up unless the transistor is switched on by an output from port B. Thus by switching between PB0, PB1, PB2 and PB3 the output from port A can be switched to the appropriate display. To maintain a constant display, a display is repeatedly turned on sufficiently often for the display to appear flicker-free.

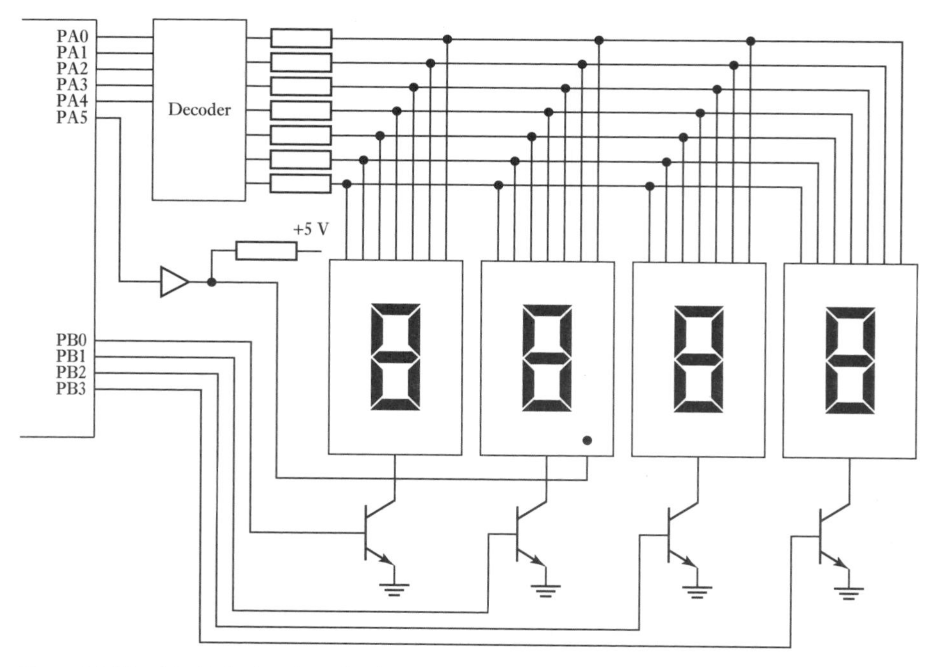

Figure 20.21 Multiplexing four displays.

Time division multiplexing can be used to enable more than one digit to be displayed at a time.

20.6.2 Interfacing analogue

Digital-to-analogue conversion is required when the output from a microprocessor or microcontroller is required to provide an analogue signal output. For example, the DAC Analog Devices AD557 can be used for this purpose. It produces an output voltage proportional to its digital input and input latches are available for microprocessor interfacing. If the latches are not required then pins 9 and 10 are connected to ground. Data is latched when there is a positive edge, i.e. a change from low to high, on the input to either pin 9 or pin 10. The data is then held until both these pins return to the low level. When this happens, the data is transferred from the latch to the DAC for conversion to an analogue voltage.

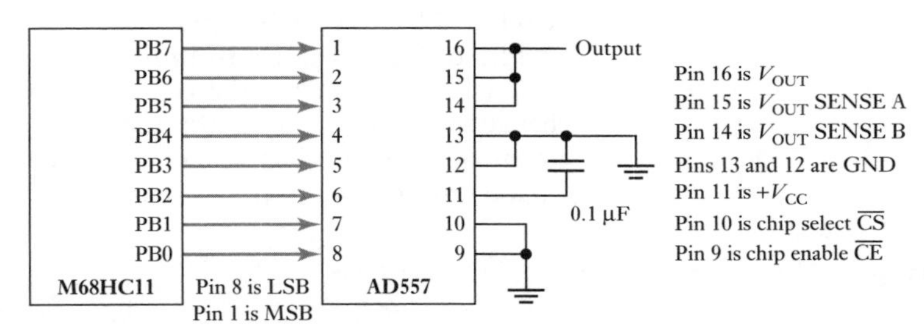

Figure 20.22 Waveform generation.

Figure 20.22 shows the AD557 with the latch not being used and connected to a Motorola M68HC11 so that, when the following program is run, it generates a sawtooth voltage output. Other voltage waveforms can readily be generated by changing the program:

```
BASE     EQU $1000      ; Base address of I/O registers
PORTB    EQU $04        ; Offset of PORTB from BASE

         ORG $C000
         LDX #BASE      ; Point X to register base
         CLR PORTB,X    ; Send 0 to the DAC
AGAIN    INC PORTB,X    ; Increment by 1
         BRA AGAIN      ; Repeat
         END
```

Summary

Interface requirements often mean electrical buffering/isolation, timing control, code conversion, changing the number of lines, serial to parallel and vice versa, conversion from analogue to digital and vice versa. Unless two devices can send and receive data at identical rates, **handshaking** is necessary.

Polling is the program control of inputs/outputs in which a loop is used continuously to read inputs and update outputs, with jumps to service routines as required, i.e. a process of repeatedly checking each peripheral device to see if it is ready to send or accept a new byte of data. An alternative to program control is **interrupt control.** An interrupt involves a peripheral device activating a separate interrupt request line. The reception of an interrupt results in the microprocessor suspending execution of its main program and jumping to the service routine for the peripheral. After the interrupt service routine, the contents of the memory are restored and the microprocessor can continue executing the main program from where it was interrupted.

There are two basic types of serial data transfer: asynchronous and synchronous. With **asynchronous transmission**, the receiver and the transmitter each use their own clock signals so it is not possible for a receiver to know when a word starts or stops. Thus it is necessary for each transmitted data word to carry its own start and stop bits so that it is possible for the receiver to tell where one word stops and another starts. With **synchronous transmission**, the transmitter and receiver have a common clock signal and thus transmission and reception can be synchronised.

Peripheral interface adapters (PIAs) are programmable input/output interface devices which permit various different input and output options to be selected by means of software.

The **universal asynchronous receiver/transmitter** (UART) is the essential element of a serial communication system, the function being to change serial data to parallel for input and parallel data to serial for output. A common programmable form of a UART is the **asynchronous communications interface adapter** (ACIA).

Problems

20.1 Describe the functions that can be required of an interface.

20.2 Explain the difference between a parallel and a serial interface.

20.3 Explain what is meant by a memory-mapped system for inputs/outputs.

20.4 What is the function of a peripheral interface adapter?

20.5 Describe the architecture of the Motorola MC6821 PIA.

20.6 Explain the function of an initialisation program for a PIA.

20.7 What are the advantages of using external interrupts rather than software polling as a means of communication with peripherals?

20.8 For a Motorola MC6821 PIA, what value should be stored in the control register if CA1 is to be disabled, CB1 is to be an enabled interrupt input and set by a low-to-high transition, CA2 is to be enabled and used as a set/reset output, and CB2 is to be enabled and go low on the first low-to-high E transition following a microprocessor write into peripheral data register B and return high by the next low-to-high E transition?

20.9 Write, in assembly language, a program to initialise the Motorola MC6821 PIA to achieve the specification given in problem 20.8.

20.10 Write, in assembly language, a program to initialise the Motorola MC6821 PIA to read 8 bits of data from port A.

Chapter twenty-one Programmable Logic Controllers

Objectives

The objectives of this chapter are that, after studying it, the reader should be able to:

- **Describe the basic structure of PLCs and their operation.**
- **Develop ladder programs for a PLC involving logic functions, latching, internal relays and sequencing.**
- **Develop programs involving timers, counters, shift registers, master relays, jumps and data handling.**

21.1 Programmable logic controller

A **programmable logic controller** (PLC) is a digital electronic device that uses a programmable memory to store instructions and to implement functions such as logic, sequencing, timing, counting and arithmetic in order to control machines and processes and has been specifically designed to make programming easy. The term logic is used because the programming is primarily concerned with implementing logic and switching operations. Input devices, e.g. switches, and output devices, e.g. motors, being controlled are connected to the PLC and then the controller monitors the inputs and outputs according to the program stored in the PLC by the operator and so controls the machine or process. Originally PLCs were designed as a replacement for hard-wired relay (e.g. Figure 9.2) and timer logic control systems. PLCs have the great advantage that it is possible to modify a control system without having to rewire the connections to the input and output devices, the only requirement being that an operator has to key in a different set of instructions. Also they are much faster than relay-operated systems. The result is a flexible system which can be used to control systems which vary quite widely in their nature and complexity. Such systems are widely used for the implementation of logic control functions because they are easy to use and program.

PLCs are similar to computers but have certain features which are specific to their use as controllers. These are:

1 They are rugged and designed to withstand vibrations, temperature, humidity and noise.
2 The interfacing for inputs and outputs is inside the controller.
3 They are easily programmed.

21.2 Basic PLC structure

Figure 21.1 shows the basic internal structure of a PLC. It consists essentially of a central processing unit (CPU), memory and input/output interfaces. The CPU controls and processes all the operations within the PLC. It

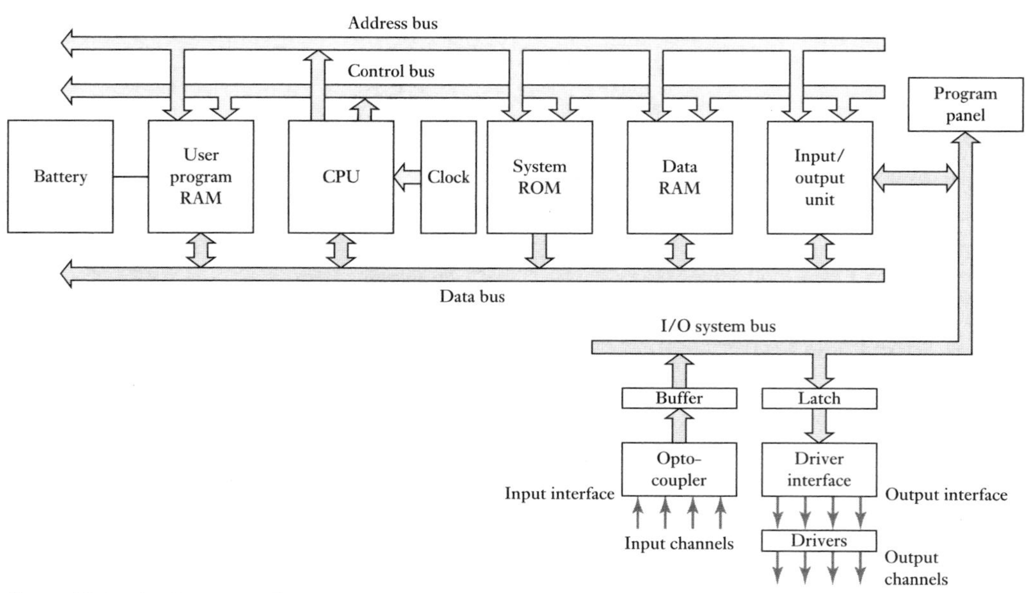

Figure 21.1 Architecture of a PLC.

is supplied with a clock with a frequency of typically between 1 and 8 MHz. This frequency determines the operating speed of the PLC and provides the timing and synchronisation for all elements in the system. A bus system carries information and data to and from the CPU, memory and input/output units. There are several memory elements: a system ROM to give permanent storage for the operating system and fixed data, RAM for the user's program, and temporary buffer stores for the input/output channels.

21.2.1 Input/output

The input and output units provide the interface between the system and the outside world and are where the processor receives information from external devices and communicates information to external devices. The input/output interfaces provide isolation and signal conditioning functions so that sensors and actuators can often be directly connected to them without the need for other circuitry. Inputs might be from limit switches which are activated when some event occurs, or other sensors such as temperature sensors, or flow sensors. The outputs might be to motor starter coils, solenoid valves, etc. Electrical isolation from the external world is usually by means of optoisolators (see Section 3.3).

Figure 21.2 shows the basic form of an input channel. The digital signal that is generally compatible with the microprocessor in the PLC is 5 V d.c. However, signal conditioning in the input channel, with isolation, enables a wide range of input signals to be supplied to it. Thus, with a larger PLC we might have possible input voltages of 5 V, 24 V, 110 V and 240 V. A small PLC is likely to have just one form of input, e.g. 24 V.

The output to the output unit will be digital with a level of 5 V. Outputs are specified as being of relay type, transistor type or triac type. With the relay type, the signal from the PLC output is used internally to operate a relay and so is able to switch currents of the order of a few amperes in an

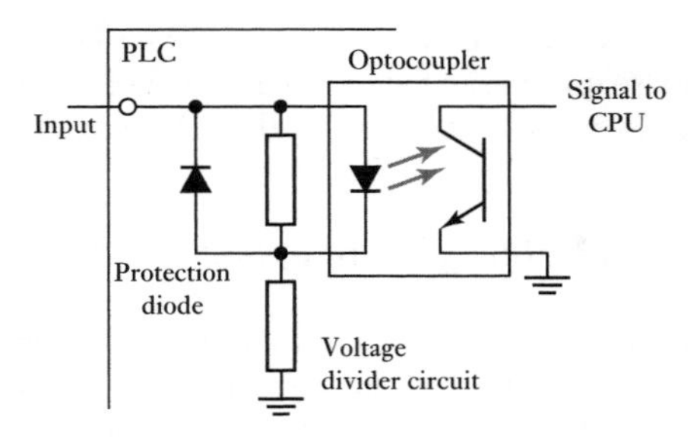

Figure 21.2 Input channel.

external circuit. The relay isolates the PLC from the external circuit and can be used for both d.c. and a.c. switching. Relays are, however, relatively slow to operate. The transistor type of output uses a transistor to switch current through the external circuit. This gives a faster switching action. Optoisolators are used with transistor switches to provide isolation between the external circuit and the PLC. The transistor output is only for d.c. switching. Triac outputs can be used to control external loads which are connected to the a.c. power supply. Optoisolators are again used to provide isolation. Thus we can have outputs from the output channel which might be a 24 V, 100 mA switching signal, a d.c. voltage of 110 V, 1 A or perhaps 240 V, 1 A a.c., or 240 V, 2 A a.c., from a triac output channel. With a small PLC, all the outputs might be of one type, e.g. 240 V a.c., 1 A. With modular PLCs, however, a range of outputs can be accommodated by selection of the modules to be used.

The terms **sourcing** and **sinking** are used to describe the way in which d.c. devices are connected to a PLC. With sourcing, using the conventional current flow direction as from positive to negative, an input device receives current from the input module (Figure 21.3(a)). If the current flows from the output module to an output load then the output module is referred to as sourcing (Figure 21.3(b)). With sinking, an input device supplies current to the input module (Figure 21.3(c)). If the current flows to the output module from an output load then the output module is referred to as sinking (Figure 21.3(d)).

The input/output unit provides the interface between the system and the outside world, allowing for connections to be made through input/output channels to input devices such as sensors and output devices such as motors and solenoids. It is also through the input/output unit that programs are entered from a program panel. Every input/output point has a unique address which can be

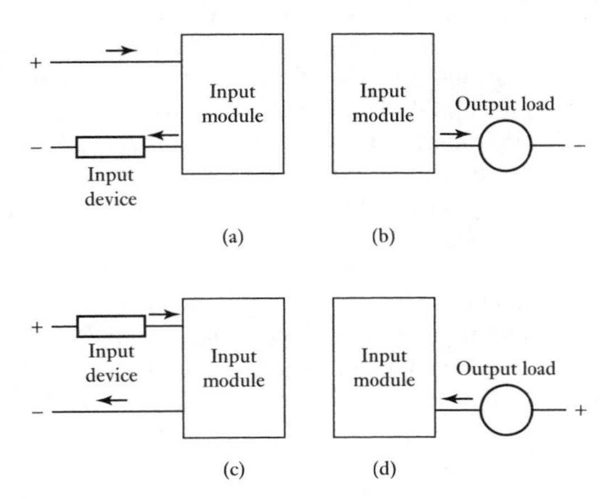

Figure 21.3 (a), (b) Sourcing, (c), (d) sinking.

used by the CPU. It is like a row of houses along a road: number 10 might be the 'house' to be used for an input from a particular sensor while number '45' might be the 'house' to be used for the output to a particular motor.

21.2.2 Inputting programs

Programs are entered into the input/output unit from small hand-held programming devices, desktop consoles with a visual display unit (VDU), keyboard and screen display, or by means of a link to a personal computer (PC) which is loaded with an appropriate software package. Only when the program has been designed on the programming device and is ready is it transferred to the memory unit of the PLC.

The programs in RAM can be changed by the user. However, to prevent the loss of these programs when the power supply is switched off, a battery is likely to be used in the PLC to maintain the RAM contents for a period of time. After a program has been developed in RAM it may be loaded into an EPROM chip and so made permanent. Specifications for small PLCs often specify the program memory size in terms of the number of program steps that can be stored. A program step is an instruction for some event to occur. A program task might consist of a number of steps and could be, for example: examine the state of switch A, examine the state of switch B, if A and B are closed then energise solenoid P which then might result in the operation of some actuator. When this happens another task might then be started. Typically the number of steps that can be handled by a small PLC is of the order of 300 to 1000, which is generally adequate for most control situations.

21.2.3 Forms of PLCs

PLCs were first conceived in 1968. They are now widely used and extend from small self-contained units, i.e. single boxes, for use with perhaps 20 digital input/outputs to rack-mounted systems which can be used for large numbers of inputs/outputs, handle digital or analogue inputs/outputs, and also carry out PID control modes. The single-box type is commonly used for small programmable controllers and is supplied as an integral compact package complete with power supply, processor, memory and input/output units. Typically such a PLC might have 6, 8, 12 or 24 inputs and 4, 8 or 16 outputs and a memory which can store some 300 to 1000 instructions. For example, the MELSEC FX3U has models which can have 6, 8, 12 or 24 inputs and 4, 8 or 16 relay outputs and a memory which can store some 300 to 1000 instructions. Some systems are able to be extended to cope with more inputs and outputs by linking input/output boxes to them.

Systems with larger numbers of inputs and outputs are likely to be modular and designed to fit in racks. These consist of separate modules for power supply, processor, input/output, etc., and are mounted on rails within a metal cabinet. The rack type can be used for all sizes of programmable controllers and has the various functional units packaged in individual modules which can be plugged into sockets in a base rack. The mix of modules required for a particular purpose is decided by the user and the appropriate ones then plugged into the rack. So the number of input/output connections can be increased by just adding more input/output modules. For example, the SIMATIC S7-300/400 PLC is rack mounted with components for the power supply, the

CPU, input/output interface modules, signal modules which can be used to provide signal conditioning for inputs or outputs and communication modules which can be used to connect PLCs to each other or to other systems.

Another example of a modular system is that provided by the Allen-Bradley SLC-500 programmable logic controller system. This is a small, chassis-based, modular family of programmable controllers having multiple processor choices, numerous power supply options and extensive I/O capacity. The SLC 500 allows the creation of a system specifically designed for an application. PLC blocks are mounted in a rack, with interconnections between the blocks being via a backplane bus. The PLC power supply is the end box in a rack with the next box containing the microprocessor. The backplane bus has copper conductors and provides the means by which the blocks slotted into the rack receive electrical power and for exchanging data between the modules and the processor. The modules slide into the rack and engage connectors on the backplane. SLC 500 series PLC racks are available to take 4, 7, 10 or 13 modules. Modules are available providing 8, 16 or current sinking d.c. inputs, 8, 16 or 32 current sourcing d.c. outputs, 8, 16 or 32 current sourcing d.c. outputs, 4, 8 or 16 relay a.c./d.c. outputs, communication module to enable additional communications with other computers or PLCs. Software is available to allow for programming from a Windows environment.

21.3 Input/output processing

A PLC is continuously running through its program and updating it as a result of the input signals. Each such loop is termed a **cycle**. There are two methods that can be used for input/output processing: continuous updating and mass input/output copying.

21.3.1 Continuous updating

Continuous updating involves the CPU scanning the input channels as they occur in the program instructions. Each input point is examined individually and its effect on the program determined. There will be a built-in delay, typically about 3 ms, when each input is examined in order to ensure that only valid input signals are read by the microprocessor. This delay enables the microprocessor to avoid counting an input signal twice, or, more frequently, if there is contact bounce at a switch. A number of inputs may have to be scanned, each with a 3 ms delay, before the program has the instruction for a logic operation to be executed and an output to occur. The outputs are latched so that they retain their status until the next updating.

21.3.2 Mass input/output copying

Because, with continuous updating, there has to be a 3 ms delay on each input, the time taken to examine several hundred input/output points can become comparatively long. To allow a more rapid execution of a program, a specific area of RAM is used as a buffer store between the control logic and the input/output unit. Each input/output has an address in this memory. At the start of each program cycle the CPU scans all the inputs and copies their status into the input/output addresses in RAM. As the program is executed the stored, input data is read, as required, from RAM and the logic operations carried out. The resulting output signals are stored in the reserved input/output section of RAM. At the end of each program cycle all the outputs are transferred from RAM to the output channels. The outputs are latched so that they retain their status until the next updating. The sequence is:

1 Scan all the inputs and copy into RAM.
2 Fetch and decode and execute all program instructions in sequence, copying output instructions to RAM.
3 Update all outputs.
4 Repeat the sequence.

A PLC takes time to complete a cycle of scanning inputs and updating outputs according to the program instructions and so the inputs are not watched all the time but only examined periodically. A typical PLC cycle time is of the order of 10 to 50 ms and so the inputs and outputs are updated every 10 to 50 ms. This means that if a very brief input appears at the wrong moment in the cycle, it could be missed. Thus for a PLC with a cycle time of 40 ms, the maximum frequency of digital impulses that can be detected will be if one pulse occurs every 40 ms. The Mitsubishi compact PLC, MELSEC FX3U, has a quoted program cycle time of 0.065 μs per logical instruction and so the more complex the program, the longer the cycle time.

21.3.3 I/O addresses

The PLC has to be able to identify each particular input and output and it does this by assigning addresses to each, rather like houses in a town have addresses to enable post to be delivered to the right families. With a small PLC the addresses are likely to be just a number preceded by a letter to indicate whether it is an input or output. For example, Mitsubishi and Toshiba have inputs identified as X400, X401, X402, etc., and outputs as Y430, Y431, etc. With larger PLCs having several racks of input and output channels and a number of modules in each rack, the racks and modules are numbered and so an input or output is identified by its rack number followed by the number of the module in that rack and then a number to show its terminal number in the module. For example, the Allen-Bradley PCL-5 has I:012/03 to indicate an input in rack 01 at module 2 and terminal 03.

21.4 Ladder programming

The form of programming commonly used with PLCs is **ladder programming**. This involves each program task being specified as though a rung of a ladder. Thus such a rung could specify that the state of switches A and B, the inputs, be examined and if A and B are both closed then a solenoid, the output, is energised. Figure 21.4 illustrates this idea by comparing it with an electric circuit.

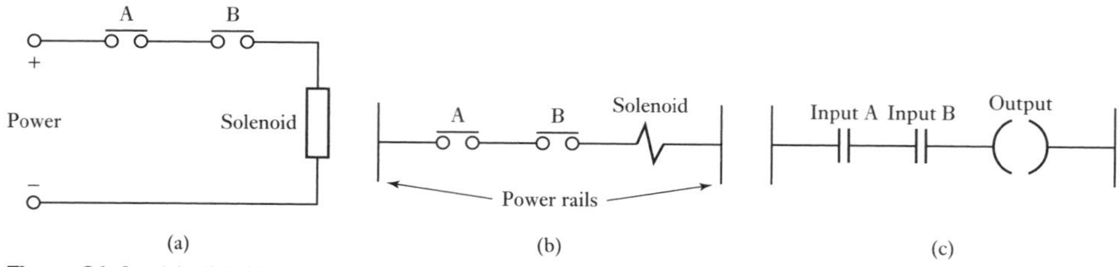

Figure 21.4 (a), (b) Alternative ways of drawing an electric circuit, (c) comparable rung in a ladder program.

The sequence followed by a PLC when carrying out a program can be summarised as:

1 Scan the inputs associated with one rung of the ladder program.
2 Solve the logic operation involving those inputs.
3 Set/reset the outputs for that rung.
4 Move on to the next rung and repeat operations 1, 2, 3.
5 Move on to the next rung and repeat operations 1, 2, 3.
6 Move on to the next rung and repeat operations 1, 2, 3.
7 And so on until the end of the program with each rung of the ladder program scanned in turn. The PLC then goes back to the beginning of the program and starts again.

PLC programming based on the use of **ladder diagrams** involves writing a program in a similar manner to drawing a switching circuit. The ladder diagram consists of two vertical lines representing the power rails. Circuits are connected as horizontal lines, i.e. the rungs of the ladder, between these two verticals. Figure 21.5 shows the basic standard symbols that are used and an example of rungs in a ladder diagram. In drawing the circuit line for a rung, inputs must always precede outputs and there must be at least one output on each line. Each rung must start with an input or a series of inputs and end with an output.

Figure 21.5 Ladder program.

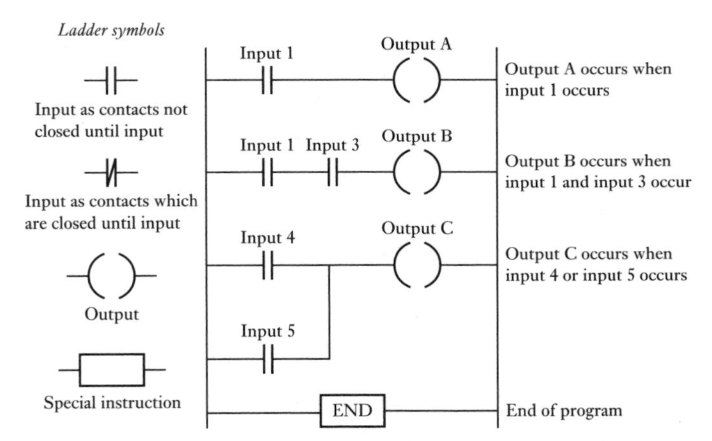

To illustrate the drawing of a ladder diagram, consider a situation where the output from the PLC is to energise a solenoid when a normally open start switch connected to the input is activated by being closed (Figure 21.6(a)). The program required is shown in Figure 21.6(b)). Starting with the input, we have the normally open symbol | |. This might have an input address X400. The line terminates with the output, the solenoid, with the symbol ( ). This might have the output address Y430. To indicate the end of the program, the end rung is marked. When the switch is closed the solenoid is activated. This might, for example, be a solenoid valve which opens to allow water to enter a vessel.

Figure 21.6 Switch controlling a solenoid.

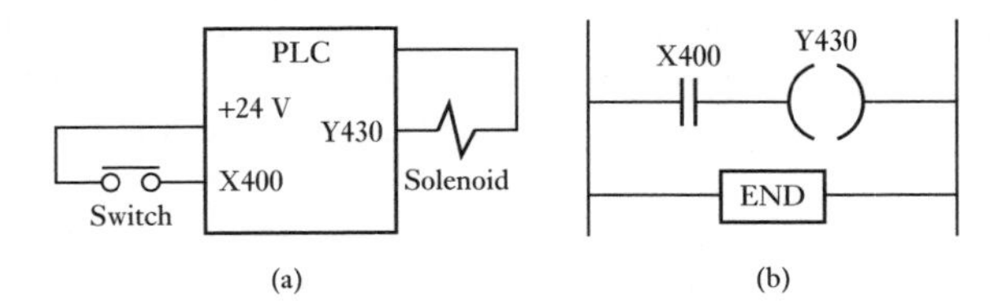

Another example might be an on/off temperature control (Figure 21.7(a)) in which the input goes from low to high when the temperature sensor reaches the set temperature. The output is then to go from on to off. The temperature sensor shown in the figure is a thermistor connected in a bridge arrangement with output to an operational amplifier connected as a comparator (see Section 3.2.7). The program (Figure 21.7(b)) shows the input as a normally closed pair of contacts, so giving the on signal and hence an output. When the contacts are opened to give the off signal then the output is switched off.

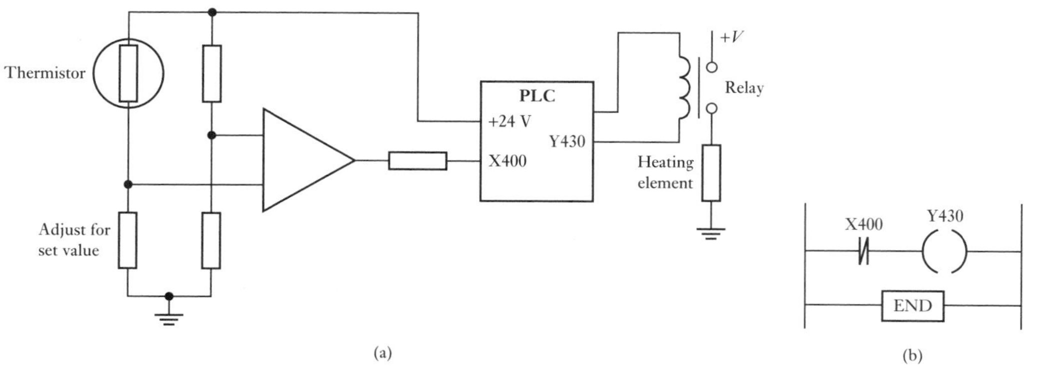

Figure 21.7 Temperature control system.

21.4.1 Logic functions

The logic functions can be obtained by combinations of switches (see Section 5.2) and the following shows how we can write ladder programs for such combinations (Figure 21.8):

1 AND
Figure 21.8(a) shows a situation where a coil is not energised unless two, normally open, switches are both closed. Switch A and switch B have both to be closed, which thus gives an AND logic situation. The equivalent ladder diagram starts with | |, labelled Input 1, to represent switch A and in series with it | |, labelled Input 2, to represent switch B. The line then terminates with ( ) to represent the output.

2 OR
Figure 21.8(b) shows a situation where a coil is not energised until either, normally open, switch A or B is closed. The situation is an OR logic gate. The equivalent ladder diagram starts with | |, labelled Input 1, to represent switch A and in parallel with it | |, labelled Input 2, to represent switch B. The line then terminates with ( ) to represent the output.

3 NOR
Figure 21.8(c) shows how we can represent the ladder program line for a NOR gate. Since there has to be an output when neither A nor B have an input, and when there is an input to A or B the output ceases, the ladder program shows Input 1 in parallel with Input 2, with both being represented by normally closed contacts.

4 NAND
Figure 21.8(d) shows a NAND gate. There is no output when both A and B have an input. Thus for the ladder program line to obtain an output we require no inputs to Input 1 and to Input 2.

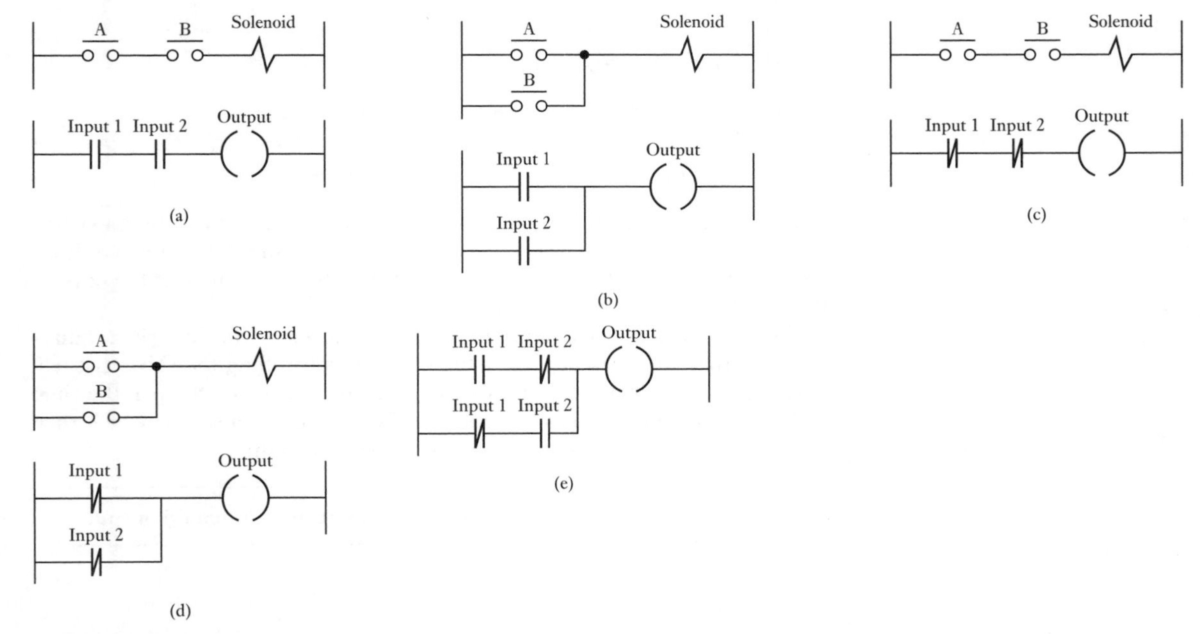

Figure 21.8 (a) AND, (b) OR, (c) NOR, (d) NAND, (e) XOR.

5 EXCLUSIVE-OR (XOR)
Figure 21.8(e) shows how we can draw the ladder program line for an XOR gate, there being no output when there is no input to Input 1 and Input 2 and when there is an input to both Input 1 and Input 2. Note that we have represented each input by two sets of contacts, one normally open and one normally closed.

Consider a situation where a normally open switch A must be activated and either of two other, normally open, switches B and C must be activated for a coil to be energised. We can represent this arrangement of switches as switch A in series with two parallel switches B and C (Figure 21.9(a)). For the coil to be energised we require A to be closed and either B or C to be closed. Switch A when considered with the parallel switches gives an AND logic situation. The two parallel switches give an OR logic situation. We thus have a combination of two gates. The truth table is:

Inputs			
A	**B**	**C**	**Output**
0	0	0	0
0	0	1	0
0	1	0	0
0	1	1	0
1	0	0	0
1	0	1	1
1	1	0	1
1	1	1	1

Figure 21.9 Switches controlling a solenoid.

For the ladder diagram, we start with | |, labelled Input 1, to represent switch A. This is in series with two | | in parallel, labelled Input 1 and Input 2, for switches B and C. The line then terminates with ( ) to represent the output, the coil. Figure 21.9(b) shows the line.

As a simple example of a program using logic gates, consider the requirement for there to be an output to the solenoid controlling the valve that will open a shop door when the shopkeeper has closed a switch to open the shop and a customer approaches the door and is detected by a sensor which then gives a high signal. The truth table for this system is thus:

Shop open switch	Customer approaching sensor	Solenoid output
Off	Off	Off
Off	On	Off
On	Off	Off
On	On	On

This truth table is that of an AND gate and thus the program for a PLC controlling the door is as shown in Figure 21.10.

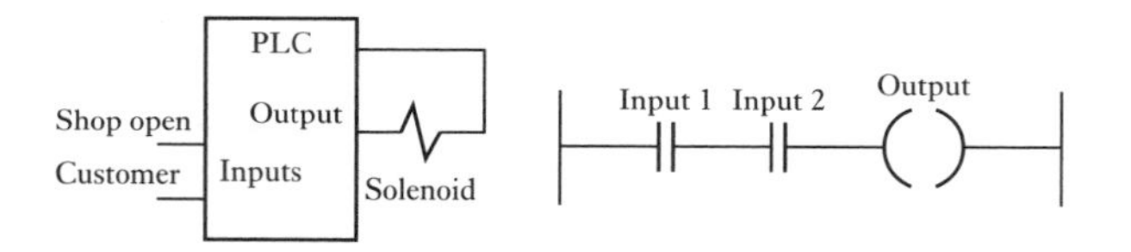

Figure 21.10 Shop door system.

21.5 Instruction lists

Each horizontal rung on the ladder in a ladder program represents a line in the program and the entire ladder gives the complete program in 'ladder language'. The programmer can enter the program into the PLC using a keyboard with the graphic symbols for the ladder elements, or using a computer screen and a mouse to select symbols, and the program panel or computer then translates these symbols into machine language that can be stored in the PLC memory. There is an alternative way of entering a program and that is to translate the ladder program into an **instruction list** and then enter this into the programming panel or computer.

Instruction lists consist of a series of instructions with each instruction being on a separate line. An instruction consists of an operator followed by one or more operands, i.e. the subjects of the operator. In terms of ladder programs, each operator in a program may be regarded as a ladder element. Thus we might have for the equivalent of an input to a ladder program:

```
LD A (*Load input A*)
```

Table 21.1 Instruction code mnemonics.

IEC 1131-3	Mitsubishi	OMRON	Siemens	Operation	Ladder diagram
LD	LD	LD	A	Load operand into result register	Start a rung with open contacts
LDN	LDI	LD NOT	AN	Load negative operand into result register	Start a rung with closed contacts
AND	AND	AND	A	Boolean AND	A series element with open contacts
ANDN	ANI	AND NOT	AN	Boolean AND with negative operand	A series element with closed contacts
OR	OR	OR	O	Boolean OR	A parallel element with open contacts
ORN	ORI	OR NOT	ON	Boolean OR with negative operand	A parallel element with closed contacts
ST	OUT	OUT	=	Store result register into operand	An output from a rung

The operator is LD for loading, the operand A as the subject being loaded and the words preceded by and concluded by * in brackets are comments explaining what the operation is and are not part of the program operation instructions to the PLC, but to aid a reader in comprehending what the program is about.

The mnemonic codes used by different PLC manufacturers differ but an international standard (IEC 1131-3) has been proposed and is widely used. Table 21.1 shows common core mnemonics. In examples discussed in the rest of this chapter, where general descriptions are not used, the Mitsubishi mnemonics will be used. However, those used by other manufacturers do not differ widely from these and the principles involved in their use are the same.

21.5.1 Instruction lists and logic functions

The following shows how individual rungs on a ladder are entered using the Mitsubishi mnemonics where logic functions are involved (Figure 21.11).

21.5.2 Instruction lists and branching

The EXCLUSIVE-OR (XOR) gate shown in Figure 21.12 has two parallel arms with an AND situation in each arm. In such a situation Mitsubishi (Figure 21.12(a)) uses an ORB instruction to indicate 'OR together parallel branches'. The first instruction is for a normally open pair of contacts X400, the next instruction for a set of normally closed contacts X401, hence ANI X401. The third instruction describes a new line, its being recognised as a new line because it starts with LDI, all new lines starting with LD or LDI. Because the first line has not been ended by an output, the PLC recognises that a parallel line is involved for the second line and reads together

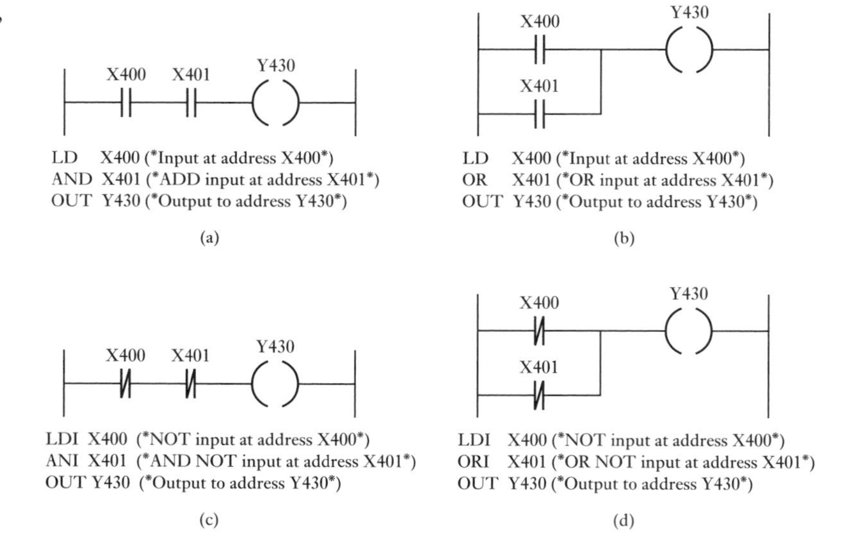

Figure 21.11 (a) AND, (b) OR, (c) NOR, (d) NAND.

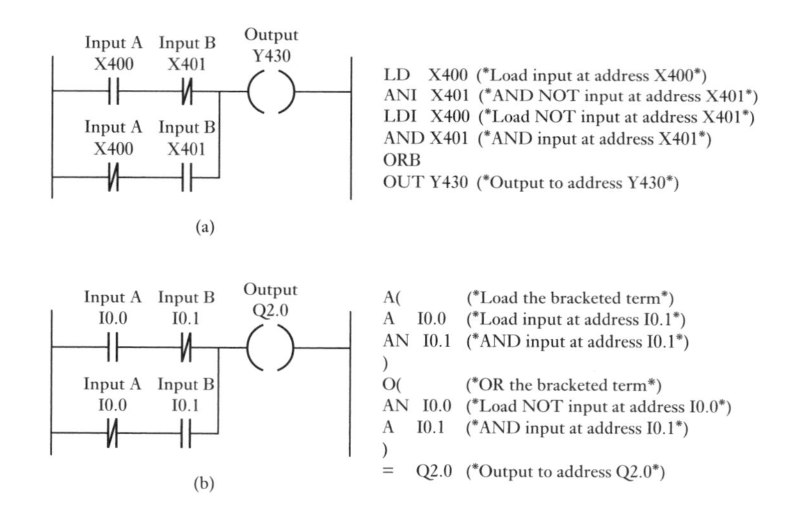

Figure 21.12 XOR.

the listed elements until the ORB instruction is reached. ORB indicates to the PLC that it should OR the results of the first and second instructions with that of the new branch with the third and fourth instructions. The list concludes with the output OUT Y430. Figure 21.12(b) shows the Siemens version of XOR gate. Brackets are used to indicate that certain instructions are to be carried out as a block and are used in the same way as brackets in any mathematical equation. For example, $(1 + 2)/4$ means that the 1 and 2 must be added before dividing by 4. Thus with the Siemens instruction list the A(means that the load instruction A is only applied after the bracketed steps have been completed and) is reached. The IEC 1131-3 standard for such programming is to use brackets in the way used in the above Siemens example.

21.6 Latching and internal relays

There are often situations where it is necessary to hold a coil energised, even when the input which energised it ceases. The term **latch circuit** is used for the circuit which carries out such an operation. It is a self-maintaining circuit in that, after being energised, it maintains that state until another input is received. It remembers its last state. An example of a latch circuit is shown in Figure 21.13. When Input 1 is energised and closes, there is an output. However, when there is an output, a set of contacts associated with the output is energised and closes. These contacts OR the Input 1 contacts. Thus, even if Input 1 contacts open, the circuit will still maintain the output energised. The only way to release the output is by operating the normally closed contact Input 2.

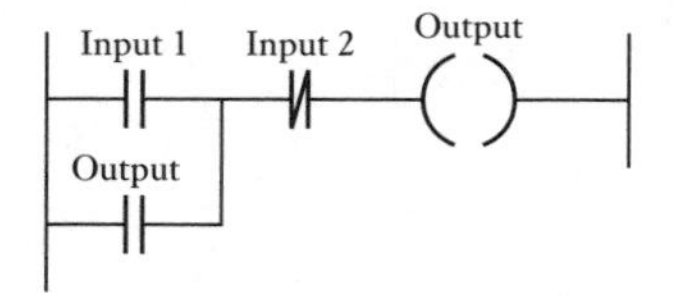

Figure 21.13 A latch circuit.

As an example of the use of a latching circuit, consider the requirement for a PLC to control a motor so that when the start signal button is momentarily pressed the motor starts and when the stop switch is used the motor switches off. Safety must be a priority in the design of a PLC system, so stop buttons must be hard-wired and not depend on the PLC software for implementation so that if there is a failure of the stop switch or PLC, the system is automatically safe. With a PLC system, a stop signal can be provided by a switch as shown in Figure 21.14(a). To start we momentarily close the press-button start switch and the motor internal control relay latches this closure and the output remains on. To stop we momentarily open the stop switch and this unlatches the start switch. However, if the stop switch cannot be operated then we cannot stop the system. Thus this system must *not* be used as it is unsafe, because if there is a fault and the switch cannot be operated, then no stop signal can be provided. What we require is a system that will still stop if a failure occurs in the stop switch. Figure 21.14(b) shows such a system. The program now has the stop switch as open contacts. However, because the hard-wired stop switch has normally closed contacts, then the program receives the signal to close the program contacts. Pressing the stop switch then opens the program contacts and stops the system.

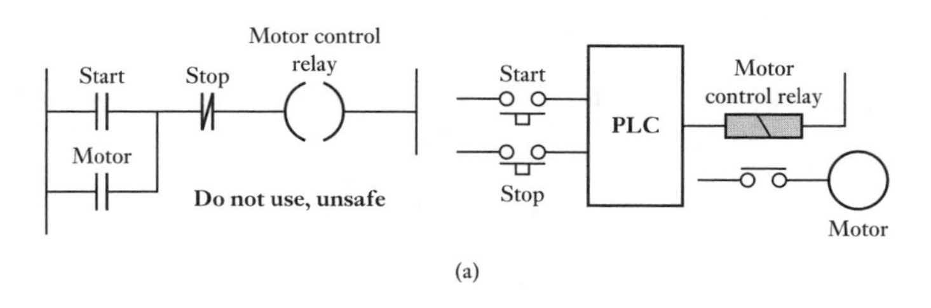

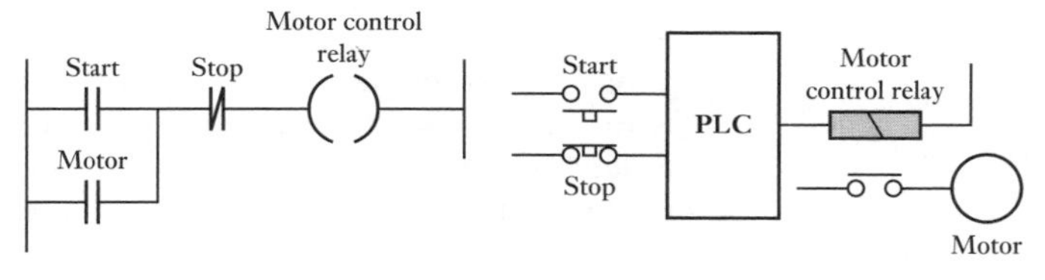

Figure 21.14 Stop system: (a) unsafe, (b) safe.

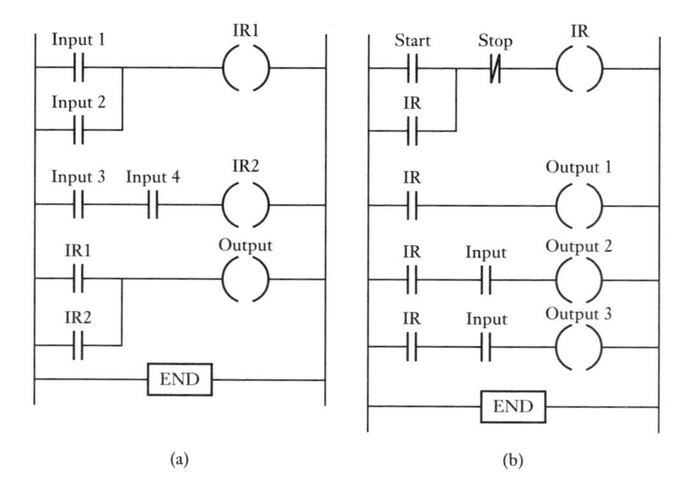

Figure 21.15 (a) An output controlled by two input arrangements, (b) starting of multiple outputs.

21.6.1 Internal relays

The term **internal relay, auxiliary relay** or **marker** is used for what can be considered as an internal relay in the PLC. These behave like relays with their associated contacts, but in reality are not actual relays but simulations by the software of the PLC. Some have battery back-up so that they can be used in circuits to ensure a safe shut-down of plant in the event of a power failure. Internal relays can be very useful aids in the implementation of switching sequences.

Internal relays are often used when there are programs with multiple input conditions. Consider the situation where the excitation of an output depends on two different input arrangements. Figure 21.15(a) shows how we can draw a ladder diagram using internal relays. The first rung shows one input arrangement being used to control the coil of internal relay IR1. The second rung shows the other input arrangement controlling the coil of internal relay IR2. The contacts of the two relays are then put in an OR situation to control the output.

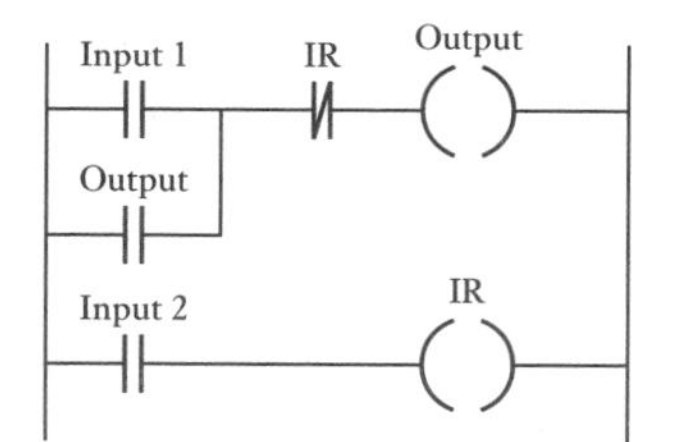

Figure 21.16 Resetting a latch.

Another use of internal relays is for the starting of multiple outputs. Figure 21.15(b) shows such a ladder program. When the start contacts are closed, the internal relay is activated and latches the input. It also starts Output 1 and makes it possible for Outputs 2 and 3 to be activated.

Another example of the use of internal relays is resetting a latch. Figure 21.16 shows the ladder diagram. When the contacts of Input 1 are momentarily pressed, the output is energised. The contacts of the output are then closed and so latch the output, i.e. keep it on even when the contacts of the input are no longer closed. The output can be unlatched by the internal relay contacts opening. This will occur if Input 2 is closed and energises the coil of the internal relay.

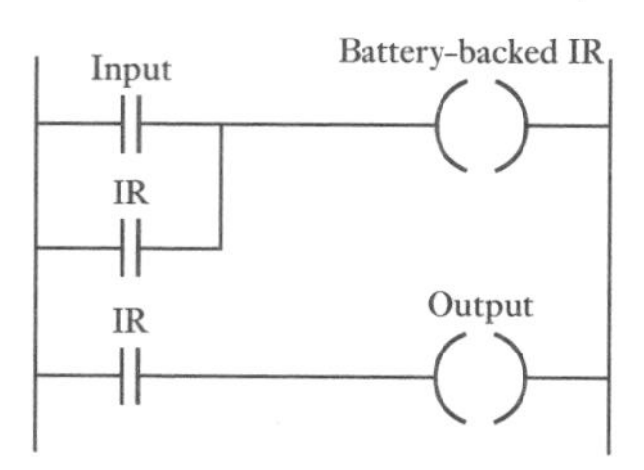

Figure 21.17 Use of a battery-backed internal relay.

An example of the use of a battery-backed internal relay is shown in Figure 21.17. When the contacts of Input 1 close, the coil of the battery-backed internal relay is energised. This closes the internal relay contacts and so even if contacts of the input open as a result of power failure, the internal relay contacts remain closed. This means that the output controlled by the internal relay remains energised, even when there is a power failure.

21.7 Sequencing

There are often control situations where sequences of outputs are required, with the switch from one output to another being controlled by sensors. Consider the requirement for a ladder program for a pneumatic system (Figure 21.18) with double-solenoid valves controlling two double-acting cylinders A and B if limit switches a−, a+, b−, b+ are used to detect the limits of the piston rod movements in the cylinders and the cylinder activation sequence A+, B+, A−, B− is required. A possible program is shown in the figure. A start switch input has been included in the first rung. Thus cylinder extension for A, i.e. the solenoid A+ energised, only occurs when the start switch is closed and the b− switch is closed, this switch indicating that the B cylinder is retracted. When cylinder A is extended, the switch a+, which indicates the extension of A, is activated. This then leads to an output to solenoid B+ which results in B extending. This closes the switch indicating the extension of B, i.e. the b+ switch, and leads to the output to solenoid A− and the retraction of cylinder A. This retraction closes limit switch a− and so gives the output to solenoid B− which results in B retracting. This concludes the program cycle and leads to the first rung again, which awaits the closure of the start switch before being repeated.

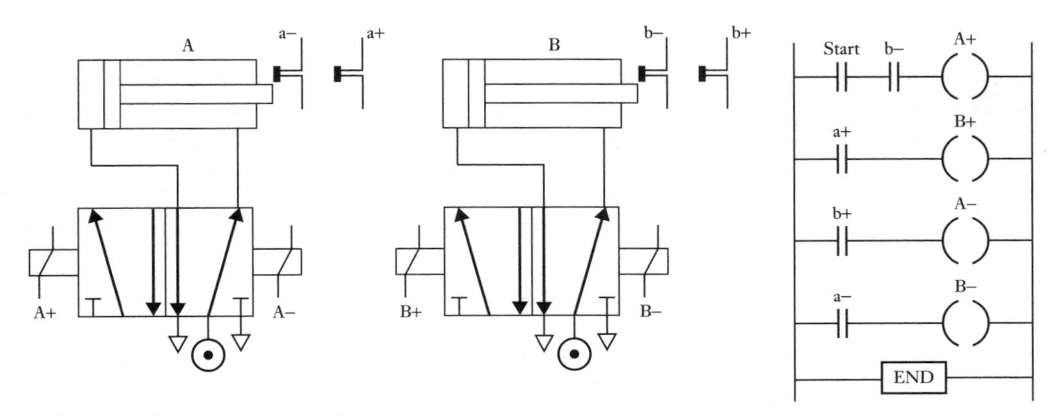

Figure 21.18 Cylinder sequencing.

As a further illustration, consider the problem of devising a ladder program to control a pneumatic system with double-solenoid-controlled valves and two cylinders A and B if limit switches a−, a+, b− and b+ are used to detect the limits of movement of the piston rod movements in the cylinders and the sequence required is for the piston rod in A to extend, followed by the piston rod in B extending, then the piston in B retracting and finally the cycle is completed by the piston in A retracting. An internal relay can be used to switch between groups of outputs to give the form of control for pneumatic cylinders, which is termed **cascade control** (see Section 7.5). Figure 21.19 shows a possible program. When the start switch is closed, the internal relay is activated. This energises solenoid A+ with the result that the piston in cylinder A extends. When extended it closes limit switch a+ and the piston in cylinder B extends. When this is extended it closes the limit switch b+. This activates the relay. As a result the B− solenoid is energised and the piston in B retracts. When this closes limit switch b−, solenoid A− is energised and the piston in cylinder A retracts.

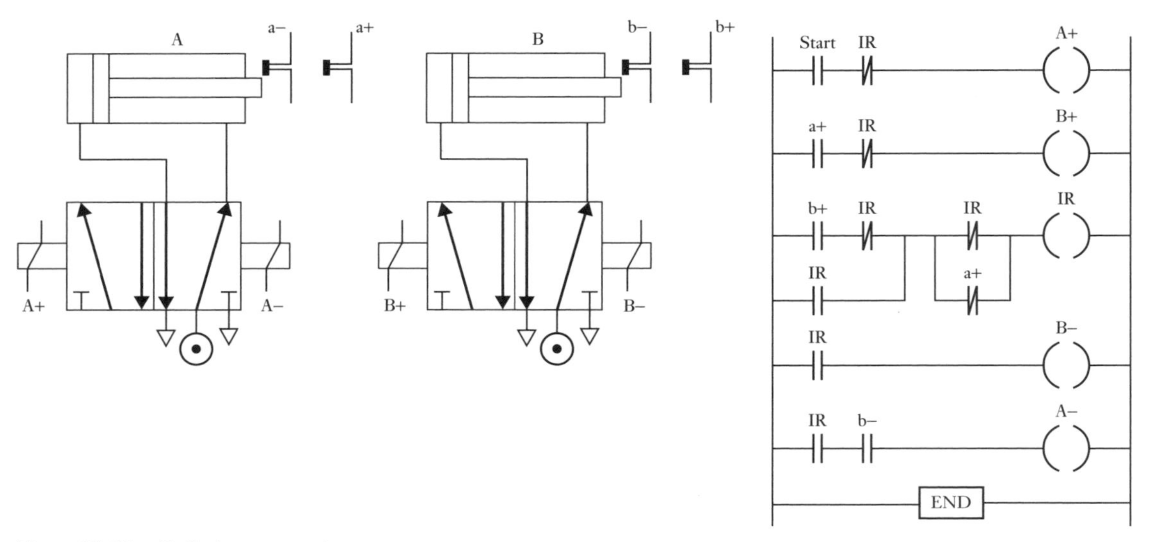

Figure 21.19 Cylinder sequencing.

21.8 Timers and counters

The previous sections in this chapter have been concerned with tasks requiring the series and parallel connections of input contacts. However, there are tasks which can involve time delays and event counting. These requirements can be met by the timers and counters which are supplied as a feature of PLCs. They can be controlled by logic instructions and represented on ladder diagrams.

21.8.1 Timers

A common approach used by PLC manufacturers is to consider timers to behave like relays with coils which when energised result in the closure or opening of contacts after some preset time. The timer is thus treated as an output for a rung with control being exercised over pairs of contacts elsewhere (Figure 21.20(a)). Others consider a timer as a delay block in a rung which delays signals in that rung reaching the output (Figure 21.20(b)).

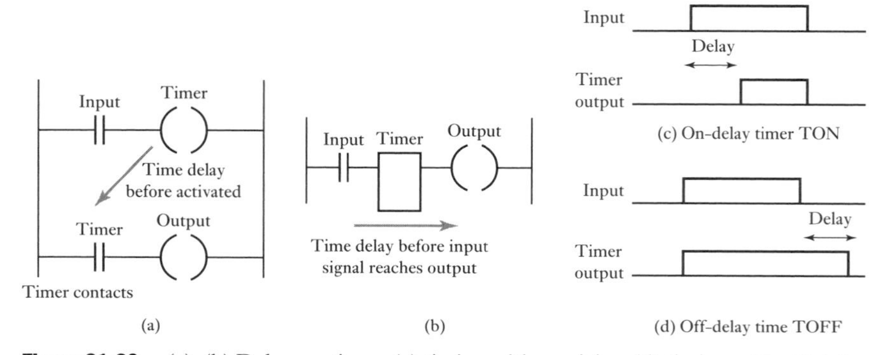

Figure 21.20 (a), (b) Delay-on timer, (c) timing with on-delay, (d) timing with off-delay.

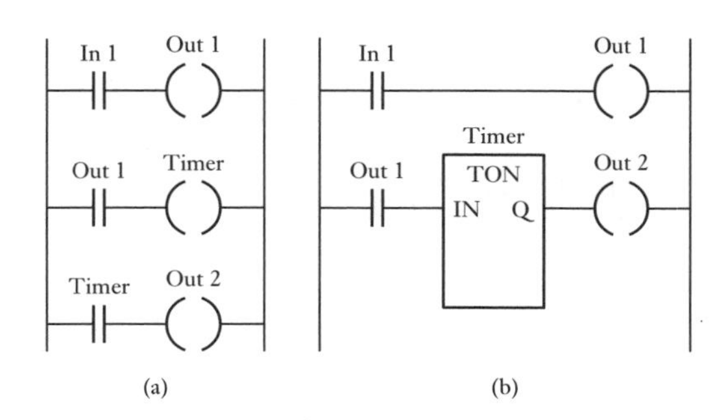

Figure 21.21 Timed sequence.

PLCs are generally provided with only a delay-on timer (TON), small PLCs possibly having only this type. Such a timer waits for a fixed delay period before turning on (Figure 21.20(c)), e.g. a period which can be set between 0.1 and 999 s in steps of 0.1 s. Other time-delay ranges and steps are possible.

As an illustration of the use of a timer for sequencing, consider the ladder diagram shown in Figure 21.21(a) or (b). When the input In 1 is on, the output Out 1 is switched on. The contacts associated with this output then start the timer. The contacts of the timer will close after the preset time delay. When this happens, output Out 2 is switched on.

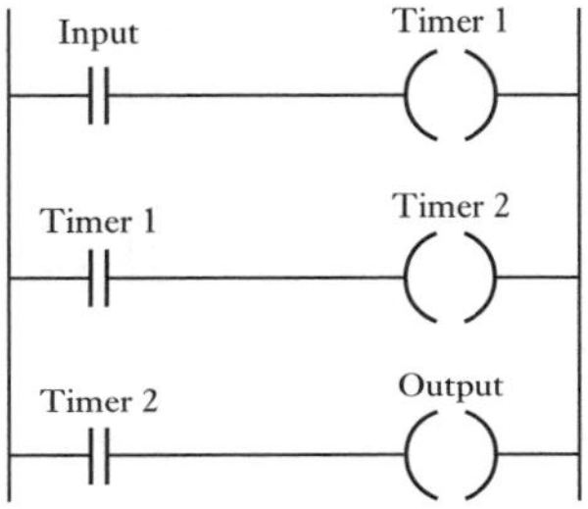

Figure 21.22 Cascaded timers.

Timers can be linked together, or **cascaded**, to give larger delay times than is possible with just one timer. Figure 21.22 shows such an arrangement. When the input contacts close, Timer 1 is started. After its time delay, its contacts close and Timer 2 is started. After its time delay, its contacts close and there is an output.

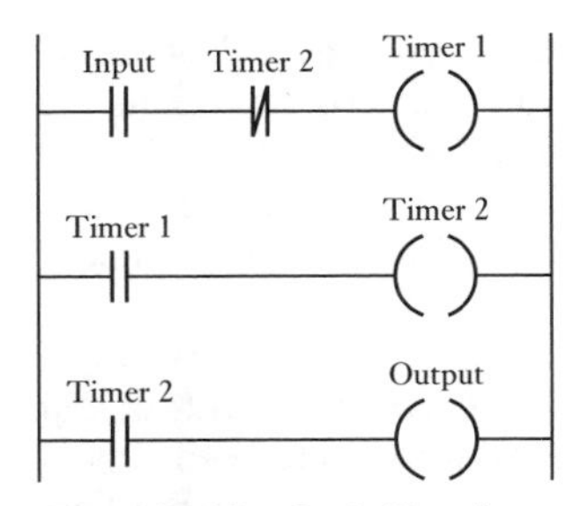

Figure 21.23 On/off cyclic timer.

Figure 21.23 shows a program that can be used to cause an output to go on for 0.5 s, then off for 0.5 s, then on for 0.5 s, then off for 0.5 s, and so on. When the input contacts close, Timer 1 is started and comes on after 0.5 s, this being the time for which it was preset. After this time the Timer 1 contacts close and start Timer 2. It comes on after 0.5 s, its preset time, and opens its contacts. This results in Timer 1 being switched off. This results in its contact opening and switching off Timer 2. This then closes its contact and so starts the entire cycle again. The result is that timer contacts for Timer 1 are switched on for 0.5 s, then off for 0.5 s, on for 0.5 s, and so on. Thus the output is switched on for 0.5 s, then off for 0.5 s, on for 0.5 s, and so on.

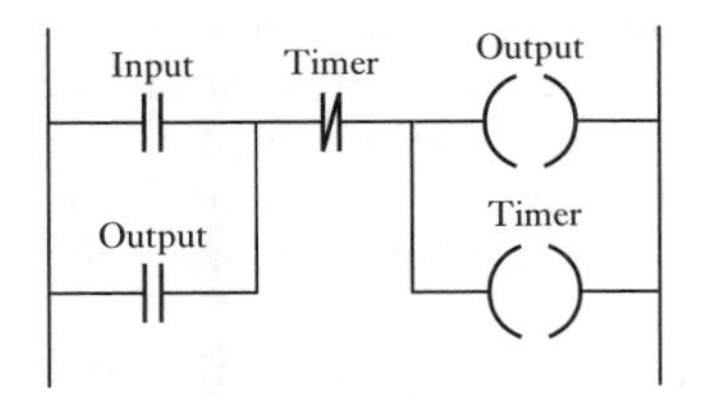

Figure 21.24 Delay-off timer.

Figure 21.24 shows how a delay-off timer, i.e. a timer which switches off an output after a time delay from being energised, can be devised. When the input contacts are momentarily closed the output is energised (Figure 21.20(d)) and the timer started. The output contacts latch the input and keep the output on. After the preset time of the timer, the timer comes on and breaks the latch circuit, so switching the output off.

21.8.2 Counters

Counters are used when there is a need to count a specified number of contact operations, e.g. where items pass along a conveyor into boxes, and when the specified number of items has passed into a box, the next item is diverted into another box. Counter circuits are supplied as an internal feature of

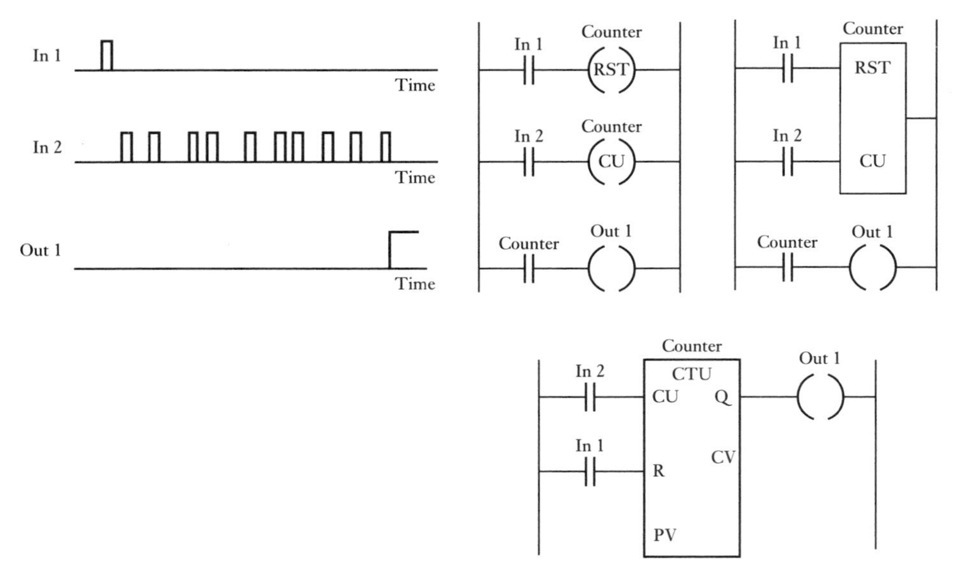

Figure 21.25 The inputs and output for a counter and various ways of representing the same program.

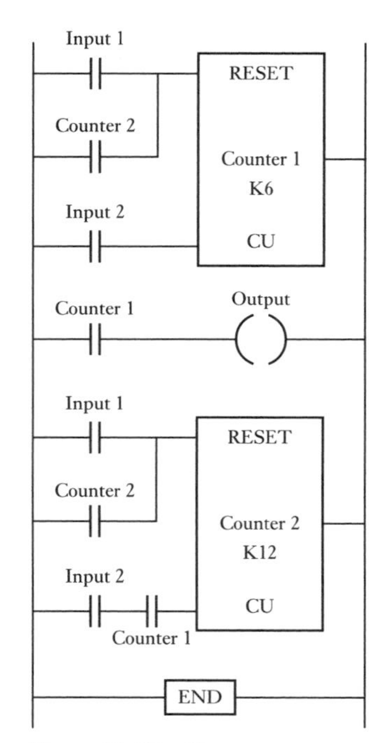

Figure 21.26 Counter.

PLCs. In most cases the counter operates as a **down-counter**. This means that the counter counts down from the present value to zero, i.e. events are subtracted from the set value. When zero is reached the counter's contact changes state. An **up-counter** would count up to the preset value, i.e. events are added until the number reaches the set value. When the set value is reached the counter's contact changes state.

Different PLC manufacturers deal with counters in different ways. Some consider the counter to consist of two basic elements: one output coil to count input pulses and one to reset the counter, the associated contacts of the counter being used in other rungs, e.g. Mitsubishi and Allen-Bradley. Others treat the counter as an intermediate block in a rung from which signals emanate when the count is attained, e.g. Siemens. As an illustration, Figure 21.25 shows a basic counting circuit. When there is a pulse input to In 1, the counter is reset. When there is an input to In 2, the counter starts counting. If the counter is set for, say, 10 pulses, then when 10 pulse inputs have been received at In 2, the counter's contacts will close and there will be an output from Out 1. If at any time during the counting there is an input to In 1, the counter will be reset and start all over again and count for 10 pulses.

As an illustration of the use of a counter, consider the problem of the control for a machine which is required to direct 6 items along one path for packaging in a box, and then 12 items along another path for packaging in another box. Figure 21.26 shows the program that could be used. It involves two counters, one preset to count 6 and the other to count 12. Input 1 momentarily closes its contacts to start the counting cycle, resetting both counters. Input 2 contacts could be activated by a microswitch which is activated every time an item passes up to the junction in the paths. Counter 1 counts 6 items and then closes its contact. This activates the output, which might be a solenoid used to activate a flap which closes one path and opens another. Counter 1 also has contacts which close and enables Counter 2 to start counting. When Counter 2 has counted 12 items it resets both the counters and opens the Counter 1 contacts, which then results in the output becoming deactivated and items no longer directed towards the box to contain 12 items.

21.9 Shift registers

A number of internal relays can be grouped together to form a register which can provide a storage area for a series sequence of individual bits. A 4-bit register would be formed by using four internal relays, an 8-bit using eight. The term **shift register** is used because the bits can be shifted along by 1 bit when there is a suitable input to the register. For example, with an 8-bit register we might initially have:

1	0	1	1	0	1	0	1

Then there is an input of a 0 shift pulse:

0 → | 0 | 1 | 0 | 1 | 1 | 0 | 1 | 0 | → 1

with the result that all the bits shift along one place and the last bit overflows.

The grouping together of a number of auxiliary registers to form a shift register is done automatically by a PLC when the shift register function is selected at the control panel. With the Mitsubishi PLC, this is done by using the programming function SFT (shift) against the auxiliary relay number that is to be the first in the register array. This then causes the block of relays, starting from that initial number, to be reserved for the shift register. Thus, if we select M140 to be the first relay then the shift register will consist of M140, M141, M142, M143, M144, M145, M146 and M147.

Shift registers have three inputs: one to load data into the first element of the register (OUT), one as the shift command (SFT) and one for resetting (RST). With OUT, a logic level 0 or 1 is loaded into the first element of the shift register. With SFT, a pulse moves the contents of the register along 1 bit at a time, the final bit overflowing and being lost. With RST, a pulse of the closure of a contact resets the register contents to all zeros.

Figure 21.27 gives a ladder diagram involving a shift register when Mitsubishi notation is used; the principle is, however, the same with other manufacturers. M140 has been designated as the first relay of the register. When X400 is switched on, a logic 1 is loaded into the first element of the shift register, i.e. M140. We thus have the register with 10000000. The circuit shows that each element of the shift register has been connected as a contact in the circuit. Thus M140 contact closes and Y430 is switched on. When contact X401 is closed, then the bits in the register are shifted along the register by one place to give 11000000, a 1 being shifted into the register because X400 is still on. Contact M141 thus closes and Y430 is switched on. As each bit is shifted along, the outputs are energised in turn. Shift registers can thus be used to sequence events.

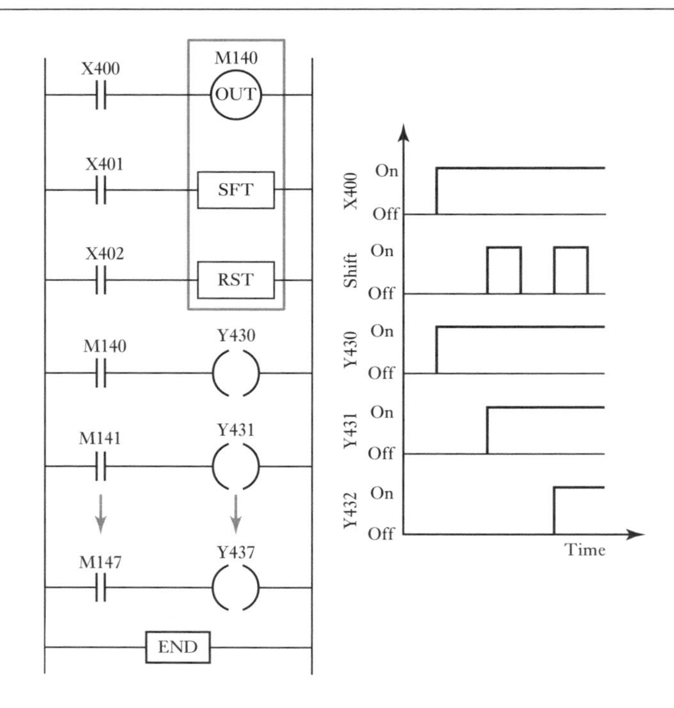

Figure 21.27 Shift register.

21.10 Master and jump controls

A whole block of outputs can be simultaneously turned off or on by using the same internal relay contacts in each output rung so that switching it on or off affects every one of the rungs. An alternative way of programming to achieve the same effect is to use a **master relay**. Figure 21.28 illustrates its use. We can think of it as controlling the power to a length of the vertical rails of the ladder. When there is an input to close Input 1 contacts, master relay MC1 is activated and then the block of program rungs controlled by that relay follows. The end of a master-relay-controlled section is indicated by the reset MCR. It is thus a branching program in that if there is Input 1 then branch to follow the MC1 controlled path; if not, follow the rest of the program and ignore the branch.

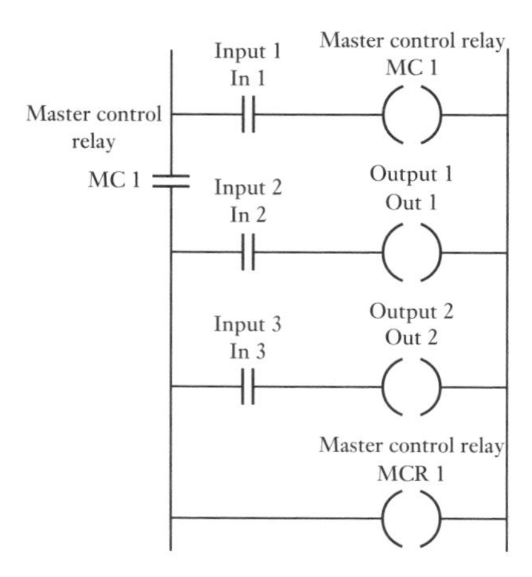

Figure 21.28 Master control relay.

With a Mitsubishi PLC, an internal relay can be designated as a master control relay by programming it accordingly. Thus to program an internal relay M100 as a master control relay, the program instruction is:

```
MC M100
```

To indicate the end of the section controlled by a master control relay, the program instruction is:

```
MCR M100
```

21.10.1 Jumps

A function which is often provided with PLCs is the **conditional jump** function. Such a function enables programs to be designed so that if a certain condition exists then a section of the program is jumped. Figure 21.29 illustrates this on a flow diagram and with a section of ladder program. Following a section of program, A, the program rung is encountered with Input 1 and the conditional jump relay CJP. If Input 1 occurs then the program jumps to the rung with the end of jump relay coil EJP and so continues with that section of the program labelled as C, otherwise it continues with the program rungs labelled as program B.

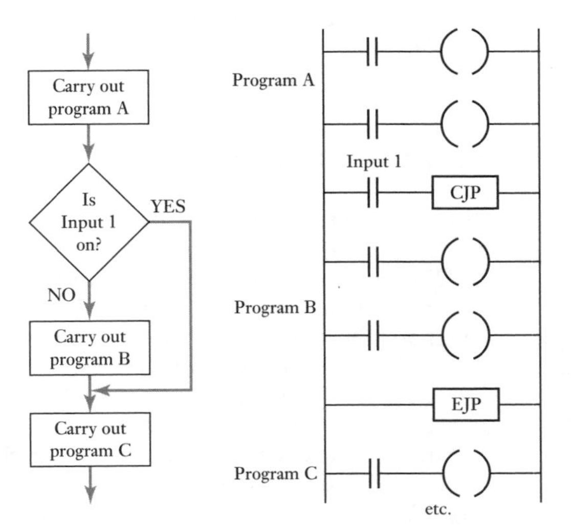

Figure 21.29 Jump.

21.11 Data handling

With the exception of the shift register, the previous parts of this chapter have been concerned with the handling of individual bits of information, e.g. a switch being closed or not. There are, however, some control tasks where it is useful to deal with related groups of bits, e.g. a block of eight inputs, and so operate on them as a data word. Such a situation can arise when a sensor supplies an analogue signal which is converted to, say, an 8-bit word before becoming an input to a PLC.

The operations that may be carried out with a PLC on data words normally include:

1 Moving data.
2 Comparison of magnitudes of data, i.e. greater than, equal to, or less than.
3 Arithmetic operations such as addition and subtraction.
4 Conversions between binary-coded decimal (BCD), binary and octal.

As discussed earlier, individual bits have been stored in memory locations specified by unique addresses. For example, for the Mitsubishi PLC, input memory addresses have been preceded by an A, outputs by a Y, timers by a T, auxiliary relays by an M, etc. Data instructions also require memory addresses and the locations in the PLC memory allocated for data are termed **data registers**. Each data register can store a binary word of, usually, 8 or 16 bits and is given an address such as D0, D1, D2, etc. An 8-bit word means that a quantity is specified to a precision of 1 in 256, a 16-bit word a precision of 1 in 65536.

Each instruction has to specify the form of the operation, the source of the data used in terms of its data register and the destination data register of the data.

21.11.1 Data movement

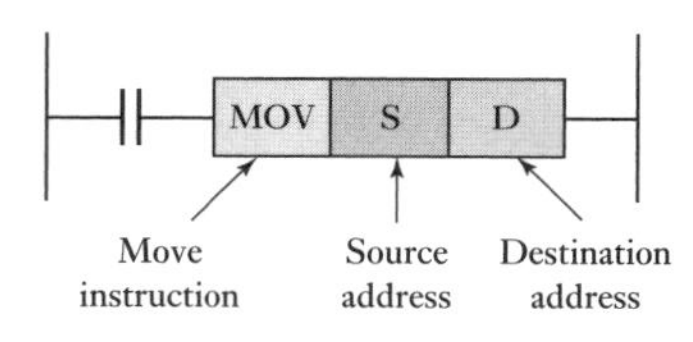

Figure 21.30 Move data.

For data movement the instruction will contain the move data instruction, the source address of the data and the destination address of the data. Thus the ladder rung could be of the form shown in Figure 21.30.

Such data transfers might be to move a constant into a data register, a time or count value to a data register, data from a data register to a timer or counter, data from a data register to an output, input data to a data register, etc.

21.11.2 Data comparison

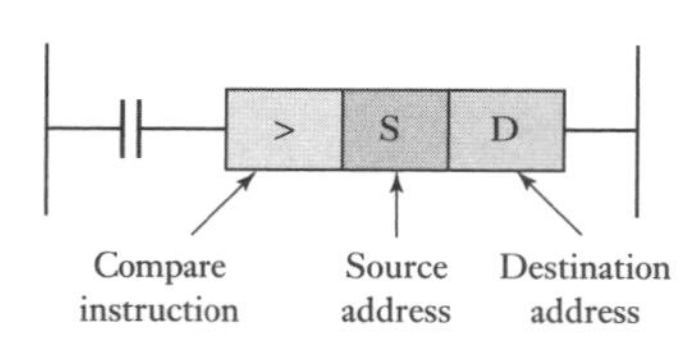

Figure 21.31 Compare data.

PLCs can generally make the data comparisons of *less than* (usually denoted by < or LES), *equal to* (= or EQU), *less than or equal to* (≤ or <= or LEQ), *greater than* (> or GRT), *greater than or equal to* (≥ , >= or GEQ) and *not equal to* (≠ or <> or NEQ). To compare data, the program instruction will contain the comparison instruction, the source address of the data and the destination address. Thus to compare the data in data register D1 to see if it is greater than data in data register D2, the ladder program rung would be of the form shown in Figure 21.31.

Such a comparison might be used when the signals from two sensors are to be compared by the PLC before action is taken. For example, an alarm might be required to be sounded if a sensor indicates a temperature above 80°C and remain sounding until the temperature falls below 70°C. Figure 21.32 shows the ladder program that could be used. The input temperature data is inputted to the source address and the destination address contains the set value. When the temperature rises to 80°C, or higher, the data value in the source address becomes ≥ the destination address value and there is an output to the alarm which latches the input. When the temperature falls to 70°C, or lower, the data value in the source address becomes ≤ the destination address value and there is an output to the relay which then opens its contacts and so switches the alarm off.

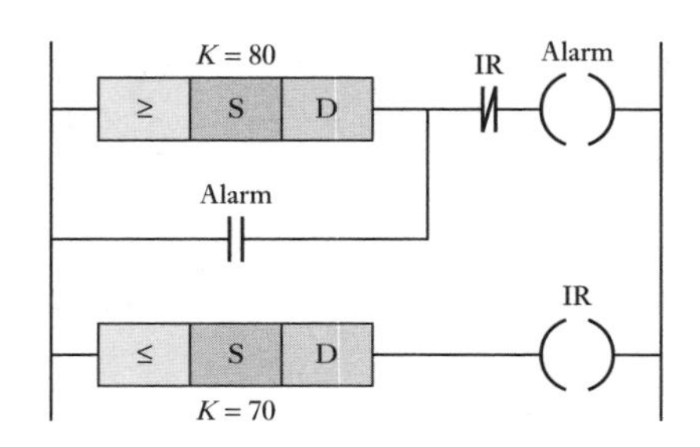

Figure 21.32 Temperature alarm.

21.11.3 Arithmetic operations

Some PLCs can carry out just the arithmetic operations of addition and subtraction, others have even more arithmetic functions. The instruction to add or subtract generally states the instruction, the register containing the address of the value to be added or subtracted, the address of the value to which the addition or from which the subtraction is to be made and the register where the result is to be stored. Figure 21.33 shows the form used for the ladder symbol for addition with OMRON.

Addition or subtraction might be used to alter the value of some sensor input value, perhaps a correction or offset term, or alter the preset values of timers or counters.

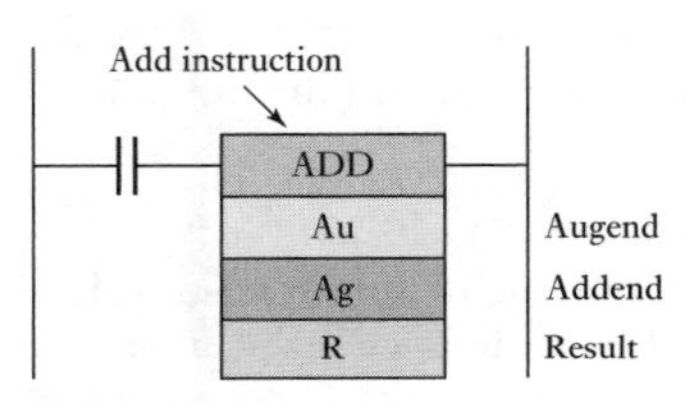

Figure 21.33 Add data.

21.11.4 Code conversions

All the internal operations in the CPU of a PLC are carried out using binary numbers. Thus, when the input is a signal which is decimal, conversion to BCD is used. Likewise, where a decimal output is required, conversion to decimal is required. Such conversions are provided with most PLCs. For example, with Mitsubishi, the ladder rung to convert BCD to binary is of the form shown in Figure 21.34. The data at the source address is in BCD and converted to binary and placed at the destination address.

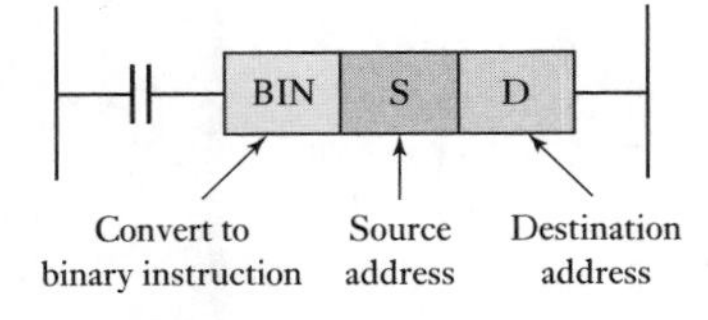

Figure 21.34 BCD to binary.

21.12 Analogue input/output

Many sensors generate analogue signals and many actuators require analogue signals. Thus, some PLCs may have an analogue-to-digital converter module fitted to input channels and a digital-to-analogue converter module fitted to output channels. An example of where such an item might be used is for the control of the speed of a motor so that its speed moves up to its steady value at a steady rate (Figure 21.35). The input is an on/off switch to start the operation. This opens the contacts for the data register and so it stores zero. Thus the output from the controller is zero and the analogue signal from the DAC is zero and hence motor speed is zero. The closing of the start contacts gives outputs to the DAC and the data register. Each time the program cycles through these rungs on the program, the data register is incremented by 1 and so the analogue signal is increased and hence the motor speed. Full speed is realised when the output from the data register is the word 11111111. The timer function of a PLC can be used to incorporate a delay between each of the output bit signals.

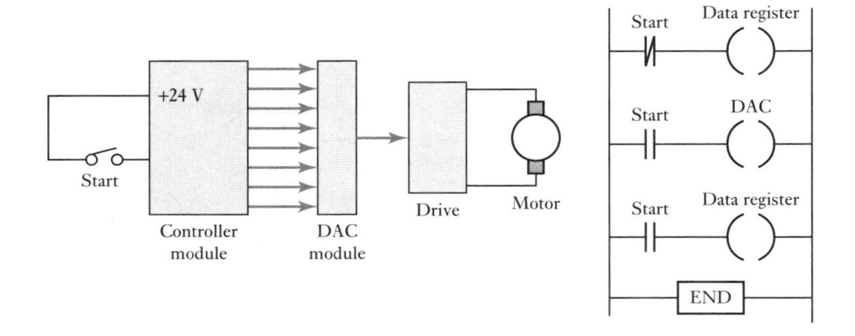

Figure 21.35 Ramping the speed of a motor.

A PLC equipped with analogue input channels can be used to carry out a continuous control function, i.e. PID control (see Section 15.7). Thus, for example, to carry out proportional control on an analogue input the following set of operations can be used:

1 Convert the sensor output to a digital signal.
2 Compare the converted actual sensor output with the required sensor value, i.e. the set point, and obtain the difference. This difference is the error.
3 Multiply the error by the proportional constant K_P.
4 Move this result to the DAC output and use the result as the correction signal to the actuator.

An example of where such a control action might be used is with a temperature controller. Figure 21.36 shows a possibility. The input could be from a thermocouple, which after amplification is fed through an ADC into the PLC. The PLC is programmed to give an output proportional to the error

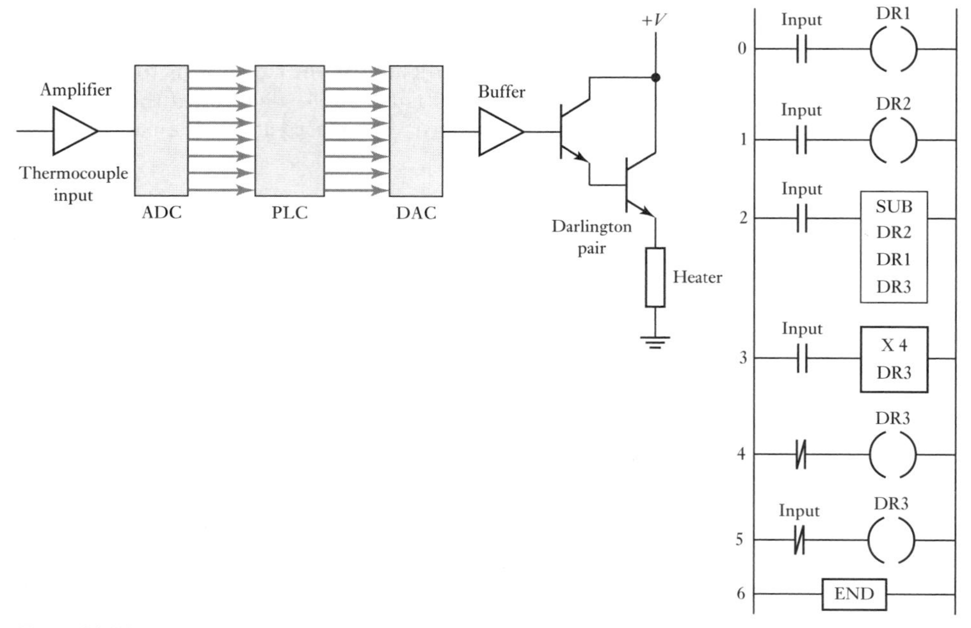

Figure 21.36 Proportional control of temperature.

between the input from the sensor and the required temperature. The output word is then fed through a DAC to the actuator, a heater, in order to reduce the error.

With the ladder program shown, rung 0 reads the ADC and stores the temperature value in data register DR1. With rung 1, the data register DR2 is used to store the set point temperature. Rung 2 uses the subtract function to subtract the values held in data registers DR1 and DR2 and store the result in data register DR3, i.e. this data register holds the error value. With rung 3 a multiply function is used, in this case to multiply the value in data register DR3 by the proportional gain of 4. Rung 4 uses an internal relay which can be programmed to switch off DR3 if it takes a negative value. With rung 5 the data register DR3 is reset to zero when the input is switched off. Some PLCs have add-on modules which more easily enable PLC control to be used, without the need to write lists of instructions in the way outlined above.

Summary

A **programmable logic controller** (PLC) is a digital electronic device that uses a programmable memory to store instructions and to implement functions such as logic, sequencing, timing, counting and arithmetic in order to control machines and processes and has been specifically designed to make programming easy.

A PLC is continuously running through its program and updating it as a result of the input signals. Each such loop is termed a **cycle**. The form of programming commonly used with PLCs is **ladder programming**. This involves each program task being specified as though a rung of a ladder. There is an alternative way of entering a program and that is to translate the ladder program into an **instruction list**. Instruction lists consist of a series of instructions with each instruction being on a separate line. An instruction consists of an operator followed by one or more operands, i.e. the subjects of the operator.

A **latch circuit** is a circuit that, after being energised, maintains that state until another input is received. The term **internal relay**, **auxiliary relay** or **marker** is used for what can be considered as an internal relay in the PLC, these behaving like relays with their associated contacts. **Timers** can be considered to behave like relays with coils which when energised result in the closure or opening of contacts after some preset time or as a delay block in a rung which delays signals in that rung reaching the output. **Counters** are used to count a specified number of contact operations, being considered to be an output coil to count input pulses with a coil to reset the counter and the associated contacts of the counter being used in other rungs or as an intermediate block in a rung from which signals emanate when the count is attained. A **shift register** is a number of internal relays which have been grouped together to form a register for a series sequence of individual bits. A **master relay** enables a whole block of outputs to be simultaneously turned off or on. The **conditional jump** function enables a section of the program to be jumped if a certain condition exists. Operations that may be carried out with **data words** include moving data, comparison of magnitudes of data, arithmetic operations and conversions between binary-coded decimal (BCD), binary and octal.

Problems

21.1 What are the logic functions used for switches (a) in series, (b) in parallel?

21.2 Draw the ladder rungs to represent:

(a) Two switches are normally open and both have to be closed for a motor to operate.

(b) Either of two, normally open, switches have to be closed for a coil to be energised and operate an actuator.

(c) A motor is switched on by pressing a spring-return push-button start switch, and the motor remains on until another spring-return push-button stop switch is pressed.

21.3 Write the program instructions corresponding to the latch program shown in Figure 21.37.

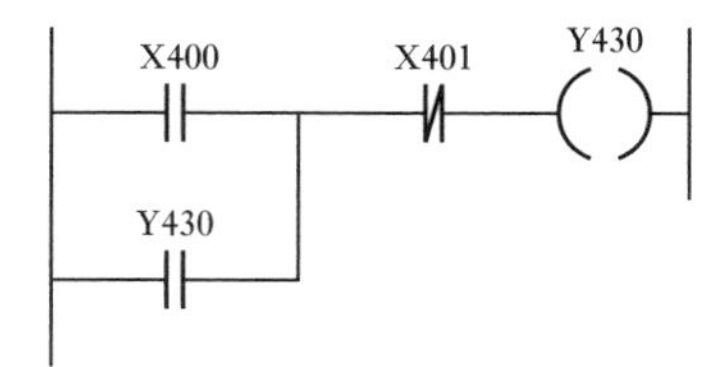

Figure 21.37 Problem 21.3.

21.4 Write the program instructions for the program in Figure 21.38 and state how the output varies with time.

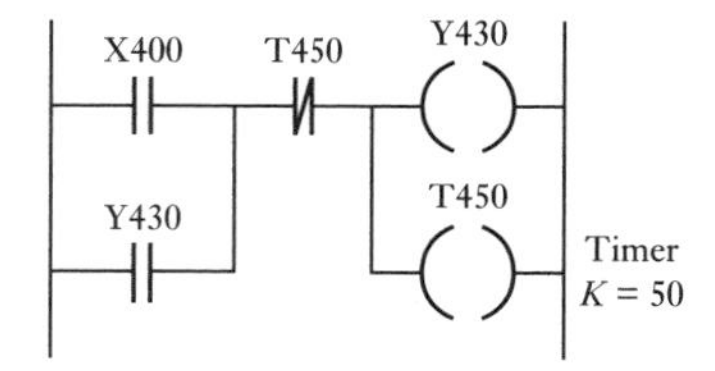

Figure 21.38 Problem 21.4.

21.5 Write the program instructions corresponding to the program in Figure 21.39 and state the results of inputs to the PLC.

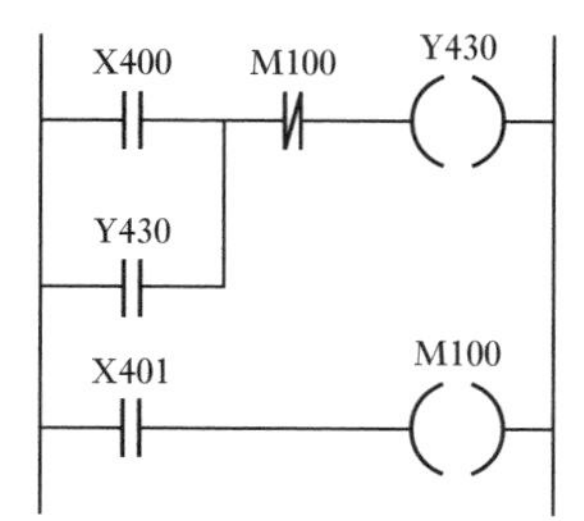

Figure 21.39 Problem 21.5.

21.6 Devise a timing circuit that will switch on an output for 1 s then off for 20 s, then on for 1 s, then off for 20 s, and so on.

21.7 Devise a timing circuit that will switch on an output for 10 s then switch it off.

21.8 Devise a circuit that can be used to start a motor and then after a delay of 100 s start a pump. When the motor is switched off there should be a delay of 10 s before the pump is switched off.

21.9 Devise a circuit that could be used with a domestic washing machine to switch on a pump to pump water for 100 s into the machine, then switch off and switch on a heater for 50 s to heat the water. The heater is then switched off and another pump is to empty the water from the machine for 100 s.

21.10 Devise a circuit that could be used with a conveyor belt which is used to move an item to a work station. The presence of the item at the work station is detected by means of breaking a contact activated by a beam of light to a photosensor. There the item stops for 100 s for an operation to be carried out before moving on and off the conveyor. The motor for the belt is started by a normally open start switch and stopped by a normally closed switch.

21.11 How would the timing pattern for the shift register in Figure 21.27 change if the data input X400 was of the form shown in Figure 21.40?

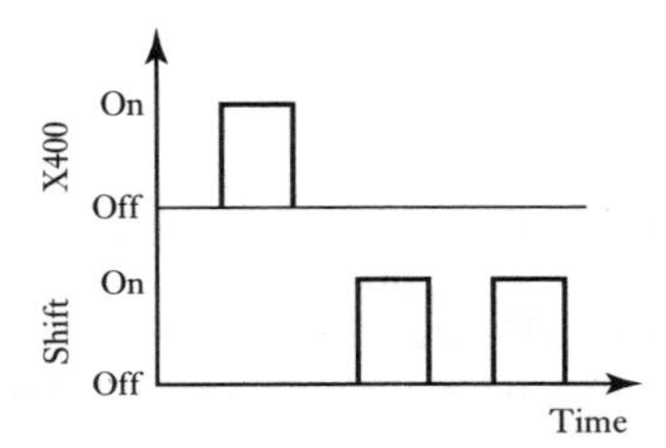

Figure 21.40 Problem 21.11.

21.12 Explain how a PLC can be used to handle an analogue input.

21.13 Devise a system, using a PLC, which can be used to control the movement of a piston in a cylinder so that when a switch is momentarily pressed, the piston moves in one direction and when a second switch is momentarily pressed, the piston moves in the other direction. Hint: you might consider using a 4/2 solenoid-controlled valve.

21.14 Devise a system, using a PLC, which can be used to control the movement of a piston in a cylinder using a 4/2 solenoid-operated pilot valve. The piston is to move in one direction when a proximity sensor at one end of the stroke closes contacts and in the other direction when a proximity sensor at the other end of the stroke indicates its arrival there.

Chapter twenty-two Communication Systems

Objectives

The objectives of this chapter are that, after studying it, the reader should be able to:

- **Describe centralised, hierarchical and distributed control systems, network configurations and methods of transmitting data and protocols used.**
- **Describe the Open Systems Interconnection communication model.**
- **Describe commonly used communication interfaces: RS-232, IEEE 488, 20 mA current loop, I^2C bus and CAN.**

22.1 Digital communications

An **external bus** is a set of signal lines that interconnects microprocessors, microcontrollers, computers and PLCs and also connects them with peripheral equipment. Thus a computer needs to have a bus connecting it with a printer if its output is to be directed to the printer and printed. Multiprocessor systems are quite common. For example, in a car there are likely to be several microcontrollers with each controlling a different part of the system, e.g. engine management, braking and instrument panel, and communication between them is necessary. In automated plant not only is there a need for data to pass between programmable logic controllers, displays, sensors and actuators and allow for data and programs to be inputted by the operator, but there can also be data communications with other computers. There may, for example, be a need to link a PLC to a control system involving a number of PLCs and computers. Computer integrated manufacturing (CIM) is an example of a large network which can involve large numbers of machines linked together. This chapter is a consideration of how such data communications between computers can take place, whether it is just simply machine-to-machine or a large network involving large numbers of machines linked together, and the forms of standard communication interfaces.

22.2 Centralised, hierarchical and distributed control

Centralised computer control involves the use of one central computer to control an entire plant. This has the problem that failure of the computer results in the loss of control of the entire plant. This can be avoided by the use of dual computer systems. If one computer fails, the other one takes over. Such centralised systems were common in the 1960s and 1970s. The development of the microprocessor and the ever reducing costs of computers have led to multi-computer systems becoming more common and the development of hierarchical and distributed systems.

With the **hierarchical system,** there is a hierarchy of computers according to the tasks they carry out. The computers handling the more routine tasks are supervised by computers which have a greater decision-making role. For example, the computers which are used for direct digital control of systems are subservient to a computer which performs supervisory control of the entire system. The work is divided between the computers according to the function involved. There is specialisation of computers with some computers only receiving some information and others different information.

With the **distributed system,** each computer system carries out essentially similar tasks to all the other computer systems. In the event of a failure of one, or overloading of a particular computer, work can be transferred to other computers. The work is spread across all the computers and not allocated to specific computers according to the function involved. There is no specialisation of computers. Each computer thus needs access to all the information in the system.

In most modern systems there is generally a mixture of distributed and hierarchical systems. For example, the work of measurement and actuation may be distributed among a number of microcontrollers/computers which are linked together and provide the database for the plant. These may be overseen by a computer used for direct digital control or sequencing and this in turn may be supervised by one used for supervisory control of the plant as a whole. Typical levels in such a scheme are:

Level 1 Measurement and actuators
Level 2 Direct digital and sequence control
Level 3 Supervisory control
Level 4 Management control and design

Distributed/hierarchical systems have the advantage of allowing the task of measurement scanning and signal conditioning in control systems to be carried out by sharing it between a number of microprocessors. This can involve a large number of signals with a high frequency of scanning. If extra measurement loops are required, it is a simple matter to increase the capacity of the system by adding microprocessors. The units can be quite widely dispersed, being located near the source of the measurements. Failure of one unit does not result in failure of the entire system.

22.2.1 Parallel and serial data transmission

Data communication can be via parallel or serial transmission links.

1 *Parallel data transmission*
Within computers, data transmission is usually by **parallel data paths.** Parallel data buses transmit 8, 16 or 32 bits simultaneously, having a separate bus wire for each data bit and the control signals. Thus, if there are eight data bits to be transmitted, e.g. 11000111, then eight data wires are needed. The entire eight data bits are transmitted in the same time as it takes to transmit one data bit because each bit is on a parallel wire. Handshaking (see Section 20.3.2) lines are also needed, handshaking being used for each character transmitted with lines needed to indicate that data is available for transmission and that the receiving terminal is ready to receive. Parallel data transmission permits high data transfer rates but is

expensive because of the cabling and interface circuitry required. It is thus normally only used over short distances or where high transfer rates are essential.

2 *Serial data transmission*
This involves the transmission of data which, together with control signals, is sent bit by bit in sequence along a single line. Only two conductors are needed, to transmit data and to receive data. Since the bits of a word are transmitted sequentially and not simultaneously, the data transfer rate is considerably less than with parallel data transmission. However, it is cheaper since far fewer conductors are required. For example, with a car when a number of microcontrollers are used, the connections between them are by serial data transmission. Without the use of serial transmission the number of wires involved would be considerable. In general, serial data transmission is used for all but the shortest peripheral connections.

Consider the problem of sending a sequence of characters along a serial link. The receiver needs to know where one character starts and stops. Serial data transmission can be either asynchronous or synchronous. **Asynchronous transmission** implies that both the transmitter and receiver computers are not synchronised, each having its own independent clock signals. The time between transmitted characters is arbitrary. Each character transmitted along the link is thus preceded by a start bit to indicate to the receiver the start of a character, and followed by a stop bit to indicate its completion. This method has the disadvantage of requiring extra bits to be transmitted along with each character and thus reduces the efficiency of the line for data transmission. With **synchronous transmission** there is no need for start and stop bits since the transmitter and receiver have a common clock signal and thus characters automatically start and stop always at the same time in each cycle.

The **rate of data transmission** is measured in bits per second. If a group of n bits form a single symbol being transmitted and the symbol has a duration of T seconds then the data rate of transmission is n/T. The **baud** is the unit used. The baud rate is only the same as the number of bits per second transmitted if each character is represented by just one symbol. Thus a system which does not use start and stop pulses has a baud rate equal to the bit rate, but this will not be the case when there are such bits.

22.2.2 Serial data communication modes

Serial data transmission occurs in one of three modes:

1 *Simplex mode*
Transmission is only possible in one direction, from device A to device B, where device B is not capable of transmitting back to device A (Figure 22.1(a)). You can think of the connection between the devices as being like a one-way road. This method is usually only used for transmission to devices such as printers which never transmit information.

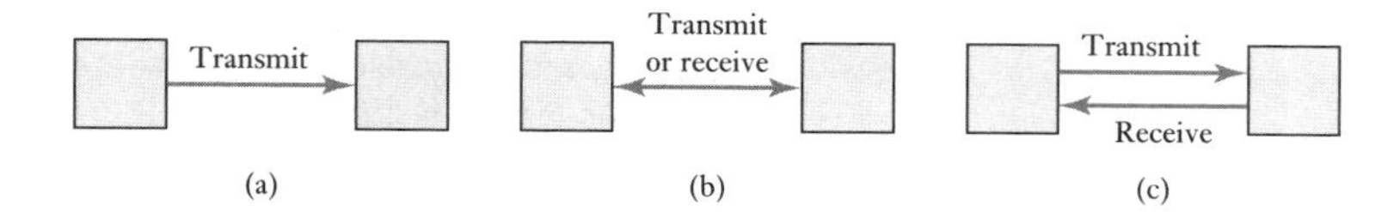

Figure 22.1 Communication modes.

2 *Half-duplex mode*
Data is transmitted in one direction at a time but the direction can be changed (Figure 22.1(b)). Terminals at each end of the link can be switched from transmit to receive. Thus device A can transmit to device B and device B to device A but not at the same time. You can think of this as being like a two-lane road under repair with traffic from one lane being stopped by a traffic control to allow the traffic from the other lane through. Citizens Band (CB) radio is an example of half-duplex mode; a person can receive or talk but not do both simultaneously.

3 *Full-duplex mode*
Data may be transmitted simultaneously in both directions between devices A and B (Figure 22.1(c)). This is like a two-lane highway in which traffic can occur in both directions simultaneously. The telephone system is an example of full-duplex mode in that a person can talk and receive at the same time.

22.3 Networks

The term **network** is used for a system which allows two or more computers/microprocessors to be linked for the interchange of data. The logical form of the links is known as the network **topology.** The term **node** is used for a point in a network where one or more communication lines terminate or a unit is connected to the communication lines. Commonly used forms are:

1 *Data bus*
This has a linear bus (Figure 22.2(a)) into which all the stations are plugged. This system is often used for multipoint terminal clusters. It is generally the preferred method for distances between nodes of more than 100 m.

2 *Star*
This has dedicated channels between each station and a central switching hub (Figure 22.2(b)) through which all communications must pass. This is the type of network used in the telephone systems (private branch exchanges (PBXs)) in many companies, all the lines passing through a central exchange. This system is also often used to connect remote and local terminals to a central mainframe computer. There is a major problem with this system in that if the central hub fails then the entire system fails.

3 *Hierarchy or tree*
This consists of a series of branches converging indirectly to a point at the head of the tree (Figure 22.2(c)). With this system there is only one transmission path between any two stations. This arrangement may be formed from a number of linked data bus systems. Like the bus method, it is often used for distances between nodes of more than 100 m.

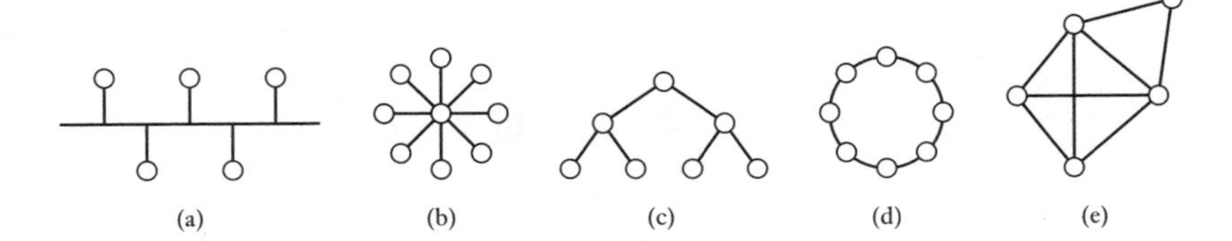

Figure 22.2 Network topologies: (a) data bus, (b) star, (c) hierarchy, (d) ring, (e) mesh.

4 *Ring*
This is a very popular method for local area networks, involving each station being connected to a ring (Figure 22.2(d)). The distances between nodes are generally less than 100 m. Data put into the ring system continues to circulate round the ring until some system removes it. The data is available to all the stations.

5 *Mesh*
This method (Figure 22.2(e)) has no formal pattern to the connections between stations and there will be multiple data paths between them.

The term **local area network** (LAN) is used for a network over a local geographic area such as a building or a group of buildings on one site. The topology is commonly bus, star or ring. A **wide area network** is one that interconnects computers, terminals and local area networks over a national or international level. This chapter is primarily concerned with local area networks.

22.3.1 Network access control

Access control methods are necessary with a network to ensure that only one user of the network is able to transmit at any one time. The following are methods used.

With ring-based local area networks, two commonly used methods are:

1 *Token passing*
With this method a token, a special bit pattern, is circulated. When a station wishes to transmit it waits until it receives the token, then transmits the data with the token attached to its end. Another station wishing to transmit removes the token from the package of data and transmits its own data with the token attached to its end.

2 *Slot passing*
This method involves empty slots being circulated. When a station wishes to transmit data it deposits it in the first empty slot that comes along.

With bus or tree networks a method that is often used is:

3 *Carrier sense multiple access with collision detection (CSMA/CD)*
This method is generally identified with the **Ethernet LAN bus.** With the CSMA/CD method, stations have to listen for other transmissions before transmitting, with any station being able to gain control of the network and transmit, hence the term multiple access. If no activity is detected then transmission can occur. If there is activity then the system has to wait until it can detect no further activity. Despite this listening before transmission, it is still possible for two or more systems to start to transmit at the same time. If such a situation is detected, both stations cease submitting and wait a random time before attempting to retransmit.

22.3.2 Broadband and baseband

The term **broadband transmission** is used for a network in which information is modulated onto a radio frequency carrier which passes through the transmission medium such as a coaxial cable. Typically the topology of broadband local area networks is a bus with branches. Broadband transmission allows a number of modulated radio frequency carriers to be simultaneously transmitted and so offers a multichannel capability. The term **baseband transmission** is used when digital information is passed directly through the transmission medium. Baseband transmission networks can only support one information signal at a time. A LAN may be either baseband or broadband.

22.4 Protocols

Transmitted data will contain two types of information. One is the data which one computer wishes to send to another, the other is information termed **protocol data** and is used by the interface between a computer and the network to control the transfer of the data into the network or from the network into the computer. A protocol is a formal set of rules governing data format, timing, sequencing, access control and error control. The three elements of a protocol are:

1 *Syntax*, which defines data format, coding and signal levels.
2 *Semantics*, which deals with synchronisation, control and error handling.
3 *Timing*, which deals with the sequencing of data and the choice of data rate.

When a sender communicates with a receiver then both must employ the same protocol, e.g. two microcontrollers with data to be serially transmitted between them. With simplex communication the data block can be just sent from sender to receiver. However, with half-duplex, each block of transmitted data, if valid, must be acknowledged (ACK) by the receiver before the next block of data can be sent (Figure 22.3(a)); if invalid a NAK, negative acknowledgement, signal is sent. Thus a continuous stream of data cannot be transmitted. The CRC bits, **cyclic redundancy check bits,** are a means of error detection and are transmitted immediately after a block of data. The data is transmitted as a binary number and at the transmitter the data is divided by a number and the remainder is used as the cyclic check code. At the receiver the incoming data, including the CRC, is divided by the same number and will give zero remainder if the signal is error-free. With full-duplex mode (Figure 22.3(b)), data can be continuously sent and received.

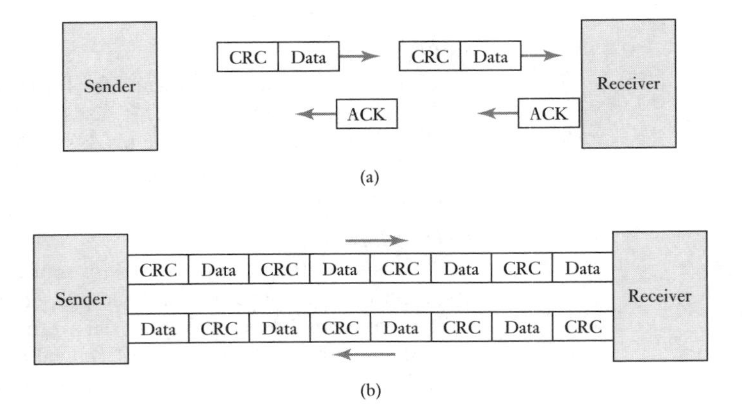

Figure 22.3 Protocols: (a) half-duplex, (b) full-duplex.

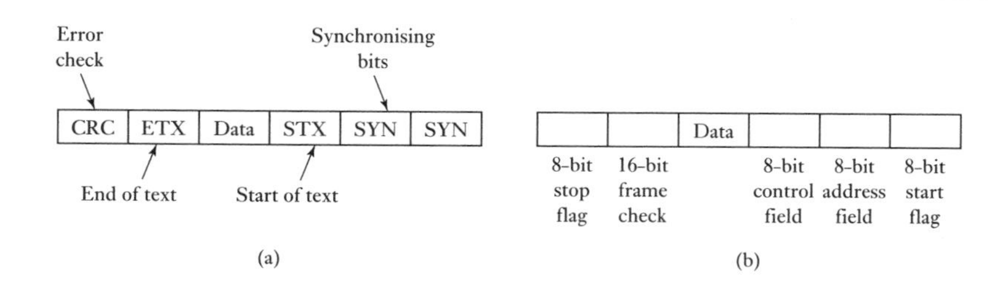

Figure 22.4 (a) Bisync, (b) HDLC.

Within a package being sent, there is a need to include protocol information. For example, with asynchronous transmission there may be characters to indicate the start and end of data. With synchronous transmission and the **Bisync Protocol,** a data block is preceded by a synchronising sequence of bits, usually the ASCII character SYN (Figure 22.4(a)). The SYN characters are used by the receiver to achieve character synchronisation, preparing the receiver to receive data in 8-bit groupings. The Motorola MC6852 is a synchronous serial data adapter (SSDA) that is designed for use with 6800 microprocessors to provide a synchronous serial communications interface using the Bisync Protocol. It is similar to the asynchronous communications interface adapter described in Section 20.5. Another protocol is the **High-level Data Link Control** (HDLC) (Figure 22.4(b)). This is a full-duplex protocol with the beginning and end of a message being denoted by the bit pattern 01111110. Address and control fields follow the start flag. The address identifies the address of the destination station, the control field defines whether the frame is supervisory, information or unnumbered. Following the message is a 16-bit frame check sequence which is used to give a CRC. The Motorola 6854 is an example of a serial interface adapter using this HDLC Protocol.

22.5 Open Systems Interconnection communication model

Communication protocols have to exist on a number of levels. The International Organization for Standardization (ISO) has defined a seven-layer standard protocol system known as the **Open Systems Interconnection** (OSI) model. The model is a framework for developing a co-ordinated system of standards. The layers are:

1 *Physical layer*
This layer describes the means for bit transmission to and from physical components of the network. It deals with hardware issues, e.g. the types of cable and connectors to be used, synchronising data transfer and signal levels. Commonly used LAN systems defined at the physical layer are Ethernet and token ring.

2 *Data link layer*
This layer defines the protocols for sending and receiving messages, error detection and correction and the proper sequencing of transmitted data. It is concerned with packaging data into packets and placing them on the cable and then taking them off the cable at the receiving end. Ethernet and token ring are also defined at this level.

3 *Network layer*
This deals with communication paths and the addressing, routing and control of messages on the network and thus making certain that the messages get to the right destinations. Commonly used network layer protocols are Internet Protocol (IP) and Novell's Internetwork Packet Exchange (IPX).

4 *Transport layer*
This provides for reliable end-to-end message transport. It is concerned with establishing and maintaining the connection between transmitter and receiver. Commonly used transport layer protocols are Internet Transmission Control Protocol (TCP) and Novell's Sequenced Packet Exchange (SPX).

5 *Session layer*
This layer is concerned with the establishment of dialogues between application processes which are connected together by the network. It is responsible for determining when to turn a communication between two stations on or off.

6 *Presentation layer*
This layer is concerned with allowing the encoded data transmitted to be presented in a suitable form for user manipulation.

7 *Application layer*
This layer provides the actual user information processing function and application-specific services. It provides such functions as file transfer or electronic mail which a station can use to communicate with other systems on the network.

22.5.1 Network standards

There are a number of network standards, based on the OSI layer model, that are commonly used. The following are examples.

In the United States, General Motors realised that the automation of its manufacturing activities posed a problem of equipment being supplied with a variety of non-standard protocols. GM thus developed a standard communication system for factory automation applications. The standard is referred to as the **Manufacturing Automation Protocol** (MAP) (Figure 22.5). The choice of protocols at the different layers reflects the requirement for the system to fit the manufacturing environment. Layers 1 and 2 are implemented in hardware electronics and levels 3 to 7 using software. For the physical layer, broadband transmission is used. The broadband method allows the system to be used for services in addition to those required for MAP communications. For the data link layer, the token system with a bus is used with logical link control (LLC) to implement such functions as error checking, etc. For the other layers, ISO standards are used. At layer 7, MAP includes manufacturing message services (MMS), an application relevant to factory floor communications which defines interactions between programmable logic controllers and numerically controlled machines or robots.

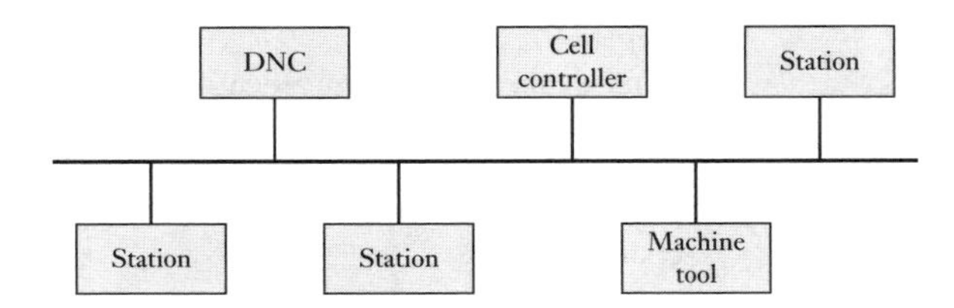

Figure 22.5 MAP.

The **Technical and Office Protocol** (TOP) is a standard that was developed by Boeing Computer Services. It has much in common with MAP but can be implemented at a lower cost because it is a baseband system. It differs from MAP in layers 1 and 2, using either the token with a ring or the CSMA/CD method with a bus network. Also, at layer 7, it specifies application protocols that concern office requirements, rather than factory floor requirements. With the CSMA/CD method, stations have to listen for other transmissions before transmitting. TOP and MAP networks are compatible and a gateway device can be used to connect TOP and MAP networks. This device carries out the appropriate address conversions and protocol changes.

Systems Network Architecture (SNA) is a system developed by IBM as a design standard for IBM products. SNA is divided into seven layers; it, however, differs to some extent from the OSI layers (Figure 22.6). The data link control layer provides support for token ring for LANs. Five of the SNA layers are consolidated in two packages: the path control network for layers 2 and 3 and the network addressable units for layers 4, 5 and 6.

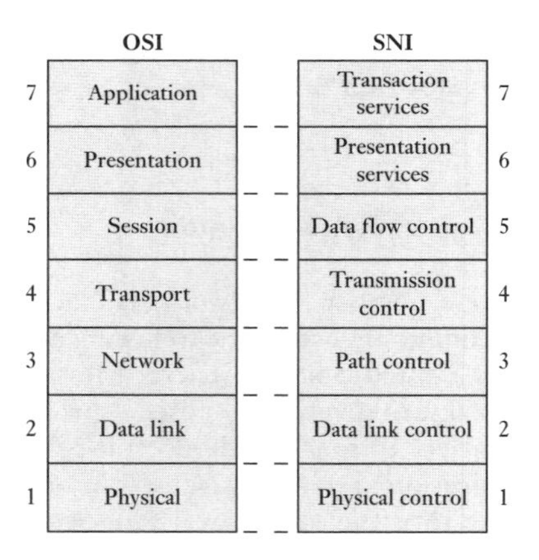

Figure 22.6 SNA.

With PLC systems, it is quite common for the system used to be that marketed by the PLC manufacturer. For example, Allen-Bradley has the **Allen-Bradley data highway** which uses token passing to control message transmission; Mitsubishi has Melsec-Net and Texas Instruments has TIWAY. A commonly used system with PLC networks is the Ethernet. This is a single-bus system with CSMA/CD used to control access and is widely used with systems involving PLCs communicating with computers. The problem with using CSMA/CD is that, though this method works well when traffic is light, as network traffic increases the number of collisions and corresponding back-off of transmitters increases. Network throughput can thus slow down quite dramatically.

22.6 Serial communication interfaces

Serial interfacing can involve synchronous or asynchronous protocols. Commonly used asynchronous interfaces are RS-232, and later versions, the 20 mA current loop, I^2C, CAN and USB.

22.6.1 RS-232

The most popular serial interface is **RS-232**; this was first defined by the American Electronic Industries Association (EIA) in 1962. The standard relates to data terminal equipment (DTE) and data circuit-terminating equipment (DCE). Data terminal equipment can send or receive data via the interface, e.g. a microcontroller. Data circuit-terminating equipment is devices which facilitate communication; a typical example is a modem. This forms an essential link between a microcomputer and a conventional analogue telephone line.

RS-232 signals can be grouped into three categories:

1 *Data*
RS-232 provides two independent serial data channels, termed primary and secondary. Both these channels are used for full-duplex operation.

2 *Handshake control*
Handshaking signals are used to control the flow of serial data over the communication path.

3 *Timing*
For synchronous operation it is necessary to pass clock signals between transmitters and receivers.

Table 22.1 gives the RS-232C connector pin numbers and signals for which each is used; not all the pins and signals are necessarily used in a particular set-up. The signal ground wire allows for a return path. The connector to a RS-232C serial port is via a 25-pin D-type connector; usually a male plug is used on cables and a female socket on the DCE or DTE.

For the simplest bidirectional link, only the two lines 2 and 3 for transmitted data and received data, with signal ground (7) for the return path of these signals, are needed (Figure 22.7(a)). Thus the minimum connection is via a three-wire cable. For a simple set-up involving a personal computer (PC) being linked with a visual display unit (VDU), pins 1, 2, 3, 4, 5, 6, 7 and 20 are involved (Figure 22.7(b)). The signals sent through pins 4, 5, 6 and 20 are used to check that the receiving end is ready to receive a signal; the transmitting end is ready to send and the data is ready to be sent.

RS-232 is limited concerning the distance over which it can be used as noise limits the transmission of high numbers of bits per second when the length of cable is more than about 15 m. The maximum data rate is about 20 kbits/s. Other standards such as RS-422 and RS-485 are similar to RS-232 and can be used for higher transmission rates and longer distances.

Table 22.1 RS-232 pin assignments.

Pin	Abbreviation	Direction: To	Signal/function
1	FG		Frame ground
2	TXD	DCE	Transmitted data
3	RXD	DTE	Received data
4	RTS	DCE	Request to send
5	CTS	DTE	Clear to send
6	DSR	DTE	DCE ready
7	SG		Signal ground/common return
8	DCD	DTE	Received line detector
12	SDCD	DTE	Secondary received line signal detector
13	SCTS	DTE	Secondary clear to send
14	STD	DCE	Secondary transmitted data
15	TC	DTE	Transmit signal timing
16	SRD	DTE	Secondary received data
17	RC	DTE	Received signal timing
18		DCE	Local loop-back
19	SRTS	DCE	Secondary request to send
20	DTR	DCE	Data terminal ready
21	SQ	DEC/DTE	Remote loop-back/signal quality detector
22	RI	DTE	Ring indicator
23		DEC/DTE	Data signal rate selector
24	TC	DCE	Transmit signal timing
25		DTE	Test mode

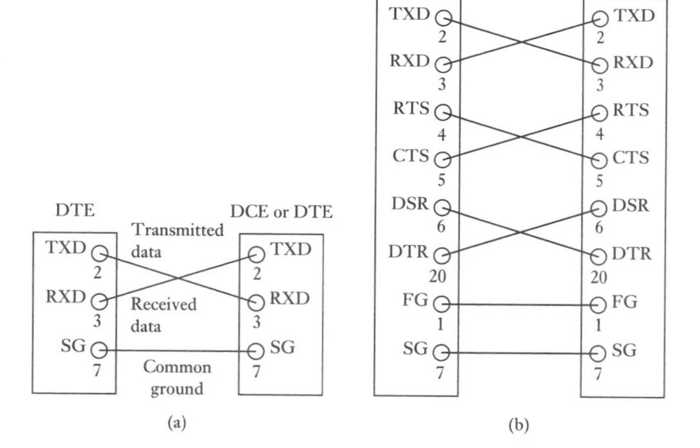

Figure 22.7 RS-232 connections: (a) minimum configuration, (b) PC connection.

RS-422 uses a pair of lines for each signal and can operate up to about 1220 m or at higher transmission speeds up to 100 bits/s and in noisier environments; maximum speed and maximum distance cannot, however, be achieved simultaneously. RS-485 can be used up to about 1220 m with speeds of 100 kbits/s.

The serial communications interface of the Motorola microcontroller MC68HC11 is capable of full-duplex communications at a variety of baud rates. However, the input and output of this system use transistor–transistor logic (TTL) for which logic 0 is 0 V and logic 1 is +5 V. The RS-232C standards are +12 V for logic 0 and −12 V for logic 1. Thus conversion in signal levels is necessary. This can be achieved by using integrated circuit devices such as MC1488 for TTL to RS-232C conversion and MC1489 for RS-232C to TTL conversion (Figure 22.8).

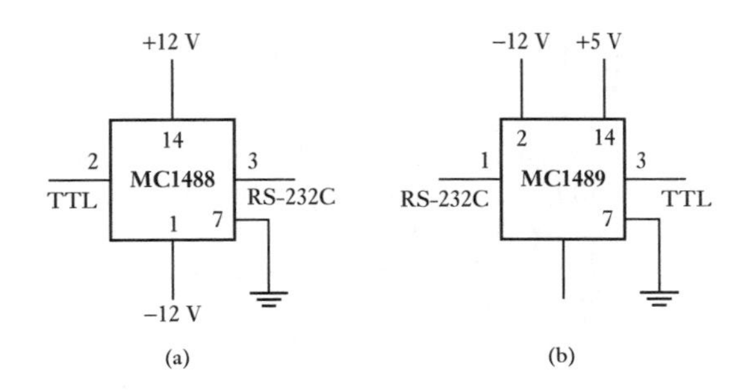

Figure 22.8 Level conversion: (a) MC1488, (b) MC1489.

22.6.2 The 20 mA current loop

Another technique, based on RS-232 but not part of the standard, is the **20 mA current loop** (Figure 22.9). This uses a current signal rather than a voltage signal. A pair of separate wires is used for the transmission and the receiver loops with a current level of 20 mA used to indicate a logic 1 and 0 mA a logic 0. The serial data is encoded with a start bit, eight data bits and two stop bits. Such current signals enable a far greater distance, a few kilometres, between transmitter and receiver than with the standard RS-232 voltage connections.

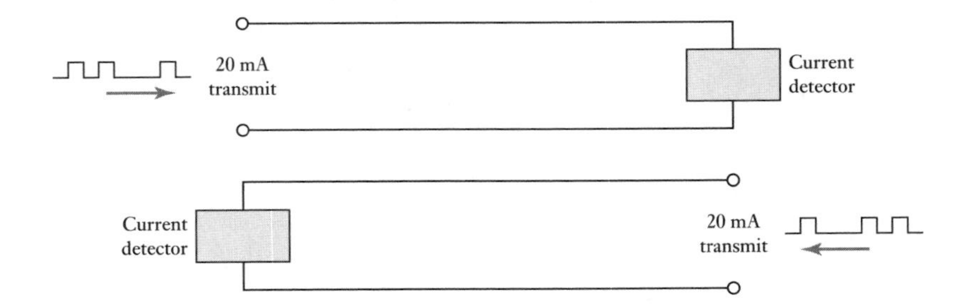

Figure 22.9 The 20 mA current loop.

22.6.3 I²C bus

The **Inter-IC Communication bus**, referred to as the I^2C bus, is a serial data bus designed by Philips for use for communications between integrated circuits or modules. The bus allows data and instructions to be exchanged between devices by means of just two wires. This results in a considerable simplification of circuits.

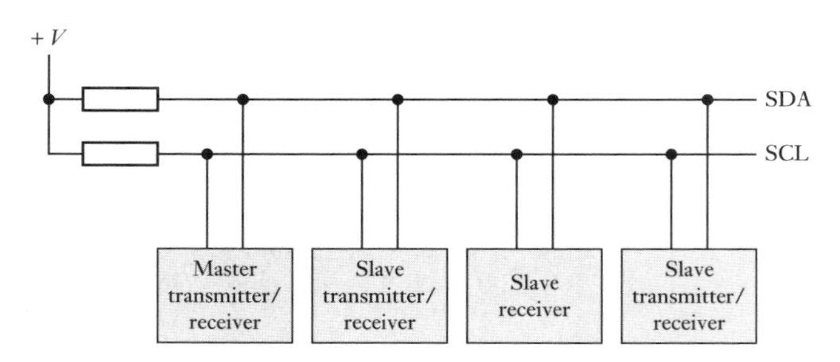

Figure 22.10 I²C bus.

The two lines are a bidirectional data line (SDA) and a clock line (SCL). Both lines are connected to the positive power supply via resistors (Figure 22.10). The device generating the message is the transmitter and the device receiving the message the receiver. The device that controls the bus operation is the master and the devices which are controlled by the master are the slaves.

The following is the protocol used: a data transfer may only be initiated when the bus is not busy and during the data transfer, when the clock line is high, the data line must remain. Changes in the data line when the clock line is high are interpreted as control signals.

1 When both the data and clock lines are high the bus is not busy.
2 A change in the state of the data line from high to low while the clock is high is used to define the start of data transfer.
3 A change in the state of the data line from low to high while the clock is high defines the stop of data transfer.
4 Data is transferred between start and stop conditions.
5 After a start of data transfer, the data line is stable for the duration of the high periods of the clock signal, being able to change during the low periods of the clock signal.
6 There is one clock pulse per data bit transferred with no limit on the number of data bytes that can be transferred between the start and stop conditions; after each byte of data the receiver acknowledges with a ninth bit.
7 The acknowledge bit is a high level put on the bus by the transmitter, a low level by a receiver.

Figure 22.11 illustrates the above by showing the form of the clock signal and the outputs by a transmitter and a receiver.

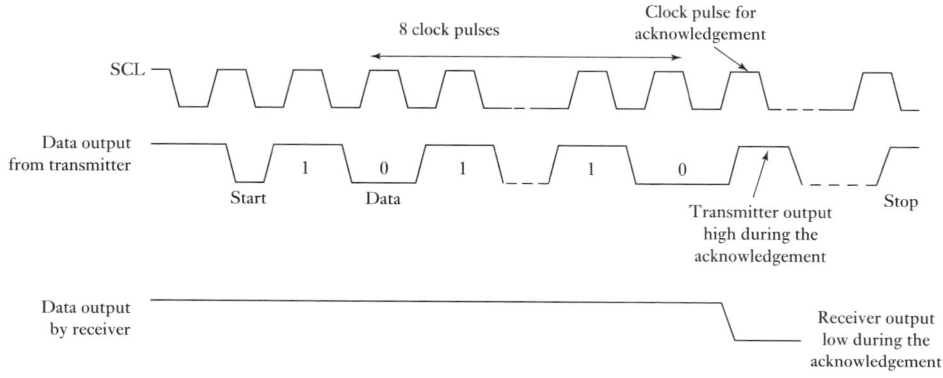

Figure 22.11 Bus conditions.

22.6.4 CAN bus

A modern automobile may have as many as seventy electronic control units (ECUs) for various subsystems, e.g. engine management systems, anti-lock brakes, traction control, active suspension, airbags, cruise control, windows, etc. This could involve a lot of wiring. However, an alternative approach is to use a common data bus with data transmitted along it and made available to all parts of the car. Bosch has thus developed a protocol known as **CAN** or **Controller Area Network**. The CAN bus is now also used as a fieldbus in other automation systems.

CAN is a multi-master serial bus standard for connecting ECUs. Each node in the system is able to both send and receive messages and requires:

1 A host processor to determine what received messages mean and which messages it wants to transmit. Sensors, actuators and control devices are not connected directly to the CAN bus but to a host processor and a CAN controller.
2 A CAN controller to store bits received serially from the bus until an entire message is available. After the CAN controller has triggered an interrupt call, the message can then be fetched by the host processor. The controller also stores messages ready for transmission serially onto the bus.
3 A transceiver, which is possibly integrated into the CAN controller, to adapt signal levels received from the bus to levels that the CAN controller expects and which has protective circuitry that protects the CAN controller. It also is used to convert the transmit-bit signal received from the CAN controller into a signal that is sent onto the bus.

Each message consists of an identification (ID) field, to identify the message-type or sender, and up to eight data bytes. Some means of arbitration is, however, needed if two or more nodes begin sending messages at the same time. A non-destructive arbitration method is used to determine which node may transmit, and it is the ID with 0s which is deemed dominant and allowed to win the conflict and transmit. Thus when a transmitting node puts a bit on the bus but detects that there is a more dominant one already on the bus, it disables its transmitter and waits until the end of the current transmission before attempting to start transmitting its own data. For example, suppose we have the 11-bit ID 11001100110 for message 1 and 10001101110 for message 2. By the time transmission has reached the fourth bit the arbitration indicates that message 1 is dominant and so message 2 ceases transmission.

The standard CAN data frame format for serial transmission consists of a message sandwiched between a start bit and a confirmation sent and end of frame bits. The message will have:

1 A 12-bit ID, the last bit being a remote transmission request bit.
2 A 6-bit control field consisting of an identifier extension bit, and a reserved bit, a 4-bit data length code to indicate the number of bytes of data.
3 The data field.
4 A 16-bit CRC field, i.e. a cyclic redundancy check for error detection.

22.6.5 USB

The *Universal Serial Bus* (USB) is designed to enable monitors, printers, modems and other input devices to be easily connected to PCs – the term plug-and-play is used. USB uses a star topology (see Section 22.3); thus only one device needs to be plugged into a PC with other devices then being able to be plugged into the resulting hub so resulting in a tiered star topology. We thus have a host hub at the PC into which other external hubs can be connected. Each port is a four-pin socket with two of the pins being for power and two for communications. The USB 1.0 and 2.0 provide a 5 V supply from which USB devices can draw power, although there is a limit of 500 mA current. USB devices requiring more power than is provided by a single port can use an external power supply.

The low-speed version USB 1.0 specification was introduced in 1996 and has a transfer rate of 12 Mbits/s and is limited to cable lengths of 3 m. The high-speed version USB 2.0 specification was introduced in April 2000 and has a data transfer rate of 480 Mbits/s and is limited to cable lengths of 5 m although, as up to five USB hubs can be used, a long chain of cables and hubs would enable distances up 30 m to be covered. A superspeed USB 3.0 specification was released by Intel and partners in August 2008 for a data transfer rate of 4.8 Gbits/s, and products using this specification are now becoming available. Data is transmitted in half-duplex mode for USB 1.0 and USB 2.0 with full-duplex being possible with USB 3.0 (see Section 22.2.2).

The root hub has complete control over all the USB ports. It initiates all communications with hubs and devices. No USB device can transfer any data onto the bus without a request from the host controller. In USB 2.0, the host controller polls the bus for traffic. For USB 3.0, connected devices are able to request service from the host. When a USB device is first connected to a USB host, an enumeration process is started by the host sending a reset signal to the USB device. After reset, the USB device's information is read by the host and the device is assigned a unique 7-bit address. If the device is supported by the host, the device driver needed for communicating with the device is loaded. The driver is used to supply the information about the device's needs, i.e. such things as speed, priority, function of the device and the size of packet needed for data transfer. When the application's software wants to send or receive some information from a device, it initiates a transfer via the device driver. The driver software then places the request in a memory location together with the requests that have been made by other device drivers. The host controller then takes all the requests and transfers it serially to the host hub ports. Since all the devices are in parallel on the USB bus, all of them hear the information. The host waits for a response. The relevant devices then respond with the appropriate information.

Packets sent out are in three basic types, namely handshaking, token and data, each having a different format and CRC (cyclic redundancy check, see Section 22.4). There are four types of token packet – start of frame, in and out packets to command a device to transmit or receive data, and set-up packet used for the initial set-up of a device.

22.6.6 Firewire

Firewire is a serial bus developed by Apple Computers, the specification being given by IEEE 1394. It offers plug-and-play capabilities and is used for applications such as disk drives, printers and cameras.

22.7 Parallel communication interfaces

For the parallel interface to a printer the Centronics parallel interface is commonly used. However, with instrumentation the most commonly used parallel interface in communications is the **General Purpose Instrument Bus** (GPIB), the IEEE 488 standard, originally developed by Hewlett Packard to link its computers and instruments and thus often referred to as the **Hewlett Packard Instrumentation Bus.** Each of the devices connected to the bus is termed a listener, talker or controller. Listeners are devices that accept data from the bus, talkers place data, on request, on the bus and controllers manage the flow of data on the bus by sending commands to talkers and listeners and carry out polls to see which devices are active (Figure 22.12(a)).

There is a total of 24 lines with the interface:

1 Eight bidirectional lines to carry data and commands between the various devices connected to the bus.
2 Five lines for control and status signals.
3 Three lines for handshaking between devices.
4 Eight lines are ground return lines.

Table 22.2 lists the functions of the lines and their pin numbers in a 25-way D-type connector. Up to 15 devices can be attached to the bus at any one time, each device having its own address.

The 8-bit parallel data bus can transmit data as one 8-bit byte at a time. Each time a byte is transferred the bus goes through a handshake cycle. Each device on the bus has its own address. Commands from the controller are signalled by taking the attention line (ATN) low. Commands are then directed to individual devices by placing addresses on the data lines; device addresses are sent via the data lines as a parallel 7-bit word with the lowest 5 bits providing the device address and the other 2 bits control information. If both these bits are 0 then the commands are sent to all addresses; if bit 6 is 1 and bit 7 is 0 the addressed device is switched to be a listener; if bit 6 is 0 and bit 7 is 1 then the device is switched to be a talker.

Handshaking uses the lines DAV, NRFD and NDAC, the three lines ensuring that the talker will only talk when it is being listened to by listeners

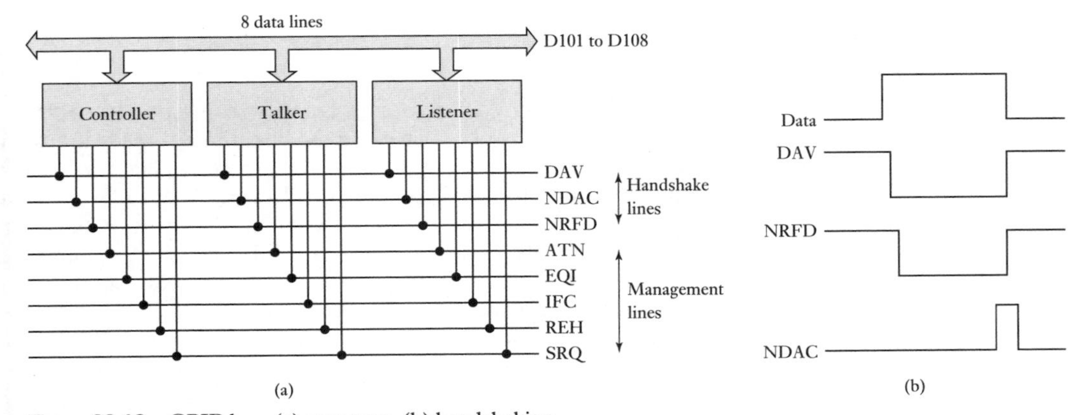

Figure 22.12 GPIB bus: (a) structure, (b) handshaking.

Table 22.2 IEEE 488 bus system.

Pin	Signal group	Abbreviation	Function
1	Data	D101	Data line 1
2	Data	D102	Data line 2
3	Data	D103	Data line 3
4	Data	D104	Data line 4
5	Management	EOI	End Or Identify. This is either used to signify the end of a message sequence from a talker device or used by the controller to ask a device to identify itself
6	Handshake	DAV	Data Valid. When the level is low on this line then the information on the data bus is valid and acceptable
7	Handshake	NRFD	Not Ready For Data. This line is used by listener devices taking it high to indicate that they are ready to accept data
8	Handshake	NDAC	Not Data Accepted. This line is used by listeners taking it high to indicate that data is being accepted
9	Management	IFC	Interface Clear. This is used by the controller to reset all the devices of the system to the start state
10	Management	SRQ	Service Request. This is used by devices to signal to the controller that they need attention
11	Management	ATN	Attention. This is used by the controller to signal that it is placing a command on the data lines
12		SHIELD	Shield
13	Data	D105	Data line 5
14	Data	D106	Data line 6
15	Data	D107	Data line 7
16	Data	D108	Data line 8
17	Management	REN	Remote Enable. This enables a device to indicate that it is to be selected for remote control rather than by its own control panel
18		GND	Ground/common (twisted pair with DAV)
19		GND	Ground/common (twisted pair with NRFD)
20		GND	Ground/common (twisted pair with NDAC)
21		GND	Ground/common (twisted pair with IFC)
22		GND	Ground/common (twisted pair with SRG)
23		GND	Ground/common (twisted pair with ATN)
24		GND	Signal ground

(Figure 22.12(b)). When a listener is ready to accept data, NRFD is made high. When data has been placed on the line, DAV is made low to notify devices that data is available. When a device accepts a data word it sets NDAC high to indicate that it has accepted the data and NRFD low to indicate that it is now not ready to accept data. When all the listeners have set NDAC high, then the talker cancels the data valid signal, DAV going high. This then results in NDAC being set low. The entire process can then be repeated for another word being put on the data bus.

The GPIB is a bus which is used to interface a wide range of instruments, e.g. digital multimeters and digital oscilloscopes, via plug-in boards (Figure 22.13) to computers with standard cables used to link the board with the instruments via interfaces.

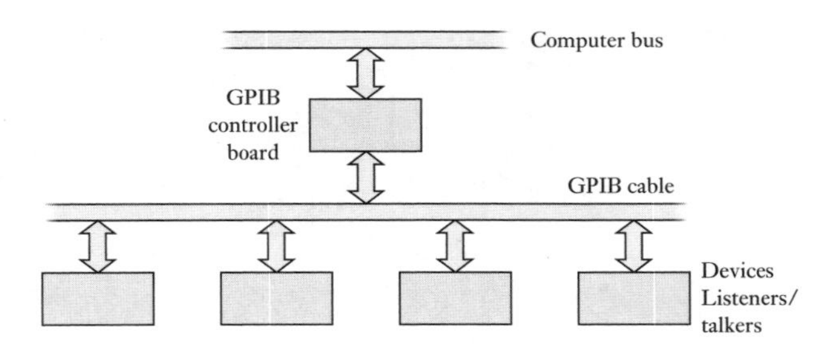

Figure 22.13 GPIB hardware.

22.7.1 Other buses

Buses used to connect the CPU to the input/output ports or other devices include:

1 The *XT computer bus* was introduced in 1983 for 8-bit data transfers with IBM PC/XT and compatible computers.
2 The *AT bus*, also referred to as the *industry standard architecture* (ISA) *bus*, was later introduced for use with 16-bit transfers with IBM PC and other compatible computers using 80286 and 80386 microprocessors. The AT bus is compatible with the XT bus so that plug-in XT boards can be used in AT bus slots.
3 The *extended industry standard architecture* (EISA) *bus* was developed to cope with 32-bit data transfers with IBM PC and other compatible computers using 80386 and 80486 microprocessors.
4 The *Micro Channel Architecture* (MCA) *bus* is a 16-bit or 32-bit data transfer bus designed for use with IBM Personal System/1 (PS/2) computers. Boards for use with this bus are not compatible with PC/XT/AT boards.
5 The *NuBus* is the 32-bit bus used in Apple's Macintosh II computers.
6 The *S-bus* is the 32-bit bus used in Sun Microsystem's SPARC stations.
7 The *TURBOchannel* is the 32-bit bus used in DECstation 5000 work stations.
8 The *VME bus* is the bus designed by Motorola for use with its 32-bit 68000-microprocessor-based system. Such a bus is now, however, widely used with other computer systems as the bus for use with instrumentation systems.

The above are called **backplane buses,** the term backplane being for the board (Figure 22.14) on which connectors are mounted and into which printed circuit boards containing a particular function, e.g. memory, can be plugged. The backplane provides the data, address and control bus signals to each board, so enabling systems to be easily expanded by the use of off-the-shelf boards. It is into such computer buses that data acquisition boards and boards used for interfacing instruments and other peripherals have to interface. Data acquisition and instrument boards are usually available in various configurations, depending on the computer with which they are to be used.

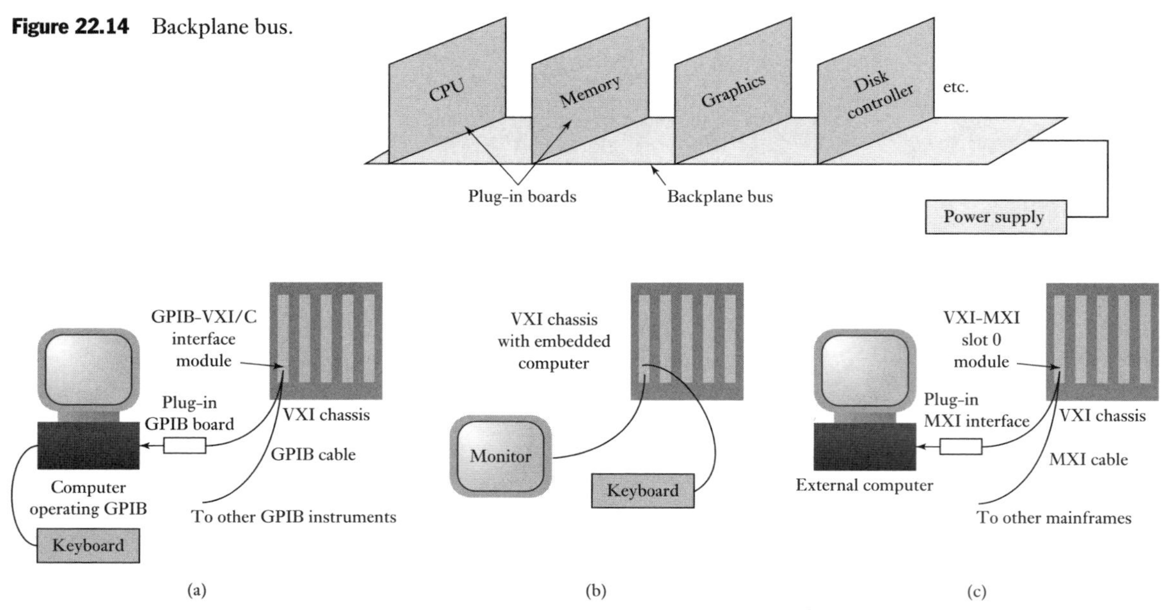

Figure 22.14 Backplane bus.

Figure 22.15 VXI options.

The VXIbus (VME Extensions for Instrumentation) is an extension of the specification of the VMEbus which has been designed for instrumentation applications such as automatic test equipment where higher speed communications are required than possible with the GPIB bus. It also provides better synchronisation and triggering and has been developed by a consortium of instrument manufacturers so that interoperability is possible between the products of different companies. The system involves VXI boards plugging into a mainframe. Figure 22.15 shows a number of possible system configurations that can be used. In Figure 22.15(a) a VXI mainframe is linked to an external controller, a computer, via a GPIB link. The controller talks across this link using a GPIB protocol to an interface board in the chassis which translates the GPIB Protocol into the VXI Protocol. This makes VXI instruments appear to the controller to be GPIB instruments and enables them to be programmed using GPIB methods. Figure 22.15(b) shows the complete computer embedded in the VXI chassis. This option offers the smallest possible physical size for the system and allows the computer to use directly the VXI backplane bus. Figure 22.15(c) uses a special high-speed system-bus-on-cable, the MXIbus, to link a computer and the VXI chassis, the MXI being 20 times faster than the GPIB.

22.8 Wireless protocols

IEEE 802.11 is a proposed standard for wireless LANs, specifying both the physical (PHY) and medium access control (MAC) layers of the network. The MAC layer specifies a carrier sense multiple access collision avoidance (CSMA/CA) protocol. With this, when a node has a packet ready for transmission, it first listens to ensure no other node is transmitting and if clear then transmits. Otherwise, it waits and then tries again. When a packet is transmitted, the transmitting node first sends out a ready-to-send (RTS) packet containing information about the length of the packet and then sends its packet. When the packet is successfully received, the receiving node transmits an acknowledgement (ACK) packet.

Bluetooth is a global standard for short-range radio transmission. When two Bluetooth-equipped devices are within 10 m of each other, a connection can be established. It is widely used for mobile phones and PCs.

Summary

An **external bus** is a set of signal lines that interconnects microprocessors, microcontrollers, computers and PLCs and also connects them with peripheral equipment.

Centralised computer control involves the use of one central computer to control an entire plant. With the **hierarchical system**, there is a hierarchical system of computers according to the tasks they carry out. With the **distributed system,** each computer system carries out essentially similar tasks to all the other computer systems.

Data communication can be via **parallel** or **serial transmission** links. Serial data transmission can be either asynchronous or synchronous transmission. **Asynchronous transmission** implies that both the transmitter and receiver computers are not synchronised, each having its own independent clock signals. Serial data transmission occurs in one of three modes: simplex, half-duplex and full-duplex.

The term **network** is used for a system which allows two or more computers/microprocessors to be linked for the interchange of data. Commonly used forms are data bus, star, hierarchy/tree, ring and mesh. Network access control is necessary to ensure that only one user is able to transmit at any one time; with ring-based networks, methods used are token passing and slot passing, while carrier sense multiple access with collision detection is used with bus or hierarchy networks. A **protocol** is a formal set of rules governing data format, timing, sequencing, access control and error control.

The International Organization for Standardization (ISO) has defined a seven-layer standard protocol system known as the **Open Systems Interconnection** (OSI) model.

Serial communication interfaces include RS-232 and its later versions, I^2C and CAN. **Parallel communication interfaces** include the General Purpose Instrument Bus (GPIB).

Problems

22.1 Explain the difference between centralised and distributed communication systems.

22.2 Explain the forms of bus/tree and ring networks.

22.3 A LAN is required with a distance between nodes of more than 100 m. Should the choice be bus or ring topology?

22.4 A multichannel LAN is required. Should the choice be broadband or baseband transmission?

22.5 What are MAP and TOP?

22.6 Explain what is meant by a communication protocol.

22.7 Briefly explain the two types of multiple-access control used with LANs.

22.8 A microcontroller M68HC11 is a 'listener' to be connected to a 'talker' via a GPIB bus. Indicate the connections to be made if full handshaking is to be used.

22.9 What problem has to be overcome before the serial data communications interface of the microcontroller M68HC11 can output data through an RS-232C interface?

22.10 What is a backplane bus?

Chapter twenty-three Fault Finding

Objectives

The objectives of this chapter are that, after studying it, the reader should be able to:

- **Recognise the techniques used to identify faults in microprocessor-based systems, including both hardware and software.**
- **Explain the use of emulation and simulation.**
- **Explain how fault finding can be achieved with PLC systems.**

23.1 Fault-detection techniques

This chapter is a brief consideration of the problems of fault detection with measurement, control and data communication systems. For details of the fault-finding checks required for specific systems or components, the manufacturer's manuals should be used.

There are a number of techniques that can be used to detect faults:

1. *Replication checks*
   This involves duplicating or replicating an activity and comparing the results. In the absence of faults it is assumed that the results should be the same. It could mean, with transient errors, just repeating an operation twice and comparing the results or it could involve having duplicate systems and comparing the results given by the two. This can be an expensive option.

2. *Expected value checks*
   Software errors are commonly detected by checking whether an expected value is obtained when a specific numerical input is used. If the expected value is not obtained then there is a fault.

3. *Timing checks*
   This involves the use of timing checks that some function has been carried out within a specified time. These checks are commonly referred to as **watchdog timers**. For example, with a PLC, when an operation starts, a timer is also started and if the operation is not completed within the specified time a fault is assumed to have occurred. The watchdog timer trips, sets off an alarm and closes down part or the entire plant.

4. *Reversal checks*
   Where there is a direct relationship between input and output values, the value of the output can be taken and the input which should have caused it computed. This can then be compared with the actual input.

5 *Parity and error coding checks*
This form of checking is commonly used for detecting memory and data transmission errors. Communication channels are frequently subject to interference which can affect data being transmitted. To detect whether data has been corrupted, a parity bit is added to the transmitted data word. The parity bit is chosen to make the resulting number of ones in the group either odd (odd parity) or even (even parity). If odd parity then the word can be checked after transmission to see if it is still odd. Other forms of checking involve codes added to transmitted data in order to detect corrupt bits.

6 *Diagnostic checks*
Diagnostic checks are used to test the behaviour of components in a system. Inputs are applied to a component and the outputs compared with those which should occur.

23.2 Watchdog timer

A watchdog timer is basically a timer that the system must reset before it times out. If the timer is not reset in time then an error is assumed to have occurred.

As an illustration of such a timer, Figure 23.1 shows a simple ladder program that can be used to provide a PLC with a watchdog timer for an operation involving the movement of a piston in a cylinder. When the start switch is closed, the solenoid A+ is activated and starts the movement of the piston in the cylinder. It also starts a timer. When the piston is fully extended it opens the limit switch a+. This stops the timer. However, if a+ is not opened before the timer has timed out then the timer contacts close and an alarm is sounded. Thus the timer might be set for, say, 4 s on the assumption that the piston will be fully extended within that time. If, however, it sticks and fails to meet this deadline then the alarm sounds.

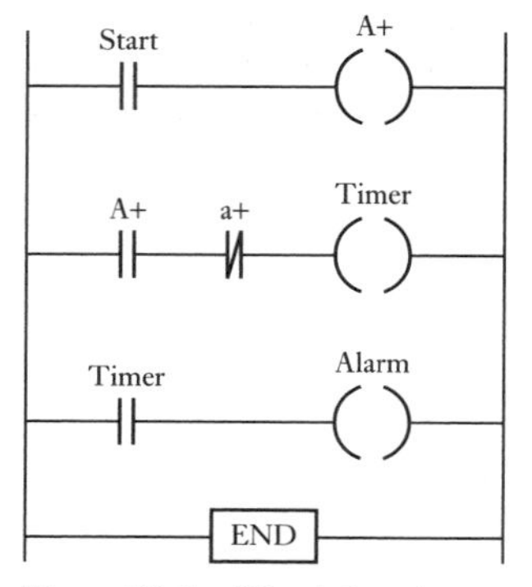

Figure 23.1 Watchdog timer.

When a microprocessor executes instructions from its memory, a nearby electrical disturbance might momentarily upset the processor data bus and the wrong byte is accessed. Alternatively a software bug can result in the processor getting into problems when it returns from a subroutine. The consequence of such faults is that the system may crash with possibly dangerous consequences for actuators controlled by the microprocessor. To avoid this happening with critical systems, a watchdog timer is used to reset the microprocessor.

As an illustration of the provision in microprocessor-based systems of internal watchdog timers, consider the microcontroller MC68HC11 which includes an internal watchdog timer, called **computer operating properly** (COP), to detect software processing errors. When the COP timer has been started it is necessary for the main program to reset the COP periodically before it times out. If the watchdog timer times out before being set to start timing all over again then a COP failure reset occurs. The COP timer can be reset to zero time by writing a \$55 (0x55 in C language) to the COP reset register (COPRST) at address \$103A (0x103A) followed later in the program by writing a \$AA (0xAA) to clear the COP timer. If the program hangs in between the two instructions then COP times out and results in the COP failure reset routine being executed. The program, in assembly language, thus has the following lines:

```
LDAA    #$55     ; reset timer
STAA    $103A    ; writing $55 to COPRST
                 ; other program lines
```

```
LDAA    #$AA     ; clearing timer
STAA    $103A    ; writing $AA to COPRST
```

The COP operating period is set by setting CR1 and CR2 in the OPTION register, address $1039 (0x1039), to either 0 or 1. For example, with CR1 set to 0 and CR2 set to 0 a time out of 16.384 ms might be given, whereas with CR1 set to 1 and CR2 set to 0 a time out of 262.14 ms is given.

23.3 Parity and error coding checks

In order to try and detect when a data signal has been corrupted and has an error as a result of noise, error detection techniques such as **parity checks** are used.

In Section 4.5.2 a brief account was given of the parity method for error detection. With such a method an extra bit is added to a message to make the total number of ones an even number when even parity is used or an odd number when odd parity is used. For example, the character 1010000 would have a parity bit placed after the most significant bit of a 0 with an even-parity system, i.e. 01010000, or a 1 with odd parity, i.e. 11010000.

Such a method can detect the presence of a single error in the message but not the presence of two errors which result in no change in parity, e.g. with even parity and the above number a single error in, say, the third bit would be detected in 1101100 because the parity check bit would be wrong, but not if also there was an error in the first bit since 1101110 would have the correct parity bit.

If no error is detected the signal is acknowledged as being error-free by the return of the ACK character to the sending terminal; if an error is detected the signal NAK is used. This is called an **automatic repeat request** (ARQ). The NAK signal then results in the retransmission of the message.

The efficiency of error detection can be increased by the use of **block parity**. The message is divided into a number of blocks and each block has a block check character added at the end of the block. For example, with the following block, a check bit for even parity is placed at the end of each row and a further check bit at the foot of each column:

	Information bits				Check bit
First symbol	0	0	1	1	0
Second symbol	0	1	0	0	1
Third symbol	1	0	1	1	1
Fourth symbol	0	0	0	0	0
Block check bits	1	1	0	0	0

At the receiver the parity of each row and each column is checked and any single error is detected by the intersection of the row and column containing the error check bit.

Another form of error detection is the **cyclic redundancy check** (CRC). At the transmitting terminal the binary number representing the data being transmitted is divided by a predetermined number using modulo-2 arithmetic. The remainder from the division is the CRC character which is transmitted with the data. At the receiver the data plus the CRC character are divided by the same number. If no transmission errors have occurred there is no remainder.

A common CRC code is CRC-16, 16 bits being used for the check sequence. The 16 bits are considered to be the coefficients of a polynomial with the number of bits equal to the highest power of the polynomial. The data block is first multiplied by the highest power of the polynomial, i.e. x^{16}, and then divided by the CRC polynomial

$$x^{16} + x^{12} + x^5 + 1$$

using modulo-2 arithmetic, i.e. $x = 2$ in the polynomial. The CRC polynomial is thus 10001000000100001. The remainder of the division by this polynomial is the CRC.

As an illustration, suppose we have the data 10110111 or polynomial

$$x^7 + x^5 + x^4 + x^2 + x^1 + 1$$

and a CRC polynomial of

$$x^5 + x^4 + x^1 + 1$$

or 110011. The data polynomial is first multiplied by x^5 to give

$$x^{12} + x^{10} + x^9 + x^7 + x^6 + x^5$$

and so 1011011100000. Dividing this by the CRC polynomial gives

```
          11010111
110011 |1011011100000
        110011
         110011
         110011
          100100
          110011
           101110
           110011
            111010
            110011
             01001
```

and so a remainder of 01001 which then becomes the CRC code transmitted with the data.

23.4 Common hardware faults

The following are some of the commonly encountered faults that can occur with specific types of components and systems.

23.4.1 Sensors

If there are faults in a measurement system then the sensor might be at fault. A simple test is to substitute the sensor with a new one and see what effect this has on the results given by the system. If the results change then it is likely that the original sensor was faulty; if the results do not change then the fault is elsewhere in the system. It is also possible to check that the voltage/current sources are supplying the correct voltages/currents, whether there is electrical continuity in connecting wires, that the sensor is correctly mounted and used under the conditions specified by the manufacturer's data sheet, etc.

23.4.2 Switches and relays

Dirt and particles of waste material between switch contacts are a common source of incorrect functioning of mechanical switches. A voltmeter used across a switch should indicate the applied voltage when the contacts are open and very nearly zero when they are closed. Mechanical switches used to detect the position of some item, e.g. the presence of a workpiece on a conveyor, can fail to give the correct responses if the alignment is incorrect or if the actuating lever is bent.

Inspection of a relay can disclose evidence of arcing or contact welding. The relay should then be replaced. If a relay fails to operate then a check can be made for the voltage across the coil. If the correct voltage is present then coil continuity can be checked with an ohmmeter. If there is no voltage across the coil then the fault is likely to be the switching transistor used with the relay.

23.4.3 Motors

Maintenance of both d.c. and a.c. motors involves correct lubrication. With d.c. motors the brushes wear and can require changing. Setting of new brushes needs to be in accordance with the manufacturer's specification. A single-phase capacitor start a.c. motor that is sluggish in starting probably needs a new starting capacitor. The three-phase induction motor has no brushes, commutator, slip rings or starting capacitor and, short of a severe overload, the only regular maintenance that is required is periodic lubrication.

23.4.4 Hydraulic and pneumatic systems

A common cause of faults with hydraulic and pneumatic systems is dirt. Small particles of dirt can damage seals, block orifices, cause valve spools to jam, etc. Thus filters should be regularly checked and cleaned, components should only be dismantled in clean conditions, and oil should be regularly checked and changed. With an electric circuit a common method of testing the circuit is to measure the voltages at a number of test points. Likewise, with a hydraulic and pneumatic system there needs to be points at which pressures can be measured. Damage to seals can result in hydraulic and pneumatic cylinders leaking, beyond that which is normal, and result in a drop in system pressure when the cylinder is actuated. This can be remedied by replacing the seals in the cylinders. The vanes in vane-type motors are subject to wear and can then fail to make a good seal with the motor housing, with the result of a loss of motor power. The vanes can be replaced. Leaks in hoses, pipes and fittings are common faults.

23.5 Microprocessor systems

Typical faults in microprocessor systems are:

1 *Chip failure*
Chips are fairly reliable but occasionally there can be failure.

2 *Passive component failure*
Microprocessor systems will usually include passive components such as resistors and capacitors. Failure of any of these can cause system malfunction.

3 *Open circuits*
Open circuits can result in a break in a signal path or in a power line. Typical reasons for such faults are unsoldered or faulty soldered joints, fracture of a printed circuit track, a faulty connection on a connector and breaks in cables.

4 *Short circuits*
Short circuits between points on a board which should not be connected often arise as a result of surplus solder bridging the gaps between neighbouring printed circuit tracks.

5 *Externally introduced interference*
Externally induced pulses will affect the operation of the system since they will be interpreted as valid digital signals. Such interference can originate from the mains supply having 'spikes' as a result of other equipment sharing the same mains circuit and being switched on or off. Filters in the mains supply to the system can be used to remove such 'spikes'.

6 *Software faults*
Despite extensive testing it is still quite feasible for software to contain bugs and under particular input or output conditions cause a malfunction.

23.5.1 Fault-finding techniques

Fault-finding techniques that are used with microprocessor-based systems include:

1 *Visual inspection*
Just carefully looking at a faulty system may reveal the source of a fault, e.g. an integrated circuit which is loose in its holder or surplus solder bridging tracks on a board.

2 *Multimeter*
This is of limited use with microprocessor systems but can be used to check for short- or open-circuit connections and the power supplies.

3 *Oscilloscope*
The oscilloscope is essentially limited to where repetitive signals occur and the most obvious such signal is the clock signal. Most of the other signals with a microprocessor system are not repetitive and depend on the program being executed.

4 *Logic probe*
The logic probe is a hand-held device (Figure 23.2(a)), shaped like a pen, which can be used to determine the logic level at any point in the

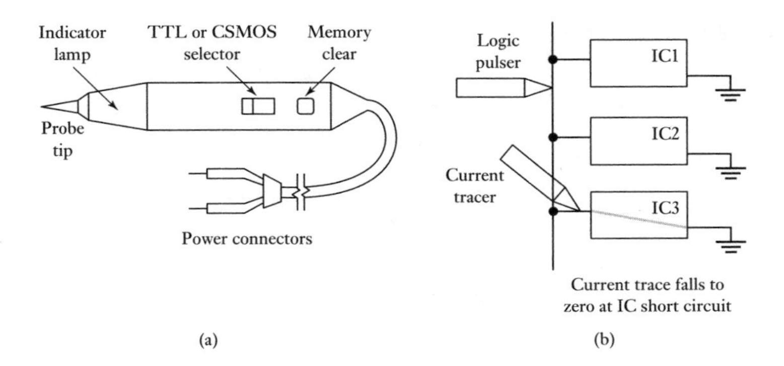

Figure 23.2 (a) Logic probe, (b) current tracer.

circuit to which it is connected. The selector switch is set for TTL or CSMOS operations and when the probe tip is touched to the point in question, the indicator lamp indicates whether it is below the logic level 0 threshold, above the logic 1 threshold or a pulsating signal. A pulse stretching circuit is often included with the probe in order to stretch the duration of a pulse to allow sufficient time for it to operate the indicator lamp and the effect to be noticed. A memory circuit can be used for detecting a single pulse, the memory clear button being pressed to turn off the indicator lamp and then any change in logic level is registered by the lamp.

5 *Logic pulser*
The logic pulser is a hand-held pulse generator, shaped like a pen, that is used to inject controlled pulses into circuits. The pulser probe tip is pressed against a node in the circuit and the button on the probe pressed to generate a pulse. It is often used with the logic probe to check the functions of logic gates.

6 *Current tracer*
The current tracer is similar to the logic probe but it senses pulsing current in a circuit rather than voltage levels. The tip of the current tracer is magnetically sensitive and is used to detect the changing magnetic field near a conductor carrying a pulsing current. The current tracer tip is moved along printed circuit tracks to trace out the low-impedance paths along which current is flowing (Figure 23.2(b)).

7 *Logic clip*
A logic clip is a device which clips to an integrated circuit and makes contact with each of the integrated circuit pins. The logic state of each pin is then shown by LED indicators, there being one for each pin.

8 *Logic comparator*
The logic comparator tests integrated circuits by comparing them to a good, reference, integrated circuit (Figure 23.3). Without removing the integrated circuit being tested from its circuit, each input pin is connected in parallel with the corresponding input pin on the reference integrated circuit; likewise each output pin is connected with the corresponding output pin on the reference integrated circuit. The two outputs are compared with an EXCLUSIVE-OR gate, which then gives an output when the two outputs differ. The pulse stretcher is used to extend the duration of the

Figure 23.3 Logic comparator.

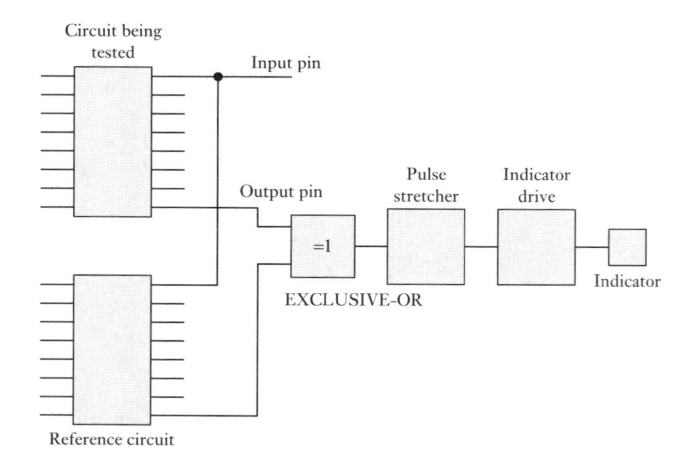

signal fed to the indicator so that very short-duration pulses will result in the indicator being on for a noticeable period.

9 *Signature analyser*
With analogue systems, fault finding usually involves tracing through the circuitry and examining the waveforms at various nodes, comparison of the waveforms with what would be expected enabling faults to be identified and located. With digital systems the procedure is more complex since trains of pulses at nodes all look very similar. To identify whether there is a fault, the sequence of pulses is converted into a more readily identifiable form, e.g. 258F, this being termed the **signature**. The signature obtained at a node can then be compared with that which should occur. When using the signature analyser with a circuit, it is often necessary for the circuit to have been designed so that data bus feedback paths can be broken easily for the test to stop faulty digital sequences being fed back during the testing. A short program, which is stored in ROM, is activated to stimulate nodes and enable signatures to be obtained. The microprocessor itself can be tested if the data bus is broken to isolate it from memory and it is then made to 'free-run' and give a 'no operation' (NO) instruction to each of its addresses in turn. The signatures for the microprocessor bus in this state can then be compared with those expected.

10 *Logic analyser*
The logic analyser is used to sample and store simultaneously in a 'first-in–first-out' (FIFO) memory the logic levels of bus and control signals in a unit under test. The point in the program at which the data capture starts or finishes is selected by the use of a 'trigger word'. The analyser compares its trigger word with the incoming data and only starts to store data when the word occurs in the program. Data capture then continues for a predetermined number of clock pulses and is then stopped. The stored data may then be displayed as a list of binary, octal, decimal or hexadecimal codes, or as a time display in which the waveforms are displayed as a function of time, or as a mnemonic display.

23.5.2 Systematic fault-location methods

Systematic fault-location methods are:

1 *Input to output*
A suitable input signal is injected into the first block of the system and then measurements are made in sequence, starting from the first block, at the output of each block in turn until the faulty block is found.

2 *Output to input*
A suitable input signal is injected into the first block of the system and then measurements are made in sequence, starting from the last block, at the output of each block in turn until the faulty block is found.

3 *Half-split*
A suitable input signal is injected into the first block of the system. The blocks constituting the system are split in half and each half tested to determine in which half the fault lies. The faulty half is then split into half and the procedure repeated.

23.5.3 Self-testing

Software can be used by a microprocessor-based system to institute a self-test program for correct functioning. Such programs are often initiated during the start-up sequence of a system when it is first switched on. For example, printers include microprocessors in their control circuits and generally the control program stored in ROM also includes test routines. Thus when first switched on, it goes through these test routines and is not ready to receive data until all tests indicate the system is fault-free.

A basic ROM test involves totalling all the data bytes stored in each location in ROM and comparing the sum against that already stored (the so-called **checksum test**). If there is a difference then the ROM is faulty; if they agree it is considered to be fault-free. A basic RAM test involves storing data patterns in which adjacent bits are at opposite logic levels, i.e. hex 55 and AA, into every memory location and then reading back each value stored to check that it corresponds to the data sent (the so-called **checker board test**).

23.6 Emulation and simulation

An emulator is a test board which can be used to test a microcontroller and its program. The board contains:

1 The microcontroller.
2 Memory chips for the microcontroller to use as data and program memory.
3 An input/output port to enable connections to be made with the system under test.
4 A communications port to enable program code to be downloaded from a computer and the program operation to be monitored.

The program code can be written in a host computer and then downloaded through either a serial or a parallel link into the memory on the board. The microcontroller then operates as though this program was contained within its own internal memory. Figure 23.4 shows the general arrangement.

Figure 23.4 Using an emulator.

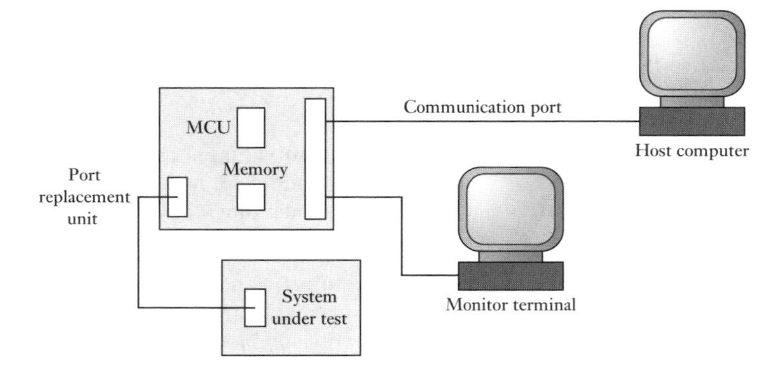

The input/output lines of the microcontroller are connected via an input/output port on the board to a plug-in device on the system under test so that it operates as though it had the microcontroller plugged into it. The board is already programmed with a monitor system which enables the operation of the program to be monitored and the contents of memory, registers, input/output ports to be checked and modified.

Figure 23.5 shows the basic elements of the evaluation board MC68HC11EVB that is provided by Motorola. This uses a monitor program called Bit User Fast Friendly Aid to Logical Operations (Buffalo). The 8K EPROM contains the Buffalo monitor. An MC6850 asynchronous communications interface adapter (ACIA) (see Section 20.5) is used for interfacing parallel and serial lines. A partial RS-232 interface is supplied with each of the two serial ports for connection to the host computer and the monitoring terminal.

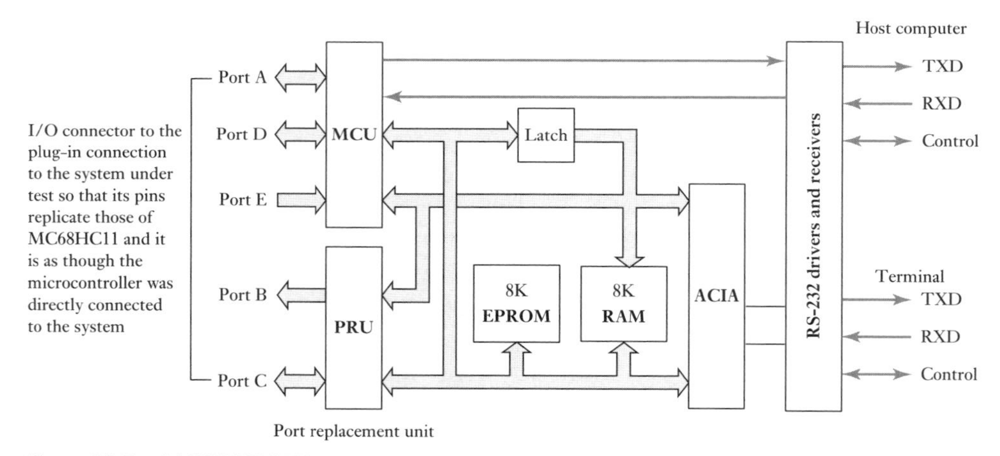

Figure 23.5 MC68HC11EVB.

23.6.1 Simulation

Instead of testing a program by running it with an actual microcontroller, one can test it by running it with a computer program that *simulates* the microcontroller. Such simulation can assist in the debugging of the program code. The display screen can be divided into a number of windows in which information is displayed such as the source code as it is executed, the CPU registers and flags with their current states, the input/output ports, registers and timers, and the memory situation.

23.7 PLC systems

PLCs have a high reliability. They are electrically isolated by optoisolators or relays from potentially damaging voltages and currents at input/output ports; battery-backed RAM protects the application software from power failures or corruption; and the construction is so designed that the PLCs can operate reliably in industrial conditions for long periods of time. PLCs generally have several built-in fault procedures. Critical faults cause the CPU to stop, while other less critical faults cause the CPU to continue running but display a fault code on a display. The PLC manual will indicate the remedial action required when a fault code is displayed.

23.7.1 Program testing

The software checking program checks through a ladder program for incorrect device addresses, and provides a list on a screen, or as a printout, of all the input/output points used, counter and timer settings, etc., with any errors detected. Thus the procedure might involve:

1 Opening and displaying the ladder program concerned.
2 Selecting from the menu on the screen Ladder Test.
3 The screen might then display the message: Start from beginning of program (Y/N)?
4 Type Y and press Enter.
5 Any error message is then displayed or the message of 'No errors found'.

For example, there might be a message for a particular output address that it is used as an output more than once in the program, a timer or counter is being used without a preset value, a counter is being used without a reset, there is no END instruction, etc. As a result of such a test, there may be a need to make changes to the program. Changes needed to rectify the program might be made by selecting exchange from the menu displayed on screen and following through the set of displayed screen messages.

23.7.2 Testing inputs and outputs

Most PLCs have the facility for testing inputs and outputs by what is termed forcing. Software is used to 'force' inputs and outputs on or off. To force inputs or outputs, a PLC has to be switched into the forcing or monitor mode by perhaps pressing a key marked FORCE or selecting the MONITOR mode on a screen display. As a result of forcing an input we can check that the consequential action of that input being on occurs. The installed

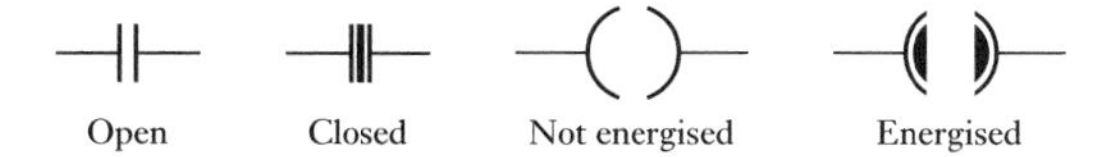

Figure 23.6 Monitor mode symbols.

program can thus be run and inputs and outputs simulated so that they, and all preset values, can be checked. However, care must be exercised with forcing in that an output might be forced that can result in a piece of hardware moving in an unexpected and dangerous manner.

As an illustration of the type of display obtained when forcing, Figure 23.6 shows how inputs might appear in the ladder program display when open and closed, and outputs when not energised and energised, and Figure 23.7(a) a selected part of a displayed ladder program and Figure 23.7(b) what happens when forcing occurs. Initially, Figure 23.7(a) shows rung 11, with inputs to X400, X401 and M100, but not X402, and with no output from Y430. For rung 12, the timer T450 contacts are closed, the display at the bottom of the screen indicating that there is no time left to run on T450. Because Y430 is not energised, the Y430 contacts are open and so there is no output from Y431. If we now force an input to X402 then the screen display changes to that shown in Figure 23.7(b) and Y430 is energised and consequently Y431 energised.

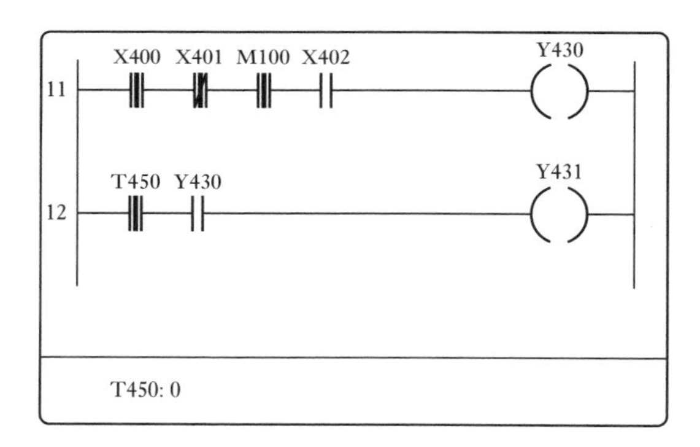

(a)

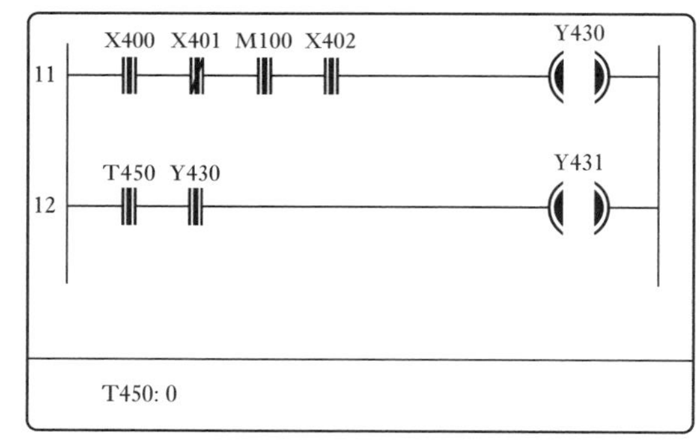

(b)

Figure 23.7 Forcing an input.

23.7.3 PLC as a monitor of systems

The PLC can also be used as a monitor of the system being controlled. It can be used to sound an alarm or light up a red light if inputs move outside prescribed limits, using the greater than, equal to, or less than functions, or its operations take longer than a prescribed time. See Figure 23.1 for an illustration of how a PLC ladder program might be used as a watchdog timer for some operation.

Often with PLC systems, status lamps are used to indicate the last output that has been set during a process and so, if the system stops, where the fault has occurred. The lamps are built into the program so that as each output occurs a lamp comes on and turns off the previous output status lamps. Figure 23.8 illustrates this.

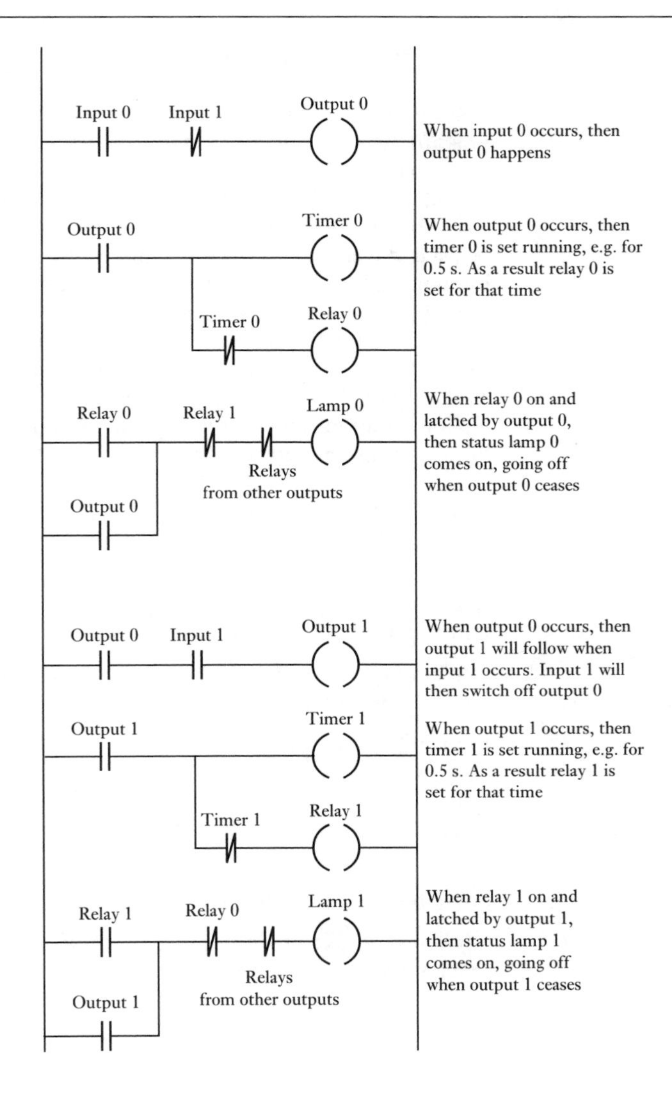

Figure 23.8 Last output set diagnostic program.

Summary

Techniques used to detect faults are replication checks, expected value checks, timing checks, i.e. watchdog timers, reversal checks, parity and error coding checks and diagnostic checks.

A **watchdog timer** is basically a timer that the system must reset before it times out. If the timer is not reset in time then an error is assumed to have occurred. **Parity checks** involve an extra bit being added to a message to make the total number of ones an even number when even parity is used or an odd number when odd parity is used. The efficiency of error detection can be increased by the use of **block parity** in which the message is divided into a number of blocks and each block has a check character added at the end of the block. The **cyclic redundancy check** (CRC) involves the binary number representing the data being transmitted being divided by a predetermined number using modulo-2 arithmetic. The remainder from the division is the CRC character which is transmitted with the data. At the receiver the data plus the CRC character are divided by the same number. If no transmission errors have occurred there is no remainder.

Software can be used by a microprocessor-based system to institute a self-test program for correct functioning. An **emulator** is a test board which can be used to test a microcontroller and its program. Instead of testing a program by running it with an actual microcontroller, one can test it by running it with a computer program that *simulates* the microcontroller.

PLCs generally have several built-in fault procedures. Critical faults cause the CPU to stop, while other less critical faults cause the CPU to continue running but display a fault code on a display. Most PLCs have the facility for testing inputs and outputs by what is termed **forcing**. Software is used to 'force' inputs and outputs on or off.

Problems

23.1 Explain what is meant by (a) replication checks, (b) expected value checks, (c) reversal checks, (d) parity checks.

23.2 Explain how a watchdog timer can be used with a PLC-controlled plant in order to indicate the presence of faults.

23.3 Explain the function of COP in the microcontroller MC68HC11.

23.4 The F2 series Mitsubishi PLC is specified as having:

Diagnosis: Programmable check (sum, syntax, circuit check), watchdog timer, battery voltage, power supply voltage

Explain the significance of the terms.

23.5 Explain how self-testing can be used by a microprocessor-based system to check its ROM and RAM.

Part VI
Conclusion

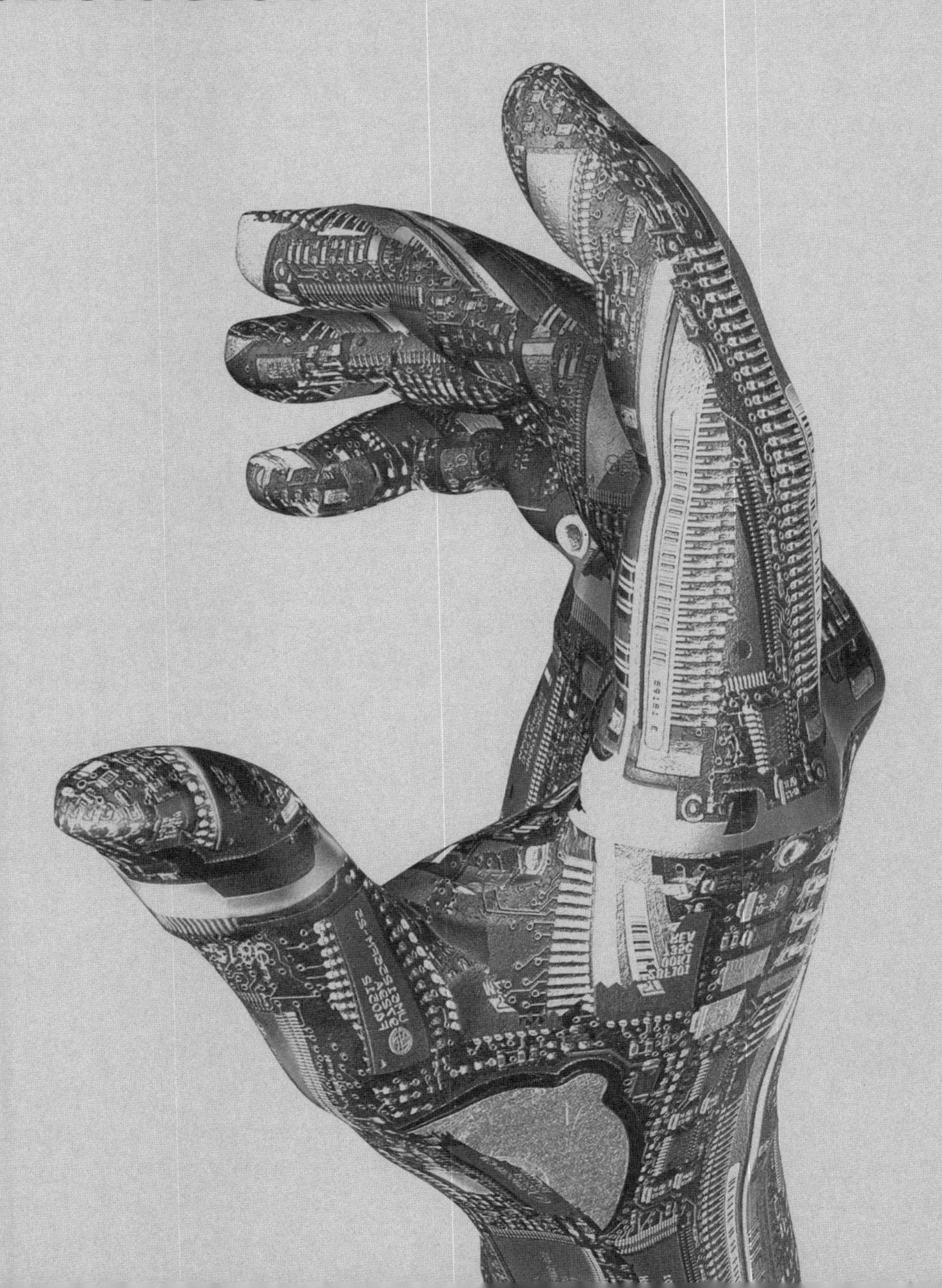

Chapter twenty-four Mechatronic Systems

Objectives

The objectives of this chapter are that, after studying it, the reader should be able to:

- **Develop possible solutions to design problems when considered from the mechatronics point of view.**
- **Analyse case studies of mechatronics solutions.**

24.1 Mechatronic designs

This chapter brings together many of the topics discussed in this book in the consideration of mechatronics solutions to design problems and gives case studies.

24.1.1 Timed switch

Consider a simple requirement for a device which switches on some actuator, e.g. a motor, for some prescribed time. Possible solutions might involve:

1 A rotating cam
2 A PLC
3 A microprocessor
4 A microcontroller
5 A timer, e.g. 555

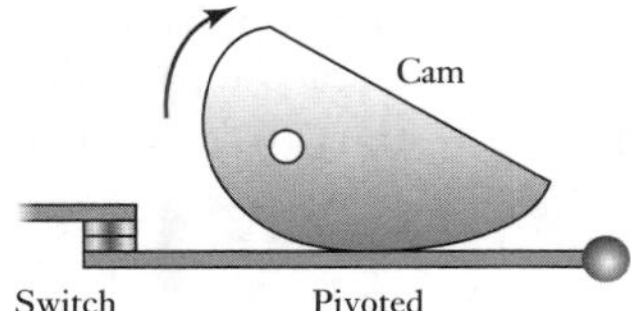

Figure 24.1 Cam-operated switch.

A mechanical solution could involve a rotating cam (Figure 24.1) (see Section 8.4). The cam would be rotated at a constant rate and the cam follower used to actuate a switch, the length of time for which the switch is closed depending on the shape of the cam. This is a solution that has widely been used in the past.

A PLC solution could involve the arrangement shown in Figure 24.2 with the given ladder program. This would have the advantage over the rotating cam of having off and on times which can be adjusted by purely changing the timer preset values in the program, whereas a different cam is needed if the times have to be changed with the mechanical solution. The software solution is much easier to implement than the hardware one.

A microprocessor-based solution could involve a microprocessor combined with a memory chip and input/output interfaces. The program is then used to switch an output on and then off after some time delay, with the time delay being produced by a block of program in which there is a timing loop. This generates a time delay by branching round a loop the number of

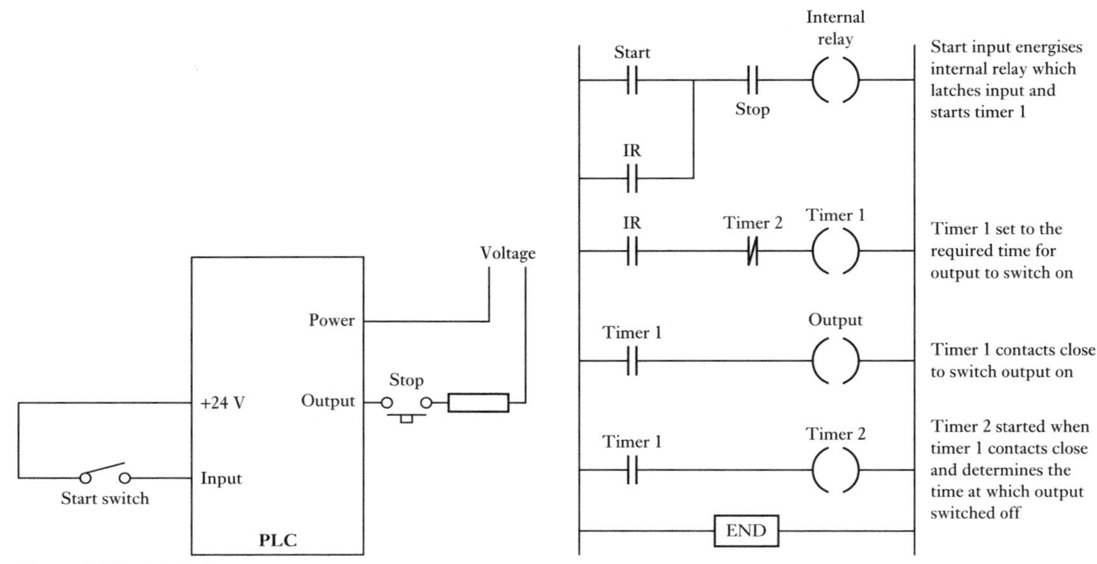

Figure 24.2 PLC timer system.

cycles required to generate the requisite time. Thus, in assembly language we might have:

```
DELAY   LDX   #F424   ; F424 is number of loops
LOOP    DEX
        BNE   LOOP
        RTS
```

DEX decrements the index register, and this and BNE, branch if not equal, each take 4 clock cycles. The loop thus takes 8 cycles and there will be n such loops until $8n + 3 + 5$ gives the number F424 (LDX takes 3 cycles and RTS takes 5 cycles). In C we could write the program lines using the while function.

Another possibility is to use the timer system in a microcontroller such as MC68HC11. This is based on a 16-bit counter TCNT operating from the system E-clock signal (Figure 24.3(a)). The system E-clock can be prescaled by setting bits in the timer interrupt mask register 2 (TMSK2), address

Figure 24.3 (a) Generating 2 MHz internal clock, (b) prescale factor.

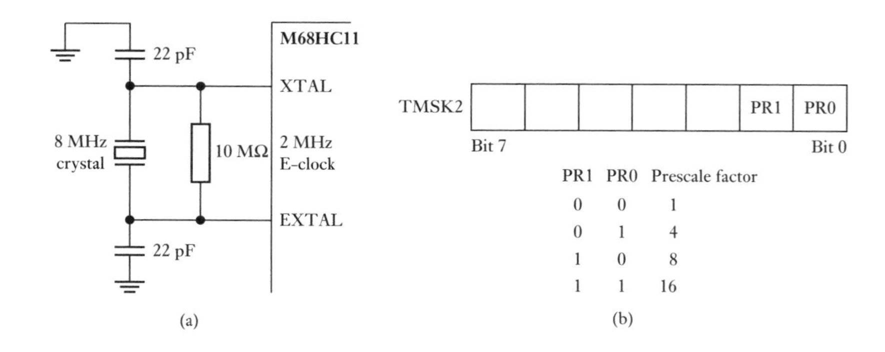

PR1	PR0	Prescale factor
0	0	1
0	1	4
1	0	8
1	1	16

$1024 (Figure 24.3(b)). The TCNT register starts at $0000 when the processor is reset and counts continuously until it reaches the maximum count of $FFFF. On the next pulse it overflows and reads $0000 again. When it overflows, it sets the timer overflow flag TOF (bit 7 in miscellaneous timer interrupt flag register 2, TFLG2 at address $1025). Thus with a prescale factor of 1 and an E-clock of 2 MHz, overflow occurs after 32.768 ms.

One way of using this for timing is for the TOF flag to be watched by polling. When the flag is set, the program increments its counter. The program then resets the flag, by writing a 1 to bit 7 in the TFLG2 register. Thus the timing operation just consists of the program waiting for the required number of overflag settings.

A better way of timing involves the use of the output-compare function. Port A of the microcontroller can be used for general inputs or outputs or for timing functions. The timer has output pins, OC1, OC2, OC3, OC4 and OC5, with internal registers TOC1, TOC2, TOC3, TOC4 and TOC5. We can use the output-compare function to compare the values in the TOC1 to TOC5 registers with the value in the free-running counter TCNT. This counter starts at 0000 when the CPU is reset and then runs continuously. When a match occurs between a register and the counter, then the corresponding OCx flag bit is set and output occurs through the relevant output pin. Figure 24.4 illustrates this. Thus by programming the TOCx registers, so the times at which outputs occur can be set. The output-compare function can generate timing delays with much higher accuracy than the timer overflow flag.

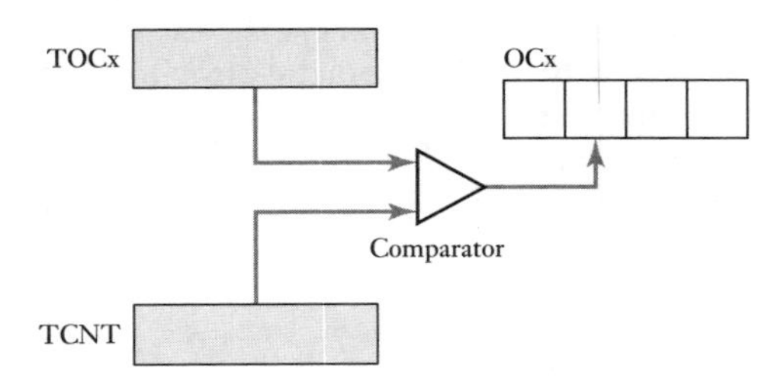

Figure 24.4 Output compare.

The following program illustrates how output compare can be used to produce a time delay. The longest delay that can be generated in one output-compare operation is 32.7 ms when the E-clock is 2 MHz. In order to generate longer delays, multiple output-compare operations are required. Thus we might have each output-compare operation producing a delay of 25 ms and repeating this 40 times to give a total delay of 1 s.

```
REGBAS   EQU   $1000   ; Base address of registers
TOC2     EQU   $18     ; Offset of TOC2 from REGBAS
TCNT     EQU   $0E     ; Offset of TCNT from REGBAS
TFLG1    EQU   $23     ; Offset of TFLGI from REGBAS
OC1      EQU   $40     ; Mask to clear OC1 pin and OC1F flag
CLEAR    EQU   $40     ; Clear OC2F flag
D25MS    EQU   50000   ; Number of E-clock cycles to generate a 25 ms delay
NTIMES   EQU   40      ; Number of output-compare operations needed to give 1 s delay
         ORG   $1000
COUNT    RMB   1       ; Memory location to keep track of the number of
                       ; output-compare operations still to be carried out
```

```
         ORG     $C000          ; Starting address of the program
         LDX     #REGBAS
         LDAA    #OC1           ; Clear OC1 flag
         STAA    TFLG1,X
         LDAA    #NTIMES        ; Initialise the output-compare count
         STAA    COUNT
         LDD     TCNT,X
WAIT     ADDD    #D25MS         ; Add 25 ms delay
         STD     TOC2,X         ; Start the output-compare operation
         BRCLR   TFLG1,X OC1    ; Wait until the OC1F flag is set
         LDAA    #OC1           ; Clear the OC1F flag
         STAA    TFLG1,X
         DEC     COUNT          ; Decrement the output-compare counter
         BEQ     OTHER          ; Branch to OTHER if 1 s elapsed
         LDD     TOC2,X         ; Prepare to start the next compare operation
         BRA     WAIT
OTHER                           ; The other operations of the program which occur after the 1 s
                                  delay
```

Another possible method of producing a timed output signal is to use a timer module, e.g. 555. With the 555 timer, the timing intervals are set by external resistors and capacitors. Figure 24.5 shows the timer and the external circuitry needed to give an on output when triggered, the duration of the on output being $1.1RC$. Large times need large values of R and C. R is limited to about 1 MΩ, otherwise leakage becomes a problem, and C is limited to about 10 μF if electrolytic capacitors with the problems of leakage and low accuracy are to be avoided. Thus the circuit shown is limited to times less than about 10 s. The lower limit is about $R = 1$ kΩ and $C = 100$ pF, i.e. times of a fraction of a millisecond. For longer times, from 16 ms to days, an alternative timer such as the ZN1034E can be used.

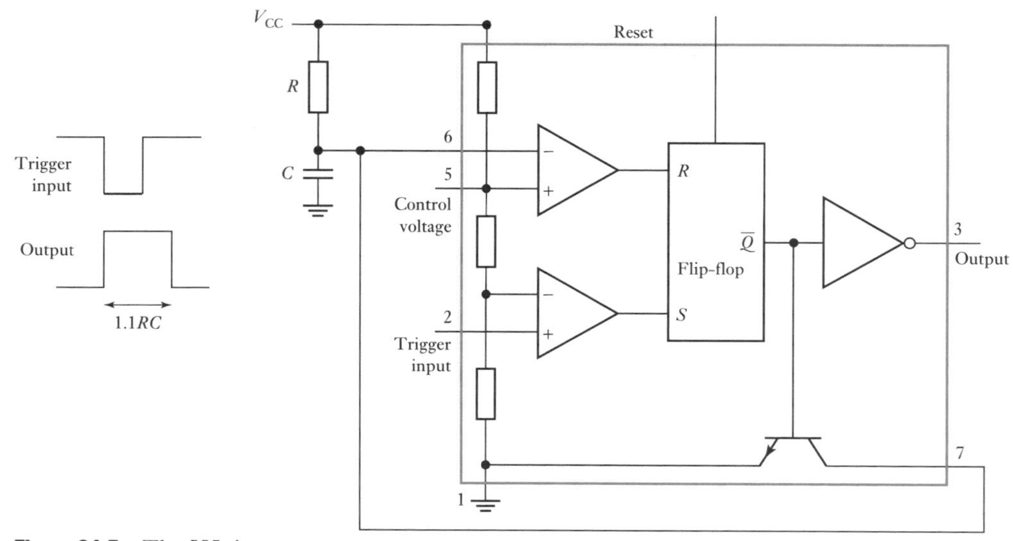

Figure 24.5 The 555 timer.

24.1.2 Windscreen-wiper motion

Consider a requirement for a device which will oscillate an arm back and forth in an arc like a windscreen wiper. Possible solutions might be:

1 Mechanical linkage and a d.c. motor
2 A stepper motor

A mechanical solution is shown in Figure 24.6. Rotation of arm 1 by a motor causes arm 2 to impart an oscillatory motion to arm 3. Car windscreen wipers generally use such a mechanism with a d.c. permanent magnet motor.

Figure 24.6 Wiper mechanism.

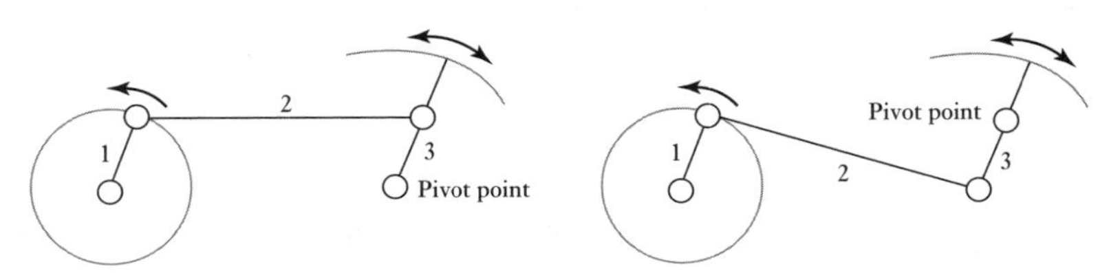

An alternative solution is to use a stepper motor. Figure 24.7 shows how a microprocessor with a PIA, or a microcontroller, might be used with a stepper. The input to the stepper is required to cause it to rotate a number of steps in one direction and then reverse to rotate the same number of steps in the other direction.

Figure 24.7 Interfacing a stepper.

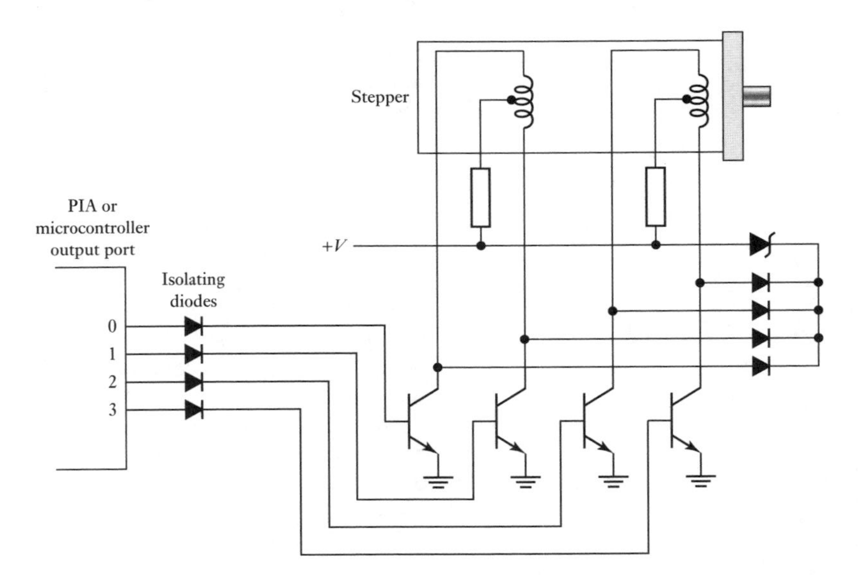

If the stepper is to be in the 'full-step' configuration then the outputs need to be as shown in Table 24.1(a). Thus to start and rotate the motor in a forward direction involves the sequence A, 9, 5, 6 and then back to the beginning with 1 again. To reverse we would use the sequence 6, 5, 9, A and then back to begin with 6 again. If 'half-step' configuration is used then the outputs need to be as shown in Table 24.1(b). Forward motion then involves the sequence A, 8, 9, 1, 5, 4, 6, 2 and then back to A, with reverse requiring 2, 6, 4, 5, 1, 9, 8, A and back to 2.

Table 24.1 (a) Full-step, (b) half-step configuration.

(a)

Step	Bit 3	Bit 2	Bit 1	Bit 0	Code
1	1	0	1	0	A
2	1	0	0	1	9
3	0	1	0	1	5
4	0	1	1	0	6
1	1	0	1	0	A

(b)

Step	Bit 3	Bit 2	Bit 1	Bit 0
1	1	0	1	0
2	1	0	0	0
3	1	0	0	1
4	0	0	0	1
5	0	1	0	1
6	0	1	0	0
7	0	1	1	0
8	0	0	1	0
1	1	0	1	0

The basic elements of a program could be:

Advance a step
Jump to time-delay routine to give time for the step to be completed
Loop or repeat the above until the requisite number of steps completed in the forward direction
Reverse direction
Repeat the above for the same number of steps in reverse direction

Such a program in C might, for three half-steps forward and three back, and following the inclusion of an appropriate header file, have the following elements:

```
main ( )
{
   portB = 0xa; /*first step*/
   delay ( ); /*incorporate delay program for, say, 20 ms*/
   portB = 0x8; /*second step*/
   delay ( ); /*incorporate delay program for 20 ms*/
   port B = 0x9; /*third step*/
   delay ( ); /*incorporate delay program for 20 ms*/
   port B = 0x8; /*reverse a step*/
   delay ( ); /*incorporate delay program for 20 ms*/
   port B = 0xa; /*reverse a further step*/
   delay ( ); /*incorporate delay program for 20 ms*/
   port B = 0x2; /*reverse back to where motor started*/
   delay ( ); /*incorporate delay program for 20 ms*/
}
```

Where there are many steps involved, a simpler program is to increment a counter with each step and loop until the counter value reaches the required number. Such a program would have the basic form of:

Advance a step
Jump to time-delay routine to give time for the step to be completed
Increment the counter

Loop or repeat the above with successive steps until the counter indicates the requisite number of steps completed in the forward direction
Reverse direction
Repeat the above for the same number of steps in reverse direction

Integrated circuits are available for step motor control and their use can simplify the interfacing and the software. Figure 24.8 shows how such a circuit can be used. All that is then needed is the requisite number of input pulses to the trigger, the motor stepping on the low-to-high transition of a high–low–high pulse. A high on the rotation input causes the motor to step anti-clockwise, while a low gives clockwise rotation. Thus we just need one output from the microcontroller for output pulses to the trigger and one output to rotation. An output to set is used to reset the motor back to its original position.

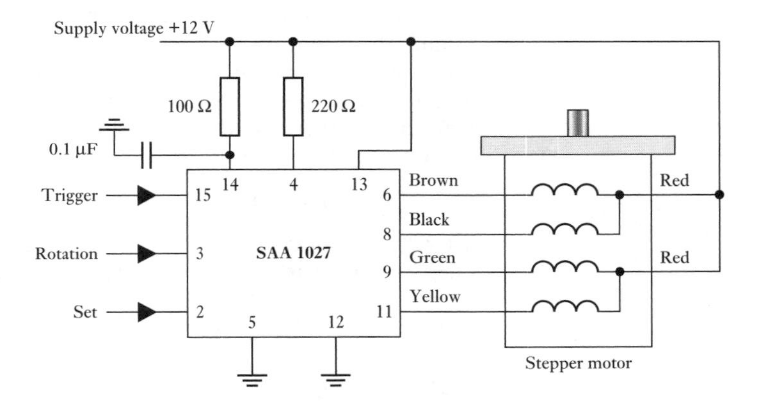

Figure 24.8 Integrated circuit SAA 1027 for stepper motor.

The above has indicated how we might use a stepper motor to give an angular rotation. But how will a stepper motor behave when given a voltage signal input? Can we expect it to rotate directly to the angle concerned with no overshoot and no oscillations before settling down to the required angle? As an illustration of how we can develop a model for the stepper motor system and so predict its behaviour, consider the following simplified analysis (for a more detailed analysis, *see Stepping Motors and their Microprocessor Controls* by T. Kenjo and A. Sugawara (Clarenden Press 1995)).

The system involving a stepper motor being driven by pulses from a microcontroller is an open-loop control system. The permanent magnet stepper motor (see Section 9.7) has a stator with a number of poles, the poles being energised by current being passed through coils wound on them. We can determine a model for how the rotor will rotate when there is a voltage pulse input to it by considering, for simplicity, a stepper with just a pair of poles and treat it in the same manner as the d.c. motor that was analysed in Section 11.3.2. If v is the voltage supplied to the motor pair of coils and v_b the back e.m.f. then

$$v - v_b = L\frac{di}{dt} + Ri$$

where L is the inductance of the circuit, R the resistance and i the circuit current. We will make the simplifying assumption that the inductance does not significantly change and so treat L as a constant.

The back e.m.f will be proportional to the rate at which the magnetic flux is changing for the pair of coils. This will depend on the angle θ of the rotor relative to the poles concerned. Thus we can write

$$v_b = -k_b \frac{d}{dt} \cos\theta = k_b \sin\theta \frac{d\theta}{dt}$$

where k_b is a constant. Thus

$$v - k_b \sin\theta \frac{d\theta}{dt} = L\frac{di}{dt} + Ri$$

Taking the Laplace transform of this equation gives

$$V(s) - k_b s \sin\theta\, \theta(s) = sL\, I(s) + R\, I(s) = (sL + R)\, I(s)$$

As with the d.c. motor, the current through a pair of coils will generate a torque (the torque on the magnet, i.e. the rotor, being the reaction resulting from the torque exerted on the coils – Newton's third law). The torque is proportional to the product of the flux density at the coil turns and the current through them. The flux density will depend on the angular position of the rotor and thus we can write

$$T = k_t i \sin\theta$$

where k_t is a constant. This torque will cause an angular acceleration α and since $T = J\alpha$, where J is the moment of inertia of the rotor,

$$T = J\frac{d^2\theta}{dt^2} = k_t i \sin\theta$$

Taking the Laplace transform of this equation gives

$$s^2 J\theta(s) = k_t \sin\theta\, I(s)$$

and so we can write

$$V(s) - k_b s \sin\theta\, \theta(s) = (sL + R)(s^2 J\theta(s)/k_t \sin\theta)$$

The transfer function between the input voltage and the resulting angular displacement is

$$G(s) = \frac{\theta(s)}{V(s)} = \frac{k_t \sin\theta}{J(sL + R)s^2 + k_b k_t s \sin^2\theta}$$

$$= \frac{1}{s} \times \frac{k_t \sin\theta}{JLs^2 + JRs + k_b k_t \sin^2\theta}$$

When there is a voltage impulse supplied to the motor coils, since for a unit impulse $V(s) = 1$,

$$\theta(s) = \frac{1}{s} \times \frac{k_t \sin\theta}{JLs^2 + JRs + k_b k_t \sin^2\theta}$$

$$= \frac{1}{s} \times \frac{(k_t \sin\theta)/JL}{s^2 + (R/L)s + (k_b k_t \sin^2\theta)/JL}$$

The quadratic equation in s is of the form $s^2 + 2\zeta\omega_n s + \omega_n^2$ (see Section 13.3.1) and thus has a natural frequency ω_n of $\sqrt{(k_b k_t \sin^2\theta / JL)}$ and a damping factor ζ of $(R/L)/2\omega_n$. The rotor will rotate to some angle and gives oscillations about that angle with the oscillations dying away with time.

24.1.3 Bathroom scales

Consider the design of a simple weighing machine, i.e. bathroom scales. The main requirements are that a person can stand on a platform and the weight of that person will be displayed on some form of readout. The weight should be given with reasonable speed and accuracy and be independent of where on the platform the person stands. Possible solutions can involve:

1 A purely mechanical system based on a spring and gearing.
2 A load cell and a microprocessor/microcontroller system.

One possible solution is to use the weight of the person on the platform to deflect an arrangement of two parallel leaf springs (Figure 24.9(a)). With such an arrangement the deflection is virtually independent of where on the platform the person stands. The deflection can be transformed into movement of a pointer across a scale by using the arrangement shown in Figure 24.9(b). A rack-and-pinion is used to transform the linear motion into a circular motion about a horizontal axis. This is then transformed into a rotation about a vertical axis, and hence movement of a pointer across a scale, by means of a bevel gear.

Figure 24.9 Bathroom scales.

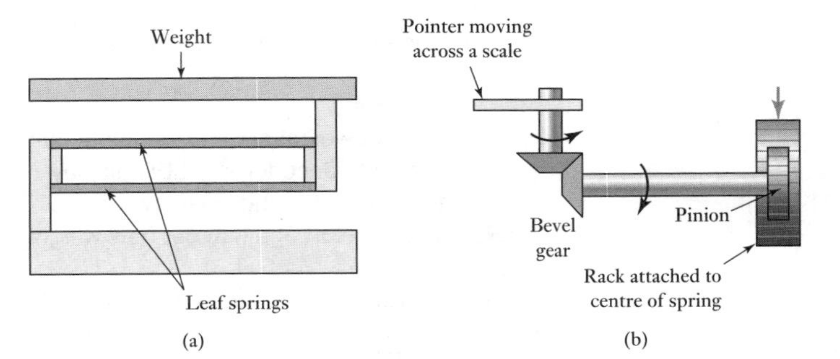

Another possible solution involves the use of a microprocessor. The platform can be mounted on load cells employing electrical resistance strain gauges. When the person stands on the platform the gauges suffer strain and change resistance. If the gauges are mounted in a four-active-arm Wheatstone bridge then the out-of-balance voltage output from the bridge is a measure of the weight of the person. This can be amplified by a differential operational amplifier. The resulting analogue signal can then be fed through a latched analogue-to-digital converter for inputting to the microprocessor, e.g. the Motorola 6820. Figure 24.10 shows the input interface. There will also be a need to provide a non-erasable memory and this can be provided by an EPROM chip, e.g. Motorola 2716. The output to the display can then be taken through a PIA, e.g. Motorola 6821.

However, if a microcontroller is used then memory is present within the single microprocessor chip, and by a suitable choice of microcontroller,

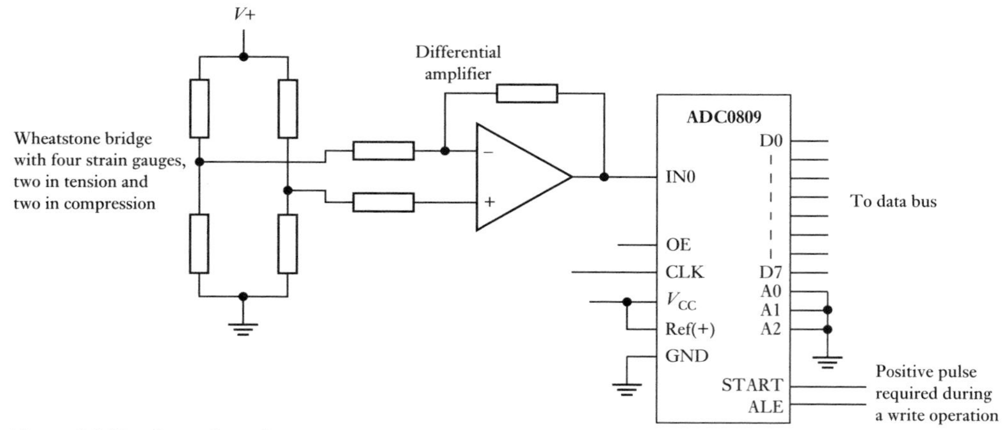

Figure 24.10 Input interface.

e.g. M68HC11, we can obtain analogue-to-digital conversion for the inputs. The system then becomes: strain gauges feeding through an operational amplifier a voltage to the port E (the ADC input) of the microcontroller, with the output passing through suitable drives to output through ports B and C to a decoder and hence an LED display (Figure 24.11).

The program structure might be:

```
Initialisation by clearing LED displays and memory

Start
  Is someone on the scales? If not display 000
  If yes
    input data
    convert weight data into suitable output form
    output to decoder and LED display
    time delay to retain display
Repeat from start again to get new weight
```

In considering the design of the mechanical parts of the bathroom scales we need to consider what will happen when someone stands on the scales. We have a spring–damper–mass system comparable with that described in Figure 10.3(a) (see Section 10.2.2) and so can describe its behaviour by

$$m\frac{d^2x}{dt^2} + c\frac{dx}{dt} + kx = F$$

where x is the vertical deflection of the platform when a force F is applied. Taking the Laplace transform gives

$$ms^2 X(s) + cs X(s) + kX(s) = F(s)$$

and so the system can be described by a transfer function of the form

$$G(s) = \frac{X(s)}{F(s)} = \frac{1}{ms^2 + cs + k}$$

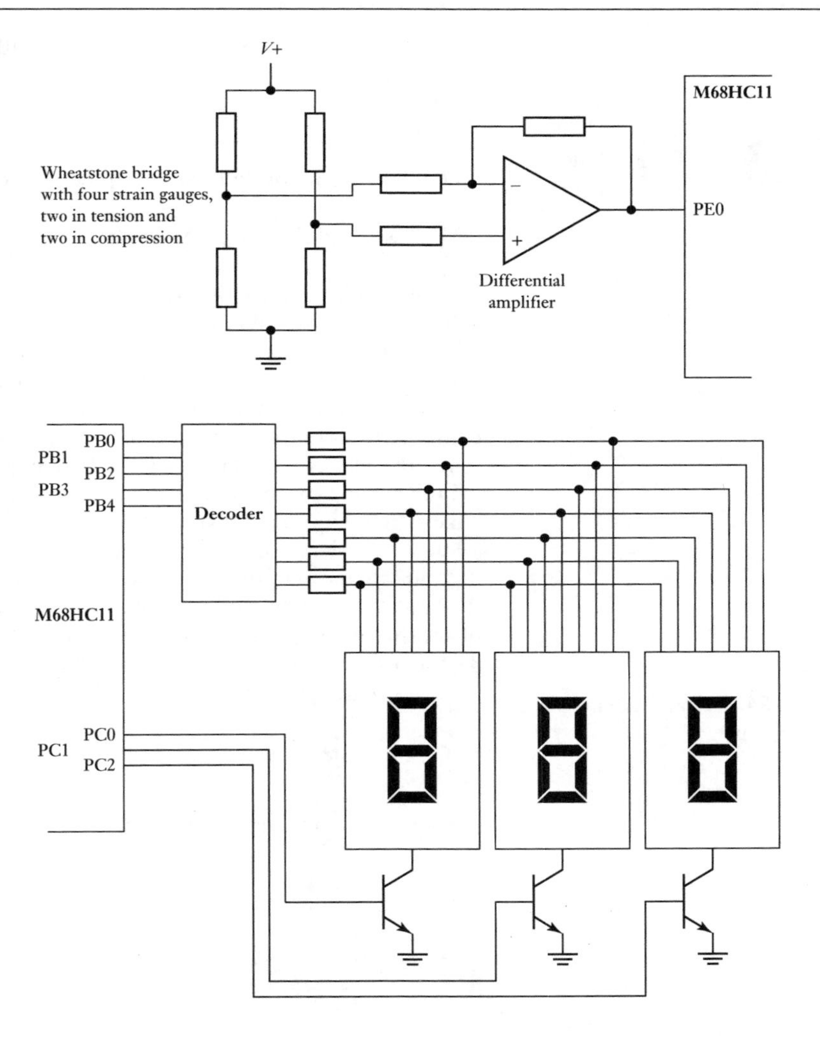

Figure 24.11 Bathroom scales.

We can consider a person of weight W standing on the platform as a step input and so

$$X(s) = \frac{1}{ms^2 + cs + k} \times \frac{W}{s}$$

The quadratic term is of the form $s^2 + 2\zeta\omega_n s + \omega_n^2$ (see Section 13.3.1) and thus has a natural frequency ω_n of $\sqrt{(k/m)}$ and a damping factor ζ of $c/(2\sqrt{(mk)})$.

When a person stands on the scales he or she wants the scales to indicate quickly the weight value and not oscillate for a long time about the value. If the damping was adjusted to be critical it would take too long to reach the value and so the damping needs to be adjusted to allow some oscillations which are rapidly damped away. We might decide that a 2% settling time t_s (see Section 12.5) of, say, 4 s was desirable. Since $t_s = 4/\zeta\omega_n$ then we require $\zeta\omega_n = 1$ and so $\zeta = \sqrt{(m/k)}$. A simple way of altering the damping is thus by changing the mass.

The above indicates how we can use a mathematical model to predict the behaviour of a system and what factors we can then change to improve its performance.

24.2 Case studies

The following are outlines of examples of mechatronic systems.

24.2.1 A pick-and-place robot

Figure 24.12(a) shows the basic form of a pick-and-place robot unit. The robot has three axes about which motion can occur: rotation in a clockwise or anti-clockwise direction of the unit on its base, arm extension or contraction and arm up or down; also the gripper can open or close. These movements can be actuated by the use of pneumatic cylinders operated by solenoid-controlled valves with limit switches to indicate when a motion is completed. Thus the clockwise rotation of the unit might result from the piston in a cylinder being extended and the anti-clockwise direction by its retraction. Likewise the upward movement of the arm might result from the piston in a linear cylinder being extended and the downward motion from it retracting; the extension of the arm by the piston in another cylinder extending and its return movement by the piston retracting. The gripper can be opened or closed by the piston in a linear cylinder extending or retracting, Figure 24.12(b) showing a basic mechanism that could be used.

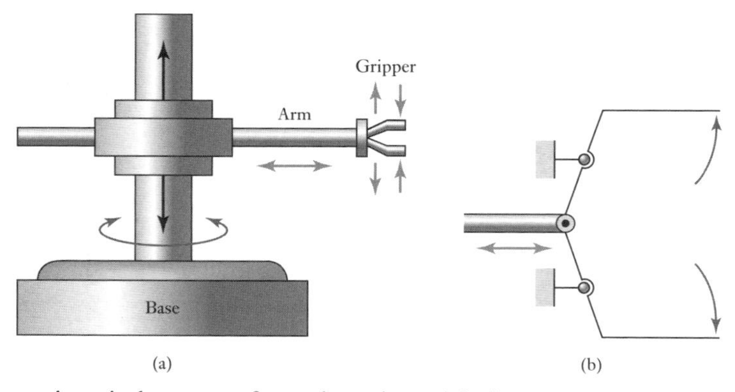

Figure 24.12 (a) Pick-and-place, (b) a gripper.

A typical program for such a robot might be:

1 Close an upright gripper on a component hanging from an overhead feeder.
2 Contract the arm so that the component is withdrawn from the feeder.
3 Rotate the arm in a horizontal plane so that it points in the direction of the workpiece.
4 Extend the arm so that the gripper is over the workpiece.
5 Rotate the wrist so that the component hangs downwards from the gripper.
6 Release the gripper so that the component falls into the required position.
7 Rotate the gripper into an upright position.
8 Contract the arm.
9 Rotate the arm to point towards the feeder.

Repeat the sequence for the next component.

Figure 24.13 shows how a microcontroller could be used to control solenoid valves and hence the movements of the robot unit.

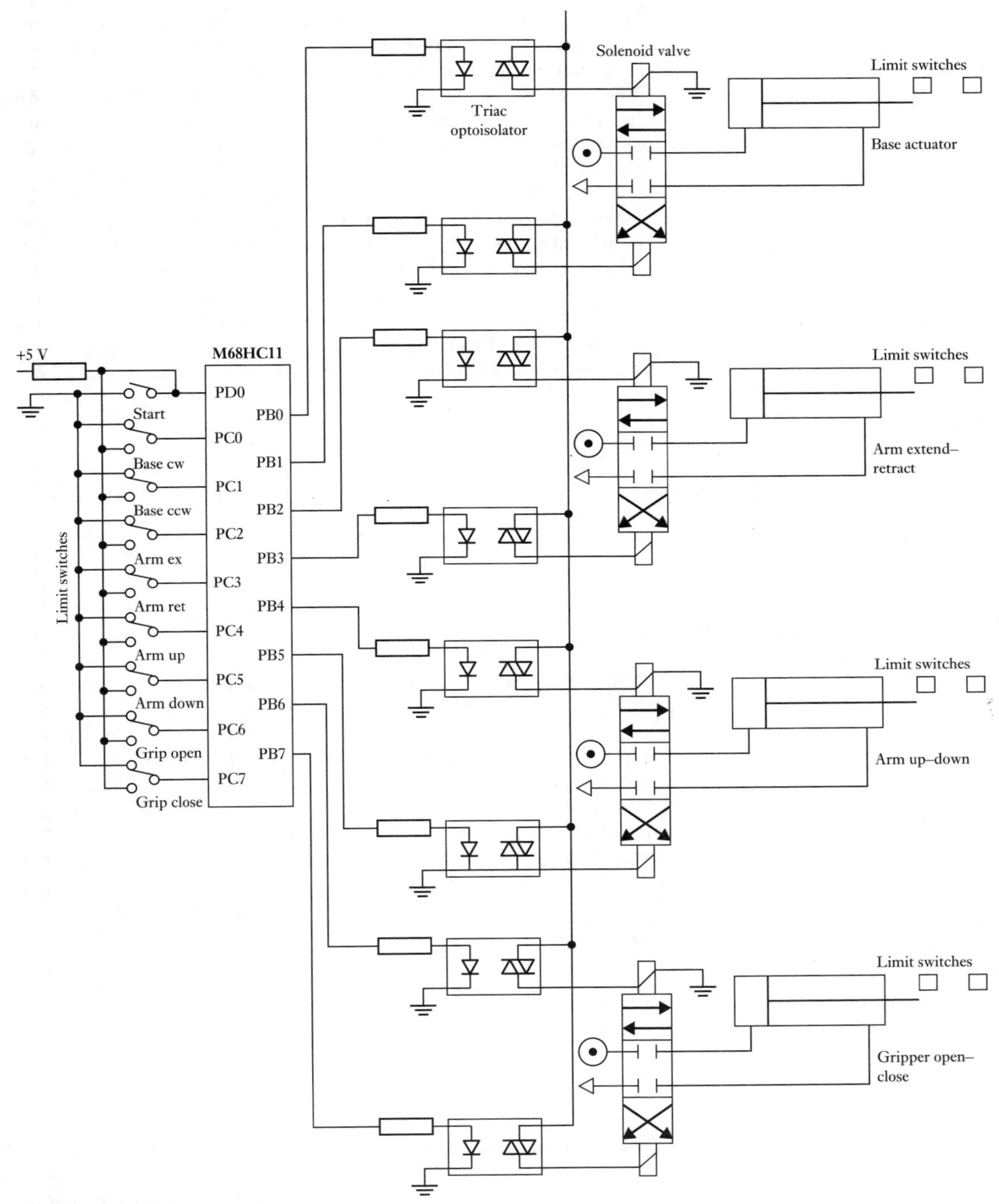

Figure 24.13 Robot control.

Hydraulic and pneumatic rams are widely used to drive robot arms since they can easily be controlled to move limbs at a relatively slow speed, while electric motors would need to operate through a gearbox.

The positions of the arm and gripper in Figure 24.13 are determined by limit switches. This means that only two positions can be accurately attained with each actuator and the positions cannot be readily changed without physically moving the positions of the switches. The arrangement is an open-loop control system. In some applications this may not be a problem.

However, it is more common to use closed-loop control with the positions of an arm and gripper being monitored by sensors and fed back to be compared in the controller with the positions required. When there is a difference from the required positions, the controller operates actuators to reduce the error. The angular position of a joint is often monitored by using an encoder (see Section 2.3.7), this being capable of high precision. Figure 24.14 shows a closed-loop arrangement that might be used for linear motion of a robot arm.

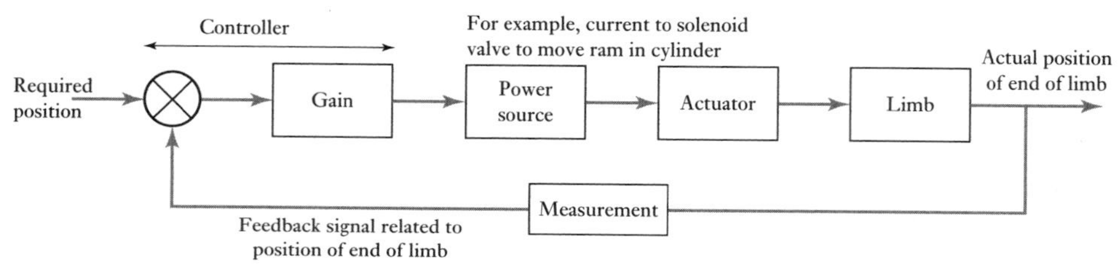

Figure 24.14 Closed-loop control for limb.

The output from the actuator is a force F applied to move the end of the limb. For a set position of y_s and an actual position y, the error signal will be $y_s - y$, assuming for simplicity that the measurement system has a gain of 1. If we consider the controller to have a gain of G_c and G_a to be that of the actuator assembly, then $F = G_c G_a(y_s - y)$. The masses to be accelerated by this force are the mass of the load that the arm is carrying, the mass of the arm and the mass of the moving parts of the actuator. If this is a total mass of m, then Newton's law gives $F = ma$, where the acceleration a can be written as $\mathrm{d}^2y/\mathrm{d}t^2$. However, this does not take account of friction and since we can take the friction force to be proportional to the velocity, the frictional force is $k\,\mathrm{d}y/\mathrm{d}t$. Thus we can write

$$F = G_c G_a(y_s - y) = m\frac{\mathrm{d}^2y}{\mathrm{d}t^2} + k\frac{\mathrm{d}y}{\mathrm{d}t}$$

and so

$$y_s = \frac{m}{G_c G_a}\frac{\mathrm{d}^2y}{\mathrm{d}t^2} + \frac{k}{G_c G_a}\frac{\mathrm{d}y}{\mathrm{d}t} + y$$

This is a second-order differential equation and so the deflection y will be as described in Section 13.3.1 and the form it will take will depend on the damping factor. An under-damped system will have a natural angular frequency ω_n given by

$$\omega_n = \sqrt{\frac{G_c G_a}{m}}$$

This angular frequency will determine how fast the system responds to a change (see Section 12.5): the larger the angular frequency, the faster the system responds (the rise time is inversely proportional to the angular frequency). This means that increasing the controller gain or decreasing the mass can increase the speed of response. The damping factor ζ is given from the differential equation as

$$\zeta = \frac{k}{2\sqrt{G_c G_a m}}$$

The time taken for the oscillations to die away, i.e. the settling time (see Section 12.5), is inversely proportional to the damping factor and so, for example, increasing any part of the mass will result in a decrease in the damping factor and so the oscillations take longer to die away.

24.2.2 Car park barriers

As an illustration of the use of a PLC, consider the coin-operated barriers for a car park. The in-barrier is to open when the correct money is inserted in the collection box and the out-barrier is to open when a car is detected at the car park side of the barrier. Figure 24.15 shows the types of valve systems that can be used to lift and lower the pivoted barriers.

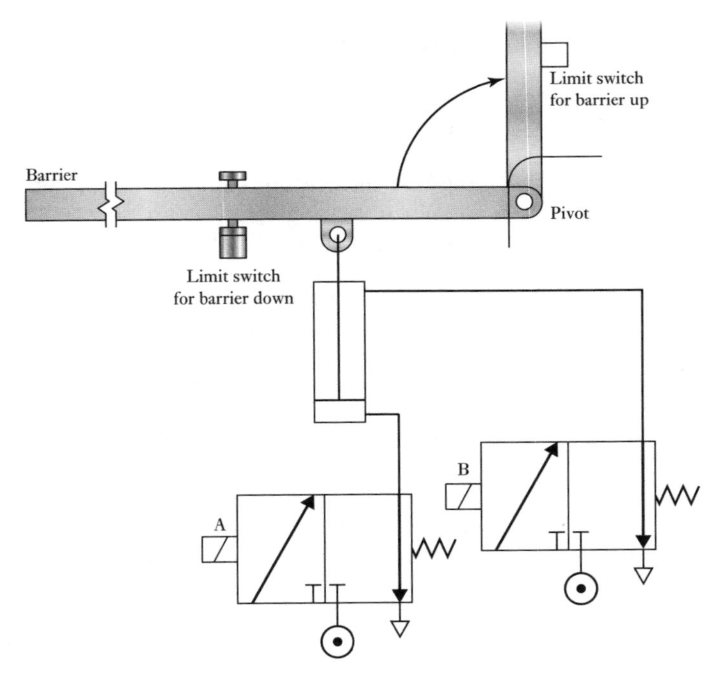

Figure 24.15 System for raising and lowering a barrier.

When a current flows through the solenoid of valve A, the piston in a cylinder moves upwards and causes the barrier to rotate about its pivot and raise to let a car through. When the current through the solenoid of valve A ceases, the return spring of the valve results in the valve position changing back to its original position. When the current is switched through the solenoid of valve B, the pressure is applied to lower the barrier. Limit switches are used to detect when the barrier is down and also when fully up.

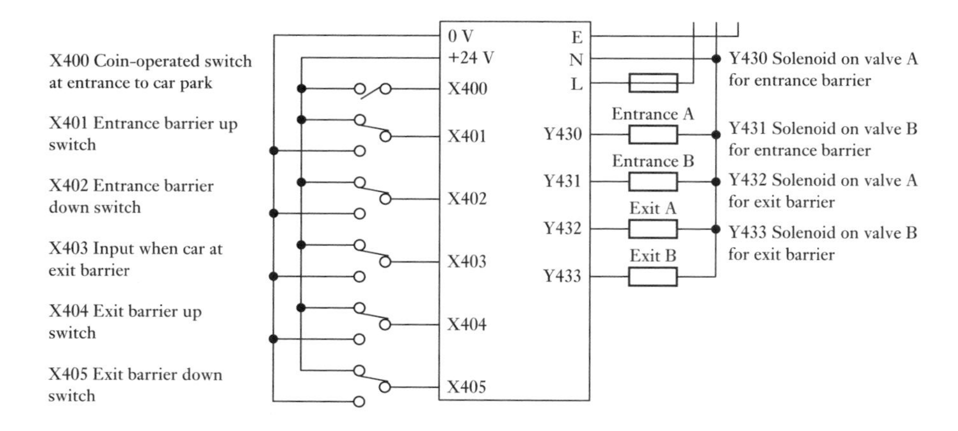

Figure 24.16 PLC connections.

With two of the systems shown in Figure 24.15, one for the entrance barrier and one for the exit barrier, and the connections to PLC inputs and outputs shown in Figure 24.16, the ladder program can be of the form shown in Figure 24.17.

24.2.3 Digital camera

A digital camera is one that captures images and stores them in a digital format in a memory card, unlike the earlier film cameras where the image was stored in an analogue form as a chemical change on film. Figure 24.18 shows the basic elements of a less expensive digital camera.

When the photographer presses the shutter button to its first position, that of being partially depressed, a microcontroller calculates the shutter speed and aperture settings from the input from the metering sensor and displays them on the LCD screen. At the same time, the microcontroller processes the input from the range sensor and sends signals to drive a motor to adjust the focusing of the lens. When the photographer presses the shutter button to its second position, that of being fully depressed, the microcontroller issues signals to change the aperture to that required, open the shutter for the required exposure time, and then, when the shutter has closed, process the image received at the image sensor and store it on the memory card. Also when the shutter button is partially depressed, the automatic focus control system is used to move the lens so that the image will be in focus (see Section 1.7.1 for details of autofocus mechanisms and later in this section for a discussion of the motor used to move the lens).

The light from the object being photographed passes through a lens system and is focused onto an image sensor. This is typically a charge-coupled device (CCD) (see Section 2.10) consisting of an array of many small light-sensitive cells, termed pixels, which are exposed to the light passing through the lens when the electromechanical shutter is opened for a brief interval of time. The light falling on a cell is converted into a small amount of electric charge which, when the exposure has been completed, is read and stored in a register before being processed and stored on the memory card.

The sensors are colour-blind and so, in order that colour photographs can be produced, a colour filter matrix is situated prior to the array of cells.

Figure 24.17 Ladder program.

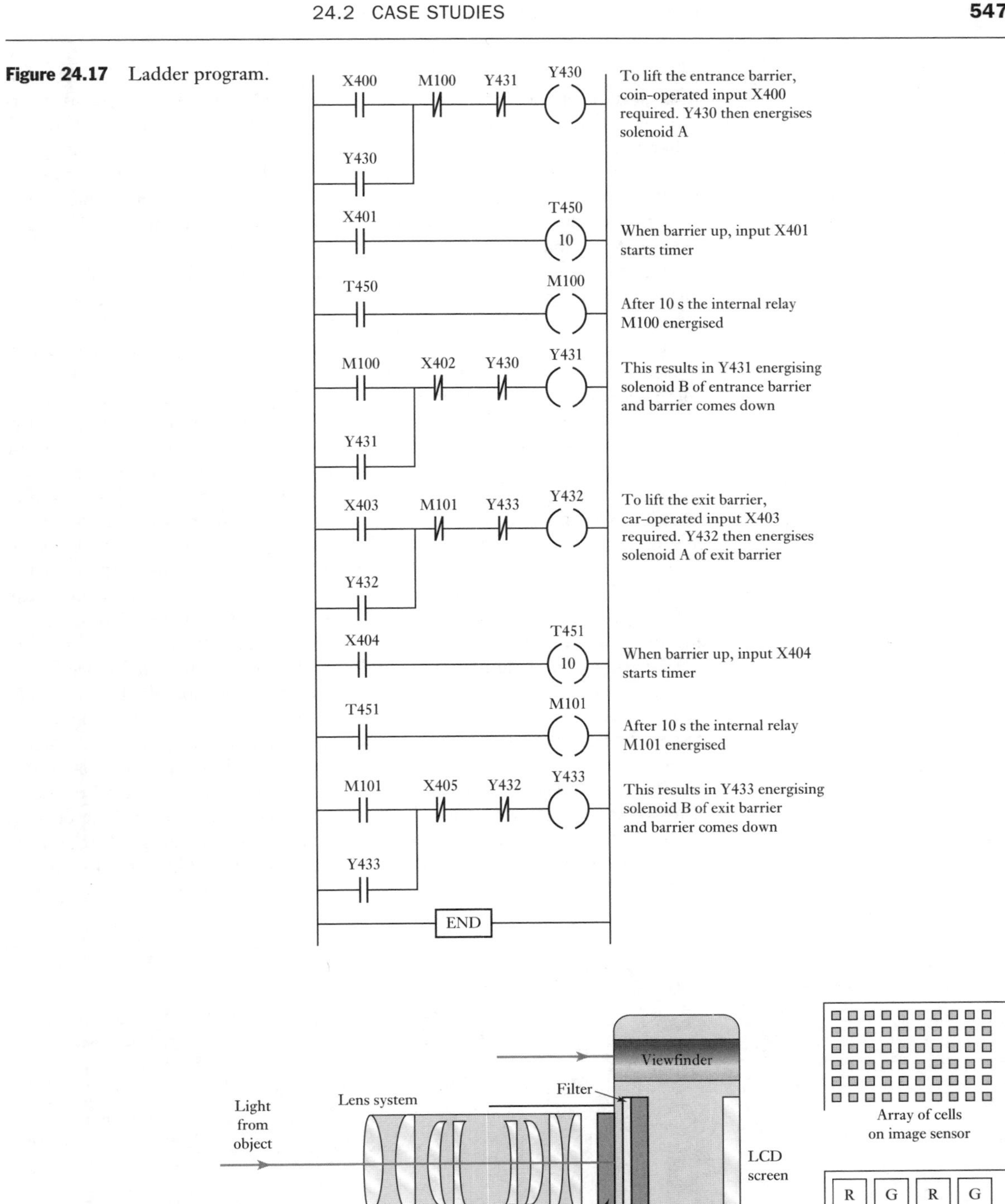

Figure 24.18 The basic elements of a digital camera.

There are separate filters, blue, green or red, for each cell. The most common design for the matrix is the Bayer array. This has the three colours arranged in a pattern so that no two filters of the same colour are next to each other and there are twice as many green filters as either red or blue, this being because green is roughly in the centre of the visible spectrum and gives more detail. The result at this stage is a mosaic of red, green and blue pixels. The files of the results for the pixels at this stage are termed RAW files in that no processing has been done to them. In order to give the full range of colours for a particular pixel, an algorithm is used in which the colour to be allocated to a particular pixel is determined by taking into account the intensities of the colours of neighbouring pixels.

The next stage in processing the signal is to compress the files so that they take up as little memory as possible. This way, more can be stored on a memory card than would be the case with RAW files. Generally, the compressed file format is JPEG, short for Joint Photographic Experts Group. JPEG compression uses the principle that in many photographs, many of the pixels in the same area are identical and so instead of storing the same information for each it can effectively store one and tell the others to just repeat it.

The exposure required is determined by a camera microcontroller in response to the output from a sensor such as a photodiode detecting the intensity of the light. It gives outputs which are used to control the aperture and the shutter speed. The aperture drive system with a digital camera can be a stepper motor which opens or closes a set of diaphragm blades according to the signal received from the microcontroller. The shutter mechanism used with a digital camera is generally of the form shown in Figure 24.19. The shutter involves two sets of curtains, each being controlled by a spring-loaded latch. In the absence of a current to the electromagnet, the spring forces the latch over to a position which has the upper set of curtains down to overlap with the lower set. When a current is passed through an electromagnet it causes the latch to rotate and in doing so lifts the upper set of curtains. The lower set of curtains is initially held down at the bottom by a current through its electromagnet holding the latch. When the current through the lower curtain latch is switched off, the curtains rise. Thus the opening of

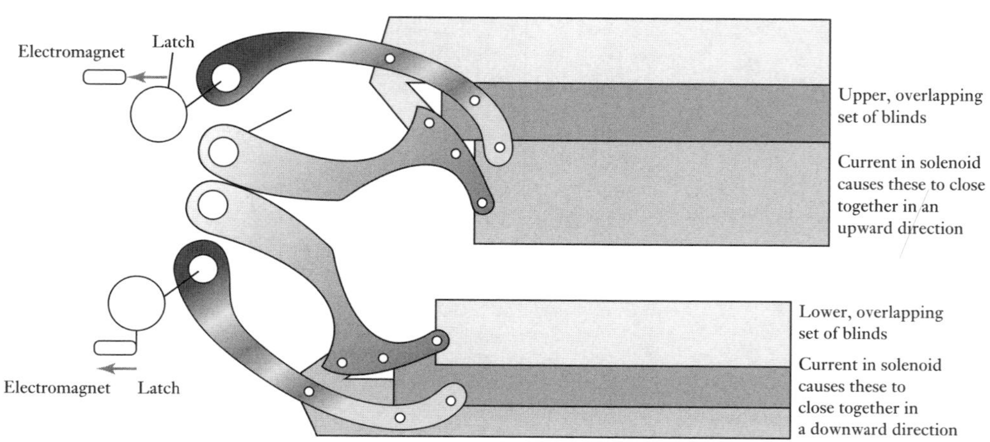

Figure 24.19 Shutter mechanism.

the aperture through to the image sensor is determined by the time between switching a current to the upper latch and switching it off at the lower latch.

The focusing requires a mechanism to move the lens. This is often an ultrasonic motor which consists of a series of piezoelectric elements, such as lead zirconium titanate (PZT). When a current is supplied to such a piezoelectric element it expands or contracts according to the polarity of the current (Figure 24.20(a)). PZT elements are bonded to both sides of a thin strip of spring steel and then, when a potential difference is applied across the strip, the only way the PZT elements can expand or contract is by bending the metal strip (Figure 24.20(b)). When opposite polarity is applied to alternating elements, they are made to bend in opposite directions (Figure 24.20(c)). Thus by using an alternating current with a sequence of such elements round a ring, a displacement wave can be made to travel around the piezoelectric ring of elements in either a clockwise or anti-clockwise direction. The amplitude of this displacement wave is only about 0.001 mm. There is a strip of material with minute cogs attached to the outside of the PZT elements and when the displacement wave moves round the PZT elements they are able to push against the lens mount (Figure 24.20(d)) and thus drive the focusing element.

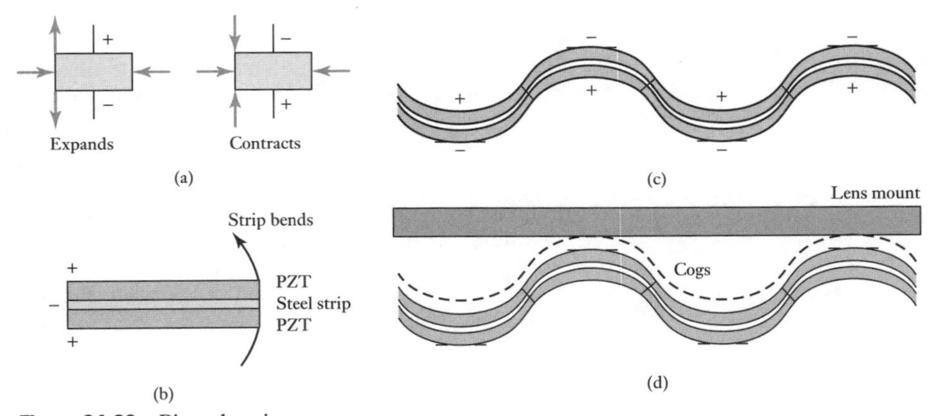

Figure 24.20 Piezoelectric motor.

As an illustration of the use of the modelling techniques discussed in earlier chapters of this book, consider this ultrasonic motor. The torque T generated by the motor is required to rotate the motor ring to some angular position θ. The ring is very light and so we neglect its inertia in comparison with the friction between the rings. Assuming the frictional force is proportional to the angular velocity ω, then $T = c\omega = c\,\mathrm{d}\theta/\mathrm{d}t$, where c is the friction constant. Integration then gives

$$\theta = \frac{1}{c}\int \mathrm{d}t$$

and so a transfer function $G(s)$ of $1/cs$.

The control system for the ultrasonic motor is of the form shown in Figure 24.21. y_n is the nth input pulse and x_n the nth output pulse. With the microprocessor exercising proportional control gain K, the input to it is

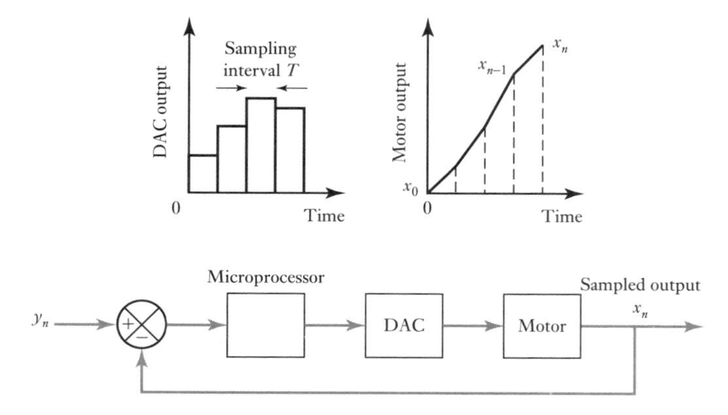

Figure 24.21 Control system.

$y_n - x_n$ and the output is $K(y_n - x_n)$. This then passes through the DAC to give an analogue output consisting of a number of steps (Figure 24.21). The motor acts as an integrator and so its output will be $1/c$ times the progressive sum of the areas under the steps (Figure 24.21). Each step has an area of (DAC change in output for the step) $\times T$. Thus

$$x_n - x_{n-1} = (\text{DAC output for } x_{n-1})T/c = K(y_{n-1} - x_{n-1})T/c$$

Hence

$$x_n = [1 - (KT/c)]x_{n-1} + (KT/c)y_{n-1}$$

Suppose we have $K/c = 5$ and a sampling interval of 0.1 s. Thus

$$x_n = 0.5y_{n-1} + 0.5x_{n-1}$$

If there is an input to the control system for the focusing of a sequence of pulses of constant size 1, prior to that there being no input, i.e. $y_0 = 1, y_1 = 1, y_2 = 1, \ldots$, then

$$x_0 = 0$$

$$x_1 = 0.5 \times 0 + 0.5 \times 1 = 0.5$$

$$x_2 = 0.5 \times 0.5 + 0.5 \times 1 = 0.75$$

$$x_3 = 0.5 \times 0.75 + 0.5 \times 1 = 0.875$$

$$x_4 = 0.5 \times 0.875 + 0.5 \times 1 = 0.9375$$

$$x_5 = 0.5 \times 0.9375 + 0.5 \times 1 = 0.968\,75$$

$$x_6 = 0.5 \times 0.96875 + 0.5 \times 1 = 0.984\,375$$

$$x_7 = 0.5 \times 0.984365 + 0.5 \times 1 = 0.992\,187\,5$$

and so on

The output thus takes about seven sampling periods, i.e. 0.7 s, for the focusing to be achieved. This is too long. Suppose, however, we choose values so that $KT/c = 1$. The difference equation then becomes $x_n = y_{n-1}$. Then we have

$$x_0 = 0$$

$$x_1 = 1$$

$$x_2 = 1$$

$$x_3 = 1$$

...

This means that the output will reach the required position after just one sample. This is a much faster response. By using a high sampling rate a very fast response can be achieved. This form of response is termed a **deadbeat response**.

24.2.4 Car engine management

The modern car is likely to include many electronic control systems involving microcontrollers, the engine control system being one, its aim being to ensure that the engine is operated at its optimum settings. Figure 24.22 shows a generalised block diagram of such a system. The system consists of sensors supplying, after suitable signal conditioning, the input signals to the microcontroller and its providing output signals via drivers to actuate actuators. Figure 24.23 shows some of these elements in relation to an engine, only one cylinder being shown.

The engine speed sensor is an inductive sensor and consists of a coil for which the inductance changes as the teeth of the sensor wheel pass it and so gives an oscillating voltage. The temperature sensor is usually a thermistor. The mass air flow sensor may be a hot wire sensor; as air passes over a heated wire it will be cooled, the amount of cooling depending on the mass rate of

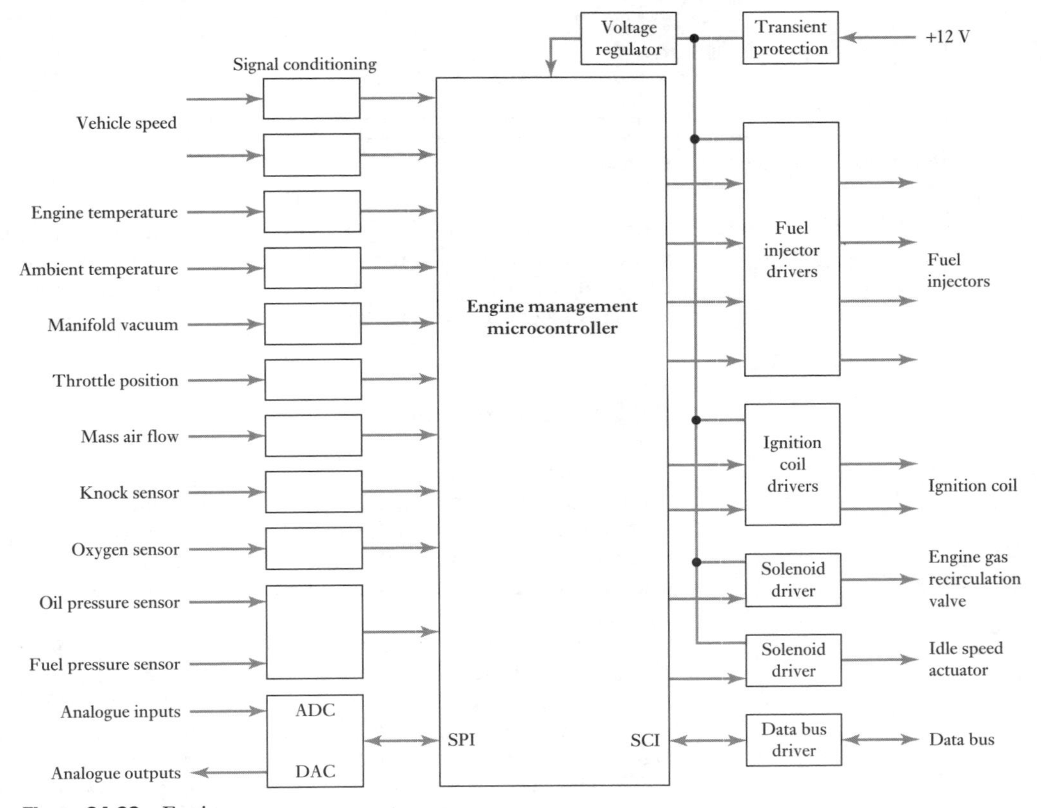

Figure 24.22 Engine management system.

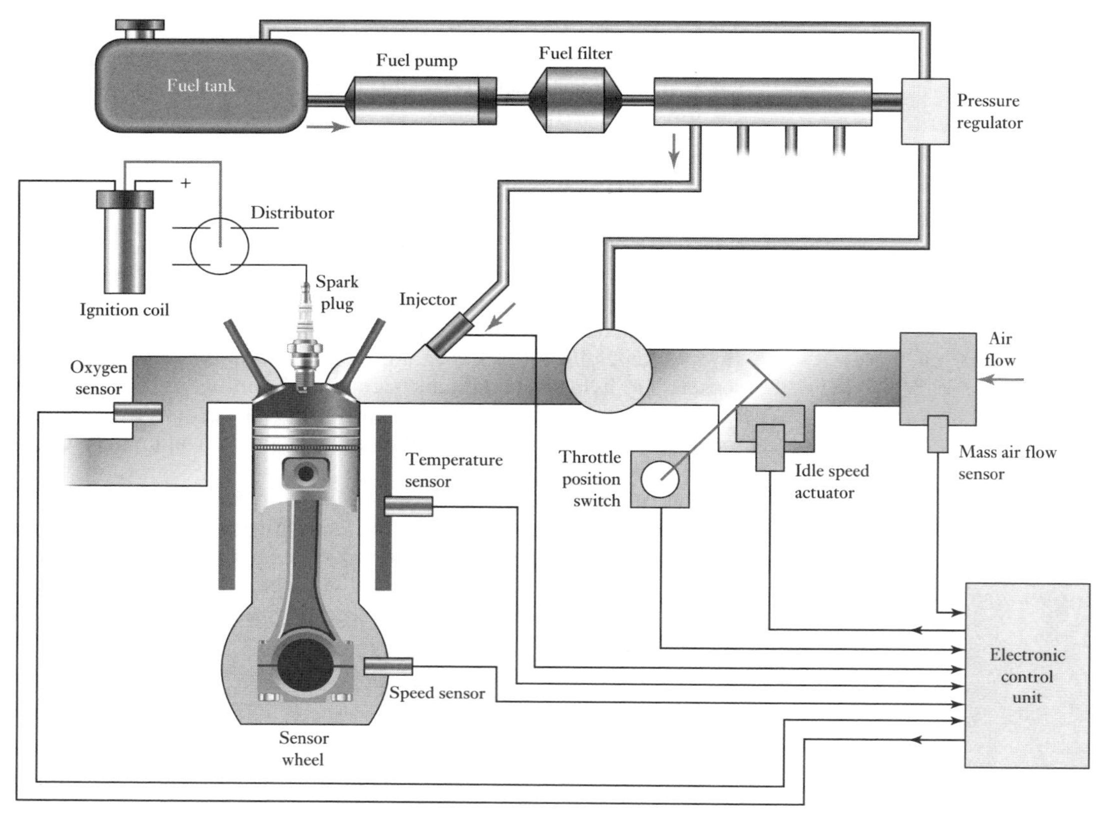

Figure 24.23 An engine management system.

flow. The oxygen sensor is generally a closed-end tube made of zirconium oxide with porous platinum electrodes on the inner and outer surfaces. Above about 300°C the sensor becomes permeable to oxygen ions and so gives a voltage between the electrodes.

24.2.5 Bar code reader

The familiar scene at the check-out of a supermarket is of the purchases being passed in front of a light beam or a hand-held wand being passed over the goods so that the bar code can be read and the nature of the purchase and hence its price automatically determined. The code consists of a series of black and white bars of varying widths. For example, there is such a bar code on the back of this book.

Figure 24.24 shows the basic form of the bar code used in the retail trade. The bar code represents a series of numbers. There is a prefix which identifies the coding scheme being used; this is a single digit for the regular Universal Product Coding (UPC) scheme used in the United States and two digits for the European Article Number (EAN) scheme used in Europe. The UPC code uses a 0 prefix for grocery and a 3 for pharmaceuticals. The EAN prefix is from 00 to 09 and is such that the UPC code can be read within the

Figure 24.24 Bar code.

EAN code. This is followed by five digits to represent the manufacturer, each manufacturer having been assigned a unique number. This brings up the centre of the code pattern which is identified by two taller bar patterns. The five-digit number that then follows represents the product. The final number is a check digit which is used to check that the code has been correctly read. A guard pattern of two taller bars at the start and end of the bar pattern is used to frame the bars.

Each number is coded as seven 0 or 1 digits. The codes used on either side of the centre line are different so that the direction of the scan can be determined. To the right the characters have an even number of ones and so even parity and, for UPC, to the left an odd number of ones and so odd parity, the EAN coding for the left being a mixture. Table 24.2 shows the UPC and EAN codings, UPC being the left A coding and the EAN using both left A and left B character codes.

Table 24.2 UPC and EAN codings.

Decimal number	Left A characters	Left B characters	Right characters
0	0001101	0100111	1110010
1	0011001	0110011	1100110
2	0010011	0011011	1101100
3	0111101	0100001	1000010
4	0100011	0011101	0011100
5	0110001	0111001	0001110
6	0101111	0000101	1010000
7	0111011	0010001	1000100
8	0110111	0001001	1001000
9	0001011	0010111	1110100

Each 1 is entered as a dark bar and thus the right-hand character 2 would be represented 1101100 and, with the adjacent dark bars run together, it appears as a double-width dark bar followed by a narrow space and then another double-width dark wide bar followed by a double-width space. This is illustrated in Figure 24.25. The guard pattern at the ends of the code represents 101 and the central band of bars is 01010.

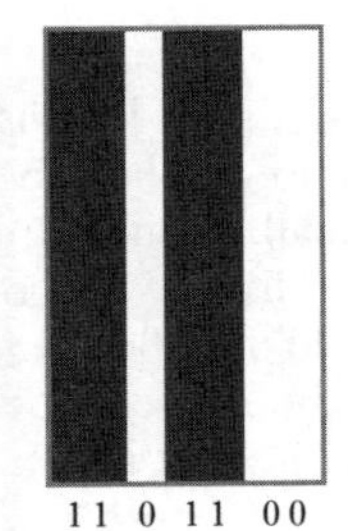

Figure 24.25 Bar code for right-hand 2.

The bar code shown in Figure 24.24 was that for the first edition of this book. It uses the EAN code and has the prefix 97 to identify it as a publication, 80582 to identify the publisher, 25634 to identify the particular book and a check digit of 7. Note that the bar code contains the relevant parts of the ISBN number, this also being a number to identify the publisher and the book concerned.

The procedure for using the check code digit is:

1 Starting at the left, sum all the characters, excluding the check digit, in the odd positions, i.e. first, third, fifth, etc., and then multiply the sum by 3.
2 Starting at the left, sum all the characters in the even positions.
3 Add the results of steps 1 and 2. The check character is the smallest number which when added to this sum produces a multiple of 10.

As an illustration of the use of the check digit, consider the bar code for the book where we have 9780582256347. For the odd characters we have $9 + 8 + 5 + 2 + 5 + 3 = 32$ and when multiplied by 3 we have 96. For the even characters we have $7 + 0 + 8 + 2 + 6 + 4 = 27$. The sum is 123 and thus the check digit should be 7.

Reading the bar code involves determining the widths of the dark and light bands. This can involve a solid-state laser being used to direct an intense, narrow, beam of light at the code and detecting the reflected light by means of a photocell. Usually with the supermarket version the scanner is fixed and a spinning mirror is used to direct the light across the bar code and so scan all the bars. Signal conditioning involves amplification of the output of the photocell using operational amplifiers and then using an operational amplifier circuit as a comparator in order to give a high, i.e. 1, output when a black bar is scanned and a low, i.e. 0, output when a white space is scanned. This sequence of zeros and ones can then be an input to, say, a PIA connected to a Motorola 6800 microprocessor. The overall form of the microprocessor program is:

1 Initialisation to clear the various memory locations.
2 Recovering the data from the input. This involves continually testing the input to determine whether it is 0 or 1.
3 Processing the data to obtain the characters in binary format. The input is a serial signal consisting of different-duration zeros and ones depending on the width of the spaces of black bars. The microprocessor system is programmed to find the module time width by dividing the time of scan between the end marker bars by the number of modules, a module being a light or dark band to represent a single 0 or 1. The program can then determine whether a dark or light band is a single digit or more than one and hence interpret the scanner signal.
4 Process the binary outcome into a statement of the item purchased and cost.

24.2.6 Hard disk drive

Figure 24.26(a) shows the basic form of a hard disk drive. It consists of a disk coated with a metal layer which can be magnetised. The gap between the write/read head and the disk surface is very small, about 0.1 μm. The data is stored in the metal layer as a sequence of bit cells (see Section 6.3.1). The disk is rotated by a motor at typically 3600, 5400 or 7200 rev/min and an actuator arm has to be positioned so that the relevant concentric track and relevant part of the track come under the read/write head at the end of that arm. The head is controlled by a closed-loop system (Figure 24.26(b)) in order to position it. Control information is written onto a disk during the formatting process, this enabling each track and sector of track to be identified.

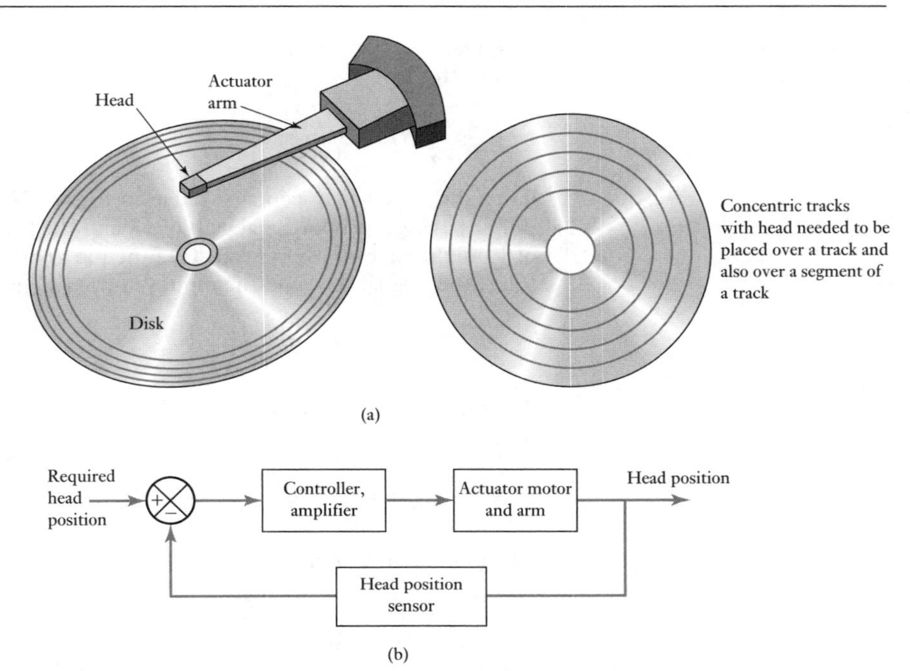

Figure 24.26 Hard disk: (a) basic form, (b) basic closed-loop control system for positioning of the read/write head.

The control process then involves the head using this information to go to the required part of the disk.

The actuator movement generally involves a voice coil actuator (Figure 24.27) to rotate the arm. This voice coil actuator is essentially a coil mounted in an iron core so that when a current is passed through the coil it moves, the arrangement being like that of a moving coil loudspeaker, and so is able to move the actuator arm to position the head over the required track. The head element reads the magnetic field on the disk and provides a feedback signal to the control amplifier.

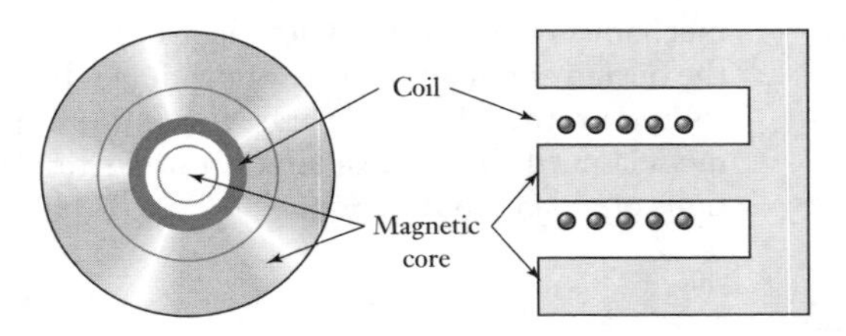

Figure 24.27 Voice coil actuator.

The voice coil actuator is a form of field-controlled permanent magnet d.c. motor and has a transfer function of the same form (see Section 13.5). Thus, since we are concerned with the transfer function relating displacement with time, i.e. the integral of the velocity time function given in Section 13.5, the voice coil actuator has a transfer function of the form

$$G(s) = \frac{k}{s(Ls + R)(Is + c)} = \frac{k/Rc}{s(\tau_L s + 1)(\tau s + 1)}$$

The $(\tau s + 1)$ term is generally close to 1 and so the transfer function approximates to

$$G(s) = \frac{k/Rc}{s(\tau_L s + 1)}$$

Thus the closed-loop control system in Figure 24.26(a), with a control amplifier having a proportional gain of K_a and the head position transfer a gain of 1, might have an overall transfer function giving the relationship between the output signal $X(s)$ and the input required signal $R(s)$ of

$$\frac{X(s)}{R(s)} = \frac{K_a G(s)}{1 + K_a G(S)}$$

Thus if we have, say, $G(s) = 0.25/s(0.05s + 1) = 5/s(s + 20)$ and $K_a = 40$, then

$$X(s) = \frac{200}{s^2 + 20s + 200} R(s)$$

Thus for a unit step input, i.e. $R(s) = 1/s$, the output will be described by

$$X(s) = \frac{200}{s(s^2 + 20s + 200)}$$

The quadratic term is of the form $s^2 + 2\zeta\omega_n s + \omega_n^2$ (see Section 13.3.1) and thus has a natural frequency ω_n of $\sqrt{(200)}$ and a damping factor ζ of $10/\sqrt{(200)}$. Thus we can work out what the response of this second-order system will be to step input signals and how long the system will need to settle down; for example, the 2% settling time (see Section 12.5) is $4/\zeta\omega_n$ and so $4/10 = 0.4$ s. This is rather a long time and so we would need to consider how it can be reduced to perhaps milliseconds. We might consider replacing the amplifier with its proportional gain by one exercising PD control.

Summary

Mechatronics is a co-ordinated, and concurrently developed, integration of mechanical engineering with electronics and intelligent computer control in the design and manufacture of products. It involves developing an integrated solution rather than a separate discipline approach. In developing solutions, models need to be considered in order to make predictions as to how solutions are likely to function.

Problems

24.1 Present outline solutions of possible designs for the following:

(a) A temperature controller for an oven.

(b) A mechanism for sorting small-, medium- and large-size objects moving along a conveyor belt so that they each are diverted down different chutes for packaging.

(c) An x–y plotter (such a machine plots graphs showing how an input to x varies as the input to y changes).

Research assignments

24.2 Research the anti-lock braking system used in cars and describe the principles of its operation.

24.3 Research the mechanism used in the dot matrix printer and describe the principles of its operation.

24.4 Research the control area network (CAN) protocol used with cars.

Design assignments

24.5 Design a digital thermometer system which will display temperatures between 0 and 99°C. You might like to consider a solution based on the use of a microprocessor with RAM and ROM chips or a microcontroller solution.

24.6 Design a digital ohmmeter which will give a display of the resistance of a resistor connected between its terminals. You might like to consider basing your solution on the use of a monostable multivibrator, e.g. 74121, which will provide an impulse with a width related to the time constant *RC* of the circuit connected to it.

24.7 Design a digital barometer which will display the atmospheric pressure. You might like to base your solution on the use of the MPX2100AP pressure sensor.

24.8 Design a system which can be used to control the speed of a d.c. motor. You might like to consider using the M68HC11 evaluation board.

24.9 Design a system involving a PLC for the placing on a conveyor belt of boxes in batches of four.

Appendices

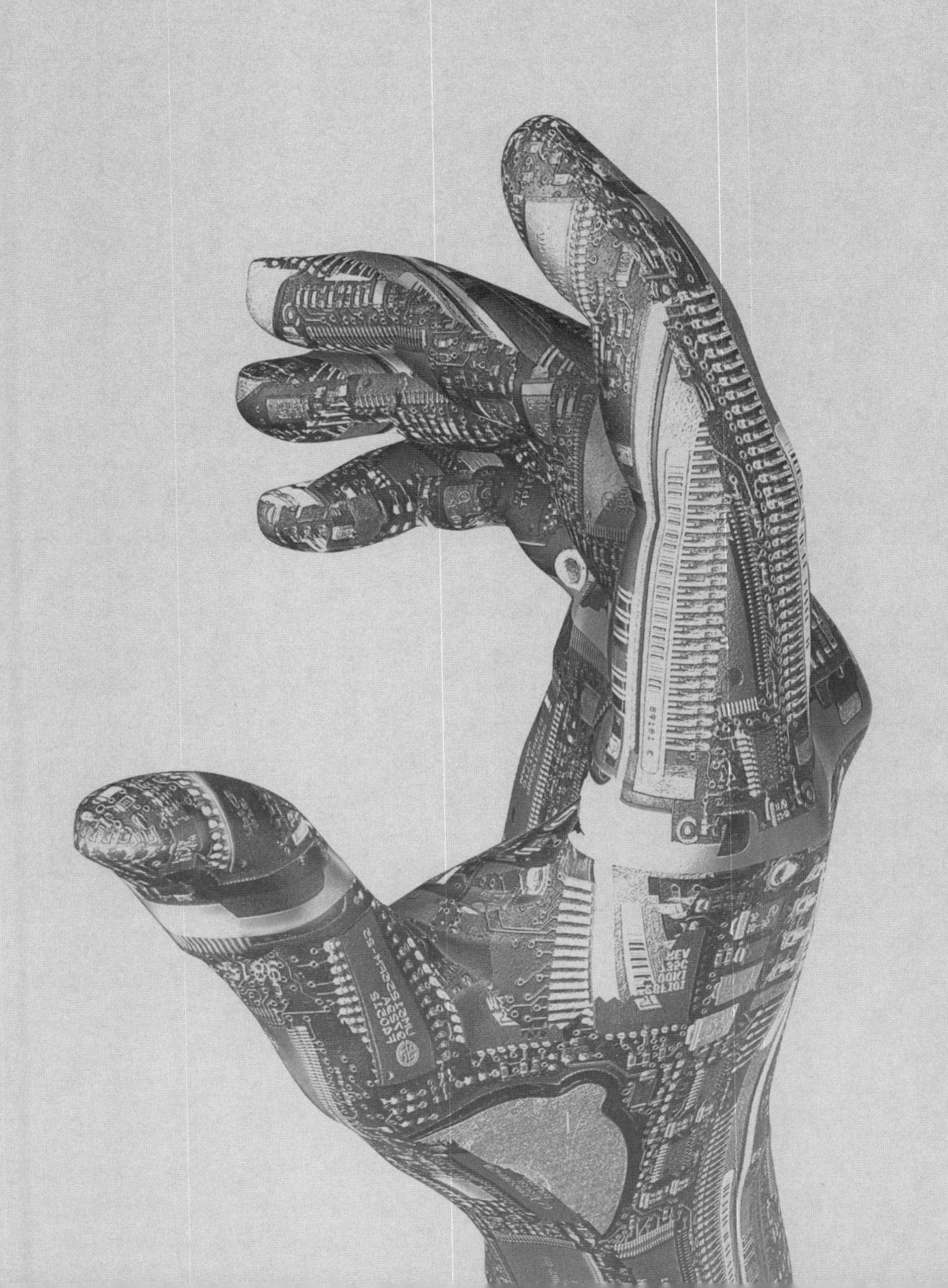

Appendix A: The Laplace Transform

A.1 The Laplace transform

Consider a quantity which is a function of time. We can talk of this quantity being in the **time domain** and represent such a function as $f(t)$. In many problems we are only concerned with values of time greater than or equal to 0, i.e. $t \geq 0$. To obtain the Laplace transform of this function we multiply it by e^{-st} and then integrate with respect to time from zero to infinity. Here s is a constant with the unit of 1/time. The result is what we now call the **Laplace transform** and the equation is then said to be in the **s-domain**. Thus the Laplace transform of the function of time $f(t)$, which is written as $\mathcal{L}\{f(t)\}$, is given by

$$\mathcal{L}\{f(t)\} = \int_0^\infty e^{-st} f(t)\, dt$$

The transform is **one-sided** in that values are only considered between 0 and $+\infty$, and not over the full range of time from $-\infty$ to $+\infty$.

We can carry out algebraic manipulations on a quantity in the s-domain, i.e. adding, subtracting, dividing and multiplying, in the normal way we do on any algebraic quantities. We could not have done this on the original function, assuming it to be in the form of a differential equation, when in the time domain. By this means we can obtain a considerably simplified expression in the s-domain. If we want to see how the quantity varies with time in the time domain then we have to carry out the inverse transformation. This involves finding the time domain function that could have given the simplified s-domain expression.

When in the s-domain a function is usually written, since it is a function of s, as $F(s)$. It is usual to use a capital letter F for the Laplace transform and a lower case letter f for the time-varying function $f(t)$. Thus

$$\mathcal{L}\{f(t)\} = F(s)$$

For the inverse operation, when the function of time is obtained from the Laplace transform, we can write

$$f(t) = \mathcal{L}^{-1}\{F(s)\}$$

This equation thus reads as: $f(t)$ is the inverse transform of the Laplace transform $F(s)$.

A.1.1 The Laplace transform from first principles

To illustrate the transformation of a quantity from the time domain into the s-domain, consider a function that has the constant value of 1 for all values of

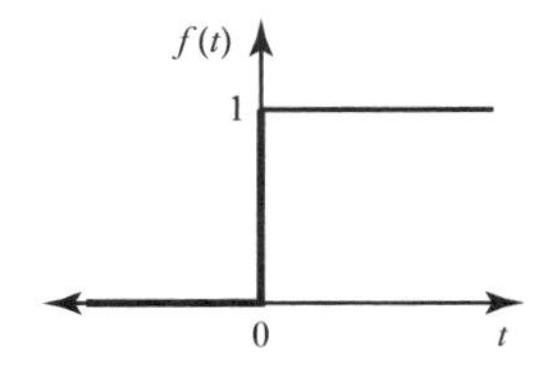

Figure A.1 Unit step function.

time greater than 0, i.e. $f(t) = 1$ for $t \geq 0$. This describes a **unit step** function and is shown in Figure A.1.

The Laplace transform is then

$$\mathcal{L}\{f(t)\} = F(s) = \int_0^\infty 1\mathrm{e}^{-st}\,\mathrm{d}t = -\frac{1}{s}[\mathrm{e}^{-st}]_0^\infty$$

Since with $t = \infty$ the value of e is 0 and with $t = 0$ the value of e^{-0} is -1, then

$$F(s) = \frac{1}{s}$$

As another example, the following shows the determination, from first principles, of the Laplace transform of the function e^{at}, where a is a constant. The Laplace transform of $f(t) = \mathrm{e}^{at}$ is thus

$$F(s) = \int_0^\infty \mathrm{e}^{at}e^{-st}\,\mathrm{d}t = \int_0^\infty \mathrm{e}^{-(s-a)t}\,\mathrm{d}t = -\frac{1}{s-a}[\mathrm{e}^{-(s-a)t}]_0^\infty$$

When $t = \infty$ the term in the square brackets becomes 0 and when $t = 0$ it becomes -1. Thus

$$F(s) = \frac{1}{s-a}$$

A.2 Unit steps and impulses

Common input functions to systems are the unit step and the impulse. The following indicates how their Laplace transforms are obtained.

A.2.1 The unit step function

Figure A.1 shows a graph of a unit step function. Such a function, when the step occurs at $t = 0$, has the equation

$f(t) = 1$ for all values of t greater than 0
$f(t) = 0$ for all values of t less than 0

The step function describes an abrupt change in some quantity from zero to a steady value, e.g. the change in the voltage applied to a circuit when it is suddenly switched on.

The unit step function thus cannot be described by $f(t) = 1$ since this would imply a function that has the constant value of 1 at all values of t, both positive and negative. The unit step function that switches from 0 to $+1$ at $t = 0$ is conventionally described by the symbol $u(t)$ or $H(t)$, the H being after the originator O. Heaviside. It is thus sometimes referred to as the **Heaviside function**.

The Laplace transform of this step function is, as derived in the previous section,

$$F(s) = \frac{1}{s}$$

The Laplace transform of a step function of height a is

$$F(s) = \frac{a}{s}$$

A.2.2 Impulse function

Consider a rectangular pulse of size $1/k$ that occurs at time $t = 0$ and which has a pulse width of k, i.e. the area of the pulse is 1. Figure A.2(a) shows such a pulse. The pulse can be described as

$$f(t) = \frac{1}{k} \text{ for } 0 \leq t < k$$

$$f(t) = 0 \text{ for } t > k$$

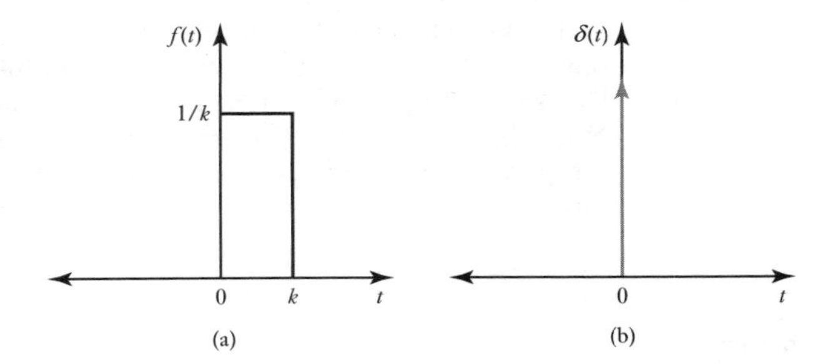

Figure A.2 (a) Rectangular pulse, (b) impulse.

If we maintain this constant pulse area of 1 and then decrease the width of the pulse (i.e. reduce k), the height increases. Thus, in the limit as $k \to 0$ we end up with just a vertical line at $t = 0$, with the height of the graph going off to infinity. The result is a graph that is zero except at a single point where there is an infinite spike (Figure A.2(b)). Such a graph can be used to represent an impulse. The impulse is said to be a unit impulse because the area enclosed by it is 1. This function is represented by $\delta(t)$, the **unit impulse function** or the **Dirac delta function**.

The Laplace transform for the unit area rectangular pulse in Figure A.2(a) is given by

$$F(s) = \int_0^\infty f(t)\mathrm{e}^{-st}\,\mathrm{d}t = \int_0^k \frac{1}{k}\mathrm{e}^{-st}\,\mathrm{d}t + \int_k^\infty 0\,\mathrm{e}^{-st}\mathrm{d}t$$

$$= \left[-\frac{1}{sk}\mathrm{e}^{-st}\right]_0^k = -\frac{1}{sk}(\mathrm{e}^{-sk} - 1)$$

To obtain the Laplace transform for the unit impulse we need to find the value of the above in the limit as $k \to 0$. We can do this by expanding the exponential term as a series. Thus

$$\mathrm{e}^{-sk} = 1 - sk + \frac{(-sk)^2}{2!} + \frac{(-sk)^3}{3!} + \cdots$$

and so we can write

$$F(s) = 1 - \frac{sk}{2!} + \frac{(sk)^2}{3!} + \cdots$$

Thus in the limit as $k \to 0$ the Laplace transform tends to the value 1:

$$\mathcal{L}\{\delta(t)\} = 1$$

Since the area of the above impulse is 1 we can define the size of such an impulse as being 1. Thus the above equation gives the Laplace transform for a unit impulse. An impulse of size a is represented by $a\delta(t)$ and the Laplace transform is

$$\mathcal{L}\{a\delta(t)\} = a$$

A.3 Standard Laplace transforms

In determining the Laplace transforms of functions it is not usually necessary to evaluate integrals since tables are available that give the Laplace transforms of commonly occurring functions. These, when combined with a knowledge of the properties of such transforms (see the next section), enable most commonly encountered problems to be tackled. Table A.1 lists some of the commoner time functions and their Laplace transforms. Note that in the table $f(t) = 0$ for all negative values of t and the $u(t)$ terms have been omitted from most of the time functions and have to be assumed.

Table A.1 Laplace transforms.

	Time function $f(t)$	Laplace transform $F(s)$
1	$\delta(t)$, unit impulse	1
2	$\delta(t - T)$, delayed unit impulse	e^{-sT}
3	$u(t)$, a unit step	$\dfrac{1}{s}$
4	$u(t - T)$, a delayed unit step	$\dfrac{e^{-sT}}{s}$
5	t, a unit ramp	$\dfrac{1}{s^2}$
6	t^n, nth-order ramp	$\dfrac{n!}{s^{n+1}}$
7	e^{-at}, exponential decay	$\dfrac{1}{s + a}$
8	$1 - e^{-at}$, exponential growth	$\dfrac{a}{s(s + a)}$
9	te^{-at}	$\dfrac{1}{(s + a)^2}$
10	$t^n e^{-at}$	$\dfrac{n!}{(s + a)^{n+1}}$

(Continued)

Table A.1 (*Continued*)

	Time function $f(t)$	Laplace transform $F(s)$
11	$t - \dfrac{1 - e^{-at}}{a}$	$\dfrac{a}{s^2(s + a)}$
12	$e^{-at} - e^{-bt}$	$\dfrac{b - a}{(s + a)(s + b)}$
13	$(1 - at)e^{-at}$	$\dfrac{s}{(s + a)^2}$
14	$1 - \dfrac{b}{b - a}e^{-at} + \dfrac{a}{b - a}e^{-bt}$	$\dfrac{ab}{s(s + a)(s + b)}$
15	$\dfrac{e^{-at}}{(b - a)(c - a)} + \dfrac{e^{-bt}}{(c - a)(a - b)} + \dfrac{e^{-ct}}{(a - c)(b - c)}$	$\dfrac{1}{(s + a)(s + b)(s + c)}$
16	$\sin \omega t$, a sine wave	$\dfrac{\omega}{s^2 + \omega^2}$
17	$\cos \omega t$, a cosine wave	$\dfrac{s}{s^2 + \omega^2}$
18	$e^{-at} \sin \omega t$, a damped sine wave	$\dfrac{\omega}{(s + a)^2 + \omega^2}$
19	$e^{-at} \cos \omega t$, a damped cosine wave	$\dfrac{s + a}{(s + a)^2 + \omega^2}$
20	$1 - \cos \omega t$	$\dfrac{\omega^2}{s(s^2 + \omega^2)}$
21	$t \cos \omega t$	$\dfrac{s^2 - \omega^2}{(s^2 + \omega^2)^2}$
22	$t \sin \omega t$	$\dfrac{2\omega s}{(s^2 + \omega^2)^2}$
23	$\sin(\omega t + \theta)$	$\dfrac{\omega \cos \theta + s \sin \theta}{s^2 + \omega^2}$
24	$\cos(\omega t + \theta)$	$\dfrac{s \cos \theta - \omega \sin \theta}{s^2 + \omega^2}$
25	$\dfrac{\omega}{\sqrt{1 - \zeta^2}} e^{-\zeta\omega t} \sin \omega\sqrt{1 - \zeta^2}\, t$	$\dfrac{\omega^2}{s^2 + 2\zeta\omega s + \omega^2}$
26	$1 - \dfrac{1}{\sqrt{1 - \zeta^2}} e^{-\zeta\omega t} \sin(\omega\sqrt{1 - \zeta^2}\, t + \phi), \cos \phi = \zeta$	$\dfrac{\omega^2}{s(s^2 + 2\zeta\omega s + \omega^2)}$

A.3.1 Properties of Laplace transforms

In this section the basic properties of the Laplace transform are outlined. These properties enable the table of standard Laplace transforms to be used in a wide range of situations.

Linearity property

If two separate time functions, e.g. $f(t)$ and $g(t)$, have Laplace transforms then the transform of the sum of the time functions is the sum of the two separate Laplace transforms:

$$\mathcal{L}\{af(t) + bg(t)\} = a\mathcal{L}f(t) + b\mathcal{L}g(t)$$

a and b are constants. Thus, for example, the Laplace transform of $1 + 2t + 4t^2$ is given by the sum of the transforms of the individual terms in the expression. Thus, using items 1, 5 and 6 in Table A.1,

$$F(s) = \frac{1}{s} + \frac{2}{s^2} + \frac{8}{s^3}$$

The s-domain shifting property

This property is used to determine the Laplace transform of functions that have an exponential factor and is sometimes referred to as the **first shifting property**. If $F(s) = \mathcal{L}\{f(t)\}$ then

$$\mathcal{L}\{e^{at}f(t)\} = F(s - a)$$

For example, the Laplace transform of $e^{at}t^n$ is, since the Laplace transform of t^n is given by item 6 in Table A.1 as $n!/s^{n+1}$, given by

$$\mathcal{L}\{e^{at}t^n\} = \frac{n!}{(s - a)^{n+1}}$$

Time domain shifting property

If a signal is delayed by a time T then its Laplace transform is multiplied by e^{-sT}. If $F(s)$ is the Laplace transform of $f(t)$ then

$$\mathcal{L}\{f(t - T)u(t - T)\} = e^{-sT}F(s)$$

This delaying of a signal by a time T is referred to as the **second shift theorem**.

The time domain shifting property can be applied to all Laplace transforms. Thus for an impulse $\delta(t)$ which is delayed by a time T to give the function $\delta(t - T)$, the Laplace transform of $\delta(t)$, namely 1, is multiplied by e^{-sT} to give $1e^{-sT}$ as the transform for the delayed function.

Periodic functions

For a function $f(t)$ which is a periodic function of period T, the Laplace transform of that function is

$$\mathcal{L}f(t) = \frac{1}{1 - e^{-sT}}F_1(s)$$

where $F_1(s)$ is the Laplace transform of the function for the first period. Thus, for example, consider the Laplace transform of a sequence of periodic rectangular pulses of period T, as shown in Figure A.3. The Laplace transform

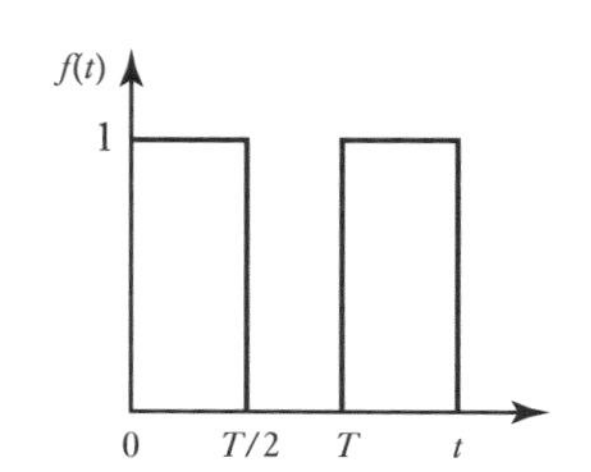

Figure A.3 Rectangular pulses.

of a single rectangular pulse is given by $(1/s)(1 - e^{-sT/2})$. Hence, using the above equation, the Laplace transform is

$$\frac{1}{1 - e^{-sT}} \times \frac{1}{s}(1 - e^{-sT/2}) = \frac{1}{s(1 + e^{-sT/2})}$$

Initial- and final-value theorems

The initial-value theorem can be stated as: if a function of time $f(t)$ has a Laplace transform $F(s)$ then in the limit as the time tends to zero the value of the function is given by

$$\lim_{t \to 0} f(t) = \lim_{s \to \infty} sF(s)$$

For example, the initial value of the function giving the Laplace transform $3/(s + 2)$ is the limiting value of $3s/(s + 2) = 3/(1 + 2/s)$ as s tends to infinity and so is 3.

The final-value theorem can be stated as: if a function of time $f(t)$ has a Laplace transform $F(s)$ then in the limit as the time tends to infinity the value of the function is given by

$$\lim_{t \to \infty} f(t) = \lim_{s \to 0} sF(s)$$

Derivatives

The Laplace transform of a derivative of a function $f(t)$ is given by

$$\mathcal{L}\left\{\frac{d}{dt}f(t)\right\} = sF(s) - f(0)$$

where $f(0)$ is the value of the function when $t = 0$. For example, the Laplace transform of $2(dx/dt) + x = 4$ is $2[sX(s) - x(0)] + X(s) = 4/s$ and if we have $x = 0$ at $t = 0$ then it is $2sX(s) + X(s) = 4/s$ or $X(s) = 4/[s(2s + 1)]$.

For a second derivative

$$\mathcal{L}\left\{\frac{d^2}{dt^2}f(t)\right\} = s^2F(s) - sf(0) - \frac{d}{dt}f(0)$$

where $df(0)/dt$ is the value of the first derivative at $t = 0$.

Integrals

The Laplace transform of the integral of a function $f(t)$ which has a Laplace transform $F(s)$ is given by

$$\mathcal{L}\left\{\int_0^t f(t)\,dt\right\} = \frac{1}{s}F(s)$$

For example, the Laplace transform of the integral of the function e^{-t} between the limits 0 and t is given by

$$\mathcal{L}\left\{\int_0^t e^{-t}\,dt\right\} = \frac{1}{s}\mathcal{L}\{e^{-t}\} = \frac{1}{s(s + 1)}$$

A.4 The inverse transform

The inverse Laplace transformation is the conversion of a Laplace transform $F(s)$ into a function of time $f(t)$. This operation can be written as

$$\mathcal{L}^{-1}\{F(s)\} = f(t)$$

The inverse operation can generally be carried out by using Table A.1. The linearity property of Laplace transforms means that if we have a transform as the sum of two separate terms then we can take the inverse of each separately and the sum of the two inverse transforms is the required inverse transform:

$$\mathcal{L}^{-1}\{aF(s) + bG(s)\} = a\mathcal{L}^{-1}F(s) + b\mathcal{L}^{-1}G(s)$$

Thus, to illustrate how rearrangement of a function can often put it into the standard form shown in the table, the inverse transform of $3/(2s + 1)$ can be obtained by rearranging it as

$$\frac{3(1/2)}{s + (1/2)}$$

The table (item 7) contains the transform $1/(s + a)$ with the inverse of e^{-at}. Thus the inverse transformation is just this multiplied by the constant $(3/2)$ with $a = (1/2)$, i.e. $(3/2)\,e^{-t/2}$.

As another example, consider the inverse Laplace transform of $(2s + 2)/(s^2 + 1)$. This expression can be rearranged as

$$2\left[\frac{s}{s^2 + 1} + \frac{1}{s^2 + 1}\right]$$

The first term in the square brackets has the inverse transform of $\cos t$ (item 17 in Table A.1) and the second term $\sin t$ (item 16 in Table A.1). Thus the inverse transform of the expression is $2\cos t + 2\sin t$.

A.4.1 Partial fractions

Often $F(s)$ is a ratio of two polynomials and cannot be readily identified with a standard transform in Table A.1. It has to be converted into simple fraction terms before the standard transforms can be used. The process of converting an expression into simple fraction terms is called decomposing into **partial fractions**. This technique can be used provided the degree of the numerator is less than the degree of the denominator. The degree of a polynomial is the highest power of s in the expression. When the degree of the numerator is equal to or higher than that of the denominator, the denominator must be divided into the numerator until the result is the sum of terms with the remainder fractional term having a numerator of lower degree than the denominator.

We can consider there to be basically three types of partial fractions:

1 The denominator contains factors which are only of the form $(s + a)$, $(s + b)$, $(s + c)$, etc. The expression is of the form

$$\frac{f(s)}{(s + a)(s + b)(s + c)}$$

and has the partial fractions of

$$\frac{A}{(s+a)} + \frac{B}{(s+b)} + \frac{C}{(s+c)}$$

2 There are repeated $(s + a)$ factors in the denominator, i.e. the denominator contains powers of such a factor, and the expression is of the form

$$\frac{f(s)}{(s+a)^n}$$

This then has partial fractions of

$$\frac{A}{(s+a)^1} + \frac{B}{(s+a)^2} + \frac{C}{(s+a)^3} + \cdots + \frac{N}{(s+a)^n}$$

3 The denominator contains quadratic factors and the quadratic does not factorise without imaginary terms. For an expression of the form

$$\frac{f(s)}{(as^2 + bs + c)(s+d)}$$

the partial fractions are

$$\frac{As + B}{as^2 + bs + c} + \frac{C}{s+d}$$

The values of the constants A, B, C, etc. can be found by making use of the fact either that the equality between the expression and the partial fractions must be true for all values of s or that the coefficients of s^n in the expression must equal those of s^n in the partial fraction expansion. The use of the first method is illustrated by the following example where the partial fractions of

$$\frac{3s+4}{(s+1)(s+2)}$$

are

$$\frac{A}{s+1} + \frac{B}{s+2}$$

Then, for the expressions to be equal, we must have

$$\frac{3s+4}{(s+1)(s+2)} = \frac{A(s+2) + B(s+1)}{(s+1)(s+2)}$$

and consequently $3s + 4 = A(s + 2) + B(s + 1)$. This must be true for all values of s. The procedure is then to pick values of s that will enable some of the terms involving constants to become zero and so enable other constants to be determined. Thus if we let $s = -2$ then we have $3(-2) + 4 = A(-2 + 2) + B(-2 + 1)$ and so $B = 2$. If we now let $s = -1$ then $3(-1) + 4 = A(-1 + 2) + B(-1 + 1)$ and so $A = 1$. Thus

$$\frac{3s+4}{(s+1)(s+2)} = \frac{1}{s+1} + \frac{2}{s+2}$$

Problems

A.1 Determine the Laplace transforms of (a) $2t$, (b) $\sin 2t$, (c) a unit impulse at time $t = 2$ s, (d) $4\, dx/dt$ when $x = 2$ at $t = 0$, (e) $3\, d^2x/dt^2$ when $x = 0$ and $dx/dt = 0$ at $t = 0$, (f) the integral between t and 0 of e^{-t}.

A.2 Determine the inverses of the Laplace transforms (a) $1/s^2$, (b) $5s/(s^2 + 9)$, (c) $(3s - 1)/[s(s - 1)]$, (d) $1/(s + 3)$.

A.3 Determine the initial value of the function with the Laplace transform $5/(s + 2)$.

Appendix B: Number Systems

B.1 Number systems

The **decimal system** is based on the use of 10 symbols or digits: 0, 1, 2, 3, 4, 5, 6, 7, 8, 9. When a number is represented by this system, the digit position in the number indicates that the weight attached to each digit increases by a factor of 10 as we proceed from right to left:

...	10^3	10^2	10^1	10^0
	thousands	hundreds	tens	units

The **binary system** is based on just two symbols or states: 0 and 1. These are termed *bi*nary digi*ts* or **bits**. When a number is represented by this system, the digit position in the number indicates that the weight attached to each digit increases by a factor of 2 as we proceed from right to left:

...	2^3	2^2	2^1	2^0
	bit 3	bit 2	bit 1	bit 0

For example, the decimal number 15 in the binary system is 1111. In a binary number the bit 0 is termed the **least significant bit** (LSB) and the highest bit the **most significant bit** (MSB).

The **octal system** is based on eight digits: 0, 1, 2, 3, 4, 5, 6, 7. When a number is represented by this system, the digit position in the number indicates that the weight attached to each digit increases by a factor of 8 as we proceed from right to left:

$$\ldots \quad 8^3 \quad 8^2 \quad 8^1 \quad 8^0$$

For example, the decimal number 15 in the octal system is 17.

The **hexadecimal system** is based on 16 digits/symbols: 0, 1, 2, 3, 4, 5, 6, 7, 8, 9, A, B, C, D, E, F. When a number is represented by this system, the digit position in the number indicates that the weight attached to each digit increases by a factor of 16 as we proceed from right to left:

$$\ldots \quad 16^3 \quad 16^2 \quad 16^1 \quad 16^0$$

For example, the decimal number 15 is F in the hexadecimal system. This system is generally used in the writing of programs for microprocessor-based systems since it represents a very compact method of entering data.

The **Binary-Coded Decimal system** (BCD system) is a widely used system with computers. Each decimal digit is coded separately in binary. For example, the decimal number 15 in BCD is 0001 0101. This code is useful for outputs from microprocessor-based systems where the output has to drive decimal displays, each decimal digit in the display being supplied by the microprocessor with its own binary code.

Table B.1 Number systems.

Decimal	Binary	BCD	Octal	Hexadecimal
0	0000	0000 0000	0	0
1	0001	0000 0001	1	1
2	0010	0000 0010	2	2
3	0011	0000 0011	3	3
4	0100	0000 0100	4	4
5	0101	0000 0101	5	5
6	0110	0000 0110	6	6
7	0111	0000 0111	7	7
8	1000	0000 1000	10	8
9	1001	0000 1001	11	9
10	1010	0001 0000	12	A
11	1011	0001 0001	13	B
12	1100	0001 0010	14	C
13	1101	0001 0011	15	D
14	1110	0001 0100	16	E
15	1111	0001 0101	17	F

Table B.1 gives examples of numbers in the decimal, binary, BCD, octal and hexadecimal systems.

B.2 Binary mathematics

Addition of binary numbers follows the following rules:

$0 + 0 = 0$
$0 + 1 = 1 + 0 = 1$
$1 + 1 = 10$ i.e. $0 +$ carry 1
$1 + 1 + 1 = 11$ i.e. $1 +$ carry 1

In decimal numbers the addition of 14 and 19 gives 33. In binary numbers this addition becomes

Augend	01110
Addend	10111
Sum	100001

For bit 0, $0 + 1 = 1$. For bit 1, $1 + 1 = 10$ and so we have 0 with 1 carried to the next column. For bit 3, $1 + 0 +$ carried $1 = 10$. For bit 4, $1 + 0 +$ carried $1 = 10$. We continue this through the various bits and end up with the sum plus a carry 1. The final number is thus 100001. When adding binary numbers A and B to give C, i.e. $A + B = C$, then A is termed the **augend**, B the **addend** and C the **sum**.

Subtraction of binary numbers follows the following rules:

$0 - 0 = 0$
$1 - 0 = 1$
$1 - 1 = 0$
$0 - 1 = 10 - 1 +$ borrow $= 1 +$ borrow

When evaluating $0 - 1$, a 1 is borrowed from the next column on the left containing a 1. The following example illustrates this. In decimal numbers the subtraction of 14 from 27 gives 13.

Minuend	11011
Subtrahend	01110
Difference	01101

For bit 0 we have $1 - 0 = 1$. For bit 1 we have $1 - 1 = 0$. For bit 2 we have $0 - 1$. We thus borrow 1 from the next column and so have $10 - 1 = 1$. For bit 3 we have $0 - 1$; remember that we borrowed the 1. Again borrowing 1 from the next column, we then have $10 - 1 = 1$. For bit 4 we have $0 - 0 = 0$; remember that we borrowed the 1. When subtracting binary numbers A and B to give C, i.e. we have $A - B = C$, then A is termed the **minuend**, B the **subtrahend** and C the **difference**.

The subtraction of binary numbers is more easily carried out electronically when an alternative method of subtraction is used. The subtraction example above can be considered to be the addition of a positive number and a negative number. The following techniques indicate how we can specify negative numbers and so turn subtraction into addition. It also enables us to deal with negative numbers in any circumstances.

The numbers used so far are referred to as **unsigned**. This is because the number itself contains no indication whether it is negative or positive. A number is said to be **signed** when the most significant bit is used to indicate the sign of the number, a 0 being used if the number is positive and a 1 if it is negative. When we have a positive number then we write it in the normal way with a 0 preceding it. Thus a positive binary number of 10010 would be written as 010010. A negative number of 10010 would be written as 110010. However, this is not the most useful way of representing negative numbers for ease of manipulation by computers.

A more useful way of representing negative numbers is to use the two's complement method. A binary number has two complements, known as the **one's complement** and the **two's complement**. The one's complement of a binary number is obtained by changing all the ones in the unsigned number into zeros and the zeros into ones. The two's complement is then obtained by adding 1 to the one's complement. When we have a negative number then we obtain the two's complement and then sign it with a 1, the positive number being signed by a 0. Consider the representation of the decimal number -3 as a signed two's complement number. We first write the binary number for the unsigned 3 as 0011, then obtain the one's complement of 1100, add 1 to give the unsigned two's complement of 1101, and finally sign it with a 1 to indicate it is negative. The result is thus 11101. The following is another example, the signed two's complement being obtained as an 8-bit number for -6:

Unsigned binary number	000 0110
One's complement	111 1001
Add 1	1
Unsigned two's complement	111 1010
Signed two's complement	1111 1010

Table B.2 Signed numbers.

Denary number	Signed number		Denary number	Signed number	
+127	0111 1111	Just the binary number signed with a 0	−1	1111 1111	The two's complement signed with a 1
...			−2	1111 1110	
+6	0000 0110		−3	1111 1101	
+5	0000 0101		−4	1111 1100	
+4	0000 0101		−5	1111 1011	
+3	0000 0011		−6	1111 1010	
+2	0000 0010		...		
+1	0000 0001		−127	1000 0000	
+0	0000 0000				

When we have a positive number then we write it in the normal way with a 0 preceding it. Thus a positive binary number of 100 1001 would be written as 01001001. Table B.2 shows some examples of numbers on this system.

Subtraction of a positive number from a positive number involves obtaining the signed two's complement of the subtrahend and then adding it to the signed minuend. Hence, for the subtraction of the decimal number 6 from the decimal number 4 we have

Signed minuend	0000 0100
Subtrahend, signed two's complement	1111 1010
Sum	1111 1110

The most significant bit of the outcome is 1 and so the result is negative. This is the signed two's complement for −2.

Consider another example, the subtraction of 43 from 57. The signed positive number of 57 is 00111001. The signed two's complement for −43 is given by

Unsigned binary number for 43	010 1011
One's complement	101 0100
Add 1	1
Unsigned two's complement	101 0101
Signed two's complement	1101 0101

Thus we obtain by the addition of the signed positive number and the signed two's complement number

Signed minuend	0011 1001
Subtrahend, signed two's complement	1101 0101
Sum	0000 1110 + carry 1

The carry 1 is ignored. The result is thus 0000 1110 and since the most significant bit is 0, the result is positive. The result is the decimal number 14.

If we wanted to add two negative numbers then we would obtain the signed two's complement for each number and then add them. Whenever a number is negative, we use the signed two's complement; when positive, just the signed number.

B.3 Floating numbers

In the decimal number system, large numbers such as 120 000 are often written in **scientific notation** as 1.2×10^5 or perhaps 120×10^3 and small numbers such as 0.000 120 as 1.2×10^{-4} rather than as a number with a fixed location for the decimal point. Numbers in this form of notation are written in terms of 10 raised to some power. Likewise we can use such notation for binary numbers but with them written in terms of 2 raised to some power. For example, we might have 1010 written as 1.010×2^3 or perhaps 10.10×2^2. Because the binary point can be moved to different locations by a choice of the power to which the 2 is raised, this notation is termed **floating point**.

A floating-point number is in the form $a \times r^e$, where a is termed the **mantissa**, r the **radix** or **base** and e the **exponent** or **power**. With binary numbers the base is understood to be 2, i.e. we have $a \times 2^e$. The advantage of using floating-point numbers is that, compared with fixed-point representation, a much wider range of numbers can be represented by a given number of digits.

Because with floating-point numbers it is possible to store a number in a number of different ways, e.g. 0.1×10^2 and 0.01×10^3, with computing systems such numbers are **normalised**, i.e. they are all put in the form $0.1 \times r^e$. Hence, with binary numbers we have 0.1×2^e and so if we had 0.00001001 it would become 0.1001×2^{-4}. In order to take account of the sign of a binary number we then add a sign bit of 0 for a positive number and 1 for a negative number. Thus the number 0.1001×2^{-4} becomes 1.1001×2^{-4} if negative and 0.1001×2^{-4} if positive.

If we want to add 2.01×10^3 and 10.2×10^2 we have to make the power (the term exponent is generally used) the same for each. Thus we can write $2.01 \times 10^3 + 1.02 \times 10^3$. We can then add them digit by digit, taking account of any carry, to give 2.03×10^3. We adopt a similar procedure for binary floating-point numbers. Thus if we want to add 0.101100×2^4 and 0.111100×2^2 we first adjust them to have the same exponents, e.g. 0.101100×2^4 and 0.001111×2^4, and then add them digit by digit to give 0.111011×2^4.

Likewise for subtraction, digit-by-digit subtraction of floating-point numbers can only occur between two numbers when they have the same exponent. Thus 0.1101100×2^{-4} minus 0.1010100×2^{-5} can be written as $0.01010100 \times 2^{-4} - 0.101010 \times 2^{-4}$ and the result given as 0.1000010×2^{-4}.

B.4 Gray code

Consider two successive numbers in binary code 0001 and 0010 (denary 2 and 3); 2 bits have changed in the code group in going from one number to the next. Thus if we had, say, an absolute encoder (see Section 2.3.7) and assigned successive positions to successive binary numbers then two changes have to be made in this case. This can present problems in that both changes must be made at exactly the same instant; if one occurs fractionally before the other then there can momentarily be another number indicated. Thus in going from 0001 to 0010 we might momentarily have 0011 or 0000. Thus an alternative method of coding is likely to be used.

The **Gray code** is such a code: only 1 bit in the code group changes in going from one number to the next. The Gray code is unweighted in that

Table B.3 Gray code.

Decimal number	Binary code	Gray code	Decimal number	Binary code	Gray code
0	0000	0000	8	1000	1100
1	0001	0001	9	1001	1101
2	0010	0011	10	1010	1111
3	0011	0010	11	1011	1110
4	0100	0110	12	1100	1010
5	0101	0111	13	1101	1011
6	0110	0101	14	1110	1001
7	0111	0100	15	1111	1000

the bit positions in the code group do not have any specific weight assigned to them. It is thus not suited to arithmetic operations but is widely used for input/output devices such as absolute encoders. Table B.3 lists decimal numbers and their values in the binary code and in Gray code.

Problems

B.1 What is the largest decimal number that can be represented by the use of an 8-bit binary number?

B.2 Convert the following binary numbers to decimal numbers: (a) 1011, (b) 10 0001 0001.

B.3 Convert the decimal numbers (a) 423, (b) 529 to hex.

B.4 Convert the BCD numbers (a) 0111 1000 0001, (b) 0001 0101 011 1 to decimal.

B.5 What are the two's complement representations of the decimal numbers (a) −90, (b) −35?

B.6 What even-parity bits should be attached to (a) 100 1000, (b) 100 1111?

B.7 Subtract the following decimal numbers using two's complements: (a) 21 − 13, (b) 15 − 3.

Appendix C: Boolean Algebra

C.1 Laws of Boolean algebra

Boolean algebra involves the binary digits 1 and 0 and the operations $\cdot$, $+$ and the inverse. The laws of this algebra are:

1 Anything ORed with itself is equal to itself: $A + A = A$.
2 Anything ANDed with itself is equal to itself: $A \cdot A = A$.
3 It does not matter in which order we consider inputs for OR and AND gates, e.g.

$$A + B = B + A \quad \text{and} \quad A \cdot B = B \cdot A$$

4 As the following truth table indicates:

$$A + (B \cdot C) = (A + B) \cdot (A + C)$$

A	B	C	$B \cdot C$	$A + B \cdot C$	$A + B$	$A + C$	$(A + B) \cdot (A + C)$
0	0	0	0	0	0	0	0
0	0	1	0	0	0	1	0
0	1	0	0	0	1	0	0
0	1	1	1	1	1	1	1
1	0	0	0	1	1	1	1
1	0	1	0	1	1	1	1
1	1	0	0	1	1	1	1
1	1	1	1	1	1	1	1

5 Likewise we can use a truth table to show that we can treat bracketed terms in the same way as in ordinary algebra, e.g.

$$A \cdot (B + C) = A \cdot B + A \cdot C$$

6 Anything ORed with its own inverse equals 1:

$$A + \overline{A} = 1$$

7 Anything ANDed with its own inverse equals 0:

$$A \cdot \overline{A} = 0$$

8 Anything ORed with a 0 is equal to itself; anything ORed with a 1 is equal to 1. Thus $A + 0 = A$ and $A + 1 = 1$.
9 Anything ANDed with a 0 is equal to 0; anything ANDed with a 1 is equal to itself. Thus $A \cdot 0 = 0$ and $A \cdot 1 = A$.

As an illustration of the use of the above to simplify Boolean expressions, consider simplifying

$$(A + B) \cdot \overline{C} + A \cdot C$$

Using item 5 for the first term gives

$$A \cdot \overline{C} + B \cdot \overline{C} + A \cdot C$$

We can regroup this and use item 6 to give

$$A \cdot (\overline{C} + C) + B \cdot \overline{C} = A \cdot 1 + B \cdot \overline{C}$$

Hence, using item 9 the simplified expression becomes

$$A + B \cdot \overline{C}$$

C.2 De Morgan's laws

As illustrated above, the laws of Boolean algebra can be used to simplify Boolean expressions. In addition we have what are known as **De Morgan's laws:**

1 The inverse of the outcome of ORing A and B is the same as when the inverses of A and B are separately ANDed. The following truth table shows the validity of this:

$$\overline{A + B} = \overline{A} \cdot \overline{B}$$

A	B	$A + B$	$\overline{A + B}$	$\overline{A}$	$\overline{B}$	$\overline{A} \cdot \overline{B}$
0	0	0	1	1	1	1
0	1	1	0	1	0	0
1	0	1	0	0	1	0
1	1	1	0	0	0	0

2 The inverse of the outcome of ANDing A and B is the same as when the inverses of A and B are separately ORed. The following truth table shows the validity of this:

$$\overline{A \cdot B} = \overline{A} + \overline{B}$$

A	B	$A \cdot B$	$\overline{A \cdot B}$	$\overline{A}$	$\overline{B}$	$\overline{A} + \overline{B}$
0	0	0	1	1	1	1
0	1	0	1	1	0	1
1	0	0	1	0	1	1
1	1	1	0	0	0	0

As an illustration of the use of De Morgan's laws, consider the simplification of the logic circuit shown in Figure C.1.

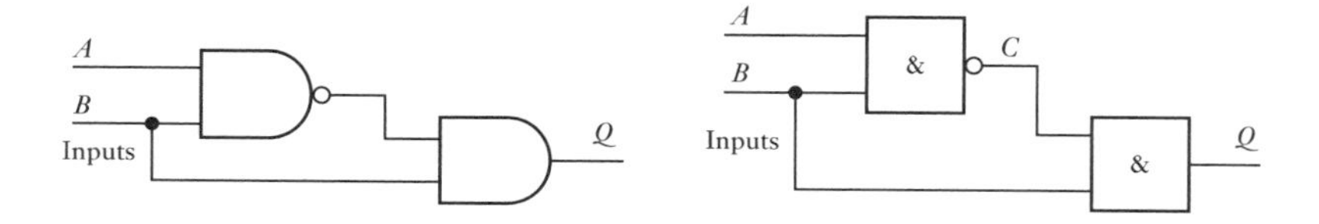

Figure C.1 Circuit simplification.

The Boolean equation for the output in terms of the input is

$$Q = \overline{A \cdot B} \cdot B$$

Applying the second law from above gives

$$Q = (\overline{A} + \overline{B}) \cdot B$$

We can write this as

$$Q = \overline{A} \cdot B + \overline{B} \cdot B = \overline{A} \cdot B + 0 = \overline{A} \cdot B$$

Hence the simplified circuit is as shown in Figure C.2.

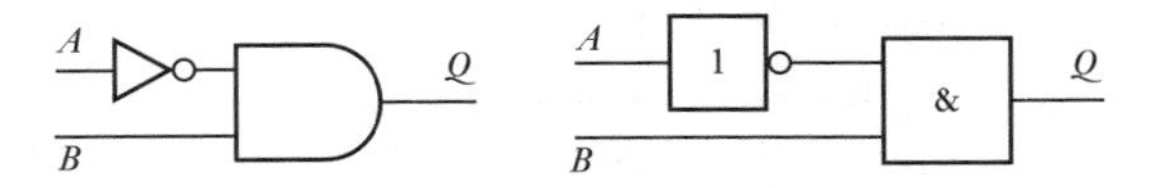

Figure C.2 Circuit simplification.

C.3 Boolean function generation from truth tables

Given a situation where the requirements of a system can be specified in terms of a truth table, how can a logic gate system using the minimum number of gates be devised to give that truth table?

Boolean algebra can be used to manipulate switching functions into many equivalent forms, some of which take many more logic gates than others; the form, however, to which most are minimised is AND gates driving a single OR gate or vice versa. Two AND gates driving a single OR gate (Figure C.3(a)) give

$$A \cdot B + A \cdot C$$

This is termed the **sum of products** form.

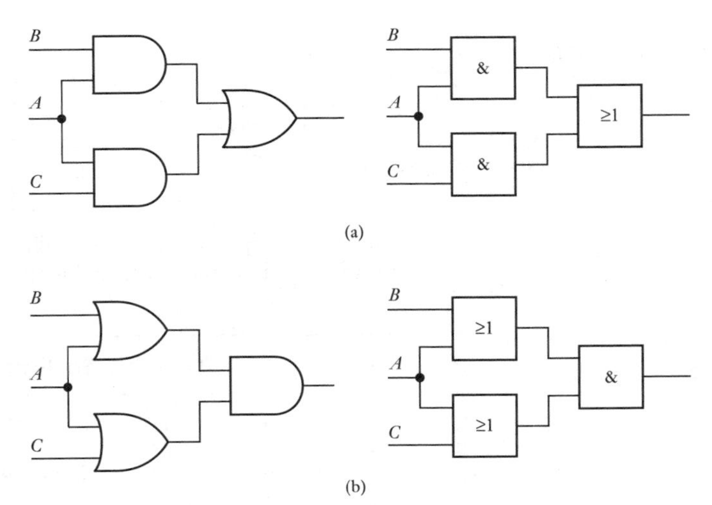

Figure C.3 (a) Sum of products, (b) product of sums.

For two OR gates driving a single AND gate (Figure C.3(b)), we have

$$(A + B) \cdot (A + C)$$

This is known as the **product of sums** form. Thus in considering what minimum form might fit a given truth table, the usual procedure is to find the sum of products or the product of sums form that fits the data. Generally the sum of products form is used. The procedure used is to consider each row of the truth table in turn and find the product that would fit a row. The overall result is then the sum of all these products.

Suppose we have a row in a truth table of

$$A = 1, B = 0 \text{ and output } Q = 1$$

When A is 1 and B is not 1 then the output is 1, thus the product which fits this is

$$Q = A \cdot \overline{B}$$

We can repeat this operation for each row of a truth table, as the following table indicates:

A	B	Output	Products
0	0	0	$\overline{A} \cdot \overline{B}$
0	1	0	$\overline{A} \cdot B$
1	0	1	$A \cdot \overline{B}$
1	1	0	$A \cdot B$

However, only the row of the truth table that has an output of 1 need be considered, since the rows with 0 output do not contribute to the final expression; the result is thus

$$Q = A \cdot \overline{B}$$

The logic gate system that will give this truth table is thus that shown in Figure C.4.

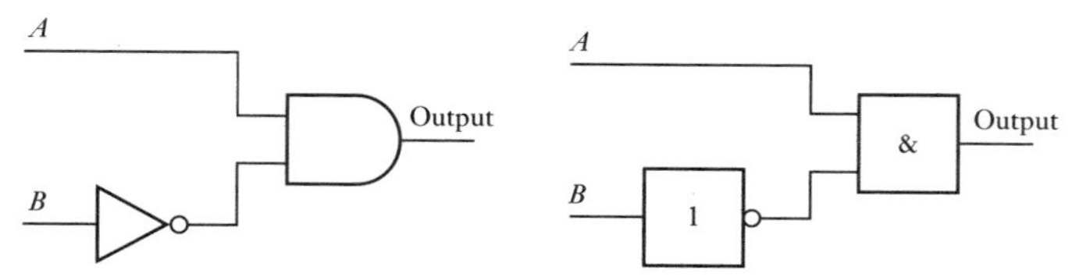

Figure C.4 Logic gates for truth table.

As a further example, consider the following truth table, only the products terms giving a 1 output being included:

A	B	C	Output	Products
0	0	0	1	$\overline{A} \cdot \overline{B} \cdot \overline{C}$
0	0	1	0	
0	1	0	1	$\overline{A} \cdot B \cdot \overline{C}$
0	1	1	0	
1	0	0	0	
1	0	1	0	
1	1	0	0	
1	1	1	0	

Thus the sum of products which fits this table is

$$Q = \overline{A} \cdot \overline{B} \cdot \overline{C} + \overline{A} \cdot B \cdot \overline{C}$$

This can be simplified to give

$$Q = \overline{A} \cdot \overline{C} \cdot (\overline{B} + B) = \overline{A} \cdot \overline{C}$$

The truth table can thus be generated by just a NAND gate.

C.4 Karnaugh maps

The **Karnaugh map** is a graphical method that can be used to produce simplified Boolean expressions from sums of products obtained from truth tables. The truth table has a row for the value of the output for each combination of input values. With two input variables there are four lines in the truth table, with three input variables six lines and with four input variables sixteen lines. Thus with two input variables there are four product terms, with three input variables there are six and with four input variables sixteen. The Karnaugh map is drawn as a rectangular array of cells, with each cell corresponding to a particular product value. Thus with two input variables there are four cells, with three input variables six cells and with four input variables sixteen cells. The output values for the rows are placed in their cells in the Karnaugh map, though it is usual to indicate only the 1 output values and leave the cells having 0 output as empty.

Figure C.5(a) shows the map for two input variables. The cells are given the output values for the following products:

the upper left cell $\overline{A} \cdot \overline{B}$,
the lower left cell $A \cdot \overline{B}$,
the upper right cell $\overline{A} \cdot B$,
the lower right cell $A \cdot B$.

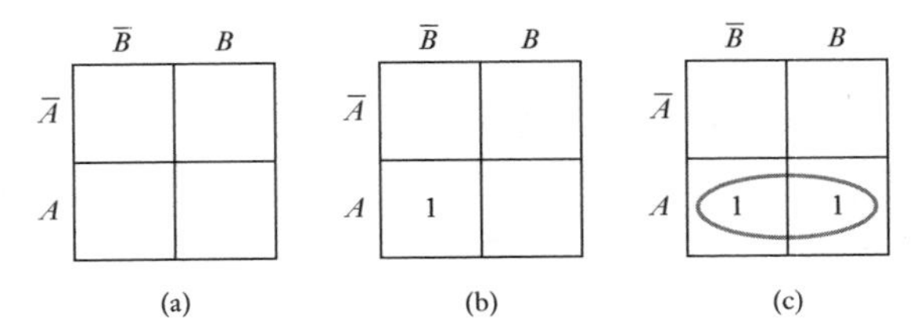

Figure C.5 Two input variable map.

The arrangement of the map squares is such that horizontally adjacent squares differ only in one variable and, likewise, vertically adjacent squares differ in only one variable. Thus horizontally with our two-variable map the variables differ only in A and vertically they differ only in B.

For the following truth table, if we put the values given for the products in the Karnaugh map, only indicating where a cell has a 1 value and leaving blank those with a 0 value, then the map shown in Figure C.5(b) is obtained:

A	B	Output	Products
0	0	0	$\overline{A} \cdot \overline{B}$
0	1	0	$\overline{A} \cdot B$
1	0	1	$A \cdot \overline{B}$
1	1	0	$A \cdot B$

Because the only 1 entry is in the lower right square, the truth table can be represented by the Boolean expression

$$\text{output} = A \cdot \overline{B}$$

As a further example, consider the following truth table:

A	B	Output	Products
0	0	0	$\overline{A} \cdot \overline{B}$
0	1	0	$\overline{A} \cdot B$
1	0	1	$A \cdot \overline{B}$
1	1	1	$A \cdot B$

It gives the Karnaugh map shown in Figure C.5(c). This has an output given by

$$\text{output} = A \cdot \overline{B} + A \cdot B$$

We can simplify this to

$$A \cdot \overline{B} + A \cdot B = A \cdot (\overline{B} + B) = A$$

When two cells containing a 1 have a common vertical edge we can simplify the Boolean expression to just the common variable. We can do this by inspection of a map, indicating which cell entries can be simplified by drawing loops round them, as in Figure C.5(c).

Figure C.6(a) shows the Karnaugh map for the following truth table having three input variables:

A	B	C	Output	Products
0	0	0	1	$\overline{A} \cdot \overline{B} \cdot \overline{C}$
0	0	1	0	$\overline{A} \cdot \overline{B} \cdot C$
0	1	0	1	$\overline{A} \cdot B \cdot \overline{C}$
0	1	1	0	$\overline{A} \cdot B \cdot C$
1	0	0	0	$A \cdot \overline{B} \cdot \overline{C}$
1	0	1	0	$A \cdot \overline{B} \cdot C$
1	1	0	0	$A \cdot B \cdot \overline{C}$
1	1	1	0	$A \cdot B \cdot C$

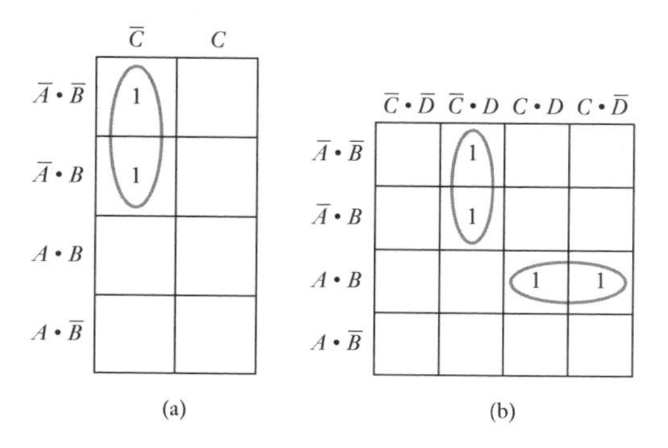

Figure C.6 (a) Three-input-variable map, (b) four-input-variable map.

As before we can use looping to simplify the resulting Boolean expression to just the common variable. The result is

$$\text{output} = \overline{A} \cdot \overline{C}$$

Figure C.6(b) shows the Karnaugh map for the following truth table having four input variables. Looping simplifies the resulting Boolean expression to give

$$\text{output} = \overline{A} \cdot \overline{C} \cdot D + A \cdot B \cdot C$$

A	B	C	D	Output	Products
0	0	0	0	0	
0	0	0	1	1	$\overline{A} \cdot \overline{B} \cdot \overline{C} \cdot D$
0	0	1	0	0	
0	0	1	1	0	
0	1	0	0	0	
0	1	0	1	1	$\overline{A} \cdot B \cdot \overline{C} \cdot D$
0	1	1	0	0	
0	1	1	1	0	
1	0	0	0	0	
1	0	0	1	0	
1	0	1	0	0	
1	0	1	1	0	
1	1	0	0	0	
1	1	0	1	0	
1	1	1	0	1	$A \cdot B \cdot C \cdot \overline{D}$
1	1	1	1	1	$A \cdot B \cdot C \cdot D$

The above represents just some simple examples of Karnaugh maps and the use of looping. Note that, in looping, adjacent cells can be considered to be those in the top and bottom rows of the left- and right-hand columns. Think of opposite edges of the map being joined together. Looping a pair of adjacent ones in a map eliminates the variable that appears in complemented and uncomplemented form. Looping a quad of adjacent ones eliminates the two variables that appear in both complemented and uncomplemented form. Looping an octet of adjacent ones eliminates the three variables that appear in both complemented and uncomplemented form.

As a further illustration, consider an automated machine that will only start when two of three sensors A, B and C give signals. The following truth table fits this requirement and Figure C.7(a) shows the resulting three-variable Karnaugh diagram. The Boolean expression which fits the map and thus describes the outcome from the machine is

$$\text{outcome} = A \cdot B + B \cdot C + A \cdot C$$

Figure C.7(b) shows the logic gates that could be used to generate this Boolean expression. $A \cdot B$ describes an AND gate for the inputs A and B. Likewise $B \cdot C$ and $A \cdot C$ are two more AND gates. The + signs indicate that the outputs from the three AND gates are then the inputs to an OR gate.

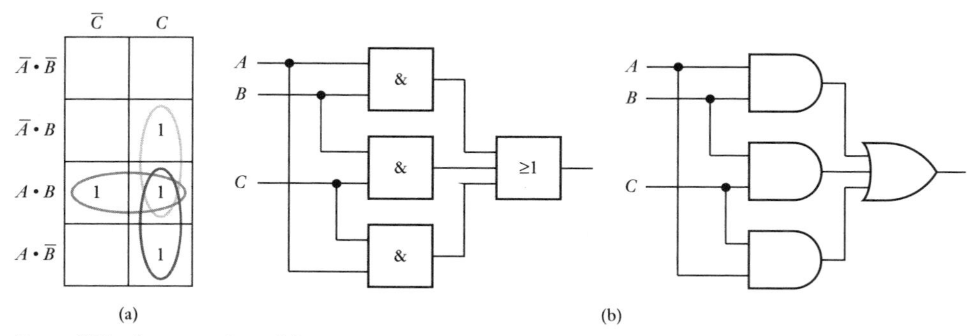

Figure C.7 Automated machine.

A	B	C	Output	Products
0	0	0	0	
0	0	1	0	
0	1	0	0	
0	1	1	1	$\overline{A} \cdot B \cdot C$
1	0	0	0	
1	0	1	1	$A \cdot \overline{B} \cdot C$
1	1	0	1	$A \cdot B \cdot \overline{C}$
1	1	1	1	$A \cdot B \cdot C$

In some logic systems there are some input variable combinations for which outputs are not specified. They are termed 'don't care states'. When entering these on a Karnaugh map, the cells can be set to either 1 or 0 in such a way that the output equations can be simplified.

Problems

C.1 State the Boolean functions that can be used to describe the following situations:

(a) There is an output when switch A is closed and either switch B or switch C is closed.

(b) There is an output when either switch A or switch B is closed and either switch C or switch D is closed.

(c) There is an output when either switch A is opened or switch B is closed.

(d) There is an output when switch A is opened and switch B is closed.

C.2 State the Boolean functions for each of the logic circuits shown in Figure C.8.

C.3 Construct a truth table for the Boolean equation $Q = (A \cdot C + B \cdot C) \cdot (A + C)$.

C.4 Simplify the following Boolean equations:

(a) $Q = A \cdot C + A \cdot C \cdot D + C \cdot D$

(b) $Q = A \cdot \overline{B} \cdot D + A \cdot \overline{B} \cdot \overline{D}$

(c) $Q = A \cdot B \cdot C + C \cdot D + C \cdot D \cdot E$

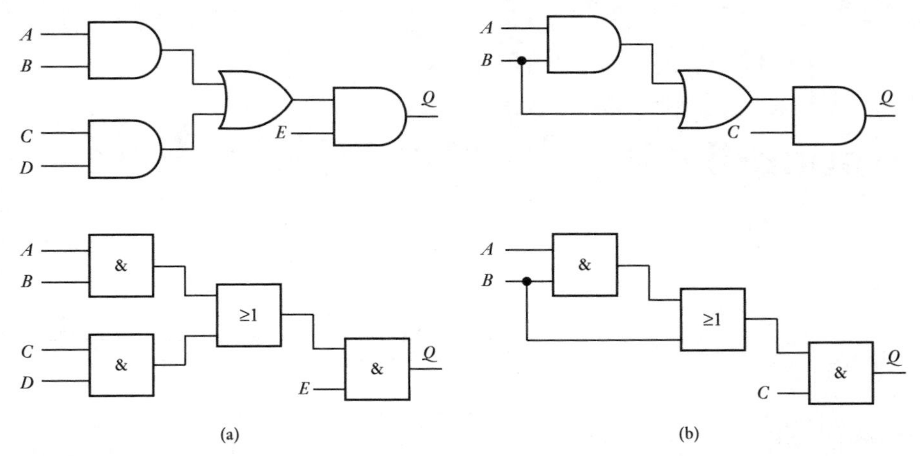

Figure C.8 Problem C.2.

C.5 Use De Morgan's laws to show that a NOR gate with inverted inputs is equivalent to an AND gate.

C.6 Draw the Karnaugh maps for the following truth tables and hence determine the simplified Boolean equation for the outputs:

(a)

A	B	Q
0	0	1
0	1	1
1	0	1
1	1	1

(b)

A	B	C	Q
0	0	0	0
0	0	1	1
0	1	0	1
0	1	1	1
1	0	0	0
1	0	1	1
1	1	0	0
1	1	1	1

C.7 Simplify the following Boolean equations by the use of Karnaugh maps:

(a) $Q = \overline{A} \cdot \overline{B} \cdot C + \overline{A} \cdot \overline{B} \cdot \overline{C} + \overline{A} \cdot B \cdot \overline{C}$

(b) $Q = \overline{A} \cdot B \cdot \overline{C} \cdot D + A \cdot \overline{B} \cdot \overline{C} \cdot D + \overline{A} \cdot \overline{B} \cdot \overline{C} \cdot D$
$+ A \cdot B \cdot \overline{C} \cdot D + A \cdot B \cdot \overline{C} \cdot \overline{D} + A \cdot B \cdot C \cdot D$

C.8 Devise a system which will allow a door to be opened only when the correct combination of four push-buttons is pressed, any incorrect combination sounding an alarm.

Appendix D: Instruction Sets

The following are the instructions used with the Motorola M68HC11, Intel 8051 and PIC16Cxx microcontrollers.

M68HC11

Instruction	Mnemonic	Instruction	Mnemonic
Loading		*Rotate/shift*	
Load accumulator A	LDAA	Rotate bits in memory left	ROL
Load accumulator B	LDAB	Rotate bits in accumulator A left	ROLA
Load double accumulator	LDD	Rotate bits in accumulator B left	ROLB
Load stack pointer	LDS	Rotate bits in memory right	ROR
Load index register X	LDX	Rotate bits in accumulator A right	RORA
Load index register Y	LDY	Rotate bits in accumulator B right	RORB
Pull data from stack and load acc. A	PULA	Arithmetic shift bits in memory left	ASL
Pull data from stack and load acc. B	PULB	Arithmetic shift bits in acc. A left	ASLA
Pull index register X from stack	PULX	Arithmetic shift bits in acc. B left	ASLB
Pull index register Y from stack	PULY	Arithmetic shift bits in memory right	ASR
Transfer registers		Arithmetic shift bits in acc. A right	ASRA
Transfer from acc. A to acc. B	TAB	Arithmetic shift bits in acc. B right	ASRB
Transfer from acc. B to acc. A	TBA	Logical shift bits in memory left	LSL
From stack pointer to index reg. X	TSX	Logical shift bits in acc. A left	LSLA
From stack pointer to index reg. Y	TSY	Logical shift bits in acc. B left	LSLB
From index reg. X to stack pointer	TXS	Logical shift bits in acc. D left	LSLD
From index reg. Y to stack pointer	TYS	Logical shift bits in memory right	LSR
Exchange double acc. and index reg. X	XGDX	Logical shift bits in acc. A right	LSRA
Exchange double acc. and index reg. Y	XGDY	Logical shift bits in acc. B right	LSRB
Decrement/increment		Logical shift bits in acc. C right	LSRD
Subtract 1 from contents of memory	DEC	*Data test with setting of condition codes*	
Subtract 1 from contents of acc. A	DECA	Logical test AND between acc. A & memory	BITA
Subtract 1 from contents of acc. B	DECB		
Subtract 1 from stack pointer	DES	Logical test AND between acc. B & memory	BITB
Subtract 1 from index register X	DEX		
Subtract 1 from index register Y	DEY	Compare accumulator A to accumulator B	CBA
Add 1 to contents of memory	INC	Compare accumulator A and memory	CMPA
Add 1 to contents of accumulator A	INCA	Compare accumulator B and memory	CMPB
Add 1 to contents of accumulator B	INCB	Compare double accumulator with memory	CPD
Add 1 to stack pointer	INS	Compare index register X with memory	CPX
Add 1 to index register X	INX	Compare index register Y with memory	CPY
Add 1 to index register Y	INY	Subtract $00 from memory	TST

(Continued)

Instruction	Mnemonic	Instruction	Mnemonic
Subtract $00 from accumulator A	TSTA	Subtract mem. from acc. B with carry	SBCB
Subtract $00 from accumulator B	TSTB	Subtract mem. from accumulator A	SUBA
Interrupt		Subtract mem. from accumulator B	SUBB
Clear interrupt mask	CLI	Subtract mem. from double acc.	SUBD
Set interrupt mask	SEI	Replace acc. A with two' complement	NEGA
Software interrupt	SWI	Replace acc. B with two' complement	NEGB
Return from interrupt	RTI	Multiply unsigned acc. A by acc. B	MUL
Wait for interrupt	WAI	Unsigned integer divide D by index reg. X	IDIV
Complement and clear		Unsigned fractional divide D by index reg. X	FDIV
Clear memory	CLR	*Conditional branch*	
Clear A	CLRA	Branch if minus	BMI
Clear B	CLRB	Branch if plus	BPL
Clear bits in memory	BCLR	Branch if overflow set	BVS
Set bits in memory	BSET	Branch if overflow clear	BVC
Store registers		Branch if less than zero	BLT
Store contents of accumulator A	STAA	Branch if greater than or equal to zero	BGE
Store contents of accumulator B	STAB	Branch if less than or equal to zero	BLE
Store contents of double acc.	STD	Branch if greater than zero	BGT
Store stack pointer	STS	Branch if equal	BEQ
Store index register X	STX	Branch if not equal	BNE
Store index register Y	STY	Branch if higher	BHI
Push data from acc. A onto stack	PSHA	Branch if lower or same	BLS
Push data from acc. B onto stack	PSHB	Branch if higher or same	BHS
Push index reg. X contents onto stack	PSHX	Branch if lower	BLO
Push index reg. Y contents onto stack	PSHY	Branch if carry clear	BCC
Logic		Branch if carry set	BCS
AND with contents of accumulator A	ANDA	*Jump and branch*	
AND with contents of accumulator B	ANDB	Jump to address	JMP
EXCLUSIVE-OR with contents of acc. A	EORA	Jump to subroutine	JSR
EXCLUSIVE-OR with contents of acc. B	EORB	Return from subroutine	RTS
OR with contents of accumulator A	ORAA	Branch to subroutine	BSR
OR with contents of accumulator B	ORAB	Branch always	BRA
Replace memory with one's complement	COM	Branch never	BRN
Replace acc. A with one's complement	COMA	Branch bits set	BRSET
Replace acc. B with one's complement	COMB	Branch bits clear	BRCLR
Arithmetic		*Condition code*	
Add contents of acc. A to acc. B	ABA	Clear carry	CLC
Add contents of acc. B to index reg. X	ABX	Clear overflow	CLV
Add contents of acc. B to index reg. Y	ABY	Set carry	SEC
Add memory to acc. A without carry	ADDA	Set overflow	SEV
Add memory to acc. B without carry	ADDB	Transfer from acc. A to condition code reg.	TAP
Add mem. to double acc. without carry	ADDD	Transfer from condition code reg. to acc. A	TPA
Add memory to acc. A with carry	ADCA	*Miscellaneous*	
Add memory to acc. B with carry	ADCB	No operation	NOP
Decimal adjust	DAA	Stop processing	STOP
Subtract contents of acc. B from acc. A	SBA	Special test mode	TEST
Subtract mem. from acc. A with carry	SBCA		

Note: The number of bits in a register depends on the processor. An 8-bit microprocessor generally has 8-bit registers. Sometimes two of the data registers may be used together to double the number of bits. Such a combined register is referred to as a doubled register.

Intel 8051

Instruction	Mnemonic
Data transfer	
Move data to accumulator	MOV A, #data
Move register to accumulator	MOV A, Rn
Move direct byte to accumulator	MOV A, direct
Move indirect RAM to accumulator	MOV A, @Ri
Move accumulator to direct byte	MOV direct, A
Move accumulator to external RAM	MOVX @Ri, A
Move accumulator to register	MOV Rn, A
Move direct byte to indirect RAM	MOV @Ri, direct
Move immediate data to register	MOV Rn, #data
Move direct byte to direct byte	MOV direct, direct
Move indirect RAM to direct byte	MOV direct, @Ri
Move register to direct byte	MOV direct, Rn
Move immediate data to direct byte	MOV direct, #data
Move immediate data to indirect RAM	MOV @Ri, #data
Load data pointer with a 16-bit constant	MOV DPTR, #data16
Move code byte relative to DPTR to acc.	MOV A, @A+DPTR
Move external RAM, 16-bit addr., to acc.	MOVX A, @DPTR
Move acc. to external RAM, 16-bit addr.	MOVX @DPTR, A
Exchange direct byte with accumulator	XCH A, direct
Exchange indirect RAM with acc.	XCH A, @Ri
Exchange register with accumulator	XCH A, Rn
Push direct byte onto stack	PUSH direct
Pop direct byte from stack	POP direct
Branching	
Absolute jump	AJMP addr 11
Long jump	LJMP addr 16
Short jump, relative address	SJMP rel
Jump indirect relative to the DPTR	JMP @A+DPTR
Jump if accumulator is zero	JZ rel
Jump if accumulator is not zero	JNZ rel
Compare direct byte to acc. and jump if not equal	CJNE A, direct, rel
Compare immediate to acc. and jump if not equal	CJNE A, #data, rel
Compare immediate to register and jump if not equal	CJNE Rn, #data, rel
Compare immediate to indirect and jump if not equal	CJNE @Ri, #data, rel
Decrement register and jump if not zero	DJNZ Rn, rel
Decrement direct byte, jump if not zero	DJNZ A, direct, rel
Jump if carry is set	JC rel
Jump if carry not set	JNC rel
Jump if direct bit is set	JB bit, rel
Jump if direct bit is not set	JNB bit, rel
Jump if direct bit is set and clear bit	JBC bit, rel
Subroutine call	
Absolute subroutine call	ACALL addr 11
Long subroutine call	LCALL addr 16
Return from subroutine	RET
Return from interrupt	RETI
Bit manipulation	
Clear carry	CLR C
Clear bit	CLR bit
Set carry but	SETB C
Set bit	SETB bit
Complement carry	CPL C
AND bit to carry bit	ANL C,bit
AND complement of bit to carry bit	ANL C,/bit
OR bit to carry bit	ORL C,bit
OR complement of bit to carry bit	ORL C,/bit
Move bit to carry	MOV C,bit
Move carry bit to bit	MOV bit,C
Logical operations	
AND accumulator to direct byte	ANL direct, A
AND immediate data to direct byte	ANL direct, #data
AND immediate data to acc.	ANL A, #data
AND direct byte to accumulator	ANL A, direct
AND indirect RAM to accumulator	ANL A, @Ri
AND register to accumulator	ANL A, Rn
OR accumulator to direct byte	ORL direct, A
OR immediate data to direct byte	ORL direct, #data
OR immediate data to accumulator	ORL A, #data
OR direct byte to accumulator	ORL A, direct
OR indirect RAM to accumulator	ORL A, @Ri
OR register to accumulator	ORL A, Rn
XOR accumulator to direct byte	XRL direct, A
XOR immediate data to acc.	XRL direct, #data
XOR immediate data to acc.	XRL A, #data
XOR direct byte to accumulator	XRL A, direct
XOR indirect RAM to accumulator	XRL A, @Ri
XOR register to accumulator	XRL A, Rn
Addition	
Add immediate data to acc.	ADD A, #data

(Continued)

Instruction	Mnemonic	Instruction	Mnemonic
Add direct byte to accumulator	ADD A, direct	Swap nibbles within the acc.	SWAP A
Add indirect RAM to accumulator	ADD A, @Ri	Decimal adjust accumulator	DA A
Add register to accumulator	ADD A, Rn	*Increment and decrement*	
Add immediate data to acc. with carry	ADDC A, #data	Increment accumulator	INC A
		Increment direct byte	INC direct
Add direct byte to acc. with carry	ADDC A, direct	Increment indirect RAM	INC @Ri
Add indirect RAM to acc. with carry	ADDC A, @Ri	Increment register	INC Rn
		Decrement accumulator	DEC A
Add register to acc. with carry	ADDC A, Rn	Decrement direct byte	DEC direct
Subtraction		Decrement indirect RAM	DEC @Ri
Subtract immediate data from acc. with borrow	SUBB A, #data	Decrement register	DEC Rn
		Increment data pointer	INC DPTR
Subtract direct byte from acc. with Borrow	SUBB A, 29	*Clear and complement operations*	
		Complement accumulator	CPL A
Subtract indirect RAM from acc. with borrow	SUBB A, @Ri	Clear accumulator	CLR A
		Rotate operations	
Multiplication and division		Rotate accumulator right	RR A
Multiply A and B	MUL AB	Rotate accumulator right thro. C	RRC A
Divide A by B	DIV AB	Rotate accumulator left	RL A
Decimal maths operations		Rotate accumulator left through C	RLC A
Exchange low-order digit indirect		*No operation*	
RAM with accumulator	XCHD A, @Ri	No operation	NOP

Note: A value preceded by # is a number, #data16 is a 16-bit constant; Rn refers to the contents of a register; @Ri refers to the value in memory where the register points; DPTR is the data pointer; direct is the memory location where data used by an instruction can be found.

PIC16Cxx

Instruction	Mnemonic
Add a number with number in working reg.	addlw number
Add number in working reg. to number in file register and put number in file register	addwf FileReg,f
Add number in working reg. to number in file register and put number in working reg.	addwf FileReg,w
AND a number with the number in the working reg. and put result in working reg.	andlw number
AND a number in the working reg. with the number in file reg., and put result in file reg.	andwf FileReg,f
Clear a bit in a file reg., i.e. make it 0	bcf FileReg,bit
Set a bit in a file reg., i.e. make it 1	bsf FileReg,bit
Test a bit in a file reg. and skip the next instruction if the bit is 0	btfsc FileReg,bit
Test a bit in a file reg. and skip the next instruction if the bit is 1	btfss FileReg,bit
Call a subroutine, after which return to where it left off	call AnySub
Clear, i.e. make 0, the number in file reg.	clrf FileReg
Clear, i.e. make 0, the no. in working reg.	clrw
Clear the number in the watchdog timer	clrwdt
Complement the number in file reg. and leave result in file register	comf FileReg,f
Decrement a file reg., result in file reg.	decf FileReg,f
Decrement a file reg. and if result zero skip the next instruction	decfsz FileReg,f
Go to point in program labelled	gotot label
Increment file reg. and put result in file reg.	Incf FileReg,f
OR a number with number in working reg.	iorlw number
OR the number in working reg. with the number in file reg., put result in file reg.	iorwf FileReg,f
Move (copy) the number in a file reg. into the working reg.	movf FileReg,w
Move (copy) number into working reg.	movlw number
Move (copy) the number in file reg. into the working reg.	movwf FileReg
No operation	nop
Return from a subroutine and enable global interrupt enable bit	refie
Return from a subroutine with a number in the working register	retlw number
Return from a subroutine	return
Rotate bits in file reg. to the left through the carry bit	rlf FileReg,f
Rotate bits in file reg. to the right through the carry bit	rrf FileReg,f
Send the PIC to sleep, a low-power-consumption mode	sleep
Subtract the number in working reg. from a number	sublw number
Subtract the no. in working reg. from number in file reg., put result in file reg.	subwf FileReg,f
Swap the two halves of the 8 bit no. in a file reg, leaving result in file reg.	swapf FileReg, f
Use the number in working reg. to specify which bits are input or output	tris PORTX
XOR a number with number in working register	xorlf number
XOR the number in working reg. with number in file reg. and put result in the file reg.	xorwf FileReg,f

Note: f is used for the file register, w for the working register and b for bit. The mnemonics indicate the types of operand involved, e.g. movlw indicates the move operation with the lw indicating that a literal value, i.e. a number, is involved in the working register w; movwf indicates the move operation when the working register and a file register are involved.

Appendix E: C Library Functions

The following are some common C library functions. This is not a complete list of all the functions within any one library or a complete list of all the libraries that are likely to be available with any one compiler.

<ctype.h>

isalnum	int isalnum(int ch)	Tests for alphanumeric characters, returning non-zero if argument is either a letter or a digit or a 0 if it is not alphanumeric.
isalpha	int isalpha(int ch)	Tests for alphabetic characters, returning non-zero if a letter of the alphabet, otherwise 0.
iscntrl	int iscntrl(int ch)	Tests for control character, returning non-zero if between 0 and 0x1F or is equal to 0x7F (DEL), otherwise 0.
isdigit	int isdigit(int ch)	Tests for decimal digit character, returning non-zero if a digit (0 to 9), otherwise 0.
isgraph	int isgraph(int ch)	Tests for a printable character (except space), returning non-zero if printable, otherwise 0.
islower	int islower(int ch)	Tests for lower case character, returning non-zero if lower case, otherwise 0.
isprint	int isprint(int ch)	Tests for printable character (including space), returning non-zero if printable, otherwise 0.
ispunct	int ispunct(int ch)	Tests for punctuation character, returning non-zero if a punctuation character, otherwise 0.
isspace	int isspace(int ch)	Tests for space character, returning non-zero if a space, tab, form feed, carriage return or new-line character, otherwise 0.
isupper	int isupper(int ch)	Tests for upper case character, returning non-zero if upper case, otherwise 0.
isxdigit	int isxdigit(int ch)	Tests for hexadecimal character, returning non-zero if a hexadecimal digit, otherwise 0.

<math.h>

acos	double acos(double arg)	Returns the arc cosine of the argument.
asin	double asin(double arg)	Returns the arc sine of the argument.
atan	double atan(double arg)	Returns the arc tangent of the argument. Requires one argument.

atan2	double atan2(double y, double x)	Returns the arc tangent of y/x.
ceil	double ceil(double num)	Returns the smallest integer that is not less than num.
cos	double cos(double arg)	Returns the cosine of arg. The value of arg must be in radians.
cosh	double cosh(double arg)	Returns the hyperbolic cosine of arg.
exp	double exp(double arg)	Returns e^x where x is arg.
fabs	double fabs(double num)	Returns the absolute value of num.
floor	double floor(double num)	Returns the largest integer not greater than num.
fmod	double fmod(double x, double y)	Returns the floating-point remainder of x/y.
ldexp	double ldexp(double x, int y)	Returns x times 2^y.
log	double log(double num)	Returns the natural logarithm of num.
log10	double log10(double num)	Returns the base 10 logarithm of num.
pow	double pow(double base, double exp)	Returns base raised to the exp power.
sin	double sin(double arg)	Returns the sine of arg.
sinh	double sinh(double arg)	Returns the hyperbolic sine of arg.
sqrt	double sqrt(double num)	Returns the square root of num.
tan	double tan(double arg)	Returns the tangent of arg.
tanh	double tanh(double arg)	Returns the hyperbolic tangent of arg.

<stdio.h>

getchar	int getchar(void)	Returns the next character typed on the keyboard.
gets	char gets(char *str)	Reads characters entered at the keyboard until a carriage return is read and stores them in the array pointed to by str.
printf	int printf(char *str, ...)	Outputs the string pointed to by str.
puts	int puts(char *str)	Outputs the string pointed to by str.
scanf	int scanf(char *str, ...)	Reads information into the variables pointed to by the arguments following the control string.

<stdlib.h>

abort	void abort(void)	Causes immediate termination of a program.
abs	int abs(int num)	Returns the absolute value of the integer num.
bsearch	void bsearch(const void *key, const void *base, size_t num, size_t size, int(*compare)(const void *, const void *))	Performs a binary search on the sorted array pointed to by base and returns a pointer to the first member that matches the key pointed to by key. The number of the elements in the array is specified by num and the size in bytes of each element by size.

calloc	void *calloc(size_t num, size_t size)	Allocates sufficient memory for an array of num objects of size given by size, returning a pointer to the first byte of the allocated memory.
exit	void exit(int status)	Causes immediate normal termination of a program. The value of the status is passed to the calling process.
free	void free(void *ptr)	Frees the allocated memory pointed to by ptr.
labs	long labs(long num)	Returns the absolute value of the long int num.
malloc	void *malloc(size_t size)	Returns a pointer to the first byte of memory of size given by size that has been allocated.
qsort	void qsort(void *base, size_t num, size_t size, int(*compare)(const void*, const void*))	Sorts the array pointed to by base. The number of elements in the array is given by num and the size in bytes of each element by size.
realloc	void *realloc(void *ptr, size_t size)	Changes the size of the allocated memory pointed to by ptr to that specified by size.

Note: size_t is the type for 'size of' variables and usually represents the size of another parameter or object.

<time.h>

asctime	char *asctime(const struct tm *ptr)	Converts time from a structure form to a character string appropriate for display, returning a pointer to the string.
clock	clock_t clock(void)	Returns the number of clock cycles that have occurred since the program began execution.
ctime	char *ctime(const time_t *time)	Returns a pointer to a string of the form day month date hours:minutes:seconds year\n\0 given a pointer to the numbers of seconds elapsed since 00:00:00 Greenwich Mean Time.
difftime	double difftime(time_t time 2, time_t time 1)	Returns the difference in seconds between time 1 and time 2.
gmtime	struct tm *gmtime (const time_t *time)	Returns a pointer to time converted from long inter form to a structure form.
localtime	struct tm *localtime (const time_t *time)	Returns a pointer to time converted from long inter form to structure form in local time.
time	time_t time(time_t *system)	Returns the current calendar time of the system.

Note: time_t and clock_t are used as the type for 'time of' and 'number of cycles of' variables.

Appendix F: MATLAB and SIMULINK

F.1 MATLAB

Computer software can be used to aid computation and modelling of systems; a program that is often used is MATLAB. The following is a brief introduction to MATLAB (registered trademark of the Mathworks Inc.) version 4.0 or later. For additional information you are referred to the user guide or textbooks such as *The MATLAB Handbook* by Eva Pärt-Enander, Anders Sjöberg, Bo Melin and Pernilla Isaksson (Addison-Wesley 1996) and *Using MATLAB to Analyse and Design Control Systems*, 2nd edition, by Naomi Ehrich Leonard and William S. Levine (Addison-Wesley 1995).

Commands are entered by typing them in after the prompt and then pressing the enter or return key in order that the command can be executed. In the discussion of the commands that follow, this pressing of the enter or return key will not be repeated but should be assumed in all cases. To start MATLAB, in Windows or the Macintosh systems, click on the MATLAB icon, otherwise type matlab. The screen will then produce the MATLAB prompt ≫. To quit MATLAB type quit or exit after the prompt. Because MATLAB is case sensitive, lower case letters should be used throughout for commands.

Typing help after the prompt, or selecting help from the menu bar at the top of the MATLAB window, displays a list of MATLAB broad help topics. To get help on a particular topic in the list, e.g. exponentials, type help exp. Typing lookfor plus some topic will instruct MATLAB to search for information on that topic, e.g. lookfor integ will display a number of commands which could be considered for integration.

In general, mathematical operations are entered into MATLAB in the same way as they would be written on paper. For example,

```
≫ a = 4/2
```

yields the response

```
a =
    2
```

and

```
≫ a = 3*2
```

yields the response

```
a =
    6
```

Operations are carried out in the following order: ^ power operation, * multiplication, / division, + addition, − subtraction. Precedence of operators is from left to right but brackets ( ) can be used to affect the order. For example,

```
≫ a = 1 + 2^3/4*5
```

yields the response

```
a =

    11
```

because we have $2^3/4$ which is multiplied by 5 and then added to 1, whereas

```
≫ a = 1 + 2^3/(4*5)
```

yields the response

```
a =

    1.4
```

because we have 2^3 divided by the product of 4 and 5, and then added to 1.

The following are some of the mathematical functions available with MATLAB:

abs(x)	Gives the absolute value of x, i.e. $\lvert x \rvert$
exp(x)	Gives the exponential of x, i.e. e^x
log(x)	Gives the natural logarithm of x, i.e. $\ln x$
log10(x)	Gives the base 10 logarithm of x, i.e. $\log_{10} x$
sqrt(x)	Gives the square root of x, i.e. $\sqrt{x}$
sin(x)	Gives $\sin x$ where x is in radians
cos(x)	Gives $\cos x$ where x is in radians
tan(x)	Gives $\tan x$ where x is in radians
asin(x)	Gives arcsin x, i.e. $\sin^{-1} x$
acos(x)	Gives arccos x, i.e. $\cos^{-1} x$
atan(x)	Gives arctan x, i.e. $\tan^{-1} x$
csc(x)	Gives $1/\sin x$
sec(x)	Gives $1/\cos x$
cot(x)	Gives $1/\tan x$

π is entered by typing pi.

Instead of writing a series of commands at the prompt, a text file can be written and then the commands executed by referring MATLAB to that file. The term M-file is used since these text files, containing a number of consecutive MATLAB commands, have the suffix .m. In writing such a file, the first line must begin with the word function followed by a statement identifying the name of the function and the input and output in the form

function [output] = function name [input]

e.g. function y = cotan(x) which is the file used to determine the value of y given by cotan x. Such a file can be called up in some MATLAB sequence of commands by writing the name followed by the input, e.g. cotan(x). It is in fact already included in MATLAB and is used when the cotangent of x is required. However, the file could have been user written. A function that has multiple inputs should list all of them in the function statement.

Likewise, a function that is to return more than one value should list all the outputs.

Lines that start with % are comment lines; they are not interpreted by MATLAB as commands. For example, supposing we write a program to determine the root-mean-square values of a single column of data points, the program might look like

```
function y=rms(x)
% rms Root mean square
% rms(x) gives the root mean square value of the
% elements of column vector x.
xs=x^2;
 s=size(x);
y=sqrt(sum(xs)/s);
```

We have let xs be the square values of each x value. The command s=size(x) obtains the size, i.e. number of entries, in the column of data. The command y=sqrt(sum(xs)/s (1)) obtains the square root of the sum of all the xs values divided by s. The ; command is used at the end of each program line.

MATLAB supplies a number of toolboxes containing collections of M-files. Of particular relevance to this book is the Control System toolbox. It can be used to carry out time responses of systems to impulses, steps, ramps, etc., along with Bode and Nyquist analysis, root locus, etc. For example, to carry out a Bode plot of a system described by a transfer function $4/(s^2 + 2s + 3)$, the program is

```
%Generate Bode plot for G(s)=4/(s^2 + 2s + 3)
num=4
den=[1 2 3];
bode(num,den)
```

The command bode (num,den) produces the Bode plot of gain in dB against frequency in rad/s on a log scale and phase in degrees against frequency in rad/s on a log scale.

F.1.1 Plotting

Two-dimensional linear plots can be produced by using the plot(x,y) command; this plots the values of x and y. For example, we might have

```
x=[0 1 2 3 4 5];
y=[0 1 4 9 16 25];
plot(x,y)
```

To plot a function, whether standard or user defined, we use the command fplot(function name,lim), where lim determines the plotting interval, i.e. the minimum and maximum values of x.

The command semilogx(x,y) generates a plot of the values of x and y using a logarithmic scale for x and a linear scale for y. The command semilogy(x,y) generates a plot of the values of x and y using a linear scale for x and a logarithmic scale for y. The command loglog(x,y) generates a plot of the values of x and y using logarithmic scales for both x and y. The command polar(theta,r) plots in polar co-ordinates with theta being the argument in radians and r the magnitude.

The subplot command enables the graph window to be split into subwindows and plots to be placed in each. For example, we might have

```
x=(0 1 2 3 4 5 6 7);
y=expx;
subplot(2,1,1);plot(x,y);
subplot(2,1,2);semilogy(x,y);
```

Three integers m, n, p are given with the subplot command; the digits m and n indicate the graph window is to be split into an $m \times n$ grid of smaller windows, where m is the number of rows and n is the number of columns, and the digit p specifies the window to be used for the plot. The subwindows are numbered by row from left to right and top to bottom. Thus the above sequence of commands divides the window into two, with one plot above the other; the top plot is a linear plot and the lower plot is a semilogarithmic plot.

The number and style of grid lines, the plot colour and the adding of text to a plot can all be selected. The command print is used to print a hard copy of a plot, either to a file or a printer. This can be done by selecting the file menu-bar item in the figure window and then selecting the print option.

F.1.2 Transfer functions

The following lines in a MATLAB program illustrate how a transfer function can be entered and displayed on screen:

```
% G(s)=4(s+10)/(s+5)(s+15)
num=4*[1 10];
den=conv([1 5],[1 15]);
printsys(num,den,'s')
```

The command num is used to indicate the numerator of the transfer function, in descending powers of s. The command den is used to indicate the denominator in descending powers of s for each of the two polynomials in the denominator. The command conv multiplies two polynomials, in this case they are $(s + 5)$ and $(s + 15)$. The printsys command displays the transfer function with the numerator and denominator specified and written in the s-domain.

Sometimes we may be presented with a transfer function as the ratio of two polynomials and need to find the poles and zeros. For this we can use

```
% Finding poles and zeros for the transfer function
% G(s)=(5s^2 + 3s + 4)/(s^3 + 2s^2 + 4s + 7)
num=[5 3 4];
den=[1 2 4 7];
[z,p,k]=tf2zp(num,den)
```

[z,p,k]=tf2zp(num,den) is the command to determine and display the zeros (z), poles (p) and gain (k) of the zero–pole–gain transfer function entered.

MATLAB can be used to give graphs showing the response of a system to different inputs. For example, the following program will give the response of the system to a unit step input $u(t)$ with a specified transfer function:

```
% Display of response to a step input for a system with
% transfer function G(s )=5/(s^2 + 3s + 12)
```

```
num=5;
den=[1 3 12];
step(num,den)
```

F.1.3 Block diagrams

Control systems are often represented as a series of interconnected blocks, each block having specific characteristics. MATLAB allows systems to be built up from interconnected blocks. The commands used are cloop when a block with a given open-loop transfer function has unity feedback. If the feedback is not unity the command feedback is used, e.g. with Figure F.1 we have the program

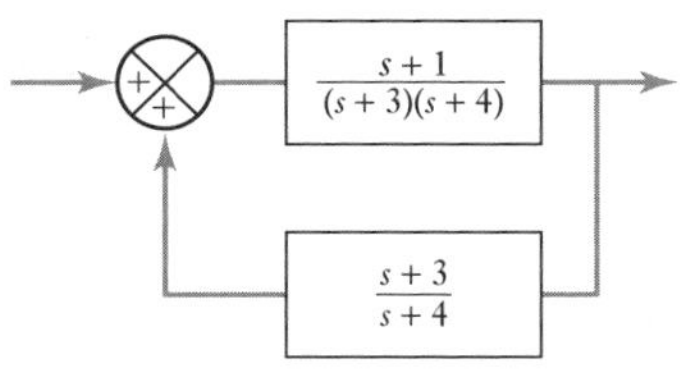

Figure F.1 Block diagram.

```
% System with feedback loop
ngo=[1 1];
dgo=conv([1 3],[1 4]);
nh=[1 3];
dh=[1 4];
[ngc2,dgc2]=feedback(ngo,dgo,nh,dh)
printsys(ngc2,dgc2,'s')
```

ngo and dgo indicate the numerator and denominator of open-loop transfer function $G_0(s)$, nh and dh the numerator and denominator of the feedback loop transfer function $H(s)$. The program results in the display of the transfer function for the system as a whole.

The command series is used to indicate that two blocks are in series in a particular path; the command parallel indicates that they are in parallel.

F.2 SIMULINK

SIMULINK is used in conjunction with MATLAB to specify systems by 'connecting' boxes on the screen rather than, as above, writing a series of commands to generate the description of the block diagram. Once MATLAB has been started, SIMULINK is selected using the command ≫ simulink. This opens the SIMULINK control window with its icons and pull-down menus in its header bar. Click on file, then click on new from the drop-down menu. This opens a window in which a system can be assembled.

To start assembling the blocks required, go back to the control window and double click on the linear icon. Click and then drag the transfer Fcn icon into the untitled window. If you require a gain block, click and drag the gain icon into the untitled window. Do the same for a sum icon and perhaps an integrator icon. In this way, drag all the required icons into the untitled window. Then double click on the Sources icon and select the appropriate source from its drop-down menu, e.g. step input, and drag it into the untitled window. Now double click on the sinks icon and drag the graph icon into the untitled window. To connect the icons, depress the mouse button while the mouse arrow is on the output symbol of an icon and drag to it the input symbol of the icon to which it is to be connected. Repeat this for all the icons until the complete block diagram is assembled.

To give the transfer Fcn box a transfer function, double click in the box. This will give a dialogue box in which you can use MATLAB commands

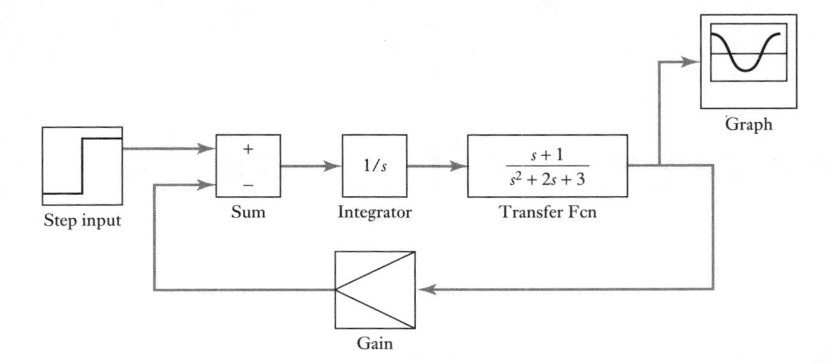

Figure F.2 Example of use of SIMULINK.

for numerator and denominator. Click on the numerator and type in [1 1] if $(s + 1)$ is required. Click on the denominator and type in [1 2 3] if $(s^2 + 2s + 3)$ is required. Then click the done icon. Double click on the gain icon and type in the gain value. Double click on the sum icon and set the signs to + or − according to whether positive or negative feedback is required. Double click on the graph icon and set the parameters for the graph. You then have the complete simulation diagram on screen. Figure F.2 shows the form it might take. To delete any block or connection, select them by clicking and then press the <DEL> key.

To simulate the behaviour of the system, click on Simulation to pull down its menu. Select Parameters and set the start and stop times for the simulation. From the Simulation menu, select Start. SIMULINK will then create a graph window and display the output of the system. The file can be saved by selecting File and clicking on SAVE AS in the drop-down menu. Insert a file name in the dialogue box then click on Done.

Appendix G: Electrical Circuit Analysis

G.1 D.C. circuits

The fundamental laws used in circuit analysis are Kirchhoff's laws:

1. **Kirchhoff's current law** states that at any junction in an electrical circuit, the current entering it equals the current leaving it.
2. **Kirchhoff's voltage law** states that around any closed path in a circuit, the sum of the voltage drops across all the components is equal to the sum of the applied voltage rises.

While circuits containing combinations of series- and parallel-connected resistors can often be reduced to a simple circuit by systematically determining the equivalent resistance of series or parallel connected resistors and reducing the analysis problem to a very simple circuit, the following techniques are likely to be needed for more complex circuits.

G.1.1 Node analysis

A **node** is a point in a circuit where two or more devices are connected together, i.e. it is a junction at which we have current entering and current leaving. A **principal node** is a point where three or more elements are connected together. Thus in Figure G.1, just b and d are principal nodes. One of the principal nodes is chosen to be a reference node so that the potential differences at the other nodes are then considered with reference to it. For the following analysis with Figure G.1, d has been taken as the reference node. Kirchhoff's current law is then applied to each non-reference node. The procedure is thus:

1. Draw a labelled circuit diagram and mark the principal nodes.
2. Select one of the principal nodes as a reference node.
3. Apply Kirchhoff's current law to each of the non-reference nodes, using Ohm's law to express the currents through resistors in terms of node voltages.
4. Solve the resulting simultaneous equations. If there are n principal nodes there will be $(n - 1)$ equations.
5. Use the derived values of the node voltages to determine the currents in each branch of the circuit.

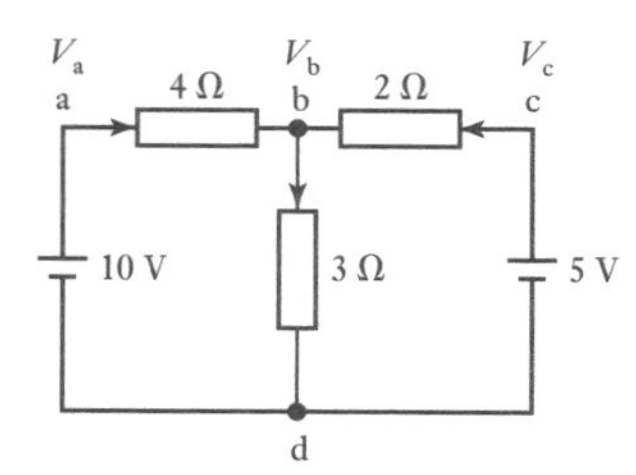

Figure G.1 Circuit for node analysis.

As an illustration, consider Figure G.1. The nodes are a, b, c and d with b and d being principal nodes. Take node d as the reference node.

If V_a, V_b and V_c are the node voltages relative to node d then the potential difference across the 4 Ω resistor is $(V_a - V_b)$, that across the 3 Ω resistor is V_b and that across the 2 Ω resistor is $(V_c - V_b)$. Thus the current through the 4 Ω resistor is $(V_a - V_b)/4$, that through the 3 Ω resistor is $V_b/3$ and that through the 2 Ω resistor is $(V_c - V_b)/2$. Thus, applying Kirchhoff's current law to node b gives:

$$\frac{V_a - V_b}{4} + \frac{V_c - V_b}{2} = \frac{V_b}{3}$$

However, $V_a = 10$ V and $V_c = 5$ V and so:

$$\frac{10 - V_b}{4} + \frac{5 - V_b}{2} = \frac{V_b}{3}$$

Thus $V_b = 4.62$ V. The potential difference across the 4 Ω resistor is thus $10 - 4.62 = 5.38$ V and so the current through it is $5.38/4 = 1.35$ A. The potential difference across the 3 Ω resistor is 4.62 V and so the current through it is $4.62/3 = 1.54$ A. The potential difference across the 2 Ω resistor is $5 - 4.62 = 0.38$ V and so the current through it is $0.38/2 = 0.19$ A.

G.1.2 Mesh analysis

The term **loop** is used for a sequence of circuit elements that form a closed path. A **mesh** is a circuit loop which does not contain any other loops within it. Mesh analysis involves defining a current as circulating round each mesh. The same direction must be chosen for each mesh current and the usual convention is to make all the mesh currents circulate in a clockwise direction. Having specified mesh currents, Kirchhoff's voltage law is then applied to each mesh. The procedure is thus:

1 Label each of the meshes with clockwise mesh currents.
2 Apply Kirchhoff's voltage law to each of the meshes, the potential differences across each resistor being given by Ohm's law in terms of the current through it and in the opposite direction to the current. The current through a resistor which borders just one mesh is the mesh current; the current through a resistor bordering two meshes is the algebraic sum of the mesh currents through the two meshes.
3 Solve the resulting simultaneous equations to obtain the mesh currents. If there are n meshes there will be n equations.
4 Use the results for the mesh currents to determine the currents in each branch of the circuit.

As an illustration, for the circuit shown in Figure G.2 there are three loops – ABCF, CDEF and ABCDEF – but only the first two are meshes. We can define currents I_1 and I_2 as circulating in a clockwise direction in these meshes.

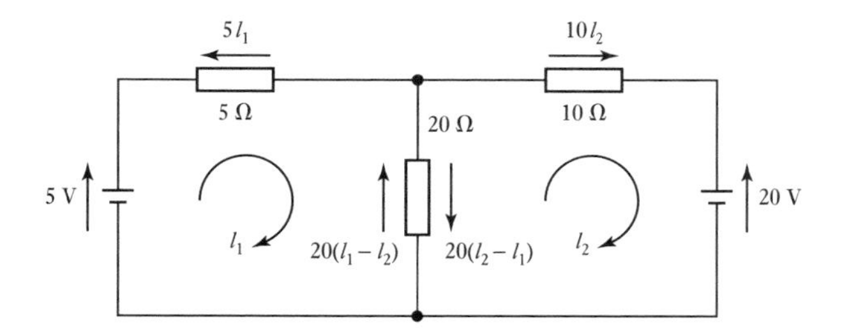

Figure G.2 Circuit illustrating mesh analysis.

For mesh 1, applying Kirchhoff's voltage law gives $5 - 5I_1 - 20(I_1 - I_2) = 0$. This can be rewritten as:

$$5 = 25I_1 - 20I_2$$

For mesh 2, applying Kirchhoff's voltage law gives $-10I_2 - 20 - 20(I_2 - I_1) = 0$. This can be rewritten as:

$$20 = 20I_1 - 30I_2$$

We now have a pair of simultaneous equations and so $I_2 = -1.14$ A and $I_1 = -0.71$ A. The minus signs indicate that the currents are in the opposite directions to those indicated in the figure. The current through the 20 Ω resistor is thus in the direction of I_1 and $-0.71 + 1.14 = 0.43$ A.

G.1.3 Thévenin's theorem

The equivalent circuit for any two-terminal network containing a voltage or current source is given by **Thévenin's theorem**:

> Any two-terminal network (Figure G.3(a)) containing voltage or current sources can be replaced by an equivalent circuit consisting of a voltage equal to the open-circuit voltage of the original circuit in series with the resistance measured between the terminals when no load is connected between them and all independent sources in the network are set equal to zero (Figure G.3(b)).

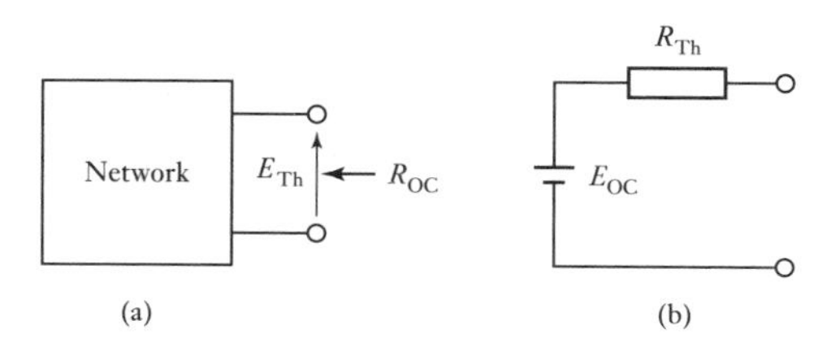

Figure G.3 (a) The network, (b) its equivalent.

If we have a linear circuit, to use Thévenin's theorem, we have to divide it into two circuits, A and B, connected at a pair of terminals. We can then use Thévenin's theorem to replace, say, circuit A by its equivalent circuit. The open-circuit Thévenin voltage for circuit A is that given when circuit B is disconnected and the Thévenin resistance for A is the resistance looking into the terminals of A with all its independent sources set equal to zero. Figure G.4 illustrates this sequence of steps.

Figure G.4 Step-by-step approach for circuit analysis.

1. Identify the two parts A and B of the circuit and separate them by terminals.
2. Separate part A from part B.

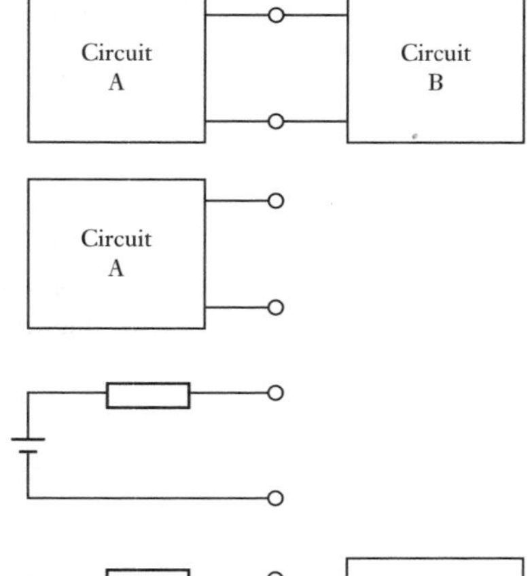

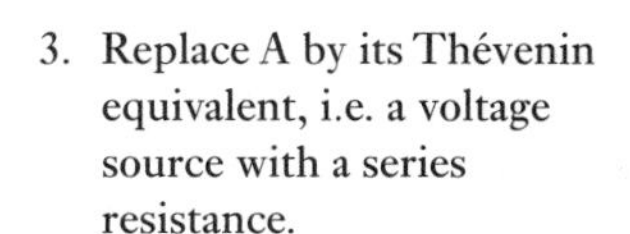

3. Replace A by its Thévenin equivalent, i.e. a voltage source with a series resistance.

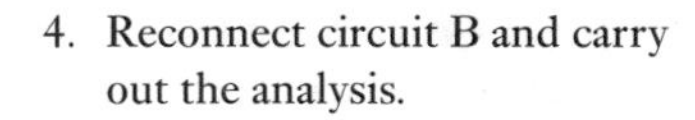

4. Reconnect circuit B and carry out the analysis.

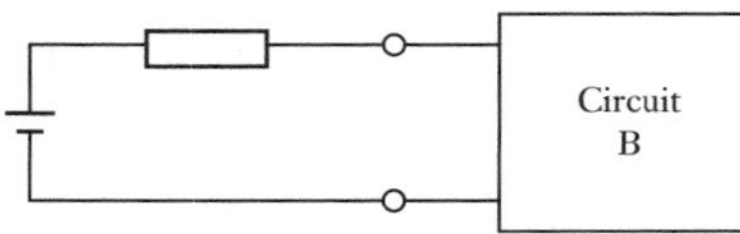

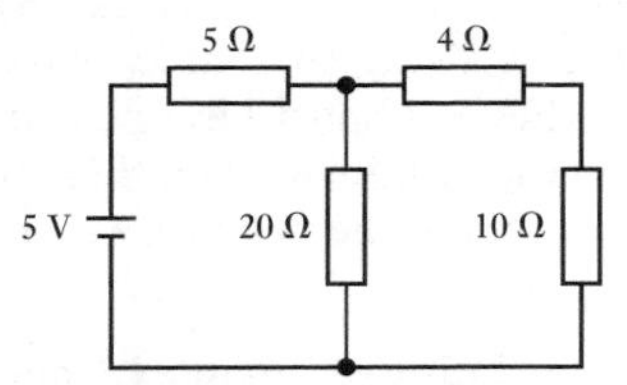

Figure G.5 Example circuit illustrating the use of Thévenin's theorem.

As an illustration, consider the use of Thévenin's theorem to determine the current through the 10 Ω resistor in the circuit given in Figure G.5.

Since we are interested in the current through the 10 Ω resistor, we identify it as network B and the rest of the circuit as network A, connecting them by terminals (Figure G.6(a)). We then separate A from B (Figure G.6(b)) and determine the Thévenin equivalent for it.

The open-circuit voltage is that across the 20 Ω resistor, i.e. the fraction of the total voltage drop across the 20 Ω:

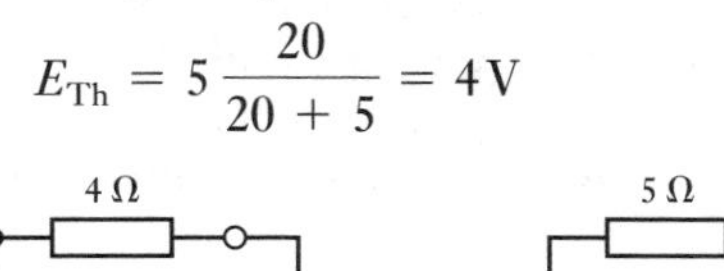

$$E_{Th} = 5\frac{20}{20 + 5} = 4\text{V}$$

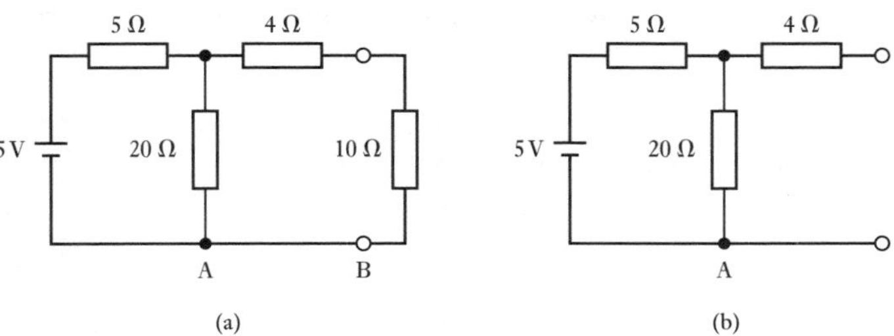

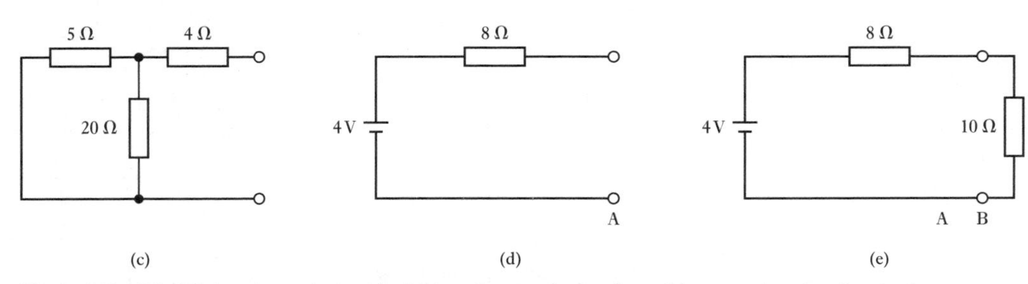

Figure G.6 The Thévenin analysis: (a) picking the terminal points; (b) separating the circuit elements; (c) resistance looking in the terminals; (d) equivalents circuit; (e) the complete circuit.

The resistance looking into the terminals when voltage source is equal to zero is that of 4 Ω in series with a parallel arrangement of 5 Ω and 20 Ω (Figure G.6(c)):

$$R_{\text{Th}} = 4 + \frac{20 \times 5}{20 + 5} = 8\,\Omega$$

Thus the equivalent Thévenin circuit is as shown in Figure G.6(d) and when network B is connected to it we have the circuit shown in Figure G.6(e). Hence the current through the 10 Ω resistor is $I_{10} = 4/(8 + 10) = 0.22$ A.

G.1.4 Norton's theorem

In a manner similar to Thévenin's theorem, we can have an equivalent circuit for any two-terminal network containing voltage or current sources in terms of an equivalent network of a current source shunted by a resistance. This is known as **Norton's theorem**:

> Any two-terminal network containing voltage or current sources can be replaced by an equivalent network consisting of a current source, equal to the current between the terminals when they are short-circuited, in parallel with the resistance measured between the terminals when there is no load between them and all independent sources in the network are set equal to zero.

If we have a linear circuit we have to divide it into two circuits, A and B, connected at a pair of terminals (Figure G.7). We can then use Norton's theorem to replace, say, circuit A by its equivalent circuit. The short-circuit Norton current for circuit A is that given when circuit B is disconnected and the Norton resistance for A is the resistance looking into the terminals of A with all its independent sources set equal to zero.

1. Identify the two parts A and B of the circuit and separate by terminals.
2. Separate A from B.
3. Replace A by its Norton equivalent.
4. Reconnect circuit B and carry out the analysis.

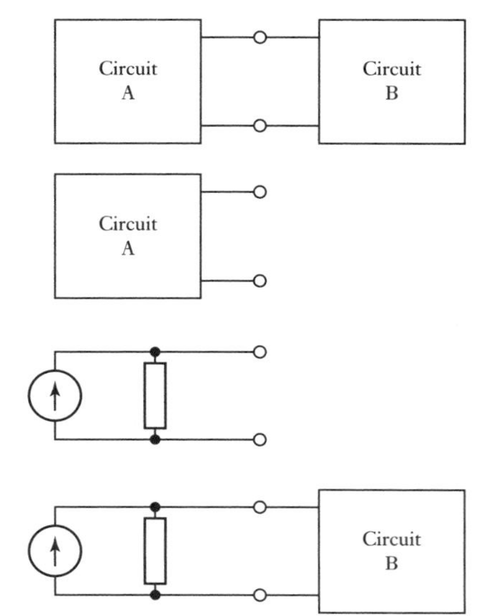

Figure G.7 Step-by-step approach for circuit analysis using Norton's theorem.

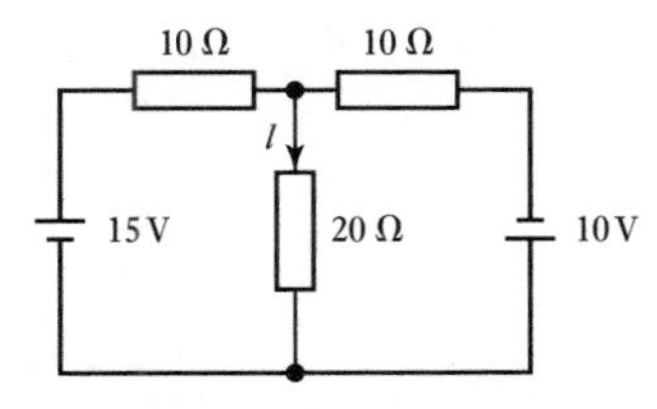

Figure G.8 Circuit for analysis using Norton's theorem.

As an illustration of the use of Norton's theorem, consider the determination of the current I through the 20 Ω resistor in Figure G.8.

We can redraw the circuit in the form shown in Figure G.9(a) as two connected networks A and B with network B selected to be the 20 Ω resistor through which we require the current. We then determine the Norton equivalent circuit for network A (Figure G.9(b)). Short-circuiting the terminals of network A gives the circuit shown in Figure G.9(c). The short circuit current will be the sum, taking into account directions, of the currents from the two branches of the circuits containing the voltage sources, i.e. $I_{sc} = I_1 - I_2$. The current $I_1 = 15/10 = 1.5$ A, since the other branch of the network is short-circuited, and $I_2 = 10/10 = 1.0$ A. Thus $I_{sc} = 0.5$ A. The Norton resistance is given by that across the terminals when all the sources are set to zero (Figure G.9(d)). It is thus:

$$R_N = \frac{10 \times 10}{10 + 10} = 5\ \Omega$$

Thus the Norton equivalent circuit is that shown in Figure G.9(e). Hence when we put this with network B (Figure G.9(f)), we can obtain the current I. The p.d. across the resistors is $0.5 \times R_{total}$ and so the current I is this p.d. divided by 20. Hence:

$$I = 0.5 \times \frac{5}{5 + 20} = 0.1\ \text{A}$$

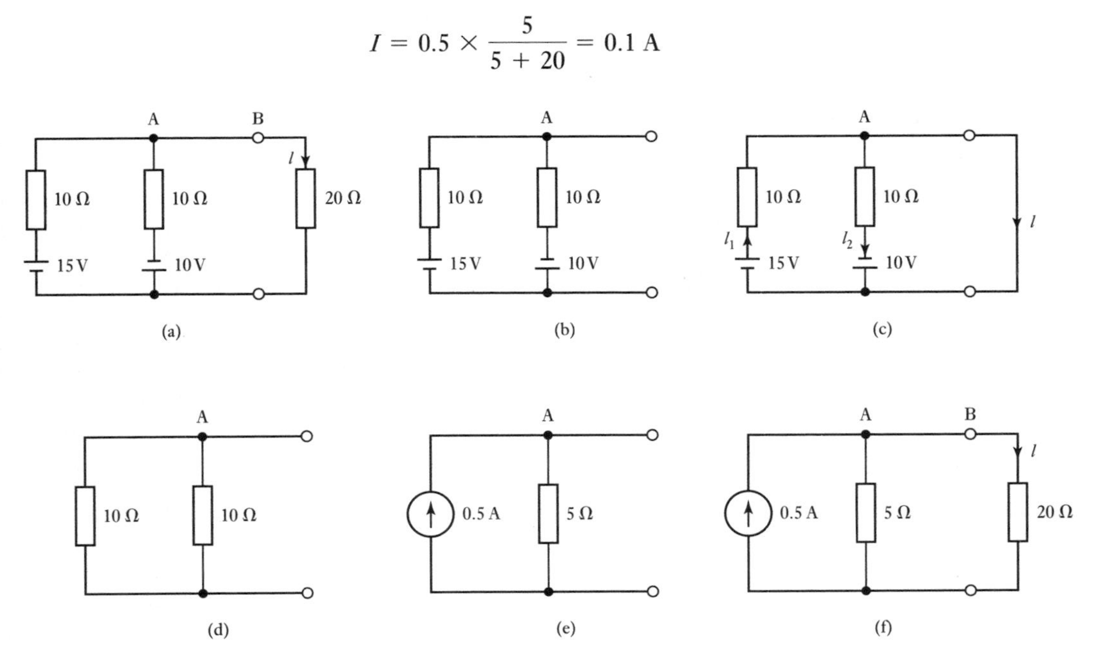

Figure G.9 The Norton analysis: (a) redrawing the circuit; (b) network A; (c) short-circuiting the terminals; (d) sources set to zero; (e) Norton equivalent; (f) the combined parts of the circuits.

G.2 A.C. circuits

We can generate a sinusoidal waveform by rotating a radius line OA at a constant angular velocity ω (Figure G.10(a)), the vertical projection of the line AB varying with time in a sinusoidal manner. The angle θ of the line AB at

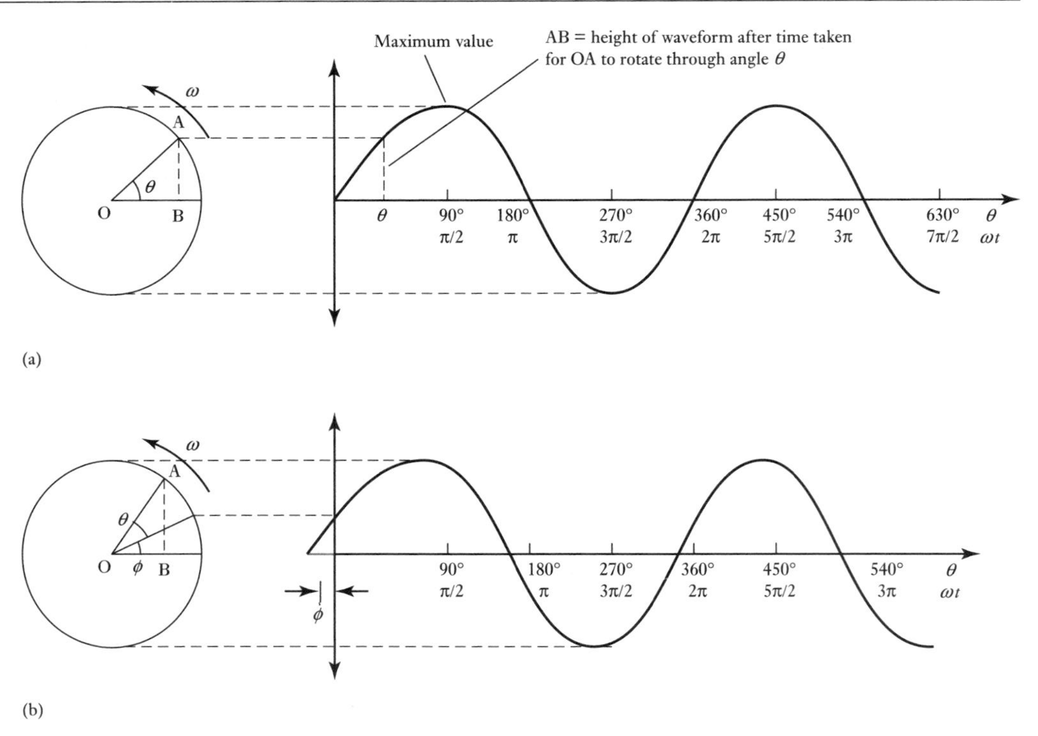

Figure G.10 Generating sinusoidal waveforms: (a) zero value at time $t = 0$, (b) an initial value at $t = 0$.

a time t is ωt. The frequency f of rotation is $1/T$, where T is the time taken for one complete rotation, and so $\omega = 2\pi f$. In Figure G.10(a) the rotating line OA was shown as starting from the horizontal position at time $t = 0$. Figure G.10(b) shows that the line OA at $t = 0$ is already at some angle ϕ. As the line OA rotates with an angular velocity ω, then in a time t the angle swept out is ωt and thus at time t the angle with respect to the horizontal is $\omega t + \phi$. Sinusoidal alternating currents and voltages can be described by such rotating lines and hence by the equations $i = I_m \sin \omega t$ and $v = V_m \sin \omega t$ for currents and voltages with zero values at time $t = 0$ and for those starting at some initial angle ϕ by $i = I_m \sin(\omega t + \phi)$ and $v = V_m \sin(\omega t + \phi)$. Lowercase symbols are used for the current and voltage terms that change with time, uppercase being reserved for the non-varying terms.

With alternating current circuits there is a need to consider the relationship between an alternating current through a component and the alternating voltage across it. If we take the alternating current as the reference for a series circuit and consider it to be represented by $i = I_m \sin \omega t$, then the voltage may be represented by $v = V_m \sin(\omega t + \phi)$. There is said to be a **phase difference** of ϕ between the current and the voltage. If ϕ has a positive value then the voltage is said to be **leading** the current (as in Figure G.10 if (a) represents the current and (b) the voltage); if it has a negative value then it is said to be **lagging** the current.

We can describe a sinusoidal alternating current by just specifying the rotating line in terms of its length and its initial angle relative to a horizontal reference line. The term **phasor**, being an abbreviation of the term phase vector, is used for such lines. The length of the phasor can represent the maximum value of the sinusoidal waveform or the root-mean-square (r.m.s.) value, since the maximum value is proportional to the r.m.s value. Because currents and voltages in the same circuit will have the same frequency and thus the phasors used to represent them will rotate with the same angular velocity and maintain the same phase angles between them at all times, we do not need to bother about drawing the effects of their rotation but can just draw phasor diagrams giving the relative angular positions of the phasors and ignore their rotations.

The following summarises the main points about phasors:

1 A phasor has a length that is directly proportional to the maximum value of the sinusoidally alternating quantity or, because the maximum value is proportional to the r.m.s. value, a length proportional to the r.m.s. value.
2 Phasors are taken to rotate anti-clockwise and have an arrow-head at the end which rotates.
3 The angle between two phasors shows the phase angle between their waveforms. The phasor which is at a larger anti-clockwise angle is said to be leading, the one at the lesser anti-clockwise angle lagging.
4 The horizontal line is taken as the reference axis and one of the phasors is given that direction; the others have their phase angles given relative to this reference axis.

G.2.1 Resistance, inductance and capacitance in a.c. circuits

Consider a sinusoidal current $i = I_m \sin \omega t$ passing through a **pure resistance**. A pure resistance is one that has only resistance and no inductance or capacitance. Since we can assume Ohm's law to apply, then the voltage v across the resistance must be $v = Ri$ and so $v = RI_m \sin \omega t$. The current and the voltage are thus in phase. The maximum voltage will be when $\sin \omega t = 1$ and so $V_m = RI_m$.

Consider a sinusoidal current $i = I_m \sin \omega t$ passing through a **pure inductance**. A pure inductance is one which has only inductance and no resistance or capacitance. With an inductance, a changing current produces a back e.m.f. $L\, di/dt$, where L is the inductance. The applied e.m.f. must overcome this back e.m.f. for a current to flow. Thus the voltage v across the inductance is $L\, di/dt$ and so

$$v = L\frac{di}{dt} = L\frac{d}{dt}(I_m \sin \omega t) = \omega L I_m \cos \omega t$$

Since $\cos \omega t = \sin(\omega t + 90°)$, the current and the voltage are out of phase with the voltage leading the current by 90°. The maximum voltage is when $\cos \omega t = 1$ and so we have $V_m = \omega L I_m$. V_m/I_m is called the **inductive reactance** X_L. Thus $X_L = V_m/I_m = \omega L$. Since $\omega = 2\pi f$ then $X_L = 2\pi f L$ and so the reactance is proportional to the frequency f. The higher the frequency the greater the opposition to the current.

Consider a circuit having just **pure capacitance** with a sinusoidal voltage $v = V_m \sin \omega t$ being applied across it. A pure capacitance is one which has only capacitance and no resistance or inductance. The charge q on the plates of a capacitor is related to the voltage v by $q = Cv$. Thus, since current is the rate of movement of charge dq/dt, we have i = rate of change of q = rate of change of $(Cv) = C \times$ (rate of change of v), i.e. $i = C\,\mathrm{d}v/\mathrm{d}t$. Thus

$$i = \frac{\mathrm{d}q}{\mathrm{d}t} = \frac{\mathrm{d}}{\mathrm{d}t}(Cv) = C\frac{\mathrm{d}}{\mathrm{d}t}(V_m \sin \omega t) = \omega C V_m \cos \omega t$$

Since $\cos \omega t = \sin(\omega t + 90°)$, the current and the voltage are out of phase, the current leading the voltage by 90°. The maximum current occurs when $\cos \omega t = 1$ and so $I_m = \omega C V_m$. V_m/I_m is called the **capacitive reactance** X_C. Thus $X_C = V_m/I_m = 1/\omega C$. The reactance has the unit of ohms and is a measure of the opposition to the current. The bigger the reactance the greater the voltage has to be to drive the current through it. Since $\omega = 2\pi f$, the reactance is inversely proportional to the frequency f and so the higher the frequency the smaller the opposition to the current. With d.c., i.e. zero frequency, the reactance is infinite and so no current flows.

In summary, Figure G.11 shows the voltage and current phasors for (a) pure resistance, (b) pure inductance, (c) pure capacitance.

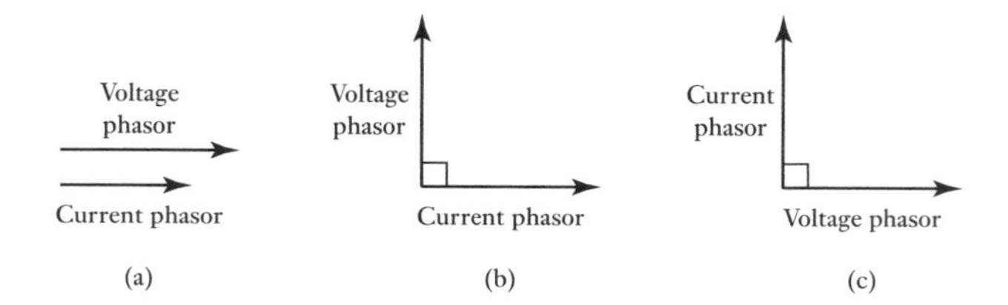

Figure G.11 Phasors with (a) pure resistance, (b) pure inductance, (c) pure capacitance.

G.2.2 Series a.c. circuits

For a series circuit, the total voltage is the sum of the p.d.s across the series components, though the p.d.s may differ in phase. This means that if we consider the phasors, they will rotate with the same angular velocity but may have different lengths and start with a phase angle between them. We can obtain the sum of two series voltages by using the **parallelogram law** of vectors to add the two phasors:

> If two phasors are represented in size and direction by adjacent sides of a parallelogram, then the diagonal of that parallelogram is the sum of the two (Figure G.12).

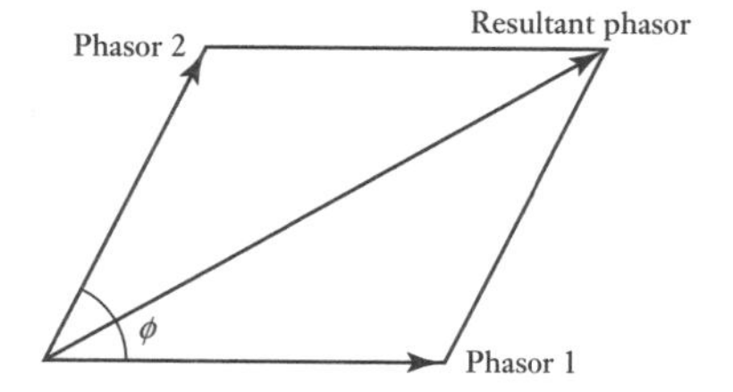

Figure G.12 Adding phasors 1 and 2 which have a phase angle ϕ between them.

If the phase angle between the two phasors of sizes V_1 and V_2 is 90°, then the resultant can be calculated by the use of Pythagoras theorem as having a size V given by $V^2 = V_1^2 + V_2^2$ and a phase angle ϕ relative to the phasor for V_1 given by $\tan \phi = V_1/V_2$.

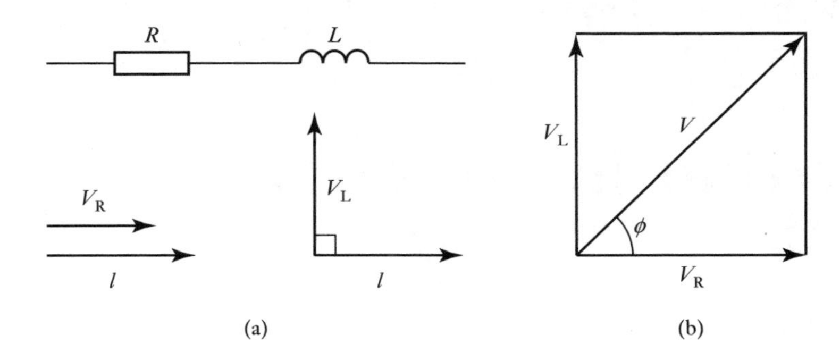

Figure G.13 RL series circuit.

As an illustration of the use of the above, consider an alternating current circuit having resistance in series with inductance (Figure G.13(a)). For such a circuit, the voltage for the resistance is in phase with the current and the voltage for the inductor leads the current by 90°. Thus the phasor for the sum of the voltage drops across the two series components is given by Figure G.13(b) as a voltage phasor with a phase angle ϕ. We can use Pythagoras theorem to give the magnitude V of the voltage, i.e. $V^2 = V_R^2 + V_L^2$, and trigonometry to give the phase angle ϕ, i.e. the angle by which the voltage leads the current as $\tan \phi = V_L/V_R$ or $\cos \phi = V_R/V$.

Since $V_R = IR$ and $V_L = IX_L$ then $V^2 = (IR)^2 + (IX_L)^2 = I^2(R^2 + X_L^2)$. The term **impedance** Z is used for the opposition of a circuit to the flow of current, being defined as $Z = V/I$ with the unit of ohms. Thus, for the resistance and inductance in series, the circuit impedance is given by

$$Z = \sqrt{R^2 + X_L^2} = \sqrt{R^2 + (\omega L)^2}$$

Further information

The following is a short list of texts which can usefully provide further information of relevance to a study of mechatronics.

Sensors and signal conditioning

Bolton, W., *Newnes Instrumentation and Measurement*, Newnes 1991, 1996, 2000
Boyes W., *Instrumentation Reference Book*, Newnes 2002
Clayton, G. B. and Winder, S., *Operational Amplifiers*, Newnes 2003
Figliola, R.S., and Beasley, D. E., *Theory and Design for Mechanical Measurements*, John Wiley 2000, 2005, 2011
Fraden, J., *Handbook of Modern Sensors*, Springer 2001, 2004, 2010
Gray, P. R., Hurst P. J., Lewis S.H. and Meyer, R. G., *Analysis and Design of Analog Integrated Circuits*, Wiley 2009
Holdsworth, B., *Digital Logic Design*, Newnes 2000
Johnson, G. W. and Jennings R., *LabVIEW Graphical Programming*, McGraw-Hill 2006
Morris, A.S., *Measurement and Instrumentation Principles*, 3rd edition, Newnes 2001
Park J. and Mackay, S., *Practical Data Acquisition for Instrumentation and Control Systems*, Elsevier 2003
Travis, J. and Kring J., *LabVIEW for Everyone*, Prentice-Hall 2006

Actuation

Bolton, W., *Mechanical Science*, Blackwell Scientific Publications 1993, 1998, 2006
Cathey, J. J., *Electric Machines: Analysis and Design Applying MATLAB*, McGraw-Hill 2001
Gottlieb, I. M., *Electric Motors and Control Techniques*, TAB Books, McGraw-Hill 1994
Kenjo, T. and Sugawara, A., *Stepping Motors and their Microprocessor Controls*, Clarenden Press 1995
Manring, N., *Hydraulic Control Systems*, Wiley 2005
Norton, R. L., *Design of Machinery*, McGraw-Hill 2003
Pinches, M. J. and Callear, B. J., *Power Pneumatics*, Prentice-Hall 1996
Wildi, T., *Electrical Machines, Drives and Power Systems*, Pearson 2005

System models

Åstrom, K. J. and Wittenmark, B., *Adaptive Control*, Dover 1994

Attaway S., *Matlab: A Practical Introduction to Programming and Problem Solving*, Butterworth-Heinemann 2009

Bennett, A., *Real-time Computer Control*, Prentice-Hall 1993

Bolton, W., *Laplace and z-Transforms*, Longman 1994

Bolton, W., *Control Engineering*, Longman 1992, 1998

Bolton, W., *Control Systems*, Newnes 2002

D'Azzo J. J., Houpis C. H. and Sheldon N., *Linear Control System Analysis and Design with Matlab*, CRC Press 2003

Dorf, R. C. and Bishop, H., *Modern Control Systems*, Pearson 2007

Fox. H. and Bolton, W., *Mathematics for Engineers and Technologists*, Butterworth-Heinemann 2002

Close C. M., Frederick, C. and Newell J. C. *Modelling and Analysis of Dynamic Systems*, Wiley 2001

Pärt-Enander, E., Sjöberg, A., Melin, B. and Isaksson, P., *The MATLAB Handbook*, Addison-Wesley 1996

Microprocessor systems

Barnett, R. H., *The 8051 Family of Microcontrollers*, Prentice-Hall 1994

Bates, M., *PIC Microcontrollers*, Newnes 2000, 2004

Bolton, W., *Microprocessor Systems*, Longman 2000

Bolton, W., *Programmable Logic Controllers*, Newnes 1996, 2003, 2006, 2009

Cady, F. M., *Software and Hardware Engineering: Motorola M68HC11*, OUP 2000

Calcutt D., Cowan F. and Parchizadeh H., *8051 Microcontrollers: An Application Based Introduction*, Newnes 2004

Ibrahim, D., *PIC Basic: Programming and Projects*, Newnes 2001

Johnsonbaugh, R. and Kalinn, M., *C for Scientists and Engineers*, Prentice Hall 1996

Lewis, R. W., *Programming Industrial Control Systems Using IEC 1131-3*, The Institution of Electrical Engineers 1998

Morton, J., *PIC: Your Personal Introductory Course*, Newnes 2001, 2005

Parr, E. A., *Programmable Controllers*, Newnes 1993, 1999, 2003

Pont, M. J., *Embedded C*, Addison-Wesley 2002

Predko, M., *Programming and Customizing the PIC Microcontroller*, Tab Electronics 2007

Rohner, P., *Automation with Programmable Logic Controllers*, Macmillan 1996

Spasov, P., *Microcontroller Technology: The 68HC11*, Prentice-Hall 1992, 1996, 2001

Vahid, F. and Givargis, T., *Embedded System Design*, Wiley 2002

Van Sickle, T., *Programming Microcontrollers in C*, Newnes 2001

Yeralan, S. and Ahluwalia A., *Programming and Interfacing the 8051 Microcontroller*, Addison-Wesley 1995

Zurrell, K., *C Programming for Embedded Systems*, Kindle Edition 2000

Answers

The following are answers to numerical problems and brief clues as to possible answers with descriptive problems.

Chapter 1

1.1 (a) Sensor, mercury; signal conditioner, fine bore stem; display, marks on the stem; (b) sensor, curved tube; signal conditioner, gears; display, pointer moving across a scale

1.2 See text

1.3 Comparison/controller, thermostat; correction, perhaps a relay; process, heat; variable, temperature; measurement, a temperature-sensitive device, perhaps a bimetallic strip

1.4 See Figure P.1

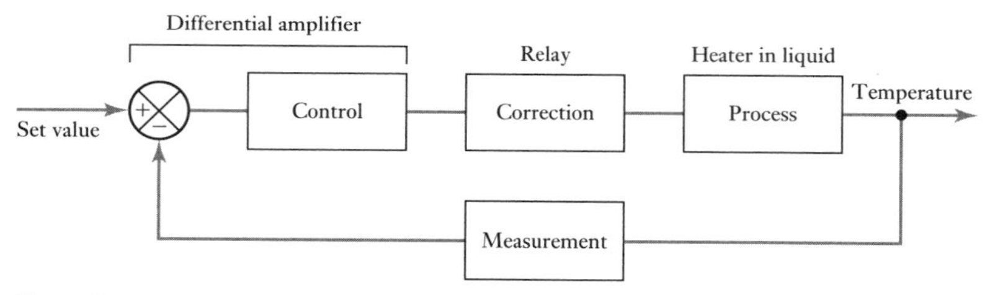

Figure P.1 Problem 1.4.

1.5 See text

1.6 See text

1.7 For example: water in, rinse, water out, water in, heat water, rinse, water out, water in, rinse, water out

1.8 Traditional: bulky, limited functions, requires rewinding. Mechatronics: compact, many functions, no rewinding, cheaper

1.9 Bimetallic element: slow, limited accuracy, simple functions, cheap. Mechatronics: fast, accurate, many functions, getting cheaper

Chapter 2

2.1 See the text for explanation of the terms

2.2 −3.9%

2.3	67.5 s
2.4	0.73%
2.5	0.105 Ω
2.6	Incremental, angle from some datum, not absolute; absolute, unique identification of an angle
2.7	162
2.8	(a)±1.2° (b) 3.3 mV
2.9	See text
2.10	2.8 kPa
2.11	19.6 kPa
2.12	−0.89%
2.13	+1.54°C
2.14	Yes
2.15	−9.81 N, −19.62 N, e.g. strain gauges
2.16	For example, orifice plate with differential pressure cell
2.17	For example, differential pressure cell
2.18	For example, LVDT displacement sensor

Chapter 3

3.1	As Figure 3.2 with $R_2/R_1 = 50$, e.g. $R_1 = 1$ kΩ, $R_2 = 50$ kΩ
3.2	200 kΩ
3.3	Figure 3.5 with two inputs, e.g. $V_A = 1$ V, $V_B = 0$ to 100 mV, $R_A = R_2 = 40$ kΩ, $R_B = 1$ kΩ
3.4	Figure 3.11 with $R_1 = 1$ kΩ and $R_2 = 2.32$ kΩ
3.5	$V = K\sqrt{I}$
3.6	100 kΩ
3.7	80 dB
3.8	Fuse to safeguard against high current, limiting resistor to reduce currents, diode to rectify a.c., Zener diode circuit for voltage and polarity protection, low-pass filter to remove noise and interference, optoisolator to isolate the high voltages from the microprocessor
3.9	0.059 V
3.10	5.25×10^{-5} V
3.11	As given in the problem

Chapter 4

4.1	24.4 mV
4.2	9
4.3	0.625 V
4.4	1, 2, 4, 8
4.5	12 μs
4.6	See text
4.7	Buffer, digital-to-analogue converter, protection
4.8	0.33 V, 0.67 V, 1.33 V, 2.67 V
4.9	32 768R
4.10	15.35 ms
4.11	Factor of 315

Chapter 5

5.1 For example: (a) ticket selected AND correct money in, correct money decided by OR gates analysis among possibilities, (b) AND with safety guards, lubricant, coolant, workpiece, power, etc., all operating or in place, (c) Figure P.2, (d) AND
5.2 (a) Q, (b) P
5.3 AND
5.4 A as 1, B as 0
5.5 See Figure P.3
5.6 See Figure P.4
5.7 As in the text, Section 5.3.1, for cross-coupled NOR gates

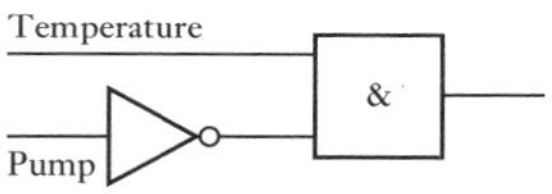

Temperature
&
Pump

Figure P.2 Problem 5.1(c).

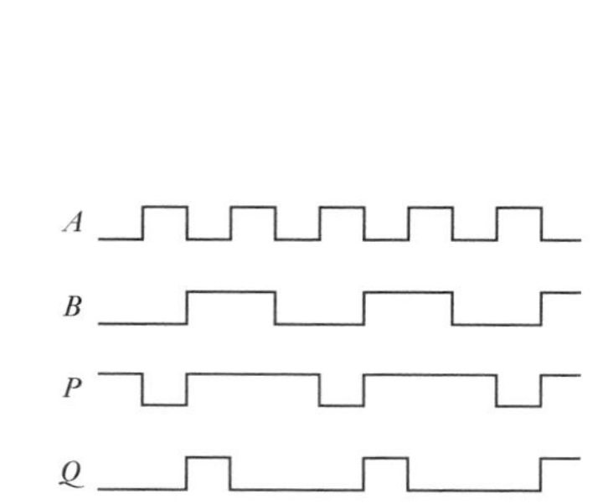

Figure P.3 Problem 5.5.

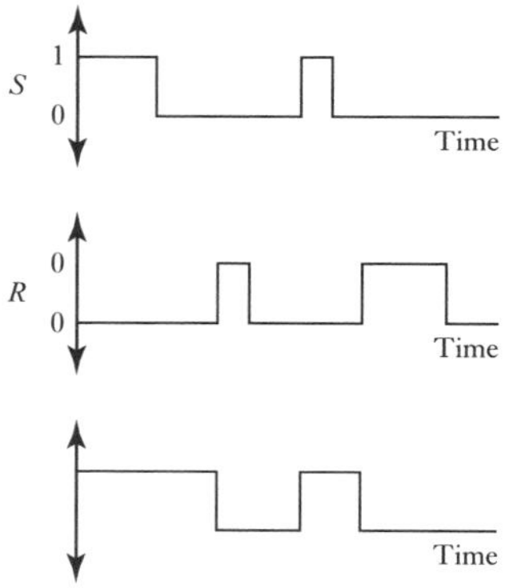

S 1 0 Time
R 0 0 Time
Time

Figure P.4 Problem 5.6.

Chapter 6

6.1 See text
6.2 See Section 6.1
6.3 For example: (a) a recorder, (b) a moving-coil meter, (c) a hard disk or CD, (d) a storage oscilloscope or a hard disk or CD
6.4 Could be four-active-arm bridge, differential operational amplifier, display of a voltmeter. The values of components will depend on the thickness chosen for the steel and the diameter of a load cell. You might choose to mount the tank on three cells
6.5 Could be as in Figure 3.8 with cold junction compensation by a bridge (see Section 3.5.2). Linearity might be achieved by suitable choice of thermocouple materials
6.6 Could be thermistors with a sample and hold element followed by an analogue-to-digital converter for each sensor. This would give a digital signal for transmission, so reducing the effects of possible interference. Optoisolators could be used to isolate high voltages/currents, followed by a multiplexer feeding digital meters
6.7 This is based on Archimedes' principle: the upthrust on the float equals the weight of fluid displaced
6.8 Could use an LVDT or strain gauges with a Wheatstone bridge
6.9 For example: (a) Bourdon gauge, (b) thermistors, galvanometric chart recorder, (c) strain-gauged load cells, Wheatstone bridge, differential amplifier, digital voltmeter, (d) tachogenerator, signal conditioning to shape pulses, counter

Chapter 7

7.1 See Section 7.3
7.2 See Section 7.3.2
7.3 See Section 7.4
7.4 See (a) Figure 7.14, (b) Figure 7.8(b), (c) Figure 7.10, (d) Figure 7.13
7.5 A+, B+, A−, B−
7.6 See Figure P.5
7.7 $0.0057\ m^2$
7.8 124 mm
7.9 1.27 MPa, $3.9 \times 10^{-5}\ m^3/s$
7.10 (a) $0.05\ m^3/s$, (b) $0.10\ m^3/s$
7.11 (a) $0.42\ m^3/s$, (b) $0.89\ m^3/s$
7.12 960 mm

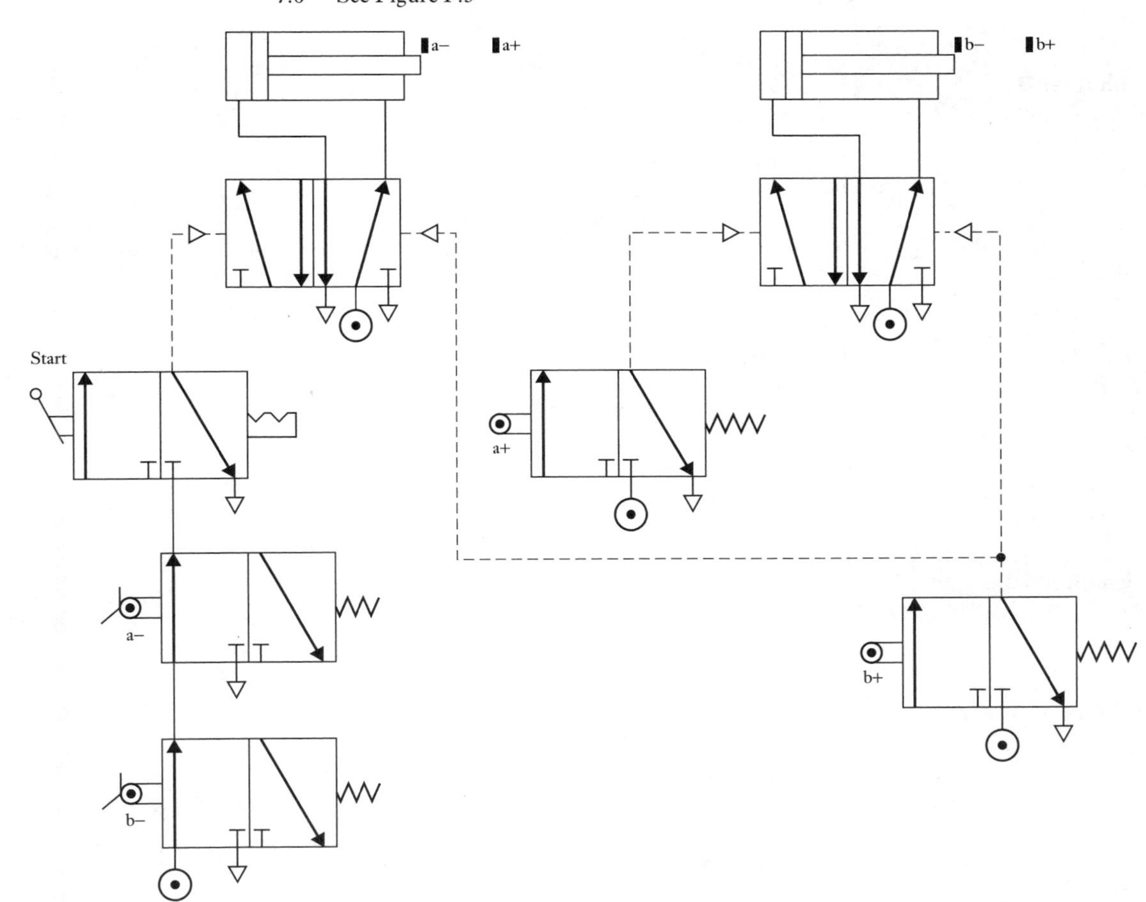

Figure P.5 Problem 7.6.

Chapter 8

8.1 (a) A system of elements arranged to transmit motion from one form to another form (b) A sequence of joints and links to provide a controlled output in response to a supplied input motion
8.2 See Section 8.3.1
8.3 (a) 1, (b) 2, (c) 1, (d) 1, (e) 3

8.4 (a) Pure translation, (b) pure translation, (c) pure rotation, (d) pure rotation, (e) translation plus rotation
8.5 Quick-return
8.6 Sudden drop in displacement followed by a gradual rise back up again
8.7 60 mm
8.8 Heart-shaped with distance from axis of rotation to top of heart 40 mm and to base 100 mm. See Figure 8.14a
8.9 For example: (a) cams on a shaft, (b) quick-return mechanism, (c) eccentric cam, (d) rack-and-pinion, (e) belt drive, (f) bevel gears
8.10 1/24

Chapter 9

9.1 It acts as a flip-flop
9.2 See text and Figure 9.7
9.3 (a) Series wound, (b) shunt wound
9.4 (a) D.C. shunt wound, (b) induction or synchronous motor with an inverter, (c) d.c., (d) a.c.
9.5 See Section 9.5.4
9.6 See Section 9.7
9.7 480 pulses/s
9.8 9°
9.9 (a) 4 kW, (b) 800 W, (c) 31.8 N m
9.10 0.65 N m
9.11 2
9.12 3.6 N m

Chapter 10

10.1 (a) $m\frac{d^2x}{dt^2} + c\frac{dx}{dt} = F$, (b) $m\frac{d^2x}{dt^2} + c\frac{dx}{dt} + (k_1 + k_2)x = F$

10.2 As in Figure 10.3(a)

10.3 $c\frac{d\theta_i}{dt} = c\frac{d\theta_o}{dt} + k\theta_o$

10.4 Two torsional springs in series with a moment of inertia block,

$$T = I\frac{d^2\theta}{dt^2} + k_1(\theta_1 - \theta_2) = m\frac{d^2\theta}{dt^2} + \frac{k_1 k_2}{k_1 + k_2}\theta_1$$

10.5 $v = v_R + \frac{1}{RC}\int v_R\,dt$

10.6 $v = \frac{L}{R}\frac{dv_R}{dt} + \frac{1}{CR}\int v_R\,dt + v_R$

10.7 $v = R_1 C\frac{dv_C}{dt} + \left(\frac{R_1}{R_2} + 1\right)v_C$

10.8 $RA_2 \frac{dh_2}{dt} + h_2\rho g = h_1$

10.9 $RC \frac{dT}{dt} + T = T_r$. Charged capacitor discharging through a resistor

10.10 $RC \frac{dT_1}{dt} = Rq - 2T_1 + T_2 + T_3$, $RC \frac{dT_2}{dt} = T_1 - 2T_2 + T_3$

10.11 $pA = m \frac{d^2x}{dt^2} + R \frac{dx}{dt} + \frac{1}{C}x$, $R =$ resistance to stem movement, $c =$ capacitance of spring

10.12 $T = \left(\frac{I_1}{n} + n\right)\frac{d^2\theta}{dt^2} + \left(\frac{c_1}{n} + nc_2\right)\frac{d\theta}{dt} + \left(\frac{k_1}{n} + nk_2\right)\theta$

Chapter 11

11.1 $\frac{IR}{k_1k_2} \frac{d\omega}{dt} + \omega = \frac{1}{k_2}v$

11.2 $(L_a + L_L)\frac{di_a}{dt} + (R_a + R_L)i_a - k_1 \frac{d\theta}{dt} = 0$, $I \frac{d^2\theta}{dt^2} + B \frac{d\theta}{dt} + k_2 i_a = T$

11.3 Same as armature-controlled motor

Chapter 12

12.1 $4 \frac{dx}{dt} + x = 6y$

12.2 (a) 59.9°C, (b) 71.9°C

12.3 (a) $i = \frac{V}{R}(1 - e^{-Rt/L})$, (b) L/R, (c) V/R

12.4 (a) Continuous oscillations, (b) under-damped, (c) critically damped, (d) over-damped

12.5 (a) 4 Hz, (b) 1.25, (c) $i = I(\frac{1}{3}e^{-8t} - \frac{4}{3}e^{-2t} + 1)$

12.6 (a) 5 Hz, (b) 1.0, (c) $x = (-32 + 6t)e^{-5t} + 6$

12.7 (a) 9.5%, (b) 0.020 s

12.8 (a) 4 Hz, (b) 0.625, (c) 1.45 Hz, (d) 0.5 s, (e) 8.1%, (f) 1.4 s

12.9 (a) 0.59, (b) 0.87

12.10 2.4

12.11 0.09

12.12 3.93 rad/s, 0.63 Hz

Chapter 13

13.1 (a) $\frac{1}{As + \rho g/R}$, (b) $\frac{1}{ms^2 + cs + k}$, (c) $\frac{1}{LCs^2 + RCs + 1}$

13.2 (a) 3 s, (b) 0.67 s

13.3 (a) $1 + e^{-2t}$, (b) $2 + 2\,e^{-5t}$

13.4 (a) Over-damped, (b) under-damped, (c) critically damped, (d) under-damped

13.5 $t\,e^{-3t}$

13.6 $2e^{-4t} - 2e^{-3t}$

13.7 (a) $\frac{4s}{s^2(s + 1) + 4}$, (b) $\frac{2(s + 2)}{(s + 1)(s + 2) + 2}$,

(c) $\frac{4}{(s + 2)(s + 3) + 20}$, (d) $\frac{2}{s(s + 2) + 20}$

13.8 $5/(s + 53)$

13.9 $5s/(s^2 + s + 10)$

13.10 $2/(3s + 1)$

13.11 $-1, -2$

13.12 (a) Stable, (b) unstable, (c) unstable, (d) stable, (e) unstable

Chapter 14

14.1 (a) $\frac{5}{\sqrt{\omega^2 + 4}}, \frac{\omega}{2}$, (b) $\frac{2}{\sqrt{\omega^4 + \omega^2}}, \frac{1}{\omega}$,

(c) $\frac{1}{\sqrt{4\omega^6 - 3\omega^4 + 3\omega^2 + 1}}, \frac{\omega(3 - 2\omega^2)}{1 - 3\omega^2}$

14.2 $0.56 \sin(5t - 38°)$

14.3 $1.18 \sin(2t + 25°)$

14.4 (a) (i) ∞, 90°, (ii) 0.44, 450°, (iii) 0.12, 26.6°, (iv) 0, 0°,

(b) (i) 1, 0°, (ii) 0.32, −71.6°, (iii) 0.16, −80.5°, (iv) 0, −90°.

14.5 See Figure P.6

14.6 (a) $1/s$, (b) $3.2/(1 + s)$, (c) $2.0/(s^2 + 2\zeta s + 1)$,

(d) $3.2/[(1 + s)(0.01s^2 + 0.2\zeta s + 1)]$

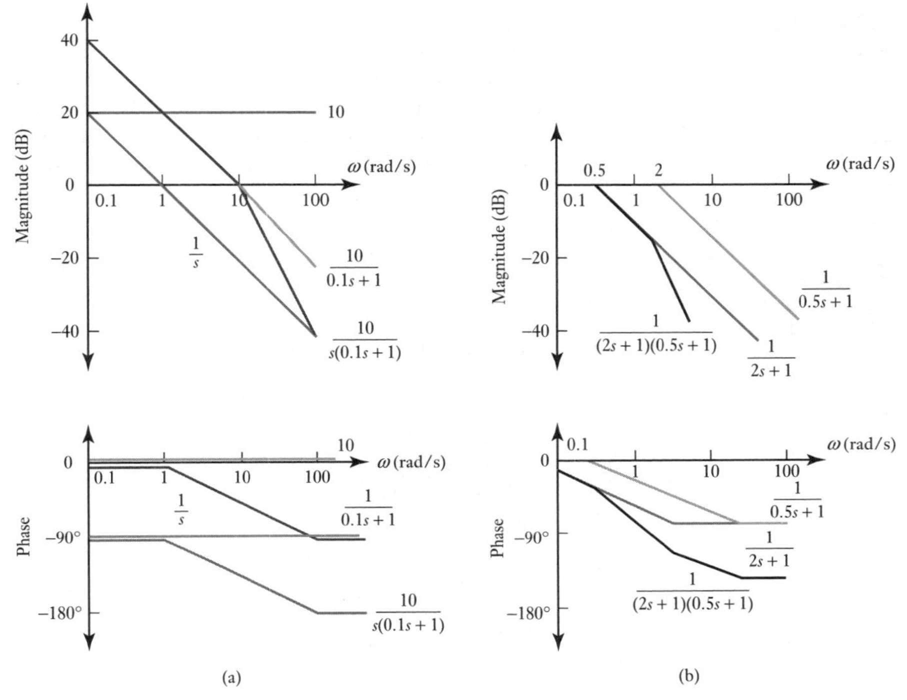

Figure P.6 Problem 14.5.

Chapter 15

15.1 See Section 15.3
15.2 (a) 8 minutes, (b) 20 minutes
15.3 (a) 12 s, (b) 24 s
15.4 5
15.5 See the text. In particular P offset, PI and PID no offset
15.6 3, 666 s, 100 s
15.7 3, 100 s, 25 s
15.8 See Sections (a) 15.12.1, (b) 15.12.2, (c) 15.12.3
15.9 1.6
15.10 First-order response with time constant c/K_P

Chapter 16

16.1 For example, try diameter and degree of redness. You might also consider weight. Your results need to be able to distinguish clearly between denominations of coins, whatever their condition
16.2 (a) 1/6, (b) 1/36, (c) 1/10
16.3 0.99
16.4 0.002
16.5 0.625

16.6 0.761

16.7 For example, if room temperature < 20°C and timer ON, then boiler ON; if boiler ON, then pump ON; if pump ON and room temperature < 20°C, then valve ON; if timer NOT ON, then boiler NOT ON; if room temperature NOT < 20°C, then valve NOT ON; if boiler NOT ON, then pump NOT ON. You might also refine this by considering there to be a restriction in that the boiler is restricted to operating below 60°C

Chapter 17

17.1 See Section 17.2

17.2 256

17.3 64K × 8

17.4 See Section 17.3

17.5 See Figure 17.9 and associated text

17.6 (a) E, (b) C, (c) D, (d) B

17.7 256

17.8 (a) 0, (b) 1

17.9 See Section 17.3.1, item 6

17.10 See Section 17.3.2, item 5

17.11 High to reset pin

17.12 (a)
```
IF A
      THEN
        BEGIN B
        END B
  ELSE
        BEGIN C
        END C
  ENDIF A
```
(b)
```
WHILE A
      BEGIN B
        END B
  ENDWHILE A
```

Chapter 18

18.1 (a) 89, (b) 99

18.2 No address has to be specified since the address is implied by the mnemonic

18.3 (a) CLRA, (b) STAA, (c) LDAA, (d) CBA, (e) LDX

18.4 (a) LDAA $20, (b) DECA, (c) CLR $0020, (d) ADDA $0020

18.5 (a) Store accumulator B value at address 0035, (b) load accumulator A with data F2, (c) clear the carry flag, (d) add 1 to value in accumulator A, (e) compare C5 to value in accumulator A, (f) clear address 2000, (g) jump to address given by index register plus 05

18.6 (a)

```
DATA1   EQU    $0050
DATA2   EQU    $0060
DIFF    EQU    $0070
        ORG    $0010
        LDAA   DATA1    ; Get minuend
        SUBA   DATA2    ; Subtract subtrahend
        STAA   DIFF     ; Store difference
        SWI             ; Program end
```

(b)

```
MULT1   EQU    $0020
MULT2   EQU    $0021
PROD    EQU    $0022
        ORG    $0010
        CLR    PROD     ; Clear product address
        LDAB   MULT1    ; Get first number
SUM     LDAA   MULT2    ; Get multiplicand
        ADDA   PROD     ; Add multiplicand
        STAA   PROD     ; Store result
        DECB            ; Decrement acc. B
        BNE    SUM      ; Branch if adding not complete
        WAI             ; Program end
```

(c)

```
FIRST   EQU    $0020
        ORG    $0000
        CLRA            ; Clear accumulator
        LDX    #0
MORE    STAA   $20,X
        INX             ; Increment index reg.
        INCA            ; Increment accumulator
        CMPA   #$10     ; Compare with number 10
        BNE    MORE     ; Branch if not zero
        WAI             ; Program end
```

(d)

```
        ORG    $0100
        LDX    #$2000   ; Set pointer
LOOP    LDA A  $00,X    ; Load data
        STA A  $50,X    ; Store data
        INX             ; Increment index register
        CPX    $3000    ; Compare
        BNE    LOOP     ; Branch
        SWI             ; Program end
```

18.7

```
YY      EQU    $??      ; Value chosen to give required time delay
SAVEX   EQU    $0100
        ORG    $0010
        STA    SAVEX    ; Save accumulator A
        LDAA   YY       ; Load accumulator A
LOOP    DECA            ; Decrement acc. A
        BNE    LOOP     ; Branch if not zero
        LDA    SAVEX    ; Restore accumulator
        RTS             ; Return to calling program
```

18.8

```
LDA      $2000    ; Read input data
AND A    #$01     ; Mask off all bits but bit 0
BEQ      $03      ; If switch low, branch over JMP which is 3
                  ; program lines
JMP      $3000    ; If switch high no branch and so execute JMP
Continue
```

Chapter 19

19.1 (a) The variable counter is an integer, (b) the variable num is assigned the value 10, (c) the word name will be displayed, (d) the display is Number 12, (e) include the file stdio.h

19.2 (a) Calls up the library necessary for the printf( ) function, (b) indicates the beginning and end of a group of statements, (c) starts a new line, (d) problem 3

19.3 The number is 12

19.4 # include <stdio.h>

```
int main(void);
{
    int len, width;
    printf("Enter length: ");
    scanf("%d", &len);
    printf("Enter width: ");
    scanf("%d", &width);
    printf("Area is %d", lens * width);
    return 0;
{
```

19.5 Similar to program given in Section 19.3, item 4

19.6 Divides first number by second number unless second is 0

Chapter 20

20.1 See Section 20.3

20.2 See Section 20.3. A parallel interface has the same number of input/output lines as the microprocessor. A serial interface has just a single input/output line

20.3 See Section 20.2

20.4 See Section 20.4

20.5 See Section 20.4 and Figure 20.10

20.6 See Section 20.4.1

20.7 See Section 20.3.3. Polling involves the interrogation of all peripherals at frequent intervals, even when some are not activated. It is thus wasteful of time. Interrupts are only initiated when a peripheral requests it and so is more efficient

20.8 CRA 00110100, CRB 00101111

20.9 As the program in 18.4.2 with LDAA #$05 replaced by LDAA #$34 and LDAA #$34 replaced by LDAA #$2F

20.10 As the program in Section 20.4.2 followed by
READ LDAA $2000 ; Read port A
Perhaps after some delay program there may then be
BRA READ

Chapter 21

21.1 (a) AND, (b) OR
21.2 (a) Figure 21.9(b), (b) Figure 21.10(b), (c) a latch circuit, Figure 21.16, with Input 1 the start and Input 2 the stop switches
21.3 0 LD X400, 1 LD Y430, 2 ORB, 3 ANI X401, 4 OUT Y430
21.4 0 LD X400, 1 OR Y430, 3 OUT Y430, 4 OUT T450, 5 K 50; delay-on timer
21.5 0 LD X400, 1 OR Y430, 2 ANI M100, 3 OUT Y430, 4 LD X401, 5 OUT M100; reset latch
21.6 As in Figure 21.28 with Timer 1 having $K = 1$ for 1 s and Timer 2 with $K = 20$ for 20s
21.7 Figure P.7
21.8 Figure P.8

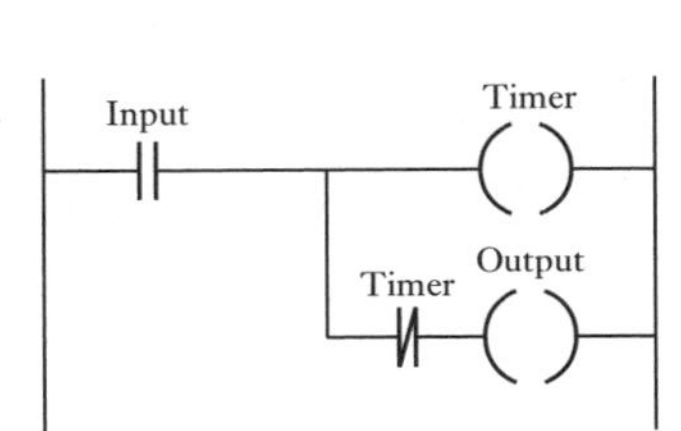

Figure P.7 Problem 21.7.

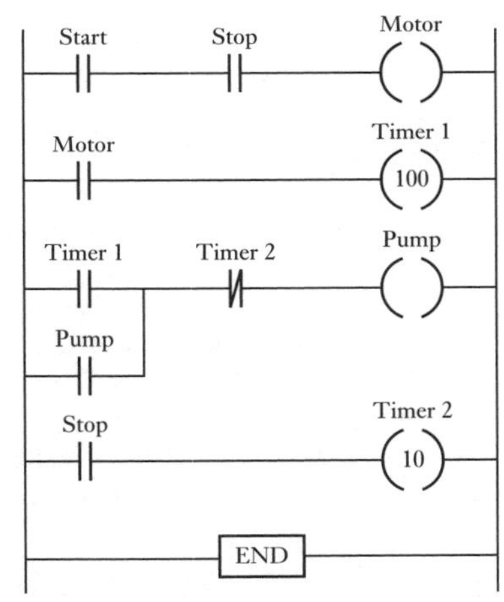

Figure P.8 Problem 21.8.

21.9 Figure P.9
21.10 Figure P.10
21.11 An output would come on, as before, but switch off when the next input occurs
21.12 See Section 21.10
21.13 Two latch circuits, as in Figure P.11
21.14 Figure P.12

Chapter 22

22.1 See Section 22.2
22.2 See Section 22.3
22.3 Bus
22.4 Broadband

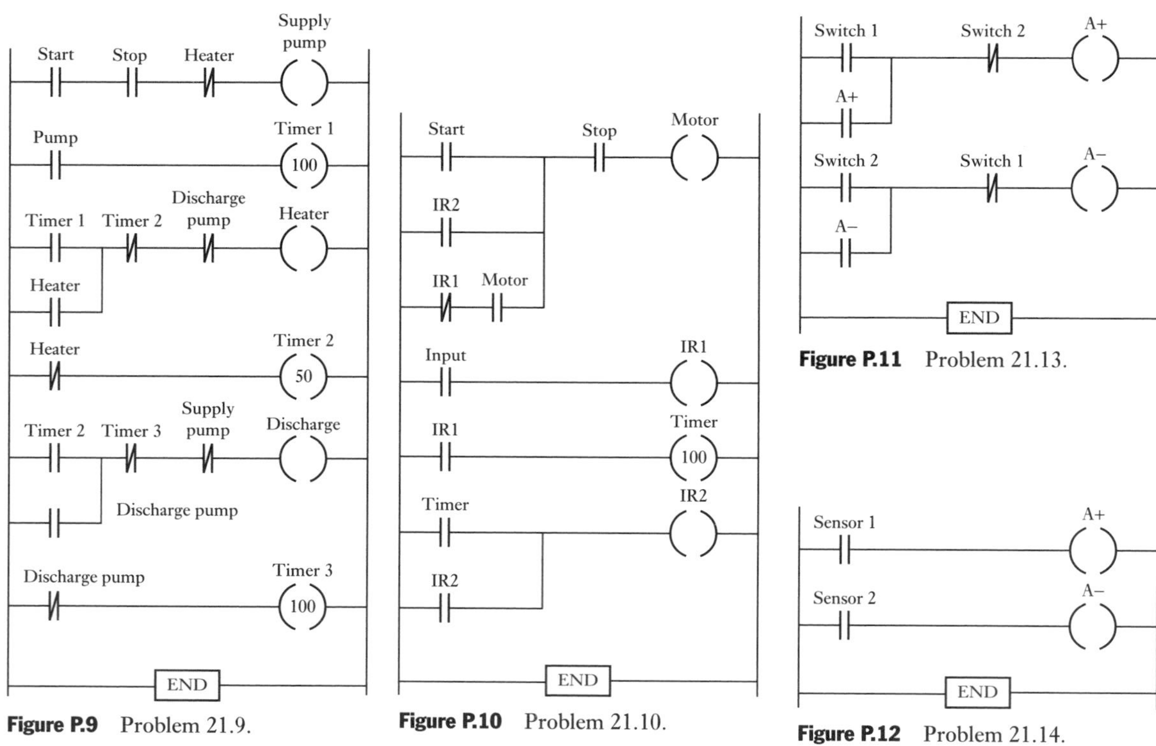

Figure P.9 Problem 21.9.

Figure P.10 Problem 21.10.

Figure P.11 Problem 21.13.

Figure P.12 Problem 21.14.

22.5 See Section 22.5.1
22.6 See Section 22.4
22.7 See Section 22.3.1
22.8 NRFD to PD0, DAV to STRA and IRQ, NDAC to STRB, data to Port C
22.9 TTL to RS-232C signal-level conversion
22.10 See Section 22.7.1

Chapter 23

23.1 See Section 23.1
23.2 See Section 23.2
23.3 See Section 23.2
23.4 See Section 23.5.3 for programmable checks and checksum and Section 23.2 for watchdog timer
23.5 See Section 23.5.3

Chapter 24

24.1. Possible solutions might be: (a) thermocouple, cold junction compensation, amplifier, ADC, PIA, microprocessor, DAC, thyristor unit to control the oven heating element, (b) light beam sensors, PLC, solenoid-operated delivery chute deflectors, (c) closed-loop control with, for movement in each direction, a d.c. motor as actuator for movement of pen, microprocessor as comparator and controller, and feedback from an optical encoder.

Research assignments

The following are brief indications of the type of information which might be contained in an answer:

24.2 A typical ABS system has sensors, inductance types, sensing the speeds of each of the car wheels, signal conditioning to convert the sensor signals into 5 V pulses, a microcontroller with a program to calculate the wheels' speed and rate of deceleration during braking so that when a set limit is exceeded the microcontroller gives an output to solenoid valves in the hydraulic modulator unit either to prevent an increase in braking force or if necessary to reduce it.

24.3 The carriage motor moves the printer head sideways while the print head prints the characters. After printing a line the paper feed motor advances the paper. The print head consists of solenoids driving pins, typically a row of nine, to impact on an ink ribbon. A microcontroller can be used to control the outputs. For more details, see *Microcontroller Technology: The 68HC11* by P. Spasov (Prentice Hall 1992, 1996).

24.4 The CAN bus operates with signals which have a start bit, followed by the name which indicates the message destination and its priority, followed by control bits, followed by the data being sent, followed by CRC bits, followed by confirmation of reception bits, and concluding with end bits.

Design assignments

The following are brief indications of possible solutions:

24.5 A digital thermometer using a microprocessor might have a temperature sensor such as LM35, an ADC, a ROM chip such as the Motorola MCM6830 or Intel 8355, a RAM chip such as the Motorola MCM6810 or Intel 8156, a microprocessor such as Motorola M6800 or Intel 8085A and a driver with LED display. With a microcontroller such as the Motorola MC68HC11 or Intel 8051 there might be just the temperature sensor, with perhaps signal conditioning, the microcontroller and the driver with LED display.

24.6 A digital ohmmeter might involve a monostable multivibrator which provides an impulse with a duration of $0.7RC$. A range of different fixed capacitors could be used to provide different resistance ranges. The time interval might then be determined using a microcontroller or a microprocessor plus memory, and then directed through a suitable driver to an LED display.

24.7 This might involve a pressure sensor, e.g. the semiconductor transducer Motorola MPX2100AP, signal conditioning to convert the small differential signal from the sensor to the appropriate level, e.g. an instrumentation amplifier using operational amplifiers, a microcontroller, e.g. MC68HC11, an LCD driver, e.g. MC145453, and a four-digit LCD display.

24.8 This could be tackled by using the M68HC11EVM board with a PWM output to the motor. Where feedback is wanted an optical encoder might be used.

24.9 The arrangement might be for each box to be loaded by current being supplied to a solenoid valve to operate a pneumatic cylinder to move a flap and allow a box down a chute. The box remains in the chute which is closed by a flap. Its presence is detected by a sensor which then indicates the next box can be allowed into the chute. This continues until four boxes are counted as being in the chute. The flap at the end of the chute might then be activated by another solenoid valve operating a cylinder and so allowing the boxes onto the belt. The arrival of the boxes on the belt might be indicated by a sensor mounted on the end of the chute. This can then allow the entire process to be repeated.

Appendix A

A.1 (a) $2/s^2$, (b) $2/(s^2 + 4)$, (c) e^{-2s}, (d) $sX(s) - 2$, (e) $3s^2X(s)$, (f) $1/[s(s + 1)]$

A.2 (a) t, (b) $5 \cos 3t$, (c) $1 + 2e^t$, (d) e^{-3t}

A.3 5

Appendix B

B.1 255

B.2 (a) 11, (b) 529

B.3 (a) 1A7, (b) 211

B.4 (a) 781, (b) 157

B.5 (a) 1010 0110, (b) 1101 1101

B.6 (a) 0, (b) 1

B.7 (a) 8, (b) 12

Appendix C

C.1 (a) $A \cdot (B + C)$, (b) $(A + B) \cdot (C + D)$, (c) $\overline{A} + B$, (d) $\overline{A} \cdot B$

C.2 (a) $Q = (A \cdot B + C \cdot D) \cdot E$, (b) $Q = (A \cdot B + B) \cdot C$

C.3

A	B	C	Q
0	0	0	0
0	0	1	0
0	1	0	0
0	1	1	1
1	0	0	0
1	0	1	1
1	1	0	0
1	1	1	1

C.4 (a) $Q = C \cdot (A + D)$, (b) $Q = A \cdot B$, (c) $Q = A \cdot \overline{B} \cdot C + C \cdot D$

C.5 As given in the problem

C.6 (a) $Q = A + B$, (b) $Q = C + \overline{A} \cdot C$

C.7 (a) $Q = \overline{A} \cdot \overline{B} + \overline{A} \cdot \overline{C}$, (b) $Q = A \cdot B \cdot D + A \cdot B \cdot \overline{C} + \overline{C} \cdot D$

C.8 Four input AND gates with two NOT gates if correct combination is 1, 1, 0, 0: $Q = A \cdot B \cdot \overline{C} \cdot \overline{D}$

Index

DISCOVERING STATISTICS USING SAS

(and sex and drugs and rock 'n' roll)

ANDY FIELD
and JEREMY MILES

Los Angeles | London | New Delhi
Singapore | Washington DC

SAGE Publications Ltd
1 Oliver's Yard
55 City Road
London EC1Y 1SP

SAGE Publications Inc.
2455 Teller Road
Thousand Oaks, California 91320

SAGE Publications India Pvt Ltd
B 1/I 1 Mohan Cooperative Industrial Area
Mathura Road
New Delhi 110 044

SAGE Publications Asia-Pacific Pte Ltd
33 Pekin Street #02-01
Far East Square
Singapore 048763

Library of Congress Control Number Available

British Library Cataloguing in Publication data

A catalogue record for this book is available from the British Library

ISBN 978-1-84920-091-2
ISBN 978-1-84920-092-9 (pbk)

Typeset by C&M Digitals (P) Ltd, Chennai, India
Printed by CPI Antony Rowe, Chippenham, Wiltshire
Printed on paper from sustainable resources

DISCOVERING STATISTICS USING SAS

Student comments about *Discovering Statistics Using SPSS*

'This book is amazing, I love it! It is responsible for raising my degree classification to a 1st class. I want to thank you for making a down to earth, easy to follow no nonsense book about the stuff that really matters in statistics.' **Tim Kock**

'I wanted to tell you; I LOVE YOUR SENSE OF HUMOUR. Statistics make me cry usually but with your book I almost mastered it. All I can say is keep writing such books that make our life easier and make me love statistics.' **Naïlah Moussa**

'Just a quick note to say how fantastic your stats book is. I was very happy to find a stats book which had sensible and interesting (sex, drugs and rock and roll) worked examples.' **Josephine Booth**

'I am deeply in your debt for your having written *Discovering Statistics Using SAS* (2nd edition). Thank you for a great contribution that has made life easier for so many of us.' **Bill Jervis Groton**

'I love the way that you write. You make this twoddle so lively. I am no longer crying, swearing, sweating or threatening to jack in this stupid degree just because I can't do the statistics. I am elated and smiling and jumping and grinning that I have come so far and managed to win over the interview panel using definitions and phrases that I read in your book!!! Bring it on you nasty exams. This candidate is Field Trained ...' **Sara Chamberlain**

'I just wanted to thank you for your book. I am working on my thesis and making sense of the statistics. Your book is wonderful!' **Katya Morgan**

'Sitting in front of a massive pile of books, in the midst of jamming up revision for exams, I cannot help distracting myself to tell you that your book keeps me smiling (or mostly laughing), although I am usually crying when I am studying. Thank you for your genius book. You have actually made a failing math student into a first class honors student, all with your amazing humor. Moreover, you have managed to convert me from absolutely detesting statistics to 'actually' enjoying them. For this, I thank you immensely. At university we have a great laugh on your jokes ... till we finish our degrees your book will keep us going!' **Amber Atif Ghani**

'Your book has brought me out of the darkness to a place where I feel I might see the light and get through my exam. Stats is by far not my strong point but you make me feel like I could live with it! Thank you.' **Vicky Learmont**

'I just wanted to email you and thank you for writing your book, *Discovering Statistics Using SAS*. I am a graduate student at the University of Victoria, Canada, and have found your book invaluable over the past few years. I hope that you will continue to write more in-depth stats books in the future! Thank you for making my life better!' **Leila Scannell**

'For a non-math book, this book is the best stat book that I have ever read.' **Dvir Kleper**

CONTENTS

PREFACE

Karma Police, arrest this man, he talks in maths, he buzzes like a fridge, he's like a detuned radio.

Radiohead *(1997)*

Introduction

Many social science students (and researchers for that matter) despise statistics. For one thing, most of us have a non-mathematical background, which makes understanding complex statistical equations very difficult. Nevertheless, the evil goat-warriors of Satan force our non-mathematical brains to apply themselves to what is, essentially, the very complex task of becoming a statistics expert. The end result, as you might expect, can be quite messy. The one weapon that we have is the computer, which allows us to neatly circumvent the considerable disability that is not understanding mathematics. The computer to a goat-warrior of Satan is like catnip to a cat: it makes them rub their heads along the ground and purr and dribble ceaselessly. The only downside of the computer is that it makes it really easy to make a complete idiot of yourself if you don't really understand what you're doing. Hence this book. Well, actually, hence a book called 'Discovering Statistics Using SPSS'.

I wrote 'Discovering statistics using SPSS' just as I was finishing off my Ph.D. in Psychology. The advent of computer programs like SAS, SPSS, R and the like provided the unique opportunity to teach statistics at a conceptual level without getting too bogged down in equations. However, some books based on statistical computer packages concentrate on 'doing the test' at the expense of theory. Using a computer without any statistical knowledge at all can be a dangerous thing. My main aim, therefore, was to write a book that attempted to strike a good balance between theory and practice: I want to use the computer as a tool for teaching statistical concepts in the hope that you will gain a better understanding of both theory and practice. If you want theory and you like equations then there are certainly better books: Howell (2006), Stevens (2002) and Tabachnick and Fidell (2007) are peerless as far as I am concerned and have taught me (and continue to teach me) more about statistics than you could possibly imagine. (I have an ambition to be cited in one of these books but I don't think that will ever happen.) However, if you want a book that incorporates digital rectal stimulation then you have just spent your money wisely. (I should probably clarify that the stimulation is in the context of an example, you will not find any devices attached to the inside cover for you to stimulate your rectum while you read. Please feel free to get your own device if you think it will help you to learn.)

A second, not in any way ridiculously ambitious, aim was to make this the only statistics textbook that anyone ever needs to buy. As such, it's a book that I hope will become your friend from first year right through to your professorship. I've tried, therefore, to write a book that can be read at several levels (see the next section for more guidance). There are chapters for first-year undergraduates (1, 2, 3, 4, 5, 6, 9 and 15), chapters for second-year undergraduates (5, 7, 10, 11, 12, 13 and 14) and chapters on more advanced topics that postgraduates might use (8, 16, 17, 18 and 19). All of these chapters should be accessible to everyone, and I hope to achieve this by flagging the level of each section (see the next section).

My third, final and most important aim is make the learning process fun. I have a sticky history with maths because I used to be terrible at it:

MATHEMATICS ADDL. MATHS.	43	59	27=	D	C	His work shows lack of discipline in thought and presentation. I hope it will mature next year.	[illegible]

Above is an extract of my school report at the age of 11. The '27' in the report is to say that I came equal 27th with another student out of a class of 29. That's almost bottom of the class. The 43 is my exam mark as a percentage! Oh dear. Four years later (at 15) this was my school report:

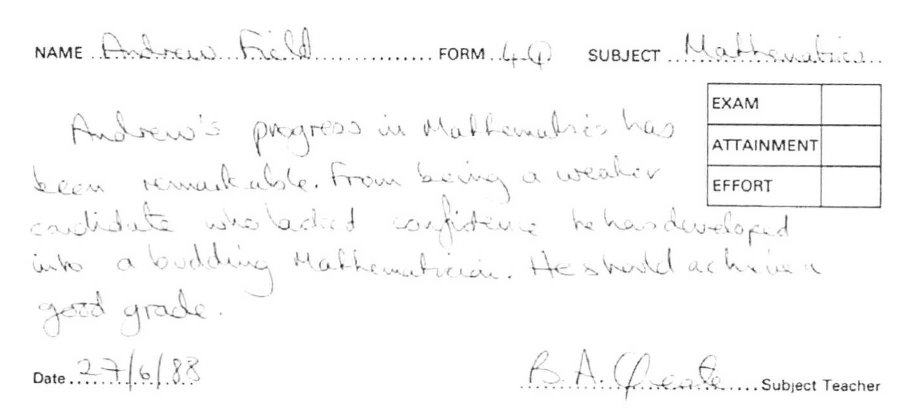
NAME Andrew Field FORM 4(?) SUBJECT Mathematics

EXAM	
ATTAINMENT	
EFFORT	

Andrew's progress in Mathematics has been remarkable. From being a weaker candidate who lacked confidence he has developed into a budding Mathematician. He should achieve a good grade.

Date 27/6/88

B. A. [illegible] Subject Teacher

What led to this remarkable change? It was having a good teacher: my brother, Paul. In fact I owe my life as an academic to Paul's ability to do what my maths teachers couldn't: teach me stuff in an engaging way. To this day he still pops up in times of need to teach me things (a crash course in computer programming some Christmases ago springs to mind). Anyway, the reason he's a great teacher is because he's able to make things interesting and relevant to me. Sadly he seems to have got the 'good teaching' genes in the family (and he doesn't even work as a bloody teacher, so they're wasted), but his approach inspires my lectures and books. One thing that I have learnt is that people appreciate the human touch, and so I tried to inject a lot of my own personality and sense of humour (or lack of) into 'Discovering Statistics Using SPSS', the book upon which this SAS version is based. Many of the examples in this book, although inspired by some of the craziness that you find in the real world, are designed to reflect topics that play on the minds of the average student (i.e. sex, drugs, rock and roll, celebrity, people doing crazy stuff). There are also some examples that are there just because they made me laugh. So, the examples are light-hearted (some have said 'smutty' but I prefer 'light-hearted') and by the end, for better or worse, I think you will have some idea of what goes on in my head on a daily basis! I apologise to those who think it's crass, hate it, or think that I'm undermining the seriousness of science, but, come on, what's not funny about a man putting an eel up his anus?

Did I succeed in these aims? Well, maybe I did, maybe I didn't, but the book has certainly been popular and I enjoy the rare luxury of having many complete strangers emailing me to tell me how wonderful I am. (Admittedly, occassionally people email to tell me that they think I'm an annoying idiot but you have to take the rough with the smooth.) It also won the British Psychological Society book award in 2007. I must have done something right. However, 'Discovering statistics using SPSS' has one very large flaw: not everybody uses SPSS. Some people use SAS instead. It occurred to me that it would be great to have a version of the book that was basically identical to the SPSS one except that all of the

SPSS stuff was replaced by stuff about SAS. Genius. Genius except that I know absolutely nothing about SAS and have never used it. Not even once. I don't even know what it looks like. I had two choices: I could either learn all about SAS and re-write bits of my book, or I could find someone else who already knew a lot about SAS and get them to re-write it for me. Fortunately, many moons ago I met a very nice guy called Jeremy (a man who, as you will see, likes to put eels in his CD player rather than anywhere else). Jeremy knows a lot more than I do about both statistics and SAS. He is also one of the most engaging writers about statistics that I know; but how could I persuade him to re-write sections of my book for me? Armed with some gaffa tape, a bag of jelly babies, a small puppy called Eric, a bear-cage and some chloroform, I went to Santa Monica where he lives. Fortunately for me Jeremy agreed to re-write the necessary sections of this book, and fortunately for Jeremy I agreed to unlock the cage and return his children.

What are you getting for your money?

As I have hinted, the SPSS incarnation of this book takes you through a journey not just of statistics but of the weird and wonderful contents of the world and my brain. In short, it's full of stupid examples, bad jokes, smut and filth. Aside from the smut, I have been forced reluctantly to include some academic content.Over many editions of the SPSS book many people have emailed me with suggestions, so, in theory, what you currently have in your hands should answer any question anyone has asked me over the past four years. It won't, but it should, and I'm sure you can find some new questions to ask. It has some other unusual features:

- **Everything you'll ever need to know:** I want this to be good value for money so the book guides you from complete ignorance (Chapter 1 tells you the basics of doing research) to being an expert on multilevel modelling (Chapter 19). Of course no book that you can actually lift off of the floor will contain everything, but I think this one has a fair crack at taking you from complete ignorance to postgraduate level expertise. It's pretty good for developing your biceps also.
- **Stupid faces:** You'll notice that the book is riddled with stupid faces, some of them my own. You can find out more about the pedagogic function of these 'characters' in the next section, but even without any useful function they're still nice to look at.
- **Data sets:** There are about 100 data files associated with this book on the companion website. Not unusual in itself for a statistics book, but my datasets contain more sperm (not literally) than other books. I'll let you judge for yourself whether this is a good thing.
- **My life story:** each chapter is book-ended by a chronological story from my life. Does this help you to learn about statistics? Probably not, but hopefully it provides some light relief between chapters.
- **SAS tips:** SAS does weird things sometimes. In each chapter, there are boxes containing tips, hints and pitfalls related to SAS.
- **Self-test questions:** Everyone loves an exam, don't they? Well, everyone that is apart from people who breathe. Given how much everyone hates tests, I thought the best way to commit commercial suicide was to liberally scatter tests throughout each chapter. These range from simple questions to test out what you have just learned to going back to a technique that you read about several chapters before and applying it in a new context. All of these questions have answers to them on the companion website. They are there so that you can check on your progress.

- **Additional material:** Enough trees have died in the name of this book, but still it gets longer and still people want to know more. Therefore, I've written nearly 300 pages, yes, three hundred, of additional material for the book. So for some more technical topics and help with tasks in the book the material has been provided electronically so that (1) the planet suffers a little less, and (2) you can actually lift the book.

The book also has some more conventional features:

- **Reporting your analysis:** Every single chapter has a guide to writing up your analysis. Obviously, how one writes up an analysis varies a bit from one discipline to another and being a psychologist these sections are quite psychology-based. Nevertheless though, they should get you heading in the right direction.
- **Glossary:** Writing the glossary was so horribly painful that it made me stick a vacuum cleaner into my ear to suck out my own brain. You can find my brain in the bottom of the vacuum cleaner in my house.
- **Real-world data:** Lots of people like to have 'real data' to play with. The trouble is that real research can be quite boring. However, just for you, I trawled the world for examples of research on really fascinating topics (in my opinion). I then stalked the authors of the research until they gave me their data. Every chapter has a real research example.

Goodbye

The first edition of the SPSS version of this book was the result of two years (give or take a few weeks to write up my Ph.D.) of trying to write a statistics book that I would enjoy reading. The second edition was another two years of work and I was terrified that all of the changes would be the death of it. The third edition was another 6 months of solid work and I was still hugely anxious that I'd just ruined the only useful thing that I've ever done with my life. I can hear the cries of lecturers around the world refusing to use the book because of cruelty to eels. Luckily, if the SAS version of the book dies on its arse I can blame Jeremy; after all, the SPSS one has done well and the only things that are different in this book are the bits Jeremy wrote. It's a win–win situation for me.

This book in its various forms has been part of my life now for over 12 years; it began and continues to be a labour of love. I'm delighted that Jeremy has done an SAS-update of the third edition and I hope that his efforts are richly rewarded with enough cash to finish building his swimming pool. The book isn't perfect, and I still love to have feedback (good or bad) from the people who matter most: you.(Unless of course it's about SAS and then you should send it to Jeremy because I won't have a clue what you're talking about.)

Andy
(My contact details are at www.statisticshell.com.)

HOW TO USE THIS BOOK

When the publishers asked me to write a section on 'How to use this book' it was obviously tempting to write 'Buy a large bottle of Olay anti-wrinkle cream (which you'll need to fend off the effects of ageing while you read), find a comfy chair, sit down, fold back the front cover, begin reading and stop when you reach the back cover.' However, I think they wanted something more useful☺

What background knowledge do I need?

In essence, I assume you know nothing about statistics, but I do assume you have some very basic grasp of computers (I won't be telling you how to switch them on, for example) and maths (although I have included a quick revision of some very basic concepts so I really don't assume anything).

Do the chapters get more difficult as I go through the book?

In a sense they do (Chapter 16 on MANOVA is more difficult than Chapter 1), but in other ways they don't (Chapter 15 on non-parametric statistics is arguably less complex than Chapter 14, and Chapter 9 on the *t*-test is definitely less complex than Chapter 8 on logistic regression). Why have I done this? Well, I've ordered the chapters to make statistical sense (to me, at least). Many books teach different tests in isolation and never really give you a grip of the similarities between them; this, I think, creates an unnecessary mystery. Most of the tests in this book are the same thing expressed in slightly different ways. So, I wanted the book to tell this story. To do this I have to do certain things such as explain regression fairly early on because it's the foundation on which nearly everything else is built!

However, to help you through I've coded each section with an icon. These icons are designed to give you an idea of the difficulty of the section. It doesn't necessarily mean you can skip the sections (but see Smart Alex in the next section), but it will let you know whether a section is at about your level, or whether it's going to push you. I've based the icons on my own teaching so they may not be entirely accurate for everyone (especially as systems vary in different countries!):

① This means 'level 1' and I equate this to first-year undergraduate in the UK. These are sections that everyone should be able to understand.

② This is the next level and I equate this to second-year undergraduate in the UK. These are topics that I teach my second years and so anyone with a bit of background in statistics should be able to get to grips with them. However, some of these sections will be quite challenging even for second years. These are intermediate sections.

③ This is 'level 3' and represents difficult topics. I'd expect third-year (final-year) UK undergraduates and recent postgraduate students to be able to tackle these sections.

④ This is the highest level and represents very difficult topics. I would expect these sections to be very challenging to undergraduates and recent postgraduates, but postgraduates with a reasonable background in research methods shouldn't find them too much of a problem.

Why do I keep seeing stupid faces everywhere?

Brian Haemorrhage: Brian's job is to pop up to ask questions and look permanently confused. It's no surprise to note, therefore, that he doesn't look entirely different from the first author. As the book progresses he becomes increasingly despondent. Read into that what you will.

Curious Cat: He also pops up and asks questions (because he's curious). Actually the only reason he's here is because I wanted a cat in the book … and preferably one that looks like mine. Of course the educational specialists think he needs a specific role, and so his role is to look cute and make bad cat-related jokes.

Cramming Sam: Samantha hates statistics. In fact, she thinks it's all a boring waste of time and she just wants to pass her exam and forget that she ever had to know anything about normal distributions. So, she appears and gives you a summary of the key points that you need to know. If, like Samantha, you're cramming for an exam, she will tell you the essential information to save you having to trawl through hundreds of pages of my drivel.

Jane Superbrain: Jane is the cleverest person in the whole universe (she makes Smart Alex look like a bit of an imbecile). The reason she is so clever is that she steals the brains of statisticians and eats them. Apparently they taste of sweaty tank tops, but nevertheless she likes them. As it happens she is also able to absorb the contents of brains while she eats them. Having devoured some top statistics brains she knows all the really hard stuff and appears in boxes to tell you really advanced things that are a bit tangential to the main text. (Readers should note that Jane wasn't interested in eating my brain. That tells you all that you need to know about my statistics ability.)

Labcoat Leni: Leni is a budding young scientist and he's fascinated by real research. He says, 'Andy, man, I like an example about using an eel as a cure for constipation as much as the next man, but all of your examples are made up. Real data aren't like that, we need some real examples, dude!' So off Leni went; he walked the globe, a lone data warrior in a thankless quest for real data. He turned up at universities, cornered academics, kidnapped their families and threatened to put them in a bath of crayfish unless he was given real data. The generous ones relented, but others? Well, let's just say their families are sore. So, when you see Leni you know that you will get some real data, from a real research study to analyse. Keep it real.

Oliver Twisted: With apologies to Charles Dickens, Oliver, like his more famous fictional London urchin, is always asking 'Please sir, can I have some more?' Unlike Master Twist though, our young Master Twisted, always wants more statistics information. Of course he does, who wouldn't? Let us not be the ones to disappoint a young, dirty, slightly smelly boy who dines on gruel, so when Oliver appears you can be certain of one thing: there is additional information to be found on the companions website. (Don't be shy; download it and bathe in the warm asp's milk of knowledge.)

Satan's Assistant Statistician: Satan is a busy boy – he has all of the lost souls to torture in hell; then there are the fires to keep fuelled, not to mention organizing enough carnage on the planet's surface to keep Norwegian black metal bands inspired. Like many of us, this leaves little time for him to analyse data, and this makes him very sad. So, he has his own assistant statistician, who, also like some of us, spends all day dressed in a gimp mask and tight leather pants in front of SAS analysing Satan's data. Consequently, he knows a thing or two about SAS, and when Satan's busy spanking a goat, he pops up in a box with SAS tips.

Smart Alex: Alex is a very important character because he appears when things get particularly difficult. He's basically a bit of a smart alec and so whenever you see his face you know that something scary is about to be explained. When the hard stuff is over he reappears to let you know that it's safe to continue. Now, this is not to say that all of the rest of the material in the book is easy, he just let's you know the bits of the book that you can skip if you've got better things to do with your life than read all 800 pages! So, if you see Smart Alex then you can *skip the section* entirely and still understand what's going on. You'll also find that Alex pops up at the end of each chapter to give you some tasks to do to see whether you're as smart as he is.

What is on the companion website?

In this age of downloading, CD-ROMs are for losers (at least that's what the 'kids' tell me) so this time around I've put my cornucopia of additional funk on that worldwide interweb thing. This has two benefits: (1) The book is *slightly* lighter than it would have been, and (2) rather than being restricted to the size of a CD-ROM, there is no limit to the amount of fascinating extra material that I can give you (although Sage have had to purchase a new server to fit it all on). To enter my world of delights, go to **www.sagepub.co.uk/fieldandmilesSAS**.

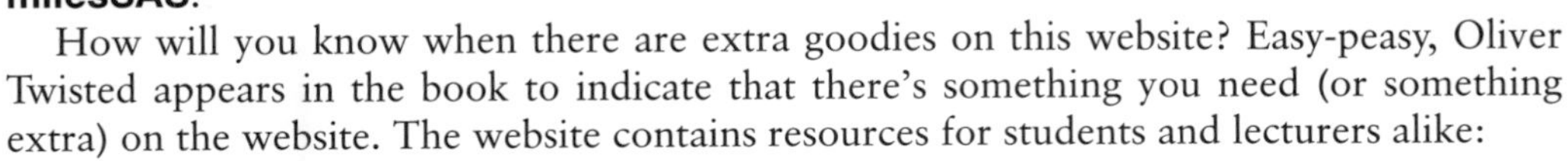

How will you know when there are extra goodies on this website? Easy-peasy, Oliver Twisted appears in the book to indicate that there's something you need (or something extra) on the website. The website contains resources for students and lecturers alike:

- **Data files:** You need data files to work through the examples in the book and they are all on the companion website.We did this so that you're forced to go there and once you're there Sage will flash up subliminal messages that make you buy more of their books. There are data files here for a range of students, including those studying psychology, business and health sciences.
- **Podcast:** My publishers think that watching a film of me explaining what this book is all about is going to get people flocking to the bookshop. I think it will have people flocking to the medicine cabinet. Either way, if you want to see how truly uncharismatic I am, watch and cringe.
- **Self-Assessment Multiple-Choice Questions:** Organized by chapter, these will allow you to test whether wasting your life reading this book has paid off so that you can walk confidently into an examination much to the annoyance of your friends. If you fail said exam, you can employ a good lawyer and sue Jeremy.
- **Flashcard Glossary:** As if a printed glossary wasn't enough, my publishers insisted that you'd like one in electronic format too. Have fun here flipping about between

terms and definitions that are covered in the textbook, it's better than actually learning something.

- **Additional material:** Enough trees have died in the name of this book and still people want to know more. Therefore, we've written nearly 300 pages, yes, three hundred, of additional material for the book. So for some more technical topics and help with tasks in the book the material has been provided electronically so that (1) the planet suffers a little less, and (2) you can actually lift the book.
- **Answers:** each chapter ends with a set of tasks for you to test your newly acquired expertise. The Chapters are also littered with self-test questions. How will you know if you get these correct? Well, the companion website contains around 300 hundred pages (that's a different three hundred pages to the three hundred above) of detailed answers. Will we ever stop writing?
- **Links:** every website has to have links to other useful websites and the companion website is no exception.
- **Cyberworms of knowledge:** I have used nanotechnology to create cyberworms that crawl down your broadband connection, pop out of the USB port of your computer then fly through space into your brain. They re-arrange your neurons so that you understand statistics. You don't believe me? Well, you'll never know for sure unless you visit the companion website …

Happy reading, and don't get sidetracked by Facebook.

ACKNOWLEDGEMENTS

This book (in the SPSS or SAS version) wouldn't have happened if it hadn't been for Dan Wright, who not only had an unwarranted faith in a then-postgraduate to write the first edition, but also read and commented on draft chapters in all three SPSS editions.

The SPSS versions of the book has been through a few editions and various people have contributed to those editions. This SAS version is based on edition 3 of the SPSS book and so benefits from the many people who have made comments and suggestions over the years for the other version of the book. I really appreciate the following people's contributions: John Alcock, Aliza Berger-Cooper, Sanne Bongers, Thomas Brügger, Woody Carter, Brittany Cornell, Peter de Heus, Edith de Leeuw, Sanne de Vrie, Jaap Dronkers, Anthony Fee, Andy Fugard, Massimo Garbuio, Ruben van Genderen, David Hitchin, Daniel Hoppe, Tilly Houtmans, Joop Hox, Suh-Ing (Amy) Hsieh, Don Hunt, Laura Hutchins-Korte, Mike Kenfield, Laura Murray, Zoë Nightingale, Ned Palmer, Jim Parkinson, Nick Perham, Thusha Rajendran, Paul Rogers, Alf Schabmann, Mischa Schirris, Mizanur Rashid Shuvra, Lynne Slocombe, Nick Smith, Craig Thorley, Paul Tinsley, Keith Tolfrey, Frederico Torracchi, Djuke Veldhuis, Jane Webster, Gareth Williams, and Enrique Woll.

I have incorporated data sets from real research papers. All of these research papers are studies that I find fascinating and it's an honour for me to have these researchers' data in my book: Hakan Çetinkaya, Tomas Chamorro-Premuzic, Graham Davey, Mike Domjan, Gordon Gallup, Eric Lacourse, Sarah Marzillier, Geoffrey Miller, Peter Muris, Laura Nichols and Achim Schüetzwohl.

Kate Lester read every single chapter of the SPSS version, but also kept my research laboratory ticking over while my mind was on writing that book. I literally could not have done it without her support and constant offers to take on extra work that she did not have to do so that I could be a bit less stressed. I am very lucky to have her in my research team. All of these people have taken time out of their busy lives to help me out. I'm not sure what that says about their mental states, but they are all responsible for a great many improvements. May they live long and their data sets be normal.

Not all contributions are as tangible as those above. With the possible exception of them not understanding why sometimes I don't answer my phone, I could not have asked for more loving and proud parents – a fact that I often take for granted. Also, very early in my career Graham Hole made me realise that teaching research methods didn't have to be dull.My whole approach to teaching has been to steal all of his good ideas and I'm pleased that he has had the good grace not to ask for them back! He is also a rarity in being brilliant, funny and nice. I also thank my Ph.D. students Carina Ugland, Khanya Price-Evans and Saeid Rohani for their patience for the three months that I was physically away in Rotterdam, and for the three months that I was mentally away upon my return.

The people at Sage are less hardened drinkers than they used to be, but I have been very fortunate to work with Michael Carmichael. Mike, despite his failings on the football field(!), has provided me with some truly memorable nights out and he also read some of my chapters this time around which, as an editor, made a pleasant change. Mike takes a lot of crap from me (especially when I am tired and stressed) and I'm grateful for his constant positive demeanor in the face of my grumpiness. Alex Lee did a fantastic job of turning the characters in my head into characters on the page.

I wrote much of the book that Jeremy converted to SAS while on sabbatical at the Department of Psychology at the Erasmus University, Rotterdam, The Netherlands. I'm grateful to the clinical research group (especially the white ape posse!) who so unreservedly made me part of the team. Sorry that I keep coming back and making a nuisance of myself when you probably hoped that I'd gone for good. Mostly, though, I thank Peter (Muris), Birgit (Mayer), Jip and Kiki who made me part of their family while in Rotterdam. I'm grateful for their kindness, hospitality, and for not getting annoyed when I was still in their kitchen having drunk all of their wine after the last tram home had gone. Mostly, I thank them for the wealth of happy memories that they gave me.

I always write listening to music, over the various incarnations of the 'Discovering statistics using …' books I owe my sanity to: 1349, Abba, AC/DC, Air, Angantyr, Arvo Pärt, Audrey Horne, Beck, The Beyond, Blondie, Busta Rhymes, Cardiacs, Cradle of Filth, Cobalt, Danzig, Dark Angel, Darkthrone, Death Angel, Deathspell Omega, DJ Shadow, Elliott Smith, Emperor, Exodus, Frank Black and the Catholics, Fugazi, Genesis, Hefner, High on Fire, Iron Maiden, Jane's Addiction, Love, Manowar, The Mars Volta, Massive Attack, Mastodon, Megadeth, Meshuggah, Mercury Rev, Metallica, Morrissey, Muse, Nevermore, Nick Cave, Nusrat Fateh Ali Khan, Opeth, Peter Gabriel, Porcupine Tree, Placebo, Quasi, Radiohead, Rush, Serj Tankian, Sevara Nazarkhan, She Said!, Slayer, Slipknot, Soundgarden, Supergrass, Taake, Tool, the Wedding Present and The White Stripes.

All this book-writing nonsense requires many lonely hours (mainly late at night) of typing. Without some wonderful friends to drag me out of my dimly lit room from time to time I'd be even more of a gibbering cabbage than I already am. My eternal gratitude goes to Graham Davey, Benie MacDonald, Ben Dyson, Martin Watts, Paul Spreckley, Darren Hayman, Helen Liddle, Sam Cartwright-Hatton, Karina Knowles and Mark Franklin for reminding me that there is more to life than work. Also, my eternal gratitude to my brothers of metal Doug Martin and Rob Mepham for letting me deafen them with my drumming on a regular basis. Finally, thanks to Leonora for her support while writing editions 2 and 3 of the SPSS book.

My greatest thanks go to Jeremy Miles for turning my book into an SAS one. As if this wasn't enough he also stopped me making a complete and utter fool of myself (in the book – sadly his powers don't extend to everyday life) by pointing out some glaring errors; he's also been a very nice person to know over the past few years (apart from when he's saying that draft sections of my books are, and I quote, 'bollocks'!). He has done a brilliant job on this SAS version of the book and I'm absolutely delighted to finally achieve my ambition of seeing both of our names on the same book cover.

Dedication

This book is dedicated to my brother Paul and my cat Fuzzy, because one of them is a constant source of intellectual inspiration and the other wakes me up in the morning by sitting on me and purring in my face until I give him cat food: mornings will be considerably more pleasant when my brother gets over his love of cat food for breakfast.☺

Andy Field

The reader might want to know who is to blame for the existence of this book. In 1985, I was persuaded by Joss Griffiths (of Newcastle College) to take psychology. Joss thereby set me off on my career path in psychology and that led eventually to this book; he was also the first person who tried to teach me about research methods and statistics, and through him I learned that statistics might be more interesting than it first appears. I'd like to dedicate this to him. You, dear reader, can blame him.

Jeremy Miles

SYMBOLS USED IN THIS BOOK

Mathematical operators

Σ	This symbol (called sigma) means 'add everything up'. So, if you see something like Σx_i it just means 'add up all of the scores you've collected'.
Π	This symbol means 'multiply everything'. So, if you see something like Πx_i it just means 'multiply all of the scores you've collected'.
$\sqrt{x}$	This means 'take the square root of x'.

Greek symbols

α	The probability of making a Type I error
β	The probability of making a Type II error
β_i	Standardized regression coefficient
χ^2	Chi-square test statistic
χ^2_F	Friedman's ANOVA test statistic
ε	Usually stands for 'error'
η^2	Eta squared
μ	The mean of a population of scores
ρ	The correlation in the population
σ^2	The variance in a population of data
σ	The standard deviation in a population of data
$\sigma_{\bar{x}}$	The standard error of the mean
τ	Kendall's tau (non-parametric correlation coefficient)
ω^2	Omega squared (an effect size measure). This symbol also means 'expel the contents of your intestine immediately into your trousers'; you will understand why in due course

English symbols

b_i	The regression coefficient (unstandardized)
df	Degrees of freedom
e_i	The error associated with the *i*th person
F	*F*-ratio (test statistic used in ANOVA)
H	Kruskal–Wallis test statistic
k	The number of levels of a variable (i.e. the number of treatment conditions), or the number of predictors in a regression model
ln	Natural logarithm
MS	The mean squared error (Mean Square). The average variability in the data
N, n, n_i	The sample size. *N* usually denotes the total sample size, whereas *n* usually denotes the size of a particular group
P	Probability (the probability value, *p*-value or significance of a test are usually denoted by *p*)
r	Pearson's correlation coefficient
r_s	Spearman's rank correlation coefficient
R	The multiple correlation coefficient
R^2	The coefficient of determination (i.e. the proportion of data explained by the model)
s^2	The variance of a sample of data
s	The standard deviation of a sample of data
SS	The sum of squares, or sum of squared errors to give it its full title
SS_A	The sum of squares for variable *A*
SS_M	The model sum of squares (i.e. the variability explained by the model fitted to the data)
SS_R	The residual sum of squares (i.e. the variability that the model can't explain – the error in the model)
SS_T	The total sum of squares (i.e. the total variability within the data)
t	Test statistic for Student's *t*-test
T	Test statistic for Wilcoxon's matched-pairs signed-rank test
U	Test statistic for the Mann–Whitney test
W_s	Test statistic for Wilcoxon's rank-sum test
$\bar{X}$ or $\bar{x}$	The mean of a sample of scores
z	A data point expressed in standard deviation units

SOME MATHS REVISION

1 **Two negatives make a positive**: Although in life two wrongs don't make a right, in mathematics they do! When we multiply a negative number by another negative number, the result is a positive number. For example, $-2 \times -4 = 8$.

2 **A negative number multiplied by a positive one make a negative number**: If you multiply a positive number by a negative number then the result is another negative number. For example, $2 \times -4 = -8$, or $-2 \times 6 = -12$.

3 **BODMAS**: This is an acronym for the order in which mathematical operations are performed. It stands for Brackets, Order, Division, Multiplication, Addition, Subtraction and this is the order in which you should carry out operations within an equation. Mostly these operations are self-explanatory (e.g. always calculate things within brackets first) except for order, which actually refers to power terms such as squares. Four squared, or 4^2, used to be called four raised to the order of 2, hence the reason why these terms are called 'order' in BODMAS (also, if we called it power, we'd end up with BPDMAS, which doesn't roll off the tongue quite so nicely). Let's look at an example of BODMAS: what would be the result of $1 + 3 \times 5^2$? The answer is 76 (not 100 as some of you might have thought). There are no brackets so the first thing is to deal with the order term: 5^2 is 25, so the equation becomes $1 + 3 \times 25$. There is no division, so we can move on to multiplication: 3×25, which gives us 75. BODMAS tells us to deal with addition next: $1 + 75$, which gives us 76 and the equation is solved. If I'd written the original equation as $(1 + 3) \times 5^2$, then the answer would have been 100 because we deal with the brackets first: $(1 + 3) = 4$, so the equation becomes 4×5^2. We then deal with the order term, so the equation becomes $4 \times 25 = 100$!

4 **http://www.easymaths.com** is a good site for revising basic maths.

Why is my evil lecturer forcing me to learn statistics?

1

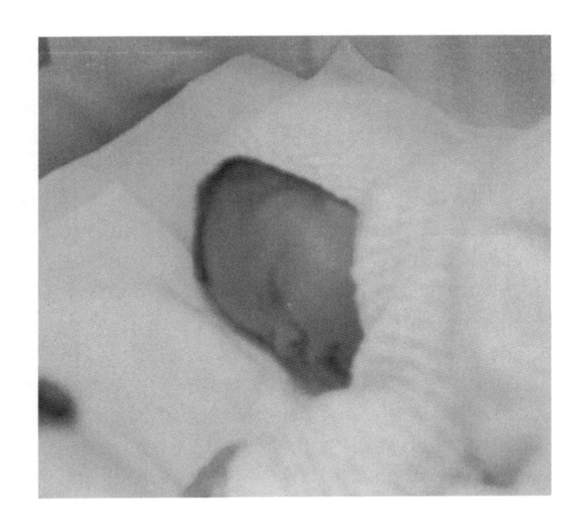

FIGURE 1.1
When I grow up, please don't let me be a statistics lecturer

1.1. What will this chapter tell me? ①

I was born on 21 June 1973. Like most people, I don't remember anything about the first few years of life and like most children I did go through a phase of driving my parents mad by asking 'Why?' every five seconds. 'Dad, why is the sky blue?', 'Dad, why doesn't mummy have a willy?', etc. Children are naturally curious about the world. I remember at the age of 3 being at a party of my friend Obe (this was just before he left England to return to Nigeria, much to my distress). It was a hot day, and there was an electric fan blowing cold air around the room. As I said, children are natural scientists and my little scientific brain was working through what seemed like a particularly pressing question: 'What happens when you stick your finger in a fan?' The answer, as it turned out, was that it hurts – a lot.[1] My point is this: my curiosity to explain the world never went away, and that's why

[1] In the 1970s fans didn't have helpful protective cages around them to prevent idiotic 3 year olds sticking their fingers into the blades.

I'm a scientist, and that's also why your evil lecturer is forcing you to learn statistics. It's because you have a curious mind too and you want to answer new and exciting questions. To answer these questions we need statistics. Statistics is a bit like sticking your finger into a revolving fan blade: sometimes it's very painful, but it does give you the power to answer interesting questions. This chapter is going to attempt to explain why statistics are an important part of doing research. We will overview the whole research process, from why we conduct research in the first place, through how theories are generated, to why we need data to test these theories. If that doesn't convince you to read on then maybe the fact that we discover whether Coca-Cola kills sperm will. Or perhaps not.

1.2. What the hell am I doing here? I don't belong here ①

You're probably wondering why you have bought this book. Maybe you liked the pictures, maybe you fancied doing some weight training (it *is* heavy), or perhaps you need to reach something in a high place (it *is* thick). The chances are, though, that given the choice of spending your hard-earned cash on a statistics book or something more entertaining (a nice novel, a trip to the cinema, etc.) you'd choose the latter. So, why have you bought the book (or downloaded an illegal pdf of it from someone who has way too much time on their hands if they can scan a 700-page textbook)? It's likely that you obtained it because you're doing a course on statistics, or you're doing some research, and you need to know how to analyse data. It's possible that you didn't realize when you started your course or research that you'd have to know this much about statistics but now find yourself inexplicably wading, neck high, through the Victorian sewer that is data analysis. The reason that you're in the mess that you find yourself in is because you have a curious mind. You might have asked yourself questions like why do people behave the way they do (psychology) or why do behaviours differ across cultures (anthropology), how do businesses maximize their profit (business), how did the dinosaurs die? (palaeontology), does eating tomatoes protect you against cancer (medicine, biology), is it possible to build a quantum computer (physics, chemistry), is the planet hotter than it used to be and in what regions (geography, environmental studies)? Whatever it is you're studying or researching, the reason you're studying it is probably because you're interested in answering questions. Scientists are curious people, and you probably are too. However, you might not have bargained on the fact that to answer interesting questions, you need two things: data and an explanation of those data.

The answer to 'what the hell are you doing here?' is, therefore, simple: to answer interesting questions you need data. Therefore, one of the reasons why your evil statistics lecturer is forcing you to learn about numbers is because they are a form of data and are vital to the research process. Of course there are forms of data other than numbers that can be used to test and generate theories. When numbers are involved the research involves **quantitative methods**, but you can also generate and test theories by analysing language (such as conversations, magazine articles, media broadcasts and so on). This involves **qualitative methods** and it is a topic for another book not written by me. People can get quite passionate about which of these methods is *best*, which is a bit silly because they are complementary, not competing, approaches and there are much more important issues in the world to get upset about. Having said that, all qualitative research is rubbish.[2]

[2] This is a joke. I thought long and hard about whether to include it because, like many of my jokes, there are people who won't find it remotely funny. Its inclusion is also making me fear being hunted down and forced to eat my own entrails by a hoard of rabid qualitative researchers. However, it made me laugh, a lot, and despite being vegetarian I'm sure my entrails will taste lovely.

1.2.1. The research process ①

How do you go about answering an interesting question? The research process is broadly summarized in Figure 1.2. You begin with an observation that you want to understand, and this observation could be anecdotal (you've noticed that your cat watches birds when they're on TV but not when jellyfish are on[3]) or could be based on some data (you've got several cat owners to keep diaries of their cat's TV habits and have noticed that lots of them watch birds on TV). From your initial observation you generate explanations, or theories, of those observations, from which you can make predictions (hypotheses). Here's where the data come into the process because to test your predictions you need data. First you collect some relevant data (and to do that you need to identify things that can be measured) and then you analyse those data. The analysis of the data may support your theory or give you cause to modify the theory. As such, the processes of data collection and analysis and generating theories are intrinsically linked: theories lead to data collection/analysis and data collection/analysis informs theories! This chapter explains this research process in more detail.

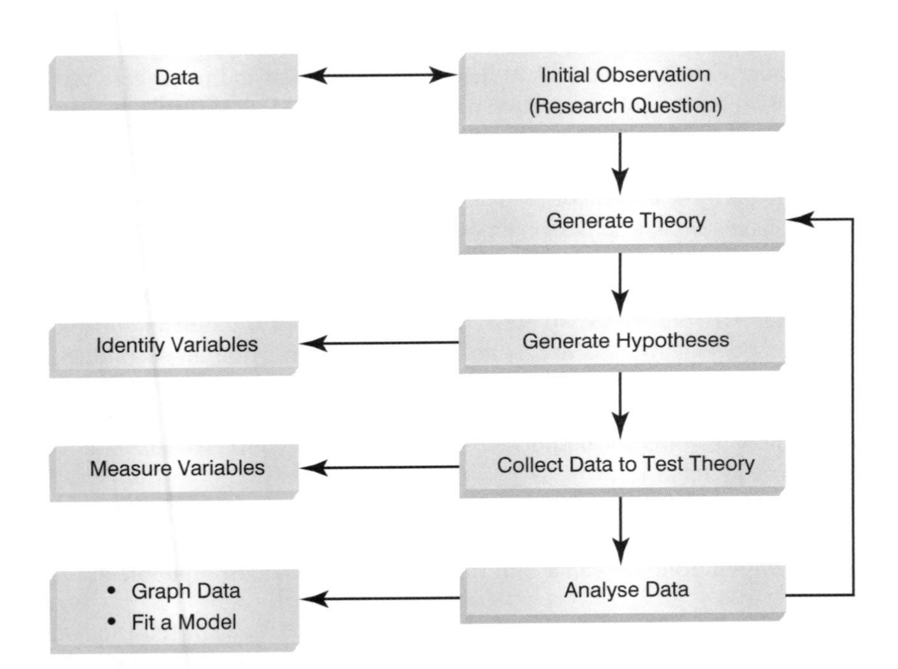

FIGURE 1.2 The research process

1.3. Initial observation: finding something that needs explaining ①

The first step in Figure 1.2 was to come up with a question that needs an answer. I spend rather more time than I should watching reality TV. Every year I swear that I won't get hooked on *Big Brother*, and yet every year I find myself glued to the TV screen waiting for

[3] My cat does actually climb up and stare at the TV when it's showing birds flying about.

the next contestant's meltdown (I am a psychologist, so really this is just research – honestly). One question I am constantly perplexed by is why every year there are so many contestants with really unpleasant personalities (my money is on narcissistic personality disorder[4]) on the show. A lot of scientific endeavour starts this way: not by watching *Big Brother*, but by observing something in the world and wondering why it happens.

Having made a casual observation about the world (*Big Brother* contestants on the whole have profound personality defects), I need to collect some data to see whether this observation is true (and not just a biased observation). To do this, I need to define one or more **variables** that I would like to measure. There's one variable in this example: the personality of the contestant. I could measure this variable by giving them one of the many well-established questionnaires that measure personality characteristics. Let's say that I did this and I found that 75% of contestants did have narcissistic personality disorder. These data support my observation: a lot of *Big Brother* contestants have extreme personalities.

1.4. Generating theories and testing them ①

The next logical thing to do is to explain these data (Figure 1.2). One explanation could be that people with narcissistic personality disorder are more likely to audition for *Big Brother* than those without. This is a **theory**. Another possibility is that the producers of *Big Brother* are more likely to select people who have narcissistic personality disorder to be contestants than those with less extreme personalities. This is another theory. We verified our original observation by collecting data, and we can collect more data to test our theories. We can make two predictions from these two theories. The first is that the number of people turning up for an audition that have narcissistic personality disorder will be higher than the general level in the population (which is about 1%). A prediction from a theory, like this one, is known as a **hypothesis** (see Jane Superbrain Box 1.1). We could test this hypothesis by getting a team of clinical psychologists to interview each person at the *Big Brother* audition and diagnose them as having narcissistic personality disorder or not. The prediction from our second theory is that if the *Big Brother* selection panel are more likely to choose people with narcissistic personality disorder then the rate of this disorder in the final contestants will be even higher than the rate in the group of people going for auditions. This is another hypothesis. Imagine we collected these data; they are in Table 1.1.

In total, 7662 people turned up for the audition. Our first hypothesis is that the percentage of people with narcissistic personality disorder will be higher at the audition than the general level in the population. We can see in the table that of the 7662 people at the audition,

TABLE 1.1 A table of the number of people at the *Big Brother* audition split by whether they had narcissistic personality disorder and whether they were selected as contestants by the producers

	No Disorder	Disorder	Total
Selected	3	9	12
Rejected	6805	845	7650
Total	6808	854	7662

[4] This disorder is characterized by (among other things) a grandiose sense of self-importance, arrogance, lack of empathy for others, envy of others and belief that others envy them, excessive fantasies of brilliance or beauty, the need for excessive admiration and exploitation of others.

854 were diagnosed with the disorder; this is about 11% (854/7662 × 100) which is much higher than the 1% we'd expect. Therefore, the first hypothesis is supported by the data. The second hypothesis was that the *Big Brother* selection panel have a bias to chose people with narcissistic personality disorder. If we look at the 12 contestants that they selected, 9 of them had the disorder (a massive 75%). If the producers did not have a bias we would have expected only 11% of the contestants to have the disorder. The data again support our hypothesis. Therefore, my initial observation that contestants have personality disorders was verified by data, then my theory was tested using specific hypotheses that were also verified using data. Data are *very* important!

JANE SUPERBRAIN 1.1

When is a hypothesis not a hypothesis? ①

A good theory should allow us to make statements about the state of the world. Statements about the world are good things: they allow us to make sense of our world, and to make decisions that affect our future. One current example is global warming. Being able to make a definitive statement that global warming is happening, and that it is caused by certain practices in society, allows us to change these practices and, hopefully, avert catastrophe. However, not all statements are ones that can be tested using science. Scientific statements are ones that can be verified with reference to empirical evidence, whereas non-scientific statements are ones that cannot be empirically tested. So, statements such as 'The Led Zeppelin reunion concert in London in 2007 was the best gig ever',[5] 'Lindt chocolate is the best food' and 'This is the worst statistics book in the world' are all non-scientific; they cannot be proved or disproved. Scientific statements can be confirmed or disconfirmed empirically. 'Watching *Curb Your Enthusiasm*' makes you happy', 'having sex increases levels of the neurotransmitter dopamine' and 'Velociraptors ate meat' are all things that can be tested empirically (provided you can quantify and measure the variables concerned). Non-scientific statements can sometimes be altered to become scientific statements, so 'The Beatles were the most influential band ever' is non-scientific (because it is probably impossible to quantify 'influence' in any meaningful way) but by changing the statement to 'The Beatles were the best-selling band ever' it becomes testable (we can collect data about worldwide record sales and establish whether The Beatles have, in fact, sold more records than any other music artist). Karl Popper, the famous philosopher of science, believed that non-scientific statements were nonsense, and had no place in science. Good theories should, therefore, produce hypotheses that are scientific statements.

I would now be smugly sitting in my office with a contented grin on my face about how my theories and observations were well supported by the data. Perhaps I would quit while I was ahead and retire. It's more likely, though, that having solved one great mystery, my excited mind would turn to another. After another few hours (well, days probably) locked up at home watching *Big Brother* I would emerge triumphant with another profound observation, which is that these personality-disordered contestants, despite their obvious character flaws, enter the house convinced that the public will love them and that they will win.[6] My hypothesis would, therefore, be that if I asked the contestants if they thought that they would win, the people with a personality disorder would say yes.

[5] It was pretty awesome actually.

[6] One of the things I like about *Big Brother* in the UK is that year upon year the winner tends to be a nice person, which does give me faith that humanity favours the nice.

Let's imagine I tested my hypothesis by measuring their expectations of success in the show, by just asking them, 'Do you think you will win *Big Brother*?'. Let's say that 7 of 9 contestants with personality disorders said that they thought that they will win, which confirms my observation. Next, I would come up with another theory: these contestants think that they will win because they don't realize that they have a personality disorder. My hypothesis would be that if I asked these people about whether their personalities were different from other people they would say 'no'. As before, I would collect some more data and perhaps ask those who thought that they would win whether they thought that their personalities were different from the norm. All 7 contestants said that they thought their personalities were different from the norm. These data seem to contradict my theory. This is known as **falsification**, which is the act of disproving a hypothesis or theory.

It's unlikely that we would be the only people interested in why individuals who go on *Big Brother* have extreme personalities and think that they will win. Imagine these researchers discovered that: (1) people with narcissistic personality disorder think that they are more interesting than others; (2) they also think that they deserve success more than others; and (3) they also think that others like them because they have 'special' personalities.

This additional research is even worse news for my theory: if they didn't realize that they had a personality different from the norm then you wouldn't expect them to think that they were more interesting than others, and you certainly wouldn't expect them to think that others will like their unusual personalities. In general, this means that my theory sucks: it cannot explain all of the data, predictions from the theory are not supported by subsequent data, and it cannot explain other research findings. At this point I would start to feel intellectually inadequate and people would find me curled up on my desk in floods of tears wailing and moaning about my failing career (no change there then).

At this point, a rival scientist, Fester Ingpant-Stain, appears on the scene with a rival theory to mine. In his new theory, he suggests that the problem is not that personality-disordered contestants don't realize that they have a personality disorder (or at least a personality that is unusual), but that they falsely believe that this special personality is perceived positively by other people (put another way, they believe that their personality makes them likeable, not dislikeable). One hypothesis from this model is that if personality-disordered contestants are asked to evaluate what other people think of them, then they will overestimate other people's positive perceptions. To test this hypothesis, Fester Ingpant-Stain collected yet more data. When each contestant came to the diary room they had to fill out a questionnaire evaluating all of the other contestants' personalities, and also answer each question as if they were each of the contestants responding about them. (So, for every contestant there is a measure of what they thought of every other contestant, and also a measure of what they believed every other contestant thought of them.) He found out that the contestants with personality disorders did overestimate their housemates' view of them; in comparison the contestants without personality disorders had relatively accurate impressions of what others thought of them. These data, irritating as it would be for me, support the rival theory that the contestants with personality disorders know they have unusual personalities but believe that these characteristics are ones that others would feel positive about. Fester Ingpant-Stain's theory is quite good: it explains the initial observations and brings together a range of research findings. The end result of this whole process (and my career) is that we should be able to make a general statement about the state of the world. In this case we could state: '*Big Brother* contestants who have personality disorders overestimate how much other people like their personality characteristics'.

SELF-TEST Based on what you have read in this section, what qualities do you think a scientific theory should have?

1.5. Data collection 1: what to measure ①

We have seen already that data collection is vital for testing theories. When we collect data we need to decide on two things: (1) what to measure, (2) how to measure it. This section looks at the first of these issues.

1.5.1. Variables ①

1.5.1.1. Independent and dependent variables ①

To test hypotheses we need to measure variables. Variables are just things that can change (or vary); they might vary between people (e.g. IQ, behaviour) or locations (e.g. unemployment) or even time (e.g. mood, profit, number of cancerous cells). Most hypotheses can be expressed in terms of two variables: a proposed cause and a proposed outcome. For example, if we take the scientific statement 'Coca-Cola is an effective spermicide'[7] then proposed cause is 'Coca-Cola' and the proposed effect is dead sperm. Both the cause and the outcome are variables: for the cause we could vary the type of drink, and for the outcome, these drinks will kill different amounts of sperm. The key to testing such statements is to measure these two variables.

A variable that we think is a cause is known as an **independent variable** (because its value does not depend on any other variables). A variable that we think is an effect is called a **dependent variable** because the value of this variable depends on the cause (independent variable). These terms are very closely tied to experimental methods in which the cause is actually manipulated by the experimenter (as we will see in section 1.6.2). In cross-sectional research we don't manipulate any variables and we cannot make causal statements about the relationships between variables, so it doesn't make sense to talk of dependent and independent variables because all variables are dependent variables in a sense. One possibility is to abandon the terms dependent and independent variable and use the terms **predictor variable** and **outcome variable.** In experimental work the cause, or independent variable, is a predictor, and the effect, or dependent variable, is simply an outcome. This terminology also suits cross-sectional work where, statistically at least, we can use one or more variables to make predictions about the other(s) without needing to imply causality.

CRAMMING SAM'S TIPS **Some Important Terms**

When doing research there are some important generic terms for variables that you will encounter:

- **Independent variable:** A variable thought to be the cause of some effect. This term is usually used in experimental research to denote a variable that the experimenter has manipulated.
- **Dependent variable:** A variable thought to be affected by changes in an independent variable. You can think of this variable as an outcome.
- **Predictor variable:** A variable thought to predict an outcome variable. This is basically another term for independent variable (although some people won't like me saying that; I think life would be easier if we talked only about predictors and outcomes).
- **Outcome variable:** A variable thought to change as a function of changes in a predictor variable. This term could be synonymous with 'dependent variable' for the sake of an easy life.

[7] Actually, there is a long-standing urban myth that a post-coital douche with the contents of a bottle of Coke is an effective contraceptive. Unbelievably, this hypothesis has been tested and Coke does affect sperm motility, and different types of Coke are more or less effective – Diet Coke is best apparently (Umpierre, Hill, & Anderson, 1985). Nevertheless, a Coke douche is ineffective at preventing pregnancy.

1.5.1.2. Levels of measurement ①

As we have seen in the examples so far, variables can take on many different forms and levels of sophistication. The relationship between what is being measured and the numbers that represent what is being measured is known as the **level of measurement**. Broadly speaking, variables can be categorical or continuous, and can have different levels of measurement.

A **categorical variable** is made up of categories. A categorical variable that you should be familiar with already is your species (e.g. human, domestic cat, fruit bat, etc.). You are a human or a cat or a fruit bat: you cannot be a bit of a cat and a bit of a bat, and neither a batman nor (despite many fantasies to the contrary) a catwoman (not even one in a nice PVC suit) exist. A categorical variable is one that names distinct entities. In its simplest form it names just two distinct types of things, for example male or female. This is known as a **binary variable**. Other examples of binary variables are being alive or dead, pregnant or not, and responding 'yes' or 'no' to a question. In all cases there are just two categories and an entity can be placed into only one of the two categories.

When two things that are equivalent in some sense are given the same name (or number), but there are more than two possibilities, the variable is said to be a **nominal variable**. It should be obvious that if the variable is made up of names it is pointless to do arithmetic on them (if you multiply a human by a cat, you do not get a hat). However, sometimes numbers are used to denote categories. For example, the numbers worn by players in a rugby or football (soccer) team. In rugby, the numbers of shirts denote specific field positions, so the number 10 is always worn by the fly-half (e.g. England's Jonny Wilkinson),[8] and the number 1 is always the hooker (the ugly-looking player at the front of the scrum). These numbers do not tell us anything other than what position the player plays. We could equally have shirts with FH and H instead of 10 and 1. A number 10 player is not necessarily better than a number 1 (most managers would not want their fly-half stuck in the front of the scrum!). It is equally as daft to try to do arithmetic with nominal scales where the categories are denoted by numbers: the number 10 takes penalty kicks, and if the England coach found that Jonny Wilkinson (his number 10) was injured he would not get his number 4 to give number 6 a piggy-back and then take the kick. The only way that nominal data can be used is to consider frequencies. For example, we could look at how frequently number 10s score tries compared to number 4s.

JANE SUPERBRAIN 1.2

Self-report data ①

A lot of self-report data are ordinal. Imagine if two judges at our beauty pageant were asked to rate Billie's beauty on a 10-point scale. We might be confident that a judge who gives a rating of 10 found Billie more beautiful than one who gave a rating of 2, but can we be certain that the first judge found her five times more beautiful than the second? What about if both judges gave a rating of 8, could we be sure they found her equally beautiful? Probably not: their ratings will depend on their subjective feelings about what constitutes beauty. For these reasons, in any situation in which we ask people to rate something subjective (e.g. rate their preference for a product, their confidence about an answer, how much they have understood some medical instructions) we should probably regard these data as ordinal although many scientists do not.

[8] Unlike, for example, NFL American football where a quarterback could wear any number from 1 to 19.

So far the categorical variables we have considered have been unordered (e.g. different brands of Coke with which you're trying to kill sperm), but they can be ordered too (e.g. increasing concentrations of Coke with which you're trying to skill sperm). When categories are ordered, the variable is known as an **ordinal variable.** Ordinal data tell us not only that things have occurred, but also the order in which they occurred. However, these data tell us nothing about the differences between values. Imagine we went to a beauty pageant in which the three winners were Billie, Freema and Elizabeth. The names of the winners don't provide any information about where they came in the contest; however, labelling them according to their performance does – first, second and third. These categories are ordered. In using ordered categories we now know that the woman who won was better than the women who came second and third. We still know nothing about the differences between categories, though. We don't, for example, know how much better the winner was than the runners-up: Billie might have been an easy victor, getting much higher ratings from the judges than Freema and Elizabeth, or it might have been a very close contest that she won by only a point. Ordinal data, therefore, tell us more than nominal data (they tell us the order in which things happened) but they still do not tell us about the differences between points on a scale.

The next level of measurement moves us away from categorical variables and into continuous variables. A **continuous variable** is one that gives us a score for each person and can take on any value on the measurement scale that we are using. The first type of continuous variable that you might encounter is an **interval variable.** Interval data are considerably more useful than ordinal data and most of the statistical tests in this book rely on having data measured at this level. To say that data are interval, we must be certain that equal intervals on the scale represent equal differences in the property being measured. For example, on www.ratemyprofessors.com students are encouraged to rate their lecturers on several dimensions (some of the lecturers' rebuttals of their negative evaluations are worth a look). Each dimension (i.e. helpfulness, clarity, etc.) is evaluated using a 5-point scale. For this scale to be interval it must be the case that the difference between helpfulness ratings of 1 and 2 is the same as the difference between say 3 and 4, or 4 and 5. Similarly, the difference in helpfulness between ratings of 1 and 3 should be identical to the difference between ratings of 3 and 5. Variables like this that look interval (and are treated as interval) are often ordinal – see Jane Superbrain Box 1.2.

Ratio variables go a step further than interval data by requiring that in addition to the measurement scale meeting the requirements of an interval variable, the ratios of values along the scale should be meaningful. For this to be true, the scale must have a true and meaningful zero point. In our lecturer ratings this would mean that a lecturer rated as 4 would be twice as helpful as a lecturer rated with a 2 (who would also be twice as helpful as a lecturer rated as 1!). The time to respond to something is a good example of a ratio variable. When we measure a reaction time, not only is it true that, say, the difference between 300 and 350 ms (a difference of 50 ms) is the same as the difference between 210 and 260 ms or 422 and 472 ms, but also it is true that distances along the scale are divisible: a reaction time of 200 ms is twice as long as a reaction time of 100 ms and twice as short as a reaction time of 400 ms.

Continuous variables can be, well, continuous (obviously) but also discrete. This is quite a tricky distinction (Jane Superbrain Box 1.3). A truly continuous variable can be measured to any level of precision, whereas a **discrete variable** can take on only certain values (usually whole numbers) on the scale. What does this actually mean? Well, our example above of rating lecturers on a 5-point scale is an example of a discrete variable. The range of the scale is 1–5, but you can enter only values of 1, 2, 3, 4 or 5; you cannot enter a value of 4.32 or 2.18. Although a continuum exists underneath the scale (i.e. a rating of 3.24 makes sense), the actual values that the variable takes on are limited. A continuous variable would be something like age, which can be measured at an infinite level of precision (you could be 34 years, 7 months, 21 days, 10 hours, 55 minutes, 10 seconds, 100 milliseconds, 63 microseconds, 1 nanosecond old).

JANE SUPERBRAIN 1.3

Continuous and discrete variables ①

The distinction between discrete and continuous variables can be very blurred. For one thing, continuous variables can be measured in discrete terms; for example, when we measure age we rarely use nanoseconds but use years (or possibly years and months). In doing so we turn a continuous variable into a discrete one (the only acceptable values are years). Also, we often treat discrete variables as if they were continuous. For example, the number of boyfriends/girlfriends that you have had is a discrete variable (it will be, in all but the very weird cases, a whole number). However, you might read a magazine that says 'the average number of boyfriends that women in their 20s have has increased from 4.6 to 8.9'. This assumes that the variable is continuous, and of course these averages are meaningless: no one in their sample actually had 8.9 boyfriends.

CRAMMING SAM'S TIPS Levels of Measurement

Variables can be split into categorical and continuous, and within these types there are different levels of measurement:

- **Categorical (entities are divided into distinct categories):**
  - **Binary variable:** There are only two categories (e.g. dead or alive).
  - **Nominal variable:** There are more than two categories (e.g. whether someone is an omnivore, vegetarian, vegan, or fruitarian).
  - **Ordinal variable:** The same as a nominal variable but the categories have a logical order (e.g. whether people got a fail, a pass, a merit or a distinction in their exam).
- **Continuous (entities get a distinct score):**
  - **Interval variable:** Equal intervals on the variable represent equal differences in the property being measured (e.g. the difference between 6 and 8 is equivalent to the difference between 13 and 15).
  - **Ratio variable:** The same as an interval variable, but the ratios of scores on the scale must also make sense (e.g. a score of 16 on an anxiety scale means that the person is, in reality, twice as anxious as someone scoring 8).

1.5.2. Measurement error ①

We have seen that to test hypotheses we need to measure variables. Obviously, it's also important that we measure these variables accurately. Ideally we want our measure to be calibrated such that values have the same meaning over time and across situations. Weight is one example: we would expect to weigh the same amount regardless of who weighs us, or where we take the measurement (assuming it's on Earth and not in an anti-gravity chamber). Sometimes variables can be directly measured (profit, weight, height) but in other cases we are forced to use indirect measures such as self-report, questionnaires and computerized tasks (to name a few).

Let's go back to our Coke as a spermicide example. Imagine we took some Coke and some water and added them to two test tubes of sperm. After several minutes, we measured the motility (movement) of the sperm in the two samples and discovered no difference. A few years passed and another scientist, Dr Jack Q. Late, replicated the study but found that sperm motility was worse in the Coke sample. There are two measurement-related issues that could explain his success and our failure: (1) Dr Late might have used more Coke in the test tubes (sperm might need a critical mass of Coke before they are affected); (2) Dr Late measured the outcome (motility) differently to us.

The former point explains why chemists and physicists have devoted many hours to developing standard units of measurement. If you had reported that you'd used 100 ml of Coke and 5 ml of sperm, then Dr Late could have ensured that he had used the same amount – because millilitres are a standard unit of measurement we would know that Dr Late used exactly the same amount of Coke that we used. Direct measurements such as the millilitre provide an objective standard: 100 ml of a liquid is known to be twice as much as only 50 ml.

The second reason for the difference in results between the studies could have been to do with how sperm motility was measured. Perhaps in our original study we measured motility using absorption spectrophotometry, whereas Dr Late used laser light-scattering techniques.[9] Perhaps his measure is more sensitive than ours.

There will often be a discrepancy between the numbers we use to represent the thing we're measuring and the actual value of the thing we're measuring (i.e. the value we would get if we could measure it directly). This discrepancy is known as **measurement error.** For example, imagine that you know as an absolute truth that you weigh 83 kg. One day you step on the bathroom scales and it says 80 kg. There is a difference of 3 kg between your actual weight and the weight given by your measurement tool (the scales): there is a measurement error of 3 kg. Although properly calibrated bathroom scales should produce only very small measurement errors (despite what we might want to believe when it says we have gained 3 kg), self-report measures do produce measurement error because factors other than the one you're trying to measure will influence how people respond to our measures. Imagine you were completing a questionnaire that asked you whether you had stolen from a shop. If you had, would you admit it, or might you be tempted to conceal this fact?

1.5.3. Validity and reliability ①

One way to try to ensure that measurement error is kept to a minimum is to determine properties of the measure that give us confidence that it is doing its job properly. The first property is **validity,** which is whether an instrument actually measures what it sets out to measure. The second is **reliability,** which is whether an instrument can be interpreted consistently across different situations.

Validity refers to whether an instrument measures what it was designed to measure; a device for measuring sperm motility that actually measures sperm count is not valid. Things like reaction times and physiological measures are valid in the sense that a reaction time does in fact measure the time taken to react and skin conductance does measure the conductivity of your skin. However, if we're using these things to infer other things (e.g. using skin conductance to measure anxiety) then they will be valid only if there are no other factors other than the one we're interested in that can influence them.

Criterion validity is whether the instrument is measuring what it claims to measure (does your lecturer's helpfulness rating scale actually measure lecturers' helpfulness?). In an ideal world, you could assess this by relating scores on your measure to real-world observations.

[9] In the course of writing this chapter I have discovered more than I think is healthy about the measurement of sperm.

For example, we could take an objective measure of how helpful lecturers were and compare these observations to student's ratings on ratemyprofessor.com. This is often impractical and, of course, with attitudes you might not be interested in the reality so much as the person's perception of reality (you might not care whether they are a psychopath but whether they think they are a psychopath). With self-report measures/questionnaires we can also assess the degree to which individual items represent the construct being measured, and cover the full range of the construct (**content validity**).

Validity is a necessary but not sufficient condition of a measure. A second consideration is reliability, which is the ability of the measure to produce the same results under the same conditions. To be valid the instrument must first be reliable. The easiest way to assess reliability is to test the same group of people twice: a reliable instrument will produce similar scores at both points in time (**test–retest reliability**). Sometimes, however, you will want to measure something that does vary over time (e.g. moods, blood-sugar levels, productivity). Statistical methods can also be used to determine reliability (we will discover these in Chapter 17).

SELF-TEST What is the difference between reliability and validity?

1.6. Data collection 2: how to measure ①

1.6.1. Correlational research methods ①

So far we've learnt that scientists want to answer questions, and that to do this they have to generate data (be they numbers or words), and to generate good data they need to use accurate measures. We move on now to look briefly at how the data are collected. If we simplify things quite a lot then there are two ways to test a hypothesis: either by observing what naturally happens, or by manipulating some aspect of the environment and observing the effect it has on the variable that interests us.

The main distinction between what we could call **correlational** or **cross-sectional research** (where we observe what naturally goes on in the world without directly interfering with it) and **experimental research** (where we manipulate one variable to see its effect on another) is that experimentation involves the direct manipulation of variables. In correlational research we do things like observe natural events or we take a snapshot of many variables at a single point in time. As some examples, we might measure pollution levels in a stream and the numbers of certain types of fish living there; lifestyle variables (smoking, exercise, food intake) and disease (cancer, diabetes); workers' job satisfaction under different managers; or children's school performance across regions with different demographics. Correlational research provides a very natural view of the question we're researching because we are not influencing what happens and the measures of the variables should not be biased by the researcher being there (this is an important aspect of **ecological validity**).

At the risk of sounding like I'm absolutely obsessed with using Coke as a contraceptive (I'm not, but my discovery that people in the 1950s and 1960s actually tried this has, I admit, intrigued me), let's return to that example. If we wanted to answer the question

'Is Coke an effective contraceptive?' we could administer questionnaires about sexual practices (quantity of sexual activity, use of contraceptives, use of fizzy drinks as contraceptives, pregnancy, etc.). By looking at these variables we could see which variables predict pregnancy, and in particular whether those reliant on coca-cola as a form of contraceptive were more likely to end up pregnant than those using other contraceptives, and less likely than those using no contraceptives at all. This is the only way to answer a question like this because we cannot manipulate any of these variables particularly easily. Even if we could, it would be totally unethical to insist on some people using Coke as a contraceptive (or indeed to do anything that would make a person likely to produce a child that they didn't intend to produce). However, there is a price to pay, which relates to causality.

1.6.2. Experimental research methods ①

Most scientific questions imply a causal link between variables; we have seen already that dependent and independent variables are named such that a causal connection is implied (the dependent variable *depends* on the independent variable). Sometimes the causal link is very obvious in the research question 'Does low self-esteem cause dating anxiety?' Sometimes the implication might be subtler, such as 'Is dating anxiety all in the mind?' The implication is that a person's mental outlook causes them to be anxious when dating. Even when the cause–effect relationship is not explicitly stated, most research questions can be broken down into a proposed cause (in this case mental outlook) and a proposed outcome (dating anxiety). Both the cause and the outcome are variables: for the cause some people will perceive themselves in a negative way (so it is something that varies); and for the outcome, some people will get anxious on dates and others won't (again, this is something that varies). The key to answering the research question is to uncover how the proposed cause and the proposed outcome relate to each other; is it the case that the people who have a low opinion of themselves are the same people that get anxious on dates?

David Hume (see Hume, 1739–40; 1748 for more detail),[10] an influential philosopher, said that to infer cause and effect: (1) cause and effect must occur close together in time (contiguity); (2) the cause must occur before an effect does; and (3) the effect should never occur without the presence of the cause. These conditions imply that causality can be inferred through corroborating evidence: cause is equated to high degrees of correlation between contiguous events. In our dating example, to infer that low self-esteem caused dating anxiety, it would be sufficient to find that whenever someone had low self-esteem they would feel anxious when on a date, that the low self-esteem emerged before the dating anxiety did, and that the person should never have dating anxiety if they haven't been suffering from low self-esteem.

In the previous section on correlational research, we saw that variables are often measured simultaneously. The first problem with doing this is that it provides no information about the contiguity between different variables: we might find from a questionnaire study that people with low self-esteem also have dating anxiety but we wouldn't know whether the low self-esteem or the dating anxiety came first.

Let's imagine that we find that there are people who have low self-esteem but do not get dating anxiety. This finding doesn't violate Hume's rules: he doesn't say anything about the cause happening without the effect. It could be that both low self-esteem and dating anxiety are caused by a third variable (e.g., poor social skills which might make you feel generally worthless but also put pressure on you in dating situations). This illustrates a second problem

[10] Both of these can be read online at http://www.utilitarian.net/hume/ or by doing a Google search for David Hume.

with correlational evidence: the ***tertium quid*** ('a third person or thing of indeterminate character'). For example, a correlation has been found between having breast implants and suicide (Koot, Peeters, Granath, Grobbee, & Nyren, 2003). However, it is unlikely that having breast implants causes you to commit suicide – presumably, there is an external factor (or factors) that causes both; for example, low self-esteem might lead you to have breast implants and also attempt suicide. These extraneous factors are sometimes called **confounding variables** or confounds for short.

What's the difference between experimental and correlational research?

The shortcomings of Hume's criteria led John Stuart Mill (1865) to add a further criterion: that all other explanations of the cause–effect relationship be ruled out. Put simply, Mill proposed that, to rule out confounding variables, an effect should be present when the cause is present and that when the cause is absent the effect should be absent also. Mill's ideas can be summed up by saying that the only way to infer causality is through comparison of two controlled situations: one in which the cause is present and one in which the cause is absent. This is what *experimental methods* strive to do: to provide a comparison of situations (usually called *treatments* or *conditions*) in which the proposed cause is present or absent.

As a simple case, we might want to see what the effect of positive encouragement has on learning about statistics. I might, therefore, randomly split some students into three different groups in which I change my style of teaching in the seminars on the course:

- **Group 1 (positive reinforcement)**: During seminars I congratulate all students in this group on their hard work and success. Even when they get things wrong, I am supportive and say things like 'that was very nearly the right answer, you're coming along really well' and then give them a nice piece of chocolate.
- **Group 2 (negative reinforcement)**: This group receives seminars in which I give relentless verbal abuse to all of the students even when they give the correct answer. I demean their contributions and am patronizing and dismissive of everything they say. I tell students that they are stupid, worthless and shouldn't be doing the course at all.
- **Group 3 (no reinforcement)**: This group receives normal university style seminars (some might argue that this is the same as group 2!). Students are not praised or punished and instead I give them no feedback at all.

The thing that I have manipulated is the teaching method (positive reinforcement, negative reinforcement or no reinforcement). As we have seen earlier in this chapter, this variable is known as the independent variable and in this situation it is said to have three *levels*, because it has been manipulated in three ways (i.e. reinforcement has been split into three types: positive, negative and none). Once I have carried out this manipulation I must have some kind of outcome that I am interested in measuring. In this case it is statistical ability, and I could measure this variable using a statistics exam after the last seminar. We have also already discovered that this outcome variable is known as the dependent variable because we assume that these scores will depend upon the type of teaching method used (the independent variable). The critical thing here is the inclusion of the 'no reinforcement' group because this is a group where our proposed cause (reinforcement) is absent, and we can compare the outcome in this group against the two situations where the proposed cause is present. If the statistics scores are different in each of the reinforcement groups (cause is present) compared to the group for which no reinforcement was given (cause is absent) then this difference can be attributed to the style of reinforcement. In other words, the type of reinforcement caused a difference in statistics scores (Jane Superbrain Box 1.4).

JANE SUPERBRAIN 1.4

Causality and statistics ①

People sometimes get confused and think that certain statistical procedures allow causal inferences and others don't. This isn't true, it's the fact that in experiments we manipulate the causal variable systematically to see its effect on an outcome (the effect). In correlational research we observe the co-occurrence of variables; we do not manipulate the causal variable first and then measure the effect, therefore we cannot compare the effect when the causal variable is present against when it is absent. In short, we cannot say which variable causes a change in the other; we can merely say that the variables co-occur in a certain way. The reason why some people think that certain statistical tests allow causal inferences is because historically certain tests (e.g. ANOVA, *t*-tests, etc.) have been used to analyse experimental research, whereas others (e.g. regression, correlation) have been used to analyse correlational research (Cronbach, 1957). As you'll discover, these statistical procedures are, in fact, mathematically identical.

1.6.2.1. Two methods of data collection ①

When we collect data in an experiment, we can choose between two methods of data collection. The first is to manipulate the independent variable using different participants. This method is the one described above, in which different groups of people take part in each experimental condition (a **between-groups**, **between-subjects**, or **independent design**). The second method is to manipulate the independent variable using the same participants. Simplistically, this method means that we give a group of students positive reinforcement for a few weeks and test their statistical abilities and then begin to give this same group negative reinforcement for a few weeks before testing them again, and then finally giving them no reinforcement and testing them for a third time (a **within-subject** or **repeated-measures design**). As you will discover, the way in which the data are collected determines the type of test that is used to analyse the data.

1.6.2.2. Two types of variation ①

Imagine we were trying to see whether you could train chimpanzees to run the economy. In one training phase they are sat in front of a chimp-friendly computer and press buttons which change various parameters of the economy; once these parameters have been changed a figure appears on the screen indicating the economic growth resulting from those parameters. Now, chimps can't read (I don't think) so this feedback is meaningless. A second training phase is the same except that if the economic growth is good, they get a banana (if growth is bad they do not) – this feedback is valuable to the average chimp. This is a repeated-measures design with two conditions: the same chimps participate in condition 1 *and* in condition 2.

Let's take a step back and think what would happen if we did *not* introduce an experimental manipulation (i.e. there were no bananas in the second training phase so condition 1 and condition 2 were identical). If there is no experimental manipulation then we expect a chimp's behaviour to be similar in both conditions. We expect this because external factors such as age, gender, IQ, motivation and arousal will be the same for both conditions (a chimp's gender etc. will not change from when they are tested in condition 1 to when they are tested in condition 2). If the performance measure is reliable (i.e. our test of how well they run the economy), and the variable or characteristic that we are measuring (in this case ability to run an economy) remains stable over time, then a participant's performance in condition 1 should be very highly related to their performance in condition 2. So, chimps who score highly in condition 1 will also score highly in condition 2, and those who have low scores for condition 1 will have low scores in condition 2. However, performance won't be *identical*, there will be small differences in performance created by unknown factors. This variation in performance is known as **unsystematic variation.**

If we introduce an experimental manipulation (i.e. provide bananas as feedback in one of the training sessions), then we do something different to participants in condition 1 to what we do to them in condition 2. So, the *only* difference between conditions 1 and 2 is the manipulation that the experimenter has made (in this case that the chimps get bananas as a positive reward in one condition but not in the other). Therefore, any difference between the means of the two conditions is probably due to the experimental manipulation. So, if the chimps perform better in one training phase than the other then this *has* to be due to the fact that bananas were used to provide feedback in one training phase but not the other. Differences in performance created by a specific experimental manipulation are known as **systematic variation.**

Now let's think about what happens when we use different participants – an independent design. In this design we still have two conditions, but this time different participants participate in each condition. Going back to our example, one group of chimps receives training without feedback, whereas a second group of different chimps does receive feedback on their performance via bananas.[11] Imagine again that we didn't have an experimental manipulation. If we did nothing to the groups, then we would still find some variation in behaviour between the groups because they contain different chimps who will vary in their ability, motivation, IQ and other factors. In short, the type of factors that were held constant in the repeated-measures design are free to vary in the independent measures design. So, the unsystematic variation will be bigger than for a repeated-measures design. As before, if we introduce a manipulation (i.e. bananas) then we will see additional variation created by this manipulation. As such, in both the repeated-measures design and the independent-measures design there are always two sources of variation:

- **Systematic variation**: This variation is due to the experimenter doing something to all of the participants in one condition but not in the other condition.
- **Unsystematic variation**: This variation results from random factors that exist between the experimental conditions (such as natural differences in ability, the time of day, etc.).

The role of statistics is to discover how much variation there is in performance, and then to work out how much of this is systematic and how much is unsystematic.

In a repeated-measures design, differences between two conditions can be caused by only two things: (1) the manipulation that was carried out on the participants, or (2) any other factor that might affect the way in which a person performs from one time to the next. The latter factor is likely to be fairly minor compared to the influence of the experimental manipulation. In an independent design, differences between the two conditions can also be caused by one of two things: (1) the manipulation that was carried out on the participants, or (2) differences between the characteristics of the people allocated to each of the groups. The latter factor in this instance is likely to create considerable random variation both within each condition and between them. Therefore, the effect of our experimental manipulation is likely to be more apparent in a repeated-measures design than in a between-groups design,

[11] When I say 'via' I don't mean that the bananas developed little banana mouths that opened up and said 'well done old chap, the economy grew that time' in chimp language. I mean that when they got something right they received a banana as a reward for their correct response.

because in the former unsystematic variation can be caused only by differences in the way in which someone behaves at different times. In independent designs we have differences in innate ability contributing to the unsystematic variation. Therefore, this error variation will almost always be much larger than if the same participants had been used. When we look at the effect of our experimental manipulation, it is always against a background of 'noise' caused by random, uncontrollable differences between our conditions. In a repeated-measures design this 'noise' is kept to a minimum and so the effect of the experiment is more likely to show up. This means that, other things being equal, repeated-measures designs have more power to detect effects than independent designs.

1.6.3. Randomization ①

In both repeated measures and independent measures designs it is important to try to keep the unsystematic variation to a minimum. By keeping the unsystematic variation as small as possible we get a more sensitive measure of the experimental manipulation. Generally, scientists use the **randomization** of participants to treatment conditions to achieve this goal. Many statistical tests work by identifying the systematic and unsystematic sources of variation and then comparing them. This comparison allows us to see whether the experiment has generated considerably more variation than we would have got had we just tested participants without the experimental manipulation. Randomization is important because it eliminates most other sources of systematic variation, which allows us to be sure that any systematic variation between experimental conditions is due to the manipulation of the independent variable. We can use randomization in two different ways depending on whether we have an independent or repeated-measures design.

Let's look at a repeated-measures design first. When the same people participate in more than one experimental condition they are naive during the first experimental condition but they come to the second experimental condition with prior experience of what is expected of them. At the very least they will be familiar with the dependent measure (e.g. the task they're performing). The two most important sources of systematic variation in this type of design are:

- **Practice effects**: Participants may perform differently in the second condition because of familiarity with the experimental situation and/or the measures being used.
- **Boredom effects**: Participants may perform differently in the second condition because they are tired or bored from having completed the first condition.

Although these effects are impossible to eliminate completely, we can ensure that they produce no systematic variation between our conditions by **counterbalancing** the order in which a person participates in a condition.

We can use randomization to determine in which order the conditions are completed. That is, we randomly determine whether a participant completes condition 1 before condition 2, or condition 2 before condition 1. Let's look at the teaching method example and imagine that there were just two conditions: no reinforcement and negative reinforcement. If the same participants were used in all conditions, then we might find that statistical ability was higher after the negative reinforcement condition. However, if every student experienced the negative reinforcement after the no reinforcement then they would enter the negative reinforcement condition already having a better knowledge of statistics than when they began the no reinforcement condition. So, the apparent improvement after negative reinforcement would not be due to the experimental manipulation (i.e. it's not because negative reinforcement works), but because participants had attended more statistics seminars by the end of the negative reinforcement condition compared to the no reinforcement one. We can use randomization to ensure that the number of statistics seminars does not introduce a systematic bias by randomly assigning students to have the negative reinforcement seminars first or the no reinforcement seminars first.

If we turn our attention to independent designs, a similar argument can be applied. We know that different participants participate in different experimental conditions and that these participants will differ in many respects (their IQ, attention span, etc.). Although we know that these confounding variables contribute to the variation between conditions, we need to make sure that these variables contribute to the unsystematic variation and *not* the systematic variation. The way to ensure that confounding variables are unlikely to contribute systematically to the variation between experimental conditions is to randomly allocate participants to a particular experimental condition. This should ensure that these confounding variables are evenly distributed across conditions.

A good example is the effects of alcohol on personality. You might give one group of people 5 pints of beer, and keep a second group sober, and then count how many fights each person gets into. The effect that alcohol has on people can be very variable because of different tolerance levels: teetotal people can become very drunk on a small amount, while alcoholics need to consume vast quantities before the alcohol affects them. Now, if you allocated a bunch of teetotal participants to the condition that consumed alcohol, then you might find no difference between them and the sober group (because the teetotal participants are all unconscious after the first glass and so can't become involved in any fights). As such, the person's prior experiences with alcohol will create systematic variation that cannot be dissociated from the effect of the experimental manipulation. The best way to reduce this eventuality is to randomly allocate participants to conditions.

SELF-TEST Why is randomization important?

1.7. Analysing data ①

The final stage of the research process is to analyse the data you have collected. When the data are quantitative this involves both looking at your data graphically to see what the general trends in the data are, and also fitting statistical models to the data.

1.7.1. Frequency distributions ①

Once you've collected some data a very useful thing to do is to plot a graph of how many times each score occurs. This is known as a **frequency distribution**, or **histogram**, which is a graph plotting values of observations on the horizontal axis, with a bar showing how many times each value occurred in the data set. Frequency distributions can be very useful for assessing properties of the distribution of scores. We will find out how to create these types of charts in Chapter 4.

Frequency distributions come in many different shapes and sizes. It is quite important, therefore, to have some general descriptions for common types of distributions. In an ideal world our data would be distributed symmetrically around the centre of all scores. As such, if we drew a vertical line through the centre of the distribution then it should look the same on both sides. This is known as a **normal distribution** and is characterized by the bell-shaped curve with which you might already be familiar. This shape basically implies that the majority of scores lie around the centre of the distribution (so the largest bars on the histogram are all around the central value).

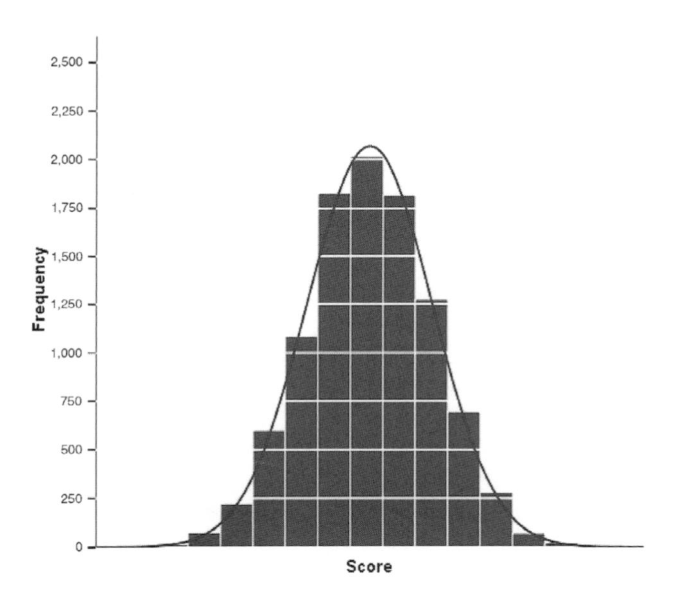

FIGURE 1.3 A 'normal' distribution (the curve shows the idealized shape)

Also, as we get further away from the centre the bars get smaller, implying that as scores start to deviate from the centre their frequency is decreasing. As we move still further away from the centre our scores become very infrequent (the bars are very short). Many naturally occurring things have this shape of distribution. For example, most men in the UK are about 175 cm tall,[12] some are a bit taller or shorter but most cluster around this value. There will be very few men who are really tall (i.e. above 205 cm) or really short (i.e. under 145 cm). An example of a normal distribution is shown in Figure 1.3.

There are two main ways in which a distribution can deviate from normal: (1) lack of symmetry (called **skew**) and (2) pointyness (called **kurtosis**). Skewed distributions are not symmetrical and instead the most frequent scores (the tall bars on the graph) are clustered at one end of the scale. So, the typical pattern is a cluster of frequent scores at one end of the scale and the frequency of scores tailing off towards the other end of the scale. A skewed distribution can be either *positively skewed* (the frequent scores are clustered at the lower end and the tail points towards the higher or more positive scores) or *negatively skewed* (the frequent scores are clustered at the higher end and the tail points towards the lower or more negative scores). Figure 1.4 shows examples of these distributions.

Distributions also vary in their kurtosis. Kurtosis, despite sounding like some kind of exotic disease, refers to the degree to which scores cluster at the ends of the distribution (known as the *tails*) and how pointy a distribution is (but there are other factors that can affect how pointy the distribution looks – see Jane Superbrain Box 2.3). A distribution with *positive kurtosis* has many scores in the tails (a so-called heavy-tailed distribution) and is pointy. This is known as a **leptokurtic** distribution. In contrast, a distribution with *negative kurtosis* is relatively thin in the tails (has light tails) and tends to be flatter than normal. This distribution is called **platykurtic**. Ideally, we want our data to be normally distributed (i.e. not too skewed, and not too many or too few scores at the extremes!). For everything there is to know about kurtosis read DeCarlo (1997).

In a normal distribution the values of skew and kurtosis are 0 (i.e. the tails of the distribution are as they should be). If a distribution has values of skew or kurtosis above or below 0 then this indicates a deviation from normal: Figure 1.5 shows distributions with kurtosis values of +1 (left panel) and –4 (right panel).

[12] I am exactly 180 cm tall. In my home country this makes me smugly above average. However, I'm writing this in The Netherlands where the average male height is 185 cm (a massive 10 cm higher than the UK), and where I feel like a bit of a dwarf.

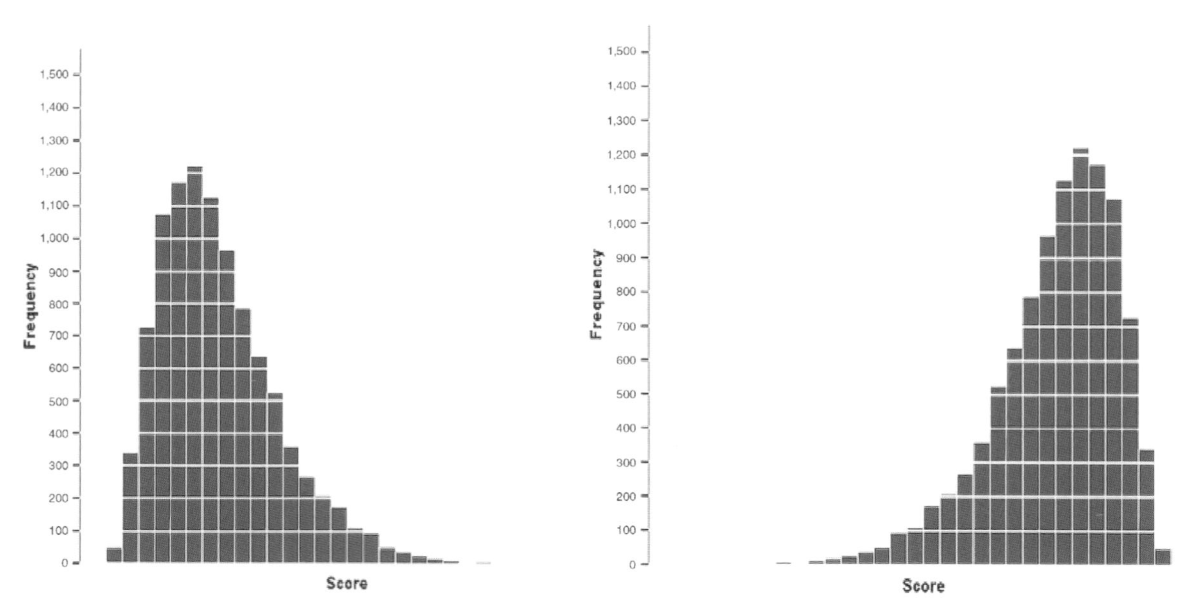

FIGURE 1.4 A positively (left figure) and negatively (right figure) skewed distribution

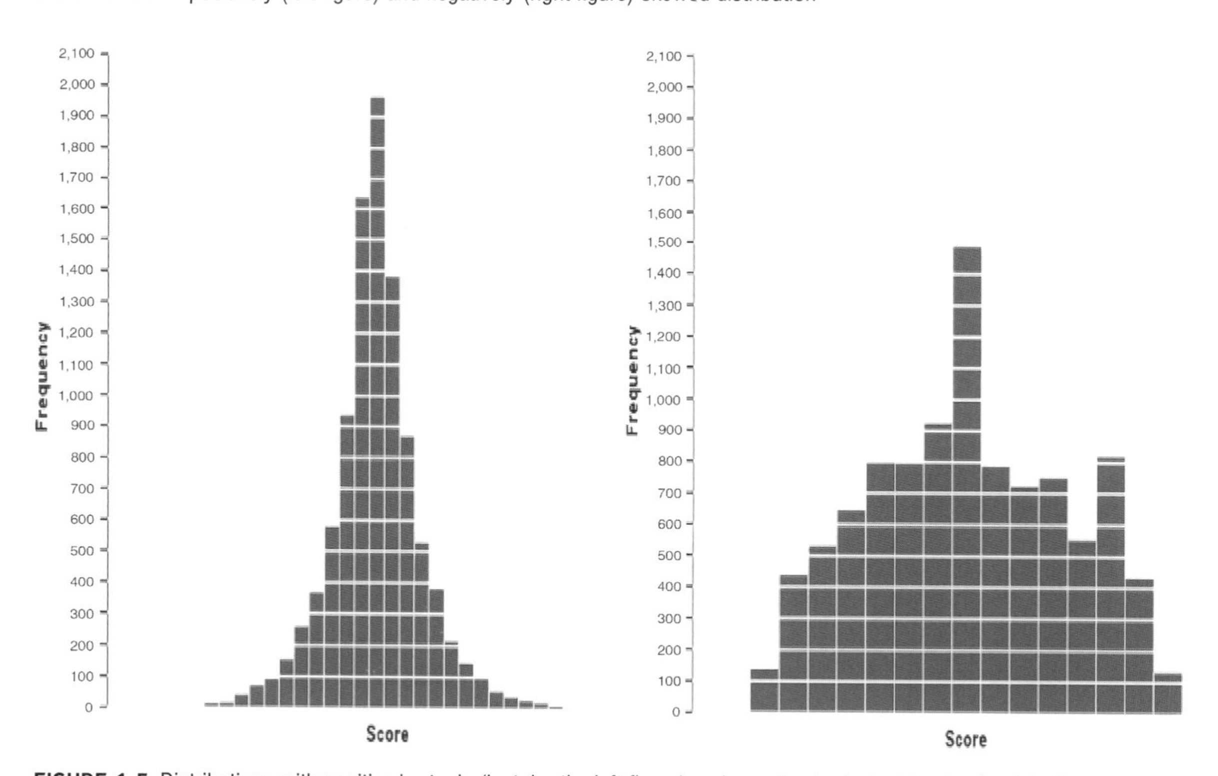

FIGURE 1.5 Distributions with positive kurtosis (leptokurtic, left figure) and negative kurtosis (platykurtic, right figure)

1.7.2. The centre of a distribution ①

We can also calculate where the centre of a frequency distribution lies (known as the **central tendency**). There are three measures commonly used: the mean, the mode and the median.

1.7.2.1. The mode ①

The **mode** is simply the score that occurs most frequently in the data set. This is easy to spot in a frequency distribution because it will be the tallest bar! To calculate the mode, simply place the data in ascending order (to make life easier), count how many times each score occurs, and the score that occurs the most is the mode! One problem with the mode is that it can often take on several values. For example, Figure 1.6 shows an example of a distribution with two modes (there are two bars that are the highest), which is said to be **bimodal**. It's also possible to find data sets with more than two modes (**multimodal**). Also, if the frequencies of certain scores are very similar, then the mode can be influenced by only a small number of cases.

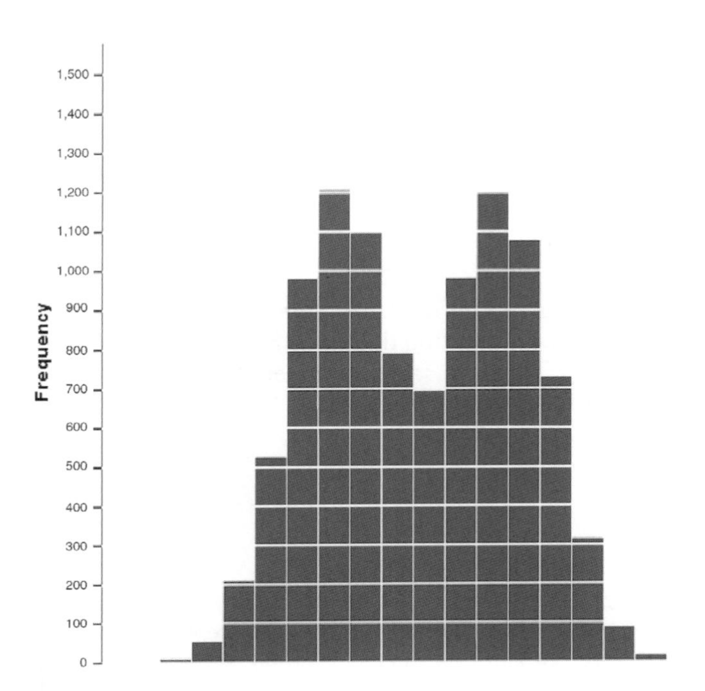

FIGURE 1.6
A bimodal distribution

1.7.2.2. The median ①

Another way to quantify the centre of a distribution is to look for the middle score when scores are ranked in order of magnitude. This is called the **median**. For example, Facebook is a popular social networking website, in which users can sign up to be 'friends' of other users. Imagine we looked at the number of friends that a selection (actually, some of my friends) of 11 Facebook users had. Number of friends: 108, 103, 252, 121, 93, 57, 40, 53, 22, 116, 98.

To calculate the median, we first arrange these scores into ascending order: 22, 40, 53, 57, 93, 98, 103, 108, 116, 121, 252.

Next, we find the position of the middle score by counting the number of scores we have collected (n), adding 1 to this value, and then dividing by 2. With 11 scores, this gives us $(n + 1)/2 = (11 + 1)/2 = 12/2 = 6$. Then, we find the score that is positioned at the location we have just calculated. So, in this example we find the sixth score:

22, 40, 53, 57, 93, 98, 103, 108, 116, 121, 252

Median

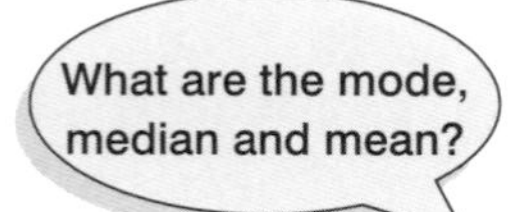

This works very nicely when we have an odd number of scores (as in this example) but when we have an even number of scores there won't be a middle value. Let's imagine that we decided that because the highest score was so big (more than twice as large as the next biggest number), we would ignore it. (For one thing, this person is far too popular and we hate them.) We have only 10 scores now. As before, we should rank-order these scores: 22, 40, 53, 57, 93, 98, 103, 108, 116, 121. We then calculate the position of the middle score, but this time it is $(n + 1)/2 = 11/2 = 5.5$. This means that the median is halfway between the fifth and sixth scores. To get the median we add these two scores and divide by 2. In this example, the fifth score in the ordered list was 93 and the sixth score was 98. We add these together (93 + 98 = 191) and then divide this value by 2 (191/2 = 95.5). The median number of friends was, therefore, 95.5.

The median is relatively unaffected by extreme scores at either end of the distribution: the median changed only from 98 to 95.5 when we removed the extreme score of 252. The median is also relatively unaffected by skewed distributions and can be used with ordinal, interval and ratio data (it cannot, however, be used with nominal data because these data have no numerical order).

1.7.2.3. The mean ①

The **mean** is the measure of central tendency that you are most likely to have heard of because it is simply the average score and the media are full of average scores.[13] To calculate the mean we simply add up all of the scores and then divide by the total number of scores we have. We can write this in equation form as:

$$\overline{X} = \frac{\sum_{i=1}^{n} x_i}{n} \tag{1.1}$$

This may look complicated, but the top half of the equation simply means 'add up all of the scores' (the x_i just means 'the score of a particular person'; we could replace the letter i with each person's name instead), and the bottom bit means divide this total by the number of scores you have got (n). Let's calculate the mean for the Facebook data. First, we first add up all of the scores:

$$\begin{aligned}\sum_{i=1}^{n} x_i &= 22 + 40 + 53 + 57 + 93 + 98 + 103 + 108 + 116 + 121 + 252 \\ &= 1063\end{aligned}$$

We then divide by the number of scores (in this case 11):

$$\overline{X} = \frac{\sum_{i=1}^{n} x_i}{n} = \frac{1063}{11} = 96.64$$

The mean is 96.64 friends, which is not a value we observed in our actual data (it would be ridiculous to talk of having 0.64 of a friend). In this sense the mean is a statistical model – more on this in the next chapter.

[13] I'm writing this on 15 February 2008, and to prove my point the BBC website is running a headline about how PayPal estimates that Britons will spend an average of £71.25 each on Valentine's Day gifts, but uSwitch.com said that the average spend would be £22.69!

SELF-TEST Compute the mean but excluding the score of 252.

If you calculate the mean without our extremely popular person (i.e. excluding the value 252), the mean drops to 81.1 friends. One disadvantage of the mean is that it can be influenced by extreme scores. In this case, the person with 252 friends on Facebook increased the mean by about 15 friends! Compare this difference with that of the median. Remember that the median hardly changed if we included or excluded 252, which illustrates how the median is less affected by extreme scores than the mean. While we're being negative about the mean, it is also affected by skewed distributions and can be used only with interval or ratio data.

If the mean is so lousy then why do we use it all of the time? One very important reason is that it uses every score (the mode and median ignore most of the scores in a data set). Also, the mean tends to be stable in different samples.

1.7.3. The dispersion in a distribution ①

It can also be interesting to try to quantify the spread, or dispersion, of scores in the data. The easiest way to look at dispersion is to take the largest score and subtract from it the smallest score. This is known as the **range** of scores. For our Facebook friends data, if we order these scores we get 22, 40, 53, 57, 93, 98, 103, 108, 116, 121, 252. The highest score is 252 and the lowest is 22; therefore, the range is 252 – 22 = 230. One problem with the range is that because it uses only the highest and lowest score it is affected dramatically by extreme scores.

SELF-TEST Compute the range but excluding the score of 252.

If you have done the self-test task you'll see that without the extreme score the range drops dramatically from 230 to 99 – less than half the size!

One way around this problem is to calculate the range when we exclude values at the extremes of the distribution. One convention is to cut off the top and bottom 25% of scores and calculate the range of the middle 50% of scores – known as the **interquartile range.** Let's do this with the Facebook data. First we need to calculate what are called **quartiles.** Quartiles are the three values that split the sorted data into four equal parts. First we calculate the median, which is also called the **second quartile**, which splits our data into two equal parts. We already know that the median for these data is 98. The **lower quartile** is the median of the lower half of the data and the **upper quartile** is the median of the upper half of the data. One rule of thumb is that the median is not included in the two halves when they are split (this is convenient if you have an odd number of values), but you can include it (although which half you put it in is another question). Figure 1.7 shows how we would calculate these values for the Facebook data. Like the median, the upper and lower quartile need not be values that actually appear in the data (like the median, if each half of the data had an even number of values in it then the upper and lower quartiles would be the average

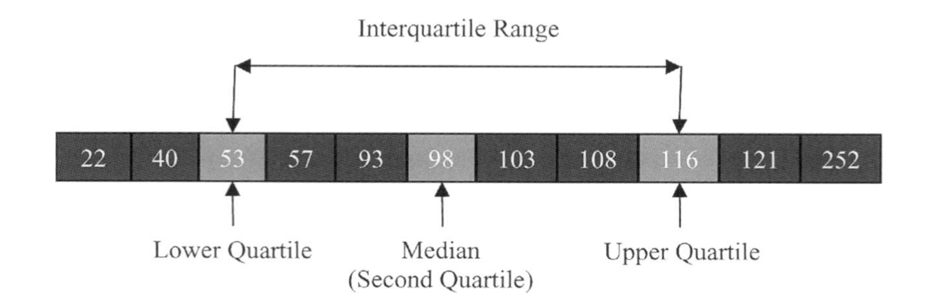

FIGURE 1.7 Calculating quartiles and the interquartile range

of two values in the data set). Once we have worked out the values of the quartiles, we can calculate the interquartile range, which is the difference between the upper and lower quartile. For the Facebook data this value would be 116–53 = 63. The advantage of the interquartile range is that it isn't affected by extreme scores at either end of the distribution. However, the problem with it is that you lose a lot of data (half of it in fact!).

SELF-TEST Twenty-one heavy smokers were put on a treadmill at the fastest setting. The time in seconds was measured until they fell off from exhaustion: 18, 16, 18, 24, 23, 22, 22, 23, 26, 29, 32, 34, 34, 36, 36, 43, 42, 49, 46, 46, 57

Compute the mode, median, mean, upper and lower quartiles, range and interquartile range

1.7.4. Using a frequency distribution to go beyond the data ①

Another way to think about frequency distributions is not in terms of how often scores actually occurred, but how likely it is that a score would occur (i.e. probability). The word 'probability' induces suicidal ideation in most people (myself included) so it seems fitting that we use an example about throwing ourselves off a cliff. Beachy Head is a large, windy cliff on the Sussex coast (not far from where I live) that has something of a reputation for attracting suicidal people, who seem to like throwing themselves off it (and after several months of rewriting this book I find my thoughts drawn towards that peaceful chalky cliff top more and more often). Figure 1.8 shows a frequency distribution of some completely made up data of the number of suicides at Beachy Head in a year by people of different ages (although I made these data up, they are roughly based on general suicide statistics such as those in Williams, 2001). There were 172 suicides in total and you can see that the suicides were most frequently aged between about 30 and 35 (the highest bar). The graph also tells us that, for example, very few people aged above 70 committed suicide at Beachy Head.

I said earlier that we could think of frequency distributions in terms of probability. To explain this, imagine that someone asked you 'how likely is it that a 70 year old committed suicide at Beach Head?' What would your answer be? The chances are that if you looked at the frequency distribution you might respond 'not very likely' because you can see that only 3 people out of the 172 suicides were aged around 70. What about if someone asked you 'how likely is it that a 30 year old committed suicide?' Again, by looking at the graph, you might say 'it's actually quite likely' because 33 out of the 172 suicides were by people aged around 30 (that's more than 1 in every 5 people who committed suicide). So based

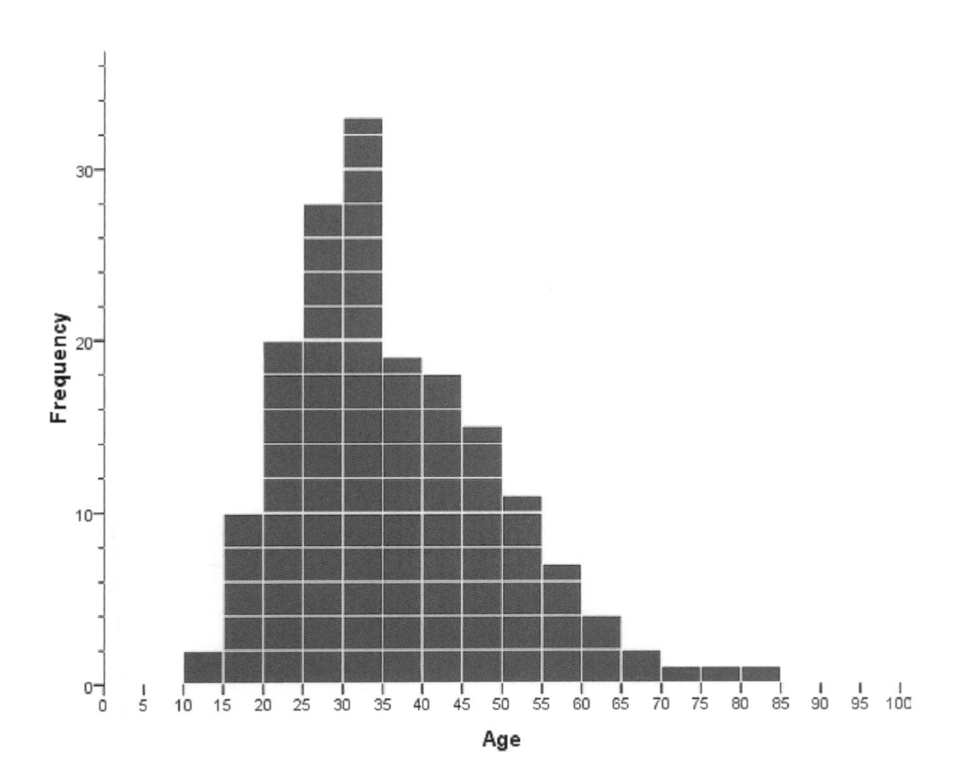

FIGURE 1.8 Frequency distribution showing the number of suicides at Beachy Head in a year by age

on the frequencies of different scores it should start to become clear that we could use this information to estimate the probability that a particular score will occur. We could ask, based on our data, 'what's the probability of a suicide victim being aged 16–20?' A probability value can range from 0 (there's no chance whatsoever of the event happening) to 1 (the event will definitely happen). So, for example, when I talk to my publishers I tell them there's a probability of 1 that I will have completed the revisions to this book by April 2008. However, when I talk to anyone else, I might, more realistically, tell them that there's a .10 probability of me finishing the revisions on time (or put another way, a 10% chance, or 1 in 10 chance that I'll complete the book in time). In reality, the probability of my meeting the deadline is 0 (not a chance in hell) because I never manage to meet publisher's deadlines! If probabilities don't make sense to you then just ignore the decimal point and think of them as percentages instead (i.e. .10 probability that something will happen = 10% chance that something will happen).

I've talked in vague terms about how frequency distributions can be used to get a rough idea of the probability of a score occurring. However, we can be precise. For any distribution of scores we could, in theory, calculate the probability of obtaining a score of a certain size – it would be incredibly tedious and complex to do it, but we could. To spare our sanity, statisticians have identified several common distributions. For each one they have worked out mathematical formulae that specify idealized versions of these distributions (they are specified in terms of a curved line). These idealized distributions are known as **probability distributions** and from these distributions it is possible to calculate the probability of getting particular scores based on the frequencies with which a particular score occurs in a distribution with these common shapes. One of these 'common' distributions is the normal distribution, which I've already mentioned in section 1.7.1. Statisticians have calculated the probability of certain scores occurring in a normal distribution with a mean of 0 and a standard deviation of 1. Therefore, if we have any data that are shaped like a normal distribution, then if the mean and standard deviation

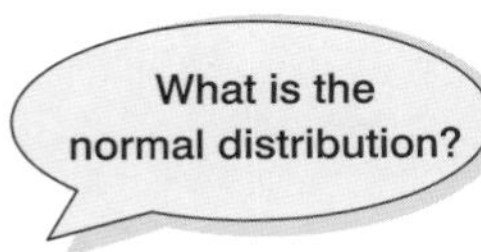

are 0 and 1 respectively we can use the tables of probabilities for the normal distribution to see how likely it is that a particular score will occur in the data (I've produced such a table in the Appendix to this book).

The obvious problem is that not all of the data we collect will have a mean of 0 and standard deviation of 1. For example, we might have a data set that has a mean of 567 and a standard deviation of 52.98. Luckily any data set can be converted into a data set that has a mean of 0 and a standard deviation of 1. First, to centre the data around zero, we take each score and subtract from it the mean of all. Then, we divide the resulting score by the standard deviation to ensure the data have a standard deviation of 1. The resulting scores are known as **z-scores** and in equation form, the conversion that I've just described is:

$$z = \frac{X - \overline{X}}{s} \tag{1.2}$$

The table of probability values that have been calculated for the standard normal distribution is shown in the Appendix. Why is this table important? Well, if we look at our suicide data, we can answer the question 'What's the probability that someone who threw themselves off of Beachy Head was 70 or older?' First we convert 70 into a z-score. Say, the mean of the suicide scores was 36, and the standard deviation 13; then 70 will become (70 – 36)/13 = 2.62. We then look up this value in the column labelled 'Smaller Portion' (i.e. the area above the value 2.62). You should find that the probability is .0044, or put another way, only a 0.44% chance that a suicide victim would be 70 years old or more. By looking at the column labelled 'Bigger Portion' we can also see the probability that a suicide victim was aged 70 or less. This probability is .9956, or put another way, there's a 99.56% chance that a suicide victim was less than 70 years old.

Hopefully you can see from these examples that the normal distribution and z-scores allow us to go a first step beyond our data in that from a set of scores we can calculate the probability that a particular score will occur. So, we can see whether scores of a certain size are likely or unlikely to occur in a distribution of a particular kind. You'll see just how useful this is in due course, but it is worth mentioning at this stage that certain z-scores are particularly important. This is because their value cuts off certain important percentages of the distribution. The first important value of z is 1.96 because this cuts off the top 2.5% of the distribution, and its counterpart at the opposite end (–1.96) cuts off the bottom 2.5% of the distribution. As such, taken together, this value cuts of 5% of scores, or put another way, 95% of z-scores lie between –1.96 and 1.96. The other two important benchmarks are ±2.58 and ±3.29, which cut off 1% and 0.1% of scores respectively. Put another way, 99% of z-scores lie between –2.58 and 2.58, and 99.9% of them lie between –3.29 and 3.29. Remember these values because they'll crop up time and time again.

SELF-TEST Assuming the same mean and standard deviation for the Beachy Head example above, what's the probability that someone who threw themselves off Beachy Head was 30 or younger?

1.7.5. Fitting statistical models to the data ①

Having looked at your data (and there is a lot more information on different ways to do this in Chapter 4), the next step is to fit a statistical model to the data. I should really just write 'insert the rest of the book here', because most of the remaining chapters discuss the various models that you can fit to the data. However, I do want to talk here briefly about two very important types of hypotheses that are used when analysing the data. Scientific statements, as we have seen, can be split into testable hypotheses. The hypothesis or prediction that comes from your theory is usually saying that an effect will be present. This hypothesis is called the **alternative hypothesis** and is denoted by H_1. (It is sometimes also called the *experimental hypothesis* but because this term relates to a specific type of methodology it's probably best to use 'alternative hypothesis'.) There is another type of hypothesis, though, and this is called the **null hypothesis** and is denoted by H_0. This hypothesis is the opposite of the alternative hypothesis and so would usually state that an effect is absent. Taking our *Big Brother* example from earlier in the chapter we might generate the following hypotheses:

- **Alternative hypothesis:** *Big Brother* contestants will score higher on personality disorder questionnaires than members of the public.
- **Null hypothesis:** *Big Brother* contestants and members of the public will not differ in their scores on personality disorder questionnaires.

The reason that we need the null hypothesis is because we cannot prove the experimental hypothesis using statistics, but we can reject the null hypothesis. If our data give us confidence to reject the null hypothesis then this provides support for our experimental hypothesis. However, be aware that even if we can reject the null hypothesis, this doesn't prove the experimental hypothesis – it merely supports it. So, rather than talking about accepting or rejecting a hypothesis (which some textbooks tell you to do) we should be talking about 'the chances of obtaining the data we've collected assuming that the null hypothesis is true'.

Using our *Big Brother* example, when we collected data from the auditions about the contestants' personalities we found that 75% of them had a disorder. When we analyse our data, we are really asking, 'Assuming that contestants are no more likely to have personality disorders than members of the public, is it likely that 75% or more of the contestants would have personality disorders?' Intuitively the answer is that the chances are very low: if the null hypothesis is true, then most contestants would not have personality disorders because they are relatively rare. Therefore, we are very unlikely to have got the data that we did if the null hypothesis were true.

What if we found that only 1 contestant reported having a personality disorder (about 8%)? If the null hypothesis is true, and contestants are no different in personality to the general population, then only a small number of contestants would be expected to have a personality disorder. The chances of getting these data if the null hypothesis is true are, therefore, higher than before.

When we collect data to test theories we have to work in these terms: we cannot talk about the null hypothesis being true or the experimental hypothesis being true, we can only talk in terms of the probability of obtaining a particular set of data if, hypothetically speaking, the null hypothesis was true. We will elaborate on this idea in the next chapter.

Finally, hypotheses can also be directional or non-directional. A directional hypothesis states that an effect will occur, but it also states the direction of the effect. For example, 'readers will know more about research methods after reading this chapter' is a one-tailed hypothesis because it states the direction of the effect (readers will know more). A non-directional hypothesis states that an effect will occur, but it doesn't state the direction of the effect. For example, 'readers' knowledge of research methods will change after they have read this chapter' does not tell us whether their knowledge will improve or get worse.

What have I discovered about statistics? ①

Actually, not a lot because we haven't really got to the statistics bit yet. However, we have discovered some stuff about the process of doing research. We began by looking at how research questions are formulated through observing phenomena or collecting data about a 'hunch'. Once the observation has been confirmed, theories can be generated about why something happens. From these theories we formulate hypotheses that we can test. To test hypotheses we need to measure things and this leads us to think about the variables that we need to measure and how to measure them. Then we can collect some data. The final stage is to analyse these data. In this chapter we saw that we can begin by just looking at the shape of the data but that ultimately we should end up fitting some kind of statistical model to the data (more on that in the rest of the book). In short, the reason that your evil statistics lecturer is forcing you to learn statistics is because it is an intrinsic part of the research process and it gives you enormous power to answer questions that are interesting; or it could be that they are a sadist who spends their spare time spanking politicians while wearing knee-high PVC boots, a diamond-encrusted leather thong and a gimp mask (that'll be a nice mental image to keep with you throughout your course). We also discovered that I was a curious child (you can interpret that either way). As I got older I became more curious, but you will have to read on to discover what I was curious about.

Key terms that I've discovered

Alternative hypothesis
Between-group design
Between-subject design
Bimodal
Binary variable
Boredom effect
Categorical variable
Central tendency
Confounding variable
Content validity
Continuous variable
Correlational research
Counterbalancing
Criterion validity
Cross-sectional research
Dependent variable
Discrete variable
Ecological validity
Experimental hypothesis
Experimental research
Falsification
Frequency distribution
Histogram
Hypothesis
Independent design
Independent variable
Interquartile range
Interval variable
Kurtosis
Leptokurtic
Level of measurement
Lower quartile
Mean
Measurement error
Median
Mode
Multimodal
Negative skew
Nominal variable
Normal distribution
Null hypothesis
Ordinal variable
Outcome variable
Platykurtic
Positive skew
Practice effect

Predictor variable
Probability distribution
Qualitative methods
Quantitative methods
Quartile
Randomization
Range
Ratio variable
Reliability
Repeated-measures design
Second quartile
Skew
Systematic variation
Tertium quid
Test–retest reliability
Theory
Unsystematic variance
Upper quartile
Validity
Variables
Within-subject design
z-scores

Smart Alex's tasks

Smart Alex knows everything there is to know about statistics and SAS. He also likes nothing more than to ask people stats questions just so that he can be smug about how much he knows. So, why not really annoy him and get all of the answers right!

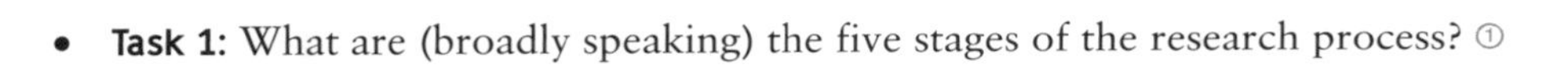

- **Task 1:** What are (broadly speaking) the five stages of the research process? ①

- **Task 2:** What is the fundamental difference between experimental and correlational research? ①
- **Task 3:** What is the level of measurement of the following variables? ①
  a. The number of downloads of different bands' songs on iTunes.
  b. The names of the bands that were downloaded.
  c. The position in the iTunes download chart.
  d. The money earned by the bands from the downloads.
  e. The weight of drugs bought by the bands with their royalties.
  f. The type of drugs bought by the bands with their royalties.
  g. The phone numbers that the bands obtained because of their fame.
  h. The gender of the people giving the bands their phone numbers.
  i. The instruments played by the band members.
  j. The time they had spent learning to play their instruments.
- **Task 4:** Say I own 857 CDs. My friend has written a computer program that uses a webcam to scan the shelves in my house where I keep my CDs and measure how many I have. His program says that I have 863 CDs. Define measurement error. What is the measurement error in my friends CD-counting device? ①
- **Task 5:** Sketch the shape of a normal distribution, a positively skewed distribution and a negatively skewed distribution. ①

Answers can be found on the companion website.

Further reading

Field, A. P., & Hole, G. J. (2003). *How to design and report experiments*. London: Sage. (I am rather biased, but I think this is a good overview of basic statistical theory and research methods.)

Miles, J. N. V., & Banyard, P. (2007). *Understanding and using statistics in psychology: a practical introduction*. London: Sage. (A fantastic and amusing introduction to statistical theory.)
Wright, D. B., & London, K. (2009). *First steps in statistics* (2nd ed.). London: Sage. (This book is a very gentle introduction to statistical theory.)

Interesting real research

Umpierre, S. A., Hill, J. A., & Anderson, D. J. (1985). Effect of Coke on sperm motility. *New England Journal of Medicine*, *313*(21), 1351.

Everything you ever wanted to know about statistics (well, sort of)

2

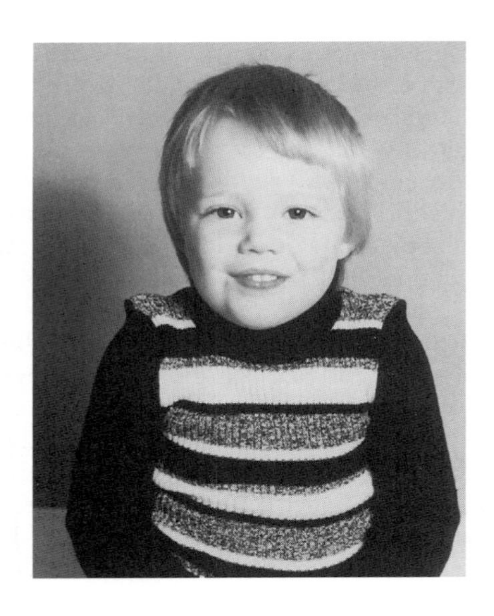

FIGURE 2.1
The face of innocence ... but what are the hands doing?

2.1. What will this chapter tell me? ①

As a child grows, it becomes important for them to fit models to the world: to be able to reliably predict what will happen in certain situations. This need to build models that accurately reflect reality is an essential part of survival. According to my parents (conveniently I have no memory of this at all), while at nursery school one model of the world that I was particularly enthusiastic to try out was 'If I get my penis out, it will be really funny'. No doubt to my considerable disappointment, this model turned out to be a poor predictor of positive outcomes. Thankfully for all concerned, I soon learnt that the model 'If I get my penis out at nursery school the teachers and mummy and daddy are going to be quite annoyed' was a better 'fit' of the observed data. Fitting models that accurately reflect the observed data is important to establish whether a theory is true. You'll be delighted to know that this chapter is all about fitting statistical models (and not about my penis). We edge sneakily away from the frying pan of research methods and trip accidentally into the fires of statistics hell. We begin by discovering what a statistical model is by using the mean as a straightforward example. We then see how we can use the properties of data to go beyond the data we have collected and to draw inferences about the world at large. In a nutshell then, this chapter lays the foundation for the whole of the rest of the book, so it's quite important that you read it or nothing that comes later will make any sense. Actually, a lot of what comes later probably won't make much sense anyway because *I've* written it, but there you go.

2.2. Building statistical models ①

We saw in the previous chapter that scientists are interested in discovering something about a phenomenon that we assume actually exists (a 'real-world' phenomenon). These real-world phenomena can be anything from the behaviour of interest rates in the economic market to the behaviour of undergraduates at the end-of-exam party. Whatever the phenomenon we desire to explain, we collect data from the real world to test our hypotheses about the phenomenon. Testing these hypotheses involves building statistical models of the phenomenon of interest.

The reason for building statistical models of real-world data is best explained by analogy. Imagine an engineer wishes to build a bridge across a river. That engineer would be pretty daft if she just built any old bridge, because the chances are that it would fall down. Instead, an engineer collects data from the real world: she looks at bridges in the real world and sees what materials they are made from, what structures they use and so on (she might even collect data about whether these bridges are damaged!). She then uses this information to construct a model. She builds a scaled-down version of the real-world bridge because it is impractical, not to mention expensive, to build the actual bridge itself. The model may differ from reality in several ways – it will be smaller for a start – but the engineer will try to build a model that best fits the situation of interest based on the data available. Once the model has been built, it can be used to predict things about the real world: for example, the engineer might test whether the bridge can withstand strong winds by placing the model in a wind tunnel. It seems obvious that it is important that the model is an accurate representation of the real world. Social scientists do much the same thing as engineers: they build models of real-world processes in an attempt to predict how these processes operate under certain conditions (see Jane Superbrain Box 2.1 below). We don't have direct access to the processes, so we collect data that represent the processes and then use these data to build statistical models (we reduce the process to a statistical model). We then use this statistical model to make predictions about the real-world phenomenon. Just like the engineer, we want our models to be as accurate as possible so that we can be confident that the predictions we make are also accurate. However, unlike engineers we don't have access to the real-world situation and so we can only ever *infer* things about psychological, societal, biological or economic processes based upon the models we build. If we want our inferences to be accurate then the statistical model we build must represent the data collected (the *observed data*) as closely as possible. The degree to which a statistical model represents the data collected is known as the **fit** of the model.

Figure 2.2 illustrates the kinds of models that an engineer might build to represent the real-world bridge that she wants to create. The first model (a) is an excellent representation of the real-world situation and is said to be a *good fit* (i.e. there are a few small differences but the model is basically a very good replica of reality). If this model is used to make predictions about the real world, then the engineer can be confident that these predictions will be very accurate, because the model so closely resembles reality. So, if the model collapses in a strong

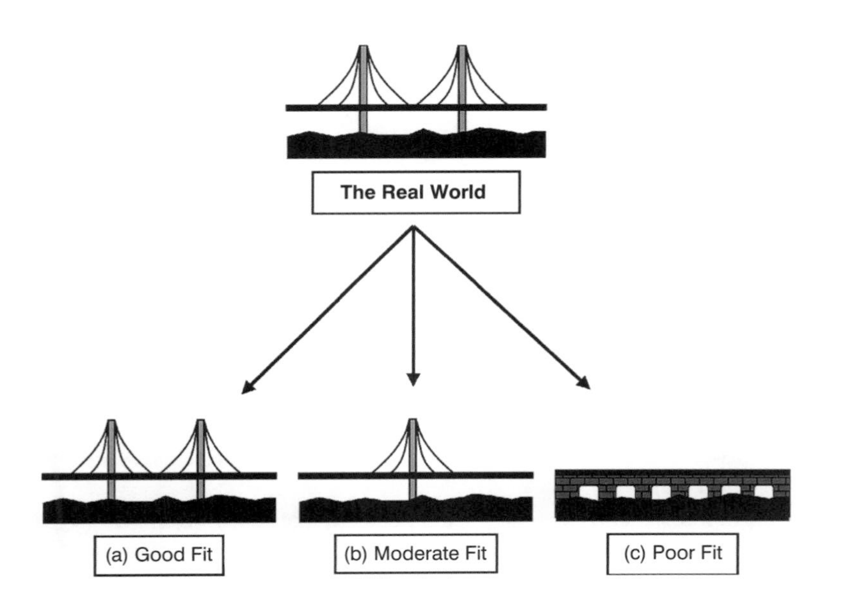

FIGURE 2.2 Fitting models to real-world data (see text for details)

wind, then there is a good chance that the real bridge would collapse also. The second model (b) has some similarities to the real world: the model includes some of the basic structural features, but there are some big differences from the real-world bridge (namely the absence of one of the supporting towers). This is what we might term a *moderate fit* (i.e. there are some differences between the model and the data but there are also some great similarities). If the engineer uses this model to make predictions about the real world then these predictions may be inaccurate and possibly catastrophic (e.g. the model predicts that the bridge will collapse in a strong wind, causing the real bridge to be closed down, creating 100-mile tailbacks with everyone stranded in the snow; all of which was unnecessary because the real bridge was perfectly safe – the model was a bad representation of reality). We can have some confidence, but not complete confidence, in predictions from this model. The final model (c) is completely different to the real-world situation; it bears no structural similarities to the real bridge and is a poor fit (in fact, it might more accurately be described as an abysmal fit!). As such, any predictions based on this model are likely to be completely inaccurate. Extending this analogy to the social sciences we can say that it is important when we fit a statistical model to a set of data that this model fits the data well. If our model is a poor fit of the observed data then the predictions we make from it will be equally poor.

JANE SUPERBRAIN 2.1

Types of statistical models ①

As behavioural and social scientists, most of the models that we use to describe data tend to be **linear models**. For example, analysis of variance (ANOVA) and regression are identical systems based on linear models (Cohen, 1968), yet they have different names and, in psychology at least, are used largely in different contexts due to historical divisions in methodology (Cronbach, 1957).

A linear model is simply a model that is based upon a straight line; this means that we are usually trying to summarize our observed data in terms of a straight line. Suppose we measured how many chapters of this book a person had read, and then measured their spiritual enrichment. We could represent these hypothetical data in the form of a scatterplot in which each dot represents an individual's score on both variables (see section 4.7). Figure 2.3 shows two versions of such a graph that summarize the pattern of these data with either a straight

FIGURE 2.3 A scatterplot of the same data with a linear model fitted (left), and with a non-linear model fitted (right)

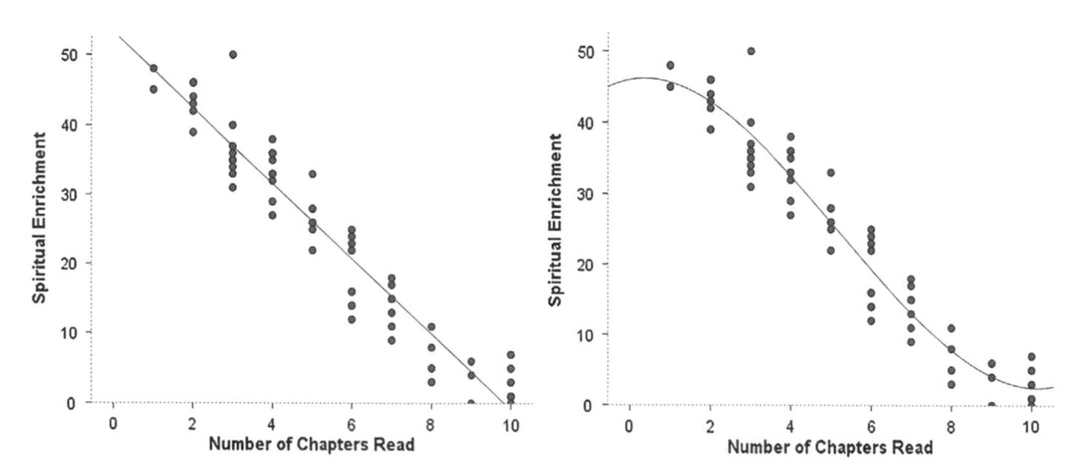

(left) or curved (right) line. These graphs illustrate how we can fit different types of models to the same data. In this case we can use a straight line to represent our data and it shows that the more chapters a person reads, the less their spiritual enrichment. However, we can also use a curved line to summarize the data and this shows that when most, or all, of the chapters have been read, spiritual enrichment seems to increase slightly (presumably because once the book is read everything suddenly makes sense – yeah, as if!). Neither of the two types of model is necessarily correct, but it will be the case that one model fits the data better than another and this is why when we use statistical models it is important for us to assess how well a given model fits the data.

It's possible that many scientific disciplines are progressing in a biased way because most of the models that we tend to fit are linear (mainly because books like this tend to ignore more complex curvilinear models). This could create a bias because most published scientific studies are ones with statistically significant results and there may be cases where a linear model has been a poor fit of the data (and hence the paper was not published), yet a non-linear model would have fitted the data well. This is why it is useful to plot your data first: plots tell you a great deal about what models should be applied to data. If your plot seems to suggest a non-linear model then investigate this possibility (which is easy for me to say when I don't include such techniques in this book!).

2.3. Populations and samples ①

As researchers, we are interested in finding results that apply to an entire population of people or things. For example, psychologists want to discover processes that occur in all humans, biologists might be interested in processes that occur in all cells, economists want to build models that apply to all salaries, and so on. A population can be very general (all human beings) or very narrow (all male ginger cats called Bob). Usually, scientists strive to infer things about general populations rather than narrow ones. For example, it's not very interesting to conclude that psychology students with brown hair who own a pet hamster named George recover more quickly from sports injuries if the injury is massaged (unless, like René Koning,[1] you happen to be a psychology student with brown hair who has a pet hamster named George). However, if we can conclude that *everyone's* sports injuries are aided by massage this finding has a much wider impact.

Scientists rarely, if ever, have access to every member of a population. Psychologists cannot collect data from every human being and ecologists cannot observe every male ginger cat called Bob. Therefore, we collect data from a small subset of the population (known as a **sample**) and use these data to infer things about the population as a whole. The bridge-building

[1] A brown-haired psychology student with a hamster called Sjors (Dutch for George, apparently), who, after reading one of my web resources, emailed me to weaken my foolish belief that this is an obscure combination of possibilities.

engineer cannot make a full-size model of the bridge she wants to build and so she builds a small-scale model and tests this model under various conditions. From the results obtained from the small-scale model the engineer infers things about how the full-sized bridge will respond. The small-scale model may respond differently to a full-sized version of the bridge, but the larger the model, the more likely it is to behave in the same way as the full-size bridge. This metaphor can be extended to scientists. We never have access to the entire population (the real-size bridge) and so we collect smaller samples (the scaled-down bridge) and use the behaviour within the sample to infer things about the behaviour in the population. The bigger the sample, the more likely it is to reflect the whole population. If we take several random samples from the population, each of these samples will give us slightly different results. However, on average, large samples should be fairly similar.

2.4. Simple statistical models ①

2.4.1. The mean: a very simple statistical model ①

One of the simplest models used in statistics is the mean, which we encountered in section 1.7.2.3. In Chapter 1 we briefly mentioned that the mean was a statistical model of the data because it is a hypothetical value that doesn't have to be a value that is actually observed in the data. For example, if we took five statistics lecturers and measured the number of friends that they had, we might find the following data: 1, 2, 3, 3 and 4. If we take the mean number of friends, this can be calculated by adding the values we obtained, and dividing by the number of values measured: $(1 + 2 + 3 + 3 + 4)/5 = 2.6$. Now, we know that it is impossible to have 2.6 friends (unless you chop someone up with a chainsaw and befriend their arm, which frankly is probably not beyond your average statistics lecturer) so the mean value is a *hypothetical* value. As such, the mean is a model created to summarize our data.

2.4.2. Assessing the fit of the mean: sums of squares, variance and standard deviations ①

With any statistical model we have to assess the fit (to return to our bridge analogy we need to know how closely our model bridge resembles the real bridge that we want to build). With most statistical models we can determine whether the model is accurate by looking at how different our real data are from the model that we have created. The easiest way to do this is to look at the difference between the data we observed and the model fitted. Figure 2.4 shows the number of friends that each statistics lecturer had, and also the mean number that we calculated earlier on. The line representing the mean can be thought of as our model, and the circles are the observed data. The diagram also has a series of vertical lines that connect each observed value to the mean value. These lines represent the **difference** between the observed data and our model and can be thought of as the error in the model. We can calculate the magnitude of these deviances by simply subtracting the mean value ($\bar{x}$) from each of the observed values (x_i).[2] For example, lecturer 1 had only 1 friend (a glove puppet of an ostrich called Kevin) and so the difference is $x_1 - \bar{x} = 1 - 2.6 = -1.6$. You might notice that the deviance is a negative number, and this represents the fact that our model *overestimates* this lecturer's popularity: it

[2] The x_i simply refers to the observed score for the *i*th person (so, the *i* can be replaced with a number that represents a particular individual). For these data: for lecturer 1, $x_i = x_1 = 1$; for lecturer 3, $x_i = x_3 = 3$; for lecturer 5, $x_i = x_5 = 4$.

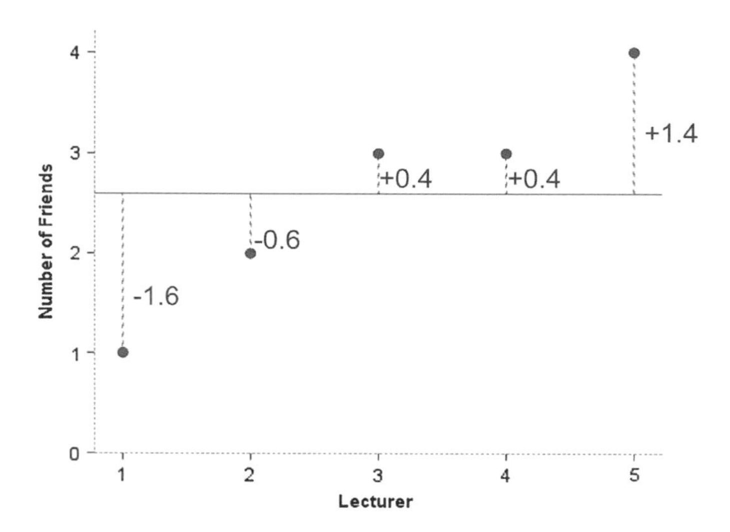

FIGURE 2.4 Graph showing the difference between the observed number of friends that each statistics lecturer had, and the mean number of friends

predicts that he will have 2.6 friends yet in reality he has only 1 friend (bless him!). Now, how can we use these deviances to estimate the accuracy of the model? One possibility is to add up the deviances (this would give us an estimate of the total error). If we were to do this we would find that (don't be scared of the equations, we will work through them step by step – if you need reminding of what the symbols mean there is a guide at the beginning of the book):

$$\begin{aligned}\text{total error} &= \text{sum of deviances}\\ &= \sum(x_i - \bar{x}) = (-1.6) + (-0.6) + (0.4) + (0.4) + (1.4) = 0\end{aligned}$$

So, in effect the result tells us that there is no total error between our model and the observed data, so the mean is a perfect representation of the data. Now, this clearly isn't true: there were errors but some of them were positive, some were negative and they have simply cancelled each other out. It is clear that we need to avoid the problem of which direction the error is in and one mathematical way to do this is to square each error,[3] that is, multiply each error by itself. So, rather than calculating the sum of errors, we calculate the sum of squared errors. In this example:

$$\begin{aligned}\text{sum of squrared errors (SS)} &= \sum(x_i - \bar{x})(x_i - \bar{x})\\ &= (-1.6)^2 + (-0.6)^2 + (0.4)^2 + (0.4)^2 + (1.4)^2\\ &= 2.56 + 0.36 + 0.16 + 0.16 + 1.96\\ &= 5.20\end{aligned}$$

The **sum of squared errors** (SS) is a good measure of the accuracy of our model. However, it is fairly obvious that the sum of squared errors is dependent upon the amount of data that has been collected – the more data points, the higher the SS. To overcome this problem we calculate the average error by dividing the SS by the number of observations (*N*). If we are interested only in the average error for the sample, then we can divide by *N* alone. However, we are generally interested in using the error in the sample to estimate the error in the population and so we divide the SS by the number of observations minus 1 (the reason why is explained in Jane Superbrain Box 2.2). This measure is known as the **variance** and is a measure that we will come across a great deal:

[3] When you multiply a negative number by itself it becomes positive.

JANE SUPERBRAIN 2.2

Degrees of freedom ②

Degrees of freedom (*df*) is a very difficult concept to explain. I'll begin with an analogy. Imagine you're the manager of a rugby team and you have a team sheet with 15 empty slots relating to the positions on the playing field. There is a standard formation in rugby and so each team has 15 specific positions that must be held constant for the game to be played. When the first player arrives, you have the choice of 15 positions in which to place this player. You place his name in one of the slots and allocate him to a position (e.g. scrum-half) and, therefore, one position on the pitch is now occupied. When the next player arrives, you have the choice of 14 positions but you still have the freedom to choose which position this player is allocated. However, as more players arrive, you will reach the point at which 14 positions have been filled and the final player arrives. With this player you have no freedom to choose where they play – there is only one position left. Therefore there are 14 degrees of freedom; that is, for 14 players you have some degree of choice over where they play, but for 1 player you have no choice. The degrees of freedom is one less than the number of players.

In statistical terms the degrees of freedom relate to the number of observations that are free to vary. If we take a sample of four observations from a population, then these four scores are free to vary in any way (they can be any value). However, if we then use this sample of four observations to calculate the standard deviation of the population, we have to use the mean of the sample as an estimate of the population's mean. Thus we hold one parameter constant. Say that the mean of the sample was 10; then we assume that the population mean is 10 also and we keep this value constant. With this parameter fixed, can all four scores from our sample vary? The answer is no, because to keep the mean constant only three values are free to vary. For example, if the values in the sample were 8, 9, 11, 12 (mean = 10) and we changed three of these values to 7, 15 and 8, then the final value *must* be 10 to keep the mean constant. Therefore, if we hold one parameter constant then the degrees of freedom must be one less than the sample size. This fact explains why when we use a sample to estimate the standard deviation of a population, we have to divide the sums of squares by $N - 1$ rather than N alone.

$$\text{variance } (s^2) = \frac{\text{SS}}{N-1} = \frac{\sum(x_i - \bar{x})^2}{N-1} = \frac{5.20}{4} = 1.3 \tag{2.1}$$

The variance is, therefore, the average error between the mean and the observations made (and so is a measure of how well the model fits the actual data). There is one problem with the variance as a measure: it gives us a measure in units squared (because we squared each error in the calculation). In our example we would have to say that the average error in our data (the variance) was 1.3 friends squared. It makes little enough sense to talk about 1.3 friends, but it makes even less to talk about friends squared! For this reason, we often take the square root of the variance (which ensures that the measure of average error is in the same units as the original measure). This measure is known as the standard deviation and is simply the square root of the variance. In this example the **standard deviation** is:

$$\begin{aligned} s &= \sqrt{\frac{\sum(x_i - \bar{x})^2}{N-1}} \\ &= \sqrt{1.3} \\ &= 1.14 \end{aligned} \tag{2.2}$$

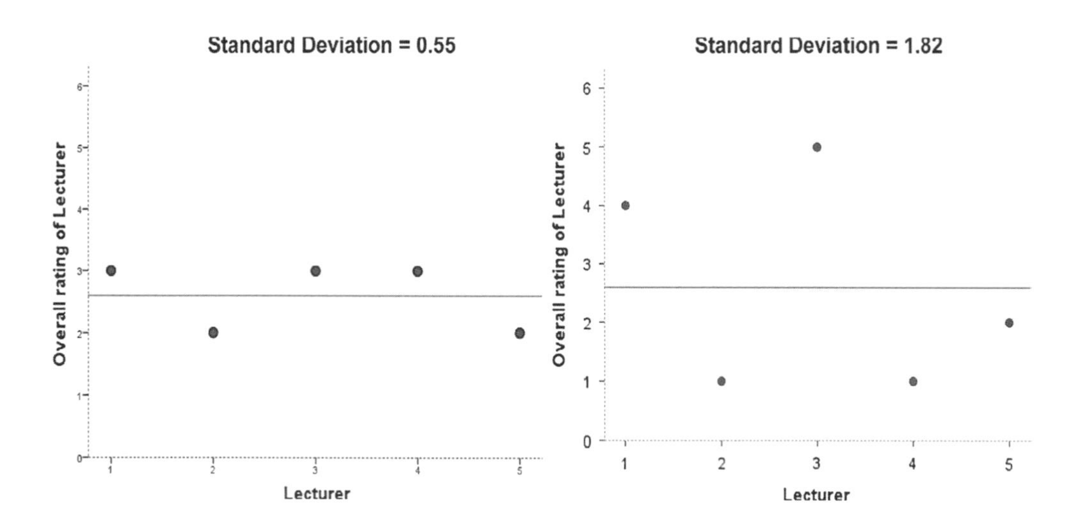

FIGURE 2.5 Graphs illustrating data that have the same mean but different standard deviations

The sum of squares, variance and standard deviation are all, therefore, measures of the 'fit' (i.e. how well the mean represents the data). Small standard deviations (relative to the value of the mean itself) indicate that data points are close to the mean. A large standard deviation (relative to the mean) indicates that the data points are distant from the mean (i.e. the mean is not an accurate representation of the data). A standard deviation of 0 would mean that all of the scores were the same. Figure 2.5 shows the overall ratings (on a 5-point scale) of two lecturers after each of five different lectures. Both lecturers had an average rating of 2.6 out of 5 across the lectures. However, the first lecturer had a standard deviation of 0.55 (relatively small compared to the mean). It should be clear from the graph that ratings for this lecturer were consistently close to the mean rating. There was a small fluctuation, but generally his lectures did not vary in popularity. As such, the mean is an accurate representation of his ratings. The mean is a good fit to the data. The second lecturer, however, had a standard deviation of 1.82 (relatively high compared to the mean). The ratings for this lecturer are clearly more spread from the mean; that is, for some lectures he received very high ratings, and for others his ratings were appalling. Therefore, the mean is not such an accurate representation of his performance because there was a lot of variability in the popularity of his lectures. The mean is a poor fit to the data. This illustration should hopefully make clear why the standard deviation is a measure of how well the mean represents the data.

SELF-TEST In section 1.7.2.2 we came across some data about the number of friends that 11 people had on Facebook (22, 40, 53, 57, 93, 98, 103, 108, 116, 121, 252). We calculated the mean for these data as 96.64. Now calculate the sums of squares, variance and standard deviation.

SELF-TEST Calculate these values again but excluding the extreme score (252).

2.4.3. Expressing the mean as a model ②

The discussion of means, sums of squares and variance may seem a side track from the initial point about fitting statistical models, but it's not: the mean is a simple statistical model

JANE SUPERBRAIN 2.3

The standard deviation and the shape of the distribution ①

As well as telling us about the accuracy of the mean as a model of our data set, the variance and standard deviation also tell us about the shape of the distribution of scores. As such, they are measures of dispersion like those we encountered in section 1.7.3. If the mean represents the data well then most of the scores will cluster close to the mean and the resulting standard deviation is small relative to the mean. When the mean is a worse representation of the data, the scores cluster more widely around the mean (think back to Figure 2.5) and the standard deviation is larger. Figure 2.6 shows two distributions that have the same mean (50) but different standard deviations. One has a large standard deviation relative to the mean (*SD* = 25) and this results in a flatter distribution that is more spread out, whereas the other has a small standard deviation relative to the mean (*SD* = 15) resulting in a more pointy distribution in which scores close to the mean are very frequent but scores further from the mean become increasingly infrequent. The main message is that as the standard deviation gets larger, the distribution gets fatter. This can make distributions look platykurtic or leptokurtic when, in fact, they are not.

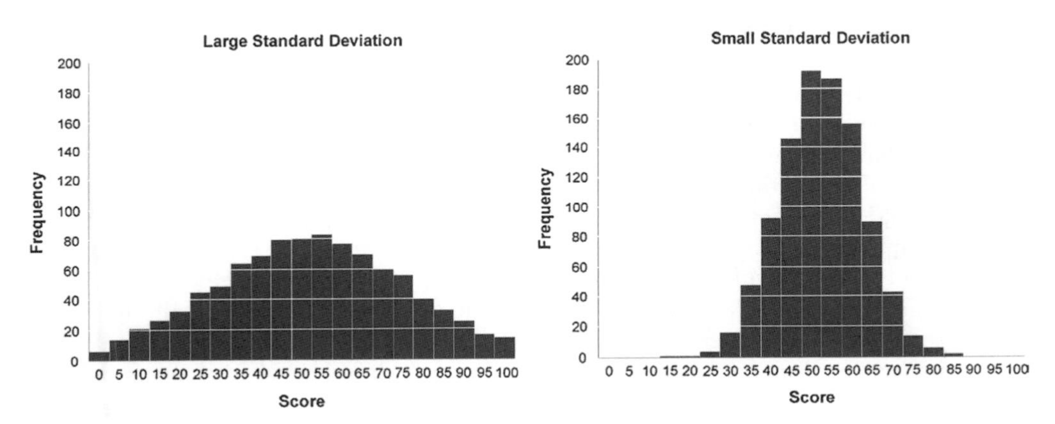

FIGURE 2.6 Two distributions with the same mean, but large and small standard deviations

that can be fitted to data. What do I mean by this? Well, everything in statistics essentially boils down to one equation:

$$\text{outcome}_i = (\text{model}) + \text{error}_i \tag{2.3}$$

This just means that the data we observe can be predicted from the model we choose to fit to the data plus some amount of error. When I say that the mean is a simple statistical model, then all I mean is that we can replace the word 'model' with the word 'mean' in that equation. If we return to our example involving the number of friends that statistics lecturers have and look at lecturer 1, for example, we observed that they had one friend and the mean of all lecturers was 2.6. So, the equation becomes:

$$\text{outcome}_{\text{lecturer1}} = \overline{X} + \varepsilon_{\text{lecturer1}}$$

$$1 = 2.6 + \varepsilon_{\text{lecturer1}}$$

From this we can work out that the error is 1 – 2.6, or –1.6. If we replace this value in the equation we get 1 = 2.6 – 1.6 or 1 = 1. Although it probably seems like I'm stating the obvious, it is worth bearing this general equation in mind throughout this book because if you do you'll discover that most things ultimately boil down to this one simple idea!

Likewise, the variance and standard deviation illustrate another fundamental concept: how the goodness of fit of a model can be measured. If we're looking at how well a model fits the data (in this case our model is the mean) then we generally look at deviation from the model, we look at the sum of squared error, and in general terms we can write this as:

$$\text{deviation} = \sum (\text{observed} - \text{model})^2 \tag{2.4}$$

Put another way, we assess models by comparing the data we observe to the model we've fitted to the data, and then square these differences. Again, you'll come across this fundamental idea time and time again throughout this book.

2.5. Going beyond the data ①

Using the example of the mean, we have looked at how we can fit a statistical model to a set of observations to summarize those data. It's one thing to summarize the data that you have actually collected but usually we want to go beyond our data and say something general about the world (remember in Chapter 1 that I talked about how good theories should say something about the world). It is one thing to be able to say that people in our sample responded well to medication, or that a sample of high-street stores in Brighton had increased profits leading up to Christmas, but it's more useful to be able to say, based on our sample, that all people will respond to medication, or that all high-street stores in the UK will show increased profits. To begin to understand how we can make these general inferences from a sample of data we can first look not at whether our model is a good fit to the sample from which it came, but whether it is a good fit to the **population** from which the sample came.

2.5.1. The standard error ①

We've seen that the standard deviation tells us something about how well the mean represents the sample data, but I mentioned earlier on that usually we collect data from samples because we don't have access to the entire population. If you take several samples from a population, then these samples will differ slightly; therefore, it's also important to know how well a particular sample represents the population. This is where we use the **standard error**. Many students get confused about the difference between the standard deviation and the standard error (usually because the difference is never explained clearly). However, the standard error is an important concept to grasp, so I'll do my best to explain it to you.

We have already learnt that social scientists use samples as a way of estimating the behaviour in a population. Imagine that we were interested in the ratings of all lecturers (so lecturers in general were the population). We could take a sample from this population. When someone takes a sample from a population, they are taking one of many possible

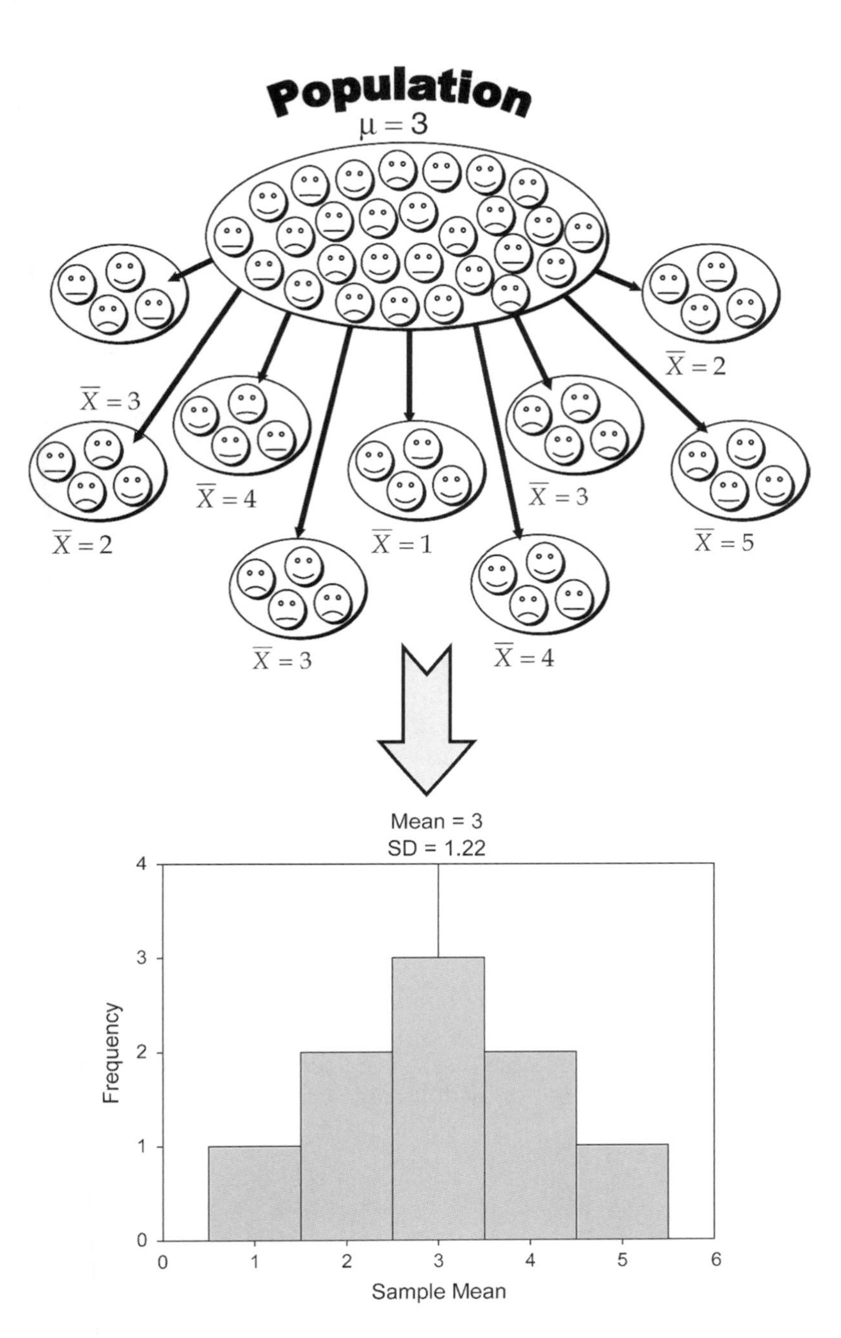

FIGURE 2.7 Illustration of the standard error (see text for details)

samples. If we were to take several samples from the same population, then each sample has its own mean, and some of these sample means will be different.

Figure 2.7 illustrates the process of taking samples from a population. Imagine that we could get ratings of all lecturers on the planet and that, on average, the rating is 3 (this is the *population mean*, μ). Of course, we can't collect ratings of all lecturers, so we use a sample. For each of these samples we can calculate the average, or *sample mean*. Let's imagine we took nine different samples (as in the diagram); you can see that some of the samples have the same mean as the population but some have different means: the first sample of lecturers were rated, on average, as 3, but the second sample were, on average, rated as only 2. This illustrates **sampling variation**: that is, samples will vary because they contain different members of the population; a sample that by chance includes some very good lecturers will have a higher average than a sample that, by chance, includes some awful lecturers! We can actually plot the sample means as a frequency distribution, or histogram,[4] just like I have done in the diagram. This distribution shows that there were three samples that had a mean of 3, means of 2 and 4 occurred in two samples each, and means of 1 and 5 occurred in only one sample each. The end result is a nice symmetrical distribution known as a **sampling distribution**. A sampling distribution is simply the frequency distribution of sample means from the same population. In theory you need to imagine that we're taking hundreds or thousands of samples to construct a sampling distribution, but I'm just using nine to keep the diagram simple.[5] The sampling distribution tells us about the behaviour of samples from the population, and you'll notice that it is centred at the same value as the mean of the population (i.e. 3). This means that if we took the average of all sample means we'd get the value of the population mean. Now, if the average of the sample means is the same value as the population mean, then if we knew the accuracy of that average we'd know something about how likely it is that a given sample is representative of the population. So how do we determine the accuracy of the population mean?

Think back to the discussion of the standard deviation. We used the standard deviation as a measure of how representative the mean was of the observed data. Small standard deviations represented a scenario in which most data points were close to the mean, a large standard deviation represented a situation in which data points were widely spread from the mean. If you were to calculate the standard deviation between *sample means* then this too would give you a measure of how much variability there was between the means of different samples. The standard deviation of sample means is known as the **standard error of the mean (SE)**. Therefore, the standard error could be calculated by taking the difference between each sample mean and the overall mean, squaring these differences, adding them up, and then dividing by the number of samples. Finally, the square root of this value would need to be taken to get the standard deviation of sample means, the standard error.

Of course, in reality we cannot collect hundreds of samples and so we rely on approximations of the standard error. Luckily for us some exceptionally clever statisticians have demonstrated that as samples get large (usually defined as greater than 30), the sampling distribution has a normal distribution with a mean equal to the population mean, and a standard deviation of:

$$\sigma_{\overline{X}} = \frac{s}{\sqrt{N}} \quad (2.5)$$

This is known as the **central limit theorem** and it is useful in this context because it means that if our sample is large we can use the above equation to approximate the standard error (because, remember, it is the standard deviation of the sampling distribution).[6] When the sample is relatively small (fewer than 30) the sampling distribution has a different shape, known as a *t*-distribution, which we'll come back to later.

[4] This is just a graph of each sample mean plotted against the number of samples that have that mean – see section 1.7.1 for more details.

[5] It's worth pointing out that I'm talking hypothetically. We don't need to *actually* collect these samples because clever statisticians have worked out what these sampling distributions would look like and how they behave.

[6] In fact it should be the *population* standard deviation (σ) that is divided by the square root of the sample size; however, for large samples this is a reasonable approximation.

CRAMMING SAM'S TIPS **The standard error**

The standard error is the standard deviation of sample means. As such, it is a measure of how representative a sample is likely to be of the population. A large standard error (relative to the sample mean) means that there is a lot of variability between the means of different samples and so the sample we have might not be representative of the population. A small standard error indicates that most sample means are similar to the population mean and so our sample is likely to be an accurate reflection of the population.

2.5.2. Confidence intervals ②

2.5.2.1. Calculating confidence intervals ②

Remember that usually we're interested in using the sample mean as an estimate of the value in the population. We've just seen that different samples will give rise to different values of the mean, and we can use the standard error to get some idea of the extent to which sample means differ. A different approach to assessing the accuracy of the sample mean as an estimate of the mean in the population is to calculate boundaries within which we believe the true value of the mean will fall. Such boundaries are called **confidence intervals.** The basic idea behind confidence intervals is to construct a range of values within which we think the population value falls.

Let's imagine an example: Domjan, Blesbois, and Williams (1998) examined the learnt release of sperm in Japanese quail. The basic idea is that if a quail is allowed to copulate with a female quail in a certain context (an experimental chamber) then this context will serve as a cue to copulation and this in turn will affect semen release (although during the test phase the poor quail were tricked into copulating with a terry cloth with an embalmed female quail head stuck on top).[7] Anyway, if we look at the mean amount of sperm released in the experimental chamber, there is a true mean (the mean in the population); let's imagine it's 15 million sperm. Now, in our actual sample, we might find the mean amount of sperm released was 17 million. Because we don't know the true mean, we don't really know whether our sample value of 17 million is a good or bad estimate of this value. What we can do instead is use an interval estimate: we use our sample value as the mid-point, but set a lower and upper limit as well. So, we might say, we think the true value of the mean sperm release is somewhere between 12 million and 22 million spermatozoa (note that 17 million falls exactly between these values). Of course, in this case the true value (15 million) does falls within these limits. However, what if we'd set smaller limits, what if we'd said we think the true value falls between 16 and 18 million (again, note that 17 million is in the middle)? In this case the interval does not contain the true value of the mean. Let's now imagine that you were particularly fixated with Japanese quail sperm, and you repeated the experiment 50 times using different samples. Each time you did the experiment again you constructed an interval around the sample mean as I've just described. Figure 2.8 shows this scenario: the circles represent the mean for each sample with the lines sticking out of them representing the intervals for these means. The true value of the mean (the mean in the population) is 15 million and is shown by a vertical line. The first thing to note is that most of the sample means are different from

What is a confidence interval?

[7] This may seem a bit sick, but the male quails didn't appear to mind too much, which probably tells us all we need to know about male mating behaviour.

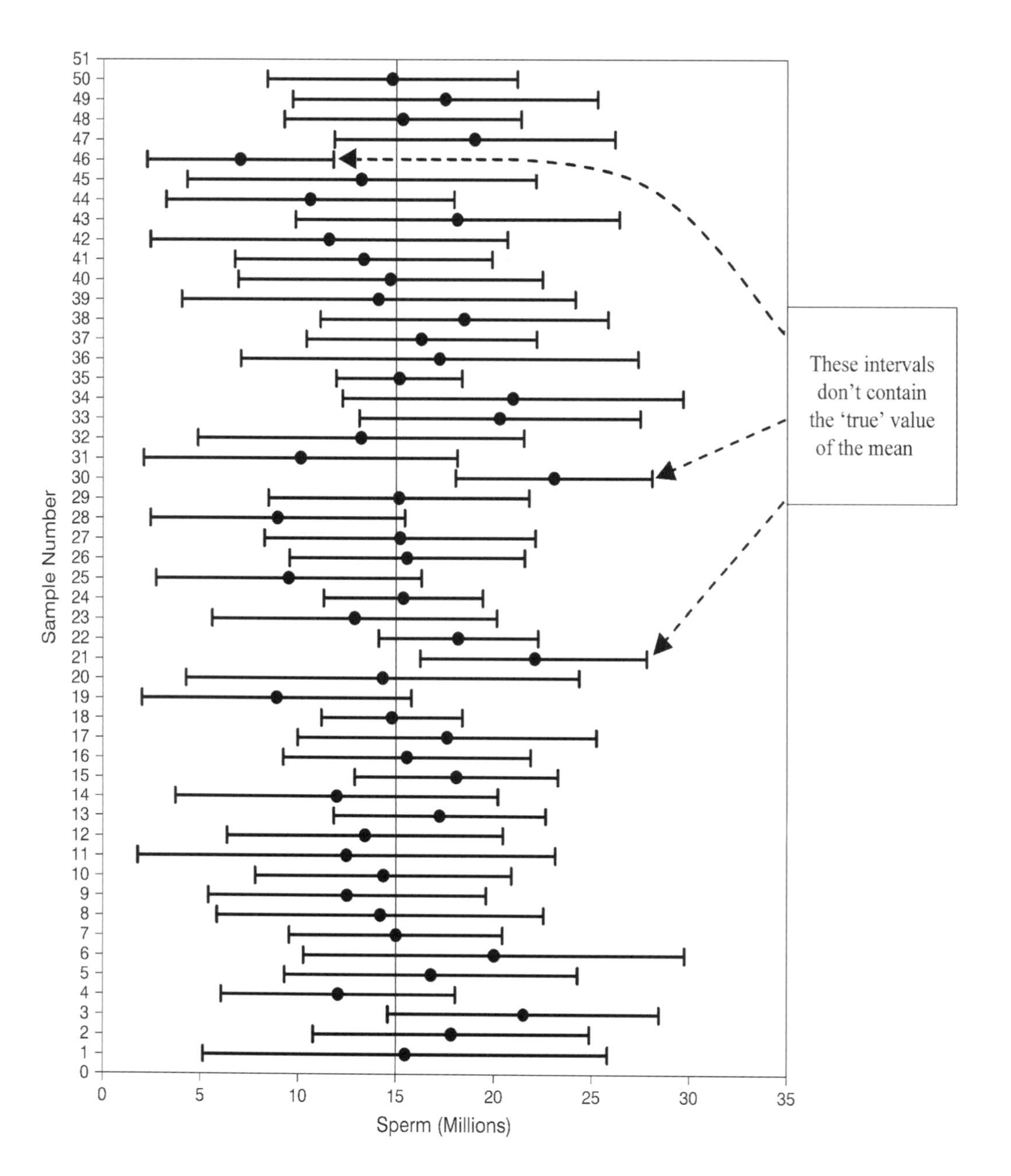

FIGURE 2.8 The confidence intervals of the sperm counts of Japanese quail (horizontal axis) for 50 different samples (vertical axis)

the true mean (this is because of sampling variation as described in the previous section). Second, although most of the intervals do contain the true mean (they cross the vertical line, meaning that the value of 15 million spermatozoa falls somewhere between the lower and upper boundaries), a few do not.

Up until now I've avoided the issue of how we might calculate the intervals. The crucial thing with confidence intervals is to construct them in such a way that they tell us something useful. Therefore, we calculate them so that they have certain properties: in particular they tell us the likelihood that they contain the true value of the thing we're trying to estimate (in this case, the mean).

Typically we look at 95% confidence intervals, and sometimes 99% confidence intervals, but they all have a similar interpretation: they are limits constructed such that for a certain percentage of the time (be that 95% or 99%) the true value of the population mean will fall within these limits. So, when you see a 95% confidence interval for a mean, think of it like this: if we'd collected 100 samples, calculated the mean and then calculated a confidence interval for that mean (a bit like in Figure 2.8) then for 95 of these samples, the confidence intervals we constructed would contain the true value of the mean in the population.

To calculate the confidence interval, we need to know the limits within which 95% of means will fall. How do we calculate these limits? Remember back in section 1.7.4 that I said that 1.96 was an important value of z (a score from a normal distribution with a mean of 0 and standard deviation of 1) because 95% of z-scores fall between −1.96 and 1.96. This means that if our sample means were normally distributed with a mean of 0 and a standard error of 1, then the limits of our confidence interval would be −1.96 and +1.96. Luckily we know from the central limit theorem that in large samples (above about 30) the sampling distribution will be normally distributed (see section 2.5.1). It's a pity then that our mean and standard deviation are unlikely to be 0 and 1; except not really because, as you might remember, we can convert scores so that they do have a mean of 0 and standard deviation of 1 (z-scores) using equation (1.2):

$$z = \frac{X - \bar{X}}{s}$$

If we know that our limits are −1.96 and 1.96 in z-scores, then to find out the corresponding scores in our raw data we can replace z in the equation (because there are two values, we get two equations):

$$1.96 = \frac{X - \bar{X}}{s} \qquad -1.96 = \frac{X - \bar{X}}{s}$$

We rearrange these equations to discover the value of X:

$$1.96 \times s = X - \bar{X} \qquad -1.96 \times s = X - \bar{X}$$

$$(1.96 \times s) + \bar{X} = X \qquad (-1.96 \times s) + \bar{X} = X$$

Therefore, the confidence interval can easily be calculated once the standard deviation (s in the equation above) and mean ($\bar{X}$ in the equation) are known. However, in fact we use the standard error and not the standard deviation because we're interested in the variability of sample means, not the variability in observations within the sample. The lower boundary of the confidence interval is, therefore, the mean minus 1.96 times the standard error, and the upper boundary is the mean plus 1.96 standard errors.

$$\text{lower boundary of confidence interval} = \bar{X} - (1.96 \times \text{SE})$$

$$\text{upper boundary of confidence interval} = \bar{X} + (1.96 \times \text{SE})$$

As such, the mean is always in the centre of the confidence interval. If the mean represents the true mean well, then the confidence interval of that mean should be small. We know that 95% of confidence intervals contain the true mean, so we can assume this confidence interval contains the true mean; therefore, if the interval is small, the sample mean must be very close to the true mean. Conversely, if the confidence interval is very wide then the sample mean could be very different from the true mean, indicating that it is a bad representation of the population You'll find that confidence intervals will come up time and time again throughout this book.

2.5.2.2. Calculating other confidence intervals ②

The example above shows how to compute a 95% confidence interval (the most common type). However, we sometimes want to calculate other types of confidence interval such as a 99% or 90% interval. The 1.96 and –1.96 in the equations above are the limits within which 95% of z-scores occur. Therefore, if we wanted a 99% confidence interval we could use the values within which 99% of z-scores occur (–2.58 and 2.58). In general then, we could say that confidence intervals are calculated as:

$$\text{lower boundary of confidence interval} = \overline{X} - \left(z_{\frac{1-p}{2}} \times \text{SE}\right)$$

$$\text{upper boundary of confidence interval} = \overline{X} + \left(z_{\frac{1-p}{2}} \times \text{SE}\right)$$

in which p is the probability value for the confidence interval. So, if you want a 95% confidence interval, then you want the value of z for (1 – .95)/2 = .025. Look this up in the 'smaller portion' column of the table of the standard normal distribution (see the Appendix) and you'll find that z is 1.96. For a 99% confidence interval we want z for (1 – .99)/2 = .005, which from the table is 2.58. For a 90% confidence interval we want z for (1 – .90)/2 = .05, which from the table is 1.65. These values of z are multiplied by the standard error (as above) to calculate the confidence interval. Using these general principles we could work out a confidence interval for any level of probability that takes our fancy.

2.5.2.3. Calculating confidence intervals in small samples ②

The procedure that I have just described is fine when samples are large, but for small samples, as I have mentioned before, the sampling distribution is not normal, it has a t-distribution. The t-distribution is a family of probability distributions that change shape as the sample size gets bigger (when the sample is very big, it has the shape of a normal distribution). To construct a confidence interval in a small sample we use the same principle as before but instead of using the value for z we use the value for t:

$$\text{lower boundary of confidence interval} = \overline{X} - (t_{n-1} \times \text{SE})$$

$$\text{upper boundary of confidence interval} = \overline{X} + (t_{n-1} \times \text{SE})$$

The n – 1 in the equations is the degrees of freedom (see Jane Superbrain Box 2.3) and tells us which of the t-distributions to use. For a 95% confidence interval we find the value of t for a two-tailed test with probability of .05, for the appropriate degrees of freedom.

SELF-TEST In section 1.7.2.2 we came across some data about the number of friends that 11 people had on Facebook. We calculated the mean for these data as 96.64 and standard deviation as 61.27. Calculate a 95% confidence interval for this mean.

SELF-TEST Recalculate the confidence interval assuming that the sample size was 56

2.5.2.4. Showing confidence intervals visually ②

Confidence intervals provide us with very important information about the mean, and, therefore, you often see them displayed on graphs. (We will discover more about how to create these graphs in Chapter 4.) The confidence interval is usually displayed using something called an error bar, which just looks like the letter 'I'. An error bar can represent the standard deviation, or the standard error, but more often than not it shows the 95% confidence interval of the mean. So, often when you see a graph showing the mean, perhaps displayed as a bar (Section 4.6) or a symbol (section 4.7), it is often accompanied by this funny I-shaped bar. Why is it useful to see the confidence interval visually?

What's an error bar?

We have seen that the 95% confidence interval is an interval constructed such that in 95% of samples the true value of the population mean will fall within its limits. We know that it is possible that any two samples could have slightly different means (and the standard error tells us a little about how different we can expect sample means to be). Now, the confidence interval tells us the limits within which the population mean is likely to fall (the size of the confidence interval will depend on the size of the standard error). By comparing the confidence intervals of different means we can start to get some idea about whether the means came from the same population or different populations.

Taking our previous example of quail sperm, imagine we had a sample of quail and the mean sperm release had been 9 million sperm with a confidence interval of 2 to 16. Therefore, we know that the population mean is probably between 2 and 16 million sperm. What if we now took a second sample of quail and found the confidence interval ranged from 4 to 15? This interval overlaps a lot with our first sample:

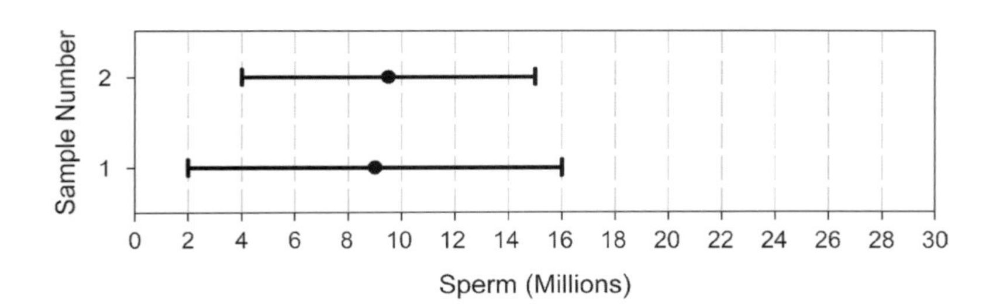

The fact that the confidence intervals overlap in this way tells us that these means could plausibly come from the same population: in both cases the intervals are likely to contain the true value of the mean (because they are constructed such that in 95% of studies they will), and both intervals overlap considerably, so they contain many similar values. What if the confidence interval for our second sample ranges from 18 to 28? If we compared this to our first sample we'd get:

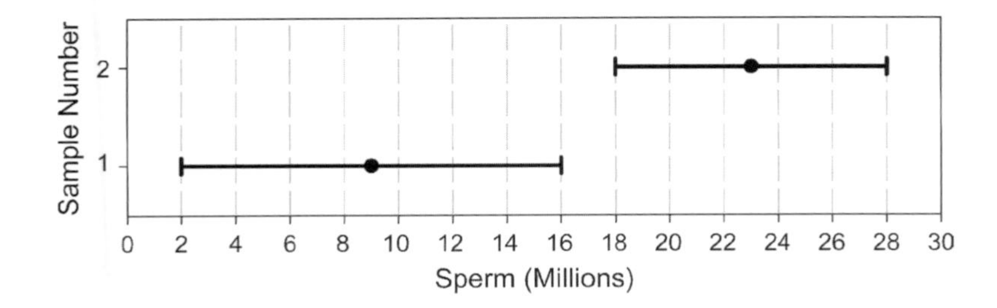

Now, these confidence intervals don't overlap at all. So, one confidence interval, which is likely to contain the population mean, tells us that the population mean is somewhere between 2 and 16 million, whereas the other confidence interval, which is also likely to contain the population mean, tells us that the population mean is somewhere between 18 and 28. This suggests that either our confidence intervals both do contain the population mean, but they come from different populations (and, therefore, so do our samples), or both samples come from the same population but one of the confidence intervals doesn't contain the population mean. If we've used 95% confidence intervals then we know that the second possibility is unlikely (this happens only 5 times in 100 or 5% of the time), so the first explanation is more plausible.

OK, I can hear you all thinking 'so what if the samples come from a different population?' Well, it has a very important implication in experimental research. When we do an experiment, we introduce some form of manipulation between two or more conditions (see section 1.6.2). If we have taken two random samples of people, and we have tested them on some measure (e.g. fear of statistics textbooks), then we expect these people to belong to the same population. If their sample means are so different as to suggest that, in fact, they come from different populations, why might this be? The answer is that our experimental manipulation has induced a difference between the samples.

To reiterate, when an experimental manipulation is successful, we expect to find that our samples have come from different populations. If the manipulation is unsuccessful, then we expect to find that the samples came from the same population (e.g. the sample means should be fairly similar). Now, the 95% confidence interval tells us something about the likely value of the population mean. If we take samples from two populations, then we expect the confidence intervals to be different (in fact, to be sure that the samples were from different populations we would not expect the two confidence intervals to overlap). If we take two samples from the same population, then we expect, if our measure is reliable, the confidence intervals to be very similar (i.e. they should overlap completely with each other).

This is why error bars showing 95% confidence intervals are so useful on graphs, because if the bars of any two means do not overlap then we can infer that these means are from different populations – they are significantly different.

CRAMMING SAM'S TIPS **Confidence intervals**

A confidence interval for the mean is a range of scores constructed such that the population mean will fall within this range in 95% of samples.

The confidence interval is not an interval within which we are 95% confident that the population mean will fall.

2.6. Using statistical models to test research questions ①

In Chapter 1 we saw that research was a five-stage process:

1. Generate a research question through an initial observation (hopefully backed up by some data).
2. Generate a theory to explain your initial observation.
3. Generate hypotheses: break your theory down into a set of testable predictions.

4. Collect data to test the theory: decide on what variables you need to measure to test your predictions and how best to measure or manipulate those variables.
5. Analyse the data: fit a statistical model to the data – this model will test your original predictions. Assess this model to see whether or not it supports your initial predictions.

This chapter has shown that we can use a sample of data to estimate what's happening in a larger population to which we don't have access. We have also seen (using the mean as an example) that we can fit a statistical model to a sample of data and assess how well it fits. However, we have yet to see how fitting models like these can help us to test our research predictions. How do statistical models help us to test complex hypotheses such as 'is there a relationship between the amount of gibberish that people speak and the amount of vodka jelly they've eaten?', or 'is the mean amount of chocolate I eat higher when I'm writing statistics books than when I'm not?' We've seen in section 1.7.5 that hypotheses can be broken down into a null hypothesis and an alternative hypothesis.

SELF-TEST What are the null and alternative hypotheses for the following questions:

- ✓ 'Is there a relationship between the amount of gibberish that people speak and the amount of vodka jelly they've eaten?'
- ✓ 'Is the mean amount of chocolate eaten higher when writing statistics books than when not?'

Most of this book deals with *inferential statistics*, which tell us whether the alternative hypothesis is likely to be true – they help us to confirm or reject our predictions. Crudely put, we fit a statistical model to our data that represents the alternative hypothesis and see how well it fits (in terms of the variance it explains). If it fits the data well (i.e. explains a lot of the variation in scores) then we assume our initial prediction is true: we gain confidence in the alternative hypothesis. Of course, we can never be completely sure that either hypothesis is correct, and so we calculate the probability that our model would fit if there were no effect in the population (i.e. the null hypothesis is true). As this probability decreases, we gain greater confidence that the alternative hypothesis is actually correct and that the null hypothesis can be rejected. This works provided we make our predictions before we collect the data (see Jane Superbrain Box 2.4).

To illustrate this idea of whether a hypothesis is likely, Fisher (1925/1991) (Figure 2.9) describes an experiment designed to test a claim by a woman that she could determine, by tasting a cup of tea, whether the milk or the tea was added first to the cup. Fisher thought that he should give the woman some cups of tea, some of which had the milk added first and some of which had the milk added last, and see whether she could correctly identify them. The woman would know that there are an equal number of cups in which milk was added first or last but wouldn't know in which order the cups were placed. If we take the simplest situation in which there are only two cups then the woman has a 50% chance of guessing correctly. If she did guess correctly we wouldn't be that confident in concluding that she can tell the difference between cups in which the milk was added first from those in which it was added last, because even by guessing she would be correct half of the time. However, what about if we complicated things by having six cups? There are 20 orders in which these cups can be arranged and the woman would guess the correct order only 1 time in 20 (or 5% of the time). If she got the correct order we would be much more

JANE SUPERBRAIN 2.4

Cheating in research ①

The process I describe in this chapter works only if you generate your hypotheses and decide on your criteria for whether an effect is significant before collecting the data. Imagine I wanted to place a bet on who would win the Rugby World Cup. Being an Englishman, I might want to bet on England to win the tournament. To do this I'd: (1) place my bet, choosing my team (England) and odds available at the betting shop (e.g. 6/4); (2) see which team wins the tournament; (3) collect my winnings (if England do the decent thing and actually win).

To keep everyone happy, this process needs to be equitable: the betting shops set their odds such that they're not paying out too much money (which keeps them happy), but so that they do pay out sometimes (to keep the customers happy). The betting shop can offer any odds before the tournament has ended, but it can't change them once the tournament is over (or the last game has started). Similarly, I can choose any team before the tournament, but I can't then change my mind halfway through, or after the final game!

The situation in research is similar: we can choose any hypothesis (rugby team) we like before the data are collected, but we can't change our minds halfway through data collection (or after data collection). Likewise we have to decide on our probability level (or betting odds) before we collect data. *If* we do this, the process works. However, researchers sometimes cheat. They don't write down their hypotheses before they conduct their experiments, sometimes they change them when the data are collected (like me changing my team after the World Cup is over), or worse still decide on them after the data are collected! With the exception of some complicated procedures called *post hoc* tests, this is cheating. Similarly, researchers can be guilty of choosing which significance level to use after the data are collected and analysed, like a betting shop changing the odds after the tournament.

Every time you change your hypothesis or the details of your analysis you appear to increase the chance of finding a significant result, but in fact you are making it more and more likely that you will publish results that other researchers can't reproduce (which is very embarrassing!). If, however, you follow the rules carefully and do your significance testing at the 5% level you at least know that in the long run at most only 1 result out of every 20 will risk this public humiliation.

(With thanks to David Hitchin for this box, and with apologies to him for turning it into a rugby example!)

confident that she could genuinely tell the difference (and bow down in awe of her finely tuned palette). If you'd like to know more about Fisher and his tea-tasting antics see David Salsburg's excellent book *The lady tasting tea* (Salsburg, 2002). For our purposes the take-home point is that only when there was a very small probability that the woman could complete the tea task by luck alone would we conclude that she had genuine skill in detecting whether milk was poured into a cup before or after the tea.

It's no coincidence that I chose the example of six cups above (where the tea-taster had a 5% chance of getting the task right by guessing), because Fisher suggested that 95% is a useful threshold for confidence: only when we are 95% certain that a result is genuine (i.e. not a chance finding) should we accept it as being true.[8] The opposite way to look at this is to say that if there is only a 5% chance (a probability of .05) of something occurring by chance then we can accept that it is a genuine effect: we say it is a **statistically significant** finding (see Jane Superbrain Box 2.5 to find out how the criterion of .05 became popular!).

[8] Of course, in reality, it might not be true – we're just prepared to believe that it is!

FIGURE 2.9 Sir Ronald A. Fisher, the cleverest person ever ($p < .0001$)

JANE SUPERBRAIN 2.5

Why do we use .05? ①

This criterion of 95% confidence, or a .05 probability, forms the basis of modern statistics and yet there is very little justification for it. How it arose is a complicated mystery to unravel. The significance testing that we use today is a blend of Fisher's idea of using the probability value p as an index of the weight of evidence against a null hypothesis, and Jerzy Neyman and Egon Pearson's idea of testing a null hypothesis *against* an alternative hypothesis. Fisher objected to Neyman's use of an alternative hypothesis (among other things), and Neyman objected to Fisher's exact probability approach (Berger, 2003; Lehmann, 1993). The confusion arising from both parties' hostility to each other's ideas led scientists to create a sort of bastard child of both approaches.

This doesn't answer the question of why we use .05. Well, it probably comes down to the fact that back in the days before computers, scientists had to compare their test statistics against published tables of 'critical values' (they did not have SAS to calculate exact probabilities for them). These critical values had to be calculated by exceptionally clever people like Fisher. In his incredibly influential textbook *Statistical methods for research workers* (Fisher, 1925)[9] Fisher produced tables of these critical values, but to save space produced tables for particular probability values (.05, .02 and .01). The impact of this book should not be underestimated (to get some idea of its influence 25 years after publication see Mather, 1951; Yates, 1951) and these tables were very frequently used – even Neyman and Pearson admitted the influence that these tables had on them (Lehmann, 1993). This disastrous combination of researchers confused about the Fisher and Neyman–Pearson approaches and the availability of critical values for only certain levels of probability led to a trend to report test statistics as being significant at the now infamous $p < .05$ and $p < .01$ (because critical values were readily available at these probabilities).

However, Fisher acknowledged that the dogmatic use of a fixed level of significance was silly: 'no scientific worker has a fixed level of significance at which from year to year, and in all circumstances, he rejects hypotheses; he rather gives his mind to each particular case in the light of his evidence and his ideas' (Fisher, 1956).

The use of effect sizes (section 2.6.4) strikes a balance between using arbitrary cut-off points such as $p < .05$ and assessing whether an effect is meaningful within the research context. The fact that we still worship at the shrine of $p < .05$ and that research papers are more likely to be published if they contain significant results does make me wonder about a parallel universe where Fisher had woken up in a $p < .10$ kind of mood. My filing cabinet full of research with p just bigger than .05 are published and I am Vice-Chancellor of my university (although, if this were true, the parallel universe version of my university would be in utter chaos, but it would have a campus full of cats).

[9] You can read this online at http://psychclassics.yorku.ca/Fisher/Methods/.

2.6.1. Test statistics ①

We have seen that we can fit statistical models to data that represent the hypotheses that we want to test. Also, we have discovered that we can use probability to see whether scores are likely to have happened by chance (section 1.7.4). If we combine these two ideas then we can test whether our statistical models (and therefore our hypotheses) are significant fits of the data we collected. To do this we need to return to the concepts of systematic and unsystematic variation that we encountered in section 1.6.2.2. Systematic variation is variation that can be explained by the model that we've fitted to the data (and, therefore, due to the hypothesis that we're testing). Unsystematic variation is variation that cannot be explained by the model that we've fitted. In other words, it is error, or variation not attributable to the effect we're investigating. The simplest way, therefore, to test whether the model fits the data, or whether our hypothesis is a good explanation of the data we have observed, is to compare the systematic variation against the unsystematic variation. In doing so we compare how good the model/hypothesis is at explaining the data against how bad it is (the error):

$$\text{test statistic} = \frac{\text{variance explained by the model}}{\text{variance not explained by the model}} = \frac{\text{effect}}{\text{error}}$$

This ratio of systematic to unsystematic variance or effect to error is a **test statistic**, and you'll discover later in the book there are lots of them: t, F and χ^2 to name only three. The exact form of this equation changes depending on which test statistic you're calculating, but the important thing to remember is that they all, crudely speaking, represent the same thing: the amount of variance explained by the model we've fitted to the data compared to the variance that can't be explained by the model (see Chapters 7 and 9 in particular for a more detailed explanation). The reason why this ratio is so useful is intuitive really: if our model is good then we'd expect it to be able to explain more variance than it can't explain. In this case, the test statistic will be greater than 1 (but not necessarily significant).

A test statistic is a statistic that has known properties; specifically we know how frequently different values of this statistic occur. By knowing this, we can calculate the probability of obtaining a particular value (just as we could estimate the probability of getting a score of a certain size from a frequency distribution in section 1.7.4). This allows us to establish how likely it would be that we would get a test statistic of a certain size if there were no effect (i.e. the null hypothesis were true). Field and Hole (2003) use the analogy of the age at which people die. Past data have told us the distribution of the age of death. For example, we know that on average men die at about 75 years old, and that this distribution is top heavy; that is, most people die above the age of about 50 and it's fairly unusual to die in your twenties. So, the frequencies of the age of demise at older ages are very high but are lower at younger ages. From these data, it would be possible to calculate the probability of someone dying at a certain age. If we randomly picked someone and asked them their age, and it was 53, we could tell them how likely it is that they will die before their next birthday (at which point they'd probably punch us!). Also, if we met a man of 110, we could calculate how probable it was that he would have lived that long (it would be a very small probability because most people die before they reach that age). The way we use test statistics is rather similar: we know their distributions and this allows us, once we've calculated the test statistic, to discover the probability of having found a value as big as we have. So, if we calculated a test statistic and its value was 110 (rather like our old man) we can then calculate the probability of obtaining a value that large. The more variation our model explains (compared to the variance it can't explain), the bigger the test statistic will be, and the more unlikely it is to occur by chance (like our 110 year old man). So, as test statistics get bigger, the probability of them occurring becomes smaller. When this probability falls below .05 (Fisher's criterion), we accept this as giving us enough confidence to assume that the test statistic is as large as it is because our model explains a sufficient amount of variation to reflect what's genuinely happening in the real world (the population). The test statistic is said to be *significant* (see Jane Superbrain Box 2.6 for a discussion of what statistically significant actually means). Given that the statistical model that we fit to the data reflects the hypothesis that we set out to test, then a significant test statistic tells us that the model would be unlikely to fit this well if the there was no effect in the population (i.e. the null hypothesis was true). Therefore, we can reject our null hypothesis and gain confidence that the alternative hypothesis is true (but, remember, we don't accept it – see section 1.7.5).

JANE SUPERBRAIN 2.6

What we can and can't conclude from a significant test statistic ②

- **The importance of an effect:** We've seen already that the basic idea behind hypothesis testing involves us generating an experimental hypothesis and a null hypothesis, fitting a statistical model to the data, and assessing that model with a test statistic. If the probability of obtaining the value of our test statistic by chance is less than .05 then we generally accept the experimental hypothesis as true: there is an effect in the population. Normally we say 'there is a *significant* effect of ...'. However, don't be fooled by that word 'significant', because even if the probability of our effect being a chance result is small (less than .05) it doesn't necessarily follow that the effect is important. Very small and unimportant effects can turn out to be statistically significant just because huge numbers of people have been used in the experiment (see Field & Hole, 2003: 74).
- **Non-significant results:** Once you've calculated your test statistic, you calculate the probability of that test statistic occurring by chance; if this probability is greater than .05 you reject your alternative hypothesis. However, this does *not* mean that the null hypothesis is true. Remember that the null hypothesis is that there is no effect in the population. All that a non-significant result tells us is that the effect is not big enough to be anything other than a chance finding – it doesn't tell us that the effect is zero. As Cohen (1990) points out, a non-significant result should never be interpreted (despite the fact that it often is) as 'no difference between means' or 'no relationship between variables'. Cohen also points out that the null hypothesis is *never* true because we know from sampling distributions (see section 2.5.1) that two random samples will have slightly different means, and even though these differences can be very small (e.g. one mean might be 10 and another might be 10.00001) they are nevertheless different. In fact, even such a small difference would be deemed as statistically significant if a big enough sample were used. So, significance testing can never tell us that the null hypothesis is true, because it never is!
- **Significant results:** OK, we may not be able to accept the null hypothesis as being true, but we can at least conclude that it is false when our results are significant, right? Wrong! A significant test statistic is based on probabilistic reasoning, which severely limits what we can conclude. Again, Cohen (1994), who was an incredibly lucid writer on statistics, points out that formal reasoning relies on an initial statement of fact followed by a statement about the current state of affairs, and an inferred conclusion. This syllogism illustrates what I mean:
  - If a man has no arms then he can't play guitar:
  - This man plays guitar.
  - Therefore, this man has arms.

  The syllogism starts with a statement of fact that allows the end conclusion to be reached because you can deny the man has no arms (the antecedent) by denying that he can't play guitar (the consequent).[10] A comparable version of the null hypothesis is:

- If the null hypothesis is correct, then this test statistic cannot occur:
- This test statistic has occurred.
- Therefore, the null hypothesis is false.

This is all very nice except that the null hypothesis is not represented in this way because it is based on probabilities. Instead it should be stated as follows:

- If the null hypothesis is correct, then this test statistic is highly unlikely:
- This test statistic has occurred.
- Therefore, the null hypothesis is highly unlikely.

If we go back to the guitar example we could get a similar statement:

- If a man plays guitar then he probably doesn't play for Fugazi (this is true because there are thousands of people who play guitar but only two who play guitar in the band Fugazi):
- Guy Picciotto plays for Fugazi:
- Therefore, Guy Picciotto probably doesn't play guitar.

This should hopefully seem completely ridiculous – the conclusion is wrong because Guy Picciotto does play guitar. This illustrates a common fallacy in hypothesis testing. In fact significance testing allows us to say very little about the null hypothesis.

2.6.2. One- and two-tailed tests ①

We saw in section 1.7.5 that hypotheses can be directional (e.g. 'the more someone reads this book, the more they want to kill its author') or non-directional (i.e. 'reading more of this book could increase or decrease the reader's desire to kill its author'). A statistical model that tests a directional hypothesis is called a **one-tailed test**, whereas one testing a non-directional hypothesis is known as a **two-tailed test**.

Imagine we wanted to discover whether reading this book increased or decreased the desire to kill me. We could do this either (experimentally) by taking two groups, one who had read this book and one who hadn't, or (correlationally) by measuring the amount of this book that had been read and the corresponding desire to kill me. If we have no directional hypothesis then there are three possibilities. (1) People who read this book want to kill me more than those who don't so the difference (the mean for those reading the book minus the mean for non-readers) is positive. Correlationally, the more of the book you read, the more you want to kill me – a positive relationship. (2) People who read this book want to kill me less than those who don't so the difference (the mean for those reading the book minus the mean for non-readers) is negative. Correlationally, the more of the book you read, the less you want to kill me – a negative relationship. (3) There is no difference between readers and non-readers in their desire to kill me – the mean for readers minus the mean for non-readers is exactly zero. Correlationally, there is no relationship between reading this book and wanting to kill me. This final option is the null hypothesis. The direction of the test statistic (i.e. whether it is positive or negative) depends on whether the difference is positive or negative. Assuming there is a positive difference or relationship (reading this book makes you want to kill me), then to detect this difference we have to take account of the fact that the mean for readers is bigger than for non-readers (and so derive a positive test statistic). However, if we've predicted incorrectly and actually reading this book makes readers want to kill me less then the test statistic will actually be negative.

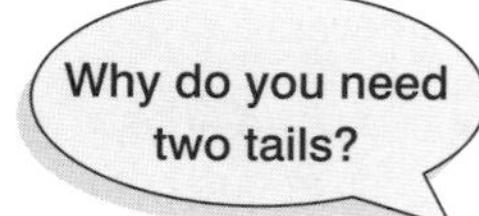

[10] Thanks to Philipp Sury for unearthing footage that disproves my point (http://www.parcival.org/2007/05/22/when-syllogisms-fail/).

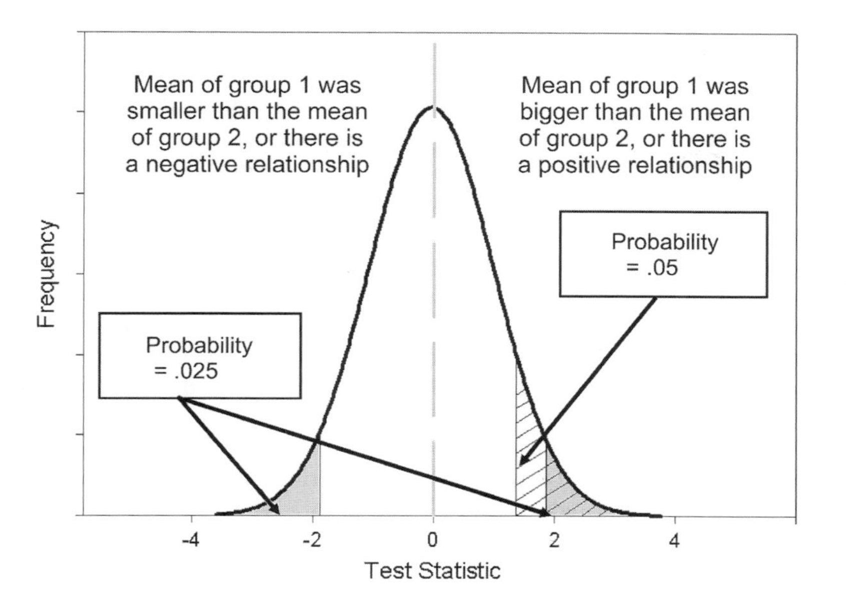

FIGURE 2.10 Diagram to show the difference between one- and two-tailed tests

What are the consequences of this? Well, if at the .05 level we needed to get a test statistic bigger than say 10 and the one we get is actually –12, then we would reject the hypothesis even though a difference does exist. To avoid this we can look at both ends (or tails) of the distribution of possible test statistics. This means we will catch both positive and negative test statistics. However, doing this has a price because to keep our criterion probability of .05 we have to split this probability across the two tails: so we have .025 at the positive end of the distribution and .025 at the negative end. Figure 2.10 shows this situation – the tinted areas are the areas above the test statistic needed at a .025 level of significance. Combine the probabilities (i.e. add the two tinted areas together) at both ends and we get .05, our criterion value. Now if we have made a prediction, then we put all our eggs in one basket and look only at one end of the distribution (either the positive or the negative end depending on the direction of the prediction we make). So, in Figure 2.10, rather than having two small tinted areas at either end of the distribution that show the significant values, we have a bigger area (the lined area) at only one end of the distribution that shows significant values. Consequently, we can just look for the value of the test statistic that would occur by chance with a probability of .05. In Figure 2.10, the lined area is the area above the positive test statistic needed at a .05 level of significance. Note on the graph that the value that begins the area for the .05 level of significance (the lined area) is smaller than the value that begins the area for the .025 level of significance (the tinted area). This means that if we make a specific prediction then we need a smaller test statistic to find a significant result (because we are looking in only one tail of the distribution), but if our prediction happens to be in the wrong direction then we'll miss out on detecting the effect that does exist! In this context it's important to remember what I said in Jane Superbrain Box 2.4: you can't place a bet or change your bet when the tournament is over. If you didn't make a prediction of direction before you collected the data, you are too late to predict the direction and claim the advantages of a one-tailed test.

2.6.3. Type I and Type II errors ①

We have seen that we use test statistics to tell us about the true state of the world (to a certain degree of confidence). Specifically, we're trying to see whether there is an effect in our population. There are two possibilities in the real world: there is, in reality, an effect in the population, or there is, in reality, no effect in the population. We have no way of knowing which of these possibilities is true; however, we can look at test statistics and their associated probability to tell us which of the two is more likely. Obviously, it is important that we're as accurate as possible, which is why Fisher originally said that we should be very conservative and only believe that a result is genuine when we are 95% confident that it is – or when there is only a 5% chance that the results could occur if there was not an effect (the null hypothesis is true). However, even if we're 95% confident there is still a small chance that we get it wrong. In fact there are two mistakes we can make: a Type I and a Type II error. A **Type I error** occurs when we believe that there is a genuine effect in our population, when in fact there isn't. If we use Fisher's criterion then the probability of this error is .05 (or 5%) when there is no effect in the population – this value is known as the **α-level**. Assuming there is no effect in our population, if we replicated our data collection 100 times we could expect that on five occasions we would obtain a test statistic large enough to make us think that there was a genuine effect in the population even though there isn't. The opposite is a **Type II error**, which occurs when we believe that there is no effect in the population when, in reality, there is. This would occur when we obtain a small test statistic (perhaps because there is a lot of natural variation between our samples). In an ideal world, we want the probability of this error to be very small (if there is an effect in the population then it's important that we can detect it). Cohen (1992) suggests that the maximum acceptable probability of a Type II error would be .2 (or 20%) – this is called the **β-level**. That would mean that if we took 100 samples of data from a population in which an effect exists, we would fail to detect that effect in 20 of those samples (so we'd miss 1 in 5 genuine effects).

There is obviously a trade-off between these two errors: if we lower the probability of accepting an effect as genuine (i.e. make α smaller) then we increase the probability that we'll reject an effect that does genuinely exist (because we've been so strict about the level at which we'll accept that an effect is genuine). The exact relationship between the Type I and Type II error is not straightforward because they are based on different assumptions: to make a Type I error there has to be no effect in the population, whereas to make a Type II error the opposite is true (there has to be an effect that we've missed). So, although we know that as the probability of making a Type I error decreases, the probability of making a Type II error increases, the exact nature of the relationship is usually left for the researcher to make an educated guess (Howell, 2006, gives a great explanation of the trade-off between errors).

2.6.4. Effect sizes ②

The framework for testing whether effects are genuine that I've just presented has a few problems, most of which have been briefly explained in Jane Superbrain Box 2.6. The first problem we encountered was knowing how important an effect is: just because a test statistic is significant doesn't mean that the effect it measures is meaningful or important. The solution to this criticism is to measure the size of the effect that we're testing in a standardized way. When we measure the size of an effect (be that an experimental manipulation or the strength of a relationship between variables) it is known as an **effect size**. An effect size is simply an objective and (usually) standardized measure of the magnitude of observed effect. The fact that the measure is standardized just means that we can compare effect sizes across different studies that have measured different variables, or have used different scales of measurement (so an effect size based on speed in milliseconds could be compared to an effect size based on heart rates). Such is the utility of effect size estimates that the American Psychological Association is now recommending that all psychologists report these effect sizes in the results of any published work. So, it's a habit well worth getting into.

Many measures of effect size have been proposed, the most common of which are Cohen's *d*, Pearson's correlation coefficient *r* (Chapter 6) and the odds ratio (Chapter 18). Many of you will be familiar with the correlation coefficient as a measure of the strength of relationship between two variables (see Chapter 6 if you're not); however, it is also a very versatile measure of the strength of an experimental effect. It's a bit difficult to reconcile how the humble correlation coefficient can also be used in this way; however, this is only because students are typically taught about it within the context of non-experimental research. I don't want to get into it now, but as you read through Chapters 6, 9 and 10 it will (I hope!) become clear what I mean. Personally, I prefer Pearson's correlation coefficient, *r*, as an effect size measure because it is constrained to lie between 0 (no effect) and 1 (a perfect effect).[11] However, there are situations in which *d* may be favoured; for example, when group sizes are very discrepant *r* can be quite biased compared to *d* (McGrath & Meyer, 2006).

Can we measure how important an effect is?

Effect sizes are useful because they provide an objective measure of the importance of an effect. So, it doesn't matter what effect you're looking for, what variables have been measured, or how those variables have been measured – we know that a correlation coefficient of 0 means there is no effect, and a value of 1 means that there is a perfect effect. Cohen (1988, 1992) has also made some widely used suggestions about what constitutes a large or small effect:

- ***r* = .10 (small effect)**: In this case the effect explains 1% of the total variance.
- ***r* = .30 (medium effect)**: The effect accounts for 9% of the total variance.
- ***r* = .50 (large effect)**: The effect accounts for 25% of the variance.

It's worth bearing in mind that *r* is not measured on a linear scale so an effect with *r* = .6 isn't twice as big as one with *r* = .3! Although these guidelines can be a useful rule of thumb to assess the importance of an effect (regardless of the significance of the test statistic), it is worth remembering that these 'canned' effect sizes are no substitute for evaluating an effect size within the context of the research domain that it is being used (Baguley, 2004; Lenth, 2001).

A final thing to mention is that when we calculate effect sizes we calculate them for a given sample. When we looked at means in a sample we saw that we used them to draw inferences about the mean of the entire population (which is the value in which we're actually interested). The same is true of effect sizes: the size of the effect in the population is the value in which we're interested, but because we don't have access to this value, we use the effect size in the sample to estimate the likely size of the effect in the population. We can also combine effect sizes from different studies researching the same question to get better estimates of the population effect sizes. This is called **meta-analysis** – see Field (2001, 2005b).

[11] The correlation coefficient can also be negative (but not below –1), which is useful when we're measuring a relationship between two variables because the sign of *r* tells us about the direction of the relationship, but in experimental research the sign of *r* merely reflects the way in which the experimenter coded their groups (see Chapter 6).

2.6.5. Statistical power ②

Effect sizes are an invaluable way to express the importance of a research finding. The effect size in a population is intrinsically linked to three other statistical properties: (1) the sample size on which the sample effect size is based; (2) the probability level at which we will accept an effect as being statistically significant (the α-level); and (3) the ability of a test to detect an effect of that size (known as the statistical **power**, not to be confused with statistical powder, which is an illegal substance that makes you understand statistics better). As such, once we know three of these properties, then we can always calculate the remaining one. It will also depend on whether the test is a one- or two-tailed test (see section 2.6.2). Typically, in psychology we use an α-level of .05 (see earlier) so we know this value already. The power of a test is the probability that a given test will find an effect assuming that one exists in the population. If you think back you might recall that we've already come across the probability of failing to detect an effect when one genuinely exists (β, the probability of a Type II error). It follows that the probability of detecting an effect if one exists must be the opposite of the probability of not detecting that effect (i.e. $1 - \beta$). I've also mentioned that Cohen (1988, 1992) suggests that we would hope to have a .2 probability of failing to detect a genuine effect, and so the corresponding level of power that he recommended was 1 – .2, or .8. We should aim to achieve a power of .8, or an 80% chance of detecting an effect if one genuinely exists. The effect size in the population can be estimated from the effect size in the sample, and the sample size is determined by the experimenter anyway so that value is easy to calculate. Now, there are two useful things we can do knowing that these four variables are related:

1. **Calculate the power of a test**: Given that we've conducted our experiment, we will have already selected a value of α, we can estimate the effect size based on our sample, and we will know how many participants we used. Therefore, we can use these values to calculate $1 - \beta$, the power of our test. If this value turns out to be .8 or more we can be confident that we achieved sufficient power to detect any effects that might have existed, but if the resulting value is less, then we might want to replicate the experiment using more participants to increase the power.
2. **Calculate the sample size necessary to achieve a given level of power**: Given that we know the value of α and β, we can use past research to estimate the size of effect that we would hope to detect in an experiment. Even if no one had previously done the exact experiment that we intend to do, we can still estimate the likely effect size based on similar experiments. We can use this estimated effect size to calculate how many participants we would need to detect that effect (based on the values of α and β that we've chosen).

The latter use is the more common: to determine how many participants should be used to achieve the desired level of power. The actual computations are very cumbersome, but fortunately there are now computer programs available that will do them for you (one example is G*Power, which is free and can be downloaded from a link on the companion website; another is nQuery Adviser but this has to be bought!). Also, Cohen (1988) provides extensive tables for calculating the number of participants for a given level of power (and vice versa). Based on Cohen (1992) we can use the following guidelines: if we take the standard α-level of .05 and require the recommended power of .8, then we need 783 participants to detect a small effect size ($r = .1$), 85 participants to detect a medium effect size ($r = .3$) and 28 participants to detect a large effect size ($r = .5$).

What have I discovered about statistics? ①

OK, that has been your crash course in statistical theory! Hopefully your brain is still relatively intact. The key point I want you to understand is that when you carry out research you're trying to see whether some effect genuinely exists in your population (the effect you're interested in will depend on your research interests and your specific predictions). You won't be able to collect data from the entire population (unless you want to spend your entire life, and probably several after-lives, collecting data) so you use a sample instead. Using the data from this sample, you fit a statistical model to test your predictions, or, put another way, detect the effect you're looking for. Statistics boil down to one simple idea: observed data can be predicted from some kind of model and an error associated with that model. You use that model (and usually the error associated with it) to calculate a test statistic. If that model can explain a lot of the variation in the data collected (the probability of obtaining that test statistic is less than .05) then you infer that the effect you're looking for genuinely exists in the population. If the probability of obtaining that test statistic is more than .05, then you conclude that the effect was too small to be detected. Rather than rely on significance, you can also quantify the effect in your sample in a standard way as an *effect size* and this can be helpful in gauging the importance of that effect. We also discovered that I managed to get myself into trouble at nursery school. It was soon time to move on to primary school and to new and scary challenges. It was a bit like using SAS for the first time!

Key terms that I've discovered

α-level
β-level
Central limit theorem
Confidence interval
Degrees of freedom
Deviance
Effect size
Fit
Linear model
Meta-analysis
One-tailed test
Population
Power
Sample
Sampling distribution
Sampling variation
Standard deviation
Standard error
Standard error of the mean (SE)
Sum of squared errors (SS)
Test statistic
Two-tailed test
Type I error
Type II error
Variance

Smart Alex's tasks

- **Task 1:** Why do we use samples? ①
- **Task 2:** What is the mean and how do we tell if it's representative of our data? ①

- **Task 3:** What's the difference between the standard deviation and the standard error? ①
- **Task 4:** In Chapter 1 we used an example of the time taken for 21 heavy smokers to fall off a treadmill at the fastest setting (18, 16, 18, 24, 23, 22, 22, 23, 26, 29, 32, 34, 34, 36, 36, 43, 42, 49, 46, 46, 57). Calculate the sums of squares, variance, standard deviation, standard error and 95% confidence interval of these data. ①
- **Task 5:** What do the sum of squares, variance and standard deviation represent? How do they differ? ①
- **Task 6:** What is a test statistic and what does it tell us? ①
- **Task 7:** What are Type I and Type II errors? ①
- **Task 8:** What is an effect size and how is it measured? ②
- **Task 9:** What is statistical power? ②

Answers can be found on the companion website.

Further reading

Cohen, J. (1990). Things I have learned (so far). *American Psychologist*, *45*(12), 1304–1312.
Cohen, J. (1994). The earth is round ($p < .05$). *American Psychologist*, *49*(12), 997–1003. (A couple of beautiful articles by the best modern writer of statistics that we've had.)
Field, A. P., & Hole, G. J. (2003). *How to design and report experiments*. London: Sage. (I am rather biased, but I think this is a good overview of basic statistical theory.)
Miles, J. N. V., & Banyard, P. (2007). *Understanding and using statistics in psychology: a practical introduction*. London: Sage. (A fantastic and amusing introduction to statistical theory.)
Wright, D. B., & London, K. (2009). *First steps in statistics* (2nd ed.). London: Sage. (This book has very clear introductions to sampling, confidence intervals and other important statistical ideas.)

Interesting real research

Domjan, M., Blesbois, E., & Williams, J. (1998). The adaptive significance of sexual conditioning: Pavlovian control of sperm release. *Psychological Science*, *9*(5), 411–415.

3 The SAS environment

FIGURE 3.1
All I want for Christmas is ... some tasteful wallpaper

3.1. What will this chapter tell me? ①

At about 5 years old I moved from nursery (note that I moved, I was not 'kicked out' for showing my ...) to primary school. Even though my older brother was already there, I remember being really scared about going. None of my nursery school friends were going to the same school and I was terrified about meeting all of these new children. I arrived in my classroom, and as I'd feared, it was full of scary children. In a fairly transparent ploy to make me think that I'd be spending the next 6 years building sand castles, the teacher told me to play in the sand pit. While I was nervously trying to discover whether I could build a pile of sand high enough to bury my head in it, a boy came and joined me. He was Jonathan Land, and he was really nice. Within an hour he was my new best friend (5 year olds are fickle ...) and I loved school. Sometimes new environments seem more scary than they really are. This chapter introduces you to a scary new environment: SAS. The SAS environment is a generally more unpleasant environment in which to spend time than your normal environment; nevertheless, we have to spend time there if we are to analyse our data. The purpose of this chapter is, therefore, to put you in a sand pit with a 5 year old called Jonathan. I will orient you in your new home and everything will be fine. We will explore the key windows in SAS (the *Enhanced Editor*, *the Log*, the *data window* and the *Output window*) and also look at how to create variables, enter data and adjust the properties of your variables. We finish off by looking at some alternatives.

3.2. Versions of SAS ①

This book is based primarily on version 9.2 of SAS (at least in terms of the diagrams); however, don't be fooled too much by version numbers because SAS doesn't change too much between versions. Most of the things that we'll talk about in this book worked just fine with version 6, and that was released in the 1980s. The most recent version before 9.2 was 9.1.3, which was released in 2005, 9.2 was released in 2008, and you'd have to look pretty hard to find things that differed.[1] On the rare occasions that we mention something new, we'll flag that up for users of older versions.

You should also be aware that SAS is a vast, vast, VAST program. SAS has so many parts and components and add-ons that I would be surprised if anyone could even list them all, never mind know what they do. If you read something about SAS that you don't understand, and doesn't make any sense at all, that's OK. It happens to me too (today I found out that SAS has a module for helping you decide how to lay out a supermarket; someone else was having trouble getting SAS to write to their Twitter account).

3.3. Getting started ①

You seem like a pretty computer and technology literate kind of person. I bet you don't read a manual very often. I mean, manuals are for wimps, aren't they? When you get a new iPod you didn't read the manual, you just plugged it in, and off you went. I bet you didn't even read the manual to program the video recorder. Let's face it, you don't read manuals. I mean, who does?[2]

When you start to use a computer program, you don't read the manual either. I mean, there is probably a manual for Word or Excel or iTunes, but really, who's ever seen it? You either start typing, or you click File… Open… and choose a file. SAS isn't like that. You can click File, but then there isn't an Open option. There's an Open program option, but we don't want a program, we want some data – we're trying to use SAS to analyse data, so why can't we open data? In the old days, SAS produced manuals. There were something like 20 of them, and together they took up about four feet of shelf. One of them was called (something like) 'Getting Started', it had two volumes, each of which were a couple of hundred pages. I don't think they print the manuals any more (if they do, I don't know anyone who has them), but you can find whatever you need on the web at http://support.sas.com/. However, you don't want to have to wade through all that, so instead, we're going to help you here.

[1] That's not to say that there weren't a lot of new things, it's just that almost all of them were way beyond the level of this book. People like Jeremy got pretty excited about something called PROC TCALIS, for example.

[2] Actually, Jeremy had to read the manual for his automatic goldfish feeder. But that was really complicated, and if he'd got it wrong it would have been very sad for his goldfish.

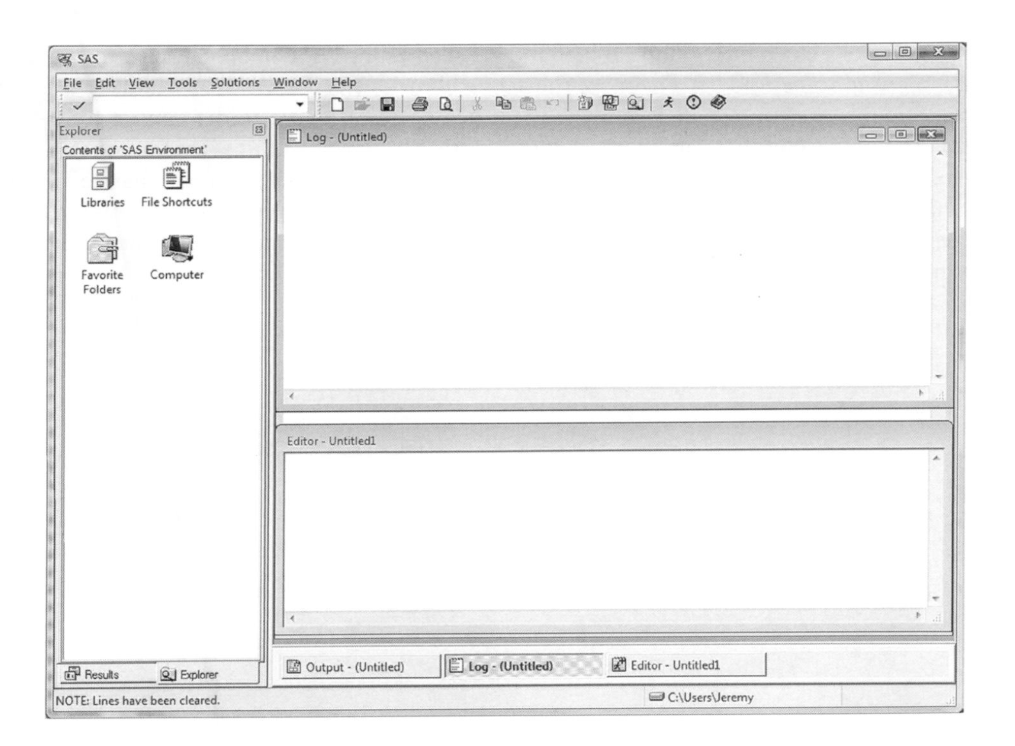

FIGURE 3.2 Opening SAS for the first time

We're going to assume that you are running SAS on a Windows computer. SAS will run on a Macintosh, but it is only version 6.12, which is pretty old. SAS will also run on some flavours of Linux and Unix, but if you are using Linux or Unix for your work, you are possibly cleverer than us, and won't be needing our advice.[3]

We want to warn you now that the learning curve for SAS is steep. SAS is not like most other computer programs (you've already seen that, if you tried to choose File…Open). But although the learning curve is steep, it's also short. Once you understand how SAS works, it's pretty straightforward.

Really.

Let's open SAS and take a look.

The first time you open SAS, it should look something like Figure 3.2. SAS has 4 windows that we need to worry about. The one on the left is called the Explorer, and contains libraries. Libraries are where SAS stores its data. In the main window, you can see the Log. The Log is where SAS tells you what's going on – when things went right, and when things went wrong, it tells you in the Log. It's easy to spot when things go wrong, because they are in a different colour (the publisher didn't think you wanted to spend the extra money on a book with nice colours in it, so ours are just black and white. But you'll soon see.)

Below the Log is the Editor. This is the place where you type commands for SAS – SAS calls them programs. Behind the Log and the Editor is the Output. (You can see that the Output has a button at the bottom of the screen). The results of your statistical procedures appear in the Output.

3.4. Libraries ①

Libraries are very important in SAS. Data (and formats, more on those later) are kept in libraries. And a library in SAS is just a folder on your disk. In the Explorer window on the left, one

[3] If you do need my advice, it's basically the same. You just need to change the LIBNAME statement.

of the icons looks like a little filing cabinet and is called Libraries. Double-click on that, and it opens, to reveal some folders. These are the active libraries (see Figure 3.3). Libraries are just like the folders that you use in Windows, with the exception that a folder cannot have subfolders. The only thing that can go in a folder is a file.

When you clicked on the library, you might have noticed that the toolbar changed at the top of the window. These are just like the toolbar in Windows Explorer, they do things like change the way that the folders are shown to you, or copy and paste. The only one that really matters is the first one, which takes us back to the library level.

FIGURE 3.3
The libraries

SAS has already created four libraries for us. The only one that we will use is the one called Work. As we will soon see, Work is a special library, for two reasons. First, if you don't tell SAS which library to put your data into, then SAS will put it into Work. Second, and really, really important: when you shut down SAS, all of the files in Work are deleted. If you want to save your work, you need to save it somewhere else. Anything that you put in a folder that is not Work will be saved.

If you want to keep your work organized, you'll need to create more libraries. To do that, right-click with your mouse in an empty space in the Explorer, and then select New… from the menu (if you clicked on an empty space, New… will be the only thing you can select).

When you have selected that, the *New Library* dialog box will appear – as shown in Figure 3.4. Type a name for the new library in the Name box – you are limited to 8 characters. Chapter3 has eight characters, and is suitable descriptive for this chapter. Then click on the Browse… button, and find the folder on your disk that you want to use for that library.

FIGURE 3.3
The New Library dialog

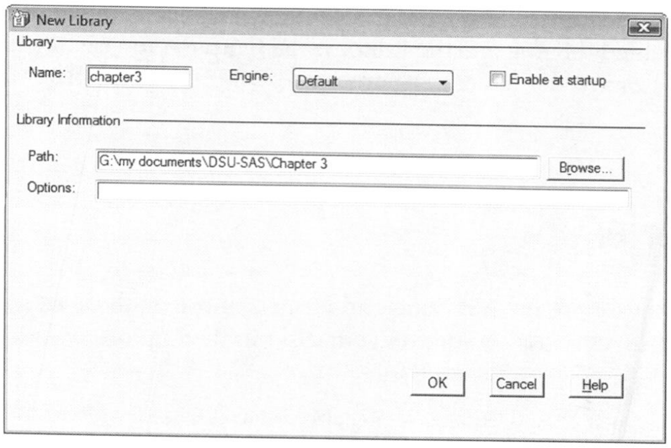

If you select 'Enable at startup' that library will always be there when you start SAS. If you don't select that, you'll have to go through this procedure again. I tend not to enable the library at startup, because then I have too many libraries, and I get in a mess. I also use SAS Tip 3.1.

SAS TIP 3.1 **Entering data**

Creating the library each time you start SAS is a little tedious sometimes – you might forget where you put it last time, for example. But if you enable all libraries at startup, you'll get in a mess, because you have so many libraries. The alternative is to type the equivalent command into the Editor and save it.

To create a library we'll use the LIBNAME command, and type:

```
LIBNAME chapter3 "g:\my documents\DSU-SAS\Chapter 3";
```

3.5. The program editor: Getting data into SAS ①

OMG! Have I got to write programs?

In short, no. SAS calls them programs but they're not really programs; at least the ones we are going to write are not programs.[4] They are just a series of commands that you type into SAS. The great thing about typing them using syntax is that you know exactly what you typed, and it's really easy to fix it when you've made a mistake, and run it again. If you enter commands by clicking the mouse, you might not recall what you did to get the output you got, and you might not be able to tell someone else.[5]

The program editor in SAS is called the Enhanced Editor. The main enhancement is that it knows what SAS commands are supposed to look like, and it helps you to write them. When you type a command, the command should turn blue. If you've mistyped it, it doesn't turn blue, and you know you've got it wrong (there are a few other things too, but that's the big one).

There are a few things that you need to know about the program editor.

1 Commands always end with a semicolon. The most common mistake to make in your code is to forget to put a semicolon. It's been said that SAS actually stands for 'Semi-colons, Always Semi-colons'.

2 SAS (usually) doesn't care about white space – that is tabs, spaces, returns. It ignores them all. As far as SAS is concerned, a command ends on a semi colon. If you were SAS, this:

```
This is a set of commands; This is another;
```

would be considered to be exactly the same as:

```
This
    is
    a set of        commands
    ;
    This
    is
    another          ;
```

As we will see later on, we can use this to our advantage.

[4] Given that it's a statistics program, I'm sometimes amazed by what people can persuade SAS to do. The thing that amazes me more than anything else is Wolfensäs – a first person shooter game. No, really. Go to http://wolfensas.subshock.net/.

[5] Sometimes people come to me with printouts from statistics packages, asking for help in understanding them. I ask them what they did to get those results. They don't remember. If they had the syntax, they could show me.

3 SAS is not case sensitive. As far as SAS is concerned

```
A COMMAND;
a command;
```

are the same.

And there's one thing you need to know about how we set out this book. We always put SAS commands in CAPITAL LETTERS. Things that you need to type which are not SAS commands (names of variables, names of files) will be in lower case.

SAS TIP 3.2 Entering data

There is a simple rule for how variables should be placed into SAS: data from different things go in different rows of the data viewer, whereas data from the same things go in different columns. As such, each person (or mollusc, goat, organization, or whatever you have measured) is represented in a different row. Data within each person (or mollusc, etc.) go in different columns. So, if you've prodded your mollusc, or human, several times with a pencil and measured how much it twitches as an outcome, then each prod will be represented by a column.

In experimental research this means that any variable measured with the same participants (a repeated measure) should be represented by several columns (each column representing one level of the repeated-measures variable). However, any variable that defines different groups of things (such as when a between-group design is used and different participants are assigned to different levels of the independent variable) is defined using a single column. This idea will become clearer as you learn about how to carry out specific procedures. (This golden rule is broken in mixed models but until Chapter 19 we can overlook this annoying anomaly.)

3.5.1. Entering words and numbers into SAS with the program editor ①

For a simple data set, the easiest way to get data into SAS is to use the program editor, and to type a short series of commands. The approach we're going to use is called a *DATA step*.

The first thing to do is to tell SAS what we want to call the data that we've created. If we decide to call the file myfirstdata we could write:

```
DATA myfirstdata;
```

(Notice that the command ends with a semicolon, and that when you finished typing the word 'data' it turned blue?)

However, if we just call it myfirstdata, SAS will put it into the Work folder, and it will be deleted when we shut down SAS. Instead, let's put it into the chapter3 library that we created earlier.

```
DATA chapter3.myfirstdata;
```

(Remember that we're using capital letters, but you don't have to. We're just showing that DATA is an SAS keyword. SAS is always pronounced SAS, to rhyme with gas, not ess-ay-ess. Hence a sas word).

Then we'll tell SAS the names of the variables we intend to input. We put the names for each variable in a row. This name will appear at the top of the corresponding column in the data viewer, and helps you to identify variables in the data view. In current versions of SAS you can more or less write what you like, but there are certain symbols you can't use (mainly symbols that have other uses in SAS such as +, –, $, &, .) , and you can't use spaces. (Many people use a 'hard' space in variable names, which replaces the space with an underscore, for example, Andy_Field instead of Andy Field.) You are allowed to put numbers in variable names, but you can't start a variable name with a number, so question6a is OK, but 6a is not OK. If you use a character that SAS doesn't like you'll get an error message saying that the variable name is invalid when you look at the Log.

```
INPUT name numberoffriends;
```

(Notice that 'input' changed colour when you had typed it correctly – we can't show that, because the publisher thinks that you don't want to pay the extra that it would cost for a book that was printed in colour).

We are going to have a slight problem here, because SAS is expecting numbers, and we're going to give it letters or words under name – technically, this is called a string (because it's a string of characters) and is represented with a $ sign. We need to tell SAS that the name variable is a string.

TABLE 3.1 Some data with which to play

Name	*Birth Date*	*Job*	*No. of Friends*	*Alcohol (Units)*	*Income (p.a.)*	*Neuroticism*
Leo	17-Feb-1977	Lecturer	5	10	20,000	10
Martin	24-May-1969	Lecturer	2	15	40,000	17
Andy	21-Jun-1973	Lecturer	0	20	35,000	14
Paul	16-Jul-1970	Lecturer	4	5	22,000	13
Graham	10-Oct-1949	Lecturer	1	30	50,000	21
Carina	05-Nov-1983	Student	10	25	5,000	7
Karina	08-Oct-1987	Student	12	20	100	13
Doug	16-Sep-1989	Student	15	16	3,000	9
Mark	20-May-1973	Student	12	17	10,000	14
Mark	28-Mar-1990	Student	17	18	10	13

SAS TIP 3.3 Naming variables

Why is it a good idea to take the time and effort to type in long variable names for your variables? Surely it's a waste of my time? In a way, I can understand why it would seem to be so, but as you go through your course accumulating data files, you will be grateful that you did. Imagine you had a variable called 'number of times I wanted to shoot myself during Andy Field's statistics lecture'; then you might have called the column in SAS 'shoot'. If you don't add a label, SAS will use this variable name in all of the output from an analysis. That's all well and good, but what happens in three weeks' time when you look at your data and output again? The chances are that you'll probably think 'What did shoot stand for? Number of shots at goal? Number of shots I drank?' Imagine the chaos you could get into if you had used an acronym for the variable 'workers attending new kiosk'. I have many data sets with variables called things like 'sftg45c', and if I didn't give them proper labels I would be in all sorts of trouble. Get into a good habit and label all of your variables!

```
INPUT name $ friends;
```

Then we tell SAS the data are coming, with the CARDS command.

```
CARDS;
```

And we enter the data.

```
Leo   5
Martin 2
Andy 0
Paul 4
Graham 1
Carina 10
Karina 12
Doug 15
Mark 12
;
```

Recall that SAS doesn't mind where you put spaces. We can use this to our advantage, by adding spaces to line up the numbers – this makes it easier to see that we haven't made a mistake.

Also notice that we don't need a semi colon until the end.

Finally, we finish the statement with a RUN command.

```
RUN;
```

The whole thing is shown in SAS Syntax 3.1.

```
DATA chapter3.myfirstdata;
INPUT name $ numberoffriends;
CARDS;
Leo   5
Martin 2
Andy 0
Paul 4
Graham 1
Carina 10
Karina 12
Doug 15
Mark 12
;
RUN;
```

SAS Syntax 3.1

3.5.1.1 Running the program ①

There are three ways to submit the set of commands that we've written.

1 On the toolbar, there is a picture of a small person running. You can click that icon.

2 You can click on the Run menu, and then choose Submit.

3 You can press and hold down the Alt key, and then press R. (That's what I do, because it's the fastest.)

If you have not selected any text, choosing Run submits all of the commands in the program editor to SAS. If you select some text, it only submits the text that you have selected. It's a good habit to get into selecting the text you want – you can do that with the mouse, or by pressing the arrow keys, while Shift is pressed. Then run the syntax using one of the methods described above.

Whenever you run a command you should have a look at the Log window, and check it went OK. Figure 3.5 shows the Log window. The commands that we typed (minus the data) are shown in black. SAS then gives us the results, in blue. It tells us we created 9 observations and 2 variables (which sounds good), and that it took a total of 0.00 seconds.

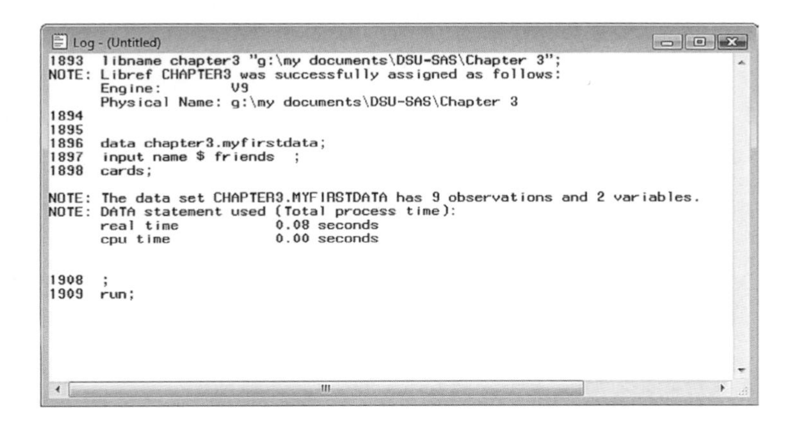

FIGURE 3.5
The Log window

We can also have a look at the data set, and make sure it was created properly. Go to the Explorer on the left of the screen, open the library called Chapter3 by double-clicking on it. It should look like Figure 3.6.

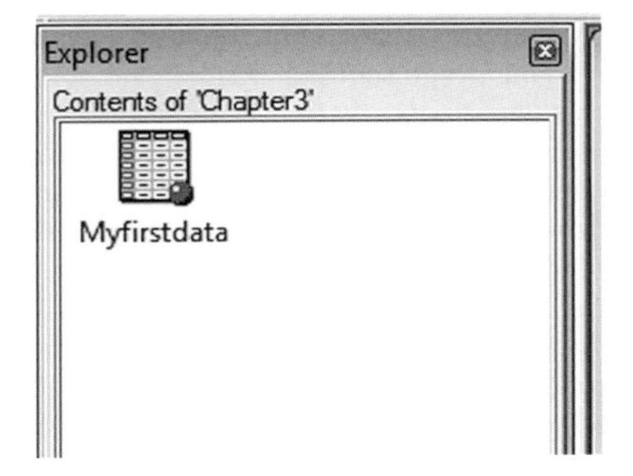

FIGURE 3.6
The library Chapter3 should contain myfirstdata

Open the data by double-clicking on that, and the data window will open, as shown in Figure 3.7. Note that this is a viewtable, you can view the data here, but you can't change it.

3.5.2. Entering dates ①

Let's expand that program to include the rest of the data. We've another string in there (the job type), and we've also got a date. We need to tell SAS that that variable will be a date, and tell it what kind. Handling dates can get a little bit complicated in many computer programs, because there are so many different ways of writing a date (e.g. 17 June, 2006; June 17 06; 6/17/06; 16/6/06; 200606017). SAS calls these different types of date formats. If you are really, really interested, you can use your favourite web search engine to search for date formats in SAS. Right now, all you need to know (and all you'll probably ever need to know) is that the format we've used in the table is called date9.

This is our first introduction to formats in SAS. We'll talk about them more soon, but right now, you also need to know that a format type in SAS ends in a dot, it's date9., not

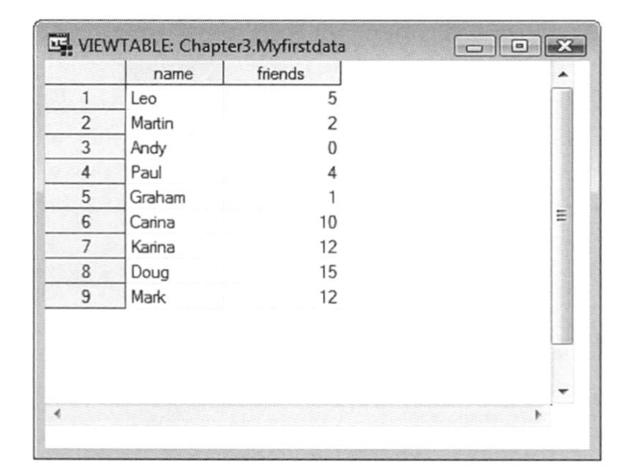
VIEWTABLE: Chapter3.Myfirstdata

	name	friends
1	Leo	5
2	Martin	2
3	Andy	0
4	Paul	4
5	Graham	1
6	Carina	10
7	Karina	12
8	Doug	15
9	Mark	12

FIGURE 3.7
Myfirst data table

date9 without a dot. In the previous example, we put a string symbol after a variable name to indicate to SAS that a variable was a string, we do the same to indicate that a variable is a date, except we put the date statement.

3.5.2.1. Creating coding variables ①

A coding variable (also known as a grouping variable) is a variable that uses numbers to represent different groups of data. As such, it is a *numeric variable*, but these numbers represent names (i.e. it is a nominal variable). These groups of data could be levels of a treatment variable in an experiment, different groups of people (men or women, an experimental group or a control group, ethnic groups, etc.), different geographic locations, different organizations, etc.

In experiments, coding variables represent independent variables that have been measured between groups (i.e. different participants were assigned to different groups). If you were to run an experiment with one group of participants in an experimental condition and a different group of participants in a control group, you might assign the experimental group a code of 1 and the control group a code of 0. When you come to put the data into the data editor you would create a variable (which you might call **group**) and type in the value 1 for any participants in the experimental group, and 0 for any participant in the control group. These codes tell SAS that all of the cases that have been assigned the value 1 should be treated as belonging to the same group, and likewise for the cases assigned the value 0. In situations other than experiments, you might simply use codes to distinguish naturally occurring groups of people (e.g. you might give students a code of 1 and lecturers a code of 0).

We have a coding variable in our data: the one describing whether a person was a lecturer or student. To create this coding variable, we follow the steps for creating a normal variable, we just put 1 or 2 in place of 'student' or 'lecturer' (in section 3.8 we will tell SAS which numeric codes have been assigned to which groups). (It's a good habit to follow to use alphabetical order for the codes – that way you will remember what they were).

Remember we said that SAS usually doesn't care about spaces? Well, sometimes it does care about spaces in the CARDS statement, and that's why we can't use spaces to make it quite as nice as we did before. The full syntax to enter our data is in SAS Syntax 3.2.

```
DATA chapter3.myfirstdata;
    INPUT name $ dateofbirth DATE9. job numberoffriends alcohol
    income neuroticism ;
    CARDS;
    Leo 17Feb1977 1 5 10 20000 10
    Martin 24May1969 1 2 15 40000 17
    Andy 21Jun1973 1 0 20 35000 14
    Paul 16Jul1970 1 4 5 22000 13
    Graham 10Oct1949 1 1 30 50000 21
    Carina 05Nov1983 2 10 25 5000 7
```

```
    Karina 08Oct1987 2 12 20 100 13
    Doug 16Sep1989 2 15 16 3000 9
    Mark 20May1973 2 12 17 10000 14
    Mark 28Mar1990 2 17 18  10 13
    ;
RUN;
```
SAS Syntax 3.2

3.6. Entering data with Excel ①

With a large data set, entering data into SAS with the CARDS command can get a bit arduous, and so if we can't get someone else to do it for us it would be good to find an easier way. The easier way that we usually use is to enter the data into a spreadsheet, like Excel or OpenOffice Calc, save the data, and then import it into SAS.

	A	B	C	D	E	F	G	H
1	Name	date	Job	friends	alcohol	Income	Neuroticism	
2	Leo	17-Feb-77	1	5	10	20,000	10	
3	Martin	24-May-69	1	2	15	40,000	17	
4	Andy	21-Jun-73	1	0	20	35,000	14	
5	Paul	16-Jul-70	1	4	5	22,000	13	
6	Graham	10-Oct-49	1	1	30	50,000	21	
7	Carina	5-Nov-83	2	10	25	5,000	7	
8	Karina	8-Oct-87	2	12	20	100	13	
9	Doug	16-Sep-89	2	15	16	3,000	9	
10	Mark	20-May-73	2	12	17	10,000	14	
11	Mark	28-Mar-90	2	17	18	10	13	
12								

Sheet1 Sheet2 Sheet3

FIGURE 3.8 Entering data in Excel

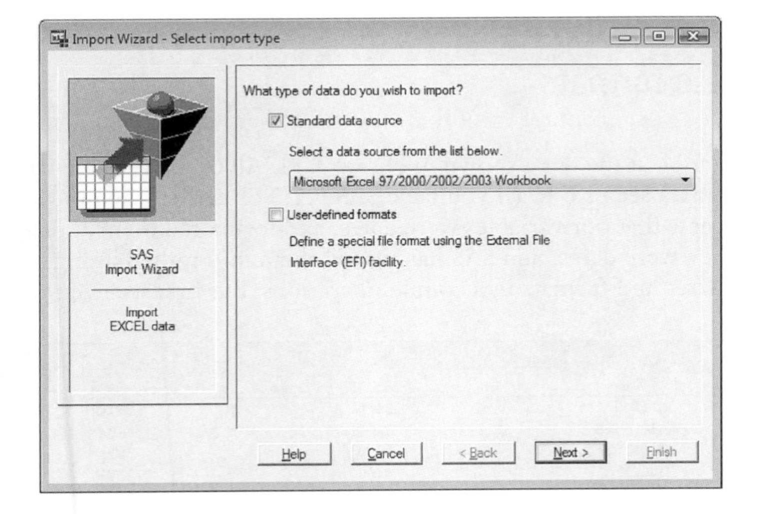

FIGURE 3.9

We enter the data into Excel in the same way. A person is a row, and a column is a variable. The first row is used as the variable name that you want SAS to give to that column. Figure 3.8 shows the data in Excel. We'll save that data set – let's call it **myfirstdata.xls**, and then we'll import it into SAS.

In SAS, click on File, and then Import Data Source. This starts the Import Wizard, which takes you through the steps of importing data. Figure 3.9 shows the steps that SAS takes you through. First, SAS wants to know what sort of data you want; the default is Excel, with a .xls extension. (Notice that if you are using Excel 2007 SAS can't open that kind of file, so you need to save in the older, .xls format, not the new .xlsx format). SAS then asks you to find the file. You Click Browse, and then tell SAS which sheet in the Excel file you want to import – it's almost always Sheet1, and that's what SAS guesses. Finally, you tell SAS which library you want to save the file in, and what you want the SAS data file to be called. We'll call it myfirstdata, and we'll save it in the Chapter3 library.

3.6.1. Saving the data ①

You might have noticed another difference between SAS and other programs. We have never saved the data file. In fact, there isn't an option to save the file. The file is always saved in the folder that the library is attached to. It's worth going and having a look in Windows Explorer, just to convince ourselves that that is true. Figure 3.10 shows Windows Explorer, and there's the file, called **myfirstdata.sas7bdat**. (SAS data files get the somewhat unwieldy extension **.sas7bdat.**)

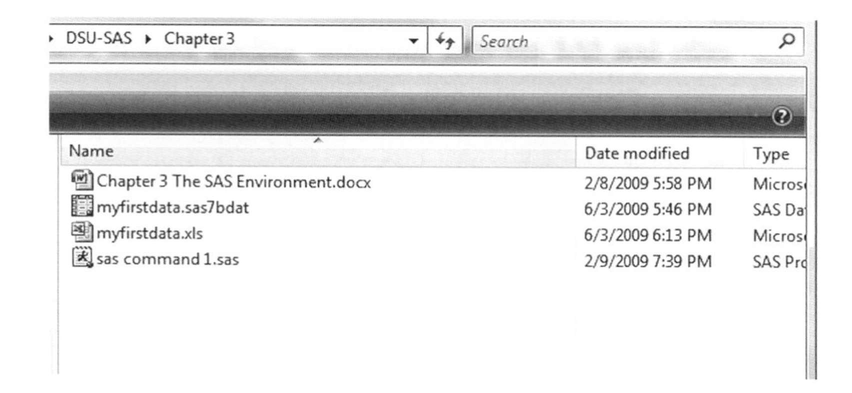

FIGURE 3.10 Windows Explorer, showing SAS file

3.6.2. Date formats ①

Let's also take a look at the data, shown in Figure 3.11. All of the variable names seem OK, and most of the data seems OK. (If you have SAS 9.1.3 or earlier, then the date will not be correct. Excel knew that our variables were dates, and formatted them appropriately. Excel told SAS that they were dates, and SAS has turned them into rather strange numbers. Let's have a look at dates and formats in a couple of sections, but first we need to digress.)

VIEWTABLE: Chapter3.Myfirstformatteddata

	Name	date	Job	friends	alcohol	Income	Neuroticism
1	Leo	17FEB1977	1	5.000	10	20000	10
2	Martin	24MAY1969	1	2.000	15	40000	17
3	Andy	21JUN1973	1	0.000	20	35000	14
4	Paul	16JUL1970	1	4.000	5	22000	13
5	Graham	10OCT1949	1	1.000	30	50000	21
6	Carina	05NOV1983	2	10.000	25	5000	7
7	Karina	08OCT1987	2	12.000	20	100	13
8	Doug	16SEP1989	2	15.000	16	3000	9
9	Mark	20MAY1973	2	12.000	17	10000	14
10	Mark	28MAR1990	2	17.000	18	10	13

FIGURE 3.11 Checking the data in SAS

3.6.3. Missing values ①

Although as researchers we strive to collect complete sets of data, it is often the case that we have missing data. Missing data can occur for a variety of reasons: in long questionnaires participants accidentally (or, depending on how paranoid you're feeling, deliberately just to piss you off) miss out questions; in experimental procedures mechanical faults can lead to a datum not being recorded; and in research on delicate topics (e.g. sexual behaviour) participants may exert their right not to answer a question. However, just because we have missed out on some data for a participant doesn't mean that we have to ignore the data we do have (although it sometimes creates statistical difficulties). Nevertheless, we do need to tell SAS that a value is missing for a particular case.

To specify a missing value, we use a dot. If, for example, Carina had declined to answer the question about the number of friends that she had (or had put an answer which was obviously false, like 37,342 or –9.3), we would replace the number with a dot, and SAS would understand that that piece of information is not known.

```
Graham 1
Carina .
Karina 12
```

SAS TIP 3.4 SAS and missing data

The way that SAS stores a missing value is a little strange, and can get you in a tangle if you forget. We'll mention it again when it comes up, but it is so easy to forget that we're going to mention it here as well. SAS stores missing values as the lowest value that it can represent. Imagine you ask two questions: What age did you first smoke a cigarette (called agecigarette)? What age did you first drink beer (called agebeer)? If a person had never smoked a cigarette, they might leave that question blank. However, if you ask SAS which number was smaller, the age they smoked a cigarette, or the age the smoked a beer, SAS will give you an answer – the cigarette value is missing, hence it is lower than the beer value.

3.7. The DATA step ①

To copy data from one file to another, or to manipulate data, we can use a command called a *data step* in SAS. If we have got a data set, called got, and we want a data set called want, the DATA step has the following format:

```
DATA want; SET got;
RUN;
```

It takes the data set called got, and copies it into a new data set called want.

Let's give an example. We have got a data set – myfirstdata – and it's in the library called chapter3. And we want a copy of it. SAS Syntax 3.3 shows the syntax. If you run that, you will create a copy of the data set, and the copy will be in the library **chapter3**.

```
DATA chapter3.myfirstdatacopy; SET chapter3.myfirstdata;
RUN;
```

SAS Syntax 3.3

SAS TIP 3.5 DATA want; SET got!

It is very important that you remember 'data want; set got'. I remember that data and want have 4 letters, and set and got have 3 letters. Getting those the wrong way around can lead to an awful lot of trouble. Here's why. Say you swap over those two file names, and you type

```
DATA chapter3.myfirstdata; SET chapter3.myfirstdatacopy;
```

What will SAS do? It will try to find the file called myfirstdatacopy, and it will find that that file does not exist, it's empty. So it will take an empty file, and put it on top of myfirstdata. Your original data will be deleted. This could range between slightly irritating and disastrous. One way to avoid it is to never work with your only copy of the data. Save the file somewhere else, write-protect it, and leave it alone. A useful strategy is to copy the file into the Work library, mess with it there, and when you are finished copy it back to where you were using it (but don't forget to copy it back – anything in the work library is deleted at the end of the day.)

You might have guessed that DATA steps can do a whole lot more than just copy files around, as we'll see in the next section.

3.8. SAS formats ①

If you have SAS 9.2 (or later) then the date of birth in Figure 3.11 will look fine. If you have an earlier version of SAS then the dates of birth in the data viewer won't look much like dates of birth (but even if you are using SAS 9.2 or later, you need to understand this, so don't stop reading). That's because SAS stores a date as the number of days after or before 1 January 1960. (SAS was first written in the 1960s, and the start of that decade seemed like a nice round number to use.) That's all very well if you happen to be a computer running SAS, it doesn't take long for you to work out that 1 June 1972 was 4920 days after 1 January 1960. But we are humans, not computers, and we would like it presented in a way that is more, well, human. To do that we need to tell SAS that dateofbirth is actually a date, and to do that we need to apply a format, and to apply a format, we use a DATA step.

3.8.1. Built-in formats ①

We can use formats in two ways. We can take the formats that SAS has built into it, or if SAS doesn't have one we like we can create new formats.

3.8.1.1. Date formats ①

First, we'll use one of the formats that SAS has built in – we'll tell it that the dateofbirth variable is actually a date, and we'd like it to look like a date.

SAS has a *lot* of built-in formats for dates. You can easily find lists of them by typing something like "SAS date formats" into your favourite web search engine. The format we entered the data in is called DATE9, so let's use a different one – we'll change the format to DDMMYY10.

To apply a format, we use a DATA step. We put a format command into the DATA step, and tell SAS which variable we'd like formatted, and which format to apply to it. SAS knows that the name we are using is the name of a format, because formats always end with a dot. SAS Syntax 3.4 shows the data step to add the variable format to the date variable. We create a data set called myfirstdataformatted, and we tell SAS that the format for the variable date is date9. Don't forget the dot at the end of the format, and the semi-colon at the end of the line.

```
DATA chapter3.myfirstformatteddata; SET chapter3.myfirstdata;
    FORMAT date DDMMYY10.;
    RUN;
```
SAS Syntax 3.4

(Notice that when you put the dot at the end of the date9, it went a sort of turquoise colour. That shows you that SAS knows it is a format.)

If we open the data and take a look, the dates are formatted differently, as shown in Figure 3.12. SAS hasn't changed what it is storing – it still keeps the number of seconds since 1960 in its memory, but it is showing something different to us.

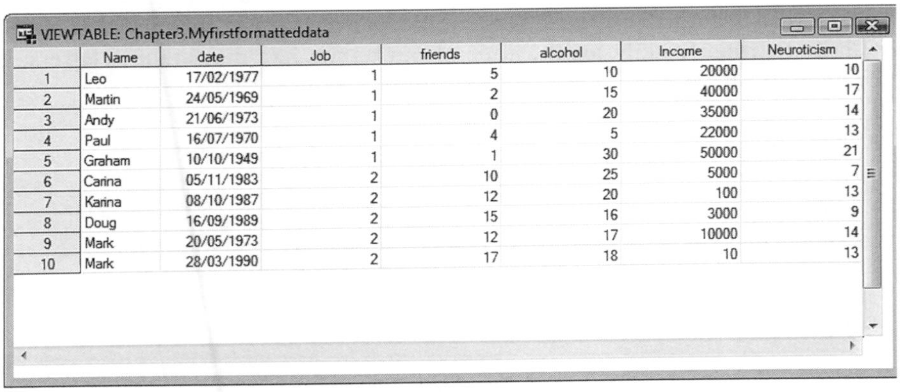

VIEWTABLE: Chapter3.Myfirstformatteddata

	Name	date	Job	friends	alcohol	Income	Neuroticism
1	Leo	17/02/1977	1	5	10	20000	10
2	Martin	24/05/1969	1	2	15	40000	17
3	Andy	21/06/1973	1	0	20	35000	14
4	Paul	16/07/1970	1	4	5	22000	13
5	Graham	10/10/1949	1	1	30	50000	21
6	Carina	05/11/1983	2	10	25	5000	7
7	Karina	08/10/1987	2	12	20	100	13
8	Doug	16/09/1989	2	15	16	3000	9
9	Mark	20/05/1973	2	12	17	10000	14
10	Mark	28/03/1990	2	17	18	10	13

FIGURE 3.12 Data with date9. format applied to date variable

3.8.1.2. Numeric formats ①

Numeric formats don't change anything about the numbers that SAS stores, they just change the way that the numbers are shown to you. The format tells SAS how wide the format should be, and how many decimal places to show. (Note that the width has to be wide enough to show the decimals you have asked for.) Numeric formats are written as *w.d*, where *w* is the width, and *d* is the number of decimal places, so 6.3 would have a width of 6 characters, and have 3 decimal places.

Currently, number of friends is displayed with no decimal places. Let's change the format to 6.3, as shown in SAS Syntax 3.5.

```
DATA chapter3.myfirstformatteddata; SET chapter3.myfirstdata;
    FORMAT date DATE9.;
    FORMAT friends 6.3 ;
    RUN;
```
SAS Syntax 3.5

And if we look at the data viewer, shown in Figure 3.13, we can see that friends is now formatted with three decimal places.

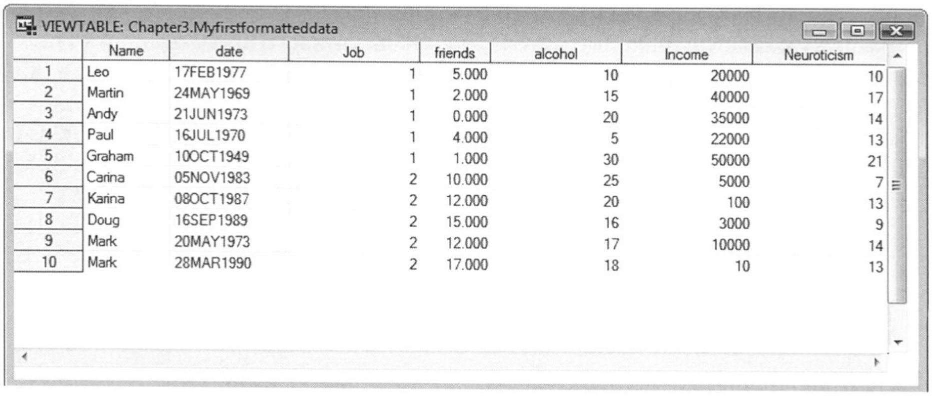

FIGURE 3.13 Myfirstformatted dataset in the data viewer

VIEWTABLE: Chapter3.Myfirstformatteddata

	Name	date	Job	friends	alcohol	Income	Neuroticism
1	Leo	17FEB1977	1	5.000	10	20000	10
2	Martin	24MAY1969	1	2.000	15	40000	17
3	Andy	21JUN1973	1	0.000	20	35000	14
4	Paul	16JUL1970	1	4.000	5	22000	13
5	Graham	10OCT1949	1	1.000	30	50000	21
6	Carina	05NOV1983	2	10.000	25	5000	7
7	Karina	08OCT1987	2	12.000	20	100	13
8	Doug	16SEP1989	2	15.000	16	3000	9
9	Mark	20MAY1973	2	12.000	17	10000	14
10	Mark	28MAR1990	2	17.000	18	10	13

3.8.1.3. String formats ①

String formats are used to define the length of strings. A standard string in SAS is 8 letters, and when we told SAS that a variable was going to be a string, it assumed that we wanted 8 letters in that string. Luckily for us, none of our names had more than 8 letters; if they had, they wouldn't have fitted. Let's try that – in SAS Syntax 3.6.

```
DATA smalldata;
     INPUT name $ ;
     CARDS;
     Emily-Elizabeth
     ;
     RUN;
```

SAS Syntax 3.6

Open the data set, and you'll notice that Emily-Elizabeth's name didn't fit – we reached the second L and SAS discarded the rest. We need to tell SAS that we need more space for the name variable. All we need to do is put a number (and a dot) after the $ symbol, to tell SAS the width of the variable. SAS Syntax 3.6 shows that syntax in action.

```
DATA smalldata;
     INPUT name $20. ;
     CARDS;
     Emily-Elizabeth
     ;
     RUN;
```

SAS Syntax 3.7

3.8.2. User-defined formats ①

When we put people into the group of students or lecturers, we used 1 to represent lecturers, and 2 to represent students. We'd like to label those, so that SAS knows that

1 means student, and 2 means lecturer, and we do that with a format. It's unlikely that SAS has a built-in format that uses those numbers and labels, so we need to create our own.

SAS uses *procedures* to do many things. These are shortened to PROC in SAS syntax, and hence are called procs, which rhymes with socks. We want to create a format, and so we will use PROC FORMAT.[6]

SAS Syntax 3.8 shows the use of PROC FORMAT. The command starts with a PROC FORMAT line. The next line is the VALUE line to tell SAS we want to apply labels to different values, and that we want to call the format **group**. We then provide the value, and the label. The label should be in quotes (single or double, it doesn't matter), and ends with a semicolon. (Notice the way that I have spaced things out, to make it easier to read – we could have put everything on one line if we'd wanted to.) Also notice that if you put something inside quotes, it turns pink – if everything is pink, you've missed out a quote.

```
PROC FORMAT;
    VALUE group
          1= 'Lecturer'
          2= 'Student'
          ;
    RUN;
```
SAS Syntax 3.8

When we have created the format, we need to apply it in a DATA step, shown in SAS Syntax 3.9. Notice that we need to add the dot after the second occurrence of the word job, to show SAS that this is a format. We gave the variable and the format the same name – I sometimes do this to make it easier to remember which variable applies to which format, but there is no need to do that. You might have many variables with the same format (for example, a series of yes–no questions), in which case you would give the format a more generic name.

```
DATA chapter3.myfirstformatteddata; SET chapter3.myfirstdata;
    FORMAT date date9.;
    FORMAT friends 6.3 ;
    FORMAT job job. ;
    RUN;
```
SAS Syntax 3.9

3.9. Variable labels ①

As well as a name, a variable can have a label. There are two reasons to label your variables in SAS. First, the label can be longer, and more descriptive than the name, for example, if you have given people a survey, you can call the variable that represents Question 28 Q28, but you can give it the label 'How much do you like your statistics classes?' When you can't remember what Q28 was, you can look at the label to remember. Second, when you produce tables or graphs, using the variable label, such as 'Number of Friends' makes your work look much nicer than if it says 'numberoffriends'.

Variable labels are added in a DATA step, as shown in SAS Syntax 3.10. Notice that you don't put a semicolon until the very end of the block of text for the labels.

[6] The names of procs aren't always that easy to guess, but they often are.

```
  DATA chapter3.myfirstformatteddata; SET
chapter3.myfirstdata;
  FORMAT dateofbirth DATE9.;
  FORMAT friends 6.3 ;
  FORMAT job job. ;
  LABEL date        = "Date of Birth"
      job           = "Position"
      friends       = "Number of Friends"
      alcohol       = "Alcohol Units Consumed"
      income        = "Income"
      neuroticism = "Neuroticism Score"
                    ;
      RUN;
```

SAS Syntax 3.10

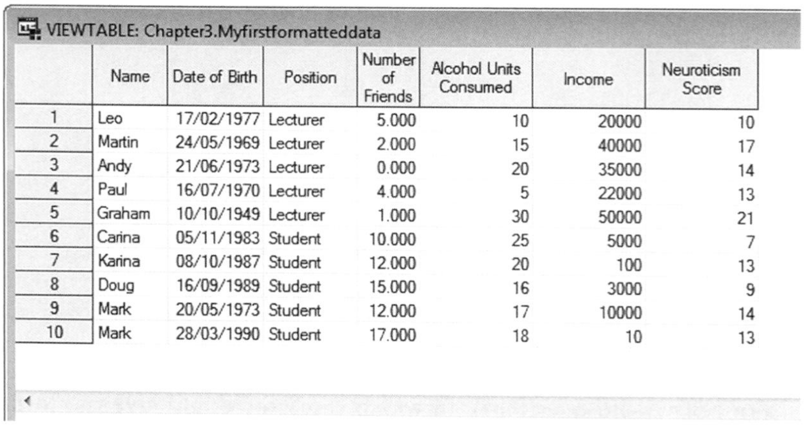

VIEWTABLE: Chapter3.Myfirstformatteddata

	Name	Date of Birth	Position	Number of Friends	Alcohol Units Consumed	Income	Neuroticism Score
1	Leo	17/02/1977	Lecturer	5.000	10	20000	10
2	Martin	24/05/1969	Lecturer	2.000	15	40000	17
3	Andy	21/06/1973	Lecturer	0.000	20	35000	14
4	Paul	16/07/1970	Lecturer	4.000	5	22000	13
5	Graham	10/10/1949	Lecturer	1.000	30	50000	21
6	Carina	05/11/1983	Student	10.000	25	5000	7
7	Karina	08/10/1987	Student	12.000	20	100	13
8	Doug	16/09/1989	Student	15.000	16	3000	9
9	Mark	20/05/1973	Student	12.000	17	10000	14
10	Mark	28/03/1990	Student	17.000	18	10	13

FIGURE 3.14 Data viewer, with labels, rather than names

When we open the data file from the Explorer, the variable labels are now displayed at the top of the columns, rather than the variable names (Figure 3.14).

If you double click on the label at the top of the column, a dialog box opens which tells you the name of the variable, the label, and the format that is applied. For example, double-clicking on Position gives the dialog shown in Figure 3.15.

Column Attributes

General | Colors | Fonts

Close | Apply | Help

Name: Job

Label: Position

Length: 8

Format: GROUP8.

Informat: 8.

Type: Character / Numeric

FIGURE 3.15 Dialog box for 'Position' showing label and format

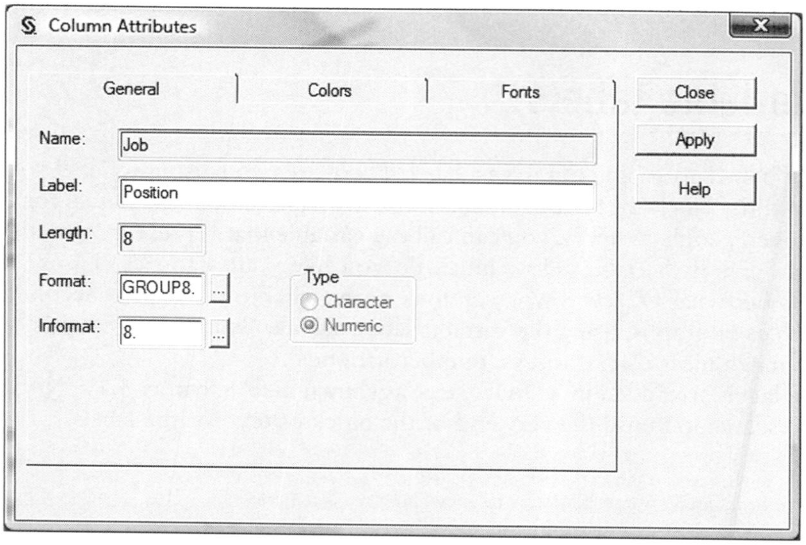

3.10. More on DATA steps ①

DATA steps can be used to manipulate your data in many ways. We can calculate new variables from old variables, we can use conditional statements, and we can remove variables.

3.10.1. Calculating a new variable ①

We might want to calculate the value of a new variable based on an old value. For example, we might decided to represent income as 1000s, to make the numbers a more sensible size. The DATA step to do this is shown in SAS Syntax 3.11.

```
DATA chapter3.mycalculateddata; SET chapter3.myfirstdata;
    income_in_thousands = income / 1000;
    RUN;
```

SAS Syntax 3.11

Figure 3.16 shows the data, along with our newly created variable.

VIEWTABLE: Chapter3.Mycalculateddata

	Name	date	Job	friends	alcohol	Income	Neuroticism	income_in_thousands
1	Leo	17FEB1977	1	5	10	20000	10	20
2	Martin	24MAY1969	1	2	15	40000	17	40
3	Andy	21JUN1973	1	0	20	35000	14	35
4	Paul	16JUL1970	1	4	5	22000	13	22
5	Graham	10OCT1949	1	1	30	50000	21	50
6	Carina	05NOV1983	2	10	25	5000	7	5
7	Karina	08OCT1987	2	12	20	100	13	0.1
8	Doug	16SEP1989	2	15	16	3000	9	3
9	Mark	20MAY1973	2	12	17	10000	14	10
10	Mark	28MAR1990	2	17	18	10	13	0.01

FIGURE 3.16 Check the calculated data, after running SAS Syntax 3.11

3.10.2. Conditional (if ...then...) statements ①

We can calculate a new variable based on a question about a variable that we already have. For example, we might decide that anyone consuming more than 21 units of alcohol is consuming too much.

SAS lets us use abbreviations for greater than and less than (rather than < and >, although we can use them if we want to). The abbreviations are:

GT – Greater Than
LT – Less Than
GE – Greater than or Equal to
LE – Less than or Equal to
EQ – EQual to[7]
NE – Not Equal to

[7] You might wonder why we need to use EQ for 'equal to' when we already have '='. The reason is that the symbol = means two things. If we say 'X = Y' we might be asking 'Is X equal to Y?' (geeky types call that a comparison operator) or we might be saying 'Make X equal to Y' (geeky types call that the assignment operator). If we always use EQ for comparison, and = for assignment, we're less likely to get in a tangle.

We can add a couple of lines to create a variable which tells us if a person is drinking too much:

```
IF alcohol GT 21 THEN too_much_alcohol = 1;
IF alcohol LE 21 THEN too_much_alcohol = 0;
```

Remember what we said about missing values (SAS Tip 3.4)? Of course you do, you're that kind of person. If alcohol is missing, then it will be less than 21, but that doesn't mean that the person doesn't drink 21 units or less. We just don't know – so the result needs to be missing. We have two choices. We could create the variable first, and then set it to missing.

```
IF alcohol GT 21 THEN too_much_alcohol = 1;
IF alcohol LE 21 THEN too_much_alcohol = 0;
IF alcohol EQ . THEN too_much_alcohol = .;
```

Or (and this is my preference) we could check it at the same time, with an AND. The advantage of this is it doesn't matter if we forget to do the check at the end.

```
IF alcohol GT 21 AND alcohol NE . THEN too_much_alcohol = 1;
IF alcohol LE 21 AND alcohol NE .THEN too_much_alcohol = 0;
```

We end up with SAS Syntax 3.12.

```
DATA chapter3.mycalculateddata; SET chapter3.myfirstdata;
    income_in_thousands = income / 1000;
    IF alcohol GT 21 AND alcohol NE . THEN too_much_alcohol = 1;
    IF alcohol LE 21 AND alcohol NE . THEN too_much_alcohol = 0;
    run;
```

SAS Syntax 3.12

SAS TIP **Some more on comparisons**

We can shorten the code a little if we want to type less. We can think of a comparison as if we ask SAS a question, and if the answer is Yes (or true) it gives 1 as the answer. If the answer is No (or false) it gives 0 as the answer. So, instead of writing:

```
IF alcohol GT 21 THEN too_much_alcohol = 1;
IF alcohol LE 21 THEN too_much_alcohol = 0;
```

we could have put:

```
too_much_alcohol = alcohol GT 21;
```

[8]

Of course, we need to check for missingness too, so we put:

```
IF alcohol NE . THEN too_much_alcohol = alcohol GT 21;
```

[8] See why we need to distinguish between EQ and =? SAS will understand if you write:
a = b = c;
But we might not.

3.10.3. Getting rid of variables ①

Sometimes we have so many variables we are at risk of getting in a tangle with them, in which case we might want to get rid of them. If we are getting rid of a small number of variables, we can add a DROP statement to the DATA step. We just ask to drop the variables we don't want any more.

```
DATA chapter3.mycalculateddata; SET chapter3.myfirstdata;
    income_in_thousands = income / 1000;
    IF alcohol GT 21 AND alcohol NE . THEN too_much_alcohol = 1;
    IF alcohol LE 21 AND alcohol NE . THEN too_much_alcohol = 0;
    DROP alcohol;
    RUN;
```
SAS Syntax 3.13

If we want to get rid of a larger number of variables, we use KEEP. All of the variables we don't keep will be deleted. SAS Syntax 3.14 drops all of the variables except for name and too_much_alcohol.

```
DATA chapter3.mycalculateddata; SET chapter3.myfirstdata;
    IF alcohol GT 21 AND alcohol NE . THEN too_much_alcohol = 1;
    IF alcohol LE 21 AND alcohol NE . then too_much_alcohol = 0;
    DROP alcohol;
    KEEP name too_much_alcohol;
    RUN;
```
SAS Syntax 3.14

Be very careful with the DROP command. If you overwrite your data with a DROP command, your data are lost – there is no chance to undo, because the data are saved as soon as you have run the command. Also, if you misspell the variable, then the variable you wanted to keep will be dropped, and a variable that doesn't exist will be kept.

3.10.4. Getting rid of cases ①

There are some people that you might like to get rid of in real life. That can be hard. Getting rid of them in SAS is much easier. We might decide that we don't want to keep any lecturers (who, remember, have a 1 on the variable group) in our data set. SAS Syntax 3.15 shows you how to do it.

```
DATA chapter3.mystudentdata; SET chapter3.myfirstdata;
    IF group EQ 1 THEN DELETE;
    RUN;
```
SAS Syntax 3.15

3.11. Procs for checking data ①

So far, when we've wanted to look at our data, we've been double-clicking on the file, and opening the data viewer, but there are a couple of other things we could have done instead. Three procedures are particularly useful – PROC PRINT; PROC MEANS and PROC FREQ. PROC PRINT prints your data to the Output window, PROC MEANS prints means (and standard deviations, and a couple of other things) to the Output window, and PROC FREQ prints frequencies to the Output window. The syntax for using these are shown in SAS Syntax 3.16–3.18. We'll run these, and then take a look at the results.

```
PROC PRINT DATA=chapter3.myfirstformatteddata;
    RUN;
```
SAS Syntax 3.16

```
PROC FREQ DATA=chapter3.myfirstformatteddata;
    RUN;
```
SAS Syntax 3.17

```
PROC MEANS DATA=chapter3.myfirstformatteddata;
    RUN;
```
SAS Syntax 3.18

3.12. Output and results ①

When you run procs that produce statistical results, those results are written to the Output window. There is nothing very clever about the Output window; new results are written to the bottom of it. If you change your data, the results in the Output window don't change. (The results are sometimes called 'dead', unlike, say, Excel or OpenOffice Calc, where the results will change if you change your data – these are called 'live').

The window on the left is called 'Results' and is like a table of contents, showing which procs you have run. This is a little like Windows Explorer, in that there are folders – you click on a folder to open it, and it either contains a table of results or more folders. The Results and Output windows are shown in Figure 3.17.

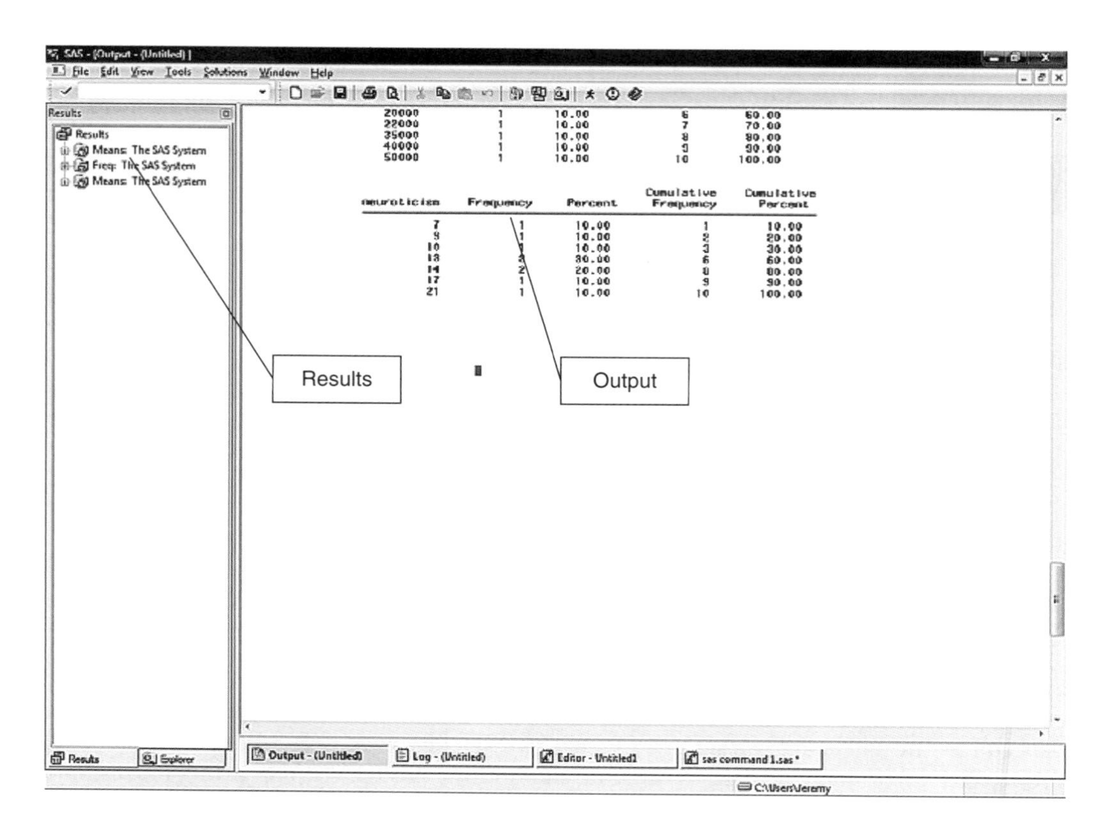

FIGURE 3.17 SAS showing Results and Output windows

You can navigate through the Output window by scrolling up and down with the mouse, the arrows, or the Page Up and Page Down keys. Alternatively, you can find the output you are interested in by using the Results window, and opening folders to find the table that you want.

If we wanted to look at the table of frequencies of number_of_friends, we would click on the folder called Freq, then on the folder called Table Numberoffriends, then on One-Way Frequencies. That will take you to the table of frequencies for Numberoffriends, shown in Figure 3.13

Scroll through the rest of the output, and see what it shows you.

SAS TIP 3.6 Funny numbers

You might notice that SAS sometimes reports numbers with the letter 'E' placed in the mix just to confuse you. For example, you might see a value such as 5.172173E-2, and many students find this notation confusing. Well, this notation means 5.172×10^{-2} (which might be a more familiar notation, or could be even more confusing). OK, some of you are still confused. Well, think of E-02 as meaning 'move the decimal place 2 places to the left', so 5.172E-2 becomes 0.05172. If the notation read 5.172E-1, then that would be 0.5172, and if it read 5.172E-3, that would be 0.005172. Likewise, think of E+02 (notice the minus sign has changed) as meaning 'move the decimal place 2 places to the right'. So 5.172+02 becomes 517.2.

You can copy and paste text from the Output window to your report, or presentation. If you do that, you'll find it doesn't look very good. Jane Superbrain knows how to make it nicer.

JANE SUPERBRAIN 3.1

Redirecting output with the Output Delivery System

You don't have to send your output to the SAS Output window, instead, you can ask SAS to send it to another program, or to do something else with it. You do this with the Output Delivery System, or ODS for short. (We're going to have a very quick look at ODS – the whole of ODS is extraordinarily complex – some people have suggested that they called it ODS because it is so similar to ODiouS. But the people at SAS say that's not true. There is a book (Haworth, 2001) called ***Output Delivery System: The Basics***, and its 302 pages long.)

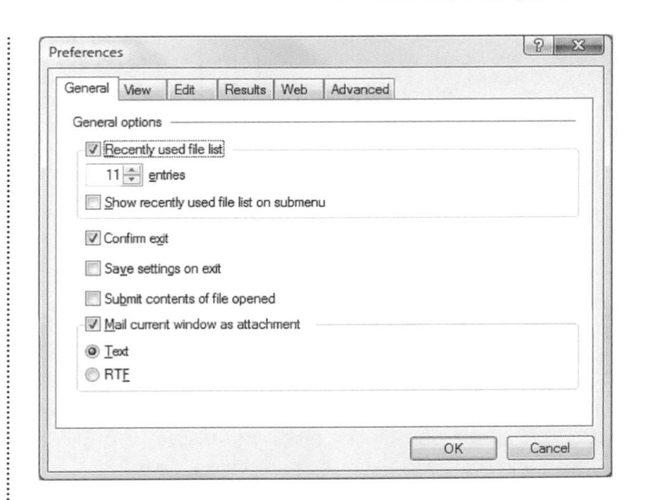

The first thing to do is to tell SAS that we don't want it to try to show our files (it's not vital, but if you're going to send your data out of SAS, you might as well send them right out). Click on the Tools menu, then on Options, and then choose Preferences.

Choose the Results tab, and then choose Preferred Web Browser.

You can send output to an HTML file (for viewing in Internet Explorer, Firefox, Opera, Safari, etc.), an RTF file (which means a word processor file – such as Word), a PDF for reading in Acrobat, straight to the printer (you probably don't want to do that) or to a new data file.

Let's have a look at those.

To send data to an HTML file, you type the command:

```
ODS HTML;
```

and then the command you want:

```
PROC MEANS DATA=chapter3.myfirstformat
teddata;
     RUN;
```

SAS then sends the results to your default viewer for web pages, and it will look something like this: which we

Variable	Label	N	Mean	Std Dev	Minimum	Maximum
dateofbirth	Date of Birth	10	6035.60	4495.95	-3735.00	11044.00
job	Position	10	1.5000000	0.5270463	1.0000000	2.0000000
numberoffriends	Number of Friends	10	7.8000000	6.1427464	0	17.0000000
alcohol	Alcohol Units Consumed	10	17.6000000	7.0427267	5.0000000	30.0000000
income	Income	10	18511.00	18001.35	10.0000000	50000.00
neuroticism	Neuroticism Score	10	13.1000000	3.9846929	7.0000000	21.0000000

think looks a lot nicer than the SAS output in the Output window.

SAS will keep sending output to HTML until you ask it to stop, by giving the command: ODS HTML CLOSE;

You can send your data to a word processor, using ODS RTF (RTF stands for Rich Text Format). The slight difference is that you have to give a file name for the RTF file (in HTML, SAS makes one up) and you don't get to see the RTF file until you tell SAS to stop sending output to RTF.

```
ODS RTF FILE= 'means.rtf';
PROC MEANS
DATA=chapter3.myfirstformatteddata;
     RUN;
ODS RTF CLOSE;
```

Your word processor will open, with a file that looks like:

Variable	N	Mean	Std Dev	Minimum	Maximum
dateofbirth	10	6035.60	4495.95	–3735.00	11044.00
job	10	1.5000000	0.5270463	1.0000000	2.0000000
numberoffriends	10	7.8000000	6.1427464	0	17.0000000
alcohol	10	17.6000000	7.0427267	5.0000000	30.0000000
income	10	18511.00	18001.35	10.0000000	50000.00
neuroticism	10	13.1000000	3.9846929	7.0000000	21.0000000

ODS PDF is exactly the same as ODS RTF, so we're not going to show you that.

There's one more trick that's worth knowing. One problem with SAS output is that it has rather a lot of decimal places. It would be nice if we could reduce that easily. Some programs can format numbers easily – like Excel, but sending output to Excel isn't an option. Or is it?

We can trick SAS (and Windows) into sending its data to Excel. To do this, we tell it to output an HTML file, but we give it a file name that ends in .xls. That way, SAS creates an HTML file, and gives it to Windows. Windows says 'Oh, that ends .xls, it must be an Excel file, and gives it to Excel.

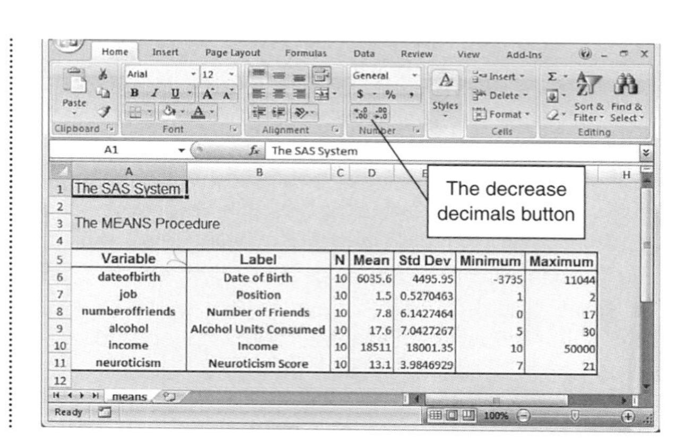

```
ODS HTML FILE= 'means.xls';
PROC MEANS
DATA=chapter3.myfirstformatteddata;
    RUN;
ODS HTML CLOSE;
```

(Note that Windows Vista knows that you are trying to trick it, and will give you a warning. You can click 'Yes' to say that's OK.

In Excel, you can use the Decrease Decimal button (that's this one:) to reduce the number of decimal places to something more sensible (two is usually good).

Almost everywhere in this book, we use output in RTF format.

3.13. Looking at data with PROC CONTENTS ①

Although we can open data and look at it with the data viewer, it's often quicker to get a summary of the contents of the data using PROC CONTENTS. The syntax for PROC CONTENTS is shown in SAS Syntax 3.19. Running that syntax causes SAS to write a summary of the data to the Log, and the part that we are interested in is shown in SAS Output 3.1.

```
PROC CONTENTS DATA=chapter3.myfirstformatteddata;
    RUN;
```

SAS Syntax 3.19

SAS OUTPUT 3.1

Alphabetic List of Variables and Attributes

#	Variable	Type	Len	Format	Label
5	alcohol	Num	8		Alcohol Units Consumed
2	dateofbirth	Num	8	DATE9.	Date of Birth
6	income	Num	8		Income
3	job	Num	8	JOB.	Position
1	name	Char	8		
7	neuroticism	Num	8		Neuroticism Score
4	numberoffriends	Num	8	6.3	Number of Friends

What have I discovered about statistics? ①

This chapter has provided a basic introduction to the SAS environment. We've seen that SAS uses three main windows: the Editor (or Enhanced Editor) the Log, and the Output. We also learned about how SAS works with files, which it keeps in libraries.

We also discovered that I was scared of my new school. However, with the help of Jonathan Land my confidence grew. With this new confidence I began to feel comfortable not just at school but in the world at large. It was time to explore.

Key terms that I've discovered

Data table
Date variable
Enhanced Editor
Library
Numeric variable
String variable
Variable format
Variable label

Smart Alex's tasks

- **Task 1**: Your first task is to enter the data shown in the figure below. These data show the scores (out of 20) for 20 different students, some of whom are male and some female, and some of whom were taught using positive reinforcement (being nice) and others who were taught using punishment (electric shock). Just to make it hard, the data should not be entered in the same way that they are laid out below:

Male		Female	
Electric Shock	Being Nice	Electric Shock	Being Nice
15	12	6	10
14	10	7	9
20	7	5	8
13	8	4	8
13	13	8	7

- **Task 2**: Research has looked at emotional reactions to infidelity and found that men get homicidal and suicidal and women feel undesirable and insecure (Shackelford, LeBlanc, & Drass, 2000). Let's imagine we did some similar research: we took some men and women and got their partners to tell them they had slept with someone else. We then took each person to two shooting galleries and each time gave them a gun and 100 bullets. In one gallery was a human-shaped target with a picture of their own face on it, and in the other was a target with their partner's face on it. They were left alone with each target for 5 minutes and the number of bullets used was measured. The data are below; enter them into SAS and, call them **Infidelity** (clue: they are not entered in the format in the table!).

Male		Female	
Partner's Face	Own Face	Partner's Face	Own Face
69	33	70	97
76	26	74	80
70	10	64	88
76	51	43	100
72	34	51	100
65	28	93	58
82	27	48	95

71	9	51	83
71	33	74	97
75	11	73	89
52	14	41	69
34	46	84	82

Answers can be found on the companion website.

Further reading

SAS is such a big thing, with so many uses, that it can be hard to find a good book that tells you what you want to know without getting badly bogged down in details. One that we like, which tells you what you need to know without too much detail, is *SAS programming: the one day course*, by Neil Spencer. If you want more detail on the kind of thing we have talked about in this chapter, you can look at *The Little SAS Book: A Primer (3rd Edition)*, by Lora Delwiche and Susan Slaughter (although it is not very little – it's around 300 pages). One book that takes a very different approach from us is *SAS for Dummies* – although the title makes it tempting, we didn't think it was helpful for the sort of work we do. (Jeremy tried to give his copy away, and couldn't find anyone that wanted it.)

There is lots and lots of useful information on the web. UCLA (University of California, Los Angeles) has a great web site at: http://www.ats.ucla.edu/stat/sas/; you can often find information about how to do something in SAS by typing SAS [thing I want to know about] into your favourite search engine, for example [SAS formats]. Another useful site is http://www.sascommunity.org, which includes the SASopedia, although again there's a lot of detail there, which makes it hard to find the thing that you want. One word of advice – if the search engine sends you to an SAS.com page, the information will be correct, but so detailed it might be hard to find the what you actually want.

4 Exploring data with graphs

FIGURE 4.1 Explorer Field borrows a bike and gets ready to ride it recklessly around a caravan site

4.1. What will this chapter tell me? ①

As I got a bit older I used to love exploring. At school they would teach you about maps and how important it was to know where you were going and what you were doing. I used to have a more relaxed view of exploration and there is a little bit of a theme of me wandering off to whatever looked most exciting at the time. I got lost at a holiday camp once when I was about 3 or 4. I remember nothing about this but apparently my parents were frantically running around trying to find me while I was happily entertaining myself (probably by throwing myself head first out of a tree or something). My older brother, who was supposed to be watching me, got a bit of flak for that but he was probably working out equations to bend time and space at the time. He did that a lot when he was 7. The careless explorer in me hasn't really gone away: in new cities I tend to just wander off and hope for the best, and usually get lost and fortunately usually don't die (although I tested my luck once by wandering through part of New Orleans where apparently tourists get mugged a lot – it seemed fine to me). When exploring data you can't afford not to have a map; to explore data in the way that the 6 year old me used to explore the world is to spin around 8000 times while drunk and then run along the edge of a cliff. Wright (2003) quotes Rosenthal who said that researchers should 'make friends with their data'. This wasn't meant to imply that people who use statistics may as well befriend their data because the data are the only friend they'll have; instead Rosenthal meant that researchers often rush their analysis. Wright makes the analogy of a fine wine: you should savour the bouquet and delicate flavours to truly enjoy the experience. That's perhaps overstating the joys of data analysis, but rushing your analysis is, I suppose, a bit like gulping down a bottle of wine: the outcome is likely to be messy and incoherent. To negotiate your way around your data you need a map, maps of data are called graphs, and it is into this tranquil and tropical ocean that we now dive (with a compass and ample supply of oxygen, obviously).

4.2. The art of presenting data ①

4.2.1. What makes a good graph? ①

Before we get down to the nitty-gritty of how to draw graphs in SAS, I want to begin by talking about some general issues when presenting data. SAS (and other packages) make it possible to produce snazzy-looking graphs, and you may find yourself losing consciousness at the excitement of colouring your graph bright pink (really, it's amazing how excited my undergraduate psychology students get at the prospect of bright pink graphs – personally I'm not a fan of pink). Much as pink graphs might send a twinge of delight down your spine, I want to urge you to remember why you're doing the graph – it's not to make yourself (or others) purr with delight at the pinkness of your graph, it's to present information (dull, perhaps, but true).

Tufte (2001) wrote an excellent book about how data should be presented. He points out that graphs should do, among other things:

- ✓ Show the data.
- ✓ Induce the reader to think about the data being presented (rather than some other aspect of the graph, like how pink it is).
- ✓ Avoid distorting the data.
- ✓ Present many numbers with minimum ink.
- ✓ Make large data sets (assuming you have one) coherent.
- ✓ Encourage the reader to compare different pieces of data.
- ✓ Reveal data.

However, graphs often don't do these things (see Wainer, 1984, for some examples; or the Junk Charts blog, at http://junkcharts.typepad.com).

Let's look at an example of a bad graph. When searching around for the worst example of a graph that I have ever seen, it turned out that I didn't need to look any further than myself – it's in the first edition of *Discovering Statistics Using SPSS* (Field, 2000). Overexcited by my computer's ability to put all sorts of useless crap on graphs (like 3-D effects, fill effects and so on – Tufte calls these *chartjunk*) I literally went into some weird orgasmic state and produced an absolute abomination (I'm surprised Tufte didn't kill

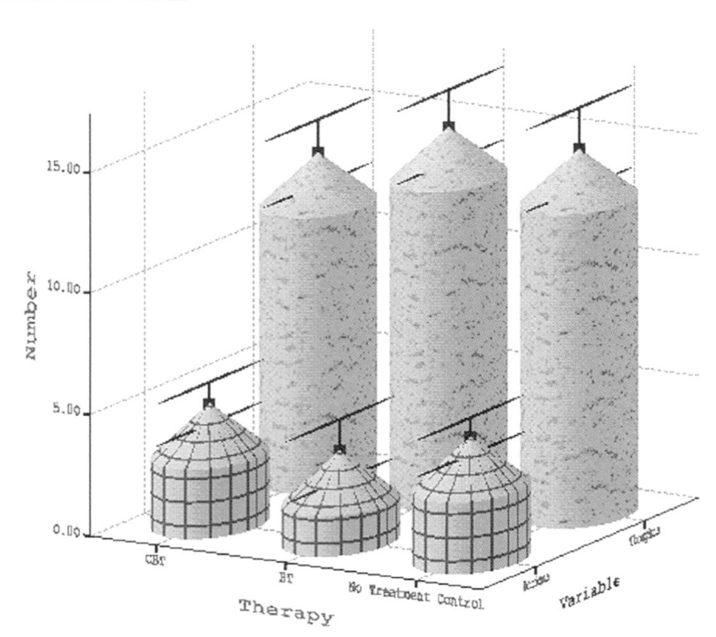

FIGURE 4.2
A cringingly bad example of a graph from Field (2000)

himself just so he could turn in his grave at the sight of it). The only consolation was that because the book was published in black and white, it's not bloody pink! The graph is reproduced in Figure 4.2 (you should compare this to the more sober version in this edition, Figure 16.3). What's wrong with this graph?

- ✗ The bars have a 3-D effect: Never use 3-D plots for a graph plotting two variables because it obscures the data.[1] In particular it makes it hard to see the values of the bars. This graph is a great example because the 3-D effect makes the error bars almost impossible to read.
- ✗ Patterns: The bars also have patterns, which, although very pretty, merely distract the eye from what matters (namely the data). These are completely unnecessary!
- ✗ Cylindrical bars: What's that all about eh? Again, they muddy the data and distract the eye from what is important.
- ✗ Badly labelled *y*-axis: 'Number' of what? Delusions? Fish? Cabbage-eating sea lizards from the eighth dimension? Idiots who don't know how to draw graphs?

Now, take a look at the alternative version of this graph (Figure 4.3). Can you see what improvements have been made?

- ✓ A 2-D plot: The completely unnecessary third dimension is gone, making it much easier to compare the values across therapies and thoughts/behaviours.
- ✓ The *y*-axis has a more informative label: We now know that it was the number of obsessive thoughts or actions per day that was being measured.

[1] If you do 3-D plots when you're plotting only two variables then a bearded statistician will come to your house, lock you in a room and make you write I μυστ νοτ δο 3–Δ γραπησ 75,172 times on the blackboard. Really, they will.

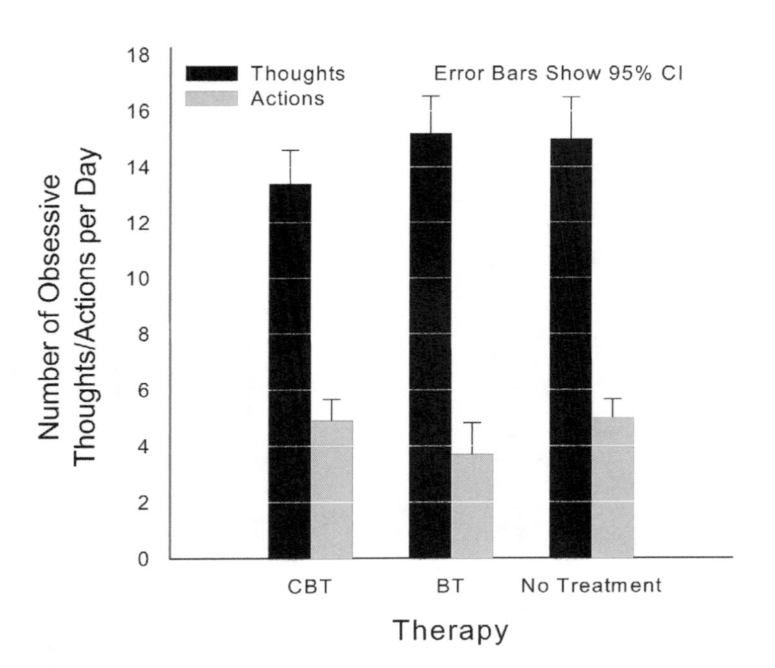

FIGURE 4.3 Figure 4.2 drawn properly

- ✓ Distractions: There are fewer distractions like patterns, cylindrical bars and the like!
- ✓ Minimum ink: I've got rid of superfluous ink by getting rid of the axis lines and by using lines on the bars rather than grid lines to indicate values on the *y*-axis. Tufte would be pleased.

4.2.2. Lies, damned lies, and ... erm ... graphs ①

Governments lie with statistics, but scientists shouldn't. How you present your data makes a huge difference to the message conveyed to the audience. As a big fan of cheese, I'm often curious about whether the urban myth that it gives you nightmares is true. Shee (1964) reported the case of a man who had nightmares about his workmates: 'He dreamt of one, terribly mutilated, hanging from a meat-hook.[2] Another he dreamt of falling into a bottomless abyss. When cheese was withdrawn from his diet the nightmares ceased.' This would not be good news if you were the minister for cheese in your country.

Figure 4.4 shows two graphs that, believe it or not, display exactly the same data: the number of nightmares had after eating cheese. The first panel shows how the graph should probably be scaled. The *y*-axis reflects the maximum of the scale, and this creates the correct impression: that people have more nightmares about colleagues hanging from meathooks if they eat cheese before bed. However, as minister for cheese, you want people to think the opposite; all you have to do is rescale the graph (by extending the *y*-axis way beyond the average number of nightmares) and there suddenly seems to be a little difference. Tempting as it is, don't do this (unless, of course, you plan to be a politician at some point in your life).

[2] I have similar dreams, but that has more to do with some of my workmates than cheese.

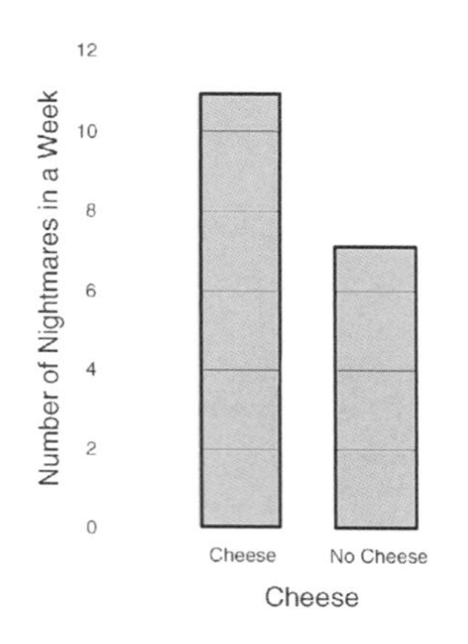

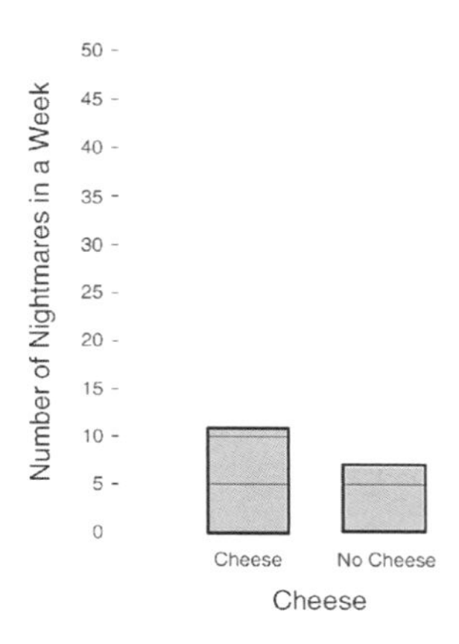

FIGURE 4.4 Two graphs about cheese

CRAMMING SAM'S TIPS **Graphs**

- ✓ The vertical axis of a graph is known as the *y*-axis (or *ordinate*) of the graph.
- ✓ The horizontal axis of a graph is known as the *x*-axis (or *abscissa*) of the graph.

If you want to draw a good graph follow the cult of Tufte:

- ✓ Don't create false impressions of what the data actually show (likewise, don't hide effects!) by scaling the *y*-axis in some weird way.
- ✓ Abolish chartjunk: Don't use patterns, 3-D effects, shadows, pictures of spleens, photos of your Uncle Fred or anything else.
- ✓ Avoid excess ink: This is a bit radical, and difficult to achieve on SAS, but if you don't need the axes, then get rid of them.

4.3. Charts in SAS ①

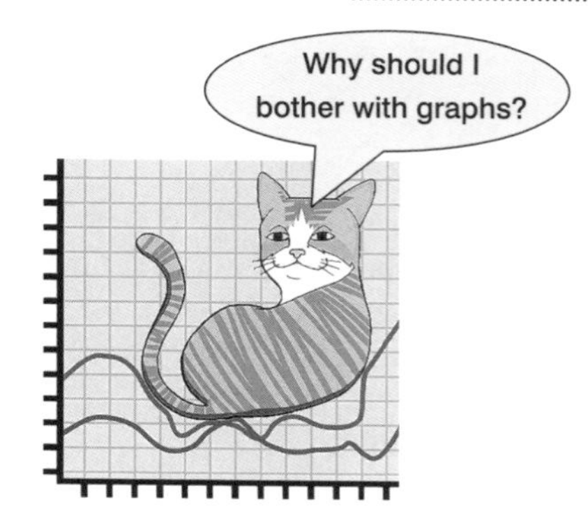

Graphs are a really useful way to look at your data before you get to the nitty-gritty of actually analysing them. You might wonder why you should bother drawing graphs – after all, you are probably drooling like a rabid dog to get into the statistics and to discover the answer to your really interesting research question. Graphs are just a waste of your precious time, right? Data analysis is a bit like Internet dating (actually it's not, but bear with me), you can scan through the vital statistics and find a perfect match (good IQ, tall, physically fit, likes arty French films, etc.) and you'll think you have found the perfect answer to your question. However, if you haven't looked at a picture, then you don't really know how to interpret this information – your perfect match might turn out to be Rimibald the Poisonous, King of the Colorado River Toads, who has genetically combined himself with a human to further his plan to start up a lucrative rodent farm (they like to eat small rodents).[3] Data analysis is much the same: inspect your data with a picture, see how it looks and only then can you interpret the more vital statistics.

We should warn you that SAS isn't really known for being able to produce the most beautiful charts at the click of a mouse. Whilst it can produce beautiful charts, they need to be wrestled out of SAS, kicking and screaming.

4.4. Histograms: a good way to spot obvious problems ①

In this section we'll look at how we can use frequency distributions to screen our data. We'll use an example to illustrate what to do. A biologist was worried about the potential health effects of music festivals. So, one year she went to the Download Music Festival[4] (for those of you outside the UK, you can pretend it is Roskilde Festival, Ozzfest, Lollopalooza, Waken or something) and measured the hygiene of 810 concert-goers over the three days of the festival. In theory each person was measured on each day but because it was difficult to track people down, there were some missing data on days 2 and 3. Hygiene was measured using a standardized technique (don't worry, it *wasn't* licking the person's armpit) that results in a score ranging between 0 (you smell like a corpse that's been left to rot up a skunk's arse) and 4 (you smell of sweet roses on a fresh spring day). We also know the gender of each person, and we have their ticket number (`ticknumb`). Now I know from bitter experience that sanitation is not always great at these places (the Reading Festival seems particularly bad) and so this researcher predicted that personal hygiene would go down dramatically over the three days of the festival. The data file, **DownloadFestival.sas7bdat**, can be found on the companion website. You could run PROC CONTENTS to confirm which variables are in the file – see section 3.13.

We encountered histograms (frequency distributions) in Chapter 1; we will now learn how to create one in SAS using these data.

SELF-TEST What does a histogram show?

Histograms are drawn using PROC UNIVARIATE in SAS. This proc can do lots of things, and so we need to specify that we want it to draw a histogram for us. SAS Syntax 4.1 shows the use of PROC UNIVARIATE to draw a histogram.

```
PROC UNIVARIATE DATA=chapter4.downloadfestival;
    VAR day1;
    HISTOGRAM day1;
    RUN;
```

SAS Syntax 4.1

The resulting histogram is shown in Figure 4.5. The first thing that should leap out at

[3] On the plus side, he would have a long sticky tongue and if you smoke his venom (which, incidentally, can kill a dog) you'll hallucinate (if you're lucky, you'd hallucinate that he wasn't a Colorado River Toad–Human hybrid).

[4] http://www.downloadfestival.co.uk

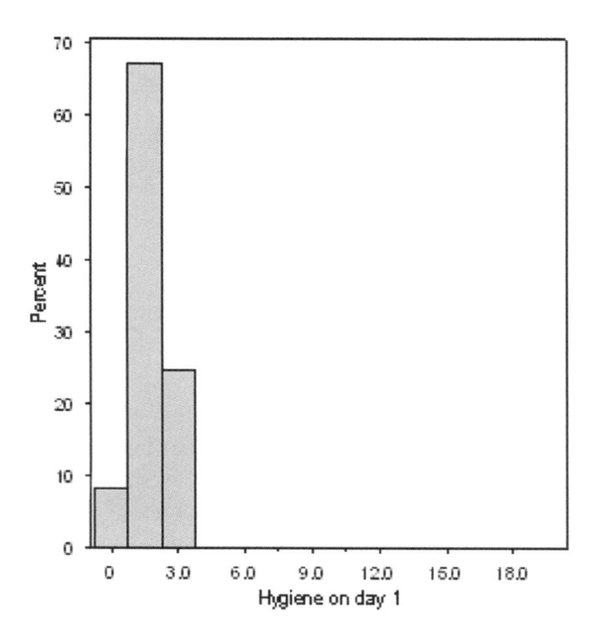

FIGURE 4.5 Histogram of hygiene scores on day 1 of the Download Festival

you is that there appears to be one case that is very different to the others. All of the scores appear to be squashed up at one end of the distribution because they are all less than 5 (yielding a very pointy distribution!) except for one, which has a value of 20! This is an **outlier**: a score very different to the rest (Jane Superbrain Box 4.1). Outliers bias the mean and inflate the standard deviation (you should have discovered this from the self-test tasks in Chapters 1 and 2) and screening data is an important way to detect them. You can look for outliers in two ways: (1) graph the data with a histogram (as we have done here) or a boxplot (as we will do in the next section), or (2) look at z-scores (this is quite complicated but if you want to know how, see Jane Superbrain Box 4.3).

The outlier shown on the histogram is particularly odd because it has a score of 20, which is above the top of our scale (remember our hygiene scale ranged only from 0 to 4) and so it must be a mistake (or the person had obsessive compulsive disorder and had washed themselves into a state of extreme cleanliness). However, with 810 cases, how on earth do we find out which case it was? You could just look through the data, but that would certainly give you a headache and so instead we can use a **boxplot** which is another very useful graph for spotting outliers.

JANE SUPERBRAIN 4.1

What is an outlier?

An outlier is a score very different from the rest of the data. When we analyse data we have to be aware of such values because they bias the model we fit to the data. A

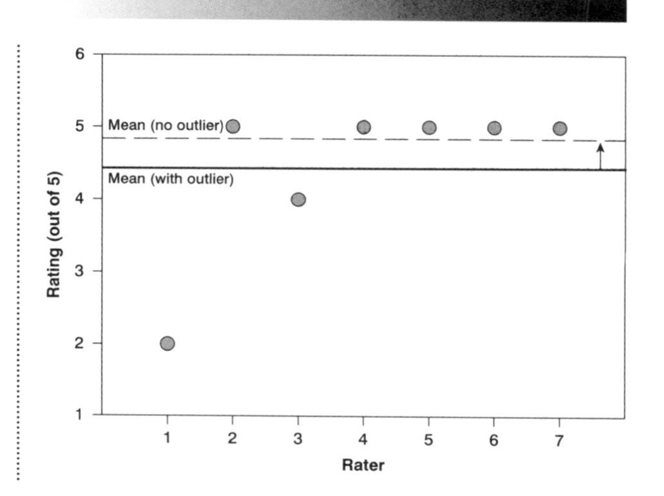

good example of this bias can be seen by looking at the mean. When the first edition of *Discovering Statistics Using SPSS* came out in 2000, I was quite young and became very excited about obsessively checking the book's ratings on Amazon.co.uk. These ratings can range from 1 to 5 stars. Back in 2002, the first edition of this book had seven ratings (in the order given) of 2, 5, 4, 5, 5, 5, 5. All but one of these ratings are fairly similar (mainly 5 and 4) but the first rating was quite different from the rest – it was a rating of 2 (a mean and horrible rating). The graph plots seven reviewers on the horizontal axis and their ratings on the vertical axis and there is also a horizontal line that represents the mean rating (4.43 as it happens). It should be clear that all of the scores except one lie close to this line. The score of 2 is very different and lies some way below the mean. This score is an example of an outlier – a weird and unusual person (sorry, I mean score) that deviates from the rest of humanity (I mean, data set). The dashed horizontal line represents the mean of the scores when the outlier is not included (4.83). This line is higher than the original mean indicating that by ignoring this score the mean increases (it increases by 0.4). This example shows how a single score, from some mean-spirited badger turd, can bias the mean; in this case the first rating (of 2) drags the average down. In practical terms this had a bigger implication because Amazon rounded off to half numbers, so that single score made a difference between the average rating reported by Amazon as a generally glowing 5 stars and the less impressive 4.5 stars. (Nowadays Amazon sensibly produces histograms of the ratings and has a better rounding system.) Although I am consumed with bitterness about this whole affair, it has at least given me a great example of an outlier! (Data for this example were taken from http://www.amazon.co.uk/ in about 2002.)

SAS TIP 4.1 Histogram options

The histogram we just saw was OK, as histograms go, but you might not want to put it in your report without making it a little more aesthetically pleasing. In SAS, options are often given following a slash at the end of the line, so we can add some histogram options.

One useful option is to add a normal curve, with the same mean and standard deviation as our data.

We can add some colours. We can colour the bars with the CFILL option, and the frame with the CFRAME option.

```
PROC UNIVARIATE DATA=chapter4.downloadfestival;
     TITLE "Histogram for Day 1 Hygiene";
     VAR day1;
     HISTOGRAM day1 /CFILL=GREY CFRAME=WHITE NORMAL;
     run;
TITLE;
```

(We have to clear the title at the end, or all analyses will be given the same title.)

4.5. Boxplots (box–whisker diagrams) ①

Boxplots or box–whisker diagrams are really useful ways to display your data. At the centre of the plot is the *median*, which is surrounded by a box the top and bottom of which are the limits within which the middle 50% of observations fall (the interquartile range). Sticking out of the top and bottom of the box are two whiskers which extend to the extreme scores at each end.

To draw a boxplot, we use (and this might surprise you) PROC BOXPLOT. We can produce a boxplot for a set of variables, or for a variable split across a group – we know the gender of the respondents, so we could draw a separate box for the males and the females. However, PROC BOXPLOT requires that the two groups be sorted, so all the males are together and all the females are together. SAS Syntax 4.2 shows the syntax for PROC SORT, which we used to sort the data, followed by PROC BOXPLOT. In this syntax, we've specified an option for the plot, which is to have a schematic id plot – this is the most common kind of plot, but if you don't specify SAS will instead produce a skeletal plot, which is not as useful. In addition, we've included an ID command which tells SAS to label any outliers with their ticket numbers. (I also included `IDHEIGHT=3`, which tells SAS to make the label on the ID for the outliers a bit bigger than normal, but that's just so that you can read it when we put it in this book; you won't normally need to do that.)

```
PROC SORT DATA=chapter4.downloadfestival;
 BY gender;
 RUN;
PROC BOXPLOT DATA=chapter4.downloadfestival;
 PLOT day1*gender /BOXSTYLE=schematicid IDHEIGHT=3;
 ID ticknumb;
RUN;
```
SAS Syntax 4.2

The resulting boxplot is shown in Figure 4.6. It shows a separate boxplot for the men and women in the data. You may remember that the whole reason that we got into this boxplot malarkey was to help us to identify an outlier from our histogram (if you have skipped straight to this section then you might want to backtrack a bit). The important thing to note is that the outlier that we detected in the histogram is shown up as a square on the boxplot and next to it is the ticket number of the case (4158) that's producing this outlier. We can use PROC PRINT, as shown in SAS Syntax 4.3, command to have a look at this person.

```
PROC PRINT DATA=chapter4.downloadfestival;
    WHERE ticknumb=4158;
    RUN;
```
SAS Syntax 4.3

This gives us SAS Output 4.1

Obs	ticknumb	gender	day1	day2	day3
367	4158	Female	20.02	2.44	.

SAS Output 4.1

Looking at this case reveals a score of 20.02, which is probably a mistyping of 2.02. We'd have to go back to the raw data and check. We'll assume we've checked the raw data and it should be 2.02, so we will replace the value 20.02 with the value 2.02 before we continue this example.

We'll use a DATA step to replace the data point, shown in SAS Syntax 4.4, we'll create a new data file called downloadfestival2, so we don't overwrite our old file.

```
DATA chapter4.downloadfestival2; SET chapter4.downloadfestival;
    IF day1 = 20.02 THEN day1 = 2.02;
    RUN;
```
SAS Syntax 4.4

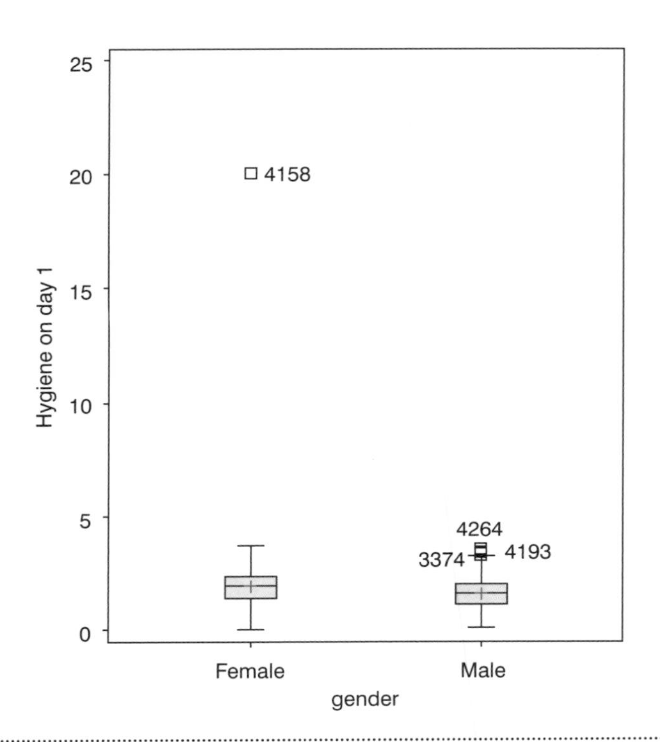

FIGURE 4.6 Boxplot of hygiene scores on day 1 of the Download Festival split by gender

SELF-TEST Now we have removed the outlier in the data, try replotting the boxplot. The resulting graph should look like Figure 4.7.

Figure 4.7 shows the boxplots for the hygiene scores on day 1 after the outlier has been corrected. Let's look now in more detail at what the boxplot represents. First, it shows us the lowest score (the bottom horizontal line on each plot) and the highest (the top horizontal line of each plot). Comparing the males and females, we can see they both had similar low scores (0, or very smelly) but the women had a slightly higher top score (i.e. the most fragrant female was more hygienic than the cleanest male). The lowest edge of the tinted box is the lower quartile (see section 1.7.3); therefore, the distance between the lowest horizontal line and the lowest edge of the tinted box is the range between which the lowest 25% of scores fall. This range is slightly larger for women than for men, which means that if we take the most unhygienic 25% females then there is more variability in their hygiene scores than the lowest 25% of males. The box (the tinted area) shows the interquartile range (see section 1.7.3): that is, 50% of the scores are bigger than the lowest part of the tinted area but smaller than the top part of the tinted area. These boxes are of similar size in the males and females.

The top edge of the tinted box shows the value of the upper quartile (see section 1.7.3); therefore, the distance between the top edge of the shaded box and the top horizontal line shows the range between which the top 25% of scores fall. In the middle of the tinted box is a slightly thicker horizontal line. This represents the value of the median (see section 1.7.2). The median for females is higher than for males, which tells us that the middle female scored higher, or was more hygienic, than the middle male. The plus sign shows the position of the mean, which is very near the median in the case of our two boxplots.

Boxplots show us the range of scores, the range in which the middle 50% of scores fall, and the median, the upper quartile and lower quartile scores. Like histograms, they also tell us whether the distribution is symmetrical or skewed. If the whiskers are the same length then the distribution is symmetrical (the range of the top and bottom 25% of scores is the same); however, if the top or bottom whisker is much longer than the opposite whisker then

FIGURE 4.7 Boxplot of hygiene scores on day 1 of the Download Festival split by gender

the distribution is asymmetrical (the range of the top and bottom 25% of scores is different). Finally, you'll notice some squares above the male boxplot. These are cases that we might consider to be outliers. Each square has a number next to it that tells us in which row of the data editor to find that case. In Chapter 5 we'll see what can be done about these outliers.

SELF-TEST Produce boxplots for the day 2 and day 3 hygiene scores and interpret them.

JANE SUPERBRAIN 4.2

Boxplots of one variable ③

Sometimes we want to produce a boxplot of one variable, not broken down by groups. Strangely, SAS doesn't want do to this – you have to have a variable that defines groups. There is a way around this though: we create a variable where everyone has the same score, and use that to define the groups. We'll create that variable, and call it **a**.

```
DATA chapter4.downloadfestival2; SET
chapter4.downloadfestival2;
  a = 1;
  RUN;
```

Then we can use **a** as the grouping variable, and SAS will happily draw the plot we want.

```
PROC BOXPLOT DATA=chapter4.
downloadfestival2;
  PLOT day1*a /BOXSTYLE=schematicid;
  RUN;
```

JANE SUPERBRAIN 4.3

Using z-scores to find outliers ③

To check for outliers we can look at z-scores. We saw in section 1.7.4 that z-scores are simply a way of *standardizing* a data set by expressing the scores in terms of a distribution with a mean of 0 and a standard deviation of 1. In doing so we can use benchmarks that we can apply to any data set (regardless of what its original mean and standard deviation were). We also saw in this section that to convert a set of scores into z-scores we simply take each score (X) and subtract the mean of all scores ($\bar{X}$) from it and divide by the standard deviation of all scores (s).

We do this in a DATA step, or SAS can do the conversion for us using PROC STANDARD. The syntax is shown below – there are a couple of things to note about this. First, in the PROC line, we have to include the value of the mean and standard deviation that we want – we almost always want a mean of 0 and standard deviation of 1, but SAS wants to be sure. Second, we include an OUT = statement. This tells SAS to create an output file with the name that we specify – in this case dlfstandard (also notice that we didn't specify a library for that file, so it will go into the Work library, and be deleted when we shut down SAS – see section 3.4.)

```
PROC STANDARD DATA=chapter4.download-
festival2 MEAN=0 STD=1 OUT=dlfstandard;
      VAR day2;
      RUN;
```

To look for outliers we could use these z-scores and count how many fall within certain important limits. If we take the absolute value (i.e. we ignore whether the z-score is positive or negative) then in a normal distribution we'd expect about 5% to have absolute values greater than 1.96 (we often use 2 for convenience), and 1% to have absolute values greater than 2.58, and none to be greater than about 3.29.

Alternatively, you could use a DATA step, followed by PROC FREQ.

```
DATA dlfstandardoutliers; SET dlf
standard;
      IF day2 NE . then outlier = 0;
      IF abs(day2) GT 1.96 THEN outlier=1;
      IF abs(day2) GT 2.58 THEN outlier=2;
      IF abs(day2) GT 3.29 THEN outlier=3;
      RUN;
```

This creates a new variable in the data editor called *outlier1*, which contains a 1 if the absolute value of the z-score that we just created is greater than 1.96, and 2 if the absolute value is greater than 2.58. (An absolute value just means ignore the minus sign if there is one.)

Then we use PROC FREQ to produce a table.

```
PROC FREQ DATA= dlfstandardoutliers;
 TABLES outlier;
 run;
```

The table produced by this syntax is shown below. Look at the column labelled 'Percent'. We would expect to see 95% of cases with absolute value less than 1.96, 5% (or less) with an absolute value greater than 1.96, and 1% (or less) with an absolute value greater than 2.58. Finally, we'd expect no cases above 3.29 (well, these cases are significant outliers). For hygiene scores on day 2 of the festival, 93.2% of values had z-scores less than 1.96; put another way, 6.8% were above (looking at the table we get this figure by adding 4.5% + 1.5% + 0.8%). This is slightly more than the 5% we would expect in a normal distribution. Looking at values above 2.58, we would expect to find only 1%, but again here we have a higher value of 2.3% (1.5% + 0.8%). Finally, we find that 0.8% of cases were above 3.29 (so 0.8% are significant outliers). This suggests that there may be too many outliers in this data set and we might want to do something about them!

outlier	Frequency	Percent	Cumulative Frequency	Cumulative Percent
0	245	93.18	246	93.18
1	12	4.55	258	97.73
2	4	1.52	262	99.24
3	2	0.76	264	100.00

Frequency Missing = 546

4.6. Graphing means: bar charts and error bars ①

Bar charts are the usual way for people to display means. How you create these graphs in SAS depends largely on how you collected your data (whether the means come from independent cases and are, therefore, independent, or came from the same cases and so are related). Now maybe we shouldn't tell you this, but bar charts in SAS are fairly straightforward, unless you want to make them look nice, and then they are rather a lot of effort. When we want to draw a bar chart in SAS, we usually just copy the means (or whatever) to Excel and draw it there. But you should make your own mind up.

4.6.1. Simple bar charts for independent means

To begin with, imagine that a film company director was interested in whether there was really such a thing as a 'chick flick' (a film that typically appeals to women more than men). He took 20 men and 20 women and showed half of each sample a film that was supposed to be a 'chick flick' (*Bridget Jones's Diary*), and the other half of each sample a film that didn't fall into the category of 'chick flick' (*Memento*, a brilliant film by the way). In all cases he measured their physiological arousal as an indicator of how much they enjoyed the film. The data are in a file called **ChickFlick.sasb7dat** on the companion website. Create a library called chapter4, and put the file into the folder called chapter4. The variable **arousal** represents the level of arousal of the person watching the film, and the variable **film** contains the film they watched.

First of all, let's just plot the mean rating of the two films. We have just one grouping variable (the film) and one outcome (the arousal); therefore, we want a simple bar chart. We use PROC GCHART for this – the syntax is shown in SAS Syntax 4.5. VBAR is used because we want a vertical bar chart (we can use HBAR for horizontal bar chart). SUMVAR is short for summary variable, SAS which variable we would like summarized, and TYPE tells SAS what sort of summary we would like.

```
PROC GCHART DATA=chapter4.chickflick2;
    VBAR film /SUMVAR = arousal
    TYPE=MEAN;
    RUN;
```

SAS Syntax 4.5

The resulting chart is shown on the left-hand side of **Figure 4.8**.

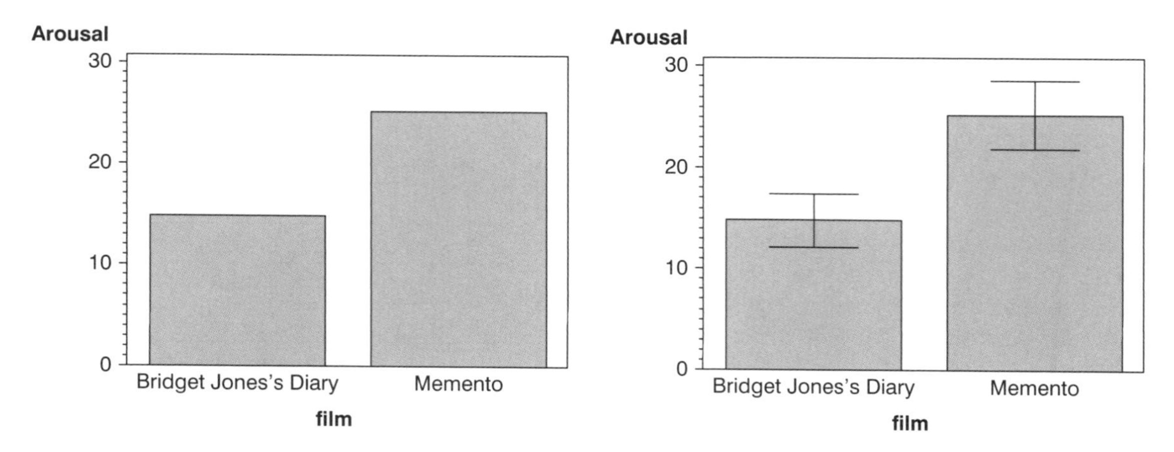

FIGURE 4.8
Vertical chart of mean arousal by film, with and without confidence limits

This being SAS, there are many other things you can do with your chart. One of the most useful is to add error bars to your bar chart to create an *error bar chart* by adding ERRORBAR=BOX as an option. The syntax is shown in SAS Syntax 4.6 is on the right hand side of Figure 4.8.

```
PROC GCHART DATA=chapter4.chickflick2;
    VBAR film /SUMVAR = arousal
    TYPE=MEAN
GROUP=gender
PATTERNID=MIDPOINT
;
RUN;
```
SAS Syntax 4.6

You have a choice of what your error bars represent. Normally, error bars show the 95% confidence interval (see section 2.5.2), and this option is selected if you don't specify something different.[5] Note, though, that you can change the width of the confidence interval displayed by adding CLM=xx to the options. If you don't put anything, SAS assumes you wanted CLM=95, but you can change this by changing the '95' to a different value.

We used the ERRORBAR=BOTH option, but you can use ERRORBAR=TOP (try it, they look like skyscrapers with antennae on) or ERRORBAR=BOX.

These graphs display the mean (and the confidence interval of the mean) and show us that on average, people were more aroused by *Memento* than they were by *Bridget Jones's Diary*. However, we originally wanted to look for gender effects, so this graph isn't really telling us what we need to know. The graph we need is a *clustered graph*.

4.6.2. Clustered bar charts for independent means ①

To do a clustered bar chart for means that are independent (i.e. have come from different groups) we need to extend our previous syntax, but only a little, by adding two options: First, GROUP=gender, to tell SAS we have a separate grouping variable, and second PATTERNID=MIDPOINT – this command makes the bars a different colour (we'll admit that we don't actually understand this, but it works). The complete syntax, with error bars, is shown in SAS Syntax 4.7. (Note that for the sake of variety we used ERRORBAR=BOTH, to get a different shaped error bar).

```
PROC GCHART DATA=chapter4.chickflick2;
          VBAR film /SUMVAR = arousal
          TYPE=MEAN
GROUP=gender
PATTERNID=MIDPOINT
;
    RUN;
```
SAS Syntax 4.7

Figure 4.9 shows the resulting bar chart. Like the simple bar chart, this graph tells us that arousal was overall higher for *Memento* than *Bridget Jones's Diary*, but it also splits this information by gender. The mean arousal for *Bridget Jones's Diary* shows that males were actually more aroused during this film than females. This indicates they enjoyed the film more than the women did. Contrast this with *Memento*, for which arousal levels are comparable in males and females. On the face of it, this contradicts the idea of a 'chick

[5] It's also worth mentioning at this point that error bars from SAS are suitable only for normally distributed data (see section 5.4).

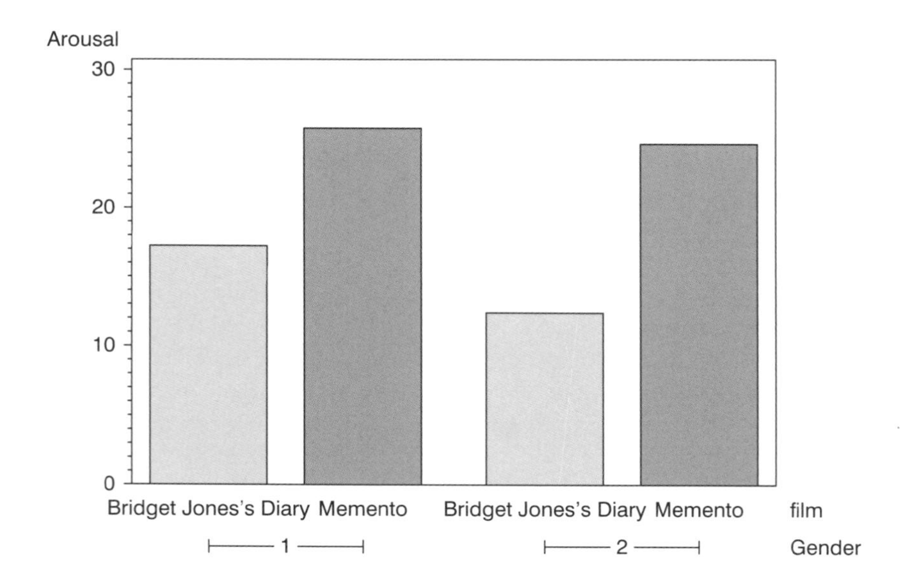

FIGURE 4.9 Bar chart of the mean arousal for each of the two films

flick': it actually seems that men enjoy chick flicks more than chicks do (probably because it's the only help we get to understand the complex workings of the female mind).

4.6.3. Simple bar charts for related means ①

Hiccups can be a serious problem: Charles Osborne apparently got a case of hiccups while slaughtering a hog (well, who wouldn't?) that lasted 67 years. People have many methods for stopping hiccups (a surprise; holding your breath), but actually medical science has put its collective mind to the task too. The official treatment methods include tongue-pulling manoeuvres, massage of the carotid artery, and, believe it or not, digital rectal massage (Fesmire, 1988). I don't know the details of what the digital rectal massage involved, but I can probably imagine. Let's say we wanted to put digital rectal massage to the test (erm, as a cure of hiccups I mean). We took 15 hiccup sufferers, and during a bout of hiccups administered each of the three procedures (in random order and at intervals of 5 minutes) after taking a baseline of how many hiccups they had per minute. We counted the number of hiccups in the minute after each procedure. Download the file **Hiccups.sas7bdat** from the companion website and place it into your chapter4 folder. Note that these data are laid out in different columns as shown in Figure 4.10.

FIGURE 4.10 Hiccups data

VIEWTABLE: Chapter4.Hiccups

	Baseline	Tongue	Carotid	Rectum	id
1	15	9	7	2	1
2	13	18	7	4	2
3	9	17	5	4	3
4	7	15	10	5	4
5	11	18	7	4	5
6	14	8	10	3	6
7	20	3	7	3	7
8	9	16	12	3	8
9	17	10	9	4	9
10	19	10	8	4	10
11	3	14	11	4	11
12	13	22	6	4	12
13	20	4	13	4	13
14	14	16	11	2	14
15	13	12	8	3	15

For repeated-measures data we need to transpose the data, to make it into a format that SAS can use to draw the chart; transposing the data is just a clever sounding way of saying that the variables need to be stacked on top of one another, with a grouping variable to tell us which treatment the person had for that measurement. This is also sometimes called going from wide (the way that the data are now) to long (the way that we want them).

Don't worry, this sounds a lot harder than it is. SAS Syntax 4.8 shows the syntax for PROC TRANSPOSE. On the first line, we put the name of the data set to be transposed (which we're used to by now), but we also need to put the name of the data set we are going to produce – in this case we are going to call that data set longhiccups (and note that we are not specifying a library to keep it in, which means that SAS will put it in the Work library, and will delete it when we close the program).

```
PROC TRANSPOSE DATA=chapter4.hiccups4 OUT=longhiccups PREFIX=
  hiccups;
  BY id;
  RUN;
```

SAS Syntax 4.8

Figure 4.11 shows the data in the viewer, transposed to long format. The id is on the left, and the person with id number 1 has four measures – baseline, tongue, carotid and rectum; SAS has called the grouping variable **_name_**. For each measure, each person has a score, which SAS has called **hiccups1** (it used the prefix that we provided, and added a 1 after it).

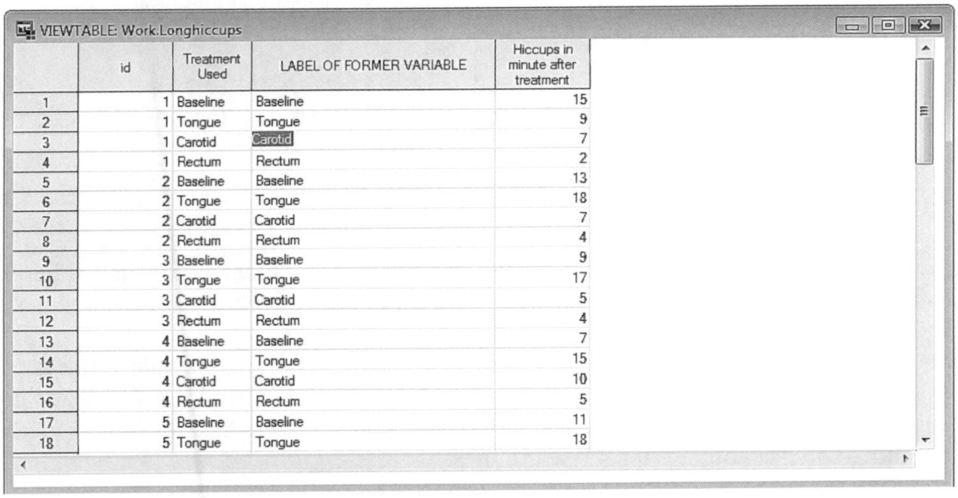

VIEWTABLE: Work.Longhiccups

	id	Treatment Used	LABEL OF FORMER VARIABLE	Hiccups in minute after treatment
1	1	Baseline	Baseline	15
2	1	Tongue	Tongue	9
3	1	Carotid	Carotid	7
4	1	Rectum	Rectum	2
5	2	Baseline	Baseline	13
6	2	Tongue	Tongue	18
7	2	Carotid	Carotid	7
8	2	Rectum	Rectum	4
9	3	Baseline	Baseline	9
10	3	Tongue	Tongue	17
11	3	Carotid	Carotid	5
12	3	Rectum	Rectum	4
13	4	Baseline	Baseline	7
14	4	Tongue	Tongue	15
15	4	Carotid	Carotid	10
16	4	Rectum	Rectum	5
17	5	Baseline	Baseline	11
18	5	Tongue	Tongue	18

FIGURE 4.11 Hiccups data, transposed to long format

We can go ahead and use PROC GCHART, as before, but first we could do a little renaming, to make our graph a bit nicer. The syntax for the renaming and labels is shown in SAS Syntax 4,9. One slightly weird thing is that the variables don't get renamed until the RUN statement, so even though we've renamed the variable **_name_** to **treatment,** we still apply the label to _name_.

```
DATA longhiccups; SET longhiccups;
    RENAME _name_ = treatment;
    LABEL _name_ = "Treatment Used";
    LABEL hiccups1 = "Hiccups in minute after treatment";
    RUN;
```

SAS Syntax 4.9

We can then run the syntax shown in SAS Syntax 4.10 to produce the graph shown in Figure 4.12.

```
PROC GCHART DATA=longhiccups;
    VBAR treatment /SUMVAR = hiccups
    TYPE=MEAN
    PATTERNID=MIDPOINT
    ;
    RUN;
```

SAS Syntax 4.10

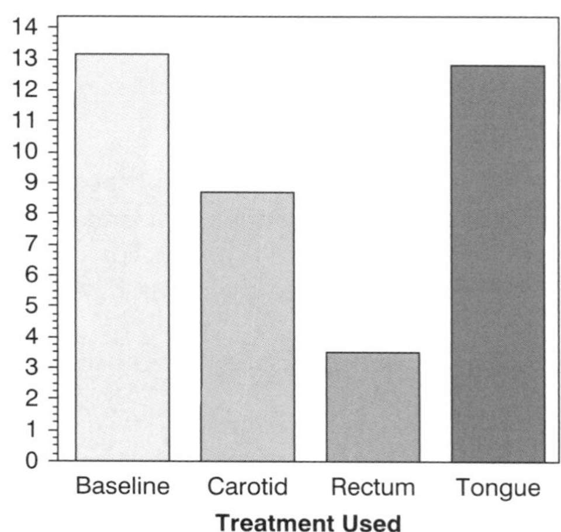

FIGURE 4.12 Graph of mean number of hiccups following different treatments

The resulting bar chart in Figure 4.12 displays the mean number of hiccups at baseline and after the three interventions. The error bars on graphs of repeated-measures designs aren't actually correct as we will see in Chapter 9.

We can conclude that the amount of hiccups after tongue pulling was about the same as at baseline; however, carotid artery massage reduced hiccups, but not by as much as a good old-fashioned digital rectal massage. The moral here is: if you have hiccups, find something digital and go amuse yourself for a few minutes.

4.6.4. Clustered bar charts for 'mixed' designs ①

We can also graph what is known as a mixed design (see Chapter 14). This is a design in which you have one or more independent variables measured using different groups, and one or more independent variables measured using the same sample. Basically, we can easily produce a graph provided we have only one variable that was a repeated measure. (We can produce a graph with more than one, but it's harder, and we're not going to cover it here).

We all like to text-message (especially students in my lectures who feel the need to text-message the person next to them to say 'Bloody hell, this guy is so boring I need to poke out my own eyes'). What will happen to the children, though? Not only will they develop super-sized thumbs, they might not learn correct written English. Imagine we conducted an experiment in which a group of 25 children was encouraged to send text messages on their mobile phones over a six-month period – we'll call that group *text messagers*. A second group of 25 children was forbidden from sending text messages for the same period – we will call that group *controls*. To ensure that kids in this latter group didn't use their phones, this group was given armbands that administered painful shocks in the presence of microwaves (like those emitted from phones).[6] The outcome was a score on a grammatical test (as a percentage) that was measured both before and after the intervention. The first independent variable was, therefore, text message use (text messagers versus controls) and the second independent variable was the time at which grammatical ability was assessed (baseline or after six months). The data are in the file **TextMessages.sas7bdat**, which you will need to download to the folder that is represented in SAS by the library chapter4.

To graph these data we need to follow the procedure for graphing related means in section 4.6.3. Our repeated-measures variable is time (whether grammatical ability was measured at baseline or six months) and is represented in the data file by two columns, one for the baseline data (**time1**) and the other for the follow-up data (**time2**). We're going to use PROC TRANSPOSE again, but this time we've got two variables that we need to keep – **id**, as before, but we also need **group**, to know if the person was in the intervention or control group. The syntax to transpose the data, rename and relabel the variables, and draw the graph is shown in SAS Syntax 4.11. Note that this time we've drawn a vertical bar chart.

```
  PROC TRANSPOSE DATA=chapter4.textmessages OUT=longtextmessages
PREFIX=grammar ;
    VAR time1 time2;
    BY id group2;
    RUN;

  DATA longtextmessages; SET longtextmessages;
      RENAME _name_ = time;
      LABEL _name_ = "Time";
      RUN;

  PROC GCHART DATA=longtextmessages;
   HBAR group2 / DISCRETE
   TYPE=MEAN       SUMVAR=grammar1
   GROUP=time
   ERRORBAR=BOTH;
   RUN;
```
SAS Syntax 4.11

Figure 4.13 shows the resulting bar chart. It shows that at baseline (before the intervention) the grammar scores were comparable in our two groups; however, after the intervention, the grammar scores were lower in the text messagers than in the controls. Also, if you compare the text messagers' grammar scores, you can see that they have fallen over the six months; compare this to the controls whose grammar scores are fairly similar over time. We could, therefore, conclude that text messaging has a detrimental effect on children's understanding of English grammar and civilization will crumble, with Abaddon rising cackling from his bottomless pit to claim our wretched souls. Maybe.

You will also have noticed that in the vertical bar chart, the mean and number of people in each group is added to the end of each bar. You can get rid of this, if you don't like it, by adding NOSTAT to the syntax, after ERRORBAR=BOTH.

[6] Although this punished them for any attempts to use a mobile phone, because other people's phones also emit microwaves, an unfortunate side effect was that these children acquired a pathological fear of anyone talking on a mobile phone.

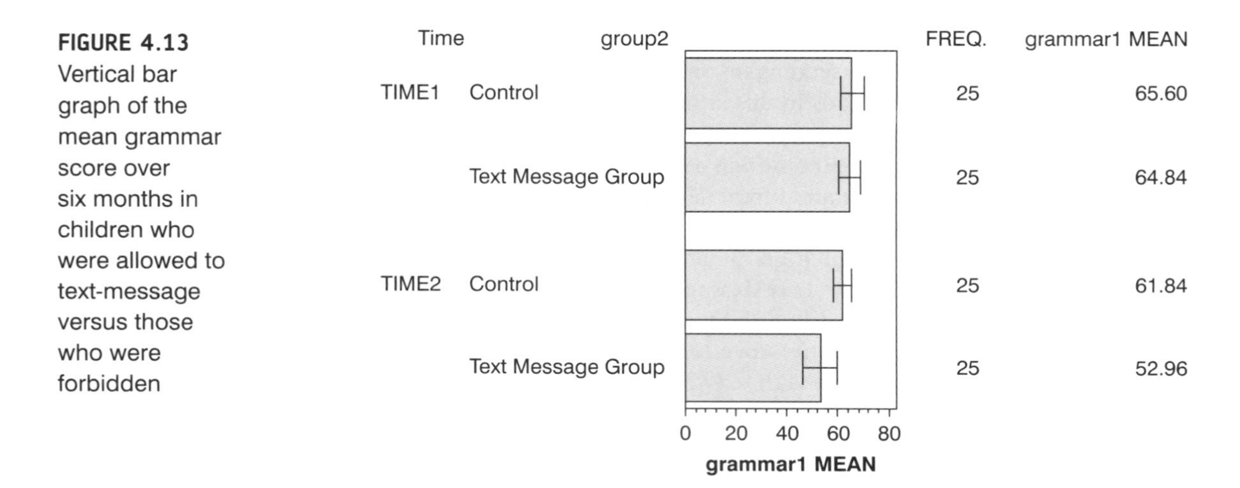

FIGURE 4.13 Vertical bar graph of the mean grammar score over six months in children who were allowed to text-message versus those who were forbidden

4.7. Graphing relationships: the scatterplot ①

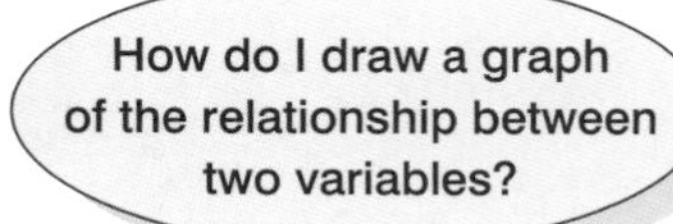

Sometimes we need to look at the relationships between variables (rather than their means, or frequencies). A *scatterplot* is simply a graph that plots each person's score on one variable against their score on another. A scatterplot tells us several things about the data, such as whether there seems to be a relationship between the variables, what kind of relationship it is and whether any cases are markedly different from the others. We saw earlier that a case that differs substantially from the general trend of the data is known as an *outlier* and such cases can severely bias statistical procedures (see Jane Superbrain Box 4.1 and section 7.6.1.1 for more detail). We can use a scatterplot to show us if any cases look like outliers.

We can draw several kinds of scatterplot:

- *Simple scatter*: Use this option when you want to plot values of one continuous variable against another.
- *Grouped scatter*: This is like a simple scatterplot except that you can display points belonging to different groups in different colours (or symbols).
- *Simple 3-D scatter*: Use this option to plot values of one continuous variable against values of two others.
- *Grouped 3-D scatter*: Use this option if you want to plot values of one continuous variable against two others but differentiating groups of cases with different-coloured dots.
- *Summary point plot*: This graph is the same as a bar chart (see section 4.6) except that a dot is used instead of a bar.
- *Scatterplot matrix*: This option produces a grid of scatterplots showing the relationships between multiple pairs of variables.

Drawing a scatterplot using SAS is dead easy using PROC GPLOT.

4.7.1. Simple scatterplot ①

This type of scatterplot is for looking at just two variables. For example, a psychologist was interested in the effects of exam stress on exam performance. So, she devised and validated a questionnaire to assess state anxiety relating to exams (called the Exam Anxiety Questionnaire, or EAQ). This scale produced a measure of anxiety scored out of 100. Anxiety was measured before an exam, and the percentage mark of each student on the exam was used to assess the exam performance. The first thing that the psychologist should do is draw a scatterplot of the two variables (her data are in the file **ExamAnxiety.sas7bdat** and you should load this file into your chapter4 folder). We'll draw a plot with exam **anxiety** on the *x*-axis, and exam performance (**exam**) on the *y*-axis, using SAS Syntax 4.12, which produces the graph shown in Figure 4.14

```
PROC GPLOT DATA=chapter4.examanxiety;
     PLOT exam*anxiety;
     RUN;
```
SAS Syntax 4.12

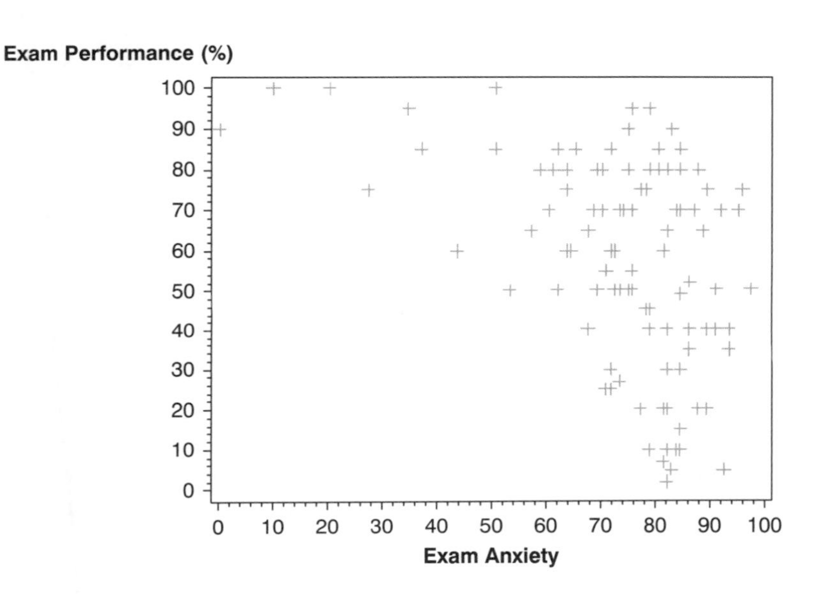

FIGURE 4.14 Scatterplot of exam anxiety and exam performance

The scatterplot tells us that the majority of students suffered from high levels of anxiety (there are very few cases that had anxiety levels below 60). Also, there are no obvious outliers in that most points seem to fall within the vicinity of other points. There also seems to be some general trend in the data, shown by the line, such that higher levels of anxiety are associated with lower exam scores and low levels of anxiety are almost always associated with high examination marks. Another noticeable trend in these data is that there were no cases having low anxiety and low exam performance – in fact, most of the data are clustered in the upper region of the anxiety scale.

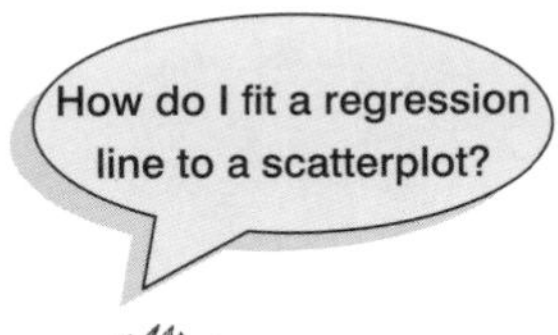

Often when you plot a scatterplot it is useful to plot a line that summarizes the relationship between variables (this is called a *regression line* and we will discover more about it in Chapter 7). To add a regression line, we need to interpolate (that's a posh way of saying *add a line*). In SAS we do this before we draw the scatterplot, using the SYMBOL1 command. SYMBOL1 tells SAS what we'd like the first set of symbols in the graph to look like – writing INTERPOL=R tells SAS we'd like to interpolate with a regression line. I've also added V=CIRCLE, to change the symbol used from a plus to a circle.

```
SYMBOL1 V=circle INTERPOL=R;
PROC GPLOT DATA=chapter4.examanxiety;
     PLOT exam*anxiety;
     RUN;
```
SAS Syntax 4.13

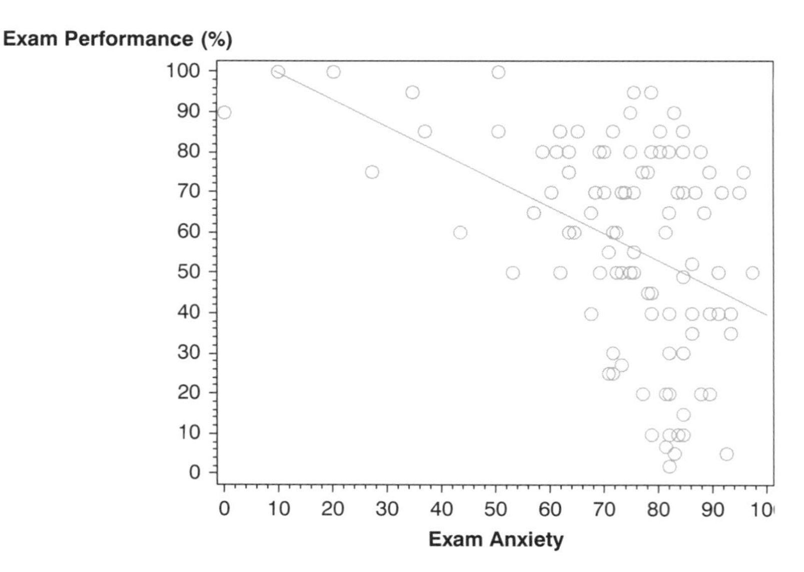

FIGURE 4.15 Scatterplot of anxiety and performance, with regression line

The scatterplot should now look like Figure 4.15. What if we want to see whether male and female students had different reactions to exam anxiety? To do this, we need a grouped scatterplot. (If you want to draw a scatterplot without interpolation, you need to turn off the option, by typing: `SYMBOL1 INTERPOL=;`.)

4.7.2. Grouped scatterplot ①

This type of scatterplot is for looking at two continuous variables, but when you want to colour data points by a third categorical variable. Sticking with our previous example, we could look at the relationship between exam anxiety and exam performance in males and females (our grouping variable).

We now have two sets of symbols, so we should supply two symbol commands, we can also add a colour to the symbol command. Then we put the grouping variable onto the plot line, following an equal sign. This is shown in SAS Syntax 4.14.

```
SYMBOL1 INTERPOL=R V=CIRCLE COLOR=RED;
SYMBOL2 INTERPOL=R V=DIAMOND COLOR=BLACK;
PROC GPLOT DATA=chapter4.examanxiety;
    PLOT exam*anxiety=gender;
    RUN;
```

SAS Syntax 4.14

Figure 4.16 shows the resulting scatterplot; as before I have added regression lines, but this time I have added different lines for each group. These lines tell us that the relationship between exam anxiety and exam performance was slightly stronger in males (the line is steeper) indicating that men's exam performance was more adversely affected by anxiety than women's exam anxiety. (Whether this difference is significant is another issue – see section 6.7.1.)

4.7.3. Simple and grouped 3-D scatterplots ①

I'm now going to show you one of the few times when you can use a 3-D graph without a bearded statistician locking you in a room. Having said that, even in this situation it

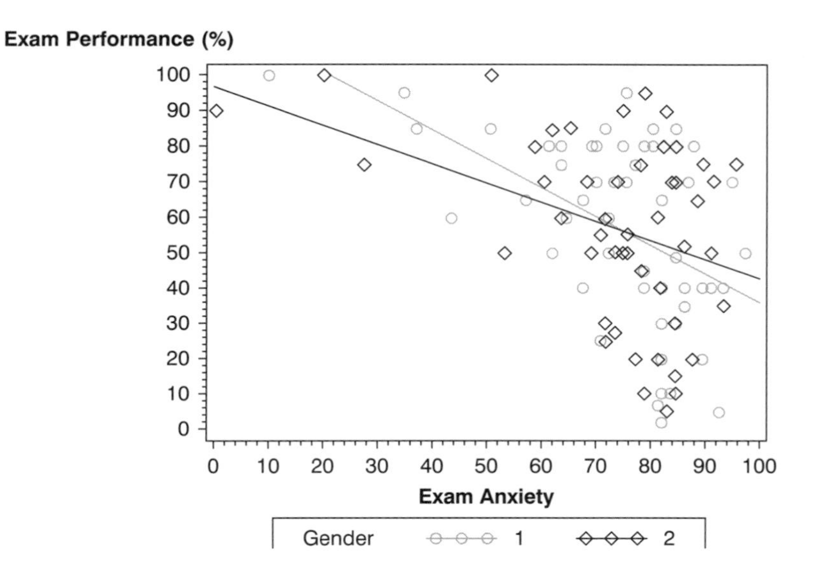

FIGURE 4.16 Scatterplot of exam anxiety and exam performance split by gender

is, arguably, still not a clear way to present data. A 3-D scatterplot is used to display the relationship between three variables. The reason why it's alright to use a 3-D graph here is because the third dimension is actually telling us something useful (and isn't just there to look pretty). As an example, imagine our researcher decided that exam anxiety might not be the only factor contributing to exam performance. So, she also asked participants to keep a revision diary from which she calculated the number of hours spent revising for the exam. She wanted to look at the relationships between these variables simultaneously, using a 3D scatterplot. The syntax for such a plot is shown in SAS Syntax 4.15.

```
PROC G3D DATA = chapter4.examanxiety;
    SCATTER anxiety*revise=exam ;
    RUN;
```

SAS Syntax 4.15

A 3-D scatterplot is great for displaying data concisely; however, as the resulting scatterplot in Figure 4.17 shows, it can be quite difficult to interpret (can you really see what the relationship between exam revision and exam performance is?).

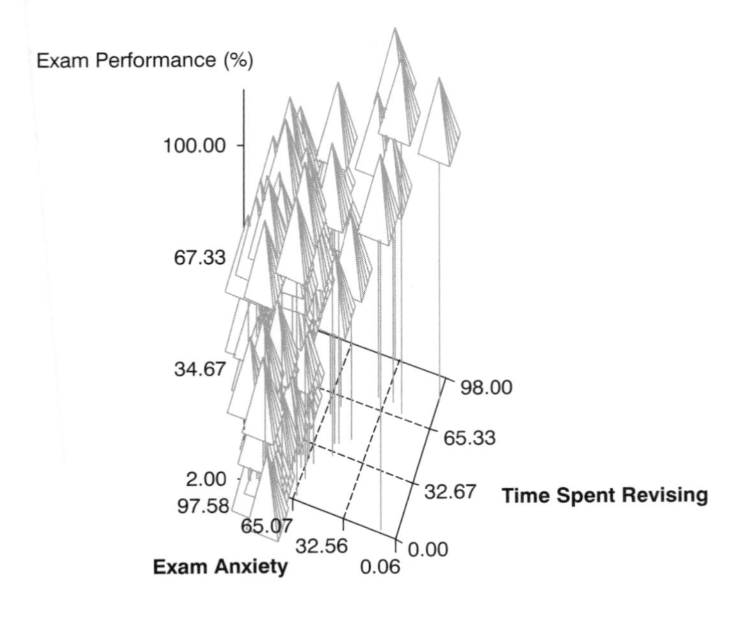

FIGURE 4.17 3D scatterplot of anxiety, revision and performance

4.7.4. Matrix scatterplot ①

Instead of plotting several variables on the same axes on a 3-D scatterplot (which, as we have seen, can be difficult to interpret), it is possible to plot a matrix of 2-D scatterplots. This type of plot allows you to see the relationship between all combinations of many different pairs of variables. We'll use the same data set as with the other scatterplots in this chapter.

A scatterplot matrix is drawn using PROC SGSCATTER. The syntax is very similar to the other scatterplot procs that we've seen.

```
PROC SGSCATTER DATA=chapter4.examanxiety;
 MATRIX revise exam anxiety  ;
 RUN;
```

SAS Syntax 4.16

The six scatterplots in Figure 4.18 represent the various combinations of each variable plotted against each other variable. So, the grid references represent the following plots:

Top middle: revision time (*Y*) vs. exam performance (*X*)
Top right: revision time (*Y*) vs. anxiety (*X*)
Middle left: exam performance (*Y*) vs. revision time (*X*)
Middle right: exam performance (*Y*) vs. anxiety (*X*)
Middle bottom: anxiety (*Y*) vs. exam performance (*X*)
Bottom right: anxiety (*Y*) vs. revision time (*X*)

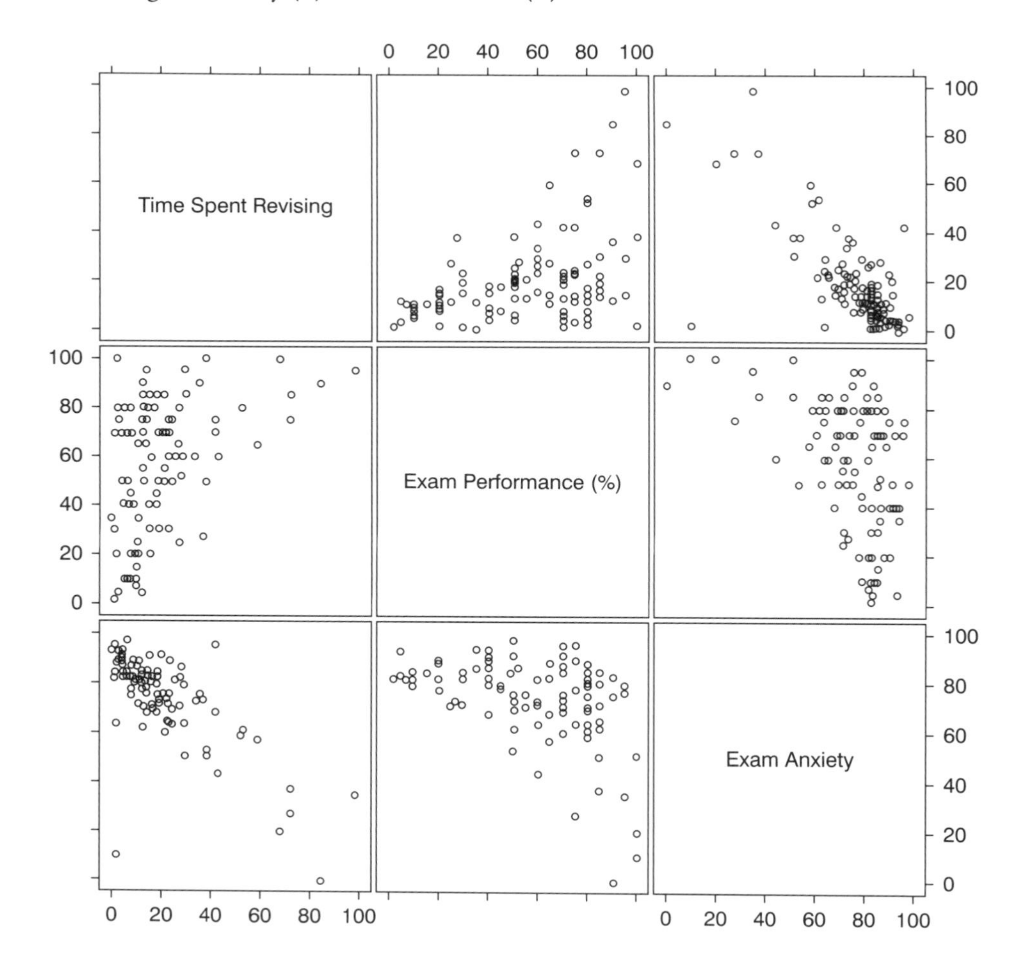

FIGURE 4.18 Matrix scatterplot of exam performance, exam anxiety and revision time

Thus, the three scatterplots below the diagonal of the matrix are the same plots as the ones above the diagonal but with the axes reversed. From this matrix we can see that revision time and anxiety are negatively related (so the more time spent revising the less anxiety the participant had about the exam). Also, in the scatterplot of revision time against anxiety there looks as though there is one possible outlier – there is a single participant who spent very little time revising yet suffered very little anxiety about the exam. As all participants who had low anxiety scored highly on the exam, we can deduce that this person also did well on the exam (don't you just hate a smart alec!). We could choose to examine this case more closely if we believed that their behaviour was caused by some external factor (such as taking brain-pills!). Matrix scatterplots are very convenient for examining pairs of relationships between variables (see SAS Tip 4.2). However, I don't recommend plotting them for more than three or four variables because they become very confusing indeed.

SAS TIP 4.2 Regression lines on a scatterplot matrix ①

You can add regression lines to each scatterplot in the matrix as you did for a simple scatterplot. However, we need to tell SAS which variables we'd like plotted on the *x*-axes, and which variables on the *y*-axes. To do this, we use:

```
PROC sgscatter DATA=chapter4.examanxiety;
 COMPARE y=(revise exam anxiety)
      x=(revise exam anxiety)   / REG ;
 RUN;
```

What have I discovered about statistics? ①

This chapter has looked at how to inspect your data using graphs. We've covered a lot of different graphs. We began by covering some general advice on how to draw graphs and we can sum that up as 'minimal is best': no pink, no 3-D effects, no pictures of Errol your pet ferret superimposed on the graph – oh, and did I mention no pink? We have looked at graphs that tell you about the distribution of your data (histograms and boxplots), that show summary statistics about your data (bar charts, error bar charts) and that show relationships between variables (scatterplots).

We also discovered that I liked to explore as a child. I was constantly dragging my dad (or was it the other way around?) over piles of rocks along any beach we happened to be on. However, at this time I also started to explore great literature, although unlike my cleverer older brother who was reading Albert Einstein's papers (well, Isaac Asimov) as an embryo, my literary preferences were more in keeping with my intellect, as we will see.

Key terms that I've discovered

Bar chart
Boxplot (box–whisker plot)
Chartjunk
Error bar chart
Outlier
Regression line
Scatterplot

Smart Alex's tasks

- **Task 1**: Using the data from Chapter 2 (which you should have saved, but if you didn't then re-enter it from Section 3.5.1) plot and interpret the following graphs:
  - An error bar chart showing the mean number of friends for students and lecturers.
  - An error bar chart showing the mean alcohol consumption for students and lecturers.
  - A scatterplot with regression lines of alcohol consumption and neuroticism grouped by lecturer/student.
  - A scatterplot matrix with regression lines of alcohol consumption, neuroticism and number of friends.

- **Task 2**: Using the **Infidelity.sas7bdat** data from Chapter 3 (see Smart Alex's task) plot a clustered error bar chart of the mean number of bullets used against the self and the partner for males and females.

Answers can be found on the companion website.

Further reading

Tufte, E. R. (2001). *The Visual Display of Quantitative Information* (2nd ed.). Cheshire, CT: Graphics Press.
Wainer, H. (1984). How to display data badly. *American Statistician*, *38*, 137–147.
Wright, D. B., & Williams, S. (2003). Producing bad results sections. *The Psychologist*, *16*, 646–648. (This is a very accessible article on how to present data. Dan usually has this article on his website but he's just moved institution so Google Dan Wright to find where his web pages are located!)
http.//junkcharts.typepad.com/ is an amusing look at bad graphs.

Interesting real research

Fesmire, F. M. (1988). Termination of intractable hiccups with digital rectal massage. *Annals of Emergency Medicine*, *17*(8), 872.

Exploring assumptions 5

FIGURE 5.1
I came first in the competition for who has the smallest brain

5.1. What will this chapter tell me? ①

When we were learning to read at primary school, we used to read versions of stories by the famous storyteller Hans Christian Andersen. One of my favourites was the story of the ugly duckling. This duckling was a big ugly grey bird, so ugly that even a dog would not bite him. The poor duckling was ridiculed, ostracized and pecked by the other ducks. Eventually, it became too much for him and he flew to the swans, the royal birds, hoping that they would end his misery by killing him because he was so ugly. As he stared into the water, though, he saw not an ugly grey bird but a beautiful swan. Data are much the same. Sometimes they're just big, grey and ugly and don't do any of the things that they're supposed to do. When we get data like these, we swear at them, curse them, peck them and hope that they'll fly away and be killed by the swans. Alternatively, we can try to force our data into becoming beautiful swans. That's what this chapter is all about: assessing how much of an ugly duckling of a data set you have, and discovering how to turn it into a swan. Remember, though, a swan can break your arm.[1]

5.2. What are assumptions? ①

Some academics tend to regard assumptions as rather tedious things about which no one really need worry. When I mention statistical assumptions to my fellow psychologists they tend to give me that raised eyebrow, 'good grief, get a life' look and then ignore me. However, there are good reasons for taking assumptions seriously. Imagine that I go over to a friend's house, the lights are on and it's obvious that someone is at home. I ring the doorbell and no one answers. From that experience, I conclude that my friend hates me and that I am a terrible, unlovable, person. How tenable is this conclusion? Well, there is a reality that I am trying to tap (i.e. whether my friend likes or hates me), and I have collected data about that reality (I've gone to his house, seen that he's at home, rung the doorbell and got no response). Imagine that in reality my friend likes me (he never was a good judge of character!); in this scenario, my conclusion is false. Why have my data led me to the wrong conclusion? The answer is simple: I had assumed that my friend's doorbell was working and under this assumption the conclusion that I made from my data was accurate (my friend heard the bell but chose to ignore it because he hates me). However, this assumption was not true – his doorbell was not working, which is why he didn't answer the door – and as a consequence the conclusion I drew about reality was completely false.

Enough about doorbells, friends and my social life: the point to remember is that when assumptions are broken we stop being able to draw accurate conclusions about reality. Different statistical models assume different things, and if these models are going to reflect reality accurately then these assumptions need to be true. This chapter is going to deal with some particularly ubiquitous assumptions so that you know how to slay these particular beasts as we battle our way through the rest of the book. However, be warned: some tests have their own unique two-headed, fire-breathing, green-scaled assumptions and these will jump out from behind a mound of blood-soaked moss and try to eat us alive when we least expect them to. Onward into battle …

5.3. Assumptions of parametric data ①

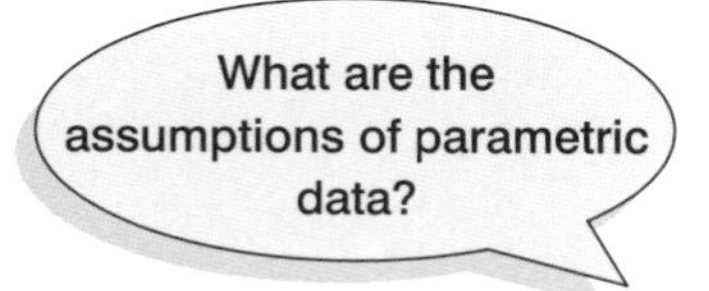

Many of the statistical procedures described in this book are **parametric tests** based on the normal distribution (which is described in section 1.7.4). A parametric test is one that requires data from one of the large catalogue of distributions that statisticians have described, and for data to be parametric certain assumptions must be true. If you use a parametric test when your data are not parametric then the results are likely to be inaccurate. Therefore, it is very important that you check the assumptions before deciding which statistical test is appropriate. Throughout this book you will become aware of my obsession with assumptions and checking them. Most parametric tests based on the normal distribution have four basic assumptions that must be met for the test to be accurate. Many students find checking assumptions a pretty tedious affair, and often get confused about how to tell whether or not an assumption has been met. Therefore, this chapter is designed to take you on a step-by-step tour of the world of parametric assumptions (wow, how exciting!). Now, you may think that

[1] Although it is theoretically possible, apparently you'd have to be weak boned, and swans are nice and wouldn't do that sort of thing.

assumptions are not very exciting, but they can have great benefits: for one thing, you can impress your supervisor/lecturer by spotting all of the test assumptions that they have violated throughout their careers. You can then rubbish, on statistical grounds, the theories they have spent their lifetime developing – and they can't argue with you[2] – but they can poke your eyes out! The assumptions of parametric tests are:

1 **Normally distributed data**: This is a tricky and misunderstood assumption because it means different things in different contexts. For this reason I will spend most of the chapter discussing this assumption! In short, the rationale behind hypothesis testing relies on having something that is normally distributed (in some cases it's the sampling distribution, in others the errors in the model) and so if this assumption is not met then the logic behind hypothesis testing is flawed (we came across these principles in Chapters 1 and 2).

2 **Homogeneity of variance**: This assumption means that the variances should be the same throughout the data. In designs in which you test several groups of participants this assumption means that each of these samples comes from populations with the same variance. In correlational designs, this assumption means that the variance of one variable should be stable at all levels of the other variable (see section 5.6).

3 **Interval data:** Data should be measured at least at the interval level. This assumption is tested by common sense and so won't be discussed further (but do reread section 1.5.1.2 to remind yourself of what we mean by interval data).

4 **Independence**: This assumption, like that of normality, is different depending on the test you're using. In some cases it means that data from different participants are independent, which means that the behaviour of one participant does not influence the behaviour of another. In repeated-measures designs (in which participants are measured in more than one experimental condition), we expect scores in the experimental conditions to be non-independent for a given participant, but behaviour between different participants should be independent. As an example, imagine that two people, Paul and Julie, were participants in an experiment where they had to indicate whether they remembered having seen particular photos earlier on in the experiment. If Paul and Julie were to confer about whether they'd seen certain pictures then their answers would *not* be independent: Julie's response to a given question would depend on Paul's answer, and this would violate the assumption of independence. If Paul and Julie were unable to confer (if they were locked in different rooms) then their responses should be independent (unless they're telepathic): Paul's responses should not be influenced by Julie's. In regression, however, this assumption also relates to the errors in the regression model being uncorrelated, but we'll discuss that more in Chapter 7.

We will, therefore, focus in this chapter on the assumptions of normality and homogeneity of variance.

5.4. The assumption of normality ①

We encountered the normal distribution back in Chapter 1, we know what it looks like and we (hopefully) understand it. You'd think then that this assumption would be easy to

[2] When I was doing my Ph.D., we were set a task by our statistics lecturer in which we had to find some published papers and criticize the statistical methods in them. I chose one of my supervisor's papers and proceeded to slag off every aspect of the data analysis (and I was being *very* pedantic about it all). Imagine my horror when my supervisor came bounding down the corridor with a big grin on his face and declared that, unbeknownst to me, he was the second marker of my essay. Luckily, he had a sense of humour and I got a good mark.☺

understand – it just means that our data are normally distributed, right? Actually, no. In many statistical tests (e.g. the *t*-test) we assume that the sampling distribution is normally distributed. This is a problem because we don't have access to this distribution – we can't simply look at its shape and see whether it is normally distributed. However, we know from the central limit theorem (section 2.5.1) that if the sample data are approximately normal then the sampling distribution will be also. Therefore, people tend to look at their sample data to see if they are normally distributed. If so, then they have a little party to celebrate and assume that the sampling distribution (which is what actually matters) is also. We also know from the central limit theorem that in big samples the sampling distribution tends to be normal anyway – regardless of the shape of the data we actually collected (and remember that the sampling distribution will tend to be normal regardless of the population distribution in samples of 30 or more). As our sample gets bigger then, we can be more confident that the sampling distribution is normally distributed (but see Jane Superbrain Box 5.1).

The assumption of normality is also important in research using regression (or general linear models). General linear models, as we will see in Chapter 7, assume that errors in the model (basically, the deviations we encountered in section 2.4.2) are normally distributed.

In both cases it might be useful to test for normality, and that's what this section is dedicated to explaining. Essentially, we can look for normality visually, look at values that quantify aspects of a distribution (i.e. skew and kurtosis) and compare the distribution we have to a normal distribution to see if it is different.

5.4.1. Oh no, it's that pesky frequency distribution again: checking normality visually ②

We discovered in section 1.7.1 that frequency distributions are a useful way to look at the shape of a distribution. In addition, we discovered how to plot these graphs in section 4.4. Therefore, we are already equipped to look for normality in our sample using a graph. Let's return to the Download Festival data from Chapter 4. Remember that a biologist had visited the Download Festival (a rock and heavy metal festival in the UK) and assessed people's hygiene over the three days of the festival using a standardized technique that results in a score ranging between 0 (you smell like a corpse that's been left to rot up a skunk's anus) and 4 (you smell of sweet roses on a fresh spring day). The data file can be downloaded from the companion website (**DownloadFestival.sas7bdat**) – remember to use the version of the data for which the outlier has been corrected (if you haven't a clue what I mean then read section 4.4 or your graphs will look very different to mine!).

SELF-TEST Using what you learnt in section 4.4 plot histograms for the hygiene scores for the three days of the Download Festival.

There is another useful graph that we can inspect to see if a distribution is normal called a **P–P plot** (probability-probability plot). This graph plots the cumulative probability of a variable against the cumulative probability of a particular distribution (in this case we would specify a normal distribution). What this means is that the data are ranked and sorted. Then for each rank the corresponding *z*-score is calculated. This is the expected value that the score should have in a normal distribution. Next the score itself is converted to a *z*-score (see section 1.7.4). The actual *z*-score is plotted against the expected *z*-score. If the data are normally distributed then the actual *z*-score will be the same as the expected

z-score and you'll get a lovely straight diagonal line. This ideal scenario is helpfully plotted on the graph and your job is to compare the data points to this line. If values fall on the diagonal of the plot then the variable is normally distributed, but deviations from the diagonal show deviations from normality.

```
PROC UNIVARIATE DATA=chapter4.downloadfestival2;
PPPLOT day1 ;
RUN;
```
SAS Syntax 5.1

Figure 5.2 shows the histograms (from the three hygiene scores) and the corresponding P–P plots. The first thing to note is that the data from day 1 look a lot more healthy since we've removed the data point that was mistyped back in section 4.5. In fact the distribution is amazingly normal looking: it is nicely symmetrical and doesn't seem too pointy or flat – these are good things! This is echoed by the P–P plot: note that the data points all fall very close to the 'ideal' diagonal line.

However, the distributions for days 2 and 3 are not nearly as symmetrical. In fact, they both look positively skewed. Again, this can be seen in the P–P plots by the data values deviating away from the diagonal. In general, what this seems to suggest is that by days 2 and 3, hygiene scores were much more clustered around the low end of the scale. Remember that the lower the score, the less hygienic the person is, so this suggests that generally people became smellier as the festival progressed. The skew occurs because a substantial minority insisted on upholding their levels of hygiene (against all odds!) over the course of the festival (baby wet-wipes are indispensable, I find). However, these skewed distributions might cause us a problem if we want to use parametric tests. In the next section we'll look at ways to try to quantify the skewness and kurtosis of these distributions.

5.4.2. Quantifying normality with numbers ②

It is all very well to look at histograms, but they are subjective and open to abuse (I can imagine researchers sitting looking at a completely distorted distribution and saying 'yep, well Bob, that looks normal to me', and Bob replying 'yep, sure does'). Therefore, having inspected the distribution of hygiene scores visually, we can move on to look at ways to quantify the shape of the distributions and to look for outliers.

When we ran PROC UNIVARIATE in the previous example, we looked at the charts. But PROC UNIVARIATE produces a lot of other information, which it writes to the Output window. The first part of each of the outputs is labelled Moments.[3]

SAS Output 5.1 shows the table of moments for the three variables in this example. From this table, we can see that, on average, hygiene scores were 1.77 (out of 4) on day 1 of the festival, but went down to 0.96 and 0.98 on days 2 and 3 respectively. The other important measures for our purposes are the skewness and the kurtosis (see section 1.7.1). The values of skewness and kurtosis should be zero in a normal distribution. Positive values of skewness indicate a pile-up of scores on the left of the distribution, whereas negative values indicate a pile-up on the right. Positive values of kurtosis indicate a pointy and heavy-tailed distribution, whereas negative values indicate a flat and light-tailed distribution. The further the value is from zero, the more likely it is that the data are not normally distributed. For day 1 the skew value is very close to zero (which is good) and kurtosis is a little negative. For days 2 and 3, though, there is a skewness of around 1 (positive skew).

[3] That's a weird name, we know. We'd explain why, but it's really complicated – mathematics stole the word from physics, where it's used in things like levers. If you are really interested, you can look it up on Wikipedia at http://en.wikipedia.org/wiki/Moment_(mathematics) or on Mathworld at http://mathworld.wolfram.com/Moment.html. But you will almost certainly regret it.

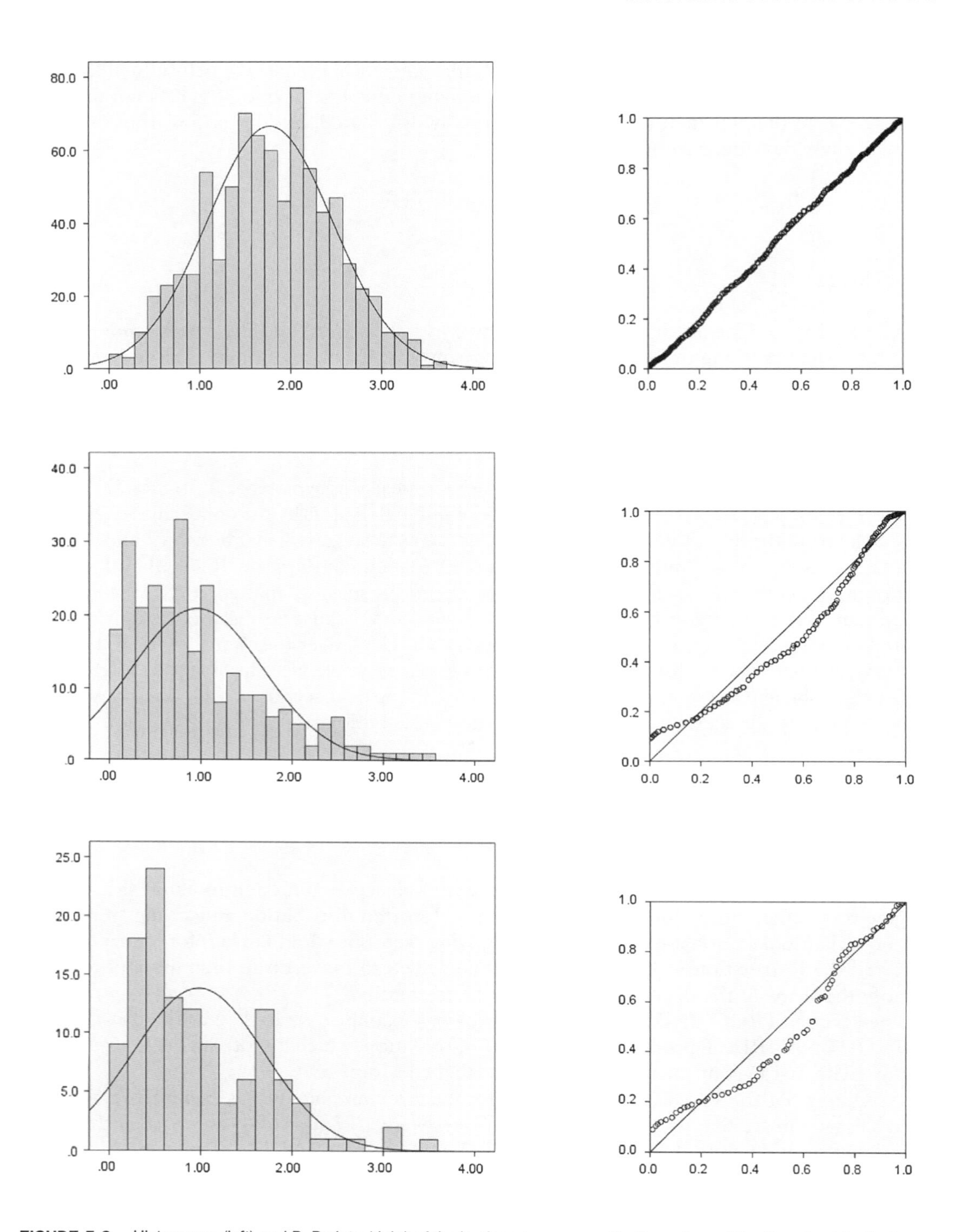

FIGURE 5.2 Histograms (left) and P–P plots (right) of the hygiene scores over the three days of the Download Festival

Although the values of skew and kurtosis are informative, we can convert these values to *z*-scores. We saw in section 1.7.4 that a *z*-score is simply a score from a distribution that has a mean of 0 and a standard deviation of 1. We also saw that this distribution has known properties that we can use. Converting scores to a *z*-score is useful then because (1) we can compare skew and kurtosis values in different samples that used different measures, and (2) we can see how likely our values of skew and kurtosis are to occur. To transform any score to a *z*-score you simply subtract the mean of the distribution (in this case zero) and then divide by the standard deviation of the distribution (in this case we use the standard error).

SAS OUTPUT 5.1

Moments			
N	810	Sum Weights	810
Mean	1.7711358	Sum Observations	1434.62
Std Deviation	0.69353892	Variance	0.48099624
Skewness	–0.0044448	Kurtosis	–0.4103462
Uncorrected SS	2930.0328	Corrected SS	389.125955
Coeff Variation	39.1578625	Std Error Mean	0.02436847

Moments			
N	264	Sum Weights	264
Mean	0.96090909	Sum Observations	253.68
Std Deviation	0.72078005	Variance	0.51952389
Skewness	1.09522552	Kurtosis	0.82220569
Uncorrected SS	380.3982	Corrected SS	136.634782
Coeff Variation	75.0102231	Std Error Mean	0.04436095

Moments			
N	123	Sum Weights	123
Mean	0.97650407	Sum Observations	120.11
Std Deviation	0.710277	Variance	0.50449342
Skewness	1.0328681	Kurtosis	0.73150026
Uncorrected SS	178.8361	Corrected SS	61.5481967
Coeff Variation	72.7367171	Std Error Mean	0.06404352

Skewness and kurtosis are converted to *z*-scores in exactly this way:

$$Z_{\text{skewness}} = \frac{S-0}{SE_{\text{skewness}}}, \quad Z_{\text{kurtosis}} = \frac{K-0}{SE_{\text{kurtosis}}}$$

In the above equations, the values of S (skewness) and K (kurtosis) are produced by SAS, but their standard errors are not.

SAS does not usually calculate these for you. However, it's possible to create new functions in SAS, which didn't exist – these are called macros. Since you look like the kind of person who doesn't solve equations for fun, we've written an SAS macro which does the calculations. You can find it on the website, were it's called **Calculate standard error of skew and kurtosis.sas.** To run it, just run the whole file. The only parts you need to change are on the last line, where you need to say what the data is that contain the variable of interest, and which variable for which you want to calculate the standard error of skew and kurtosis. Then run the whole file, and it will write the results to the Log (not the Output window).
The final line of the syntax file **Calculate standard error of skew and kurtosis.sas** is:

```
%seskewkurt(data, variable);
```

We change it to:

```
%seskewkurt(chapter4.downloadfestival2 ,day1);
```

SAS Output 5.2

```
SE of skew is 0.08590748701603
Z score of skew is -0.05173936701431
```

```
Kurtosis is -0.410346161
SE of kurtosis is 0.17160520561361
Z score of Kurtosis is -2.39122210502136

N is         264
Skew is 1.0952255244
SE of skew is 0.14990849688396
Z score of skew is 7.30596028354406
Kurtosis is 0.8222056945
SE of kurtosis is 0.2987196408297
Z score of Kurtosis is 2.75243265630712

N is         123
Skew is 1.0328680983
SE of skew is 0.21823243373229
Z score of skew is 4.73288081260661
Kurtosis is 0.7315002614
SE of kurtosis is 0.43317042299182
Z score of Kurtosis is 1.68871239256751
```

These z-scores can be compared against values that you would expect to get by chance alone (i.e. known values for the normal distribution shown in the Appendix). So, an absolute value greater than 1.96 is significant at $p < .05$, above 2.58 is significant at $p < .01$ and absolute values above about 3.29 are significant at $p < .001$. Large samples will give rise to small standard errors and so when sample sizes are big, significant values arise from even small deviations from normality. In smallish samples it's OK to look for values above 1.96; however, in large samples this criterion should be increased to the 2.58 one and in very large samples, because of the problem of small standard errors that I've described, no criterion should be applied! If you have a large sample (200 or more) it is more important to look at the shape of the distribution visually and to look at the value of the skewness and kurtosis statistics rather than calculate their significance.

For the hygiene scores, the z-score of skewness = –0.052 on day 1, 7.31 on day 2 and 4.73 on day 3. It is pretty clear that although on day 1 scores are not at all skewed, on days 2 and 3 there is a very significant positive skew (as was evident from the histogram) – however, bear in mind what I just said about large samples! The kurtosis z-scores are: −2.39 on day 1, 2.75 on day 2 and 1.69 on day 3. These values indicate significant kurtosis (at $p < .05$) for two of the three days; however, because of the large sample, this isn't surprising and so we can take comfort in the fact that all values of kurtosis are below our upper threshold of 3.29.

CRAMMING SAM'S TIPS Skewness and kurtosis

- To check that the distribution of scores is approximately normal, we need to look at the values of *skewness* and *kurtosis* in the SAS output.
- Positive values of skewness indicate too many low scores in the distribution, whereas negative values indicate a build-up of high scores.
- Positive values of kurtosis indicate a pointy and heavy-tailed distribution, whereas negative values indicate a flat and light-tailed distribution.
- The further the value is from zero, the more likely it is that the data are not normally distributed.
- You can convert these scores to z-scores by dividing by their standard error. If the resulting score (when you ignore the minus sign) is greater than 1.96 then it is significant ($p < .05$).
- Significance tests of skew and kurtosis should not be used in large samples (because they are likely to be significant even when skew and kurtosis are not too different from normal).

5.4.3. Exploring groups of data ②

Sometimes we have data in which there are different groups of people (men and women, different universities, people with depression and people without, for example). There are several ways to produce basic descriptive statistics for separate groups of people (and we will come across some of these methods in section 5.5.1). However, I intend to use this opportunity to introduce you to the BY option. This option allows you to specify a grouping variable (remember, these variables are used to specify categories of cases). Any time you use the BY option in a procedure in SAS, you will find that the procedure is then carried out on each category of cases separately.

You're probably getting sick of the hygiene data from the Download Festival so let's use the data in the file **SASExam.sas7bdat**. This file contains data regarding students' performance on an SAS exam. Four variables were measured: **exam** (first-year SAS exam scores as a percentage), **computer** (percentage measure of computer literacy), **lecture** (percentage of SAS lectures attended) and **numeracy** (a measure of numerical ability out of 15). There is a variable called **uni** indicating whether the student attended Sussex University (where I work) or Duncetown University. Let's begin by looking at the data as a whole.

5.4.3.1. Running the analysis for all data ③

To see the distribution of the variables, we can use the PROC FREQ and PROC UNIVARIATE commands, which we came across in section 3.11. We'll also show you a couple of cool things, in SAS Syntax 5.2. Well, we think they are cool. First look at PROC FREQ in SAS Syntax 5.2; we give the PROC line, and then run. We don't have a tables line, which says the variables we'd like to have. That's because if we don't specify the variables for PROC FREQ, it just assumes that we want all of the variables, which is OK in this case, because we do want all of the variables. We could do the same thing in PROC UNIVARIATE – if we don't say which variables we want, it will use them all; but in this case it might cause a problem if we did that, one of the variables (uni) is a string, and hence we can't have things like the mean of that variable. Instead, we can use the keyword `_numeric_`. To SAS, this means use all of the variables in the data set which contain numbers, not strings.

```
PROC FREQ DATA=chapter5.sasexam;
     RUN;
PROC UNIVARIATE DATA=chapter5.sasexam;
     HISTOGRAM _numeric_       /normal;
     RUN;
```

SAS Syntax 5.2

SAS Output 5.3 shows the table of descriptive statistics for the four variables in this example. From this table, we can see that, on average, students attended nearly 60% of lectures, obtained 58% in their SAS exam, scored only 51% on the computer literacy test, and only 5 out of 15 on the numeracy test. In addition, the standard deviation for computer literacy was relatively small compared to that of the percentage of lectures attended and exam scores. These latter two variables had several modes (multimodal).

The other important measures are the skewness and the kurtosis, both of which have an associated standard error. We came across these measures earlier on and found that we can convert these values to *z*-scores by dividing using the macro we provided. For the SAS exam scores, the *z*-score of skewness is $-0.107/0.241 = -0.44$. For numeracy, the *z*-score of skewness is $0.961/0.241 = 3.99$. It is pretty clear then that the numeracy scores are significantly positively skewed ($p < .05$) because the *z*-score is greater than 1.96, indicating a pile-up of scores on the left of the distribution (so most students got low scores).

SELF-TEST Calculate and interpret the *z*-scores for skewness of the other variables (computer literacy and percentage of lectures attended).

SELF-TEST Calculate and interpret the *z*-scores for kurtosis of all of the variables.

The output provides tabulated frequency distributions of each variable (not reproduced here). These tables list each score and the number of times that it is found within the data set. In addition, each frequency value is expressed as a percentage of the sample (in this case the frequencies and percentages are the same because the sample size was 100). Also, the cumulative percentage is given, which tells us how many cases (as a percentage) fell below a certain score. So, for example, 66% of numeracy scores were 5 or less, 74% were 6 or less, and so on. Looking in the other direction, only 8% (100 – 92%) got scores greater than 8.

SAS OUTPUT 5.3

Variable: EXAM (Percentage on SAS exam)

Moments			
N	100	Sum Weights	100
Mean	58.1	Sum Observations	5810
Std Deviation	21.3155703	Variance	454.353535
Skewness	–0.1069926	Kurtosis	–1.1051309
Uncorrected SS	382542	Corrected SS	44981
Coeff Variation	36.6877285	Std Error Mean	2.13155703

Basic Statistical Measures			
Location		Variability	
Mean	58.10000	Std Deviation	21.31557
Median	60.00000	Variance	454.35354
Mode	72.00000	Range	84.00000
	382542	Interquartile Range	37.00000
	36.6877285		

Note: The mode displayed is the smallest of 2 modes with a count of 4.

Variable: COMPUTER (Computer literacy)

Moments			
N	100	Sum Weights	100
Mean	50.71	Sum Observations	5071
Std Deviation	8.26003522	Variance	68.2281818
Skewness	–0.1742601	Kurtosis	0.36355167
Uncorrected SS	263905	Corrected SS	6754.59
Coeff Variation	16.2887699	Std Error Mean	0.82600352

Variable: LECTURES (Percentage of lectures attended)

Moments			
N	100	Sum Weights	100
Mean	59.765	Sum Observations	5976.5
Std Deviation	21.6847774	Variance	470.229571
Skewness	–0.4224335	Kurtosis	–0.1789184
Uncorrected SS	403738.25	Corrected SS	46552.7275
Coeff Variation	36.2834057	Std Error Mean	2.16847774

Basic Statistical Measures			
Location		Variability	
Mean	59.76500	Std Deviation	21.68478
Median	62.00000	Variance	470.22957
Mode	48.50000	Range	92.00000
		Interquartile Range	28.50000

Variable: NUMERACY (Numeracy)

Moments			
N	100	Sum Weights	100
Mean	4.85	Sum Observations	485
Std Deviation	2.70568052	Variance	7.32070707
Skewness	0.96136378	Kurtosis	0.94595294
Uncorrected SS	3077	Corrected SS	724.75
Coeff Variation	55.7872272	Std Error Mean	0.27056805

Basic Statistical Measures			
Location		Variability	
Mean	4.850000	Std Deviation	2.70568
Median	4.000000	Variance	7.32071
Mode	4.000000	Range	13.00000
		Interquartile Range	4.00000

Finally, we are given histograms of each variable with the normal distribution overlaid. These graphs are displayed in Figure 5.3 and show us several things. The exam scores are very interesting because this distribution is quite clearly not normal; in fact, it looks suspiciously bimodal (there are two peaks indicative of two modes). This observation corresponds with the earlier information from the table of descriptive statistics. It looks as though computer literacy is fairly normally distributed (a few people are very good with computers and a few are very bad, but the majority of people have a similar degree of knowledge) as is the lecture attendance. Finally, the numeracy test has produced very positively skewed data (i.e. the majority of people did very badly on this test and only a few did well). This corresponds to what the skewness statistic indicated.

Descriptive statistics and histograms are a good way of getting an instant picture of the distribution of your data. This snapshot can be very useful: for example, the bimodal distribution

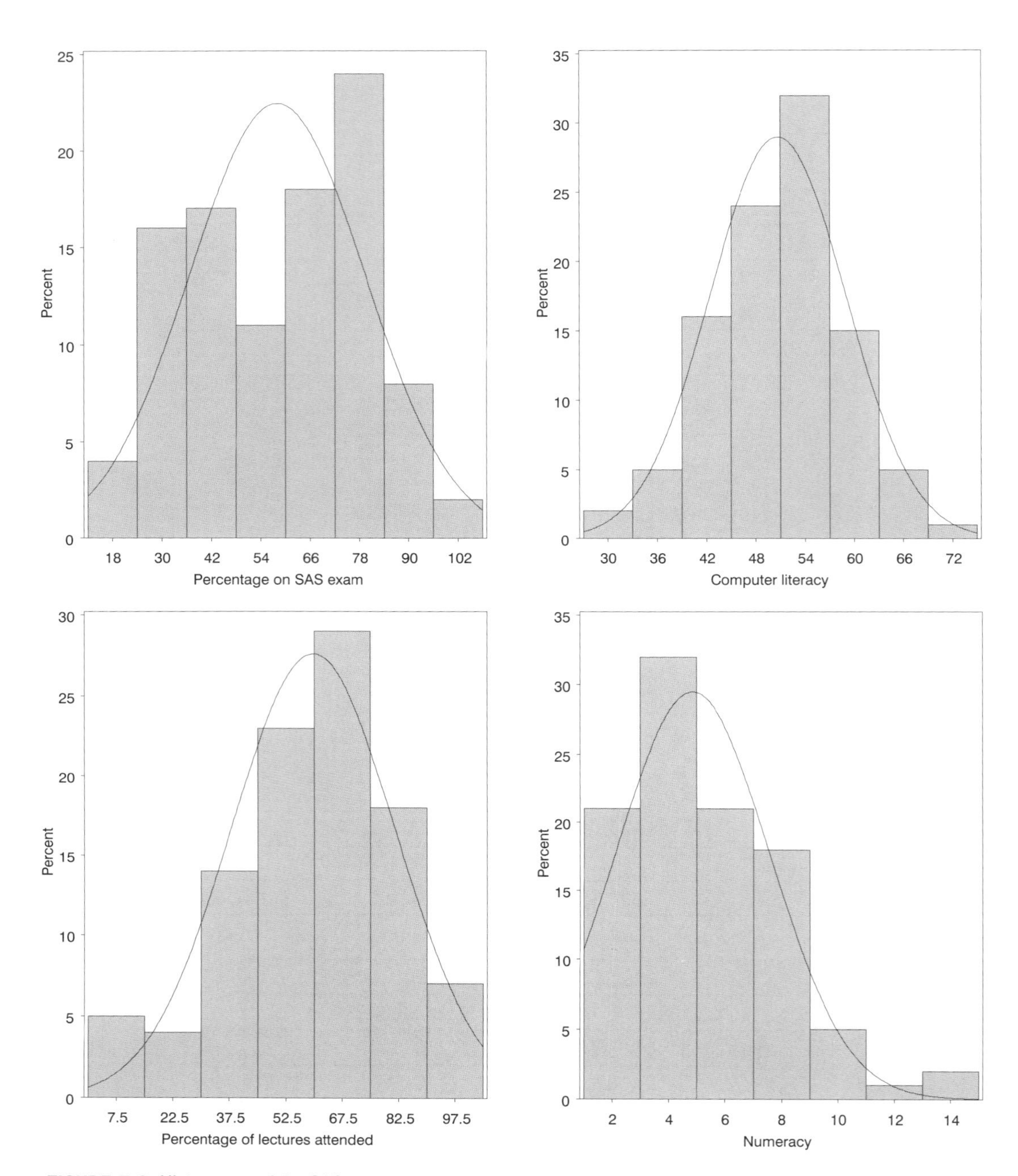

FIGURE 5.3 Histograms of the SAS exam data

of SAS exam scores instantly indicates a trend that students are typically either very good at statistics or struggle with it (there are relatively few who fall in between these extremes). Intuitively, this finding fits with the nature of the subject: statistics is very easy once everything falls into place, but before that enlightenment occurs it all seems hopelessly difficult!

5.4.3.2. Running the analysis for different groups ③

All SAS commands have the option of splitting the file into different groups, and running the analysis separately. To do this, we use the BY option in the command. However, SAS will get upset if the data are not first sorted by the variable that you want to split by, so before you can run the procedure you are interested in, you need to run PROC SORT.

We'll run PROC UNIVARIATE separately for each university, looking at the variable exam. The syntax is shown in SAS Syntax 5.3. Notice that we first use PROC SORT to sort the file by university, and then we run PROC UNIVARIATE. Because we only want statistics for one variable, we need to specify this in the VAR statement, and we'll also get a histogram for that variable.

```
PROC SORT DATA=chapter5.sasexam;
    BY univ;
PROC UNIVARIATE data=chapter5.sasexam;
    BY univ;
    VAR exam;
    HISTOGRAM exam /normal;
    RUN;
```

SAS Syntax 5.3

SAS OUTPUT 5.4

univ=Duncetown

Moments			
N	50	Sum Weights	50
Mean	40.18	Sum Observations	2009
Std Deviation	12.5887705	Variance	158.477143
Skewness	0.30896685	Kurtosis	–0.5666551
Uncorrected SS	88487	Corrected SS	7765.38
Coeff Variation	31.330937	Std Error Mean	1.780321

Basic Statistical Measures			
Location		Variability	
Mean	40.18000	Std Deviation	12.58877
Median	38.00000	Variance	158.47714
Mode	34.00000	Range	51.00000
		Interquartile Range	17.00000

univ=Sussex

Moments			
N	50	Sum Weights	50
Mean	76.02	Sum Observations	3801
Std Deviation	10.2050208	Variance	104.142449
Skewness	0.27209464	Kurtosis	–0.2643288
Uncorrected SS	294055	Corrected SS	5102.98
Coeff Variation	13.4241262	Std Error Mean	1.44320788

Basic Statistical Measures			
Location		Variability	
Mean	76.02000	Std Deviation	10.20502
Median	75.00000	Variance	104.14245
Mode	72.00000	Range	43.00000
		Interquartile Range	12.00000

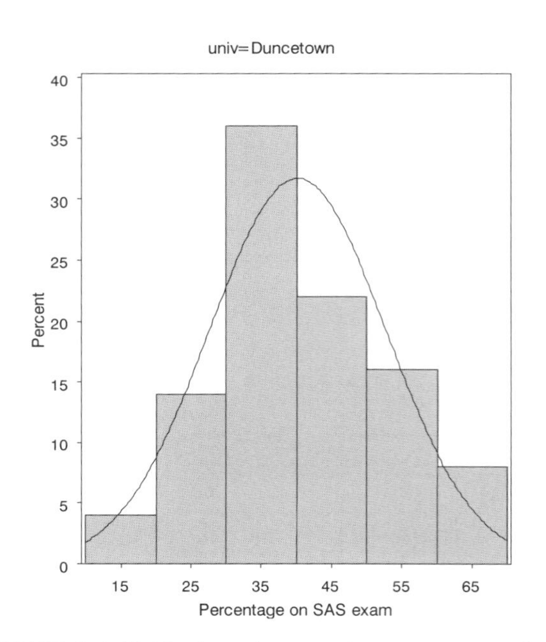

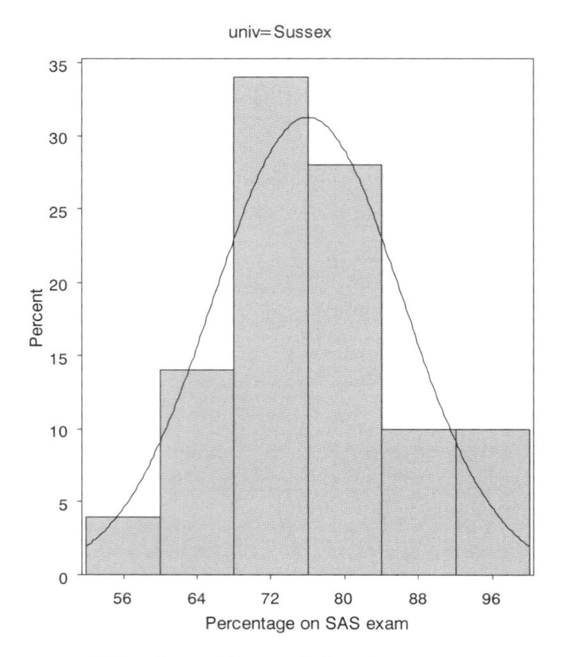

FIGURE 5.4 Distributions of exam and numeracy scores for Duncetown University and Sussex University students

The SAS output is split into two sections: first the results for students at Duncetown University, then the results for those attending Sussex University. SAS Output 5.4 shows the two main summary tables. From these tables it is clear that Sussex students scored higher on both their SAS exam and the numeracy test than their Duncetown counterparts. In fact, looking at the means reveals that, on average, Sussex students scored an amazing 36% more on the SAS exam than Duncetown students, and had higher numeracy scores too (what can I say, my students are the best).

Figure 5.4 shows the histograms of these variables split according to the university attended. The first interesting thing to note is that for exam marks, the distributions are both fairly normal. This seems odd because the overall distribution was bimodal. However, it starts to make sense when you consider that for Duncetown the distribution is centred around a mark of about 40%, but for Sussex the distribution is centred around a mark of about 76%. This illustrates how important it is to look at distributions within groups. If we were interested in comparing Duncetown to Sussex it wouldn't matter that overall the distribution of scores was bimodal; all that's important is that each group comes from a normal distribution, and in this case it appears to be true. When the two samples are combined, these two normal distributions create a bimodal one (one of the modes being around the centre of the Duncetown distribution, and the other being around the centre of the Sussex data!). For numeracy scores, the distribution is slightly positively skewed in the Duncetown group (there is a larger concentration at the lower end of scores). Therefore, the overall positive skew observed before is due to the mixture of universities (the Duncetown students contaminate Sussex's normally distributed scores!).

SELF-TEST Repeat these analyses for the computer literacy and percentage of lectures attended and interpret the results.

FIGURE 5.5 Andrei Kolmogorov, wishing he had a Smirnov

5.5. Testing whether a distribution is normal ①

Another way of looking at the problem is to see whether the distribution as a whole deviates from a comparable normal distribution. The **Kolmogorov–Smirnov test** and **Shapiro–Wilk test** do just this: they compare the scores in the sample to a normally distributed set of scores with the same mean and standard deviation. If the test is non-significant ($p > .05$) it tells us that the distribution of the sample is not significantly different from a normal distribution (i.e. it is probably normal). If, however, the test is significant ($p < .05$) then the distribution in question is significantly different from a normal distribution (i.e. it is non-normal). These tests seem great: in one easy procedure they tell us whether our scores are normally distributed (nice!). However, they have their limitations because with large sample sizes it is very easy to get significant results from small deviations from normality, and so a significant test doesn't necessarily tell us whether the deviation from normality is enough to bias any statistical procedures that we apply to the data. I guess the take-home message is: by all means use these tests, but plot your data as well and try to make an informed decision about the extent of non-normality.

Did someone say Smirnov? Great, I need a drink after all this data analysis!

5.5.1. Doing the Kolmogorov–Smirnov test on SAS ②

The Kolmogorov–Smirnov (K–S from now on; Figure 5.5) test is part of the PROC UNIVARIATE output (one of the parts that we have been ignoring until now). We can also ask SAS PROC UNIVARIATE to produce some graphs called normal Q–Q plots. A Q–Q plot is very similar to the P–P plot that we encountered in section 5.4.1 except that it plots the quantiles of the data set instead of every individual score in the data. Quantiles are just values that split a data set into equal portions. We have already used quantiles without knowing it because quartiles (as in the interquartile range in section 1.7.3) are a special case of quantiles that split the data into four equal parts. However, you can have other quantiles such as percentiles (points that split the data into 100 equal parts), noniles (points that split the data into nine equal parts) and so on. In short, then, the Q–Q plot can be interpreted in the same way as a P–P plot but it will have fewer points on it because rather than plotting every single data point it plots only values that divide the data into equal parts (so they can be easier to interpret if you have a lot of scores).

PROC UNIVARIATE can produce Q–Q plots with the option. This is very similar to the HISTOGRAM option, and is shown in SAS Syntax 5.4. However, note that we need to put in an extra option of MU=EST SIGMA=EST. Mu and sigma are the mean and standard deviation of the distribution. With this option there, SAS draws a reference line of a normal distribution with mean and standard deviation estimated from the data. (We could give a different value, but that would be, well, weird.)

```
PROC UNIVARIATE DATA=chapter5.sasexam NORMAL;
      VAR exam numeracy;
      QQPLOT exam numeracy /NORMAL (MU=EST SIGMA=EST);
      RUN;
```

SAS Syntax 5.4

5.5.2. Distribution output from the PROC UNIVARIATE ②

Now we are going to look at the part of the UNIVARIATE output that contains the goodness of fit. There are three tests, the K-S test, and also the Cramér–von Mises and Anderson–Darling. We're not going to worry about anything other than the K–S test (SAS Output 5.5). This table includes the test statistic itself, the degrees of freedom (which should equal the sample size) and the significance value of this test. Remember that a significant value ($p<.05$) indicates a deviation from normality. For both numeracy and SAS exam scores, the K–S test is significant, indicating that both distributions are not normal. This result is likely to reflect the bimodal distribution found for exam scores, and the positively skewed distribution observed in the numeracy scores. However, these tests confirm that these deviations were statistically *significant*. (But bear in mind that the sample is fairly big.)

As a final point, bear in mind that when we looked at the exam scores for separate groups, the distributions seemed quite normal, now if we'd asked for separate tests for the two universities (by using BY univ, in SAS Syntax 5.5) the K–S test might not have been significant. In fact if you try this out, you'll find that the SAS exam scores are not significantly different from normal within the two groups (the values in the *p* Value. column are greater than .05). This is important because if our analysis involves comparing groups, then what's important is not the overall distribution but the distribution in each group.

SAS also produces a normal Q–Q plot for any variables specified (see Figure 5.6). The normal Q–Q chart plots the values you would expect to get if the distribution were normal (expected values) against the values actually seen in the data set (observed values). The expected values are a straight diagonal line, whereas the observed values are plotted as individual points. If the data are exactly normally distributed, then the observed values (the dots on the chart) should fall exactly along the straight line (meaning that the observed

SAS OUTPUT 5.5

Variable: EXAM (Percentage on SAS exam)

Tests for Normality				
Test	Statistic		p Value	
Shapiro-Wilk	W	0.961309	Pr < W	0.0050
Kolmogorov-Smirnov	D	0.1021	Pr > D	0.0112
Cramer-von Mises	W-Sq	0.24529	Pr > W-Sq	<0.0050
Anderson-Darling	A-Sq	1.403126	Pr > A-Sq	<0.0050

Variable: NUMERACY

Tests for Normality				
Test	Statistic		p Value	
Shapiro-Wilk	W	0.924387	Pr < W	<0.0001
Kolmogorov-Smirnov	D	0.153298	Pr > D	<0.0100
Cramer-von Mises	W-Sq	0.3388	Pr > W-Sq	<0.0050
Anderson-Darling	A-Sq	2.063862	Pr > A-Sq	<0.0050

values are the same as you would expect to get from a normally distributed data set). Any deviation of the dots from the line represents a deviation from normality. So, if the Q–Q plot looks like a straight line with a wiggly snake wrapped around it then you have some deviation from normality! Specifically, when the line sags consistently below the diagonal, or consistently rises above it, then this shows that the kurtosis differs from a normal distribution, and when the curve is S-shaped, the problem is skewness.

In both of the variables analysed we already know that the data are not normal, and these plots confirm this observation because the dots deviate substantially from the line. It is noteworthy that the deviation is greater for the numeracy scores, and this is consistent with the higher significance value of this variable on the K–S test.

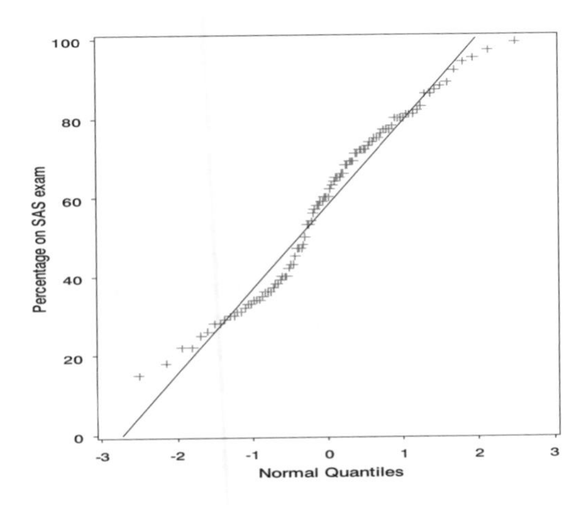

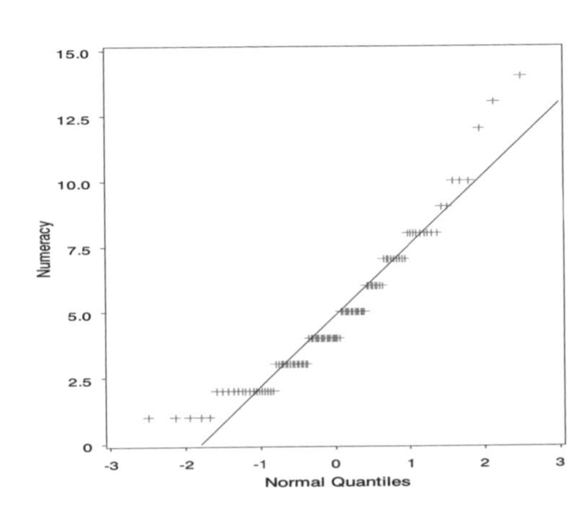

FIGURE 5.6 Normal Q–Q plots of numeracy and SAS exam scores

5.5.3. Reporting the K–S test ①

The test statistic for the K–S test is denoted by D and we must also report the degrees of freedom (df) from the table in brackets after the D. We can report the results in SAS Output 5.4 in the following way:

✓ The percentage on the SAS exam, $D(100) = 0.10$, $p < .05$, and the numeracy scores, $D(100) = 0.15$, $p < .01$, were both significantly non-normal.

CRAMMING SAM'S TIPS **Normality tests**

- The K–S test can be used to see if a distribution of scores significantly differs from a normal distribution.
- If the K–S test is significant (*p*-value. in the SAS table is less than .05) then the scores are significantly different from a normal distribution.
- Otherwise, scores are approximately normally distributed.
- **Warning**: In large samples these tests can be significant even when the scores are only slightly different from a normal distribution. Therefore, they should always be interpreted in conjunction with histograms, P–P or Q–Q plots, and the values of skew and kurtosis.

5.6. Testing for homogeneity of variance ①

So far I've concentrated on the assumption of normally distributed data; however, at the beginning of this chapter I mentioned another assumption: homogeneity of variance. This assumption means that as you go through levels of one variable, the variance of the other should not change. If you've collected groups of data then this means that the variance of your outcome variable or variables should be the same in each of these groups. If you've collected continuous data (such as in correlational designs), this assumption means that the variance of one variable should be stable at all levels of the other variable. Let's illustrate this with an example. An audiologist was interested in the effects of loud concerts on people's hearing. So, she decided to send 10 people on tour with the loudest band she could find, Motörhead. These people went to concerts in Brixton (London), Brighton, Bristol, Edinburgh, Newcastle, Cardiff and Dublin and after each concert the audiologist measured the number of hours after the concert that these people had ringing in their ears.

Figure 5.7 shows the number of hours that each person had ringing in their ears after each concert (each person is represented by a circle). The horizontal lines represent the average number of hours that there was ringing in the ears after each concert and these means are connected by a line so that we can see the general trend of the data. Remember that for each concert, the circles are the scores from which the mean is calculated. Now, we can see in both graphs that the means increase as the people go to more concerts. So, after the first concert their ears ring for about 12 hours, but after the second they ring for about 15–20 hours, and by the final night of the tour, they ring for about 45–50 hours (2 days). So, there is a cumulative effect of the concerts on ringing in the ears. This pattern is found in both graphs; the difference between the graphs is not in terms of the means (which are roughly the same), but in terms of the spread of scores around

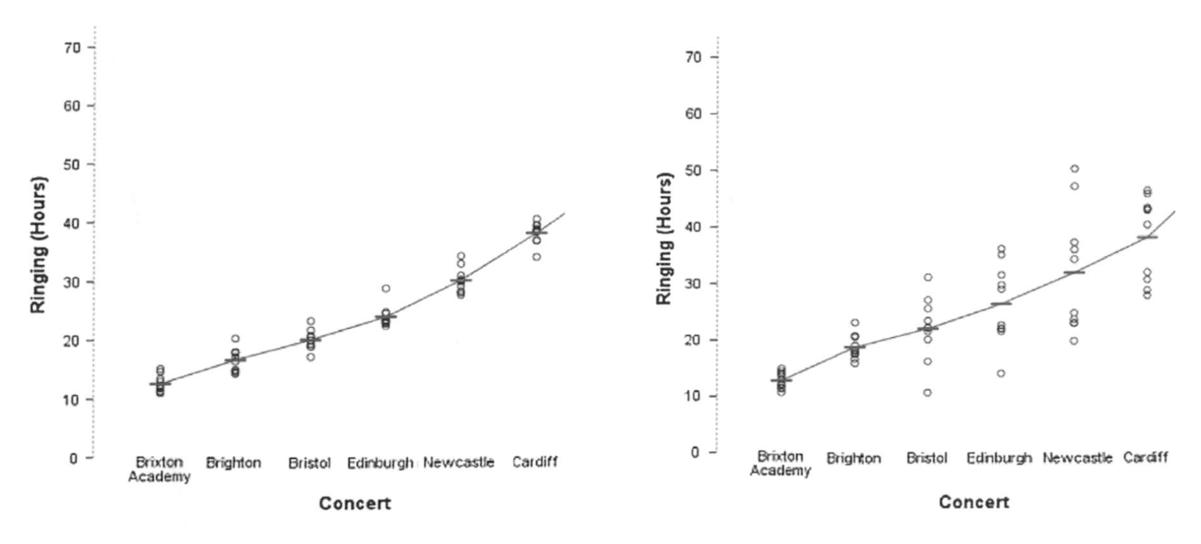

FIGURE 5.7 Graphs illustrating data with homogeneous (left) and heterogeneous (right) variances

the mean. If you look at the left-hand graph, the spread of scores around the mean stays the same after each concert (the scores are fairly tightly packed around the mean). To put it another way, if you measured the vertical distance between the lowest score and the highest score after the Brixton concert, and then did the same after the other concerts, all of these distances would be fairly similar. Although the means increase, the spread of scores for hearing loss is the same at each level of the concert variable (the spread of scores is the same after Brixton, Brighton, Bristol, Edinburgh, Newcastle, Cardiff and Dublin). This is what we mean by *homogeneity of variance*. The right-hand graph shows a different picture: if you look at the spread of scores after the Brixton concert, they are quite tightly packed around the mean (the vertical distance from the lowest score to the highest score is small), but after the Dublin show (for example) the scores are very spread out around the mean (the vertical distance from the lowest score to the highest score is large). This is an example of *heterogeneity of variance*: that is, at some levels of the concert variable the variance of scores is different to other levels (graphically, the vertical distance from the lowest to highest score is different after different concerts).

5.6.1. Levene's test ②

Hopefully you've got a grip on what homogeneity of variance actually means. Now, how do we test for it? Well, we could just look at the values of the variances and see whether they are similar. However, this approach would be very subjective and probably prone to academics thinking 'Ooh look, the variance in one group is only 3000 times larger than the variance in the other: that's roughly equal'. Instead, in correlational analysis such as regression we tend to use graphs (see section 7.8.6) and for groups of data we tend to use a test called **Levene's test** (Levene, 1960). Levene's test tests the null hypothesis that the variances in different groups are equal (i.e. the difference between the variances is zero). It's a very simple and elegant test that works by doing a one-way ANOVA (see Chapter 10) conducted on the deviation scores; that is, the absolute difference between each score and the mean of the group from which it came (see Glass, 1966, for a very readable explanation).[4] For now,

[4] We haven't covered ANOVA yet, so this explanation won't make much sense to you now, but in Chapter 10 we will look in more detail at how Levene's test works.

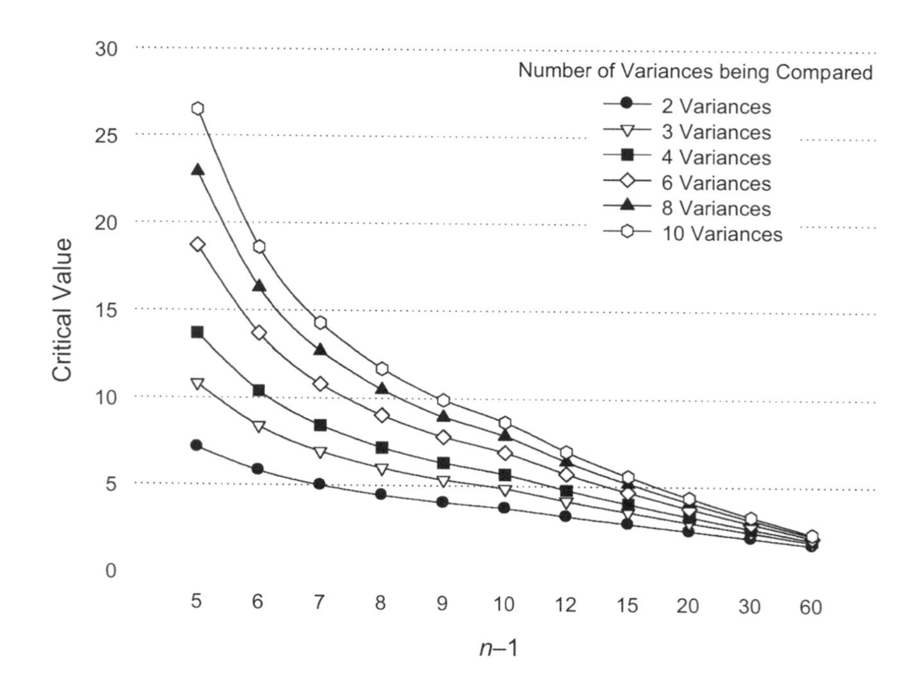

FIGURE 5.8 Selected critical values for Hartley's F_{Max} test

all we need to know is that if Levene's test is significant at $p \leq .05$ then we can conclude that the null hypothesis is incorrect and that the variances are significantly different – therefore, the assumption of homogeneity of variances has been violated. If, however, Levene's test is non-significant (i.e. $p > .05$) then the variances are roughly equal and the assumption is tenable. Although Levene's test can be selected as an option in many of the statistical tests that require it, it can also be examined when you're exploring data (and strictly speaking it's better to examine Levene's test now than wait until your main analysis).

As with the K–S test (and other tests of normality), when the sample size is large, small differences in group variances can produce a Levene's test that is significant (because, as we saw in Chapter 1, the power of the test is improved). A useful double check, therefore, is to look at **Hartley's F_{Max}**, also known as the **variance ratio** (Pearson & Hartley, 1954). This is the ratio of the variances between the group with the biggest variance and the group with the smallest variance. This ratio was compared to critical values in a table published by Hartley. Some of the critical values (for a .05 level of significance) are shown in Figure 5.8 (see Oliver Twisted); as you can see, the critical values depend on the number of cases per group (well, $n - 1$ actually), and the number of variances being compared. From this graph you can see that with sample sizes (n) of 10 per group, an F_{Max} of less than 10 is more or less always going to be non-significant, with 15–20 per group the ratio needs to be less than about 5, and with samples of 30–60 the ratio should be below about 2 or 3.

OLIVER TWISTED

Please, Sir, can I have some more … Hartley's F_{Max}?

Oliver thinks that my graph of critical values is stupid. 'Look at that graph,' he laughed, 'it's the most stupid thing I've ever seen since I was at Sussex University and I saw my statistics lecturer, Andy Fie…'. Well, go choke on your gruel you Dickensian bubo because the full table of critical values is in the additional material for this chapter on the companion website.

Levene's test is calculated in many different procedures – when it is needed. We will not worry about how to do it yet, but we will worry about it when we get to those procedures.

5.7. Correcting problems in the data ①

The previous section showed us various ways to explore our data; we saw how to look for problems with our distribution of scores and how to detect heterogeneity of variance. In Chapter 4 we also discovered how to spot outliers in the data. The next question is what to do about these problems?

5.7.1. Dealing with outliers ②

If you detect outliers in the data there are several options for reducing the impact of these values. However, before you do any of these things, it's worth checking that the data have been entered correctly for the problem cases. If the data are correct then the three main options you have are:

1. *Remove the case*: This entails deleting the data for the person who contributed the outlier. However, this should be done only if you have good reason to believe that this case is not from the population that you intended to sample. For example, if you were investigating factors that affected how much cats purr and one cat didn't purr at all, this would likely be an outlier (all cats purr). Upon inspection, if you discovered that this cat was actually a dog wearing a cat costume (hence why it didn't purr), then you'd have grounds to exclude this case because it comes from a different population (dogs who like to dress as cats) than your target population (cats).
2. *Transform the data*: Outliers tend to skew the distribution and, as we will see in the next section, this skew (and, therefore, the impact of the outliers) can sometimes be reduced by applying **transformations** to the data.
3. *Change the score*: If transformation fails, then you can consider replacing the score. This, on the face of it may seem like cheating (you're changing the data from what was actually collected); however, if the score you're changing is very unrepresentative and biases your statistical model anyway then changing the score is the lesser of two evils! There are several options for how to change the score:

    (a) *The next highest score plus one*: Change the score to be one unit above the next highest score in the data set.

    (b) *Convert back from a z-score*: A z-score of 3.29 constitutes an outlier (see Jane Superbrain Box 4.1) so we can calculate what score would give rise to a z-score of 3.29 (or perhaps 3) by rearranging the z-score equation in section 1.7.4, which gives us $X = (z \times s) + \bar{X}$. All this means is that we calculate the mean ($\bar{X}$) and standard deviation (s) of the data; we know that z is 3 (or 3.29 if you want to be exact) so we just add three times the standard deviation to the mean, and replace our outliers with that score.

    (c) *The mean plus two standard deviations*: A variation on the above method is to use the mean plus two times the standard deviation (rather than three times the standard deviation).

5.7.2. Dealing with non-normality and unequal variances ②

5.7.2.1. Transforming data ③

The next section is quite hair raising so don't worry if it doesn't make much sense – many undergraduate courses won't cover transforming data so feel free to ignore this section if you want to!

We saw in the previous section that you can deal with outliers by transforming the data and these transformations are also useful for correcting problems with normality and the assumption of homogeneity of variance. The idea behind transformations is that you do something to every score to correct for distributional problems, outliers or unequal variances. Although some students often (understandably) think that transforming data sounds dodgy (the phrase 'fudging your results' springs to some people's minds), in fact it isn't because you do the same thing to all of your scores.[5] As such, transforming the data won't change the *relationships* between variables (the relative differences between people for a given variable stay the same), but it does change the *differences* between different variables (because it changes the units of measurement). Therefore, even if you only have one variable that has a skewed distribution, you should still transform any other variables in your data set if you're going to compare differences between that variable and others that you intend to transform.

Let's return to our Download Festival data (**DownloadFestival.sas7bdat**) from earlier in the chapter. These data were not normal on days 2 and 3 of the festival (section 5.4). Now, we might want to look at how hygiene levels changed across the three days (i.e. compare the mean on day 1 to the means on days 2 and 3 to see if people got smellier). The data for days 2 and 3 were skewed and need to be transformed, but because we might later compare the data to scores on day 1, we would also have to transform the day 1 data (even though scores were not skewed). If we don't change the day 1 data as well, then any differences in hygiene scores we find from day 1 to day 2 or 3 will be due to us transforming one variable and not the others.

There are various transformations that you can do to the data that are helpful in correcting various problems.[6] However, whether these transformations are necessary or useful is quite a complex issue (see Jane Superbrain Box 5.1). Nevertheless, because they *are* used by researchers Table 5.1 shows some common transformations and their uses.

5.7.2.2. Choosing a transformation ③

Given that there are many transformations that you can do, how can you decide which one is best? The simple answer is trial and error: try one out and see if it helps and if it doesn't then try a different one. Remember that you must apply the same transformation to all variables (you cannot, for example, apply a log transformation to one variable and a square root transformation to another if they are to be used in the same analysis). This can be quite time consuming. However, for homogeneity of variance we can see the effect of a transformation quite quickly.

[5] Although there aren't statistical consequences of transforming data, there may be empirical or scientific implications that outweigh the statistical benefits (see Jane Superbrain Box 5.1).

[6] You'll notice in this section that I keep writing X_i. We saw in Chapter 1 that this refers to the observed score for the *i*th person (so, the *i* could be replaced with the name of a particular person, thus for Graham, $X_i = X_{Graham}$ = Graham's score, and for Carol, $X_i = X_{Carol}$ = Carol's score).

TABLE 5.1 Data transformations and their uses

Data Transformation	Can Correct For
Log transformation ($\log(X_i)$): Taking the logarithm of a set of numbers squashes the right tail of the distribution. As such it's a good way to reduce positive skew. However, you can't get a log value of zero or negative numbers, so if your data tend to zero or produce negative numbers you need to add a constant to all of the data before you do the transformation. For example, if you have zeros in the data then do $\log(X_i + 1)$, or if you have negative numbers add whatever value makes the smallest number in the data set positive.	Positive skew, unequal variances
Square root transformation ($\sqrt{X_i}$): Taking the square root of large values has more of an effect than taking the square root of small values. Consequently, taking the square root of each of your scores will bring any large scores closer to the centre – rather like the log transformation. As such, this can be a useful way to reduce positive skew; however, you still have the same problem with negative numbers (negative numbers don't have a square root).	Positive skew, unequal variances
Reciprocal transformation ($1/X_i$): Dividing 1 by each score also reduces the impact of large scores. The transformed variable will have a lower limit of 0 (very large numbers will become close to 0). One thing to bear in mind with this transformation is that it reverses the scores: scores that were originally large in the data set become small (close to zero) after the transformation, but scores that were originally small become big after the transformation. For example, imagine two scores of 1 and 10; after the transformation they become $1/1 = 1$, and $1/10 = 0.1$: the small score becomes bigger than the large score after the transformation. However, you can avoid this by reversing the scores before the transformation, by finding the highest score and changing each score to the highest score minus the score you're looking at. So, you do a transformation $1/(X_{\text{Highest}} - X_i)$.	Positive skew, unequal variances
Reverse score transformations: Any one of the above transformations can be used to correct negatively skewed data, but first you have to reverse the scores. To do this, subtract each score from the highest score obtained, or the highest score + 1 (depending on whether you want your lowest score to be 0 or 1). If you do this, don't forget to reverse the scores back afterwards, and remember that the interpretation of the variable is reversed: big scores have become small and small scores have become big!	Negative skew

JANE SUPERBRAIN 5.1

To transform or not to transform, that is the question ③

Not everyone agrees that transforming data is a good idea; for example, Glass, Peckham, and Sanders (1972) in a very extensive review commented that 'the payoff of normalizing transformations in terms of more valid probability statements is low, and they are seldom considered to be worth the effort' (p. 241). In which case, should we bother?

The issue is quite complicated (especially for this early in the book), but essentially we need to know whether the statistical models we apply perform better on transformed data than they do when applied to data that violate the assumption that the transformation corrects. If a statistical model is still accurate even when its assumptions are broken it is said to be a **robust test** (section 5.7.4). I'm not going to discuss whether particular tests are robust here, but I will discuss the issue for particular tests in their respective chapters. The question of whether to transform is linked to this issue of robustness (which in turn is linked to what test you are performing on your data).

A good case in point is the *F*-test in ANOVA (see Chapter 10), which is often claimed to be robust (Glass et al., 1972). Early findings suggested that *F* performed as it should in skewed distributions and that transforming the data helped as often as it hindered the accuracy of *F* (Games & Lucas, 1966). However, in a lively but informative exchange Levine and Dunlap (1982) showed that transformations of skew did improve the performance of *F*; however, in a response Games (1983) argued that their conclusion was incorrect, which Levine and Dunlap (1983) contested in a response to the response. Finally, in a response to the response to the response, Games (1984) pointed out several important questions to consider:

1. The central limit theorem (section 2.5.1) tells us that in big samples the sampling distribution will be normal regardless, and this is what's actually important so the debate is academic in anything other than small samples. Lots of early research did indeed show that with samples of 40 the normality of the sampling distribution was, as predicted, normal. However, this research focused on distributions with light tails and subsequent work has shown that with heavy-tailed distributions larger samples would be necessary to invoke the central limit theorem (Wilcox, 2005). This research suggests that transformations might be useful for such distributions.
2. By transforming the data you change the hypothesis being tested (when using a log transformation and comparing means you change from comparing arithmetic means to comparing geometric means). Transformation also means that you're now addressing a different construct to the one originally measured, and this has obvious implications for interpreting that data (Grayson, 2004).
3. In small samples it is tricky to determine normality one way or another (tests such as K–S will have low power to detect deviations from normality and graphs will be hard to interpret with so few data points).
4. The consequences for the statistical model of applying the 'wrong' transformation could be worse than the consequences of analysing the untransformed scores.

As we will see later in the book, there is an extensive library of robust tests that can be used and which have considerable benefits over transforming data. The definitive guide to these is Wilcox's (2005) outstanding book.

5.7.3. Transforming the data using SAS ②

5.7.3.1. The DATA step ③

To do transformations we use a DATA step which we first came across in section 3.5.1. The DATA step enables us to carry out various functions on columns of data. Some typical functions are adding scores across several columns, taking the square root of the scores in a column or calculating the mean of several variables. We'll use the download festival data **(downloadfestival.sas7bdat)** for this example.

Let's first look at some of the simple functions that you can use in a DATA step:

Operator	Description
+	**Addition**: You can add two scores together to get a total score for a person. For example, with our hygiene data, 'day1 + day2' creates a column in which each row contains the hygiene score from the column labelled *day1* added to the score from the column labelled *day2* (e.g. for participant 1: 2.65 + 1.35 = 4).
-	**Subtraction**: We can subtract one score from another. For example, if we wanted to calculate the change in hygiene from day 1 to day 2 we could type 'day2 – day1'. This creates a column in which each row contains the score from the column labelled *day1* subtracted from the score from the column labelled *day2* (e.g. for participant 1: 1.35 – 2.65 = –1.30). Therefore, this person's hygiene went down by 1.30 (on our 4-point scale) from day 1 to day 2 of the festival.
*	**Multiply**: Notice that to multiply you need to use an asterisk. For example, 'day1 * day2' creates a column that contains the score from the column labelled *day1* multiplied by the score from the column labelled *day2* (e.g. for participant 1: 2.65 × 1.35 = 3.58).
/	**Divide**: You can divide one score by another. For example, 'day1/day2' creates a column that contains the score from the column labelled *day1* divided by the score from the column labelled *day2* (e.g. for participant 1: 2.65/1.35 = 1.96).
**	**Exponentiation**: This raises the preceding term by the power of the succeeding term. So, 'day1**2' creates a column that contains the scores in the *day1* column raised to the power of 2 (i.e. the square of each number in the *day1* column: for participant 1, $(2.65)^2 = 7.02$). Likewise, 'day1**3' creates a column with values of *day1* cubed.

< LT	**Less than:** You can use the < symbol for less than, or you can use LT. This is often used to select particular people for an analysis, or to create a variable only using some people. For example, you could type:
	`if day1 LT 1 then day1highlow = "low";`
	Note that SAS also uses 1 to mean 'yes' or 'true' and 0 to mean 'no' or 'false'. You can write:
	`day1low = day1 LT 1;`
	If the day1 score is less than 1, the answer is 'yes' and the person gets a 1. If the score is equal to or greater than 1, the answer is no, and the score gets a 0.
	But before you use less than, you need to understand SAS Tip 5.1.

<= or LE	**Less than or equal to**: This operation is the same as above except that in the example above, cases that are exactly 1 would be included as well.
> or GT	**Greater than**: This operation is used to include cases above a certain value. This could be used to exclude people who were already smelly at the start of the festival. We might want to exclude them because these people will contaminate the data (not to mention our nostrils) because they reek of putrefaction to begin with so the festival cannot further affect their hygiene!
>= or GE	**Greater than or equal to**: This operation is the same as above but will include cases that are exactly 1 as well.
= or EQ	**Equal to**: You can use this operation to include cases for which participants have a specific value. For example, if you type 'day1 = 1' or 'day1 EQ 1' then only cases that have a value of exactly 1 for the day1 variable are included. This is most useful when you have a coding variable and you want to look at only one of the groups. For example, if we wanted to look only at females at the festival we could type 'gender = 1', then the analysis would be carried out on only females (who are coded as 1 in the data). Some people prefer to always use 'EQ' when making a comparison, and '=' when assigning a variable. It makes it much easier to understand things like: `Day1one = day1 EQ 1;`
>< or <> or ~= or NE	**Not equal to**: This operation will include all cases except those with a specific value. So, 'gender ~= 1' (as in Figure 5.9) will include all cases except those that were female (have a 1 in the gender column). In other words, the output dataset will only contain the males.

Some of the most useful functions are listed in Table 5.2, which shows the standard form of the function, the name of the function, an example of how the function can be used and what SAS would output if that example were used. There are several basic functions for calculating means, standard deviations and sums of sets of columns. There are also functions such as the square root and logarithm that are useful for transforming data that are skewed and we will use these functions now.

SAS TIP 5.1 Storing missing data ③

SAS has a nasty trick that you need to be aware of, and which can make things go horribly wrong if you don't know about it. The problem is the way that SAS stores missing values. SAS stores a missing value as the lowest possible number it can store.

Imagine what would happen if we use the day2 scores from the download festival data set.

We might be interested in marking the people who scored one or less on day 2 of the download festival, so we will use:

```
day2scorelow = day2 LT 1;
```

A person who scores 0.5 on the day2 hygiene score has a score less than 1, and so the answer for them is 'yes' and they get a value in the variable day2scorelow of 1. Similarly, for a person who scores 1, the answer is 'no' and they get a zero. But what happens to a person who we couldn't track down on day 2? We probably want the answer to be that they should be missing. However, if we use the syntax above SAS will answer yes – a missing value is a very low score, and it is less than 1. This can cause you all kinds of trouble (and has caused trouble in published research.)

If you need to use less than, you should always use a check to see if the value is missing, so write:

```
If day2 NE . then day2low = day2 LT 1;
```

If that seems a little convoluted, an easier thing is to just avoid 'less than', and always use 'greater than'.

5.7.3.2. The log transformation on SAS ③

Now we've found out some basic information about the functions available in DATA steps, we can compute the log of day1 in a DATA step, as shown in SAS Syntax 5.5. This creates a new data set in the Work library (remember that it goes into the Work library because we didn't specify another library).

```
DATA downloadfestival2; SET chapter4.downloadfestival;
      logday1 = LOG(day1);
      RUN;
```

SAS Syntax 5.5

For the day 2 hygiene scores there is a value of 0 in the original data, and there is no logarithm of the value 0 (or of any number less than zero). To overcome this we should add a constant to our original scores before we take the log of those scores. Any constant will do, provided that it makes all of the scores greater than 0. In this case our lowest score is 0 in the data set so we can simply add 1 to all of the scores and that will ensure that all scores are greater than zero. In addition, the log of 1 is zero, so the zeros will still be zeros when we have the new scores.

We can add 1 to the day 1 hygiene scores before we do the log transformation, as shown in SAS Syntax 5.6.

TABLE 5.2 Some useful data step functions

Function	Name	Example Input	Output
MEAN(?,?, ..)	Mean	Mean(day1, day2, day3)	For each row, SAS calculates the average hygiene score across the three days of the festival
SD(?,?, ..)	Standard deviation	SD(day1, day2, day3)	Across each row, SAS calculates the standard deviation of the values in the columns labelled *day1, day2* and *day3*
SUM(?,?, ..)	Sum	SUM(day1, day2)	For each row, SAS adds the values in the columns labelled *day1* and *day2*
SQRT(?)	Square root	SQRT(day2)	Produces a column containing the square root of each value in the column labelled *day2*
ABS(?)	Absolute value	ABS(day1)	Produces a variable that contains the absolute value of the values in the column labelled *day1* (absolute values are ones where the signs are ignored: so −5 becomes +5 and +5 stays as +5)
LOG10(?)	Base 10 logarithm	LOG10(day1)	Produces a variable that contains the logarithmic (to base 10) values of the variable *day1*
LOG(?)	Base e logarithm	LOG(day1)	Produces a variable that contains the logarithmic (to base e) values of the variable *day1*. For reasons we don't want to go into now, statisticians prefer e over 10 to use as a base for logs. (e, in case you didn't remember, is 2.71828183).
NORMAL (seed)	Normal random numbers	Normal(seed)	Produces a variable of pseudo-random numbers from a normal distribution with a mean of 0 and a standard deviation of 1. *seed* is a random number seed that helps SAS generate different numbers each time (and the same numbers, with the same seed).

```
DATA downloadfestival2; SET chapter4.downloadfestival;
    logday1 = LOG(day1 + 1);
    RUN;
```
SAS Syntax 5.6

SELF-TEST Have a go at creating similar variables **logday2** and **logday3** for the day 2 and day 3 data. Plot histograms of the transformed scores for all three days.

5.7.3.3. The square root transformation on SAS ③

To do a square root transformation, we run through the same process, by creating a variable such as sqrtday1, as shown in SAS Syntax 5.7.

```
DATA downloadfestival2; SET chapter4.downloadfestival;
    sqrtday1 = SQRT(day1);
    RUN;
```
SAS Syntax 5.7

SELF-TEST Repeat this process for **day2** and **day3** to create variables called **sqrtday2** and **sqrtday3**. Plot histograms of the transformed scores for all three days.

5.7.3.4. The effect of transformations ③

Figure 5.9 shows the distributions for days 1 and 2 of the festival after the three different transformations. Compare these to the untransformed distributions in Figure 5.2. Now, you can see that all three transformations have cleaned up the hygiene scores for day 2: the positive skew is reduced (the square root transformation in particular has been useful). However, because our hygiene scores on day 1 were more or less symmetrical to begin with, they have now become slightly negatively skewed for the log and square root transformation, and positively skewed for the reciprocal transformation![7] If we're using scores from day 2 alone then we could use the transformed scores; however, if we wanted to look at the change in scores then we'd have to weigh up whether the benefits of the transformation for the day 2 scores outweigh the problems it creates in the day 1 scores – data analysis can be frustrating sometimes!

5.7.4. When it all goes horribly wrong ②

It's very easy to think that transformations are the answer to all of your broken assumption prayers. However, as we have seen, there are reasons to think that transformations are not necessarily a good idea (see Jane Superbrain Box 5.1) and even if you think that they are they do not always solve the problem, and even when they do solve the problem they often create different problems in the process. This happens more frequently than you might imagine (messy data are the norm).

If you find yourself in the unenviable position of having irksome data then there are some other options available to you (other than sticking a big samurai sword through your head). The first is to use a test that does not rely on the assumption of normally distributed data and as you go through the various chapters of this book I'll point out these tests – there is also a whole chapter dedicated to them later on.[8] One thing that you will quickly discover about non-parametric tests is that they have been developed for only a fairly limited range of situations. So, happy days if you want to compare two means, but sad lonely days listening to Joy Division if you have a complex experimental design.

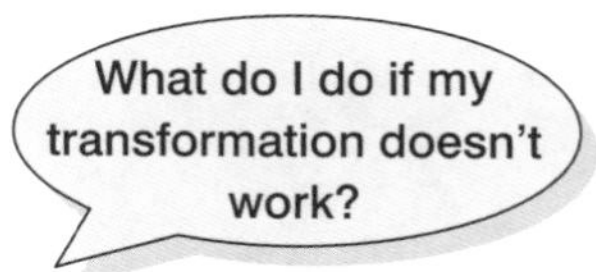

A much more promising approach is to use robust methods (mentioned in Jane Superbrain Box 5.1). These tests have developed as computers have got more sophisticated (doing these tests without computers would be only marginally less painful than ripping off your skin and diving into a bath of salt). How these tests work is beyond the scope of this book (and my brain) but two simple concepts will give you the general idea. Some of these procedures use a **trimmed mean.** A trimmed mean is simply a mean based on the distribution of scores after some percentage of scores has been removed from each extreme

[7] The reversal of the skew for the reciprocal transformation is because, as I mentioned earlier, the reciprocal has the effect of reversing the scores.

[8] For convenience a lot of textbooks refer to these tests as *non-parametric distribution-free* or *assumption-free* tests and stick them in a separate chapter. Actually neither of these terms are particularly accurate (none of these tests is assumption-free) but in keeping with tradition I've put them in Chapter 16 a on their own, ostracized from their 'parametric' counterparts and feeling lonely.

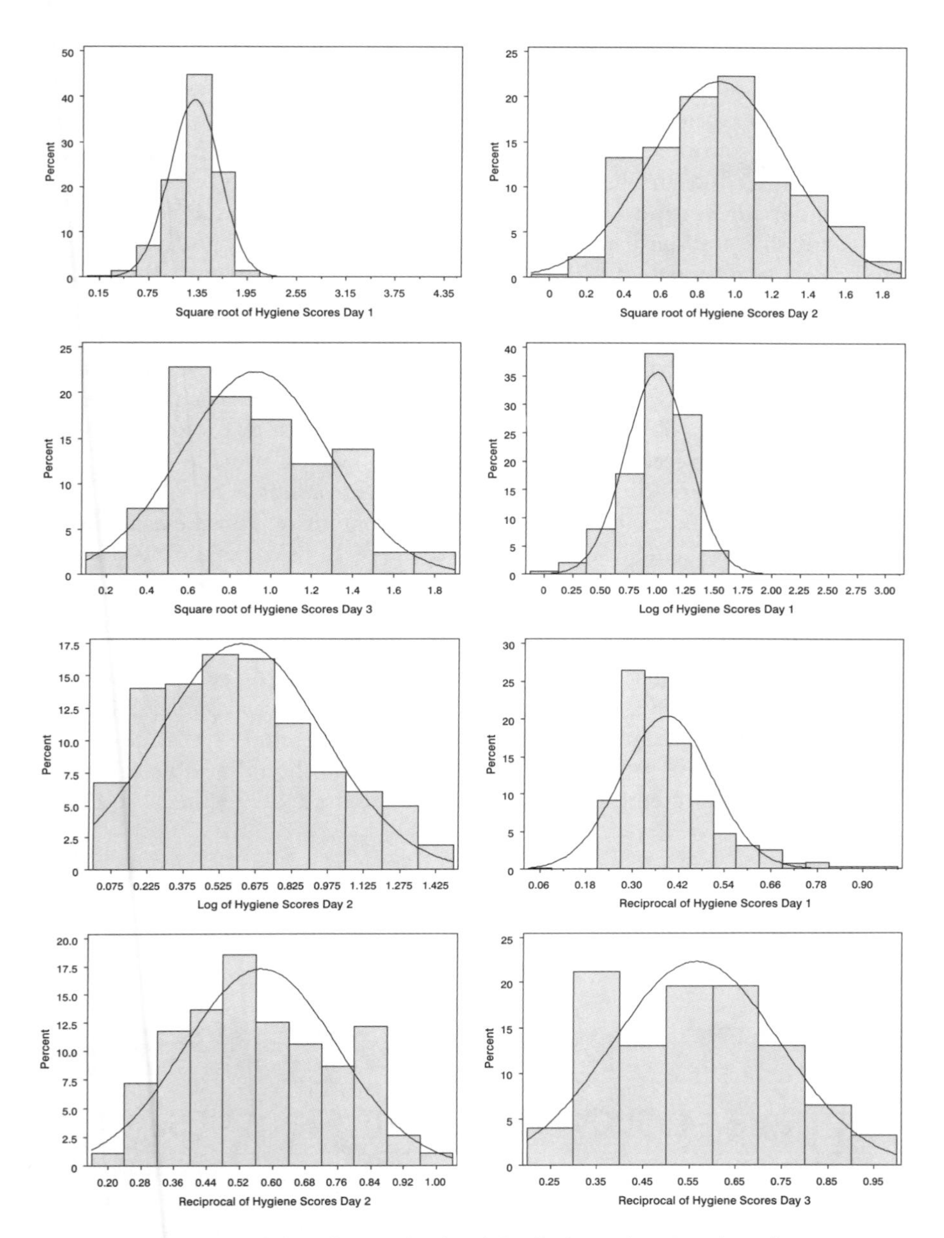

FIGURE 5.9 Distributions of the hygiene data on day 1 and day 2 after various transformations

of the distribution. So, a 10% trimmed mean will remove 10% of scores from the top and bottom before the mean is calculated. We saw in Chapter 2 that the accuracy of the mean depends on a symmetrical distribution, but a trimmed mean produces accurate results even when the distribution is not symmetrical, because by trimming the ends of the distribution we remove outliers and skew that bias the mean. Some robust methods work by taking advantage of the properties of the trimmed mean.

The second general procedure is the **bootstrap** (Efron & Tibshirani, 1993). The idea of the bootstrap is really very simple and elegant. The problem that we have is that we don't know the shape of the sampling distribution, but normality in our data allows us to infer that the sampling distribution is normal (and hence we can know the probability of a particular test statistic occurring). Lack of normality prevents us from knowing the shape of the sampling distribution unless we have big samples (but see Jane Superbrain Box 5.1). Bootstrapping gets around this problem by estimating the properties of the sampling distribution from the sample data. In effect, the sample data are treated as a population from which smaller samples (called bootstrap samples) are taken (putting the data back before a new sample is drawn). The statistic of interest (e.g. the mean) is calculated in each sample, and by taking many samples the sampling distribution can be estimated (rather like in Figure 2.7). The standard error of the statistic is estimated from the standard deviation of this sampling distribution created from the bootstrap samples. From this standard error, confidence intervals and significance tests can be computed. This is a very neat way of getting around the problem of not knowing the shape of the sampling distribution.

These techniques sound pretty good, don't they? It might seem a little strange then that I haven't written a chapter on them. The reason why is that SAS does not do most of them directly, which is something that I hope it will correct sooner rather than later. However, thanks to Rand Wilcox you can do them using a free statistics program called R (www.r-project.org) and a non-free program called S-Plus. Wilcox provides a very comprehensive review of robust methods in his excellent book *Introduction to robust estimation and hypothesis testing* (Wilcox, 2005) and has written programs to run these methods using R. Among many other things, he has files to run robust versions of many tests discussed in this book: ANOVA, ANCOVA, correlation and multiple regression. If there is a robust method, it is likely to be in his book, and he will have written a macro procedure to run it! You can also download these macros from his website. There are several good introductory books to tell you how to use R also (e.g. Dalgaard, 2002). By the time you read this book, it is likely that SAS will have developed a way of integrating R into SAS – it was released in the summer of 2009 (http://support.sas.com/rnd/app/studio/Rinterface2.html) – that allows you to use R commands from the SAS program editor. With this you should be able to analyse an SAS data file using the robust methods of R. These analyses are quite technical so I don't discuss them in the book.

What have I discovered about statistics? ①

'You promised us swans,' I hear you cry, 'and all we got was normality this, homosomethingorother that, transform this – it's all a waste of time that. Where were the bloody swans?!' Well, the Queen owns them all so I wasn't allowed to have them. Nevertheless, this chapter did negotiate Dante's eighth circle of hell (Malebolge), where data of deliberate and knowing evil dwell. That is, data that don't conform to all of those pesky assumptions that make statistical tests work properly. We began by seeing what assumptions need to be met for parametric tests to work, but we mainly focused on the assumptions of normality and homogeneity of variance. To look for normality we rediscovered the joys of frequency distributions, but also encountered some other graphs that tell us about deviations from normality (P–P and Q–Q plots). We saw how we can use skew and kurtosis values to assess normality and that there are statistical tests that we can use (the Kolmogorov–Smirnov test). While negotiating these evildoers, we discovered what homogeneity of variance is, and how to test it with Levene's test and Hartley's F_{Max}. Finally, we discovered redemption for our data. We saw we can cure their sins, make them good, with transformations (and on the way we discovered some of the uses of the *transform* function of SAS data step). Sadly, we also saw that some data are destined to always be evil.

We also discovered that I had started to read. However, reading was not my true passion; it was music. One of my earliest memories is of listening to my dad's rock and soul records (back in the days of vinyl) while waiting for my older brother to come home from school, so I must have been about 3 at the time. The first record I asked my parents to buy me was 'Take on the world' by Judas Priest which I'd heard on *Top of the Pops* (a now defunct UK TV show) and liked. This record came out in 1978 when I was 5. Some people think that this sort of music corrupts young minds. Let's see if it did ...

Key terms that I've discovered

Bootstrap
Hartley's F_{Max}
Heterogeneity of variance
Homogeneity of variance
Independence
Kolmogorov–Smirnov test
Levene's test
Noniles
Normally distributed data
P–P plot
Parametric test
Percentiles
Q–Q plot
Quantiles
Robust test
Transformation
Trimmed mean
Variance ratio

Smart Alex's tasks

- **Task 1**: Using the **ChickFlick.sas7bdat** data from Chapter 4, check the assumptions of normality and homogeneity of variance for the two films (ignore gender): are the assumptions met? ①
- **Task 2**: Remember that the numeracy scores were positively skewed in the **SASExam.sas7bdat** data (see Figure 5.3)? Transform these data using one of the transformations described in this chapter: do the data become normal? ②

Answers can be found on the companion website.

Further reading

Tabachnick, B. G., & Fidell, L. S. (2007). *Using multivariate statistics* (5th edition). Boston: Allyn & Bacon. (Chapter 4 is the definitive guide to screening data!)
Wilcox, R. R. (2005). *Introduction to robust estimation and hypothesis testing* (2nd ed.). Burlington, MA: Elsevier. (Quite technical, but this is the definitive book on robust methods.)

6 Correlation

FIGURE 6.1
I don't have a photo from Christmas 1981, but this was taken about that time at my grandparents' house. I'm trying to play an E by the looks of it, no doubt because it's in 'Take on the world'

6.1. What will this chapter tell me? ①

When I was 8 years old, my parents bought me a guitar for Christmas. Even then, I'd desperately wanted to play the guitar for years. I could not contain my excitement at getting this gift (had it been an *electric* guitar I think I would have actually exploded with excitement). The guitar came with a 'learn to play' book and after a little while of trying to play what was on page 1 of this book, I readied myself to unleash a riff of universe-crushing power onto the world (well, 'skip to my Lou' actually). But, I couldn't do it. I burst into tears and ran upstairs to hide.[1] My dad sat with me and said 'Don't worry, Andy, everything is hard to begin with, but the more you practise the easier it gets.'

[1] This is not a dissimilar reaction to the one I have when publishers ask me for new editions of statistics textbooks.

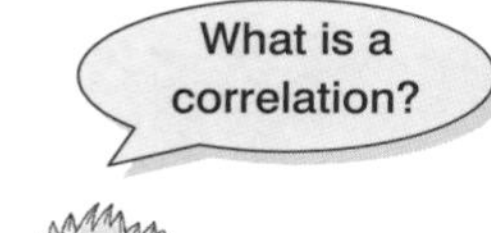

In his comforting words, my dad was inadvertently teaching me about the relationship, or correlation, between two variables. These two variables could be related in three ways: (1) *positively related*, meaning that the more I practised my guitar, the better a guitar player I would become (i.e. my dad was telling me the truth); (2) *not related* at all, meaning that as I practise the guitar my playing ability remains completely constant (i.e. my dad has fathered a cretin); or (3) *negatively related*, which would mean that the more I practised my guitar the worse a guitar player I became (i.e. my dad has fathered an indescribably strange child). This chapter looks first at how we can express the relationships between variables statistically by looking at two measures: *covariance* and the *correlation coefficient*. We then discover how to carry out and interpret correlations in SAS. The chapter ends by looking at more complex measures of relationships; in doing so it acts as a precursor to the chapter on multiple regression.

6.2. Looking at relationships ①

In Chapter 4 I stressed the importance of looking at your data graphically before running any other analysis on them. I just want to begin by reminding you that our first starting point with a correlation analysis should be to look at some scatterplots of the variables we have measured. I am not going to repeat how to get SAS to produce these graphs, but I am going to urge you (if you haven't done so already) to read section 4.7 before embarking on the rest of this chapter.

6.3. How do we measure relationships? ①

6.3.1. A detour into the murky world of covariance ①

The simplest way to look at whether two variables are associated is to look at whether they *covary*. To understand what **covariance** is, we first need to think back to the concept of variance that we met in Chapter 2. Remember that the variance of a single variable represents the average amount that the data vary from the mean. Numerically, it is described by:

$$\text{variance}(s^2) = \frac{\sum(x_i - \bar{x})^2}{N-1} = \frac{\sum(x_i - \bar{x})(x_i - \bar{x})}{N-1} \tag{6.1}$$

The mean of the sample is represented by $\bar{x}$, x_i is the data point in question and N is the number of observations (see section 2.4.1). If we are interested in whether two variables are related, then we are interested in whether changes in one variable are met with similar changes in the other variable. Therefore, when one variable deviates from its mean we would expect the other variable to deviate from its mean in a similar way. To illustrate what I mean, imagine we took five people and subjected them to a certain number of advertisements promoting toffee sweets, and then measured how many packets of those sweets each person bought during the next week. The data are in Table 6.1 as well as the mean and standard deviation (s) of each variable.

TABLE 6.1

Subject:	1	2	3	4	5	Mean	s
Adverts Watched	5	4	4	6	8	5.4	1.67
Packets Bought	8	9	10	13	15	11.0	2.92

If there were a relationship between these two variables, then as one variable deviates from its mean, the other variable should deviate from its mean in the same or the directly opposite way. Figure 6.2 shows the data for each participant (hollow circles represent the number of packets bought and solid circles represent the number of adverts watched); the dashed line is the average number of packets bought and the solid line is the average number of adverts watched. The vertical lines represent the differences (remember that these differences are called *deviations*) between the observed values and the mean of the relevant variable. The first thing to notice about Figure 6.2 is that there is a very similar pattern of deviations for both variables. For the first three participants the observed values are below the mean for both variables, for the last two people the observed values are above the mean for both variables. This pattern is indicative of a potential relationship between the two variables (because it seems that if a person's score is below the mean for one variable then their score for the other will also be below the mean).

So, how do we calculate the exact similarity between the pattern of differences of the two variables displayed in Figure 6.2? One possibility is to calculate the total amount of deviation, but we would have the same problem as in the single-variable case: the positive and negative deviations would cancel out (see section 2.4.1). Also, by simply adding the deviations, we would gain little insight into the *relationship* between the variables. Now, in the single-variable case, we squared the deviations to eliminate the problem of positive and negative deviations cancelling out each other. When there are two variables, rather than squaring each deviation, we can multiply the deviation for one variable by the corresponding deviation for the second variable. If both deviations are positive or negative then this will give us a positive value (indicative of the deviations being in the same direction), but if one deviation is positive and one negative then the resulting product will be negative

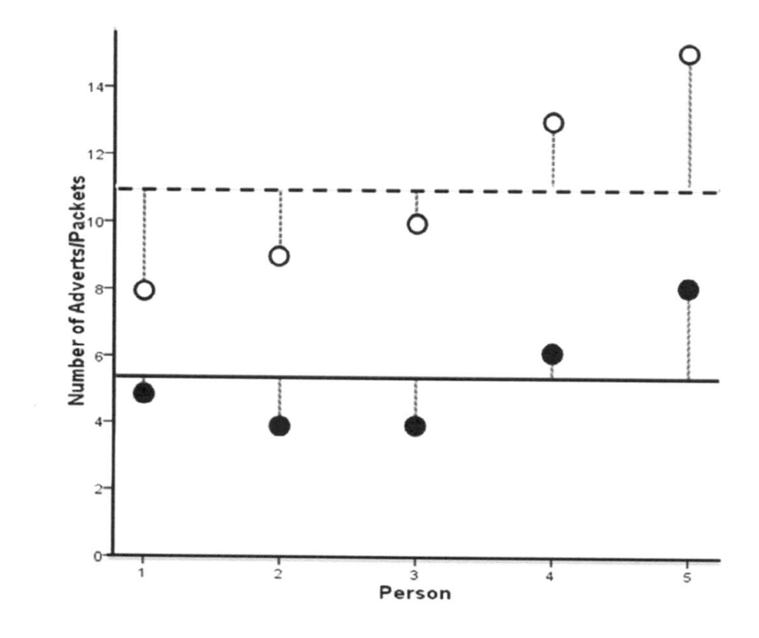

FIGURE 6.2 Graphical display of the differences between the observed data and the means of two variables

(indicative of the deviations being opposite in direction). When we multiply the deviations of one variable by the corresponding deviations of a second variable, we get what is known as the **cross-product deviations.** As with the variance, if we want an average value of the combined deviations for the two variables, we must divide by the number of observations (we actually divide by $N - 1$ for reasons explained in Jane Superbrain Box 2.2). This averaged sum of combined deviations is known as the *covariance*. We can write the covariance in equation form as in equation (6.2) – you will notice that the equation is the same as the equation for variance, except that instead of squaring the differences, we multiply them by the corresponding difference of the second variable:

$$\mathrm{cov}(x,y) = \frac{\sum(x_i - \bar{x})(y_i - \bar{y})}{N-1} \tag{6.2}$$

For the data in Table 6.1 and Figure 6.2 we reach the following value:

$$\begin{aligned}
\mathrm{cov}(x,y) &= \frac{\sum(x_i - \bar{x})(y_i - \bar{y})}{N-1} \\
&= \frac{(-0.4)(-3) + (-1.4)(-2) + (-1.4)(-1) + (0.6)(2) + (2.6)(4)}{4} \\
&= \frac{1.2 + 2.8 + 1.4 + 1.2 + 10.4}{4} \\
&= \frac{17}{4} \\
&= 4.25
\end{aligned}$$

Calculating the covariance is a good way to assess whether two variables are related to each other. A positive covariance indicates that as one variable deviates from the mean, the other variable deviates in the same direction. On the other hand, a negative covariance indicates that as one variable deviates from the mean (e.g. increases), the other deviates from the mean in the opposite direction (e.g. decreases).

There is, however, one problem with covariance as a measure of the relationship between variables and that is that it depends upon the scales of measurement used. So, covariance is not a standardized measure. For example, if we use the data above and assume that they represented two variables measured in miles then the covariance is 4.25 (as calculated above). If we then convert these data into kilometres (by multiplying all values by 1.609) and calculate the covariance again then we should find that it increases to 11. This dependence on the scale of measurement is a problem because it means that we cannot compare covariances in an objective way – so, we cannot say whether a covariance is particularly large or small relative to another data set unless both data sets were measured in the same units.

6.3.2. Standardization and the correlation coefficient ①

To overcome the problem of dependence on the measurement scale, we need to convert the covariance into a standard set of units. This process is known as **standardization.** A very basic form of standardization would be to insist that all experiments use the same units of measurement, say metres – that way, all results could be easily compared.

However, what happens if you want to measure attitudes – you'd be hard pushed to measure them in metres! Therefore, we need a unit of measurement into which any scale of measurement can be converted. The unit of measurement we use is the *standard deviation*. We came across this measure in section 2.4.1 and saw that, like the variance, it is a measure of the average deviation from the mean. If we divide any distance from the mean by the standard deviation, it gives us that distance in standard deviation units. For example, for the data in Table 6.1, the standard deviation for the number of packets bought is approximately 3.0 (the exact value is 2.92). In Figure 6.2 we can see that the observed value for participant 1 was 3 packets less than the mean (so there was an error of −3 packets of sweets). If we divide this deviation, −3, by the standard deviation, which is approximately 3, then we get a value of −1. This tells us that the difference between participant 1's score and the mean was −1 standard deviation. So, we can express the deviation from the mean for a participant in standard units by dividing the observed deviation by the standard deviation.

It follows from this logic that if we want to express the covariance in a standard unit of measurement we can simply divide by the standard deviation. However, there are two variables and, hence, two standard deviations. Now, when we calculate the covariance we actually calculate two deviations (one for each variable) and then multiply them. Therefore, we do the same for the standard deviations: we multiply them and divide by the product of this multiplication. The standardized covariance is known as a *correlation coefficient* and is defined by equation (6.3) in which s_x is the standard deviation of the first variable and s_y is the standard deviation of the second variable (all other letters are the same as in the equation defining covariance):

$$r = \frac{\text{cov}_{xy}}{s_x s_y} = \frac{\sum(x_i - \bar{x})(y_i - \bar{y})}{(N-1)s_x s_y} \tag{6.3}$$

The coefficient in equation (6.3) is known as the *Pearson product-moment correlation coefficient* or **Pearson correlation coefficient** (for a really nice explanation of why it was originally called the 'product-moment' correlation see Miles & Banyard, 2007) and was invented by Karl Pearson (see Jane Superbrain Box 6.1).[2] If we look back at Table 6.1 we see that the standard deviation for the number of adverts watched (s_x) was 1.67, and for the number of packets of crisps bought (s_y) was 2.92. If we multiply these together we get $1.67 \times 2.92 = 4.88$. Now, all we need to do is take the covariance, which we calculated a few pages ago as being 4.25, and divide by these multiplied standard deviations. This gives us $r = 4.25/4.88 = .87$.

By standardizing the covariance we end up with a value that has to lie between −1 and +1 (if you find a correlation coefficient less than −1 or more than +1 you can be sure that something has gone hideously wrong!). A coefficient of +1 indicates that the two variables are perfectly positively correlated, so as one variable increases, the other increases by a proportionate amount. Conversely, a coefficient of −1 indicates a perfect negative relationship: if one variable increases, the other decreases by a proportionate amount. A coefficient of zero indicates no linear relationship at all. We also saw in section 2.6.4 that because the correlation coefficient is a standardized measure of an observed effect, it is a commonly used measure of the size of an effect and that values of ±.1 represent a small

[2] You will find Pearson's product-moment correlation coefficient denoted by both r and R. Typically, the upper-case form is used in the context of regression because it represents the multiple correlation coefficient; however, for some reason, when we square r (as in section 6.5.2.3) an upper case R is used. Don't ask me why – it's just to confuse me, I suspect.

JANE SUPERBRAIN 6.1

Who said statistics was dull? ①

Students often think that statistics is dull, but back in the early 1900s it was anything but dull with various prominent figures entering into feuds on a soap opera scale. One of the most famous was between Karl Pearson and Ronald Fisher (whom we met in Chapter 2). It began when Pearson published a paper of Fisher's in his journal but made comments in his editorial that, to the casual reader, belittled Fisher's work. Two years later Pearson's group published work following on from Fisher's paper without consulting him. The antagonism persisted, with Fisher turning down a job to work in Pearson's, group and publishing 'improvements' on Pearson's ideas. Pearson for his part wrote in his own journal about apparent errors made by Fisher.

Another prominent statistician, Jerzy Neyman, criticized some of Fisher's most important work in a paper delivered to the Royal Statistical Society on 28 March 1935 at which Fisher was present. Fisher's discussion of the paper at that meeting directly attacked Neyman. Fisher more or less said that Neyman didn't know what he was talking about and didn't understand the background material on which his work was based. Relations soured so much that while they both worked at University College London, Neyman openly attacked many of Fisher's ideas in lectures to his students. The two feuding groups even took afternoon tea (a common practice in the British academic community of the time) in the same room but at different times! The truth behind who fuelled these feuds is, perhaps, lost in the mists of time, but Zabell (1992) makes a sterling effort to unearth it.

Basically, then, the founders of modern statistical methods were a bunch of squabbling children. Nevertheless, these three men were astonishingly gifted individuals. Fisher, in particular, was a world leader in genetics, biology and medicine as well as possibly the most original mathematical thinker ever (Barnard, 1963; Field, 2005d; Savage, 1976).

effect, $\pm.3$ a medium effect and $\pm.5$ a large effect (although I re-emphasize my caveat that these canned effect sizes are no substitute for interpreting the effect size within the context of the research literature).

6.3.3. The significance of the correlation coefficient ③

Although we can directly interpret the size of a correlation coefficient, we have seen in Chapter 2 that scientists like to test hypotheses using probabilities. In the case of a correlation coefficient we can test the hypothesis that the correlation is different from zero (i.e. different from 'no relationship'). If we find that our observed coefficient was very unlikely to happen if there was no effect in the population then we can gain confidence that the relationship that we have observed is statistically meaningful.

There are two ways that we can go about testing this hypothesis. The first is to use our trusty z-scores that keep cropping up in this book. As we have seen, z-scores are useful because we know the probability of a given value of z occurring, if the distribution from which it comes is normal. There is one problem with Pearson's r, which is that it is known to have a sampling distribution that is not normally distributed. This is a bit of a nuisance, but luckily thanks to our friend Fisher we can adjust r so that its sampling distribution *is* normal as follows (Fisher, 1921):

$$z_r = \frac{1}{2}\log_e\left(\frac{1+r}{1-r}\right) \tag{6.4}$$

The resulting z_r has a standard error of:

$$\mathrm{SE}_{z_r} = \frac{1}{\sqrt{N-3}} \tag{6.5}$$

For our advert example, our $r = .87$ becomes 1.33 with a standard error of 0.71.

We can then transform this adjusted r into a z-score just as we have done for raw scores, and for skewness and kurtosis values in previous chapters. If we want a z-score that represents the size of the correlation relative to a particular value, then we simply compute a z-score using the value that we want to test against and the standard error. Normally we want to see whether the correlation is different from 0, in which case we can subtract 0 from the observed value of r and divide by the standard error (in other words, we just divide z_r by its standard error):

$$z = \frac{z_r}{\mathrm{SE}z_r} \tag{6.6}$$

For our advert data this gives us 1.33/0.71 = 1.87. We can look up this value of z (1.87) in the table for the normal distribution in the Appendix and get the one-tailed probability from the column labelled 'Smaller Portion'. In this case the value is .0307. To get the two-tailed probability we simply multiply the one-tailed probability value by 2, which gives us .0614. As such the correlation is significant, $p < .05$ one-tailed, but not two-tailed.

In fact, the hypothesis that the correlation coefficient is different from 0 is usually (SAS, for example, does this) tested not using a z-score, but using a t-statistic with $N - 2$ degrees of freedom, which can be directly obtained from r:

$$t_r = \frac{r\sqrt{N-2}}{\sqrt{1-r^2}} \tag{6.7}$$

So, you might wonder then why I told you about z-scores. Partly it was to keep the discussion framed in concepts with which you are already familiar (we don't encounter the t-test properly for a few chapters), but also it is useful background information for the next section.

6.3.4. Confidence intervals for *r* ③

I was moaning on earlier about how SAS doesn't make tea for you. Another thing that it doesn't do is compute confidence intervals for r. This is a shame because as we have seen in Chapter 2 these intervals tell us something about the likely value (in this case of the correlation) in the population. However, we can calculate these manually (or if you search the web you will also find SAS macros that will do this). To do this we need to take advantage of what we learnt in the previous section about converting r to z_r (to make the sampling distribution normal), and using the associated standard errors. We can then construct a confidence interval in the usual way. For a 95% confidence interval we have (see section 2.5.2.1):

$$\text{lower boundary of confidence interval} = \overline{X} - (1.96 \times \text{SE})$$
$$\text{upper boundary of confidence interval} = \overline{X} + (1.96 \times \text{SE})$$

In the case of our transformed correlation coefficients these equations become:

$$\text{lower boundary of confidence interval} = z_r - (1.96 \times \text{SE}_{z_r})$$
$$\text{upper boundary of confidence interval} = z_r + (1.96 \times \text{SE}_{z_r})$$

For our advert data this gives us 1.33 – (1.96 × 0.71) = –0.062, and 1.33 + (1.96 × 0.71) = 2.72. Remember that these values are in the z_r metric and so we have to convert back to correlation coefficients using:

$$r = \frac{e^{(2z_r)} - 1}{e^{(2z_r)} + 1} \tag{6.8}$$

This gives us an upper bound of r = .991 and a lower bound of –.062 (because this value is so close to zero the transformation to z has no impact).

CRAMMING SAM'S TIPS

- A crude measure of the relationship between variables is the *covariance*.
- If we standardize this value we get *Pearson's correlation coefficient, r*.
- The correlation coefficient has to lie between –1 and +1.
- A coefficient of +1 indicates a perfect positive relationship, a coefficient of –1 indicates a perfect negative relationship, a coefficient of 0 indicates no linear relationship at all.
- The correlation coefficient is a commonly used measure of the size of an effect: values of ±.1 represent a small effect, ±.3 is a medium effect and ±.5 is a large effect.

6.3.5. A word of warning about interpretation: causality ①

Considerable caution must be taken when interpreting correlation coefficients because they give no indication of the direction of *causality*. So, in our example, although we can conclude that as the number of adverts watched increases, the number of packets of toffees bought increases also, we cannot say that watching adverts *causes* you to buy packets of toffees. This caution is for two reasons:

- **The third-variable problem**: We came across this problem in section 1.6.2. To recap, in any correlation, causality between two variables cannot be assumed because there may be other measured or unmeasured variables affecting the results. This is known as the *third-variable* problem or the *tertium quid* (see section 1.6.2 and Jane Superbrain Box 1.1).
- **Direction of causality**: Correlation coefficients say nothing about which variable causes the other to change. Even if we could ignore the third-variable problem described above, and we could assume that the two correlated variables were the only important ones, the correlation coefficient doesn't indicate in which direction causality operates. So, although it is intuitively appealing to conclude that watching adverts causes us to buy packets of toffees, there is no *statistical* reason why buying packets of toffees cannot cause us to watch more adverts. Although the latter conclusion makes less intuitive sense, the correlation coefficient does not tell us that it isn't true.

6.4. Data entry for correlation analysis using SAS ①

Data entry for correlation, regression and multiple regression is straightforward because each variable is entered in a separate column. So, for each variable you have measured, create a variable in the data editor with an appropriate name, and enter a participant's scores across one row of the data editor. There may be occasions on which you have one or more categorical variables (such as gender). As an example, if we wanted to calculate the correlation between the two variables in Table 6.1 we would enter each variable in a separate column, and each row represents a single individual's data (so the first consumer saw 5 adverts and bought 8 packets).

SELF-TEST Enter the advert data (you might want to look at a DATA step, in SAS Syntax 3-1 to remind yourself how to do this) and use the gplot to produce a scatterplot (number of packets bought on the *y*-axis, and adverts watched on the *x*-axis) of the data.

6.5. Bivariate correlation ①

6.5.1. General procedure for running correlations on SAS ①

There are two types of correlation: *bivariate* and *partial*. A **bivariate correlation** is a correlation between two variables (as described at the beginning of this chapter) whereas a

partial correlation looks at the relationship between two variables while 'controlling' the effect of one or more additional variables. Pearson's product-moment correlation coefficient (described earlier) and Spearman's rho (see section 6.5.3) are examples of bivariate correlation coefficients.

Let's return to the example from Chapter 4 about exam scores. Remember that a psychologist was interested in the effects of exam stress and revision on exam performance. She had devised and validated a questionnaire to assess state anxiety relating to exams (called the Exam Anxiety Questionnaire, or EAQ). This scale produced a measure of anxiety scored out of 100. Anxiety was measured before an exam, and the percentage mark of each student on the exam was used to assess the exam performance. She also measured the number of hours spent revising. These data are in **ExamAnxiety.sas7bdat** on the companion website. We have already created scatterplots for these data (section 4.7) so we don't need to do that again.

To conduct bivariate correlations we use PROC CORR. In the first line we list the data set, and then on the VAR line we list the variables that we want to correlate. SAS Syntax 6.1 shows how PROC CORR is used.

```
PROC CORR data=chapter6.examanxiety;
    VAR revise exam anxiety;
    RUN;
```

SAS Syntax 6.1

The default setting is Pearson's product-moment correlation, but you can also calculate Spearman's correlation and Kendall's correlation—we will see the differences between these correlation coefficients in due course.

6.5.2. Pearson's correlation coefficient ①

6.5.2.1. Assumptions of Pearson's r ③

Pearson's (Figure 6.3) correlation coefficient was described in full at the beginning of this chapter. Pearson's correlation requires only that data are interval (see section 1.5.1.2) for it to be an accurate measure of the linear relationship between two variables. However, if you want to establish whether the correlation coefficient is significant, then more assumptions are required: for the test statistic to be valid the sampling distribution has to be normally distributed and, as we saw in Chapter 5, we assume that it is if our sample data are normally distributed (or if we have a large sample). Although, typically, to assume that the sampling distribution is normal, we would want both variables to be normally distributed, there is one exception to this rule: one of the variables can be a categorical variable provided there are only two categories (this is the same as doing a t-test, but I'm jumping the gun a bit). In any case, if your data are non-normal (see Chapter 5) or are not measured at the interval level then you should not use a Pearson correlation.

OLIVER TWISTED

Please, Sir, can I have some more ... options?

Oliver is so excited to get onto analysing his data that he doesn't want me to spend pages waffling on about options that you will probably never use. 'Stop writing, you waffling fool,' he says. 'I want to analyse my data.' Well, he's got a point. If you want to find out more about what other options are available in SAS PROC CORR, then the additional material for this chapter on the companion website will tell you.

SAS TIP 6.1 **Pairwise or listwise?** ①

As we run through the various analyses in this book, many of them have additional options. One common option is to choose whether you do 'pairwise', 'analysis by analysis' or 'listwise'. First, we can exclude cases listwise, which means that if a case has a missing value for any variable, then they are excluded from the whole analysis. So, for example, in our exam anxiety data if one of our students had reported their anxiety and we knew their exam performance but we didn't have data about their revision time, then their data would not be used to calculate any of the correlations: *they would be completely excluded from the analysis*. Another option is to excluded cases on a pairwise or analysis-by-analysis basis, which means that if a participant has a score missing for a particular variable or analysis, then their data are excluded only from calculations involving the variable for which they have no score. For our student about whom we don't have any revision data, this means that their data would be excluded when calculating the correlation between exam scores and revision time, and when calculating the correlation between exam anxiety and revision time; however, the student's scores would be *included* when calculating the correlation between exam anxiety and exam performance because for this pair of variables we have both of their scores.

SAS PROC CORR does pairwise deletion by default – that is, a person is included wherever possible. If you want to exclude people who have missing data for one variable, use the nomiss option on the PROC CORR line. You would write:

```
PROC CORR DATA=chapter6.examanxiety nomiss;
```

6.5.2.2. Running Pearson's r on SAS ①

We have already seen the syntax for PROC CORR (in SAS Syntax 6.1). SAS produces Pearson's correlation coefficient by default. Our researcher predicted that (1) as anxiety increases, exam performance will decrease, and (2) as the time spent revising increases, exam performance will increase.

SAS Output 6.1 first provides descriptive statistics – mean, standard deviation, sum, minimum and maximum – and then provides a matrix of the correlation coefficients for the three

FIGURE 6.3
Karl Pearson

variables. Underneath each correlation coefficient the significance value of the correlation is displayed (if there were missing data, it would also show the sample size for each correlation – as there are no missing data, the sample size is always the same and it tells us in the title to the table). Each variable is perfectly correlated with itself (obviously) and so $r = 1$ along the diagonal of the table. Exam performance is negatively related to exam anxiety with a Pearson correlation coefficient of $r = -.441$ and the significance value is a less than .0001. This significance value tells us that the probability of getting a correlation coefficient this big in a sample of 103 people if the null hypothesis were true (there was no relationship between these variables) is very low (close to zero in fact). Hence, we can gain confidence that there is a genuine relationship between exam performance and anxiety. (Remember that our criterion for statistical significance is usually less than 0.05.) The output also shows that exam performance is positively related to the amount of time spent revising, with a coefficient of $r = .397$, which is also significant at $p < .0001$. Finally, exam anxiety appears to be negatively related to the time spent revising, $r = -.709$, $p < .0001$.

In psychological terms, this all means that people who have higher anxiety about an exam obtain a lower percentage mark in that exam. Conversely, people who spend more time revising obtain higher marks in the exam. Finally, people who spend more time revising, have lower anxiety about the exam. So there is a complex interrelationship between the three variables.

6.5.2.3. Using R^2 for interpretation ③

Although we cannot make direct conclusions about causality from a correlation, there is still more that it can tell us. The square of the correlation coefficient (known as the

SAS OUTPUT 6.1

Simple Statistics							
Variable	N	Mean	Std Dev	Sum	Minimum	Maximum	Label
REVISE	103	19.85437	18.15910	2045	0	98.00000	Time Spent Revising
EXAM	103	56.57282	25.94058	5827	2.00000	100.00000	Exam Performance (%)
ANXIETY	103	74.34367	17.18186	7657	0.05600	97.58200	Exam Anxiety

Pearson Correlation Coefficients, N = 103 Prob > |r| under H0: Rho=0			
	REVISE	EXAM	ANXIETY
REVISE Time Spent Revising	1.00000	0.39672 <.0001	−0.70925 <.0001
EXAM Exam Performance (%)	0.39672 <.0001	1.00000	−0.44099 <.0001
ANXIETY Exam Anxiety	−0.70925 <.0001	−0.44099 <.0001	1.00000

coefficient of determination, R^2) is a measure of the amount of variability in one variable that is shared by the other. For example, we may look at the relationship between exam anxiety and exam performance. Exam performances vary from person to person because of any number of factors (different ability, different levels of preparation and so on). If we add up all of this variability (rather like when we calculated the sum of squares in section 2.4.1) then we would have an estimate of how much variability exists in exam performances. We can then use R^2 to tell us how much of this variability is shared by exam anxiety. These two variables had a correlation of −0.4410 and so the value of R^2 will be $(-0.4410)^2 = 0.194$. This value tells us how much of the variability in exam performance is shared by exam anxiety.

If we convert this value into a percentage (multiply by 100) we can say that exam anxiety shares 19.4% of the variability in exam performance. So, although exam anxiety was highly correlated with exam performance, it can account for only 19.4% of variation in exam scores. To put this value into perspective, this leaves 80.6% of the variability still to be accounted for by other variables. I should note at this point that although R^2 is an extremely useful measure of the substantive importance of an effect, it cannot be used to infer causal relationships. Although we usually talk in terms of 'the variance in *y* *accounted for* by *x*', or even the variation in one variable *explained* by the other, this still says nothing about which way causality runs. So, although exam anxiety can account for 19.4% of the variation in exam scores, it does not necessarily cause this variation – it may be that people who do well feel anxious because they are under pressure to do well. If Andy released a CD tomorrow he would be under less pressure to sell one million copies than if Metallica released a CD tomorrow.

6.5.3. Spearman's correlation coefficient ①

Spearman's correlation coefficient (Spearman, 1910; Figure 6.4), r_s, is a non-parametric statistic and so can be used when the data have violated parametric assumptions such as non-normally distributed data (see Chapter 5). You'll sometimes hear the test referred to as Spearman's rho (pronounced 'row', as in 'row your boat gently down the stream'), which does make it difficult for some people to distinguish from the London lap-dancing club Spearmint Rhino.[3] Spearman's test works by first ranking the data (see section 15.3.1), and then applying Pearson's equation (6.3) to those ranks.

I was born in England, which has some bizarre traditions. One such oddity is the World's Biggest Liar Competition held annually at the Santon Bridge Inn in Wasdale (in the Lake District). The contest honours a local publican, 'Auld Will Ritson', who in the nineteenth century was famous in the area for his far-fetched stories (one such tale being that Wasdale turnips were big enough to be hollowed out and used as garden sheds). Each year locals are encouraged to attempt to tell the biggest lie in the world (lawyers and politicians are apparently banned from the competition). Over the years there have been tales of mermaid farms, giant moles, and farting sheep blowing holes in the ozone layer. (I am thinking of entering next year and reading out some sections of this book.)

Imagine I wanted to test a theory that more creative people will be able to create taller tales. I gathered together 68 past contestants from this competition and asked them where they were placed in the competition (first, second, third, etc.) and also gave them a creativity questionnaire (maximum score 60). The position in the competition is an ordinal variable (see section 1.5.1.2) because the places are categories but have a meaningful order (first place is better than second place and so on). Therefore, Spearman's correlation coefficient should be used (Pearson's r requires interval or ratio data). The data for this study are in the file **TheBiggestLiar.sas7bdat.** The data are in two columns: one labelled **Creativity** and one labelled **Position** (there's actually a third variable in there but we will ignore it for the time being). For the **Position** variable, each of the categories described above has been coded with a numerical value. First place has been coded with the value 1, with positions being labelled 2, 3 and so on. Note that for each numeric code I have provided a value label (just like we did for coding variables).

The procedure for doing a Spearman correlation is the same as for a Pearson correlation except on the options line we add the word SPEARMAN. This is shown in SAS Syntax 6.2.

```
PROC CORR data=chapter6.thebiggestliar SPEARMAN;
    VAR creativity position;
    RUN;
```
SAS Syntax 6.2

SAS Output 6.2 shows the output for a Spearman correlation on the variables **Creativity** and **Position.** The output is very similar to that of the Pearson correlation: a matrix is

[3]Seriously, a colleague of mine asked a student what analysis she was thinking of doing and she responded 'a Spearman's Rhino'.

FIGURE 6.4 Charles Spearman, ranking furiously

SAS OUTPUT 6.2

Simple Statistics							
Variable	N	Mean	Std Dev	Median	Minimum	Maximum	Label
CREATIVITY	68	39.98529	8.11759	39.00000	21.00000	56.00000	Creativity
POSITION	68	2.22059	1.38052	2.00000	1.00000	6.00000	Position in Best Liar Competition

Spearman Correlation Coefficients, N = 68 Prob > |r| under H0: Rho=0		
	CREATIVITY	POSITION
CREATIVITY Creativity	1.00000	–0.30022 0.0017
POSITION Position in Best Liar Competition	–0.37322 0.0017	1.00000

displayed giving the correlation coefficient between the two variables (–.373), underneath is the significance value of this coefficient (.0017). The significance value for this correlation coefficient is less than .05; therefore, it can be concluded that there is a significant relationship between creativity scores and how well someone did in the World's Biggest Liar Competition. Note that the relationship is negative: as creativity increased, position decreased. This might seem contrary to what we predicted until you remember that a low number means that you did well in the competition (a low number such as 1 means you came first, and a high number like 4 means you came fourth). Therefore, our hypothesis is supported: as creativity increased, so did success in the competition.

SELF-TEST Did creativity cause success in the World's Biggest Liar Competition?

6.5.4. Kendall's tau (non-parametric) ②

Kendall's tau, τ, is another non-parametric correlation and it should be used rather than Spearman's coefficient when you have a small data set with a large number of tied ranks. This means that if you rank all of the scores and many scores have the same rank, then Kendall's tau should be used. Although Spearman's statistic is the more popular of the two coefficients, there is much to suggest that Kendall's statistic is actually a better estimate of the correlation in the population (see Howell, 1997, p. 293). As such, we can draw more accurate generalizations from Kendall's statistic than from Spearman's. To carry out Kendall's correlation on the world's biggest liar data simply follow the same steps as for Pearson and Spearman correlations but write KENDALL on the PROC CORR line, instead of SPEARMAN. The output is much the same as for Spearman's correlation.

SAS OUTPUT 6.3

Kendall Tau b Correlation Coefficients, N = 68
Prob > |tau| under H0: Tau=0

	CREATIVITY	POSITION
CREATIVITY Creativity	1.00000	–0.30024 0.0013
POSITION Position in Best Liar Competition	–0.30024 0.0013	1.00000

You'll notice from SAS Output 6.3 that the actual value of the correlation coefficient is closer to zero than the Spearman correlation (it has increased from –.373 to –.300). Despite the difference in the correlation coefficients we can still interpret this result as being a highly significant relationship (because the significance value of .001 is less than .05). However, Kendall's value is a more accurate gauge of what the correlation in the population would be. As with the Pearson correlation, we cannot assume that creativity caused success in the World's Best Liar Competition.

SELF-TEST Conduct a Pearson correlation analysis of the advert data from the beginning of the chapter.

CRAMMING SAM'S TIPS

- We can measure the relationship between two variables using *correlation coefficients*.
- These coefficients lie between −1 and +1.
- *Pearson's correlation coefficient, r,* is a parametric statistic and requires interval data for both variables. To test its significance we assume normality too.
- *Spearman's correlation coefficient, r_s,* is a non-parametric statistic and requires only ordinal data for both variables.
- *Kendall's correlation coefficient, τ,* is like Spearman's r_s but probably better for small samples.

6.6. Partial correlation ②

6.6.1. The theory behind part and partial correlation ②

I mentioned earlier that there is a type of correlation that can be done that allows you to look at the relationship between two variables when the effects of a third variable are held constant. For example, analyses of the exam anxiety data (in the file **ExamAnxiety.sas 7bdat**) showed that exam performance was negatively related to exam anxiety, but positively related to revision time, and revision time itself was negatively related to exam anxiety. This scenario is complex, but given that we know that revision time is related to both exam anxiety and exam performance, then if we want a pure measure of the relationship between exam anxiety and exam performance we need to take account of the influence of revision time. Using the values of R^2 for these relationships, we know that exam anxiety accounts for 19.4% of the variance in exam performance, that revision time accounts for 15.7% of the variance in exam performance, and that revision time accounts for 50.2% of the variance in exam anxiety. If revision time accounts for half of the variance in exam anxiety, then it seems feasible that at least some of the 19.4% of variance in exam performance that is accounted for by anxiety is the same variance that is accounted for by revision time. As such, some of the variance in exam performance explained by exam anxiety is not *unique* and can be accounted for by revision time. A correlation between two variables in which the effects of other variables are held constant is known as a *partial correlation*.

Figure 6.7 illustrates the principle behind partial correlation. In part 1 of the diagram there is a box for exam performance that represents the total variation in exam scores (this value would be the variance of exam performance). There is also a box that represents the variation in exam anxiety (again, this is the variance of that variable). We know already that exam anxiety and exam performance share 19.4% of their variation (this value is the correlation coefficient squared). Therefore, the variations of these two variables overlap (because they share variance) creating a third box (the one with diagonal lines). The overlap of the boxes representing exam performance and exam anxiety is the common variance. Likewise, in part 2 of the diagram the shared variation between exam performance and revision time is illustrated. Revision time shares 15.7% of the variation in exam scores. This shared variation is represented by the area of overlap (filled with diagonal lines). We know that revision time and exam anxiety also share 50% of their variation; therefore, it is very probable that some of

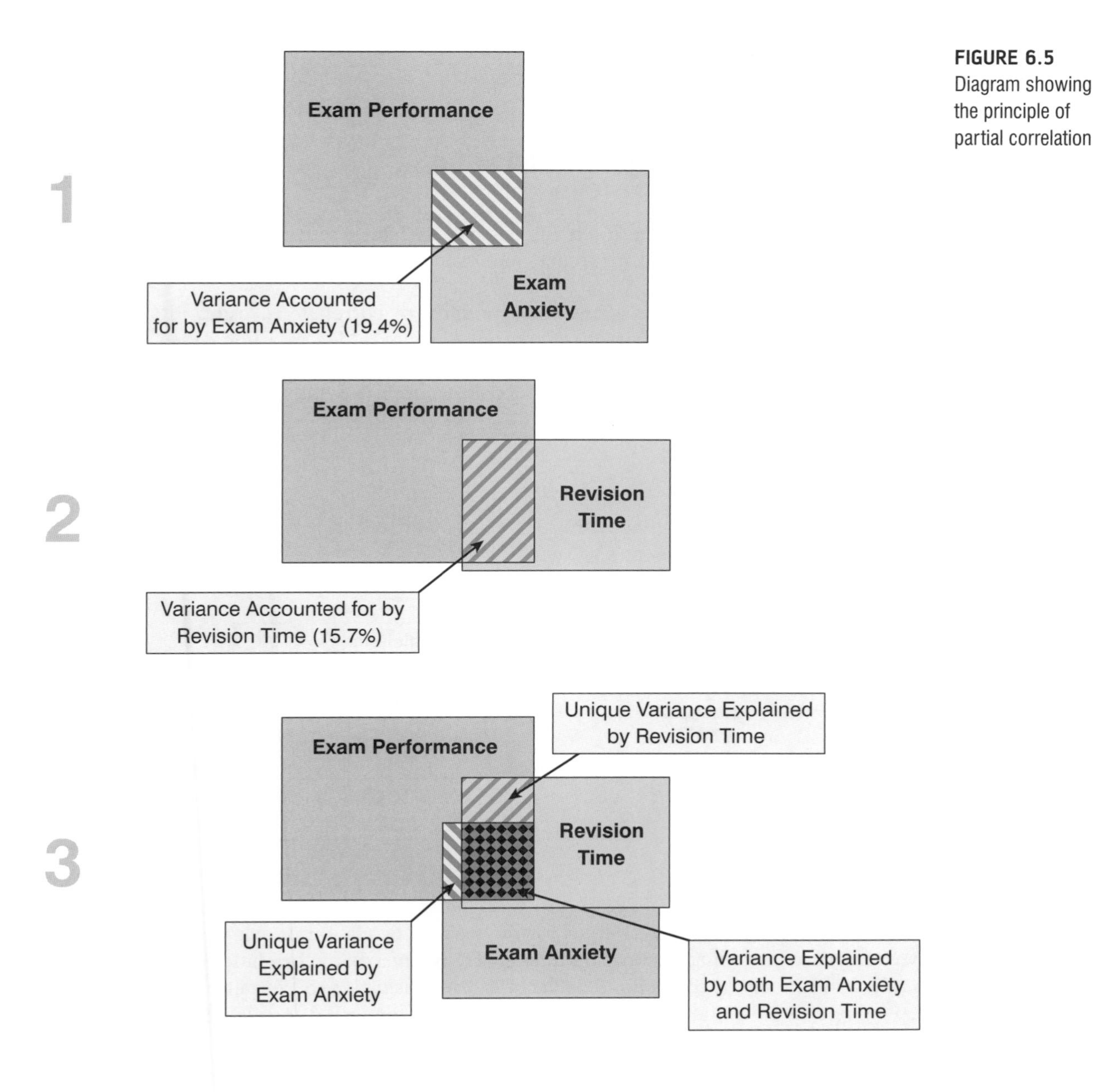

FIGURE 6.5 Diagram showing the principle of partial correlation

the variation in exam performance shared by exam anxiety is the same as the variance shared by revision time.

Part 3 of the diagram shows the complete picture. The first thing to note is that the boxes representing exam anxiety and revision time have a large overlap (this is because they share 50% of their variation). More important, when we look at how revision time and anxiety contribute to exam performance we see that there is a portion of exam performance that is shared by both anxiety and revision time (the dotted area). However, there are still small chunks of the variance in exam performance that are unique to the other two variables. So, although in part 1 exam anxiety shared a large chunk of variation in exam performance, some of this overlap is also shared by revision time. If we remove the portion of variation that is also shared by revision time, we get a measure of the unique relationship between exam performance and exam anxiety. We use partial correlations to find out the size of the unique portion of variance. Therefore, we could conduct a partial correlation between exam anxiety and exam performance while 'controlling' for the effect of revision time. Likewise, we could carry out a partial correlation between revision time and exam performance while 'controlling' for the effects of exam anxiety.

6.6.2. Partial correlation Using SAS ②

Let's use the **ExamAnxiety.sas7bdat** file so that, as I suggested above, we can conduct a partial correlation between exam anxiety and exam performance while 'controlling' for the effect of revision time.

We calculate partial correlations very similarly to correlations, using PROC CORR. Variables that we want to control for (partial out) are added in a line called (you might guess) PARTIAL. So to examine the relationship between test scores and test anxiety, controlling for revision time, we use the syntax shown in SAS Syntax 6.3.

```
PROC CORR data=chapter6.examanxiety;
  VAR exam anxiety;
  PARTIAL revise;
  RUN;
```

SAS Syntax 6.3

SAS OUTPUT 6.4

Pearson Partial Correlation Coefficients, N = 103 Prob > |r| under H0: Partial Rho=0		
	EXAM	ANXIETY
EXAM Exam Performance (%)	1.00000	–0.24667 0.0124
ANXIETY Exam Anxiety	–0.24667 0.0124	1.00000

SAS Output 6.4 shows the output for the partial correlation of exam anxiety and exam performance controlling for revision time. There is a matrix of correlations for the variables **anxiety** and **exam** but controlling for the effect of revision. In this instance we have controlled for one variable and so this is known as a first-order partial correlation. It is possible to control for the effects of two variables at the same time (a second-order partial correlation) or control three variables (a third-order partial correlation) and so on. First, notice that the partial correlation between exam performance and exam anxiety is −.247, which is considerably less than the correlation when the effect of revision time is not controlled for ($r = -.441$). In fact, the correlation coefficient is only about half what it was before. Although this correlation is still statistically significant (its *p*-value is still below .05), the relationship is diminished. In terms of variance, the value of R^2 for the partial correlation is .06, which means that exam anxiety can now account for only 6% of the variance in exam performance. When the effects of revision time were not controlled for, exam anxiety shared 19.4% of the variation in exam scores and so the inclusion of revision time has severely diminished the amount of variation in exam scores shared by anxiety. As such, a truer measure of the role of exam anxiety has been obtained. Running this analysis has shown us that exam anxiety alone does explain some of the variation in exam scores, but there is a complex relationship between anxiety, revision and exam performance that might otherwise have been ignored. Although causality is still not certain, because relevant variables are being included, the third-variable problem is, at least, being addressed in some form.

These partial correlations can be done when variables are dichotomous (including the 'third' variable). So, for example, we could look at the relationship between bladder relaxation (did the person wet themselves or not?) and the number of large tarantulas crawling up your leg, controlling for fear of spiders (the first variable is dichotomous, but the second variable and 'controlled for' variables are continuous). Also, to use an earlier example, we could examine the relationship between creativity and success in the World's Greater Liar Contest controlling for whether someone had previous experience in the competition (and therefore had some idea of the type of tale that would win) or not. In this latter case the 'controlled for' variable is dichotomous.[4]

6.6.3. Semi-partial (or part) correlations ②

In the next chapter, we will come across another form of correlation known as a **semi-partial correlation** (also referred to as a part correlation). While I'm babbling on about partial correlations it is worth my explaining the difference between this type of correlation and a semi-partial correlation. When we do a partial correlation between two variables, we control for the effects of a third variable. Specifically, the effect that the third variable has on *both* variables in the correlation is controlled. In a semi-partial correlation we control for the effect that the third variable has on only one of the variables in the correlation. Figure 6.6 illustrates this principle for the exam performance data. The partial correlation that we calculated took account not only of the effect of revision on exam performance, but also of the effect of revision on anxiety. If we were to calculate the semi-partial correlation for the same data, then this would control for only the effect of revision on exam performance (the effect of revision on exam anxiety is ignored). Partial correlations are most useful for looking at the unique relationship between two variables when other variables are ruled out. Semi-partial correlations are, therefore, useful when trying to explain the variance in one particular

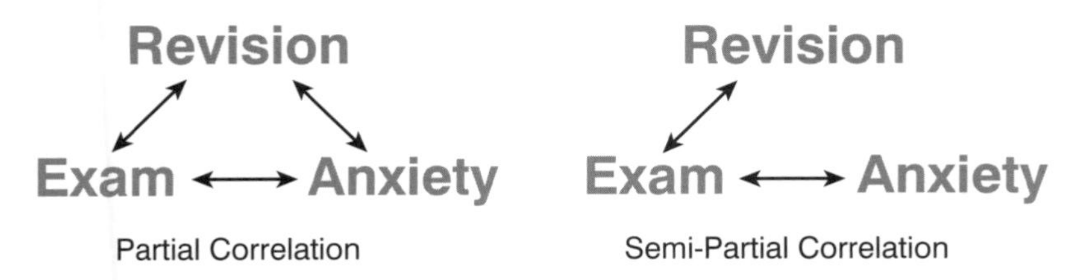

FIGURE 6.6 The difference between a partial and a semi-partial correlation

[4] Both these examples are, in fact, simple cases of hierarchical regression (see the next chapter) and the first example is also an example of analysis of covariance. This may be confusing now, but as we progress through the book I hope it'll become clearer that virtually all of the statistics that you use are actually the same things dressed up in different names.

CRAMMING SAM'S TIPS

- A *partial correlation* quantifies the relationship between two variables while controlling for the effects of a third variable on *both* variables in the original correlation.
- A *semi-partial correlation* quantifies the relationship between two variables while controlling for the effects of a third variable on only *one* of the variables in the original correlation.

variable (an outcome) from a set of predictor variables. (Bear this in mind when you read Chapter 7.)

6.7. Comparing correlations ③

6.7.1. Comparing independent *r*s ③

Sometimes we want to know whether one correlation coefficient is bigger than another. For example, when we looked at the effect of exam anxiety on exam performance, we might have been interested to know whether this correlation was different in men and women. We could compute the correlation in these two samples, but then how would we assess whether the difference was meaningful?

SELF-TEST Use the BY command to compute the correlation coefficient between exam anxiety and exam performance in men and women. (Remember to sort by gender with PROC SORT, and then add: BY gender; to the PROC CORR command.

If we did this, we would find that the correlations were $r_{Male} = -.506$ and $r_{Female} = -.381$. These two samples are independent; that is, they contain different entities. To compare these correlations we can again use what we discovered in section 6.3.3 to convert these coefficients to z_r (just to remind you, we do this because it makes the sampling distribution normal and we know the standard error). If you do the conversion, then we get z_r (males) = $-.557$ and z_r (females) = $-.401$. We can calculate a z-score of the differences between these correlations as:

$$z_{Difference} = \frac{z_{r_1} - z_{r_2}}{\sqrt{\frac{1}{N_1 - 3} + \frac{1}{N_2 + 3}}} \qquad (6.9)$$

We had 52 men and 51 women so we would get:

$$z_{Difference} = \frac{-.557 - (-.401)}{\sqrt{\frac{1}{49} + \frac{1}{48}}} = \frac{-.156}{0.203} = -0.768$$

We can look up this value of z (0.768, we can ignore the minus sign) in the table for the normal distribution in the Appendix and get the one-tailed probability from the column labelled 'Smaller Portion'. In this case the value is .221. To get the two-tailed probability we simply multiply the one-tailed probability value by 2, which gives us .442. As such the correlation between exam anxiety and exam performance is not significantly different in men and women.

6.7.2. Comparing dependent *r*s ②

If you want to compare correlation coefficients that come from the same entities then things are a little more complicated. You can use a *t*-statistic to test whether a difference between two dependent correlations from the same sample is significant. For example, in our exam anxiety data we might want to see whether the relationship between exam anxiety (*x*) and exam performance (*y*) is stronger than the relationship between revision (*z*) and exam performance. To calculate this, all we need are the three *r*s that quantify the relationships between these variables: r_{xy}, the relationship between exam anxiety and exam performance (–.441); r_{zy}, the relationship between revision and exam performance (.397); and r_{xz}, the relationship between exam anxiety and revision (–.709). The *t*-statistic is computed as (Chen & Popovich, 2002):

$$t_{Difference} = (r_{xy} - r_{zy})\sqrt{\frac{(n-3)(1+r_{xz})}{2\left(1 - r_{xy}^2 - r_{xz}^2 - r_{zy}^2 + 2r_{xy}r_{xz}r_{zy}\right)}} \quad (6.10)$$

Admittedly that equation looks hideous, but really it's not too bad: it just uses the three correlation coefficients and the sample size *N*. Place the numbers from the exam anxiety example in it (*N* was 103) and you should end up with:

$$t_{Difference} = (-.838)\sqrt{\frac{29.1}{2(1 - .194 - .503 - .158 + 0.248}} = -5.09$$

This value can be checked against the appropriate critical value in the Appendix with $N - 3$ degrees of freedom (in this case 100). The critical values in the table are 1.98 ($p < .05$) and 2.63 ($p < .01$), two-tailed. As such we can say that the correlation between exam anxiety and exam performance was significantly higher (more towards the positive) than the correlation between revision time and exam performance (this isn't a massive surprise given that these relationships went in the opposite directions to each other).

6.8. Calculating the effect size ①

Calculating effect sizes for correlation coefficients couldn't be easier because, as we saw earlier in the book, correlation coefficients *are* effect sizes! So, no calculations (other than those you have already done) necessary! However, I do want to point out one caveat when using non-parametric correlation coefficients as effect sizes. Although the Spearman and Kendall correlations are comparable in many respects (their power, for example, is similar under parametric conditions), there are two important differences (Strahan, 1982).

First, we saw for Pearson's r that we can square this value to get the proportion of shared variance, R^2. For Spearman's r_s we can do this too because it uses the same equation as Pearson's r. However, the resulting R_s^2 needs to be interpreted slightly differently: it is the proportion of variance in the *ranks* that two variables share. Having said this, R_s^2 is usually a good approximation of R^2 (especially in conditions of near-normal distributions). Kendall's τ, however, is not numerically similar to either r or r_s and so τ^2 does not tell us about the proportion of variance shared by two variables (or the ranks of those two variables).

Second, Kendall's τ is 66–75% smaller than both Spearman's r_s and Pearson's r, but r and r_s are generally similar sizes (Strahan, 1982). As such, if τ is used as an effect size it should be borne in mind that it is not comparable to r and r_s and should not be squared. More generally, when using correlations as effect sizes you should remember (both when reporting your own analysis and when interpreting others) that the choice of correlation coefficient can make a substantial difference to the apparent size of the effect.

6.9. How to report correlation coefficents ①

Reporting correlation coefficients is pretty easy: you just have to say how big they are and what their significance value was (although the significance value isn't *that* important because the correlation coefficient is an effect size in its own right!). Four things to note are that: (1) coefficients are reported to 2 decimal places; (2) if you are quoting a one-tailed probability, you should say so; (3) each correlation coefficient is represented by a different letter (and some of them are Greek!); and (4) there are standard criteria of probabilities that we use (.05, .01 and .001). Let's take a few examples from this chapter:

- ✓ There was a significant relationship between the number of adverts watched and the number of packets of sweets purchased, $r = .87$, p (one-tailed) $< .05$.
- ✓ Exam performance was significantly correlated with exam anxiety, $r = -.44$, and time spent revising, $r = .40$; the time spent revising was also correlated with exam anxiety, $r = -.71$ (all ps $< .001$).
- ✓ Creativity was significantly related to how well people did in the World's Biggest Liar Competition, $r_s = -.37$, $p < .001$.

✓ Creativity was significantly related to how well people did in the World's Biggest Liar Competition, $\tau = -.30$, $p < .001$. (Note that I've quoted Kendall's τ here.)

Scientists, rightly or wrongly, tend to use several *standard* levels of statistical significance. Primarily, the most important criterion is that the significance value is less than .05; however, if the exact significance value is much lower then we can be much more confident about the strength of the experimental effect. In these circumstances we like to make a big song and dance about the fact that our result isn't just significant at .05, but is significant at a much lower level as well (hooray!). The values we use are .05, .01, and .001.

TABLE 6.2 An example of reporting a table of correlations

	Exam Performance	Exam Anxiety	Revision Time
Exam Performance	1	−.44***	.40***
Exam Anxiety	103	1	−.71***
Revision Time	103	103	1

Ns = not significant ($p > .05$), $*p < .05$, $** p < .01$, $*** p < .001$

When we have lots of correlations we sometimes put them into a table. For example, our exam anxiety correlations could be reported as in Table 6.2. Note that above the diagonal I have reported the correlation coefficients and used symbols to represent different levels of significance. Under the table there is a legend to tell readers what symbols represent. (Actually, none of the correlations were non-significant or had p bigger than .001 so most of these are here simply to give you a reference point – you would normally include symbols that you had actually used in the table in your legend.) Finally, in the lower part of the table I have reported the sample sizes. These are all the same (103) but sometimes when you have missing data it is useful to report the sample sizes in this way because different values of the correlation will be based on different sample sizes. For some more ideas on how to report correlations have a look at Labcoat Leni's Real Research 6.1.

LABCOAT LENI'S REAL RESEARCH 6.1

Why do you like your lecturers? ①

As students you probably have to rate your lecturers at the end of the course. There will be some lecturers you like and others that you hate. As a lecturer I find this process horribly depressing (although this has a lot to do with the fact that I tend focus on negative feedback and ignore the good stuff). There is some evidence that students tend to pick courses of lecturers who they perceive to be enthusastic and good communicators. In a fascinating study, Tomas Chamorro-Premuzic and his colleagues (Chamorro-Premuzic, Furnham, Christopher, Garwood, & Martin, 2008) tested a slightly different hypothesis, which was that students tend to like lecturers who are like themselves. (This hypothesis will have the students on my course who like my lectures screaming in horror.)

First of all the authors measured students' own personalities using a very well-established measure (the NEO-FFI) which gives rise to scores on five fundamental personality traits: Neuroticism, Extroversion, Openness to experience, Agreeableness and Conscientiousness. They also gave students a questionnaire that asked them to rate how much they wanted their lecturer to have each of a list of characteristics. For example, they would be given the description 'warm: friendly, warm, sociable, cheerful, affectionate, outgoing' and asked to rate how much they wanted to see this in a lecturer from −5 (they don't want this characteristic at all) through 0 (the characteristic is not important) to +5 (I really want this characteristic in my lecturer). The characteristics on the questionnaire all related to personality characteristics measured by the NEO-FFI. As such, the authors had a measure of how much a student had each of the five core personality characteristics, but also a measure of how much they wanted to see those same characteristics in their lecturer.

In doing so, Tomas and his colleagues could test whether, for instance, extroverted students want extrovert lecturers. The data from this study (well, for the variables that I've mentioned) are in the file **Chamorro Premuzic.sas7bdat**. Run some Pearson correlations on these variables to see if students with certain personality characteristics want to see those characteristics in their lecturers. What conclusions can you draw?

Answers are in the additional material on the companion website (or look at Table 3 in the original article, which will also show you how to report a large number of correlations).

CHAMORRO-PREMUZIC, T., ET AL. (2008). *PERSONALITY AND INDIVIDUAL DIFFERENCES, 44*, 965–976.

What have I discovered about statistics? ①

This chapter has looked at ways to study relationships between variables. We began by looking at how we might measure relationships statistically by developing what we already know about variance (from Chapter 1) to look at variance shared between variables. This shared variance is known as *covariance*. We then discovered that when data are parametric we can measure the strength of a relationship using Pearson's correlation coefficient, r. When data violate the assumptions of parametric tests we can use Spearman's r_s, or for small data sets Kendall's τ may be more accurate. We also saw that correlations can be calculated between two variables when one of those variables is a dichotomy (i.e. composed of two categories). Finally, we looked at the difference between *partial correlations*, in which the relationship between two variables is measured controlling for the effect that one or more variables has on both of those variables, and *semi-partial correlations*, in which the relationship between two variables is measured controlling for the effect that one or more variables has on only one of those variables. We also discovered that I had a guitar and, like my favourite record of the time, I was ready to 'Take on the world'. Well, Wales at any rate …

Key terms that I've discovered

Bivariate correlation
Coefficient of determination
Covariance
Cross-product deviations
Kendall's tau
Partial correlation
Pearson correlation coefficient
Semi-partial correlation
Spearman's correlation coefficient
Standardization

Smart Alex's tasks

- **Task 1**: A student was interested in whether there was a positive relationship between the time spent doing an essay and the mark received. He got 45 of his friends and timed how long they spent writing an essay (**hours**) and the percentage they got in the essay (**essay**). He also translated these grades into their degree classifications (**grade**): first, upper second, lower second and third class. Using the data in the file **EssayMarks.sas7bdat** find out what the relationship was between the time spent doing an essay and the eventual mark in terms of percentage and degree class (draw a scatterplot too!). ①
- **Task 2**: Using the **ChickFlick.sas7bdat.** data from Chapter 3, is there a relationship between gender and arousal? Using the same data, is there a relationship between the film watched and arousal? ①
- **Task 3**: As a statistics lecturer I am always interested in the factors that determine whether a student will do well on a statistics course. One potentially important factor is their previous expertise with mathematics. Imagine I took 25 students and looked at their degree grades for my statistics course at the end of their first year at university. In the UK, a student can get a first-class mark (the best), an upper-second-class mark, a lower second, a third, a pass or a fail (the worst). I also asked these students what grade they got in their GCSE maths exams. In the UK GCSEs are school exams taken at age 16 that are graded A, B, C, D, E or F (an A grade is better than all of the lower grades). The data for this study are in the file **grades.sas7bdat.** Carry out the appropriate analysis to see if GCSE maths grades correlate with first-year statistics grades. ①

Answers can be found on the companion website.

Further reading

Chen, P. Y., & Popovich, P. M. (2002). *Correlation: Parametric and nonparametric measures*. Thousand Oaks, CA: Sage.
Howell, D. C. (2006). *Statistical methods for psychology* (6th ed.). Belmont, CA: Duxbury. (Or you might prefer his *Fundamental statistics for the behavioral sciences,* also in its 6th edition, 2007. Both are excellent texts that are a bit more technical than this book so they are a useful next step.)
Miles, J. N. V., & Banyard, P. (2007). *Understanding and using statistics in psychology: a practical introduction*. London: Sage. (A fantastic and amusing introduction to statistical theory.)

Wright, D. B., & London, K. (2009). *First steps in statistics* (2nd ed.). London: Sage. (This book is a very gentle introduction to statistical theory.)

Interesting real research

Chamorro-Premuzic, T., Furnham, A., Christopher, A. N., Garwood, J., & Martin, N. (2008). Birds of a feather: Students' preferences for lecturers' personalities as predicted by their own personality and learning approaches. *Personality and Individual Differences*, *44*, 965–976.

Regression 7

FIGURE 7.1 Me playing with my ding-a-ling in the Holimarine Talent Show. Note the groupies queuing up at the front

7.1. What will this chapter tell me? ①

Although none of us can know the future, predicting it is so important that organisms are hard wired to learn about predictable events in their environment. We saw in the previous chapter that I received a guitar for Christmas when I was 8. My first foray into public performance was a weekly talent show at a holiday camp called 'Holimarine' in Wales (it doesn't exist anymore because I am old and this was 1981). I sang a Chuck Berry song called 'My ding-a-ling'[1] and to my absolute amazement I won the competition.[2] Suddenly other 8 year olds across the land (well, a ballroom in Wales) worshipped me (I made lots of friends after the competition). I had tasted success, it tasted like praline chocolate, and so I wanted to enter the competition in the second week of our holiday. To ensure success, I needed to know why I had won in the first week. One way to do this would have been to collect data and to use these data to predict people's evaluations of children's performances in the contest

[1] It appears that even then I had a passion for lowering the tone of things that should be taken seriously.

[2] I have a very grainy video of this performance recorded by my dad's friend on a video camera the size of a medium-sized dog that had to be accompanied at all times by a 'battery pack' the size and weight of a tank. Maybe I'll put it up on the companion website …

from certain variables: the age of the performer, what type of performance they gave (singing, telling a joke, magic tricks), and maybe how cute they looked. A regression analysis on these data would enable us to predict future evaluations (success in next week's competition) based on values of the predictor variables. If, for example, singing was an important factor in getting a good audience evaluation, then I could sing again the following week; however, if jokers tended to do better then I could switch to a comedy routine. When I was 8 I wasn't the sad geek that I am today, so I didn't know about regression analysis (nor did I wish to know); however, my dad thought that success was due to the winning combination of a cherub-looking 8 year old singing songs that can be interpreted in a filthy way. He wrote me a song to sing in the competition about the keyboard player in the Holimarine Band 'messing about with his organ', and first place was mine again. There's no accounting for taste.

7.2. An introduction to regression ①

In the previous chapter we looked at how to measure relationships between two variables. These correlations can be very useful but we can take this process a step further and predict one variable from another. A simple example might be to try to predict levels of stress from the amount of time until you have to give a talk. You'd expect this to be a negative relationship (the smaller the amount of time until the talk, the larger the anxiety). We could then extend this basic relationship to answer a question such as 'if there's 10 minutes to go until someone has to give a talk, how anxious will they be?' This is the essence of regression analysis: we fit a model to our data and use it to predict values of the dependent variable from one or more independent variables.[3] Regression analysis is a way of predicting an **outcome variable** from one **predictor variable** (**simple regression**) or several predictor variables (**multiple regression**). This tool is incredibly useful because it allows us to go a step beyond the data that we collected.

In section 2.4.3 I introduced you to the idea that we can predict any data using the following general equation:

$$\text{outcome}_i = (\text{model}) + \text{error}_i \tag{7.1}$$

This just means that the outcome we're interested in for a particular person can be predicted by whatever model we fit to the data plus some kind of error. In regression, the model we fit is linear, which means that we summarize a data set with a straight line (think back to Jane Superbrain Box 2.1). As such, the word 'model' in the equation above simply gets replaced by 'things that define the line that we fit to the data' (see the next section).

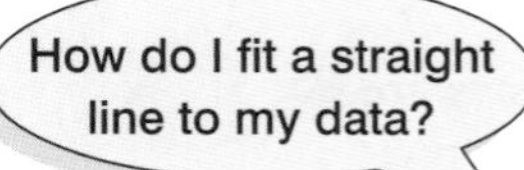

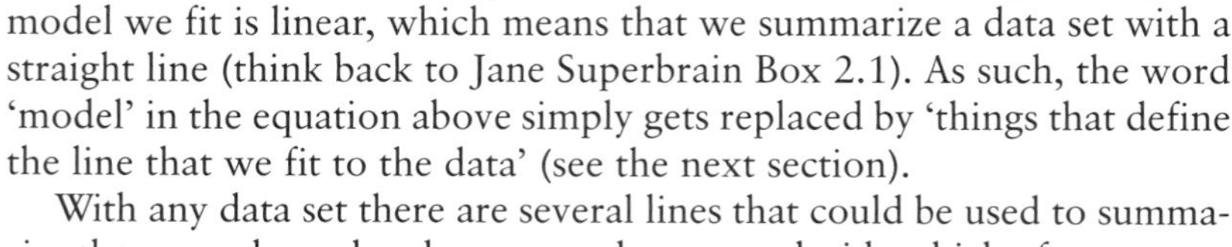

With any data set there are several lines that could be used to summarize the general trend and so we need a way to decide which of many possible lines to chose. For the sake of making accurate predictions we want to fit a model that *best* describes the data. The simplest way to do this would be to use your eye to gauge a line that looks as though it summarizes the data well. You don't need to be a genius to realize that the 'eyeball' method is very subjective and so offers no assurance that the model is the best one that could have been chosen. Instead, we use a mathematical technique called the *method of least squares* to establish the line that best describes the data collected.

[3] I want to remind you here of something I discussed in Chapter 1: SAS refers to regression variables as dependent and independent variables (as in controlled experiments). However, correlational research by its nature seldom controls the independent variables to measure the effect on a dependent variable and so I will talk about 'independent variables' as *predictors*, and the 'dependent variable' as the *outcome*.

7.2.1. Some important information about straight lines ①

I mentioned above that in our general equation the word 'model' gets replaced by 'things that define the line that we fit to the data'. In fact, any straight line can be defined by two things: (1) the slope (or gradient) of the line (usually denoted by b_1); and (2) the point at which the line crosses the vertical axis of the graph (known as the *intercept* of the line, b_0). In fact, our general model becomes equation (7.2) below in which Y_i is the outcome that we want to predict and X_i is the *i*th participant's score on the predictor variable.[4] Here b_1 is the gradient of the straight line fitted to the data and b_0 is the intercept of that line. These parameters b_1 and b_0 are known as the *regression coefficients* and will crop up time and time again in this book, where you may see them referred to generally as *b* (without any subscript) or $\mathbf{b}_i$ (meaning the *b* associated with variable *i*). There is a residual term, ε_i, which represents the difference between the score predicted by the line for participant *i* and the score that participant *i* actually obtained. The equation is often conceptualized without this residual term (so ignore it if it's upsetting you); however, it is worth knowing that this term represents the fact that our model will not fit the data collected perfectly:

$$Y_i = (b_0 + b_1 X_i) + \varepsilon_i \tag{7.2}$$

A particular line has a specific intercept and gradient. Figure 7.2 shows a set of lines that have the same intercept but different gradients, and a set of lines that have the same gradient but different intercepts. Figure 7.2 also illustrates another useful point: the gradient of the line tells us something about the nature of the relationship being described. In Chapter 6 we saw how relationships can be either positive or negative (and I don't mean the difference between getting on well with your girlfriend and arguing all the time!). A line that has a gradient with a positive value describes a positive relationship, whereas a line with a negative gradient describes a negative relationship. So, if you look at the graph in Figure 7.2

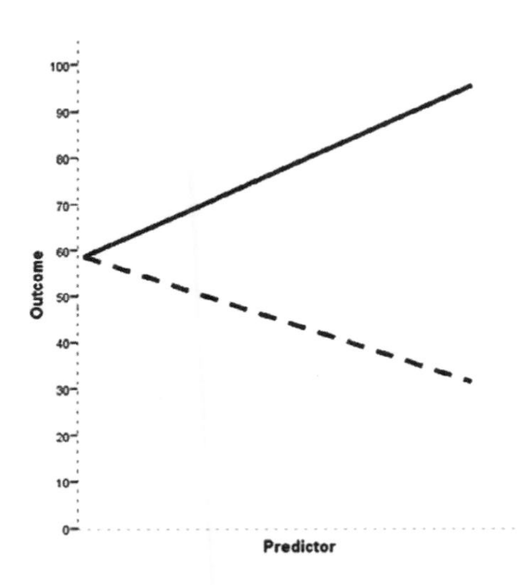

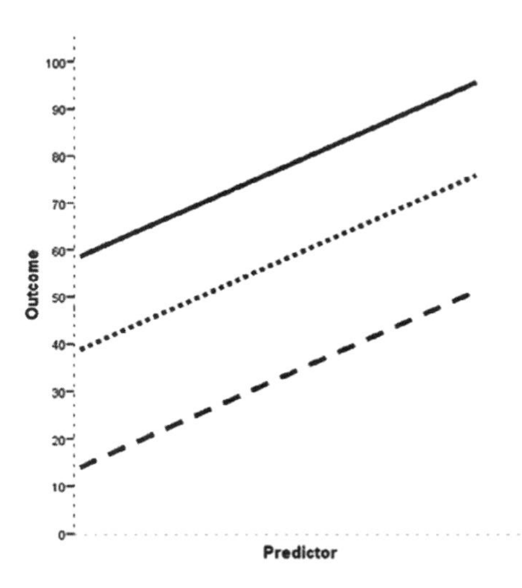

FIGURE 7.2 Lines with the same gradients but different intercepts, and lines that share the same intercept but have different gradients

[4] You'll sometimes see this equation written as:

$$Y_i = (\beta_0 + \beta_1 X_i) + \varepsilon_i$$

The only difference is that this equation has got *β*s in it instead of *b*s and in fact both versions are the same thing, they just use different letters to represent the coefficients.

in which the gradients differ but the intercepts are the same, then the dashed line describes a positive relationship whereas the solid line describes a negative relationship. Basically, then, the gradient (b_1) tells us what the model looks like (its shape) and the intercept (b_0) tells us where the model is (its location in geometric space).

If it is possible to describe a line knowing only the gradient and the intercept of that line, then we can use these values to describe our model (because in linear regression the model we use is a straight line). So, the model that we fit to our data in linear regression can be conceptualized as a straight line that can be described mathematically by equation (7.2). With regression we strive to find the line that best describes the data collected, then estimate the gradient and intercept of that line. Having defined these values, we can insert different values of our predictor variable into the model to estimate the value of the outcome variable.

7.2.2. The method of least squares ①

I have already mentioned that the method of least squares is a way of finding the line that best fits the data (i.e. finding a line that goes through, or as close to, as many of the data points as possible). This 'line of best fit' is found by ascertaining which line, of all of the possible lines that could be drawn, results in the least amount of difference between the observed data points and the line. Figure 7.3 shows that when any line is fitted to a set of data, there will be small differences between the values predicted by the line and the data that were actually observed.

Back in Chapter 2 we saw that we could assess the fit of a model (the example we used was the mean) by looking at the deviations between the model and the actual data collected. These deviations were the vertical distances between what the model predicted and each data point that was actually observed. We can do exactly the same to assess the fit of a regression line (which, like the mean, is a statistical model). So, again we are interested in the vertical differences between the line and the actual data because the line is our model: we use it to predict values of Y from values of the X variable. In regression these differences are usually called *residuals* rather than deviations, but they are the same thing. As with the mean, data points fall both above (the model underestimates their value) and

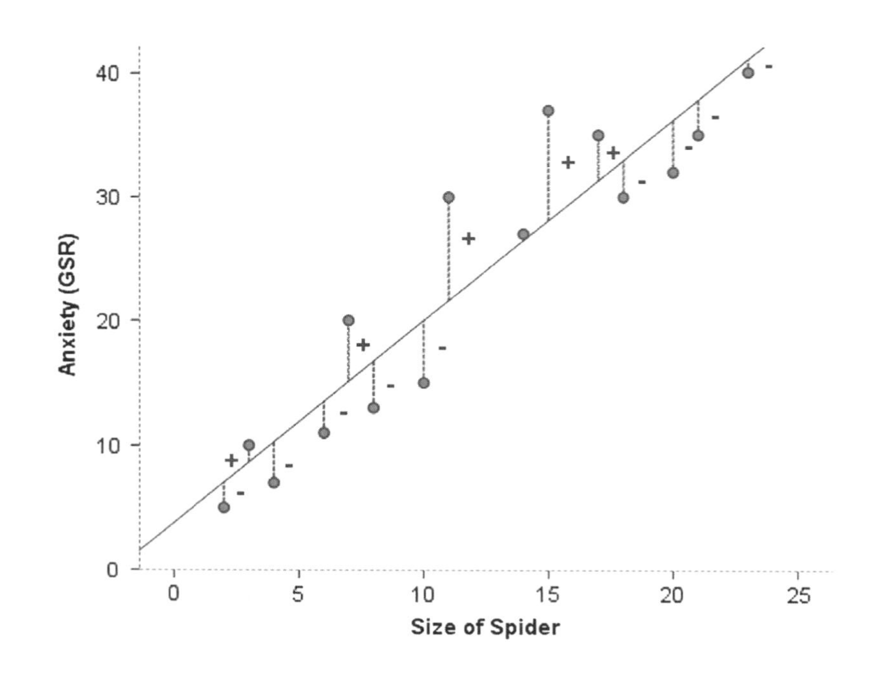

FIGURE 7.3 This graph shows a scatterplot of some data with a line representing the general trend. The vertical lines (dotted) represent the differences (or residuals) between the line and the actual data

below (the model overestimates their value) the line, yielding both positive and negative differences. In the discussion of variance in section 2.4.2 I explained that if we sum positive and negative differences then they cancel each other out and that to circumvent this problem we square the differences before adding them up. We do the same thing here. The resulting squared differences provide a gauge of how well a particular line fits the data: if the squared differences are large, the line is not representative of the data; if the squared differences are small, the line is representative.

You could, if you were particularly bored, calculate the sum of squared differences (or SS for short) for every possible line that is fitted to your data and then compare these 'goodness-of-fit' measures. The one with the lowest SS is the line of best fit. Fortunately we don't have to do this because the method of least squares does it for us: it selects the line that has the lowest sum of squared differences (i.e. the line that best represents the observed data). How exactly it does this is by using a mathematical technique for finding maxima and minima and this technique is used to find the line that minimizes the sum of squared differences. I don't really know much more about it than that, to be honest, so I tend to think of the process as a little bearded wizard called Nephwick the Line Finder who just magically finds lines of best fit. Yes, he lives inside your computer. The end result is that Nephwick estimates the value of the slope and intercept of the 'line of best fit' for you. We tend to call this line of best fit a *regression line*.

7.2.3. Assessing the goodness of fit: sums of squares, R and R^2 ①

Once Nephwick the Line Finder has found the line of best fit it is important that we assess how well this line fits the actual data (we assess the **goodness of fit** of the model). We do this because even though this line is the best one available, it can still be a lousy fit to the data! In section 2.4.2 we saw that one measure of the adequacy of a model is the sum of squared differences (or more generally we assess models using equation (7.3) below). If we want to assess the line of best fit, we need to compare it against something, and the thing we choose is the most basic model we can find. So we use equation (7.3) to calculate the fit of the most basic model, and then the fit of the best model (the line of best fit), and basically if the best model is any good then it should fit the data significantly better than our basic model:

$$\text{deviation} = \sum(\text{observed} - \text{model})^2 \tag{7.3}$$

This is all quite abstract so let's look at an example. Imagine that I was interested in predicting record sales (*Y*) from the amount of money spent advertising that record (*X*). One day my boss came into my office and said 'Andy, I know you wanted to be a rock star and you've ended up working as my stats-monkey, but how many records will we sell if we spend £100,000 on advertising?' If I didn't have an accurate model of the relationship between record sales and advertising, what would my best guess be? Well, probably the best answer I could give would be the mean number of record sales (say, 200,000) because on average that's how many records we expect to sell. This response might well satisfy a brainless record company executive (who didn't offer my band a record contract). However, what if he had asked 'How many records will we sell if we spend £1 on advertising?' Again, in the absence of any accurate information, my best guess would be to give the average number of sales (200,000). There is a problem: whatever amount of money is spent on advertising I

always predict the same level of sales. As such, the mean is a model of 'no relationship' at all between the variables. It should be pretty clear then that the mean is fairly useless as a model of a relationship between two variables – but it is the simplest model available.

So, as a basic strategy for predicting the outcome, we might choose to use the mean, because on average it will be a fairly good guess of an outcome, but that's all. Using the mean as a model, we can calculate the difference between the observed values, and the values predicted by the mean (equation (7.3)). We saw in section 2.4.1 that we square all of these differences to give us the sum of squared differences. This sum of squared differences is known as the **total sum of squares** (denoted SS_T) because it is the total amount of differences present when the most basic model is applied to the data. This value represents how good the mean is as a model of the observed data. Now, if we fit the more sophisticated model to the data, such as a line of best fit, we can again work out the differences between this new model and the observed data (again using equation (7.3)). In the previous section we saw that the method of least squares finds the best possible line to describe a set of data by minimizing the difference between the model fitted to the data and the data themselves. However, even with this optimal model there is still some inaccuracy, which is represented by the differences between each observed data point and the value predicted by the regression line. As before, these differences are squared before they are added up so that the directions of the differences do not cancel out. The result is known as the **sum of squared residuals** or **residual sum of squares** (SS_R). This value represents the degree of inaccuracy when the best model is fitted to the data. We can use these two values to calculate how much better the regression line (the line of best fit) is than just using the mean as a model (i.e. how much better is the best possible model than the worst model?). The improvement in prediction resulting from using the regression model rather than the mean is calculated by calculating the difference between SS_T and SS_R. This difference shows us the reduction in the inaccuracy of the model resulting from fitting the regression model to the data. This improvement is the *model sum of squares* (SS_M). Figure 7.4 shows each sum of squares graphically.

If the value of SS_M is large then the regression model is very different from using the mean to predict the outcome variable. This implies that the regression model has made a big improvement to how well the outcome variable can be predicted. However, if SS_M is small then using the regression model is little better than using the mean (i.e. the regression model is no better than taking our 'best guess'). A useful measure arising from these sums of squares is the proportion of improvement due to the model. This is easily calculated by dividing the sum of squares for the model by the total sum of squares. The resulting value is called R^2 and to express this value as a percentage you should multiply it by 100. R^2 represents the amount of variance in the outcome explained by the model (SS_M) relative to how much variation there was to explain in the first place (SS_T). Therefore, as a percentage, it represents the percentage of the variation in the outcome that can be explained by the model:

$$R^2 = \frac{SS_M}{SS_T} \tag{7.4}$$

This R^2 is the same as the one we met in Chapter 6 (section 6.5.2.3) and you might have noticed that it is interpreted in the same way. Therefore, in simple regression we can take the square root of this value to obtain Pearson's correlation coefficient. As such, the correlation coefficient provides us with a good estimate of the overall fit of the regression model, and R^2 provides us with a good gauge of the substantive size of the relationship.

A second use of the sums of squares in assessing the model is through the *F*-test. I mentioned way back in Chapter 2 that test statistics (like *F*) are usually the amount of systematic variance divided by the amount of unsystematic variance, or, put another way, the model compared against the error in the model. This is true here: *F* is based upon the ratio of the improvement due to the model (SS_M) and the difference between the model and the observed data (SS_R). Actually, because the sums of squares depend on the number

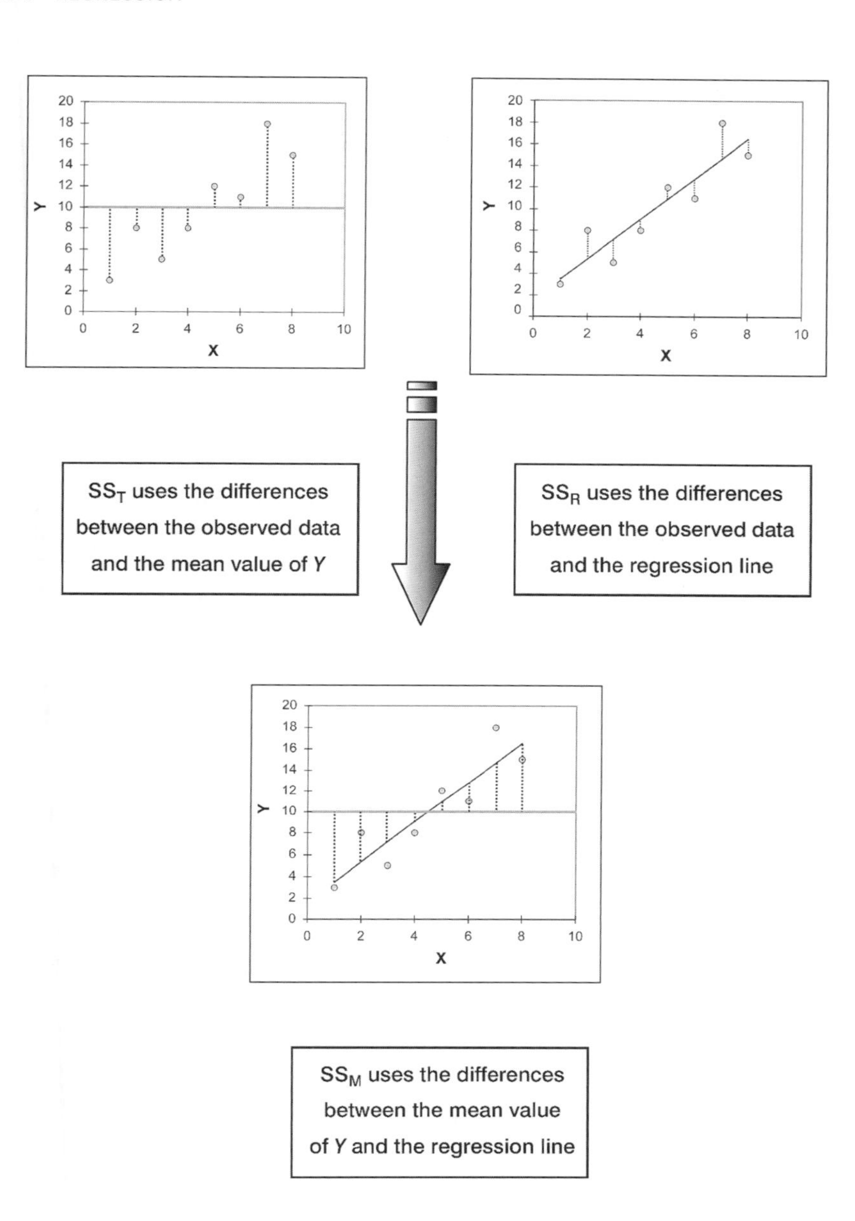

FIGURE 7.4 Diagram showing from where the regression sums of squares derive

of differences that we have added up, we use the average sums of squares (referred to as the **mean squares** or MS). To work out the mean sums of squares we divide by the degrees of freedom (this is comparable to calculating the variance from the sums of squares – see section 2.4.2). For SS_M the degrees of freedom are simply the number of variables in the model, and for SS_R they are the number of observations minus the number of parameters being estimated (i.e. the number of beta coefficients including the constant). The result is the mean squares for the model (MS_M) and the residual mean squares (MS_R). At this stage it isn't essential that you understand how the mean squares are derived (it is explained in Chapter 10). However, it is important that you understand that the ***F*-ratio** (equation (7.5)) is a measure of how much the model has improved the prediction of the outcome compared to the level of inaccuracy of the model:

$$F = \frac{MS_M}{MS_R} \tag{7.5}$$

If a model is good, then we expect the improvement in prediction due to the model to be large (so MS_M will be large) and the difference between the model and the observed data to be small (so MS_R will be small). In short, a good model should have a large *F*-ratio (greater than 1 at least) because the top of equation (7.5) will be bigger than the bottom. The exact magnitude of this *F*-ratio can be assessed using critical values for the corresponding degrees of freedom (as in the Appendix).

7.2.4. Assessing individual predictors ①

We've seen that the predictor in a regression model has a coefficient (b_1), which in simple regression represents the gradient of the regression line. The value of *b* represents the change in the outcome resulting from a unit change in the predictor. If the model was useless at predicting the outcome, then if the value of the predictor changes, what might we expect the change in the outcome to be? Well, if the model is very bad then we would expect the change in the outcome to be zero. Think back to Figure 7.4 (see the panel representing SS_T) in which we saw that using the mean was a very bad way of predicting the outcome. In fact, the line representing the mean is flat, which means that as the predictor variable changes, the value of the outcome does *not* change (because for each level of the predictor variable, we predict that the outcome will equal the mean value). The important point here is that a bad model (such as the mean) will have regression coefficients of 0 for the predictors. A regression coefficient of 0 means: (1) a unit change in the predictor variable results in no change in the predicted value of the outcome (the predicted value of the outcome does not change at all); and (2) the gradient of the regression line is 0, meaning that the regression line is flat. Hopefully, you'll see that it logically follows that if a variable significantly predicts an outcome, then it should have a *b*-value significantly different from zero. This hypothesis is tested using a *t*-test (see Chapter 9). The **t-statistic** tests the null hypothesis that the value of *b* is 0: therefore, if it is significant we gain confidence in the hypothesis that the *b*-value is significantly different from 0 and that the predictor variable contributes significantly to our ability to estimate values of the outcome.

Like *F*, the *t*-statistic is also based on the ratio of explained variance to unexplained variance or error. Well, actually, what we're interested in here is not so much variance but whether the *b* we have is big compared to the amount of error in that estimate. To estimate how much error we could expect to find in *b* we use the standard error. The standard error tells us something about how different *b*-values would be across different samples. We could take lots and lots of samples of data regarding record sales and advertising budgets and calculate the *b*-values for each sample. We could plot a frequency distribution of these samples to discover whether the *b*-values from all samples would be relatively similar, or whether they would be very different (think back to section 2.5.1). We can use the standard deviation of this distribution (known as the *standard error*) as a measure of the similarity of *b*-values across samples. If the standard error is very small, then it means that most samples are likely to have a *b*-value similar to the one in our sample (because there is little variation across samples). The *t*-test tells us whether the *b*-value is different from 0 relative to the variation in *b*-values across samples. When the standard error is small even a small deviation from zero can reflect a meaningful difference because *b* is representative of the majority of possible samples.

Equation (7.6) shows how the *t*-test is calculated and you'll find a general version of this equation in Chapter 9 (equation (9.1)). The b_{expected} is simply the value of *b* that we would expect to obtain if the null hypothesis were true. I mentioned earlier that the null hypothesis is that b is 0 and so this value can be replaced by 0. The equation simplifies to become the observed value of b divided by the standard error with which it is associated:

$$\begin{aligned} t &= \frac{b_{\text{observed}} - b_{\text{expected}}}{\text{SE}_b} \\ &= \frac{b_{\text{observed}}}{\text{SE}_b} \end{aligned} \tag{7.6}$$

The values of t have a special distribution that differs according to the degrees of freedom for the test. In regression, the degrees of freedom are $N - p - 1$, where N is the total sample size and p is the number of predictors. In simple regression when we have only one predictor, this reduces down to $N - 2$. Having established which t-distribution needs to be used, the observed value of t can then be compared to the values that we would expect to find if there was no effect (i.e. $b = 0$): if t is very large then it is unlikely to have occurred when there is no effect (these values can be found in the Appendix). SAS provides the exact probability that the observed value (or a larger one) of t would occur if the value of b was, in fact, 0. As a general rule, if this observed significance is less than .05, then scientists assume that b is significantly different from 0; put another way, the predictor makes a significant contribution to predicting the outcome.

7.3. Doing simple regression on SAS ①

So far, we have seen a little of the theory behind regression, albeit restricted to the situation in which there is only one predictor. To help clarify what we have learnt so far, we will go through an example of a simple regression on SAS. Earlier on I asked you to imagine that I worked for a record company and that my boss was interested in predicting record sales from advertising. There are some data for this example in the file **Record1.sas7bdat.** This data file has 200 rows, each one representing a different record. There are also two columns, one representing the sales of each record in the week after release and the other representing the amount (in pounds) spent promoting the record before release. This is the format for entering regression data: the outcome variable and any predictors should be entered in different columns, and each row should represent independent values of those variables.

The pattern of the data is shown in Figure 7.5 and it should be clear that a positive relationship exists: so, the more money spent advertising the record, the more it is likely to sell. Of course there are some records that sell well regardless of advertising (top left of scatterplot), but there are none that sell badly when advertising levels are high (bottom right of scatterplot). The scatterplot also shows the line of best fit for these data: bearing in mind that the mean would be represented by a flat line at around the 200,000 sales mark, the regression line is noticeably different.

To find out the parameters that describe the regression line, and to see whether this line is a useful model, we need to run a regression analysis. To do the analysis you need to use PROC REG.

PROC REG syntax is very straightforward, and is shown in SAS Syntax 7.1. The MODEL statement is written in the form of the equation: we think that sales are a function of adverts, so we write sales = adverts. Notice that PROC REG is slightly different to other procedures, because we need to write QUIT; after the RUN statement.

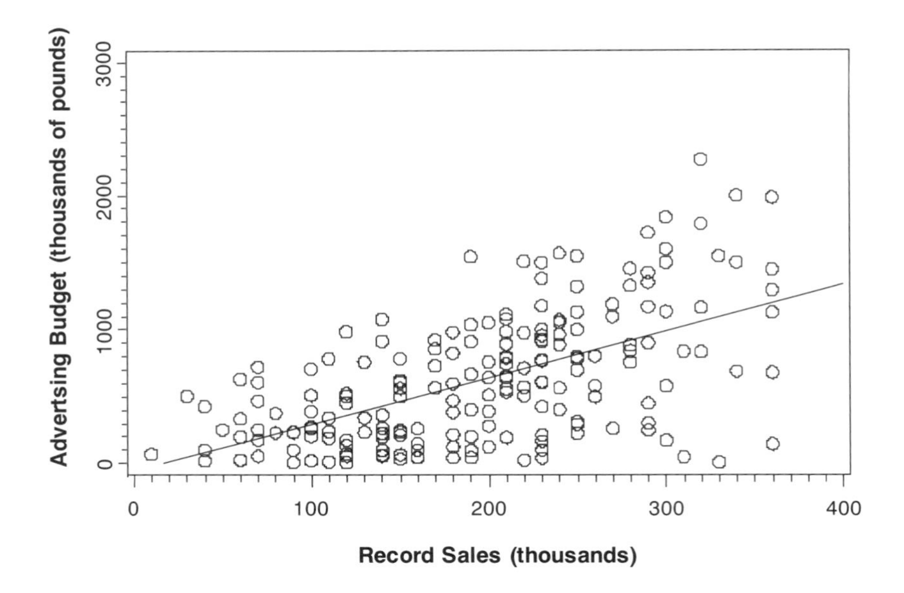

FIGURE 7.5 Scatterplot showing the relationship between record sales and the amount spent promoting the record

```
PROC REG DATA=chapter6.record1;
    MODEL sales = adverts;
    RUN;
QUIT;
```

SAS Syntax 7.1

7.4. Interpreting a simple regression ①

7.4.1. Overall fit of the model ①

The output from the regression is shown in SAS Output 7.1. The first part of the output reports an analysis of variance (ANOVA – see Chapter 10). The summary table shows the various sums of squares described in Figure 7.4 and the degrees of freedom associated with each. From these two values, the average sums of squares (the mean squares) can be calculated by dividing the sums of squares by the associated degrees of freedom. The most important part of the table is the *F*-ratio, which is calculated using equation (7.5), and the associated significance value of that *F*-ratio. For these data, *F* is 99.59, which is significant at $p < .001$ (because the value in the column labelled Sig. is less than .001). Researchers usually dont report p-values below 0.001 though. This result tells us that there is less than a 0.1% chance that an *F*-ratio at least this large would happen if the null hypothesis were true. Therefore, we can conclude that our regression model results in significantly better prediction of record sales than if we used the mean value of record sales. In short, the regression model overall predicts record sales significantly well.

The second part of the output is a summary of the model. This summary table provides the value of R and R^2 for the model that has been derived (as well as some other things we are not going to worry about for now). For these data, R has a value of .578 and because there is only one predictor, this value represents the simple correlation between advertising

SAS OUTPUT 7.1

The REG Procedure
Model: MODEL1
Dependent Variable: SALES Record Sales (thousands)

Number of Observations Read	200
Number of Observations Used	200

Analysis of Variance					
Source	DF	Sum of Squares	Mean Square	F Value	Pr > F
Model	1	433688	433688	99.59	<.0001
Error	198	862264	4354.86953		
Corrected Total	199	1295952			

Root MSE	65.99144	R-Square	0.3346
Dependent Mean	193.20000	Adj R-Sq	0.3313
Coeff Var	34.15706		

Parameter Estimates						
Variable	Label	DF	Parameter Estimate	Standard Error	t Value	Pr > |t|
Intercept	Intercept	1	134.13994	7.53657	17.80	<.0001
ADVERTS	Advertising Budget (thousands of pounds)	1	0.09612	0.00963	9.98	<.0001

and record sales (you can confirm this by running a correlation using what you were taught in Chapter 6). The value of R^2 is .335, which tells us that advertising expenditure can account for 33.5% of the variation in record sales. In other words, if we are trying to explain why some records sell more than others, we can look at the variation in sales of different records. There might be many factors that can explain this variation, but our model, which includes only advertising expenditure, can explain approximately 33% of it. This means that 67% of the variation in record sales cannot be explained by advertising alone. Therefore, there must be other variables that have an influence also.

7.4.2. Model parameters ①

The ANOVA tells us whether the model, overall, results in a significantly good degree of prediction of the outcome variable. However, the ANOVA doesn't tell us about the individual contribution of variables in the model (although in this simple case there is only one variable in the model and so we can infer that this variable is a good predictor. The third part of the output provides details of the model parameter estimates (the beta values) and the significance of these values. We saw in equation (7.2) that b_0 was the Y intercept and this value is the value labelled *Parameter Estimate* (in the SAS output) for the constant. So, from the table, we can say that b_0 is 134.14, and this can be interpreted as meaning that when no money is spent on advertising (when X = 0), the model predicts that 134,140 records will be sold (remember that our unit of measurement was thousands of records). We can also read off the value of b_1 from the table and this value represents the gradient of the regression line. It is 0.096. Although this value is the slope of the regression line, it is more useful to think of this value as representing *the change in the outcome associated with a unit change in the predictor*. Therefore, if our predictor variable is increased by one unit (if the advertising budget is increased by 1), then our model predicts that 0.096 extra records will be sold. Our units of measurement were thousands of pounds and thousands of records sold, so we can say that for an increase in advertising of £1000 the model predicts 96 (0.096 × 1000 = 96) extra record sales. As you might imagine, this investment is pretty bad for the record company: it invests £1000 and gets only 96 extra sales! Fortunately, as we already know, advertising accounts for only one-third of record sales.

We saw earlier that, in general, values of the regression coefficient *b* represent the change in the outcome resulting from a unit change in the predictor and that if a predictor is having a significant impact on our ability to predict the outcome then this *b* should be different from 0 (and big relative to its standard error). We also saw that the *t*-test tells us whether the *b*-value is different from 0. SAS provides the exact probability that the observed value of *t* would occur if the value of *b* in the population were 0. If this observed significance is less than .05, then scientists agree that the result reflects a genuine effect (see Chapter 2). For these two values, the probabilities are <.0001 and so we can say that the probability of these *t* values (or larger) occurring if the values of *b* in the population were 0 is less than .0001. Therefore, the *b*s are different from 0 and we can conclude that the advertising budget makes a significant contribution ($p < .0001$) to predicting record sales.

SELF-TEST How is the *t* in SAS Output 7.1 calculated? Use the values in the table to see if you can get the same value as SAS.

7.4.3. Using the model ①

So far, we have discovered that we have a useful model, one that significantly improves our ability to predict record sales. However, the next stage is often to use that model to make some predictions. The first stage is to define the model by replacing the *b*-values in equation (7.2) with the values from SAS Output 7.1. In addition, we can replace the *X* and *Y* with the variable names so that the model becomes:

$$\begin{aligned}\text{record sales}_i &= b_0 + b_1\text{advertising budget}_i \\ &= 134.14 + (0.096 \times \text{advertising budget}_i)\end{aligned} \quad (7.7)$$

It is now possible to make a prediction about record sales, by replacing the advertising budget with a value of interest. For example, imagine a record executive wanted to spend £100,000 on advertising a new record. Remembering that our units are already in thousands of pounds; we can simply replace the advertising budget with 100. He would discover that record sales should be around 144,000 for the first week of sales:

$$\begin{aligned}\text{record sales}_i &= 134.14 + (0.096 \times \text{advertising budget}_i)\\ &= 134.14 + (0.096 \times 100) \\ &= 143.74\end{aligned} \tag{7.8}$$

SELF-TEST How many records would be sold if we spent £666,000 on advertising the latest CD by black metal band Abgott?

CRAMMING SAM'S TIPS **Simple regression**

- Simple regression is a way of predicting values of one variable from another.
- We do this by fitting a statistical model to the data in the form of a straight line.
- This line is the line that best summarizes the pattern of the data.
- We have to assess how well the line fits the data using:
  - R^2 which tells us how much variance is explained by the model compared to how much variance there is to explain in the first place. It is the proportion of variance in the outcome variable that is shared by the predictor variable.
  - F, which tells us how much variability the model can explain relative to how much it can't explain (i.e. it's the ratio of how good the model is compared to how bad it is).
- The b-value tells us the gradient of the regression line and the strength of the relationship between a predictor and the outcome variable. If it is significant (*Sig.* < .05 in the SAS table) then the predictor variable significantly predicts the outcome variable.

7.5. Multiple regression: the basics ①

To summarize what we have learnt so far, in simple linear regression the outcome variable Y is predicted using the equation of a straight line (equation (7.2)). Given that we have collected several values of Y and X, the unknown parameters in the equation can be calculated. They are calculated by fitting a model to the data (in this case a straight line) for which the sum of the squared differences between the line and the actual data points is minimized. This method is called the method of least squares. Multiple regression is a logical extension of these principles to situations in which there are several predictors. Again, we still use our basic equation of:

$$\text{outcome}_i = (\text{model}) + \text{error}_i$$

but this time the model is slightly more complex. It is basically the same as for simple regression except that for every extra predictor you include, you have to add a coefficient;

so, each predictor variable has its own coefficient, and the outcome variable is predicted from a combination of all the variables multiplied by their respective coefficients plus a residual term (see equation (7.9) – the brackets aren't necessary, they're just to make the connection to the general equation above):

$$Y_i = (b_0 + b_1 X_{1i} + b_2 X_{2i} + \ldots + b_n X_{ni}) + \varepsilon_i \tag{7.9}$$

Y is the outcome variable, b_1 is the coefficient of the first predictor (X_1), b_2 is the coefficient of the second predictor (X_2), b_n is the coefficient of the nth predictor (X_n), and ε_i is the difference between the predicted and the observed value of Y for the ith participant. In this case, the model fitted is more complicated, but the basic principle is the same as simple regression. That is, we seek to find the linear combination of predictors that correlate maximally with the outcome variable. Therefore, when we refer to the regression model in multiple regression, we are talking about a model in the form of equation (7.9).

7.5.1. An example of a multiple regression model ②

Imagine that our record company executive was interested in extending his model of record sales to incorporate another variable. We know already that advertising accounts for 33% of variation in record sales, but a much larger 67% remains unexplained. The record executive could measure a new predictor in an attempt to explain some of the unexplained variation in record sales. He decides to measure the number of times the record is played on Radio 1 (the UK's biggest national radio station) during the week prior to release. The existing model that we derived using SAS (see equation (7.7)) can now be extended to include this new variable (**airplay**):

$$\text{record sales}_i = b_0 + b_1 \text{advertising budget}_i + b_2 \text{airplay}_i + \varepsilon_i \tag{7.10}$$

The new model is based on equation (7.9) and includes a b-value for both predictors (and, of course, the constant). If we calculate the b-values, we could make predictions about record sales based not only on the amount spent on advertising but also in terms of radio play. There are only two predictors in this model and so we could display this model graphically in three dimensions (Figure 7.6).

Equation (7.9) describes the tinted trapezium in the diagram (this is known as the regression *plane*) and the dots represent the observed data points. Like simple regression, the plane fitted to the data aims to best predict the observed data. However, there are invariably some differences between the model and the real-life data (this fact is evident because some of the dots do not lie exactly on the tinted area of the graph). The b-value for advertising describes the slope of the left and right sides of the regression plane, whereas the b-value for airplay describes the slope of the top and bottom of the regression plane. Just like simple regression, knowledge of these two slopes tells us about the shape of the model (what it looks like) and the intercept locates the regression plane in space.

It is fairly easy to visualize a regression model with two predictors, because it is possible to plot the regression plane using a 3-D scatterplot. However, multiple regression can be used with three, four or even ten or more predictors. Although you can't immediately visualize what such complex models look like, or visualize what the b-values represent, you should be able to apply the principles of these basic models to more complex scenarios.

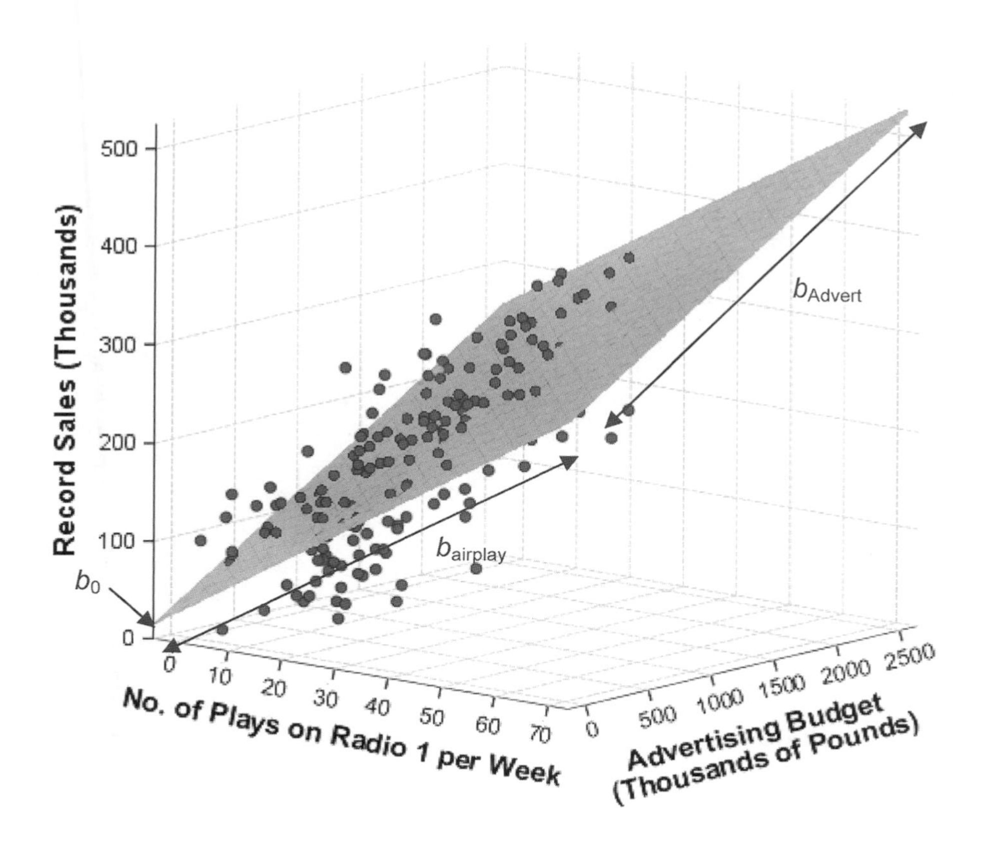

FIGURE 7.6 Scatterplot of the relationship between record sales, advertising budget and radio play

7.5.2. Sums of squares, R and R^2 ②

When there are several predictors it does not make sense to look at the simple correlation coefficient and instead SAS produces the squared multiple correlation coefficient R^2 (labelled *R-Square*). R^2 is the square of the correlation between the observed values of Y and the values of Y predicted by the multiple regression model. Therefore, large values of R^2 represent a large correlation between the predicted and observed values of the outcome. A R^2 of 1 represents a situation in which the model perfectly predicts the observed data. As such, R^2 is a gauge of how well the model predicts the observed data. It follows that the resulting R^2 can be interpreted in the same way as simple regression: it is the amount of variation in the outcome variable that is accounted for by the model.

When there are several predictors it does not make sense to look at the simple correlation coefficient and instead SAS produces a multiple correlation coefficient (labelled *Multiple R*). Multiple R is the correlation between the observed values of Y and the values of Y predicted by the multiple regression model. Therefore, large values of multiple R represent a large correlation between the predicted and observed values of the outcome. A multiple R of 1 represents a situation in which the model perfectly predicts the observed data. As such, multiple R is a gauge of how well the model predicts the observed data. It follows that the resulting R^2 can be interpreted in the same way as simple regression: it is the amount of variation in the outcome variable that is accounted for by the model.

7.5.3. Methods of regression ②

If we are interested in constructing a complex model with several predictors, how do we decide which predictors to use? A great deal of care should be taken in selecting predictors for a model because the values of the regression coefficients depend upon the variables in the model. Therefore, the predictors included and the way in which they are entered into the model can have a great impact. In an ideal world, predictors should be selected based on past research.[5] If new predictors are being added to existing models then select these new variables based on the substantive *theoretical* importance of these variables. One thing *not* to do is select hundreds of random predictors, bung them all into a regression analysis and hope for the best. In addition to the problem of selecting predictors, there are several ways in which variables can be entered into a model. When predictors are all completely uncorrelated the order of variable entry has very little effect on the parameters calculated; however, we rarely have uncorrelated predictors and so the method of predictor selection is crucial.

7.5.3.1. Hierarchical (blockwise entry) ②

In **hierarchical regression** predictors are selected based on past work and the experimenter decides in which order to enter the predictors into the model. As a general rule, known predictors (from other research) should be entered into the model first in order of their importance in predicting the outcome. After known predictors have been entered, the experimenter can add any new predictors into the model. New predictors can be entered either all in one go, in a stepwise manner, or hierarchically (such that the new predictor suspected to be the most important is entered first).

7.5.3.2. Forced entry ②

Forced entry is a method in which all predictors are forced into the model simultaneously. Like hierarchical, this method relies on good theoretical reasons for including the chosen predictors, but unlike hierarchical the experimenter makes no decision about the order in which variables are entered. Some researchers believe that this method is the only appropriate method for theory testing (Studenmund & Cassidy, 1987) because stepwise techniques are influenced by random variation in the data and so seldom give replicable results if the model is retested.

7.5.3.3. Stepwise methods ②

In **stepwise regressions** decisions about the order in which predictors are entered into the model are based on a purely mathematical criterion. In the *forward* method, an initial model is defined that contains only the constant (b_0). The computer then searches for the predictor (out of the ones available) that best predicts the outcome variable – it does this by selecting the predictor that has the highest simple correlation with the outcome. If this predictor significantly improves the ability of the model to predict the outcome, then this predictor is retained in the model and the computer searches for a second predictor. The criterion used for selecting this second predictor is that it is the variable that has the

[5] I might cynically qualify this suggestion by proposing that predictors be chosen based on past research that has utilized good methodology. If basing such decisions on regression analyses, select predictors based only on past research that has used regression appropriately and yielded reliable, generalizable models!

largest semi-partial correlation with the outcome. Let me explain this in plain English. Imagine that the first predictor can explain 40% of the variation in the outcome variable; then there is still 60% left unexplained. The computer searches for the predictor that can explain the biggest part of the remaining 60% (so it is not interested in the 40% that is already explained). As such, this semi-partial correlation gives a measure of how much 'new variance' in the outcome can be explained by each remaining predictor (see section 6.6). The predictor that accounts for the most new variance is added to the model and, if it makes a significant contribution to the predictive power of the model, it is retained and another predictor is considered.

The *stepwise* method in SAS is the same as the forward method, except that each time a predictor is added to the equation, a removal test is made of the least useful predictor. As such the regression equation is constantly being reassessed to see whether any redundant predictors can be removed.

The *backward* method is the opposite of the forward method in that the computer begins by placing all predictors in the model and then calculating the contribution of each one by looking at the significance value of the *t*-test for each predictor. This significance value is compared against a removal criterion (which can be either an absolute value of the test statistic or a probability value for that test statistic). If a predictor meets the removal criterion (i.e. if it is not making a statistically significant contribution to how well the model predicts the outcome variable) it is removed from the model and the model is re-estimated for the remaining predictors. The contribution of the remaining predictors is then reassessed.

If you do decide to use a stepwise method then the backward method is preferable to the forward method. This is because of **suppressor effects**, which occur when a predictor has a significant effect but only when another variable is held constant. Forward selection is more likely than backward elimination to exclude predictors involved in suppressor effects. As such, the forward method runs a higher risk of making a Type II error (i.e. missing a predictor that does in fact predict the outcome).

7.5.3.4. Choosing a method ②

SAS allows you to opt for any one of these methods and it is important to select an appropriate one. The forward, backward and stepwise methods all come under the general heading of *stepwise methods* because they all rely on the computer selecting variables based upon mathematical criteria. Many writers argue that this takes many important methodological decisions out of the hands of the researcher. What's more, the models derived by computer often take advantage of random sampling variation and so decisions about which variables should be included will be based upon slight differences in their semi-partial correlation. However, these slight statistical differences may contrast dramatically with the theoretical importance of a predictor to the model. There is also the danger of over-fitting (having too many variables in the model that essentially make little contribution to predicting the outcome) and under-fitting (leaving out important predictors) the model. For this reason stepwise methods are best avoided except for exploratory model building. If you must do a stepwise regression then it is advisable to cross-validate your model by splitting the data (see section 7.6.2.2).

When there is a sound theoretical literature available, then base your model upon what past research tells you. Include any meaningful variables in the model in their order of importance. After this initial analysis, repeat the regression but exclude any variables that were statistically redundant the first time around. There are important considerations in deciding which predictors should be included. First, it is important not to include too many predictors. As a general rule, the fewer predictors the better, and certainly include only predictors for which you have a good theoretical grounding (it is meaningless to measure hundreds of variables and then put them all into a regression model). So, be selective and remember that you should have a decent sample size – see section 7.6.2.3.

7.6. How accurate is my regression model? ②

When we have produced a model based on a sample of data there are two important questions to ask: (1) does the model fit the observed data well, or is it influenced by a small number of cases; and (2) can my model generalize to other samples? These questions are vital to ask because they affect how we use the model that has been constructed. These questions are also, in some sense, hierarchical because we wouldn't want to generalize a bad model. However, it is a mistake to think that because a model fits the observed data well we can draw conclusions beyond our sample. **Generalization** is a critical additional step and if we find that our model is not generalizable, then we must restrict any conclusions based on the model to the sample used. First, we will look at how we establish whether a model is an accurate representation of the actual data, and in section 7.6.2 we move on to look at how we assess whether a model can be used to make inferences beyond the sample of data that has been collected.

7.6.1. Assessing the regression model I: diagnostics ②

To answer the question of whether the model fits the observed data well, or if it is influenced by a small number of cases, we can look for outliers and influential cases (the difference is explained in Jane Superbrain Box 7.2). We will look at these in turn.

7.6.1.1. Outliers and residuals ②

An outlier is a case that differs substantially from the main trend of the data (see Jane Superbrain Box 4.1). Figure 7.7 shows an example of such a case in regression. Outliers can cause your model to be biased because they affect the values of the estimated regression coefficients. For example, Figure 7.7 uses the same data as Figure 7.3 except that the score of one participant has been changed to be an outlier (in this case a person who was very calm in the presence of a very big spider). The change in this one point has had a dramatic effect on the regression model chosen to fit the data. With the outlier present, the regression model changes: its gradient is reduced (the line becomes flatter) and the intercept increases (the new line will cross the *Y*-axis at a higher point). It should be clear from this diagram that it is important to try to detect outliers to see whether the model is biased in this way.

How do you think that you might detect an outlier? Well, we know that an outlier, by its nature, is very different from all of the other scores. This being true, do you think that the model will predict that person's score very accurately? The answer is *no*: looking at Figure 7.7 it is evident that even though the outlier has biased the model, the model still predicts that one value very badly (the regression line is long way from the outlier). Therefore, if we were to work out the differences between the data values that were collected, and the values predicted by the model, we could detect an outlier by looking for large differences. This process is the same as looking for cases that the model predicts inaccurately. The differences between the values of the outcome predicted by the model

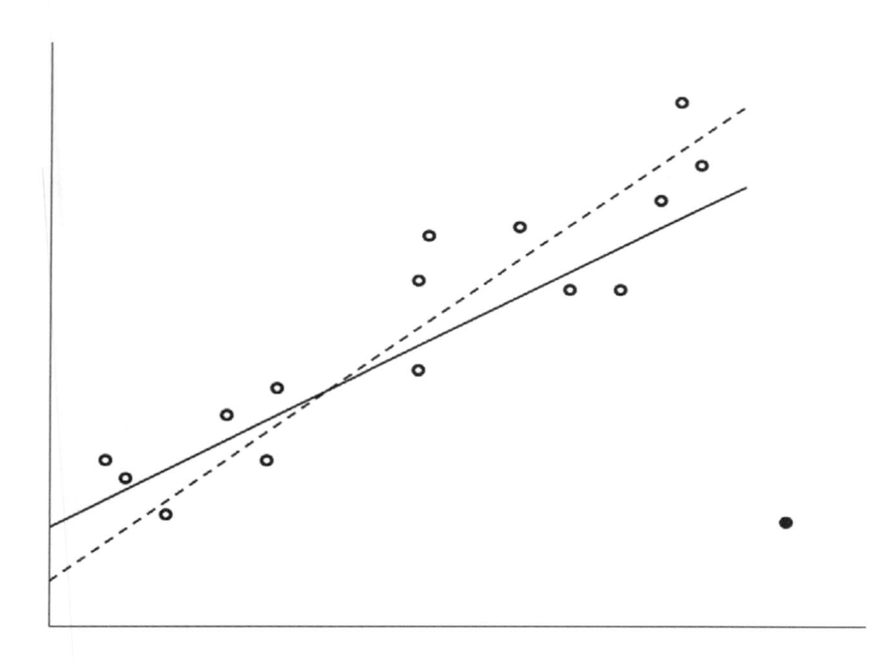

FIGURE 7.7 Graph demonstrating the effect of an outlier. The dashed line represents the original regression line for these data (see Figure 7.3), whereas the solid line represents the regression line when an outlier is present

and the values of the outcome observed in the sample are known as *residuals*. These residuals represent the error present in the model. If a model fits the sample data well then all residuals will be small (if the model was a perfect fit of the sample data – all data points fall on the regression line – then all residuals would be zero). If a model is a poor fit of the sample data then the residuals will be large. Also, if any cases stand out as having a large residual, then they could be outliers.

The **normal or unstandardized residuals** described above are measured in the same units as the outcome variable and so are difficult to interpret across different models. What we can do is to look for residuals that stand out as being particularly large. However, we cannot define a universal cut-off point for what constitutes a large residual. To overcome this problem, we use **studentized residuals** (see Jane Superbrain 7.1), which are the residuals divided by an estimate of their standard deviation. We came across standardization in section 6.3.2 as a means of converting variables into a standard unit of measurement (the standard deviation); we also came across *z*-scores (see section 1.7.4) in which variables are converted into standard deviation units (i.e. they're converted into scores that are distributed around a mean of 0 with a standard deviation of 1). Studentized residuals don't follow a *z*-distribution, instead they follow (something close to) a *t*-distribution. The *t*-distribution is actually very similar to the *z*-distribution, and becomes more and more similar as the sample size increases – if your sample is more than about 30, it is probably similar enough, and if your sample is larger than 100, it's definitely similar enough. By converting residuals into *t*-scores (studentized residuals) we can compare residuals from different models and use what we know about the properties of *t*-scores to devise universal guidelines for what constitutes an acceptable (or unacceptable) value. For example, we know from Chapter 1 that in a large, normally distributed sample, 95% of *z*-scores should lie between −1.96 and +1.96 (if your sample is around 30, this value becomes 2.04 and –2.04, so not far off 2), 99% should lie between −2.6 and +2.6, and 99.9% (i.e. nearly all of them) should lie between −3.3 and +3.3. Some general rules for studentized residuals are derived from these facts: (1) studentized residuals with an absolute value greater than 3.3 (we can use 3 as an approximation) are cause for concern because in an average sample a value this high is unlikely to happen by chance; (2) if more than 1% of our sample cases have studentized residuals with an absolute value greater than 2.6 (we usually just say 2.5) there is evidence that the level of error within our model is unacceptable (the model is a fairly poor fit of the sample data); and (3) if more than 5% of cases have studentized residuals with an absolute value greater than 1.96 (we can use 2 for convenience) then there is also evidence that the model is a poor representation of the actual data.

JANE SUPERBRAIN 7.1

Standardized residuals and standardized residuals. ③

Standardizing residuals is surprisingly tricky. One way to do it (the obvious way) is to find the standard deviation of the residuals and divide each residual by that standard deviation (we don't need to subtract the mean, because the mean is already 0). Dividing by the standard deviation sounds right, but it's not really right, because each residual has a different standard deviation. Specifically, residuals near to the high and low predicted scores have smaller standard deviations than residuals with middling predicted scores.

However, some people (and some software packages) do standardize the residuals by dividing by the overall standard deviation, and they call these the standardized residuals. Other people (and software packages) divide by the corrected standard deviation, and they call these the standardized residuals, while other people call them studentized residuals. So we have the same name for two (slightly) different things. Which is a little confusing. To make life slightly easier we don't ever talk about standardized residuals, and nor does SAS, instead, we only talk about studentized residuals. If you're really interested in learning more about different kinds of residuals and how confusing the different names are, you could consult Miles (2005).

7.6.1.2. Influential cases ③

As well as testing for outliers by looking at the error in the model, it is also possible to look at whether certain cases exert undue influence over the parameters of the model. So, if we were to delete a certain case, would we obtain different regression coefficients? This type of analysis can help to determine whether the regression model is stable across the sample, or whether it is biased by a few influential cases. Again, this process will unveil outliers.

There are several residual statistics that can be used to assess the influence of a particular case. One statistic is the **adjusted predicted value** for a case when that case is excluded from the analysis. In effect, the computer calculates a new model without a particular case and then uses this new model to predict the value of the outcome variable for the case that was excluded. If a case does not exert a large influence over the model then we would expect the adjusted predicted value to be very similar to the predicted value when the case is included. Put simply, if the model is stable then the predicted value of a case should be the same regardless of whether or not that case was used to calculate the model. The difference between the adjusted predicted value and the original predicted value is known as **DFFit** (see below). We can also look at the residual based on the adjusted predicted value: that is, the difference between the adjusted predicted value and the original observed value. This is the **PRESS residual** (PRESS stands for Prediction Sum of Squares). The PRESS residual can be divided by the standard error to give a standardized value known as the **Studentized deleted residual**, which SAS calls Rstudent. This residual can be compared across different regression analyses because it is measured in standard units and is exactly t-distributed.

The PRESS residuals are very useful to assess the influence of a case on the ability of the model to predict that case. However, they do not provide any information about how a case influences the model as a whole (i.e. the impact that a case has on the model's ability to predict *all* cases). One statistic that does consider the effect of a single case on the model as a whole is **Cook's distance.** Cook's distance is a measure of the overall influence of a case on the model and Cook and Weisberg (1982) have suggested that values greater than 1 may be cause for concern.

A second measure of influence is **leverage** (sometimes called **hat value**), which gauges the influence of the observed value of the outcome variable over the predicted values. The average leverage value is defined as $(k+1)/n$ in which k is the number of predictors in the model and n is the number of participants.[6] Leverage values can lie between 0 (indicating that the case has no influence whatsoever) and 1 (indicating that the case has complete influence over prediction). If no cases exert undue influence over the model then we would expect all of the leverage values to be close to the average value $((k+1)/n)$. Hoaglin and Welsch (1978) recommend investigating cases with values greater than twice the average $(2(k + 1)/n)$ and Stevens (2002) recommends using three times the average $(3(k + 1)/n)$ as a cut-off point for identifying cases having undue influence. We will see how to use these cut-off points later. However, cases with large leverage values will not necessarily have a large influence on the regression coefficients because they are measured on the outcome variables rather than the predictors.

Related to the leverage values are the **Mahalanobis distances** (Figure 7.8), which measure the distance of cases from the mean(s) of the predictor variable(s). You need to look for the cases with the highest values. It is not easy to establish a cut-off point at which to worry, although Barnett and Lewis (1978) have produced a table of critical values dependent on the number of predictors and the sample size. From their work it is clear that even with large samples ($N = 500$) and 5 predictors, values above 25 are cause for concern. In smaller samples ($N = 100$) and with fewer predictors (namely 3) values greater than 15 are problematic, and in very small samples ($N = 30$) with only 2 predictors values greater than 11 should be examined. However, for more specific advice, refer to Barnett and Lewis's table.

FIGURE 7.8 Prasanta Chandra Mahalanobis staring into his distances

It is possible to run the regression analysis with a case included and then rerun the analysis with that same case excluded. If we did this, undoubtedly there would be some difference between the b coefficients in the two regression equations. This difference would tell us how

[6] You may come across the average leverage denoted as p/n in which p is the number of parameters being estimated. In multiple regression, we estimate parameters for each predictor and also for a constant and so p is equivalent to the number of predictors plus one $(k + 1)$.

much influence a particular case has on the parameters of the regression model. To take a hypothetical example, imagine two variables that had a perfect negative relationship except for a single case (case 30). If a regression analysis was done on the 29 cases that were perfectly linearly related then we would get a model in which the predictor variable X perfectly predicts the outcome variable Y, and there are no errors. If we then ran the analysis but this time include the case that didn't conform (case 30), then the resulting model has different parameters. Some data are stored in the file **dfbeta.sas7bdat** which illustrate such a situation. Try running a simple regression first with all the cases included and then with case 30 deleted. The results are summarized in Table 7.1, which shows: (1) the parameters for the regression model when the extreme case is included or excluded; (2) the resulting regression equations; and (3) the value of Y predicted from participant 30's score on the X variable (which is obtained by replacing the X in the regression equation with participant 30's score for X, which was 1).

When case 30 is excluded, these data have a perfect negative relationship; hence the coefficient for the predictor (b_1) is −1 (remember that in simple regression this term is the same as Pearson's correlation coefficient), and the coefficient for the constant (the intercept, b_0) is 31. However, when case 30 is included, both parameters are reduced[7] and the difference between the parameters is also displayed. The difference between a parameter estimated using all cases and estimated when one case is excluded is known as the **DFBeta**. DFBeta is calculated for every case and for each of the parameters in the model. So, in our hypothetical example, the DFBeta for the constant is −2, and the DFBeta for the predictor variable is 0.1. By looking at the values of DFBeta, it is possible to identify cases that have a large influence on the parameters of the regression model. Again, the units of measurement used will affect these values and so SAS produces a **standardized DFBeta**. These standardized values are easier to use because universal cut-off points can be applied. In this case absolute values above 1 indicate cases that substantially influence the model parameters (although Stevens, 2002, suggests looking at cases with absolute values greater than 2).

TABLE 7.1 The difference in the parameters of the regression model when one case is excluded

Parameter (b)	Case 30 Included	Case 30 Excluded	Difference
Constant (intercept)	29.00	31.00	−2.00
Predictor (gradient)	−0.90	−1.00	0.10
Model (regression line):	$Y = (-0.9)X + 29$	$Y = (-1)X + 31$	
Predicted Y	28.10	30.00	−1.90

A related statistic is the **DFFit**, which is the difference between the predicted value for a case when the model is calculated including that case and when the model is calculated excluding that case: in this example the value is −1.90 (see Table 7.1). If a case is not influential then its DFFit should be zero – hence, we expect non-influential cases to have small DFFit values. However, we have the problem that this statistic depends on the units of measurement of the outcome and so a DFFit of 0.5 will be very small if the outcome ranges from 1 to 100, but very large if the outcome varies from 0 to 1; because of this problem SAS standardizes the DFFit statistic before presenting it to you.

A final measure is that of the **covariance ratio (CVR)**, which is a measure of whether a case influences the variance of the regression parameters. A description of the computation of this

[7] The value of b_1 is reduced because the data no longer have a perfect linear relationship and so there is now variance that the model cannot explain.

statistic would leave most readers dazed and confused, so suffice to say that when this ratio is close to 1 the case is having very little influence on the variances of the model parameters. Belsey, Kuh and Welsch (1980) recommend the following:

- If $CVR_i > 1 + [3(k + 1)/n]$ then deleting the *i*th case will damage the precision of some of the model's parameters.
- If $CVR_i < 1 - [3(k + 1)/n]$ then deleting the *i*th case will improve the precision of some of the model's parameters.

In both equations, k is the number of predictors, CVR_i is the covariance ratio for the *i*th participant, and n is the sample size.

JANE SUPERBRAIN 7.2

The difference between residuals and influence statistics ③

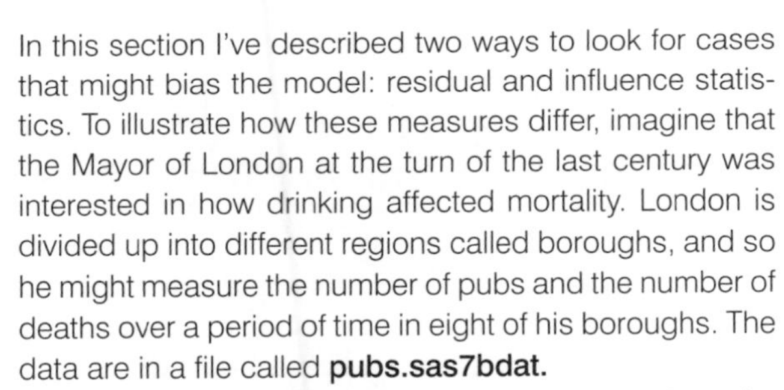

In this section I've described two ways to look for cases that might bias the model: residual and influence statistics. To illustrate how these measures differ, imagine that the Mayor of London at the turn of the last century was interested in how drinking affected mortality. London is divided up into different regions called boroughs, and so he might measure the number of pubs and the number of deaths over a period of time in eight of his boroughs. The data are in a file called **pubs.sas7bdat.**

The scatterplot of these data reveals that without the last case there is a perfect linear relationship (the dashed straight line). However, the presence of the last case (case 8) changes the line of best fit dramatically (although this line is still a significant fit of the data – do the regression analysis and see for yourself).

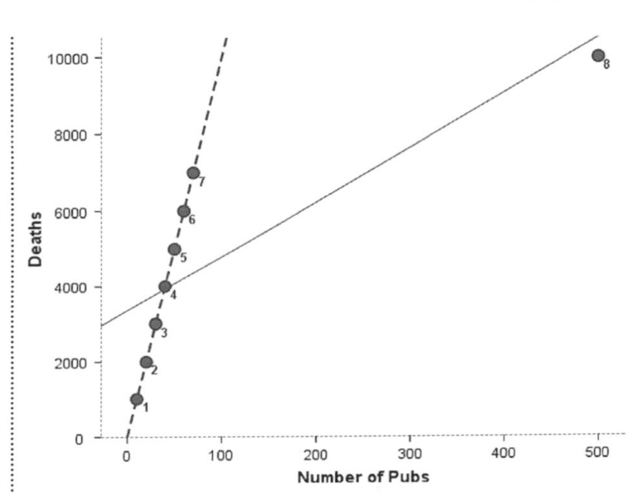

What's interesting about these data is when we look at the residuals and influence statistics. The standardized residual for case 8 is the second *smallest*: this outlier produces a very small residual (most of the non-outliers have larger residuals) because it sits very close to the line that has been fitted to the data. How can this be? Look at the influence statistics below and you'll see that they're massive for case 8: it exerts a huge influence over the model.

			Hat Diag	Cov		-------DFBETAS-------	
Obs	Residual	RStudent	H	Ratio	DFFITS	Intercept	PUBS
1	−2508	−1.6773	0.1650	0.7062	−0.7455	−0.7445	0.3669
2	−1637	−0.9467	0.1571	1.2285	−0.4087	−0.4067	0.1847
3	−780.7819	−0.4211	0.1494	1.5789	−0.1764	−0.1744	0.0712
4	75.8957	0.0401	0.1427	1.6786	0.0163	0.0160	−0.0058
5	932.5732	0.5030	0.1371	1.5119	0.2005	0.1934	−0.0595
6	1789	1.0346	0.1325	1.1262	0.4044	0.3831	−0.0965
7	2646	1.7658	0.1291	0.6272	0.6798	0.6290	−0.1206
8	−516.9374	−138.2589	0.9872	0.0000	−1215.56	245.6972	−1136.00

As always when you see a statistical oddity you should ask what was happening in the real world. The last data point represents the City of London, a tiny area of only 1 square mile in the centre of London where very few people lived but where thousands of commuters (even then) came to work and had lunch in the pubs. Hence the pubs didn't rely on the resident population for their business and the residents didn't consume all of their beer! Therefore, there was a massive number of pubs.

This illustrates that a case exerting a massive influence can produce a small residual – so look at both! (I'm very grateful to David Hitchin for this example, and he in turn got it from Dr Richard Roberts.)

7.6.1.3. A final comment on diagnostic statistics ②

There are a lot of diagnostic statistics that should be examined after a regression analysis, and it is difficult to summarize this wealth of material into a concise conclusion. However, one thing I would like to stress is a point made by Belsey et al. (1980) who noted the dangers inherent in these procedures. The point is that diagnostics are tools that enable you to see how good or bad your model is in terms of fitting the sampled data. They are a way of assessing your model. They are *not*, however, a way of justifying the removal of data points to effect some desirable change in the regression parameters (e.g. deleting a case that changes a non-significant *b*-value into a significant one). Stevens (2002), as ever, offers excellent advice:

> If a point is a significant outlier on Y, but its Cook's distance is < 1, there is no real need to delete that point since it does not have a large effect on the regression analysis. However, one should still be interested in studying such points further to understand why they did not fit the model. (p. 135)

7.6.2. Assessing the regression model II: generalization ②

When a regression analysis is done, an equation can be produced that is correct for the sample of observed values. However, in the social sciences we are usually interested in generalizing our findings outside of the sample. So, although it can be useful to draw conclusions about a particular sample of people, it is usually more interesting if we can then assume that our conclusions are true for a wider population. For a regression model to generalize we must be sure that underlying assumptions have been met, and to test whether the model does generalize we can look at cross-validating it.

7.6.2.1. Checking assumptions ②

To draw conclusions about a population based on a regression analysis done on a sample, several assumptions must be true (see Berry, 1993):

- **Variable types**: All predictor variables must be quantitative or categorical (with two categories), and the outcome variable must be quantitative, continuous and unbounded. By quantitative I mean that they should be measured at the interval level and by unbounded I mean that there should be no constraints on the variability of the outcome. If the outcome is a measure ranging from 1 to 10 yet the data collected vary between 3 and 7, then these data are constrained.

- **Non-zero variance:** The predictors should have some variation in value (i.e. they should not have variances of 0).
- **No perfect multicollinearity:** There should be no perfect linear relationship between two or more of the predictors. So, the predictor variables should not correlate too highly (see section 7.6.2.4).
- **Predictors are uncorrelated with 'external variables':** *External variables* are variables that haven't been included in the regression model which influence the outcome variable.[8] These variables can be thought of as similar to the 'third variable' that was discussed with reference to correlation. This assumption means that there should be no external variables that correlate with any of the variables included in the regression model. Obviously, if external variables do correlate with the predictors, then the conclusions we draw from the model become unreliable (because other variables exist that can predict the outcome just as well).
- **Homoscedasticity:** At each level of the predictor variable(s), the variance of the residual terms should be constant. This just means that the residuals at each level of the predictor(s) should have the same variance (**homoscedasticity**); when the variances are very unequal there is said to be **heteroscedasticity** (see section 5.6 as well).
- **Independent errors:** For any two observations the residual terms should be uncorrelated (or independent). This eventuality is sometimes described as a lack of **autocorrelation**. This assumption can be tested with the **Durbin–Watson test**, which tests for serial correlations between errors. Specifically, it tests whether adjacent residuals are correlated. The test statistic can vary between 0 and 4, with a value of 2 meaning that the residuals are uncorrelated. A value greater than 2 indicates a negative correlation between adjacent residuals, whereas a value below 2 indicates a positive correlation. The size of the Durbin–Watson statistic depends upon the number of predictors in the model and the number of observations. For accuracy, you should look up the exact acceptable values in Durbin and Watson's (1951) original paper. As a very conservative rule of thumb, values less than 1 or greater than 3 are definitely cause for concern; however, values closer to 2 may still be problematic depending on your sample and model.
- **Normally distributed errors:** It is assumed that the residuals in the model are random, normally distributed variables with a mean of 0. This assumption simply means that the differences between the model and the observed data are most frequently zero or very close to zero, and that differences much greater than zero happen only occasionally. Some people confuse this assumption with the idea that predictors have to be normally distributed. In fact, predictors do not need to be normally distributed (see section 7.11).
- **Independence:** It is assumed that all of the values of the outcome variable are independent (in other words, each value of the outcome variable comes from a separate entity).
- **Linearity:** The mean values of the outcome variable for each increment of the predictor(s) lie along a straight line. In plain English this means that it is assumed that the relationship we are modelling is a linear one. If we model a non-linear relationship using a linear model then this obviously limits the generalizability of the findings.

[8] Some authors choose to refer to these external variables as part of an error term that includes any random factor in the way in which the outcome varies. However, to avoid confusion with the residual terms in the regression equations I have chosen the label 'external variables'. Although this term implicitly washes over any random factors, I acknowledge their presence here!

This list of assumptions probably seems pretty daunting but as we saw in Chapter 5, assumptions are important. When the assumptions of regression are met, the model that we get for a sample can be accurately applied to the population of interest (the coefficients and parameters of the regression equation are said to be *unbiased*). Some people assume that this means that when the assumptions are met the regression model from a sample is always identical to the model that would have been obtained had we been able to test the entire population. Unfortunately, this belief isn't true. What an unbiased model does tell us is that *on average* the regression model from the sample is the same as the population model. However, you should be clear that even when the assumptions are met, it is possible that a model obtained from a sample may not be the same as the population model – but the likelihood of them being the same is increased.

7.6.2.2. Cross-validation of the model ③

Even if we can't be confident that the model derived from our sample accurately represents the entire population, there are ways in which we can assess how well our model can predict the outcome in a different sample. Assessing the accuracy of a model across different samples is known as cross-validation. If a model can be generalized, then it must be capable of accurately predicting the same outcome variable from the same set of predictors in a different group of people. If the model is applied to a different sample and there is a severe drop in its predictive power, then the model clearly does *not* generalize. As a first rule of thumb, we should aim to collect enough data to obtain a reliable regression model (see the next section). Once we have a regression model there are two main methods of cross-validation:

- **Adjusted R^2**: In SAS not only is the value of R^2 calculated, but also an **adjusted R^2**. This adjusted value indicates the loss of predictive power or **shrinkage**. Whereas R^2 tells us how much of the variance in Y is accounted for by the regression model from our sample, the adjusted value tells us how much variance in Y would be accounted for if the model had been derived from the population from which the sample was taken. However, the adjusted value has been criticized because it tells us nothing about how well the regression model would predict an entirely different set of data (how well can the model predict scores of a different sample of data from the same population?). One version of R^2 that does tell us how well the model cross-validates uses Stein's formula (see Stevens, 2002) which is given by:

$$\text{adjusted } R^2 = 1 - \left[\left(\frac{n-1}{n-k-1}\right)\left(\frac{n-2}{n-k-2}\right)\left(\frac{n+1}{n}\right)\right](1-R^2) \quad (7.11)$$

  In Stein's equation, R^2 is the unadjusted value, n is the number of participants and k is the number of predictors in the model. For the more mathematically minded of you, it is worth using this equation to cross-validate a regression model.

- **Data splitting**: This approach involves randomly splitting your data set, computing a regression equation on both halves of the data and then comparing the resulting models. When using stepwise methods, cross-validation is a good idea; you should run the stepwise regression on a random selection of about 80% of your cases. Then force this model on the remaining 20% of the data. By comparing values of the R^2 and b-values in the two samples you can tell how well the original model generalizes (see Tabachnick & Fidell, 2007, for more detail).

7.6.2.3. Sample size in regression ③

In the previous section I said that it's important to collect enough data to obtain a reliable regression model. Well, how much is enough? You'll find a lot of rules of thumb floating about, the two most common being that you should have 10 cases of data for each predictor in the model, or 15 cases of data per predictor. So, with five predictors, you'd need 50 or 75 cases respectively (depending on the rule you use). These rules are very pervasive (even I used the 15 cases per predictor rule in the first edition of this book) but they oversimplify the issue considerably. In fact, the sample size required will depend on the size of effect that we're trying to detect (i.e. how strong the relationship is that we're trying to measure) and how much power we want to detect these effects. The simplest rule of thumb is: the bigger the sample size, the better! The reason is that the estimate of R that we get from regression is dependent on the number of predictors, k, and the sample size, N. In fact the expected R for random data is $k/(N-1)$ and so with small sample sizes random data can appear to show a strong effect: for example, with six predictors and 21 cases of data, $R = 6/(21-1) = .3$ (a medium effect size by Cohen's criteria described in section 6.3.2). Obviously for random data we'd want the expected R to be 0 (no effect) and for this to be true we need large samples (to take the previous example, if we had 100 cases not 21, then the expected R would be a more acceptable .06).

It's all very well knowing that larger is better, but researchers usually need some more concrete guidelines (much as we'd all love to collect 1000 cases of data it isn't always practical!). Green (1991) makes two rules of thumb for the *minimum* acceptable sample size, the first based on whether you want to test the overall fit of your regression model (i.e. test the R^2), and the second based on whether you want to test the individual predictors within the model (i.e. test b-values of the model). If you want to test the model overall, then he recommends a minimum sample size of $50 + 8k$, where k is the number of predictors. So, with five predictors, you'd need a sample size of $50 + 40 = 90$. If you want to test the individual predictors then he suggests a minimum sample size of $104 + k$, so again taking the example of five predictors you'd need a sample size of $104 + 5 = 109$. Of course, in most cases we're interested both in the overall fit and in the contribution of individual predictors, and in this situation Green recommends you calculate both of the minimum sample sizes I've just described, and use the one that has the largest value (so in the five-predictor example, we'd use 109 because it is bigger than 90).

Now, these guidelines are all right as a rough and ready guide, but they still oversimplify the problem. As I've mentioned, the sample size required actually depends on the size of the effect (i.e. how well our predictors predict the outcome) and how much statistical power we want to detect these effects. Miles and Shevlin (2001) produce some extremely useful graphs that illustrate the sample sizes needed to achieve different levels of power, for different effect sizes, as the number of predictors vary. For precise estimates of the sample size you should be using, I recommend using these graphs. I've summarized some of the general findings in Figure 7.9. This diagram shows the sample size required to achieve a high level of power (I've taken Cohen's, 1988, benchmark of .8) depending on the number of predictors and the size of expected effect. To summarize the graph very broadly: (1) if you expect to find a large effect then a sample size of 80 will always suffice (with up to 20 predictors) and if there are fewer predictors then you can afford to have a smaller sample; (2) if you're expecting a medium effect, then a sample size of 200 will always suffice (up to 20 predictors), you should always have a sample size above 60, and with six or fewer predictors you'll be fine with a sample of 100; and (3) if you're expecting a small effect size then just don't bother unless you have the time and resources to collect at least 600 cases of data (and many more if you have six or more predictors!).

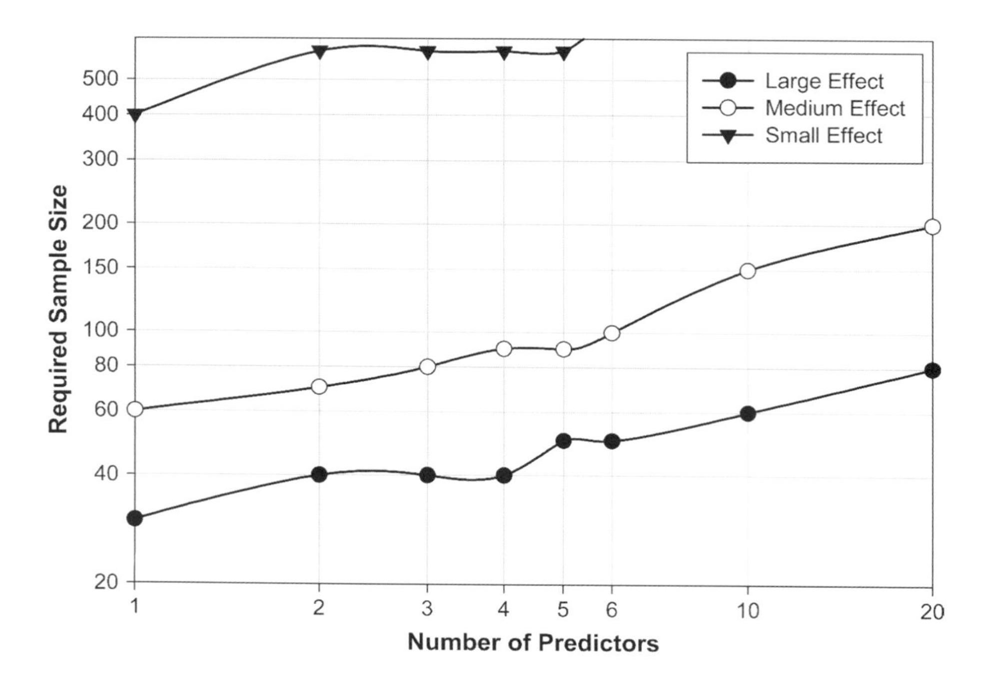

FIGURE 7.9 Graph to show the sample size required in regression depending on the number of predictors and the size of expected effect

7.6.2.4. Multicollinearity ②

Multicollinearity exists when there is a strong correlation between two or more predictors in a regression model. Multicollinearity poses a problem only for multiple regression because (without wishing to state the obvious) simple regression requires only one predictor. **Perfect collinearity** exists when at least one predictor is a perfect linear combination of the others (the simplest example being two predictors that are perfectly correlated – they have a correlation coefficient of 1). If there is perfect collinearity between predictors it becomes impossible to obtain unique estimates of the regression coefficients because there are an infinite number of combinations of coefficients that would work equally well. Put simply, if we have two predictors that are perfectly correlated, then the values of *b* for each variable are interchangeable. The good news is that perfect collinearity is rare in real-life data. The bad news is that less than perfect collinearity is virtually unavoidable. Low levels of collinearity pose little threat to multiple regression models, but as collinearity increases there are three problems that arise:

- **Untrustworthy *b*s**: As collinearity increases so do the standard errors of the *b* coefficients. If you think back to what the standard error represents, then big standard errors for *b* coefficients means that these *b*s are more variable across samples. Therefore, it means that the *b* coefficient in our sample is less likely to represent the population. Crudely put, multicollinearity means that the *b*-values are less trustworthy. Don't lend them money and don't let them go for dinner with your boy- or girlfriend. Of course if the *b*s are variable from sample to sample then the resulting predictor equations will be unstable across samples too.
- **It limits the size of *R***: Remember that *R* is a measure of the multiple correlation between the predictors and the outcome and that R^2 indicates the variance in the outcome for which the predictors account. Imagine a situation in which a single variable predicts the outcome variable fairly successfully (e.g. $R = .80$) and a second predictor variable is then added to the model. This second variable might account for a lot of the variance in the outcome (which is why it is included in the model), but the variance it accounts for is the same variance accounted for by the first variable. In other words, once the variance

accounted for by the first predictor has been removed, the second predictor accounts for very little of the remaining variance (the second variable accounts for very little *unique variance*). Hence, the overall variance in the outcome accounted for by the two predictors is little more than when only one predictor is used (so R might increase from .80 to .82). This idea is connected to the notion of partial correlation that was explained in Chapter 6. If, however, the two predictors are completely uncorrelated, then the second predictor is likely to account for different variance in the outcome to that accounted for by the first predictor. So, although in itself the second predictor might account for only a little of the variance in the outcome, the variance it does account for is different to that of the other predictor (and so when both predictors are included, R is substantially larger, say .95). Therefore, having uncorrelated predictors is beneficial.

- **Importance of predictors:** Multicollinearity between predictors makes it difficult to assess the individual importance of a predictor. If the predictors are highly correlated, and each accounts for similar variance in the outcome, then how can we know which of the two variables is important? Quite simply we can't tell which variable is important – the model could include either one, interchangeably.

One way of identifying multicollinearity is to scan a correlation matrix of all of the predictor variables and see if any correlate very highly (by very highly I mean correlations of above .80 or .90). This is a good 'ball park' method but misses more subtle forms of multicollinearity. Luckily, SAS produces various collinearity diagnostics, one of which is the **variance inflation factor (VIF)**. The VIF indicates whether a predictor has a strong linear relationship with the other predictor(s). Although there are no hard and fast rules about what value of the VIF should cause concern, Myers (1990) suggests that a value of 10 is a good value at which to worry. What's more, if the average VIF is greater than 1, then multicollinearity may be biasing the regression model (Bowerman & O'Connell, 1990). Related to the VIF is the **tolerance** statistic, which is its reciprocal (1/VIF). As such, values below 0.1 indicate serious problems although Menard (1995) suggests that values below 0.2 are worthy of concern.

Other measures that are useful in discovering whether predictors are dependent are the *eigenvalues of the scaled, uncentred cross-products matrix*, the *condition indexes* and the *variance proportions*. These statistics are extremely complex and will be covered as part of the interpretation of SAS output (see section 7.8.5). If none of this has made any sense then have a look at Hutcheson and Sofroniou (1999, pp. 78–85) who give a really clear explanation of multicollinearity.

7.7. How to do multiple regression using SAS ②

7.7.1. Some things to think about before the analysis ②

A good strategy to adopt with regression is to measure predictor variables for which there are sound theoretical reasons for expecting them to predict the outcome. Run a regression analysis in which all predictors are entered into the model and examine the output to see which predictors contribute substantially to the model's ability to predict the outcome. Once you have established which variables are important, rerun the analysis including only the important predictors and use the resulting parameter estimates to define your regression model. If the initial analysis reveals that there are two or more significant predictors then you could consider running a forward stepwise analysis (rather than forced entry) to find out the individual contribution of each predictor.

I have spent a lot of time explaining the theory behind regression and some of the diagnostic tools necessary to gauge the accuracy of a regression model. It is important to remember that SAS may appear to be very clever, but in fact it is not. Admittedly, it can do lots of complex calculations in a matter of seconds, but what it can't do is control the quality of the model that is generated – to do this requires a human brain (and preferably a trained one). SAS will happily generate output based on any garbage you decide to feed into it and SAS will not judge the results or give any indication of whether the model can be generalized or if it is valid. However, SAS provides the statistics necessary to judge these things, and at this point our brains must take over the job – which is slightly worrying (especially if your brain is as small as mine).

7.7.2. Main options ②

Imagine that the record company executive was now interested in extending the model of record sales to incorporate other variables. He decides to measure two new variables: (1) the number of times the record is played on Radio 1 during the week prior to release; and (2) the attractiveness of the band (**attract**). Before a record is released, the executive notes the amount spent on advertising, the number of times the record is played on radio the week before release, and the attractiveness of the band. He does this for 200 different records (each made by a different band). Attractiveness was measured by asking a random sample of the target audience to rate the attractiveness of each band on a scale from 0 (hideous potato-heads) to 10 (gorgeous sex objects). The modal value of attractiveness given by the sample was used in the regression (because he was interested in what the majority of people thought, rather than the average of people's opinions).

These data are in the file **Record2.sas7bdat** and you should note that each variable has its own column (the same layout as for correlation) and each row represents a different record. So, the first record had £10,260 spent advertising it, sold 330,000 copies, received 43 plays on Radio 1 the week before release, and was made by a band that the majority of people rated as gorgeous sex objects (Figure 7.10).

FIGURE 7.10

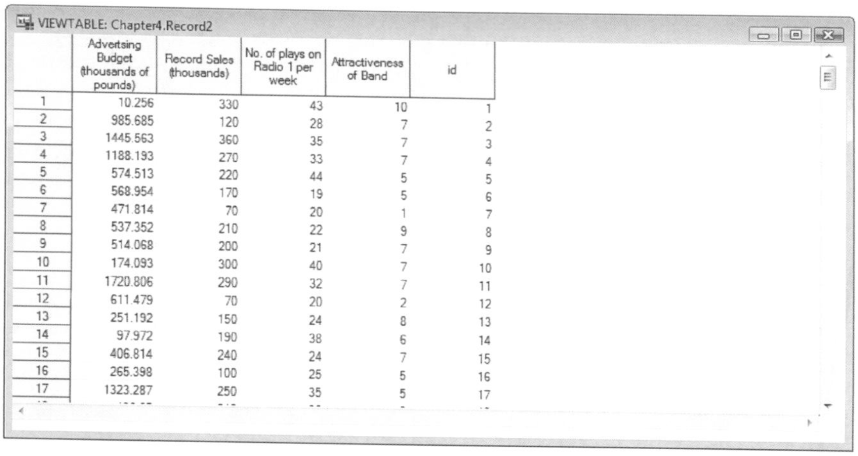

VIEWTABLE: Chapter4.Record2

	Advertising Budget (thousands of pounds)	Record Sales (thousands)	No. of plays on Radio 1 per week	Attractiveness of Band	id
1	10.256	330	43	10	1
2	985.685	120	28	7	2
3	1445.563	360	35	7	3
4	1188.193	270	33	7	4
5	574.513	220	44	5	5
6	568.954	170	19	5	6
7	471.814	70	20	1	7
8	537.352	210	22	9	8
9	514.068	200	21	7	9
10	174.093	300	40	7	10
11	1720.806	290	32	7	11
12	611.479	70	20	2	12
13	251.192	150	24	8	13
14	97.972	190	38	6	14
15	406.814	240	24	7	15
16	265.398	100	25	5	16
17	1323.287	250	35	5	17

The executive has past research indicating that advertising budget is a significant predictor of record sales, and so he should include this variable in the model first. His new variables (**airplay** and **attract**) should, therefore, be entered into the model *after* advertising budget. This method is hierarchical (the researcher decides in which order to enter variables into the model based on past research). To do a hierarchical regression in SAS we have to enter the variables in blocks (each block representing one step in the hierarchy).

First we could run a regression model with sales predicted by the advertising budget. SAS Syntax 7.2 shows this.

```
PROC REG DATA=chapter6.record2;
    MODEL sales = adverts;
    RUN; QUIT;
```

SAS Syntax 7.2

Then we could run a second regression, with sales predicted by adverts, airplay and attractiveness. This is shown in SAS Syntax 7.3.

```
PROC REG DATA=chapter6.record2;
    MODEL sales = adverts airplay attract;
    RUN; QUIT;
```

SAS Syntax 7.3

However, SAS will let us do both models in one step. All we need to do is to add a second model line, as shown in SAS Syntax 7.4.

```
PROC REG DATA=chapter6.record2;
    MODEL sales = adverts;
    MODEL sales = adverts airplay attract;
    RUN; QUIT;
```

SAS Syntax 7.4

SAS is going to run both models and give us the statistics for both. However, we want to know if adding the two additional predictors had a significant impact on the value of R^2, and this is done with a TEST statement, as shown in SAS Syntax 7.5.

```
PROC REG DATA=chapter6.record2;
    MODEL sales = adverts;
    MODEL sales = adverts airplay attract;
TEST airplay=0, attract=0;
    RUN; QUIT;
```

SAS Syntax 7.5

7.7.3. Statistics ②

If you just run the regression syntax as we've shown it above, you just get the basic output, including the estimated coefficients of the regression model (i.e. the estimated *b*-values), test statistics and their significance for each regression coefficient, and a t-test is used to see whether each *b* differs significantly from zero (see section 7.2.4). But there are many other things you can ask for, and a lot of them are useful. These are obtained by either putting an option on either the PROC REG line or the MODEL line. Options on the MODEL line are preceded by a slash /. There are many, many options available, and we are not going to try to go through them all, instead we will list the most important and useful options. (For a full listing check the help file, or search the web for SAS PROC REG.)

- STB. Added to the MODEL line, this gives standardized estimates. Just as we saw with covariances and correlations, it can be useful to standardize parameter estimates. The standardized estimates are the estimate that you would get if you standardized the variables before you did the regression. We saw in Section 7.4.3 that increasing the budget by £1000 increased the predicted number of records sold by 96. If these were not meaningful units, we could talk instead in standard deviations, and say that an increase of 1 standard deviation in the predictor variable increases the outcome variable by a certain number of standard deviations.

- CLB. This option, added to the MODEL line, produces confidence limits (also called intervals) for each of the unstandardized regression coefficients. Confidence intervals can be a very useful tool in assessing the likely value of the regression coefficients in the population. I will describe their exact interpretation later.

- CORR. Added to the PROG REG line, this option will display between the variables in the model. This option is extremely useful because the correlation matrix can be used to assess whether predictors are interrelated (which can be used to establish whether there is multicollinearity).
- SIMPLE. By adding this option to the PROG REG line, you can obtain a table of simple descriptive statistics including the mean, standard deviation and number of observations of all of the variables included in the analysis.
- VIF, TOL, COLLIN. You can add options to the MODEL line to obtain collinearity diagnostics. Adding VIF gives the variance inflation factor, and adding TOL will give you the tolerance. If you use the term COLLIN you will be given eigenvalues of the scaled, uncentred cross-products matrix and condition indexes (see section 7.6.2.3).
- *Collinearity diagnostics*: You can add options to the model line to obtain collinearity diagnostics. Adding vif, and adding tol will give you the tolerance. If you use the term collin you will be given eigenvalues of the scaled, uncentred cross-products matrix and condition indexes (see section 7.6.2.3).
- INFLUENCE. Adding this to the MODEL line produces the influence statistics for every case – that is, the residual, studentized residual, leverage (hat value), covariance ratio, DFFit and DFbetas. This makes for rather a lot of output, and it's often better to send it to a data set to be analysed. We'll see how to do that soon.

Let's recap and show where all those different options are put (see SAS Syntax 7.6).

```
PROC REG DATA=chapter6.record2 CORR SIMPLE ;
MODEL sales = adverts /STB CLB COLLIN TOL VIF ;
MODEL sales = adverts airplay attract /STB CLB COLLIN TOL VIF ;
TEST airplay=0, attract=0;
OUTPUT OUT=temp;
RUN; QUIT;
```

SAS Syntax 7.6

7.7.4. Saving regression diagnostics ②

In section 7.6 we met two types of regression diagnostics: those that help us assess how well our model fits our sample and those that help us detect cases that have a large influence on the model generated. In SAS we can choose to save these diagnostic a new data file.

To save regression diagnostics you need use the OUTPUT option. First, we give SAS a data file to write output to, e.g.:

OUTPUT OUT = myoutputfile;

Notice that we haven't specified a library to put this into – that's OK, because SAS will place it in the Work library, and will delete it next time we close SAS. We probably want it to be deleted, so we don't find it next year, and worry about whether or not it was important. However, if that's all we write then myoutputfile will just contain the input file that we used. We need to tell SAS what else we want in there. For example, if we want the residuals, and we want them to be saved in a variable called advertresiduals, we type:

OUTPUT OUT = myoutputfile RESIDUAL = advertresiduals;

- R or **Residual:** Unstandardized residuals.
- P or **predicted:** unstandardized predicted value.

- **Student** : Studentized residual.
- **Rstudent**: Studentized deleted residual.
- **dffits**: Standardized change in predicted value associated with that case.
- **Cookd**: Cook's distance.
- **H**: leverage value.

As we shall see later, producing these files and then plotting them can be very useful for testing assumptions of your regression analyses.

There is another way of getting many of these statistics, and that is to use the INFLUENCE option on the MODEL line. This sends the influence statistics (Residual, Studentized residual, leverage, covariance ratio, DFFit and DFBetas) to your output. We can ask instead to send them to a data file, so we can look at them more easily – this is done by adding the line:

ODS OUTPUT OUTPUTSTATISTICS=advertdfbetas;

I tend not to use that very often, because it's not very often that I want to look at the DFbetas, and it's easier to specify what you want.

Our final syntax is shown in SAS Syntax 7.7.

```
PROC REG DATA=chapter6.record2 CORR SIMPLE ;
    MODEL sales = adverts /STB CLB COLLIN TOL VIF ;
    MODEL sales = adverts airplay attract /STB CLB COLLIN TOL VIF ;
    TEST airplay=0, attract=0;
    OUTPUT OUT = myoutputfile
          RESIDUAL = advertresiduals
          PREDICTED = advertpredicted
          STUDENT = advertstudent
          RSTUDENT = advertrstudent
          DFFITS = advertdffits
          COOKD = advertcooks
          H = advertleverage;
ODS output OUTPUTSTATISTICS=advertdfbetas;
    RUN; QUIT;
```

SAS Syntax 7.7

The sharper eyed amongst you will have noticed that we never calculated the Mahalanobis distance. It's not an option in SAS, so we have to calculate it manually. Luckily for us, it's very easy, because it is given by:

$$md = (N-1)(h-1/N)$$

where *md* is the Mahalanobis distance, *H* is the leverage for that case, and *N* is the sample size. We can ask SAS to do that for us in a DATA step:

```
DATA myoutputfile2; SET myoutputfile;
    md = (200-1)*(advertleverage-1/200);
    RUN;
```

SAS Syntax 7.8

7.8. Interpreting multiple regression ②

When we run our analysis, SAS will spew out copious amounts of output, and we now turn to look at how to make sense of this information. We'll break it into parts, to make it easier.

We ran a hierarchical regression, with two MODEL statements. SAS produces one model output and then the second model output, but I have rearranged it to make it easier to talk about – specifically, I've put the same parts of the model output together.

7.8.1. Simple statistics ②

We asked for simple statistics, with the SIMPLE option, and these are shown in SAS Output 7.2. We start with the number of cases that were in the dataset, and the number that were used in the analysis. Then SAS gives us the sum, mean, uncorrected sum of squares, variance, standard deviation and the label for the variable. (The uncorrected sum of squares is the sum of the squared scores for the variable, not the sum of the squares of the differences between the mean and each score, which is how we would normally refer to the sum of squares).

Notice that there's a variable there called Intercept? In order to do the regression analysis, SAS creates a variable called Intercept for us, which is equal to 1.00 for every person. Sometimes the intercept is called the constant – that's because it is constant, it's always the same. We only get one set of descriptive statistics for both of the models.

SAS OUTPUT 7.2

Number of Observations Read	200
Number of Observations Used	200

Descriptive Statistics						
Variable	Sum	Mean	Uncorrected SS	Variance	Standard Deviation	Label
Intercept	200.00000	1.00000	200.00000	0	0	Intercept
ADVERTS	122882	614.41226	122436819	235861	485.65521	Advertsing Budget (thousands of pounds)
SALES	38640	193.20000	8761200	6512.32161	80.69896	Record Sales (thousands)
AIRPLAY	5500.00000	27.50000	181208	150.54271	12.26958	No. of plays on Radio 1 per week
ATTRACT	1354.00000	6.77000	9554.00000	1.94683	1.39529	Attractiveness of Band

SAS OUTPUT 7.3

Correlation					
Variable	Label	ADVERTS	SALES	AIRPLAY	ATTRACT
ADVERTS	Advertsing Budget (thousands of pounds)	1.0000	0.5785	0.1019	0.0808
SALES	Record Sales (thousands)	0.5785	1.0000	0.5989	0.3261
AIRPLAY	No. of plays on Radio 1 per week	0.1019	0.5989	1.0000	0.1820
ATTRACT	Attractiveness of Band	0.0808	0.3261	0.1820	1.0000

Next come the correlations (SAS Output 7.3), which we obtained with the CORR option on the PROC REG line. You might notice that along the diagonal of the matrix the values for the correlation coefficients are all 1.00 (i.e. a perfect positive correlation). The reason for this is that these values represent the correlation of each variable with itself, so obviously the resulting values are 1. The correlation matrix is extremely useful for getting a rough idea of the relationships between predictors and the outcome, and for a preliminary look for multicollinearity. If there is no multicollinearity in the data then there should be no substantial correlations ($r > .9$) between predictors.

If we look only at the predictors (ignore record sales) then the highest correlation is between the attractiveness of the band and the amount of airplay which is $r = .182$; the coefficient is small and so it looks as though our predictors are measuring different things (there is no collinearity). We can see also that of all of the predictors the number of plays on Radio 1 correlates best with the outcome ($r = .599$) and so it is likely that this variable will best predict record sales.

CRAMMING SAM'S TIPS

Use the descriptive statistics to check the correlation matrix for multicollinearity; that is, predictors that correlate too highly with each other, $R > .9$.

SAS OUTPUT 7.4

Analysis of Variance (Model 1)					
Source	DF	Sum of Squares	Mean Square	F Value	Pr > F
Model	1	433688	433688	99.59	<.0001
Error	198	862264	4354.86953		
Corrected Total	199	1295952			

Analysis of Variance (Model 2)					
Source	DF	Sum of Squares	Mean Square	F Value	Pr > F
Model	3	861377	287126	129.50	<.0001
Error	196	434575	2217.21725		
Corrected Total	199	1295952			

SAS Output 7.4 shows the next part of the output, which contains an ANOVA that tests whether the model is significantly better at predicting the outcome than using the mean as a 'best guess' (I've added 'Model 1' and 'Model 2' to the titles). Specifically, the F-ratio represents the ratio of the improvement in prediction that results from fitting the model, relative to the inaccuracy that still exists in the model (see section 7.2.3). We are told the value of the sum of squares for the model (this value is SS_M in section 7.2.3 and represents the improvement in prediction resulting from fitting a regression line to the data rather than using the mean as an estimate of the outcome). We are also told the error sum of squares (this value is SS_R in section 7.2.3 and represents the total difference between the model and the observed data). We are also told the degrees of freedom (df) for each term. In the case of the improvement due to the model, this value is equal to the number of predictors (one), and for SS_R it is the number of observations (200) minus the number of coefficients in the regression model. The model has two coefficients (one for the predictor and one for the constant) therefore model 1 has 198 degrees of freedom. The average sum of squares (MS) is then calculated for each term by dividing the SS by the df. The F-ratio is calculated by dividing the average improvement in prediction by the model (MS_M) by the average difference between the model and the observed data (MS_R). If the improvement due to fitting the regression model is much greater than the inaccuracy within the model then the value of F will be greater than 1 and SAS calculates the exact probability of obtaining the value of F by chance. For the initial model the F-ratio is 99.59, which is very unlikely to have happened by chance ($p < .0001$). For the second model the value of F is even higher (129.50), which is also highly significant ($p < .0001$). We can interpret these results as meaning that the initial model significantly improved our ability to predict the outcome variable, but that the new model (with the extra predictors) was even better (because the F-ratio is more significant).

The next section of output, shown in SAS Output 7.5, describes the overall model (so it tells us whether the model is successful in predicting record sales). This section provides us with some very important information about the model: the values of R^2 (R-Square) and the adjusted R^2 (Adj R-Sq). (And some other things which we are not going to worry about.)

R^2 is the squared value of the multiple correlation coefficient between the predictor(s) and the outcome. When only advertising budget is used as a predictor, this is the square of the simple correlation between advertising and record sales ($.578^2$=.3346). R^2 is also a measure of how much of the variability in the outcome is accounted for by the predictors. For the first model its value is .335, which means that advertising budget accounts for 33.5% of the variation in record sales. However, when the other two predictors are included as well (model 2), this value increases to .665 or 66.5% of the variance in record sales. Therefore, if advertising accounts for 33.5%, we can tell that attractiveness and radio play account for an additional 66.5.% – 33.5% = 33%. So, the inclusion of the two new predictors has explained quite a large amount of the variation in record sales.

SAS OUTPUT 7.5

Model 1			
Root MSE	65.99144	R-Square	0.3346
Dependent Mean	193.20000	Adj R-Sq	0.3313
Coeff Var	34.15706		

Model 2			
Root MSE	47.08734	R-Square	0.6647
Dependent Mean	193.20000	Adj R-Sq	0.6595
Coeff Var	24.37233		

The adjusted R^2 gives us some idea of how well our model generalizes and ideally we would like its value to be the same, or very close to, the value of R^2. In this example the difference for the final model is small (in fact the difference between the values is .665 – .660 = .005 (about 0.5%). This shrinkage means that if the model were derived from the population rather than a sample it would account for approximately 0.5% less variance in the outcome. Advanced students might like to apply Stein's formula to the R^2 to get some idea of its likely value in different samples. Stein's formula was given in equation (7.11) and can be applied by replacing n with the sample size (200) and k with the number of predictors (3):

$$\begin{aligned}\text{adjusted } R^2 &= 1 - \left[\left(\frac{200-1}{200-3-1}\right)\left(\frac{200-2}{200-3-2}\right)\left(\frac{200+1}{200}\right)\right](1-0.665)\\ &= 1 - [(1.015)(1.015)(1.005)](0.335)\\ &= 1 - 0.347\\ &= 0.653\end{aligned}$$

This value is very similar to the observed value of R^2 (.665) indicating that the cross-validity of this model is very good.

The change statistics tell us whether the change in R^2 is significant. The significance of R^2 can actually be tested using an F-ratio, and this F is calculated from the following equation (in which N is the number of cases or participants, and k is the number of predictors in the model):

$$F = \frac{(N - k - 1)R^2}{k(1 - R^2)} \tag{7.12}$$

CRAMMING SAM'S TIPS

The fit of the regression model can be assessed using the **Analysis of Variance** and **Fit Statistics** tables from SAS (note that SAS doesn't label the Fit Statistics). Look for the R^2 to tell you the proportion of variance explained by the model. If you have done a hierarchical regression then you can assess the improvement of the model at each stage of the analysis by looking at the change in R^2 and whether this change is significant (look for values less than .05 in the column labelled *Sig F Change*.). The ANOVA also tells us whether the model is a significant fit of the data overall (look for values less than .05 in the column labelled *Sig.*).

7.8.2. Model parameters ②

SAS OUTPUT 7.6

Parameter Estimates (Model 1)

Variable	DF	Parameter Estimate	Standard Error	t Value	Pr > |t|	Standardized Estimate	Tolerance	Variance Inflation
Intercept	1	134.13994	7.53657	17.80	<.0001	0	.	0
ADVERTS	1	0.09612	0.00963	9.98	<.0001	0.57849	1.00000	1.00000

Parameter Estimates (Model 1)

Variable	Label	DF	95% Confidence Limits	
Intercept	Intercept	1	119.27768	149.00219
ADVERTS	Advertising Budget (thousands of pounds)	1	0.07713	0.11512

Parameter Estimates (Model 2)

Variable	DF	Parameter Estimate	Standard Error	t Value	Pr > |t|	Standardized Estimate	Tolerance	Variance Inflation
Intercept	1	–26.61296	17.35000	–1.53	0.1267	0	.	0
ADVERTS	1	0.08488	0.00692	12.26	<.0001	0.51085	0.98562	1.01459
AIRPLAY	1	3.36743	0.27777	12.12	<.0001	0.51199	0.95923	1.04250
ATTRACT	1	11.08634	2.43785	4.55	<.0001	0.19168	0.96297	1.03845

Parameter Estimates (model 2)				
Variable	Label	DF	95% Confidence Limits	
Intercept	Intercept	1	-60.82961	7.60369
ADVERTS	Advertsing Budget (thousands of pounds)	1	0.07123	0.09854
AIRPLAY	No. of plays on Radio 1 per week	1	2.81962	3.91523
ATTRACT	Attractiveness of Band	1	6.27855	15.89412

So far we have looked at several summary statistics telling us whether or not the model has improved our ability to predict the outcome variable. The next part of the output is concerned with the parameters of the model. SAS Output 7.6 shows the model parameters for both steps in the hierarchy. Now, the first step in our hierarchy was to include advertising budget (as we did for the simple regression earlier in this chapter) and so the parameters for the first model are identical to the parameters obtained in SAS Output 7.1. Therefore, we will be concerned only with the parameters for the final model (in which all predictors were included). The format of the table of coefficients will depend on the options specified in the syntax. The standardized estimates, confidence interval for the *b*-values, and the collinearity diagnostics will be present only if specified in the syntax shown in SAS Syntax 7.6. (Standardized estimates were obtained with STB, confidence intervals with CLB, tolerance with TOL and variance inflation facts with VIF.) Also notice that the table was so wide that SAS has split it across two lines.

Remember that in multiple regression the model takes the form of equation (7.9) and in that equation there are several unknown quantities (the *b*-values or parameter estimates). The first part of the table gives us estimates for these *b*-values and these values indicate the individual contribution of each predictor to the model. If we replace the *b*-values in equation (7.9) we find that we can define the model as follows:

$$\begin{aligned} \text{sales}_i &= b_0 + b_1\text{advertising}_i + b_2\text{airplay}_i + b_3\text{attractiveness}_i \\ &= -26.61 + (0.08\text{advertising}_i) + (3.37\text{airplay}_i) \\ &\quad + (11.09\text{attractiveness}_i) \end{aligned} \tag{7.13}$$

The *b*-values tell us about the relationship between record sales and each predictor. If the value is positive we can tell that there is a positive relationship between the predictor and the outcome, whereas a negative coefficient represents a negative relationship. For these data all three predictors have positive *b*-values indicating positive relationships. So, as advertising budget increases, record sales increase; as plays on the radio increase, so do record sales; and finally more attractive bands will sell more records. The *b*-values tell us more than this, though. They tell us to what degree each predictor affects the outcome *if the effects of all other predictors are held constant*:

- **Advertising budget** ($b = 0.085$): This value indicates that as advertising budget increases by one unit, record sales increase by 0.085 units. Both variables were measured in thousands; therefore, for every £1000 more spent on advertising, an extra 0.085 thousand records (85 records) are sold. This interpretation is true only if the effects of attractiveness of the band and airplay are held constant.

- **Airplay** (b = 3.367): This value indicates that as the number of plays on radio in the week before release increases by one, record sales increase by 3.367 units. Therefore, every additional play of a song on radio (in the week before release) is associated with an extra 3.367 thousand records (3367 records) being sold. This interpretation is true only if the effects of attractiveness of the band and advertising are held constant.
- **Attractiveness** (b = 11.086): This value indicates that a band rated one unit higher on the attractiveness scale can expect additional record sales of 11.086 units. Therefore, every unit increase in the attractiveness of the band is associated with an extra 11.086 thousand records (11,086 records) being sold. This interpretation is true only if the effects of radio airplay and advertising are held constant.

Each of these parameter estimates (or beta values) values has an associated standard error indicating to what extent these values would vary across different samples, and these standard errors are used to determine whether or not the b-value differs significantly from zero. As we saw in section 7.4.2, a t-statistic can be derived that tests whether a b-value is significantly different from 0. In simple regression, a significant value of t indicates that the slope of the regression line is significantly different from horizontal, but in multiple regression, it is not so easy to visualize what the value tells us. Well, it is easiest to conceptualize the t-tests as measures of whether the predictor is making a significant contribution to the model. Therefore, if the t-test associated with a b-value is significant (if the value in the column labelled Pr > |t|. is less than .05) then the predictor is making a significant contribution to the model. The smaller the significance value (and the larger the value of t), the greater the contribution of that predictor. For this model, the advertising budget ($t(196) = 12.26$, $p < .0001$), the amount of radio play prior to release ($t(196) = 12.12$, $p < .0001$) and attractiveness of the band ($t(196) = 4.55$, $p < .0001$) are all significant predictors of record sales.[9] From the magnitude of the t-statistics we can see that the advertising budget and radio play had a similar impact, whereas the attractiveness of the band had less impact.

The b-values and their significance are important statistics to look at; however, the standardized versions of the b-values are in many ways easier to interpret (because they are not dependent on the units of measurement of the variables). The standardized estimates are provided by SAS and they tell us the number of standard deviations that the outcome will change as a result of one standard deviation change in the predictor. The standardized estimates are all measured in standard deviation units and so are directly comparable: therefore, they provide a better insight into the 'importance' of a predictor in the model. The standardized estimates for airplay and advertising budget are virtually identical (.512 and .511 respectively) indicating that both variables have a comparable degree of importance in the model (this concurs with what the magnitude of the t-statistics told us). To interpret these values literally, we need to know the standard deviations of all of the variables and these values can be found in SAS Output 7.2.

- **Advertising budget** (*standardized* β = .511): This value indicates that as advertising budget increases by one standard deviation (£485,655), record sales increase by 0.511 standard deviations. The standard deviation for record sales is 80,699 and so this constitutes a change of 41,240 sales (0.511 × 80,699). Therefore, for every £485,655 more spent on advertising, an extra 41,240 records are sold. This interpretation is true only if the effects of attractiveness of the band and airplay are held constant.

[9] For all of these predictors I wrote t(196). The number in brackets is the degrees of freedom. We saw in section 8.2.4 that in regression the degrees of freedom are N–p–1, where N is the total sample size (in this case 200) and p is the number of predictors (in this case 3). For these data we get 200–3–1 = 196.

- **Airplay** (*standardized* β = .512): This value indicates that as the number of plays on radio in the week before release increases by 1 standard deviation (12.27), record sales increase by 0.512 standard deviations. The standard deviation for record sales is 80,699 and so this constitutes a change of 41,320 sales (0.512 × 80,699). Therefore, if Radio 1 plays the song an extra 12.27 times in the week before release, 41,320 extra record sales can be expected. This interpretation is true only if the effects of attractiveness of the band and advertising are held constant.
- **Attractiveness** (*standardized* β = .192): This value indicates that a band rated one standard deviation (1.40 units) higher on the attractiveness scale can expect additional record sales of 0.192 standard deviations units. This constitutes a change of 15,490 sales (0.192 × 80,699). Therefore, a band with an attractiveness rating 1.40 higher than another band can expect 15,490 additional sales. This interpretation is true only if the effects of radio airplay and advertising are held constant.

SELF-TEST Think back to what the confidence interval of the mean represented (section 2.5.2). Can you guess what the confidence interval for *b* represents?

Imagine that we collected 100 samples of data measuring the same variables as our current model. For each sample we could create a regression model to represent the data. If the model is reliable then we hope to find very similar parameters in all samples. Therefore, each sample should produce approximately the same *b*-values. The confidence intervals of the unstandardized estimates are boundaries constructed such that in 95% of these samples these boundaries will contain the true value of *b* (see section 2.5.2). Therefore, if we'd collected 100 samples, and calculated the confidence intervals for *b*, we are saying that 95% of these confidence intervals would contain the true value of *b*. Therefore, we can be fairly confident that the confidence interval we have constructed for this sample will contain the true value of *b* in the population. This being so, a good model will have a small confidence interval, indicating that the value of *b* in this sample is close to the true value of *b* in the population. The sign (positive or negative) of the *b*-values tells us about the direction of the relationship between the predictor and the outcome. Therefore, we would expect a very bad model to have confidence intervals that cross zero, indicating that in some samples the predictor has a negative relationship to the outcome whereas in others it has a positive relationship. In this model, the two best predictors (advertising and airplay) have very tight confidence intervals indicating that the estimates for the current model are likely to be representative of the true population values. The interval for attractiveness is wider (but still does not cross zero) indicating that the parameter for this variable is less representative, but nevertheless significant.

CRAMMING SAM'S TIPS

The individual contribution of variables to the regression model can be found in the **Coefficients** table from SAS. If you have done a hierarchical regression then look at the values for the final model. For each predictor variable, you can see if it has made a significant contribution to predicting the outcome by looking at the column labelled *Sig.* (values less than .05 are significant). You should also look at the standardized estimates because these tell you the importance of each predictor (bigger absolute value = more important). The tolerance and VIF values will also come in handy later on, so make a note of them!

7.8.3. Comparing the models ①

In SAS Output 7.5, the R^2 and F values are reported for both of our models – each representing a block of the hierarchy. So, model 1 causes R^2 to change from 0 to .335, and this change in the amount of variance explained gives rise to an F-ratio of 99.59, which is significant with a probability less than .0001. Bearing in mind for this first model that we have only one predictor (so $k = 1$) and 200 cases ($N = 200$), this F comes from the equation above (7.12).

The addition of the new predictors (model 2) causes R^2 to increase of .330 (see above). We can calculate the F-ratio for this change using the same equation, but because we're looking at the change in the model R^2 we use the change in R^2 – that is the difference between the two values of R^2. The difference is often called ΔR^2 (where Δ is the Greek letter delta, which is used to mean 'difference'). We also use the change in the number of predictors, k_{Change} (model 1 had one predictor and model 2 had three predictors, so the change in the number of predictors is $3-1 = 2$),

As such, the change in the amount of variance that can be explained gives rise to an F-ratio of 96.45, which is again significant ($p < .0001$). The change statistics therefore tell us about the difference made by adding new predictors to the model.

We can ask SAS to calculate this for us, but we do it in a slightly backwards manner. Instead of asking what the significance of the increase in R^2 we ask SAS how much worse the model would be if we removed the variables of interest. We did this with the line following the second MODEL statement in SAS Syntax 7.6, which said:

```
TEST airplay=0, attract=0;
```

Adding that line gives us the output shown in SAS Output 7.7, which shows the value for the change that we found above.

SAS OUTPUT 7.7

Test 1 Results for Dependent Variable SALES				
Source	DF	Mean Square	F Value	Pr > F
Numerator	2	213845	96.45	<.0001
Denominator	196	2217.21725		

7.8.4. Assessing the issue of multicollinearity ①

SAS Output 7.6 provides some measures of whether there is collinearity in the data. Specifically, it provides the VIF and tolerance statistics (with tolerance being 1 divided by the VIF). There are a few guidelines from section 7.6.2.3 that can be applied here:

- If the largest VIF is greater than 10 then there is cause for concern (Bowerman & O'Connell, 1990; Myers, 1990).
- If the average VIF is substantially greater than 1 then the regression may be biased (Bowerman & O'Connell, 1990).
- Tolerance below 0.1 indicates a serious problem.
- Tolerance below 0.2 indicates a potential problem (Menard, 1995).

For our current model the VIF values are all well below 10 and the tolerance statistics all well above 0.2; therefore, we can safely conclude that there is no collinearity within our data. To calculate the average VIF we simply add the VIF values for each predictor and divide by the number of predictors (k):

$$F_{Model1} = \frac{(200-1-1)0.334648}{1(1-0.334648)} = 99.587$$

$$\begin{aligned} F_{Change} &= \frac{(N-k_2-1)R^2_{Change}}{k_{Change}(1-R^2_2)} \\ &= \frac{(200-3-1)\times 0.330}{2(1-0.664668)} \\ &= 96.44 \end{aligned}$$

The average VIF is very close to 1 and this confirms that collinearity is not a problem for this model.

If we add the option COLLIN to the model line, SAS also produces a table of eigenvalues of the scaled, uncentred cross-products matrix, condition indexes and variance proportions (see Jane Superbrain Box 7.3). There is a lengthy discussion, and example, of collinearity in section 8.8.1 and how to detect it using variance proportions, so I will limit myself now to saying that we are looking for large variance proportions on the same *small* eigenvalues. Therefore, in SAS Output 7.8 we look at the bottom few rows of the table (these are the small eigenvalues) and look for any variables that both have high variance proportions for that eigenvalue. The variance proportions vary between 0 and 1, and for each predictor should be distributed across different dimensions (or eigenvalues). For this model, you can see that each predictor has most of its variance loading onto a different dimension (advertising has 96% of variance on dimension 2, airplay has 93% of variance on dimension 3 and attractiveness has 92% of variance on dimension 4). These data represent a classic example of no multicollinearity. For an example of when collinearity exists in the data and some suggestions about what can be done, see Chapters 8 (section 8.8.1) and 17 (section 17.3.3.3).

SAS OUTPUT 7.8

Collinearity Diagnostics (Model 1)

Number	Eigenvalue	Condition Index	Proportion of Variation	
			Intercept	ADVERTS
1	1.78527	1.00000	0.10737	0.10737
2	0.21473	2.88340	0.89263	0.89263

Collinearity Diagnostics (Model 2)

Number	Eigenvalue	Condition Index	Proportion of Variation			
			Intercept	ADVERTS	AIRPLAY	ATTRACT
1	3.56209	1.00000	0.00277	0.02259	0.01122	0.00293
2	0.30803	3.40061	0.00638	0.95988	0.05348	0.00763
3	0.10949	5.70369	0.05369	0.01522	0.93240	0.06854
4	0.02039	13.21880	0.93716	0.00231	0.00289	0.92090

CRAMMING SAM'S TIPS

To check for multicollinearity, use the VIF values from the table labelled *Parameter Estimates*. If these values are less than 10 then that indicates there probably isn't cause for concern. If you take the average of VIF values, and this average is not substantially greater than 1, then that also indicates that there's no cause for concern.

JANE SUPERBRAIN 7.3

What are eigenvectors and eigenvalues? ④

The definitions and mathematics of eigenvalues and eigenvectors are very complicated and most of us need not worry about them (although they do crop up again in Chapters 16 and 17). However, although the mathematics is hard, they are quite easy to visualize! Imagine we have two variables: the salary a supermodel earns in a year, and how attractive she is. Also imagine these two variables are normally distributed and so can be considered together as a bivariate normal distribution. If these variables are correlated, then their scatterplot forms an ellipse. This is shown in the scatterplots below: if we draw a dashed line around the outer values of the scatterplot we get something oval shaped. Now, we can draw two lines to measure the length and height of this ellipse. These lines are the *eigenvectors* of the original correlation matrix for these two variables (a vector is just a set of numbers that tells us the location of a line in geometric space). Note that the two lines we've drawn (one for height and one for width of the oval) are perpendicular; that is, they are at 90 degrees, which means that they are independent of one another). So, with two variables, eigenvectors are just lines measuring the length and height of the ellipse that surrounds the scatterplot of data for those variables. If we add a third variable (e.g. experience of the supermodel) then all that happens is our scatterplot gets a third dimension, the ellipse turns into something shaped like a rugby ball (or American football), and because we now have a third dimension (height, width and depth) we get an extra eigenvector to measure this extra dimension. If we add a fourth variable, a similar logic applies (although it's harder to visualize): we get an extra dimension, and an eigenvector to measure that dimension. Now, each eigenvector has an *eigenvalue* that tells us its length (i.e. the distance from one end of the eigenvector to the other). So, by looking at all of the eigenvalues for a data set, we know the dimensions of the ellipse or rugby ball: put more generally, we know the dimensions of the data. Therefore, the eigenvalues show how evenly (or otherwise) the variances of the matrix are distributed.

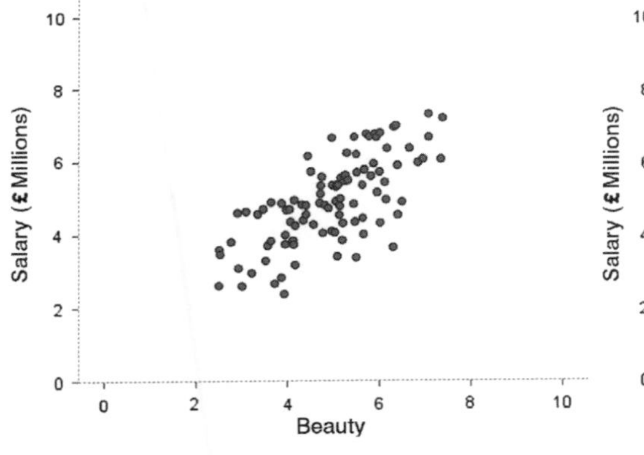

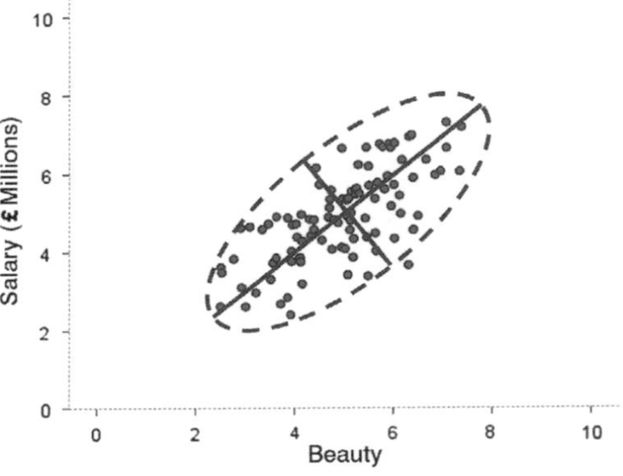

In the case of two variables, the *condition* of the data is related to the ratio of the larger eigenvalue to the smaller. Let's look at the two extremes: when there is no relationship at all between variables, and when there is a perfect relationship. When there is no relationship, the scatterplot will, more or less, be contained within a circle (or a sphere if we have three variables). If we again draw lines that measure the height and width of this circle we'll find that these lines are the same length. The eigenvalues measure the length, therefore the eigenvalues will also be the same. So, when we divide the largest eigenvalue by the smallest we'll get a value of 1 (because the eigenvalues are the same). When the variables are perfectly correlated (i.e. there is perfect collinearity) then the scatterplot forms a straight line and the ellipse surrounding it will also collapse to a straight line. Therefore, the height of the ellipse will be very small indeed (it will approach zero). Therefore, when we divide the largest eigenvalue by the smallest we'll get a value that tends to infinity (because the smallest eigenvalue is close to zero). Therefore, an infinite condition index is a sign of deep trouble.

7.8.5. Casewise diagnostics ①

You may remember that in section 7.7.4 we asked SAS to save various diagnostic statistics in a new dataset called myoutputfile.

One useful strategy is to use PROC UNIVARIATE to identify residuals that you want to investigate further. We can investigate the residuals, and see if any might be considered to be outliers, using the PROC UNIVARIATE syntax. (We came across PROC UNIVARIATE in Section 4.4.) The syntax to look at the residuals and studentized residuals is shown in SAS Syntax 7.7.

```
PROC UNIVARIATE DATA=myoutputfile PLOTS ;
 VAR advertstudent;
 HISTOGRAM advertstudent;
 RUN;
```

SAS Syntax 7.9

PROC UNIVARIATE produces quite a lot of output, but one of the most useful parts is the extreme observations – this prints the cases with the highest and lowest values for the studentized residual, and is shown in SAS Output 7.9.

SAS OUTPUT 7.9

Extreme Observations			
Lowest		Highest	
Value	Obs	Value	Obs
–2.66958	164	2.11803	52
–2.49354	47	2.12282	61
–2.48822	55	2.14988	10
–2.39184	68	2.19860	1
–2.34972	2	3.16362	169

These look like cases we might want to consider in more detail – perhaps we should consider all of the variables for these people. To do this, we can use PROC PRINT, is a simple procedure that just prints all of the variables, for all of the cases, to the Output window. This isn't going to be very useful in our case, because we are only really interested in those 10. However, we can ask PROC PRINT to only print those cases for us, by using the WHERE option. In addition, we need to know a little trick that SAS has – the case number (referred to as Obs in SAS Output 7.9) is a special variable, with the name _N_. We'll first create a shorter data file, with only those cases of interest in it, and then we'll print it. The syntax to do this is shown in SAS Syntax 7.10.

```
DATA myshortoutputfile; SET myoutputfile;
    casenum = _N_;
    IF casenum IN ( 164, 47, 55, 68, 2, 52, 61, 10, 1, 169) EQ 0
    THEN DELETE;
    RUN;
PROC PRINT DATA=myshortoutputfile;
    VAR advertresiduals advertpredicted
          advertstudent advertrstudent
          advertdffits advertcooks advertleverage          ;
    RUN;
```

SAS Syntax 7.10

Obs	Casenum	advertresiduals	advertpredicted	advertstudent	advertrstudent	advertdffits	advertcooks	advertleverage
1	1	100.080	229.920	2.17740	2.19860	0.48929	0.058704	0.047191
2	2	−108.949	228.949	−2.32308	−2.34972	−0.21110	0.010889	0.008007
3	10	99.534	200.466	2.13029	2.14988	0.26896	0.017756	0.015410
4	47	−114.970	154.970	−2.46100	−2.49354	−0.31469	0.024115	0.015677
5	52	97.403	92.597	2.09945	2.11803	0.36742	0.033159	0.029213
6	55	−114.123	304.123	−2.45591	−2.48822	−0.40736	0.040416	0.026104
7	61	98.810	201.190	2.10408	2.12282	0.15562	0.005948	0.005346
8	68	−110.416	180.416	−2.36355	−2.39184	−0.30216	0.022289	0.015709
9	164	−121.324	241.324	−2.62881	−2.66958	−0.54029	0.070766	0.039349
10	169	144.132	215.868	3.09333	3.16362	0.46132	0.050867	0.020821

SAS OUTPUT 7.10

SAS Output 7.10 shows the influence statistics for the 10 cases that I selected. The average leverage can be calculated as 0.02 ($(k + 1)/n = 4/200$) and so we are looking for values either twice as large as this (0.04) or three times as large (0.06) depending on which statistician you trust most (see section 7.6.1.2). All cases are within the boundary of three times the average and only case 1 is over two times the average. The evidence suggests that there are no influential cases within our data (although all cases would need to be examined to confirm this fact).

We can look also at the DFBeta statistics to see whether any case would have a large influence on the regression parameters. To do this, we need to look at the *other* output file we created, the one we called advertdfbetas. This one already has a variable called observation, and so we can use that with our PROC PRINT command. Using SAS Syntax 7.11 gives us SAS Output 7.11 (I've actually cheated a little, and removed some of the things we've already seen).

```
PROC PRINT DATA=advertdfbetas;
      WHERE observation IN ( 164, 47, 55, 68, 2, 52, 61, 10,
      1, 169)      ;
      RUN;
```

SAS Syntax 7.11

SAS OUTPUT 7.11

Obs	CovRatio	DFFITS	DFB_Intercept	DFB_ADVERTS	DFB_AIRPLAY	DFB_ATTRACT
1	0.9713	0.4893	−0.3155	−0.2423	0.1577	0.3533
2	0.9202	−0.2111	0.0126	−0.1264	0.0094	−0.0187
10	0.9439	0.2690	−0.0126	−0.1561	0.1677	0.0067
47	0.9146	−0.3147	0.0664	0.1960	0.0483	−0.1786
52	0.9600	0.3674	0.3529	−0.0288	−0.1367	−0.2697
55	0.9249	−0.4074	0.1743	−0.3265	−0.0231	−0.1243
61	0.9365	0.1556	0.0008	−0.0154	0.0279	0.0205
68	0.9237	−0.3022	−0.0028	0.2115	−0.1477	−0.0176
164	0.9204	−0.5403	0.1798	0.2899	−0.4009	−0.1171
169	0.8532	0.4613	−0.1682	−0.2576	0.2574	0.1697

For the DFBeta statistics, an absolute value greater than 1 is a problem and in all cases the values lie within ±1, which shows that these cases have no undue influence over the regression parameters. There is also a column for the covariance ratio. We saw in section 7.6.1.2 that we need to use the following criteria:

- $CVR_i > 1 + [3(k + 1)/n] = 1 + [3(3 + 1)/200] = 1.06$
- $CVR_i < 1 - [3(k + 1)/n] = 1 - [3(3 + 1)/200] = 0.94$.

Therefore, we are looking for any cases that deviate substantially from these boundaries. Most of our 12 potential outliers have CVR values within or just outside these boundaries. The only case that causes concern is case 169 whose CVR is some way below the bottom limit. However, given the other statistics for this case, there is probably little cause for alarm.

You would have requested other diagnostic statistics and from what you know from the earlier discussion of them you would be well advised to glance over them in case of any unusual cases in the data. However, from this minimal set of diagnostics we appear to have a fairly reliable model that has not been unduly influenced by any subset of cases.

CRAMMING SAM'S TIPS

You need to look for cases that might be influencing the regression model:

- Look at studentized residuals and check that no more than 5% of cases have absolute values above 2, and that no more than about 1% have absolute values above 2.5. Any case with a value above about 3, could be an outlier.
- Look in the data editor for the values of Cook's distance: any value above 1 indicates a case that might be influencing the model.
- Calculate the average leverage (the number of predictors plus 1, divided by the sample size) and then look for values greater than twice or three times this average value.
- For Mahalanobis distance, a crude check is to look for values above 25 in large samples (500) and values above 15 in smaller samples (100). However, Barnett and Lewis (1978) should be consulted for more detailed analysis.
- Look for absolute values of DFBeta greater than 1.
- Calculate the upper and lower limit of acceptable values for the covariance ratio, CVR. The upper limit is 1 plus three times the average leverage, whereas the lower limit is 1 minus three times the average leverage. Cases that have a CVR that fall outside of these limits may be problematic.

7.8.6. Checking assumptions ①

As a final stage in the analysis, you should check the assumptions of the model. Two useful plots to produce are a histogram and normal probability plot of the residuals, and a scatterplot of the residuals and the predicted values. The scatterplot of residuals against predicted values should look like a random array of dots evenly dispersed around zero. If this graph funnels out, then the chances are that there is heteroscedasticity in the data. If there is any sort of curve in this graph then the chances are that the data have broken the assumption of linearity. Figure 7.12 shows several examples of the plot of studentized residuals against standardized predicted values. Panel (a) shows the graph for the data in our record sales example. Note how the points are randomly and evenly dispersed throughout the plot. This pattern is indicative of a situation in which the assumptions of linearity and homoscedasticity have been met. Panel (b) shows a similar plot for a data set that violates the assumption of homoscedasticity. Note that the points form the shape of a funnel so they become more spread out across the graph. This funnel shape is typical of heteroscedasticity and indicates increasing variance across the residuals. Panel (c) shows a plot of some data in which there is a non-linear relationship between the outcome and the predictor. This pattern is shown up by the residuals. A line illustrating the curvilinear relationship has been drawn over the top of the graph to illustrate the trend in the data. Finally, panel (d) represents a situation in which the data not only represent a non-linear relationship, but also show heteroscedasticity. Note first the curved trend in the data, and then also note that at one end of the plot the points are very close together whereas at the other end they are widely dispersed. When these assumptions have been violated you will not see these exact patterns, but hopefully these plots will help you to understand the types of anomalies you should look out for.

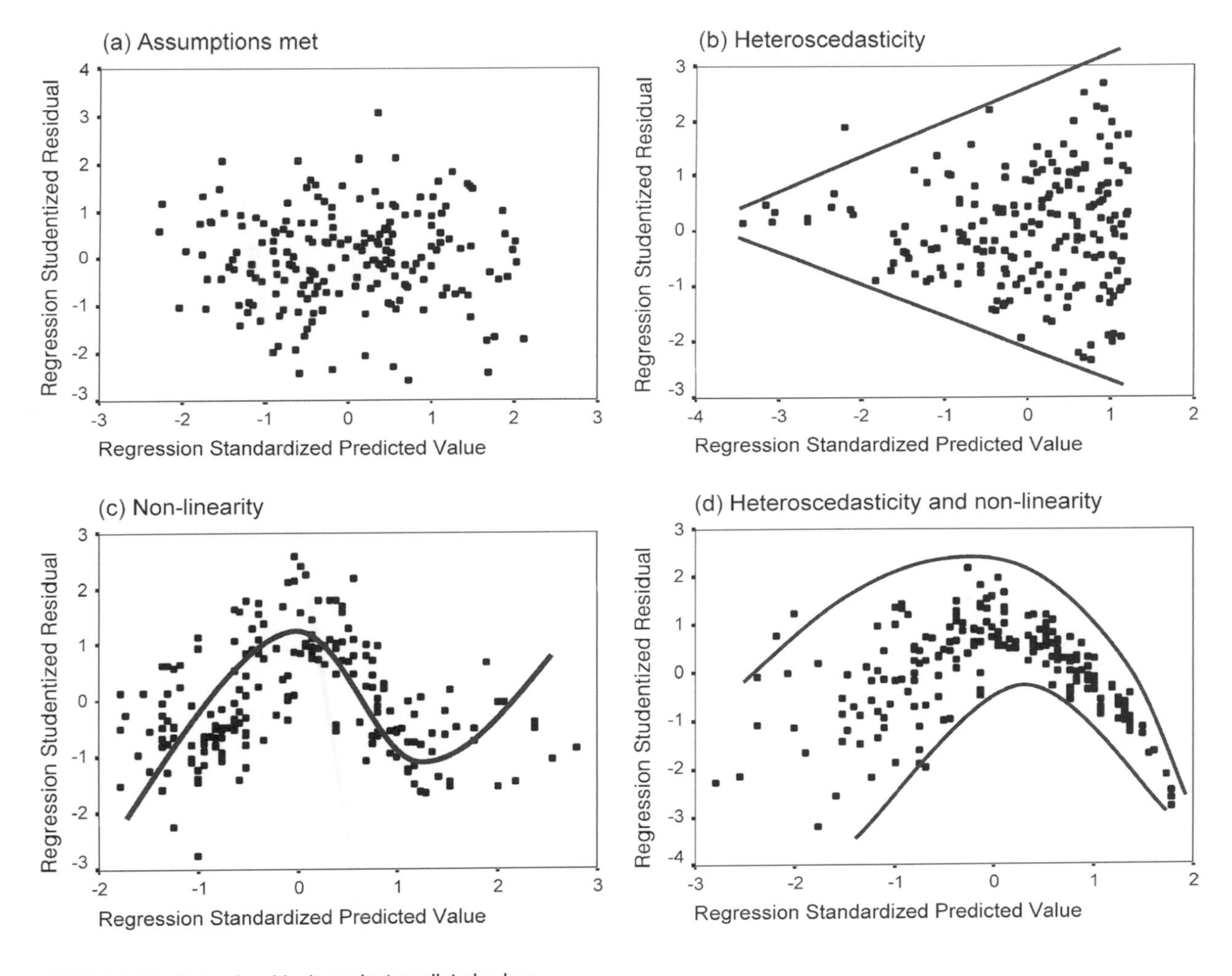

FIGURE 7.12 Plots of residuals against predicted values

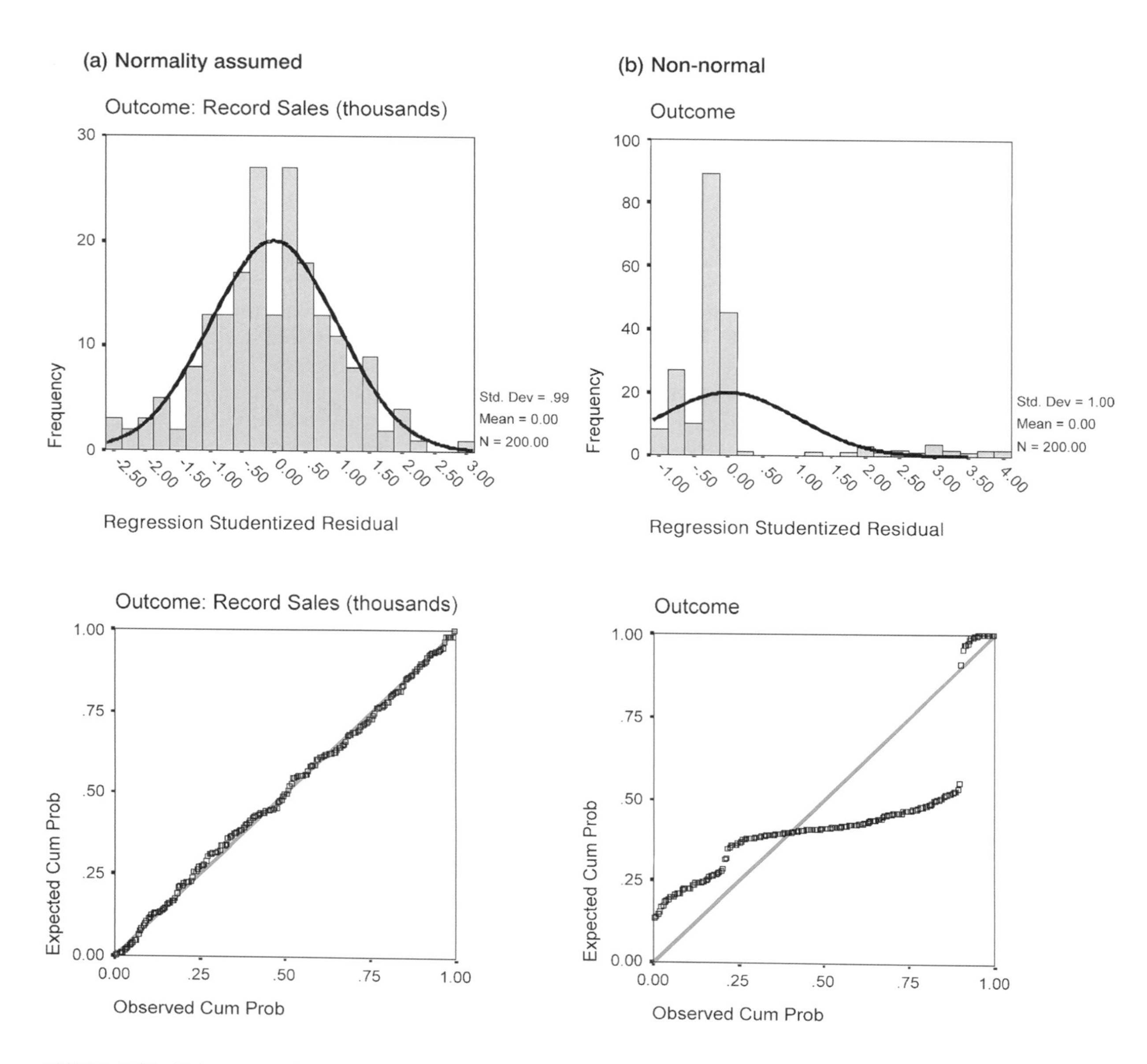

FIGURE 7.13 Histograms and normal P–P plots of normally distributed residuals (left-hand side) and non-normally distributed residuals (right-hand side)

To test the normality of residuals, we must look at the histogram and normal probability plots in Figure 7.13. Figure 7.13 shows the histogram and normal probability plot of the data for the current example (left-hand side). The histogram should look like a normal distribution (a bell-shaped curve). SAS can draw a curve on the histogram to show the shape of the distribution (as we saw in Chapter 5). For the record company data, the distribution is roughly normal (although there is a slight deficiency of residuals exactly on zero). Compare this histogram to the extremely non-normal histogram next to it and it should be clear that the non-normal distribution is extremely skewed. So, you should look for a curve that has the same shape as the one for the record sales data: any deviation from this curve is a sign of non-normality and the greater the deviation, the more non-normally distributed the residuals. The normal probability plot also shows up deviations from normality (see Chapter 5). The straight line in this plot represents a normal distribution, and the points represent the observed residuals. Therefore, in a perfectly normally distributed data set, all points will lie on the line. This is pretty much what we see for the record sales data. However, next to the normal probability plot of the record sales data is an example of an extreme deviation from normality. In this plot, the dots are very distant from the line, which indicates a large deviation from normality. For both plots, the non-normal data are extreme cases and you should be aware that the deviations from normality are likely to be subtler. Of course, you can use what you learnt in Chapter 5 to do a K–S test on the studentized residuals to see whether they deviate significantly from normality.

There are two ways to draw these graphs. The first is to use the datasets that we created when we ran the regression, and then follow the instructions shown in Chapter 5. However, if you are anything like me, you'd much rather do it the easy way. In version 9.1.3 the nice people at SAS added some features to draw these graphs automatically – they called them experimental, and in SAS 9.2 they are no longer called experimental. All you do is put either ODS HTML; or ODS RTF; then ODS GRAPHICS ON; the regression commands, ODS GRAPHICS OFF; and ODS RTF/HTML CLOSE; . If you use ODS RTF, your results will appear in your word processor. If you use ODS HTML, they will appear in your web browser (or possibly in SAS, depending on how you have SAS set up). These graphs are shown in Figures 7.14 and 7.15

```
ODS RTF;
ODS GRAPHICS ON;
PROC REG DATA=chapter6.record2 CORR SIMPLE ;
    MODEL sales = adverts airplay attract ;
    RUN; QUIT;
ODS RTF CLOSE;
ODS GRAPHICS OFF;
```

SAS Syntax 7.12

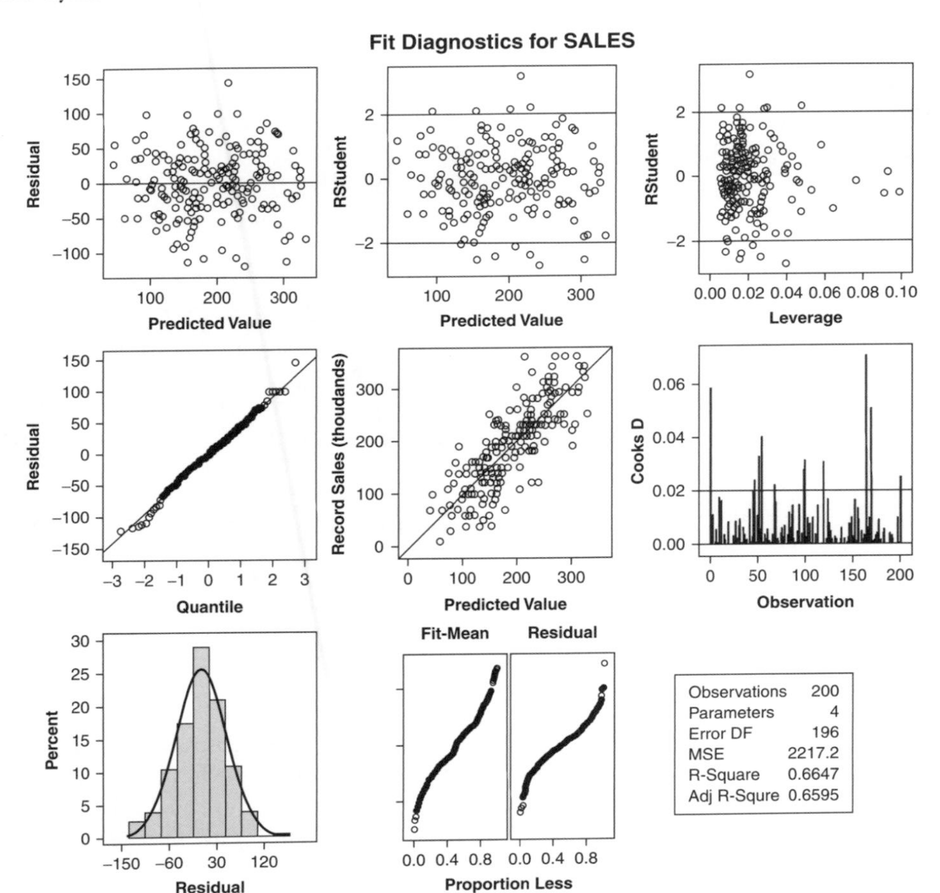

FIGURE 7.14 Plots from ODS Graphics

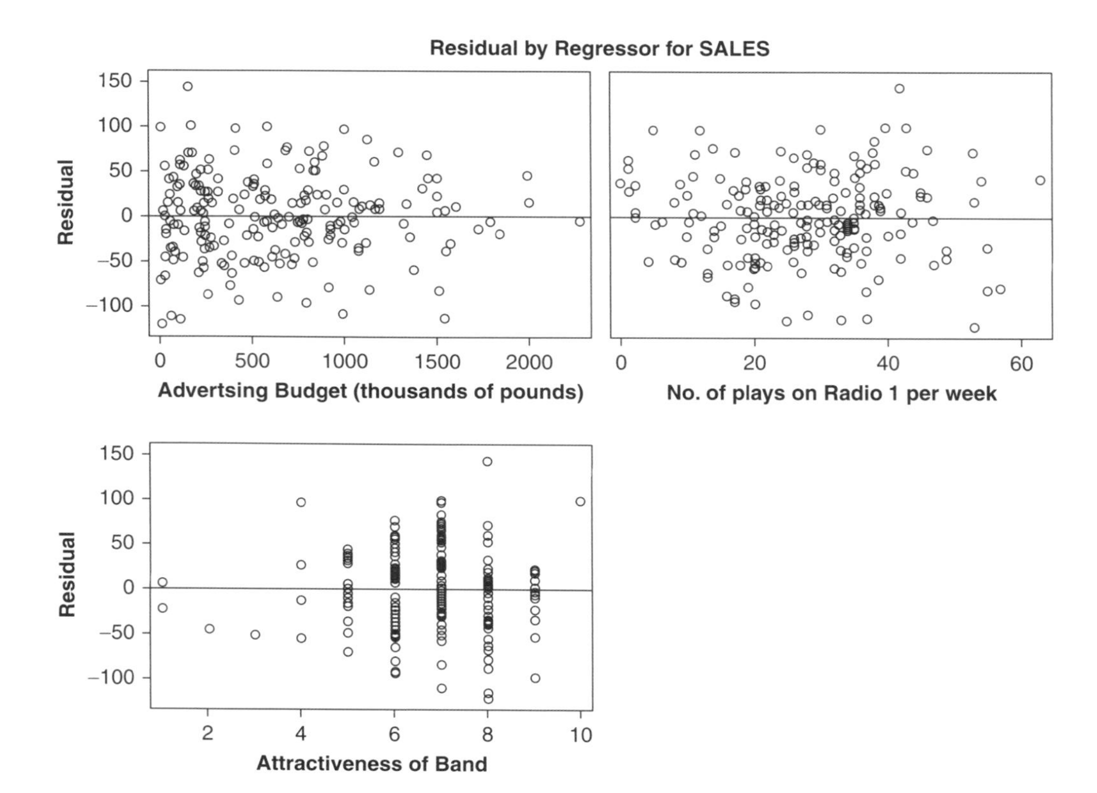

FIGURE 7.15 More plots from ODS Graphics

CRAMMING SAM'S TIPS

You need to check some of the assumptions of regression to make sure your model generalizes beyond your sample:

- Look at the graph of residuals plotted against predicted values. If it looks like a random array of dots then this is good. If the dots seem to get more or less spread out over the graph (look like a funnel) then this is probably a violation of the assumption of homogeneity of variance. If the dots have a pattern to them (i.e. a curved shape) then this is probably a violation of the assumption of linearity. If the dots seem to have a pattern and are more spread out at some points on the plot than others then this probably reflects violations of both homogeneity of variance *and* linearity. Any of these scenarios puts the validity of your model into question. Repeat the above for all partial plots too.
- Look at histograms and P–P plots. If the histograms look like normal distributions (and the P–P plot looks like a diagonal line), then all is well. If the histogram looks non-normal and the P–P plot looks like a wiggly snake curving around a diagonal line then things are less good! Be warned, though: distributions can look very non-normal in small samples even when they are normal!

We could summarize by saying that the model appears, in most senses, to be both accurate for the sample and generalizable to the population. The only slight glitch is some concern over whether attractiveness ratings had violated the assumption of homoscedasticity. Therefore, we could conclude that in our sample advertising budget and airplay are fairly equally important in predicting record sales. Attractiveness of the band is a significant predictor of record sales but is less important than the other two predictors (and probably needs verification because of possible heteroscedasticity). The assumptions seem to have been met and so we can probably assume that this model would generalize to any record being released.

7.9 What if I violate an assumption? ①

It's worth remembering that you can have a perfectly good model for your data (no outliers, influential cases, etc.) and you can use that model to draw conclusions about your sample, even if your assumptions are violated. However, it's much more interesting to generalize your regression model and this is where assumptions become important. If they have been violated then you cannot generalize your findings beyond your sample. The options for correcting for violated assumptions are a bit limited. If residuals show problems with heteroscedasticity or non-normality you could try transforming the raw data – but this won't necessarily affect the residuals! If you have a violation of the linearity assumption then you could see whether you can do logistic regression instead (described in the next chapter). Finally, there are a series of robust regression techniques (see section 5.7.4), which are described extremely well by Rand Wilcox in Chapter 10 of his book. SAS can't do these methods directly (well, technically it can do robust parameter estimates but it's not easy), but you can attempt a robust regression using some of Wilcox's files for the software R (Wilcox, 2005).

7.10 How to report multiple regression ①

If you follow the American Psychological Association guidelines for reporting multiple regression then the implication seems to be that tabulated results are the way forward. The APA also seem in favour of reporting, as a bare minimum, the standardized estimates, their significance value and some general statistics about the model (such as the R^2). If you do decide to do a table then the parameter estimates and their standard errors are also very useful. Personally I'd like to see the constant as well because then readers of your work can construct the full regression model if they need to. Also, if you've done a hierarchical regression you should report these values at each stage of the hierarchy. So, basically, you want to reproduce the table labelled *Estimates* from the SAS output and omit some of the non-essential information. For the example in this chapter we might produce a table like that in Table 7.2.

TABLE 7.2 How to report multiple regression

	B	SE *B*	β
Step 1			
Constant	134.14	7.54	
Advertising Budget	0.10	0.01	.58*
Step 2			
Constant	–26.61	17.35	
Advertising Budget	0.09	0.01	.51*
Plays on BBC radio 1	3.37	0.28	.51*
Attractiveness	11.09	2.44	.19*

Note: R^2 = .34 for Step 1: ΔR^2 = .33 for Step 2 (p < .001). * $p < .001$.

See if you can look back through the SAS output in this chapter and work out from where the values came. Things to note are: (1) I've rounded off to 2 decimal places throughout; (2) for the standardized regression estimates there is no zero before the decimal point (because these values cannot exceed 1) but for all other values less than 1 the zero is present.

LABCOAT LENI'S REAL RESEARCH 7.1

Why do you like your lecturers? ①

In the previous chapter we encountered a study by Chamorro-Premuzic et al. in which they measured students' personality characteristics and asked them to rate how much they wanted these same characteristics in their lecturers (see Labcoat Leni's Real Research 6.1 for a full description). In that chapter we correlated these scores; however, we could go a step further and see whether students' personality characteristics predict the characteristics that they would like to see in their lecturers.

The data from this study are in the file **Chamorro Premuzic.sas7bdat**. Labcoat Leni wants you to carry out five multiple regression analyses: the outcome variable in each of the five analyses is the ratings of how much students want to see Neuroticism, Extroversion, Openness to experience, Agreeableness and Conscientiousness. For each of these outcomes, force Age and Gender into the analysis in the first step of the hierarchy, then in the second block force in the five student personality traits (Neuroticism, Extroversion, Openness to experience, Agreeableness and Conscientiousness). For each analysis create a table of the results.

Answers are in the additional material on the companion website (or look at Table 4 in the original article).

CHAMORRO-PREMUZIC, T., ET AL. (2008). *PERSONALITY AND INDIVIDUAL DIFFERENCES*, 44, 965–976.

7.11. Categorical predictors and multiple regression ①

Often in regression analysis you'll collect data about groups of people (e.g. ethnic group, gender, socio-economic status, diagnostic category). You might want to include these groups as predictors in the regression model; however, we saw from our assumptions that variables need to be continuous or categorical with only two categories. It shouldn't be too inconceivable that we could then extend this model to incorporate several predictors that had two categories. All that is important is that we code the two categories with the values of 0 and 1. Why is it important that there are only two categories and that they're coded 0 and 1? Actually, I don't want to get into this here because this chapter is already too long, the publishers are going to break my legs if it gets any longer, and I explain it anyway later in the book (sections 9.7 and 10.2.3), so, for the time being, just trust me!

7.11.1. Dummy coding ①

The obvious problem with wanting to use categorical variables as predictors is that often you'll have more than two categories. For example, if you'd measured religiosity you might have categories of Muslim, Jewish, Hindu, Catholic, Buddhist, Protestant, Jedi (for those of you not in the UK, we had a census here a few years back in which a significant portion of people put down Jedi as their religion). Clearly these groups cannot be distinguished using a single variable coded with zeros and ones. In these cases we have to use what are called **dummy variables.** Dummy coding is a way of representing groups of people using only zeros and ones. To do it, we have to create several variables; in fact, the number of variables we need is one less than the number of groups we're recoding. There are eight basic steps:

1. Count the number of groups you want to recode and subtract 1.
2. Create as many new variables as the value you calculated in step 1. These are your dummy variables.
3. Choose one of your groups as a baseline (i.e. a group against which all other groups should be compared). This should usually be a control group, or, if you don't have a specific hypothesis, it should be the group that represents the majority of people (because it might be interesting to compare other groups against the majority).
4. Having chosen a baseline group, assign that group values of 0 for all of your dummy variables.
5. For your first dummy variable, assign the value 1 to the first group that you want to compare against the baseline group. Assign all other groups 0 for this variable.
6. For the second dummy variable assign the value 1 to the second group that you want to compare against the baseline group. Assign all other groups 0 for this variable.
7. Repeat this until you run out of dummy variables.
8. Place all of your dummy variables into the regression analysis!

Let's try this out using an example. In Chapter 4 we came across an example in which a biologist was worried about the potential health effects of music festivals. She collected some data at the Download Festival, which is a music festival specialising in heavy metal. The biologist was worried that her findings were a function of the fact that she had tested only one type of person: metal fans. Perhaps it's not the festival that makes people smelly, maybe it's only metal fans at festivals who get smellier (as a metal fan, I would at this point sacrifice the biologist to Satan for being so prejudiced). Anyway, to answer this question she went to another festival that had a more eclectic clientele. The Glastonbury Music Festival attracts all sorts of people because it has many styles of music there. Again, she measured the hygiene of concert-goers over the three days of the festival using a technique that results in a score ranging between 0 (you smell like you've bathed in sewage) and 4 (you smell of freshly baked bread). Now, in Chapters 4 and 5, we just looked at the distribution of scores for the three days of the festival, but now the biologist wanted to look at whether the type of music you like (your cultural group) predicts whether hygiene decreases over the festival. The data are in the file called **GlastonburyFestivalRegression.sas7bdat**. This file contains the hygiene scores for each of three days of the festival, but it also contains a variable called **change** which is the change in hygiene over the three days of the festival (so it's the change from day 1 to day 3).[10] Finally, the biologist categorized people according to their musical affiliation: if they mainly liked alternative music she called them 'indie kid', if they mainly liked heavy metal she called them a 'metaller' and if they mainly liked sort of hippy/folky/ambient type of stuff then she labelled them a 'crusty'. Anyone not falling into these categories was labelled 'no musical affiliation'. In the data file she coded these groups 1, 2, 3 and 4 respectively.

[10] Not everyone could be measured on day 3, so there is a change score only for a subset of the original sample.

The first thing we should do is set up the dummy variables. We have four groups, so there will be three dummy variables (one less than the number of groups). Next we need to choose a baseline group. We're interested in comparing those that have different musical affiliations against those that don't, so our baseline category will be 'no musical affiliation'. We give this group a code of 0 for all of our dummy variables. For our first dummy variable, we could look at the 'crusty' group, and to do this we give anyone that was a crusty a code of 1, and everyone else a code of 0. For our second dummy variable, we could look at the 'metaller' group, and to do this we give anyone that was a metaller a code of 1, and everyone else a code of 0. We have one dummy variable left and this will have to look at our final category: 'indie kid'. To do this we give anyone that was an indie kid a code of 1, and everyone else a code of 0. The resulting coding scheme is shown in Table 7.3. The thing to note is that each group has a code of 1 on only one of the dummy variables (except the base category that is always coded as 0).

TABLE 7.3 Dummy coding for the Glastonbury Festival data

	Dummy Variable 1	*Dummy Variable 2*	*Dummy Variable 3*
No Affiliation	0	0	0
Indie Kid	0	0	1
Metaller	0	1	0
Crusty	1	0	0

As I said, we'll look at why dummy coding works in sections 9.7 and 10.2.3, but for the time being let's look at how to recode our grouping variable into these dummy variables using SAS. To recode variables you need to use a DATA step.

First, we'll create a variable called **crusty**. This will have the value 1 if the person is in the crusty group, and the value 0 if they are in some other group. In the original variable **music**, a the crusty group was labelled with the value 3. The simplest way is to use two if... then statements:

```
IF music EQ 3 then crusty = 1;
IF music NE 3 then crusty = 0;
```

We can write that in one line, instead of two:

```
crusty = music EQ 3;
```

Remember that music EQ **3** is treated as a question by SAS, and the answer is either 'yes' (or 1) or the answer is 'no' or zero.

However, we'll encounter a problem with this if we have missing data – if we don't know a person's classification and they are missing on the music variable, they will be given a 0 on the crusty variable, when they should be given a missing value. The way around this is to check at the same time:

```
IF music NE . THEN crusty = music EQ 3;
```

We would then repeat that for the other two classes we need.

If we want to be a little more elegant, we can use a DO statement to test for missingness and then run the commands. We might as well do the other recodes in the same block. A DO statement looks like an IF... THEN statement:

```
IF music NE . THEN do;
     crusty = music EQ 3;
END;
```

The whole DATA step is shown in SAS Syntax 7.13. (Notice how I've used spaces to line things up – that's a handy trick, because it makes it much easier to see if you've made some sort of mistake, like missing out an equals sign).

```
DATA   GlastonburyDummies;   SET   chapter7.GlastonburyFestival
Regression;
     IF music NE . THEN DO;
          indiekid = music EQ 1;
          metaller = music EQ 2;
          crusty   = music EQ 3;
          END;
     RUN;
```

SAS Syntax 7.13

Whenever you created dummies, it's worth doing a check to make sure that it worked as you expected. You can do this using the PROC FREQ syntax shown in SAS Syntax 7.12. (There are several options on the TABLES line. The first three – NOCOL, NOROW and NOPERCENT – tell SAS not to display percentages, the last one – MISSING – tells SAS that we want to see missing values, if there are any).

```
PROC FREQ DATA=GlastonburyDummies;
     TABLES music*indiekid music*metaller music*crusty
          /NOCOL NOROW NOPERCENT MISSING;
     RUN;
```

SAS Syntax 7.14

This produces a series of tables; I've shown the first one in SAS Output 7.12. We can see that everyone who scored a 1 on **music** has scored a 1 on **indiekid**, and everyone who had another value on music has a 0 on **indiekid**.

SAS OUTPUT 7.12

Table of MUSIC by indiekid			
MUSIC(Musical Affiliation)	indiekid		
Frequency	0	1	Total
1	0	102	102
2	177	0	177
3	194	0	194
4	337	0	337
Total	708	102	810

7.11.2. SAS output for dummy variables ①

Let's assume you've successfully created the three dummy coding variables (if you're stuck there is a data file called **GlastonburyDummy.sas7bdat** (the 'Dummy' refers to the fact

it has dummy variables in it – I'm not implying that if you need to use this file you're a dummy!). With dummy variables, you have to enter all related dummy variables in the same block. So, in this case we have to enter our dummy variables in the same block; however, if we'd had another variable (e.g. socio-economic status) that had been transformed into dummy variables, we could enter these dummy variables in a different block (so it's only dummy variables that have recoded the same variable that need to be entered in the same block).

SELF-TEST Use what you've learnt in this chapter to run a multiple regression using the change scores as the outcome, and the three dummy variables (entered in the same block) as predictors.

Let's have a look at at the output. SAS Output 7.13 shows the model statistics. This shows that by entering the three dummy variables we can explain 7.6% of the variance in the change in hygiene scores (the R^2 value × 100). In other words, 7.6% of the variance in the change in hygiene can be explained by the musical affiliation of the person. The ANOVA tells us that the model is significantly better at predicting the change in hygiene scores than having no model (or, put another way, the 7.6% of variance that can be explained is a significant amount). Most of this should be clear from what you've read in this chapter already; what's more interesting is how we interpret the individual dummy variables.

SAS OUTPUT 7.13

Analysis of Variance					
Source	DF	Sum of Squares	Mean Square	F Value	Pr > F
Model	3	4.64647	1.54882	3.27	0.0237
Error	119	56.35780	0.47359		
Corrected Total	122	61.00427			

Root MSE	0.68818	R-Square	0.0762
Dependent Mean	−0.67504	Adj R-Sq	0.0529
Coeff Var	−101.94680		

SAS OUTPUT 7.14

Parameter Estimates								
Variable	Label	DF	Parameter Estimate	Standard Error	t Value	Pr > |t|	95% Confidence Limits	
Intercept	Intercept	1	−0.55431	0.09036	−6.13	<.0001	−0.73324	-0.37538
crusty		1	−0.41152	0.16703	−2.46	0.0152	−0.74226	-0.08079
indiekid		1	−0.40998	0.20492	−2.00	0.0477	−0.81574	-0.00421
metaller		1	0.02838	0.16033	0.18	0.8598	−0.28909	0.34586

SAS Output 7.14 shows a basic *Parameter Estimates* table for the dummy variables (I didn't ask for the standardized estimate, nor the collinearity diagnostics). The first dummy variable (crusty) shows the difference between the change in hygiene scores for the no affiliation group and the crusty group. Remember that the parameter estimate tells us the change in the outcome due to a unit change in the predictor. In this case, a unit change in the predictor is the change from 0 to 1. As such it shows the shift in the change in hygiene scores that results from the dummy variable changing from 0 to 1 (crusty). By including all three dummy variables at the same time, our baseline category is always zero, so this actually represents the difference in the change in hygiene scores if a person has no musical affiliation, compared to someone who is a crusty. This difference is the difference between the two group means.

To illustrate this fact, I've produced a table (SAS Output 7.15) of the group means for each of the four groups and also the difference between the means for each group and the *no affiliation* group. These means represent the average change in hygiene scores for the three groups (i.e. the mean of each group on our outcome variable). If we calculate the difference in these means for the no affiliation group and the crusty group we get, crusty − no affiliation = (−0.966) − (−0.554) = −0.412. In other words, the change in hygiene scores is greater for the crusty group than it is for the no affiliation group (crusties' hygiene decreases more over the festival than those with no musical affiliation). This value is the same as the parameter estimate in SAS Output 7.14! So, the estimate values tell us the relative difference between each group and the group that we chose as a baseline category. This estimate is converted to a *t*-statistic and the significance of this *t* reported. This *t*-statistic is testing, as we've seen before, whether the parameter estimate is 0 and when we have two categories coded with 0 and 1, that means it's testing whether the difference between group means is 0. If it is significant then it means that the group coded with 1 is significantly different from the baseline category – so, it's testing the difference between two means, which is the context in which students are most familiar with the *t*-statistic (see Chapter 9). For our first dummy variable, the *t*-test is significant, and the parameter estimate has a negative value so we could say that the change in hygiene scores goes down as a person changes from having no affiliation to being a crusty. Bear in mind that a decrease in hygiene scores represents more change (you're becoming smellier) so what this actually means is that hygiene decreased significantly more in crusties compared to those with no musical affiliation!

SAS OUTPUT 7.15

Musical Affiliation	N	Mean
Indie Kid	14	–0.964
Metaller	27	–0.526
Crusty	24	–0.966
No Affiliation	58	–0.554

Moving on to our next dummy variable, this compares metallers to those that have no musical affiliation. The regression estimate again represents the shift in the change in hygiene scores if a person has no musical affiliation, compared to someone who is a metaller. If we calculate the difference in the group means for the no affiliation group and the metaller group we get, metaller − no affiliation = (−0.526) − (−0.554) = 0.028. This value is again the same as the value of the parameter estimate in SAS Output 7.14. For this second dummy variable, the *t*-test is not significant, so we could say that the change in hygiene scores is the same if a person changes from having no affiliation to being a metaller. In other words, the change in hygiene scores is not predicted by whether someone is a metaller compared to if they have no musical affiliation.

For the final dummy variable, we're comparing indie kids to those that have no musical affiliation. The parameter estimate again represents the shift in the change in hygiene scores if a person has no musical affiliation, compared to someone who is an indie kid. If we calculate the difference in the group means for the no affiliation group and the indie kid group we get, indie kid − no affiliation = (−0.964) − (−0.554) = −0.410. It should be no surprise to you by now that this is the parameter estimate value in SAS Output 7.14. The *t*-test is significant, and the estimate has a negative value so, as with the first dummy variable, we could say that the change in hygiene scores goes down if we compare a person having no affiliation to an indie kid. Bear in mind that a decrease in hygiene scores represents more change (you're becoming smellier) so what this actually means is that hygiene decreased significantly more in indie kids compared to those with no musical affiliation.

So, overall this analysis has shown that compared to having no musical affiliation, crusties and indie kids get significantly smellier across the three days of the festival, but metallers don't. This section has introduced some really complex ideas that I expand upon in Chapters 9 and 10. It might all be a bit much to take in, and so if you're confused or want to know more about why dummy coding works in this way I suggest reading sections 9.7 and 10.2.3 and then coming back here. Alternatively, read Hardy's (1993) excellent monograph.

What have I discovered about statistics? ①

This chapter is possibly the longest book chapter ever written, and if you feel like you aged several years while reading it then, well, you probably have (look around, there are cobwebs in the room, you have a long beard, and when you go outside you'll discover a second ice age has been and gone leaving only you and a few woolly mammoths to populate the planet). However, on the plus side, you now know more or less everything you'll ever need to know about statistics. Really, it's true; you'll discover in the coming chapters that everything else we discuss is basically a variation on the theme of regression. So, although you may be near death having spent your life reading this chapter (and I'm certainly near death having written it) you are a stats genius – it's official.

We started the chapter by discovering that at 8 years old I could have really done with regression analysis to tell me which variables are important in predicting talent competition success. Unfortunately I didn't have regression, but fortunately I had my dad instead (and he's better than regression). We then looked at how we could use statistical models to make similar predictions by looking at the case when you have one predictor and one outcome. This allowed us to look at some basic principles such as the equation of a straight line, the method of least squares, and how to assess how well our model fits the data using some important quantities that you'll come across again in future chapters: the model sum of squares, SS_M, the residual sum of squares, SS_R, and the total sum of squares, SS_T. We used these values to calculate several important statistics such as R^2 and the *F*-ratio. We also learnt how to do a regression on SAS, and how we can plug the resulting parameter estimates into the equation of a straight line to make predictions about our outcome.

Next, we saw that the question of a straight line can be extended to include several predictors and looked at different methods of placing these predictors in the model (hierarchical, forced entry, stepwise). Then, we looked at factors that can affect the accuracy of a model (outliers and influential cases) and ways to identify these factors. We then moved on to look at the assumptions necessary to generalize our model beyond the sample of data we've collected before discovering how to do the analysis on SAS, and how to interpret the output, create our multiple regression model and test its reliability and generalizability. I finished the chapter by looking at how we can use categorical predictors in regression. In general, multiple regression is a long process and should be done with care and attention to detail. There are a lot of important things to consider and you should approach the analysis in a systematic fashion. I hope this chapter helps you to do that!

So, I was starting to get a taste for the rock-idol lifestyle: I had friends, a fortune (well, two gold-plated winner's medals), fast cars (a bike) and dodgy-looking 8 year olds were giving me suitcases full of lemon sherbet to lick off mirrors. However, my parents and teachers were about to impress reality upon my young mind …

Key terms that I've discovered

Adjusted predicted value
Adjusted r^2
Autocorrelation
b_i
β_i
Cook's distance
Covariance ratio (CVR)
Cross-validation
Deleted residual
DFBeta
DFFit
Dummy variables
Durbin-Watson test
F-ratio
Generalization
Goodness of fit
Hat values
Heteroscedasticity
Hierarchical regression
Homoscedasticity
Independent errors
Leverage
Mahalanobis distances
Mean squares
Model sum of squares
Multicollinearity
Multiple *R*
Multiple regression
Outcome variable
Perfect collinearity
Predictor variable
Residual
Residual sum of squares
Shrinkage
Simple regression
Standardized DFBeta
Standardized DFFit
Stepwise regression
Standardized residuals
Studentized deleted residuals
Studentized residuals
Suppressor effects
t-statistic
Tolerance
Total sum of squares
Unstandardized residuals
Variance inflation factor (VIF)

Smart Alex's tasks

- **Task 1**: A fashion student was interested in factors that predicted the salaries of catwalk models. She collected data from 231 models. For each model she asked them their salary per day on days when they were working (**salary**), their age (**age**), how many years they had worked as a model (**years**), and then got a panel of experts from modelling agencies to rate the attractiveness of each model as a percentage with 100% being perfectly attractive (**beauty**). The data are in the file **Supermodel.sas7bdat.** Unfortunately, this fashion student bought some substandard statistics text and so doesn't know how to analyse her data. Can you help her out by conducting a multiple regression to see which variables predict a model's salary? How valid is the regression model?

- **Task 2**: Using the Glastonbury data from this chapter (with the dummy coding in **GlastonburyDummy.sas7bdat**), which you should've already analysed, comment on whether you think the model is reliable and generalizable.
- **Task 3**: A study was carried out to explore the relationship between **Aggression** and several potential predicting factors in 666 children who had an older sibling. Variables measured were **Parenting_Style** (high score = bad parenting practices), **Computer_Games** (high score = more time spent playing computer games), **Television** (high score = more time spent watching television), **Diet** (high score = the child has a good diet low in additives), and **Sibling_Aggression** (high score = more aggression seen in their older sibling). Past research indicated that parenting style and sibling aggression were good predictors of the level of aggression in the younger child. All other variables were treated in an exploratory fashion. The data are in the file **Child Aggression.sas7bdat.** Analyse them with multiple regression.

Answers can be found on the companion website.

Further reading

Bowerman, B. L., & O'Connell, R. T. (1990). *Linear statistical models: an applied approach* (2nd ed.). Belmont, CA: Duxbury. (This text is only for the mathematically minded or postgraduate students but provides an extremely thorough exposition of regression analysis.)

Hardy, M. A. (1993). *Regression with dummy variables*. Sage University Paper Series on Quantitative Applications in the Social Sciences, 07–093. Newbury Park, CA: Sage.

Howell, D. C. (2006). *Statistical methods for psychology* (6th ed.). Belmont, CA: Duxbury. (Or you might prefer his *Fundamental statistics for the behavioral sciences,* also in its 6th edition, 2007. Both are excellent introductions to the mathematics behind regression analysis.)

Littell, R.C., Stroup, W.W., Freund, R. J. (2002). *SAS for linear models*. Cary, NC: SAS Publishing. (This is another guide produced by SAS – it tells you everything there is to know about PROC REG (and a few other procs too), but it's fairly hard going if you're new to regression.)

Miles, J. N. V. & Shevlin, M. (2001). *Applying regression and correlation: a guide for students and researchers*. London: Sage. (This is an extremely readable text that covers regression in loads of detail but with minimum pain – highly recommended.)

Stevens, J. (2002). *Applied multivariate statistics for the social sciences* (4th ed.). Hillsdale, NJ: Erlbaum. (See Chapter 3.)

Interesting real research

Chamorro-Premuzic, T., Furnham, A., Christopher, A. N., Garwood, J., & Martin, N. (2008). Birds of a feather: Students' preferences for lecturers' personalities as predicted by their own personality and learning approaches. *Personality and Individual Differences*, *44*, 965–976.

Logistic regression

8

FIGURE 8.1
Practising for my career as a rock star by slaying the baying throng of Grove Primary School at the age of 10. (Note the girl with her hands covering her ears.)

8.1. What will this chapter tell me? ①

We saw in the previous chapter that I had successfully conquered the holiday camps of Wales with my singing and guitar playing (and the Welsh know a thing or two about good singing). I had jumped on a snowboard called oblivion and thrown myself down the black run known as world domination. About 10 metres after starting this slippery descent I hit the lumpy patch of ice called 'adults'. I was 9, life was fun, and yet every adult that I encountered seemed to be was obsessed with my future. 'What do you want to be when you grow up?' they would ask. I was 9 and 'grown-up' was a lifetime away; all I knew was that I was going to marry Clair Sparks (more on her in the next chapter) and that I was a rock legend who didn't need to worry about such adult matters as having a job. It was a difficult question, but adults require answers and I wasn't going to let them know that I didn't care about 'grown-up' matters.

We saw in the previous chapter that we can use regression to predict future outcomes based on past data, when the outcome is a continuous variable, but this question had a categorical outcome (e.g. would I be a fireman, a doctor, a pimp?). Luckily, though, we can use an extension of regression called logistic regression to deal with these situations. What a result; bring on the rabid wolves of categorical data. To make a prediction about a categorical outcome then, as with regression, I needed to draw on past data: I hadn't tried conducting brain surgery, neither had I experience of sentencing psychopaths to prison sentences for eating their husbands, nor had I taught anyone. I had, however, had a go at singing and playing guitar; 'I'm going to be a rock star' was my prediction. A prediction can be accurate (which would mean that I am a rock star) or it can be inaccurate (which would mean that I'm writing a statistics textbook). This chapter looks at the theory and application of **logistic regression**, an extension of regression that allows us to predict categorical outcomes based on predictor variables.

8.2. Background to logistic regression ①

In a nutshell, logistic regression is multiple regression but with an outcome variable that is a categorical variable and predictor variables that are continuous or categorical. In its simplest form, this means that we can predict which of two categories a person is likely to belong to given certain other information. A trivial example is to look at which variables predict whether a person is male or female. We might measure laziness, pig-headedness, alcohol consumption and number of burps that a person does in a day. Using logistic regression, we might find that all of these variables predict the gender of the person, but the technique will also allow us to predict whether a certain person is likely to be male or female. So, if we picked a random person and discovered they scored highly on laziness, pig-headedness, alcohol consumption and the number of burps, then the regression model might tell us that, based on this information, this person is likely to be male. Admittedly, it is unlikely that a researcher would ever be interested in the relationship between flatulence and gender (it is probably too well established by experience to warrant research), but logistic regression can have life-saving applications. In medical research logistic regression is used to generate models from which predictions can be made about the likelihood that a tumour is cancerous or benign (for example). A database of patients can be used to establish which variables are influential in predicting the malignancy of a tumour. These variables can then be measured for a new patient and their values placed in a logistic regression model, from which a probability of malignancy could be estimated. If the probability of the tumour being malignant is suitably low then the doctor may decide not to carry out expensive and painful surgery that in all likelihood is unnecessary. We might not face such life-threatening decisions but logistic regression can nevertheless be a very useful tool. When we are trying to predict membership of only two categorical outcomes the analysis is known as **binary logistic regression**, but when we want to predict membership of more than two categories we use **multinomial** (polychotomous) **logistic regression.**

8.3. What are the principles behind logistic regression? ③

I don't wish to dwell on the underlying principles of logistic regression because they aren't necessary to understand the test. However, I do wish to draw a few parallels to normal regression so that you can get the gist of what's going on using a framework that will be familiar to you already (what do you mean you haven't read the regression chapter yet!).

To keep things simple I'm going to explain binary logistic regression, but most of the principles extend easily to when there are more than two outcome categories. Now would be a good time for the equation-phobes to look away. In simple linear regression, we saw that the outcome variable Y is predicted from the equation of a straight line:

$$Y_i = b_0 + b_1 X_{1i} + \varepsilon_i \tag{8.1}$$

in which b_0 is the Y intercept, b_1 is the gradient of the straight line, X_1 is the value of the predictor variable and ε is a residual term. Given the values of Y and X_1, the unknown parameters in the equation can be estimated by finding a solution for which the squared distance between the observed and predicted values of the dependent variable is minimized (the method of least squares).

This stuff should all be pretty familiar by now. In multiple regression, in which there are several predictors, a similar equation is derived in which each predictor has its own coefficient. As such, Y is predicted from a combination of each predictor variable multiplied by its respective regression coefficient.

$$Y_i = b_0 + b_1 X_{1i} + b_2 X_{2i} + \ldots + b_n X_{ni} + \varepsilon_i \tag{8.2}$$

in which b_n is the regression coefficient of the corresponding variable X_n. In logistic regression, instead of predicting the value of a variable Y from a predictor variable X_1 or several predictor variables (Xs), we predict the *probability* of Y occurring given known values of X_1 (or Xs). The logistic regression equation bears many similarities to the regression equations just described. In its simplest form, when there is only one predictor variable X_1, the logistic regression equation from which the probability of Y is predicted is given by:

$$P(Y) = \frac{1}{1 + e^{-(b_0 + b_1 X_{1i})}} \tag{8.3}$$

in which $P(Y)$ is the probability of Y occurring, e is the base of natural logarithms, and the other coefficients form a linear combination much the same as in simple regression. In fact, you might notice that the bracketed portion of the equation is identical to the linear regression equation in that there is a constant (b_0), a predictor variable (X_1) and a coefficient (or weight) attached to that predictor (b_1). Just like linear regression, it is possible to extend this equation so as to include several predictors. When there are several predictors the equation becomes:

$$P(Y) = \frac{1}{1 + e^{-(b_0 + b_1 X_{1i} + b_2 X_{2i} + \ldots + b_n X_{ni})}} \tag{8.4}$$

Equation (8.4) is the same as the equation used when there is only one predictor except that the linear combination has been extended to include any number of predictors. So, whereas the one-predictor version of the logistic regression equation contained the simple linear regression equation within it, the multiple-predictor version contains the multiple regression equation.

Despite the similarities between linear regression and logistic regression, there is a good reason why we cannot apply linear regression directly to a situation in which the outcome variable is categorical. The reason is that one of the assumptions of linear regression is that the relationship between variables is linear (see section 7.6.2.1). In that section we saw how important it is

that the assumptions of a model are met for it to be accurate. Therefore, for linear regression to be a valid model, the observed data should contain a linear relationship. When the outcome variable is categorical, this assumption is violated (Berry, 1993). One way around this problem is to transform the data using the logarithmic transformation (see Berry & Feldman, 1985, and Chapter 5). This transformation is a way of expressing a non-linear relationship in a linear way. The logistic regression equation described above is based on this principle: it expresses the multiple linear regression equation in logarithmic terms (called the *logit*) and thus overcomes the problem of violating the assumption of linearity.

The exact form of the equation can be arranged in several ways but the version I have chosen expresses the equation in terms of the probability of Y occurring (i.e. the probability that a case belongs in a certain category). The resulting value from the equation, therefore, varies between 0 and 1. A value close to 0 means that Y is very unlikely to have occurred, and a value close to 1 means that Y is very likely to have occurred. Also, just like linear regression, each predictor variable in the logistic regression equation has its own coefficient. When we run the analysis we need to estimate the value of these coefficients so that we can solve the equation. These parameters are estimated by fitting models, based on the available predictors, to the observed data. The chosen model will be the one that, when values of the predictor variables are placed in it, results in values of Y closest to the observed values. Specifically, the values of the parameters are estimated using **maximum-likelihood estimation**, which selects coefficients that make the observed values most likely to have occurred. So, as with multiple regression, we try to fit a model to our data that allows us to estimate values of the outcome variable from known values of the predictor variable or variables.

8.3.1. Assessing the model: the log-likelihood statistic ③

We've seen that the logistic regression model predicts the probability of an event occurring for a given person (we would denote this as $P(Y_i)$, the probability that Y occurs for the ith person), based on observations of whether or not the event did occur for that person (we could denote this as Y_i, the actual outcome for the ith person). So, for a given person, Y will be either 0 (the outcome didn't occur) or 1 (the outcome did occur), and the predicted value, $P(Y)$, will be a value between 0 (there is no chance that the outcome will occur) and 1 (the outcome will certainly occur). We saw in multiple regression that if we want to assess whether a model fits the data we can compare the observed and predicted values of the outcome (if you remember, we use R^2, which is the Pearson correlation between observed values of the outcome and the values predicted by the regression model). Likewise, in logistic regression, we can use the observed and predicted values to assess the fit of the model. The measure we use is the **log-likelihood**:

$$\text{log-likelihood} = \sum_{i=1}^{N} [Y_i \ln(P(Y_i)) + (1 - Y_i)\ln(1 - P(Y_i))] \tag{8.5}$$

The log-likelihood is based on summing the probabilities associated with the predicted and actual outcomes (Tabachnick & Fidell, 2007). The log-likelihood statistic is analogous to the residual sum of squares in multiple regression in the sense that it is an indicator of how much unexplained information there is after the model has been fitted. It, therefore, follows that large values of the log-likelihood statistic indicate poorly fitting statistical models, because the larger the value of the log-likelihood, the more unexplained observations there are.

Now, it's possible to calculate a log-likelihood for different models and to compare these models by looking at the difference between their log-likelihoods. One use of this is to compare the state of a logistic regression model against some kind of baseline state. The baseline state that's usually used is the model when only the constant is included. In multiple regression, the baseline model we use is the mean of all scores (this is our best guess of the outcome when we have no other information). In logistic regression, if we want to predict the outcome, what would our best guess be? Well, we can't use the mean score because our outcome is made of zeros and ones and so the mean is meaningless! However, if we know the frequency of zeros and ones, then the best guess will be the category with the largest number of cases. So, if the outcome occurs 107 times, and doesn't occur only 72 times, then our best guess of the outcome will be that it occurs (because it occurs 107 times compared to only 72 times when it doesn't occur). As such, like multiple regression, our baseline model is the model that gives us the best prediction when we know nothing other than the values of the outcome: in logistic regression this will be to predict the outcome that occurs most often. This is the logistic regression model when only the constant is included. If we then add one or more predictors to the model, we can compute the improvement of the model as follows:

$$\begin{aligned} \chi^2 &= 2[\text{LL(new)} - \text{LL(baseline)}] \\ &(df = k_{\text{new}} - k_{\text{baseline}}) \end{aligned} \tag{8.6}$$

So, we merely take the new model and subtract from it the baseline model (the model when only the constant is included). You'll notice that we multiply this value by 2; this is because it gives the result a chi-square distribution (see Chapter 18 and the Appendix) and so makes it easy to calculate the significance of the value. The chi-square distribution we use has degrees of freedom equal to the number of parameters, k, in the new model minus the number of parameters in the baseline model. The number of parameters in the baseline model will always be 1 (the constant is the only parameter to be estimated); any subsequent model will have degrees of freedom equal to the number of predictors plus 1 (i.e. the number of predictors plus one parameter representing the constant).

8.3.2. Assessing the model: R, R^2 and c ③

When we talked about linear regression, we saw that the multiple correlation coefficient R and the corresponding R^2-value were useful measures of how well the model fits the data. We've also just seen that the likelihood ratio is similar in the respect that it is based on the level of correspondence between predicted and actual values of the outcome. However, you can calculate a more literal version of the multiple correlation in logistic regression known as the R-statistic. This R-statistic is the partial correlation between the outcome variable and each of the predictor variables and it can vary between –1 and 1. A positive value indicates that as the predictor variable increases, so does the likelihood of the event occurring. A negative value implies that as the predictor variable increases, the likelihood of the outcome occurring decreases. If a variable has a small value of R then it contributes only a small amount to the model.

The equation for R is given in equation (8.7). The **−2LL** is –2 times the log-likelihood for the original model, the Wald statistic is calculated as described in the next section, and the degrees of freedom can be read from the summary table for the variables in the equation. However, because this value of R is dependent upon the Wald statistic it is by no means an accurate measure (we'll see in the next section that the Wald statistic can be inaccurate under certain circumstances). For this reason the value of R should be treated with some caution, and it is invalid to square this value and interpret it as you would in linear regression:

$$R = \pm\sqrt{\frac{\text{Wald} - (2 \times df)}{-2\text{LL(baseline)}}} \tag{8.7}$$

There is some controversy over what would make a good analogue to the R^2-value in linear regression, but one measure described by Hosmer and Lemeshow (1989) can be easily calculated. In SAS terminology, **Hosmer and Lemeshow's** R_L^2 measure is calculated as:

$$R_L^2 = \frac{-2\text{LL(model)}}{-2\text{LL(original)}} \tag{8.8}$$

As such, R_L^2 is calculated by dividing the model chi-square (based on the log-likelihood) by the *original* −2LL (the log-likelihood of the model before any predictors were entered). R_L^2 is the proportional reduction in the absolute value of the log-likelihood measure and as such it is a measure of how much the badness of fit improves as a result of the inclusion of the predictor variables. It can vary between 0 (indicating that the predictors are useless at predicting the outcome variable) and 1 (indicating that the model predicts the outcome variable perfectly).

However, this is not the measure used by SAS. SAS uses **Cox and Snell's** R_{CS}^2 (1989), which is based on the log-likelihood of the model (LL(new)) and the log-likelihood of the original model (LL(baseline)), and the sample size, n:

$$R_{CS}^2 = 1 - e^{\left[-\frac{2}{n}(\text{LL(new)}) - (\text{LL(baseline)})\right]} \tag{8.9}$$

However, this statistic never reaches its theoretical maximum of 1. Therefore, Nagelkerke (1991) suggested the following amendment (Nagelkerke's R_N^2):

$$R_N^2 = \frac{R_{CS}^2}{1 - e^{\left[\frac{2(\text{LL(baseline)})}{n}\right]}} \tag{8.10}$$

which SAS refers to as the rescaled chi-squared. Although all of these measures differ in their computation (and the answers you get), conceptually they are somewhat the same. So, in terms of interpretation they can be seen as similar to the R^2 in linear regression in that they provide a gauge of the substantive significance of the model.

However, a better summary of the model is the *c*-statistic. The *c*-statistic is a little complex to calculates; it's based on the receiver operating characteristic (ROC) curve, which goes beyond this book. It gives a measure of how well people are classified according to your model, compared with how well they would have been classified if you did something like flipping a coin. If the model is very poor, the *c*-statistic is 0.5, which means that you would have classified half of the people correctly just by flipping a coin.[1] If the *c*-statistic is 1.0 this means that your model was capable of classifying everyone correctly.

[1] Which would have been a lot easier than doing a logistic regression, so you shouldn't be happy about that.

8.3.3. Assessing the contribution of predictors: the Wald statistic ②

As in linear regression, we want to know not only how well the model overall fits the data, but also the individual contribution of predictors. In linear regression, we used the estimated regression coefficients (*b*) and their standard errors to compute a *t*-statistic. In logistic regression there is an analogous statistic known as the **Wald statistic**, which has a special distribution known as the **chi-square distribution**. Like the *t*-test in linear regression, the Wald statistic (Figure 8.2) tells us whether the *b* coefficient for that predictor is significantly different from zero. If the coefficient is significantly different from zero then we can assume that the predictor is making a significant contribution to the prediction of the outcome (*Y*):

$$\text{Wald} = \frac{b}{\text{SE}_b} \tag{8.11}$$

FIGURE 8.2 Abraham Wald writing 'I must not devise test statistics prone to having inflated standard errors' on the blackboard 100 times

Equation (8.11) shows how the Wald statistic is calculated and you can see it's basically identical to the *t*-statistic in linear regression (see equation (7.6)): it is the value of the regression coefficient divided by its associated standard error. The Wald statistic is usually used to ascertain whether a variable is a significant predictor of the outcome; however, it is probably more accurate to examine the likelihood ratio statistics. The reason why the Wald statistic should be used cautiously is because, when the regression coefficient (*b*) is large, the standard error tends to become inflated, resulting in the Wald statistic being underestimated (see Menard, 1995). The inflation of the standard error increases the probability of rejecting a predictor as being significant when in reality it is making a significant contribution to the model (i.e. you are more likely to make a Type II error).

8.3.4. The odds ratio ③

More crucial to the *interpretation* of logistic regression is the value of the odds ratio, which is an indicator of the change in odds resulting from a unit change in the predictor. As such, it is similar to the b coefficient in logistic regression but easier to understand (because it doesn't require a logarithmic transformation). When the predictor variable is categorical the odds ratio is easier to explain, so imagine we had a simple example in which we were trying to predict whether or not someone got pregnant from whether or not they used a condom last time they had sex. The **odds** of an event occurring are defined as the probability of an event occurring divided by the probability of that event not occurring (see equation (8.12)) and should not be confused with the more colloquial usage of the word to refer to probability. So, for example, the odds of becoming pregnant are the probability of becoming pregnant divided by the probability of not becoming pregnant:

$$\begin{aligned} \text{odds} &= \frac{P(\text{event})}{P(\text{no event})} \\ P(\text{event } Y) &= \frac{1}{1+e^{-(b_0+b_1X_1)}} \\ P(\text{no event } Y) &= 1 - P(\text{event } Y) \end{aligned} \tag{8.12}$$

To calculate the change in odds that results from a unit change in the predictor, we must first calculate the odds of becoming pregnant given that a condom *wasn't* used. We then calculate the odds of becoming pregnant given that a condom *was* used. Finally, we calculate the proportionate change in these two odds.

To calculate the first set of odds, we need to use equation (8.3) to calculate the probability of becoming pregnant given that a condom wasn't used. If we had more than one predictor we would use equation (8.4). There are three unknown quantities in this equation: the coefficient of the constant (b_0), the coefficient for the predictor (b_1) and the value of the predictor itself (X). We'll know the value of X from how we coded the condom use variable (chances are we would've used 0 = condom wasn't used and 1 = condom was used). The values of b_1 and b_0 will be estimated for us. We can calculate the odds as in equation (8.12).

Next, we calculate the same thing after the predictor variable has changed by one unit. In this case, because the predictor variable is dichotomous, we need to calculate the odds of getting pregnant, given that a condom *was* used. So, the value of X is now 1 (rather than 0).

We now know the odds before and after a unit change in the predictor variable. It is a simple matter to calculate the proportionate change in odds by dividing the odds after a unit change in the predictor by the odds before that change:

$$\Delta\text{odds} = \frac{\text{odds after a unit change in the predictor}}{\text{original odds}} \tag{8.13}$$

This proportionate change in odds is the odds ratio, and we can interpret it in terms of the change in odds: if the value is greater than 1 then it indicates that as the predictor increases, the odds of the outcome occurring increase. Conversely, a value less than 1 indicates that as the predictor increases, the odds of the outcome occurring decrease. We'll see how this works with a real example shortly.

8.4. Assumptions and things that can go wrong ④

8.4.1. Assumptions ②

Logistic regression shares some of the assumptions of normal regression:

1 **Linearity**: In ordinary regression we assumed that the outcome had linear relationships with the predictors. In logistic regression the outcome is categorical and so this assumption is violated. As I explained before, this is why we use the log of the odds (or *logit*). The assumption of linearity in logistic regression, therefore, means that there is a linear relationship between any continuous predictors and the logit of the outcome variable. This assumption can be tested by looking at whether the interaction term between the predictor and its log transformation is significant (Hosmer & Lemeshow, 1989). We will go through an example in section 8.8.1.
2 **Independence of errors**: This assumption is the same as for ordinary regression (see section 7.6.2.1). Basically it means that cases of data should not be related; for example, you cannot measure the same people at different points in time. Violating this assumption produces overdispersion.
3 **Multicollinearity**: Although not really an assumption as such, multicollinearity is a problem as it was for ordinary regression (see section 7.6.2.1). In essence, predictors should not be too highly correlated. As with ordinary regression, this assumption can be checked with tolerance and VIF statistics, the eigenvalues of the scaled, uncentred cross-products matrix, the condition indexes and the variance proportions. We go through an example in section 8.8.1.

Logistic regression also has some unique problems of its own (not assumptions, but things that can go wrong). SAS solves logistic regression problems by an iterative procedure (SAS Tip 8.1). Sometimes, instead of pouncing on the correct solution quickly, you'll notice nothing happening: SAS begins to move infinitely slowly, or appears to have just got fed up with you asking it to do stuff and has gone on strike. If it can't find a correct solution, then sometimes it actually does give up, quietly offering you (without any apology) a result which is completely incorrect. Usually this is revealed by implausibly large standard errors. Two situations can provoke this situation, both of which are related to the ratio of cases to variables: incomplete information and complete separation.

SAS TIP 8.1 **Error messages about 'failure to converge'** ③

Many statistical procedures use an *iterative process*, which means that SAS attempts to estimate the parameters of the model by finding successive approximations of those parameters. Essentially, it starts by estimating the parameters with a 'best guess'. It then attempts to approximate them more accurately (known as an *iteration*). It then tries again, and then again, and so on through many iterations. It stops either when the approximations of parameters converge (i.e. at each new attempt the 'approximations' of parameters are the same or very similar to the previous attempt), or it reaches the maximum number of attempts (iterations).

Sometimes you will get an error message in the log that says something like '***convergence criteria not met***'. What this means is that SAS has attempted to estimate the parameters the maximum number of times (as specified in the options) but they are not converging (i.e. at each iteration SAS is getting quite different estimates).

This certainly means that you should ignore any output that SAS has produced, and it might mean that your data are beyond help. You can try increasing the number of iterations that SAS attempts, by including MAXITER=XXX on the model line (after the /), where XXX is the number of iterations you want to do – SAS tries 25 by default, so increase the number over 25 (say, to 50).

8.4.2. Incomplete information from the predictors ④

Imagine you're trying to predict lung cancer from smoking (a foul habit believed to increase the risk of cancer) and whether or not you eat tomatoes (which are believed to reduce the risk of cancer). You collect data from people who do and don't smoke, and from people who do and don't eat tomatoes; however, this isn't sufficient unless you collect data from all combinations of smoking and tomato eating. Imagine you ended up with the following data:

Do you smoke?	*Do you eat tomatoes?*	*Do you have cancer?*
Yes	No	Yes
Yes	Yes	Yes
No	No	Yes
No	Yes	??????

Observing only the first three possibilities does not prepare you for the outcome of the fourth. You have no way of knowing whether this last person will have cancer or not based on the other data you've collected. Therefore, SAS will have problems unless you've collected data from all combinations of your variables. This should be checked before you run the analysis using a frequency table, and I describe how to do this in Chapter 18. While you're checking these tables, you should also look at the expected frequencies in each cell of the table to make sure that they are greater than 1 and no more than 20% are less than 5 (see section 18.4). This is because the goodness-of-fit tests in logistic regression make this assumption.

This point applies not only to categorical variables, but also to continuous ones. Suppose that you wanted to investigate factors related to human happiness. These might include age, gender, sexual orientation, religious beliefs, levels of anxiety and even whether a person is right-handed. You interview 1000 people, record their characteristics, and whether they are happy ('yes' or 'no'). Although a sample of 1000 seems quite large, is it likely to include an 80 year old, highly anxious, Buddhist left-handed lesbian? If you found one such person and she was happy, should you conclude that everyone else in the same category is happy? It would, obviously, be better to have several more people in this category to confirm that this combination of characteristics causes happiness. One solution is to collect more data.

As a general point, whenever samples are broken down into categories and one or more combinations are empty it creates problems. These will probably be signalled by coefficients that have unreasonably large standard errors. Conscientious researchers produce and check multiway crosstabulations of all categorical independent variables. Lazy but cautious ones don't bother with crosstabulations, but look carefully at the standard errors. Those who don't bother with either should expect trouble.

8.4.3. Complete separation ④

A second situation in which logistic regression collapses might surprise you: it's when the outcome variable can be perfectly predicted by one variable or a combination of variables! This is known as **complete separation.**

FIGURE A

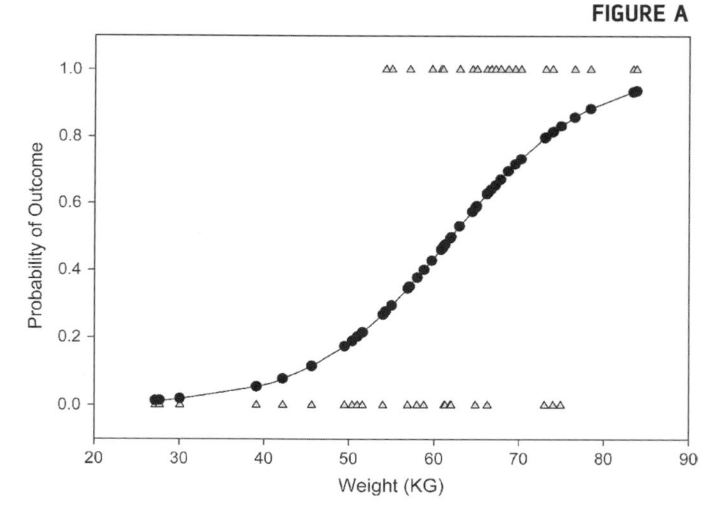

Let's look at an example: imagine you placed a pressure pad under your door mat and connected it to your security system so that you could detect burglars when they creep in at night. However, because your teenage children (which you would have if you're old enough and rich enough to have security systems and pressure pads) and their friends are often coming home in the middle of the night, when they tread on the pad you want it to work out the probability that the person is a burglar and not your teenager. Therefore, you could measure the weight of some burglars and some teenagers and use logistic regression to predict the outcome (teenager or burglar) from the weight. The graph shown in Figure A would show a line of triangles at zero (the data points for all of the teenagers you weighed) and a line of triangles at 1 (the data points for burglars you weighed). Note that these lines of triangles overlap (some teenagers are as heavy as burglars). We've seen that in logistic regression, SAS tries to predict the probability of the outcome given a value of the predictor. In this case, at low weights the fitted probability follows the bottom line of the plot, and at high weights it follows the top line. At intermediate values it tries to follow the probability as it changes.

FIGURE B

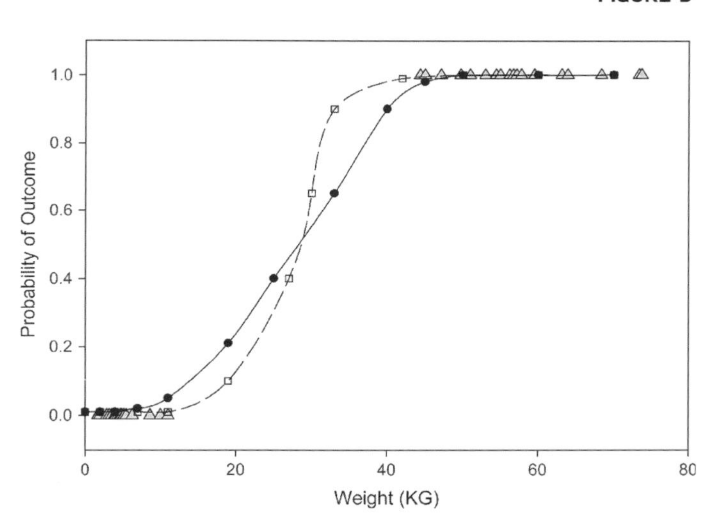

Imagine that we had the same pressure pad, but our teenage children had left home to go to university. We're now interested in distinguishing burglars from our pet cat based on weight. Again, we can weigh some cats and weigh some burglars. This time the graph shown in Figure B still has a row of triangles at zero (the cats we weighed) and a row at 1 (the burglars) but this time the rows of triangles do not overlap: there is no burglar who weighs the same as a cat – obviously there were no cat burglars in the sample (groan now at that sorry excuse for a joke!). This is known as perfect separation: the outcome (cats and burglars) can be perfectly predicted from weight (anything less than 15 kg is a cat, anything more than 40 kg is a burglar). If we try to calculate the probabilities of the outcome given a certain weight then we run into trouble. When the weight is low, the probability is 0, and when the weight is high, the probability is 1, but what happens in between? We have no data between 15 and 40 kg on which to base these probabilities. The figure shows two possible probability curves that we could fit to these data: one much steeper than the other. Either one of these curves is valid based on the data we have available. The lack of data means that SAS will be uncertain about how steep it should make the intermediate slope and it will try to bring the centre as close to vertical as possible, but its estimates veer unsteadily towards infinity (hence large standard errors).

This problem often arises when too many variables are fitted to too few cases. Often the only satisfactory solution is to collect more data, but sometimes a neat answer is found by adopting a simpler model.

CRAMMING SAM'S TIPS **Issues in logistic regression**

- In logistic regression, like ordinary regression, we assume linearity, no multicollinearity and independence of errors.
- The linearity assumption is that each predictor has a linear relationship with the log of the outcome variable.
- If we created a table that combined all possible values of all variables then we should ideally have some data in every cell of this table. If you don't then watch out for big standard errors.
- If the outcome variable can be predicted perfectly from one predictor variable (or a combination of predictor variables) then we have *complete separation*. This problem creates large standard errors too.

8.5. Binary logistic regression: an example that will make you feel eel ②

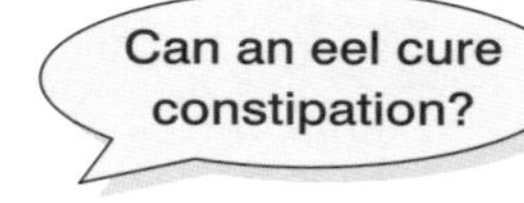

It's amazing what you find in academic journals sometimes. It's a bit of a hobby of mine trying to unearth bizarre academic papers (really, if you find any email them to me). I believe that science should be fun, and so I like finding research that makes me laugh. A research paper by Lo and colleagues is the one that (so far) has made me laugh the most (Lo, Wong, Leung, Law, & Yip, 2004). Lo and colleagues report the case of a 50 year old man who reported to an Accident and Emergency Department (ED for the Americans) with abdominal pain. A physical examination revealed peritonitis so they took an X-ray of the man's abdomen. Although it somehow slipped the patient's mind to mention this to the receptionist upon arrival at the hospital, the X-ray revealed the shadow of an eel. The authors don't directly quote the man's response to this news, but I like to imagine it was something to the effect of 'Oh, that! Erm, yes, well I didn't think it was terribly relevant to my abdominal pain so I didn't mention it, but I did insert an eel into my anus. Do you think that's the problem?'. Whatever he *did* say, the authors report that he admitted inserting an eel into his anus to 'relieve constipation'.

I can have a lively imagination at times, and when I read this article I couldn't help thinking about the poor eel. There it was, minding its own business swimming about in a river (or fish tank possibly), thinking to itself 'Well, today seems like a nice day, there are no eel-eating sharks about, the sun is out, the water is nice, what could possibly go wrong?' The next thing it knows, it's being shoved up the anus of a man from Hong Kong. 'Well, I didn't see that coming', thinks the eel. Putting myself in the mindset of an eel for a moment, he has found himself in a tight dark tunnel, there's no light, there's a distinct lack of water compared to his usual habitat, and he's probably fearing for his life. His day has gone *very* wrong. How can he escape this horrible fate? Well, doing what any self-respecting eel would do, he notices that his prison cell is fairly soft and decides 'bugger this,[2] I'll *eat* my way out of here'. Unfortunately he didn't make it, but he went out with a fight (there's a fairly unpleasant photograph in the article of the eel biting the splenic flexure). The authors conclude that 'Insertion of a live animal into the rectum causing rectal perforation has never been reported. This may be related to a bizarre healthcare belief, inadvertent sexual behaviour, or criminal assault. However, the true reason may never be known.' Quite.

[2] Literally.

OK, so this is a really grim tale.[3] It's not really very funny for the man or the eel, but it was so unbelievably bizarre that I did laugh. Of course my instant reaction was that sticking an eel up your anus to 'relieve constipation' is the poorest excuse for bizarre sexual behaviour I have ever heard. But upon reflection I wondered if I was being harsh on the man – maybe an eel up the anus really can cure constipation. If we wanted to test this, we could collect some data. Our outcome might be 'constipated' vs. 'not constipated', which is a dichotomous variable that we're trying to predict. One predictor variable would be intervention (eel up the anus) vs. waiting list (no treatment). We might also want to factor how many days the patient had been constipated before treatment. This scenario is perfect for logistic regression (but not for eels). The data are in **Eel.sas7bdat**.

I'm quite aware that many statistics lecturers would rather not be discussing eel-created rectal perforations with their students, so I have named the variables in the file more generally:

- **Outcome (dependent variable): Cured** (cured or not cured).
- **Predictor (independent variable): Intervention** (intervention or no treatment).
- **Predictor (independent variable): Duration** (the number of days before treatment that the patient had the problem).

In doing so, your tutor can adapt the example to something more palatable if they wish to, but you will secretly know that it's all about having eels up your bum.

8.5.1. The main analysis ②

Logistic regression is carried out (usually) using PROC LOGISTIC. The basic structure of PROC LOGISTIC is very similar to PROC REG, but there are one or two little things to watch out for. We are going to do a logistic regression with predictors **intervention** and **duration**, and we're also going to introduce the interaction of **intervention** and **duration**.

The basic syntax for logistic regression is shown in SAS Syntax 8.1. However, there are a couple of ways that SAS tries to trick you with logistic regression, and we need to deal with one of those first.

```
PROC LOGISTIC DATA=chapter8.eel;
  MODEL cured = intervention duration;
  RUN;
```

SAS Syntax 8.1

The problem with PROC LOGISTIC is that it will analyse a variable with two categories, and these can be any two values, it doesn't have to be 0 and 1, it can be numeric values 17 and 43, text strings 'yes' and 'no' or 'mars and 'venus'. SAS has to do the same thing we did in section 7.11.1 – it has to pick a reference category and create a dummy. The way

[3] As it happens, it isn't an isolated grim tale. Through this article I found myself hurtling down a road of morbid curiosity that was best left untravelled. Although the eel was my favourite example, I could have chosen from a very large stone (Sachdev, 1967), a test tube (Hughes, Marice, & Gathright, 1976), a baseball (McDonald & Rosenthal, 1977), an aerosol deodorant can, hose pipe, iron bar, broomstick, penknife, marijuana, bank notes, blue plastic tumbler, vibrator and primus stove (Clarke, Buccimazza, Anderson, & Thomson, 2005), or (a close second place to the eel) a toy pirate ship, with or without pirates I'm not sure (Bemelman & Hammacher, 2005). So, although I encourage you to send me bizarre research, if it involves objects in the rectum then probably don't, unless someone has managed to put Buckingham Palace up there.

that SAS does this is to put the values into either numeric or alphabetical order, and use the last one as the reference category – that is, it makes whichever value comes last into 0 and whichever comes first into 1. That's all well and good, unless your values are 0 and 1, which ours happen to be. If that is the case, then SAS takes the highest value (which comes last) and makes it 0, so it takes our 1, and makes it into a 0, and the first category (which is 0 for us) and makes it into a 1. That's not very useful, and we need to stop it from doing that, by telling it which value to use as the reference category. There are two ways to do that. First, we can tell SAS not to use the highest one as the reference category, but to use the lowest. We do that by putting the option descending on the PROC LOGISTIC line.

```
PROC LOGISTIC DATA=chapter8.eel descending;
```

The second way to do it is to tell SAS which value we want for the reference category. We do that on the model line, by putting:

```
MODEL cured (REF='0') = intervention duration;
```

Both these will give the same result.

8.5.1.1. Confidence intervals

As with PROC REG, we're not going to get confidence intervals unless we ask for them. As with PROC REG it's an option on the MODEL line – in this case the option to put is CL.

8.5.1.2. Odds ratios

It's easier to interpret odds ratios than logits, and so it's worth asking for the results in odds ratio format. The odds ratios are e^b, which we can also write as exp(b), where exp is short for exponentiate, which means to take the 'anti-log'. This is also an option on the MODEL line: EXPB.

8.5.1.3. Obtaining R^2

If you want R^2, you need to put the RSQ option on the MODEL line.

8.5.2. Obtaining predicted probabilities and residuals

As with linear regression, it is possible to save a set of residuals (see section 7.6.1.1) as variables in a new data set. Although the outcome for a logistic regression is either a 0 or a 1, the predicted values are not – they are the probability (according to our model) that a person would score a 1.

Residuals are a little curious in logistic regression – your actual score is either a 0 or a 1. But your predicted score is a probability, so the residual is not (quite) the difference between the two. We aren't going to worry about exactly what it is for now, but we'll note that there are two slightly different ways to calculate the residual – these give us the Pearson residual and the deviance residual.

To reiterate a point from the previous chapter, running a regression without checking how well the model fits the data is like buying a new pair of trousers without trying them on – they might look fine on the hanger but get them home and you find you're Johnny-tight-pants. The trousers do their job (they cover your legs and keep you warm) but they have no real-life value (because they cut off the blood circulation to your legs, which then have to be amputated). Likewise, regression does its job regardless of the data – it will create a model – but the real-life value of the model may be limited (see section 7.6).

We use the same approach that we used in linear regression (in Section 7.7.3) to obtain the new data file, that is we use the OUTPUT option, however, in SAS 9.1.3 or later it is easier to use the ODS GRAPHICS option, which we saw in Section 7.8.6, together with the INFLUENCE option.

8.5.3. Final syntax

Putting all of that together gives us the syntax shown in SAS Syntax 8.2.

```
ODS RTF;
ODS GRAPHICS ON;
PROC LOGISTIC DATA=chapter8.eel DESCENDING;
    MODEL cured = intervention duration/EXPB CL INFLUENCE RSQ;
    OUTPUT OUT=res PRED=pred ;
    RUN;
ODS GRAPHICS OFF;
ODS RTF CLOSE;
```

SAS Syntax 8.2

8.6. Interpreting logistic regression

8.6.1. The initial output

The first thing SAS tells us about is the basics of our model and data, as shown in SAS Output 8.1. The first table tells us about the data, the second about the number of observations (we should check this is what we expect) , the third table how many people were cured or not cured, and tells us how SAS decided to code the dummy outcome variable – we need to make sure we get that right, or we'll have our results backwards. The final table is a new one – this tells us that the model converged, that is, it managed to find a solution. If the model did not converge, SAS would still give us results, but we would not be able to interpret them.

SAS OUTPUT 8.1

Model Information		
Data Set	**CHAPTER8.EEL**	
Response Variable	CURED	Cured?
Number of Response Levels	2	
Model	binary logit	
Optimization Technique	Fisher's scoring	

Number of Observations Read	113
Number of Observations Used	113

Response Profile		
Ordered Value	**CURED**	**Total Frequency**
1	1	65
2	0	48

SAS OUTPUT 8.1
(Continued)

Probability modeled is CURED=1.

Model Convergence Status
Convergence criterion (GCONV=1E-8) satisfied.

8.6.2. Intervention

SAS Output 8.2 tells us about the fit of the model. It actually tells us about two models, the first is the intercept only model, and contains no predictors. It has only the intercept, or the constant.[4] This represents the fit of the most basic model to the data.

We are given the AIC (Akaike information criterion) and SC (Schwartz criterion; also known as **Schwartz's Bayesian Criterion, or the BIC**), and the -2 log likelihood –2LL for both the intercept only model and the model with our predictor variables in it. We have come across the –2LL before, and the other two statistics are based on it. They are a way of deciding if the predictor variables that you added to the model were worth it: –2LL will always go down when you add predictors, but we want to know if it went down enough, and so we have two corrections, which are based on the number of predictors (AIC is easy, it's just –2LL + 2 times the number of predictors; SC is more difficult, so we won't worry about it). Lower values of AIC and SC indicate better fit, and in our case, both are lower for the fitted model than for the intercept only model, indicating that the model with predictors is better than the indicator only model. The AIC and SC are rarely used, unless you want to compare models with different sets of predictors, because, as we'll see in a minute, we have a better way of knowing.

Following the fit statistics we get two values of R^2 – the first is the regular (or as regular as you get in logistic regression) R^2, the second is the rescaled R^2. I will reiterate that these are not without their problems, and the *c*-statistic is a better summary (that comes later).

Finally in this section, we get the chi-square statistics. The chi-square is equivalent to the *F*-test in linear regression: it tests the overall significance of our model, to see if we did better than chance. The likelihood ratio chi-square is the difference between the –2LL for the intercept only model and the model with predictors. The degrees of freedom are equal to the number of predictors in the model, and as we can see, the result is significant at $p = 0.007$. Thus our model is significantly ($p < 0.05$) better than the intercept only model. We weren't very interested in the AIC and SC, because in this situation we have a significance test that does the same thing – tells us which model is better. In the situation where we had two sets of predictors, we would not have a significance test, and so we would need to use the AIC and SC.

SAS OUTPUT 8.2

Model Fit Statistics		
Criterion	Intercept Only	Intercept and Covariates
AIC	156.084	148.158
SC	158.811	153.613
–2 Log L	154.084	144.158

[4] We don't need to know the fit of the intercept only model for linear regression – we know that R^2 will be zero. Although we do calculate the intercept only model to get the total sum of squares. (SS_T).

SAS OUTPUT 8.2 (*Continued*)

R-Square	0.0841	Max-rescaled R-Square	0.1130

Testing Global Null Hypothesis: BETA=0			
Test	Chi-Square	DF	Pr > ChiSq
Likelihood Ratio	9.9262	2	0.0016
Score	9.7714	2	0.0018
Wald	9.4465	2	0.0021

SAS OUTPUT 8.3

Analysis of Maximum Likelihood Estimates						
Parameter	DF	Estimate	Standard Error	Wald Chi-Square	Pr > ChiSq	Exp(Est)
Intercept	1	−0.2877	0.2700	1.1350	0.2867	0.750
INTERVENTION	1	1.2287	0.3998	9.4465	0.0021	3.417

Odds Ratio Estimates			
Effect	Point Estimate	95% Wald Confidence Limits	
INTERVENTION	3.417	1.561	7.480

The next part of the output (SAS Output 8.3) is crucial because it tells us the estimates for the coefficients for the predictors included in the model. This section of the output gives us the coefficients and statistics for the variables that have been included in the model at this point (namely **intervention, duration** and the constant). The parameter estimate is the same as the parameter estimate in linear regression: they are the values that we need to replace in equation (8.4) to establish the probability that a case falls into a certain category. We saw in linear regression that the value of b represents the change in the outcome resulting from a unit change in the predictor variable. The interpretation of this coefficient in logistic regression is very similar in that it represents the change in the logit of the outcome variable associated with a one-unit change in the predictor variable. The logit of the outcome is simply the natural logarithm of the odds of Y occurring.

The crucial statistic is the Wald chi-square[5] which tells us whether the b coefficient for that predictor is significantly different from zero. If the coefficient is significantly different from zero then we can assume that the predictor is making a significant contribution to the prediction of the outcome (Y). We came across the Wald statistic in section 8.3.2 and saw that it should be used cautiously because when the regression coefficient (b) is large, the standard error tends to become inflated, resulting in the Wald statistic being underestimated (see Menard, 1995). However, for these data it seems to indicate that having the intervention (or not) is a significant predictor of whether the patient is cured (note that the significance of the Wald statistic is less than .05).

[5] As we have seen, this is simply b divided by its standard error (1.229/0.40 = 3.0725); however, SAS actually quotes the Wald statistic squared.

In section 8.3.2 we saw that we could calculate an analogue of R using equation (8.7). For these data, the Wald statistic and its df can be read from SAS Output 8.3 (8.85 and 1 respectively), and the original −2LL was 154.08. Therefore, R can be calculated as:

$$\begin{aligned} R &= \pm\sqrt{\frac{8.85-(2\times 1)}{154.08}} \\ &= .21 \end{aligned} \tag{8.14}$$

Earlier on in SAS Output 8.2, SAS gave us one other measure of R^2 that was described in section 8.3.2. This is what SAS calls R^2, and is Cox and Snell's measure, which SAS reports as .084.

The final thing we need to look at is the *odds ratio* estimates, which were described in section 8.3.4. To calculate the change in odds that results from a unit change in the predictor for this example, we must first calculate the odds of a patient being cured given that they *didn't* have the intervention. We then calculate the odds of a patient being cured given that they *did* have the intervention. Finally, we calculate the proportionate change in these two odds.

To calculate the first set of odds, we need to use equation (8.12) to calculate the probability of a patient being cured given that they *didn't* have the intervention (to make our lives easier, we're going to ignore the duration estimate, as it wasn't statistically significant). The parameter coding was done so that patients who did not have the intervention were coded with a 0, so we can use this value in place of X. The value of b_1 has been estimated for us as 1.2335 and the coefficient for the constant can be taken from the same table and is −0.2347. We can calculate the odds as:

$$\begin{aligned} P(\text{Cured}) &= \frac{1}{1+e^{-(b_0+b_1X_1)}} \\ &= \frac{1}{1+e^{-[-0.2347+(1.2335\times 0)]}} \\ &= 0.442 \\ P(\text{Not Cured}) &= 1-P(\text{Cured}) \\ &= 1-0.442 \\ &= 0.558 \\ \text{odds} &= \frac{0.442}{0.558} \\ &= 0.792 \end{aligned} \tag{8.15}$$

Now, we calculate the same thing after the predictor variable has changed by one unit. In this case, because the predictor variable is dichotomous, we need to calculate the odds of a patient being cured, given that they have had the intervention. So, the value of the intervention variable, X, is now 1 (rather than 0). The resulting calculations are as follows:

$$\begin{aligned}
P(\text{Cured}) &= \frac{1}{1+e^{-(b_0+b_1X_1)}} \\
&= \frac{1}{1+e^{-[-0.2347+(1.2335\times 1)]}} \\
&= 0.731 \\
P(\text{Not Cured}) &= 1 - P(\text{Cured}) \\
&= 1 - 0.719 \\
&= 0.269 \\
\text{odds} &= \frac{0.719}{0.281} \\
&= 2.715
\end{aligned} \tag{8.16}$$

We now know the odds before and after a unit change in the predictor variable. It is now a simple matter to calculate the proportionate change in odds by dividing the odds after a unit change in the predictor by the odds before that change:

$$\begin{aligned}
\Delta\text{odds} &= \frac{\text{odds after a unit change in the predictor}}{\text{original odds}} \\
&= \frac{2.715}{0.792} \\
&= 3.43
\end{aligned} \tag{8.17}$$

You should notice that the value of the proportionate change in odds is the same as the value that SAS reports for the odds ratio estimate. We can interpret the odds ratio in terms of the change in odds. If the value is greater than 1 then it indicates that as the predictor increases, the odds of the outcome occurring increase. Conversely, a value less than 1 indicates that as the predictor increases, the odds of the outcome occurring decrease. In this example, we can say that the odds of a patient who is treated being cured are 3.43 times higher than those of a patient who is not treated.

We also requested a confidence interval for the odds ratio and it can also be found in the output. The way to interpret this confidence interval is the same as any other confidence interval (section 2.5.2): if we calculated confidence intervals for the value of the odds ratio in 100 different samples, then these intervals would encompass the actual value of the odds ratio in the population (rather than the sample) in 95 of those samples. In this case, we can be fairly confident that the population value of the odds ratio lies between 1.56 and 7.48. However, our sample could be one of the 5% that produces a confidence interval that 'misses' the population value.

The important thing about this confidence interval is that it doesn't cross 1 (both values are greater than 1). This is important because values greater than 1 mean that as the predictor variable increases, so do the odds of (in this case) being cured. Values less than 1 mean the opposite: as the predictor variable increases, the odds of being cured decrease. The fact that both limits of our confidence interval are above 1 gives us confidence that the direction of the relationship that we have observed is true in the population (i.e. it's likely that having an intervention compared to not increases the odds of being cured). If the lower limit had been below 1 then it would tell us that there is a chance that in the population the direction of the relationship is the opposite to what we have observed. This would mean that we could not trust that our intervention increases the odds of being cured.

CRAMMING SAM'S TIPS

- The overall fit of the final model is shown by the −2 log-likelihood statistic and its associated chi-square statistic. If the significance of the chi-square statistic is less than .05, then the model is a significant fit of the data.
- Check the table labelled *Analysis of maximum* to see which variables significantly predict the outcome.
- For each variable in the model, look at the Wald statistic and its significance (which again should be below .05). More important though, use the odds ratio for interpretation. If the value is greater than 1 then as the predictor increases, the odds of the outcome occurring increase. Conversely, a value less than 1 indicates that as the predictor increases, the odds of the outcome occurring decrease. For the aforementioned interpretation to be reliable the confidence interval of the odds ratio should not cross 1!

8.6.3. Listing predicted probabilities ②

When we ran SAS Syntax 8.2 we asked to create file called res, which contained the predicted probabilities of being cured, for each case in our model (and which will be placed in the WORK library).

We can use the PROC PRINT function in SAS to create a table for the first 15 cases in the file **res.** showing the values of **cured, intervention, duration,** and the predicted probability (we called that **pred**). Rather than showing everyone, we'll display just the first 15 cases, by using the OBS option. SAS Syntax 8.3 shows how this is done.

```
PROC PRINT DATA=res (OBS = 15);
    RUN;
```

SAS Syntax 8.3

SAS Output 8.4 shows a selection of the predicted probabilities (that result from running PROC PRINT. It is also worth listing the predictor variables as **well** to clarify from where the predicted probabilities come.

SAS OUTPUT 8.4

Obs	CURED	INTERVENTION	DURATION	_LEVEL_	pred
1	0	0	7	1	0.42812
2	0	0	7	1	0.42812
3	0	0	6	1	0.43004
4	1	0	8	1	0.42621
5	1	1	7	1	0.71991
6	1	0	6	1	0.43004

(Continued)

Obs	CURED	INTERVENTION	DURATION	_LEVEL_	pred
7	0	1	7	1	0.71991
8	1	1	7	1	0.71991
9	1	0	8	1	0.42621
10	0	0	7	1	0.42812
11	1	1	7	1	0.71991
12	1	0	7	1	0.42812
13	1	0	5	1	0.43197
14	0	1	9	1	0.71674
15	0	0	6	1	0.43004

We found from the model that the only significant predictor of being cured was having the intervention. This could have a value of either 1 (have the intervention) or 0 (no intervention). If these two values are placed into equation (8.4) with the respective regression coefficients, then the two probability values are derived. In fact, we calculated these values as part of equation (8.15) and (8.16) and you should note that the calculated probabilities – *P*(Cured) in these equations – correspond to the values in **pred**. These values tells us that when a patient is not treated (**intervention** = 0, no treatment), there is a probability of about .43 that they will be cured – basically, about 43% of people get better without any treatment! However, if the patient does have the intervention (**intervention** = 1, yes), there is a probability of about .72 that they will get better – about 72% of people treated get better (note that the values differ slightly because duration still had a small effect, even though it wasn't statistically significant. When you consider that a probability of 0 indicates no chance of getting better, and a probability of 1 indicates that the patient will definitely get better, the values obtained provide strong evidence that having the intervention increases your chances of getting better (although the probability of recovery without the intervention is still not bad).

Assuming we are content that the model is accurate and that the intervention has some substantive significance, then we could conclude that our intervention (which, to remind you, was putting an eel up the anus) is the single best predictor of getting better (not being constipated). Furthermore, the duration of the constipation pre-intervention and its interaction with the intervention did not significantly predict whether a person got better.

8.6.4. Interpreting residuals ②

Our conclusions so far are fine in themselves, but to be sure that the model is a good one, it is important to examine the residuals. In SAS Syntax 8.2 we saw how to get SAS to produce tables of residuals in the output. The first 15 cases are shown in SAS Output 8.5. We can now interpret them.

Regression Diagnostics									
	Covariates								
Case Number	Intervention	Number of Days with Problem before Treatment	Pearson Residual	Deviance Residual	Hat Matrix Diagonal	Intercept DfBeta	INTERVENTION DfBeta	DURATION DfBeta
1	0	7.0000	−0.8652	−1.0572	0.0183	−0.00858	0.0814	−0.0179
2	0	7.0000	−0.8652	−1.0572	0.0183	−0.00858	0.0814	−0.0179
3	0	6.0000	−0.8686	−1.0604	0.0223	−0.0841	0.0617	0.0593
4	0	8.0000	1.1603	1.3060	0.0293	−0.0898	−0.1379	0.1283
5	1.0000	7.0000	0.6237	0.8107	0.0185	0.0191	0.0650	-0.0196
6	0	6.0000	1.1512	1.2991	0.0223	0.1115	−0.0817	−0.0786
7	1.0000	7.0000	−1.6032	−1.5954	0.0185	−0.0492	−0.1671	0.0504
8	1.0000	7.0000	0.6237	0.8107	0.0185	0.0191	0.0650	−0.0196
9	0	8.0000	1.1603	1.3060	0.0293	−0.0898	−0.1379	0.1283
10	0	7.0000	−0.8652	−1.0572	0.0183	−0.00858	0.0814	−0.0179
11	1.0000	7.0000	0.6237	0.8107	0.0185	0.0191	0.0650	−0.0196
12	0	7.0000	1.1558	1.3026	0.0183	0.0115	−0.1088	0.0239
13	0	5.0000	1.1467	1.2957	0.0416	0.2150	−0.0555	−0.1842
14	1.0000	9.0000	−1.5907	−1.5883	0.0339	0.2048	−0.0999	−0.2100
15	0	6.0000	−0.8686	−1.0604	0.0223	−0.0841	0.0617	0.0593

SAS OUTPUT 8.5

We saw in the previous chapter that the main purpose of examining residuals in any regression is to (1) isolate points for which the model fits poorly, and (2) isolate points that exert an undue influence on the model. To assess the former we examine the residuals, especially the Pearson residual and deviance statistics. To assess the latter we use influence statistics such as the leverage (which in SAS is called the Hat Matrix Diagonal), and DFBeta. The leverage and DFBeta were explained in detail in section 7.6 and their interpretation in logistic regression is the same; therefore, Table 8.1 summarizes the main statistics that you should look at and what to look for, but for more detail consult the previous chapter.

TABLE 8.1 Summary of residual statistics saved by SAS

Name	*Comment*
Hat Matrix Diagonal (Leverage)	Lies between 0 (no influence) and 1 (complete influence). The expected leverage is $(k+1)/N$, where k is the number of predictors and N is the sample size. In this case it would be 2/113 = 0.018
Deviance	Only 5% should lie outside ±1.96, and about 1% should lie outside ±2.58. Cases above 3 are cause for concern and cases close to 3 warrant inspection
DFBeta for the Constant	Should be less than 1
DFBeta for the first predictor (Intervention)	Should be less than 1

The basic residual statistics for this example (leverage), standardized residuals and DFBeta values) are pretty good: note that all cases have DFBetas less than 1, and leverage statistics are very close to the calculated expected value of 0.018. The residuals all have values of less than ± 2 and so there seems to be very little here to concern us.

You should note that these residuals are slightly unusual because they are based on a single predictor that is categorical. This is why there isn't a lot of variability in the values of the residuals. Also, if substantial outliers or influential cases had been isolated, you are not justified in eliminating these cases to make the model fit better. Instead these cases should be inspected closely to try to isolate a good reason why they were unusual. It might simply be an error in inputting data, or it could be that the case was one which had a special reason for being unusual: for example, there were other medical complications that might contribute to the constipation that were noted during the patient's assessment. In such a case, you may have good reason to exclude the case and duly note the reasons why.

CRAMMING SAM'S TIPS **Diagnostic statistics**

You need to look for cases that might be influencing the logistic regression model:

- Look at residuals and check that no more than 5% of cases have absolute values above 2, and that no more than about 1% have absolute values above 2.5. Any case with a value above about 3 could be an outlier.
- Calculate the average leverage (the number of predictors plus 1, divided by the sample size) and then look for values greater than twice or three times this average value.
- Look for absolute values of DFBeta greater than 1.

8.6.5. Calculating the effect size ②

We've already seen (section 8.3.2) that SAS produces a value of *r* for each predictor, based on the Wald statistic. This can be used as your effect size measure for a predictor; however, it's worth bearing in mind what I've already said about it: it won't be accurate when the Wald statistic is inaccurate! The better effect size to use is the odds ratio (see section 18.5.6).

8.7. How to report logistic regression ②

Logistic regression is fairly rarely used in my discipline of psychology, so it's difficult to find any concrete guidelines about how to report one. My personal view is that you should report it much the same as linear regression (see section 7.9). I'd be inclined to tabulate the results, unless it's a very simple model. As a bare minimum, report the beta values and their standard errors and significance value and some general statistics about the model (such as the R^2 and goodness-of-fit statistics). I'd also highly recommend reporting the odds ratio and its confidence interval. I'd include the constant so that readers of your work can construct the full regression model if they need to. You might also consider reporting the variables that were not significant predictors because this can be as valuable as knowing about which predictors were significant.

For the example in this chapter we might produce a table like that in Table 8.2. Hopefully you can work out from where the values came by looking back through the chapter so far. As with multiple regression, I've rounded off to 2 decimal places throughout; for the R^2 and p-values there is no zero before the decimal point (because these values cannot exceed 1) but for all other values less than 1 the zero is present; the significance of the variable is denoted by an asterisk with a footnote to indicate the significance level being used.

TABLE 8.2 How to report logistic regression

		95% CI for Odds Ratio		
	B (SE)	Lower	Odds Ratio	Upper
Included				
Constant	−0.29 (0.27)			
Intervention	1.23* (0.40)	1.56	3.43	7.48

Note: R^2 = .08 (Cox & Snell), .11 (Nagelkerke). Model $\chi^2(1)$ = 9.93, p < .01. * p < .01.

8.8. Testing assumptions: another example ②

This example was originally inspired by events in the soccer World Cup of 1998 (a long time ago now, but such crushing disappointments are not easily forgotten). Unfortunately for me (being an Englishman), I was subjected to watching England get knocked out of the competition by losing a penalty shootout. Reassuringly, six years later I watched England get knocked out of the European Championship in another penalty shootout. Even more reassuring, a few months ago I saw them fail to even qualify for the European Championship (not a penalty shootout this time, just playing like fools).

Now, if I were the England coach, I'd probably shoot the spoilt overpaid prima donnas, or I might be interested in finding out what factors predict whether or not a player will score a penalty. Those of you who hate football can read this example as being factors that predict success in a free, throw in basketball or netball, a penalty in hockey, a penalty kick in rugby[6] or a field goal in American football. Now, this research question is perfect for logistic regression because our outcome variable is a dichotomy: a penalty can be either scored or missed. Imagine that past research (Eriksson, Beckham, & Vassell, 2004; Hoddle, Batty, & Ince, 1998) had shown that there are two factors that reliably predict whether a penalty kick will be missed or scored. The first factor is whether the player taking the kick is a worrier (this factor can be measured using a measure such as the Penn State Worry Questionnaire, PSWQ). The second factor is the player's past success rate at scoring (so whether the player has a good track record of scoring penalty kicks). It is fairly well accepted that anxiety has detrimental effects on the performance of a variety of tasks and so it was also predicted that state anxiety might be able to account for some of the unexplained variance in penalty success.

This example is a classic case of building on a well-established model, because two predictors are already known and we want to test the effect of a new one. So, 75 football players

[6] Although this would be an unrealistic example because our rugby team, unlike their football counterparts, have Jonny Wilkinson who is the lord of penalty kicks and we bow at his great left foot in wonderment (well, I do).

were selected at random and before taking a penalty kick in a competition they were given a state anxiety questionnaire to complete (to assess anxiety before the kick was taken). These players were also asked to complete the PSWQ to give a measure of how much they worried about things generally, and their past success rate was obtained from a database. Finally, a note was made of whether the penalty was scored or missed. The data can be found in the file **penalty.sas7bdat**, which contains four variables – each in a separate column:

- **Scored**: This variable is our outcome and it is coded such that 0 = penalty missed and 1 = penalty scored.
- **PSWQ**: This variable is the first predictor variable and it gives us a measure of the degree to which a player worries.
- **Previous**: This variable is the percentage of penalties scored by a particular player in their career. As such, it represents previous success at scoring penalties.
- **Anxious**: This variable is our third predictor and it is a variable that has not previously been used to predict penalty success. It is a measure of state anxiety before taking the penalty.

SELF-TEST We learnt how to do hierarchical regression in the previous chapter. Try to conduct a hierarchical logistic regression analysis on these data. Enter **Previous** and **PSWQ** in the first model and then add **Anxious** in the second. Note that you can't put two model lines in one PROC LOGISTIC statement, so you need to run it twice.

8.8.1. Testing for linearity of the logit ③

In this example we have three continuous variables, therefore we have to check that each one is linearly related to the log of the odds of the outcome variable (**Scored**). I mentioned earlier in this chapter that to test this assumption we need to run the logistic regression but include predictors that are the interaction between each predictor and the log of itself (Hosmer & Lemeshow, 1989). To create these interaction terms, we need to include a DATA step (see section 5.7.3). For each variable create a new variable that is the log of the original variable. For example, for **PSWQ**, create a new variable called **LnPSWQ** by using

```
lnpwsq = LOG(pwsq);
```

SELF-TEST Try creating two new variables that are the natural logs of **Anxious** and **Previous**.

To test the assumption we need to redo the analysis exactly the same as before except that we put in three new interaction terms of each predictor and their logs. To add interaction terms, we put the two variables we want to test on the model line, with * between them.

SAS Syntax 8.4 shows the logistic regression statement (and the DATA step to create the variables – you didn't think I'd miss that out, did you?).

```
DATA penalty; SET chapter8.penalty;
    lnPSWQ = LOG(pswq);
    lnPrevious = LOG(previous);
    lnAnxious = LOG(anxious);
    RUN;

PROC LOGISTIC DATA=penalty DESCENDING;
    MODEL scored = pswq previous anxious
          pswq*lnpswq
          previous*lnprevious
          anxious*lnanxious;
    RUN;
```

SAS Syntax 8.4

SAS Output 8.6 shows the part of the output that tests the assumption. We're interested only in whether the interaction terms are significant. Any interaction that is significant indicates that the main effect has violated the assumption of linearity of the logit. All three interactions have *p*-values greater than .05, indicating that the assumption of linearity of the logit has been met for PSWQ, Anxious and Previous.

SAS OUTPUT 8.6

Analysis of Maximum Likelihood Estimates

Parameter	DF	Estimate	Standard Error	Wald Chi-Square	Pr > ChiSq
Intercept	1	−3.8789	14.9241	0.0676	0.7949
PSWQ	1	−0.4223	1.1027	0.1467	0.7017
PREVIOUS	1	1.6660	1.4820	1.2637	0.2610
ANXIOUS	1	−2.6448	2.7970	0.8941	0.3444
PSWQ*lnPSWQ	1	0.0439	0.2967	0.0219	0.8823
PREVIOUS*lnPrevious	1	−0.3186	0.3173	1.0078	0.3154
ANXIOUS*lnAnxious	1	0.6808	0.6528	1.0876	0.2970

SELF-TEST Using what you learned in Chapter 6, carry out a Pearson correlation between all of the variables in this analysis. Can you work out why we have a problem with collinearity?

8.8.2. Testing the assumption of linearity of the logit ③

If you have identified collinearity then, unfortunately, there's not much that you can do about it. One obvious solution is to omit one of the variables (so for example, we might stick with the model that ignored state anxiety). The problem with this should be obvious: there is no way of knowing which variable to omit. The resulting theoretical conclusions are meaningless because, statistically speaking, any of the collinear variables could be omitted. There are no statistical grounds for omitting one variable over another. Even if a predictor is removed, Bowerman and O'Connell (1990) recommend that another equally important predictor that does not have such strong multicollinearity replaces it. They also suggest collecting more data to see whether the multicollinearity can be lessened. Another possibility when there are several predictors involved in the multicollinearity is to run a factor analysis on these predictors and to use the resulting factor scores as a predictor (see Chapter 17). The safest (although unsatisfactory) remedy is to acknowledge the unreliability of the model. So, if we were to report the analysis of which factors predict penalty success, we might acknowledge that previous experience significantly predicted penalty success in the first model, but propose that this experience might affect penalty taking by increasing state anxiety. This statement would be highly speculative because the correlation between **Anxious** and **Previous** tells us nothing of the direction of causality, but it would acknowledge the inexplicable link between the two predictors. I'm sure that many of you may find the lack of remedy for collinearity grossly unsatisfying – unfortunately statistics is frustrating sometimes!

LABCOAT LENI'S REAL RESEARCH 8.1

Mandatory suicide? ②

Although I have fairly eclectic tastes in music, my favourite kind of music is heavy metal. One thing that is mildly irritating about liking heavy music is that everyone assumes that you're a miserable or aggressive bastard. When not listening (and often while listening) to heavy metal, I spend most of my time researching clinical psychology: I research how anxiety develops in children. Therefore, I was literally beside myself with excitement when a few years back I stumbled on a paper that combined these two interests: Lacourse, Claes, and Villeneuve (2001) carried out a study to see whether a love of heavy metal could predict suicide risk. Fabulous stuff!

Eric Lacourse and his colleagues used questionnaires to measure several variables: suicide risk (yes or no), marital status of parents (together or divorced/separated), the extent to which the person's mother and father were neglectful, self-estrangement/powerlessness (adolescents who have negative self-perceptions, are bored with life, etc.), social isolation (feelings of a lack of support), normlessness (beliefs that socially disapproved behaviours can be used to achieve certain goals), meaninglessness (doubting that school is relevant to gain employment) and drug use. In addition, the authors measured liking of heavy metal; they included the sub-genres of classic (Black Sabbath, Iron Maiden), thrash metal (Slayer, Metallica), death/black metal (Obituary, Burzum) and gothic (Marilyn Manson). As well as liking they measured behavioural manifestations of worshipping these bands (e.g. hanging posters, hanging out with other metal fans) and vicarious music listening (whether music was used when angry or to bring out aggressive moods). They used logistic regression to predict suicide risk from these predictors for males and females separately.

The data for the female sample are in the file **Lacourseetal2001Females.sas7bdat**. Labcoat Leni wants you to carry out a logistic regression predicting **Suicide_Risk** from all of the predictors (forced entry). (To make your results easier to compare to the published results, enter the predictors in the same order as Table 3 in the paper: **Age**, **Marital_Status**, **Mother_Negligence**, **Father_Negligence**, **Self_Estrangement**, **Isolation**, **Normlessness**, **Meaninglessness**, **Drug_Use**, **Metal**, **Worshipping**, **Vicarious**). Create a table of the results. Does listening to heavy metal predict girls' suicide? If not, what does?

Answers are in the additional material on the companion website (or look at Table 3 in the original article).

LACOURSE, E. ET AL. (2001). *JOURNAL OF YOUTH AND ADOLESCENCE, 30*, 321–332.

8.9. Predicting several categories: multinomial logistic regression ③

I mentioned earlier that it is possible to use logistic regression to predict membership of more than two categories and that this is called multinomial logistic regression. Essentially, this form of logistic regression works in the same way as binary logistic regression so there's no need for any additional equations to explain what is going on (hooray!). The analysis breaks the outcome variable down into a series of comparisons between two categories (which helps explain why no extra equations are really necessary). For example, if you have three outcome categories (A, B and C), then the analysis will consist of two comparisons. The form that these comparisons take depends on how you specify the analysis: you can compare everything against your first category (e.g. A vs. B and A vs. C), or your last category (e.g. A vs. C and B vs. C), or a custom category, for example category B (e.g. B vs. A and B vs. C). In practice, this means that you have to select a baseline category. The important parts of the analysis and output are much the same as we have just seen for binary logistic regression.

Let's look at an example. There has been some recent work looking at how men and women evaluate chat-up lines (Bale, Morrison, & Caryl, 2006; Cooper, O'Donnell, Caryl, Morrison, & Bale, 2007). This research has looked at how the content (e.g. whether the chat-up line is funny, has sexual content, or reveals desirable personality characteristics) affects how favourably the chat-up line is viewed. To sum up this research, it has found that men and women like different things in chat-up lines: men prefer chat-up lines with a high sexual content, and women prefer chat-up lines that are funny and show good moral fibre!

Imagine that we wanted to assess how *successful* these chat-up lines were. We did a study in which we recorded the chat-up lines used by 348 men and 672 women in a night-club. Our outcome was whether the chat-up line resulted in one of the following three events: the person got no response or the recipient walked away, the person obtained the recipient's phone number, or the person left the night-club with the recipient. Afterwards, the chat-up lines used in each case were rated by a panel of judges for how funny they were (0 = not funny at all, 10 = the funniest thing that I have ever heard), sexuality (0 = no sexual content at all, 10 = very sexually direct) and whether the chat-up line reflected good moral values (0 = the chat-up line does not reflect good characteristics, 10 = the chat-up line is very indicative of good characteristics). For example, 'I may not be Fred Flintstone, but I bet I could make your bed rock' would score high on sexual content, low on good characteristics and medium on humour; 'I've been looking all over for YOU, the woman of my dreams' would score high on good characteristics, low on sexual content and low on humour (but high on cheese, had it been measured). We predict based on past research that the success of different types of chat-up line will interact with gender.

This situation is perfect for multinomial regression. The data are in the file **Chat-Up Lines.sas7bdat**. There is one outcome variable (**Success**) with three categories (no response, phone number, go home with recipient) and four predictors: funniness of the chat-up line (**Funny**), sexual content of the chat-up line (**Sex**), degree to which the chat-up line reflects good characteristics (**Good_Mate**) and the gender of the person being chatted up (**Gender**).

8.9.1. Running multinomial logistic regression in SAS ③

To run multinomial logistic regression in SAS, we use PROC LOGISTIC again, but we need to make a couple of tweaks.

The first thing we need to do is tell SAS about our gender variable. This is categorical, with two categories – Male and Female. We saw in section 7.11.1 how to dummy-code a variable, and we could do the same procedure again; however, we can do this automatically in PROC LOGISTIC, provided that we declare that the variable is a classification variable, using a CLASS line. To make sure it dummy-codes, we need to add the option PARAM=REF (param is short for parameterizaton, ref means reference – dummy coding is sometimes called reference coding).[7] SAS will pick the last category to be the reference category – in this case it is the last category, when the categories are put in alphabetical order, and so the last one is male. Male seems OK, but we can force it to be male by specifying a reference category. We write:

```
CLASS gender (REF="Male") /PARAM =REF;
```

SELF-TEST Think about the three categories that we have as an outcome variable. Which of these categories do you think makes most sense to use as a baseline category?

By default SAS uses the last category, (No response/walk off – remember they are alphabetical again), and this makes sense for us because this category represents failure (the chat-up line did not have the desired effect) whereas the other two categories represent some form of success (getting a phone number or leaving the club together).

Next we have to specify the predictor variables, which we do on the MODEL line. The variables are **Gender**, (which is continuous) **Funny**, **Sex** and **Good_Mate** (which are continuous).

8.9.1.1. Adding interactions

In this example, the main effects are not particularly interesting: based on past research we don't necessarily expect funny chat-up lines to be successful, but we do expect them to be more successful when used on women than on men. What this prediction implies is that the *interaction* of **Gender** and **Funny** will be significant. Similarly, chat-up lines with a high sexual content might not be successful overall, but we expect them to be relatively successful when used on men. Again, this means that we might not expect the **Sex** main effect to be significant, but we do expect the **Sex×Gender** interaction to be significant. As such, we need to enter some interaction terms into the model. We specify these on the model line, using **Funny×Sex**, **Funny×Gender** and **Sex×Gender**). We could keep going, adding all interactions with two variables (**Funny×Sex, Funny×Gender, Funny×Good_Mate, Sex×Gender, Sex×Good_Mate, Gender×Good_Mate**), all interactions with three variables (**Funny×Sex×Gender, Funny×Sex×Good_Mate, Good_Mate×Sex×Gender, Funny×Good_Mate×Gender**) and the interaction of all four variables (**Funny×Sex×Gender×Good_Mate**). In this scenario, we want to specify interactions between the ratings of the chat-up lines and gender only (we're not interested in any interactions involving three variables, or all four variables).

[7] We'll see later that when we tell SAS that a predictor variable is a classification variable, it usually dummy-codes by default. PROC LOGISTIC is different from the other procs in that it does not dummy-code by default.

8.9.2. The final touches ③

We can request all of the different options and statistics that we requested last time (in SAS Syntax 8.2). We'll ask for EXPB (to get odds ratios), CL (to get confidence limits) and RSQ (to obtain the two pseudo-R^2 values).

```
PROC LOGISTIC data=chapter8.chatuplines ;
    CLASS gender (REF="Male") /PARAM=REF;
    MODEL success = gender good_mate sex funny gender*funny
    gender*sex
          /LINK=GLOGIT EXPB CL RSQ;
    RUN;
```

SAS Syntax 8.5

8.9.3. Interpreting the multinomial logistic regression output ③

The first thing we see in our output is some model summary information, which tells us how many cases were in the data set, how many of each outcome occurred and what the reference category was, for both the outcome (Success) and the categorical predictor (Gender). We are also told that the model converged successfully. (We haven't printed those here.)

The chi-square statistics for each of these steps are highly significant, indicating that these interactions have a significant effect on predicting whether a chat-up line was significant. Also note that the AIC gets smaller as these terms are added to the model, indicating that the fit of the model is getting better as these terms are added (the SC changes less, but still shows that having the interaction terms in the model results in a better fit than when just the main effects are present).

Next come the model fit statistics, shown in SAS Output 8.7. The first table shows the AIC, SC and –2 log likelihood. The output also shows us the two other measures of R^2 that were described in section 8.3.2. The first is Cox and Snell's measure, which SAS reports as .24, and the second is Nagelkerke's adjusted value, which SAS reports as .28. As you can see, they are reasonably similar values and represent relatively decent-sized effects, and the final table shows the overall tests of the model.

SAS OUTPUT 8.7

Model Fit Statistics		
Criterion	Intercept Only	Intercept and Covariates
AIC	2019.998	1765.473
SC	2029.853	1834.459
–2 Log L	2015.998	1737.473

R-Square	0.2390	Max-rescaled R-Square	0.2774

Testing Global Null Hypothesis: BETA=0			
Test	Chi-Square	DF	Pr > ChiSq
Likelihood Ratio	278.5249	12	<.0001
Score	242.4315	12	<.0001
Wald	189.6931	12	<.0001

SELF-TEST What does the log-likelihood measure

Remember that the log-likelihood is a measure of how much unexplained variability there is in the data; therefore, the difference or change in log-likelihood indicates how much new variance has been explained by the model. The chi-square test tests the decrease in unexplained variance from the baseline model (1149.53) to the final model (871.00), which is a difference of 1149.53−871 = 278.53. This change is significant, which means that our final model explains a significant amount of the original variability (in other words, it's a better fit than the original model).

SAS Output 8.8 shows the results of the likelihood ratio tests and these can be used to ascertain the significance of predictors to the model. This table tells us, though, that gender had a significant main effect on success rates of chat-up lines, $\chi^2(2) = 17.98$, $p = .001$, as did whether the chat-up lined showed evidence of being a good partner, ($\chi^2(2) = 6.31$, $p < .043$) the sexual content ($\chi^2(2) = 14.08$, $p = 0.009$) and the humour ($\chi^2(2) = 7.09$, $p = 0.029$). Most interesting are the interactions which showed that the humour in the chat-up line interacted with gender to predict success at getting a date, ($\chi^2(2) = 34.65$, $p < .001$); also the sexual content of the chat-up line interacted with the gender of the person being chatted up in predicting their reaction, ($\chi^2(2) = 13.42$, $p < .001$). These χ^2 statistics can be seen as sorts of overall statistics that tell us which predictors significantly enable us to predict the outcome category, but they don't really tell us specifically what the effect is. To see this we have to look at the individual parameter estimates.

SAS OUTPUT 8.8

Type 3 Analysis of Effects			
Effect	DF	Wald Chi-Square	Pr > ChiSq
gender	2	17.9802	0.0001
good_mate	2	6.3096	0.0426
sex	2	14.0833	0.0009
funny	2	7.0938	0.0288
funny*gender	2	34.6513	<.0001
sex*gender	2	13.4246	0.0012

SAS Output 8.9 shows the individual parameter estimates. Note that each predictor variable in the table is split into two halves. This is because these parameters compare pairs of outcome categories. We specified the first category as our reference category; therefore, the part of the table labelled *Get Phone Number* is comparing this category against the 'No response/walked away' category. Let's look at the effects one by one; because we are just comparing two categories the interpretation is the same as for binary logistic regression (so if you don't understand my conclusions reread the start of this chapter):

- **Good_Mate**: Whether the chat-up line showed signs of good moral fibre significantly predicted whether you got a phone number or no response/walked away, $b = 0.13$, Wald $\chi^2(1) = 6.02$, $p = 0.141$. The odds ratio (Exp(Est)) tells us that as this variable increases, so as chat-up lines show one more unit of moral fibre, the change in the odds of getting a phone number (rather than no response/walked away) is 1.14. In short, you're more likely to get a phone number than not if you use a chat-up line that demonstrates good moral fibre. (Note that this effect is superseded by the interaction with gender below.)

- **Funny**: Whether the chat-up line was funny did not significantly predict whether you got a phone number or no response, $b = 0.14$, Wald $\chi^2(1) = 1.60$, $p = 0.206$. Note that although this predictor is not significant, the odds ratio is approximately the same as for the previous predictor (which was significant). So, the effect size is comparable, but the non-significance stems from a relatively higher standard error.
- **Gender**: The gender of the person being chatted up significantly predicted whether they gave out their phone number or gave no response, $b = -1.65$, Wald $\chi^2(1) = 4.27$, $p = .039$. Remember that 0 = female and 1 = male, so this is the effect of females compared to males. The odds ratio tells us that as gender changes from female (0) to male (1) the change in the odds of giving out a phone number compared to not responding is 0.19. In other words, the odds of a man giving out his phone number compared to not responding are 1/0.19 = 5.26 times more than for a woman. (Men are cheap.)
- **Sex**: The sexual content of the chat-up line significantly predicted whether you got a phone number or no response/walked away, $b = 0.28$, Wald $\chi^2(1) = 9.59$, $p = .002$. The odds ratio tells us that as the sexual content increased by a unit, the change in the odds of getting a phone number (rather than no response) is 1.32. In short, you're more likely to get a phone number than not if you use a chat-up line with high sexual content. (But this effect is superseded by the interaction with gender.)
- **Funny×Gender**: The success of funny chat-up lines depended on whether they were delivered to a man or a woman because in interaction these variables predicted whether or not you got a phone number, $b = 0.49$, Wald $\chi^2(1) = 12.37$, $p = .004$. Bearing in mind how we interpreted the effect of gender above, the odds ratio tells us that as gender changes from female (0) to male (1) in combination with funniness increasing, the change in the odds of giving out a phone number compared to not responding was 1.64. In other words, as funniness increases, women become more likely to hand out their phone number than men. Funny chat-up lines are more successful when used on women than men.
- **Sex×Gender**: The success of chat-up lines with sexual content depended on whether they were delivered to a man or a woman because in interaction these variables predicted whether or not you got a phone number, $b = -0.35$, Wald $\chi^2(1) = 10.82$, $p = .001$. Bearing in mind how we interpreted the interaction above (note that b is negative here but positive above), the odds ratio tells us that as gender changes from female (0) to male (1) in combination with the sexual content increasing, the change in the odds of giving out a phone number compared to not responding is 0.71. In other words, as sexual content increases, women become *less* likely than men to hand out their phone number. Chat-up lines with a high sexual content are more successful when used on men than women.

Analysis of Maximum Likelihood Estimates

Parameter	success	DF	Estimate	Standard Error	Wald Chi-Square	Pr > ChiSq	Exp(Est)	
Intercept		Get Phone Number	1	−1.7831	0.6698	7.0873	0.0078	0.168
Intercept		Go Home with Person	1	−4.2864	0.9414	20.7317	<.0001	0.014
gender	Female	Get Phone Number	1	−1.6462	0.7962	4.2744	0.0387	0.193
gender	Female	Go Home with Person	1	−5.6257	1.3285	17.9309	<.0001	0.004
good_mate		Get Phone Number	1	0.1318	0.0537	6.0217	0.0141	1.141
good_mate		Go Home with Person	1	0.1300	0.0835	2.4235	0.1195	1.139
sex		Get Phone Number	1	0.2762	0.0892	9.5888	0.0020	1.318

SAS OUTPUT 8.9 (Continued)

Analysis of Maximum Likelihood Estimates								
Parameter	**success**	**DF**	**Estimate**	**Standard Error**	**Wald χ**	**Pr > ChiSq**	**Exp(Est)**	
sex		Go Home with Person	1	0.4173	0.1221	11.6830	0.0006	1.518
funny		Get Phone Number	1	0.1394	0.1101	1.6020	0.2056	1.150
funny		Go Home with Person	1	0.3185	0.1253	6.4593	0.0110	1.375
funny*gender	Female	Get Phone Number	1	0.4924	0.1400	12.3735	0.0004	1.636
funny*gender	Female	Go Home with Person	1	1.1723	0.1992	34.6216	<.0001	3.229
sex*gender	Female	Get Phone Number	1	–0.3483	0.1059	10.8238	0.0010	0.706
sex*gender	Female	Go Home with Person	1	–0.4766	0.1634	8.5053	0.0035	0.621

SAS OUTPUT 8.9

The second half of SAS Output 8.9 shows the individual parameter estimates for the *Go Home with Person* category compared to the 'No response/walked away' category. We can interpret these effects as follows:

- **Good_Mate**: Whether the chat-up line showed signs of good moral fibre did not significantly predict whether you went home with the date or got a slap in the face, $b = 0.13$, Wald $\chi^2(1) = 2.42$, $p = .120$. In short, you're not significantly more likely to go home with the person if you use a chat-up line that demonstrates good moral fibre.
- **Funny**: Whether the chat-up line was funny significantly predicted whether you went home with the date or no response, $b = 0.32$, Wald $\chi^2(1) = 6.46$, $p = .011$. The odds ratio tells us that as chat-up lines are one more unit funnier, the change in the odds of going home with the person (rather than no response) is 1.38. In short, you're more likely to go home with the person than get no response if you use a chat-up line that is funny. (This effect, though, is superseded by the interaction with gender below.)
- **Gender**: The gender of the person being chatted up significantly predicted whether they went home with the person or gave no response, $b = -5.63$, Wald $\chi^2(1) = 17.93$, $p < .0001$. The odds ratio tells us that as gender changes from female (0) to male (1) the change in the odds of going home with the person compared to not responding is 0.004. In other words, the odds of a man going home with someone compared to not responding are 1/0.004 = 250 times more likely than for a woman. Men are *really* cheap.
- **Sex**: The sexual content of the chat-up line significantly predicted whether you went home with the date or got a slap in the face, $b = 0.42$, Wald $\chi^2(1) = 11.68$, $p = .0.0006$. The odds ratio tells us that as the sexual content increased by a unit, the change in the odds of going home with the person (rather than no response) is 1.52: you're more likely to go home with the person than not if you use a chat-up line with high sexual content.
- **Funny×Gender**: The success of funny chat-up lines depended on whether they were delivered to a man or a woman because in interaction these variables predicted whether or not you went home with the date, $b = 1.17$, Wald $\chi^2(1) = 34.63$, $p < .0001$. The odds ratio tells us that as gender changes from female (0) to male (1) in combination with funniness increasing, the change in the odds of going home with the person compared to not responding is 3.23. As funniness increases, women become more likely to go home with the person than men. Funny chat-up lines are more successful when used on women compared to men.

- **Sex×Gender**: The success of chat-up lines with sexual content depended on whether they were delivered to a man or a woman because in interaction these variables predicted whether or not you went home with the date, $b = -0.48$, Wald $\chi^2(1) = 8.51$, $p = .004$. The odds ratio tells us that as gender changes from female (0) to male (1) in combination with the sexual content increasing, the change in the odds of going home with the date compared to not responding is 0.62. As sexual content increases, women become less likely than men to go home with the person. Chat-up lines with sexual content are more successful when used on men than women.

SELF-TEST Use what you learnt earlier in this chapter to check the assumptions of multicollinearity and linearity of the logit.

8.9.4. Reporting the results ③

We can report the results as with binary logistic regression using a table (see Table 8.3). Note that I have split the table by the outcome categories being compared, but otherwise it is the same as before. These effects are interpreted as in the previous section.

TABLE 8.3 How to report multinomial logistic regression

		95% CI for Odds Ratio		
	B (SE)	*Lower*	*Odds Ratio*	*Upper*
Phone Number vs. No Response				
Intercept	−1.78 (0.67)**			
Good Mate	0.13 (0.05)*	1.03	1.14	1.27
Funny	0.14 (0.11)	0.93	1.15	1.43
Gender	−1.65 (0.80)*	0.04	0.19	0.92
Sexual Content	0.28 (0.09)**	1.11	1.32	1.57
Gender × Funny	0.49 (0.14)***	1.24	1.64	2.15
Gender × Sex	−0.35 (0.11)*	0.57	0.71	0.87
Going Home vs. No Response				
Intercept	−4.29 (0.94)***			
Good Mate	0.13 (0.08)	0.97	1.14	1.34
Funny	0.32 (0.13)*	1.08	1.38	1.76
Gender	−5.63 (1.33)***	0.00	0.00	0.05
Sexual Content	0.42 (0.12)**	1.20	1.52	1.93
Gender × Funny	01.17 (0.20)***	2.19	3.23	4.77
Gender × Sex	−0.48 (0.16)**	0.45	0.62	0.86

Note: $R^2 = .24$ (Cox & Snell), .28 (Nagelkerke). Model $\chi^2(12) = 278.53$, $p < .001$. * $p < .05$, ** $p < .01$, *** $p < .001$.

What have I discovered about statistics? ①

At the age of 10 I thought I was going to be a rock star. Such was my conviction about this that even today (many years on) I'm still not entirely sure how I ended up *not* being a rock star (lack of talent, not being a very cool person, inability to write songs that don't make people want to throw rotting vegetables at you, are all possible explanations). Instead of the glitzy and fun life that I anticipated I am instead reduced to writing chapters about things that I don't even remotely understand.

We began the chapter by looking at why we can't use linear regression when we have a categorical outcome, but instead have to use binary logistic regression (two outcome categories) or multinomial logistic regression (several outcome categories). We then looked into some of the theory of logistic regression by looking at the regression equation and what it means. Then we moved onto assessing the model and talked about the log-likelihood statistic and the associated chi-square test. I talked about different methods of obtaining equivalents to R^2 in regression. We also discovered the Wald statistic and odds ratio. The rest of the chapter looked at three examples using SAS to carry out various logistic regressions. So, hopefully, you should have a pretty good idea of how to conduct and interpret a logistic regression by now.

Having decided that I was going to be a rock star I put on my little denim jacket with Iron Maiden patches sewn onto it and headed off down the rocky road of stardom. The first stop was ... my school.

Key terms that I've discovered

AIC (Akaike's Information Criterion)
−2LL
Binary logistic regression
Chi-square distribution
Complete separation
Hosmer and Lemeshow's R^2_L
Interaction effect
Likelihood
Logistic regression
Log-likelihood
Main effect
Maximum-likelihood estimation
Multinomial logistic regression
Nagelkerke's R^2_N
Odds
Odds ratio
Polychotomous logistic regression
SC (Schwartz Criterion; also known as the Bayesian Information Criterion; BIC)
Wald statistic

Smart Alex's tasks

- **Task 1**: A psychologist was interested in whether children's understanding of display rules can be predicted from their age, and whether the child possesses a theory of mind. A display rule is a convention of displaying an appropriate emotion in a given situation. For example, if you receive a Christmas present that you don't like, the appropriate emotional display is to smile politely and say 'Thank you Auntie Kate, I've always wanted a rotting cabbage.' The inappropriate emotional display is to start crying and scream 'Why did you buy me a rotting cabbage, you selfish old bag?' Using appropriate display rules has been linked to having a theory of mind (the ability to understand what another person might be thinking). To test this theory, children were given a false belief task (a task used to measure whether someone has a theory of mind) and a display rule task (which they could either pass or fail), and their age in months was measured. The data are in **Display.sas7bdat**. Run a logistic regression to see whether possession of display rule understanding (did the child pass the test: Yes/No?) can be predicted from possession of a theory of mind (did the child pass the false belief task: Yes/No?), age in months and their interaction. ③

- **Task 2**: Recent research has shown that lecturers are among the most stressed workers. A researcher wanted to know exactly what it was about being a lecturer that created this stress and subsequent burnout. She took 467 lecturers and administered several questionnaires to them that measured: **Burnout** (burnt out or not), **Perceived Control** (high score = low perceived control), **Coping Style** (high score = high ability to cope with stress), **Stress from Teaching** (high score = teaching creates a lot of stress for the person), **Stress from Research** (high score = research creates a lot of stress for the person) and **Stress from Providing Pastoral Care** (high score = providing pastoral care creates a lot of stress for the person). The outcome of interest was burnout, and Cooper, Sloan, and Williams's (1988) model of stress indicates that perceived control and coping style are important predictors of this variable. The remaining predictors were measured to see the unique contribution of different aspects of a lecturer's work to their burnout. Can you help her out by conducting a logistic regression to see which factors predict burnout? The data are in **Burnout.sas7bdat**. ③
- **Task 3**: A health psychologist interested in research into HIV wanted to know the factors that influenced condom use with a new partner (relationship less than 1 month old). The outcome measure was whether a condom was used (**Use**: condom used = 1, not used = 0). The predictor variables were mainly scales from the Condom Attitude Scale (CAS) by Sacco, Levine, Reed, and Thompson (1991): **Gender** (gender of the person); **Safety** (relationship safety, measured out of 5, indicates the degree to which the person views this relationship as 'safe' from sexually transmitted disease); **Sexexp** (sexual experience, measured out of 10, indicates the degree to which previous experience influences attitudes towards condom use); **Previous** (a measure not from the CAS, this variable measures whether or not the couple used a condom in their previous encounter, 1 = condom used, 0 = not used, 2 = no previous encounter with this partner); **selfcon** (self-control, measured out of 9, indicates the degree of self-control that a person has when it comes to condom use, i.e. do they get carried away with the heat of the moment, or do they exert control?); **Perceive** (perceived risk, measured out of 6, indicates the degree to which the person feels at risk from unprotected sex). Previous research (Sacco, Rickman, Thompson, Levine, & Reed, 1993) has shown that gender, relationship safety and perceived risk predict condom use. Carry out an appropriate analysis to verify these previous findings, and to test whether self-control, previous usage and sexual experience can predict any of the remaining variance in condom use. (1) Interpret all important parts of the SAS output. (2) How reliable is the final model? (3) What are the probabilities that participants 12, 53 and 75 will use a condom? (4) A female who used a condom in her previous encounter with her new partner scores 2 on all variables except perceived risk (for which she scores 6). Use the model to estimate the probability that she will use a condom in her next encounter. Data are in the file **condom.sas7bdat**. ③

Answers can be found on the companion website.

Further reading

Allison, P. (2001). *Logistic regression using the SAS system*. Cary, NC: SAS Publishing. (This book is a little more about logistic regression, and a little less about SAS than some of the other SAS books, it's also a little lighter on the mathematics.)

Hutcheson, G., & Sofroniou, N. (1999). *The multivariate social scientist*. London: Sage. Chapter 4.

Menard, S. (1995). *Applied logistic regression analysis*. Sage University Paper Series on Quantitative Applications in the Social Sciences, 07-106. Thousand Oaks, CA: Sage. (This is a fairly advanced text, but great nevertheless. Unfortunately, few basic-level texts include logistic regression so you'll have to rely on what I've written!)

Miles, J. & Shevlin, M. (2001). *Applying regression and correlation: a guide for students and researchers*. London: Sage. (Chapter 6 is a nice introduction to logistic regression).

Interesting real research

Bale, C., Morrison, R., & Caryl, P. G. (2006). Chat-up lines as male sexual displays. *Personality and Individual Differences*, *40*(4), 655–664.

Bemelman, M., & Hammacher, E. R. (2005). Rectal impalement by pirate ship: A case report. *Injury Extra*, *36*, 508–510.

Cooper, M., O'Donnell, D., Caryl, P. G., Morrison, R., & Bale, C. (2007). Chat-up lines as male displays: Effects of content, sex, and personality. *Personality and Individual Differences*, *43*(5), 1075–1085.

Lacourse, E., Claes, M., & Villeneuve, M. (2001). Heavy metal music and adolescent suicidal risk. *Journal of Youth and Adolescence*, *30*(3), 321–332.

Lo, S. F., Wong, S. H., Leung, L. S., Law, I. C., & Yip, A. W. C. (2004). Traumatic rectal perforation by an eel. *Surgery*, *135*(1), 110–111.

9 Comparing two means

FIGURE 9.1
My (probably) 8th birthday. L–R: My brother Paul (who still hides behind cakes rather than have his photo taken), Paul Spreckley, Alan Palsey, Clair Sparks and me

9.1. What will this chapter tell me? ①

Having successfully slain audiences at holiday camps around the country, my next step towards global domination was my primary school. I had learnt another Chuck Berry song ('Johnny B. Goode'), but also broadened my repertoire to include songs by other artists (I have a feeling 'Over the edge' by Status Quo was one of them).[1] Needless to say, when the opportunity came to play at a school assembly I jumped at it. The headmaster tried to have me banned,[2] but the show went on. It was a huge success (I want to reiterate my

[1] This would have been about 1982, so just before they became the most laughably bad band on the planet. Some would argue that they were *always* the most laughably bad band on the planet, but they were the first band that I called my favourite band.

[2] Seriously! Can you imagine, a headmaster banning a 10 year old from assembly? By this time I had an electric guitar and he used to play hymns on an acoustic guitar; I can assume only that he somehow lost all perspective on the situation and decided that a 10 year old blasting out some Quo in a squeaky little voice was subversive or something.

earlier point that 10 year olds are very easily impressed). My classmates carried me around the playground on their shoulders. I was a hero. Around this time I had a childhood sweetheart called Clair Sparks. Actually, we had been sweethearts since before my new-found rock legend status. I don't think the guitar playing and singing impressed her much, but she rode a motorbike (really, a little child's one) which impressed *me* quite a lot; I was utterly convinced that we would one day get married and live happily ever after. I was utterly convinced, that is, until she ran off with Simon Hudson. Being 10, she probably literally did run off with him – across the playground. To make this important decision of which boyfriend to have, Clair had needed to compare two things (Andy and Simon) to see which one was better; sometimes in science we want to do the same thing, to compare two things to see if there is evidence that one is different to the other. This chapter is about the process of comparing two means using a *t*-test.

9.2. Looking at differences ①

Rather than looking at relationships between variables, researchers are sometimes interested in looking at differences between groups of people. In particular, in experimental research we often want to manipulate what happens to people so that we can make causal inferences. For example, if we take two groups of people and randomly assign one group a programme of dieting pills and the other group a programme of sugar pills (which they think will help them lose weight) then if the people who take the dieting pills lose more weight than those on the sugar pills we can infer that the diet pills caused the weight loss. This is a powerful research tool because it goes one step beyond merely observing variables and looking for relationships (as in correlation and regression).[3] This chapter is the first of many that look at this kind of research scenario, and we start with the simplest scenario: when we have two groups, or, to be more specific, when we want to compare two means. As we have seen (Chapter 1), there are two different ways of collecting data: we can either expose different people to different experimental manipulations (*between-group* or *independent* design), or take a single group of people and expose them to different experimental manipulations at different points in time (a *repeated-measures design*).

9.3. The *t*-test ①

We have seen in previous chapters that the *t*-test is a very versatile statistic: it can be used to test whether a correlation coefficient is different from 0; it can also be used to test whether a regression coefficient, *b*, is different from 0. However, it can also be used to test whether two group means are different. It is to this use that we now turn.

The simplest form of experiment that can be done is one with only one independent variable that is manipulated in only two ways and only one outcome is measured. More often than not the manipulation of the independent variable involves having an

What's the difference between the independent and dependent *t*-test?

[3] People sometimes get confused and think that certain statistical procedures allow causal inferences and others don't (see Jane Superbrain Box 1.4).

experimental condition and a control group (see Field & Hole, 2003). Some examples of this kind of design are:

- Is the movie *Scream 2* scarier than the original *Scream*? We could measure heart rates (which indicate anxiety) during both films and compare them.
- Does listening to music while you work improve your work? You could get some people to write an essay (or book!) listening to their favourite music, and then write a different essay when working in silence (this is a control group). You could then compare the essay grades!
- Does listening to Andy's favourite music improve your work? You could repeat the above but rather than letting people work with their favourite music, you could play them some of my favourite music (as listed in the acknowledgements) and watch the quality of their work plummet!

The *t*-test can analyse these sorts of scenarios. Of course, there are more complex experimental designs and we will look at these in subsequent chapters. There are, in fact, two different *t*-tests and the one you use depends on whether the independent variable was manipulated using the same participants or different:

- **Independent-means *t*-test**: This test is used when there are two experimental conditions and different participants were assigned to each condition (this is sometimes called the *independent-measures* or *independent-samples t*-test).
- **Dependent-means *t*-test**: This test is used when there are two experimental conditions and the same participants took part in both conditions of the experiment (this test is sometimes referred to as the *matched-pairs* or *paired-samples t*-test).

9.3.1. Two example data sets ①

We're going to look at an example that I use throughout this chapter (not because I am too lazy to think up different data sets, but because it allows me to illustrate various things). The example relates to whether arachnophobia (fear of spiders) is specific to real spiders or whether pictures of spiders can evoke similar levels of anxiety. Twenty-four arachnophobes were used in all. Twelve were asked to play with a big hairy tarantula spider with big fangs and an evil look in its eight eyes. Their subsequent anxiety was measured. The remaining twelve were shown only pictures of the same big hairy tarantula and again their anxiety was measured. The data are in Table 9.1 (and **spiderBG.sas7bdat** if you're having difficulty entering them into SAS yourself). Remember that each row in the data editor represents a different participant's data. Therefore, you need a column representing the group to which they belonged and a second column representing their anxiety.

OK, now let's imagine that we'd collected these data using the same participants; that is, all participants had their anxiety rated after seeing the real spider, but also after seeing the picture (in counterbalanced order obviously!). The data would now be arranged differently in SAS. Instead of having a coding variable, and a single column with anxiety scores in, we would arrange the data in two columns (one representing the **picture** condition and one representing the **real** condition). The data are displayed in Table 9.2 (and **spiderRM.sas7bdat** if you're having difficulty entering them into SAS yourself). Note that the anxiety scores are identical to the between-group data (Table 9.1) — it's just that we're pretending that they came from the same people rather than different people.

TABLE 9.1 Data from **spiderBG.sas7bdat**

Participant	*Group*	*Anxiety*
1	Picture	30
2	Picture	35
3	Picture	45
4	Picture	40
5	Picture	50
6	Picture	35
7	Picture	55
8	Picture	25
9	Picture	30
10	Picture	45
11	Picture	40
12	Picture	50
13	Real	40
14	Real	35
15	Real	50
16	Real	55
17	Real	65
18	Real	55
19	Real	50
20	Real	35
21	Real	30
22	Real	50
23	Real	60
24	Real	39

9.3.2. Rationale for the *t*-test ①

Both *t*-tests have a similar rationale, which is based on what we learnt in Chapter 2 about hypothesis testing:

- Two samples of data are collected and the sample means calculated. These means might differ by either a little or a lot.
- If the samples come from the same population, then we expect their means to be roughly equal (see section 2.5.1). Although it is possible for their means to differ by chance alone, we would expect large differences between sample means to occur very infrequently. Under the null hypothesis we assume that the experimental manipulation has no effect on the participants: therefore, we expect the sample means to be very similar.

TABLE 9.2 Data from **spiderRM.sas7bdat**

Subject	*Picture (Anxiety score)*	*Real (Anxiety Score)*
1	30	40
2	35	35
3	45	50
4	40	55
5	50	65
6	35	55
7	55	50
8	25	35
9	30	30
10	45	50
11	40	60
12	50	39

- We compare the difference between the sample means that we collected to the difference between the sample means that we would expect to obtain if there were no effect (i.e. if the null hypothesis were true). We use the standard error (see section 2.5.1) as a gauge of the variability between sample means. If the standard error is small, then we expect most samples to have very similar means. When the standard error is large, large differences in sample means are more likely. If the difference between the samples we have collected is larger than what we would expect based on the standard error then we can assume one of two things:
  - There is no effect and sample means in our population fluctuate a lot and we have, by chance, collected two samples that are atypical of the population from which they came.
  - The two samples come from different populations but are typical of their respective parent population. In this scenario, the difference between samples represents a genuine difference between the samples (and so the null hypothesis is incorrect).
- As the observed difference between the sample means gets larger, the more confident we become that the second explanation is correct (i.e. that the null hypothesis should be rejected). If the null hypothesis is incorrect, then we gain confidence that the two sample means differ because of the different experimental manipulation imposed on each sample.

I mentioned in section 2.6.1 that most test statistics can be thought of as the 'variance explained by the model' divided by the 'variance that the model can't explain'. In other words, effect/error. When comparing two means the 'model' that we fit to the data (the effect) is the difference between the two group means. We saw also in Chapter 2 that means vary from sample to sample (sampling variation) and that we can use the standard error as a measure of how much means fluctuate (in other words, the error in the estimate of the mean). Therefore, we can also use the standard error of the differences between the two means as an estimate of the error in our model (or the error in the difference between means). Therefore, we calculate the

t-test using equation (9.1) below. The top half of the equation is the 'model' (our model being that the difference between means is bigger than the expected difference, which in most cases will be 0 – we expect the difference between means to be different to zero). The bottom half is the 'error'. So, just as I said in Chapter 2, we're basically getting the test statistic by dividing the model (or effect) by the error in the model. The exact form that this equation takes depends on whether the same or different participants were used in each experimental condition:

$$t = \frac{\begin{array}{c}\text{observed difference}\\ \text{between sample means}\end{array} - \begin{array}{c}\text{expected difference}\\ \text{between population means}\\ \text{(if null hypothesis is true)}\end{array}}{\begin{array}{c}\text{estimate of the standard error of the}\\ \text{difference between two sample means}\end{array}} \tag{9.1}$$

9.3.3. Assumptions of the *t*-test ①

Both the **independent *t*-test** and the **dependent *t*-test** are *parametric tests* based on the normal distribution (see Chapter 5). Therefore, they assume:

- The sampling distribution is normally distributed. In the dependent t-test this means that the sampling distribution of the *differences* between scores should be normal, not the scores themselves (see section 9.4.3).
- Data are measured at least at the interval level.

The independent t-test, because it is used to test different groups of people, also assumes:

- Variances in these populations are roughly equal (*homogeneity of variance*).
- Scores in different treatment conditions are independent (because they come from different people).

These assumptions were explained in detail in Chapter 5 and, in that chapter, I emphasized the need to check these assumptions before you reach the point of carrying out your statistical test. As such, I won't go into them again, but it does mean that if you have ignored my advice and haven't checked these assumptions then you need to do it now! SAS also incorporates some procedures into the t-test (e.g. Levene's test, discussed in section 5.6.1, can be done at the same time as the t-test). Let's now look at each of the two t-tests in turn.

9.4. The dependent *t*-test ①

If we stay with our repeated-measures data for the time being we can look at the dependent t-test, or paired-samples t-test. The dependent t-test is easy to calculate. In effect, we use a numeric version of equation (9.1):

$$t = \frac{\overline{D} - \mu_D}{s_D/\sqrt{N}} \tag{9.2}$$

Equation (9.2) compares the mean difference between our samples ($\overline{D}$) to the difference that we would expect to find between population means (μ_D), and then takes into account the standard error of the differences ($s_D/\sqrt{N}$). If the null hypothesis is true, then we expect there to be no difference between the population means (hence $\mu_D = 0$).

9.4.1. Sampling distributions and the standard error ①

In equation (9.1) I referred to the lower half of the equation as the standard error of differences. The standard error was introduced in section 2.5.1 and is simply the standard deviation of the sampling distribution. Have a look back at this section now to refresh your memory about sampling distributions and the standard error. Sampling distributions have several properties that are important. For one thing, if the population is normally distributed then so is the sampling distribution; in fact, if the samples contain more than about 50 scores the sampling distribution should be normally distributed. The mean of the sampling distribution is equal to the mean of the population, so the average of all possible sample means should be the same as the population mean. This property makes sense because if a sample is representative of the population then you would expect its mean to be equal to that of the population. However, sometimes samples are unrepresentative and their means differ from the population mean. On average, though, a sample mean will be very close to the population mean and only rarely will the sample mean be substantially different from that of the population. A final property of a sampling distribution is that its standard deviation is equal to the standard deviation of the population divided by the square root of the number of observations in the sample. As I mentioned before, this standard deviation is known as the standard error.

We can extend this idea to look at the *differences* between sample means. If you were to take several pairs of samples from a population and calculate their means, then you could also calculate the difference between their means. I mentioned earlier that *on average* sample means will be very similar to the population mean: as such, most samples will have very similar means. Therefore, most of the time the difference between sample means from the same population will be zero, or close to zero. However, sometimes one or both of the samples could have a mean very deviant from the population mean and so it is possible to obtain large differences between sample means by chance alone. However, this would happen less frequently.

In fact, if you plotted these differences between sample means as a histogram, you would again have a sampling distribution with all of the properties previously described. The standard deviation of this sampling distribution is called the **standard error of differences.** A small standard error tells us that most pairs of samples from a population will have very similar means (i.e. the difference between sample means should normally be very small). A large standard error tells us that sample means can deviate quite a lot from the population mean and so differences between pairs of samples can be quite large by chance alone.

9.4.2. The dependent *t*-test equation explained ①

In an experiment, a person's score in condition 1 will be different to their score in condition 2, and this difference could be very large or very small. If we calculate the differences between each person's score in each condition and add up these differences we would get the total amount of difference. If we then divide this total by the number of participants we get the average difference (thus how much, on average, a person's score differed in condition 1 compared to condition 2). This average difference is $\bar{D}$ in equation (9.2) and it is an indicator of the systematic variation in the data (i.e. it represents the experimental effect). We need to compare this systematic variation against some kind of measure of the 'systematic variation that we could naturally expect to find'. In Chapter 2 we saw that the standard deviation was a measure of the 'fit' of the mean to the observed data (i.e. it measures the error in the model when the model is the mean), but it is does not measure the fit of the mean to the population. To do this we need the standard error (see the previous section, where we revised this idea).

The standard error is a measure of the error in the mean as a model of the population. In this context, we know that if we had taken two random samples from a population (and not done anything to these samples) then the means could be different just by chance. The standard error tells us by how much these samples could differ. A small standard error means that sample means should be quite similar, so a big difference between two sample means is unlikely. In contrast, a large standard error tells us that big differences between the means of two random samples are more likely. Therefore it makes sense to compare the average difference between means against the standard error of these differences. This gives us a test statistic that, as I've said numerous times in previous chapters, represents model/error. Our model is the average difference between condition means, and we divide by the standard error which represents the error associated with this model (i.e. how similar two random samples are likely to be from this population).

To clarify, imagine that an alien came down and cloned me millions of times. This population is known as Landy of the Andys (this would be possibly the most dreary and strangely terrifying place I could imagine). Imagine the alien were interested in spider phobia in this population (because I am petrified of spiders). Everyone in this population (my clones) will be the same as me, and would behave in an identical way to me. If you took two samples from this population and measured their spider fear, then the means of these samples would be the same (we are clones), so the difference between sample means would be zero. Also, because we are all identical, then all samples from the population will be perfect reflections of the population (the standard error would be zero also). Therefore, if we were to get two samples that differed even very slightly then this would be very unlikely indeed (because our population is full of cloned Andys). Therefore, a difference between samples must mean that they have come from different populations. Of course, in reality we don't have samples that perfectly reflect the population, but the standard error gives an idea of how well samples reflect the population from which they came.

Therefore, by dividing by the standard error we are doing two things: (1) standardizing the average difference between conditions (this just means that we can compare values of t without having to worry about the scale of measurement used to measure the outcome variable); and (2) contrasting the difference between means that we have against the difference that we could *expect* to get based on how well the samples represent the populations from which they came. If the standard error is large, then large differences between samples are more common (because the distribution of differences is more spread out). Conversely, if the standard error is small, then large differences between sample means are uncommon (because the distribution is very narrow and centred around zero). Therefore, if the average difference between our samples is large, and the standard error of differences is small, then we can be confident that the difference we observed in our sample is not a chance result. If the difference is not a chance result then it must have been caused by the experimental manipulation.

In a perfect world, we could calculate the standard error by taking all possible pairs of samples from a population, calculating the differences between their means, and then working out the standard deviation of these differences. However, in reality this is impossible. Therefore, we estimate the standard error from the standard deviation of differences obtained within the sample (s_D) and the sample size (N). Think back to section 2.5.1 where we saw that the standard error is simply the standard deviation divided by the square root of the sample size; likewise the standard error of differences ($\sigma_{\bar{D}}$) is simply the standard deviation of differences divided by the square root of the sample size:

$$\sigma_{\bar{D}} = \frac{s_D}{\sqrt{N}}$$

If the standard error of differences is a measure of the unsystematic variation within the data, and the sum of difference scores represents the systematic variation, then it should be clear that the t-statistic is simply the ratio of the systematic variation in the experiment to the unsystematic variation. If the experimental manipulation creates any kind of effect, then we would expect the systematic variation to be much greater than the unsystematic variation (so at the very least, t should be greater than 1). If the experimental manipulation is unsuccessful then we might expect the variation caused by individual differences to be much greater than that caused by the experiment (so t will be less than 1). We can compare the obtained value of t against the maximum value we would expect to get by chance alone in a t-distribution with the same degrees of freedom (these values can be found in the Appendix); if the value we obtain exceeds this critical value we can be confident that this reflects an effect of our independent variable.

9.4.3. The dependent *t*-test and the assumption of normality ①

We talked about the assumption of normality in Chapter 5 and discovered that parametric tests (like the dependent t-test) assume that the sampling distribution is normal. This should be true in large samples, but in small samples people often check the normality of their data because if the data themselves are normal then the sampling distribution is likely to be also. With the dependent t-test we analyse the *differences* between scores because we're interested in the sampling distribution of these differences (not the raw data). Therefore, if you want to test for normality before a dependent t-test then what you should do is compute the differences between scores, and then check if this new variable is normally distributed (or use a big sample and not worry about normality!). It is possible to have two measures that are highly non-normal that produce beautifully distributed differences!

SELF-TEST Using the **spiderRM.sas7bdat** data, compute the differences between the picture and real condition and check the assumption of normality for these differences.

9.4.4. Dependent *t*-tests using SAS ①

Using our spider data (**spiderRM.sas7bdat**), we have 12 arachnophobes who were exposed to a picture of a spider (**picture**) and on a separate occasion a real live tarantula (**real**). Their anxiety was measured in each condition (half of the participants were exposed to the picture before the real spider while the other half were exposed to the real spider first). I have already described how the data are arranged, and so we can move straight on to doing the test itself.

In SAS, when we want to do a t-test, we used PROC TTEST (notice no hyphen). The syntax for the dependent t-test is very straightforward, and is shown in SAS Syntax 9.1. We first put the PROC TTEST line, and declare our data, then we have a PAIRED line, and we list the pairs of variables we are interested in comparing. In this case, the paired variables are the picture of a spider (**picture**) and a real spider (**real**).

```
PROC TTEST DATA=chapter9.spiderrm;
      PAIRED picture*real     ;
      RUN;
```
SAS Syntax 9.1

9.4.5. Output from the dependent *t*-test ①

The resulting output produces three tables, shown in SAS Output 9.1. First, the output tells us the mean difference between scores (this value, i.e. $\bar{D}$ in equation (9.2), is the difference between the mean scores of each condition: 40 − 47 = −7). The table also reports the standard deviation of the differences between the means and more important the standard error of the differences between participants' scores in each condition (see section 9.4.1).

Second, this output provides a 95% confidence interval for the mean difference. Imagine we took 100 samples from a population of difference scores and calculated their means ($\bar{D}$) and a confidence interval for that mean. In 95 of those samples the constructed confidence interval contains the true value of the mean difference. The confidence interval tells us the boundaries within which the true mean difference is likely to lie.[4] So, assuming this sample's confidence interval is one of the 95 out of 100 that contains the population value, we can say that the true mean difference lies between −13.23 and −0.77. The importance of this interval is that it does not contain zero (i.e. both limits are negative) because this tells us that the true value of the mean difference is unlikely to be zero. Crucially, if we were to compare pairs of random samples from a population we would expect most of the differences between sample means to be zero. This interval tells us that, based on our two samples, the true value of the difference between means is unlikely to be zero. Therefore, we can be confident that our two samples do not represent random samples from the same population. Instead they represent samples from different populations induced by the experimental manipulation. (We are also told the 95% confident intervals for the standard deviation, but that's not as interesting).

The third table shows the test statistic, *t*, which is calculated by dividing the mean of differences by the standard error of differences (see equation (9.2): $t = -7/2.8311 = -2.47$). The size of *t* is compared against known values based on the degrees of freedom. When the same participants have been used, the degrees of freedom are simply the sample size minus 1 ($df = N - 1 = 11$). SAS uses the degrees of freedom to calculate the exact probability that a value of *t* as big as the one obtained could occur if the null hypothesis were true (i.e. there was no difference between these means). This probability value is in the column labelled Pr > |t|. By default, SAS provides the two-tailed probability, which is the probability when no prediction was made about the direction of group differences. If a specific prediction was made (e.g. we might predict that anxiety will be higher when a real spider is used) then the one-tailed probability should be reported and this value is obtained by dividing the two-tailed probability by 2 (see SAS Tip 9.1). The two-tailed probability for the spider data is very low ($p = .031$) and in fact it tells us that there is only a 3.1% chance that a value of *t* this big could happen if the null hypothesis were true. We saw in Chapter 2 that we generally accept a $p < .05$ as statistically meaningful; therefore, this *t* is significant because .031 is smaller than .05. The fact that the *t*-value is a negative number tells us that the first condition (the picture condition) had a smaller mean than the second (the real condition) and so the real spider led to greater anxiety than the picture. Therefore, we can conclude that exposure to a real spider caused significantly more reported anxiety in arachnophobes than exposure to a picture, $t(11) = -2.47, p < .05$.

[4] We saw in section 3.5.2 that these intervals represent the value of two (well, 1.96 to be precise) standard errors either side of the mean of the sampling distribution. For these data, in which the mean difference was −7 and the standard error was 2.8311, these limits will be $-7 \pm (1.96 \times 2.8311)$. However, because we're using the *t*-distribution, not the normal distribution, we use the critical value of *t* to compute the confidence intervals. This value is (with $df = 11$ in this example) 2.201 (two-tailed), which gives us $-7 \pm (2.201 \times 2.8311)$.

SAS OUTPUT 9.1

N	Mean	Std Dev	Std Err	Minimum	Maximum
12	-7.0000	9.8072	2.8311	-20.0000	11.0000

Mean	95% CL Mean		Std Dev	95% CL Std Dev	
-7.0000	-13.2312	-0.7688	9.8072	6.9474	16.6515

DF	t Value	Pr > |t|
11	-2.47	0.0310

SAS TIP 9.1 One- and two-tailed significance in SAS ①

Some students get a bit upset by the fact that SAS produces only the two-tailed significance much of the time and are confused by why there isn't an option that can be selected to produce the one-tailed significance. The answer is simple: there is no need for an option because the one-tailed probability can be ascertained by dividing the two-tailed significance value by 2. For example, if the two-tailed probability is .107, then the one-tailed probability is .107/2 = .054.

However, we should warn you that some people (e.g. Miles & Banyard, 2007) have suggested that one-tailed tests should be used extremely rarely.

9.4.6. Calculating the effect size ②

Even though our *t*-statistic is statistically significant, this doesn't mean our effect is important in practical terms. To discover whether the effect is substantive we need to use what we know about effect sizes (see section 2.6.4). I'm going to stick with the effect size *r* because it's widely understood, frequently used, and yes, I'll admit it, I actually like it! Converting a *t*-value into an *r*-value is actually really easy; we can use the following equation (e.g. Rosenthal, 1991; Rosnow & Rosenthal, 2005):[5]

$$r = \sqrt{\frac{t^2}{t^2 + df}}$$

We know the value of *t* and the *df* from the SAS output and so we can compute *r* as follows:

$$r = \sqrt{\frac{-2.473^2}{-2.473^2 + 11}} = \sqrt{\frac{6.116}{17.116}} = .60$$

[5] Actually, this will overestimate the effect size because of the correlation between the two conditions. This is quite a technical issue and I'm trying to keep things simple here, but bear this in mind and if you're interested read Dunlap, Cortina, Vaslow, and Burke (1996).

If you think back to our benchmarks for effect sizes this represents a very large effect (it is above .5, the threshold for a large effect). Therefore, as well as being statistically significant, this effect is large and so represents a substantive finding.

JANE SUPERBRAIN 9.1

Calculating automatically ②

We can use the SAS ODS system (which I introduced you to in Jane Superbrain Box 3.1) to automatically calculate *r* for us. We can send the output to a data file, and then manipulate it.

First, we use ODS output before our PROC TTEST statement, to tell SAS to send some of the output to a data file. Then we run PROC TTEST and close the output. We can then calculate *r* in a DATA step. Finally, we use PROC PRINT to print the dataset, and see the result.

```
ODS OUTPUT TTESTS=myttest;
PROC TTEST DATA=chapter9.spiderrm;
     PAIRED picture*real   ;
     RUN;
     ODS OUTPUT CLOSE;
DATA myttest; SET myttest;
     r=SQRT(tvalue**2/(tvalue**2+df));
     RUN;
PROC PRINT DATA=myttest;
     VAR r;
     RUN;
```

The result is:

```
Obs       r
 1     0.59769
```

9.4.7. Reporting the dependent *t*-test ①

There is a fairly standard way to report any test statistic: you usually state the finding to which the test relates and then report the test statistic, its degrees of freedom and the probability value of that test statistic. There has also been a recent move (by the American Psychological Association among others) to recommend that an estimate of the effect size is routinely reported. Although effect sizes are still rather sporadically used, I want to get you into good habits so we'll start thinking about effect sizes now. In this example the SAS output tells us that the value of t was −2.47, based on 11 degrees of freedom, and that it was significant at $p = .031$. We can also see the means for each group. We could write this as:

✓ On average, participants experienced significantly greater anxiety with real spiders ($M = 47.00$, $SE = 3.18$) than with pictures of spiders ($M = 40.00$, $SE = 2.68$), $t(11) = -2.47$, $p = 0.031$, $r = .60$.

Note how we've reported the means in each group (and standard errors) in the standard format. For the test statistic, note that we've used an italic *t* to denote the fact that we've calculated a *t*-statistic, then in brackets we've put the degrees of freedom and then stated the value of the test statistic. The probability can be expressed in several ways: here I've reported the exact significance. Finally, note that I've reported the effect size at the end – you won't always see this in published papers but that's no excuse for you not to report it!

Try to avoid writing vague, unsubstantiated things like this:

× People were more scared of real spiders ($t = -2.47$).

More scared than what? Where are the *df*? Was the result statistically significant? Was the effect important (what was the effect size)?

CRAMMING SAM'S TIPS

- The dependent *t*-test compares two means, when those means have come from the same entities; for example, if you have used the same participants in each of two experimental conditions.
- Look at the column labelled *Pr* > |*t*|. If the value is less than .05 then the means of the two conditions are significantly different.
- Look at the values of the means to tell you how the conditions differ.
- Report the *t*-statistic, the degrees of freedom and the significance value. Also report the means and their corresponding standard errors (or draw an error bar chart).
- If you're feeling brave, calculate and report the effect size too!

9.5. The independent *t*-test ①

9.5.1. The independent *t*-test equation explained ①

The independent *t*-test is used in situations in which there are two experimental conditions and different participants have been used in each condition. There are two different equations that can be used to calculate the *t*-statistic, depending on whether the samples contain an equal number of people. As with the dependent *t*-test we can calculate the *t*-statistic by using a numerical version of equation (9.1); in other words, we are comparing the model or effect against the error. With the dependent *t*-test we could look at differences between pairs of scores, because the scores came from the same participants and so individual differences between conditions were eliminated. Hence, the difference in scores should reflect only the effect of the experimental manipulation. Now, when different participants participate in different conditions then pairs of scores will differ not just because of the experimental manipulation, but also because of other sources of variance (such as individual differences between participants' motivation, IQ, etc.). If we cannot investigate differences between conditions on a *per participant* basis (by comparing pairs of scores as we did for the dependent *t*-test) then we must make comparisons on a *per condition* basis (by looking at the overall effect in a condition):

$$t = \frac{(\overline{X}_1 - \overline{X}_2) - (\mu_1 - \mu_2)}{\text{estimate of the standard error}} \tag{9.3}$$

Instead of looking at differences between pairs of scores, we now look at differences between the overall means of the two samples and compare them to the differences we would expect to get between the means of the two populations from which the samples come. If the null hypothesis is true then the samples have been drawn from the same population. Therefore, under the null hypothesis $\mu_1 = \mu_2$ and therefore, $\mu_1 - \mu_2 = 0$. Therefore, under the null hypothesis the equation becomes:

$$t = \frac{\overline{X}_1 - \overline{X}_2}{\text{estimate of the standard error}} \tag{9.4}$$

In the dependent *t*-test we divided the mean difference between pairs of scores by the standard error of these differences. For the independent *t*-test we are looking at differences between groups and so we need to divide by the standard deviation of differences between groups. We can still apply the logic of sampling distributions to this situation. Now, imagine we took several pairs of samples – each pair containing one sample from the two different populations – and compared the means of these samples. From what we have learnt about sampling distributions, we know that the majority of samples from a population will have fairly similar means. Therefore, if we took several pairs of samples (from different populations), the differences between the sample means will be similar across pairs. However, often the difference between a pair of sample means will deviate by a small amount and very occasionally it will deviate by a large amount. If we could plot a sampling distribution of the differences between every pair of sample means that could be taken from two populations, then we would find that it had a normal distribution with a mean equal to the difference between population means ($\mu_1 - \mu_2$). The sampling distribution would tell us by how much we can expect the means of two (or more) samples to differ. As before, the standard deviation of the sampling distribution (the standard error) tells us how variable the differences between sample means are by chance alone. If the standard deviation is high then large differences between sample means can occur by chance; if it is small then only small differences between sample means are expected. It, therefore, makes sense that we use the standard error of the sampling distribution to assess whether the difference between two sample means is statistically meaningful or simply a chance result. Specifically, we divide the difference between sample means by the standard deviation of the sampling distribution.

So, how do we obtain the standard deviation of the sampling distribution of differences between sample means? Well, we use the **variance sum law**, which states that the variance of a difference between two independent variables is equal to the sum of their variances (see, for example, Howell, 2006). This statement means that the variance of the sampling distribution is equal to the sum of the variances of the two populations from which the samples were taken. We saw earlier that the standard error is the standard deviation of the sampling distribution of a population. We can use the sample standard deviations to calculate the standard error of each population's sampling distribution:

$$\text{SE of sampling distribution of population 1} = \frac{s_1}{\sqrt{N_1}}$$

$$\text{SE of sampling distribution of population 2} = \frac{s_2}{\sqrt{N_2}}$$

Therefore, remembering that the variance is simply the standard deviation squared, we can calculate the variance of each sampling distribution:

$$\text{variance of sampling distribution of population } 1 = \left(\frac{s_1}{\sqrt{N_1}}\right)^2 = \frac{s_1^2}{N_1}$$

$$\text{variance of sampling distribution of population } 2 = \left(\frac{s_2}{\sqrt{N_2}}\right)^2 = \frac{s_2^2}{N_2}$$

The variance sum law means that to find the variance of the sampling distribution of differences we merely add together the variances of the sampling distributions of the two populations:

$$\text{variance of sampling distribution of differences} = \frac{s_1^2}{N_1} + \frac{s_2^2}{N_2}$$

To find out the standard error of the sampling distribution of differences we merely take the square root of the variance (because variance is the standard deviation squared):

$$\text{SE of the sampling distribution of differences} = \sqrt{\frac{s_1^2}{N_1} + \frac{s_2^2}{N_2}}$$

Therefore, equation (9.4) becomes:

$$t = \frac{\overline{X}_1 - \overline{X}_2}{\sqrt{\frac{s_1^2}{N_1} + \frac{s_2^2}{N_2}}} \tag{9.5}$$

Equation (9.5) is true only when the sample sizes are equal. Often in the social sciences it is not possible to collect samples of equal size (because, for example, people may not complete an experiment). When we want to compare two groups that contain different numbers of participants then equation (9.5) is not appropriate. Instead the pooled variance estimate *t*-test is used which takes account of the difference in sample size by *weighting* the variance of each sample. We saw in Chapter 1 that large samples are better than small ones because they more closely approximate the population; therefore, we weight the variance by the size of sample on which it's based (we actually weight by the number of degrees of freedom, which is the sample size minus 1). Therefore, the pooled variance estimate is:

$$s_p^2 = \frac{(n_1 - 1)s_1^2 + (n_2 - 1)s_2^2}{n_1 + n_2 - 2}$$

This is simply a weighted average in which each variance is multiplied (weighted) by its degrees of freedom, and then we divide by the sum of weights (or sum of the two degrees of freedom). The resulting weighted average variance is then just replaced in the *t*-test equation:

$$t = \frac{\overline{X}_1 - \overline{X}_2}{\sqrt{\frac{s_p^2}{n_1} + \frac{s_p^2}{n_2}}}$$

As with the dependent *t*-test we can compare the obtained value of *t* against the maximum value we would expect to get by chance alone in a *t*-distribution with the same degrees of freedom (these values can be found in the Appendix); if the value we obtain exceeds this critical value we can be confident that this reflects an effect of our independent variable. One thing that should be apparent from the equation for t is that to compute it you don't actually need any data! All you need are the means, standard deviations and sample sizes (see SAS Tip 9.2).

The derivation of the t-statistic is merely to provide a conceptual grasp of what we are doing when we carry out a t-test on SAS. Therefore, if you don't know what on earth I'm babbling on about then don't worry about it (just spare a thought for my cat: he has to listen to this rubbish all the time!) because SAS knows how to do it and that's all that matters!

SAS TIP 9.2 Computing *t* from means, *SDs* and *Ns* ③

Using a DATA step you can compute a t-test in SAS from only the two group means, the two group standard deviations and the two group sizes.

We'll create a dataset called **ttest**, then we'll assign values to six new variables: **x1** (mean of group 1), **x2** (mean of group 2), **sd1** (standard deviation of group 1), **sd2** (standard deviation of group 2), **n1** (sample size of group 1) and **n2** (sample size of group 2).

```
DATA ttest;
     x1 = 10; x2 = 15; n1 = 20;  n2 = 25; sd1 = 7; sd2 = 8;
     df = (n1+n2-2) ;
     poolvar = (((n1-1)*(sd1 ** 2))+((n2-1)*(sd2 ** 2)))/df;
     t = (x1-x2)/SQRT(poolvar*((1/n1)+(1/n2)));
     sig = 2*(1-(PROBT(ABS(t),df))) ;
     RUN;

PROC PRINT DATA=ttest;
     RUN;
```

These commands will produce a table of the variables **x1, x2, df, t** and **sig** so you'll see the means of the two groups, the degrees of freedom, the value of t and its two-tailed significance.

LABCOAT LENI'S REAL RESEARCH 9.1

You don't have to be mad here, but it helps ③

BOARD, B. J., & FRITZON, K. (2005). *PSYCHOLOGY, CRIME & LAW, 11*, 17–32.

In the UK you often see the 'humorous' slogan 'You don't have to be mad to work here, but it helps' stuck up in work places. Well, Board and Fritzon (2005) took this a step further by measuring whether 39 senior business managers and chief executives from leading UK companies were mad (well, had personality disorders, PDs). They gave them the Minnesota Multiphasic Personality Inventory Scales for DSM III Personality Disorders (MMPI-PD), which is a well-validated measure of 11 personality disorders: Histrionic, Narcissistic, Antisocial, Borderline, Dependent, Compulsive, Passive-aggressive, Paranoid, Schizotypal, Schizoid and Avoidant. They needed a comparison group, and what better one to choose than 317 legally classified psychopaths at Broadmoor Hospital (a famous high-security psychiatric hospital in the UK).

The authors report the means and SDs for these two groups in Table 2 of their paper. Using these we can run t-tests on these means. The data from Board and Fritzon's Table 2 are in the file **BoardandFritzon2005.sas7bdat**. Use this file and the syntax file to run t-tests to see whether managers score higher on personality disorder questionnaires than legally classified psychopaths. Report these results. What do you conclude?

Answers are in the additional material on the companion website (or look at Table 2 in the original article).

9.5.2. The independent *t*-test using SAS ①

I have probably bored most of you to the point of wanting to eat your own legs by now. Equations are boring and that is why SAS was invented to help us minimize our contact with them. Using our spider data again (**spiderBG.sas7bdat**), we have 12 arachnophobes who were exposed to a picture of a spider and 12 different arachnophobes who were exposed to a real-life tarantula (the groups are coded using the variable **group**). Their anxiety was measured in each condition (**anxiety**). I have already described how the data are arranged (see section 9.2), so we can move straight on to doing the test itself.

We use PROC TTEST again, with a slightly different format to tell SAS that we want an independent groups t-test, not a repeated measures *t*-test.

First, we select an independent variable – this should be a categorical variable with two values – such as **group** (which has the two values 'real' and 'picture'). We tell SAS that this is a categorical variable with a CLASS statement. Then we tell SAS which variables we'd like to be compared on, with the VAR statement – in our case, we'd like to compare the mean anxiety. SAS Syntax 9.2 shows the PROC TTEST syntax.

```
PROC TTEST DATA=chapter9.spiderbg;
    CLASS group;
    VAR anxiety;
    RUN;
```

SAS Syntax 9.2

JANE SUPERBRAIN 9.1

Are median splits the devil's work? ②

Often in research papers you see that people have analysed their data using a 'median split'. In our spider phobia example, this means that you measure scores on a spider phobia questionnaire and calculate the median. You then classify anyone with a score above the median as a 'phobic', and those below the median as 'non-phobic'. In doing this you 'dichotomize' a continuous variable. This practice is quite common, but is it sensible?

MacCallum, Zhang, Preacher, and Rucker (2002) wrote a splendid paper pointing out various problems on turning a perfectly decent continuous variable into a categorical variable:

1. Imagine there are four people: Peter, Birgit, Jip and Kiki. We measure how scared of spiders they are as a percentage and get Jip (100%), Kiki (60%), Peter (40%) and Birgit (0%). If we split these four people at the median (50%) then we're saying that Jip and Kiki are the same (they get a score of 1 = phobic) and Peter and Birgit are the same (they both get a score of 0 = not phobic). In reality, Kiki and Peter are the most similar of the four people, but they have been put in different groups. So, median splits change the original information quite dramatically (Peter and Kiki are originally very similar but become very different after the split, Jip and Kiki are relatively dissimilar originally but become identical after the split).
2. Effect sizes get smaller: if you correlate two continuous variables then the effect size will be larger than if you correlate the same variables after one of them has been dichotomized. Effect sizes also get smaller in ANOVA and regression.
3. There is an increased chance of finding spurious effects.

So, if your supervisor has just told you to do a median split, have a good think about whether it is the right thing to do (and read MacCallum et al.'s paper). One of the rare situations in which dichotomizing a continuous variable is justified, according to MacCallum et al., is when there is a clear theoretical rationale for distinct categories of people based on a meaningful break point (i.e. not the median); for example, phobic versus not phobic based on diagnosis by a trained clinician would be a legitimate dichotomization of anxiety.

9.5.3. Output from the independent *t*-test ①

The output from the independent *t*-test contains only two tables. The first table (SAS Output 9.2) provides summary statistics for the two experimental conditions. From this table, we can see that both groups had 12 participants (column labelled *N*). The group who saw the picture of the spider had a mean anxiety of 40, with a standard deviation of 9.29. What's more, the standard error of that group (the standard deviation of the sampling distribution) is 2.68 ($SE = 9.293/\sqrt{12} = 9.293/3.464 = 2.68$). In addition, the table tells us that the average anxiety level in participants who were shown a real spider was 47, with a standard deviation of 11.03 and a standard error of 3.18 ($SE = 11.029/\sqrt{12} = 11.029/3.464 = 3.18$).

SAS OUTPUT 9.2

group	N	Mean	Std Dev	Std Err	Minimum	Maximum
Picture	12	40.0000	9.2932	2.6827	25.0000	55.0000
Real	12	47.0000	11.0289	3.1838	30.0000	65.0000
Diff (1-2)		−7.0000	10.1980	4.1633		

SAS Output 9.3 contains the main test statistics. The first thing to notice is that there are two rows containing values for the test statistics: one row is labelled Pooled, while the other is labelled Satterthwaite. In Chapter 5, we saw that parametric tests assume that the variances in experimental groups are roughly equal. Well, in reality there are adjustments that can be made in situations in which the variances are not equal, one of which was proposed by Satterthwaite, and so SAS produces a pooled variance *t*-test, and a Satterthwaite corrected *t*-test (sometimes it's called the Welch–Satterthwaite correction, because Welch also worked on a similar correction).[6] The rows of the table relate to whether or not this assumption has been broken. How do we know whether this assumption has been broken?

We saw in section 5.6.1 that we can use Levene's test to see whether variances are different in different groups, and SAS produces this test for us (actually, in PROC TTEST SAS produces a slightly different version, called the Folded F, but we won't worry about that).[7] Remember that Levene's test is similar to a *t*-test in that it tests the hypothesis that the variances in the two groups are equal (i.e. the difference between the variances is zero). Therefore, if this test is significant at $p \leq .05$, we can gain confidence in the hypothesis that the variances are significantly different and that the assumption of homogeneity of variances has been violated. If, however, the test of homogeneity of variance is non-significant (i.e. $p > .05$) then we do not have sufficient evidence to reject the null hypothesis that the difference between the variances is zero – in other words, we can assume that the variances are roughly equal and the assumption is tenable. For these data, Levene's test is non-significant (because $p = .580$, which is greater than .05) and so we should read the test statistics in the row labelled pooled. Had the equality of variances test been significant, then we would have read the test statistics from the row labelled *Satterthwaite*.

[6] It's an interesting thing in statistics ... well, we think it's interesting ... that different people often come to the same way of doing things from different directions, and so don't realize, at first, that they are doing the same thing. This is one example, but there are many others. The same correction exists in many forms, and has different many different names – one of our favorites is a sandwich estimate.

[7] If you are worried about this, you can do a Levene's test with PROC GLM. We'll cover that in Chapter 10.3.3.

SAS OUTPUT 9.3

group	Method	Mean	95% CL Mean		Std Dev	95% CL Std Dev	
Picture		40.0000	34.0954	45.9046	9.2932	6.5833	15.7787
Real		47.0000	39.9926	54.0074	11.0289	7.8128	18.7257
Diff (1-2)	Pooled	–7.0000	–15.6342	1.6342	10.1980	7.8871	14.4338
Diff (1-2)	Satterthwaite	–7.0000	–15.6486	1.6486			

Method	Variances	DF	t Value	Pr > |t|
Pooled	Equal	22	–1.68	0.1068
Satterthwaite	Unequal	21.385	–1.68	0.1072

Equality of Variances				
Method	Num DF	Den DF	F Value	Pr > F
Folded F	11	11	1.41	0.5797

Having established that the assumption of homogeneity of variances is met, we can move on to look at the *t*-test itself. We are told the mean difference $(\overline{X}_1 - \overline{X}_2 = 40 - 47 = -7)$ and the standard error of the sampling distribution of differences, which is calculated using the lower half of equation (9.5):

$$\begin{aligned}\sqrt{\frac{s_1^2}{N_1}+\frac{s_2^2}{N_2}} &= \sqrt{\frac{9.29^2}{12}+\frac{11.03^2}{12}}\\ &= \sqrt{7.19+10.14}\\ &= \sqrt{17.33}\\ &= 4.16\end{aligned}$$

The *t*-statistic is calculated by dividing the mean difference by the standard error of the sampling distribution of differences ($t = -7/4.16 = -1.68$). The value of *t* is then assessed against the value of *t* you might expect to get by chance when you have certain degrees of freedom. For the independent *t*-test, degrees of freedom are calculated by adding the two sample sizes and then subtracting the number of samples ($df = N_1 + N_2 - 2 = 12 + 12 - 2 = 22$). SAS produces the exact significance value of *t*, and we are interested in whether this value is less than or greater than .05. In this case the two-tailed value of *p* is .107, which is greater than .05, and so we would have to conclude that there was no significant difference between the means of these two samples. In terms of the experiment, we can infer that arachnophobes are not made less anxious by pictures of spiders than they are by the real thing.

9.5.4. Calculating the effect size ②

To discover whether our effect is substantive we can use the same equation as in section 9.4.6 to convert the *t*-statistics into a value of *r*. We know the value of *t* and the *df* from the SAS output and so we can compute *r* as follows:

$$r = \sqrt{\frac{-1.681^2}{-1.681^2 + 22}} = \sqrt{\frac{2.826}{24.826}} = .34$$

If you think back to our benchmarks for effect sizes this represents a medium effect (it is around .3, the threshold for a medium effect). Therefore, even though the effect was non-significant, it still represented a fairly substantial effect. You may also notice that the effect has shrunk, which may seem slightly odd given that we used exactly the same data (but see section 9.6)!

Jane Superbrain Box 9.1 showed us how to use the SAS ODS system to calculate r automatically. We can do the same thing again, with the syntax shown in SAS Syntax 9.3, which produces the output shown in SAS Output 9.4.

```
ODS OUTPUT TTESTS=bgttest;
PROC TTEST DATA=chapter9.spiderbg;
      CLASS group;
      VAR anxiety;
      RUN;
ODS OUTPUT CLOSE;
DATA bgttest; SET bgttest;
      r=SQRT(tvalue**2/(tvalue**2+df));
      RUN;
PROC PRINT DATA=bgttest;
      RUN;
ODS RTF CLOSE;
```

SAS Syntax 9.3

Obs	Variable	Method	Variances	tValue	DF	Probt	r
1	anxiety	Pooled	Equal	−1.68	22	0.1068	0.33744
2	anxiety	Satterthwaite	Unequal	−1.68	21.385	0.1072	0.34170

SAS OUTPUT 9.4

9.5.5. Reporting the independent *t*-test ①

The rules that I reported for the dependent *t*-test pretty much apply for the independent *t*-test. The SAS output tells us that the value of t was 1.68, based on 22 degrees of freedom, and that it was not significant at $p < .05$. We can also see the means for each group. We could write this as:

- ✓ On average, participants experienced greater anxiety with real spiders (M = 47.00, SE = 3.18), than with pictures of spiders (M = 40.00, SE = 2.68). This difference was not significant, $t(22) = 1.68$, $p = .337$; however, it did represent a medium-sized effect, $r = .34$.

Note how we've reported the means in each group (and standard errors) as before. For the test statistic everything is much the same as before except that I've had to report that p was greater than (>) .05 rather than less than (<). Finally, note that I've commented on the effect size at the end.

CRAMMING SAM'S TIPS

- The independent *t*-test compares two means, when those means have come from different groups of entities; for example, if you have used different participants in each of two experimental conditions.
- Look at the table labelled *Homogeneity Equality of Variance*. If the *Pr* > *F*. value is less than .05 then the assumption of homogeneity of variance has been broken and you should look at the row in the table labelled Equal *Satterthwaite*. If the significance value of the homogeneity of variance test is bigger than .05 then you should look at the row in the table labelled *Pooled*.
- Look at the column labelled *Pr* > *t*. If the value is less than .05 then the means of the two groups are significantly different.
- Look at the values of the means to tell you how the groups differ.
- Report the *t*-statistic, the degrees of freedom and the significance value. Also report the means and their corresponding standard errors (or draw an error bar chart).
- Calculate and report the effect size. Go on, you can do it!

9.6. Between groups or repeated measures? ①

The two examples in this chapter are interesting (honestly!) because they illustrate the difference between data collected using the same participants and data collected using different participants. The two examples use the same scores in each condition. When analysed as though the data came from the same participants the result was a significant difference between means, but when analysed as though the data came from different participants there was no significant difference between group means. This may seem like a puzzling finding – after all the numbers were identical in both examples. What this illustrates is the relative *power* of repeated-measures designs. When the same participants are used across conditions the unsystematic variance (often called the error variance) is reduced dramatically, making it easier to detect any systematic variance. It is often assumed that the way in which you collect data is irrelevant, but I hope to have illustrated that it can make the difference between detecting a difference and not detecting one. In fact, researchers have carried out studies using the same participants in experimental conditions, then repeated the study using different participants in experimental conditions, and then used the method of data collection as an independent variable in the analysis. Typically, they have found that the method of data collection interacts significantly with the results found (see Erlebacher, 1977).

9.7. The *t*-test as a general linear model ②

A lot of you might think it's odd that I've chosen to represent the effect size for my *t*-tests using *r*, the correlation coefficient. In fact you might well be thinking 'but correlations show relationships, not differences between means'. I used to think this too until I read a fantastic paper by Cohen (1968), which made me realize what I'd been missing; the complex, thorny, weed-infested and large Andy-eating tarantula-inhabited world of statistics suddenly turned into a beautiful meadow filled with tulips and little bleating lambs all jumping for joy at the wonder of life. Actually, I'm still a bumbling fool trying desperately to avoid having the blood sucked from my flaccid corpse by the tarantulas of statistics, but it was a good paper! What I'm about to say will either make no sense at all, or might help you to appreciate what I've said in most of the chapters so far: all statistical procedures are basically the same, they're just more or less elaborate versions of the correlation coefficient!

In Chapter 7 we saw that the *t*-test was used to test whether the regression coefficient of a predictor was equal to zero. The experimental design for which the independent *t*-test is used can be conceptualized as a regression equation (after all, there is one independent variable (predictor) and one dependent variable (outcome)). If we want to predict our outcome, then we can use the general equation that I've mentioned at various points:

$$\text{outcome}_i = (\text{model}) + \text{error}_i$$

If we want to use a linear model, then we saw that this general equation becomes equation (7.2) in which the model is defined by the slope and intercept of a straight line. Equation (9.6) shows a very similar equation in which A_i is the dependent variable (outcome), b_0 is the intercept, b_1 is the weighting of the predictor and G_i is the independent variable (predictor). Now, I've also included the same equation but with some of the letters replaced with what they represent in the spider experiment (so A = **anxiety**, G = **group**). When we run an experiment with two conditions, the independent variable has only two values (group 1 or group 2). There are several ways in which these groups can be coded (in the spider example we coded group 1 with the value 0 and group 2 with the value 1). This coding variable is known as a *dummy variable* and values of this variable represent groups of entities. We have come across this coding in section 7.11:

$$\begin{aligned} A_i &= b_0 + b_1 G_i + \varepsilon_i \\ \text{anxiety}_i &= b_0 + b_1 \text{group}_i + \varepsilon_i \end{aligned} \qquad (9.6)$$

Using the spider example, we know that the mean **anxiety** of the picture group was 40, and that the **group** variable is equal to 0 for this condition. Look at what happens when the **group** variable is equal to 0 (the picture condition): equation 9.6) becomes (if we ignore the residual term):

$$\begin{aligned} \overline{X}_{\text{Picture}} &= b_0 + (b_1 \times 0) \\ b_0 &= \overline{X}_{\text{Picture}} \\ b_0 &= 40 \end{aligned}$$

Therefore, b_0 (the intercept) is equal to the mean of the picture group (i.e. it is the mean of the group coded as 0). Now let's look at what happens when the **group** variable is equal to 1. This condition is the one in which a real spider was used, therefore and the mean **anxiety** ($\overline{X}_{\text{Real}}$) of this condition was 47. Remembering that we have just found out that b_0 is equal to the mean of the picture group ($\overline{X}_{\text{Picture}}$), equation (9.6) becomes:

$$\begin{aligned} \overline{X}_{\text{Real}} &= b_0 + (b_1 \times 1) \\ \overline{X}_{\text{Real}} &= \overline{X}_{\text{Picture}} + b_1 \\ b_1 &= \overline{X}_{\text{Real}} - \overline{X}_{\text{Picture}} \\ &= 47 - 40 \\ &= 7 \end{aligned}$$

b_1, therefore, represents the difference between the group means. As such, we can represent a two-group experiment as a regression equation in which the coefficient of the independent variable (b_1) is equal to the difference between group means, and the intercept (b_0) is equal to the mean of the group coded as 0. In regression, the *t*-test is used to ascertain whether the regression coefficient (b_1) is equal to 0, and when we carry out a *t*-test on grouped data we, therefore, test whether the difference between group means is equal to 0.

SELF-TEST To prove that I'm not making it up as I go along, run a regression on the data in **spiderBG.sas7bdat** with **group** as the predictor and **anxiety** as the outcome. **Group** is coded using zeros and ones and represents the dummy variable described above.

The resulting SAS output should contain the regression summary table shown in SAS Output 9.5. The first thing to notice is the value of the constant (b_0): its value is 40, the same as the mean of the base category (the picture group). The second thing to notice is that the value of the regression coefficient b_1 is 7, which is the difference between the two group means (47 – 40 = 7). Finally, the *t*-statistic, which tests whether b_1 is significantly different from zero, is the same as for the independent *t*-test (see SAS Output 9.3) and so is the significance value (apart from a charge of sign; see SAS Output 9.3).

SAS OUTPUT 9.5 Regression analysis of between-group spider data

Parameter Estimates					
Variable	DF	Parameter Estimate	Standard Error	t Value	Pr > |t|
Intercept	1	40.00000	2.94392	13.59	<.0001
condition	1	7.00000	4.16333	1.68	0.1068

This section has demonstrated that differences between means can be represented in terms of linear models, and this concept is essential in understanding the following chapters on the general linear model.

9.8. What if my data are not normally distributed? ②

We've seen in this chapter that there are adjustments that can be made to the *t*-test when the assumption of homogeneity of variance is broken, but what about when you have non-normally distributed data? The first thing to note is that although a lot of early evidence suggested that *t* was accurate when distributions were skewed, the *t*-test can be biased when the assumption of normality is not met (Wilcox, 2005). Second, we need to remember that it's the shape of the sampling distribution that matters, not the sample data. One option then is to use a big sample and rely on the central limit theorem (section 2.5.1) which says that the sampling distribution should be normal when samples are big. You could also try to correct the distribution using a transformation (but see Jane Superbrain Box 5.1). Another useful solution is to use one of a group of tests commonly referred to as *non-parametric tests*. These tests have fewer assumptions than their parametric counterparts and so are useful when your data violate the assumptions of parametric data described in Chapter 5. Some of these tests are described in Chapter 15. The non-parametric counterpart of the dependent t-test is called the Wilcoxon signed-rank test (section 15.4), and the independent t-test has two non-parametric counterparts (both extremely similar) called the Wilcoxon rank-sum test and the Mann–Whitney test (section 15.3). I'd recommend reading these sections before moving on.

A final option is to use robust methods (see section 5.7.4). There are various robust ways to test differences between means that involve using trimmed means or a bootstrap. However, they are a little fiddly and go beyond this book.

What have I discovered about statistics? ①

We started this chapter by looking at my relative failures as a human being compared to Simon Hudson before investigating some problems with the way SAS produces error bars for repeated-measures designs. We then had a look at some general conceptual features of the t-test, a parametric test that's used to test differences between two means. After this general taster, we moved on to look specifically at the dependent t-test (used when your conditions involve the same entities). I explained how it was calculated, how to do it on SAS and how to interpret the results. We then discovered much the same for the independent t-test (used when your conditions involve different entities). After this I droned on excitedly about how a situation with two conditions can be conceptualized as a general linear model, by which point those of you who have a life had gone to the pub for a stiff drink. My excitement about things like general linear models could explain why Clair Sparks chose Simon Hudson all those years ago. Perhaps she could see the writing on the wall! Fortunately, I was a ruthless pragmatist at the age of 10, and the Clair Sparks episode didn't seem to concern me unduly; I just set my sights elsewhere during the obligatory lunchtime game of kiss chase. These games were the last I would see of women for quite some time ...

Key terms that I've discovered

Dependent t-test
Independent t-test
Standard error of differences
Variance sum law

Smart Alex's tasks

These scenarios are taken from Field and Hole (2003). In each case analyse the data on SAS:

- **Task 1**: One of my pet hates is 'pop psychology' books. Along with banishing Freud from all bookshops, it is my avowed ambition to rid the world of these rancid putrefaction-ridden wastes of trees. Not only do they give psychology a very bad name by stating the bloody obvious and charging people for the privilege, but they are also considerably less enjoyable to look at than the trees killed to produce them (admittedly the same could be said for the turgid tripe that I produce in the name of education, but let's not go there just for now!). Anyway, as part of my plan to rid the world of popular psychology I did a little experiment. I took two groups of people who were in relationships and randomly assigned them to one of two conditions. One group read the famous popular psychology book *Women are from Bras and men are from Penis*, whereas another group read *Marie Claire*. I tested only 10 people in each of these groups, and the dependent variable was an objective measure of their happiness with their relationship after reading the book. I didn't make any specific prediction about which reading material would improve relationship happiness. The data are in the file **Penis.sas7bdat.** Analyse them with the appropriate *t*-test. ①

- **Task 2**: Imagine Twaddle and Sons, the publishers of *Women are from Bras and men are from Penis*, were upset about my claims that their book was about as useful as a paper umbrella. They decided to take me to task and design their own experiment in which participants read their book and one of my books (Field and Hole) at different times. Relationship happiness was measured after reading each book. To maximize their chances of finding a difference they used a sample of 500 participants, but got each participant to take part in both conditions (they read both books). The order in which books were read was counterbalanced and there was a delay of six months between reading the books. They predicted that reading their wonderful contribution to popular psychology would lead to greater relationship happiness than reading some dull and tedious book about experiments. The data are in **FieldHole.sas7bdat.** Analyse them using the appropriate *t*-test. ①

Answers can be found on the companion website (or for more detail see Field and Hole, 2003).

Further reading

Field, A. P., & Hole, G. (2003). *How to design and report experiments*. London: Sage. (In my completely unbiased opinion this is a useful book to get some more background on experimental methods.)

Miles, J. N. V., & Banyard, P. (2007). *Understanding and using statistics in psychology: a practical introduction*. London: Sage. (A fantastic and amusing introduction to statistical theory.)

Rosnow, R. L., & Rosenthal, R. (2005). *Beginning behavioral research: A conceptual primer* (5th ed.). Englewood Cliffs, NJ: Pearson/Prentice Hall.

Wright, D. B., & London, K. (2009). *First steps in statistics* (2nd ed.) London: Sage. (This book has very clear introductions to the *t*-test.)

Interesting real research

Board, B. J., & Fritzon, K. (2005). Disordered personalities at work. *Psychology, Crime & Law*, *11*(1), 17–32.

Comparing several means: ANOVA (GLM 1)

10

FIGURE 10.1 My brother Paul (left) and I (right) in our very fetching school uniforms

10.1. What will this chapter tell me? ①

There are pivotal moments in everyone's life, and one of mine was at the age of 11. Where I grew up in England there were three choices when leaving primary school and moving on to secondary school: (1) state school (where most people go); (2) grammar school (where clever people who pass an exam called the 11+ go); and (3) private school (where rich people go). My parents were not rich and I am not clever and consequently I failed my 11+, so private school and grammar school (where my clever older brother had gone) were out. This left me to join all of my friends at the local state school. I could not have been happier. Imagine everyone's shock when my parents received a letter saying that some extra spaces had become available at the grammar school; although the local authority could scarcely believe it and had remarked the 11+ papers several million times to confirm their findings, I was next on their list. I could not have been unhappier. So, I waved goodbye to all of my friends and trundled off to join my brother at Ilford County High School for Boys (a school that still hit students with a cane if they were particularly bad and that, for some considerable time and with good reason, had 'H.M. Prison' painted in huge white letters on its roof). It was goodbye to normality, and hello to six years of learning how not to function in society. I often wonder how my life would have turned out had I not gone to this school; in the parallel universes where the letter didn't arrive and the Andy went to state school, or where my parents were rich and the Andy went to private school, what became of him? If we wanted to compare these three situations we couldn't use a *t*-test because there are more than two conditions.[1] However, this chapter tells us all about the statistical models that we use to analyse situations in which we want to compare more than two conditions: **analysis of variance** (or ANOVA to its friends). This chapter will begin by explaining the theory of ANOVA when different participants are used (**independent ANOVA**). We'll then look at how to carry out the analysis on SAS and interpret the results.

10.2. The theory behind ANOVA ②

10.2.1. Inflated error rates ②

Before explaining how ANOVA works, it is worth mentioning why we don't simply carry out several *t*-tests to compare all combinations of groups that have been tested. Imagine a situation in which there were three experimental conditions and we were interested in differences between these three groups. If we were to carry out *t*-tests on every pair of groups, then we would have to carry out three separate tests: one to compare groups 1 and 2, one to compare groups 1 and 3, and one to compare groups 2 and 3. If each of these *t*-tests uses a .05 level of significance then for each test the probability of falsely rejecting the null hypothesis (known as a Type I error) is only 5%. Therefore, the probability of no Type I errors is .95 (95%) for each test. If we assume that each test is independent (hence, we can multiply the probabilities) then the overall probability of no Type I errors is $(.95)^3 = .95 \times .95 \times .95 = .857$, because the probability of no Type I errors is .95 for each test and there are three tests. Given that the probability of no Type I errors is .857, then we can calculate the probability of making at least one Type I error by subtracting this number from 1 (remember that the maximum probability of any event occurring is 1). So, the probability of at least one Type I error is $1 - .857 = .143$, or 14.3%. Therefore, across this group of tests, the probability of making a Type I error has increased from 5% to 14.3%, a value greater than the criterion accepted by social scientists. This error rate across statistical tests conducted on the same experimental data is known as the **familywise** or **experimentwise error rate**. An experiment with three conditions is a relatively simple design, and so the effect of carrying out several tests is not severe. If you imagine that we now increase the number of experimental conditions from three to five (which is only two more groups) then the number of *t*-tests that would need to done increases to 10.[2] The familywise error rate can be calculated using the general equation

[1] Really, this is the least of our problems: there's the small issue of needing access to parallel universes.

[2] These comparisons are group 1 vs. 2, 1 vs. 3, 1 vs. 4, 1 vs. 5, 2 vs. 3, 2 vs. 4, 2 vs. 5, 3 vs. 4, 3 vs. 5 and 4 vs. 5. The number of tests required is calculated using this equation:

$$\text{number of comparisons, } C = \frac{k!}{2(k-2)!}$$

in which k is the number of experimental conditions. The ! symbol stands for *factorial*, which means that you multiply the value preceding the symbol by all of the whole numbers between one and that value (so $5! = 5 \times 4 \times 3 \times 2 \times 1 = 120$). Thus, with five conditions we find that:

$$C = \frac{5!}{2(5-2)!} = \frac{120}{2 \times 6} = 10$$

$$\text{familywise error} = 1 - (.95)^n \tag{10.1}$$

in which n is the number of tests carried out on the data. With 10 tests carried out, the familywise error rate is .40 ($1 - .95^{10} = .40$), which means that there is a 40% chance of having made at least one Type I error. For this reason we use ANOVA rather than conducting lots of t-tests.

10.2.2. Interpreting *F* ②

When we perform a t-test, we test the hypothesis that the two samples have the same mean. Similarly, ANOVA tells us whether three or more means are the same, so it tests the null hypothesis that all group means are equal. An ANOVA produces an *F-statistic* or *F-ratio*, which is similar to the t-statistic in that it compares the amount of systematic variance in the data to the amount of unsystematic variance. In other words, F is the ratio of the model to its error.

What does an ANOVA tell me?

ANOVA is an *omnibus* test, which means that it tests for an overall experimental effect: so, there are things that ANOVA cannot tell us. Although ANOVA tells us whether the experimental manipulation was generally successful, it does not provide specific information about which groups were affected. Assuming an experiment was conducted with three different groups, the F-ratio tells us that the means of these three samples are not equal (i.e. that $\bar{X}_1 = \bar{X}_2 = \bar{X}_3$ is *not* true). However, there are several ways in which the means can differ. The first possibility is that all three sample means are significantly different ($\bar{X}_1 \neq \bar{X}_2 \neq \bar{X}_3$). A second possibility is that the means of group 1 and 2 are the same but group 3 has a significantly different mean from both of the other groups ($\bar{X}_1 = \bar{X}_2 \neq \bar{X}_3$). Another possibility is that groups 2 and 3 have similar means but group 1 has a significantly different mean ($\bar{X}_1 \neq \bar{X}_2 = \bar{X}_3$). Finally, groups 1 and 3 could have similar means but group 2 has a significantly different mean from both ($\bar{X}_1 = \bar{X}_3 \neq \bar{X}_2$). So, in an experiment, the F-ratio tells us only that the experimental manipulation has had some effect, but it doesn't tell us specifically what the effect was.

10.2.3. ANOVA as regression ②

I've hinted several times that all statistical tests boil down to variants on regression. In fact, ANOVA is just a special case of regression. This surprises many scientists because ANOVA and regression are usually used in different situations. The reason is largely historical in that two distinct branches of methodology developed in the social sciences: correlational research and experimental research. Researchers interested in controlled experiments adopted ANOVA as their method of choice whereas those looking for real-world relationships adopted multiple regression. As we all know, scientists are intelligent, mature and rational people and so neither group was tempted to slag off the other and claim that their own choice of methodology was far superior to the other (yeah, right!). With the divide in methodologies came a chasm between the statistical methods adopted by the two opposing camps (Cronbach, 1957, documents this divide in a lovely article). This divide has lasted many decades to the extent that now students are generally taught regression and ANOVA in very different contexts and many textbooks teach ANOVA in an entirely different way to regression.

Although many considerably more intelligent people than me have attempted to redress the balance (notably the great Jacob Cohen, 1968), I am passionate about making my own small, feeble-minded attempt to enlighten you (and I set the ball rolling in sections 7.11 and 9.7). There are several good reasons why I think ANOVA should be taught within the context of regression. First, it provides a familiar context: I wasted many trees trying to explain regression, so why not use this base of knowledge to explain a new concept (it should make it easier to understand)? Second, the traditional method of teaching ANOVA (known as the variance-ratio method) is fine for simple designs, but becomes impossibly cumbersome in more complex situations (such as analysis of covariance). The regression model extends very logically to these more complex designs without anyone needing to get bogged down in mathematics. Finally, the variance-ratio method becomes extremely unmanageable in unusual circumstances such as when you have unequal sample sizes.[3] The regression method makes these situations considerably simpler. Although these reasons are good enough, it is also the case that SAS has moved away from the variance-ratio method of ANOVA and progressed towards solely using the regression model (known as the general linear model, or GLM).

I have mentioned that ANOVA is a way of comparing the ratio of systematic variance to unsystematic variance in an experimental study. The ratio of these variances is known as the *F*-ratio. However, any of you who have read Chapter 7 should recognize the *F*-ratio (see section 7.2.3) as a way to assess how well a regression model can predict an outcome compared to the error within that model. If you haven't read Chapter 7 (surely not!), have a look before you carry on (it should only take you a couple of weeks to read). How can the *F*-ratio be used to test differences between means *and* whether a regression model fits the data? The answer is that when we test differences between means we *are* fitting a regression model and using *F* to see how well it fits the data, but the regression model contains only categorical predictors (i.e. grouping variables). So, just as the *t*-test could be represented by the linear regression equation (see section 9.7), ANOVA can be represented by the multiple regression equation in which the number of predictors is one less than the number of categories of the independent variable.

Let's take an example. There was a lot of controversy, a few years ago, surrounding the drug Viagra. Admittedly there's less controversy now, but the controversy has been replaced by an alarming number of spam emails on the subject (for which I'll no doubt be grateful in 20 years' time), so I'm going to stick with the example. Viagra is a sexual stimulant (used to treat impotence) that broke into the black market under the belief that it will make someone a better lover (oddly enough, there was a glut of journalists taking the stuff at the time in the name of 'investigative journalism' … hmmm!). Suppose we tested this belief by taking three groups of participants and administering one group with a placebo (such as a sugar pill), one group with a low dose of Viagra and one with a high dose. The dependent variable was an objective measure of libido (I will tell you only that it was measured over the course of a week – the rest I will leave to your imagination). The data can be found in the file **Viagra.sas7bdat** (which is described in detail later in this chapter) and are in Table 10.1.

TABLE 10.1 Data in **Viagra.sas7bdat**

	Placebo	Low Dose	High Dose
	3	5	7
	2	2	4
	1	4	5
	1	2	3
	4	3	6
$\bar{X}$	**2.20**	**3.20**	**5.00**
s	**1.30**	**1.30**	**1.58**
s^2	**1.70**	**1.70**	**2.50**
Grand Mean = **3.467**	Grand SD = **1.767**	Grand Variance = **3.124**	

[3] Having said this, it is well worth the effort to try to obtain equal sample sizes in your different conditions because unbalanced designs do cause statistical complications (see section 10.2.10).

If we want to predict levels of libido from the different levels of Viagra then we can use the general equation that keeps popping up:

$$\text{outcome}_i = (\text{model}) + \text{error}_i$$

If we want to use a linear model, then we saw in section 9.7 that when there are only two groups we could replace the 'model' in this equation with a linear regression equation with one dummy variable to describe two experimental groups. This dummy variable was a categorical variable with two numeric codes (0 for one group and 1 for the other). With three groups, however, we can extend this idea and use a multiple regression model with two dummy variables. In fact, as a general rule we can extend the model to any number of groups and the number of dummy variables needed will be one less than the number of categories of the independent variable. In the two-group case, we assigned one category as a base category (remember that in section 9.7 we chose the picture condition to act as a base) and this category was coded with 0. When there are three categories we also need a base category and you should choose the condition to which you intend to compare the other groups. Usually this category will be the control group. In most well-designed social science experiments there will be a group of participants who act as a baseline for other categories. This baseline group should act as the reference or base category, although the group you choose will depend upon the particular hypotheses that you want to test. In unbalanced designs (in which the group sizes are unequal) it is important that the base category contains a fairly large number of cases to ensure that the estimates of the regression coefficients are reliable. In the Viagra example, we can take the placebo group as the base category because this group was a placebo control. We are interested in comparing both the high- and low-dose groups to the group which received no Viagra at all. If the placebo group is the base category then the two dummy variables that we have to create represent the other two conditions: so, we should have one dummy variable called High and one called Low). The resulting equation is described as:

$$\text{Libido}_i = b_0 + b_1\text{Low}_i + b_2\text{High}_i + \varepsilon_i \tag{10.2}$$

In equation (10.2), a person's libido can be predicted from knowing their group code (i.e. the code for the High and Low dummy variables) and the intercept (b_0) of the model. The dummy variables in equation (10.2) can be coded in several ways, but the simplest way is to use a similar technique to that of the t-test. The base category is always coded as 0. If a participant was given a high dose of Viagra then they are coded with a 1 for the High dummy variable and 0 for all other variables. If a participant was given a low dose of Viagra then they are coded with the value 1 for the Low dummy variable and coded with 0 for all other variables (this is the same type of scheme we used in section 7.11). Using this coding scheme we can express each group by combining the codes of the two dummy variables (see Table 10.2).

TABLE 10.2 Dummy coding for the three-group experimental design

Group	*Dummy Variable 1 (Low)*	*Dummy Variable 2 (High)*
Placebo	0	0
Low Dose Viagra	0	1
High Dose Viagra	1	0

Placebo Group: Let's examine the model for the placebo group. In the placebo group both the High and Low dummy variables are coded as 0. Therefore, if we ignore the error term (ε_i), the regression equation becomes:

$$\text{Libido}_i = b_0 + (b_1 \times 0) + (b_2 \times 0)$$
$$\text{Libido}_i = b_0$$
$$\overline{X}_{\text{Placebo}} = b_0$$

This is a situation in which the high- and low-dose groups have both been excluded (because they are coded with 0). We are looking at predicting the level of libido when both doses of Viagra are ignored, and so the predicted value will be the mean of the placebo group (because this group is the only one included in the model). Hence, the intercept of the regression model, b_0, is always the mean of the base category (in this case the mean of the placebo group).

High-dose group: If we examine the high-dose group, the dummy variable High will be coded as 1 and the dummy variable Low will be coded as 0. If we replace the values of these codes into equation (10.2) the model becomes:

$$\text{Libido}_i = b_0 + (b_1 \times 0) + (b_2 \times 1)$$
$$\text{Libido}_i = b_0 + b_2$$

We know already that b_0 is the mean of the placebo group. If we are interested in only the high-dose group then the model should predict that the value of Libido for a given participant equals the mean of the high-dose group. Given this information, the equation becomes:

$$\text{Libido}_i = b_0 + b_2$$
$$\overline{X}_{\text{High}} = \overline{X}_{\text{Placebo}} + b_2$$
$$b_2 = \overline{X}_{\text{High}} - \overline{X}_{\text{Placebo}}$$

Hence, b_2 represents the difference between the means of the high-dose group and the placebo group.

Low-dose group: Finally, if we look at the model when a low dose of Viagra has been taken, the dummy variable Low is coded as 1 (and hence High is coded as 0). Therefore, the regression equation becomes:

$$\text{Libido}_i = b_0 + (b_1 \times 1) + (b_2 \times 0)$$
$$\text{Libido}_i = b_0 + b_1$$

We know that the intercept is equal to the mean of the base category and that for the low-dose group the predicted value should be the mean libido for a low dose. Therefore the model can be reduced down to:

$$\text{Libido}_i = b_0 + b_1$$
$$\overline{X}_{\text{Low}} = \overline{X}_{\text{Placebo}} + b_1$$
$$b_1 = \overline{X}_{\text{Low}} - \overline{X}_{\text{Placebo}}$$

Hence, b_1 represents the difference between the means of the low-dose group and the placebo group. This form of dummy variable coding is the simplest form, but as we will see later, there are other ways in which variables can be coded to test specific hypotheses. These alternative coding schemes are known as *contrasts* (see section 10.2.11.2). The idea behind contrasts is that you code the dummy variables in such a way that the *b*-values represent differences between groups that you are interested in testing.

SELF-TEST To illustrate exactly what is going on I have created a file called **dummy.sas7bdat**. This file contains the Viagra data but with two additional variables (**dummy1** and **dummy2**) that specify to which group a data point belongs (as in Table 10.2). Access this file and run multiple regression analysis using libido as the outcome and **dummy1** and **dummy2** as the predictors. If you're stuck on how to run the regression then read Chapter 7 again (see, these chapters are ordered for a reason)!

The resulting analysis is shown in SAS Output 10.1. It might be a good idea to remind yourself of the group means from Table 10.1. The first thing to notice is that just as in the regression chapter, an ANOVA has been used to test the overall fit of the model. This test is significant, $F(2, 12) = 5.12$, $p = .0.025$. Given that our model represents the group differences, this ANOVA tells us that using group means to predict scores is significantly better than using the overall mean: in other words, the group means are significantly different.

Turning to the regression coefficients, *b*s, the constant is equal to the mean of the base category (the placebo group). The regression coefficient for the first dummy variable (b_2) is equal to the difference between the means of the high-dose group and the placebo group ($5.0 - 2.2 = 2.8$). Finally, the regression coefficient for the second dummy variable (b_1) is equal to the difference between the means of the low-dose group and the placebo group ($3.2 - 2.2 = 1$). This analysis demonstrates how the regression model represents the three-group situation. We can see from the significance values of the *t*-tests that the difference between the high-dose group and the placebo group (b_2) is significant because $p < .05$. The difference between the low-dose and the placebo group is not, however, significant ($p = .282$).

A four-group experiment can be described by extending the three-group scenario. I mentioned earlier that you will always need one less dummy variable than the number of groups in the experiment: therefore, this model requires three dummy variables. As before, we need to specify one category that is a base category (a control group). This base category should have a code of 0 for all three dummy variables. The remaining three conditions will have a code of 1 for the dummy variable that describes that condition and a code of 0 for the other two dummy variables. Table 10.3 illustrates how the coding scheme would work.

SAS OUTPUT 10.1

Analysis of Variance					
Source	DF	Sum of Squares	Mean Square	F Value	Pr > F
Model	2	20.13333	10.06667	5.12	0.0247
Error	12	23.60000	1.96667		
Corrected Total	14	43.73333			

Parameter Estimates						
Variable	Label	DF	Parameter Estimate	Standard Error	t Value	Pr > |t|
Intercept	Intercept	1	2.20000	0.62716	3.51	0.0043
DUMMY1	Dummy Variable 1	1	2.80000	0.88694	3.16	0.0083
DUMMY2	Dummy Variable 2	1	1.00000	0.88694	1.13	0.2816

TABLE 10.3 Dummy coding for the four-group experimental design

	Dummy Variable 1	Dummy Variable 2	Dummy Variable 3
Group 1	1	0	0
Group 2	0	1	0
Group 3	0	0	1
Group 4 (base)	0	0	0

10.2.4. Logic of the *F*-ratio ②

In Chapter 7 we learnt a little about the *F*-ratio and its calculation. To recap, we learnt that the *F*-ratio is used to test the overall fit of a regression model to a set of observed data. In other words, it is the ratio of how good the model is compared to how bad it is (its error). I have just explained how ANOVA can be represented as a regression equation, and this should help you to understand what the *F*-ratio tells you about your data. Figure 10.2 shows the Viagra data in graphical form (including the group means, the overall mean and the difference between each case and the group mean). In this example, there were three groups; therefore, we want to test the hypothesis that the means of three groups are different (so the null hypothesis is that the group means are the same). If the group means were all the same, then we would not expect the placebo group to differ from the low-dose group or the high-dose group, and we would not expect the low-dose group to differ from the high-dose group. Therefore, on the diagram, the three solid horizontal lines would be in the same vertical position (the exact position would be the grand mean – the dotted line in the figure). We can see from the diagram that the group means are actually different because the lines (the group means) are in different vertical positions. We have just found out that in the regression model, b_2 represents the difference between the means of the placebo and the high-dose group, and b_1 represents the difference in means between the low-dose and placebo groups. These two distances are represented in Figure 10.2 by the vertical arrows. If the null hypothesis is true

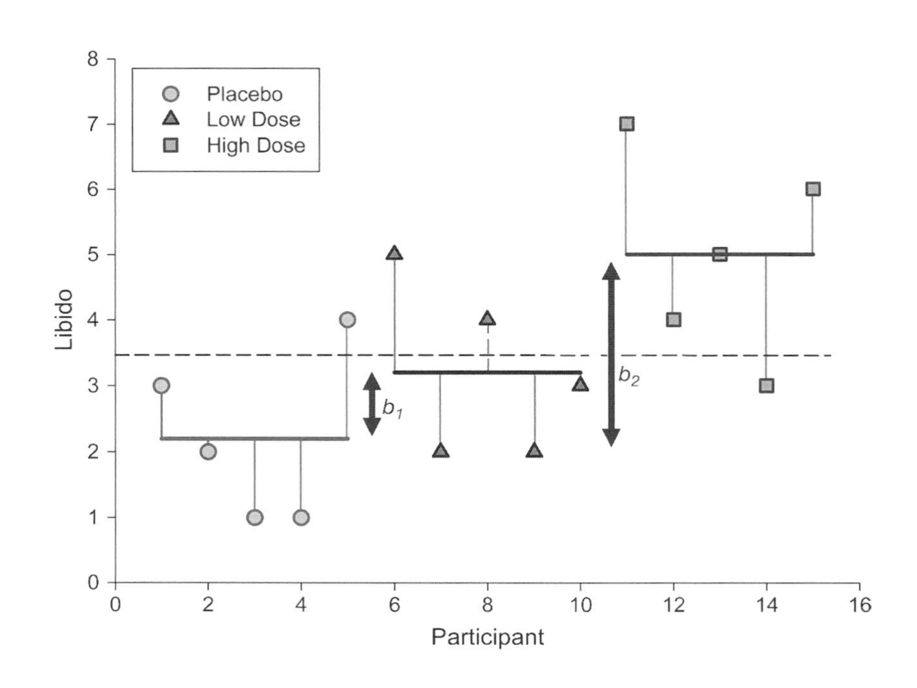

FIGURE 10.2 The Viagra data in graphical form. The solid horizontal lines represent the mean libido of each group. The shapes represent the libido of individual participants (different shapes indicate different experimental groups). The dashed horizontal line is the average libido of all participants

and all the groups have the same means, then these b coefficients should be zero (because if the group means are equal then the difference between them will be zero).

The logic of ANOVA follows from what we understand about regression:

- The simplest model we can fit to a set of data is the grand mean (the mean of the outcome variable). This basic model represents 'no effect' or 'no relationship between the predictor variable and the outcome'.
- We can fit a different model to the data collected that represents our hypotheses. If this model fits the data well then it must be better than using the grand mean. Sometimes we fit a linear model (the line of best fit) but in experimental research we often fit a model based on the means of different conditions.
- The intercept and one or more regression coefficients can describe the chosen model.
- The regression coefficients determine the shape of the model that we have fitted; therefore, the bigger the coefficients, the greater the deviation between the line and the grand mean.
- In correlational research, the regression coefficients represent the slope of the line, but in experimental research they represent the differences between group means.
- The bigger the differences between group means, the greater the difference between the model and the grand mean.
- If the differences between group means are large enough, then the resulting model will be a better fit of the data than the grand mean.
- If this is the case we can infer that our model (i.e. predicting scores from the group means) is better than not using a model (i.e. predicting scores from the grand mean). Put another way, our group means are significantly different.

Just like when we used ANOVA to test a regression model, we can compare the improvement in fit due to using the model (rather than the grand mean) to the error that still remains. Another way of saying this is that when the grand mean is used as a model, there will be a certain amount of variation between the data and the grand mean. When a model is fitted it will explain some of this variation but some will be left unexplained. The F-ratio is the ratio of the explained to the unexplained variation. Look back at section 7.2.3 to refresh your memory on these concepts before reading on. This may all sound quite complicated, but actually most of it boils down to variations on one simple equation (see Jane Superbrain Box 10.1).

JANE SUPERBRAIN 10.1

You might be surprised to know that ANOVA boils down to one equation (well, sort of) ②

At every stage of the ANOVA we're assessing variation (or deviance) from a particular model (be that the most basic model, or the most sophisticated model). We saw back in section 2.4.1 that the extent to which a model deviates from the observed data can be expressed, in general, in the form of equation (10.3). So, in ANOVA, as in regression, we use equation (10.3) to calculate the fit of the most basic model, and then the fit of the best model (the line of best fit). If the best model is any good then it should fit the data significantly better than our basic model:

$$\text{deviation} = \sum(\text{observed} - \text{model})^2 \quad (10.3)$$

The interesting point is that all of the sums of squares in ANOVA are variations on this one basic equation. All that changes is what we use as the model, and what the corresponding observed data are. Look through the various sections on the sums of squares and compare the resulting equations to equation (10.3); hopefully, you can see that they are all basically variations on this general form of the equation!

10.2.5. Total sum of squares (SS_T) ②

To find the total amount of variation within our data we calculate the difference between each observed data point and the grand mean. We then square these differences and add them together to give us the total sum of squares (SS_T):

$$SS_T = \sum (x_i - \bar{x}_{\text{grand}})^2 \tag{10.4}$$

We also saw in section 2.4.1 that the variance and the sums of squares are related such that variance, $s^2 = SS/(N - 1)$, where N is the number of observations. Therefore, we can calculate the total sums of squares from the variance of all observations (the **grand variance**) by rearranging the relationship ($SS = s^2(N - 1)$). The grand variance is the variation between all scores, regardless of the experimental condition from which the scores come. Therefore, in Figure 10.2 it would be the sum of the squared distances between each point and the dashed horizontal line. The grand variance for the Viagra data is given in Table 10.1, and if we count the number of observations we find that there were 15 in all. Therefore, SS_T is calculated as follows:

$$\begin{aligned} SS_T &= s^2_{\text{grand}}(n - 1) \\ &= 3.124(15 - 1) \\ &= 3.124 \times 14 \\ &= 43.74 \end{aligned}$$

Before we move on, it is important to understand degrees of freedom, so have a look back at Jane Superbrain Box 2.2 to refresh your memory. We saw before that when we estimate population values, the degrees of freedom are typically one less than the number of scores used to calculate the population value. This is because to get these estimates we have to hold something constant in the population (in this case the mean), which leaves all but one of the scores free to vary (see Jane Superbrain Box 2.2). For SS_T, we used the entire sample (i.e. 15 scores) to calculate the sums of squares and so the total degrees of freedom (df_T) are one less than the total sample size ($N - 1$). For the Viagra data, this value is 14.

10.2.6. Model sum of squares (SS_M) ②

So far, we know that the total amount of variation within the data is 43.74 units. We now need to know how much of this variation the regression model can explain. In the ANOVA scenario, the model is based upon differences between group means and so the model sums of squares tell us how much of the total variation can be explained by the fact that different data points come from different groups.

In section 7.2.3 we saw that the model sum of squares is calculated by taking the difference between the values predicted by the model and the grand mean (see Figure 7.4). In ANOVA, the values predicted by the model are the group means (therefore, in Figure 10.2 the coloured horizontal lines represented the values of libido predicted by the model). For each participant the value predicted by the model is the mean for the group to which the participant belongs. In the Viagra example, the predicted value for the five participants in the placebo group will be 2.2, for the five participants in the low-dose condition it will be 3.2, and for the five participants in the high-dose condition it will be 5. The model sum of squares requires us to calculate the differences between each participant's predicted value and the grand mean. These differences are then squared and added together (for reasons that should be clear in your mind by now). We know that the predicted value for participants in a particular group is the mean of that group. Therefore, the easiest way to calculate SS_M is to:

1 Calculate the difference between the mean of each group and the grand mean.
2 Square each of these differences.
3 Multiply each result by the number of participants within that group (n_k).
4 Add the values for each group together.

The mathematical expression of this process is:

$$SS_M = \sum n_k (\bar{x}_k - \bar{x}_{grand})^2 \tag{10.5}$$

Using the means from the Viagra data, we can calculate SS_M as follows:

$$\begin{aligned} SS_M &= 5(2.200 - 3.467)^2 + 5(3.200 - 3.467)^2 + 5(5.000 - 3.467)^2 \\ &= 5(-1.267)^2 + 5(-0.267)^2 + 5(1.533)^2 \\ &= 8.025 + 0.355 + 11.755 \\ &= 20.135 \end{aligned}$$

For SS_M, the degrees of freedom (df_M) will always be one less than the number of parameters estimated. In short, this value will be the number of groups minus one (which you'll see denoted as $k - 1$). So, in the three-group case the degrees of freedom will always be 2 (because the calculation of the sums of squares is based on the group means, two of which will be free to vary in the population if the third is held constant).

10.2.7. Residual sum of squares (SS_R) ②

We now know that there are 43.74 units of variation to be explained in our data, and that our model can explain 20.14 of these units (nearly half). The final sum of squares is the residual sum of squares (SS_R), which tells us how much of the variation cannot be explained by the model. This value is the amount of variation caused by extraneous factors such as individual differences in weight, testosterone or whatever. Knowing SS_T and SS_M already, the simplest way to calculate SS_R is to subtract SS_M from SS_T ($SS_R = SS_T - SS_M$); however, telling you to do this provides little insight into what is being calculated and, of course, if you've messed up the calculations of either SS_M or SS_T (or indeed both!) then SS_R will be incorrect also. We saw in section 7.2.3 that the residual sum of squares is the difference between what the model predicts and what was actually observed. We already know that for a given participant, the model predicts the mean of the group to which that person belongs. Therefore, SS_R is calculated by looking at the difference between the score obtained by a person and the mean of the group to which the person belongs. In graphical terms the vertical lines in Figure 10.2 represent this sum of squares. These distances between each data point and the group mean are squared and then added together to give the residual sum of squares, SS_R, thus:

$$SS_R = \sum (x_{ik} - \bar{x}_k)^2 \tag{10.6}$$

Now, the sum of squares for each group represents the sum of squared differences between each participant's score in that group and the group mean. Therefore, we can express SS_R as $SS_R = SS_{group1} + SS_{group2} + SS_{group3} \ldots$ and so on. Given that we know the relationship between the variance and the sums of squares, we can use the variances for each group of the Viagra data to create an equation like we did for the total sum of squares. As such, SS_R can be expressed as:

$$SS_R = \sum s_k^2 (n_k - 1) \tag{10.7}$$

This just means take the variance from each group (s_k^2) and multiply it by one less than the number of people in that group ($n_k - 1$). When you've done this for each group, add them all up. For the Viagra data, this gives us:

$$\begin{aligned} SS_R &= s^2_{\text{group1}}(n_1 - 1) + s^2_{\text{group2}}(n_2 - 1) + s^2_{\text{group3}}(n_3 - 1) \\ &= 1.70(5 - 1) + 1.70(5 - 1) + 2.50(5 - 1) \\ &= (1.70 \times 4) + (1.70 \times 4) + (2.50 \times 4) \\ &= 6.8 + 6.8 + 10 \\ &= 23.60 \end{aligned}$$

The degrees of freedom for SS_R (df_R) are the total degrees of freedom minus the degrees of freedom for the model ($df_R = df_T - df_M = 14 - 2 = 12$). Put another way, it's $N - k$: the total sample size, N, minus the number of groups, k.

10.2.8. Mean squares ②

SS_M tells us the *total* variation that the regression model (e.g. the experimental manipulation) explains and SS_R tells us the *total* variation that is due to extraneous factors. However, because both of these values are summed values they will be influenced by the number of scores that were summed; for example, SS_M used the sum of only 3 different values (the group means) compared to SS_R and SS_T, which used the sum of 12 and 14 values respectively. To eliminate this bias we can calculate the average sum of squares (known as the *mean squares*, MS), which is simply the sum of squares divided by the degrees of freedom. The reason why we divide by the degrees of freedom rather than the number of parameters used to calculate the SS is because we are trying to extrapolate to a population and so some parameters within that populations will be held constant (this is the same reason that we divide by $N - 1$ when calculating the variance, see Jane Superbrain Box 2.2). So, for the Viagra data we find the following mean squares.

$$MS_M = \frac{SS_M}{df_M} = \frac{20.135}{2} = 10.067$$

$$MS_R = \frac{SS_R}{df_R} = \frac{23.60}{12} = 1.967$$

MS_M represents the average amount of variation explained by the model (e.g. the systematic variation), whereas MS_R is a gauge of the average amount of variation explained by extraneous variables (the unsystematic variation).

10.2.9. The *F*-ratio ②

The *F*-ratio is a measure of the ratio of the variation explained by the model and the variation explained by unsystematic factors. In other words, it is the ratio of how good the model is against how bad it is (how much error there is). It can be calculated by dividing the model mean squares by the residual mean squares.

$$F = \frac{MS_M}{MS_R} \tag{10.8}$$

As with the independent *t*-test, the *F*-ratio is, therefore, a measure of the ratio of systematic variation to unsystematic variation. In experimental research, it is the ratio of the experimental effect to the individual differences in performance. An interesting point about the *F*-ratio is that because it is the ratio of systematic variance to unsystematic variance, if its value is less than 1 then it must, by definition, represent a non-significant effect. The reason why this statement is true is because if the *F*-ratio is less than 1 it means that MS_R is greater than MS_M, which in real terms means that there is more unsystematic than systematic variance. You can think of this in terms of the effect of natural differences in ability being greater than differences brought about by the experiment. In this scenario, we can, therefore, be sure that our experimental manipulation has been unsuccessful (because it has bought about less change than if we left our participants alone!). For the Viagra data, the *F*-ratio is:

$$F = \frac{MS_M}{MS_R} = \frac{10.067}{1.967} = 5.12$$

This value is greater than 1, which indicates that the experimental manipulation had some effect above and beyond the effect of individual differences in performance. However, it doesn't yet tell us whether the *F*-ratio is large enough not to be a chance result. To discover this we can compare the obtained value of *F* against the maximum value we would expect to get by chance if the group means were equal in an *F*-distribution with the same degrees of freedom (these values can be found in the Appendix); if the value we obtain exceeds this critical value we can be confident that this reflects an effect of our independent variable (because this value would be very unlikely if there were no effect in the population). In this case, with 2 and 12 degrees of freedom the critical values are 3.89 ($p = .05$) and 6.93 ($p = .01$). The observed value, 5.12, is, therefore, significant at the .05 level of significance but not significant at the .01 level. The exact significance produced by SAS should, therefore, fall somewhere between .05 and .01 (which, incidentally, it does).

10.2.10. Assumptions of ANOVA ③

The assumptions under which the *F*-statistic is reliable are the same as for all parametric tests based on the normal distribution (see section 5.2). That is, the variances in each experimental condition need to be fairly similar, observations should be independent and the dependent variable should be measured on at least an interval scale. In terms of normality, what matters is that distributions *within groups* are normally distributed.

You often hear people say 'ANOVA is a robust test', which means that it doesn't matter much if we break the assumptions of the test: the *F* will still be accurate. There is some truth to this statement, but it is also an oversimplification of the situation. For one thing, the term *ANOVA* covers many different situations and the performance of *F* has been investigated in only some of those situations. There are two issues to consider: (1) does the *F* control the Type I error rate or is it significant even when there are no differences between means; and (2) does the *F* have enough power (i.e. is it able to detect differences when they are there)? Let's have a look at the evidence.

Looking at normality first, Glass et al. (1972) reviewed a lot of evidence that suggests that *F* controls the Type I error rate well under conditions of skew, kurtosis and non-normality. Skewed distributions seem to have little effect on the error rate and power for two-tailed tests (but can have serious consequences for one-tailed tests). However, some of this evidence has been questioned (see Jane

Superbrain Box 5.1). In terms of kurtosis, leptokurtic distributions make the Type I error rate too low (too many null effects are significant) and consequently the power is too high; platykurtic distributions have the opposite effect. The effects of kurtosis seem unaffected by whether sample sizes are equal or not. One study that is worth mentioning in a bit of detail is by Lunney (1970) who investigated the use of ANOVA in about the most non-normal situation you could imagine: when the dependent variable is binary (it could have values of only 0 or 1). The results showed that when the group sizes were equal, ANOVA was accurate when there were at least 20 degrees of freedom and the smallest response category contained at least 20% of all responses. If the smaller response category contained less than 20% of all responses then ANOVA performed accurately only when there were 40 or more degrees of freedom. The power of F also appears to be relatively unaffected by non-normality (Donaldson, 1968). This evidence suggests that *when group sizes are equal* the F-statistic can be quite robust to violations of normality.

However, when group sizes are not equal the accuracy of F is affected by skew, and non-normality also affects the power of F in quite unpredictable ways (Wilcox, 2005). One situation that Wilcox describes shows that when means are equal the error rate (which should be 5%) can be as high as 18%. If you make the differences between means bigger you should find that power increases, but actually he found that initially power *decreased* (although it increased when he made the group differences bigger still). As such F can be biased when normality is violated.

Turning to violations of the assumption of homogeneity of variance, ANOVA is fairly robust in terms of the error rate when sample sizes are equal. However, when sample sizes are unequal, ANOVA is not robust to violations of homogeneity of variance (this is why earlier on I said it's worth trying to collect equal-sized samples of data across conditions!). When groups with larger sample sizes have larger variances than the groups with smaller sample sizes, the resulting F-ratio tends to be conservative. That is, it's more likely to produce a non-significant result when a genuine difference does exist in the population. Conversely, when the groups with larger sample sizes have smaller variances than the groups with smaller samples sizes, the resulting F-ratio tends to be liberal. That is, it is more likely to produce a significant result when there is no difference between groups in the population (put another way, the Type I error rate is not controlled) – see Glass et al. (1972) for a review. When variances are proportional to the means then the power of F seems to be unaffected by the heterogeneity of variance and trying to stabilize variances does not substantially improve power (Budescu, 1982; Budescu & Appelbaum, 1981). Problems resulting from violations of homogeneity of variance assumption can be corrected (see Jane Superbrain Box 10.2).

Violations of the assumption of independence are very serious indeed. Scariano and Davenport (1987) showed that when this assumption is broken (i.e. observations across groups are correlated) then the Type I error rate is substantially inflated. For example, using the conventional .05 Type I error rate when observations are independent, if these observations are made to correlate moderately (say, with a Pearson coefficient of .5), when comparing three groups of 10 observations per group the actual Type I error rate is .74 (a substantial inflation!). Therefore, if observations are correlated you might think that you are working with the accepted .05 error rate (i.e. you'll incorrectly find a significant result only 5% of the time) when in fact your error rate is closer to .75 (i.e. you'll find a significant result on 75% of occasions when, in reality, there is no effect in the population)!

10.2.11. Planned contrasts ②

The F-ratio tells us only whether the model fitted to the data accounts for more variation than extraneous factors, but it doesn't tell us where the differences between groups lie.

So, if the *F*-ratio is large enough to be statistically significant, then we know only that one or more of the differences between means is statistically significant (e.g. either b_2 or b_1 is statistically significant). It is, therefore, necessary after conducting an ANOVA to carry out further analysis to find out which groups differ. In multiple regression, each *b* coefficient is tested individually using a *t*-test and we could do the same for ANOVA. However, we would need to carry out two *t*-tests, which would inflate the familywise error rate (see section 10.2.1). Therefore, we need a way to contrast the different groups without inflating the Type I error rate. There are two ways in which to achieve this goal. The first is to break down the variance accounted for by the model into component parts, and the second is to compare every group (as if conducting several *t*-tests) but to use a stricter acceptance criterion such that the familywise error rate does not rise above .05. The first option can be done using planned comparisons (also known as **planned contrasts**)[4] whereas the latter option is done using *post hoc* comparisons (see next section). The difference between planned comparisons and **post hoc tests** can be likened to the difference between one- and two-tailed tests in that planned comparisons are done when you have specific hypotheses that you want to test, whereas *post hoc* tests are done when you have no specific hypotheses. Let's first look at planned contrasts.

10.2.11.1. Choosing which contrasts to do ②

In the Viagra example we could have had very specific hypotheses. For one thing, we would expect any dose of Viagra to change libido compared to the placebo group. As a second hypothesis we might believe that a high dose should increase libido more than a low dose. To do planned comparisons, these hypotheses must be derived *before* the data are collected. It is fairly standard in social sciences to want to compare experimental conditions to the control conditions as the first contrast, and then to see where the differences lie between the experimental groups. ANOVA is based upon splitting the total variation into two component parts: the variation due to the experimental manipulation (SS_M) and the variation due to unsystematic factors (SS_R) (see Figure 10.3).

Planned comparisons take this logic a step further by breaking down the variation due to the experiment into component parts (see Figure 10.4). The exact comparisons that are carried out depend upon the hypotheses you want to test. Figure 10.4 shows a situation in which the experimental variance is broken down to look at how much variation is created by the two drug conditions compared to the placebo condition (*contrast 1*). Then the variation explained by taking Viagra is broken down to see how much is explained by taking a high dose relative to a low dose (*contrast 2*).

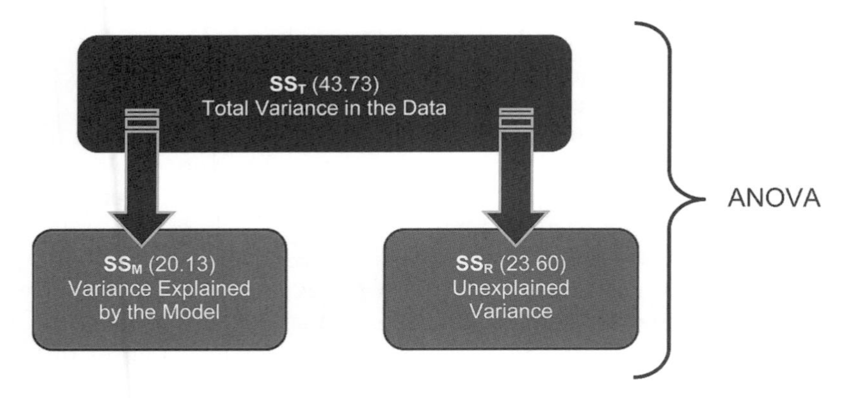

FIGURE 10.3 Partitioning variance for ANOVA

[4] The terms *comparison* and *contrast* are used interchangeably.

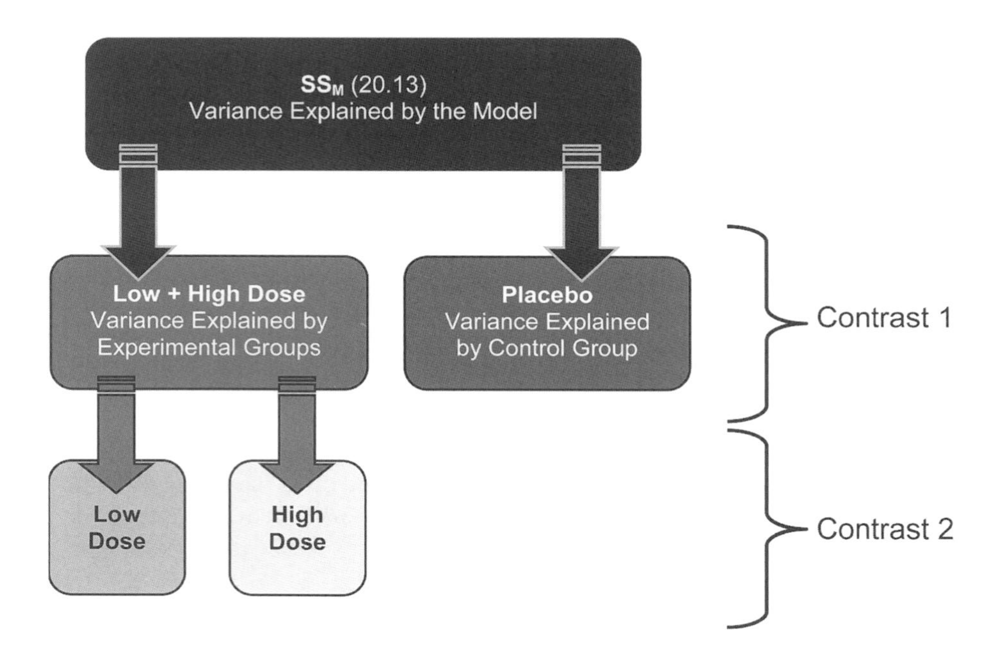

FIGURE 10.4 Partitioning of experimental variance into component comparisons

Typically, students struggle with the notion of planned comparisons, but there are several rules that can help you to work out what to do. The important thing to remember is that we are breaking down one chunk of variation into smaller independent chunks. This means several things. First, if a group is singled out in one comparison, then it should not reappear in another comparison. So, in Figure 10.4 contrast 1 involved comparing the placebo group to the experimental groups; because the placebo group is singled out, it should not be incorporated into any other contrasts. You can think of partitioning variance as being similar to slicing up a cake. You begin with a cake (the total sum of squares) and you then cut this cake into two pieces (SS_M and SS_R). You then take the piece of cake that represents SS_M and divide this up into smaller pieces. Once you have cut off a piece of cake you cannot stick that piece back onto the original slice, and you cannot stick it onto other pieces of cake, but you can divide it into smaller pieces of cake. Likewise, once a slice of variance has been split from a larger chunk, it cannot be attached to any other pieces of variance, it can only be subdivided into smaller chunks of variance. Now, all of this talk of cake is making me hungry, but hopefully it illustrates a point.

Each contrast must compare only two chunks of variance. This rule is so that we can draw firm conclusions about what the contrast tells us. The *F*-ratio tells us that some of our means differ, but not which ones, and if we were to perform a contrast on more than two chunks of variance we would have the same problem. By comparing only two chunks of variance we can be sure that a significant result represents a difference between these two portions of experimental variation.

If you follow the independence of contrasts rule that I've just explained (the cake slicing!), and always compare only two pieces of variance, then you should always end up with one less contrast than the number of groups; there will be $k - 1$ contrasts (where k is the number of conditions you're comparing).

In most social science research we use at least one control condition, and in the vast majority of experimental designs we predict that the experimental conditions will differ from the control condition (or conditions). As such, the biggest hint that I can give you is that when planning comparisons the chances are that your first contrast should be one that compares all of the experimental groups with the control group (or groups). Once you have done this first comparison, any remaining comparisons will depend upon which of the experimental groups you predict will differ.

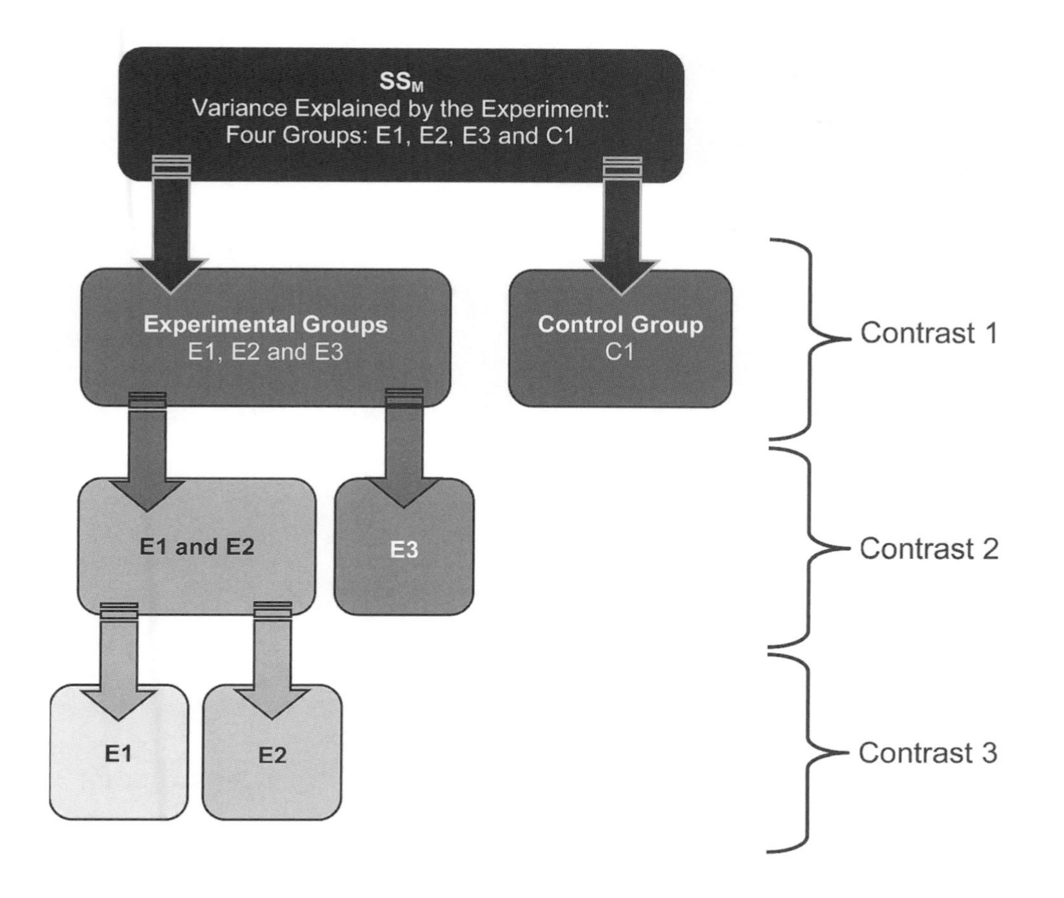

FIGURE 10.5 Partitioning variance for planned comparisons in a four-group experiment using one control group

To illustrate these principles, Figure 10.5 and Figure 10.6 show the contrasts that might be done in a four-group experiment. The first thing to notice is that in both scenarios there are three possible comparisons (one less than the number of groups). Also, every contrast compares only two chunks of variance. What's more, in both scenarios the first contrast is the same: the experimental groups are compared against the control group or groups. In Figure 10.5 there was only one control condition and so this portion of variance is used only in the first contrast (because it cannot be broken down any further). In Figure 10.6 there were two control groups, and so the portion of variance due to the control conditions (contrast 1) can be broken down again so as to see whether or not the scores in the control groups differ from each other (contrast 3).

In Figure 10.5, the first contrast contains a chunk of variance that is due to the three experimental groups and this chunk of variance is broken down by first looking at whether groups E1 and E2 differ from E3 (contrast 2). It is equally valid to use contrast 2 to compare groups E1 and E3 to E2, or to compare groups E2 and E3 to E1. The exact comparison that you choose depends upon your hypotheses. For contrast 2 in Figure 10.5 to be valid we need to have a good reason to expect group E3 to be different from the other two groups. The third comparison in Figure 10.5 depends on the comparison chosen for contrast 2. Contrast 2 necessarily had to involve comparing two experimental groups against a third, and the experimental groups chosen to be combined must be separated in the final comparison. As a final point, you'll notice that in Figure 10.5 and Figure 10.6, once a group has been singled out in a comparison, it is never used in any subsequent contrasts.

What does a planned contrast tell me?

When we carry out a planned contrast, we compare 'chunks' of variance and these chunks often consist of several groups. It is

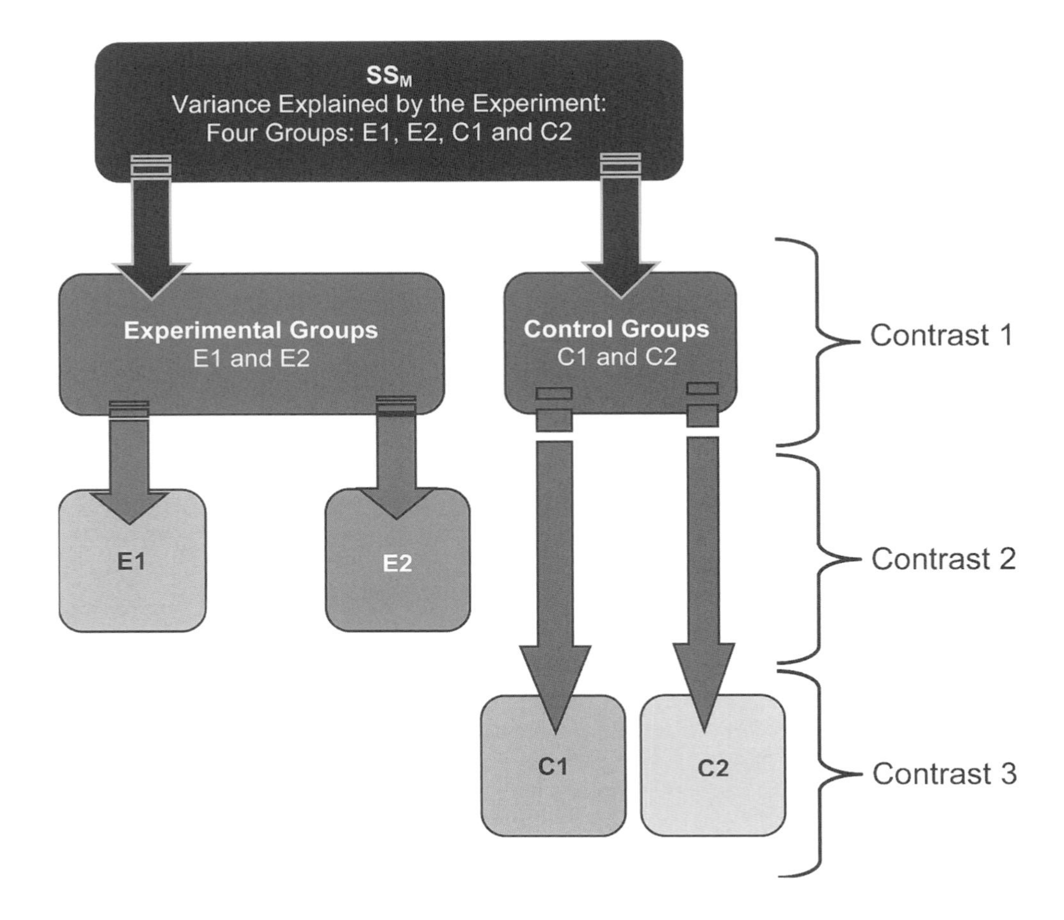

FIGURE 10.6 Partitioning variance for planned comparisons in a four-group experiment using two control groups

perhaps confusing to understand exactly what these contrasts tell us. Well, when you design a contrast that compares several groups to one other group, you are comparing the means of the groups in one chunk with the mean of the group in the other chunk. As an example, for the Viagra data I suggested that an appropriate first contrast would be to compare the two dose groups with the placebo group. The means of the groups are 2.20 (placebo), 3.20 (low dose) and 5.00 (high dose) and so the first comparison, which compared the two experimental groups to the placebo, is comparing 2.20 (the mean of the placebo group) to the average of the other two groups ($(3.20 + 5.00)/2 = 4.10$). If this first contrast turns out to be significant, then we can conclude that 4.10 is significantly greater than 2.20, which in terms of the experiment tells us that the average of the experimental groups is significantly different to the average of the controls. You can probably see that logically this means that, if the standard errors are the same, the experimental group with the highest mean (the high-dose group) will be significantly different from the mean of the placebo group. However, the experimental group with the lower mean (the low-dose group) might not necessarily differ from the placebo group; we have to use the final comparison to make sense of the experimental conditions. For the Viagra data the final comparison looked at whether the two experimental groups differ (i.e. is the mean of the high-dose group significantly different from the mean of the low-dose group?). If this comparison turns out to be significant then we can conclude that having a high dose of Viagra significantly affected libido compared to having a low dose. If the comparison is non-significant then we have to conclude that the dosage of Viagra made no significant difference to libido. In this latter scenario it is likely that both doses affect libido more than placebo, whereas the former case implies that having a low dose may be no different to having a placebo. However, the word *implies* is important here: it is possible that the low-dose group might not differ from the placebo. To be completely sure we must carry out *post hoc* tests.

10.2.11.2. Defining contrasts using weights ②

Hopefully by now you have got some idea of how to plan which comparisons to do (i.e. if your brain hasn't exploded by now). Much as I'd love to tell you that all of the hard work is now over and SAS will magically carry out the comparisons that you've selected, it won't. To get SAS to carry out planned comparisons we need to tell it which groups we would like to compare and doing this can be quite complex. In fact, when we carry out contrasts we assign values to certain variables in the regression model (sorry, I'm afraid that I have to start talking about regression again) – just as we did when we used dummy coding for the main ANOVA. To carry out contrasts we assign certain values to the dummy variables in the regression model. Whereas before we defined the experimental groups by assigning the dummy variables values of 1 or 0, when we perform contrasts we use different values to specify which groups we would like to compare. The resulting coefficients in the regression model (b_2 and b_1) represent the comparisons in which we are interested. The values assigned to the dummy variables are known as **weights**.

This procedure is horribly confusing, but there are a few basic rules for assigning values to the dummy variables to obtain the comparisons you want. I will explain these simple rules before showing how the process actually works. Remember the previous section when you read through these rules, and remind yourself of what I mean by a 'chunk' of variation!

- **Rule 1**: Choose sensible comparisons. Remember that you want to compare only two chunks of variation and that if a group is singled out in one comparison, that group should be excluded from any subsequent contrasts.
- **Rule 2**: Groups coded with positive weights will be compared against groups coded with negative weights. So, assign one chunk of variation positive weights and the opposite chunk negative weights.
- **Rule 3**: The sum of weights for a comparison should be zero. If you add up the weights for a given contrast the result should be zero.
- **Rule 4**: If a group is not involved in a comparison, automatically assign it a weight of 0. If we give a group a weight of 0 then this eliminates that group from all calculations
- **Rule 5**: For a given contrast, the weights assigned to the group(s) in one chunk of variation should be equal to the number of groups in the opposite chunk of variation.

OK, let's follow some of these rules to derive the weights for the Viagra data. The first comparison we chose was to compare the two experimental groups against the control:

Therefore, the first chunk of variation contains the two experimental groups, and the second chunk contains only the placebo group. Rule 2 states that we should assign one chunk positive weights, and the other negative. It doesn't matter which way round we do this, but for convenience let's assign chunk 1 positive weights, and chunk 2 negative weights:

Using rule 5, the weight we assign to the groups in chunk 1 should be equivalent to the number of groups in chunk 2. There is only one group in chunk 2 and so we assign each group in chunk 1 a weight of 1. Likewise, we assign a weight to the group in chunk 2 that is equal to the number of groups in chunk 1. There are two groups in chunk 1 so we give the placebo group a weight of 2. Then we combine the sign of the weights with the magnitude to give us weights of –2 (placebo), 1 (low dose) and 1 (high dose):

Rule 3 states that for a given contrast, the weights should add up to zero, and by following rules 2 and 5 this rule will always be followed (if you haven't followed these rules properly then this will become clear when you add the weights). So, let's check by adding the weights: sum of weights = 1 + 1 – 2 = 0.

The second contrast was to compare the two experimental groups and so we want to ignore the placebo group. Rule 4 tells us that we should automatically assign this group a weight of 0 (because this will eliminate this group from any calculations). We are left with two chunks of variation: chunk 1 contains the low-dose group and chunk 2 contains the high-dose group. By following rules 2 and 5 it should be obvious that one group is assigned a weight of +1 while the other is assigned a weight of –1. The control group is ignored (and so given a weight of 0). If we add the weights for contrast 2 we should find that they again add up to zero: sum of weights = 1 – 1 + 0 = 0.

The weights for each contrast are codings for the two dummy variables in equation (10.2). Hence, these codings can be used in a multiple regression model in which b_2 represents contrast 1 (comparing the experimental groups to the control), b_1 represents contrast 2 (comparing the high-dose group to the low-dose group), and b_0 is the grand mean:

$$\text{Libido}_i = b_0 + b_1\text{Contrast}_1 + b_2\text{Contrast}_2 \tag{10.9}$$

Each group is specified now not by the 0 and 1 coding scheme that we initially used, but by the coding scheme for the two contrasts. A code of –2 for contrast 1 and a code of 0 for contrast 2 identify participants in the placebo group. Likewise, the high-dose group is identified by a code of 1 for both variables, and the low-dose group has a code of 1 for one contrast and a code of –1 for the other (see Table 10.4).

It is important that the weights for a comparison sum to zero because it ensures that you are comparing two unique chunks of variation. Therefore, SAS can perform a *t*-test. A more important consideration is that when you multiply the weights for a particular group, these products should also add up to zero (see final column of Table 10.4). If the products add to zero then we can be sure that the contrasts are *independent* or **orthogonal**. It is important for interpretation that contrasts are orthogonal. When we used dummy variable coding and ran a regression on the Viagra data, I commented that we couldn't look at the individual *t*-tests done on the regression coefficients because the familywise error rate is inflated (see section 10.2.11 and SAS Output 10.1). However, if the contrasts are independent then the *t*-tests done on the *b* coefficients are independent also and so the resulting *p*-values are uncorrelated. You might think that it is very difficult to ensure that the weights you choose for your contrasts conform to the requirements for independence but, provided you follow the rules I have laid out, you should always derive a set of *orthogonal* comparisons. You should double-check by looking at the sum of the multiplied weights and if this total is not zero then go back to the rules and see where you have gone wrong (see last column of Table 10.4).

TABLE 10.4 Orthogonal contrasts for the Viagra data

Group	Dummy Variable 1 (Contrast_1)	Dummy Variable 2 (Contrast_2)	Product $\text{Contrast}_1 \times \text{Contrast}_2$
Placebo	–2	0	0
Low Dose	1	–1	–1
High Dose	1	1	1
Total	0	0	0

Earlier on, I mentioned that when you used contrast codings in dummy variables in a regression model the *b*-values represented the differences between the means that the contrasts were designed to test. Although it is reasonable for you to trust me on this issue, for the more advanced students I'd like to take the trouble to show you how the regression model works (this next part is not for the faint-hearted and so equation-phobes should move on to the next section!). When we do planned contrasts, the intercept b_0 is equal to the grand mean (i.e. the value predicted by the model when group membership is not known), which when group sizes are equal is:

$$b_0 = \text{grand mean} = \frac{\overline{X}_{\text{High}} + \overline{X}_{\text{Low}} + \overline{X}_{\text{Placebo}}}{3}$$

Placebo group: If we use the contrast codings for the placebo group (see Table 10.4), the predicted value of libido equals the mean of the placebo group. The regression equation can, therefore, be expressed as:

$$\text{Libido}_i = b_0 + b_1\text{Contrast}_1 + b_2\text{Contrast}_2$$

$$\overline{X}_{\text{Placebo}} = \left(\frac{\overline{X}_{\text{High}} + \overline{X}_{\text{Low}} + \overline{X}_{\text{Placebo}}}{3}\right) + (-2b_1) + (b_2 \times 0)$$

Now, if we rearrange this equation and then multiply everything by 3 (to get rid of the fraction) we get:

$$2b_1 = \left(\frac{\overline{X}_{\text{High}} + \overline{X}_{\text{Low}} + \overline{X}_{\text{Placebo}}}{3}\right) - \overline{X}_{\text{Placebo}}$$

$$6b_1 = \overline{X}_{\text{High}} + \overline{X}_{\text{Low}} + \overline{X}_{\text{Placebo}} - 3\overline{X}_{\text{Placebo}}$$

$$6b_1 = \overline{X}_{\text{High}} + \overline{X}_{\text{Low}} - 2\overline{X}_{\text{Placebo}}$$

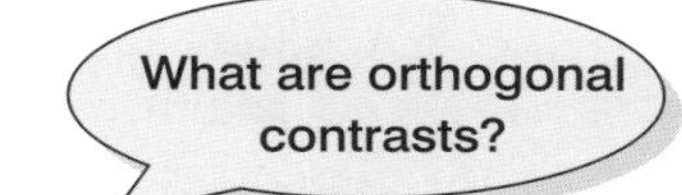

We can then divide everything by 2 to reduce the equation to its simplest form:

$$3b_1 = \left(\frac{\overline{X}_{\text{High}} + \overline{X}_{\text{Low}}}{2}\right) - \overline{X}_{\text{Placebo}}$$

$$b_1 = \frac{1}{3}\left[\left(\frac{\overline{X}_{\text{High}} + \overline{X}_{\text{Low}}}{2}\right) - \overline{X}_{\text{Placebo}}\right]$$

This equation shows that b_1 represents the difference between the average of the two experimental groups and the control group:

$$3b_1 = \left(\frac{\overline{X}_{\text{High}} + \overline{X}_{\text{Low}}}{2}\right) - \overline{X}_{\text{Placebo}}$$

$$= \frac{5 + 3.2}{2} - 2.2$$

$$= 1.9$$

We planned contrast 1 to look at the difference between the average of the experimental groups and the control and so it should now be clear how b_1 represents this difference. The observant among you will notice that rather than being the true value of the difference between experimental and control groups, b_1 is actually a third of this difference ($b_1 = 1.9/3 = 0.633$). The reason for this division is that the familywise error is controlled by making the regression coefficient equal to the actual difference divided by the number of groups in the contrast (in this case 3).

High-dose group: For the situation in which the codings for the high-dose group (see Table 10.4) are used, the predicted value of libido is the mean for the high-dose group, and so the regression equation becomes:

$$\text{Libido}_i = b_0 + b_1\text{Contrast}_{1i} + b_2\text{Contrast}_{2i}$$

$$\overline{X}_{\text{High}} = b_0 + (b_1 \times 1) + (b_2 \times 1)$$

$$b_2 = \overline{X}_{\text{High}} - b_1 - b_0$$

We know already what b_1 and b_0 represent, so we place these values into the equation and then multiply by 3 to get rid of some of the fractions:

$$b_2 = \overline{X}_{\text{High}} - b_1 - b_0$$

$$b_2 = \overline{X}_{\text{High}} - \left\{\frac{1}{3}\left[\left(\frac{\overline{X}_{\text{High}} + \overline{X}_{\text{Low}}}{2}\right) - \overline{X}_{\text{Placebo}}\right]\right\} - \left(\frac{\overline{X}_{\text{High}} + \overline{X}_{\text{Low}} + \overline{X}_{\text{Placebo}}}{3}\right)$$

$$3b_2 = 3\overline{X}_{\text{High}} - \left[\left(\frac{\overline{X}_{\text{High}} + \overline{X}_{\text{Low}}}{2}\right) - \overline{X}_{\text{Placebo}}\right] - \left(\overline{X}_{\text{High}} + \overline{X}_{\text{Low}} + \overline{X}_{\text{Placebo}}\right)$$

If we multiply everything by 2 to get rid of the other fraction, expand all of the brackets and then simplify the equation we get:

$$6b_2 = 6\overline{X}_{\text{High}} - \left(\overline{X}_{\text{High}} + \overline{X}_{\text{Low}} - 2\overline{X}_{\text{Placebo}}\right) - 2\left(\overline{X}_{\text{High}} + \overline{X}_{\text{Low}} + \overline{X}_{\text{Placebo}}\right)$$

$$6b_2 = 6\overline{X}_{\text{High}} - \overline{X}_{\text{High}} - \overline{X}_{\text{Low}} + 2\overline{X}_{\text{Placebo}} - 2\overline{X}_{\text{High}} - 2\overline{X}_{\text{Low}} - 2\overline{X}_{\text{Placebo}}$$

$$6b_2 = 3\overline{X}_{\text{High}} - 3\overline{X}_{\text{Low}}$$

Finally, we can divide the equation by 6 to find out what b_2 represents (remember that 3/6 = 1/2):

$$b_2 = \frac{1}{2}\left(\overline{X}_{\text{High}} - \overline{X}_{\text{Low}}\right)$$

We planned contrast 2 to look at the difference between the experimental groups:

$$\overline{X}_{\text{High}} - \overline{X}_{\text{Low}} = 5 - 3.2 = 1.8$$

It should now be clear how b_2 represents this difference. Again, rather than being the absolute value of the difference between the experimental groups, b_2 is actually half of this difference (1.8/2 = 0.9). The familywise error is again controlled, by making the regression coefficient equal to the actual difference divided by the number of groups in the contrast (in this case 2).

SELF-TEST To illustrate these principles, I have created a file called **Contrast.sas7bdat** in which the Viagra data are coded using the contrast coding scheme used in this section. Run multiple regression analyses on these data using libido as the outcome and using **dummy1** and **dummy2** as the predictor variables (leave all default options).

SAS Output 10.2 shows the result of this regression. The main ANOVA for the model is the same as when dummy coding was used (compare it to SAS Output 10.1) showing that the model fit is the same (it should be because the model represents the group means and these have not changed); however, the regression coefficients have now changed. The first thing to notice is that the intercept is the grand mean, 3.467 (see, I wasn't telling lies). Second, the regression coefficient for contrast 1 is one-third of the difference between the average of the experimental conditions and the control condition (see above). Finally, the regression coefficient for contrast 2 is half of the difference between the experimental groups (see above). So, when a planned comparison is done in ANOVA a *t*-test is conducted comparing the mean of one chunk of variation with the mean of a different chunk. From the significance values of the *t*-tests we can see that our experimental groups were significantly different from the control ($p < .05$) but that the experimental groups were not significantly different ($p > .05$).

SAS OUTPUT 10.2

Parameter Estimates						
Variable	Label	DF	Parameter Estimate	Standard Error	t Value	Pr > |t|
Intercept	Intercept	1	3.46667	0.36209	9.57	<.0001
DUMMY1	Dummy Variable 1	1	0.63333	0.25604	2.47	0.0293
DUMMY2	Dummy Variable 2	1	0.90000	0.44347	2.03	0.0652

CRAMMING SAM'S TIPS **Planned contrasts**

- After an ANOVA you need more analysis to find out which groups differ.
- When you have generated specific hypotheses before the experiment use *planned contrasts*.
- Each contrast compares two 'chunks' of variance. (A chunk can contain one or more groups.)
- The first contrast will usually be experimental groups vs. control groups.
- The next contrast will be to take one of the chunks that contained more than one group (if there were any) and divide it in to two chunks.
- You then repeat this process: if there are any chunks in previous contrasts that contained more than one group that haven't already been broken down into smaller chunks, then create a new contrast that breaks it down into smaller chunks.
- Carry on creating contrasts until each group has appeared in a chunk on its own in one of your contrasts.
- You should end up with one less contrast than the number of experimental conditions. If not, you've done it wrong.
- In each contrast assign a 'weight' to each group that is the value of the number of groups in the opposite chunk in that contrast.
- For a given contrast, randomly select one chunk, and for the groups in that chunk change their weights to be negative numbers.
- Breathe a sigh of relief.

10.2.11.3. Non-orthogonal comparisons ②

I have spent a lot of time on how to design appropriate orthogonal comparisons without mentioning the possibilities that non-orthogonal contrasts provide. Non-orthogonal contrasts are comparisons that are in some way related and the best way to get them is to disobey rule 1 in the previous section. Using my cake analogy again, non-orthogonal comparisons are where you slice up your cake and then try to stick slices of cake together again! So, for the Viagra data a set of non-orthogonal contrasts might be to have the same initial contrast (comparing experimental groups against the placebo), but then to compare the high-dose group to the placebo. This disobeys rule 1 because the placebo group is singled out in the first contrast but used again in the

TABLE 10.5 Non-orthogonal contrasts for the Viagra data

Group	Dummy Variable 1 (Contrast$_1$)	Dummy Variable 2 (Contrast$_2$)	Product Contrast$_1$ × Contrast$_2$
Placebo	−2	−1	2
Low Dose	1	0	0
High Dose	1	1	1
Total	0	0	3

second contrast. The coding for this set of contrasts is shown in Table 10.5 and by looking at the last column it is clear that when you multiply and add the codings from the two contrasts the sum is not zero. This tells us that the contrasts are not orthogonal.

There is nothing intrinsically wrong with performing non-orthogonal contrasts. However, if you choose to perform this type of contrast you must be very careful as to how you interpret the results. With non-orthogonal contrasts, the comparisons you do are related and so the resulting test statistics and *p*-values will be correlated to some extent. For this reason you should use a more conservative probability level to accept that a given contrast is statistically meaningful (see section 10.2.12).

10.2.11.4. Polynomial contrasts: trend analysis ②

One additional type of contrast is the **polynomial contrast**. This contrast tests for trends in the data and in its most basic form it looks for a linear trend (i.e. that the group means increase proportionately). However, there are more complex trends such as quadratic, cubic and quartic trends that can be examined. Figure 10.7 shows examples of the types of trend that can exist in data sets. The *linear* trend should be familiar to you all by now and represents a simply proportionate change in the value of the dependent variable across ordered categories (the diagram shows a positive linear trend but of course it could be negative). A *quadratic trend* is where there is one change in the direction of the line (e.g. the line is curved in one place). An example of this might be a situation in which a drug enhances performance on a task at first but then as the dose increases the performance drops again. To find a quadratic trend you need at least three groups (because in the two-group situation there are not enough categories of the independent variable for the means of the dependent variable to change one way and then another). A *cubic trend* is where there are two changes in the direction of the trend. So, for example, the mean of the dependent variable at first goes up across the first couple of categories of the independent variable, then across the succeeding categories the means go down, but then across the last few categories the means rise again. To have two changes in the direction of the mean you must have at least four categories of the independent variable. The final trend that you are likely to come across is the *quartic trend*, and this trend has three changes of direction (so you need at least five categories of the independent variable).

Polynomial trends should be examined in data sets in which it makes sense to order the categories of the independent variable (so, for example, if you have administered five doses of a drug it makes sense to examine the five doses in order of magnitude). For the Viagra data there are only three groups and so we can expect to find only a linear or quadratic trend (and it would be pointless to test for any higher-order trends).

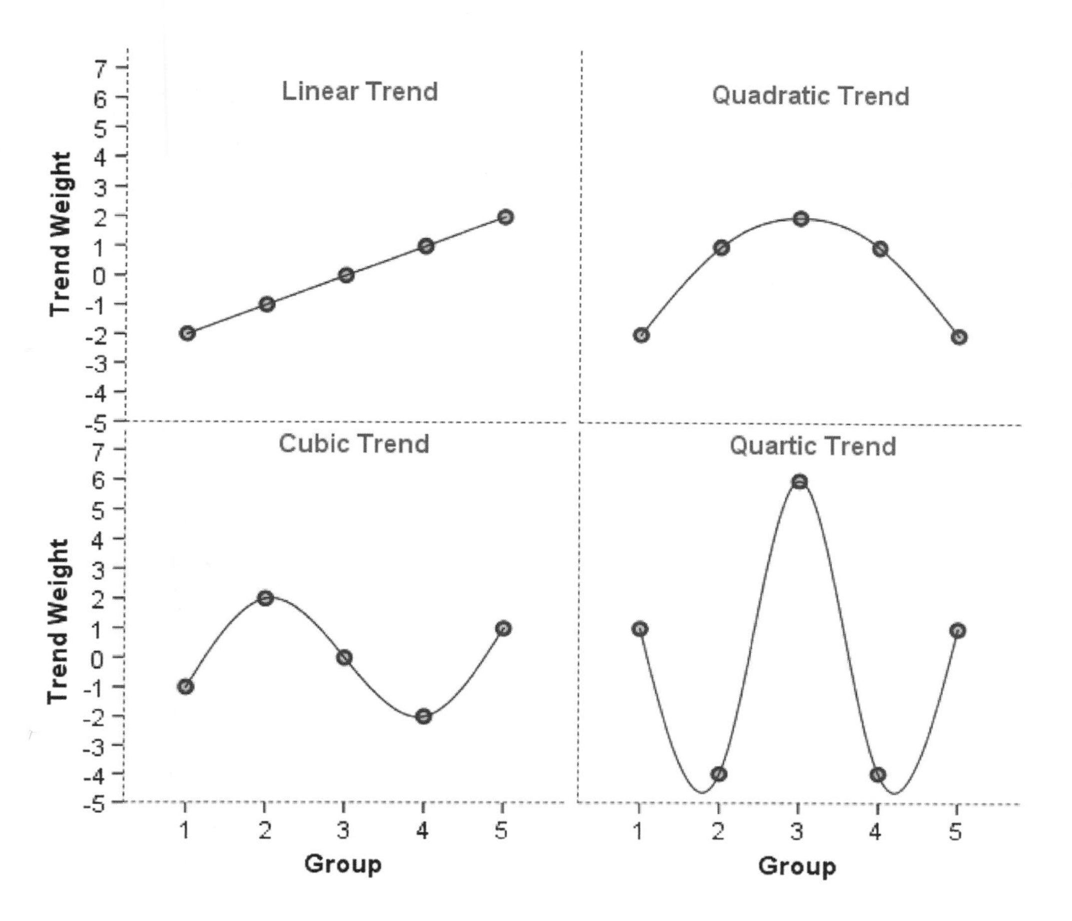

FIGURE 10.7 Linear, quadratic, cubic and quartic trends across five groups

Each of these trends has a set of codes for the dummy variables in the regression model, so we are doing the same thing that we did for planned contrasts except that the codings have already been devised to represent the type of trend of interest. In fact, the graphs in Figure 10.7 have been constructed by plotting the coding values for the five groups. Also, if you add the codes for a given trend the sum will equal zero and if you multiply the codes you will find that the sum of the products also equals zero. Hence, these contrasts are orthogonal. The great thing about these contrasts is that you don't need to construct your own coding values to do them, because the codings already exist.

10.2.12. *Post hoc* procedures ②

Often it is the case that you have no specific a priori predictions about the data you have collected and instead you are interested in exploring the data for any between-group differences between means that exist. This procedure is sometimes called *data mining* or *exploring data*. Now, personally I have always thought that these two terms have certain 'rigging the data' connotations to them and so I prefer to think of these procedures as 'finding the differences that I should have predicted if only I'd been clever enough'.

Post hoc tests consist of **pairwise comparisons** that are designed to compare all different combinations of the treatment groups. So, it is rather like taking every pair of groups and then performing a *t*-test on each pair of groups. Now, this might seem like a particularly stupid thing to say (but then again, I am particularly stupid) in the light of what I have

FIGURE 10.8
Carlo Bonferroni before the celebrity of his correction led to drink, drugs and statistics groupies

already told you about the problems of inflated familywise error rates. However, pairwise comparisons control the familywise error by correcting the level of significance for each test such that the overall Type I error rate (α) across all comparisons remains at .05. There are several ways in which the familywise error rate can be controlled. The most popular (and easiest) way is to divide α by the number of comparisons, thus ensuring that the cumulative Type I error is below .05. Therefore, if we conduct 10 tests, we use .005 as our criterion for significance. This method is known as the **Bonferroni correction** (Figure 10.8). There is a trade-off for controlling the familywise error rate and that is a loss of statistical power. This means that the probability of rejecting an effect that does actually exist is increased (this is called a Type II error). By being more conservative in the Type I error rate for each comparison, we increase the chance that we will miss a genuine difference in the data.

Therefore, when considering which *post hoc* procedure to use we need to consider three things: (1) does the test control the Type I error rate; (2) does the test control the Type II error rate (i.e. does the test have good statistical power); and (3) is the test reliable when the test assumptions of ANOVA have been violated?

Although I would love to go into tedious details about how all of the various *post hoc* tests work, there is really very little point. For one thing, there are some excellent texts already available for those who wish to know (Klockars & Sax, 1986; Toothaker, 1993), for another, SAS provides no fewer than nine *post hoc* procedures and so it would use up several square miles of rainforest to explain them. By far the best reason, though, is that to explain them I would have to learn about them first, and I may be a nerd but even I draw the line at reading up on nine different *post hoc* tests. However, it *is* important that you know which *post hoc* tests perform best according to the aforementioned criteria.

10.2.12.1. *Post hoc* procedures and Type I (α) and Type II error rates ②

The Type I error rate and the statistical power of a test are linked. Therefore, there is always a trade-off: if a test is conservative (the probability of a Type I error is small) then it is likely to lack statistical power (the probability of a Type II error will be high). Therefore, it is important that multiple comparison procedures control the Type I error rate but without a substantial loss in power. If a test is too conservative then we are likely to reject differences between means that are, in reality, meaningful.

The least-significant difference (LSD) pairwise comparison makes no attempt to control the Type I error and is equivalent to performing multiple *t*-tests on the data. The only difference is that the LSD requires the overall ANOVA to be significant. The studentized Newman–Keuls (SNK) procedure is also a very liberal test and lacks control over the familywise error rate. *Bonferroni's* and *Tukey's* tests both control the Type I error rate very well but are conservative tests (they lack statistical power). Of the two, Bonferroni has more power when the number of comparisons is small, whereas Tukey is more powerful when testing large numbers of means. Tukey generally has greater power than *Dunn* and *Scheffé*. The Ryan, Einot, Gabriel and Welsch *Q* procedure (REGWQ) has good power and tight control of the Type I error rate. In fact, when you want to test all pairs of means this procedure is probably the best. However, when group sizes are different this procedure should not be used.

10.2.12.2. *Post hoc* procedures and violations of test assumptions ②

Most research on *post hoc* tests has looked at whether the test performs well when the group sizes are different (an unbalanced design), when the population variances are very different, and when data are not normally distributed. The good news is that most multiple comparison procedures perform relatively well under small deviations from normality. The bad news is that they perform badly when group sizes are unequal and when population variances are different.

Gabriel's pairwise test procedure was designed to cope with situations in which sample sizes are different. Gabriel's procedure is generally more powerful but can become too liberal when the sample sizes are very different. Also, *Hochberg's GT2* is very unreliable when the population variances are different and so should be used only when you are sure that this is not the case.

10.2.12.3. Summary of *post hoc* procedures ②

The choice of comparison procedure will depend on the exact situation you have and whether it is more important for you to keep strict control over the familywise error rate or to have greater statistical power. However, some general guidelines can be drawn (Toothaker, 1993). When you have equal sample sizes and you are confident that your population variances are similar then use REGWQ or Tukey as both have good power and tight control over the Type I error rate. Bonferroni is generally conservative, but if you want guaranteed control over the Type I error rate then this is the test to use. If sample sizes are slightly different then use Gabriel's procedure because it has greater power, but if sample sizes are very different use Hochberg's GT2.

CRAMMING SAM'S TIPS ***Post hoc* tests**

- After an ANOVA you need a further analysis to find out which groups differ.
- When you have no specific hypotheses before the experiment use *post hoc tests*.
- When you have equal sample sizes and group variances are similar use REGWQ or Tukey.
- If you want guaranteed control over the Type I error rate then use Bonferroni.
- If sample sizes are slightly different then use Gabriel's procedure but if sample sizes are very different use Hochberg's GT2.

10.3. Running one-way ANOVA on SAS ②

Hopefully you should all have some appreciation for the theory behind ANOVA, so let's put that theory into practice by conducting an ANOVA test on the Viagra data. As with the independent t-test we need to enter the data into the data editor using a coding variable to specify to which of the three groups the data belong. So, the data must be entered in two columns (one called **dose** which specifies how much Viagra the participant was given and one called **libido** which indicates the person's libido over the following week). The data are in the file **Viagra.sas7bdat** but I recommend entering them using a DATA step with the CARDS statement to gain practice in data entry. I have coded the grouping variable so that 1 = placebo, 2 = low dose and 3 = high dose (see section 3.4.2.3).

To do a one-way ANOVA in SAS, we use PROC GLM (GLM, you'll recall, stands for general linear model).[5] PROC GLM is very similar to PROC REG. The syntax for our model is shown in SAS Syntax 10.1. We use a CLASS statement to tell SAS that dose is a categorical variable, and then write a MODEL line, just as in PROC REG.

We will want descriptive statistics for the three groups, and to get this we add a MEANS statement to the syntax, telling SAS which predictors we would like to see the means for.

```
PROC GLM DATA=chap10.viagra;
    CLASS dose;
    MODEL libido = dose;
    MEANS dose;
    run;
```

SAS Syntax 10.1

10.3.1. Planned comparisons using SAS ②

Planned comparisons are done in PROC GLM using one of two options – we can use an ESTIMATE line, or we can use a CONTRAST line. We're going to use an ESTIMATE line, which carries out a *t*-test to estimate the difference between the groups (if we used a CONTRAST line, it would give the same result though).

First, we'll do a trend analysis, looking for a linear effect. Now, it is important from the point of view of trend analysis that we have coded the grouping variable in a meaningful order. Also, we expect libido to be smallest in the placebo group, to increase in the low-dose group and then to increase again in the high-dose group. To detect a meaningful trend, we need to have coded these groups in ascending order. We have done this by coding the placebo group with the lowest value 1, the low-dose group with the middle value 2 and the high-dose group with the highest coding value of 3. If we coded the groups differently, this would influence both whether a trend is detected and, if a trend is detected, whether it is statistically meaningful.

For a linear trend, we would use SAS Syntax 10.2. We first put the estimate statement, then a name for the estimate, which needs to be in speech marks. Next we specify the predictor variable that we would like to test for estimates, and finally we put the weights – because we want linear estimate the weights are –1, 0 and 1.

```
PROC GLM DATA=chap10.viagra;
      CLASS dose;
      MODEL libido = dose;
      MEANS dose;
      ESTIMATE 'linear' dose -1 0 1;
      RUN;
```

SAS Syntax 10.2

[5] SAS does have a procedure called PROC ANOVA. However, it assumes that you have equal numbers of people in each group. As that's often not the case, we're not going to trouble ourselves with it.

We can also add a quadratic estimate, but we can't add any more, because we only had three groups. The quadratic estimate is shown in SAS Syntax 10.3. – notice we've given it the name 'quad'.

```
PROC GLM DATA=chap10.viagra;
     CLASS dose;
     MODEL libido = dose;
     MEANS dose;
     ESTIMATE 'linear' dose -1 0 1;
     ESTIMATE 'quad' dose 1 -2 1;
     RUN;
```

SAS Syntax 10.3

We can also specify planned comparisons. To conduct planned comparisons we need to tell SAS what weights to assign to each group. The first step is to decide which comparisons you want to do and then what weights must be assigned to each group for each of the estimates. We have already gone through this process in section 10.2.11.2, so we know that the weights for estimate 1 were −2 (placebo group), +1 (low-dose group) and +1 (high-dose group). We will specify this estimate first.

Planned comparisons are very similar to trends, so we use the ESTIMATE option again.[6] It is important to make sure that you enter the correct weight for each group, so you should remember that the first weight that you enter should be the weight for the first group (i.e. the group coded with the lowest value in the data editor). For the Viagra data, the group coded with the lowest value was the placebo group (which had a code of 1) so we should enter the weighting for this group first. The placebo group is weighted –2, so we enter this value first. Next, we need to input the weight for the second group, which for the Viagra data is the low-dose group (because this group was coded with the second-highest value). The next weight is therefore 1. Finally, we need to input the weight for the last group, which for the Viagra data is the high-dose group (because this group was coded with the highest value in the data editor). This is also coded 1. We then enter an ESTIMATE line, giving it a name, and entering the weights, as shown in SAS Syntax 10.4.

The second planned comparison that we want to make is between the two doses of Viagra. Specifically, we want to see if the high dose of Viagra is significantly different from the low dose. Now we don't want the control group to be included in this model, so we weight them at 0, and we weight the other two doses –1 and 1, as shown in SAS Syntax 10.4.

```
PROC GLM DATA=chap10.viagra;
      CLASS dose;
      MODEL libido = dose ;
      MEANS dose;
      ESTIMATE 'linear' dose -1 0 1;
      ESTIMATE 'quad' dose 1 -2 1 /;
      ESTIMATE 'placebo vs other' dose -2 1 1 /;
      ESTIMATE low vs high' dose 0 -1 1 /;
      RUN;
```

SAS Syntax 10.4

[6] Although we could use CONTRAST.

10.3.2. *Post hoc* tests in SAS ②

Having told SAS which planned comparisons to do, we can choose to do *post hoc* tests. In theory, if we have done planned comparisons we shouldn't need to do *post hoc* tests (because we have already tested the hypotheses of interest). Likewise, if we choose to conduct *post hoc* tests then we should not need to do planned contrasts (because we have no hypotheses to test). However, we will conduct some *post hoc* tests on the Viagra data.

In section 11.2.12.3, I recommended various *post hoc* procedures for various situations. For the Viagra data there are equal sample sizes and so we need not use Gabriel's test, instead we should use Tukey's test and REGWQ.

We specify the *post-hoc* tests with another MEANS statement. Then we specify that we'd like those means compared using post hoc tests, with an option. SAS Syntax 11 5 shows the current syntax, with the MEANS line specified, and the options placed after the /.

```
PROC GLM DATA=chap10.viagra ;
  CLASS dose;
  MODEL libido = dose /SOLUTION ;
  MEANS dose;
  ESTIMATE 'linear' dose -1 0 1;
  ESTIMATE 'linear' dose -1 0 1;
  ESTIMATE 'placebo vs other' dose -2 1 1 /;
  ESTIMATE low vs high' dose 0 -1 1 /;
  MEANS dose /TUKEY REGWQ;
  RUN;
```
SAS Syntax 10.5

10.3.3. Other options ②

We've almost finished, but there are a few other things we want to specify.

The options for one-way ANOVA are fairly straightforward. A vital option to select is the homogeneity of variance tests. As with the t-test, there is an assumption that the variances of the groups are equal and selecting this option tests this assumption. SAS offers several options to test the homogeneity of variance – the default is the Levene's test (you can also request some other kinds of HOVTEST). If the homogeneity of variance assumption is broken, then SAS offers us an alternative version of the F-ratio: the Welch's F (1951). These tests are added as options on the MEANS line. The test of homogeneity of variance is requested with the HOVTEST option, if you want a Levene's test, just leave it alone, if you want a Browne-Forsythe test instead (which some statisticians argue is more powerful) you can use HOVTEST=BF.

We can request the Welch corrected F test, also on the MEANS line, by adding the option WELCH.

Finally, if you've read Chapter 7, you'll see that PROC GLM is very similar to PROC REG. One curious difference is that PROC GLM doesn't give you the parameter estimates, unless you ask for them. We do this by adding the SOLUTION option to the MODEL statement. The latest syntax is shown in SAS Syntax 10.6.

```
PROC GLM DATA=chap10.viagra ;
      CLASS dose;
```

```
    MODEL libido = dose /SOLUTION ;
    MEANS dose;
    ESTIMATE 'linear' dose -1 0 1;
    ESTIMATE 'linear' dose -1 0 1;
    ESTIMATE 'placebo vs other' dose -2 1 1 /;
    ESTIMATE low vs high' dose 0 -1 1 /;
    MEANS dose /TUKEY REGWQ HOVTEST WELCH ;
    run;
```

SAS Syntax 10.6

JANE SUPERBRAIN 10.2

What do I do in ANOVA when the homogeneity of variance assumption is broken? ③

In section 10.2.10 I mentioned that when group sizes are unequal, violations of the assumption of homogeneity of variance can have quite serious consequences. SAS has an option for an alternative *F*-ratios, which has been derived to be robust when homogeneity of variance has been violated, called Welch's (1951) *F*. Welch's *F* adjusts *F* and the residual degrees of freedom to combat problems arising from violations of the homogeneity of variance assumption. There is a lengthy explanation about Welch's *F* in the additional material available on the companion website.

10.3.4. ODS graphics ②

As with PROC REG and PROC LOGISTIC, we can request that SAS produces some diagnostic plots for us, using ODS graphics. We can specify what graphs we'd like, or we can just let SAS choose what it thinks is the most appropriate, given our model. Note that if you let SAS do the choosing, it will produce different kinds of plots, depending on your model.

To turn on ODS graphics, we use add ODS GRAPHICS ON; before the PROC GLM statement. If we want to tell SAS what graphics we would like to produce we add a PLOTS= option to the PROC GLM line. It will be easier to understand if we show you, it's in SAS Syntax 10.7.

```
ODS GRAPHICS ON ;
PROC GLM DATA=chap10.viagra PLOTS=all ;
    CLASS dose;
    MODEL libido = dose /SOLUTION ;
    MEANS dose;
    ESTIMATE 'linear' dose -1 0 1;
    ESTIMATE 'linear' dose -1 0 1;
    ESTIMATE 'placebo vs other' dose -2 1 1 /;
    ESTIMATE low vs high' dose 0 -1 1 /;
    MEANS dose /TUKEY REGWQ HOVTEST WELCH ;
    RUN; QUIT;
ODS RTF CLOSE;
```

SAS Syntax 10.7

10.4. Output from one-way ANOVA ②

If you load up the Viagra data (or enter it in by hand) and select all of the options I have suggested, you should find that the output looks the same as what follows. If your output is different we should panic because one of us has done it wrong – hopefully not me, or a lot of trees have died for nothing.

SELF-TEST After you have run the syntax, change the ESTIMATE statements to CONTRAST statements, and run it again. What effect does that have?

FIGURE 10.9 Error bar chart of the Viagra data

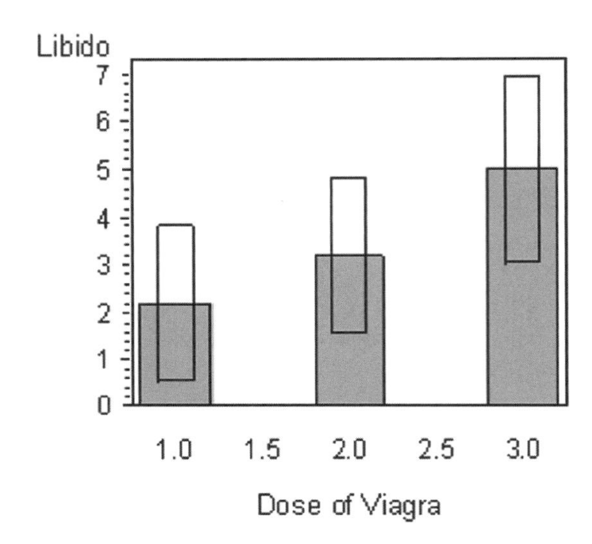

Figure 10.9 shows a line chart with error bars of the Viagra data. It's clear from this chart that all of the error bars overlap, indicating that, on face value, there are no between-group differences (although you can't always trust this). The fact that the means are increasing seems to indicate a linear trend in that, as the dose of Viagra increases, so does the mean level of libido.

10.4.1. Output for the main analysis ②

The first part of the output shows how the classification variable was categorized, and the number of options. We won't reproduce them here, but check that they agree with what you think you have.

As we go through the output, you need to remember that the order can change according to where you put different statements in the model. If your output isn't in the same order as mine, don't worry, just find the table that looks the same.

SAS Output 10.3 shows the main ANOVA summary table. The table is divided into model sums of squares (between-group effects due to the model – the experimental effect) and error sums of squares (within-group effects – this is the unsystematic variation in the data). The model effect is the overall experimental effect. In this row we are told the sums of squares for the model (SS_M = 20.13) and this value corresponds to the value calculated in section 10.2.6. The degrees of freedom are equal to 2 and the mean squares for the model corresponds to the value calculated in section 10.2.8 (10.067). The sum of squares and mean squares represent the experimental effect. The row labelled Error gives details of the unsystematic variation within the data (the variation due to natural individual differences in libido and different reactions to Viagra). The table tells us how much unsystematic variation exists (the residual sum of squares, SS_R) and this value (23.60) corresponds to the value calculated in section 10.2.7. The table then gives the average amount of unsystematic variation, the mean squares (MS_R), which corresponds to the value (1.967) calculated in section 10.2.8. The test of whether the group means are the same is represented by the *F*-ratio for the combined between-group effect. The value of this ratio is 5.12, which is the same as was calculated in section 10.2.9. Finally, SAS tells us whether this value is likely to have happened by chance. The final column labelled *Pr > F*. indicates the likelihood of an *F*-ratio the size of the one obtained occurring if there was no effect in the population (see also SAS Tip 10.1). In this case, there is a probability of .025 that an *F*-ratio of this size would occur if in reality there was no effect (that's only a 2.5% chance!). We have seen in previous chapters that we use a cut-off point of .05 as a criterion for statistical significance. Hence, because the observed significance value is less than .05 we can say that there was a significant effect of Viagra. However, at this stage we still do not know exactly what the effect of Viagra was (we don't know which groups differed). One thing that is interesting here is that we obtained a significant experimental effect yet our error bar plot indicated that no significant difference would be found. This contradiction illustrates how the error bar chart can act only as a rough guide to the data.

The next table shows us the value of R^2, which is a measure of effect size for the overall model, and can be calculated as the model sum of squares divided by the total sum of squares.

Finally, we get the sum of squares broken down by each predictor variable. We only have one predictor variable and so the sum of squares accounted for by the variable is equal to the sum of squares accounted for by the model. You'll see that this table repeats – it gives the Type I sum of squares, and the Type III sum of squares. Right now these are the same, because we only have one predictor. If they are not the same, you should look at the Type III sum of squares, and (for now) ignore the Type I.[7]

SAS OUTPUT 10.3

Source	DF	Sum of Squares	Mean Square	F Value	Pr > F
Model	2	20.13333333	10.06666667	5.12	0.0247
Error	12	23.60000000	1.96666667		
Corrected Total	14	43.73333333			

R-Square	Coeff Var	Root MSE	LIBIDO Mean
0.460366	40.45324	1.402379	3.466667

Source	DF	Type I SS	Mean Square	F Value	Pr > F
DOSE	2	20.13333333	10.06666667	5.12	0.0247

Source	DF	Type III SS	Mean Square	F Value	Pr > F
DOSE	2	20.13333333	10.06666667	5.12	0.0247

[7] There are also Type II and Type IV sums of squares. These are something else not to worry about (for now).

The next part of the output (SAS Output 10.4) gives the parameter estimates from the regression. We requested this when we added the SOLUTION option to the model line. This is very similar to the output that we could have obtained from PROC REG, and is not especially useful here (but that won't stop us working through it). We covered this in more detail in Section 7.11.

SAS has dummy coded the dose variable, and treated the high dose (with the value of 3) as the reference category, hence the intercept is the mean of the high dose category, and is equal to 5.0. The mean libido score for the control group was 2.2, this is 2.8 lower than the mean for the high dose group, and so the estimate for DOSE 1 is –2.8. Similarly, the mean for the low dose group was 3.2, this is 1.8 lower than the mean for the high dose group, and so the estimate is –1.8.

SAS OUTPUT 10.4

Parameter	Estimate		Standard Error	t Value	Pr > |t|
Intercept	5.000000000	B	0.62716292	7.97	<.0001
DOSE 1	–2.800000000	B	0.88694231	–3.16	0.0083
DOSE 2	–1.800000000	B	0.88694231	–2.03	0.0652
DOSE 3	0.000000000	B	.	.	.

The means and standard deviations come in the next table (SAS Output 10.5). These were requested by the first MEANS statement. They show the mean and standard deviation for each group. You should be pretty familiar with these statistics by now.

SAS OUTPUT 10.5

Level of DOSE	N	LIBIDO	
		Mean	Std Dev
1	5	2.20000000	1.30384048
2	5	3.20000000	1.30384048
3	5	5.00000000	1.58113883

The next part of the output (SAS Output 10.6) is a summary table of Levene's test (see section 5.6.1). This test is designed to test the null hypothesis that the variances of the groups are the same. It is an ANOVA test conducted on the absolute differences between the observed data and the mean from which the data came (see Oliver Twisted). In this case, Levene's test is, therefore, testing whether the variances of the three groups are significantly different. If Levene's test is significant (i.e. the significance value is less than .05) then we can say that the variances are significantly different. This would mean that we had violated one of the assumptions of ANOVA and we would have to take steps to rectify this matter. As we saw in Chapter 5, one common way to rectify differences between group variances is to transform all of the data and then reanalyse these transformed values. However, given the apparent utility of Welch's F, or the Brown-Forsythe F, and the fact that transformations often don't help at all, you can instead report Welch's F (and I'd probably suggest reporting this instead of the Brown–Forsythe F unless you have an extreme mean that is also causing the problem with the variances). Luckily, for these data the variances are very similar (hence the high probability value); in fact, if you look at SAS Output 10.5 you'll see that the variances of the placebo and low-dose groups are identical.

The next table shows the Welch's ANOVA, which has corrected for departures from homogeneity fo vvariance. As it turned out we didn't need these because our Levene's test was not significant, indicating that our variances were equal. However, when homogeneity of variance has been violated you should look at these F-ratios instead of the ones in the main table. If you're interested in how these values are calculated then look at Jane Superbrain Box 10.2, but to be honest it's just bloody confusing if you ask me; you're much better off just looking at the values in SAS Output 10.6 and trusting that they do what they're supposed to do (you should also note that the error degrees of freedom have been adjusted and you should remember this when you report the values!).

SAS OUTPUT 10.6

Levene's Test for Homogeneity of LIBIDO Variance ANOVA of Squared Deviations from Group Means					
Source	DF	Sum of Squares	Mean Square	F Value	Pr > F
DOSE	2	1.3653	0.6827	0.32	0.7323
Error	12	25.6160	2.1347		

Welch's ANOVA for LIBIDO			
Source	DF	F Value	Pr > F
DOSE	2.0000	4.32	0.0537
Error	7.9434		

OLIVER TWISTED

Please, Sir, can I have some more … Levene's test?

'Liar! Liar! Pants on fire!' screams Oliver, his cheeks red and eyes about to explode, 'You promised in Chapter 5 to explain Levene's test properly and you haven't, you spatula head'. True enough, Oliver, I do have a spatula for a head. I also have a very nifty little demonstration of Levene's test in the additional material for this chapter on the companion website. It will tell you more than you could possibly want to know. Let's go fry an egg …

10.4.2. Output for trends and planned comparisons ②

Knowing that the overall effect of Viagra was significant, we can now look at the trend analysis (SAS Output 10.7). The trend analysis breaks down the experimental effect to see whether it can explained by either a linear or a quadratic relationship in the data. First, let's look at the linear component. This comparison tests whether the means increase across groups in a linear way. The estimate for the linear trend is the average increase when the dose increases – this is 2.8, and just as with regression, it has a standard error, and an estimate, t-value (which is the estimate divided by the standard error) and an associated probability value. The estimate for the linear trend is 2.8, indicating that average libido increases by 2.8 each time the dose is increased, and this value is significant at the .008 level. Therefore, we can say that as the dose of Viagra increased from nothing to a low dose to a high dose, libido increased proportionately. Moving onto the quadratic trend, this comparison is testing whether the pattern of means is curvilinear (i.e. is represented by a curve that has one bend). The error bar graph of the data suggests that the means cannot be represented by a curve and the results for the quadratic trend bear this out. The t-value for the quadratic trend is non-significant (in fact, the value of t is less than 1 which immediately indicates that this contrast will not be significant).

SAS OUTPUT 10.7

Parameter	Estimate	Standard Error	t Value	Pr > |t|
linear	2.80000000	0.88694231	3.16	0.0083
quadratic	0.80000000	1.53622915	0.52	0.6120
placebo vs other	3.80000000	1.53622915	2.47	0.0293
low vs high	1.80000000	0.88694231	2.03	0.0652

In section 10.3.1 we told SAS to conduct two planned comparisons: one to test whether the control group was different to the two groups which received Viagra, and one to see whether the two doses of Viagra made a difference to libido. The third and fourth rows of the table in SAS Output 10.7 shows the results of the planned comparisons that we requested for the Viagra data.

The table tells us the value of the contrast itself, which is the weighted sum of the group means. This value is obtained by taking each group mean, multiplying it by the weight for the contrast of interest, and then adding these values together.[8] The table also gives the standard error of each contrast and a t-statistic. The t-statistic is derived by dividing the contrast value by the standard error ($t = 3.8/1.5362 = 2.47$) and is compared against critical values of the t-distribution. The significance value of the contrast is given in the final column and this value is two-tailed. Hence, for contrast 1, we can say that taking Viagra significantly increased libido compared to the control group ($p = 0.029$). For contrast 2 we had the hypothesis that a high dose of Viagra would increase libido significantly more than a low dose and the means bear this hypothesis out. However, significance of contrast 2 tells us that a high dose of Viagra did not increase libido significantly more than a low dose ($p = .065$).

In summary, there is an overall effect of Viagra on libido. Furthermore, the planned contrasts revealed that having Viagra significantly increased libido compared to a control group, $t(12) = 2.47$, $p = .029$, and having a high dose did not significantly increase libido compared to a low dose, $t(12) = 2.03$, $p = 0.065$ (although it was close).

SAS TIP 10.1 One- and two-tailed tests in ANOVA ②

A question I get asked a lot by students is 'is the significance of the ANOVA one- or two-tailed, and if it's two-tailed can I divide by 2 to get the one-tailed value?' The answer is that to do a one-tailed test you have to be making a directional hypothesis (i.e. the mean for cats is greater than for dogs). ANOVA is a non-specific test, so it just tells us generally whether there is a difference or not and because there are several means you can't possibly make a directional hypothesis. As such, it's invalid to halve the significance value.

10.4.3. Output for *post hoc* tests ②

If we had no specific hypotheses about the effect that Viagra might have on libido then we could carry out *post hoc* tests to compare all groups of participants with each other. In fact, we asked SAS to do this (see section 10.3.2) and the results of this analysis are shown in SAS Output 10.8. This table shows the results of Tukey's test (known as Tukey's HSD,[9]) the and the Ryan–Einot–Gabriel–Welsch multiple range test (REGWQ is a lot easier to say), which were all specified earlier on. If we look at Tukey's test first (because we have no reason to doubt that the population variances are unequal) it first shows that we are using an alpha value (that's the probability cut-off) of .05 – that's good, because that's what we expect to be using. We have 12

[8]For the first contrast this value is:

$$\Sigma(\bar{X}W) = [(2.2\times-2)+(3.2\times1)+(5.0\times1)] = 3.8$$

[9]The HSD stands for 'honestly significant difference', which has a slightly dodgy ring to it, if you ask me!

degrees of freedom (the number of people – the number of groups), and the error mean square is shown. It then gives a critical range – for two groups to be significantly different, when we are comparing three groups, the difference in the means must be larger than 2.36. That might be a little hard for us to work out, so it shows us in the next table which groups were significantly different, and which were not. It divides the three doses (control, low and high) into groups. Doses levels in the same group are not significantly different from one another. The high dose (dose = 3) and low dose (dose = 2) are both in group A, and so these two groups are not significantly different. The low dose (group = 2) and control (group = 1) are both in group B, so these two are not significantly different from one another. However, the high dose and low dose are not in the same group, and so we can conclude that the means of these two groups are significantly different from one another.

You might like to examine the REGWQ output, and see if the results are in agreement with the Tukey test.

SELF-TEST Our planned comparison showed that any dose of Viagra produced a significant increase in libido, yet the *post hoc* tests indicate that a low dose does not. Why is there this contradiction?

SAS OUTPUT 10.8

Ryan-Einot-Gabriel-Welsch Multiple Range Test for LIBIDO

Alpha	0.05
Error Degrees of Freedom	12
Error Mean Square	1.966667

Number of Means	2	3
Critical Range	1.9324813	2.3662412

Means with the same letter are not significantly different.				
REGWQ Grouping		Mean	N	DOSE
	A	5.0000	5	3
	A			
B	A	3.2000	5	2
B				
B		2.2000	5	1

Tukey's Studentized Range (HSD) Test for LIBIDO

Alpha	0.05
Error Degrees of Freedom	12
Error Mean Square	1.966667
Critical Value of Studentized Range	3.77293
Minimum Significant Difference	2.3662

In section 10.2.11.2, I explained that the first planned comparison would compare the experimental groups to the placebo group. Specifically, it would compare the average of the two group means of the experimental groups ((3.2 + 5.0)/2 = 4.1) to the mean of the placebo group (2.2). So, it was assessing whether the difference between these values (4.1 − 2.2 = 1.9) was significant. In the *post hoc* tests, when the low dose is compared to the placebo, the contrast is testing whether the difference between the means of these two groups is significant. The difference in this case is only 1, compared to a difference of 1.9 for the planned comparison. This explanation illustrates how it is possible to have apparently contradictory results from planned contrasts and *post hoc* comparisons. More important, it illustrates how careful we must be in interpreting planned contrasts.

10.4.4. Graph output for one-way ANOVA using PROC GLM ②

We requested PLOTS=all, and so SAS produced several plots for us. The most useful is the boxplot (see section 4.5), which plots the means and medians, and helps us to see the distribution of each group.

Boxplot showing distribution of libido for three Viagra doses

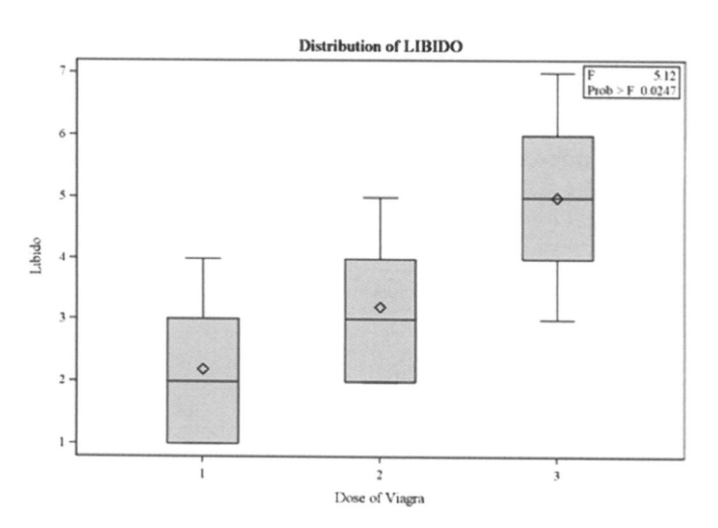

CRAMMING SAM'S TIPS One-way ANOVA

- The one-way independent ANOVA compares several means, when those means have come from different groups of people; for example, if you have several experimental conditions and have used different participants in each condition.
- When you have generated specific hypotheses before the experiment use *planned comparisons*, but if you don't have specific hypotheses use *post hoc* tests.
- There are lots of different *post hoc* tests: when you have equal sample sizes and homogeneity of variance is met, use *REGWQ* or *Tukey's* HSD. If sample sizes are slightly different then use *Gabriel's* procedure, but if sample sizes are very different use *Hochberg's GT2*.
- Test for homogeneity of variance using *Levene's test*. Find the table with this label: if the value in the column labelled significance. is less than .05 then the assumption is violated. If this is the case go to the table labelled **Welch's ANOVA**. If homogeneity of variance has been met (the significance of Levene's test is greater than *.05)* go to the table labelled **ANOVA**.
- In the model results summary (or **Welch's ANOVA** – see above), look at the column labelled **Pr > F**. If the value is less than .05 then the means of the groups are significantly different.
- For contrasts and *post hoc* tests, again look to the columns labelled **Pr < |*t*|**. to discover if your comparisons are significant (they will be if the significance value is less than .05).

LABCOAT LENI'S REAL RESEARCH 10.1

Scraping the barrel? ①

Evolution has endowed us with many beautiful things (cats, dolphins, the Great Barrier Reef, etc.), all selected to fit their ecological niche. Given evolution's seemingly limitless capacity to produce beauty, it's something of a wonder how it managed to produce such a monstrosity as the human penis. One theory is that the penis evolved into the shape that it is because of sperm competition. Specifically, the human penis has an unusually large glans (the 'bell-end' as it's affectionately known) compared to other primates, and this may have evolved so that the penis can displace seminal fluid from other males by 'scooping it out' during intercourse. To put this idea to the test, Gordon Gallup and his colleagues came up with an ingenious study (Gallup et al., 2003). Armed with various female masturbatory devices from Hollywood Exotic Novelties, an artificial vagina from California Exotic Novelties, and some water and cornstarch to make fake sperm, they loaded the artificial vagina with 2.6 ml of fake sperm and inserted one of three female sex toys into it before withdrawing it. Over several trials, three different female sex toys were used: a control phallus that had no coronal ridge (i.e. no bell-end), a phallus with a minimal coronal ridge (small bell-end) and a phallus with a coronal ridge.

They measured sperm displacement as a percentage using the following equation (included here because it is more interesting than all of the other equations in this book):

$$\frac{\text{weight of vagina with semen} - \text{weight of vagina following insertion and removal of phallus}}{\text{weight of vagina with semen} - \text{weight of empty vagina}} \times 100$$

As such, 100% means that all of the sperm was displaced by the phallus, and 0% means that none of the sperm was displaced. If the human penis evolved as a sperm displacement device then Gallup et al. predicted: (1) that having a bell-end would displace more sperm than not; and (2) the phallus with the larger coronal ridge would displace more sperm than the phallus with the minimal coronal ridge. The conditions are ordered (no ridge, minimal ridge, normal ridge) so we might also predict a linear trend. The data can be found in the file **Gallupetal.sas7bdat**. Draw an error bar graph of the means of the three conditions. Conduct a one-way ANOVA with planned comparisons to test the two hypotheses above. What did Gallup et al. find?

Answers are in the additional material on the companion website (or look at pages 280–281 in the original article).

GALLUP, G.G.J. ET AL. (2003). *EVOLUTION AND HUMAN BEHAVIOR*, 24, 277–289.

10.5. Calculating the effect size ②

SAS provides a measure of R^2 for GLM models – although for some bizarre reason it's usually called eta squared, η^2. It is then a simple matter to take the square root of this value to give us the effect size r for our libido example:

$$r^2 = \eta^2$$

$$r = \sqrt{\frac{20.13}{43.73}} = \sqrt{.46} = .68$$

Using the benchmarks for effect sizes this represents a large effect (it is above the .5 threshold for a large effect). Therefore, the effect of Viagra on libido is a substantive finding.

However, this measure of effect size is slightly biased because it is based purely on sums of squares from the sample and no adjustment is made for the fact that we're trying to estimate the effect size in the population. Therefore, we often use a slightly more complex measure called **omega squared** (ω^2). This effect size estimate is still based on the sums of squares that we've met in this chapter, but like the F-ratio it uses the variance explained by the model, and the error variance (in both cases the average variance, or mean squared error, is used):

$$\omega^2 = \frac{SS_M - (df_M)MS_R}{SS_T + MS_R}$$

The df_M in the equation is the degrees of freedom for the effect, which you can get from the SAS output (in the case of the main effect this is the number of experimental conditions minus one). So, in this example we'd get:

$$\begin{aligned}\omega^2 &= \frac{20.13-(2)1.97}{43.73+1.97}\\ &= \frac{16.19}{45.70}\\ &= .35\\ \omega &= .60\end{aligned}$$

As you can see, this has led to a slightly lower estimate to using r, and in general ω is a more accurate measure. Although in the sections on ANOVA I will use ω as my effect size measure, think of it as you would r (because it's basically an unbiased estimate of r anyway). People normally report ω^2 and it has been suggested that values of .01, .06 and .14 represent small, medium and large effects respectively (Kirk, 1996). Remember, though, that these are rough guidelines and that effect sizes need to be interpreted within the context of the research literature.

Most of the time it isn't that interesting to have effect sizes for the overall ANOVA because it's testing a general hypothesis. Instead, we really want effect sizes for the contrasts (because these compare only two things, so the effect size is considerably easier to interpret). Planned comparisons are tested with the t-statistic and, therefore, we can use the same equation as in section 9.4.6:

$$r_{\text{contrast}} = \sqrt{\frac{t^2}{t^2+df}}$$

We know the value of t and the df from SAS Output 10.4 and so we can compute r as follows:

$$\begin{aligned}r_{\text{contrast1}} &= \sqrt{\frac{2.474^2}{2.474^2+12}}\\ &= \sqrt{\frac{6.12}{18.12}}\\ &= 0.58\end{aligned}$$

If you think back to our benchmarks for effect sizes this represents a large effect (it is above .5, the threshold for a large effect). Therefore, as well as being statistically significant, this effect is large and so represents a substantive finding. For contrast 2 we get:

$$\begin{aligned}r_{\text{contrast2}} &= \sqrt{\frac{2.029^2}{2.029^2+12}}\\ &= \sqrt{\frac{4.12}{16.12}}\\ &= 0.51\end{aligned}$$

This too is a substantive finding and represents a large effect size.

10.6. Reporting results from one-way independent ANOVA ②

When we report an ANOVA, we have to give details of the F-ratio and the degrees of freedom from which it was calculated. For the experimental effect in these data the F-ratio was derived by dividing the mean squares for the effect by the mean squares for the residual. Therefore, the degrees of freedom used to assess the F-ratio are the degrees of freedom for the effect of the model ($df_M = 2$) and the degrees of freedom for the residuals of the model ($df_R = 12$). Therefore, the correct way to report the main finding would be:

✓ There was a significant effect of Viagra on levels of libido, $F(2, 12) = 5.12$, $p = .025$, $\omega = .60$.

Notice that the value of the F-ratio is preceded by the values of the degrees of freedom for that effect. The linear contrast can be reported in much the same way:

✓ There was a significant linear trend, $F(1, 12) = 9.97$, $p < .01$, $\omega = .62$, indicating that as the dose of Viagra increased, libido increased proportionately.

Notice that the degrees of freedom have changed to reflect how the F-ratio was calculated. I've also included an effect size measure (have a go at calculating this as we did for the main F-ratio and see if you get the same value). Also, we have now reported that the F-value was significant at a value less than the criterion value of .01. We can also report our planned contrasts:

✓ Planned contrasts revealed that having any dose of Viagra significantly increased libido compared to having a placebo, $t(12) = 2.47$, $p = .029$, $r = .58$, and that having a high dose was associated with a marginally non-significant increase in libido compared to having a low dose, $t(12) = 2.03$, $p = .065$ (one-tailed), $r = .51$.

Note that in both cases I've stated that we used a one-tailed probability.

10.7. Violations of assumptions in one-way independent ANOVA ②

I've mentioned several times in this chapter that ANOVA can be robust to violations of its assumptions, but not always. We also saw that there are measures that can be taken when you have heterogeneity of variance (Jane Superbrain Box 10.2). However, there is another alternative. There are a group of tests (often called assumption-free, distribution-free or non-parametric tests, none of which are particularly accurate names!). Well, the one-way independent ANOVA also has a non-parametric counterpart called the Kruskal–Wallis test. If you have non-normally distributed data, or have violated some other assumption, then this test can be a useful way around the problem. This test is described in Chapter 15.

There are also robust methods available (see section 5.7.4) to compare independent means (and even medians) that involve, for example, using 20% trimmed means or a bootstrap. SAS doesn't do any of them so I advise investigating Wilcox's Chapter 7 and the associated files for the software R (Wilcox, 2005). SAS are planning to release a plugin that will allow you to run R code from within SAS, but at the time of writing it doesn't exist, so we can't tell you about it (yet – they say it will come out in 2010). When it does come out, we'll try to put something on the website about it (as long as I can actually work out how to do it).

What have I discovered about statistics? ①

This chapter has introduced you to analysis of variance (ANOVA), which is the topic of the next few chapters also. One-way independent ANOVA is used in situations when you want to compare several means, and you've collected your data using different participants in each condition. I started off explaining that if we just do lots of *t*-tests on the same data then our Type I error rate becomes inflated. Hence we use ANOVA instead. I looked at how ANOVA can be conceptualized as a general linear model (GLM) and so is in fact the same as multiple regression. Like multiple regression, there are three important measures that we use in ANOVA: the total sum of squares, SS_T (a measure of the variability in our data), the model sum of squares, SS_M (a measure of how much of that variability can be explained by our experimental manipulation), and SS_R (a measure of how much variability can't be explained by our experimental manipulation). We discovered that, crudely speaking, the *F*-ratio is just the ratio of variance that we can explain against the variance that we can't. We also discovered that a significant *F*-ratio tells us only that our groups differ, not how they differ. To find out where the differences lie we have two options: specify specific contrasts to test hypotheses (*planned contrasts*), or test every group against every other group (*post hoc tests*). The former are used when we have generated hypotheses before the experiment, whereas the latter are for exploring data when no hypotheses have been made. Finally we discovered how to implement these procedures on SAS.

We also saw that my life was changed by a letter that popped through the letterbox one day saying only that I could go to the local grammar school if I wanted to. When my parents told me, rather than being in celebratory mood, they were very downbeat; they knew how much it meant to me to be with my friends and how I had got used to my apparent failure. Sure enough, my initial reaction was to say that I wanted to go to the local school. I was unwavering in this view. Unwavering, that is, until my brother convinced me that being at the same school as him would be really cool. It's hard to measure how much I looked up to him, and still do, but the fact that I willingly subjected myself to a lifetime of social dysfunction just to be with him is a measure of sorts. As it turned out, being at school with him was not always cool – he was bullied for being a boffin (in a school of boffins) and being the younger brother of a boffin made me a target. Luckily, unlike my brother, I was stupid and played football, which seemed to be good enough reasons for them to leave me alone. Most of the time.

Key terms that I've discovered

Analysis of variance (ANOVA)
Bonferroni correction
Brown–Forsythe *F*
Cubic trend
Deviation contrast
Eta squared, η^2
Experimentwise error rate
Familywise error rate
Grand variance
Independent ANOVA
Omega squared
Orthogonal
Pairwise comparisons
Planned contrasts
Polynomial contrast
Post hoc tests
Quadratic trend
Quartic trend
Weights
Welch's *F*

Smart Alex's tasks

- **Task 1**: Imagine that I was interested in how different teaching methods affected students' knowledge. I noticed that some lecturers were aloof and arrogant in their teaching style and humiliated anyone who asked them a question, while others were encouraging and supporting of questions and comments. I took three statistics courses where I taught the same material. For one group of students I wandered around with a large cane and beat anyone who asked daft questions or got questions wrong (*punish*). In the second group I used my normal teaching style, which is to encourage students to discuss things that they find difficult and to give anyone working hard a nice sweet (*reward*). The final group I remained indifferent to and neither punished nor rewarded students' efforts (*indifferent*). As the dependent measure I took the students' exam marks (*percentage*). Based on theories of operant conditioning, we expect punishment to be a very unsuccessful way of reinforcing learning, but we expect reward to be very successful. Therefore, one prediction is that reward will produce the best learning. A second hypothesis is that punishment should actually retard learning such that it is worse than an indifferent approach to learning. The data are in the file **Teach.sas7bdat**. Carry out a one-way ANOVA and use planned comparisons to test the hypotheses that (1) reward results in better exam results than either punishment or indifference; and (2) indifference will lead to significantly better exam results than punishment. ②
- **Task 2**: In Chapter 15 (section 15.5) there are some data looking at whether eating soya meals reduces your sperm count. Have a look at this section, access the data for that example, but analyse them with ANOVA. What's the difference between what you find and what is found in section 15.5.4? Why do you think this difference has arisen? ②
- **Task 3**: Students (and lecturers for that matter) love their mobile phones, which is rather worrying given some recent controversy about links between mobile phone use and brain tumours. The basic idea is that mobile phones emit microwaves, and so holding one next to your brain for large parts of the day is a bit like sticking your brain in a microwave oven and hitting the 'cook until well done' button. If we wanted to test this experimentally, we could get six groups of people and strap a mobile phone on their heads (that they can't remove). Then, by remote control, we turn the phones on for a certain amount of time each day. After six months, we measure the size of any tumour (in mm^3) close to the site of the phone antenna (just behind the ear). The six groups experienced 0, 1, 2, 3, 4 or 5 hours per day of phone microwaves for six months. The data are in **Tumour.sas7bdat**. (From Field & Hole, 2003, so there is a very detailed answer in there.) ②
- **Task 4**: Labcoat Leni's Real Research 15.2 describes an experiment (Çetinkaya & Domjan, 2006) on quails with fetishes for terrycloth objects (really, it does). In this example, you are asked to analyse two of the variables that they measured with a Kruskal–Wallis test. However, there were two other outcome variables (time spent near the terrycloth object and copulatory efficiency). These data can be analysed with one-way ANOVA. Read Labcoat Leni's Real Research 15.2 to get the full story, then carry out two one-way ANOVAs and Bonferroni *post hoc* tests on the aforementioned outcome variables. ②

Answers can be found on the companion website.

Further reading

Howell, D. C. (2006). *Statistical methods for psychology* (6th ed.). Belmont, CA: Duxbury. (Or you might prefer his *Fundamental statistics for the behavioral sciences,* also in its 6th edition, 2007. Both are excellent texts that provide very detailed coverage of the standard variance approach to ANOVA but also the GLM approach that I have discussed.)

Iversen, G. R., & Norpoth, H. (1987). *ANOVA* (2nd ed.). Sage University Paper Series on Quantitative Applications in the Social Sciences, 07-001. Newbury Park, CA: Sage. (Quite high-level, but a good read for those with a mathematical brain.)

Klockars, A. J., & Sax, G. (1986). *Multiple comparisons.* Sage University Paper Series on Quantitative Applications in the Social Sciences, 07-061. Newbury Park, CA: Sage. (High-level but thorough coverage of multiple comparisons – in my view this book is better than Toothaker for planned comparisons.)

Rosenthal, R., Rosnow, R. L., & Rubin, D. B. (2000). *Contrasts and effect sizes in behavioural research: a correlational approach.* Cambridge: Cambridge University Press. (Fantastic book on planned comparisons by three of the great writers on statistics.)

Rosnow, R. L., & Rosenthal, R. (2005). *Beginning behavioural research: a conceptual primer* (5th ed.). Englewood Cliffs, NJ: Pearson/Prentice Hall. (Look, they wrote another great book!)

Toothaker, L. E. (1993). *Multiple comparison procedures.* Sage University Paper Series on Quantitative Applications in the Social Sciences, 07-089. Newbury Park, CA: Sage. (Also high-level, but gives an excellent precis of *post hoc* procedures.)

Wright, D. B., & London, K. (2009). *First steps in statistics* (2nd ed.). London: Sage. (If this chapter is too complex then this book is a very readable basic introduction to ANOVA.)

Interesting real research

Gallup, G. G. J., Burch, R. L., Zappieri, M. L., Parvez, R., Stockwell, M., & Davis, J. A. (2003). The human penis as a semen displacement device. *Evolution and Human Behavior*, *24*, 277–289.

Analysis of covariance, ANCOVA (GLM 2)

11

FIGURE 11.1 Davey Murray (guitarist from Iron Maiden) and me backstage in London in 1986; my grimace reflects the utter terror I was feeling at meeting my hero

11.1. What will this chapter tell me? ②

My road to rock stardom had taken a bit of a knock with my unexpected entry to an all-boys grammar school (rock bands and grammar schools really didn't go together). I needed to be inspired and I turned to the masters: Iron Maiden. I first heard Iron Maiden at the age of 11 when a friend of mine lent me *Piece of Mind* and told me to listen to 'The Trooper'. It was, to put it mildly, an epiphany. I became their smallest (I was 11) biggest fan and started to obsess about them in the most unhealthy way possible. I started stalking the man who ran their fan club with letters, and, bless him, he replied. Eventually this stalking paid off and he arranged for me to go backstage when they played the Hammersmith Odeon in London (now the Carling Apollo Hammersmith) on 5 November 1986 (*Somewhere on Tour,* in case you're interested). Not only was it the first time that I had seen them live, but I got to meet them. It's hard to put into words how bladder-splittingly exciting this was. I was so utterly awestruck that I managed to say precisely no words to them. As usual, then, a social situation provoked me to make an utter fool of myself.[1] When it was over I was in no doubt that this was the best day of my life. In fact, I thought, I should just kill myself there and then because nothing would ever be as good as that again. This may be true, but I have subsequently had many other very nice experiences, so who is to say that they were not better? I could compare experiences to see which one is the best, but there is an important confound: my age. At the age of 13, meeting Iron Maiden was bowel-weakeningly exciting, but adulthood (sadly) dulls your capacity for this kind of unqualified joy of life. Therefore, to really see which experience was best, I would have to take account of the variance in enjoyment that is attributable to my age at the time. This will give me a purer measure of how much variance in my enjoyment is attributable to the event itself. This chapter describes **analysis of covariance**, which extends the basic idea of ANOVA from the previous chapter to situations when we want to factor in other variables that influence the outcome variable.

11.2. What is ANCOVA? ②

In the previous chapter we saw how one-way ANOVA could be characterized in terms of a multiple regression equation that used dummy variables to code group membership. In addition, in Chapter 7. we saw how multiple regression could incorporate several continuous predictor variables. It should, therefore, be no surprise that the regression equation for ANOVA can be extended to include one or more continuous variables that predict the outcome (or dependent variable). Continuous variables such as these, that are not part of the main experimental manipulation but have an influence on the dependent variable, are known as *covariates* and they can be included in an ANOVA analysis. When we measure covariates and include them in an analysis of variance we call it analysis of covariance (or ANCOVA for short). This chapter focuses on this technique.

In the previous chapter, we used an example about looking at the effects of Viagra on libido. Let's think about things other than Viagra that might influence libido: well, the obvious one is the libido of the participant's sexual partner (after all, 'it takes two to tango'!), but there are other things too such as other medication that suppresses libido (such as antidepressants or the contraceptive pill) and fatigue. If these variables (the **covariates**) are measured, then it is possible to control for the influence they have on the dependent variable by including them in the regression model. From what we know of hierarchical regression (see Chapter 7) it should be clear that if we enter the covariate into the regression model first, and then enter the dummy variables representing the experimental manipulation, we can see what effect an independent variable has *after* the effect of the covariate. As such, we **partial out** the effect of the covariate. There are two reasons for including covariates in ANOVA:

- **To reduce within-group error variance**: In the discussion of ANOVA and *t*-tests we got used to the idea that we assess the effect of an experiment by comparing the amount of variability in the data that the experiment can explain against the variability that it cannot explain. If we can explain some of this 'unexplained' variance (SS_R) in terms of other variables (covariates), then we reduce the error variance, allowing us to more accurately assess the effect of the independent variable (SS_M).

[1] In my teens I stalked many bands and Iron Maiden are by far the nicest of the bands I've met.

- **Elimination of confounds:** In any experiment, there may be unmeasured variables that confound the results (i.e. variables that vary systematically with the experimental manipulation). If any variables are known to influence the dependent variable being measured, then ANCOVA is ideally suited to remove the bias of these variables. Once a possible confounding variable has been identified, it can be measured and entered into the analysis as a covariate.

There are other reasons for including covariates in ANOVA but because I do not intend to describe the computation of ANCOVA in any detail I recommend that the interested reader consult my favourite sources on the topic (Stevens, 2002; Wildt & Ahtola, 1978).

Imagine that the researcher who conducted the Viagra study in the previous chapter suddenly realized that the libido of the participants' sexual partners would affect the participants' own libido (especially because the measure of libido was behavioural). Therefore, they repeated the study on a different set of participants, but this time took a measure of the partner's libido. The partner's libido was measured in terms of how often they tried to initiate sexual contact. In the previous chapter, we saw that this experimental scenario could be characterized in terms of equation (10.2). Think back to what we know about multiple regression (Chapter 7) and you can hopefully see that this equation can be extended to include this covariate as follows:

$$\begin{aligned} \text{Libido}_i &= b_0 + b_3\text{Covariate}_i + b_2\text{High}_i + b_1\text{Low}_i + \varepsilon_i \\ \text{Libido}_i &= b_0 + b_3\text{Partner's Libido}_i + b_2\text{High}_i + b_1\text{Low}_i + \varepsilon_i \end{aligned} \tag{11.1}$$

11.3. Assumptions and issues in ANCOVA ③

ANCOVA has the same assumptions as ANOVA except that there are two important additional considerations: (1) independence of the covariate and treatment effect, and (2) homogeneity of regression slopes.

11.3.1. Independence of the covariate and treatment effect ③

I said in the previous section that one use of ANCOVA is to reduce within-group error variance by allowing the covariate to explain some of this error variance. However, for this to be true the covariate must be independent of the experimental effect.

Figure 11.2 shows three different scenarios. Part A shows a basic ANOVA and is similar to Figure 10.3; it shows that the experimental effect (in our example, libido) can be partitioned into two parts that represent the experimental or treatment effect (in this case the administration of Viagra) and the error or unexplained variance (i.e. factors that affect libido that we haven't measured). Part B shows the ideal scenario for ANCOVA in which the covariate shares its variance only with the bit of libido that is currently unexplained. In other words, it is completely independent of the treatment effect (it does not overlap with the effect of Viagra at all). This scenario is the only one in which ANCOVA is appropriate. Part C shows a situation in which people often use ANCOVA when they should not. In this situation the effect of the covariate overlaps with the experimental effect. In other words, the experimental effect is confounded with the effect of the covariate. In this situation, the covariate will reduce (statistically speaking) the experimental effect because it explains some of the variance that would otherwise be attributable to the experiment. When the covariate and the experimental effect (independent variable) are

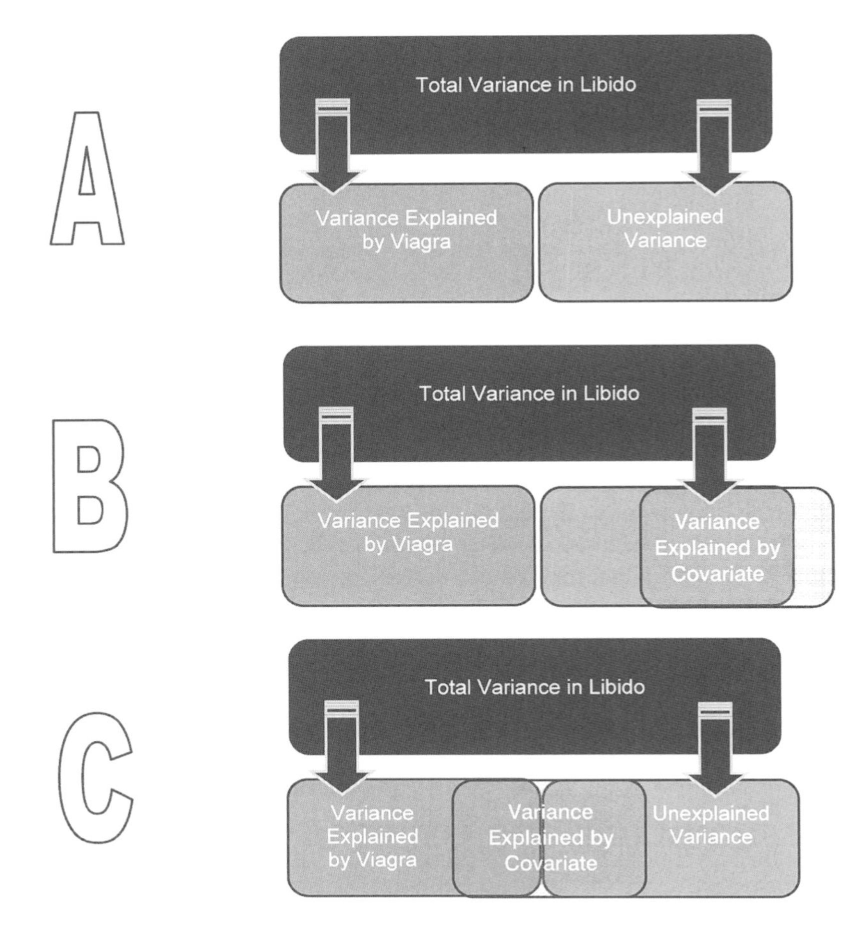

FIGURE 11.2 The role of the covariate in ANCOVA (see text for details)

not independent, the treatment effect is obscured, spurious treatment effects can arise and at the very least the interpretation of the ANCOVA is seriously compromised (Wildt & Ahtola, 1978).

The problem of the covariate and treatment sharing variance is common and is ignored or misunderstood by many people (Miller & Chapman, 2001). Miller and Chapman in a very readable review cite many situations in which people misapply ANCOVA and I recommend reading this paper. To summarize the main issue, when treatment groups differ on the covariate then putting the covariate into the analysis will not 'control for' or 'balance out' those differences (Lord, 1967, 1969). This situation arises mostly when participants are not randomly assigned to experimental treatment conditions. For example, anxiety and depression are closely correlated (anxious people tend to be depressed) so if you wanted to compare an anxious group of people against a non-anxious group on some task, the chances are that the anxious group would also be more depressed than the non-anxious group. You might think that by adding depression as a covariate into the analysis you can look at the 'pure' effect of anxiety, but you can't. This would be the situation in part C of Figure 11.2; the effect of the covariate (depression) would contain some of the variance from the effect of anxiety. Statistically speaking, all that we know is that anxiety and depression share variance; we cannot separate this shared variance into 'anxiety variance' and 'depression variance', it will always just be 'shared'. Another common example is if you happen to find that your experimental groups differ in their ages. Placing age into the analysis as a covariate will not solve this problem – it is still confounded with the experimental manipulation. ANCOVA is not a magic solution to this problem.

This problem can be avoided by randomizing participants to experimental groups, by matching experimental groups on the covariate (in our anxiety example, you could try to find participants for the low-anxiety-group who score high on depression) or by something called propensity scoring (which goes beyond this book). We can check whether this problem is likely to be an issue by checking whether experimental groups differ on the covariate before we run the ANCOVA. To use our anxiety example again, we could test whether our high- and low-anxiety groups differ on levels of depression (with a *t*-test or ANOVA). If the groups do not significantly differ then we can use depression as a covariate.

11.3.2. Homogeneity of regression slopes ③

When an ANCOVA is conducted we look at the overall relationship between the outcome (dependent variable) and the covariate: we fit a regression line to the entire data set, ignoring to which group a person belongs. In fitting this overall model we, therefore, assume that this overall relationship is true for all groups of participants. For example, if there's a positive relationship between the covariate and the outcome in one group, we assume that there is a positive relationship in all of the other groups too. If, however, the relationship between the outcome (dependent variable) and covariate differs across the groups then the overall regression model is inaccurate (it does not represent all of the groups). This assumption is very important and is called the assumption of **homogeneity of regression slopes.** The best way to think of this assumption is to imagine plotting a scatterplot for each experimental condition with the covariate on one axis and the outcome on the other. If you then calculated, and drew, the regression line for each of these scatterplots you should find that the regression lines look more or less the same (i.e. the values of *b* in each group should be equal). We will have a look at an example of this assumption and how to test it in section 11.7.

11.4. Conducting ANCOVA on SAS ②

11.4.1. Inputting data ①

The data for this example are in Table 11.1 and can be found in the file **ViagraCovariate.sas7bdat.** Table 11.1 shows the participant's libido and their partner's libido and Table 11.2 shows the means and standard deviations of these data. For practice, let's use a DATA step and enter these data by hand. This can be done in much the same way as the Viagra data from the previous chapter except that an extra variable must be created in which to place the values of the covariate.

The data should be laid out in the data editor as they are in Table 11.1. Remember that each row contains information about one person. We'll also add an id variable, which is often a good idea to keep track. We'll keep the dose as a string variable – to do that, we put a $ sign after its name, and it will default to 89 characters, which is OK.

The syntax to enter the data is shown in SAS Syntax 11.1.

```
DATA chap11.ViagraCovariate;
   INPUT id group $ libido partner_libido;
   DATALINES;
1 Control 3     4
2 Control 2     1
3 Control 5     5
4 Control 2     1
5 Control 2     2
```

TABLE 11.1 Data from **ViagraCovariate.sas7bdat**

Dose	Participant's Libido	Partner's Libido
Placebo	3	4
	2	1
	5	5
	2	1
	2	2
	2	2
	7	7
	2	4
	4	5
Low Dose	7	5
	5	3
	3	1
	4	2
	4	2
	7	6
	5	4
	4	2
High Dose	9	1
	2	3
	6	5
	3	4
	4	3
	4	3
	4	2
	6	0
	4	1
	6	3
	2	0
	8	1
	5	0

```
6 Control 2     2
7 Control 7     7
8 Control 2     4
9 Control 4     5
10 Low    7     5
11 Low    5     3
12 Low    3     1
13 Low    4     2
14 Low    4     2
15 Low    7     6
16 Low    5     4
17 Low    4     2
18 High   9     1
19 High   2     3
20 High   6     5
21 High   3     4
22 High   4     3
```

```
23 High    4     3
24 High    4     2
25 High    6     0
26 High    4     1
27 High    6     3
28 High    2     0
29 High    8     1
30 High    5     0
;
run;
```
SAS Syntax 11.1

SELF-TEST Use SAS to find out the means and standard deviations of both the participant's libido and that of their partner in the three groups. (Answers are in Table 11.2).

TABLE 11.2 Means (and standard deviations) from **ViagraCovariate.sas7bdat**

Dose	*Participant's Libido*	*Partner's Libido*
Placebo	3.22 (1.79)	3.44 (2.07)
Low Dose	4.88 (1.46)	3.12 (1.73)
High Dose	4.85 (2.12)	2.00 (1.63)

11.4.2. Initial considerations: testing the independence of the independent variable and covariate ②

In section 11.3.1, I mentioned that before including a covariate in an analysis we should check that it is independent of the experimental manipulation. In this case, the proposed covariate is partner's libido, and we need to check that this variable was roughly equal across levels of our independent variable. In other words, is the mean level of partner's libido roughly equal across our three Viagra groups? We can test this by running an ANOVA with **Partner_Libido** as the outcome and **Dose** as the predictor.

SELF-TEST Conduct an ANOVA to test whether partner's libido (our covariate) is independent of the dose of Viagra (our independent variable).

SAS OUTPUT 11.1

Source	DF	Sum of Squares	Mean Square	F Value	Pr > F
Model	2	12.76944444	6.38472222	1.98	0.1577
Error	27	87.09722222	3.22582305		
Corrected Total	29	99.86666667			

SAS Output 11.1 shows the results of such an ANOVA. The main effect of dose is not significant, $F(2, 27) = 1.98$, $p = .16$, which shows that the average level of partner's libido was roughly the same in the three Viagra groups. In other words, the means for partner's libido in Table 11.2 are not significantly different in the placebo, low- and high-dose groups. This result means that it is appropriate to use partner's libido as a covariate in the analysis.

11.4.3. The main analysis ②

The analysis for ANCOVA is very similar to the analysis for ANOVA. We just add an additional variable to the MODEL line, and if that variable is continuous (and SAS will assume it is continuous if we don't say that it's a class variable) it will be treated as a covariate. SAS Syntax 11.2 shows the basic setup for PROC GLM. As you can see, it is very similar to the ANOVA setup. The only difference is that we've added partner_libido to the MODEL line.

```
PROC GLM DATA=chap11.viagracovariate;
      CLASS dose;
      MODEL libido = dose partner_libido ;
      RUN;
```
SAS Syntax 11.2

11.4.4. Contrasts and other options ②

The first thing to note is that if a covariate is selected, the *post hoc* tests are still possible, but they rapidly get tricky and a little difficult to interpret. Because of that, if you're not doing one-way ANOVA, it's best to steer clear of them. Instead, some comparisons should be still be done using contrasts or estimates.

Contrasts and Estimates are done in the same way that they were for one-way ANOVA (and if you didn't try it then, I'll tell you now that they do the same thing, and give the same results, at this point). To write the CONTRAST or ESTIMATE line properly we need to know the order of the groups – SAS puts them in alphabetical order, so they go: Control, High, Low. Because keeping track of these contrasts can be difficult, we also add an option to the contrast line to tell SAS to report the weights that we are using for the different groups /E.

We will contrast both the low dose group and the high-dose group with the reference group – controlling for partner libido. To do this, we add two CONTRAST lines, one for each comparison, with appropriate weights. This is shown in SAS Syntax 11.3.

```
PROC GLM DATA=chap11.viagracovariate;
    CLASS dose ;
    MODEL libido = dose partner_libido /SOLUTION;
    CONTRAST 'High dose vs control' dose  -1 1 0 / E ;
    CONTRAST 'Low dose vs control' dose  -1 0 1 / E ;
    RUN;
```
SAS Syntax 11.3

The other way to do *post hoc* tests is to use LSMEANS. LSMEANS is short for 'least squares means'; we need to use least squares means, because we have a covariate – we are comparing means, adjusting for the covariate. If someone asked 'what is the mean **libido** score adjusting for the covariate', we would have to answer 'it depends – what do you want the value of the covariate to be?'. SAS solves this problem by estimating the value of the means at the mean of the covariate. Using an LSMEANS statement on its own will give us the means;[2] however, we can test the differences between the means, and get confidence intervals of those differences if we use the ADJUST option. If we just specify ADJUST, SAS won't actually do an adjustment at all – it will do the the default, which is to have no adjustment and simply perform a Tukey LSD *post hoc* test (this option is not recommended). Another option is to ask for a Bonferroni correction (recommended; use ADJUST=BON). The final option is to have a **Šidák correction** (ADJUST=SIDAK). The Šidák correction is similar to the Bonferroni correction but is less conservative and so should be selected if you are concerned about the loss of power associated with Bonferroni corrected values. For this example use the Šidák correction (we will use Bonferroni later in the book).

Our final syntax is now shown in SAS Syntax 11.4.

```
PROC GLM DATA=chap11.viagracovariate;
    CLASS dose ;
    MODEL libido = dose partner_libido /SOLUTION;
    CONTRAST 'High dose vs control' dose  -1 1 0 / E ;
    CONTRAST 'Low dose vs control' dose  -1 0 1 / E ;
    LSMEANS dose ;
    LSMEANS dose /ADJUST=SIDAK ;
    RUN;
    ODS RTF CLOSE;
```
SAS Syntax 11.4

11.5. Interpreting the output from ANCOVA ②

11.5.1. What happens when the covariate is excluded? ②

SELF-TEST Run a one-way ANOVA to see whether the three groups differ in their levels of libido.

[2] Note that it won't give us the standard deviations though. The standard deviation doesn't really exist for an adjusted mean.

SAS Output 11.2 shows (for illustrative purposes) the ANOVA table for these data when the covariate is not included. It is clear from the significance value, which is greater than .05, that Viagra seems to have no significant effect on libido. It should also be noted that the total amount of variation to be explained (SS_T) was 110.97 (Corrected Total), of which the experimental manipulation accounted for 16.84 units (SS_M), while 94.12 were unexplained (SS_E).

SAS OUTPUT 11.2

Source	DF	Sum of Squares	Mean Square	F Value	Pr > F
Model	2	16.8438034	8.4219017	2.42	0.1083
Error	27	94.1228632	3.4860320		
Corrected Total	29	110.9666667			

11.5.2. The main analysis ②

The order of the output for this analysis is a little curious, so please forgive me for not describing it in the order that SAS produces it. We'll start with the overall model results – SAS Output 11.3 shows these. We see the sum of squares for the model, error and total in the first table, along with the mean squares (SS/*df*), *F* and the probability value. The amount of variation accounted for by the model (SS_M) has increased to 31.92 units (corrected model) of which Viagra accounts for 25.19 units. Most important, the large amount of variation in libido that is accounted for by the covariate has meant that the unexplained variance (SS_R) has been reduced to 79.05 units. Notice that SS_T has not changed; all that has changed is how that total variation is explained.

We can see from this table that the overall model is statistically significant – that is, it is explaining more variance than we would expect by chance alone. The second table shows R^2, which in ANOVA we can refer to as η^2.

The format of these tables is largely the same as without the covariate, except that there is an additional row of information about the covariate (partner_libido). Looking first at the significance values, it is clear that the covariate significantly predicts the dependent variable, because the significance value is less than .05. Therefore, the person's libido is influenced by their partner's libido. What's more interesting is that when the effect of partner's libido is removed, the effect of Viagra becomes significant (*p* is .027 which is less than .05).

This example illustrates how ANCOVA can help us to exert stricter experimental control by taking account of confounding variables to give us a 'purer' measure of effect of the experimental manipulation. Without taking account of the libido of the participants' partners we would have concluded that Viagra had no effect on libido, yet it does. Looking back at the group means from Table 11.1 for the libido data, it seems pretty clear that the significant ANOVA reflects a difference between the placebo group and the two experimental groups (because the low- and high-dose groups have very similar means – 4.88 and 4.85 – whereas the placebo group mean is much lower at 3.22). However, we'll need to check some contrasts to verify this.

Source	DF	Sum of Squares	Mean Square	F Value	Pr > F
Model	3	31.9195511	10.6398504	3.50	0.0295
Error	26	79.0471155	3.0402737		
Corrected Total	29	110.9666667			

SAS OUTPUT 11.3

R-Square	Coeff Var	Root MSE	libido Mean
0.287650	39.93064	1.743638	4.366667

Source	DF	Type I SS	Mean Square	F Value	Pr > F
dose	2	16.84380342	8.42190171	2.77	0.0812
partner_libido	1	15.07574771	15.07574771	4.96	0.0348

SAS OUTPUT 11.4

Source	DF	Type III SS	Mean Square	F Value	Pr > F
dose	2	25.18519421	12.59259710	4.14	0.0274
partner_libido	1	15.07574771	15.07574771	4.96	0.0348

SAS Output 11.5 shows the parameter estimates selected with SOLUTION. These estimates are calculated using a regression analysis with **dose** split into two dummy coding variables (see sections 10.2.3 and 11.6). SAS codes the two dummy variables such that the last category alphabetically (the category coded with the highest value in the data editor – in this case the low-dose group) is the reference category. This reference category is coded with a 0 for both dummy variables (see section 10.2.3. for a reminder of how dummy coding works). The dose High line, therefore, shows the difference between high dose and the reference category (low dose), and dose Control represents the difference between the control group (placebo) and the reference category (low dose). The *b*-values represent the differences between the means of these groups and so the significances of the *t*-tests tell us whether the group means differ significantly. The degrees of freedom for these *t*-tests can be calculated as in normal regression (see section 7.2.4) as $N-p-1$ in which N is the total sample size (in this case 30) and p is the number of predictors (in this case 3, the two dummy variables and the covariate). For these data, $df = 30-3-1 = 26$. From these estimates we could conclude that the high-dose group differs significantly from the placebo group but not from the low-dose group.

The final thing to notice is the value of *b* for the covariate (0.416). This value tells us that, other things being equal, if a partner's libido is one unit higher, then the person's libido should be just under half a unit higher (although there is nothing to suggest a causal link between the two). The sign of this coefficient tells us the direction of the relationship between the covariate and the outcome. So, in this example, because the coefficient is positive it means that partner's libido has a positive relationship with the participant's libido: as one increases so does the other. A negative coefficient would mean the opposite: as one increases, the other decreases.

SAS OUTPUT 11.5

Parameter	Estimate		Standard Error	t Value	Pr > |t|
Intercept	3.574868442	B	0.84906986	4.21	0.0003
dose Control	−1.785680115	B	0.84935531	−2.10	0.0454
dose High	0.439201207	B	0.81122140	0.54	0.5928
dose Low	0.000000000	B	.	.	.
partner_libido	0.416042099		0.18683326	2.23	0.0348

11.5.3. Contrasts ②

SAS Output 11.6 shows the result of the contrast and compares level 2 (low dose) against level 1 (placebo) as a first comparison, and level 3 (high dose) against level 1 (placebo) as a second comparison. These contrasts are consistent with what was specified: all groups are compared to the first group. These results show that both the low-dose group (p = .045) and high-dose group (contrast 2, p = .010) had significantly different libidos than the placebo group. These results are consistent with the regression parameter estimates (in fact, note that Low dose vs control is identical to the regression parameters for dose Control in the previous section).

SAS OUTPUT 11.6

Contrast	DF	Contrast SS	Mean Square	F Value	Pr > F
High dose vs control	1	23.35069722	23.35069722	7.68	0.0102
Low dose vs control	1	13.43820291	13.43820291	4.42	0.0454

These contrasts and parameter estimates tell us that there were group differences, but to interpret them we need to know the means. We produced the means in Table 11.2 so surely we can just look at these values? Actually we can't because these group means have not been adjusted for the effect of the covariate. These original means tell us nothing about the group differences reflected by the significant ANCOVA. SAS Output 11.7 gives the adjusted values of the group means and it is these values that should be used for interpretation (this is the main reason we used the LSMEANS statement). The adjusted means (and our contrasts) show that levels of libido were significantly higher in the low- and high-dose groups compared to the placebo group. The regression parameters also told us that the high- and low-dose groups did not significantly differ (p = .593). These conclusions can be verified with the *post hoc* tests specified in the options menu but normally you would do only contrasts or *post hoc* tests, not both.

SAS Output 11.8 shows the results of the Šidák corrected *post hoc* comparisons that were requested using/ADJUST=SIDAK: dialog box. The significant difference between the

SAS OUTPUT 11.7

dose	libido LSMEAN	LSMEAN Number
Control	2.92637006	1
High	5.15125139	2
Low	4.71205018	3

SAS OUTPUT 11.8

Least Squares Means for effect dose
Pr > |t| for H0: LSMean(i)=LSMean(j)

Dependent Variable: libido

i/j	1	2	3
1		0.0302	0.1300
2	0.0302		0.9325
3	0.1300	0.9325	

high-dose and placebo group remains (p = .030), and the high-dose and low-dose groups do not significantly differ (p = .93). However, it is interesting that the significant difference between the low-dose and placebo groups shown by the regression parameters and contrasts (SAS Output 11.5) is gone (p is only .13).

SELF-TEST Why do you think that the results of the *post hoc* test differ from the contrasts for the comparison of the low-dose and placebo groups?

11.5.4. Interpreting the covariate ②

I've already mentioned that the parameter estimates (SAS Output 11.5) tell us how to interpret the covariate. If the b-value for the covariate is positive then it means that the covariate and the outcome variable have a positive relationship (as the covariate increases, so does the outcome). If the b-value is negative it means the opposite: that the covariate and the outcome variable have a negative relationship (as the covariate increases, the outcome decreases). For these data the b-value was positive, indicating that as the partner's libido increases, so does the participant's libido. Another way to discover the same thing is simply to draw a scatterplot of the covariate against the outcome. We came across scatterplots in section 4.7. so have a look back there to find out how to produce one. Figure 11.4 shows the resulting scatterplot for these data and confirms what we already know: the effect of the covariate is that as partner's libido increases, so does the participant's libido (as shown by the slope of the regression line).

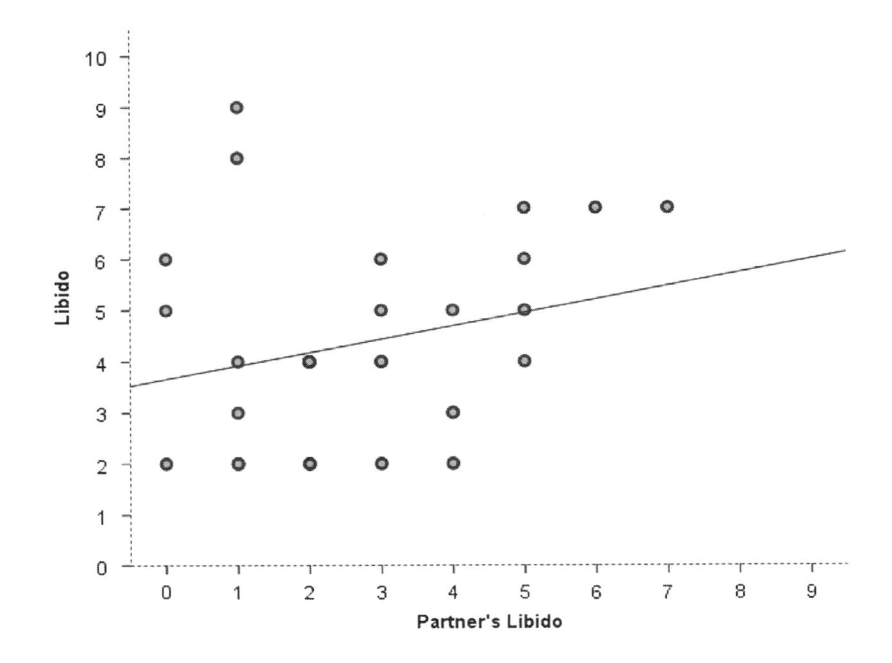

FIGURE 11.4 Scatterplot of libido against partner's libido

LABCOAT LENI'S REAL RESEARCH 11.1

Space invaders ②

Anxious people tend to interpret ambiguous information in a negative way. For example, being highly anxious myself, if I overheard a student saying 'Andy Field's lectures are really different' I would assume that 'different' meant rubbish, but it could also mean 'refreshing' or 'innovative'. One current mystery is how these interpretational biases develop in children. Peter Muris and his colleagues addressed this issue in an ingenious study (Muris, Huijding, Mayer, & Hameetman, 2008). Children did a task in which they imagined that they were astronauts who had discovered a new planet. Although the planet was similar to Earth, some things were different. They were given some scenarios about their time on the planet (e.g. 'On the street, you encounter a spaceman. He has a sort of toy handgun and he fires at you …') and the child had to decide which of two outcomes occurred. One outcome was positive ('You laugh: it is a water pistol and the weather is fine anyway') and the other negative ('Oops, this hurts! The pistol produces a red beam which burns your skin!'). After each response the child was told whether their choice was correct. Half of the children were always told that the negative interpretation was correct, and the remainder were told that the positive interpretation was correct.

Over 30 scenarios children were trained to interpret their experiences on the planet as negative or positive. Muris et al. then gave children a standard measure of interpretational biases in everyday life to see whether the training had created a bias to interpret things negatively. In doing so, they could ascertain whether children learn interpretational biases through feedback (e.g. from parents) about how to disambiguate ambiguous situations.

The data from this study are in the file **Muris et al 2008.sas7bdat.** The independent variable is **Training** (positive or negative) and the outcome was the child's interpretational bias score (**Interpretational_Bias**) – a high score reflects a tendency to interpret situations negatively. It is important to factor in the **Age** and **Gender** of the child and also their natural anxiety level (which they measured with a standard questionnaire of child anxiety called the **SCARED**) because these things affect interpretational biases also. Labcoat Leni wants you to carry out a one-way ANCOVA on these data to see whether **Training** significantly affected children's **Interpretational_Bias** using **Age**, **Gender** and **SCARED** as covariates. What can you conclude?

Answers are in the additional material on the companion website (or look at pages 475 to 476 in the original article).

MURIS P. ET AL. (2008). *CHILD PSYCHIATRY AND HUMAN DEVELOPMENT*, 39, 469–480.

11.6. ANCOVA run as a multiple regression ②

Although the ANCOVA is essentially done, it is a useful educational exercise to rerun the analysis as a hierarchical multiple regression (it will, I hope, help you to understand what's happening when we do an ANCOVA).

SELF-TEST Add two dummy variables to the file **ViagraCovariate.sas7bdat** that compare the low dose to the placebo **(Low_Placebo)** and the high dose to the placebo **(High_Placebo)** – see section 10.2.3. for help. If you get stuck then download **ViagraCovariateDummy.sas7bdat.**

SELF-TEST Run a hierarchical regression analysis with **Libido** as the outcome. In the first block enter partner's libido **(Partner_Libido)** as a predictor, and then in a second model block enter both dummy variables see section 7.7 for help. or help.

To summarize, you don't have to run ANCOVA through PROC REG; I have done this merely to illustrate that when we do ANCOVA we are using a regression model. In other words, we could just ignore the PROC GLM and run the analysis as regression. However, you wouldn't do both.

11.7. Testing the assumption of homogeneity of regression slopes ③

We saw earlier that when we conduct ANCOVA we assume homogeneity of regression slopes. This just means that we assume that the relationship between the outcome (dependent variable) and the covariate is the same in each of our treatment groups. Figure 11.5 shows scatterplots that display the relationship between partner's libido (the covariate) and the outcome (participant's libido) for each of the three experimental conditions (different colours and symbols). Each symbol represents the data from a particular participant, and the type of symbol tells us the group (circles = placebo, triangles = low dose, squares = high dose). The lines are the regression slopes for the particular group, they summarize the relationship between libido and partner's libido shown by the dots (black solid = placebo group, black dashed = low-dose group, grey dashed = high-dose group). It should be clear that there is a positive relationship (the regression line slopes upwards from left to right) between partner's libido and participant's libido in both the placebo and low-dose conditions. In fact, the slopes of the lines for these two groups (the two black lines) are very similar, showing that the relationship between libido and partner's libido is very similar in these two groups. This situation is an example of homogeneity of regression slopes (the regression slopes in the two groups are similar). However, in the high-dose condition there appears to be no relationship at all between participant's libido and that of their partner (the squares are fairly randomly scattered and the regression line is very flat and shows a slightly negative relationship). The slope of this line is very different to the other two, and this difference gives us cause to doubt whether there is homogeneity of regression slopes (because the relationship between participant's libido and that of their partner is different in the high-dose group to the other two groups).

To test the assumption of homogeneity of regression slopes, we need to specify a model that includes the interaction between the covariate and independent variable. Ordinarily, the

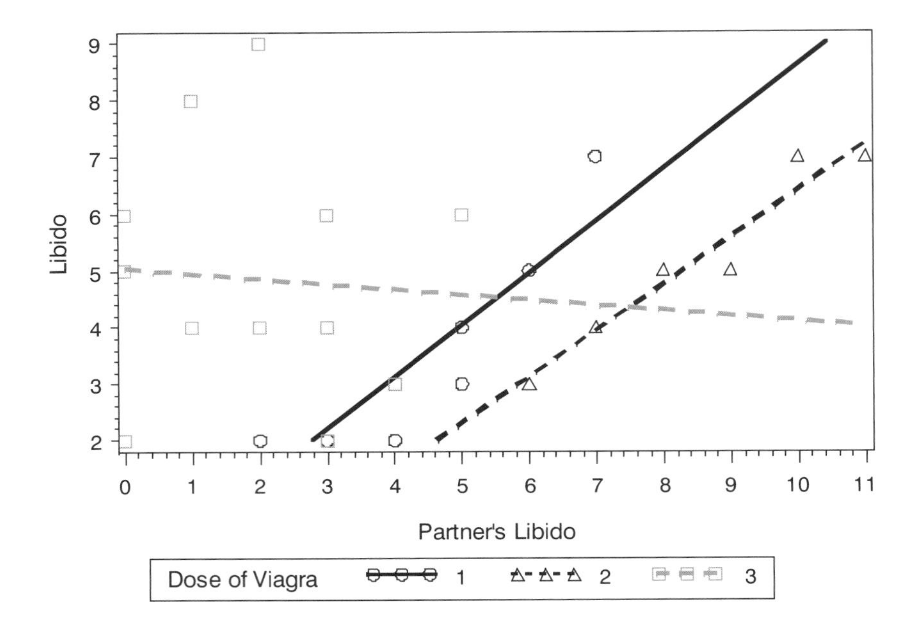

FIGURE 11.5 Scatterplot and regression lines of libido against partner's libido for each of the experimental conditions

ANCOVA includes only the main effect of dose and partner's libido and does not include this interaction term. To test this interaction term it's important to still include the main effects of dose and partner so that the interaction term is tested controlling for these main effects. If we don't include the main effects then variance in libido may become attributed to the interaction term that would otherwise be attributed to main effects. To add the interaction, we add dose*partner_libido to the model statement, as shown in SAS Syntax 11.5.

```
  PROC GLM DATA=chap11.viagracovariate ;
      CLASS dose ;
      MODEL libido = dose partner_libido dose*partner_libido /
SOLUTION ;
      RUN;
```

SAS Syntax 11.5

SAS OUTPUT 11.9

Source	DF	Type III SS	Mean Square	F Value	Pr > F
dose	2	36.55755997	18.27877998	7.48	0.0030
partner_libido	1	17.18222420	17.18222420	7.03	0.0139
partner_libido*dose	2	20.42659366	10.21329683	4.18	0.0277

SAS Output 11.9 shows the main summary table for the ANCOVA including the interaction term. The effects of the dose of Viagra and the partner's libido are still significant, but the main thing in which we're interested is the interaction term, so look at the significance value of the covariate by outcome interaction **(dose×partner_libido).** If this effect is significant then the assumption of homogeneity of regression slopes has been broken. The effect here is significant (p = 0.028); therefore the assumption is not tenable. Although this finding is not surprising given the pattern of relationships shown in Figure 11.5, it does raise concern about the main analysis. This example illustrates why it is important to test assumptions and not to just blindly accept the results of an analysis.

CRAMMING SAM'S TIPS **ANCOVA**

- Analysis of covariance (ANCOVA) compares several means, but adjusting for the effect of one or more other variables (called *covariates*); for example, if you have several experimental conditions and want to adjust for the age of the participants.
- Before the analysis you should check that the independent variables and covariate(s) are independent. You can do this using ANOVA or a *t*-test to check that levels of the covariate do not differ significantly across groups.
- In the ANOVA table, look at the column labelled *Pr* > *F* for both the covariate and the independent variable. If the value is less than .05 then for the covariate it means that this variable has a significant relationship with the outcome variable; for the independent variable it means that the means are significantly different across the experimental conditions after partialling out the effect that the covariate has on the outcome.
- As with ANOVA, if you have generated specific hypotheses before the experiment use planned comparisons.
- For contrasts and *post hoc* tests, again look to the probability values to discover if your comparisons are significant (they will be if the significance value is less than .05).
- Test the same assumptions as for ANOVA, but in addition you should test the assumption of homogeneity of regression slopes. This has to be done by customizing the ANCOVA model in SAS to look at the independent variable × covariate interaction.

11.8. Calculating the effect size ②

We saw in the previous chapter that we can use eta squared, η^2, as an effect size measure in ANOVA. This effect size is just r^2 by another name and is calculated by dividing the effect of interest, SS_M, by the total amount of variance in the data, SS_T. As such, it is the proportion of total variance explained by an effect. In ANCOVA (and some of the more complex ANOVAs that we'll encounter in future chapters), we have more than one effect; therefore, we could calculate eta squared for each effect. However, we can also use an effect size measure called partial eta squared (partial η^2). This differs from eta squared in that it looks not at the proportion of total variance that a variable explains, but at the proportion of variance that a variable explains that *is not explained by other variables in the analysis*. Let's look at this with our example; say we want to know the effect size of the dose of Viagra. Partial eta squared is the proportion of variance in libido that the dose of Viagra shares that is not attributed to partner's libido (the covariate). If you think about the variance that the covariate cannot explain, there are two sources: it cannot explain the variance attributable to the dose of Viagra, SS_{Viagra} and it cannot explain the error variability, SS_R. Therefore, we use these two sources of variance instead of the total variability, SS_T, in the calculation. The difference between eta squared and partial eta squared is shown as:

$$\eta^2 = \frac{SS_{Effect}}{SS_{Total}} \qquad \text{Partial } \eta^2 = \frac{SS_{Effect}}{SS_{Effect} + SS_{Residual}} \tag{11.2}$$

We can't ask SAS to produce partial eta squared for us directly (but see Jane Superbrain 11.1 if you want to know how to use SAS to calculate it). To illustrate its calculation let's look at our Viagra example. We need to use the sums of squares in SAS Output 11.4 for the effect of dose (25.19), the covariate (15.08) and the error (79.05):

$$\begin{aligned}\text{Partial } \eta^2_{\text{Dose}} &= \frac{SS_{\text{Dose}}}{SS_{\text{Dose}} + SS_{\text{Residual}}} \\ &= \frac{25.19}{25.19 + 79.05} \\ &= \frac{25.19}{104.24} \\ &= .24\end{aligned}$$

$$\begin{aligned}\text{Partial } \eta^2_{\text{PartnerLibido}} &= \frac{SS_{\text{PartnerLibido}}}{SS_{\text{PartnerLibido}} + SS_{\text{Residual}}} \\ &= \frac{15.08}{15.08 + 79.05} \\ &= \frac{15.08}{94.13} \\ &= .16\end{aligned}$$

These values show that **dose** explained a bigger proportion of the variance not attributable to other variables than **partner_libido**.

As with ANOVA, you can also use omega squared (ω^2). However, as we saw in Section 10.5 this measure can be calculated only when we have equal numbers of participants in each group (which is not the case in this example!). So, we're a bit stumped!

However, not all is lost because, as I've said many times already, the overall effect size is not nearly as interesting as the effect size for more focused comparisons. These are easy to calculate because we selected regression parameters (by putting/SOLUTION on the MODEL line (see SAS Output 11.5) and so we have *t*-statistics for the covariate and comparisons between the low- and high-dose groups and the placebo and high-dose group. These *t*-statistics have 26 degrees of freedom (see section 11.5.1). We can use the same equation as in section Section 10.5:[3]

$$r_{\text{contrast}} = \sqrt{\frac{t^2}{t^2 + df}}$$

Therefore we get (with *t* from SAS Output 11.5):

$$\begin{aligned}r_{\text{Covariate}} &= \sqrt{\frac{2.23^2}{2.23^2 + 26}} \\ &= \sqrt{\frac{4.97}{30.97}} \\ &= .40\end{aligned}$$

$$\begin{aligned}r_{\text{High Dose vs. Placebo}} &= \sqrt{\frac{-2.77^2}{-2.77^2 + 26}} \\ &= \sqrt{\frac{7.67}{33.67}} \\ &= .48\end{aligned}$$

$$\begin{aligned}r_{\text{High vs. Low Dose}} &= \sqrt{\frac{-0.54^2}{-0.54^2 + 26}} \\ &= \sqrt{\frac{0.29}{26.29}} \\ &= .11\end{aligned}$$

[3] Strictly speaking, we have to use a slightly more elaborate procedure when groups are unequal. It's a bit beyond the scope of this book but Rosnow, Rosenthal, and Rubin (2000) give a very clear account.

If you think back to our benchmarks for effect sizes, the effect of the covariate and the difference between the high dose and the placebo both represent medium to large effect sizes (they're all between .4 and .5). Therefore, as well as being statistically significant, these effects are substantive findings. The difference between the high- and low-dose groups was a fairly small effect.

Jane Superbrain Box 11.1:

Using ODS to calculate effect sizes. ③

We don't want to have to do all those calculations by hand, especially when there's an easier way – which there is, and that's to use ODS (the Output Delivery System – see Jane Superbrain 3.1). We can ask SAS to send the output to a data set, rather than to the listing window, and once it's in a data set, we can do that calculations with SAS, instead of doing them by. Here's what to do.

First we need to modify our PROC GLM command, to add an ODS statement. We're also going to add the SS3 option to the model line, so that we only get type III sums of squares.

```
PROC GLM DATA=chap11.viagracovariate ;
       CLASS dose ;
       MODEL libido = dose partner_libido/
       SOLUTION SS3;
       ODS    OUTPUT    OVERALLANOVA=anova1
MODELANOVA=anova2
       RUN;
```

Now we have two new files. The file called anova1 contains the model, error and corrected sums of squares, the file called anova2 contains the sums of squares for the independent variables. We need to get the error sum of squares into the data set with the sums of squares for the model. We do that with two DATA steps. The first DATA step removes everything except the error sum of squares from the DATA set anova1, the second DATA step merges that with the data set anova2, and calculates partial eta-squared. Finally, we run PROC PRINT to show us the results.

```
DATA anova1; SET anova1;
       IF source NE 'Error' THEN DELETE;
       DROP ms Fvalue probF source;
       RENAME ss = sse;
       RUN;
DATA bothanovas; MERGE anova1 anova2;
       BY dependent;
       partial_eta2 = ss / (ss + sse);
       RUN;
       PROC PRINT DATA=bothanovas;
       VAR source partial_eta2;
       RUN;
```

Obs	Source	partial_eta2
1	DOSE	0.24163
2	PARINER_LIBIDO	0.16017

11.9. Reporting results ②

Reporting ANCOVA is much the same as reporting ANOVA except we now have to report the effect of the covariate as well. For the covariate and the experimental effect we give details of the *F*-ratio and its degrees of freedom. In both cases, the *F*-ratio was derived from dividing the mean squares for the effect by the mean squares for the residual. Therefore, the degrees of freedom used to assess the *F*-ratio are the degrees of freedom for the effect of the model (df_M = 1 for the covariate and 2 for the experimental effect) and the degrees of freedom for the residuals of the model (df_R = 26 for both the covariate and the experimental effect) – see SAS Output 11.4. Therefore, the correct way to report the main findings would be:

✓ The covariate, partner's libido, was significantly related to the participant's libido, $F(1, 26) = 4.96$, $p = .035$, $r = .40$. There was also a significant effect of Viagra on levels of libido after controlling for the effect of partner's libido, $F(2, 26) = 4.14$, $p = .027$, *partial* $\eta^2 = .24$.

We can also report some contrasts (see SAS Output 11.6):

✓ Planned contrasts revealed that having a high dose of Viagra significantly increased libido compared to having a placebo, $t(26) = -2.77$, $p < .05$, $r = .48$, but not compared to having a low dose, $t(26) = -0.54$, $p > .05$, $r = .11$.

11.10. What to do when assumptions are violated in ANCOVA ③

In previous chapters we have seen that when the assumptions of a test have been violated, there is often a non-parametric test to which we can turn (Chapter 15). However, as we start to discover more complicated procedures, we will also see that the squid of despair squirts its ink of death on our data when they violate assumptions. For complex analyses, there are no non-parametric counterparts that are easily run on SAS. ANCOVA is the first such example of a test that does not have an SAS-friendly non-parametric test. As such, if our data violate the assumptions of normality or homogeneity of variance the only real solutions are robust methods (see section 5.7.4) such as those described in Wilcox's Chapter 11 and the associated files for the software R (Wilcox, 2005). You could use R to do these tests (directly from SAS when this is possible). You could also use the robust regression procedures in Wilcox's book because, as we have seen, ANCOVA is simply a version of regression. Also, if you have violated the assumption of homogeneity of regression slopes then you can explicitly model this variation using multilevel linear models (see Chapter 19). If you run ANCOVA as a multilevel model you can also bootstrap the parameters to get robust estimates.

What have I discovered about statistics? ②

This chapter has shown you how the general linear model that was described in Chapter 10 can be extended to include additional variables. The advantages of doing so are that we can remove the variance in our outcome that is attributable to factors other than our experimental manipulation. This gives us tighter experimental control, and may also help us to explain some of our error variance, and, therefore, give us a purer measure of the experimental manipulation. We didn't go into too much theory about ANCOVA, just looked conceptually at how the regression model can be expanded to include these additional variables (covariates). Instead we jumped straight into an example, which was to look at the effect of Viagra on libido (as in Chapter 10) but including partner's libido as a covariate. I explained how to do the analysis on SAS and interpret the results but also showed how the same output could be obtained by running the analysis as a regression. This was to try to get the message home that ANOVA and ANCOVA are merely forms of regression! Anyway, we finished off by looking at an additional assumption that has to be considered when doing ANCOVA: the assumption of homogeneity of regression slopes. This just means that the relationship between the covariate and the outcome variable should be the same in all of your experimental groups.

Having seen Iron Maiden in all of their glory, I was inspired. Although I had briefly been deflected from my destiny by the shock of grammar school, I was back on track. I had to form a band. There was just one issue: no one else played a musical instrument. The solution was easy: through several months of covert subliminal persuasion I convinced my two best friends (both called Mark oddly enough) that they wanted nothing more than to start learning the drums and bass guitar. A power trio was in the making!

Key terms that I've discovered

Adjusted mean
Analysis of covariance (ANCOVA)
Covariate
Homogeneity of regression slopes
Partial eta squared (partial η^2)
Partial out
Šidák correction

Smart Alex's tasks

- **Task 1:** Stalking is a very disruptive and upsetting (for the person being stalked) experience in which someone (the stalker) constantly harasses or obsesses about another person. It can take many forms, from sending intensely disturbing letters threatening to boil your cat if your undying love isn't reciprocated, to following the stalker around your local area in a desperate attempt to see which CD they buy on a Saturday (as if it would be anything other than Fugazi!). A psychologist, who'd had enough of being stalked by people, decided to try two different therapies on different groups of stalkers (25 stalkers in each group – this variable is called **Group**). To the first group of stalkers he gave what he termed cruel-to-be-kind therapy. This therapy was based on punishment for stalking behaviours; in short, every time the stalkers followed him around, or sent him a letter, the psychologist attacked them with a cattle prod until they stopped their stalking behaviour. It was hoped that the stalkers would learn an aversive reaction to anything resembling stalking. The second therapy was psychodyshamic therapy, which is a recent development on Freud's psychodynamic therapy that acknowledges what a sham this kind of treatment is (so you could say it's based on Fraudian theory). The stalkers were hypnotized and regressed into their childhood, the therapist would also discuss their penis (unless it was a woman, in which case they discussed their lack of penis), the penis of their father, their dog's penis, the penis of the cat down the road and anyone else's penis that sprang to mind. At the end of therapy, the psychologist measured the number of hours in the week that the stalker spent stalking their prey (this variable is called **stalk2**). Now, the therapist believed that the success of therapy might well depend on how bad the problem was to begin with, so before therapy the therapist measured the number of hours that the patient spent stalking as an indicator of how much of a stalker the person was (this variable is called **stalk1**). The data are in the file **Stalker.sas7bdat.** Analyse the effect of therapy on stalking behaviour after therapy, controlling for the amount of stalking behaviour before therapy. ②

- **Task 2:** A marketing manager for a certain well-known drinks manufacturer was interested in the therapeutic benefit of certain soft drinks for curing hangovers. He took 15 people out on the town one night and got them drunk. The next morning as they awoke, dehydrated and feeling as though they'd licked a camel's sandy feet clean with their tongue, he gave five of them water to drink, five of them Lucozade (in case this isn't sold outside of the UK, it's a very nice glucose-based drink) and the remaining five a leading brand of cola (this variable is called **drink**). He then measured how well they felt (on a scale from 0 = I feel like death to 10 = I feel really full of beans and healthy) two hours later (this variable is called **well**). He wanted to know which drink produced the greatest level of wellness. However, he realized it was important to control for how drunk the person got the night before, and so he measured this on a scale of 0 = as sober as a nun to 10 = flapping about like a haddock out of water on the floor in a puddle of their own vomit. The data are in the file **HangoverCure.sas7bdat.** Conduct an ANCOVA to see whether people felt better after different drinks when controlling for how drunk they were the night before. ②

The answers are on the companion website and task 1 has a full interpretation in Field & Hole (2003).

Further reading

Howell, D. C. (2006). *Statistical methods for psychology* (6th ed.). Belmont, CA: Duxbury. (Or you might prefer his *Fundamental statistics for the behavioral sciences,* also in its 6th edition, 2007.)

Miller, G. A., & Chapman, J. P. (2001). Misunderstanding analysis of covariance. *Journal of Abnormal Psychology*, *110*, 40–48.

Rutherford, A. (2000). *Introducing ANOVA and ANCOVA: A GLM approach*. London: Sage.

Wildt, A. R. & Ahtola, O. (1978). *Analysis of covariance*. Sage University Paper Series on Quantitative Applications in the Social Sciences, 07-012. Newbury Park, CA: Sage. (This text is pretty high level but very comprehensive if you want to know the maths behind ANCOVA.)

Interesting real research

Muris, P., Huijding, J., Mayer, B. and Hameetman, M. (2008). A space odyssey: Experimental manipulation of threat perception and anxiety-related interpretation bias in children. *Child Psychiatry and Human Development, 39*(4), 469–480.

Factorial ANOVA (GLM 3) 12

FIGURE 12.1 Andromeda coming to a living room near you in 1988 (L-R: Malcolm, me and the two Marks)

12.1. What will this chapter tell me? ②

After persuading my two friends (Mark and Mark) to learn the bass and drums, I took the rather odd decision to *stop* playing the guitar. I didn't stop, as such, but I focused on singing instead. In retrospect, I'm not sure why because I am *not* a good singer. Mind you, I'm not a good guitarist either. The upshot was that a classmate, Malcolm, ended up as our guitarist. I really can't remember how or why we ended up in this configuration, but we called ourselves Andromeda, we learnt several Queen and Iron Maiden songs and we were truly awful. I have some tapes somewhere to prove just what a cacophony of tuneless drivel we produced, but the chances of their appearing on the companion website are slim at best. Suffice it to say, you'd be hard pushed to recognize *which* Iron Maiden and Queen songs we were trying to play. I try to comfort myself with the fact that we were only 14 or 15 at the time, but even youth does not excuse the depths of ineptitude to which we sank. Still, we garnered a reputation for being too loud in school assembly and we did a successful tour of our friends' houses (much to their parents' amusement, I'm sure). We even started to write a few songs (I wrote one called 'Escape From Inside' about the film *The Fly* that contained the wonderful rhyming couplet of 'I am a fly, I want to die': genius!). The only thing that we did that resembled the activities of a 'proper' band was to split up due to 'musical differences', these differences being that Malcolm wanted to write 15-part symphonies about a boy's journey to worship electricity pylons and discover a mythical beast called the cuteasaurus, whereas I wanted to write songs about flies, and dying. When we could not agree on a musical direction the split became inevitable. We could have tested empirically the best musical direction for the band if Malcolm and I had both written a 15-part symphony and a 3-minute song about a fly. If we played these songs to various people and measured their screams of agony then we could ascertain the best musical direction to gain popularity. We have two variables that predict screams: whether I or Malcolm wrote the song (songwriter), and whether the song was a 15-part symphony or a song about a fly (song type). The one-way ANOVA that we encountered in Chapter 10 cannot deal with two predictor variables – this is a job for factorial ANOVA!

12.2. Theory of factorial ANOVA (between groups) ②

In the previous two chapters we have looked at situations in which we've tried to test for differences between groups when there has been a single independent variable (i.e. one variable has been manipulated). However, at the beginning of Chapter 10 I said that one of the advantages of ANOVA was that we could look at the effects of more than one independent variable (and how these variables interact). This chapter extends what we already know about ANOVA to look at situations where there are two independent variables. We've already seen in the previous chapter that it's very easy to incorporate a second variable into the ANOVA framework when that variable is a continuous variable (i.e. not split into groups), but now we'll move on to situations where there is a second independent variable that has been systematically manipulated by assigning people to different conditions.

12.2.1. Factorial designs ②

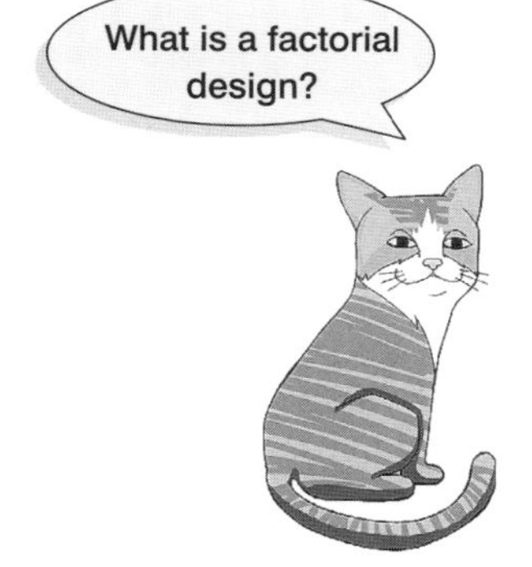

In the previous two chapters we have explored situations in which we have looked at the effects of a single independent variable on some outcome. However, independent variables often get lonely and want to have friends. Scientists are obliging individuals and often put a second (or third) independent variable into their designs to keep the others company. When an experiment has two or more independent variables it is known as a *factorial design* (this is because variables are sometimes referred to as *factors*). There are several types of factorial design:

- **Independent factorial design**: In this type of experiment there are several independent variables or predictors and each has been measured using different participants (between groups). We discuss this design in this chapter.
- **Repeated-measures (related) factorial design**: This is an experiment in which several independent variables or predictors have been measured, but the same participants have been used in all conditions. This design is discussed in Chapter 13.
- **Mixed design**: This is a design in which several independent variables or predictors have been measured; some have been measured with different participants whereas others used the same participants. This design is discussed in Chapter 14.

As you might imagine, it can get quite complicated analysing these types of experiments. Fortunately, we can extend the ANOVA model that we encountered in the previous two chapters to deal with these more complicated situations. When we use ANOVA to analyse a situation in which there are two or more independent variables it is sometimes called **factorial ANOVA**; however, the specific names attached to different ANOVAs reflect the experimental design that they are being used to analyse (see Jane Superbrain Box 12.1). This section extends the one-way ANOVA model to the factorial case (specifically when there are two independent variables). In subsequent chapters we will look at repeated-measures designs, factorial repeated-measures designs and finally mixed designs.

JANE SUPERBRAIN 12.1

Naming ANOVAs ②

ANOVAs can be quite confusing because there appear to be lots of them. When you read research articles you'll quite often come across phrases like 'a two-way independent ANOVA was conducted', or 'a three-way repeated-measures ANOVA was conducted'. These names may look confusing but they are quite easy if you break them down. All ANOVAs have two things in common: they involve some quantity of independent variables and these variables can be measured using either the same or different participants. If the same participants are used we typically use the term *repeated measures* and if different participants are used we use the term *independent*. When there are two or more independent variables, it's possible that some variables use the same participants whereas others use different participants. In this case we use the term *mixed*. When we name an ANOVA, we are simply telling the reader how many independent variables we used and how they were measured. In general terms we could write the name of an ANOVA as:

- A (number of independent variables)-way (how these variables were measured) ANOVA.

By remembering this you can understand the name of any ANOVA you come across. Look at these examples and try to work out how many variables were used and how they were measured:

- One-way independent ANOVA
- Two-way repeated-measures ANOVA
- Two-way mixed ANOVA
- Three-way independent ANOVA

The answers you should get are:

- One independent variable measured using different participants.
- Two independent variables both measured using the same participants.
- Two independent variables: one measured using different participants and the other measured using the same participants.
- Three independent variables all of which are measured using different participants.

12.2.2. An example with two independent variables ②

Throughout this chapter we'll use an example that has two independent variables. This is known as a two-way ANOVA (see Jane Superbrain Box 12.1). I'll look at an example with two independent variables because this is the simplest extension of the ANOVAs that we have already encountered.

TABLE 12.1 Data for the beer-goggles effect

Alcohol	None		2 Pints		4 Pints	
Gender	Female	Male	Female	Male	Female	Male
	65	50	70	45	55	30
	70	55	65	60	65	30
	60	80	60	85	70	30
	60	65	70	65	55	55
	60	70	65	70	55	35
	55	75	60	70	60	20
	60	75	60	80	50	45
	55	65	50	60	50	40
Total	485	535	500	535	460	285
Mean	60.625	66.875	62.50	66.875	57.50	35.625
Variance	24.55	106.70	42.86	156.70	50.00	117.41

An anthropologist was interested in the effects of alcohol on mate selection at night-clubs. Her rationale was that after alcohol had been consumed, subjective perceptions of physical attractiveness would become more inaccurate (the well-known **beer-goggles effect**). She was also interested in whether this effect was different for men and women. She picked 48 students: 24 male and 24 female. She then took groups of eight participants to a night-club and gave them no alcohol (participants received placebo drinks of alcohol-free lager), 2 pints of strong lager, or 4 pints of strong lager. At the end of the evening she took a photograph of the person that the participant was chatting up. She then got a pool of independent judges to assess the attractiveness of the person in each photograph (out of 100). The data are in Table 12.1 and **Goggles.sas7bdat**.

12.2.3. Total sums of squares (SS_T) ②

Two-way ANOVA is conceptually very similar to one-way ANOVA. Basically, we still find the total sum of squared errors (SS_T) and break this variance down into variance that can be explained by the experiment (SS_M) and variance that cannot be explained (SS_R). However, in two-way ANOVA, the variance explained by the experiment is made up of not one experimental manipulation but two. Therefore, we break the model sum of squares down into variance explained by the first independent variable (SS_A), variance explained by the second independent variable (SS_B) and variance explained by the interaction of these two variables ($SS_{A \times B}$) – see Figure 12.2.

We start off in the same way as we did for one-way ANOVA. That is, we calculate how much variability there is between scores when we ignore the experimental condition from which they came. Remember from one-way ANOVA (equation (10.4)) that SS_T is calculated using the following equation:

$$SS_T = s^2_{grand}(N-1)$$

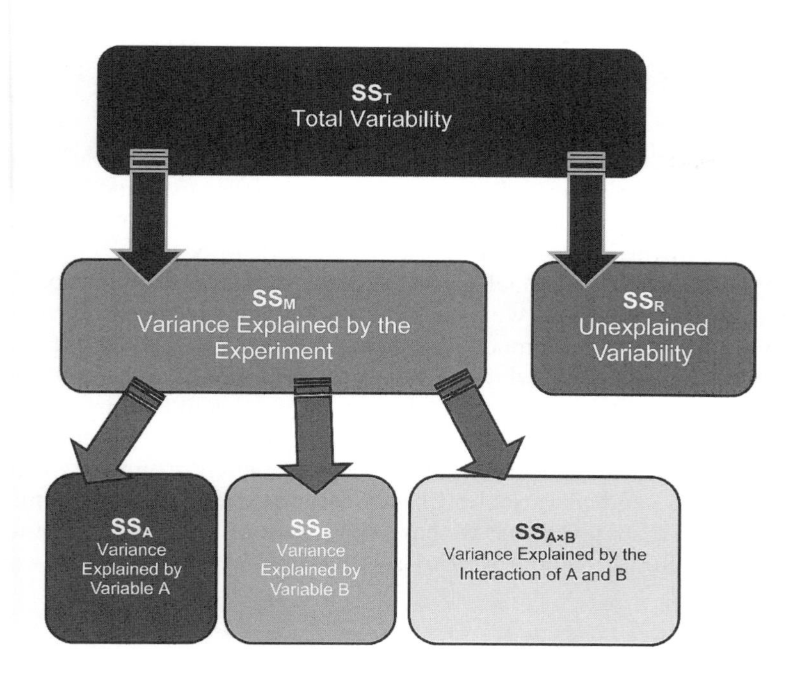

FIGURE 12.2 Breaking down the variance in two-way ANOVA

The grand variance is simply the variance of all scores when we ignore the group to which they belong. So if we treated the data as one big group it would look as follows:

65	50	70	45	55	30
70	55	65	60	65	30
60	80	60	85	70	30
60	65	70	65	55	55
60	70	65	70	55	35
55	75	60	70	60	20
60	75	60	80	50	45
55	65	50	60	50	40
		Grand Mean = 58.33			

If we calculate the variance of all of these scores, we get 190.78 (try this on your calculators if you don't trust me). We used 48 scores to generate this value, and so N is 48. As such the equation becomes:

$$\begin{aligned} SS_T &= s^2_{\text{grand}}(N-1) \\ &= 190.78(48-1) \\ &= 8966.66 \end{aligned}$$

The degrees of freedom for this SS will be $N - 1$, or 47.

12.2.4. The model sum of squares (SS_M) ②

The next step is to work out the model sum of squares. As I suggested earlier, this sum of squares is then further broken into three components: variance explained by the first independent variable (SS_A), variance explained by the second independent variable (SS_B) and variance explained by the interaction of these two variables ($SS_{A \times B}$).

Before we break down the model sum of squares into its component parts, we must first calculate its value. We know we have 8966.66 units of variance to be explained and our first step is to calculate how much of that variance is explained by our experimental manipulations overall (ignoring which of the two independent variables is responsible). When we did one-way ANOVA we worked out the model sum of squares by looking at the difference between each group mean and the overall mean (see section 10.2.6). We can do the same here. We effectively have six experimental groups if we combine all levels of the two independent variables (three doses for the male participants and three doses for the females). So, given that we have six groups of different people we can then apply the equation for the model sum of squares that we used for one-way ANOVA (equation (10.5)):

$$SS_M = \sum n_k \left(\bar{x}_k - \bar{x}_{\text{grand}}\right)^2$$

The grand mean is the mean of all scores (we calculated this above as 58.33) and *nk* is the number of scores in each group (i.e. the number of participants in each of the six experimental groups; eight in this case). Therefore, the equation becomes:

$$\begin{aligned} SS_M &= 8(60.625 - 58.33)^2 + 8(66.875 - 58.33)^2 + 8(62.5 - 58.33)^2 + \ldots \\ &\quad + 8(66.875 - 58.33)^2 + 8(57.5 - 58.33)^2 + 8(35.625 - 58.33)^2 \\ &= 8(2.295)^2 + 8(8.545)^2 + 8(4.17)^2 + 8(8.545)^2 + 8(-0.83)^2 + 8(-22.705)^2 \\ &= 42.1362 + 584.1362 + 139.1112 + 584.1362 + 5.5112 + 4124.1362 \\ &= 5479.167 \end{aligned}$$

The degrees of freedom for this SS will be the number of groups used, k, minus 1. We used six groups and so $df = 5$.

At this stage we know that the model (our experimental manipulations) can explain 5479.167 units of variance out of the total of 8966.66 units. The next stage is to further break down this model sum of squares to see how much variance is explained by our independent variables separately.

12.2.4.1. The main effect of gender (SS_A) ②

To work out the variance accounted for by the first independent variable (in this case, gender) we need to group the scores in the data set according to which gender they belong to. So, basically we ignore the amount of drink that has been drunk, and we just place all of the male scores into one group and all of the female scores into another. So, the data will look like this (note that the first box contains the three female columns from our original table and the second box contains the male columns):

A_1: Female		
65	70	55
70	65	65
60	60	70
60	70	55
60	65	55
55	60	60
60	60	50
55	50	50
Mean Female = 60.21		

A_2: Male		
50	45	30
55	60	30
80	85	30
65	65	55
70	70	35
75	70	20
75	80	45
65	60	40
Mean Male = 56.46		

We can then apply the equation for the model sum of squares that we used to calculate the overall model sum of squares:

$$SS_A = \sum n_k (\bar{x}_k - \bar{x}_{grand})^2$$

The grand mean is the mean of all scores (above) and n is the number of scores in each group (i.e. the number of males and females; 24 in this case). Therefore, the equation becomes:

$$\begin{aligned} SS_{Gender} &= 24(60.21 - 58.33)^2 + 24(56.46 - 58.33)^2 \\ &= 24(1.88)^2 + 24(-1.87)^2 \\ &= 84.8256 + 83.9256 \\ &= 168.75 \end{aligned}$$

The degrees of freedom for this SS will be the number of groups used, k, minus 1. We used two groups (males and females) and so $df = 1$. To sum up, the main effect of gender compares the mean of all males against the mean of all females (regardless of which alcohol group they were in).

12.2.4.2. The main effect of alcohol (SS_B) ②

To work out the variance accounted for by the second independent variable (in this case, alcohol) we need to group the scores in the data set according to how much alcohol was consumed. So, basically we ignore the gender of the participant, and we just place all of the scores after no drinks in one group, the scores after 2 pints in another group and the scores after 4 pints in a third group. So, the data will look like this:

B_1: None	
65	50
70	55
60	80
60	65
60	70
55	75
60	75
55	65
Mean None = 63.75	

B_2: 2 Pints	
70	45
65	60
60	85
70	65
65	70
60	70
60	80
50	60
Mean 2 Pints = 64.6875	

B_3: 4 Pints	
55	30
65	30
70	30
55	55
55	35
60	20
50	45
50	40
Mean 4 Pints = 46.5625	

We can then apply the same equation for the model sum of squares that we used for the overall model sum of squares and for the main effect of gender:

$$SS_B = \sum n_k (\bar{x}_k - \bar{x}_{grand})^2$$

The grand mean is the mean of all scores (58.33 as before) and n is the number of scores in each group (i.e. the number of scores in each of the boxes above, in this case 16). Therefore, the equation becomes:

$$\begin{aligned} SS_{alcohol} &= 16(63.75 - 58.33)^2 + 16(64.6875 - 58.33)^2 + 16(46.5625 - 58.33)^2 \\ &= 16(5.42)^2 + 16(6.3575)^2 + 16(-11.7675)^2 \\ &= 470.0224 + 646.6849 + 2215.5849 \\ &= 3332.292 \end{aligned}$$

The degrees of freedom for this SS will be the number of groups used minus 1 (see section 10.2.6). We used three groups and so $df = 2$. To sum up, the main effect of alcohol compares the means of the no alcohol, 2 pint and 4 pint groups (regardless of whether the scores come from men or women).

12.2.4.3. The interaction effect ($SS_{A \times B}$) ②

The final stage is to calculate how much variance is explained by the **interaction** of the two variables. The simplest way to do this is to remember that the SS_M is made up of three components (SS_A, SS_B and $SS_{A \times B}$). Therefore, given that we know SS_A and SS_B we can calculate the interaction term using subtraction:

$$SS_{A \times B} = SS_M - SS_A - SS_B$$

Therefore, for these data, the value is:

$$\begin{aligned} SS_{A \times B} &= SS_M - SS_A - SS_B \\ &= 5479.167 - 168.75 - 3322.292 \\ &= 1978.125 \end{aligned}$$

The degrees of freedom can be calculated in the same way, but are also the product of the degrees of freedom for the main effects (either method works):

$$\begin{aligned} df_{A \times B} &= df_M - df_A - df_B \\ &= 5 - 1 - 2 \\ &= 2 \end{aligned} \qquad \begin{aligned} df_{A \times B} &= df_A \times df_B \\ &= 1 \times 2 \\ &= 2 \end{aligned}$$

12.2.5. The residual sum of squares (SS_R) ②

The residual sum of squares is calculated in the same way as for one-way ANOVA (see section 10.2.7) and again represents individual differences in performance or the variance that can't be explained by factors that were systematically manipulated. We saw in one-way

ANOVA that the value is calculated by taking the squared error between each data point and its corresponding group mean. An alternative way to express this was as (see equation (10.7)):

$$SS_R = s^2_{\text{group 1}}(n_1 - 1) + s^2_{\text{group 2}}(n_2 - 1) + s^2_{\text{group 3}}(n_3 - 1) + \ldots + s^2_{\text{group }n}(n_n - 1)$$

So, we use the individual variances of each group and multiply them by one less than the number of people within the group (n). We have the individual group variances in our original table of data (Table 12.1) and there were eight people in each group (therefore, $n = 8$) and so the equation becomes:

$$\begin{aligned}
SS_R &= s^2_{\text{group 1}}(n_1 - 1) + s^2_{\text{group 2}}(n_2 - 1) + s^2_{\text{group 3}}(n_3 - 1) + s^2_{\text{group 4}}(n_4 - 1) + \ldots \\
&\quad + s^2_{\text{group 5}}(n_5 - 1) + s^2_{\text{group 6}}(n_6 - 1) \\
&= (24.55)(8-1) + (106.7)(8-1) + (42.86)(8-1) + (156.7)(8-1) + \cdots \\
&\quad + (50)(8-1) + (117.41)(8-1) \\
&= (24.55 \times 7) + (106.7 \times 7) + (42.86 \times 7) + (156.7 \times 7) + (50 \times 7) + \cdots \\
&\quad + (117.41 \times 7) \\
&= 171.85 + 746.9 + 300 + 1096.9 + 350 + 821.87 \\
&= 3487.52
\end{aligned}$$

The degrees of freedom for each group will be one less than the number of scores per group (i.e. 7). Therefore, if we add the degrees of freedom for each group, we get a total of $6 \times 7 = 42$.

12.2.6. The *F*-ratios ②

Each effect in a two-way ANOVA (the two main effects and the interaction) has its own *F*-ratio. To calculate these we have to first calculate the mean squares for each effect by taking the sum of squares and dividing by the respective degrees of freedom (think back to section 10.2.8). We also need the mean squares for the residual term. So, for this example we'd have four mean squares calculated as follows:

$$\begin{aligned}
MS_A &= \frac{SS_A}{df_A} = \frac{168.75}{1} = 168.75 \\
MS_B &= \frac{SS_B}{df_B} = \frac{3332.292}{2} = 1666.146 \\
MS_{A \times B} &= \frac{SS_{A \times B}}{df_{A \times B}} = \frac{1978.125}{2} = 989.062 \\
MS_R &= \frac{SS_R}{df_R} = \frac{3487.52}{42} = 83.036
\end{aligned}$$

The *F*-ratios for the two independent variables and their interactions are then calculated by dividing their mean squares by the residual mean squares. Again, if you think back to one-way ANOVA this is exactly the same process!

$$F_A = \frac{MS_A}{MS_R} = \frac{168.75}{83.036} = 2.032$$

$$F_B = \frac{MS_B}{MS_R} = \frac{1666.146}{83.036} = 20.065$$

$$F_{A \times B} = \frac{MS_{A \times B}}{MS_R} = \frac{989.062}{83.036} = 11.911$$

Each of these *F*-ratios can be compared against critical values (based on their degrees of freedom, which can be different for each effect) to tell us whether these effects are likely to reflect data that have arisen by chance, or reflect an effect of our experimental manipulations (these critical values can be found in the Appendix). If an observed *F* exceeds the corresponding critical values then it is significant. SAS will calculate these *F*-ratios and exact significance for each, but what I hope to have shown you in this section is that two-way ANOVA is basically the same as one-way ANOVA except that the model sum of squares is partitioned into three parts: the effect of each of the independent variables and the effect of how these variables interact.

12.3. Factorial ANOVA using SAS ②

12.3.1. Exploring the data: PROC MEANS ②

Before jumping into an ANOVA, it is a good idea to explore your data, to make sure nothing too strange is going on, and to give some idea of what you are looking for. The easiest way to do this is using PROC MEANS. SAS Syntax 12.1 shows the syntax to do this, and SAS Output 12.1 shows the output. Looking at the output, the values all look reasonable. The standard deviations are all similar to each other, and the minimum and maximum scores are within acceptable bounds (no one has scored –73, or 3846).

```
PROC MEANS DATA=chap12.goggles;
      CLASS gender alcohol;
      VAR attract;
      RUN;
```
SAS Syntax 12.1

SAS OUTPUT 12.1

Analysis Variable : ATTRACT Attractiveness of Date							
gender	alcohol	N Obs	N	Mean	Std Dev	Minimum	Maximum
female	2 pints	8	8	62.5000000	6.5465367	50.0000000	70.0000000
	4 pints	8	8	57.5000000	7.0710678	50.0000000	70.0000000
	no alcohol	8	8	60.6250000	4.9551560	55.0000000	70.0000000
male	2 pints	8	8	66.8750000	12.5178444	45.0000000	85.0000000
	4 pints	8	8	35.6250000	10.8356225	20.0000000	55.0000000
	no alcohol	8	8	66.8750000	10.3293963	50.0000000	80.0000000

Factorial ANOVA is carried out using PROC GLM, and it looks very like the ANOVA that we met in Chapter 10. All we need to do is put more independent variables on the MODEL line. In a factorial ANOVA, we are also interested in the interaction effects between the variables, and so we should specify them as well. The interactions can be expressed in two ways: the first is to specify them as we have been doing so far, as shown in SAS Syntax 12.2.

```
PROC GLM DATA=chap12.goggles;
      CLASS alcohol gender;
      MODEL attract= gender alcohol gender * alcohol;
      RUN;
```

SAS Syntax 12.2

This way of specifying interactions is fine, until we happen to have more than a couple of independent variables. If we have two independent variables (we'll call them **a** and **b**) we just need to write:

```
a·b·a·*·b
```

If we have three, then we have:

```
a b c a*b a*c b*c a*b*c
```

And as we add more, it gets more and more complicated (and easier and easier to make a mistake). Instead, SAS lets us use the | symbol, which means 'put these variables in, and all of their interactions'. So we can write:

```
a | b
```

or

```
a | b | c
```

or

```
a | b | c | d | e | f
```

Although I would try quite hard to discourage you from trying the last one.

12.3.2. Contrasts and estimates ②

We saw in Chapter 10 that it's useful to follow up ANOVA with contrasts that break down the main effects and tell us where the differences between groups lie. When we have two predictor variables, we do the contrasts in much the same way – however, you should be aware before you continue that contrasts in PROC GLM are really rather fiddly. In fact, if you don't really, really want the specific contrast, you should try to do it a different way (which we'll look at in the next couple of sections). But if you are feeling strong, here we go.

We looked at contrasts when we had one categorical predictor, in section 10.2.11. Alcohol has three categories, and when we place them in alphabetical order, they are '2 pints', '4 pints', 'no alcohol'. We set the weights for the contrast in the same way as before. To compare the no alcohol group with the two alcohol groups, we use:

ESTIMATE "No alcohol vs alcohol" alcohol 1 1 -2;

If we want to compare the two alcohol groups, we use:

ESTIMATE "2 pints vs 4 pints" alcohol 1 -1 0;

However, we might want to do more complex contrasts involving the interaction. We can compare the two pint group with the four pint group, in the males only. To do this more complex contrast (or estimate) we need to think about the six groups that we have. The order of the six groups is determined by the order that we put them in the CLASS statement. We put alcohol first, and so the order goes:

2 pints female
2 pints male
4 pints female
4 pints male
no alcohol female
no alcohol male

To compare the males with the females when both groups had consumed four pints, we add an ESTIMATE statement with two sections. The first part tells SAS that we want to compare the males with the females, the second part which cells we want to compare:

```
ESTIMATE "Gender difference for 4 pints"
         gender -1 1
         alcohol*gender 0 0 -1 1 0 0 ;
```

To compare the females who had 4 pints with the females who had no alcohol, we would use:

```
ESTIMATE 'females 4 pints vs no alcohol'
      alcohol 0 1 -1
      gender*alcohol 0 0 1 0 -1 0;
```

It's very, very easy to get these wrong, and you need to be very, very careful when you use contrasts or estimates, to make sure that you are contrasting (or estimating) what you think you are contrasting (or estimating).

Let's put those into our syntax (SAS Syntax 12.3).

```
PROC GLM DATA=chap12.goggles;
      CLASS  alcohol gender;
      MODEL attract= gender alcohol gender * alcohol ;
      ESTIMATE "No alcohol vs alcohol" alcohol  1 1 -2;
      ESTIMATE "2 pints vs 4 pints " alcohol  1 -1 0;
      ESTIMATE  "Gender difference for 4 pints"
            gender  -1 1
            alcohol*gender 0 0 -1 1  0  0 ;
      ESTIMATE 'females 4 pints vs no alcohol'
            alcohol 0 1 -1
            gender*alcohol 0 0 1 0 -1 0;
      RUN; QUIT;
```

SAS Syntax 12.3

12.3.3. *Post hoc* tests ②

We can do *post hoc* tests using the LSMEANS statement, just as we did with ANCOVA. However, as you might expect, things are now a little bit more complex, because there are several different sorts of *post hoc* comparisons that can be made. As before, we can use a correction for multiple comparisons.

First, we can compare the mean of each group on one independent variable at at time. So, to compare the males with the females we would use:

```
LSMEANS gender  / ADJUST=sidak ;
```

To compare the three levels of alcohol, we can use:

```
LSMEANS alcohol / ADJUST=sidak ;
```

We can also compare all six different groups (male no alcohol; male 2 pints; male 4 pints; female no alcohol; female 2 pints; female 4 pints), using a *post hoc* test. We do this by putting an interaction on the LSMEANS line:

```
LSMEANS alcohol*gender / ADJUST=sidak ;
```

12.3.4. Simple effects ②

You might be thinking that nothing in PROC GLM might be described as a 'simple effect.' But you'd be wrong. A simple effect is when we look at the effect of one independent variable, in only one group of the other independent variable. For example, we might ask whether there is a statistically significant effect of alcohol in males only.

In SAS, we do this by slicing the data. We can 'slice' the data by one variable, and do comparisons of groups within those slices. For example, if we wanted to test whether males and females differed for each level of alcohol, we would put:

```
LSMEANS gender * alcohol / slice = alcohol ADJUST=SIDAK ;
```

If we want to test the effect of alcohol for the two different genders, we could put:

```
LSMEANS gender *alcohol / slice = gender ADJUST=SIDAK;
```

Our syntax is now as shown in SAS Syntax 12.4.

```
PROC GLM DATA=chap12.goggles;
      CLASS  alcohol gender;
      MODEL attract= gender alcohol gender * alcohol ;
      MODEL attract= gender alcohol gender * alcohol ;
      ESTIMATE "No alcohol vs alcohol" alcohol  1 1 -2;
      ESTIMATE "2 pints vs 4 pints " alcohol  1 -1 0;
      ESTIMATE  "Gender difference for 4 pints"
            gender  -1 1
            alcohol*gender 0 0 -1 1  0  0 ;
      ESTIMATE 'females 4 pints vs no alcohol'
            alcohol 0 1 -1
            gender*alcohol  0  0  1  0  -1  0;  LSMEANS  gender  /
            ADJUST=SIDAK ;
      LSMEANS alcohol / ADJUST=sidak ;
      LSMEANS alcohol*gender / ADJUST=SIDAK ;
      LSMEANS gender *alcohol / SLICE = alcohol;
      LSMEANS gender *alcohol / SLICE = gender;
      RUN; QUIT;
```

SAS Syntax 12.4.

12.3.5. ODS graphics

If we add a request for ODS graphics, SAS will automatically produce several charts for us (some of which are useful, and some of which are less useful). Before you run the PROC GLM syntax, run: ODS GRAPHICS ON;

12.4. Output from factorial ANOVA ②

12.4.1. The main ANOVA tables ②

SAS Output 12.2 is the most important part of the output because it tells us whether any of the independent variables have had an effect on the dependent variable. The first table tells us about the overall significance of the model – the *F* value is 13.2, with 5, 42 df, and this is highly significant, $p < .0001$, and so our model is doing better than chance. The second table tells us about R^2, which is 0.61, and is therefore large. The important things to look at in the third table are the significance values of the independent variables. The first thing to notice is that there is a significant main effect of alcohol (because the significance value is less than .05). The *F*-ratio is highly significant, indicating that the amount of alcohol consumed significantly affected who the participant would try to chat up. This means that overall, when we ignore whether the participant was male or female, the amount of alcohol influenced their mate selection. The best way to see what this means is to look at a bar chart of the average attractiveness at each level of alcohol (ignore gender completely). (remember to look at the type III SS, not the type I SS).

SELF-TEST Plot error bar graphs of the main effects of alcohol and gender.

Source	DF	Sum of Squares	Mean Square	F Value	Pr > F
Model	5	5479.166667	1095.833333	13.20	<.0001
Error	42	3487.500000	83.035714		
Corrected Total	47	8966.666667			

R-Square	Coeff Var	Root MSE	ATTRACT Mean
0.611059	15.62125	9.112393	58.33333

Source	DF	Type I SS	Mean Square	F Value	Pr > F
gender	1	168.750000	168.750000	2.03	0.1614
alcohol	2	3332.291667	1666.145833	20.07	<.0001
alcohol*gender	2	1978.125000	989.062500	11.91	<.0001

SAS OUTPUT 12.2

Source	DF	Type III SS	Mean Square	F Value	Pr > F
gender	1	168.750000	168.750000	2.03	0.1614
alcohol	2	3332.291667	1666.145833	20.07	<.0001
alcohol*gender	2	1978.125000	989.062500	11.91	<.0001

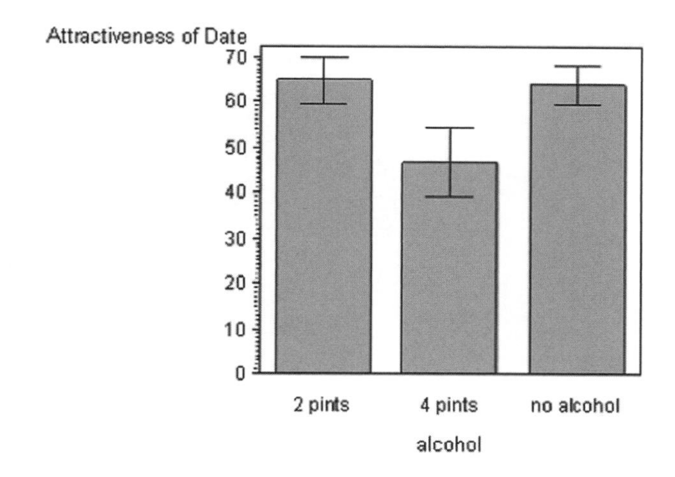

FIGURE 12.3 Graph showing the main effect of alcohol

Figure 12.3 clearly shows that when you ignore gender the overall attractiveness of the selected mate is very similar when no alcohol has been drunk and when 2 pints have been drunk (the means of these groups are approximately equal). Hence, this significant main effect is *likely* to reflect the drop in the attractiveness of the selected mates when 4 pints have been drunk. This finding seems to indicate that a person is willing to accept a less attractive mate after 4 pints.

SAS Output 12.2 also tells us about the main effect of gender. This time the *F*-ratio is not significant (p = .161, which is larger than .05). This effect means that overall, when we ignore how much alcohol had been drunk, the gender of the participant did not influence the attractiveness of the partner that the participant selected. In other words, other things being equal, males and females selected equally attractive mates. The bar chart (that you have hopefully produced from the self-test) of the average attractiveness of mates for men and women (ignoring how much alcohol had been consumed) reveals the meaning of this main effect. Figure 12.4 plots the means that we calculated in section 12.2.4.1. This graph shows that the average attractiveness of the partners of male and female participants was fairly similar. Therefore, this non-significant effect reflects the fact that the mean attractiveness was similar. We can conclude from this that, *other things being equal,* men and women chose equally attractive partners.

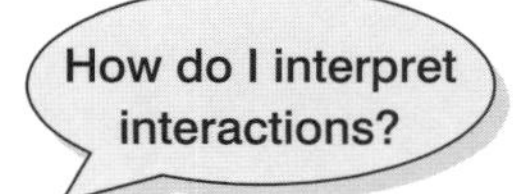

Finally, SAS Output 12.2 tells us about the interaction between the effect of gender and the effect of alcohol. The *F*-value is highly significant (because the *p*-value is less than .05). What this actually means is that the effect of alcohol on mate selection was different for male participants than it was for females.

Figure 12.5 was produced by the ODS GRAPHICS ON statement, and clearly shows that for women, alcohol has very little effect: the attractiveness of their selected partners is quite stable across the three conditions (as shown by the near-horizontal line). However, for the men, the

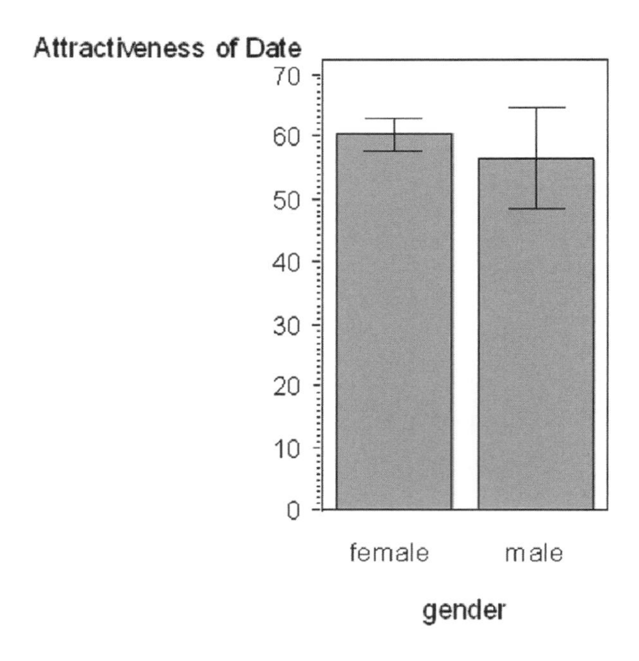

FIGURE 12.4 Graph to show the main effect of gender on mate selection

attractiveness of their partners is stable when only a small amount has been drunk, but rapidly declines when more is drunk. Non-parallel lines often indicate a significant interaction effect. In this particular graph the lines actually cross, which indicates a fairly large interaction between independent variables. The interaction tells us that alcohol has little effect on mate selection until 4 pints have been drunk and that the effect of alcohol is prevalent only in male participants. In short, the results show that women maintain high standards in their mate selection regardless of alcohol, whereas men have a few beers and then try to get off with anything on legs. One interesting point that these data demonstrate is that we earlier concluded that alcohol significantly affected how attractive a mate was selected (the **Alcohol** main effect); however, the interaction effect tells us that this is true only in males (females appear unaffected). This shows how misleading main effects can be: it is usually the interactions between variables that are most interesting in a factorial design.

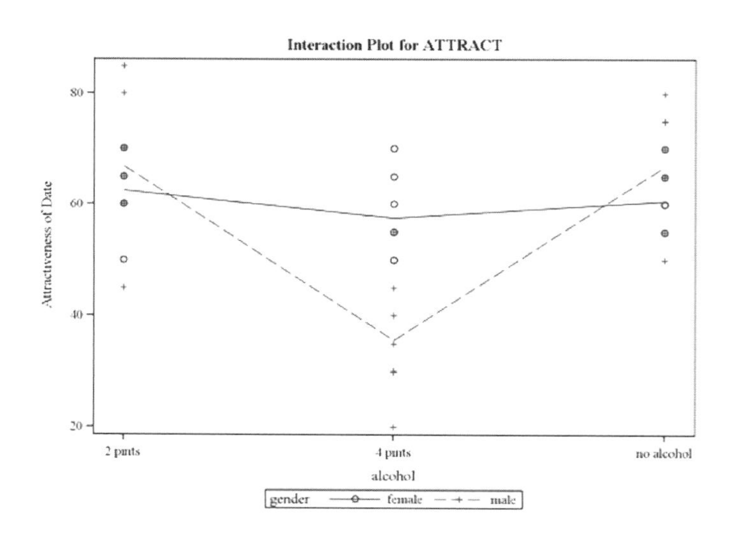

FIGURE 12.5 Graph of the interaction of gender and alcohol consumption in mate selection

12.4.2. Contrasts

SAS Output 12.3 shows the results of our contrast estimates. This helps us to break down the effect of alcohol. The first row of the table shows the contrast for No alcohol vs alcohol, which in this case means the no alcohol group compared to the two alcohol groups. This tests whether the mean of the no alcohol group (63.75) is different to the mean of the 2 pint and 4 pint groups combined ((64.69 + 46.56)/2 = 55.625). This is a difference of 8.125 (63.75 − 55.63). The important thing to look at is the value of the significance test, Pr > |*t*|, which tells us if this difference is significant. It is, because this value is .006, which is smaller than .05. So we could conclude that the effect of alcohol is that any amount of alcohol reduces the attractiveness of the dates selected compared to when no alcohol is drunk. Of course this is misleading because, in fact, the means for the no alcohol and 2 pint groups are fairly similar (63.75 and 64.69), so 2 pints of alcohol don't reduce the attractiveness of selected dates! The reason why the comparison is significant is because it's testing the combined effect of 2 and 4 pints, and because 4 pints has such a drastic effect it drags down the overall mean. This shows why you need to be careful about how you interpret these contrasts: you need to have a look at the remaining contrast as well.

The second row of the table shows the contrast for 2 pints vs 4 pints – the contrast for the 2 pint group compared with the 4 pint group. This tests whether the mean of the 2 pint groups (64.69) is different to the mean of the 4 pint groups (46.56). This is a difference of 18.13 (64.69 − 46.56). Again, the important thing to look at is the value of Pr > |*t*| which tells us if this difference is significant. It is, because this value is <0.0001, which is smaller than .05. This tells us that having 4 pints significantly reduced the attractiveness of selected dates compared to having only 2 pints.

The next contrast broke peered a little further into the results. We compare the mean scores for the males and females who had 4 pints. The males who had consumed 4 pints had a mean score of 35.625, the females who had consumed 4 pints had a considerable higher mean score, of 57.5. The difference between these two scores is –21.875, and this difference is highly statistically significant, $p < 0.0001$.

The final row of the table compares the females who had 4 pints (mean 57.5) with the females who had no alcohol (mean 60.625). The difference between these two means is –3.125, and when we look at the Pr > |*t*| value, we can see that it is 0.497, higher than 0.05 and therefore this difference is not statistically significant.

SAS OUTPUT 12.3

Parameter	Estimate	Standard Error	t Value	Pr > |t|
No alcohol vs alcohol	–8.1250000	2.79008928	–2.91	0.0057
2 pints vs 4 pints	18.1250000	3.22171760	5.63	<.0001
Gender difference for 4 pints	–21.8750000	4.55619672	–4.80	<.0001
females 4 pints vs no alcohol	–3.1250000	4.55619672	–0.69	0.4966

12.4.3. Least squares means and *post hoc* analysis ②

The *post hoc* tests (see section 10.2.12) are requested with the LSMEANS statement and the ADJUST option. These tests break down the main effect of alcohol and can be interpreted as if a one-way ANOVA had been conducted on the **Gender** variable, the **Alcohol** variable (i.e. the reported effects for alcohol are collapsed with regard to gender), or each combination is treated as a separate group and compared. Several corrections for multiple testing are available, including Šidák, Tukey and Bonferroni. We used the Šidák correction for each of these *post hoc* comparisons.

The first comparison, shown in SAS Output 12.4 compares the mean for males with the mean for females. The overall mean difference is not statistically significant – however, we knew that already – because there are only two groups (males and females) there is only one comparison, and no correction takes place. The *p*-value is therefore exactly the same as the *p*-value for the effect of gender that we saw in SAS Output 12.2.

SAS OUTPUT 12.4

gender	ATTRACT LSMEAN	H0: LSMean1=LSMean2 Pr > |t|
female	60.2083333	0.1614
male	56.4583333	

The second *post hoc* test we requested was for a comparison amongst the alcohol groups. SAS Output 12.5 shows the means and *p*-values for three comparisons However, after 4 pints had been consumed, participants selected significantly less attractive mates than after both 2 pints ($p < .0001$) and no alcohol ($p < .0001$). It is interesting to note that the mean attractiveness of partners after no alcohol and 2 pints were so similar that the probability of the obtained difference between those means is very nearly 1. The test confirms that the means of the placebo and 2 pint conditions were not different, whereas the mean of the 4 pint group was different. It should be noted that these *post hoc* tests ignore the interactive effect of gender and alcohol.

SAS OUTPUT 12.5

alcohol	ATTRACT LSMEAN	LSMEAN Number
2 pints	64.6875000	1
4 pints	46.5625000	2
no alcohol	63.7500000	3

Least Squares Means for effect alcohol Pr > |t| for H0: LSMean(i)=LSMean(j) Dependent Variable: ATTRACT			
i/j	1	2	3
1		<.0001	0.9882
2	<.0001		<.0001
3	0.9882	<.0001	

SELF-TEST We could have compared the no alcohol group and the 2 pint group with a contrast or estimate statement, rather than a *post hoc* test. If we did that, would the *p*-value be higher, lower, or the same? (Try it.)

The final set of *post hoc* tests is shown in SAS Output 12.6. This compares each of the six groups defined by the two independent variables with each of the other groups. The groups are numbered, and their means are given in the first table, and then the second table presents the *p*-values for the differences between them. Looking at this table, we can see that all of the significant effects involve group 4 – the males who drank 4 pints. In fact, this group is significantly different from every other group, and no other groups are significantly different from one another.

SAS OUTPUT 12.6

alcohol	gender	ATTRACT LSMEAN	LSMEAN Number
2 pints	female	62.5000000	1
2 pints	male	66.8750000	2
4 pints	female	57.5000000	3
4 pints	male	35.6250000	4
no alcohol	female	60.6250000	5
no alcohol	male	66.8750000	6

Least Squares Means for effect alcohol*gender
Pr > |t| for H0: LSMean(i)=LSMean(j)

Dependent Variable: ATTRACT

i/j	1	2	3	4	5	6
1		0.9981	0.9926	<.0001	1.0000	0.9981
2	0.9981		0.5055	<.0001	0.9466	1.0000
3	0.9926	0.5055		0.0003	1.0000	0.5055
4	<.0001	<.0001	0.0003		<.0001	<.0001
5	1.0000	0.9466	1.0000	<.0001		0.9466
6	0.9981	1.0000	0.5055	<.0001	0.9466	

12.4.4. Simple effects

The final type of comparison that we can make is to look at simple effects, by 'slicing' the data with the LSMEANS command. SAS Output 12.7 shows the results of the analysis that we requested with the two LSMEANS statements and the SLICE option.

The first table shows the effect of gender, at different levels of alcohol. By now it shouldn't really surprise us to see that for those who had no alcohol, and for those who had a low dose of alcohol, there is no statistically significant effect of gender. However, for those who had 4 pints, there is a highly significant effect of gender.

The second table shows us the effect of alcohol, sliced by gender. For the females, the effect of alcohol is not statistically significant, $p = 0.342$. For males, though, the effect is highly significant, $p < 0.0001$.

SAS OUTPUT 12.7

alcohol*gender Effect Sliced by alcohol for ATTRACT					
alcohol	DF	Sum of Squares	Mean Square	F Value	Pr > F
2 pints	1	76.562500	76.562500	0.92	0.3424
4 pints	1	1914.062500	1914.062500	23.05	<.0001
no alcohol	1	156.250000	156.250000	1.88	0.1774

alcohol*gender Effect Sliced by gender for ATTRACT					
gender	DF	Sum of Squares	Mean Square	F Value	Pr > F
female	2	102.083333	51.041667	0.61	0.5456
male	2	5208.333333	2604.166667	31.36	<.0001

12.4.5. Summary

In summary, we should conclude that alcohol has an effect on the attractiveness of selected mates. Overall, after a relatively small dose of alcohol (2 pints) humans are still in control of their judgements and the attractiveness levels of chosen partners are consistent with a control group (no alcohol consumed). However, after a greater dose of alcohol, the attractiveness of chosen mates decreases significantly. This effect is what is referred to as the 'beer-goggles effect'. More interesting, the interaction shows a gender difference in the beer-goggles effect. Specifically, it looks as though men are significantly more likely to pick less attractive mates when drunk. Women, in comparison, manage to maintain their standards despite being drunk. What we still don't know is whether women will become susceptible to the beer-goggles effect at higher doses of alcohol.

CRAMMING SAM'S TIPS

- Two-way independent ANOVA compares several means when there are two independent variables and different participants have been used in all experimental conditions. For example, if you wanted to know whether different teaching methods worked better for different subjects, you could take students from four courses (Psychology, Geography, Management and Statistics) and assign them to either lecture-based or book-based teaching. The two variables are course and method of teaching. The outcome might be the end of year mark (as a percentage).
- In the overall ANOVA tables labelled, look at the column labelled *Pr > F* for the overall model, and for all three of your effects: there should be a main effect of each variable and an effect of the interaction between the two variables; if the value is less than .05 then the effect is significant. For main effects consult *post hoc* tests to see which groups differ, and for the interaction look at an interaction graph or conduct a simple effects analysis.
- For *post hoc* tests, again look at the columns labelled $Pr > |t|$ to discover if your comparisons are significant (they will be if the significance value is less than .05).
- Choose planned comparisons and simple effects to analyse, to help you to understand interaction effects.
- Test the same assumptions as for one-way independent ANOVA (see Chapter 10).

12.5. Interpreting interaction graphs ②

We've already had a look at one interaction graph when we interpreted the analysis in this chapter. However, interactions are very important, and the key to understanding them is being able to interpret interaction graphs. In the example in this chapter we used Figure 12.5 to conclude that the interaction probably reflected the fact that men and women chose equally attractive dates after no alcohol and 2 pints, but that at 4 pints men's standards dropped significantly more than women's. Imagine we'd got the profile of results shown in Figure 12.6; do you think we would've still got a significant interaction effect?

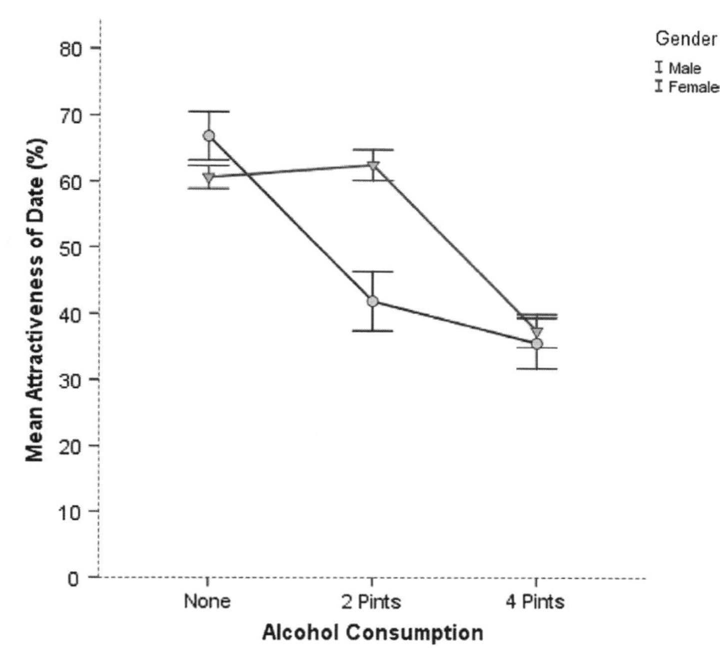

FIGURE 12.6 Another interaction graph

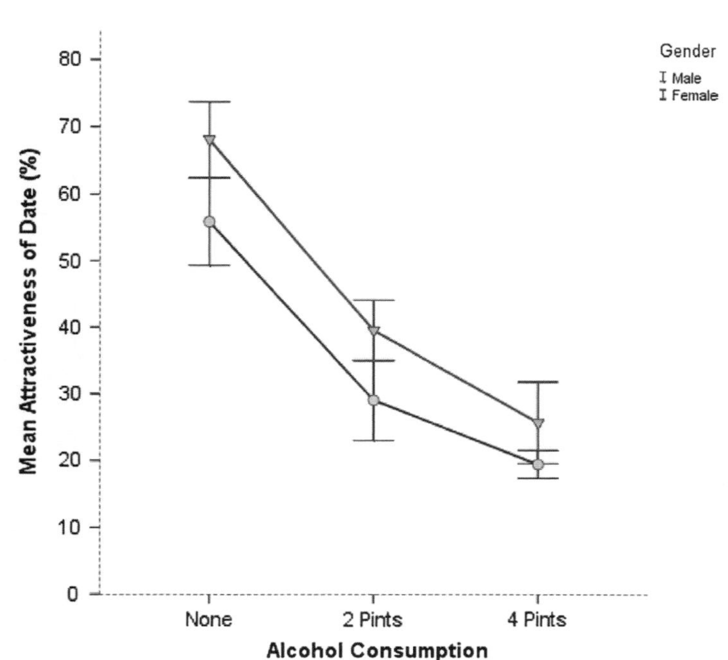

FIGURE 12.7 A 'lack of' interaction graph

This profile of data probably would also give rise to a significant interaction term because, although the attractiveness of men and women's dates is similar after no alcohol and 4 pints of alcohol, there is a big difference after 2 pints. This reflects a scenario in which the beer goggles effect is equally big in men and women after 4 pints (and doesn't exist after no alcohol) but kicks in quicker for men: the attractiveness of their dates plummets after 2 pints, whereas women maintain their standards until 4 pints (at which point they'd happily date an unwashed skunk). Let's try another example. Is there a significant interaction in Figure 12.7?

For the data in Figure 12.7 there is unlikely to be a significant interaction because the effect of alcohol is the same for men and women. So, for both men and women, the attractiveness of their dates after no alcohol is quite high, but after 2 pints all types drop by a similar amount (the slope of the male and female lines is about the same). After 4 pints there is a further drop and, again, this drop is about the same in men and women

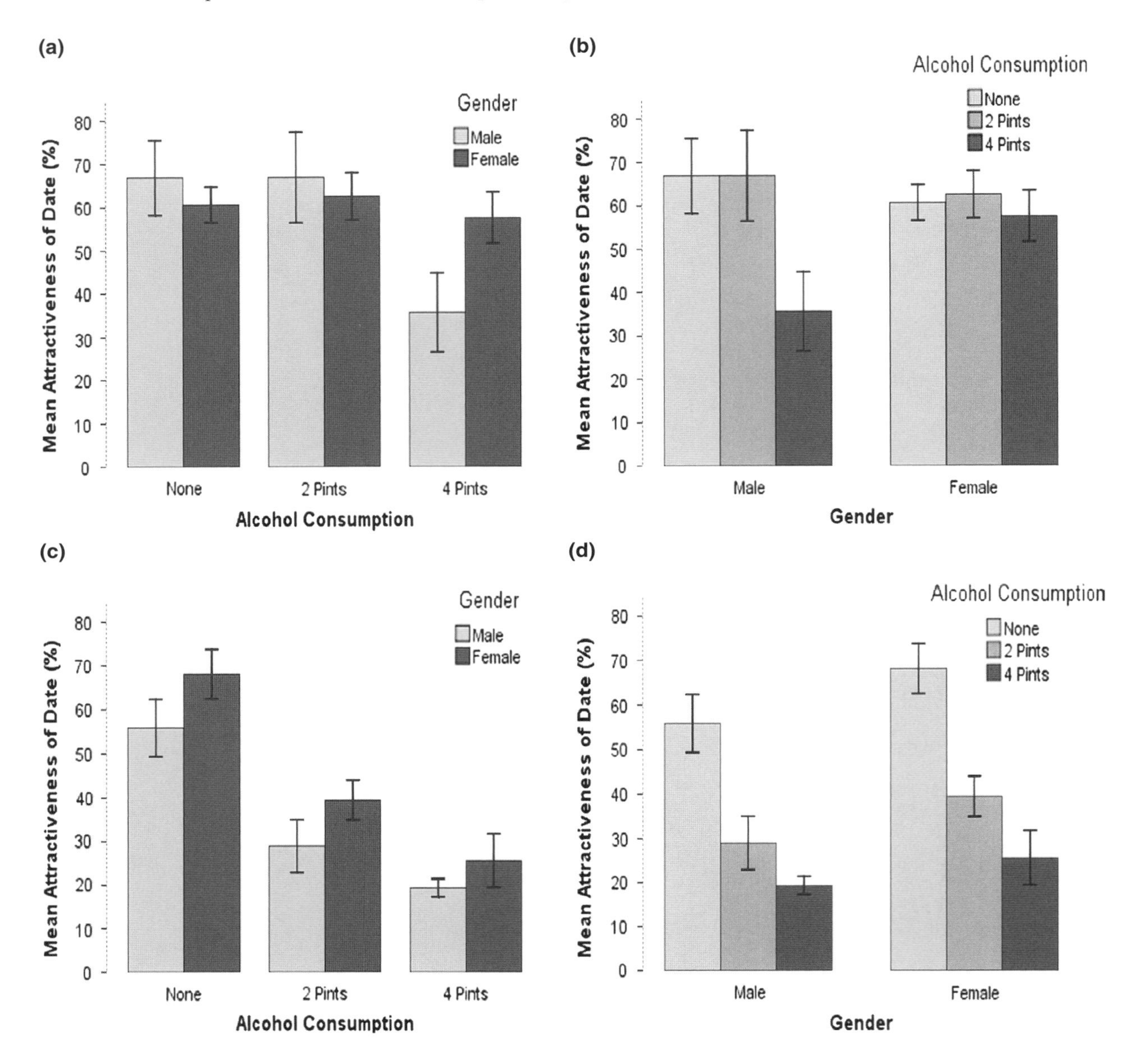

FIGURE 12.8 Bar charts showing interactions between two variables

(the lines again slope at about the same angle). The fact that the line for males is lower than for females just reflects the fact that across all conditions, men have lower standards than their female counterparts: this reflects a main effect of gender (i.e. males generally chose less attractive dates than females at all levels of alcohol). Two general points that we can make from these examples are that:

- Significant interactions are shown up by non-parallel lines on an interaction graph. However, it's important to remember that this doesn't mean that non-parallel lines automatically mean that the interaction is significant: whether the interaction is significant will depend on the degree to which the lines are not parallel!
- If the lines on an interaction graph cross then obviously they are not parallel and this can be a dead giveaway that you have a possible significant interaction. However, contrary to popular belief it isn't *always* the case that if the lines of the interaction graph cross then the interaction is significant.

A further complication is that sometimes people draw bar charts rather than line charts. Figure 12.8 shows some bar charts of interactions between two independent variables. Panels (a) and (b) actually display the data from the example used in this chapter (in fact, why not have a go at plotting them!). As you can see, there are two ways to present the same data: panel (a) shows the data when levels of alcohol are placed along the x-axis and different-coloured bars are used to show means for males and females, and panel (b) shows the opposite scenario where gender is plotted on the x-axis and different colours distinguish the dose of alcohol. Both of these graphs show an interaction effect. What you're looking for is the differences between coloured bars to be different at different points along the x-axis.

So, for panel (a) you'd look at the difference between the light and dark blue bars for no alcohol, and then look to 2 pints and ask, 'Is the difference between the bars different to when I looked at no alcohol?' In this case the dark and light bars look the same at no alcohol as they do at 2 pints: hence, no interaction. However, we'd then move on to look at 4 pints, and we'd again ask, 'Is the difference between the light and dark bars different to what it has been in any of the other conditions?' In this case the answer is yes: for no alcohol and 2 pints, the light and dark bars were about the same height, but at 4 pints the dark bar is much higher than the light one. This shows an interaction: the pattern of responses changes at 4 pints. Panel (b) shows the same thing but plotted the other way around. Again we look at the pattern of responses. So, first we look at the men and see that the pattern is that the first two bars are the same height, but the last bar is much shorter. The interaction effect is shown up by the fact that for the women there is a different pattern: all three bars are about the same height.

SELF-TEST What about panels (c) and (d): do you think there is an interaction?

Again, they display the same data in two different ways, but it's different data to what we've used in this chapter. First let's look at panel (c): for the no alcohol data, the dark bar is a little bit bigger than the light one; moving on to the 2 pint data, the dark bar is also a little bit taller than the light bar; and finally for the 4 pint data, the dark bar is again higher than the light one. In all conditions the same pattern is shown – the dark bar is a bit higher than the light one (i.e. females pick more attractive dates than men regardless of alcohol consumption) – therefore, there is no interaction. Looking at panel (d) we see a similar result. For men, the pattern is that attractiveness ratings fall as more alcohol is drunk (the bars decrease in height) and then for the women we see the same pattern: ratings fall as more is drunk. This again is indicative of no interaction: the change in attractiveness due to alcohol is similar in men and women.

12.6. Calculating effect sizes ③

As we saw in previous chapters (e.g. section 11.8), we can calculate partial eta squared, η^2. However, you're well advised, for reasons explained in these other sections, to use omega squared (ω^2). The calculation of omega squared becomes somewhat more cumbersome in factorial designs ('somewhat' being one of my characteristic understatements!). Howell (2006), as ever, does a wonderful job of explaining the complexities of it all (and has a nice table summarizing the various components for a variety of situations). Condensing all of this down, I'll just say that we need to first compute a variance component for each of the effects (the two main effects and the interaction term) and the error, and then use these to calculate effect sizes for each. If we call the first main effect A, the second main effect B and the interaction effect A × B, then the variance component for each of these is based on the mean squares of each effect and the sample sizes on which they're based:

$$\hat{\sigma}^2_A = \frac{(a-1)(\text{MS}_\text{A} - \text{MS}_\text{R})}{nab}$$

$$\hat{\sigma}^2_B = \frac{(b-1)(\text{MS}_\text{B} - \text{MS}_\text{R})}{nab}$$

$$\hat{\gamma}^2_{A\times B} = \frac{(a-1)(b-1)(\text{MS}_{\text{A}\times\text{B}} - \text{MS}_\text{R})}{nab}$$

In these equations, a is the number of levels of the first independent variable, b is the number of levels of the second independent variable and n is the number of people per condition. Let's calculate these for our data. We need to look at SAS Output 12.2 to find out the mean squares for each effect, and for the error term. Our first independent variable was alcohol. This had three levels (hence $a = 3$) and had a mean square of 1666.146. Our second independent variable was gender, which had two levels (hence $b = 2$) and a mean square of 168.75. The number of people in each group was 8 and the residual mean square was 83.036. Therefore, our equations become:

$$\hat{\sigma}^2_A = \frac{(3-1)(1666.146 - 83.036)}{8 \times 3 \times 2} = 65.96$$

$$\hat{\sigma}^2_B = \frac{(2-1)(168.75 - 83.036)}{8 \times 3 \times 2} = 1.79$$

$$\hat{\sigma}^2_{A\times B} = \frac{(3-1)(2-1)(989.062 - 83.036)}{8 \times 3 \times 2} = 37.75$$

We also need to estimate the total variability and this is just the sum of these other variables plus the residual mean square:

$$\begin{aligned}\hat{\sigma}^2_{\text{total}} &= \hat{\sigma}^2_A + \hat{\sigma}^2_B + \hat{\sigma}^2_{A\times B} + \text{MS}_\text{R}\\ &= 65.96 + 1.79 + 37.75 + 83.04\\ &= 188.54\end{aligned}$$

The effect size is then simply the variance estimate for the effect in which you're interested divided by the total variance estimate:

$$\omega^2_{\text{effect}} = \frac{\hat{\sigma}^2_{\text{effect}}}{\hat{\sigma}^2_{\text{total}}}$$

As such, for the main effect of alcohol we get:

$$\omega^2_{\text{alcohol}} = \frac{\hat{\sigma}^2_{\text{alcohol}}}{\hat{\sigma}^2_{\text{total}}} = \frac{65.96}{188.54} = .35$$

For the main effect of gender we get:

$$\omega^2_{\text{gender}} = \frac{\hat{\sigma}^2_{\text{gender}}}{\hat{\sigma}^2_{\text{total}}} = \frac{1.79}{188.54} = .009$$

For the interaction of gender and alcohol we get:

$$\omega^2_{\text{alcohol}\times\text{gender}} = \frac{\hat{\sigma}^2_{\text{alcohol}\times\text{gender}}}{\hat{\sigma}^2_{\text{total}}} = \frac{37.75}{188.54} = .20$$

To make these values comparable to r we can take the square root, which gives us effect sizes of .59 for alcohol, .09 for gender and .45 for the interaction term. As such, the effects of alcohol and the interaction are fairly large, but the effect of gender, which was non-significant in the main analysis, is small (close to zero in fact).

It's also possible to calculate effect sizes for our simple effects analysis (if you read section 12.4.4). These effects have 1 degree of freedom for the model (which means they're comparing only two things) and in these situations F can be converted to r using the following equation (which just uses the F-ratio and the residual degrees of freedom):[1]

$$r = \sqrt{\frac{F(1, df_\text{R})}{F(1, df_\text{R}) + df_\text{R}}}$$

Looking at the simple effects, we can see that we got F-ratios of 1.88, 0.92 and 23.05 for the effects of gender at no alcohol, 2 pints and 4 pints respectively. For each of these, the degrees of freedom were 1 for the model and 44 (given by $N - g - 1$; where N is the number of people, 48, and g is the number of groups, 3) for the residual. Therefore, we get the following effect sizes:

[1] If your F compares more than two things then a different equation is needed (see Rosenthal et al., 2000, p. 44), but I prefer to try to keep effect sizes to situations in which only two things are being compared because interpretation is easier.

$$r_{\text{Gender (no alcohol)}} = \sqrt{\frac{1.88}{1.88 + 44}} = 0.20$$

$$r_{\text{Gender (no alcohol)}} = \sqrt{\frac{0.92}{0.92 + 44}} = 0.14$$

$$r_{\text{Gender (no alcohol)}} = \sqrt{\frac{23.05}{23.05 + 44}} = 0.59$$

Therefore, the effect of gender is small at both no alcohol and 2 pints, but becomes large at 4 pints of alcohol.

12.7. Reporting the results of two-way ANOVA ②

As with the other ANOVAs we've encountered, we have to report the details of the *F*-ratio and the degrees of freedom from which it was calculated. For the various effects in these data the *F*-ratio will be based on different degrees of freedom: it was derived from dividing the mean squares for the effect by the mean squares for the residual. For the effects of alcohol and the alcohol × gender interaction, the model degrees of freedom were 2 ($df_M = 2$), but for the effect of gender the degrees of freedom were only 1 ($df_M = 1$). For all effects, the degrees of freedom for the residuals were 42 ($df_R = 42$). We can, therefore, report the three effects from this analysis as follows:

✓ There was a significant main effect of the amount of alcohol consumed at the nightclub, on the attractiveness of the mate selected, $F(2, 42) = 20.07$, $p < .001$, $\omega^2 = .35$. The Šidák *post hoc* test revealed that the attractiveness of selected dates was significantly lower after 4 pints than both after 2 pints and no alcohol (both *ps* < .001). The attractiveness of dates after 2 pints and no alcohol was not significantly different ($p = 0.988$).

✓ There was a non-significant main effect of gender on the attractiveness of selected mates, $F(1, 42) = 2.03$, $p = .161$, $\omega^2 = .009$.

✓ There was a significant interaction effect between the amount of alcohol consumed and the gender of the person selecting a mate, on the attractiveness of the partner selected, $F(2, 42) = 11.91$, $p < .001$, $\omega^2 = .20$. This indicates that male and female genders were affected differently by alcohol. Specifically, the attractiveness of partners was similar in males ($M = 66.88$, $SD = 10.33$) and females ($M = 60.63$, $SD = 4.96$) after no alcohol; the attractiveness of partners was also similar in males ($M = 66.88$, $SD = 12.52$) and females ($M = 62.50$, $SD = 6.55$) after 2 pints; however, attractiveness of partners selected by males ($M = 35.63$, $SD = 10.84$) was significantly lower than those selected by females ($M = 57.50$, $SD = 7.07$) after 4 pints.

12.8. Factorial ANOVA as regression ③

We saw in section 10.2.3. that one-way ANOVA could be conceptualized as a regression equation (a general linear model). In this section we'll consider how we extend this linear model to incorporate two independent variables. To keep things as simple as possible I just

LABCOAT LENI'S REAL RESEARCH 12.1

Don't forget your toothbrush? ②

We have all experienced that feeling after we have left the house of wondering whether we locked the door, or if we remembered to close the window, or if we remembered to remove the bodies from the fridge in case the police turn up. This behaviour is normal; however, people with obsessive compulsive disorder (OCD) tend to check things excessively. They might, for example, check whether they have locked the door so often that it takes them an hour to leave their house. It is a very debilitating problem.

One theory of this checking behaviour in OCD suggests that it is caused by a combination of the mood you are in (positive or negative) interacting with the rules you use to decide when to stop a task (do you continue until you feel like stopping, or until you have done the task as best as you can?). Davey, Startup, Zara, MacDonald and Field (2003) tested this hypothesis by inducing a negative, positive or no mood in different people and then asking them to imagine that they were going on holiday and to generate as many things as they could that they should check before they went away. Within each mood group, half of the participants were instructed to generate as many items as they could (known as an 'as many as can' stop rule), whereas the remainder were asked to generate items for as long as they felt like continuing the task (known as a 'feel like continuing' stop rule). The data are in the file **Davey2003.sas7bdat**.

Davey et al. hypothesized that people in negative moods, using an 'as many as can' stop rule, would generate more items than those using a 'feel like continuing' stop rule. Conversely, people in a positive mood would generate more items when using a 'feel like continuing' stop rule compared to an 'as many as can' stop rule. Finally, in neutral moods, the stop rule used shouldn't affect the number of items generated. Draw an error bar chart of the data and then conduct the appropriate analysis to test Davey et al.'s hypotheses.

Answers are in the additional material on the companion website (or look at pages 148–149 in the original article).

DAVEY, G. C. L. ET AL. (2003). *JOURNAL OF BEHAVIOR THERAPY & EXPERIMENTAL PSYCHIATRY, 34*, 141–160.

want you to imagine that we have only two levels of the alcohol variable in our example (none and 4 pints). As such, we have two variables, each with two levels. All of the general linear models we've considered in this book take the general form of:

$$\text{outcome}_i = (\text{model}) + \text{error}_i$$

For example, when we encountered multiple regression in Chapter 7. we saw that this model was written as (see equation (7.9)):

$$Y_i = (b_0 + b_1 X_{1i} + b_2 X_{2i} + \ldots + b_n X_{ni}) + \varepsilon_i$$

Also, when we came across one-way ANOVA, we adapted this regression model to conceptualize our Viagra example as (see equation 10.2)):

$$\text{Libido}_i = (b_0 + b_2 \text{High}_i + b_2 \text{Low}_i) + \varepsilon_i$$

In this model, the High and Low variables were dummy variables (i.e. variables that can take only values of 0 or 1). In our current example, we have two variables: gender (male or female) and alcohol (none and 4 pints). We can code each of these with zeros and ones, for example, we could code gender as male = 0, female = 1; and we could code the alcohol variable as 0 = none, 1 = 4 pints. We could then directly copy the model we had in one-way ANOVA:

$$\text{Attractiveness}_i = (b_0 + b_1 \text{Gender}_i + b_2 \text{Alcohol}_i) + \varepsilon_i$$

Now the astute among you might say, 'Where has the interaction term gone?' Well, of course, we have to include this too, and so the model simply extends to become (first expressed generally and then in terms of this specific example):

$$\text{Attractive}_i = (b_0 + b_1\text{A}_i + b_2B_i + b_3AB_i) + \varepsilon_i$$
$$\text{Attractive}_i = (b_0 + b_1\text{Gender}_i + b_2\text{Alcohol}_i + b_3\text{Interaction}_i) + \varepsilon_i \tag{12.1}$$

The question is: how do we code the interaction term? The interaction term represents the combined effect of alcohol and gender, and in fact to get any interaction term in regression you simply multiply the variables involved in that interaction term. This is why you see interaction terms written as gender × alcohol, because in regression terms the interaction variable literally is the two variables multiplied by each other. Table 12.2 shows the resulting variables for the regression (note that the interaction variable is simply the value of the gender dummy variable multiplied by the value of the alcohol dummy variable). So, for example, a male receiving 4 pints of alcohol would have a value of 0 for the gender variable, 1 for the alcohol variable and 0 for the interaction variable. The group means for the various combinations of gender and alcohol are also included because they'll come in useful in due course.

TABLE 12.2 Coding scheme for factorial ANOVA

Gender	Alcohol	Dummy (Gender)	Dummy (Alcohol)	Interaction	Mean
Male	None	0	0	0	66.875
Male	4 Pints	0	1	0	35.625
Female	None	1	0	0	60.625
Female	4 Pints	1	1	1	57.500

To work out what the b-values represent in this model we can do the same as we did for the t-test and one-way ANOVA; that is, look at what happens when we insert values of our predictors (gender and alcohol)! To begin with, let's see what happens when we look at men who had no alcohol. In this case, the value of gender is 0, the value of alcohol is 0 and the value of the interaction is also 0. The outcome we predict (as with one-way ANOVA) is the mean of this group (66.875), so our model becomes:

$$\text{Attractive}_i = (b_0 + b_1\text{Gender}_i + b_2\text{Alcohol}_i + b_3\text{Interaction}_i) + \varepsilon_i$$
$$\overline{X}_{\text{Men, None}} = b_0 + (b_1 \times 0) + (b_2 \times 0) + (b_3 \times 0)$$
$$b_0 = \overline{X}_{\text{Men, None}}$$
$$b_0 = 66.875$$

So, the constant b_0 in the model represents the mean of the group for which all variables are coded as 0. As such it's the mean value of the base category (in this case men who had no alcohol).

Now, let's see what happens when we look at females who had no alcohol. In this case, the gender variable is 1 and the alcohol and interaction variables are still 0. Also remember that b_0 is the mean of the men who had no alcohol. The outcome is the mean for women who had no alcohol. Therefore, the equation becomes:

$$\overline{X}_{\text{Women, None}} = b_0 + (b_1 \times 1) + (b_2 \times 0) + (b_3 \times 0)$$
$$\overline{X}_{\text{Women, None}} = b_0 + b_1$$
$$\overline{X}_{\text{Women, None}} = \overline{X}_{\text{Men, None}} + b_1$$
$$b_1 = \overline{X}_{\text{Women, None}} - \overline{X}_{\text{Men, None}}$$
$$b_1 = 60.625 - 66.875$$
$$b_1 = -6.25$$

So, b_1 in the model represents the difference between men and women who had no alcohol. More generally we can say it's the effect of gender for the base category of alcohol (the base category being the one coded with 0, in this case no alcohol).

Now let's look at males who had 4 pints of alcohol. In this case, the gender variable is 0, the alcohol variable is 1 and the interaction variable is still 0. We can also replace b_0 with the mean of the men who had no alcohol. The outcome is the mean for men who had 4 pints. Therefore, the equation becomes:

$$\overline{X}_{\text{Men, 4 Pints}} = b_0 + (b_1 \times 0) + (b_2 \times 1) + (b_3 \times 0)$$
$$\overline{X}_{\text{Men, 4 Pints}} = b_0 + b_2$$
$$\overline{X}_{\text{Men, 4 Pints}} = \overline{X}_{\text{Men, None}} + b_2$$
$$b_2 = \overline{X}_{\text{Men, 4 Pints}} - \overline{X}_{\text{Men, None}}$$
$$b_2 = 35.625 - 66.875$$
$$b_2 = -31.25$$

So, b_2 in the model represents the difference between having no alcohol and 4 pints in men. Put more generally, it's the effect of alcohol in the base category of gender (i.e. the category of gender that was coded with 0, in this case men).

Finally, we can look at females who had 4 pints of alcohol. In this case, the gender variable is 1, the alcohol variable is 1 and the interaction variable is also 1. We can also replace b_0, b_1, and b_2, with what we now know they represent. The outcome is the mean for women who had 4 pints. Therefore, the equation becomes:

$$\overline{X}_{\text{Women, 4 Pints}} = b_0 + (b_1 \times 1) + (b_2 \times 1) + (b_3 \times 1)$$
$$\overline{X}_{\text{Women, 4 Pints}} = b_0 + b_1 + b_2 + b_3$$
$$\overline{X}_{\text{Women, 4 Pints}} = \overline{X}_{\text{Men, None}} + \left(\overline{X}_{\text{Women, None}} - \overline{X}_{\text{Men, None}}\right) + \left(\overline{X}_{\text{Men, 4 Pints}} - \overline{X}_{\text{Men, None}}\right) + b_3$$
$$\overline{X}_{\text{Women, 4 Pints}} = \overline{X}_{\text{Women, None}} + \overline{X}_{\text{Men, 4 Pints}} - \overline{X}_{\text{Men, None}} + b_3$$
$$b_3 = \overline{X}_{\text{Men, None}} - \overline{X}_{\text{Women, None}} + \overline{X}_{\text{Women, 4 Pints}} - \overline{X}_{\text{Men, 4 Pints}}$$
$$b_3 = 66.875 - 60.625 + 57.500 - 35.625$$
$$b_3 = 28.125$$

So, b_3 in the model really compares the difference between men and women in the no alcohol condition to the difference between men and women in the 4 pint condition. Put another way, it compares the effect of gender after no alcohol to the effect of gender after

4 pints.[2] If you think about it in terms of an interaction graph this makes perfect sense. For example, the top left-hand side of Figure 12.9 shows the interaction graph for these data. Now imagine we calculated the difference between men and women for the no alcohol groups. This would be the difference between the lines on the graph for the no alcohol group (the difference between group means, which is 6.25). If we then do the same for the 4 pints group, we find that the difference between men and women is −21.875. If we plotted these two values as a new graph we'd get a line connecting 6.25 to −21.875 (see the bottom left-hand side of Figure 12.9). This reflects the difference between the effect of gender after no alcohol compared to after 4 pints. We know that beta values represent gradients of lines and in fact b_3 in our model is the gradient of this line! (This is 6.25 − (−21.875) = 28.125.) Let's also see what happens if there isn't an interaction effect: the right-hand side of Figure 12.9 shows the same data except that the mean for the females who had 4 pints has been changed to 30. If we calculate the difference between men and women after no alcohol we get the same as before: 6.25. If we calculate the difference between men and women after 4 pints we now get 5.625. If we again plot these differences on a new graph, we find a virtually horizontal line. So, when there's no interaction, the line connecting the effect of gender after no alcohol and after 4 pints is flat and the resulting b_3 in our model would be close to 0 (remember that a zero gradient means a flat line). In fact its actual value would be 6.25 − 5.625 = 0.625.

SELF-TEST The file **GogglesRegression.sas7bdat** contains the dummy variables used in this example. Just to prove that all of this works, use this file and run a multiple regression on the data.

The resulting table of coefficients is in SAS Output 12.8. The important thing to note is that the various beta values are the same as we've just calculated, which should hopefully convince you that factorial ANOVA is (and this should be familiar by now) just regression dressed up in a different costume!

What I hope to have shown you in this example is how even complex ANOVAs are just forms of regression (a general linear model). You'll be pleased to know (I'll be pleased to know for that matter) that this is the last I'm going to say about ANOVA as a general linear model. I hope I've given you enough background so that you get a sense of the fact that we can just keep adding in independent variables into our model. All that happens is these new variables just get added into a multiple regression equation with an associated beta value (just like the regression chapter). Interaction terms can also be added simply by multiplying the variables that interact. These interaction terms will also have an associated beta value. So, any ANOVA (no matter how complex) is just a form of multiple regression.

12.9. What to do when assumptions are violated in factorial ANOVA ③

There is not a simple non-parametric counterpart of factorial ANOVA. As such, if our data violate the assumption of normality the only real solution is robust methods (see

[2] In fact, if you re arrange the terms in the equation you'll see that you can also phrase the interaction the opposite way around: it represents the effect of alcohol in men compared to women.

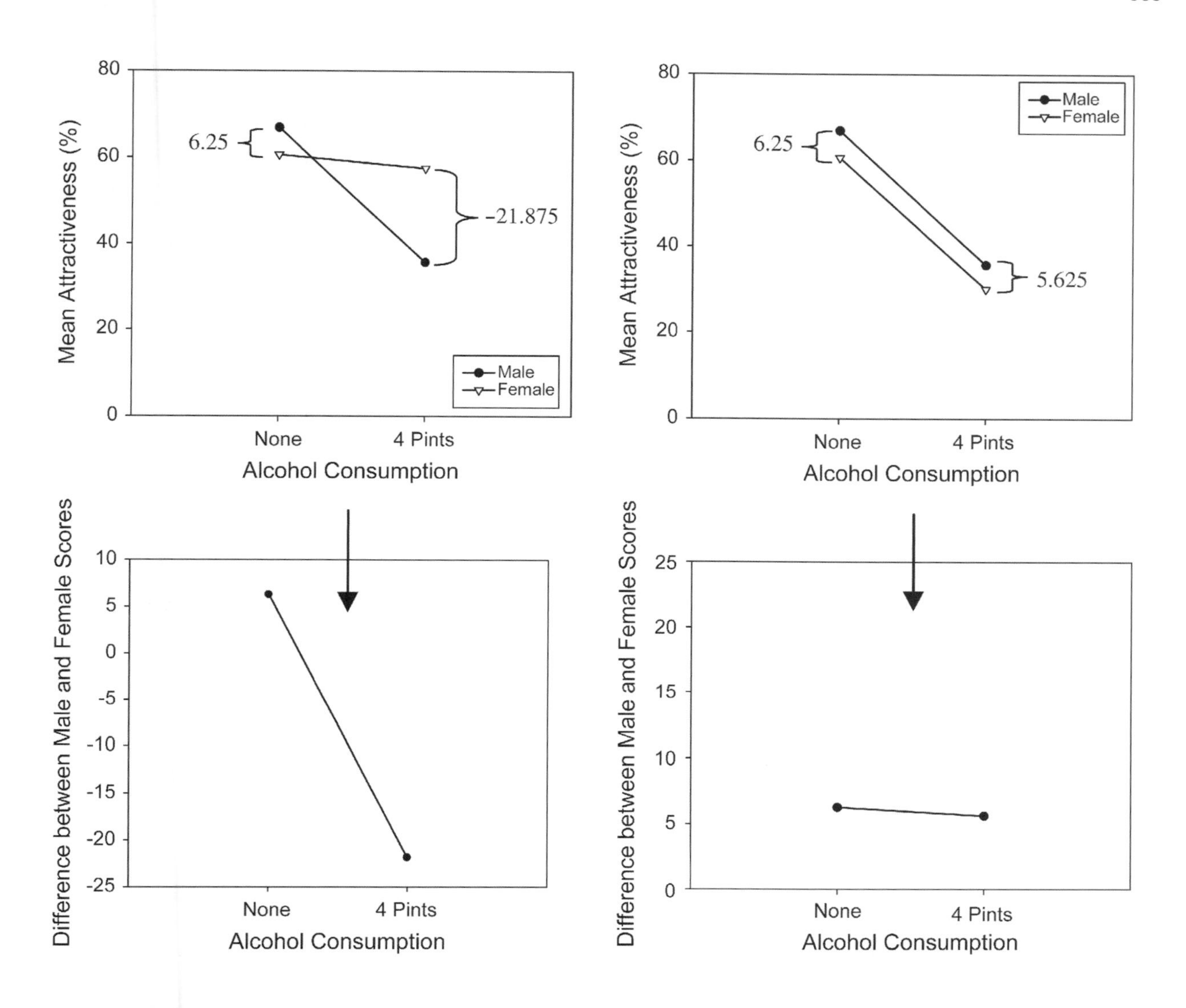

FIGURE 12.9 Breaking down what an interaction represents

section 5.7.4) such as those described in Wilcox's Chapter 7 and the associated files for the software R (Wilcox, 2005). Also, if you have violated the assumption of homogeneity of variance then you can try to implement corrections based on the Welch procedure that was described in the previous chapter. However, this is quite technical, and if you have anything more complicated than a 2 × 2 design then, really, it would be less painful to cover your body in paper cuts and then bathe in chilli sauce (see Algina & Olejnik, 1984).

SAS OUTPUT 12.8

Parameter Estimates						
Variable	Label	DF	Parameter Estimate	Standard Error	t Value	Pr > |t|
Intercept	Intercept	1	66.87500	3.05502	21.89	<.0001
GENDER	Gender	1	–6.25000	4.32045	–1.45	0.1591
ALCOHOL	Alcohol Consumption	1	–31.25000	4.32045	–7.23	<.0001
INTERACT	Interaction	1	28.12500	6.11004	4.60	<.0001

What have I discovered about statistics? ②

This chapter has been a whistle-stop tour of factorial ANOVA. In fact we'll come across more factorial ANOVAs in the next two chapters, but for the time being we've just looked at the situation where there are two independent variables, and different people have been used in all experimental conditions. We started off by looking at how to calculate the various sums of squares in this analysis, but most importantly we saw that we get three effects: two main effects (the effect of each of the independent variables) and an interaction effect. We moved on to see how this analysis is done on SAS and how the output is interpreted. Much of this was similar to the ANOVAs we've come across in previous chapters, but one big difference was the interaction term. We spent a bit of time exploring interactions (and especially interaction graphs) to see what an interaction looks like and how to spot it. The brave readers also found out how to follow up an interaction with simple effects analysis and also discovered that even complex ANOVAs are simply regression analyses in disguise. Finally, we discovered that calculating effect sizes in factorial designs is a complete headache and should be attempted only by the criminally insane. So far we've steered clear of repeated-measures designs, but in the next chapter I have to resign myself to the fact that I can't avoid explaining them for the rest of my life.☹

We also discovered that no sooner had I started my first band than it disintegrated. I went with drummer Mark to sing in a band called the Outlanders, who were much better musically but were not, if the truth be told, metal enough for me. They also sacked me after a very short period of time for not being able to sing like Bono (an insult at the time, but in retrospect ...).

Key terms that I've discovered

Beer-goggles effect
Factorial ANOVA
Independent factorial design
Interaction graph
Mixed design
Related factorial design
Simple effects analysis

Smart Alex's tasks

- **Task 1**: People's musical tastes tend to change as they get older (my parents, for example, after years of listening to relatively cool music when I was a kid in the 1970s, subsequently hit their mid-forties and developed a worrying obsession with country and western music – or maybe it was the stress of having me as a teenage son!). Anyway, this worries me immensely as the future seems incredibly bleak if it is spent listening to Garth Brooks and thinking 'oh boy, did I underestimate

Garth's immense talent when I was in my 20s'. So, I thought I'd do some research to find out whether my fate really was sealed, or whether it's possible to be old and like good music too. First, I got myself two groups of people (45 people in each group): one group contained young people (which I arbitrarily decided was under 40 years of age) and the other group contained more mature individuals (above 40 years of age). This is my first independent variable, **age**, and it has two levels (less than or more than 40 years old). I then split each of these groups of 45 into three smaller groups of 15 and assigned them to listen to either Fugazi (who everyone knows are the coolest band on the planet),[3] ABBA or Barf Grooks (who is a lesser known country and western musician not to be confused with anyone who has a similar name and produces music that makes you want to barf). This is my second independent variable, **music**, and has three levels (Fugazi, ABBA or Barf Grooks). There were different participants in all conditions, which means that of the 45 under-forties, 15 listened to Fugazi, 15 listened to ABBA and 15 listened to Barf Grooks; likewise, of the 45 over-forties, 15 listened to Fugazi, 15 listened to ABBA and 15 listened to Barf Grooks. After listening to the music I got each person to rate it on a scale ranging from –100 (I hate this foul music of Satan) through 0 (I am completely indifferent) to +100 (I love this music so much I'm going to explode). This variable is called **liking**. The data are in the file **Fugazi.sas7bdat**. Conduct a two-way independent ANOVA on them. ②

- **Task 2**: In Chapter 4 we used some data that related to men and women's arousal levels when watching either *Bridget Jones's Diary* or *Memento* (**ChickFlick.sas7bdat**). Analyse these data to see whether men and women differ in their reactions to different types of films. ②
- **Task 3**: At the start of this chapter I described a way of empirically researching whether I wrote better songs than my old band mate Malcolm, and whether this depended on the type of song (a symphony or a song about flies). The outcome variable would be the number of screams elicited by audience members during the songs. These data are in the file **Escape From Inside.sas7bdat**. Draw an error bar graph (lines) and analyse and interpret these data. ②

The answers are on the companion website. Task 1 is an example from Field and Hole (2003) and so them is a more detailed answer in there if you feel like you want it.

12.13. Further reading

Howell, D. C. (2006). *Statistical methods for psychology* (6th ed.). Belmont, CA: Duxbury. (Or you might prefer his *Fundamental statistics for the behavioral sciences,* also in its 6th edition, 2007.)

Rosenthal, R., Rosnow, R. L., & Rubin, D. B. (2000). *Contrasts and effect sizes in behavioural research: a correlational approach*. Cambridge: Cambridge University Press. (This is quite advanced but really cannot be bettered for contrasts and effect size estimation.)

Rosnow, R. L., & Rosenthal, R. (2005). *Beginning behavioral research: A conceptual primer* (5th ed.). Englewood, Cliffs, NJ: Pearson/Prentice Hall. (Has some wonderful chapters on ANOVA, with a particular focus on effect size estimation, and some very insightful comments on what interactions actually mean.)

[3] See http://www.dischord.com.

Interesting real research

Davey, G. C. L., Startup, H. M., Zara, A., MacDonald, C. B., & Field, A. P. (2003). Perseveration of checking thoughts and mood-as-input hypothesis. *Journal of Behavior Therapy & Experimental Psychiatry*, *34*, 141–160.

Repeated-measures designs (GLM 4)

13

FIGURE 13.1 Scansion in the early days; I used to stare a lot (L–R: me, Mark and Mark)

13.1. What will this chapter tell me? ②

At the age of 15, I was on holiday with my friend Mark (the drummer) in Cornwall. I had a pretty decent mullet by this stage (nowadays I just wish I had enough hair to grow a mullet … or perhaps not) and had acquired a respectable collection of heavy metal T-shirts from going to various gigs. We were walking along the cliff tops one evening at dusk reminiscing about our times in Andromeda. We came to the conclusion that the only thing we hadn't enjoyed about that band was Malcolm and that maybe we should reform it with a different guitarist.[1] As I was wondering who we could get to play guitar, Mark pointed

[1] I feel bad about saying this because Malcolm was a very nice guy and, to be honest, at that age (and some would argue beyond) I could be a bit of a cock.

out the blindingly obvious: I played guitar. So, when we got home Scansion was born.[2] As the singer, guitarist and songwriter I set about writing some songs. I moved away from writing about flies and set my sights on the pointlessness of existence, death, betrayal and so on. We had the dubious honour of being reviewed in the music magazine *Kerrang!* (in a live review they called us 'twee', which is really not what you want to be called if you're trying to make music so heavy that it ruptures the bowels of Satan himself). Our highlight, however, was playing a gig at the famous Marquee Club in London (this club has closed now, not as a result of us playing there I hasten to add, but in its day it started the careers of people like Jimi Hendrix, the Who, Iron Maiden and Led Zeppelin).[3] This was the biggest gig of our career and it was essential that we played like we never had before. As it turned out, we did: I ran on stage, fell over and in the process de-tuned my guitar beyond recognition and broke the zip on my trousers. I spent the whole gig out of tune and spread-eagle to prevent my trousers falling down. Like I said, I'd never played like *that* before. We used to get quite obsessed with comparing how we played at different gigs. I didn't know about statistics then (happy days) but if I had I would have realized that we could rate ourselves and compare the mean ratings for different gigs; because we would always be the ones doing the rating, this would be a repeated-measures design, so we would need a repeated-measures ANOVA to compare these means. That's what this chapter is about.

13.2. Introduction to repeated-measures designs ②

Over the last three chapters we have looked at a procedure called ANOVA which is used for testing differences between several means. So far we've concentrated on situations in which different people contribute to different means; put another way, different people take part in different experimental conditions. Actually, it doesn't have to be different people (I tend to say people because I'm a psychologist and so spend my life torturing, I mean testing, children in the name of science), it could be different plants, different companies, different plots of land, different viral strains, different goats or even different duck-billed platypuses (or whatever the plural is). Anyway, the point is I've completely ignored situations in which the same people (plants, goats, hamsters, seven-eyed green galactic leaders from space, or whatever) contribute to the different means. I've put it off long enough, and now I'm going to take you through what happens when we do ANOVA on repeated-measures data.

SELF-TEST What is a repeated-measures design? (Clue: it is described in Chapter 1.)

Repeated-measures is a term used when the same participants participate in all conditions of an experiment. For example, you might test the effects of alcohol on enjoyment of a party. Some people can drink a lot of alcohol without really feeling the consequences,

[2] Scansion is a term for the rhythm of poetry. We got the name by searching through a dictionary until we found a word that we liked. Originally we didn't think it was 'metal' enough, and we decided that any self-respecting heavy metal band needed to have a big spiky 'X' in their name. So, for the first couple of years we spelt it 'Scanxion'. Like I said, I could be a bit of a cock back then.

[3] http://www.themarqueeclub.net.

whereas others, like myself, have only to sniff a pint of lager and they start flapping around on the floor waving their arms and legs around shouting 'Look at me, I'm Andy, King of the lost world of the Haddocks.' Therefore, it is important to control for individual differences in tolerance to alcohol and this can be achieved by testing the same people in all conditions of the experiment: participants could be given a questionnaire assessing their enjoyment of the party after they had consumed 1 pint, 2 pints, 3 pints and 4 pints of lager.

We saw in Chapter 1 that this type of design has several advantages; however, there is a big disadvantage. In Chapter **10** we saw that the accuracy of the *F*-test in ANOVA depends upon the assumption that scores in different conditions are independent (see section 10.2.10). When repeated measures are used this assumption is violated: scores taken under different experimental conditions are likely to be related because they come from the same participants. As such, the conventional *F*-test will lack accuracy. The relationship between scores in different treatment conditions means that an additional assumption has to be made and, put simplistically, we assume that the relationship between pairs of experimental conditions is similar (i.e. the level of dependence between experimental conditions is roughly equal). This assumption is called the assumption of **sphericity**, which, trust me, is a pain in the neck to try to pronounce when you're giving statistics lectures at 9 a.m.

13.2.1. The assumption of sphericity ②

The assumption of sphericity can be likened to the assumption of homogeneity of variance in between-group ANOVA. Sphericity (denoted by ε and sometimes referred to as *circularity*) is a more general condition of **compound symmetry**. Compound symmetry holds true when both the variances across conditions are equal (this is the same as the homogeneity of variance assumption in between-group designs) and the covariances between pairs of conditions are equal. So, we assume that the variation within experimental conditions is fairly similar and that no two conditions are any more dependent than any other two. Although compound symmetry has been shown to be a sufficient condition for ANOVA using repeated-measures data, it is not a necessary condition. Sphericity is a less restrictive form of compound symmetry (in fact much of the early research into repeated-measures ANOVA confused compound symmetry with sphericity). Sphericity refers to the equality of variances of the *differences* between treatment levels. So, if you were to take each pair of treatment levels, and calculate the differences between each pair of scores, then it is necessary that these differences have approximately equal variances. As such, *you need at least three conditions for sphericity to be an issue.*

What is sphericity?

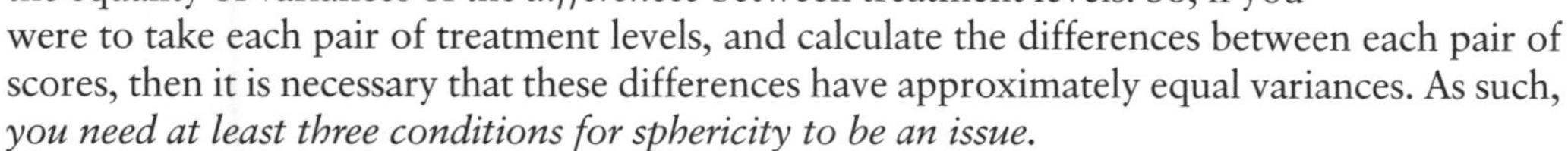

13.2.2. How is sphericity measured? ②

If we were going to check the assumption of sphericity by hand rather than getting SAS to do it for us then we could start by calculating the differences between pairs of scores in all combinations of the treatment levels. Once this has been done, we could calculate the variance of these differences. Table 13.1 shows data from an experiment with three conditions. The differences between pairs of scores are computed for each participant and the variance for each set of differences is calculated. We saw above that sphericity is met when these variances are roughly equal. For these data, sphericity will hold when:

$$\text{variance}_{A-B} \approx \text{variance}_{A-C} \approx \text{variance}_{B-C}$$

TABLE 13.1 Hypothetical data to illustrate the calculation of the variance of the differences between conditions

Group A	Group B	Group C	A–B	A–C	B–C
10	12	7	−2	2	5
15	15	12	0	3	3
25	30	20	−5	5	10
35	30	28	5	7	2
30	27	20	3	10	7
		Variance:	**15.7**	**10.3**	**10.3**

In these data there is some deviation from sphericity because the variance of the differences between conditions A and B (15.7) is greater than the variance of the differences between A and C and between B and C (10.3). However, these data have *local circularity* (or local sphericity) because two of the variances of differences are identical. Therefore, the sphericity assumption has been met for any multiple comparisons involving these conditions (for a discussion of local circularity see Rouanet & Lépine, 1970). The deviation from sphericity in the data in Table 13.1 does not seem too severe (all variances are *roughly* equal), but can we assess whether a deviation is severe enough to warrant action?

13.2.3. Assessing the severity of departures from sphericity ②

SAS produces a test known as **Mauchly's test**, which tests the hypothesis that the variances of the differences between conditions are equal. Therefore, if Mauchly's test statistic is significant (i.e. has a probability value less than .05) we should conclude that there are significant differences between the variances of differences and, therefore, the condition of sphericity is not met. If, however, Mauchly's test statistic is non-significant (i.e. $p > .05$) then it is reasonable to conclude that the variances of differences are not significantly different (i.e. they are roughly equal). So, in short, if Mauchly's test is significant then we must be wary of the *F*-ratios produced by the computer. However, like any significance test it is dependent on sample size: in big samples small deviations from sphericity can be significant, and in small samples large violations can be non-significant.

13.2.4. What is the effect of violating the assumption of sphericity? ③

Rouanet and Lépine (1970) provided a detailed account of the validity of the *F*-ratio under violations of the sphericity assumption. They argued that there are two different *F*-ratios that can be used to assess treatment comparisons, labelled F' and F'' respectively. F' refers to an *F*-ratio derived from the mean squares of the comparison in question and the specific error term for the comparison of interest – this is the *F*-ratio normally used. F'' is derived not from the specific error mean square but from the total error mean squares for *all* repeated-measures comparisons. Rouanet and Lépine (1970) showed that for F'' to be valid, *overall* sphericity must hold (i.e. the whole data set must be spherical), but for F' to be valid, sphericity must hold for the *specific comparison in question* (see also Mendoza, Toothaker, & Crain, 1976). F' is the statistic generally used and the effect of violating sphericity is a loss of power (compared to when F'' is used) and a test statistic (F-ratio) that simply cannot be compared to tabulated values of the F-distribution (see Oliver Twisted).

OLIVER TWISTED

Please, Sir, can I have some more … sphericity?

'Balls …,' says Oliver, '… are spherical, and I like balls. Maybe I'll like sphericity too if only you could explain it to me in more detail.' Be careful what you wish for, Oliver. In my youth I wrote an article called 'A bluffer's guide to sphericity', which I used to cite in previous version of this book, roughly on this page. A few people ask me for it, so I thought I might as well reproduce it in the additional material for this chapter.

13.2.5. What do you do if you violate sphericity? ②

If data violate the sphericity assumption there are several corrections that can be applied to produce a valid F-ratio. SAS produces two corrections based upon the estimates of sphericity advocated by Greenhouse and Geisser (1959) and Huynh and Feldt (1976). Both of these estimates give rise to a correction factor that is applied to the degrees of freedom used to assess the observed F-ratio. The calculation of these estimates is beyond the scope of this book (interested readers should consult Girden, 1992); we need know only that the three estimates differ. The **Greenhouse–Geisser correction** (usually denoted as $\hat{\varepsilon}$) varies between $1/k - 1$ (where k is the number of repeated-measures conditions) and 1. The closer that $\hat{\varepsilon}$ is to 1, the more homogeneous the variances of differences, and hence the closer the data are to being spherical. For example, in a situation in which there are five conditions the lower limit of $\hat{\varepsilon}$ will be $1/(5 - 1)$, or 0.25 (known as the **lower-bound estimate** of sphericity).

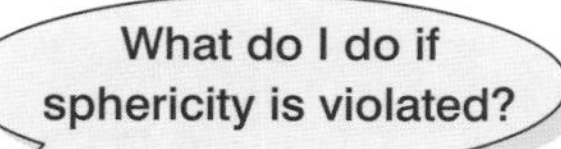

Huynh and Feldt (1976) reported that when the Greenhouse–Geisser estimate is greater than 0.75 too many false null hypotheses fail to be rejected (i.e. the correction is too conservative) and Collier, Baker, Mandeville, and Hayes (1967) showed that this was also true when the sphericity estimate was as high as 0.90. Huynh and Feldt, therefore, proposed their own less conservative correction (usually denoted as $\tilde{\varepsilon}$). However, Maxwell and Delaney (1990) report that $\tilde{\varepsilon}$ overestimates sphericity. Stevens (2002) therefore recommends taking an average of the two and adjusting *df* by this averaged value. Girden (1992) recommends that when estimates of sphericity are greater than 0.75 then the **Huynh–Feldt correction** should be used, but when sphericity estimates are less than 0.75 or nothing is known about sphericity at all, then the Greenhouse–Geisser correction should be used instead. We will see how these values are used in due course.

A final option, when you have data that violate sphericity, is to use multivariate test statistics (MANOVA – see Chapter **16**), because they are not dependent upon the assumption of sphericity (see O'Brien & Kaiser, 1985). MANOVA is covered in depth in Chapter **16**, but the repeated-measures procedure in SAS automatically produces multivariate test statistics. However, there may be trade-offs in power between these univariate and multivariate tests (see Jane Superbrain Box 13.1).

JANE SUPERBRAIN 13.1

Power in ANOVA and MANOVA ③

There is a trade-off in test power between univariate and multivariate approaches (although some authors argue that this can be overcome with suitable mastery of the techniques – O'Brien & Kaiser, 1985). Davidson (1972) compared the power of adjusted univariate techniques with those of Hotelling's T^2 (a MANOVA test statistic) and found that the univariate technique was relatively powerless to detect small reliable changes between highly correlated conditions when other less correlated conditions were also present. Mendoza, Toothaker, and Nicewander (1974) conducted a Monte Carlo study comparing univariate and multivariate techniques under violations of compound symmetry and normality and found that 'as the degree of violation of compound symmetry increased, the empirical power for the multivariate tests also increased. In contrast, the power for the univariate tests generally decreased' (p. 174). Maxwell and Delaney (1990) noted that the univariate test is relatively more powerful than the multivariate test as n decreases and proposed that 'the multivariate approach should probably not be used if n is less than $a + 10$ (a is the number of levels for repeated measures)' (p. 602). As a rule it seems that when you have a large violation of sphericity ($\varepsilon < 0.7$) and your sample size is greater than $a + 10$ then multivariate procedures are more powerful, but with small sample sizes or when sphericity holds ($\varepsilon > 0.7$) the univariate approach is preferred (Stevens, 2002). It is also worth noting that the power of MANOVA increases and decreases as a function of the correlations between dependent variables (see Jane Superbrain Box 16.1) and so the relationship between treatment conditions must be considered also.

13.3. Theory of one-way repeated-measures ANOVA ②

In a **repeated-measures ANOVA** the effect of our experiment is shown up in the within-participant variance (rather than in the between-group variance). Remember that in independent ANOVA (section 10.2) the within-participant variance is our residual variance (SS_R); it is the variance created by individual differences in performance. This variance is not contaminated by the experimental effect, because whatever manipulation we've carried out has been done on different people. However, when we carry out our experimental manipulation on the same people then the within-participant variance will be made up of two things: the effect of our manipulation and, as before, individual differences in performance. So, some of the within-subjects variation comes from the effects of our experimental manipulation: we did different things in each experimental condition to the participants, and so variation in an individual's scores will partly be due to these manipulations. For example, if everyone scores higher in one condition than another, it's reasonable to assume that this happened not by chance, but because we did something different to the participants in one of the conditions compared to any other one. *Because* we did the *same* thing to everyone within a particular condition, any variation that cannot be explained by the manipulation we've carried out must be due to random factors outside our control, unrelated to our experimental manipulations (we could call this 'error'). As in independent ANOVA, we use an F-ratio that compares the size of the variation due to our experimental manipulations to the size of the variation due to random factors, the only difference being how we calculate these variances. If the variance due to our manipulations is big relative to the variation due to random factors, we get a big value of F, and we can conclude that the observed results are unlikely to have occurred if there was no effect in the population.

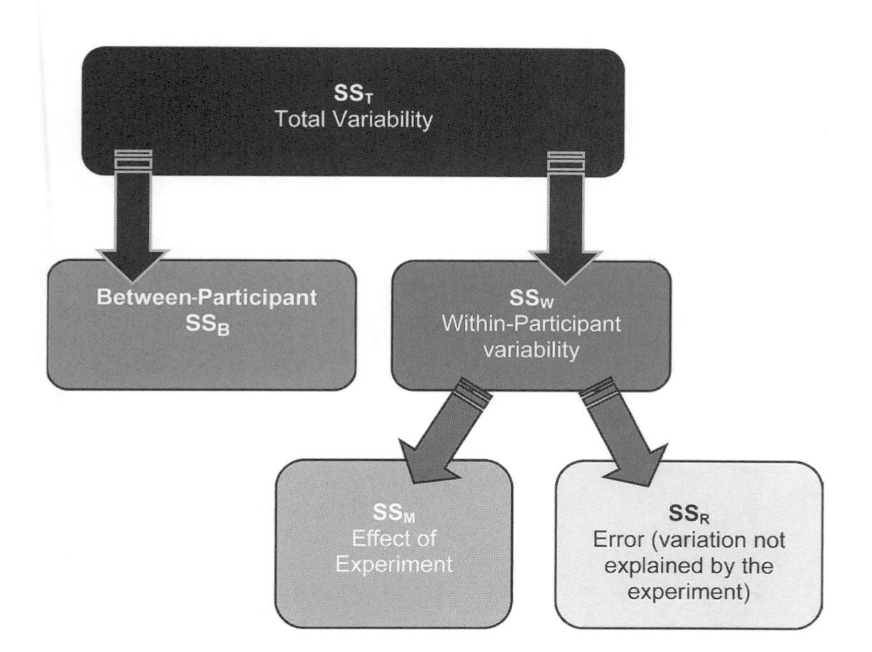

FIGURE 13.2 Partitioning variance for repeated-measures ANOVA

Figure 13.2 shows how the variance is partitioned in a repeated-measures ANOVA. The important thing to note is that we have the same types of variances as in independent ANOVA: we have a total sum of squares (SS_T), a model sum of squares (SS_M) and a residual sum of squares (SS_R). The *only* difference between repeated-measures and independent ANOVA is from where those sums of squares come: in repeated-measures ANOVA the model and residual sums of squares are both part of the within-participant variance. Let's have a look at an example.

I'm a celebrity, get me out of here! is a TV show in the UK in which celebrities (well, they're not really celebrities as such, more like ex-celebrities), in a pitiful attempt to salvage their careers (or just have careers in the first place), go and live in the jungle in Australia for a few weeks. During the show these contestants have to do various humiliating and degrading tasks to win food for their camp mates. These tasks invariably involve creepy-crawlies in places where creepy-crawlies shouldn't go; for example, you might be locked in a coffin full of rats, forced to put your head in a bowl of large spiders, or have eels and cockroaches poured onto you. It's cruel, voyeuristic, gratuitous, car crash TV, and I love it. As a vegetarian, a particular favourite task for me is the bushtucker trials in which the celebrities have to eat things like live stick insects, witchetty grubs, fish eyes and kangaroo testicles/penises. Honestly, your mental image of someone is forever scarred by seeing a fish eye exploding in their mouth (here's praying that Angela Gossow (Ed – who?) never goes on the show, although she'd probably just eat the other contestants which could enhance rather than detract from her appeal). I've often wondered (perhaps a little too much) which of the bushtucker foods is the most revolting. Imagine that I tested this by getting eight celebrities, and forcing them to eat four different animals (the aforementioned stick insect, kangaroo testicle, fish eye and witchetty grub) in counterbalanced order. On each occasion I measured the time it took the celebrity to retch, in seconds. This design is repeated measures because every celebrity eats every food. The independent variable was the type of food eaten and the dependent variable was the time taken to retch.

Table 13.2 shows the data for this example. There were four foods, each eaten by 8 different celebrities. Their times taken to retch are shown in the table. In addition, the mean amount of time to retch for each celebrity is shown in the table (and the variance in the time taken to retch), and also the mean time to retch for each animal. The total variance in retching time will, in part, be caused by the fact that different animals are more or less palatable (the manipulation), and will, in part, be caused by the fact that the celebrities themselves will differ in their constitution (individual differences).

TABLE 13.2 Data for the bushtucker example

Celebrity	Stick Insect	Kangaroo Testicle	Fish Eye	Witchetty Grub	Mean	s^2
1	8	7	1	6	5.50	9.67
2	9	5	2	5	5.25	8.25
3	6	2	3	8	4.75	7.58
4	5	3	1	9	4.50	11.67
5	8	4	5	8	6.25	4.25
6	7	5	6	7	6.25	0.92
7	10	2	7	2	5.25	15.58
8	12	6	8	1	6.75	20.92
Mean	8.13	4.25	4.13	5.75		

13.3.1. The total sum of squares (SS_T) ②

Remember from one-way independent ANOVA that SS_T is calculated using the following equation (see equation **10.4**):

$$SS_T = s^2_{\text{grand}}(N-1)$$

Well, in repeated-measures designs the total sum of squares is calculated in exactly the same way. The grand variance in the equation is simply the variance of all scores when we ignore the group to which they belong. So if we treated the data as one big group it would look as follows:

8	7	1	6
9	5	2	5
6	2	3	8
5	3	1	9
8	4	5	8
7	5	6	7
10	2	7	2
12	6	8	1

Grand Mean = 5.56
Grand Variance = 8.19

The variance of these scores is 8.19 (try this on your calculator). We used 32 scores to generate this value, so *N* is 32. As such the equation becomes:

$$
\begin{aligned}
SS_T &= s^2_{\text{grand}}(N-1) \\
&= 8.19(32-1) \\
&= 253.88
\end{aligned}
$$

The degrees of freedom for this sum of squares, as with the independent ANOVA, will be $N-1$, or 31.

13.3.2. The within-participant sum of squares (SS_W) ②

The crucial difference in this design is that there is a variance component called the within-participant variance (this arises because we've manipulated our independent variable within each participant). This is calculated using a sum of squares. Generally speaking, when we calculate any sum of squares we look at the squared difference between the mean and individual scores. This can be expressed in terms of the variance across scores and the number of scores on which that variance is based. For example, when we calculated the residual sum of squares in independent ANOVA (SS_R) we used the following equation (look back to equation**10.7**):

$$
SS_R = \sum_{i=1}^{n}(x_i - \bar{x}_i)^2
$$

$$
SS_R = s^2(n-1)
$$

This equation gave us the variance between individuals within a particular group, and so is an estimate of individual differences within a particular group. Therefore, to get the total value of individual differences we have to calculate the sum of squares within each group and then add them up:

$$
SS_R = s^2_{\text{group 1}}(n_1-1) + s^2_{\text{group 2}}(n_2-1) + s^2_{\text{group 3}}(n_3-1)\ldots
$$

This is all well and good when we have different people in each group, but in repeated-measures designs we've subjected people to more than one experimental condition, and, therefore, we're interested in the variation not within a group of people (as in independent ANOVA) but within an actual person. That is, how much variability is there within an individual? To find this out we actually use the same equation but we adapt it to look at people rather than groups. So, if we call this sum of squares SS_W (for within-participant SS) we could write it as:

$$
SS_W = s^2_{\text{Person 1}}(n_1-1) + s^2_{\text{Person 2}}(n_2-1) + s^2_{\text{Person 3}}(n_3-1) + \ldots + s^2_{\text{Person } n}(n_n-1)
$$

This equation simply means that we are looking at the variation in an individual's scores and then adding these variances for all the people in the study. The *n*s simply represent the number of scores on which the variances are based (i.e. the number of experimental conditions, or in this case the number of foods). All of the variances we need are in Table 13.2, so we can calculate SS_W as:

$$
\begin{aligned}
SS_W &= s^2_{\text{Celebrity 1}}(n_1-1) + s^2_{\text{Celebrity 2}}(n_2-1) + \ldots + s^2_{\text{Celebrity } n}(n_n-1) \\
&= 9.67(4-1) + 8.25(4-1) + 7.58(4-1) + 11.67(4-1) \\
&\quad + 4.25(4-1) + 0.92(4-1) + 15.58(4-1) + 20.92(4-1) \\
&= 29 + 24.75 + 22.75 + 35 + 12.75 + 2.75 + 46.75 + 62.75 \\
&= 236.50
\end{aligned}
$$

The degrees of freedom for each person are $n-1$ (i.e. the number of conditions minus 1). To get the total degrees of freedom we add the *df*s for all participants. So, with eight participants (celebrities) and four conditions (i.e. $n = 4$), there are 3 degrees of freedom for each celebrity and $8 \times 3 = 24$ degrees of freedom in total.

13.3.3. The model sum of squares (SS_M) ②

So far, we know that the total amount of variation within the data is 253.58 units. We also know that 236.50 of those units are explained by the variance created by individuals' (celebrities') performances under different conditions. Now some of this variation is the result of our experimental manipulation and some of this variation is simply random fluctuation. The next step is to work out how much variance is explained by our manipulation and how much is not.

In independent ANOVA, we worked out how much variation could be explained by our experiment (the model SS) by looking at the means for each group and comparing these to the overall mean. So, we measured the variance resulting from the differences between group means and the overall mean (see equation **10.5**). We do exactly the same thing with a repeated-measures design. First we calculate the mean for each level of the independent variable (in this case the mean time to retch for each food) and compare these values to the overall mean of all foods.

So, we calculate this SS in the same way as for independent ANOVA:

1. Calculate the difference between the mean of each group and the grand mean.
2. Square each of these differences.
3. Multiply each result by the number of participants that contribute to that mean (n_i).
4. Add the values for each group together:

$$SS_M = \sum_{i=1}^{k} n_i \left(\bar{x}_i - \bar{x}_{grand}\right)^2$$

Using the means from the bushtucker data (see Table 13.2), we can calculate SS_M as follows:

$$\begin{aligned} SS_M &= 8(8.13 - 5.56)^2 + 8(4.25 - 5.56)^2 + 8(4.13 - 5.56)^2 + 8(5.75 - 5.56)^2 \\ &= 8(2.57)^2 + 8(-1.31)^2 + 8(-1.44)^2 + 8(0.196)^2 \\ &= 83.13 \end{aligned}$$

For SS_M, the degrees of freedom (df_M) are again one less than the number of things used to calculate the sum of squares. For the model sums of squares we calculated the sum of squared errors between the four means and the grand mean. Hence, we used four things to calculate these sums of squares. Therefore, the degrees of freedom will be 3. So, as with independent ANOVA the model degrees of freedom are always the number of conditions (k) minus 1:

$$df_M = k - 1 = 3$$

13.3.4. The residual sum of squares (SS_R) ②

We now know that there are 253.58 units of variation to be explained in our data, and that the variation across our conditions accounts for 236.50 units. Of these 236.50 units, our experimental manipulation can explain 83.13 units. The final sum of squares is the residual sum of squares (SS_R), which tells us how much of the variation cannot be explained by the model. This value is the amount of variation caused by extraneous factors outside of experimental control. Knowing SS_W and SS_M already, the simplest way to calculate SS_R is to subtract SS_M from SS_W ($SS_R = SS_W - SS_M$):

$$\begin{aligned} SS_R &= SS_W - SS_M \\ SS_R &= 236.50 - 83.13 \\ &= 153.37 \end{aligned}$$

The degrees of freedom are calculated in a similar way:

$$\begin{aligned} df_R &= df_W - df_M \\ &= 24 - 3 \\ &= 21 \end{aligned}$$

13.3.5. The mean squares ②

SS_M tells us how much variation the model (e.g. the experimental manipulation) explains and SS_R tells us how much variation is due to extraneous factors. However, because both of these values are summed values the number of scores that were summed influences them. As with independent ANOVA we eliminate this bias by calculating the average sum of squares (known as the *mean squares*, MS), which is simply the sum of squares divided by the degrees of freedom:

$$MS_M = \frac{SS_M}{df_M} = \frac{83.13}{3} = 27.71$$

$$MS_R = \frac{SS_R}{df_R} = \frac{153.37}{21} = 7.30$$

MS_M represents the average amount of variation explained by the model (e.g. the systematic variation), whereas MS_R is a gauge of the average amount of variation explained by extraneous variables (the unsystematic variation).

13.3.6. The *F*-ratio ②

The *F*-ratio is a measure of the ratio of the variation explained by the model and the variation explained by unsystematic factors. It can be calculated by dividing the model mean squares by the residual mean squares. You should recall that this is exactly the same as for independent ANOVA:

$$F = \frac{MS_M}{MS_R}$$

So, as with the independent ANOVA, the *F*-ratio is still the ratio of systematic variation to unsystematic variation. As such, it is the ratio of the experimental effect to the effect on performance of unexplained factors. For the bushtucker data, the *F*-ratio is:

$$F = \frac{\text{MS}_\text{M}}{\text{MS}_\text{R}} = \frac{27.71}{7.30} = 3.79$$

This value is greater than 1, which indicates that the experimental manipulation had some effect above and beyond the effect of extraneous factors. As with independent ANOVA this value can be compared against a critical value based on its degrees of freedom (df_M and df_R), which are 3 and 21 in this case.

13.3.7. The between-participant sum of squares ②

I mentioned that the total variation is broken down into a within-participant variation and a between-participant variation. We sort of forgot about the between-participant variation because we didn't need it to calculate the *F*-ratio. However, I will just briefly mention what it represents. The easiest way to calculate this term is by subtraction, because we know from Figure 13.2 that:

$$\text{SS}_\text{T} = \text{SS}_\text{B} + \text{SS}_\text{W}$$

Now, we have already calculated SS_W and SS_T so by rearranging the equation and replacing the values of these terms, we get:

$$\begin{aligned}\text{SS}_\text{B} &= \text{SS}_\text{T} - \text{SS}_\text{W}\\ \text{SS}_\text{B} &= 253.89 - 236.50\\ &= 17.39\end{aligned}$$

This term represents individual differences between cases. So, in this example, different celebrities will have different tolerances of eating these sorts of food. This is shown by the means for the celebrities in Table 13.2. For example, celebrity 4 (M = 4.50) was, on average, more than 2 seconds quicker to retch than participant 8 (M = 6.75). Celebrity 8 just had a better constitution than celebrity 4. The between-participant sum of squares reflects these differences between individuals. In this case only 17.08 units of variation in the times to retch can be explained by individual differences between our celebrities.

13.4. One-way repeated measures ANOVA using SAS

13.4.1. The main analysis ②

Sticking with the bushtucker example, we know that *each row of the data editor should represent data from one entity while each column represents a level of a variable*

(SAS Tip 3.2). Therefore, separate columns represent levels of a repeated-measures variables. As such, there is no need for a coding variable (as with between-group designs). The data are in Table 13.2 and can be entered into SAS in the same format as this table (you don't need to include the columns labelled Celebrity, Mean or s^2 as they were included only to clarify that the celebrities ate the same food and to help explain how this ANOVA is calculated). Alternatively, these data can also be found in the file **Bushtucker.sas7bdat.**

To conduct an ANOVA using a repeated-measures design we use PROC GLM, just as we did with the one-way and two-way ANOVA. We have four outcome variables for each person (stick, ball, eye, and witchetty) so we put all four of these on the left-hand side of the equals sign. So that SAS doesn't think we want a univariate test of each of these, we also put NOUNI in the options.

We have a slightly weird situation here in that we have an independent variable (as before) but now we have not got a variable in the dataset that represents the independent variable. We need to tell SAS that these four measures are levels of a variable, and we do that, with the REPEATED statement, where we label the variable – we'll call it **animal.** We'd also like a test of sphericity – to get that, we add the /PRINTE option to the REPEATED line. At this stage, you don't need to tell SAS how many different groups make up the animal variable, but it doesn't hurt, so we'll put a 4 in there.

We end up with SAS Syntax 13.1.

```
PROC GLM DATA=chap13.bushtucker;
    MODEL stick ball eye witchetty = /NOUNI;
    REPEATED animal 4 /PRINTE;
RUN;
```

SAS Syntax 13.1

13.4.2. Defining contrasts for repeated measures

This being SAS, there are many different ways to incorporate contrasts into your repeated measures. We will have a look at two of them,

The first way to do contrasts is to add a statement to the REPEATED line that tells SAS you want to contrast each variable against one of the other variables. We haven't got an obvious control variable to select, so it is not obvious which one is best to use – we'll use **stick,** just because it was the first variable in the list. Because **stick** was the first variable on the list, we write CONTRAST(1) on the REPEATED line, if we'd wanted the second variable, we'd have written CONTRAST(2), and so no.

The second way to do contrasts is using a MANOVA statement. The MANOVA statement is very similar to the CONTRAST or ESTIMATE statement that we saw in Section 10.2.11. We set weights for the four variables, which sum to zero and set up the contrasts that we want. We will compare **stick** (which was the first variable in the list) with **ball** (the second variable). Because we don't have a between-subjects measure (just wait a chapter or two if you are eager) we specify that we are using the intercept only with H=intercept, and then specify the weights. For each of these different ways of setting up the contrast, we need to request a summary. The syntax is shown in SAS Syntax 13.2.

```
PROC GLM DATA=chap13.bushtucker;
    MODEL stick ball eye witchetty = /NOUNI;
```

```
      REPEATED animal 4 CONTRAST(1) /SUMMARY;
      MANOVA H=intercept M=(1 -1 0 0 ) /SUMMARY ;
RUN;
```

SAS Syntax 13.2

Each line only allows one contrast, but we can repeat the lines with different contrasts. For example, to contrast each food with each other type of food, we can use:

```
      REPEATED animal 4 CONTRAST(1) /SUMMARY;
      REPEATED animal 4 CONTRAST(2) /SUMMARY;
      REPEATED animal 4 CONTRAST(3) /SUMMARY;
      REPEATED animal 4 CONTRAST(4) /SUMMARY;
```

In the above syntax, the first line compares each food to the first (stick), the second line compares each food to the second (ball), the third line compares each food to the third (eye) and the fourth line compares each food to the fourth (witchetty).

Similarly, we can repeat the MANOVA line:

```
      MANOVA H=intercept M=(1 -1 0 0 ) /SUMMARY ;
      MANOVA H=intercept M=(1 0 -1 0 ) /SUMMARY ;
      MANOVA H=intercept M=(1 0 0 -1 ) /SUMMARY ;
```

Here, the first line compares stick and ball, the second line compares stick and eye, the third line compares stick and witchetty. Using the MANOVA approach means that we take three lines to do what we can do with one line in a REPEATED statement. The advantage of the MANOVA approach is that it allows us to make other theoretically derived comparisons. For example, we can compare the insects (witchetty grub[4] and stick insect) with the non-insects (fish eye and kangaroo testicle). The stick insect was the first variable and the witchetty grub was the fourth, so we use:

```
      MANOVA H=intercept M=(1 -1 -1 1 ) /summary ;
```

Putting that all together gives SAS Syntax 13.3.

```
PROC GLM DATA=chap13.bushtucker;
      MODEL stick ball eye witchetty = /NOUNI;
      REPEATED animal 4 CONTRAST(1) /SUMMARY PRINTE;
      REPEATED animal 4 CONTRAST(2) /SUMMARY;
      REPEATED animal 4 CONTRAST(3) /SUMMARY;
      REPEATED animal 4 CONTRAST(4) /SUMMARY;
      MANOVA H=intercept M= (1 -1  0  0) / SUMMARY ;
      MANOVA H=intercept M= (1  0 -1  0) / SUMMARY;
      MANOVA H=intercept M= (1  0  0 -1) / SUMMARY;
      MANOVA H=intercept M= (1 -1 -1  1) / SUMMARY;
RUN;
```

SAS Syntax 13.3

[4] A witchetty grub is a larva of a moth, and is therefore an insect, even if it doesn't look much like one.

JANE SUPERBRAIN 13.2

Sphericity and post hoc tests ③

The issue of sphericity makes life horribly complicated for the world of repeated measures. The violation of sphericity has implications for multiple comparisons which give rise to problems. Because of these issues, SAS does not allow you to do *post hoc* tests. Instead, you have two options. The first is to use the contrasts available to make all possible contrasts, and then use Bonferroni correction (which is available in PROC MULTTEST) to correct the *p*-values. The second, which is preferable, but harder, is to use a multilevel modelling approach to repeated measures, which is done using PROC MIXED, and which we look at in Chapter 19. The multilevel modelling approach is a lot more useful, because we don't treat the failure of sphericity as a violation of an assumption which needs to be corrected, instead we try to model the covariance structure and incorporate it into our model. A second advantage of PROC MIXED is that all of the CONTRAST and TEST features that are in PROC GLM are also available in PROC MIXED.

The issue of modelling repeated-measures ANOVA has, to a large extent, disappeared from modern statistics, because of the multilevel modelling approach.

13.5. Output for one-way repeated-measures ANOVA ②

SAS gives a lot of output for a repeated-measures ANOVA. We're not going to worry about all of it here – some of it we'll cover in more detail when we get to Chapter 16, on MANOVA.

13.5.1. Model description ①

The first part of the output (SAS output 13.1) shows a summary of the model. It's worth checking this to make sure that you are using the correct data – make sure the sample size is right, and check that the contrasts are doing what you think they are doing.

Number of Observations Read	8
Number of Observations Used	8

Repeated Measures Level Information				
Dependent Variable	stick	ball	eye	witchetty
Level of food	1	2	3	4

SAS OUTPUT 13.1

13.5.2. Assessing and correcting for sphericity: Mauchly's test

In section 13.2.3 you were told that SAS produces a test of whether the data violate the assumption of sphericity. Mauchly's test should be non-significant if we are to assume that the condition of sphericity has been met. SAS Output 13.2 shows Mauchly's test for the bushtucker data; the important row is the one labelled Orthogonal Components, and the important column is the one containing the significance value. The significance value (.044) is less than the critical value of .05, so we reject the assumption that the variances of the differences between levels are equal. In other words, the assumption of sphericity has been violated. Knowing that we have violated this assumption a pertinent question is: how should we proceed?

SAS OUTPUT 13.2

Sphericity Tests				
Variables	DF	Mauchly's Criterion	Chi-Square	Pr > ChiSq
Transformed Variates	5	0.1308828	11.635872	0.0401
Orthogonal Components	5	0.136248	11.405981	0.0439

We discovered in section 13.2.5 that SAS produces two corrections based upon the estimates of sphericity advocated by Greenhouse and Geisser (1959) and Huynh and Feldt (1976). Both of these estimates give rise to a correction factor that is applied to the degrees of freedom used to assess the observed *F*-ratio. The closer the *Greenhouse–Geisser correction*, $\hat{\varepsilon}$, is to 1, the more homogeneous the variances of differences, and hence the closer the data are to being spherical. In a situation in which there are four conditions (as with our data) the lower limit of $\hat{\varepsilon}$ will be $1/(4-1)$, or 0.33 (the lower-bound estimate in the table). The second table in SAS Output 13.3 shows that the calculated value of $\hat{\varepsilon}$ is 0.533. This is closer to the lower limit of 0.33 than it is to the upper limit of 1 and it therefore represents a substantial deviation from sphericity. We will see how these values are used in the next section.

SAS Tip 13.1 My Mauchly's test looks weird

Sometimes the SAS output for Mauchly's test is missing. Naturally, you fear that SAS has gone crazy and is going to break into your bedroom at night and tattoo the equation for the Greenhouse–Geisser correction on your face. The reason that this happens is that (as I mentioned in section 13.2.1) you need at least three conditions for sphericity to be an issue (read that section if you want to know why). Therefore, if you have a repeated-measures variable that has only two levels then sphericity is met. Hence, the estimates computed by SAS are 1 (perfect sphericity) and the resulting significance test cannot be computed (hence the reason why the table isn't there). This might confuse you if you remember to look for it, and can't find it. Maybe it should just print in big letters 'Hooray! Hooray! Sphericity has gone away!' We can dream.[5]

[5] Or we can do multilevel models.

13.5.3. The main ANOVA ②

SAS Output 13.3 shows the results of the ANOVA for the within-subject variable. This table can be read much the same as for one-way between-group ANOVA (see Chapter 10). There is a sum of squares for the repeated-measures effect of **food**, which tells us how much of the total variability is explained by the experimental effect. Note the value of 83.13, which is the model sum of squares (SS_M) that we calculated in section 13.3.3. There is also an error term, which is the amount of unexplained variation across the conditions of the repeated-measures variable. This is the residual sum of squares (SS_R) that was calculated in section 13.3.4 and note that the value is 153.38 (which is the same value as calculated). As I explained earlier, these sums of squares are converted into mean squares by dividing by the degrees of freedom. As we saw before, the *df* for the effect of **Animal** is simply $k-1$, where k is the number of levels of the independent variable. The error *df* is $(n-1)(k-1)$, where n is the number of participants (in this case, the number of celebrities) and k is as before. The F-ratio is obtained by dividing the mean squares for the experimental effect (27.71) by the error mean squares (7.30). As with between-group ANOVA, this test statistic represents the ratio of systematic variance to unsystematic variance. The value of $F = 3.79$ (the same as we calculated earlier) is then compared against a critical value for 3 and 21 degrees of freedom. SAS displays the exact significance level for the F-ratio. The significance of F is .0256, which is significant because it is less than the criterion value of .05. We can, therefore, conclude that there was a significant difference between the four animals in their capacity to induce retching when eaten. However, this main test does not tell us which animals differed from each other.

						Adj Pr > F	
Source	DF	Type III SS	Mean Square	F Value	Pr > F	G - G	H - F
food	3	83.1250000	27.7083333	3.79	0.0256	0.0626	0.0483
Error(food)	21	153.3750000	7.3035713				

Greenhouse-Geisser Epsilon	0.5328
Huynh-Feldt Epsilon	0.6658

SAS OUTPUT 13.3

Although this result seems very plausible, we have learnt that the violation of the sphericity assumption makes the F-test inaccurate. We know from the second table that these data were non-spherical and so we need to make allowances for this violation. The table in SAS Output 13.3 shows the F-ratio and associated degrees of freedom when sphericity is assumed (in the F-value and Pr > F columns) and the significant F-statistic indicated some difference(s) between the mean time to retch after eating the four animals. This table also contains corrected p-values for the for the two different types of adjustment (Greenhouse–Geisser, labelled as G-G, Huynh–Feldt, labelled as H-F).

The new degrees of freedom are then used to ascertain the significance of F. For these data the corrections result in the observed F being non-significant when using the Greenhouse–Geisser correction (because $p > .05$). However, it was noted earlier that this correction is quite conservative, and so can miss effects that genuinely exist. It is, therefore, useful to consult the Huynh–Feldt corrected F-statistic. Using this correction, the F-value is still significant because the probability value of .048 is just below the criterion value of .05. So, by this correction we would accept the hypothesis that the lecturers differed in their marking. However, it was also noted earlier that this correction is quite liberal and so tends to accept values as significant when, in reality, they are not significant. This leaves us with

SAS Tip 13.2 How are df adjusted?

The degrees of freedom have been adjusted using the estimates of ε (epsilon) shown in SAS Output 13.3. The adjustment is made by multiplying the degrees of freedom by the estimate of sphericity (see the previous Oliver Twisted). However, SAS does not print the corrected degrees of freedom – if you want to see them, you have to calculate them.

We can calculate them by setting up a dataset with variables F, equal to 4.79, df1, equal to 3, df2 equal to 21, and epsilon, equal to 0.5328. We'll then calculate the *p*-value using the CDF (cumulative distribution function) function. The cumulative distribution function gives the probability of a value lower than the value of F; we want to know the probability of a value higher than F, so we need to subtract it from 1. The code is shown below.

```
DATA sphericity_corrected;
      df1 = 3;
      df2 = 21;
      F = 3.79 ;
      p_spher = 1 - cdf("F", F, df1, df2);
      epsilon_gg = 0.5328;
      df1_gg=df1 * epsilon_gg;
      df2_gg=df2 * epsilon_gg;
      p_gg = 1 - CDF("F", F, df1_gg, df2_gg);
      RUN;
PROC PRINT data=sphericity_corrected;
     RUN;
```

This produces the following. p_spher gives the *p*-value when sphericity is assumed, and df1_gg and df2_gg give the Greenhouse–Geisser corrected *df*, and p_gg gives the Greenhouse–Geisser corrected *p*-value.

```
df1  df2   F      p_spher   gg       df1_gg  df2_gg   p_gg
3    21   3.79   0.026    0.533   1.598    11.189   0.0627
```

the puzzling dilemma of whether or not to accept this *F*-statistic as significant (and also illustrates how ridiculous it is to have a fixed criterion like .05 against which to determine significance). I mentioned earlier that Stevens (2002) recommends taking an average of the two estimates, and certainly when the two corrections give different results (as is the case here) this can be useful. If the two corrections give rise to the same conclusion it makes little difference which you choose to report (although if you accept the *F*-statistic as significant you might as well report the more conservative Greenhouse–Geisser estimate to avoid criticism). Although it is easy to calculate the average of the two correction factors and to correct the degrees of freedom accordingly, it is not so easy to then calculate an exact probability for those degrees of freedom. Therefore, should you ever be faced with this perplexing situation (and to be honest that's fairly unlikely) I suggest taking an average of the two significance values to give you a rough idea of which correction is giving the most accurate answer. In this case, the average of the two *p*-values is $(.063 + .048)/2 = .056$. Therefore, we should probably go with the Greenhouse–Geisser correction and conclude that the *F*-ratio is non-significant.

These data illustrate how important it is to use a valid critical value of *F*: it can potentially mean the difference between making a Type I error and not. However, it also highlights how arbitrary it is that we use a .05 level of significance. These two corrections produce significance values that differ by only .015 and yet they lead to completely opposite conclusions. The decision about 'significance' has, in some ways, become rather arbitrary. The *F*, and hence the size of effect, is unaffected by these corrections and so whether the *p* falls slightly above or slightly below .05 is less important than how big the effect is. We might be well advised to look at an effect size to see whether the effect is substantive regardless of its significance.

We also saw earlier that a final option, when you have data that violate sphericity, is to use multivariate test statistics (MANOVA – see Chapter 16), because they do not make this assumption (see O'Brien and Kaiser, 1985). MANOVA is covered in depth in Chapter 16, but the repeated-measures procedure in SAS automatically produces multivariate test statistics. SAS Output 13.4 shows the multivariate test statistics for this example (details of these test statistics can be found in section 16.4.4.). The column displaying the significance values shows that the multivariate tests are significant (because *p* is .002, which is less than the criterion value of .05). This result supports a decision to conclude that there are significant differences between the time taken to retch after eating different animals.

SAS OUTPUT 13.4

MANOVA Test Criteria and Exact F Statistics for the Hypothesis of no food Effect H = Type III SSCP Matrix for food E = Error SSCP Matrix S=1 M=0.5 N=1.5					
Statistic	Value	F Value	Num DF	Den DF	Pr > F
Wilks' Lambda	0.05823014	26.96	3	5	0.0016
Pillai's Trace	0.94176986	26.96	3	5	0.0016
Hotelling-Lawley Trace	16.17323850	26.96	3	5	0.0016
Roy's Greatest Root	16.17323850	26.96	3	5	0.0016

13.5.4. Contrasts ②

We requested contrasts in two differents ways (section 13.4.2). First, we used the CONTRAST() option on the REPEATED line, and second, we used the MANOVA statement. Both of these statements produce a lot of output, and there's a lot of repetition, so we won't go through it all.

First, we'll look at the output that we requested by using

```
REPEATED animal 4 CONTRAST(1) /SUMMARY PRINTE;
```

The output from this statement is shown in SAS Output 13.5. Each of the tables is a test of the first level of food compared to a later level. We need to check SAS Output 13.1 to make sure we know which is which – to remind you, stick insect is number 1, kangaroo testicle is number 2, fish eye is number 3 and witchetty grub is number 4.

The first contrast is a comparison of stick insect, and kangaroo testicle. The value of *F* is 22.8, with 1, 7 *df*, is significant at $p = .002$, and as this is below 0.05, this is a significant difference. The second contrast compares the stick insect with the fish eye – the *p*-value here is .001, and is less than 0.05 and is significant. The third, and final, contrast is the comparison of stick insect with witchetty grub – the value of *F* of 1.76, with 1, 9 df, has a *p*-value of .227, and as this is greater than the 0.05 cut-off, this is not a statistically significant difference.

food_N represents the contrast between the nth level of food and the 1st

Contrast Variable: food_2

SAS OUTPUT 13.5

Source	DF	Type III SS	Mean Square	F Value	Pr > F
Mean	1	120.1250000	120.1250000	22.80	0.0020
Error	7	36.8750000	5.2678571		

Contrast Variable: food_3

Source	DF	Type III SS	Mean Square	F Value	Pr > F
Mean	1	128.0000000	128.0000000	29.87	0.0009
Error	7	30.0000000	4.2857143		

Contrast Variable: food_4

Source	DF	Type III SS	Mean Square	F Value	Pr > F
Mean	1	45.1250000	45.1250000	1.76	0.2267
Error	7	179.8750000	25.6964286		

This repeats itself a further three times, once for each time we had a repeated line with a contrast statement on it. We won't go through them all, as the interpretation is pretty straightforward.

We also asked for contrasts with a MANOVA statement:

```
MANOVA H=intercept M= (1 -1 0 0) / SUMMARY ;
```

Here, we had to put one contrast per line rather than getting three contrasts per line. The output from this statement starts off with the M matrix, shown in SAS Output 13.6. It's worth checking this, to ensure that we have coded everything correctly. It appears that we have a contrast between stick insect and kangaroo testicle for our first contrast – this is what we thought we put, so that's good.

SAS OUTPUT 13.6

M Matrix Describing Transformed Variables				
	stick	ball	eye	Witchetty
MVAR1	1	–1	0	0

Source	DF	Type III SS	Mean Square	F Value	Pr > F
Intercept	1	120.1250000	120.1250000	22.80	0.0020
Error	7	36.8750000	5.2678571		

Again, this is a MANOVA table and so it appears a little scary, but it will seem less scary when we've covered it in Chapter 16.[6] You can see that although there are four different tests, they are all giving the same result – an F of 22.8 with 1, 7 df, and a probability value of .002, hence this test suggests that there is a significant difference in the time to retch for the stick insect and fish eyeball. You might also notice that this result is exactly the same as the result we obtained in the equivalent contrast. The second table in the output also provides us with the sums of squares and mean squares.

[6] Well, that's what I hope, anyway.

SAS OUTPUT 13.7

MANOVA Test Criteria and Exact F Statistics for the Hypothesis of No Overall Intercept Effect on the Variables Defined by the M Matrix Transformation H = Type III SSCP Matrix for Intercept E = Error SSCP Matrix S=1 M= -0.5 N=2.5					
Statistic	Value	F Value	Num DF	Den DF	Pr > F
Wilks' Lambda	0.23487261	22.80	1	7	0.0020
Pillai's Trace	0.76512739	22.80	1	7	0.0020
Hotelling-Lawley Trace	3.25762712	22.80	1	7	0.0020
Roy's Greatest Root	3.25762712	22.80	1	7	0.0020

Source	DF	Type III SS	Mean Square	F Value	Pr > F
Intercept	1	120.1250000	120.1250000	22.80	0.0020
Error	7	36.8750000	5.2678571		

However, it's worth remembering that by some criteria our main effect of the type of animal eaten was not significant, and if this is the case then the significant contrasts should be ignored. We have to make some kind of decision about whether we think there really is an effect of eating different animals or not before we can look at further tests. Personally, given the multivariate tests, I would be inclined to conclude that the main effect of animal was significant and proceed with further tests. The important point to note is that the sphericity in our data has raised the issue that statistics is not a recipe book and that sometimes we have to use our own discretion to interpret data (it's comforting to know that the computer does not have all of the answers – but it's alarming to realize that this means that we have to know some of the answers ourselves).

CRAMMING SAM'S TIPS

- The one-way repeated-measures ANOVA compares several means, when those means have come from the same participants; for example, if you measured people's statistical ability each month over a year-long course.
- In repeated-measures ANOVA there is an additional assumption: sphericity. This assumption needs to be considered only when you have three or more repeated-measures conditions. Test for sphericity using *Mauchly's* test. Find the table with this label: value in the column labelled Pr > ChiSq is less than .05 then the assumption is violated. If the significance of Mauchly's test is greater than .05 then the assumption of sphericity has been met.
- Look at the table with the main result of your ANOVA. If the assumption of sphericity has been met then look at the **Pr** > ***F*** column to obtain the significance value. If the assumption was violated then read the adjusted probability from the column labelled G-G (*Greenhouse–Geisser*) (you can also look at H-F – *Huynh–Feldt* but you'll have to read this chapter to find out the relative merits of the two procedures). If the value is less than .05 then the means of the groups are significantly different.
- For contrasts tests, again look to the columns labelled Pr > F to discover if your comparisons are significant (they will be if the significance value is less than .05).

13.6. Effect sizes for repeated-measures ANOVA ③

As with independent ANOVA the best measure of the overall effect size is omega squared (ω^2). However, just to make life even more complicated than it already is, the equations we've previously used for omega squared can't be used for repeated-measures data! If you do use the same equation on repeated-measures data it will slightly overestimate the effect size. For the sake of simplicity some people do use the same equation for one-way independent and repeated-measures ANOVAs (and I'm guilty of this in another book), but I'm afraid that in this book we're going to hit simplicity in the face with Stingy the particularly poison-ridden jellyfish, and embrace complexity like a particularly hot date.

In repeated-measures ANOVA, the equation for omega squared is (hang onto your hats):

$$\omega^2 = \frac{\left[\frac{k-1}{nk}(\mathrm{MS_M} - \mathrm{MS_R})\right]}{\mathrm{MS_R} + \frac{\mathrm{MS_B} - \mathrm{MS_R}}{k} + \left[\frac{k-1}{nk}(\mathrm{MS_M} - \mathrm{MS_R})\right]} \tag{13.1}$$

I know what you're thinking and it's something along the lines of 'are you having a bloody laugh?'. Well, no, I'm not, but really the equation isn't too bad if you break it down. First, there are some mean squares that we've come across before (and calculated before). There's the mean square for the model ($\mathrm{MS_M}$) and the residual mean square ($\mathrm{MS_R}$) both of which can be obtained from the ANOVA table that SAS produces. There's also k, the number of conditions in the experiment, which for these data would be 4 (there were four animals), and there's n, the number of people that took part (in this case, the number of celebrities, 8). The main problem is this term $\mathrm{MS_B}$. Back at the beginning of section 13.3 (Figure 13.2) I mentioned that the total variation is broken down into a within-participant variation and a between-participant variation. In section 13.3.7 we saw that we could calculate this term from:

$$\mathrm{SS_T} = \mathrm{SS_B} + \mathrm{SS_W}$$

The problem is that SAS doesn't give us $\mathrm{SS_W}$ in the output, but we know that this is made up of $\mathrm{SS_M}$ and $\mathrm{SS_R}$, which we are given. By substituting these terms and rearranging the equation we get:

$$\mathrm{SS_T} = \mathrm{SS_B} + \mathrm{SS_M} + \mathrm{SS_R}$$

$$\mathrm{SS_B} = \mathrm{SS_T} - \mathrm{SS_M} - \mathrm{SS_R}$$

The next problem is that SAS, which is clearly trying to hinder us at every step, doesn't give us $\mathrm{SS_T}$ and I'm afraid (unless I've missed something in the output) you're just going to have to calculate it by hand (see section 13.3.1). From the values we calculated earlier, you should get:

$$\begin{aligned}\mathrm{SS_B} &= 253.89 - 83.13 - 153.38 \\ &= 17.38\end{aligned}$$

The next step is to convert this to a mean squares by dividing by the degrees of freedom, which in this case are the number of people in the sample minus 1 ($N - 1$):

$$\begin{aligned}\mathrm{MS_B} &= \frac{\mathrm{SS_B}}{df_\mathrm{B}} = \frac{\mathrm{SS_B}}{N-1} \\ &= \frac{17.38}{8-1} \\ &= 2.48\end{aligned}$$

Having done all this and probably died of boredom in the process, we must now resurrect our corpses with renewed vigour for the effect size equation, which becomes:

$$\omega^2 = \frac{\left[\frac{4-1}{8\times 4}(27.71-7.30)\right]}{7.30+\frac{2.48-7.30}{4}+\left[\frac{4-1}{8\times 4}(27.71-7.30)\right]}$$
$$= \frac{1.91}{8.01}$$
$$= .24$$

So, we get an omega squared of .24.

I've mentioned at various other points that it's actually more useful to have effect size measures for focused comparisons anyway (rather than the main ANOVA), and so a slightly easier approach to calculating effect sizes is to calculate them for the contrasts we did (see 10.5). For these we can use the equation that we've seen before to convert the *F*-values (because they all have 1 degree of freedom for the model) to *r*:

$$r = \sqrt{\frac{F(1, df_R)}{F(1, df_R) + df_R}}$$

For the three comparisons we did, we would get:

$$r_{\text{Stick insect vs. kangaroo testicle}} = \sqrt{\frac{20.80}{20.80+7}} = .86$$

$$r_{\text{kangaroo testicle vs. fish eyeball}} = \sqrt{\frac{0.01}{0.01+7}} = .04$$

$$r_{\text{Fish eyeball vs. witchetty grub}} = \sqrt{\frac{0.80}{0.80+7}} = .32$$

The difference between the stick insect and the testicle was a large effect, between the fish eye and witchetty grub a medium effect, but between the testicle and eyeball a very small effect.

13.7. Reporting one-way repeated-measures ANOVA ②

When we report repeated-measures ANOVA, we give the same details as for an independent ANOVA. The only additional thing we should concern ourselves with is reporting the corrected degrees of freedom if sphericity was violated. Personally, I'm also keen on reporting the results of sphericity tests as well. As with the independent ANOVA the degrees of freedom used to assess the *F*-ratio are the degrees of freedom for the effect of the model ($df_M = 1.60$) and the degrees of freedom for the residuals of the model ($df_R = 11.19$). Remember that in this example we corrected both using the Greenhouse–Geisser estimates of sphericity, which is why the degrees of freedom are as they are. Therefore, we could report the main finding as:

- The results show that the time to retch was not significantly affected by the type of animal eaten, $F(1.60, 11.19) = 3.79$, $p > .05$.

However, as I mentioned earlier, because the multivariate tests were significant we should probably be confident that the differences between the animals is significant. We could report these multivariate tests. There are four different test statistics, but in most situations you should probably report Pillai's trace, V (see Chapter 16). You should report the value of V as well as the associated F and its degrees of freedom (all from SAS Output 13.6). If you choose to report the sphericity test as well, you should report the chi-square approximation, its degrees of freedom and the significance value. It's also nice to report the degree of sphericity by reporting the epsilon value. We'll also report the effect size in this improved version:

- ✓ Mauchly's test indicated that the assumption of sphericity had been violated, $\chi^2(5) = 11.41$, $p < .05$, therefore multivariate tests are reported ($\varepsilon = .53$). The results show that the time to retch was significantly affected by the type of animal eaten, $V = 0.94$, $F(3, 5) = 26.96$, $p < .01$, $\omega^2 = .24$.
- ✓ Mauchly's test indicated that the assumption of sphericity had been violated, $\chi^2(5) = 11.41$, $p < .05$, therefore degrees of freedom were corrected using Huynh–Feldt estimates of sphericity ($\varepsilon = .67$). The results show that the time to retch was significantly affected by the type of animal eaten, $F(2, 13.98) = 3.79$, $p < .05$, $\omega^2 = .24$.

LABCOAT LENI'S REAL RESEARCH 13.1

Who's afraid of the big bad wolf? ②

I'm going to let my ego get the better of me and talk about some of my own research. When I'm not scaring my students with statistics, I scare small children with Australian marsupials. There is a good reason for doing this, which is to try to discover how children develop fears (which will help us to prevent them). Most of my research looks at the effect of giving children information about animals or situations that are novel to them (rather like a parent, teacher or TV show would do). In one particular study (Field, 2006), I used three novel animals (the quoll, quokka and cuscus) and children were told negative things about one of the animals, positive things about another, and were given no information about the third (our control). I then asked the children to place their hands in three wooden boxes each of which they believed contained one of the aforementioned animals. My hypothesis was that they would take longer to place their hand in the box containing the animal about which they had heard negative information.

The data from this part of the study are in the file **Field2006.sas7bdat.** Labcoat Leni wants you to carry out a one-way repeated-measures ANOVA on the times taken for children to place their hands in the three boxes (negative information, positive information, no information). First, draw an error bar graph of the means, then do some normality tests on the data, then do a log-transformation on the scores, and do the ANOVA on these log-transformed scores (if you read the paper you'll notice that I found that the data were not normal, so I log-transformed them before doing the ANOVA). Do children take longer to put their hands in a box that they believe contains an animal about which they have been told nasty things?

Answers are in the additional material on the companion website (or look at page 748 in the original article).

FIELD, A. P. (2006). *JOURNAL OF ABNORMAL PSYCHOLOGY, 115*(4), 742–752.

13.8. Repeated-measures with several independent variables ②

We have already seen that simple between-group designs can be extended to incorporate a second (or third) independent variable. It is equally easy to incorporate a second, third or even fourth independent variable into a repeated-measures analysis. As an example, some social scientists were asked to research whether imagery could influence public attitudes towards alcohol. There is evidence that attitudes towards stimuli can be changed using positive and negative imagery (Field, 2005c; Stuart, Shimp, & Engle, 1987) and these researchers were interested in answering two questions. On the one hand, the government had funded them to look at whether negative imagery in advertising could be used to change attitudes towards alcohol. Conversely, an alcohol company had provided funding to see whether positive imagery could be used to improve attitudes towards alcohol. The scientists designed a study to address both issues. Table 13.3 illustrates the experimental design and contains the data for this example (each row represents a single participant).

TABLE 13.3 Data from **Attitude.sas7bdat**

Drink	***Beer***			***Wine***			***Water***		
Image	**+ve**	**−ve**	**Neut**	**+ve**	**−ve**	**Neut**	**+ve**	**−ve**	**Neut**
Male	1	6	5	38	−5	4	10	−14	−2
	43	30	8	20	−12	4	9	−10	−13
	15	15	12	20	−15	6	6	−16	1
	40	30	19	28	−4	0	20	−10	2
	8	12	8	11	−2	6	27	5	−5
	17	17	15	17	−6	6	9	−6	−13
	30	21	21	15	−2	16	19	−20	3
	34	23	28	27	−7	7	12	−12	2
	34	20	26	24	−10	12	12	−9	4
	26	27	27	23	−15	14	21	−6	0
Female	1	−19	−10	28	−13	13	33	−2	9
	7	−18	6	26	−16	19	23	−17	5
	22	−8	4	34	−23	14	21	−19	0
	30	−6	3	32	−22	21	17	−11	4
	40	−6	0	24	−9	19	15	−10	2
	15	−9	4	29	−18	7	13	−17	8
	20	−17	9	30	−17	12	16	−4	10
	9	−12	−5	24	−15	18	17	−4	8
	14	−11	7	34	−14	20	19	−1	12
	15	−6	13	23	−15	15	29	−1	10

Participants viewed a total of nine mock adverts over three sessions. In one session, they saw three adverts: (1) a brand of beer (Brain Death) presented with a negative image (a dead body with the slogan 'Drinking Brain Death makes your liver explode'); (2) a brand of wine (Dangleberry) presented in the context of a positive image (a sexy naked man or woman – depending on the participant's preference – and the slogan 'Drinking Dangleberry wine makes you a horny stud muffin'); and (3) a brand of water (Puritan) presented alongside a neutral image (a person watching television accompanied by the slogan 'Drinking Puritan water makes you behave completely normally'). In a second session (a week later), the participants saw the same three brands, but this time Brain Death was accompanied by the positive imagery, Dangleberry by the neutral image and Puritan by the negative. In a third session, the participants saw Brain Death accompanied by the neutral image, Dangleberry by the negative image and Puritan by the positive. After each advert participants were asked to rate the drinks on a scale ranging from −100 (dislike very much) through 0 (neutral) to 100 (like very much). The order of adverts was randomized, as was the order in which people participated in the three sessions. This design is quite complex. There are two independent variables: the type of drink (beer, wine or water) and the type of imagery used (positive, negative or neutral). These two variables completely cross over, producing nine experimental conditions.

13.8.1. Calculating and comparing means ②

Getting descriptive statistics will help us to understand the effects in this analysis. We can't get the effects directly out of SAS PROC GLM, and in order to get the means we need to make the data long – this means that we need to break our golden rule that a person goes on a row and taking up a lot of columns (which is called wide). Instead, we need to put a person on multiple rows, and only have one column.

The first step is to add an ID variable, in a DATA step.

```
DATA attitude; SET chap13.attitude;
      id = _n_;
      RUN;
```

Then we transpose the data, to make them long:

```
PROC TRANSPOSE DATA=attitude OUT=attitudelong;
      VAR beerpos - waterneu ;
      BY id;
      RUN;
```

Finally, we label the drink and imagery variables:

```
DATA attitudelong; set attitudelong;
      IF _name_ = "BEERPOS" OR
         _name_ = "BEERNEG" OR
         _name_ = "BEERNEUT" then drink = "BEER";
      IF _name_ = "WINEPOS" OR
         _name_ = "WINENEG" OR
         _name_ = "WINENEUT" then drink = "WINE";
      IF _name_ = "WATERPOS" OR
         _name_ = "WATERNEG" OR
         _name_ = "WATERNEU" then drink = "WATER";
      IF _name_ = "BEERPOS" OR
         _name_ = "WATERPOS" OR
         _name_ = "WINEPOS" then imagery = "POS";
      IF _name_ = "BEERNEG" OR
         _name_ = "WATERNEG" OR
         _name_ = "WINENEG" then imagery = "NEG";
      IF _name_ = "BEERNEUT" OR
         _name_ = "WATERNEU" OR
         _name_ = "WINENEUT" then imagery = "NEUT";
      RUN;
```

Then we can use PROC MEANS to produce a table of means and standard deviations. We want to examine the main effects, and the interactions, so we need to call PROC MEANS three times, one time for the means for the three drinks, one time for the means for the three types of imagery, and one time for the interaction.

```
PROC MEANS DATA=attitudelong MEAN STD;
      VAR col1;
      CLASS drink;
      RUN;
PROC MEANS DATA=attitudelong MEAN STD;
      VAR col1;
      CLASS imagery ;
      RUN;
PROC MEANS DATA=attitudelong MEAN STD;
      VAR col1;
      CLASS imagery drink;
      RUN;
```

13.8.2. The main analysis ②

To enter these data into SAS we need to remember that each row represents a single participant's data. If a person participates in all experimental conditions (in this case (s)he sees all types of stimuli presented with all types of imagery) then each experimental condition must be represented by a column in the data editor. In this experiment there are nine experimental conditions and so the data need to be entered in nine columns (so the format is identical to Table 13.3). You should create the following nine variables in a DATA step, and for each one, you should also create a label name (see section XXXX) for clarity in the output:

beerpos	Beer	+	Sexy Person
beerneg	Beer	+	Corpse
beerneut	Beer	+	Person in Armchair
winepos	Wine	+	Sexy Person
wineneg	Wine	+	Corpse
wineneut	Wine	+	Person in Armchair
waterpos	Water	+	Sexy Person
waterneg	Water	+	Corpse
waterneut	Water	+	Person in Armchair

SELF-TEST Enter the data as in Table 13.3. If you have problems entering the data then use the file **Attitude.sas7bdat.**

We need to be careful when we are setting up a GLM with two repeated measures factors, because we have to tell PROC GLM that there are two sets of predictors, and we have to get them into the correct order.

The MODEL statement looks just like it did before.

```
MODEL beerpos beerneg beerneut winepos wineneg wineneut waterpos
waterneg waterneu = /NOUNI;
```

That's a long list of variable names – it would be handy if there were an easier way to list them, so we didn't have to do all that typing, and risk making a mistake.

SAS Tip 13.3 Using variable lists?

SAS has a couple of handy ways of referring to a long string of variables.

The first of these is to just name the first and last variable. If your variables all end with numbers and are in order you can use a single hyphen, for example for x1, x2, x3, x4, x5 you can use: `x1-x5`.

If they do not end in a number, you can use a double hyphen. Instead of writing `beerpos beerneg beerneut winepos wineneg wineneut waterpos waterneg waterneu`, we can write: `beerpos -- waterneu`.

That's pretty cool of SAS, I hear you say, but wait! There is more!

```
DATA sphericity_corrected;
```

If we have many variables with the same intial letters, we can use a colon as a 'wildcard'. If we want variables q1 q2 q3 q4 q5a q5b, and if these variables are not in order *and* if they are the only variables that being with the letter Q, we can use q: The colon means choose all variables that start with that name.

Finally, we can choose all of the variables by using `_ALL_`, and all of the numeric variables by using `_NUMERIC_`.

I can think of three ways to refer to the list of variables in the current example. Can you?

Hey! Thanks for the tip. So, we can write:

```
MODEL beerpos -- waterneu = /NOUNI;
```

(Or, because we are using all of the variables in the dataset, we could write `_ALL_`, and finally, because all of our variables are numeric, we could write `_NUMERIC_`.)

On the REPEATED line, we need to tell SAS which variables go together to make which factor. SAS assumes that the variables are listed in an order so that the first set of variables are the first level of the first variable, the second set are the second set of the first variable, and so on. Within each of these sets, it expects the second set to be in the same order. Let's try to explain that with an example. We will put display `REPEATED drink 3, imagery 3` So SAS knows that drink has three levels – SAS doesn't know that those levels are beer, wine and water, but we do. So SAS expects the first three variables to be beer, the next three to be wine, and the final three to be water. That's good, because that's what we have.

Then SAS knows that imagery has three levels. These are positive, negative, and neutral. SAS doesn't know what the labels are, but it expects them to be in the same order within each drink. This is really easy to get wrong, and luckily SAS produces a table in the output that means we can check.

The syntax for the repeated measures model is:

```
PROC GLM DATA=chap13.attitude;
      MODEL BEERPOS -- WATERNEU = /NOUNI;
      REPEATED drink 3, imagery 3 /PRINTE SUMMARY;
      RUN;
```

13.9. Output for factorial repeated-measures ANOVA ②

13.9.1. Descriptive statistics ②

SAS Output 13.8 shows the output from PROC MEANS in the form of three tables, and provides the mean and standard deviation for each of the nine conditions. (Note that the standard deviations in the drink and imagery tables are a combination of between-variable and within variable variance, and are not really interpretable. You should only look at the standard deviations in the third table.)

SAS OUTPUT 13.7

Analysis Variable : COL1			
drink	N Obs	Mean	Std Dev
BEER	60	11.8333333	15.2795051
WATE	60	3.5166667	12.9110276
WINE	60	8.3333333	16.7783826

Analysis Variable : COL1			
imagery	N Obs	Mean	Std Dev
NEG	60	–5.5833333	13.2719810
NEU	60	8.0000000	8.8566742
POS	60	21.2666667	9.7959983

Analysis Variable : COL1				
imagery	drink	N Obs	Mean	Std Dev
NEG	BEER	20	4.4500000	17.3037112
	WATE	20	–9.2000000	6.8024763
	WINE	20	–12.0000000	6.1814664
NEU	BEER	20	10.0000000	10.2956301
	WATE	20	2.3500000	6.8385517
	WINE	20	11.6500000	6.2431015
POS	BEER	20	21.0500000	13.0079935
	WATE	20	17.4000000	7.0740445
	WINE	20	25.3500000	6.7377569

The descriptive statistics are interesting in that they tell us that the variability among scores was greatest when beer was used as a product (compare the standard deviations of the beer variables against the others). Also, when a corpse image was used, the ratings given to the products were negative (as expected) for wine and water but not for beer (so for some reason, negative imagery didn't seem to work when beer was used as a stimulus). The values in this table will help us later to interpret the main effects of the analysis.

13.9.2. Main analysis ②

The first table produced by PROC GLM is shown in SAS Output 13.8. This table is important, because it shows that SAS has interpreted our repeated measures input correctly.[7]

SAS OUTPUT 13.8

Repeated Measures Level Information		
Dependent Variable	WATERNEG	WATERNEU
Level of drink	3	3
Level of imagery	2	3

Repeated Measures Level Information							
Dependent Variable	BEERPOS	BEERNEG	BEERNEUT	WINEPOS	WINENEG	WINENEUT	WATERPOS
Level of drink	1	1	1	2	2	2	3
Level of imagery	1	2	3	1	2	3	1

Repeated Measures Level Information		
Dependent Variable	WATERNEG	WATERNEU
Level of drink	3	3
Level of imagery	2	3

The order of the output gets a little confusing now, so you might need to hunt around for what you need.

First, we will consider the main effect of drink. We need to look at the Mauchly sphericity test, and the univariate effects – these are in different places, you'll find the sphericity test first, and then the univariate tests later.

SAS Output 13.9 shows the results of the Mauchly's sphericity tests (see section 13.2.3) for each of the three effects in the model (two main effects and one interaction). The significance values of these tests indicate that both the main effects of **drink** and **imagery** have violated this assumption and so the *F*-values should be corrected (see section 13.5.2).

[7] Actually what it usually shows is that we have set up our input incorrectly and we need to do it again.

For the interaction the assumption of sphericity is met (because $p > .05$) and so we need not correct the F-ratio for this effect.

Drink

SAS OUTPUT 13.9

Sphericity Tests				
Variables	DF	Mauchly's Criterion	Chi-Square	Pr > ChiSq
Transformed Variates	2	0.2444113	25.36025	<.0001
Orthogonal Components	2	0.2672411	23.752875	<.0001

Imagery

Sphericity Tests				
Variables	DF	Mauchly's Criterion	Chi-Square	Pr > ChiSq
Transformed Variates	2	0.6459192	7.8674566	0.0196
Orthogonal Components	2	0.6621013	7.4220619	0.0245

Drink * Imagery Interaction

Sphericity Tests				
Variables	DF	Mauchly's Criterion	Chi-Square	Pr > ChiSq
Transformed Variates	9	0.2438851	24.575927	0.0035
Orthogonal Components	9	0.595044	9.0413389	0.4335

SAS Output 13.10 shows the results of the ANOVA (with corrected F-values). The output is split into sections that refer to each of the effects in the model and the error terms associated with these effects. By looking at the significance values it is clear that there is a significant effect of the type of drink used as a stimulus, a significant main effect of the type of imagery used and a significant interaction between these two variables. I will examine each of these effects in turn.

SAS OUTPUT 13.10

						Adj Pr > F	
Source	DF	Type III SS	Mean Square	F Value	Pr > F	G–G	H–F
drink	2	2092.344444	1046.172222	5.11	0.0109	0.0298	0.0288
Error(drink)	38	7785.877778	204.891520				

Greenhouse-Geisser Epsilon	0.5771
Huynh-Feldt Epsilon	0.5907

						Adj Pr > F	
Source	DF	Type III SS	Mean Square	F Value	Pr > F	G–G	H–F
imagery	2	21628.67778	10814.33889	122.56	<.0001	<.0001	<.0001
Error(imagery)	38	3352.87778	88.23363				

Greenhouse-Geisser Epsilon	0.7474
Huynh-Feldt Epsilon	0.7968

						Adj Pr > F	
Source	DF	Type III SS	Mean Square	F Value	Pr > F	G–G	H–F
drink*imagery	4	2624.422222	656.105556	17.15	<.0001	<.0001	<.0001
Error(drink*imagery)	76	2906.688889	38.245906				

SAS OUTPUT 13.10

Greenhouse-Geisser Epsilon	0.7984
Huynh-Feldt Epsilon	0.9786

13.9.3. The effect of drink ②

The first part of SAS Output 13.10 tells us the effect of the type of drink used in the advert. For this effect we must look at one of the corrected significance values because sphericity was violated (see above). All of the corrected values are significant and so we should report the conservative Greenhouse–Geisser *p*-value. This effect tells us that if we ignore the type of imagery that was used, participants still rated some types of drink significantly differently.

If we look at the first table in SAS Output 13.7 again, it is clear that beer and wine were rated higher than water (with beer being rated most highly). To see the nature of this effect we can look at contrasts (see section 13.9.6).

13.9.4. The effect of imagery ②

SAS Output 13.10 also indicates that the effect of the type of imagery used in the advert had a significant influence on participants' ratings of the stimuli. Again, we must look at one of the corrected significance values because sphericity was violated (see above). All of the corrected values are highly significant and so we can again report the Greenhouse–Geisser *p*-value. This effect tells us that if we ignore the type of drink that was used, participants' ratings of those drinks were different according to the type of imagery that was used. Looking again at SAS Output 13.7 we can see that positive imagery resulted in very positive ratings (compared to the neutral imagery) and negative imagery resulted in negative ratings (especially compared to the effect of neutral imagery). To see the nature of this effect we can look at contrasts (see section 13.9.6).

13.9.5. The interaction effect (drink × imagery) ②

SAS Output 13.10 indicates that imagery interacted in some way with the type of drink used as a stimulus. From that table we should report that there was a significant interaction between the type of drink used and imagery associated with it, $F(4, 76) = 17.16$, $p < .0001$. This effect tells us that the type of imagery used had a different effect depending on which type of drink it was presented alongside. As before, we can use the means from SAS Output 13 .7 to determine the nature of this interaction).

The ideal way to look at an interaction effect is in a line graph. SAS doesn't make it especially easy to draw a line graph of these data. We often prefer to copy the means into another program (like Excel or OpenOffice) and draw the graph that way. But this is a book about SAS, so we'll show you the syntax to do it in SAS Syntax 13.4, and the graph that it produces in Figure 13.3.

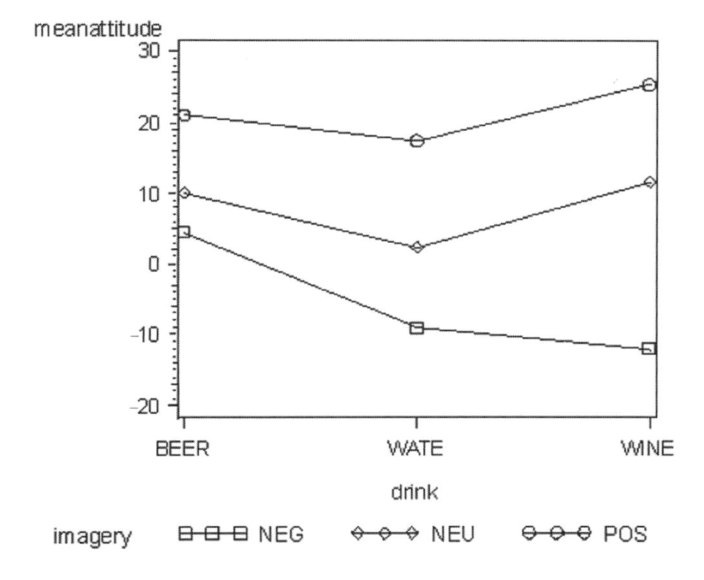

FIGURE 13.3 Interaction plot for the effect of imagery and drink on attitude

Figure 13.3 shows the interaction graph, and we are looking for non-parallel lines. The graph shows that the pattern of responding across drinks was similar when positive and neutral imagery were used. That is, ratings were positive for beer, slightly higher for wine and then went down slightly for water. The fact that the line representing positive imagery is higher than the neutral line indicates that positive imagery gave rise to higher ratings than neutral imagery across all drinks. The bottom line (representing negative imagery) shows a different effect: ratings were lower for wine and water but not for beer. Therefore, negative imagery had the desired effect on attitudes towards wine and water, but for some reason attitudes towards beer remained fairly neutral. Therefore, the interaction is likely to reflect the fact that negative imagery has a different effect to both positive and neutral imagery (because it decreases ratings rather than increasing them). This interaction is completely in line with the experimental predictions. To verify the interpretation of the interaction effect, we need to look at the contrasts.

13.9.6. Contrasts for repeated-measures variables ②

SAS produces a set of contrasts for the repeated-measures effects, comparing each variable to the last one. I set up the data so that the last of the three drinks was water, and the last of the three imagery types was neutral, because these provide appropriate controls. SAS Output 13.11 shows the summary results for these contrasts. The table is split up into main effects and interactions, and each effect is split up into components of the contrast. So, for the main effect of drink, the first contrast compares level 1 (beer) against the base category (in this case, the last category: water). This result is significant, $F(1, 19) = 6.22$, $p < .05$, which contradicts what was found using the *post hoc* tests.

SELF-TEST Why do you think that this contradiction has occurred?

The next contrast compares level 2 (wine) with the base category (water) and confirms the significant difference found with the *post hoc* tests, $F(1, 19) = 18.61$, $p < .001$. For the imagery main effect, the first contrast compares level 1 (positive) to the base category (the last category: neutral) and verifies the significant difference found with the *post hoc* tests, $F(1, 19) = 142.19$, $p < .0001$. The second contrast confirms the significant difference in ratings found in the negative imagery condition compared to the neutral, $F(1, 19) = 47.07$, $p < .0001$. These contrasts are all very well, but they tell us only what we already knew (although note the increased statistical power with these tests, shown by the higher significance values). The contrasts become much more interesting when we look at the interaction term.

SAS OUTPUT 13.11

drink_N represents the contrast between the nth level of drink and the last
Contrast Variable: drink_1

Source	DF	Type III SS	Mean Square	F Value	Pr > F
Mean	1	12450.05000	12450.05000	6.22	0.0220
Error	19	38040.95000	2002.15526		

*Contrast Variable: drink_*

Source	DF	Type III SS	Mean Square	F Value	Pr > F
Mean	1	4176.050000	4176.050000	18.61	0.0004
Error	19	4262.950000	224.365789		

imagery_N represents the contrast between the nth level of imagery and the last
Contrast Variable: drink_1

Source	DF	Type III SS	Mean Square	F Value	Pr > F
Mean	1	31680.80000	31680.80000	142.19	<.0001
Error	19	4233.20000	222.80000		

Contrast Variable: drink_2

Source	DF	Type III SS	Mean Square	F Value	Pr > F
Mean	1	33211.25000	33211.25000	47.07	<.0001
Error	19	13405.75000	705.56579		

drink_N represents the contrast between the nth level of drink and the last
imagery_N represents the contrast between the nth level of imagery and the last

*Contrast Variable: drink_1*imagery_1*

Source	DF	Type III SS	Mean Square	F Value	Pr > F
Mean	1	320.000000	320.000000	1.58	0.2246
Error	19	3858.000000	203.052632		

*Contrast Variable: drink_1*imagery_2*

Source	DF	Type III SS	Mean Square	F Value	Pr > F
Mean	1	720.000000	720.000000	6.75	0.0176
Error	19	2026.000000	106.631579		

*Contrast Variable: drink_2*imagery_1*

Source	DF	Type III SS	Mean Square	F Value	Pr > F
Mean	1	36.450000	36.450000	0.24	0.6334
Error	19	2946.550000	155.081579		

*Contrast Variable: drink_2*imagery_2*

Source	DF	Type III SS	Mean Square	F Value	Pr > F
Mean	1	2928.200000	2928.200000	26.91	<.0001
Error	19	2067.800000	108.831579		

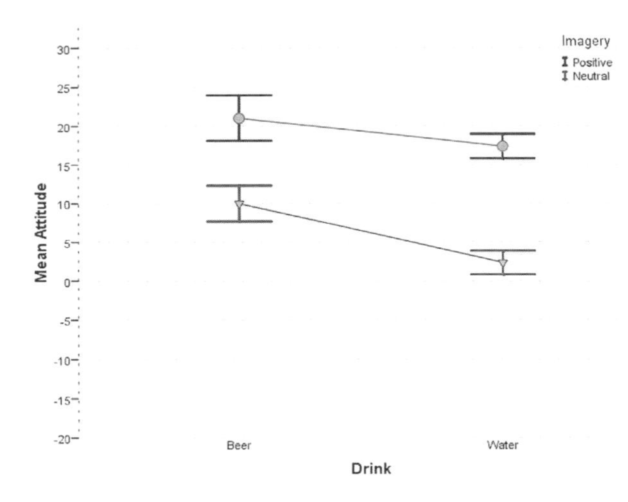

13.9.6.1. Beer vs. water, positive vs. neutral imagery ②

The first The contrasts are particularly useful for interpreting the interactions, which are the final four tables in SAS Output 14.11. The first interaction term looks at level 1 of drink (beer) compared to level 3 (water), when positive imagery (level 1) is used compared to neutral (level 3). This contrast is non-significant. This result tells us that the increased liking found when positive imagery is used (compared to neutral imagery) is the same for both beer and water. In terms of the interaction graph (Figure 13.3) it means that the distance between the circle and the diamond in the beer condition is the same as the distance between the circle and the diamond in the water condition. It's easy to see that the lines are approximately parallel, indicating no significant interaction effect. We could conclude that the improvement of ratings due to positive imagery compared to neutral is not affected by whether people are evaluating beer or water.

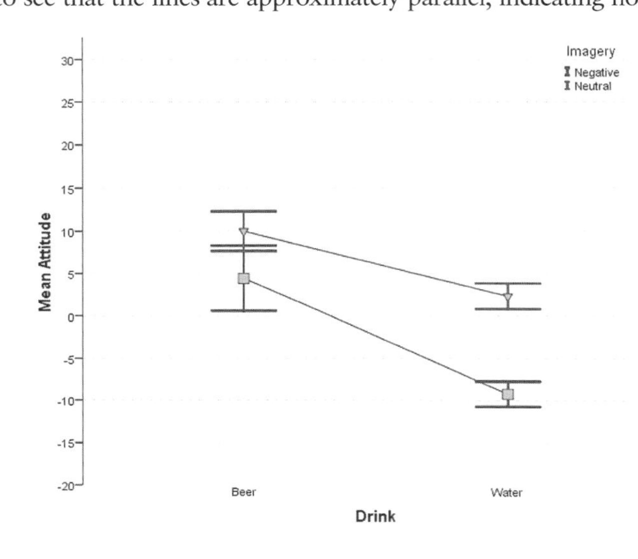

13.9.6.2. Beer vs. water, negative vs. neutral imagery ②

The second interaction term looks at level 1 of drink (beer) compared to level 3 (water), when negative imagery (level 2) is used compared to neutral (level 3). This contrast is significant, $F(1, 19) = 6.75$, $p < .05$. This result tells us that the decreased liking found when negative imagery is used (compared to neutral imagery) is different when beer is used compared to when water is used. In terms of the interaction graph (Figure 13.3) it means that the distance between the square and the diamond in the beer condition (a small difference) is significantly smaller than the distance between the square and the triangle in the water condition (a larger difference). We could conclude that the decrease in ratings due to negative imagery (compared to neutral) found when water is used in the advert is smaller than when beer is used.

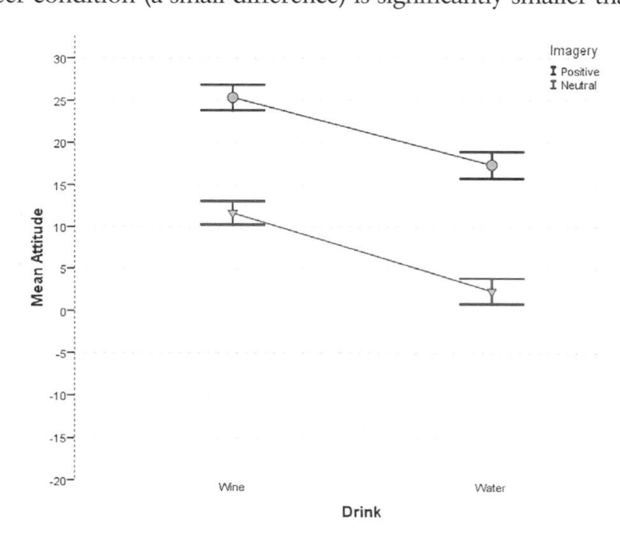

13.9.6.3. Wine vs. water, positive vs. neutral imagery ②

The third interaction term looks at level 2 of drink (wine) compared to level 3 (water), when positive imagery (level 1) is used compared to neutral (level 3). This contrast is non-significant, indicating that the increased liking found when positive imagery is used (compared to neutral imagery) is the same for both wine and water.

same as the distance between the circle and the diamond in the water condition. Again, we can see that the lines are close to parallel, indicating no interaction effect. We could conclude that the improvement of ratings due to positive imagery compared to neutral is not affected by whether people are evaluating wine or water.

13.9.6.4. Wine vs. water, negative vs. neutral imagery ②

The final interaction term looks at level 2 of drink (wine) compared to level 3 (water), when negative imagery (level 2) is used compared to neutral (level 3). This contrast is significant, $F(1, 19) = 26.91, p < .0001$. This result tells us that the decreased liking found when negative imagery is used (compared to neutral imagery) is different when wine is used compared to when water is used. In terms of the interaction graph (Figure 13.3) it means that the distance between the square and the diamond in the wine condition (a big difference) is significantly larger than the distance between the square and the diamond in the water condition (a smaller difference). We could conclude that the decrease in ratings due to negative imagery (compared to neutral) is significantly greater when wine is advertised than when water is advertised.

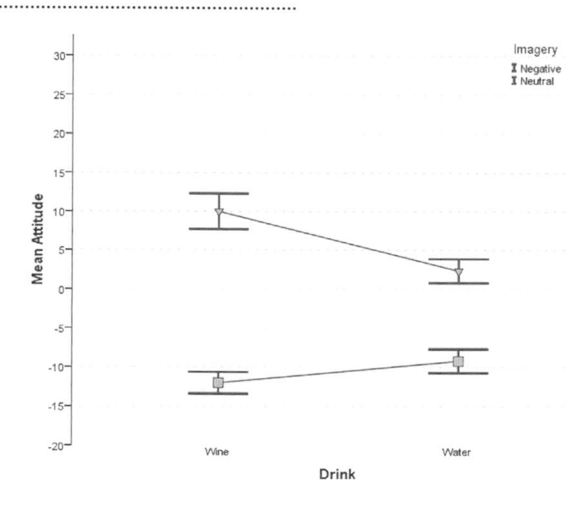

13.9.6.5. Limitations of these contrasts ②

These contrasts, by their nature, tell us nothing about the differences between the beer and wine conditions (or the positive and negative conditions) and different contrasts would have to be run to find out more. However, what is clear so far is that, relative to the neutral condition, positive images increased liking for the products more or less regardless of the product; however, negative imagery had a greater effect on wine and a lesser effect on beer. These differences were not predicted. Although it may seem tiresome to spend so long interpreting an analysis so thoroughly, you are well advised to take such a systematic approach if you want to truly understand the effects that you obtain. Interpreting interaction terms is complex, and I can think of a few well-respected researchers who still struggle with them, so don't feel disheartened if you find them hard. Try to be thorough, and break each effect down as much as possible using contrasts and hopefully you will find enlightenment.

CRAMMING SAM'S TIPS

- Two-way repeated-measures ANOVA compares several means when there are two independent variables, and the same participants have been used in all experimental conditions.
- Test the assumption of *sphericity* when you have three or more repeated-measures conditions. Test for sphericity using *Mauchly's test.* Find the table with labelled Sphericity Tests: if the value for the *Orthogonal Components* in the column labelled *Pr > ChiSq.* is less than .05 then the assumption is violated. If the significance of Mauchly's test is greater than

.05 then the assumption of sphericity has been met. You should test this assumption for all effects (in a two-way ANOVA this means you test it for the effect of both variables and the interaction term).

- The tables in the section labelled *Univariate Tests of Hypotheses for Within Subject Effects* shows the main result of your ANOVA. In a two-way ANOVA you will have three effects: a main effect of each variable and the interaction between the two. For each effect, if the assumption of sphericity has been met then look at the *Pr* > *F* column to obtain the significance value. If the assumption was violated then read the *Adj Pr F* and the G-G (Greenhouse-Geisser) column. (you can also look at H-F [*Huynh–Feldt*] but you'll have to read this chapter to find out the relative merits of the two procedures). If the value is less than .05 then the means of the groups are significantly different.
- Break down the main effects and interaction terms using contrasts. These contrasts appear in the section labelled *Analysis of Variance of Contrast Variables*; again look to the columns labelled *Pr* > *F* to discover if your comparisons are significant (they will be if the significance value is less than .05)

13.10. Effect sizes for factorial repeated-measures ANOVA ③

Calculating omega squared for a one-way repeated-measures ANOVA was hair-raising enough, and as I keep saying, effect sizes are really more useful when they describe a focused effect, so I'd advise calculating effect sizes for your contrasts when you've got a factorial design (and any main effects that compare only two groups). SAS Output 13.11 shows the values for several contrasts, all of which have 1 degree of freedom for the model (i.e. they represent a focused and interpretable comparison) and have 19 residual degrees of freedom. We can use these *F*-ratios and convert them to an effect size *r*, using a formula we've come across before:

$$r = \sqrt{\frac{F(1, df_R)}{F(1, df_R) + df_R}}$$

For the two comparisons we did for the drink variable we would get:

$$r_{\text{Beer vs. Water}} = \sqrt{\frac{6.22}{6.22 + 19}} = .50$$

$$r_{\text{Wine vs. Water}} = \sqrt{\frac{18.61}{18.61 + 19}} = .70$$

Therefore, both comparisons yielded very large effect sizes. For the two comparisons we did for the imagery variable we would get:

$$r_{\text{Positive vs. Neutral}} = \sqrt{\frac{142.19}{142.19 + 19}} = .94$$

$$r_{\text{Negative vs. Neutral}} = \sqrt{\frac{47.07}{47.07 + 19}} = .84$$

Again, both comparisons yield very large effect sizes. For the interaction term, we had four contrasts, but again we can convert them to r because they all have 1 degree of freedom for the model:

$$r_{\text{Beer vs. Water, Positive vs. Neutral}} = \sqrt{\frac{1.58}{1.58 + 19}} = .28$$

$$r_{\text{Beer vs. Water, Negative vs. Neutral}} = \sqrt{\frac{6.75}{6.75 + 19}} = .51$$

$$r_{\text{Wine vs. Water, Positive vs. Neutral}} = \sqrt{\frac{0.24}{0.24 + 19}} = .11$$

$$r_{\text{Wine vs. Water, Negative vs. Neutral}} = \sqrt{\frac{26.91}{26.91 + 19}} = .77$$

As such, the two effects that were significant (beer vs. water, negative vs. neutral and wine vs. water, negative vs. neutral) yield large effect sizes. The two effects that were not significant yielded a medium effect size (beer vs. water, positive vs. neutral) and a small effect size (wine vs. water, positive vs. neutral).

13.11. Reporting the results from factorial repeated-measures ANOVA ②

We can report a factorial repeated-measures ANOVA in much the same way as any other ANOVA. Remember that we've got three effects to report, and these effects might have different degrees of freedom. For the main effects of drink and imagery, the assumption of sphericity was violated so we'd have to report the Greenhouse–Geisser corrected degrees of freedom. We can, therefore, begin by reporting the violation of sphericity:

- ✓ Mauchly's test indicated that the assumption of sphericity had been violated for the main effects of drink, $\chi^2(2) = 23.75, p < .0001$, and imagery, $\chi^2(2) = 7.42, p < .05$. Therefore degrees of freedom were corrected using Greenhouse–Geisser estimates of sphericity ($\varepsilon = .58$ for the main effect of drink and .75 for the main effect of imagery).

We can then report the three effects from this analysis as follows:

- ✓ All effects are reported as significant at $p < .05$. There was a significant main effect of the type of drink on ratings of the drink, $F(1.15, 21.93) = 5.11$. Contrasts revealed that ratings of beer, $F(1, 19) = 6.22$, $r = .50$, and wine, $F(1, 19) = 18.61$, $r = .70$, were significantly higher than water.

✓ There was also a significant main effect of the type of imagery on ratings of the drinks, $F(1.50, 28.40) = 122.57$. Contrasts revealed that ratings after positive imagery were significantly higher than after neutral imagery, $F(1, 19) = 142.19$, $r = .94$. Conversely, ratings after negative imagery were significantly lower than after neutral imagery, $F(1, 19) = 47.07$, $r = .84$.

✓ There was a significant interaction effect between the type of drink and the type of imagery used, $F(4, 76) = 17.16$. This indicates that imagery had different effects on people's ratings depending on which type of drink was used. To break down this interaction, contrasts were performed comparing all drink types to their baseline (water) and all imagery types to their baseline (neutral imagery). These revealed significant interactions when comparing negative imagery to neutral imagery both for beer compared to water, $F(1, 19) = 6.75$, $r = .51$, and wine compared to water, $F(1, 19) = 26.91$, $r = .77$. Looking at the interaction graph, these effects reflect that negative imagery (compared to neutral) lowered scores significantly more in water than it did for beer, and lowered scores significantly more for wine than it did for water. The remaining contrasts revealed no significant interaction term when comparing positive imagery to neutral imagery both for beer compared to water, $F(1, 19) = 1.58$, $r = .28$, and wine compared to water, $F(1, 19) < 1$, $r = .11$. However, these contrasts did yield small to medium effect sizes.

13.12. What to do when assumptions are violated in repeated-measures ANOVA ③

When you have only one independent variable then you can use a non-parametric test called Friedman's ANOVA (see Chapter 15) if you find that you your assumptions are being irksome. However, for factorial repeated measures designs there is not a non-parametric counterpart. At the risk of sounding like a broken record, this means that if our data violate the assumption of normality then the solution is to read Rand Wilcox's book. I know, I say this in every chapter, but it really is the definitive source. So, to get a robust method (see section 5.7.4) for factorial repeated-measures ANOVA designs, read Wilcox's Chapter 8, get the associated files for the software R (Wilcox, 2005) to implement these procedures, and away you go to statistics oblivion.

What have I discovered about statistics? ②

This chapter has helped us to walk through the murky swamp of repeated-measures designs. We discovered that is was infested with rabid leg-eating crocodiles. The first thing we learnt was that with repeated-measures designs there is yet another assumption to worry about: *sphericity*. Having recovered from this shock revelation, we were fortunate to discover that this assumption, if violated, can be easily remedied. Sorted! We then moved on to look at the theory of repeated-measures ANOVA for one independent variable. Although not essential by any stretch of the imagination, this was a useful exercise to demonstrate that basically it's exactly the same as when we have an independent design (well, there are a few subtle differences but I was trying to emphasize the similarities). We then worked through an example on SAS, before tackling the particularly foul-tempered, starving hungry, and mad as 'Stabby' the mercury-sniffing hatter, piranha fish of omega squared. That's a road I kind of regretted going down after I'd started, but, stubborn as ever, I persevered. This led us ungracefully on to factorial repeated-measures designs and specifically the situation where we have two independent variables. We learnt that as with other factorial designs we have to worry about interaction terms. But, we also discovered some useful ways to break these terms down using contrasts.

By 16 I had started my first 'serious' band. We actually stayed together for about 7 years (with the same line-up, and we're still friends now) before Mark (drummer) moved to Oxford, I moved to Brighton to do my Ph.D., and rehearsing became a mammoth feat of organization. We had a track on a CD, some radio play and transformed from a thrash metal band to a blend of Fugazi, Nirvana and metal. I never split my trousers during a gig again (although I did once split my head open). Why didn't we make it? Well, Mark was an astonishingly good drummer so it wasn't his fault, the other Mark was an extremely good bassist too (of the three of us he is the one that has always been in a band since we split up), so the weak link was me. This was especially unfortunate given that I had three roles in the band (guitar, singing, songs) – my poor band mates never stood a chance.☺ I stopped playing music for quite a few years after we split. I still wrote songs (for personal consumption) but the three of us were such close friends that I couldn't bear the thought of playing with other people. At least not for a few years ...

Key terms that I've discovered

Compound symmetry
Greenhouse–Geisser correction
Huynh–Feldt correction
Lower bound
Mauchly's test
Repeated-measures ANOVA
Sphericity

Smart Alex's tasks

- **Task 1**: There is often concern among students as to the consistency of marking between lecturers. It is common that lecturers obtain reputations for being 'hard' or 'light' markers (or to use the students' terminology, 'evil manifestations from Beelzebub's bowels' and 'nice people') but there is often little to substantiate these reputations. A group of students investigated the consistency of marking by submitting the same essays to four different lecturers. The mark given by each lecturer was recorded for each of the eight essays. It was important that the same essays were used for all lecturers because this eliminated any individual differences in the standard of work that each lecturer marked. This design is repeated measures because every lecturer marked every essay.

The independent variable was the lecturer who marked the report and the dependent variable was the percentage mark given. The data are in the file **TutorMarks.sas7bdat**. Conduct a one-way ANOVA on these data by hand. ②

- **Task 2**: Repeat the analysis above on SAS and interpret the results. ②
- **Task 3**: Imagine I wanted to look at the effect alcohol has on the roving eye. The 'roving eye' effect is the propensity of people in relationships to 'eye up' members of the opposite sex. I took 20 men and fitted them with incredibly sophisticated glasses that could track their eye movements and record both the movement and the object being observed (this is the point at which it should be apparent that I'm making it up as I go along). Over four different nights I plied these poor souls with 1, 2, 3 or 4 pints of strong lager in a night-club. Each night I measured how many different women they eyed up (a woman was categorized as having been eyed up if the man's eye moved from her head to her toe and back up again). To validate this measure we also collected the amount of dribble on the man's chin while looking at a woman. The data are in the file **RovingEye.sas7bdat.** Analyse them with a one-way ANOVA. ②
- **Task 4**: In the previous chapter we came across the beer-goggles effect, a severe perceptual distortion after imbibing several pints of beer. The specific visual distortion is that previously unattractive people suddenly become the hottest thing since Spicy Gonzalez' extra-hot Tabasco-marinated chillies. In short, one minute you're standing in a zoo admiring the orang-utans, and the next you're wondering why someone would put Angela Gossow in a cage. Anyway, in that chapter, a blatantly fabricated data set demonstrated that the beer-goggles effect was much stronger for men than women, and took effect only after 2 pints. Imagine we wanted to follow this finding up to look at what factors mediate the beer-goggles effect. Specifically, we thought that the beer goggles effect might be made worse by the fact that it usually occurs in clubs which have dim lighting. We took a sample of 26 men (because the effect is stronger in men) and gave them various doses of alcohol over four different weeks (0 pints, 2 pints, 4 pints and 6 pints of lager). This is our first independent variable, which we'll call alcohol consumption, and it has four levels. Each week (and, therefore, in each state of drunkenness) participants were asked to select a mate in a normal club (that had dim lighting) and then select a second mate in a specially designed club that had bright lighting. As such, the second independent variable was whether the club had dim or bright lighting. The outcome measure was the attractiveness of each mate as assessed by a panel of independent judges. To recap, all participants took part in all levels of the alcohol consumption variable, and selected mates in both brightly and dimly lit clubs. The data are in the file **BeerGogglesLighting.sas7bdat.** Analyse them with a two-way repeated-measures ANOVA. ②

Answers can be found on the companion website.

Further reading

Field, A. P. (1998). A bluffer's guide to sphericity. *Newsletter of the Mathematical, Statistical and Computing section of the British Psychological Society*, 6(1), 13–22. (Available in the additional material for this chapter.)

Howell, D. C. (2006). *Statistical methods for psychology* (6th ed.). Belmont, CA: Duxbury. (Or you might prefer his *Fundamental statistics for the behavioral sciences,* also in its 6th edition, 2007.)

Rosenthal, R., Rosnow, R. L., & Rubin, D. B. (2000). *Contrasts and effect sizes in behavioural research: a correlational approach*. Cambridge: Cambridge University Press. (This is quite advanced but really cannot be bettered for contrasts and effect size estimation.)

Interesting real research

Field, A. P. (2006). The behavioral inhibition system and the verbal information pathway to children's fears. *Journal of Abnormal Psychology*, *115*(4), 742–752.

14 Mixed design ANOVA (GLM 5)

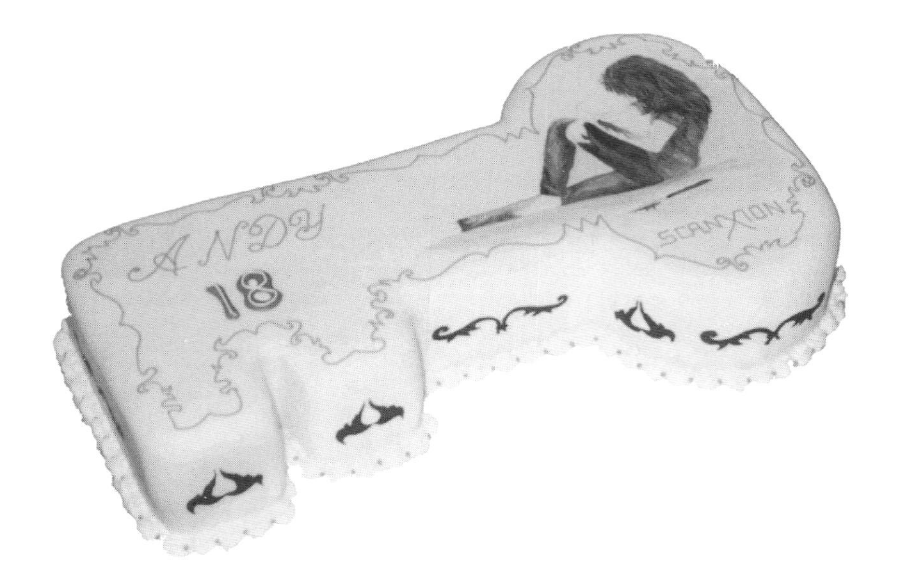

FIGURE 14.1
My 18th birthday cake

14.1. What will this chapter tell me? ①

Most teenagers are anxious and depressed, but I probably had more than my fair share. The parasitic leech that was the all boys grammar school that I attended had feasted on my social skills leaving in its wake a terrified husk. Although I had no real problem with playing my guitar and shouting in front of people, speaking to them was another matter entirely. In the band I felt at ease, in the real world I did not. Your 18th birthday is a time of great joy, where (in the UK at any rate) you cast aside the shackles of childhood and embrace the exciting new world of adult life. Your birthday cake might symbolize this happy transition by reflecting one of your great passions. Mine had a picture on it of a long-haired person who looked somewhat like me, slitting his wrists. That pretty much sums it up. Still, you can't lock yourself in your bedroom with your Iron Maiden CDs for ever and soon enough I tried to integrate with society. Between the ages of 16 and 18 this pretty much involved getting drunk. I quickly discovered that getting drunk made it much easier to speak to people, and getting *really* drunk made you unconscious and then the problem of speaking to people went away entirely. This situation was exacerbated by the sudden presence of girls in my social circle. I hadn't seen a girl since Clair Sparks; they were particularly problematic because not only did you have to talk to them, but what you said had to be really impressive because then they might become your girlfriend. Also, in 1990, girls didn't like to talk about Iron Maiden – they probably still don't. Speed dating[1] didn't exist back then, but if it had it would have been a sick and twisted manifestation of hell on earth for me. The idea of having a highly pressured social situation where you *have* to think of something witty and amusing to say or be thrown to the baying vultures of eternal loneliness would have had me injecting pure alcohol into my eyeballs; at least that way I could be in a coma and unable to see the disappointment on the faces of those forced to spend 3 minutes in my company. That's what this chapter is all about: speed dating, oh, and mixed ANOVA too, but if I mention that you'll move swiftly on to the next chapter when the bell rings.

14.2. Mixed designs ②

If you thought that the previous chapter was bad, well, I'm about to throw an added complication into the mix. We can combine repeated measures and independent designs, and this chapter looks at this situation. As if this wasn't bad enough, I'm also going to use this as an excuse to show you a design with three independent variables (at this point you should imagine me leaning back in my chair, cross-eyed, dribbling and laughing maniacally). A mixture of between-group and repeated-measures variables is called a **mixed design**. It should be obvious that you need at least two independent variables for this type of design to be possible, but you can have more complex scenarios too (e.g. two between-group and one repeated measures, one between-group and two repeated measures, or even two of each). SAS allows you to test almost any design you might want to, and of virtually any degree of complexity. However, interaction terms are difficult enough to interpret with only two variables, so imagine how difficult they are if you include four! The best advice I can offer is to stick to three or fewer independent variables if you want to be able to interpret your interaction terms,[2] and certainly don't exceed four unless you want to give yourself a migraine.

This chapter will go through an example of a **mixed ANOVA**. There won't be any theory because really and truly you've probably had enough ANOVA theory by now to have a good idea of what's going on (you can read this as 'it's too complex for me and I'm going to cover up my own incompetence by pretending you don't need to know about it'). So, we look at an example using SAS and then interpret the output. In the process you'll hopefully develop your understanding of interactions and how to break them down using contrasts.

Also be aware that 'mixed' is used to mean different things in statistics – we are going to discuss a mixed ANOVA, but that's not (necessarily) the same as a mixed model, which we would analyse using PROC MIXED (and which we'll cover in Chapter 19).

[1] In case speed dating goes out of fashion and no one knows what I'm going on about, the basic idea is that lots of men and women turn up to a venue (or just men or just women if it's a gay night), one-half of the group sit individually at small tables and the remainder choose a table, get 3 minutes to impress the other person at the table with their tales of heteroscedastic data, then a bell rings and they get up and move to the next table. Having worked around all of the tables, the end of the evening is spent either stalking the person whom you fancied or avoiding the hideous mutant who was going on about hetero…something or other.

[2] Fans of irony will enjoy the four-.ay ANOVAs that I conducted in Field and Davey (1999) and Field and Moore (2005), to name but two examples!

14.3. What do men and women look for in a partner? ②

The example we're going to use in this chapter stays with the dating theme. It seems that lots of magazines go on all the time about how men and women want different things from relationships (or perhaps it's just my girlfriend's copies of *Marie Claire*, which I don't read – honestly). The big question to which we all want to know the answer is: are looks or personality more important? Imagine you wanted to put this to the test. You devised a cunning plan whereby you'd set up a speed-dating night. Little did the people who came along know that you'd got some of your friends to act as the dates. Each date varied in their attractiveness (attractive, average or ugly) and their charisma (charismatic, average and dull) and by combining these characteristics you get nine different combinations. Each combination was represented by one of your stooge dates. As such, your stooge dates were made up of nine different people. Three were extremely attractive people but differed in their personality: one had tons of charisma,[3] one had some charisma and the other was as dull as this book. Another three people were of average attractiveness, and again differed in their personality: one was highly charismatic, one had some charisma and the third was a dullard. The final three were, not wishing to be unkind in any way, pig-ugly, and again one was charismatic, one had some charisma and the final poor soul was mind-numbingly tedious. Obviously you had two sets of stooge dates: one set was male and the other female so that your participants could match up with dates of the appropriate sex.

The participants themselves were not these nine stooges, but 10 men and 10 women who came to the speed-dating event that you had set up. Over the course of the evening they speed dated all nine members of the opposite sex that you'd set up for them. After their 5-minute date, they rated how much they'd like to have a proper date with the person as a percentage (100% = 'I'd pay large sums of money for your phone number', 0% = 'I'd pay a large sum of money for a plane ticket to get me as far away from you as possible'). As such, each participant rated nine different people who varied in their attractiveness and personality. So, there are two repeated-measures variables: **Looks** (with three levels because the person could be attractive, average or ugly) and **Personality** (again with three levels because the person could have lots of charisma, have some charisma or be a dullard). The people giving the ratings could be male or female, so we should also include the gender of the person making the ratings (male or female), and this, of course, will be a between-group variable. The data are in Table 14.1.

14.4. Mixed ANOVA on SAS ②

To enter these data into SAS we use the same procedure as the two-way repeated-measures ANOVA that we came across in the previous chapter. Remember that each row in the data represents a single participant's data. If a person participates in all experimental conditions (in this case they date all of the people who differ in attractiveness and all of the people who differ in their charisma) then each experimental condition must be represented by a column. In this experiment there are nine experimental conditions and so the data need to be entered in nine columns (the format is identical

[3] The highly attractive people with tons of charisma were, of course, taken to a remote cliff top and shot after the experiment because life is hard enough without having people like that floating around making you feel inadequate.

TABLE 14.1 Data from **LooksOrPersonality.sas7bdat** (Att = Attractive, Av = Average, Ug = Ugly)

	High Charisma			*Some Charisma*			*Dullard*		
Looks	***Att***	***Av***	***Ug***	***Att***	***Av***	***Ug***	***Att***	***Av***	***Ug***
Male	86	84	67	88	69	50	97	48	47
	91	83	53	83	74	48	86	50	46
	89	88	48	99	70	48	90	45	48
	89	69	58	86	77	40	87	47	53
	80	81	57	88	71	50	82	50	45
	80	84	51	96	63	42	92	48	43
	89	85	61	87	79	44	86	50	45
	100	94	56	86	71	54	84	54	47
	90	74	54	92	71	58	78	38	45
	89	86	63	80	73	49	91	48	39
Female	89	91	93	88	65	54	55	48	52
	84	90	85	95	70	60	50	44	45
	99	100	89	80	79	53	51	48	44
	86	89	83	86	74	58	52	48	47
	89	87	80	83	74	43	58	50	48
	80	81	79	86	59	47	51	47	40
	82	92	85	81	66	47	50	45	47
	97	69	87	95	72	51	45	48	46
	95	92	90	98	64	53	54	53	45
	95	93	96	79	66	46	52	39	47

to Table 14.1). Therefore, create the following ten variables in the data editor with the names as given.

at_high	Attractive	+	High Charisma
av_high	Average Looks	+	High Charisma
ug_high	Ugly	+	High Charisma
at_some	Attractive	+	Some Charisma
av_some	Average Looks	+	Some Charisma
ug_some	Ugly	+	Some Charisma
at_none	Attractive	+	Dullard
av_none	Average Looks	+	Dullard
ug_none	Ugly	+	Dullard
gender			

SELF-TEST Once these variables have been created, enter the data as in Table 14.1. If you have problems entering the data then use the file **LooksOrPersonality.sas7bdat.**

Using PROC GLM for a mixed design is very straightforward, if you've already worked through the previous two chapters. We have a categorical independent variable, gender, which we add as a CLASS statement. The MODEL statement includes the list of dependent variables, as with a repeated measures ANOVA, and after the equals sign, the categorical predictor variable.

The REPEATED statement is formatted as it was for repeated measures (see section 13.4).

We would also like to get the means for each group, and to do this we need to run PROC TRANSPOSE, followed by PROC MEANS as we did in Section 13.4. We need to recode the _name_variable, which we'll do this time with the SUBSTR() function (see SAS Tip 14.1). We'll also use LOWCASE() to make sure everything is in lower case.

```
PROC GLM DATA=chap14.looksorpersonality;
      CLASS gender;
      MODEL at_high at_none at_some ug_high
            ug_none ug_some av_high av_none av_some = gender /
            NOUNI ;
      REPEATED  looks 3, charisma 3 /PRINTE SUMMARY;
      MEANS gender /HOVTEST;
      RUN;

*Add an id variable;
DATA looksorpersonality; SET chap14.looksorpersonality;
      id = _N_;
      RUN;

*Transpose data;
PROC TRANSPOSE DATA=looksorpersonality OUT=looksorpersonality long;
      BY id gender;
      VAR at_high at_none at_some ug_high ug_none ug_some av_high
          av_none av_some  ;
      RUN;
*Label measurements;
DATA looksorpersonalitylong; SET looksorpersonalitylong;
      looks = LOWCASE(SUBSTR(_name_, 1, 2));
      charisma = LOWCASE(SUBSTR(_name_, 4, 4));
      RUN;
*Obtain means;
PROC MEANS DATA=looksorpersonalitylong;
      CLASS looks charisma;
      VAR col1;
      RUN;
```
SAS Syntax 14.1

SAS TIP 14.1 **Using SUBSTR()** ①

After we have transposed, we need to tell SAS about the relationship between the categorical variables. SAS doesn't know that at_high, at_some and at_none represent levels of looks, so we need to tell it that this is the case. Last time we ran PROC TRANSPOSE we used nine lines in a DATA step to code the variables. We could do that again, but we are kind of lazy, and want a shorter way. The way you or I would do it would be to look at the variables, see what was there, and use that to decide what the groups are. SAS can do exactly the same thing, with the SUBSTR() command – SUBSTR finds substrings in a character string. SUBSTR needs to know three things: the character string you want to look inside, where you want to start, and how many characters to look for. So SUBSTR("Andy Field", 1, 4) would give 'Andy', and SUBSTR("Jeremy Miles", 8, 5) would give 'Miles'.

14.5. Output for mixed factorial ANOVA: main analysis ③

The initial output is the same as the two-way ANOVA example: there is a table listing the repeated-measures variables from the data editor and the level of each independent variable that they represent – you should check these and make sure they are correct.

Repeated Measures Level Information							
Dependent Variable	ATT_HIGH	ATT_NONE	ATT_SOME	UG_HIGH	UG_NONE	UG_SOME	AV_HIGH
Level of looks	1	1	1	2	2	2	3
Level of charisma	1	2	3	1	2	3	1

SAS OUTPUT 14.1

Looks

Sphericity Tests				
Variables	DF	Mauchly's Criterion	Chi-Square	Pr > ChiSq
Transformed Variates	2	0.5954943	8.8121783	0.0122
Orthogonal Components	2	0.9602054	0.690337	0.7081

SAS OUTPUT 14.2

Charisma

Sphericity Tests				
Variables	DF	Mauchly's Criterion	Chi-Square	Pr > ChiSq
Transformed Variates	2	0.6321467	7.7967732	0.0203
Orthogonal Components	2	0.9293298	1.2459569	0.5363

Looks * Charisma

Sphericity Tests				
Variables	DF	Mauchly's Criterion	Chi-Square	Pr > ChiSq
Transformed Variates	9	0.1034439	37.24491	<.0001
Orthogonal Components	9	0.6133545	8.0246674	0.5317

SELF-TEST SAS Output 14.2 shows the results of Mauchly's sphericity test. Based on what you have already learnt, was sphericity violated?

SAS Output 14.2 shows the results of Mauchly's sphericity tests for each of the three repeated-measures effects in the model. None of the effects violate the assumption of sphericity because all of the values in the column labelled *Pr > ChiSq.* in the Orthogonal Components line are above .05; therefore, we can assume sphericity when we look at our *F*-statistics.

SAS Output 14.3 shows the ANOVA table of the repeated-measures effects in the ANOVA with corrected *F*-values. As with factorial repeated-measures ANOVA, the output is split into sections for each of the effects in the model and their associated error terms. The only difference is that the interactions between our between-group variable of gender and the repeated-measures effects are also included.

Again, we need to look at the column labelled *Pr > F*; and if the values in this column are less than .05 for a particular effect then it is statistically significant. Working down from the top of the table we find a significant effect of looks, which means that if we ignore whether the date was charismatic, and whether the rating was from a man or a woman, then the attractiveness of a person significantly affected the ratings they received. The looks × gender interaction is also significant, which means that although the ratings were affected by whether the date was attractive, average or ugly, the way in which ratings were affected by attractiveness was different in male and female raters.

Next, we find a significant effect of charisma, which means that if we ignore whether the date was attractive, and whether the rating was from a man or a woman, then the charisma of a person significantly affected the ratings they received. The charisma × gender interaction is also significant, indicating that this effect of charisma differed in male and female raters.

There is a significant interaction between looks and charisma, which means that if we ignore the gender of the rater, the profile of ratings across different levels of attractiveness was different for highly charismatic dates, charismatic dates and dullards. (It is equally true to say this the opposite way around: the profile of ratings across different levels of charisma was different for attractive, average and ugly dates.) Just to add to the mounting confusion, the looks × charisma × gender interaction is also significant, meaning that the looks × charisma interaction was significantly different in men and women participants!

This is all a lot to take in so we'll look at each of these effects in turn in subsequent sections. First, though, we need to see what has happened to our main effect of gender.

SAS OUTPUT 14.3

Univariate Tests of Hypotheses for Within Subject Effects						Adj Pr > F	
Source	DF	Type III SS	Mean Square	F Value	Pr > F	G–G	H–F
looks	2	20779.63333	10389.81667	423.73	<.0001	<.0001	<.0001
looks*gender	2	3944.10000	1972.05000	80.43	<.0001	<.0001	<.0001
Error(looks)	36	882.71111	24.51975				

Greenhouse-Geisser Epsilon	0.9617
Huynh-Feldt Epsilon	1.1342

						Adj Pr > F	
Source	DF	Type III SS	Mean Square	F Value	Pr > F	G–G	H–F
charisma	2	23233.60000	11616.80000	328.25	<.0001	<.0001	<.0001
charisma*gender	2	4420.13333	2210.06667	62.45	<.0001	<.0001	<.0001
Error(charisma)	36	1274.04444	35.39012				

Greenhouse-Geisser Epsilon	0.9340
Huynh-Feldt Epsilon	1.0960

						Adj Pr > F	
Source	DF	Type III SS	Mean Square	F Value	Pr > F	G–G	H–F
looks*charisma	4	4055.266667	1013.816667	36.63	<.0001	<.0001	<.0001
looks*charisma*gender	4	2669.666667	667.416667	24.12	<.0001	<.0001	<.0001
Error(looks*charisma)	72	1992.622222	27.675309				

Greenhouse-Geisser Epsilon	0.7994
Huynh-Feldt Epsilon	1.0462

SELF-TEST What is the difference between a main effect and an interaction?

14.5.1. The main effect of gender ②

The main effect of gender is listed separately from the repeated-measures effects in the section labelled **Tests of Hypotheses for Between Subjects Effects.** Before looking at this table it is important to check the assumption of homogeneity of variance using Levene's test (see section 6.6.1). SAS produces a table of Levene's tests for each of the variables, and as we had nine variables, it requires nine tests. You'd think that whoever decided what this should look like really hates trees. For the sake of not making this book so heavy you can't carry it, I've only reproduced the first two tables, in SAS Output 14.4.

SAS OUTPUT 14.4

Levene's Test for Homogeneity of at_high Variance ANOVA of Squared Deviations from Group Means					
Source	DF	Sum of Squares	Mean Square	F Value	Pr > F
gender	1	543.9	543.9	0.33	0.5721
Error	18	29563.6	1642.4		

Levene's Test for Homogeneity of at_none Variance ANOVA of Squared Deviations from Group Means					
Source	DF	Sum of Squares	Mean Square	F Value	Pr > F
gender	1	1256.1	1256.1	1.67	0.2120
Error	18	13502.5	750.1		

SELF-TEST Was the assumption of homogeneity of variance met (for the first two variables, if you haven't run the analysis yourself – see SAS Output 14.4)?

The tables showing the Levene's tests indicate that variances are homogeneous for all levels of the repeated-measures variables (because all significance values are greater than .05; you'll have to believe me about the rest of them if you haven't run the analysis yourself). If any values were significant, then this would compromise the accuracy of the *F*-test for gender, and we would have to consider transforming all of our data to stabilize the variances between groups (see Chapter 6). Fortunately, in this example a transformation is unnecessary. The table in the section labelled *Tests of Hypotheses for Between-Subjects Effects* shows the ANOVA summary table for the main effect of gender, shown in SAS Output 14.5, and this reveals a non-significant effect (because the significance of .946 is greater than the standard cut-off point of .05).

We can report that there was a non-significant main effect of gender, $F(1, 18) = 0.00$, $p = 0.946$. This effect tells us that if we ignore all other variables, male participants' ratings were basically the same as females'. We can use PROC MEANS (on the long data that we created earlier) to obtain a table of means for the main effect of gender. The syntax is shown in SAS Syntax 14.2, and the output (SAS Output 14.6) reveals that the means for the males and females (ignoring everything else) are very similar.

SAS OUTPUT 14.5

Source	DF	Type III SS	Mean Square	F Value	Pr > F
gender	1	0.2000000	0.2000000	0.00	0.9459
Error	18	760.2222222	42.2345679		

```
PROC MEANS DATA=looksorpersonalitylong MEAN;
     CLASS gender ;
     VAR col1;
     RUN;
```

SAS Syntax 14.2

SAS OUTPUT 14.6

Analysis Variable : COL1		
gender	N Obs	Mean
Female	90	68.5333333
Male	90	68.6000000

14.5.2. The main effect of looks ②

We came across the main effect of looks in SAS Output 14.3. Now we're going to have a look at what this effect means. We can report that there was a significant main effect of looks, $F(2, 36) = 423.73$, $p < .0001$. This effect tells us that if we ignore all other variables, ratings were different for attractive, average and unattractive dates.

We can use PROC MEANS to obtain the mean for each of the three levels of looks, using the transposed data, as shown in SAS Syntax 14.3, which produces the table shown in SAS Output 14.7. This shows that the attractive group (at) had the highest scores, followed by the average group, and then followed by the ugly group.

```
PROC MEANS DATA=looksorpersonalitylong MEAN;
      CLASS looks ;
      VAR col1;
      OUTPUT OUT=looksorpersonalitymeans  MEAN=meanscore;
      RUN;
```

SAS Syntax 14.3

SAS OUTPUT 14.7

Analysis Variable : COL1		
looks	N Obs	Mean
at	60	82.1000000
av	60	67.7833333
ug	60	55.8166667

SAS OUTPUT 14.8

looks_N represents the contrast between the nth level of looks and the last
Contrast Variable: looks_1

Source	DF	Type III SS	Mean Square	F Value	Pr > F
Mean	1	36894.05000	36894.05000	226.99	<.0001
gender	1	7031.25000	7031.25000	43.26	<.0001
Error	18	2925.70000	162.53889		

Contrast Variable: looks_2

Source	DF	Type III SS	Mean Square	F Value	Pr > F
Mean	1	25776.20000	25776.20000	160.07	<.0001
gender	1	4867.20000	4867.20000	30.22	<.0001
Error	18	2898.60000	161.03333		

Contrast Variable: charisma_1

Source	DF	Type III SS	Mean Square	F Value	Pr > F
Mean	1	29491.20000	29491.20000	109.94	<.0001
gender	1	7296.20000	7296.20000	27.20	<.0001
Error	18	4828.60000	268.25556		

Contrast Variable: charisma_2

Source	DF	Type III SS	Mean Square	F Value	Pr > F
Mean	1	40500.00000	40500.00000	227.94	<.0001
gender	1	5985.80000	5985.80000	33.69	<.0001
Error	18	3198.20000	177.67778		

looks_N represents the contrast between the nth level of looks and the last
charisma_N represents the contrast between the nth level of charisma and the last
*Contrast Variable: looks_1*charisma_1*

Source	DF	Type III SS	Mean Square	F Value	Pr > F
Mean	1	3976.200000	3976.200000	21.94	0.0002
gender	1	168.200000	168.200000	0.93	0.3481
Error	18	3261.600000	181.200000		

*Contrast Variable: looks_1*charisma_2*

Source	DF	Type III SS	Mean Square	F Value	Pr > F
Mean	1	441.800000	441.800000	4.09	0.0582
gender	1	6552.200000	6552.200000	60.67	<.0001
Error	18	1944.000000	108.000000		

*Contrast Variable: looks_2*charisma_1*

Source	DF	Type III SS	Mean Square	F Value	Pr > F
Mean	1	911.250000	911.250000	6.23	0.0225
gender	1	1711.250000	1711.250000	11.70	0.0030
Error	18	2632.500000	146.250000		

*Contrast Variable: looks_2*charisma_2*

Source	DF	Type III SS	Mean Square	F Value	Pr > F
Mean	1	7334.450000	7334.450000	88.60	<.0001
gender	1	110.450000	110.450000	1.33	0.2632
Error	18	1490.100000	82.783333		

SAS Output 14.8 shows the contrasts. For the time being, just look at the tables labelled looks_1 and looks_2. As SAS tells us, each variable is compared with the last, so we get a contrast comparing level 1 to level 3, and then comparing level 2 to level 3; because of the order in which we entered the variables, these contrasts represent attractive compared to average (level 1 vs. level 3) and ugly compared to average (level 2 vs. level 3). Looking at the values of F for each contrast, and their related significance values, tells us that the effect of attractiveness represents the fact that attractive dates were rated significantly higher than average dates, $F(1, 18) = 226.99$, $p < .0001$, and average dates were rated significantly higher than ugly ones, $F(1, 18) = 160.07$, $p < .0001$.

14.5.3. The main effect of charisma ②

The main effect of charisma is in SAS Output 14.3. We can report that there was a significant main effect of charisma, $F(2, 36) = 328.25$, $p < .0001$. This effect tells us that if we ignore all other variables, ratings were different for highly charismatic, a bit charismatic and dullard people. We can use PROC MEANS to display the means for the three different levels of charisma, shown in SAS Output 14.9. Again, the levels of charisma are labelled simply 1, 2 and 3. You can see that as charisma declines, the mean rating falls too. So this main effect seems to reflect that the raters were more likely to express a greater interest in going out with charismatic people than average people or dullards.

SAS OUTPUT 14.9

Analysis Variable: COL1		
charisma	N Obs	Mean
High	60	82.1000000
None	60	54.3000000
Some	60	69.3000000

Look at the contrasts which are shown in SAS Output 14.8. Looking at the table labelled charisma contrast variable: 1 and contrast variable: charisma 2 and remembering that we get a contrast comparing level 1 to level 3, and then comparing level 2 to level 3. How we interpret these contrasts depends on the order in which we entered the repeated-measures variables: in this case these contrasts represent high charisma compared to some charisma (level 1 vs. level 3) and no charisma compared to some charisma (level 2 vs. level 3). The contrasts tell us that the effect of charisma represents the fact that highly charismatic dates were rated significantly higher than dates with some charisma, $F(1, 18) = 109.94$, $p < .0001$, and dates with some charisma were rated significantly higher than dullards, $F(1, 18) = 227.94$, $p < .0001$.

14.5.4. The interaction between gender and looks ②

We saw in SAS Output 14.3 that gender interacted in some way with the attractiveness of the date. From the summary table we can report that there was a significant interaction between the attractiveness of the date and the gender of the participant, $F(2, 36) = 80.43$, $p < .0001$. This effect tells us that the profile of ratings across dates of different attractiveness was different for men and women. We can use PROC MEANS to produce the means for this interaction, shown in SAS Syntax 14.4. I've also used those means to draw a graph, which helps to interpret the interaction. The means and interaction graph (Figure 14.2 and SAS Output 14.10) show the meaning of this result. The graph shows the average male ratings of dates of different attractiveness ignoring how charismatic the date was (circles). The women's scores are shown as squares. The graph clearly shows that male and female ratings are very similar for average-looking dates, but men give higher ratings (i.e. they're really keen to go out with these people) than women for attractive dates, while women express more interest in going out with ugly people than men. In general this interaction seems to suggest than men's interest in dating a person is more influenced by their looks than for females. Although both males' and females' interest decreases as attractiveness decreases, this decrease is more pronounced for men. This interaction can be clarified using the contrasts in SAS Output 14.8.

```
PROC MEANS DATA=looksorpersonalitylong MEAN;
      CLASS looks gender ;
      VAR col1;
      RUN;
```
SAS Syntax 14.4

SAS OUTPUT 14.10

Analysis Variable: COL1			
looks	gender	N Obs	Mean
at	Female	30	76.1666667
	Male	30	88.0333333
av	Female	30	68.1000000
	Male	30	67.4666667
ug	Female	30	61.3333333
	Male	30	50.3000000

FIGURE 14.2

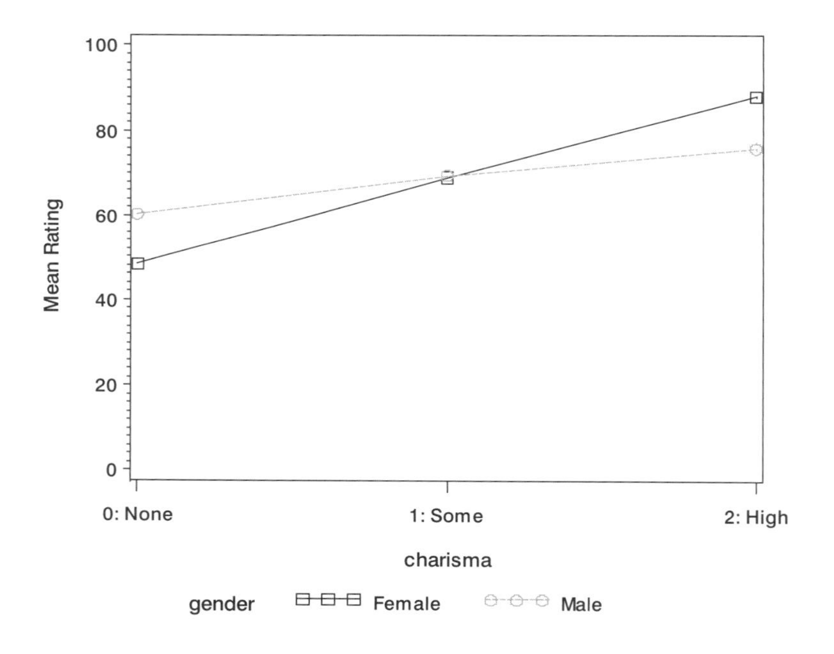

14.5.4.1. Looks x gender interaction 1: attractive vs. average, male vs. female ②

The contrast for the first interaction term looks at level 1 of looks (attractive) compared to level 3 (average), comparing male and female scores. This contrast is highly significant, $F(1, 18) = 43.26$, $p < .0001$. This result tells us that the increased interest in attractive dates compared to average-looking dates found for men is significantly more than for women. So, in Figure 14.2 the slope of the line between the circles representing male ratings of attractive dates and average dates is steeper than the line joining the squares representing female ratings of attractive dates and average dates. We can conclude that the preferences for attractive dates, compared to average-looking dates, are greater for males than females.

14.5.4.2. Looks × gender interaction 2: ugly vs. average, male vs. female ②

The second contrast compares level 2 of looks (ugly) to level 3 (average), comparing male and female scores. This contrast is highly significant, $F(1, 18) = 30.23$, $p < .0001$. This tells us that the decreased interest in ugly dates compared to average-looking dates found for men is significantly more than for women. So, in Figure 14.2 the slope of the line between the circles representing male ratings of ugly dates and average dates is steeper than the line joining the squares representing female ratings of ugly dates and average dates. We can conclude that the preferences for average-looking dates, compared to ugly dates, are greater for males than females.

14.5.5. The interaction between gender and charisma ②

We also saw in SAS Output 14.3 that gender interacted in some way with how charismatic the date was. From the summary table we should report that there was a significant interaction between the attractiveness of the date and the gender of the participant, $F(2, 36) = 62.45$, $p < .0001$. This effect tells us that the profile of ratings across dates of different levels of charisma was different for men and women.) We can use PROC MEANS to tell us the meaning of this interaction, and provide a graph to help the interpretation (see SAS Syntax 14.5 for syntax, SAS Output 14.11 for the table of means, and Figure 14.3). The graph shows the average male ratings of dates of different levels of charisma ignoring how attractive they were (circles). The women's scores are shown as squares. The graph shows almost the reverse pattern as for the attractiveness data; again male and female ratings are very similar for dates with normal amounts of charisma, but this time men show more interest in dates who are dullards than women do, and women show slightly more interest in very charismatic dates than men do. In general this interaction seems to suggest than women's interest in dating a person is more influenced by their charisma than for men. Although both male's and female's interest decreases as charisma decreases, this decreases is more pronounced for females. This interaction can be clarified using the contrasts in SAS Output 14.8.

```
PROC MEANS DATA=looksorpersonalitylong MEAN;
      CLASS charisma gender ;
      VAR col1;
      RUN;
```
SAS Syntax 14.5

SAS OUTPUT 14.11

Analysis Variable : COL1			
charisma	gender	N Obs	Mean
High	Female	30	88.2333333
	Male	30	75.9666667
None	Female	30	48.3000000
	Male	30	60.3000000
Some	Female	30	69.0666667
	Male	30	69.5333333

FIGURE 14.3

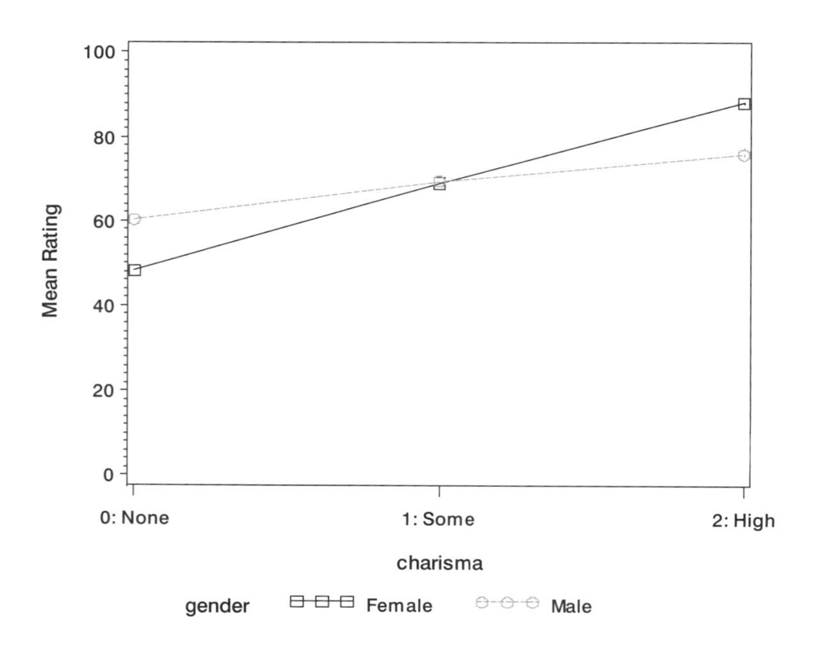

14.5.5.1. Charisma × gender interaction 1: high vs. some charisma, male vs. female ②

The first contrast for this interaction term looks at level 1 of charisma (high charisma) compared to level 3 (some charisma), comparing male and female scores. This contrast is highly significant, $F(1, 18) = 27.20$, $p < .0001$. This result tells us that the increased interest in highly charismatic dates compared to averagely charismatic dates found for women is significantly more than for men. So, in Figure 14.3 the slope of the line between the squares representing female ratings of very charismatic dates and dates with some charisma is steeper than the line joining the circles representing male ratings of very charismatic dates and dates with some charisma. We can conclude that the preferences for very charismatic dates, compared to averagely charismatic dates, are greater for females than males.

14.5.5.2. Charisma × gender interaction 2: dullard vs. some charisma, male vs. female ②

The second contrast for this interaction term looks at level 2 of charisma (dullard) compared to level 3 (some charisma), comparing male and female scores. This contrast is highly significant, $F(1, 18) = 33.69$, $p < .0001$. This result tells us that the decreased interest in dullard dates compared to averagely charismatic dates found for women is significantly more than for men. So, in Figure 14.3 the slope of the line between the squares representing female ratings of dates with some charisma and dullard dates is steeper than the line joining the circles representing male ratings of dates with some charisma and dullard dates. We can conclude that the preferences for dates with some charisma over dullards are greater for females than males.

14.5.6. The interaction between attractiveness and charisma ②

SAS Output 14.3 indicated that the attractiveness of the date interacted in some way with how charismatic the date was. From the summary table we should report that there was a significant interaction between the attractiveness of the date and the charisma of the date, $F(4, 72) = 36.63$, $p < .0001$. This effect tells us that the profile of ratings across dates of different levels of charisma was different for attractive, average and ugly dates. We can use PROC MEANS to tell us the meaning of this interaction (see SAS Syntax 14.6 for syntax, SAS Output 14.11 for the table of means, and Figure 14.4).

```
PROC MEANS DATA=looksorpersonalitylong MEAN;
      CLASS charisma looks ;
      VAR col1;
      RUN;
```
SAS Syntax 14.6

SAS OUTPUT 14.11

Analysis Variable: COL1			
charisma	looks	N Obs	Mean
hig	at	20	88.9500000
	av	20	85.6000000
	ug	20	71.7500000
non	at	20	69.5500000
	av	20	47.4000000
	ug	20	45.9500000
som	at	20	87.8000000
	av	20	70.3500000
	ug	20	49.7500000

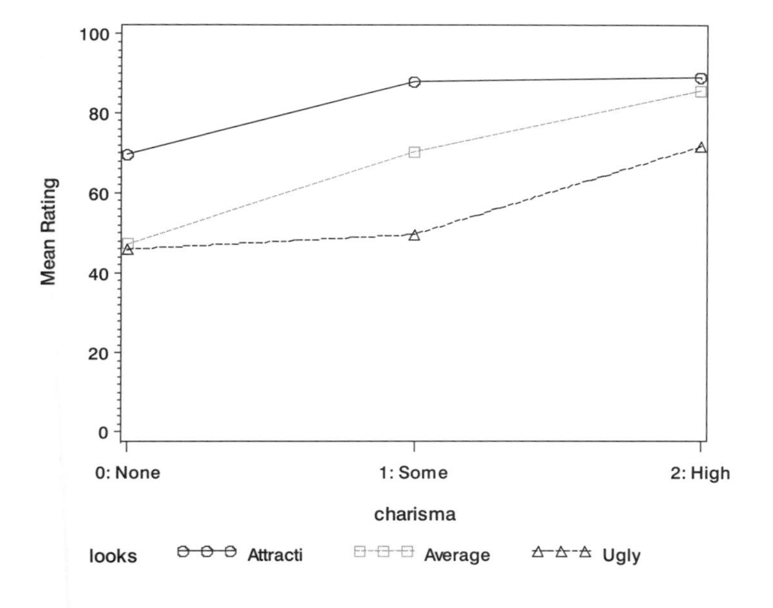

FIGURE 14.4

The graph shows the average ratings of dates of different levels of attractiveness when the date also had high levels of charisma (circles), some charisma (squares) and no charisma (triangles). Look first at the difference between attractive and average-looking dates. The interest in highly charismatic dates doesn't change (the line is more or less flat between these two points), but for dates with some charisma or no charisma interest levels decline. So, if you have lots of charisma you can get away with being average looking and people will still want to date you. Now, if we look at the difference between average-looking and ugly dates, a different pattern is observed. For dates with no charisma (triangles) there is no difference between ugly and average people (so if you're a dullard you have to be really attractive before people want to date you). However, for those with charisma, there is a decline in interest if you're ugly (so if you're ugly, having charisma won't help you much). This interaction is very complex, but we can break it down using the contrasts in SAS Output 14.8.

14.5.6.1. Looks × charisma interaction 1: attractive vs. average, high charisma vs. some charisma ②

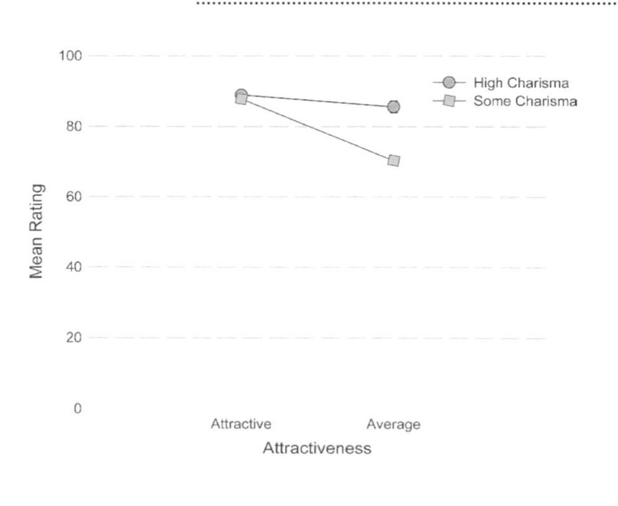

The first contrast for this interaction term investigates level 1 of looks (attractive) compared to level 3 (average looking), comparing level 1 of charisma (high charisma) to level 3 of charisma (some charisma). This is like asking 'is the difference between high charisma and some charisma the same for attractive people and average-looking people?' The best way to understand what this contrast is testing is to extract the relevant bit of the interaction graph in Figure 14.4. If you look at this you can see that the interest (as indicated by high ratings) in attractive dates was the same regardless of whether they had high or average charisma. However, for average-looking dates, there was more interest when that person had high charisma rather than average. The contrast is highly significant, $F(1, 18) = 21.94$, $p < .001$, and tells us that as dates become less attractive there is a greater decline in interest when charisma is average compared to when charisma is high.

14.5.6.2. Looks × charisma interaction 2: attractive vs. average, dullard vs. some charisma ②

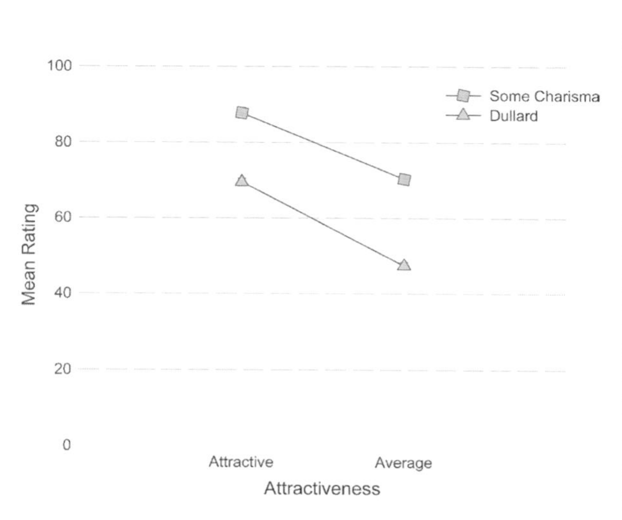

The second contrast for this interaction term investigates level 1 of looks (attractive) compared to level 3 (average looking), when comparing level 2 of charisma (dullard) to level 3 of charisma (some charisma). This is like asking 'is the difference between no charisma and some charisma the same for attractive people and average-looking people?' Again, the best way to understand what this contrast is testing is to extract the relevant bit of the interaction graph in Figure 14.4. If you look at this you can see that the interest (as indicated by high ratings) in attractive dates was higher when they had some charisma than when they were a dullard. The same is also true for average-looking dates. In fact the two lines are fairly parallel. The contrast is not significant, $F(1, 18) = 4.09$, $p = .058$, and tells us that as dates become less attractive there is a decline in interest both when charisma is low and when there is no charisma at all.

14.5.6.3. Looks × charisma interaction 3: ugly vs. average, high charisma vs. some charisma ②

The third contrast for this interaction term investigates level 2 of looks (ugly) compared to level 3 (average looking), comparing level 1 of charisma (high charisma) to level 3 of charisma (some charisma). This is like asking 'is the difference between high charisma and some charisma the same for ugly people and average-looking people?' If we again extract the relevant bit of the interaction graph in Figure 14.4 you can see that the interest (as indicated by high ratings) decreases from average-looking dates to ugly ones in both high- and some-charisma dates; however, this fall is slightly greater in the low-charisma dates (the line connecting the squares is slightly steeper). The contrast is significant, $F(1, 18) = 6.23$, $p = .023$, and tells us that as dates become less attractive there is a greater decline in interest when charisma is low compared to when charisma is high.

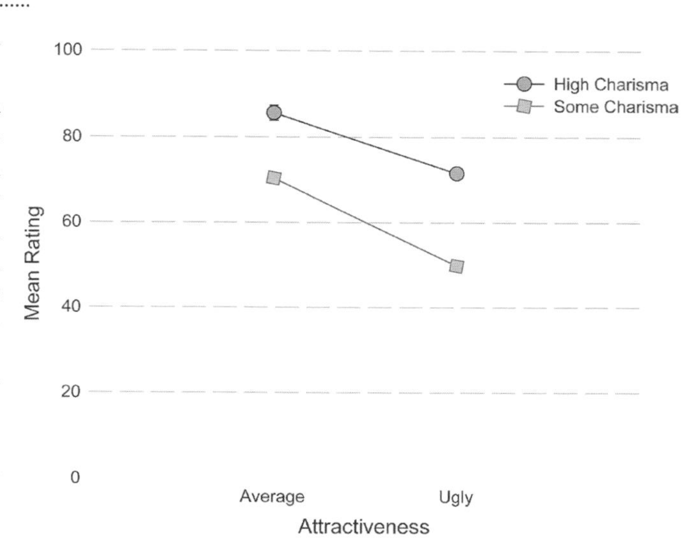

14.5.6.4. Looks × charisma interaction 4: ugly vs. average, dullard vs. some charisma ②

The final contrast for this interaction term investigates level 2 of looks (ugly) compared to level 3 (average looking), when comparing level 2 of charisma (dullard) to level 3 of charisma (some charisma). This is like asking 'is the difference between no charisma and some charisma the same for ugly people and average-looking people?' If we extract the relevant bit of the interaction graph in Figure 14.4 you can see that the interest (as indicated by high ratings) in average-looking dates was higher when they had some charisma than when they were a dullard, but for ugly dates the ratings were roughly the same regardless of the level of charisma. This contrast is highly significant, $F(1, 18) = 88.60$, $p < .0001$, and tells us that as dates become less attractive the decline in interest in dates with a bit of charisma is significantly greater than for dullards.

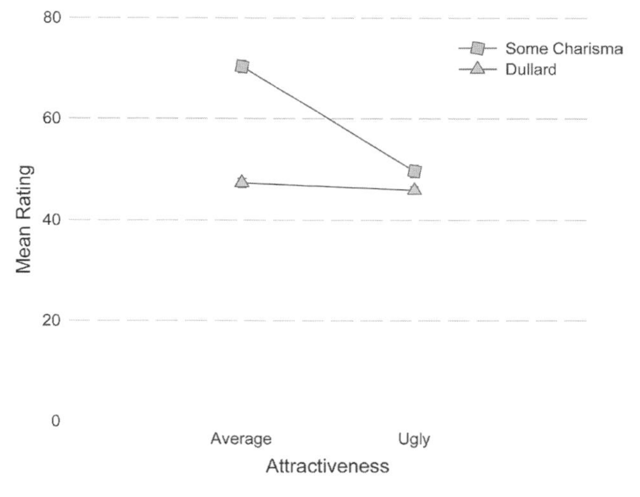

14.5.7. The interaction between looks, charisma and gender ③

The three-way interaction tells us whether the looks × charisma interaction described above is the same for men and women (i.e. whether the combined effect of attractiveness of the date and their level of charisma is the same for male as for female participants). SAS Output 14.3 tells us that there is a significant three-way looks × charisma × gender interaction, $F(4, 72) = 24.12$, $p < .0001$. The nature of this interaction is revealed in Figure 14.5,

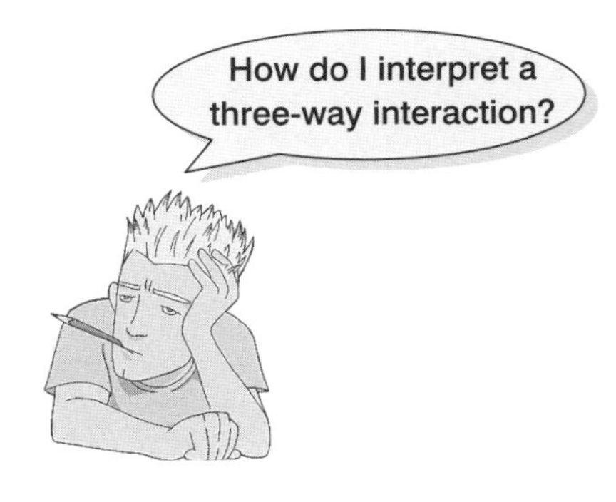

which shows the looks × charisma interaction for men and women separately (the means on which this graph is based appear in SAS Output 14.12 and were produced by SAS Syntax 14.7). The male graph shows that when dates are attractive, men will express a high interest regardless of charisma levels (the circle, square and triangle all overlap). At the opposite end of the attractiveness scale, when a date is ugly, regardless of charisma men will express very little interest (ratings are all low). The only time charisma makes any difference to a man is if the date is average-looking, in which case high charisma boosts interest, being a dullard reduces interest, and having a bit of charisma leaves things somewhere in between. The take-home message is that men are superficial fools who are more interested in physical attributes. The picture for women is very different. If someone has high levels of charisma then it doesn't really matter what they look like, women will express an interest in them (the line of circles is relatively flat). At the other extreme, if the date is a dullard, then they will express no interest in them, regardless of how attractive they are (the line of triangles is relatively flat). The only time attractiveness makes a difference is when someone has an average amount of charisma, in which case being attractive boosts interest, and being ugly reduces it. Put another way, women prioritize charisma over physical appearance. Again, we can look at some contrasts to further break this interaction down (SAS Output 14.8). These contrasts are similar to those for the looks × charisma interaction, but they now also take into account the effect of gender as well!

```
PROC MEANS DATA=looksorpersonalitylong MEAN;
      CLASS charisma looks gender  ;
      VAR col1;
      RUN;
```
SAS Syntax 14.7

SAS OUTPUT 14.12

Analysis Variable : COL1				
charisma	looks	gender	N Obs	Mean
high	at	Female	10	89.6000000
		Male	10	88.3000000
	av	Female	10	88.4000000
		Male	10	82.8000000
	ug	Female	10	86.7000000
		Male	10	56.8000000
none	at	Female	10	51.8000000
		Male	10	87.3000000
	av	Female	10	47.0000000
		Male	10	47.8000000
	ug	Female	10	46.1000000
		Male	10	45.8000000
some	at	Female	10	87.1000000
		Male	10	88.5000000
	av	Female	10	68.9000000
		Male	10	71.8000000
	ug	Female	10	51.2000000
		Male	10	48.3000000

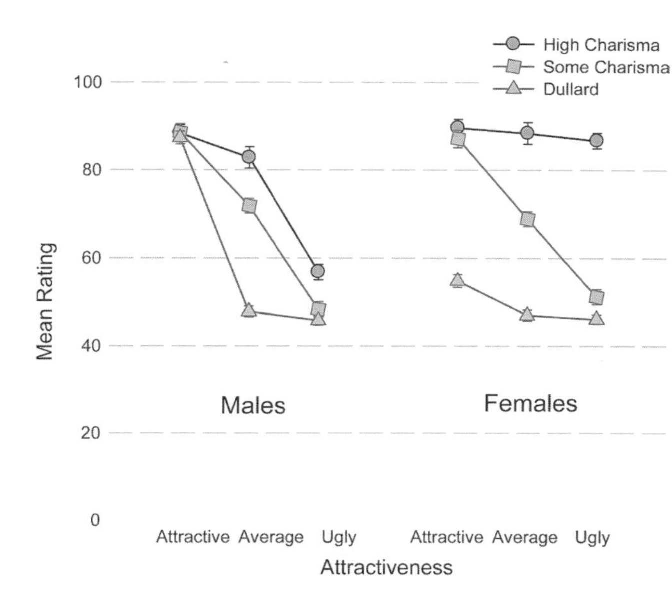

FIGURE 14.5 Graphs showing the looks by charisma interaction for men and women. Lines represent high charisma (circles), some charisma (squares) and no charisma (triangles)

14.5.7.1. Looks × charisma × gender interaction 1: attractive vs. average, high charisma vs. some charisma, male vs. female ③

The first contrast for this interaction term compares level 1 of looks (attractive) to level 3 (average looking), when level 1 of charisma (high charisma) is compared to level 3 of charisma (some charisma) in males compared to females, $F(1, 18) = 0.93$, $p = 0.348$. If we extract the relevant bits of the interaction graph in Figure 14.5, we can see that interest (as indicated by high ratings) in attractive dates was the same regardless of whether they had high or average charisma. However, for average-looking dates, there was more interest when that person had high charisma rather than some charisma. Importantly, this pattern of results is the same in males and females and this is reflected in the non-significance of this contrast.

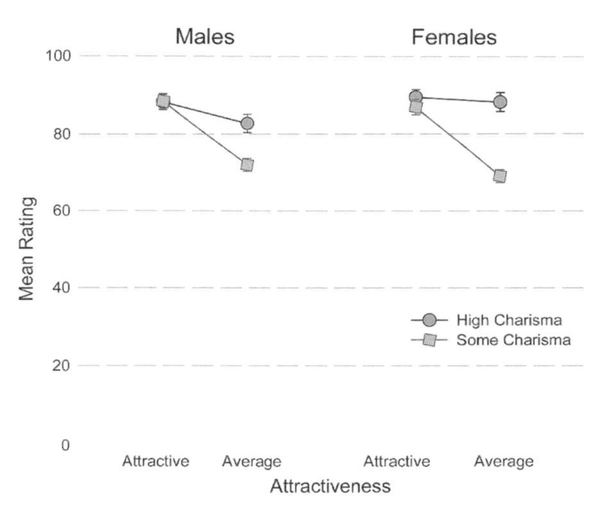

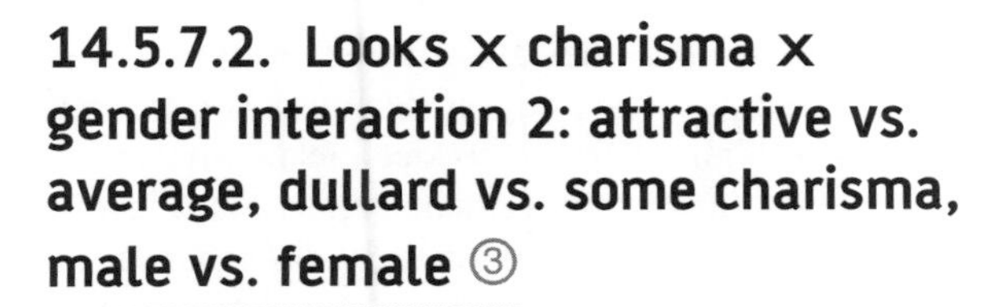

14.5.7.2. Looks × charisma × gender interaction 2: attractive vs. average, dullard vs. some charisma, male vs. female ③

The second contrast for this interaction term compares level 1 of looks (attractive) to level 3 (average looking), when level 2 of charisma (dullard) is compared to level 3 of charisma (some charisma), in men compared to women. Again, we extract the relevant

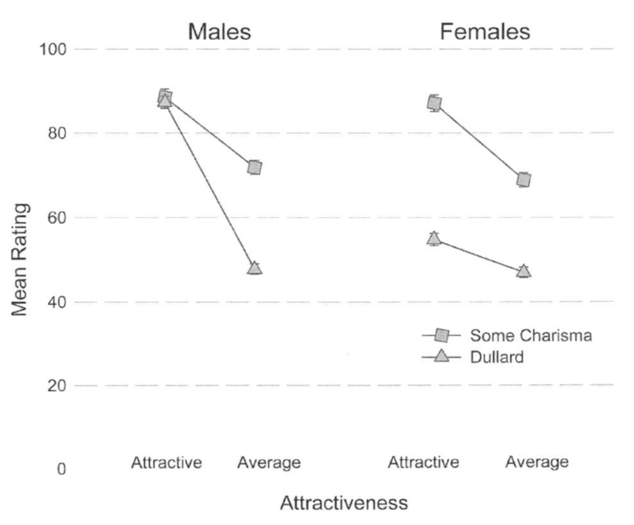

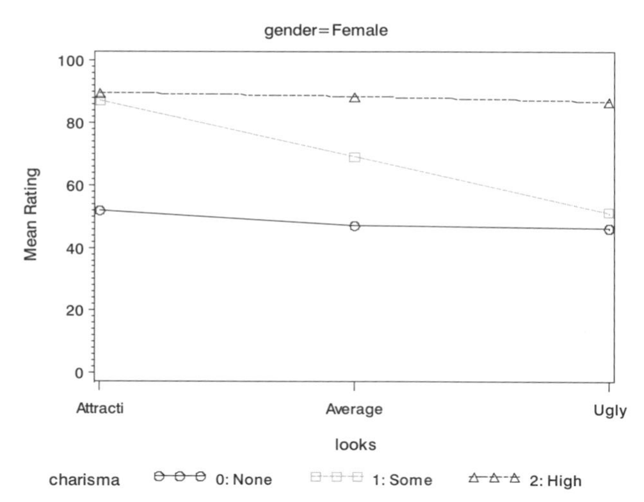

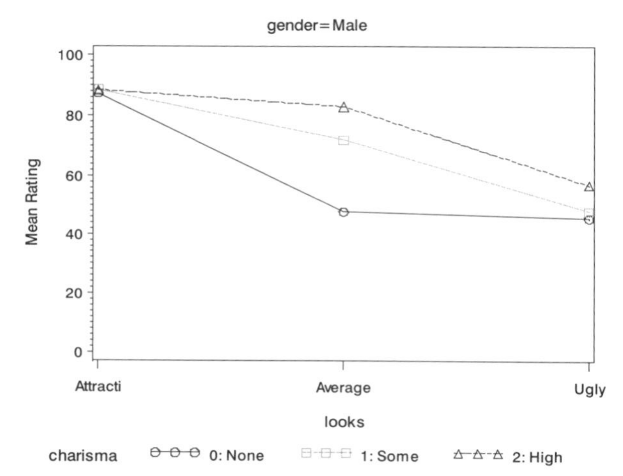

bit of the interaction graph in Figure 14.5 and you can see that the patterns are different for men and women. This is reflected by the fact that the contrast is significant, $F(1, 18) = 60.67$, $p < .0001$. To unpick this we need to look at the graph. First, if we look at average-looking dates, for both men and women more interest is expressed when the date has some charisma than when they have none (and the distance between the square and the triangle is about the same). So the difference doesn't appear to be here. If we now look at attractive dates, we see that men are equally interested in their dates regardless of their charisma, but women are much less interested in an attractive person if they are a dullard. Put another way, for attractive dates, the distance between the square and the triangle is much smaller for men than it is for women. Another way to look at it is that for dates with some charisma, the reduction in interest as attractiveness goes down is about the same in men and women (the lines with squares have the same slope). However, for dates who are dullards, the decrease in interest if these dates are average-looking rather than attractive is much more dramatic in men than women (the line with triangles is much steeper for men than it is for women).

14.5.7.3. Looks × charisma × gender interaction 3: ugly vs. average, high charisma vs. some charisma, males vs. females ③

The third contrast for this interaction term compares level 2 of looks (ugly) to level 3 (average looking), when level 1 of charisma (high charisma) is compared to level 3 of charisma (some charisma), in men compared to women. Again, we extract the relevant bit of the interaction graph in Figure 14.5 and you can see that the patterns are different for men and women. This is reflected by the fact that the contrast is significant, $F(1, 18) = 11.70$, $p = 0.003$. To unpick this we need to look at the graph. First, let's look at the men. For men, as attractiveness goes down, so does interest when the date has high charisma and when they have some charisma. In fact the lines are parallel. So, regardless of charisma, there is a similar reduction in interest as attractiveness declines. For women the picture is quite different. When charisma is high, there is no decline in interest as attractiveness falls (the line connecting the circles is flat); however, when charisma is lower, the attractiveness of the date does matter and interest is lower in an ugly date than in an average-looking date. Another way to look at it is that for dates with some charisma, the reduction in interest as attractiveness goes down is about the same in men and women (the lines with squares have the same slope). However, for dates who have high charisma, the decrease in interest if these dates are ugly rather than average-looking is much more dramatic in men than women (the line with circles is much steeper for men than it is for women). This is what the significant contrast tells us.

14.5.7.4. Looks × charisma × gender interaction 4: ugly vs. average, dullard vs. some charisma, male vs. female ③

The final contrast for this interaction term compares level 2 of looks (ugly) to level 3 (average looking), when comparing level 2 of charisma (dullard) to level 3 of charisma (some charisma), in men compared to women. If we extract the relevant bits of the interaction graph in Figure 14.5, we can see that interest (as indicated by high ratings) in ugly dates was the same regardless of whether they had some charisma or were a dullard. However, for average-looking dates, there was more interest when that person had some charisma rather than if they were a dullard. Importantly, this pattern of results is the same in males and females and this is reflected in the non-significance of this contrast, $F(1, 18) = 1.33$, $p = 0.263$.

14.5.8. Conclusions ③

These contrasts tell us nothing about the differences between the attractive and ugly conditions, or the high-charisma and dullard conditions, because these were never compared. We could rerun the analysis and specify our contrasts differently to get these effects. However, what is clear from our data is that differences exist between men and women in terms of how they're affected by the looks and personality of potential dates. Men appear to be enthusiastic about dating anyone who is attractive regardless of how awful their personality. Women are almost completely the opposite: they are enthusiastic about dating anyone with a lot of charisma, regardless of how they look (and are unenthusiastic about dating people without charisma regardless of how attractive they look). The only consistency between men and women is when there is some charisma (but not lots), in which case for both genders the attractiveness influences how enthusiastic they are about dating the person.

What should be even clearer from this chapter is that when more than two independent variables are used in ANOVA, it yields complex interaction effects that require a great deal of concentration to interpret (imagine interpreting a four-way interaction!). Therefore, it is essential to take a systematic approach to interpretation, and plotting graphs is a particularly useful way to proceed. It is also advisable to think carefully about the appropriate contrasts to use to answer the questions you have about your data. It is these contrasts that will help you to interpret interactions, so make sure you select sensible ones! (You will find in Chapter 19 that the PROC MIXED approach to repeated measures gives a great deal more flexibility to look at contrasts and interactions).

CRAMMING SAM'S TIPS

- Mixed ANOVA compares several means when there are two or more independent variables, and at least one of them has been measured using the same participants and at least one other has been measured using different participants.
- Test the assumption of *sphericity* for the repeated-measures variable(s) when they have three or more conditions using *Mauchly's test*. If the value in the column labelled *PR* ± *Chisq* is less than .05 then the assumption is violated. You should test this assumption for all effects (if there are two or more repeated-measures variables this means you test the assumption for all variables and the corresponding interaction terms).
- The section labelled **Univariate Tests of Hypotheses for Within Subjects Effects** shows the results of your ANOVA for the repeated-measures variables and all of the interaction effects. For each effect, if the assumption of sphericity has been met then look

at the column labelled Pr > F. Otherwise, read the column labelled G-G or H-F (read the previous chapter to find out the relative merits of the two procedures). If the value is less than .05 then the means are significantly different.

- The table labelled **Tests of Hypotheses for Between-Subjects Effects** shows the results of your ANOVA for the between-group variables. Look at the column labelled *Sig.* If the value is less than .05 then the means of the groups are significantly different.
- Break down the main effects and interaction terms using contrasts. These contrasts appear in the section labelled **Analysis of Variance of Contrast Variables**. Again look to the columns labelled *Pr > F.* to discover if your comparisons are significant (they will be if the significance value is less than .05).
- Look at the means, or better still draw graphs, to help you interpret the contrasts.

14.6. Calculating effect sizes ③

I keep emphasizing the fact that effect sizes are really more useful when they summarize a focused effect. This also gives me a useful excuse to circumvent the complexities of omega squared in mixed designs (it's the road to madness, I assure you). Therefore, just calculate effect sizes for your contrasts when you've got a factorial design (and any main effects that compare only two groups). SAS Output 14.8 shows the values for several contrasts, all of which have 1 degree of freedom for the model (i.e. they represent a focused and interpretable comparison) and have 18 residual degrees of freedom. We can use these *F*-ratios and convert them to an effect size *r*, using a formula we've come across before:

$$r = \sqrt{\frac{F(1, df_R)}{F(1, df_R) + df_R}}$$

First, we can deal with the main effect of gender because this compares only two groups:

$$r_{\text{Gender}} = \sqrt{\frac{0.005}{0.005 + 18}} = .02$$

For the two comparisons we did for the looks variable, we would get:

$$r_{\text{Attractive vs. Average}} = \sqrt{\frac{226.99}{226.99 + 18}} = .96$$

$$r_{\text{Ugly vs. Average}} = \sqrt{\frac{160.07}{160.07 + 18}} = .95$$

Therefore, both comparisons yielded massive effect sizes. For the two comparisons we did for the charisma variable (SAS Output 14.7), we would get:

$$r_{\text{High vs. Some}} = \sqrt{\frac{109.94}{109.94 + 18}} = .93$$

$$r_{\text{Dullard vs. Some}} = \sqrt{\frac{227.94}{227.94 + 18}} = .96$$

Again, both comparisons yield massive effect sizes. For the looks × gender interaction, we again had two contrasts:

$$r_{\text{Attractive vs. Average, Male vs. Female}} = \sqrt{\frac{43.26}{43.26 + 18}} = .84$$

$$r_{\text{Ugly vs. Average, Male vs. Female}} = \sqrt{\frac{30.23}{30.23 + 18}} = .79$$

Again, these are massive effects. For the charisma × gender interaction, the two contrasts give us:

$$r_{\text{High vs. Some, Male vs. Female}} = \sqrt{\frac{27.20}{27.20 + 18}} = .78$$

$$r_{\text{Dullard vs. Some, Male vs. Female}} = \sqrt{\frac{33.69}{33.69 + 18}} = .81$$

Yet again massive effects (can't you tell the data are fabricated!). Moving on to the looks × charisma interaction, we get four contrasts:

$$r_{\text{Attractive vs. Average, High vs. Some}} = \sqrt{\frac{21.94}{21.94 + 18}} = .74$$

$$r_{\text{Attractive vs. Average, Dullard vs.Some}} = \sqrt{\frac{4.09}{4.09 + 18}} = .43$$

$$r_{\text{Ugly vs. Average, High vs. Some}} = \sqrt{\frac{6.23}{6.23 + 18}} = .51$$

$$r_{\text{Ugly vs. Average, Dullard vs. Some}} = \sqrt{\frac{88.60}{88.60 + 18}} = .91$$

All of these effects are in the medium to massive range. Finally, for the looks × charisma × gender interaction we had four contrasts:

$$r_{\text{Attractive vs. Average, High vs. Some, Male vs. Female}} = \sqrt{\frac{0.93}{0.93 + 18}} = .22$$

$$r_{\text{Attractive vs. Average, Dullard vs. Some, Male vs. Female}} = \sqrt{\frac{60.67}{60.67 + 18}} = .88$$

$$r_{\text{Ugly vs. Average, High vs. Some, Male vs. Female}} = \sqrt{\frac{11.70}{11.70 + 18}} = .63$$

$$r_{\text{Ugly vs. Average, Dullard vs. Some, Male vs. Female}} = \sqrt{\frac{1.33}{1.33 + 18}} = .26$$

As such, the two effects that were significant (attractive vs. average, dullard vs. some, male vs. female, and ugly vs. average, high vs. some, male vs. female) yielded large effect sizes. The two effects that were not significant yielded close to medium effect sizes.

14.7. Reporting the results of mixed ANOVA ②

As you've probably gathered, when you have more than two independent variables there's an awful lot of information to report. You have to report all of the main effects, all of the interactions and any contrasts you may have done. This can take up a lot of space and one good tip is: reserve the most detail for the effects that actually matter (e.g. main effects are usually not that interesting if you've got a significant interaction that includes that variable). I'm a big fan of giving brief explanations of results in the results section to really get the message across about what a particular effect is telling us and so I tend to not just report results, but offer some interpretation as well. Having said that, some journal editors enjoy telling me my results sections are too long. So, you should probably ignore everything I say. Assuming we want to report all of our effects, we could do it something like this (although not as a list!):

✓ All effects are reported as significant at $p < .05$. There was a significant main effect of the attractiveness of the date on interest expressed by participant, $F(2, 36) = 423.73$. Contrasts revealed that attractive dates were more desirable than average-looking ones,

$F(1, 18) = 226.99$, $r = .96$, and ugly dates were less desirable than average-looking ones $F(1, 18) = 160.07$, $r = .95$.

- ✓ There was also a significant main effect of the amount of charisma the date possessed on the interest expressed in dating them, $F(2, 36) = 328.25$. Contrasts revealed that dates with high charisma were more desirable than dates with some charisma, $F(1, 18) = 109.94$, $r = .93$, and dullards were less desirable than dates with some charisma, $F(1, 18) = 227.94$, $r = .96$.
- ✓ There was no significant effect of gender, indicating that ratings from male and female participants were in general the same, $F(1, 18) < 1$, $r = .02$.
- ✓ There was a significant interaction effect between the attractiveness of the date and the gender of the participant, $F(2, 36) = 80.43$. This indicates that the desirability of dates of different levels of attractiveness differed in men and women. To break down this interaction, contrasts were performed comparing each level of attractiveness to average-looking across male and female participants. These revealed significant interactions when comparing male and female scores to attractive dates compared to average-looking dates, $F(1, 18) = 43.26$, $r = .84$, and to ugly dates compared to average dates, $F(1, 18) = 30.23$, $r = .79$. Looking at the interaction graph, this suggests that male and female ratings are very similar for average-looking dates, but men rate attractive dates higher than women, whereas women rate ugly dates higher than men. Although both males' and females' interest decreases as attractiveness decreases, this decrease is more pronounced for men, suggesting that when charisma is ignored, men's interest in dating a person is more influenced by their looks than for females.
- ✓ There was a significant interaction effect between the level of charisma of the date and the gender of the participant, $F(2, 36) = 62.45$. This indicates that the desirability of dates of different levels of charisma differed in men and women. To break down this interaction, contrasts were performed comparing each level of charisma to the middle category of 'some charisma' across male and female participants. These revealed significant interactions when comparing male and female scores to highly charismatic dates compared to dates with some charisma, $F(1, 18) = 27.20$, $r = .78$, and to dullards compared to dates with some charisma, $F(1, 18) = 33.69$, $r = .81$. The interaction graph reveals that men show more interest in dates who are dullards than women do, and women show slightly more interest in very charismatic dates than men do. Both males' and females' interest decrease as charisma decreases, but this decrease is more pronounced for females, suggesting women's interest in dating a person is more influenced by their charisma than for men.
- ✓ There was a significant interaction effect between the level of charisma of the date and the attractiveness of the date, $F(4, 72) = 36.63$. This indicates that the desirability of dates of different levels of charisma differed according to their attractiveness. To break down this interaction, contrasts were performed comparing each level of charisma to the middle category of 'some charisma' across each level of attractiveness compared to the category of average attractiveness. The first contrast revealed a significant interaction when comparing attractive dates to average-looking dates when the date had high charisma compared to some charisma, $F(1, 18) = 21.94$, $r = .74$, and tells us that as dates become less attractive there is a greater decline in interest when charisma is low compared to when charisma is high. The second contrast compared attractive dates to average-looking dates when the date was a dullard compared to when they had some charisma. This was not significant, $F(1, 18) = 4.09$, $r = .43$, and tells us that as dates become less attractive there is decline in interest both when charisma is low and when there is no charisma at all. The third contrast compared ugly dates to average-looking dates when the date had high charisma compared to some charisma. This was significant, $F(1, 18) = 6.23$, $r = .51$, and tells us that as dates become less attractive there is a greater decline in interest when charisma is low compared to when charisma is high. The final contrast compared ugly dates to average-looking dates when the date was a dullard compared to when they had some charisma. This contrast was highly significant, $F(1, 18) = 88.60$, $r = .91$, and tells us that as dates become less attractive the decline in interest in dates with a bit of charisma is significantly greater than for dullards.

✓ Finally, the looks × charisma × gender interaction was significant $F(4, 72) = 24.12$. This indicates that the looks × charisma interaction described previously was different in male and female participants. Again, contrasts were used to break down this interaction; these contrasts compared male and females scores at each level of charisma compared to the middle category of 'some charisma' across each level of attractiveness compared to the category of average attractiveness. The first contrast revealed a non-significant difference between male and female responses when comparing attractive dates to average-looking dates when the date had high charisma compared to some charisma, $F(1, 18) < 1$, $r = .22$, and tells us that for both males and females, as dates become less attractive there is a greater decline in interest when charisma is low compared to when charisma is high. The second contrast investigated differences between males and females when comparing attractive dates to average-looking dates when the date was a dullard compared to when they had some charisma. This was significant, $F(1, 18) = 60.67$, $r = .88$, and tells us that for dates with some charisma, the reduction in interest as attractiveness goes down is about the same in men and women, but for dates who are dullards, the decrease in interest if these dates are average-looking rather than attractive is much more dramatic in men than women. The third contrast looked for differences between males and females when comparing ugly dates to average-looking dates when the date had high charisma compared to some charisma. This was significant, $F(1, 18) = 11.70$, $r = .63$, and tells us that for dates with some charisma, the reduction

LABCOAT LENI'S REAL RESEARCH 14.1

Keep the faith(ful)? ③

People can be jealous. People can be especially jealous when they think that their partner is being unfaithful. An evolutionary view of jealousy suggests that men and women have evolved distinctive types of jealousy because male and female reproductive success is threatened by different types of infidelity. Specifically, a woman's sexual infidelity deprives her mate of a reproductive opportunity and in some cases burdens him with years investing in a child that is not his. Conversely, a man's sexual infidelity does not burden his mate with unrelated children, but may divert his resources from his mate's progeny. This diversion of resources is signalled by emotional attachment to another female. Consequently, men's jealousy mechanism should have evolved to prevent a mate's *sexual* infidelity, whereas in women it has evolved to prevent emotional infidelity. If this is the case then men and women should divert their attentional resources towards different cues to infidelity: women should be 'on the look-out' for *emotional* infidelity, whereas men should be watching out for sexual infidelity.

Achim Schützwohl put this theory to the test in a unique study in which men and women saw sentences presented on a computer screen (Schützwohl, 2008). On each trial, participants saw a target sentence that was always emotionally neutral (e.g. 'The gas station is at the other side of the street'). However, the trick was that before each of these targets, a distractor sentence was presented that could also be affectively neutral, or could indicate sexual infidelity (e.g. 'Your partner suddenly has difficulty becoming sexually aroused when he and you want to have sex') or emotional infidelity (e.g. 'Your partner doesn't say "I love you" to you anymore'). The idea was that if these distractor sentences grabbed a person's attention then (1) they would remember them, and (2) they would not remember the target sentence that came afterwards (because their attentional resources were still focused on the distractor). These effects should show up only in people currently in a relationship. The outcome was the number of sentences that a participant could remember (out of 6), and the predictors were whether the person had a partner or not (**Relationship**), whether the trial used a neutral distractor, an emotional infidelity distractor or a sexual infidelity distractor, and whether the sentence was a distractor or the target following the distractor. Schützwohl analysed men and women's data separately. The predictions are that women should remember more emotional infidelity sentences (distractors) but fewer of the targets that followed those sentences (target). For men, the same effect should be found but for sexual infidelity sentences.

The data from this study are in the file **Schutzwohl 2008.sas7bdat**. Labcoat Leni wants you to carry out two three-way Mixed ANOVAs (one for men and the other for women) to test these hypotheses. Answers are in the additional material on the companion website (or look at pages 638–642 in the original article).

SCHÜTZWOHL, A. (2008). *PERSONALITY AND INDIVIDUAL DIFFERENCES, 44*, 633–644.

in interest as attractiveness goes down is about the same in men and women, but for dates who have high charisma, the decrease in interest if these dates are ugly rather than average-looking is much more dramatic in men than women. The final contrast looked for differences between men and women when comparing ugly dates to average-looking dates when the date was a dullard compared to when they had some charisma. This contrast was not significant, $F(1, 18) = 1.33$, $r = .26$, and tells us that for both men and women, as dates become less attractive the decline in interest in dates with a bit of charisma is significantly greater than for dullards.

14.8. What to do when assumptions are violated in mixed ANOVA ③

If I had £1 (or $1, €1, or whatever currency you fancy) for every time someone told me with 100% confidence that there was no 'non-parametric' equivalent of mixed ANOVA, then I'd have a nice shiny new drum kit. As with other factorial ANOVAs, there is not a nonparametric counterpart of mixed ANOVA, as such, but there are robust methods that can be used (see section 5.7.4). As in previous chapters, my advice is to read Rand Wilcox's book (Wilcox, 2005), then use the R plugin for SAS to run the analysis (see Oliver Twisted).

What have I discovered about statistics? ②

Three-way ANOVA is a confusing nut to crack. I've probably done hundreds of three-way ANOVAs in my life and still I getting confused throughout writing this chapter (and so if you're confused after reading it it's not your fault, it's mine). Hopefully, what you should have discovered is that ANOVA is flexible enough that you can mix and match independent variables that are measured using the same or different participants. In addition, we've looked at how ANOVA is also flexible enough to go beyond merely including two independent variables. Hopefully, you've also started to realize why there are good reasons to limit the number of independent variables that you include (for the sake of interpretation).

Of course far more interesting than that is that you've discovered that men are superficial creatures who value looks over charisma, and that women are prepared to date the hunchback of Notre Dame provided he has a sufficient amount of charisma. This is why as a 16–18 year old my life was so complicated, because where on earth do you discover your hidden charisma? Luckily for me, some girls find alcoholics appealing. The girl I was particularly keen on at 16 was, as it turned out, keen on me too. I refused to believe this for at least a month. All of our friends were getting bored of us declaring our undying love for each other to them but then not speaking to each other; they eventually intervened. There was a party one evening and all of her friends had spent hours convincing me to ask her on a date, guaranteeing me that she would say 'yes'. I had psyched myself up, I was going to do it, I was actually going to ask a girl out on a date. My whole life had been leading up to this moment and I must not do anything to ruin it. By the time she arrived my nerves had got the better of me and she had to step over my paralytic corpse to get into the house. Later on, my friend Paul Spreckley (see Figure 9.1) physically carried the girl in question from another room and put her next to me and then said something to the effect of 'Andy, I'm going to sit here until you ask her out.' He had a long wait but eventually, miraculously, the words came out of my mouth. What happened next is the topic for another book, not about statistics.

Key terms that I've discovered

Mixed ANOVA
Mixed design

Smart Alex's tasks

- **Task 1**: I am going to extend the example from the previous chapter (advertising and different imagery) by adding a between-group variable into the design.[4] To recap, in case you haven't read the previous chapter, participants viewed a total of nine mock adverts over three sessions. In these adverts there were three products (a brand of beer, Brain Death, a brand of wine, Dangleberry, and a brand of water, Puritan). These could be presented alongside positive, negative or neutral imagery. Over the three sessions and nine adverts, each type of product was paired with each type of imagery (read the previous chapter if you need more detail). After each advert participants rated the drinks on a scale ranging from −100 (dislike very much) through 0 (neutral) to 100 (like very much). The design, thus far, has two independent variables: the type of drink (beer, wine or water) and the type of imagery used (positive, negative or neutral). These two variables completely cross over, producing nine experimental conditions. Now imagine that I also took note of each person's gender. Subsequent to the previous analysis it occurred to me that men and women might respond differently to the products (because, in keeping with stereotypes, men might mostly drink lager whereas women might drink wine). Therefore, I wanted to reanalyse the data taking this additional variable into account. Now, gender is a between-group variable because a participant can be only male or female: they cannot participate as a male and then change into a female and participate again! The data are the same as in the previous chapter (Table 13.3) and can be found in the file **MixedAttitude.s7bdat.** Run a mixed ANOVA on these data. ③

- **Task 2**: Text messaging is very popular among mobile phone owners, to the point that books have been published on how to write in text speak (BTW, hope u no wat I mean by txt spk). One concern is that children may use this form of communication so much that it will hinder their ability to learn correct written English. One concerned researcher conducted an experiment in which one group of children was encouraged to send text messages on their mobile phones over a six-month period. A second group was forbidden from sending text messages for the same period. To ensure that kids in this latter group didn't use their phones, this group were given armbands that administered painful shocks in the presence of microwaves (like those emitted from phones). There were 50 different participants: 25 were encouraged to send text messages, and 25 were forbidden. The outcome was a score on a grammatical test (as a percentage) that was measured both before and after the experiment. The first independent variable was, therefore, text message use (text messagers versus controls) and the second independent variable was the time at which grammatical ability was assessed (before or after the experiment). The data are in the file **TextMessages.sas7bdat.** ③

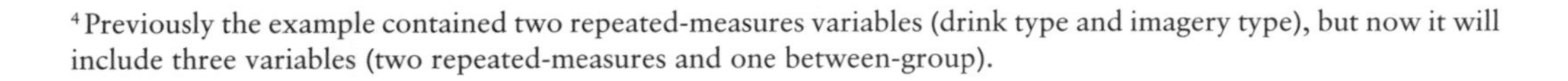

[4] Previously the example contained two repeated-measures variables (drink type and imagery type), but now it will include three variables (two repeated-measures and one between-group).

- **Task 3**: A researcher was interested in the effects on people's mental health of participating in *Big Brother* (see Chapter 1 if you don't know what *Big Brother* is). The researcher hypothesized that they start off with personality disorders that are exacerbated by being forced to live with people as attention-seeking as themselves. To test this hypothesis, she gave eight contestants a questionnaire measuring personality disorders before they entered the house, and again when they left the house. A second group of eight people acted as a waiting list control. These were people short listed to go into the house, but never actually made it. They too were given the questionnaire at the same points in time as the contestants. The data are in **BigBrother.sas7bdat.** Conduct a mixed ANOVA on the data. ②

Answers can be found on the companion website. Some more detailed comments about Task 2 can be found in Field and Hole (2003).

Further reading

Field, A. P. (1998). A bluffer's guide to sphericity. *Newsletter of the Mathematical, Statistical and Computing section of the British Psychological Society*, 6(1), 13–22. (Available in the additional material on the companion website.)

Howell, D. C. (2006). *Statistical methods for psychology* (6th ed.). Belmont, CA: Duxbury. (Or you might prefer his *Fundamental statistics for the behavioral sciences,* also in its 6th edition, 2007.)

Interesting real research

Schützwohl, A. (2008). The disengagement of attentive resources from task-irrelevant cues to sexual and emotional infidelity. *Personality and Individual Differences*, *44*, 633–644.

Non-parametric tests 15

FIGURE 15.1 In my office during my Ph.D., probably preparing some teaching. Yes, as if my life isn't embarrassing enough, I had quite long hair back then

15.1. What will this chapter tell me? ①

After my psychology degree (at City University, London) I went to the University of Sussex to do my Ph.D. (also in psychology) and like many people I had to teach to survive. Much to my dread, I was allocated to teach second-year undergraduate statistics. This was possibly the worst combination of events that I could imagine. I was still very shy at the time, and I didn't have a clue about research methods. Standing in front of a room full of strangers and trying to teach them ANOVA was only marginally more appealing than dislocating my knees and running a marathon – with broken glass in my trainers (sneakers). I obsessively prepared for my first session so that it would go well; I created handouts, I invented examples, I rehearsed what I would say. I went in terrified but at least knowing that if preparation was any predictor of success then I would be OK. About half way through the first session as I was mumbling on to a room of bored students, one of them rose majestically from their chair. She walked slowly towards me and I'm convinced that she was surrounded by an aura of bright white light and dry ice. Surely she had been chosen by her peers to impart a message of gratitude for the hours of preparation I had done and the skill with which I was unclouding their brains of the mysteries of ANOVA. She stopped beside me. We stood inches apart and my eyes raced around the floor looking for the reassurance of my shoelaces: 'No one in this room has a rabbit[1] clue what you're going on about', she spat before storming out. Scales have not been

[1] She didn't say rabbit, but she did say a word that describes what rabbits do a lot; it begins with an 'f' and the publishers think that it will offend you.

invented yet to measure how much I wished I'd run the dislocated-knees marathon that morning and then taken the day off. I was absolutely mortified. To this day I have intrusive thoughts about groups of students in my lectures walking zombie-like towards the front of the lecture theatre chanting 'No one knows what you're going on about' before devouring my brain in a rabid feeding frenzy. The point is that sometimes our lives, like data, go horribly, horribly wrong. This chapter is about data that are as wrong as dressing a cat in a pink tutu.

15.2. When to use non-parametric tests ①

We've seen in the last few chapters how we can use various techniques to look for differences between means. However, all of these tests rely on parametric assumptions (see Chapter 5). Data are often unfriendly and don't always turn up in nice normally distributed packages! Just to add insult to injury, it's not always possible to correct for problems with the distribution of a data set – so, what do we do in these cases? The answer is that we have to use special kinds of statistical procedures known as **non-parametric tests**. Non-parametric tests are sometimes known as assumption-free tests because they make fewer assumptions about the type of data on which they can be used.[2] Most of these tests work on the principle of **ranking** the data: that is, finding the lowest score and giving it a rank of 1, then finding the next highest score and giving it a rank of 2, and so on. This process results in high scores being represented by large ranks, and low scores being represented by small ranks. The analysis is then carried out on the ranks rather than the actual data. This process is an ingenious way around the problem of using data that break the parametric assumptions. Some people believe that non-parametric tests have less power than their parametric counterparts, but as we will see in Jane Superbrain Box 15.1 this is not always true. In this chapter we'll look at four of the most common non-parametric procedures: the Mann–Whitney test, the Wilcoxon signed-rank test, Friedman's test and the Kruskal–Wallis test. For each of these we'll discover how to carry out the analysis on SAS and how to interpret and report the results.

15.3. Comparing two independent conditions: the Wilcoxon rank-sum test and Mann–Whitney test ①

When you want to test differences between two conditions and different participants have been used in each condition then you have two choices: the **Mann-Whitney test** (Mann & Whitney, 1947) and **Wilcoxon's rank-sum test** (Wilcoxon, 1945; Figure 15.2). These tests are the non-parametric equivalent of the independent *t*-test. In fact both tests are equivalent, and there's another, more famous, Wilcoxon test, so it gets extremely confusing for most of us.

To make life slightly more confusing, there are two Wilcoxon tests – the Wilcoxon rank-sum test, and the Wilcoxon signed-rank test. As you can imagine, these similar names mean that it's very easy to confuse things – many people prefer to never mention the rank-sum test, always calling it the Mann–Whitney test. However SAS calls them both Wilcoxon tests. To make it easier to tell which test I am talking about, I'm going to refer to the

[2] Non-parametric tests sometimes get referred to as distribution-free tests, with an explanation that they make *no* assumptions about the distribution of the data. Technically, this isn't true: they do make distributional assumptions (e.g. the ones in this chapter all assume a continuous distribution), but they are less restrictive ones than their parametric counterparts.

FIGURE 15.2
Frank Wilcoxon

Wilcoxon two-sample test when we have independent groups (i.e. two samples), and the Wilcoxon signed-rank test when we have repeated measures (i.e. one sample, measured twice) – because that's what SAS calls it.

For example, a neurologist might collect data to investigate the depressant effects of certain recreational drugs. She tested 20 clubbers in all: 10 were given an ecstasy tablet to take on a Saturday night and 10 were allowed to drink only alcohol. Levels of depression were measured using the Beck Depression Inventory (BDI) the day after and midweek. The data are in Table 15.1.

TABLE 15.1 Data for drug experiment

Participant	Drug	BDI (Sunday)	BDI (Wednesday)
1	Ecstasy	15	28
2	Ecstasy	35	35
3	Ecstasy	16	35
4	Ecstasy	18	24
5	Ecstasy	19	39
6	Ecstasy	17	32
7	Ecstasy	27	27
8	Ecstasy	16	29
9	Ecstasy	13	36
10	Ecstasy	20	35
11	Alcohol	16	5
12	Alcohol	15	6
13	Alcohol	20	30

(Continued)

TABLE 15.1 *(Continued)*

Participant	Drug	BDI (Sunday)	BDI (Wednesday)
14	Alcohol	15	8
15	Alcohol	16	9
16	Alcohol	13	7
17	Alcohol	14	6
18	Alcohol	19	17
19	Alcohol	18	3
20	Alcohol	18	10

15.3.1. Theory ②

The logic behind the Wilcoxon two-sample test is incredibly elegant. First, let's imagine a scenario in which there is no difference in depression levels between ecstasy and alcohol users. If we were to rank the data *ignoring the group to which a person belonged* from lowest to highest (i.e. give the lowest score a rank of 1 and the next lowest a rank of 2, etc.), then what should you find? Well, if there's no difference between the groups then you expect to find a similar number of high and low ranks in each group; specifically, if you added up the ranks, then you'd expect the summed total of ranks in each group to be about the same. Now think about what would happen if there was a difference between the groups. Let's imagine that the ecstasy group is more depressed than the alcohol group. If you rank the scores as before, then you would expect the higher ranks to be in the ecstasy group and the lower ranks to be in the alcohol group. Again, if we summed the ranks in each group, we'd expect the sum of ranks to be higher in the ecstasy group than in the alcohol group.

The Mann–Whitney and Wilcoxon two-sample tests both work on this principle. In fact, when the groups have unequal numbers of participants in them then the test statistic (W_s) for the Wilcoxon two-sample test is simply the sum of ranks in the group that contains the fewer people; when the group sizes are equal it's the value of the smaller summed rank. Let's have a look at how ranking works in practice.

Figure 15.3 shows the ranking process for both the Wednesday and Sunday data. To begin with, let's use our data for Wednesday, because it's more straightforward. First, just arrange the scores in ascending order, attach a label to remind you which group they came from (I've used A for alcohol and E for ecstasy), then starting at the lowest score assign potential ranks starting with 1 and going up to the number of scores you have. The reason why I've called these potential ranks is because sometimes the same score occurs more than once in a data set (e.g. in these data a score of 6 occurs twice, and a score of 35 occurs three times). These are called *tied ranks* and these values need to be given the same rank, so all we do is assign a rank that is the average of the potential ranks for those scores. So, with our two scores of 6, because they would've been ranked as 3 and 4, we take an average of these values (3.5) and use this value as a rank for both occurrences of the score! Likewise, with the three scores of 35, we have potential ranks of 16, 17 and 18; we actually use the average of these three ranks, (16 + 17 + 18)/3 = 17. When we've ranked the data, we add up all of

Wednesday Data

Score	3	5	6	6	7	8	9	10	17	24	27	28	29	30	32	35	35	35	36	39
Potential Rank	1	2	3	4	5	6	7	8	9	10	11	12	13	14	15	16	17	18	19	20
Actual Rank	1	2	3.5	3.5	5	6	7	8	9	10	11	12	13	14	15	17	17	17	19	20
Group	A	A	A	A	A	A	A	A	A	E	E	E	E	A	E	E	E	E	E	E

Sum of Ranks for Alcohol (A) = 59 **Sum of Ranks for Ecstasy (E) = 151**

Sunday Data

Score	13	13	14	15	15	15	16	16	16	16	17	18	18	18	19	19	20	20	27	35
Potential Rank	1	2	3	4	5	6	7	8	9	10	11	12	13	14	15	16	17	18	19	20
Actual Rank	1.5	1.5	3	5	5	5	8.5	8.5	8.5	8.5	11	13	13	13	15.5	15.5	17.5	17.5	19	20
Group	A	E	A	A	A	E	A	A	E	E	E	E	A	A	E	A	E	A	E	E

Sum of Ranks for Alcohol (A) = 90.5 **Sum of Ranks for Ecstasy (E) = 119.5**

FIGURE 15.3 Ranking the depression scores for Wednesday and Sunday the ranks for the two groups. So, add the ranks for the scores that came from the alcohol group (you should find the sum is 59) and then add the ranks for the scores that came from the ecstasy group (this value should be 151). We take the lowest of these sums to be our test statistic, therefore the test statistic for the Wednesday data is $W_s = 59$.

SELF-TEST Based on what you have just learnt, try ranking the Sunday data. (The answers are in Figure 15.3 – there are lots of tied ranks and the data are generally horrible.)

You should find that when you've ranked the data, and added the ranks for the two groups, the sum of ranks for the alcohol group is 90.5 and for the ecstasy group it is 119.5. The lowest of these sums is our test statistic; therefore the test statistic for the Sunday data is $W_s = 90.5$.

The next issue is: how do we determine whether this test statistic is significant? It turns out that the mean ($\overline{W}_s$) and standard error of this test statistic ($SE_{\overline{W}_s}$) can be easily calculated from the sample sizes of each group (n_1 is the sample size of group 1 and n_2 is the sample size of group 2):

$$\overline{W}_s = \frac{n_1(n_1 + n_2 + 1)}{2}$$

$$\mathrm{SE}_{\overline{W}_s} = \sqrt{\frac{n_1 n_2(n_1 + n_2 + 1)}{12}}$$

For our data, we actually have equal-sized groups and there are 10 people in each, so n_1 and n_2 are both 10. Therefore, the mean and standard deviation are:

$$\overline{W}_s = \frac{10(10 + 10 + 1)}{2} = 105$$

$$\mathrm{SE}_{\overline{W}_s} = \sqrt{\frac{(10 \times 10)(10 + 10 + 1)}{12}} = 13.23$$

If we know the test statistic, the mean of test statistics and the standard error, then we can easily convert the test statistic to a z-score using the equation that we came across way back in Chapter 1:

$$z = \frac{\mathrm{X} - \overline{\mathrm{X}}}{s} = \frac{W_s - \overline{W}_s}{\mathrm{SE}_{\overline{W}_s}}$$

If we calculate this value for the Sunday and Wednesday depression scores we get:

$$z_{\text{Sunday}} = \frac{W_s - \overline{W}_s}{\mathrm{SE}_{\overline{W}_s}} = \frac{90.5 - 105}{13.23} = -1.10$$

$$z_{\text{Wednesday}} = \frac{W_s - \overline{W}_s}{\mathrm{SE}_{\overline{W}_s}} = \frac{59 - 105}{13.23} = -3.48$$

If these values are bigger than 1.96 (ignoring the minus sign) then the test is significant at $p < .05$. So, it looks as though there is a significant difference between the groups on Wednesday, but not on Sunday.

The procedure I've actually described is the Wilcoxon two-sample (rank-sum) test. The Mann–Whitney test, with which many of you may be more familiar, is basically the same. It is based on a test statistic U, which is derived in a fairly similar way to the Wilcoxon two sample procedure (in fact there's a direct relationship between the two). If you're interested, U is calculated using an equation in which n_1 and n_2 are the sample sizes of groups 1 and 2 respectively, and R_1 is the sum of ranks for group 1:

$$U = n_1 n_2 + \frac{n_1(n_1+1)}{2} - R_1$$

So, for our data we'd get the following (remember we have 10 people in each group and the sum of ranks for group 1, the ecstasy group, were 119.5 for the Sunday data and 151 for the Wednesday data):

$$U_{\text{Sunday}} = (10 \times 10) + \frac{10(11)}{2} - 119.50 = 35.50$$

$$U_{\text{Wednesday}} = (10 \times 10) + \frac{10(11)}{2} - 151.00 = 4.00$$

SAS produces the Wilcoxon statistic, but not the Mann–Whitney U – they are equivalent so this doesn't matter. In fact because the tests are so similar, many people refer to the Mann–Whitney–Wilcoxon (or Wilcoxon–Mann–Whitney) test.

15.3.2. Inputting data and provisional analysis ①

SELF-TEST See whether you can use what you have learnt about data entry to enter the data in Table 15.1 into SAS using a CARDS statement, in a DATA step. I'm going to call the dataset ecs_alc.

When the data are collected using different participants in each group, we need to input the data using a coding variable. So, the data editor will have three columns of data. The first column is a coding variable (called something like **drug**) which will contain either 'ecstasy' or 'alcohol'. The second column will have values for the dependent variable (BDI) measured the day after (call this variable **Sunday_BDI**) and the third will have the midweek scores on the same questionnaire (call this variable **Wednesday_BDI**).

There were no specific predictions about which drug would have the most effect so the analysis should be two-tailed. First, we would run some exploratory analyses on the data and because we're going to be looking for group differences we need to run these exploratory analyses for each group.

SELF-TEST Carry out some analyses to test for normality and homogeneity of variance in these data (see sections 5.5 and 5.6.).

The results of the exploratory analysis are shown in SAS Output 15.1. These tables show first of all that for the Sunday data the distributions for ecstasy, $D(10) = 0.28$, $p < .05$, appears to be non-normal whereas the alcohol data, $D(10) = 0.17$, $n > 0.25$, were normal; we can tell this by whether the significance of the K–S and Shapiro–Wilk tests are less than .05 (and, therefore, significant) or greater than .05 (and, therefore, non-significant, *ns*). For the Wednesday data, although the data for ecstasy were normal, $D(10) = 0.24$, *ns*, the data for alcohol appeared to be significantly non-normal, $D(10) = 0.31$, $p < .01$. This finding would alert us to the fact that the sampling distribution might also be non-normal for the Sunday and Wednesday data and that a non-parametric test should be used. You should note that the Shapiro–Wilk statistic yields exact significance values whereas the K–S test sometimes gives an approximation for the significance (see the Sunday data for the alcohol group) because SAS cannot calculate exact significances. This finding highlights an important difference between the K–S test and the Shapiro–Wilk test: in general the Shapiro–Wilk test is more accurate (see Chapter 5).

SAS OUTPUT 15.1

BDI Sunday: Alcohol

Tests for Normality				
Test	***Statistic***		***p Value***	
Shapiro-Wilk	**W**	0.959466	**Pr < W**	0.7798
Kolmogorov-Smirnov	**D**	0.169918	**Pr > D**	>0.1500
Cramer-von Mises	**W-Sq**	0.043071	**Pr > W-Sq**	>0.2500
Anderson-Darling	**A-Sq**	0.241479	**Pr > A-Sq**	>0.2500

BDI Wednesday: Alcohol

Tests for Normality				
Test	***Statistic***		***p Value***	
Shapiro-Wilk	**W**	0.753466	**Pr < W**	0.0039
Kolmogorov-Smirnov	**D**	0.305018	**Pr > D**	<0.0100
Cramer-von Mises	**W-Sq**	0.188985	**Pr > W-Sq**	0.0056
Anderson-Darling	**A-Sq**	1.05232	**Pr > A-Sq**	0.0052

BDI Sunday: Ecstasy

Tests for Normality				
Test	***Statistic***		***p Value***	
Shapiro-Wilk	**W**	0.81064	**Pr < W**	0.0195
Kolmogorov-Smirnov	**D**	0.275848	**Pr > D**	0.0297
Cramer-von Mises	**W-Sq**	0.153433	**Pr > W-Sq**	0.0184
Anderson-Darling	**A-Sq**	0.842705	**Pr > A-Sq**	0.0202

BDI Sunday: Alcohol

Tests for Normality				
Test	***Statistic***		***p Value***	
Shapiro-Wilk	**W**	0.941141	**Pr < W**	0.5658
Kolmogorov-Smirnov	**D**	0.23469	**Pr > D**	0.1183
Cramer-von Mises	**W-Sq**	0.065011	**Pr > W-Sq**	>0.2500
Anderson-Darling	**A-Sq**	0.353869	**Pr > A-Sq**	>0.2500

15.3.3 Running the analysis ①

Most non-parametric tests are run using PROC NPAR1WAY. This procedure is very straightforward – we put the categorical variable that defines the groups (**drug** in our case) in a CLASS statement, and the test variables in a VAR statement. Finally, SAS includes a continuity correction to the formulae by default[3] – we don't want to have that, so we stop it by writing CORRECT=NO on the NPAR1WAY line. The syntax is shown in SAS Syntax 15.1.

```
PROC NPAR1WAY DATA=ecs_alc WILCOXON CORRECT=NO;
   CLASS drug;
   VAR bdi_sunday bdi_wednesday;
   RUN;

PROC NPAR1WAY DATA=ecs_alc WILCOXON CORRECT=NO ;
   CLASS drug;
   VAR  bdi_sunday bdi_wednesday;
EXACT WILCOXON;
   RUN;
```

SAS Syntax 15.1

By default SAS calculates the significance of the Wilcoxon two-sample test using a method that is accurate with large samples (called the *asymptotic method*); however, when samples are smaller, or the data are particularly poorly distributed, then more accurate methods are available. The most accurate method is an *exact* test, which calculates the significance of the Wilcoxon two-sample test exactly. However, to get this precision, there is a price, and because of the complexities of the computation SAS can take some time to find a solution – especially in large samples. The second block of syntax in SAS Syntax 15.1 shows the use of the EXACT option.

SAS TIP 15.1 An alternative to the Wilcoxon–Mann–Whitney test ②

In Chapter 5 we met a Kolmogorov–Smirnov test that tested whether a sample was from a normally distributed population. This is a different test! In fact, it tests whether two groups have been drawn from the same population (regardless of what that population may be). In effect, this means it does much the same as the Wilcoxon test! However, this test tends to have better power than the Mann–Whitney test when sample sizes are less than about 25 per group, and so is worth selecting if that's the case. To implement the K–S test, just write K–S instead of WILCOXON on the PROC NPAR1WAY line.

[3]We are assuming that we have a normal distribution of test statistics, to calculate the *p*-value. That can't be true, because the test statistic can't take on any value – it has to go up or down in 0.5 jumps. If we wanted to be very cautious we would subtract 0.5 from the score, to ensure that it was always conservative (that is, less likely to be significant). However, this is generally seen as being overcautious and is rarely recommended. If you are worried, use an exact test.

15.3.4. Output from the Wilcoxon two-sample test ①

I explained earlier that the Wilcoxon test works by looking at differences in the ranked positions of scores in different groups. Therefore, the first part of the output summarizes the data after they have been ranked. Specifically, SAS tells us the total ranks in each condition (see SAS Output 15.2), and the total rank we would expect under the null hypothesis, the standard deviation of the ranks and the mean rank. Remember that the Wilcoxon two-sample test relies on scores being ranked from lowest to highest; therefore, the group with the lowest mean rank is the group with the greatest number of lower scores in it. Similarly, the group that has the highest mean rank should have a greater number of high scores within it. Therefore, this initial table can be used to ascertain which group had the highest scores, which is useful in case we need to interpret a significant result. You should note that the sums of ranks are the same as those calculated in section 15.3.1 (which is something of a relief to me!).

SAS Output 15.3 provides the actual test statistics for the Wilcoxon procedure and the corresponding *z*-score. The output has a table for each variable (one for **Sunday_BDI** and one for **Wednesday_BDI**) and in each table there is the value of the value of Wilcoxon's statistic (which most people call *W*, but SAS calls *S*) and the associated

SAS OUTPUT 15.2

Wilcoxon Scores (Rank Sums) for Variable bdi_sunday Classified by Variable drug					
drug	***N***	***Sum of Scores***	***Expected Under H0***	***Std Dev Under H0***	***Mean Score***
Alcohol	10	90.50	105.0	13.123903	9.050
Ecstacy	10	119.50	105.0	13.123903	11.950
Average scores were used for ties.					

Wilcoxon Scores (Rank Sums) for Variable bdi_wednesday Classified by Variable drug					
drug	***N***	***Sum of Scores***	***Expected Under H0***	***Std Dev Under H0***	***Mean Score***
Alcohol	10	59.0	105.0	13.203867	5.90
Ecstacy	10	151.0	105.0	13.203867	15.10
Average scores were used for ties.					

z approximation (there's also a *t* approximation, but we won't worry about that). Note that the values of *W* and the associated *z*-score are the same as we calculated in section 15.3.1.

The important part of each table is the significance value of the test, which gives the two-tailed probability that a test statistic of at least that magnitude is a chance result. This significance value can be used as it is when no prediction has been made about which group will differ from which. However, if a prediction has been made (e.g. if we said that ecstasy users would be more depressed than alcohol users the day after taking the drug) then we can use the one-tailed probability. For these data, the Wilcoxon two-sample test is non-significant (two-tailed) for the depression scores taken on the Sunday. This finding indicates that ecstasy is no more of a depressant, the day after taking it, than alcohol: both groups report comparable levels of depression. However, for the midweek measures the results are highly significant ($p < .0001$).[4] The value of the mean rankings indicates that the ecstasy group had significantly higher levels of depression midweek than the alcohol group. This conclusion is reached by noting that for the Wednesday scores, the average rank is higher in the ecstasy users (15.10) than in the alcohol users (5.90).

SAS OUTPUT 15.3

Wilcoxon Two-Sample Test	
Statistic (S)	90.5000
Normal Approximation	
Z	−1.1049
One-Sided Pr < Z	0.1346
Two-Sided Pr > |Z|	0.2692
t Approximation	
One-Sided Pr < Z	0.1415
Two-Sided Pr > |Z|	0.2830
Exact Test	
One-Sided Pr <= S	0.1441
Two-Sided Pr > = |S – Mean|	0.2882

Wilcoxon Two-Sample Test	
Statistic (S)	59.0000
Normal Approximation	
Z	-3.4838
One-Sided Pr < Z	0.0002
Two-Sided Pr > |Z|	0.0005
t Approximation	
One-Sided Pr < Z	0.0012
Two-Sided Pr > |Z|	0.0025
Exact Test	
One-Sided Pr < = S	6.495E-05
Two-Sided Pr > = |S – Mean|	1.299E-04

[4] Remember that when SAS writes something like 6.495E-05, it means ‘a really small number’,

Also included in SAS Output 15.3 are the exact probability values. These don't actually change our conclusions, but be aware that you should probably consult these values in preference to the asymptotic value, especially when sample sizes are small.

15.3.5. Calculating an effect size ②

As we've seen throughout this book, it's important to report effect sizes so that people have a standardized measure of the size of the effect you observed, which they can compare to other studies. SAS doesn't calculate an effect size for us, but we can calculate approximate effect sizes really easily thanks to the fact that SAS converts the test statistics into a *z*-score. The equation to convert a *z*-score into the effect size estimate, *r*, is as follows (from Rosenthal, 1991, p. 19):

$$r = \frac{Z}{\sqrt{N}}$$

in which *z* is the *z*-score that SAS produces and *N* is the size of the study (i.e. the number of total observations) on which *z* is based. In this case SAS Output 15.3 tells us that *z* is –1.10 for the Sunday data and –3.48 for the Wednesday data. In both cases we had 10 ecstasy users and 10 alcohol users and so the total number of observations was 20. The effect sizes are therefore:

$$r_{\text{Sunday}} = \frac{-1.11}{\sqrt{20}} = -0.25$$

$$r_{\text{Wednesday}} = \frac{-3.48}{\sqrt{20}} -0.78$$

This represents a small to medium effect for the Sunday data (it is below the .3 criterion for a medium effect size) and a huge effect for the Wednesday data (the effect size is well above the .5 threshold for a large effect). The Sunday data show how a fairly large effect size can still be non-significant in a small sample.

15.3.6. Writing the results ①

For the Mann–Whitney test, we need to report only the test statistic (which is denoted by *U*) and its significance. Of course, we really ought to include the effect size as well. So, we could report something like:

✓ Depression levels in ecstasy users (*Mdn* = 17.50) did not differ significantly from alcohol users (*Mdn* = 16.00) the day after the drugs were taken, $U = 35.50$, $z = -1.10$, *ns*, $r = -.25$. However, by Wednesday, ecstasy users (*Mdn* = 33.50) were significantly more depressed than alcohol users (*Mdn* = 7.50), $U = 4.00$, $z = -3.48$, $p < .001$, $r = -.78$.

Note that I've reported the median for each condition – this statistic is more appropriate than the mean for non-parametric tests. We could also choose to report Wilcoxon's test rather than Mann–Whitney's *U* statistic and this would be as follows:

✓ Depression levels in ecstasy users (*Mdn* = 17.50) did not significantly differ from alcohol users (*Mdn* = 16.00) the day after the drugs were taken, $W_s = 90.50$, $z = -1.10$, *ns*, $r = -.25$.

However, by Wednesday, ecstasy users (Mdn = 33.50) were significantly more depressed than alcohol users (Mdn = 7.50), $W_s = 59.00$, $z = -3.48$, $p < .001$, $r = -.78$.

CRAMMING SAM'S TIPS

- The Mann–Whitney test and Wilcoxon rank-sum test compare two conditions when different participants take part in each condition and the resulting data violate any assumption of the independent *t*-test.
- Look at the row labelled *Two Sided > Pr |Z|*. If the value is less than .05 then the two groups are significantly different. (If you opted for exact tests then look at the row labelled *Exact Test*.)
- The values of the ranks tell you how the groups differ (the group with the highest scores will have the highest ranks).
- Report the *S* statistic (or W_s if you prefer), the corresponding *z* and the significance value. Also report the medians and their corresponding ranges (or draw a boxplot).
- You should calculate the effect size and report this too!

JANE SUPERBRAIN 15.1

Non-parametric tests and statistical power ②

Ranking the data is a useful way around the distributional assumptions of parametric tests but there is a price to pay: by ranking the data we lose some information about the magnitude of differences between scores. Consequently, non-parametric tests can be less powerful than their parametric counterparts. Statistical power (section 2.6.5) refers to the ability of a test to find an effect that genuinely exists. So, by saying that non-parametric tests are less powerful, we mean that if there is a genuine effect in our data then a parametric test is more likely to detect it than a non-parametric one. However, this statement is true only *if the assumptions of the parametric test are met*. So, if we use a parametric test and a non-parametric test on the same data, and those data meet the appropriate assumptions, then the parametric test will have greater power to detect the effect than the non-parametric test.

The problem is that to define the power of a test we need to be sure that it controls the Type I error rate (the number of times a test will find a significant effect when in reality there is no effect to find – see section 2.6.2). We saw in Chapter 2 that this error rate is normally set at 5%. We know that when the sampling distribution is normally distributed then the Type I error rate of tests based on this distribution is indeed 5%, and so we can work out the power. However, when data are not normal the Type I error rate of tests based on this distribution won't be 5% (in fact we don't know what it is for sure as it will depend on the shape of the distribution) and so we have no way of calculating power (because power is linked to the Type I error rate – see section 2.6.5). So, although you often hear of non-parametric tests having an increased chance of a Type II error (i.e. more chance of accepting that there is no difference between groups when, in reality, a difference exists), this is true only if the sampling distribution is normally distributed.

15.4. Comparing two related conditions: the Wilcoxon signed-rank test ①

The **Wilcoxon signed-rank test** (Wilcoxon, 1945), not to be confused with the two-sample, or rank-sum test in the previous section, is used in situations in which there are two sets of scores to compare, but these scores come from the same participants. As such, think of it as the non-parametric equivalent of the dependent *t*-test. Imagine the experimenter in the previous section was now interested in the *change* in depression levels, within people, for each of the two drugs. We now want to compare the BDI scores on Sunday to those on Wednesday. We still have to use a non-parametric test because the distributions of scores for both drugs were non-normal on one of the two days (see SAS Output 15.1).

15.4.1. Theory of the Wilcoxon signed-rank test ②

The Wilcoxon signed-rank test works in a fairly similar way to the dependent *t*-test (Chapter 9) in that it is based on the differences between scores in the two conditions you're comparing. Once these differences have been calculated they are ranked (just like in section 15.3.1) but the sign of the difference (positive or negative) is assigned to the rank. If we use the same data as before we can compare depression scores on Sunday to those on Wednesday for the two drugs separately.

Table 15.2 shows the ranking for these data. Remember that we're ranking the two drugs separately. First, we calculate the difference between Sunday and Wednesday (that's just Sunday's score subtracted from Wednesday's). If the difference is zero (i.e. the scores are the same on Sunday and Wednesday) then we exclude these data from the ranking. We make a note of the sign of the difference (was it positive or negative) and then rank the differences (starting with the smallest) ignoring whether they are positive or negative. The ranking process is the same as in section 15.3.1, and we deal with tied scores in exactly the same way. Finally, we collect together the ranks that came from a positive difference between the conditions, and add them up to get the sum of positive ranks (r_+). We also add up the ranks that came from negative differences between the conditions to get the sum of negative ranks (r_-). So, for ecstasy, $r_+ = 36$ and $r_- = 0$ (in fact there were no negative ranks), and for alcohol, $r_+ = 8$ and $r_- = 47$. The test statistic, *S*, is computed using:

$$S = r_+ - \frac{n(n+1)}{4}$$

where *n* is the number of cases which did not have the same score on both occasions. For the ecstasy group, this is:

$$S = 36 - \frac{8(8+1)}{4}$$

$$S = 18$$

For the alcohol group:

$$S = 47 - \frac{10(10+1)}{4}$$

$$S = 19.5$$

TABLE 15.2 Ranking data in the Wilcoxon signed-rank test

BDI Sunday	BDI Wednesday	Difference	Sign	Rank	Positive Ranks	Negative Ranks
			Ecstasy			
15	28	13	+	2.5	2.5	
35	35	0	Exclude			
16	35	19	+	6	6	
18	24	6	+	1	1	
19	39	20	+	7	7	
17	32	15	+	4.5	4.5	
27	27	0	Exclude			
16	29	13	+	2.5	2.5	
13	36	23	+	8	8	
20	35	15	+	4.5	4.5	
				Total =	**36**	**0**
			Alcohol			
16	5	−11	−	9		9
15	6	−9	−	7		7
20	30	10	+	8	+8	
15	8	−7	−	3.5		3.5
16	9	−7	−	3.5		3.5
13	7	−6	−	2		2
14	6	−8	−	5.5		5.5
19	17	−2	−	1		1
18	3	−15	−	10		10
18	10	−8	−	5.5		5.5
				Total =	**8**	**47**

When the sample is less than 20, SAS uses a very sophisticated method to calculate the *p*-values, which we are not going to concern ourselves with.[5] When the sample is over 20, SAS converts *S* to a *t*-distributed variable, using:

$$t = s\sqrt{\frac{n-1}{nV - S^2}}$$

with *n*–1 degrees of freedom (*N* is still the number of non-tied cases) You'll notice there's an extra symbol called V in that formula, which we'll need to calculate.

If there are no ties, then *V* is calculated using:

[5] Although if you do want to know the details it uses a convolution of scaled binomial distributions. I'm not exactly sure what that is either.

$$V = \frac{1}{24}n(n+1)(2n+1)$$

If there are ties, then SAS uses a more complex formula, which has a very small effect on the result, and which we are going to ignore.

If we plug the numbers in for the ecstasy group, first we'll calculate V:

$$V = \frac{1}{24}8(8+1)(2\times 8+1)$$

$$V = \frac{1}{24}\times 72\times 17$$

$$V = 51$$

$$t = 10\sqrt{\frac{(8-1)}{8\times 51 - 18^2}}$$

$$t = 18\sqrt{\frac{7}{408-324}}$$

$$t = 5.2$$

The degrees of freedom for the t-statistic are given by $n-1 = 7$. We can calculate the probability of finding a value of t as high as 5.2 when the null hypothesis is false, and it is 0.001. (However, bear in mind that this was an example only – our sample was too small to use this formula.)

JANE SUPERBRAIN 15.2

Different sorts of Wilcoxon tests

Everyone knows how to do a t-test, or an analysis of variance, but there is surprisingly little agreement on how to handle some of the trickier aspects of the Wilcoxon test. We've described how SAS does a Wilcoxon test, but different programs approach the problem of ties and small samples differently, so don't be surprised if you see a slightly different result. (As we'll see in a minute SAS gives a p-value of 0.0076 for the ecstasy data. When I ran the same analysis in some other widely used programs, I got: Stata, $p = 0.0076$; SAS, $p = 0.012$; and R, $p = 0.014$.

15.4.2. Running the analysis ①

There isn't any way to do a Wilcoxon signed-rank test on two variables in SAS directly. Instead, we need to create a difference score first in a DATA step, and then use PROC UNIVARIATE to test this against zero. This is shown in SAS Syntax 15.2.

```
DATA diff; SET ecs_alc;
    diff = bdi_wednesday-bdi_sunday;
    RUN;
PROC UNIVARIATE DATA=diff;
    BY drug;
    VAR diff;
    RUN;
```
SAS Syntax 15.2

15.4.3 Output for the alcohol group ①

The first set of results is for the alcohol group. The first table shown in SAS Output 15.4 gives descriptive information about the difference scores. Because we subtracted the Wednesday score from the Sunday score, a negative number means that a person's BDI score has decreased, and a positive score means that the BDI score has increased. The median score is –7.5, so the median score has decreased.

The second table shows the location statistics, which tests the null hypothesis that the difference score is equal to zero. Of particular interest to us is the signed-rank test, which has a value of 19.5 (the same as we calculated earlier) and a *p*-value of .045 (not the same as we calculated earlier, because of the complex method SAS uses when the sample is fewer than 20). Therefore, we can conclude that for ecstasy users there was a significant increase in depression from the next day to midweek (S = 19.5, n = 10, p = .022).

SAS OUTPUT 15.4

Basic Statistical Measures				
Location		**Variability**		
Mean	–6.30000	Std Deviation	6.63409	
Median	–7.50000	Variance	44.01111	
Mode	–8.00000	Range	25.00000	
		Interquartile Range	3.00000	

Tests for Location: Mu0=0				
Test	**Statistic**		**p Value**	
Student's t	t	–3.00303	Pr > |t|	0.0149
Sign	M	–4	Pr >= |M|	0.0215
Signed Rank	S	–19.5	Pr >= |S|	0.0449

15.4.4. Output for the ecstasy group ①

The output for the ecstasy group is shown in SAS Output 15.5. The median score here is positive, showing that depression scores have *increased* from Sunday to Wednesday. The signed-rank statistic, *S*, has a value of –19.5 (we can ignore the minus sign, if there is one), and this is the same as the value that we calculated. The *p*-value is .008, and as this is less than .05 we can conclude that the difference is statistically significant.

From the results of the two different groups, we can see that there is an opposite effect when alcohol is taken to when ecstasy is taken. Alcohol makes you slightly depressed the morning after but this depression has dropped by midweek. Ecstasy also causes some depression the morning after consumption; however, this depression increases towards the middle of the week, because the ecstasy users have a positive median score. Of course, to see the true effect of the morning after we would have had to take measures of depression before the drugs were administered. This opposite effect between groups of people is known as an interaction (i.e. you get one effect under certain circumstances and a different effect under other circumstances) and we came across these in Chapters 10 to 14.

SAS OUTPUT 15.5

Basic Statistical Measures			
Location		**Variability**	
Mean	12.40000	**Std Deviation**	8.00278
Median	14.00000	**Variance**	64.04444
Mode	0.00000	**Range**	23.00000
		Interquartile Range	13.00000

Tests for Location: Mu0=0				
Test		**Statistic**		**p Value**
Student's t	t	4.899829	Pr > |t|	0.0008
Sign	M		Pr >= |M|	0.0078
Signed Rank	S		Pr >= |S|	0.0078

15.4.5. Writing the results ①

For the Wilcoxon test, we need to report only the test statistic (which is denoted by the letter *S* and the smallest of the two sum of ranks), its significance and preferably an effect size. So, we could report something like:

✓ For ecstasy users, depression levels were significantly higher on Wednesday (*Mdn* = 33.50) than on Sunday (*Mdn* = 17.50), $S = 18$, $p = .008$. However, for alcohol users the opposite was true: depression levels were significantly lower on Wednesday (*Mdn* = 7.50) than on Sunday (*Mdn* = 16.0), $S = 19.5$, $p = .045$.

CRAMMING SAM'S TIPS

- The Wilcoxon signed-rank test compares two conditions when the same participants take part in each condition and the resulting data violate an assumption of the dependent *t*-test.
- Look at the value for Pr > |*S*|. If the value is less than .05 then the two groups are significantly different.
- Look at the median to determine the direction of the effect.
- Report the *S* statistic and the significance value. Also report the medians and their corresponding ranges (or draw a boxplot).

LABCOAT LENI'S REAL RESEARCH 15.1

Having a quail of a time? ①

We encountered some research in Chapter 2 in which we discovered that you can influence aspects of male quail's sperm production through 'conditioning'. The basic idea is that the male is granted access to a female for copulation in a certain chamber (e.g. one that is coloured green) but gains no access to a female in a different context (e.g. a chamber with a tilted floor). The male, therefore, learns that when he is in the green chamber his luck is in, but if the floor is tilted then frustration awaits. For other males the chambers will be reversed (i.e. they get sex only when in the chamber with the tilted floor). The human equivalent (well, sort of) would be if you always managed to pull in the Pussycat Club but never in the Honey Club.[6] During the test phase, males get to mate in both chambers; the question is: after the males have learnt that they will get a mating opportunity in a certain context, do they produce more sperm or better-quality sperm when mating in that context compared to the control context? (i.e. are you more of a stud in the Pussycat club? OK, I'm going to stop this analogy now.)

Mike Domjan and his colleagues predicted that if conditioning evolved because it increases reproductive fitness then males who mated in the context that had previously signalled a mating opportunity would fertilize a significantly greater number of eggs than quails that mated in their control context (Matthews, Domjan, Ramsey, & Crews, 2007). They put this hypothesis to the test in an experiment that is utter genius. After training, they allowed 14 females to copulate with two males (counterbalanced): one male copulated with the female in the chamber that had previously signalled a reproductive opportunity (**Signalled**), whereas the second male copulated with the same female but in the chamber that had not previously signalled a mating opportunity (**Control**). Eggs were collected from the females for 10 days after the mating and a genetic analysis was used to determine the father of any fertilized eggs.

The data from this study are in the file **Matthewsetal2007.sas7bdat**. Labcoat Leni wants you to carry out a Wilcoxon signed-rank test to see whether more eggs were fertilized by males mating in their signalled context compared to males in their control context.

Answers are in the additional material on the companion website (or look at page 760 in the original article).

MATTHEWS, R. C. ET AL. (2007). *PSYCHOLOGICAL SCIENCE, 18*(9), 758–762.

15.5. Differences between several independent groups: the Kruskal–Wallis test ①

In Chapter 10 we discovered a technique called one-way independent ANOVA that could be used to test for differences between several independent groups. I mentioned several times in that chapter that the *F* statistic can be robust to violations of its assumptions (section 10.2.10). We also saw that there are measures that can be taken when you have heterogeneity of variance (Jane Superbrain Box 10.2). However, there is another alternative: the one-way independent ANOVA has a non-parametric counterpart called the **Kruskal–Wallis test** (Kruskal & Wallis, 1952). If you have data that have violated an assumption then this test can be a useful way around the problem.

I read a story in a newspaper claiming that scientists had discovered that the chemical genistein, which is naturally occurring in soya, was linked to lowered sperm counts in western males. In fact, when you read the actual study, it had been conducted on rats, it found no link to lowered sperm counts, but there was evidence of abnormal sexual development in male rats (probably because this chemical acts like oestrogen). The journalist naturally interpreted this as a clear link to apparently declining sperm counts in western males. Anyway, as a vegetarian who eats lots of soya products and probably would like to have

[6] These are both clubs in Brighton that I don't go to because although my social skills are marginally better than they used to be, they're not that good.

kids one day, I might want to test this idea in humans rather than rats. I took 80 males and split them into four groups that varied in the number of soya meals they ate per week over a year-long period. The first group was a control group and had no soya meals at all per week (i.e. none in the whole year); the second group had one soya meal per week (that's 52 over the year); the third group had four soya meals per week (that's 208 over the year); and the final group had seven soya meals a week (that's 364 over the year). At the end of the year, all of the participants were sent away to produce some sperm that I could count (when I say 'I', I mean someone else in a laboratory as far away from me as humanly possible.)[7]

FIGURE 15.4 Joseph Kruskal spotting some more errors in his well-thumbed first edition of *Discovering Statistics*... by that idiot Field

15.5.1. Theory of the Kruskal–Wallis test ②

The theory for the Kruskal–Wallis test is very similar to that of the Mann–Whitney (and Wilcoxon rank-sum) test, so before reading on look back at section 15.3.1. Like the Mann–Whitney test, the Kruskal–Wallis test is based on ranked data. So, to begin with, you simply order the scores from lowest to highest, ignoring the group to which the score belongs, and then assign the lowest score a rank of 1, the next highest a rank of 2 and so on (see section 15.3.1 for more detail). When you've ranked the data you collect the scores back into their groups and simply add up the ranks for each group. The sum of ranks for each group is denoted by R_i (where i is used to denote the particular group).

Table 15.3 shows the raw data for this example along with the ranks.

SELF-TEST Have a go at ranking the data and see if you get the same results as me.

Once the sum of ranks has been calculated for each group, the test statistic, H, is calculated as:

$$H = \frac{12}{N(N-1)} \sum_{i=1}^{k} \frac{R_i^2}{n_i} - 3(N+1) \qquad (15.1)$$

In this equation, R_i is the sum of ranks for each group, N is the total sample size (in this case 80) and n_i is the sample size of a particular group (in this case we have equal sample sizes and they are all 20). Therefore, all we really need to do for each group is square the sum of ranks and divide this value by the sample size for that group. We then add up these

[7] In case any medics are reading this chapter, these data are made up and, because I have absolutely no idea what a typical sperm count is, they're probably ridiculous. I apologize and you can laugh at my ignorance.

values. That deals with the middle part of the equation; the rest of it involves calculating various values based on the total sample size. For these data we get:

$$\begin{aligned}
H &= \frac{12}{80(81)}\left(\frac{972^2}{20}+\frac{883^2}{20}+\frac{833^2}{20}+\frac{547^2}{20}\right)-3(81)\\
&= \frac{12}{6480}(42{,}966.45+38{,}984.45+38{,}984.45+14{,}960.45)-243\\
&= 0.0019(135{,}895.8)-243\\
&= 251.66-243\\
&= 8.659
\end{aligned}$$

This test statistic has a special kind of distribution known as the chi-square distribution (see Chapter 18) and for this distribution there is one value for the degrees of freedom, which is one less than the number of groups (k–1): in this case 3.

TABLE 15.3 Data for the soya example with ranks

No Soya		*1 Soya Meal*		*4 Soya Meals*		*7 Soya Meals*	
Sperm (Millions)	*Rank*	*Sperm (Millions)*	*Rank*	*Sperm (Millions)*	*Rank*	*Sperm (Millions)*	*Rank*
0.35	4	0.33	3	0.40	6	0.31	1
0.58	9	0.36	5	0.60	10	0.32	2
0.88	17	0.63	11	0.96	19	0.56	7
0.92	18	0.64	12	1.20	21	0.57	8
1.22	22	0.77	14	1.31	24	0.71	13
1.51	30	1.53	32	1.35	27	0.81	15
1.52	31	1.62	34	1.68	35	0.87	16
1.57	33	1.71	36	1.83	37	1.18	20
2.43	41	1.94	38	2.10	40	1.25	23
2.79	46	2.48	42	2.93	48	1.33	25
3.40	55	2.71	44	2.96	49	1.34	26
4.52	59	4.12	57	3.00	50	1.49	28
4.72	60	5.65	61	3.09	52	1.50	29
6.90	65	6.76	64	3.36	54	2.09	39
7.58	68	7.08	66	4.34	58	2.70	43
7.78	69	7.26	67	5.81	62	2.75	45
9.62	72	7.92	70	5.94	63	2.83	47
10.05	73	8.04	71	10.16	74	3.07	51
10.32	75	12.10	77	10.98	76	3.28	53
21.08	80	18.47	79	18.21	78	4.11	56
Total (R_i)	**927**		**883**		**883**		**547**

15.5.2. Inputting data and provisional analysis ①

SELF-TEST See whether you can enter the data in Table 15.3 into SAS (you don't need to enter the ranks). Then conduct some exploratory analyses on the data (see sections 5.5. and 5.6).

You can find the data in **soya.sas7bdat**.

First, we would run some exploratory analyses on the data and because we're going to be looking for group differences we need to run these exploratory analyses for each group. If you do these analyses you should find the same tables shown in SAS Output 15.6. The first table shows that the Kolmogorov–Smirnov test (see section 5.5) was not significant for the control group ($D = .181, p = .085$) but the Shapiro–Wilk test is significant and this test is actually more accurate (though less widely reported) than the Kolmogorov–Smirnov test (see Chapter 5), especially for small samples. Data for the group that ate one soya meal per week were significantly different from normal ($D = .207, p = .023$), as were the data for those that ate 4 ($D = .267, p < .010$) and 7 ($D = .204, p = .027$).

SAS OUTPUT 15.6

Number of soya meals=No soya meals

Tests for Normality				
Test	***Statistic***		***p Value***	
Shapiro-Wilk	**W**	0.805231	**Pr < W**	0.0010
Kolmogorov-Smirnov	**D**	0.180951	**Pr > D**	0.0850
Cramer-von Mises	**W-Sq**	0.175862	**Pr > W-Sq**	0.0095
Anderson-Darling	**A-Sq**	1.150504	**Pr > A-Sq**	<0.0050

Number of soya meals=1 per week

Tests for Normality				
Test	***Statistic***		***p Value***	
Shapiro-Wilk	**W**	0.82591	**Pr < W**	0.0022
Kolmogorov-Smirnov	**D**	0.207445	**Pr > D**	0.0231
Cramer-von Mises	**W-Sq**	0.172822	**Pr > W-Sq**	0.0104
Anderson-Darling	**A-Sq**	1.102828	**Pr > A-Sq**	0.0053

Number of soya meals=4 per week

Tests for Normality				
Test	***Statistic***		***p Value***	
Shapiro-Wilk	**W**	0.742552	**Pr < W**	0.0001
Kolmogorov-Smirnov	**D**	0.267145	**Pr > D**	<0.0100
Cramer-von Mises	**W-Sq**	0.337002	**Pr > W-Sq**	<0.0050
Anderson-Darling	**A-Sq**	1.849512	**Pr > A-Sq**	<0.0050

SAS OUTPUT 15.6

Number of soya meals=7 per week				
Tests for Normality				
Test		***Statistic***		***p Value***
Shapiro-Wilk	**W**	0.912299	**Pr < W**	0.0705
Kolmogorov-Smirnov	**D**	0.204278	**Pr > D**	0.0269
Cramer-von Mises	**W-Sq**	0.12514	**Pr > W-Sq**	0.0474
Anderson-Darling	**A-Sq**	0.69088	**Pr > A-Sq**	0.0629

15.5.3. Doing the Kruskal–Wallis test on SAS ①

The Kruskal Wallis test is done in SAS in a very similar to the Wilcoxon–Mann–Whitney test, using PROC NPAR1WAY. The syntax is shown in SAS Syntax 15.3.

```
PROC NPAR1WAY DAYA=chap16.soya;
   CLASS soya;
   VAR sperm;
   RUN;
```

SAS Syntax 15.3

As with the Mann–Whitney test you can ask for an exact test to be carried out, with the EXACT option. However, you should be aware that is is going to take a *long* time to run, even with a relatively small sample of 80.

15.5.4. Output from the Kruskal–Wallis test ①

SAS Output 15.7 shows the output from the Kruskal–Wallis test. The first table is a summary of the ranked data in each condition and we'll need these for interpreting any effects.

SAS OUTPUT 15.7

Wilcoxon Scores (Rank Sums) for Variable SPERM Classified by Variable soya					
soya	***N***	***Sum of Scores***	***Expected Under H0***	***Std Dev Under H0***	***Mean Score***
1 per week	20	883.0	810.0	90.0	44.150
4 per week	20	883.0	810.0	90.0	44.150
7 per week	20	547.0	810.0	90.0	27.350
No soya meals	20	927.0	810.0	90.0	46.350

Kruskal-Wallis Test	
Chi-Square	8.6589
DF	3
Pr > Chi-Square	0.0342

The second table shows the test statistic H for the Kruskal–Wallis test (although SAS labels it chi-square, because of its distribution, rather than H), its associated degrees of freedom (in this case we had 4 groups so there are 3 degrees of freedom) and the significance. The crucial thing to look at is the significance value, which is .034; because this value is less than .05 we could conclude that the amount of soya meals eaten per week does significantly affect sperm counts. Like a one-way ANOVA, though, this test tells us only that a difference exists; it doesn't tell us exactly where the differences lie.

SELF-TEST Use PROC BOXPLOT to draw a boxplot of these data

One way to see which groups differ is to look at a boxplot (see section 4.5) of the groups (see Figure 15.5). The first thing to note is that there are some outliers (note the circles and asterisks that lie above the top whiskers) – these are men who produced a particularly rampant amount of sperm. Using the control as our baseline, the medians of the first three groups seem quite similar; however, the median of the group which ate seven soya meals per week does seem a little lower, so perhaps this is where the difference lies. However, these conclusions are subjective. What we really need are some contrasts or *post hoc* tests like we used in ANOVA (see sections 10.2.11 and 10.2.12).

15.5.5. *Post hoc* tests for the Kruskal–Wallis test ②

There are two ways to do non-parametric *post hoc* procedures, the first being to use Wilcoxon two-sample tests (section 15.3). However, if we use lots of Wilcoxon two-sample tests we will inflate the Type I error rate (section 10.2.1) and this is precisely why we don't begin by doing lots of tests! However, if we want to use lots of Wilcoxon two sample tests to follow up a Kruskal–Wallis test, we can if we make some kind of adjustment to ensure that the Type I errors

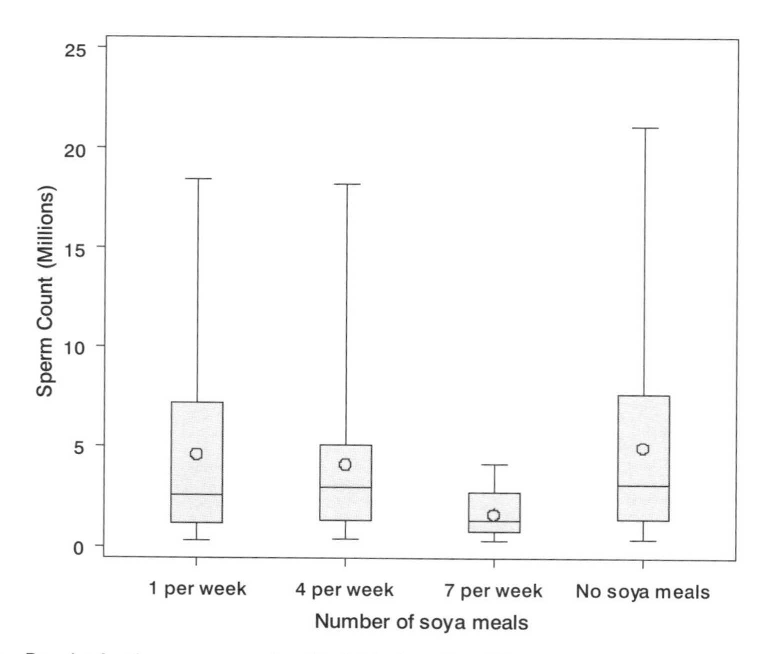

FIGURE 15.5 Boxplot for the sperm counts of individuals eating different numbers of soya meals per week

don't build up to more than .05. The easiest method is to use a Bonferroni correction, which in its simplest form just means that instead of using .05 as the critical value for significance for each test, you use a critical value of .05 divided by the number of tests you've conducted. If you do this, you'll soon discover that you quickly end up using a critical value for significance that is so small that it is very restrictive. Therefore, it's a good idea to be selective about the comparisons you make. In this example, we have a control group which had no soya meals. As such, a nice succinct set of comparisons would be to compare each group against the control:

Can I do non-parametric *post hoc* tests?

- Test 1: one soya meal per week compared to no soya meals
- Test 2: four soya meals per week compared to no soya meals
- Test 3: seven soya meals per week compared to no soya meals

This results in three tests, so rather than use .05 as our critical level of significance, we'd use .05/3 = .0167. If we didn't use focused tests and just compared all groups to all other groups we'd end up with six tests rather than three (no soya vs. 1 meal, no soya vs. 4 meals, no soya vs. 7 meals, 1 meal vs. 4 meals, 1 meal vs. 7 meals, 4 meals vs. 7 meals), meaning that our critical value would fall to .05/6 = .0083.

SELF-TEST Carry out the three Wilcoxon tests suggested above.

SAS Output 15.8 shows the test statistics from doing Wilcoxon two sample tests on the three focused comparisons that I suggested. Remember that we are now using a critical value of .0167, so the only comparison that is significant is when comparing those that had seven soya meals a week to those that had none (because the observed significance value of .009 is less than .0167). The other two comparisons produce significance values that are greater than .0167 so we'd have to say they're non-significant. So the effect we got seems to mainly reflect the fact that eating soya seven times per week lowers (I know this from the medians in Figure 15.5) sperm counts compared to eating no soya. However, eating some soya (one meal or four meals) doesn't seem to affect sperm counts significantly.

SAS OUTPUT 15.8

No Soya vs. 1 Meal per

Wilcoxon Two-Sample Test	
Statistic	401.0000
Normal Approximation	
Z	–0.2435
One-Sided Pr < Z	0.4038
Two-Sided Pr > |Z|	0.8077
t Approximation	
One-Sided Pr < Z	0.4045
Two-Sided Pr > |Z|	0.8089

No Soya vs. 4 Meals per week

Wilcoxon Two-Sample Test	
Statistic	398.0000
Normal Approximation	
Z	–0.3246
Two-Sided Pr > |Z|	0.7455
One-Sided Pr < Z	0.3727
t Approximation	
One-Sided Pr < Z	0.3736
Two-Sided Pr > |Z|	0.7472

No Soya vs. 7 Meals per week

Wilcoxon Two-Sample Test	
Statistic	314.0000
Normal Approximation	
Z	−2.5968
One-Sided Pr < Z	0.0047
Two-Sided Pr > |Z|	0.0094
t Approximation	
One-Sided Pr < Z	0.0066
Two-Sided Pr > |Z|	0.0132

The second way to do *post hoc* tests is essentially the same as doing Wilcoxon two-sample tests on all possible comparisons, but for the sake of completeness I'll run you through it! It is described by Siegel and Castellan (1988) and involves taking the difference between the mean ranks of the different groups and comparing this to a value based on the value of z (corrected for the number of comparisons being done) and a constant based on the total sample size and the sample size in the two groups being compared. The inequality is:

$$\left|\bar{R}_u - \bar{R}_v\right| \geq z_{\alpha/k(k-1)}\sqrt{\frac{N(N+1)}{12}\left(\frac{1}{n_u}+\frac{1}{n_v}\right)} \quad (15.2)$$

The left-hand side of this inequality is just the difference between the mean rank of the two groups being compared, but ignoring the sign of the difference (so the two vertical lines that enclose the difference between mean ranks just indicate that if the difference is negative then we ignore the negative sign and treat it as positive). For the rest of the expression, k is the number of groups (in the soya example, 4), N is the total sample size (in this case 80), n_u is the number of people in the first group that's being compared (we have equal group sizes in the soya example so it will be 20 regardless of which groups we compare), and n_v is the number of people in the second group being compared (again this will be 20 regardless of which groups we compare because we have equal group sizes in the soya example). The only other thing we need to know is $z_{\alpha/k(k-1)}$, and to get this value we need to decide a level for α, which is the level of significance at which we want to work. You should know by now that in the social sciences we traditionally work at a .05 level of significance, so α will be .05. We then calculate $k(k - 1)$, which for these data will be $4(4 - 1) = 12$. Therefore, $\alpha/k(k - 1) = .05/12 = .00417$. So, $z_{\alpha/k(k-1)}$ just means 'the value of z for which only $\alpha/k(k - 1)$ other values of z are bigger' (or in this case 'the value of z for which only .00417 other values of z are bigger'). In practical terms this means we go to the table in the Appendix, look at the column labelled *Smaller Portion* and find the number .00417 (or the nearest value to this, which if you look at the table is .00415), and we then look in the same row at the column labelled z. In this case, you should find that the value of z is 2.64. The next thing to do is to calculate the right-hand side of inequality 15.2:

$$\begin{aligned}\text{critical difference} &= z_{\alpha/k(k-1)}\sqrt{\frac{N(N+1)}{12}\left(\frac{1}{n_u}+\frac{1}{n_v}\right)}\\ &= 2.64\sqrt{\frac{80(80+1)}{2}\left(\frac{1}{20}+\frac{1}{20}\right)}\\ &= 2.64\sqrt{540(0.1)}\\ &= 2.64\sqrt{54}\\ &= 19.40\end{aligned}$$

For this example, because the sample sizes across groups are equal, this critical difference can be used for all comparisons. However, when sample sizes differ across groups, the critical difference will have to be calculated for each comparison individually. The next step is simply to calculate all of the differences between the mean ranks of all of the groups (the mean ranks can be found in SAS Output 15.7), as in Table 15.4.

Inequality (15.2) basically means that if the difference between mean ranks is bigger than or equal to the critical difference for that comparison, then that difference is significant. In this case, because we have only one critical difference, it means that if any difference is bigger than 19.40, then it is significant. As you can see, all differences are below this value so we would have to conclude that none of the groups were significantly different! This contradicts our earlier findings where the Mann–Whitney test for the no-meals group compared to the seven-meals group was deemed significant; why do you think that is? Well, for our Wilcoxon two-sample tests, we did only three comparisons and so only corrected the significance value for the three tests we'd done (.05/3 = .0167). Earlier on in this section I said that if we compared all groups against all other groups, that would be six comparisons, so we could accept a difference as being significant only if the significance value was less than .05/6 = .0083. If we go back to our one significant Mann–Whitney test (SAS Output 15.8) the significance value was .009; therefore, if we had done all six comparisons this would've been non-significant (because .009 is bigger than .0083). This illustrates what I said earlier about the benefits of choosing selective comparisons.

TABLE 15.4 Differences between mean ranks for the soya data

Comparison	$\bar{R}_u$	$\bar{R}_v$	$\bar{R}_u - \bar{R}_v$	$\lvert\bar{R}_u - \bar{R}_v\rvert$
No Meals–1 Meal	46.35	44.15	2.20	2.20
No Meals–4 Meals	46.35	44.15	2.20	2.20
No-Meals–7 Meals	46.35	27.35	19.00	19.00
1 Meal–4 Meals	44.15	44.15	0.00	0.00
1 Meal–7 Meals	44.15	27.35	16.80	16.80
4 Meals–7 Meals	44.15	27.35	16.80	16.80

15.5.6. Writing and interpreting the results ①

For the Kruskal–Wallis test, we need only report the test statistic (which we saw earlier is denoted by *H*), its degrees of freedom and its significance. So, we could report something like:

✓ Sperm counts were significantly affected by eating soya meals, $H(3) = 8.66$, $p < .05$.

However, we need to report the follow-up tests as well (including their effect sizes):

✓ Sperm counts were significantly affected by eating soya meals, $H(3) = 8.66$, $p < .05$. Wilcoxon two-sample tests were used to follow up this finding. A Bonferroni correction was applied and so all effects are reported at a .0167 level of significance. It appeared that sperm counts were no different when one soya meal ($U = 191$) or four soya meals ($U = 188$) were eaten per week compared to none. However, when seven soya meals were eaten per week sperm counts were significantly lower than when no soya was eaten ($U = 104$). We can conclude that if soya is eaten every day it significantly reduces sperm counts compared to eating none; however, eating soya less than every day has no significant effect on sperm counts ('phew!' says the vegetarian man!).

CRAMMING SAM'S TIPS

- The Kruskal–Wallis test compares several conditions when different participants take part in each condition and the resulting data violate an assumption of one-way independent ANOVA.
- Look at the row labelled *Pr > ChiSq* If the value is less than .05 then the groups are significantly different.
- You can follow up the main analysis with Wilcoxon two-sample tests between pairs of conditions, but only accept them as significant if they're significant below .05/number of tests.
- Report the *H*-statistic, the degrees of freedom and the significance value for the main analysis. For any *post hoc* tests, report the *U*-statistic and an effect size if possible (you can also report the corresponding *z* and the significance value). Also report the medians and their corresponding ranges (or draw a boxplot).

LABCOAT LENI'S REAL RESEARCH 15.2

Eggs-traordinary! ①

There seems to be a lot of sperm in this book (not literally, I hope) – it's possible that I have a mild obsession. We saw in Labcoat Leni's Real Research 15.1 that male quail fertilized more eggs if they had been trained to be able to predict when a mating opportunity would arise. However, some quail develop fetishes. Really. In the previous example the type of compartment acted as a predictor of an opportunity to mate, but in studies where a terrycloth object acts as a sign that a mate will shortly become available, some quail start to direct their sexuial behaviour towards the terrycloth object. (I may regret this anology but in human terms if you imagine that every time you were going to have sex with your boyfriend you gave him a green towel a few moments before seducing him, then after enough seductions he would start rubbing his crotch against any green towel he saw. If you've ever wondered why you boyfriend rubs his crotch on green towels, then I hope this explanation has been enlightening.)

ÇETINKAYA, H., & DOMJAN, M. (2006). *JOURNAL OF COMPARATIVE PSYCHOLOGY, 120*(4), 427–432.

In evolutionary terms, this fetishistic behaviour seems counterproductive because sexual behaviour becomes directed towards something that cannot provide reproductive success. However, perhaps this behaviour serves to prepare the organism for the 'real' mating behaviour.

Hakan Çetinkaya and Mike Domjan conducted a brilliant study in which they sexually conditioned male quail (Çetinkaya & Domjan, 2006). All quail experienced the terrycloth stimulus and an opportunity to mate, but for some the terrycloth stimulus immediately preceded the mating opportunity (paired group) whereas for others they experienced it 2 hours after the mating opportunity (this was the control group because the terrycloth stimulus did not predict a mating opportuinity). In the paired group, quail were classified as fetishistic or not depending on whether they engaged in sexual behaviour with the terrycloth object.

During a test trial the quail mated with a female and the researchers measured the percentage of eggs fertilized, the time spent near the terrycloth object, the latency to initiate copulation, and copulatory efficiency. If this fetishistic behaviour provides an evolutionary advantage then we would expect the fetishistic quail to fertilize more eggs, initiate copulation faster and be more efficient in their copulations.

The data from this study are in the file **CetinkayaandDomjan2006.sas7bdat**. Labcoat Leni wants you to carry out a Kruskal–Wallis test to see whether fetishist quail produced a higher percentage of fertilized eggs and initiated sex more quickly.

Answers are in the additional material on the companion website (or look at pages 429–430 in the original article).

15.6. Differences between several related groups: Friedman's ANOVA ①

In Chapter 13 we discovered a technique called one-way related ANOVA that could be used to test for differences between several related groups. Although, as we've seen, ANOVA can be robust to violations of its assumptions, there is another alternative to the repeated-measures case: **Friedman's ANOVA** (Friedman, 1937). As such, it is used for testing differences between conditions when there are more than two conditions and the same participants have been used in all conditions (each case contributes several scores to the data). If you have violated some assumption of parametric tests then this test can be a useful way around the problem.

Young people (women especially) can become obsessed with body weight and diets, and because the media are insistent on ramming ridiculous images of stick-thin celebrities down our throats (should that be 'into our eyes'?) and brainwashing us into believing that these emaciated corpses are actually attractive, we all end up terribly depressed that we're not perfect (because we don't have a couple of slugs stuck to our faces instead of lips). Then corporate parasites jump on our vulnerability by making loads of money on diets that will help us attain the body beautiful! Well, not wishing to miss out on this great opportunity to exploit people's insecurities, I came up with my own diet called the Andikins diet.[8] The principle is that you follow my lifestyle: you eat no meat, drink lots of Darjeeling tea, eat shed-loads of lovely European cheese, lots of fresh crusty bread, pasta, chocolate at every available opportunity (especially when writing books), then enjoy a few beers at the weekend, play football and rugby twice a week and play your drum kit for an hour a day or until your neighbour threatens to saw your arms off and beat you around the head with them for making so much noise. To test the efficacy of my wonderful new diet, I took 10 women who considered themselves to be in need of losing weight and put them on this diet for two months. Their weight was measured in kilograms at the start of the diet and then after one month and two months.

[8] Not to be confused with the Atkins diet obviously.☺

15.6.1. Theory of Friedman's ANOVA ②

The theory for Friedman's ANOVA is much the same as the other tests we've seen in this chapter: it is based on ranked data. To begin with, you simply place your data for different conditions into different columns (in this case there were three conditions so we have three columns). The data for the diet example are in Table 15.5; note that the data are in different columns and so each row represents the weight of a different person. The next thing we have to do is rank the data *for each person*. So, we start with person 1, we look at their scores (in this case person 1 weighed 63.75 kg at the start, 65.38 kg after one month on the diet, and 81.34 kg after two months on the diet), and then we give the lowest one a rank of 1, the next highest a rank of 2 and so on (see section 15.3.1 for more detail). When you've ranked the data for the first person, you move on to the next person, and starting at 1 again, rank their lowest score, then rank the next highest as 2 and so on. You do this for all people from whom you've collected data. You then simply add up the ranks for each condition (R_i, where i is used to denote the particular group).

SELF-TEST Have a go at ranking the data and see if you get the same results as in Table 15.5.

TABLE 15.5 Data for the diet example with ranks

	Weight				Weight	
	Start	**Month 1**	**Month 2**	**Start (Ranks)**	**Month 1 (Ranks)**	**Month 2 (Ranks)**
Person 1	63.75	65.38	81.34	1	2	3
Person 2	62.98	66.24	69.31	1	2	3
Person 3	65.98	67.70	77.89	1	2	3
Person 4	107.27	102.72	91.33	3	2	1
Person 5	66.58	69.45	72.87	1	2	3
Person 6	120.46	119.96	114.26	3	2	1
Person 7	62.01	66.09	68.01	1	2	3
Person 8	71.87	73.62	55.43	2	3	1
Person 9	83.01	75.81	71.63	3	2	1
Person 10	76.62	67.66	68.60	3	1	2
			R_i	19	20	21

Once the sum of ranks has been calculated for each group, the test statistic, F_r, is calculated as:

$$F_r = \left[\frac{12}{Nk(k+1)}\sum_{i=1}^{k} R_i^2\right] - 3N(k+1) \qquad (15.3)$$

In this equation, R_i is the sum of ranks for each group, N is the total sample size (in this case 10) and k is the number of conditions (in this case 3). This equation is very similar to that of the Kruskal–Wallis test (compare equations (15.1) and (15.3)). All we need to do for each condition is square the sum of ranks and then add up these values. That deals with the middle part of the equation; the rest of it involves calculating various values based on the total sample size and the number of conditions. For these data we get:

$$\begin{aligned} F_r &= \left[\frac{12}{(10\times 3)(3+1)}(19^2+20^2+21^2)\right] - (3\times 10)(3+1) \\ &= \frac{12}{120}(361+400+441) - 120 \\ &= 0.1(1202) - 120 \\ &= 120.2 - 120 \\ &= 0.2 \end{aligned}$$

When the number of people tested is large (bigger than about 10) this test statistic, like the Kruskal–Wallis test in the previous section, has a chi-square distribution (see Chapter 18) and for this distribution there is one value for the degrees of freedom, which is one fewer than the number of groups (k–1): in this case 2.

15.6.2. Inputting data and provisional analysis ①

SELF-TEST Using what you know about inputting data, try to enter these data into SAS and run some exploratory analyses (see Chapter 5).

When the data are collected using the same participants in each condition, the data are entered using different columns. So, the data editor will have three columns of data. The first column is for the data from the start of the diet (called something like **start**), the second column will have values for the weights after one month (called **month1**) and the final column will have the weights at the end of the diet (called **month2**). The data can be found in the file **Diet.sas7bdat.**

First, we run some exploratory analyses on the data. With a bit of luck you'll get the same table shown in SAS Output 15.9, which shows that the Kolmogorov–Smirnov test (see section 5.5.1.). was not significant for the initial weights at the start of the diet ($D(10) = .23$, $p > .05$), but the Shapiro–Wilk test is significant and this test is actually more accurate than the Kolmogorov–Smirnov test. The data one month into the diet were significantly different from normal ($D(10) = .34$, $p < .01$). The data at the end of the diet do appear to be normal, though ($D(10) = .20$, $p > .05$). Some of these data are not normally distributed.

SAS OUTPUT 15.9

Variable: START (Weight at Start (Kg))

Tests for Normality				
Test	**Statistic**		**p Value**	
Shapiro-Wilk	**W**	0.784385	**Pr < W**	0.0094
Kolmogorov-Smirnov	**D**	0.228184	**Pr > D**	0.1436
Cramer-von Mises	**W-Sq**	0.164983	**Pr > W-Sq**	0.0124
Anderson-Darling	**A-Sq**	0.93358	**Pr > A-Sq**	0.0110

Variable: MONTH1 (Weight after 1 month (Kg))

Tests for Normality				
Test	**Statistic**		**p Value**	
Shapiro-Wilk	**W**	0.684879	**Pr < W**	0.0006
Kolmogorov-Smirnov	**D**	0.335373	**Pr > D**	<0.0100
Cramer-von Mises	**W-Sq**	0.279429	**Pr > W-Sq**	<0.0050
Anderson-Darling	**A-Sq**	1.447031	**Pr > A-Sq**	<0.0050

Variable: MONTH2 (Weight after 2 month (Kg))

Tests for Normality				
Test	**Statistic**		**p Value**	
Shapiro-Wilk	**W**	0.877213	**Pr < W**	0.1212
Kolmogorov-Smirnov	**D**	0.202795	**Pr > D**	>0.1500
Cramer-von Mises	**W-Sq**	0.103356	**Pr > W-Sq**	0.0897
Anderson-Darling	**A-Sq**	0.595546	**Pr > A-Sq**	0.0907

15.6.3. Doing Friedman's ANOVA on SAS

We have to do a little bit of fiddling to get SAS to do a Friedman's ANOVA for us. SAS doesn't actually do the test, but with a bit of tweaking, it can do another test which is equivalent (a little like the way that a *t*-test can be thought of as an ANOVA with only two groups).

First, we need to make the data long, using PROC TRANSPOSE. Remember that PROC TRANSPOSE calls the outcome variable **col1**, and the names of the variables are put into a variable called **_name_** (notice the underscore at the beginning and end of that variable).

When we have transposed the data, we used PROC FREQ to carry out a Cochran–Mantel–Haensel test, with the CMH2 option in the TABLES statement. This will produce a large table that we're not really interested in, and so it's worth turning that off with the NOPRINT option. This is shown in SAS Syntax 15.4.

```
PROC TRANSPOSE DATA=chap16.diet OUT=dietlong;
  VAR start month1 month2;
  BY id;
  RUN;
```

```
PROC FREQ DATA=dietlong NOPRINT;
  TABLES id*_name_*col1 / CMH2 SCORES=RANK EXACT;
  RUN;
```
SAS Syntax 15.4

However we are also going to need to interpret the result. When we used PROC NPAR1WAY to carry out non-parametric tests, we were also shown the mean rank for each group. PROC FREQ is not going to show this, instead we need to rank the data, and then ask SAS to show the mean of those ranks. We rank data with PROC RANK, and then use PROC MEANS to display the mean ranks. SAS Syntax 15.5 shows the use of these two procedures.

```
PROC RANK DATA=dietlong OUT=dietlongrank;
  VAR col1;
  RUN;

PROC MEANS data=dietlongrank MEAN;
  CLASS _name_;
  VAR col1;
  RUN;
```
SAS Syntax 15.5

15.6.4. Output for Friedman's ANOVA

The output is shown in SAS Output 15.10. The first table, produced by PROC FREQ, shows the test statistics for the Cochran–Mantel–Haenszel test. This is equivalent to the chi-square statistic from the Friedman test. The column labeled *value* contains the value for chi-square, and the *p*-value is in the column labeled *Prob*. (Also, the column labeled *DF* has the degrees of freedom, but you probably didn't need me to tell you that.) We are interested in the second row, labelled *Row Mean Scores Differ.* The result of our test is non-significant, as the *p*-value is greater than 0.05. However, if the result were significant, we would want to know which variable had the higher scores. To determine this, we would look at the second table in SAS Output 15.10, which contains the mean ranks for each of the three variables. We can see that the scores seem to be increasing – **start** has the lowest ranked scores, followed by **month1**, and followed by **month2**; however, we should not interpret these differences, as they are not significantly different.

SAS OUTPUT 15.10

Cochran-Mantel-Haenszel Statistics (Based on Rank Scores)				
Statistic	**Alternative Hypothesis**	**DF**	**Value**	**Prob**
1	Nonzero Correlation	1	0.0500	0.8231
2	Row Mean Scores Differ	2	0.2000	0.9048

Analysis Variable : COL1 Values of COL1 Were Replaced by Ranks		
NAME OF FORMER VARIABLE	**N Obs**	**Mean**
START	10	14.3000000
MONTH1	10	15.0000000
MONTH2	10	17.2000000

15.6.5. *Post hoc* tests for Friedman's ANOVA ②

In normal circumstances we wouldn't do any follow-up tests because the overall effect from Friedman's ANOVA was not significant. However, in case you get a result that is significant we will have a look at what options you have. As with the Kruskal–Wallis test, there are two ways to do non-parametric *post hoc* procedures, which are in essence the same. The first is to use Wilcoxon signed-rank tests (section 15.4) but correcting for the number of tests we do (see sections 2.6.3. and 15.6.5 for the reasons why). The way we correct for the number of tests is to accept something as significant only if its significance is less than α/number of comparisons (the Bonferroni correction). In the social sciences this usually means .05/ number of comparisons. In this example, we have only three groups, so if we compare all of the groups we simply get three comparisons:

- Test 1: Weight at the start of the diet compared to at one month.
- Test 2: Weight at the start of the diet compared to at two months.
- Test 3: Weight at one month compared to at two months.

Therefore, rather than use .05 as our critical level of significance, we'd use .05/3 = .0167. In fact we wouldn't bother with *post hoc* tests at all for this example because the main ANOVA was non-significant, but I'll go through the motions to illustrate what to do.

SELF-TEST Carry out the three Wilcoxon one-sample tests suggested above.

SAS OUTPUT 15.11

Start – Month1

Signed Rank	S	0.5	Pr >= |S|	1.0000

Start – Month 2

Signed Rank	S	2.5	Pr >= |S|	0.8457

Month 1 – Month 2

Signed Rank	S	1.5	Pr >= |S|	0.9219

SAS Output 15.11 shows the Wilcoxon signed-rank test statistics from doing the three comparisons (I only put the important line from the table). Remember that we are now using a critical value of .0167, and in fact none of the comparisons are significant because they have significance values of 1.00, .846 and .922 (this isn't surprising because the main analysis was non-significant).

The second way to do *post hoc* tests is very similar to what we did for the Kruskal–Wallis test in section 15.6.5 and is, likewise, described by Siegel and Castellan (1988). Again, we take the difference between the mean ranks of the different groups and compare these differences to a value based on the value of z (corrected for the number of comparisons being done) and a constant based on the total sample size, N (10 in this example) and the number of conditions, k (3 in this case). The inequality is:

$$|\bar{R}_u - \bar{R}_v| \geq z_{\alpha/k(k-1)}\sqrt{\frac{k(k+1)}{6N}} \quad (15.4)$$

The left-hand side of this inequality is just the difference between the mean rank of the two groups being compared, but ignoring the sign of the difference. As with Kruskal–Wallis, we need to know $z_{\alpha/k(k-1)}$, and if we stick to tradition and use an α level of .05, knowing that k is 3, we get $\alpha/k(k-1) = .05/3(3-1) = .05/6 = .00833$. So, $z_{\alpha/k(k-1)}$ just means 'the value of z for which only $\alpha/k(k-1)$ other values of z are bigger' (or in this case 'the value of z for which only .00833 other values of z are bigger'). Therefore, we go to the table in the Appendix, look at the column labelled *Smaller Portion* and find the number .00833 and then find the value in the same row in the column labelled z. In this case there are values of .00842 and .00820, which give z-values of 2.39 and 2.40 respectively; because .00833 lies about midway between the values we found, we could just take the midpoint of the two z-values, 2.395, or we could err on the side of caution and use 2.40. I'll err on the cautious side and use 2.40. We can now calculate the right-hand side of inequality (15.4):

$$\begin{aligned}\text{critical difference} &= z_{\alpha/k(k-1)}\sqrt{\frac{k(k+1)}{6N}}\\ &= 2.40\sqrt{\frac{3(3+1)}{6(10)}}\\ &= 2.40\sqrt{\frac{12}{60}}\\ &= 2.40\sqrt{0.2}\\ &= 1.07\end{aligned}$$

When the same people have been used, the same critical difference can be used for all comparisons. The next step is simply to calculate all of the differences between the mean ranks of all of the groups, as in Table 15.6.

Inequality (15.4) means that if the differences between mean ranks is bigger than or equal to the critical difference, then that difference is significant. In this case, it means that if any difference is bigger than 1.07, then it is significant. All differences are below this value so we could conclude that none of the groups were significantly different and this is consistent with the non-significance of the initial ANOVA.

TABLE 15.6 Differences between mean ranks for the diet data

Comparison	$\bar{R}_u$	$\bar{R}_v$	$\bar{R}_u - \bar{R}_v$	$\lvert\bar{R}_u - \bar{R}_v\rvert$
Start–1 Month	1.90	2.00	−0.10	0.10
Start–2 Months	1.90	2.10	−0.20	0.20
1 Month–2 Months	2.00	2.10	−0.10	0.10

15.6.6. Writing and interpreting the results ①

For Friedman's ANOVA we need only report the test statistic (which we saw earlier is denoted by χ^2),[9] its degrees of freedom and its significance. So, we could report something like:

✓ The weight of participants did not significantly change over the two months of the diet, $\chi^2(2) = 0.20$, $p > .05$.

Although with no significant initial analysis we wouldn't report *post hoc* tests for these data, in case you need to, you should say something like this (remember that the test statistic T is the smaller of the two sums of ranks for each test and these values are in SAS Output 15.10):

✓ The weight of participants did not significantly change over the two months of the diet, $\chi^2(2) = 0.20$, $p > .05$. Wilcoxon tests were used to follow up this finding. A Bonferroni correction was applied and so all effects are reported at a .0167 level of significance. It appeared that weight didn't significantly change from the start of the diet to one month, $T = 27$, $r = -.01$, from the start of the diet to two months, $T = 25$, $r = -.06$, or from one month to two months, $T = 26$, $r = -.03$. We can conclude that the Andikins diet, like its creator, is a complete failure.

CRAMMING SAM'S TIPS

- Friedman's ANOVA compares several conditions when the same participants take part in each condition and the resulting data violate an assumption of one-way repeated-measures ANOVA.
- Look at the row labelled *Row Mean Scores Differ,* in the column labelled *Prob.* If the value is less than .05 then the conditions are significantly different.
- You can follow up the main analysis with Wilcoxon signed-rank tests between pairs of conditions, but only accept them as significant if they're significant below .05/number of tests.
- Report the χ^2 statistic, its degrees of freedom and significance. For any *post hoc* tests report the *S*-statistic and its significance value.
- Report the medians and their ranges (or draw a boxplot).

What have I discovered about statistics? ①

This chapter has dealt with an alternative approach to violations of parametric assumptions, which is to use tests based on ranking the data. We started with the Wilcoxon rank-sum test and the Mann–Whitney test, which is used for comparing two independent groups. This test allowed us to look in some detail at the process of ranking data. We then moved on to look at the Wilcoxon signed-rank test, which is used to compare two related conditions. We moved on to more complex situations in which there are several conditions (the Kruskal–Wallis test for independent conditions and Friedman's ANOVA for related conditions). For each of these tests we looked at the theory of the test (although these sections could be ignored) and then focused on how to conduct them on SAS, how to interpret the results and how to report the results of the test. In the process we discovered that drugs make you depressed, soya reduces your sperm count, and my lifestyle is not conducive to losing weight.

[9] The test statistic is often denoted as χ^2_F but the official APA style guide doesn't recognize this term.

We also discovered that my teaching career got off to an inauspicious start. As it turned out, one of the reasons why the class did not have a clue what I was talking about was because I hadn't been shown their course handouts and I was trying to teach them ANOVA using completely different equations to their lecturer (there are many ways to compute an ANOVA). The other reason was that I was a rubbish teacher. This event did change my life, though, because the experience was so awful that I did everything in my power to make sure that it didn't happen again. After years of experimentation I can now pass on the secret of avoiding students telling you how awful your ANOVA classes are: the more penis jokes you tell, the less likely you are to be emotionally crushed by dissatisfied students.

Key terms that I've discovered

Friedman's ANOVA
Kolmogorov–Smirnov *Z* test
Kruskal–Wallis test
Mann–Whitney test
Non-parametric tests
Ranking
Sign test
Wilcoxon rank-sum test
Wilcoxon signed-rank test
Wilcoxon two-sample test

Smart Alex's tasks

- **Task 1**: A psychologist was interested in the cross-species differences between men and dogs. She observed a group of dogs and a group of men in a naturalistic setting (20 of each). She classified several behaviours as being dog-like (urinating against trees and lamp posts, attempts to copulate with anything that moved, and attempts to lick their own genitals). For each man and dog she counted the number of dog-like behaviours displayed in a 24 hour period. It was hypothesized that dogs would display more dog-like behaviours than men. The data are in the file **MenLikeDogs.sas7bdat.** Analyse them with a Mann–Whitney test. ①
- **Task 2**: There's been much speculation over the years about the influence of subliminal messages on records. To name a few cases, both Ozzy Osbourne and Judas Priest have been accused of putting backward masked messages on their albums that subliminally influence poor unsuspecting teenagers into doing things like blowing their heads off with shotguns. A psychologist was interested in whether backward masked messages really did have an effect. He took the master tapes of Britney Spears's 'Baby one more time' and created a second version that had the masked message 'deliver your soul to the dark lord' repeated in the chorus. He took this version, and the original, and played one version (randomly) to a group of 32 people. He took the same group six months later and played them whatever version they hadn't heard the time before. So each person heard both the original, and the version with the masked message, but at different points in time. The psychologist measured the number of goats that were sacrificed in the week after listening to each version. It was hypothesized that the backward message would lead to more goats being sacrificed. The data are in the file **DarkLord.sas7dbat.** Analyse them with a Wilcoxon signed-rank test. ①
- **Task 3**: A psychologist was interested in the effects of television programmes on domestic life. She hypothesized that through 'learning by watching', certain programmes might

actually encourage people to behave like the characters within them. This in turn could affect the viewer's own relationships (depending on whether the programme depicted harmonious or dysfunctional relationships). She took episodes of three popular TV shows and showed them to 54 couples, after which the couple were left alone in the room for an hour. The experimenter measured the number of times the couple argued. Each couple viewed all three of the TV programmes at different points in time (a week apart) and the order in which the programmes were viewed was counterbalanced over couples. The TV programmes selected were *EastEnders* (which typically portrays the lives of extremely miserable, argumentative, London folk who like nothing more than to beat each other up, lie to each other, sleep with each other's wives and generally show no evidence of any consideration to their fellow humans!), *Friends* (which portrays a group of unrealistically considerate and nice people who love each other oh so very much – but for some reason I love it anyway!), and a National Geographic programme about whales (this was supposed to act as a control). The data are in the file **Eastenders.sas7bdat.** Access them and conduct Friedman's ANOVA on the data. ①

- **Task 4**: A researcher was interested in trying to prevent coulrophobia (fear of clowns) in children. She decided to do an experiment in which different groups of children (15 in each) were exposed to different forms of positive information about clowns. The first group watched some adverts for McDonald's in which their mascot Ronald McDonald is seen cavorting about with children going on about how they should love their mums. A second group was told a story about a clown who helped some children when they got lost in a forest (although what on earth a clown was doing in a forest remains a mystery). A third group was entertained by a real clown, who came into the classroom and made balloon animals for the children.[10] A final group acted as a control condition and they had nothing done to them at all. The researcher took self-report ratings of how much the children liked clowns resulting in a score for each child that could range from 0 (not scared of clowns at all) to 5 (very scared of clowns). The data are in the file **coulrophobia.sas7dbat.** Access them and conduct a Kruskal–Wallis test. ①

Answers can be found on the companion website and because these examples are used in Field and Hole (2003), you could steal this book or photocopy Chapter 7 to get some very detailed answers.

Further reading

Siegel, S., & Castellan, N. J. (1988). *Nonparametric statistics for the behavioral sciences* (2nd ed.). New York: McGraw-Hill. (This has become the definitive text on non-parametric statistics, and is the only book seriously worth recommending as 'further' reading. It is probably not a good book for stats-phobes, but if you've coped with my chapter then this book will be an excellent next step.)

Wilcox, R. R. (2005). *Introduction to robust estimation and hypothesis testing* (2nd ed.). Burlington, MA: Elsevier. (This book is quite technical, compared to this one, but really is a wonderful resource. Wilcox describes how to use an astonishing range of robust tests that can't be done directly in SAS!)

Interesting real research

Çetinkaya, H., & Domjan, M. (2006). Sexual fetishism in a quail (*Coturnix japonica*) model system: Test of reproductive success. *Journal of Comparative Psychology*, *120*(4), 427–432.

Matthews, R. C., Domjan, M., Ramsey, M., & Crews, D. (2007). Learning effects on sperm competition and reproductive fitness. *Psychological Science*, *18*(9), 758–762.

[10] Unfortunately, the first time they attempted the study the clown accidentally burst one of the balloons. The noise frightened the children and they associated that fear response with the clown. All 15 children are currently in therapy for coulrophobia!

Multivariate analysis of variance (MANOVA)

16

FIGURE 16.1
Fuzzy doing some light reading

16.1. What will this chapter tell me? ②

Having had what little confidence I had squeezed out of me by my formative teaching experiences, I decided that I could either kill myself, or get a cat. I'd wanted to do both for years but when I was introduced to a little four-week old bundle of gingerness the choice was made. Fuzzy (as I named him) was born on 8 April 1996 and has been my right-hand feline ever since. He is like the Cheshire cat in Lewis Carroll's *Alice's adventures in Wonderland*[1] in that he seemingly vanishes and reappears at will: I go to find clothes in my wardrobe and notice

[1] This is one of my favourite books from my childhood. For those that haven't read it, the Cheshire cat is a big fat cat mainly remembered for vanishing and reappearing out of nowhere; on one occasion it vanished leaving only its smile behind.

a ginger face peering out at me, I put my pants in the laundry basket and he looks up at me from a pile of smelly socks, I go to have a bath and he's sitting in it, and I shut the bedroom door yet wake up to find him asleep next to me. His best vanishing act was a few years ago when I moved house. He'd been locked up in his travel basket (which he hates) during the move, so once we were in our new house I thought I'd let him out as soon as possible. I found a quiet room, checked the doors and windows to make sure he couldn't escape, opened the basket, gave him a cuddle and left him to get to know his new house. When I returned five minutes later, he was gone. The door had been shut, the windows closed and the walls were solid (I checked). He had literally vanished into thin air and he didn't even leave behind his smile. Before his dramatic disappearance, Fuzzy had stopped my suicidal tendencies, and there is lots of research showing that having a pet is good for your mental health. If you wanted to test this you could compare people with pets against those without to see if they had better mental health. However, the term *mental health* covers a wide range of concepts including (to name a few) anxiety, depression, general distress and psychosis. As such, we have four outcome measures and all the tests we have encountered allow us to look at one. Fear not, when we want to compare groups on several outcome variables we can extend ANOVA to become a MANOVA. That's what this chapter is all about.

16.2. When to use MANOVA ②

Over Chapters 9–14, we have seen how the general linear model (GLM) can be used to detect group differences on a single dependent variable. However, there may be circumstances in which we are interested in several dependent variables and in these cases the simple ANOVA model is inadequate. Instead, we can use an extension of this technique known as multivariate analysis of variance (or MANOVA). *MA*NOVA can be thought of as ANOVA for situations in which there are several dependent variables. The principles of ANOVA extend to MANOVA in that we can use MANOVA when there is only one independent variable or when there are several, we can look at interactions between independent variables, and we can even do contrasts to see which groups differ from each other. ANOVA can be used only in situations in which there is one dependent variable (or outcome) and so is known as a *univariate* test (univariate quite obviously means 'one variable'); MANOVA is designed to look at several dependent variables (outcomes) simultaneously and so is a *multivariate* test (multivariate means 'many variables'). This chapter will explain some basics about MANOVA for those of you who want to skip the fairly tedious theory sections and just get on with the test. However, for those who want to know more there is a fairly lengthy theory section to try to explain the workings of MANOVA. We then look at an example using SAS and see how the output from MANOVA can be interpreted. This leads us to look at another statistical test known as *discriminant function analysis*.

16.3. Introduction: similarities and differences to ANOVA ②

If we have collected data about several dependent variables then we could simply conduct a separate ANOVA for each dependent variable (and if you read research articles you'll find that it is not unusual for researchers to do this!). Think back to Chapter 10 and you should remember that a similar question was posed regarding why ANOVA was used in preference to multiple *t*-tests. The answer to why MANOVA is used instead of multiple ANOVAs is the same: the more tests we conduct on the same data, the more we inflate the familywise error rate (see section 10.2.1). The more dependent variables that have been measured, the more ANOVAs would need to be conducted and the greater the chance of making a Type I error. However, there are other reasons for preferring MANOVA to several ANOVAs. For one thing, there is important additional information that is gained from a MANOVA. If separate ANOVAs are conducted on each dependent variable, then any relationship between dependent variables is ignored. As such, we lose information about any correlations that might exist between the dependent variables. MANOVA, by including all dependent variables in the same analysis, takes account of the relationship between outcome variables. Related to this point, ANOVA can tell us only whether groups differ along a single dimension whereas MANOVA has the power to detect whether groups differ along a combination of dimensions. For example, ANOVA tells us how scores on a single dependent variable distinguish groups of participants (so, for example, we might be able to distinguish drivers, non-drivers and drunk drivers by the number of pedestrians they kill). MANOVA incorporates information about several outcome measures and, therefore, informs us of whether groups of participants can be distinguished by a combination of scores on several dependent measures. For example, it may not be possible to distinguish drivers, non-drivers and drunk drivers only by the number of pedestrians they kill, but they might be distinguished by *a combination* of the number of pedestrians they kill, the number of lamp posts they hit, and the number of cars they crash into. So, in this sense MANOVA has greater power to detect an effect, because it can detect whether groups differ along a combination of variables, whereas ANOVA can detect only if groups differ along a single variable (see Jane Superbrain Box 16.1). For these reasons, MANOVA is preferable to conducting several ANOVAs.

JANE SUPERBRAIN 16.1

The power of MANOVA ③

I mentioned in the previous section that MANOVA had greater power than ANOVA to detect effects because it could take account of the correlations between dependent variables (Huberty & Morris, 1989). However, the issue of power is more complex than alluded to by my simple statement. Ramsey (1982) found that as the correlation between dependent variables increased, the power of MANOVA decreased. This led Tabachnick and Fidell (2007) to recommend that MANOVA 'works best with highly negatively correlated [dependent variables DVs], and acceptably well with moderately correlated DVs in either direction' and that 'MANOVA also is wasteful when DVs are uncorrelated' (p. 268). In contrast, Stevens's (1980) investigation of the effect of dependent variable correlations on test power revealed that 'the power with high intercorrelations is in most cases greater than that for moderate intercorrelations, and in some cases it is dramatically higher' (p. 736). These findings are slightly contradictory, which leaves us with the puzzling conundrum of what, exactly, is the relationship between power and intercorrelation of the dependent variables? Luckily, Cole, Maxwell, Arvey, and Salas (1994) have done a great deal to illuminate this relationship. They found that the power of MANOVA depends on a combination of the correlation between dependent variables and the effect size to be detected. In short, if you are expecting to find a large effect, then MANOVA will have greater power if the measures are somewhat different (even negatively correlated) and if the group differences are in the same direction for each measure. If you have two dependent variables, one of which exhibits a large group difference, and one of which exhibits a small, or no, group difference, then power will be increased if these variables are highly correlated. The take-home message from Cole et al.'s work is that if you are interested in how powerful the MANOVA is likely to be you should consider not just the intercorrelation of dependent variables but also the size and pattern of group differences that you expect to get. However, it should be noted that Cole et al.'s work is limited to the case where two groups are being compared and power considerations are more complex in multiple-group situations.

16.3.1. Words of warning ②

From my description of MANOVA it is probably looking like a pretty groovy little test that allows you to measure hundreds of dependent variables and then just sling them into the analysis. This is not the case. It is not a good idea to lump all of your dependent variables together in a MANOVA unless you have a good theoretical or empirical basis for doing so. I mentioned way back at the beginning of this book that statistical procedures are just a way of number crunching and so even if you put rubbish into an analysis you will still reach conclusions that are statistically meaningful, but are unlikely to be empirically meaningful. In circumstances where there is a good theoretical basis for including some but not all of your dependent variables, you should run separate analyses: one for the variables being tested on a heuristic basis and one for the theoretically meaningful variables. The point to take on board here is not to include lots of dependent variables in a MANOVA just because you have measured them.

16.3.2. The example for this chapter ②

Throughout the rest of this chapter we're going to use a single example to look at how MANOVA works and then how to conduct one on SAS. Imagine that we were interested in the effects of cognitive behaviour therapy on obsessive compulsive disorder (OCD). OCD is a disorder characterized by intrusive images or thoughts that the sufferer finds abhorrent (in my case this might be the thought of someone carrying out a *t*-test on data that are not normally distributed, or imagining my parents have died). These thoughts lead the sufferer to engage in activities to neutralize the unpleasantness of these thoughts (these activities can be mental, such as doing a MANOVA in my head to make me feel better about the *t*-test thought, or physical, such as touching the floor 23 times so that my parents won't die). Now, we could compare a group of OCD sufferers after cognitive behaviour therapy (CBT) and after behaviour therapy (BT) with a group of OCD sufferers who are still awaiting treatment (a no-treatment condition, NT).[2] Now, most psychopathologies have both behavioural and cognitive elements to them. For example, in OCD if someone had an obsession with germs and contamination, this disorder might manifest itself in obsessive hand-washing and would influence not just how many times they actually wash their hands (behaviour), but also the number of times they think about washing their hands (cognitions). Similarly, someone with an obsession about bags won't just think about bags a lot, but they might carry out bag-related behaviours (such as saying 'bag' repeatedly, or buying lots of bags). If we are interested in seeing how successful a therapy is, it is not enough to look only at behavioural outcomes (such as whether obsessive behaviours are reduced); it is important to establish whether cognitions are being changed also. Hence, in this example two dependent measures were taken: the occurrence of obsession-related behaviours (**Actions**) and the occurrence of obsession-related cognitions (**Thoughts**). These dependent variables were measured on a single day and so represent the number of obsession-related behaviours/thoughts in a normal day.

The data are in Table 16.1 and can be found in the file **OCD.sas7bdat**. Participants belonged to group 1 (CBT), group 2 (BT) or group 3 (NT) and within these groups all participants had both actions and thoughts measured.

[2] The non-psychologists out there should note that behaviour therapy works on the basis that if you stop the maladaptive behaviours the disorder will go away, whereas cognitive therapy is based on the idea that treating the maladaptive cognitions will stop the disorder.

TABLE 16.1 Data from **OCD.sas7bdat**

	DV 1: Actions			*DV 2: Thoughts*		
Group:	*CBT (1)*	*BT (2)*	*NT (3)*	*CBT (1)*	*BT (2)*	*NT (3)*
	5	4	4	14	14	13
	5	4	5	11	15	15
	4	1	5	16	13	14
	4	1	4	13	14	14
	5	4	6	12	15	13
	3	6	4	14	19	20
	7	5	7	12	13	13
	6	5	4	15	18	16
	6	2	6	16	14	14
	4	5	5	11	17	18
$\bar{X}$	**4.90**	**3.70**	**5.00**	**13.40**	**15.20**	**15.00**
s	**1.20**	**1.77**	**1.05**	**1.90**	**2.10**	**2.36**
s^2	**1.43**	**3.12**	**1.11**	**3.60**	**4.40**	**5.56**
	$\bar{X}_{\text{grand (Actions)}} = 4.53$			$\bar{X}_{\text{grand (Thoughts)}} = 14.53$		
	$s^2_{\text{grand (Actions)}} = 2.1195$			$s^2_{\text{grand (Thoughts)}} = 4.8780$		

16.4. Theory of MANOVA ③

The theory of MANOVA is very complex to understand without knowing matrix algebra, and frankly matrix algebra is way beyond the scope of this book (those with maths brains can consult Namboodiri, 1984; Stevens, 2002). However, I intend to give a flavour of the conceptual basis of MANOVA, using matrices, without requiring you to understand exactly how those matrices are used. Those interested in the exact underlying theory of MANOVA should read Bray and Maxwell's (1985) superb monograph.

16.4.1. Introduction to matrices ③

A matrix is simply a collection of numbers arranged in columns and rows. In fact, throughout this book you have been using a matrix without even realizing it: the SAS data table. In the data editor we have numbers arranged in columns and rows and this is exactly what a matrix is. A matrix can have many columns and many rows and we usually specify the dimensions of the matrix using numbers. So, a 2×3 matrix is a matrix with two rows and three columns, and a 5×4 matrix is one with five rows and four columns (examples below):

$$\begin{pmatrix} 2 & 5 & 6 \\ 3 & 5 & 8 \end{pmatrix} \qquad \begin{pmatrix} 2 & 4 & 6 & 8 \\ 3 & 4 & 6 & 7 \\ 4 & 3 & 5 & 8 \\ 2 & 5 & 7 & 9 \\ 4 & 6 & 6 & 9 \end{pmatrix}$$

2 × 3 matrix   5 × 4 matrix

You can think of these matrices in terms of each row representing the data from a single participant and each column as representing data relating to a particular variable. So, for the 5 × 4 matrix we can imagine a situation where five participants were tested on four variables: so, the first participant scored 2 on the first variable and 8 on the fourth variable. The values within a matrix are typically referred to as *components* or *elements*.

A **square matrix** is one in which there are an equal number of columns and rows. In this type of matrix it is sometimes useful to distinguish between the diagonal components (i.e. the values that lie on the diagonal line from the top left component to the bottom right component) and the off-diagonal components (the values that do not lie on the diagonal). In the matrix below, the diagonal components are 5, 12, 2 and 6 because they lie along the diagonal line. The off-diagonal components are all of the other values. A square matrix in which the diagonal elements are equal to 1 and the off-diagonal elements are equal to 0 is known as an **identity matrix**:

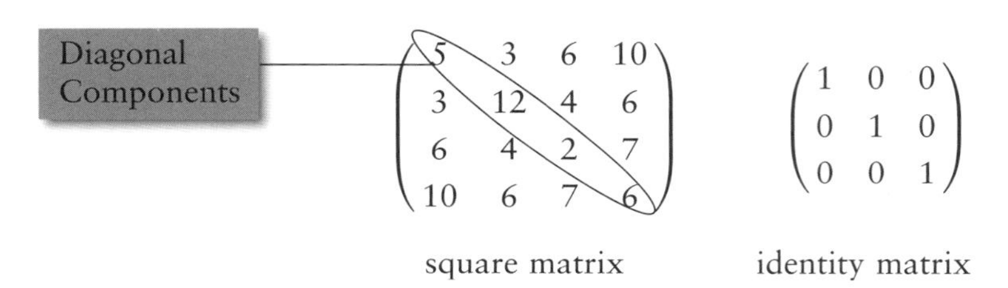

Hopefully, the concept of a matrix should now be slightly less scary than it was previously: it is not some magical mathematical entity, merely a way of representing a data set – just like a spreadsheet.

Now, there is a special case of a matrix where there are data from only one person, and this is known as a *row vector*. Likewise, if there is only one column in a matrix this is known as a *column vector*. In the examples below, the row vector can be thought of as a single person's score on four different variables, whereas the column vector can be thought of as five participants' scores on one variable:

$$\begin{pmatrix} 2 & 6 & 4 & 8 \end{pmatrix} \qquad \begin{pmatrix} 8 \\ 6 \\ 10 \\ 15 \\ 6 \end{pmatrix}$$

row vector   column vector

Armed with this knowledge of what vectors are, we can have a brief look at how they are used to conduct MANOVA.

16.4.2. Some important matrices and their functions ③

As with ANOVA, we are primarily interested in how much variance can be explained by the experimental manipulation (which in real terms means how much variance is explained by the fact that certain scores appear in certain groups). Therefore, we need to know the sum of squares due to the grouping variable (the systematic variation, SS_M), the sum of squares due to natural differences between participants (the residual variation, SS_R) and of course the total amount of variation that needs to be explained (SS_T); for more details about these sources of variation reread Chapters 7 and 10. However, I mentioned that MANOVA also takes into account several dependent variables simultaneously and it does this by using a matrix that contains information about the variance accounted for by each dependent variable. For the univariate *F*-test (e.g. ANOVA) we calculated the ratio of systematic variance to unsystematic variance for a single dependent variable. In MANOVA, the test statistic is derived by comparing the ratio of systematic to unsystematic variance for several dependent variables. This comparison is made by using the ratio of a matrix representing the systematic variance of all dependent variables to a matrix representing the unsystematic variance of all dependent variables. To sum up, the test statistic in both ANOVA and MANOVA represents the ratio of the effect of the systematic variance to the unsystematic variance; in ANOVA these variances are single values, but in MANOVA each is a matrix containing many variances and covariances.

The matrix that represents the systematic variance (or the model sum of squares for all variables) is denoted by the letter *H* and is called the **hypothesis sum of squares and cross-products matrix** (or **hypothesis SSCP**). The matrix that represents the unsystematic variance (the residual sums of squares for all variables) is denoted by the letter *E* and is called the **error sum of squares and cross-products matrix** (or **error SSCP**). Finally, there is a matrix that represents the total amount of variance present for each dependent variable (the total sums of squares for each dependent variable) and this is denoted by *T* and is called the **total sum of squares and cross-products matrix** (or **total SSCP**).

Later, I will show how these matrices are used in exactly the same way as the simple sums of squares (SS_M, SS_R and SS_T) in ANOVA to derive a test statistic representing the ratio of systematic to unsystematic variance in the model. The observant among you may have noticed that the matrices I have described are all called *sum of squares and cross-products (SSCP) matrices*. It should be obvious why these matrices are referred to as sum of squares matrices, but why is there a reference to cross-products in their name?

SELF-TEST Can you remember (from Chapter 6) what a cross-product is?

Cross-products represent a total value for the combined error between two variables (so in some sense they represent an unstandardized estimate of the total correlation between two variables). As such, whereas the sum of squares of a variable is the total squared difference between the observed values and the mean value, the cross-product is the total combined error between two variables. I mentioned earlier that MANOVA had the power to account for any correlation between dependent variables and it does this by using these cross-products.

16.4.3. Calculating MANOVA by hand: a worked example ③

To begin with let's carry out univariate ANOVAs on each of the two dependent variables in our OCD example (see Table 16.1). A description of the ANOVA model can be found in Chapter 10 and I will draw heavily on the assumption that you have read this chapter; if you are hazy on the details of Chapter 10, now would be a good time to (re)read sections 10.2.5–10.2.9.

16.4.3.1. Univariate ANOVA for DV 1 (Actions) ②

There are three sums of squares that need to be calculated. First we need to assess how much variability there is to be explained within the data (SS_T), next we need to see how much of this variability can be explained by the model (SS_M), and finally we have to assess how much error there is in the model (SS_R). From Chapter 10 we can calculate each of these values:

- $SS_{T(Actions)}$: The total sum of squares is obtained by calculating the difference between each of the 20 scores and the mean of those scores, then squaring these differences and adding these squared values up. Alternatively, you can get SAS to calculate the variance for the action data (regardless of which group the score falls into) and then multiplying this value by the number of scores minus 1:

$$\begin{aligned} SS_T &= s^2_{grand}(n-1) \\ &= 2.1195(30-1) \\ &= 2.1195 \times 29 \\ &= 61.47 \end{aligned}$$

- $SS_{M(Actions)}$: This value is calculated by taking the difference between each group mean and the grand mean and the squaring them. Multiply these values by the number of scores in the group and then add them together:

$$\begin{aligned} SS_M &= 10(4.90-4.53)^2 + 10(3.70-4.53)^2 + 10(5.00-4.53)^2 \\ &= 10(0.37)^2 + 10(-0.83)^2 + 10(0.47)^2 \\ &= 1.37 + 6.89 + 2.21 \\ &= 10.47 \end{aligned}$$

- $SS_{R(Actions)}$: This value is calculated by taking the difference between each score and the mean of the group from which it came. These differences are then squared and then added together. Alternatively, we can get SAS to calculate the variance within each group, multiply each group variance by the number of scores minus 1 and then add them together:

$$\begin{aligned} SS_R &= s^2_{CBT}(n_{CBT}-1) + s^2_{BT}(n_{BT}-1) + s^2_{NT}(n_{NT}-1) \\ &= (1.433)(10-1) + (3.122)(10-1) + (1.111)(10-1) \\ &= (1.433 \times 9) + (3.122 \times 9)(1.111 \times 9) \\ &= 12.9 + 28.1 + 10.0 \\ &= 51.00 \end{aligned}$$

The next step is to calculate the average sums of squares (the mean square) of each by dividing by the degrees of freedom (see section 10.2.8):

SS	df	MS
$SS_{M(Actions)} = 10.47$	2	5.235
$SS_{R(Actions)} = 51.00$	27	1.889

The final stage is to calculate F by dividing the mean squares for the model by the mean squares for the error in the model:

$$F = \frac{MS_M}{MS_R} = \frac{5.235}{1.889} = 2.771$$

This value can then be evaluated against critical values of F. The point to take home here is the calculation of the various sums of squares and what each one relates to.

16.4.3.2. Univariate ANOVA for DV 2 (Thoughts) ②

As with the data for dependent variable 1, there are three sums of squares that need to be calculated as before:

- $SS_{T(Thoughts)}$:

$$\begin{aligned} SS_T &= s^2_{grand}(n-1) \\ &= 4.878(30-1) \\ &= 4.878 \times 29 \\ &= 141.46 \end{aligned}$$

- $SS_{M(Thoughts)}$:

$$\begin{aligned} SS_M &= 10(13.40-14.53)^2 + 10(15.2-14.53)^2 + 10(15.0-14.53)^2 \\ &= 10(-1.13)^2 + 10(0.67)^2 + 10(0.47)^2 \\ &= 12.77 + 4.49 + 2.21 \\ &= 19.47 \end{aligned}$$

- $SS_{R(Thoughts)}$:

$$\begin{aligned} SS_R &= s^2_{CBT}(n_{CBT}-1) + s^2_{BT}(n_{BT}-1) + s^2_{NT}(n_{NT}-1) \\ &= (3.6)(10-1) + (4.4)(10-1) + (5.56)(10-1) \\ &= (3.6 \times 9) + (4.4 \times 9)(5.56 \times 9) \\ &= 32.4 + 39.6 + 50.0 \\ &= 122 \end{aligned}$$

The next step is to calculate the average sums of squares (the mean square) of each by dividing by the degrees of freedom (see section 10.2.8):

SS	df	MS
$SS_{M(Thoughts)} = 19.47$	2	9.735
$SS_{R(Thoughts)} = 122.00$	27	4.519

The final stage is to calculate *F* by dividing the mean squares for the model by the mean squares for the error in the model:

$$F = \frac{MS_M}{MS_R} = \frac{9.735}{4.519} = 2.154$$

This value can then be evaluated against critical values of *F*. Again, the point to take home here is the calculation of the various sums of squares and what each one relates to.

16.4.3.3. The relationship between DVs: cross-products ②

We know already that MANOVA uses the same sums of squares as ANOVA, and in the next section we will see exactly how it uses these values. However, I have also mentioned that MANOVA takes account of the relationship between dependent variables by using the cross-products. There are three different cross-products that are of interest and these relate to the three sums of squares that we calculated for the univariate ANOVAs: that is, there is a total cross-product, a cross-product due to the model and a residual cross-product. Let's look at the total cross-product (CP_T) first.

I mentioned in Chapter 6 that the cross-product was the difference between the scores and the mean in one group multiplied by the difference between the scores and the mean in the other group. In the case of the total cross-product, the mean of interest is the grand mean for each dependent variable (see Table 16.2). Hence, we can adapt the cross-product equation described in Chapter 6 using the two dependent variables. The resulting equation for the total cross-product is described as in equation (16.1). Therefore, for each dependent variable you take each score and subtract from it the grand mean for that variable. This leaves you with two values per participant (one for each dependent variable) which should be multiplied together to get the cross-product for each participant. The total can then be found by adding the cross-products of all participants. Table 16.2 illustrates this process:

$$CP_T = \sum \left(x_{i(\text{Actions})} - \overline{X}_{\text{grand(Actions)}}\right)\left(x_{i(\text{Thoughts})} - \overline{X}_{\text{grand(Thoughts)}}\right) \tag{16.1}$$

The total cross-product is a gauge of the overall relationship between the two variables. However, we are also interested in how the relationship between the dependent variables is influenced by our experimental manipulation and this relationship is measured by the model cross-product (CP_M). The CP_M is calculated in a similar way to the model sum of squares. First, the difference between each group mean and the grand mean is calculated for each dependent variable. The cross-product is calculated by multiplying the differences found for each group. Each product is then multiplied by the number of scores within the group (as was done with the sum of squares). This principle is illustrated in the following equation and Table 16.3:

$$CP_M = \sum n\left[\left(\overline{x}_{\text{group(Actions)}} - \overline{X}_{\text{grand(Actions)}}\right)\left(\overline{x}_{\text{group(Thoughts)}} - \overline{X}_{\text{grand(Thoughts)}}\right)\right] \tag{16.2}$$

TABLE 16.2 Calculation of the total cross-product

Group	Actions	Thoughts	Actions $-\overline{X}_{\text{grand(Actions)}}$ (D_1)	Thoughts $-\overline{X}_{\text{grand(Thoughts)}}$ (D_2)	$D_1 \times D_2$
CBT	5	14	0.47	−0.53	−0.25
	5	11	0.47	−3.53	−1.66
	4	16	−0.53	1.47	−0.78
	4	13	−0.53	−1.53	0.81
	5	12	0.47	−2.53	−1.19
	3	14	−1.53	−0.53	0.81
	7	12	2.47	−2.53	−6.25
	6	15	1.47	0.47	0.69
	6	16	1.47	1.47	2.16
	4	11	−0.53	−3.53	1.87
BT	4	14	−0.53	−0.53	0.28
	4	15	−0.53	0.47	−0.25
	1	13	−3.53	−1.53	5.40
	1	14	−3.53	−0.53	1.87
	4	15	−0.53	0.47	−0.25
	6	19	1.47	4.47	6.57
	5	13	0.47	−1.53	−0.72
	5	18	0.47	3.47	1.63
	2	14	−2.53	−0.53	1.34
	5	17	0.47	2.47	1.16
NT	4	13	−0.53	−1.53	0.81
	5	15	0.47	0.47	0.22
	5	14	0.47	−0.53	−0.25
	4	14	−0.53	−0.53	0.28
	6	13	1.47	−1.53	−2.25
	4	20	−0.53	5.47	−2.90
	7	13	2.47	−1.53	−3.78
	4	16	−0.53	1.47	−0.78
	6	14	1.47	−0.53	−0.78
	5	18	0.47	3.47	1.63
$\overline{X}_{\text{grand}}$	4.53	14.53		$CP_T = \Sigma(D_1 \times D_2) = 5.47$	

Finally, we also need to know how the relationship between the two dependent variables is influenced by individual differences in participants' performances. The residual cross-product (CP_R) tells us about how the relationship between the dependent variables is affected by individual differences, or error in the model. The CP_R is calculated in a similar way to the total cross-product except that the group means are used rather than the grand mean (see equation (16.3). So, to calculate each of the difference scores, we take each score and subtract from it the mean of the group to which it belongs (see Table 16.4):

$$\text{CP}_\text{R} = \sum\left(x_{i(\text{Actions})} - \overline{X}_{\text{group(Actions)}}\right)\left(x_{i(\text{Thoughts})} - \overline{X}_{\text{group(Thoughts)}}\right) \quad (16.3)$$

TABLE 16.3 Calculating the model cross-product

	$\bar{X}_{group}$ Actions	$\bar{X}_{group} - \bar{X}_{grand}$ (D_1)	$\bar{X}_{group}$ Thoughts	$\bar{X}_{group} - \bar{X}_{grand}$ (D_2)	$D_1 \times D_2$	$N(D_1 \times D_2)$
CBT	4.9	0.37	13.4	−1.13	−0.418	−4.18
BT	3.7	−0.83	15.2	0.67	−0.556	−5.56
NT	5.0	0.47	15.0	0.47	0.221	2.21
$\bar{X}_{grand}$	4.53		14.53	$CP_M = \Sigma N(D_1 \times D_2) = -7.53$		

The observant among you may notice that the residual cross-product can also be calculated by subtracting the model cross-product from the total cross-product:

$$\begin{aligned} CP_R &= CP_T - CP_M \\ &= 5.47 - (-7.53) = 13 \end{aligned}$$

However, it is useful to calculate the residual cross-product manually in case of mistakes in the calculation of the other two cross-products. The fact that the residual and model cross-products should sum to the value of the total cross-product can be used as a useful double-check.

Each of the different cross-products tells us something important about the relationship between the two dependent variables. Although I have used a simple scenario to keep the maths relatively simple, these principles can be easily extended to more complex scenarios. For example, if we had measured three dependent variables then the cross-products between pairs of dependent variables would be calculated (as they were in this example) and entered into the appropriate SSCP matrix (see next section). As the complexity of the situation increases, so does the amount of calculation that needs to be done. At times such as these the benefit of software like SAS becomes ever more apparent!

16.4.3.4. The total SSCP matrix (*T*) ③

In this example we have only two dependent variables and so all of the SSCP matrices will be 2 × 2 matrices. If there had been three dependent variables then the resulting matrices would all be 3 × 3 matrices. The total SSCP matrix, *T*, contains the total sums of squares for each dependent variable and the total cross-product between the two dependent variables. You can think of the first column and first row as representing one dependent variable and the second column and row as representing the second dependent variable:

	Column 1 Actions	Column 2 Thoughts
Row 1 Actions	$SS_{T(Actions)}$	CP_T
Row 1 Thoughts	CP_T	$SS_{T(Thoughts)}$

TABLE 16.4 Calculation of CP_R

Group	Actions	Actions $-\bar{X}_{group(Actions)}$ (D_1)	Thoughts	Thoughts $-\bar{X}_{group(Thoughts)}$ (D_2)	$D_1 \times D_2$
CBT	5	0.10	14	0.60	0.06
	5	0.10	11	−2.40	−0.24
	4	−0.90	16	2.60	−2.34
	4	−0.90	13	−0.40	0.36
	5	0.10	12	−1.40	−0.14
	3	−1.90	14	0.60	−1.14
	7	2.10	12	−1.40	−2.94
	6	1.10	15	1.60	1.76
	6	1.10	16	2.60	2.86
	4	−0.90	11	−2.40	2.16
$\bar{X}_{CBT}$	**4.9**		**13.4**		Σ = **0.40**
BT	4	0.30	14	−1.20	−0.36
	4	0.30	15	−0.20	−0.06
	1	−2.70	13	−2.20	5.94
	1	−2.70	14	−1.20	3.24
	4	0.30	15	−0.20	−0.06
	6	2.30	19	3.80	8.74
	5	1.30	13	−2.20	−2.86
	5	1.30	18	2.80	3.64
	2	−1.70	14	−1.20	2.04
	5	1.30	17	1.80	2.34
$\bar{X}_{BT}$	**3.7**		**15.2**		Σ = **22.60**
NT	4	−1.00	13	−2.00	2.00
	5	0.00	15	0	0.00
	5	0.00	14	−1.00	0.00
	4	−1.00	14	−1.00	1.00
	6	1.00	13	−2.00	−2.00
	4	−1.00	20	5.00	−5.00
	7	2.00	13	−2.00	−4.00
	4	−1.00	16	1.00	−1.00
	6	1.00	14	−1.00	−1.00
	5	0.00	18	3.00	0.00
$\bar{X}_{NT}$	**5**		**15**		Σ = **−10.00**
				$CP_R = \Sigma(D_1 \times D_2) = 13$	

We calculated these values in the previous sections and so we can simply place the appropriate values in the appropriate cell of the matrix:

$$T = \begin{pmatrix} 61.47 & 5.47 \\ 5.47 & 141.47 \end{pmatrix}$$

From the values in the matrix (and what they represent) it should be clear that the total SSCP represents both the total amount of variation that exists within the data and the total co-dependence that exists between the dependent variables. You should also note that the off-diagonal components are the same (they are both the total cross-product) because this value is equally important for both of the dependent variables.

16.4.3.5. The residual SSCP matrix (*E*) ③

The residual (or error) sum of squares and cross-product matrix, *E*, contains the residual sums of squares for each dependent variable and the residual cross-product between the two dependent variables. This SSCP matrix is similar to the total SSCP except that the information relates to the error in the model:

	Column 1 Actions	Column 2 Thoughts
Row 1 Actions	$SS_{R(Actions)}$	CP_R
Row 1 Thoughts	CP_R	$SS_{R(Thoughts)}$

We calculated these values in the previous sections and so we can simply place the appropriate values in the appropriate cell of the matrix:

$$E = \begin{pmatrix} 51 & 13 \\ 13 & 122 \end{pmatrix}$$

From the values in the matrix (and what they represent) it should be clear that the residual SSCP represents both the unsystematic variation that exists for each dependent variable and the co-dependence between the dependent variables that is due to chance factors alone. As before, the off-diagonal elements are the same (they are both the residual cross-product).

16.4.3.6. The model SSCP matrix (*H*) ③

The model (or hypothesis) sum of squares and cross-product matrix, *H*, contains the model sums of squares for each dependent variable and the model cross-product between the two dependent variables:

	Column 1 Actions	Column 2 Thoughts
Row 1 Actions	$SS_{M(Actions)}$	CP_M
Row 1 Thoughts	CP_M	$SS_{M(Thoughts)}$

These values were calculated in the previous sections and so we can simply place the appropriate values in the appropriate cell of the matrix (see below). From the values in the matrix (and what they represent) it should be clear that the model SSCP represents both the systematic variation that exists for each dependent variable and the co-dependence between the dependent variables that is due to the model (i.e. is due to the experimental manipulation). As before, the off-diagonal components are the same (they are both the model cross-product):

$$H = \begin{pmatrix} 10.47 & -7.53 \\ -7.53 & 19.47 \end{pmatrix}$$

Matrices are additive, which means that you can add (or subtract) two matrices together by adding (or subtracting) corresponding components. Now, when we calculated univariate ANOVA we saw that the total sum of squares was the sum of the model sum of squares and the residual sum of squares (i.e. $SS_T = SS_M + SS_R$). The same is true in MANOVA except that we are adding matrices rather than single values:

$$\begin{aligned} T &= H + E \\ T &= \begin{pmatrix} 10.47 & -7.53 \\ -7.53 & 19.47 \end{pmatrix} + \begin{pmatrix} 51 & 13 \\ 13 & 122 \end{pmatrix} \\ &= \begin{pmatrix} 10.47+51 & -7.53+13 \\ -7.53+13 & 19.47+122 \end{pmatrix} \\ &= \begin{pmatrix} 61.47 & 5.47 \\ 5.47 & 141.47 \end{pmatrix} \end{aligned}$$

The demonstration that these matrices add up should (hopefully) help you to understand that the MANOVA calculations are conceptually the same as for univariate ANOVA – the difference is that matrices are used rather than single values.

16.4.4. Principle of the MANOVA test statistic ④

In univariate ANOVA we calculate the ratio of the systematic variance to the unsystematic variance (i.e. we divide SS_M by SS_R).[3] The conceptual equivalent would therefore be to divide the matrix H by the matrix E. There is, however, a problem in that matrices are not divisible by other matrices! However, there is a matrix equivalent to division, which is to multiply by what's known as the inverse of a matrix. So, if we want to divide H by E we have to multiply H by the inverse of E (denoted as E^{-1}). So, therefore, the test statistic is based upon the matrix that results from multiplying the model SSCP with the inverse of the residual SSCP. This matrix is called HE^{-1}.

Calculating the inverse of a matrix is incredibly difficult and there is no need for you to understand how it is done because SAS will do it for you. However, the interested reader should consult either Stevens (2002) or Namboodiri (1984) – these texts provide very accessible accounts of how to derive an inverse matrix. For readers who do consult these sources, see Oliver Twisted. For the uninterested reader, you'll have to trust me on the following:

[3] In reality we use the mean squares but these values are merely the sums of squares corrected for the degrees of freedom.

$$E^{-1} = \begin{pmatrix} 0.0202 & -0.0021 \\ -0.0021 & 0.0084 \end{pmatrix}$$

$$HE^{-1} = \begin{pmatrix} 0.2273 & -0.0852 \\ -0.1930 & 0.1794 \end{pmatrix}$$

Remember that HE^{-1} represents the ratio of systematic variance in the model to the unsystematic variance in the model and so the resulting matrix is conceptually the same as the *F*-ratio in univariate ANOVA. There is another problem, though. In ANOVA, when we divide the systematic variance by the unsystematic variance we get a single figure: the *F*-ratio. In MANOVA, when we divide the systematic variance by the unsystematic variance we get a matrix containing several values. In this example, the matrix contains four values, but had there been three dependent variables the matrix would have had nine values. In fact, the resulting matrix will always contain p^2 values, where p is the number of dependent variables. The problem is how to convert these matrix values into a meaningful single value. This is the point at which we have to abandon any hope of understanding the maths behind the test and talk conceptually instead.

16.4.4.1. Discriminant function variates ④

The problem of having several values with which to assess statistical significance can be simplified considerably by converting the dependent variables into underlying dimensions or factors (this process will be discussed in more detail in Chapter 17). In Chapter 7, we saw how multiple regression worked on the principle of fitting a linear model to a set of data to predict an outcome variable (the dependent variable in ANOVA terminology). This linear model was made up of a combination of predictor variables (or independent variables) each of which had a unique contribution to this linear model. We can do a similar thing here, except that we are interested in the opposite problem (i.e. predicting an independent variable from a set of dependent variables). So, it is possible to calculate underlying linear dimensions of the dependent variables. These linear combinations of the dependent variables are known as *variates* (or sometimes called *latent variables* or *factors*). In this context we wish to use these linear variates to predict which group a person belongs to (i.e. whether they were given CBT, BT or no treatment), so we are using them to discriminate groups of people. Therefore, these variates are called *discriminant functions* or *discriminant function variates*. Although I have drawn a parallel between these discriminant functions and the model in multiple regression, there is a difference in that we can extract several discriminant functions from a set of dependent variables, whereas in multiple regression all independent variables are included in a single model.

That's the theory in simplistic terms, but how do we discover these discriminant functions? Well, without going into too much detail, we use a mathematical procedure of maximization, such that the first discriminant function (V_1) is the linear combination of dependent variables that maximizes the differences between groups.

It follows from this that the ratio of systematic to unsystematic variance (SS_M/SS_R) will be maximized for this first variate, but subsequent variates will have smaller values of this ratio. Remember that this ratio is an analogue of what the *F*-ratio represents in univariate ANOVA, and so in effect we obtain the maximum possible value of the *F*-ratio when we look at the first discriminant function. This variate can be described in terms of a linear regression equation (because it is a linear combination of the dependent variables):

$$\begin{aligned} Y &= b_0 + b_1X_1 + b_2X_2 \\ V_1 &= b_0 + b_1\text{DV}_1 + b_2\text{DV}_2 \\ V_1 &= b_0 + b_1\text{Actions} + b_2\text{Thoughts} \end{aligned} \tag{16.4}$$

Equation (16.4) shows the multiple regression equation for two predictors and then extends this to show how a comparable form of this equation can describe discriminant functions. The *b*-values in the equation are weights (just as in regression) that tell us something about the contribution of each dependent variable to the variate in question. In regression, the values of *b* are obtained by the method of least squares; in discriminant function analysis the values of *b* are obtained from the *eigenvectors* (see Jane Superbrain 7.3) of the matrix HE^{-1}. We can actually ignore b_0 as well because this serves only to locate the variate in geometric space, which isn't necessary when we're using it to discriminate groups.

In a situation in which there are only two dependent variables and two groups for the independent variable, there will be only one variate. This makes the scenario very simple: by looking at the discriminant function of the dependent variables, rather than looking at the dependent variables themselves, we can obtain a single value of SS_M/SS_R for the discriminant function, and then assess this value for significance. However, in more complex cases where there are more than two dependent variables or more than three levels of the independent variable (as is the case in our example), there will be more than one variate. The number of variates obtained will be the smaller of *p* (the number of dependent variables) and $k-1$ (where *k* is the number of levels of the independent variable). In our example, both *p* and $k-1$ are 2, so we should be able to find two variates. I mentioned earlier that the *b*-values that describe the variates are obtained by calculating the eigenvectors of the matrix HE^{-1}, and in fact, there will be two eigenvectors derived from this matrix: one with the *b*-values for the first variate, and one with the *b*-values of the second variate. Conceptually speaking, eigenvectors are the vectors associated with a given matrix that are unchanged by transformation of that matrix to a diagonal matrix (look back to Jane Superbrain 7.3. for a visual explanation of eigenvectors and eigenvalues). A diagonal matrix is simply a matrix in which the off-diagonal elements are zero and by changing HE^{-1} to a diagonal matrix we eliminate all of the off-diagonal elements (thus reducing the number of values that we must consider for significance testing). Therefore, by calculating the eigenvectors and eigenvalues, we still end up with values that represent the ratio of systematic to unsystematic variance (because they are unchanged by the transformation), but there are considerably farer of them. The calculation of eigenvectors is extremely complex (insane students can consider reading Namboodiri, 1984), so you can trust me that for the matrix HE^{-1} the eigenvectors obtained are:

$$\text{eigenvector}_1 = \begin{pmatrix} 0.603 \\ -0.335 \end{pmatrix}$$

$$\text{eigenvector}_2 = \begin{pmatrix} 0.425 \\ 0.339 \end{pmatrix}$$

Replacing these values into the two equations for the variates and bearing in mind we can ignore b_0, we obtain the models described in the following equation:

$$\begin{aligned} V_{1i} &= b_0 + 0.603\text{Actions}_i - 0.335\text{Thoughts}_i \\ V_{2i} &= b_0 + 0.425\text{Actions}_i - 0.339\text{Thoughts}_i \end{aligned} \qquad (16.5)$$

It is possible to use the equations for each variate to calculate a score for each person on the variate. For example, the first participant in the CBT group carried out 5 obsessive actions, and had 14 obsessive thoughts. Therefore, this participant's score on variate 1 would be −1.675:

$$V_1 = (0.603 \times 5) - (0.335 \times 14) = -1.675$$

The score for variate 2 would be 6.871:

$$V_2 = (0.425 \times 5) - (0.339 \times 14) = 6.871$$

If we calculated these variate scores for each participant and then calculated the SSCP matrices (e.g. *H*, *E*, *T* and HE^{-1}) that we used previously, we would find that all of them have cross-products of zero. The reason for this is because the variates extracted from the data are orthogonal, which means that they are uncorrelated. In short, the variates extracted are independent dimensions constructed from a linear combination of the dependent variables that were measured.

This data reduction has a very useful property in that if we look at the matrix HE^{-1} calculated from the variate scores (rather than the dependent variables) we find that all of the off-diagonal elements (the cross-products) are zero. The diagonal elements of this matrix represent the ratio of the systematic variance to the unsystematic variance (i.e. SS_M/SS_R) for each of the underlying variates. So, for the data in this example, this means that instead of having four values representing the ratio of systematic to unsystematic variance, we now have only two. This reduction may not seem a lot. However, in general if we have *p* dependent variables, then ordinarily we would end up with p^2 values representing the ratio of systematic to unsystematic variance; by looking at discriminant functions, we reduce this number to *p*. If there were four dependent variables we would end up with four values rather than 16 (which highlights the benefit of this process).

For the data in our example, the matrix HE^{-1} calculated from the variate scores is:

$$HE^{-1}_{\text{variates}} = \begin{pmatrix} 0.335 & 0.000 \\ 0.000 & 0.073 \end{pmatrix}$$

It is clear from this matrix that we have two values to consider when assessing the significance of the group differences. It probably seems like a complex procedure to reduce the data down in this way: however, it transpires that the values along the diagonal of the matrix for the variates (namely 0.335 and 0.073) are the *eigenvalues* of the original HE^{-1} matrix. Therefore, these values can be calculated directly from the data collected without first forming the eigenvectors. If you have lost all sense of rationality and want to see how these eigenvalues are calculated then see Oliver Twisted. These eigenvalues are conceptually equivalent to the *F*-ratio in ANOVA and so the final step is to assess how large these values are compared to what we would expect by chance alone. There are four ways in which the values are assessed.

OLIVER TWISTED

Please Sir, can I have some more … maths?

'You are a bit stupid. I think it would be fun to check your maths so that we can see exactly how much of a village idiot you are', mocks Oliver. Luckily you can. Never one to shy from public humiliation on a mass scale, I have provided the matrix calculations for this example on the companion website. Find a mistake, go on, you know that you can …

16.4.4.2. Pillai–Bartlett trace (*V*) ④

The *Pillai–Bartlett trace* (also known as Pillai's trace) is given by equation (16.6) in which λ represents the eigenvalues for each of the discriminant variates and *s* represents the number of variates. Pillai's trace is the sum of the proportion of explained variance on the discriminant functions. As such, it is similar to the ratio of SS_M/SS_T, which is known as R^2:

$$V = \sum_{i=1}^{s} \frac{\lambda_i}{1+\lambda_i} \tag{16.6}$$

For our data, Pillai's trace turns out to be 0.319, which can be transformed to a value that has an approximate F-distribution:

$$V = \frac{0.335}{1+0.335} + \frac{0.073}{1+0.073} = 0.319$$

16.4.4.3. Hotelling's T^2 ④

The *Hotelling–Lawley trace* (also known as Hotelling's T^2; Figure 16.2) is simply the sum of the eigenvalues for each variate (see equation 16.7) and so for these data its value is 0.408 (0.335 + 0.073). This test statistic is the sum of SS_M/SS_R for each of the variates and so it compares directly to the F-ratio in ANOVA:

$$T = \sum_{i=1}^{s} \lambda_i \tag{16.7}$$

FIGURE 16.2 Harold Hotelling enjoying my favourite activity of drinking tea

16.4.4.4. Wilks's lambda (Λ) ④

Wilks's lambda is the product of the *unexplained* variance on each of the variates (see equation (16.8) – the Π symbol is similar to the summation symbol (Σ) that we have encountered already except that it means *multiply* rather than add up). So, Wilks's lambda represents the ratio of error variance to total variance (SS_R/SS_T) for each variate:

$$\Lambda = \prod_{i=1}^{s} \frac{1}{1+\lambda_i} \tag{16.8}$$

For the data in this example the value is 0.698, and it should be clear that large eigenvalues (which in themselves represent a large experimental effect) lead to small values of Wilks's lambda – hence statistical significance is found when Wilks's lambda is small:

$$\Lambda = \left(\frac{1}{1+0.335}\right)\left(\frac{1}{1+0.073}\right) = 0.698$$

16.4.4.5. Roy's greatest root ④

Roy's greatest root always makes me think of some bearded statistician with a garden spade digging up a fantastic parsnip (or similar root vegetable); however, it isn't a parsnip but, as the name suggests, is simply the eigenvalue for the first variate. So, in a sense it is the same as the Hotelling–Lawley trace but for the first variate only, that is:

$$\Theta = \lambda_{\text{Largest}} \tag{16.9}$$

As such, Roy's greatest root represents the proportion of explained variance to unexplained variance (SS_M/SS_R) for the first discriminant function.[4] For the data in this example, the value of Roy's greatest root is simply 0.335 (the eigenvalue for the first variate). So, this value is conceptually the same as the *F*-ratio in univariate ANOVA. It should be apparent, from what we have learnt about the maximizing properties of these discriminant variates, that Roy's root represents the maximum possible between-group difference given the data collected. Therefore, this statistic should in many cases be the most powerful.

16.5. Practical issues when conducting MANOVA ③

There are three main practical issues to be considered before running MANOVA. First of all, as always we have to consider the assumptions of the test. Next, for the main analysis there are four commonly used ways of assessing the overall significance of a MANOVA and debate exists about which method is best in terms of power and sample size considerations. Finally, we also need to think about what analysis to do *after* the MANOVA: like ANOVA, MANOVA is a two-stage test in which an overall (or omnibus) test is first performed before more specific procedures are applied to tease apart group differences. As you will see, there is substantial debate over how best to further analyse and interpret group differences when the overall MANOVA is significant. We will look at these issues in turn.

16.5.1. Assumptions and how to check them ③

MANOVA has similar assumptions to ANOVA but extended to the multivariate case:

- **Independence**: Observations should be statistically independent.
- **Random sampling**: Data should be randomly sampled from the population of interest and measured at an interval level.
- **Multivariate normality:** In ANOVA, we assume that our dependent variable is normally distributed within each group. In the case of MANOVA, we assume that the dependent variables (collectively) have multivariate normality within groups.
- **Homogeneity of covariance matrices**: In ANOVA, it is assumed that the variances in each group are roughly equal (homogeneity of variance). In MANOVA we must assume that this is true for each dependent variable, but also that the correlation between any two dependent variables is the same in all groups. This assumption is examined by testing whether the population variance–covariance matrices of the different groups in the analysis are equal.[5]

Most of the assumptions can be checked in the same way as for univariate tests (see Chapter 10); the additional assumptions of multivariate normality and equality of covariance matrices require different procedures. The assumption of multivariate normality is difficult to test (you need to

[4] This statistic is sometimes characterized as $\lambda_{\text{largest}}/(1 + \lambda_{\text{largest}})$ but this is not the statistic reported by SAS.

[5] For those of you who read about SSCP matrices, if you think about the relationship between sums of squares and variance, and cross-products and correlations, it should be clear that a variance–covariance matrix is basically a standardized form of an SSCP matrix.

get a macro from the SAS website) and so the only practical solution is to check the assumption of univariate normality for each dependent variable in turn (see Chapter 5). This solution is practical (because it is easy to implement) and useful (because univariate normality is a necessary condition for multivariate normality), but it does not *guarantee* multivariate normality – but it's rare for univariate normality to hold, and multivariate normality not to hold. So, although this approach is the best we can do, I urge interested readers to consult Stevens (2002) who provides some alternative solutions.

The assumption of equality of covariance matrices is more easily checked. First, for this assumption to be true the univariate tests of equality of variances between groups should be met. This assumption is easily checked using Levene's test (see section 5.6.1). As a preliminary check, Levene's test should not be significant for any of the dependent variables. However, Levene's test does not take account of the covariances and so the variance–covariance matrices should be compared between groups using Box's test (see Section 16.8). This test should be non-significant if the matrices are the same. The effect of violating this assumption is unclear, except that Hotelling's T^2 is robust in the two-group situation when sample sizes are equal (Hakstian, Roed, & Lind, 1979).

Box's test is susceptible to deviations from multivariate normality and so can be non-significant not because the matrices are similar, but because the assumption of multivariate normality is not tenable. Hence, it is vital to have some idea of whether the data meet the multivariate normality assumption before interpreting the result of Box's test. Also, as with any significance test, in large samples Box's test could be significant even when covariance matrices are relatively similar. As a general rule, if sample sizes are equal then disregard Box's test, because (1) it is unstable, and (2) in this situation we can assume that Hotelling's and Pillai's statistics are robust (see section 16.5.2). However, if group sizes are different, then robustness cannot be assumed (especially if Box's test is significant at $p < .001$). The more dependent variables you have measured, and the greater the differences in sample sizes, the more distorted the probability values produced by SAS become. Tabachnick and Fidell (2007) suggest that if the larger samples produce greater variances and covariances then the probability values will be conservative (and so significant findings can be trusted). However, if it is the smaller samples that produce the larger variances and covariances then the probability values will be liberal and so significant differences should be treated with caution (although non-significant effects can be trusted). Therefore, the variance–covariance matrices for samples should be inspected to assess whether the printed probabilities for the multivariate test statistics are likely to be conservative or liberal. In the event that you cannot trust the printed probabilities, there is little you can do except equalize the samples by randomly deleting cases in the larger groups (although with this loss of information comes a loss of power).

16.5.2. Choosing a test statistic ③

Only when there is one underlying variate will the four test statistics necessarily be the same. Therefore, it is important to know which test statistic is best in terms of test power and robustness. A lot of research has investigated the power of the four MANOVA test statistics (Olson, 1974, 1976, 1979; Stevens, 1980). Olson (1974) observed that for small and moderate sample sizes the four statistics differ little in terms of power. If group differences are concentrated on the first variate (as will often be the case in social science research) Roy's statistic should prove most powerful (because it takes account of only that first variate), followed by the Hotelling–Lawley trace, Wilks's lambda and Pillai's trace. However, when groups differ along more than one variate, the power ordering is the reverse (i.e. Pillai's trace is most powerful and Roy's root is least). One final issue pertinent to test power is that of sample size and the number of dependent variables. Stevens (1980) recommends using fewer than 10 dependent variables unless sample sizes are large.

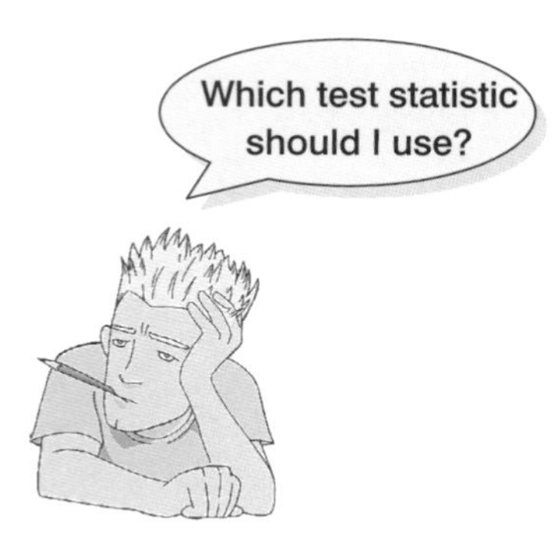

In terms of robustness, all four test statistics are relatively robust to violations of multivariate normality (although Roy's root is affected by platykurtic distributions – see Olson, 1976). Roy's root is also not robust when the homogeneity of covariance matrix assumption is untenable (Stevens, 1979). The work of Olson and Stevens led Bray and Maxwell (1985) to conclude that when sample sizes are equal the Pillai–Bartlett trace is the most robust to violations of assumptions. However, when sample sizes are unequal this statistic is affected by violations of the assumption of equal covariance matrices. As a rule, with unequal group sizes, check the assumption of homogeneity of covariance matrices using Box's test; if this test is non-significant, *and if the assumption of multivariate normality is tenable* (which allows us to assume that Box's test is accurate), then assume that Pillai's trace is accurate.

16.5.3. Follow-up analysis ③

There is some controversy over how best to follow up the main MANOVA. The traditional approach is to follow a significant MANOVA with separate ANOVAs on each of the dependent variables. If this approach is taken, you might well wonder why we bother with the MANOVA in the first place (earlier on I said that multiple ANOVAs were a bad thing to do). Well, the ANOVAs that follow a significant MANOVA are said to be 'protected' by the initial MANOVA (Bock, 1975). The idea is that the overall multivariate test protects against inflated Type I error rates because if that initial test is non-significant (i.e. the null hypothesis is true) then any subsequent tests are ignored (any significance must be a Type I error because the null hypothesis is true). However, the notion of protection is somewhat fallacious because a significant MANOVA, more often than not, reflects a significant difference for one, but not all, of the dependent variables. Subsequent ANOVAs are then carried out on all of the dependent variables, but the MANOVA protects only the dependent variable for which group differences genuinely exist (see Bray and Maxwell, 1985, pp. 40–41). Therefore, you might want to consider applying a Bonferroni correction to the subsequent ANOVAs (Harris, 1975).

By following up a MANOVA with ANOVAs you assume that the significant MANOVA is not due to the dependent variables representing a set of underlying dimensions that differentiate the groups. Therefore, some researchers advocate the use of discriminant analysis, which finds the linear combination(s) of the dependent variables that best *separates* (or discriminates) the groups. This procedure is more in keeping with the ethos of MANOVA because it embraces the relationships that exist between dependent variables and it is certainly useful for illuminating the relationship between the dependent variables and group membership. The major advantage of this approach over multiple ANOVAs is that it reduces and explains the dependent variables in terms of a set of underlying dimensions thought to reflect substantive theoretical dimensions. By default PROC GLM in SAS provides univariate ANOVAs, but not the discriminant analysis – for that we need PROC CANDISC.

16.6. MANOVA on SAS ②

In the remainder of this chapter we will use the OCD data to illustrate how MANOVA is done (those of you who skipped the theory section should refer to Table 16.1).

16.6.1. The main analysis ②

MANOVA on SAS looks very like ANOVA on SAS. Our classification variable is **group**, so we place that on the CLASS line. The MODEL line has two outcome variables and one predictor, so we put the two outcomes (**actions** and **thoughts**) on the left-hand side of the equals sign, and one predictor (**group**) on the right-hand side. We put the E option on the MODEL line, so that SAS shows us how the model was parameterized (E stands for Estimable functions).

For this analysis there are no covariates; however, you can apply the principles of ANCOVA to the multivariate case and conduct multivariate analysis of covariance (MANCOVA): just put continuous variables on this line, as before.

Next, we put a MANOVA statement, which tells SAS we'd like to do a MANOVA. We use H= to tell SAS which hypotheses we'd like to test, and we'd like to test them all. The PRINTE option displays the *E* matrix (or SSCP error matrix) and the PRINTH option displays the *H* matrix (or SSCP model matrix – H stands for Hypothesis).

We'd like the means of the groups, so we use the MEANS statement, which will display the means for each dependent variable, for each group. Putting HOVTEST will give us a test of homogeneity of variance. SAS Syntax 16.1 shows the syntax.

```
PROC GLM DATA=chap16.ocd;
    CLASS group;
    MODEL actions thoughts = group / E;
    MANOVA H=_all_ / PRINTE;
    MEANS group /HOVTEST;
    RUN;
```
SAS Syntax 16.1

16.6.2. Multiple comparisons in MANOVA ②

The default way to follow up a MANOVA is to look at individual univariate ANOVAs for each dependent variable. We can use contrasts, just as we did with ANOVA. The no-treatment control group will be coded as the last category (it is the last alphabetically, but if we're not sure we can use the E option on the MODEL line to check). We will therefore use two contrasts (using ESTIMATE statements), one comparing behavioral therapy with control, and one comparing cognitive behavioral therapy with control.

```
PROC GLM DATA=chap16.ocd;
    CLASS group;
    MODEL actions thoughts = group / E;
    MANOVA H=_all_ / PRINTE;
    MEANS group /HOVTEST;
    ESTIMATE "BT vs Control" group 1 0 -1;
    ESTIMATE "CBT vs Control" group 0 1 -1;
    RUN;
```
SAS Syntax 16.2

The sharper eyed amongst you might have noticed that we have not carried out Box's test, to examine the assumption of homogeneity of covariance matrices. PROC GLM doesn't allow that test, but we'll be coming to it.

16.7. Output from MANOVA ②

As we're getting used to, the output from PROC GLM isn't necessarily in the order in which we'd like to think about it. In fact, it starts with the univariate ANOVAs, which we'd only like to look after we've looked at the multivariate ANOVAs, so please forgive me for tackling the output in a different order.

16.7.1. Preliminary analysis and testing assumptions ③

SAS Output 16.1 shows an initial table of descriptive statistics that is produced by the MEANS option. This table contains group means and standard deviations for each dependent variable in turn. These values correspond to those calculated by hand in Table 16.1, and by looking at that table it should be clear what this part of the output tells us. It is clear from the means that participants had many more obsession-related thoughts than behaviours.

The second and third tables in SAS Output 16.1 shows a summary table of Levene's test of equality of variances for each of the dependent variables. These tests are the same as would be found if a one-way ANOVA had been conducted on each dependent variable in turn (see section 5.6.1). Levene's test should be non-significant for all dependent variables if the assumption of homogeneity of variance has been met. The results for these data clearly show that the assumption has been met. This finding not only gives us confidence in the reliability of the univariate tests to follow, but also strengthens the case for assuming that the multivariate test statistics are robust.

SAS OUTPUT 16.1

		ACTIONS		THOUGHTS	
Level of group	N	Mean	Std Dev	Mean	Std Dev
BT	10	3.70000000	1.76698110	15.2000000	2.09761770
CBT	10	4.90000000	1.19721900	13.4000000	1.89736660
NT	10	5.00000000	1.05409255	15.0000000	2.35702260

Levene's Test for Homogeneity of ACTIONS Variance ANOVA of Squared Deviations from Group Means					
Source	DF	Sum of Squares	Mean Square	F Value	Pr > F
group	2	18.9020	9.4510	2.43	0.1070
Error	27	105.0	3.8886		

Levene's Test for Homogeneity of THOUGHTS Variance ANOVA of Squared Deviations from Group Means					
Source	DF	Sum of Squares	Mean Square	F Value	Pr > F
group	2	15.6587	7.8293	0.28	0.7571
Error	27	752.0	27.8530		

16.7.2. MANOVA test statistics ③

SAS Output 16.2 shows the main table of results. Test statistics are quoted for the group variable (we could also get the statistics for the intercept of the model – even MANOVA can be characterized as a regression model, although how this is done is beyond the scope of my brain – by specifying INTERCEPT on the MANOVA line). For our purposes, the group effects are of interest because they tell us whether or not the therapies had an effect on the OCD clients. You'll see that SAS lists the four multivariate test statistics, and their values correspond to those calculated in sections 16.4.4.2–16.4.4.5. In the next column these values are transformed into an *F*-ratio with 2 degrees of freedom. The column of real interest, however, is the one containing the significance values of these *F*-ratios. For these data, Pillai's trace (p = .049), Wilks's lambda (p = .050) and Roy's greatest root (p = .020) all reach the criterion for significance of .05. However, the Hotelling–Lawley trace (p = .055) is non-significant by this criterion. This scenario is interesting, because the test statistic we choose determines whether or not we reject the null hypothesis that there are no between-group differences. However, given what we know about the robustness of Pillai's trace when sample sizes are equal, we might be well advised to trust the result of that test statistic, which indicates a significant difference. This example highlights the additional power associated with Roy's greatest root (you should note how this statistic is considerably more significant than all others) when the test assumptions have been met and when the group differences are focused on one variate (which they are in this example, as we will see later).

SAS OUTPUT 16.2

MANOVA Test Criteria and F Approximations for the Hypothesis of No Overall group Effect H = Type III SSCP Matrix for group E = Error SSCP Matrix

S=2 M= -0.5 N=12

Statistic	Value	F Value	Num DF	Den DF	Pr > F
Wilks' Lambda	0.69850905	2.55	4	52	0.0497
Pillai's Trace	0.31845458	2.56	4	54	0.0490
Hotelling-Lawley Trace	0.40733521	2.62	4	30.19	0.0546
Roy's Greatest Root	0.33479736	4.52	2	27	0.0203

NOTE: F Statistic for Roy's Greatest Root is an upper bound.

NOTE: F Statistic for Wilks' Lambda is exact.

From this result we should probably conclude that the type of therapy employed had a significant effect on OCD. The nature of this effect is not clear from the multivariate test statistic. First, it tells us nothing about which groups differed from which; and second, it tells us nothing about whether the effect of therapy was on the obsession-related thoughts, the obsession-related behaviours, or a combination of both. To determine the nature of the effect, SAS provides us with univariate tests.

16.7.3. Univariate test statistics ②

The next part of the output that we will look at (but actually the first part of the output) contains the ANOVA summary table for the dependent variables. The row of interest in the first table is that labelled *Model,* which tests the overall significance of the model. The second table gives us R^2 as a measure of effect size. The third and fourth tables break the model effects into the effects of each independent variable. We only have one independent variable, and so the effect of *Group* is the same as the effect of the whole model. In addition, because we have only one measure independent variable, the Type I and Type III sums of squares are the same – if this is not the case, and you have more than one independent variable, you should look at the Type III sums of squares (if you need to look at Type I, you will know). The row labelled *Group* contains an ANOVA summary table for each of the dependent variables, and values are given for the sums of squares for both actions and thoughts (these values correspond to the values of SS_M calculated in sections 16.4.3.1 and 16.4.3.2 respectively). The row labelled *Error* contains information about the residual sums of squares and mean squares for each of the dependent variables: these values of SS_R were calculated in sections 16.4.3.1 and 16.4.3.2 and I urge you to look back to these sections to consolidate what these values mean.. The important parts of this table are the columns labelled *F Value* and *Pr > F* in which the *F*-ratios for each univariate ANOVA and their significance values are listed. What should be clear from SAS Output 16.3 and the calculations made in sections 16.4.3.1 and 16.4.3.2 is that the values associated with the univariate ANOVAs conducted after the MANOVA are *identical* to those obtained if one-way ANOVA was conducted on each dependent variable. This fact illustrates that MANOVA offers only hypothetical protection of inflated Type I error rates: there is no real-life adjustment made to the values obtained.

The values of Pr > *F* in SAS Output 16.3 indicate that there was a non-significant difference between therapy groups in terms of both obsession-related thoughts (p = .136) and obsession-related behaviours (p = .081). These two results should lead us to conclude that the type of therapy has had no significant effect on the levels of OCD experienced by clients. Those of you that are still awake may have noticed something odd about this example: the multivariate test statistics led us to conclude that therapy had had a significant impact on OCD, yet the univariate results indicate that therapy has not been successful.

SELF-TEST Why might the univariate tests be non-significant when the multivariate tests were significant?

SAS OUTPUT 16.3

Dependent Variable: ACTIONS Number of obsession-related behaviours

Source	DF	Sum of Squares	Mean Square	F Value	Pr > F
Model	2	10.46666667	5.23333333	2.77	0.0805
Error	27	51.00000000	1.88888889		
Corrected Total	29	61.46666667			

R-Square	Coeff Var	Root MSE	ACTIONS Mean
0.170282	30.31695	1.374369	4.533333

Source	DF	Type I SS	Mean Square	F Value	Pr > F
group	2	10.46666667	5.23333333	2.77	0.0805

Source	DF	Type III SS	Mean Square	F Value	Pr > F
group	2	10.46666667	5.23333333	2.77	0.0805

Dependent Variable: THOUGHTS Number of obsession-related thoughts

Source	DF	Sum of Squares	Mean Square	F Value	Pr > F
Model	2	19.4666667	9.7333333	2.15	0.1355
Error	27	122.0000000	4.5185185		
Corrected Total	29	141.4666667			

R-Square	Coeff Var	Root MSE	THOUGHTS Mean
0.137606	14.62624	2.125681	14.53333

Source	DF	Type I SS	Mean Square	F Value	Pr > F
group	2	19.46666667	9.73333333	2.15	0.1355

Source	DF	Type III SS	Mean Square	F Value	Pr > F
group	2	19.46666667	9.73333333	2.15	0.1355

The reason for the anomaly in these data is simple: the multivariate test takes account of the correlation between dependent variables and so for these data it has more power to detect group differences. With this knowledge in mind, the univariate tests are not particularly useful for interpretation, because the groups differ along a combination of the dependent variables. To see how the dependent variables interact we need to carry out a discriminant function analysis, which will be described in due course.

16.7.4. SSCP matrices ③

If you added the PRINTH and PRINTE options to the MANOVA statement, then SAS will produce the tables in SAS Output 16.4. The first table displays the error SSCP (E), and the second table model SSCP (H). The matrix for the intercept is displayed also, but this matrix is not important for our purposes. It should be pretty clear that the values in the model and error matrices displayed in correspond to the values we calculated in sections 16.4.3.6 and 16.4.3.5 respectively. These matrices are useful, therefore, for gaining insight into the pattern of the data, and especially in looking at the values of the cross-products to indicate the relationship between dependent variables. In this example, the sums of squares for the error SSCP matrix are substantially bigger than in the model (or group) SSCP matrix, whereas the absolute value of the cross-products is fairly similar. This pattern suggests that if the MANOVA is significant then it might be the relationship between dependent variables that is important rather than the individual dependent variables themselves.

SAS OUTPUT 16.4

E = Error SSCP Matrix		
	ACTIONS	THOUGHTS
ACTIONS	51	13
THOUGHTS	13	122

H = Type III SSCP Matrix for group		
	ACTIONS	THOUGHTS
ACTIONS	10.466666667	−7.533333333
THOUGHTS	−7.533333333	19.466666667

16.7.5. Contrasts ③

I need to begin this section by reminding you that because the univariate ANOVAs were both non-significant we should not interpret these contrasts. However, purely to give you an example to follow for when your main analysis is significant, we'll look at this part of the output anyway. In section 16.6.2 I suggested carrying out a contrast that compares each of the therapy groups to the no-treatment control group. SAS Output 16.5 shows the results of these contrasts. The table is divided into two sections conveniently labelled *Thoughts* and *Actions*.

SAS OUTPUT 16.5

Dependent Variable: ACTIONS Number of obsession-related behaviours

Parameter	Estimate	Standard Error	t Value	Pr > |t|
BT vs Control	−1.30000000	0.61463630	−2.12	0.0438
CBT vs Control	−0.10000000	0.61463630	−0.16	0.8720

Dependent Variable: THOUGHTS Number of obsession-related thoughts

Parameter	Estimate	Standard Error	t Value	Pr > |t|
BT vs Control	0.20000000	0.95063332	0.21	0.8349
CBT vs Control	−1.60000000	0.95063332	−1.68	0.1039

The first thing that you might notice (from the values of $Pr > |t|$) is that when we compare CBT to control there are no significant differences in thoughts ($p = .104$) or behaviours ($p = .872$) because both values are above the .05 threshold. However, comparing BT to control, there is no significant difference in thoughts ($p = .835$) but there is a significant difference in behaviours between the groups ($p = .044$, which is less than .05). This is a little unexpected because the univariate ANOVA for behaviours was non-significant and so we would not expect there to be group differences.

16.8. Box's test ③

You might have noticed that we haven't seen any results for Box's test (sometimes called Box's *M*-test). That's because PROC GLM doesn't have an option to produce Box's test. Instead, we need to use PROC DISCRIM to produce this test. SAS Syntax 16.3 shows how this is done. We use PROC DISCRIM with the same data as we used for the MANOVA, but we also put POOL=test on the PROC line – this tests whether we can pool the covariance matrices. We put the independent variable on the CLASS line, and the VAR line contains the dependent variables.

```
PROC DISCRIM DATA=chap16.ocd POOL=test ;
    CLASS group;
    VAR thoughts actions;
    RUN;
```

SAS Syntax 16.3

SAS OUTPUT 16.6

Test of Homogeneity of Within Covariance Matrices

Chi-Square	DF	Pr > ChiSq
8.893217	6	0.1797

PROC DISCRIM produces a lot of output, but the part that we want to look at is labelled *Test of homogeneity of within covariance matrices*, and is shown in SAS Output 16.6. This statistic tests the null hypothesis that the variance–covariance matrices are the same in all three groups. Therefore, if the matrices are equal (and therefore the assumption of homogeneity is met) this statistic should be *non-significant*. For these data $p = .18$ (which is greater than .05): hence, the covariance matrices are roughly equal and the assumption is tenable.

If the value of Box's test is significant ($p < .05$) then the covariance matrices are significantly different and so the homogeneity assumption would have been violated. Bartlett's test of sphericity tests whether the assumption of sphericity has been met and is useful only in univariate repeated-measures designs because MANOVA does not require this assumption.

CRAMMING SAM'S TIPS MANOVA

- MANOVA is used to test the difference between groups across several dependent variables simultaneously.
- Box's test looks at the assumption of equal covariance matrices. This test can be ignored when sample sizes are equal because when they are some MANOVA test statistics are robust to violations of this assumption. If group sizes differ this test should be inspected. If the significance value is less than .001 then the results of the analysis should not be trusted (see section 16.7.1).
- The table labelled **Multivariate Tests** gives us the results of the MANOVA. There are four test statistics (*Pillai's trace, Wilks's lambda, Hotelling–Lawley Trace and Roy's greatest root*). I recommend using Pillai's trace. If the significance value for this statistic is less than .05 then the groups differ significantly with respect to the dependent variables.
- ANOVAs can be used to follow up the MANOVA (a different ANOVA for each dependent variable). The results of these are listed in the table entitled *Tests of Between-Subjects Effects*. These ANOVAs can in turn be followed up using contrasts (see Chapters 10–14). Personally I don't recommend this approach and suggest conducting a *discriminant function analysis*.

16.9 Reporting results from MANOVA ②

Reporting a MANOVA is much like reporting an ANOVA. As you can see in SAS Output 16.2, the multivariate tests are converted into approximate *F*s, and people often just report these *F*s just as they would for ANOVA (i.e. they give details of the *F*-ratio and the degrees of freedom from which it was calculated). For our effect of group, we would report the hypothesis *df* and the error *df*. Therefore, we could report these analyses as:

✓ There was a significant effect of therapy on the number of obsessive thoughts and behaviours, $F(4, 54) = 2.56, p < .05$.

However, personally, I think the multivariate test statistic should be quoted as well. There are four different multivariate tests reported in sections 16.4.4.2–16.4.4.5; I'll report each one in turn (note that the degrees of freedom and value of *F* change), but in reality you would just report one of the four:

✓ Using Pillai's trace, there was a significant effect of therapy on the number of obsessive thoughts and behaviours, $V = 0.32$, $F(4, 54) = 2.56$, $p = .049$.

✓ Using Wilks's statistic, there was a significant effect of therapy on the number of obsessive thoughts and behaviours, $\Lambda = 0.70$, $F(4, 52) = 2.56$, $p = .050$.

✓ Using the Hotelling–Lawley trace statistic, there was no significant effect of therapy on the number of obsessive thoughts and behaviours, $T = 0.41$, $F(4, 50) = 2.55$, $p = .055$.

✓ Using Roy's greatest root, there was a significant effect of therapy on the number of obsessive thoughts and behaviours, $\Theta = 0.35$, $F(2, 27) = 4.52$, $p = .020$.

We can also report the follow-up ANOVAs in the usual way (SAS Output 16.2):

✓ Using Pillai's trace, there was a significant effect of therapy on the number of obsessive thoughts and behaviours, $V = 0.32$, $F(4, 54) = 2.56$, $p < .05$. However, separate univariate ANOVAs on the outcome variables revealed non-significant treatment effects on obsessive thoughts, $F(2, 27) = 9.73$, $p > .05$, and behaviours, $F(2, 27) = 5.23$, $p > .05$.

LABCOAT LENI'S REAL RESEARCH 16.1

A lot of hot air! ④

MARZILLIER, S. L., & DAVEY, G. C. L. (2005). *COGNITION AND EMOTION*, 19, 729–750.

Have you ever wondered what researchers do in their spare time? Well, some of them spend it tracking down the sounds of people burping and farting! It has long been established that anxiety and disgust are linked. Anxious people are, typically, easily disgusted. Throughout this book I have talked about how you cannot infer causality from relationships between variables. This has been a bit of a conundrum for anxiety researchers: does anxiety cause feelings of digust or does a low threshold for being disgusted cause anxiety?

Two colleagues of mine at Sussex addressed this in an unusual study in which they induced feelings of anxiety, feelings of disgust, or a neutral mood, and they looked at the effect that these induced moods had on feelings of anxiety, sadness, happiness, anger, disgust and contempt. To induce these moods, they used three different types of manipulation: vignettes (e.g. 'You're swimming in a dark lake and something brushes your leg' for anxiety, and 'You go into a public toilet and find it has not been flushed. The bowl of the toilet is full of diarrhoea' for disgust), music (e.g. some scary music for anxiety, and a tape of burps, farts and vomitting for disgust), videos (e.g. a clip from *Silence of the lambs* for anxiety and a scene from *Pink flamingos* in which Divine eats dog faeces) and memory (remembering events from the past that had made the person anxious, disgusted or neutral).

Different people underwent anxious, disgust and neutral mood inductions. Within these groups, the induction was done using either vignettes and music, videos, or memory recall and music for different people. The outcome variables were the change (from before to after the induction) in six moods: anxiety, sadness, happiness, anger, disgust and contempt.

The data are in the file **MarzillierandDav2005**.sas7bdat. Draw an error bar graph of the changes in moods in the different conditions, then conduct a 3 (Mood: anxiety, disgust, neutral) × 3 (Induction: vignettes + music, videos, memory recall + music) MANOVA on these data.

Whatever you do, don't imagine what their fart tape sounded like while you do the analysis!

Answers are in the additional material on the companion website (or look at page 738 of the original article).

16.10 Following up MANOVA with discriminant function analysis ③

I mentioned earlier on that a significant MANOVA could be followed up using either univariate ANOVA or discriminant function analysis (sometimes just called discriminant analysis). In the example in this chapter, the univariate ANOVAs were not a useful way of looking at what the multivariate tests showed because the relationship between dependent variables is obviously having an effect. However, these data were designed especially to illustrate how the univariate ANOVAs should be treated cautiously and in real life a significant MANOVA is likely to be accompanied by at least one significant ANOVA. However, this does not mean that the relationship between dependent variables is not important, and it is still vital to investigate the nature of this relationship. Discriminant function analysis is the best way to achieve this, and I strongly recommend that you follow up a MANOVA with both univariate tests and discriminant function analysis if you want to fully understand your data.

We've already come across discriminant function analysis when we used PROC DISCRIM to produce Box's test. We could use PROC DISCRIM, but it's easier to use PROC CANDISC instead. (That stands for Canonical Discriminant, and if you want to know why that is, you need more help than we are able to provide.)

```
PROC CANDISC DATA=chap16.ocd;
     CLASS group;
     VAR thoughts actions;
     RUN;
```
SAS Syntax 16.4

The syntax is shown in SAS Syntax 16.4 – you'll notice that it looks very like SAS Syntax 16.3, with the exception that we cannot request the Box's test in PROC CANDISC. We then put the classification, group, on the CLASS line, and the continuous variables on the VAR line.

16.11. Output from the discriminant function analysis ③

SAS Output 16.7 shows the initial statistics from the discriminant function analysis. We are told the eigenvalues for each variate and you should note that the values correspond to the values of the diagonal elements of the matrix HE^{-1} (for the calculation see Oliver Twisted). These eigenvalues are converted into proportion of variance accounted for, and the first variate accounts for 0.8219 (=82.2%) of the variance compared to the second variate, which accounts for only 17.8%. This table also shows the canonical correlation, which we can square to use as an effect size (just like R^2 which we have encountered in regression).

	Canonical Correlation	Adjusted Canonical Correlation	Approximate Standard Error	Squared Canonical Correlation	Eigenvalues of Inv(E)*H = CanRsq/(1–CanRsq)			
					Eigenvalue	Difference	Proportion	Cumulative
1	0.500822	0.441441	0.139119	0.250823	0.3348	0.2623	0.8219	0.8219
2	0.260061	.	0.173136	0.067632	0.0725		0.1781	1.0000

	Test of H0: The canonical correlations in the current row and all that follow are zero				
	Likelihood Ratio	Approximate F Value	Num DF	Den DF	Pr > F
1	0.69850905	2.55	4	52	0.0497
2	0.93236803	1.96	1	27	0.1731

SAS OUTPUT 16.7

The next part of the output shows the significance tests of the variates. These show the significance of both variates (row 1), and the significance after the first variate has been removed (row 2). So, effectively we test the model as a whole, and then peel away variates one at a time to see whether what's left is significant. In this case with only two variates we get only two steps: the whole model, and then the model after the first variate is removed (which leaves only the second variate). When both variates are tested in combination the likelihood ratio has the same value (0.699), degrees of freedom (4), approximate F and significance value (.05) as in the Wilks's lambda in the MANOVA (see SAS Output 16.2). The important point to note from this table is that the two variates significantly discriminate the groups in combination ($p = .05$), but the second variate alone is non-significant, $p = .173$. Therefore, the group differences shown by the MANOVA can be explained in terms of *two* underlying dimensions in combination.

SAS OUTPUT 16.8

Pooled Within Canonical Structure			
Variable	Label	Can1	Can2
THOUGHTS	Number of obsession-related thoughts	–0.576045	0.817418
ACTIONS	Number of obsession-related behaviours	0.711304	0.702885

Pooled Within-Class Standardized Canonical Coefficients			
Variable	Label	Can1	Can2
THOUGHTS	Number of obsession-related thoughts	–0.7126297132	0.7211649822
ACTIONS	Number of obsession-related behaviours	0.8287506239	0.5840312004

The tables in SAS Output 16.8 are the most important for interpretation. The first table shows the pooled within canonical structure. The values in this matrix are the canonical variate correlation coefficients. These values are comparable to factor loadings and indicate the substantive nature of the variates (see Chapter 17). Bargman (1970) argues that when some dependent variables have high canonical variate correlations while others have low ones, then the ones with high correlations contribute most to group separation. As such they represent the relative contribution of each dependent variable to group separation (see Bray and Maxwell, 1985, pp. 42–45). The second table shows the standardized discriminant function coefficients for the two variates. These values are standardized versions of the values in the eigenvectors calculated in section 16.4.4.1. Recall that if the variates can be expressed in terms of a linear regression equation (see equation (16.4)), the standardized discriminant function coefficients are equivalent to the standardized betas in regression. Hence, the coefficients in these tables tell us the relative contribution of each variable to the variates.

Looking at the first canonical variate, labelled Can1, thoughts and behaviours have the opposite effect (behaviour has a positive relationship with this variate whereas thoughts have a negative relationship). Given that these values (in both tables) can vary between 1 and −1, we can also see that both relationships are strong (although behaviours have slightly larger contribution to the first variate). The first variate, then, could be seen as one that differentiates thoughts and behaviours (it affects thoughts and behaviours in the opposite way). Both thoughts and behaviours have a strong positive relationship with the second variate. This tells us that this variate represents something that affects thoughts and behaviours in a similar way. Remembering that ultimately these variates are used to differentiate groups, we could say that the first variate differentiates groups by some factor that affects thoughts and behaviours differently, whereas the second variate differentiates groups on some dimension that affects thoughts and behaviours in the same way.

SAS OUTPUT 16.9

Raw Canonical Coefficients			
Variable	Label	Can1	Can2
THOUGHTS	Number of obsession-related thoughts	−0.3352477665	0.3392630774
ACTIONS	Number of obsession-related behaviours	0.6030046517	0.4249451167

Class Means on Canonical Variables		
group	Can1	Can2
BT	−0.7260023874	−0.1279455457
CBT	0.6010491743	−0.2286849449
NT	0.1249532131	0.3566304906

SAS Output 16.9 tells us first the canonical discriminant function coefficients, which are the unstandardized versions of the standardized coefficients described above. These values are the values of *b* in equation (16.4) and you'll notice that these values correspond to the values in the eigenvectors derived in section 16.4.4.1 and used in equation (16.5). The values are less useful than the standardized versions, but do demonstrate from where the standardized versions come.

The class means are simply the mean variate scores for each group. For interpretation we should look at the sign of the class means (positive or negative). The tabulated values of the class means tell us that canonical variate 1 discriminates the BT group from the CBT (look at the difference between them on Can1, it's about 1.3). The second variate differentiates the no-treatment group from the two interventions (the differences on Can2 – the two treatment groups differ from each other by about 0.1, but they differ from the control group by about 0.50 and 0.60), but this difference is not as dramatic as for the first variate. Remember that the variates significantly discriminate the groups in combination (i.e. when both are considered).

CRAMMING SAM'S TIPS

- Discriminant function analysis (DFA) can be used after MANOVA to see how the dependent variables discriminate the groups.
- DFA identifies variates (combinations of the dependent variables) and to find out how many variates are significant look at the *F*-ratio: if the degrees of freedom and significance value is less than .05 then the variate is significantly discriminating the groups.
- Once the significant variates have been identified, use the table labelled Pooled Within-Class **Standardized Canonical Coefficients** to find out how the dependent variables contribute to the variates. High scores indicate that a dependent variable is important for a variate, and variables with positive and negative coefficients are contributing to the variate in opposite ways.
- Finally, to find out which groups are discriminated by a variate look at the table labelled **Class Means on Canonical Variates:** for a given variate, groups with values opposite in sign are being discriminated by that variate.

16.12. Reporting results from discriminant function analysis ②

The guiding principle (for the APA, whose guidelines, as a psychologist, are the ones that I try to follow) in presenting data is to give the readers enough information to be able to judge for themselves what your data mean. The APA does not have specific guidelines for what needs to be reported for discriminant function analysis. Personally, I would suggest reporting percentage of variance explained (which gives the readers the same information as the eigenvalue but in a more palatable form) and the squared canonical correlation for each variate (this is the appropriate effect size measure for discriminant analysis). I would also report the chi-square significance tests of the variates. All of these values can be found in SAS Output 16.7 (although remember to square the canonical correlation). It is probably also useful to quote the values in the structure matrix in SAS Output 16.8 (which will tell the reader about how the outcome variables relate to the underlying variates). We could, therefore, write something like this:

✓ The MANOVA was followed up with discriminant analysis, which revealed two discriminant functions. The first explained 82.2% of the variance, canonical $R^2 = .25$, whereas the second explained only 17.8%, canonical $R^2 = .07$. In combination these discriminant functions significantly differentiated the treatment groups, $\Lambda = 0.70$, $\chi^2(4) = 9.51$, $p = .05$, but removing the first function indicated that the second function did not significantly differentiate the treatment groups, $\Lambda = 0.93$, $\chi^2(1) = 1.86$, $p > .05$. The correlations between outcomes and the discriminant functions revealed that obsessive behaviours loaded fairly evenly highly onto both functions ($r = .71$ for the first function and $r = .70$ for the second); obsessive thoughts loaded more highly on the second function ($r = .82$) than the first function ($r = -.58$). The discriminant function plot showed that the first function discriminated the BT group from the CBT group, and the second function differentiated the no-treatment group from the two interventions.

16.13. Some final remarks ④

16.13.1. The final interpretation ④

So far we have gathered an awful lot of information about our data, but how can we bring all of it together to answer our research question: can therapy improve OCD and if so which therapy is best? Well, the MANOVA tells us that therapy can have a significant effect on OCD symptoms, but the non-significant univariate ANOVAs suggested that this improvement is not simply in terms of either thoughts or behaviours. The discriminant analysis suggests that the group separation can be best explained in terms of one underlying dimension. In this context the dimension is likely to be OCD itself (which we can realistically presume is made up of both thoughts and behaviours). So, therapy doesn't necessarily change behaviours or thoughts *per se*, but it does influence the underlying dimension of OCD. So, the answer to the first question seems to be: yes, therapy can influence OCD, but the nature of this influence is unclear.

The next question is more complex: which therapy is best? Figure 16.3 shows graphs of the relationships between the dependent variables and the group means of the original data. The graph of the means shows that for actions, BT reduces the number of obsessive behaviours, whereas CBT and NT do not. For thoughts, CBT reduces the number of obsessive thoughts, whereas BT and NT do not (check the pattern of the bars). Looking now at the relationships between thoughts and actions, in the BT group there is a positive relationship between thoughts and actions, so the more obsessive thoughts a person has, the more obsessive behaviours they carry out. In the CBT group there is no relationship at all (thoughts and actions vary quite independently). In the no-treatment group there is a negative (and non-significant incidentally) relationship between thoughts and actions.

What we have discovered from the discriminant analysis is that BT and CBT can be differentiated from the control group based on variate 2, a variate that has a similar effect on both thoughts and behaviours. We could say then that BT and CBT are both better than a no-treatment group at changing obsessive thoughts and behaviours. We also discovered that BT and CBT could be distinguished by variate 1, a variate that had the opposite effects on thoughts and behaviours. Combining this information with that in Figure 16.3, we could conclude that BT is better at changing behaviours and CBT is better at changing thoughts. So, the NT group can be distinguished from the CBT and BT graphs using a variable that affects both thoughts and behaviours. Also, the CBT and BT groups can be distinguished by a variate that has opposite effects on thoughts and behaviours. So, some therapy is better than none, but the choice of CBT or BT depends on whether you think it's more important to target thoughts (CBT) or behaviours (BT).

16.13.2. Univariate ANOVA or discriminant function analysis? ④

This example should have made clear that univariate ANOVA and discriminant function analysis are ways of answering different questions arising from a significant MANOVA. If univariate ANOVAs are chosen, Bonferroni corrections should be applied to the level at which you accept significance. The truth is that you should run both analyses to get a full picture of what is happening in your data. The advantage of discriminant function analysis is that it tells you something about the underlying dimensions within your data (which is especially useful if you have employed several dependent measures in an attempt to capture some social or psychological construct). Even if univariate ANOVAs are significant, the discriminant function analysis provides useful insight into your data and should be used. I hope that this chapter will convince you of this recommendation!

16.14. What to do when assumptions are violated in MANOVA ④

SAS doesn't offer a non-parametric version of MANOVA; however, some ideas have been put forward based on ranked data (much like the non-parametric tests we saw in

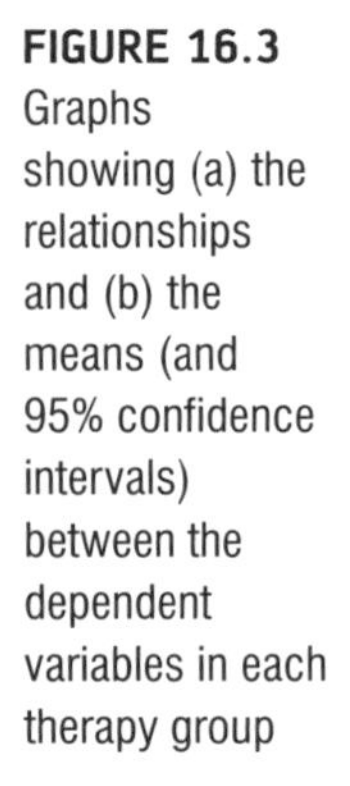
FIGURE 16.3 Graphs showing (a) the relationships and (b) the means (and 95% confidence intervals) between the dependent variables in each therapy group

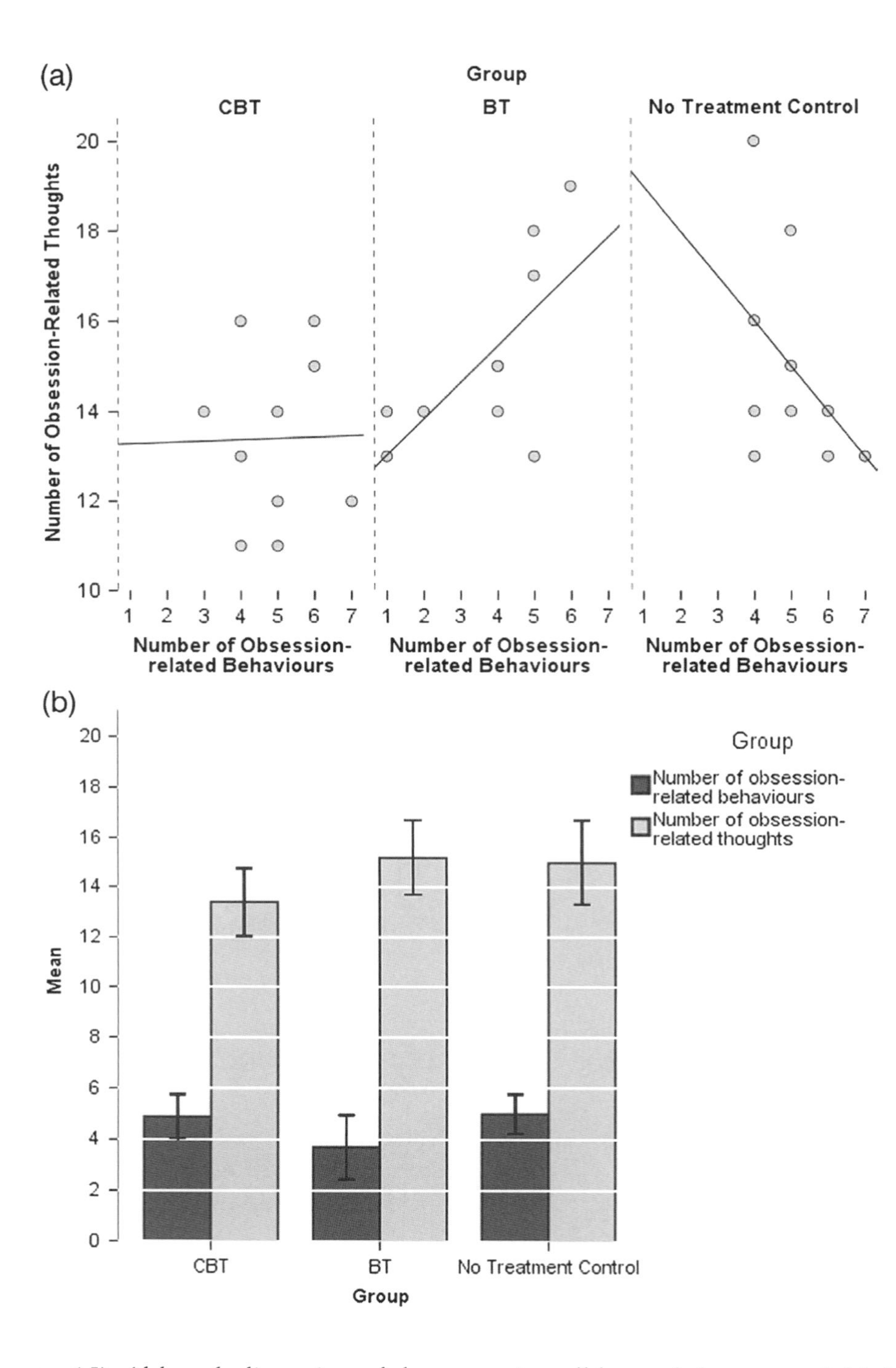

Chapter 15). Although discussion of these tests is well beyond the scope of this book, there are some techniques that can be beneficial when multivariate normality or homogeneity of covariance matrices cannot be assumed (Zwick, 1985). In addition, there are robust methods (see section 5.7.4) described by Wilcox (2005) for fairly straightforward designs with multiple outcome variables (for example, the Munzel–Brunner method). R can be used to run Wilcox's files.

What have I discovered about statistics? ④

In this chapter we've cackled maniacally in the ear of MANOVA, force-fed discriminant function analysis cod-liver oil, and discovered to our horror that Roy has a great root. There are sometimes situations in which several outcomes have been measured in different groups and we discovered that in these situations the ANOVA technique can be extended and is called MANOVA (multivariate analysis of variance). The reasons for using this technique rather than running lots of ANOVAs is that we retain control over the Type I error rate, and we can incorporate the relationships between outcome variables into the analysis. Some of you will have then discovered that MANOVA works in very similar ways to ANOVA, but just with matrices rather than single values. Others will have discovered that it's best to ignore the theory sections of this book. We had a look at an example of MANOVA on SAS and discovered that just to make life as confusing as possible you get four test statistics relating to the same effect! Of these, I tried to convince you that Pillai's trace was the safest option. Finally, we had a look at the two options for following up MANOVA: running lots of ANOVAs, or discriminant function analysis. Of these, discriminant function analysis gives us the most information, but can be a bit of nightmare to interpret.

We also discovered that pets can be therapeutic. I left the whereabouts of Fuzzy a mystery. Now admit it, how many of you thought he was dead? He's not: he is lying next to me as I type this sentence. After frantically searching the house I went back to the room that he had vanished from to check again whether there was a hole that he could have wriggled through. As I scuttled around on my hands and knees tapping the walls, a little ginger (and sooty) face popped out from the fireplace with a look as if to say 'have you lost something?' (see the picture). Yep, freaked out by the whole moving experience, he had done the only sensible thing and hidden up the chimney. Cats, you gotta love 'em.

Key terms that I've discovered

Bartlett's test of sphericity
Box's test
Discriminant function analysis (*DFA*)
Discriminant function variates
Discriminant scores
Error SSCP (*E*)
HE^{-1}
Homogeneity of covariance matrices
Hotelling–Lawley trace (T^2)
Hypothesis SSCP (*H*)
Identity matrix
Matrix
Multivariate
Multivariate analysis of variance (*MANOVA*)
Multivariate normality
Pillai–Bartlett trace (*V*)
Roy's greatest root
Square matrix
Sum of squares and cross-products matrix (*SSCP*)
Total SSCP (*T*)
Univariate
Variance–covariance matrix
Wilks's lambda (Λ)

Smart Alex's tasks

- **Task 1**: A clinical psychologist noticed that several of his manic psychotic patients did chicken impersonations in public. He wondered whether this behaviour could be used to diagnose this disorder and so decided to compare his patients against a normal sample. He observed 10 of his patients as they went through a normal day. He also needed to observe 10 of the most normal people he could find: naturally he chose to observe lecturers at the University of Sussex. He measured all participants using two dependent variables: first, how many chicken impersonations they did in the streets of Brighton over the course of a day, and, second, how good their impersonations were (as scored out of 10 by an independent farmyard noise expert). The data are in the file **chicken.sas7bdat**. Use MANOVA and DFA to find out whether these variables could be used to distinguish manic psychotic patients from those without the disorder. ③
- **Task 2**: I was interested in whether students' knowledge of different aspects of psychology improved throughout their degree. I took a sample of first years, second years and third years and gave them five tests (scored out of 15) representing different aspects of psychology: **Exper** (experimental psychology such as cognitive and neuropsychology, etc.); **Stats** (statistics); **Social** (social psychology); **Develop** (developmental psychology); **Person** (personality). Your task is to: (1) carry out an appropriate general analysis to determine whether there are overall group differences along these five measures; (2) look at the scale-by-scale analyses of group differences produced in the output and interpret the results accordingly; (3) select contrasts that test the hypothesis that second and third years will score higher than first years on all scales; (4) select tests that compare all groups to each other and briefly compare these results with the contrasts; and (5) carry out a separate analysis in which you test whether a combination of the measures can successfully discriminate the groups (comment only briefly on this analysis). Include only those scales that revealed group differences for the contrasts. How do the results help you to explain the findings of your initial analysis? The data are in the file **psychology.sas7bdat**. ④

Answers can be found on the companion website.

Further reading

Bray, J. H., & Maxwell, S. E. (1985). *Multivariate analysis of variance*. Sage University Paper Series on Quantitative Applications in the Social Sciences, 07-054. Newbury Park, CA: Sage.) (This monograph on MANOVA is superb: I cannot recommend anything better.)

Huberty, C. J., & Morris, J. D. (1989). Multivariate analysis versus multiple univariate analysis. *Psychological Bulletin*, *105*, 302–308.

Interesting real research

Marzillier, S. L., & Davey, G. C. L. (2005). Anxiety and disgust: Evidence for a unidirectional relationship. *Cognition and Emotion*, *19*(5), 729–750.

Exploratory factor analysis 17

FIGURE 17.1 Me at Niagara Falls in 1998. I was in the middle of writing the first edition of this book at the time. Note how fresh faced I look

17.1. What will this chapter tell me? ①

I was a year or so into my Ph.D., and thanks to my initial terrible teaching experiences I had developed a bit of an obsession with over-preparing for classes. I wrote detailed handouts and started using funny examples. Through my girlfriend at the time I met Dan Wright (a psychologist, who was in my department but sadly moved recently to Florida). He had published a statistics book of his own and was helping his publishers to sign up new authors. On the basis that my handouts were quirky and that I was too young to realize that writing a textbook at the age of 23 was academic suicide (really, textbooks take a long time to write and they are not at all valued compared to research articles) I was duly signed up. The commissioning editor was a man constantly on the verge of spontaneously combusting with intellectual energy. He can start a philosophical debate about literally anything: should he ever be trapped in a elevator he will be compelled to attempt to penetrate the occupants' minds with probing arguments that the elevator doesn't exist, that they don't exist, and that their entrapment is an illusory construct generated by their erroneous beliefs in the physical world. Ultimately though, he'd still be a man trapped in an elevator (with several exhausted corpses). A combination of his unfaltering self-confidence, my fear of social interactions with people that I don't know, and my utter bemusement that anyone would want me to write a book made me incapable of saying anything sensible to him. Ever. He must have thought that he had signed up an imbecile. He was probably right. (I find him less intimidating since thinking up the elevator scenario.) The trouble with agreeing to write books is that you then have to write them. For the next two years or so I found myself trying to juggle my research, a lectureship at the University of London, and writing a book. Had I been writing a book on heavy metal it would have been fine because all of the information was moshing away in my memory waiting to stage-dive out. Sadly, however, I had agreed to write a book on something that I knew nothing about: statistics. I soon discovered that writing the book was like doing a **factor analysis**: in factor analysis we take a lot of information (variables) and SAS effortlessly reduces this mass of confusion to a simple message (fewer variables) that is easier to digest. SAS does this (sort of) by filtering out the bits of the information overload that we don't need to know about. It takes a few seconds. Similarly, my younger self took a mass of information about statistics that I didn't understand and filtered it down into a simple message that I *could* understand: I became a living, breathing factor analysis … except that, unlike SAS, it took me two years and some considerable effort.

17.2. When to use factor analysis ②

In the social sciences we are often trying to measure things that cannot directly be measured (so-called **latent variables**). For example, management researchers (or psychologists even) might be interested in measuring 'burnout', which is when someone who has been working very hard on a project (a book, for example) for a prolonged period of time suddenly finds themselves devoid of motivation, inspiration, and wants to repeatedly headbutt their computer screaming 'please Mike, unlock the door, let me out of the basement, I need to feel the soft warmth of sunlight on my skin'. You can't measure burnout directly: it has many facets. However, you can measure different aspects of burnout: you could get some idea of motivation, stress levels, whether the person has any new ideas and so on. Having done this, it would be helpful to know whether these differences really do reflect a single variable. Put another way, are these different variables driven by the same underlying variable? This chapter will look at factor analysis (and principal component analysis) – a technique for identifying groups or clusters of variables. This technique has three main uses: (1) to understand the structure of a set of variables (e.g. pioneers of intelligence such as Spearman and Thurstone used factor analysis to try to understand the structure of the latent variable 'intelligence'); (2) to construct a questionnaire to measure an underlying variable (e.g. you might design a questionnaire to measure burnout); and (3) to reduce a data set to a more manageable size while retaining as much of the original information as possible (for example, we saw in Chapter 7 that multicollinearity can be a problem in multiple regression, and factor analysis can be used to solve this problem by combining variables that are collinear). Through this chapter we'll discover what factors are, how we find them, and what they tell us (if anything) about the relationship between the variables we've measured.

17.3. Factors ②

If we measure several variables, or ask someone several questions about themselves, the correlation between each pair of variables (or questions) can be arranged in what's known as an *R-matrix*. An *R*-matrix is just a correlation matrix: a table of correlation coefficients between variables (in fact, we saw small versions of these matrices in Chapter 6). The diagonal elements of an *R*-matrix are all 1s because each variable will correlate perfectly with itself. The off-diagonal elements are the correlation coefficients between pairs of variables, or questions.[1] The existence of clusters of large correlation coefficients between subsets of variables suggests that those variables could be measuring aspects of the same underlying dimension. These underlying dimensions are known as factors (or *latent variables*). By reducing a data set from a group of interrelated variables in to a smaller set of factors, factor analysis achieves

[1] This matrix is called an *R*-matrix, or *R*, because it contains correlation coefficients and *r* usually denotes Pearson's correlation (see Chapter 6) – the *r* turns into a capital letter when it denotes a matrix.

parsimony by explaining the maximum amount of common variance in a correlation matrix using the smallest number of explanatory constructs.

There are numerous examples of the use of factor analysis in the social sciences. The trait theorists in psychology used factor analysis endlessly to assess personality traits. Most readers will be familiar with the extraversion–introversion and neuroticism traits measured by Eysenck (1953). Most other personality questionnaires are based on factor analysis – notably Cattell's (1966a) 16 personality factors questionnaire – and these inventories are frequently used for recruiting purposes in industry (and even by some religious groups). However, although factor analysis is probably most famous for being adopted by psychologists, its use is by no means restricted to measuring dimensions of personality. Economists, for example, might use factor analysis to see whether productivity, profits and workforce can be reduced down to an underlying dimension of company growth, and Jeremy Miles told me of a biochemist who used it to analyse urine samples!

Let's put some of these ideas into practice by imagining that we wanted to measure different aspects of what might make a person popular. We could administer several measures that we believe tap different aspects of popularity. So, we might measure a person's social skills (Social Skills), their selfishness (Selfish), how interesting others find them (Interest), the proportion of time they spend talking about the other person during a conversation (Talk 1), the proportion of time they spend talking about themselves (Talk 2), and their propensity to lie to people (the Liar scale). We can then calculate the correlation coefficients for each pair of variables and create an *R*-matrix. Table 17.1 shows this matrix. Any significant correlation coefficients are shown in bold type. It is clear that there are two clusters of interrelating variables. Therefore, these variables might be measuring some common underlying dimension. The amount that someone talks about the other person during a conversation seems to correlate highly with both the level of social skills and how interesting the other finds that person. Also, social skills correlate well with how interesting others perceive a person to be. These relationships indicate that the better your social skills, the more interesting and talkative you are likely to be. However, there is a second cluster of variables. The amount that people talk about themselves within a conversation correlates with how selfish they are and how much they lie. Being selfish also correlates with the degree to which a person tells lies. In short, selfish people are likely to lie and talk about themselves.

In factor analysis we strive to reduce this *R*-matrix down in to its underlying dimensions by looking at which variables seem to cluster together in a meaningful way. This data reduction is achieved by looking for variables that correlate highly with a group of other variables, but do not correlate with variables outside of that group. In this example, there appear to be two clusters that fit the bill. The first factor seems to relate to general sociability, whereas the second factor seems to relate to the way in which a person treats others socially (we might call it consideration). It might, therefore, be assumed that popularity depends not only on your ability to socialize, but also on whether you are genuine towards others.

TABLE 17.1 An *r*-matrix

	Talk 1	*Social Skills*	*Interest*	*Talk 2*	*Selfish*	*Liar*
Talk 1	1.000					
Social Skills	Factor 2 .772	1.000				
Interest	**.646**	**.879**	1.000			
Talk 2	.074	−.120	.054	1.000		
Selfish	−.131	.031	−.101	Factor 2 .441	1.000	
Liar	.068	.012	.110	**.361**	**.277**	1.000

17.3.1. Graphical representation of factors ②

Factors (not to be confused with independent variables in factorial ANOVA) are statistical entities that can be visualized as classification axes along which measurement variables can be plotted. In plain English, this statement means that if you imagine factors as being the axis of a graph, then we can plot variables along these axes. The coordinates of variables along each axis represent the strength of relationship between that variable and each factor. Figure 17.2 shows such a plot for the popularity data (in which there were only two factors). The first thing to notice is that for both factors, the axis goes from –1 to 1, which are the outer limits of a correlation coefficient. Therefore, the position of a given variable depends on its correlation to the two factors. The circles represent the three variables that correlate highly with factor 1 (sociability: horizontal axis) but have a low correlation with factor 2 (consideration: vertical axis). Conversely, the triangles represent variables that correlate highly with consideration to others but have a low correlation to sociability. From this plot, we can tell that selfishness, the amount a person talks about themselves and their propensity to lie all contribute to a factor which could be called consideration of others. Conversely, how much a person takes an interest in other people, how interesting they are and their level of social skills contribute to a second factor, sociability. This diagram therefore supports the structure that was apparent in the *R*-matrix. Of course, if a third factor existed within these data it could be represented by a third axis (creating a 3-D graph). It should also be apparent that if more than three factors exist in a data set, then they cannot all be represented by a 2-D graph.

If each axis on the graph represents a factor, then the variables that go to make up a factor can be plotted according to the extent to which they relate to a given factor. The coordinates of a variable, therefore, represent its relationship to the factors. In an ideal world a variable should have a large coordinate for one of the axes, and low coordinates for any other factors. This scenario would indicate that this particular variable related to only one factor. Variables that have large coordinates on the same axis are assumed to measure different aspects of some common underlying dimension. The coordinate of a variable along

FIGURE 17.2 Example of a factor plot

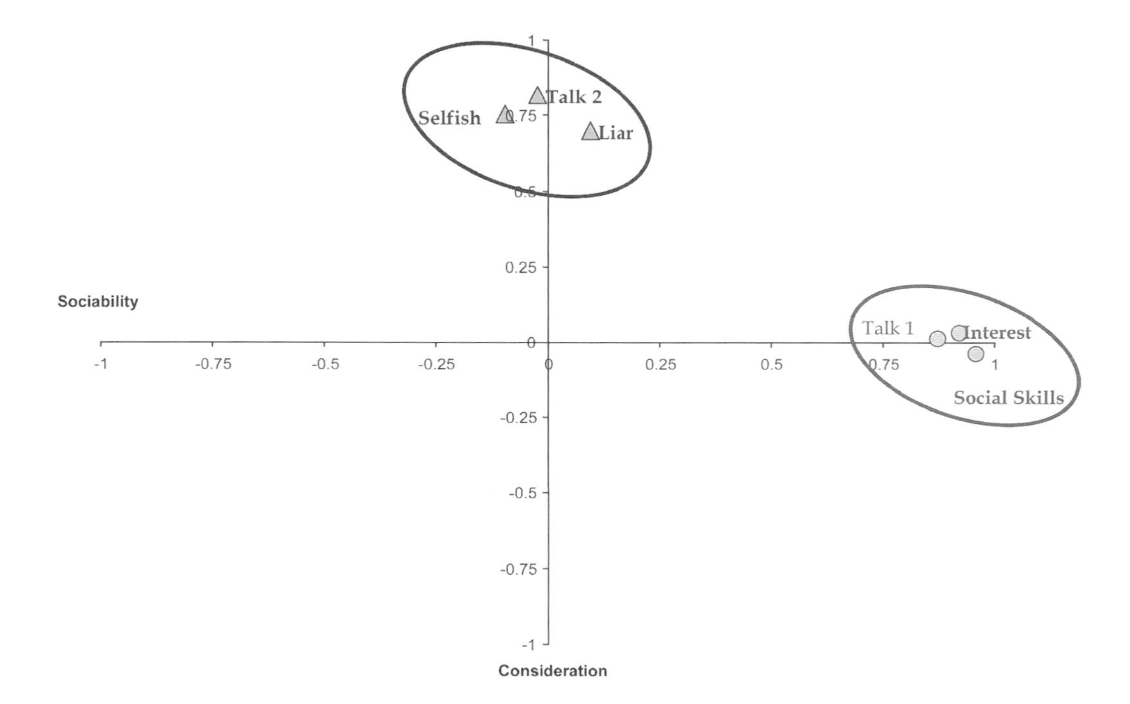

a classification axis is known as a **factor loading.** The factor loading can be thought of as the Pearson correlation between a factor and a variable (see Jane Superbrain Box 17.1). From what we know about interpreting correlation coefficients (see section 6.5.2.3) it should be clear that if we square the factor loading we obtain a measure of the substantive importance of a particular variable to a factor.

JANE SUPERBRAIN 17.1

What's the difference between a pattern matrix and a structure matrix? ③

Throughout my discussion of factor loadings I've been quite vague. Sometimes I've said that these loadings can be thought of as the correlation between a variable and a given factor, then at other times I've described these loadings in terms of regression coefficients (b). Now, it should be obvious from what we discovered in Chapters 6 and 7 that correlation coefficients and regression coefficients are quite different things, so what the hell am I going on about: shouldn't I make up my mind what the factor loadings actually are?

Well, in vague terms (the best terms for my brain) both correlation coefficients and regression coefficients represent the relationship between a variable and linear model in a broad sense, so the key take-home message is that factor loadings tell us about the relative contribution that a variable makes to a factor. As long as you understand that much, you have no problems.

However, the factor loadings in a given analysis can be both correlation coefficients and regression coefficients. In a few sections' time we'll discover that the interpretation of factor analysis is helped greatly by a technique known as *rotation*. Without going into details, there are two types: orthogonal and oblique rotation (see section 17.4.6). When orthogonal rotation is used, any underlying factors are assumed to be independent, and the factor loading *is* the correlation between the factor and the variable, but is also the regression coefficient. Put another way, the values of the correlation coefficients are the same as the values of the regression coefficients. However, there are situations in which the underlying factors are assumed to be related or correlated to each other. In these situations, oblique rotation is used and the resulting correlations between variables and factors will differ from the corresponding regression coefficients. In this case, there are, in effect, two different sets of factor loadings: the correlation coefficients between each variable and factor (which are put in the *factor* **structure matrix**) and the regression coefficients for each variable on each factor (which are put in the *factor* **pattern matrix**). These coefficients can have quite different interpretations (see Graham, Guthrie, & Thompson, 2003).

17.3.2. Mathematical representation of factors ②

The axes drawn in Figure 17.2 are straight lines and so can be described mathematically by the equation of a straight line. Therefore, factors can also be described in terms of this equation.

SELF-TEST What is the equation of a straight line?

Equation (17.1) reminds us of the equation describing a linear model and then applies this to the scenario of describing a factor.

$$
\begin{aligned}
Y_i &= b_1X_{1i} + b_2X_{2i} + \cdots + b_nX_{ni} + \varepsilon_i \\
\text{Factor}_i &= b_1\text{Variable}_{1i} + b_2\text{Variable}_{2i} + \cdots + b_n\text{Variable}_{ni} + \varepsilon_i
\end{aligned}
\tag{17.1}
$$

You'll notice that there is no intercept in the equation, the reason being that the lines intersect at zero (hence the intercept is also zero). The *b*s in the equation represent the factor loadings.

Sticking with our example of popularity, we found that there were two factors underlying this construct: general sociability and consideration. We can, therefore, construct an equation that describes each factor in terms of the variables that have been measured. The equations are as follows:

$$
\begin{aligned}
Y_i &= b_1X_{1i} + b_2X_{2i} + \cdots + b_nX_{ni} + \varepsilon_i \\
\text{Sociability}_i &= b_1\text{Talk 1}_i + b_2\text{ Social Skills}_i + b_3\text{Interest}_i \\
&\quad + b_4\text{Talk 2}_i + b_5\text{Selfish}_i + b_6\text{Liar}_i + \varepsilon_i \\
\text{Consideration}_i &= b_1\text{Talk 1}_i + b_2\text{ Social Skills}_i + b_3\text{Interest}_i \\
&\quad + b_4\text{Talk 2}_i + b_5\text{Selfish}_i + b_6\text{Liar}_i + \varepsilon_i
\end{aligned}
\tag{17.2}
$$

First, notice that the equations are identical in form: they both include all of the variables that were measures. However, the values of *b* in the two equations will be different (depending on the relative importance of each variable to the particular factor). In fact, we can replace each value of *b* with the coordinate of that variable on the graph in Figure 17.2 (i.e. replace the values of *b* with the factor loading). The resulting equations are as follows:

$$
\begin{aligned}
Y_i &= b_1X_{1i} + b_2X_{2i} + \cdots + b_nX_{ni} + \varepsilon_i \\
\text{Sociability}_i &= 0.87\text{Talk 1}_i + 0.96\text{Social Skills}_i + 0.92\text{Interest}_i \\
&\quad + 0.00\text{Talk 2}_i - 0.10\text{Selfish}_i + 0.09\text{Liar}_i + \varepsilon_i \\
\text{Consideration}_i &= 0.01\text{Talk 1}_i - 0.03\text{Social Skills}_i + 0.04\text{Interest}_i \\
&\quad + 0.82\text{Talk 2}_i + 0.75\text{Selfish}_i + 0.70\text{Liar}_i + \varepsilon_i
\end{aligned}
\tag{17.3}
$$

Notice that, for the sociability factor, the values of *b* are high for Talk 1, Social Skills and Interest. For the remaining variables (Talk 2, Selfish and Liar) the values of *b* are very low (close to 0). This tells us that three of the variables are very important for that factor (the ones with high values of *b*) and three are very unimportant (the ones with low values of *b*). We saw that this point is true because of the way that three variables clustered highly on the factor plot. The point to take on board here is that the factor plot and these equations represent the same thing: the factor loadings in the plot are simply the *b*-values in these equations (but see Jane Superbrain Box 17.1). For the second factor, consideration to others, the opposite pattern can be seen in that Talk 2, Selfish and Liar all have high values of *b* whereas the remaining three variables have *b*-values close to 0. In an ideal world, variables would have very high *b*-values for one factor and very low *b*-values for all other factors.

These factor loadings can be placed in a matrix in which the columns represent each factor and the rows represent the loadings of each variable on each factor. For the popularity data this matrix would have two columns (one for each factor) and six rows (one for each variable). This matrix, usually denoted A, is given below. To understand what the matrix means, try relating the elements to the loadings in equation (17.3). For example, the top row represents the first variable, Talk 1, which had a loading of .87 for the first factor (sociability) and a loading of .01 for the second factor (consideration). This matrix is called the **factor matrix** or **component matrix** (if doing **principal component analysis**) – see Jane Superbrain Box 17.1 to find out about the different forms of this matrix:

$$A = \begin{pmatrix} 0.87 & 0.01 \\ 0.96 & -0.03 \\ 0.92 & 0.04 \\ 0.00 & 0.82 \\ -0.10 & 0.75 \\ 0.09 & 0.70 \end{pmatrix}$$

The major assumption in factor analysis is that these algebraic factors represent real-world dimensions, the nature of which must be *guessed at* by inspecting which variables have high loads on the same factor. So, psychologists might believe that factors represent dimensions of the psyche, education researchers might believe they represent abilities, and sociologists might believe they represent races or social classes. However, it is an extremely contentious point whether this assumption is tenable and some believe that the dimensions derived from factor analysis are real only in the statistical sense – and are real-world fictions.

17.3.3. Factor scores ②

A factor can be described in terms of the variables measured and their relative importance for that factor (represented by the value of b). Therefore, having discovered which factors exist, and estimated the equation that describes them, it should be possible to also estimate a person's score on a factor, based on their scores for the constituent variables. As such, if we wanted to derive a score of sociability for a particular person, we could place their scores on the various measures into equation (17.3). This method is known as a *weighted average*. In fact, this method is overly simplistic and rarely used, but it is probably the easiest way to explain the principle. For example, imagine the six scales all range from 1 to 10 and that someone scored the following: Talk 1 (4), Social Skills (9), Interest (8), Talk 2 (6), Selfish (8), and Liar (6). We could replace these values into equation (17.3) to get a score for this person's sociability and their consideration to others (see equation (17.3). The resulting scores of 19.22 and 15.21 reflect the degree to which this person is sociable and their consideration to others respectively. This person scores higher on sociability than inconsideration.

$$\begin{aligned}
\text{Sociability} &= 0.87\text{Talk 1} + 0.96\text{Social Skills} + 0.92\text{Interest} + 0.00\text{Talk 2} \\
&\quad - 0.10\text{Selfish} + 0.09\text{Liar} \\
&= (0.87 \times 4) + (0.96 \times 9) + (0.92 \times 8) + (0.00 \times 6) - (0.10 \times 8) + (0.09 \times 6) \\
&= 19.22 \\
\text{Consideration} &= 0.01\text{Talk 1} - 0.03\text{Social Skills} + 0.04\text{Interest} + 0.82\text{Talk 2} \\
&\quad + 0.75\text{Selfish} + 0.70\text{Liar} \\
&= (0.01 \times 4) - (0.03 \times 9) + (0.04 \times 8) + (0.82 \times 6) + (0.75 \times 8) + (0.70 \times 6) \\
&= 15.21
\end{aligned} \tag{17.4}$$

There are several uses of factor scores. First, if the purpose of the factor analysis is to reduce a large set of data to a smaller subset of measurement variables, then the factor scores tell us an individual's score on this subset of measures. Therefore, any further analysis can be carried out on the factor scores rather than the original data. For example, we could carry out a *t*-test to see whether females are significantly more sociable than males using the factor scores for sociability. A second use is in overcoming collinearity problems in regression. If, following a multiple regression analysis, we have identified sources of multicollinearity then the interpretation of the analysis is questioned (see section 7.6.2.3). In this situation, we can simply carry out a factor analysis on the predictor variables to reduce them down to a subset of uncorrelated factors. The variables causing the multicollinearity will combine to form a factor. If we then rerun the regression but using the factor scores as predictor variables then the problem of multicollinearity should vanish (because the variables are now combined into a single factor). There are ways in which we can ensure that the factors are uncorrelated. By using uncorrelated factor scores as predictors in the regression we can be confident that there will be no correlation between predictors: hence, no multicollinearity!

17.4. Discovering factors ②

By now, you should have some grasp of the concept of what a factor is, how it is represented graphically, how it is represented algebraically, and how we can calculate composite scores representing an individual's 'performance' on a single factor. I have deliberately restricted the discussion to a conceptual level, without delving into how we actually find these mythical beasts known as factors. This section will look at how we find factors. Specifically we will examine different types of method, look at the maths behind one method (principal components), investigate the criteria for determining whether factors are important, and discover how to improve the interpretation of a given solution.

17.4.1. Choosing a method ②

The first thing you need to know is that there are several methods for unearthing factors in your data. The method you choose will depend on what you hope to do with the analysis. Tinsley and Tinsley (1987) give an excellent account of the different methods available. There are two things to consider: whether you want to generalize the findings from your sample to a population and whether you are exploring your data or testing a specific hypothesis. This chapter describes techniques for exploring data using factor analysis. Testing hypotheses about the structures of latent variables and their relationships to each other requires PROC CALIS or PROC TCALIS (TCALIS is an update of CALIS). Those interested in hypothesis testing techniques (known as **confirmatory factor analysis**) are advised to read Pedhazur and Schmelkin (1991: Chapter 23) for an introduction. Assuming we want to explore our data, we then need to consider whether we want to apply our findings to the sample collected (descriptive method) or to generalize our findings to a population (inferential methods). When factor analysis was originally developed it was assumed that it would be used to explore data to generate future hypotheses. As such, it was assumed that the technique would be applied to the entire population of interest. Therefore, certain techniques assume that the sample used is the population, and so results cannot be extrapolated beyond that particular sample. Principal component analysis is an example of one of these techniques, as are principal factors analysis (*principal axis factoring*) and image covariance analysis (*image factoring*). Of these, principal component analysis and principal factors analysis are the preferred methods and usually result in similar solutions (see section 17.4.3). When these methods are used, conclusions are restricted to the sample collected and generalization of the results can be achieved only if analysis using different samples reveals the same factor structure.

Another approach has been to assume that participants are randomly selected and that the variables measured constitute the population of variables in which we're interested. By assuming this, it is possible to develop techniques from which the results can be generalized from the sample participants to a larger population. However, a constraint is that any findings hold true only for the set of variables measured (because we've assumed this set constitutes the entire population of variables). Techniques in this category include the maximum-likelihood method (see Harman, 1976) and Kaiser's alpha factoring. The choice of method depends largely on what generalizations, if any, you want to make from your data.[2]

17.4.2. Communality ②

Before continuing it is important that you understand some basic things about the variance within an *R*-matrix. It is possible to calculate the variability in scores (the variance) for any given measure (or variable). You should be familiar with the idea of variance by now and comfortable with how it can be calculated (if not, see Chapter 2). The total variance for a particular variable will have two components: some of it will be shared with other variables or measures (**common variance**) and some of it will be specific to that measure (**unique variance**). We tend to use the term *unique variance* to refer to variance that can be reliably attributed to only one measure. However, there is also variance that is specific to one measure but not reliably so; this variance is called *error* or **random variance**. The proportion of common variance present in a variable is known as the **communality**. As such, a variable that has no specific variance (or random variance) would have a communality of 1; a variable that shares none of its variance with any other variable would have a communality of 0.

In factor analysis we are interested in finding common underlying dimensions within the data and so we are primarily interested only in the common variance. Therefore, when

[2] It's worth noting at this point that principal component analysis is not in fact the same as factor analysis. This doesn't stop idiots like me from discussing them as though they are, but more on that later.

we run a factor analysis it is fundamental that we know how much of the variance present in our data is common variance. This presents us with a logical impasse: to do the factor analysis we need to know the proportion of common variance present in the data, yet the only way to find out the extent of the common variance is by carrying out a factor analysis! There are two ways to approach this problem. The first is to assume that all of the variance is common variance. As such, we assume that the communality of every variable is 1. By making this assumption we merely transpose our original data into constituent linear components (known as principal component analysis). The second approach is to estimate the amount of common variance by estimating communality values for each variable. There are various methods of estimating communalities but the most widely used (including **alpha factoring**) is to use the squared multiple correlation (SMC) of each variable with all others. So, for the popularity data, imagine you ran a multiple regression using one measure (Selfish) as the outcome and the other five measures as predictors: the resulting multiple R^2 (see section 7.5.2) would be used as an estimate of the communality for the variable Selfish. This second approach is used in factor analysis. These estimates allow the factor analysis to be done. Once the underlying factors have been extracted, new communalities can be calculated that represent the multiple correlation between each variable and the factors extracted. Therefore, the communality is a measure of the proportion of variance explained by the extracted factors.

17.4.3. Factor analysis vs. principal component analysis ②

I have just explained that there are two approaches to locating underlying dimensions of a data set: factor analysis and principal component analysis. These techniques differ in the communality estimates that are used. Simplistically, though, factor analysis derives a mathematical model from which factors are estimated, whereas principal component analysis merely decomposes the original data into a set of linear variates (see Dunteman, 1989: Chapter 8, for more detail on the differences between the procedures). As such, only factor analysis can estimate the underlying factors, and it relies on various assumptions for these estimates to be accurate. Principal component analysis is concerned only with establishing which linear components exist within the data and how a particular variable might contribute to that component. In terms of theory, this chapter is dedicated to principal component analysis rather than factor analysis. The reasons are that principal component analysis is a psychometrically sound procedure, it is conceptually less complex than factor analysis, and it bears numerous similarities to discriminant factor analysis (described in the previous chapter).

However, we should consider whether the techniques provide different solutions to the same problem. Based on an extensive literature review, Guadagnoli and Velicer (1988) concluded that the solutions generated from principal component analysis differ little from those derived from factor analytic techniques. In reality, there are some circumstances for which this statement is untrue. Stevens (2002) summarizes the evidence and concludes that with 30 or more variables and communalities greater than .7 for all variables, different solutions are unlikely; however, with fewer than 20 variables and any low communalities ($< .4$) differences can occur.

The flip-side of this argument is eloquently described by Cliff (1987) who observed that proponents of factor analysis 'insist that components analysis is at best a common factor analysis with some error added and at worst an unrecognizable hodgepodge of things from which nothing can be determined' (p. 349). Indeed, feeling is strong on this issue with some arguing that when principal component analysis is used it should not be described as a factor analysis and that you should not impute substantive meaning to the resulting components. However, to non-statisticians the difference between a principal component and a factor may be difficult to conceptualize (they are both linear models), and the differences arise largely from the calculation.[3]

17.4.4. Theory behind principal component analysis ③

Principal component analysis works in a very similar way to MANOVA and discriminant function analysis (see previous chapter). Although it isn't necessary to understand the mathematical principles in any detail, readers of the previous chapter may benefit from some comparisons between the two techniques. For those who haven't read that chapter, I suggest you flick through it before moving ahead!

In MANOVA, various sum of squares and cross-product matrices were calculated that contained information about the relationships between dependent variables. I mentioned before that these SSCP matrices could be easily converted to variance–covariance matrices, which represent the same information but in averaged form (i.e. taking account of the number of observations). I also said that by dividing each element by the relevant standard deviation the variance–covariance matrix becomes standardized. The result is a correlation matrix. In principal component analysis we usually deal with correlation matrices (although it is possible to analyse a variance–covariance matrix too) and the point to note is that this matrix pretty much represents the same information as an SSCP matrix in MANOVA. The difference is just that the correlation matrix is an averaged version of the SSCP that has been standardized.

In MANOVA, we used several SSCP matrices that represented different components of experimental variation (the model variation and the residual variation). In principal component analysis the covariance (or correlation) matrix cannot be broken down in this way (because all data come from the same group of participants). In MANOVA, we ended up looking at the variates or components of the SSCP matrix that represented the ratio of the model variance to the error variance. These variates were linear dimensions that separated the groups tested, and we saw that the dependent variables mapped onto these underlying components. In short, we looked at whether the groups could be separated by some linear combination of the dependent variables. These variates were found by calculating the eigenvectors of the SSCP. The number of variates obtained was the smaller of p (the number of dependent variables) and $k - 1$ (where k is the number of groups). In component analysis we do something similar (I'm simplifying things a little, but it will give you the basic idea). That is, we take a correlation matrix and calculate the variates. There are no groups of observations, and so the number of variates calculated will always equal the number of variables measured (p). The variates are described, as for MANOVA, by the eigenvectors associated with the correlation matrix. The elements of the eigenvectors are the weights of each variable on the variate (see equation (16.5)). These values are the factor loadings described earlier. The largest eigenvalue associated with each of the eigenvectors provides a single indicator of the substantive importance of each variate (or component). The basic idea is that we retain factors with relatively large eigenvalues and ignore those with relatively small eigenvalues.

In summary, component analysis works in a similar way to MANOVA. We begin with a matrix representing the relationships between variables. The linear components (also called variates, or factors) of that matrix are then calculated by determining the eigenvalues of the matrix. These eigenvalues are used to calculate eigenvectors, the elements of which

[3] For this reason I have used the terms *components* and *factors* interchangeably throughout this chapter. Although this use of terms will reduce some statisticians (and psychologists) to tears, I'm banking on these people not needing to read this book! I acknowledge the methodological differences, but I think it's easier for students if I dwell on the similarities between the techniques and not the differences.

provide the loading of a particular variable on a particular factor (i.e. they are the b-values in equation (17.1)). The eigenvalue is also a measure of the substantive importance of the eigenvector with which it is associated.

17.4.5. Factor extraction: eigenvalues and the scree plot ②

Not all factors are retained in an analysis, and there is debate over the criterion used to decide whether a factor is statistically important. I mentioned above that eigenvalues associated with a variate indicate the substantive importance of that factor. Therefore, it seems logical that we should retain only factors with large eigenvalues. How do we decide whether or not an eigenvalue is large enough to represent a meaningful factor? Well, one technique advocated by Cattell (1966b) is to plot a graph of each eigenvalue (Y-axis) against the factor with which it is associated (X-axis). This graph is known as a **scree plot** (because it looks like a rock face with a pile of debris, or scree, at the bottom). I mentioned earlier that it is possible to obtain as many factors as there are variables and that each has an associated eigenvalue. By graphing the eigenvalues, the relative importance of each

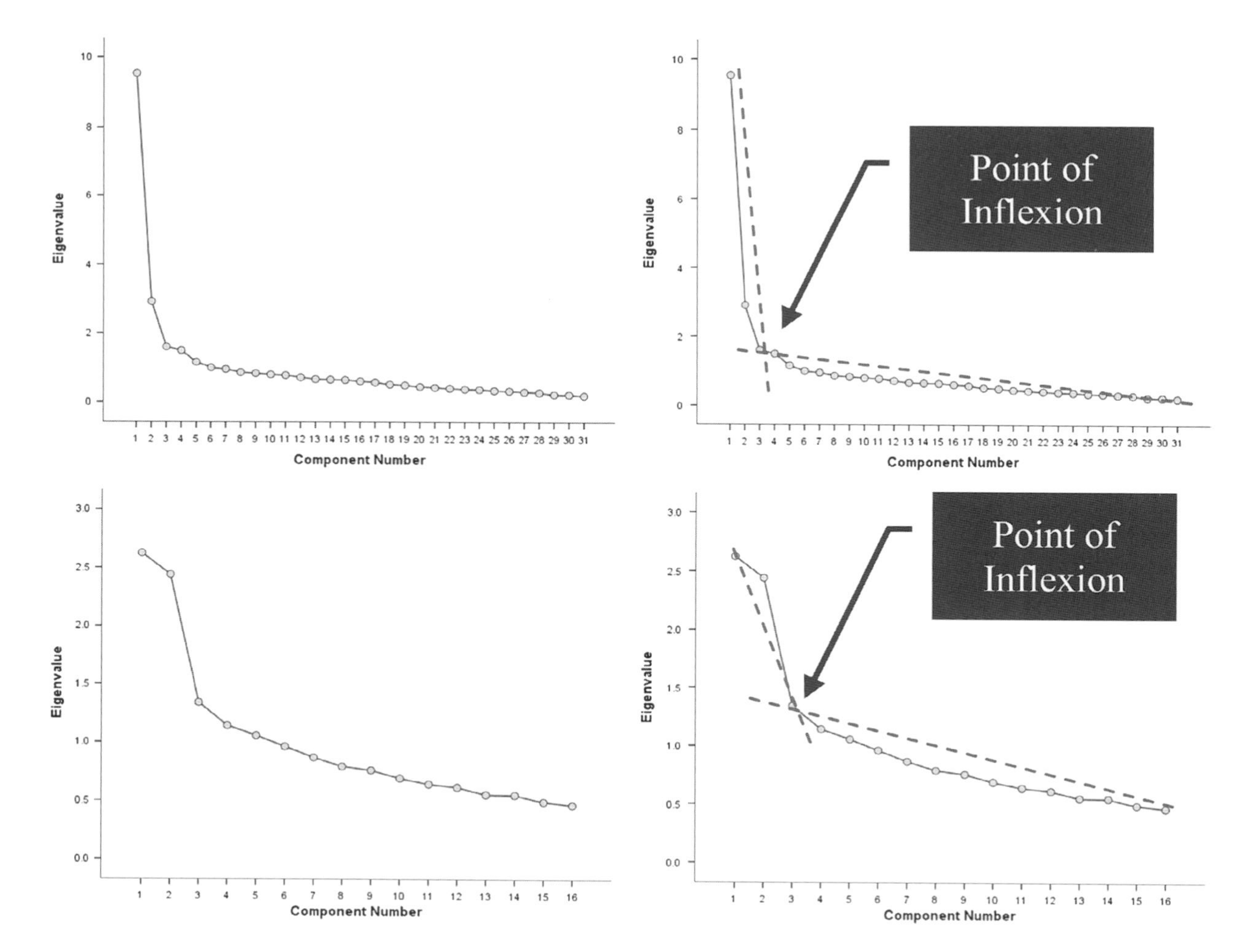

FIGURE 17.3 Examples of scree plots for data that probably have two underlying factors

factor becomes apparent. Typically there will be a few factors with quite high eigenvalues, and many factors with relatively low eigenvalues and so this graph has a very characteristic shape: there is a sharp descent in the curve followed by a tailing off (see Figure 17.3). Cattell (1966b) argued that the cut-off point for selecting factors should be at the point of inflexion of this curve. The point of inflexion is where the slope of the line changes dramatically: so, in Figure 17.3, imagine drawing a straight line that summarizes the vertical part of the plot and another that summarizes the horizontal part (the red dashed lines); then the point of inflexion is the data point at which these two lines meet. In both examples in Figure 17.3 the point of inflexion occurs at the third data point (factor); therefore, we would extract two factors. Thus, you retain (or extract) only factors to the left of the point of inflexion (and do not include the factor at the point of inflexion itself).[4] With a sample of more than 200 participants, the scree plot provides a fairly reliable criterion for factor selection (Stevens, 2002).

Although scree plots are very useful, factor selection should not be based on this criterion alone. Kaiser (1960) recommended retaining all factors with eigenvalues greater than 1. This criterion is based on the idea that the eigenvalues represent the amount of variation explained by a factor and that an eigenvalue of 1 represents a substantial amount of variation. Jolliffe (1972, 1986) reports that **Kaiser's criterion** is too strict and suggests the third option of retaining all factors with eigenvalues more than .7. The difference between how many factors are retained using Kaiser's methods and Jolliffe's can be dramatic.

You might well wonder how the methods compare. Generally speaking, Kaiser's criterion overestimates the number of factors to retain (see Jane Superbrain Box 17.2) but there is some evidence that it is accurate when the number of variables is less than 30 and the resulting communalities (after extraction) are all greater than .7. Kaiser's criterion can also be accurate when the sample size exceeds 250 and the average communality is greater than or equal to .6. In any other circumstances you are best advised to use a scree plot provided the sample size is greater than 200 (see Stevens, 2002, for more detail). By default, SAS uses Kaiser's criterion to extract factors. Therefore, if you use the scree plot to determine how many factors are retained you may have to rerun the analysis specifying that SAS extracts the number of factors you require.

However, as is often the case in statistics, the three criteria often provide different solutions! In these situations the communalities of the factors need to be considered. In principal component analysis we begin with communalities of 1 with all factors retained (because we assume that all variance is common variance). At this stage all we have done is to find the linear variates that exist in the data – so we have just transformed the data without discarding any information. However, to discover what common variance *really* exists between variables we must decide which factors are meaningful and discard any that are too trivial to consider. Therefore, we discard some information. The factors we retain will not explain all of the variance in the data (because we have discarded some information) and so the communalities after extraction will always be less than 1. The factors retained do not map perfectly onto the original variables – they merely reflect the common variance present in the data. If the communalities represent a loss of information then they are important statistics. The closer the communalities are to 1, the better our factors are at explaining the original data. It is logical that the greater the number of factors retained, the greater the communalities will be (because less information is discarded); therefore, the communalities are good indices of whether too few factors have been retained. In fact, with *generalized least-squares factor analysis* and *maximum-likelihood factor analysis* you can get a statistical measure of the goodness of fit of the factor solution (see the next chapter for more on goodness-of-fit tests). This basically measures the proportion of variance that the factor solution explains (so can be thought of as comparing communalities before and after extraction).

[4] Actually if you read Cattell's original paper he advised including the factor at the point of inflexion as well because it is 'desirable to include at least one common error factor as a "garbage can"'. The idea is that the point of inflexion represents an error factor. However, in practice this garbage can factor is rarely retained; also Thurstone argued that it is better to retain too few rather than too many factors so most people do *not* to retain the factor at the point of inflexion.

JANE SUPERBRAIN 17.2

How many factors do I retain? ③

The discussion of factor extraction in the text is somewhat simplified. In fact, there are fundamental problems with Kaiser's criterion (Nunnally & Bernstein, 1994). For one thing, an eigenvalue of 1 means different things in different analyses: with 100 variables it means that a factor explains 1% of the variance, but with 10 variables it means that a factor explains 10% of the variance. Clearly, these two situations are very different and a single rule that covers both is inappropriate. An eigenvalue of 1 also means only that the factor explains as much variance as a variable, which rather defeats the original intention of the analysis to reduce variables down to 'more substantive' underlying factors (Nunnally & Bernstein, 1994). Consequently, Kaiser's criterion often overestimates the number of factors. On this basis Jolliffe's criterion is even worse (a factor explains less variance than a variable!).

There are more complex ways to determine how many factors to retain, but they are not easy to do on SAS (which is why I'm discussing them outside of the main text). The best is probably parallel analysis (Horn, 1965). Essentially each eigenvalue (which represents the size of the factor) is compared against an eigenvalue for the corresponding factor in many randomly generated data sets that have the same characteristics as the data being analysed. In doing so, each eigenvalue is being compared to an eigenvalue from a data set that has no underlying factors. This is a bit like asking whether our observed factor is bigger than a non-existing factor. Factors that are bigger than their 'random' counterparts are retained. Of parallel analysis, the scree plot and Kaiser's criterion, Kaiser's criterion is, in general, worst and parallel analysis best (Zwick & Velicer, 1986).

As a final word of advice, your decision on how many factors to extract will depend also on why you're doing the analysis; for example, if you're trying to overcome multicollinearity problems in regression, then it might be better to extract too many factors than too few.

17.4.6. Improving interpretation: factor rotation ③

Once factors have been extracted, it is possible to calculate to what degree variables load onto these factors (i.e. calculate the loading of the variable on each factor). Generally, you will find that most variables have high loadings on the most important factor and small loadings on all other factors. This characteristic makes interpretation difficult, and so a technique called factor rotation is used to discriminate between factors. If a factor is a classification axis along which variables can be plotted, then factor **rotation** effectively rotates these factor axes such that variables are loaded maximally to only one factor. Figure 17.4 demonstrates how this process works using an example in which there are only two factors. Imagine that a sociologist was interested in classifying university lecturers as a demographic group. She discovered that two underlying dimensions best describe this group: alcoholism and achievement (go to any academic conference and you'll see that academics drink heavily!). The first factor, alcoholism, has a cluster of variables associated with it (hollow circles) and these could be measures such as the number of units drunk in a week, dependency and obsessive personality. The second factor, achievement, also has a cluster of variables associated with it (solid circles) and these could be measures relating to salary, job status and number of research publications. Initially, the full lines represent the factors, and by looking at the coordinates it should be clear that the solid circles have high loadings for factor 2 (they are a long way up this axis) and medium loadings for factor 1 (they are not very far up this axis). Conversely, the light grey circles have high loadings for factor 1 and medium loadings for factor 2. By rotating the axes (dashed lines), we ensure that both clusters of variables are intersected by the factor to which they relate most. So, after rotation, the loadings of the variables are maximized onto one factor (the factor that intersects the cluster) and minimized on the remaining factor(s). If an axis passes through a cluster of variables, then these variables will have a loading of approximately zero on the opposite axis. If this idea is confusing, then look at Figure 17.4 and think about the values of the coordinates before and after rotation (this is best achieved by turning the book when you look at the rotated axes).

There are two types of rotation that can be done. The first is **orthogonal rotation**, and the left-hand side of Figure 17.4 represents this method. In Chapter 10 we saw that the term *orthogonal* means unrelated, and in this context it means that we rotate factors while keeping them independent, or unrelated. Before rotation, all factors are independent (i.e. they do not correlate at all) and orthogonal rotation ensures that the factors remain uncorrelated. That is why in Figure 17.4 the axes are turned while remaining perpendicular.[5] The other form of rotation is **oblique rotation**. The difference with oblique rotation is that the factors are allowed to correlate (hence, the axes of the right-hand diagram of Figure 17.4 do not remain perpendicular).

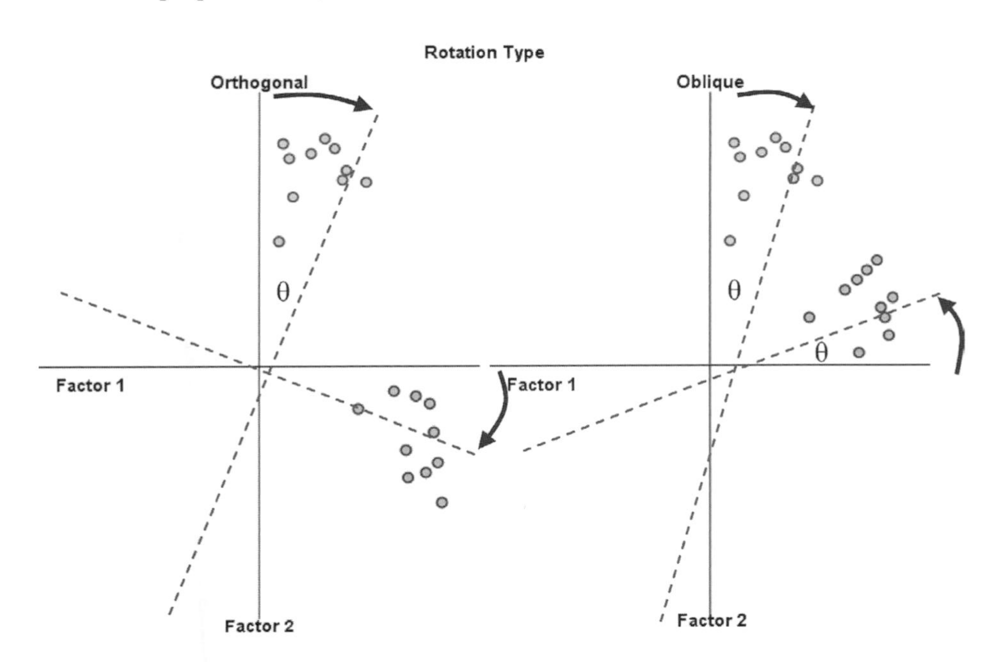

FIGURE 17.4 Schematic representations of factor rotation. The left graph displays orthogonal rotation whereas the right graph displays oblique rotation (see text for more details). θ is the angle through which the axes are rotated

The choice of rotation depends on whether there is a good theoretical reason to suppose that the factors should be related or independent (but see my later comments on this), and also how the variables cluster on the factors before rotation. On the first point, we might not expect alcoholism to be completely independent of achievement (after all, high achievement leads to high stress, which can lead to the drinks cabinet!). Therefore, on theoretical grounds, we might choose oblique rotation. On the second point, Figure 17.4 demonstrates how the positioning of clusters is important in determining how successful the rotation will be (note the position of the light grey circles). Specifically, if an orthogonal rotation was carried out on the right-hand diagram it would be considerably less successful in maximizing loadings than the oblique rotation that is displayed. One approach is to run the analysis using both types of rotation. Pedhazur and Schmelkin (1991) suggest that if the

[5] This term means that the axes are at right angles to one another.

oblique rotation demonstrates a negligible correlation between the extracted factors then it is reasonable to use the orthogonally rotated solution. If the oblique rotation reveals a correlated factor structure, then the orthogonally rotated solution should be discarded. In any case, an oblique rotation should be used only if there are good reasons to suppose that the underlying factors *could* be related in theoretical terms.

The mathematics behind factor rotation is complex (especially oblique rotation). However, in oblique rotation, because each factor can be rotated by different amounts a **factor transformation matrix,** Λ, is needed. The factor transformation matrix is a square matrix and its size depends on how many factors were extracted from the data. If two factors are extracted then it will be a 2×2 matrix, but if four factors are extracted then it becomes a 4×4 matrix. The values in the factor transformation matrix consist of sines and cosines of the angle of axis rotation (θ). This matrix is multiplied by the matrix of unrotated factor loadings, A, to obtain a matrix of rotated factor loadings.

For the case of two factors the factor transformation matrix would be:

$$\Lambda = \begin{pmatrix} \cos\theta & -\sin\theta \\ \sin\theta & \cos\theta \end{pmatrix}$$

Therefore, you should think of this matrix as representing the angle through which the axes have been rotated, or the degree to which factors have been rotated. The angle of rotation necessary to optimize the factor solution is found in an iterative way (see SAS Tip 8.1) and different methods can be used.

17.4.6.1. Choosing a Method of Factor Rotation ③

SAS has three methods of orthogonal rotation (**varimax**, **quartimax**, and **equamax**), along with some less used methods (which have names like orthogonal generalized Crawford–Ferguson rotation), and two methods of oblique rotation (**direct oblimin** and **promax**). These methods differ in how they rotate the factors and, therefore, the resulting output depends on which method you select. Quartimax rotation attempts to maximize the spread of factor loadings for a variable across all factors. Therefore, interpreting variables becomes easier. However, this often results in lots of variables loading highly onto a single factor. Varimax is the opposite in that it attempts to maximize the dispersion of loadings within factors. Therefore, it tries to load a smaller number of variables highly onto each factor, resulting in more interpretable clusters of factors. Equamax is a hybrid of the other two approaches and is reported to behave fairly erratically (see Tabachnick & Fidell, 2007). For a first analysis, you should probably select varimax because it is a good general approach that simplifies the interpretation of factors.

The case with oblique rotations is more complex because correlation between factors is permitted. Promax is a faster procedure designed for very large data sets.

In theory, the exact choice of rotation will depend largely on whether or not you think that the underlying factors should be related. If you expect the factors to be independent then you should choose one of the orthogonal rotations (I recommend varimax). If, however, there are theoretical grounds for supposing that your factors might correlate, then direct oblimin should be selected. In practice, there are strong grounds to believe that orthogonal rotations are a complete nonsense for naturalistic data, and certainly for any data involving humans (can you think of any psychological construct that is not in any way correlated with some other psychological construct?) As such, some argue that orthogonal rotations should never be used.

17.4.6.2. Substantive importance of factor loadings ②

Once a factor structure has been found, it is important to decide which variables make up which factors. Earlier I said that the factor loadings were a gauge of the substantive importance of a given variable to a given factor. Therefore, it makes sense that we use these values to place variables with factors. It is possible to assess the statistical significance of a factor loading (after all, it is simply a correlation coefficient or regression coefficient); however, there are various reasons why this option is not as easy as it seems (see Stevens, 2002, p. 393). Typically, researchers take a loading of an absolute value of more than .3 to be important. However, the significance of a factor loading will depend on the sample size. Stevens (2002) produced a table of critical values against which loadings can be compared. To summarize, he recommends that for a sample size of 50 a loading of .722 can be considered significant, for 100 the loading should be greater than .512, for 200 it should be greater than .364, for 300 it should be greater than .298, for 600 it should be greater than .21, and for 1000 it should be greater than .162. These values are based on an alpha level of .01 (two-tailed), which allows for the fact that several loadings will need to be tested (see Stevens, 2002, for further detail). Therefore, in very large samples, small loadings can be considered statistically meaningful. SAS does not provide significance tests of factor loadings but by applying Stevens's guidelines you should gain some insight into the structure of variables and factors.

The significance of a loading gives little indication of the substantive importance of a variable to a factor. This value can be found by squaring the factor loading to give an estimate of the amount of variance in a factor accounted for by a variable (like R^2). In this respect Stevens (2002) recommends interpreting only factor loadings with an absolute value greater than .4 (which explain around 16% of the variance in the variable).

17.5. Research example ②

One of the uses of factor analysis is to develop questionnaires: after all, if you want to measure an ability or trait, you need to ensure that the questions asked relate to the construct that you intend to measure. I have noticed that a lot of students become very stressed about learning SAS. Therefore I wanted to design a questionnaire to measure various aspects of a trait that I termed 'SAS anxiety'. I generated questions based on interviews with anxious and non-anxious students and came up with 23 possible questions to include. Each question was a statement followed by a 5-point Likert scale ranging from 'strongly disagree' through 'neither agree nor disagree' to 'strongly agree'. The questionnaire is printed in Figure 17.5.

The questionnaire was designed to predict how anxious a given individual would be about learning how to use SAS. What's more, I wanted to know whether anxiety about SAS could be broken down into specific forms of anxiety. In other words, what latent variables contribute to anxiety about SAS? With a little help from a few lecturer friends I collected 2571 completed questionnaires (at this point it should become apparent that this example is fictitious!). The data are stored in the file **SAQ.sas7bdat.** Open this file viewer and have a look at the variables. The first thing to note is that each question (variable) is represented by a different column. We know that in SAS, cases (or people's data) are stored in rows and variables are stored in columns and so this layout is consistent with past chapters. The second thing to notice is that there are 23 variables labelled **q1** to **q23** and that each has a label indicating the question. By labelling my variables I can be very clear about what each variable represents (this is the value of giving your variables full titles rather than just using restrictive column headings).

FIGURE 17.5 The SAS anxiety questionnaire (SAQ)

SD = Strongly Disagree, D = Disagree, N = Neither, A = Agree, SA = Strongly Agree						
		SD	**D**	**N**	**A**	**SA**
1	Statistics makes me cry	O	O	O	O	O
2	My friends will think I'm stupid for not being able to cope with SAS	O	O	O	O	O
3	Standard deviations excite me	O	O	O	O	O
4	I dream that Pearson is attacking me with correlation coefficients	O	O	O	O	O
5	I don't understand statistics	O	O	O	O	O
6	I have little experience of computers	O	O	O	O	O
7	All computers hate me	O	O	O	O	O
8	I have never been good at mathematics	O	O	O	O	O
9	My friends are better at statistics than me	O	O	O	O	O
10	Computers are useful only for playing games	O	O	O	O	O
11	I did badly at mathematics at school	O	O	O	O	O
12	People try to tell you that SAS makes statistics easier to understand but it doesn't	O	O	O	O	O
13	I worry that I will cause irreparable damage because of my incompetence with computers	O	O	O	O	O
14	Computers have minds of their own and deliberately go wrong whenever I use them	O	O	O	O	O
15	Computers are out to get me	O	O	O	O	O
16	I weep openly at the mention of central tendency	O	O	O	O	O
17	I slip into a coma whenever I see an equation	O	O	O	O	O
18	SAS always crashes when I try to use it	O	O	O	O	O
19	Everybody looks at me when I use SAS	O	O	O	O	O
20	I can't sleep for thoughts of eigenvectors	O	O	O	O	O
21	I wake up under my duvet thinking that I am trapped under a normal distribution	O	O	O	O	O
22	My friends are better at SAS than I am	O	O	O	O	O
23	If I am good at statistics people will think I am a nerd	O	O	O	O	O

OLIVER TWISTED

Please, Sir, can I have some more ... questionnaires?

'I'm going to design a questionnaire to measure one's propensity to pick a pocket or two,' says Oliver, 'but how would I go about doing it?' You'd read the useful information about the dos and don'ts of questionnaire design in the additional material for this chapter on the companion website, that's how. Rate how useful it is on a Likert scale from 1 = not useful at all, to 5 = very useful.

17.5.1. Before you begin ②

17.5.1.1. Sample size ②

Correlation coefficients fluctuate from sample to sample, much more so in small samples than in large. Therefore, the reliability of factor analysis is also dependent on sample size. Much has been written about the necessary sample size for factor analysis, resulting in many 'rules of thumb'. The common rule is to suggest that a researcher has at least 10–15 participants per variable. Although I've heard this rule bandied about on numerous occasions its empirical basis is unclear (although Nunnally, 1978, did recommend having 10 times as many participants as variables). Kass and Tinsley (1979) recommended having between 5 and 10 participants per variable up to a total of 300 (beyond which test parameters tend to be stable regardless of the participant to variable ratio). Indeed, Tabachnick and Fidell (2007) agree that 'it is comforting to have at least 300 cases for factor analysis' (p. 613) and Comrey and Lee (1992) class 300 as a good sample size, 100 as poor and 1000 as excellent.

Fortunately, recent years have seen empirical research done in the form of experiments using simulated data (so-called Monte Carlo studies). Arrindell and van der Ende (1985) used real-life data to investigate the effect of different participant to variable ratios. They concluded that changes in this ratio made little difference to the stability of factor solutions. Guadagnoli and Velicer (1988) found that the most important factors in determining reliable factor solutions was the absolute sample size and the absolute magnitude of factor loadings. In short, they argue that if a factor has four or more loadings greater than .6 then it is reliable regardless of sample size. Furthermore, factors with 10 or more loadings greater than .40 are reliable if the sample size is greater than 150. Finally, factors with a few low loadings should not be interpreted unless the sample size is 300 or more. MacCallum, Widaman, Zhang, and Hong (1999) have shown that the minimum sample size or sample to variable ratio depends on other aspects of the design of the study. In short, their study indicated that as communalities become lower the importance of sample size increases. With all communalities above .6, relatively small samples (less than 100) may be perfectly adequate. With communalities in the .5 range, samples between 100 and 200 can be good enough provided there are relatively few factors each with only a small number of indicator variables. In the worst scenario of low communalities (well below .5) and a larger number of underlying factors they recommend samples above 500.

What's clear from this work is that a sample of 300 or more will probably provide a stable factor solution, but that a wise researcher will measure enough variables to adequately measure all of the factors that theoretically they would expect to find.

Another alternative is to use the **Kaiser–Meyer–Olkin measure of sampling adequacy (KMO)** (Kaiser, 1970). The KMO can be calculated for individual and multiple variables and represents the ratio of the squared correlation between variables to the squared partial correlation between variables. The KMO statistic varies between 0 and 1. A value of 0 indicates that the sum of partial correlations is large relative to the sum of correlations, indicating diffusion in the pattern of correlations (hence, factor analysis is likely to be inappropriate). A value close to 1 indicates that patterns of correlations are relatively compact and so factor analysis should yield distinct and reliable factors. Kaiser (1974) recommends accepting values greater than .5 as barely acceptable (values below this should lead you to either collect more data or rethink which variables to include). Furthermore, values between .5 and .7 are mediocre, values between .7 and .8 are good, values between.8 and .9 are great, and values above .9 are superb (Hutcheson & Sofroniou, 1999).

17.5.1.2. Correlations between variables ③

When I was an undergraduate, my statistics lecturer always used to say 'if you put garbage in, you get garbage out'. This saying applies particularly to factor analysis because SAS will always find a factor solution to a set of variables. However, the solution is unlikely to have any real meaning if the variables analysed are not sensible. The first thing to do when conducting a factor analysis or principal component analysis is to look at the intercorrelation between variables. There are essentially two potential problems: (1) correlations that are not high enough; and (2) correlations that are too high. The correlations between variables can be checked using PROC CORR (see Chapter 6) to create a correlation matrix of all variables. This matrix can also be created as part of the main factor analysis. In both cases the remedy is to remove variables from the analysis. We will look at each problem in turn.

If our test questions measure the same underlying dimension (or dimensions) then we would expect them to correlate with each other (because they are measuring the same thing). Even if questions measure different aspects of the same things (e.g. we could measure overall anxiety in terms of sub-components such as worry, intrusive thoughts and physiological arousal), there should still be high intercorrelations between the variables relating to these sub-traits. We can test for this problem first by visually scanning the correlation matrix and looking for correlations below about .3 (you could use the significance of correlations but, given the large sample sizes normally used with factor analysis, this approach isn't helpful because even very small correlations will be significant in large samples). If any variables have lots of correlations below .3 then consider excluding them. It should be immediately clear that this approach is very subjective: I've used fuzzy terms such as 'about .3' and 'lots of', but I have to because every data set is different. Analysing data really is a skill, it's not like following a recipe book!

A problem that can arise is when variables correlate too highly. Although mild multicollinearity is not a problem for factor analysis it is important to avoid extreme multicollinearity (i.e. variables that are very highly correlated) and **singularity** (variables that are perfectly correlated). As with regression, multicollinearity causes problems in factor analysis because it becomes impossible to determine the unique contribution to a factor of the variables that are highly correlated (as was the case for multiple regression). Multicollinearity does not cause a problem for principal component analysis. Therefore, as well as scanning the correlation matrix for low correlations, we could also look out for very high correlations ($r > .8$). The problem with a heuristic such as this is that the effect of two variables correlating with $r = .9$ might be less than the effect of, say, three variables that all correlate at $r = .6$. In other words, eliminating such highly correlating variables might not be getting at the cause of the multicollinearity (Rockwell, 1975).

Multicollinearity can be detected by looking at the determinant of the *R*-matrix, denoted $|R|$ (see Jane Superbrain Box 17.3). One simple heuristic is that the determinant of the *R*-matrix should be greater than 0.00001. However, SAS doesn't print the determinant when using PROC FACTOR, and it's not especially easy to get the determinant of a correlation matrix out of SAS. If your determinant is zero, or negative, SAS will print an error in the log, so you will know.

If you have reason to believe that the correlation matrix has multicollinearity then you could look through the correlation matrix for variables that correlate very highly ($r > 0.8$) and consider eliminating one of the variables (or more depending on the extent of the problem) before proceeding. You may have to try some trial and error to work out which variables are creating the problem (it's not always the two with the highest correlation, it could be a larger number of variables with correlations that are not obviously too large).

JANE SUPERBRAIN 17.3

What is the determinant? ③

The determinant of a matrix is an important diagnostic tool in factor analysis, but the question of what it is is not easy to answer because it has a mathematical definition and I'm not a mathematician. Rather than pretending that I understand the maths, all I'll say is that a good explanation of how the determinant is derived can be found at mathworld.wolfram.com. However, we can bypass the maths and think about the determinant conceptually. The way that I think of the determinant is as describing the 'area' of the data. Here are two scatterplots:

At the time I used these to describe eigenvectors and eigenvalues (which describe the shape of the data). The determinant is related to eigenvalues and eigenvectors but instead of describing the height and width of the data it describes the overall area. So, in the left diagram above, the determinant of those data would represent the area inside the red dashed elipse. These variables have a low correlation so the determinant (area) is big; the biggest value it can be is 1. In the right diagram, the variables are perfectly correlated or singular, and the elipse (dashed line) has been squashed down to a straight line. In other words, the opposite sides of the ellipse have actually met each other and there is no distance between them at all. Put another way, the area, or determinant, is 0. Therefore, the determinant tells us whether the correlation matrix is singular (determinant is 0), or if all variables are completely unrelated (determinant is 1), or somewhere in between!

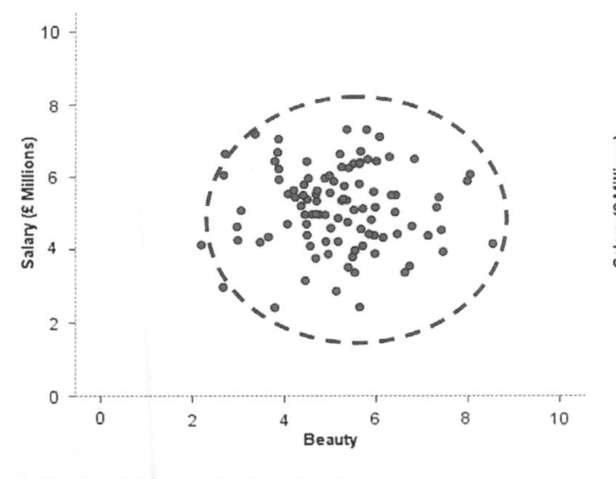

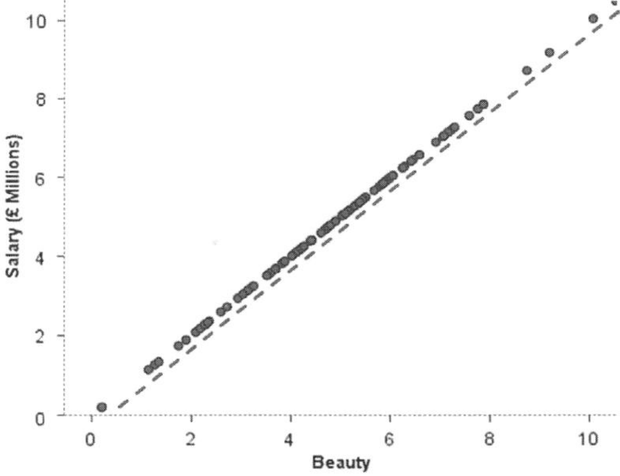

17.5.1.3. The distribution of data ②

As well as looking for interrelations, you should ensure that variables have roughly normal distributions and are measured at an interval level (which Likert scales are, perhaps wrongly, assumed to be). The assumption of normality is most important if you wish to generalize the results of your analysis beyond the sample collected. You can do factor analysis on non-continuous data; for example, if you had dichotomous variables, it's possible (using syntax) to do the factor analysis direct from the correlation matrix, but you should construct the correlation matrix from tetrachoric correlation coefficients (http://ourworld.compuserve.com/homepages/jsuebersax/tetra.htm). This keeps the analysis close to the familiar framework of factor analysis, and the only hassle is computing the correlations (but see the website for software options). Alternatively you can use software other than SAS, such as Mplus (http://www.statmodel.com/), to analyse the raw data.

17.6. Running the analysis ②

Factor analysis in SAS is run using PROC FACTOR. PROC FACTOR is slightly different from other Procs in that most of the information goes on the PROC FACTOR line, not in statements – so make sure you don't put any semicolons there which shoudn't be there.

The basic PROC FACTOR syntax is remarkably straightforward, and is shown in SAS Syntax 17.1. Because we haven't told SAS which variables to use, it will use them all – that's OK in this case, but it usually isn't, in which case we should put in a VAR statement.

You would probably already have examined the descriptive statistics for your variables, but if you haven't (or if you want to check again) you can add the CORR option to the PROC FACTOR line to obtain the correlation matrix, and the SIMPLE option to obtain the means and standard deviations. Adding the MSA option to the PROC FACTOR line gives the Kaiser–Meyer–Olkin measure of sampling adequacy. With a sample of 2571 we shouldn't have cause to worry about the sample size (see section 17.5.1.1). We have already stumbled across KMO (see section 17.5.1.1) and have seen the various criteria for adequacy.

The RES option compares the correlation matrix based on the model (rather than the real data) with the actual correlation matrix. Differences between the matrix based on the model and the matrix based on the observed data indicate the residuals of the model (i.e. differences). We want relatively few of these values to be greater than .05.

The syntax with these options is shown in SAS Syntax 17.2.

```
PROC FACTOR DATA=chap18.saq ;
      RUN;
```
SAS Syntax 17.1

```
PROC FACTOR DATA=chap18.saq SIMPLE CORR MSA RES;
      VAR q01—q23;
      RUN;
```
SAS Syntax 17.2

17.6.1. Factor extraction on SAS ②

To choose the extraction method, we put METHOD= on the PROC FACTOR line. There are several ways of conducting a factor analysis (see section 17.4.1), and when and where you use the various methods depend on numerous things. For our purposes we will use principal component analysis, which strictly speaking isn't factor analysis; however, the two procedures may often yield similar results (see section 17.4.3). For principal component analysis, we write METHOD=PRINCIPAL (or METHOD=P, or METHOD=PRIN).

We also need to choose the number of factors to extract. The default is Kaiser's recommendation of eigenvalues over 1. It is probably best to run a primary analysis with this option, draw a scree plot and compare the results. To obtain a scree plot, we add the option SCREE to the PROC FACTOR line.

If looking at the scree plot and the eigenvalues over 1 leads you to retain the same number of factors then continue with the analysis and be happy. If the two criteria give different results then examine the eigenvalues and decide for yourself which of the two criteria to believe. If you decide to use the scree plot then you may want to redo the analysis specifying the number of factors to extract. The number of factors to be extracted can be specified by specifying NFACT= and then typing the appropriate number (e.g. 4).

Our syntax with these options is shown in SAS Syntax 17.3 (notice there is no semicolon until SCREE).

```
PROC FACTOR DATA=chap18.saq SIMPLE CORR MSA RES METHOD=PRINCIPAL
     NFACT=4 SCREE;
     VAR q01-q23;
     RUN;
```

SAS Syntax 17.3

17.6.2. Rotation ②

We have already seen that the interpretability of factors can be improved through rotation. Rotation maximizes the loading of each variable on one of the extracted factors while minimizing the loading on all other factors. This process makes it much clearer which variables relate to which factors. Rotation works through changing the absolute values of the variables while keeping their differential values constant. To choose a rotation method, we use the ROTATE= option on the PROC FACTOR line. I've discussed the various rotation options in section 17.4.6.1, but, to summarize, the exact choice of rotation will depend on whether or not you think that the underlying factors should be related. If there are theoretical grounds to think that the factors are independent (unrelated) then you should choose one of the orthogonal rotations (I recommend varimax). However, if theory suggests that your factors might correlate then one of the oblique rotations (direct oblimin or promax) should be selected. In this example I've selected varimax.

The syntax including the rotation option is shown in SAS Syntax 17.4.

```
PROC FACTOR DATA=chap18.saq SIMPLE CORR MSA RES METHOD=PRINCIPAL
     NFACT=4 SCREE ROTATE=VARIMAX;
     VAR q01-q23;
     RUN;
```

SAS Syntax 17.4

17.6.3. Saving factor scores

You can save a new dataset containing factor scores for the variables. SAS creates a new column for each factor extracted and then places the factor score for each case within that column. These scores can then be used for further analysis, or simply to identify groups of participants who score highly on particular factors. You create a new dataset using the OUT=dataset option (see SAS Syntax 17.5), on the PROC FACTOR line (note that you can only use OUT= when you specify the number of factors, with NFACT, otherwise SAS doesn't know how many factors to create scores for).

```
PROC FACTOR DATA=chap18.saq SIMPLE CORR MSA RES METHOD=PRINCIPAL
      NFACT=4 SCREE ROTATE=VARIMAX OUT=saqscores;
     VAR q01-q23;
     RUN;
```

SAS Syntax 17.5

17.7. Interpreting output from SAS

You should now run the factor analysis using the syntax we have descried.

SELF-TEST Having done this, choose the OBLIMIN option and repeat the analysis. You should obtain two outputs identical in all respects except that one used an orthogonal rotation and the other an oblique.

Sometimes the factor analysis doesn't work, the KMO test and determinant are nowhere to be found and SAS spits out an error message about a 'singular correlation matrix' (see SAS Tip 17.1).

SAS TIP 17.1 Error messages about a 'Singular definite matrix' ④

What is a non-positive definite matrix?: As we have seen, factor analysis works by looking at your correlation matrix. This matrix has to be 'positive definite' for the analysis to work. What does that mean in plain English? It means lots of horrible things mathematically (e.g. the eigenvalues and determinant of the matrix have to be positive) and about the best explanation I've seen is at http://www2.gsu.edu/~mkteer/npdmatri.html. In more basic terms, factors are like lines floating in space, and eigenvalues measure the length of those lines. If your eigenvalue is negative then it means that the length of your line/factor is negative too. It's a bit like me asking you how tall you are, and you responding 'I'm minus 175 cm tall'. That would be nonsense. By analogy, if a factor has negative length, then that too is nonsense. When SAS decomposes the correlation matrix to look for factors, if it comes across a negative eigenvalue it starts thinking 'oh dear, I've entered some weird parallel universe where the usual rules of maths no longer apply and things can have negative lengths, and this probably means that time runs backwards, my mum is my dad, my sister is a dog, my head is a fish, and my toe is a frog called Gerald.' As you might well imagine, it does the sensible thing and decides not to proceed.

Things like the KMO test and the determinant rely on a positive definite matrix; if you don't have one they can't be computed.

Why have I got a non-positive definite matrix?: The most likely answer is that you have too many variables and too few cases of data, which makes the correlation matrix a bit unstable. It could also be that you have too many highly correlated items in your matrix. In any case it means that your data are bad, naughty data, and not to be trusted; if you let them loose then you have only yourself to blame for the consequences.

What can I do?: Other than cry, there's not that much you can do. You could try to limit your items, or selectively remove items (especially highly correlated ones) to see if that helps. Collecting more data can help too. There are some mathematical fudges you can do, but they're not as tasty as vanilla fudge and they are hard to implement easily.

SAS OUTPUT 17.1

	Q01	Q02	Q03	Q04	Q05	Q06	Q07	Q08	Q09	Q10	Q11	Q12	Q13	Q14	Q15	Q16	Q17	Q18	Q19	Q20	Q21	Q22	Q23
Q01	1	−.10	−.34	.44	.40	.22	.31	.33	−.09	.21	.36	.35	.35	.34	.25	.50	.37	.35	−.19	.21	.33	−.10	.00
Q02	−.10	1	.32	−.11	−.12	−.07	−.16	−.05	.31	−.08	−.14	−.19	−.14	−.16	−.17	−.17	−.09	−.16	.20	−.20	−.20	.23	.10
Q03	−.34	.32	1	−.38	−.31	−.23	−.38	−.26	.30	−.19	−.35	−.41	−.32	−.37	−.31	−.42	−.33	−.38	.34	−.32	−.42	.20	.15
Q04	.44	−.11	−.38	1	.40	.28	.41	.35	−.12	.22	.37	.44	.34	.35	.33	.42	.38	.38	−.19	.24	.41	−.10	−.03
Q05	.40	−.12	−.31	.40	1	.26	.34	.27	−.10	.26	.30	.35	.30	.32	.26	.39	.31	.32	−.17	.20	.33	−.13	−.04
Q06	.22	−.07	−.23	.28	.26	1	.51	.22	−.11	.32	.33	.31	.47	.40	.36	.24	.28	.51	−.17	.10	.27	−.17	−.07
Q07	.31	−.16	−.38	.41	.34	.51	1	.30	−.13	.28	.34	.42	.44	.44	.39	.39	.39	.50	−.27	.22	.48	−.17	−.07
Q08	.33	−.05	−.26	.35	.27	.22	.30	1	.02	.16	.63	.25	.31	.28	.30	.32	.59	.28	−.16	.18	.30	−.08	−.05
Q09	−.09	.31	.30	−.12	−.10	−.11	−.13	.02	1	−.13	−.12	−.17	−.17	−.12	−.19	−.19	−.04	−.15	.25	−.16	−.14	.26	.17
Q10	.21	−.08	−.19	.22	.26	.32	.28	.16	−.13	1	.27	.25	.30	.25	.30	.29	.22	.29	−.13	.08	.19	−.13	−.06
Q11	.36	−.14	−.35	.37	.30	.33	.34	.63	−.12	.27	1	.34	.42	.33	.36	.37	.59	.37	−.20	.26	.35	−.16	−.09
Q12	.35	−.19	−.41	.44	.35	.31	.42	.25	−.17	.25	.34	1	.49	.43	.33	.41	.33	.49	−.27	.30	.44	−.17	−.05
Q13	.35	−.14	−.32	.34	.30	.47	.44	.31	−.17	.30	.42	.49	1	.45	.34	.36	.41	.53	−.23	.20	.37	−.20	−.05
Q14	.34	−.16	−.37	.35	.32	.40	.44	.28	−.12	.25	.33	.43	.45	1	.38	.42	.35	.50	−.25	.23	.40	−.17	−.05
Q15	.25	−.17	−.31	.33	.26	.36	.39	.30	−.19	.30	.36	.33	.34	.38	1	.45	.37	.34	−.21	.21	.30	−.17	−.06
Q16	.50	−.17	−.42	.42	.39	.24	.39	.32	−.19	.29	.37	.41	.36	.42	.45	1	.41	.42	−.27	.27	.42	−.16	−.08
Q17	.37	−.09	−.33	.38	.31	.28	.39	.59	−.04	.22	.59	.33	.41	.35	.37	.41	1	.38	−.16	.21	.36	−.13	−.09
Q18	.35	−.16	−.38	.38	.32	.51	.50	.28	−.15	.29	.37	.49	.53	.50	.34	.42	.38	1	−.26	.24	.43	−.16	−.08
Q19	−.19	.20	.34	−.19	−.17	−.17	−.27	−.16	.25	−.13	−.20	−.27	−.23	−.25	−.21	−.27	−.16	−.26	1	−.25	−.27	.23	.12
Q20	.21	−.20	−.32	.24	.20	.10	.22	.18	−.16	.08	.26	.30	.20	.23	.21	.27	.21	.24	−.25	1	.47	−.10	−.03
Q21	.33	−.20	−.42	.41	.33	.27	.48	.30	−.14	.19	.35	.44	.37	.40	.30	.42	.36	.43	−.27	.47	1	−.13	−.07
Q22	−.10	.23	.20	−.10	−.13	−.17	−.17	−.08	.26	−.13	−.16	−.17	−.20	−.17	−.17	−.16	−.13	−.16	.23	−.10	−.13	1	.23
Q23	.00	.10	.15	−.03	−.04	−.07	−.07	−.05	.17	−.06	−.09	−.05	−.05	−.05	−.06	−.08	−.09	−.08	.12	−.03	−.07	.23	1

17.7.1. Preliminary analysis ②

The first body of output concerns data screening, assumption testing and sampling adequacy. You'll find several large tables (or matrices) that tell us interesting things about our data. If you selected the SIMPLE option then the first table will contain descriptive statistics for each variable (the mean, standard deviation and number of cases). This table is not included here, but you should have enough experience to be able to interpret it. The table also includes the number of missing cases; this summary is a useful way to determine the extent of missing data.

SAS Output 17.1 shows the *R*-matrix (or correlation matrix) produced using the CORR option. You should be comfortable with the idea that to do a factor analysis we need to have variables that correlate fairly well, but not perfectly. Also, any variables that correlate with no others should be eliminated. Therefore, we can use this correlation matrix to check the pattern of relationships. First, scan the matrix for correlations greater than .3, then look for variables that only have a small number of correlations greater than this value. Then scan the correlation coefficients themselves and look for any greater than .9. If any are found then you should be aware that a problem could arise because of multicollinearity in the data.

We came across the KMO statistic in section 17.5.1.1 and saw that Kaiser (1974) recommends a bare minimum of .5 and that values between .5 and .7 are mediocre, values between .7 and .8 are good, values between .8 and .9 are great, and values above .9 are superb (Hutcheson & Sofroniou, 1999). For these data the overall MSA is .93, which rates as superb, so we should be confident that the sample size is adequate for factor analysis.

SAS OUTPUT 17.2

Kaiser's Measure of Sampling Adequacy: Overall MSA = 0. 93022450								
Q01	Q02	Q03	Q04	Q05	Q06	Q07	Q08	Q09
0.92976103	0.87477544	0.95103784	0.95534035	0.96008925	0.89133139	0.94167998	0.87130545	0.83372947

Kaiser's Measure of Sampling Adequacy: Overall MSA = 0.98989259								
Q10	Q11	Q12	Q13	Q14	Q15	Q16	Q17	Q18
0.94868576	0.90593384	0.95483238	0.94822699	0.96717218	0.94044020	0.93364394	0.93062054	0.94795084

Kaiser's Measure of Sampling Adequacy: Overall MSA = 0.98989259				
Q19	Q20	Q21	Q22	Q23
0.94070207	0.88905143	0.92933694	0.87845085	0.76639941

I mentioned that KMO can be calculated for multiple and individual variables. The KMO values for individual variables are also shown in SAS Output 17.2. As well as checking the overall KMO statistic, it is important to examine the KMO value for each variable. Each of these should be above the bare minimum of .5 for all variables (and preferably higher). For these data all values are well above .5, which is good news! If you find any variables with values below .5 then you should consider excluding them from the analysis (or run the analysis with and without that variable and note the difference). Removal of a variable affects the KMO statistics, so if you do remove a variable be sure to re-examine the new KMO statistics.

CRAMMING SAM'S TIPS **Preliminary analysis**

- Scan the **correlation matrix;** look for variables that don't correlate with any other variables, or correlate very highly (r = .9) with one or more other variables. In factor analysis, check that the determinant of this matrix is bigger than 0.00001; if it is then multicollinearity isn't a problem.
- In the table labelled *Kaiser's Measure of Sampling Adequacy* the KMO statistic should be greater than .5 as a bare minimum; if it isn't, collect more data. You can also check the KMO statistic for individual variables by looking at the same table, again, these values should be above .5 (this is useful for identifying problematic variables if the overall KMO is unsatisfactory).

17.7.2. Factor extraction ②

The first part of the factor extraction process is to determine the linear components within the data set (the eigenvectors) by calculating the eigenvalues of the R-matrix (see section 17.4.4). We know that there are as many components (eigenvectors) in the R-matrix as there are variables, but most will be unimportant. To determine the importance of a particular vector we look at the magnitude of the associated eigenvalue. We can then apply criteria to determine which factors to retain and which to discard. By default SAS uses Kaiser's criterion of retaining factors with eigenvalues greater than 1.

Prior Communality Estimates: ONE
SAS Output 17.3

SAS Output 17.3 shows the initial communalities. Principal component analysis works on the initial assumption that all variance is common; therefore, before extraction the communalities are all 1. In effect, all of the variance associated with a variable is assumed to be common variance. Since the communalities are all equal to 1.00, SAS doesn't bother listing them for us. If we'd used a different method, where they were not equal to 1.00, SAS would provide a table here.

SAS Output 17.4 lists the eigenvalues associated with each linear component (factor) before extraction. SAS has identified 23 linear components within the data set (we know that there should be as many eigenvectors as there are variables and so there will be as many factors as variables – see section 17.4.4). The eigenvalues associated with each factor represent the variance explained by that particular linear component and SAS also displays the eigenvalue in terms of the percentage of variance explained (so factor 1 explains 31.70% of total variance). It should be clear that the first few factors explain relatively large amounts of variance (especially factor 1) whereas subsequent factors explain only small amounts of variance. SAS then extracts all factors with eigenvalues greater than 1, which leaves us with four factors.

SAS OUTPUT 17.4

Eigenvalues of the Correlation Matrix: Total = 23 Average = 1				
	Eigenvalue	Difference	Proportion	Cumulative
1	7.29004706	5.55121832	0.3170	0.3170
2	1.73882875	0.42207722	0.0756	0.3926
3	1.31675153	0.08955337	0.0573	0.4498
4	1.22719815	0.23932026	0.0534	0.5032
5	0.98787789	0.09254749	0.0430	0.5461
6	0.89533041	0.08977002	0.0389	0.5850
7	0.80556039	0.02274045	0.0350	0.6201
8	0.78281994	0.03184875	0.0340	0.6541
9	0.75097119	0.03401347	0.0327	0.6868
10	0.71695772	0.03336999	0.0312	0.7179
11	0.68358773	0.01408618	0.0297	0.7476
12	0.66950156	0.05750398	0.0291	0.7768
13	0.61199758	0.03425984	0.0266	0.8034
14	0.57773774	0.02855020	0.0251	0.8285
15	0.54918754	0.02603716	0.0239	0.8524
16	0.52315038	0.01475421	0.0227	0.8751
17	0.50839618	0.05245632	0.0221	0.8972
18	0.45593985	0.03213628	0.0198	0.9170
19	0.42380357	0.01601264	0.0184	0.9355
20	0.40779093	0.02831107	0.0177	0.9532
21	0.37947986	0.01545760	0.0165	0.9697
22	0.36402226	0.03096046	0.0158	0.9855
23	0.33306180		0.0145	1.0000

SAS Output 17.5 shows the table of communalities after extraction. Remember that the communality is the proportion of common variance within a variable (see section 17.4.1). All methods of factor analysis have to have an estimate of the communality before extraction (and principal component analysis assumes they are all equal to 1.00). Once factors have been extracted, we have an idea of how much variance is, in reality, common. The communalities reflect this common variance. So, for example, we can say that 43.5% of the variance associated with question 1 is common, or shared, variance. Another way to look at these communalities is in terms of the proportion of variance explained by the underlying factors. Before extraction, there are as many factors as there are variables, so all variance is explained by the factors and communalities are all 1. However, after extraction some of the factors are discarded and so some information is lost. The retained factors cannot explain all of the variance present in the data, but they can explain some. The amount of variance in each variable that can be explained by the retained factors is represented by the communalities after extraction.

Final Communality Estimates: Total = 11.572825								
Q01	Q02	Q03	Q04	Q05	Q06	Q07	Q08	Q09
0.43464771	0.41375251	0.52971603	0.46858901	0.34304982	0.65393170	0.54529426	0.73946353	0.48448046

Q10	Q11	Q12	Q13	Q14	Q15	Q16	Q17	Q18
0.33477263	0.68960485	0.51332808	0.53582844	0.48826489	0.37799178	0.48708221	0.68280849	0.59733779

Q19	Q20	Q21	Q22	Q23
0.34324231	0.48399646	0.54990692	0.46354431	0.41219129

SAS OUTPUT 17.5

SAS Output 17.6 shows the component matrix before rotation. This matrix contains the loadings of each variable onto each factor. This matrix is not particularly important for interpretation, but it is interesting to note that before rotation most variables load highly onto the first factor (that is why this factor accounts for most of the variance in SAS Output 17.4).

At this stage SAS has extracted four factors. Factor analysis is an exploratory tool and so it should be used to guide the researcher to make various decisions: you shouldn't leave the computer to make them. One important decision is the number of factors to extract. In section 17.4.5 we saw various criteria for assessing the importance of factors. By Kaiser's criterion we should extract four factors and this is what SAS has done. However, this criterion is accurate when there are fewer than 30 variables and communalities after extraction are greater than .7 or when the sample size exceeds 250 and the average communality is greater than .6. The communalities are shown in 17.4, and only one exceeds .7. The average of the communalities can be found by adding them up and dividing by the number of communalities (11.573/23 = .503). So, on both grounds Kaiser's rule may not be accurate. However, you should consider the huge sample that we have, because the research into Kaiser's criterion gives recommendations for much smaller samples. By Jolliffe's criterion (retain factors with eigenvalues greater than .7) we should retain 10 factors, but there is little to recommend this criterion over Kaiser's. As a final guide we can use the scree plot which we asked SAS to produce by using the SCREE option. The scree plot in SAS is not very attractive, and can lose some information, because SAS tries to draw the plot with text. Instead, I've drawn a graph of the eigenvalues in Excel (I just pasted them in, and clicked on line graph). Both scree plots are shown in SAS Output 17.5. This curve is difficult to interpret because it begins to tail off after three factors, but there is another drop after four factors before a stable plateau is reached. Therefore, we could probably justify retaining either two or four factors. Given the large sample, it is probably safe to assume Kaiser's criterion; however, you might like to rerun the analysis specifying that SAS extract only two factors and compare the results.

SAS Output 17.8 shows (some of) the residual correlation matrix that was requested using the RES option. The diagonal of this matrix contains the communalities after extraction for each variable (you can check the values against SAS Output 17.7).

The values in the residual matrix are the differences between the actual correlations that we have, and the correlations that we would have if our model were perfectly correct. If the model were a perfect fit of the data then we would expect the reproduced correlation coefficients to be the same as the original correlation coefficients. Therefore, to assess the fit of the model we can look at the differences between the observed correlations and the correlations based on the model. For example, if we take the correlation between questions 1 and 2, the correlation based on the observed data is −0.099 (taken from SAS Output 17.1). We can calculate the correlation based on the model – if the model were correct, the only reason why question 1 and question 2 would correlated would be because of the factors that they load on. The correlation between them that is due to the factor is calculated by multiplying the loadings on that factor. So the correlation between question 1 and question 2 that is due to factor 1 is equal to the product of the loadings:

SAS OUTPUT 17.6

Factor Pattern					
		Factor1	Factor2	Factor3	Factor4
Q01	Statistics makes me cry	0.58608	0.17514	–0.21528	0.11893
Q02	My friends will think I'm stupid for not being able to cope with SAS	–0.30259	0.54833	0.14639	0.00997
Q03	Standard deviations excite me	–0.62894	0.29016	0.21310	–0.06738
Q04	I dream that Pearson is attacking me with correlation coefficients	0.63446	0.14351	–0.14877	0.15271
Q05	I don't understand statistics	0.55555	0.10088	–0.07414	0.13691
Q06	I have little experience of computers	0.56185	0.09741	0.57134	–0.04829
Q07	All computers hate me	0.68518	0.03914	0.25215	0.10347
Q08	I have never been good at mathematics	0.54891	0.40064	–0.32276	–0.41650
Q09	My friends are better at statistics than me	–0.28385	0.62703	–0.00831	0.10334
Q10	Computers are useful only for playing games	0.43708	0.03453	0.36310	–0.10345
Q11	I did badly at mathematics at school	0.65247	0.24539	–0.20893	–0.40002
Q12	People try to tell you that SAS makes statistics easier to understand but it doesn't	0.66872	–0.04770	0.05065	0.24759
Q13	I worry that I will cause irreparable damage because of my incompetence with computers	0.67299	0.07579	0.27769	–0.00759
Q14	Computers have minds of their own and deliberately go wrong whenever I use them	0.65581	0.02296	0.19832	0.13533
Q15	Computers are out to get me	0.59285	0.01026	0.11716	–0.11262
Q16	I weep openly at the mention of central tendency	0.67930	0.01417	–0.13814	0.07967
Q17	I slip into a coma whenever I see an equation	0.64311	0.32951	–0.20962	–0.34162
Q18	SAS always crashes when I try to use it	0.70121	0.03340	0.29812	0.12514
Q19	Everybody looks at me when I use SAS	–0.42681	0.38961	0.09545	–0.01307
Q20	I can't sleep for thoughts of eigenvectors	0.43577	–0.20524	–0.40447	0.29729
Q21	I wake up under my duvet thinking that I am trapped under a normal distribtion	0.65752	–0.05530	–0.18700	0.28205
Q22	My friends are better at SAS than I am	–0.30161	0.46540	–0.11590	0.37754
Q23	If I'm good at statistics my friends will think I'm a nerd	–0.14394	0.36646	–0.02115	0.50669

SAS OUTPUT 17.7

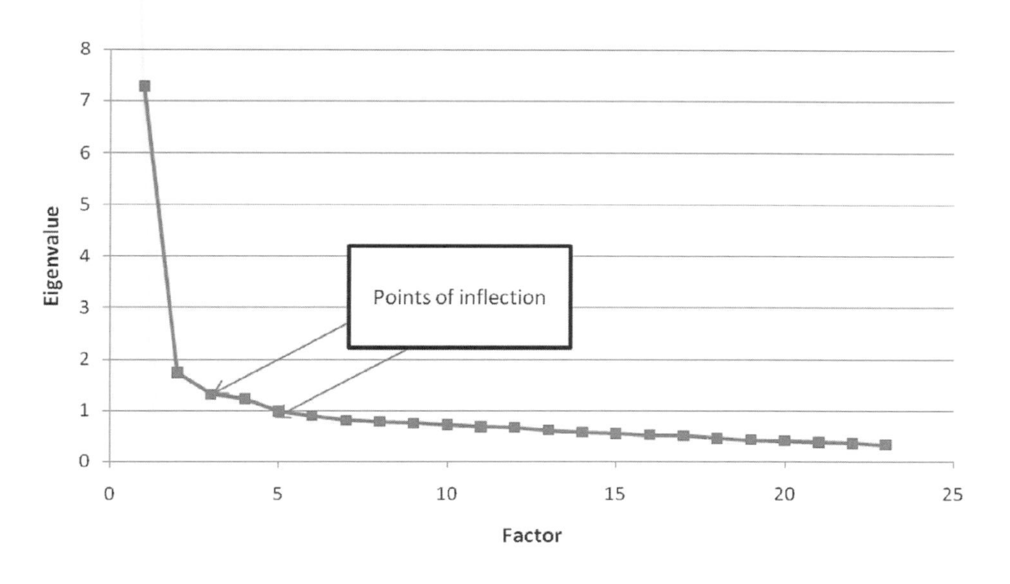

$$0.58608 \times -0.30259 = -0.177341947$$

For factors 2, 3, and 4:

$$0.17514 \times 0.54833 = 0.096034516$$
$$-0.21528 \times 0.14639 = -0.031514839$$
$$0.11893 \times 0.00997 = 0.001185732$$

And we can add these correlations together to get the total correlation between the variables, if our model is perfectly correct:

$$-0.177341947 + 0.096034516 - 0.031514839 + 0.001185732 = -0.111636538$$

The correlation based on the model is -0.112, which is slightly higher. We can calculate the difference as follows:

$$\begin{aligned} \text{residual} &= r_{\text{observed}} - r_{\text{from model}} \\ \text{residual}_{Q_1Q_2} &= (-0.099) - (-0.112) \\ &= 0.013 \end{aligned}$$

CRAMMING SAM'S TIPS Factor extraction

- To decide how many factors to extract look at the table labelled **Final Communality Estimates.** If these values are all .7 or above and you have less than 30 variables then the SAS default option for extracting factors is fine (Kaiser's criterion of retaining factors with eigenvalues greater than 1). Likewise, if your sample size exceeds 250 and the average of the communalities is .6 or greater then the default option is fine. Alternatively, with 200 or more participants the scree plot can be used.
- Check the table labelled ***Residual Correlations with Uniqueness on the Diagonal*** for the percentage of non-redundant residuals with absolute values > .05. This percentage should be less than 50% and the smaller it is, the better.

SAS OUTPUT 17.8

Residual Correlations With Uniqueness on the Diagonal										
		Q01	Q02	Q03	Q04	Q05	Q06	Q07	Q08	Q09
Q01	Statistics makes me cry	0.56535	0.01291	0.03503	−0.01130	0.02694	−0.00087	−0.06108	−0.08108	−0.04988
Q02	My friends will think I'm stupid for not being able to cope with SAS	0.01291	0.58625	−0.06155	0.02169	0.00293	−0.04077	−0.01125	−0.05181	−0.11488
Q03	Standard deviations excite me	0.03503	−0.06155	0.47028	0.01893	0.03485	−0.02664	−0.00913	0.01106	−0.05193
Q04	I dream that Pearson is attacking me with correlation coefficients	−0.01130	0.02169	0.01893	0.53141	0.00178	0.00012	−0.01001	−0.04074	−0.05146
Q05	I don't understand statistics	0.02694	0.00293	0.03485	0.00178	0.65695	−0.01553	−0.04068	−0.04364	−0.01603
Q06	I have little experience of computers	−0.00087	−0.04077	−0.02664	0.00012	−0.01553	0.34607	−0.01427	0.03970	−0.00450
Q07	All computers hate me	−0.06108	−0.01125	−0.00913	−0.01001	−0.04068	−0.01427	0.45471	0.03020	0.03305
Q08	I have never been good at mathematics	−0.08108	−0.05181	0.01106	−0.04074	−0.04364	0.03970	0.03020	0.26054	−0.03932
Q09	My friends are better at statistics than me	−0.04988	−0.11488	−0.05193	−0.05146	−0.01603	−0.00450	0.03305	−0.03932	0.51552
Q10	Computers are useful only for playing games	0.04194	−0.02281	−0.01286	0.00336	0.05299	−0.13916	−0.09796	−0.02103	−0.01807
Q11	I did badly at mathematics at school	−0.06599	−0.04638	0.00609	−0.05052	−0.05013	0.03763	−0.01785	−0.06121	−0.04459
Q12	People try to tell you that SAS makes statistics easier to understand but it doesn't	−0.05673	0.02375	0.03036	−0.00606	−0.05009	−0.07555	−0.07173	0.02351	0.02717
Q13	I worry that I will cause irreparable damage because of my incompetence with computers	0.00763	−0.02124	0.02367	−0.05110	−0.05807	−0.07812	−0.09120	0.00094	−0.02085
Q14	Computers have minds of their own and deliberately go wrong whenever I use them	−0.02390	−0.00923	0.00191	−0.05974	−0.05514	−0.07524	−0.07356	0.03179	0.03791
Q15	Computers are out to get me	−0.06488	−0.00726	0.02494	−0.00876	−0.04492	−0.04658	−0.03314	−0.03894	−0.01211
Q16	I weep openly at the mention of central tendency	0.05880	0.04966	0.03929	−0.04987	−0.00504	−0.05594	−0.05087	−0.06846	−0.01431
Q17	I slip into a coma whenever I see an equation	−0.06857	−0.03899	0.00315	−0.05159	−0.04887	−0.00790	0.02540	−0.10482	−0.02733
Q18	SAS always crashes when I try to use it	−0.02039	−0.01492	0.00100	−0.04244	−0.06586	−0.04819	−0.06901	0.02981	0.01806
Q19	Everybody looks at me when I use SAS	0.01500	−0.15332	−0.06114	0.04510	0.04136	−0.02006	−0.01464	−0.05593	−0.11399
Q20	I can't sleep for thoughts of eigenvectors	−0.12798	0.09905	0.11502	−0.10968	−0.09241	0.12152	0.00163	0.01145	0.05965
Q21	I wake up under my duvet thinking that I am trapped under a normal distribtion	−0.12032	0.04922	0.07125	−0.06983	−0.07757	0.02875	0.05261	0.01407	0.05467
Q22	My friends are better at SAS than I am	−0.07900	−0.10238	−0.07095	−0.04871	−0.07221	0.04344	0.01040	0.01976	−0.16056
Q23	If I'm good at statistics my friends will think I'm a nerd	−0.04911	−0.14677	−0.00801	−0.07561	−0.06960	0.01304	−0.03310	0.08616	−0.15240

You should notice that this difference is the value quoted in the residual correlation matrix for questions 1 and 2. Therefore, residual matrix contains the differences between the observed correlation coefficients and the ones predicted from the model. For a good model these values will all be small. In fact, we want most values to be less than 0.05. For these data there are 91 residuals (35%) that are greater than 0.05. There are no hard and fast rules about what proportion of residuals should be below 0.05; however, if more than 50% are greater than 0.05 you probably have grounds for concern.

17.7.3. Factor rotation ②

The first analysis I asked you to run was using an orthogonal rotation. However, you were asked to rerun the analysis using oblique rotation too. In this section the results of both analyses will be reported so as to highlight the differences between the outputs. This comparison will also be a useful way to show the circumstances in which one type of rotation might be preferable to another.

17.7.3.1. Orthogonal rotation (varimax)

SAS Output 17.9 shows the factor transformation matrix (see section 17.4.6). This matrix provides information about the degree to which the factors were rotated to obtain a solution. If no rotation were necessary, this matrix would be an identity matrix. If orthogonal rotation were completely appropriate then we would expect a symmetrical matrix (same values above and below the diagonal). However, in reality the matrix is not easy to interpret, although very unsymmetrical matrices might be taken as a reason to try oblique rotation. For the inexperienced factor analyst you are probably best advised to ignore the factor transformation matrix.

SAS OUTPUT 17.9

Orthogonal Transformation Matrix

	1	2	3	4
1	0.63469	0.58503	0.44289	–0.24240
2	0.13673	–0.16761	0.48821	0.84550
3	0.75761	–0.51345	–0.40290	0.00834
4	0.06710	0.60500	–0.63496	0.47572

SAS Output 17.10 shows the rotated component matrix (also called the rotated factor matrix in factor analysis) which is a matrix of the factor loadings for each variable onto each factor. This matrix contains the same information as the component matrix in SAS Output 17.6 except that it is calculated *after* rotation. There are several things to consider about the format of this matrix.

Compare this matrix to the unrotated solution. Before rotation, most variables loaded highly on the first factor and the remaining factors didn't really get a look-in. However, the rotation of the factor structure has clarified things considerably: there are four factors and variables load very highly on only one factor (with the exception of one question). I have highlighted loadings that have an absolute value greater than .4.

The next step is to look at the content of questions that load on the same factor to try to identify common themes. If the mathematical factor produced by the analysis represents

Rotated Factor Pattern					
		Factor1	Factor2	Factor3	Factor4
Q01	Statistics makes me cry	0.24080	**0.49600**	0.35630	0.06079
Q02	My friends will think I'm stupid for not being able to cope with SAS	–0.00550	–0.33806	0.06838	**0.54292**
Q03	Standard deviations excite me	–0.20258	**–0.56676**	–0.17997	0.36751
Q04	I dream that Pearson is attacking me with correlation coefficients	0.31985	**0.51590**	0.31403	0.03895
Q05	I don't understand statistics	0.31941	**0.42900**	0.23824	0.01514
Q06	I have little experience of computers	**0.79954**	–0.01019	0.09686	–0.07204
Q07	All computers hate me	**0.63820**	0.32743	0.15528	–0.08167
Q08	I have never been good at mathematics	0.13069	0.16771	**0.83321**	0.00486
Q09	My friends are better at statistics than me	–0.09378	–0.20437	0.11814	**0.64804**
Q10	Computers are useful only for playing games	**0.55028**	0.00090	0.12983	–0.12293
Q11	I did badly at mathematics at school	0.26254	0.20585	**0.74695**	–0.14272
Q12	People try to tell you that SAS makes statistics easier to understand but it doesn't	**0.47289**	**0.52300**	0.09526	–0.08422
Q13	I worry that I will cause irreparable damage because of my incompetence with computers	**0.64737**	0.23384	0.22800	–0.10034
Q14	Computers have minds of their own and deliberately go wrong whenever I use them	**0.57871**	0.35987	0.13583	–0.07352
Q15	Computers are out to get me	**0.45889**	0.21683	0.29188	–0.18763
Q16	I weep openly at the mention of central tendency	0.33377	**0.51416**	0.31285	–0.11594
Q17	I slip into a coma whenever I see an equation	0.27149	0.22196	**0.74707**	–0.04155
Q18	SAS always crashes when I try to use it	**0.68387**	0.32727	0.12729	–0.07972
Q19	Everybody looks at me when I use SAS	–0.14619	–0.37192	–0.02898	**0.42745**
Q20	I can't sleep for thoughts of eigenvectors	–0.03797	**0.67687**	0.06700	–0.14111
Q21	I wake up under my duvet thinking that I am trapped under a normal distribtion	0.28702	**0.66059**	0.16047	–0.07352
Q22	My friends are better at SAS than I am	–0.19027	0.03346	–0.09940	**0.64525**
Q23	If I'm good at statistics my friends will think I'm a nerd	–0.02328	0.17177	–0.19804	**0.58560**

SAS OUTPUT 17.10

some real-world construct then common themes among highly loading questions can help us identify what the construct might be. The questions that load highly on factor 1 seem to all relate to using computers or SAS. Therefore we might label this factor *fear of computers*. The questions that load highly on factor 2 all seem to relate to different aspects of statistics; therefore, we might label this factor *fear of statistics*. The three questions that load highly on factor 3 all seem to relate to mathematics; therefore, we might label this factor *fear of mathematics*. Finally, the questions that load highly on factor 4 all contain some component of social evaluation from friends; therefore, we might label this factor *peer evaluation*. This analysis seems to reveal that the initial questionnaire, in reality, is composed of four subscales: fear of computers, fear of statistics, fear of maths and fear of negative peer evaluation. There are two possibilities here. The first is that the SAQ failed to measure what it set out to (namely, SAS anxiety) but does measure some related constructs. The second is that these four constructs are sub-components of SAS anxiety; however, the factor analysis does not indicate which of these possibilities is true.

Finally, we are presented with SAS Output 17.11 on the variance explained by each factor. This is equivalent to the eigenvalues for each factor, after the factors have been rotated. The eigenvalues of the four factors we extracted added up to 11.57, and if you add these eigenvalues, they still sum to 11.57 (indicating that the factors explain 11.57/23 = 50.3% of the variance. We cannot explain more variance by rotating the factors, but the variance that we have explained is divided up amongst the factors, rather than the first factor getting everything it can.

SAS OUTPUT 17.11

Variance Explained by Each Factor			
Factor1	Factor2	Factor3	Factor4
3.7304611	3.3402529	2.5529277	1.9491838

17.7.3.2. Oblique rotation

When an oblique rotation is conducted the factor matrix is split into two matrices: the *pattern matrix* and the *structure matrix* (see Jane Superbrain Box 17.1). For orthogonal rotation these matrices are the same. The pattern matrix contains the factor loadings and is comparable to the factor matrix that we interpreted for the orthogonal rotation. The structure matrix takes into account the relationship between factors (in fact it is a product of the pattern matrix and the matrix containing the correlation coefficients between factors). Most researchers interpret the pattern matrix, because it is usually simpler; however, there are situations in which values in the pattern matrix are suppressed because of relationships between the factors. Therefore, the structure matrix is a useful double-check and Graham et al. (2003) recommend reporting both (with some useful examples of why this can be important).

For the pattern matrix for these data (SAS Output 17.12) the same four factors seem to have emerged. Factor 1 represents fear of computers and factor 2 represents fear of mathematics, factor 3 seems to represent fear of statistics, and factor 4 represents fear of peer evaluation. The structure matrix (SAS Output 17.13) differs in that shared variance is not ignored. The picture becomes more complicated because with the exception of factor 4, several variables load highly on more than one factor. This has occurred because of the relationship between factors 1 and 2 and factors 2 and 3. This example should highlight why the pattern matrix is preferable for interpretative reasons: because it contains information about the *unique* contribution of a variable to a factor.

The final part of the output that we will worry about is the matrix of correlation coefficient between the factors (SAS Output 17.14). As predicted from the structure matrix, factor 4 has little or no relationship with any other factors (correlation coefficients are low),

Rotated Factor Pattern (Standardized Regression Coefficients)					
		Factor1	Factor2	Factor3	Factor4
Q01	Statistics makes me cry	0.136	0.338	0.396	0.136
Q02	My friends will think I'm stupid for not being able to cope with SAS	0.045	0.100	–0.370	**0.502**
Q03	Standard deviations excite me	–0.104	–0.136	**–0.515**	0.286
Q04	I dream that Pearson is attacking me with correlation coefficients	0.235	0.273	**0.404**	0.124
Q05	I don't understand statistics	0.262	0.190	0.327	0.091
Q06	I have little experience of computers	**0.885**	–0.057	–0.229	–0.012
Q07	All computers hate me	**0.653**	0.032	0.155	0.008
Q08	I have never been good at mathematics	–0.098	**0.902**	–0.001	0.015
Q09	My friends are better at statistics than me	–0.075	0.173	–0.218	**0.617**
Q10	Computers are useful only for playing games	**0.585**	0.029	–0.159	–0.083
Q11	I did badly at mathematics at school	0.063	**0.774**	0.022	–0.116
Q12	People try to tell you that SAS makes statistics easier to understand but it doesn't	**0.462**	–0.006	**0.412**	0.019
Q13	I worry that I will cause irreparable damage because of my incompetence with computers	**0.650**	0.114	0.043	–0.025
Q14	Computers have minds of their own and deliberately go wrong whenever I use them	**0.588**	0.022	0.208	0.016
Q15	Computers are out to get me	**0.411**	0.222	0.063	–0.132
Q16	I weep openly at the mention of central tendency	0.240	0.265	**0.401**	–0.031
Q17	I slip into a coma whenever I see an equation	0.079	**0.774**	0.036	–0.011
Q18	SAS always crashes when I try to use it	**0.713**	–0.009	0.148	0.014
Q19	Everybody looks at me when I use SAS	–0.095	0.018	–0.352	0.372
Q20	I can't sleep for thoughts of eigenvectors	–0.133	0.062	**0.708**	–0.059
Q21	I wake up under my duvet thinking that I am trapped under a normal distribtion	0.219	0.102	**0.592**	0.031
Q22	My friends are better at SAS than I am	–0.146	–0.058	0.092	**0.643**
Q23	If I'm good at statistics my friends will think I'm a nerd	0.056	–0.207	0.213	**0.617**

SAS OUTPUT 17.12

Factor Structure (Correlations)					
		Factor1	Factor2	Factor3	Factor4
Q01	Statistics makes me cry	**0.413**	**0.501**	**0.519**	0.024
Q02	My friends will think I'm stupid for not being able to cope with SAS	−0.136	−0.022	**−0.403**	**0.543**
Q03	Standard deviations excite me	**−0.407**	−0.350	**−0.635**	0.394
Q04	I dream that Pearson is attacking me with correlation coefficients	**0.487**	**0.486**	**0.548**	−0.002
Q05	I don't understand statistics	**0.453**	0.396	**0.462**	−0.022
Q06	I have little experience of computers	**0.777**	0.290	0.081	−0.133
Q07	All computers hate me	**0.723**	0.377	**0.402**	−0.137
Q08	I have never been good at mathematics	0.317	**0.855**	0.214	−0.036
Q09	My friends are better at statistics than me	−0.187	0.030	−0.292	**0.651**
Q10	Computers are useful only for playing games	**0.556**	0.262	0.076	−0.167
Q11	I did badly at mathematics at school	**0.451**	**0.818**	0.280	−0.190
Q12	People try to tell you that SAS makes statistics easier to understand but it doesn't	**0.607**	0.322	**0.576**	−0.128
Q13	I worry that I will cause irreparable damage because of my incompetence with computers	**0.723**	**0.429**	0.316	−0.158
Q14	Computers have minds of their own and deliberately go wrong whenever I use them	**0.671**	0.351	**0.426**	−0.125
Q15	Computers are out to get me	**0.561**	**0.440**	0.296	−0.233
Q16	I weep openly at the mention of central tendency	**0.516**	**0.491**	**0.568**	−0.157
Q17	I slip into a coma whenever I see an equation	**0.453**	**0.821**	0.283	−0.090
Q18	SAS always crashes when I try to use it	**0.761**	0.362	**0.405**	−0.138
Q19	Everybody looks at me when I use SAS	−0.283	−0.153	**−0.439**	**0.442**
Q20	I can't sleep for thoughts of eigenvectors	0.165	0.202	**0.685**	−0.148
Q21	I wake up under my duvet thinking that I am trapped under a normal distribution	**0.477**	0.366	**0.695**	−0.108
Q22	My friends are better at SAS than I am	−0.256	−0.149	−0.076	**0.660**
Q23	If I'm good at statistics my friends will think I'm a nerd	−0.075	−0.169	0.080	**0.590**

SAS OUTPUT 17.13

but all other factors are interrelated to some degree (notably factors 1 and 3 and factors 3 and 4). The fact that these correlations exist tells us that the constructs measured can be interrelated. If the constructs were independent then we would expect oblique rotation to provide an identical solution to an orthogonal rotation and the component correlation matrix should be an identity matrix (i.e. all factors have correlation coefficients of 0). Therefore, this final matrix gives us a guide to whether it is reasonable to assume independence between factors: for these data it appears that we cannot assume independence. Therefore, the results of the orthogonal rotation should not be trusted: the obliquely rotated solution is probably more meaningful.

On a theoretical level the dependence between our factors does not cause concern; we might expect a fairly strong relationship between fear of maths, fear of statistics and fear of computers. Generally, the less mathematically and technically minded people struggle with statistics. However, we would not expect these constructs to correlate with fear of peer evaluation (because this construct is more socially based). In fact, this factor is the one that correlates fairly badly with all others – so on a theoretical level, things have turned out rather well!

SAS OUTPUT 17.14

Inter-Factor Correlations				
	Factor1	Factor2	Factor3	Factor4
Factor1	1.00000	0.46333	0.36598	–0.18173
Factor2	0.46333	1.00000	0.27995	–0.07659
Factor3	0.36598	0.27995	1.00000	–0.15421
Factor4	–0.18173	–0.07659	–0.15421	1.00000

CRAMMING SAM'S TIPS **Interpretation**

- If you've conduced orthogonal rotation then look at the table labelled ***Rotated Factor Pattern.*** For each variable, note the component for which the variable has the highest loading. Also, for each component, note the variables that load highly onto it (by high I'd say loadings should be above .4 when you ignore the plus or minus sign). Try to make sense of what the factors represent by looking for common themes in the items that load on them.
- If you've conducted oblique rotation then look at the table labelled ***Rotated Factor Pattern.*** For each variable, note the component for which the variable has the highest loading. Also, for each component, note the variables that load highly on it (by high I'd say loadings should be above .4 when you ignore the plus or minus sign). Double-check what you find by doing the same thing for the ***Factor Structure.*** Try to make sense of what the factors represents by looking for common themes in the items that load on them.

17.7.4. Factor scores ②

In the original analysis we asked for scores to be calculated based and placed into a dataset called saqscores. (Remember that because we didn't specify a folder this dataset will go into the Work folder). The dataset should contain all of the old variables, plus four new variables, called *Factor1, Factor2, Factor3 and Factor4.*

SAS Output 17.15 shows the factor scores for the first 10 participants. It should be pretty clear that participant 9 scored highly on all four factors and so this person is very anxious about statistics, computing and maths, but less so about peer evaluation (factor 4). Factor scores can be used in this way to assess the relative fear of one person compared to another, or we could add the scores up to obtain a single score for each participant (that we might assume represents SAS anxiety as a whole). We can also use factor scores in regression when groups of predictors correlate so highly that there is multicollinearity.

SAS OUTPUT 17.15

Obs	Factor1	Factor2	Factor3	Factor4
1	−0.42251	−1.88832	−0.92242	−0.39041
2	−0.63645	−0.23119	−0.28401	0.33861
3	−0.37223	0.08048	0.11543	−0.93551
4	0.78470	−0.30334	0.77901	−0.22384
5	−0.08056	−0.66183	−0.60246	0.69073
6	1.66867	−0.21471	−0.04352	−0.63805
7	−0.70292	−0.39223	−1.53628	0.94297
8	−0.62907	−0.37617	−1.07495	1.06890
9	2.40502	3.63610	1.32896	0.55794
10	−0.96251	−0.18978	−0.98200	1.54315

17.7.5. Summary

To sum up, the analyses revealed four underlying scales in our questionnaire that may, or may not, relate to genuine sub-components of SAS anxiety. It also seems as though an obliquely rotated solution was preferred due to the interrelationships between factors. The use of factor analysis is purely exploratory; it should be used only to guide future hypotheses, or to inform researchers about patterns within data sets. A great many decisions are left to the researcher using factor analysis and I urge you to make informed decisions, rather than basing decisions on the outcomes you would like to get. The next question is whether or not our scale is reliable.

17.8. How to report factor analysis ②

As with any analysis, when reporting factor analysis we need to provide our readers with enough information to form an informed opinion about our data. As a bare minimum we should be very clear about our criteria for extracting factors and the method of rotation used. We must also produce a table of the rotated factor loadings of all items and flag (in bold) values above a criterion level (I would personally choose .40, but I discussed the various criteria you could use in section 17.4.6.2). You should also report the percentage of variance that each factor explains and possibly the eigenvalue too. Table 17.2 shows an example of such a table for the SAQ data; note that I have also reported the sample size in the title.

TABLE 17.2 Summary of exploratory factor analysis results for the SAS anxiety questionnaire ($N = 2571$)

	Rotated Factor Loadings			
Item	**Fear of Computers**	**Fear of Statistics**	**Fear of Maths**	**Peer Evaluation**
I have little experience of computers	**.80**	−.01	.10	−.07
SAS always crashes when I try to use it	**.68**	.33	.13	−.08
I worry that I will cause irreparable damage because of my incompetence with computers	**.65**	.23	.23	−.10
All computers hate me	**.64**	.33	.16	−.08
Computers have minds of their own and deliberately go wrong whenever I use them	**.58**	.36	.14	−.07
Computers are useful only for playing games	**.55**	.00	.13	−.12
Computers are out to get me	**.46**	.22	.29	−.19
I can't sleep for thoughts of eigenvectors	−.04	**.68**	.08	−.14
I wake up under my duvet thinking that I am trapped under a normal distribution	.29	**.66**	.16	−.07
Standard deviations excite me	−.20	**−.57**	−.18	.37
People try to tell you that SAS makes statistics easier to understand but it doesn't	**.47**	**.52**	.10	−.08
I dream that Pearson is attacking me with correlation coefficients	.32	**.52**	.31	.04
I weep openly at the mention of central tendency	.33	**.51**	.31	−.12
Statistics makes me cry	.24	**.50**	.36	.06
I don't understand statistics	.32	**.43**	.24	.02
I have never been good at mathematics	.13	.17	**.83**	.01
I slip into a coma whenever I see an equation	.27	.22	**.75**	−.04
I did badly at mathematics at school	.26	.21	**.75**	−.14
My friends are better at statistics than me	−.09	−.20	.12	**.65**
My friends are better at SAS than I am	−.19	.03	−.10	**.65**
If I'm good at statistics my friends will think I'm a nerd	−.02	.17	−.20	**.59**
My friends will think I'm stupid for not being able to cope with SAS	−.01	−.34	.07	**.54**
Everybody looks at me when I use SAS	−.15	−.37	−.03	**.43**
Eigenvalues	3.73	3.34	2.55	1.95
% of variance	16.2%	14.5%	11.1%	8.5%
α	.82	.82	.82	.57

Note: Factor loadings over .40 appear in bold.

In my opinion, a table of factor loadings and a description of the analysis are a bare minimum, though. You could consider (if it's not too large) including the table of correlations from which someone could reproduce your analysis (should they want to). You could also consider including some information on sample size adequacy. For this example we might write something like this (although obviously you don't have to cite this book as much as I have):

✓ A principal component analysis (PCA) was conducted on the 23 items with orthogonal rotation (varimax). The Kaiser–Meyer–Olkin measure verified the sampling adequacy for the analysis, KMO = .93 ('superb' according to Field, 2009), and all KMO values for individual items were > .77, which is well above the acceptable limit of .5 (Field, 2009). An initial analysis was run to obtain eigenvalues for each component in the data. Four components had eigenvalues over Kaiser's criterion of 1 and in combination explained 50.32% of the variance. The scree plot was slightly ambiguous and showed inflexions that would justify retaining both components 2 and 4. Given the large sample size, and the convergence of the scree plot and Kaiser's criterion on four components, this is the number of components that were retained in the final analysis. Table 17.2 shows the factor loadings after rotation. The items that cluster on the same components suggest that component 1 represents a fear of computers, component 2 a fear of statistics, component 3 a fear of maths and component 4 peer evaluation concerns.

Finally, if you have used oblique rotation you should consider reporting a table of both the structure and pattern matrix because the loadings in these tables have different interpretations (see Jane Superbrain Box 17.1).

LABCOAT LENI'S REAL RESEARCH 17.1

World wide addiction? ②

The Internet is now a houshold tool. In 2007 it was estimated that around 179 million people worldwide used the Internet (over 100 million of those were in the USA and Canada). From the increasing populatrity (and usefulness) of the Internet has emerged a new phenomenon: Internet addiction. This is now a serious and recognized problem, but until very recently it was very difficult to research this topic because there was not a psychometrically sound measure of Internet addition. That is, until Laura Nichols and Richard Nicki developed the Internet Addiction Scale, IAS (Nichols & Nicki, 2004). (Incidentally, while doing some research on this topic I encountered an Internet addiction recovery website that I won't name but that offered a whole host of resources that would keep you online for ages, such as questionnaires, an online support group, videos, articles, a recovery blog and podcasts. It struck me that that this was a bit like having a recovery centre for heroin addiction where the addict arrives to be greeted by a nice-looking counsellor who says 'there's a huge pile of heroin in the corner over there, just help yourself'.)

Anyway, Nichols and Nicki developed a 36-item questionnaire to measure internet addiction. It contained items such as 'I have stayed on the Internet longer than I intended to' and 'My grades/work have suffered because of my Internet use' which could be responded to on a 5-point scale (Never, Rarely, Sometimes, Frequently, Always). They collected data from 207 people to validate this measure.

The data from this study are in the file **NicholsandNicki.sas7bdat**. The authors dropped two items because they had low means and variances, and dropped three others because of relatively low correlations with other items. They performed a principal component analysis on the remaining 31 items. Labcoat Leni wants you to run some descriptive statistics to work out which two items were dropped for having low means/variances, then inspect a correlation matrix to find the three items that were dropped for having low correlations. Finally, he wants you to run a principal component analysis on the data.

Answers are in the additional material on the companion website (or look at the original article).

NICHOLS, L.A., & NICKI, R. (2004). *PSYCHOLOGY OF ADDICTIVE BEHAVIORS, 18*(4), 381–384.

17.9. Reliability analysis ②

17.9.1. Measures of reliability ②

If you're using factor analysis to validate a questionnaire, it is useful to check the reliability of your scale.

SELF-TEST Thinking back to Chapter 1, what are reliability and test–retest reliability?

Reliability means that a measure (or in this case questionnaire) should consistently reflect the construct that it is measuring. One way to think of this is that, other things being equal, a person should get the same score on a questionnaire if they complete it at two different points in time (we have already discovered that this is called test–retest reliability). So, someone who is terrified of statistics and who scores highly on our SAQ should score similarly highly if we tested them a month later (assuming they hadn't gone into some kind of statistics-anxiety therapy in that month). Another way to look at reliability is to say that two people who are the same in terms of the construct being measured should get the same score. So, if we took two people who were equally statistics-phobic, then they should get more or less identical scores on the SAQ. Likewise, if we took two people who loved statistics, they should both get equally low scores. It should be apparent that if we took someone who loved statistics and someone who was terrified of it, and they got the same score on our questionnaire, then it wouldn't be an accurate measure of statistics anxiety! In statistical terms, the usual way to look at reliability is based on the idea that individual items (or sets of items) should produce results consistent with the overall questionnaire. So, if we take someone scared of statistics, then their overall score on the SAQ will be high; if the SAQ is reliable then if we randomly select some items from it the person's score on those items should also be high.

The simplest way to do this is in practice is to use split-half reliability. This method randomly splits the data set into two. A score for each participant is then calculated based on each half of the scale. If a scale is very reliable a person's score on one half of the scale should be the same as (or similar to) their score on the other half: therefore, across several participants, scores from the two halves of the questionnaire should correlate perfectly (well, very highly). The correlation between the two halves is the statistic computed in the split-half method, with large correlations being a sign of reliability. The problem with this method is that there are several ways in which a set of data can be split into two and so the results could be a product of the way in which the data were split.

To overcome this problem, Cronbach (1951) came up with a measure that is loosely equivalent to splitting data in two in every possible way and computing the correlation coefficient for each split. The average of these values is equivalent to Cronbach's alpha, α, which is the most common measure of scale reliability:[6]

$$\alpha = \frac{N^2\overline{\text{Cov}}}{\sum s^2_{\text{item}} + \sum \text{Cov}_{\text{item}}} \tag{17.7}$$

which may look complicated, but actually isn't. The first thing to note is that for each item on our scale we can calculate two things: the variance within the item, and the covariance between a particular item and any other item on the scale. Put another way, we can construct a variance–covariance matrix of all items. In this matrix the diagonal elements will be the variance within a particular item, and the off-diagonal elements will be covariances between pairs of items. The top half of the equation is simply the number of items (*N*) squared multiplied by the average covariance between items (the average of the off-diagonal elements in the aforementioned variance–covariance matrix). The bottom half is just the sum of all the item variances and item covariances (i.e. the sum of everything in the variance–covariance matrix).

There is a standardized version of the coefficient too, which essentially uses the same equation except that correlations are used rather than covariances, and the bottom half of the equation uses the sum of the elements in the correlation matrix of items (including the ones that appear on the diagonal of that matrix). The normal alpha is appropriate when items on a scale are summed to produce a single score for that scale (the standardized alpha is not appropriate in these cases). The standardized alpha is useful, though, when items on a scale are standardized before being summed.

17.9.2. Interpreting Cronbach's α (some cautionary tales ...) ②

You'll often see in books or journal articles, or be told by people, that a value of .7 to .8 is an acceptable value for Cronbach's α; values substantially lower indicate an unreliable scale. Kline (1999) notes that although the generally accepted value of .8 is appropriate for cognitive tests such as intelligence tests, for ability tests a cut-off point of .7 is more suitable. He goes on to say that when dealing with psychological constructs values below even .7 can, realistically, be expected because of the diversity of the constructs being measured.

However, Cortina (1993) notes that such general guidelines need to be used with caution because the value of α depends on the number of items on the scale. You'll notice that the top half of the equation for α includes the number of items squared. Therefore, as the number of items on the scale increases, α will increase. Therefore, it's possible to get a large value of α because you have a lot of items on the scale, and not because your scale is reliable! For example, Cortina reports data from two scales, both of which have α = .8. The first scale has only three items, and the average correlation between items was a respectable

[6] Although this is the easiest way to conceptualize Cronbach's α, whether or not it is exactly equal to the average of all possible split-half reliabilities depends on exactly how you calculate the split-half reliability (see the Glossary for computational details). If you use the Spearman–Brown formula, which takes no account of item standard deviations, then Cronbach's α will be equal to the average split-half reliability only when the item standard deviations are equal; otherwise α will be smaller than the average. However, if you use a formula for split-half reliability that does account for item standard deviations (such as Flanagan, 1937; Rulon, 1939) then α will always equal the average split-half reliability (see Cortina, 1993).

.57; however, the second scale had 10 items with an average correlation between these items of a less respectable .28. Clearly the internal consistency of these scales differs enormously, yet according to Cronbach's α they are both equally reliable!

A second common interpretation of alpha is that it measures 'unidimensionality', or the extent to which the scale measures one underlying factor or construct. This interpretation stems from the fact that when there is one factor underlying the data, α is a measure of the strength of that factor (see Cortina, 1993). However, Grayson (2004) demonstrates that data sets with the same α can have very different structures. He showed that $\alpha = .8$ can be achieved in a scale with one underlying factor, with two moderately correlated factors and with two uncorrelated factors. Cortina (1993) has also shown that with more than 12 items, and fairly high correlations between items ($r > .5$), α can reach values around and above .7 (.65 to .84). These results compellingly show that α should not be used as a measure of 'unidimensionality'. Indeed, Cronbach (1951) suggested that if several factors exist then the formula should be applied separately to items relating to different factors. In other words, if your questionnaire has subscales, α should be applied separately to these subscales.

The final warning is about items that have a reverse phrasing. For example, in our SAQ that we used in the factor analysis part of this chapter, we had one item (question 3) that was phrased the opposite way around to all other items. The item was 'standard deviations excite me'. Compare this to any other item and you'll see it requires the opposite response. For example, item 1 is 'statistics makes me cry'. Now, if you don't like statistics then you'll strongly agree with this statement and so will get a score of 5 on our scale. For item 3, if you hate statistics then standard deviations are unlikely to excite you so you'll strongly disagree and get a score of 1 on the scale. These reverse-phrased items are important for reducing response bias; participants will actually have to read the items in case they are phrased the other way around. For factor analysis, this reverse phrasing doesn't matter, all that happens is you get a negative factor loading for any reversed items (in fact, look at SAS Output 17.10 and you'll see that item 3 has a negative factor loading). However, in reliability analysis these reverse-scored items do make a difference. To see why, think about the equation for Cronbach's α. In this equation, the top half incorporates the average covariance between items. If an item is reverse-phrased then it will have a negative relationship with other items, hence the covariances between this item and other items will be negative. The average covariance is obviously the sum of covariances divided by the number of covariances, and by including a bunch of negative values we reduce the sum of covariances, and hence we also reduce Cronbach's α, because the top half of the equation gets smaller. In extreme cases, it is even possible to get a negative value for Cronbach's α, simply because the magnitude of negative covariances is bigger than the magnitude of positive ones! A negative Cronbach's α doesn't make much sense, but it does happen, and if it does, ask yourself whether you included any reverse-phrased items!

If you have reverse-phrased items then you have to also reverse the way in which they're scored before you conduct reliability analysis. This is quite easy. To take our SAQ data, we have one item which is currently scored as 1 = strongly disagree, 2 = disagree, 3 = neither, 4 = agree and 5 = strongly agree. This is fine for items phrased in such a way that agreement indicates statistics anxiety, but for item 3 (standard deviations excite me), disagreement indicates statistics anxiety. To reflect this numerically, we need to reverse the scale such that 1 = strongly agree, 2 = agree, 3 = neither, 4 = disagree and 5 = strongly disagree. This way, an anxious person still gets 5 on this item (because they'd strongly disagree with it).

To reverse the scoring find the maximum value of your response scale (in this case 5) and add 1 to it (so you get 6 in this case). Then for each person, you take this value and subtract from it the score they actually got. Therefore, someone who scored 5 originally now scores $6-5 = 1$, and someone who scored 1 originally now gets $6-1 = 5$. Someone in the middle of the scale with a score of 3 will still get $6-3 = 3$! Obviously it would take a long time to do this for each person, but we can get SAS to do it for us.

SELF-TEST Using what you learnt in Chapter 5, use a DATA step to reverse-score item 3. (Clue: Remember that you are simply changing the variable to 6 minus its original value.)

17.9.3. Reliability analysis on SAS ②

Let's test the reliability of the SAQ using the data in **SAQ.sas7bdat.** Now, you should have reverse scored item 3 (see above), but if you can't be bothered then load up the file **SAQItem3Reversed.sas7bdat** instead. Remember also that I said we should conduct reliability analysis on any subscales individually. If we use the results from our orthogonal rotation (look back atSAS Output 17.9), then we have four subscales:

1. Subscale 1 (*Fear of computers*): items 6, 7, 10, 13, 14, 15, 18
2. Subscale 2 (*Fear of statistics*): items 1, 3, 4, 5, 12, 16, 20, 21
3. Subscale 3 (*Fear of mathematics*): items 8, 11, 17
4. Subscale 4 (*Peer evaluation*): items 2, 9, 19, 22, 23

To conduct each reliability analysis on these data we use PROC CORR, which we first encountered in Chapter 6. We add ALPHA to the PROC CORR line, and then put the variables we are interested in on the VAR line. Because we have already seen the correlations between item items, we might consider turning them off by adding the NOCORR option to the PROC CORR line.

```
PROC CORR DATA=chap18.saqitem3reversed ALPHA NOCORR;
VAR q06 q07 q10 q13 q14 q15 q18;
      RUN;
```

17.9.4. Interpreting the output ②

SAS Output 17.16 shows the results of this basic reliability analysis for the fear of computing subscale.

First, and perhaps most important, we have the value of Cronbach's α: the overall reliability of the scale. To reiterate, we're looking for values in the range of .7 to .8 (or thereabouts) bearing in mind what we've already noted about effects from the number of items.

SAS OUTPUT 17.16

Cronbach Coefficient Alpha	
Variables	**Alpha**
Raw	0.823360
Standardized	0.821413

Cronbach Coefficient Alpha with Deleted Variable					
Deleted Variable	**Raw Variables**		**Standardized Variables**		
	Correlation with Total	**Alpha**	**Correlation with Total**	**Alpha**	**Label**
Q06	0.618726	0.790593	0.616204	0.788542	I have little experience of computers
Q07	0.619032	0.790475	0.614746	0.788792	All computers hate me
Q10	0.399922	0.823902	0.399945	0.824088	Computers are useful only for playing games
Q13	0.606744	0.793677	0.604642	0.790520	I worry that I will cause irreparable damage because of my incompetence with computers
Q14	0.576825	0.797975	0.574940	0.795561	Computers have minds of their own and deliberately go wrong whenever I use them
Q15	0.491267	0.811871	0.491876	0.809349	Computers are out to get me
Q18	0.647389	0.785533	0.644668	0.783635	SAS always crashes when I try to use it

We are shown two different values of alpha – the raw value, and the standardized value – we should use the raw value. In this case α is equal to 0.82 – slightly above .8, and is certainly in the region indicated by Kline (1999), so this probably indicates good reliability.

The next table gives some items statistics when each item is removed. Again, we should look at the *Raw Variables* columns. The values in the column labelled *Correlation with Total* are the correlations between each item and the total score from the questionnaire (without that item) these are sometimes called corrected item total correlations, or item–rest correlations. In a reliable scale all items should correlate with the total. So, we're looking for items that don't correlate with the overall score from the scale: if any of these values are less than about .3 then we've got problems, because it means that a particular item does not correlate very well with the scale overall. Items with low correlations may have to be dropped. For these data, all data have item-total correlations above .3, which is encouraging.

The values in the column labelled *Alpha* are the values of the overall α if that item isn't included in the calculation. As such, they reflect the change in Cronbach's α that would be seen if a particular item were deleted. The overall α is .823, and so all values in this column should be either less than or around that same value.[7] What we're actually looking for is values of alpha greater than the overall α. If you think about it, if the deletion of an item increases Cronbach's α then this means that the deletion of that item improves reliability. Therefore, any items that result in substantially greater values of α than the overall α may need to be deleted from the scale to improve its reliability. None of the items here would substantially affect reliability if they were deleted. The worst offender is question 10: deleting this question would increase the α from .823 to .824. Nevertheless this increase is negligible and both values reflect a good degree of reliability.

As a final point, it's worth noting that if items do need to be removed at this stage then you should rerun your factor analysis as well to make sure that the deletion of the item has not affected the factor structure.

[7] Recall that the the reliability of a scale increases when more items are added. If all of the items are equally good, then removing an item should always lower alpha.

OK, let's move on to to do the fear of statistics subscale (items 1, 3, 4, 5, 12, 16, 20 and 21). I won't show the SAS syntax again, but SAS Output 17.17 shows the output from the analysis). The overall α is .821, and none of the items here would increase the reliability if they were deleted. In the second table, *Cronbach Coefficient Alpha with Deleted Variable*, the values in the column labelled *Correlation with Total* are again all above .3, which is good. The values in the column labelled *Alpha* are the values of the overall α if that item isn't included in the calculation. This indicates that all items are positively contributing to the overall reliability. The overall α is also excellent (.821) because it is above .8, and indicates good reliability.

SAS OUTPUT 17.17

Cronbach Coefficient Alpha	
Variables	Alpha
Raw	0.820836
Standardized	0.823419

Cronbach Coefficient Alpha with Deleted Variable					
Deleted Variable	Raw Variables		Standardized Variables		
	Correlation with Total	Alpha	Correlation with Total	Alpha	Label
Q01	0.536123	0.801701	0.540145	0.803666	Statistics makes me cry
Q03	0.549230	0.799600	0.548622	0.802500	Standard deviations excite me
Q04	0.575007	0.795502	0.579366	0.798239	I dream that Pearson is attacking me with correlation coefficients
Q05	0.494449	0.806645	0.499927	0.809153	I don't understand statistics
Q12	0.571532	0.796235	0.570491	0.799474	People try to tell you that SAS makes statistics easier to understand but it doesn't
Q16	0.597286	0.792774	0.601908	0.795085	I weep openly at the mention of central tendency
Q20	0.418519	0.818455	0.414226	0.820582	I can't sleep for thoughts of eigenvectors
Q21	0.606126	0.790807	0.601509	0.795141	I wake up under my duvet thinking that I am trapped under a normal distribution

Moving swiftly on to the fear of maths subscale (items 8, 11 and 17), SAS Output 17.18 shows the output from the analysis. As with the previous two subscales, the overall α is around .8, which indicates good reliability The values in the column labelled *Correlation with Total* are again all above .3, which is good, and the values in the column labelled *Alpha* indicate that none of the items here would increase the reliability if they were deleted because all values in this column are less than the overall reliability of .819.

SAS OUTPUT 17.18

Cronbach Coefficient Alpha	
Variables	**Alpha**
Raw	0.819409
Standardized	0.819475

Cronbach Coefficient Alpha with Deleted Variable					
Deleted Variable	**Raw Variables**		**Standardized Variables**		
	Correlation with Total	**Alpha**	**Correlation with Total**	**Alpha**	**Label**
Q08	0.684466	0.739626	0.684508	0.739629	I have never been good at mathematics
Q11	0.681786	0.742210	0.681943	0.742249	I did badly at mathematics at school
Q17	0.651997	0.772458	0.652007	0.772477	I slip into a coma whenever I see an equation

SAS OUTPUT 17.19

Cronbach Coefficient Alpha	
Variables	**Alpha**
Raw	0.569918
Standardized	0.572442

Cronbach Coefficient Alpha with Deleted Variable					
Deleted Variable	**Raw Variables**		**Standardized Variables**		
	Correlation with Total	**Alpha**	**Correlation with Total**	**Alpha**	**Label**
Q02	0.338909	0.515281	0.332108	0.516254	My friends will think I'm stupid for not being able to cope with SAS
Q09	0.390653	0.476482	0.396910	0.478791	My friends are better at statistics than me
Q19	0.316246	0.521817	0.314738	0.526031	Everybody looks at me when I use SAS
Q22	0.377606	0.487048	0.378682	0.489488	My friends are better at SAS than I am
Q23	0.238881	0.562830	0.235943	0.569003	If I'm good at statistics my friends will think I'm a nerd

Finally, if you run the analysis for the final subscale of peer evaluation, you should get the output in SAS Output 17.19. Unlike the previous subscales, the overall α is quite low and although this is in keeping with what Kline says we should expect for this kind of social science data, it is well below the other scales. The scale has five items, compared to seven, eight and three on the other scales, so its reduced reliability is not going to be dramatically affected by the number of items (in fact, it has more items than the fear of maths subscale). If you look at the items on this subscale, they cover quite diverse themes of peer evaluation, and this might explain the relative lack of consistency. The values in the column labelled *Correlation* with total are all around .3, and in fact for item 23 the value is below .3. This indicates fairly bad internal consistency and identifies item 23 as a potential problem. The values in the column labelled *Alpha* indicate that none of the items here would increase the reliability if they were deleted

CRAMMING SAM'S TIPS **Reliability**

- Reliability is really the consistency of a measure.
- Reliability analysis can be used to measure the consistency of a questionnaire.
- Remember to reverse-score any items that were reverse-phrased on the original questionnaire before you run the analysis.
- Run separate reliability analyses for all subscales of your questionnaire.
- Cronbach's α indicates the overall reliability, of a questionnaire, and values around .8 are good (or 0.7 for ability tests and such like).
- The *Correlation with Total* tells you whether removing an item will improve the overall reliability: values greater than the overall reliability indicate that removing that item will improve the overall reliability of the scale. Look for items that dramatically increase the value of α.
- If you do remove items, rerun your factor analysis to check that the factor structure still holds!

because all values in this column are less than the overall reliability of .57. This might lead us to rethink this subscale.

17.10. How to report reliability analysis ②

You can report the reliabilities in the text using the symbol α and remembering that because Cronbach's α can't be larger than 1 then we drop the zero before the decimal place (if we are following APA style):

✓ The fear of computers, fear of statistics and fear of maths subscales of the SAQ all had high reliabilities, all Cronbach's $\alpha = .82$. However, the fear of negative peer evaluation subscale had relatively low reliability, Cronbach's $\alpha = .57$.

However, the most common way to report reliability analysis when it follows a factor analysis is to report the values of Cronbach's α as part of the table of factor loadings. For example, in Table 17.2 notice that in the last row of the table I have quoted the value of Cronbach's α for each subscale in turn.

What have I discovered about statistics? ②

This chapter has made us tiptoe along the craggy rockface that is factor analysis. This is a technique for identifying clusters of variables that relate to each other. One of the difficult things with statistics is realizing that they are subjective: many books (this one included, I suspect) create the impression that statistics are like a cook book and if you follow the instructions you'll get a nice tasty chocolate cake (yum!). Factor analysis perhaps more than any other test in this book illustrates how incorrect this is. The world of statistics is full of arbitrary rules that we probably shouldn't follow (.05 being the classic example) and nearly all of the time, whether you realize it or not, we should act upon our own discretion. So, if nothing else I hope you've discovered enough to give you sufficient discretion about factor analysis to act upon! We saw that the first stage of factor analysis is to scan your variables to check that they relate to each other to some degree but not too strongly. The factor analysis itself has several stages: check some initial issues (e.g. sample size adequacy), decide how many factors to retain, and finally decide which items load on which factors (and try to make sense of the meaning of the factors). Having done all that, you can consider whether the items you have are reliable measures of what you're trying to measure.

We also discovered that at the age of 23 I took it upon myself to become a living homage to the digestive system. I furiously devoured articles and books on statistics (some of them I even understood), I mentally chewed over them, I broke them down with the stomach acid of my intellect, I stripped them of their goodness and nutrients, I compacted them down, and after about two years I forced the smelly brown remnants of those intellectual meals out of me in the form of a book. I was mentally exhausted at the end of it; 'It's a good job I'll never have to do that again', I thought.

Key terms that I've discovered

Alpha factoring
Common variance
Communality
Component matrix
Confirmatory factor analysis (CFA)
Cronbach's α
Direct oblimin
Extraction
Equamax
Factor analysis
Factor loading
Factor matrix
Factor scores
Factor transformation matrix, Λ
Kaiser's criterion
Kaiser–Meyer–Olkin (KMO) measure of sampling adequacy
Latent variable
Oblique rotation
Orthogonal rotation
Pattern matrix
Principal component analysis (PCA)
Promax
Quartimax
Random variance
Rotation
Scree plot
Singularity
Split-half reliability
Structure matrix
Unique variance
Varimax

Smart Alex's tasks

- **Task 1**: The University of Sussex is constantly seeking to employ the best people possible as lecturers (no, really, it is). Anyway, they wanted to revise a questionnaire based on Bland's theory of research methods lecturers. This theory predicts that good research methods lecturers should have four characteristics: (1) a profound love of statistics; (2) an enthusiasm for experimental design; (3) a love of teaching; and (4) a complete absence of normal interpersonal skills. These characteristics should be related (i.e. correlated). The 'Teaching of Statistics for Scientific Experiments' (TOSSE) questionnaire it already existed, but the university revised and it became the 'Teaching of Statistics for Scientific Experiments – Revised' (TOSSE–R). They gave this questionnaire to 239 research methods lecturers around the world to see if it supported Bland's theory. The questionnaire is in Figure 17.6, and the data are in **TOSSE_R.sas7bdat.** Conduct a factor analysis (with appropriate rotation) to see the factor structure of the data.

SD = Strongly Disagree, D = Disagree, N = Neither, A = Agree, SA = Strongly Agree						
		SD	**D**	**N**	**A**	**SA**
1	I once woke up in the middle of a vegetable patch hugging a turnip that I'd mistakenly dug up thinking it was Roy's greatest root	O	O	O	O	O
2	If I had a big gun I'd shoot all the students I have to teach	O	O	O	O	O
3	I memorize probability values for the *F*-distribution	O	O	O	O	O
4	I worship at the shrine of Pearson	O	O	O	O	O
5	I still live with my mother and have little personal hygiene	O	O	O	O	O
6	Teaching others makes me want to swallow a large bottle of bleach because the pain of my burning oesophagus would be light relief in comparison	O	O	O	O	O
7	Helping others to understand sums of squares is a great feeling	O	O	O	O	O
8	I like control conditions	O	O	O	O	O
9	I calculate 3 ANOVAs in my head before getting out of bed every morning	O	O	O	O	O
10	I could spend all day explaining statistics to people	O	O	O	O	O
11	I like it when people tell me I've helped them to understand factor rotation	O	O	O	O	O
12	People fall asleep as soon as I open my mouth to speak	O	O	O	O	O
13	Designing experiments is fun	O	O	O	O	O
14	I'd rather think about appropriate dependent variables than go to the pub	O	O	O	O	O
15	I soil my pants with excitement at the mere mention of factor analysis	O	O	O	O	O
16	Thinking about whether to use repeated or independent measures thrills me	O	O	O	O	O
17	I enjoy sitting in the park contemplating whether to use participant observation in my next experiment	O	O	O	O	O
18	Standing in front of 300 people in no way makes me lose control of my bowels	O	O	O	O	O

19	I like to help students	O	O	O	O	O
20	Passing on knowledge is the greatest gift you can bestow on an individual	O	O	O	O	O
21	Thinking about Bonferroni corrections gives me a tingly feeling in my groin	O	O	O	O	O
22	I quiver with excitement when thinking about designing my next experiment	O	O	O	O	O
23	I often spend my spare time talking to the pigeons ... and even they die of boredom	O	O	O	O	O
24	I tried to build myself a time machine so that I could go back to the 1930s and follow Fisher around on my hands and knees licking the floor on which he'd just trodden	O	O	O	O	O
25	I love teaching	O	O	O	O	O
26	I spend lots of time helping students	O	O	O	O	O
27	I love teaching because students have to pretend to like me or they'll get bad marks	O	O	O	O	O
28	My cat is my only friend	O	O	O	O	O

FIGURE 17.6 The Teaching of Statistics for Scientific Experiments – Revised (TOSSE–R)

- **Task 2**: Dr Sian Williams (University of Brighton) devised a questionnaire to measure organizational ability. She predicted five factors to do with organizational ability: (1) preference for organization; (2) goal achievement; (3) planning approach; (4) acceptance of delays; and (5) preference for routine. These dimensions are *theoretically independent*. Williams's questionnaire contains 28 items using a 7-point Likert scale (1 = strongly disagree, 4 = neither, 7 = strongly agree). She gave it to 239 people. Run a principal component analysis on the data in **Williams.sas7bdat.** ②

1	I like to have a plan to work to in everyday life
2	I feel frustrated when things don't go to plan
3	I get most things done in a day that I want to
4	I stick to a plan once I have made it
5	I enjoy spontaneity and uncertainty
6	I feel frustrated if I can't find something I need
7	I find it difficult to follow a plan through
8	I am an organized person
9	I like to know what I have to do in a day
10	Disorganized people annoy me
11	I leave things to the last minute

12	I have many different plans relating to the same goal
13	I like to have my documents filed and in order
14	I find it easy to work in a disorganized environment
15	I make 'to do' lists and achieve most of the things on it
16	My workspace is messy and disorganized
17	I like to be organized
18	Interruptions to my daily routine annoy me
19	I feel that I am wasting my time
20	I forget the plans I have made
21	I prioritize the things I have to do
22	I like to work in an organized environment
23	I feel relaxed when I don't have a routine
24	I set deadlines for myself and achieve them
25	I change rather aimlessly from one activity to another during the day
26	I have trouble organizing the things I have to do
27	I put tasks off to another day
28	I feel restricted by schedules and plans

Answers can be found on the companion website.

Further reading

Cortina, J. M. (1993). What is coefficient alpha? An examination of theory and applications. *Journal of Applied Psychology*, *78*, 98–104. (A very readable paper on Cronbach's α.)

Dunteman, G. E. (1989). *Principal components analysis*. Sage University Paper Series on Quantitative Applications in the Social Sciences, 07-069. Newbury Park, CA: Sage. (This monograph is quite high level but comprehensive.)

Hatcher, L. (1994). *A step-by-step approach to using the SAS system for factor analysis and structural equation modeling*. Cary, NC: SAS Publishing. (Very comprehensive and detailed guide to what you can do with factor analysis in SAS.)

Pedhazur, E., & Schmelkin, L. (1991). *Measurement, design and analysis*. Hillsdale, NJ: Erlbaum. (Chapter 22 is an excellent introduction to the theory of factor analysis.)

Tabachnick, B. G. & Fidell, L. S. (2007). *Using multivariate statistics* (5th ed.) Boston: Allyn & Bacon. (Chapter 13 is a technical but wonderful overview of factor analysis.)

Interesting real research

Nichols, L. A., & Nicki, R. (2004). Development of a psychometrically sound internet addiction scale: a preliminary step. *Psychology of Addictive Behaviors*, *18*(4), 381–384.

18 Categorical data

FIGURE 18.1
Midway through writing the SAS version of this book, things had gone a little strange

18.1. What will this chapter tell me? ①

We discovered in the previous chapter that I wrote a book. A bit like this one. There are a lot of good things about writing books. The main benefit is that your parents are impressed. Well, they're not *that* impressed actually because they think that a good book sells as many copies as *Harry Potter* and that people should queue outside bookshops for the latest enthralling instalment of *Discovering statistics* … . My parents are, consequently, quite baffled about how this book is seen as successful, yet I don't get invited to dinner by the Queen. Nevertheless, given that my family don't really understand what I do, books are tangible proof that I do *something*. The size of this book and the fact it has equations in it are an added bonus because it makes me look cleverer than I actually am. However, there is a price to pay, which is immeasurable mental anguish. In England we don't talk about our emotions, because we fear that if they get out into the open, civilization as we know it will collapse, so I definitely will not mention that the writing process for the second edition of the SPSS version of this book was so stressful that I came within one of Fuzzy's whiskers of a total meltdown. It took me two years to recover, just in time to start thinking about this third edition. Still, it was worth it because the feedback suggests that some people found the book vaguely useful. Of course, the publishers don't care about helping people, they care only about raking in as much cash as possible to feed their cocaine habits and champagne addictions. Therefore, they are obsessed with sales figures and comparisons with other books. They have databases that have sales figures of this book and its competitors in different 'markets' (you are not a person, you are a 'consumer' and you don't live in a country, you live in a 'market') and they gibber and twitch at their consoles creating frequency distributions (with 3-D effects) of these values. The data they get are frequency data (the number of books sold in a certain timeframe). Therefore, if they wanted to compare sales of this book to its competitors, in different countries, they would need to read this chapter because it's all about analysing data, for which we know only the frequency with which events occur. Of course, they won't read this chapter, but they should ...

18.2. Analysing categorical data ①

Sometimes, we are interested not in test scores, or continuous measures, but in *categorical variables*. These are not variables involving cats (although the examples in this chapter might convince you otherwise), but are what we have mainly used as grouping variables. They are variables that describe categories of entities (see section 1.5.1.2). We've come across these types of variables in virtually every chapter of this book. There are different types of categorical variable, but in theory a person, or case, should fall into only one category. Good examples of categorical variables are gender (with few exceptions people can be only biologically male or biologically female),[1] pregnancy (a woman can be only pregnant or not pregnant) and voting in an election (as a general rule you are allowed to vote for only one candidate). In all cases (except logistic regression) so far, we've used such categorical variables to predict some kind of continuous outcome, but there are times when we want to look at relationships between lots of categorical variables. This chapter looks at two techniques for doing this. We begin with the simple case of two categorical variables and discover the chi-square statistic (which we're not really discovering because we've unwittingly come across it countless times before). We then extend this model to look at relationships between several categorical variables.

18.3. Theory of analysing categorical data ①

We will begin by looking at the simplest situation that you could encounter; that is, analysing two categorical variables. If we want to look at the relationship between two categorical variables then we can't use the mean or any similar statistic because we don't have any variables that have been measured continuously. Trying to calculate the mean of a categorical variable is completely meaningless because the numeric values you attach to different categories are arbitrary, and the mean of those numeric values will depend on how many members each category has. Therefore, when we've measured only categorical variables, we analyse frequencies. That is, we analyse the number of things that fall into each combination

[1] Before anyone rips my arms from their sockets and beats me around the head with them, I am aware that numerous chromosomal and hormonal conditions exist that complicate the matter. Also, people can have a different gender identity to their biological gender.

TABLE 18.1 Contingency table showing how many cats will line-dance after being trained with different rewards

		Training		
		Food as Reward	**Affection as Reward**	**Total**
Could they dance?	Yes	28	48	76
	No	10	114	124
	Total	38	162	200

of categories. If we take an example, a researcher was interested in whether animals could be trained to line-dance. He took 200 cats and tried to train them to line-dance by giving them either food or affection as a reward for dance-like behaviour. At the end of the week they counted how many animals could line-dance and how many could not. There are two categorical variables here: **training** (the animal was trained using either food or affection, not both) and **dance** (the animal either learnt to line-dance or it did not). By combining categories, we end up with four different categories. All we then need to do is to count how many cats fall into each category. We can tabulate these frequencies as in Table 18.1 (which shows the data for this example) and this is known as a contingency table.

18.3.1. Pearson's chi-square test ①

If we want to see whether there's a relationship between two categorical variables (i.e. does the number of cats that line-dance relate to the type of training used?) we can use Pearson's chi-square test (Fisher, 1922; Pearson, 1900). This is an extremely elegant statistic based on the simple idea of comparing the frequencies you observe in certain categories to the frequencies you might expect to get in those categories by chance. All the way back in Chapters 2, 7 and 10 we saw that if we fit a model to any set of data we can evaluate that model using a very simple equation (or some variant of it):

$$\text{deviation} = \sum(\text{observed} - \text{model})^2$$

This equation was the basis of our sums of squares in regression and ANOVA. Now, when we have categorical data we can use the same equation. There is a slight variation in that we divide by the model scores as well, which is actually much the same process as dividing the sum of squares by the degrees of freedom in ANOVA. So, basically, what we're doing is standardizing the deviation for each observation. If we add all of these standardized deviations together the resulting statistic is Pearson's chi-square (χ^2) given by:

$$\chi^2 = \sum \frac{(\text{observed}_{ij} - \text{model}_{ij})^2}{\text{model}_{ij}}, \qquad (18.1)$$

in which i represents the rows in the contingency table and j represents the columns. The observed data are, obviously, the frequencies in Table 18.1, but we need to work out what the model is. In ANOVA the model we use is group means, but as I've mentioned we can't work with means when we have only categorical variables so we work with frequencies instead. Therefore, we use 'expected frequencies'. One way to estimate the expected frequencies would be to say 'well, we've got 200 cats in total, and four categories, so the expected value is simply 200/4 = 50'. This would be fine if, for example, we had the same number of cats that had affection as a reward and food as a reward; however, we didn't: 38 got food and 162 got affection as a reward. Likewise there are not equal numbers that could and couldn't dance. To take account of this, we calculate expected frequencies for each of the cells in the table (in this case there are four cells) and we use the column and row totals for a particular cell to calculate the expected value:

$$\text{model}_{ij} = E_{ij} = \frac{\text{row total}_i \times \text{column total}_j}{n}$$

n is simply the total number of observations (in this case 200). We can calculate these expected frequencies for the four cells within our table (row total and column total are abbreviated to RT and CT respectively):

$$\text{model}_{\text{Food, Yes}} = \frac{\text{RT}_{\text{Yes}} \times \text{CT}_{\text{Food}}}{n} = \frac{76 \times 38}{200} = 14.44$$

$$\text{model}_{\text{Food, No}} = \frac{\text{RT}_{\text{No}} \times \text{CT}_{\text{Food}}}{n} = \frac{124 \times 38}{200} = 23.56$$

$$\text{model}_{\text{Affection, Yes}} = \frac{\text{RT}_{\text{Yes}} \times \text{CT}_{\text{Affection}}}{n} = \frac{76 \times 162}{200} = 61.56$$

$$\text{model}_{\text{Affection, No}} = \frac{\text{RT}_{\text{No}} \times \text{CT}_{\text{Affection}}}{n} = \frac{124 \times 162}{200} = 100.44$$

Given that we now have these model values, all we need to do is take each value in each cell of our data table, subtract from it the corresponding model value, square the result, and then divide by the corresponding model value. Once we've done this for each cell in the table, we just add them up!

$$\begin{aligned}
\chi^2 &= \frac{(28-14.44)^2}{14.44} + \frac{(10-23.56)^2}{23.56} + \frac{(48-61.56)^2}{61.56} + \frac{(114-100.44)^2}{100.44} \\
&= \frac{(13.56)^2}{14.44} + \frac{(-13.56)^2}{23.56} + \frac{(-13.568)^2}{61.56} + \frac{(13.56)^2}{100.44} \\
&= 12.73 + 7.80 + 2.99 + 1.83 \\
&= 25.35
\end{aligned}$$

This statistic can then be checked against a distribution with known properties. All we need to know is the degrees of freedom, and these are calculated as $(r-1)(c-1)$ in which r is the number of rows and c is the number of columns. Another way to think of it is the number of levels of each variable minus one multiplied. In this case we get $df = (2-1)(2-1) = 1$. If you were doing the test by hand, you would find a critical value for the chi-square distribution with $df = 1$ and if the observed value was bigger than this critical value you would say that there was a significant relationship between the two variables. These critical values are produced in the Appendix, and for $df = 1$ the critical values are 3.84 ($p = .05$) and 6.63 ($p = .01$) and so because the observed chi-square is bigger than these values it is significant at $p < .01$. However, if you use SAS, it will simply produce an estimate of the precise probability of obtaining a chi-square statistic at least as big as (in this case) 25.35 if there were no association in the population between the variables.

18.3.2. Fisher's exact test ①

There is one problem with the chi-square test, which is that the sampling distribution of the test statistic has an *approximate* chi-square distribution. The larger the sample is, the better this approximation becomes and in large samples the approximation is good enough not to worry about the fact that it is an approximation. However, in small samples, the approximation is not good enough, making significance tests of the chi-square distribution inaccurate. This is why you often read that to use the chi-square test the expected frequencies in each cell must be greater than 5 (see section 18.4). When the expected frequencies are greater than 5, the sampling distribution is probably close enough to a perfect chi-square distribution for us not to worry. However, when the expected frequencies are too low, it probably means that the sample size is too small and that the sampling distribution of the test statistic is too deviant from a chi-square distribution to be of any use.

Fisher came up with a method for computing the exact probability of the chi-square statistic that is accurate when sample sizes are small. This method is called Fisher's exact test (Fisher, 1922) even though it's not so much of a test as a way of computing the exact probability of the chi-square statistic. This procedure is normally used on 2 × 2 contingency tables (i.e. two variables each with two options) and with small samples. However, it can be used on larger contingency tables and with large samples, but on larger contingency tables it becomes computationally intensive and you might find SAS taking a long time to give you an answer. In large samples there is really no point because it was designed to overcome the problem of small samples, so you don't need to use it when samples are large.

18.3.3. The likelihood ratio ②

An alternative to Pearson's chi-square is the likelihood ratio statistic, which is based on maximum-likelihood theory. The general idea behind this theory is that you collect some data and create a model for which the probability of obtaining the observed set of data is maximized, then you compare this model to the probability of obtaining those data under the null hypothesis. The resulting statistic is, therefore, based on comparing observed frequencies with those predicted by the model:

$$L\chi^2 = 2\sum \text{observed}_{ij} \ln\left(\frac{\text{observed}_{ij}}{\text{model}_{ij}}\right) \qquad (18.2)$$

in which i and j are the rows and columns of the contingency table and ln is the natural logarithm (this is the standard mathematical function that we came across in Chapter 8 and you can find it on your calculator usually labelled as ln or $\log_e$). Using the same model and observed values as in the previous section, this would give us:

$$\begin{aligned}
L\chi^2 &= 2\left[28 \times \ln\left(\frac{28}{14.44}\right) + 10 \times \ln\left(\frac{10}{23.56}\right) + 48 \times \ln\left(\frac{48}{61.56}\right) + 114 \times \ln\left(\frac{114}{100.44}\right)\right] \\
&= 2[(28 \times 0.662) + (10 \times -0.857) + (48 \times -0.249) + (114 \times 0.0.127)] \\
&= 2[18.54 - 8.57 - 11.94 + 14.44] \\
&= 24.94
\end{aligned}$$

As with Pearson's chi-square, this statistic has a chi-square distribution with the same degrees of freedom (in this case 1). As such, it is tested in the same way: we could look up the critical value of chi-square for the number of degrees of freedom that we have. As before, the value we have here will be significant because it is bigger than the critical values of 3.84 ($p = .05$) and 6.63 ($p = .01$). For large samples this statistic will be roughly the same as Pearson's chi-square, but is preferred when samples are small.

18.3.4. Yates's correction ②

When you have a 2 × 2 contingency table (i.e. two categorical variables each with two categories) then Pearson's chi-square tends to produce significance values that are too small (in other words, it tends to make a Type I error). Therefore, Yates suggested a correction to the Pearson formula (usually referred to as Yates's continuity correction). The basic idea is that when you calculate the deviation from the model (the $\text{observed}_{ij} - \text{model}_{ij}$ in equation (18.1)) you subtract 0.5 from the absolute value of this deviation before you square it. In plain English this means you calculate the deviation, ignore whether it is positive or negative, subtract 0.5 from the value and then square it. Pearson's equation then becomes:

$$\chi^2 = \sum \frac{\left(|\text{observed}_{ij} - \text{model}_{ij}| - 0.5\right)^2}{\text{model}_{ij}}$$

For the data in our example this just translates into:

$$\begin{aligned}\chi^2 &= \frac{(13.56 - 0.5)^2}{14.44} + \frac{(13.56 - 0.5)^2}{23.56} + \frac{(13.56 - 0.5)^2}{61.56} + \frac{(13.56 - 0.5)^2}{100.44} \\ &= 11.81 + 7.24 + 2.77 + 1.70 \\ &= 23.52\end{aligned}$$

The key thing to note is that it lowers the value of the chi-square statistic and, therefore, makes it less significant. Although this seems like a nice solution to the problem, there is a fair bit of evidence that this overcorrects and produces chi-square values that are too small! Howell (2006) provides an excellent discussion of the problem with Yates's correction for continuity if you're interested; all I will say is that although it's worth knowing about, it's probably best ignored!

18.4. Assumptions of the chi-square test ①

It should be obvious that the chi-square test does not rely on assumptions such as having continuous normally distributed data like most of the other tests in this book (categorical data cannot be normally distributed because they aren't continuous). However, the chi-square test still has two important assumptions:

1 Pretty much all of the tests we have encountered in this book have made an assumption about the independence of data and the chi-square test is no exception. For the chi-square test to be meaningful it is imperative that each person, item or entity contributes to only one cell of the contingency table. Therefore, you cannot use a chi-square test on a repeated-measures design (e.g. if we had trained some cats with food to see if they would dance and then trained the same cats with affection to see if they would dance we couldn't analyse the resulting data with Pearson's chi-square test).

2. The expected frequencies should be greater than 5. Although it is acceptable in larger contingency tables to have up to 20% of expected frequencies below 5, the result is a loss of statistical power (so the test may fail to detect a genuine effect). Even in larger contingency tables no expected frequencies should be below 1. Howell (2006) gives a nice explanation of why violating this assumption creates problems. If you find yourself in this situation consider using Fisher's exact test (section 18.3.2).

Finally, although it's not an assumption, it seems fitting to mention in a section in which a gloomy and foreboding tone is being used that proportionately small differences in cell frequencies can result in statistically significant associations between variables if the sample is large enough (although it might need to be very large indeed). Therefore, we must look at row and column percentages to interpret any effects we get. These percentages will reflect the patterns of data far better than the frequencies themselves (because these frequencies will be dependent on the sample sizes in different categories).

18.5. Doing chi-square on SAS ①

There are two ways in which categorical data can be entered: enter the raw scores, or enter weighted cases. We'll look at both in turn.

18.5.1. Entering data: raw scores ①

If we input the raw scores, it means that every row of the data editor represents each entity about which we have data (in this example, each row represents a cat). So, you would create two coding variables (**Training** and **Dance**) and specify appropriate numeric codes for each. **Training** could be coded with 0 to represent a food reward and 1 to represent affection, and **Dance** could coded with 0 to represent an animal that danced and 1 to represent one that did not. For each animal, you put the appropriate numeric code into each column. So a cat that was trained with food that did not dance would have 0 in the training column and 1 in the dance column. The data in the file **Cats.sas7bdat** are entered in this way and you should be able to identify the variables described. There were 200 cats in all and so there are 200 rows of data.

18.5.2. Entering data: weight cases ①

An alternative method of data entry is to create the same coding variables as before, but to have a third variable that represents the number of animals that fell into each combination of categories. In other words we input the frequency data (the number of cases that fall into a particular category). We could call this variable **Frequency**. Now, instead of having 200 rows, each one representing a different animal, we have one row representing each combination of categories and a variable telling us how many animals fell into this category combination. So, the first row represents cats that had food as a reward and then danced. The variable **Frequency** tells us that there were 28 cats that had food as a reward and then danced. This information was previously represented by 28 different rows in the file **Cats.sas7bdat** and so you can see how this method of data entry saves you a lot of time! Extending this principle, we can see that when affection was used as a reward 114 cats did not dance.

Entering data using a variable representing the number of cases that fall into a combination of categories can be quite labour saving. However, to analyze data entered in this way we must tell the computer that the variable **Frequency** represents the number of cases that fell into a particular combination of categories.

18.5.3. Running the analysis ①

The analysis is run using PROC FREQ. We want to crosstabulate the two variables, so in the TABLES statement we put training × dance. We also add the CHISQ option to the tables statement to get the tests. For a 2 × 2 table we get Fisher's exact test automatically. For larger tables, we need to request an exact test using the EXACT option. We'll put the exact option here, so you can see how it's done – just remember that you don't actually need it here.

We show the syntax twice in SAS Syntax 18.1 – the first time we use the raw data (**cats**), the second time we use the weighted data (**catsweight**), and add the WEIGHT statement. Note: Be very (very) careful with the WEIGHT statement, because it has different effects in different procs, so don't use it unless you are sure you know what it is doing.

```
PROC FREQ DATA=chap17.cats;
    TABLES training*dance /CHISQ EXACT EXPECTED;
    RUN;

PROC FREQ DATA=chap17.catsweight;
    TABLES training*dance /CHISQ EXACT EXPECTED;
    WEIGHT frequent;
    RUN;
```
SAS Syntax 18.1

18.5.4. Output for the chi-square test ①

The crosstabulation table produced by SAS (SAS Output 18.1) contains the number of cases that fall into each combination of categories and is rather like our original contingency table. We can see that in total 76 cats danced (38% of the total) and of these 28 were trained using food (36.8% of the total that danced) and 48 were trained with affection (63.2% of the total that danced). Further, 124 cats didn't dance at all (62% of the total) and of those that didn't dance, 10 were trained using food as a reward (8.1% of the total that didn't dance) and a massive 114 were trained using affection (91.9% of the total that didn't dance). The numbers of cats can be read from the rows labelled *Frequency* (the first row) and the percentages are read from the rows labelled *Col Pct*. We can also look at the percentages within the training categories by looking at the rows labelled *Percent*. This tells us, for example, that of those trained with food as a reward, 73.7% danced and 26.3% did not. Similarly, for those trained with affection only 29.6% danced compared to 70.4% that didn't. In summary, when food was used as a reward most cats would dance, but when affection was used most cats refused to dance.

SAS OUTPUT 18.1

Table of training by dance			
training	dance		
Frequency Expected Percent Row Pct Col Pct	No	Yes	Total
Affection	48 61.56 24.00 29.63 63.16	114 100.44 57.00 70.37 91.94	162 81.00
Food	28 14.44 14.00 73.68 36.84	10 23.56 5.00 26.32 8.06	38 19.00
Total	76 38.00	124 62.00	200 100.00

Before moving on to look at the test statistic itself it is vital that we check that the assumption for chi-square has been met. The assumption is that in 2 × 2 tables (which is what we have here), all expected frequencies should be greater than 5. If you look at the expected counts in the crosstabulation table (in the second row in each cell) which incidentally are the same as we calculated earlier), it should be clear that the smallest expected count is 14.4 (for cats that were trained with food and did dance). This value exceeds 5 and so the assumption has been met. If you found an expected count lower than 5 the best remedy is to collect more data to try to boost the proportion of cases falling into each category.

SAS TIP 18.1 Choosing your statistics ②

We are getting rather a lot of information, and some of it is getting confusing. For example, we have the percentage of cats that were rewarded with affection that danced, the percentage of cats that danced which were rewarded with food, and the percentage of cats who danced *and* were rewarded with food. Then we have the actual frequencies, and the expected frequencies under the null hypothesis. This all gets a bit confusing. There are two things we can do to reduce the amount of confusion. First, we can turn off some of the statistics – we can remove the percentage for the row by putting NOROW, the percentage for the column by putting NOCOL, the percentage of the total by putting NOPERCENT, and we can turn off the frequency count by putting NOFREQ.

The second thing we can do is have more than one TABLES statement in a PROC FREQ. This way we can choose what we want. We might put:

```
TABLES training*dance /CHISQ EXACT NOCOL NOPERCENT ;

TABLES training*dance / NOCOL NOROW NOFREQ NOPERCENT EXPECTED;
```

The first TABLES statement will produce a table with just the frequencies and the row percentages (which is the percentage we are interested in). The second TABLES statement will produce a table with only the expected frequencies.

As we saw earlier, Pearson's chi-square test examines whether there is an association between two categorical variables (in this case the type of training and whether the animal danced or not). As part of PROC FREQ, SAS produces a table that includes the chi-square statistic and its significance value (SAS Output 18.2). The Pearson chi-square statistic tests whether the two variables are independent. If the significance value is small enough (conventionally it must be less than .05) then we reject the hypothesis that the variables are independent and gain confidence in the hypothesis that they are in some way related. The value of the chi-square statistic is given in the table (and the degrees of freedom) as is the significance value. The value of the chi-square statistic is 25.356, which is within rounding error of what we calculated in section 18.3.1. This value is highly significant ($p < .0001$), indicating that the type of training used had a significant effect on whether an animal would dance.

A series of other statistics are also included in the table *Continuity Adj Chi-Square* is Yates's continuity correction (see section 18.3.4) and its value is the same as the value we calculated earlier (23.52). As I mentioned earlier, this test is probably best ignored anyway, but it does confirm the result from the main chi-square test. The *Likelihood Ratio Chi-Square* is the statistic we encountered in section 18.3.3 (and is again within rounding error of the value we calculated: 24.93). Again this confirms the main chi-square result, but this statistic would be preferred in smaller samples.

Statistic	DF	Value	Prob
Chi-Square	1	25.3557	<.0001
Likelihood Ratio Chi-Square	1	24.9316	<.0001
Continuity Adj. Chi-Square	1	23.5203	<.0001
Mantel-Haenszel Chi-Square	1	25.2289	<.0001
Phi Coefficient		–0.3561	
Contingency Coefficient		0.3354	
Cramer's V		–0.3561	

SAS OUTPUT 18.2

The highly significant result indicates that there is an association between the type of training and whether the cat danced or not. What we mean by an association is that the pattern of responses (i.e. the proportion of cats that danced to the proportion that did not) in the two training conditions is significantly different. This significant finding reflects the fact that when food is used as a reward, about 74% of cats learn to dance and 26% do not, whereas when affection is used, the opposite is true (about 70% refuse to dance and 30% do dance). Therefore, we can conclude that the type of training used significantly influences the cats: they will dance for food but not for love! Having lived with a lovely cat for many years now, this supports my cynical view that they will do nothing unless there is a bowl of cat-food waiting for them at the end of it!

Underneath the chi-square statistics are three more measures – these are based on modifying the chi-square statistic to take account of sample size and degrees of freedom and they try to restrict the range of the test statistic from 0 to 1 (to make them similar to the correlation coefficient described in Chapter 7).

- *Phi*: This statistic is accurate for 2 × 2 contingency tables. However, for tables with more than two dimensions the value of phi may not lie between 0 and 1 because the chi-square value can exceed the sample size. Therefore, Pearson suggested the use of the contingency coefficient.

- *Contingency Coefficient*: This coefficient ensures a value between 0 and 1 but, unfortunately, it seldom reaches its upper limit of 1 and for this reason Cramér devised Cramer's *V*.
- *Cramer's V*: When both variables have only two categories, phi and Cramér's *V* are identical. However, when variables have more than two categories Cramér's statistic can attain its maximum of – unlike the other two – and so it is the most useful.

For these data, Cramér's statistic is 0.36 out of a possible maximum value of 1. This represents a medium association between the type of training and whether the cats danced or not (if you think of it like a correlation coefficient then this represents a medium effect size). This value is highly significant ($p < .001$) indicating that a value of the test statistic that is this big is unlikely to have happened by chance, and therefore the strength of the relationship is significant. These results confirm what the chi-square test already told us but also give us some idea of the size of effect.

18.5.5. Breaking down a significant chi-square test with standardized residuals ②

Although in a 2 × 2 contingency table like the one we have in this example, the nature of the association can be quite clear from just the cell percentages or counts, in larger contingency tables it can be useful to do a finer-grained investigation of the table. In a way, you can think of a significant chi-square test in much the same way as a significant interaction in ANOVA: it is an effect that needs to be broken down further. One very easy way to break down a significant chi-square test is to use data that we already have – the standardized residual.

Just like regression, the residual is simply the error between what the model predicts (the expected frequency) and the data actually observed (the observed frequency):

$$\text{residual}_{ij} = \text{observed}_{ij} - \text{model}_{ij}$$

in which i and j represent the two variables (i.e. the rows and columns in the contingency table). This is the same as every other residual or deviation that we have encountered in this book (compare this equation to, for example, equation (2.4)). To standardize this equation, we simply divide by the square root of the expected frequency:

$$\text{standardized residual} = \frac{\text{observed}_{ij} - \text{model}_{ij}}{\sqrt{\text{model}_{ij}}}$$

Does this equation look familiar? Well, it's basically part of equation (18.1). The only difference is that rather than looking at squared deviations, we're looking at the pure deviation. Remember that the rationale for squaring deviations in the first place is simply to make them positive so that they don't cancel out when we add them. The chi-square statistic is based on adding together values, so it is important that the deviations are squared so that they don't cancel out. However, if we're not planning to add up the deviations or residuals then we can inspect them in their unsquared form. There are two important things about these standardized residuals:

1. Given that the chi-square statistic is the sum of these standardized residuals (sort of), then if we want to decompose what contributes to the overall association that the

chi-square statistic measures, then looking at the individual standardized residuals is a good idea because they have a direct relationship with the test statistic.

2 These standardized residuals behave like any other (see section 7.6.1.1) in the sense that each one is a *z*-score. This is very useful because it means that just by looking at a standardized residual we can assess its significance (see section 1.7.4). As we have learnt many times before, if the value lies outside of ± 1.96 then it is significant at $p < .05$, if it lies outside ± 2.58 then it is significant at $p < .01$ and if it lies outside ± 3.29 then it is significant at $p < .001$.

Getting the standardized residuals out of SAS is a little difficult. I'm going to show you how to do it, but I'm not going to explain to you why it works. First, we need the weighted data, with the counts in, not the raw data. Then we use PROC GENMOD. We model the **frequency** variable as the outcome, and we don't tell SAS that it is a frequency value. The syntax is shown in SAS Syntax 18.2.

```
PROC GENMOD DATA=chap17.catsweight;
    CLASS training dance;
    MODEL frequent = dance training /DIST=poi LINK=log RESIDUALS;
    RUN;
```

SAS Syntax 18.2

SAS OUTPUT 18.3

Observation Statistics

Observation	Raw Residual	Pearson Residual	Deviance Residual	Std Deviance Residual	Std Pearson Residual	Likelihood Residual
1	13.559997	3.5684201	3.156474	4.4541401	5.0354423	4.7524084
2	−13.56	−2.793651	−3.159225	−5.694377	−5.035445	−5.247087
3	−13.56	−1.728265	−1.798365	−5.239684	−5.035443	−5.059931
4	13.56	1.3530266	1.3241769	4.9280757	5.0354432	5.0277681

FIGURE 18.2

Observation Statistics

Observation	Raw Residual	Pearson Residual	Deviance Residual	Std Deviance Residual	Std Pearson Residual	Likelihood Residual
1	13.559997	3.5684201	3.156474	4.4541401	5.0354423	4.7524084
2	−13.56	−2.793651	−3.159225	−5.694377	−5.035445	−5.247087
3	−13.56	−1.728265	−1.798365	−5.239684	−5.035443	−5.059931
4	13.56	1.3530266	1.3241769	4.9280757	5.0354432	5.0277681

SAS Output 18.3 shows the residual part of the output from PROC GENMOD. Notice that it shows us the residuals for each of the four 'observations' in the dataset, but it doesn't tell us which observation was which. The ressduals that we are interested in are those labelled 'Pearson Residual' by SAS. We need to go and look at the dataset, shown in Figure 18.2, where we find that observation 1 is *food yes*, 2 is *food no*, 3 is *affection yes* and 4 is *affection no*.

There are four residuals: one for each combination of the type of training and whether the cats danced. When food was used as a reward the standardized residual was significant for both those that danced (observation 1, $z = 3.6$) and those that didn't dance (observation 2, $z = -2.8$) because both values are bigger than 1.96 (when you ignore the minus sign). The plus or minus sign (and the counts and expected counts within the cells) tells us that when food was used as a reward significantly more cats than expected danced, and significantly fewer cats than expected did not dance. When affection was used as a reward the standardized residual was not significant for both those that danced ($z = -1.7$) and those that didn't dance ($z = 1.4$) because they are both smaller than 1.96 (when you ignore the minus sign). This tells us that when affection was used a reward as many cats as expected danced and did not dance. In a nutshell, the cells for when food was used as a reward both significantly contribute to the overall chi-square statistic. Put another way, the association between the type of reward and dancing is mainly driven by when food is a reward.

SAS TIP 18.2 Creating the weights file ②

This is all very clever, but what if you don't have a dataset with the counts in – it might be rather a lot of effort to creat it, especially if your sample size is large.

Never fear, for our trusty assistant is here to help us out. We can use a PROC SUMMARY command to produce the weights file:

```
PROC SUMMARY data=chap17.cats NWAY;
   CLASS dance training;
   OUTPUT OUT=catfreqs (drop=_type_ ) / LEVELS;
   RUN;
```

18.5.6. Calculating an effect size ②

Although Cramér's *V* is an adequate effect size (in the sense that it is constrained to fall between 0 and 1 and is, therefore, easily interpretable), a more common and possibly more useful measure of effect size for categorical data is the *odds ratio*, which we encountered in Chapter 9. Odds ratios are most interpretable in 2 × 2 contingency tables and are probably not useful for larger contingency tables. However, this isn't as restrictive as you might think because, as I've said more times than I care to recall in the GLM chapters, effect sizes are only ever useful when they summarize a focused comparison. A 2 × 2 contingency table is the categorical data equivalent of a focused comparison!

The odds ratio is simple enough to calculate. If we look at our example, we can first calculate the odds that a cat danced given that they had food as a reward. This is simply the number of cats that were given food and danced, divided by the number of cats given food that didn't dance:

$$\begin{aligned}\text{Odds}_{\text{dancing after food}} &= \frac{\text{Number that had food and danced}}{\text{Number that had food but didn't dance}}\\ &= \frac{28}{10}\\ &= 2.8\end{aligned}$$

Next we calculate the odds that a cat danced given that they had affection as a reward. This is simply the number of cats that were given affection and danced, divided by the number of cats given affection that didn't dance:

$$\text{Odds}_{\text{dancing after affection}} = \frac{\text{Number that had affection and danced}}{\text{Number that had affection but didn't dance}} = \frac{48}{114} = 0.421$$

The odds ratio is simply the odds of dancing after food divided by the odds of dancing after affection:

$$\text{Odds ratio} = \frac{\text{Odds}_{\text{dancing after food}}}{\text{Odds}_{\text{dancing after affection}}} = \frac{2.8}{0.421} = 6.65$$

What this tells us is that if a cat was trained with food the odds of their dancing were 6.65 times higher than if they had been trained with affection. As you can see, this is an extremely elegant and easily understood metric for expressing the effect you've got.

However, there is one thing to watch out for – and I'll illustrate it by comparing with a mean difference. When we have a mean difference, it doesn't really matter which way around we express the difference. If group A has a mean that is 10 points higher than group B, we could also say that group B is 10 points lower than group A. With an odds ratio, if one group's odds are 6.65 times the other group's odds, we can't reverse it and say that the other group's odds are –6.65 times the first group's odds. Instead, because they are ratios, we take the inverse (that's a posh way of saying divide by 1). So the odds of dancing given food as a reward are 6.65 times the odds of dancing when affection is given as a reward. We could also say that the odds of dancing when given affection are 1/6.65 = 0.15 times the odds of dancing when given food.

SAS doesn't provide the odds ratio directly, but we can persuade it to, by adding the CMH option to the TABLES statement in PROC FREQ (see SAS Syntax 18.3).

```
PROC FREQ DATA=catsweight;
     TABLES dance*training /CHISQ CMH;
     WEIGHT frequent;
     RUN;
```

SAS Syntax 18.3

18.5.7. Reporting the results of chi-square ①

When reporting Pearson's chi-square we simply report the value of the test statistic with its associated degrees of freedom and the significance value. The test statistic, as we've seen, is denoted by χ^2. The SAS output tells us that the value of χ^2 was 25.36, that the degrees of freedom on which this was based were 1, and that it was significant at $p < .0001$. It's also useful to reproduce the contingency table, and my vote would go to quoting the odds ratio too. As such, we could report:

- ✓ There was a significant association between the type of training and whether or not cats would dance $\chi^2(1) = 25.36$, $p < .0001$. This seems to represent the fact that, based on the odds ratio, the odds of cats dancing were 6.65 times higher if they were trained with food than if trained with affection.

CRAMMING SAM'S TIPS

- If you want to test the relationship between two categorical variables you can do this with *Pearson's chi-square test* or the *likelihood ratio statistic*.
- Look at the table labelled **Statistics for Table of Dance by Training** (or whatever your variables are called); if the *Prob.* value is less than .05 for the row labelled *Chi-Square* then there is a significant relationship between your two variables.
- Check the expected frequencies to make sure that none are less than 5.
- Look at the crosstabulation table to work out what the relationship between the variables is. Better still, calculate standardized residuals (using PROC GENMOD) and look out for significant standardized residuals (values outside of ±1.96), and calculate the *odds ratio*.
- Report the χ^2 statistic, the degrees of freedom and the significance value. Also report the contingency table.

LABCOAT LENI'S REAL RESEARCH 18.1

Is the black American happy? ①

When I was doing my psychology degree I spent a lot of time reading about the civil rights movement in the USA. Although I was supposed to be reading psychology, I became more interested in Malcolm X and Martin Luther King Jr. This is why I find Beckham's (1929) study of black Americans such an interesting piece of research. Beckham was a black American academic who founded the psychology laboratory at Howard University, Washington, DC, and his wife Ruth was the first black woman ever to be awarded a Ph.D. (also in psychology) at the University of Minnesota. The article needs to be placed within the era in which it was published. To put some context on the study, it was published 36 years before the Jim Crow laws were finally overthrown by the Civil Rights Act of 1964, and in a time when black Americans were segregated, openly discriminated against and were victims of the most abominable violations of civil liberties and human rights. For a richer context I suggest reading James Baldwin's superb novel *The fire next time*. Even the language of the study and the data from it are an uncomfortable reminder of the era in which it was conducted.

Beckham sought to measure the psychological state of black Americans with three questions put to 3443 black Americans from different walks of life. He asked them whether they thought black Americans were happy, whether they personally were happy as a black American, and whether black Americans *should* be happy. They could answer only *yes* or *no* to each question. By today's standards the study is quite simple, and he did no formal statistical analysis of his data (Fisher's article containing the popularized version of the chi-square test was published only 7 years earlier in a statistics journal that would not have been read by psychologists). I love this study, though, because it demonstrates that you do not need elaborate methods to answer important and far-reaching questions; with just three questions, Beckham told the world an enormous amount about very real and important psychological and sociological phenomena.

The frequency data (number of yes and no responses within each employment category) from this study are in the file **Beckham1929.sas7bdat**. Labcoat Leni wants you to carry out three chi-square tests (one for each question that was asked). What conclusions can you draw?

Answers are in the additional material on the companion website.

BECKHAM, A. S. (1929). *JOURNAL OF ABNORMAL AND SOCIAL PSYCHOLOGY, 24*, 186–190.

18.6. Several categorical variables: loglinear analysis ①

So far we've looked at situations in which there are only two categorical variables. However, often we want to analyse more complex contingency tables in which there are three or more variables. For example, what about if we took the example we've just used but also collected data from a sample of 70 dogs? We might want to compare the behaviour in dogs to that in cats. We would now have three variables: **Animal** (dog or cat), **Training** (food as reward or affection as reward) and **Dance** (did they dance or not?). This couldn't be analysed with the Pearson chi-square and instead has to be analysed with a technique called loglinear analysis.

18.6.1. Chi-square as regression ①

To begin with, let's have a look at how our simple chi-square example can be expressed as a regression model. Although we already know about as much as we need to about the chi-square test, if we want to understand more complex situations life becomes considerably easier if we consider our model as a general linear model (i.e. regression). All of the general linear models we've considered in this book take the general form of:

$$\text{outcome}_i = (\text{model}) + \text{error}_i$$

For example, when we encountered multiple regression in Chapter 8 we saw that this model was written as (see equation (7.9)):

$$Y_i = (b_0 + b_1X_{1i} + b_2X_{2i} + \ldots + b_nX_{ni}) + \varepsilon_i$$

Also, when we came across one-way ANOVA, we adapted this regression model to conceptualize our Viagra example, as (see equation (11.2)):

$$\text{Libido}_i = b_0 + b_2\text{High}_i + b_1\text{Low}_i + \varepsilon_i$$

The *t*-test was conceptualized in a similar way. In all cases the same basic equation is used; it's just the complexity of the model that changes. With categorical data we can use the same model in much the same way as with regression to produce a linear model. In our current example we have two categorical variables: training (food or affection) and dance (yes they did dance or no they didn't dance). Both variables have two categories and so we can represent each one with a single dummy variable (see section 7.11.1) in which one category is coded as 0 and the other as 1. So for training, we could code 'food' as 0 and 'affection' as 1, and we could code the dancing variable as 0 for 'yes' and 1 for 'no' (see Table 18.2).

TABLE 18.2 Coding scheme for dancing cats

Training	Dance	Dummy (Training)	Dummy (Dance)	Interaction	Frequency
Food	Yes	0	0	0	28
Food	No	0	1	0	10
Affection	Yes	1	0	0	48
Affection	No	1	1	1	114

This situation might be familiar if you think back to factorial ANOVA (section 12.8) in which we also had two variables as predictors. In that situation we saw that when there are two variables the general linear model became (think back to equation (12.1)):

$$\text{outcome}_i = (b_0 + b_1 A_i + b_2 B_i + b_3 AB_i) + \varepsilon_i$$

in which A represents the first variable, B represents the second and AB represents the interaction between the two variables. Therefore, we can construct a linear model using these dummy variables that is exactly the same as the one we used for factorial ANOVA (above). The interaction term will simply be the training variable multiplied by the dance variable (look at Table 18.2 and if it doesn't make sense look back to section 12.8 because the coding is exactly the same as this example):

$$\begin{aligned}\text{outcome}_i &= (\text{model}) + \text{Error}_i \\ \text{outcome}_{ij} &= (b_0 + b_1\text{Training}_i + b_2\text{Dance}_j + b_3\text{Interaction}_{ij})\end{aligned} \tag{18.3}$$

However, because we're using categorical data, to make this model linear we have to actually use log values (see Chapter 9) and so the actual model becomes:[2]

$$\begin{aligned}\ln(O_i) &= \ln(\text{model}) + \ln(\varepsilon_i) \\ \ln(O_{ij}) &= (b_0 + b_1\text{Training}_i + b_2\text{Dance}_j + b_3\text{Interaction}_{ij}) + \ln(\varepsilon_i)\end{aligned} \tag{18.4}$$

The training and dance variables and the interaction can take the values 0 and 1, depending on which combination of categories we're looking at (Table 18.2). Therefore, to work out what the b-values represent in this model we can do the same as we did for the t-test and ANOVA and look at what happens when we replace training and dance with values of 0 and 1. To begin with, let's see what happens when we look at when training and dance are both zero. This represents the category of cats that got food reward and did line-dance. When we used this sort of model for the t-test and ANOVA the outcomes we used were taken from the observed data: we used the group means (e.g. see sections 9.7 and 10.2.3). However, with categorical variables, means are rather meaningless because we haven't measured anything on an ordinal or interval scale, instead we merely have frequency data. Therefore, we use the observed frequencies (rather than observed means) as our outcome instead. In Table 18.1 we saw that there were 28 cats that had food for a reward and did line-dance. If we use this as the observed outcome then the model can be written as (if we ignore the error term for the time being):

$$\ln(O_{ij}) = b_0 + b_1\text{Training}_i + b_2\text{Dance}_j + b_3\text{Interaction}_{ij}$$

For cats that had food reward and did dance, the training and dance variables and the interaction will all be 0 and so the equation reduces down to:

$$\begin{aligned}\ln(O_{\text{Food, Yes}}) &= b_0 + (b_1 \times 0) + (b_2 \times 0) + (b_3 \times 0) \\ \ln(O_{\text{Food, Yes}}) &= b_0 \\ \ln(28) &= b_0 \\ b_0 &= 3.332\end{aligned}$$

[2] Actually, the convention is to denote b_0 as θ and the b-values as λ, but I think these notational changes serve only to confuse people so I'm sticking with b because I want to emphasize the similarities to regression and ANOVA.

Therefore, b_0 in the model represents the log of the observed value when all of the categories are zero. As such it's the log of the observed value of the base category (in this case cats that got food and danced). Now, let's see what happens when we look at cats that had affection as a reward and danced. In this case, the training variable is 1 and the dance variable and the interaction are still 0. Also, our outcome now changes to be the observed value for cats that received affection and danced (from Table 18.1 we can see the value is 48). Therefore, the equation becomes:

$$\ln(O_{\text{Affection, Yes}}) = b_0 + (b_1 \times 1) + (b_2 \times 0) + (b_3 \times 0)$$
$$\ln(O_{\text{Affection, Yes}}) = b_0 + b_1$$
$$b_1 = \ln(O_{\text{Affection, Yes}}) - b_0$$

Remembering that b_0 is the expected value for cats that had food and danced, we get:

$$\begin{aligned} b_1 &= \ln(O_{\text{Affection, Yes}}) - \ln(O_{\text{Food, Yes}}) \\ &= \ln(48) - \ln(28) \\ &= 3.871 - 3.332 \\ &= 0.539 \end{aligned}$$

The important thing is that b_1 is the difference between the log of the observed frequency for cats that received affection and danced, and the log of the observed values for cats that received food and danced. Put another way, within the group of cats that danced it represents the difference between those trained using food and those trained using affection.

Now, let's see what happens when we look at cats that had food as a reward and did not dance. In this case, the training variable is 0, the dance variable is 1 and the interaction is again 0. Our outcome now changes to be the observed frequency for cats that received food but did not dance (from Table 18.1 we can see the value is 10). Therefore, the equation becomes:

$$\ln(O_{\text{Food, No}}) = b_0 + (b_1 \times 0) + (b_2 \times 1) + (b_3 \times 0)$$
$$\ln(O_{\text{Food, No}}) = b_0 + b_2$$
$$b_2 = \ln(O_{\text{Food, No}}) - b_0$$

Remembering that b_0 is the expected value for cats that had food and danced, we get:

$$\begin{aligned} b_2 &= \ln(O_{\text{Food, No}}) - \ln(O_{\text{Food, Yes}}) \\ &= \ln(10) - \ln(28) \\ &= 2.303 - 3.332 \\ &= -1.029 \end{aligned}$$

The important thing is that b_2 is the difference between the log of the observed frequency for cats that received food and danced, and the log of the observed frequency for cats that received food and didn't dance. Put another way, within the group of cats that received food as a reward it represents the difference between cats that didn't dance and those that did.

Finally, we can look at cats that had affection and danced. In this case, the training and dance variables are both 1 and the interaction (which is the value of training multiplied by the value of dance) is also 1. We can also replace b_0, b_1, and b_2, with what we now know they represent. The outcome is the log of the observed frequency for cats that received affection but didn't dance (this expected value is 114 – see Table 18.1). Therefore, the equation becomes (I've used the shorthand of A for affection, F for food, Y for yes, and N for no):

$$\ln(O_{A,N}) = b_0 + (b_1 \times 1) + (b_2 \times 1) + (b_3 \times 1)$$
$$\ln(O_{A,N}) = b_0 + b_1 + b_2 + b_3$$
$$\ln(O_{A,N}) = \ln(O_{F,Y}) + (\ln(O_{A,Y}) - \ln(O_{F,Y})) + (\ln(O_{F,N}) - \ln(O_{F,Y})) + b_3$$
$$\ln(O_{A,N}) = \ln(O_{A,Y}) + \ln(O_{F,N}) - \ln(O_{F,Y}) + b_3$$
$$\begin{aligned} b_3 &= \ln(O_{A,N}) - \ln(O_{F,N}) + \ln(O_{F,Y}) - \ln(O_{A,Y}) \\ &= \ln(114) - \ln(10) + \ln(28) - \ln(48) \\ &= 1.895 \end{aligned}$$

So, b_3 in the model really compares the difference between affection and food when the cats didn't dance to the difference between food and affection when the cats did dance. Put another way, it compares the effect of training when cats didn't dance to the effect of training when they did dance.

The final model is therefore:

$$\ln(O_{ij}) = 3.332 + 0.539\text{Training} - 1.029\text{Dance} + 1.895\text{Interaction} + \ln(\varepsilon_{ij})$$

The important thing to note here is that everything is exactly the same as factorial ANOVA except that we dealt with log-transformed values (in fact compare this section to section 12.8 to see just how similar everything is). In case you still don't believe me that this works as a general linear model, I've prepared a file called **CatRegression.sas7bdat** which contains the two variables **Dance** and **Training** (both dummy coded with 0 and 1 as described above) and the interaction (**Interaction**). There is also a variable called **Observed** which contains the observed frequencies in Table 18.1 for each combination of **Dance** and **Training**. Finally, there is a variable called **LnObserved**, which is the natural logarithm of these observed frequencies (remember that throughout this section we've dealt with the log observed values).

Run a multiple regression analysis using **CatsRegression.sas7bdat** with **LnObserved** as the outcome, and **Training**, **Dance** and **Interaction** as your three predictors.

SAS Output 18.4 shows the resulting coefficients table from this regression. The important thing to note is that the constant, b_0, is 3.332 as calculated above, the beta value for type of training, b_1, is 0.539 and for dance, b_2, is −1.030, both of which are within rounding error of what was calculated above. Also the coefficient for the interaction, b_3, is 1.895 as predicted. There is one interesting point, though: all of the standard errors are zero, or put differently there is *no* error at all in this model (which is also why there are no significance tests). This is because the various combinations of coding variables completely explain the observed values. This is known as a *saturated model* and I will return to this point later, so bear it in mind. For the time being, I hope this convinces you that chi-square can be conceptualized as a linear model.

SAS OUTPUT 18.4

Parameter Estimates						
Variable	Label	DF	Parameter Estimate	Standard Error	t Value	Pr > |t|
Intercept	Intercept	1	3.33220	0	Infty	<.0001
TRAINING	Type of Training	1	0.53900	0	Infty	<.0001
DANCE	Did they dance?	1	–1.02962	0	–Infty	<.0001
INT	Interaction	1	1.89462	0	Infty	<.0001

OK, this is all very well, but the heading of this section did rather imply that I would show you how the chi-square test can be conceptualized as a linear model. Well, basically, the chi-square test looks at whether two variables are independent; therefore, it has no interest in the combined effect of the two variables, only their unique effect. Thus, we can conceptualize chi-square in much the same way as the saturated model, except that we don't include the interaction term. If we remove the interaction term, our model becomes:

$$\ln(\text{model}_{ij}) = b_0 + b_1\text{Training}_i + b_2\text{Dance}_j$$

With this new model, we cannot predict the observed values like we did for the saturated model because we've lost some information (namely, the interaction term). Therefore, the outcome from the model changes, and therefore the beta values change too. We saw earlier that the chi-square test is based on 'expected frequencies'. Therefore, if we're conceptualizing the chi-square test as a linear model, our outcomes will be these expected values. If you look back to the beginning of this chapter you'll see we already have the expected frequencies based on this model. We can recalculate the beta values based on these expected values:

$$\ln(E_{ij}) = b_0 + b_1\text{Training}_i + b_2\text{Dance}_j$$

For cats that had food reward and did dance, the training and dance variables will be 0 and so the equation reduces down to:

$$\begin{aligned}
\ln(E_{\text{Food, Yes}}) &= b_0 + (b_1 \times 0) + (b_2 \times 0)\\
\ln(E_{\text{Food, Yes}}) &= b_0 + b_1\\
b_0 &= \ln(14.44)\\
&= 2.67
\end{aligned}$$

Therefore, b_0 in the model represents the log of the expected value when all of the categories are zero.

When we look at cats that had affection as a reward and danced, the training variable is 1 and the dance variable is still 0. Also, our outcome now changes to be the expected value for cats that received affection and danced:

$$\begin{aligned}
\ln(E_{\text{Affection, Yes}}) &= b_0 + (b_1 \times 1) + (b_2 \times 0)\\
\ln(E_{\text{Affection, Yes}}) &= b_0 + b_1\\
b_1 &= \ln(E_{\text{Affection, Yes}}) - b_0\\
&= \ln(E_{\text{Affection, Yes}}) - \ln(E_{\text{Food, Yes}})\\
&= \ln(61.56) - \ln(14.44)\\
&= 1.45
\end{aligned}$$

The important thing is that b_1 is the difference between the log of the expected frequency for cats that received affection and danced and the log of the expected values for cats that received food and danced. In fact, the value is the same as the column marginal, that is, the difference between the total number of cats getting affection and the total number of cats getting food: ln(162)−ln(38) = 1.45. Put simply, it represents the main effect of the type of training.

When we look at cats that had food as a reward and did not dance, the training variable is 0 and the dance variable is 1. Our outcome now changes to be the expected frequency for cats that received food but did not dance:

$$\begin{aligned}
\ln(E_{\text{Food, No}}) &= b_0 + (b_1 \times 0) + (b_2 \times 1) \\
\ln(E_{\text{Food, No}}) &= b_0 + b_2 \\
b_2 &= \ln(O_{\text{Food, No}}) - b_0 \\
&= \ln(O_{\text{Food, No}}) - \ln(O_{\text{Food, yes}}) \\
&= \ln(23.56) - \ln(14.44) \\
&= 0.49
\end{aligned}$$

Therefore, b_2 is the difference between the log of the expected frequencies for cats that received food and didn't or did dance. In fact, the value is the same as the row marginal, that is, the difference between the total number of cats that did and didn't dance: ln(124)−ln(76) = 0.49. In simpler terms, it is the main effect of whether or not the cat danced.

We can double-check all of this by looking at the final cell:

$$\begin{aligned}
\ln(E_{\text{Affection, No}}) &= b_0 + (b_1 \times 1) + (b_2 \times 1) \\
\ln(E_{\text{Affection, No}}) &= b_0 + b_1 + b_2 \\
\ln(100.44) &= 2.67 + 1.45 + 0.49 \\
4.61 &= 4.61
\end{aligned}$$

The final chi-square model is therefore:

$$\begin{aligned}
\ln(O_i) &= \text{model}_{ij} \cdot \ln ln(\varepsilon_i) \\
\ln(O_i) &= 2.67 + 1.45\text{Training} + 0.49\text{Dance} + \ln(\varepsilon_i)
\end{aligned}$$

We can rearrange this to get some residuals (the error term):

$$\ln(\varepsilon_i) = \ln(O_i) - (\text{model})$$

In this case, the model is merely the expected frequencies that were calculated for the chi-square test, so the residuals are the differences between the observed and expected frequencies.

SELF-TEST To show that this all actually works, run another multiple regression analysis using **CatsRegression.sas7bdat.** This time the outcome is the log of expected frequencies (**LnExpected**) and **Training** and **Dance** are the predictors (the interaction is not included).

This demonstrates how chi-square can work as a linear model, just like regression and ANOVA, in which the beta values tell us something about the relative differences in frequencies across categories of our two variables. If nothing else made sense I want you to leave this section aware that chi-square (and analysis of categorical data generally) can be expressed as a linear model (although we have to use log values). We can express categories of a variable using dummy variables, just as we did with regression and ANOVA, and the resulting beta values can be calculated in exactly the same way as for regression and ANOVA. In ANOVA, these beta values represented differences between the means of a particular category compared against a baseline category. With categorical data, the beta values represent the same thing, the only difference being that rather than dealing with means, we're dealing with predicted probabilities. Grasping this idea (that regression, *t*-tests, ANOVAs and categorical data analysis are basically the same) will help (me) considerably in the next section.

18.6.2. Loglinear analysis ①

In the previous section, after nearly reducing my brain to even more of a rotting vegetable than it already is trying to explain how categorical data analysis is just another form of regression, I ran the data through an ordinary regression on SAS to prove that I wasn't talking complete gibberish. At the time I rather glibly said 'oh, by the way, there's no error in the model, that's odd isn't it?' and sort of passed this off by telling you that it was a 'saturated' model and not to worry too much about it because I'd explain it all later just as soon as I'd worked out what the hell was going on. That seemed like a good avoidance tactic at the time but unfortunately I now have to explain what I was going on about.

To begin with, I hope you're now happy with the idea that categorical data can be expressed in the form of a linear model provided that we use log values (this, incidentally, is why the technique we're discussing is called loglinear analysis). From what you hopefully already know about ANOVA and linear models generally, you should also be cosily tucked up in bed with the idea that we can extend any linear model to include any amount of predictors and any resulting interaction terms between predictors. Therefore, if we can represent a simple two-variable categorical analysis in terms of a linear model, then it shouldn't amaze you to discover that if we have more than two variables this is no problem: we can extend the simple model by adding whatever variables and the resulting interaction terms. This is all you really need to know. So, just as in multiple regression and ANOVA, if we think of things in terms of a linear model, then conceptually it becomes very easy to understand how the model expands to incorporate new variables. So, for example, if we have three predictors (*A*, *B* and *C*) in ANOVA (think back to section 14.4) we end up with three two-way interactions (*AB*, *AC*, *BC*) and one-three way interaction (*ABC*). Therefore, the resulting linear model of this is just:

$$\text{outcome}_i = (b_0 + b_1A + b_2B + b_3B + b_4AB + b_5AC + b_6BC + b_7ABC + \varepsilon_i$$

In exactly the same way, if we have three variables in a categorical data analysis we get an identical model, but with an outcome in terms of logs:

$$\ln(\text{O}_{ijk} = (b_0 + b_1A_i + b_2B_i + b_3C_k + b_4AB_{ij} + b_5AC_{ik} + b_6BC_{jk} + b_7ABC_{ijk} + \ln(\varepsilon_{ijk})$$

Obviously the calculation of beta values and expected values from the model becomes considerably more cumbersome and confusing, but that's why we invented computers – so that we don't have to worry about it! Loglinear analysis works on these principles. However, as we've seen in the two-variable case, when our data are categorical and we include all of the available terms (main effects and interactions) we get no error: our predictors can perfectly predict our outcome (the expected values). So, if we start with the most complex model possible, we will get no error. The job of loglinear analysis is to try to fit a simpler model to the data without any substantial loss of predictive power. Therefore, loglinear analysis typically works on a principle of backward elimination (yes, the same kind of backward elimination that we can use in multiple regression – see section 7.5.3.3). So we begin with the saturated model, and then we remove a predictor from the model and using this new model we predict our data (calculate expected frequencies, just like the chi-square test) and then see how well the model fits the data (i.e. are the expected frequencies close to the observed frequencies?). If the fit of the new model is not very different from the more complex model, then we abandon the complex model in favour of the new one. Put another way, we assume the term we removed was not having a significant impact on the ability of our model to predict the observed data.

However, the analysis doesn't just remove terms randomly, it does it hierarchically. So, we start with the saturated model and then remove the highest-order interaction, and assess the effect that this has. If removing the interaction term has no effect on the model, we get rid of it and move on to remove any lower-order interactions. If removing these interactions has no effect then we carry on to any main effects until we find an effect that does affect the fit of the model if it is removed.

To put this in more concrete terms, at the beginning of the section on loglinear analysis I asked you to imagine we'd extended our training and line-dancing example to incorporate a sample of dogs. So, we now have three variables: animal (dog or cat), training (food or affection) and dance (did they dance or not?). Just as in ANOVA this results in three main effects:

- Animal
- Training
- Dance

three interactions involving two variables:

- Animal × Training
- Animal × Dance
- Training × Dance

and one interaction involving all three variables:

- Animal × Training × Dance

I mentioned that the likelihood ratio statistic (see section 18.3.3) is used to assess each model. It should be clear how equation (18.2) can be adapted to fit any model: the observed values are the same throughout, and the model frequencies are simply the expected frequencies from the model being tested. For the saturated model, this statistic will always be 0 (because the observed and model frequencies are the same so the ratio of observed to model frequencies will be 1, and $\ln(1) = 0$), but as we've seen, in other cases it will provide a measure of how well the model fits the observed frequencies. To test whether a new model has changed the likelihood ratio, all we need do is to take the likelihood ratio for a model and subtract from it the likelihood statistic for the previous model (provided the models are hierarchically structured):

$$L\chi^2_{\text{Change}} = L\chi^2_{\text{Current Model}} - L\chi^2_{\text{Previous Model}} \quad (18.5)$$

I've tried in this section to give you a flavour of how loglinear analysis works, without actually getting too much into the nitty-gritty of the calculations. I've tried to show you how we can conceptualize a chi-square analysis as a linear model and then relied on what I've previously told you about ANOVA to hope that you can extrapolate these conceptual ideas to understand roughly what's going on. The curious among you might want to know exactly how everything is calculated and to these people I have two things to say: 'I don't know' and 'I know a really good place where you can buy a straitjacket'. If you're that interested then Tabachnick and Fidell (2007) have, as ever, written a wonderfully detailed and lucid chapter on the subject which frankly puts this feeble attempt to shame. Still, assuming you're happy to live in relative ignorance, we'll now have a look at how to do a loglinear analysis.

18.7. Assumptions in loglinear analysis ②

Loglinear analysis is an extension of the chi-square test and so has similar assumptions; that is, an entity should fall into only one cell of the contingency table (i.e. cells of the table must be independent) and the expected frequencies should be large enough for a reliable analysis. In loglinear analysis with more than two variables it's all right to have up to 20% of cells with expected frequencies less than 5; however, all cells must have expected frequencies greater than 1. If this assumption is broken the result is a radical reduction in test power – so dramatic in fact that it may not be worth bothering with the analysis at all. Remedies for problems with expected frequencies are: (1) collapse the data across one of the variables (preferably the one you least expect to have an effect!); (2) collapse levels of one of the variables; (3) collect more data; or (4) accept the loss of power.

If you want to collapse data across one of the variables then certain things have to be considered:

1 The highest-order interaction should be non-significant.
2 At least one of the lower-order interaction terms involving the variable to be deleted should be non-significant.

Let's take the example we've been using. Say we wanted to delete the animal variable; then for this to be valid, the animal × training × dance variable should be non-significant, and either the animal × training or the animal × dance interaction should also be non-significant. You can also collapse categories within a variable. So, if you had a variable of 'season' relating to spring, summer, autumn and winter, and you had very few observations in winter, you could consider reducing the variable to three categories: spring, summer, autumn/winter perhaps. However, you should really only combine categories that it makes theoretical sense to combine.

Finally, some people overcome the problem by simply adding a constant to all cells of the table, but there really is no point in doing this because it doesn't address the issue of power.

18.8. Loglinear analysis using SAS ②

18.8.1. Initial considerations ②

Data are entered for loglinear analysis in the same way as for the chi-square test (see sections 18.5.1 and 18.5.2). The data for the cat and dog example are in the file **CatsandDogs.**

sas7bdat; open this file. Notice that it has three variables (**Animal**, **Training** and **Dance**) and each one contains codes representing the different categories of these variables. To begin with, we should use PROC FREQ to produce a contingency table of the data. The syntax I have used for this is shown in SAS Syntax 18.4. Notice I have two TABLES statements, the first to get frequencies and row and column percentages, the second to get only the expected values.

```
PROC FREQ DATA=catsanddogs;
    TABLES animal*training*dance /NOPERCENT;
    TABLES animal*training*dance /NOPERCENT NOROW NOCOL NOFREQ
    EXPECTED;
    RUN;
```
SAS Syntax 18.4

The crosstabulation table produced by SAS (SAS Output 18.5) contains the number of cases that fall into each combination of categories. The top half of this table is the same as SAS Output 18.1 because the data are the same (we've just added some dogs) and if you look back in this chapter there's a summary of what this tells us. For the dogs we can summarize the data in a similar way. In total 49 dogs danced (70% of the total) and of these 20 were trained using food (40.8% of the total that danced) and 29 were trained with affection (59.2% of the total that danced). Further, 21 dogs didn't dance at all (30% of the total) and of those that didn't dance, 14 were trained using food as a reward (66.7% of the total that didn't dance) and 7 were trained using affection (33.3% of the total that didn't dance). In summary, a lot more dogs danced (70%) than didn't (30%). About half of those that danced were trained with affection and about half with food as a reward. In short, dogs seem more willing to dance than cats (70% compared to 38%), and they're not too worried what training method is used.

Before moving on to look at the test statistics it is vital that we check that the assumptions of loglinear analysis have been met: specifically, there should be no expected counts less than 1, and no more than 20% less than 5. If you look at the expected counts in the crosstabulation table, it should be clear that the smallest expected count is 10.2 (for dogs that were trained with food but didn't dance). This value still exceeds 5 and so the assumption has been met.

SAS OUTPUT 18.5

Table 1 of training by dance			
Controlling for animal=cat			
training	dance		
Frequency Row Pct Col Pct	No	Yes	Total
Affection	114	48	162
	70.37	29.63	
	91.94	63.16	
Food	10	28	38
	26.32	73.68	
	8.06	36.84	
Total	124	76	200

Table 2 of training by dance			
Controlling for animal=dog			
training	**dance**		
Frequency Row Pct Col Pct	**No**	**Yes**	**Total**
Affection	7	29	36
	19.44	80.56	
	33.33	59.18	
Food	14	20	34
	41.18	58.82	
	66.67	40.82	
Total	21	49	70

Table 1 of training by dance			
Controlling for animal=cat			
training	**dance**		
Expected	**No**	**Yes**	**Total**
Affection	100.44	61.56	
Food	23.56	14.44	
Total	124	76	200

Table 2 of training by dance			
Controlling for animal=dog			
training	**dance**		
Expected	**No**	**Yes**	**Total**
Affection	10.8	25.2	
Food	10.2	23.8	
Total	21	49	70

18.8.2. The loglinear analysis ②

Having established that the assumptions have been met we can move on to the main analysis.

Loglinear models are run in PROC CATMOD (CATegorical MODels) in SAS. (You can also run them in PROC GENMOD, and in a couple of other procs, but if you're anything like me, you only want to know the easiest way, and that's PROC CATMOD.) We need to be a little bit careful with PROC CATMOD, because it's a very powerful procedure that can do all kinds of very clever things, but that means we can get into quite a big tangle with it.

PROC CATMOD looks a lot like PROC GLM. We have a MODEL statement, which defines the model. Then we have a LOGLIN statement, in which we define the parameters that we want in the model – a little like the ESTIMATE statement in PROC GLM. However, in the LOGLIN statement, we *remove* parameters from the model and see if the model still fits.

The syntax for the first model is shown in SAS Syntax 18.5. The MODEL statement looks a bit like a GLM MODEL statement (see Chapter 10), except we have three variables on the left-hand side of the equals sign, and on the right we have _response_, which isn't a variable at all.

Then we have the LOGLIN statement, where we specify the model that we want to test. Because we have three variables, we have three main effects (animal, training and dance), three two-way interaction effects (animal × training, animal × dance, training × dance), and one three-way effect (animal × training × dance). We could write each of those out in full, but we're kind of lazy and would rather have a simpler way. SAS provides it with the | symbol – putting animal|training|dance tells SAS that we'd like all the main effects and interactions.

```
PROC CATMOD DATA=chap17.catsanddogs;
   MODEL animal*training*dance = _response_;
   LOGLIN animal|training|dance ;
   RUN;
```

SAS Syntax 18.5

Next, we remove the three-way interaction, leaving only the main effects and two-way interactions. Again, we could write them out in full, but thankfully SAS gives us a lazy way to do it – by putting @2 at the end of the LOGLIN line, we include all interactions up to two-way. When we have run that model, we might want to include a model with only main effects – we can do this by putting @1 at the end of the LOGLIN line. Both of these are shown in SAS Syntax 18.6.

```
PROC CATMOD DATA=chap17.catsanddogs;
    MODEL animal*training*dance = _response_;
    LOGLIN animal|training|dance @2;
    RUN;

PROC CATMOD DATA=chap17.catsanddogs;
    MODEL animal*training*dance = _response_;
    LOGLIN animal|training|dance @1;
    RUN;
```

SAS Syntax 18.6

18.9. Output from loglinear analysis ①

SAS Output 18.6 shows the initial output from the loglinear analysis. The first table tells us that we have 270 cases (remember thatwe had 200 cats and 70 dogs and this is a useful check that no cats or dogs have been lost – they do tend to wander off). SAS then provides parameter estimates for the model that we requested (all terms are in the model including the highest-order interaction, in this case the animal × training × dance interaction) – the table labelled Maximum Likelihood Analysis of Variance contains the parameter estimates for each effect, along with standard error, chi-square and probability, shown in the first table in SAS Output 18.7.

SAS OUTPUT 18.6

Data Summary			
Response	**animal*training*dance**	**Response Levels**	**8**
Weight Variable	None	Populations	1
Data Set	CATSANDDOGS	Total Frequency	270
Frequency Missing	0	Observations	270

Population Profiles	
Sample	**Sample Size**
1	270

Response Profiles			
Response	**animal**	**training**	**dance**
1	cat	Affection	No
2	cat	Affection	Yes
3	cat	Food	No
4	cat	Food	Yes
5	dog	Affection	No
6	dog	Affection	Yes
7	dog	Food	No
8	dog	Food	Yes

The second table shows the chi-square statistic for each variable in the model – because all of our variables had only two categories, these two tables are very similar to one another. However, the final line in this second table gives us the goodness-of-fit statistic in the form of the likelihood ratio statistic. In this context these tests are testing the hypothesis that the frequencies predicted by the model (the expected frequencies) are significantly different from the actual frequencies in our data (the observed frequencies). Now, obviously, if our model is a good fit of the data then the observed and expected frequencies should be very similar (i.e. not significantly different). Therefore, we want these statistics to be non-significant. A significant result would mean that our model was significantly different from our data (i.e. the model is a bad fit of the data). In this example, the chi-square statistic is 0 and yields a probability value, p, of '.', which is a rather confusing way of saying that the probability cannot be computed. The reason why it cannot be computed is because at this stage the model *perfectly* predicts the data. If you read the theory section this shouldn't surprise you because I showed there that the saturated model is a perfect fit of the data and I also mentioned that the resulting likelihood ratio would be zero. What's interesting in loglinear analysis is what bits of the model we can then remove without significantly affecting the fit of the model.

SAS OUTPUT 18.7

Analysis of Maximum Likelihood Estimates					
Parameter		Estimate	Standard Error	Chi-Square	Pr > ChiSq
animal	cat	0.4118	0.0852	23.38	<.0001
training	Affection	0.3314	0.0852	15.14	<.0001
animal*training	cat Affection	0.4118	0.0852	23.38	<.0001
dance	No	–0.2428	0.0852	8.13	0.0043
animal*dance	cat No	0.2017	0.0852	5.61	0.0179
training*dance	Affection No	0.1037	0.0852	1.48	0.2231
animal*training*dance	cat Affection No	0.3699	0.0852	18.87	<.0001

Maximum Likelihood Analysis of Variance			
Source	DF	Chi-Square	Pr > ChiSq
animal	1	23.38	<.0001
training	1	15.14	<.0001
animal*training	1	23.38	<.0001
dance	1	8.13	0.0043
animal*dance	1	5.61	0.0179
training*dance	1	1.48	0.2231
animal*training*dance	1	18.87	<.0001
Likelihood Ratio	0	.	.

So we run the second piece of syntax, with the highest level effect removed (SAS Syntax 18.6).

The next part of the output (SAS Output 18.8) tells us something about which components of the model can be removed. Here we have fitted the same model, except we have removed the highest-order interaction. This model is now highly significant, which, in a reverse of our usual thinking, means that it does *not* fit the data. In other words, the three-way interaction is required for the model, and we should retain it.

SAS OUTPUT 18.8

Maximum Likelihood Analysis of Variance			
Source	DF	Chi-Square	Pr > ChiSq
animal	1	31.48	<.0001
training	1	27.54	<.0001
animal*training	1	13.86	0.0002
dance	1	4.24	0.0396
animal*dance	1	13.18	0.0003
training*dance	1	8.46	0.0036
Likelihood Ratio	1	20.30	<.0001

Normally we would stop there. We know that the three-way interaction is significant, and it is therefore required. We can't interpret three-way interaction effects in the absence of two-way interaction effects, so there is little point in testing a model without the two-way effects. If we were to test it (we've got the syntax in SAS Syntax 18.6) we would find a likelihood ratio chi-square of 72.3, with 4 degrees of freedom, $p < 0.001$.

What this is actually telling us is that the three-way interaction is significant: removing it from the model has a significant effect on how well the model fits the data. We also know that removing all two-way interactions has a significant effect on the model, but you have to remember that loglinear analysis should be done hierarchically and so these two-way interactions aren't of interest to us because the three-way interaction is significant (we'd look only at these effects if the three-way interaction were non-significant).

Now we know which model we should be looking at, we can look at the parameter estimates, found in the table labelled Analysis of Maximum Likelihood Estimates, and shown in SAS Output 18.7. This simply breaks down the table that we've just looked at into its component parts. So, for example, although we know from the previous output that removing all of the two-way interactions significantly affects the model, we don't know which of the two-way interactions is having the effect. This table tells us. We get a chi-square test for each of the two-way interactions and the main effects and the column labelled *Pr > ChiSq* tells us which of these effects is significant (values less than .05 are significant). We can tell from this that the animal × dance, and the animal × training are both significant. Likewise, we saw in the previous output that removing the one-way effects (the main effects of animal, training and dance) also significantly affected the fit of the model, and these findings are confirmed here because the main effects of animal and training are both significant. However, the main effect of dance is not (the probability value is greater than .05). Interesting as these findings are, we should ignore them because of the hierarchical nature of loglinear analysis: these effects are all confounded with the higher-order interaction of animal × training × dance.

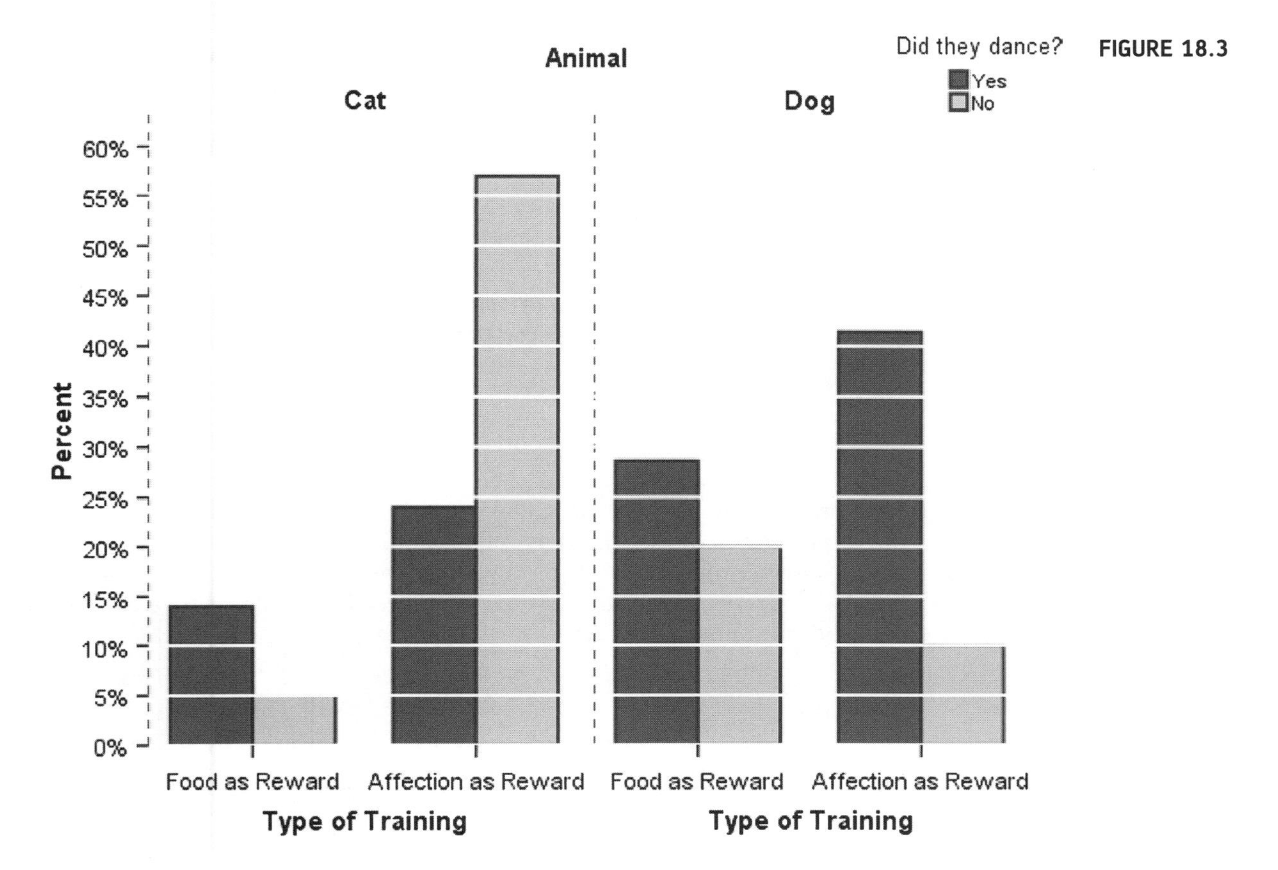

FIGURE 18.3

The next step is to try to interpret this interaction. The first useful thing we can do is to plot the frequencies across all of the different categories. You should plot the frequencies in terms of the percentage of the total (you can calculate these using PROC FREQ). The resulting graph is shown in Figure 18.3 and this shows what we already know about cats: they will dance (or do anything else for that matter) when there is food involved but if you train them with affection they're not interested. Dogs on the other hand will dance when there's affection involved (actually more dogs danced than didn't dance regardless of the type of reward, but the effect is more pronounced when affection was the training method). In fact, both animals show similar responses to food training, it's just that cats won't do anything for affection. So cats are sensible creatures that only do stupid stuff when there's something in it for them (i.e. food), whereas dogs are just plain stupid!

SELF TEST Can you produce a chart which replicates that shown in Figure 18.3?

18.10. Following up loglinear analysis ②

An alternative way to interpret a three-way interaction is to conduct chi-square analysis at different levels of one of your variables. For example, to interpret our animal × training × dance interaction, we could perform a chi-square test on training and dance but do this separately for dogs and cats (in fact the analysis for cats will be the same as the example we used for chi-square). You can then compare the results in the different animals.

SELF TEST Use the BY command (see section 5.4.3) to run a chi-square test on **Dance** and **Training** for dogs and cats.

The results and interpretation for cats are in SAS Output 18.2 and for dogs the output is shown in SAS Output 18.9. For dogs there is still a significant relationship between the types of training and whether they danced but it is weaker (the chi-square is 3.93 compared to 25.2 for the cats).[3] This reflects the fact that dogs are more likely to dance if given affection than if given food, the opposite of cats!

[3] The chi-square statistic depends on the sample size, so really you need to calculate effect sizes and compare them to make this kind of statement (unless you had equal numbers of dogs and cats!).

SAS OUTPUT 18.9

Statistic	DF	Value	Prob
Chi-Square	1	3.9325	0.0474
Likelihood Ratio Chi-Square	1	3.9839	0.0459
Continuity Adj. Chi-Square	1	2.9657	0.0850
Mantel-Haenszel Chi-Square	1	3.8763	0.0490
Phi Coefficient		–0.2370	
Contingency Coefficient		0.2306	
Cramer's V		–0.2370	

18.11. Effect sizes in loglinear analysis ②

As with Pearson's chi-square, one of the most elegant ways to report your effects is in terms of odds ratios. Odds ratios are easiest to understand for 2 × 2 contingency tables and so if you have significant higher-order interactions, or your variables have more than two categories, it is worth trying to break these effects down into logical 2 × 2 tables and calculating odds ratios that reflect the nature of the interaction. So, for example, in this example we could calculate odds ratios for dogs and cats separately. We have the odds ratios for cats already (section 18.5.5), and for dogs we would get:

$$\begin{aligned}\text{odds}_{\text{dancing after food}} &= \frac{\text{number that had food and danced}}{\text{number that had food but didn't dance}}\\ &= \frac{20}{14}\\ &= 1.43\end{aligned}$$

$$\begin{aligned}\text{odds}_{\text{dancing after affection}} &= \frac{\text{number that had affection and danced}}{\text{number that had affection but didn't dance}}\\ &= \frac{29}{7}\\ &= 4.14\end{aligned}$$

$$\begin{aligned}\text{odds ratio} &= \frac{\text{odds}_{\text{dancing after food}}}{\text{odds}_{\text{dancing after affection}}}\\ &= \frac{1.43}{4.14}\\ &= 0.35\end{aligned}$$

This tells us is that if a dog was trained with food the odds of their dancing were 0.35 times the odds if they were rewarded with affection (i.e. they were less likely to dance). Another way to say this is that the odds of their dancing were 1/0.35 = 2.90 times lower if they were trained with food instead of affection. Compare this to cats where the odds of dancing were 6.65 higher if they were trained with food rather than affection. As you can see, comparing the odds ratios for dogs and cats is an extremely elegant way to present the three-way interaction term in the model.

18.12. Reporting the results of loglinear analysis ②

When reporting loglinear analysis you need to report the likelihood ratio statistic for the final model, usually denoted just by χ^2. For any terms that are significant you should report the chi-square change, or you could consider reporting the z-score for the effect and its associated confidence interval. If you break down any higher-order interactions in subsequent analyses then obviously you need to report the relevant chi-square statistics (and odds ratios). For this example we could report:

✓ The three-way loglinear analysis produced a final model that retained all effects. The likelihood ratio of this model was $\chi^2(0) = 0, p = 1$. This indicated that the highest-order interaction (the animal × training × dance interaction) was significant, $\chi^2(1) = 20.30$, $p < .0001$. To break down this effect, separate chi-square tests on the training and dance variables were performed separately for dogs and cats. For cats, there was a significant association between the type of training and whether or not cats would dance, $\chi^2(1) = 25.36, p < .0001$; this was true in dogs also, $\chi^2(1) = 3.93, p < .05$. Odds ratios indicated that the odds of dancing were 6.65 higher after food than affection in cats, but only 0.35 in dogs (i.e. in dogs, the odds of dancing were 2.90 times lower if trained with food compared to affection). Therefore, the analysis seems to reveal a fundamental difference between dogs and cats: cats are more likely to dance for food rather than affection, whereas dogs are more likely to dance for affection than food.

CRAMMING SAM'S TIPS

- If you want to test the relationship between more than two categorical variables you can do this with *loglinear analysis*.
- Loglinear analysis is hierarchical: the initial model contains all main effects and interactions. Starting with the highest-order interaction, terms are removed to see whether their removal significantly affects the fit of the model. If it does then this term is not removed and all lower-order effects are ignored.
- Run a model with all effects in, then remove effects, and check the likelihood ratio Chi-Square. When this is significant (when Pr > ChiSq is less than 0.05), stop removing effects.
- Look at the crosstabulation table to interpret any significant effects (% of total for cells is the best thing to look at).

What have I discovered about statistics? ①

When I wrote the first edition of SPSS book I had always intended to do a chapter on loglinear analysis, but by the time I got to that chapter I had already written 300 pages more than I was contracted to do, and had put so much effort into the rest of it that, well, the thought of that extra chapter was making me think of large cliffs and jumping. When the second edition needed to be written, I wanted to make sure that at the very least I did a loglinear chapter. However, when I came to it, I'd already written 200 pages more than I was supposed to for the new edition, and with deadlines fading into the distance, history was repeating itself. It won't surprise you to know then that I was really happy to have written the damn thing!

This chapter has taken a very brief look at analysing categorical data. What I've tried to do is to show you how really we approach categorical data in much the same way as any other kind of data: we fit a model, we calculate the deviation between our model and the observed data, and we use that to evaluate the model we've fitted. I've also tried to show that the model we fit is the same one that we've come across throughout this book: it's a linear model (regression). When we have only two variables we can use Pearson's chi-square test or the likelihood ratio test to look at whether those two variables are associated. In more complex situations, we simply extend these models into something known as a loglinear model. This is a bit like ANOVA for categorical data: for every variable we have, we get a main effect but we also get interactions between variables. Loglinear analysis simply evaluates all of these effects hierarchically to tell us which ones best predict our outcome.

Fortunately the experience of writing this loglinear chapter taught me a valuable lesson, which is never to agree to write a chapter about something that you know very little about, and if you do then definitely don't leave it until the very end of the writing process when you're under pressure and mentally exhausted. It's lucky that we learn from our mistakes, isn't it …?

Key terms that I've discovered

Chi-square test
Contingency table
Cramér's *V*
Fisher's exact test
Loglinear analysis
Odds ratio
Phi
Saturated model
Yates's continuity correction

Smart Alex's tasks

- **Task 1**: Certain editors at Sage like to think they're a bit of a whiz at football (soccer if you prefer). To see whether they are better than Sussex lecturers and postgraduates we invited various employees of Sage to join in our football matches (oh, sorry, I mean we invited them down for important meetings about books). Every player was only allowed to play in one match. Over many matches, we counted the number of players that scored goals. The data are in the file **SageEditorsCantPlayFootball.sas7bdat.** Do a chi-square test to see whether more publishers or academics scored goals. We predict that Sussex people will score more than Sage people. ③
- **Task 2**: I wrote much of this update while on sabbatical in the Netherlands (I have a real soft spot for Holland). However, living there for three months did enable me to notice certain cultural differences to England. The Dutch are famous for travelling by bike; they do it much more than the English. However, I noticed that many more Dutch people cycle while steering with only one hand. I pointed this out to one of my friends, Birgit Mayer, and she said that I was being a crazy English fool and that Dutch people did not cycle one-handed. Several weeks of me pointing at one-handed cyclists

and her pointing at two-handed cyclists ensued. To put it to the test I counted the number of Dutch and English cyclists who ride with one or two hands on the handlebars (**Handlebars.sas7bdat**). Can you work out which one of us is right? ①

- **Task 3**: I was interested in whether horoscopes are just a figment of people's minds. Therefore, I got 2201 people, made a note of their star sign (this variable, obviously, has 12 categories: Capricorn, Aquarius, Pisces, Aries, Taurus, Gemini, Cancer, Leo, Virgo, Libra, Scorpio and Sagittarius) and whether they believed in horoscopes (this variable has two categories: believer or unbeliever). I then sent them a horoscope in the post of what would happen over the next month: everybody, regardless of their star sign, received the same horoscope, which read 'August is an exciting month for you. You will make friends with a tramp in the first week of the month and cook him a cheese omelette. Curiosity is your greatest virtue, and in the second week, you'll discover knowledge of a subject that you previously thought was boring, statistics perhaps. You might purchase a book around this time that guides you towards this knowledge. Your new wisdom leads to a change in career around the third week, when you ditch your current job and become an accountant. By the final week you find yourself free from the constraints of having friends, your boy/girlfriend has left you for a Russian ballet dancer with a glass eye, and you now spend your weekends doing loglinear analysis by hand with a pigeon called Hephzibah for company.' At the end of August I interviewed all of these people and I classified the horoscope as having come true, or not, based on how closely their lives had matched the fictitious horoscope. The data are in the file **Horoscope.sas7bdat.** Conduct a loglinear analysis to see whether there is a relationship between the person's star sign, whether they believe in horoscopes and whether the horoscope came true. ③
- **Task 4**: On my statistics course students have weekly SAS classes in a computer laboratory. These classes are run by postgraduate tutors but I often pop in to help out. I've noticed in these sessions that many students are studying Facebook rather more than they are studying the very interesting statistics assignments that I have set them. I wanted to see the impact that this behaviour had on their exam performance. I collected data from all 260 students on my course. First I checked their **Attendance** and classified them as having attended either more or less than 50% of their lab classes. Next, I classified them as being either someone who looked at **Facebook** during their lab class, or someone who never did. Lastly, after the Research Methods in Psychology (RMiP) exam, I classified them as having either passed or failed (**Exam**). The data are in **Facebook.sas7bdat.** Do a loglinear analysis on the data to see if there is an association between studying Facebook and failing your exam. ③

Answers can be found on the companion website.

Further reading

Hutcheson, G., & Sofroniou, N. (1999). *The multivariate social scientist*. London: Sage.
Tabachnick, B. G., & Fidell, L. S. (2007). *Using multivariate statistics* (4th ed.). Boston: Allyn & Bacon. (Chapter 16 is a fantastic account of loglinear analysis.)

Interesting real research

Beckham, A. S. (1929). Is the Negro happy? A psychological analysis. *Journal of Abnormal and Social Psychology*, *24*, 186–190.

Multilevel linear models 19

FIGURE 19.1
Having a therapy session in 2007

19.1. What will this chapter tell me? ①

Over the last couple of chapters we saw that I had gone from a child having dreams and aspirations of being a rock star, to becoming a living (barely) statistical test. A more dramatic demonstration of my complete failure to achieve my life's ambitions I can scarcely imagine. Having devoted far too much of my life to statistics it was time to unlock the latent rock star once more. The second edition of the book had left me in desperate need for some therapy and, therefore, at the age of 29 I decided to start playing the drums (there's a joke in there somewhere about it being the perfect instrument for a failed musician, but really they're much harder to play than people think). A couple of years later I had a call from an old friend of mine, Doug, who used to be in a band that my old band Scansion used to play with a lot: 'Remember the last time I saw you we talked about you coming and having a jam with us?' I had absolutely no recollection whatsoever of him saying this so I responded

'Yes'. 'Well, how about it then?' he said. 'OK,' I said, 'you arrange it and I'll bring my guitar.' 'No, you whelk,' he said, 'we want you to drum and maybe you could learn some of the songs on the CD I gave you last year?' I'd played his band's CD and I liked it, but there was no way on this earth that I could play the drums as well as their drummer. 'Sure, no problem,' I lied. I spent the next two weeks playing along to this CD as if my life depended on it and when the rehearsal came, much as I'd love to report that I drummed like a lord, I didn't. I did, however, nearly have a heart attack and herniate everything in my body that it's possible to herniate (really, the music is pretty fast!). Still, we had another rehearsal, and then another and, well, three years down the line we're still having them. The only difference is that now I can play the songs at a speed that makes their old recordings seem as though a sedated snail was on the drums (www.myspace.com/fracturepattern). The point is that it's never too late to learn something new. This is just as well because, as a man who clearly doesn't learn from his mistakes, I agreed to write a chapter on multilevel linear models, a subject about which I know absolutely nothing. I'm writing it last, when I feel mentally exhausted and stressed. Hopefully at some point between now and the end of writing the chapter I will learn something. With a bit of luck you will too.

19.2. Hierarchical data ②

In all of the analyses in this book so far we have treated data as though they are organized at a single level. However, in the real world, data are often hierarchical. This just means that some variables are clustered or *nested* within other variables. For example, when I'm not writing statistics books I spend most of my time researching how anxiety develops in children below the age of 10. This typically involves my running experiments in schools. When I run research in a school, I test children who have been assigned to different classes, and who are taught by different teachers. The classroom that a child is in could conceivably affect my results. Let's imagine I test in two different classrooms. The first class is taught by Mr Nervous. Mr Nervous is very anxious and often when he supervises children he tells them to be careful, or that things that they do are dangerous, or that they might hurt themselves. The second class is taught by Little Miss Daredevil.[1] She is very carefree and she believes that children in her class should have the freedom to explore new experiences. Therefore, she is always telling them not to be scared of things and to explore new situations. One day I go into the school to test the children. I take in a big animal carrier, which I tell them has an animal inside. I measure whether they will put their hand in the carrier to stroke the animal. Children taught by Mr Nervous have grown up in an environment where their teacher reinforces caution, whereas children taught by Miss Daredevil have been encouraged to embrace new experiences. Therefore, we might expect Mr Nervous's children to be more reluctant to put their hand in the box because of the classroom experiences that they have had. The classroom is, therefore, known as a *contextual variable*. In reality, as an experimenter I would be interested in a much more complicated situation. For example, I might tell some of the children that the animal is a bloodthirsty beast, whereas I tell others that the animal is friendly. Now obviously I'm expecting the information I give the children to affect their enthusiasm for stroking the animal. However, it's also possible that their classroom has an effect. Therefore, my manipulation of the information that I give the children also has to be placed within the context of the classroom to which the

[1] Those of you who don't spot the Mr Men references here, check out http://www.mrmen.com. Mr Nervous used to be called Mr Jelly and was a pink jelly-shaped blob, which in my humble opinion was better than his current incarnation.

child belongs. My threat information is likely to have more impact on Mr Nervous's children than it will on Miss Daredevil's children. One consequence of this is that children in Mr Nervous's class will be more similar to each other than they are to children in Miss Daredevil's class and vice versa.

Figure 19.2 illustrates this scenario more generally. In a big data set, we might have collected data from lots of children. This is the bottom of the hierarchy and is known as a *level 1* variable. So, children (or cases) are our level 1 variable. However, these children are organized by classroom (children are said to be *nested* within classes). The class to which a child belongs is a level up from the participant in the hierarchy and is said to be a *level 2* variable.

The situation that I have just described is the simplest hierarchy that you can have because there are just two levels. However, you can have other layers to your hierarchy. The easiest way to explain this is to stick with our example of my testing children in different classes and then to point out the obvious fact that classrooms are themselves nested within schools. Therefore, if I ran a study incorporating lots of different schools, as well as different classrooms within those schools, then I would have to add another level to the hierarchy. We can apply the same logic as before, in that children in particular schools will be more similar to each other than to children in different schools. This is because schools tend to reflect their social demographic (which can differ from school to school) and they may differ in their policies also. Figure 19.3 shows this scenario. There are now three levels in the hierarchy: the child (level 1), the class to which the child belongs (level 2) and the school within which that class exists (level 3). In this situation we have two contextual variables: school and classroom.

Hierarchical data structures need not apply only to between-participant situations. We can also think of data as being nested within people. In this situation the case, or person, is not at the bottom of the hierarchy (level 1), but is further up. A good example is memory. Imagine that after giving children threat information about my caged animal I asked them a week later to recall everything they could about the animal. For each child there are many facts that they could recall. Let's say that I originally gave them 15 pieces of information; some children might recall all 15 pieces of information, but others might remember only 2 or 3 bits of information. The bits of information, or memories, are nested within the person and their recall depends on the person. The probability of a given memory being recalled depends on what other memories are available, and the recall of one memory may have knock-on effects for what other memories are recalled. Therefore, memories are not independent units. As such, the person acts as a context within which memories are recalled (Wright, 1998).

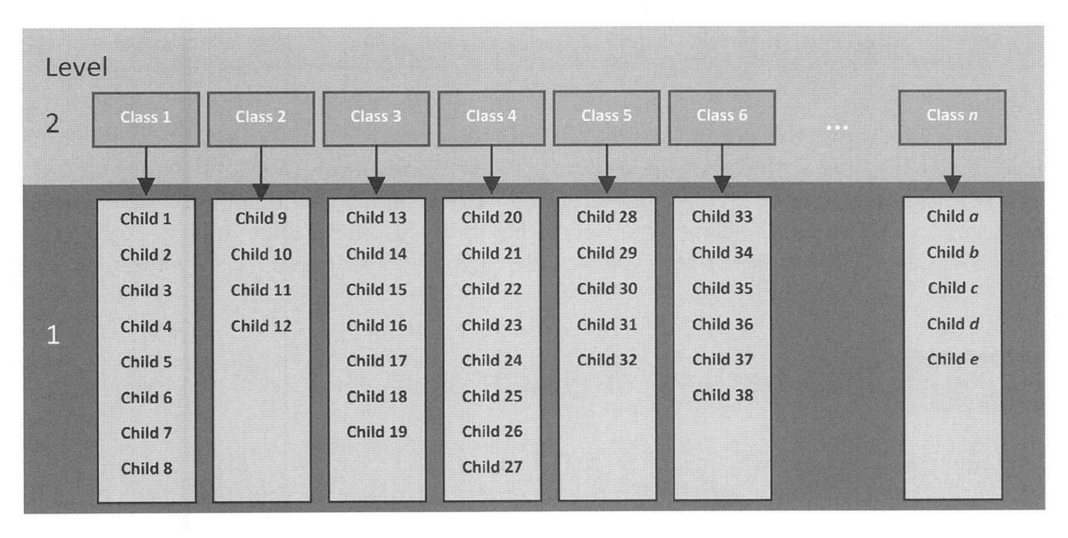

FIGURE 19.2 An example of a two-level hierarchical data structure. Children (level 1) are organized within classrooms (level 2)

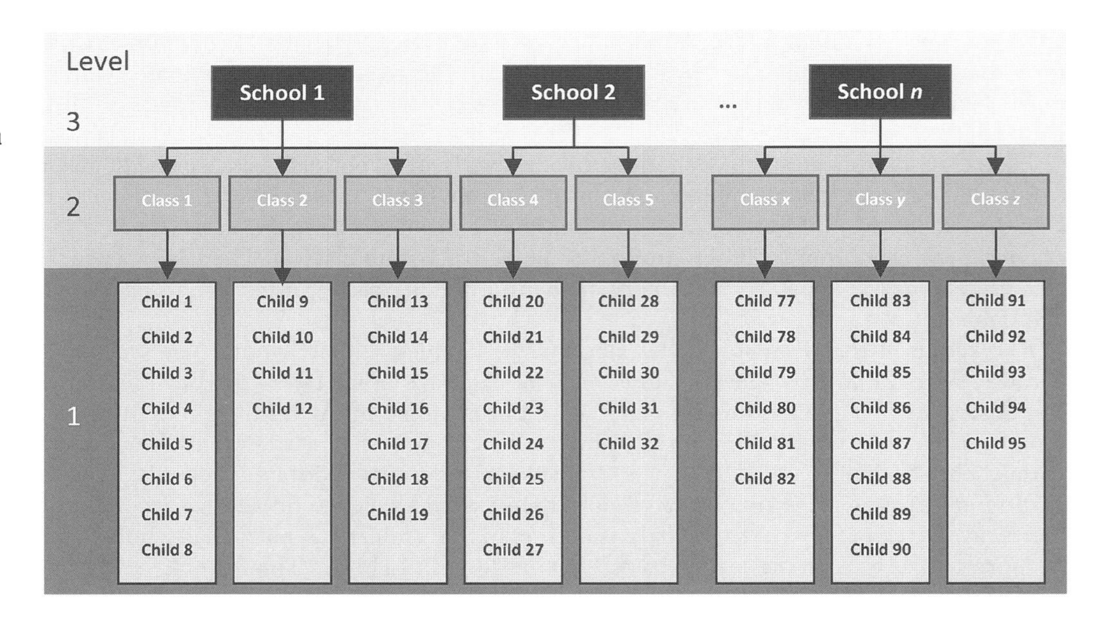

FIGURE 19.3 An example of a three-level hierarchical data structure

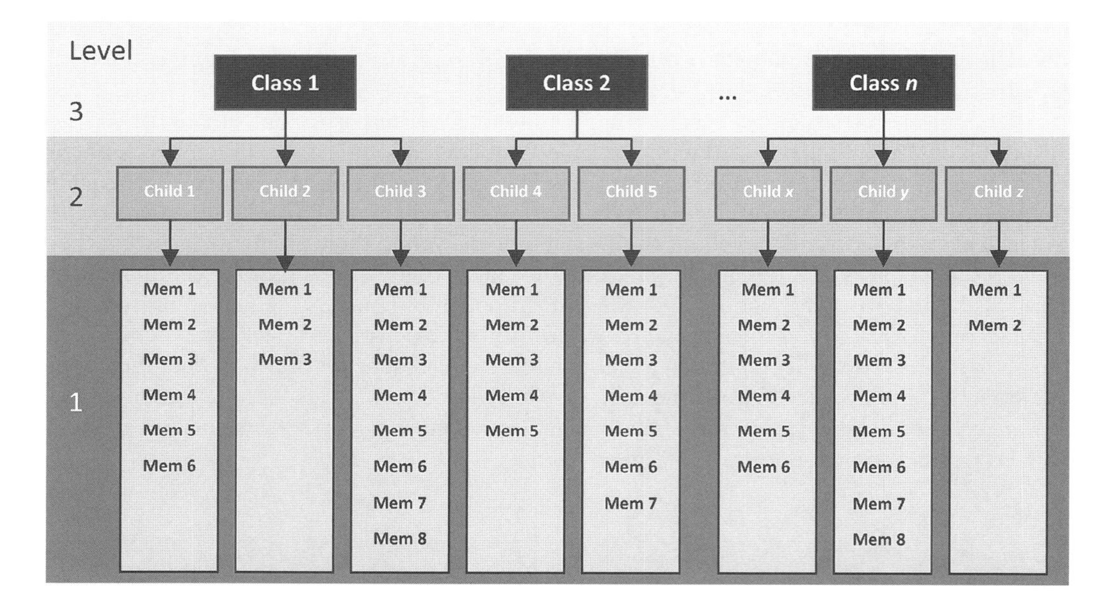

FIGURE 19.4 An example of a three-level hierarchical data structure, where the level 1 variable is a repeated measure (memories recalled)

Figure 19.4 shows the structure of the situation that I have just described. The child is our level 2 variable, and within each child there are several memories (our level 1 variable). Of course we can also have levels of the hierarchy above the child. So, we could still, for example, factor in the context of the class from which they came (as I have done in Figure 19.4) as a level 3 variable. Indeed, we could even include the school again as a level 4 variable!

19.2.1. The intraclass correlation ②

You might well wonder why it matters that data are hierarchical (or not). The main problem is that the contextual variables in the hierarchy introduce dependency in the data. In plain English this means that residuals will be correlated. I have alluded to this fact already when I noted that children in Mr Nervous's class would be more similar to each other than to children in Miss Daredevil's class. In some sense, having the same teacher makes children more similar to each other. This similarity is a problem because in nearly every test we have covered in this book we assume that cases are independent. In other words, there is absolutely no correlation between residual scores of one child and another. However, when people are sampled from similar contexts, this independence is unlikely to be true. For example, Charlotte and Emily's responses to the animal in the carrier have both been influenced by Mr Nervous's cautious manner, so their behaviour will be similar. Likewise, Kiki and Jip's responses to the animal in the box have both been influenced by Miss Daredevil's carefree manner, so their behaviour will be similar too. We have seen before that in ANOVA, for example, a lack of independence between cases is a huge problem that really affects the resulting test statistic – and not in a good way! (See section 10.2.10.)

By thinking about contextual variables and factoring them into the analysis we can overcome this problem of non-independent observations. One way that we can do this is to use the intraclass correlation (ICC). We'll skip the formalities of calculating the ICC (but see Oliver Twisted if you're keen to know), and we'll just give a conceptual grasp of what it represents. In our two-level example of children within classes, the ICC represents the proportion of the total variability in the outcome that is attributable to the classes. It follows that if a class has had a big effect on the children within it then the variability within the class will be small (the children will behave similarly). As such, variability in the outcome within classes is minimized, and variability in the outcome between classes is maximized; therefore, the ICC is large. Conversely, if the class has little effect on the children then the outcome will vary a lot within classes, which will make differences between classes relatively small. Therefore, the ICC is small too. Thus, the ICC tells us that variability within levels of a contextual variable (in this case the class to which a child belongs) is small, but between levels of a contextual variable (comparing classes) is large. As such the ICC is a good gauge of whether a contextual variable has an effect on the outcome.

OLIVER TWISTED

Please, Sir, can I have some more ... ICC?

'I have a dependency on gruel,' whines Oliver. 'Maybe I could measure this dependency if I knew more about the ICC.' We'll you're so high on gruel Oliver that you have rather missed the point. Still, I did write an article on the ICC once upon a time (Field, 2005a) and it's reproduced in the additional web material for your delight and amusement.

19.2.2. Benefits of multilevel models ②

Multilevel linear models have numerous uses. To convince you that trawling through this chapter is going to reward you with statistical possibilities beyond your wildest dreams, here are just a few (slightly overstated) benefits of multilevel models:

- **Cast aside the assumption of homogeneity of regression slopes:** We saw in Chapter 11 that when we use analysis of covariance we have to assume that the relationship between our covariate and our outcome is the same across the different groups that make up our predictor variable. However, this doesn't always happen. Luckily, in multilevel models we can explicitly model this variability in regression slopes, thus overcoming this inconvenient problem.

- **Say 'bye bye' to the assumption of independence**: We saw in Chapter 10 that when we use independent ANOVA we have to assume that the different cases of data are independent. If this is not true, little lizards climb out of your mattress while you're asleep and eat you. Again, multilevel models are specifically designed to allow you to model these relationships between cases. Also, in Chapter 7 we saw that multiple regression relies on having independent observations. However, there are situations in which you might want to measure someone on more than one occasion (i.e. over time). Ordinary regression turns itself into cheese and hides in the fridge at the prospect of cases of data that are related. Multilevel models eat these data for breakfast, with a piece of regression-flavoured cheese.
- **Laugh in the face of missing data**: I've spent a lot of this book extolling the virtues of balanced designs and not having missing data. Regression, ANOVA, ANCOVA and most of the other tests we have covered do strange things when data are missing or the design is not balanced. This can be a real pain. Multilevel models open the door to missing data, invite them to sit by the fire and make them a cup of tea. Multilevel models expect missing data, they love them in fact. So, if you have some kind of ANOVA or regression (of any variety) for which you have missing data, fear not, just do a multilevel model.

I think you'll agree that multilevel models are pretty funky. 'Is there anything they can't do?' I hear you cry. Well, no, not really.

19.3. Theory of multilevel linear models ③

The underlying theory of multilevel models is very complicated indeed – far too complicated for my little peanut of a brain to comprehend. Fortunately, the advent of computers and software like SAS makes it possible for feeble-minded individuals such as myself to take advantage of this wonderful tool without actually needing to know the maths. Better still, this means I can get away with not explaining the maths (and really, I'm not kidding, I don't understand any of it). What I will do though is try to give you a flavour of what multilevel models are and what they do by describing the key concepts within the framework of linear models that has permeated this whole book.

19.3.1. An example ②

Throughout the first part of the chapter we will use an example to illustrate some of the concepts in multilevel models. Cosmetic surgery is on the increase at the moment. In the USA, there was a 1600% increase in cosmetic surgical and non-surgical treatments between 1992 and 2002, and in 2004, 65,000 people in the UK underwent privately and publicly funded operations (Kellett, Clarke, & McGill, 2008). With the increasing popularity of this surgery, many people are starting to question the motives of those who want to go under the knife. There are two main reasons to have cosmetic surgery: (1) to help a physical problem, such as having breast reduction surgery to relieve back ache; and (2) to change your external appearance, for example by having a face lift. Related to this second point, there is even some case for arguing that cosmetic surgery could be performed as a psychological intervention: to improve self-esteem (Cook, Rosser, & Salmon, 2006; Kellett et al., 2008). The main example for this chapter looks at the effects of cosmetic surgery on quality of life. The variables in the data file are (CosmeticSurgery.sas7bdat):

- **Post_QoL**: This is a measure of quality of life after the cosmetic surgery. This is our outcome variable.
- **Base_QoL**: We need to adjust our outcome for quality of life before the surgery.
- **Surgery**: This variable is a dummy variable that specifies whether the person has undergone cosmetic surgery (1) or whether they are on the waiting list (0), which acts as our control group.
- **Clinic**: This variable specifies which of 10 clinics the person attended to have their surgery.
- **Age**: this variable tells us the person's age in years.
- **BDI**: It is becoming increasingly apparent that people volunteering for cosmetic surgery (especially when the surgery is purely for vanity) might have very different personality profiles than the general public (Cook, Rosser, Toone, James, & Salmon, 2006). In particular, these people might have low self-esteem or be depressed. When looking at quality of life it is important to assess natural levels of depression and this variable used the Beck Depression Inventory (BDI) to do just that.
- **Reason**: This dummy variable specifies whether the person had/is waiting to have surgery purely to change their appearance (0), or because of a physical reason (1).
- **Gender**: This variable simply specifies whether the person was a man (1) or a woman (0).

When conducting hierarchical models we generally work up from a very simple model to more complicated models and we will take that approach in this chapter. In doing so I hope to illustrate multilevel modelling by attaching it to frameworks that you already understand, such as ANOVA and ANCOVA.

Figure 19.5 shows the hierarchical structure of the data. Essentially, people being treated in the same surgeries are not independent of each other because they will have had surgery from the same surgeon. Surgeons will vary in how good they are, and quality of life will to some extent depend on how well the surgery went (if they did a nice neat job then quality of life should be higher than if they left you with unpleasant scars). Therefore, people within clinics will be more similar to each other than people in different clinics. As such, the person undergoing surgery is the level 1 variable, but there is a level 2 variable, a variable higher in the hierarchy, which is the clinic attended.

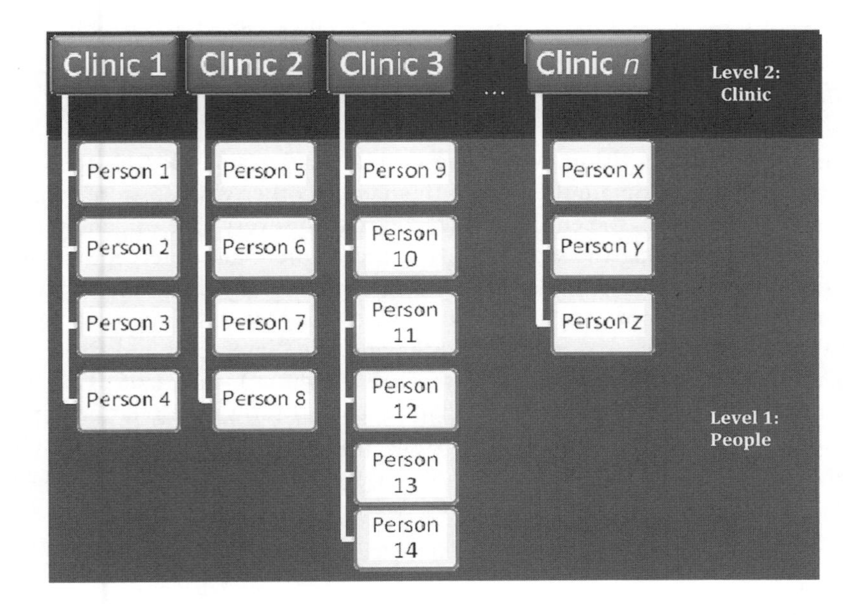

FIGURE 19.5 Diagram to show the hierarchical structure of the cosmetic surgery data set. People are clustered within clinics. Note that for each person there would be a series of variables measured: surgery, BDI, age, gender, reason and pre-surgery quality of life

19.3.2. Fixed and random coefficients ③

Throughout this book we have discussed effects and variables and these concepts should be very familiar to you by now. However, we have viewed these effects and variables in a relatively simple way: we have not distinguished between whether something is fixed or random.

What we mean by 'fixed' and 'random' can be a bit confusing because the terms are used in a variety of contexts. You hear people talk about fixed effects and random effects. An effect in an experiment is said to be a fixed effect if all possible treatment conditions that a researcher is interested in are present in the experiment. An effect is said to be random if the experiment contains only a random sample of possible treatment conditions. This distinction is important because fixed effects can be generalized only to the situations in your experiment, whereas random effects can be generalized beyond the treatment conditions in the experiment (provided that the treatment conditions are representative). For example, in our Viagra example from Chapter 10, the effect is fixed if we say that we are interested only in the three conditions that we had (placebo, low dose and high dose) and we can generalize our findings only to the situation of a placebo, low dose and high dose. However, if we were to say that the three doses were only a sample of possible doses (maybe we could have tried a very high dose), then it is a random effect and we can generalize beyond just placebos, low doses and high doses. All of the effects in this book so far we have treated as fixed effects. The vast majority of academic research that you read will treat variables as fixed effects.

People also talk about fixed variables and random variables. A fixed variable is one that is not supposed to change over time (e.g. for most people their gender is a fixed variable – it never changes), whereas a random one varies over time (e.g. your weight is likely to fluctuate over time).

In the context of multilevel models we need to make a distinction between **fixed coefficients** and **random coefficients.** In the regressions, ANOVAs and ANCOVAs throughout this book we have assumed that the regression parameters are fixed. We have seen numerous times that a linear model is characterized by two things – the intercept, b_0, and the slope, b_1:

$$Y_i = b_0 + b_1 X_{1i} + \varepsilon_i$$

Note that the outcome (*Y*), the predictor (*X*) and the error (ε) all vary as a function of *i*, which normally represents a particular case of data. In other words, it represents the level 1 variable. If, for example, we wanted to predict Sam's score, we could replace the *i*s with her name:

$$Y_{\text{Sam}} = b_0 + b_1 X_{1\text{Sam}} + \varepsilon_{\text{Sam}}$$

This is just some basic revision. Now, when we do a regression like this we assume that the *b*s are fixed and we estimate them from the data. In other words, we're assuming that the model holds true across the entire sample and that for every case of data in the sample we can predict a score using the same values of the gradient and intercept. However, we can also conceptualize these parameters as being random.[2] If we say that a parameter is random then we assume not that it is a fixed value, but that its value can vary. Up until now we have thought of regression models as having fixed intercepts and fixed slopes, but this opens up three new possibilities for us that are shown in Figure 19.6. This figure uses the data from our ANCOVA example in Chapter 11 and shows the relationship between a person's libido and that of their partner overall (the dashed line) and separately for the three groups in the study (a placebo group, a group that had a low dose of Viagra and a group that had a high dose).

[2] In a sense random isn't an intuitive term for us non-statisticians because it implies that values are plucked out of thin air (randomly selected). However, this is not the case, they are carefully estimated just as fixed parameters are.

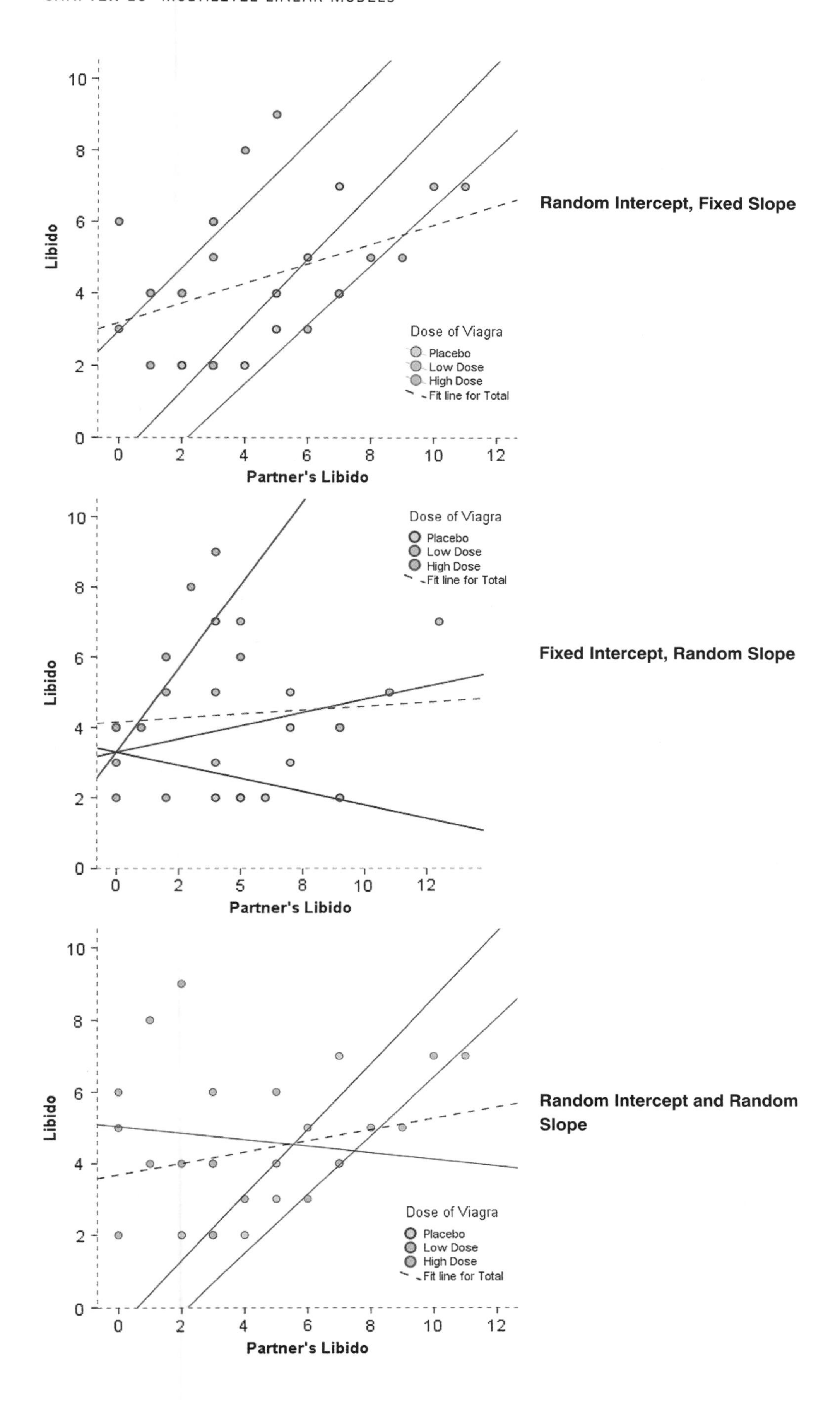

FIGURE 19.6 Data sets showing an overall model (dashed line) and the models for separate contexts within the data (i.e. groups of cases)

19.3.2.1. The random intercept model ③

The simplest way to introduce random parameters into the model is to assume that the intercepts vary across contexts (or groups) – because the intercepts vary, we call them random intercepts. For our libido data this is like assuming that the relationship between libido and partner's libido is the same in the placebo, low- and high-dose groups (i.e. the slope is the same), but that the models for each group are in different locations (i.e. the intercepts are different). This is shown in the diagram in which the models within the different contexts have the same shape (slope) but are located in different geometric space (they have different intercepts – top panel of Figure 19.6).

19.3.2.2. Random slope model ③

We can also assume that the slopes vary across contexts – i.e. we assume random slopes. For our libido data this is like assuming that the relationship between libido and partner's libido is different in the placebo, low- and high-dose groups (i.e. the slopes are different), but that the models for each group are fixed at the same geometric location (i.e. the intercepts are the same). This is what happens when we violate the assumption of homogeneity of regression slopes in ANCOVA. Homogeneity of regression slopes is the assumption that regression slopes are the same across contexts. If this assumption is not tenable than we can use a multilevel model to explicitly estimate that variability in slopes. This is shown in the diagram in which the models within the different contexts converge on a single intercept but have different slopes (middle panel of Figure 19.6).

19.3.2.3. The random intercept and slope model ③

The most realistic situation is to assume that both intercepts and slopes vary around the overall model. This is shown in the diagram in which the models within the different contexts have different slopes but are also located in different geometric space and so have different intercepts (bottom panel of Figure 19.6).

19.4. The multilevel model ④

We have seen conceptually what a random intercept, random slope and random intercept and slope model looks like. Now let's look at how we actually represent the models. To keep things concrete, let's use our example. For the sake of simplicity, let's imagine first that we wanted to predict someone's quality of life (QoL) after cosmetic surgery. We can represent this as a linear model as follows:

$$\text{QoL After Surgery}_i = b_0 + b_1\text{Surgery}_i + \varepsilon_i \tag{19.1}$$

We have seen equations like this many times and it represents a linear model: regression, a *t*-test (in this case) and ANOVA. In this example, we had a contextual variable, which was the clinic in which the cosmetic surgery was conducted. We might expect the effect of surgery on quality of life to vary as a function of which clinic the surgery was conducted at because surgeons will differ in their skill. This variable is a level 2 variable. As such we could allow the model that represents the effect of surgery on quality of life to vary across the different contexts (clinics). We can do this by allowing the intercepts to vary across clinics, or by allowing the slopes to vary across clinics or by allowing both to vary across clinics.

To begin with, let's say we want to include a random intercept for quality of life. All we do is add a component to the intercept that measures the variability in intercepts, u_{0j}. Therefore, the intercept changes from b_0 to become $b_0 + u_{0j}$. This term estimates the intercept of the overall model fitted to the data, b_0, and the variability of intercepts around that overall model, u_{0j}. The overall model becomes:[3]

$$Y_{ij} = (b_0 + u_{0j}) + b_1 X_{ij} + \varepsilon_{ij} \tag{19.2}$$

The js in the equation reflect levels of the variable over which the intercept varies (in this case the clinic) – the level 2 variable. Another way that we could write this is to take out the error terms so that it looks like an ordinary regression equation except that the intercept has changed from a fixed, b_0, to a random one, b_{0j}, which is defined in a separate equation:

$$\begin{aligned} Y_{ij} &= b_{0j} + b_1 X_{ij} + \varepsilon_{ij} \\ b_{0j} &= b_0 + u_{0j} \end{aligned} \tag{19.3}$$

Therefore, if we want to know the estimated intercept for clinic 7, we simply replace the j with 'clinic 7' in the second equation:

$$b_{0\text{Clinic7}} = b_0 + u_{0\text{Clinic7}}$$

If we want to include random slopes for the effect of surgery on quality of life, then all we do is add a component to the slope of the overall model that measures the variability in slopes, u_{1j}. Therefore, the gradient changes from b_1 to become $(b_1 + u_{1j})$. This term estimates the slope of the overall model fitted to the data, b_1, and the variability of slopes in different contexts around that overall model, u_{1j}. The overall model becomes (compare to the random intercept model above):

$$Y_{ij} = b_0 + (b_1 + u_{1j}) X_{ij} + \varepsilon_{ij} \tag{19.4}$$

Again we can take the error terms out into a separate equation to make the link to a familiar linear model even clearer. It now looks like an ordinary regression equation except that the slope has changed from a fixed, b_1, to a random one, b_{1j}, which is defined in a separate equation:

$$\begin{aligned} Y_{ij} &= b_{0i} + b_{1j} X_{ij} + \varepsilon_{ij} \\ b_{1j} &= b_1 + u_{1j} \end{aligned} \tag{19.5}$$

If we want to model a situation with random slopes *and* intercepts, then we combine the two models above. We still estimate the intercept and slope of the overall model (b_0 and b_1) but we also include the two terms that estimate the variability in intercepts, u_{0j}, and slopes, u_{1j}. The overall model becomes (compare to the two models above):

$$Y_{ij} = (b_0 + u_{0j}) + (b_1 + u_{1j}) X_{ij} + \varepsilon_{ij} \tag{19.6}$$

We can link this more directly to a simple linear model if we take some of these extra terms out into separate equations. We could write this model as a basic linear model, except

[3] Some people use gamma (γ), not b, to represent the parameters, but I prefer b because it makes the link to the other linear models that we have used in this book clearer.

we've replaced our fixed intercept and slope (b_0 and b_1) with their random counterparts (b_{0j} and b_{1j}):

$$\begin{aligned} Y_{ij} &= b_{0j} + b_{1j}X_{ij} + \varepsilon_{ij} \\ b_{0j} &= b_0 + u_{0j} \\ b_{1j} &= b_1 + u_{1j} \end{aligned} \tag{19.7}$$

The take-home point is that we're not doing anything terribly different from the rest of the book: it's basically just a posh regression.

Now imagine we wanted to add in another predictor, for example quality of life before surgery. Knowing what we do about multiple regression we shouldn't be invading the personal space of the idea that we can simply add this variable in with an associated beta:

$$\text{QoL After Surgery}_i = b_0 + b_1\text{Surgery}_i + b_2\text{QoL Before Surgery}_i + \varepsilon_i \tag{19.8}$$

This is all just revision of ideas from earlier in the book. Remember also that the *i* represents the level 1 variable, in this case the people we tested. Therefore, we can predict a given person's quality of life after surgery by replacing the *i* with their name:

$$\text{QoL After}_{\text{Sam}} = b_0 + b_1\text{Surgery}_{\text{Sam}} + b_2\text{QoL Before}_{\text{Sam}} + \varepsilon_{\text{Sam}}$$

Now, if we want to allow the intercept of the effect of surgery on quality of life after surgery to vary across contexts then we simply replace b_0 with b_{0j} . If we want to allow the slope of the effect of surgery on quality of life after surgery to vary across contexts then we replace b_1 with b_{1j}. So, even with a random intercept and slope, our model stays much the same:

$$\begin{aligned} \text{QoL After}_{ij} &= b_{0j} + b_{1j}\text{Surgery}_{ij} + b_2\text{QoL Before}_{ij} + \varepsilon_{ij} \\ b_{0j} &= b_0 + u_{0j} \\ b_{1j} &= b_1 + u_{1j} \end{aligned} \tag{19.9}$$

Remember that the *j* in the equation relates to the level 2 contextual variable (clinic in this case). So, if we wanted to predict someone's score we wouldn't just do it from their name, but also from the clinic they attended. Imagine our guinea pig Sam had her surgery done at clinic 7, then we could replace the *i*s and *j*s as follows:

$$\begin{aligned} \text{QoL After Surgery}_{\text{Sam, Clinic7}} = {} & b_{0\text{Clinic7}} + b_{1\text{Clinic7}}\text{Surgery}_{\text{Sam, Clinic7}} \\ & + b_2\text{QoL Before Surgery}_{\text{Sam, Clinic7}} + \varepsilon_{\text{Sam, Clinic7}} \end{aligned}$$

I want to sum up by just reiterating that all we're really doing in a multilevel model is a fancy regression in which we allow either the intercepts or slopes, or both, to vary across different contexts. All that really changes is that for every parameter that we allow to be random, we get an estimate of the variability of that parameter as well as the parameter itself. So, there isn't anything terribly complicated; we can add new predictors to the model and for each one decide whether its regression parameter is fixed or random.

19.4.1. Assessing the fit and comparing multilevel models ④

As in logistic regression (Chapter 8) the overall fit of a multilevel model is tested using a chi-square likelihood ratio test (see section 18.3.3) and just as in logistic regression, SAS reports the −2 log-likelihood (−2LL; see section 8.3.2). Essentially, the smaller the value of the log-likelihood, the better. SAS also produces three adjusted versions of the log-likelihood value. All of these can be interpreted in the same way as the log-likelihood, but they have been corrected for various things:

- *Akaike's information criterion* (*AIC*): This is basically a goodness-of-fit measure that is corrected for model complexity. That just means that it takes into account how many parameters have been estimated.
- *Hurvich and Tsai's criterion* (*AICC*): This is the same as AIC but is designed for small samples.
- *Schwarz's Bayesian information criterion* (*BIC*): This statistic is again comparable to the AIC, although it is slightly more conservative (it corrects more harshly for the number of parameters being estimated). It should be used when sample sizes are large and the number of parameters is small.

All of these measures are similar but the AIC and BIC are the most commonly used. None of them are intrinsically interpretable (it's not meaningful to talk about their values being large or small *per se*); however, they are all useful as a way of comparing models. The value of AIC, AICC and BIC can all be compared to their equivalent values in other models. In all cases smaller values mean better-fitting models.

Many writers recommend building up multilevel models starting with a 'basic' model in which all parameters are fixed and then adding in random coefficients as appropriate and exploring confounding variables (Raudenbush & Bryk, 2002; Twisk, 2006). One advantage of doing this is that you can compare the fit of the model as you make parameters random, or as you add in variables. To compare models we simply subtract the log-likelihood of the new model from the value for the old:

$$\begin{aligned} \chi^2_{\text{Change}} &= -2\text{LL}_{\text{Old}} - 2\text{LL}_{\text{New}} \\ df_{\text{Change}} &= k_{\text{Old}} - k_{\text{New}} \end{aligned} \tag{19.10}$$

This equation is the same as equations (18.5) and (8.6), but written in a way that uses the names of the actual values that SAS produces. There are two caveats to this equation: (1) it works only if full maximum-likelihood estimation is used (and not restricted maximal likelihood – see SAS Tip 19.1); and (2) the new model contains all of the effects of the older model.

19.4.2. Types of covariance structures ④

If you have any random effects or repeated measures in your multilevel model then you have to decide upon the *covariance structure* of your data. If you have random effects and repeated measures then you can specify different covariance structures for each. The covariance structure simply specifies the form of the variance–covariance matrix (a matrix in which the diagonal elements are variances and the off-diagonal elements are covariances). There are various forms that this matrix could take and we have to tell SAS what form we think it *does* take. Of course we might not know what form it takes (most of the time we'll be taking an educated guess), so it is sometimes useful to run the model with different covariance structures defined and use the goodness-of-fit indices (the AIC, AICC and BIC) to see whether changing the covariance structure improves the fit of the model (remember that a smaller value of these statistics means a better-fitting model).

The covariance structure is important because SAS uses it as a starting point to estimate the model parameters. As such, you will get different results depending on which covariance structure you choose. If you specify a covariance structure that is too simple then you are more likely to make a Type I error (finding a parameter is significant when in reality it is not), but if you specify one that is too complex then you run the risk of a Type II error (finding parameters to be non-significant when in reality they are). SAS has 23 different covariance structures that you can use. We will look at four of the commonest covariance structures to give you a feel for what they are and when they should be used. In each case I use a representation of the variance–covariance matrix to illustrate. With all of these matrices you could imagine that the rows and columns represents four different clinics in our cosmetic surgery data:

$\begin{pmatrix} 1 & 0 & 0 & 0 \\ 0 & 1 & 0 & 0 \\ 0 & 0 & 1 & 0 \\ 0 & 0 & 0 & 1 \end{pmatrix}$	*Variance Components*: This covariance structure is very simple and assumes that all random effects are independent (this is why all of the covariances in the matrix are 0). Variances of random effects are assumed to be the same (hence why they are 1 in the matrix) and sum to the variance of the outcome variable. In SAS this is the default covariance structure for random effects and it sometimes called the independence model.
$\begin{pmatrix} \sigma_1^2 & 0 & 0 & 0 \\ 0 & \sigma_1^2 & 0 & 0 \\ 0 & 0 & \sigma_1^2 & 0 \\ 0 & 0 & 0 & \sigma_1^2 \end{pmatrix}$	*Diagonal*: This variance structure is like variance components except that variances are assumed to be heterogeneous (this is why the diagonal of the matrix is made up of different variance terms). This structure again assumes that variances are independent and, therefore, that all of the covariances are 0. In SAS this is the default covariance structure for repeated measures.
$\begin{pmatrix} 1 & \rho & \rho^2 & \rho^3 \\ \rho & 1 & \rho & \rho^2 \\ \rho^2 & \rho & 1 & \rho \\ \rho^3 & \rho^2 & \rho & 1 \end{pmatrix}$	*AR(1)*: This stands for first-order autoregressive structure. In layman's terms this means that the relationship between variances changes in a systematic way. If you imagine the rows and columns of the matrix to be points in time, then it assumes that the correlations between repeated measurements are highest at adjacent time points. So, in the first column, the correlation between time points 1 and 2 is ρ; let's assume that this value is .3. As we move to time point 3, the correlation between time point 1 and 3 is ρ^2, or .09. In other words, it has decreased: scores at time point 1 are more related to scores at time 2 than they are to scores at time 3. At time 4, the correlation goes down again to ρ^3 or .027. So, the correlations between time points next to each other are assumed to be ρ, scores two intervals apart are assumed to have correlations of ρ^2, and scores three intervals apart are assumed to have correlations of ρ^3. So the correlation between scores gets smaller over time. Variances are assumed to be homogeneous, but there is a version of this covariance structure where variance can be heterogeneous. This structure is often used for repeated-measures data (especially when measurements are taken over time such as in growth models)
$\begin{pmatrix} \sigma_1^2 & \sigma_{21} & \sigma_{31} & \sigma_{41} \\ \sigma_{21} & \sigma_2^2 & \sigma_{32} & \sigma_{42} \\ \sigma_{31} & \sigma_{32} & \sigma_3^2 & \sigma_{43} \\ \sigma_{41} & \sigma_{42} & \sigma_{43} & \sigma_4^2 \end{pmatrix}$	*Unstructured*: This covariance structure is completely general. Covariances are assumed to be completely unpredictable: they do not conform to a systematic pattern.

CRAMMING SAM'S TIPS Multilevel models

- Multilevel models should be used to analyse data that have a hierarchical structure. For example, you might measure depression after psychotherapy. In your sample, patients will see different therapists within different clinics. This is a three-level hierarchy with depression scores from patients (level 1), nested within therapists (level 2) who are themselves nested within clinics (level 3).
- Hierarchical models are just like regression, except that you can allow parameters to vary (this is called a random effect). In ordinary regression, parameters generally are a fixed value estimated from the sample (a fixed effect).
- If we estimate a linear model within each context (e.g. the therapist or clinic, to use the example above) rather than the sample as a whole, then we can assume that the intercepts of these models vary (a random intercepts model), or that the slopes of these models differ (a random slopes model) or that both vary.
- We can compare different models (assuming that they differ in only one additional parameter) by looking at the difference in the −2 log-likelihood. Usually we would do this when we have changed only one parameter (added one new thing to the model).
- For any model we have to assume a covariance structure. For random intercepts models the default of *variance components* is fine, but when slopes are random an *unstructured* covariance structure is often assumed. When data are measured over time an autoregressive structure (AR(1)) is often assumed.

19.5. Some practical issues ③

19.5.1. Assumptions ③

Multilevel linear models are an extension of regression so all of the assumptions for regression apply to multilevel models (see section 7.6.2). There is a caveat, though, which is that the assumptions of independence and independent errors can sometimes be relaxed by a multilevel model because the purpose of this model is to factor in the correlations between cases caused by higher-level variables. As such, if a lack of independence is being caused by a level 2 or level 3 variable then a multilevel model should make this problem go away (although not always). As such, try to check the usual assumptions in the usual way.

There are two additional assumptions in multilevel models that relate to the random coefficients. These coefficients are assumed to be normally distributed around the overall model. So, in a random intercepts model the intercepts in the different contexts are assumed to be normally distributed around the overall model. Similarly, in a random slopes model, the slopes of the models in different contexts are assumed to be normally distributed.

Also it's worth mentioning that multicollinearity can be a particular problem in multilevel models if you have interactions that cross levels in the data hierarchy (cross-level interactions). However, centring predictors can help matters enormously (Kreft & de Leeuw, 1998), and we will see how to centre predictors in section 19.5.3.

19.5.2. Sample size and power ③

As you might well imagine, the situation with power and sample size is very complex indeed. One complexity is that we are trying to make decisions about our power to detect both fixed and random effects coefficients. Kreft and de Leeuw (1998) do a tremendous job of making sense of things for us. Essentially, the take-home message is the more data, the better. As more levels are introduced into the model, more parameters need to be estimated and the larger the sample sizes need to be. Kreft and de Leeuw conclude that if you are looking for cross-level interactions then you should aim to have more than 20 contexts (groups) in the higher-level variable, and that group sizes 'should not be too small'. They conclude by saying that there are so many factors involved in multilevel analysis that it is impossible to produce any meaningful rules of thumb.

Twisk (2006) agrees that the number of contexts relative to individuals within those contexts is important. He also points out that standard sample size and power calculations can be used but then 'corrected' for the multilevel component of the analysis (by factoring, among other things, the intraclass correlation). However, there are two corrections that he discusses that yield very different sample sizes! He recommends using sample size calculations with caution.

The easiest option is to get a computer to do it for you. HLM (http://www.ssicentral.com/hlm/index.html) will do power calculations for multilevel models, and for two-level models you could try Tom Snijders' PinT program (http://stat.gamma.rug.nl/multilevel.htm).

19.5.3. Centring variables ④

Centring refers to the process of transforming a variable into deviations around a fixed point. This fixed point can be any value that you choose, but typically we use the grand mean. We have already come across a form of centring way back in Chapter 1, when we discovered how to compute z-scores. When we calculate a z-score we take each score and subtract from it the mean of all scores (this centres the values at 0), and then divide by the standard deviation (this changes the units of measurement to standard deviations). When we centre a variable around the mean we simply subtract the mean from all of the scores: this centres the variables around 0.

There are two forms of centring that are typically used in multilevel modelling: *grand mean centring* and *group mean centring*. Grand mean centring means that for a given variable we take each score and subtract from it the mean of all scores (for that variable). Group mean centring means that for a given variable we take each score and subtract from it the mean of the scores (for that variable) within a given group. In both cases it is usually only level 1 predictors that are centred (in our cosmetic surgery example this would be predictors such as age, BDI and pre-surgery quality of life). If group mean centring is used then a level 1 variable is typically centred around means of a level 2 variable (in our cosmetic surgery data this would mean that, for example, the age of a person would be centred around the mean of age for the clinic at which the person had their surgery).

Centring can be used in ordinary multiple regression too, and because this form of regression is already familiar to you I'd like to begin by looking at the effects of centring in regression. In multiple regression the intercept represents the value of the outcome when all of the predictors take a value of 0. There are some predictors for which a value of 0 makes little sense. For example, if you were using heart rate as a predictor variable then a value of 0 would be meaningless (no one will have a heart rate of 0 unless they are dead). As such, the intercept in this case has no real-world use: why would you want to know the value of the outcome when heart rate was 0 given than no alive person would have a heart rate that low? Centring heart rate around its mean changes the meaning of the intercept. The intercept becomes the value of the outcome when heart rate is its average value. In more general terms, if all predictors are centred around their mean then the intercept is the value of the outcome when all predictors are the value of their mean. Centring can, therefore, be a useful tool for interpretation when a value of 0 for the predictor is meaningless.

The effect of centring in multilevel models, however, is much more complicated. There are some excellent reviews that look in detail at the effects of centring on multilevel models (Kreft & de Leeuw, 1998; Kreft, de Leeuw, & Aiken, 1995), and here I will just give a very basic précis of what they say. Essentially if you fit a multilevel model using the raw score predictors and then fit the same model but with grand mean centred predictors then the resulting models are equivalent. By this, I mean that they will fit the data equally well, have the same predicted values, and the residuals will be the same. The parameters themselves (the *b*s) will, of course, be different but there will be a direct relationship between the parameters from the two models (i.e. they can be directly transformed into each other). Therefore, grand mean centring doesn't change the model, but it would change your interpretation of the parameters (you can't interpret them as though they are raw scores). When group mean centring is used the picture is much more complicated. In this situation the raw score model is not equivalent to the centred model in either the fixed part or the random part. One exception is when only the intercept is random (which arguably is an unusual situation), and the group means are reintroduced into the model as level 2 variables (Kreft & de Leeuw, 1998).

The decision about whether to centre or not is quite complicated and you really need to make the decision yourself in a given analysis. Centring can be a useful way to combat multicollinearity between predictor variables. It's also helpful when predictors do not have a meaningful zero point. Finally, multilevel models with centred predictors tend to be more stable, and estimates from these models can be treated as more or less independent of each other, which might be desirable. If group mean centring is used then the group means should be reintroduced as a level 2 variable unless you want to look at the effect of your 'group' or level 2 variable uncorrected for the mean effect of the centred level 1 predictor, such as when fitting a model when time is your main explanatory variable (Kreft & de Leeuw, 1998).

OLIVER TWISTED

Please, Sir, can I have some more ... centring?

'Recentgin', babbles Oliver as he stumbles drunk out of Mrs Moonshine's alcohol emporium. 'I need some more recent gin.' I think you mean *centring* Oliver, not *recentgin*. If you want to know how to centre your variables using SAS, then the additional material for this chapter on the companion website will tell you.

19.6. Multilevel modelling on SAS ④

Multilevel modelling is unusual amongst statistical procedures in that there are many statistical programs out there which just do multilevel models – such as MLwiN or HLM.

We saw in section 19.4.1 that it is useful to build up models starting with a 'basic' model in which all parameters are fixed and then add random coefficients as appropriate before exploring confounding variables. We will take this approach to look at an example of conducting a multilevel model on SAS.

19.6.1. Entering the data ②

Data entry depends a bit on the type of multilevel model that you wish to run: the data layout is slightly different when the same variables are measured at several points in time. However, we will look at the case of repeated-measures data in a second example. In this first example, the situation we have is very much like multiple regression in that data from each person who had surgery are not measured over multiple time points. Figure 19.7 shows the data layout. Each row represents a case of data (in this case a person who had surgery). Their scores on the various variables are simply entered in different columns. So, for example, the first person was 31 years old, had a BDI score of 12, they were in the waiting list control group at clinic 1, were female and were waiting for surgery for a physical reason.

FIGURE 19.7 Data layout for multilevel modelling with no repeated measure

VIEWTABLE: Chap18.Cosmeticsurgery

	Baseline Quality of Life	Cosmetic Surgery	Clinic	Age	Beck Depression Inventory	Reason for Surgery	Gender
1	73	Waiting List	1	31	12	Physical reason	Female
2	74	Waiting List	1	32	16	Physical reason	Female
3	80	Waiting List	1	33	13	Physical reason	Female
4	76	Waiting List	1	59	11	Physical reason	Male
5	71	Waiting List	1	61	11	Physical reason	Male
6	72	Waiting List	1	32	10	Change Appearance	Female
7	71	Waiting List	1	33	11	Change Appearance	Female
8	73	Waiting List	1	35	15	Change Appearance	Female
9	80	Cosmetic Surgery	1	25	30	Physical reason	Female
10	64	Waiting List	1	55	36	Physical reason	Male
11	71	Waiting List	1	57	37	Physical reason	Male
12	72	Waiting List	1	29	34	Physical reason	Female
13	68	Waiting List	1	31	30	Change Appearance	Female
14	65	Waiting List	1	32	31	Change Appearance	Female
15	66	Waiting List	1	43	41	Physical reason	Female
16	76	Waiting List	1	45	34	Change Appearance	Female
17	69	Waiting List	1	46	36	Physical reason	Female

19.6.2. Ignoring the data structure: ANOVA ②

First of all, let's ground the example in something very familiar to us: ANOVA. Let's say for the time being that we were interested only in the effect that surgery has on post-operative quality of life. We could analyse this with a simple one-way independent ANOVA (or indeed a *t*-test), and the model is described by equation (19.1).

SELF-TEST Using what you know about ANOVA, conduct a one-way ANOVA using **Surgery** as the predictor and **Post_QoL** as the outcome.

In reality we wouldn't do an ANOVA, I'm just using it as a way of showing you that multilevel models are not big and scary, but are simply extensions of what we have done before.

SAS Output 19.1 shows the results of the ANOVA that you should get if you did the self-test. We find a non-significant effect of surgery on quality of life, $F(1, 274) = 0.33, p > 0.5$.

SAS OUTPUT 19.1

Source	DF	Sum of Squares	Mean Square	F Value	Pr > F
Model	1	28.62030	28.62030	0.33	0.5660
Error	274	23747.88329	86.67111		
Corrected Total	275	23776.50359			

PROC MIXED is very similar to PROC GLM but, as you might expect, has a whole lot of additional options. One option we'll find useful later on is to ask for the regression equation – to request this, put /SOLUTION on the MODEL line.

A second option is the method of estimation. By default, PROC MIXED uses restricted maximum likelihood (REML). (If you look at the top of the output in your SAS output window, you'll notice it says: Estimation Method REML.) Because we want to compare models (see SAS Tip 19.1) we want to use maximum likelihood – to tell SAS, we put METHOD=ML on the PROC MIXED line. The syntax with these options is shown in SAS Syntax 19.1.

The third useful option is to request tests of the covariance parameter estimates, which will give us a significance test of each of the covariance estimates in the model (i.e. the values of u in equations (19.3), (19.5) and (19.7)). These estimates tell us about the variability of intercepts or slopes across our contextual variable and so significance testing them can be useful (we can then say that there was significant, or not, variability in intercepts or slopes).[4] These are requested with the COVTEST option on the PROC MIXED line.

```
PROC MIXED DATA=chap19.cosmeticsurgery METHOD=ML COVTEST;
    CLASS surgery;
    MODEL post_qol = surgery /SOLUTION;
    RUN;
```

SAS Syntax 19.1

SAS TIP 19.1 Estimation ③

SAS gives you the choice of two methods for estimating the parameters in the analysis: maximum likelihood (ML), which we have encountered before, and restricted maximum likelihood (REML). The conventional wisdom seems to be that ML produces more accurate estimates of fixed regression parameters, whereas REML produces more accurate estimates of random variances (Twisk, 2006). As such, the choice of estimation procedure depends on whether your hypotheses are focused on the fixed regression parameters or on estimating variances of the random effects. However, in many situations the choice of ML or REML will make only a small difference to parameter estimates. Also, if you want to compare models you must use ML.

[4] This is a slightly curious thing to test – we're asking if the variance parameters are significantly different from zero, but we *know* that they cannot be below zero, because they are variances (recall that variances are sums of squares, and if you square a negative number, it becomes positive). We're going to ask for them anyway.

SAS Output 19.2 shows the main table for the model. Compare this table with SAS Output 19.1 and you'll see that there is basically no difference: we get a non-significant effect of surgery with an *F* of 0.33, and a *p* of .565 (as opposed to 0.566.)[5] The point I want you to absorb here is that if we ignore the hierarchical structure of the data then what we are left with is something very familiar: an ANOVA/regression. The numbers are more or less exactly the same; all that has changed is that we have used a different procedure get to the same end point.

SAS OUTPUT 19.2

Type 3 Tests of Fixed Effects				
Effect	**Num DF**	**Den DF**	**F Value**	**Pr > F**
SURGERY	1	274	0.33	0.5646

19.6.3. Ignoring the data structure: ANCOVA ②

We have seen that there is no effect of cosmetic surgery on quality of life, but we did not take into account the quality of life before surgery. Let's, therefore, extend the example a little to look at the effect of the surgery on quality of life while taking into account the quality of life scores before surgery. Our model is now described by equation (19.8). You would normally do this analysis with an ANCOVA, through the univariate GLM menu. As in the previous section we'll run the analysis both ways, just to illustrate that we're doing the same thing when we run a hierarchical model.

SELF-TEST Using what you know about ANCOVA, conduct a one-way ANCOVA using **Surgery** as the predictor, **Post_QoL** as the outcome and **Base_QoL** as the covariate.

As before, we probably wouldn't do an ANCOVA using the mixed model procedure, but there's no reason that we can't, and it's a useful illustration. If you need a reminder, SAS Syntax 19.2 shows how to run this ANCOVA using PROC GLM.

```
PROC GLM DATA=chap19.cosmeticsurgery ;
    MODEL post_qol = surgery base_qol/SOLUTION;
    RUN;
```

SAS Syntax 19.2

SAS Output 19.3 shows the results of the ANCOVA that you should get if you did the self-test. With baseline quality of life included, we find a significant effect of surgery on quality of life, $F(1, 273) = 4.04$, $p = .045$. Baseline quality of life also predicted quality of life after surgery, $F(1, 273) = 214.89$, $p < .0001$.

[5] If you'd used REML, the default, you would have found that *p* was 0.566.

SAS OUTPUT 19.3

Source	DF	Sum of Squares	Mean Square	F Value	Pr > F
Model	2	10488.25317	5244.12659	107.74	<.0001
Error	273	13288.25042	48.67491		
Corrected Total	275	23776.50359			

R-Square	Coeff Var	Root MSE	POST_QOL Mean
0.441118	11.70402	6.976741	59.60978

Source	DF	Type III SS	Mean Square	F Value	Pr > F
SURGERY	1	196.81587	196.81587	4.04	0.0453
BASE_QOL	1	10459.63287	10459.63287	214.89	<.0001

Parameter	Estimate		Standard Error	t Value	Pr > |t|
Intercept	18.14702471	B	2.90766570	6.24	<.0001
SURGERY Cosmetic Surgery	1.69723334	B	0.84404203	2.01	0.0453
SURGERY Waiting List	0.00000000	B	.	.	.
BASE_QOL	0.66503584		0.04536693	14.66	<.0001

We can run the equivalent model using PROC MIXED. SAS Syntax 19.3 shows how the model is run using PROC MIXED. It is very similar to PROC GLM.

```
PROC MIXED DATA=chap19.cosmeticsurgery METHOD=ML COVTEST;
    MODEL post_qol = surgery base_qol/SOLUTION;
    RUN;
```

SAS Syntax 19.3

SAS Output 19.4 shows the main table for the model. Compare this table with SAS Output 19.3 and you'll see that again there is (almost) no difference: we get a significant effect of surgery with an F of 4.09, $p = .044$, and a significant effect of baseline quality of life with an F of 217.25, $p < .0001$. We can also see that the regression coefficient for surgery is −1.70. Again, the results are pretty similar to when we ran the analysis as ANCOVA (the values are slightly different because here we're using maximum-likelihood methods to estimate the parameters of the model but in ANCOVA we us ordinary least squares methods).

Hopefully this has convinced you that we're just doing a regression here, something you have been doing throughout this book. This technique isn't radically different, and if you think about it as just an extension of what you already know, then it's really relatively easy to understand. So, having shown you that we can do basic analyses through the mixed models syntax, let's now use its power to factor in the hierarchical structure of the data.

SAS OUTPUT 19.4

Covariance Parameter Estimates				
Cov Parm	Estimate	Standard Error	Z Value	Pr > Z
Residual	48.1458	4.0984	11.75	<.0001

Fit Statistics	
-2 Log Likelihood	1852.5
AIC (smaller is better)	1860.5
AICC (smaller is better)	1860.7
BIC (smaller is better)	1875.0

Solution for Fixed Effects						
Effect	Cosmetic Surgery	Estimate	Standard Error	DF	t Value	Pr > |t|
Intercept		18.1470	2.8918	273	6.28	<.0001
SURGERY	Cosmetic Surgery	−1.6972	0.8394	273	2.02	0.0442
SURGERY	Waiting List	0	.	.	.	.
BASE_QOL		0.6650	0.04512	273	14.74	<.0001

Type 3 Tests of Fixed Effects				
Effect	Num DF	Den DF	F Value	Pr > F
SURGERY	1	273	4.09	0.0442
BASE_QOL	1	273	217.25	<.0001

19.6.4. Factoring in the data structure: random intercepts ③

We have seen that when we factor in the pre-surgery quality of life scores, which themselves significantly predict post-surgery quality of life scores, surgery seems to positively affect quality of life. However, at this stage we have ignored that fact that our data have a hierarchical structure. Essentially we have violated the independence assumption because scores from people who had their surgery at the same clinic are likely to be related to each other (and certainly more related than with people at different clinics). We have seen that violating the assumption of independence can have some quite drastic consequences (see section 10.2.10). However, rather than just panic and gibber about our *F*-ratio being inaccurate, we can model this covariation within clinics explicitly by including the hierarchical data structure in our analysis.

To begin with, we will include the hierarchy in a fairly crude way by assuming simply that intercepts vary across clinics. Our model is now described by:

$$\text{QoL After Surgery}_{ij} = b_{0j} + b_1\text{Surgery}_{ij} + b_2\text{QoL Before Surgery}_{ij} + \varepsilon_{ij}$$
$$b_{0j} = b_0 + u_{0j}$$

What we need to do is to tell PROC MIXED that our individuals are grouped within clinics, and that we'd like to have random intercepts (please). To do this, we add a RANDOM statement to PROC MIXED. We specify what we'd like to be random (in this first instance, it's only the intercept). In the options to the RANDOM statement, we need to tell SAS what the clustering variable is – curiously, SAS calls this the SUBJECT variable, and in our case that is clinic.[5] In addition, clinic is a categorical variable, so we need to add it to the CLASS statement.

```
PROC MIXED DATA=chap19.cosmeticsurgery METHOD=ML COVTEST;
    CLASS clinic;
    MODEL post_qol = surgery base_qol/SOLUTION;
    RANDOM intercept /SUBJECT=clinic;
    RUN;
```
SAS Syntax 19.4

The output of this analysis is shown in SAS Output 19.5. The first issue is whether allowing the intercepts to vary has made a difference to the model. We can test this from the change in the −2 log-likelihood (equation (19.10)). In our new model the −2LL is 1837.5 (SAS Output 19.5) based on a total of five parameters. In the old model (SAS Output 19.4) the −2LL was 1852.54, based on four parameters. Therefore:

$$\chi^2_{\text{Change}} = 1852.5 - 1837.5 = 15$$
$$df_{\text{Change}} = 5 - 4 = 1$$

If we look at the critical values for the chi-square statistic with 1 degree of freedom in the Appendix, they are 3.84 ($p < .05$) and 6.63 ($p < .01$); therefore, this change is highly significant. Put another way, it is important that we modelled this variability in intercepts because when we do our model is significantly improved. We can conclude then that the intercepts for the relationship between surgery and quality of life (when controlling for baseline quality of life) vary significantly across the different clinics.

The variance estimate for the intercept is equal to 9.24. You will also notice that the significance of the variance estimate for the intercept is tested using a standard z-score in this case ($z = 1.69$), this is just about statistically significant, $p = .045$. (Notice that this is a one-tailed test, because the variance can only be positive.) You should be cautious in interpreting the Wald statistic because, for random parameters especially, it can be quite unpredictable (for fixed effects it should be OK). The change in the −2LL is much more reliable, and you should use this to assess the significance of changes to the model – just like with logistic regression (Chapter 8).

By allowing the intercept to vary we also have a new regression parameter for the effect of surgery, which is −0.31 compared to −1.70 when the intercept was fixed (SAS Output 19.4). In other words, by allowing the intercepts to vary over clinics, the effect of surgery has decreased dramatically. In fact, it is not significant any more, $F(1, 264) = 0.14$, $p = 0.709$.

[5] This is because multilevel models are often used when we have multiple measures from the same person.

This shows how, had we ignored the hierarchical structure in our data, we would have reached very different conclusions than we have found here.

SAS OUTPUT 19.5

Covariance Parameter Estimates					
Cov Parm	Subject	Estimate	Standard Error	Z Value	Pr > Z
Intercept	CLINIC	9.2326	5.4576	1.69	0.0454
Residual		42.4978	3.7040	11.47	<.0001

Fit Statistics	
–2 Log Likelihood	1837.5
AIC (smaller is better)	1847.5
AICC (smaller is better)	1847.7
BIC (smaller is better)	1849.0

Solution for Fixed Effects						
Effect	Cosmetic Surgery	Estimate	Standard Error	DF	t Value	Pr > |t|
Intercept		29.5622	3.4529	9	8.56	<.0001
SURGERY	Cosmetic Surgery	0.3132	0.8386	264	0.37	0.7091
SURGERY	Waiting List	0	.	.	.	.
BASE_QOL		0.4787	0.05249	264	9.12	<.0001

Type 3 Tests of Fixed Effects				
Effect	Num DF	Den DF	F Value	Pr > F
SURGERY	1	264	0.14	0.7091
BASE_QOL	1	264	83.17	<.0001

19.6.5. Factoring in the data structure: random intercepts and slopes ④

We have seen that including a random intercept is important for this model (it changes the log-likelihood significantly). However, we could now look at whether adding a random slope will also be beneficial by adding this term to the model. The model is now described by equation (19.9), which we saw earlier on; it can be specified in SAS with only minor modifications. All we are doing is adding another random term to the model; therefore, the only changes we need to make are in the RANDOM statement. As well as the intercept varying across clinics, we'd like the effect of surgery to vary across clinics, so we add that as a random effect. However, now we have two random effects, and so they can have a structure (see Section 19.4.2). Because we are adding one parameter at a time, we'll tell SAS that these two random effects should be uncorrelated. To keep them uncorrelated, we tell SAS to use a variance components structure, shortened to VC. SAS Syntax 19.5 shows the syntax – compare this with

```
PROC MIXED DATA=chap19.cosmeticsurgery METHOD=REML COVTEST;
    CLASS clinic;
    MODEL post_qol = surgery base_qol/SOLUTION;
    RANDOM intercept surgery /SUBJECT=clinic TYPE=VC;
    RUN;
```

SAS Syntax 19.5

All we're interested in at this stage is estimating the effect of including the variance in intercepts. SAS Output 19.6 gives us the −2LL for the new model and the value of the variance in slopes (29.63). To find the significance of the variance in slopes, we subtract this value from the −2LL for the previous model. This gives us a chi-square statistic with df = 1 (because we have added only one new parameter to the model: the variance in slopes). In our new model the −2LL is 1816 (SAS Output 19.6) based on a total of six parameters. In the old model (SAS Output 19.5) the −2LL was 1837.5, based on five parameters. Therefore:

$$\chi^2_{\text{Change}} = 1837.5 - 1816 = 21.5$$
$$df_{\text{Change}} = 6 - 5 = 1$$

Comparing this value to the same critical values as before for the chi-square statistic with $df = 1$ (i.e. 3.84 and 6.63) shows that this change is highly significant because 21.5 is much larger than these two values. Put another way, the fit of our model significantly improved when the variance of slopes was included: there is significant variability in slopes.

SAS OUTPUT 19.6

Covariance Parameter Estimates					
Cov Parm	Subject	Estimate	Standard Error	Z Value	Pr > Z
Intercept	CLINIC	33.1815	16.9008	1.96	0.0248
SURGERY	CLINIC	29.6269	16.4948	1.80	0.0362
Residual		35.0086	3.1329	11.17	<.0001

Fit Statistics	
-2 Log Likelihood	1816.0
AIC (smaller is better)	1828.0
AICC (smaller is better)	1828.3
BIC (smaller is better)	1829.8

Solution for Fixed Effects					
Effect	Estimate	Standard Error	DF	t Value	Pr > |t|
Intercept	39.9811	3.8328	9	10.43	<.0001
SURGERY	–0.2292	1.8941	9	-0.12	0.9063
BASE_QOL	0.3121	0.05355	255	5.83	<.0001

Type 3 Tests of Fixed Effects				
Effect	Num DF	Den DF	F Value	Pr > F
SURGERY	1	9	0.01	0.9063
BASE_QOL	1	255	33.97	<.0001

Now that we know that there is significant variability in slopes, we can look to see whether the slopes and intercepts are correlated (or covary). By selecting TYPE=VC in the previous analysis, we assumed that the covariances between the intercepts and slopes were zero. Therefore, SAS estimated only the variance of slopes. This was a useful thing to do because it allowed us to look at the effect of the variance of slopes in isolation. If we now want to include the covariance between random slopes and random intercepts we do this by changing TYPE=VC to TYPE = UN, where UN is short for Unstructured. We remove the assumption that the covariances between slopes and intercepts are zero, and so SAS will estimate this covariance. As such, by changing to unstrcutured, we add a new term to the model that estimates the covariance between random slopes and intercepts. This is shown in SAS Syntax 19.6.

```
PROC MIXED DATA=chap19.cosmeticsurgery METHOD=ML COVTEST;
    CLASS clinic;
    MODEL post_qol = surgery base_qol /SOLUTION;
    RANDOM intercept surgery /SUBJECT=clinic TYPE=UN;
    RUN;
```

SAS Syntax 19.6

SAS OUTPUT 19.7

Covariance Parameter Estimates					
Cov Parm	Subject	Estimate	Standard Error	Z Value	Pr Z
UN(1,1)	CLINIC	37.6084	18.7251	2.01	0.0223
UN(2,1)	CLINIC	–36.6798	18.7632	–1.95	0.0506
UN(2,2)	CLINIC	38.4081	19.2091	1.90	0.0287
Residual		34.9557	3.1167	11.22	<.0001

Fit Statistics	
-2 Log Likelihood	1798.6
AIC (smaller is better)	1812.6
AICC (smaller is better)	1813.0
BIC (smaller is better)	1814.7

Solution for Fixed Effects					
Effect	Estimate	Standard Error	DF	t Value	Pr > |t|
Intercept	40.1025	3.8717	9	10.36	<.0001
SURGERY	–0.6545	2.0994	9	–0.31	0.7623
BASE_QOL	0.3102	0.05321	255	5.83	<.0001

Type 3 Tests of Fixed Effects				
Effect	Num DF	Den DF	F Value	Pr > F
SURGERY	1	9	0.10	0.7623
BASE_QOL	1	255	33.98	<.0001

The output of this analysis is shown in SAS Output 19.7. The first issue is whether adding the covariance between slopes and intercepts has made a difference to the model using the change in the −2LL (equation (19.10)). In our new model the −2LL is 1798.6 (SAS Output 19.7) based on a total of seven parameters. In the old model (SAS Output 19.6) the −2LL was 1816, based on six parameters. Therefore:

$$\chi^2_{\text{Change}} = 1816 - 1798.6 = 17.4$$
$$df_{\text{Change}} = 7 - 6 = 1$$

This change is highly significant at $p < .01$ because 17.4 is bigger than the critical value of 6.63 for the chi-square statistic with 1 degree of freedom (as before). Put another way, our model is significantly improved when the covariance term is included in the model. We now have three random parameters: the intercept (37.61), which SAS calls UN(1, 1); and we referred to as μ_{0j} the slope (38.41), which we called μ_{1j}, and SAS calls UN(2, 2); and the covariance of the slope and intercept (−36.68), and SAS calls UN(2, 1). Each of these parameters has an associated significance based again on the

Wald test, all three estimates are significant, or close to (although I reiterate my earlier point that this Wald statistic should be interpreted with caution).

The covariance of the slope and intercept is a new parameter in the model. The reason why there are two values is that we changed from a covariance structure of variance components, which forces the parameters to be uncorrelated, to unstructured, which does not and, therefore, the covariance is estimated too. The first of these values is the covariance between the random slope and random intercept, and the second is the variance of the random slopes. We encountered covariance in Chapter 6 and saw that it is an unstandardized measure of the relationship between variables. In other words, it's like a correlation. Therefore, the covariance term tells us whether there is a relationship or interaction between the random slope and the random intercept within the model. The actual size of this value is not terribly important because it is unstandardized (so we can't compare the size of covariances measured across different variables), but its direction is. In this case the covariance is negative (−36.68) indicating a negative relationship between the intercepts and the slopes. Remember that we are looking at the effect of surgery on quality of life in 10 different clinics, so this means that, across these clinics, as the intercept for the relationship between surgery and quality of life increases, the value of the slope decreases.

There are two ways to examine these effects. First, we can ask for the parameters of the random effects from SAS, by putting the SOLUTION option on the RANDOM line, as shown in SAS Syntax 19.7. If we do that, SAS will produce the table shown in SAS Output 19.8. SAS has dummy coded, so clinic 10 is indicated with an asterisk. Clinic 1, for example, has a highly positive intercept (6.62) and a highly negative slope (−6.67). In contrast, clinic 7 has a high negative intercept, and a high positive slope. If you check you will find that every clinic that has a positive intercept has a negative slope, and every clinic that has a negative intercept has a positive slope – it's no surprise that these are negatively correlated then.

```
PROC MIXED DATA=chap19.cosmeticsurgery METHOD=ML COVTEST;
    CLASS clinic;
    MODEL post_qol = surgery base_qol /SOLUTION outp=outp
outpm=outpm;
    RANDOM intercept surgery /SUBJECT=clinic TYPE=UN SOLUTION;
    RUN;
```

SAS Syntax 19.7

SAS OUTPUT 19.8

Solution for Random Effects						
Effect	Clinic	Estimate	Std Err Pred	DF	t Value	Pr > |t|
Intercept	*	−7.1327	2.3609	255	−3.02	0.0028
SURGERY	*	7.4015	2.6200	255	2.83	0.0051
Intercept	1	6.6290	2.3999	255	2.76	0.0062
SURGERY	1	−6.6799	2.7392	255	−2.44	0.0154
Intercept	2	9.7314	2.4233	255	4.02	<.0001
SURGERY	2	−7.7595	2.7230	255	−2.85	0.0047
Intercept	3	3.6661	2.5774	255	1.42	0.1561

Solution for Random Effects						
Effect	Clinic	Estimate	Std Err Pred	DF	t Value	Pr > |t|
SURGERY	3	−4.6380	2.7933	255	−1.66	0.0981
Intercept	4	5.8521	2.3990	255	2.44	0.0154
SURGERY	4	−6.3475	2.6498	255	−2.40	0.0173
Intercept	5	0.8451	2.5689	255	0.33	0.7424
SURGERY	5	−1.9864	2.7825	255	−0.71	0.4759
Intercept	6	−1.9863	2.4580	255	−0.81	0.4198
SURGERY	6	3.3130	2.6855	255	1.23	0.2185
Intercept	7	−7.6701	2.4109	255	−3.18	0.0016
SURGERY	7	7.5267	2.6155	255	2.88	0.0043
Intercept	8	−4.0100	2.5795	255	−1.55	0.1213
SURGERY	8	2.1728	2.7704	255	0.78	0.4336
Intercept	9	−5.9245	2.5307	255	−2.34	0.0200
SURGERY	9	6.9973	2.7361	255	2.56	0.0111

OLIVER TWISTED

You fool! Those estimates in SAS Output 19.8 are wrong!

Ah, Oliver, you've run a GLM with BY clinic, haven't you? You thought that the estimates from each clinic from PROC GLM should match the estimates from each clinic in the SOLUTION that PROC MIXED gave. I can see why you think that – after all, we are freely estimating the slope and intercept in each clinic, so why shouldn't they match? The reason is actually a little complex; however, it does involve the statistic with my favourite name – the EBLUP (EBLUP is pronounced ee-blup; and stands for empirical best linear unbiased prediction). Sadly, we don't have space to go into it here, but you'll find more on the website.

The other thing we can do is draw a graph. First, we need to find the predicted values from the mixed model. We do this by adding OUTP=dataset to the MODEL line, then we can plot a line graph of the observed values of quality of life after surgery compared with the predicted values. This is shown in SAS Syntax 19.8.

```
PROC MIXED DATA=chap19.cosmeticsurgery METHOD=ML COVTEST;
    CLASS clinic;
    MODEL post_qol = surgery base_qol /SOLUTION OUTP=outp ;
    RANDOM intercept surgery /SUBJECT=clinic TYPE=UN SOLUTION;
    RUN;
```

```
  SYMBOL1 COLOR=BLACK INTERPOL=rl VALUE=NONE REPEAT=1000;
PROC GPLOT DATA=outp;
  PLOT pred*post_qol=clinic / NOLEGEND;
  RUN;
```

SAS Syntax 19.8

Running SAS Syntax 19.8 should produce a chart like that shown in Figure 19.8. The chart shows the observed values of quality of life after surgery plotted against those predicted by our model. In this diagram each line represents a different clinic. We can see that the 10 clinics differ: those with low intercepts (low values on the y-axis) have quite steep positive slopes. However, as the intercept increases (as we go from the line that crosses the y-axis at the lowest point up to the line that hits the y-axis at the highest point) the slopes of the lines get flatter (the slope decreases). The negative covariance between slope and intercept reflects this relationship. Had it been positive it would have meant the opposite: as intercepts increase, the slopes increase also.

The second term that we get with the random slope is its variance (in this case 38.41). This tells us how much the slopes vary around a single slope fitted to the entire data set (i.e. ignoring the clinic from which the data came). This confirms what our chi-square test showed us; that the slopes across clinics are significantly different.

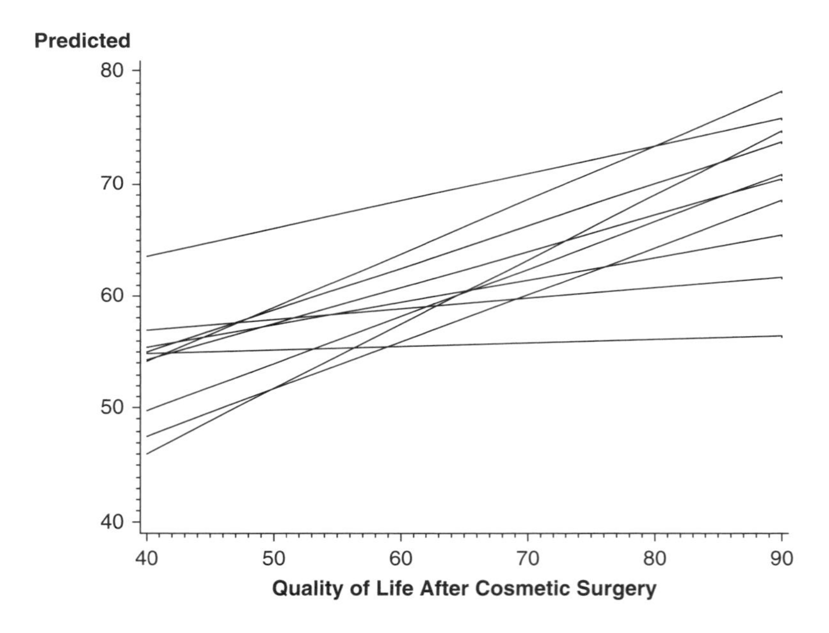

FIGURE 19.8 Predicted values from the model (surgery predicting quality of life after controlling for baseline quality of life) plotted against the observed values

We can conclude, then, that the intercepts and slopes for the relationship between surgery and quality of life (when controlling for baseline quality of life) vary significantly across the different clinics. By allowing the intercept and slopes to vary we also have a new regression parameter for the effect of surgery, which is -0.65 compared to -0.31 when the slopes were fixed (SAS Output 19.5). In other words, by allowing the intercepts to vary over clinics, the effect of surgery has increased slightly, although it is still nowhere near significant, $F(1, 9) = 0.10$, $p > .05$. This shows how, had we ignored the hierarchical structure in our data, we would have reached very different conclusions to what we have found here.

19.6.6. Adding an interaction to the model ④

We can now build up the model by adding in another variable. One of the variables we measured was the reason for the person having cosmetic surgery: was it to resolve a physical problem or was it purely for vanity? We can add this variable to the model, and also look at whether it interacts with surgery in predicting quality of life.[6] Our model will simply expand to incorporate these new terms, and each term will have a regression coefficient (which we select to be fixed). Therefore, our new model can be described as in the equation below (note that all that has changed is that there are two new predictors):

$$\begin{aligned} \text{QoL After}_{ij} &= b_{0j} + b_{1j}\text{Surgery}_{ij} + b_2\text{QoL Before Surgery}_{ij} + b_3\text{Reason}_{ij} \\ &\quad + b_4\,(\text{Reason} \times \text{Surgery})_{ij} + \varepsilon_{ij} \\ b_{0j} &= b_0 + u_{0j} \\ b_{1j} &= b_1 + u_{1j} \end{aligned} \tag{19.11}$$

To set up this model in SAS is very easy to do and just requires some minor changes to the model. We add **reason** and **reason*surgery** to the model line, as shown in SAS Syntax 19.9.

```
PROC MIXED DATA=temp METHOD=ML COVTEST;
  CLASS clinic;
  MODEL post_qol = surgery base_qol reason reason*surgery /SOLUTION
  OUTP=outp;
  RANDOM intercept surgery /SUBJECT=clinic TYPE=UN ;
  RUN;
```

SAS Syntax 19.9

SAS Output 19.9 shows the resulting output, which is similar to the previous output except that we now have two new fixed effects. The first issue is whether these new effects make a difference to the model. We can use the log-likelihood statistics again:

$$\chi^2_{\text{Change}} = 1798.6 - 1789.0 = 9.6$$
$$df_{\text{Change}} = 9 - 7 = 2$$

If we look at the critical values for the chi-square statistic in the Appendix, it is 5.99 ($p < .05$, $df = 2$); therefore, this change is significant. We can look at the effects individually in the table of fixed effects. This tells us that quality of life before surgery significantly predicted quality of life after surgery, $t(253) = 5.80$, $p < .0001$, surgery still did not significantly predict quality of life, $t(9) = -1.47$, $p = .175$, but the reason for surgery, $t(253) = 3.11$, $p = 0.002$, and the interaction of the reason for surgery and surgery, $t(253) = 2.51$, $p = 0.013$, both did significantly predict quality of life.

[7] In reality, because we would use the change in the –2LL to see whether effects are significant, we would build this new model up a term at a time. Therefore, we would first include only **Reason** in the model, then in a separate analyse we would add the interaction. By doing so we can calculate the change in –2LL for each effect. To save space I'm going to put both into the model in a single step.

SAS OUTPUT 19.9

Covariance Parameter Estimates					
Cov Parm	**Subject**	**Estimate**	**Standard Error**	**Z Value**	**Pr Z**
UN(1,1)	CLINIC	30.0557	15.4440	1.95	0.0258
UN(2,1)	CLINIC	–28.0831	15.1952	–1.85	0.0646
UN(2,2)	CLINIC	29.3489	16.4041	1.79	0.0368
Residual		33.8598	3.0244	11.20	<.0001

Fit Statistics	
−2 Log Likelihood	1789.0
AIC (smaller is better)	1807.0
AICC (smaller is better)	1807.7
BIC (smaller is better)	1809.8

Solution for Fixed Effects					
Effect	**Estimate**	**Standard Error**	**DF**	**t Value**	**Pr > |t|**
Intercept	22.5178	3.8400	9	5.86	0.0002
SURGERY	–3.1877	2.1655	9	–1.47	0.1751
BASE_QOL	0.3054	0.05264	253	5.80	<.0001
REASON	–3.5152	1.1306	253	–3.11	0.0021
SURGERY*REASON	4.2213	1.6848	253	2.51	0.0129

The values of the variance for the intercept (30.06) and the slope (29.35) are lower than the previous model but still significant (one-tailed). Also the covariance between the slopes and intercepts is still negative (−28.08). As such our conclusions about our random parameters stay much the same as in the previous model.

The effect of the reason for surgery is easy to interpret. Given that we coded this predictor as 1 = physical reason and 0 = change appearance, the negative coefficient tells us that as reason increases (i.e. for surgery = 0, as a person goes from changing their appearance to a physical reason) quality of life decreases. However, this effect in isolation isn't that interesting because it includes both people who had surgery and the waiting list controls. More interesting is the interaction term, because this takes account of whether or not the person had surgery. To break down this interaction we could rerun the analysis separately for the two 'reason groups'. Obviously we would remove the interaction term and the main effect of **Reason** from this analysis (because we are analysing the physical reason group separately from the group that wanted to change their appearance). As such, you need to fit the model in the previous section, but first split the file by **Reason**.

SELF-TEST Use a BY statement[8] to split the file by **Reason** and then run a multilevel model predicting **Post_QoL** with a random intercept, and random slopes for **Surgery**, and including **Base_QoL** and **Surgery** as predictors.

Surgery to Change Appearance:

SAS OUTPUT 19.10

Solution for Fixed Effects					
Effect	Estimate	Standard Error	DF	t Value	Pr > |t|
Intercept	18.0207	4.6661	9	3.86	0.0038
SURGERY	1.1965	2.0820	9	0.57	0.5796
BASE_QOL	0.3177	0.06888	157	4.61	<.0001

Surgery for a Physical Problem:

Solution for Fixed Effects					
Effect	Estimate	Standard Error	DF	t Value	Pr > |t|
Intercept	21.7843	5.4877	8	3.97	0.0041
SURGERY	–4.3074	2.2396	6	–1.92	0.1028
BASE_QOL	0.3385	0.07903	81	4.28	<.0001

SAS Output 19.10 shows the parameter estimates from these analyses. It shows that for those operated on only to change their appearance, surgery almost significantly predicted quality of life after surgery, $b = -4.31$, $t(6) = -1.92$, $p = 0.103$. The negative gradient shows that in these people, quality of life was lower after surgery compared to the control group. However, for those that had surgery to solve a physical problem surgery did not significantly predict quality of life, $b = 1.20$, $t(9) = 0.57$, $p = .580$. However, the slope was positive indicating that people who had surgery scored higher on quality of life than those on the waiting list (although not significantly so!). The interaction effect, therefore, reflects the difference in slopes for surgery as a predictor of quality of life in those who had surgery for physical problems (slight positive slope) and those who had surgery purely for vanity (a negative slope).

We could sum up these results by saying that quality of life after surgery, after controlling for quality of life before surgery, was lower for those who had surgery to change their appearance than those that had surgery for a physical reason. This makes sense because for those having surgery to correct a physical problem, the surgery has probably bought relief and so their quality of life will improve. However, those having surgery for vanity might well discover that having a different appearance wasn't actually at the root of their unhappiness, so their quality of life is lower.

[8] You might need to run PROC SORT first.

CRAMMING SAM'S TIPS **Multilevel models SAS Output**

- The **Fit Statistics** table can be used to assess the overall fit of the model. The value of –2LL can be significance tested with *df* = the number of parameters being estimated. It is mainly used, though, to compare models that are the same in all but one parameter by testing the difference in –2LL in the two models against *df* = 1 (if only one parameter has been changed). The AIC, AICC and BIC can also be compared across models (but not significance tested).
- The table of **Type 3 Tests of Fixed Effects** tells you whether your predictors significantly predict the outcome: look in the column labelled *Pr > F Sig.* If the value is less than .05 then the effect is significant.
- The table of **Solution for Fixed Effects** gives us the regression coefficient for each effect The direction of these coefficients tells us whether the relationship between each predictor and the outcome is positive or negative.
- The table labelled **Covariance Parameters Estmates** tells us about any random effects in the model. These values can tell us how much intercepts and slopes varied over our level 1 variable. The significance of these estimates should be treated cautiously. The exact labelling of these effects depends on which covariance structure you selected for the analysis.

19.7. Growth models ④

Growth models are extremely important in many areas of science including psychology, medicine, physics, chemistry and economics. In a growth model the aim is to look at the rate of change of a variable over time: for example, we could look at white blood cell counts, attitudes, radioactive decay or profits. In all cases we're trying to see which model best describes the change over time.

19.7.1. Growth curves (polynomials) ④

Figure 19.9 gives some examples of possible growth curves. This diagram shows three polynomials representing a linear trend (labelled) otherwise known as a first-order polynomial, a quadratic trend (labelled) otherwise known as a second-order polynomial, and a cubic trend (labelled) otherwise known as a third-order polynomial. Notice first that the linear trend is a straight line, but as the polynomials increase they get more and more curved, indicating more rapid growth over time. Also, as polynomials increase, the change in the curve is quite dramatic (so dramatic that I adjusted the scale of the graph to fit all three curves on the same diagram). This observation highlights the fact that any growth curve higher than a quadratic (or possibly cubic) trend is very unrealistic in real data. By fitting a growth model to the data we can see which trend best describes the growth of an outcome variable over time (although no one will believe that a significant fifth-order polynomial is telling us anything meaningful about the real world!).

The growth curves that we have described might seem familiar to you: they are the same as the trends that we described for ordered means in section 10.2.11.4. What we are discussing now is really no different. There are just two important things to remember when fitting growth curves: (1) you can fit polynomials of order up to one less than the number of time points that you have; and (2) a polynomial is defined by a simple power function. On the first point, this means that with three time points you can fit a linear and

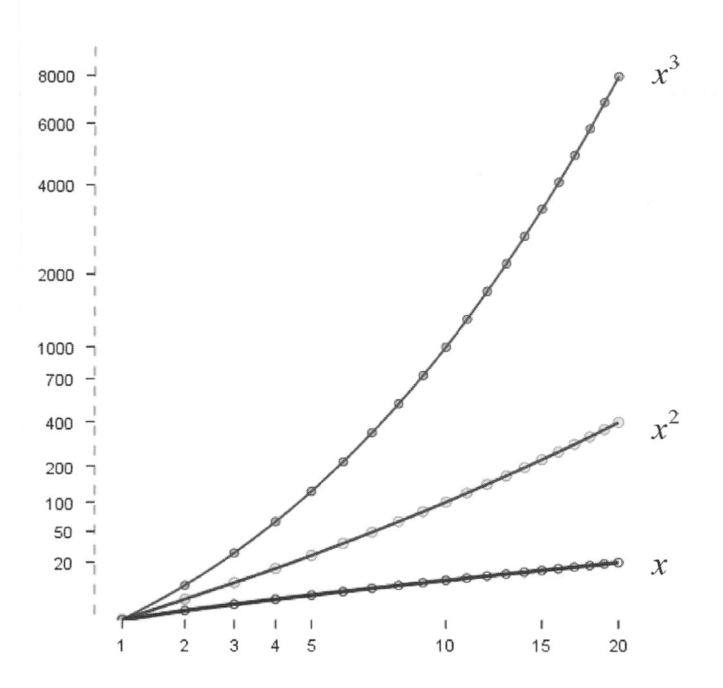

FIGURE 19.9 Illustration of a first-order (linear, x), second-order (quadratic, x^2) and third-order (cubic, x^3) polynomial

quadratic growth curve (or a first- and second-order polynomial), but you cannot fit any higher-order growth curves. Similarly, if you have six time points you can (in theory) fit up to a fifth-order polynomial. This is the same basic idea as having one less contrast than the number of groups in ANOVA (see section 10.2.11).

On the second point, we have to define growth curves manually in multilevel models in SAS: there is not a convenient option that we can select to do it for us. However, this is quite easy to do. If *time* is our predictor variable, then a linear trend is tested by including this variable alone. A quadratic or second-order polynomial is tested by including a $time^2$ predictor, a cubic or third-order polynomial is tested by including a $time^3$ predictor and so on. So any polynomial is tested by including a variable that is the predictor to the power of the order of polynomial that you want to test: for a fifth-order polynomial we need a predictor of $time^5$ and for an n-order polynomial we would have to include $time^n$ as a predictor. Hopefully you get the general idea.

19.7.2. An example: the honeymoon period ②

I recently saw a brilliant talk given by Professor Daniel Kahneman, who won the 2002 Nobel Prize for Economics. In this talk Kahneman bought together an enormous amount of research on life satisfaction (he explored questions such as whether people are happier if they are richer). There was one graph in this talk that particularly grabbed my attention. It showed that leading up to marriage people reported greater life satisfaction, but by about two years after marriage this life satisfaction decreased back to its baseline level. This graph perfectly illustrated what people talk about as the 'honeymoon period': a new relationship/marriage is great at first (no matter how ill suited you may be) but after six months or so the cracks start to appear and everything turns to elephant dung. Kahneman argued that people adapt to marriage; it does not make them happier in the long run (Kahneman & Krueger,

2006).[9] This got me thinking about relationships not involving marriage (is it marriage that makes you happy, or just being in a long-term relationship?). Therefore, in a completely fictitious parallel world where I don't research child anxiety, but instead concern myself with people's life satisfaction, I collected some data. I organized a massive speed-dating event (see Chapter 14). At the start of the night I measured everyone's life satisfaction (**Satisfaction_Baseline**) on a 10-point scale (0 = completely dissatisfied, 10 = completely satisfied) and their gender (**Gender**). After the speed dating I noted all of the people who had found dates. If they ended up in a relationship with the person that they met on the speed-dating night then I stalked these people over the next 18 months of that relationship. As such, I had measures of their life satisfaction at 6 months (**Satisfaction_6_Months**), 12 months (**Satisfaction_12_Months**) and 18 months (**Satisfaction_18_Months**), after they entered the relationship. None of the people measured were in the same relationship (i.e. I measured only life satisfaction from one of the people in the couple).[10] Also, as is often the case with longitudinal data, I didn't have scores for all people at all time points because not everyone was available at the follow-up sessions. One of the benefits of a multilevel approach is that these missing data do not pose a particular problem. The data are in the file **HoneymoonPeriod.sas7bdat**.

Figure 19.10 shows the data. Each dot is a data point and the line shows the average life satisfaction over time. Basically, from baseline, life satisfaction rises slightly at time 2 (6 months) but then starts to decrease over the next 12 months. There are two things to note about the data. First, time 0 is before the people enter into their new relationship yet already there is a lot of variability in their responses (reflecting the fact that people will vary in their satisfaction due to other reasons such as finances, personality and so on). This suggests that intercepts for life satisfaction differ across people. Second, there is also a lot of variability in life satisfaction after the relationship has started (time 1) and at all subsequent time points, which suggests that the slope of the relationship between time and life satisfaction might vary across people also. If we think of the time points as a level 1 variable that is nested with people (a level 2 variable) then we can easily model this variability in intercepts and slopes within people. We have a situation similar to Figure 19.4 (except with two levels instead of three, although we could add in the location of the speed-dating event as a level three variable if we had that information!).

19.7.3. Restructuring the data ③

The first problem with having data measured over time is that to do a multilevel model the data need to be in a different format to that which we are used to. Figure 19.11 shows how we would normally set up the data editor for a repeated-measures design: each row represents a person, and notice that the repeated-measures variable of time is represented by four different columns. If we were going to run an ordinary repeated-measures ANOVA this data layout would be fine; however, for a multilevel model we need the variable **Time** to be represented by a single column. We could enter all of the data again, but that would be a pain; luckily we don't have to do this because SAS has a a procedure that will do it for us – PROC TRANSPOSE, which is also a pain, but not as much as retyping the data.[11] This command enables you to take your data set and create a new data set that is organized differently.

[9] The romantics among you might be relieved to know that others have used the same data to argue the complete opposite: that married people are happier than non-married people in the long term (Easterlin, 2003).

[10] However, I could have measured both people in the couple because using a multilevel model I could have treated people as being nested within 'couples' to take account of the dependency in their data.

[11] If you are a better SAS programmer than I am, you can also do it in a DATA step.

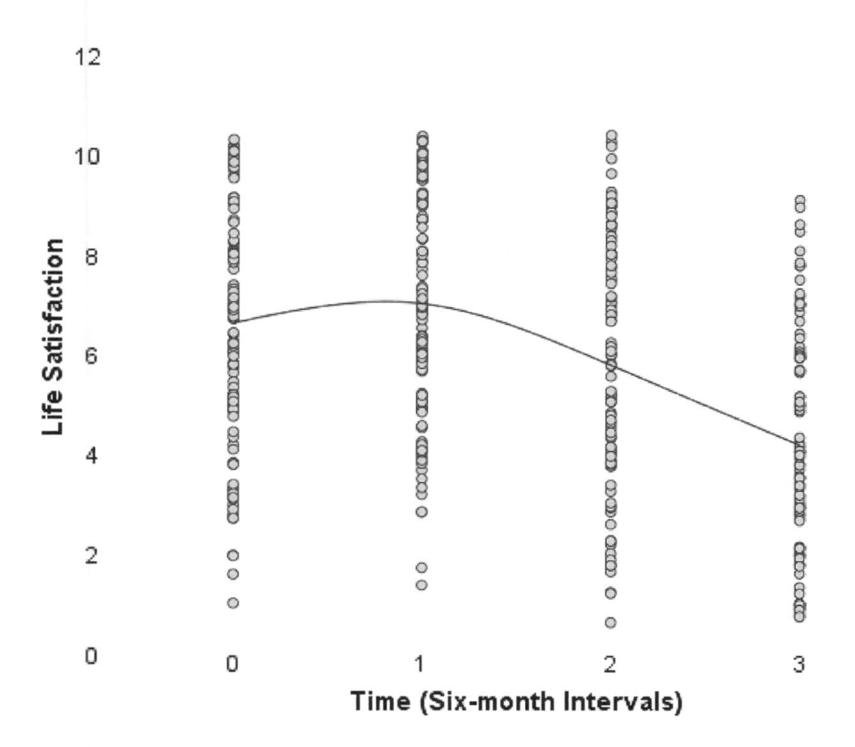

FIGURE 19.10 Life satisfaction over time

VIEWTABLE: Written by SAS

	Participant Number	Life Satisfaction: Baseline	Life Satisfaction: 6 Months	Life Satisfaction: 12 Months	Life Satisfaction: 18 Months	Gender
1	1	6	6	5	2	Female
2	2	7	7	8	4	Male
3	3	4	6	2	2	Male
4	4	6	9	4	1	Female
5	5	6	7	6	6	Female
6	6	5	10	4	2	Male
7	7	6	6	4	2	Female
8	8	2	5	4	.	Female
9	9	10	9	5	6	Female
10	10	10	10	10	9	Female
11	11	8	8	10	9	Female
12	12	6	10	9	9	Female
13	13	7	8	9	6	Male
14	14	6	7	9	5	Female
15	15	9	10	8	6	Male
16	16	10	10	8	6	Male
17	17	1	2	1	.	Male
18	18	5	6	7	3	Female
19	19	6	10	10	6	Male
20	20	5	6	.	.	Female

FIGURE 19.11 The view table for a normal repeated-measures data set

The syntax for PROC TRANSPOSE is shown in SAS Syntax 19.10. The DATA option refers to the data that we want to use, and the OUT option tells SAS the name of the dataset we want to create. We want to transpose BY **person** and **gender** (we don't really want to transpose by **gender**, but it works if we tell SAS we do). Finally, the variables we want to transpose are salled **satisfaction_base, satisfaction_6_months, satisfaction_12_months, satisfaction_18_months**. Rather than typing out all those variable names, and risking making a mistake, we can type **sat:** and the colon tells to SAS to make a list of all the variables that start with **sat**. (We did not specify a library for the output file, and so this file will go into the Work folder, and it will be deleted when we close SAS).

We need to do a little bit of tidying up of the file, and we'll do that in a DATA step, also shown in SAS Syntax 19.10. First, we'd like the periods to be labelled 0, 1, 2 and 3 – there are some automatic ways to do this if you had a large number of categories, but it's easier to just do it by hand, with four IF ... THEN statements.

```
  PROC    TRANSPOSE    DATA=chap19.honeymoon_period
OUT=honeymoonperiodlong;
      BY person gender;
      VAR sat: ;
      RUN;

  DATA honeymoonperiodlong; SET honeymoonperiodlong;
      IF _name_ = "SATISFACTION_BASE"    THEN time = 0;
      IF _name_ = "SATISFACTION_6_MONTHS" THEN time = 1;
      IF _name_ = "SATISFACTION_12_MONTHS" THEN time = 2;
      IF _name_ = "SATISFACTION_18_MONTHS" THEN time = 3;
      RENAME col1 = life_satisfaction;
      DROP _name_ _label_;
      RUN;
```

SAS Syntax 19.10

The restructured data are shown in Figure 19.12; it's useful to compare the restructured data with the old data file in Figure 19.11. Notice that each person is now represented by four rows (one for each time point) and that variables such as gender that are invariant over the time points have the same value within each person. However, our outcome variable (life satisfaction) does change over the four time points (the four rows for each person).

FIGURE 19.12 Data entry for a repeated-measures multilevel model

VIEWTABLE: Work.Honeymoonperiodlong

	Participant Number	Gender	life_satisfaction	time
1	1	Female	6	0
2	1	Female	6	1
3	1	Female	5	2
4	1	Female	2	3
5	2	Male	7	0
6	2	Male	7	1
7	2	Male	8	2
8	2	Male	4	3
9	3	Male	4	0
10	3	Male	6	1
11	3	Male	2	2
12	3	Male	2	3
13	4	Female	6	0
14	4	Female	9	1
15	4	Female	4	2
16	4	Female	1	3
17	5	Female	6	0
18	5	Female	7	1
19	5	Female	6	2

19.7.4. Running a growth model on SAS ④

Now that we have set up our data, we can run the analysis. Essentially, we can set up this analysis in a very similar way to the previous example.

We need to add the potential growth curves (see section 19.7.1) that we want to test as fixed effects to our model. With four time points we can fit up to a third-order polynomial. One way to do this would be to start with just the linear effect (**Time**), then run a new model with the linear and quadratic (**Time**^{2}) polynomials to see if the quadratic trend improves the model. Finally, run a third model with the linear, quadratic and cubic (**Time**^{3}) polynomial in, and see if the cubic trend adds to the model. So, basically, we add in polynomials one at a time and assess the change in −2LL.

I mentioned earlier on that we expected the relationship between time and life satisfaction to have both a random intercept and a random slope, so we'll make sure that **time** is on both the MODEL line, and the RANDOM line. Finally, we saw in section 19.4.2 that when we have repeated measures over time it can be useful to specify a covariance structure that assumes that scores become less correlated over time. Therefore, let's choose an autoregressive covariance structure, AR(1), and let's also assume that variances will be heterogeneous. Therefore, we'll use TYPE=ARH(1) on the random line. The final syntax is shown in SAS Syntax 19.11.

```
PROC MIXED DATA=honeymoonperiodlong METHOD=ML COVTEST;
    CLASS person;
    MODEL life_satisfaction = time /SOLUTION  ;
    RANDOM intercept time /SUBJECT=person TYPE=ARH(1);
    RUN;
```
SAS Syntax 19.11

SAS Output 19.11 shows the preliminary tables from the output. We can see that the linear trend was significant, $F(1, 114) = 134.27$, $p < .0001$. For evaluating the improvement in the model when we add in new polynomials, we also need to note the value −2LL, which is 1862.6, and we need to work out the degrees of freedom, which are 6 (there are four parameters in the table labelled **Covariance Parameter Estimates**, and 2 parameters in the table labelled **Solution for Fixed Effects**).

Now, let's add the quadratic trend. This is easier than you might guess – we want to add **time**^{2}, and **time**^{2} is equal to **time*time**, so we just add that to the MODEL line. This is shown in SAS Syntax 19.12.

SAS OUTPUT 19.11

Fit Statistics	
-2 Log Likelihood	1862.6
AIC (smaller is better)	1874.6
AICC (smaller is better)	1874.8
BIC (smaller is better)	1891.1

Solution for Fixed Effects					
Effect	Estimate	Standard Error	DF	t Value	Pr > |t|
Intercept	7.2078	0.2137	114	33.72	<.0001
time	-0.8757	0.07557	114	-11.59	<.0001

Type 3 Tests of Fixed Effects				
Effect	Num DF	Den DF	F Value	Pr > F
time	1	114	134.27	<.0001

```
PROC MIXED DATA=honeymoonperiodlong METHOD=ML COVTEST;
    CLASS person;
    MODEL life_satisfaction = time time*time/SOLUTION  ;
    RANDOM intercept time /SUBJECT=person TYPE=ARH(1);
    RUN;
```

SAS Syntax 19.12

The output will now include the quadratic polynomial. To see whether this quadratic trend has improved the model we need to compare the −2LL for this new model, to the value when only the linear polynomial was included. The value of −2LL is shown in SAS Output 19.12, and it is 1802.0. We have added only one term to the model so the new degrees of freedom will have risen by 1, from 6 to 7 (because we've added one parameter). We can compute the change in −2LL as a result of the quadratic term by subtracting the −2LL for this model from the −2LL for the model with only the linear trend:

$$\chi^2_{\text{Change}} = 1862.6 - 1802.0 = 60.6$$
$$df_{\text{Change}} = 7 - 6 = 1$$

If we look at the critical values for the chi-square statistic for $df = 1$ in the Appendix, they are 3.84 ($p < .05$) and 6.63 ($p < .01$); therefore, this change is highly significant because 60.6 is bigger than these values.

SAS OUTPUT 19.12

Fit Statistics	
-2 Log Likelihood	1802.0
AIC (smaller is better)	1816.0
AICC (smaller is better)	1816.3
BIC (smaller is better)	1835.2

Finally, let's add the cubic trend. I hardly need to tell you how to do this, but I will, because I'm nice like that. We add **time**^{3}, which is written as **time*time*time**, to the model line. This is shown in SAS Syntax 19.13.

```
PROC MIXED DATA=honeymoonperiodlong METHOD=ML COVTEST;
    CLASS person;
    MODEL life_satisfaction = time time*time time*time*time /
    SOLUTION  ;
    RANDOM intercept time/SUBJECT=person TYPE=ARH(1);
    RUN;
```
SAS Syntax 19.13

The output will now include the cubic polynomial. To see whether this cubic trend has improved the model we again compare the −2LL for this new model, to the value in the previous model. The value of −2LL is shown in SAS Output 19.12 , and it is 1798.9. We have added only one term to the model so the new degrees of freedom will have risen by 1, from 7 to 8. We can compute the change in −2LL as a result of the cubic term by subtracting the −2LL for this model from the −2LL for the model with only the linear trend:

$$\chi^2_{\text{Change}} = 1802.0 - 1798.9 = 3.1$$
$$df_{\text{Change}} = 8 - 7 = 1$$

Using the same critical values for the chi-square statistic as before, we can conclude that this change is not significant, because 3.1 is less than the critical value of 3.84.

We will look at the SAS output for this final model in a little more detail (SAS Output 19.12). First, we will look at the fit indices (the −2LL, AIC, AICC and BIC). As we have seen, these are useful mainly for comparing models, so we have used the log-likelihood, for example, to test whether the addition of a polynomial significantly affects the fit of the model.

The main part of the output is the table of fixed effects and the parameter estimates. These tell us that the linear, $F(1, 114) = 10.01$, $p = 0.002$, and quadratic, $F(1, 114) = 9.41$, $p = 0.003$, trends both significantly described the pattern of the data over time; however, the cubic trend, $F(1, 114) = 3.19$, $p = 0.077$, does not. This confirms what we already know from comparing the fit of successive models. The trend in the data is best described by a second-order polynomial, or a quadratic trend. This reflects the initial increase in life satisfaction 6 months after finding a new partner but a subsequent reduction in life satisfaction at 12 and 18 months after the start of the relationship (Figure 19.10). The parameter estimates tell us much the same thing. It's worth remembering that this quadratic trend is only an *approximation*: if it were completely accurate then we would predict from the model that couples who had been together for 10 years would have negative life satisfaction, which is impossible given the scale we used to measure it.

SAS OUTPUT 19.12

Covariance Parameter Estimates					
Cov Parm	Subject	Estimate	Standard Error	Z Value	Pr Z
Var(1)	PERSON	3.8893	0.7002	5.55	<.0001
Var(2)	PERSON	0.2444	0.09686	2.52	0.0058
ARH(1)	PERSON	−0.3826	0.1514	−2.53	0.0115
Residual		1.8343	0.1789	10.25	<.0001

Fit Statistics	
–2 Log Likelihood	1798.9
AIC (smaller is better)	1814.9
AICC (smaller is better)	1815.2
BIC (smaller is better)	1836.8

Solution for Fixed Effects					
Effect	Estimate	Standard Error	DF	t Value	Pr > |t|
Intercept	6.6348	0.2231	114	29.74	<.0001
time	1.5447	0.4882	114	3.16	0.0020
time*time	–1.3236	0.4315	114	–3.07	0.0027
time*time*time	0.1703	0.09537	114	1.79	0.0769

Type 3 Tests of Fixed Effects				
Effect	Num DF	Den DF	F Value	Pr > F
time	1	114	10.01	0.0020
time*time	1	114	9.41	0.0027
time*time*time	1	114	3.19	0.0769

The final part of the output to look at tells us about the random parameters in the model. First of all, the variance of the random intercepts was = 3.89. This suggests that we were correct to assume that life satisfaction at baseline varied significantly across people. Also, the variance of the people's slopes varied significantly, = 0.24. This suggests also that the change in life satisfaction over time varied significantly across people. Finally, the covariance between the slopes and intercepts (−0.38) suggests that as intercepts increased, the slope decreased. (Ideally, all of these terms should have been added in individually so that we could calculate the chi-square statistic for the change in the −2LL for each of them.)

19.7.5. Further analysis ④

It's worth pointing out that I've kept this growth curve analysis simple to give you the basic tools. In the example I allowed only the linear term to have a random intercept and slopes, but given that we discovered that a second-order polynomial described the change in responses, we could redo the analysis and allow random intercepts and slopes for the second-order polynomial also. To do these we would just have to add these terms to the RANDOM line. If we were to do this it would make sense to add the random components one at a time and test whether they have a significant impact on the model by comparing the log-likelihood values or other fit indices.

Also, the polynomials I have described are not the only ones that can be used. You could test for a logarithmic trend over time, or even an exponential one.

CRAMMING SAM'S TIPS Growth models

- Growth models are multilevel models in which changes in an outcome over time are modelled using potential growth patterns.
- These growth patterns can be linear, quadratic, cubic, logarithmic, exponential, or anything you like really.
- The hierarchy in the data is that time points are nested within people (or other entities). As such, it's a way of analysing repeated-measures data that have a hierarchical structure.
- The **Fit Statistics** table can be used to assess the overall fit of the model. The −2LL can be significance tested with *df* = the number of parameters being estimated. It is mainly used, though, to compare models that are the same in all but one parameter by testing the difference in −2LL in the two models against *df* = 1 (if only one parameter has been changed). The AIC, AICC and BIC can also be compared across models (but not significance tested).
- The table of **Type 3 Tests of Fixed Effects** tells you whether the growth functions that you have entered into the model significantly predict the outcome the *F*-value is less than .05 then the effect is significant.
- The table labelled **Covariance Parameter Estimates** tells us about any random effects in the model. These values can tell us how much intercepts and slopes varied over our level 1 variable. The significance of these estimates should be treated cautiously. The exact labelling of these effects depends on which covariance structure you selected for the analysis.
- An autoregressive covariance structure, AR(1), is often assumed in time course data such as that in growth models.

LABCOAT LENI'S REAL RESEARCH 19.1

A fertile gesture ③

Most female mammals experience a phase of 'estrus' during which they are more sexually receptive, proceptive, selective and attractive. As such, the evolutionary benefit to this phase is believed to be to attract mates of superior genetic stock. However, some people have argued that this important phase became uniquely lost or hidden in human females. Testing these evolutionary ideas is exceptionally difficult but Geoffrey Miller and his colleagues came up with an incredibly elegant piece of research that did just that (Miller, Tybur, & Jordon, 2007). They reasoned that if the 'hidden-estrus' theory is incorrect then men should find women most attractive during the fertile phase of their menstrual cycle compared to the pre-fertile (menstrual) and post-fertile (luteal) phase.

To measure how attractive men found women in an ecologically valid way, they came up with the ingeneous idea of collecting data from women working at lap-dancing clubs. These women maximize their tips from male visitors by attracting more dances. In effect the men 'try out' several dancers before choosing a dancer for a prolonged dance. For each dance the male pays a 'tip', therefore the greater the number of men who choose a particular woman, the more her earnings will be. As such, each dancer's earnings are a good index of how attractive the male customers have found her. Miller et al. argued, therefore, that if women do have an estrus phase then they will be more attractive during this phase and therefore earn more money. This study is a brilliant example of using a real-world phenomenon to address an important scientific question in an ecologically valid way.

MILLER, G. TYBUR, J.M. & JORDAN, B.D. (2007). *EVOLUTION AND HUMAN BEHAVIOR*, 28, 375–381.

The data for this study are in the file **Milleretal2007.sas7-bdat** The researchers collected data via a website from several dancers (**ID**), who provided data for multiple lap-dancing shifts (so for each person there are several rows of data). They also measured what phase of the menstrual cycle the women were in at a given shift (**Cyclephase**), and whether they were using hormonal contraceptives (**Contraceptive**) because this would affect their cycle. The outcome was their earnings on a given shift in dollars (**Tips**).

A multilevel model can be used here because the data are unbalanced: each woman differed in the number of shifts she provided data for (the range was 9 to 29 shifts); multilevel models can handle this problem.

Labcoat Leni wants you to carry out a multilevel model to see whether **Tips** can be predicted from **Cyclephase**, **Contraceptive** and their interaction. Is the 'estrus-hidden' hypothesis supported? Answers are in the additional material on the companion website (or look at page 378 in the original article).

19.8. How to report a multilevel model ③

Specific advice on reporting multilevel models is hard to come by. Also, the models themselves can take on so many forms that giving standard advice is hard. If you have built up your model from one with only fixed parameters to one with a random intercept, and then random slope, it is advisable to report all stages of this process (or at the very least report the fixed-effects-only model and the final model). For any model you need to say something about the random effects. For the final model of the cosmetic surgery example you could write something like:

✓ The relationship between surgery and quality of life showed significant variance in intercepts across participants, $\text{Var}(u_{0j}) = 30.06$, $\chi^2(1) = 15.05$, $p < .01$. In addition, the slopes varied across participants, $\text{Var}(u_{1j}) = 29.35$, $\chi^2(1) = 21.49$, $p < .01$, and the slopes and intercepts negatively and significantly covaried, $\text{Cov}(u_{0j}, u_{1j}) = -28.08$, $\chi^2(1) = 17.38$, $p < .01$.

For the model itself, you have two choices. The first is to report the results rather like an ANOVA, with the *F*s and degrees of freedom for the fixed effects, and then report the parameters for the random effects in the text as well. The second is to produce a table of parameters as you would for regression. For example, we might report our cosmetic surgery example as follows:

✓ Quality of life before surgery significantly predicted quality of life after surgery, $F(1, 268.92) = 33.65$, $p < .001$. Surgery did not significantly predict quality of life, $F(1, 15.86) = 2.17$, $p = .161$, but the reason for surgery, $F(1, 259.89) = 9.67$, $p < .01$, and the interaction of the reason for surgery and surgery, $F(1, 217.09) = 6.28$, $p < .05$, both did significantly predict quality of life. This interaction was broken down by conducting separate multilevel models on the 'physical reason' and 'attractiveness reason'. The models specified were the same as the main model but excluded the main effect and interaction term involving the reason for surgery. These analyses showed that for those operated on only to change their appearance, surgery almost significantly predicted quality of life after surgery, $b = -4.31$, $t(7.72) = -1.92$, $p = .09$: quality of life was lower after surgery compared to the control group. However, for those who had surgery to solve a physical problem, surgery did not significantly predict quality of life, $b = 1.20$, $t(7.61) = 0.58$, $p = .58$. The interaction effect, therefore, reflects the difference in slopes for surgery as a predictor of quality of life in those who had surgery for physical problems (slight positive slope) and those who had surgery purely for vanity (a negative slope).

Alternatively we could present parameter information in a table:

	b	SE b	95% CI
Baseline QoL	0.31	0.05	0.20, 0.41
Surgery	−3.19	2.17	−7.78, 1.41
Reason	−3.51	1.13	−5.74, −1.29
Surgery × Reason	4.22	1.68	0.90, 7.54

What have I discovered about statistics? ②

Writing this chapter was quite a steep learning curve for me. I've been meaning to learn about multilevel modelling for ages, and now I finally feel like I know something. This is pretty amazing considering that the bulk of the reading and writing was done between 11 p.m. and 3 a.m. over many nights. However, despite now feeling as though I understand them, I don't, and if you feel like you now understand them then you're wrong. This sounds harsh, but sadly multilevel modelling is very complicated and we have scratched only the surface of what there is to know. Multilevel models often fail to converge with no apology or explanation, and trying to fathom out what's happening can feel like hammering nails into your head.

Needless to say I didn't mention any of this at the start of the chapter because I wanted you to read it. Instead, I lulled you into a false sense of security by looking gently at how data can be hierarchical and how this hierarchical structure can be important. Most of the tests in this book simply ignore the hierarchy. We also saw that hierarchical models are just basically a fancy regression in which you can estimate the variability in the slopes and intercepts within entities. We saw that you should start with a model that ignores the hierarchy and then add in random intercepts and slopes to see if they improve the fit of the model. Having submerged ourselves in the warm bath of standard multilevel models we moved on to the icy lake of growth curves. We saw that there are ways to model trends in the data over time (and that these trends can also have variable intercepts and slopes). We also discovered that these trends have long confusing names like fourth-order polynomial. We asked ourselves why they couldn't have a sensible name, like Kate. In fact, we decided to ourselves that we'd secretly call a linear trend Kate, a quadratic trend Benjamin, a cubic trend Zoë, and a fourth-order trend Doug. 'That will show the statisticians' we thought to ourselves, and felt a little bit self-satisfied too.

We also saw that after years of denial, my love of making a racket got the better of me. This brings my life story up to date. Admittedly I left out some of the more colourful bits, but only because I couldn't find an extremely tenuous way to link them to statistics. We saw that over my life I managed to completely fail to achieve any of my childhood dreams. It's OK, I have other ambitions now (a bit smaller scale than 'rock star') and I'm looking forward to failing to achieve them too. The question that remains is whether there is life after *Discovering Statistics*. What effect does writing a statistics book have on your life?

Key terms that I've discovered

AIC
AICC
AR(1)
BIC
CAIC
Centring
Diagonal
Fixed coefficient
Fixed effect
Fixed intercept
Fixed slope
Fixed variable
Grand mean centring
Group mean centring
Growth curve
Multilevel linear model
Polynomial
Random coefficient
Random effect
Random intercept
Random slope
Random variable
Unstructured
Variance components

Smart Alex's tasks

- **Task 1**: Using the cosmetic surgery example, run the analysis described in section 19.6.5 but also including BDI, age and gender as fixed effect predictors. What differences does including these predictors make? ④
- **Task 2**: Using our growth model example in this chapter, analyse the data but include **Gender** as an additional covariate. Does this change your conclusions? ④
- **Task 3: Getting kids to exercise (Hill, Abraham, & Wright, 2007)**: The purpose of this research was to examine whether providing children with a leaflet based on the 'theory of planned behaviour' increases children's exercise. There were four different interventions (**Intervention**): a control group, a leaflet, a leaflet and quiz, and a leaflet and plan. A total of 503 children from 22 different classrooms were sampled (**Classroom**). It was not practical to have children in the same classrooms in different conditions, therefore the 22 classrooms were randomly assigned to the four different conditions. Children were asked 'On average over the last three weeks, I have exercised energetically for at least 30 minutes ______ times per week' after the intervention (**Post_Exercise**). Run a multilevel model analysis on these data (**Hilletal.sas7bdat**) to see whether the intervention affected the children's exercise levels (the hierarchy in the data is: children within classrooms within interventions). ④
- Repeat the above analysis but include the pre-intervention exercise scores (**Pre_Exercise**) as a covariate. What difference does this make to the results? ④

Answers can be found on the companion website.

Further reading

Kreft, I., & de Leeuw, J. (1998). *Introducing multilevel modeling*. London: Sage. (This is a fantastic book that is easy to get into but has a lot of depth too.)

Little, R. C. et al. (2006). *SAS for mixed models* (2nd ed.) Cary, NC: SAS Publishing. (This book is pretty hard going, but it's very detailed, and tells you everything there is to know about PROC MIXED, and a couple of other procs as well.)

Tabachnick, B. G., & Fidell, L. S. (2007). *Using multivariate statistics* (5th ed.). Boston: Allyn & Bacon. (Chapter 15 is a fantastic account of multilevel linear models that goes a bit more in depth than I do.)

Twisk, J. W. R. (2006). *Applied multilevel analysis: A practical guide.* Cambridge: Cambridge University Press. (An absolutely superb introduction to multilevel modelling. This book is exceptionally clearly written and is aimed at novices. Without question, this is the best beginner's guide that I have read.)

Interesting real research

Cook, S. A., Rosser, R., & Salmon, P. (2006). Is cosmetic surgery an effective psychotherapeutic intervention? A systematic review of the evidence. *Journal of Plastic, Reconstructive & Aesthetic Surgery*, *59*, 1133–1151.

Miller, G., Tybur, J. M., & Jordan, B. D. (2007). Ovulatory cycle effects on tip earnings by lap dancers: economic evidence for human estrus? *Evolution and Human Behavior*, *28*, 375–381.

EPILOGUE: LIFE AFTER DISCOVERING STATISTICS

'Here's some questions that the writer sent
Can an observer be a participant?
Have I seen too much?
Does it count if it doesn't touch?
If the view is all I can ascertain,
Pure understanding is out of range'

(Fugazi, 'Ex Spectator', *The Argument*, 2001)

When I wrote the first edition of the SPSS book my main ambition was to write a statistics book that I would enjoy reading. Pretty selfish I know. I thought that if I had a reference book that had a few examples that amused me then it would make life a lot easier when I needed to look something up. I honestly didn't think anyone would buy the thing (well, apart from my mum and dad) and I anticipated a glut of feedback along the lines of 'the whole of Chapter X is completely wrong and you're an arrant fool', or 'you should be ashamed of how many trees have died in the name of this rubbish, you brainless idiot'. In fact, even the publishers didn't think it would sell (they have only revealed this subsequently I might add). There are several other things that I didn't expect to happen:

1 *Nice emails*: I didn't expect to receive hundreds of extremely nice emails from people who liked the book. To this day it still absolutely amazes me that anyone reads it, let alone takes the time to write me a nice email and knowing that the book has helped people always puts a huge smile on my face. When the nice comments are followed by four pages of statistics questions the smile fades a bit ...
2 *Everybody thinks that I'm a statistician*: I should have seen this one coming really, but since writing a statistics textbook everyone assumes that I'm a statistician. I'm not, I'm a psychologist. Consequently, I constantly disappoint people by not being able to answer their statistics questions. In fact, this book is the sum total of my knowledge about statistics; there is nothing else (statistics-wise) in my brain that isn't in this book. Actually, that's a lie: there is more in this book about statistics than in my brain. For example, in the logistic regression chapter there is a new example on multinomial logistic regression. To write this new section I read a lot about multinomial logistic regression because I'd never used it. I wrote that new section about four months ago, and I've now forgotten everything that I wrote. Should I ever need to do a multinomial logistic regression I will read the chapter in this book and think to myself 'wow, it really sounds as though I know what I'm talking about'. Jeremy is a statistician so he's much less disappointing than I am in every respect.

3 *Craziness on a grand scale*: The nicest thing about life after discovering statistics is the effort that people go to to demonstrate that they are even stranger than me. All of these people have made life after 'Discovering Statistics' a profoundly enjoyable experience.

- *Catistics*: I've had quite a few photographs of people's cats (and dogs) reading my book (check out my 'discovering catistics' website at http://www.statisticshell.com/catistics.html). There has been many a week where one of these in my inbox has turned what was going to be a steaming turd of a day into a fragrant romp through fields of tulips. How can you not get a big stupid grin on your face when you see these? SAS users I think you now need to rise to the challenge set by the SPSS users.
- *Facebook*: Two particularly strange people from Exeter (UK) whom I have never met set up an 'Andy Field appreciation society' on Facebook. I don't go there much because it scares me a bit. But secretly I think it is quite cool. It's almost like being the rock star that I always wanted to be, except that when people join a rock star's appreciation society they mean it, but people join mine because it's funny. Nevertheless, beggars can't be choosers and I'm happy to overlook a technicality such as the truth if it means that I can believe that I'm popular. With this book, I think Jeremy deserves an appreciatation society too. Go on, you know it makes sense.
- *Films*: Possibly the strangest thing to have happened is Julie-Renée Kabriel and her bonkers friends from Washburn University producing a video homage to 'Discovering Stats' (http://www.youtube.com/watch?v=oLsrt594Xxc). I was in equal parts crippled with laughter and utterly bemused watching this video. My parents liked it too. (Oddly enough, it's to the tune of Sweet Home Alabama by LynyrdSkynyrd; I once gave a talk at Aberdeen University (Scotland) after which I got taken to a bar and ended up (quite unexpectedly) playing drums to that song with a makeshift band of complete strangers.) I re-iterate my point about SAS users rising to the challenge.
- *Invitation to an autopsy*: : I got invited to an autopsy. Really! Some (very nice) forensic scientists in Leicester loved this book so much that they felt that I needed to be rewarded for my efforts. They felt that the most appropriate reward would be to offer to take me to see a dead body being carved up (or to spend a day visiting crime scenes). In a strange way, I can see their logic. I haven't been because I'm slightly scared that it's a cruel trick and that it will turn out to be my body on the slab. However, in the interests of having a good story for the next edition I might just go …

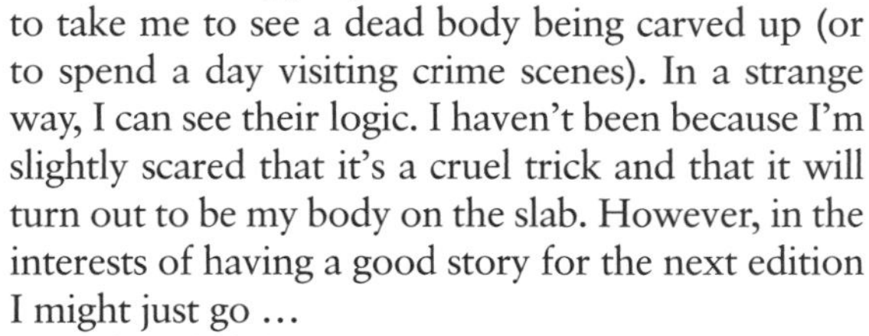

- *Befriended by Satan*: I got an email from the manager of a black metal band from London who, while using my book for her studies, was impressed to see that I like black metal bands. My band was playing the next week in London and never one to miss an opportunity, I invited her to come along. She not only turned up, but bought some of the band and some free CDs. They're called Abgott, they rock, and they renamed me 'The Evil Statistic'. I've subsequently spent many a happy night in London listening to deafening music and drinking too much with them. Buy their albums, buy their albums, buy their albums …

Life after *Discovering Statistics* … never ceases to amuse me. I never dreamed for a second that I'd be writing third editions or kidnapping my friend's children to persuade him to write a SAS version. It would have seemed utterly insane to contemplate at that time that this book would become such a huge part of my life. I would recommend writing a statistics book to anyone: it changes your life. You get a constant warm fuzzy feeling from being told that you've helped people, strangers send you photos of their pets, they make films about you, they give you CDs, you get an appreciation society, you can go to see corpses being cut up, join a black metal band (well, maybe not, but if my drumming improves and their drummer's arms and legs fall off, who knows?) and have people constantly overestimate your intelligence. It's a great life and long may the craziness continue.

GLOSSARY

0: the amount of a clue that Sage have about how much effort I put into writing this book.

−2LL: the *log-likelihood* multiplied by minus 2. This version of the likelihood is used in *logistic regression*.

α-level: the probability of making a *Type I error* (usually this value is .05).

A life: what you don't have when writing statistics textbooks.

Adjusted mean: in the context of *analysis of covariance* this is the value of the group mean adjusted for the effect of the *covariate*.

Adjusted predicted value: a measure of the influence of a particular case of data. It is the predicted value of a case from a model estimated without that case included in the data. The value is calculated by re-estimating the model without the case in question, then using this new model to predict the value of the excluded case. If a case does not exert a large influence over the model then its predicted value should be similar regardless of whether the model was estimated including or excluding that case. The difference between the predicted value of a case from the model when that case was included and the predicted value from the model when it was excluded is the *DFFit*.

Adjusted R^2: a measure of the loss of predictive power or *shrinkage* in regression. The adjusted R^2 tells us how much variance in the outcome would be accounted for if the model had been derived from the population from which the sample was taken.

AIC (Akaike's information criterion): a *goodness-of-fit* measure that is corrected for model complexity. That just means that it takes into account how many parameters have been estimated. It is not intrinsically interpretable, but can be compared in different models to see how changing the model affects the fit. A small value represents a better fit of the data.

AICC (Hurvich and Tsai's criterion): a *goodness-of-fit* measure that is similar to *AIC* but is designed for small samples. It is not intrinsically interpretable, but can be compared in different models to see how changing the model affects the fit. A small value represents a better fit of the data.

Alpha factoring: a method of *factor analysis*.

Alternative hypothesis: the prediction that there will be an effect (i.e. that your experimental manipulation will have some effect or that certain variables will relate to each other).

Analysis of covariance: a statistical procedure that uses the *F*-ratio to test the overall fit of a linear model controlling for the effect that one or more *covariates* have on the *outcome variable*. In experimental research this linear model tends to be defined in terms of group means and the resulting ANOVA is therefore an overall test of whether group means differ after the variance in the outcome variable explained by any *covariates* has been removed.

Analysis of variance: a statistical procedure that uses the *F*-ratio to test the overall fit of a linear model. In experimental research this linear model tends to be defined in terms of group means and the resulting ANOVA is therefore an overall test of whether group means differ.

ANCOVA: acronym for *analysis of covariance*.

ANOVA: acronym for *analysis of variance*.

AR(1): this stands for first-order autoregressive structure. It is a covariance structure used in *multilevel models* in which the relationship between scores changes in a systematic way. It is assumed that the correlation between scores gets smaller over time and variances are assumed to be homogeneous. This structure is often used for repeated-measures data (especially when measurements are taken over time such as in growth models).

Autocorrelation: when the *residuals* of two observations in a regression model are correlated.

b_i: unstandardized regression coefficient. Indicates the strength of relationship between a given predictor, *i*, and an outcome in the units of measurement of the predictor. It is the change in the outcome associated with a unit change in the predictor.

β_i: standardized regression coefficient. Indicates the strength of relationship between a given predictor, *i*, and an outcome in a *standardized* form. It is the change in the outcome (in standard deviations) associated with a one standard deviation change in the predictor.

β-level: the probability of making a *Type II error* (Cohen, 1992, suggests a maximum value of .2).

Bar chart: a graph in which a summary statistic (usually the mean) is plotted on the *y*-axis against a categorical variable on the *x*-axis (this categorical variable could represent, for example, groups of people, different times or different experimental conditions). The value of the mean for each category is shown by a bar. Different-coloured bars may be used to represent levels of a second categorical variable.

Bartlett's test of sphericity: unsurprisingly this is a test of the assumption of *sphericity*. This test examines whether a *variance–covariance matrix* is proportional to an *identity matrix*. Therefore, it effectively tests whether the diagonal elements of the variance–covariance matrix are equal (i.e. group variances are the same), and that the off-diagonal elements are approximately zero (i.e. the *dependent variables* are not *correlated*). Jeremy Miles, who does a lot of multivariate stuff, claims he's never ever seen a matrix that reached non-significance using this test and, come to think of it, I've never seen one either (although I do less multivariate stuff) so you've got to wonder about it's practical utility.

Beer-goggles effect: the phenomenon that people of the opposite gender (or the same depending on your sexual orientation) appear much more attractive after a few alcoholic drinks.

Between-group design: another name for *independent design*.

Between-subject design: another name for *independent design*.

BIC (Schwarz's Bayesian criterion): a *goodness-of-fit* statistic comparable to the AIC, although it is slightly more conservative (it corrects more harshly for the number of parameters being estimated). It should be used when sample sizes are large and the number of parameters is small. It is not intrinsically interpretable, but can be compared in different models to see how changing the model affects the fit. A small value represents a better fit of the data.

Bimodal: a description of a distribution of observations that has two *modes*.

Binary logistic regression: *logistic regression* in which the outcome variable has exactly two categories.

Binary variable: a *categorical variable* that has only two mutually exclusive categories (e.g. being dead or alive).

Bivariate correlation: a correlation between two variables.

Blockwise regression: another name for *hierarchical regression*.

Bonferroni correction: a correction applied to the *α-level* to control the overall *Type I error rate* when multiple significance tests are carried out. Each test conducted should use a criterion of significance of the α-level (normally .05) divided by the number of tests conducted. This is a simple but effective correction, but tends to be too strict when lots of tests are performed.

Bootstrap: a technique for estimating the sampling distribution of a statistic by taking repeated samples (with replacement) from the data set (so in effect, treating the data as a population from which smaller samples are taken). The statistic of interest (e.g. the *mean*, or *b* coefficient) is calculated for each sample, from which the sampling distribution of the statistic is estimated. The standard error of the statistic is estimated as the standard deviation of the sampling distribution created from the bootstrap samples. From this, confidence intervals and significance tests can be computed.

Boredom effect: refers to the possibility that performance in tasks may be influenced (the assumption is a negative influence) by boredom/lack of concentration if there are many tasks, or the task goes on for a long period of time. In short, what you are experiencing reading this glossary is a boredom effect.

Box's test: a test of the assumption of *homogeneity of covariance matrices*. This test should be non-significant if the matrices are roughly the same. Box's test is very susceptible to deviations from *multivariate normality* and so can be non-significant, not because the *variance–covariance matrices* are similar across groups, but because the assumption of multivariate normality is not tenable. Hence, it is vital to have some idea of whether the data meet the multivariate normality assumption (which is extremely difficult) before interpreting the result of Box's test.

Boxplot (or box–whisker diagram): a graphical representation of some important characteristics of a set of observations. At the centre of the plot is the *median*, which is surrounded by a box the top and bottom of which are the limits within which the middle 50% of observations fall (the *interquartile range*). Sticking out of the top and bottom of the box are two whiskers which extend to the most and least extreme scores respectively.

Box–whisker plot: see *Boxplot*.

Brown–Forsythe *F*: *a* version of the *F*-ratio designed to be accurate when the assumption of *homogeneity of variance* has been violated.

Categorical variable: any variable made up of categories of objects/entities. The UK degree classifications are a good example because degrees are classified as 1, 2:1, 2:2, 3, pass or fail. Therefore, graduates form a categorical variable because they will fall into only one of these categories (hopefully the category of students receiving a first!).

Central limit theorem: this theorem states that when samples are large (above about 30) the *sampling distribution* will take the shape of a *normal distribution* regardless of the shape of the population from which the sample was drawn. For small samples the *t*-distribution better approximates the shape of the sampling distribution. We also know from this theorem that the *standard deviation* of the sampling distribution (i.e. the *standard error* of the sample *mean*) will be equal to the standard deviation of the sample (*s*) divided by the square root of the sample size (*N*).

Central tendency: a generic term describing the centre of a *frequency distribution* of observations as measured by the *mean*, *mode* and *median*.

Centring: the process of transforming a variable into deviations around a fixed point. This fixed point can be any value that is chosen, but typically a mean is used. To centre a variable the mean is subtracted from each score. See *Grand mean centring*, *Group mean centring*.

Chartjunk: superfluous material that distracts from the data being displayed on a graph.

Chi-squared distribution: a *probability distribution* of the sum of squares of several normally distributed variables. It tends to be used to (1) test hypotheses about categorical data, and (2) test the fit of models to the observed data.

Chi-square test: although this term can apply to any *test statistic* having a *chi-square distribution*, it generally refers to Pearson's chi-square test of the independence of two categorical variables. Essentially it tests whether two categorical variables forming a *contingency table* are associated.

Cocaine: the drug of Sage. They inject it into their eyeballs, you know.

Coefficient of determination: the proportion of variance in one variable explained by a second variable. It is the *Pearson correlation coefficient* squared.

Common variance: variance shared by two or more variables.

		Glossary		
		Author made to write glossary	No glossary	Total
Mental state	Normal	5	423	428
	Sobbing uncontrollably	23	46	69
	Utterly psychotic	127	2	129
	Total	155	471	626

Communality: the proportion of a variable's variance that is *common variance*. This term is used primarily in *factor analysis*. A variable that has no *unique variance* (or *random variance*) would have a communality of 1, whereas a variable that shares none of its variance with any other variable would have a communality of 0.

Complete separation: a situation in *logistic regression* when the outcome variable can be perfectly predicted by one predictor or a combination of predictors! Suffice it to say this situation makes your computer have the equivalent of a nervous breakdown: it'll start gibbering, weeping and saying it doesn't know what to do.

Component matrix: general term for the *structure matrix* in SAS *principal component analysis*.

Compound symmetry: a condition that holds true when both the variances across conditions are equal (this is the same as the *homogeneity of variance* assumption) and the *covariances* between pairs of conditions are also equal.

Confidence interval: for a given statistic calculated for a sample of observations (e.g. the mean), the confidence interval is a range of values around that statistic that are believed to contain, with a certain probability (e.g. 95%), the true value of that statistic (i.e. the population value).

Confirmatory factor analysis (CFA): a version of *factor analysis* in which specific hypotheses about structure and relations between the *latent variables* that underlie the data are tested.

Confounding variable: a variable (that we may or may not have measured) other than the *predictor variables* in which we're interested that potentially affects an *outcome variable*.

Content validity: evidence that the content of a test corresponds to the content of the construct it was designed to cover.

Contingency table: a table representing the cross-classification of two or more *categorical variables*. The levels of each variable are arranged in a grid, and the number of observations falling into each category is noted in the cells of the table. For example, if we took the categorical variables of glossary (with two categories: whether an author was made to write a glossary or not), and **mental state** (with three categories: normal, sobbing uncontrollably and utterly psychotic), we could construct a table as below. This instantly tells us that 127 authors who were made to write a glossary ended up as utterly psychotic, compared to only 2 who did not write a glossary.

Continuous variable: a variable that can be measured to any level of precision. (Time is a continuous variable, because there is in principle no limit on how finely it could be measured.)

Cook's distance: a measure of the overall influence of a case on a model. Cook and Weisberg (1982) have suggested that values greater than 1 may be cause for concern.

Correlation coefficient: a measure of the strength of association or relationship between two variables. See *Pearson's correlation coefficient*, *Spearman's correlation coefficient*, *Kendall's tau*.

Correlational research: a form of research in which you observe what naturally goes on in the world without directly interfering with it. This term implies that data will be analysed so as to look at relationships between naturally-occurring variables rather than making statements about cause and effect. Compare with *cross-sectional research* and *experimental research*.

Counterbalancing: a process of systematically varying the order in which experimental conditions are conducted. In the simplest case of there being two conditions (A and B), counterbalancing simply implies that half of the participants complete condition A followed by condition B, whereas the remainder do condition B followed by condition A. The aim is to remove systematic bias caused by *practice effects* or *boredom effects*.

Covariance: a measure of the 'average' relationship between two variables. It is the average *cross-product deviation* (i.e. the cross-product divided by one less than the number of observations).

Covariance ratio (CVR): a measure of whether a case influences the variance of the parameters in a *regression model*. When this ratio is close to 1 the case is having very little influence on the variances of the model parameters. Belsey et al. (1980) recommend the following: if the CVR of a case is greater than $1 + [3(k + 1)/n]$ then deleting that case will damage the precision of some of the model's parameters, but if it is less than $1 - [3(k + 1)/n]$ then deleting the case will improve the precision of some of the model's parameters (k is the number of predictors and n is the sample size).

Covariate: a variable that has a relationship with (in terms of *covariance*), or has the potential to be related to, the *outcome variable* we've measured.

Cox and Snell's R^2_{CS}: a version of the *coefficient of determination* for logistic regression. It is based on the log-likelihood of a model (*LL(new)*) and the log-likelihood of the original model (*LL(baseline)*), and the sample size, *n*. However, it is notorious for not reaching its maximum value of 1 (see *Nagelkerke's* R_N^2).

Cramér's V: a measure of the strength of association between two *categorical variables* used when one of these variables has more than two categories. It is a variant of *phi* used because when one or both of the categorical variables contain more than two categories, phi fails to reach its minimum value of 0 (indicating no association).

Criterion validity: evidence that scores from an instrument correspond with or predict concurrent external measures conceptually related to the measured construct.

Cronbach's α: a measure of the reliability of a scale defined by:

$$\alpha = \frac{N^2\overline{\text{Cov}}}{\sum s^2_{\text{item}} + \sum \text{Cov}_{\text{item}}}$$

in which the top half of the equation is simply the number of items (*N*) squared multiplied by the average covariance between items (the average of the off-diagonal elements in the *variance–covariance matrix*). The bottom half is the sum of all the elements in the *variance–covariance matrix*.

Cross-product deviations: a measure of the 'total' relationship between two variables. It is the deviation of one variable from its mean multiplied by the other variable's deviation from its mean.

Cross-sectional research: a form of research in which you observe what naturally goes on in the world without directly interfering with it. This term specifically implies that data come from people at different age points with different people representing each age point. See also *correlational research*.

Cross-validation: assessing the accuracy of a model across different samples. This is an important step in *generalization*. In a *regression model* there are two main methods of cross-validation: *adjusted* R^2 or data splitting, in which the data are split randomly into two halves, and a regression model is estimated for each half and then compared.

Crying: what you feel like doing after writing statistics textbooks.

Cubic trend: if you connected the means in ordered conditions with a line then a cubic trend is shown by two changes in the direction of this line. You must have at least four ordered conditions.

Currency variable: a variable containing values of money.

Date variable: a variable containing dates. The data can take forms such as dd-mmm-yyyy (e.g. 21-Jun-1973), dd-mmm-yy (e.g. 21-Jun-73), mm/dd/yy (e.g. 06/21/73), dd.mm.yyyy (e.g. 21.06.1973).

Degrees of freedom: a impossible thing to define in a few pages, let alone a few lines. Essentially it is the number of 'entities' that are free to vary when estimating some kind of statistical parameter. In a more practical sense, it has a bearing on significance tests for many commonly used *test statistics* (such as the *F-ratio*, *t-test*, *chi-square statistic*) and determines the exact form of the *probability distribution* for these *test statistics*. The explanation involving rugby players in Chapter 8 is far more interesting...

Deleted residual: a measure of the influence of a particular case of data. It is the difference between the *adjusted predicted value* for a case and the original observed value for that case.

Density plot: similar to a *histogram* except that rather than having a summary bar representing the frequency of scores, it shows each individual score as a dot. They can be useful for looking at the shape of a distribution of scores.

Dependent *t*-test: a test using the *t-statistic* that establishes whether two means collected from the same sample (or related observations) differ significantly.

Dependent variable: another name for *outcome variable*. This name is usually associated with experimental methodology (which is the only time it really makes sense) and is used because it is the variable that is not manipulated by the experimenter and so its value depends on the variables that have been manipulated. To be honest I just use the term outcome variable all the time – it makes more sense (to me) and is less confusing.

Deviance: the difference between the observed value of a variable and the value of that variable predicted by a statistical model.

Deviation contrast: a non-orthogonal *planned contrast* that compares the mean of each group (except first or last depending on how the contrast is specified) to the overall mean.

DFA: acronym for *discriminant function analysis*.

DFBeta: a measure of the influence of a case on the values of b_i in a *regression model*. If we estimated a regression parameter b_i and then deleted a particular case and re-estimated the same regression parameter b_i, then the difference between these two estimates would be the DFBeta for the case that was deleted. By looking at the values of the DFBetas, it is possible to identify cases that have a large influence on the parameters of the regression model; however, the size of DFBeta will depend on the units of measurement of the regression parameter.

DFFit: a measure of the influence of a case. It is the difference between the *adjusted predicted value* and the original predicted value of a particular case. If a case is not influential then its DFFit should be zero – hence, we expect non-influential cases to have small DFFit values. However, we have the problem that this statistic depends on the units of measurement of the outcome and so a DFFit of 0.5 will be very small if the outcome ranges from 1 to 100, but very large if the outcome varies from 0 to 1.

Dichotomous: description of a variable that consists of only two categories (e.g. the variable gender is dichotomous because it consists of only two categories: male and female).

Difference contrast: a non-orthogonal *planned contrast* that compares the mean of each condition (except the first) to the overall mean of all previous conditions combined.

Direct oblimin: a method of *oblique rotation*.

Discrete variable: a variable that can only take on certain values (usually whole numbers) on the scale.

Discriminant function analysis: also known as discriminant analysis. This analysis identifies and describes the *discriminant function variates* of a set of variables and is useful as a follow-up test to *MANOVA* as a means of seeing how these variates allow groups of cases to be discriminated.

Discriminant function variate: a linear combination of variables created such that the differences between group means on the transformed variable are maximized. It takes the general form:

$$Variate_1 = b_1X_1 + b_2X_2 + \ldots + b_nX_n.$$

Discriminant score: a score for an individual case on a particular *discriminant function variate* obtained by replacing that case's scores on the measured variables into the equation that defines the variate in question.

Dummy variables: a way of recoding a categorical variable with more than two categories into a series of variables all of which are *dichotomous* and can take on values of only 0 or 1. There are seven basic steps to create such variables: (1) count the number of groups you want to recode and subtract 1; (2) create as many new variables as the value you calculated in step 1 (these are your dummy variables); (3) choose one of your groups as a baseline (i.e. a group against which all other groups should be compared, such as a control group); (4) assign that baseline group values of 0 for all of your dummy variables; (5) for your first dummy variable, assign the value 1 to the first group that you want to compare against the baseline group (assign all other groups 0 for this variable); (6) for the second dummy variable assign the value 1 to the second group that you want to compare against the baseline group (assign all other groups 0 for this variable); (7) repeat this process until you run out of dummy variables.

Ecological validity: evidence that the results of a study, experiment or test can be applied, and allow inferences, to real-world conditions.

Eel: long, snakelike, scaleless fish that lacks pelvic fins. From the order Anguilliformes or Apodes, they should probably not be inserted into your anus to cure constipation (or for any other reason).

Effect size: an objective and (usually) standardized measure of the magnitude of an observed effect. Measures include Cohen's *d*, Glass's *g* and Pearson's correlations coefficient, *r*.

Equamax: a method of *orthogonal rotation* that is a hybrid of *quartimax* and *varimax*. It is reported to behave fairly erratically (see Tabachnick & Fidell, 2007) and so is probably best avoided.

Error bar chart: a graphical representation of the mean of a set of observations that includes the 95% confidence interval of the mean. The mean is usually represented as a circle, square or rectangle at the value of the mean (or a bar extending to the value of the mean). The confidence interval is represented by a line protruding from the mean (upwards, downwards or both) to a short horizontal line representing the limits of the confidence interval. Error bars can be drawn using the standard error or standard deviation instead of the 95% confidence interval.

Error SSCP (*E*): the error sum of squares and cross-product matrix. This is a *sum of squares and cross-product matrix* for the error in a predictive *linear model* fitted to *multivariate* data. It represents the *unsystematic variance* and is the multivariate equivalent of the *residual sum of squares*.

Eta squared (η^2): an *effect size* measure that is the ratio of the *model sum of squares* to the *total sum of squares*. So, in essence, *the coefficient of determination* by another name. It doesn't have an awful lot going for it: not only is it biased, but it typically measures the overall effect of an ANOVA and effect sizes are more easily interpreted when they reflect specific comparisons (e.g. the difference between two means).

Experimental hypothesis: synonym for *alternative hypothesis*.

Experimental research: a form of research in which one or more variables is systematically manipulated to see their effect (alone or in combination) on an *outcome variable*. This term implies that data will be able to be used to make statements about cause and effect. Compare with *cross-sectional research* and *correlational research*.

Experimentwise error rate: the probability of making a *Type I error* in an experiment involving one or more statistical comparisons when the null hypothesis is true in each case.

Extraction: a term used for the process of deciding whether a *factor* in *factor analysis* is statistically important enough to 'extract' from the data and interpret. The decision is based on the magnitude of the eigenvalue associated with the factor. See *Kaiser's criterion*, *scree plot*.

F_{Max}: see *Hartley's F_{Max}*.

***F*-ratio:** a test statistic with a known *probability distribution* (the *F*-distribution). It is the ratio of the average variability in the data that a given model can explain to the average variability unexplained by that same model. It is used to test the overall fit of the model in *simple regression* and *multiple regression*, and to test for overall differences between group means in experiments.

Factor: another name for an *independent variable* or *predictor* that's typically used when describing experimental designs. However, to add to the confusion, it is also used synonymously with *latent variable* in factor analysis.

Factor analysis: a *multivariate* technique for identifying whether the correlations between a set of observed variables stem from their relationship to one or more *latent variables* in the data, each of which takes the form of a *linear model*.

Factor loading: the *regression coefficient* of a variable for the *linear model* that describes a *latent variable* or *factor* in *factor analysis*.

Factor scores: a single score from an individual entity representing their performance on some *latent variable*. The score can be crudely conceptualized as follows: take an entity's score on each of the variables that make up the factor and multiply it by the corresponding *factor loading* for the variable, then add these values up (or average them).

Factor transformation matrix, Λ: a matrix used in *factor analysis*. It can be thought of as containing the angles through which factors are rotated in factor *rotation*.

Factorial ANOVA: an analysis of variance involving two or more *independent variables* or *predictors*.

Falsification: the act of disproving a hypothesis or theory.

Familywise error rate: the probability of making a *Type I error* in any family of tests when the null hypothesis is true in each case. The 'family of tests' can be loosely defined as a set of tests conducted on the same data set and addressing the same empirical question.

Fisher's exact test: Fisher's exact test (Fisher, 1922) is not so much a test as a way of computing the exact probability of a statistic. It was designed originally to overcome the problem that with small samples the sampling distribution of the chi-square statistic deviates substantially from a chi-square distribution. It should be used with small samples.

Fit: how sexually attractive you find a statistical test. Alternatively, it's the degree to which a statistical model is an accurate representation of some observed data. (Incidentally, it's just plain *wrong* to find statistical tests sexually attractive).

Fixed coefficient: a coefficient or model parameter that is fixed; that is, it cannot vary over situations or contexts (cf. *Random coefficient*).

Fixed effect: An effect in an experiment is said to be a fixed effect if all possible treatment conditions that a researcher is interested in are present in the experiment. Fixed effects can be generalized only to the situations in the experiment. For example, the effect is fixed if we say that we are interested only in the conditions that we had in our experiment (e.g. placebo, low dose and high dose) and we can generalize our findings only to the situation of a placebo, low dose and high dose.

Fixed intercept: A term used in *multilevel modelling* to denote when the intercept in the model is fixed. That is, it is not free to vary across different groups or contexts (cf. *Random intercept*).

Fixed slope: A term used in *multilevel modelling* to denote when the slope of the model is fixed. That is, it is not free to vary across different groups or contexts (cf. *Random slope*).

Fixed variable: A fixed variable is one that is not supposed to change over time (e.g. for most people their gender is a fixed variable – it never changes).

Frequency distribution: a graph plotting values of observations on the horizontal axis, and the frequency with which each value occurs in the data set on the vertical axis (a.k.a. *histogram*).

Friedman's ANOVA: a non-parametric test of whether more than two related groups differ. It is the non-parametric version of one-way *repeated-measures ANOVA*.

Generalization: the ability of a statistical model to say something beyond the set of observations that spawned it. If a model generalizes it is assumed that predictions from that model can be applied not just to the sample on which it is based, but to a wider population from which the sample came.

Glossary: a collection of grossly inaccurate definitions (written late at night when you really ought to be asleep) of things that you thought you understood until some evil book publisher forced you to try to define them.

Goodness of fit: an index of how well a model fits the data from which it was generated. It's usually based on how well the data predicted by the model correspond to the data that were actually collected.

Grand mean: the *mean* of an entire set of observations.

Grand mean centring: the transformation of a variable by taking each score and subtracting the mean of all scores (for that variable) from it (cf. *Group mean centring*).

Grand variance: the *variance* within an entire set of observations.

Greenhouse–Geisser correction: an estimate of the departure from *sphericity*. The maximum value is 1 (the data completely meet the assumption of sphericity) and minimum is the *lower bound*. Values below 1 indicate departures from sphericity and are used to correct the *degrees of freedom* associated with the corresponding *F-ratios* by multiplying them by the value of the estimate. Some say the Greenhouse–Geisser correction is too conservative (strict) and recommend the *Huynh–Feldt correction* instead.

Group mean centring: the transformation of a variable by taking each score and subtracting from it the mean of the scores (for that variable) for the group to which that score belongs (cf. *Grand mean centring*).

Growth curve: a curve that summarizes the change in some outcome over time. See *Polynomial*.

Harmonic mean: a weighted version of the *mean* that takes account of the relationship between variance and sample size. It is calculated by summing the reciprocal of all observations, then dividing by the number of observations. The reciprocal of the end product is the harmonic mean:

$$H = \frac{1}{\frac{1}{n}\sum_{i=1}^{n}\frac{1}{x_i}}$$

Hartley's F_{Max}: also known as the *variance ratio*, is the ratio of the variances between the group with the biggest variance and the group with the smallest variance. This ratio is compared to critical values in a table published by Hartley as a test of *homogeneity of variance*. Some general rules are that with sample sizes (n) of 10 per group, an F_{Max} less than 10 is more or less always going to be non-significant, with 15–20 per group the ratio needs to be less than about 5, and with samples of 30–60 the ratio should be below about 2 or 3.

Hat values: another name for *leverage*.

***HE*$^{-1}$:** this is a matrix that is functionally equivalent to the *hypothesis SSCP* divided by the *error SSCP* in *MANOVA*. Conceptually it represents the ratio of *systematic* to *unsystematic variance*, so is a *multivariate* analogue of the *F-ratio*.

Helmert contrast: a non-orthogonal *planned contrast* that compares the mean of each condition (except the last) to the overall mean all subsequent conditions combined.

Heterogeneity of variance: the opposite of *homogeneity of variance*. This term means that the variance of one variable varies (i.e. is different) across levels of another variable.

Heteroscedasticity: the opposite of *homoscedasticity*. This occurs when the residuals at each level of the predictor variables(s) have unequal variances. Put another way, at each point along any predictor variable, the spread of residuals is different.

Hierarchical regression: a method of *multiple regression* in which the order in which predictors are entered into the regression model is determined by the researcher based on previous research: variables already known to be predictors are entered first, new variables are entered subsequently.

Histogram: a *frequency distribution*.

Homogeneity of covariance matrices: an assumption of some *multivariate* tests such as *MANOVA*. It is an extension of the *homogeneity of variance assumption* in *univariate* analyses. However, as well as assuming that *variances* for each *dependent variable* are the same across groups, it also assumes that relationships (*covariances*) between these dependent variables are roughly equal. It is tested by comparing the population *variance–covariance matrices* of the different groups in the analysis.

Homogeneity of regression slopes: an assumption of *analysis of covariance*. This is the assumption that the relationship between the *covariate* and *outcome variable* is constant across different treatment levels. So, if we had three treatment conditions, if there's a positive relationship between the covariate and the outcome in one group, we assume that there is a similar-sized positive relationship between the covariate and outcome in the other two groups too.

Homogeneity of variance: the assumption that the variance of one variable is stable (i.e. relatively similar) at all levels of another variable.

Homoscedasticity: an assumption in regression analysis that the residuals at each level of the predictor variables(s) have similar variances. Put another way, at each point along any predictor variable, the spread of residuals should be fairly constant.

Hosmer and Lemeshow's R^2_L: a version of the *coefficient of determination* for logistic regression. It is a fairly literal translation in that it is the −2LL for the model divided by the original −2LL, in other words, it's the ratio of what the model can explain compared to what there was to explain in the first place!

Hotelling–Lawley trace (T^2): a *test statistic* in *MANOVA*. It is the sum of the eigenvalues for each *discriminant function variate* of the data and so is conceptually the same as the *F-ratio* in *ANOVA* it is the sum of the ratio of *systematic* and *unsystematic variance* (SS_M/SS_R) for each of the variates.

Huynh–Feldt correction: an estimate of the departure from *sphericity*. The maximum value is 1 (the data completely meet the assumption of sphericity). Values below this indicate departures from sphericity and are used to correct the *degrees of freedom* associated with the corresponding *F-ratios* by multiplying them by the value of the estimate. It is less conservative than the *Greenhouse–Geisser estimate*, but some say it is too liberal.

Hypothesis: a prediction about the state of the world (see *experimental hypothesis* and *null hypothesis*).

Hypothesis SSCP (*H*): the hypothesis sum of squares and cross product matrix. This is a *sum of squares and cross-product matrix* for a predictive *linear model* fitted to *multivariate* data. It represents the *systematic variance* and is the multivariate equivalent of the *model sum of squares*.

Identity matrix: a square matrix (i.e. with the same number of rows and columns) in which the diagonal elements are equal to 1, and the off-diagonal elements are equal to 0. The following are all examples:

$$\begin{pmatrix} 1 & 0 \\ 0 & 1 \end{pmatrix} \begin{pmatrix} 1 & 0 & 0 \\ 0 & 1 & 0 \\ 0 & 0 & 1 \end{pmatrix} \begin{pmatrix} 1 & 0 & 0 & 0 \\ 0 & 1 & 0 & 0 \\ 0 & 0 & 1 & 0 \\ 0 & 0 & 0 & 1 \end{pmatrix}$$

Independence: the assumption that one data point does not influence another. When data come from people, it basically means that the behaviour of one person does not influence the behaviour of another.

Independent ANOVA: *analysis of variance* conducted on any design in which all *independent variables* or *predictors* have been manipulated using different participants (i.e. all data come from different entities).

Independent design: an experimental design in which different treatment conditions utilize different organisms (e.g. in psychology, this would mean using different people in different treatment conditions) and so the resulting data are independent (a.k.a. between-group or between-subject designs).

Independent errors: for any two observations in regression the *residuals* should be uncorrelated (or independent).

Independent factorial design: an experimental design incorporating two or more *predictors* (or *independent variables*) all of which have been manipulated using different participants (or whatever entities are being tested).

Independent *t*-test: a test using the *t-statistic* that establishes whether two means collected from independent samples differ significantly.

Independent variable: another name for a *predictor variable*. This name is usually associated with experimental methodology (which is the only time it makes sense) and is used because it is the variable that is manipulated by the experimenter and so its value does not depend on any other variables (just on the experimenter). I just use the term *predictor variable* all the time because the meaning of the term is not constrained to a particular methodology.

Interaction effect: the combined effect of two or more *predictor variables* on an *outcome variable*.

Interaction graph: a graph showing the means of two or more *independent variables* in which means of one variable are shown at different levels of the other variable. Unusually the means are connected with lines, or are displayed as bars. These graphs are used to help understand *interaction effects*.

Interquartile range: the limits within which the middle 50% of an ordered set of observations falls. It is the difference between the value of the *upper quartile* and *lower quartile*.

Interval variable: data measured on a scale along the whole of which intervals are equal. For example, people's ratings of this book on Amazon.com can range from 1 to 5; for these data to be interval it should be true that the increase in appreciation for this book represented by a change from 3 to 4 along the scale should be the same as the change in appreciation represented by a change from 1 to 2, or 4 to 5.

Intraclass correlation (ICC): a *correlation coefficient* that assess the consistency between measures of the same class (i.e. measures of the same thing). (Cf. *Pearson product-moment correlation* which measures the relationship between variables of a different class.) Two common uses are in comparing paired data (such as twins) on the same measure, and assessing the consistency between judges' ratings of a set of objects.

Kaiser–Meyer–Olkin measure of sampling adequacy (KMO): the KMO can be calculated for individual and multiple variables and represents the ratio of the squared correlation between variables to the squared *partial correlation* between variables. It varies between 0 and 1: a value of 0 indicates that the sum of partial correlations is large relative to the sum of correlations, indicating diffusion in the pattern of correlations (hence, *factor analysis* is likely to be inappropriate); a value close to 1 indicates that patterns of correlations are relatively compact and so factor analysis should yield distinct and reliable factors. Values between .5 and .7 are mediocre, values between .7 and .8 are good, values between .8 and .9 are great and values above .9 are superb (see Hutcheson & Sofroniou, 1999).

Kaiser's criterion: a method of *extraction* in *factor analysis* based on the idea of retaining factors with associated eigenvalues greater than 1. This method appears to be accurate when the number of variables in the analysis is less than 30 and the resulting *communalities* (after *extraction*) are all greater than .7, or when the sample size exceeds 250 and the average communality is greater than or equal to .6.

Kendall's tau: a non-parametric correlation coefficient similar to *Spearman's correlation coefficient*, but should be used in preference for a small data set with a large number of tied ranks.

Kolmogorov–Smirnov test: a test of whether a distribution of scores is significantly different from a *normal distribution*. A significant value indicates a deviation from normality, but this test is notoriously affected by large samples in which small deviations from normality yield significant results.

Kolmogorov–Smirnov Z: not to be confused with the *Kolmogorov–Smirnov test* that tests whether a sample comes from a normally distributed population. This tests whether two groups have been drawn from the same population (regardless of what that population may be). It does much the same as the *Mann–Whitney test* and *Wilcoxon rank-sum test*! This test tends to have better power than the Mann–Whitney test when sample sizes are less than about 25 per group.

Kruskal–Wallis test: non-parametric test of whether more than two independent groups differ. It is the non-parametric version of one-way *independent ANOVA*.

Kurtosis: this measures the degree to which scores cluster in the tails of a frequency distribution. A distribution with positive kurtosis (*leptokurtic*, kurtosis > 0) has too many scores in the tails and is too peaked, whereas a distribution with negative kurtosis (*platykurtic*, kurtosis < 0) has too few scores in the tails and is quite flat.

Latent variable: a variable that cannot be directly measured, but is assumed to be related to several variables that can be measured.

Leptokurtic: see *Kurtosis*.

Levels of measurement: the relationship between what is being measured and the numbers obtained on a scale.

Levene's test: tests the hypothesis that the variances in different groups are equal (i.e. the difference between the variances is zero). It basically does a one-way ANOVA on the deviations (i.e. the absolute value of the difference between each score and the mean of its group). A significant result indicates that the variances are significantly different – therefore, the assumption of *homogeneity of variances* has been violated. When samples sizes are large, small differences in group variances can produce a significant Levene's test and so the *variance ratio* is a useful double-check.

Leverage: leverage statistics (or hat values) gauge the influence of the observed value of the outcome variable over the predicted values. The average leverage value is $(k+1)/n$ in which k is the number of predictors in the model and n is the number of participants. Leverage values can lie between 0 (the case has no influence whatsoever) and 1 (the case has complete influence over prediction). If no cases exert undue influence over the model then we would expect all of the leverage values to be close to the average value. Hoaglin and Welsch (1978) recommend investigating cases with values greater than twice the average $(2(k + 1)/n)$ and Stevens (2002) recommends using three times the average $(3(k + 1)/n)$ as a cut-off point for identifying cases having undue influence.

Likelihood: the probability of obtaining a set of observations given the parameters of a model fitted to those observations.

Linear model: a model that is based upon a straight line.

Line chart: a graph in which a summary statistic (usually the mean) is plotted on the y-axis against a categorical variable on the x-axis (this categorical variable could represent, for example, groups of people, different times or different experimental conditions). The value of the mean for each category is shown by a symbol and means across categories are connected by a line. Different-coloured lines may be used to represent levels of a second categorical variable.

Logistic regression: a version of *multiple regression* in which the outcome is a *categorical variable*. If the categorical variable has exactly two categories the analysis is called *binary logistic regression*, and when the outcome has more than two categories it is called *multinomial logistic regression*.

Log-likelihood: a measure of error, or unexplained variation, in categorical models. It is based on summing the probabilities associated with the predicted and actual outcomes and is analogous to the *residual sum of squares* in multiple regression in that it is an indicator of how much unexplained information there is after the model has been fitted. Large values of the log-likelihood statistic indicate poorly fitting statistical models, because the larger the value of the log-likelihood, the more unexplained observations there are. The log-likelihood is the logarithm of the *likelihood*.

Loglinear analysis: a procedure used as an extension of the *chi-squared test* to analyse situations in which we have more than two *categorical variables* and we want to test for relationships between these variables. Essentially, a *linear model* is fitted to the data that predicts expected frequencies (i.e. the number of cases expected in a given category). In this respect it is much the same as *analysis of variance* but for entirely categorical data.

Lower bound: the name given to the lowest possible value of the *Greenhouse–Geisser estimate* of *sphericity*. Its value is $1(k-1)$, in which k is the number of treatment conditions.

Lower quartile: the value that cuts off the lowest 25% of the data. If the data are ordered and then divided into two halves at the median, then the lower quartile is the median of the lower half of the scores.

Mahalanobis distances: these measure the influence of a case by examining the distance of cases from the mean(s) of the predictor variable(s). One needs to look for the cases with the highest values. It is not easy to establish a cut-off point at which to worry, although Barnett and Lewis (1978) have produced a table of critical values dependent on the number of predictors and the sample size. From their work it is clear that even with large samples ($N = 500$) and five predictors, values above 25 are cause for concern. In smaller samples ($N = 100$) and with fewer predictors (namely three) values greater than 15 are problematic, and in very small samples ($N = 30$) with only two predictors values greater than

11 should be examined. However, for more specific advice, refer to Barnett and Lewis's (1978) table.

Main effect: the unique effect of a *predictor variable* (or *independent variable*) on an *outcome variable*. The term is usually used in the context of *ANOVA*.

Mann–Whitney test: a *non-parametric test* that looks for differences between two independent samples. That is, it tests whether the populations from which two samples are drawn have the same location. It is functionally the same as *Wilcoxon's rank-sum test*, and both tests are non-parametric equivalents of the *independent t-test*.

MANOVA: acronym for *multivariate analysis of variance*.

Matrix: a collection of numbers arranged in columns and rows. The values within a matrix are typically referred to as *components* or *elements*.

Mauchly's test: a test of the assumption of *sphericity*. If this test is significant then the assumption of *sphericity* has not been met and an appropriate correction must be applied to the *degrees of freedom* of the *F-ratio* in *repeated-measures ANOVA*. The test works by comparing the *variance–covariance matrix* of the data to an *identity matrix;* if the variance–covariance matrix is a scalar multiple of an *identity matrix* then sphericity is met.

Maximum-likelihood estimation: a way of estimating statistical parameters by choosing the parameters that make the data most likely to have happened. Imagine for a set of parameters that we calculated the probability (or likelihood) of getting the observed data; if this probability was high then these particular parameters yield a good fit of the data, but conversely if the probability was low, these parameters are a bad fit of our data. Maximum-likelihood estimation chooses the parameters that maximize the probability.

Mean: a simple statistical model of the centre of a distribution of scores. A hypothetical estimate of the 'typical' score.

Mean squares: a measure of average variability. For every *sum of squares* (which measure the total variability) it is possible to create mean squares by dividing by the number of things used to calculate the sum of squares (or some function of it).

Measurement error: the discrepancy between the numbers used to represent the thing that we're measuring and the actual value of the thing we're measuring (i.e. the value we would get if we could measure it directly).

Median: the middle score of a set of ordered observations. When there is an even number of observations the median is the average of the two scores that fall either side of what would be the middle value.

Meta-analysis: this is a statistical procedure for assimilating research findings. It is based on the simple idea that we can take effect sizes from individual studies that research the same question, quantify the observed effect in a standard way (using *effect sizes*) and then combine these effects to get a more accurate idea of the true effect in the population.

Mixed ANOVA: *analysis of variance* used for a *mixed design*.

Mixed design: an experimental design incorporating two or more *predictors* (or *independent variables*) at least one of which has been manipulated using different participants (or whatever entities are being tested) and at least one of which has been manipulated using the same participants (or entities). Also known as a split-plot design because Fisher developed ANOVA for analysing agricultural data involving 'plots' of land containing crops.

Mode: the most frequently occurring score in a set of data.

Model sum of squares: a measure of the total amount of variability for which a model can account. It is the difference between the *total sum of squares* and the *residual sum of squares*.

Monte Carlo method: a term applied to the process of using data simulations to solve statistical problems. Its name comes from the use of Monte Carlo roulette tables to generate 'random' numbers in the pre-computer age. Karl Pearson, for example, purchased copies of *Le Monaco*, a weekly Paris periodical that published data from the Monte Carlo casinos' roulette wheels. He used these data as pseudo-random numbers in his statistical research.

Multicollinearity: a situation in which two or more variables are very closely linearly related.

Multilevel linear model: A linear model (just like regression, ANCOVA, ANOVA, etc.) in which the hierarchical structure of the data is explicitly considered. In this analysis regression parameters can be fixed (as in regression and ANOVA) but also random (i.e. free to vary across different contexts at a higher level of the hierarchy). This means that for each regression parameter there is a fixed component but also an estimate of how much the parameter varies across contexts (see *Fixed coefficient, Random coefficient*).

Multimodal: description of a distribution of observations that has more than two *modes*.

Multinomial logistic regression: *logistic regression* in which the outcome variable has more than two categories.

Multiple *R*: the multiple correlation coefficient. It is the correlation between the observed values of an outcome and the values of the outcome predicted by a multiple regression model.

Multiple regression: an extension of *simple regression* in which an outcome is predicted by a linear combination of two or more predictor variables. The form of the model is: $Y_i = (b_0 + b_1X_{1i} + b_2X_{2i} + \ldots + b_nX_{ni}) + \varepsilon_i$ in which the outcome is denoted as Y, and each predictor is denoted as X. Each predictor has a regression coefficient b associated with it, and b_0 is the value of the outcome when all predictors are zero.

Multivariate: means 'many variables' and is usually used when referring to analyses in which there is more than one *outcome variable* (*MANOVA*, *principal component analysis*, etc.).

Multivariate analysis of variance: family of tests that extend the basic *analysis of variance* to situations in which more than one *outcome variable* has been measured.

Multivariate normality: an extension of a normal distribution to multiple variables. It is a *probability distribution* of a set of variables $v' = [v_1, v_2, \ldots, v_n]$ given by:

$$f(v_1, v_2, \ldots, v_n) = 2\pi^{n/2}|\Sigma|^{1/2} \exp\left[-\frac{1}{2}(v-\mu)'\Sigma^{-1}(v-\mu)\right]$$

in which μ is the vector of means of the variables, and Σ is the *variance–covariance* matrix. If that made any sense to you then you're cleverer than I am.

Nagelkerke's R^2_N: a version of the *coefficient of determination* for logistic regression. It is a variation on *Cox and Snell's* R^2_{CS} which overcomes the problem that this statistic has of not being able to reach its maximum value.

Negative skew: see *Skew.*

Nominal variable: where numbers merely represent names. For example, the numbers on sports players shirts: a player with the number 1 on her back is not necessarily worse than a player with a 2 on her back. The numbers have no meaning other than denoting the type of player (i.e. full back, centre forward, etc.).

Noniles: a type of *quantile*; they are values that split the data into nine equal parts. They are comonly used in educational research.

Non-parametric tests: a family of statistical procedures that do not rely on the restrictive assumptions of parametric tests. In particular, they do not assume that the sampling distribution is normally distributed.

Normal distribution: a *probability distribution* of a random variable that is known to have certain properties. It is perfectly symmetrical (has a *skew* of 0), and has a *kurtosis* of 0.

Null hypothesis: reverse of the *experimental hypothesis* that your prediction is wrong and the predicted effect doesn't exist.

Numeric variables: variables involving numbers.

Oblique rotation: a method of *rotation* in *factor analysis* that allows the underlying factors to be correlated.

Odds: the probability of an event occurring divided by the probability of that event not occurring.

Odds ratio: the ratio of the *odds* of an event occurring in one group compared to another. So, for example, if the odds of dying after writing a glossary are 4, and the odds of dying after not writing a glossary are 0.25, then the odds ratio is 4/0.25 = 16. This means that the *odds* of dying if you write a glossary are 16 times higher than if you don't. An odds ratio of 1 would indicate that the *odds* of a particular outcome are equal in both groups.

Omega squared: an *effect size* measure associated with ANOVA that is less biased than *eta squared*. It is a (sometimes hideous) function of the *model sum of squares* and the *residual sum of squares* and isn't actually much use because it measures the overall effect of the ANOVA and so can't be interpreted in a meaningful way. In all other respects it's great though.

One-tailed test: a test of a directional hypothesis. For example, the hypothesis 'the longer I write this glossary, the more I want to place my editor's genitals in a starved crocodile's mouth' requires a one-tailed test because I've stated the direction of the relationship (see also *two-tailed test*).

Ordinal variable: data that tell us not only that things have occurred, but also the order in which they occurred. These data tell us nothing about the differences between values. For example, gold, silver and bronze medals are ordinal: they tell us that the gold medallist was better than the silver medallist, but they don't tell us how much better (was gold a lot better than silver, or were gold and silver very closely competed?).

Orthogonal: means perpendicular (at right angles) to something. It tends to be equated to *independence* in statistics because of the connotation that perpendicular *linear models* in geometric space are completely independent (one is not influenced by the other).

Orthogonal rotation: a method of *rotation* in *factor analysis* that keeps the underlying factors independent (i.e. not correlated).

Outcome variable: a variable whose values we are trying to predict from one or more *predictor variables*.

Outlier: an observation very different from most others. Outliers can bias statistics such as the mean.

Pairwise comparisons: comparisons of pairs of means.

Parametric test: a test that requires data from one of the large catalogue of distributions that statisticians have described. Normally this term is used for parametric tests based on the *normal distribution*, which require four basic assumptions that must be met for the test to be accurate: a normally distributed sampling distribution (see *normal distribution*), *homogeneity of variance*, *interval* or *ratio data*, and *Independence*.

Part correlation: another name for a *semi-partial correlation*.

Partial correlation: a measure of the relationship between two variables while 'controlling' the effect of one or more additional variables has on both.

Partial eta squared (partial η^2): a version of *eta squared* that is the proportion of variance that a variable explains when excluding other variables in the analysis. Eta squared is the proportion of total variance explained by a variable, whereas partial eta squared is the proportion of variance that a variable explains that is not explained by other variables.

Partial out: to partial out the effect of a variable is to remove the variance that the variable shares with other variables in the analysis before looking at their relationships (see *partial correlation*).

Pattern matrix: a matrix in *factor analysis* containing the *regression coefficients* for each variable on each *factor* in the data. See also *Structure matrix*.

Pearson's correlation coefficient: or Pearson's product-moment correlation coefficient to give it its full name, is a *standardized* measure of the strength of relationship between two variables. It can take any value from −1 (as one variable changes, the other changes in the opposite direction by the same amount), through 0 (as one variable changes the other doesn't change at all), to +1 (as one variable changes, the other changes in the same direction by the same amount).

Percentiles: are a type of *quantile*; they are values that split the data into 100 equal parts.

Perfect collinearity: exists when at least one predictor in a *regression model* is a perfect linear combination of the others (the simplest example being two predictors that are perfectly correlated – they have a correlation coefficient of 1).

Phi: a measure of the strength of association between two *categorical variables*. Phi is used with 2 × 2 *contingency tables* (tables which have two categorical variables and each variable has only two categories). Phi is a variant of the *chi-square test*, χ^2:

$$\phi = \sqrt{\frac{\chi^2}{n}},$$

in which n is the total number of observations.

Pillai–Bartlett trace (*V*): a *test statistic* in *MANOVA*. It is the sum of the proportion of explained variance on the *discriminant function variates* of the data. As such, it is similar to the ratio of SS_M/SS_T.

Planned comparisons: another name for *planned contrasts*.

Planned contrasts: a set of comparisons between group means that are constructed before any data are collected. These are theory-led comparisons and are based on the idea of partitioning the variance created by the overall effect of group differences into gradually smaller portions of variance. These tests have more power than *post hoc tests*.

Platykurtic: see *Kurtosis*.

Polychotomous logistic regression: another name for *multinomial logistic regression*.

Polynomial: a posh name for a *growth curve* or trend over time. If *time* is our predictor variable, then any polynomial is tested by including a variable that is the predictor to the power of the order of polynomial that we want to test: a linear trend is tested by *time* alone, a quadratic or second-order polynomial is tested by including a predictor that is $time^2$, for a fifth-order polynomial we need a predictor of $time^5$ and for an nth-order polynomial we would have to include $time^n$ as a predictor.

Polynomial contrast: a contrast that tests for trends in the data. In its most basic form it looks for a linear trend (i.e. that the group means increase proportionately).

Population: in statistical terms this usually refers to the collection of units (be they people, plankton, plants, cities, suicidal authors, etc.) to which we want to generalize a set of findings or a statistical model.

Positive skew: see *skew*.

***Post hoc* tests:** a set of comparisons between group means that were not thought of before data were collected. Typically these tests involve comparing the means of all combinations of pairs of groups. To compensate for the number of tests conducted, each test uses a strict criterion for significance. As such, they tend to have less power than *planned contrasts*. They are usually used for exploratory work for which no firm hypotheses were available on which to base planned contrasts.

Power: the ability of a test to detect an effect of a particular size (a value of .8 is a good level to aim for).

P–P plot: Short for a probability–probability plot. A graph plotting the cumulative probability of a variable against the cumulative probability of a particular distribution (often a normal distribution). Like a *Q–Q plot*, if values fall on the diagonal of the plot then the variable shares the same distribution as the one specified. Deviations from the diagonal show deviations from the distribution of interest.

Practice effect: refers to the possibility that participants' performance in a task may be influenced (positively or negatively) if they repeat the task because of familiarity with the experimental situation and/or the measures being used.

Predictor variable: a variable that is used to try to predict values of another variable known as an *outcome variable*.

Principal component analysis (PCA): a *multivariate* technique for identifying the linear components of a set of variables.

Probability distribution: a curve describing an idealized *frequency distribution* of a particular variable from which it is possible to ascertain the probability with which specific values of that variable will occur. For categorical variables it is simply a formula yielding the probability with which each category occurs.

Promax: a method of *oblique rotation* that is computationally faster than *direct oblimin* and so useful for large data sets.

Q–Q plot: short for a quantile–quantile plot. A graph plotting the *quantiles* of a variable against the quantiles of a particular distribution (often a normal distribution). Like a *P–P plot*, if values fall on the diagonal of the plot then the variable shares the same distribution as the one specified. Deviations from the diagonal show deviations from the distribution of interest.

Quadratic trend: if the means in ordered conditions are connected with a line then a quadratic trend is shown by one change in the direction of this line (e.g. the line is curved in one place); the line is, therefore, U-shaped. There must be at least three ordered conditions.

Qualitative methods: extrapolating evidence for a theory from what people say or write (contrast with *quantitative methods*).

Quantiles: values that split a data set into equal portions. *Quartiles*, for example, are a special case of quantiles that split the data into four equal parts. Similarly, *percentiles* are points that split the data into 100 equal parts and *noniles* are points that split the data into nine equal parts (you get the general idea).

Quantitative methods: inferring evidence for a theory through measurement of variables that produce numeric outcomes (contrast with *qualitative methods*).

Quartic trend: if the means in ordered conditions are connected with a line then a quartic trend is shown by three changes in the direction of this line. There must be at least five ordered conditions.

Quartiles: generic term for the three values that cut an ordered data set into four equal parts. The three quartiles are known as the *lower quartile*, the second quartile (or *median*) and the *upper quartile*.

Quartimax: a method of *orthogonal rotation*. It attempts to maximize the spread of factor loadings for a variable across all *factors*. This often results in lots of variables loading highly onto a single *factor*.

Random coefficient: a coefficient or model parameter that is free to vary over situations or contexts (cf. *Fixed coefficient*).

Random effect: an effect is said to be random if the experiment contains only a sample of possible treatment conditions. Random effects can be generalized beyond the treatment conditions in the experiment. For example, the effect is random if we say that the conditions in our experiment (e.g. placebo, low dose and high dose) are only a sample of possible conditions (maybe we could have tried a very high dose). We can generalize this random effect beyond just placebos, low doses and high doses.

Random intercept: A term used in *multilevel modelling* to denote when the intercept in the model is free to vary across different groups or contexts (cf. *Fixed intercept*).

Random slope: A term used in *multilevel modelling* to denote when the slope of the model is free to vary across different groups or contexts (cf. *Fixed slope*).

Random variable: a random variable is one that varies over time (e.g.

your weight is likely to fluctuate over time).

Random variance: variance that is unique to a particular variable but not reliably so.

Randomization: the process of doing things in an unsystematic or random way. In the context of experimental research the word usually applies to the random assignment of participants to different treatment conditions.

Range: the range of scores is the value of the smallest score subtracted from the highest score. It is a measure of the dispersion of a set of scores. See also *variance*, *standard deviation*, and *interquartile range*.

Ranking: the process of transforming raw scores into numbers that represent their position in an ordered list of those scores. i.e. the raw scores are ordered from lowest to highest and the lowest score is assigned a rank of 1, the next highest score is assigned a rank of 2, and so on.

Ratio variable: an *interval variable* but with the additional property that ratios are meaningful. For example, people's ratings of this book on Amazon.com can range from 1 to 5; for these data to be ratio not only must they have the properties of *interval variables*, but in addition a rating of 4 should genuinely represent someone who enjoyed this book twice as much as someone who rated it as 2. Likewise, someone who rated it as 1 should be half as impressed as someone who rated it as 2.

Regression coefficient: see b_i and β_i.

Regression model: see *Multiple regression* and *Simple regression*.

Regression line: a line on a scatterplot representing the *regression model* of the relationship between the two variables plotted.

Related design: another name for a *repeated-measures design*.

Related factorial design: an experimental design incorporating two or more *predictors* (or *independent variables*) all of which have been manipulated using the same participants (or whatever entities are being tested).

Reliability: the ability of a measure to produce consistent results when the same entities are measured under different conditions.

Repeated contrast: a non-orthogonal *planned contrast* that compares the mean in each condition (except the first) to the mean of the preceding condition.

Repeated-measures ANOVA: an *analysis of variance* conducted on any design in which the *independent variable* (*predictor*) or *variables* (*predictors*) have all been measured using the same participants in all conditions.

Repeated-measures design: an experimental design in which different treatment conditions utilize the same organisms (i.e. in psychology, this would mean the same people take part in all experimental conditions) and so the resulting data are related (a.k.a. related design or within-subject design).

Residual: the difference between the value a model predicts and the value observed in the data on which the model is based. When the residual is calculated for each observation in a data set the resulting collection is referred to as the *residuals*.

Residuals: see *Residual*.

Residual sum of squares: a measure of the variability that cannot be explained by the model fitted to the data. It is the total squared *deviance* between the observations, and the value of those observations predicted by whatever model is fitted to the data.

Reverse Helmert contrast: another name for a *difference contrast*.

Robust test: A term applied to a family of procedures to estimate statistics that are reliable even when the normal assumptions of the statistic are not met.

Rotation: a process in *factor analysis* for improving the interpretability of factors. In essence, an attempt is made to transform the *factors* that emerge from the analysis in such a way as to maximize *factor loadings* that are already large, and minimize factor loadings that are already small. There are two general approaches: *orthogonal rotation* and *oblique rotation*.

Roy's greatest root: a *test statistic* in *MANOVA*. It is the eigenvalue for the first *discriminant function variate* of a set of observations. So, it is the same as the *Hotelling–Lawley trace* but for the first variate only. It represents the proportion of explained variance to unexplained variance (SS_M/SS_R) for the first discriminant function.

Sample: a smaller (but hopefully representative) collection of units from a *population* used to determine truths about that population (e.g. how a given population behaves in certain conditions).

Sampling distribution: the *probability distribution* of a statistic. We can think of this as follows: if we take a *sample* from a *population* and calculate some statistic (e.g. the *mean*), the value of this statistic will depend somewhat on the sample we took. As such the statistic will vary slightly from sample to sample. If, hypothetically, we took lots and lots of samples from the population and calculated the statistic of interest we could create a frequency distribution of the values we get. The resulting distribution is what the sampling distribution represents: the distribution of possible values of a given statistic that we could expect to get from a given population.

Sampling variation: the extent to which a statistic (e.g. the mean, median, *t*, *F*, etc.) varies in samples taken from the same population.

Saturated model: a model that perfectly fits the data and, therefore, has no error. It contains all possible *main effects* and *interactions* between variables.

Scatterplot: a graph that plots values of one variable against the corresponding value of another variable (and the corresponding value of a third variable can also be included on a 3-D scatterplot).

Scree plot: a graph plotting each *factor* in a *factor analysis* (*X*-axis) against its associated eigenvalue (*Y*-axis). It shows the relative importance of each factor. This graph has a very characteristic shape (there is a sharp descent in the curve followed by a tailing off) and the point of inflexion of this curve is often used as a means of *extraction*. With a sample of more than 200 participants, this provides a fairly reliable criterion for *extraction* (Stevens, 2002)

Second quartile: another name for the *median*.

Semi-partial correlation: a measure of the relationship between two variables while 'controlling' the effect that one or more additional variables has on one of those variables. If we call our variables *x* and *y*, it gives us a measure of the variance in *y* that *x* alone shares.

Shapiro–Wilk test: a test of whether a distribution of scores is significantly different from a *normal distribution*. A significant value indicates a deviation from normality, but this test is notoriously affected by large samples in which small deviations from normality yield significant results.

Shrinkage: the loss of predictive power of a regression model if the model had been derived from the population from which the sample was taken, rather than the sample itself.

Šidák correction: slightly less conservative variant of a *Bonferroni correction*.

Sign test: tests whether two related samples are different. It does the same thing as the *Wilcoxon signed-rank test*. Differences between the conditions are calculated and the sign of this difference (positive or negative) is analysed because it indicates the direction of differences. The magnitude of change is completely ignored (unlike in Wilcoxon's test where the rank tells us something about the relative magnitude of change), and for this reason it lacks *power*. However, its computational simplicity makes it a nice party trick if ever anyone drunkenly accosts you needing some data quickly analysed without the aid of a computer … doing a sign test in your head really impresses people. Actually it doesn't, they just think you're a sad gimboid.

Simple contrast: a non-orthogonal *planned contrast* that compares the mean in each condition to the mean of either the first or last condition, depending on how the contrast is specified.

Simple effects analysis: this analysis looks at the effect of one *independent variable* (categorical *predictor variable*) at individual levels of another independent variable.

Simple regression: a *linear model* in which one variable or outcome is predicted from a single predictor variable. The model takes the form:

$$Y_i = (b_0 + b_1 X_i) + \varepsilon_i$$

in which *Y* is the outcome variable, *X* is the predictor, b_1 is the regression coefficient associated with the predictor and b_0 is the value of the outcome when the predictor is zero.

Singularity: a term used to describe variables that are perfectly correlated (i.e. the *correlation coefficient* is 1 or −1).

Skew: a measure of the symmetry of a *frequency distribution*. Symmetrical distributions have a skew of 0. When the frequent scores are clustered at the lower end of the distribution and the tail points towards the higher or more positive scores, the value of skew is positive. Conversely, when the frequent scores are clustered at the higher end of the distribution and the tail points towards the lower more negative scores, the value of skew is negative.

Spearman's correlation coefficient: a standardized measure of the strength of relationship between two variables that does not rely on the assumptions of a *parametric test*. It is *Pearson's correlation coefficient* performed on data that have been converted into ranked scores.

Sphericity: a less restrictive form of *compound symmetry* which assumes that the variances of the differences between data taken from the same participant (or other entity being tested) are equal. This assumption is most commonly found in *repeated-measures ANOVA* but applies only where there are more than two points of data from the same participant. (See also *Greenhouse–Geisser correction*, *Huynh–Feldt correction*.)

Split-half reliability: a measure of *reliability* obtained by splitting items on a measure into two halves (in some random fashion) and obtaining a score from each half of the scale. The correlation between the two scores, corrected to take account of the fact the correlations are based on only half of the items, is used as a measure of reliability. There are two popular ways to do this. Spearman (1910) and Brown (1910) developed a formula that takes no account of the standard deviation of items:

$$r_{sh} = \frac{2r_{12}}{1 + r_{12}}$$

in which r_{12} is the correlation between the two halves of the scale. Flanagan (1937) and Rulon (1939), however, proposed a measure that does account for item variance:

$$r_{sh} = \frac{4r_{12} \times s_1 \times s_2}{s_T^2}$$

in which s_1 and s_2 are the standard deviations of each half of the scale, and s_T^2 is the variance of the whole test. See Cortina (1993) for more detail.

Square matrix: a *matrix* that has an equal number of columns and rows.

Standard deviation: an estimate of the average variability (spread) of a set of data measured in the same units of measurement as the original data. It is the square root of the variance.

Standard error: the standard deviation of the *sampling distribution* of a statistic. For a given statistic (e.g. the *mean*) it tells us how much variability there is in this statistic across *samples* from the same *population*. Large values, therefore, indicate that a statistic from a given sample may not be an accurate reflection of the population from which the sample came.

Standard error of differences: if we were to take several pairs of samples from a population and calculate their means, then we could also calculate the difference between their means. If we plotted these differences between sample means as a *frequency distribution*, we would have the *sampling distribution* of differences. The standard deviation of this sampling distribution is the *standard error of differences*. As such it is a measure of the variability of differences between sample means.

Standard error of the mean (SE): the full name of the *standard error*.

Standardization: the process of converting a variable into a standard unit of measurement. The unit of measurement typically used is *standard deviation* units (see also *z-scores*). Standardization allows us to compare data when different units of measurement have been used (we could compare weight measured in kilograms to height measured in inches).

Standardized: see *Standardization*.

Standardized DFBeta: a *standardized* version of *DFBeta*. These standardized values are easier to use than DFBeta because universal cut-off points can be applied. Stevens (2002) suggests looking at cases with absolute values greater than 2.

Standardized DFFit: a *standardized* version of *DFFit*.

Standardized residuals: the *residuals* of a model expressed in standard deviation units. Standardized residuals with an absolute value greater than 3.29 (actually we usually just use 3) are cause for concern because in an average sample a value this high is unlikely to happen by chance; if more than 1% of our observations have standardized residuals with an absolute value greater than 2.58 (we usually just say 2.5) there is evidence that the level of error within our model is unacceptable (the model is a fairly poor fit of the sample data); and if more than 5% of observations have standardized residuals with an absolute value greater than 1.96 (or 2 for convenience) then there is also evidence that the model is a poor representation of the actual data.

Stepwise regression: a method of multiple regression in which variables are entered into the model based on a statistical criterion (the *semi-partial correlation* with the *outcome variable*). Once a new variable is entered into the model, all variables in the model are assessed to see whether they should be removed.

String variables: variables involving words (i.e. letter strings). Such variables could include responses to open-ended questions such as 'how much do you like writing glossary entries?'; the response might be 'about as much as I like placing my gonads on hot coals'.

Structure matrix: a matrix in *factor analysis* containing the *correlation coefficients* for each variable on each *factor* in the data. When *orthogonal rotation* is used this is the same as the *pattern matrix*, but when oblique rotation is used these matrices are different.

Studentized deleted residual: a measure of the influence of a particular case of data. This is a standardized version of the *deleted residual*.

Studentized residuals: a variation on *standardized residuals*. Studentized residuals are the *unstandardized residual* divided by an estimate of its standard deviation that varies point by point. These residuals have the same properties as the *standardized residuals* but usually provide a more precise estimate of the error variance of a specific case.

Sum of squared errors: another name for the *sum of squares*.

Sum of squares (SS): an estimate of total variability (spread) of a set of data. First the *deviance* for each score is calculated, and then this value is squared. The SS is the sum of these squared deviances.

Sum of squares and cross-products matrix (SSCP matrix): a *square matrix* in which the diagonal elements represent the *sum of squares* for a particular variable, and the off-diagonal elements represent the *cross-products* between pairs of variables. The SSCP matrix is basically the same as the *variance–covariance matrix*, except the SSCP matrix expresses variability and between-variable relationships as total values, whereas the variance–covariance matrix expresses them as average values.

Suppressor effects: when a predictor has a significant effect but only when another variable is held constant.

Systematic variation: variation due to some genuine effect (be that the effect of an experimenter doing something to all of the participants in one sample but not in other samples, or natural variation between sets of variables). We can think of this as variation that can be explained by the model that we've fitted to the data.

t-statistic: Student's *t* is a *test statistic* with a known *probability distribution* (the *t*-distribution). In the context of regression it is used to test whether a regression coefficient *b* is significantly different from zero; in the context of experimental work it is used to test whether the differences between two means are significantly different from zero. See also *Dependent t-test* and *Independent t-test*.

Tertium quid: the possibility that an apparent relationship between two variables is actually caused by the effect of a third variable on them both (often called *the third-variable problem*).

Test–retest reliability: the ability of a measure to produce consistent results when the same entities are tested at two different points in time.

Test statistic: a statistic for which we know how frequently different values occur. The observed value of such a statistic is typically used to test *hypotheses*.

Theory: although it can be defined more formally, a theory is a hypothesized general principle or set of principles that explain known findings about a topic and from which new hypotheses can be generated.

Tolerance: tolerance statistics measure *multicollinearity* and are simply the reciprocal of the *variance inflation factor* (1/VIF). Values below 0.1 indicate serious problems, although Menard (1995) suggests that values below 0.2 are worthy of concern.

Total SSCP (*T*): the total sum of squares and cross-product matrix. This is a *sum of squares and cross-product matrix* for an entire set of observations. It is the *multivariate* equivalent of the *total sum of squares*.

Total sum of squares: a measure of the total variability within a set of observation. It is the total squared *deviance* between each observation and the overall mean of all observations.

Transformation: the process of applying a mathematical function to all observations in a data set, usually to correct some distributional abnormality such as *skew* or *kurtosis*.

Trimmed mean: a statistic used in many *robust tests*. Imagine we had 20 scores representing the annual income of students (in thousands, rounded to the nearest thousand: 2, 2, 2, 2, 3, 3, 3, 3, 3, 4, 4, 4, 4, 4, 4, 4, 4, 4, 6, 35. The mean income is 5 (£5000). This value is biased by an outlier. A trimmed mean is simply a mean based on the distribution of scores after some percentage of scores has been removed from each extreme of the distribution. So, a 10% trimmed mean will remove 10% of scores from the top and bottom of ordered scores before the mean is calculated. With 20 scores, removing 10% of scores involves removing the top and bottom 2 scores. This gives us: 2, 2, 3, 3, 3, 3, 3, 4, 4, 4, 4, 4, 4, 4, 4, 4, the mean of which is 3.44. The mean depends on a symmetrical distribution to be accurate, but a trimmed mean produces accurate results even when the distribution is not symmetrical. There are more complex examples of robust methods such as the *bootstrap*.

Two-tailed test: a test of a non-directional hypothesis. For example, the hypothesis 'writing this glossary has some effect on what I want to do with my editor's genitals' requires a two-tailed test because it doesn't suggest the direction of the relationship (see also *One-tailed test*).

Type I error: occurs when we believe that there is a genuine effect in our population, when in fact there isn't.

Type II error: occurs when we believe that there is no effect in the population when, in reality, there is.

Unique variance: variance that is specific to a particular variable (i.e. is not shared with other variables). We tend to use the term 'unique variance' to refer to variance that can be reliably attributed to only one measure, otherwise it is called *random variance*.

Univariate: means 'one variable' and is usually used to refer to situations in which only one *outcome variable* has been measured (*ANOVA*, *t-tests*, *Mann–Whitney tests,* etc.).

Unstructured: a covariance structure used in *multilevel models*. This covariance structure is completely general and is, therefore, the default option used in random effects in SAS. Covariances are assumed to be completely unpredictable: they do not conform to a systematic pattern.

Unstandardized residuals: the *residuals* of a model expressed in the units in which the original outcome variable was measured.

Unsystematic variation: this is variation that isn't due to the effect in which we're interested (so could be due to natural differences between people in different samples such as differences in intelligence or motivation). We can think of this as variation that can't be explained by whatever model we've fitted to the data.

Upper quartile: the value that cuts off the highest 25% of ordered scores. If the scores are ordered and then divided into two halves at the median, then the upper quartile is the median of the top half of the scores.

Validity: evidence that a study allows correct inferences about the question it was aimed to answer or that a test measures what it set out to measure conceptually (see also *Content validity*, *Criterion validity*).

Variables: anything that can be measured and can differ across entities or across time.

Variance: an estimate of average variability (spread) of a set of data. It is the sum of squares divided by the number of values on which the sum of squares is based minus 1.

Variance components: a covariance structure used in *multilevel models*. This covariance structure is very simple and assumes that all random effects are independent and variances of random effects are assumed to be the same and sum to the variance of the outcome variable. In SAS this is the default covariance structure for random effects.

Variance–covariance matrix: a square matrix (i.e. same number of columns and rows) representing the variables measured. The diagonals represent the *variances* within each variable, whereas the off-diagonals represent the *covariances* between pairs of variables.

Variance inflation factor (VIF): a measure of *multicollinearity*. The VIF indicates whether a predictor has a strong linear relationship with the other predictor(s). Myers (1990) suggests that a value of 10 is a good value at which to worry. Bowerman and O'Connell (1990) suggest that if the average VIF is greater than 1, then multicollinearity may be biasing the regression model.

Variance ratio: see *Hartley's* F_{max}.

Variance sum law: states that the variance of a difference between two independent variables is equal to the sum of their variances.

Varimax: a method of *orthogonal rotation*. It attempts to maximize the dispersion of *factor loadings* within *factors*. Therefore, it tries to load a smaller number of variables highly onto each factor, resulting in more interpretable clusters of factors.

VIF: see *variance inflation factor*.

Wald statistic: a *test statistic* with a known *probability distribution* (a *chi-square distribution*) that is used to test whether the *b* coefficient for a predictor in a *logistic* regression model is significantly different from zero. It is analogous to the *t-statistic* in a *regression model* in that it is simply the *b* coefficient divided by its standard error. The Wald statistic is inaccurate when the regression coefficient (*b*) is large, because the standard error tends to become inflated, resulting in the Wald statistic being underestimated.

Weights: a number by which something (usually a variable in statistics) is multiplied. The weight assigned to a variable determines the influence that variable has within a mathematical equation: large weights give the variable a lot of influence.

Welch's *F*: a version of the *F*-ratio designed to be accurate when the assumption of *homogeneity of variance* has been violated. Not to be confused with the squelch test which is where you shake your head around after writing statistics books to see if you still have a brain.

Wilcoxon's rank-sum test: a *non-parametric test* that looks for differences between two independent samples. That is, it tests whether the populations from which two samples are drawn have the same location. It is functionally the same as the *Mann–Whitney test*, and both tests are non-parametric equivalents of the *independent t-test*.

Wilcoxon signed-rank test: a *non-parametric test* that looks for differences between two related samples. It is the non-parametric equivalent of the *related t-test*.

Wilks's lambda (Λ): a *test statistic* in *MANOVA*. It is the product of the unexplained variance on each of the *discriminant function variates* so it represents the ratio of error variance to total variance (SS_R/SS_T) for each variate.

Within-subject design: another name for a *repeated-measures design*.

Writer's block: something I suffered from a lot while writing this book. It's when you can't think of any decent examples and so end up talking about sperm the whole time. Seriously, look at this book, it's all sperm this, sperm that, quail sperm, human sperm. Frankly, I'm amazed donkey sperm didn't get in there somewhere. Oh, it just did.

Yates's continuity correction: an adjustment made to the *chi-square* *test* when the *contingency table* is 2 rows by 2 columns (i.e. there are two categorical variables both of which consist of only two categories). In large samples the adjustment makes little difference and is slightly dubious anyway (see Howell, 2006).

z-score: the value of an observation expressed in standard deviation units. It is calculated by taking the observation, subtracting from it the mean of all observations, and dividing the result by the standard deviation of all observations. By converting a distribution of observations into *z*-scores a new distribution is created that has a mean of 0 and a standard deviation of 1.

APPENDIX

A.1. Table of the standard normal distribution

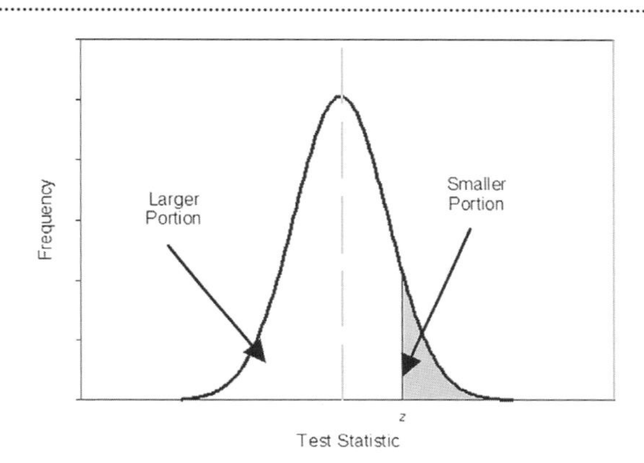

z	Larger Portion	Smaller Portion	y
.00	.50000	.50000	.3989
.01	.50399	.49601	.3989
.02	.50798	.49202	.3989
.03	.51197	.48803	.3988
.04	.51595	.48405	.3986
.05	.51994	.48006	.3984
.06	.52392	.47608	.3982
.07	.52790	.47210	.3980
.08	.53188	.46812	.3977
.09	.53586	.46414	.3973
.10	.53983	.46017	.3970
.11	.54380	.45620	.3965
.12	.54776	.45224	.3961
.13	.55172	.44828	.3956
.14	.55567	.44433	.3951
.15	.55962	.44038	.3945
.16	.56356	.43644	.3939
.17	.56749	.43251	.3932
.18	.57142	.42858	.3925
.19	.57535	.42465	.3918
.20	.57926	.42074	.3910
.21	.58317	.41683	.3902
.22	.58706	.41294	.3894
.23	.59095	.40905	.3885

(Continued)

(Continued)

z	Larger Portion	Smaller Portion	y
.24	.59483	.40517	.3876
.25	.59871	.40129	.3867
.26	.60257	.39743	.3857
.27	.60642	.39358	.3847
.28	.61026	.38974	.3836
.29	.61409	.38591	.3825
.30	.61791	.38209	.3814
.31	.62172	.37828	.3802
.32	.62552	.37448	.3790
.33	.62930	.37070	.3778
.34	.63307	.36693	.3765
.35	.63683	.36317	.3752
.36	.64058	.35942	.3739
.37	.64431	.35569	.3725
.38	.64803	.35197	.3712
.39	.65173	.34827	.3697
.40	.65542	.34458	.3683
.41	.65910	.34090	.3668
.42	.66276	.33724	.3653
.43	.66640	.33360	.3637
.44	.67003	.32997	.3621
.45	.67364	.32636	.3605
.46	.67724	.32276	.3589
.47	.68082	.31918	.3572
.48	.68439	.31561	.3555
.49	.68793	.31207	.3538
.50	.69146	.30854	.3521
.51	.69497	.30503	.3503
.52	.69847	.30153	.3485
.53	.70194	.29806	.3467

z	Larger Portion	Smaller Portion	y
.54	.70540	.29460	.3448
.55	.70884	.29116	.3429
.56	.71226	.28774	.3410
.57	.71566	.28434	.3391
.58	.71904	.28096	.3372
.59	.72240	.27760	.3352
.60	.72575	.27425	.3332
.61	.72907	.27093	.3312
.62	.73237	.26763	.3292
.63	.73565	.26435	.3271
.64	.73891	.26109	.3251
.65	.74215	.25785	.3230
.66	.74537	.25463	.3209
.67	.74857	.25143	.3187
.68	.75175	.24825	.3166
.69	.75490	.24510	.3144
.70	.75804	.24196	.3123
.71	.76115	.23885	.3101
.72	.76424	.23576	.3079
.73	.76730	.23270	.3056
.74	.77035	.22965	.3034
.75	.77337	.22663	.3011
.76	.77637	.22363	.2989
.77	.77935	.22065	.2966
.78	.78230	.21770	.2943
.79	.78524	.21476	.2920
.80	.78814	.21186	.2897
.81	.79103	.20897	.2874
.82	.79389	.20611	.2850
.83	.79673	.20327	.2827

z	Larger Portion	Smaller Portion	y
.84	.79955	.20045	.2803
.85	.80234	.19766	.2780
.86	.80511	.19489	.2756
.87	.80785	.19215	.2732
.88	.81057	.18943	.2709
.89	.81327	.18673	.2685
.90	.81594	.18406	.2661
.91	.81859	.18141	.2637
.92	.82121	.17879	.2613
.93	.82381	.17619	.2589
.94	.82639	.17361	.2565
.95	.82894	.17106	.2541
.96	.83147	.16853	.2516
.97	.83398	.16602	.2492
.98	.83646	.16354	.2468
.99	.83891	.16109	.2444
1.00	.84134	.15866	.2420
1.01	.84375	.15625	.2396
1.02	.84614	.15386	.2371
1.03	.84849	.15151	.2347
1.04	.85083	.14917	.2323
1.05	.85314	.14686	.2299
1.06	.85543	.14457	.2275
1.07	.85769	.14231	.2251
1.08	.85993	.14007	.2227
1.09	.86214	.13786	.2203
1.10	.86433	.13567	.2179
1.11	.86650	.13350	.2155
1.12	.86864	.13136	.2131
1.13	.87076	.12924	.2107

z	Larger Portion	Smaller Portion	y
1.14	.87286	.12714	.2083
1.15	.87493	.12507	.2059
1.16	.87698	.12302	.2036
1.17	.87900	.12100	.2012
1.18	.88100	.11900	.1989
1.19	.88298	.11702	.1965
1.20	.88493	.11507	.1942
1.21	.88686	.11314	.1919
1.22	.88877	.11123	.1895
1.23	.89065	.10935	.1872
1.24	.89251	.10749	.1849
1.25	.89435	.10565	.1826
1.26	.89617	.10383	.1804
1.27	.89796	.10204	.1781
1.28	.89973	.10027	.1758
1.29	.90147	.09853	.1736
1.30	.90320	.09680	.1714
1.31	.90490	.09510	.1691
1.32	.90658	.09342	.1669
1.33	.90824	.09176	.1647
1.34	.90988	.09012	.1626
1.35	.91149	.08851	.1604
1.36	.91309	.08691	.1582
1.37	.91466	.08534	.1561
1.38	.91621	.08379	.1539
1.39	.91774	.08226	.1518
1.40	.91924	.08076	.1497
1.41	.92073	.07927	.1476
1.42	.92220	.07780	.1456
1.43	.92364	.07636	.1435

(Continued)

(Continued)

z	Larger Portion	Smaller Portion	y
1.44	.92507	.07493	.1415
1.45	.92647	.07353	.1394
1.46	.92785	.07215	.1374
1.47	.92922	.07078	.1354
1.48	.93056	.06944	.1334
1.49	.93189	.06811	.1315
1.50	.93319	.06681	.1295
1.51	.93448	.06552	.1276
1.52	.93574	.06426	.1257
1.53	.93699	.06301	.1238
1.54	.93822	.06178	.1219
1.55	.93943	.06057	.1200
1.56	.94062	.05938	.1182
1.57	.94179	.05821	.1163
1.58	.94295	.05705	.1145
1.59	.94408	.05592	.1127
1.60	.94520	.05480	.1109
1.61	.94630	.05370	.1092
1.62	.94738	.05262	.1074
1.63	.94845	.05155	.1057
1.64	.94950	.05050	.1040
1.65	.95053	.04947	.1023
1.66	.95154	.04846	.1006
1.67	.95254	.04746	.0989
1.68	.95352	.04648	.0973
1.69	.95449	.04551	.0957
1.70	.95543	.04457	.0940
1.71	.95637	.04363	.0925
1.72	.95728	.04272	.0909
1.73	.95818	.04182	.0893

z	Larger Portion	Smaller Portion	y
1.74	.95907	.04093	.0878
1.75	.95994	.04006	.0863
1.76	.96080	.03920	.0848
1.77	.96164	.03836	.0833
1.78	.96246	.03754	.0818
1.79	.96327	.03673	.0804
1.80	.96407	.03593	.0790
1.81	.96485	.03515	.0775
1.82	.96562	.03438	.0761
1.83	.96638	.03362	.0748
1.84	.96712	.03288	.0734
1.85	.96784	.03216	.0721
1.86	.96856	.03144	.0707
1.87	.96926	.03074	.0694
1.88	.96995	.03005	.0681
1.89	.97062	.02938	.0669
1.90	.97128	.02872	.0656
1.91	.97193	.02807	.0644
1.92	.97257	.02743	.0632
1.93	.97320	.02680	.0620
1.94	.97381	.02619	.0608
1.95	.97441	.02559	.0596
1.96	.97500	.02500	.0584
1.97	.97558	.02442	.0573
1.98	.97615	.02385	.0562
1.99	.97670	.02330	.0551
2.00	.97725	.02275	.0540
2.01	.97778	.02222	.0529
2.02	.97831	.02169	.0519
2.03	.97882	.02118	.0508

z	Larger Portion	Smaller Portion	y
2.04	.97932	.02068	.0498
2.05	.97982	.02018	.0488
2.06	.98030	.01970	.0478
2.07	.98077	.01923	.0468
2.08	.98124	.01876	.0459
2.09	.98169	.01831	.0449
2.10	.98214	.01786	.0440
2.11	.98257	.01743	.0431
2.12	.98300	.01700	.0422
2.13	.98341	.01659	.0413
2.14	.98382	.01618	.0404
2.15	.98422	.01578	.0396
2.16	.98461	.01539	.0387
2.17	.98500	.01500	.0379
2.18	.98537	.01463	.0371
2.19	.98574	.01426	.0363
2.20	.98610	.01390	.0355
2.21	.98645	.01355	.0347
2.22	.98679	.01321	.0339
2.23	.98713	.01287	.0332
2.24	.98745	.01255	0325
2.25	.98778	.01222	.0317
2.26	.98809	.01191	.0310
2.27	.98840	.01160	.0303
2.28	.98870	.01130	.0297
2.29	.98899	.01101	.0290
2.30	.98928	.01072	.0283
2.31	.98956	.01044	.0277
2.32	.98983	.01017	.0270
2.33	.99010	.00990	.0264

z	Larger Portion	Smaller Portion	y
2.34	.99036	.00964	.0258
2.35	.99061	.00939	.0252
2.36	.99086	.00914	.0246
2.37	.99111	.00889	.0241
2.38	.99134	.00866	.0235
2.39	.99158	.00842	.0229
2.40	.99180	.00820	.0224
2.41	.99202	.00798	.0219
2.42	.99224	.00776	.0213
2.43	.99245	.00755	.0208
2.44	.99266	.00734	.0203
2.45	.99286	.00714	.0198
2.46	.99305	.00695	.0194
2.47	.99324	.00676	.0189
2.48	.99343	.00657	.0184
2.49	.99361	.00639	.0180
2.50	.99379	.00621	.0175
2.51	.99396	.00604	.0171
2.52	.99413	.00587	.0167
2.53	.99430	.00570	.0163
2.54	.99446	.00554	.0158
2.55	.99461	.00539	.0154
2.56	.99477	.00523	.0151
2.57	.99492	.00508	.0147
2.58	.99506	.00494	.0143
2.59	.99520	.00480	.0139
2.60	.99534	.00466	.0136
2.61	.99547	.00453	.0132
2.62	.99560	.00440	.0129
2.63	.99573	.00427	.0126

(Continued)

(Continued)

z	Larger Portion	Smaller Portion	y
2.64	.99585	.00415	.0122
2.65	.99598	.00402	.0119
2.66	.99609	.00391	.0116
2.67	.99621	.00379	.0113
2.68	.99632	.00368	.0110
2.69	.99643	.00357	.0107
2.70	.99653	.00347	.0104
2.71	.99664	.00336	.0101
2.72	.99674	.00326	.0099
2.73	.99683	.00317	.0096
2.74	.99693	.00307	.0093
2.75	.99702	.00298	.0091
2.76	.99711	.00289	.0088
2.77	.99720	.00280	.0086
2.78	.99728	.00272	.0084
2.79	.99736	.00264	.0081
2.80	.99744	.00256	.0079
2.81	.99752	.00248	.0077
2.82	.99760	.00240	.0075
2.83	.99767	.00233	.0073
2.84	.99774	.00226	.0071
2.85	.99781	.00219	.0069

z	Larger Portion	Smaller Portion	y
2.86	.99788	.00212	.0067
2.87	.99795	.00205	.0065
2.88	.99801	.00199	.0063
2.89	.99807	.00193	.0061
2.90	.99813	.00187	.0060
2.91	.99819	.00181	.0058
2.92	.99825	.00175	.0056
2.93	.99831	.00169	.0055
2.94	.99836	.00164	.0053
2.95	.99841	.00159	.0051
2.96	.99846	.00154	.0050
2.97	.99851	.00149	.0048
2.98	.99856	.00144	.0047
2.99	.99861	.00139	.0046
3.00	.99865	.00135	.0044
⋮	⋮	⋮	⋮
3.25	.99942	.00058	.0020
⋮	⋮	⋮	⋮
3.50	.99977	.00023	.0009
⋮	⋮	⋮	⋮
4.00	.99997	.00003	.0001

A.2. Critical values of the *t*-distribution

	Two-Tailed Test		One-Tailed Test	
df	0.05	0.01	0.05	0.01
1	12.71	63.66	6.31	31.82
2	4.30	9.92	2.92	6.96
3	3.18	5.84	2.35	4.54
4	2.78	4.60	2.13	3.75
5	2.57	4.03	2.02	3.36
6	2.45	3.71	1.94	3.14
7	2.36	3.50	1.89	3.00
8	2.31	3.36	1.86	2.90
9	2.26	3.25	1.83	2.82
10	2.23	3.17	1.81	2.76
11	2.20	3.11	1.80	2.72
12	2.18	3.05	1.78	2.68
13	2.16	3.01	1.77	2.65
14	2.14	2.98	1.76	2.62
15	2.13	2.95	1.75	2.60
16	2.12	2.92	1.75	2.58
17	2.11	2.90	1.74	2.57
18	2.10	2.88	1.73	2.55
19	2.09	2.86	1.73	2.54
20	2.09	2.85	1.72	2.53
21	2.08	2.83	1.72	2.52
22	2.07	2.82	1.72	2.51
23	2.07	2.81	1.71	2.50
24	2.06	2.80	1.71	2.49
25	2.06	2.79	1.71	2.49
26	2.06	2.78	1.71	2.48
27	2.05	2.77	1.70	2.47
28	2.05	2.76	1.70	2.47
29	2.05	2.76	1.70	2.46
30	2.04	2.75	1.70	2.46
35	2.03	2.72	1.69	2.44
40	2.02	2.70	1.68	2.42
45	2.01	2.69	1.68	2.41
50	2.01	2.68	1.68	2.40
60	2.00	2.66	1.67	2.39
70	1.99	2.65	1.67	2.38
80	1.99	2.64	1.66	2.37
90	1.99	2.63	1.66	2.37
100	1.98	2.63	1.66	2.36
∞ (z)	1.96	2.58	1.64	2.33

All values computed by the author using SPSS.

A.3. Critical values of the *F*-distribution

df (Denominator)	p	df (Numerator) 1	2	3	4	5	6	7	8	9	10
1	.05	161.45	199.50	215.71	224.58	230.16	233.99	236.77	238.88	240.54	241.88
	.01	4052.18	4999.50	5403.35	5624.58	5763.65	5858.99	5928.36	5981.07	6022.47	6055.85
2	.05	18.51	19.00	19.16	19.25	19.30	19.33	19.35	19.37	19.38	19.40
	.01	98.50	99.00	99.17	99.25	99.30	99.33	99.36	99.37	99.39	99.40
3	.05	10.13	9.55	9.28	9.12	9.01	8.94	8.89	8.85	8.81	8.79
	.01	34.12	30.82	29.46	28.71	28.24	27.91	27.67	27.49	27.35	27.23
4	.05	7.71	6.94	6.59	6.39	6.26	6.16	6.09	6.04	6.00	5.96
	.01	21.20	18.00	16.69	15.98	15.52	15.21	14.98	14.80	14.66	14.55
5	.05	6.61	5.79	5.41	5.19	5.05	4.95	4.88	4.82	4.77	4.74
	.01	16.26	13.27	12.06	11.39	10.97	10.67	10.46	10.29	10.16	10.05
6	.05	5.99	5.14	4.76	4.53	4.39	4.28	4.21	4.15	4.10	4.06
	.01	13.75	10.92	9.78	9.15	8.75	8.47	8.26	8.10	7.98	7.87
7	.05	5.59	4.74	4.35	4.12	3.97	3.87	3.79	3.73	3.68	3.64
	.01	12.25	9.55	8.45	7.85	7.46	7.19	6.99	6.84	6.72	6.62
8	.05	5.32	4.46	4.07	3.84	3.69	3.58	3.50	3.44	3.39	3.35
	.01	11.26	8.65	7.59	7.01	6.63	6.37	6.18	6.03	5.91	5.81
9	.05	5.12	4.26	3.86	3.63	3.48	3.37	3.29	3.23	3.18	3.14
	.01	10.56	8.02	6.99	6.42	6.06	5.80	5.61	5.47	5.35	5.26
10	.05	4.96	4.10	3.71	3.48	3.33	3.22	3.14	3.07	3.02	2.98
	.01	10.04	7.56	6.55	5.99	5.64	5.39	5.20	5.06	4.94	4.85
11	.05	4.84	3.98	3.59	3.36	3.20	3.09	3.01	2.95	2.90	2.85
	.01	9.65	7.21	6.22	5.67	5.32	5.07	4.89	4.74	4.63	4.54
12	.05	4.75	3.89	3.49	3.26	3.11	3.00	2.91	2.85	2.80	2.75
	.01	9.33	6.93	5.95	5.41	5.06	4.82	4.64	4.50	4.39	4.30
13	.05	4.67	3.81	3.41	3.18	3.03	2.92	2.83	2.77	2.71	2.67
	.01	9.07	6.70	5.74	5.21	4.86	4.62	4.44	4.30	4.19	4.10
14	.05	4.60	3.74	3.34	3.11	2.96	2.85	2.76	2.70	2.65	2.60
	.01	8.86	6.51	5.56	5.04	4.69	4.46	4.28	4.14	4.03	3.94
15	.05	4.54	3.68	3.29	3.06	2.90	2.79	2.71	2.64	2.59	2.54
	.01	8.68	6.36	5.42	4.89	4.56	4.32	4.14	4.00	3.89	3.80
16	.05	4.49	3.63	3.24	3.01	2.85	2.74	2.66	2.59	2.54	2.49
	.01	8.53	6.23	5.29	4.77	4.44	4.20	4.03	3.89	3.78	3.69
17	.05	4.45	3.59	3.20	2.96	2.81	2.70	2.61	2.55	2.49	2.45
	.01	8.40	6.11	5.18	4.67	4.34	4.10	3.93	3.79	3.68	3.59
18	.05	4.41	3.55	3.16	2.93	2.77	2.66	2.58	2.51	2.46	2.41
	.01	8.29	6.01	5.09	4.58	4.25	4.01	3.84	3.71	3.60	3.51

		df (Numerator)									
df (Denominator)	p	1	2	3	4	5	6	7	8	9	10
19	.05	4.38	3.52	3.13	2.90	2.74	2.63	2.54	2.48	2.42	2.38
	.01	8.18	5.93	5.01	4.50	4.17	3.94	3.77	3.63	3.52	3.43
20	.05	4.35	3.49	3.10	2.87	2.71	2.60	2.51	2.45	2.39	2.35
	.01	8.10	5.85	4.94	4.43	4.10	3.87	3.70	3.56	3.46	3.37
22	.05	4.30	3.44	3.05	2.82	2.66	2.55	2.46	2.40	2.34	2.30
	.01	7.95	5.72	4.82	4.31	3.99	3.76	3.59	3.45	3.35	3.26
24	.05	4.26	3.40	3.01	2.78	2.62	2.51	2.42	2.36	2.30	2.25
	.01	7.82	5.61	4.72	4.22	3.90	3.67	3.50	3.36	3.26	3.17
26	.05	4.23	3.37	2.98	2.74	2.59	2.47	2.39	2.32	2.27	2.22
	.01	7.72	5.53	4.64	4.14	3.82	3.59	3.42	3.29	3.18	3.09
28	.05	4.20	3.34	2.95	2.71	2.56	2.45	2.36	2.29	2.24	2.19
	.01	7.64	5.45	4.57	4.07	3.75	3.53	3.36	3.23	3.12	3.03
30	.05	4.17	3.32	2.92	2.69	2.53	2.42	2.33	2.27	2.21	2.16
	.01	7.56	5.39	4.51	4.02	3.70	3.47	3.30	3.17	3.07	2.98
35	.05	4.12	3.27	2.87	2.64	2.49	2.37	2.29	2.22	2.16	2.11
	.01	7.42	5.27	4.40	3.91	3.59	3.37	3.20	3.07	2.96	2.88
40	.05	4.08	3.23	2.84	2.61	2.45	2.34	2.25	2.18	2.12	2.08
	.01	7.31	5.18	4.31	3.83	3.51	3.29	3.12	2.99	2.89	2.80
45	.05	4.06	3.20	2.81	2.58	2.42	2.31	2.22	2.15	2.10	2.05
	.01	7.23	5.11	4.25	3.77	3.45	3.23	3.07	2.94	2.83	2.74
50	.05	4.03	3.18	2.79	2.56	2.40	2.29	2.20	2.13	2.07	2.03
	.01	7.17	5.06	4.20	3.72	3.41	3.19	3.02	2.89	2.78	2.70
60	.05	4.00	3.15	2.76	2.53	2.37	2.25	2.17	2.10	2.04	1.99
	.01	7.08	4.98	4.13	3.65	3.34	3.12	2.95	2.82	2.72	2.63
80	.05	3.96	3.11	2.72	2.49	2.33	2.21	2.13	2.06	2.00	1.95
	.01	6.96	4.88	4.04	3.56	3.26	3.04	2.87	2.74	2.64	2.55
100	.05	3.94	3.09	2.70	2.46	2.31	2.19	2.10	2.03	1.97	1.93
	.01	6.90	4.82	3.98	3.51	3.21	2.99	2.82	2.69	2.59	2.50
150	.05	3.90	3.06	2.66	2.43	2.27	2.16	2.07	2.00	1.94	1.89
	.01	6.81	4.75	3.91	3.45	3.14	2.92	2.76	2.63	2.53	2.44
300	.05	3.87	3.03	2.63	2.40	2.24	2.13	2.04	1.97	1.91	1.86
	.01	6.72	4.68	3.85	3.38	3.08	2.86	2.70	2.57	2.47	2.38
500	.05	3.86	3.01	2.62	2.39	2.23	2.12	2.03	1.96	1.90	1.85
	.01	6.69	4.65	3.82	3.36	3.05	2.84	2.68	2.55	2.44	2.36
1000	.05	3.85	3.00	2.61	2.38	2.22	2.11	2.02	1.95	1.89	1.84
	.01	6.66	4.63	3.80	3.34	3.04	2.82	2.66	2.53	2.43	2.34

(Continued)

(Continued)

df (Denominator)	p	15	20	25	30	40	50	1000
1	.05	245.95	248.01	249.26	250.10	251.14	251.77	254.19
	.01	6157.31	6208.74	6239.83	6260.65	6286.79	6302.52	6362.70
2	.05	19.43	19.45	19.46	19.46	19.47	19.48	19.49
	.01	99.43	99.45	99.46	99.47	99.47	99.48	99.50
3	.05	8.70	8.66	8.63	8.62	8.59	8.58	8.53
	.01	26.87	26.69	26.58	26.50	26.41	26.35	26.14
4	.05	5.86	5.80	5.77	5.75	5.72	5.70	5.63
	.01	14.20	14.02	13.91	13.84	13.75	13.69	13.47
5	05	4.62	4.56	4.52	4.50	4.46	4.44	4.37
	.01	9.72	9.55	9.45	9.38	9.29	9.24	9.03
6	.05	3.94	3.87	3.83	3.81	3.77	3.75	3.67
	.01	7.56	7.40	7.30	7.23	7.14	7.09	6.89
7	.05	3.51	3.44	3.40	3.38	3.34	3.32	3.23
	.01	6.31	6.16	6.06	5.99	5.91	5.86	5.66
8	.05	3.22	3.15	3.11	3.08	3.04	3.02	2.93
	.01	5.52	5.36	5.26	5.20	5.12	5.07	4.87
9	.05	3.01	2.94	2.89	2.86	2.83	2.80	2.71
	.01	4.96	4.81	4.71	4.65	4.57	4.52	4.32
10	.05	2.85	2.77	2.73	2.70	2.66	2.64	2.54
	.01	4.56	4.41	4.31	4.25	4.17	4.12	3.92
11	.05	2.72	2.65	2.60	2.57	2.53	2.51	2.41
	.01	4.25	4.10	4.01	3.94	3.86	3.81	3.61
12	.05	2.62	2.54	2.50	2.47	2.43	2.40	2.30
	.01	4.01	3.86	3.76	3.70	3.62	3.57	3.37
13	.05	2.53	2.46	2.41	2.38	2.34	2.31	2.21
	.01	3.82	3.66	3.57	3.51	3.43	3.38	3.18
14	.05	2.46	2.39	2.34	2.31	2.27	2.24	2.14
	.01	3.66	3.51	3.41	3.35	3.27	3.22	3.02
15	.05	2.40	2.33	2.28	2.25	2.20	2.18	2.07
	.01	3.52	3.37	3.28	3.21	3.13	3.08	2.88
16	.05	2.35	2.28	2.23	2.19	2.15	2.12	2.02
	.01	3.41	3.26	3.16	3.10	3.02	2.97	2.76
17	.05	2.31	2.23	2.18	2.15	2.10	2.08	1.97
	.01	3.31	3.16	3.07	3.00	2.92	2.87	2.66
18	.05	2.27	2.19	2.14	2.11	2.06	2.04	1.92
	.01	3.23	3.08	2.98	2.92	2.84	2.78	2.58

Column headers 15–1000: df (Numerator)

df (Denominator)	p	df (Numerator) 15	20	25	30	40	50	1000
19	0.05	2.23	2.16	2.11	2.07	2.03	2.00	1.88
	0.01	3.15	3.00	2.91	2.84	2.76	2.71	2.50
20	0.05	2.20	2.12	2.07	2.04	1.99	1.97	1.85
	0.01	3.09	2.94	2.84	2.78	2.69	2.64	2.43
22	0.05	2.15	2.07	2.02	1.98	1.94	1.91	1.79
	0.01	2.98	2.83	2.73	2.67	2.58	2.53	2.32
24	0.05	2.11	2.03	1.97	1.94	1.89	1.86	1.74
	0.01	2.89	2.74	2.64	2.58	2.49	2.44	2.22
26	0.05	2.07	1.99	1.94	1.90	1.85	1.82	1.70
	0.01	2.81	2.66	2.57	2.50	2.42	2.36	2.14
28	0.05	2.04	1.96	1.91	1.87	1.82	1.79	1.66
	0.01	2.75	2.60	2.51	2.44	2.35	2.30	2.08
30	0.05	2.01	1.93	1.88	1.84	1.79	1.76	1.63
	0.01	2.70	2.55	2.45	2.39	2.30	2.25	2.02
35	0.05	1.96	1.88	1.82	1.79	1.74	1.70	1.57
	0.01	2.60	2.44	2.35	2.28	2.19	2.14	1.90
40	0.05	1.92	1.84	1.78	1.74	1.69	1.66	1.52
	0.01	2.52	2.37	2.27	2.20	2.11	2.06	1.82
45	0.05	1.89	1.81	1.75	1.71	1.66	1.63	1.48
	0.01	2.46	2.31	2.21	2.14	2.05	2.00	1.75
50	0.05	1.87	1.78	1.73	1.69	1.63	1.60	1.45
	0.01	2.42	2.27	2.17	2.10	2.01	1.95	1.70
60	0.05	1.84	1.75	1.69	1.65	1.59	1.56	1.40
	0.01	2.35	2.20	2.10	2.03	1.94	1.88	1.62
80	0.05	1.79	1.70	1.64	1.60	1.54	1.51	1.34
	0.01	2.27	2.12	2.01	1.94	1.85	1.79	1.51
100	0.05	1.77	1.68	1.62	1.57	1.52	1.48	1.30
	0.01	2.22	2.07	1.97	1.89	1.80	1.74	1.45
150	0.05	1.73	1.64	1.58	1.54	1.48	1.44	1.24
	0.01	2.16	2.00	1.90	1.83	1.73	1.66	1.35
300	0.05	1.70	1.61	1.54	1.50	1.43	1.39	1.17
	.01	2.10	1.94	1.84	1.76	1.66	1.59	1.25
500	.05	1.69	1.59	1.53	1.48	1.42	1.38	1.14
	.01	2.07	1.92	1.81	1.74	1.63	1.57	1.20
1000	.05	1.68	1.58	1.52	1.47	1.41	1.36	1.11
	.01	2.06	1.90	1.79	1.72	1.61	1.54	1.16

All values computed by the author using SPSS.

A.4. Critical values of the chi-square distribution

df	p 0.05	p 0.01
1	3.84	6.63
2	5.99	9.21
3	7.81	11.34
4	9.49	13.28
5	11.07	15.09
6	12.59	16.81
7	14.07	18.48
8	15.51	20.09
9	16.92	21.67
10	18.31	23.21
11	19.68	24.72
12	21.03	26.22
13	22.36	27.69
14	23.68	29.14
15	25.00	30.58
16	26.30	32.00
17	27.59	33.41
18	28.87	34.81
19	30.14	36.19
20	31.41	37.57
21	32.67	38.93
22	33.92	40.29
23	35.17	41.64
24	36.42	42.98

df	p 0.05	p 0.01
25	37.65	44.31
26	38.89	45.64
27	40.11	46.96
28	41.34	48.28
29	42.56	49.59
30	43.77	50.89
35	49.80	57.34
40	55.76	63.69
45	61.66	69.96
50	67.50	76.15
60	79.08	88.38
70	90.53	100.43
80	101.88	112.33
90	113.15	124.12
100	124.34	135.81
200	233.99	249.45
300	341.40	359.91
400	447.63	468.72
500	553.13	576.49
600	658.09	683.52
700	762.66	789.97
800	866.91	895.98
900	970.90	1001.63
1000	1074.68	1106.97

All values computed by author using SPSS.

REFERENCES

Algina, J., & Olejnik, S. F. (1984). Implementing the Welch-James procedure with factorial designs. *Educational and Psychological Measurement, 44*, 39–48.

Arrindell, W. A., & van der Ende, J. (1985). An empirical test of the utility of the observer-to-variables ratio in factor and components analysis. *Applied Psychological Measurement, 9*, 165–178.

Baguley, T. (2004). Understanding statistical power in the context of applied research. *Applied Ergonomics, 35*(2), 73–80.

Bale, C., Morrison, R., & Caryl, P. G. (2006). Chat-up lines as male sexual displays. *Personality and Individual Differences, 40*(4), 655–664.

Bargman, R. E. (1970). *Interpretation and use of a generalized discriminant function*. In R. C. Bose et al. (Eds.), *Essays in probability and statistics*. Chapel Hill: University of North Carolina Press.

Barnard, G. A. (1963). Ronald Aylmer Fisher, 1890–1962: Fisher's contributions to mathematical statistics. *Journal of the Royal Statistical Society, Series A, 126*, 162–166.

Barnett, V., & Lewis, T. (1978). *Outliers in statistical data*. New York: Wiley.

Beckham, A. S. (1929). Is the Negro happy? A psychological analysis. *Journal of Abnormal and Social Psychology, 24*, 186–190.

Belsey, D. A., Kuh, E., & Welsch, R. (1980). *Regression diagnostics: identifying influential data and sources of collinearity*. New York: Wiley.

Bemelman, M., & Hammacher, E. R. (2005). Rectal impalement by pirate ship: A case report. *Injury Extra, 36*, 508–510.

Berger, J. O. (2003). Could Fisher, Jeffreys and Neyman have agreed on testing? *Statistical Science, 18*(1), 1–12.

Berry, W. D. (1993). *Understanding regression assumptions*. Sage University Paper Series on Quantitative Applications in the Social Sciences, 07-092. Newbury Park, CA: Sage.

Berry, W. D., & Feldman, S. (1985). *Multiple regression in practice*. Sage University Paper Series on Quantitative Applications in the Social Sciences, 07-050. Beverly Hills, CA: Sage.

Board, B. J., & Fritzon, K. (2005). Disordered personalities at work. *Psychology, Crime & Law, 11*(1), 17–32.

Bock, R. D. (1975). *Multivariate statistical methods in behavioural research*. New York: McGraw-Hill.

Bowerman, B. L., & O'Connell, R. T. (1990). *Linear statistical models: An applied approach* (2nd ed.). Belmont, CA: Duxbury.

Bray, J. H., & Maxwell, S. E. (1985). *Multivariate analysis of variance*. Sage University Paper Series on Quantitative Applications in the Social Sciences, 07-054. Newbury Park, CA: Sage.

Brown, M. B., & Forsythe, A. B. (1974). The small sample behaviour of some statistics which test the equality of several means. *Technometrics, 16*, 129–132.

Brown, W. (1910). Some experimental results in the correlation of mental abilities. *British Journal of Psychology, 3*, 296–322.

Budescu, D. V. (1982). The power of the F test in normal populations with heterogeneous variances. *Educational and Psychological Measurement, 42*, 609–616.

Budescu, D. V., & Appelbaum, M. I. (1981). Variance stabilizing transformations and the power of the F test. *Journal of Educational Statistics, 6*(1), 55–74.

Cattell, R. B. (1966a). *The scientific analysis of personality*. Chicago: Aldine.

Cattell, R. B. (1966b). The scree test for the number of factors. *Multivariate Behavioral Research, 1*, 245–276.

Çetinkaya, H., & Domjan, M. (2006). Sexual fetishism in a quail (*Coturnix japonica*) model system: Test of reproductive success. *Journal of Comparative Psychology, 120*(4), 427–432.

Chamorro-Premuzic, T., Furnham, A., Christopher, A. N., Garwood, J., & Martin, N. (2008). Birds of a feather: Students' preferences for lecturers' personalities as predicted by their own personality and learning approaches. *Personality and Individual Differences, 44*, 965–976.

Chen, P. Y., & Popovich, P. M. (2002). *Correlation: Parametric and nonparametric measures*. Thousand Oaks, CA: Sage.

Clarke, D. L., Buccimazza, I., Anderson, F. A., & Thomson, S. R. (2005). Colorectal foreign bodies. *Colorectal Disease, 7*(1), 98–103.

Cliff, N. (1987). *Analyzing multivariate data*. New York: Harcourt Brace Jovanovich.

Cohen, J. (1968). Multiple regression as a general data-analytic system. *Psychological Bulletin, 70*(6), 426–443.

Cohen, J. (1988). *Statistical power analysis for the behavioural sciences* (2nd ed.). New York: Academic Press.

Cohen, J. (1990). Things I have learned (so far). *American Psychologist, 45*(12), 1304–1312.

Cohen, J. (1992). A power primer. *Psychological Bulletin, 112*(1), 155–159.

Cohen, J. (1994). The earth is round ($p < .05$). *American Psychologist, 49*(12), 997–1003.

Cole, D. A., Maxwell, S. E., Arvey, R., & Salas, E. (1994). How the power of MANOVA can both increase and decrease as a function of the intercorrelations among the dependent variables. *Psychological Bulletin, 115*(3), 465–474.

Collier, R. O., Baker, F. B., Mandeville, G. K., & Hayes, T. F. (1967). Estimates of test size for several test procedures based on conventional variance ratios in the repeated measures design. *Psychometrika, 32*(2), 339–352.

Comrey, A. L., & Lee, H. B. (1992). *A first course in factor analysis* (2nd ed.). Hillsdale, NJ: Erlbaum.

Cook, R. D., & Weisberg, S. (1982). *Residuals and influence in regression*. New York: Chapman & Hall.

Cook, S. A., Rosser, R., & Salmon, P. (2006). Is cosmetic surgery an effective psychotherapeutic intervention? A systematic review of the evidence. *Journal of Plastic, Reconstructive & Aesthetic Surgery, 59*, 1133–1151.

Cook, S. A., Rosser, R., Toone, H., James, M. I., & Salmon, P. (2006). The psychological and social characteristics of patients referred for NHS cosmetic surgery: Quantifying clinical need. *Journal of Plastic, Reconstructive & Aesthetic Surgery 59*, 54–64.

Cooper, C. L., Sloan, S. J., & Williams, S. (1988). *Occupational Stress Indicator Management Guide*. Windsor: NFER-Nelson.

Cooper, M., O'Donnell, D., Caryl, P. G., Morrison, R., & Bale, C. (2007). Chat-up lines as male displays: Effects of content, sex, and personality. *Personality and Individual Differences, 43*(5), 1075–1085.

Cortina, J. M. (1993). What is coefficient alpha? An examination of theory and applications. *Journal of Applied Psychology, 78*, 98–104.

Cox, D. R., & Snell, D. J. (1989). *The analysis of binary data* (2nd ed.). London: Chapman & Hall.

Cronbach, L. J. (1951). Coefficient alpha and the internal structure of tests. *Psychometrika, 16*, 297–334.

Cronbach, L. J. (1957). The two disciplines of scientific psychology. *American Psychologist, 12*, 671–684.

Dalgaard, P. (2002). *Introductory Statistics with R.* New York: Springer.

Davey, G. C. L., Startup, H. M., Zara, A., MacDonald, C. B., & Field, A. P. (2003). Perseveration of checking thoughts and mood-as-input hypothesis. *Journal of Behavior Therapy & Experimental Psychiatry, 34*, 141–160.

Davidson, M. L. (1972). Univariate versus multivariate tests in repeated-measures experiments. *Psychological Bulletin, 77*, 446–452.

DeCarlo, L. T. (1997). On the meaning and use of kurtosis. *Psychological Methods, 2*(3), 292–307.

Domjan, M., Blesbois, E., & Williams, J. (1998). The adaptive significance of sexual conditioning: Pavlovian control of sperm release. *Psychological Science, 9*(5), 411–415.

Donaldson, T. S. (1968). Robustness of the *F*-test to errors of both kinds and the correlation between the numerator and denominator of the *F*-ratio. *Journal of the American Statistical Association, 63*, 660–676.

Dunlap, W. P., Cortina, J. M., Vaslow, J. B., & Burke, M. J. (1996). Meta-analysis of experiments with matched groups or repeated measures designs. *Psychological Methods, 1*(2), 170–177.

Dunteman, G. E. (1989). *Principal components analysis*. Sage University Paper Series on Quantitative Applications in the Social Sciences, 07-069. Newbury Park, CA: Sage.

Durbin, J., & Watson, G. S. (1951). Testing for serial correlation in least squares regression, II. *Biometrika, 30*, 159–178.

Easterlin, R. A. (2003). Explaining happiness. *Proceedings of the National Academy of Sciences., 100*(19), 11176–11183.

Efron, B., & Tibshirani, R. (1993). *An introduction to the bootstrap.* New York: Chapman & Hall.

Eriksson, S.-G., Beckham, D., & Vassell, D. (2004). Why are the English so shit at penalties? A review. *Journal of Sporting Ineptitude, 31*, 231–1072.

Erlebacher, A. (1977). Design and analysis of experiments contrasting the within- and between-subjects manipulations of the independent variable. *Psychological Bulletin, 84*, 212–219.

Eysenck, H. J. (1953). *The structure of human personality*. New York: Wiley.

Fesmire, F. M. (1988). Termination of intractable hiccups with digital rectal massage. *Annals of Emergency Medicine, 17*(8), 872.

Field, A. P. (2000). *Discovering statistics using SPSS for Windows: Advanced techniques for the beginner*. London: Sage.

Field, A. P. (2001). Meta-analysis of correlation coefficients: A Monte Carlo comparison of fixed- and random-effects methods. *Psychological Methods, 6*(2), 161–180.

Field, A. P. (2005a). Intraclass correlation. In B. Everitt & D. C. Howell (Eds.), *Encyclopedia of statistics in behavioral science* (Vol. 2, pp. 948–954). Hoboken, NJ: Wiley.

Field, A. P. (2005b). Is the meta-analysis of correlation coefficients accurate when population correlations vary? *Psychological Methods, 10*(4), 444–467.

Field, A. P. (2005c). Learning to like (and dislike): Associative learning of preferences. In A. J. Wills (Ed.), *New directions in human associative learning* (pp. 221–252). Mahwah, NJ: Erlbaum.

Field, A. P. (2005d). Sir Ronald Aylmer Fisher. In B. S. Everitt & D. C. Howell (Eds.), *Encyclopedia of statistics in behavioral science* (Vol. 2, pp. 658–659). Hoboken, NJ: Wiley.

Field, A. P. (2006). The behavioral inhibition system and the verbal information pathway to children's fears. *Journal of Abnormal Psychology, 115*(4), 742–752.

Field, A. P. (2009). *Discovering statistics using SPSS (3rd ed.)*. London: Sage.

Field, A. P., & Davey, G. C. L. (1999). Reevaluating evaluative conditioning: A nonassociative explanation of conditioning effects in the visual evaluative conditioning paradigm. *Journal of Experimental Psychology – Animal Behavior Processes, 25*(2), 211–224.

Field, A. P., & Hole, G. J. (2003). *How to design and report experiments*. London: Sage.

Field, A. P., & Moore, A. C. (2005). Dissociating the effects of attention and contingency awareness on evaluative conditioning effects in the visual paradigm. *Cognition and Emotion, 19*(2), 217–243.

Fisher, R. A. (1921). On the probable error of a coefficient of correlation deduced from a small sample. *Metron, 1*, 3–32.

Fisher, R. A. (1922). On the interpretation of chi square from contingency tables, and the calculation of P. *Journal of the Royal Statistical Society, 85*, 87–94.

Fisher, R. A. (1925). *Statistical methods for research workers*. Edinburgh: Oliver & Boyd.

Fisher, R. A. (1925/1991). *Statistical methods, experimental design, and scientific inference*. Oxford: Oxford University Press. (This reference is for the 1991 reprint.).

Fisher, R. A. (1956). *Statistical methods and scientific inference*. New York: Hafner.

Flanagan, J. C. (1937). A proposed procedure for increasing the efficiency of objective tests. *Journal of Educational Psychology, 28*, 17–21.

Friedman, M. (1937). The use of ranks to avoid the assumption of normality implicit in the analysis of variance. *Journal of the American Statistical Association, 32*, 675–701.

Gallup, G. G. J., Burch, R. L., Zappieri, M. L., Parvez, R., Stockwell, M., & Davis, J. A. (2003). The human penis as a semen displacement device. *Evolution and Human Behavior, 24*, 277–289.

Games, P. A. (1983). Curvilinear transformations of the dependent variable. *Psychological Bulletin, 93*(2), 382–387.

Games, P. A. (1984). Data transformations, power, and skew: A rebuttal to Levine and Dunlap. *Psychological Bulletin, 95*(2), 345–347.

Games, P. A., & Lucas, P. A. (1966). Power of the analysis of variance of independent groups on non-normal and normally transformed data. *Educational and Psychological Measurement, 26*, 311–327.

Girden, E. R. (1992). *ANOVA: Repeated measures*. Sage University Paper Series on Quantitative Applications in the Social Sciences, 07-084. Newbury Park, CA: Sage.

Glass, G. V. (1966). Testing homogeneity of variances. *American Educational Research Journal, 3*(3), 187–190.

Glass, G. V., Peckham, P. D., & Sanders, J. R. (1972). Consequences of failure to meet assumptions underlying the fixed effects analyses of variance and covariance. *Review of Educational Research, 42*(3), 237–288.

Graham, J. M., Guthrie, A. C., & Thompson, B. (2003). Consequences of not interpreting structure coefficients in published CFA research: A reminder. *Structural Equation Modeling, 10*(1), 142–153.

Grayson, D. (2004). Some myths and legends in quantitative psychology. *Understanding Statistics, 3*(1), 101–134.

Green, S. B. (1991). How many subjects does it take to do a regression analysis? *Multivariate Behavioral Research, 26*, 499–510.

Greenhouse, S. W., & Geisser, S. (1959). On methods in the analysis of profile data. *Psychometrika, 24*, 95–112.

Guadagnoli, E., & Velicer, W. F. (1988). Relation of sample size to the stability of component patterns. *Psychological Bulletin, 103*(2), 265–275.

Hakstian, A. R., Roed, J. C., & Lind, J. C. (1979). Two-sample T^2 procedure and the assumption of homogeneous covariance matrices. *Psychological Bulletin, 86*, 1255–1263.

Hardy, M. A. (1993). *Regression with dummy variables*. Sage University Paper Series on Quantitative Applications in the Social Sciences, 07-093. Newbury Park, CA: Sage.

Harman, B. H. (1976). *Modern factor analysis* (3rd ed., revised.). Chicago: University of Chicago Press.

Harris, R. J. (1975). *A primer of multivariate statistics*. New York: Academic Press.

Haworth, L. E. (2001) *Output delivery system: the basics*. Cary, NC: SAS Publishing.

Hill, C., Abraham, C., & Wright, D. B. (2007). Can theory-based messages in combination with cognitive prompts promote exercise in classroom settings? *Social Science & Medicine, 65*, 1049–1058.

Hoaglin, D., & Welsch, R. (1978). The hat matrix in regression and ANOVA. *American Statistician, 32*, 17–22.

Hoddle, G., Batty, D., & Ince, P. (1998). How not to take penalties in important soccer matches. *Journal of Cretinous Behaviour, 1*, 1–2.

Horn, J. L. (1965). A rationale and test for the number of factors in factor analysis. *Psychometrika, 30*, 179–185.

Hosmer, D. W., & Lemeshow, S. (1989). *Applied logistic regression*. New York: Wiley.

Howell, D. C. (1997). *Statistical methods for psychology* (4th ed.). Belmont, CA: Duxbury.

Howell, D. C. (2006). *Statistical methods for psychology* (6th ed.). Belmont, CA: Thomson.

Huberty, C. J., & Morris, J. D. (1989). Multivariate analysis versus multiple univariate analysis. *Psychological Bulletin, 105*(2), 302–308.

Hughes, J. P., Marice, H. P., & Gathright, J. B. (1976). Method of removing a hollow object from the rectum. *Diseases of the Colon & Rectum, 19*(1), 44–45.

Hume, D. (1739–40). *A treatise of human nature* (ed. L. A. Selby-Bigge). Oxford: Clarendon Press, 1965.

Hume, D. (1748). *An enquiry concerning human understanding*. Chicago: Open Court, 1927.

Hutcheson, G., & Sofroniou, N. (1999). *The multivariate social scientist*. London: Sage.

Huynh, H., & Feldt, L. S. (1976). Estimation of the Box correction for degrees of freedom from sample data in randomised block and split-plot designs. *Journal of Educational Statistics, 1*(1), 69–82.

Jolliffe, I. T. (1972). Discarding variables in a principal component analysis, I: Artificial data. *Applied Statistics, 21*, 160–173.

Jolliffe, I. T. (1986). *Principal component analysis*. New York: Springer.

Kahneman, D., & Krueger, A. B. (2006). Developments in the measurement of subjective well-being. *Journal of Economic Perspectives, 20*(1), 3–24.

Kaiser, H. F. (1960). The application of electronic computers to factor analysis. *Educational and Psychological Measurement, 20*, 141–151.

Kaiser, H. F. (1970). A second-generation little jiffy. *Psychometrika, 35*, 401–415.

Kaiser, H. F. (1974). An index of factorial simplicity. *Psychometrika, 39*, 31–36.

Kass, R. A., & Tinsley, H. E. A. (1979). Factor analysis. *Journal of Leisure Research, 11*, 120–138.

Kellett, S., Clarke, S., & McGill, P. (2008). Outcomes from psychological assessment regarding recommendations for cosmetic surgery. *Journal of Plastic, Reconstructive & Aesthetic Surgery, 61*, 512–517.

Kirk, R. E. (1996). Practical significance: A concept whose time has come. *Educational and Psychological Measurement, 56*(5), 746–759.

Kline, P. (1999). *The handbook of psychological testing* (2nd ed.). London: Routledge.

Klockars, A. J., & Sax, G. (1986). *Multiple comparisons*. Sage University Paper Series on Quantitative Applications in the Social Sciences, 07-061. Newbury Park, CA: Sage.

Koot, V. C. M., Peeters, P. H. M., Granath, F., Grobbee, D. E., & Nyren, O. (2003). Total and cause specific mortality among Swedish women with cosmetic breast implants: Prospective study. *British Medical Journal, 326*(7388), 527–528.

Kreft, I. G. G., & de Leeuw, J. (1998). *Introducing multilevel modeling*. London: Sage.

Kreft, I. G. G., de Leeuw, J., & Aiken, L. S. (1995). The effect of different forms of centering in hierarchical linear models. *Multivariate Behavioral Research, 30*, 1–21.

Kruskal, W. H., & Wallis, W. A. (1952). Use of ranks in one-criterion variance analysis. *Journal of the American Statistical Association, 47*, 583–621.

Lacourse, E., Claes, M., & Villeneuve, M. (2001). Heavy metal music and adolescent suicidal risk. *Journal of Youth and Adolescence, 30*(3), 321–332.

Lehmann, E. L. (1993). The Fisher, Neyman-Pearson theories of testing hypotheses: One theory or two? *Journal of the American Statistical Association, 88*, 1242–1249.

Lenth, R. V. (2001). Some practical guidelines for effective sample size determination. *American Statistician, 55*(3), 187–193.

Levene, H. (1960). Robust tests for equality of variances. In I. Olkin, S. G. Ghurye, W. Hoeffding, W. G. Madow & H. B. Mann (Eds.), *Contributions to Probability and statistics: essays in honor of Harold Hotelling* (pp. 278–292). Stanford, CA: Stanford University Press.

Levine, D. W., & Dunlap, W. P. (1982). Power of the F test with skewed data: Should one transform or not? *Psychological Bulletin, 92*(1), 272–280.

Levine, D. W., & Dunlap, W. P. (1983). Data transformation, power, and skew: A rejoinder to Games. *Psychological Bulletin, 93*(3), 596–599.

Lo, S. F., Wong, S. H., Leung, L. S., Law, I. C., & Yip, A. W. C. (2004). Traumatic rectal perforation by an eel. *Surgery, 135*(1), 110–111.

Lord, F. M. (1967). A paradox in the interpretation of group comparisons. *Psychological Bulletin, 68*(5), 304–305.

Lord, F. M. (1969). Statistical adjustments when comparing preexisting groups. *Psychological Bulletin, 72*(5), 336–337.

Lunney, G. H. (1970). Using analysis of variance with a dichotomous dependent variable: An empirical study. *Journal of Educational Measurement, 7*(4), 263–269.

MacCallum, R. C., Widaman, K. F., Zhang, S., & Hong, S. (1999). Sample size in factor analysis. *Psychological Methods, 4*(1), 84–99.

MacCallum, R. C., Zhang, S., Preacher, K. J., & Rucker, D. D. (2002). On the practice of dichotomization of quantitative variables. *Psychological Methods, 7*(1), 19–40.

Mann, H. B., & Whitney, D. R. (1947). On a test of whether one of two random variables is stochastically larger than the other. *Annals of Mathematical Statistics, 18*, 50–60.

Marzillier, S. L., & Davey, G. C. L. (2005). Anxiety and disgust: Evidence for a unidirectional relationship. *Cognition and Emotion, 19*(5), 729–750.

Mather, K. (1951). R. A. Fisher's *Statistical methods for research workers*: An appreciation. *Journal of the American Statistical Association, 46*, 51–54.

Matthews, R. C., Domjan, M., Ramsey, M., & Crews, D. (2007). Learning effects on sperm competition and reproductive fitness. *Psychological Science, 18*(9), 758–762.

Maxwell, S. E., & Delaney, H. D. (1990). *Designing experiments and analyzing data*. Belmont, CA: Wadsworth.

McDonald, P. T., & Rosenthal, D. (1977). An unusual foreign body in the rectum – A baseball: Report of a case. *Diseases of the Colon & Rectum, 20*(1), 56–57.

McGrath, R. E., & Meyer, G. J. (2006). When effect sizes disagree: The case of *r* and *d*. *Psychological Methods, 11*(4), 386–401.

Menard, S. (1995). *Applied logistic regression analysis*. Sage University Paper Series on Quantitative Applications in the Social Sciences, 07-106. Thousand Oaks, CA: Sage.

Mendoza, J. L., Toothaker, L. E., & Crain, B. R. (1976). Necessary and sufficient conditions for F ratios in the L * J * K factorial design with two repeated factors. *Journal of the American Statistical Association, 71*, 992–993.

Mendoza, J. L., Toothaker, L. E., & Nicewander, W. A. (1974). A Monte Carlo comparison of the univariate and multivariate methods for the groups by trials repeated measures design. *Multivariate Behavioural Research, 9*, 165–177.

Miles, J. N. V. (2005) Regression Residuals, in D. Howell & B. Everitt (Eds) *The Encyclopedia of Statistics in Behavioural Science*. Chichester: Wiley.

Miles, J. M. V., & Banyard, P. (2007). *Understanding and using statistics in psychology: A practical introduction*. London: Sage.

Miles, J. M. V., & Shevlin, M. (2001). *Applying regression and correlation: A guide for students and researchers*. London: Sage.

Mill, J. S. (1865). *A system of logic: atiocinative and inductive*. London: Longmans, Green.

Miller, G., Tybur, J. M., & Jordan, B. D. (2007). Ovulatory cycle effects on tip earnings by lap dancers: economic evidence for human estrus? *Evolution and Human Behavior, 28*, 375–381.

Miller, G. A., & Chapman, J. P. (2001). Misunderstanding analysis of covariance. *Journal of Abnormal Psychology, 110*(1), 40–48.

Muris, P., Huijding, J., Mayer, B., & Hameetman, M. (2008). A space odyssey: Experimental manipulation of threat perception and anxiety-related interpretation bias in children. *Child Psychiatry and Human Development 39*(4), 469–480.

Myers, R. (1990). *Classical and modern regression with applications* (2nd ed.). Boston, MA: Duxbury.

Nagelkerke, N. J. D. (1991). A note on a general definition of the coefficient of determination. *Biometrika, 78*, 691–692.

Namboodiri, K. (1984). *Matrix algebra: an introduction*. Sage University Paper Series on Quantitative Applications in the Social Sciences, 07-38. Beverly Hills, CA: Sage.

Nichols, L. A., & Nicki, R. (2004). Development of a psychometrically sound internet addiction scale: A preliminary step. *Psychology of Addictive Behaviors, 18*(4), 381–384.

Nunnally, J. C. (1978). *Psychometric theory*. New York: McGraw-Hill.

Nunnally, J. C., & Bernstein, I. H. (1994). *Psychometric theory* (3rd ed.). New York: McGraw-Hill.

O'Brien, M. G., & Kaiser, M. K. (1985). MANOVA method for analyzing repeated measures designs: An extensive primer. *Psychological Bulletin, 97*(2), 316–333.

Olson, C. L. (1974). Comparative robustness of six tests in multivariate analysis of variance. *Journal of the American Statistical Association, 69*, 894–908.

Olson, C. L. (1976). On choosing a test statistic in multivariate analysis of variance. *Psychological Bulletin, 83*, 579–586.

Olson, C. L. (1979). Practical considerations in choosing a MANOVA test statistic: A rejoinder to Stevens. *Psychological Bulletin, 86*, 1350–1352.

Pearson, E. S., & Hartley, H. O. (1954). *Biometrika tables for statisticians, Volume I*. New York: Cambridge University Press.

Pearson, K. (1900). On the criterion that a given system of deviations from the probable in the case of a correlated system of variables is such that it can be reasonably supposed to have arisen from random sampling. *Philosophical Magazine, 50*(5), 157–175.

Pedhazur, E., & Schmelkin, L. (1991). *Measurement, design and analysis: an integrated approach*. Hillsdale, NJ: Erlbaum.

Ramsey, P. H. (1982). Empirical power of procedures for comparing two groups on *p* variables. *Journal of Educational Statistics, 7*, 139–156.

Raudenbush, S. W., & Bryk, A. S. (2002). *Hierarchical linear models* (2nd ed.). Thousand Oaks, CA: Sage.

Rockwell, R. C. (1975). Assessment of multicollinearity: The Haitovsky test of the determinant. *Sociological Methods and Research, 3*(4), 308–320.

Rosenthal, R. (1991). *Meta-analytic procedures for social research* (2nd ed.). Newbury Park, CA: Sage.

Rosenthal, R., Rosnow, R. L., & Rubin, D. B. (2000). *Contrasts and effect sizes in behavioural research: a correlational approach*. Cambridge: Cambridge University Press.

Rosnow, R. L., & Rosenthal, R. (2005). *Beginning behavioral research: A conceptual primer* (5th ed.). Englewood Cliffs, NJ: Pearson/Prentice Hall.

Rosnow, R. L., Rosenthal, R., & Rubin, D. B. (2000). Contrasts and correlations in effect-size estimation. *Psychological Science, 11*, 446–453.

Rouanet, H., & Lépine, D. (1970). Comparison between treatments in a repeated-measurement design: ANOVA and multivariate methods. *British Journal of Mathematical and Statistical Psychology, 23*, 147–163.

Rulon, P. J. (1939). A simplified procedure for determining the reliability of a test by split-halves. *Harvard Educational Review, 9*, 99–103.

Sacco, W. P., Levine, B., Reed, D., & Thompson, K. (1991). Attitudes about condom use as an AIDS-relevant behavior: Their factor structure and relation to condom use. *Psychological Assessment: A Journal of Consulting and Clinical Psychology, 3(2)*, 265–272.

Sacco, W. P., Rickman, R. L., Thompson, K., Levine, B., & Reed, D. L. (1993). Gender differences in AIDS-relevant condom attitudes and condom use. *AIDS Education and Prevention, 5*(4), 311–326.

Sachdev, Y. V. (1967). An unusual foreign body in the rectum. *Diseases of the Colon & Rectum, 10*(3), 220–221.

Salsburg, D. (2002). *The lady tasting tea: How statistics revolutionized science in the twentieth century*. New York: Owl Books.

Savage, L. J. (1976). On re-reading R. A. Fisher. *Annals of Statistics, 4*, 441–500.

Scariano, S. M., & Davenport, J. M. (1987). The effects of violations of independence in the one-way ANOVA. *American Statistician, 41*(2), 123–129.

Schützwohl, A. (2008). The disengagement of attentive resources from task-irrelevant cues to sexual and emotional infidelity. *Personality and Individual Differences, 44*, 633–644.

Shackelford, T. K., LeBlanc, G. J., & Drass, E. (2000). Emotional reactions to infidelity. *Cognition & Emotion, 14*(5), 643–659.

Shee, J. C. (1964). Pargyline and the cheese reaction. *British Medical Journal, 1*(539), 1441.

Siegel, S., & Castellan, N. J. (1988). *Nonparametric statistics for the behavioral sciences* (2nd ed.). New York: McGraw-Hill.

Spearman, C. (1910). Correlation calculated with faulty data. *British Journal of Psychology, 3*, 271–295.

Stevens, J. P. (1979). Comment on Olson: choosing a test statistic in multivariate analysis of variance. *Psychological Bulletin, 86*, 355–360.

Stevens, J. P. (1980). Power of the multivariate analysis of variance tests. *Psychological Bulletin, 88*, 728–737.

Stevens, J. P. (2002). *Applied multivariate statistics for the social sciences* (4th ed.). Hillsdale, NJ: Erlbaum.

Strahan, R. F. (1982). Assessing magnitude of effect from rank-order correlation coeffients. *Educational and Psychological Measurement, 42*, 763–765.

Stuart, E. W., Shimp, T. A., & Engle, R. W. (1987). Classical-conditioning of consumer attitudes – 4. Experiments in an advertising context. *Journal of Consumer Research, 14*(3), 334–349.

Studenmund, A. H., & Cassidy, H. J. (1987). *Using econometrics: a practical guide*. Boston: Little Brown.

Tabachnick, B. G., & Fidell, L. S. (2001). *Using multivariate statistics* (4th ed.). Boston: Allyn & Bacon.

Tabachnick, B. G., & Fidell, L. S. (2007). *Using multivariate statistics* (5th ed.). Boston: Allyn & Bacon.

Tinsley, H. E. A., & Tinsley, D. J. (1987). Uses of factor analysis in counseling psychology research. *Journal of Counseling Psychology, 34*, 414–424.

Toothaker, L. E. (1993). *Multiple comparison procedures*. Sage University Paper Series on Quantitative Applications in the Social Sciences, 07–089. Newbury Park, CA: Sage.

Tufte, E. R. (2001). *The visual display of quantitative information* (2nd ed.). Cheshire, CT: Graphics Press.

Twisk, J. W. R. (2006). *Applied multilevel analysis: A practical guide*. Cambridge: Cambridge University Press.

Umpierre, S. A., Hill, J. A., & Anderson, D. J. (1985). Effect of Coke on sperm motility. *New England Journal of Medicine, 313*(21), 1351.

Wainer, H. (1984). How to display data badly. *American Statistician, 38*(2), 137–147.

Welch, B. L. (1951). On the comparison of several mean values: An alternative approach. *Biometrika, 38*, 330–336.

Wilcox, R. R. (2005). *Introduction to robust estimation and hypothesis testing* (2nd ed.). Burlington, MA: Elsevier.

Wilcoxon, F. (1945). Individual comparisons by ranking methods. *Biometrics, 1*, 80–83.

Wildt, A. R., & Ahtola, O. (1978). *Analysis of covariance*. Sage University Paper Series on Quantitative Applications in the Social Sciences, 07-012. Newbury Park, CA: Sage.

Williams, J. M. G. (2001). *Suicide and attempted suicide*. London: Penguin.

Wright, D. B. (1998). Modeling clustered data in autobiographical memory research: The multilevel approach. *Applied Cognitive Psychology, 12*, 339–357.

Wright, D. B. (2003). Making friends with your data: Improving how statistics are conducted and reported. *British Journal of Educational Psychology, 73*, 123–136.

Yates, F. (1951). The influence of Statistical methods for research workers on the development of the science of statistics. *Journal of the American Statistical Association, 46*, 19–34.

Zabell, S. L. (1992). R. A. Fisher and fiducial argument. *Statistical Science, 7*(3), 369–387.

Zwick, R. (1985). Nonparametric one-way multivariate analysis of variance: A computational approach based on the Pillai-Bartlett trace. *Psychological Bulletin, 97*(1), 148–152.

Zwick, W. R., & Velicer, W. F. (1986). Comparison of five rules for determining the number of components to retain. *Psychological Bulletin, 99*(3), 432–442.

INDEX

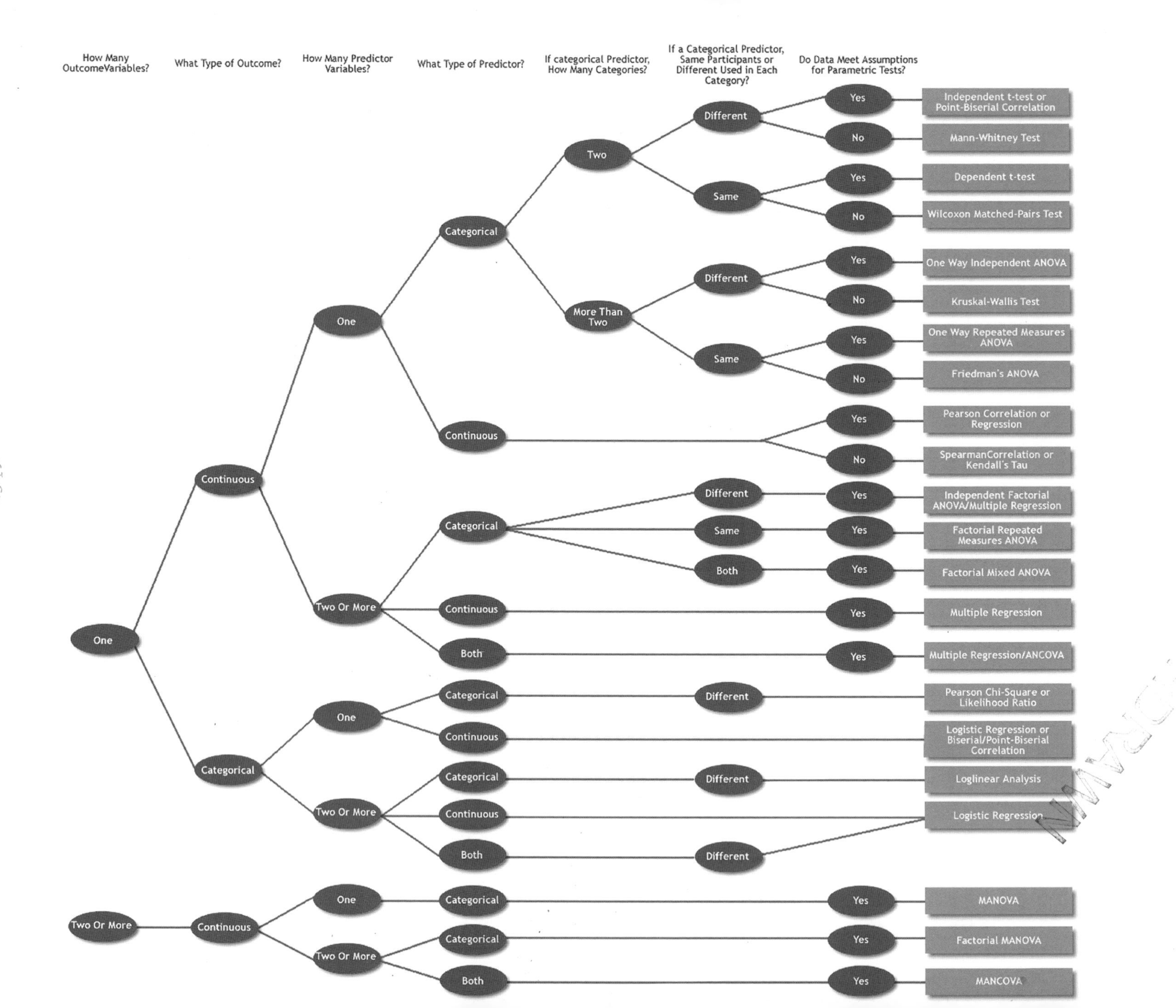
How Many OutcomeVariables?
What Type of Outcome?
How Many Predictor Variables?
What Type of Predictor?
If categorical Predictor, How Many Categories?
If a Categorical Predictor, Same Participants or Different Used in Each Category?
Do Data Meet Assumptions for Parametric Tests?
One
Two Or More
Continuous
Categorical
One
Two Or More
Categorical
Continuous
Both
Two
More Than Two
Different
Same
Both
Yes
No
Independent t-test or Point-Biserial Correlation
Mann-Whitney Test
Dependent t-test
Wilcoxon Matched-Pairs Test
One Way Independent ANOVA
Kruskal-Wallis Test
One Way Repeated Measures ANOVA
Friedman's ANOVA
Pearson Correlation or Regression
SpearmanCorrelation or Kendall's Tau
Independent Factorial ANOVA/Multiple Regression
Factorial Repeated Measures ANOVA
Factorial Mixed ANOVA
Multiple Regression
Multiple Regression/ANCOVA
Pearson Chi-Square or Likelihood Ratio
Logistic Regression or Biserial/Point-Biserial Correlation
Loglinear Analysis
Logistic Regression
MANOVA
Factorial MANOVA
MANCOVA